Fisseler

STEAM

its generation and use

4th
EDITION

Babcock & Wilcox
a McDermott company

Edited by S.C. Stultz and J.B. Kitto

Disclaimer

The information contained within this book has been obtained by The Babcock &Wilcox Company from sources believed to be reliable. However, neither The Babcock & Wilcox Company nor its authors make any guarantee or warranty, expressed or implied, about the accuracy, completeness or usefulness of the information, product, process or apparatus discussed within this book, nor shall The Babcock & Wilcox Company or any of its authors be liable for error, omission, losses or damages of any kind or nature. This book is published with the understanding that The Babcock & Wilcox Company and its authors are supplying general information and neither attempting to render engineering or professional services nor offering a product for sale. If services are desired, an appropriate professional should be consulted.

Steam/its generation and use. 40th edition.
Editors: Steven C. Stultz and John B. Kitto.
The Babcock & Wilcox Company, Barberton, Ohio, U.S.A.
1992

Includes bibliographic references and index.
Subject areas: 1. Steam boilers.
2. Combustion — Fossil fuels.
3. Nuclear power.

Library of Congress Catalog Number: 92-74123
ISBN 0-9634570-0-4

Printed in the United States of America

Steam 40

This 40th edition is for the men and women who do the work of the world every day in power plants, paper mills, oil refineries, factories, and every other institution that uses a safety water tube boiler.

125 YEARS

Babcock & Wilcox

McDERMOTT

Preface/Dedication

Dear Reader:

One hundred and twenty five years ago, George Babcock and Stephen Wilcox formed a partnership and patented a product that accelerated worldwide industrial development — the safety water tube boiler.

Today, Babcock & Wilcox continues its commitment to the safe and dependable generation of power. While B&W has added new products and services over the years, few companies have remained as dedicated to their original products and services for so long a period of time. And few products and services have provided such a steady increase in their usefulness, efficiency and economy over the years.

The common forces throughout these 125 years have been our employees, our customers and our partners. Without their support and ingenuity, we would not be celebrating this historic event. Babcock & Wilcox has been able to continue its commitment to excellence through technology and through the values of quality, service and integrity held by its people, and we will continue to do so in the future.

I would like to dedicate this 40th edition of *Steam / its generation and use* to the creativity and technical strength of all Babcock & Wilcox employees, and to our customers and partners who recognize these talents and support our commitment to excellence. For more than a century we have worked together, pursuing all opportunities to further advance the art and science of steam generation with effective control of environmental emissions.

We look forward to working together in the future to apply energy conversion technologies to do the work of the world in an environmentally safe and dependable manner.

Joe J. Stewart
President and Chief Operating Officer
The Babcock & Wilcox Company

Table of Contents

Acknowledgments

Steam / its generation and use is the culmination of the work of hundreds of B&W employees who have contributed directly and indirectly to this edition and to the technology upon which it is based. Particular recognition goes to individuals who formally committed to preparing and completing this expanded 40th edition.

Project Manager/Editor

S.C. Stultz

Technical Advisor/Editor

J.B. Kitto

Art Director/Illustrator

C.H. Rahn

Lead Authors

M.J. Albrecht
K.C. Alexander
J.B. Andrews
J.D. Blue
G.H. Branigan
E.J. Campobenedetto
J.D. Carlton
P.C. Childress
P.L. Cioffi
G.A. Clark
J.L. Clement
R.A. Clocker
T.A. Coleman
W. Downs
D.D. Dueck
S.J. Elmiger
J.L. Esakov
J.R. Farr
D.P. Finn
W.A. Fiveland
L.R. Flais
R.W. Ganthner
D.R. Gibbs
J.R. Gloudemans
M. Gold
J.E. Granger
G.H. Harth
T.C. Heil
T.P. Hoosic
D.A. Huston
D.E. James
D.W. Johnson
T.O. Johnson
J.B. Kitto
A.D. LaRue
E.H. Mayer
D.K. McDonald
T.E. Moskal
N.J. Mravich
N.D. Nelson
M.W. Parker
R.R. Piepho
M.R. Rechner
K.E. Redinger
N.E. Reeling
K.P. Sabol
D.P. Sanders
T.A. Saari
D.M. Shetler
W.R. Stirgwolt
S.C. Stultz
E.P. Szmania
D.P. Tonn
G.T. Urquhart
S.J. Vecci
R.C. Vetterick
P.W. Waanders
R.J. Waltz
J.J. Warchol
J.C. Waung

Primary Support Authors

W.E. Allmon
J.M. Bloom
S.A. Bryk
W.P. Dal Pio
N.A. Exconde
D.S. Fedock
J.D. Graham
J.M. Jevec
C.L. Jones
R.W. Kronenberger
D.C. Langley
J.A. Larose
C.J. Lenore
K.S. Lester
J.W. Locke
G.H. McClellan
B.P. Miglin
M.G. Milobowski
D.T.K. Murray
G.J. Nakoneczny
L.D. Paul
D.B. Pearson
C.E. Phelps
E.J. Piaskowski
E.F. Radke
K.J. Rogers
S.A. Scavuzzo
J. Schlichting
R.E. Spada
J.R. Strempek
R.J. Warrick
F.H. Wenderoth
J.M. Wennerstrom
L.C. Westfall
L.P. Williams

Executive Steering Committee

G.J. Clessuras
P.P. Koenderman
J.S. Kulig
J.E. Pollock
C.W. Pryor
P.E. Ralston
R.J. Thibeault
E.A. Womack

Technical Advisory Committee

T.O. Johnson
R.L. Killion
J.B. Kitto
R.B. Mady
E.R. Michaud
M.G. Morash
J.B. Rogan
J.W. Smith
S.C. Stultz

Production Group

J.L. Basar
L.A. Brower
J.L. Hanson
K.A. Labbate
R.R. McEntee
T.W. Secrest

Outside Support

J.C. Edwards (Text)
P.C. Lutjen (Art)

System of Units: English and Système International

To recognize the globalization of the power industry, the 40th edition of *Steam* has been extensively revised to incorporate the Système International d'Unitès (SI) along with the continued use of English or customary U.S. units. Upon recommendation of the Technical Advisory Committee for *Steam*, English units continue to be the primary system of units with SI provided as secondary units in parentheses. In some instances, SI units alone have been provided where these units are common usage. In selected figures and tables where dual units could detract from clarity (logarithmic scales, for example) SI conversions are provided within the figure titles or as a table footnote.

Extensive English-SI conversion tables are provided in Appendix I. This appendix also contains a complete SI set of the Steam Tables, Mollier diagram, pressure-enthalpy diagram and psychrometric chart.

The decision was made to provide exact conversions rounded to an appropriate number of figures. This was done to avoid confusion about the original source values.

Two exceptions to these conversion practices have been adopted. As is general practice in the power industry worldwide, *bar* has been used as the unit for steam pressure above atmospheric, in place of the pure SI unit of *megapascal* or *MPa* (1 MPa = 10 bar). Absolute pressure is denoted by *psi* or *bar* and gauge pressure by *psig* or *bar gauge*. The difference between *absolute pressure* and *pressure difference* is identified by the context. Finally, in Chapters 9 and 21, as well as selected other areas of *Steam* which provide extensive numerical examples, only English units have been provided for clarity.

For reference and clarity, power in British thermal units per hour (Btu/h) has typically been converted to megawatts-thermal and is denoted by MW_t while megawatts-electric in both systems of units has been denoted by MW.

It is hoped that these conversion practices will make *Steam* easily usable by the broadest possible audience.

Editors' Foreword

"While making known the character and quality of our manufactures, we have endeavored at the same time to present to our friends and customers a variety of useful information, not readily accessible to them in other ways."

So began the first edition of *Steam / its generation and use* more than a century ago. The text has been revised and updated often, and some editions have been more extensively modified than others. We, as editors of the 40th edition, feel a particular affiliation with those who undertook the project after the first ten editions, at least in their overview: *"In preparing the new edition of* Steam, *we have revised the whole, and added much new and valuable material."* This new 40th edition has been such a task, and only through the dedicated efforts of many B&W employees has this complete revision been possible.

Steam / its generation and use is accepted worldwide as an authoritative handbook on steam generation. Its principal purpose is to help professionals design, procure, construct and maintain B&W equipment in a way that will provide decades of reliable performance.

The focus of the 40th edition of *Steam* has been to analyze and incorporate industry developments since the last edition, while looking well into the 21st century and continuing the 115 year tradition of this publication.

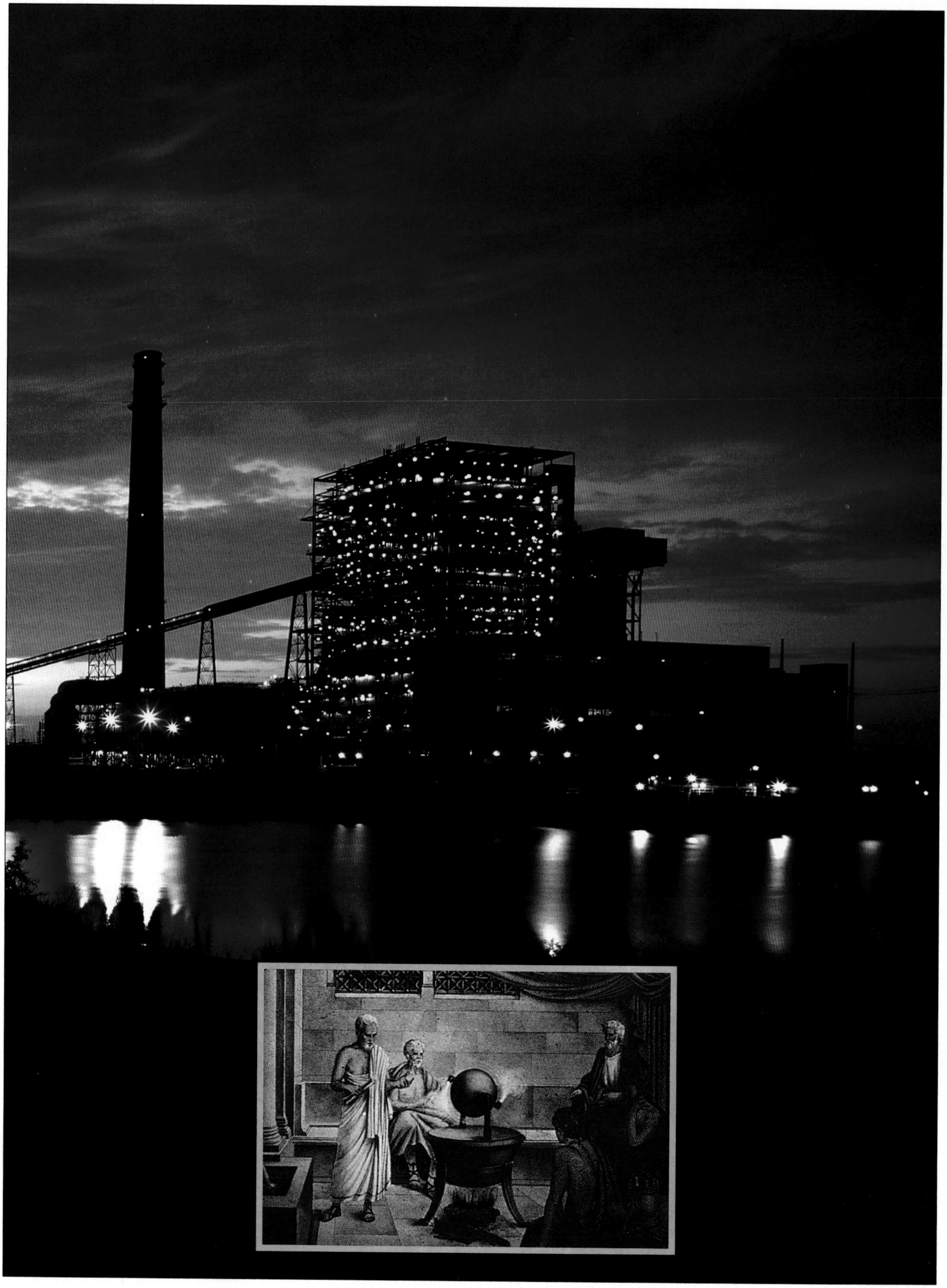

Introduction to *Steam*

Throughout history, mankind has reached beyond the acceptable to pursue a challenge, achieving significant events and developing new technology. The process is both scientific and creative. Entire civilizations, organizations and, most notably, individuals have achieved success, doing what mankind has never done before. A prime example is the safe and efficient use of steam.

The most significant series of events shaping today's world is the industrial revolution that began in the late 17th century. The ability to generate steam on demand sparked this revolution, and technical advances in steam generation allowed it to continue. Without these advancements, the industrial revolution as we know it would not have taken place.

It is therefore appropriate to say that few technologies that the ingenuity of man have produced have done so much to advance mankind as the safe and dependable generation of steam.

Fig. 1 Hero's aelopile.

Steam as a resource

In 200 B.C., a Greek named Hero designed a simple machine that used steam as a power source (Fig. 1). He began with a cauldron of water, placed above a fire. As the fire heated the cauldron, the cauldron shell transferred the heat to the water and the water reached the boiling point of 212F (100C). It then changed form and turned into steam. The steam passed through two pipes into a hollow sphere, pivoted at both sides. As the steam escaped the sphere through two tubes, each bent at an angle, the sphere moved, rotating on its axis.

Hero, a mathematician and scientist, labeled the device *aelopile*, meaning rotary steam engine.

The invention was only a novelty. However, even though Hero made no suggestion for how to use the device, the idea of generating steam to do useful work was born.

To this day, the basic idea has remained the same — generate heat, transfer the heat to water, and produce steam.

Intimately related to steam generation is the steam turbine, a device to change the energy of steam into mechanical work. In the early 1600s, an Italian named Branca produced a unique invention (Fig. 2). He first produced steam, based on Hero's aelopile. Next, he channeled the steam to a wheel and the steam pressure caused the wheel to rotate. This marked the beginning of steam turbine development.

The primary use of steam turbines today is the production of electric power. In one of the most complex systems designed by mankind, superheated high pressure steam is produced in a boiler and channeled to turbine-generators to produce electricity.

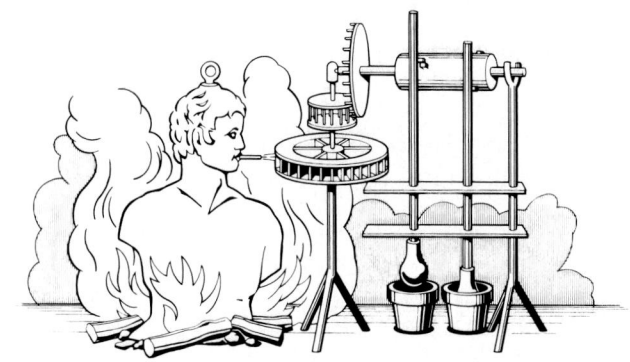

Fig. 2 Branca's steam turbine.

Today's steam plants are a complex and highly sophisticated combination of engineered elements. They must first obtain heat, usually in the form of coal, oil or natural gas, referred to as fossil fuels, or in the form of uranium, called nuclear fuel. Other forms of energy include waste heat and exhaust gases, bagasse and biomass, spent chemicals and municipal waste, and geothermal and solar energy.

Each fuel contains potential energy, or a heating value measured in Btu/lb (J/kg). The first goal is to release this energy, most often by a controlled combustion process or, with uranium, through fission. The heat is then transferred through tube walls and other components or liquids to water. The heated water then changes form, turning into steam. The steam is normally heated further to specific temperatures and pressures.

Steam is also a vital resource in industry. It drives pumps and valves, helps produce paper and wood products, prepares foods, and heats and cools large buildings and institutions. Steam also propels much of the world's naval fleets and a high percentage of commercial marine transport. In some countries, steam plays a continuing role in railway transportation.

Steam generators, commonly referred to as *boilers*, range in size from those needed to heat a small building to those used individually to produce 1300 megawatts of electricity in a power generating station — enough power for more than one million people. These larger units deliver more than ten million pounds of superheated steam per hour with steam temperatures exceeding 1000F (538C) and pressures exceeding 3800 psi (262 bar).

Today's steam generating systems owe their dependability and safety to more than 125 years of experience in the design, fabrication and operation of water tube boilers, first patented by George Babcock and Stephen Wilcox in 1867 (Fig. 3).

Steam is a resource for mankind. It is our challenge and responsibility to further develop and use this resource safely, efficiently and dependably, in an environmentally friendly manner.

The early use of steam

Steam generation as an industry began almost two thousand years after Hero's invention, in the 17th century. Many conditions began to stimulate the development of steam use in a power cycle. Mining for ores and minerals had expanded greatly and large quantities of fuel were needed for ore refining. Fuels were needed for space heating and cooking and for general industrial and military growth. Forests were being stripped and coal was becoming an important fuel. Coal mining was emerging as a major industry.

As mines became deeper, they were often flooded with underground water. The English in particular were faced with a very serious curtailment of their industrial growth if they could not find some economical way to pump water from the mines. Many people began working on the problem and numerous patents were issued for machines to pump water from the mines using *the expansive power of steam*. The early machines used wood and charcoal for fuel, but coal eventually became the dominant fuel.

The most common source of steam at the time was a *shell* boiler, little more than a large kettle filled with water and heated at the bottom (Fig. 4).

Not all early developments in steam were directed toward pumps and engines. In 1680 Dr. Denis Papin, a Frenchman, invented a steam digester for food processing, using a boiler under heavy pressure. To avoid explosion, Papin added a device which is the first safety valve on record. Papin also invented a boiler with an internal firebox, the earliest record of such construction.

Many experiments were concentrating on using steam pressure or atmospheric pressure combined with a vacuum. The result was the first commercially successful steam engine, patented by Thomas Savery in 1698 to pump water by direct displacement (Fig. 5). The patent credits Savery with an engine for raising water by the impellant force of fire, meaning steam. The mining industry needed the invention, but the engine had a limited pumping height set by the pressure the boiler and other vessels could withstand. Before its replacement by Thomas Newcomen's engine (described below), Desaguliers improved the Savery engine, adding the Papin safety valve and using an internal jet for the condensing part of the cycle.

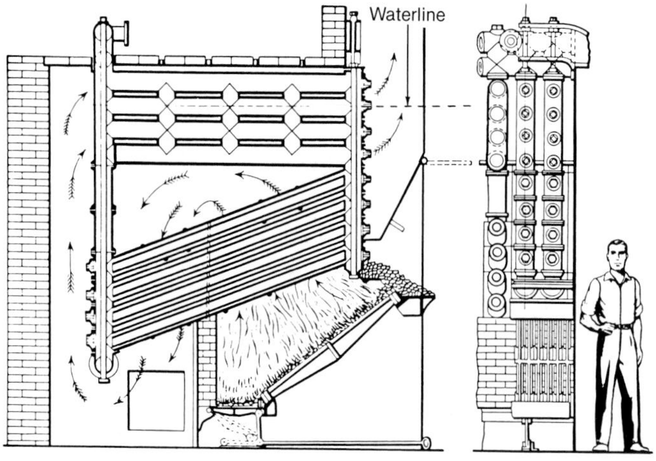

Fig. 3 First Babcock & Wilcox boiler, patented in 1867.

Fig. 4 Haycock shell boiler, 1720.

Fig. 5 Savery's engine, 1700.

Steam engine developments continued and the earliest cylinder-and-piston unit was based on Papin's suggestion in 1690 that the condensation of steam should be used to make a vacuum beneath a piston, after the piston had been raised by expanding steam. Newcomen's atmospheric pressure engine made practical use of this principle.

While Papin neglected his own ideas of a steam engine to develop Savery's invention, Thomas Newcomen and his assistant John Cawley adapted Papin's suggestions in a practical engine. Years of experimentation ended with success in 1711 (Fig. 6). Steam admitted from the boiler to a cylinder raised a piston by expansion and assistance from a counterweight on the other end of a beam, actuated by the piston. The steam valve was then closed and the steam in the cylinder was condensed by a spray of cold water. The vacuum which was formed caused the piston to be forced downward by atmospheric pressure, doing work on a pump. Condensed water in the cylinder was expelled through a valve by the entry of steam which was at a pressure slightly above atmospheric. A 25 ft (7.6 m) oak beam, used to transmit power from the cylinder to the water pump, was a dominant feature of what came to be called the *beam engine*. The boiler used by Newcomen, a plain copper brewer's kettle, was known as the Haycock type. (See Fig. 4.)

Fig. 6 Newcomen's beam engine, 1711.

The key technical challenge remained the need for higher pressures, which meant a more reliable and stronger boiler. Basically, evolution of the steam boiler paralleled evolution of the steam engine.

During the late 1700s the inventor James Watt pursued developments of the steam engine, now physically separated from the boiler. Evidence indicates that he helped introduce the first *waggon boiler*, so named because of its shape (Fig. 7). Watt concentrated on the engine and developed the separate steam condenser to create the vacuum and also replaced atmospheric pressure with steam pressure, improving the engine's efficiency. He also established the measurement of horsepower, calculating that one horse could raise 550 lb (249 kg) of weight a distance of 1 ft (0.3 m) in one second, or 33,000 lb (14,969 kg) a distance of one foot in one minute.

Fire tube boilers

The next outstanding inventor and builder was Richard Trevithick, who had observed many pumping stations at his father's mines. He realized that the problem with many pumping systems was the boiler capacity. Whereas copper was the only material previously available, hammered wrought iron plates could now be used although the maximum length was 2 ft (0.6 m). Rolled iron plates became available in 1875.

In 1804 Trevithick designed a higher pressure engine, made possible by the successful construction of a high pressure boiler (Fig. 8). Trevithick's boiler design featured a cast iron cylindrical shell and dished end.

As demand grew further, it became necessary to either build larger boilers with more capacity or put up with the inconveniences of operating many smaller units. Engineers knew that the longer the hot gases were in contact with the shell and the greater the exposed surface area, the greater the capacity and efficiency.

While a significant advance, Newcomen's engine and boiler were so thermally inefficient that they were frequently only practical at coal mine sites. To make the system more widely applicable, developers of steam engines

Fig. 7 Waggon boiler, 1769.

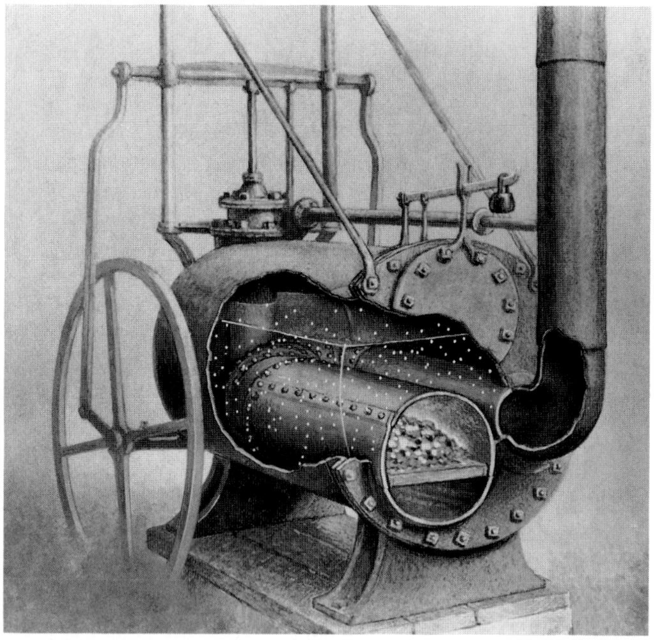

Fig. 8 Trevithick boiler, 1804.

began to think in terms of fuel economy. Noting that nearly half the heat from the fire was lost because of short contact time between the hot gases and the boiler heating surface, Dr. John Allen may have made the first calculation of boiler efficiency in 1730. To reduce heat loss, Allen developed an internal furnace with a smoke flue winding through the water, like a coil in a still. Then, to prevent a deficiency of combustion air, he suggested the use of bellows to force the gases through the flue. This probably represents the first use of forced draft.

Later developments saw the single pipe flue replaced by many gas tubes, which increased the amount of heating surface. These *fire tube* boilers were essentially the design of about 1870. However, they were limited in capacity and pressure and could not meet the needs that were developing for higher pressures and larger unit sizes. Also, there was the ominous record of explosions and personal injury because of direct heating of the pressure shell, which contained large volumes of water and steam at high temperature and pressure.

The following appeared in an 1898 edition of *Steam*: *That the ordinary forms of boilers* (fire tube boilers) *are liable to explode with disastrous effect is conceded. That they do so explode is witnessed by the sad list of casualties from this cause every year, and almost every day. In the year 1880, there were 170 explosions reported in the United States, with a loss of 259 lives, and 555 persons injured. In 1887 the number of explosions recorded was 198, with 652 persons either killed or badly wounded. The average reported for ten years past has been about the same as the two years given, while doubtless many occur which are not recorded.*

Inventors recognized the need for a new design — one which could increase capacity and limit the consequences of pressure part rupture at high pressure and temperature. *Water tube* boiler development began.

Early water tube design

A patent granted to William Blakey in 1766, covering an improvement in Savery's steam engine, includes a form of steam generator (Fig. 9). This probably was the first step in the development of the water tube boiler. However, the first successful use of a water tube design was by James Rumsey, an American inventor who patented several types of boilers in 1788 — some of which were water tube designs.

At about this time John Stevens, also an American, invented a water tube boiler consisting of a group of small tubes closed at one end and connected at the other to a central reservoir (Fig. 10). Patented in the United States (U.S.) in 1803, this boiler was used on a Hudson River steam boat. However, the design was short lived due to basic engineering problems in construction and operation.

Blakey had gone to England to obtain his patents, as there were no similar laws in North America. Stevens, a lawyer, petitioned the U.S. Congress for a patent law to protect his invention and such a law was enacted in 1790. It may be said that part of the basis of present U.S. patent laws grew out of the need to protect a water tube boiler design.

Fig. 11 shows another form of water tube boiler, this one patented in 1805 by John Cox Stevens. A boiler of this type can now be seen at the Smithsonian Institution, Washington, D.C.

In 1822, Jacob Perkins built a water tube boiler which is the predecessor of the once-through steam generator. A number of cast iron bars with longitudinal holes were arranged over the fire in three tiers by connecting the ends outside of the furnace with a series of bent pipes. Water was fed to the top tier by a feed pump and superheated steam was discharged from the lower tier to a collecting chamber.

Babcock & Wilcox

It was not until 1856, however, that a truly successful water tube boiler emerged. In that year, Stephen Wilcox introduced his version of the water tube design, with improved water circulation and increased heating surface (Fig. 12). Wilcox had designed a boiler with inclined water

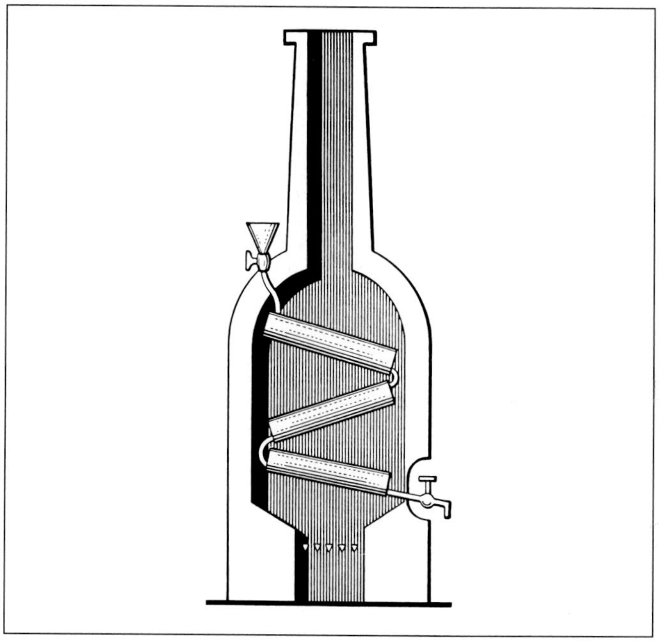

Fig. 9 William Blakey boiler, 1766.

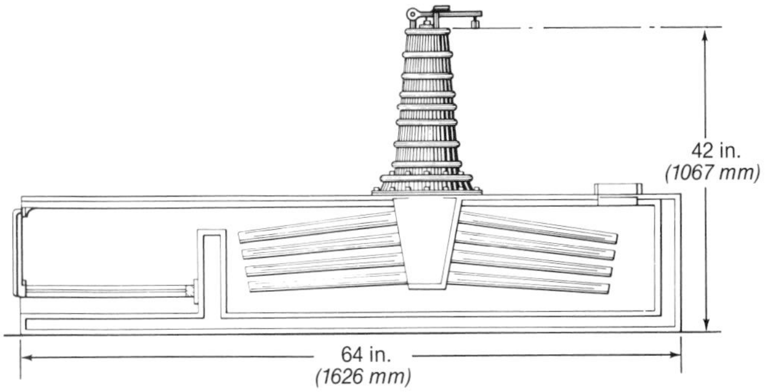

Fig. 10 John Stevens water tube boiler, 1803.

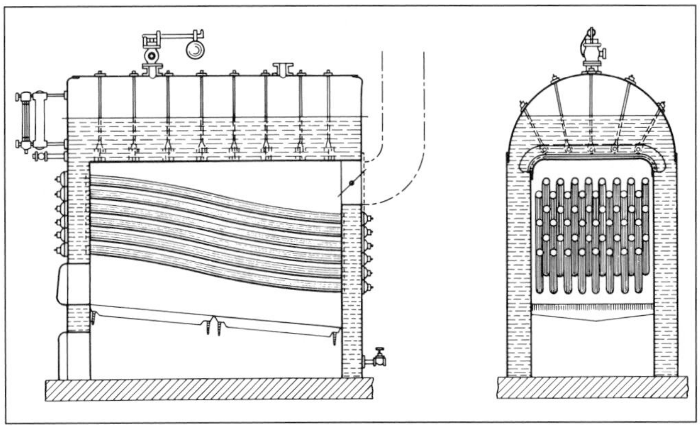

Fig. 12 Inclined water tubes connecting front and rear water spaces complete with steam space above. Stephen Wilcox, 1856.

tubes that connected water spaces at the front and rear, with a steam chamber above. Most important, as a water tube boiler, his unit was inherently *safe*. His design revolutionized the boiler industry.

In 1866, George Babcock became associated with Wilcox and the first Babcock & Wilcox (B&W) boiler was patented a year later (see Fig. 3).

Industrial progress continued. In 1876, a giant sized Corliss steam engine, a device invented in Rhode Island in 1849, went on display at the Centennial Exhibition in Philadelphia, Pennsylvania, as a symbol of worldwide industrial development. Also on prominent display was a 150 horsepower water tube boiler by George Babcock and Stephen Wilcox, two men now recognized as engineers of unusual ability (Fig. 13). Their professional reputation was high and their names carried prestige. By 1877, the B&W boiler had been modified and improved by the partners several times (Fig. 14).

At the exhibition, the public was awed by the size of the Corliss engine, which weighed 600 tons and had cylinders 3 ft (0.9 m) in diameter. But this giant size was to also mark the end of the steam engine, in favor of more efficient prime movers, such as the steam turbine. This transition would add impetus to further development of the B&W water tube boiler. By 1900 the steam turbine gained importance as the major steam powered source of rotary motion, due primarily to its lower maintenance costs, greater overloading tolerance, fewer number of moving parts and smaller size.

Perhaps the most visible technical accomplishments of the time were in Philadelphia and New York City. In 1881 in Philadelphia, the Brush Electric Light Company began operations with four boilers totaling 292 horsepower. The following year, in New York, Thomas Alva Edison threw the switch to open the Pearl Street Central station, ushering in the *age of the cities*. The boilers in Philadelphia, and the four used by Thomas Edison in New York, were built by B&W, now incorporated as a partnership. The boilers were heralded as *sturdy, safe and reliable*. When asked to comment on the units, Edison wrote: *These are the best boilers God has permitted man yet to make*. (Fig. 15).

The historic Pearl Street station opened with 59 customers using about 1300 lamps. The B&W boilers consumed 5 tons of coal and 11,500 gal (43,532 l) of water per day.

The B&W boiler of 1881 was a safe and efficient steam generator, ready for the part it would play in worldwide industrial development.

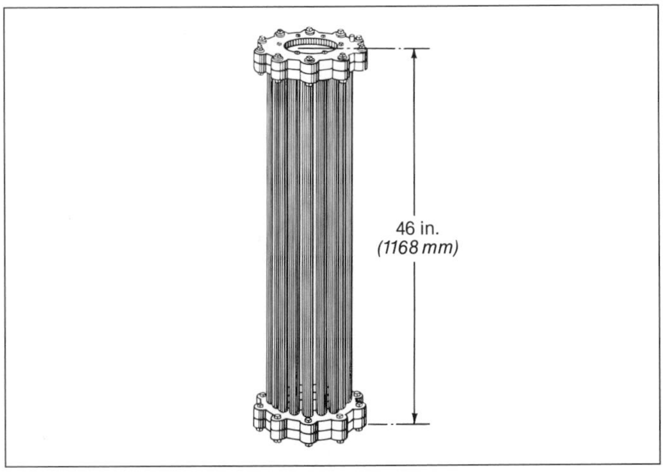

Fig. 11 Water tube boiler with tubes connecting water chamber below and steam chamber above. John Cox Stevens, 1805.

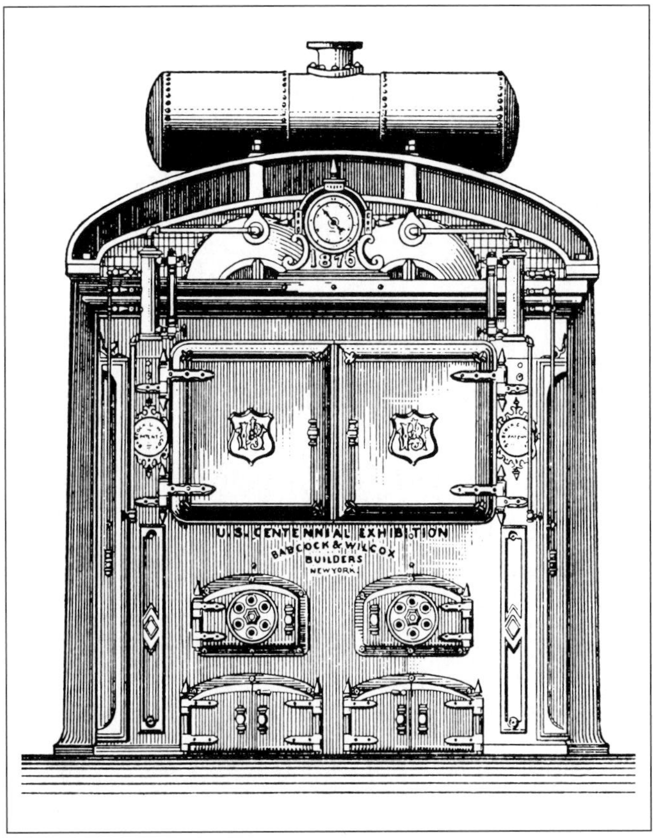

Fig. 13 Babcock & Wilcox Centennial boiler, 1876.

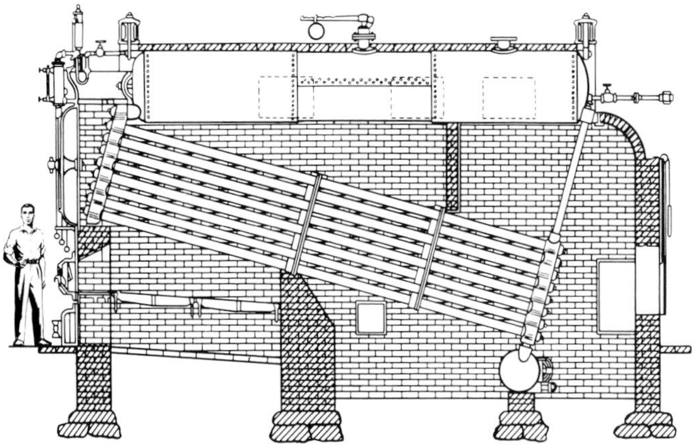

Fig. 14 Babcock & Wilcox boiler developed in 1877.

Water tube marine boilers

The first water tube marine boiler built by B&W was for the *Monroe* of the U.S. Army's Quartermaster department. A major step in water tube marine boiler design then came in 1889 with a unit for the steam yacht *Reverie*. The U.S. Navy then ordered, for three ships, a further improved design that saved about 30% in weight from previous designs. This design was further improved for a unit installed, in 1899, in the U.S. cruiser *Alert*, establishing the superiority of the water tube boiler for marine propulsion. In this installation, the firing end of the boiler was reversed, placing the firing door in what had been the rear wall of the boiler. The furnace was thereby enlarged in the direction in which combustion took place, greatly improving combustion conditions.

The development of marine boilers for naval and merchant ship propulsion has paralleled that for land use (see Fig. 16). Throughout the 20th century, dependable water tube marine boilers have contributed greatly to the excellent performance of naval and commercial ships worldwide.

Bent tube design

The success and widespread use of the inclined straight tube B&W boiler stimulated other inventors to explore new ideas. In 1880, Allan Stirling developed a design connecting the steam generating tubes directly to a steam separating drum and featuring low headroom above the furnace. The Stirling Boiler Company was formed to manufacture and market an improved Stirling design, essentially the same as shown in Fig. 17.

The merits of *bent tubes* for certain applications were soon recognized by George Babcock and Stephen Wilcox, and what had become the Stirling Consolidated Boiler Company in Barberton, Ohio was purchased by B&W in 1906. After the problems of internal tube cleaning were solved, the bent tube boiler replaced the straight tube design. The continuous and economical production of clean, dry steam, even when using poor quality feedwater, and the ability to meet sudden load swings were features of the new B&W design.

George Herman Babcock

George Herman Babcock was born June 17, 1832 near Otsego, New York. His father was a well known inventor and mechanic. When George was 12 years old, his parents moved to Westerly, Rhode Island, where he met Stephen Wilcox.

At age 19, George started the *Literary Echo*, editing the paper and running a printing business. With his father, he invented the first polychromatic printing press and he also patented a job press which won a prize at the London Crystal Palace International Exposition in 1855.

In the early 1860s, he was made chief draftsman of the Hope Iron Works at Providence, Rhode Island, where he renewed his acquaintance with Stephen Wilcox and worked with him in developing the first B&W boiler. In 1886, George Babcock became the sixth president of the American Society of Mechanical Engineers.

He was the first president of The Babcock & Wilcox Company, a position he held until his death in 1893.

Electric power

Until the late 1800s, steam was used primarily for heat and as a tool for industry. Then, with the advent of practical electric power generation and distribution, utility companies were formed to serve industrial and residential users across wide areas. The pioneer stations in the U.S. were the Brush Electric Light Company and the Commonwealth Edison Company. Both used B&W boilers exclusively.

During the first two decades of the 20th century, there was an increase in steam pressures and temperatures to 275 psi (19 bar) and 560F (293C), with 146F (81C) superheat. In 1921, the North Tess station of the Newcastle Electric Supply Company in Northern England went into operation with steam at 450 psi (31 bar) and a temperature of 650F (343C). The steam was reheated to 500F (260C) and regenerative feedwater heating was used to attain a boiler feedwater temperature of 300F (149C). Three years later, the Crawford Avenue station of the Commonwealth Edison Company and the Philo and Twin Branch stations of the present American Electric Power system were placed in service with steam at 550 psi (37.9 bar) and 725F (385C) at the turbine throttle. The steam was reheated to 700F (371C).

A station designed for much higher steam pressure, the Weymouth (later named Edgar) station of the Boston Edison Company in Massachusetts, began operation in 1925. The 3150 kW high pressure unit used steam at 1200 psi (82.7 bar) and 700F (371C), reheated to 700F (371C) for the main turbines (Fig. 18).

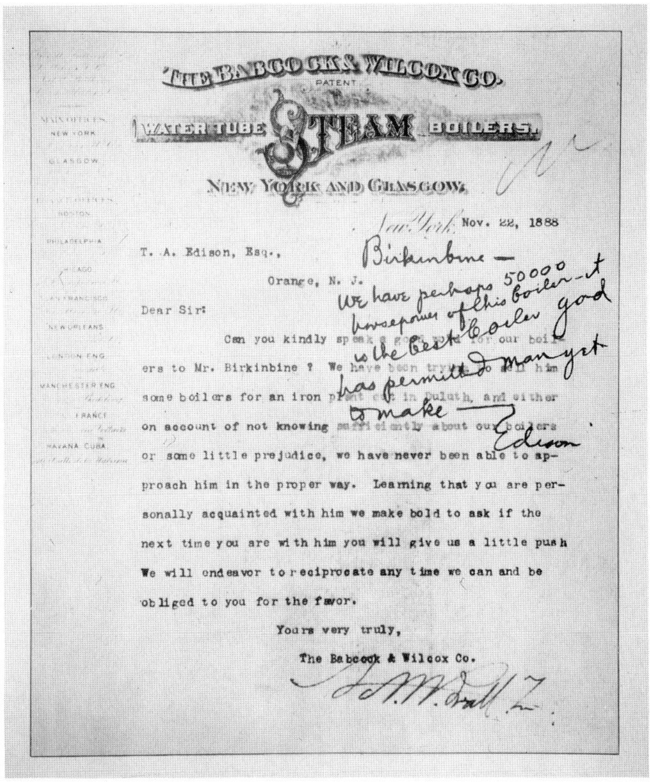
THE BABCOCK & WILCOX CO.
PATENT
WATER TUBE STEAM BOILERS.
NEW YORK AND GLASGOW.

New York, Nov. 22, 1888

T. A. Edison, Esq.,
Orange, N. J.

Dear Sir:

Can you kindly speak a good word for our boilers to Mr. Birkinbine ? We have been trying to sell him some boilers for an iron plant out in Duluth, and either on account of not knowing sufficiently about our boilers or some little prejudice, we have never been able to approach him in the proper way. Learning that you are personally acquainted with him we make bold to ask if the next time you are with him you will give us a little push. We will endeavor to reciprocate any time we can and be obliged to you for the favor.

Yours very truly,
The Babcock & Wilcox Co.

Birkinbine — We have perhaps 50000 horsepower of this boiler — it is the best Boiler god has permitted man yet to make — Edison

Fig. 15 Thomas Edison endorsement, 1888.

Stephen Wilcox

Stephen Wilcox was born February 12, 1839 at Westerly, Rhode Island.

The first definite information concerning his engineering activities locates him in Providence, Rhode Island, about 1849, trying to introduce a caloric engine. In 1853, in association with Amos Taylor of Mystic, Connecticut, he patented a letoff motion for looms. In 1856, a patent for a steam boiler was issued to Stephen Wilcox and O.M. Stillman. While this boiler differed materially from later designs, it is notable as his first recorded step into the field of steam generation.

In 1866 Wilcox, with George Babcock, developed the first B&W boiler which was patented the following year.

In 1869 he went to New York as selling agent for the Hope Iron Works and took an active part in improving the boiler and the building of the business. He was vice president of The Babcock & Wilcox Company from its incorporation in 1881 until his death in 1893.

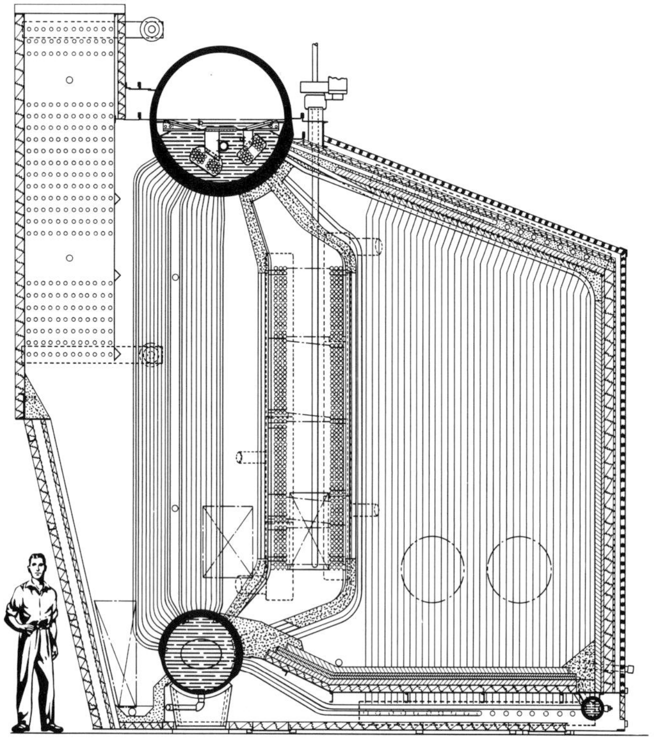

Fig. 16 Two drum Integral Furnace marine boiler.

Pulverized coal and water-cooled furnaces

The 1920s saw other major changes in boiler design and fabrication. Before that time, as power generating stations increased capacity, they increased the number of boilers. But attempts were being made to increase the size of the boilers. Soon, however, the size became such that existing furnace designs and methods of burning coal, primarily stokers, were no longer adequate.

Pulverized coal was the answer in achieving higher volumetric combustion rates and increased boiler capacity. This could not have been fully exploited without the use of water-cooled furnaces. Such furnaces eliminated the problem of rapid deterioration of the refractory walls due to slag (molten ash). Also, these designs lowered the temperature of the gases leaving the furnace and thereby reduced fouling (accumulation of ash) of convection pass heating surfaces to manageable levels.

Integral Furnace boiler

Water cooling was applied to existing boiler designs, with its circulatory system essentially independent of the boiler steam-water circulation. In the early 1930s, however, a new concept was developed in which the furnace water-cooled surface and the boiler surface were arranged together so that each was an integral part of the unit (Fig. 19).

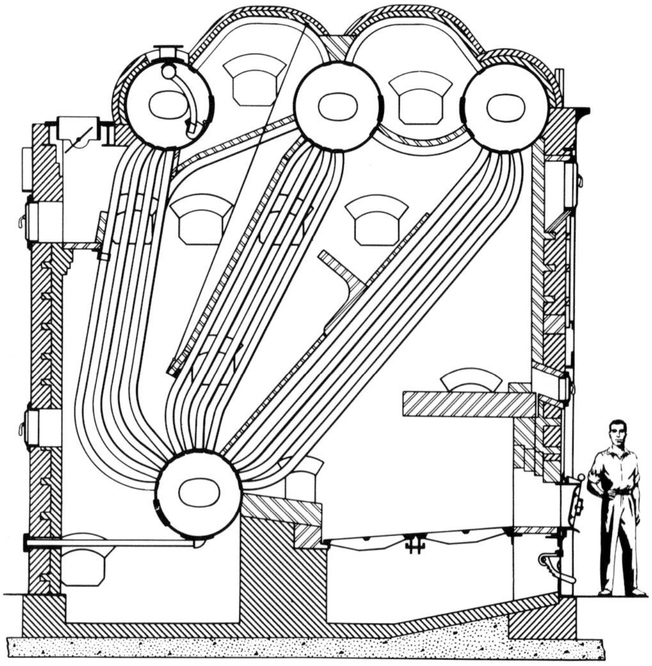

Fig. 17 Early Stirling boiler arranged for hand firing.

Requirements of a Perfect Steam Boiler - 1875

In 1875, George Babcock and Stephen Wilcox published their conception of the perfect boiler, listing twelve principles that even today generally represent good design practice:

1st. Proper workmanship and simple construction, using materials which experience has shown to be best, thus avoiding the necessity of early repairs.

2nd. A mud-drum to receive all impurities deposited from the water, and so placed as to be removed from the action of the fire.

3rd. A steam and water capacity sufficient to prevent any fluctuation in steam pressure or water level.

4th. A water surface for the disengagement of the steam from the water, of sufficient extent to prevent foaming.

5th. A constant and thorough circulation of water throughout the boiler, so as to maintain all parts at the same temperature.

6th. The water space divided into sections so arranged that, should any section fail, no general explosion can occur and the destructive effects will be confined to the escape of the contents. Large and free passages between the different sections to equalize the water line and pressure in all.

7th. A great excess of strength over any legitimate strain, the boiler being so constructed as to be free from strains due to unequal expansion, and, if possible, to avoid joints exposed to the direct action of the fire.

8th. A combustion chamber so arranged that the combustion of the gases started in the furnace may be completed before the gases escape to the chimney.

9th. The heating surface as nearly as possible at right angles to the currents of heated gases, so as to break up the currents and extract the entire available heat from the gases.

10th. All parts readily accessible for cleaning and repairs. This is a point of the greatest importance as regards safety and economy.

11th. Proportioned for the work to be done, and capable of working to its full rated capacity with the highest economy.

12th. Equipped with the very best gauges, safety valves and other fixtures.

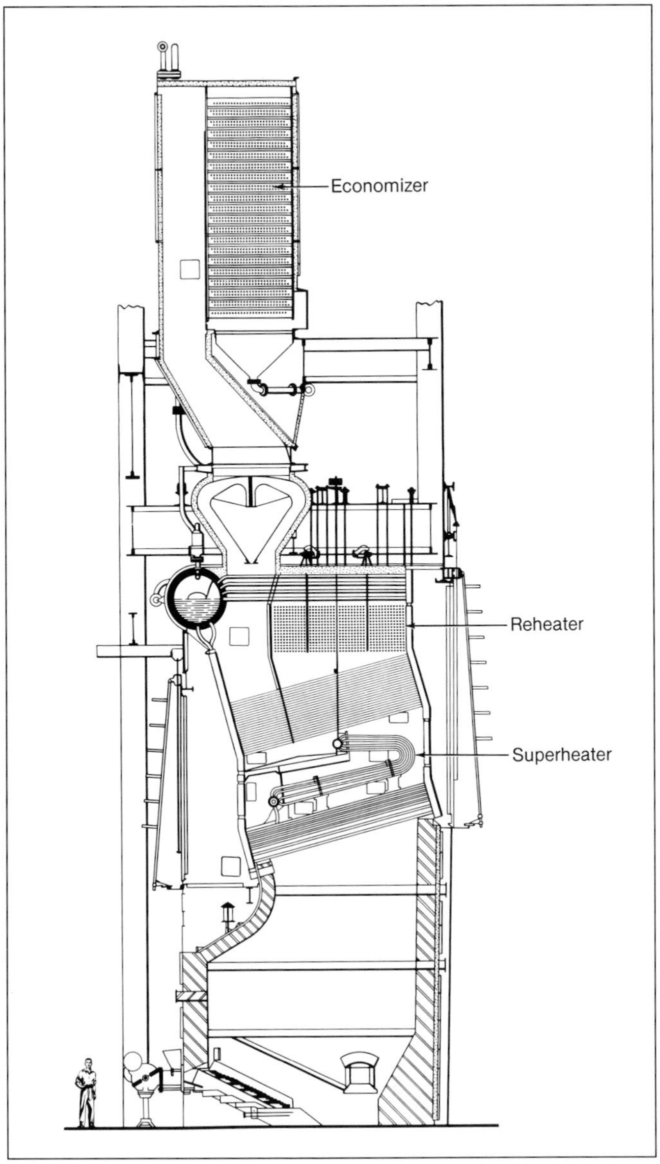

Fig. 18 High pressure reheat boiler, 1925.

Shop-assembled water tube boilers

In the late 1940s, the increasing need for industrial and heating boilers, combined with the increasing costs of field-assembled equipment, led to development of the shop-assembled *package* boiler. These units are now designed in capacities up to 600,000 lb/h (75.6 kg/s) at pressures up to 1800 psi (124 bar) and temperatures to 900F (482C).

Further developments

In addition to reducing furnace maintenance and the fouling of convection heating surfaces, water cooling also helped to generate more steam. Boiler tube bank surface was reduced because additional steam generating surface was available in the furnace. Increased feedwater and steam temperatures and increased steam pressures, for greater cycle efficiency, further reduced boiler tube bank surface, to be replaced by additional superheater surface.

As a result, units for steam pressures above 1200 psi (82.7 bar) consist essentially of furnace water wall tubes, superheaters and such heat recovery accessories as economizers and air heaters (Fig. 20). Units for lower pressures, however, have considerable steam generating surface in tube banks (boiler banks) in addition to the water-cooled furnace (Fig. 21).

Universal Pressure boilers

An important milestone in producing electricity at the lowest possible cost took place in 1957 — the first commercial operation of a boiler with steam pressure above the critical value of 3208 psi (221.2 bar). This 125 MW B&W Universal Pressure (UP) steam generator (Fig. 22), located at Ohio Power Company's Philo plant, delivered 675,000 lb/h (85 kg/s) steam at 4550 psi (313.7 bar); the steam was superheated to 1150F (621C) with two reheats to 1050 and 1000F (566 and 538C).

B&W built and tested its first once-through steam generator for 600 psi in 1916, and built an experimental 5000 psi unit in the late 1920s.

The UP boiler, so named because it can be designed for subcritical or supercritical operation, is capable of rapid load pickup. Increases in load rates up to 5% per minute are attained. If required, the unit can increase load from 25 to 90% of full load in four minutes and can go to 100% in another four minutes.

Fig. 23 shows a typical 1300 MW UP boiler rated at 9,775,000 lb/h (1232 kg/s) steam at 3845 psi (265 bar) and 1010F (543C) with reheat to 1000F (538C). In 1987, one of these B&W units, located in West Virginia, achieved 607 days of continuous operation, a record unmatched by any major coal-fired unit in the world.

Subcritical units, however, remained a standard throughout the world. Coal remained the dominant fuel because of its abundant supply in many countries. In Canada, for example, the province of Ontario was grow-

Fig. 19 Integral Furnace boiler, 1933.

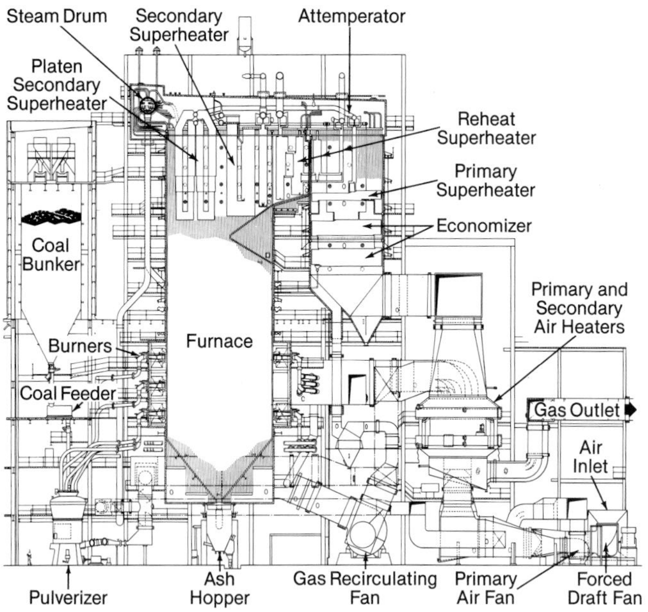

Fig. 20 Typical B&W Radiant utility boiler.

ing quickly in the 1960s and 1970s with a large demand for power. B&W supplied all eight 500 MW units at the landmark Nanticoke generating station, which remains the largest operating coal-fired plant in the world today.

Other fuels and systems

B&W has continued to produce power from an ever widening array of fuels in an increasingly clean and environmentally acceptable manner. Landmark developments by B&W have been atmospheric fluidized-bed combustion installations, both bubbling bed and circulating, for reduced emissions. Related to this technology, a pilot project for pressurized fluidized-bed combustion became operational in 1992, paving the way for larger units in the U.S.

Waste-to-energy also became a major effort worldwide, and B&W has installed both mass burn and refuse-derived fuel units to meet this growing demand for waste disposal and electric power generation. B&W installed the world's first waste-to-energy boiler in 1972. In the U.S., B&W has now installed more waste-to-energy units than any other supplier.

For the paper industry, B&W installed the first chemical recovery boiler in the U.S. in 1940. Since that time, B&W has developed a long tradition of firsts in this industry and has now installed the largest black liquor chemical recovery units operating in the world today.

Modified steam cycles

High efficiency cycles involve combinations of gas turbines and steam power in cogeneration, and direct thermal to electrical energy conversion. One direct conversion system includes using conventional fuel or char byproduct from coal gasification or liquefaction.

Despite many complex cycles devised to increase overall plant efficiency, the conventional steam cycle remains the most economical. The increasing use of high steam pressures and temperatures, reheat superheaters, economizers and air heaters has led to improved efficiency in the modern steam power cycle.

Nuclear power

Since 1942, when Enrico Fermi demonstrated a controlled self-sustaining reaction, nuclear fission has been recognized as an important source of heat for producing steam for power generation. The first significant application of this new source was the land-based prototype reactor for the U.S.S. *Nautilus* submarine (Fig. 24), operated at the National Reactor Testing Station in Idaho in the early 1950s. This prototype reactor, designed by B&W, was also the basis for land-based pressurized water reactors now being used for electric power generation worldwide. B&W has continued its active involvement in both the naval and land-based programs.

The first nuclear electric utility installation was the 90 MW unit at the Shippingport atomic power station in Pennsylvania. This plant, owned partly by Duquesne Light Company and partly by the U.S. Atomic Energy Commission, began operations in 1957.

Spurred by the trend toward larger unit capacity,

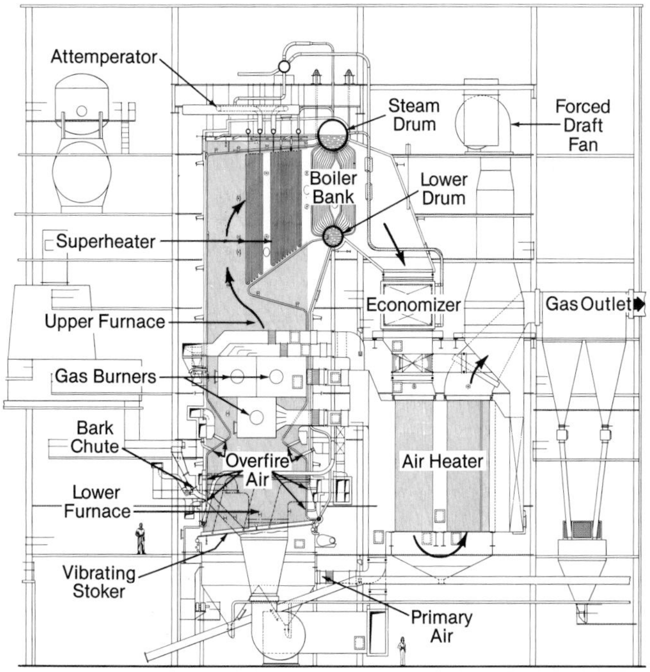

Fig. 21 Lower pressure Stirling® design.

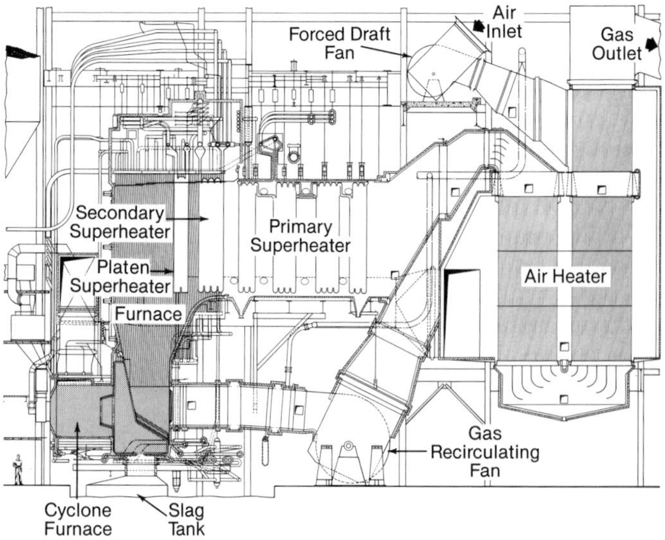

Fig. 22 125 MW B&W Universal Pressure boiler, 1957.

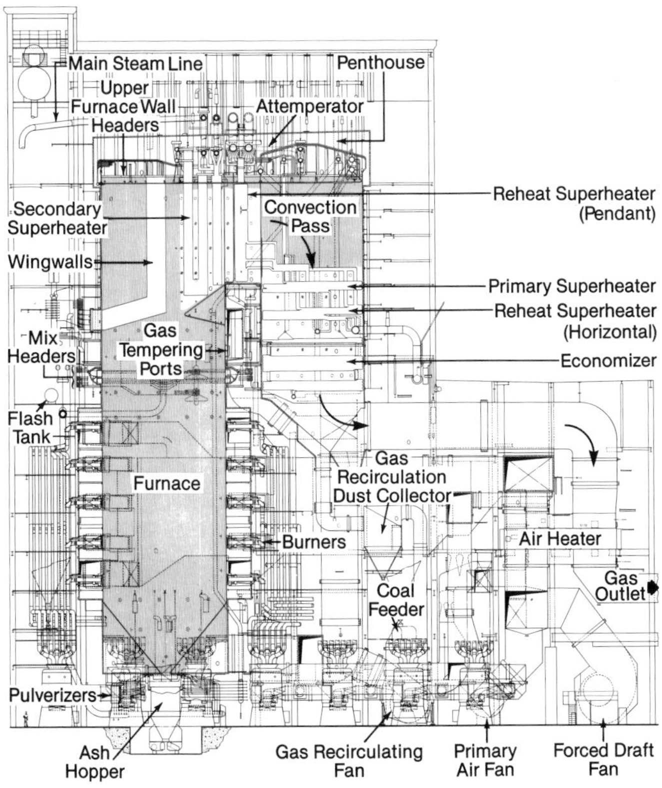

Fig. 23 1300 MW B&W Universal Pressure boiler.

developments in the use of nuclear energy for electric power reached a milestone in 1967 when, in the U.S., nuclear units constituted almost 50% of the 54,000 MW of new capacity ordered that year. Single unit capacity designs have reached 1300 MW. Activity regarding nuclear power was also active outside the U.S., especially in Europe. Development continues today, but at a slower pace. Still, by 1990 there were, in the U.S., 107 reactors operating, 18 under construction and two additional units planned. Forty-four of the operating units had net capacities greater than 1000 MW.

Throughout this period the nuclear power program in Canada continued to develop based on a design (CANDU)

Fig. 24 U.S.S. *Nautilus*, world's first nuclear-powered ship.

Fig. 25 B&W nuclear steam generator (lower section), ordered to replace a competitor's unit.

that rated high in both availability and dependability. By 1991 there were 23 units in Canada, all with B&W nuclear steam generators, and an additional four units operating outside of Canada.

The B&W recirculating steam generators in these units have continually held the best performance records in the world. Based on tube leaks, their record is a factor of 1000 times better than the next largest supplier of recirculating nuclear steam generators. (See Fig. 25.)

The future of nuclear power is somewhat uncertain as issues of plant operating safety and long term waste disposal are being resolved. However, nuclear power continues to offer one of the least polluting forms of large scale power generation available.

Materials and fabrication

Pressure parts for water tube boilers were originally made of iron and later of steel. Now, steam drums and nuclear pressure vessels are fabricated from heavy steel plates and steel forgings joined by welding. The development of the steam boiler has been necessarily concurrent with advances in metallurgy and progressive improvements in the fabrication and welding of steel and steel alloys.

The cast iron generating tubes used in the first B&W boilers were later superseded by steel tubes. Shortly after 1900, B&W developed a commercial process for the manufacture of hot finished seamless steel boiler tubes, combining strength and reliability with reasonable cost. In the midst of World War II, B&W completed a mill to manufacture tubes by the electric resistance welding process. This tubing has now been used in thousands of steam generating units throughout the world.

The cast iron tubes used for steam and water storage in the original B&W boilers were soon replaced by drums, and by 1888 drum construction was improved by changing from wrought iron to steel plates rolled into cylinders.

Before 1930, riveting was the standard method of joining boiler drum plates. Drum plate thickness was limited to about 2.75 in. (70 mm) because no satisfactory method was known to secure a tight joint in thicker plates. The only alternative available was to forge and machine a solid ingot of steel into a drum, which was an extremely expensive process. This method was only used on boilers operating at what was then considered high pressure, above 700 psi (48.3 bar).

The story behind the development of fusion welding was one of intensive research activity beginning in 1926. Welding techniques had to be improved in many respects. Equally, if not more important, an acceptable test procedure had to be found and instituted that would guarantee the drum without destroying it in the test. After extensive investigation of various testing methods, it was decided in 1929 to adapt the medical x-ray machine to production examination of welds. By x-ray examination, together with physical tests of samples of the weld material, the soundness of the weld could be determined without affecting the drum.

In 1930, the U.S. Navy adopted a specification for construction of welded boiler drums for naval vessels. In that same year the first welded drums ever accepted by an engineering authority were a part of the B&W boilers installed in several naval cruisers. Also in 1930, the Boiler Code Committee of the American Society of Mechanical Engineers (ASME) issued complete rules and specifications for the fusion welding of drums for power boilers, and in 1931 B&W shipped the first welded power boiler drum built under this code.

The x-ray examination of welded drums and the rules declared for the qualification of welders and the control of welding operations were major first steps in the development of modern methods of quality control in the boiler industry. Quality assurance has received additional stimulus from the naval nuclear propulsion program and more recently from the U.S. Nuclear Regulatory Commission in connection with the licensing of nuclear plants for power generation.

International advancement

Since 1978, B&W has been part of McDermott International, Inc., a worldwide energy services company and a leader in products and services for the oil and gas industry. This relationship has strengthened the technical and management base of both B&W and McDermott, and has opened new opportunities in the world market.

Quality and technology

B&W established its first quality program in 1867. Only through continually rededicating itself to quality has the company survived the past 125 years in a changing and increasingly competitive environment.

As mentioned earlier, the names George Babcock and Stephen Wilcox carried great esteem in the engineering field at the turn of the century. Since that beginning, B&W has remained committed to attracting and employing people who are experts at what they do, who know how their work affects others and the company, who are willing and able to be measured for their performance, and who strive to find new ways to improve their work and the company's products and services.

With McDermott, B&W has established companywide Total Quality Management to establish an improved base for future operations.

B&W licensing and joint venture activities

In 1881, shortly after the formal incorporation of The Babcock & Wilcox Company, the treasurer, Nathan W. Pratt, traveled to Scotland to extend the company's activities in Europe. Concentrating on equipment for the sugar industry (see Fig. 26), Pratt also journeyed to Belgium and France to begin business discussions.

What he set up in Europe would become licensees of the U.S. technology. The British company would in turn set up licensees in other parts of the world including Spain, Germany and Japan. However, the technology base would remain that of George Babcock and Stephen Wilcox, who now had offices in New York City.

Many licensees began using the name Babcock as part of their company identity. However, the name Wilcox was disallowed for most license agreements. Although B&W soon set up manufacturing capability in the U.S., it has continued to license parts of its technology to other companies worldwide. One result has been confusion in the world market due to the widespread use of the Babcock name. On the positive side, the use shows that the name is highly respected in the industry and has carried prestige for more than a century and a quarter.

In the 1980s, B&W launched a program to establish equity partnership joint venture companies in various countries around the world, designed to enhance and complement the company's marketing efforts. The new venture companies carried both the Babcock and the Wilcox names, the earliest being P.T. Babcock & Wilcox Indonesia, established in 1985 with facilities on Batam Island and offices in Jakarta.

The second equity venture was established in Beijing, People's Republic of China, becoming operational in 1986. This venture is called Babcock & Wilcox Beijing Company, Ltd. By 1990 two more ventures were established, in cooperation with licensees. Thermax Babcock & Wilcox Ltd. was formed in Pune, India in 1989. The following year Babcock & Wilcox Gama Kazan Teknolojisi A.S. was formed in Ankara, Turkey.

As each venture was established, the first and most important task was ASME Code approval for each manufacturing facility. All joint venture manufacturing operations are certified.

Through these efforts and through an impressive list

Fig. 26 Boilers for the sugar industry, 1874.

of past and present international contracts, B&W has assumed a strengthened direct involvement role in the worldwide steam generation industry.

Research and development

Since the founding of The Babcock & Wilcox Company a century and a quarter ago, research and development have played important roles in B&W's continuing service to the power industry. From the initial improvements of Wilcox's original safety water tube boiler to the first supercritical pressure boilers, from the first privately operated nuclear research reactor to today's advanced environmental systems, innovation and new ideas of its employees have placed B&W at the forefront of steam generation and energy conversion technology. Today, research and development activities span the entire corporate structure from the groups at various manufacturing plants and engineering offices to the Research and Development division, which focuses on tomorrow's product and process requirements.

A key to the continued success of B&W is the ability to bring together cross-disciplinary research teams of experts from the many technical specialties in the steam generation field. These are combined with state of the art test facilities and computer systems.

The Research and Development division is an important element in this effort. The division has two research centers — one in Alliance, Ohio and one in Lynchburg, Virginia. The Alliance Research Center uses equipment designed specifically for research programs in all aspects of fossil power development, nuclear steam systems, material development and evaluation, and manufacturing technology. The Lynchburg Research Center's activities include development work in such areas as ceramics and nondestructive testing. Research focuses upon areas of central importance to Babcock & Wilcox and steam power generation. However, partners in these research programs have grown to include the U.S. Department of Energy and Defense, the Environmental Protection Agency, public and private research institutes, state governments, and electric utilities.

Key areas of current research include environmental protection, fuels and combustion technology, heat transfer and fluid mechanics, materials and manufacturing technologies, structural analysis and design, and fuels and water chemistry, as well as measurement and monitoring technology.

Environmental protection

Environmental protection is a key element in all modern steam producing systems where low cost steam and electricity must be produced with a minimum impact on the environment. Air pollution control [predominantly limiting nitrogen oxides (NO_x), sulfur dioxide (SO_2) and particulate emissions] is a key issue for all combustion processes, and B&W has been a leader in this area. Several generations of low NO_x burners and combustion technology for coal-, oil- and gas-fired systems have been developed, tested and patented in Alliance. These have met all of the requirements imposed by the U.S. Clean Air Act and its revisions, reducing NO_x by 50 to 70% from uncontrolled levels. Ongoing research and testing in several combustion facilities are being combined with fundamental studies and numerical modeling to produce the ultra-low NO_x emission burners of tomorrow.

Extensive research efforts have been underway since the early 1970s to reduce SO_2 emissions. These have included both combustion modifications and postcombustion removal. Research in the 1970s aided in the development of B&W's wet SO_2 scrubbing system controlling emissions from more than 15,000 MW of boiler capacity. Current research focuses on improved removal and operational efficiency. Advanced SO_2 control systems research has focused on dry scrubbing processes where the waste product is a dry powder ready for disposal. Key technologies include slurry and water atomization and dry sorbent injection which have been developed through pilot testing, flow modeling and numerical modeling. Major pilot facilities permit the testing of in-furnace injection, in-duct injection, and dry scrubber systems as well as gas conditioning and combined SO_2, NO_x and particulate control. (See Fig. 27.)

Since 1975, B&W has been in the forefront of fluidized-bed combustion (FBC) technology which offers the ability to simultaneously control SO_2 and NO_x formation as an integral part of the combustion process. This work led to the first large scale (20 MW) bubbling-bed system installation in the U.S. The Research and Development division features three separate experimental FBC units as research tools for this technology. These units provide information on limestone utilization and performance characteristics of various fuels and sorbents. (See Fig. 28.) The larger unit (2.5 MW_t) has bridged the gap between smaller test units and commercialization of the technology.

Additional areas of ongoing environmental research include coal benefication, water purification, air toxin emissions, and the characterization and control of a variety of process waste streams. Many consumer products and process byproducts are classified as hazardous, radioactive or mixed waste. B&W is actively investigating new processes to reuse these materials or convert them into more benign products for disposal.

Fig. 27 Utility boiler with dry scrubber and baghouse.

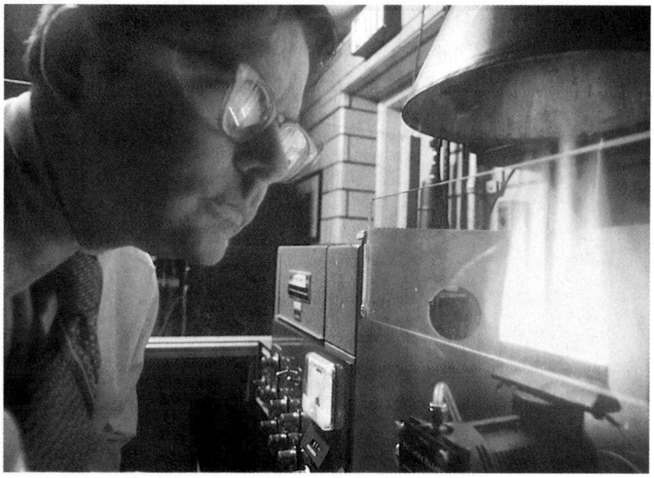

Fig. 28 Atomic absorption test for limestone utilization.

Fuels and combustion technology

A large number of fuels have been used to generate steam. This is even more true today as an ever widening and varied supply of waste and byproduct fuels, such as municipal refuse, coal mine tailings and biomass wastes, join coal, oil and gas to meet steam production needs. These fuels must be burned and their combustion products successfully handled while meeting two key trends: 1) declining fuel quality (lower heating value and poorer combustion), and 2) more restrictive emissions limits.

Major strengths of B&W and the Research and Development division have been: 1) the characterization of fuels and their ashes, 2) combustion of difficult fuels, and 3) effective heat recovery from the products of combustion (Fig. 29). B&W has earned international recognition for its fuels analysis capabilities which are based upon generally accepted as well as specialized B&W procedures. Detailed analyses include: heating value, chemical constituents, grindability, abrasion resistance, erosiveness, ignition, combustion characteristics, ash composition/viscosity/strength/fusion temperature, particle size and resistivity, among others. The results of these tests assist in pulverizer specification and design, internal boiler dimension selection, efficiency calculations, predicted unit availability, ash removal system design, sootblower placement and precipitator performance evaluation. Thousands of coal and ash samples have been analyzed and catalogued, forming part of the basis for B&W's design methods.

Combustion and fuel preparation facilities are maintained that can test a broad range of fuels at large scale. The 6×10^6 Btu/h (1.8 MW_t) Small Boiler Simulator (Fig. 30) permits a simulation of the time-temperature history of the entire combustion process. The subsystems include a vertical test furnace; fuel subsystem for pulverizing, collecting and firing solid fuels; fuel storage and feeding; gas and stack particulate analyzers for O_2, CO, CO_2 and NO_x; and instrumentation for solids grinding characterization.

This facility is supplemented by a 50×10^6 Btu/h (14.7 MW_t) boiler for combustion system scaleup.

The FBC facilities discussed above provide means to test and evaluate fuels under operating conditions. Supporting research is also conducted on pressurized fluidized-bed combustion technology. This promising approach to combined SO_2 and NO_x control can also offer improved thermodynamic cycle efficiency.

Research continues in the areas of gas side corrosion from refuse combustion, boiler fouling and cleaning characteristics, combustion of sewage sludge, advanced pulp and paper black liquor combustion, the use and application of coal-water and coal-oil fuels, and coal gasification.

Heat transfer and fluid dynamics

Heat transfer is a critical technology in the design of steam generation equipment. For many years B&W has been conducting heat transfer research from hot gases to tube walls and from the tube walls to enclosed water, steam and air. Early in the 1950s research in heat transfer and fluid mechanics was initiated in the supercritical pressure region above 3208 psi (221.2 bar). This work was the technical foundation for the large number of supercritical pressure once-through steam generators currently in service in the electric power industry.

A key advancement in steam-water flow was the invention of the ribbed tube, patented by B&W in 1960. By preventing deterioration of heat transfer under many flow conditions (called critical heat flux or departure from nucleate boiling), the internally ribbed tube made possible the use of natural circulation boilers at virtually all pressures up to the critical point. Extensive experimental studies have provided the critical heat flux data necessary for the design of boilers with both ribbed and smooth bore tubes.

Closely related to heat transfer, and of equal importance in steam generating equipment, is fluid mechanics. Both

Fig. 29 Radiant boiler flow model: one-sixteenth scale model of 650 MW boiler used at B&W research center to confirm optimum air and gas flow distribution for maximum combustion efficiency.

low pressure fluids (air and gas in ducts and flues) and high pressure fluids (water, steam-water mixtures, steam and fuel oil) must be investigated. The theories of single phase fluid flow are well understood, but the application of theory to the complex, irregular and multiple parallel path geometry of practical situations is often difficult and sometimes impossible. In these cases analytical procedures must be supplemented or replaced by experimental methods. If reliable extrapolations are possible, economical modeling techniques can be used. Where extrapolation is not feasible, large scale testing at full pressure, temperature and flow rate is needed.

Advances in numerical modeling technology have for the first time made possible the evaluation of the complex three dimensional flow, heat transfer and combustion processes in coal-fired boiler furnaces. Continuing development and validation of these models will enhance new boiler designs and expand applications. These models are also valuable tools in the design and evaluation of combustion processes, pollutant formation and environmental control equipment.

Research in these areas continues today including advanced studies in boiling heat transfer, steam-water separation, extended surface heat transfer, gas-solids flows, two-phase flow and stability, thermal/fluid system analysis, and large scale component testing. Major test facilities supporting these efforts include a 10 MW heat transfer apparatus for heat transfer and critical heat flux testing of steam-water flows as well as large, high pressure steam-water loops for performance evaluation of steam generators and equipment. Cold water flow loops [1000 and 4000 GPM (63 and 252 l/s)] are used for improving the hydraulic performance of fossil and nuclear components. Cold air and air particulate flow loops permit scale testing of components. (See Fig. 31.)

Fig. 30 Small boiler simulator.

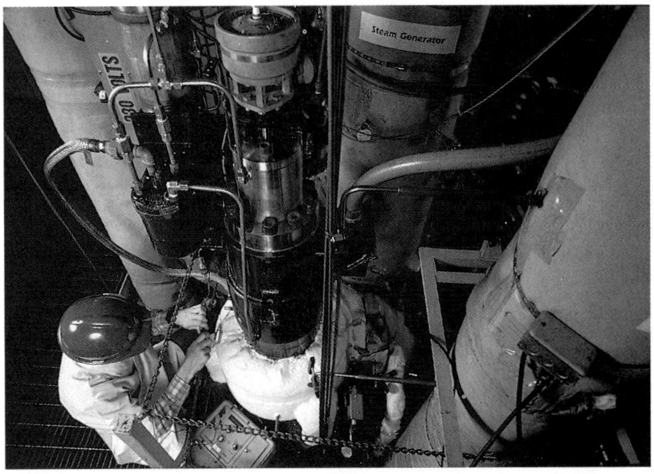

Fig. 31 Multi-loop integral systems test facility for steam-water flow.

Materials and manufacturing technologies

Because advanced steam producing and energy conversion systems require the application and fabrication of a wide variety of carbon, alloy and stainless steels, nonferrous metals and nonmetallic materials, an experienced metallurgical and materials science research staff equipped with the finest investigative tools is essential. Areas of primary interest in the metallurgical field are fabrication processes such as welding, room temperature and high temperature material properties, resistance to corrosion properties, wear resistance properties, and changes in such material properties under various operating conditions. Development of oxidation-resistant alloys, which retain strength at high temperature, and determination of short term and long term high temperature properties permitted the increase in steam temperature which has been of critical importance in increasing power plant efficiency and reducing the cost of producing electricity.

Advancements in manufacturing areas have included a process to manufacture large pressure components entirely from weld wire, designing a unique manufacturing process for bimetallic tubing, using pressure forming to produce metallic heat exchangers, developing air blown ultra-high temperature fibrous insulation, and combining sensor and control capabilities to improve quality and productivity of manufacturing processes.

Advanced materials are also studied for use in space and defense applications. Studies include ceramic composites, metal matrix composites, high density fuels and high strength metals. (See Fig. 32.)

The Research and Development division uses extensive facilities to study materials processing, joining processes, process metallurgy, analytical and physical metallurgical examination, and mechanical testing. The results are then directly applied to product improvement.

Structural analysis and design

The complex geometries and high stresses under which metals must serve in many products require careful study to allow prediction of stress distribution and intensity. Applied mechanics, a discipline with highly

Fig. 32 Lab equipment for space shuttle case material.

sophisticated analytical and experimental techniques, can provide designers with calculation methods and other information to assure the safety of structures and reduce costs by eliminating unnecessarily conservative design practices. The analytical techniques involve advanced mathematical procedures and computational tools as well as the use of advanced computers. An array of experimental tools and techniques are used to supplement these powerful analytical techniques.

Computational finite element analysis has largely displaced experimental measurement for establishing detailed local stress relationships. B&W has developed and applied some of the most advanced computer programs in the design of components for the power industry. Advanced techniques permit the evaluation of stresses resulting from component response to thermal and mechanical (including vibratory) loading.

Fracture mechanics, the evaluation of crack formation and growth, is an important area where analytical techniques and new experimental methods permit a better understanding of failure modes and the prediction of remaining component life. This branch of technology has contributed to the feasibility and safety of advanced designs in many types of equipment.

To provide part of the basis for these models, extensive computer controlled experimental facilities allow the assessment of mechanical properties for materials under environments similar to those in which they will operate. Some of the evaluations include tensile and impact testing, fatigue and corrosion fatigue, fracture toughness, as well as environmentally assisted cracking.

Fuel and water chemistry

Chemistry plays an important role in supporting the effective operation of steam generating systems. Therefore, a strong and diversified chemistry group is essential to support research, development and engineering. The design and operation of fuel burning equipment must be supported by capabilities in the analysis of a wide variety of solid, liquid and gaseous fuels and their products of combustion, and characterization of their behavior under various conditions. Long term operation of fossil and nuclear steam generating equipment requires extensive programs including high purity water analysis, water treatment and water purification. Equipment must also be chemically cleaned at intervals to remove water side deposits.

To develop customized programs to meet specific needs, B&W maintains a leadership position in these areas through extensive analytical chemistry laboratories and staff for fuels characterization, water chemistry and chemical cleaning. Studies focus on water treatment, production and measurement of ultra-high purity water (parts per billion), water side deposit analysis, corrosion product transport and on-line chemistry monitoring systems.

B&W is involved in the introduction of oxygenated water treatment to U.S. utility applications. Specialized chemical cleaning evaluations are conducted to prepare cleaning programs for utility boilers, industrial boilers and nuclear steam generators. Special analyses are frequently required to develop boiler specific cleaning solvent solutions which will remove the desired deposits without damaging the equipment. B&W has developed, in cooperation with the Electric Power Research Institute, a special solvent for the removal of magnetite from nuclear steam generators. A key element today includes developing the appropriate means to dispose of the cleaning solvents in an environmentally safe manner following use. This need frequently impacts and sometimes controls initial solvent selection.

Measurements and monitoring technology

Development, evaluation and accurate assessment of modern power systems require increasingly precise measurements in difficult to reach locations, often in hostile environments. To meet these demanding needs, B&W has maintained extensive research activities in specialized sensors, measurement and nondestructive examination. B&W's Research and Development division performs research in the areas of ultrasonic, eddy current, acoustic emissions, radiography, laser spectroscopy, optic and fiber optic technology, electro-optic systems, and electro and optic sensors. It also investigates advanced inspection techniques such as ultrasonic imaging, deep-penetrating eddy current, microfocus real time radiography, x-ray tomography and acoustic microscopy.

These techniques have been used to aid in laboratory research such as void fraction measurements for steam-water flows. They have also been applied to operating steam generating systems. New methods have been introduced by B&W to nondestructively measure oxide thicknesses on the inside of boiler tubes, detect hydrogen damage, and detect and measure corrosion fatigue cracks. Acoustic pyrometry systems have been introduced by B&W to nonintrusively measure high temperature gases in boiler furnaces. An on-line real time noncontact method to measure lignin concentration in wood pulp is being introduced. In a related area, advanced techniques are being developed to monitor and predict the life of ceramic components to enhance the application of these high temperature, erosion and corrosion resistant materials to industrial applications.

Research and development specialists have also applied fiber optic technology to solve many measurement problems. Unique advantages of this technology are freedom from electromagnetic interference, the ability to function in severe chemical and electric environments not possible with many electronic sensors, and improved safety in areas with potentially explosive atmospheres.

Among the division's fiber optic developments are a fiber optic pressure transducer, fiber optic accelerometer capable of operation with high electromagnetic interference and in potentially explosive environments, and a probe for viewing fire side boiler components.

Steam/its generation and use

This updated and expanded volume provides a broad, in-depth look at steam generating technology and equipment, plus related auxiliaries that are of interest to engineers and students in the steam power industry. These include discussions of the fundamental technologies such as thermodynamics, fluid mechanics, heat transfer, solid mechanics, materials science and fuels science. The various components of the steam generating equipment, plus their integration and performance evaluation, are covered in depth. Key elements of the balance of the steam generating system life including operation, inspection, maintenance, upgrade and life extension are also discussed.

Selected Color Plates — 40th Edition

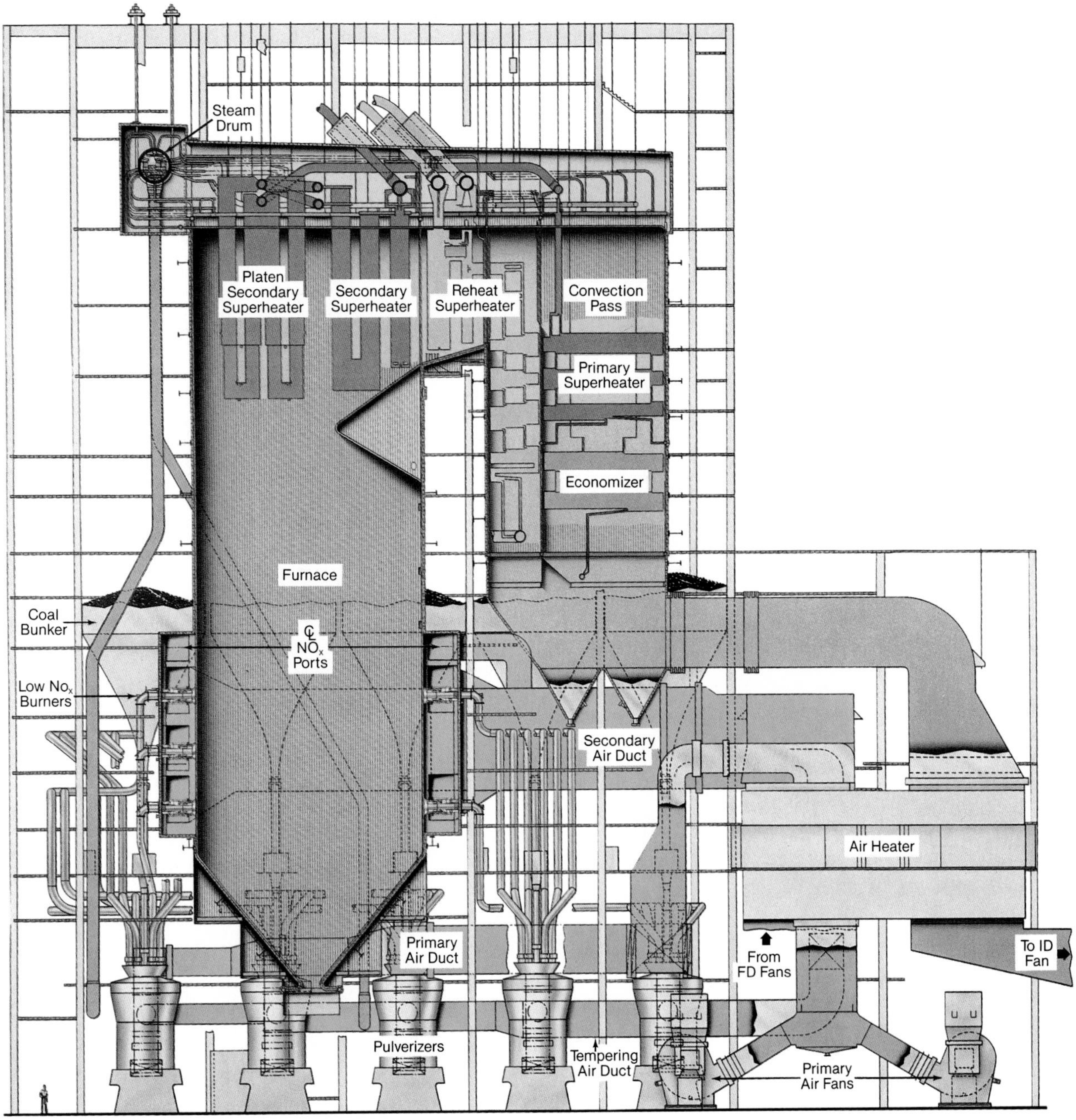

Carolina-type 455 MW Radiant boiler for pulverized coal.

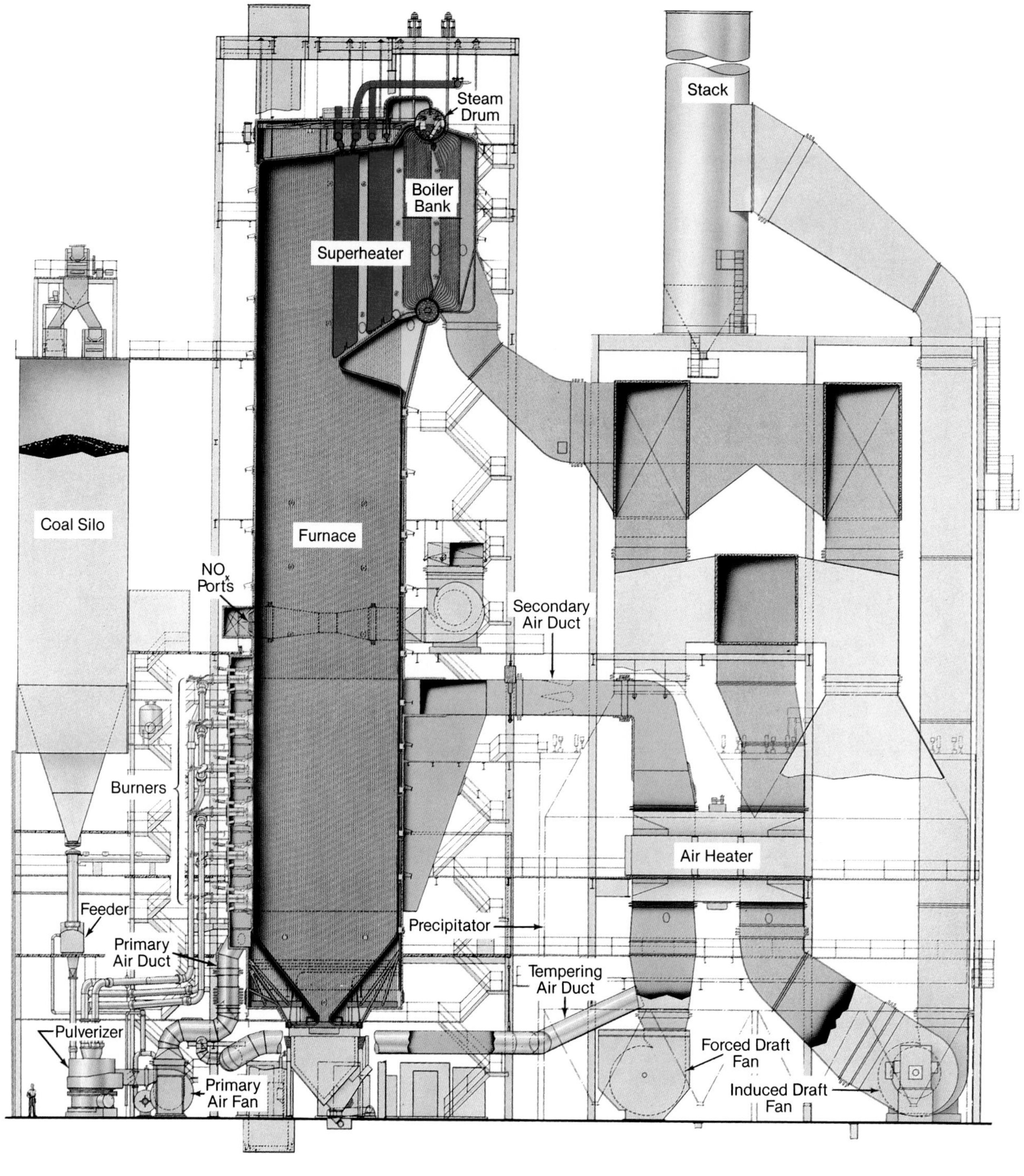

Large coal- and oil-fired two-drum Stirling® power boiler for industry, 885,000 lb/h (112 kg/s) steam flow.

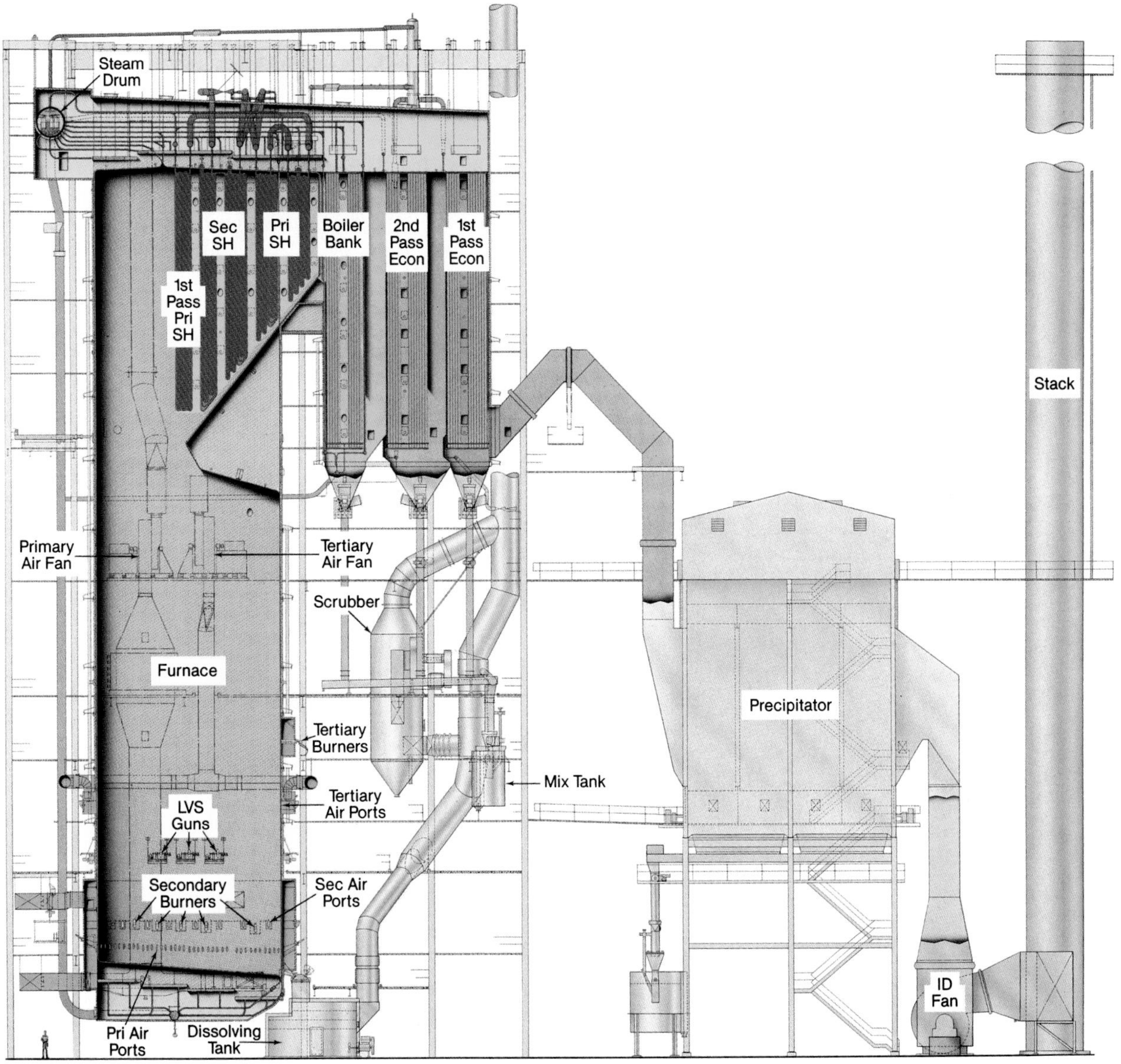

Single-drum chemical recovery boiler for the pulp and paper industry.

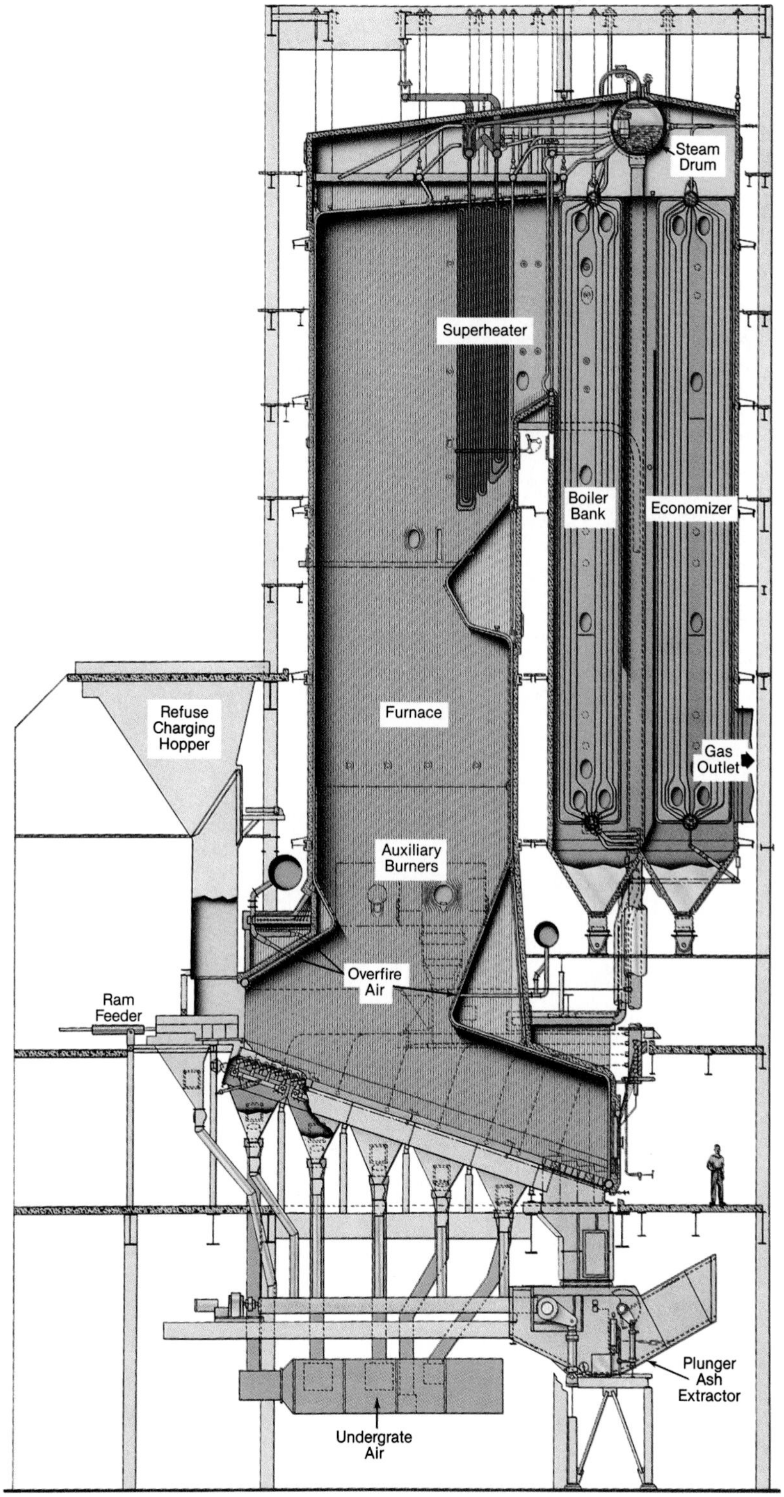

Refuse-to-energy boiler for mass burning, 1000 tons per day.

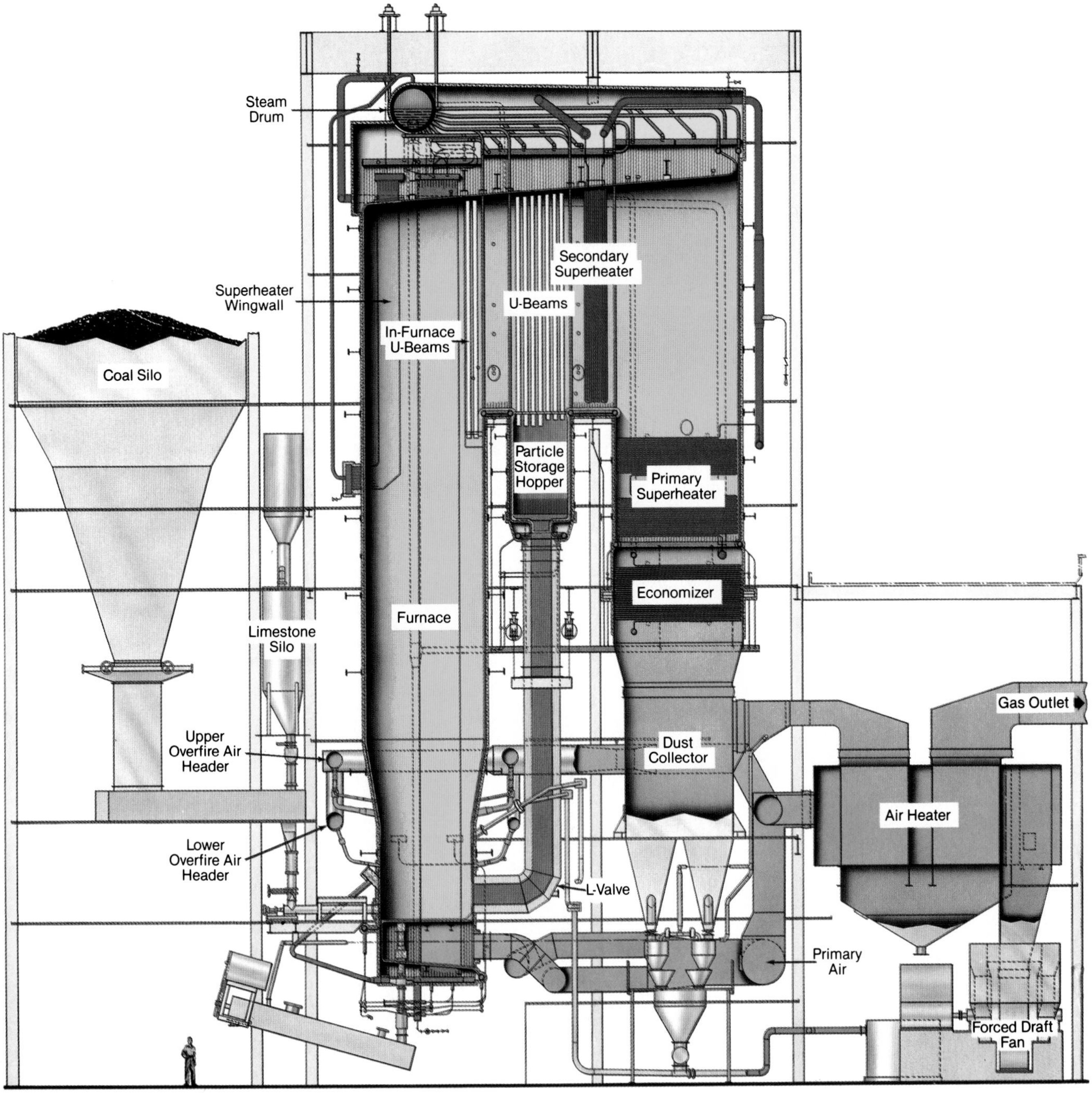

55 MW circulating fluidized-bed combustion unit for low volatile bituminous coal.

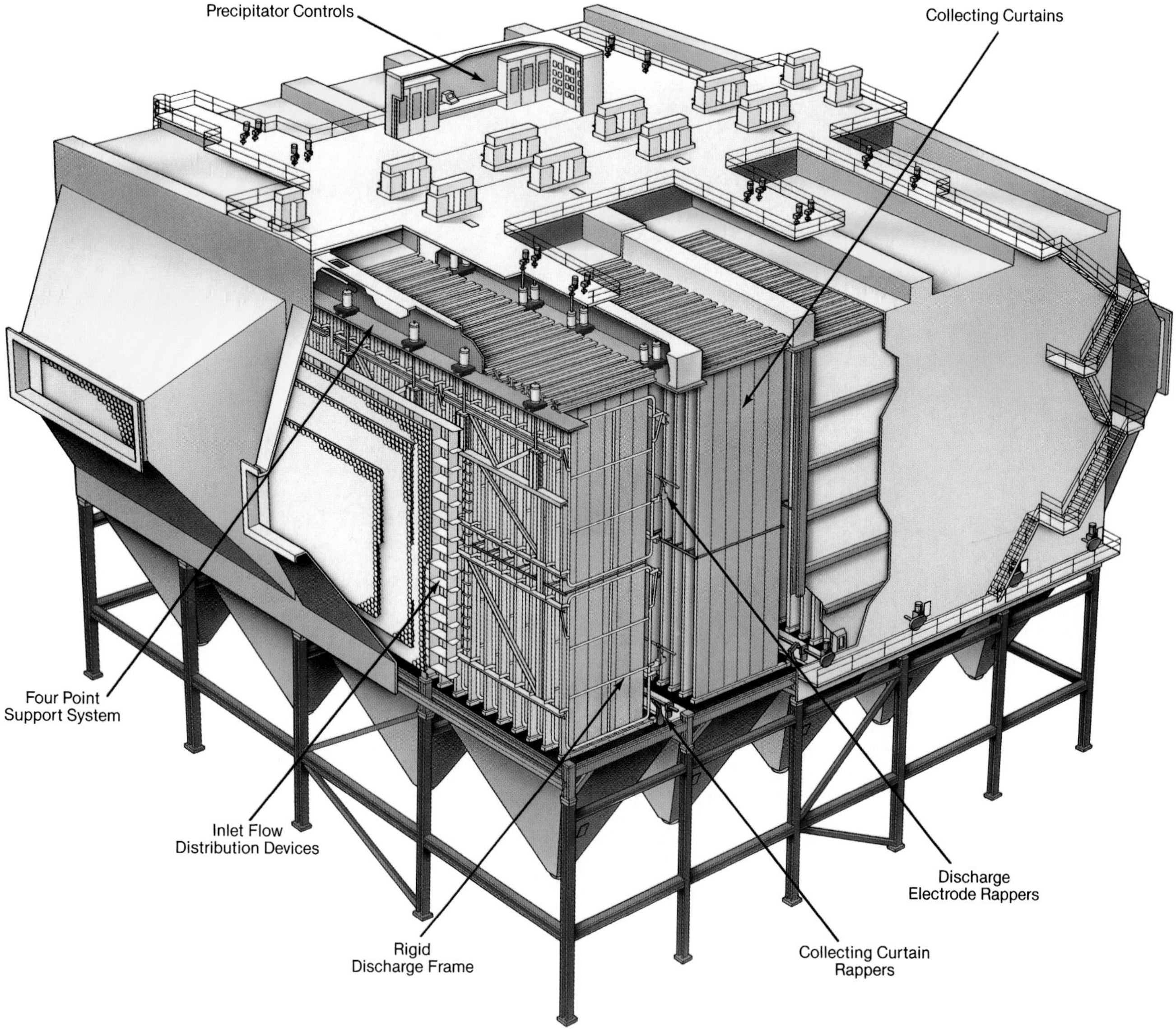

Electrostatic precipitator for particulate control.

Plate 6

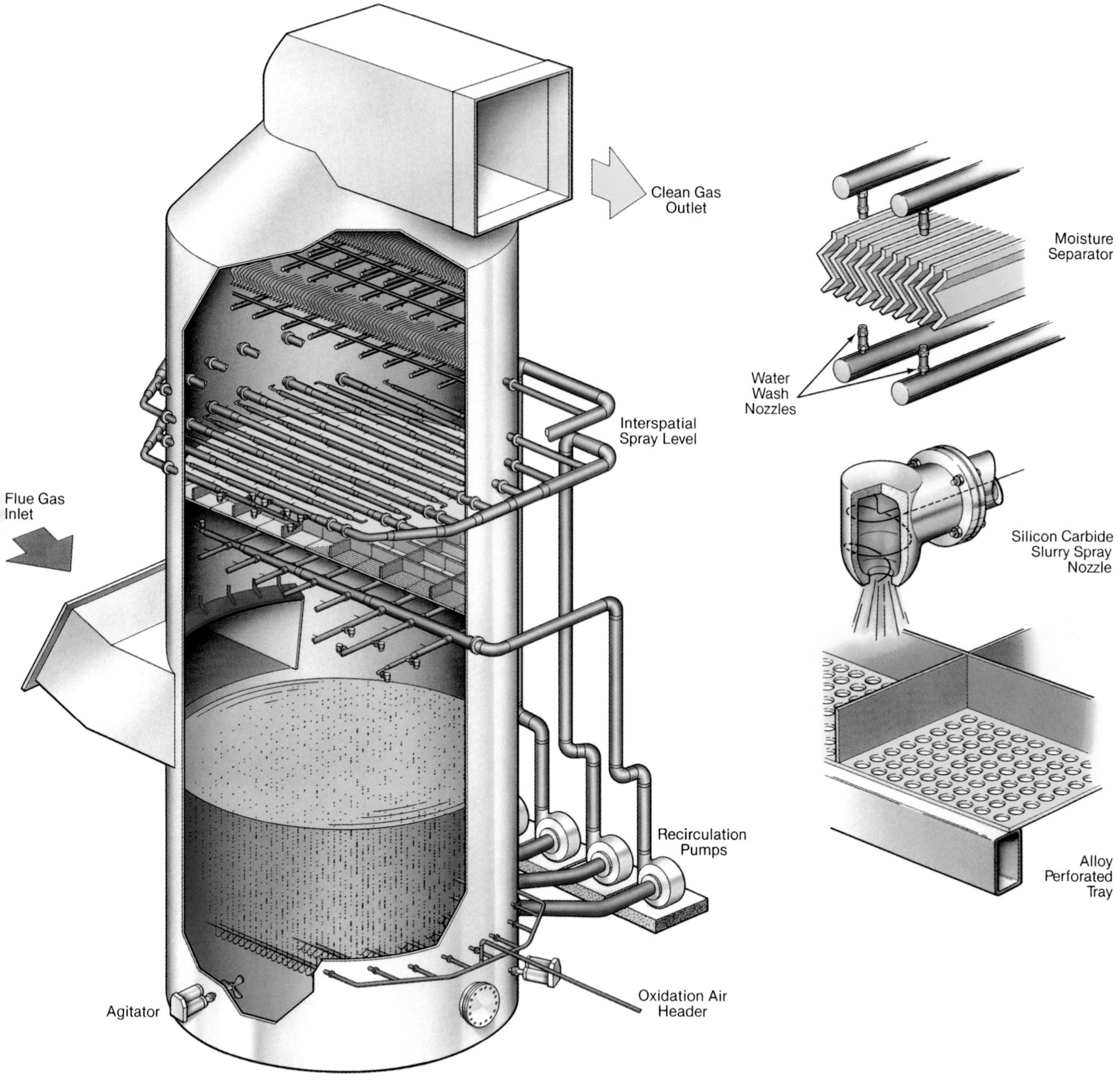

Wet flue gas desulfurization scrubber module for sulfur dioxide control.

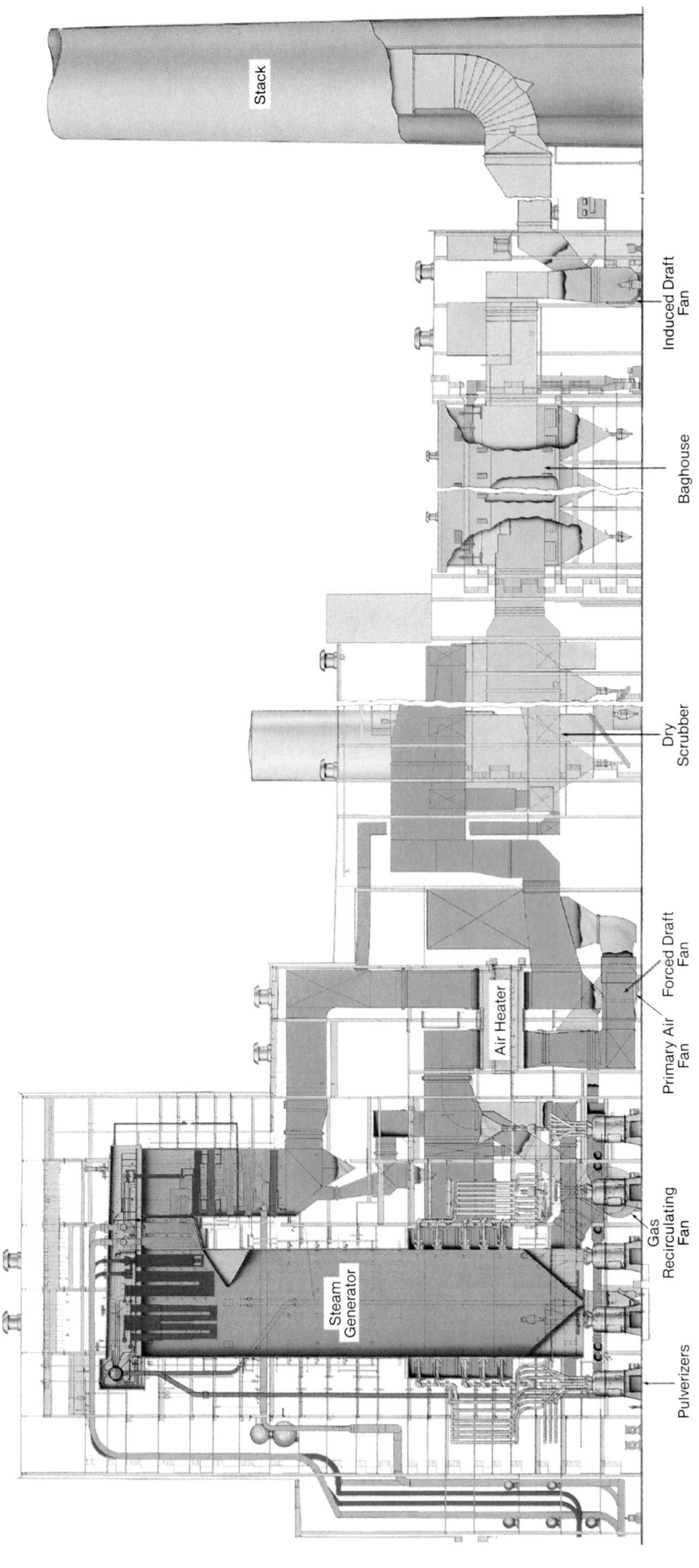

Modern 860 MW coal-fired utility boiler system with environmental control equipment.

Section I
Steam Fundamentals

Steam is uniquely adapted, by its availability and advantageous properties, for use in industrial and heating processes and in power cycles. The fundamentals of the steam generating process and the core technologies upon which performance and equipment design are based are described in this section of seven chapters. Chapter 1 has been added to this edition of *Steam* to provide an initial overview of the process, equipment and design of steam generating systems, and how they interface with other power producing and steam using processes. This is followed by fundamental discussions of thermodynamics, fluid dynamics, heat transfer, and the complexities of boiling and steam-water flow. The section concludes with key elements of material science and structural analysis which permit the safe and efficient design of the steam generating units and components.

The importance and application of computational and numerical analyses in modern steam generator design are addressed throughout this section.

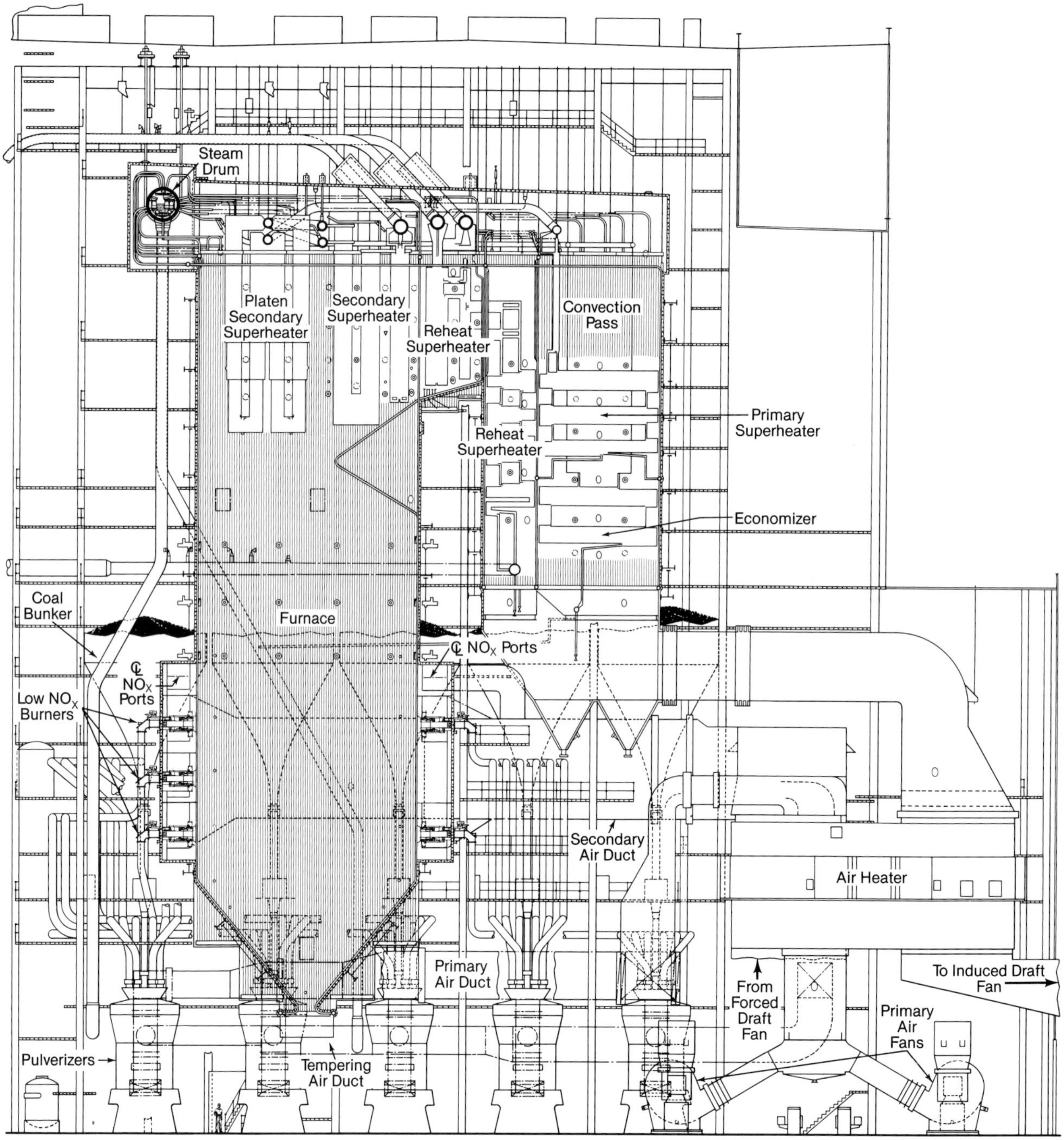

455 MW natural circulation coal-fired utility boiler.

Chapter 1
Steam Generation — An Overview

Steam generators, or boilers, use heat to convert water into steam for a variety of applications. Primary among these are electric power generation and industrial process heating. Steam has become a key resource because of its wide availability, advantageous properties and nontoxic nature. The steam flow rates and operating conditions can vary dramatically: from 1000 lb/h (0.1 kg/s) in one process use to more than 10 million lb/h (1260 kg/s) in large electric power plants; from about 14.7 psi (1 bar) and 212F (100C) in some heating applications to more than 4500 psi (310 bar) and 1100F (593C) in advanced cycle power plants.

Fuel use and handling add to the complexity and variety of steam generating systems. The fuels used in most steam generators are coal, natural gas and oil. However, during the past few decades, nuclear energy has also begun to play a major role in at least the electric power generation area. Also, an increasing variety of biomass materials and process byproducts have become heat sources for steam generation. These include peat, wood and wood wastes, bagasse, straw, coffee grounds, corn husks, coal mine wastes (culm), waste heat from steelmaking furnaces and even solar energy. The steam generating process has also been adapted to incorporate functions such as chemical recovery from paper pulping processes, volume reduction for municipal solid waste or trash, and hazardous waste destruction.

Steam generators designed to accomplish these tasks range from a small package boiler (Fig. 1) to a large, high capacity utility boiler to generate 1300 MW of electricity (Fig. 2). The former is a factory-assembled, fully automated gas-fired boiler, which can supply saturated steam perhaps for a hospital. It arrives at the site with all controls and equipment assembled. The large field-erected utility boiler will produce more than 10 million lb/h (1260 kg/s) steam at 3860 psi (266 bar) and 1010F (543C). Such a unit, or its companion nuclear option (Fig. 3), is part of some of the most complex and demanding engineering systems in operation today.

The range of combustion systems is illustrated by the 750 t/d (680 t_m/d) mass-fired refuse power boiler shown in Fig. 4 and the circulating fluidized-bed combustion boiler shown in Fig. 5.

The central job of the boiler designer in any of these applications is to combine fundamental science, technology, empirical data, and practical experience to produce a steam generating system which meets the steam supply requirements in the most economical package. Other factors in the design process include fuel characteristics, environmental protection, thermal efficiency, operations, maintenance and operating costs, regulatory requirements, and local geographic and weather conditions, among others.

Design involves balancing these complex and sometimes competing factors. For example, the reduction of pollutants such as nitrogen oxides (NO_x) may require a larger boiler volume, increasing capital costs and potentially increasing maintenance costs. Such a design activity is firmly based upon the physical and thermal sciences such as solid mechanics, thermodynamics, heat transfer, fluid mechanics and materials science. However, the real world is so complex and variable — so interrelated — that it is only by applying the art of boiler design to combine science and practice that the most economical and dependable design can be achieved.

Steam generator design must also strive to address in advance the many changes occurring in the world to provide the best possible option. Fuel prices are expected to escalate while fuel supplies become less certain, thereby enforcing the need for continued efficiency improvement and fuel flexibility. Increased environmental protection will drive improvements in combustion to reduce NO_x and in efficiency to reduce carbon dioxide (CO_2) emissions. Demand growth continues in many areas where steam generator load may have to cycle up and down more frequently and at a faster rate.

There are technologies such as pressurized fluidized-bed combustion and integrated gasification combined

Fig. 1 Small shop-assembled package boiler.

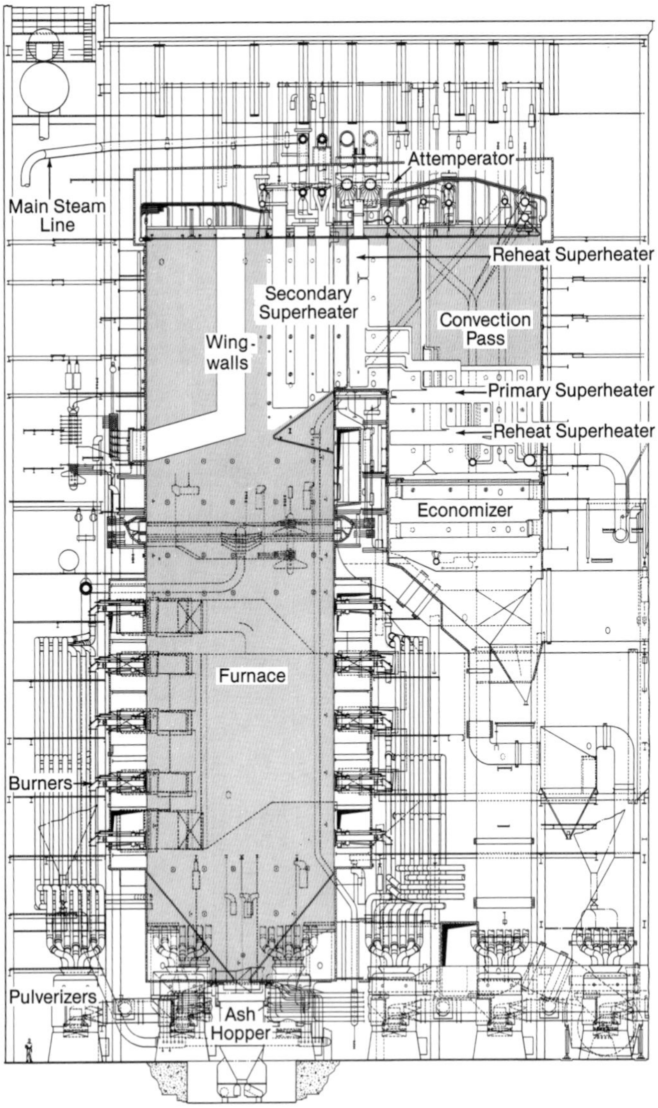

Fig. 2 1300 MW coal-fired utility steam generator.

cycle systems currently under development which actually integrate the environmental control with the entire steam generation process to reduce emissions and increase power plant thermal efficiency. Also, modularization and further standardization will help reduce fabrication and erection schedules to meet more dynamic capacity addition needs.

Fig. 3 900 MW nuclear power system.

Systematic approach

There are a variety of evaluation approaches that can be used to meet the specific steam generator performance requirements. These include the multiple iterations commonly found in thermal design where real world complexities and nonlinear, noncontinuous interactions prevent a straightforward solution. The process begins by understanding the particular application and system to define conditions such as steam flow requirements, fuel source, operating dynamics, and emissions limits, among others. From these, the designer proceeds to assess the steam generator options, interfaces, and equipment needs to achieve performance. Using a coal-fired boiler as an example, a systematic approach would include the following:

1. specify the steam supply requirements to define the overall inputs of fuel, air and water, and the steam output conditions,
2. evaluate the heat balances and heat absorption by type of steam generator surface,
3. perform combustion calculations to define heat input and gas flow requirements,
4. configure the combustion system to complete the combustion process while minimizing emissions (fuel preparation, combustion and air handling),
5. configure the furnace and other heat transfer surfaces to satisfy temperature, material and performance tradeoffs while meeting the system control needs,
6. size other water-side and steam-side components,
7. specify the back-end tradeoffs on the final heat recovery devices such as water heaters (economizers) and air heaters,
8. check the steam generating system performance to ensure that the design criteria are met,
9. verify overall unit performance,
10. repeat steps 2 through 9 until the desired steam mass flow and temperature are achieved over the specified range of load conditions,

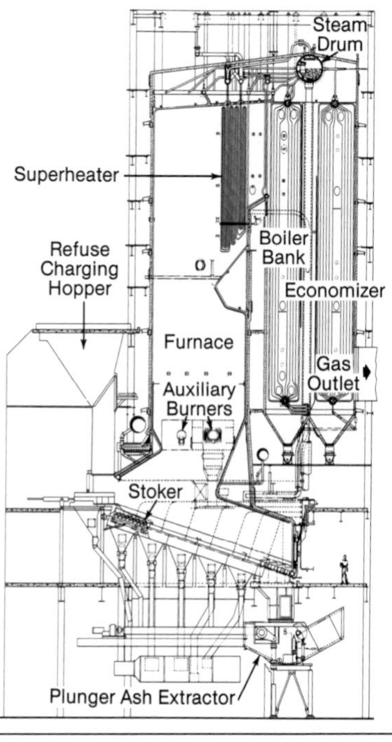

Fig. 4 Babcock & Wilcox 750 ton per day mass-fired refuse power boiler.

11. use American Society of Mechanical Engineers (ASME) Code rules to design pressure parts to meet the anticipated operating conditions and complete detailed mechanical design,
12. add environmental protection equipment to achieve prescribed emission levels, and
13. add auxiliaries as needed, such as tube surface cleaning equipment, fans, instrumentation and controls, to complete the design and assure safe and continuous operation.

A final important element of this process is to consider the life cycle of the steam generator and the plant in which it will operate. Today, some steam generators will be required to operate efficiently and reliably for up to 60 years. During this time, many components will wear out because of the aggressive environment. Routine inspection of pressure parts is needed to assure continued reliability. Unit operating procedures, such as the permitted severity and magnitude of transients, may be monitored to prevent reduced unit life. Operating practices including water treatment, cycling operation procedures, and preventive maintenance programs, among others, can significantly affect steam generator availability and reliability. Key unit components may be upgraded to improve performance. In each case, decisions made during the design phase and subsequent operation can substantially enhance the life and performance of the unit. The full life cycle of the unit should be considered from the beginning.

System arrangement and key components

Most applications of steam generators involve the production of electricity or the supply of process steam. In some cases, a combination of the two applications, called *cogeneration*, is used. In each application, the steam generator is a major part of a larger system that has many subsystems and components. Fig. 6 shows a modern coal-fired power generating facility; Fig. 7 identifies the major subsystems. Key subsystems include fuel receiving and preparation, steam generator and combustion, environmental protection, turbine-generator, and heat rejection including cooling tower.

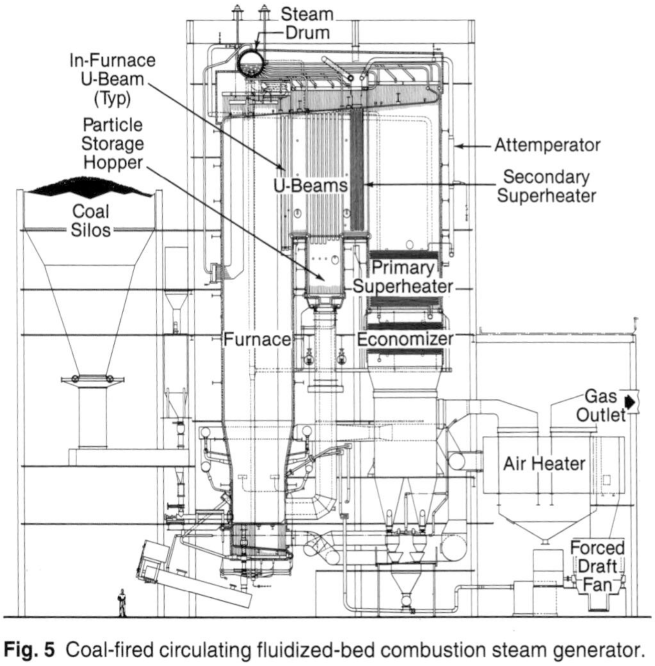

Fig. 5 Coal-fired circulating fluidized-bed combustion steam generator.

Fig. 6 Coal-fired utility power plant.

First, follow the fuel and products of combustion (flue gas) through the system. The fuel handling system stores the fuel supply (coal in this example), prepares the fuel for combustion and transports it to the steam generator. The associated air system supplies air to the burners through a forced draft fan. The steam generator subsystem, which includes the air heater, burns the fuel-air mixture, recovers the heat, and generates the controlled high pressure and high temperature steam. The flue gas leaves the air heater and passes through particulate collection and sulfur dioxide (SO_2) scrubbing systems where pollutants are collected and the ash and solid scrubber residue are removed. The remaining flue gas is then sent to the stack through an induced draft fan.

The steam generator (boiler) evaporates water and supplies high temperature, high pressure steam under carefully controlled conditions. The steam is passed to a turbine-generator set that produces the electricity. The steam may also be reheated in the steam generator, after passing through part of a multi-stage turbine system, by running the exhaust steam back to the boiler convection pass (reheater not shown). Ultimately, the steam is passed from the turbine to the condenser where the remaining waste heat is rejected. Before the water from the condenser is returned to the boiler it passes through several pumps and heat exchangers (feedwater heaters) to increase its pressure and temperature. The heat absorbed by the condenser is eventually rejected to the atmosphere by one or more cooling towers. These cooling towers are perhaps the most visible component in the power system (Fig. 6). The natural draft cooling tower shown is basically a hollow cylindrical structure which circulates air and moisture to absorb the heat rejected by the condenser. Such cooling towers exist at most modern power plant sites.

For the industrial power system, many of the same features are needed but the turbine-generator and heat rejection portions are replaced by the process application.

In a nuclear power system (Fig. 8), the fossil fuel-fired steam generator is replaced by a nuclear reactor vessel and, typically, two or more steam generators. The coal handling system is replaced by a nuclear reactor fuel bundle handling and storage facility, and the large scale air pollution control equipment is not needed.

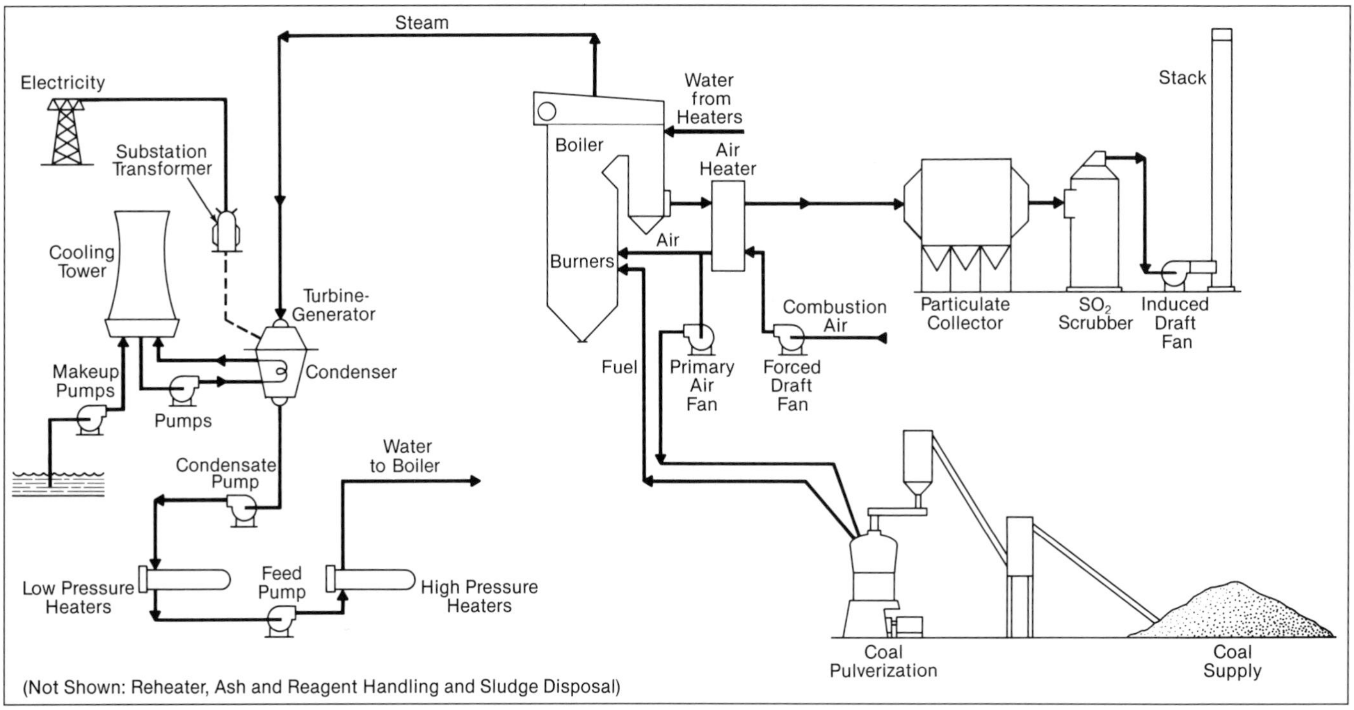

Fig. 7 Coal-fired utility power plant schematic.

Steam generator interfaces

The steam generator system's primary function is to convert chemical or nuclear energy bound in the fuel to heat and produce high temperature, high pressure steam. The variety of fuel sources, the high temperature nature of these processes, and the large number of subsystem interfaces indicate the critical and challenging design process. The initial steps in evaluating the steam generating system include establishing key interfaces with other plant systems and with the power cycle. These are typically set by the end user or consulting engineer after an in depth evaluation indicates: 1) the need for the expanded power supply or steam source, 2) the most economical fuel selection and type of steam producing system, 3) the plant location, and 4) the desired power cycle or process steam conditions. The key requirements fall into six major areas:

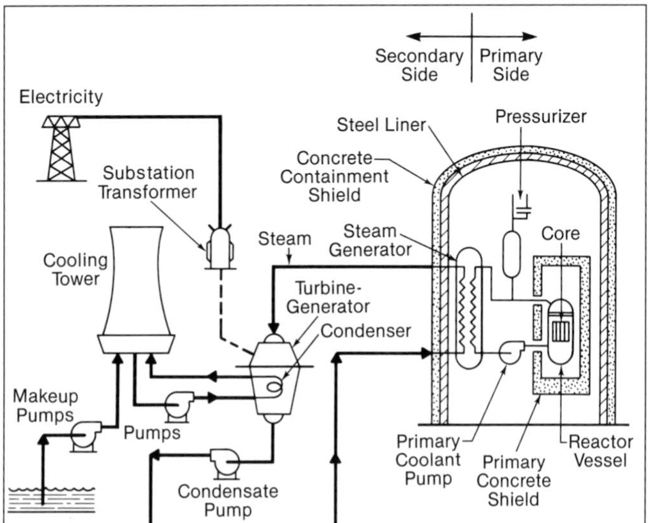

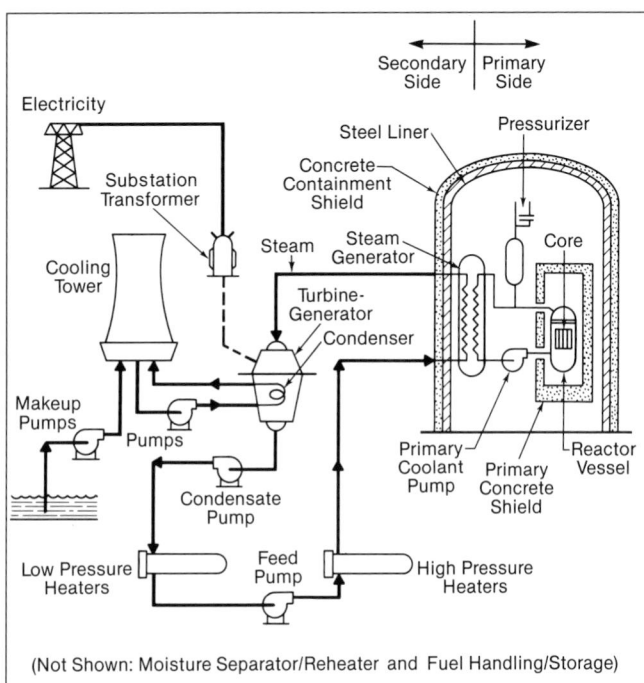

Fig. 8 Nuclear power plant schematic.

1. steam minimum, nominal, and maximum flow rates; pressure and temperature; need for one or more steam reheat stages; auxiliary equipment steam usage; and future requirements,
2. source of the steam flow makeup or replacement water supply, water chemistry and inlet temperature,
3. the type and range of fuels considered including worst case conditions, and the chemical analyses (proximate and ultimate analyses) for each fuel or mixture of fuels,
4. elevation above sea level, overall climate history and forecast, earthquake potential and space limitations,
5. emissions control requirements and applicable government regulations and standards, and
6. the types of auxiliary equipment; overall plant and boiler efficiency; access needs; evaluation penalties, e.g., power usage; planned operating modes including expected load cycling requirements, e.g., peaking, intermediate or base load; and likely future plant use.

When these interfaces are established, boiler design and evaluation may begin.

Impact of energy source

The primary fuel selected has perhaps the most significant impact on the steam generator system configuration and design. In the case of nuclear energy, a truly unique system for containing the fuel and the nuclear reaction products has been developed with an intense focus on safety and protecting the public from radiation exposure. Acceptable materials performance in the radiation environment and the long term thermal-hydraulic and mechanical performance are central to system design. When fossil, bio-

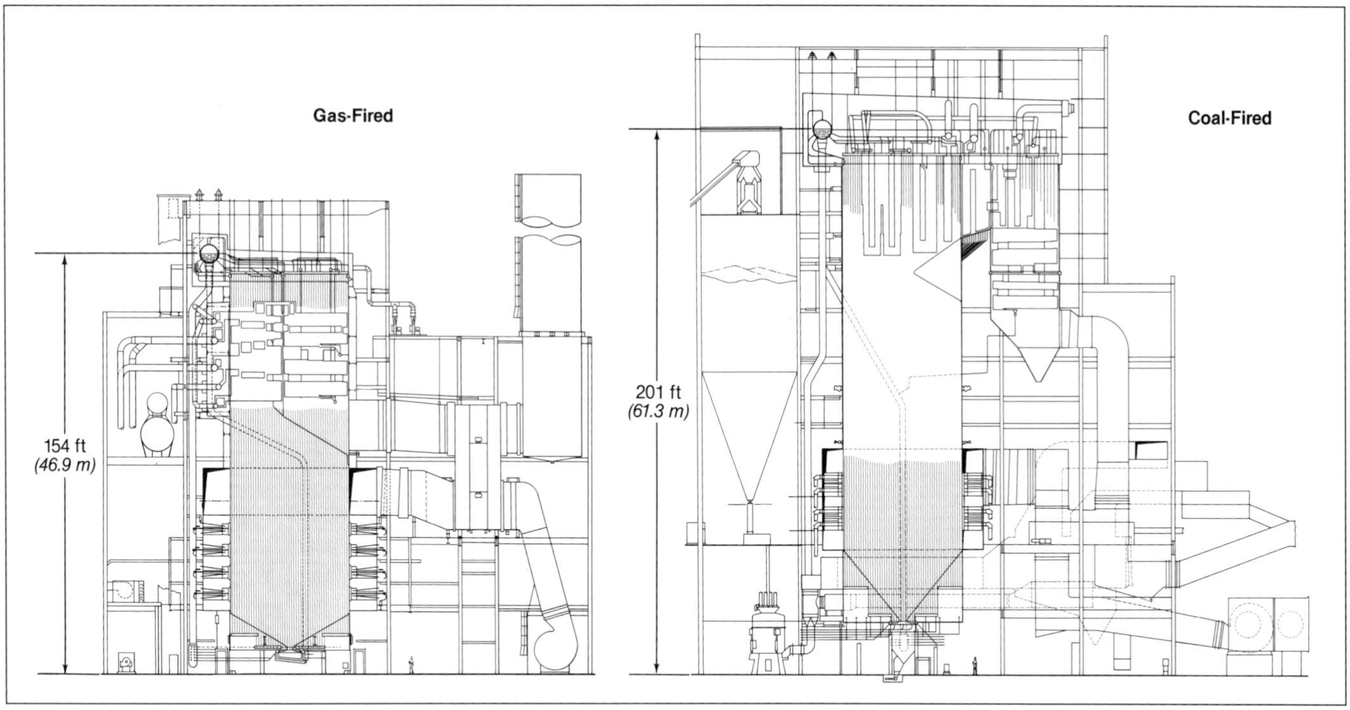

Fig. 9 Comparison of gas- and coal-fired steam generators.

mass, or byproduct fuels are burned, widely differing provisions must be made for fuel handling and preparation, fuel combustion, heat recovery, fouling of heat transfer surfaces, corrosion of materials and emissions control. For example, in a gas-fired unit (Fig. 9), there is minimal need for fuel storage and handling. Only a small furnace is needed for combustion, and closely spaced heat transfer surfaces may be used because of lack of ash deposits (fouling). The corrosion allowance is relatively small and the emissions control function is primarily for NO_x formed during the combustion process. The result is a relatively small, compact and economical design.

If a solid fuel such as coal (which has a significant level of noncombustible ash) is used, the overall system is much more complex. This system could include extensive fuel handling and preparation facilities, a much larger furnace, and more widely spaced heat transfer surfaces. Additional components could be special cleaning equipment to reduce the impact of fouling and erosion, air preheating to dry the fuel and enhance combustion, more extensive environmental equipment, and equipment to collect and remove solid wastes.

The impact of fuel on a utility boiler design alone is clearly indicated in Fig. 9 where both steam generators produce the same steam flow rate. The particular challenge when burning different solid fuels is indicated in Fig. 10 where provision is made for burning both pulverized (finely ground) coal using the burners and for wood chips and bark which are burned on the moving grate (stoker) at the bottom of the unit.

Impact of steam conditions

The steam temperature and pressure for different boiler applications can have a significant impact on design. Fig. 11 identifies several typical boiler types, as well as the relative amount of heat input needed, for water heating, evaporation (boiling), superheating and reheating, if required. The relative amount of energy needed for evaporation is dramatically reduced as operating pressure is increased. As a result, the relative amount of physical heat transfer surface (tubes) dedicated to each function can be dramatically different.

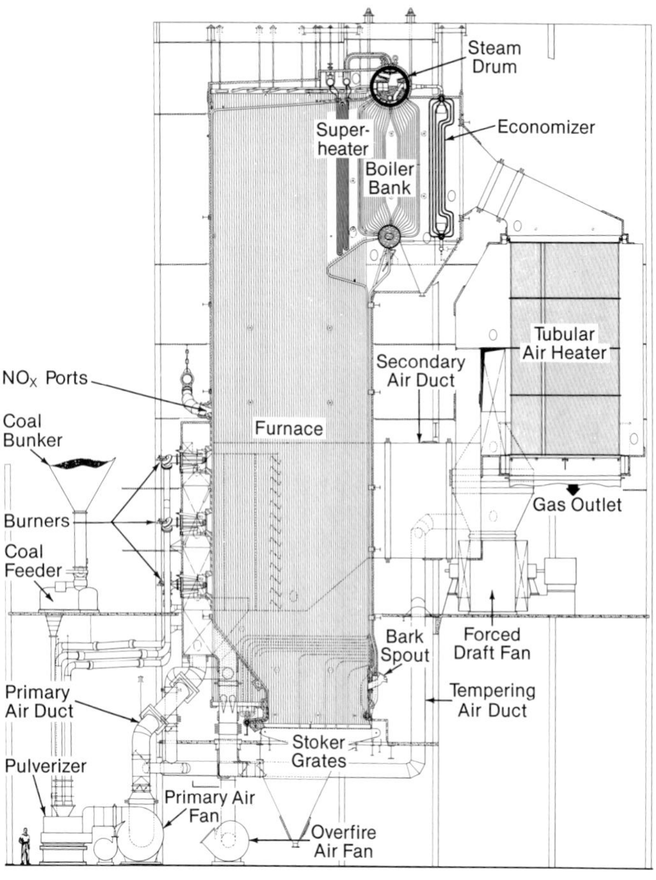

Fig. 10 Large industrial power boiler with multiple fuel capability.

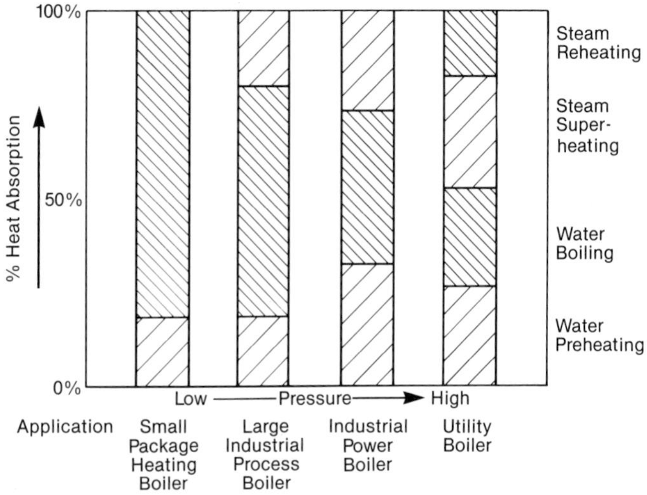

Fig. 11 Steam generator energy absorption by function.

Steam generation fundamentals

Boiling

The process of boiling water to make steam is a familiar phenomenon. Thermodynamically, instead of increasing the water temperature, the energy used results in a change of phase from a liquid to gaseous state, i.e., water to steam. A steam generating system should provide a continuous process for this conversion.

The simplest case for such a device is a kettle boiler where a fixed quantity of water is heated (Fig. 12). The applied heat raises the water temperature. Eventually, for the given pressure, the boiling (*saturation*) temperature is reached and bubbles begin to form. As heat continues to be applied, the temperature remains constant and steam escapes from the water surface. If the steam is continuously removed from the vessel, the temperature will remain constant until all of the water is evaporated. At this point, heat addition would increase the temperature of the kettle and of any steam remaining in the vessel. To provide a continuous process, all that is needed is a regulated supply of water to the vessel to equal the steam being generated and removed.

Technical and economic factors indicate that the most effective way to produce high pressure steam is to heat relatively small diameter tubes containing a continuous flow of water. This process is shown in Fig. 13a. Subcooled water (less than boiling temperature) enters the tube to which heat is applied. As the water flows through the tube, it is heated to the boiling point, bubbles are formed, and *wet* steam is generated. In most boilers, a steam-water mixture leaves the tube and enters a large vessel (steam drum) where steam is separated from water. The remaining water is then mixed with the replacement water and returned to the heated tube.

A special case of this type of boiling system is the once-through boiler. Shown in Fig. 13b, the flow rate and heat input are closely controlled so that all of the water is evaporated and only steam leaves the tube. There is no need for the steam drum.

Circulation

For the system to generate steam continuously, water must be circulated through the tubes. Two different approaches are commonly used, natural or thermal circulation, and forced or pumped circulation. Natural circulation is illustrated in Fig. 14a. In the unheated tube segment A-B, no steam is present. Heat addition generates a steam-water mixture in segment B-C. Because the steam and steam-water mixture in segment B-C are less dense than the water segment A-B, gravity will cause the water to flow downward in segment A-B and will cause the steam-water mixture (B-C) to move upward into the steam drum. The rate of water motion or circulation depends upon the difference in average density between the unheated water and the steam-water mixture.

The total circulation rate potential depends primarily

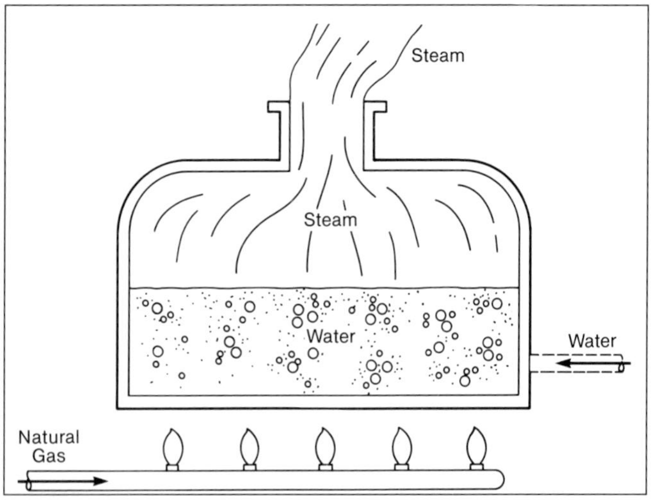

Fig. 12 Simple kettle boiler.

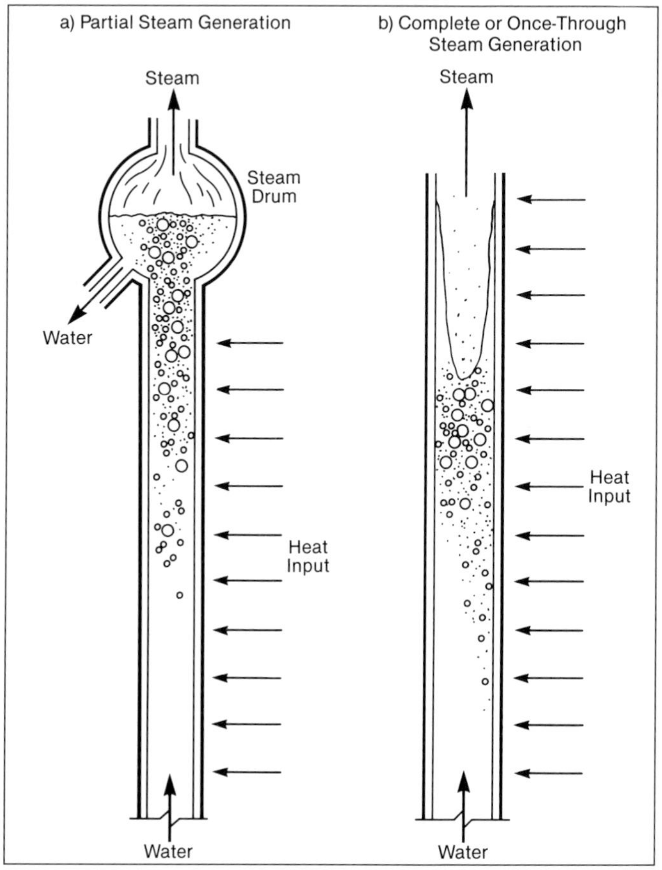

Fig. 13 Boiling process in tubular geometries.

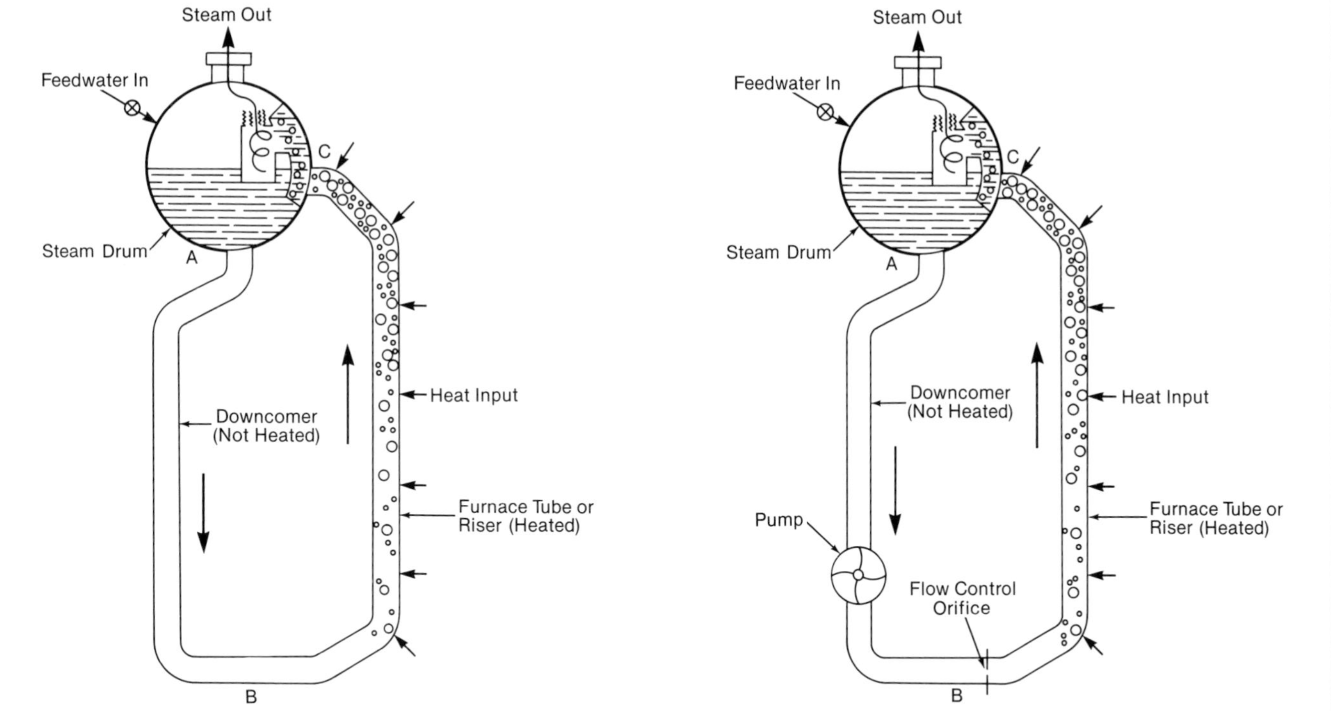

(a) Simple Natural or Thermal Circulation Loop

(b) Simple Forced or Pumped Circulation Loop

Fig. 14 Simple circulation systems.

upon four factors: 1) the height of the boiler, 2) the operating pressure, 3) the heat input rate, and 4) the free flow areas of the components. Taller boilers result in a larger total pressure difference between the heated and unheated legs and therefore can produce larger total flow rates. Higher operating pressures provide higher density steam and higher density steam-water mixtures. This reduces the total weight difference between the heated and unheated segments and tends to reduce flow rate. Higher heat input typically increases the amount of steam in the heated segments and reduces the average density of the steam-water mixture, increasing total flow rate. An increase in the cross-sectional (free flow) areas for the water or steam-water mixtures may increase the circulation rate. For each unit of steam produced, the amount of water entering the tube can vary from 3 to 25 units.

Forced circulation is illustrated in Fig. 14b. A mechanical pump is added to the simple flow loop and the pressure difference created by the pump controls the water flow rate.

The steam-water separation in the drum requires careful consideration. In small, low pressure boilers, steam-water separation can be easily accomplished with a large drum approximately half full of water. Natural gravity steam-water separation (similar to a kettle) can be sufficient. However, in today's high capacity, high pressure units, mechanical steam-water separators are needed to economically provide moisture free steam from the drum. With such devices installed in the drum, the vessel diameter and cost can be significantly reduced.

Finally, at very high pressures, a point is reached where water no longer exhibits boiling behavior. Above this critical pressure [3208 psi (221 bar)], the water temperature continuously increases with heat addition. Steam generators can be designed to operate at pressures above this critical pressure. Drums and steam-water separation are no longer required and the steam generator operates effectively on the once-through principle.

There are a large number of design methods used to evaluate the expected flow rate for a specific steam generator design and set of operating conditions. In addition, there are several criteria which establish the minimum required flow rate and maximum allowable steam content or quality in individual tubes, as well as the maximum allowable flow rates for the steam drum.

Fossil fuel systems

Fossil fuel steam generator components

Modern steam generators are a complex configuration of thermal-hydraulic (steam and water) sections which preheat and evaporate water, and superheat steam. These surfaces are arranged so that: 1) the fuel can be burned completely and efficiently while minimizing emissions, 2) the steam is generated at the required flow rate, pressure and temperature, and 3) the maximum amount of energy is recovered. A relatively simple coal-fired utility boiler is illustrated in Fig. 15. The major components in the steam generating and heat recovery system include:

1. furnace and convection pass,
2. steam superheaters (primary and secondary),
3. steam reheater,
4. boiler or steam generating bank (industrial units only),
5. economizer,
6. steam drum,
7. attemperator and steam temperature control system, and
8. air heater.

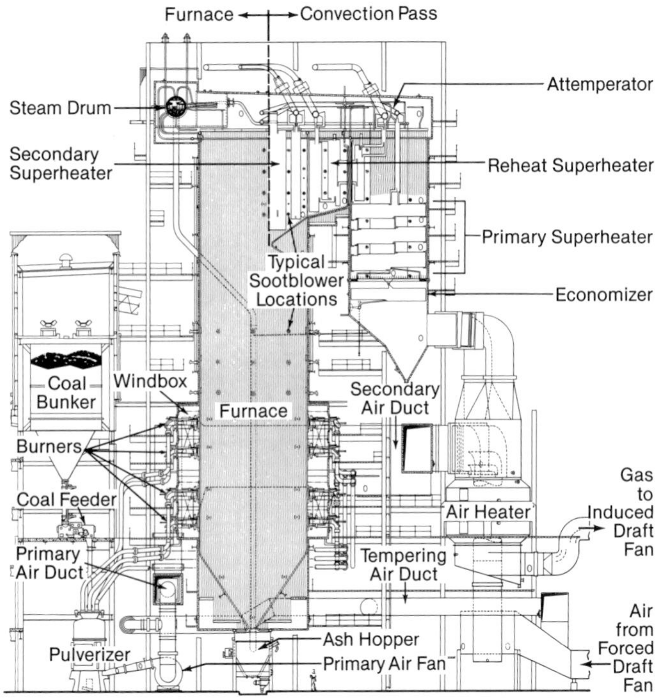

Fig. 15 Coal-fired boiler.

These components are supported by a number of subsystems and pieces of equipment such as coal pulverizers, combustion system, flues, ducts, fans, gas-side cleaning equipment and ash removal equipment. The *furnace* is a large enclosed open space for fuel combustion and for cooling of the flue gas before it enters the convection pass. Excessive gas temperatures leaving the furnace and entering the tube bundles could cause particle accumulation on the tubes or excessive tube metal temperatures. The specific geometry and dimensions of the furnace are highly influenced by the fuel and type of combustion equipment. In this case, finely ground or pulverized coal is blown into the furnace where it burns in suspension. The products of combustion then rise through the upper furnace. The superheater, reheater and economizer surfaces are typically located in the horizontal and vertical downflow sections of the boiler enclosure (*convection pass*).

In modern steam generators, the furnace and convection pass walls are composed of steam- or water-cooled carbon steel or low alloy tubes to maintain wall metal temperatures within acceptable limits. These tubes are connected at the top and bottom by headers, or manifolds. These headers distribute or collect the water, steam or steam-water mixture. The furnace wall tubes in most modern units also serve as key steam generating components or surfaces. The tubes are welded together with steel bars to provide membrane wall panels which are gas-tight, continuous and rigid. The tubes are usually prefabricated into shippable membrane panels with openings for burners, observation doors, sootblowers (boiler cleaning equipment) and gas injection ports.

Superheaters and *reheaters* are specially designed in-line tube bundles that increase the temperature of saturated steam. In general terms, they are simple single-phase heat exchangers with steam flowing inside the tubes and the flue gas passing outside, generally in crossflow. These critical components are manufactured from steel alloy material because of their high operating temperature. They are typically configured to help control steam outlet temperatures, keep metal temperatures below acceptable limits and control steam flow pressure loss.

The main difference between superheaters and reheaters is the steam pressure. In a typical drum boiler, the superheater outlet pressure might be 2700 psi (186 bar) while the reheater outlet might be only 580 psi (40 bar). The physical design and location of the surfaces depend upon the desired outlet temperatures, heat absorption, fuel ash characteristics and cleaning equipment. These surfaces can be either horizontal or vertical as shown. The superheater and sometimes reheater are often divided into multiple sections to help control steam temperature and optimize heat recovery.

The heat transfer surface in the furnace may not be sufficient to generate enough saturated steam for the particular end use. If this is the case, an additional bank of heat exchanger tubes called the *boiler bank* or *steam generating bank* is added. (See Fig. 10.) This is needed on many smaller, low pressure industrial boilers, but is not often needed in high pressure utility boilers. This boiler bank is typically composed of the *steam drum* on top, a second drum on the bottom, and a series of bent connecting tubes. The steam drum internals and tube sizes are arranged so that subcooled water travels down the tubes (farthest from the furnace) into the lower drum. The water is then distributed to the other tubes where it is partially converted to steam and returned to the steam drum. The lower drum is often called the *mud drum* because this is where sediments found in the boiler water tend to settle out and collect.

The *economizer* is a counterflow heat exchanger for recovering energy from the flue gas beyond the superheater and, if used, the reheater. It increases the temperature of the water entering the steam drum. The tube bundle is typically an arrangement of parallel horizontal serpentine tubes with the water flowing inside but in the opposite direction (counterflow) to the flue gas. Tube spacing is as tight as possible to promote heat transfer while still permitting adequate tube surface cleaning and limiting flue gas side pressure loss. By design, steam is usually not generated inside these tubes.

The steam drum is a large cylindrical vessel at the top of the boiler in which saturated steam is separated from the steam-water mixture leaving the boiler tubes. Drums can be quite large with diameters of 3 to 6 ft (0.9 to 1.8 m) and lengths approaching 100 ft (30.5 m). They are fabricated from thick steel plates rolled into cylinders with hemispherical heads. They house the steam-water separation equipment, purify the steam, mix the replacement or feedwater and chemicals, and provide limited water storage to accommodate small changes in unit load. Major connections to the steam drum are provided to receive the steam-water mixture from the boiler tubes, remove saturated steam, add replacement or makeup water, and return the near saturated water back to the inlet of the boiler tubes.

The *steam temperature control* system can be complex and include combinations of recirculating some of the flue gas to the bottom or top of the furnace, providing special gas flow passages at the back end of the steam generator, adjusting the combustion system, and adding water or low temperature steam to the high temperature steam flow (*attemperation*). The component most frequently used for the latter is called a *spray attemperator*. In large utility units, attemperators with direct injection of water or low temperature steam are used for dynamic control because of their rapid response. They are specially designed to resist thermal shock and are frequently located at the inlet of the superheater or between superheater sections to better control the superheater outlet metal temperatures. Positioning of individual superheater sections can also help maintain proper outlet steam temperatures.

The *air heater* is not a portion of the steam-water circuitry, but serves a key role in the steam generator system heat transfer and efficiency. In many cases, especially in high pressure boilers, the temperature of the flue gas leaving the economizer is still quite high. The air heater recovers much of this energy and adds it to the combustion air to reduce fuel use. Designs include tubular, flat plate, and regenerative heat exchangers, among others.

Steam-water flow system

The steam-water components are arranged for the most economical system to provide a continuous supply of steam. The circulation system (excluding reheater) for a natural circulation, subcritical pressure, drum type steam generator is shown in Fig. 16. Feedwater enters the bottom header (A) of the economizer and passes upward in the opposite direction to the flue gas. It is collected in an outlet header (B), which may also be located in the flue gas stream. The water then flows through a number of pipes which connect the economizer outlet header to the steam drum. It is sometimes appropriate to run these tubes vertically (B-C) through the convection pass to economizer outlet headers located at the top of the boiler. These tubes can then serve as water-cooled supports for the horizontal superheater and reheater when these banks span too great a distance for end support. The feedwater is injected into the steam drum (D) where it mixes with the water discharged from the steam-water separators before entering connections to the *downcomer* pipes (D-E) which exit the steam drum.

The water travels around the furnace waterwall circuit to generate steam. The water flows through the downcomer pipes (D-E) to the bottom of the furnace where *supply* tubes (E-F) route the circulating water to the individual lower furnace panel wall headers (F). The water rises through the furnace walls to an outlet header (G), absorbing energy to become a steam-water mixture. The mixture leaves the furnace wall outlet headers by means of *riser* tubes (G-D) to be discharged into the drum and steam-water separators. The separation equipment returns essentially steam-free water to the downcomer inlet connections. The residual moisture in the steam that leaves the primary steam separation devices is removed in secondary steam separators and dry steam is discharged to the superheater through a number of drum outlet connections (H-I and H-J).

The steam circuitry serves dual functions, cooling the convection pass enclosure, and generating the required superheated steam conditions. Steam from the drum passes through multiple connections to a header (I) supplying the roof tubes and, separately, to headers (J) supplying the membrane panels in the horizontal convection pass. The steam flows through these membrane panels to outlet headers (K). Steam from these headers and the roof tube outlet headers (L) then provides the cooling for the vertical convection pass enclosure (L-M). Steam flows downward through these panels and is collected in outlet headers (M) just upstream of the economizer bank.

Steam flow then rises through the primary superheater and discharges through the outlet header (N) and connecting piping equipped with a spray attemperator (O). It then enters the secondary superheater inlet header (P), flowing through the superheater sections to an outlet header (Q). A discharge pipe terminates outside of the boiler enclosure (R) where the main steam lines route the steam flow to the control valves and turbine.

Combustion system and auxiliaries

Most of the non-steam generating components and auxiliaries used in coal-fired steam generators are part of the fuel preparation and combustion systems. These include:

1. fuel preparation: feeders and coal pulverizers,
2. combustion system: burners, flame scanners, lighters, controls, windbox,
3. air/gas handling: fans, flues and ducts, dampers, control and measurement systems, silencers, and
4. other components and auxiliaries: sootblowers (heat transfer surface cleaning equipment), ash collection and handling equipment, control and monitoring equipment.

Because of their intimate relationship with the steam generation process, many of these components are supplied with the boiler. If not, careful specification and interaction with the steam generator manufacturer are critical.

The combustion system has a dramatic impact on overall furnace design. Wall mounted burners are shown in Figs. 15 and 17. These are typical for large coal-, oil-, or gas-fired units today. However, a variety of other systems are also used and are continuing to be developed to handle the variety of fuel characteristics and unit sizes. Other combustion systems include stokers (Fig. 10), Cyclone fur-

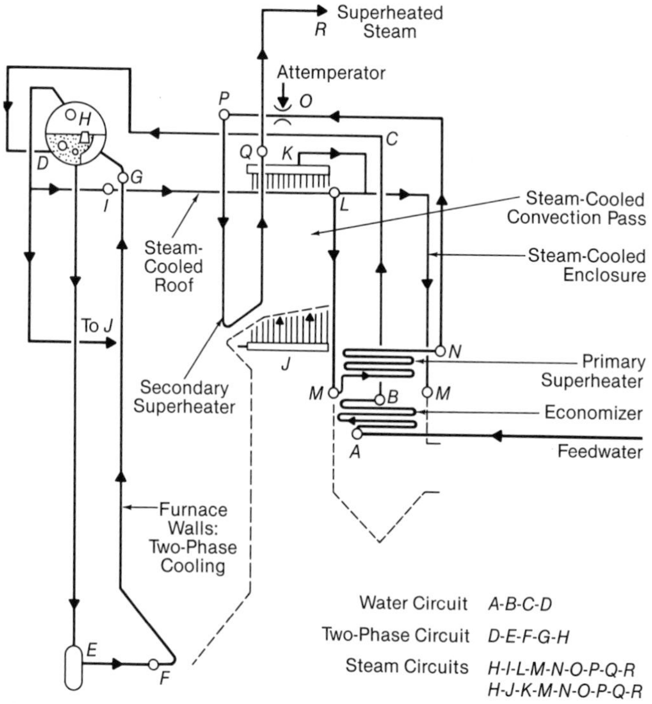

Fig. 16 Coal-fired boiler steam-water circulation system.

naces, and fluidized-bed combustion units. All have their strengths and weaknesses for particular applications. Key elements of these systems involve the need to control the formation and emission of pollutants, provide complete efficient combustion, and handle inert material found in the fuel. The fuel characteristics play a central role in how these functions are met and how the equipment is sized and designed.

Gas flow system

Many of these auxiliaries are identified in Fig. 17 along with the air/gas flow path in the large coal-fired utility boiler. Air is supplied by the *forced draft* fan (A) to the air heater (B) where it is heated to recover energy and enhance combustion. Most of the hot air (secondary air, typically 70 to 80%) passes directly to the windboxes (C) where it is distributed to individual burners. The remaining 20 to 30% passes to the booster (or primary air) fan and then to the coal pulverizers (D) where the coal is dried and ground. The hot air then pneumatically conveys the pulverized coal to the burners (E) where it is mixed with the secondary air for combustion. The coal and air are rapidly mixed and burned in the furnace (F) and the flue gas then passes up through the furnace, being cooled primarily by radiation until it reaches the furnace exit (G). The gas then progressively passes through the secondary superheater, reheater, primary superheater and economizer before leaving the steam generator enclosure (H). The gas passes through the air heater (B) and then through any pollution control equipment and *induced draft* fan (I) before being exhausted to the atmosphere.

Emissions control

A key element of fossil fuel-fired steam generator system design is environmental protection. A broad range of government regulations sets limits on primary gaseous, liquid and solid waste emissions from the steam generating process. For coal-, oil-, and gas-fired units, the primary air pollutant emissions include sulfur dioxide (SO_2), NO_x and airborne particulate or flyash. Water discharges include trace chemicals used to control corrosion and fouling as well as waste heat rejected from the condenser. Solid waste primarily involves the residual ash from the fuel and any spent sorbent from the pollution control systems.

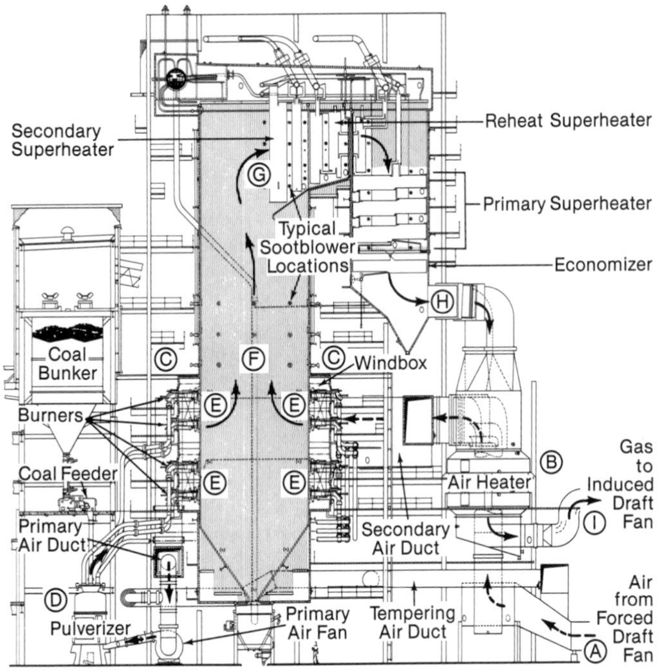

Fig. 17 Coal-fired boiler air/gas flow path.

The gaseous and solid waste from the fuel and combustion process can be minimized by fuel selection, control of the combustion process, and equipment located downstream of the steam generator. SO_2 emissions may be reduced by using fuels which contain low levels of sulfur, by fluidized-bed combustors, or by using a post-combustion scrubber system. NO_x emissions are typically controlled by using equipment such as special low NO_x burners or fluidized-bed combustors. Back-end NO_x removal systems may provide further control. Flyash or airborne particulate is collected by either a fabric filter or electrostatic precipitator (ESP) with removal efficiencies above 99%. The particulate collection equipment and SO_2 scrubbers produce solid byproduct streams which must be safely landfilled or used for some industrial applications.

The water discharges are minimized by installing recirculating cooling systems with large cooling towers which reject the waste heat from the power cycle to the air, instead of to a water source. These are used on virtually all new fossil and nuclear power plants. Chemical discharges are minimized by specially designed zero discharge systems. A set of emissions rates before and after control for a typical 500 MW power plant is shown in Table 1.

Fossil steam generator classifications

Modern steam generating systems can be classified by various criteria. These include end use, firing method, operating pressure, fuel and circulation method.

Utility steam generators are used primarily to generate electricity in large central power stations. They are designed to optimize overall thermodynamic efficiency at the highest possible availability. New units are typically characterized by large, main steam flow rates with superheated steam outlet pressures from 1800 to 3860 psi (124 to 266 bar) at about 1000F (538C). A key characteristic of newer units is the use of a reheater section to increase overall cycle efficiency.

Industrial steam generators generally supply steam to processes or manufacturing activities and are designed with particular attention to: 1) process controlled (usually lower) pressures, 2) high reliability with minimum maintenance, 3) use of one or more locally inexpensive fuels, especially process byproducts or wastes, and 4) low initial capital and minimum operating costs. On a capacity basis, the larger users of such industrial units are the pulp and paper industry, municipal solid waste reduction industry, food processing industry, petroleum/petrochemical industry, independent power producers and cogenerators, and some large manufacturing operations. Operating pressures range from 150 to 1800 psi (10 to 124 bar) with saturated or superheated steam conditions.

Nuclear steam generating systems

Overview

Nuclear steam generating systems include a series of highly specialized heat exchangers, pressure vessels, pumps and components which use the heat generated by nuclear fission reactions to efficiently and safely gener-

Table 1
Typical 500 MW Coal-Fired Steam Generator Emissions and Byproducts

Power System Characteristics
- 500 MW net
- 196 t/h (49.4 kg/s) bituminous coal
  - 2.5% sulfur
  - 16% ash
  - 12,360 Btu/lb (28,749 kJ/kg)
- 65% capacity factor

Emission	Typical Control Equipment	Discharge Rate — t/h (t_m/h) Uncontrolled		Controlled	
SO_x as SO_2	Wet limestone scrubber	9.3	(8.4)	0.9	(0.8)
NO_x as NO_2	Low NO_x burners	2.9	(2.6)	0.7	(0.7)
CO_2	Not applicable	485	(440)	485	(440)
Flyash to air*	Electrostatic precipitator or baghouse	22.9	(20.8)	0.05	(0.04)
Water stream thermal discharge	Natural draft cooling tower	2.8 x 10^9 Btu/h	(821 MW_t)	~0	(0)
Ash to landfill*	Controlled landfill	9.1	(8.3)	32	(29)
Scrubber sludge: gypsum plus water	Controlled landfill or wallboard quality gypsum	0	(0)	25	(27.7)

* As flyash emissions to the air decline, ash to landfill increases.

ate steam. The system is based upon the energy released when atoms within certain materials, such as uranium, break apart or *fission*. Fission occurs when a fissionable atom nucleus captures a free subatomic particle — a neutron. This upsets the internal forces which hold the atom nucleus together. The nucleus splits apart producing new atoms as well as an average of two to three neutrons, gamma radiation and energy.

The nuclear steam system (NSS) is designed to serve a number of functions: 1) house the nuclear fuel, 2) stimulate the controlled fission of the fuel, 3) control the nuclear reaction rate to produce the required amount of thermal energy, 4) collect the heat and generate steam, 5) safely contain the reaction products, and 6) provide backup systems to prevent release of radioactive material to the environment. Various systems have been developed to accomplish these functions. The main power producing system in commercial operation today is the pressurized water reactor, or PWR.

A key difference between the nuclear and chemical energy driven systems is the quantity of fuel. The energy released per unit mass of nuclear fuel is many orders of magnitude greater than that for chemical based fuels. For example, 1 lb (0.454 kg) of 3% enriched uranium fuel produces about the same amount of thermal energy in a commercial nuclear system as 100,000 lb (45,360 kg) of coal in a fossil-fired steam system. While a 500 MW power plant must handle approximately one million tons of coal per year, the nuclear plant will handle only 10 tons of fuel. The fossil fuel plant must be designed for a continuous fuel supply process, while most nuclear plants use a batch fuel process, where about one third of the fuel is replaced during periodic outages. However, once the steam is generated, the balance of the power producing system (turbine, condenser, cooling system, etc.) is similar to that used in the fossil fuel plant.

NSS components

A typical Babcock & Wilcox (B&W) nuclear steam system is shown in Fig. 3 and a simplified schematic is shown in Fig. 18. This nuclear system consists of two coolant loops. The primary loop cools the reactor, transports heat to two or more steam generators (only one shown), and returns coolant to the reactor by four or more primary coolant pumps (only one shown). The coolant is high purity, subcooled, single-phase water flowing at very high rates [350,000 to 450,000 GPM (22,100 to 28,400 l/s)] at around 2200 psi (152 bar) and an average temperature of about 580F (304C). The primary loop also contains a pressurizer to maintain the loop pressure at design operating levels.

The secondary loop includes the steam generation and interface with the balance of the power plant. High purity water from the last feedwater heater passes through the

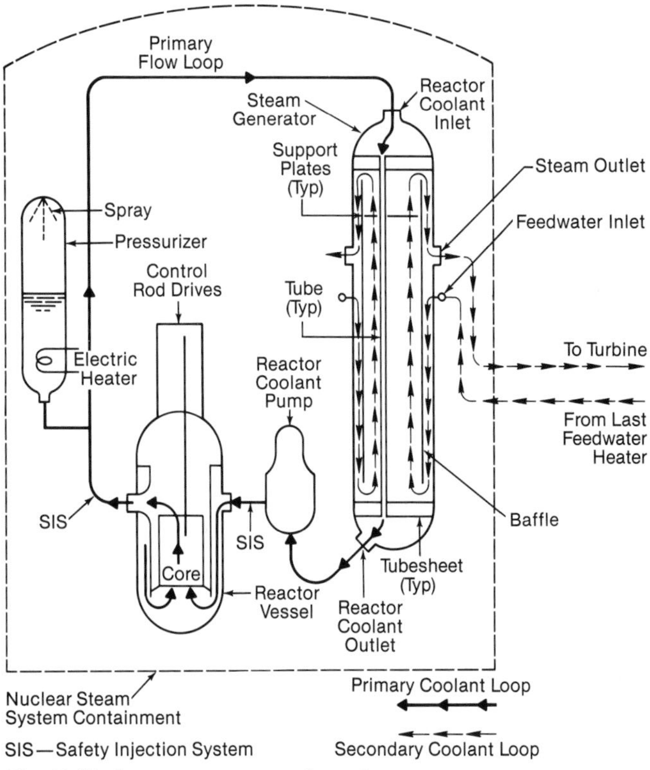

Fig. 18 Nuclear steam system schematic.

steam generator and is converted into steam. From the steam generator outlet, the saturated or superheated steam flows out of the containment building to the high pressure turbine. The operating pressure is typically around 1000 psi (69 bar). The balance of the secondary loop resembles fossil fuel-fired systems. (See Figs. 7 and 8.)

The center of the NSS is the reactor vessel and nuclear core (Fig. 19). The fuel consists of compressed pellets [for example, 0.37 in. (9.4 mm) diameter by 0.7 in. (18 mm) long] of 2.5 to 5% enriched uranium oxide. These pellets are placed in Zircaloy tubes which are sealed at both ends to protect the fuel and to contain the nuclear reaction products. The tubes are assembled into bundles with spacer and closure devices. These bundles are then assembled into the nuclear fuel core.

The reactor enclosure (Fig. 19) is a low alloy steel pressure vessel lined with stainless steel for corrosion protection. The rest of the reactor includes flow distribution devices, control rods, core support structures, thermal shielding and moderator. The moderator in this case is water which serves a dual purpose. It reduces the velocity of the neutrons thereby making the nuclear reactions more likely. It also serves as the coolant to maintain the core materials within acceptable temperature limits and transports thermal energy to the steam generators. The control rods contain neutron absorbing material and are moved into and out of the nuclear core to control the energy output.

The steam generators can be of two types, once-through (Fig. 20) and recirculating (Fig. 21). In both types, the pressure vessel is a large heat exchanger designed to generate steam for the secondary loop from heat contained in the primary coolant. The primary coolant enters a plenum and passes through several thousand small diameter [about 0.625 in. (15.9 mm)] Inconel tubes. The steam generator is a large, carbon steel pressure vessel. Specially designed tubesheets, support plates, shrouds, and baffles provide effective heat transfer, avoid thermal expansion problems and avoid flow-induced vibration.

In the once-through steam generator (OTSG), Fig. 20, the secondary loop water flows from the bottom to the top of the shell side of the tube bundle and is continuously converted from water to superheated steam. The superheated steam then passes to the high pressure turbine.

In the recirculating steam generator (RSG), Fig. 21, water moves from the bottom to the top of the shell side of the tube bundle being converted partially into steam. The steam-water mixture passes into the upper shell where steam-water separators supply saturated dry steam to the steam generator outlet. The steam is sent to the high pressure turbine. The water leaving the steam generator upper shell is mixed with feedwater and is returned to the bottom of the tube bundle.

The pressurizer is a simple cylindrical pressure vessel which contains both water and steam at equilibrium. Electrical heaters and spray systems maintain the pressure in the pressurizer vessel and the primary loop within set limits. The primary loop circulating pumps maintain high flow rates to the reactor core to control its temperature and transfer heat to the steam generators.

A number of support systems are also provided. These include reactor coolant charging systems, makeup water addition, spent fuel storage cooling, and decay heat

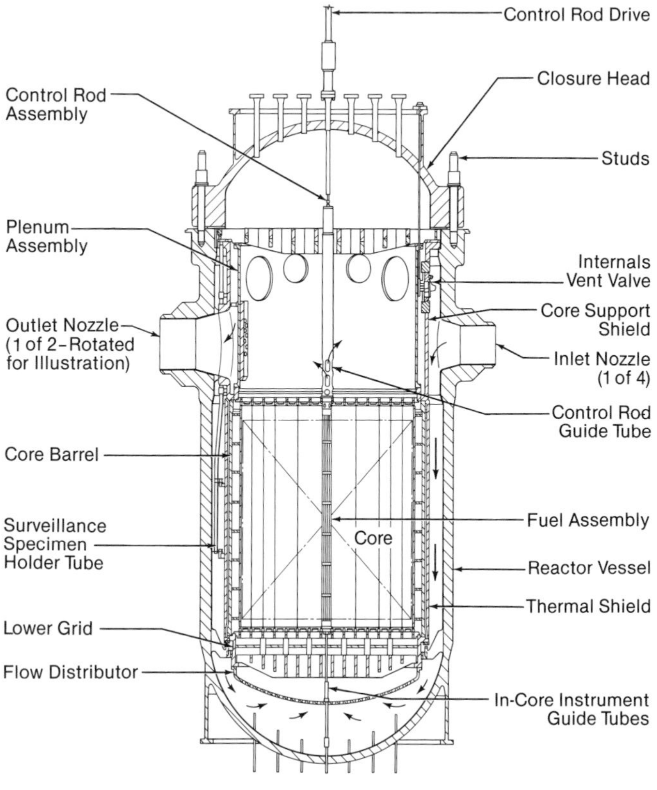

Fig. 19 Reactor vessel and internals.

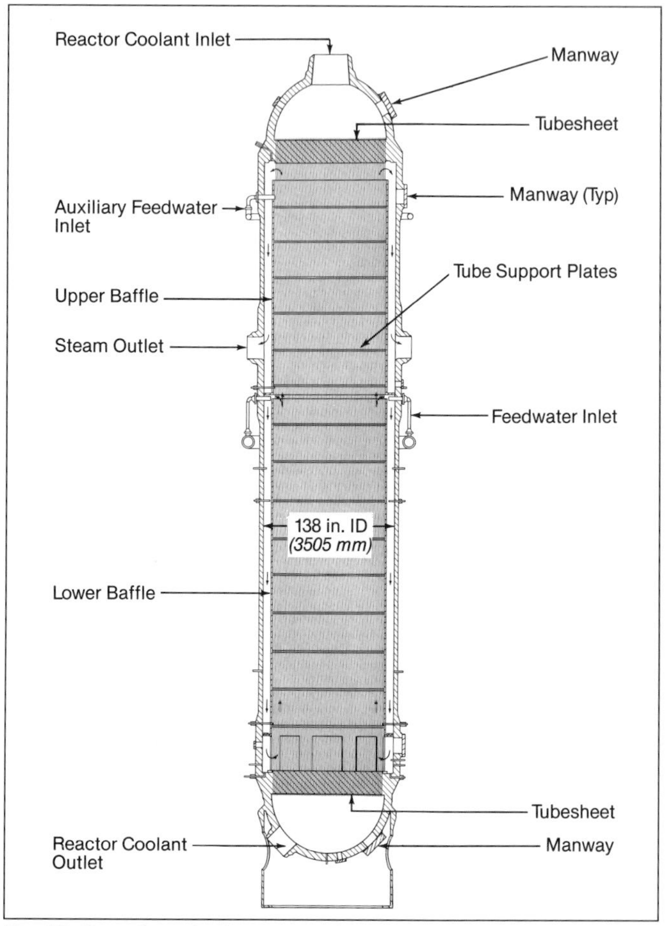

Fig. 20 Once-through steam generator.

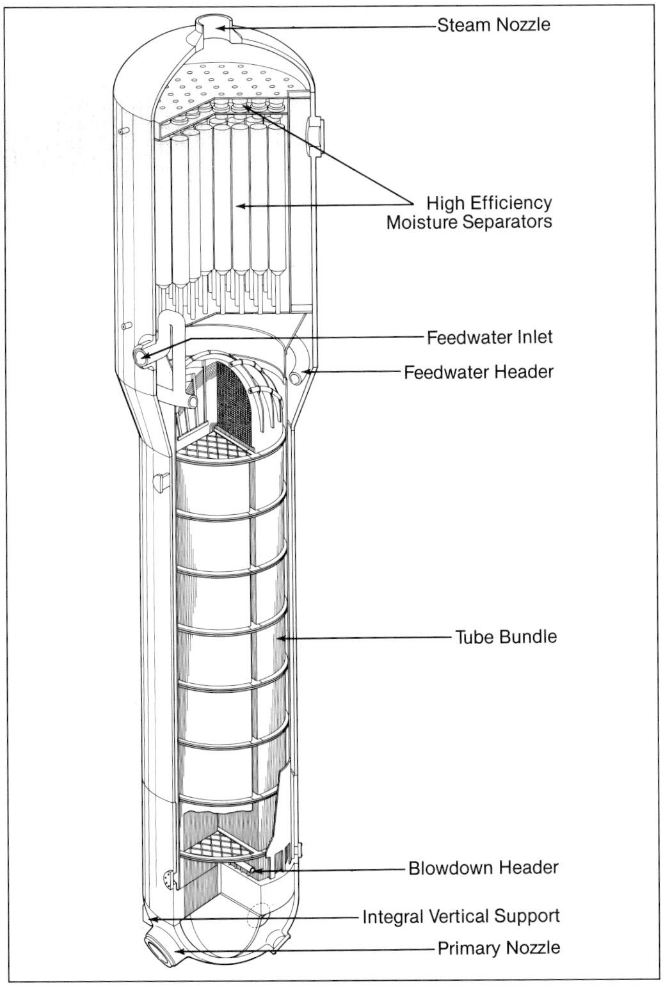

Fig. 21 Recirculating steam generator.

removal systems for when the reactor is shut down. Other specialized systems protect the reactor system in the case of a loss of coolant event. The key function of these systems is to keep the fuel bundle temperature within safe limits if the primary coolant flow is interrupted.

NSS classifications

A variety of reactor systems have been developed to recover thermal energy from nuclear fuel and to generate steam for power generation. These are usually identified by their coolant and moderator types. The principal systems for power generation include:

1. *Pressurized water reactor (PWR)* This is the system discussed above, using water as both reactor coolant and moderator, and enriched uranium oxide as the fuel.
2. *Boiling water reactor (BWR)* The steam generator is eliminated and steam is generated directly in the reactor core. A steam-water mixture cools and moderates the reactor core. Enriched uranium oxide fuel is used.
3. *CANDU (PHWR)* Heavy water (deuterium) is used as the moderator and primary loop coolant. The reactor configuration is unique but the steam generator is similar to the recirculating steam generator for the PWR. Natural (not enriched) uranium oxide is used as the fuel.
4. *Gas-cooled reactors* These are a variety of gas-cooled reactors which are typically moderated by graphite and cooled by helium or carbon dioxide.
5. *Breeder reactors* These are advanced reactor systems using sodium as the reactor coolant with no moderator. These systems are specially designed to produce more fissionable nuclear fuel than they use.

Engineered safety systems

Safety is a major concern in the design, construction and operation of nuclear power generating facilities. The focus of these efforts is to minimize the likelihood of a release of radioactive materials to the environment. Three approaches are used to accomplish this goal. First, the nuclear power industry has developed one of the most extensive and rigorous quality control programs for the design, construction and maintenance of nuclear facilities. Second, reactor systems are designed with multiple barriers to prevent radioactive material release. These include high temperature ceramic fuel pellets, sealed fuel rods, reactor vessel and primary coolant system, and the containment building including both the carbon steel reactor containment vessel and the reinforced concrete shield building. The third approach includes a series of engineered safety systems to address loss of coolant conditions and maintain the integrity of the multiple barriers. These systems include:

1. emergency reactor trip systems including rapid insertion of control rods and the addition of soluble neutron poisons in the primary coolant to shut down the nuclear reaction,
2. high and low pressure emergency core cooling systems to keep the reactor core temperature within acceptable limits and remove heat from the fuel in the event of a major loss of primary coolant or a small pipe break,
3. a heat removal system for the cooling water and containment building, and
4. spray and filtering systems to collect and remove radioactivity from the containment building.

Because of the high power densities and decay heat generation, the reactor integrity depends upon the continuous cooling of the nuclear fuel rods. Multiple independent components and backup power supplies are provided for all critical systems.

Steam system design

Now that the basic fossil fuel and nuclear steam generating systems have been described, it is appropriate to explore the general design and engineering process. While each of the many systems requires specialized evaluations, they share many common elements. To illustrate how the design process works, a small industrial B&W PFI™ gas-fired boiler has been selected for discussion. (See Figs. 22 and 23.)

Basically, the customer has one overriding need. When he turns the valve on, he expects steam to be supplied at the desired pressure, temperature and flow rate. In this example, the customer specifies 400,000 lb/h (50.5 kg/s) of superheated steam at 600 psi (41.4 bar) and 850F (454C). The customer has agreed to supply high purity feedwater at 280F (138C) and to supply natural gas as a fuel source. As with all steam generating systems, there are a number of additional constraints and requirements

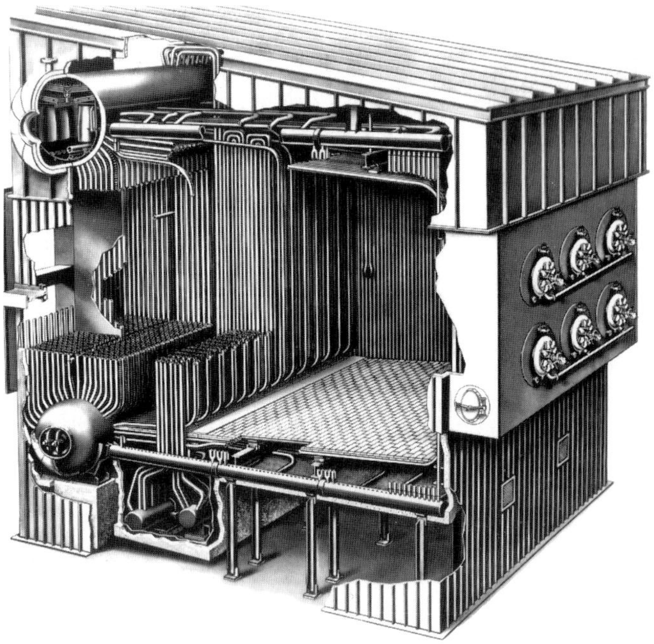

Fig. 22 Small PFI™ industrial boiler.

as discussed in *Steam Generator Interfaces*, but the major job of the steam generator or boiler is to supply steam.

Combustion of the natural gas produces a stream of combustion products or flue gas at perhaps 3600F (1982C). To maximize the steam generator thermal efficiency, it is important to cool these gases as much as possible while generating the steam. The minimum gas outlet temperature is established based upon technical and economic factors (discussed below). For now, a 310F (154C) outlet temperature to the exhaust stack is selected. The approximate steam and flue gas temperature curves are shown in Fig. 24 and define the heat transfer process. The heat transfer surface for the furnace, boiler bank, superheater and air heater is about 69,000 ft^2 (6410 m^2).

From a design perspective, the PFI boiler can be viewed as either a steam heater or gas cooler. The latter approach is most often selected for design. The design fuel heat input is calculated by dividing the steam heat output by the target steam generator thermal efficiency. Based upon the resulting fuel flow, combustion calculations define the air flow requirements and combustion products gas weight. The heat transfer surface is then configured in the most economical way to cool the gas to the temperature necessary for the target steam generator efficiency. Before proceeding to follow the gas through the cooling process, the amount of heat recovery for each of the different boiler surfaces (superheater and boiler) must be established.

Fig. 25 illustrates the water heating process from an inlet temperature of 280F (138C) to the superheater steam outlet temperature of 850F (454C). This curve indicates that about 20% of the heat absorbed is used to raise the water from its inlet temperature to the saturation temperature of 490F (254C). 60% of the energy is then used to evaporate the water to produce saturated steam. The remaining 20% of the heat input is used to superheat or raise the steam temperature to the desired outlet temperature of 850F (454C).

The fuel and the combustion process selected set the geometry of the furnace. In this case, simple circular burners are used. The objective of the burners is to mix the fuel and air rapidly to produce a stable flame and complete combustion while minimizing the formation of NO_x emissions. Burners are available in several standardized sizes. The specific size and number are selected from past experience to provide the desired heat input rate while permitting the necessary level of load range control. The windbox, which distributes the air to individual burners, is designed to provide a uniform air flow at low enough velocities to permit the burners to function properly.

The furnace volume is then set to allow complete fuel combustion. The distances between burners and between the burners and the floor, roof, and sidewalls are determined from the known characteristics of the particular burner flame. Adequate clearances are specified to prevent flame impingement on the furnace surfaces, which could overheat the tubes and cause tube failures.

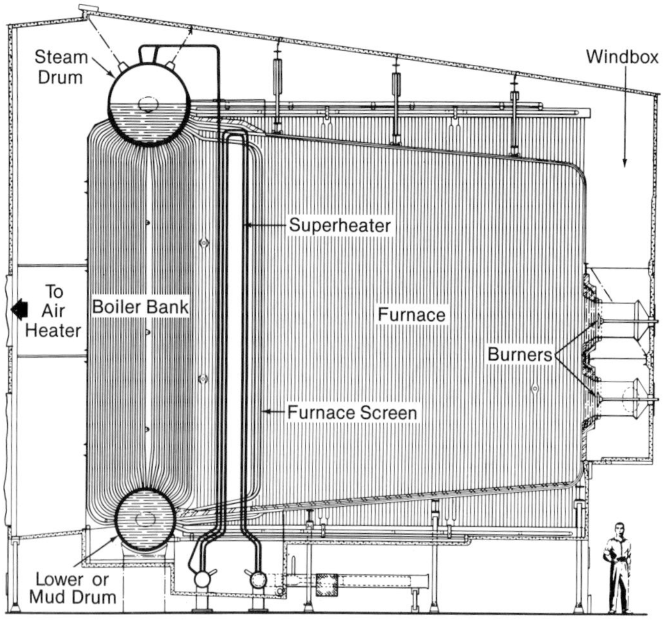

Fig. 23 Sectional view — small industrial boiler.

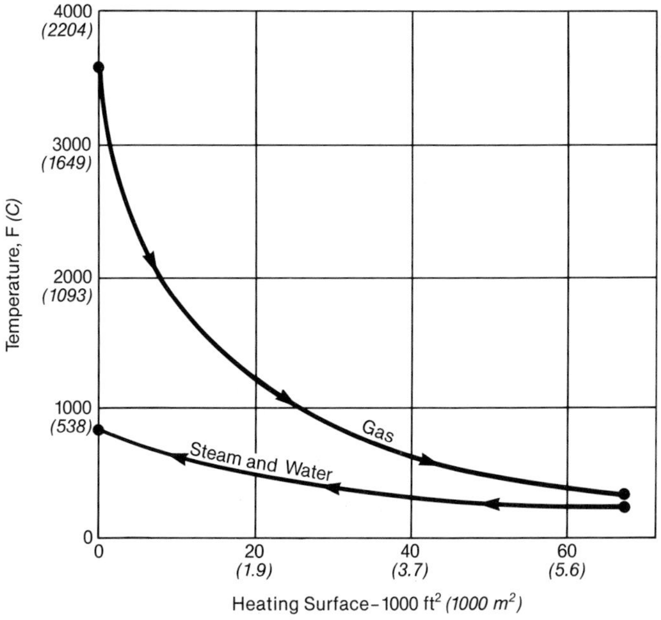

Fig. 24 Industrial boiler — temperature versus heat transfer surface.

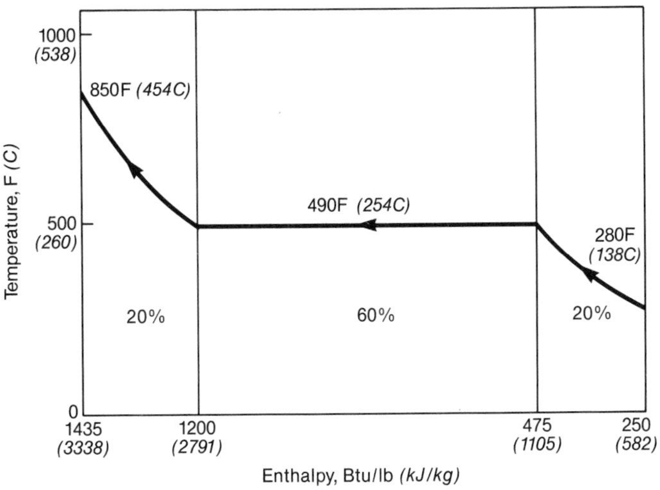

Fig. 25 Steam-water temperature profile.

Once the furnace dimensions are set, this volume is enclosed in a water-cooled membrane panel surface. This construction provides a gas-tight, all steel enclosure which minimizes energy loss, generates some steam and minimizes furnace maintenance. As shown in Fig. 23, the roof and floor tubes are inclined slightly to enhance water flow and prevent steam from collecting on the tube surface. Trapped steam could result in overheating of the tubes. Heat transfer from the flame to the furnace enclosure surfaces occurs primarily by thermal radiation. As a result, the heat input rates per unit area of surface are very high and relatively independent of the tubewall temperatures. Boiling water provides an effective means to cool the tubes and keep the tube metal temperatures within acceptable limits as long as the boiling conditions are maintained.

Fig. 26 shows the effect of the furnace on gas temperature. The gas temperature is reduced from 3600 to 2400F (1982 to 1316C), points A to B, while boiling takes place in the waterwalls (points 1 to 2). A large amount of heat transfer takes place on a small amount of surface. From the furnace, the gases pass through the furnace screen tubes shown in Fig. 23. The temperature drops a small amount [50F (28C)] from points B to C in Fig. 26 but more importantly, the superheater surface is partially shielded from the furnace thermal radiation. The furnace screen tubes are connected to the drum and contain boiling water. Next, the gas passes through the superheater where the gas temperature drops from 2350 to 1750F (1288 to 954C), points C to D. Saturated steam from the drum is passed through the superheater tubing to raise its temperature from 490F (254C) saturation temperature to the 850F (454C) desired outlet temperature (points 5 to 4).

The location of the superheater and its configuration are critical in order to keep the steam outlet temperature constant under all load conditions. This involves radiation heat transfer from the furnace with convection heat transfer from the gas passing across the surface. In addition, where dirty gases such as combustion products from coal are used, the spacing of the superheater tubes is also adjusted to accommodate the accumulation of fouling ash deposits and the use of cleaning equipment.

After the superheater, almost half of the energy in the gas stream has been recovered with only a small amount of heat transfer surface [approximately 6400 ft^2 (595 m^2)]. This is possible because of the large temperature difference between the gas and the boiling water or steam. The gas temperature has now been dramatically reduced, requiring much larger heat transfer surfaces to recover incremental amounts of energy.

The balance of the steam is generated by passing the gas through the boiler bank. (See Figs. 22 and 23.) This bank is composed of a large number of water-containing tubes that connect the steam drum to a lower (mud) drum. The temperature of the boiling water is effectively constant (points 5 to 6 in Fig. 26) while the gas temperature drops by almost 1000F (556C) to an outlet temperature of 760F (404C), points D to E. The tubes are spaced as closely as possible to increase the gas flow heat transfer rate. If a particulate laden gas stream were present, the spacing would be set to limit erosion of the tubes, reduce the heat transfer degradation due to ash deposits, and permit removal of the ash. Spacing is also controlled by the allowable pressure drop across the bank. In addition, a baffle can be used in the boiler bank bundle to force the gas to travel at higher velocity through the bundle, increase the heat transfer rate, and thereby reduce the bundle size and cost. To recover this additional 30% of the supplied energy, the boiler bank contains more than 32,000 ft^2 (3000 m^2) of surface or about nine times more surface per unit of energy than in the high temperature furnace and superheater. At this point in the process, the temperature difference between the saturated water and gas is only 270F (150C), points 6 to E in Fig. 26.

Economics and technical limits dictate the type and arrangement of additional heat transfer surfaces. An economizer or water-cooled heat exchanger could be used to heat the makeup or feedwater and cool the gas. The lowest gas exit temperature possible is the inlet temperature of the feedwater [280F (138C)]. However, the economizer would have to be infinitely large to accomplish this goal. Even if the exit gas temperature is only 310F (154C), the temperature difference at this point in the heat exchanger would only be 30F (17C), still making the heat exchanger relatively large. Instead of incorporating an economizer, an air

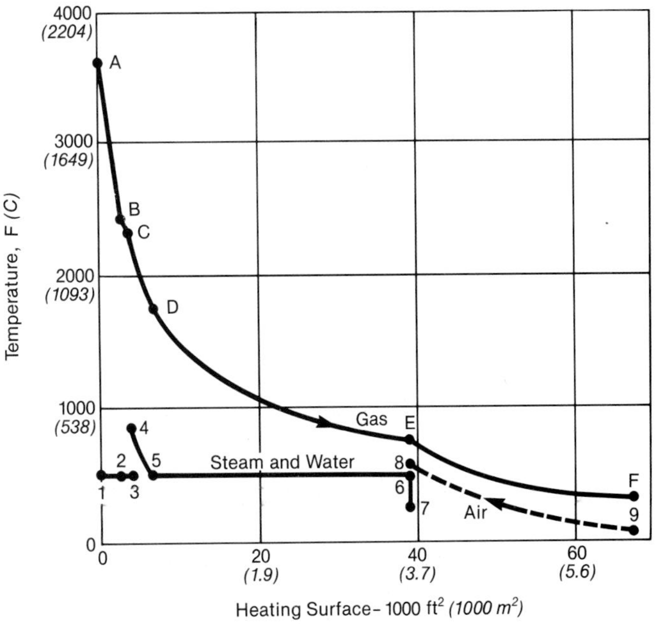

Fig. 26 Gas and steam temperature schematic.

preheater could be used to recover the remaining gas energy and preheat the combustion air. This would reduce the natural gas needed to heat the steam generator. Air heaters can be very compact. Also, air preheating can enhance the combustion of many difficult to burn fuels such as coal. All of the parameters are reviewed to select the most economical solution that meets the technical requirements.

In this case, the decision has been made to use an air heater and not an economizer. The air heater is designed to take 80F (27C) ambient temperature air (point 9) and increase the temperature to 570F (300C), point 8. This hot air is then fed to the burners. At the same time, the gas temperature is dropped from 760F (404C) to the desired 310F (154C) outlet temperature (points E to F). If a much lower gas outlet temperature than 310F (154C) is used, the heat exchanger surfaces may become uneconomically large, although this is a case by case decision. In addition, for fuels such as oil or coal which can produce acid constituents in the gas stream (such as sulfur oxides), lower exit gas temperatures may result in condensation of these constituents onto the heat transfer surfaces, and excessive corrosion damage. The gas is then exhausted through the stack to the atmosphere.

Finally, the feedwater temperature is shown jumping from 280F (138C) to saturation temperature of 490F (254C). In the absence of an economizer, the feedwater is supplied directly to the drum where it is mixed with the water flowing through the boiler bank tubes and furnace. The flow rate of this circulating water in industrial units is about 25 times higher than the feedwater flow rate. Therefore, when the feedwater is mixed in the drum, it quickly approaches the saturation temperature without appreciably lowering the temperature of the recirculating water in the boiler tubes.

Reviewing the water portion of the system, the feedwater is supplied to the drum where it mixes with the recirculating water after the steam is extracted and sent to the superheater. The drum internals are specially designed so that the now slightly subcooled water flows down through the boiler bank tubes to the lower or mud drum. This water is then distributed to the remainder of the boiler bank tubes (also called risers) and the furnace enclosure tubes where it is partially converted to steam (about 4% steam by weight). The steam-water mixture is then returned to the steam drum. Here, the steam-water mixture is passed through *separators* where the steam is separated from the water and then sent to the superheater. The steam passes through the superheater and is sent to its end use. The remaining water is mixed with the feedwater and is again distributed to the downcomer tubes.

Other steam producing systems

A variety of additional systems also produce steam for power and process applications. These systems usually take advantage of low cost or free fuels, a combination of power cycles and processes, and recovery of waste heat in order to reduce overall costs. Examples of these include:

1. *Gas turbine combined cycle (CC)* Advanced gas turbines with heat recovery steam generators as part of a bottoming cycle to use waste heat recovery and increase thermal efficiency.
2. *Integrated gasification combined cycle (IGCC)* Adds a coal gasifier to the CC to reduce fuel costs and minimize airborne emissions.
3. *Pressurized fluidized-bed combustion (PFBC)* Includes higher pressure combustion with gas cleaning and expansion of the combustion products through a gas turbine.
4. *Blast furnace hood heat recovery* Generates steam using the waste heat from a blast furnace.
5. *Solar steam generator* Uses concentrators to collect and concentrate solar radiation and generate steam.

Common technical elements

The design of steam generating systems involves the combination of scientific and technical fundamentals, empirical data, practical experience and designer insight. While technology has advanced significantly, it is still not possible to base the design of modern systems on fundamentals alone. Instead, the fundamentals provide the basis for combining field data and empirical methods.

Even given the wide variety of shapes, sizes and applications, steam generator design involves the application of a common set of technologies. The functional performance of the steam generator is established by combining thermodynamics, heat transfer, fluid mechanics, chemistry, and combustion science or nuclear science with practical knowledge of the fouling of heat transfer surfaces and empirical knowledge of the behavior of boiling water. The design and supply of the hardware are aided by structural design and advanced materials properties research combined with expertise in manufacturing technologies and erection skills to produce a quality, reliable product to meet the highly demanding system requirements.

The ASME Boiler and Pressure Vessel Code is the firm basis from which steam generator pressure parts can be safely designed. Once built, the operation and maintenance of the steam generator are critical to ensure a long life and reliable service. Water chemistry and chemical cleaning are increasingly recognized as central elements in any ongoing operating program. The impact of fuel and any residual flyash is important in evaluating the corrosion and fouling of heat transfer surfaces. The use of modern techniques to periodically inspect the integrity of the steam generator tubes leads to the ability to extend steam generator life and improve overall performance. These are accomplished by the application of engineered component modification to better meet the changing needs of the steam generating system. Finally, the control systems which monitor and operate many subsystems to optimize unit performance are important to maintain system reliability and efficiency.

All of these functions — functional performance, mechanical design, manufacture, construction, operation, maintenance, and life extension — must be fully combined to provide the best steam generating system. Long term success depends upon a complete life cycle approach to the steam generating system. Steam generator system operators routinely require their equipment to operate continuously and reliably for more than 60 years in increasingly demanding conditions. Therefore, it is important to consider later boiler life, including component replacement, in the initial phases of boiler specification. Changes in design to reduce initial capital cost must be weighed against their possible impact on future operation.

Bibliography

Aschner, F.S., *Planning Fundamentals of Thermal Power Plants*, Wiley, New York, 1977.

Axtman, W.H., Mosher, R.N., and Bahn, C.R., "The American boiler industry — a century of innovation," American Boiler Manufacturers Association, Arlington, Virginia, 1988.

Collier, J.G., and Hewitt, G.F., *Introduction to Nuclear Power*, Hemisphere, Washington, D.C., 1987.

Elliot, T.C., ed., *Standard Handbook of Powerplant Engineering*, McGraw-Hill, New York, 1989.

El-Wakil, M.M., *Powerplant Technology*, McGraw-Hill, New York, 1984.

Foster, A.R., and Wright, Jr., R.L., *Basic Nuclear Engineering*, Allyn and Bacon, Boston, 1973.

Fraas, A.P., *Heat Exchanger Design*, 2nd ed., Wiley, New York, 1989.

Gunn, D., and Horton, R., *Industrial Boilers*, Longman Scientific and Technical, London, 1989.

Jackson, A.W., "The 'how' and 'why' of boiler design," Engineers Society of Western Pennsylvania Power Symposium, Pittsburgh, February 15, 1967.

Kitto, J.B., and Albrecht, M.J., "Fossil-fuel-fired boilers: fundamentals and elements," Chapter 6 appearing in *Boilers, Evaporators, and Condensers*, S. Kakac, ed., Wiley, New York, 1991.

Li, K.W., and Priddy, A.P., *Power Plant System Design*, Wiley, New York, 1985.

Shields, C.D., *Boilers: Types, Characteristics, and Functions*, McGraw-Hill, New York, 1961.

Wiener, M., "The latest developments in natural circulation boiler design," Proceedings of the American Power Conference, Vol. 39, pp. 336-348, 1977.

"Boilers and auxiliary equipment," *Power*, Vol. 132, No. 6, pp. B-1 to B-138, June 1988.

Modern Power Station Practice, Pergamon Press, Oxford, 1991.

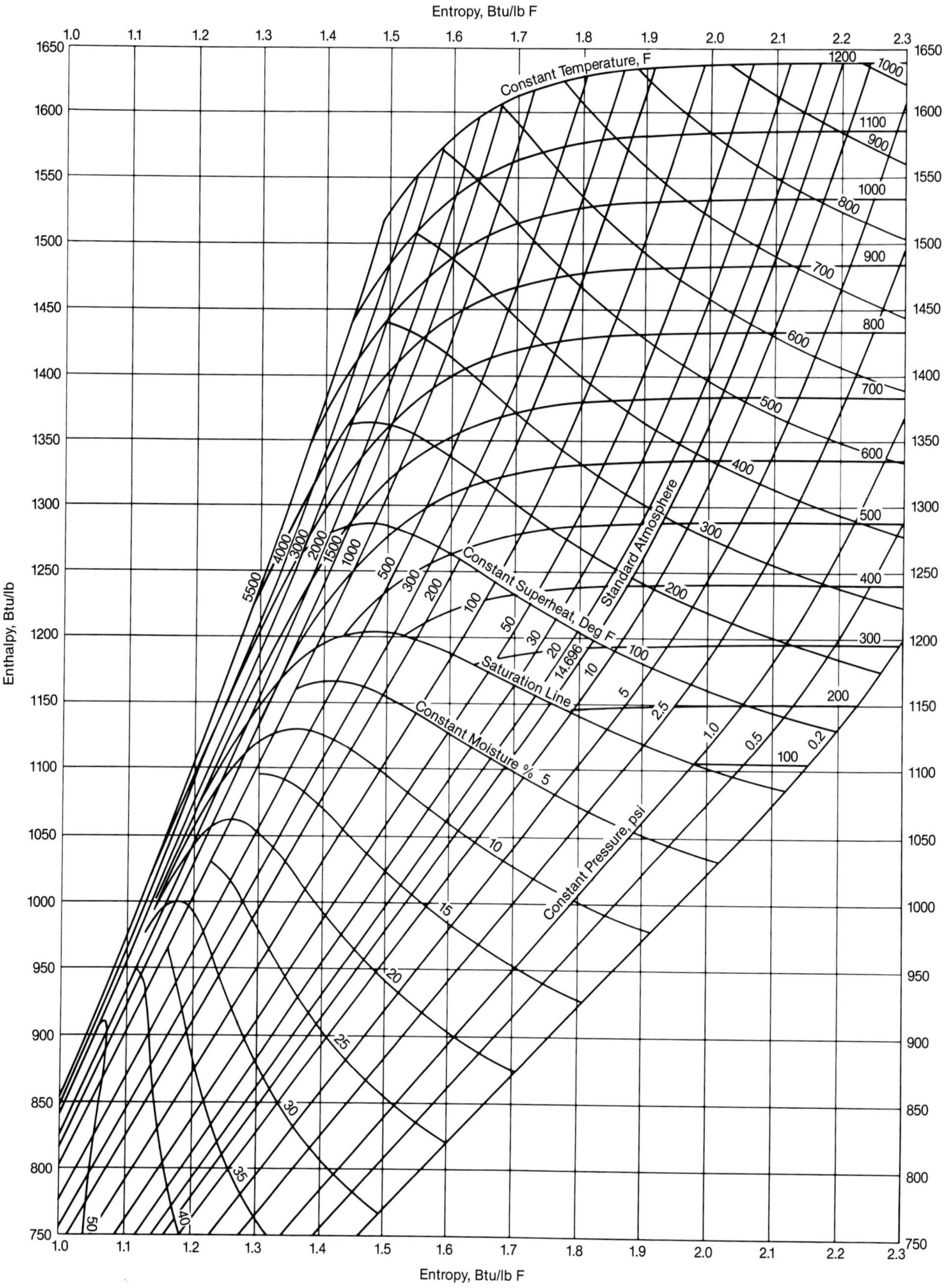

Mollier diagram (*H-s*) for steam.

Chapter 2
Thermodynamics of Steam

Thermodynamics is the science which describes and defines the transformation of one form of energy into another — chemical to thermal, thermal to mechanical and mechanical to thermal. The basic tenets include: 1) energy in all of its forms must be conserved, and 2) only a portion of *available* energy can be converted to *useful* energy or work. Generally referred to as the first and second laws of thermodynamics, these tenets evolved from the early development of the steam engine and the efforts to formalize the observations of its conversion of heat into mechanical work.

Regardless of the type of work or form of energy under consideration, the terms heat, work and energy have practical significance only when viewed in terms of systems, processes, cycles and their surroundings. In the case of expansion work, the *system* is a *fluid* capable of expansion or contraction as a result of pressure, temperature or chemical changes. The way in which these changes take place is referred to as the *process*. A *cycle* is a sequence of processes that is capable of producing net heat flow or work when placed between an energy *source* and an energy *sink*. The *surroundings* represent the sources and sinks which accommodate interchanges of mass, heat and work to or from the system.

Steam may be viewed as a thermodynamic system which is favored for power generation and heat transfer. Its unique combination of high thermal capacity (specific heat), high critical temperature, wide availability and nontoxic nature has served to maintain this dominant position. High thermal capacity of a working fluid generally results in smaller equipment for a given power output or heat transfer. The useful temperature range of water and its high thermal capacity meet the needs of many industrial processes and the temperature limitations of power conversion equipment.

Properties of steam

Before a process or cycle can be analyzed, reliable properties of the working fluid are needed. Key properties include enthalpy, entropy and specific volume. While precise definitions are provided later in this chapter, *enthalpy* is a general measure of the internally stored energy per unit mass of a flowing steam, *entropy* is a measure of the thermodynamic potential of a system in the units of energy per unit mass and *specific volume* is the volume per unit mass.

In the case of steam, a worldwide consensus of these and other thermophysical properties has been reached through the International Association for the Properties of Steam. The most frequently used tabulation of steam properties is the American Society of Mechanical Engineers (ASME) 1983 Steam Tables.[1] Selected data from this tabulation in English units are summarized in Tables 1, 2 and 3. Corresponding SI tabulations are provided in Appendix 1.

The first two columns of Tables 1 and 2 define the unique relationship between pressure and temperature referred to as saturated conditions, where liquid and vapor phases of water can coexist at thermodynamic equilibrium. For a given pressure, steam heated above the saturation temperature is referred to as superheated steam, while water cooled below the saturation temperature is referred to as subcooled or compressed water. Properties for superheated steam and compressed water are provided in Table 3. Reproduced from Reference 1, Fig. 1 shows the values of enthalpy and specific volume for steam and water over a wide range of pressure and temperature.

Under superheated or subcooled conditions, fluid properties, such as enthalpy, entropy and volume per unit mass, are unique functions of temperature and pressure. However, at saturated conditions where mixtures of steam and water coexist, the situation is more complex and requires an additional parameter for definition. For example, the enthalpy of a steam-water mixture will depend upon the relative amounts of steam and water present. This additional parameter is the thermodynamic equilibrium quality or simply quality (x) defined by convention as the mass fraction of steam:

$$x = \frac{m_s}{m_s + m_w} \quad \textbf{(1)}$$

where m_s is the mass of steam and m_w is the mass of water. The quality is frequently recorded as a percent steam by weight (% SBW) after multiplying by 100%. The mixture enthalpy (H) (see Note below), entropy (s) and

Note: To avoid confusion with the symbol for heat transfer coefficient, enthalpy (Btu/lb or kJ/kg) is denoted by H in this chapter and the balance of *Steam* unless specially noted. Enthalpy is frequently denoted by h in thermodynamic texts.

Note: The following steam tables and Fig. 1 have been abstracted from *ASME Steam Tables: Thermodynamic and Transport Properties of Steam* (copyright 1983 by the American Society of Mechanical Engineers).

Table 1
Properties of Saturated Steam and Saturated Water (Temperature)[1]

Temp F	Press. psia	Volume, ft^3/lb Water v_f	Evap v_{fg}	Steam v_g	Enthalpy,[2] Btu/lb Water H_f	Evap H_{fg}	Steam H_g	Entropy, Btu/lb F Water s_f	Evap s_{fg}	Steam s_g	Temp F
32	0.08859	0.01602	3305	3305	−0.02	1075.5	1075.5	0.0000	2.1873	2.1873	32
35	0.09991	0.01602	2948	2948	3.00	1073.8	1076.8	0.0061	2.1706	2.1767	35
40	0.12163	0.01602	2446	2446	8.03	1071.0	1079.0	0.0162	2.1432	2.1594	40
45	0.14744	0.01602	2037.7	2037.8	13.04	1068.1	1081.2	0.0262	2.1164	2.1426	45
50	0.17796	0.01602	1704.8	1704.8	18.05	1065.3	1083.4	0.0361	2.0901	2.1262	50
60	0.2561	0.01603	1207.6	1207.6	28.06	1059.7	1087.7	0.0555	2.0391	2.0946	60
70	0.3629	0.01605	868.3	868.4	38.05	1054.0	1092.1	0.0745	1.9900	2.0645	70
80	0.5068	0.01607	633.3	633.3	48.04	1048.4	1096.4	0.0932	1.9426	2.0359	80
90	0.6981	0.01610	468.1	468.1	58.02	1042.7	1100.8	0.1115	1.8970	2.0086	90
100	0.9492	0.01613	350.4	350.4	68.00	1037.1	1105.1	0.1295	1.8530	1.9825	100
110	1.2750	0.01617	265.4	265.4	77.98	1031.4	1109.3	0.1472	1.8105	1.9577	110
120	1.6927	0.01620	203.25	203.26	87.97	1025.6	1113.6	0.1646	1.7693	1.9339	120
130	2.2230	0.01625	157.32	157.33	97.96	1019.8	1117.8	0.1817	1.7295	1.9112	130
140	2.8892	0.01629	122.98	123.00	107.95	1014.0	1122.0	0.1985	1.6910	1.8895	140
150	3.718	0.01634	97.05	97.07	117.95	1008.2	1126.1	0.2150	1.6536	1.8686	150
160	4.741	0.01640	77.27	77.29	127.96	1002.2	1130.2	0.2313	1.6174	1.8487	160
170	5.993	0.01645	62.04	62.06	137.97	996.2	1134.2	0.2473	1.5822	1.8295	170
180	7.511	0.01651	50.21	50.22	148.00	990.2	1138.2	0.2631	1.5480	1.8111	180
190	9.340	0.01657	40.94	40.96	158.04	984.1	1142.1	0.2787	1.5148	1.7934	190
200	11.526	0.01664	33.62	33.64	168.09	977.9	1146.0	0.2940	1.4824	1.7764	200
210	14.123	0.01671	27.80	27.82	178.15	971.6	1149.7	0.3091	1.4509	1.7600	210
212	14.696	0.01672	26.78	26.80	180.17	970.3	1150.5	0.3121	1.4447	1.7568	212
220	17.186	0.01678	23.13	23.15	188.23	965.2	1153.4	0.3241	1.4201	1.7442	220
230	20.779	0.01685	19.364	19.381	198.33	958.7	1157.1	0.3388	1.3902	1.7290	230
240	24.968	0.01693	16.304	16.321	208.45	952.1	1160.6	0.3533	1.3609	1.7142	240
250	29.825	0.01701	13.802	13.819	218.59	945.4	1164.0	0.3677	1.3323	1.7000	250
260	35.427	0.01709	11.745	11.762	228.76	938.6	1167.4	0.3819	1.3043	1.6862	260
270	41.856	0.01718	10.042	10.060	238.95	931.7	1170.6	0.3960	1.2769	1.6729	270
280	49.200	0.01726	8.627	8.644	249.17	924.6	1173.8	0.4098	1.2501	1.6599	280
290	57.550	0.01736	7.443	7.460	259.4	917.4	1176.8	0.4236	1.2238	1.6473	290
300	67.005	0.01745	6.448	6.466	269.7	910.0	1179.7	0.4372	1.1979	1.6351	300
310	77.67	0.01755	5.609	5.626	280.0	902.5	1182.5	0.4506	1.1726	1.6232	310
320	89.64	0.01766	4.896	4.914	290.4	894.8	1185.2	0.4640	1.1477	1.6116	320
340	117.99	0.01787	3.770	3.788	311.3	878.8	1190.1	0.4902	1.0990	1.5892	340
360	153.01	0.01811	2.939	2.957	332.3	862.1	1194.4	0.5161	1.0517	1.5678	360
380	195.73	0.01836	2.317	2.335	353.6	844.5	1198.0	0.5416	1.0057	1.5473	380
400	247.26	0.01864	1.8444	1.8630	375.1	825.9	1201.0	0.5667	0.9607	1.5274	400
420	308.78	0.01894	1.4808	1.4997	396.9	806.2	1203.1	0.5915	0.9165	1.5080	420
440	381.54	0.01926	1.1976	1.2169	419.0	785.4	1204.4	0.6161	0.8729	1.4890	440
460	466.9	0.0196	0.9746	0.9942	441.5	763.2	1204.8	0.6405	0.8299	1.4704	460
480	566.2	0.0200	0.7972	0.8172	464.5	739.6	1204.1	0.6648	0.7871	1.4518	480
500	680.9	0.0204	0.6545	0.6749	487.9	714.3	1202.2	0.6890	0.7443	1.4333	500
520	812.5	0.0209	0.5386	0.5596	512.0	687.0	1199.0	0.7133	0.7013	1.4146	520
540	962.8	0.0215	0.4437	0.4651	536.8	657.5	1194.3	0.7378	0.6577	1.3954	540
560	1133.4	0.0221	0.3651	0.3871	562.4	625.3	1187.7	0.7625	0.6132	1.3757	560
580	1326.2	0.0228	0.2994	0.3222	589.1	589.9	1179.0	0.7876	0.5673	1.3550	580
600	1543.2	0.0236	0.2438	0.2675	617.1	550.6	1167.7	0.8134	0.5196	1.3330	600
620	1786.9	0.0247	0.1962	0.2208	646.9	506.3	1153.2	0.8403	0.4689	1.3092	620
640	2059.9	0.0260	0.1543	0.1802	679.1	454.6	1133.7	0.8686	0.4134	1.2821	640
660	2365.7	0.0277	0.1166	0.1443	714.9	392.1	1107.0	0.8995	0.3502	1.2498	660
680	2708.6	0.0304	0.0808	0.1112	758.5	310.1	1068.5	0.9365	0.2720	1.2086	680
700	3094.3	0.0366	0.0386	0.0752	822.4	172.7	995.2	0.9901	0.1490	1.1390	700
705.5	3208.2	0.0508	0	0.0508	906.0	0	906.0	1.0612	0	1.0612	705.5

1. SI steam tables are provided in Appendix 1.
2. In the balance of *Steam,* enthalpy is denoted by H in place of h to avoid confusion with heat transfer coefficient.

Table 2
Properties of Saturated Steam and Saturated Water (Pressure)[1]

Press. psia	Temp F	Volume, ft^3/lb			Enthalpy,[2] Btu/lb			Entropy, Btu/lb F			Energy, Btu/lb		Press. psia
		Water v_f	Evap v_{fg}	Steam v_g	Water H_f	Evap H_{fg}	Steam H_g	Water s_f	Evap s_{fg}	Steam s_g	Water u_f	Steam u_g	
0.0886	32.018	0.01602	3302.4	3302.4	0.00	1075.5	1075.5	0	2.1872	2.1872	0	1021.3	0.0886
0.10	35.023	0.01602	2945.5	2945.5	3.03	1073.8	1076.8	0.0061	2.1705	2.1766	3.03	1022.3	0.10
0.15	45.453	0.01602	2004.7	2004.7	13.50	1067.9	1081.4	0.0271	2.1140	2.1411	13.50	1025.7	0.15
0.20	53.160	0.01603	1526.3	1526.3	21.22	1063.5	1084.7	0.0422	2.0738	2.1160	21.22	1028.3	0.20
0.30	64.484	0.01604	1039.7	1039.7	32.54	1057.1	1089.7	0.0641	2.0168	2.0809	32.54	1032.0	0.30
0.40	72.869	0.01606	792.0	792.1	40.92	1052.4	1093.3	0.0799	1.9762	2.0562	40.92	1034.7	0.40
0.5	79.586	0.01607	641.5	641.5	47.62	1048.6	1096.3	0.0925	1.9446	2.0370	47.62	1036.9	0.5
0.6	85.218	0.01609	540.0	540.1	53.25	1045.5	1098.7	0.1028	1.9186	2.0215	53.24	1038.7	0.6
0.7	90.09	0.01610	466.93	466.94	58.10	1042.7	1100.8	0.3	1.8966	2.0083	58.10	1040.3	0.7
0.8	94.38	0.01611	411.67	411.69	62.39	1040.3	1102.6	0.1117	1.8775	1.9970	62.39	1041.7	0.8
0.9	98.24	0.01612	368.41	368.43	66.24	1038.1	1104.3	0.1264	1.8606	1.9870	66.24	1042.9	0.9
1.0	101.74	0.01614	333.59	333.60	69.73	1036.1	1105.8	0.1326	1.8455	1.9781	69.73	1044.1	1.0
2.0	126.07	0.01623	173.74	173.76	94.03	1022.1	1116.2	0.1750	1.7450	1.9200	94.03	1051.8	2.0
3.0	141.47	0.01630	118.71	118.73	109.42	1013.2	1122.6	0.2009	1.6854	1.8864	109.41	1056.7	3.0
4.0	152.96	0.01636	90.63	90.64	120.92	1006.4	1127.3	0.2199	1.6428	1.8626	120.90	1060.2	4.0
5.0	162.24	0.01641	73.515	73.53	130.20	1000.9	1131.1	0.2349	1.6094	1.8443	130.18	1063.1	5.0
6.0	170.05	0.01645	61.967	61.98	138.03	996.2	1134.2	0.2474	1.5820	1.8294	138.01	1065.4	6.0
7.0	176.84	0.01649	53.634	53.65	144.83	992.1	1136.9	0.2581	1.5587	1.8168	144.81	1067.4	7.0
8.0	182.86	0.01653	47.328	47.35	150.87	988.5	1139.3	0.2676	1.5384	1.8060	150.84	1069.2	8.0
9.0	188.27	0.01656	42.385	42.40	156.30	985.1	1141.4	0.2760	1.5204	1.7964	156.28	1070.8	9.0
10	193.21	0.01659	38.404	38.42	161.26	982.1	1143.3	0.2836	1.5043	1.7879	161.23	1072.3	10
14.696	212.00	0.01672	26.782	26.80	180.17	970.3	1150.5	0.3121	1.4447	1.7568	180.12	1077.6	14.696
15	213.03	0.01673	26.274	26.29	181.21	969.7	1150.9	0.3137	1.4415	1.7552	181.16	1077.9	15
20	227.96	0.01683	20.070	20.087	196.27	960.1	1156.3	0.3358	1.3962	1.7320	196.21	1082.0	20
30	250.34	0.01701	13.7266	13.744	218.9	945.2	1164.1	0.3682	1.3313	1.6995	218.8	1087.9	30
40	267.25	0.01715	10.4794	10.497	236.1	933.6	1169.8	0.3921	1.2844	1.6765	236.0	1092.1	40
50	281.02	0.01727	8.4967	8.514	250.2	923.9	1174.1	0.4112	1.2474	1.6586	250.1	1095.3	50
60	292.71	0.01738	7.1562	7.174	262.2	915.4	1177.6	0.4273	1.2167	1.6440	262.0	1098.0	60
70	302.93	0.01748	6.1875	6.205	272.7	907.8	1180.6	0.4411	1.1905	1.6316	272.5	1100.2	70
80	312.04	0.01757	5.4536	5.471	282.1	900.9	1183.1	0.4534	1.1675	1.6208	281.9	1102.1	80
90	320.28	0.01766	4.8777	4.895	290.7	894.6	1185.3	0.4643	1.1470	1.6113	290.4	1103.7	90
100	327.82	0.01774	4.4133	4.431	298.5	888.6	1187.2	0.4743	1.1284	1.6027	298.2	1105.2	100
120	341.27	0.01789	3.7097	3.728	312.6	877.8	1190.4	0.4919	1.0960	1.5879	312.2	1107.6	120
140	353.04	0.01803	3.2010	3.219	325.0	868.0	1193.0	0.5071	1.0681	1.5752	324.5	1109.6	140
160	363.55	0.01815	2.8155	2.834	336.1	859.0	1195.1	0.5206	1.0435	1.5641	335.5	1111.2	160
180	373.08	0.01827	2.5129	2.531	346.2	850.7	1196.9	0.5328	1.0215	1.5543	345.6	1112.5	180
200	381.80	0.01839	2.2689	2.287	355.5	842.8	1198.3	0.5438	1.0016	1.5454	354.8	1113.7	200
250	400.97	0.01865	1.8245	1.8432	376.1	825.0	1201.1	0.5679	0.9585	1.5264	375.3	1115.8	250
300	417.35	0.01889	1.5238	1.5427	394.0	808.9	1202.9	0.5882	0.9223	1.5105	392.9	1117.2	300
350	431.73	0.01913	1.3064	1.3255	409.8	794.2	1204.0	0.6059	0.8909	1.4968	408.6	1118.1	350
400	444.60	0.0193	1.14162	1.1610	424.2	780.4	1204.6	0.6217	0.8630	1.4847	422.7	1118.7	400
450	456.28	0.0195	1.01224	1.0318	437.3	767.5	1204.8	0.6360	0.8378	1.4738	435.7	1118.9	450
500	467.01	0.0198	0.90787	0.9276	449.5	755.1	1204.7	0.6490	0.8148	1.4639	447.7	1118.8	500
550	476.94	0.0199	0.82183	0.8418	460.9	743.3	1204.3	0.6611	0.7936	1.4547	458.9	1118.6	550
600	486.20	0.0201	0.74962	0.7698	471.7	732.0	1203.7	0.6723	0.7738	1.4461	469.5	1118.2	600
700	503.08	0.0205	0.63505	0.6556	491.6	710.2	1201.8	0.6928	0.7377	1.4304	488.9	1116.9	700
800	518.21	0.0209	0.54809	0.5690	509.8	689.6	1199.4	0.7111	0.7051	1.4163	506.7	1115.2	800
900	531.95	0.0212	0.47968	0.5009	526.7	669.7	1196.4	0.7279	0.6753	1.4032	523.2	1113.0	900
1000	544.58	0.0216	0.42436	0.4460	542.6	650.4	1192.9	0.7434	0.6476	1.3910	538.6	1110.4	1000
1100	556.28	0.0220	0.37863	0.4006	557.5	631.5	1189.1	0.7578	0.6216	1.3794	553.1	1107.5	1100
1200	567.19	0.0223	0.34013	0.3625	571.9	613.0	1184.8	0.7714	0.5969	1.3683	566.9	1104.3	1200
1300	577.42	0.0227	0.30722	0.3299	585.6	594.6	1180.2	0.7843	0.5733	1.3577	580.1	1100.9	1300
1400	587.07	0.0231	0.27871	0.3018	598.8	576.5	1175.3	0.7966	0.5507	1.3474	592.9	1097.1	1400
1500	596.20	0.0235	0.25372	0.2772	611.7	558.4	1170.1	0.8085	0.5288	1.3373	605.2	1093.1	1500
2000	635.80	0.0257	0.16266	0.1883	672.1	466.2	1138.3	0.8625	0.4256	1.2881	662.6	1068.6	2000
2500	668.11	0.0286	0.10209	0.1307	731.7	361.6	1093.3	0.9139	0.3206	1.2345	718.5	1032.9	2500
3000	695.33	0.0343	0.05073	0.0850	801.8	218.4	1020.3	0.9728	0.1891	1.1619	782.8	973.1	3000
3208.2	705.47	0.0508	0	0.0508	906.0	0	906.0	1.0612	0	1.0612	875.9	875.9	3208.2

1. See Note 1, Table 1.
2. See Note 2, Table 1.

Table 3
Properties of Superheated Steam and Compressed Water (Temperature and Pressure)[1]

Press., psia (sat. temp)		Temperature, F 100	200	300	400	500	600	700	800	900	1000	1100	1200	1300	1400	1500
1	v	0.0161	392.5	452.3	511.9	571.5	631.1	690.7								
	H	68.00	1150.2	1195.7	1241.8	1288.6	1336.1	1384.5								
(101.74)	s	0.1295	2.0509	2.1152	2.1722	2.2237	2.2708	2.3144								
5	v	0.0161	78.14	90.24	102.24	114.21	126.15	138.08	150.01	161.94	173.86	185.78	197.70	209.62	221.53	233.45
	H	68.01	1148.6	1194.8	1241.3	1288.2	1335.9	1384.3	1433.6	1483.7	1534.7	1586.7	1639.6	1693.3	1748.0	1803.5
(162.24)	s	0.1295	1.8716	1.9369	1.9943	2.0460	2.0932	2.1369	2.1776	2.2159	2.2521	2.2866	2.3194	2.3509	2.3811	2.4101
10	v	0.0161	38.84	44.98	51.03	57.04	63.03	69.00	74.98	80.94	86.91	92.87	98.84	104.80	110.76	116.72
	H	68.02	1146.6	1193.7	1240.6	1287.8	1335.5	1384.0	1433.4	1483.5	1534.6	1586.6	1639.5	1693.3	1747.9	1803.4
(193.21)	s	0.1295	1.7928	1.8593	1.9173	1.9692	2.0166	2.0603	2.1011	2.1394	2.1757	2.2101	2.2430	2.2744	2.3046	2.3337
15	v	0.0161	0.0166	29.899	33.963	37.985	41.986	45.978	49.964	53.946	57.926	61.905	65.882	69.858	73.833	77.807
	H	68.04	168.09	1192.5	1239.9	1287.3	1335.2	1383.8	1433.2	1483.4	1534.5	1586.5	1639.4	1693.2	1747.8	1803.4
(213.03)	s	0.1295	0.2940	1.8134	1.8720	1.9242	1.9717	2.0155	2.0563	2.0946	2.1309	2.1653	2.1982	2.2297	2.2599	2.2890
20	v	0.0161	0.0166	22.356	25.428	28.457	31.466	34.465	37.458	40.447	43.435	46.420	49.405	52.388	55.370	58.352
	H	68.05	168.11	1191.4	1239.2	1286.9	1334.9	1383.5	1432.9	1483.2	1534.3	1586.3	1639.3	1693.1	1747.8	1803.3
(227.96)	s	0.1295	0.2940	1.7805	1.8397	1.8921	1.9397	1.9836	2.0244	2.0628	2.0991	2.1336	2.1665	2.1979	2.2282	2.2572
40	v	0.0161	0.0166	11.036	12.624	14.165	15.685	17.195	18.699	20.199	21.697	23.194	24.689	26.183	27.676	29.168
	H	68.10	168.15	1186.6	1236.4	1285.0	1333.6	1382.5	1432.1	1482.5	1533.7	1585.8	1638.8	1992.7	1747.5	1803.0
(267.25)	s	0.1295	0.2940	1.6992	1.7608	1.8143	1.8624	1.9065	1.9476	1.9860	2.0224	2.0569	2.0899	2.1224	2.1516	2.1807
60	v	0.0161	0.0166	7.257	8.354	9.400	10.425	11.438	12.446	13.450	14.452	15.452	16.450	17.448	18.445	19.441
	H	68.15	168.20	1181.6	1233.5	1283.2	1332.3	1381.5	1431.3	1481.8	1533.2	1585.3	1638.4	1692.4	1747.1	1802.8
(292.71)	s	0.1295	0.2939	1.6492	1.7134	1.7681	1.8168	1.8612	1.9024	1.9410	1.9774	2.0120	2.0450	2.0765	2.1068	2.1359
80	v	0.0161	0.0166	0.0175	6.218	7.018	7.794	8.560	9.319	10.075	10.829	11.581	12.331	13.081	13.829	14.577
	H	68.21	168.24	269.74	1230.5	1281.3	1330.9	1380.5	1430.5	1481.1	1532.6	1584.9	1638.0	1692.0	1746.8	1802.5
(312.04)	s	0.1295	0.2939	0.4371	1.6790	1.7349	1.7842	1.8289	1.8702	1.9089	1.9454	1.9800	2.0131	2.0446	2.0750	2.1041
100	v	0.0161	0.0166	0.0175	4.935	5.588	6.216	6.833	7.443	8.050	8.655	9.258	9.860	10.460	11.060	11.659
	H	68.26	168.29	269.77	1227.4	1279.3	1329.6	1379.5	1429.7	1480.4	1532.0	1584.4	1637.6	1691.6	1746.5	1802.2
(327.82)	s	0.1295	0.2939	0.4371	1.6516	1.7088	1.7586	1.8036	1.8451	1.8839	1.9205	1.9552	1.9883	2.0199	2.0502	2.0794
120	v	0.0161	0.0166	0.0175	4.0786	4.6341	5.1637	5.6831	6.1928	6.7006	7.2060	7.7096	8.2119	8.7130	9.2134	9.7130
	H	68.31	168.33	269.81	1224.1	1277.4	1328.1	1378.4	1428.8	1479.8	1531.4	1583.9	1637.1	1691.3	1746.2	1802.0
(341.27)	s	0.1295	0.2939	0.4371	1.6286	1.6872	1.7376	1.7829	1.8246	1.8635	1.9001	1.9349	1.9680	1.9996	2.0300	2.0592
140	v	0.0161	0.0166	0.0175	3.4661	3.9526	4.4119	4.8585	5.2995	5.7364	6.1709	6.6036	7.0349	7.4652	7.8946	8.3233
	H	68.37	168.38	269.85	1220.8	1275.3	1326.8	1377.4	1428.0	1479.1	1530.8	1583.4	1636.7	1690.9	1745.9	1801.7
(353.04)	s	0.1295	0.2939	0.4370	1.6085	1.6686	1.7196	1.7652	1.8071	1.8461	1.8828	1.9176	1.9508	1.9825	2.0129	2.0421
160	v	0.0161	0.0166	0.0175	3.0060	3.4413	3.8480	4.2420	4.6295	5.0132	5.3945	5.7741	6.1522	6.5293	6.9055	7.2811
	H	68.42	168.42	269.89	1217.4	1273.3	1325.4	1376.4	1427.2	1478.4	1530.3	1582.9	1636.3	1690.5	1745.6	1801.4
(363.55)	s	0.1294	0.2938	0.4370	1.5906	1.6522	1.7039	1.7499	1.7919	1.8310	1.8678	1.9027	1.9359	1.9676	1.9980	2.0273
180	v	0.0161	0.0166	0.0174	2.6474	3.0433	3.4093	3.7621	4.1084	4.4505	4.7907	5.1289	5.4657	5.8014	6.1363	6.4704
	H	68.47	168.47	269.92	1213.8	1271.2	1324.0	1375.3	1426.3	1477.7	1529.7	1582.4	1635.9	1690.2	1745.3	1801.2
(373.08)	s	0.1294	0.2938	0.4370	1.5743	1.6376	1.6900	1.7362	1.7784	1.8176	1.8545	1.8894	1.9227	1.9545	1.9849	2.0142
200	v	0.0161	0.0166	0.0174	2.3598	2.7247	3.0583	3.3783	3.6915	4.0008	4.3077	4.6128	4.9165	5.2191	5.5209	5.8219
	H	68.52	168.51	269.96	1210.1	1269.0	1322.6	1374.3	1425.5	1477.0	1529.1	1581.9	1635.4	1689.8	1745.0	1800.9
(381.80)	s	0.1294	0.2938	0.4369	1.5593	1.6242	1.6776	1.7239	1.7663	1.8057	1.8426	1.8776	1.9109	1.9427	1.9732	2.0025
250	v	0.0161	0.0166	0.0174	0.0186	2.1504	2.4662	2.6872	2.9410	3.1909	3.4382	3.6837	3.9278	4.1709	4.4131	4.6546
	H	68.66	168.63	270.05	375.10	1263.5	1319.0	1371.6	1423.4	1475.3	1527.6	1580.6	1634.4	1688.9	1744.2	1800.2
(400.97)	s	0.1294	0.2937	0.4368	0.5667	1.5951	1.6502	1.6976	1.7405	1.7801	1.8173	1.8524	1.8858	1.9177	1.9482	1.9776
300	v	0.0161	0.0166	0.0174	0.0186	1.7665	2.0044	2.2263	2.4407	2.6509	2.8585	3.0643	3.2688	3.4721	3.6746	3.8764
	H	68.79	168.74	270.14	375.15	1257.7	1315.2	1368.9	1421.3	1473.6	1526.2	1579.4	1633.3	1688.0	1743.4	1799.6
(417.35)	s	0.1294	0.2937	0.4307	0.5665	1.5703	1.6274	1.6758	1.7192	1.7591	1.7964	1.8317	1.8652	1.8972	1.9278	1.9572
350	v	0.0161	0.0166	0.0174	0.0186	1.4913	1.7028	1.8970	2.0832	2.2652	2.4445	2.6219	2.7980	2.9730	3.1471	3.3205
	H	68.92	168.85	270.24	375.21	1251.5	1311.4	1366.2	1419.2	1471.8	1524.7	1578.2	1632.3	1687.1	1742.6	1798.9
(431.73)	s	0.1293	0.2936	0.4367	0.5664	1.5483	1.6077	1.6571	1.7009	1.7411	1.7787	1.8141	1.8477	1.8798	1.9105	1.9400
400	v	0.0161	0.0166	0.0174	0.0162	1.2841	1.4763	1.6499	1.8151	1.9759	2.1339	2.2901	2.4450	2.5987	2.7515	2.9037
	H	69.05	168.97	270.33	375.27	1245.1	1307.4	1363.4	1417.0	1470.1	1523.3	1576.9	1631.2	1686.2	1741.9	1798.2
(444.60)	s	0.1293	0.2935	0.4366	0.5663	1.5282	1.5901	1.6406	1.6850	1.7255	1.7632	1.7988	1.8325	1.8647	1.8955	1.9250
500	v	0.0161	0.0166	0.0174	0.0186	0.9919	1.1584	1.3037	1.4397	1.5708	1.6992	1.8256	1.9507	2.0746	2.1977	2.3200
	H	69.32	169.19	270.51	375.38	1231.2	1299.1	1357.7	1412.7	1466.6	1520.3	1574.4	1629.1	1684.4	1740.3	1796.9
(467.01)	s	0.1292	0.2934	0.4364	0.5660	1.4921	1.5595	1.6123	1.6578	1.6990	1.7371	1.7730	1.8069	1.8393	1.8702	1.8998

1. See Notes 1 and 2, Table 1.

Table 3
Properties of Superheated Steam and Compressed Water (Temperature and Pressure)[1]

Press., psia (sat. temp)		Temperature, F 100	200	300	400	500	600	700	800	900	1000	1100	1200	1300	1400	1500
600 (486.20)	v	0.0161	0.0166	0.0174	0.0186	0.7944	0.9456	1.0726	1.1892	1.3008	1.4093	1.5160	1.6211	1.7252	1.8284	1.9309
	H	69.58	169.42	270.70	375.49	1215.9	1290.3	1351.8	1408.3	1463.0	1517.4	1571.9	1627.0	1682.6	1738.8	1795.6
	s	0.1292	0.2933	0.4362	0.5657	1.4590	1.5329	1.5844	1.6351	1.6769	1.7155	1.7517	1.7859	1.8184	1.8494	1.8792
700 (503.08)	v	0.0161	0.0166	0.0174	0.0186	0.0204	0.7928	0.9072	1.0102	1.1078	1.2023	1.2948	1.3858	1.4757	1.5647	1.6530
	H	69.84	169.65	270.89	375.61	487.93	1281.0	1345.6	1403.7	1459.4	1514.4	1569.4	1624.8	1680.7	1737.2	1794.3
	s	0.1291	0.2932	0.4360	0.5655	0.6889	1.5090	1.5673	1.6154	1.6580	1.6970	1.7335	1.7679	1.8006	1.8318	1.8617
800 (518.21)	v	0.0161	0.0166	0.0174	0.0186	0.0204	0.6774	0.7828	0.8759	0.9631	1.0470	1.1289	1.2093	1.2885	1.3669	1.4446
	H	70.11	169.88	271.07	375.73	487.88	1271.1	1339.2	1399.1	1455.8	1511.4	1566.9	1622.7	1678.9	1735.0	1792.9
	s	0.1290	0.2930	0.4358	0.5652	0.6885	1.4869	1.5484	1.5980	1.6413	1.6807	1.7175	1.7522	1.7851	1.8164	1.8464
900 (531.95)	v	0.0161	0.0166	0.0174	0.0186	0.0204	0.5869	0.6858	0.7713	0.8504	0.9262	0.9998	1.0720	1.1430	1.2131	1.2825
	H	70.37	170.10	271.26	375.84	487.83	1260.6	1332.7	1394.4	1452.2	1508.5	1564.4	1620.6	1677.1	1734.1	1791.6
	s	0.1290	0.2929	0.4357	0.5649	0.6881	1.4659	1.5311	1.5822	1.6263	1.6662	1.7033	1.7382	1.7713	1.8028	1.8329
1000 (544.58)	v	0.0161	0.0166	0.0174	0.0186	0.0204	0.5137	0.6080	0.6875	0.7603	0.8295	0.8966	0.9622	1.0266	1.0901	1.1529
	H	70.63	170.33	271.44	375.96	487.79	1249.3	1325.9	1389.6	1448.5	1504.4	1561.9	1618.4	1675.3	1732.5	1790.3
	s	0.1289	0.2928	0.4355	0.5647	0.6876	1.4457	1.5149	1.5677	1.6126	1.6530	1.6905	1.7256	1.7589	1.7905	1.8207
1100 (556.28)	v	0.0161	0.0166	0.0174	0.0185	0.0203	0.4531	0.5440	0.6188	0.6865	0.7505	0.8121	0.8723	0.9313	0.9894	1.0468
	H	70.90	170.56	271.63	376.08	487.75	1237.3	1318.8	1384.7	1444.7	1502.4	1559.4	1616.3	1673.5	1731.0	1789.0
	s	0.1289	0.2927	0.4353	0.5644	0.6872	1.4259	1.4996	1.5542	1.6000	1.6410	1.6787	1.7141	1.7475	1.7793	1.8097
1200 (567.19)	v	0.0161	0.0166	0.0174	0.0185	0.0203	0.4016	0.4905	0.5615	0.6250	0.6845	0.7418	0.7974	0.8519	0.9055	0.9584
	H	71.16	170.78	271.82	376.20	487.72	1224.2	1311.5	1379.7	1440.9	1499.4	1556.9	1614.2	1671.6	1729.4	1787.6
	s	0.1288	0.2926	0.4351	0.5642	0.6868	1.4061	1.4851	1.5415	1.5883	1.6298	1.6679	1.7035	1.7371	1.7691	1.7996
1400 (587.07)	v	0.0161	0.0166	0.0174	0.0185	0.0203	0.3176	0.4059	0.4712	0.5282	0.5809	0.6311	0.6798	0.7272	0.7737	0.8195
	H	71.68	171.24	272.19	376.44	487.65	1194.1	1296.1	1369.3	1433.2	1493.2	1551.8	1609.9	1668.0	1726.3	1785.0
	s	0.1287	0.2923	0.4348	0.5636	0.6859	1.3652	1.4575	1.5182	1.5670	1.6096	1.6484	1.6845	1.7185	1.7508	1.7815
1600 (604.87)	v	0.0161	0.0166	0.0173	0.0185	0.0202	0.0236	0.3415	0.4032	0.4555	0.5031	0.5482	0.5915	0.6336	0.6748	0.7153
	H	72.21	171.69	272.57	376.69	487.60	616.77	1279.4	1358.5	1425.2	1486.9	1546.6	1605.6	1664.3	1723.2	1782.3
	s	0.1286	0.2921	0.4344	0.5631	0.6851	0.8129	1.4312	1.4968	1.5478	1.5916	1.6312	1.6678	1.7022	1.7344	1.7657
1800 (621.02)	v	0.0160	0.0165	0.0173	0.0185	0.0202	0.0235	0.2906	0.3500	0.3988	0.4426	0.4836	0.5229	0.5609	0.5980	0.6343
	H	72.73	172.15	272.95	376.93	487.56	615.58	1261.1	1347.2	1417.1	1480.6	1541.1	1601.2	1660.7	1720.1	1779.7
	s	0.1284	0.2918	0.4341	0.5626	0.6843	0.8109	1.4054	1.4768	1.5302	1.5753	1.6156	1.6528	1.6876	1.7204	1.7516
2000 (635.80)	v	0.0160	0.0165	0.0173	0.0184	0.0201	0.0233	0.2488	0.3072	0.3534	0.3942	0.4320	0.4680	0.5027	0.5365	0.5695
	H	73.26	172.60	273.32	377.19	487.53	614.48	1240.9	1353.4	1408.7	1474.1	1536.2	1596.9	1657.0	1717.0	1777.1
	s	0.1283	0.2916	0.4337	0.5621	0.6834	0.8091	1.3794	1.4578	1.5138	1.5603	1.6014	1.6391	1.6743	1.7075	1.7389
2500 (668.11)	v	0.0160	0.0165	0.0173	0.0184	0.0200	0.0230	0.1681	0.2293	0.2712	0.3068	0.3390	0.3692	0.3980	0.4259	0.4529
	H	74.57	173.74	274.27	377.82	487.50	612.08	1176.7	1303.4	1386.7	1457.5	1522.9	1585.9	1647.8	1709.2	1770.4
	s	0.1280	0.2910	0.4329	0.5609	0.6815	0.8048	1.3076	1.4129	1.4766	1.5269	1.5703	1.6094	1.6456	1.6796	1.7116
3000 (695.33)	v	0.0160	0.0165	0.0172	0.0183	0.0200	0.0228	0.0982	0.1759	0.2161	0.2484	0.2770	0.3033	0.3282	0.3522	0.3753
	H	75.88	174.88	275.22	378.47	487.52	610.08	1060.5	1267.0	1363.2	1440.2	1509.4	1574.8	1638.5	1701.4	1761.8
	s	0.1277	0.2904	0.4320	0.5597	0.6796	0.8009	1.1966	1.3692	1.4429	1.4976	1.5434	1.5841	1.6214	1.6561	1.6888
3200 (705.08)	v	0.0160	0.0165	0.0172	0.0183	0.0199	0.0227	0.0335	0.1588	0.1987	0.2301	0.2576	0.2827	0.3065	0.3291	0.3510
	H	76.4	175.3	275.6	378.7	487.5	609.4	800.8	1250.9	1353.4	1433.1	1503.8	1570.3	1634.8	1698.3	1761.2
	s	0.1276	0.2902	0.4317	0.5592	0.6788	0.7994	0.9708	1.3515	1.4300	1.4866	1.5335	1.5749	1.6126	1.6477	1.6806
3500	v	0.0160	0.0164	0.0172	0.0183	0.0199	0.0225	0.0307	0.1364	0.1764	0.2066	0.2326	0.2563	0.2784	0.2995	0.3198
	H	77.2	176.0	276.2	379.1	487.6	608.4	779.4	1224.6	1338.2	1422.2	1495.5	1563.3	1629.2	1693.6	1757.2
	s	0.1274	0.2899	0.4312	0.5585	0.6777	0.7973	0.9508	1.3242	1.4112	1.4709	1.5194	1.5618	1.6002	1.6358	1.6691
4000	v	0.0159	0.0164	0.0172	0.0182	0.0198	0.0223	0.0287	0.1052	0.1463	0.1752	0.1994	0.2210	0.2411	0.2601	0.2783
	H	78.5	177.2	277.1	379.8	487.7	606.9	763.0	1174.3	1311.6	1403.6	1481.3	1552.2	1619.8	1685.7	1750.6
	s	0.1271	0.2893	0.4304	0.5573	0.6760	0.7940	0.9343	1.2754	1.3807	1.4461	1.4976	1.5417	1.5812	1.6177	1.6516
5000	v	0.0159	0.0164	0.0171	0.0181	0.0196	0.0219	0.0268	0.0591	0.1038	0.1312	0.1529	0.1718	0.1890	0.2050	0.2203
	H	81.1	179.5	279.1	381.2	488.1	604.6	746.0	1042.9	1252.9	1364.6	1452.1	1529.1	1600.9	1670.0	1737.4
	s	0.1265	0.2881	0.4287	0.5550	0.6726	0.7880	0.9153	1.1593	1.3207	1.4001	1.4582	1.5061	1.5481	1.5863	1.6216
6000	v	0.0159	0.0163	0.0170	0.0180	0.0195	0.0216	0.0256	0.0397	0.0757	0.1020	0.1221	0.1391	0.1544	0.1684	0.1817
	H	83.7	181.7	281.0	382.7	488.6	602.9	736.1	945.1	1188.8	1323.6	1422.3	1505.9	1582.0	1654.2	1724.2
	s	0.1258	0.2870	0.4271	0.5528	0.6693	0.7826	0.9026	1.0176	1.2615	1.3574	1.4229	1.4748	1.5194	1.5593	1.5962
7000	v	0.0158	0.0163	0.0170	0.0180	0.0193	0.0213	0.0248	0.0334	0.0573	0.0816	0.1004	0.1160	0.1298	0.1424	0.1542
	H	86.2	184.4	283.0	384.2	489.3	601.7	729.3	901.8	1124.9	1281.7	1392.2	1482.6	1563.1	1638.6	1711.1
	s	0.1252	0.2859	0.4256	0.5507	0.6663	0.7777	0.8926	1.0350	1.2055	1.3171	1.3904	1.4466	1.4938	1.5355	1.5735

1. See Notes 1 and 2, Table 1.

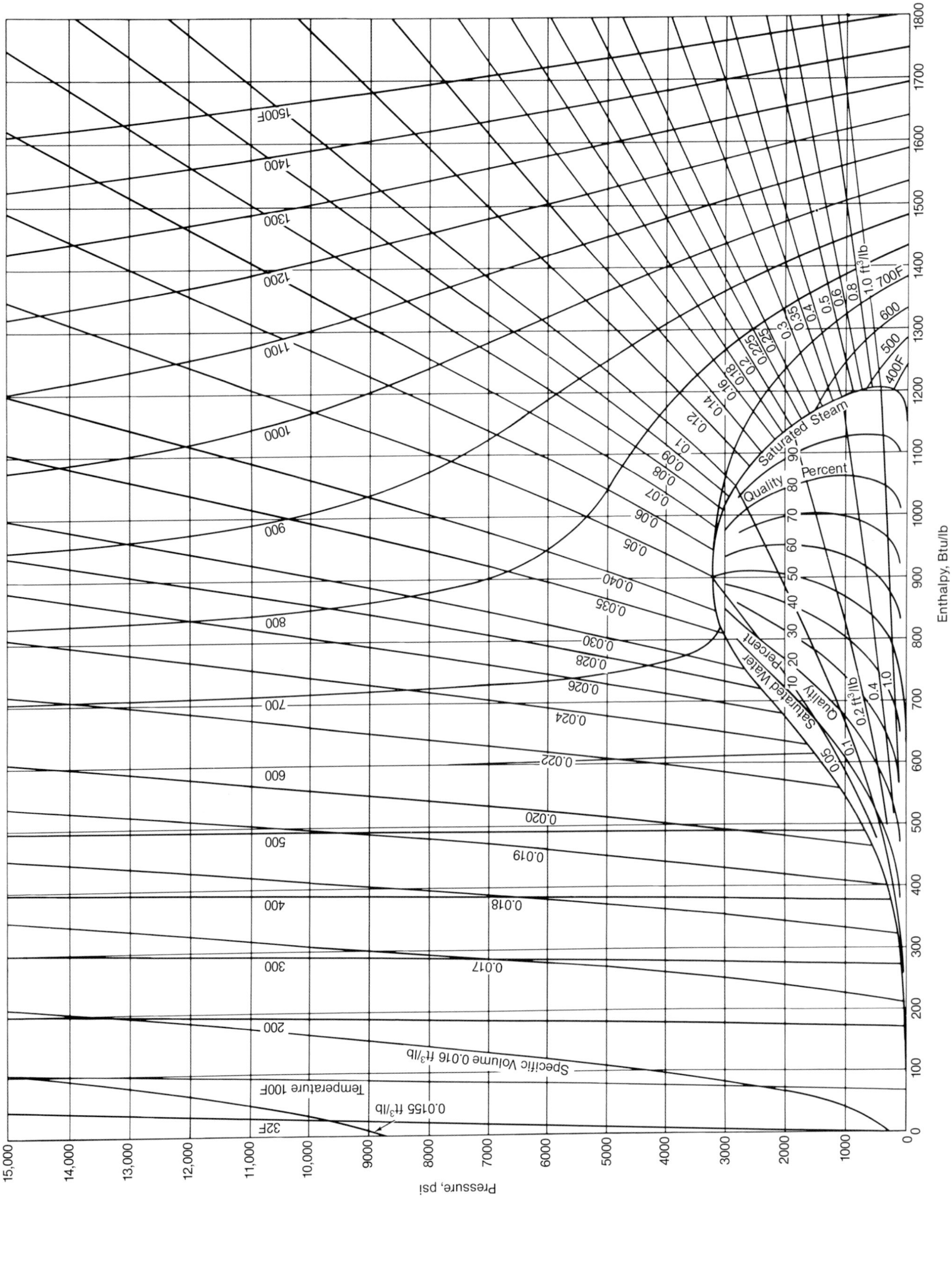

Fig. 1 Pressure-enthalpy chart for steam. (English units.)

specific volume (v) of a steam-water mixture can then be simply defined as:

$$H = H_f + x\ (H_g - H_f) \quad \textbf{(2a)}$$

$$s = s_f + x\ (s_g - s_f) \quad \textbf{(2b)}$$

$$v = v_f + x\ (v_g - v_f) \quad \textbf{(2c)}$$

where the subscripts f and g refer to properties at saturated liquid and vapor conditions respectively. The difference in a property between saturated liquid and vapor conditions is frequently denoted by the subscript fg; for example, $H_{fg} = H_g - H_f$. With these definitions, if the pressure or temperature of a steam-water mixture is known along with one of the mixture properties, the quality can then be calculated. For example, if the mixture enthalpy is known, then:

$$x = (H - H_f) / H_{fg} \quad \textbf{(3)}$$

Engineering problems deal mainly with changes or differences in enthalpy and entropy. It is not necessary to establish an absolute zero for these properties, although this may be done for entropy. The Steam Tables indicate an arbitrary zero internal energy and entropy for the liquid state of water at the triple point corresponding to a temperature of 32.018F (0.01C) and a vapor pressure of 0.08865 psi (0.6112 kPa). The triple point is a unique condition where the three states of water (solid, liquid and vapor) coexist at equilibrium.

Properties of gases

In addition to steam, air is a common working fluid for some thermodynamic cycles. As with steam, well defined properties are important in cycle analysis. Air and many common gases used in power cycle applications can usually be treated as ideal gases. An ideal gas is defined as a substance that obeys the ideal gas law:

$$Pv = RT \quad \textbf{(4)}$$

where R is a constant which varies with gas species; P and T are the pressure and temperature, respectively. R is equal to the universal gas constant, **R** [1545 ft lb/lb-mole R (8.3143 kJ/kg-mole K)], divided by the molecular weight of the gas. For dry air, R is equal to 53.34 lbf ft/lbm R (0.287 kJ/kg K). Values for other gases are summarized in Reference 2. The ideal gas law is commonly used in a first analysis of a process or cycle because it simplifies calculations. Final calculations often rely on tabulated gas properties for greater accuracy.

Tabulated gas properties are available from numerous sources. (See References 3 and 4 for examples.) Unfortunately, there is less agreement on gas properties than on those for steam. The United States (U.S.) boiler industry customarily uses 80F (27C) and 14.7 psia (1.01 bar) as the zero enthalpy of air and combustion products. A more general reference is one atmosphere pressure, 14.696 psia (1.01 bar), and 77F (25C). This is the standard reference point for heats of formation of compounds from elements in their standard states, latent heats of phase changes and free energy changes. Because of different engineering conventions, considerable care must be exercised when using tabulated properties. Selected properties for air and other gases are provided in Chapter 3, Table 3.

Conservation of mass and energy

Thermodynamic processes are governed by the laws of conservation of mass and conservation of energy except for the special case of nuclear reactions discussed in Chapter 37. These conservation laws basically state that the total mass and total energy (in any of its forms) can neither be created nor destroyed in a process. In an open flowing power system, where mass continually enters and exits a system such as Fig. 2, these laws can take the forms:

Conservation of mass

$$m_1 - m_2 = \Delta m \quad \textbf{(5)}$$

Conservation of energy

$$E_2 - E_1 + E(t) = Q - W \quad \textbf{(6)}$$

where m is the mass flow, Δm is the change in internal system mass, E is the total energy flowing into or out of the processes, $E(t)$ is the change in energy stored in the system with time, Q is the heat added to the system, W is the work removed, and the subscripts 1 and 2 refer to inlet and outlet conditions respectively. For steady-state conditions Δm and $E(t)$ are zero.

The conservation of energy states that a balance exists between energy, work and heat quantities entering and leaving the system. This balance of energy flow is also referred to as the first law of thermodynamics. The terms on the left side of Equation 6 represent stored energy entering or leaving the system as part of the mass flows and the accumulation of total stored energy within the system. The

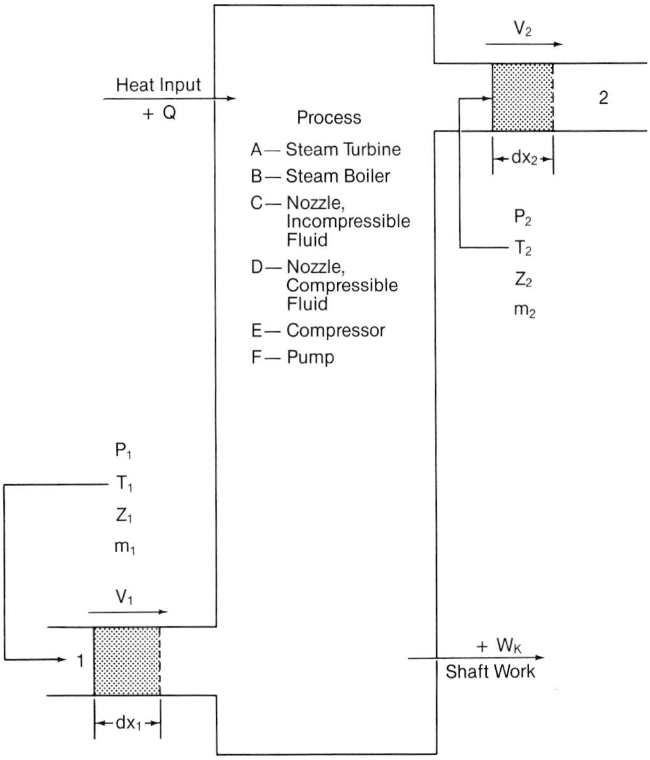

Fig. 2 Diagram illustrating thermodynamic processes.

terms on the right side are the heat transferred to the system, Q, and work done by the system, W. The stored energy components, represented by the term E, consist of the internally stored energy and the kinetic and potential energy. In an open system, there is work required to move mass into the system and work done by the system to move mass out. In each case, the total work is equal to the product of the mass, the system pressure and the specific volume. Separating this work from other work done by the system and including a breakdown of the stored energy, the energy conservation equation becomes:

$$m_2\left(u+Pv+\frac{V^2}{2g_c}+z\right)_2-m_1\left(u+Pv+\frac{V^2}{2g_c}+z\right)_1 + E(t)=Q-W_k \quad \textbf{(7)}$$

where m is the mass, u is the internal stored energy, P is the system pressure, V is the fluid velocity, v is the specific volume, z is the elevation and W_k is the sum of the work done by the system.

In this form, the work terms associated with mass moving into and out of the system (Pv) have been grouped with the stored energy crossing the system boundary. W_k represents all other work done by the system.

For many practical power applications, the energy equation can be further simplified for steady-state processes. Because the mass entering and leaving the system over any time interval is the same, dividing Equation 7 by the mass (m_2 or m_1 because they are equal) yields a simple balance between the change in stored energy due to inflow and outflow and the heat and work terms expressed on a unit mass basis. Heat and work expressed on a unit mass basis are denoted q and w, respectively. The unsteady term for system stored energy in Equation 7 is then set to zero. This yields the following form of the energy conservation equation:

$$\Delta u+\Delta(Pv)+\Delta\frac{V^2}{2g_c}+\Delta z\frac{g}{g_c}=q-w_k \quad \textbf{(8)}$$

Each Δ term on the left in Equation 8 represents the difference in the fluid property or system characteristic between the system outlet and inlet. Δu is the difference in internally stored energy associated with molecular and atomic motions and forces. Internally stored energy, or simply internal energy, accounts for all forms of energy other than the kinetic and potential energies of the collective molecule masses. This is possible because no attempt is made to absolutely define u.

The term $\Delta(Pv)$ can be viewed as externally stored energy in that it reflects the work required to move a unit mass into and out of the system. The remaining terms of externally stored energy, $\Delta(V^2/2g_c)$ and Δz, depend on physical aspects of the system. $\Delta(V^2/2g_c)$ is the difference in total kinetic energy of the fluid between two reference points (system inlet and outlet). $\Delta zg/g_c$ represents the change in potential energy due to elevation, where g is the gravitational constant 32.17 ft/s^2 (9.8 m/s^2) and g_c is a proportionality constant for English units. The value of the constant is obtained from equivalence of force and mass times acceleration:

$$\text{Force}=\frac{\text{mass}\times\text{acceleration}}{g_c} \quad \textbf{(9)}$$

In the English system, by definition, when 1 lb force (lbf) is exerted on a 1 lb mass (lbm), the mass accelerates at the rate of 32.17 ft/s^2. In the SI system, 1 N of force is exerted by 1 kg of mass accelerating at 1 m/s^2. Therefore, the values of g_c are:

$$g_c = 32.17 \text{ lbm ft / lbf s}^2 \quad \textbf{(10a)}$$

$$g_c = 1 \text{ kg m / N s}^2 \quad \textbf{(10b)}$$

Because of the numerical equivalency between g and g_c in the English system, the potential energy term in Equation 8 is frequently shown simply as Δz. When SI units are used, this term is often expressed simply as Δzg because the proportionality constant has a value of 1.

While many texts use the expression lbf to designate lb force and lbm to designate lb mass, this is not done in this text because it is believed that it is generally clear simply by using lb. As examples, the expression Btu/lb always means Btu/lb mass, and the expression ft lb/lb always means ft lb force/lb mass.

Application of the energy equation requires dimensional consistency of all terms and proper conversion constants are inserted as necessary. For example, the terms u and q, usually expressed in Btu/lb or J/kg, may be converted to ft lb/lb or N m/kg when multiplied by J, the mechanical equivalent of heat. This conversion constant, originally obtained by Joule's experiments between 1843 and 1878, is defined as:

$$J = 778.26 \text{ ft lbf / Btu} \quad \textbf{(11a)}$$

$$J = 1 \text{ N m / J} \quad \textbf{(11b)}$$

Particular attention should be given to the sign convention applied to heat and work quantities. Originating with the steam engine analysis, heat quantities are defined as positive when entering the system and work (for example, shaft work) is positive when leaving the system.

Because u and Pv of Equation 8 are system properties, their sum is also a system property. Because these properties of state can not be changed independently of one another and because the combination $(u + Pv)$ appears whenever mass enters or leaves the system, it is customary to consider the sum $(u + Pv)$ as a single property H, called enthalpy.

$$H = u + (Pv / J) \quad \textbf{(12)}$$

where Pv is divided by J to provide consistent units.

In steam applications, H is usually expressed in Btu/lb or J/kg. The examples in the following section illustrate the application of the steady-state open system energy Equation 7 and the usefulness of enthalpy in the energy balance of specific equipment.

Applications of the energy equation

Steam turbine

To apply the energy equation, each plant component is considered to be a system, as depicted in Fig. 2. In most cases, Δz, $\Delta(V^2/2g_c)$ and q from throttle (1) to exhaust (2) of the steam turbine are small compared to $(H_2 - H_1)$. This reduces Equation 8 to:

$$u_2 + (P_2 v_2 / J) - u_1 - (P_1 v_1 / J) = w_k / J \quad \textbf{(13a)}$$

or

$$H_2 - H_1 = w_k / J \quad \textbf{(13b)}$$

Equation 13 indicates that the work done by the steam turbine, w_k / J, is equal to the difference between the enthalpy of the steam entering and leaving. However, H_1 and H_2 are seldom both known and further description of the process is required for a solution of most problems.

Steam boiler

The boiler does no work, therefore $w_k = 0$. Because Δz and $\Delta(V^2/2g_c)$ from the feedwater inlet (1) to the steam outlet (2) are small compared to $(H_2 - H_1)$, the steady-state energy equation becomes:

$$q = H_2 - H_1 \quad \textbf{(14)}$$

Based on Equation 14 the heat added, q (positive), in the boiler per unit mass of flow in is equal to the difference between H_2 of the steam leaving and H_1 of the feedwater entering. Assuming that the pressure varies negligibly through the boiler and the drum pressure is known, Equation 14 can be solved knowing the temperature of the incoming feedwater.

Water flow through a nozzle

For water flowing through a nozzle, the change in specific volume is negligible. Quite commonly the change in elevation, Δz, the change in internal energy, Δu, the work done, w_k, and the heat added, q, are negligible and the energy equation reduces to:

$$(V_2^2 / 2g_c) - (V_1^2 / 2g_c) = (P_1 - P_2)v \quad \textbf{(15)}$$

The increase in kinetic energy of the water is given by Equation 15 for the pressure drop $(P_1 - P_2)$. If the approach velocity to the nozzle, V_1, is zero, Equation 15 becomes:

$$V_2 = \sqrt{2g_c(P_1 - P_2)v} \quad \textbf{(16)}$$

The quantity $(P_1 - P_2)v$ is often referred to as the static head.

Flow of a compressible fluid through a nozzle

In contrast to water flow, when steam, air or other compressible fluid flows through a nozzle, the changes in specific volume and internal energy are not negligible. In this case, assuming no change in elevation Δz, Equation 8 becomes:

$$(V_2^2 / 2g_c) - (V_1^2 / 2g_c) = (H_1 - H_2)\,J \quad \textbf{(17)}$$

If the approach velocity, V_1, is zero, this further simplifies to:

$$V_2 = \sqrt{2g_c J\,(H_1 - H_2)} \quad \textbf{(18)}$$

From this, it is evident that the velocity of a compressible fluid leaving a nozzle is a function of its entering and leaving enthalpies. Unfortunately, as with the steam turbine, H_1 and H_2 are seldom both known.

Compressor

If a compressible fluid moves through an adiabatic compressor ($q = 0$, a convenient approximation) and the change in elevation and velocity are small compared to $(H_2 - H_1)$, the energy equation reduces to:

$$-w_k / J = H_2 - H_1 \quad \textbf{(19)}$$

Note that w_k is negative because the compressor does work on the system. Therefore, the net effect of the compressor is expressed as an increase in fluid enthalpy from inlet to outlet.

Pump

The difference between a pump and a compressor is that the fluid is considered to be incompressible for the pumping process; this is a good approximation for water. For an incompressible fluid, the specific volume is the same at the inlet and outlet of the pump. If the fluid friction is negligible, then the internal energy changes, Δu, are set to zero and the energy equation can be expressed as:

$$-w_k = (P_2 - P_1)v \quad \textbf{(20)}$$

Because all real fluids are compressible, it is important to know what is implied by the term incompressible. The meaning here is that the isothermal compressibility, k_T, given by

$$k_T = -\frac{1}{v}\frac{\delta v}{\delta P} \quad \textbf{(21)}$$

is assumed to be arbitrarily small and approaching zero. Because neither v nor P is zero, δv must be zero and v must be a constant. Also, for isothermal conditions (by definition) there can be no change in internal energy, u, due to pressure changes only.

Entropy and its application to processes

The preceding examples illustrate applications of the energy balance in problems where a fluid is used for heat transfer and shaft work. They also demonstrate the usefulness of the enthalpy property. However, as was pointed out, H_1 and H_2 are seldom both known. Additional information is frequently provided by the first and second laws of thermodynamics and their consequences.

The first and second laws of thermodynamics

The first law of thermodynamics is based on the energy conservation expressed by Equation 6 and, by convention, relates the heat and work quantities of this equation to internally stored energy, u. Strictly speaking, Equation 6 is a complete form of the first law of thermodynamics. However, it is frequently useful to use the steady-state formulation provided in Equation 8 and further simplify this for the special case of 1) no change in potential energy due to gravity acting on the mass, and 2) no change in kinetic energy of the mass as a whole. In a closed system where only shaft work is permitted, these simplifying assumptions permit the energy Equation 8 for a unit mass to be reduced to:

$$\Delta u = q - (w_k / J) \quad \textbf{(22a)}$$

or in differential form

$$du = \delta q - (\delta w_k / J) \quad \textbf{(22b)}$$

The first law treats heat and work as being interchangeable, although some qualifications must apply. All forms of energy, including work, can be wholly converted to heat, but the converse is not generally true. Given a source of heat coupled with a heat-work cycle, such as heat released by high temperature combustion in a steam power plant, only a portion of this heat can be converted to work. The rest must be rejected to an energy sink, such as the atmosphere, at a lower temperature. This is essentially the *Kelvin* statement of the second law of thermodynamics. It can also be shown that it is equivalent to the *Clausius* statement wherein heat, in the absence of external assistance, can only flow from a hotter to a colder body.

Concept and definition of entropy

Heat flow is a function of temperature difference. If a quantity of heat is divided by its absolute temperature, the quotient can be considered a type of distribution property complementing the intensity factor of temperature. Such a property, proposed and named entropy by Clausius, is widely used in thermodynamics because of its close relationship to the second law.

Rather than attempt to define entropy (s) in an absolute sense, consider the significance of differences in this property given by:

$$S_2 - S_1 = \Delta S = \int_1^2 \frac{\delta q_{rev}}{T} \times \text{total system mass} \quad \mathbf{(23)}$$

where

ΔS = change in entropy, Btu/R (J/K)
q_{rev} = reversible heat flow between thermodynamic equilibrium states 1 and 2 of the system, Btu/lb (J/kg)
T = absolute temperature, R (K)

Entropy is an extensive property, i.e., a quantity of entropy, S, is associated with a finite quantity of mass, m. If the system is closed and the entire mass undergoes a change from state 1 to 2, an intensive property s is defined by S/m. The property s is also referred to as entropy, although it is actually specific entropy. If the system is open as in Fig. 2, the specific entropy is calculated by dividing by the appropriate mass.

Use of the symbol δ instead of the usual differential operator d is a reminder that q depends on the process and is not a property of the system (steam). δq represents only a small quantity, not a differential. Before Equation 23 can be integrated, q_{rev} must be expressed in terms of properties, and a reversible path between the prescribed initial and final equilibrium states of the system must be specified. For example, when heat flow is reversible and at constant pressure, $q_{rev} = c_p dT$. This may represent heat added reversibly to the system, as in a boiler, or the equivalent of internal heat flows due to friction or other irreversibilities. In these two cases, Δs is always positive.

The same qualifications for δ hold in the case of thermodynamic work. Small quantities of w similar in magnitude to differentials are expressed as δw.

Application of entropy to a reversible process

Reversible thermodynamic processes exist in theory only; however, they serve an important function of defining limiting cases for heat flow and work processes. The properties of a system undergoing a reversible process are constrained to be homogeneous because there are no variations among subregions of the system. Moreover, during interchanges of heat or work between a system and its surroundings, only corresponding potential gradients of infinitesimal magnitude may exist.

All actual processes are irreversible. To occur, they must be under the influence of a finite potential difference. A temperature difference supplies this drive and direction for heat flow. The work term, on the other hand, is more complicated, because there are as many different potentials (generalized forces) as there are forms of work. However, the main concern here is expansion work for which the potential is clearly a pressure difference.

Regardless of whether a process is to be considered reversible or irreversible, it must have specific beginning and ending points (limits) in order to be evaluated. To apply the first and second laws, the limits must be equilibrium states. Nonequilibrium thermodynamics is beyond the scope of this text. Because the limits of real processes are to be equilibrium states, any process can be approximated by a series of smaller reversible processes starting and ending at the same states as the real processes. In this way, only equilibrium conditions are considered and the substitute processes can be defined in terms of the system properties. The following lists the reversible processes for heat flow and work:

Reversible Heat Flow		Reversible Work	
Constant pressure,	$dP = 0$	Constant pressure,	$dP = 0$
Constant temperature,	$dT = 0$	Constant temperature,	$dT = 0$
Constant volume,	$dv = 0$	Constant entropy,	$ds = 0$
	$w = 0$		$q = 0$

The qualification of these processes is that each describes a path that has a continuous functional relationship on coordinate systems of thermodynamic properties.

A combined form of the first and second laws is obtained by substituting $\delta q_{rev} = Tds$ for δq in Equation 22b, yielding:

$$du = Tds - \delta w_k \quad \mathbf{(24)}$$

Because only reversible processes are to be used, δw should also be selected with this restriction. Reversible work for the limited case of expansion work can be written:

$$\delta(w_{rev}) = Pdv \quad \mathbf{(25)}$$

In this case, pressure is in complete equilibrium with external forces acting on the system and is related to v through an equation of state.

Substituting Equation 25 in 24, the combined expression for the first and second law becomes:

$$du = Tds - Pdv \quad \mathbf{(26)}$$

Equation 26, however, only applies to a system in which the reversible work is entirely shaft work. To modify this expression for an open system in which flow work $d(Pv)$ is also present, the quantity $d(Pv)$ is added to the left side of Equation 25 and added as $(Pdv + vdP)$ on the right side. The result is:

$$du + d(Pv) = Tds - Pdv + Pdv + vdP \quad \mathbf{(27a)}$$

or

$$dH = Tds + vdP \quad \mathbf{(27b)}$$

The work term vdP in Equation 27 now represents reversible shaft work in an open system, expressed on a unit mass basis.

Because Tds in Equation 26 is equivalent to δq, its value becomes zero under adiabatic or zero heat transfer conditions ($\delta q = 0$). Because T can not be zero, it follows that $ds = 0$ and s is constant. Therefore, the maximum work from stored energy in an open system during a reversible adiabatic expansion is $\int vdP$ at constant entropy. The work done is equal to the decrease in enthalpy. Likewise for the closed system, the maximum expansion work is $-\int Pdv$ at constant entropy and is equal to the decrease in internal energy. These are important cases of an adiabatic isentropic expansion.

Irreversible processes

All real processes are irreversible due to factors such as friction, heat transfer through a finite temperature difference and expansion through a process with a finite net force on the boundary. Real processes can be solved approximately, however, by substituting a series of reversible processes. An example of such a substitution is illustrated in Fig. 3, which represents the adiabatic expansion of steam in a turbine or any gas expanded from P_1 to P_2 to produce shaft work. T_1, P_1 and P_2 are known. The value of H_1 is fixed by T_1 and P_1 for a single-phase condition (vapor) at the inlet. H_1 may be found from the Steam Tables, a T-s diagram (Fig. 3) or, more conveniently, from an H-s (Mollier) diagram, shown in the chapter frontispiece. From the combined first and second laws, the maximum energy available for work in an adiabatic system is $(H_1 - H_3)$, as shown in Fig. 3, where H_3 is found by the adiabatic *isentropic* expansion (expansion at constant entropy) from P_1 to P_2. A portion of this available energy, usually about 10 to 15%, represents work lost (w_L) due to friction and form loss, limiting ΔH for shaft work to $(H_1 - H_2)$. The two reversible paths used to arrive at point b in Fig. 3 (path a to c at constant entropy, s, and path c to b at constant pressure) yield the following equation:

$$(H_1 - H_3) - (H_2 - H_3) = H_1 - H_2 \qquad \textbf{(28)}$$

Point b, identified by solving for H_2, now fixes T_2; v_1 and v_2 are available from separate tabulated values of physical properties.

Note that $\Delta H_{3\text{-}2}$ can be found from:

$$\Delta H_{2-3} = \int_3^2 Tds \qquad \textbf{(29)}$$

or, graphically, the area on the T-s diagram (Fig. 3) under the curve P_2 from points c to b. Areas bounded by reversible paths on the T-s diagram in general represent q (heat flow per unit mass) between the system and its surroundings. However, the path a to b is irreversible and the area under the curve has no significance. The area under path c to b, although it has the form of a reversible quantity q, does not represent heat added to the system but rather its equivalent in internal heat flow. A similar situation applies to the relationship between work and areas under reversible paths in a pressure-volume equation of state diagram. Because of this important distinction between reversible and irreversible paths, care must be exercised in graphically interpreting these areas in cycle analysis.

Returning to Fig. 3 and the path a to b, w_L was considered to be a percentage of the enthalpy change along the path a to c. In general, the evaluation should be handled in several smaller steps (Fig. 4) for the following reason. Point b has a higher entropy than point c and, if expansion to a pressure lower than P_2 (Fig. 3) is possible, the energy available for this additional expansion is greater than that at point c. In other words, a portion of w_L (which has the same effect as heat added to the system) for the first expansion can be recovered in the next expansion or stage. This is the basis of the reheat factor used in analyzing expansions through a multi-stage turbine. Since the pressure curves are divergent on an H-s or T-s diagram, the sum of the individual ΔH_s values (isentropic ΔH) for individual increments of ΔP (or stages in an irreversible expansion) is greater than that of the reversible ΔH_s between the initial and final pressures (Fig. 4). Therefore the shaft work that can be achieved is greater than that calculated by a simple isentropic expansion between the two pressures.

Principle of entropy increase

Although entropy has been given a quantitative meaning in previous sections, there are qualitative aspects of this property which deserve special emphasis. An increase in entropy is a measure of that portion of process heat which is unavailable for conversion to work. For example, consider the constant pressure reversible addition of heat to a working fluid with the resulting increase in steam entropy. The minimum portion of this heat flow which is unavailable for shaft work is equal to the entropy increase multiplied by the absolute temperature of the sink to which a part of the heat must be rejected (in accordance with the second law). However, because a reversible addition of heat is not possible, incremental entropy increases also occur due to internal fluid heating as a result of temperature gradients and fluid friction.

Even though the net entropy change of any portion of a fluid moving through a cycle of processes is always zero

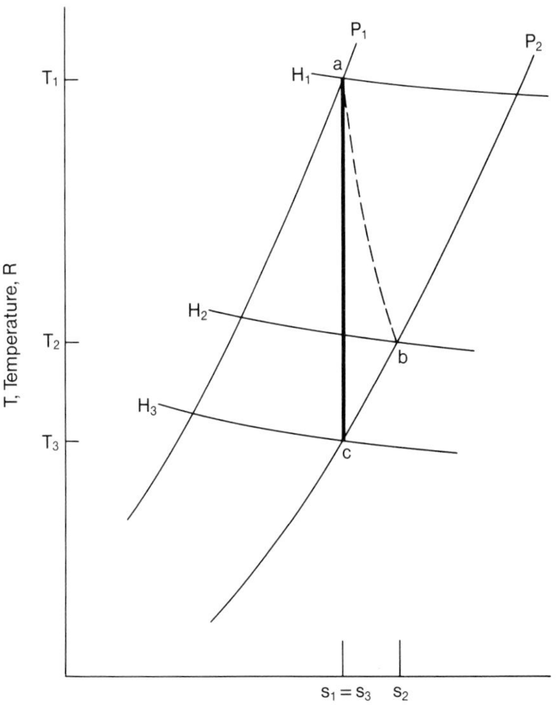

Fig. 3 Irreversible expansion, state a to state b.

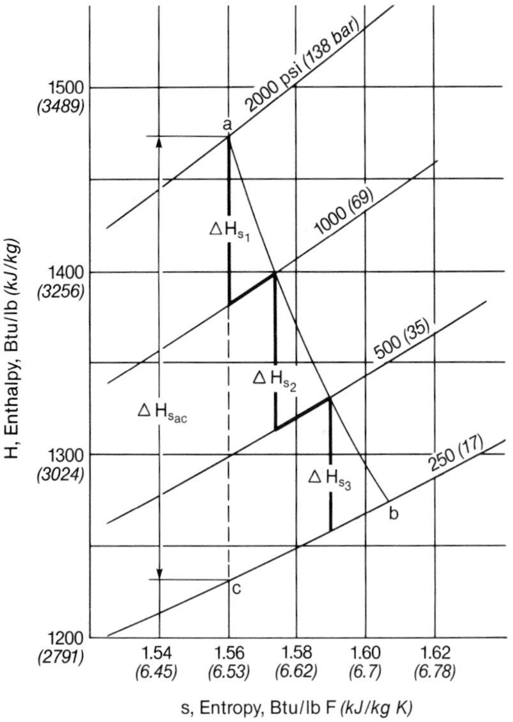

Fig.4 Three-stage irreversible expansion— $\Delta H_{s1} + \Delta H_{s2} + \Delta H_{s3} > \Delta H_{s_{ac}}$.

because the cycle requires restoration of all properties to some designated starting point, the sum of all entropy increases has a special significance. These increases in entropy, less any decreases due to recycled heat within a regenerator, multiplied by the appropriate sink absolute temperature (R or K) are equal to the heat flow to the sink. In this case, the net entropy change of the system undergoing the cycle is zero, but there is an entropy increase of the surroundings. Any thermodynamic change that takes place, whether it is a stand alone process or cycle of processes, results in a net entropy increase when both the system and its surroundings are considered.

Cycles

To this point, only thermodynamic processes have been discussed with minor references to the cycle. The next step is to couple processes so heat may be converted to work on a continuous basis. This is done by selectively arranging a series of thermodynamic processes in a cycle forming a closed curve on any system of thermodynamic coordinates. Because the main interest is steam, the following discussion emphasizes expansion or *Pdv* work. This relies on the limited differential expression for internal energy, Equation 26, and enthalpy, Equation 27. However, the subject of thermodynamics recognizes work as energy in transit under any potential other than differential temperature and electromagnetic radiation.

Carnot cycle

Sadi Carnot (1796 to 1832) introduced the concept of the cycle and reversible processes. The *Carnot cycle* is used to define heat engine performance as it constitutes a cycle in which all component processes are reversible. This cycle, on a temperature-entropy diagram, is shown in Fig. 5a for a gas and in Fig. 5b for a two-phase saturated fluid. Fig. 5c presents this cycle for a nonideal gas, such as superheated steam, on Mollier coordinates (entropy versus enthalpy).

Referring to Fig. 5, the Carnot cycle consists of the following processes:

1. Heat is added to the working medium at constant temperature ($dT = 0$) resulting in expansion work and changes in enthalpy. (For an ideal gas, changes in internal energy and pressure are zero and, therefore, changes in enthalpy are zero.)
2. Adiabatic isentropic expansion ($ds = 0$) occurs with expansion work and an equivalent decrease in enthalpy.
3. Heat is rejected to the surroundings at a constant temperature and is equivalent to the compression work and any changes in enthalpy.
4. Adiabatic isentropic compression occurs back to the starting temperature with compression work and an equivalent increase in enthalpy.

This cycle has no counterpart in practice. The only way to carry out the constant temperature processes in a one-phase system would be to approximate them through a series of isentropic expansions and constant pressure reheats for heat addition, and isentropic compressions with a series of intercoolers for heat rejections. Another serious disadvantage of a Carnot gas engine would be the small ratio of net work to gross work (net work referring to the difference between the expansion work and the compression work, and gross work being expansion work). Even a two-phase cycle, such as Fig. 5b, would be subject to the practical mechanical difficulties of wet compression and, to a lesser extent, wet expansion where a vapor-liquid mixture exists.

Nevertheless, the Carnot cycle illustrates the basic principles of thermodynamics and, because the processes are reversible, the Carnot cycle offers the maximum thermal efficiency attainable between any given temperatures of heat source and sink. The thermal efficiency of the cycle is defined as the ratio of the net work output to the total heat input. Using the *T-s* diagram for the Carnot cycle shown in Fig. 5a, the thermal efficiency depends solely on the temperatures at which heat addition and rejection occur:

$$\eta = \frac{T_1 - T_2}{T_1} = 1 - \frac{T_2}{T_1} \tag{30}$$

where

η = thermal efficiency of the conversion from heat into work
T_1 = absolute temperature of heat source, R (K)
T_2 = absolute temperature of heat sink, R (K)

The efficiency statement of Equation 30 can be extended to cover all reversible cycles where T_1 and T_2 are defined as mean temperatures found by dividing the heat added and rejected reversibly by Δs. For this reason, all reversible cycles have the same efficiencies when considered between the same mean temperature limits of heat source and heat sink.

Rankine cycle

Early thermodynamic developments were centered around the performance of the steam engine and, for comparison purposes, it was natural to select a revers-

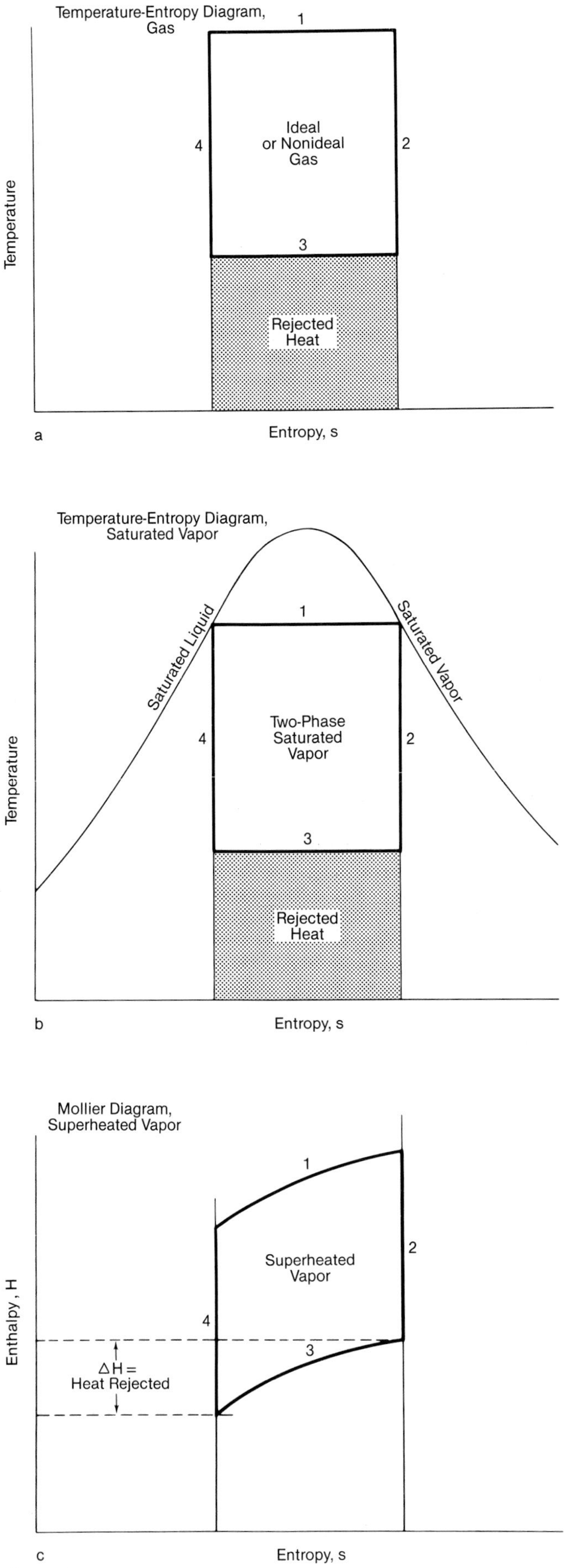

Fig. 5 Carnot cycles.

ible cycle which approximated the processes related to its operation. The *Rankine cycle* shown in Fig. 6, proposed independently by Rankine and Clausius, meets this objective. All steps are specified for the system only (working medium) and are carried out reversibly as the fluid cycles among liquid, two-phase and vapor states. Liquid is compressed isentropically from points a to b. From points b to c, heat is added reversibly in the compressed liquid, two-phase and finally superheat states. Isentropic expansion with shaft work output takes place from points c to d and unavailable heat is rejected to the atmospheric sink from points d to a.

The main feature of the Rankine cycle is that compression (pumping) is confined to the liquid phase, avoiding the high compression work and mechanical problems of a corresponding Carnot cycle with two-phase compression. This part of the cycle, from points a to b in Fig. 6, is greatly exaggerated, because the difference between the saturated liquid line and point b (where reversible heat addition begins) is too small to show in proper scale. For example, the temperature rise with isentropic compression of water from a saturation temperature of 212F (100C) and one atmosphere to 1000 psi (69.0 bar) is less than 1F (0.6C).

If the Rankine cycle is closed in the sense that the fluid repeatedly executes the various processes, it is termed a *condensing cycle*. Although the closed, condensing Rankine cycle was developed to improve steam engine efficiency, a closed cycle is essential for any toxic or hazardous working fluid. Steam has the important advantage of being inherently safe. However, the close control of water chemistry required in high pressure, high temperature power cycles also favors using a minimum of makeup water. (Makeup is the water added to the steam cycle to replace leakage and other withdrawals.) Open steam cycles are still found in small units, some special processes and heating load applications coupled with power. The condensate from process and heating loads is usually returned to the power cycle for economic reasons.

The higher efficiency of the condensing steam cycle is a result of the pressure-temperature relationship between water and its vapor state, steam. The lowest temperature at which an *open*, or *noncondensing*, steam cycle may reject heat is approximately 212F (100C), the satu-

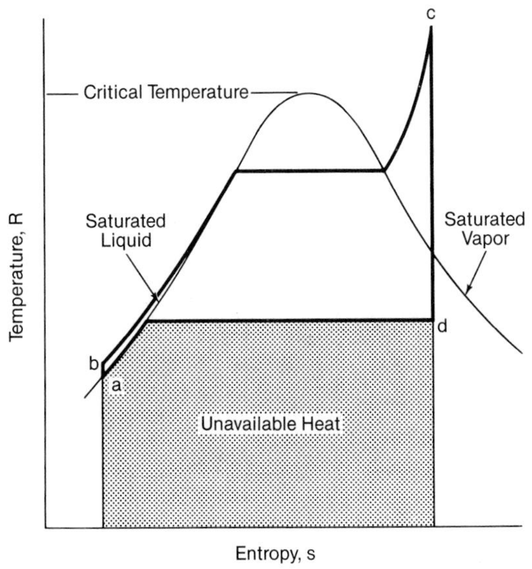

Fig. 6 Temperature-entropy diagram of the ideal Rankine cycle.

ration temperature corresponding to atmospheric pressure of 14.7 psi (1.01 bar). The pressure of the condensing fluid can be set at or below atmospheric pressure in a closed cycle. This takes advantage of the much lower sink temperature available for heat rejection in natural bodies of water and the atmosphere. Therefore, the condensing temperature in the closed cycle can be 100F (38C) or lower. Because the maximum *Pdv* work per unit mass of a compressible fluid is directly related to a function of the ratio of maximum and minimum pressures in the system available for expansion, as well as to the initial absolute temperature, an increase in this pressure ratio (higher maximum and/or lower minimum) permits an increase in the available work.

Fig. 7 illustrates the difference between an open and closed Rankine cycle. Both cycles are shown with nonideal expansion processes. Liquid compression takes place from points a to b and heat is added from points b to c. The work and heat quantities involved in each of these processes are the same for both cycles. Expansion and conversion of stored energy to work take place from points c to d' for the open cycle and from c to d for the closed cycle. Because this process is shown for the irreversible case, there is internal fluid heating and an entropy increase. From points d' to a and d to a, heat is rejected. Because this last portion of the two cycles is shown as reversible, the shaded areas are proportional to the rejected heat. The larger amount of rejected heat for the open cycle is clearly indicated.

Regenerative Rankine cycle

The reversible cycle efficiency given by Equation 30, where T_2 and T_1 are mean absolute temperatures for rejecting and adding heat respectively, indicates only three choices for improving ideal cycle efficiency: decreasing T_2, increasing T_1, or both. Little can be done to reduce T_2 in the Rankine cycle because of the limitations imposed by the temperatures of available rejected heat sinks in the general environment. Some T_2 reduction is possible by selecting variable condenser pressures for very large units with two or more exhaust hoods, because the lowest temperature in the condenser is set by the lowest temperature of the cooling water. On the other hand, there are many ways to increase T_1 even though the steam temperature may be limited by high temperature corrosion and allowable stress properties of the material.

One early improvement to the Rankine cycle was the adoption of *regenerative feedwater heating*. This is done by extracting steam from various stages in the turbine to heat the feedwater as it is pumped from the bottom of the condenser (hot well) to the boiler economizer.

Fig. 8 is a diagram of a widely used supercritical pressure steam cycle showing the arrangement of various components including the feedwater heaters. This cycle also contains one stage of steam reheat, which is another method of increasing the mean T_1. Regardless of whether the cycle is high temperature, high pressure or reheat, regeneration is used in all modern condensing steam power plants. It improves cycle efficiency and has other advantages, including lower volume flow in the final turbine stages and a convenient means of deaerating the feedwater. In the power plant heat balances shown in Fig. 8 and later in Fig. 10, several parameters require definition:

DC: In the feedwater heater blocks, this parameter is the drain cooler approach temperature or the difference between the shell-side condensate outlet (drain) temperature and the feedwater inlet temperature.

TD: In the feedwater heater blocks, this parameter is the terminal temperature difference or the difference between the shell-side steam inlet temperature and the feedwater outlet temperature.

P: In the feedwater heater blocks, this parameter is the nominal shell-side pressure.

The temperature-entropy diagram of Fig. 9 for the steam cycle of Fig. 8 illustrates the principle of regeneration in which the mean temperature level is increased for heat addition. Instead of heat input starting at the hot well temperature of 101.1F (38.4C), the water entering the boiler economizer has been raised to 502F (261C) by the feedwater heaters.

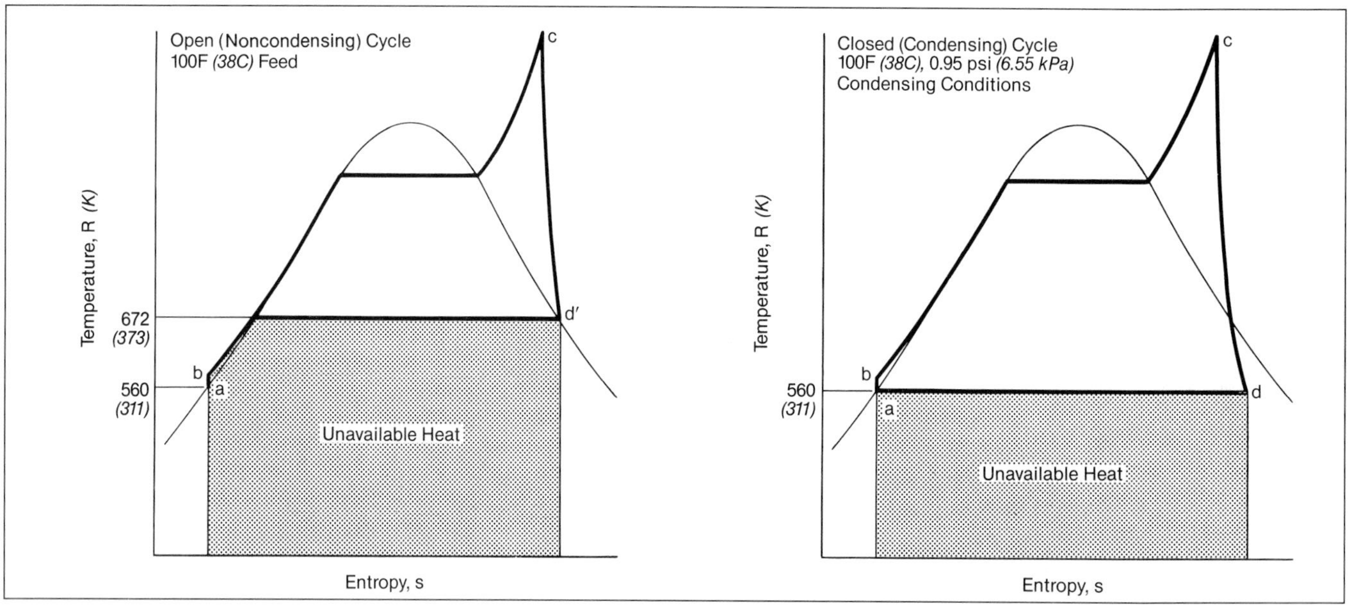

Fig. 7 Rankine cycles.

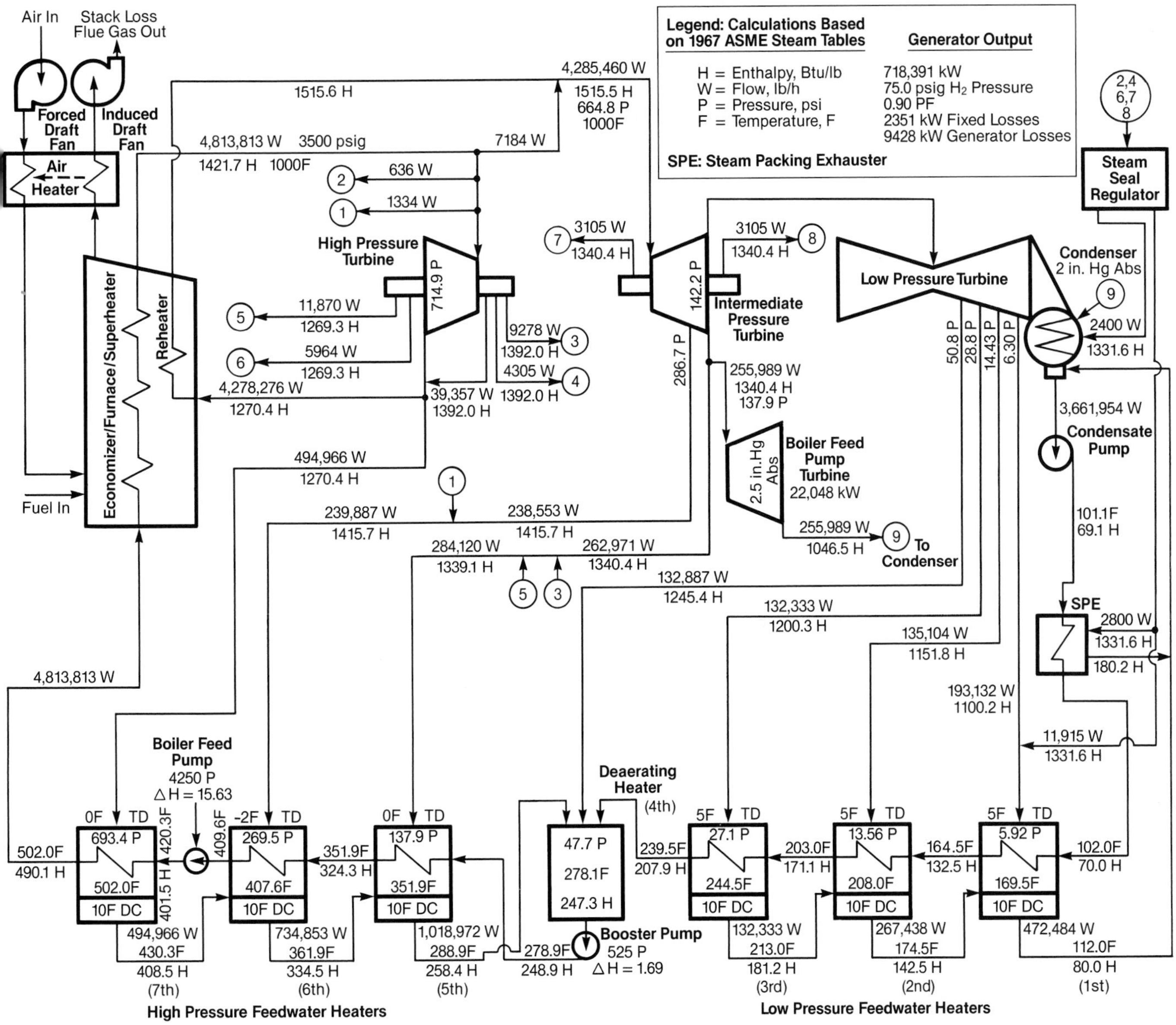

Fig. 8 Supercritical pressure, 3500 psig turbine cycle heat balance (English units).

Fig. 9 also shows that the mean temperature level for heat addition is increased by reheating the steam after a portion of the expansion has taken place. Because maximum temperatures are limited by physical or economic reasons, reheating after partial expansion of the working fluid is also effective in raising the average T_1. The hypothetical case of an infinite number of reheat and expansion stages approaches a constant temperature heat addition of the Carnot cycle, at least in the superheat region. It would appear beneficial to set the highest temperature in the superheat reheat stage at the temperature limit of the working medium or its containment. However, merely increasing T_1 may not improve efficiency. If the entropy increase accompanying reheat causes the final expansion process to terminate in superheated vapor, the mean temperature for heat rejection, T_2, has also been increased unless the superheat can be extracted in a regenerative heater, adding heat to the boiler feedwater. Such a regenerative heater would have to operate at the expense of the very effective cycle.

All of these factors, plus component design problems, must be considered in a cycle analysis where the objective is to optimize physical limits, economic limits and fuel economy. Overall cycle characteristics, including efficiency, can also be illustrated by plotting the cycle on a Mollier chart. (See chapter frontispiece and Fig. 4.)

The procedure used in preparing Fig. 9 deserves special comment because it illustrates an important function of entropy. All processes on the diagram represent total entropies divided by the high pressure steam flow rate. Total entropies at any point of the cycle are the product of the mass flowing past that point in unit time and the entropy per pound (specific entropy) corresponding to the pressure, temperature and state of the steam. Specific entropy values are provided by the Steam Tables such as those provided here in Tables 1 to 3. If a point falls in the two-phase region, entropy is calculated in the same manner as enthalpy. That is, the value for evaporation is multiplied by the steam quality (fraction of uncondensed steam) and added to the entropy value of water at saturation conditions.

Because there are different flow rates for the various cycle processes, small sections of individual T-s diagrams are superimposed in Fig. 9 on a base diagram that identifies saturated liquid and vapor parameters. However, the saturation parameters can only be compared to specific points on the T-s diagram. These points correspond to the parts of the cycle representing heat addition to high pressure steam and the expansion of this steam in a high pressure turbine. In these parts of the cycle, the specific entropy of the fluid and the value plotted in the diagram are the same. At each steam bleed point of the intermediate and low pressure turbine, the expansion line should show a decrease in entropy due to reduced flow entering the next turbine stage. However, for convenience, the individual step backs in the expansion lines have been shifted to the right to show the reheated steam expansion as one continuous process.

Feedwater heating through the regenerators and compression by the pumps (represented by the zig zag lines in Fig. 9) result in a net entropy increase. However, two factors are involved in the net increase, an entropy increase from the heat added to the feed and a decrease resulting from condensing and cooling the bleed steam and drain flows from higher pressure heaters.

Consider an example in which the feedwater heater just before the deaerating heater increases the temperature of a 3,661,954 lb/h feedwater from 203.0 to 239.5F. From Table 1, this increases the enthalpy, H, of the feed from 171.2 to 208.0 and increases the entropy, s, from 0.2985 to 0.3526. The total entropy increase per lb of high pressure steam flowing at 4,813,813 lb/h is:

$$\frac{(s_2 - s_1)\ \dot{m}_{\text{feed}}}{\dot{m}_{\text{HP steam}}} = \frac{(0.3526 - 0.2985)\ 3{,}661{,}954}{4{,}813{,}813}$$

$$= 0.0412 \ \text{Btu/lb F} \qquad \textbf{(31)}$$

The feed temperature rises 36.5F and the total heat absorbed is:

$$(H_2 - H_1)\ \dot{m}_{\text{feed}} = (208.0 - 171.2)\ 3{,}661{,}954$$

$$= 134{,}759{,}907 \ \text{Btu/h} \qquad \textbf{(32)}$$

On the heat source side of the balance, 132,333 lb/h of steam are bled from the low pressure turbine at 28.8 psig. This steam has an enthalpy of 1200.3 and an entropy of 1.7079. The steam is desuperheated and condensed according to the following equation:

$$H_2 = H_1 - \frac{\text{heat absorbed by feedwater}}{\dot{m}_{\text{LP steam}}}$$

$$= 1200.3 - \frac{134{,}759{,}907}{132{,}333} = 182.0 \ \text{Btu/lb} \qquad \textbf{(33)}$$

Interpolating Table 1, the low pressure steam is cooled to 213.0F at H_f = 181.2 Btu/lb. The corresponding entropy of the heater drain is 0.3136 Btu/lb F. Therefore, the entropy decrease is:

$$\frac{(s_1 - s_2)\ \dot{m}_{\text{LP steam}}}{\dot{m}_{\text{HP steam}}} = \frac{(1.7079 - 0.3136)\ 132{,}333}{4{,}813{,}813}$$

$$= 0.0383 \ \text{Btu/lb F} \qquad \textbf{(34)}$$

This heater shows a net entropy increase of 0.0412 – 0.0383 = 0.0029 Btu/lb F.

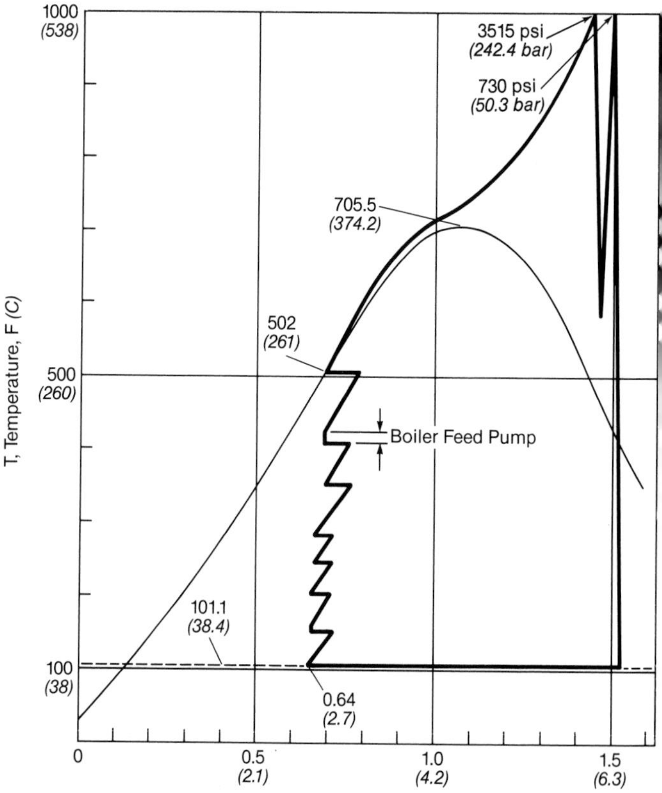

Fig. 9 Steam cycle for fossil fuel temperature-entropy diagram — single reheat, 7-stage regenerative feedwater heating—3500 psig, 1000F/1000F steam.

Recall that an increase in entropy represents heat energy that is unavailable for conversion to work. Therefore, the net entropy increase through the feedwater heater is the loss of available energy that can be attributed to the pressure drop required for flow and temperature difference. These differences are necessary for heat transfer. The quantity of heat rendered unavailable for work is the product of the entropy increase and the absolute temperature of the sink receiving the rejected heat.

Available energy

From the previous feedwater heater example, there is a derived quantity, formed by the product of the corresponding entropy and the absolute temperature of the available heat sink, which has the nature of a property. The difference between H (enthalpy) and $T_o s$ is another derived quantity called available energy.

$$e = H - T_o s \qquad \textbf{(35)}$$

where

e = available energy, Btu/lb (kJ/kg)
H = enthalpy, Btu/lb (kJ/kg)
T_o = sink temperature, R (K)
s = entropy, Btu/lb R (kJ/kg K)

Available energy is not a property because it can not be completely defined by an equation of state; rather, it is dependent on the sink temperature. However, a combined statement of the first and second laws of thermodynamics indicates that the difference in the available energy between two points in a reversible process represents the maximum amount of work (on a unit mass basis) that can be extracted from the fluid due to the change of state variables H and s between the two points. Conceptually then, differences in the value of T_0s represent energy that is unavailable for work.

The concept of available energy is useful in cycle analysis for optimizing the thermal performance of various components relative to overall cycle efficiency. In this way small, controllable changes in availability may be weighed against larger, fixed unavailable heat quantities which are inherent to the cycle. By comparing actual work to the maximum reversible work calculated from differences in available energy, the potential for improvement is obtained.

Rankine cycle efficiency

As with the Carnot cycle efficiency, the Rankine cycle efficiency (η) is defined as the ratio of the net work (w_{out}-w_{in}) produced to the energy input (Q_{in}):

$$\eta = \frac{w_{out} - w_{in}}{Q_{in}} \qquad \textbf{(36)}$$

For the simple cycle shown in Fig. 7, the work terms and the energy input are defined as:

$$w_{out} = \eta_t \; \dot{m}_w (H_c - H_d) \qquad \textbf{(37)}$$

$$w_{in} = \dot{m}_w (H_b - H_a) / \eta_p \cong \dot{m}_w v_a (P_b - P_a) / \eta_p \qquad \textbf{(38)}$$

$$Q_{in} = \dot{m}_w (H_c - H_b) \qquad \textbf{(39)}$$

where $H_{a\text{-}d}$ are the enthalpies defined in Fig. 7, $\dot{m}_w$ is the water flow rate, $P_{a\text{-}b}$ are the pressures at points a and b, v_a is the water specific volume at point a, while η_t and η_p are the efficiencies of the turbine and boiler feed pump respectively.

Substituting Equations 37 through 39 into Equation 36 and canceling the mass flow rate, $\dot{m}_w$, which is the same in all three cases provide the following overall thermal efficiency (η_{th}):

$$\eta_{th} = \frac{\eta_t (H_c - H_d) - v_a (P_b - P_a) / \eta_p}{(H_c - H_b)} \qquad \textbf{(40)}$$

In even a simple power producing facility using the Rankine cycle, several other factors must also be considered:

1. Not all of the chemical energy supplied to the boiler from the fuel is absorbed by the steam — typically 80 to 85% of the energy input is absorbed.
2. A variety of auxiliary equipment such as fans, sootblowers, environmental protection systems, water treatment equipment, and fuel handling systems, among others use part of the power produced.
3. Electrical generators and motors are not 100% efficient.

Incorporating these general factors into Equation 40 for a simple power cycle yields the net generating efficiency, η_{net}:

$$\eta_{net} = \frac{\eta_g \eta_t (H_c - H_d) - \left[v_1 (P_b - P_a) / \eta_p \eta_m \right] - w_{aux}}{(H_c - H_b) / \eta_b} \qquad \textbf{(41)}$$

where w_{aux} is the auxiliary power usage, η_b is the boiler efficiency, while η_g and η_m are the electrical generator and motor efficiencies, both typically 0.98 to 0.99. The gross power efficiency can be evaluated from Equation 41 with w_{aux} set at zero.

The evaluation of efficiency in modern high pressure steam power systems is more complex. Provision in the evaluation must be made for steam reheat or double reheat, and turbine steam extraction for regenerative feedwater heating, among others. This evaluation is based upon a steam turbine heat balance or steam cycle diagram such as that shown in Fig. 8 for a 3500 psig (241.3 bar gauge) supercritical pressure fossil fuel-fired unit, or Fig. 10 for a 2400 psig (165.5 bar gauge) subcritical pressure unit. The subcritical pressure unit shown has a single reheat, six closed feedwater heaters and one open feedwater heater.

Rankine cycle heat rate

Heat rate is a term frequently used to define various power plant efficiencies. If the electrical generation used is the net output after subtracting all auxiliary electrical power needs, then Equation 42 defines the *net heat rate* using English units. If the auxiliary electrical usage is not included, Equation 42 defines the *gross heat rate.*

$$\text{Heat rate} = \frac{\text{Total fuel heat input (Btu/h)}}{\text{Electrical generation (kW)}} \qquad \textbf{(42)}$$

Heat rate is directly related to plant efficiency, η, by the following relationships:

$$\text{Net heat rate} = \frac{3412 \text{ Btu/kWh}}{\eta_{net}} \qquad \textbf{(43a)}$$

$$\text{Gross heat rate} = \frac{3412 \text{ Btu/kWh}}{\eta_{gross}} \qquad \textbf{(43b)}$$

Steam cycle in a nuclear plant

Fig. 11 illustrates a Rankine cycle whose thermal energy source is a pressurized water nuclear steam system. High pressure cooling water is circulated from a pressurized water reactor to a steam generator. Therefore, heat produced by the fission of enriched uranium in the reactor core is transferred to feedwater supplied to the steam generator which, in turn, supplies steam for the turbine. The steam generators of a nuclear plant are shell and tube heat exchangers in which the high pressure reactor coolant flows inside the tubes and lower pressure feedwater is boiled outside of the tubes. For the pressurized water reactor system, the Rankine cycle for power generation takes place entirely in the nonradioactive water side (*secondary side*) that is boiling and circulating in the steam system; the reactor coolant system is simply the heat source for the power producing Rankine cycle.

The steam pressure at the outlet of the steam generator varies among plants due to design differences and ranges from 700 to 1000 psi (48.3 to 69.0 bar). Nominally, a Babcock & Wilcox (B&W) nuclear steam system with a once-through steam generator provides slightly superheated steam at 570F (299C) and 925 psi (63.8 bar). Steam flow from the once-through generator reaches the high pressure turbine at about 900 psi (62.1 bar) and 566F (297C). The other nuclear steam systems use a recirculating steam generator. In this design, feedwater is mixed with saturated water coming from the steam generator's separators before entering the tube bundle and boiling

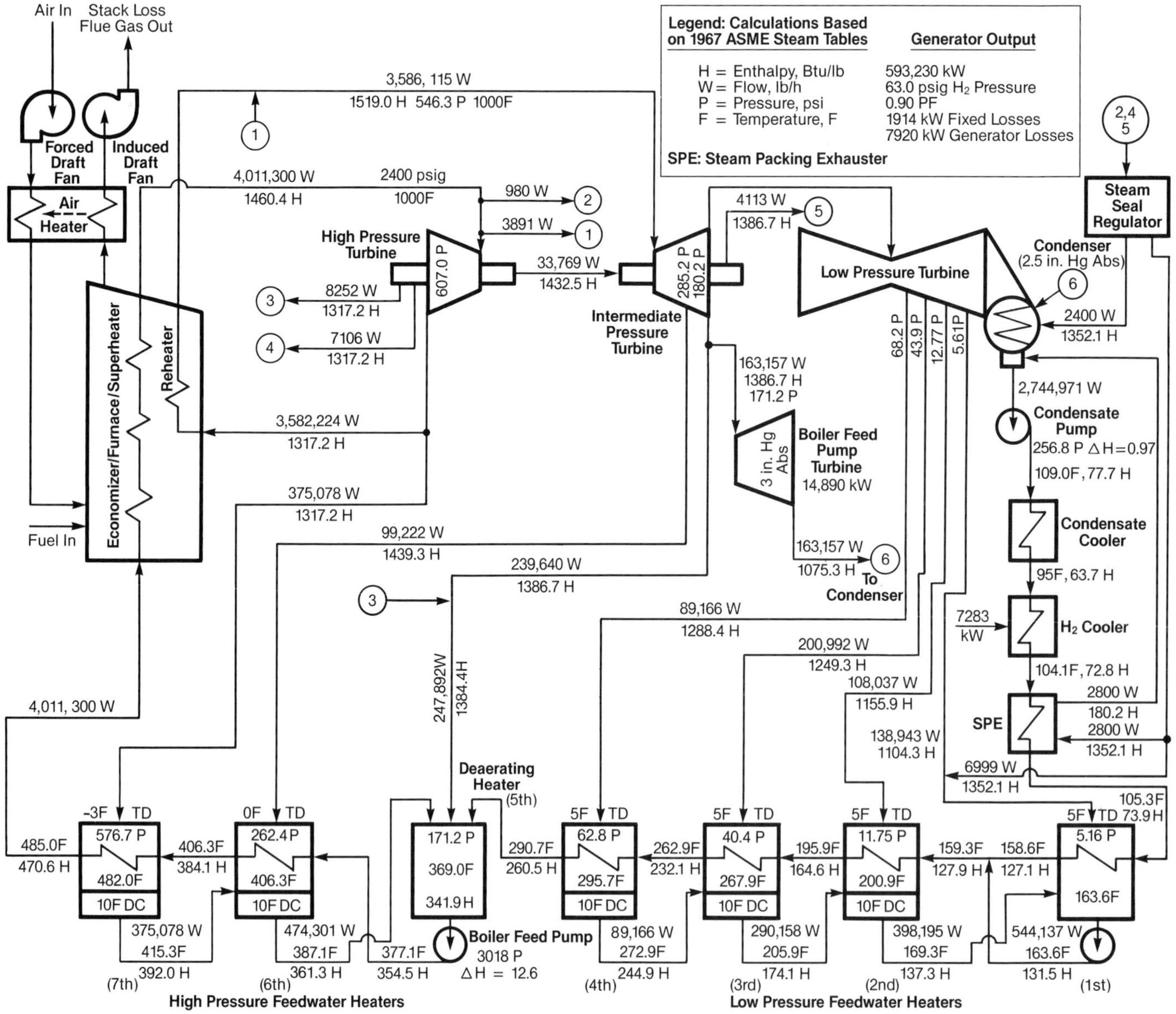

Fig. 10 Subcritical pressure, 2400 psig turbine cycle heat balance (English units).

to generate steam. This boiling steam-water mixture reaches a quality of 25 to 33% at the end of the heat exchanger and enters the steam generator's internal separators. The separators return the liquid flow to mix with incoming feedwater and direct the saturated steam flow to the outlet of the steam generator. Inevitably, a small amount of moisture is formed by the time the steam flow reaches the high pressure turbine.

Even though the once-through steam generator is capable of providing superheated steam to the turbine, the pressure and temperature limitations of nuclear plant components must be observed. As a result, the expansion lines of the power cycle lie largely in the wet steam region. This is essentially a saturated or nearly saturated steam cycle. The expansion lines for the nuclear steam system shown in Fig. 11 (featuring a once-through steam generator) are plotted on an enthalpy-entropy or *H-s* diagram in Fig. 12.

The superheated steam is delivered to the turbine at a temperature only 34F (19C) above saturation. Although this superheat improves cycle efficiency, large quantities of condensed moisture still exist in the turbine. For example, if expansion from the initial conditions shown in Fig. 12 proceed down one step to the back pressure of 2.0 in. Hg [approximately 1.0 psi (0.07 bar)] the moisture formed would exceed 20%. At best, steam turbines can accommodate about 15% moisture content. High moisture promotes erosion, especially in the turbine blades, and reduces expansion efficiency.

In addition to mechanical losses from momentum exchanges between slow moving condensate particles, high velocity steam and rotating turbine blades, there is also a thermodynamic loss resulting from the condensate in the turbine. The expansion of the steam is too rapid to permit equilibrium conditions to exist when condensation is occurring. Under this condition, the steam becomes subcooled, retaining a part of the available energy which would be released by condensation.

Fig. 11 indicates two methods of moisture removal used in this cycle and Fig. 12 shows the effect of this moisture removal on the cycle. After expansion in the high pressure turbine, the steam passes through a moisture separator, which is a low pressure drop separator external to the turbine. After passing through this separator, the steam is reheated in two stages, first by bleed steam and then by high pressure steam to 503F (262C), before entering the low pressure turbine. Here a second method of moisture removal, in which grooves on the back of the turbine blades drain the moisture from several stages of the low pressure turbine, is used. The separated moisture is carried off with the bleed steam.

Internal moisture separation reduces erosion and affords a thermodynamic advantage due to the divergence of the constant pressure lines with increasing enthalpy and entropy. This can be shown by the use of available energy, e, as follows. Consider the moisture removal stage at 10.8 psi in Fig. 12. After expansion to 10.8 psi, the steam moisture content is 8.9%. Internal separation reduces this to approximately 8.2%. Other properties are as follows:

	End of Expansion	After Moisture Extraction
P	10.8 psi	10.8 psi
H	1057.9 Btu/lb	1064.7 Btu/lb
s	1.6491 Btu/lb F	1.6595 Btu/lb F
T_o (at 2 in. Hg)	560.8 R	560.8 R
$T_o s$	924.8 Btu/lb	930.7 Btu/lb
$e = H - T_o s$	133.1 Btu/lb	134.0 Btu/lb

The increase in availability, Δe, due to moisture extraction is 134.0 - 133.1 = 0.9 Btu/lb of steam.

The values of moisture and enthalpy listed are given for equilibrium conditions without considering the nonequilibrium effects that are likely to exist within the turbine. These effects can be empirically accounted for by the isentropic efficiency of the expansion line. An important point to observe from this example is the need to retain a sufficient number of significant digits in the calculations. Frequently, the evaluation of thermodynamic processes results in working with small differences between large numbers.

Supercritical steam cycles

As previously pointed out, cycle thermal efficiency is improved by increasing the mean temperature of the heat addition process. This temperature is increased when the feedwater pressure is increased because the boiler inlet pressure sets the saturation temperature in the Rankine cycle. If the pressure is increased above the critical point of 3208.2 psi (221.2 bar), heat addition no longer results in the typical boiling process in which there is an exact division between the steam and water. Rather, the fluid is essentially a composite mixture throughout the heating process; it passes through a nondistinct point at which the properties of water change from those of a liquid to those of a gas (steam). Additional heating superheats the steam and expansion in a first stage (high pressure) turbine can occur entirely in a superheated state. This is referred to as a supercritical steam cycle, originally given the name Benson Super Pressure Plant when first proposed in the 1920s. The first commercial unit featuring the supercritical cycle and two stages of reheat was placed in service in 1957.

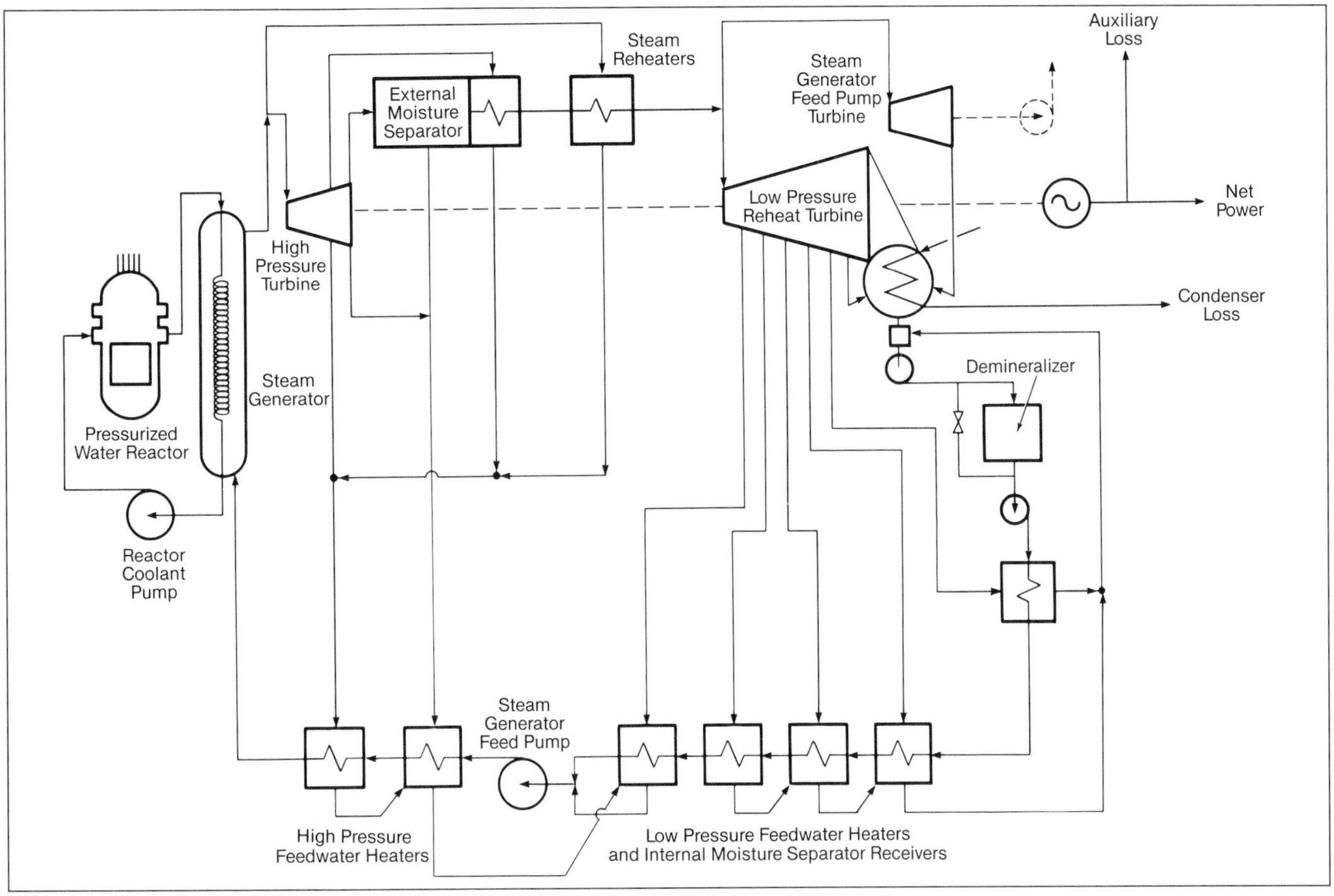

Fig. 11 Power cycle diagram, nuclear fuel: reheat by bleed and high pressure steam, moisture separation, and 6-stage regenerative feedwater heating — 900 psi, 566F/503F (62.1 bar, 297C/262C) steam.

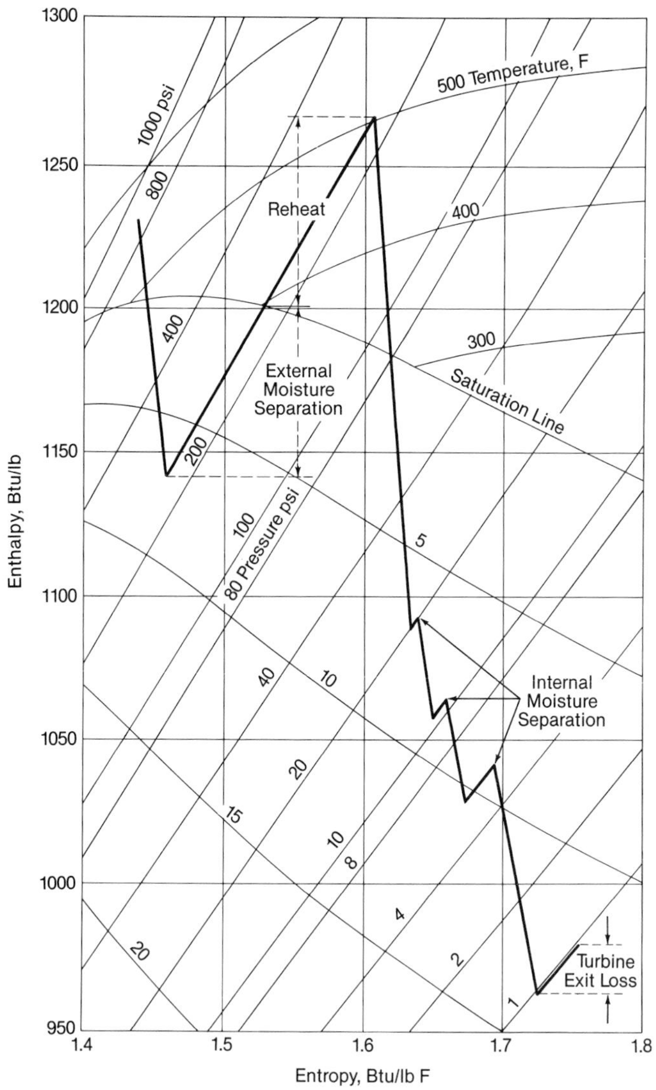

Fig. 12 Steam cycle for nuclear fuel on a Mollier chart: reheat by bleed and high pressure steam, moisture separation and 6-stage regenerative feedwater heating — 900 psi, 566F/503F steam (English units).

The steam cycle of a typical supercritical plant is shown in Fig. 13. In this *T-s* diagram, point a represents the outlet of the condensate pump. Between points a and b, the condensate is heated in the low pressure feedwater heater using saturated liquid and/or steam extracted from the steam turbines. Point b corresponds to the high pressure feedwater pump inlet. The pump increases the pressure to 4200 psi (289.6 bar), obtaining conditions of point c. Between points c and d, additional feedwater heating is provided by steam extracted from the high and low pressure turbines. Point d corresponds to the supercritical boiler inlet. Due to the nature of the fluid, the supercritical boiler is a once-through design, having no need for separation equipment. The Universal Pressure boiler design used in the supercritical unit is described further in Chapter 24. For the supercritical cycle shown, the steam arrives at the high pressure turbine at 3500 psi (241.3 bar) and 1050F (566C). Expansion in this turbine is complete at point f, which corresponds to a superheated condition. Steam exhausted from the high pressure turbine is then reheated in the boiler to approximately 1040F (560C), before entering the low pressure turbine at approximately 540 psi (37.2 bar); this corresponds to point g on the *T-s* diagram. The low pressure turbine expands the steam to point h on the diagram. The cycle is completed by condensing the exhaust from the low pressure turbine to a slightly subcooled liquid, and a condensate pump delivers the liquid to the low pressure feedwater heater, which corresponds to point a in the *T-s* diagram.

The high pressure of the feedwater in the supercritical cycle requires a substantially higher power input to the feedwater pump than that required by the saturated Rankine cycle. In a typical Rankine cycle with a steam pressure of 2400 psi (165.5 bar), the pump power input requires approximately 2.5% of the turbine output. This may increase to as much as 5% in the supercritical unit. However, this increase is justified by the improved thermal efficiency of the cycle. In general, with equivalent plant parameters (fuel type, heat sink temperature, etc.), the supercritical steam cycle generates about 4% more net power output than the subcritical pressure regenerative Rankine steam cycle.

Process steam applications

In steam power plants generating only electric power, economically justifiable thermal efficiencies range up to about 40% in fossil fuel plants (higher in combined cycle plants discussed later in this chapter) and 34% in nuclear plants. Therefore, typically more than half of the heat released from the fuel must be transferred to the environment.

Energy resources may be more efficiently used by operating multi-purpose steam plants, where steam is exhausted or extracted from the cycle at a sufficient pressure for use in an industrial process or space heating application. With these arrangements, an overall thermal utilization of 65% or greater is possible. Combination power and process installations have been common for many years, but the demand for process steam is not sufficient to permit the use of these combined cycles in most central station electric power generating plants. However, recent trends in cogeneration, biomass and waste-to-energy installations have revived interest in district heating and other process steam applications.

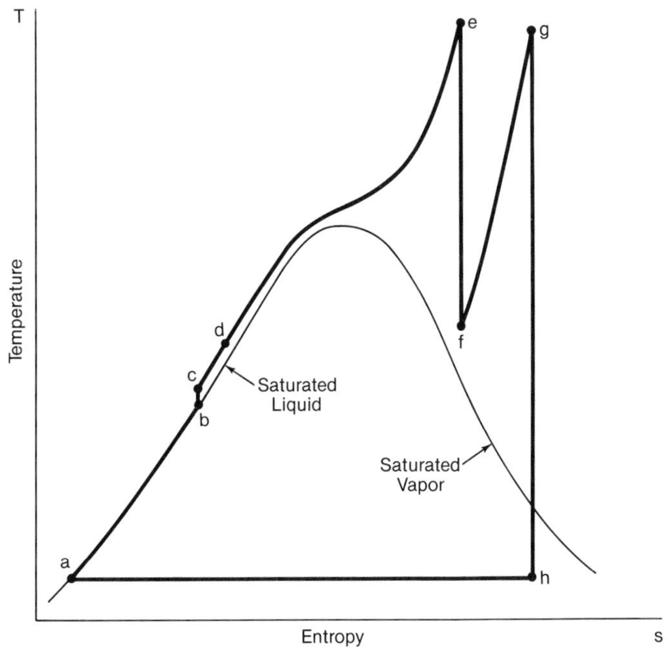

Fig. 13 Supercritical steam cycle with one reheat.

Gas turbine cycle

In the thermodynamic cycles previously described, the working fluid has been steam used in a Rankine cycle. The Rankine cycle efficiency limit is dictated by the ratio of the maximum and minimum cycle temperatures. The maximum temperature of the steam Rankine cycle is approximately 1100F (593C), which is set primarily by material constraints at the elevated pressures of the steam cycles. One means of extending the efficiency limit is to replace the working fluid with air or gas. The gas turbine system in its simplest form consists of a compressor, combustor and turbine, as shown in Fig. 14. Because of its simplicity, low capital cost and short lead time, the simple gas turbine system is being used by some utilities to add capacity in smaller increments, particularly where the capacity is needed for intermittent operation. Use of the gas turbine system in conjunction with the steam Rankine cycle is also an effective means of recovering some of the heat lost when combustion gases are released to the atmosphere at high temperatures.

In the simple gas turbine system shown in Fig. 14, air is compressed then mixed with fuel and burned in a combustor. The high temperature gaseous combustion products enter the turbine and produce work by expansion. A portion of the work produced by the turbine is used to drive the compressor and the remainder is available to produce power. The turbine exhaust gases are then vented to the atmosphere. To analyze the cycle, several simplifying assumptions are made. First, although the combustion process changes the composition of the working fluid, the fluid is treated as a gas of single composition throughout, and it is considered an ideal gas to obtain simple relationships between points in the system. Second, the combustion process is approximated as a simple heat transfer process in which the heat input to the working fluid is determined by the fuel heating values. A result of this approximation is that the mass flow rate through the system remains constant. The final approximation is to assume that each of the processes is internally reversible.

If the turbine expansion is complete with the exhaust gas at the same pressure as the compressor inlet air, the combination of processes can be viewed as a cycle. The simplifying assumptions above result in the idealized gas turbine cycle referred to as the air-standard *Brayton* cycle. Fig. 15 shows the cycle on *T-s* and *P-v* diagrams, which permit determining the state variables at the various cycle locations.

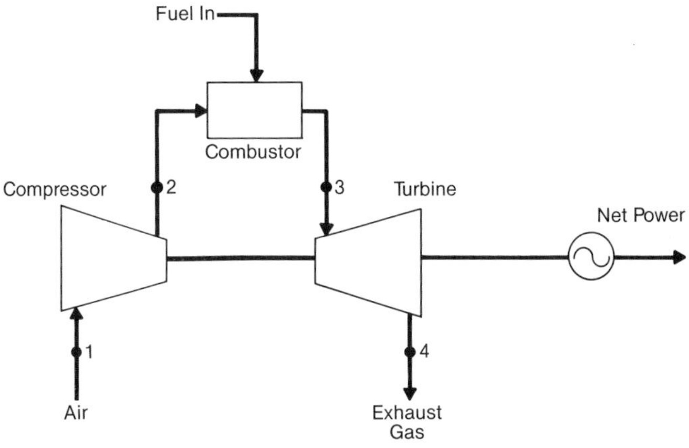

Fig. 14 Simple gas turbine system.

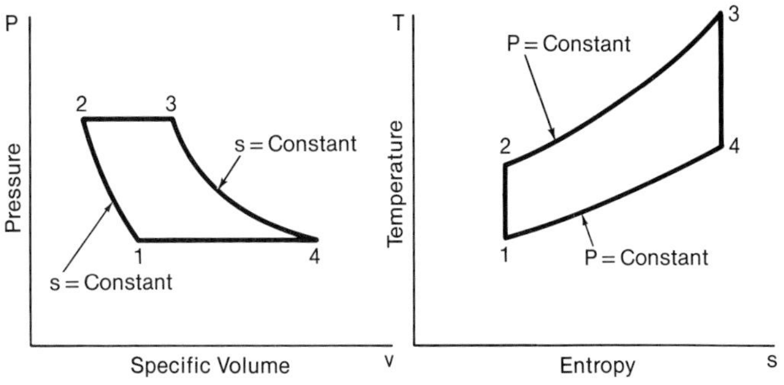

Fig. 15 Air-standard Brayton cycle.

The idealized cycle assumes an isentropic process between points 1 and 2 (compression) and between 3 and 4 (expansion work). The temperature rise between points 2 and 3 is calculated by assuming the heat addition due to combustion is at a constant pressure. In the analysis, the pressure ratio between points 1 and 2 is given by the compressor design and is assumed to be known. To determine the temperature at point 2, a relationship between the initial and final states of an isentropic ideal gas process is obtained as follows.

First, the general definitions of constant pressure and constant volume specific heats, respectively, are:

$$c_p = \left(\frac{\partial H}{\partial T}\right)_p \qquad \textbf{(44)}$$

$$c_v = \left(\frac{\partial u}{\partial T}\right)_v \qquad \textbf{(45)}$$

Strictly speaking, the specific heat values vary with temperature. In practice, however, they are assumed to be constant to facilitate the calculations. The two constants are related in that their difference equals the gas constant in the ideal gas law ($Pv = RT$):

$$c_p - c_v = R \qquad \textbf{(46)}$$

At zero pressure, the ratio of the constant pressure and constant volume specific heats is designated the specific heat ratio, k.

$$k = c_p / c_v \qquad \textbf{(47)}$$

From these definitions, changes in enthalpy and internal energy for an ideal gas can be calculated from:

$$dH = c_p dT \qquad \textbf{(48)}$$

$$du = c_v dT \qquad \textbf{(49)}$$

Although expressed to relate differential changes in enthalpy and temperature, the concept of specific heat can be used to calculate finite enthalpy changes as long as the change in temperature is not excessive. When higher accuracy is required, tabulated enthalpy values should be used.

Recalling Equation 26, the combined expression of the first and second laws of thermodynamics, setting $ds = 0$ for the isentropic process, and inserting the change in internal energy given by Equation 49, the former equation may be written:

$$Tds = du + Pdv = c_v dT + Pdv = 0 \qquad \textbf{(50)}$$

Substituting the ideal gas law (in differential form, $dT/R = Pdv + vdP$) and using the specific heat ratio definition, Equation 50 becomes:

$$\frac{dP}{P} + \frac{kdv}{v} = 0 \quad \textbf{(51)}$$

Integrating this yields:

$$Pv^k = \text{constant} \quad \textbf{(52)}$$

From Equation 52 and the ideal gas law, the following relationship between pressures and temperatures in an isentropic process is obtained, and the temperatures at points 2 and 4 are determined.

$$\frac{T_2}{T_1} = \frac{T_3}{T_4} = \left(\frac{P_2}{P_1}\right)^{(k-1)/k} \quad \textbf{(53)}$$

With this, the temperature and pressure (state variables) at all points in the cycle are determined. The turbine work output, w_t, required compressor work, w_c, and heat input to the process are calculated as:

$$w_t = c_p(T_3 - T_4) \quad \textbf{(54)}$$

$$w_c = c_p(T_2 - T_1) \quad \textbf{(55)}$$

$$q_b = c_v(T_3 - T_2) \quad \textbf{(56)}$$

As in other cycle analyses described to this point, the cycle thermal efficiency η is calculated as the net work produced divided by the total heat input to the cycle and is given by:

$$\eta = \frac{w_t - w_c}{q_b} \quad \textbf{(57)}$$

in which q_b is the heat input in the combustor (burner) per unit mass of gas (the working fluid) flowing through the system. For the ideal cycle, this can also be expressed in terms of gas temperatures by using Equation 48 to express the enthalpy change in the combustor and Equations 54 and 55 for the turbine and compressor work:

$$\eta = \left(1 - \frac{T_4 - T_1}{T_3 - T_2}\right)\frac{c_p}{c_v} \quad \textbf{(58)}$$

The actual gas turbine cycle differs from the ideal cycle due to inefficiencies in the compressor and turbine and pressure losses in the system. The effect of these irreversible aspects of the real gas turbine cycle are shown in the T-s diagram in Fig. 16. An isentropic compression would attain the point 2s, whereas the real compressor attains the pressure P_2, with an entropy corresponding to point 2 on the T-s diagram; likewise the turbine expansion attains point 4 rather than 4s. Constant-pressure lines on the diagram for pressures P_2 and P_3 illustrate the effect of pressure losses in the combustor and connecting piping, and the deviation of the process between points 4 and 1 from a constant-pressure process illustrates the effect of compressor inlet and turbine exhaust pressure losses on the cycle efficiency.

Points along the real cycle are determined by calculating the temperature T_{2s} from Equation 53 as:

$$T_{2s} = T_1\left(\frac{P_2}{P_1}\right)^{(k-1)/k} \quad \textbf{(59)}$$

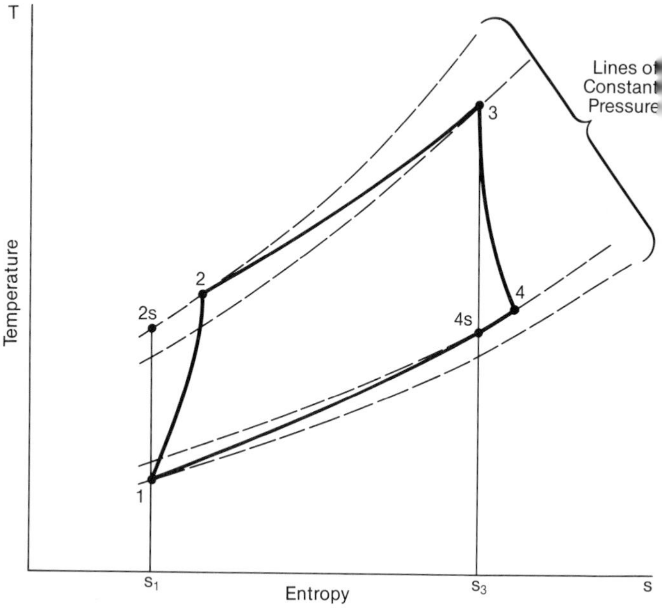

Fig. 16 *T-s* diagram of an actual gas turbine system.

Using the compressor efficiency provided by the manufacturer and solving for enthalpy or temperature at the compressor outlet yields the following:

$$\eta_c = \frac{T_{2s} - T_1}{T_2 - T_1} = \frac{H_{2s} - H_1}{H_2 - H_1} \quad \textbf{(60)}$$

In general, the cycle efficiency of a real gas turbine system is relatively low (25 to 30%) due to the high exhaust gas temperature and because a significant portion of the turbine output is used for compressor operation. The cycle efficiency may be increased by using a heat exchanger to preheat the air between the compressor and combustor. This heat is supplied by the turbine exhaust gas in a manner similar to that of the Rankine cycle regenerative heat exchangers. However, the higher efficiency is achieved in a system with a lower pressure ratio across the compressor and turbine, which in turn lowers the net work output for a given combustion system. The lower net output and extra hardware cost must be weighed in each case against the thermal efficiency improvement.

One of the key benefits of the gas turbine cycle is its ability to operate at much higher temperatures than the Rankine steam cycle. Gas turbines typically operate with an inlet temperature of 1800 to 2200F (982 to 1204C) and more recent designs have been operated as high as 2350F (1288C), raising the thermal efficiency. With the ability to operate at elevated temperatures and to use combustion gases as a working fluid, a common application of the gas turbine system is operation in conjunction with the steam Rankine cycle.

Combined cycles and cogeneration

As seen in the previous discussions of the Rankine and Brayton cycles, the gas turbine Brayton cycle efficiently uses high temperature gases from a combustion process but discharges its exhaust gas at a relatively high temperature; in the Brayton cycle, this constitutes wasted heat. On the other hand, the steam turbine Rankine cycle is unable to make full use of the highest temperatures. Combined cycles are designed to take advantage of the best features of these two cycles to improve the overall thermal efficiency of the plant. Advanced combined cycles, in which the gas turbine exhaust is used as a heat source for a steam turbine cycle, can achieve overall thermal efficiencies in excess of 50%; these are becoming increasingly popular in power generation applications. (See Chapter 31.)

Waste heat boilers

In its simplest form, the combined cycle plant is a gas turbine (Brayton cycle) plant enhanced by passing the turbine exhaust through a steam generator, as shown in Fig. 17. The steam generator uses the hot turbine exhaust as a heat source for a steam turbine Rankine cycle. Electric power is generated from the mechanical work provided by the gas turbine and the steam turbine. In concept, the steam generator in the combined cycle is recovering the wasted heat from the gas turbine exhaust, and therefore it is referred to as a heat recovery steam generator or a waste heat boiler. (See Chapter 31.) More recent applications of the combined cycle have incorporated supplemental firing in the waste heat boiler to elevate the steam temperature and, therefore, to improve the steam cycle performance. Thermodynamic efficiency of the overall system is determined by evaluating the mechanical work output (W) of the steam turbine and gas turbine cycles separately. The thermal efficiency is defined as the work output of the two cycles divided by the total heat supplied (Q_{total}):

$$\eta = \left[(W_{out} - W_{in})_{GT} + (W_{out} - W_{in})_{ST}\right] / Q_{total} \quad (61)$$

where the subscripts GT and ST refer to gas turbine and steam turbine, respectively.

Another approach to combining the gas and steam cycles, in which the steam generator serves as the combustion chamber for the gas turbine cycle, is shown in Fig. 18. In this arrangement, the principal heat source to the gas and steam cycle is the combustion process taking place in the steam generator. The gaseous combustion products are expanded in the gas turbine and the steam generated in the boiler tubes is expanded in the steam turbine. Although not shown in Fig. 18, the heat contained in the gas turbine exhaust may be recovered by using either a regenerative heat exchanger in the gas turbine cycle or a feedwater heater in the steam cycle. A pressurized fluidized-bed combustion combined cycle is a specific example of this approach to combining the gas and steam cycles. (See Chapter 29.)

Cogeneration

In the most general sense, cogeneration is the production of more than one useful form of energy (thermal, mechanical, electrical, etc.) simultaneously from a single fuel. In practice, cogeneration refers to generating electricity while principally performing an industrial function such as space heating, process heating or fuel gasification. Cogeneration systems are divided into two basic arrangements, topping and bottoming cycles.

A topping cycle is shown in Fig. 19. In this system the fuel is used for power generation in a steam boiler or gas turbine cycle combustor, and the waste heat from the power generation cycle supports an industrial process. The most common topping cycle is one in which a boiler

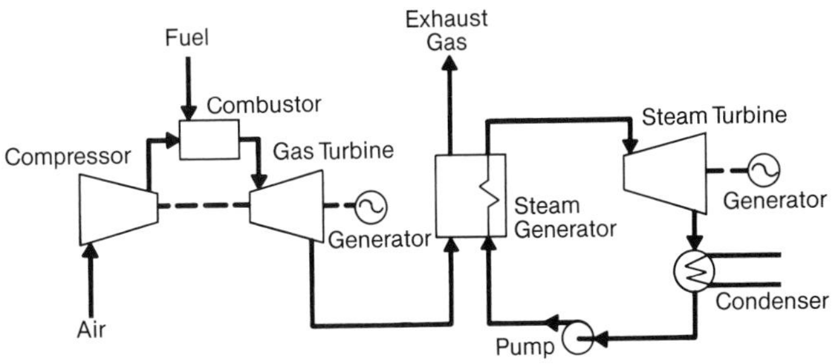

Fig. 17 Simple combined cycle plant.

generates steam at a higher pressure than that needed for the process or space conditioning application. The high pressure steam is then expanded in a turbine to a pressure that is appropriate for the application, generating electricity in the expansion process. Steam turbines, gas turbines and reciprocating engines are commonly used in topping cycles.

A bottoming cycle is most commonly associated with the recovery or waste heat boiler. In the bottoming cycle, fuel is not supplied directly to the power generating cycle. Rather, steam is generated from a waste heat source and then expanded in a turbine to produce work or to generate electricity. Steam is frequently used in the bottoming cycle because of its ability to condense at low temperatures in the closed Rankine cycle. The bottoming cycle is shown in Fig. 20. The steam Rankine cycle used as a bottoming cycle has been illustrated previously in the descriptions of combined cycle plants.

Combustion processes

To this point, cycles have been compared based on the thermal efficiency achieved, i.e., the net work produced divided by the total heat input to the cycle. To complete the evaluation of a combustion-based cycle, however, the performance must be expressed in terms of fuel consumption. In addition, the ability of the different machines to make full use of the combustion energy varies with temperatures reached in the combustion chamber and with dissociation of the combustion products.

The energy release during combustion is illustrated by considering the combustion of carbon (C) and oxygen (O_2) to form carbon dioxide (CO_2):

$$C + O_2 \rightarrow CO_2$$

If heat is removed from the combustion chamber and the reactants and products are maintained at 25C and 0.1 MPa (77F and 14.5 psi) during the process, the heat transfer from the combustion chamber would be 393,522 kJ per kmole of CO_2 formed. From the first law applied

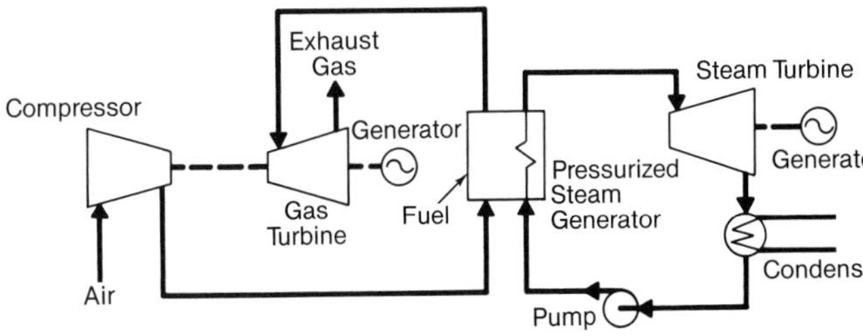

Fig. 18 Pressurized combustion combined cycle plant.

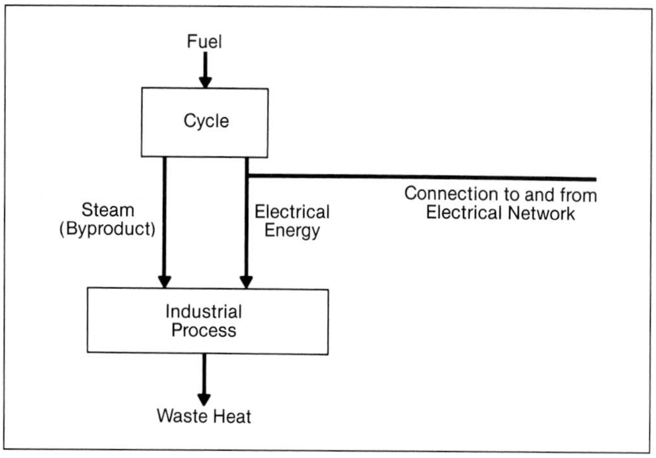

Fig. 19 Topping cycle.

to the process, the heat transfer is equal to the difference in enthalpy between the reactants and products:

$$q - w = H_P - H_R \quad \textbf{(62)}$$

The subscripts R and P refer to reactants and products, respectively. Assuming that no work is done in the combustion chamber and expressing the enthalpy of reactants and products on a per mole basis, this becomes:

$$Q = \sum n_P H_P - \sum n_R H_R \quad \textbf{(63)}$$

The number of moles of each element or molecular species entering or leaving the chamber, n_R or n_P respectively, is obtained from the chemical reaction equation. By convention, the enthalpy of elements at 25C and 0.1 MPa (77F and 14.5 psi) are assigned the value of zero. Consequently, the enthalpy of CO_2 at these conditions is -393,522 kJ/kmole (the negative sign is due to the convention of denoting heat transferred from a control volume as negative). This is referred to as the enthalpy of formation and is designated by the symbol H_f^o. The enthalpy of CO_2 (and other molecular species) at other conditions is found by adding the change in enthalpy between the desired condition and the standard state to the enthalpy of formation. [Note that some tables may not use 25C and 0.1 MPa (77F and 14.5 psi) as the standard state when listing the enthalpy of formation.] Ideal gas behavior or tabulated properties are used to determine enthalpy changes from the standard state.

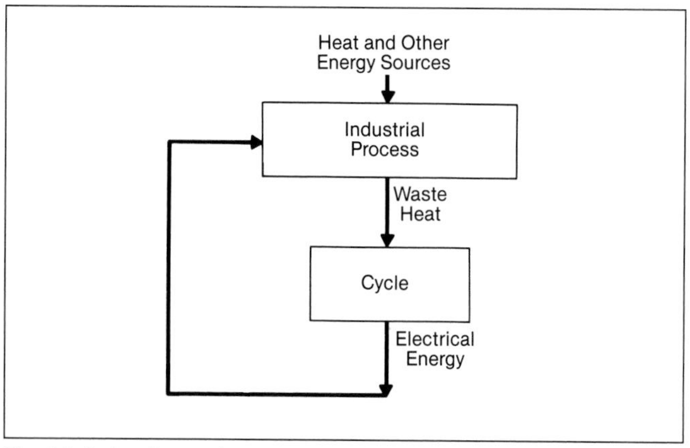

Fig. 20 Bottoming cycle.

The stoichiometrically balanced chemical reaction equation provides the relative quantities of reactants and products entering and leaving the combustion chamber. The first law analysis is usually performed on a per mole or unit mass of fuel basis. The heat transfer from a combustion process is obtained from a first law analysis of the combustion process, given the pressure and temperature of the reactants and products. Unfortunately, even in the case of complete combustion as assumed in the previous example, the temperature of the combustion products must be determined by additional calculations discussed later in this section.

The combustion of fossil or carbon based fuel is commonly accompanied by the formation of steam or water (H_2O) as in the reaction:

$$CH_4 + 2O_2 \rightarrow CO_2 + 2H_2O$$

Again, the difference in enthalpy between the reactants and products is equal to the heat transfer from the combustion process. The heat transfer per unit mass of fuel (methane in this example) is referred to as the heating value of the fuel. If the H_2O is present as liquid in the products, the heat transferred is referred to as the *higher heating value* (HHV). The term *lower heating value* (LHV) is used when the H_2O is present as a vapor. The difference between these two values is frequently small (about 4% for most hydrocarbon fuels) but still significant. When the efficiency of a cycle is expressed as a percentage of the fuel's heating value, it is important to know whether the HHV or LHV is used.

As noted, one of the difficulties in completing the first law analysis of the combustion process is determining the temperature of the products. In some applications an upper limit of the combustion temperature may be estimated. From the first law, this can occur if the combustion process takes place with no change in kinetic or potential energy, with no work and with no heat transfer (adiabatically). Under these assumptions, the first law indicates that the sum of the enthalpies of the reactants equals that of the products. The temperature of the products is then determined iteratively by successively assuming a product temperature and checking the equality of the reactant and product enthalpies. For a given fuel and reactants at the specified inlet temperature and pressure, this procedure determines the highest attainable combustion temperature, referred to as the *adiabatic combustion* or *flame temperature*.

Free energy

An important thermodynamic property derived from a combination of other properties (just as enthalpy was derived from u, P and v) is Gibbs free energy (g), which is also frequently referred to as *free energy*:

$$g = H - Ts \quad \textbf{(64)}$$

Free energy g is a thermodynamic potential similar to enthalpy and internal energy because in any thermodynamic process, reversible or irreversible, differences in this quantity depend only on initial and final states of the system.

The usefulness of free energy is particularly evident from the following expression of the combined first and second laws, expressed for a reversible process with negligible changes in kinetic and potential energy:

$$W_{rev} = \sum m_1(H_1 - T_o s_1) - \sum m_2(H_2 - T_o s_2) \quad \textbf{(65)}$$

When applied to a combustion process in which the reactants and products are in temperature equilibrium with the surroundings, this becomes:

$$W_{rev} = \sum n_R g_R - \sum n_P g_P \quad \textbf{(66)}$$

This equation indicates the maximum value of reversible work that can be obtained from the combustion of a given fuel. The reversible work is maximized when the reactants constitute a stoichiometrically balanced mixture (no excess air). The quantities n_R and n_P are obtained from the chemical reaction equation and g is expressed on a per mole basis in Equation 66.

From this, one might expect to express the efficiency of a cycle that extracts energy from a combustion process as a percentage of the Gibbs free energy decrease, rather than in terms of the heating value of the fuel. It is uncommon for this to be done, however, because the difference between the free energy decrease and the heating value of hydrocarbon fuels is small and because the use of fuel heating value is more widespread.

Free energy is more commonly used to determine the temperature reached in burning fuel, including the effects of dissociation. The problem of dissociation is illustrated by again considering the combustion of carbon and oxygen to form CO_2. If the temperature of the combustion process is high enough, the CO_2 dissociates to form CO and O_2 according to the reaction:

$$CO_2 \leftrightarrow CO + \frac{1}{2}O_2$$

As the dissociation reaction occurs from left to right (from all CO_2 to none), the sum of the reactant free energies and that of the products vary. Equilibrium of this reaction is reached when the sum of the free energies is a minimum. The equilibrium point (degree of dissociation) varies with the combustion temperature.

While the process of iteratively determining a minimum free energy point is suited for computer calculations, the equilibrium conditions of the dissociation reaction at an assumed temperature can also be determined using tabulated values of a constant relating the species involved in the reaction. This constant is known as the equilibrium constant K_{eq}, which for ideal gases is given by:

$$K_{eq} = \frac{(P_B)^b (P_C)^c}{(P_A)^a} \quad \textbf{(67)}$$

where P_A, P_B and P_C are the partial pressures, i.e., the products of total pressure and mole fractions in the mixture, of the reactants and products. The exponents represent the number of moles present for each species (A, B and C) in the stoichiometric balance equation as follows:

$$aA \leftrightarrow bB + cC \quad \textbf{(68)}$$

Equations 67 and 68 yield simultaneous equations for the mole fractions a, b and c. For nonideal gas reactions, the partial pressures are replaced by what are known as *fugacities* (the tendencies of a gas to expand or escape). Thermodynamic properties and relationships for the compounds and their elements encountered in the combustion process are available in the literature. One of the best sources for this information is the JANAF Thermochemical Tables, published by the U.S. Department of Commerce.[5] These tables include $\log_{10}$ values of the equilibrium constants for temperatures from 0 to 6000K.

To continue the carbon-oxygen combustion example, the overall chemical reaction, including dissociation, is now written as:

$$C + O_2 \rightarrow aCO_2 + bCO + cO_2$$

in which the coefficients a, b and c represent the mole fractions of the product components as determined by the solution to the dissociation reaction at the assumed combustion temperature. The overall reaction equation is now used to check the assumed temperature by adding the enthalpies of the combustion products at the assumed temperature, noting that the enthalpy per mole must be multiplied by the corresponding mole fraction a, b, or c for each product species. The combustion temperature is determined when the sum of product enthalpies minus that of the reactants equals the heat transfer to the surroundings of the combustion chamber. The convective and radiative heat transfer from the combustion products to the chamber and eventually to the working fluid of the cycle at the assumed combustion temperature are discussed in Chapter 4.

References

1. Meyer, C.A., *et al.*, *ASME Steam Tables*, 5th ed., The American Society of Mechanical Engineers, New York, 1983.

2. *CRC Handbook of Chemistry and Physics*, 70th ed., CRC Press, Inc., Boca Raton, Florida, 1989.

3. Keenan, J.H., Chao, J., and Kaye, J., *Gas Tables*, 2nd ed., Wiley, New York, 1980.

4. Vargaftik, N.B., *Tables on the Thermophysical Properties of Liquids and Gases*, 2nd ed., Wiley, New York, 1975.

5. Chase, M.W., *et al.*, *JANAF Thermochemical Tables*, 3rd ed., American Chemical Society and the American Institute of Physics, New York, 1986.

Laser velocity measurements in a steam generator flow model.

Chapter 3
Fluid Dynamics

In the production and use of steam there are many fluid dynamics considerations. Fluid dynamics addresses steam and water flow through pipes, fittings, valves, tube bundles, nozzles, orifices, pumps and turbines as well as entire circulating systems. It also considers air and gas flow through ducts, tube banks, fans, compressors and turbines plus convection flow of gases due to draft effect. The fluid may be a liquid or gas but, regardless of its state, the essential property of a fluid is that it yields under the slightest shear stress. This chapter is limited to the discussion of Newtonian liquids, gases and vapors where any shear stress is directly proportional to a velocity gradient normal to the shear force.

Liquids and gases are recognized as states of matter. In the liquid state a fluid is relatively incompressible, having a definite volume. It is also capable of forming a free surface between itself and its vapor or any other fluid with which it does not mix. On the other hand, a gas is highly compressible. It expands or diffuses indefinitely and is subject only to the limitations of gravitational forces or an enclosing vessel.

The term *vapor*, while imprecise and not universally agreed upon, generally implies a gas near saturation conditions where the liquid and the gas phase coexist at essentially the same temperature and pressure. In a similar sense the term *gas* denotes a highly superheated vapor.

Fundamental relationships

All fluid flowing systems are governed by three fundamental conservation relations: mass, momentum and energy. With the exception of nuclear reactions where minute quantities of mass are converted into energy, these laws must be satisfied in all flowing systems. Fundamental mathematical relationships for these principles are presented below and form the basis for numerical computational models. However, full analytical solutions are frequently too complex for engineering use. Simplified forms are used in engineering practice based upon various assumptions and empirical relationships in order to provide practical solutions. A more complete discussion of the relationships and vector notation may be found in References 1 and 2.

Conservation of mass

The *law of conservation of mass* simply states that the change in mass stored in a system must equal the difference in the mass flowing in to and out of the system. In its simplest form in x, y and z Cartesian coordinates, conservation of mass for a very small fixed volume can be mathematically stated as the *continuity equation*:

$$\frac{\partial}{\partial x}\rho u + \frac{\partial}{\partial y}\rho v + \frac{\partial}{\partial z}\rho w = -\frac{\partial \rho}{\partial t} \quad \textbf{(1)}$$

where u, v and w are the fluid velocities in the x, y and z coordinate directions; t is time and ρ is the fluid density. An important form of this equation is derived by assuming steady-state ($\partial / \partial t = 0$) and incompressible (constant density) flow conditions:

$$\frac{\partial u}{\partial x} + \frac{\partial v}{\partial y} + \frac{\partial w}{\partial z} = 0 \quad \textbf{(2)}$$

Although no fluid is truly incompressible, the assumption of incompressibility simplifies problem solutions and is frequently acceptable for engineering practice.

Another relationship useful in large scale pipe flow systems involves the integration of Equation 1 around the flow path for constant density, steady-state conditions. For only one inlet (subscript 1) and one outlet (subscript 2):

$$\dot{m} = \rho_1 A_1 V_1 = \rho_2 A_2 V_2 \quad \textbf{(3)}$$

where ρ is the average density, V is the average velocity, A is the cross-sectional area, and $\dot{m}$ is the mass flow rate.

Conservation of momentum

The *law of conservation of momentum* is a representation of Newton's Second Law of Motion — the mass of a particle times its acceleration is equal to the sum of all of the forces acting on the particle. In a flowing system, the equivalent relationship for a fixed (control) volume becomes: the change in momentum entering and leaving the control volume is equal to the sum of the forces acting on the control volume. This relationship is direction dependent resulting in one equation for each coordinate direction (x, y and z for Cartesian coordinates), providing three momentum equations.

The full mathematical representation of the momentum equation is complex and is of limited direct usefulness in many engineering applications except for numerical computational models. As an example, in the x coordinate direction, the full momentum equation becomes:

$$\rho\left(\frac{\partial u}{\partial t}+u\frac{\partial u}{\partial x}+v\frac{\partial u}{\partial y}+w\frac{\partial u}{\partial z}\right)=\rho f_x-\frac{\partial P}{\partial x}$$

Term 1 *Term 2* *Term 3*

$$+\frac{\partial}{\partial x}\left[\frac{2}{3}\mu\left(2\frac{\partial u}{\partial x}-\frac{\partial v}{\partial y}-\frac{\partial w}{\partial z}\right)\right]+\frac{\partial}{\partial y}\left[\mu\left(\frac{\partial v}{\partial x}+\frac{\partial u}{\partial y}\right)\right] \quad \textbf{(4)}$$

Term 4

$$+\frac{\partial}{\partial z}\left[\mu\left(\frac{\partial w}{\partial x}+\frac{\partial u}{\partial z}\right)\right]$$

Term 4 (cont'd.)

where f_x is the body force in the x direction, P is the pressure, and μ is the dynamic viscosity. This equation and the corresponding equations in the y and z Cartesian coordinates represent the Navier-Stokes equations which are valid for all compressible Newtonian fluids with variable viscosity. *Term 1* is the rate of momentum change. *Term 2* accounts for body force effects such as gravity. *Term 3* accounts for the pressure gradient. The balance of the equation accounts for momentum change due to viscous transfer. *Term 1* is sometimes abbreviated as $\rho\ (Du/Dt)$ where Du/Dt is defined as the substantial derivative of u.

For the special case of constant density and viscosity, this equation reduces to (for the x coordinate direction):

$$\frac{Du}{Dt}=f_x-\frac{1}{\rho}\frac{\partial P}{\partial x}+\frac{\mu}{\rho}\left(\frac{\partial^2 u}{\partial x^2}+\frac{\partial^2 u}{\partial y^2}+\frac{\partial^2 u}{\partial z^2}\right) \quad \textbf{(5)}$$

The y and z coordinate equations can be developed by substituting appropriate parameters for velocity u, pressure gradient $\partial P/\partial x$, and body force f_x. Where viscosity effects are negligible ($\mu = 0$), the Euler equation of momentum is produced (x direction only shown):

$$\frac{Du}{Dt}=f_x-\frac{1}{\rho}\frac{\partial P}{\partial x} \quad \textbf{(6)}$$

Energy equation (first law of thermodynamics)

The *law of conservation of energy* for nonreacting fluids basically states that the energy transferred into a system less the mechanical work done by the system must be equal to the change in stored energy, plus the energy flowing out of the system with a fluid, minus the energy flowing in with a fluid. A single overall relationship results. A general form of the energy equation for a very small volume in a flowing system using an enthalpy based formulation and vector notation is:

$$\rho\frac{DH}{Dt} = q''' + \frac{DP}{Dt} + \nabla\cdot k\nabla T + \frac{\mu}{g_c}\Phi \quad \textbf{(7)}$$

Term 1 *Term 2* *Term 3* *Term 4* *Term 5*

where ρ is the fluid density, H is the enthalpy per unit mass of a fluid, T is the fluid temperature, q''' is the internal heat generation, k is the thermal conductivity, and Φ is the dissipation function for irreversible work.[2] *Term 1* accounts for net energy convected into the system, *Term 2* accounts for internal heat generation, *Term 3* accounts for work done by the system, *Term 4* addresses heat conduction and *Term 5* accounts for viscous dissipation.

As with the momentum equations, the full energy equation is too complex for most direct engineering applications except for use in numerical models. As a result, specialized forms are based upon various assumptions and engineering approximations. As discussed in Chapter 2, the most common form of the energy equation for a simple, inviscid (i.e., frictionless) steady-state flow system with flow in at location 1 and out at location 2 is:

$$JQ-W=J(u_2-u_1)+(P_2v_2-P_1v_1)+\frac{1}{2g_c}\left(V_2^2-V_1^2\right)+(Z_2-Z_1)\frac{g}{g_c} \quad \textbf{(8a)}$$

or

$$JQ-W=J(H_2-H_1)+\frac{1}{2g_c}\left(V_2^2-V_1^2\right)+(Z_2-Z_1)\frac{g}{g_c} \quad \textbf{(8b)}$$

where

Q = heat added to the system, Btu/lbm (J/kg) (See Note below)
W = work done by the system, ft-lbf/lbm (N m/kg)
J = mechanical equivalent of heat = 778.26 ft-lbf/Btu (1 N m/J)
u = internal energy, Btu/lbm (J/kg)
P = pressure, lbf/ft^2 (N/m^2)
v = specific volume, ft^3/lbm (m^3/kg)
V = velocity, ft /s (m/s)
Z = elevation, ft (m)
H = enthalpy = $u + Pv/J$, Btu/lbm (J/kg)
g = 32.17 ft /s^2 (9.8 m /s^2)
g_c = 32.17 lbm ft /lbf s^2 (1 kg m /N s^2)

Energy equation applied to fluid flow (pressure loss without friction)

The conservation laws of mass and energy, when simplified for steady, frictionless (i.e., inviscid) flow of an incompressible fluid, result in the mechanical energy balance referred to as Bernoulli's equation:

$$P_1v+Z_1\frac{g}{g_c}+\frac{V_1^2}{2g_c}=P_2v+Z_2\frac{g}{g_c}+\frac{V_2^2}{2g_c} \quad \textbf{(9)}$$

The variables in Equation 9 are defined as follows with the subscripts referring to location 1 and location 2 in the system:

P = pressure, lbf/ft^2 (N/m^2)
v = specific volume of fluid, ft^3/lbm (m^3/kg)
Z = elevation, ft (m)
V = fluid velocity, ft/s (m/s)

Briefly, Equation 9 states that the total mechanical energy present in a flowing fluid is made up of pressure energy, gravity energy and velocity or kinetic energy; each is mutually convertible into the other forms. Furthermore, the total mechanical energy is constant along any streamtube, provided there is no friction, heat transfer or shaft

Note: Where required for clarity, the abbreviation lb is augmented by f (lbf) to indicate pound force and by m (lbm) to indicate pound mass. Otherwise lb is used with force or mass indicated by the context.

work between the points considered. This stream-tube may be an imaginary closed surface bounded by stream lines or it may be the wall of a flow channel, such as a pipe or duct, in which fluid flows without a free surface.

Applications of Equation 9 are found in flow measurements using the velocity head conversion resulting from flow channel area changes. Examples are the venturi, flow nozzle and various orifices. Also, pitot tube flow measurements depend on being able to compare the total head, $Pv + Z + (V^2/2\,g_c)$, to the static head, $Pv + Z$, at a specific point in the flow channel. Descriptions of metering instruments are found in Chapter 40. Bernoulli's equation, developed from strictly mechanical energy concepts some 50 years before any precise statement of thermodynamic laws, is a special case of the conservation of energy equation or first law of thermodynamics stated above in Equation 8.

Applications of Equation 8 to fluid flow are given in the examples on water and compressible fluid flow through a nozzle under the *Applications of the Energy Equation* section in Chapter 2. Equation 18, Chapter 2 is:

$$V_2 = \sqrt{2g_c J(H_1 - H_2)} = C\sqrt{H_1 - H_2} \tag{10}$$

where

V_2 = downstream velocity, ft/s (m/s)
g_c = 32.17 lbm ft/lbf s^2
H_2 = downstream enthalpy, Btu/lb (J/kg)
H_1 = upstream enthalpy, Btu/lb (J/kg)
J = 778.26 ft lbf/Btu
C = 223.8 $\sqrt{\text{Btu/lb}}$ × ft/s (1.414 $\sqrt{\text{J/kg}}$ × m/s)

This equation relates fluid velocity to a change in enthalpy under *adiabatic* (no heat transfer), steady, *inviscid* (no friction) flow where no work, local irreversible flow pressure losses, or change in elevation occurs. The initial velocity is assumed to be zero and compressible flow is permitted. If the temperature (T) and pressure (P) of steam or water are known at points 1 and 2, Equation 10 provides the exit velocity using the enthalpy (H) values provided in Tables 1, 2 and 3 of Chapter 2. If the pressure and temperature at point 1 are known but only the pressure at point 2 is known, the outlet enthalpy (H_2) can be evaluated by assuming constant entropy expansion from points 1 to 2, i.e., $S_1 = S_2$.

There is another method that can be used to determine velocity changes in a frictionless adiabatic expansion. This method uses the ideal gas equation of state in combination with the pressure-volume relationship for constant entropy.

From the established gas laws, the relationship between pressure, volume and temperature of an ideal gas is expressed by:

$$Pv = RT \tag{11a}$$

or

$$Pv = \frac{\mathbf{R}}{M}T \tag{11b}$$

where

P = absolute pressure, lb/ft^2 (N/m^2)
v = specific volume, ft^3/lb of gas (m^3/kg)
T = absolute temperature, R (K)
M = molecular weight of the gas, lb/lb-mole (kg/kg-mole)
R = gas constant for specific gas, ft lbf/lbm R (N m/kg K)
MR = $\mathbf{R}$ = the universal gas constant
= 1545 ft lb/lb-mole R (8.3143 kJ/kg-mole K)

The relationship between pressure and specific volume along an expansion path at constant entropy, i.e., isentropic expansion, is given by:

$$Pv^k = \text{constant} \tag{12}$$

Because P_1 and v_1 in Equation 12 are known, the constant can be evaluated from $P_1v_1^k$. The exponent k is constant and is evaluated for an ideal gas as:

$$k = c_p / c_v \tag{13}$$

where

c_p = specific heat at constant pressure, Btu/lb F (J/kg K)
c_v = specific heat at constant volume, Btu/lb F (J/kg K)
= $(u_1 - u_2)/(T_1 - T_2)$

For a steady, adiabatic flow with no work or change in elevation, Equations 8, 11, 12 and 13 can be combined to provide the following relationship:

$$V_2^2 - V_1^2 = 2g_c\left(\frac{k}{k-1}\right)P_1v_1\left\{1 - \left(\frac{P_2}{P_1}\right)^{\frac{k-1}{k}}\right\} \tag{14}$$

When V_1 is set to zero and using English units Equation 14 becomes:

$$V_2 = 8.02\sqrt{\left(\frac{k}{k-1}\right)P_1v_1\left\{1 - \left(\frac{P_2}{P_1}\right)^{\frac{k-1}{k}}\right\}}\text{, ft/s} \tag{15}$$

Equations 14 and 15 can be used for gases in pressure drop ranges where there is little change in k, provided values of k are known or can be calculated. Equation 15 is widely used in evaluating gas flow through orifices, nozzles and flow meters.

It is sufficiently accurate for most purposes to determine velocity differences caused by changes in flow area by treating a compressible fluid as incompressible. This assumption only applies when the difference in specific volumes at points 1 and 2 is small compared to the final specific volume. The accepted practice is to consider the fluid incompressible when:

$$(v_2 - v_1)/v_2 < 0.05 \tag{16}$$

Because Equation 9 represents the incompressible energy balance for frictionless adiabatic flow, it may be rearranged to solve for the velocity difference as follows:

$$V_2^2 - V_1^2 = 2g_c[\Delta(Pv) + \Delta Zg/g_c] \tag{17}$$

where

$\Delta(Pv)$ = pressure head difference between locations 1 and 2 = $(P_1 - P_2)\,v$, ft (m)
ΔZ = head (elevation) difference between locations 1 and 2, ft (m)
V = velocity at locations 1 and 2, ft/s (m/s)

When the approach velocity is approximately zero, Equation 17 in English units becomes:

$$V_2 = \sqrt{2gh} = 8.02\sqrt{h},\ \text{ft/s} \tag{18}$$

In this equation, h, in ft head of the flowing fluid, replaces $\Delta(Pv) + \Delta Z$. If the pressure difference is measured in psi, it must be converted to lb/ft^2 to obtain Pv in ft.

Pressure loss from fluid friction

So far, only pressure losses associated with changes in the kinetic energy term, $V^2/2g_c$, and static pressure term, Z, have been discussed. These losses occur at constant flow where there are variations in flow channel cross-sectional area and where inlet and outlet are at different elevations. Fluid friction and, in some cases, heat transfer with the surroundings also have important effects on pressure and velocity in a flowing fluid. The following discussion applies to fluids flowing in channels without a free surface.

When a fluid flows, molecular diffusion causes momentum interchanges between layers of the fluid that are moving at different velocities. These interchanges are not limited to individual molecules. In most flow situations there are also bulk fluid interchanges known as eddy diffusion. The net result of all inelastic momentum exchanges is exhibited in shear stresses between adjacent layers of the fluid. If the fluid is contained in a flow channel, these stresses are eventually transmitted to the walls of the channel. To counterbalance this wall shear stress, a pressure gradient proportional to the bulk kinetic energy, $V^2/2g_c$, is established in the fluid in the direction of the bulk flow. The force balance is:

$$\pi \frac{D^2}{4}(dP) = \tau_w \pi D(dx) \tag{19}$$

where

D = tube diameter or equivalent diameter D_e, ft (m)
D_e = 4 × (flow area)/(wetted perimeter) for circular or noncircular cross-sections, ft (m)
x = distance in direction of flow, ft (m)
τ_w = shear stress at the tube wall, lb/ft^2 (N/m^2)

Solving Equation 19 for the pressure gradient (dP/dx):

$$\frac{dP}{dx} = \frac{4}{D}\tau_w \tag{20}$$

This pressure gradient along the length of the flow channel can be expressed in terms of a certain number of velocity heads, f, lost in a length of pipe equivalent to one tube diameter. The symbol f is called the friction factor which has the following relationship to the shear stress at the tube wall:

$$\tau_w = \frac{f}{4}\frac{1}{v}\frac{V^2}{2g_c} \tag{21}$$

Equation 20 can be rewritten, substituting for τ_w from Equation 21 as follows:

$$\frac{dP}{dx} = \frac{4}{D}\left(\frac{f}{4}\frac{1}{v}\frac{V^2}{2g_c}\right) = \frac{f}{D}\frac{1}{v}\frac{V^2}{2g_c} \tag{22}$$

The general energy equation, Equation 8, expressed as a differential has the form:

$$du + \frac{VdV}{g_c} + d(Pv) = dQ - dW_k$$

or (23)

$$du + \frac{VdV}{g_c} + Pdv + vdP = dQ - dW_k$$

Substituting Equation 26 of Chapter 2 ($du = Tds - Pdv$) in Equation 23 yields:

$$Tds + \frac{VdV}{g_c} + vdP = dQ - dW_k \tag{24}$$

The term Tds represents heat transferred to or from the surroundings, dQ, and any heat added internally to the fluid as the result of irreversible processes. These processes include fluid friction or any irreversible pressure losses resulting from fluid flow. (See Equation 29 and explanation, Chapter 2.) Therefore:

$$Tds = dQ + dQ_F \tag{25}$$

where dQ_F is the heat equivalent of fluid friction and any *local irrecoverable pressure losses* such as those from pipe fittings, bends, expansions or contractions.

Substituting Equation 25 into Equation 24, canceling dQ on both sides of the equation, setting dW_k equal to 0 (no shaft work), and rearranging Equation 24 results in:

$$dP = -\frac{VdV}{vg_c} - \frac{dQ_F}{v} \tag{26}$$

Three significant facts should be noted from Equation 26 and its derivation. First, the general energy equation does not accommodate pressure losses due to fluid friction or geometry changes. To accommodate these losses Equation 26 must be altered based on the first and second laws of thermodynamics (Chapter 2). Second, Equation 26 does not account for heat transfer except as it may change the specific volume, v, along the length of the flow channel. Third, there is also a pressure loss as the result of a velocity change. This loss is independent of any flow area change but is dependent on specific volume changes. The pressure loss is due to acceleration which is always present in compressible fluids. It is generally negligible in incompressible flow without heat transfer because friction heating has little effect on fluid temperature and the accompanying specific volume change.

Equation 22 contains no acceleration term and applies only to friction and local pressure losses. Therefore dQ_F/v in Equation 26 is equivalent to dP of Equation 22, or:

$$\frac{dQ_F}{v} = f\frac{dx}{D}\frac{V^2}{v\,2g_c} \tag{27}$$

Substitution of Equation 27 into Equation 26 yields:

$$dP = -\frac{VdV}{vg_c} - \frac{f}{D}\frac{V^2}{v\,2g_c}dx \tag{28}$$

From Equation 3, the continuity equation permits definition of the mass flux, G, or mass flow rate per unit area [lb/h ft^2 (kg/m^2 s)] as:

$$\frac{V}{v} = G = \text{constant} \tag{29}$$

Substituting Equation 29 into Equation 28 for a flow channel of constant area:

$$dP = -2\frac{G^2}{2g_c}dv - f\frac{G^2}{2g_c}\frac{v}{D}dx \tag{30}$$

Integrating Equation 30 between points 1 and 2, located at x = 0 and x = L, respectively:

$$P_1 - P_2 = 2\frac{G^2}{2g_c}(v_2 - v_1) + f\frac{G^2}{2g_c}\frac{1}{D}\int_0^L v\,dx \tag{31}$$

The second term on the right side of Equation 31 may be integrated provided a functional relationship between v and x can be established. For example, where the heat absorption rate over the length of the flow channel is constant, temperature T is approximately linear in x, or:

$$dx = \frac{L}{T_2 - T_1}dT \tag{32}$$

and

$$\int_0^L v\,dx = \frac{L}{T_2 - T_1}\int_1^2 v\,dT = Lv_{av} \tag{33}$$

The term v_{av} is an average specific volume with respect to temperature, T.

$$v_{av} = \phi\,(v_2 + v_1) = \phi\,v_1(v_R + 1) \tag{34}$$

where

v_R = v_2/v_1
ϕ = averaging factor

In most engineering evaluations, v is almost linear in T and $\phi \approx 1/2$. Combining Equations 31 and 32, and rewriting v_2 - v_1 as v_1 (v_R - 1):

$$P_1 - P_2 = 2\frac{G^2}{2g_c}v_1(v_R - 1) + f\frac{L}{D}\frac{G^2}{2g_c}v_1\phi\,(v_R + 1) \tag{35}$$

Equation 35 is completely general. It is valid for compressible and incompressible flow in pipes of constant cross-section as long as the function $T = F(x)$ can be assigned. The only limitation is that dP/dx is negative at every point along the pipe. Equation 28 can be solved for dP/dx making use of Equation 29 and the fact that P_1v_1 can be considered equal to P_2v_2 for adiabatic flow over a short section of tube length. The result is:

$$\frac{dP}{dx} = \frac{Pf/2D}{1 - \frac{g_c Pv}{V^2}} \tag{36}$$

At any point where $V^2 = g_cPv$, the flow becomes choked because the pressure gradient is positive for velocities greater than $(g_cPv)^{0.5}$. The flow is essentially choked by excessive steam expansion due to the drop in pressure. The minimum downstream pressure that is effective in producing flow in a channel is:

$$P_2 = V^2/v_2g_c = v_2\,G^2/g_c \tag{37}$$

Dividing both sides of Equation 35 by $G^2v_1/2g_c$, the pressure loss is expressed in terms of velocity heads. One velocity head equals:

$$\Delta P\ (\text{one velocity head}) = \frac{V^2}{2g_cCv} = \frac{\rho V^2}{2g_cC} \tag{38}$$

where

ΔP= pressure drop equal to one velocity head, lb/in.^{2} (N/m^2)
V = velocity, ft/s (m/s)
v = specific volume, ft^3/lb (m^3/kg)
g_c = 32.17 lbm ft/lbf s^2 = 1 kg m/N s^2
C = 144 in.^{2}/ft^2 (1 m^2/m^2)
ρ = density, lb/ft^3 (kg/m^3)

In either case, f represents the number of velocity heads (N_{vh}) lost in each diameter length of pipe.

The dimensionless parameter defined by the pressure loss divided by twice Equation 38 is referred to as the Euler number:

$$\text{Eu} = \Delta P\,/\,(\rho V^2/g_c) \tag{39}$$

where ρ is the density, or $1/v$.

Two other examples of integrating Equation 30 have wide applications in fluid flow. First, adiabatic flow through a pipe is considered. Both H and D are constant and $P_1v_1^m$ = $P_2v_2^m$ where m is the exponent for constant enthalpy. Values of m for steam range from 0.98 to 1.0. Therefore, the assumption Pv = constant = P_1v_1 is sufficiently accurate for pressure drop calculations. This process is sometimes called isothermal pressure drop because a constant temperature ideal gas expansion also requires a constant enthalpy. For $Pv = P_1v_1$, the integration of Equation 30 reduces to:

$$P_1 - P_2 = 2\frac{G^2}{2g_c}\frac{2v_1v_2}{v_1 + v_2}\ell n\left(\frac{v_2}{v_1}\right) + f\frac{L}{D}\frac{G^2}{2g_c}\frac{2v_1v_2}{v_2 + v_1} \tag{40}$$

Neither P_2 nor v_2 are known in most cases, therefore Equation 40 is solved by iteration. Also, the term $2v_1v_2/(v_2 + v_1)$ can usually be replaced by the numerical average of the specific volumes — $v_{av} = {}^1\!/_2 v_1(P_R + 1)$ where $P_R = P_1/P_2 = v_2/v_1$. The maximum high side error at P_R = 1.10 is 0.22% and this increases to 1.3% at P_R = 1.25. It is common practice to use a numerical average for the specific volume in most fluid friction pressure drop calculations. However, where the lines are long, P_2 should be checked by Equation 37. Also, where heat transfer is taking place, P_2 is seldom constant along the flow channel and appropriate averaging factors should be used.

The second important example considering flow under adiabatic conditions assumes an almost incompressible fluid, i.e., v_1 is approximately equal to v_2. (See Equation 16.) Substituting v for v_1 and v_2 in Equation 37, the result is:

$$P_1 - P_2 = f\frac{L}{D}\frac{G^2}{2g_c}v \tag{41}$$

All terms in Equations 40 and 41 are expressed in consistent units. However, it is general practice and often more convenient to use mixed units. For example, a useful form of Equation 41 in English units is:

$$\Delta P = f\frac{L}{D_e}v\left(\frac{G}{10^5}\right)^2 \tag{42}$$

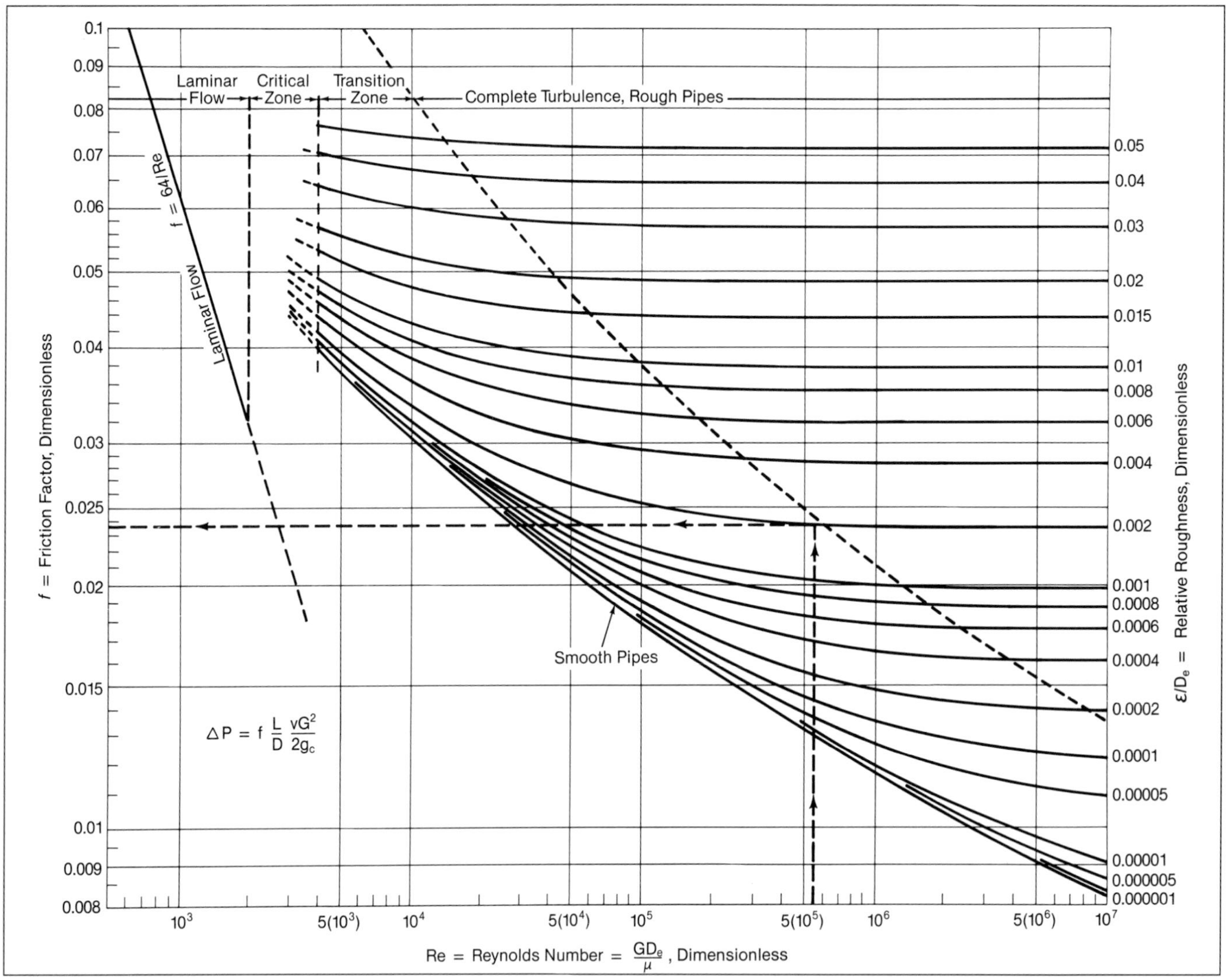

Fig. 1 Friction factor/Reynolds number relationship for determining pressure drop of fluids flowing through closed circuits (pipes and ducts).

where

ΔP = fluid pressure drop, psi
f = friction factor from Fig. 1, dimensionless
L = length, ft
D_e = equivalent diameter of flow channel, in. (note units)
v = specific volume of fluid, ft³/lb
G = mass velocity of fluid, lb/h ft² (note change from units of Equation 29)

Friction factor

The friction factor f, introduced in Equation 21, is defined as the dimensionless fluid friction loss in velocity heads per diameter length of pipe or equivalent diameter length of flow channel. Earlier correlators in this field, including Fanning, used a friction factor one fourth the magnitude indicated by Equation 21. This is because the shear stress at the wall is proportional to one fourth the velocity head. All references to f in this book combine the factor 4 in Equation 20 with f as has been done by Darcy, Blasius, Moody and others.

The friction factor is plotted in Fig. 1 as a function of the Reynolds number, a dimensionless group of variables defined as the ratio of inertial forces to viscous forces. The Reynolds number (Re) can be written:

$$\text{Re} = \frac{\rho V D_e}{\mu} \text{ or } \frac{V D_e}{\upsilon} \text{ or } \frac{G D_e}{\mu} \qquad \textbf{(43)}$$

where

ρ = density of fluid, lbm/ft³ (kg/m³)
υ = kinematic viscosity = μ / ρ, ft²/h (m²/s)
μ = absolute viscosity of fluid, lbm/ft h (kg/m s)
V = velocity of fluid, ft/h (m/s)
G = mass velocity of fluid, lb/h ft² (kg/m² s)
D_e = equivalent diameter of flow channel, ft (m)

Fluid flow inside a closed channel occurs in a viscous or laminar manner at low velocity and in a turbulent manner at high velocities. Many experiments on fluid friction pressure drop, examined by dimensional analysis and the laws of similarity, have shown that the Reynolds number can be used to characterize a flow pattern. Examination of Fig. 1 shows that flow is laminar at Reynolds numbers less than 2000, generally turbulent at values exceeding 4000 and completely turbulent at higher values. Indeterminate conditions exist in the critical zone between Reynolds numbers of 2000 and 4000.

Fluid flow can be described by a system of simultaneous partial differential equations. (See earlier *Fundamental Relationships* section.) However, due to the complexity of these equations, solutions are generally only available for the case of laminar flow, where the only momentum changes are on a molecular basis. For laminar flow, integration of the Navier-Stokes equation with velocity in the length direction only gives the following equation for friction factor:

$$f = 64/\text{Re} \quad \textbf{(44)}$$

The straight line in the laminar flow region of Fig. 1 is a plot of this equation.

It has been experimentally determined that the friction factor is best evaluated by using the Reynolds number to define the flow pattern. A factor ε/D_e is then introduced to define the relative roughness of the channel surface. The coefficient ε expresses the average height of roughness protrusions equivalent to the sand grain roughness established by Nikuradse. The friction factor values in Fig. 1 and the ε/D_e values in Fig. 2 are taken from experimental data as correlated by Moody.[3]

Laminar flow

Laminar flow is characterized by the parallel flowing of individual streams. There is no mixing between the streams except for molecular diffusion from one stream to the other. A small layer of fluid next to the boundary wall has zero velocity as a result of molecular adhesion forces. This establishes a velocity gradient normal to the main body of flow. Because the only interchanges of momentum in laminar flow are between the molecules of the fluid, the condition of the surface has no effect on the velocity gradient and therefore no effect on the friction factor. In commercial equipment, laminar flow is usually encountered only with more viscous liquids such as the heavier oils.

Turbulent flow

When turbulence exists, there are momentum interchanges between masses of fluid. These interchanges are induced through secondary velocities that are not parallel to the axis of flow. In this case, the condition of the boundary surface does have an effect on the velocity gradient near the wall, which in turn affects the friction factor. Heat transfer is substantially greater with turbulent flow (Chapter 4) and, except for viscous liquids, it is common to induce turbulent flow with steam and water without excessive friction loss. Consequently, it is customary to design for Reynolds numbers above 4000 in steam generating units.

Velocity ranges

Table 1 lists the velocity ranges generally encountered in the heat transfer equipment as well as in duct and piping systems of steam generating units. These values, plus the specific volumes from the ASME Steam Tables (see Chapter 2) and the densities listed in Tables 2 and 3 in this chapter, are used to establish mass velocities for calculating Reynolds numbers and fluid friction pressure drops. In addition, values of absolute viscosity, also required in calculating the Reynolds number, are given in Figs. 3, 4 and 5 for selected liquids and gases. Table 4 lists the relationship between various units of viscosity.

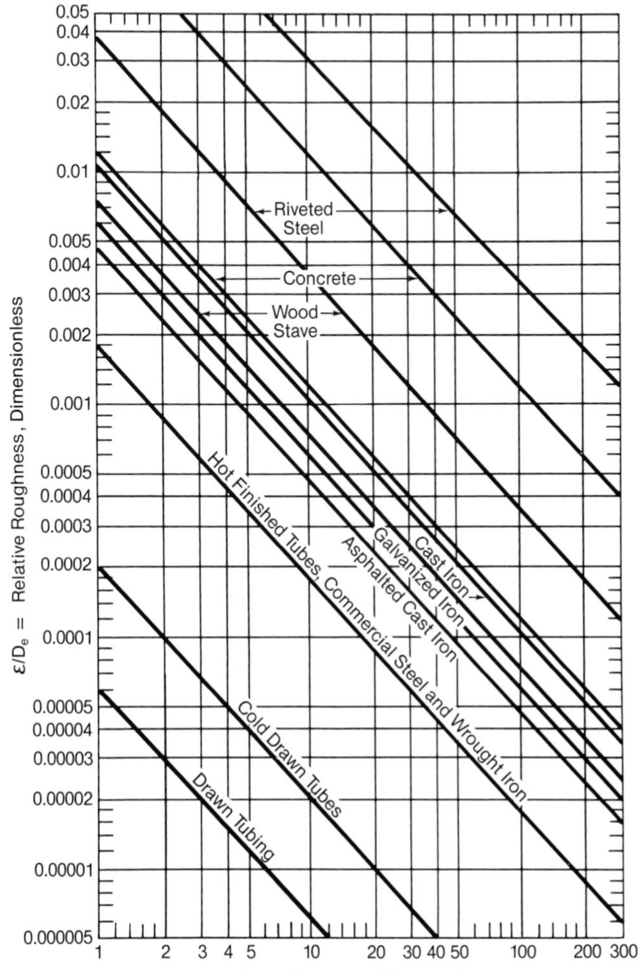

Fig. 2 Relative roughness of various conduit surfaces. (SI conversion: mm = 25.4 × in.)

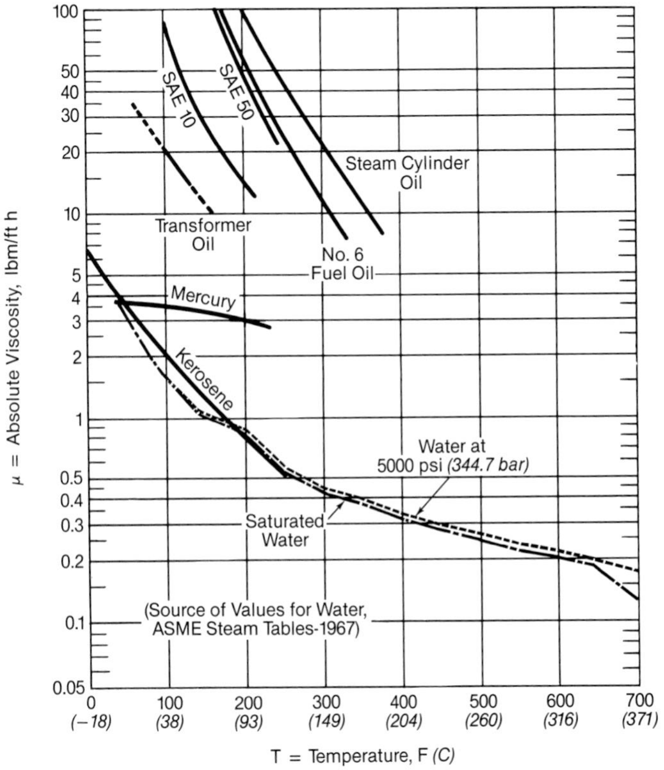

Fig. 3 Absolute viscosities of some common liquids (Pa s = 0.000413 × lbm/ft h).

Table 1
Velocities Common in Steam Generating Systems

Nature of Service	Velocity ft/min	Velocity m/s
Air:		
Air heater	1,000 to 5,000	5.1 to 25.4
Coal and air lines, pulverized coal	3,000 to 4,500	15.2 to 22.9
Compressed air lines	1,500 to 2,000	7.6 to 10.2
Forced draft air ducts	1,500 to 3,600	7.6 to 18.3
Forced draft air ducts, entrance to burners	1,500 to 2,000	7.6 to 10.2
Ventilating ducts	1,000 to 3,000	5.1 to 15.2
Crude oil lines [6 to 30 in. (152 to 762 mm)]	60 to 3,600	0.3 to 1.8
Flue gas		
Air heater	1,000 to 5,000	5.1 to 25.4
Boiler gas passes	3,000 to 6,000	15.2 to 30.5
Induced draft flues and breaching	2,000 to 3,500	10.2 to 17.8
Stacks and chimneys	2,000 to 5,000	10.2 to 25.4
Natural gas lines (large interstate)	1,000 to 1,500	5.1 to 7.6
Steam:		
Steam lines		
High pressure	8,000 to 12,000	40.6 to 61.0
Low pressure	12,000 to 15,000	61.0 to 76.2
Vacuum	20,000 to 40,000	101.6 to 203.2
Superheater tubes	2,000 to 5,000	10.2 to 25.4
Water:		
Boiler circulation	70 to 700	0.4 to 3.6
Economizer tubes	150 to 300	0.8 to 1.5
Pressurized water reactors		
Fuel assembly channels	400 to 1,300	2.0 to 6.6
Reactor coolant piping	2,400 to 3,600	12.2 to 18.3
Water lines, general	500 to 750	2.5 to 3.8

Table 3
Physical Properties of Gases at 14.7 psi (1.01 bar)**

Gas	Temperature F	Density, lb/ft^3	Instantaneous Specific Heat c_p Btu/lb F	Instantaneous Specific Heat c_v Btu/lb F	k, c_p/c_v
Air	70	0.0749	0.241	0.172	1.40
	200	0.0601	0.242	0.173	1.40
	500	0.0413	0.248	0.180	1.38
	1000	0.0272	0.265	0.197	1.34
CO_2	70	0.1148	0.202	0.155	1.30
	200	0.0922	0.216	0.170	1.27
	500	0.0634	0.247	0.202	1.22
	1000	0.0417	0.280	0.235	1.19
H_2	70	0.0052	3.440	2.440	1.41
	200	0.0042	3.480	2.490	1.40
	500	0.0029	3.500	2.515	1.39
	1000	0.0019	3.540	2.560	1.38
Flue gas*	70	0.0776	0.253	0.187	1.35
	200	0.0623	0.255	0.189	1.35
	500	0.0429	0.265	0.199	1.33
	1000	0.0282	0.283	0.217	1.30
CH_4	70	0.0416	0.530	0.406	1.30
	200	0.0334	0.575	0.451	1.27
	500	0.0230	0.720	0.596	1.21
	1000	0.0151	0.960	0.836	1.15

* From coal; 120% total air; flue gas molecular weight 30.
** SI conversions: T, C = 5/9 (F-32); ρ, kg/m^3 = 16.02 × lbm/ft^3; c_p, kJ/kg K = 4.187 × Btu/lbm F.

Resistance to flow in valves and fittings

Pipelines and duct systems contain many valves and fittings. Unless the lines are used to transport fluids over long distances, as in the distribution of process steam at a factory or the cross country transmission of oil or gas, the straight runs of pipe or duct are relatively short. Water, steam, air and gas lines in a power plant have relatively short runs of straight pipe and many valves and fittings. Consequently, the flow resistance due to valves and fittings is a substantial part of the total resistance.

Methods for estimating the flow resistance in valves and fittings are less exact than those used in establishing the friction factor for straight pipes and ducts. In the latter, pressure drop is considered to be the result of the fluid shear stress at the boundary walls of the flow channel; this leads to relatively simple boundary value evaluations. On the other hand, pressure loss associated with valves, fittings and bends are mainly the result of impacts and inelastic exchanges of momentum. These losses are frequently referred to as local losses or local nonrecoverable pressure losses. Even though momentum is conserved, kinetic energies are dissipated as heat. This means that pressure losses are influenced mainly by the

Table 2
Physical Properties of Liquids at 14.7 psi (1.01 bar)

Liquid	Temperature F (C)	Density, lb/ft^3 (Mg/m^3)	Specific Heat Btu/lb F (kJ/kg C)
Water	70 (21)	62.4 (1.000)	1.000 (4.19)
	212 (100)	59.9 (0.959)	1.000 (4.19)
Automotive oil	70 (21)		
SAE 10		55 to 57 (0.88 to 0.91)	0.435 (1.82)
SAE 50		57 to 59 (0.91 to 0.95)	0.425 (1.78)
Mercury	70 (21)	846 (13.6)	0.033 (0.138)
Fuel oil, #6	70 (21)	60 to 65 (0.96 to 1.04)	0.40 (1.67)
	180 (82)	60 to 65 (0.96 to 1.04)	0.46 (1.93)
Kerosene	70 (21)	50 to 51 (0.80 to 0.82)	0.47 (1.97)

Table 4
Relationship Between Various Units of Viscosity

Part A: Absolute (or Dynamic) Viscosity, μ

Pa s $\frac{N\,s}{m^2} = \frac{kg}{m\,s}$	Centipoise $\frac{0.01\,g}{cm\,s}$	$\frac{lbm}{ft\,s}$	$\frac{lbm}{ft\,h}$	$\frac{lbf\,s}{ft^2}$
1.0	1,000	672×10^{-3}	2420	20.9×10^{-3}
0.001	1.0	672×10^{-6}	2.42	20.9×10^{-6}
1.49	1488	1.0	3600	0.0311
413×10^{-6}	0.413	278×10^{-6}	1.0	8.6×10^{-6}
47.90	47,900	32.2	115,900	1.0

Part B: Kinematic Viscosity, υ = μ/ρ

$\frac{m^2}{s}$	Centistoke $\frac{0.01\,cm^2}{s}$	$\frac{ft^2}{s}$	$\frac{ft^2}{h}$
1.0	10^6	10.8	38,800
10^{-6}	1.0	10.8×10^{-6}	0.0389
92.9×10^{-3}	92,900	1.0	3,600
25.8×10^{-6}	25.8	278×10^{-6}	1.0

geometries of valves, fittings and bends. As with turbulent friction factors, pressure losses are determined from empirical correlations of test data. These correlations may be based on equivalent pipe lengths but are preferably defined by a multiple of velocity heads based on the connecting pipe or tube sizes. Equivalent pipe length calculations have the disadvantage of being dependent on the relative roughness (ε/D) used in the correlation. Because there are many geometries of valves and fittings,

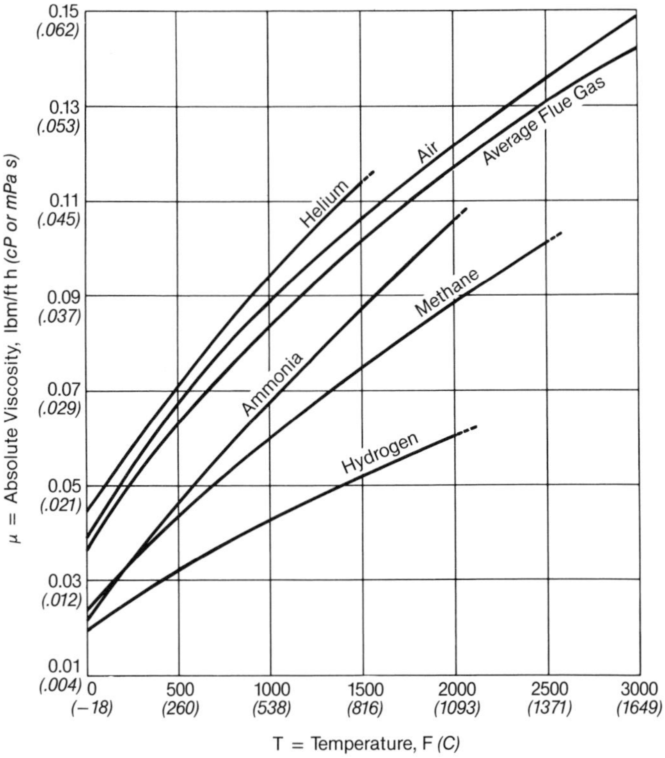

Fig. 4 Absolute viscosities of some common gases at atmospheric pressure.

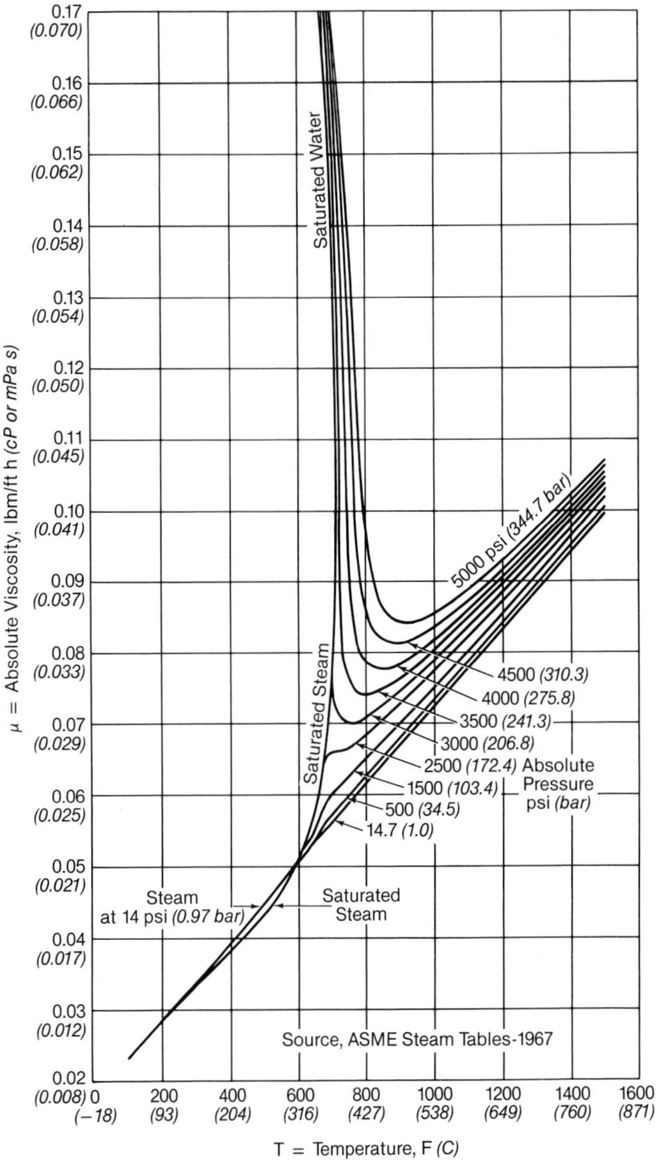

Fig. 5 Absolute viscosity of saturated and superheated steam.

it is customary to rely on manufacturers for pressure drop coefficients.

It is also customary for manufacturers to supply valve flow coefficients (C_V) for 60F (16C) water. These are expressed as ratios of weight or volume flow in the fully open position to the square root of the pressure drop. These coefficients can be used to relate velocity head losses to a connecting pipe size by the following expression:

$$N_\upsilon = k\,D^4 / C_V^{\,2} \qquad \textbf{(45)}$$

where

N_υ = number of velocity heads, dimensionless
k = units conversion factor: for C_V based upon gal/min/$(\Delta\rho)^{1/2}$, $k = 891$
D = internal diameter of connecting pipe, in. (mm)
C_V = flow coefficient in units compatible with k and D: for $k = 891$, C_V = gal/min/$(\Delta\rho)^{1/2}$

C_V and corresponding values of N_υ for valves apply only to incompressible flow. However, they may be extrapolated for compressible conditions using an average specific volume between P_1 and P_2 for ΔP values as high as 20% of P_1. This corresponds to a maximum pressure ratio of 1.25. The ΔP process for valves, bends and fittings is approximately isothermal and does not require the most stringent limits set by Equation 16.

When pressure drop can be expressed as an equivalent number of velocity heads, it can be calculated by the following formula in English units:

$$\Delta P = N_v \frac{v}{12}\left(\frac{G}{10^5}\right)^2 \tag{46}$$

where

ΔP = pressure drop, lb/in.^{2}
N_v = number of equivalent velocity heads, dimensionless
v = specific volume, ft^3/lb
G = mass velocity, lb/ft^2 h

Another convenient expression, in English units only, for pressure drop in air (or gas) flow evaluations is:

$$\Delta P = N_v \frac{30}{B} \frac{T+460}{1.73\times 10^5}\left(\frac{G}{10^3}\right)^2 \tag{47}$$

where

ΔP = pressure drop, in. wg
B = barometric pressure, in. Hg
T = air (or gas) temperature, F

Equation 47 is based on air, which has a specific volume of 25.2 ft^3/lb at 1000R and a pressure equivalent to 30 in. Hg. This equation can be used for other gases by correcting for specific volume.

The range in pressure drop through an assortment of commercial fittings is given in Table 5. This resistance to flow is presented in equivalent velocity heads based on the internal diameter of the connecting pipe. As noted, pressure drop through fittings may also be expressed as the loss in equivalent lengths of straight pipe.

Contraction and enlargement irreversible pressure loss

The simplest sectional changes in a conduit are converging or diverging boundaries. Converging boundaries can stabilize flow during the change from pressure energy to kinetic energy, and local irrecoverable flow losses (inelastic momentum exchanges) can be practically eliminated with proper design. If the included angle of the converging boundaries is 30 deg (0.52 rad) or less and the terminal junctions are smooth and tangent, any losses in mechanical energy are largely due to fluid friction. It is necessary to consider this loss as 0.05 times the velocity head, based on the smaller downstream flow area.

When the elevation change ($Z_2 - Z_1$) is zero, the mechanical energy balance for converging boundaries becomes:

$$P_1 v + \frac{V_1^2}{2g_c} = P_2 v + \frac{V_2^2}{2g_c} + N_c \frac{V_2^2}{2g_c} \tag{48}$$

Subscripts 1 and 2 identify the upstream and downstream sections. N_c, the contraction loss factor, is the number of velocity heads lost by friction and local nonrecoverable pressure loss in contraction. Fig. 6 shows values of this factor.

Table 5
Resistance to Flow of Fluids Through Commercial Fittings*

Fitting	Loss in Velocity Heads
L-shaped, 90 deg (1.57 rad) standard sweep elbow	0.3 to 0.7
L-shaped, 90 deg (1.57 rad) long sweep elbow	0.2 to 0.5
T-shaped, flow through run	0.15 to 0.5
T-shaped, flow through 90 deg (1.57 rad) branch	0.6 to 1.6
Return bend, close	0.6 to 1.7
Gate valve, open	0.1 to 0.2
Check valve, open	2.0 to 10.0
Globe valve, open	5.0 to 16.0
Angle valve, 90 deg (1.57 rad) open	3.0 to 7.0
Boiler nonreturn valve, open	1.0 to 3.0

* See Fig. 9 for loss in velocity heads for flow of fluids through pipe bends.

When there is an enlargement of the conduit section in the direction of flow, the expansion of the flow stream is proportional to the kinetic energy of the flowing fluid and is subject to a pressure loss depending on the geometry. Just as in the case of the contraction loss, this is an irreversible energy conversion to heat resulting from inelastic momentum exchanges. Because it is customary to show these losses as coefficients of the higher kinetic energy term, the mechanical energy balance for enlargement loss is:

$$P_1 v + \frac{V_1^2}{2g_c} = P_2 v + \frac{V_2^2}{2g_c} + N_e \frac{V_1^2}{2g_c} \tag{49}$$

The case of sudden enlargement [angle of divergence β = 180 deg (π rad)] yields an energy loss of $(V_1 - V_2)^2/2g_c$. This can also be expressed as:

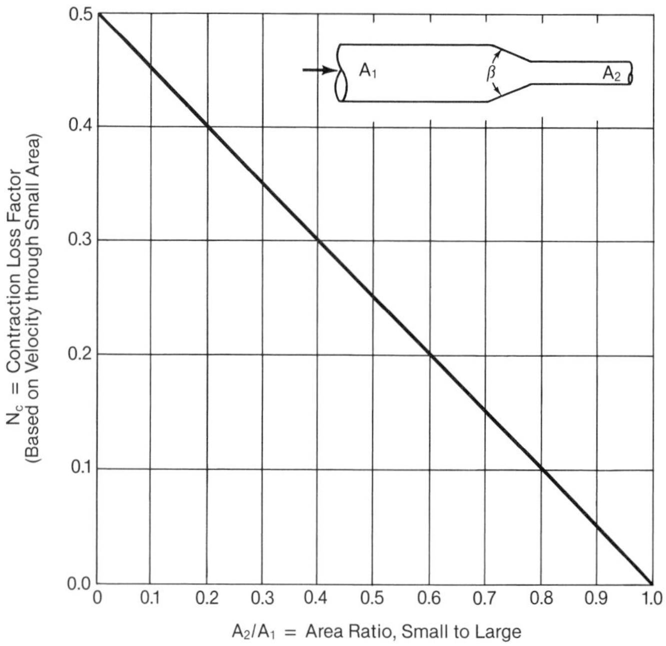

Fig. 6 Contraction loss factor for β >30 deg (N_c = 0.05 for $\beta \leq$ 30 deg).

$$N_e = \left(1 - \frac{A_1}{A_2}\right)^2 \quad \textbf{(50)}$$

where A_1 and A_2 are the upstream and downstream cross-sectional flow areas, respectively and ($A_1 < A_2$). Even this solution, based on the conservation laws, depends on qualifying assumptions regarding static pressures at the upstream and downstream faces of the enlargement.

Experimental values of the enlargement loss factor, based on different area ratios and angles of divergence, are given in Fig. 7. The differences in static pressures caused by sudden and gradual changes in section are shown graphically in Fig. 8. The pressure differences are shown in terms of the velocity head at the smaller area plotted against section area ratios.

Flow through bends

Bends in a pipeline or duct system produce pressure losses caused by both fluid friction and momentum exchanges which result from a change in flow direction. Because the axial length of the bend is normally included in the straight length friction loss of the pipeline or duct system, it is convenient to subtract a calculated equivalent straight length friction loss from experimentally determined bend pressure loss factors. These corrected data form the basis of the empirical bend loss factor, N_b.

The pressure losses for bends in round pipe in excess of straight pipe friction vary slightly with Reynolds numbers below 150,000. For Reynolds numbers above this value, they are reasonably constant and depend solely on the dimensionless ratio r/D, the ratio of the centerline radius of the bend to the internal diameter of the pipe. For commercial pipe, the effect of Reynolds number is negligible. The combined effect of radius ratio and bend angle, in terms of velocity heads, is shown in Fig. 9.

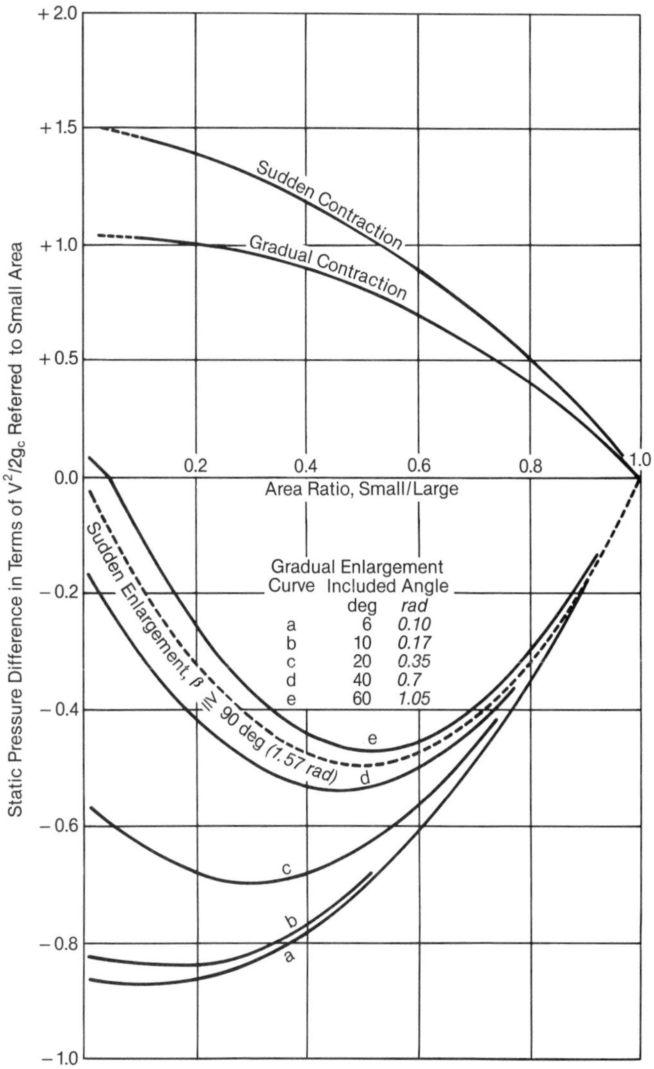

Fig. 8 Static pressure difference resulting from sudden and gradual changes in section.

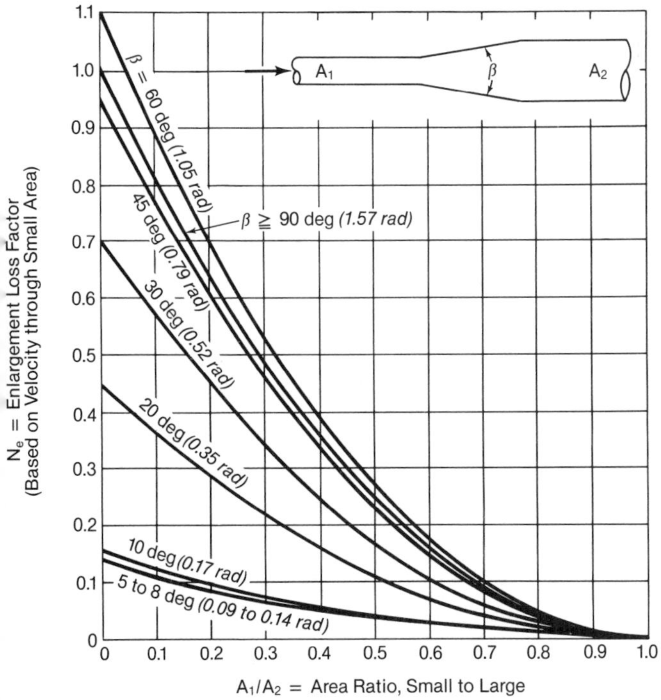

Fig. 7 Enlargement loss factor for various included angles.

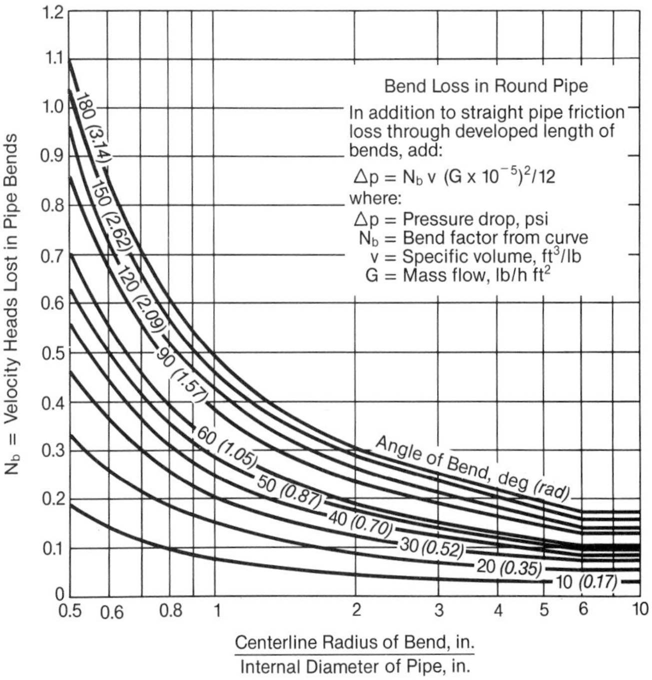

Fig. 9 Bend loss for round pipe, in terms of velocity heads.

Flow through coils

A convenient method for calculating the pressure drop in flow through coils is to apply a factor to the drop through an equivalent length of straight pipe. This factor depends upon the type of flow, laminar or turbulent, and the radius of the coil. The type of flow and the factors for laminar and turbulent flow can be determined from the curves and formulas of Fig. 10.[4]

Flow through rectangular ducts

The loss of pressure caused by a direction change in a rectangular duct system is similar to that for cylindrical pipe. However, an additional factor, the shape of the duct in relation to the direction of bend, must be taken into account. This is called the aspect ratio, which is defined as the ratio of the width to the depth of the duct, i.e., the ratio b/d in Fig. 11. The bend loss for the same radius ratio decreases as the aspect ratio increases, because of the smaller proportionate influence of secondary flows on the stream. The combined effect of radius and aspect ratios on 90 deg (1.57 rad) duct bends is given in terms of velocity heads in Fig. 11.

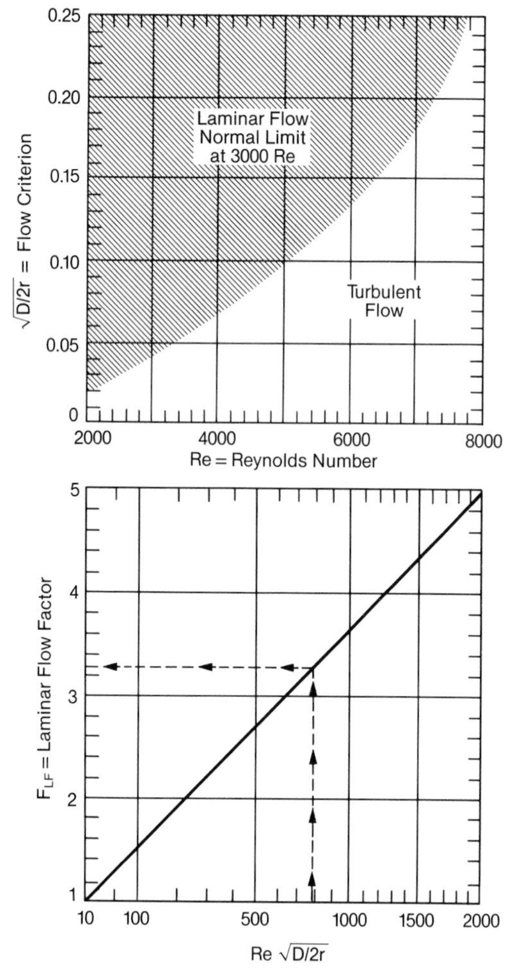

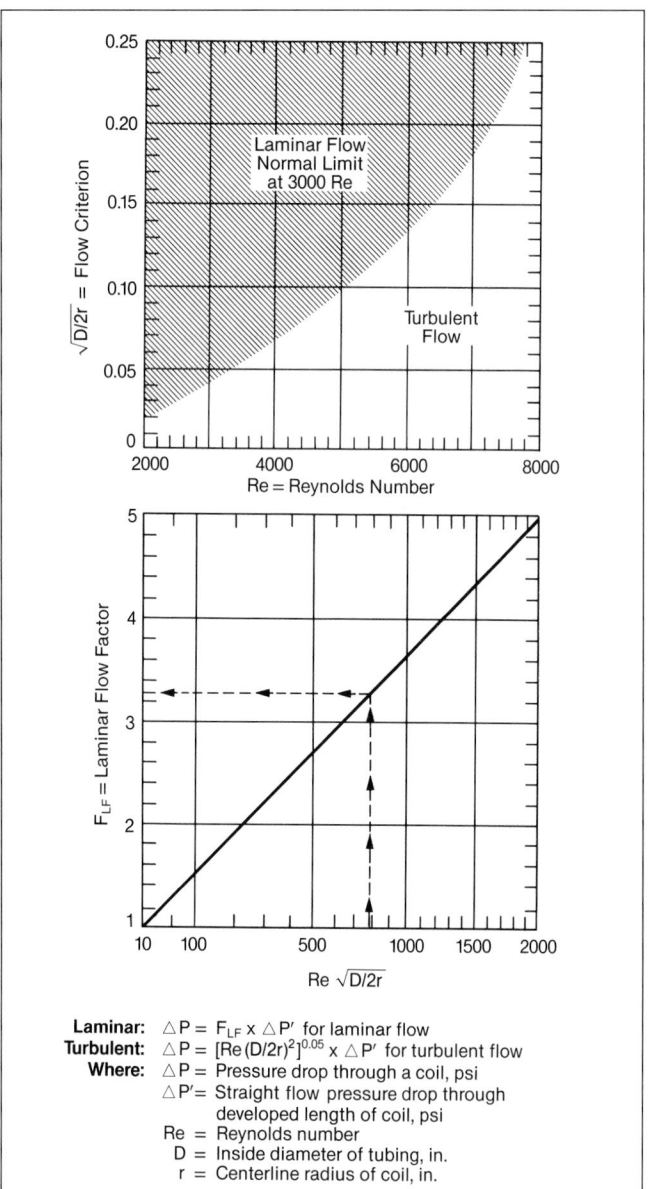

Laminar: $\Delta P = F_{LF} \times \Delta P'$ for laminar flow
Turbulent: $\Delta P = [Re(D/2r)^2]^{0.05} \times \Delta P'$ for turbulent flow
Where: ΔP = Pressure drop through a coil, psi
$\Delta P'$ = Straight flow pressure drop through developed length of coil, psi
Re = Reynolds number
D = Inside diameter of tubing, in.
r = Centerline radius of coil, in.

Fig. 10 Pressure drop for laminar flow and for turbulent flow through coils (adapted from Reference 4).

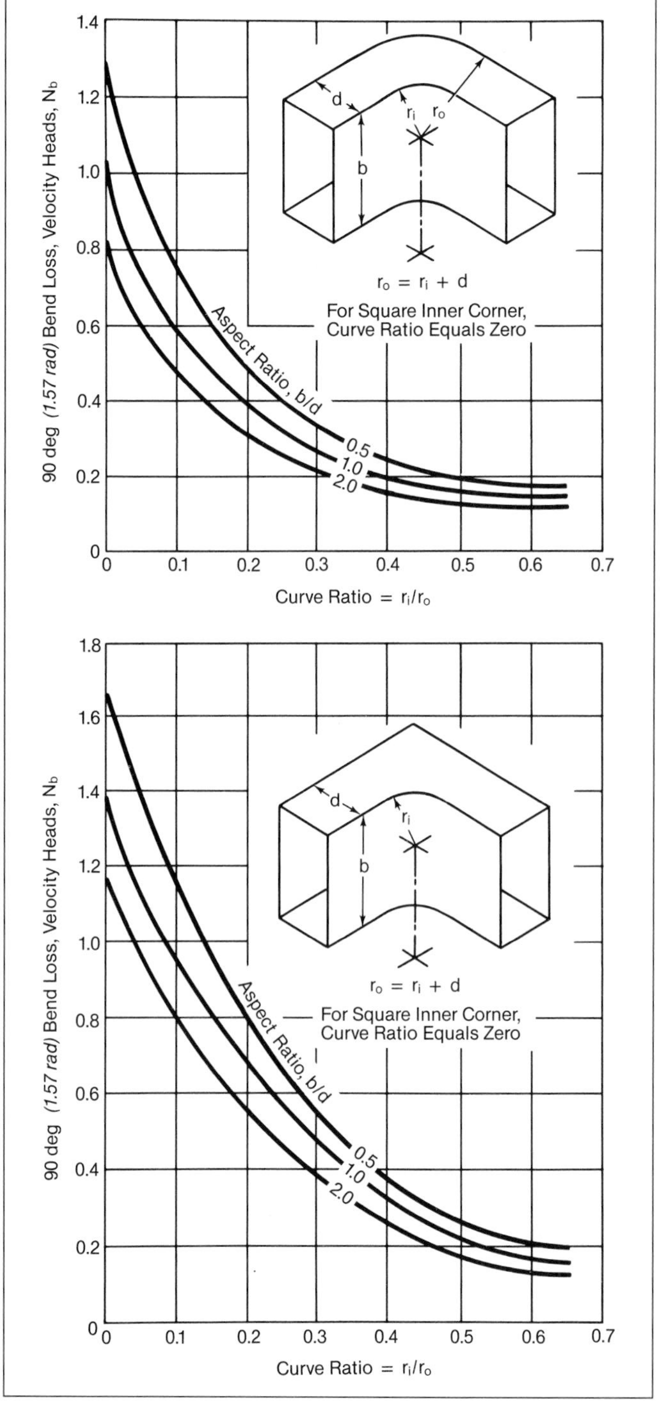

Fig. 11 Loss for 90 deg (1.57 rad) bends in rectangular ducts.

The loss factors shown in Fig. 11 are average values of test results on ducts. For the given range of aspect ratios, the losses are relatively independent of the Reynolds number. Outside this range, the variation with Reynolds number is erratic. It is therefore recommended that N_b values for b/d = 0.5 be used for all aspect ratios less than b/d = 0.5, and values for b/d = 2.0 be used for ratios greater than b/d = 2.0. Losses for bends other than 90 deg (1.57 rad) are customarily considered to be proportional to the bend angle.

Turning vanes

The losses in a rectangular elbow duct can be reduced by rounding or beveling its corners and by installing turning vanes. With rounding or beveling, the overall size of the duct can become large; however, with turning vanes, the compact form of the duct is preserved.

A number of turning vane shapes can be used in a duct. Fig. 12 shows four different arrangements. Segmented shaped vanes are shown in Fig. 12a, simple curved thin vanes are shown in Fig. 12b and concentric splitter vanes are shown in Fig. 12c. In Fig. 12c, the vanes are concentric with the radius of the duct. Fig. 12d illustrates simple vanes used to minimize flow separation from a square edged duct.

The turning vanes of identical shape and dimension, Fig. 12b, are usually mounted within the bend of an elbow. They are generally installed along a line or section of the duct and are placed from the inner corner to the outside corner of the bend. Concentric turning vanes, Fig. 12c, typically installed within the bend of the turn, are located from one end of the turn to the other end.

The purpose of the turning vanes in an elbow or turn is to deflect the flow around the bend to the inner wall of the duct. When the turning vanes are appropriately designed, the flow distribution prevents jet separation from the walls and prevents the formation of eddy zones in the downstream section of the bend. The velocity distribution over the downstream cross-section of the turn is improved (see Fig. 13), and the pressure loss of the turn or elbow is decreased.

The main factor in decreasing the pressure losses and obtaining equalization of the velocity field is the elimination of an eddy zone at the inner wall of the turn. For a uniform incoming flow field, the largest effect of decreasing the pressure losses and establishing a uniform outlet flow field for a turn or elbow is achieved by locating the turning vanes closer to the inner curvature of the bend. (See Figs. 12d and 13c.) For applications requiring a uniform velocity distribution directly after the turn, a full complement or normal arrangement of turning vanes (see Fig. 13b) is required. However, for many applications, it is sufficient to use a reduced number of vanes, as shown in Fig. 13c.

For nonuniform flow fields, the arrangement of turning vanes is more difficult to determine. Many times, numerical modeling and flow testing of the duct system must be done to determine the proper vane locations.

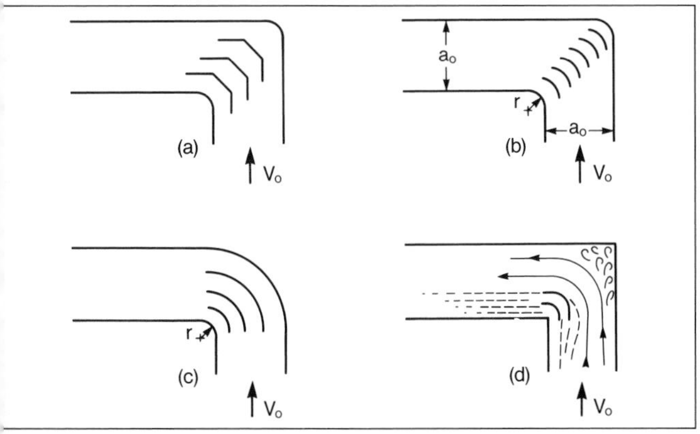

Fig. 12 Turning vanes in elbows and turns: a) segmented, b) thin concentric, c) concentric splitters, and d) slotted (adapted from Idelchik, Reference 8).

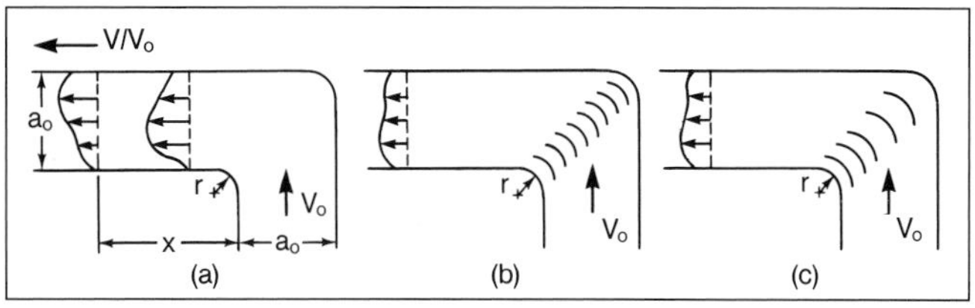

Fig. 13 Velocity profiles downstream of an elbow: a) without vanes, b) with typical vanes, and c) with optimum vanes (adapted from Idelchik, Reference 8).

Pressure loss

A convenient chart for calculating the pressure loss resulting from impact losses in duct systems conveying air (or flue gas) is shown in Fig. 14. When mass velocity and temperature are known, a base velocity head in inches of water at sea level can be obtained.

Flow over tube banks

Bare tube The transverse flow of gases across tube banks is an example of flow over repeated major cross-sectional changes. When the tubes are staggered, sectional and directional changes affect the resistance. Experimental results and the analytical conclusions of extensive research by Babcock & Wilcox (B&W) indicate that three principal variables other than mass velocity affect this resistance. The primary variable is the number of major restrictions, i.e., the number of tube rows crossed, N. The second variable is the friction factor f which is related to the Reynolds number (based on tube diameter), the tube spacing diameter ratios, and the arrangement pattern (in-line or staggered). The third variable is the depth factor, F_d (Fig. 15), which is applicable

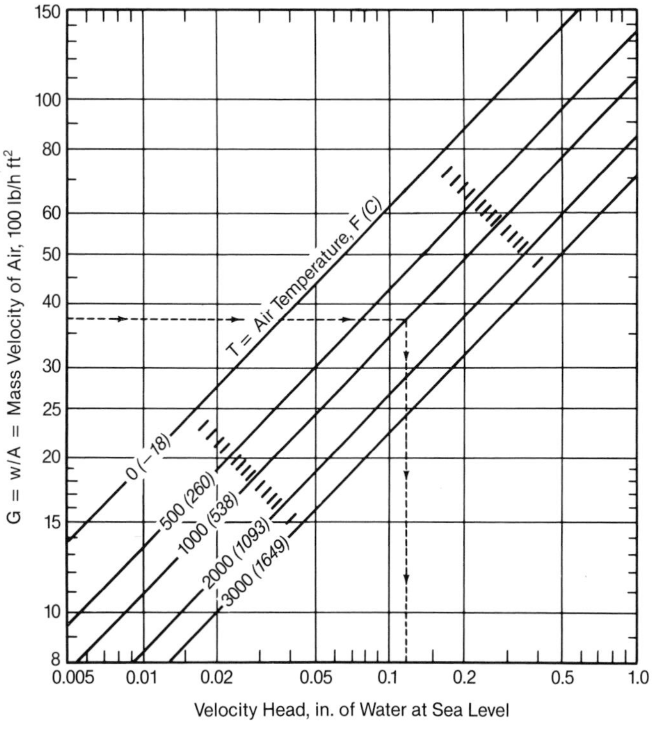

Velocity head values are for air having a specific volume of 25.2 ft^3/lb at 1000F *(538C)* absolute and 30 in. Hg. For flue gas,

$$\frac{V^2}{2g_c} = \left(\frac{V^2}{2g_c}\text{ for Air}\right)\left(\frac{\text{Flue Gas Sp Vol}}{\text{Air Sp Vol}}\right)$$

Fig. 14 Mass velocity/velocity head relationship for air.

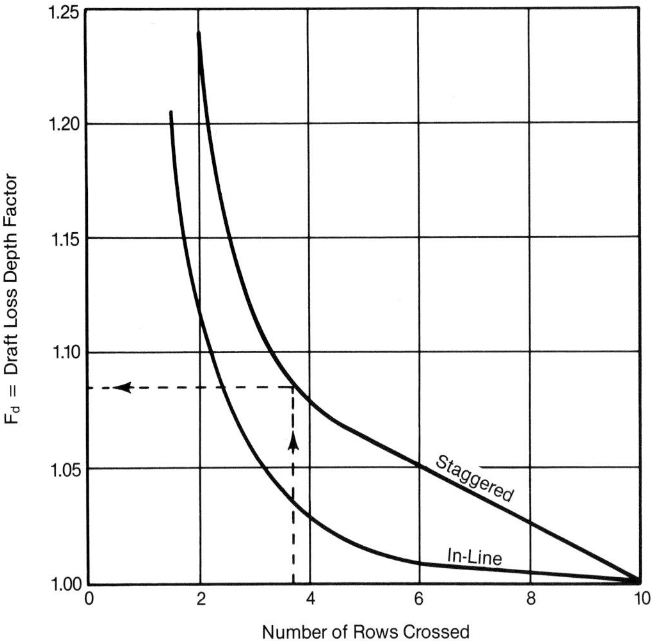

Fig. 15 Draft loss depth factor for number of tube rows crossed in convection banks.

to banks less than ten rows deep. The friction factors f for various in-line tube patterns are given in Fig. 16.

The product of the friction factor, the number of major restrictions (tube rows) and the depth factor is, in effect, the summation of velocity head losses through the tube bank.

$$N_v = f\, N\, F_d \tag{51}$$

The N_v value established by Equation 51 may be used in Equations 46 or 47 to find the tube bank pressure loss. Some test correlations indicate f values higher than the isothermal case for cooling gas and lower for heating gas.

Finned tube In some convective boiler design applications, extended surface tube banks are used. Many types of extended surface exist, i.e., solid helical fin, serrated helical fin, longitudinal fin, square fin and different types of pin studs. For furnace applications, the cleanliness of the gas or heat transfer medium dictates whether an extended surface tube bank can be used and also defines the type of extended surface.

Several different tube bank calculation methods exist for extended surface, and many are directly related to the type of extended surface that is used. Various correlations for extended surface pressure loss can be found in References 5 through 11. In all cases, a larger pressure loss per row of bank exists with an extended surface tube compared to a bare tube. For in-line tube bundles, the finned tube resistance per row of tubes is approximately 1.5 times that of the bare tube row. However, due to the increased heat transfer absorption of the extended surface, a smaller number of tube rows is required. This results in an overall bank pressure loss that can be equivalent to a larger but equally absorptive bare tube bank.

Flow through stacks or chimneys

The flow of gases through stacks or chimneys is established by the natural draft effect of the stack and/or the mechanical draft produced by a fan. The resistance to this flow, or the loss in mechanical energy between the bottom and the top of the stack, is a result of the friction and stack exit losses. Application examples of these losses are given in Chapter 23.

Pressure loss in two-phase flow

Evaluation of two-phase steam-water flows is much more complex. As with single-phase flow, pressure loss occurs from wall friction, acceleration and change in elevation. However, the relationships are more complicated. The evaluation of friction requires the assessment of the interaction of the steam and water phases. Acceleration is much more important because of the large changes in specific volume of the mixture as water is converted to steam. Finally, large changes in average mixture density at different locations significantly impact the static head. These factors are presented in detail in Chapter 5.

Entrainment by fluid flow

Collecting or transporting solid particles or a second fluid by the flow of a primary fluid at high velocity is known as entrainment. This is usually accomplished with jets using a small quantity of high pressure fluid to carry large quantities of another fluid or solid particles. The pressure energy of the high pressure fluid is converted into kinetic energy by nozzles, with a consequent reduction of pressure. The material to be transported is drawn in at the low pressure zone, where it meets and mixes with the high velocity jet. The jet is usually followed by a parallel throat section to equalize the velocity profile. The mixture then enters a diverging section where kinetic energy is partially reconverted into pressure energy. In this case, major fluid flow mechanical energy losses are an example of inelastic momentum exchanges occurring within the fluid streams.

The *injector* is a jet pump that uses condensing steam as the driving fluid to entrain low pressure water for delivery against a back pressure higher than the pressure of the steam supplied. The *ejector*, similar to the injector, is designed to entrain gases, liquids, or mixtures of solids and liquids for delivery against a pressure less than that of the primary fluid. In a water-jet *aspirator*, water is used to entrain air to obtain a partial vacuum. In the Bunsen type burner, a jet of gas entrains air for combustion. Steam jet blowers are sometimes used in the bases of the stack for small natural draft boiler plants to increase the draft for short peak loads.

In several instances, entrainment may be detrimental to the operation of steam boilers. Particles of ash entrained by the products of combustion, when deposited on heating surfaces, reduce thermal conductance, erode fan blades, and add to pollution when discharged into the atmosphere. Moisture carrying solids, either in suspension or in solution, are entrained in the stream. The solids may be carried through to the turbine and deposited on the blades, decreasing turbine capacity and efficiency. In downcomers or supply tubes, steam bubbles are entrained in the water when the drag on the bubbles is greater than the buoyant force. This reduces the density in the pumping column of natural circulation boilers.

Boiler circulation

An adequate flow of water and steam-water mixture is necessary for steam generation and control of tube metal temperatures in all circuits of a steam generating unit. At supercritical pressures this flow is produced mechanically by pumps. At subcritical pressures, circulation is produced by the force of gravity or pumps, or a combination of the two. The elements of single-phase flow discussed in this chapter, two-phase flow discussed in Chapter 5, heat input rates, and selected limiting design criteria are combined to evaluate the circulation in fossil- and nuclear-fired steam generators. The evaluation procedures and key criteria are presented in Chapter 5.

Numerical modeling

In *numerical modeling* of fluid dynamics and heat transfer, computer based or computational methods are used to predict a variety of local parameters including, but not limited to, fluid velocities, temperatures and pressures. Numerical methods permit analyses of fluid flow situations which are not practical to investigate experimentally or analytically.

Many applications for computational fluid dynamic analyses exist within the design and evaluation of boilers. Numerical models of the flue gas and steam-water flows are used to predict boiler behavior, evaluate design modifications, or investigate localized phenomena. Examples of flue gas applications include predicting temperature distributions within a furnace, evaluating fluid mixing due to

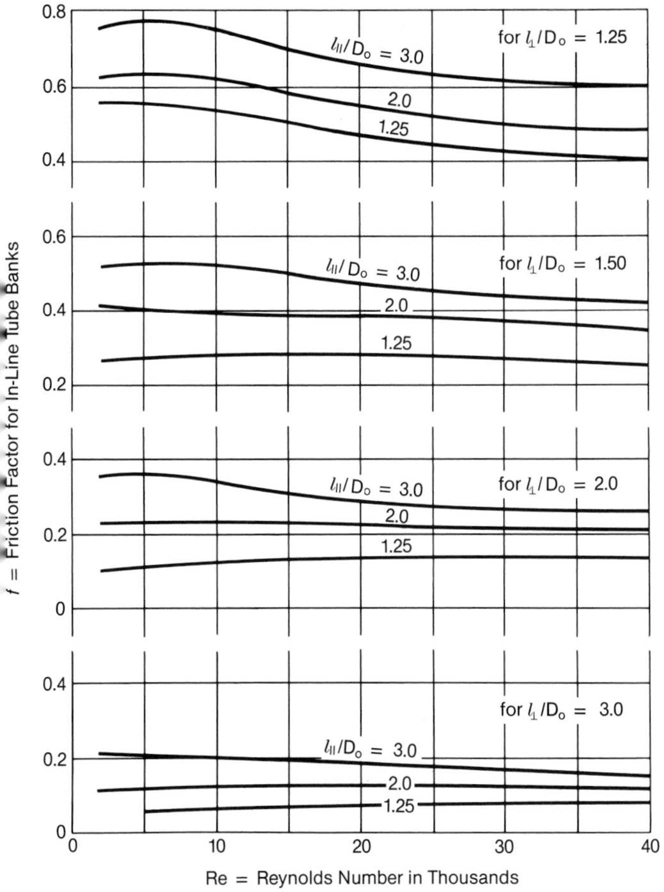

Fig. 16 Friction factor, *f*, as affected by Reynolds number for various in-line tube patterns; crossflow gas or air.

the retrofit of systems to control the emission of nitrogen oxides (NO_x), and improving air heater flow distributions to increase heat absorption. Water-side applications include determining flow rates for natural circulation systems and evaluating system stability, among others.

Two general types of numerical fluid flow models are used to aid in the analysis of boilers, *network* and *general purpose* computer programs or codes. Network programs are used to determine flow rate and pressure drop through water-side circuits and flue gas passages. They are typically composed of combinations of empirical and fundamental correlations and relationships to provide a one dimensional assessment of overall component performance. They are important where global quantities such as overall flow rate, pressure drop or thermal performance are desired.

General purpose programs are typically based upon the fundamental governing equations for mass, momentum and energy presented at the beginning of this chapter to describe a two or three dimensional flow field. These fundamental relationships are then augmented by empirical or experimental relationships where needed. Chemical reaction kinetics and thermal radiation heat transfer can be added. The result is typically a two or three dimensional matrix of local pressures, temperatures, velocities and possibly chemical species concentrations. This information can then be used to evaluate the local or overall performance parameters.

General purpose codes are commonly based on one of three numerical methods — finite difference, finite element and spectral element. The choice of a numerical method is based on its past success, ease of coding and maintenance, and geometric flexibility. All of these approaches are similar in that the same governing equations are used, and each approach divides the geometry under evaluation into smaller interconnected regions. These regions are referred to as *control volumes*, *nodes* or *elements* and must individually satisfy the governing equations. The equations are then approximated for each region by assuming algebraic forms for the partial derivatives or by assuming profiles for the primitive variables in which the constants are determined by solving the governing equations. Each method changes the governing nonlinear partial differential equations to a system of algebraic expressions which can be solved by a computer. The set of algebraic expressions are then solved simultaneously or iteratively.

If more than one conservation equation requires solving, an iterative procedure is typically used to handle the coupling between them. This is done to achieve reasonable computer run times. For example, the velocity obtained for using the conservation of momentum equations must also satisfy conservation of mass. Once a variable such as pressure has been calculated, the updated value would be used to calculate the other variables. The procedure is repeated until the difference between the updated and previous values falls within a specified tolerance. Further discussion of numerical modeling is provided in Chapters 4 and 18, as well as References 12 and 13.

Modeling process

A basic understanding of the fluid flow situation is required to determine the type of computer model best suited to the application and to determine the appropriate level of analysis detail. A multi-step process is used:

1. obtain a complete situation description including physical geometry, process flow, physical property data and the level of detail for the output,
2. define the modeling assumptions appropriate for the specific flow system and computer model selected while making appropriate tradeoffs — cost and time are balanced against level of detail and information to be obtained,
3. preprocess the input data usually using a computer model to convert the general technical inputs, such as overall geometry and input flow conditions, into the detailed inputs required by the computational code selected — verification of the preprocessing results is important,
4. run the numerical computational model until it converges on an acceptable solution, and
5. postprocess the results to verify the initial model assumptions to check the results against known trends, and to present the results in a readily usable form.

Several examples of results of this modeling process are presented in Chapters 4 and 18.

Example computational model analysis: air heater inlet distribution

To illustrate this process for using numerical computational models, the following material summarizes a relatively simple analysis of an air heater inlet flow distribution.

Step 1. Obtain a complete situation description

The air side inlet flow to a boiler air heater has been determined to be severely maldistributed and needs to be corrected. A review of the as-built design drawings showed a forced draft fan discharging into the air supply duct which goes through a 90 deg (1.57 rad) turn (without turning vanes) just upstream of the air heater. A steam coil, which is used to preheat the air, is located downstream of the turn and just upstream of the air heater. The fan, 90 deg (1.57 rad) turn, steam coil and air heater are very closely coupled.

From past experience, the fan and duct orientation would be expected to cause the flow to hug the outer radius of the turn prior to entering the steam coil and air heater. The steam coil adds resistance to the flow which helps provide a more uniform flow distribution, but apparently is insufficient to overcome the maldistribution leaving the turn in this case. Turning vanes were needed in the existing duct.

The flow distribution had to be quickly corrected during an unscheduled plant outage. A computational fluid dynamic analysis was performed to assist and validate the design of the turning vanes. Experimental modeling required too much time. Design was based on company standards for the turning vanes using a previously measured fan discharge profile, and the numerical analysis was performed to check the new turning vane performance. This allowed construction of simple flat plate vanes within the time frame available and permitted small adjustments after postinstallation testing.

Step 2. Define the modeling assumptions

The necessary modeling assumptions are the inlet velocity profile, the temperature and pressure from the fan discharge, the fluid density and viscosity, and the steam coil resistance. The inlet velocity profile was taken from previous work and adjusted to give the same average flow velocity as the stated mass flow rate. The fluid properties were assumed constant with the density calculated as an ideal gas, given the temperature and pressure. The dynamic viscosity was obtained from air tables at atmospheric pressure and actual temperature. The steam coil pressure drop was known from static pressure taps.

Based upon engineering judgement and past experience, a two dimensional model was selected to validate the turning vane design. Final adjustments could be made relatively easily in the field if needed to fine tune the design. The B&W general purpose multi-dimensional fluid flow code was used.

Step 3. Preprocess input for a given computer program

A preprocessor was used to generate the required input file from the initial geometry and flow conditions provided. Fig. 17 shows the geometry and grid used to simulate the flow field. Upstream and downstream ducting is included to provide the most accurate flow field possible near the duct 90 deg (1.57 rad) turn.

The steam coil pressure drop was specified as a local pressure loss factor:

$$k_{\text{steam coil}} = \Delta P_{\text{measured}} / (^1/_2\ \rho V^2) \tag{52}$$

The velocity used to determine the steam coil local pressure loss factor was calculated using the duct flow area:

$$V = \dot{m} / (\rho A_{\text{duct}}) \tag{53}$$

Visual inspection of the input file was performed before running the problem. The geometry boundaries were verified before running the model.

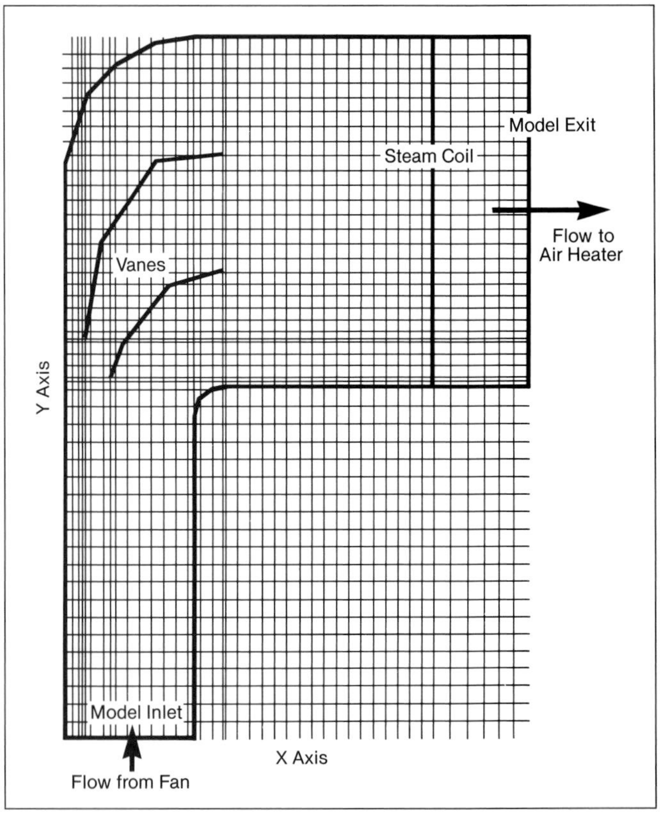

Fig. 17 Air heater fan discharge model, geometry plot.

Step 4. Run the program and obtain a converged solution

The program was run using the computational code for 1000 iterations, and the differences between runs or *residuals* were monitored. The residuals leveled off after 600 iterations to below 10^{-2} for the velocities and 10^{-3} for the pressure, an acceptable condition for this two dimensional problem based on experience. The residuals are low enough not to affect the practical problem precision (~ four significant digits) of the velocity and pressure values.

Step 5. Postprocess the results, verify modeling assumptions and present the results

The flow streamlines were plotted as shown in Fig. 18. This figure shows that the designed vanes will provide a relatively uniform flow to the steam coil and prohibit the formation of strong recirculation zones prior to the air heater inlet.

Verifying that the modeling assumptions provide realistic results was done by comparing the numerically calculated pressure drop with an empirical value for a turn containing vanes, 0.006 psi and 0.008 psi respectively. The numerical value is close in magnitude to the empirical value. The modeled turn region also contained representative velocities, pressure drop and geometry. Based on this simple check, it was concluded that the model-calculated values would be close to the real situation.

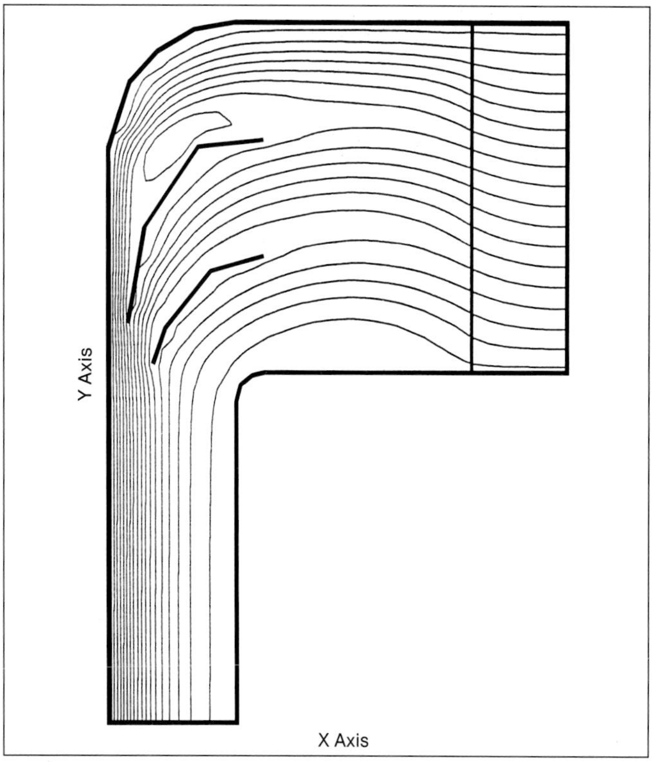

Fig. 18 Air heater fan discharge model sample analysis, streamline plot (streamline spacing is 1.250 lb/ft).

References

1. Rohsenow, W., Hartnett, J., and Ganic, E., *Handbook of Heat Transfer Fundamentals*, McGraw-Hill, New York, 1985.

2. Burmeister, L.C., *Convective Heat Transfer*, Wiley, New York, 1983.

3. Moody, L.F., "Friction factors for pipe flow," *Journal of Heat Transfer*, Vol. 66, pp. 671-684, 1944.

4. Ito, H., "Friction factors for turbulent flow in curved pipes," *Journal of Basic Engineering*, Vol. 81, pp. 123-126, 1959.

5. Briggs, D.E., and Young, E.H., "Convection heat transfer and pressure drop of air flowing across triangular pitch banks of finned tubes," *Chemical Engineering Progress Symposium Series (Heat Transfer)*, AIChE, No. 41, Vol. 59, pp. 1-10, Houston, 1963.

6. Grimison, E.D., "Correlation and utilization of new data on flow resistance and heat transfer for crossflow of gases over tube banks," *Trans. ASME*, Vol. 59, pp. 583-594, 1937.

7. Gunter, A.Y., and Shaw, W.A., "A general correlation of friction factors for various types of surfaces in crossflow," *Trans. ASME*, Vol. 67, pp. 643-660, 1945.

8. Idelchik, I.E., *Handbook of Hydraulic Resistance*, 2nd ed., Hemisphere, Washington, D.C., 1986.

9. Jakob, M., Discussion appearing in *Trans. ASME*, Vol. 60, pp. 384-386, 1938.

10. Kern, D.Q., *Process Heat Transfer*, p. 555, McGraw-Hill, New York, 1950.

11. Wimpress, R.N., *Hydrocarbon Processing and Petroleum Refiner*, Vol. 42, No. 10, pp. 115-126, 1963.

12. Patankar, S., *Numerical Heat Transfer and Fluid Flow*, Hemisphere, Washington, D.C., 1980.

13. Anderson, D.A., Tannehill, J.C., and Pletcher, R.H., *Computational Fluid Mechanics and Heat Transfer*, Hemisphere, Washington, D.C., 1984.

An upward view of a large wood-fired industrial boiler furnace showing air injection arches and auxiliary burner openings.

Chapter 4
Heat Transfer

Heat transfer deals with the transmission of thermal energy and plays a central role in most energy conversion processes. Heat transfer is important in fossil fuel combustion, chemical reaction processes, electrical systems, nuclear fission and certain fluid systems. It also occurs during everyday activities including cooking, heating and refrigeration, as well as being an important consideration in choosing clothing for different climates.

Although the fundamentals of heat transfer are simple, practical applications are complex because real systems contain irregular geometries, combined modes of heat transfer and time dependent responses.

Fundamentals

Basic modes of heat transfer

There are three modes of heat transfer: conduction, convection and radiation. One or more of these modes controls the amount of heat transfer in all applications.

Conduction Temperature is a property which indicates the index of the kinetic energy possessed by the molecules of a substance; the higher the temperature the greater the kinetic energy or molecular activity of the substance. Molecular conduction of heat is simply the transfer of energy due to a temperature difference between adjacent molecules in a solid, liquid or gas.

Conduction heat transfer is evaluated using Fourier's law:

$$q_c = -kA\frac{dT}{dx} \tag{1}$$

The flow of heat, q_c, is positive when the temperature gradient, dT/dx, is negative. This result, consistent with the second law of thermodynamics, indicates that heat flows in the direction of decreasing temperature. The heat flow, q_c, is in a direction normal (or perpendicular) to an area, A, and the gradient, dT/dx, is the change of temperature in the direction of heat flow. The thermal conductivity, k, a property of the material, quantifies its ability to conduct heat. A range of thermal conductivities is listed in the Table 1. The discrete form of the conduction law is written:

$$q = \frac{kA}{L}(T_1 - T_2) \tag{2}$$

Fig. 1 illustrates positive heat flow described by this equation and shows the effect of variable thermal conductivity on the temperature distribution. The grouping kA/L is known as the thermal conductance, K_c.

Contact resistance A special case of conduction is the thermal contact resistance across a joint between solid

Table 1
Thermal Conductivity, k, of Common Materials

Material	Btu/h ft F	W/m C
Gases at atmospheric pressure	0.004 to 0.70	0.007 to 1.2
Insulating materials	0.01 to 0.12	0.02 to 0.21
Nonmetallic liquids	0.05 to 0.40	0.09 to 0.70
Nonmetallic solids (brick, stone, concrete)	0.02 to 1.5	0.04 to 2.6
Liquid metals	5.0 to 45	8.6 to 78
Alloys	8.0 to 70	14 to 121
Pure metals	30 to 240	52 to 415

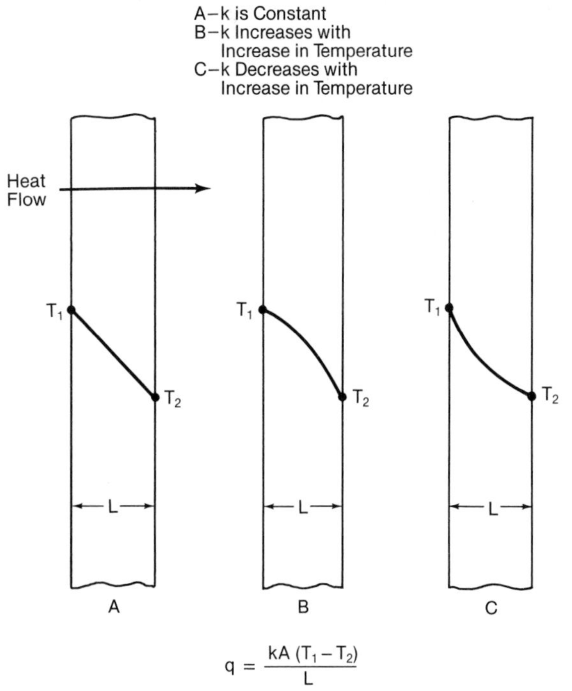

Fig. 1 Temperature-thickness relationships corresponding to different thermal conductivities, *k*.

Nomenclature

a, b, c general coefficients
A surface area, ft^2 (m^2)
c_p specific heat at constant pressure, Btu/lb F (J/kg K)
C_f cleanliness factor, dimensionless
C_t thermal capacitance, Btu/ft^3 F (J/m^3 K)
D diameter, ft (m)
D_e equivalent or hydraulic diameter, ft (m)
E_b blackbody emissive power, Btu/h ft^2 (W/m^2)
F radiation configuration factor, dimensionless
F heat exchanger arrangement factor, dimensionless
F_a crossflow arrangement factor, dimensionless
F_d tube bundle depth factor, dimensionless
F_{pp} fluid property factor, see text
F_T fluid temperature factor, dimensionless
$\mathcal{F}$ total radiation exchange factor, dimensionless
g acceleration of gravity, 32.17 ft/s^2 (9.8 m/s^2)
G incident thermal radiation, Btu/h ft^2 (W/m^2)
G mass flux or mass velocity, lb/h ft^2 (kg/m^2 s)
h heat transfer coefficient, Btu/h ft^2 F (W/m^2 K)
h_{ct} contact coefficient, Btu/h ft^2 F (W/m^2 K)
h_c crossflow heat transfer coef., Btu/h ft^2 F (W/m^2 K)
h'_c crossflow velocity and geometry factor, see text
h_l longitudinal heat transfer coef., Btu/h ft^2 F (W/m^2 K)
h'_l longitudinal flow velocity and geometry factor, see text
H enthalpy, Btu/lb (J/kg)
H_{fg} enthalpy of vaporization, Btu/lb (J/kg)
I electrical current, amperes
J radiosity, Btu/h ft^2 (W/m^2)
k thermal conductivity, Btu/h ft F (W/m K)
k_c insulation thermal conductivity, Btu/h ft F (W/m K)
K thermal conductance, Btu/h F (W/K)
K_y mass transfer coefficient, lb/ft^2 s (kg/m^2 s)
L beam length, ft (m)
L length or dimension, ft (m)
L_h fin height, ft (m)
L_t fin spacing, ft (m)
$\dot{m}$ mass flow rate, lb/h (kg/s)
N number of concentric thermocouple shields
P pressure or partial pressure, atm
P temp. ratio for surface arrgt. factor, dimensionless
q heat flow rate, Btu/h (W)
q''' volumetric heat generation rate, Btu/h ft^3 (W/m^3)
q_{rel} heat release, Btu/h ft^3 (W/m^3)
r radius, ft (m)
R temp. ratio for surface arrgt. factor, dimensionless
R thermal resistance, h ft^2 F/Btu (m^{2}K/W)
S general source term
S total exposed surface area for a finned surface, ft^2 (m^2)
S_f fin surface area; sides plus peripheral area, ft^2 (m^2)
t time, s or h, see text (s)
T temperature, F or R (C or K)
T^o temperature at initial time, F (C)
Δt time interval, s
ΔT temperature difference, F (C)
ΔT_{LMTD} log mean temperature difference, F (C)
u, v, w velocity in x, y, z coordinates respectively, ft/s (m/s)
U overall heat transfer coef., Btu/h ft^2 F (W/m^2 K)
V electrical voltage, volts
V velocity, ft/s (m/s)
V volume, ft^3 (m^3)
x dimension, ft (m)
x, y, z dimensions in Cartesian coordinate system, ft (m)
Δx change in length, ft (m)
Y Schmidt fin geometry factor, dimensionless
Y_g concentration in bulk fluid, lb/lb (kg/kg)
Y_i concentration at condensate interface, lb/lb (kg/kg)
Z Schmidt fin geometry factor, dimensionless
$\alpha = k/\rho c_p$ thermal diffusivity, ft^2/s (m^2/s)
α absorptivity, dimensionless
β volume coefficient of expansion, 1/R (1/K)
Γ effective diffusion coefficient
δ film thickness, ft (m)
ε emissivity, dimensionless
η Schmidt fin efficiency, dimensionless
μ dynamic viscosity, lbm/ft s (kg/m s)
υ kinematic viscosity, ft^2/s (m^2/s)
ρ density, lb/ft^3 (kg/m^3)
ρ reflectivity, dimensionless
σ Stefan-Boltzmann constant, 0.1713×10^{-8} Btu/h ft^2 R^4 (5.669×10^{-8} W/m^2 K^4)
τ transmissivity, dimensionless
ϕ general dependent variable

Subscripts:

b bulk
c contact or conduction
cv convection
clean unfouled or clean condition
e node point east
eq equivalent
f fluid or fin
g gas
i inside or *ith* parameter
j *jth* parameter
o outside
p node point under evaluation
p point
r radiation
sg surface to gas
t thermal
w node point west
w wall
δ liquid film surface (gas liquid interface)

Dimensionless groups:

$$\mathrm{Gr} = \frac{g\beta(T_s - T_f)\rho^3 L^3}{\mu}$$ Grashof number

$$\mathrm{Nu} = \frac{h\,L}{k}$$ Nusselt number

$$\mathrm{Pe} = \mathrm{Re}\,\mathrm{Pr}$$ Peclet number

$$\mathrm{Pr} = \frac{c_p\,\mu}{k}$$ Prandtl number

$$\mathrm{Ra} = \mathrm{Gr}\,\mathrm{Pr}$$ Rayleigh number

$$\mathrm{Re} = \frac{\rho V\,L}{\mu} = \frac{G\,L}{\mu}$$ Reynolds number

$$\mathrm{St} = \frac{\mathrm{Nu}}{\mathrm{Re}\,\mathrm{Pr}}$$ Stanton number

materials. At the interface of two solid materials the surface to surface contact is imperfect from the gap that prevails due to surface roughness. In nuclear applications with fuel pellets and fuel cladding, surface contact resistance can have a major impact on heat transfer. If one dimensional steady heat flow is assumed, the heat transfer across a gap is defined by:

$$q = \frac{T_1 - T_2}{R_{ct}} \tag{3}$$

where the quantity R_{ct} is called the thermal contact resistance, $1/h_{ct}A$, and h_{ct} is called the contact coefficient. T_1 and T_2 are the average surface temperatures on each side of the gap. Tabulated values of the contact coefficient are presented in References 1 and 2. Examples include 300 Btu/h ft^2 F (1.7 kW/m^2 K) between two sections of ground 304 stainless steel in air and 25,000 Btu/h ft^2 F (142 kW/m^2 K) between two sections of ground copper in air. The factors are usually unknown for specific applications and estimates need to be made. There are two principal contributions across the gap — solid to solid conduction at the points of contact and thermal conduction through the entrapped gases in the void spaces.

Convection Convection heat transfer within a fluid (gas or liquid) occurs by a combination of molecular conduction and macroscopic fluid motion. Convection occurs adjacent to heated surfaces as a result of fluid motion past the surface as shown in Fig. 2.

Natural convection occurs when the fluid motion is due to local density differences alone. In the top portion of Fig. 2, the fluid motion is due to heat flow from the surface to the fluid; the fluid density decreases causing the lighter fluid to rise and be replaced by cooler fluid. Forced convection results when mechanical forces from devices such as fans give motion to the fluids. The rate of heat transfer by convection, q_{cv}, is defined:

$$q_{cv} = hA\,(T_s - T_f) \tag{4}$$

where h is the local heat transfer coefficient, A is the surface area, T_s is the surface temperature and T_f is the fluid temperature. Equation 4 is known as Newton's Law of Cooling and the term hA_s is the convection conductance, K_{cv}. The heat transfer coefficient, h, is also termed the unit conductance, because it is defined as the conductance per unit area. Average heat transfer coefficients over a surface are used in most engineering applications. This convective heat transfer coefficient is a function of the thermal and hydrodynamic properties of the fluid and the surface geometry. Approximate ranges are shown in Table 2.

Radiation Radiation is the transfer of energy between bodies by electromagnetic waves. This transfer, unlike conduction or convection, requires no intervening medium. The electromagnetic radiation, in the wavelength range of 0.1 to 100 micrometers, is produced solely by the temperature of a body. Energy at the body's surface is converted into electromagnetic waves that emanate from the surface and strike another body. Some of the thermal radiation is absorbed by the receiving body and reconverted into internal energy, while the remaining energy is reflected from or transmitted through the body. The fractions of radiation reflected, transmitted and absorbed by a surface are known respectively as the reflectivity, ρ, transmissivity, τ, and absorptivity α. The sum of these fractions equals one:

$$\rho + \tau + \alpha = 1 \tag{5}$$

All surfaces whose temperatures are above absolute zero emit thermal radiation.

Thermal radiation generally passes through gases such as air with no absorption taking place. These nonabsorbing, or nonparticipating, gases do not affect the radiative transfer. Other gases, like carbon dioxide, water vapor and carbon monoxide, to a lesser degree, affect radiative transfer and are known as participating gases. These gases, prevalent in the flue gases of a boiler, affect the heat transfer to surfaces and the distribution of energy absorbed in the boiler.

All bodies continuously emit radiant energy in amounts which are determined by the temperature and the nature of the surface. A perfect radiator, or *blackbody*, absorbs all of the radiant energy reaching its surface and emits radiant energy at the maximum theoreti-

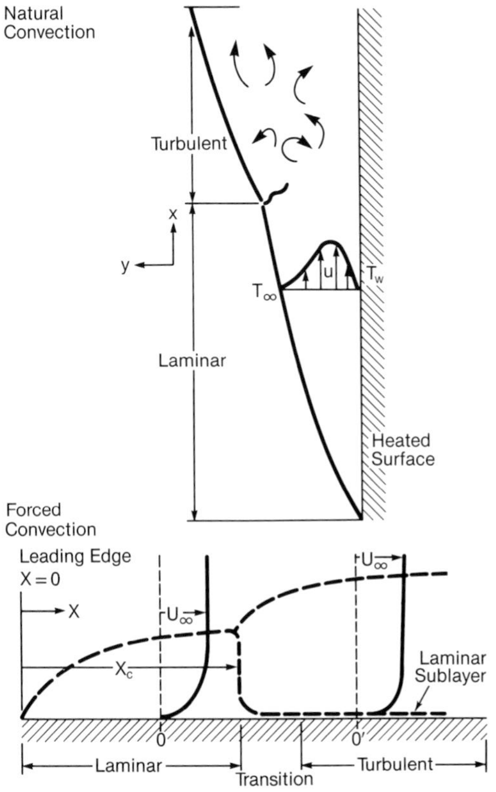

Fig. 2 Natural and forced convection. Above, boundary layer on a vertical flat plate. Below, velocity profiles for laminar and turbulent boundary layers in flow over a flat plate. (Vertical scale enlarged for clarity.)

Table 2
Typical Convective Heat Transfer Coefficients, h

Condition	Btu/h ft^2 F	W/m^2 C
Air, free convection	1 to 5	6 to 30
Air, forced convection	5 to 50	30 to 300
Steam, forced convection	300 to 800	1800 to 4800
Oil, forced convection	5 to 300	30 to 1800
Water, forced convection	50 to 2000	300 to 12,000
Water, boiling	500 to 20,000	3000 to 120,000

cal limit according to the Stefan-Boltzmann law:

$$q_r = A\,\sigma\,T_s^4 \qquad (6)$$

where σ is the Stefan-Boltzmann constant 0.1713×10^{-8} Btu/h $ft^2\,R^4$ (5.669×10^{-8} W/$m^2\,K^4$), and T_s is the absolute temperature, R (K). The product $\sigma\,T_s^4$ is also known as the blackbody emissive power, E_b.

The radiation from a blackbody extends over the whole range of wavelengths, although the bulk of it in boiler applications is concentrated in a band from 0.1 to 20 micrometers. The wavelength at which the maximum radiation intensity occurs is inversely proportional to the absolute temperature of the body; this is known as Wien's law.

A real radiator (or *graybody*) absorbs less than 100% of the energy incident on it. The heat transfer by radiation from a gray surface can be expressed by:

$$q_r = \varepsilon\,A\sigma\,T_s^4 \qquad (7)$$

where ε is the emittance of the surface, or emissivity. Table 3 shows some representative emissivity values. If the emissivity is independent of wavelength, the surface is termed a nonselective radiator, or *gray* surface. If the emissivity depends on wavelength, the surface is termed a selective or non-gray radiator. An exact analysis of non-gray conditions is generally too complicated; however, if all surfaces are assumed to be gray, a simpler treatment is possible. This treatment involves introducing a factor, $\mathcal{F}$, which depends on the configuration (geometry), the emissivities and the surface areas.[3]

The net radiation heat transfer between two blackbody surfaces which are separated by a vacuum or nonparticipating gas is written:

$$q_{12} = A_1 F_{12}\,\sigma\,(T_1^4 - T_2^4) \qquad (8)$$

A_1 is the surface area; F_{12} is the configuration factor and represents the fraction of radiant energy leaving surface 1 that directly strikes surface 2. T_1 and T_2 are the surface temperatures. Since the net energy at surface 1 must balance the net energy at surface 2, we can write:

$$q_{12} = -q_{21} \qquad (9)$$

Using Equations 8 and 9, the following results:

$$A_1 F_{12} = A_2 F_{21} \qquad (10)$$

This equation, known as the principle of reciprocity, guarantees conservation of the radiant heat transfer between surfaces. The following rule applies:

$$\sum_j F_{ij} = 1 \qquad (11)$$

stating that all energy leaving surface i to all other (j) surfaces must equal 1. Many texts include the calculation of configuration factors, commonly named shape factors or geometric factors.[1,2] Radiation balances for participating and nonparticipating media are presented later in the chapter.

Table 3
Representative Values of Emissivity

Polished metals	$0.01 < \varepsilon < 0.08$
Metals, as-received	$0.1 < \varepsilon < 0.2$
Metals oxidized	$0.25 < \varepsilon < 0.7$
Ceramic oxides	$0.4 < \varepsilon < 0.8$
Special paints	$0.9 < \varepsilon < 0.98+$

Energy balances

The solution of a heat transfer problem requires defining the system which will be analyzed. This usually involves idealizing the actual system by defining a schematic *control volume* of the modeled system. A net energy balance on the control volume reflects the first law of thermodynamics and can be stated:

$$\text{energy in} - \text{energy out} = \text{stored energy} \qquad (12)$$

For a steady flow of heat, the balance simplifies to:

$$\text{heat in} = \text{heat out} \qquad (13)$$

The laws governing the flow of heat are used to obtain equations in terms of temperature.

Electrical analogy

The basic laws of conduction, convection and radiation can be rearranged into equations of the form:

$$q = \frac{T_1 - T_2}{R_t} \qquad (14)$$

This equation can be compared to Ohm's law for electrical circuits ($I = V/R$). The heat transfer (q) compares with the current (I), the temperature difference (T_1-T_2) compares with the voltage (V), and the thermal resistance (R_t) compares with the electrical resistance (R). Thermal conductance, K_t, is defined as the reciprocal of resistance. Table 4 contains analog thermal resistances used in many applications.

For systems with changes in stored energy governed by Equation 12, an electrical analogy can be written:

Table 4
Summary of Thermal Resistances

	Rectangular Geometries and Surfaces	Cylindrical Geometries and Surfaces
Conduction, R_c	$\frac{\Delta x}{kA}$	$\frac{\ell n\,(r_2/r_1)}{2\pi kl}$
Convection, R_{cv} from surface	$\frac{1}{hA}$	$\frac{1}{2\pi r_2 lh}$
Radiation, R_r from surface	$\frac{T_2 - T_3}{\mathcal{F}_{23} A_2\,\sigma(T_2^4 - T_3^4)}$	$\frac{T_2 - T_3}{\mathcal{F}_{23}(2\pi r_2 l)\,\sigma(T_2^4 - T_3^4)}$

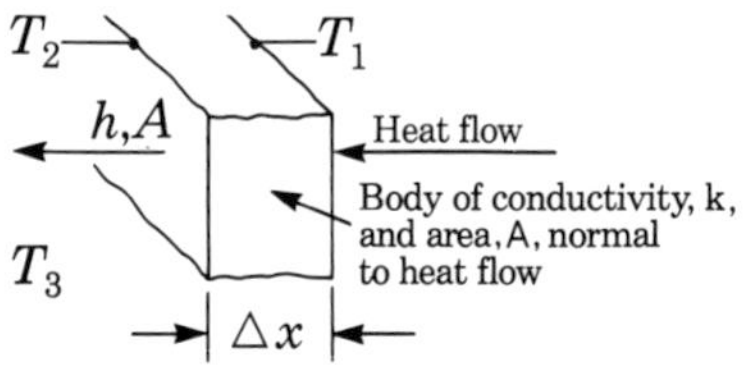

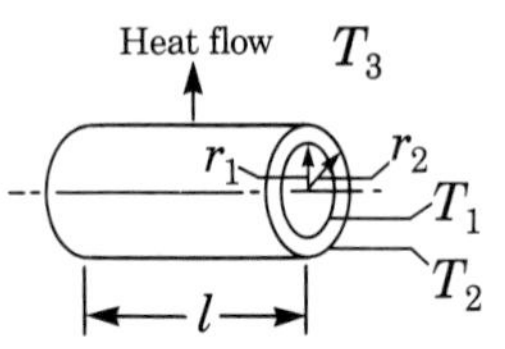

$$q = C_t \frac{dT}{dt} \quad (15)$$

where C_t is the thermal capacitance, ρc_p. This equation can be compared with its electrical equivalent:

$$I = C \frac{dV}{dt} \quad (16)$$

where C is the electrical capacitance. Kirchhoff's law for electrical circuits provides the last analogy needed. In heat transfer notation this would be:

$$\sum q = q_{stored} \quad (17)$$

This is an expression of the first law of thermodynamics which states that all heat flows into a point equal the stored energy.

Consider the composite system and the equivalent thermal circuit shown in Fig. 3. The concepts of resistance and conductance are particularly useful when more than one mode of heat transfer or more than one material or boundary is involved. When two modes of heat transfer, such as convection and radiation, occur simultaneously and independently, the combined conductance, K, is the sum of the individual conductances, K_{cv} and K_r. These individual conductances are essentially heat flows in parallel. When the heat flows are in series, the resistances, not the conductances, are additive. The total or equivalent thermal resistance can then be substituted into Equation 14 to define the energy and flow.

Governing equations

Steady-state conduction The basic laws for each heat transfer mode and the energy balance provide the tools needed to write the governing equations for rectangular and cylindrical heat transfer systems. For example, for the plane wall shown in Fig. 1, the steady flow energy balance for a slice of thickness, dx, is:

$$q_1 - q_2 = q - \left(q + \frac{dq}{dx} dx \right) = 0 \quad (18)$$

which can also be written:

$$\frac{d}{dx} kA \frac{dT}{dx} = 0 \quad (19)$$

The conditions at the boundaries, $T = T_1$ at $x = 0$ and $T = T_2$ at $x = L$, provide closure. The general symbolic form of the equation in vector notation can be represented:

$$\nabla \cdot k \nabla T = 0 \quad (20a)$$

or in x, y, z Cartesian coordinates:

$$\frac{\delta}{\delta x} k \frac{\delta T}{\delta x} + \frac{\delta}{\delta y} k \frac{\delta T}{\delta y} + \frac{\delta}{\delta z} k \frac{\delta T}{\delta z} = 0 \quad (20b)$$

This assumes there is no net heat storage or heat generation in the wall.

Unsteady-state conduction So far only steady-state conduction, where temperatures vary from point to point but do not change with time, has been discussed. All unsteady-state conduction involves heat storage. For instance, in heating a furnace, enough heat must be supplied to bring the walls to the operating temperature and also to make up for the steady-state losses of normal operation. In large power boilers that run for long periods of time, heat storage in the walls and boiler metal is an insignificant fraction of the total heat input. In small boilers with refractory settings that are operated only part time, or in furnaces that are frequently heated and cooled in batch process work, heat stored in the walls during startup may be a considerable portion of the total heat input.

Unsteady-state conduction is important when equalizing boiler drum temperature during pressure raising and reducing periods. When boiler pressure is raised, the water temperature rises. The inner surface of the steam drum is heated by contact with the water below the water line and by the condensation of steam above the water line. The inside and outside drum temperatures are increased by unsteady-state conduction. During this transient heatup period, temperature differentials across the drum wall (or thermal gradients) will be larger than during steady-state operation. Larger thermal gradients result in higher thermal stresses as discussed in Chapter 7. The rate of temperature and pressure increase must therefore be controlled to maintain the thermal stresses within acceptable levels in order to protect the drum. During pressure reducing periods, the inside of the drum below the water line is cooled by boiler water while the top of the drum is cooled by radiation to the water, by the steam flow to the outlet connections, and by unsteady-state conduction through the drum walls.

Unsteady-state conduction occurs in heating or cooling processes where temperatures change with time. Examples include heating billets, quenching steel, operating regenerative heaters, raising boiler pressure, and heating and cooling steam turbines. By introducing time as an additional variable, conduction analyses become more complicated. For unsteady heat flow, the one dimensional thermal energy equation becomes:

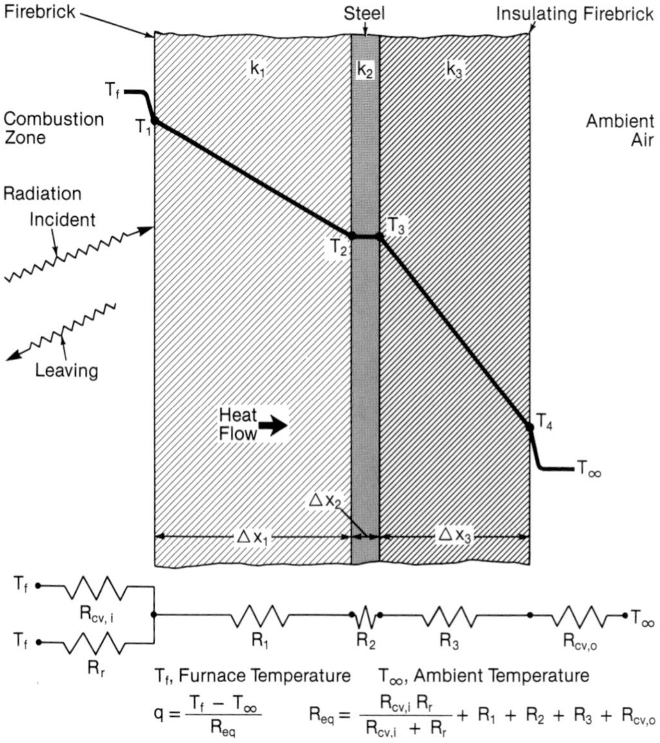

Fig. 3 Temperature distribution in composite wall with fluid films.

$$\rho c_p \frac{\partial T}{\partial t} = \frac{\partial}{\partial x} k \frac{\partial T}{\partial x} \tag{21}$$

The two boundary temperatures at $x = 0$ and L, and the initial temperature, $T = T^o$, are sufficient to find a solution. A general form of the energy equation for multi-dimensional applications is:

$$\rho c_p \frac{\partial T}{\partial t} = \nabla \cdot k \nabla T \tag{22}$$

where $\nabla \cdot k \nabla T$ is defined in Equations 20a and b.

Flowing systems Boilers have complex flow profiles. In the basic example depicted in Fig. 4, there is steady flow into and out of the system, which is lumped into a single control volume. This leads to a balance of energy written as:

$$\dot m_1 H_1 + \dot m_2 H_2 + \dot m_3 H_3 = \dot m_4 H_4 \tag{23}$$

where $\dot m$ is the mass flow rate at each inlet or outlet and H is the fluid enthalpy.

As discussed in Chapter 3, the full energy equation sets the net energy entering a system with mass flow equal to the internal heat generation plus the work done by the system plus energy conducted into the system plus a viscous dissipation term (see Equation 7, Chapter 3). Viscous dissipation and work done in the boiler system can both usually be neglected. For steady-state conditions, the energy equation in terms of enthalpy and in vector notation can then be written as:

$$\underset{\text{Convection}}{\nabla \cdot (\rho u H)} = \underset{\text{Conduction}}{\nabla \cdot \Gamma \nabla H} + \underset{\text{Internal heat generation}}{S_H} \tag{24}$$

The parameter Γ is an effective diffusion coefficient, and S_H is the energy source term. Rearranging Equation 24 for temperature and using x, y, z Cartesian coordinates, the energy equation becomes:

$$u \frac{\partial T}{\partial x} + v \frac{\partial T}{\partial y} + w \frac{\partial T}{\partial z} =$$

$$\frac{\partial}{\partial x} \Gamma_x \frac{\partial T}{\partial x} + \frac{\partial}{\partial y} \Gamma_y \frac{\partial T}{\partial y} + \frac{\partial}{\partial z} \Gamma_z \frac{\partial T}{\partial z} + S_T + q_{rel} \tag{25}$$

Again Γ is the effective diffusion coefficient, S_T is the internal heat generation term and q_{rel} is the heat release rate.

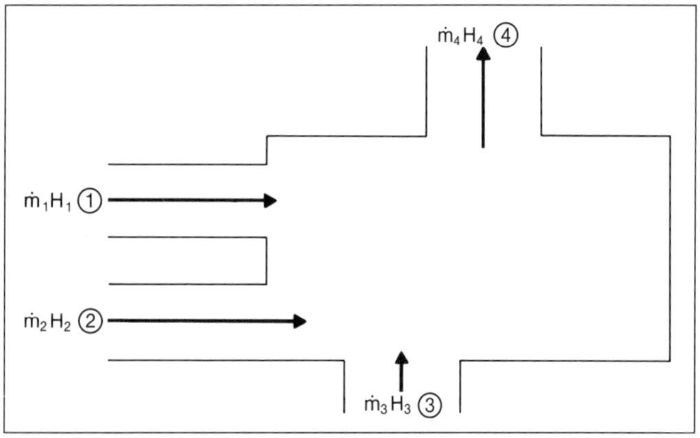

Fig. 4 Energy balance for a flowing system.

Equations 24 and 25 can be used with single phase flow or can be used with multiple phase flow (gas-solid or steam-water) by using mass averaged enthalpy. The development and application of the energy equation in rectangular, cylindrical and spherical coordinates are discussed in References 1 and 2.

Most boiler situations are too complex for direct application of the energy equation. However, the continuity and momentum equations discussed in Chapter 3 (Equations 1 to 6) are combined with the energy equation to form a fundamental part of the numerical computational furnace models discussed later in this chapter.

Radiation balances for enclosures

Nonparticipating media Referring to Equation 8, the net radiation between two black surfaces can be written:

$$q_{12} = A_1 F_{12}\, \sigma\, (T_1^4 - T_2^4) \tag{26}$$

The term F_{12} is the geometric shape factor and is shown for several common geometries in Fig. 5. Use of the tabulated values for more complex problems is demonstrated in the example problems under *Applications*. Equation 26 has limited value in boilers, because most fireside sur-

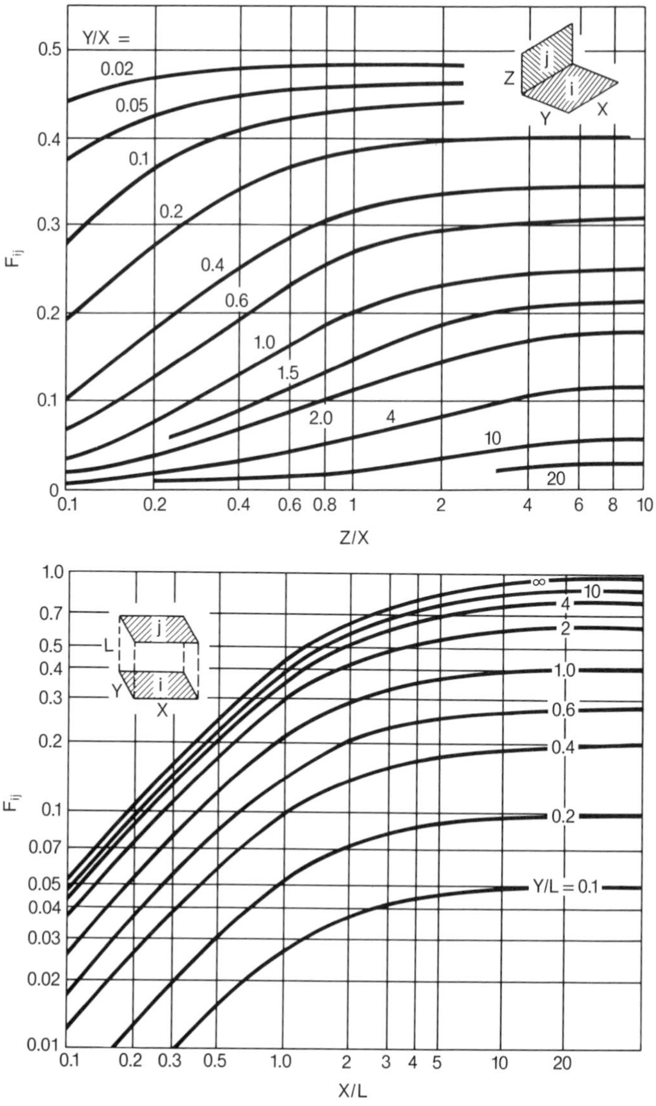

Fig. 5 Shape factors, F_{ij}, for calculating surface to surface radiation heat transfer.

faces are gray. It is better used to obtain estimates of radiation heat transfer, because it describes the maximum theoretical rate of energy transfer between two surfaces.

For the theory of radiation heat transfer in enclosures, see Reference 4. The energy striking a surface, called the *incident energy*, G, is the total energy striking a surface from all other surfaces in the enclosure. The energy leaving a surface, called the *radiosity*, J, is comprised of the energy emitted from the surface (E_b) and the reflected incident energy. These terms are related by:

$$J = \varepsilon E_b + \rho G \qquad \textbf{(27)}$$

The net radiation heat transfer from a surface is found as follows:

$$q = A\,(J - G) \qquad \textbf{(28)}$$

Combining Equations 27 and 28 leads to the electrical analogy listed first in Table 5. This circuit equivalent describes the potential difference between the surface at an emissive power, evaluated at T_s, and the surface radiosity. To evaluate the radiative heat transfer, the radiosity must first be determined. The net energy between surface i and surface j is the difference between the outgoing radiosities:

$$q_{ij} = A_i F_{ij} (J_i - J_j) \qquad \textbf{(29)}$$

The electrical analogy of the net exchange between two surfaces is listed in Table 5. The sum of similar terms for all surfaces in the enclosure yields the circuit diagram in the Table and the equation:

$$q_i = \sum_{j=1}^{N} A_i F_{ij} (J_i - J_j) \qquad \textbf{(30)}$$

The rules for electrical circuits are useful in finding the net radiation heat transfer. Consider the electrical circuit for radiation heat transfer between two gray walls shown in Fig. 6.

The rules for series circuits can be used to determine the net heat transfer:

Table 5
Network Equivalents for Radiative Exchange in Enclosures

Description	Circuit Equivalent	Resistance
Net exchange at surface		$R_i = \dfrac{\rho_i}{A_i \varepsilon_i}$
Net exchange between surfaces i and j		$R_{ij} = \dfrac{1}{A_i F_{ij}}$
Net exchange between surface i and all other surfaces	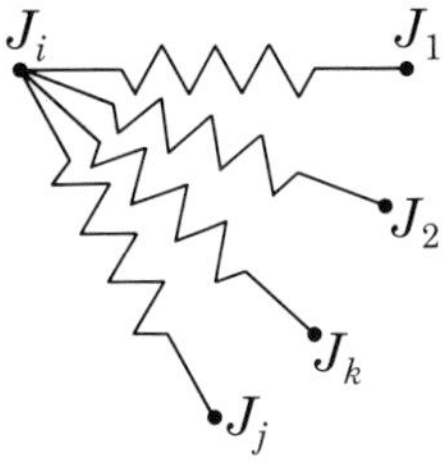	$R_{ik} = \dfrac{1}{A_i F_{ik}}$

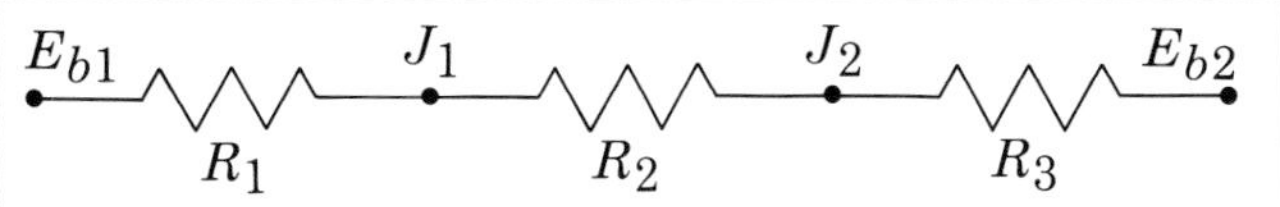

Fig. 6 Electric circuit analogy for thermal radiation.

$$q_{12} = \frac{E_{b2} - E_{b1}}{R_1 + R_2 + R_3} \qquad \textbf{(31)}$$

where

$$R_1 = \frac{\rho_1}{\varepsilon_1 A_1},\; R_2 = \frac{1}{A_1 F_{12}},\; R_3 = \frac{\rho_2}{\varepsilon_2 A_2}$$

Common two-surface geometries encountered in boiler design are listed in Table 6.

Participating media On the fire side of the boiler, the mixture of gases absorbs and emits radiant energy. When a uniform temperature-bounding surface encloses an isothermal gas volume, the radiant heat transfer can be treated as one zone. The incident radiation on the surfaces is made up of the emitted energy from the gas, $\varepsilon_g E_g$, and incoming energy from the surrounding walls, $(1 - \alpha_g) J_s$. Therefore, the incident radiation is defined by:

$$G_s = \varepsilon_g E_g + (1 - \alpha_g) J_s \qquad \textbf{(32)}$$

The energy leaving the surface is made up of direct emission, $\varepsilon_s E_s$, and reflected incident energy,

Table 6
Common Gray Two-Surface Enclosures

Geometry	Relations	$\mathcal{F}_{12}$
Large (Infinite) Parallel Planes		
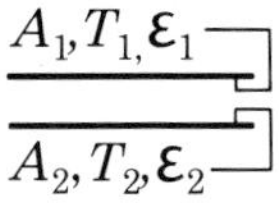	$A_1 = A_2 = A$ $F_{12} = 1$	$\mathcal{F}_{12} = \dfrac{1}{\dfrac{1}{\varepsilon_1} + \dfrac{1}{\varepsilon_2} - 1}$
Long (Infinite) Concentric Cylinders		
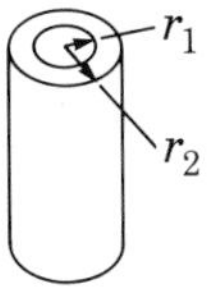	$\dfrac{A_1}{A_2} = \dfrac{r_1}{r_2}$ $F_{12} = 1$	$\mathcal{F}_{12} = \dfrac{1}{\dfrac{1}{\varepsilon_1} + \dfrac{1-\varepsilon_2}{\varepsilon_2}\left[\dfrac{r_1}{r_2}\right]}$
Concentric Spheres		
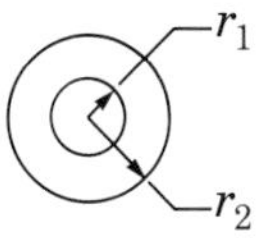	$\dfrac{A_1}{A_2} = \dfrac{r_1^2}{r_2^2}$ $F_{12} = 1$	$\mathcal{F}_{12} = \dfrac{1}{\dfrac{1}{\varepsilon_1} + \dfrac{1-\varepsilon_2}{\varepsilon_2}\left[\dfrac{r_1}{r_2}\right]^2}$
Small Convex Object in a Large Cavity		
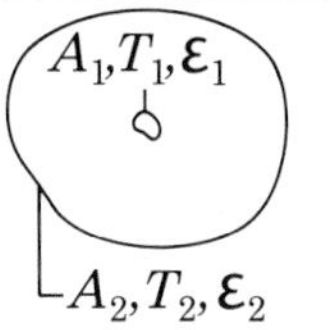	$\dfrac{A_1}{A_2} \approx 0$ $F_{12} = 1$	$\mathcal{F}_{12} = \varepsilon_1$

Note: The net heat flow is calculated from
$q_{12} = \sigma A_1 \mathcal{F}_{12} (T_1^4 - T_2^4)$

$(1 - \varepsilon_s)\ G_s$. Therefore, the radiosity is:

$$J_s = \varepsilon_s E_s + (1 - \varepsilon_s)\ G_s \qquad (33)$$

The solution of Equations 34 and 35 yields values for the incoming and the outgoing heat fluxes (q / A_s). The net heat transfer between the surface and gas becomes:

$$q_{sg} = J - G = \frac{A_s(\varepsilon_s \varepsilon_g E_g - \varepsilon_s \alpha_g E_s)}{1 - (1 - \alpha_g)(1 - \varepsilon_s)} \qquad (34)$$

The calculation of absorptivity is described in the examples at the end of the chapter. When the surfaces are radiatively black, $\varepsilon_s = 1$ and Equation 34 becomes:

$$q_{sg} = A(\varepsilon_g E_g - \alpha_g E_s) \qquad (35)$$

For slightly gray surfaces, Hottel and Sarofim[4] suggest the modification:

$$q_{sg} = A \frac{(\varepsilon_s + 1)}{2} (\varepsilon_g E_g - \alpha_g E_s) \qquad (36)$$

with errors of less than 10%.

In general, boiler enclosure wall and gas temperatures vary from wall to wall and even from point to point. A multi-zone analysis is required, because simple expressions of net surface heat transfer can not be described. Reference 4 contains procedures for calculating radiation heat transfer in multi-zone applications. However, the procedures presented here provide the basis for engineering estimates as indicated in the example at the end of the chapter.

Heat transfer properties and correlations

Thermal conductivity, specific heat and density

Thermal conductivity, k, is a material property that is expressed in Btu/h ft F (W/m K) and is dependent on the chemical composition of the substance. The relative order of magnitude of values for various substances is shown in Table 7. Thermal conductivities are generally highest for solids, lower for liquids and lower yet for gases. Insulating materials have the lowest conductivities of solid materials.

Thermal conductivities of pure metals generally decrease with an increase in temperature, while alloy conductivities may either increase or decrease. (See Fig. 7.) Conductivities of several steels and alloys are shown in Table 7. Thermal conductivities of various refractory materials are shown in Fig. 7 of Chapter 22. For many heat transfer calculations it is sufficiently accurate to assume a constant thermal conductivity that corresponds to the average temperature of the material.

The ranges of thermal conductivity and resistance for a layer of heated ash deposits may vary widely depending on the deposit's location within the furnace, on fuel type and on combustion conditions. In the available literature, the effective thermal conductivity of deposits on waterwall heating surfaces varies from 0.03 to 0.29 Btu/h ft F (0.05 to 0.5 W/m K); thermal resistances vary from 0.11 to 0.006 h ft^2 F/Btu (0.02 to 0.001 m^2 K/W).

The effective thermal conductivity of ash deposits is similar to that of carbon dioxide and air at high temperatures, and in some cases is lower. The lower limit is close to the thermal conductivity of glass wool, but does not exceed values for refractory materials. The effective thermal conductivity of the layer increases with temperature and with the amount of iron oxides present.

The thermal conductivity of water ranges from 0.33 Btu/h ft F (0.57 W/m K) at room temperature to 0.16 Btu/

Table 7
Properties of Various Substances at Room Temperature (see Note 1)

	ρ lb/ft^3	c_p Btu/lb F	k Btu/h ft F
METALS			
Copper	559	0.09	223
Aluminum	169	0.21	132
Nickel	556	0.12	52
Iron	493	0.11	42
Carbon Steel	487	0.11	25
Alloy Steel 18Cr 8Ni	488	0.11	9.4
NONMETAL SOLIDS			
Limestone	105	~0.2	0.87
Glass Pyrex®	170	~0.2	0.58
Brick K-28	27	~0.2	0.14
Plaster	140	~0.2	0.075
Kaowool	8	~0.2	0.016
GASES			
Hydrogen	0.006	3.3	0.099
Oxygen	0.09	0.22	0.014
Air	0.08	0.24	0.014
Nitrogen	0.08	0.25	0.014
Steam (see Note 2)	0.04	0.45	0.015
LIQUIDS			
Water	62.4	1.0	0.32
Sulfur dioxide (liquid)	89.8	0.33	0.12

Notes:
1. SI conversions: ρ, kg/m^3 = 16.02 × lb/ft^3; c_p, kJ/kg K = 4.1869 × Btu/lb F; k, W/m K = 1.7307 × Btu/h ft F.
2. Reference temperature equals 32F (0C) except for steam which is referenced at 212F (100C).

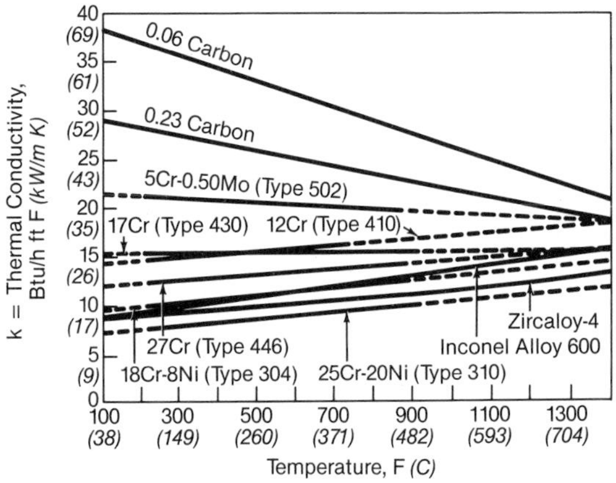

Fig. 7 Thermal conductivity, *k*, of some commonly used steels and alloys.

h ft F (0.28 W/m K) near the critical point. Water properties are relatively insensitive to pressure, particularly at pressures far from the critical point. Most other nonmetallic liquid thermal conductivities range from 0.05 to 0.15 Btu/h ft F (0.09 to 0.26 W/m K). In addition, thermal conductivities of most liquids decrease with temperature.

The thermal conductivities of gases increase with temperature and are independent of pressure at normal boiler conditions. These conductivities generally decrease with increasing molecular weight. The relatively high conductivity of hydrogen (a low molecular weight gas) makes it a good cooling medium for electric generators.

When calculating the conductivity of nonhomogeneous materials, the designer must use an apparent thermal conductivity to account for the porous or layered construction materials. In boilers and furnaces with refractory walls, thermal conductivity may vary from site to site due to variations in structure, composition, density, or porosity when the materials were installed. The thermal conductivities of these materials are strongly dependent on their apparent bulk density (mass per unit volume). For higher temperature insulations, the apparent thermal conductivity of fibrous insulations and insulating firebrick decreases as bulk density increases, because the denser material attenuates the radiation. However, there is a limit at which any increase in density increases the thermal conductivity due to conduction in the solid material.

Theory shows that specific heats of solids and liquids are generally independent of pressure. Table 7 lists specific heats of various metals, alloys and nonhomogeneous materials at 68F (20C). These values may be used at other temperatures without significant error.

The temperature dependence of the specific heat for gases is more pronounced than for solids and liquids. In boiler applications, pressure dependence may generally be neglected. Table 8 gives specific heat data for air and other gases.

In the case of steam and water, property variations (specific heat and thermal conductivity) can be significant over the ranges of temperature and pressure found in boilers. It is therefore recommended that the properties as compiled in the American Society of Mechanical Engineers (ASME) Steam Tables[5] be used.

Table 8
Properties of Selected Gases
at 14.696 psi (1.01 bar) (see Note 1)

T F	ρ lb/ft³	c_p Btu/lb F	k Btu/h ft F	ρ lb/ft³	c_p Btu/lb F	k Btu/h ft F
	Air			CO_2		
0	0.0855	0.240	0.0131	0.1320	0.184	0.0076
500	0.0408	0.248	0.0247	0.0630	0.247	0.0198
1000	0.0268	0.263	0.0334	0.0414	0.280	0.0318
1500	0.0200	0.276	0.0410	0.0308	0.298	0.042
2000	0.0159	0.287	0.0508	0.0247	0.309	0.050
2500	0.0132	0.300	0.0630	0.0122	0.311	0.055
3000	0.0113	0.314	0.0751	0.0175	0.322	0.061
	O_2			N_2		
0	0.0945	0.219	0.0133	0.0826	0.249	0.0131
500	0.0451	0.235	0.0249	0.0395	0.254	0.0236
1000	0.0297	0.252	0.0344	0.0260	0.269	0.0320
1500	0.0221	0.263	0.0435	0.0193	0.283	0.0401
2000	0.0178		0.0672	0.0156		0.0468
2500	0.0148		0.0792	0.0130		0.0528
3000	0.0127		0.0912	0.0111		
	H_2					
0	0.0059	3.421	0.1071			
500	0.0028	3.470	0.1610			
1000	0.0019	3.515	0.2206			
1500	0.0014	3.619	0.2794			
2000	0.0011	3.759	0.3444			
2500	0.0009	3.920	0.4143			
3000	0.0008	4.218	0.4880			

Thermal Conductivity of Flue Gases from Various Fuels
[Btu/h ft F: 14.696 psi (1.01 bar)] (see Note 2)

Temperature T	Natural Gas k	Fuel Oil k	Coal k
0	--	--	--
500	0.022	0.022	0.022
1000	0.030	0.029	0.029
1500	0.037	0.036	0.036
2000	0.044	0.043	0.043
2500	0.051	0.049	0.050

Notes:
1. SI conversions: T, C = 5/9 (F-32); ρ, kg/m³ = 16.02 × lb/ft³; c_p, kJ/kg K = 0.2388 × Btu/lb F; k, W/m k = 0.75778 × Btu/h ft F.
2. 115% excess air for natural gas and oil; 120% excess air for coal.

Radiation properties

Bodies that are good radiation absorbers are equally good emitters and Kirchhoff's law states that, at thermal equilibrium, their emissivities are equal to their absorptivities. A *blackbody* is one which absorbs all incident radiant energy while reflecting or transmitting none of it. The absorptivity and emissivity of a blackbody are, by definition, each equal to one. This terminology does not necessarily mean that the body appears to be black. Snow, for instance, absorbs only a small portion of the incident visible light, but to the longer wavelengths (the bulk of thermal radiation), snow is almost a blackbody. At a temperature of 2000F (1093C) a blackbody glows brightly, because a non-negligible part of its radiation is in the visible range. Bodies are never completely black, but a hole through the wall of a large enclosure can be used to approximate blackbody conditions, because radiation entering the hole undergoes multiple reflections and absorptions. As a result, most of the radiation is retained in the enclosure.

Fortunately, a number of commercial surfaces, particularly at high temperatures, have emissivities of 0.80 to 0.95 and behave much like blackbodies. Typical average emissivity values are noted in Table 9. Although emissivity depends on the surface composition and roughness and wavelength of radiation, the wavelength, or spectral, dependence is neglected in practical boiler calculations.

The emissivity of furnace ash deposits affects boiler heat transfer. The emissivity depends not only on the substrate conditions previously mentioned but also on the chemical composition, structure and porosity of the layer. Therefore, apparent emissivity describes the combined deposit and substrate emissivity; the same deposit on different surfaces yields different apparent surface emissivities. Reported values in the literature claim emissivities of 0.5 to 0.8 for most slag and ash deposits. Emissivity increases with iron content of the deposit. Apparent emissivity decreases as wall temperature increases, most likely due to the deposit becoming transparent and the effective emissivity approaching the substrate values.

Although many gases, such as oxygen and nitrogen, absorb or emit only insignificant amounts of radiation, others, such as water vapor, carbon dioxide, sulfur dioxide and carbon monoxide, substantially absorb and emit. Water vapor and carbon dioxide are important in boiler calculations because of their presence in the combustion products of hydrocarbon fuels. These gases are selective radiators. They emit and absorb radiation only in certain wavelengths that lie outside of the visible range and are consequently identified as nonluminous radiators. Whereas the radiation from a furnace wall is a surface phenomenon, a gas radiates and absorbs (within its absorption bands) at every point throughout the furnace. Furthermore, the emissivity of a gas changes with temperature, and the presence of one radiating gas has an effect on the radiating characteristics of another with which it is mixed. The energy emitted by a radiating gaseous mixture depends on gas temperature, the partial pressures, p, of the constituents and a beam length, L, that depends on the shape and dimensions of the gas volume. An estimate of the beam length is L = 3.6 V/A for radiative transfer from the gas to the surface of the enclosure, where V is the enclosure volume and A is the enclosure surface area. The factor 3.6 varies from approximately 3.4 to 3.8. For many boiler applications 3.4 is the appropriate value. Figs. 8 and 9 show the emissivity for water vapor and carbon dioxide. When using these

Table 9
Normal Emissivities, ε, for Various Surfaces[6] (see Note 1)

Material	Emissivity, ε	Temp., F	Description
Aluminum	0.09	212	Commercial sheet
Aluminum oxide	0.63 to 0.42	530 to 930	
Aluminum paint	0.27 to 0.67	212	Varying age and Al content
Brass	0.22	120 to 660	Dull plate
Copper	0.16 to 0.13	1970 to 2330	Molten
Copper	0.023	242	Polished
Cuprous oxide	0.66 to 0.54	1470 to 2012	
Iron	0.21	392	Polished, cast
Iron	0.55 to 0.60	1650 to 1900	Smooth sheet
Iron	0.24	68	Fresh emeried
Iron oxide	0.85 to 0.89	930 to 2190	
Steel	0.79	390 to 1110	Oxidized at 1100F
Steel	0.66	70	Rolled Sheet
Steel	0.28	2910 to 3270	Molten
Steel (Cr-Ni)	0.44 to 0.36	420 to 914	18-8 rough, after heating
Steel (Cr-Ni)	0.90 to 0.97	420 to 980	25-20 oxidized in service
Brick, red	0.93	70	Rough
Brick, fireclay	0.75	1832	
Carbon, lamp-black	0.945	100 to 700	0.003 in. or thicker
Water	0.95 to 0.963	32 to 212	

Note:
1. SI conversion: T, C = 5/9 (F-32).

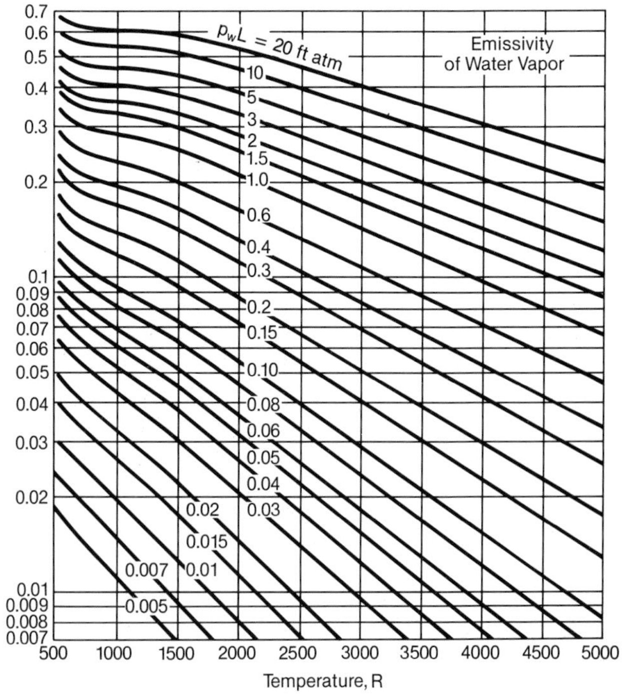

Fig. 8 Emissivity of water vapor at one atmosphere total pressure: p_wL = partial pressure in atmospheres × mean beam length in feet.[6]

figures to evaluate absorptivity, α, of a gas, Hottel[4] recommends modification of the pL product by a surface to gas temperature ratio. This is illustrated in Example 5 at the end of this chapter.

The effective surface emittance of a water-carbon dioxide mixture is calculated as follows:

$$\varepsilon = \varepsilon_{H_2O} + \varepsilon_{CO_2} - \Delta\varepsilon \quad (37)$$

where $\Delta\varepsilon$ is a correction factor that accounts for the effect of one radiating gas on the other. This equation ne-

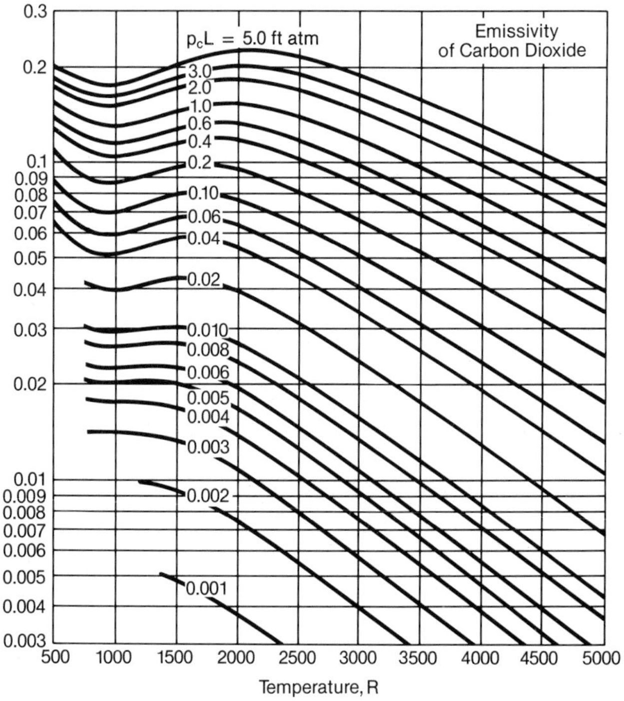

Fig. 9 Emissivity of carbon dioxide at one atmosphere total pressure: p_cL = partial pressure in atmospheres × mean beam length in feet.[6]

glects pressure corrections and considers boilers operating at 1 atm. The factors shown in Fig. 10 depend on temperature, the partial pressures, p, of the constituents and the beam length, L.

Working formulas for convection heat transfer

Heat transfer by convection between a fluid (gas or liquid) and a solid is expressed by Equation 4. This equation is a definition of the heat transfer coefficient but is inadequate in describing the details of the convective mechanisms. Only a comprehensive study of the flow and heat transfer would define the dependence of the heat transfer coefficient along the surface. In the literature, simple geometries have been modeled and predictions agree well with experimental data. However, for the more complex geometries encountered in boiler analysis, correlations are used that have been developed principally from experimental data.

Convective heat transfer near a surface takes place by a combination of conduction and mass transport. In the case of heat flowing from a heated surface to a cooler fluid, heat flows from the solid first by conduction into a fluid element, raising its internal energy. The heated element then moves to a cooler zone where heat flows from it by conduction to the cooler surrounding fluid.

Fluid motion can occur in two ways. If the fluid is set in motion due to density differences arising from temperature variations, free or natural convection occurs. If the motion is externally induced by a pump or fan, the process is referred to as forced convection.

Convective heat transfer can occur in laminar or turbulent flows. For laminar flow, the fluid moves in layers, or lamina, with each element following an orderly path. In turbulent flow, prevalent in boiler passages, the local motion of the fluid is chaotic and statistical treatment is used to establish average velocity and heat transfer values.

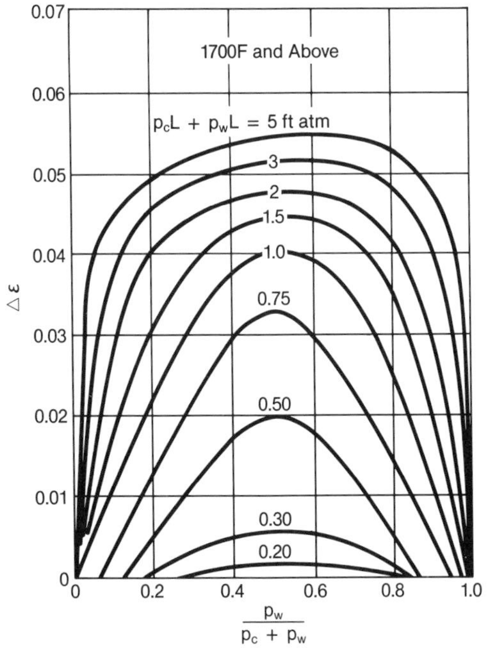

Fig. 10 Radiation heat transfer correction factor associated with mixtures of water vapor and carbon dioxide.[6]

Experimental studies have confirmed that a flow field can be divided into two zones: a viscous zone adjacent to the surface and a nonviscous zone removed from the heat transfer surface. The viscous, heated zone is termed the boundary layer region. The hydrodynamic boundary layer is defined as the flow thickness at which the local velocity reaches 99% of the velocity far from the wall.

At the entrance of a pipe or duct, the boundary layer begins to grow; this flow portion is called the developing region. Downstream, when the viscous region fills the pipe core or grows to a maximum, the flow is termed fully developed. Developing region heat transfer coefficients are larger than the fully developed values. In many applications it is sufficient to assume that the hydrodynamic and thermal boundary layers start to grow at the same location, although this is not always the case.

Flow over a body (around a circular cylinder) is termed external flow, while flow inside a confined region, like a pipe or duct, is termed internal flow.

Natural or free convection

A fluid at rest, exposed to a heated surface, will be at a higher temperature and lower density than the surrounding fluid. The differences in density, because of this difference in temperature, cause the lighter, warmer fluid elements to circulate and carry the heat elsewhere. The complex relationships governing this type of convective heat transfer are covered extensively in other texts.[1] Experimental studies have confirmed that the main dimensionless parameters governing free convection are the Grashof and Prandtl numbers:

$$\mathrm{Gr} = \frac{g\beta(T_s - T_f)\rho^3 L^3}{\mu^3} \qquad \textbf{(38)}$$

$$\mathrm{Pr} = \frac{c_p \mu}{k} \qquad \textbf{(39)}$$

The Grashof number is a ratio of the buoyant to viscous forces. The Prandtl number is a ratio of the viscous to thermal diffusivity of a gas or fluid. The product, Gr Pr, is also called the Rayleigh number, Ra.

In boiler system designs, air and flue gases are the important free convection heat transfer media. For these designs, the equation for the convective heat transfer coefficient h is:

$$h = C\ (T_s - T_f)^{1/3} \qquad \textbf{(40)}$$

This correlation is applicable when the Rayleigh number, Ra, is greater than 10^9, which is generally recognized as the transition between laminar and turbulent flow. Values of the constant C in the equation are listed below:

Geometry	Btu / h ft² F^4/3	W / m² K^4/3
Horizontal plate facing upward	0.22	1.52
Vertical plates or pipes more than 1 ft (0.3 m) high	0.19	1.31
Horizontal pipes	0.18	1.24

The correlation generally produces convective heat transfer coefficients in the range of 1 to 5 Btu/h ft² F (5.68 to 28.39 W/m² K).

Forced convection

Dimensionless numbers Forced convection implies the use of a fan, pump or natural draft stack to induce fluid motion. Studies of many heat transfer systems and numerical simulation of some simple geometries confirm that fluid flow and heat transfer data may be correlated by dimensionless numbers. Using these principles, scale models enable designers to predict field performance. For simple geometries, a minimum of dimensionless numbers is needed for modeling. More complex scaling requires more dimensionless groups to predict unit performance.

The Reynolds number is used to correlate flow and heat transfer in closed conduits. It is defined as:

$$\mathrm{Re} = \frac{\rho V L}{\mu} \tag{41}$$

This dimensionless group represents the ratio of inertial to viscous forces.

The Reynolds number is only valid for a continuous fluid filling the conduit. The use of this parameter generally assumes that gravitational and intermolecular forces are negligible compared to inertial and viscous forces.

The characteristic length, termed equivalent diameter, is different for circular and noncircular conduits. For circular conduits, the inside diameter is used. For nonround ducts, the equivalent diameter becomes:

$$D_H = 4 \times \frac{\text{Flow cross–sectional area}}{\text{Wetted perimeter}} \tag{42}$$

This approach, used to compare dynamically similar fluids in geometrically similar conduits of different size, yields equal Reynolds numbers for the flows considered.

At low velocities, the viscous forces are strong and laminar flow predominates, while at higher velocities, the inertial forces dominate and there is turbulent flow. In closed conduits, such as pipes and ducts, the transition to turbulent flow occurs near Re = 2000. The generally accepted range for transition to turbulent flow under common tube flow conditions is $2000 < \mathrm{Re} < 4000$.

For fluid flow over a flat external surface, the characteristic length for the Reynolds number is the surface length in the direction of the flow, x. Transition to turbulence is generally considered for $\mathrm{Re} \geq 10^5$. In the case of flow over a tube, the outside diameter, D, is the characteristic length. In tube bundles with crossflow, transition generally occurs at $\mathrm{Re} > 10^2$.

Experimental studies have confirmed that the convective heat transfer coefficient can be functionally characterized by the following dimensionless groups:

$$\mathrm{Nu} = f\,(\mathrm{Re}, \mathrm{Pr}) \tag{43}$$

where Nu is the Nusselt number, Re is the Reynolds number and Pr is the Prandtl number.

The Nusselt number, a ratio of the wall temperature gradient to a reference gradient is defined as follows:

$$\mathrm{Nu} = \frac{hL}{k} \tag{44}$$

The previously discussed Prandtl number, representing a ratio of the viscous to thermal diffusivity, ν/α, is a measure of the relative magnitude of momentum and thermal diffusion in the fluid. For air and flue gases, Pr < 1.0 and the thermal boundary layer is thicker than the hydrodynamic boundary layer.

In the literature, correlations are also presented using other dimensionless groups; the Peclet and Stanton numbers are the most common. The Peclet number is defined as follows:

$$\mathrm{Pe} = \mathrm{Re}\ \mathrm{Pr} \tag{45}$$

The Stanton number is defined in terms of the Nusselt, Reynolds and Prandtl numbers:

$$\mathrm{St} = \frac{\mathrm{Nu}}{\mathrm{Re}\ \mathrm{Pr}} \tag{46}$$

Laminar flow inside tubes For heating or cooling viscous liquids in horizontal or vertical tubes, the heat transfer coefficient, or film conductance, can be determined by the following equation:[6]

$$\mathrm{Nu} = 1.86\left(\mathrm{Re}\ \mathrm{Pr}\frac{\mathrm{D}}{\mathrm{L}}\right)^{1/3}\left(\frac{\mu}{\mu_s}\right)^{0.14} \tag{47}$$

or

$$h = 1.86\frac{k}{D}\left(\frac{GD}{\mu}\ \frac{c_p\mu}{k}\ \frac{D}{L}\right)^{1/3}\left(\frac{\mu}{\mu_s}\right)^{0.14} \tag{48}$$

where the parameter $G = \rho V$ is defined as the mass flux or mass flow rate per unit area. The ratio of viscosities (μ/μ_s) is a correction factor that accounts for the difference in surface and fluid temperatures.

For low viscosity fluids, such as water and gases, a more complex equation is required to account for the effects of natural convection at the heat transfer surface. This refinement is of little interest in industrial practice because it is generally impractical to use water and gases in laminar flow.

Turbulent flow Studies of turbulent flow indicate several well defined regions as shown in Fig. 11. Next to the heat transfer surface is a very thin laminar flow region, less than 0.2% of the characteristic length, where the heat flow to or from the surface is by molecular conduction. The next zone, known as the buffer layer, is less than 1% of the characteristic length and is a mixture of laminar and turbulent flow. Here the heat is transferred by a combination of convection and conduction. In the turbulent core, which comprises roughly 98% of the cross-section, heat is transferred mainly by convection.

In turbulent flow, the local but chaotic motion of the fluid causes axial and radial motion of fluid elements. This combination of motions sets up eddies, or local whirling motions, augmenting the heat transfer from the core to the laminar sublayer. The laminar flow in the sublayer and the laminar component in the buffer layer act as a barrier, or film, to the heat transfer process. Increasing the fluid velocity has been found to decrease this film

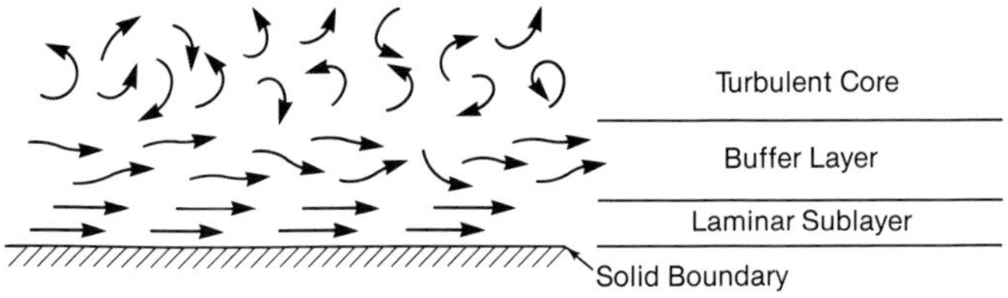

Fig. 11 Structure of a turbulent flow field near a solid boundary.

thickness, reducing the resistance to heat transfer.

Turbulent flow in tubes The distance required to obtain hydrodynamically and thermally fully developed turbulent flow is shorter than that for laminar flow. The flow length needed to achieve hydrodynamically fully developed conditions is variable and depends upon the specific Reynolds number (operating conditions) and surface geometry. It typically varies from 6 to 20 diameters (x/D). Fully developed thermal flow for gases and air, important in boiler analysis, occurs at similar x/D ratios. However, for liquids, the ratio is somewhat higher and increases with the Prandtl number.

Extensive research data using low viscosity gases and liquids have been correlated. Colburn[7] recommends the following equation in terms of a fully developed Nusselt number:

$$Nu_{fd} = 0.023\ Re^{0.8}\ Pr^{0.4} \tag{49}$$

Equation 49 applies to gases and liquids in the range $0.5 < Pr < 100$, which covers all fluids in boiler analysis. If the conditions are not fully developed, the correlation is corrected as shown below:[3]

$$Nu = Nu_{fd}\ [1 + (D/x)^{0.7}] \tag{50}$$

with the stipulation that $2 \leq x/D \leq 20$. McAdams[6] has suggested using Equation 49 with all of the properties evaluated at the film temperature, except the specific heat. The film temperature (T_f) is defined as the arithmetic mean temperature between the wall temperature (T_w) and the bulk fluid temperature (T_b) : $T_f = (T_w + T_b)/2$. Figs. 12 to 16 display factors needed to evaluate Equation 49 when it is grouped as follows with the temperature factor to account for property dependence:

$$h_l = \left[0.023 \frac{G^{0.8}}{D^{0.2}}\right] \left[\frac{c_p^{0.4} k^{0.6}}{\mu^{0.4}}\right] \left[\frac{T_b}{T_f}\right]^{0.8} \tag{51}$$

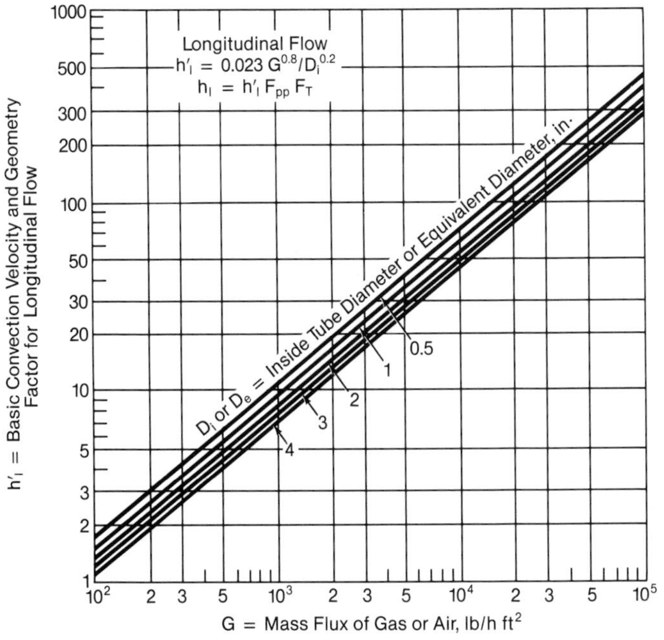

Fig. 12 Basic longitudinal flow convection velocity and geometry factor, h'_l, for air, gas or steam (English units only).

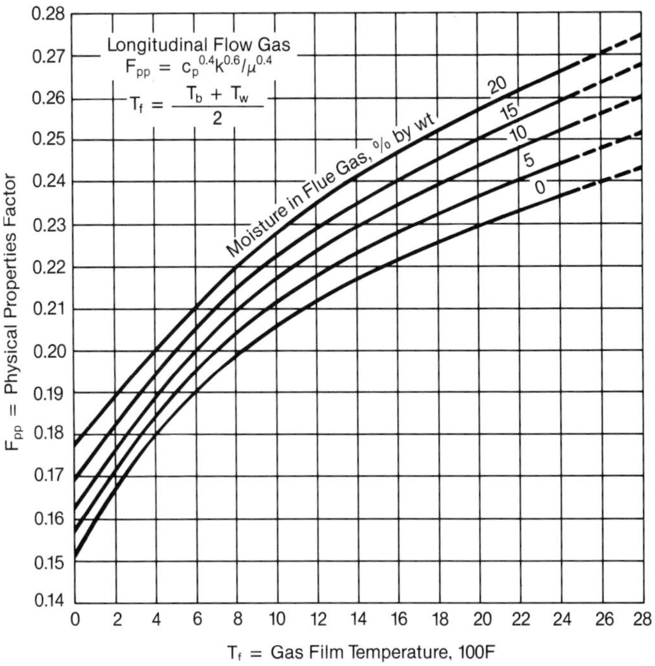

Fig. 13 Effect of film temperature, T_f, and moisture on the physical properties factor, F_{pp}, for gas in longitudinal flow (English units only).

or

$$h_l = h'_l\ F_{pp}\ F_T \tag{52}$$

Note that if F_{pp} can not be obtained from Fig. 15, it can be obtained from the ASME Steam Table[5] using c_p, k and μ at the film temperature, T_f. The temperature factor, F_T, is required to convert the properties from a bulk to film temperature basis.

Turbulent flow around tubes The most important boiler application of convection is heat transfer from the combustion gases to the tubular surfaces in the convection passes. Perhaps the most complete and authoritative research on heat transfer of tubes in crossflow was completed in an extensive Babcock & Wilcox (B&W) program. The following correlation was found:

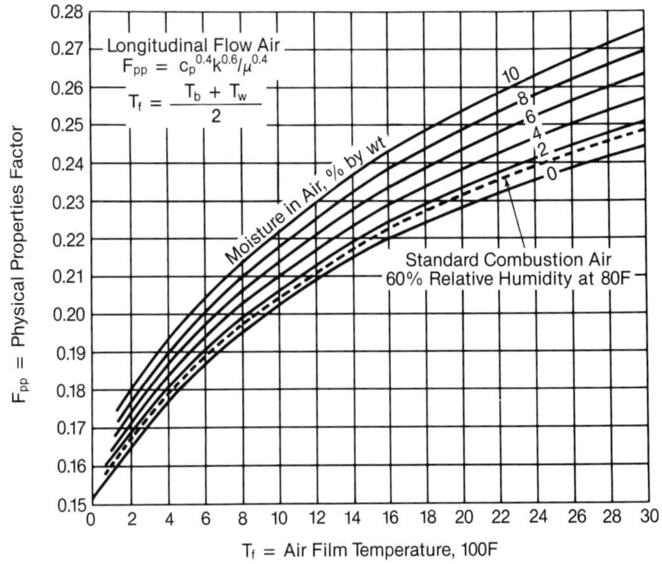

Fig. 14 Effect of film temperature, T_f, and moisture on the physical properties factor, F_{pp}, for air in longitudinal flow (English units only).

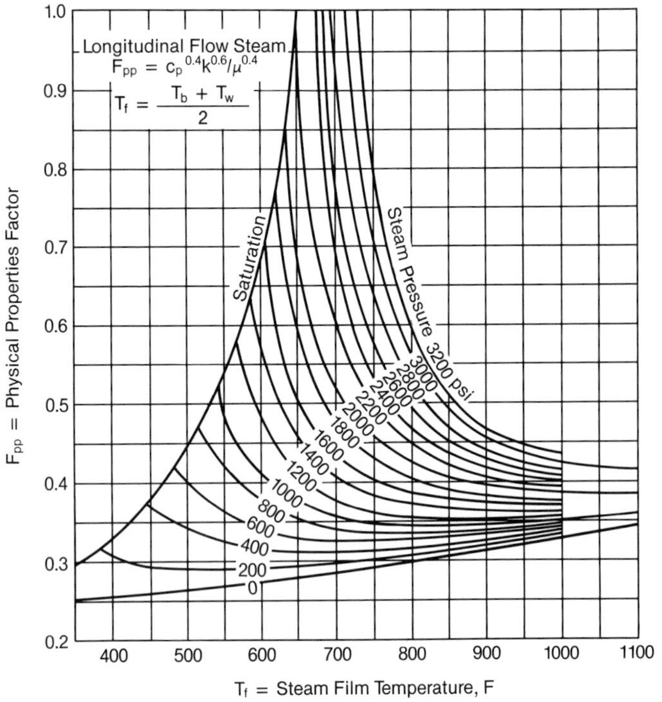

Fig. 15 Effect of film temperature, T_f, and pressure on the physical properties factor, F_{pp}, for steam in longitudinal flow (English units only).

$$\text{Nu} = 0.287\ \text{Re}^{0.61}\ \text{Pr}^{0.33}\ F_a \tag{53}$$

The last term is an arrangement factor, F_a, correcting the values from the base configuration. The equation applies to heating and cooling of fluids flowing over clean tubes in crossflow. Using the parametric groupings shown below:

$$h_c = \left[\frac{0.287\ G^{0.61}}{D^{0.39}}\right]\left[\frac{c_p^{0.33}\ k^{0.67}}{\mu^{0.28}}\right]F_a \tag{54}$$

This equation can be rewritten in the form:

$$h_c = h_c{}'F_{pp}\ F_a\ F_d \tag{55}$$

Figs. 17 to 21 define the various factors. The physical properties factor, F_{pp}, is similar to the one previously defined. The arrangement factor, F_a, depends on the tube arrangement, the ratio of tube spacing to diameter and the Reynolds number.

Values for F_a are given in Fig. 20 for various conditions. The mass flux or mass flow per unit area, G, in Equations 53 and 54 and the Reynolds numbers used in Fig. 20 are calculated using the minimum free area for fluid flow.

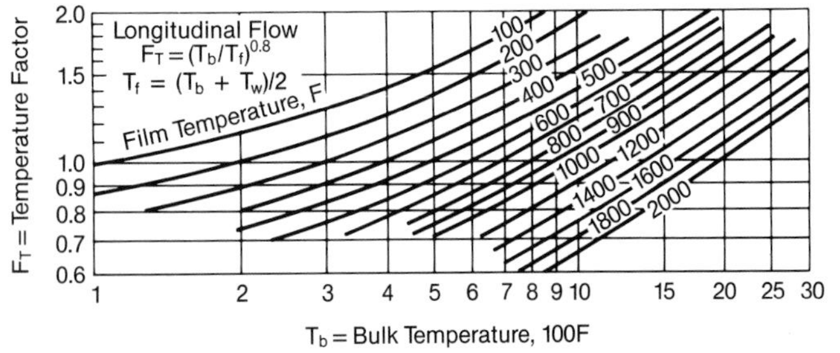

Fig. 16 Temperature factor F_T, for converting mass velocity from bulk to film basis; longitudinal flow air, gas or steam.

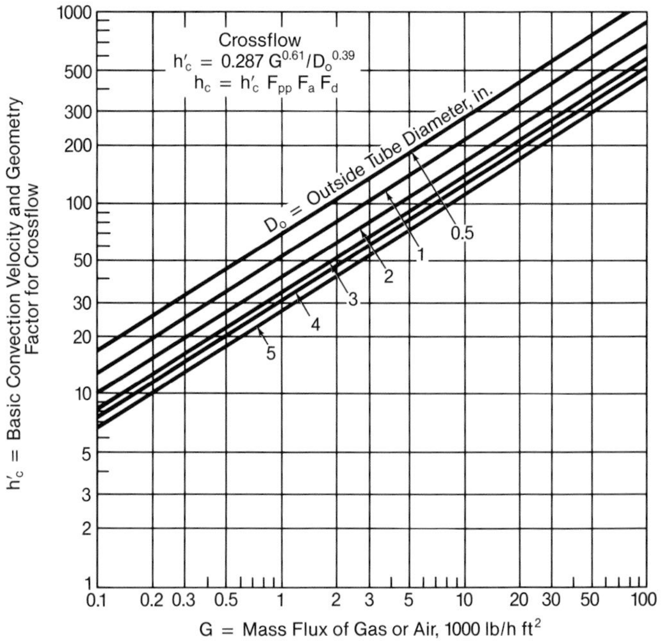

Fig. 17 Basic crossflow convection velocity and geometry factor, h'_c, for gas or air (English units only).

The value of the film conductance, h_c, in Equation 54 applies to banks of tubes which are at least ten rows deep in the direction of gas flow. For undisturbed flow (flow that is straight and uninterrupted for at least 4 ft (1.2 m) before entering a tube bank) approaching a bank of less than ten rows, the film conductance must be multiplied by a correction factor, F_d. Known as the depth factor, F_d is unity when the tube bank is preceded by a bend, screen, or damper; values are shown in Fig. 21.

Although Equations 49 and 51 were developed for flow inside tubes, they can be rewritten for external flow parallel to tubes. An equivalent diameter, D_e, for flow parallel to a bank of circular tubes arranged on rectangular spacing, is used in these equations:

$$D_e = \frac{4(L_1 L_2 - 0.785D^2)}{\pi D} \tag{56}$$

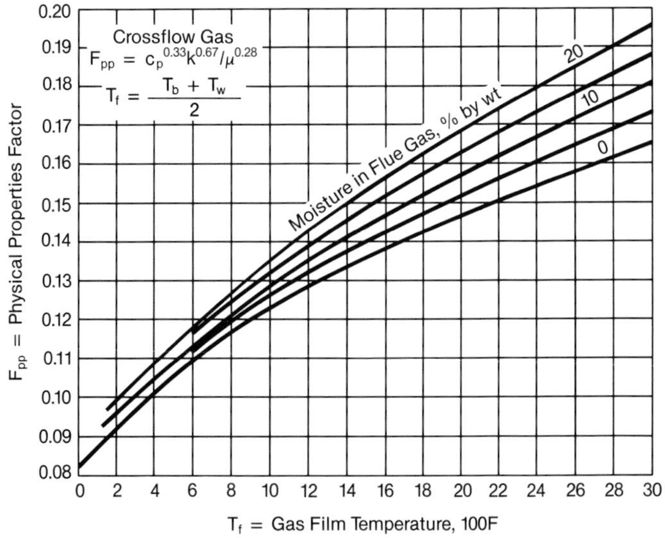

Fig. 18 Effect of film temperature, T_f, and moisture on the physical properties factor, F_{pp}, for gas in crossflow (English units only).

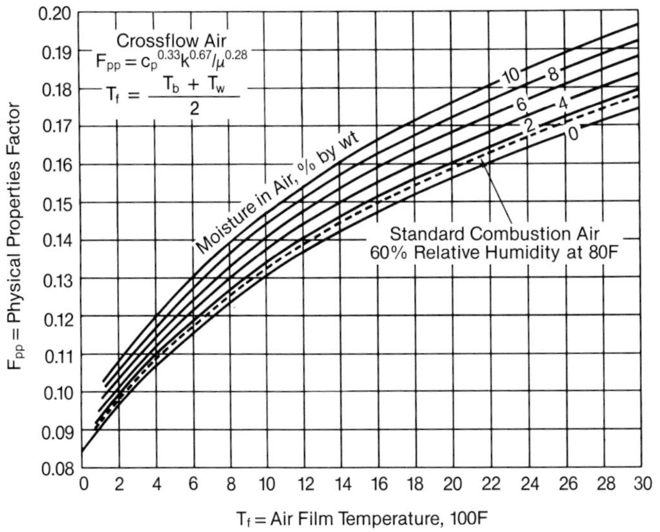

Fig. 19 Effect of film temperature, T_f, and moisture on the physical properties factor, F_{pp}, for air in crossflow (English units only).

where D is the tube outside diameter and L_1 and L_2 are the tube pitches.

General heat transfer topics

Heat exchangers

Boiler systems contain many heat exchangers. In these devices, the fluid temperature changes as the fluids pass through the equipment. With an energy balance specified between two locations, 1 and 2:

$$q = \dot{m}\, c_p (T_2 - T_1) \tag{57}$$

the change in fluid temperature can be calculated:

$$T_2 = T_1 + (q / \dot{m}\, c_p) \tag{58}$$

It is therefore appropriate to define a mean effective temperature difference governing the heat flow. This difference is determined by performing an energy balance on the energy lost by the hot fluid and that energy gained by the cold fluid. An equation of the form:

$$q = UA\, F\, \Delta T_{\text{LMTD}} \tag{59}$$

is obtained where the parameters U, A and F define the overall heat transfer coefficient, surface area, and arrangement factor, respectively. The term ΔT_{LMTD}, known as the log mean temperature difference, is defined as:

$$\Delta T_{\text{LMTD}} = \frac{\Delta T_1 - \Delta T_2}{\ell n(\Delta T_1 / \Delta T_2)} \tag{60}$$

ΔT_1 is the initial temperature difference between the hot and cold fluids (or gases), while ΔT_2 defines the final temperature difference between these media. The parameter U in Equation 59 defines the overall heat transfer coefficient for clean surfaces and represents the thermal resistance between the hot and cold fluids:

$$\frac{1}{UA_{\text{clean}}} = \frac{1}{h_i A_i} + R_w + \frac{1}{h_o A_o} \tag{61}$$

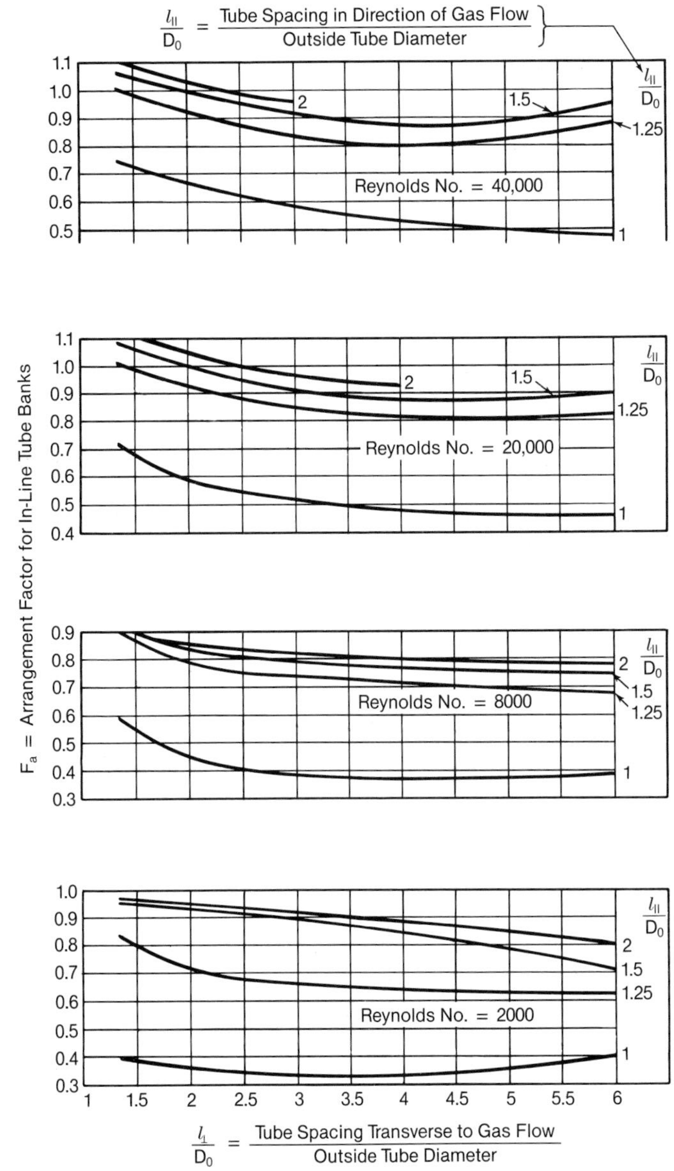

Fig. 20 Arrangement factor, F_a, as affected by Reynolds number for various in-line tube patterns, crossflow gas or air.

For surfaces that are fouled, the equation is written:

$$\frac{1}{UA} = \frac{R_{f,i}}{A_i} + \frac{1}{UA_{\text{clean}}} + \frac{R_{f,o}}{A_o} \tag{62}$$

where $R_{f,i}$ is the reciprocal effective heat transfer coefficient of the fouling on the inside surface, $(1/UA)$ is the thermal resistance and $R_{f,o}$ is the reciprocal heat transfer coefficient of the fouling on the outside surface. Estimates of overall heat transfer coefficients and fouling factors are listed in Tables 10 and 11. Actual fouling factors are site specific and depend on water chemistry and other deposition rate factors. Overall heat transfer coefficients can be predicted using: 1) the fluid conditions on each side of the heat transfer surface with either Equation 49 or 53, 2) the known materials of the heat transfer surface, and 3) the fouling factors listed in Table 11. Often the heat exchanger tube resistance is small compared to the surface resistances and can be neglected, leading to the following equation for a clean surface:

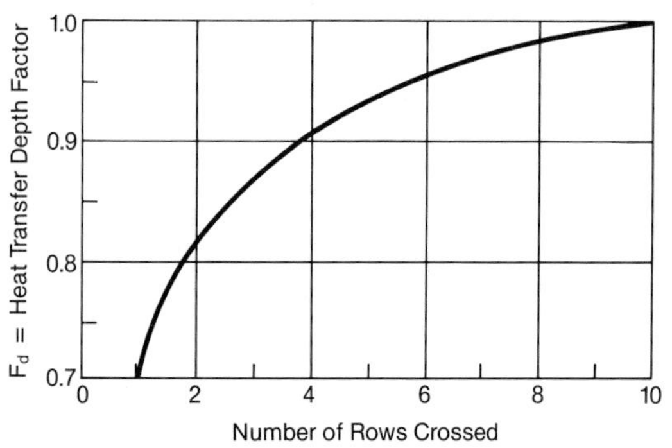

Fig. 21 Heat transfer depth factor for number of tube rows crossed in convection banks. (F_d = 1.0 if tube bank is immediately preceded by a bend, screen or damper.)

$$U = \frac{h_i h_o}{h_i + (h_o D_o / D_i)} \tag{63}$$

The difficulty in quantifying fouling factors for gas-, oil- and coal-fired units has led to use of a *cleanliness factor*. This factor provides a practical way to provide extra surface to account for the reduction in heat transfer due to fouling. In gas-fired units, experience indicates that gas side heat transfer coefficients are higher as a result of the cleanliness of the surface. In oil- and coal-fired units that are kept free of slag and deposits, a lower value is used. For units with difficult to remove deposits, values are reduced further. (See Chapter 21.)

There are three general heat transfer arrangements: parallel flow, counterflow and crossflow, as shown in Fig. 22. In parallel flow, both fluids enter at the same relative location with respect to the heat transfer surface and flow in parallel paths over the heating surface. In counterflow, the two fluids enter at opposite ends of the heat transfer surface and flow in opposite directions over the surface. This is the most efficient heat exchanger although it can also lead to the highest tube wall metal temperatures. In crossflow, the paths of the two fluids are, in general, perpendicular to one another.

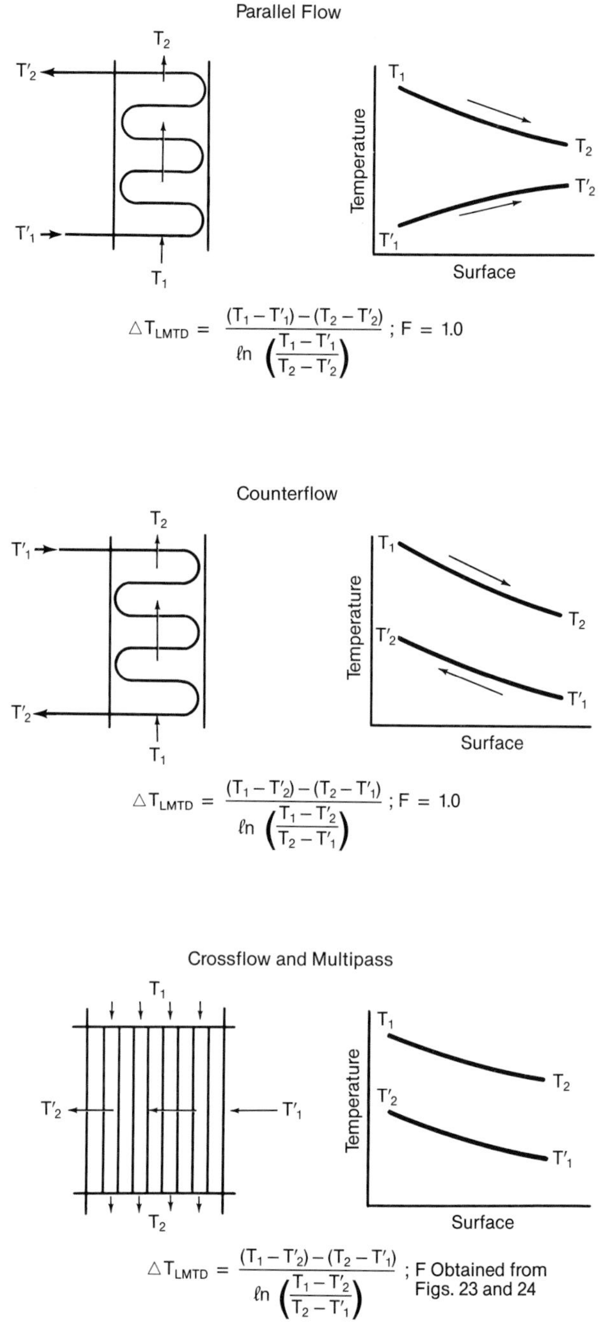

Fig. 22 Mean effective temperature difference.

Fig. 22 shows the flow arrangements and presents Equation 60 written specifically for each case. The arrangement factor, *F*, is 1.0 for parallel and counterflow cases. For crossflow and multi-pass arrangements. The correction factors are shown in Figs. 23 and 24.

Table 10
Approximate Values of Overall Heat Transfer Coefficients

Physical Situation	Btu/h ft^2 F	W/m^2 K
Plate glass window	1.10	6.20
Double plate glass window	0.40	2.30
Steam condenser	200 to 1000	1100 to 5700
Feedwater heater	200 to 1500	1100 to 8500
Water-to-water heat exchanger	150 to 300	850 to 1700
Finned tube heat exchanger, water in tubes, air across tubes	5 to 10	30 to 55
Water-to-oil heat exchanger	20 to 60	110 to 340
Steam-to-gas	5 to 50	30 to 300
Water-to-gas	10 to 20	55 to 200

Table 11
Selected Fouling Factors

Type of Fluid	h ft^2 F/Btu	m^2 K/W
Sea water above 125F (50C)	0.001	0.0002
Treated boiler feedwater above 125F (50C)	0.001	0.0002
Fuel oil	0.005	0.0010
Alcohol vapors	0.0005	0.0001
Steam, non-oil bearing	0.0005	0.0001
Industrial air	0.002	0.0004

Extended surface heat transfer

The heat absorption area in boilers can be increased using longitudinally and circumferentially finned tubes. Finned, or extended, tube surfaces are used on the flue gas

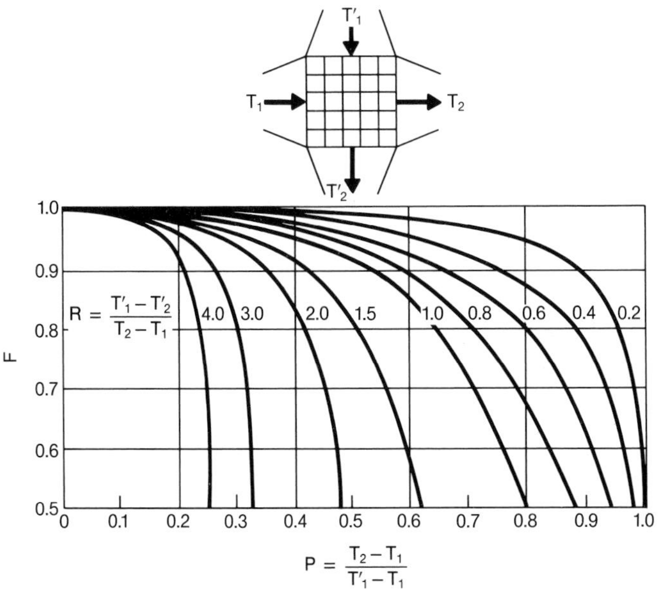

Fig. 23 Correction factors for a single-pass, crossflow heat exchanger with both fluids unmixed.

side. In regions prone to fouling, the fins must be spaced to permit cleaning. Experimental data on actual finned or extended surfaces are preferred for design purposes; the data should be collected at conditions similar to those expected to be encountered. However, in place of these data, the method by Schmidt[8] generally describes the heat transfer across finned tubes. It is based on heat transfer to the underlying bare tube configuration, and it treats the tube as if it has zero fin height. Schmidt's correlation for the gas side conductance to tubes with helical, rectangular, circular, or square fins is as follows:

$$h_f = h_c Z\left\{1-(1-\eta_f)\left(\frac{S_f}{S}\right)\right\} \tag{64}$$

where h_c is the heat transfer coefficient of the bare tubes in crossflow defined by Equations 53 and 54, and Z is the geometry factor defined as:

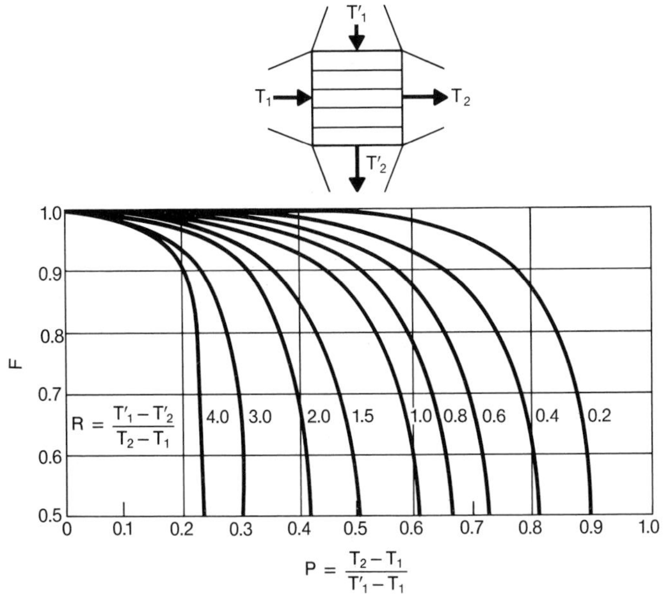

Fig. 24 Correction factors for a single-pass, crossflow heat exchanger with one fluid mixed and the other unmixed (typical tubular air heater application).

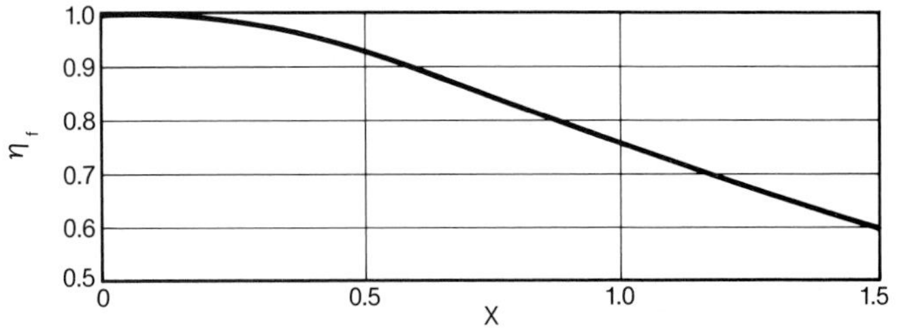

Fig. 25 Fin efficiency as a function of parameter *X*.

$$Z = 1-0.18\left(\frac{L_h}{L_t}\right)^{0.63} \tag{65}$$

S_f represents the fin surface area including both sides and the peripheral area, while S represents the exposed bare tube surface between the fins plus the fin surface, S_f. The ratio L_h/L_t is the fin height divided by the clear spacing between fins. Fin efficiency, η_f, is shown in Fig. 25 as a function of the parameter X, defined as:

$$X = L_h\sqrt{2Zh_c/(k_f L_t)} \tag{66}$$

for helical fins, and

$$X = r\,Y\sqrt{2Zh_c/(k_f L_t)} \tag{67}$$

for rectangular, square or circular fins. The parameter Y is defined in Fig. 26.

The overall conductance can be written:

$$\frac{1}{UA} = \frac{1}{C_f A_o h_{f,o}} + R_w + \frac{1}{A_i h_{c,i}} \tag{68}$$

The parameter C_f is the surface cleanliness factor as defined previously and in Chapter 21.

NTU method

There are design situations for which the performance of the heat exchanger is known, but the fluid temperatures are not. This occurs when selecting a unit for which operating flow rates are different than those previously tested. The outlet temperatures can only be found by trial and error using the methods previously presented. These applications are best handled by the net transfer unit (NTU) method that uses the heat exchanger effectiveness (see Reference 9).

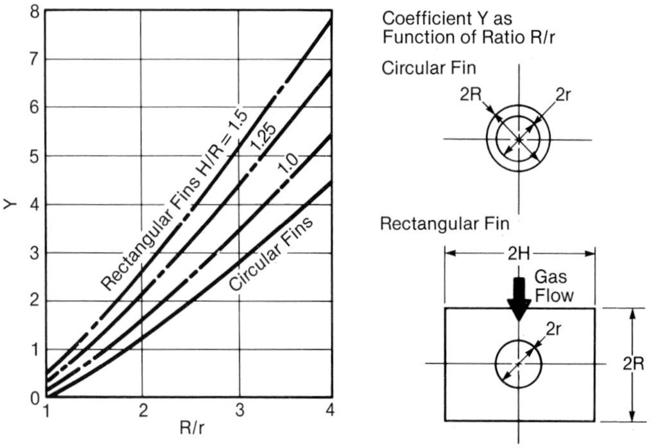

Fig. 26 Coefficient *Y* as a function of ratio *R/r* for fin efficiency.

Heat transfer in porous materials

Porosity is an important factor in evaluating the effectiveness of insulation materials. In boiler applications, porous materials are backed up by solid walls or casings, so that there is minimal flow through the pores.

Heat flow in porous insulating materials occurs by conduction through the material and by a combination of conduction and radiation through the gas-filled voids. In most refractory materials, the Grashof-Prandtl (Raleigh) number is small enough that negligible convection exists although this is not the case in low density insulations [< 2 lb/ft^3 (32 kg/m^3)]. The relative magnitudes of the heat transfer mechanisms depend, however, on various factors including porosity of the material, gas density and composition filling the voids, temperature gradient across the material, and absolute temperature of the material.

Analytical evaluation of the separate mechanisms is complex, but recent experimental studies at B&W have shown that the effective conductivity can be approximated by:

$$k_e = a + bT + cT^3 \tag{69}$$

Experimental data can be correlated through this form. The heat flow is calculated using Equation 1; k is replaced by k_e and T is the local temperature in the insulation.

In high temperature applications, heat transfer across the voids occurs mainly by radiation and the third term of Equation 69 dominates. In low temperature applications, heat flow by conduction dominates and the first two terms of Equation 69 are controlling.

Film condensation

When a pure saturated vapor strikes a surface of lower temperature, the vapor condenses and a film is formed on the surface. If the film flows along the surface because of gravity alone and a condition of laminar flow exists throughout the film thickness, then heat transfer through this film is by conduction only. As a result, the thickness of the condensate film has a direct effect on the quantity of heat transferred. The film thickness, in turn, depends on the flow rate of the condensate. On a vertical surface, because of drainage, the thickness of the film at the bottom will be greater than at the top. Film thickness increases as a plate surface is inclined from the vertical position.

As the film temperature increases, its thickness decreases primarily due to increased drainage velocity. In addition, the film thickness decreases with increasing vapor velocity in the direction of drainage.

Mass diffusion and transfer

Heat transfer can also occur by diffusion and mass transfer. When a mixture of a condensable vapor and a noncondensable gas is in contact with a surface that is below the dew point of the mixture, some condensation occurs and a film of liquid is formed on the surface. An example of this phenomenon is the condensation of water vapor on the outside of a metal container. As vapor from the main body of the mixture diffuses through the vapor-lean layer, it is condensed on the cold surface as shown in Fig. 27. The rate of condensation is therefore governed by the laws of gas diffusion. The heat transfer is controlled by the laws of conduction and convection.

The heat transferred across the liquid layer must equal

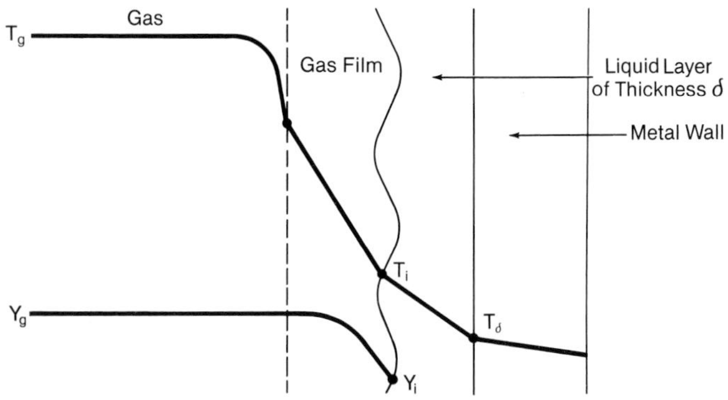

Fig. 27 Simultaneous heat and mass transfer in the dehumidification of air.

the heat transferred across the gas film plus the latent heat given up at the gas-liquid interface due to condensation of the mass transferred across the gas film. An equation relating the mass transfer is:

$$h_\delta (T_i - T_\delta) = h_g (T_g - T_i) + K_y H_{fg} (Y_g - Y_i) \tag{70}$$

where T and Y define the temperatures and concentrations respectively identified in Fig. 27, h_δ is the heat transfer coefficient across the liquid film, h_g is the heat transfer coefficient across the gas film, and K_y is the mass transfer coefficient. H_{fg} is the latent heat of vaporization.

Heat transfer due to mass transfer is important in designing cooling towers and humidifiers, where mixtures of vapors and noncondensable gases are encountered.

Evaporation or boiling

The phenomenon of boiling is discussed in Chapters 1 and 5, where the heat transfer advantages of nucleate boiling are noted. Natural-circulation fossil fuel boilers are designed to operate in the boiling range. In this range, the heat transfer coefficient varies from 5000 to 20,000 Btu/h ft^2 F (28,392 to 113,568 W/m^{2}K). This is not a limiting factor in the design of fossil fuel boilers provided scale and other deposits are prevented by proper water treatment, and provided the design avoids critical heat flux (CHF) phenomena. (See Chapter 5.)

In subcritical pressure once-through boilers, water is completely evaporated (dried out) in the furnace wall tubes which are continuous with the superheater tubes. These units must be designed for subcooled nucleate boiling, nucleate boiling, and film boiling, depending on fluid conditions and expected maximum heat absorption rates.

Fluidized-bed heat transfer

The heat transfer in gas-fluidized particle beds used in some combustion systems is complex, involving particle-to-surface contact, general convection and particle-to-surface thermal radiation. Correlations for heat transfer in fluidized-bed boilers are summarized in Chapters 16 and 29.

Numerical modeling

Advances in mainframe and workstation computers have enabled B&W to mathematically model complex heat transfer systems. These models provide a tool for analyzing thermal systems inexpensively and rapidly. Although empirical methods and extensive equipment testing continue to provide information to designers, numerical simulation of boiler components, e.g., membrane walls, will become increasingly important as computer technology evolves.

Conduction modeling using the electrical analogy

The energy equation for steady-state heat flow was previously defined as Equation 19 and is more generally written as Equation 20. Solutions of these equations for practical geometries are difficult to obtain except in idealized situations. Numerical methods permit the consideration of additional complex effects including irregular geometries, variable properties, and complex boundary conditions. Conduction heat flow through boiler membrane walls, refractory linings with several materials, and steam drum walls are several applications for these methods. The approach is to divide the heat transfer system into subvolumes called control volumes (Fig. 28). (See References 1, 2, 3, 9 and 10.) The governing equation is integrated, or averaged, over the subvolume, leading to an expression of the form:

$$\frac{T_e - T_p}{R_{pe}} + \frac{T_w - T_p}{R_{pw}} + q_p''' V_p = c_p \frac{T_p - T_p^o}{\Delta t} \qquad (71)$$

where the subscripts denote the neighbor locations as points on a compass. If the steady-state solution is desired, the right hand side of the equation, $c_p\,(T_p - T_p^o)\,/\,\Delta t$, which accounts for changes in stored energy is set to zero. A solution is then obtained numerically. Equation 71 is a discrete form of the continuous differential equation. The modeled geometry is subdivided and equations of this form are determined for each interior volume. The electrical analogy of the equation is apparent. First, each term is an expression of heat flow into a point using Fourier's law by Equation 1 and, second, Kirchhoff's law for circuits, Equation 17, is used to determine the net flow of heat into any point. The application of the electrical analogy is straightforward for any interior volume once the subvolumes are defined. At the boundaries, temperature or heat flow is defined.

For unsteady-state problems, a sequence of solutions is obtained for the time interval Δt, with T_p^o being the node temperature at the beginning of the interval and T_p being the temperature at the end of the interval. References 1, 2, 3, 9 and 10 explore these models in depth. A variety of computer codes is commercially available to perform the analysis and display the results.

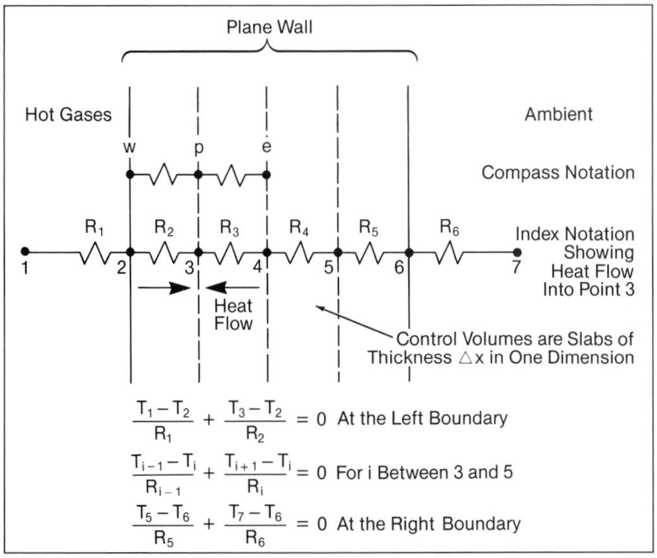

Fig. 28 Control volume layout for a plane wall with notation for heat flow to node 3 and steady-state solution.

Furnace modeling

Fossil fuel-fired boiler designers need to evaluate furnace wall temperature and heat flux, flue gas composition and temperature, and furnace exit gas temperature. These parameters are required to determine materials and their limits, to size heat transfer surface, and to ensure fuel conversion. Three approaches are generally used in combination to provide a solution:

1. empirical design methods based upon operating experience; field data, with interpolation/extrapolation based upon fundamentals,
2. semi-empirical methods, which combine the results of physical flow models and evaluate the flow field with multi-zone computations, and
3. numerical furnace modeling, which solves the fundamental equations for reaction kinetics, fluid flow, and heat transfer.

Empirical methods Considering the fuel type, firing rate and furnace configuration, empirical methods as illustrated in Fig. 29 have long been used to predict local absorption rates in the furnace. These methods, although

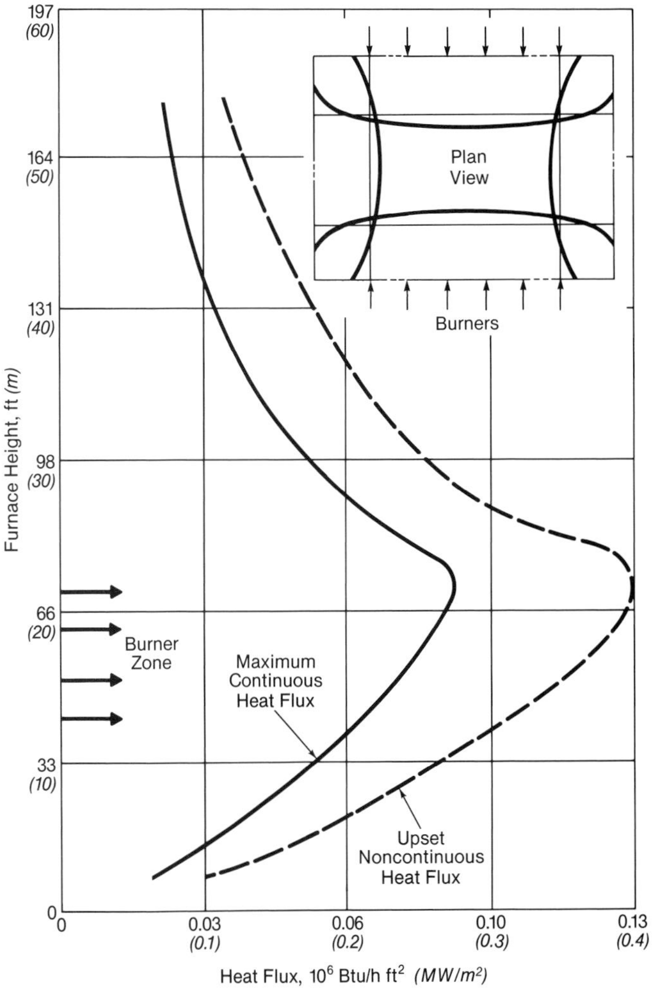

Fig. 29 Heat flux distributions for vertical and horizontal furnace walls.

largely empirical, contain engineering models which are based on fundamentals. Data and operating experience are used to tune the models used in the design envelope. Fig. 29 shows typical heat flux distributions for vertical and horizontal furnace walls.

Deviations in the heat flux distribution are caused by unbalanced firing, variations in tube surface condition, differences in slagging, load changes, sootblower operation and other variations in unit operation. A typical upset heat flux distribution is shown in Fig. 29. These upset factors are typically a function of vertical/horizontal location, firing method and fuel, and furnace configuration. They are derived from operating experience.

The heat flux applied to the tubes in the furnace wall is also nonuniform in the circumferential direction. As shown in Fig. 30, the membrane wall is exposed to the furnace on one side while the opposite side is typically insulated to minimize heat loss. The resulting heat flux distribution depends upon the tube outside diameter, wall thickness, and spacing, as well as the web thickness and materials. The fluid temperature and inside heat transfer coefficient have secondary effects. This distribution can be evaluated using electrical analogy methods.

To correlate data and calculations for different furnaces, methods for comparing the relative effectiveness of different furnace wall surfaces are needed. The effectiveness and spacing of tubes compared to a completely water-cooled surface are shown in Fig. 31. A wall of flat-studded tubes is considered completely water-cooled. The effectiveness of expected ash covering, compared with completely water-cooled surfaces, can also be estimated. The entire furnace envelope can then be evaluated in terms of equivalent cold surface.

The heat energy supplied by the fuel and by the preheated combustion air, corrected for unburned combustible loss, radiation loss, and moisture in the fuel, may be combined into a single variable, known as heat available. The heat available divided by the equivalent flat projected water-cooled furnace enclosure surface area is called the *furnace heat release rate*. The heat available divided by the furnace volume is called the *furnace liberation rate*. Furnace exit gas temperature is primarily a function of heat release rate rather than liberation rate.

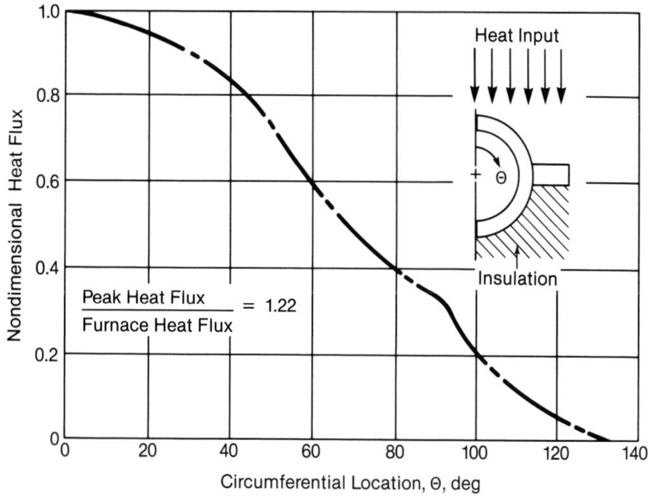

Fig. 30 Typical circumferential heat flux distribution for a furnace membrane wall panel tube.

The approximate relation of exit gas temperature to heat release rate for three typical fuels is given in Fig. 32.

Furnace exit gas temperatures and related heat absorption rates, as functions of furnace heat release rate for most pulverized coal-fired furnaces, lie within the shaded bands shown in Figs. 33 and 34. The limits indicated serve only as a general guide and may vary due to

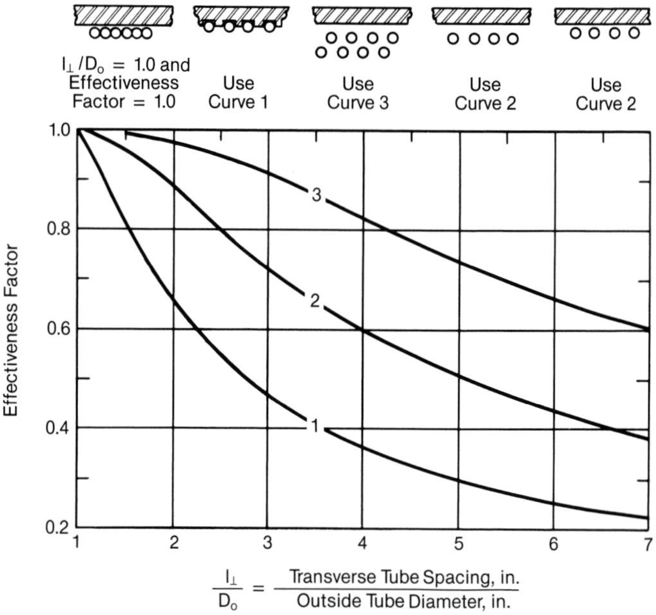

Fig. 31 Furnace wall area effectiveness factor (1.0 for completely water-cooled surface). A reduced area (equivalent cold surface) is determined from these curves for walls not completely water-cooled. (Adapted from Hottel.[4])

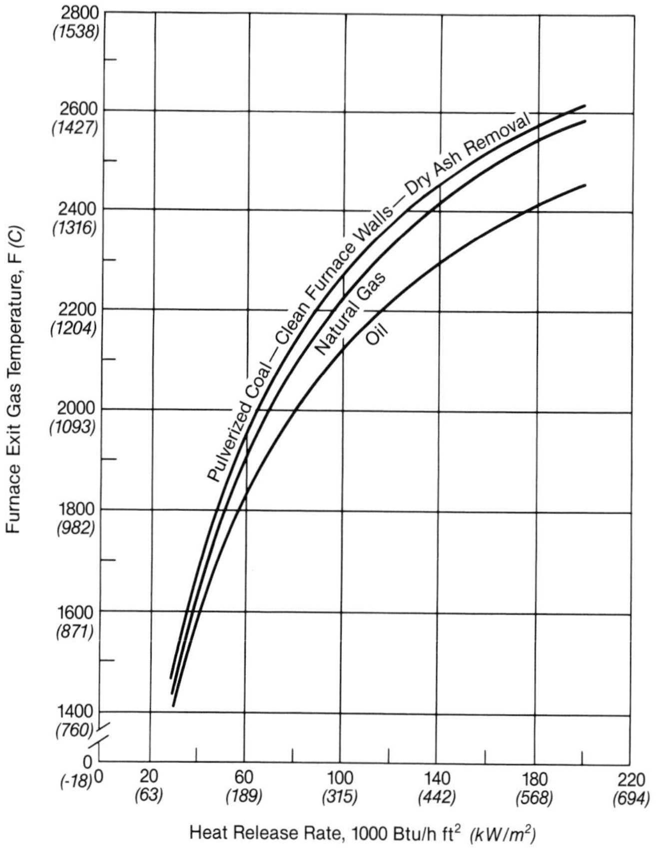

Fig. 32 Approximate relationships of furnace exit gas temperature to heat release rate for various fuels.

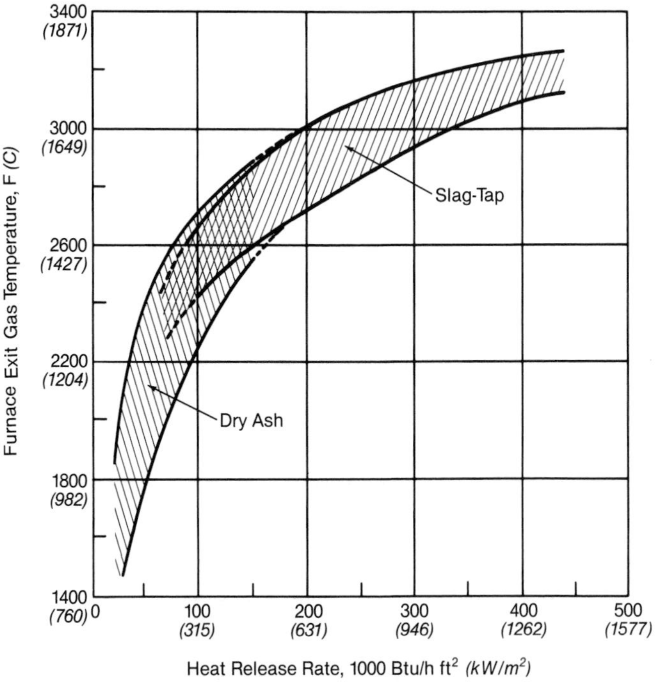

Fig. 33 General range of furnace exit gas temperature for dry ash and slag-tap pulverized coal-fired furnaces.

combustion system type, burner and air port placement, stoichiometry, fuel characteristics and cleaning cycle. The bands for dry ash and for slag-tap furnaces overlap between 100,000 and 150,000 Btu/h ft^2 (315,460 to 473,190 W/m^2), but different types of coal are involved. To be suitable for a slag-tap furnace, a bituminous coal should have an ash viscosity of 250 poises at 2450F (1343C) or lower. In the overlapping range, dry ash and slag-tap both have about the same heat absorption rate, or dirtiness factor, as shown in Fig. 34. Both bands are rather broad, but they cover a wide range of ash characteristics and a considerable diversity in waterwall construction and dirtiness.

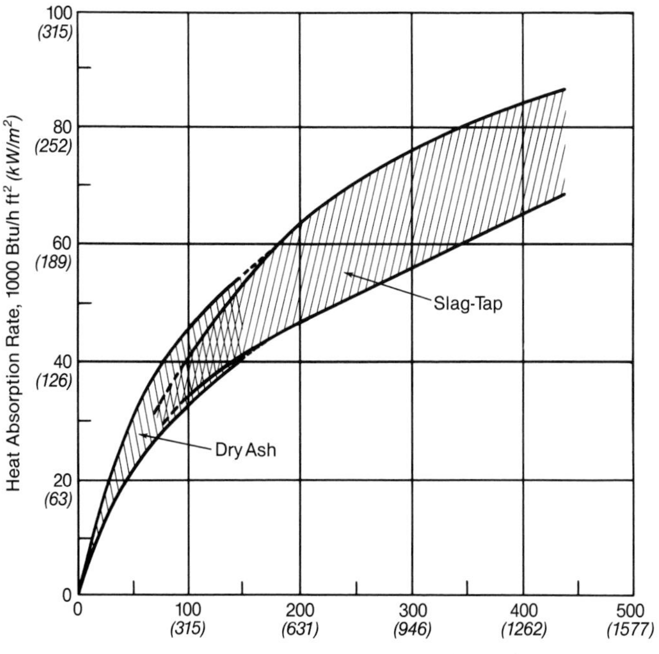

Fig. 34 General range of furnace heat absorption rates for dry ash and slag-tap pulverized coal-fired furnaces.

The heat leaving the furnace is calculated from the exiting gas flow rate (the gas enthalpy values evaluated at the furnace exit gas temperature) plus the net radiative transfer at the furnace exit. The heat absorbed in the furnace is the difference between the heat available from the fuel, including the preheated combustion air, and heat leaving the furnace.

Numerical boiler modeling The objective of B&W's combustion model development program has been to produce computational models applicable to fossil fuel-fired boilers. The models are based on a fundamental description of the interacting processes including flow, combustion, and heat transfer. While still in the development stage, these models are used to qualitatively study parametric variations on boiler performance. At some point, they will enable designers to accurately predict performance in field units. Reference 11 provides an overview of the main elements involved in a comprehensive model and shows an advanced application of a numerical model. In the B&W furnace combustion model, all transport processes are represented by the general steady-state convective transport equation:

$$\frac{\partial}{\partial x_i} \rho u_i \phi = \frac{\partial}{\partial x_i} \Gamma_\phi \frac{\partial \phi}{\partial x_i} + S_\phi \tag{72}$$

where ϕ is the dependent variable and can denote velocity, enthalpy, or chemical species; Γ_ϕ is the effective diffusion coefficient; and S_ϕ is a source term per unit volume. The subscript i indicates a summation over the coordinate directions x, y, and z. Solutions of the flow and auxiliary equations are discussed by Patankar.[10] Sources of energy include the radiant energy captured by the absorbing flue gases, e.g., carbon dioxide and water vapor; energy from the coal, char, ash, and soot particles; and the energy of the entering fuel and air. Energy losses at the boundaries provide a heat sink.

The solution of the equation is similar to the method previously described. The modeled geometry is divided into control volumes and the equation is integrated over the volume; an algebraic equation of the following form results:

$$a_p H_p = \sum_i^N a_i H_i + S_H \tag{73}$$

where the coefficients, a, have flow units,[10] N is the number of neighbor control volumes, and the source term has the form:

$$S_H = S_r + S_{cv} \tag{74}$$

The radiative absorption in each subvolume is found by either the discrete transfer method[12] or the discrete ordinates method.[13,14] These methods solve the radiation equations in discrete directions. They provide accurate estimates of radiative transfer in the absorbing and scattering media that is prevalent in the combustion and postcombustion zones of a boiler. The convective source (or sink) thermally connects the fluid to the wall, as represented by Equation 4. The wall temperature is found by creating an energy balance between the convective and radiative heat transfer to the surface and the heat loss through the wall.

The models are designed as flexible engineering tools and can handle a variety of geometries and inlet and exit conditions. One furnace performance model was created for a 560 MW supercritical steam pressure boiler firing high volatile Eastern United States bituminous coal. A schematic of the furnace is shown in Fig. 35. The sloping furnace walls, the furnace nose, and the ash hopper were modeled. Inlet fuel, inlet air and exit streams were properly located around the boundaries. An example of the predicted heat flux distribution is shown in Fig. 36. The predicted furnace exit gas temperature for this case was 2242F (1228C), while the observed average value was 2276F (1247C).

A numerical evaluation of a large unit firing Eastern bituminous coal identifies the relative magnitudes of convective and radiative heat transfer at various locations. As shown in Fig. 37, the furnace area is dominated by the radiative input while the back-end heat transfer surfaces in the direction of flow are increasingly dominated by convection.

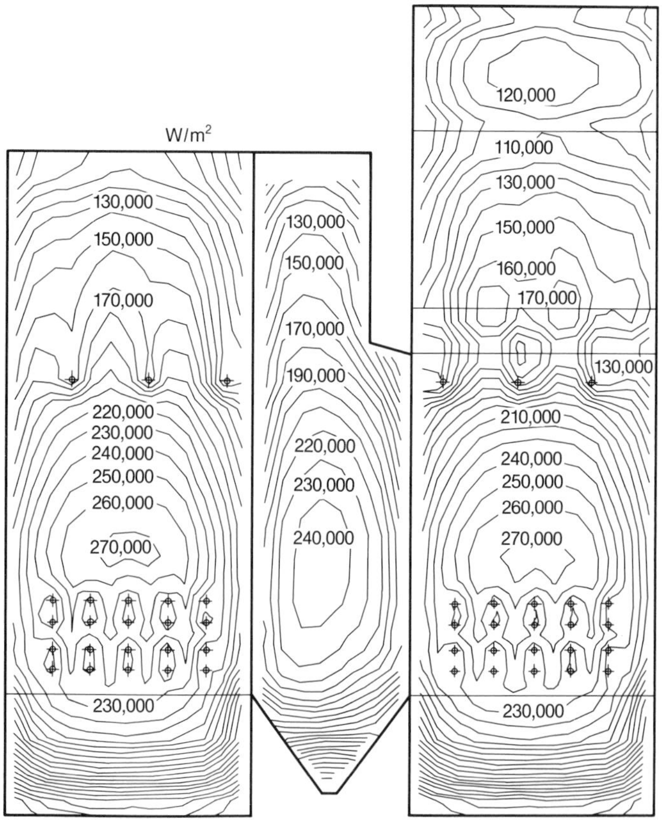

Fig. 36 Numerical model — predicted furnace wall flat projected heat flux distribution (Btu/h ft^2 = 0.317 × W/m^2).

Design considerations

Furnaces

An analytical solution for heat transfer in a steam generating furnace is extremely complex. It is not possible to calculate furnace outlet temperatures by theoretical methods alone. Nevertheless, this temperature must be correctly predicted because it determines the design of the superheater and other system components.

In a boiler, all of the principal heat transfer mechanisms take place simultaneously. These mechanisms are intersolid radiation between suspended solid particles, tubes, and refractory materials; nonluminous gas radiation from the products of combustion; convection from the gases to the furnace walls; and conduction through ash deposits on tubes.

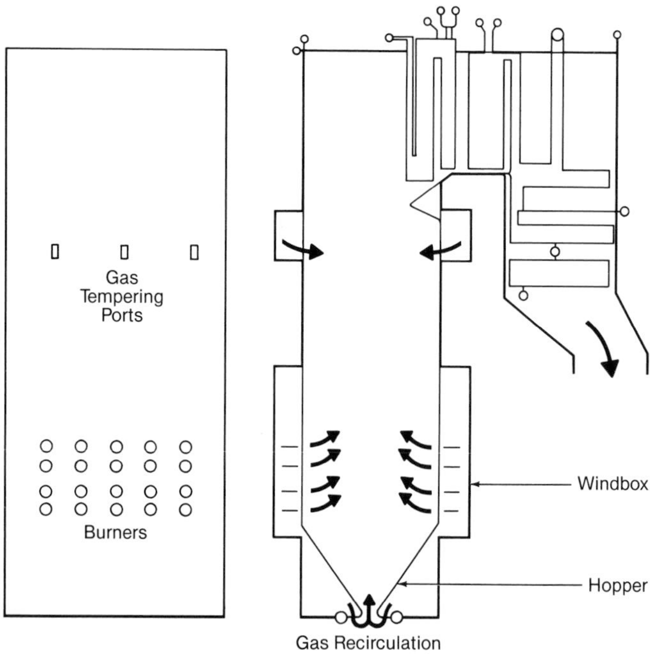

Fig. 35 560 MW utility boiler furnace schematic used for numerical model (see Fig. 36).

Fuel variation is significant. Pulverized coal, gas, oil or waste-fuel firing may be used. In addition, different types of the same fuel also cause variations. Coal, for example, may be high volatile or low volatile, and may have high or low ash and moisture contents. The ash fusion temperature may also be high or low, and may vary considerably with the oxidizing properties of the furnace atmosphere.

Furnace geometry is complex. Variations occur in the burner locations and spacing, in the fuel bed size, in the ash deposition, in the type of cooling surface, in the furnace wall tube spacing, and in the arch and hopper arrangements. Flame shape and length also affect the distribution of radiation and heat absorption in the furnace.

High intensity, high mixing burners produce bushy flames and promote large high temperature zones in the lower furnace. Lower intensity, controlled mixing burners frequently have longer flames that delay combustion while controlling pollutant formation.

Surface characteristics vary. The enclosing furnace walls may include any combination of fuel arrangements, refractory material, studded tubes, spaced tubes backed by refractory, close spaced tubes, membrane construction or tube banks. Emissivities of these surfaces are different. The water-cooled surface may be covered with fluid slag or dry ash in any thickness, or it may be clean.

Temperature varies throughout the furnace. Fuel and air enter at relatively low temperatures, reach high temperatures during combustion, and cool again as the products of combustion lose heat to the furnace enclosure. All temperatures change with load, excess air, burner adjustment and other operating conditions.

Accurate estimates of furnace exit gas temperature are

important. For example, high estimates may lead to overestimating the heat transfer surface, while low estimates may cause operational problems. These are discussed in Chapter 18.

Convection banks

Tube spacing and arrangement In addition to heat absorption and resistance to gas flow, other important factors must be considered in establishing the optimum tube spacing and arrangement for a convection surface. These are slagging or fouling of surfaces, accessibility for cleaning, and space occupied. A large longitudinal spacing relative to the transverse spacing is usually undesirable because it increases the space requirement without improving performance. These are discussed further in Chapter 20.

Tube diameter For turbulent flow, the heat transfer coefficient is inversely proportional to a power of the tube diameter. In Equations 51 and 54 the exponent for longitudinal flow is 0.20; for crossflow it is 0.39. These equations indicate that the tube diameter should be minimized for the most effective heat transfer. However, this optimum tube diameter may require an arrangement that is expensive to fabricate, difficult to install, or costly to maintain. A compromise between heat transfer effectiveness and manufacturing, erection, and service limitations is therefore necessary in selecting tube diameter.

Penetration of radiation A convection bank of tubes bordering a furnace or a cavity acts as a blackbody radiant heat absorber. Some of the impinging heat, however, radiates through the spaces between the tubes of the first row and may penetrate as far as the fourth row. The quantity of heat penetration can be established by geometric or analytical methods. The effect of this penetration is especially important in establishing tube temperatures for superheaters located close to a furnace or high temperature cavity. Consider 2.0 in. (50.8 mm) OD tubes placed in an array of tubes on a 6.0 in. (152.4 mm) pitch. Fig. 31, curve 1 can be used to estimate the remaining radiation. For a given radiant heat flux, 45% is absorbed in the first tube row, and 55% passes to the second row. 45% of this reduced amount is again absorbed in the second tube row. After the fourth row, less than 10% of the initial radiation remains.

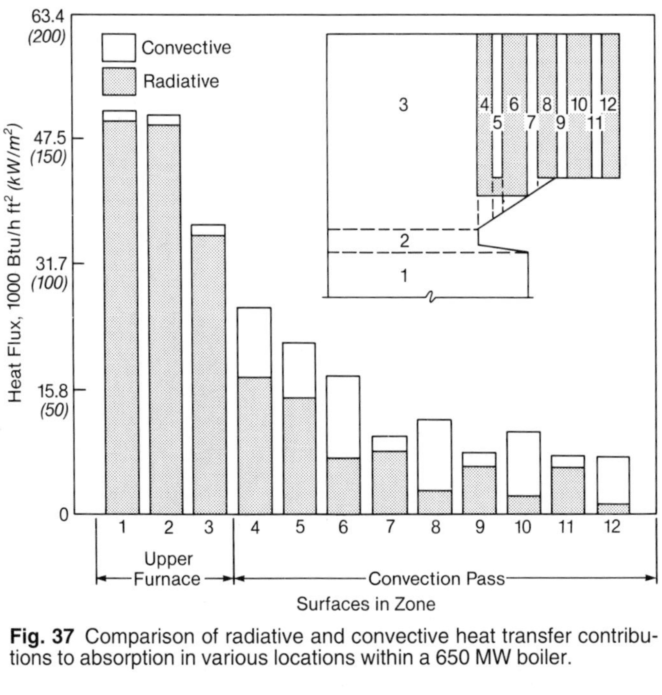

Fig. 37 Comparison of radiative and convective heat transfer contributions to absorption in various locations within a 650 MW boiler.

Effect of lanes Lanes in tube banks, formed by the omission of rows of tubes, may decrease the heat absorption considerably. These passages act as bypasses for flowing hot gases and radiation losses. Although the overall efficiency decreases, the high mass flow through the lanes increases the absorption rate of the adjacent tubes. Critical tube temperatures in superheaters or steaming conditions in economizers may develop. Whenever possible, lanes should be avoided within tube banks and between tube banks and walls; however, this is not always possible. A calculation accounting for the lanes is necessary in such cases.

Heat transfer to water

Water heat transfer coefficient The heat transfer coefficient for water in economizers is so much higher than the gas side heat transfer coefficient that it can be neglected in determining economizer surface.

Boiling water heat transfer coefficient The combined gas side heat transfer coefficient (convection plus intertube radiation) seldom exceeds 30 Btu/h ft^2 F (170 W/m^{2}K) in boiler design practice. The heat transfer coefficient for boiling water [10,000 Btu/h ft^2 F (56,784 W/m^2 K)] is so much larger that it is generally neglected in calculating the resistance to heat flow, although Equation 4 in Chapter 5 can be used to calculate this value.

Effect of oil or scale Water side and steam side scale deposits provide high resistance to heat flow. As scale thickness increases, additional heat is required to maintain a given temperature inside a furnace tube. This leads to high metal temperatures and can cause tube failure. Deposition of scale and other contaminants is prevented by good feedwater treatment and proper operating practices.

Heat transfer to steam In superheaters, the steam film constitutes a significant resistance to heat flow. Although this resistance is much lower than the gas side resistance, it can not be neglected in computing the overall heat flow resistance or the heat transfer rate. It is particularly significant in calculating superheater tube temperatures, because the mean tube wall temperature is equal to the steam temperature plus the temperature drop through the steam film plus half of the metal temperature drop.

The steam heat transfer coefficient is calculated from Equation 55 using information from Figs. 12, 15 and 16. If the steam heat transfer coefficient is designated as h, the film temperature drop, ΔT_f, is $q/(hA)$, using the outside surface area of the tube as the base in each expression.

It is imperative to prevent scale deposits in superheater tubes. Because of its high resistance to heat flow and due to the elevated temperatures, even a thin layer of scale may be sufficient to overheat and fail a tube.

Cavities

Cavities are necessary between tube banks of steam generating units for access, for sootblowers, and for possible surface addition. Hot flue gas radiates heat to the boundary surfaces while passing through the cavity. The factors involved in calculating heat transfer in cavities are as follows.

Temperature level Radiation from nonluminous gases to boundary surfaces and radiation to the gas by the surroundings increase approximately by the fourth power of their respective absolute temperatures. Remembering that $E_b = \sigma T^4$, Equations 34 to 36 illustrate this relationship.

Gas composition Carbon dioxide and water vapor are the normal constituents of flue gases which emit nonluminous radiation in steam generating units. The concentrations of these constituents depend on the fuel burned and the amount of excess air.

Particles in the gas The particles carried by flue gases receive heat from the gas by radiation, convection, and conduction, and emit heat by intersolid radiation to the furnace enclosure.

Size of cavity The heat transferred per unit of time increases with cavity size. Thick layers of gas radiate more vigorously than thin layers. The shape of the cavity can also complicate heat transfer calculations.

Receiving surface A refractory surface forming part of a cavity boundary reaches a high temperature by convection and radiation from the flue gas. It also reradiates heat to the gas and to the other walls of the enclosure. Reradiation from a clean, heat-absorbing surface is small unless the receiving surface temperature is high, as is the case with superheaters and reheaters. Ash or slag deposits on the tube reduce heat absorption and increase reradiation.

In boiler design, there are two significant effects of cavity radiation: 1) the temperature of flue gas drops, from several degrees up to 40F (22C), in passing across a cavity, and 2) gas radiation increases the heat absorption rates for the tubes forming the cavity boundaries. The second effect influences superheater tube temperatures and the selection of alloys.

Insulation

The calculation of heat transfer through insulation follows the principles outlined for conduction through a composite wall. Refer to Chapter 22 for more information regarding insulating materials.

Hot face temperature In a furnace with tube to tube walls, the hot face temperature of the insulation is the saturation temperature of the water in the tubes. If the inner face of the furnace wall is refractory, the hot face insulation temperature must be calculated using radiation and convection heat transfer principles on the gas side of the furnace wall, or estimated using empirical data.

Heat loss and cold face temperature The heat loss to the surroundings and the cold face temperature decrease as the insulation thickness increases. However, once an acceptable layer of insulation is applied, additional amounts are not cost effective. Standard commercial insulation thicknesses should be used in the composite wall.

The detailed calculation of overall heat loss by radiation and convection from the surfaces of a steam generating unit (usually called radiation loss) is tedious and time consuming. A simple approximate method is provided by the chart prepared from the American Boiler Manufacturers Association (AMBA) original. (See Fig. 11, Chapter 22.)

Ambient air conditions Low ambient air temperature and high air velocities reduce the cold face temperature. However, they have only a small effect on total heat loss, because surface film resistance is a minor part of the total insulation resistance. Combined heat loss rates (radiation plus convection) are given in Fig. 10, Chapter 22, for various temperature differences and air velocities. The effect of surface film resistance on casing temperature and on heat loss through casings is shown in Fig. 14, Chapter 22.

Temperature limits and conductivities Refractory or insulating material suitable for high temperature applications is usually more expensive and less effective than low temperature materials. It is therefore customary to use several layers of insulation. The lower cost, more effective insulation is used in the cool zones; the higher cost materials are used only where demanded by high operating temperatures. Thermal conductivities for refractory and insulating materials, and temperatures for which they are suitable, are shown in Fig. 9, Chapter 22.

Applications

Example 1 — Conduction through a plane wall

If a flat plate is heated on one side and cooled on the other, the heat flow rate in the wall, shown in Fig. 1, is given by Equation 2. The rate of heat flow through a 0.25 in. thick steel plate with 1 ft^2 surface area and ΔT = 25F may be evaluated with Equation 2 as follows:

$$q = kA\frac{\Delta T}{L} = \frac{30 \times 1 \times 25}{0.25/12} = 36{,}000 \text{ Btu/h} \quad \textbf{(75)}$$

where the thermal conductivity, k, for steel is 30 Btu/h ft F.

Example 2 — Heat flow in a composite wall with convection

The heat flow through a steel wall which is insulated on both sides is shown in Fig. 3. This example demonstrates the procedure for combining thermal resistances. In addition to the thermal resistance of the firebrick, steel, and insulation, the heat flow is impeded by the surface resistances. Consider a 600 ft^2 surface with gases at 1080F or 1540R on the inside exposed to an ambient temperature of 80F on the outside. The thermal conductivities of the firebrick, steel flue and insulation are assumed to be $k_1 = 0.09$, $k_2 = 25$, and $k_3 = 0.042$ Btu/h ft F, respectively. These assumptions are verified later. The layer thicknesses are $\Delta x_1 = 4$ in., $\Delta x_2 = 0.25$ in., and $\Delta x_3 = 3$ in. The heat transfer coefficients for convection are $h_{cv,i} = 5.0$ Btu/h ft^2 F on the inside surface $h_{cv,o} = 2.0$ Btu/h ft^2 F on the outside surface. Where the temperature difference between the radiating gas, T_g, and a surface, T_s, is small, the radiation heat transfer coefficient can be estimated by:

$$h_r \cong 4.0\,\sigma\,\varepsilon\,F[(T_g + T_s)/2]^3 \approx 4\,\sigma\,\varepsilon\,F\,T_g^3 \quad \textbf{(76a)}$$

where T_g and T_s are the absolute temperatures, R (K). In this example, the surface emissivity is assumed close to 1.0 and F = 1.0 resulting in:

$$h_r = 4.0\,(0.1713\times 10^{-8})\,(1080+460)^3$$
$$= 25 \text{ Btu/h ft}^2\text{ F} \quad \textbf{(76b)}$$

Using the R_{eq} shown in Fig. 3 and values of R evaluated using Table 4:

$$R_{eq}A = \frac{\left(\frac{1}{5}\right)\left(\frac{1}{25}\right)}{\frac{1}{5}+\frac{1}{25}} + \frac{\frac{4}{12}}{0.09} + \frac{\frac{0.25}{12}}{25} + \frac{\frac{3}{12}}{0.042} + \frac{1}{2}$$
$$= 0.033 + 3.70 + 0.000833 + 5.95 + 0.5$$
$$= 10.18\,\frac{\text{h ft}^2\text{ F}}{\text{Btu}} \quad \textbf{(77)}$$

It is clear that the firebrick and insulation control the overall resistance; the steel resistance can be neglected. If the successive material layers do not make good thermal contact with each other, there will be interface resistances due to the air space or film. These resistances may be neglected in composite walls of insulating materials. However, they must be included in calculations if the layer resistances are small compared to the interface resistances. An example of this is heat transfer through a boiler tube with internal oxide deposits.

The heat flow can be computed using Equation 2:

$$q = kA\frac{T_1 - T_2}{L} = \frac{T_1 - T_2}{R_{eq}} = A\frac{T_1 - T_2}{R_{eq}A}$$

$$= 600\left(\frac{1080 - 80}{10.18}\right) = 58,939\frac{\text{Btu}}{\text{h}} \qquad \textbf{(78)}$$

To determine if the correct thermal conductivities were assumed and if the temperature levels are within allowable operating limits of the material, it is necessary to calculate the temperatures at the material interfaces. Solving Equation 78 for temperature and substituting individual resistances and local temperatures:

$$T_1 = T_0 - \frac{q}{A}(R\,A)_f = 1080 - \frac{(58,939)(0.033)}{600} = 1077\text{F} \qquad \textbf{(79)}$$

$$T_2 = T_1 - \frac{q}{A}(R\,A)_1 = 1077 - \frac{(58,939)(3.70)}{600} = 713\text{F} \qquad \textbf{(80)}$$

$$T_3 = T_2 - \frac{q}{A}(R\,A)_2 = 713 - \frac{(58,939)(0.000833)}{600} = 713\text{F} \qquad \textbf{(81)}$$

$$T_4 = T_3 - \frac{q}{A}(R\,A)_3 = 713 - \frac{(58,939)(5.95)}{600} = 129\text{F} \qquad \textbf{(82)}$$

$$T_5 = T_4 - \frac{q}{A}(R\,A)_\infty = 129.0 - \frac{(58,939)(0.5)}{600} = 80\text{F} \qquad \textbf{(83)}$$

The negligible resistance of the steel flue is reflected in the temperature drop $T_2 - T_3 = 0$. If the calculated interface temperatures indicate the conductivity was chosen improperly, new conductivities are defined using the mean temperature of each material. For example, a new firebrick conductivity is determined using $0.5\ (T_1 + T_2)$.

Example 3 — Heat flow in an insulated pipe

Heat flow in cylindrical geometries is important in evaluating boiler heat transfer. Refer to the example steam line shown in Fig. 38. The resistances in Table 4 for cylindrical geometries must be used. The thermal analogy for the pipe in Fig. 38 can be written:

$$R_{eq} = \frac{1}{h_i(2\pi r_1 l)} + \frac{\ell n(r_2/r_1)}{2\pi k_1 l} + \frac{\ell n(r_3/r_2)}{2\pi k_2 l} + \frac{1}{h_o(2\pi r_3 l)} \qquad \textbf{(84)}$$

A 3 in. Schedule 40 steel pipe (k = 25 Btu/h ft F) is covered with 0.75 in. insulation of k = 0.10 Btu/h ft F. This pipe has a 3.07 in. ID and a 3.50 in. OD. The pipe transports fluid at 300F and is exposed to an ambient temperature of 80F. With an inside heat transfer coefficient of 50 Btu/h ft^2 F and an outside heat transfer coefficient of 4 Btu/h ft^2 F, the thermal resistance and heat flow per unit length are:

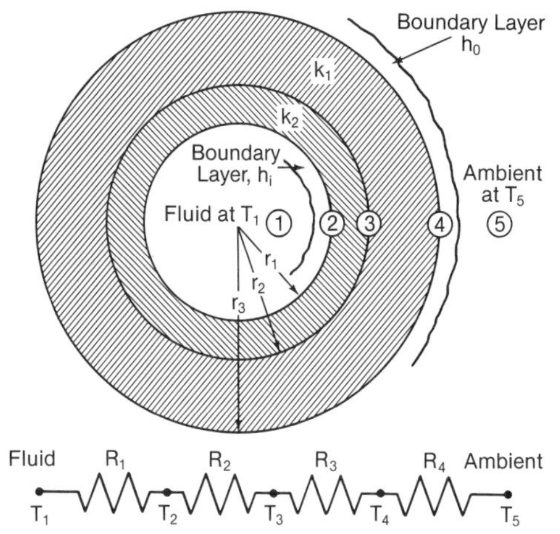

Fig. 38 Heat flow in an insulated pipe — example.

$$R_{eq}\,l = \frac{1}{50(2\pi)\left(\frac{3.07/2}{12}\right)} + \frac{\ell n\left(\frac{3.50}{3.07}\right)}{2\pi(25)} + \frac{\ell n\left(\frac{5.0}{3.5}\right)}{2\pi(0.1)} + \frac{1}{4(2\pi)\left(\frac{5.0/2}{12}\right)} \qquad \textbf{(85)}$$

$$R_{eq}\,l = 0.0249 + 0.000834 + 0.568 + 0.191$$

$$= 0.785\frac{\text{h ft F}}{\text{Btu}} \qquad \textbf{(86)}$$

The overall resistance is dominated by the insulation resistance and that of the outer film boundary layer. The resistance of the metal pipe is negligible.

$$\frac{q}{l} = \frac{300 - 80}{0.785} = 280 \ \text{Btu/h ft} \qquad \textbf{(87)}$$

Example 4 — Heat flow between a small object and a large cavity

Consider an unshielded thermocouple probe with an emissivity of 0.8 inserted in a duct at 240F carrying combustion air. If the thermocouple indicates a temperature of 540F and the surface heat transfer coefficient, h, between the thermocouple and gas is 20 Btu/h ft^2 F, the true gas temperature can be estimated. The thermocouple temperature must be below the gas temperature because heat is lost to the walls. Under steady-state conditions, an energy balance equates the radiant heat loss from the thermocouple to the wall and the rate of heat flow from the gas to the thermocouple.

Using Table 6, the heat flow between the thermocouple and the cavity becomes:

$$\frac{q}{A} = 0.8\,(0.1713 \times 10^{-8})\,[(540+460)^4 - (240+460)^4]$$
$$= 1041.9 \text{ Btu/h ft}^2 \quad \textbf{(88)}$$

The true gas temperatures becomes:

$$T_g = \frac{q/A}{h} + T_t = \frac{1041.9}{20} + 540 = 592\text{F} \quad \textbf{(89)}$$

Similar analyses can be performed for thermocouples shielded with reflective foils in high temperature environments. This practice prevents thermocouple heat losses and incorrect temperature readings. The heat flow from a shielded thermocouple is calculated as follows:

$$q_{\text{shielded}} = \frac{1}{(N+1)}\, q_{\text{no shield}} \quad \textbf{(90)}$$

where N is the number of concentric layers of material.

Example 5 — Heat flow between two surfaces

An estimate of the maximum radiant heat transfer between two surfaces can be determined using Equation 8. This approximation is valid when the walls are considered black and any intervening absorbing gases are neglected. If two 5 × 10 ft black rectangles, directly opposed, are spaced 10 ft apart with temperatures of 940 and 1040F, the energy exchange is estimated as follows:

The energy from surface 1 directly striking surface 2 is defined by the shape factor F_{12}. Referring to Fig. 5, this factor F_{12} is 0.125, indicating that 87.5% of the energy leaving surface 1 strikes a surface other than surface 2. The net heat flow is:

$$q_{12} = A_1 F_{12} \sigma (T_1^4 - T_2^4) = 50\,(0.125)\,(0.1713 \times 10^{-8})$$
$$\times [(1040+460)^4 - (940+460)^4]$$
$$q_{12} = 13{,}078 \text{ Btu/h} \quad \textbf{(91)}$$

Intervening gases and/or gray walls further reduce the net heat flow.

Example 6 — Radiation from a hot gas to furnace walls

Consider a furnace with a volume of 160,000 ft^3 and a heat transfer surface area of 19,860 ft^2. The gas traversing the furnace is at 2540F (3000R) and the furnace walls are at 1040F (1500R). The radiant heat transfer rate can be estimated using Equation 35, assuming the walls are radiatively black. If the products of combustion at one atmosphere consist of 10% carbon dioxide, 5% water vapor, and 85% nitrogen, Figs. 8 to 10 can be used to estimate the gas emissivity and absorptivity. The beam length is L = 3.6 V/A = 29.0. Then for H_2O, p_wL = (29.0) (0.05) = 1.45 and from Fig. 8 at 3000R the emissivity is found to be 0.17. For CO_2, p_cL = (29.0) (0.10)= 2.90 and from Fig. 9, at 3000R, the emissivity is found to be 0.16. The correction $\Delta\varepsilon$ is determined from Fig. 10. The total gas emissivity is then found from:

$$\varepsilon = \varepsilon_{H_2O} + \varepsilon_{CO_2} - \Delta\varepsilon \quad \textbf{(92)}$$

$$\varepsilon_g = 0.17 + 0.16 - 0.05 = 0.28 \quad \textbf{(93)}$$

Hottel[4] suggests calculating the absorptivity of the gas using modified pressure length parameters:

$$F_w = p_w L \frac{T_s}{T_g} = (0.05)(29)\frac{1040+460}{2540+460} = 0.73 \quad \textbf{(94)}$$

$$F_c = p_c L \frac{T_s}{T_g} = (0.10)(29)\frac{1040+460}{2540+460} = 1.45 \quad \textbf{(95)}$$

$$\alpha_{H_2O} = \varepsilon_{H_2O}(F_w, T_s) \times \left(\frac{T_g}{T_s}\right)^{0.45}$$
$$= 0.22\left(\frac{2540+460}{1040+460}\right)^{0.45} = 0.35 \quad \textbf{(96)}$$

$$\alpha_{CO_2} = \varepsilon_{CO_2}(F_c, T_s) \times \left(\frac{T_g}{T_s}\right)^{0.65}$$
$$= 0.16\left(\frac{2540+460}{1040+460}\right)^{0.65} = 0.25 \quad \textbf{(97)}$$

$$\Delta\alpha_g = \Delta\varepsilon\,(T_s) = 0.05 \quad \textbf{(98)}$$

$$\alpha_g = \alpha_{H_2O} + \alpha_{CO_2} - \Delta\alpha = 0.55 \quad \textbf{(99)}$$

The net rate of heat flow calculated from Equation 35:

$$q_{sg} = A\,(\varepsilon_g E_g - \alpha_g E_s)$$
$$= 19{,}860\;(0.28\;\sigma\;3000^4 - 0.55\;\sigma\;1500^4)$$
$$= 677 \times 10^6 \text{ Btu/h} \quad \textbf{(100)}$$

In estimating boiler heat transfer, the beam lengths are large, effecting large pL values. Proprietary data are used to estimate the heat transfer for these values, and extrapolation of the curves in Figs. 8 to 10 is not recommended.

Example 7 — Radiation in a cavity

Radiation in a cavity containing absorbing gases can be analyzed with the concepts previously presented. These concepts are useful in analyzing surface to surface heat transfer. Examples include boiler wall to boiler wall, platen to platen, and boiler wall to boiler enclosure heat exchanges. Table 5 contains the thermal resistances used in constructing the thermal circuit in Fig. 39. Note the resistance between surface 1 and 2 decreases as the transmission in the gas, τ_{12}, increases to a transparent condition τ_{12} = 1. At τ_{12} near zero, the gases are opaque, and the resistance is very large. As the gas emissivity decreases, the thermal circuit reduces to Equation 33. The solution of the circuit in Fig. 39 is found using Kirchhoff's rule for nodes J_1 and J_2. The equations are solved simultaneously for J_1 and J_2:

$$\frac{E_1 - J_1}{\frac{1-\varepsilon_1}{\varepsilon_1 A_1}} + \frac{J_2 - J_1}{\frac{1}{A_1 F_{12} \tau_{12}}} + \frac{E_g - J_1}{\frac{1}{A_1 \varepsilon_{g1}}} = 0 \quad \textbf{(101)}$$

$$\frac{E_2 - J_2}{\frac{1-\varepsilon_2}{\varepsilon_2 A_2}} + \frac{J_1 - J_2}{\frac{1}{A_1 F_{12} \tau_{12}}} + \frac{E_g - J_2}{\frac{1}{A_1 \varepsilon_{g2}}} = 0 \quad \textbf{(102)}$$

The net heat flow between the surfaces is:

$$q_{12} = A_1 F_{12} \tau_{12} (J_1 - J_2) \quad \textbf{(103)}$$

Hottel[4] demonstrates the procedures for finding the beam length to determine F_{12}, ε_{g1}, and ε_{g2}.

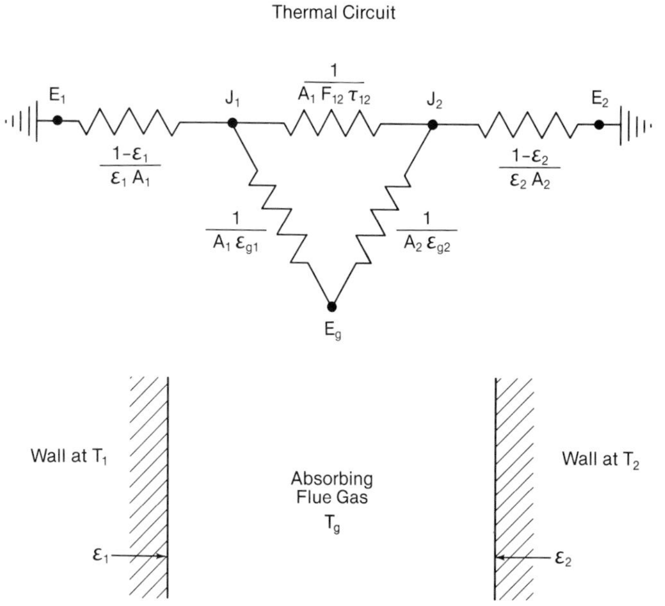

Fig. 39 Radiation in a cavity — example.

References

1. Roshenow, W., Hartnett, J., and Ganic, E., *Handbook of Heat Transfer Fundamentals*, 2nd ed., McGraw-Hill, New York, 1985.

2. Roshenow, W., Hartnett, J., and Ganic, E., *Handbook of Heat Transfer Applications*, 2nd ed., McGraw-Hill, New York, 1985.

3. Kreith, F., and Bohn, M., *Principles of Heat Transfer*, 4th ed., Harper and Row, New York, 1986.

4. Hottel, H., and Sarofim, A., *Radiative Transfer*, McGraw-Hill, New York, 1967.

5. Meyer, C.A., *et al.*, *ASME Steam Tables: Thermodynamic and Transport Properties of Steam*, American Society of Mechanical Engineers, New York, 1983.

6. McAdams, W., *Heat Transmission*, 3d ed., McGraw-Hill, New York, 1954.

7. Colburn, A., "A method of correlating forced convection heat transfer data and a comparison with fluid friction," *Trans. AIChE*, Vol. 29, p. 174, 1933.

8. Schmidt, T.F., "Wärme leistung von berippten Flächen," *Mitt. des Kältetechn. Inst. der T.H. Karlshruhn*, Vol. 4, 1949.

9. Incropera, F., and Dewitt, D., *Introduction to Heat Transfer*, Wiley, New York, 1985.

10. Patankar, S., *Numerical Heat Transfer and Fluid Flow*, McGraw-Hill, New York, 1980.

11. Fiveland, W. A., and Wessel, R. A., "A numerical model for predicting performance of three-dimensional pulverized fuel fired furnaces," *Journal of Engineering for Gas Turbines and Power*, Vol. 110, No. 1, 1988.

12. Lockwood, F.C., and Shah, N.G., "New radiation solution method for incorporation in general combustion prediction procedures," The Eighteenth Symposium (International) on Combustion, The Combustion Institute, Pittsburgh, pp. 1405-1414, 1981.

13. Fiveland, W. A., "Discrete-ordinates solutions of the radiative transport equation for rectangular enclosures," *Journal of Heat Transfer*, Vol. 106, pp. 699-706, 1984.

14. Fiveland, W. A., "Three-dimensional radiative heat transfer solutions by the discrete-ordinates method," *Journal of Thermophysics and Heat Transfer*, Vol. 2, No. 4, pp. 309-316, 1988.

Two-phase flow void fraction measurements.

Chapter 5
Boiling Heat Transfer, Two-Phase Flow and Circulation

A case of heat transfer and flow of particular interest in steam generation is the process of boiling and steam-water flow. The boiling or evaporation of water is a familiar phenomenon. In general terms, boiling is the heat transfer process where heat addition to a liquid no longer raises its temperature under constant pressure conditions; the heat is absorbed as the liquid becomes a gas. The heat transfer rates are high, making this an ideal cooling method for surfaces exposed to the high heat input rates found in fossil fuel boilers, concentrated solar energy collectors and the nuclear reactor fuel bundles. However, the boiling phenomenon poses special challenges such as: 1) the sudden breakdown of the boiling behavior at very high heat input rates, 2) the potential flow rate fluctuations which may occur in steam-water flows, and 3) the efficient separation of steam from water. An additional feature of boiling and two-phase flow is the creation of significant density differences between heated and unheated tubes. These density differences result in water flowing to the heated tubes in a well designed boiler natural circulation loop.

Most fossil fuel steam generators and all commercial nuclear steam supply systems operate in the pressure range where boiling is a key element of the heat transfer process. Therefore, a comprehensive understanding of boiling and its various related phenomena is essential in the design of these units. Even at operating conditions above the critical pressure, where water no longer boils but experiences a continuous transition from a liquid like to a gas like fluid, boiling type behavior and special heat transfer characteristics occur.

Boiling process and fundamentals

Boiling point and thermophysical properties

The boiling point, or saturation temperature, of a liquid can be defined as the temperature at which its vapor pressure is equal to the total local pressure. The saturation temperature for water at atmospheric pressure is 212F (100C). This is the point at which net vapor generation occurs and free steam bubbles are formed from a liquid undergoing continuous heating. As discussed in Chapter 2, this saturation temperature (T_{sat}) is a unique function of pressure. The American Society of Mechanical Engineers (ASME) and the International Association for the Properties of Steam (IAPS) have compiled extensive correlations of thermophysical characteristics of water. These characteristics include the enthalpy (or heat content) of water, the enthalpy of evaporation (also referred to as the latent heat of vaporization), and the enthalpy of steam. As the pressure is increased to the critical pressure [3208.2 psi (221.2 bar)], the latent heat of vaporization declines to zero and the bubble formation associated with boiling no longer occurs. Instead, a smooth transition from liquid to gaseous behavior occurs with a continuous increase in temperature as energy is applied.

Two other definitions are also helpful in discussing boiling heat transfer:

1. *Subcooling* For water below the local saturation temperature, this is the difference between the saturation temperature and the local water temperature (T_{sat}-T).
2. *Quality* This is the flowing mass fraction of steam (frequently stated as percent steam by weight or %SBW after multiplying by 100%):

$$x = \frac{\dot{m}_{steam}}{\dot{m}_{water} + \dot{m}_{steam}} \qquad \textbf{(1)}$$

where

$\dot{m}_{steam}$ = steam flow rate, lb/h (kg/s)
$\dot{m}_{water}$ = water flow rate, lb/h (kg/s)

Thermodynamically, this can also be defined as:

$$x = \frac{H - H_f}{H_{fg}} \quad or \quad \frac{H - H_f}{H_g - H_f} \qquad \textbf{(2)}$$

where

H = local average fluid enthalpy, Btu/lb (J/kg)
H_f = enthalpy of water at saturation, Btu/lb (J/kg)
H_g = enthalpy of steam at saturation, Btu/lb (J/kg)
H_{fg} = latent heat of vaporization, Btu/lb (J/kg)

When boiling is occurring at saturated, thermal equilibrium conditions, Equation 2 provides the fractional steam flow rate by mass. For subcooled conditions where $H < H_f$, quality (x) can be negative and is an indication of liquid subcooling. For conditions where $H > H_g$, this value can be greater than 100% and represents the amount of average superheat of the steam.

Boiling curve

Fig. 1 illustrates a boiling curve which summarizes the results of many investigators. This curve provides the results of a heated wire in a pool, although the characteristics are similar for most situations. The heat transfer rate per unit area, or *heat flux*, is plotted versus the temperature differential between the metal surface and the bulk fluid. From points A to B, convection heat transfer cools the wire and boiling on the surface is suppressed. Moving beyond point B, which is also referred to as the *incipient boiling point*, the temperature of the fluid immediately adjacent to the heated surface slightly exceeds the local saturation temperature of the fluid while the bulk fluid remains subcooled. Bubbles, initially very small, begin to form adjacent to the wire. The bubbles then periodically collapse as they come into contact with the cooler bulk fluid. This phenomenon, referred to as *subcooled boiling*, occurs between points B and S on the curve. The heat transfer rate is quite high, but no net steam generation occurs. From points S to C, the temperature of the bulk fluid has reached the local saturation temperature. Bubbles are no longer confined to the area immediately adjacent to the surface, but move into the bulk fluid. This region is usually referred to as the *nucleate boiling* region, and as with subcooled boiling, the heat transfer rates are quite high and the metal surface is only slightly above the saturation temperature.

As point C is approached, increasingly large surface evaporation rates occur. Eventually, the vapor generation rate becomes so large that it restricts the liquid return flow to the surface. The surface eventually becomes covered (blanketed) with an insulating layer of steam and the ability of the surface to transfer heat drops. This transition is referred to as the *critical heat flux (CHF)*, *departure from nucleate boiling (DNB)*, *burnout*, *dryout*, *peak heat flux*, or *boiling crisis*. The temperature response of the surface under this condition depends upon how the surface is being heated. In fossil fuel boiler furnaces and nuclear reactor cores, the heat input is effectively independent of surface temperature. Therefore, a reduction in the heat transfer rate results in a corresponding increase in surface temperature from point D to D′ in Fig. 1. In some cases, the elevated surface temperature is so high that the metal surface may melt. If, on the other hand, the heat input or heat transfer rate is dependent upon the surface temperature, typical of a nuclear steam generator, the average local temperature of the surface increases as the local heat transfer rate declines. This region, illustrated in Fig. 1 from points D to E, is typically referred to as *unstable film boiling* or *transition boiling*. Because a large surface temperature increase does not occur, the main consequences are a decline in heat transfer performance per unit surface area and less overall energy transfer. The actual local phenomenon in this region is quite complex and unstable as discrete areas of surface fluctuate between a wetted boiling condition and a steam blanketed, or dry patch, condition. From position E through D′ to F, the surface is effectively blanketed by an insulating layer of steam or vapor. Energy is transferred from the solid surface through this layer by radiation, conduction and microconvection to the liquid-vapor interface. From this interface, evaporation occurs and bubbles depart. This heat transfer region is frequently referred to as stable *film boiling*.

In designing steam generating systems, care must be exercised to control which of these phenomena occur. In high heat input locations, such as the furnace area of fossil fuel boilers or nuclear reactor cores, it is important to maintain nucleate or subcooled boiling to adequately cool the surface and prevent material failures. However, in low heat flux areas or in areas where the heat transfer rate is controlled by the boiling side heat transfer coefficient, stable or unstable film boiling may be acceptable. In these areas, the resultant heat transfer rate must be evaluated, any temperature limitations maintained and only allowable temperature fluctuations accepted.

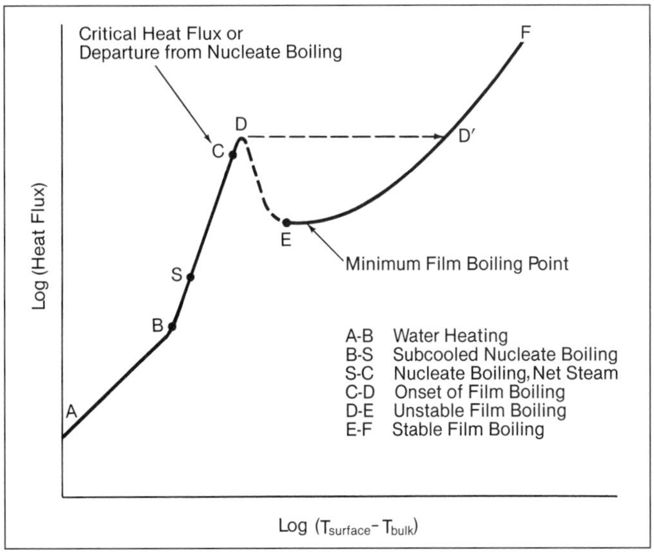

Fig. 1 Boiling curve — heat flux versus applied temperature difference.

Flow boiling

Flow or *forced convective boiling*, which is found in virtually all steam generating systems, is a more complex phenomenon involving the intimate interaction of two-phase fluid flow, gravity, material phenomena and boiling heat transfer mechanisms. Fig. 2 is a classic picture of boiling water in a long, uniformly heated, circular tube. The water enters the tube as a subcooled liquid and convection heat transfer cools the tube. The point of incipient boiling is reached (point 1 in Fig. 2). This results in the beginning of subcooled boiling and bubbly flow. The fluid temperature continues to rise until the entire bulk fluid reaches the saturation temperature and nucleate boiling occurs, point 2. At this location, flow boiling departs somewhat from the simple pool boiling model previously discussed. The steam-water mixture progresses through a series of flow structures or patterns: bubbly, intermediate and annular. This is a result of the complex interaction of surface tension forces, interfacial phenomena, pressure drop, steam-water densities and momentum effects coupled with the surface boiling behavior. While boiling heat transfer continues throughout, a point is reached in the annular flow regime where the liquid film on the wall becomes so thin that nucleation in the film is suppressed, point 3. Heat transfer then occurs through conduction and convection across the thin annular film with surface evaporation at the steam-water interface.

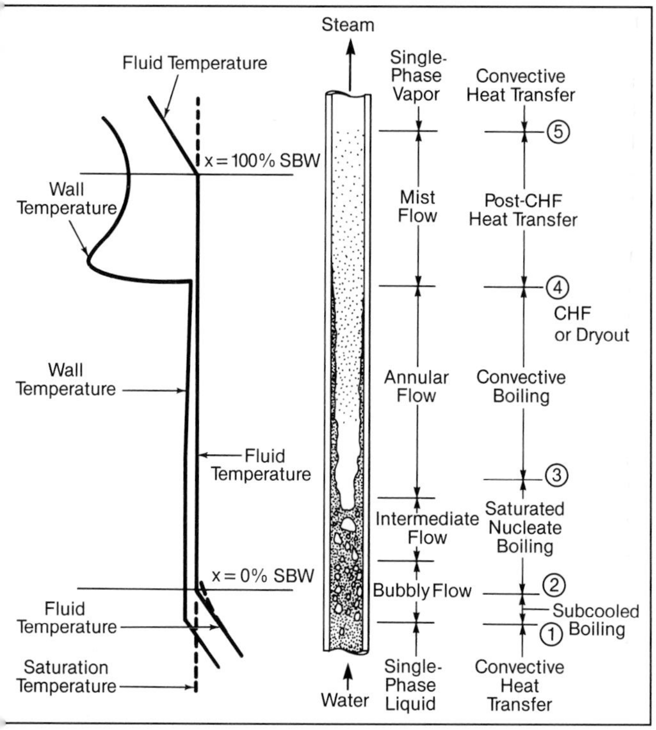

Fig. 2 Simplified flow boiling in a vertical tube (adapted from Collier[1]).

This heat transfer mechanism, called *convective boiling*, also results in high heat transfer rates. It should also be noted that not all of the liquid is on the tube wall. A portion is entrained in the steam core as dispersed droplets.

Eventually, an axial location, point 4, is reached where the tube surface is no longer wetted and CHF or dryout occurs. This is typically associated with a temperature rise. The exact tube location and magnitude of this temperature, however, depend upon a variety of parameters, such as the heat flux, mass flux, geometry and steam quality. Fig. 3 illustrates the effect of heat input rate, or heat flux, on CHF location and the associated temperature increase. From points 4 to 5 in Fig. 2, post-CHF heat transfer, which is quite complex, occurs. Beyond point 5, all of the liquid is evaporated and simple convection to steam occurs.

Boiling heat transfer evaluation

Engineering design of steam generators requires the evaluation of water and steam heat transfer rates under boiling and nonboiling conditions. In addition, the identification of the location of critical heat flux (CHF) is important where a dramatic reduction in the heat transfer rate could lead to: 1) excessive metal temperatures potentially resulting in tube failures, 2) an unacceptable loss of thermal performance, or 3) unacceptable temperature fluctuations leading to thermal fatigue failures. Data must also be available to predict the rate of heat transfer downstream of the dryout point. CHF phenomena are less important than the heat transfer rates for performance evaluation, but are more important in defining acceptable operating conditions. As discussed in Chapter 4, the heat transfer rate per unit area or heat flux is equal to the product of temperature difference and a heat transfer coefficient.

Heat transfer coefficients

Heat transfer correlations are application (surface and geometry) specific and Babcock & Wilcox (B&W) has developed extensive data for its applications through experimental testing and field experience. These detailed correlations remain proprietary to B&W. However, the following generally available correlations are provided here as representative of the heat transfer relationships.

Single-phase convection Several correlations for forced convection heat transfer are presented in Chapter 4. Forced convection is assumed to occur as long as the calculated forced convection heat flux is greater than the calculated boiling heat flux (point 1 in Fig. 2):

$$q''_{\text{Forced Convection}} > q''_{\text{Boiling}} \tag{3}$$

While not critical in most steam generator applications, correlations are available which explicitly define this onset of subcooled boiling and more accurately define the transition region.[1]

Subcooled boiling In areas where subcooled boiling occurs, several correlations are available to characterize the heat transfer process. Typical of these is the Jens and Lottes[2] correlation for water. For inputs with English units:

$$\Delta T_{sat} = 60\left(q''/10^6\right)^{1/4} e^{-P/900} \tag{4}$$

where

ΔT_{sat} = T_w - T_{sat}, F
T_w = wall temperature, F
T_{sat} = saturated water temperature, F
q'' = heat flux, Btu/h ft^2
P = pressure, psi

Another relationship frequently used is that developed by Thom.[3]

Nucleate and convective boiling Heat transfer in the saturated boiling region occurs by a complex combination of bubble nucleation at the tube surface (nucleate boiling) and direct evaporation at the steam-water interface in annular flow (convective boiling). At low steam qualities, nucleate boiling dominates while at higher qualities convective boiling dominates. While separate correlations are available for each range, the most useful relationships cover the entire saturated boiling regime. They typically involve the summation of appropriately weighted nucleate and convective boiling components as exemplified by the correlation developed by J.C. Chen and his colleagues.[4] While such correlations are frequently recommended for use in saturated boiling systems, their additional precision is not usually required in many boiler or reactor applications. For general evaluation purposes, the subcooled boiling relationship provided in Equation 4 is usually sufficient.

Post-CHF heat transfer As shown in Fig. 3, substantial increases in tube wall metal temperatures are possible if boiling is interrupted by the CHF phenomenon. The maximum temperature rise is of particular importance in establishing whether tube wall overheating may occur. In addition, the reliable estimation of the heat transfer rate may be important for an accurate assessment of thermal performance. Once the metal surface is no longer

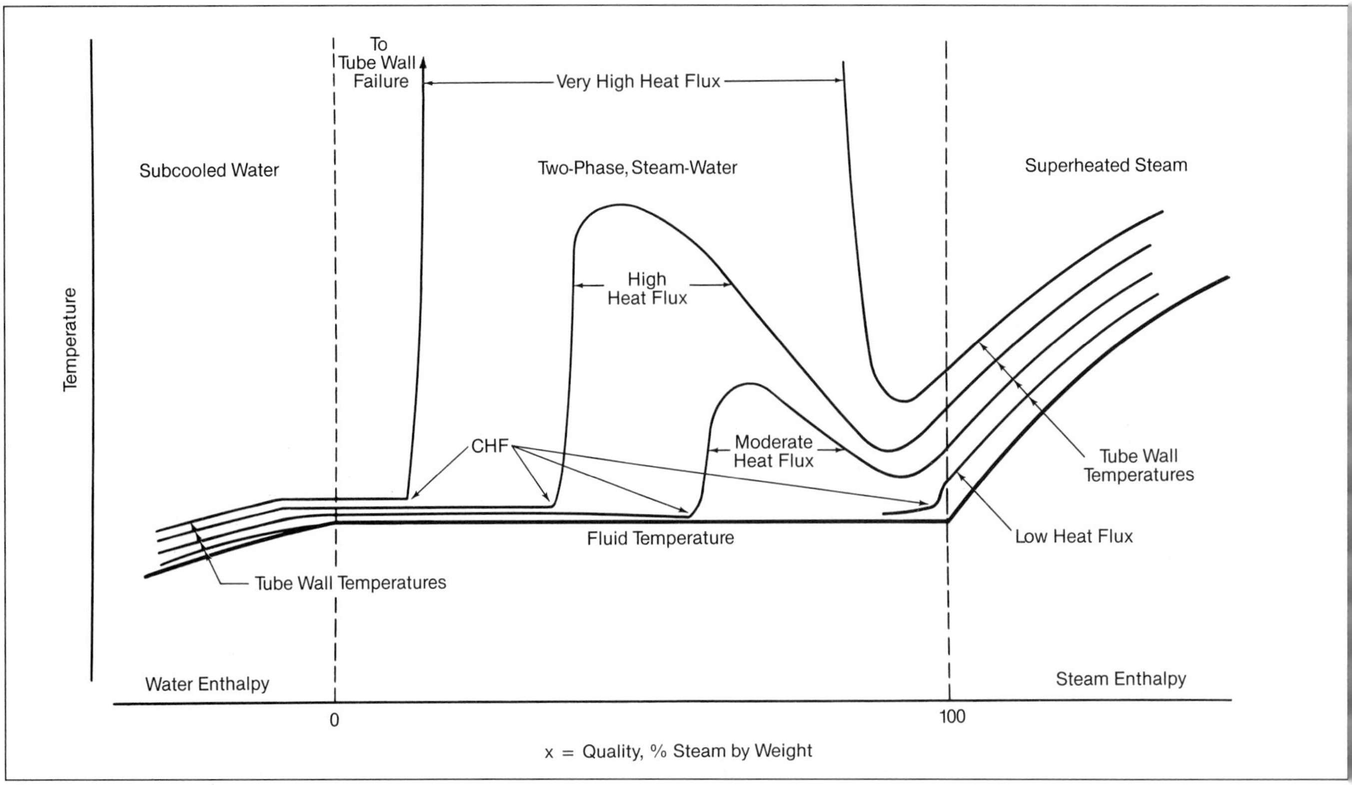

Fig. 3 Tube wall temperatures under different heat input conditions.

wetted and water droplets are carried along in the steam flow, the heat transfer process becomes more complex and includes: 1) convective heat transfer to the steam which becomes superheated, 2) heat transfer to droplets impinging on the surface from the core of the flow, 3) radiation directly from the surface to the droplets in the core flow, and 4) heat transfer from the steam to the droplets. This process results in a nonequilibrium flow featuring superheated steam mixed with water droplets. Current correlations do not provide a good estimate of the heat transfer in this region, but computer models show promise. Accurate prediction requires the use of experimental data for similar flow conditions.

Reflooding A key concept in evaluating emergency core coolant systems for nuclear power applications is *reflooding*. In a loss of coolant event, the reactor core can pass through critical heat flux conditions and can become completely dry. Reflooding is the term for the complex thermal-hydraulic phenomena involved in rewetting the fuel bundle surfaces as flow is returned to the reactor core. The fuel elements may be at very elevated temperatures so that the post-CHF, or steam blanketed, condition may continue even in the presence of returned water flow. Eventually, the surface temperature drops enough to permit a rewetting front to wash over the fuel element surface. Analysis includes transient conduction of the fuel elements and the interaction with the steam-water heat transfer processes.

Critical heat flux phenomena

Critical heat flux is one of the most important parameters in steam generator design. CHF denotes the set of operating conditions (mass flux, pressure, heat flux and steam quality) covering the transition from the relatively high heat transfer rates associated with nucleate or forced convective boiling to the lower rates resulting from transition or film boiling (Figs. 1 and 2). These operating conditions have been found to be geometry specific. CHF encompasses the phenomena of departure from nucleate boiling (DNB), burnout, dryout and boiling crisis. One objective in recirculating boiler and nuclear reactor designs is to avoid CHF conditions. In once-through steam generators, the objective is to design to accommodate the temperature increase at the CHF locations. In this process, the heat flux profile, flow passage geometry, operating pressure and inlet enthalpy are usually fixed, leaving mass flux, local quality, diameter and some surface effects as the more easily adjusted variables.

Factors affecting CHF Critical heat flux phenomena under flowing conditions found in fossil fuel and nuclear steam generators are affected by a variety of parameters.[5] The primary parameters are the operating conditions and the design geometries. The operating conditions affecting CHF are pressure, mass flux and steam quality. Numerous design geometry factors include flow passage dimensions and shape, flow path obstructions, heat flux profile, inclination and wall surface configuration. Several of these effects are illustrated in Figs. 3 through 7.

Fig. 3 illustrates the effect of increasing the heat input on the location of the temperature excursion in a uniformly heated vertical tube cooled by upward flowing water. At low heat fluxes, the water flow can be almost completely evaporated to steam before any temperature rise is observed. At moderate and high heat fluxes, the CHF location moves progressively towards the tube inlet and the maximum temperature excursion increases. At very high heat fluxes, CHF occurs at a low steam quality and the metal temperature excursion can be high enough to melt the tube. At extremely high heat input rates, CHF can occur in subcooled water. Avoiding this type of CHF

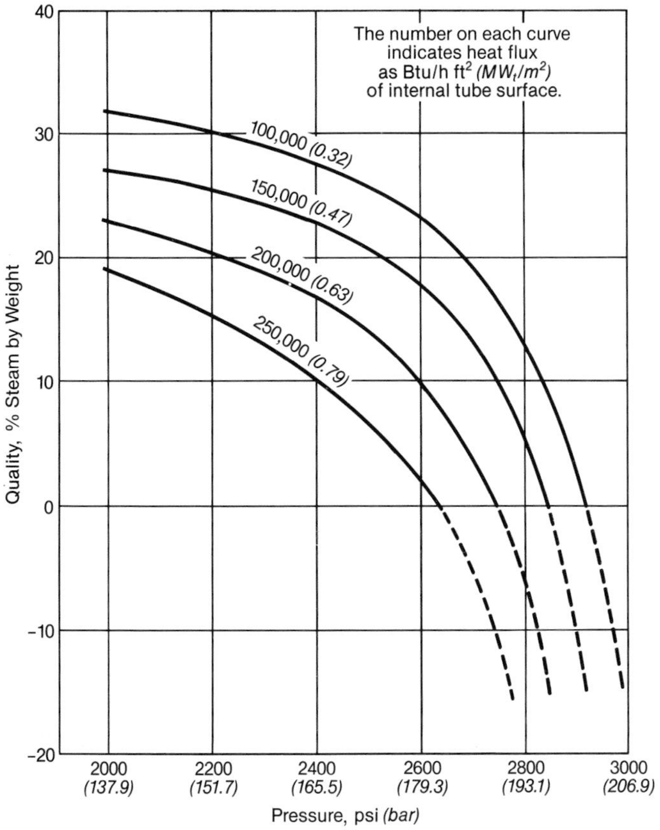

Fig. 4 Steam quality limit for CHF as a function of pressure.

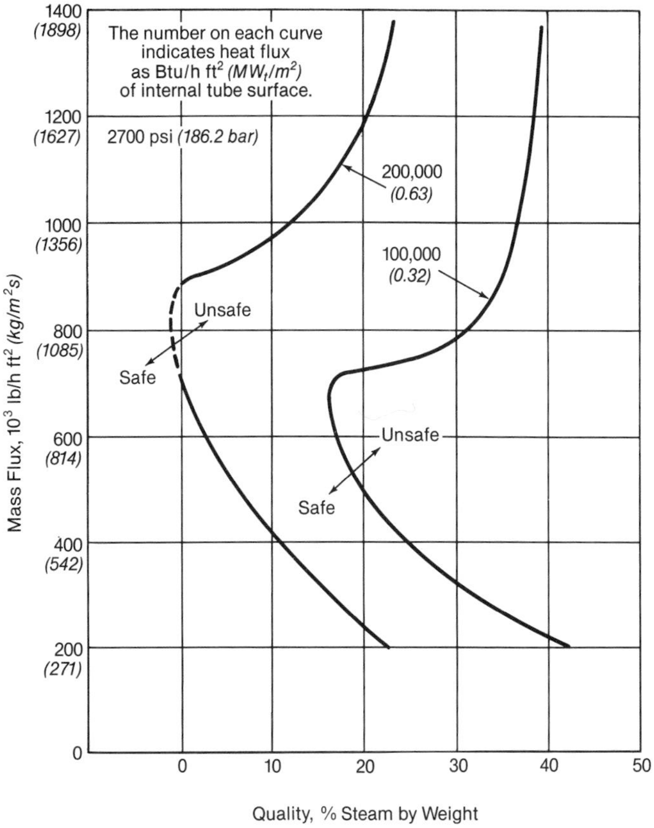

Fig. 5 Steam quality limit for CHF as a function of mass flux.

is an important design criterion for pressurized water nuclear reactors.

Many large fossil fuel boilers are designed to operate between 2000 and 3000 psi (137.9 and 206.9 bar). In this range, pressure has a very important effect, shown in Fig. 4, with the steam quality limit for CHF falling rapidly near the critical pressure; i.e., at constant heat flux, CHF occurs at lower steam qualities as pressure rises.

Many CHF correlations have been proposed and are satisfactory within certain limits of pressure, mass velocity and heat flux. Fig. 5 is an example of a correlation which is useful in the design of fossil fuel natural circulation boilers. This correlation defines safe and unsafe regimes for two heat flux levels at a given pressure in terms of steam quality and mass velocity. Additional factors must be introduced when tubes are used in membrane or tangent wall construction, are inclined from the vertical, or have different inside diameter or surface con-

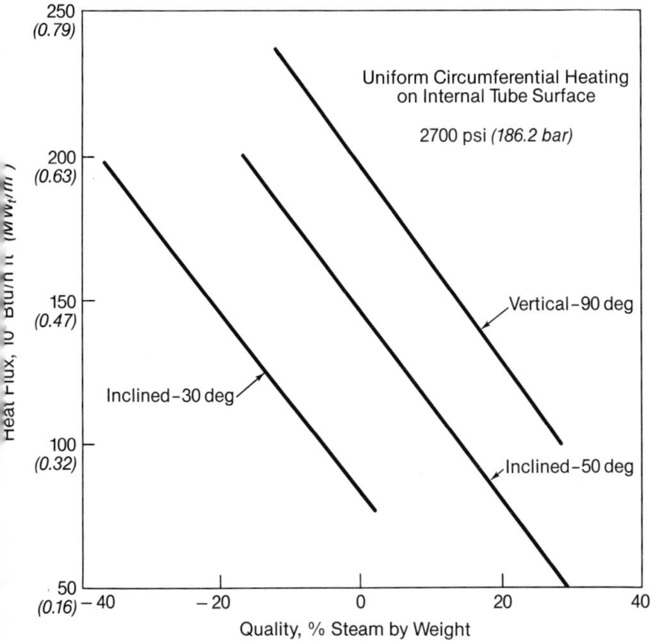

Fig. 6 Effect of inclination on CHF at 700,000 lb/h ft² (950 kg/m² s).[6]

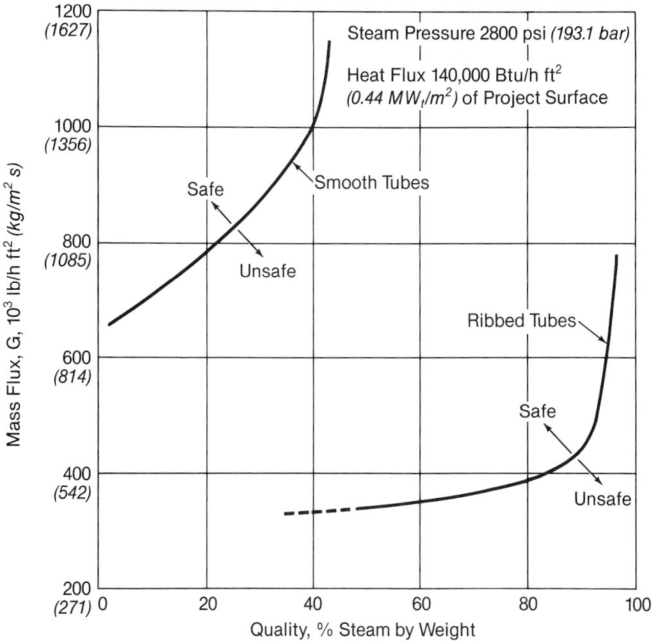

Fig. 7 Steam quality limit for CHF in smooth and ribbed bore tubes.

figuration. The inclination of the flow passage can have a particularly dramatic effect on the CHF conditions as illustrated in Fig. 6.[6]

Ribbed tubes Since the 1930s, B&W has investigated a large number of devices, including internal twisters, springs and grooved, ribbed and corrugated tubes to delay the onset of CHF. The most satisfactory overall performance was obtained with tubes having helical ribs on the inside surface.

Two rib configurations were developed:

1. single-lead ribbed tubes (Fig. 8a) for small internal diameters used in once-through subcritical pressure boilers, and
2. multi-lead ribbed tubes (Fig. 8b) for larger internal diameters used in natural circulation boilers.

Both of these ribbed tubes have shown a remarkable ability to delay the breakdown of nucleate boiling. Fig. 7 compares the effectiveness of a ribbed tube to that of a smooth tube in a membrane wall configuration. This plot is different from Fig. 5 in that heat flux is given as an average over the flat projected surface. This is more meaningful in discussing membrane wall heat absorption.

The ribbed bore tubes provide a balance of improved CHF performance at an acceptable increase in pressure drop without other detrimental effects. The ribs generate a swirl flow resulting in a centrifugal action which forces the water to the tube wall and retards entrainment of the liquid. The steam blanketing and film dryout are therefore prevented until substantially higher steam qualities or heat fluxes are reached.

Because the ribbed bore tube is more expensive than a smooth bore tube, its use involves an economic balance of several design factors. In most instances, there is less incentive to use ribbed tubes below 2200 psi (151.7 bar).

Evaluation CHF is a complex combination of thermal-hydraulic phenomena for which a comprehensive theoretical basis is not yet available. As a result, experimental data are likely to continue to be the basis for CHF evaluations. Many data and correlations define CHF well over *limited* ranges of conditions and geometries. However, some progress is being made in developing more general evaluation procedures for at least the most studied case — a uniformly heated smooth bore tube with upward flowing water.

To address this complex but critical phenomenon in the design of reliable steam generating equipment, B&W has developed an extensive proprietary database and associated correlations. A graphical example is shown in Fig. 5 for a fossil fuel boiler tube. The B&W2 correlation[7] for nuclear reactor fuel rod bundle subchannel analysis is shown in Table 1.

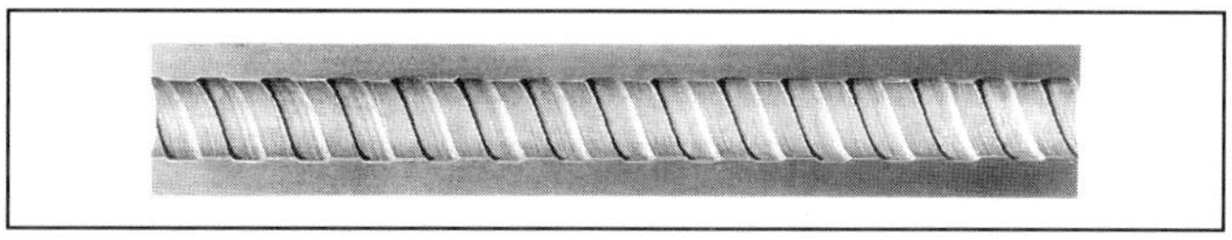

Fig. 8a Single-lead ribbed tube.

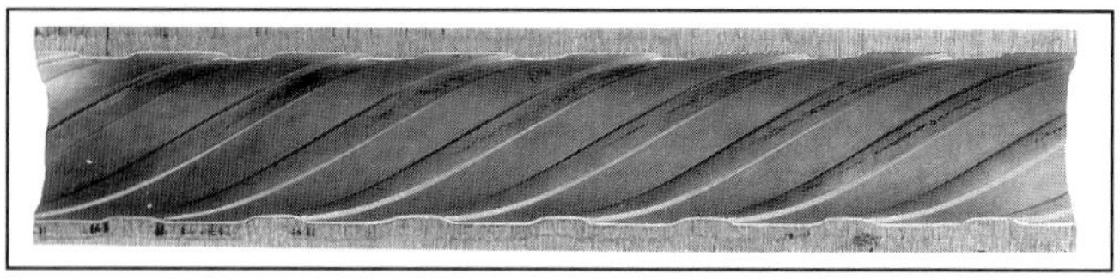

Fig. 8b Multi-lead ribbed tube.

CHF criteria A number of criteria are used to assess the CHF margins in a particular tube or tube bundle geometry.[8] These include the CHF ratio, flow ratio and quality margin, defined as follows:

1. $\text{CHF ratio} = \text{minimum value of } \dfrac{\text{CHF heat flux}}{\text{upset heat flux}}$

2. $\text{flow ratio} = \text{minimum value of } \dfrac{\text{min. design mass flux}}{\text{mass flux at CHF}}$

3. quality margin = CHF quality – max. design quality

The CHF ratios for a sample fossil fuel boiler are illustrated in Fig. 9 for a smooth bore tube (q''_B/q''_A) and a ribbed bore tube (q''_C/q''_A). The graph indicates the relative increase in local heat input which can be tolerated before the onset of CHF conditions. A similar relationship for a nuclear reactor fuel rod application is shown in Fig. 10.

Supercritical heat transfer

Unlike subcritical pressure conditions, fluids at supercritical pressures experience a continuous transition from water-like to steam-like characteristics. As a result, CHF conditions and boiling behavior would not be expected. However, at supercritical pressures, especially in the range of $1 < P/P_c < 1.15$ where P_c is the critical pressure, two types of boiling-like behavior have been observed: pseudo-boiling and pseudo-film boiling. Pseudo-boiling is an increase in heat transfer coefficient not accounted for by traditional convection relationships. In pseudo-film boiling, a dramatic reduction in the heat transfer coefficient is observed at high heat fluxes. This is similar to the critical heat flux condition at subcritical pressures.

These behaviors have been attributed to the sharp changes in fluid properties as the transition from water-like to steam-like behavior occurs.

Fluid properties In the supercritical region, the thermophysical properties important to the heat transfer pro-

Table 1
B&W2 Reactor Rod Bundle Critical Heat Flux (CHF) Correlation[7]

$$q''_{CHF} = \frac{(a - bD_i)\left[A_1(A_2G)^{A_3+A_4(P-2000)} - A_9 G x_{CHF} H_{fg}\right]}{A_5(A_6G)^{A_7+A_8(P-2000)}}$$

where

a = 1.15509		A	= area, in.^{2}
b = 0.40703		D_i	= equivalent diameter = $4A/Per$
A_1 = 0.37020 x 10^8		G	= mass flux, lb/h ft^2
A_2 = 0.59137 x 10^{-6}		H_{fg}	= latent heat of vaporization, Btu/lb
A_3 = 0.83040			
A_4 = 0.68479 x 10^{-3}		P	= pressure, psi
A_5 = 12.710		Per	= wetted perimeter, in.
A_6 = 0.30545 x 10^{-5}		x_{CHF}	= steam quality at CHF conditions, fraction steam by weight
A_7 = 0.71186			
A_8 = 0.20729 x 10^{-3}		q''_{CHF}	= heat flux at CHF conditions, Btu/h ft^2
A_9 = 0.15208			

cess, i.e., conductivity, viscosity, density and specific heat, experience radical changes as a certain pressure-dependent temperature is approached and exceeded. This is illustrated in Fig. 11. The transition temperature, referred to as the pseudo-critical temperature, is defined as the temperature where the specific heat, c_p, reaches its maximum. As the operating pressure is increased, the pseudo-critical temperature increases and the dramatic change in the thermophysical properties declines as this temperature is approached and exceeded.

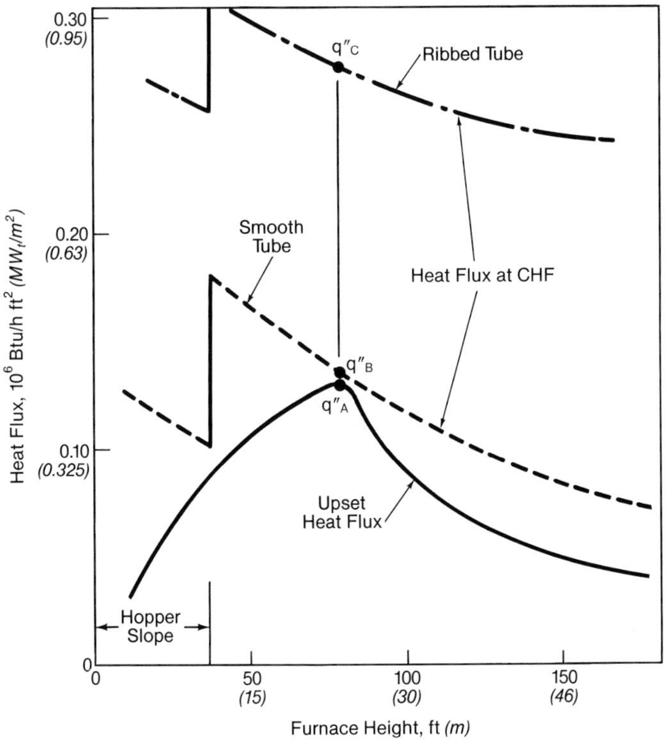

Fig. 9 Fossil boiler CHF ratio = minimum value of critical heat flux / upset heat flux.

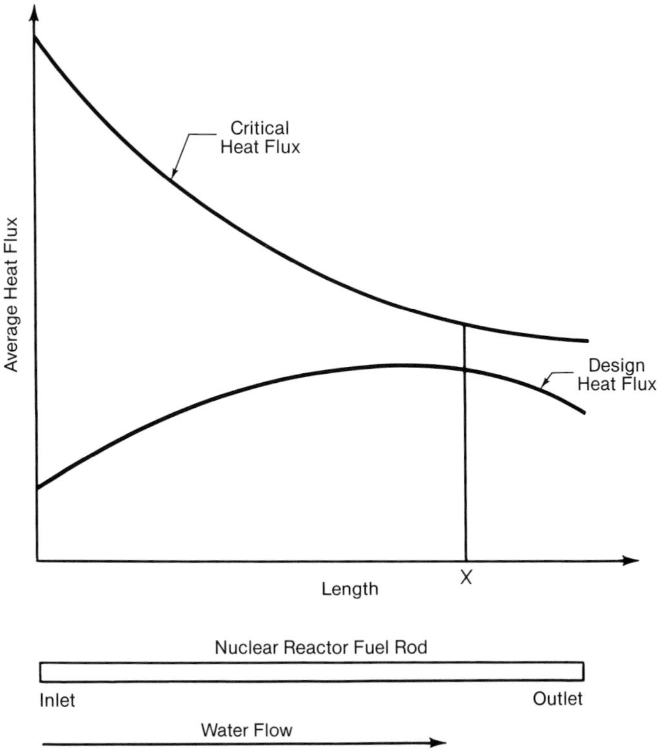

Fig. 10 Nuclear reactor CHF ratio = minimum value of critical heat flux / design heat flux.

Heat transfer rates Because of the significant changes in thermophysical properties (especially in specific heat) near the pseudo-critical temperature, a modified approach to evaluating convective heat transfer is needed. A number of correlations have been developed and a representative relationship for smooth bore tubes is:[9]

$$\frac{hD_i}{k_w} = 0.00459\left[\frac{D_i G}{\mu_w}\right]^{0.923} \times \left[\left(\frac{H_w - H_b}{T_w - T_b}\right)\left(\frac{\mu_w}{k_w}\right)\right]^{0.613}\left[\frac{v_b}{v_w}\right]^{0.231} \quad (5)$$

where

h = heat transfer coefficient, Btu/h ft^2 F (W/m^2 K)
k = thermal conductivity, Btu/h ft F (W/m K)
D_i = inside tube diameter, ft (m)
G = mass flux, lb/h ft^2 (kg/m^2 s)
μ = viscosity, lb/h ft (kg/m s)
H = enthalpy, Btu/lb (J/kg)
T = temperature F (C)
v = specific volume, ft^3/lb (m^3/kg)

The subscripts b and w refer to properties evaluated at the bulk fluid and wall temperatures respectively.

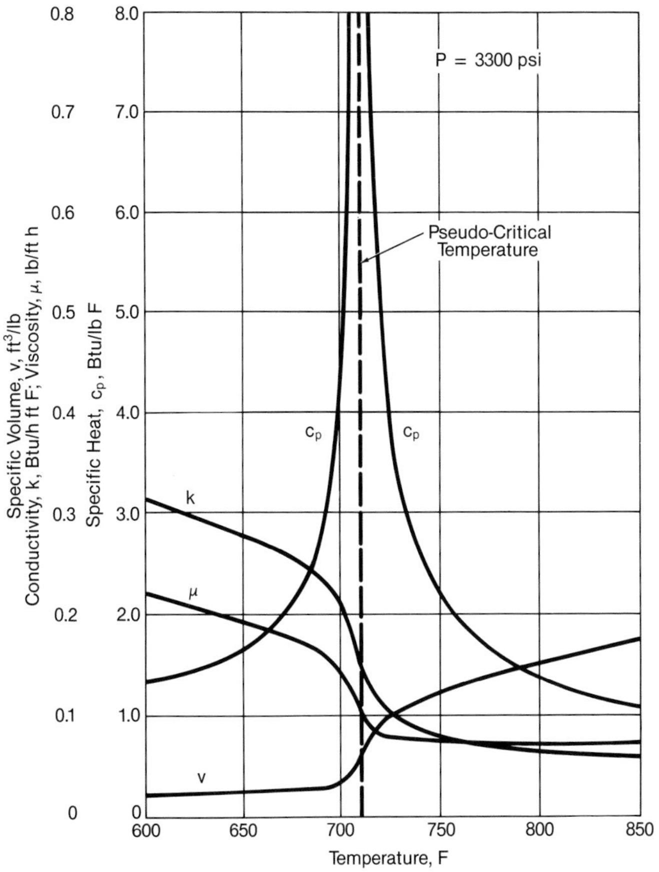

Fig. 11 Thermophysical properties of water (English units).

This correlation has demonstrated reasonable agreement with experimental data from tubes of 0.37 to 1.5 in. (9.4 to 38.1 mm) inside diameter and at low heat fluxes.

Pseudo-boiling For low heat fluxes and bulk fluid temperatures approaching the pseudo-critical temperature, an improvement in the heat transfer rate takes place. The enhanced heat transfer rate observed is sometimes referred to as pseudo-boiling. It has been attributed to the increased turbulence resulting from the interaction of the water-like and steam-like fluids near the tube wall.

Pseudo-film boiling Potentially damaging temperature excursions associated with a sharp reduction in heat transfer can be observed at high heat fluxes. This temperature behavior is similar to the CHF phenomenon observed at subcritical conditions and is referred to as pseudo-film boiling. This phenomenon has been attributed to a limited ability of the available turbulence to move the higher temperature steam-like fluid away from the tube wall into the colder higher density (water-like) fluid in the bulk stream. A phenomenon similar to steam blanketing occurs and the wall temperature increases in response to the relatively constant applied heat flux.

Single-lead ribbed bore tubes are very effective in suppressing the temperature peaks encountered in smooth bore tubes.[10]

Two-phase flow

Flow patterns

As illustrated in Fig. 2, two-phase steam-water flow may occur in many regimes or structures. The transition from one structure to another is continuous rather than abrupt, especially under heated conditions, and is strongly influenced by gravity, i.e., flow orientation. Because of the qualitative nature of flow pattern identification, there are probably as many flow pattern descriptions as there are observers. However, for vertical, heated, upward, co-current steam-water flow in a tube, four general flow patterns are generally recognized (see Fig. 12):

1. *Bubbly flow* Relatively discrete steam bubbles are dispersed in a continuous liquid water phase. Bubble size, shape and distribution are dependent upon the flow rate, local enthalpy, heat input rate and pressure.
2. *Intermediate flow* This is a range of patterns between bubbly and annular flows; the patterns are also referred to as slug or churn flow. They range from: a) large bubbles, approaching the tube size in diameter, separated from the tube wall by thin annular films and separated from each other by slugs of liquid which may also contain smaller bubbles, to b) chaotic mixtures of large nonsymmetric bubbles and small bubbles.
3. *Annular flow* A liquid layer is formed on the tube wall with a continuous steam core; most of the liquid is flowing in the annular film. At lower steam qualities, the liquid film may have larger amplitude waves adding to the liquid droplet entrainment and transport in the continuous steam core. At high qualities, the annular film becomes very thin, bubble generation is suppressed and the large amplitude waves disappear.
4. *Mist flow* A continuous steam core transports entrained water droplets which slowly evaporate until a single-phase steam flow occurs. This is also referred to as droplet or dispersed flow.

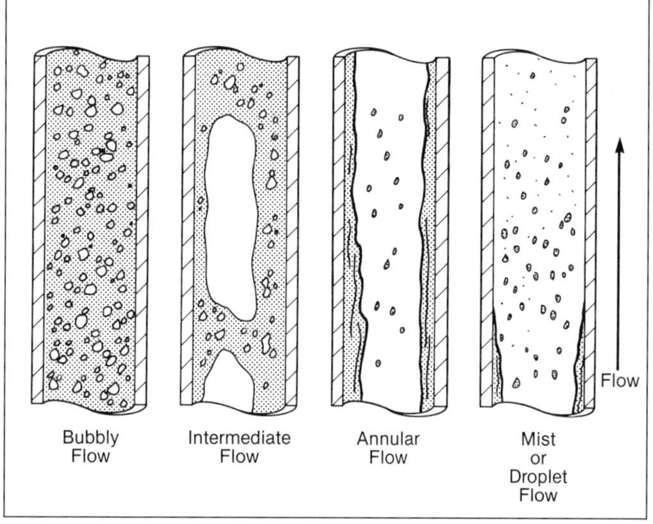

Fig. 12 Flow pattern — upward, co-current steam-water flow in a heated vertical tube.

In the case of inclined and horizontal co-current steam-water flow in heated tubes, the flow patterns are further complicated by stratification effects. At high flow rates, the flow patterns approach those of vertical tubes. At lower rates, additional distinct flow patterns (wavy, stratified and modified plug) emerge as gravity stratifies the flow with steam concentrated in the upper portion of the tube. This can be a problem where inclined tubes are heated from the top. CHF or dryout conditions occur at much lower steam qualities and lower heat input rates in such inclined or horizontal tubes.

Additional complexity in patterns is observed when two-phase flow occurs in parallel or crossflow tube bundles. The tubes, baffles, support plates and mixing devices further disrupt the flow pattern formation.

Flow maps The transitions from one flow regime to another are quite complex, with each transition representing a combination of factors. However, two dimensional flow maps provide at least a general indication of which flow pattern is likely under given operating conditions. The maps generally are functions of superficial gas and liquid velocities. An example for vertical, upward, steam-water co-current flow is provided in Fig. 13.[11] The axes in this figure represent the superficial momentum fluxes of the steam (y-axis) and water (x-axis). A sample flow line is shown beginning at nearly saturated water conditions and ending with saturated steam conditions. The tube experiences bubbly flow only near its inlet. This is followed by a brief change to intermediate flow before annular flow dominates the heated length.

Other flow maps are available for arrangements such as downflow tubes, inclined tubes and bundles. Flow maps, however, are only approximations providing guidance in determining the relevant flow structure for a given situation.

Pressure loss

The local pressure loss, ΔP [lb/ft^2 (Pa)] or gradient $\delta P / \delta l$ [lb/ft^2/ft (Pa/m)] in a two-phase steam-water system may be represented by:

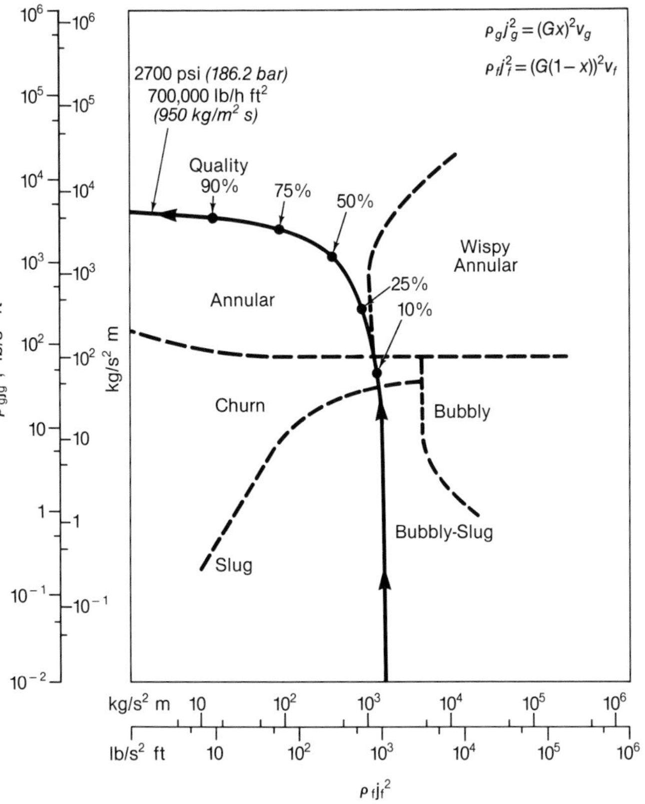

Fig. 13 Flow pattern map for vertical upward flow of water.[11]

$$\Delta P = \Delta P_f + \Delta P_a + \Delta P_g + \Delta P_l \quad \textbf{(6a)}$$

or

$$-\frac{\delta P}{\delta l} = -\left(\frac{\delta P}{\delta l}\right)_f - \left(\frac{\delta P}{\delta l}\right)_a - \left(\frac{\delta P}{\delta l}\right)_g + \Delta P_l \quad \textbf{(6b)}$$

The ΔP_f and $-(\delta P / \delta l)_f$ terms account for local wall friction losses. The ΔP_a and $-(\delta P / \delta l)_a$ terms address the momentum or acceleration loss incurred as the volume increases due to evaporation. The hydraulic or static head loss is accounted for by ΔP_g and $-(\delta P / \delta l)_g$. Finally, all of the local losses due to fittings, contractions, expansions, bends, or orifices are included in ΔP_l. The evaluation of these parameters is usually made using one of two models: homogeneous flow or separated flow.

A parameter of particular importance when evaluating the pressure loss in steam-water flows is void fraction. The void fraction can be defined by time averaged flow area ratios or local volume ratios of steam to the total flow. The area based void fraction, α, can be defined as the ratio of the time averaged steam flow cross-sectional area (A_{steam}) to the total flow area ($A_{steam} + A_{water}$):

$$\alpha = \frac{A_{steam}}{A_{steam} + A_{water}} \quad \textbf{(7)}$$

Using the simple continuity equation, the relationship between quality, x, and void fraction is:

$$\alpha = \frac{x}{x + (1-x)\dfrac{\rho_g}{\rho_f} S} \quad \textbf{(8)}$$

where

S = ratio of the average cross-sectional velocities of steam and water (referred to as *slip*)
ρ_g = saturated steam density, lb/ft³ (kg/m³)
ρ_f = saturated water density, lb/ft³ (kg/m³)

If the steam and water are moving at the same velocity, $S = 1$ (no slip). Obviously, the relationship between void fraction and quality is also a strong function of system pressure. This relationship is illustrated in Fig. 14. The difference between the homogeneous and separated flow models is illustrated by the shaded band. The upper bound is established by the homogeneous model and the lower bound by the separated flow model.

Homogeneous model The homogeneous model is the simpler approach and is based upon the premise that the two-phase flow behavior can be directly modeled after single-phase behavior (see Chapter 3) if appropriate average properties are determined. The temperature and velocities of steam and water are assumed equal. The mixed weight averaged specific volume (v) or the inverse of the homogeneous density ($1/\rho_{hom}$) is used:

$$v = v_f(1-x) + v_g x \quad \textbf{(9a)}$$

or

$$\frac{1}{\rho_{hom}} = \frac{(1-x)}{\rho_f} + \frac{x}{\rho_g} \quad \textbf{(9b)}$$

where

v_f = saturated water specific volume, ft³/lb (m³/kg)
v_g = saturated steam specific volume, ft³/lb (m³/kg)
ρ_f = saturated water density, lb/ft³ (kg/m³)
ρ_g = saturated steam density, lb/ft³ (kg/m³)
x = steam quality

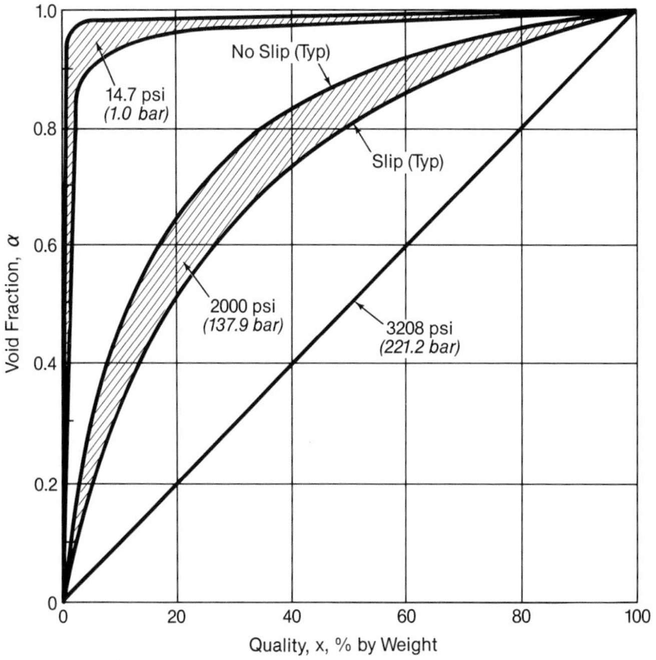

Fig. 14 Void fraction — quality relationship (homogeneous model, upper bound; separated flow model, lower bound).

This model provides reasonable results when high or low steam qualities exist, when high flow rates are present, or at higher pressures. In these cases, the flow is reasonably well mixed.

The friction pressure drop (ΔP_f) can be evaluated by the equations provided in Chapter 3 using the mixture thermophysical properties. The pressure difference due to elevation (ΔP_g) can be evaluated as:

$$\Delta P_g = \pm \rho_{hom} \left(\frac{g}{g_c} \right) L \sin \theta \qquad (10)$$

where

g = acceleration of gravity, ft/s^2 (m/s^2)
g_c = 32.17 lbm ft/lbf s^2 (1 kg m/N s^2)
L = length, ft (m)
θ = angle from the horizontal

The constant g_c is discussed at Equation 10 in Chapter 2. A pressure gain occurs in downflow and a pressure loss occurs in upflow. The acceleration loss can be evaluated by:

$$\Delta P_a = \frac{G^2}{g_c} \left(\frac{1}{\rho_{out}} - \frac{1}{\rho_{in}} \right) \qquad (11)$$

where

G = mass flux, lb/s ft^2 (kg/m^2 s)
ρ_{out} = outlet homogeneous density, lb/ft^3 (kg/m^3)
ρ_{in} = inlet homogeneous density, lb/ft^3 (kg/m^3)

Separated flow model In the steady-state separated flow model, the steam and water are treated as separate streams under the same pressure gradient but different velocities and differing properties. When the actual flow velocities of steam and water are equal, the simplest separated flow models approach the homogeneous case. Using one of several separated flow models[1] with unequal velocities, the pressure drop components (in differential form) are:

$$-\left(\frac{\delta P}{\delta l} \right)_f = -\left(\frac{\delta P}{\delta l} \right)_{LO} \phi^2_{LO} \quad \text{(friction)} \qquad (12)$$

$$-\left(\frac{\delta P}{\delta l} \right)_{LO} = \frac{f}{D_i} \frac{G^2 v_f}{2 g_c} \quad \text{(single-phase friction)} \qquad (13)$$

$$-\left(\frac{\delta P}{\delta l} \right)_a = \frac{G^2}{g_c} \frac{\delta}{\delta l} \left(\frac{x^2 v_g}{\alpha} + \frac{(1.0 - x)^2 v_f}{(1.0 - \alpha)} \right) \quad \text{(acceleration)} \qquad (14)$$

$$-\left(\frac{\delta P}{\delta l} \right)_g = \frac{g}{g_c} \sin \theta \left(\frac{\alpha}{v_g} + \frac{(1.0 - \alpha)}{v_f} \right) \quad \text{(static head)} \qquad (15)$$

$$\Delta P_l = \Phi K \frac{G^2 v_f}{2 g_c} \quad \text{(local losses)} \qquad (16)$$

where

Φ and ϕ^2_{LO} = appropriate two-phase multipliers
G = mass flux, lb/s ft^2 (kg/m^2 s)
f = fanning friction factor (see Chapter 3)
D_i = tube inside diameter, ft (m)
g = acceleration of gravity, ft/s^2 (m/s^2)
g_c = 32.17 lbm ft/lbf s^2 (1 kg m/N s^2)
v_f = liquid specific volume, ft^3/lb (m^3/kg)
v_g = vapor specific volume, ft^3/lb (m^3/kg)
x = steam quality
α = void fraction
θ = angle from the horizontal
K = loss coefficient

While ΔP_l usually represents just the irreversible pressure loss in single-phase flows, the complexity of two-phase flows results in the loss of ΔP_l typically representing the reversible and irreversible losses for fittings.

To evaluate the individual pressure losses from Equations 12 through 16 and Equation 6b, it is necessary to calculate ϕ^2_{LO}, α and Φ. Unfortunately, these factors are not well defined.

Specific correlations and evaluations can only be used where experimental data under similar conditions provide confidence in the prediction. Proprietary correlations used by B&W are based upon experimental data and practical experience.

For straight vertical tubes, generally available representative relationships include:

1. *Acceleration loss* The void fraction can frequently be evaluated with the homogeneous model ($S = 1$ in Equation 8).

2. *Friction loss and void fraction* Typical two-phase multiplier, ϕ^2_{LO}, and void fraction, α, relationships are presented by Thom,[12] Martinelli-Nelson,[13] Zuber-Findlay[14] and Chexal-Lellouche.[15] For illustration purposes the correlations of Thom are presented in Figs. 15 and 16. These curves can be approximated by:

$$\phi^2_{LO} = \left\{ \left[0.97303(1-x) + x \left(\frac{v_g}{v_f} \right) \right]^{0.5} \times \left[0.97303(1-x) + x \right]^{0.5} + 0.027(1-x) \right\}^{2.0} \qquad (17)$$

and

$$\alpha = \frac{\gamma x}{1 + x(\gamma - 1)} \qquad (18)$$

where

γ = $(v_g / v_f)^n$
n = $(0.8294 - 1.1672/P)$
P = pressure, psi
v_g = saturated steam specific volume, ft^3/lb
v_f = saturated liquid specific volume, ft^3/lb
x = steam quality

Instabilities

Instability in two-phase flow refers to the set of operating conditions under which sudden changes in flow direction, reduction in flow rate and oscillating flow rates can occur in a single flow passage. Often in manifolded multi-channel systems, the overall mass flow rate can remain constant while oscillating flows in individual channels still may occur. Such unstable conditions in steam generating systems can result in:

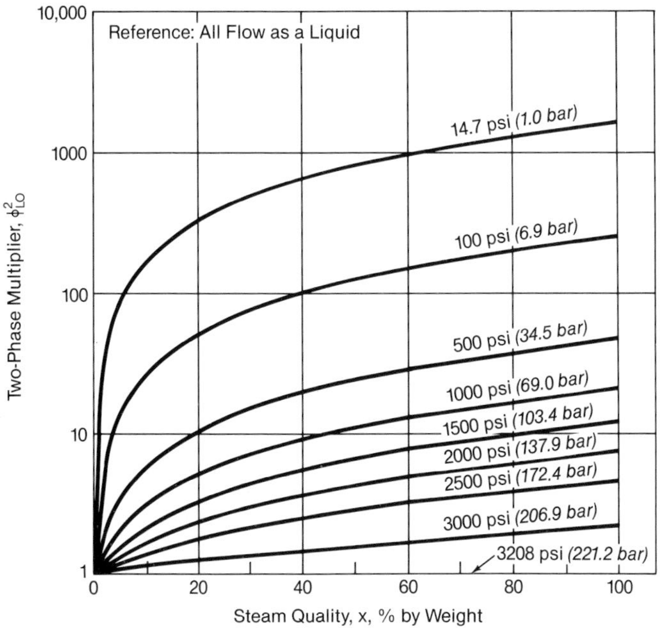

Fig. 15 Thom two-phase friction multiplier.

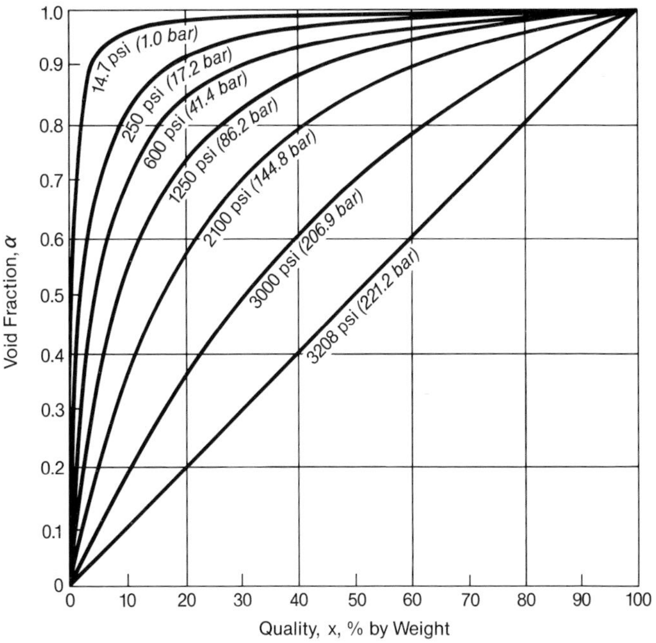

Fig. 16 Thom void fraction correlation (>3% SBW).[12]

1. unit control problems, including unacceptable variations in steam drum water level,
2. CHF,
3. tube metal temperature oscillation and thermal fatigue failure, and
4. accelerated corrosion attack.

Two of the most important types of instabilities in steam generator design are excursive instability, including Ledinegg and flow reversal, and density wave oscillations. The first is a static instability evaluated using steady-state equations while the last is dynamic in nature requiring the inclusion of time dependent factors.

Excursive and flow reversal instability evaluation The excursive instability is characterized by conditions where small perturbations in operating parameters result in a large flow rate change to a separate steady-state level. This can occur in both single channel and multi-channel manifolded systems. Excursive instabilities can be predicted by using the Ledinegg criteria.[16] Instability may occur if the slope of the pressure drop versus flow characteristic curve (internal) for the tube becomes less than the slope of the supply (or applied) curve at any intersection point:

$$\left(\frac{\delta \Delta P}{\delta G}\right)_{internal} \leq \left(\frac{\delta \Delta P}{\delta G}\right)_{applied} \tag{19}$$

The stable and unstable situations are illustrated in Fig. 17. As shown in the figure for unstable conditions, if the mass flow rate drops below point B then the flow rate continues to fall dramatically because the applied pumping head is less than that needed to move the fluid. For slightly higher mass flow rates (higher than point B), a dramatic positive flow excursion occurs because the pumping head exceeds the flow system requirement.

In most systems, the first term in Equation 19 is generally positive and the second is negative. Therefore, Equation 19 predicts stability. However, in two-phase systems, thermal-hydraulic conditions may combine to produce a local area where $(\delta \Delta P / \delta G)_{internal}$ is negative and the potential for satisfying Equation 19 and observing an instability exists. A heated tube flow characteristic showing a potential region of instability is illustrated in Fig. 18 where multiple flow rates can occur for a single applied pressure curve. Operating at point B is unstable with small disturbances resulting in a shift to point A or point C. More intense disturbances could result in flow shifts between A and C.

For the relatively small subcooling found at the entrance to tube panels in recirculating drum boilers and due to the relatively low exit steam qualities, negative slope regions in the pressure drop versus flow curves are typically not observed for positive flow cases. However, for once-through fossil fuel boilers and nuclear steam generators with high subcooling at the inlet and evaporation to dryness, negative slope regions in the upflow portion of the pressure drop characteristic may occur. Steps can be taken to avoid operation in any region where the circuit internal $\delta\Delta P/\delta G \leq 0$. General effects of operating and design parameters on the pressure drop versus mass flow curves include:

Parameter increased	Effect on ΔP	Comment
heat input	decrease	more stable
inlet ΔP	increase	more stable
pressure	increase	more stable

In situations where static instability may occur, the inlet pressure drop can be increased by adding an orifice or flow restriction to modify the overall flow characteristic as shown in Fig. 18.

Density wave instability Density wave instabilities involve kinematic wave propagation phenomena. Regenerative feedback between flow rate, vapor generation rate and pressure drop produce self sustaining alternating waves of higher and lower density mixture that travel through the tube. This dynamic instability can occur in

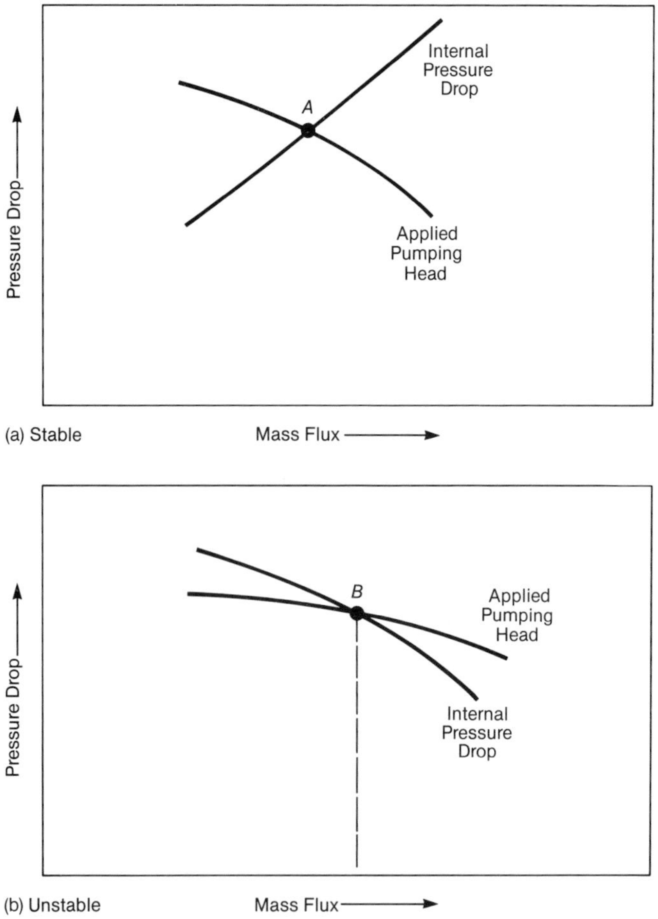

Fig. 17 Stable and unstable flow-pressure drop characteristics.

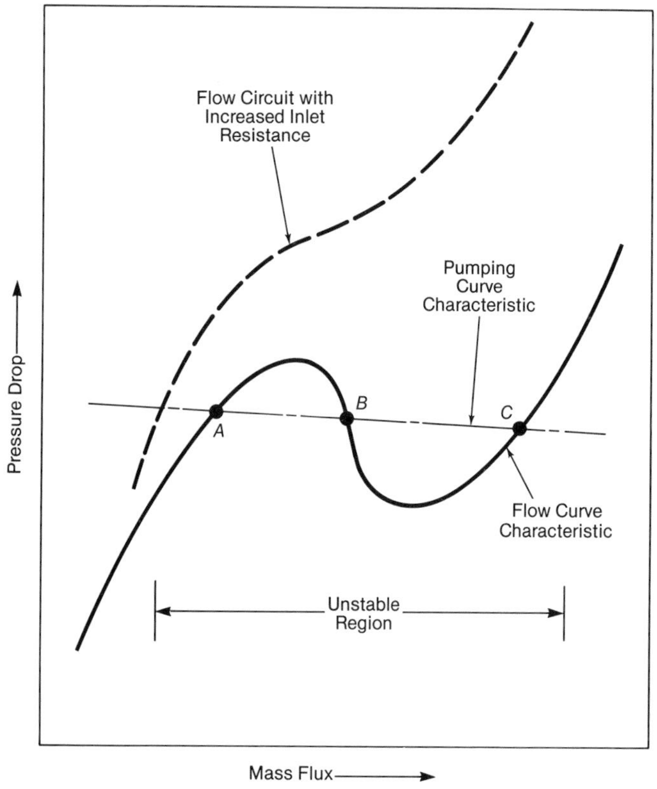

Fig. 18 Pressure drop characteristic showing unstable region.

single tubes that contain two-phase flows. In addition, when multiple tubes are connected by inlet and outlet headers, a more complex coupled channel instability, which is driven by density wave oscillations, may occur.

Density wave oscillations can be predicted by the application of feedback control theory. A number of computer codes have been developed to provide these predictions. In addition, instability criteria, which use a series of dimensionless parameters to reduce the complexity of the evaluation, have been developed.

Effects of operating and design parameters on the density wave instability include:

Parameter increased	Change in stability
mass flux	improved
heat flux	reduced
pressure	improved
inlet ΔP	improved
inlet subcooling	improved (large subcooling)
	reduced (small subcooling)

Steam-water separation

Subcritical pressure recirculating boilers and steam generators are equipped with large cylindrical vessels called steam drums. Their primary objective is to permit separation of the saturated steam from the steam-water mixture leaving the boiling heat transfer surfaces. The steam-free water is recirculated with the feedwater to the heat absorbing surfaces for further steam generation. The saturated steam is discharged through a number of outlet nozzles for direct use or further heating. The steam drum also serves to:

1. mix the feedwater with the saturated water remaining after steam separation,
2. mix the corrosion control and water treatment chemicals (if used),
3. purify the steam to remove contaminants and residual moisture,
4. remove part of the water (blowdown) to control the boiler water chemistry (solids content), and
5. provide limited water storage to accommodate rapid changes in boiler load.

However, the primary function of the steam drum is to permit the effective separation of steam and water. This may be accomplished by providing a large steam-water surface for natural gravity driven separation or by having sufficient space for mechanical separation equipment.

High efficiency separation is critical in most boiler applications in order to:

1. prevent water droplet *carryover* into the superheater where thermal damage may occur,
2. minimize steam *carryunder* in the water leaving the drum where residual steam can reduce the effective hydraulic pumping head, and
3. prevent the carryover of solids dissolved in the steam-entrained water droplets into the superheater and turbine where damaging deposits may form.

The last item is of particular importance. Boiler water may contain contaminants, principally in solution.

These arise from impurities in the makeup water, treatment chemicals and condensate system leaks, as well as from the reaction of the water and contaminants with the boiler and preboiler equipment materials. Even low levels of these solids in the steam (less than 0.6 ppm) can damage the superheater and turbine. Because the solubility of these solids is typically several orders of magnitude less in steam than in water (see Chapter 42), small amounts of water droplet carryover (>0.25% by weight) may result in dramatically increased solids carryover and unacceptable deposition in the superheater and turbine. The deposits have caused turbine damage as well as superheater tube temperature increases, distortion and burnout.

A cross-section of a horizontal steam drum found on a modern high capacity fossil fuel boiler is shown in Fig. 19. This illustrates the general arrangement of the baffle plates, primary cyclone separators, secondary separator elements (scrubbers), water discharger (downcomer) and feedwater inlets. The blowdown (water removal) connections are not shown. The steam-water separation typically takes place in two stages. The primary separation removes nearly all the steam from the water so that very little steam is recirculated from the bottom of the drum through the outlet connection (downcomer) towards the heated tubes. The steam leaving the primary separators in high pressure boilers still typically contains too much liquid in the form of contaminant containing droplets for satisfactory superheater and turbine performance. Therefore, the steam is passed through a secondary set of separators, or scrubber elements (usually closely spaced, corrugated parallel plates) for final water droplet removal. The steam is then exhausted through several connections. As this figure indicates, successful steam-water separation involves the integrated operation of primary separators, secondary scrubbers and general drum arrangement.

Factors affecting steam separation

Effective steam separation from the steam-water mixture relies on certain design and operating factors. The design factors include:

1. pressure,
2. drum length and diameter,
3. rate of steam generation,
4. average inlet steam quality,
5. type and arrangement of mechanical separators,
6. feedwater supply and steam discharge equipment arrangement, and
7. arrangement of downcomer and riser connections to the steam drum.

The operating factors include:

1. pressure,
2. boiler load (steam flow),
3. type of steam load,
4. chemical analysis of boiler water, and
5. water level.

Primary separation equipment generally takes one of three forms:

1. natural gravity driven separation,
2. baffle assisted separation, and
3. high capacity mechanical separation.

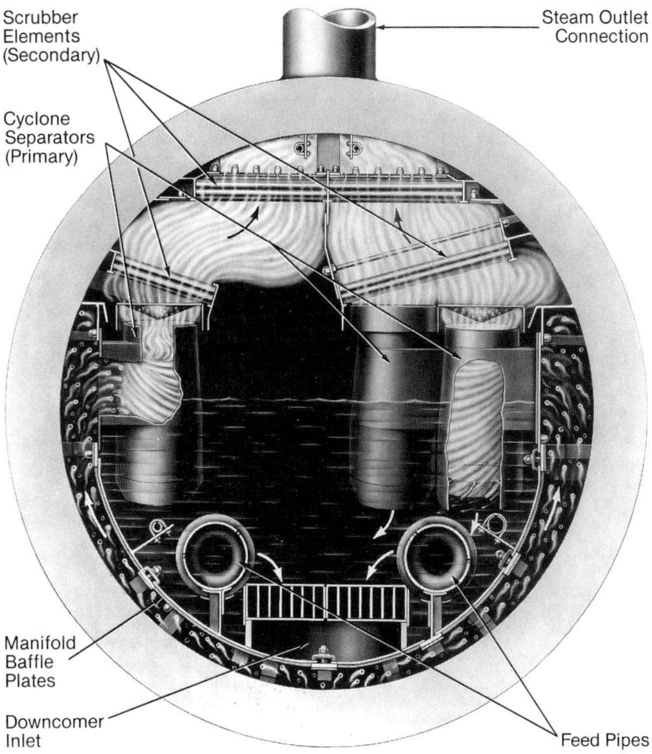

Fig. 19 Steam drum with three rows of primary cyclone separators.

Natural gravity driven separation

While simple in concept, natural steam-water separation is quite complex. It is strongly dependent upon inlet velocities and inlet locations, average inlet steam quality, water and steam outlet locations and disengagement of liquid and steam above the nominal water surface. Some of these effects are illustrated in Figs. 20 and 21.

For a low rate of steam generation, up to about 3 ft/s (0.9 m/s) velocity of steam leaving the water surface, there is sufficient time for the steam bubbles to separate from the mixture by gravity without being drawn into the discharge connections and without carrying entrained water droplets into the steam outlet (Fig. 20a). However,

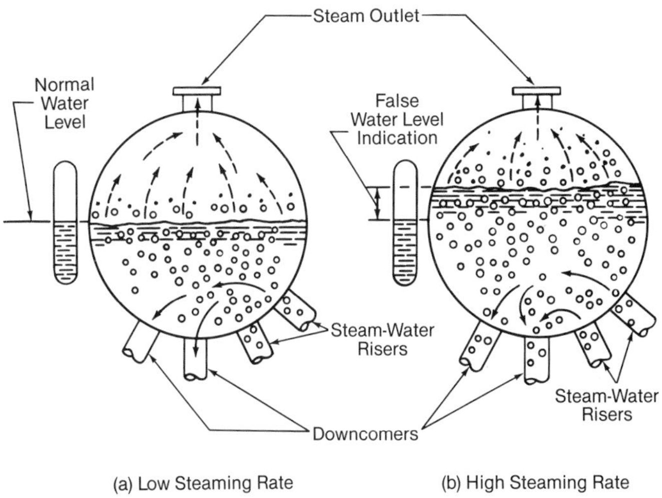

Fig. 20 Effect of rate of steam generation on steam separation in a boiler drum without separation devices.

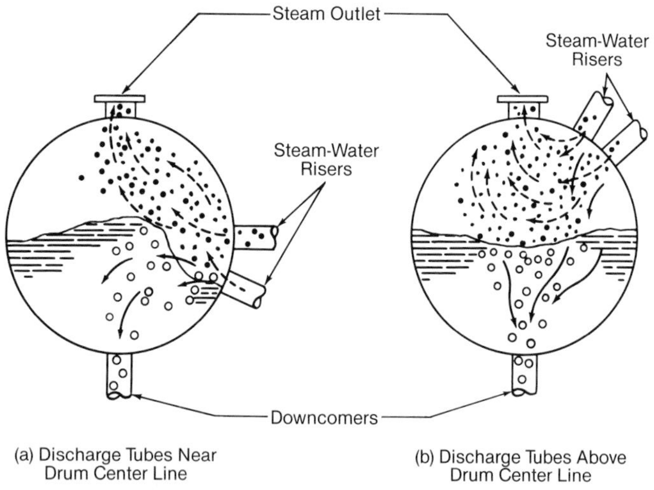

Fig. 21 Effect of location of discharge from risers on steam separation in a boiler drum without separation devices.

for the same arrangement at a higher rate of steam generation (Fig. 20b), there is insufficient time to attain either of these desirable results. Moreover, the dense upward traffic of steam bubbles in the mixture may also cause a false water level indication, as shown.

The effect of the riser or inlet connection locations in relation to the water level is illustrated in diagrams a and b of Fig. 21. Neither arrangement is likely to yield desirable results in a drum where gravity alone is used for separation.

From an economic standpoint, the diameter of a single drum may become prohibitive. To overcome this limitation, several smaller steam drums may be used, as shown in Fig. 22a, although this is no longer common. However, in most boiler applications, natural gravity driven separation alone is generally uneconomical, leading to the need for separation assistance.

Baffle assisted primary separation

Simple screens and baffle arrangements may be used to greatly improve the steam-water separation process. Three relatively common baffle arrangements are illustrated in Fig. 22. In each case, the baffles provide: 1) changes in direction, 2) more even distribution of the steam-water mixture, 3) added flow resistance, and 4) the maximum steam flow travel length to enhance the gravity driven separation process. Various combinations of perforated plates have also been used. The performance of these devices must be determined by experimental evaluations and they are typically limited to smaller, low capacity boilers.

Mechanical primary separators

Centrifugal force or radial acceleration is used almost universally for modern steam-water separators. Three types of separators are shown in Fig. 23, the conical cyclone, the curved arm and the horizontal cyclone. The B&W vertical cyclone steam separator is shown in more detail in Fig. 24. Vertical cyclones are arranged internally in rows along the length of the drum and the steam-water mixture is admitted tangentially as shown in Fig. 19. The water forms a layer against the cylinder walls and the steam moves to the core of the cylinder then upward. The water flows downward in the cylinder and is discharged through an annulus at the bottom, below the drum water level. With the water returning from drum storage to the downcomers virtually free of steam bubbles, the maximum net pumping head is available for producing flow in the circuits. The steam moving upward from the cylinder passes through a small primary corrugated scrubber at the top of the cyclone (see Fig. 24) for additional separation. Under many operating conditions, no further separation is required.

When wide load fluctuations and water analysis variations are expected, large corrugated secondary scrubbers may be installed at the top of the drum (see Fig. 19) to provide very high steam separation. These scrubbers are termed secondary separators. They provide a large surface which intercepts water droplets as the steam flows sinuously between closely fitted plates. Steam velocity through the corrugated plate assembly is very low, so that water re-entrainment is avoided. The collected water is drained from the bottom of the assembly to the water below.

One to four rows of cyclone separators are installed in boiler drums, with ample room for access. For smaller boilers at lower pressures [100 psig (6.9 bar gauge)], the separation rate of clean steam by single and double rows of cyclone separators is approximately 4000 and 6000 lb, respectively, per hour per foot of drum length (1.7 and 2.5 kg/s m). At pressures near 1050 psig (72.4 bar gauge),

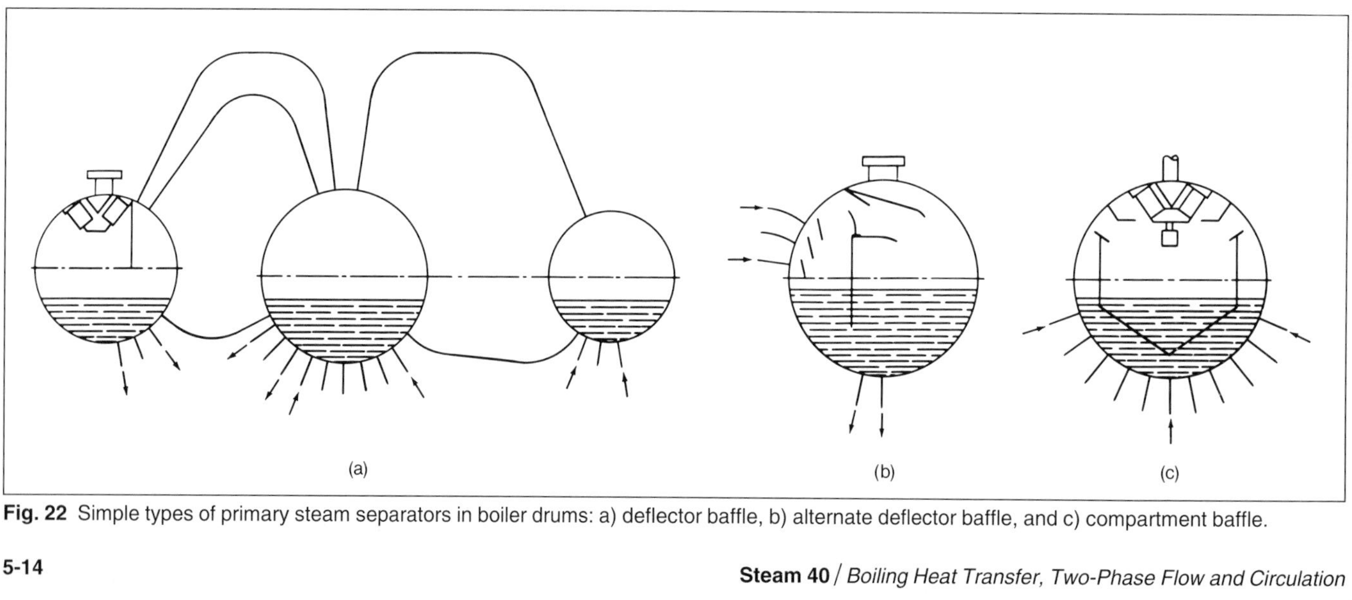

Fig. 22 Simple types of primary steam separators in boiler drums: a) deflector baffle, b) alternate deflector baffle, and c) compartment baffle.

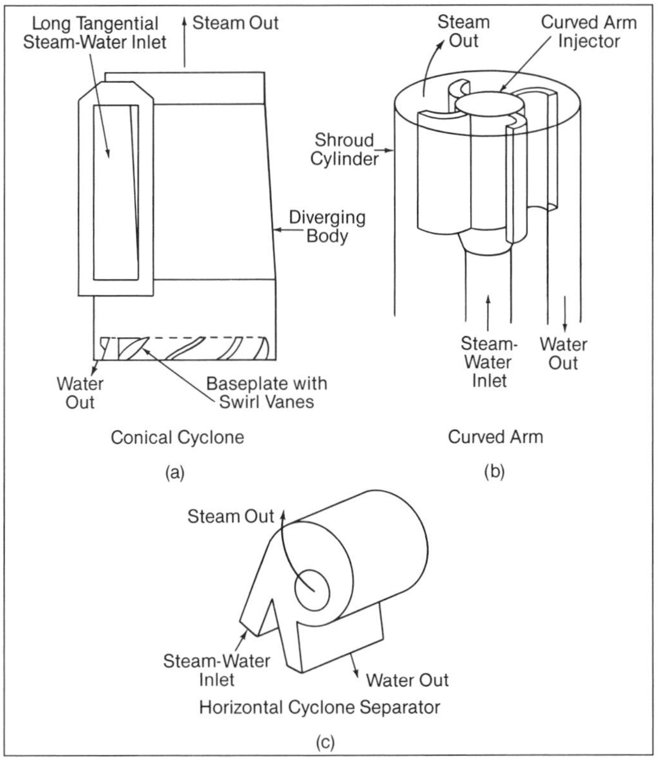

Fig. 23 Typical primary steam-water separators.

these values increase to 9000 and 15,000 lb/h ft (3.7 and 6.2 kg/s m), respectively. For large utility boilers operating at 2800 psig (193.1 bar gauge), separation can be as high as 67,000 lb/h ft (28 kg/s m) of steam with four rows of cyclone separators.

This combination of cyclone separators and scrubbers provides a steam purity of less than 1.0 ppm solids content under a wide variation of operating conditions. This purity is generally adequate in commercial practice. However, further refinement in steam purification is required where it is necessary to remove boiler water salts, such as silica, which are entrained in the steam by a vaporization or solution mechanism. Washing the steam with condensate or feedwater of acceptable purity may be used for this purpose.

Mechanical separator performance

The overall performance of mechanical separators is defined by: 1) the maximum steam flow rate at a specified average inlet quality per cyclone which meets droplet carryover limits, and 2) the predicted pressure loss. In addition, the maximum expected steam carryunder (% steam by weight) should also be known. These parameters are influenced by total flow rate, pressure, separator length, aperture sizes, drum water level, inlet steam quality, interior separator finish and overall drum arrangement. Performance characteristics are highly hardware specific. The general trends are listed in Table 2.

Steam separator evaluation To date, theoretical analyses alone do not satisfactorily predict separation performance. Therefore, extensive experimental investigations are performed to characterize individual steam-water primary separator designs.

Pressure drop of two-phase flow through a separator is extremely complex. An approximation involves using the homogeneous model two-phase multiplier, Φ, and a dimensionless loss coefficient, K_{ss}, as follows:

Table 2
Mechanical Separator Performance Trends

Moisture carryover with steam
1. increases gradually with steam flow rate until a breakaway point is reached where a sudden rise in carryover occurs,
2. increases with water level until flooding occurs, and
3. increases with steam quality.

Carryunder of steam with water
1. declines with increasing water level, and
2. declines with decreasing inlet steam quality.

Pressure drop (P_{in} - P_{drum})
1. increases with mass flow and steam quality.

$$\Delta P_{separator} = K_{ss}\,\Phi \frac{G^2 v_f}{2g_c} \qquad \textbf{(20)}$$

where

$$\Phi = 1.0 + \left(\frac{v_g - v_f}{v_f}\right) x$$

The variable K_{ss} is a unique function of pressure for each steam separator design. The other variables are defined after Equation 16.

The maximum steam flow per primary separator defines the minimum number of standard units required, while the ΔP is used in the circulation calculations. Given the unique design of each separator, B&W has acquired extensive experimental performance data under full scale, full flow and full pressure conditions for its equipment.

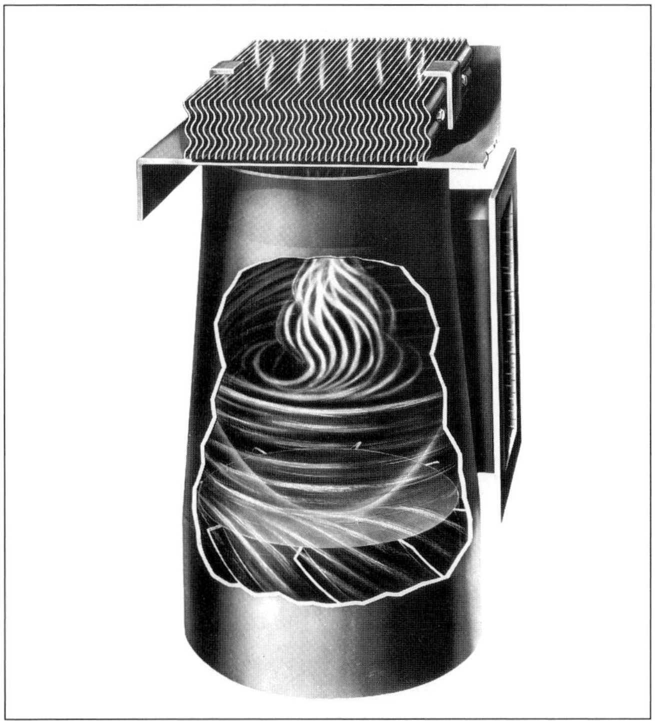

Fig. 24 Vertical cyclone separator.

Steam drum capacity

Given the flow capabilities of standardized steam-water separation equipment, the boiler drum is sized to accommodate the number of separators necessary for the largest expected boiler load (maximum steam flow rate) and to accommodate the changes in water level that occur during the expected load changes. The drum diameter, in incremental steps, and length are adjusted to meet the space requirements at a minimum cost.

An evaluation limit in steam drum design is the maximum steam carryunder into the downcomer. Carryunder, or transport of steam into the downcomers, is not desirable because it reduces the available thermal pumping force by reducing the density at the top of the downcomer. Carryunder performance is a function of physical arrangement, operating pressure, feedwater enthalpy, free water surface area, drum water level and separator efficiency. Empirical correction factors for specific designs are developed and used in the circulation calculations to account for the steam entering the downcomers. The steam is eventually completely condensed after it travels a short distance into the downcomer. However, the average density in the top portion of the downcomer is still lower than thermal equilibrium would indicate.

A rapid increase in steam demand is usually accompanied by a temporary drop in pressure until the firing rate can be sufficiently increased. During this interval, the volume of steam throughout the boiler is increased and the resulting swell raises the water level in the drum. The rise depends on the rate and magnitude of the load change and the rate at which the heat and feed inputs can be changed to meet the load demand. Steam drums are designed to provide the necessary volume, in combination with the controls and firing equipment, to prevent excessive water rise into the steam separators. This, in turn, prevents water carryover with the steam.

Circulation

The purpose of the steam-water flow circuitry is to provide the desired steam output at the specified temperature and pressure. The circuitry flow also ensures effective cooling of the tube walls under expected operating conditions, provided that the unit is properly operated and maintained. A number of methods have been developed. Four of the most common systems are illustrated in Fig. 25. These systems are typically classified as either *recirculating* or *once-through*.

In recirculating systems, water is only partially evaporated into steam in the boiler tubes. The residual water plus the makeup water supply are then recirculated to the boiler tube inlet for further heating and steam generation. A steam drum provides the space required for effective steam-water separation. Once-through systems provide for continuous evaporation of slightly subcooled water to 100% steam without steam-water separation. Steam drums are not required. These designs use forced circulation for the necessary water and steam-water flow. In some cases, a combination of these approaches is used. At low loads, recirculation maintains adequate tube wall cooling while at high loads, high pressure once-through operation enhances cycle efficiency.

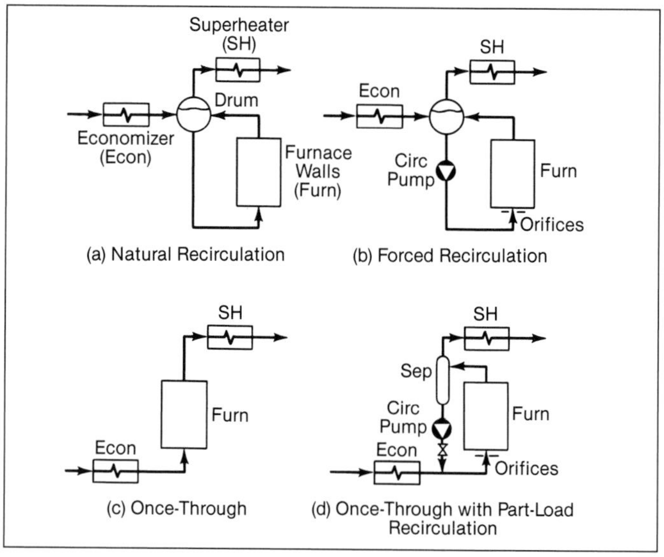

Fig. 25 Common fossil fuel boiler circulation systems.

Natural circulation

In *natural circulation*, gravity acting on the density difference between the subcooled water in the downcomer and the steam-water mixture in the tube circuits produces the driving force or pumping head to drive the flow. As shown in Fig. 26, a simplified boiler circuit consists of an unheated leg or downcomer and heated boiler tubes. The water in the downcomer is subcooled through the mixing of the low temperature feedwater from the economizer with the saturation temperature water discharged from the steam-water separators. Steam-water, two-phase flow is created in the boiler tubes as a result of the heat input. Because the steam-water mixture has a lower average density than the single-phase downcomer flow,

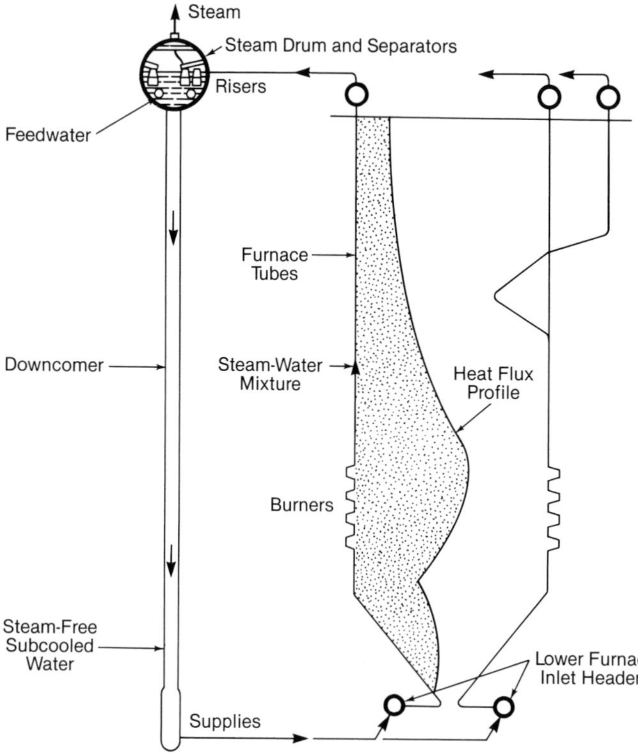

Fig. 26 Simple furnace circulation diagram.

a pressure differential or pumping pressure is created by the action of gravity and the water flows around the circuit. The flow increases or decreases until the pressure losses in all boiler circuits are balanced by the available pumping pressure. For steady-state, incompressible flow conditions, this balance takes the form:

$$\left(Z\bar{\rho}_d - \int_0^Z \rho(z)\,dz\right)\left(\frac{g}{g_c}\right) = \left(\Delta P_{friction} + \Delta P_{acceleration} + \Delta P_{local}\right) \qquad \textbf{(21)}$$

where

Z = total vertical elevation, ft (m)
z = incremental vertical elevations, ft (m)
$\rho(z)$ = heated tube local fluid density, lb/ft^3 (kg/m^3)
$\bar{\rho}_d$ = average downcomer fluid density, lb/ft^3 (kg/m^3)
g = acceleration of gravity, ft/s^2 (m/s^2)
g_c = 32.17 lbm ft/lbf s^2 (1 kg m/N s^2)
ΔP = circuitry pressure loss due to friction, fluid acceleration and local losses, lb/ft^2 (Pa)

As the heat input increases, circulation rate increases until a maximum flow rate is reached (Fig. 27). If higher heat inputs occur, they will result in larger pressure losses in the heated tubes without corresponding increases in pressure differential. As a result, the flow rate declines.

Natural circulation boilers are designed to operate in the region where increased heat input results in an increase in flow for all specified operating conditions. In this mode, a natural circulation system tends to be self compensating for numerous variations in heat absorption. These can include sudden changes in load, changes in heating surface cleanliness and changes in burner operation.

Natural circulation is most effective where there is a considerable difference in density between steam and water phases. As shown in Fig. 28, the potential for natural circulation flow remains very high even at pressures of 3100 psi (213.7 bar).

Forced circulation

In recirculating or once-through *forced circulation* systems, mechanical pumps provide the driving head to

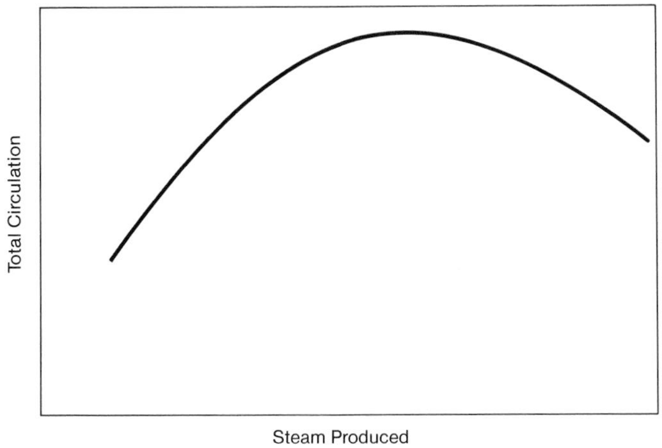

Fig. 27 Typical relationship between circulation at a given pressure and steam production (arbitrary scale).

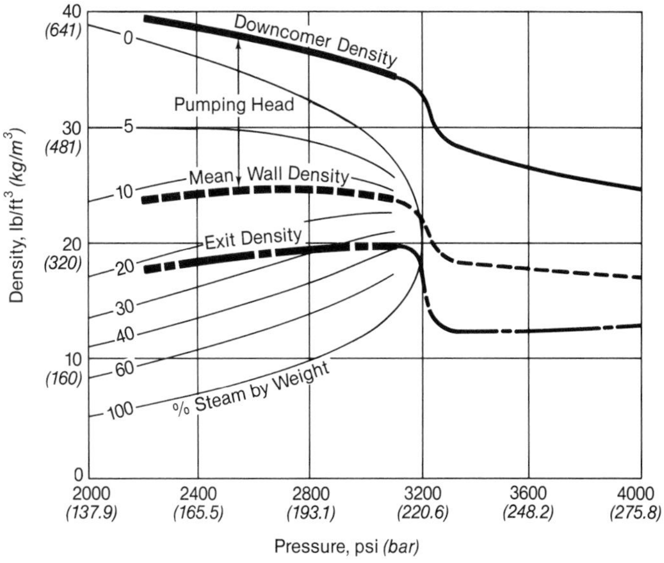

Fig. 28 Effect of pressure on pumping head.

overcome the pressure losses in the flow circuitry. Unlike natural circulation, forced circulation does not enjoy an inherent flow compensating effect when heat input changes, i.e., flow does not increase significantly with increasing heat input. This is because a large portion of the total flow resistance in the boiler tubes arises from the flow distribution devices (usually orifices) used to balance flow at the circuit inlets. The large resistance of the flow distributors prevents significant increases in flow when heat absorption is increased.

Forced circulation is, however, used where the boilers are designed to operate near or above the critical pressure [3208 psi (221.2 bar)]. There are instances in the process and waste heat fields and in some specialized boiler designs where the use of circulating pumps and forced circulation can be economically attractive. At pressures above 3100 psi (213.7 bar) a natural circulation system becomes increasingly large and costly and a pump may be more economical. In addition, the forced circulation principle can work effectively in both the supercritical and subcritical pressure ranges.

In forced recirculation there is a net thermal loss because of the separate circulating pump. While practically all the energy required to drive the pumps reappears in the water as added enthalpy, this energy originally came from the fuel at a conversion to useful energy factor of less than 1.0. If an electric motor drive is used, the net energy lost is about twice the energy supplied to the pump motor for typical fossil fuel systems.

Circulation design and evaluation

The furnace wall enclosure circuits are very important areas in a boiler. High constant heat flux conditions make uninterrupted cooling of furnace tubes essential. Inadequate cooling can result in rapid overheating, cycling thermal stress failure, or material failures from differential tube expansion. Sufficient conservatism must be engineered into the system to provide adequate cooling even during transient upset conditions. Simultaneously, the rated steam flow conditions must be maintained at the drum outlet. Any of the circulation methods discussed may be used to cool the furnace waterwall tubes. In evaluating the circulation method selected for a particular situation, the following general procedure can be used:

1. The furnace geometry is set by the fuel and combustion system selected. (See Chapters 10, 13 and 20.)
2. Standardized components (furnace walls, headers, drums, etc.) are selected to enclose the furnace arrangement as needed. (See Chapters 18 and 20.)
3. The local heat absorption is evaluated based upon the furnace geometry, fuel and firing method. Local upset factors are evaluated based upon past field experience. (See Chapter 4.)
4. Circulation calculations are performed using the pressure drop relationships.
5. The calculated circulation results (velocities, steam qualities, etc.) are compared to the design criteria.
6. The flow circuitry is modified and the circulation reevaluated until all of the design criteria are met.

Some of the design criteria include:

1. *Critical heat flux limits* For recirculating systems, CHF conditions are generally avoided. For once-through systems, the temperature excursions at CHF are accommodated as part of the design.
2. *Stability limits* These limits generally indicate acceptable pressure drop versus mass flow relationships to ensure positive flow in all circuits and to avoid oscillating flow behavior.
3. *Steam separator and steam drum limits* These indicate maximum steam and water flow rates to individual steam-water separators and maximum water flow to the drum downcomer locations to ensure that steam carryunder and water carryover will not be problems.
4. *Minimum velocity limits* Minimum circuit saturated velocities assure that solids deposition, potentially detrimental chemistry interactions and selected operating problems are minimized.
5. *Sensitivity* The system flow characteristic is checked to ensure that flow increases with heat input for all expected operating conditions.

Circulation is analyzed by dividing the boiler into individual simple circuits — groups of tubes or circuits with common end points and similar geometry and heat absorption characteristics. The balanced flow condition is the simultaneous solution of the flow characteristics of all boiler circuits.

At the heart of a B&W circulation evaluation is a circulation computer program that incorporates techniques for calculating the single- and two-phase heat transfer and flow parameters discussed above and in Chapters 3 and 4. With this program, a circulation model of the entire boiler is developed. Input into the program is a geometric description of each boiler circuit including descriptions of downcomers, supplies, risers, orifices, bends and swages, as well as individual tubes. Each of the circuits within the boiler is subjected to the local variation in heat transfer though inputs based upon the furnace heat flux distribution. (See Chapter 4.) Given the geometry description and heat absorption profile, the computer program determines the balanced steam-water flow to each circuit by solving the energy, mass and momentum equations for the model. The results of the program provide

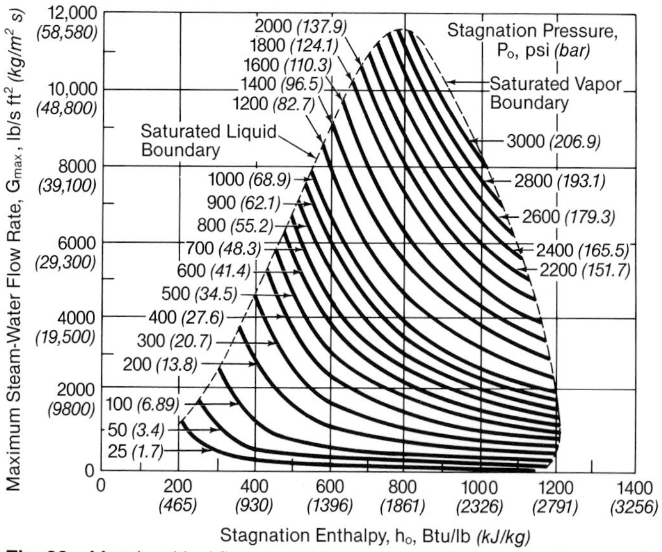

Fig. 29 Moody critical flow model for maximum steam-water flow rate.[17]

the detailed information on fluid properties, pressure drop and flow rates for each circuit so that they can be compared to the design criteria. Adjustments frequently made to improve the individual circuit circulation rates can include: changing the number of riser and supply connections, changing the number or type of steam separators in the drum, adding orifices to the inlets to individual tubes, changing the drum internal baffling, changing the operating pressure (if possible) and lowering the feedwater temperature entering the drum. Once the steam-water circuitry is finalized, the detailed mechanical design proceeds.

Critical flow

A two-phase flow parameter of particular importance in nuclear reactor safety analysis and in the operation of valves in many two-phase flow systems is the *critical flow rate*. This is the maximum possible flow rate through an opening when the flow becomes choked and further changes in upstream pressure no longer affect the rate. For single-phase flows, the critical flow rate is set by the sonic velocity. The analysis is based upon the assumption that the flow is one dimensional, homogeneous, at equilibrium and isentropic. These assumptions result in the following relationships:

$$\text{Sonic velocity} = C = \sqrt{\left(\frac{dP}{d\rho}\right)_s g_c} \qquad \textbf{(22)}$$

$$\text{Critical flow} = G_{max} = \rho\sqrt{\left(\frac{dP}{d\rho}\right) g_c} \qquad \textbf{(23)}$$

where

C = velocity, ft/s (m/s)
P = pressure, lb/ft^2 (Pa)
ρ = fluid density, lb/ft^3 (kg/m^3)
g_c = 32.17 lbm ft/lbf s^2 (1 kg m/N s^2)
G_{max} = mass flux, lb/s ft^2 (kg/m^2 s)

However, when saturated water or a two-phase steam-water mixture is present, these simplifying assumptions are no longer valid. The flow is heterogeneous and nonisentropic with strong interfacial transport and highly unstable conditions.

Moody's analysis[17] of steam-water critical flow is perhaps the most frequently used. It is based upon an annular flow model with uniform axial velocities of each phase and equilibrium between the two phases. A key element of the analysis involves maximizing the flow rate with respect to the slip ratio and the pressure. The results are presented in Fig. 29. The critical flow rate is presented as a function of the stagnation condition. Compared to experimental observations, this correlation slightly overpredicts the maximum discharge at low qualities ($x < 0.1$) and predicts reasonably accurately at moderate qualities ($0.2 < x < 0.6$), but tends to underpredict at higher qualities ($x > 0.6$).

References

1. Collier, J.G., *Convective Boiling & Condensation*, 2nd ed., McGraw-Hill, New York, 1981.

2. Jens, W.H., and Lottes, P.A., "Analysis of heat transfer, burnout, pressure drop, and density data for high pressure water," Argonne National Laboratory Report ANL-4627, May 1951.

3. Thom, J.R.S., *et al.*, "Boiling in subcooled water during flow up heated tubes or annuli," *Proceedings of Institute of Mechanical Engineers*, Vol. 180, pp. 226-246, 1966.

4. Chen, J.C., "Correlation for boiling heat transfer to saturated liquids in convective flow," *Industrial & Engineering Chemistry Process & Design Development*, Vol. 5, pp. 322-329, 1966.

5. Kitto, J.B., and Albrecht, M., "Elements of two-phase flow in fossil boilers," *Two-Phase Flow Heat Exchangers*, S. Kakac, A.E. Bergles and E.O. Fernandes, eds., Kluwer, Dordrecht, pp. 495-552, 1988.

6. Watson, G.B., Lee, R.A., and Wiener, M.,"Critical heat flux in inclined and vertical smooth and ribbed tubes," *Proceedings of the 5th International Heat Transfer Conference*, Vol. 4, Japan Society of Mechanical Engineers, Tokyo, pp. 275-279, 1974.

7. Gellerstedt, J.S., *et al.*, "Correlation of critical heat flux in a bundle cooled by pressurized water," *Two Phase Flow and Heat Transfer in Rod Bundles*, V.E. Schock, ed., ASME, New York, pp. 63-71, 1969.

8. Wiener, M., "The latest developments in natural circulation boiler design," *Proceedings of the American Power Conference*, Vol. 39, pp. 336-348, 1977.

9. Swenson, H.S., Carver, J.R., and Kakarala, C.R., "Heat transfer to supercritical water in smooth-bore tubes," *Journal of Heat Transfer*, Vol. 87, pp. 477-484, 1965.

10. Ackerman, J.W., "Pseudoboiling heat transfer to supercritical pressure water in smooth and ribbed tubes," *Journal of Heat Transfer*, Vol. 92, pp. 490-498, 1970.

11. Hewitt, G.F., and Roberts, D.W., "Studies of two-phase flow patterns by simultaneous x-ray and flash photography," Atomic Energy Research Establishment Report M2159, HMSO, London, 1969.

12. Thom, J.R.S., "Prediction of pressure drop during forced circulation boiling of water," *International Journal of Heat and Mass Transfer*, Vol. 7, pp. 709-724, 1964.

13. Martinelli, R.C., and Nelson, D.B., "Prediction of pressure drop during forced-circulation boiling of water," *Transactions of the ASME*, pp. 695-702, 1948.

14. Zuber, N., and Findlay, J.A., "Average volumetric concentration in two-phase flow systems," *Journal of Heat Transfer*, Vol. 87, pp. 453-468, 1965.

15. Chexal, B.J., Horowitz, J., and Lellouche, G.S., "An assessment of eight void fraction models for vertical flows," Electric Power Research Institute Report NSAC-107, December 1986.

16. Ledinegg, M., "Instability of flow during natural and forced circulation," *Die Warme*, Vol. 61, No. 8, pp. 891-898, 1938 (AEC-tr-1861, 1954).

17. Moody, F.J., "Maximum flow rate of a single component, two-phase mixture," *Journal of Heat Transfer*, Vol. 87, pp. 134-142, 1965.

Bibliography

Bergles, A.E., *et al.*, *Two-Phase Flow and Heat Transfer in the Power and Process Industries*, Hemisphere, Washington, D.C., 1981.

Butterworth, D., and Hewitt, G.F., eds., *Two-Phase Flow and Heat Transfer*, Oxford University Press, Oxford, UK, 1977.

Hsu, Y-Y, and Graham, R.W., *Transport Processes in Boiling and Two-Phase Systems*, Hemisphere, Washington, D.C., 1976.

Lahey, R.T., and Moody, F.J., *Thermal Hydraulics of Boiling Water Nuclear Reactors*, American Nuclear Society, Hinsdale, Illinois, 1977.

Tong, L.S., *Boiling Heat Transfer and Two-Phase Flow*, Wiley, New York, 1965.

Wallis, G.B., *One-Dimension Two-Phase Flow*, McGraw-Hill, New York, 1969.

Application of protrective coating for boiler tubes.

Chapter 6
Metallurgy, Materials and Mechanical Properties

Boilers, pressure vessels and their associated components are primarily made of metals. Most of these are various types of steels. Less common, but still important, are cast irons and nickel base alloys. Finally, ceramics and refractories, coatings, and engineered combinations are used in special applications.

Metallurgy

Crystal structure

The smallest unit of a metal is its atom. In solid structures, the atoms of metals follow an orderly arrangement, called a *lattice*. An example of a *simple point lattice* is shown in Fig. 1a and the unit cell is emphasized. The lengths of the unit cell axes are defined by a, b and c, and the angles between them are defined by α, β and γ in Fig. 1b. The steels used in boilers and pressure vessels are limited to two different lattice types: *body-centered cubic* (BCC) and *face-centered cubic* (FCC). (See Fig. 2.) Where changes in the structure or interruptions occur within a crystal, these are referred to as *defects*. *Crystal (or grain) boundaries* occur where the systematic repetition of cells for one group of atoms changes in orientation or configuration to that of a separate group of atoms. A few useful structures are composed of a single crystal in which all the unit cells have the same relationship to one another and have few defects. Some high performance jet engine turbine blades have been made of single crystals. Although these structures are difficult to make, they are attractive, because their strength is determined by close interactions of the atomic bonds in their optimum arrangement. The behavior of all other metallic structures is determined by the nature of the defects in their structures. Structures are made of imperfect assemblies of imperfect crystals, and their strengths are often orders of magnitude lower than the theoretical strengths of perfect single crystals.

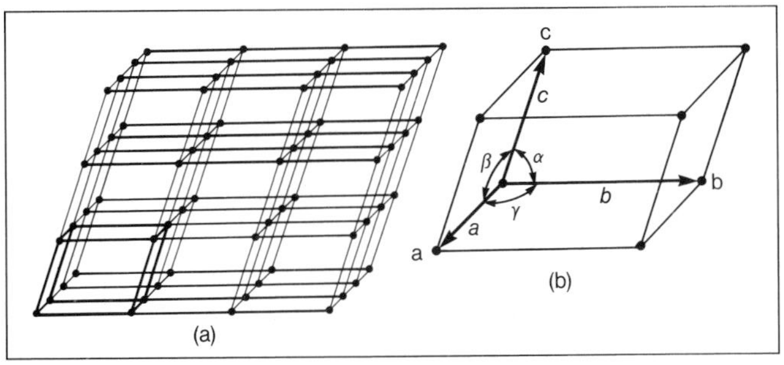

Fig. 1 Simple point lattice and unit cell (*courtesy of Addison-Wesley*).[1]

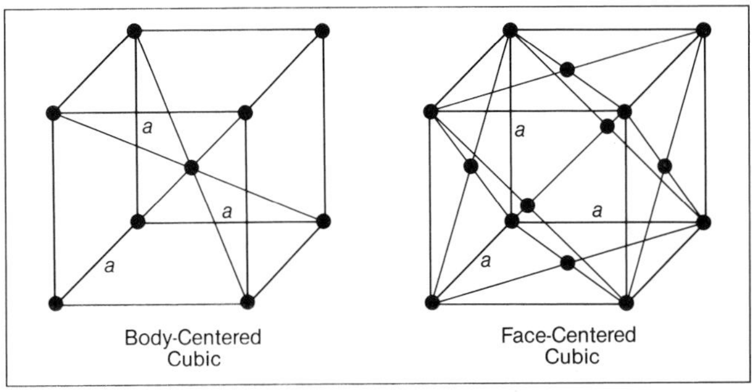

Fig. 2 Two Bravais lattices.[1]

Defects in crystals

Perfect crystals do not exist in nature. The imperfections found in metal crystals and their interactions control the material properties.

Point defects Point defects include missing atoms (vacancies), atoms of a different element occurring on crystal lattice points (substitutionals), and atoms of a different element occurring in the spaces between crystal lattice points (interstitials). Thermally created vacancies are always present, because they reduce the free energy of the crystal structure by raising its entropy. Therefore, there is an equilibrium number of thermal vacancies present; this number varies with the temperature of the crystal. The presence of such vacancies permits *diffusion* (the transport of one species of metal atom through the lattice of another) and some forms of time dependent deformation, such as *creep*. Creep is the slow deformation of continuously stressed metal over time.

Vacancies can also be created by radiation damage and plastic deformation, and the thermodynamically controlled processes of diffusion and creep can also be affected by these other processes.

When atoms of two metals are mixed in the molten state and then cooled to solidification, the atoms of one metal may take positions on the lattice of the other, forming a substitutional *alloy*. Because the atoms may be different sizes and because the bond strength between unlike atoms is different from that of like atoms, the properties of the alloy can be quite different from those of either pure metal.

Atoms of carbon, oxygen, nitrogen and boron are much smaller than metal atoms and have quite different structures. They can fit in the spaces, or interstices, between the metal atoms. The diffusion of an interstitial in a metal lattice is also affected by temperature, but it is only dependent on lattice vibration, so it is much more rapid at any given temperature. Interstitial elements are often only partly soluble in metal lattices. Certain atoms, such as carbon in iron, are nearly insoluble, so their presence in a lattice produces major defects.

Several crystal defects are illustrated in Fig. 3. This is a two dimensional schematic of a cubic iron lattice containing point defects (vacancies, substitutional foreign atoms, interstitial atoms), defects (dislocations, sub-boundaries, grain boundaries), and volume defects (voids and inclusions or precipitates of a totally different structure).[2] *Dislocations* are linear defects formed by a deformation process called *slip*, the sliding of two close-packed crystal structure planes over one another.

Grain boundaries Grain boundaries are more complex interfaces between crystals (grains) of significantly different orientations in a metal. They are arrays of dislocations between misoriented crystals. Because the atomic bonds at grain boundaries and at other planar crystal defects are different from those in the body of the more perfect crystal, they react differently to heat and chemical reagents. This difference appears as grain boundaries and other structural features on polished and etched metal surfaces under a microscope. Grain boundaries can have positive or negative effects. At lower temperatures, a steel with very small grains (fine grain size) may be stronger than the same steel with fewer large grains (coarse grain size) because the grain boundaries act as barriers to slip. At higher temperatures, where thermally activated deformation can occur, a fine grain structure material may be weaker because the irregular structure at the grain boundaries promotes local creep. This allows grains to rotate by grain boundary sliding.

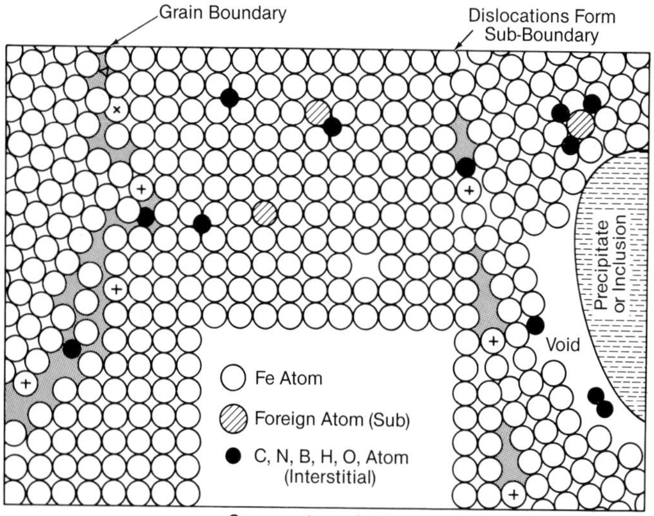

Fig. 3 Some important defects and defect complexes in metals (*courtesy of Wiley*).[3]

Volume defects Volume defects can be voids formed by coalescence of vacancies or separation of grain boundaries. More common volume defects are inclusions of oxides, sulfides and other compounds, or other phases that precipitate during solidification of complex systems.

Physical metallurgy of steel

Phases A phase is a homogeneous body of matter existing in a prescribed physical form. Metallurgists use a graph, called a *phase diagram*, to plot the stable phases for temperature versus composition. A phase diagram for a pure metal is a line because the composition does not vary.

When more than one element is involved, even for binary alloys, a variety of phases can result. One type is the binary isomorphous system, typified by only a few combinations: copper-nickel (Cu-Ni), gold-silver (Au-Ag), gold-platinum (Au-Pt) and antimony-bismuth (Sb-Bi). The phase diagram for one of these simple systems illustrates two characteristics of all solid solutions: 1) the range of composition can vary in the liquid and solid solutions, and 2) the change of phase (in these systems, from liquid to solid) takes place over a range of temperatures (unlike water and pure metals which freeze and change structure at a single temperature). Fig. 4 is a portion of the phase diagram for Cu-Ni, which shows what species precipitate out of solution when the liquid is slowly cooled.[4] (In the remainder of this chapter, chemical symbols are often used to represent the elements. See *Periodic Table*, Appendix 1.)

Alloy systems in which both species are infinitely soluble in each other are rare. More often the species are only partly soluble and mixtures of phases precipitate on cooling. Also common is the situation in which the species attract each other in a particular ratio and form a chemical compound. These intermetallic compounds may still have a range of compositions, but it is much narrower than that for solid solutions. Two systems that form such intermetallic compounds are chromium-iron (Cr-Fe) and iron-carbon (Fe-C).

Iron-carbon phase diagram Steel is an iron base alloy containing manganese (Mn), carbon and other alloying elements. Virtually all metals used in boilers and pressure vessels are steels. Mn, usually present at about 1% in carbon steels, is a substitutional solid solution element. Because its atomic size and electronic structure are similar to those of Fe, it has little effect on the Fe lattice or phase diagram in these low concentrations. Carbon, on the other hand, has significant effects; by varying the carbon content and heat treatment of Fe, an enormous range of mechanical properties can be obtained. These effects can best be understood using the Fe-C phase diagram, shown in Fig. 5. This shows that the maximum solubility of carbon in α (BCC) iron is only about 0.025%, while its solubility in γ (FCC) iron is slightly above 2.0%. Alloys of Fe-C up to 2% C are malleable and are considered steels. Iron alloys containing more than 2% C are decidedly inferior to steels in malleability, strength, toughness and ductility. They are usually used in cast form and are called *cast irons*.

Carbon atoms are substantially smaller than iron atoms, and in BCC iron, they fit at the midpoints of the cube edges and face centers. This structure is called *fer-*

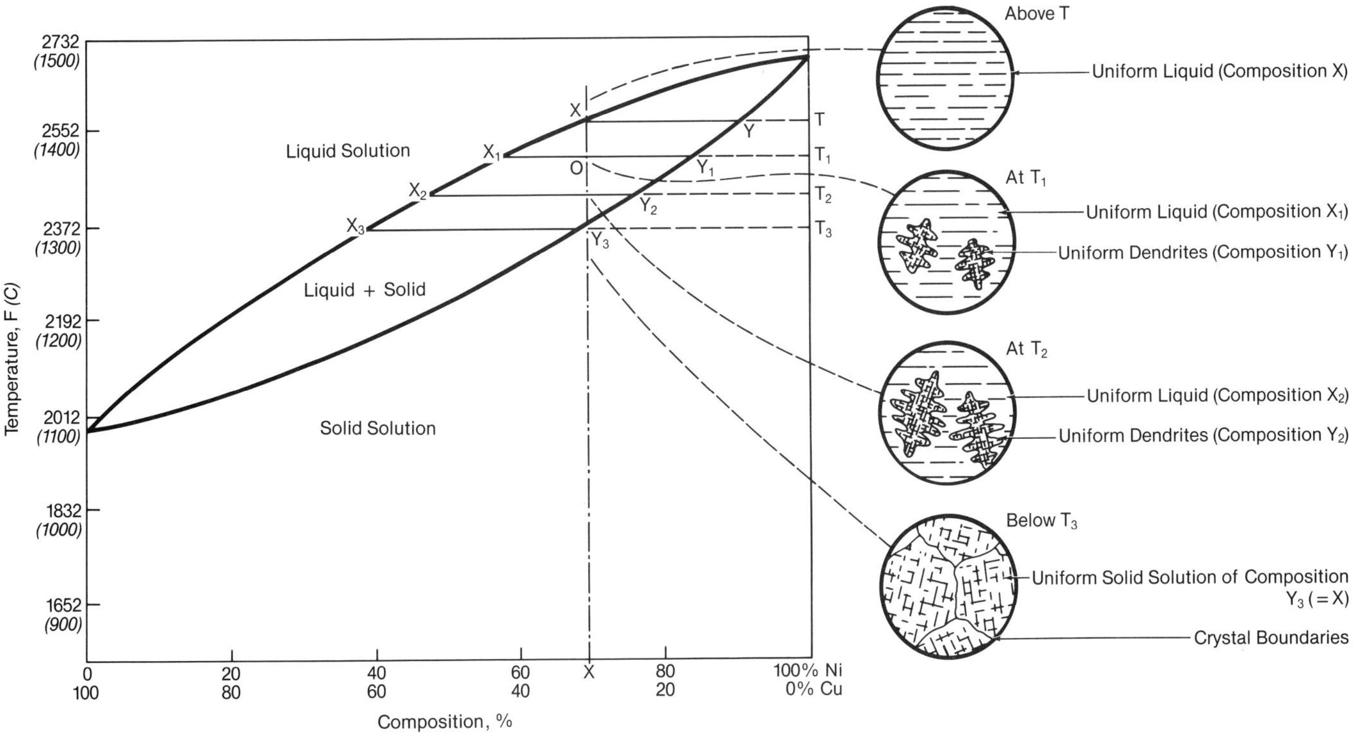

Fig. 4 The copper-nickel equilibrium diagram (*courtesy of Hodder and Staughton*).[4]

rite. In FCC iron, the carbon atoms fit at the midpoints of the cube edges at the cube center. This structure is called *austenite*. There are many more interstitial sites in ferrite than in austenite. In both structures, the interstitial spaces are much smaller than the carbon atom, leading to local distortion of the lattice and resulting in limited carbon solubility in iron. The interstices are larger in austenite than in ferrite, partly accounting for the higher solubility of carbon in austenite. If austenite containing more than 0.025% C cools slowly and transforms to ferrite, the carbon in excess of 0.025% precipitates from the solid solution. However, it is not precipitated as pure carbon (graphite) but as the intermetallic compound Fe_3C, *cementite*. As with most metallic carbides, this is a hard substance. Therefore, the hardness of steel generally increases with carbon content even without heat treatment. Cementite is not a completely stable phase; graphite is more stable.

Critical transformation temperatures The melting point of iron is reduced by the addition of carbon up to about 4.3% C. At the higher temperatures, solid and liquid coexist. The BCC δ iron range is restricted and finally eliminated as a single phase when the carbon content reaches about 0.1%. Some δ iron remains up to about 0.5% C but is in combination with other phases. Below the δ iron region, austenite absorbs carbon up to the composition limits along line S-E (Fig. 5), the limiting solid-solution solubility. The temperature at which only austenite exists decreases as the carbon increases (line G-S) to the eutectoid point: 0.80% C at 1333F (723C). Then, the temperature increases along line S-E with the carbon content because the austenite is unable to absorb additional carbon, except at higher temperatures.

Any transformation in which a single solid phase decomposes into two new phases on cooling, and in which the reverse reaction takes place on heating, is called a *eutectoid* reaction. At the eutectoid composition of 0.80% C, only austenite exists above 1333F (723C) and only ferrite and Fe_3C carbide exist below that temperature. This is the *lower critical transformation temperature*, A_1. At lower carbon contents, in the *hypoeutectoid* region, as austenite cools and reaches A_3, the *upper critical transformation temperature*, ferrite precipitates first. As the

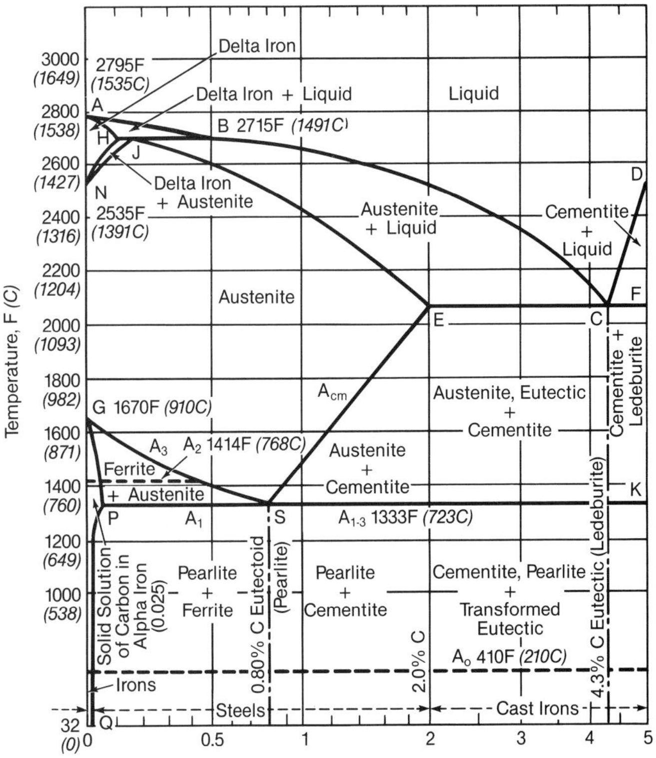

Fig. 5 Equilibrium diagram showing phase solubility limits, carbon in iron.

temperature is further reduced to 1333F (723C) at A_1, the remaining austenite is transformed to ferrite and carbide. In the *hypereutectoid* region, above 0.80% C, cementite precipitates first when austenite cools to the thermal arrest line (A_{cm}). Again, the remaining austenite transforms to ferrite and carbide when it cools to 1333F (723C). For a given steel composition, A_3, A_1 and A_{cm} represent the *critical transformation temperatures*, or critical points. A_2 is the Curie point, the temperature at which iron loses its spontaneous ferromagnetism.

At the A_1 temperature, on cooling, all the remaining austenite must transform to ferrite and carbide. Because there is not time for the carbon to go very far as it is rejected from the forming ferrite matrix, the resulting structure is one of alternating thin layers, or lamellae, of ferrite and carbide. This lamellar structure is typical of all eutectoid decomposition reactions. In steel, this structure is called *pearlite*, which always has the eutectoid composition of 0.8% C.

When pearlite is held at a moderately high temperature, such as 950F (510C), for a long time, the metastable cementite eventually decomposes to ferrite and graphite. First, the Fe_3C lamellae agglomerate into spheres. The resulting structure is considered *spheroidized*. Later, the iron atoms are rejected from the spheres, leaving a *graphitized* structure. Graphitized structures are shown in Fig. 6.

Isothermal transformation diagrams The transformation lines on the equilibrium diagram, Fig. 5, are subject to displacement when the austenite is rapidly cooled or when the pearlite and ferrite, or pearlite and cementite, are rapidly heated. This has led to the refinement of A_1 and A_3 into A_{c1} and A_{c3} on heating (c, from the French *chauffage*, heating) and into A_{r1} and A_{r3} for the displacement on cooling (r, from *refroidissement*, cooling). Because these are descriptions of dynamic effects, they distort the meaning of an equilibrium diagram which represents prevailing conditions given an infinite time. Because fabrication processes involve times ranging from seconds (laser welding) to several days (heat treatment of large vessels), the effect of time is important. Isothermal transformation experiments are used to determine phase transformation times at a particular temperature. The data are plotted on time-temperature-transformation (TTT) diagrams.

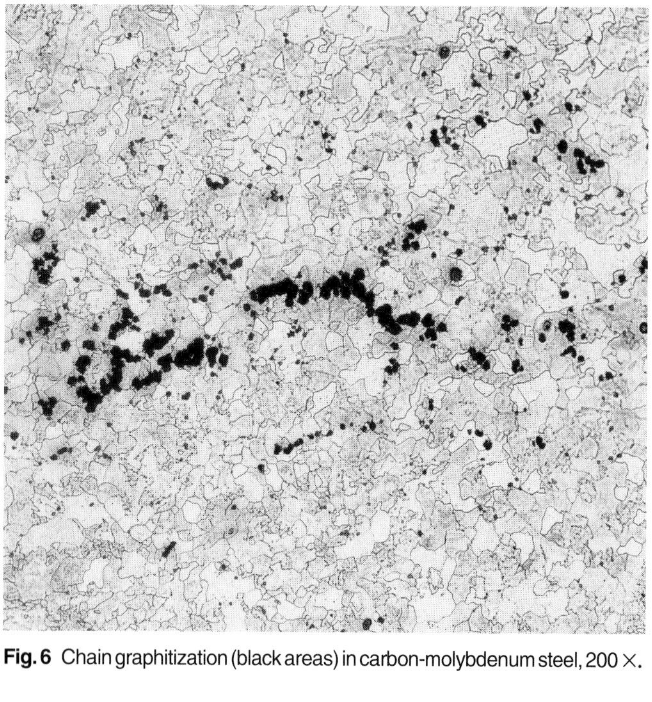

Fig. 6 Chain graphitization (black areas) in carbon-molybdenum steel, 200 ×.

The isothermal transformation (TTT) diagram in Fig. 7, for a hypoeutectoid steel, shows the time required for transformation from austenite to other constituents at the various temperature levels. The steel is heated to about 1600F (871C) and it becomes completely austenitic. It is then quickly transferred to and held in a furnace or bath at 700F (371C). Fig. 5 shows that ferrite and carbides should eventually exist at this temperature and Fig. 7 indicates how long this reaction takes. By projecting the time intervals during the transformation, as indicated in the lower portion of Fig. 7, to the top portion of the diagram, the austenite is predicted to exist for about three seconds before transformation. Then, at about 100 s, the transformation is 50% complete. At 700 s, the austenite is entirely replaced by an agglomerate of fine carbides and ferrite.

At temperatures below about 600F (316C) austenite transforms to *martensite*, the hardest constituent of heat treated steels. The temperature at which martensite starts to form is denoted M_s. It decreases with increasing austenizing temperature because M_s is sensitive to the carbon content of the austenite, and a high austenizing temperature produces a more complete solution of carbides. The nose of the left curve in Fig. 7, at about 900F (482C), is of prime significance because the transformation at this temperature is very rapid. Also, if this steel is to be quenched to form martensite (for maximum hardness), it must pass through about 900F (482C) very rapidly to prevent some of the austenite from transforming to pearlite (F + C), which is much softer.

Martensite is therefore a supercooled metastable structure that has the same composition as the austenite from which it forms. It is a solution of carbon in iron, having a *body centered tetragonal* (BCT) crystal structure. (See Reference 1.) Because martensite forms with no change of composition, diffusion is not required for the transformation to occur. It is for this reason that martensite can form at such low temperatures. Its hardness is due to the high, supersaturated carbon content, to the great lattice distortion caused by trapping excess carbon, and to the volume change of the transformation. The specific volume of martensite is greater than that of the austenite.

The formation of martensite does not occur by nucleation and growth. It can not be suppressed by quenching and it is athermal. Austenite begins to form martensite at a temperature M_s. As the temperature is lowered, the relative amount of martensite in the structure increases. Eventually, a temperature (M_f) is reached where the transformation to martensite is complete. At any intermediate temperature, the amount of martensite characteristic of that temperature forms instantly and holding at that temperature results in no further transformation. The M_s and M_f temperatures are, therefore, shown on the isothermal transformation diagram (Fig. 8) as horizontal lines. Under the microscope, martensite has the appearance of lenticular needles. Each needle is a martensite crystal.

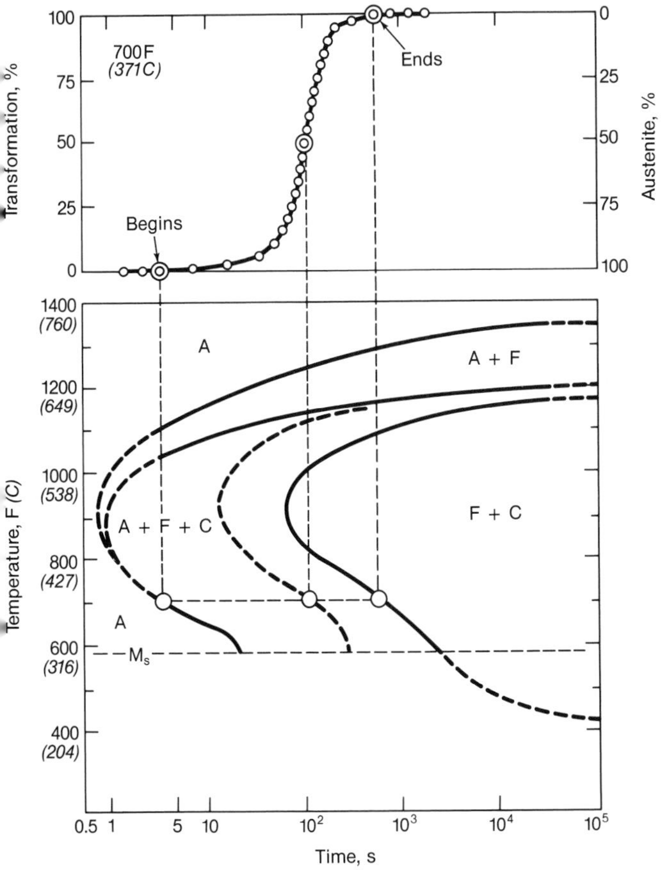

Fig. 7 Typical isothermal transformation diagram. Time required in a specific steel at 700F (371C) taken as an example.

Bainite is produced when the eutectoid (0.8% C) transformation takes place at a lower temperature (but above the M_s temperature for the alloy). The temperature regions of the TTT curve in which pearlite, bainite and martensite form are shown in Fig. 8. In the pearlite transformation, the cementite and the ferrite form in a fine lamellar pattern of alternating layers of ferrite and cementite.

Effects of alloying elements on the Fe-Fe$_3$C phase diagram Adding one or more elements to the Fe-C alloy can have significant effects on the relative size of the phase fields in the Fe-Fe$_3$C phase diagram. The elements Ni, Mn, Cu and cobalt (Co) are called *austenite formers* because their addition to the Fe-C alloy system raises the temperature at which austenite transforms to δ ferrite and greatly lowers A_3 in Fig. 5. Adding a sufficient amount of these elements increases the size of the austenite field and the FCC structure may become stable at room temperature. Because most of these elements do not form carbides, the carbon stays in solution in the austenite. Many useful material properties result, including high stability, strength and ductility, even at high temperatures. The elements Cr, molybdenum (Mo), tungsten (W), vanadium (V), aluminum (Al) and silicon (Si) have the opposite effect and are considered *ferrite formers*. They raise the A_3 temperature and some of them form very stable carbides, promoting the stability of BCC ferrite, even at very high temperatures.

Specific effect of alloying elements

Steel alloys are the chief structural materials of modern engineering because their wide range in properties suits so many applications. These properties are affected directly not only by the characteristics and the amounts of the elements which, either alone or in combination, enter into the composition of the steel, but also by their reaction as constituents under various conditions of temperature and time during fabrication and use. For example, Cr increases resistance to corrosion and scaling, Mo increases creep strength at elevated temperatures, and Ni (in adequate amounts) renders the steel austenitic. The specific effects of the most important elements found in steel are as follows.

Carbon is the most important alloying element in steel. In general, an increase in carbon content produces higher ultimate strength and hardness but lowers the ductility and toughness of steel alloys. The curves in Fig. 9 indicate the general effect of carbon on the mechanical properties of hot rolled carbon steel. Carbon also increases air hardening tendencies and weld hardness, especially in the presence of Cr. In low alloy steel for high temperature applications, the carbon content is usually restricted to a maximum of about 0.15% to assure optimum ductility for welding, expanding and bending operations, but it should be no lower than 0.07% for optimum creep strength. To minimize intergranular corrosion caused by carbide precipitation, the carbon content of austenitic stainless steel alloys is limited to 0.10%. This maximum may be reduced to 0.03% in extremely low carbon grades used in certain corrosion resistant applications. However, at least 0.04% C is required for acceptable creep strength. In plain, normalized carbon steels, the creep resistance at temperatures below 825F (441C) increases with carbon content up to 0.4% C; at higher temperatures, there is little variation of creep properties with carbon content. An increase in carbon content also lessens the thermal and electrical conductivities of steel and increases its hardness on quenching.

Manganese is infinitely soluble in austenite and up to about 10% soluble in ferrite. It combines with residual

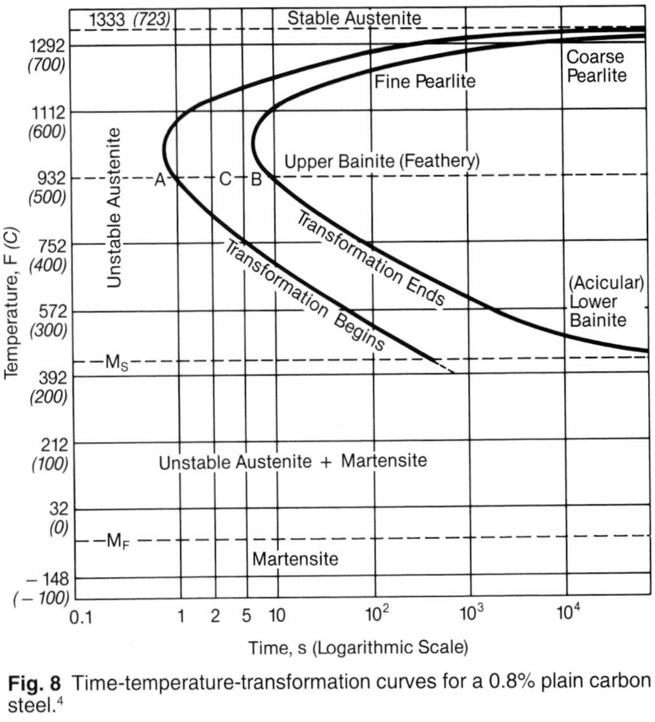

Fig. 8 Time-temperature-transformation curves for a 0.8% plain carbon steel.[4]

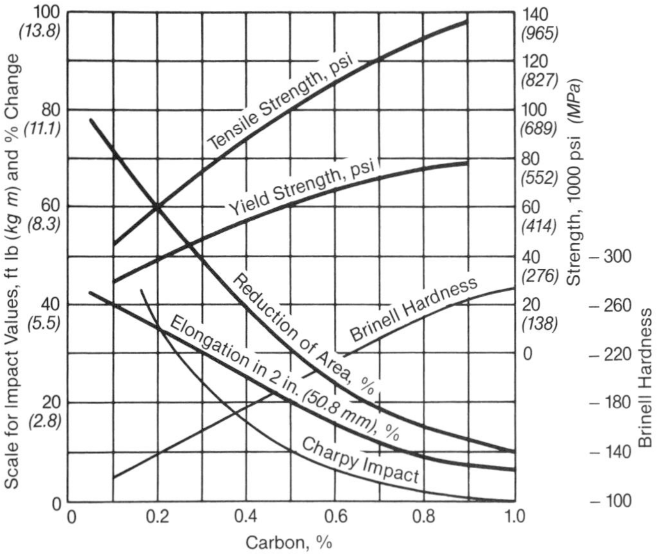

Fig. 9 General effect of carbon on the mechanical properties of hot rolled carbon steel.

sulfur while the steel is molten to form manganese sulfides, which have a much higher melting point than iron sulfides. Without the Mn, iron sulfides, which melt at about 1800F (982C), would form. This would lead to *hot-shortness*, a brittle-failure mechanism, during hot forming operations. The Mn therefore produces the malleability that differentiates steel from cast iron.

Mn forms stable carbides and its carbide forming tendency is slightly greater than Fe, although not as strong as Cr. It is a good solid solution strengthener, better than Ni and about as good as Cr. It can also be used in austenitic stainless steels to replace Ni as the austenite stabilizer at lower cost.[5]

Molybdenum, when added to steel, increases its strength, elastic limit, resistance to wear, impact qualities and hardenability. Mo contributes to high temperature strength and permits heating steel to a red hot condition without loss of hardness. It also increases the resistance to softening on tempering and restrains grain growth. Mo makes chromium steels less susceptible to temper embrittlement and it is the most effective single additive that increases high temperature creep strength.

An important use of Mo is for corrosion resistance improvement in austenitic stainless steels. It enhances the inherent corrosion resistance of these steels in reducing chemical media and it increases their passivity under mildly oxidizing conditions. Under certain conditions, molybdenum reduces the susceptibility of stainless steels to pitting.

Chromium is the essential constituent of stainless steel. While other elements are stronger oxide formers, Cr is the only one that is highly soluble in iron (about 20% in austenite and infinite in ferrite) and forms a stable, tightly adherent oxide. It is virtually irreplaceable in resisting oxidation at elevated temperatures. Cr raises the yield and ultimate strength, hardness, and toughness of steel at room temperature. It also contributes to high temperature strength. The optimum chromium content for creep strength in annealed low alloy steels is about 2.25%.

A steady improvement in resistance to atmospheric corrosion and to attack by many reagents is also noted when the chromium content is increased. A steel with 12% or more Cr is considered stainless, i.e., the Cr_2O_3 film is sufficient to prevent surface rust (hydrated iron oxide) formation. The chemical properties of the steel, however, are affected by the carbon content. Higher chromium and lower carbon levels generally promote increased corrosion resistance. The addition of sufficient Cr prevents *graphitization* during long term high temperature service.

Adding more than 1% of chromium may cause appreciable air hardening in the steel. Up to about 13.5% Cr, air hardening is a direct function of chromium and carbon content. Low carbon alloy steels containing more than 12% Cr can become nonhardening, but the impact strength is reduced and the ductility is poor. Cr lessens thermal and electrical conductivities.

Cr can be diffused into low alloy steel surfaces by a chemical vapor deposition process called *chromizing*. Very high Cr contents can be achieved, making chromized steels virtually impervious to oxidation and resistant to exfoliation.

Nickel increases toughness when added to steel, particularly in amounts over 1%. Improved resistance to corrosion by some media is attained with Ni contents over 5%. Ni dissolves in the iron matrix in all proportions and, therefore, raises the ultimate strength without impairing the ductility of the steel. Ni is particularly effective in improving impact properties, especially at low temperature.

The most important use of nickel as an alloying element in steel is its combination with chromium in amounts of 8% Ni or more. Ni is such a strong austenite former that the high chromium Fe-Ni-C alloys are austenitic at room temperature. The various combinations of chromium and nickel in iron produce alloy properties that can not be obtained with equivalent amounts of a single element. Common combinations are 18% Cr - 8% Ni, 25% Cr - 12% Ni, 25% Cr - 20% Ni and 20% Cr - 30% Ni. These steels are resistant to atmospheric corrosion and to oxidation at high temperatures. In addition, they offer greatly enhanced creep strength.

Ni is only slightly beneficial to creep properties of low alloy ferritic steels. It reduces the coefficient of thermal expansion and diminishes the electrical and thermal conductivities. It is not resistant to sulfur compounds at elevated temperatures.

Cobalt is the only element that suppresses hardenability in steels. However, when added to austenite, it is a strong solution strengthener and a carbide former. It also significantly improves creep strength. Binary Fe-Co alloys have the highest magnetic saturation induction of any known materials. Therefore, such alloys are often used in permanent magnets.

Tungsten acts similarly to molybdenum. It is a very strong carbide former and solid solution strengthener. It forms hard, abrasion resisting carbides in tool steels, develops high temperature hardness in quenched and tempered steels, and contributes to creep strength in some high temperature alloys.[5]

Vanadium is a degasifying and deoxidizing agent, but it is seldom used in that capacity because of high cost. It is applied chiefly as an alloying element in steel to increase strength, toughness and hardness. It is essentially a carbide forming element which stabilizes the structure, especially at high temperatures. Vanadium minimizes grain growth tendencies, thereby permitting much higher heat treating temperatures. It also intensifies the properties of other elements in alloy steels. Small additions of vanadium (0.1 to 0.5%), accompanied by proper heat treatment, give steels containing 0.5 to 1.0% molybdenum pronounced improvement in high temperature creep properties.

Titanium (Ti) and *columbium (niobium)* (Cb) are the most potent carbide forming elements. Ti is also a good deoxidizer and denitrider. These elements are most effective in the Cr-Ni austenitic alloys, where they react more readily with carbon than does Cr. This allows the Cr to remain in solid solution and in the concentrations necessary to maintain the corrosion resistance. Ti and Cb [or Cb plus tantalum (Ta)] are used to reduce air hardening tendencies and to increase oxidation resistance in steel containing up to 14% Cr. These elements have a beneficial effect on the long term, high temperature properties of Cr-Ni stainless steels because of the stability of their carbides, nitrides and carbonitrides. Cb and Ti have also been used in some of the super alloys to improve high temperature properties. Ti forms an intermetallic compound with Ni in these alloys, Ni_3Ti, called γ', which is a potent strengthening phase.

Copper, when added to steel in small amounts, improves its resistance to atmospheric corrosion and lowers the attack rate in reducing acids. Cu, like Ni, is not resistant to sulfur compounds at elevated temperatures. Consequently, it is not ordinarily used in low alloy steels intended for high temperature service where sulfur is a major component of the environment, as in combustion gases. Cu is added (up to 1%) in low alloy constructional steels to improve yield strength and resistance to atmospheric corrosion. Its presence in some of the high alloy steels increases corrosion resistance to sulfuric acid.

Boron (B), when combined with Mo, is a strong bainite stabilizer. Small amounts of boron in the presence of Mo suppress the formation of martensite, leading to the complete transformation to bainite before the M_s temperature is reached. This substantially improves the strength and stability of Cr-Mo pressure vessel steels. The B-10 isotope of boron has a very high neutron-capture cross-section, so it is added to steels used for containment and storage vessels of nuclear fuels and waste products.

Nitrogen (N) has two primary functions as an alloying agent in steels. In carbon and low alloy steels, it is used in case hardening, in which nascent nitrogen is diffused into the steel surface. Nitrogen and carbon are interstitial solid solution strengtheners. In the presence of Al or Ti, additional strengthening results by precipitate formations of the respective nitrides or carbonitrides. In austenitic stainless steels, nitrogen provides the same interstitial strengthening as carbon but does not deplete the austenite of chromium as does carbon. The strength of nitrogen-containing stainless steels is therefore equivalent to that of the carbon-containing stainlesses. This strength is achieved without the susceptibility to corrosive attack that results from local carbide formation at grain boundaries of these steels.

Oxygen (O) is not normally considered to be an alloying element. It is present in steel as a residual of the steel making process.[6,7] However, a few oxides are so hard and stable, notably those of Al, Ti and thorium (Th), that they are potent strengtheners when dispersed as fine particles throughout an alloy. This can be accomplished by internal oxidation in an oxygen-containing atmosphere or by powder metallurgical techniques.

Aluminum is an important minor constituent of low alloy steels. It is an efficient deoxidizer and is widely used in producing killed steel. When added to steel in appreciable quantities, Al forms tightly adhering refractory oxide scales and therefore increases resistance to scaling. It is difficult, however, to add appreciable amounts of this element without producing undesirable effects. In the amounts customarily added (0.015 to 0.080%), Al does not increase resistance to ordinary forms of corrosion. Because of their affinity for oxygen, high-aluminum steels generally contain numerous alumina inclusions which can promote pitting corrosion. Al, however, increases oxidation resistance when applied to steel as a surface coating, as in the calorizing process.

An excessive quantity of aluminum has a detrimental effect on creep properties, particularly in plain carbon steel. This is attributable to its grain refining effect and to its acceleration of spheroidization and graphitization of the carbide phase.

Silicon greatly contributes to steel quality because of its deoxidizing and degasifying properties. When added in amounts up to 2.5%, the ultimate strength of steel is increased without loss in ductility. Si in excess of 2.5% causes brittleness and amounts higher than 5% make the steel nonmalleable.

Resistance to oxidation and surface stability of steel are increased by adding silicon. These desirable effects partially compensate for its tendency to reduce creep resistance. Si increases the electrical conductivity of steel and decreases hysteresis losses. Si steels are, therefore, widely used in electrical apparatus.

Killing agents, such as Si and Al, are added to steel for deoxidation; the latter is used for grain size control. Calcium and rare earth metals, when added to the melt, have the same effects. Additionally, these elements form complex oxides or oxysulfides and can significantly improve formability by controlling the sulfide shape.

Phosphorus (P) is a surprisingly effective hardener when dissolved in quantities of up to 0.20%.[5] However, a high phosphorus content can notably decrease the resistance of carbon steel to shock and reduce ductility when the metal is cold worked. This embrittling effect, referred to as *cold-shortness*, results from an enlarged grain size which causes segregation. The detrimental effect of phosphorus increases with carbon content.

Phosphorus is effective in improving the machinability of free-cutting steels. This is related to its embrittling effect, which permits chips to break on machining. In alloy steels intended for boiler applications, the permissible phosphorus content is less than that for machining steels and its presence is objectionable for welding. Phosphorus is used as an alloying element (up to 0.15%) in proprietary low alloy, high strength steels, where increased yield strength and atmospheric corrosion resistance are primary requirements. In certain acids, however, a high phosphorus content may increase the corrosion rate.

Sulfur (S) is generally undesirable in steel and many processes have been developed to minimize its presence.

However, sulfur is sometimes added to steel to improve its machinability, as are phosphorus and other free-machining additives: calcium, lead, bismuth, selenium and tellurium. Several of these elements are virtually insoluble in steel and have low melting points, or they form low melting temperature compounds. These compounds can lead to liquid metal embrittlement or hot-shortness at even moderately elevated temperatures. Because the fastener industry favors free machining steels due to their beneficial production effects, boiler and pressure vessel manufacturers must exercise care in applying threaded fasteners at high temperatures.

Heat treating practices

Steel can be altered by modifying its microstructure through heat treatment. Various heat treatments may be used to meet hardness or ductility requirements, improve machinability, refine grain structure, remove internal stresses, or obtain high strength levels or impact properties. The more common heat treatments, annealing, normalizing, spheroidizing, hardening (quenching) and tempering, are briefly described.

Annealing is a general term applied to several distinctly different methods of heat treatment. These are full, solution, stabilization, intercritical, isothermal, and process annealing.

Full annealing is done by heating a ferritic steel above the upper critical transformation temperature (A_3 in Fig. 5), holding it there long enough to fully transform the steel to austenite, and then cooling it at a controlled rate in the furnace to below 600F (316C). A full anneal refines grain structure and provides a relatively soft, ductile material that is free of internal stresses.

Solution annealing is done by heating an austenitic stainless steel to a temperature that puts most of the carbides into solution. The steel is held at this temperature long enough to achieve grain growth. It is then quenched in water or another liquid for fast cooling, which prevents most of the carbides from reprecipitating. This process achieves optimum creep strength and corrosion resistance. For many boiler applications, austenitic stainless steels require the high creep strength of a coarse grain structure but do not require aqueous corrosion resistance, because they are only exposed to dry steam and flue gases. *Solution treatment*, used to achieve grain growth, is required for these applications, but the quenching step is not required.

Stabilization annealing is performed on austenitic stainless steels used in severe aqueous corrosion environments. The steel is first solution annealed, then reheated to about 1600F (871C) and held there. Initially, chromium carbides precipitate at the grain boundaries in the steel. Because these are mostly of the complex $M_{23}C_6$ type, which are very high in Cr, the austenite adjacent to the grain boundaries is depleted of chromium. This would normally leave the steel susceptible to corrosive attack, but holding it at 1600F (871C) permits the Cr remaining in the austenite solution to redistribute within the grains, restoring corrosion resistance, even adjacent to the grain boundaries.

Intercritical annealing and *isothermal annealing* are similar. They involve heating a hypoeutectoid ferritic steel above the lower critical transformation temperature (A_1 in Fig. 5) but below the upper critical temperature, A_3. This dissolves all the iron carbides but does not transform all the ferrite to austenite. Cooling slowly from this temperature through A_1 produces a structure of ferrite and pearlite that is free of internal stresses. In intercritical annealing, the steel continues to cool slowly in the furnace, similarly to full annealing. In isothermal annealing, cooling is stopped just below A_1, assuring complete transformation to ferrite and pearlite, and eliminating the potential for bainite formation.

Process annealing, sometimes called subcritical annealing or stress relieving, is performed at temperatures just below the lower critical temperature A_1, usually between 950 and 1300F (510 and 704C). Process annealing neither refines grains nor redissolves cementite, but it improves the ductility and decreases residual stresses in work hardened steel.

Normalizing is a variation of full annealing. Once it has been heated above the upper critical temperature, normalized steel is cooled in air rather than in a controlled furnace atmosphere. Normalizing is sometimes used as a homogenization procedure; it assures that any prior fabrication or heat treatment history of the material is eliminated. Normalizing relieves the internal stresses caused by previous working and, while it produces sufficient softness and ductility for many purposes, it leaves the steel harder and with higher tensile strength than full annealing. To remove cooling stresses, normalizing is often followed by tempering.

Spheroidizing is a type of subcritical annealing used to soften the steel and to improve its machinability. Heating fine pearlite for a long time just below the lower critical temperature of the steel, followed by very slow cooling, causes spheroidization.

Hardening (quenching) occurs when steels of the higher carbon grades are heated to produce austenite and then cooled rapidly (quenched) in a liquid such as water or oil. Upon hardening, the austenite transforms into martensite. Martensite is formed at temperatures below about 400F (204C), depending on the carbon content and the type and amount of alloying elements in the steel. It is the hardest form of heat treated steels and has high strength and abrasion resistance.

Tempering is applied after normalizing or quenching some air hardening steels. These preliminary treatments impart a degree of hardness to the steel but also make it brittle. The object of tempering, a secondary treatment, is to remove some of that brittleness by allowing certain transformations to proceed in the hardened steel. It involves heating to a predetermined point below the lower critical temperature, A_1, and is followed by any desired rate of cooling. Some hardness is lost by tempering, but toughness is increased, and stresses induced by quenching are reduced or eliminated. Higher tempering temperatures promote softer and tougher steels. Some steels may become embrittled on slow cooling from certain tempering temperatures. These steels are said to be temper brittle. To overcome this difficulty, they are quenched from the tempering temperature.

Postfabrication heat treatments are often applied to restore more stable, stress free conditions. These include postweld and postforming heat treatments and solution treatment.

Fabrication processes

Any mechanical work applied to the metal below its recrystallization temperature is *cold* work. Mechanical work performed above the recrystallization temperature is *hot* work and the simultaneous annealing that occurs at that temperature retards work-hardening. The recrystallization temperature is dependent on the rate of deformation. If a material is formed at a temperature significantly above room temperature, but below its recrystallization temperature, the process is referred to as *warm working*.

The temperature at which steel is mechanically worked has a profound effect on its properties. Cold work increases the hardness, tensile strength and yield strength of steel, but its indices of ductility — elongation and reduction of area — are decreased. The extent of the work-hardening, with progressive elongation of the grains in the direction of working, depends on the amount of cold work and on the material. If the work-hardening caused by the necessary shaping operation becomes excessive, further work can cause fracture.

Hot working variations include forging, rolling, pressing, extruding, piercing, upsetting and bending. Most of these are largely compressive operations, in which the metal is squeezed into a desired shape. They introduce some degree of orientation to the internal structure. Even if the metal experiences phase transformations or other recrystallization processes, some degree of orientation is maintained in the pattern retained by the oxides, sulfides, and other inclusions that do not dissolve during hot working or heat treatment. Depending on the application, the resultant orientation may have no effect, be useful, or be harmful. Rolled plates, for instance, often have inferior properties in the through-thickness direction due to retention of mid-plane segregated inclusions and to the predominant grain orientation in the longitudinal and transverse directions. This can result in a failure mode known as lamellar tearing if not addressed.

Hot rolling of carbon steel and low alloy steel into drum or pressure vessel sections is often done at temperatures above A_3. Temperatures and times of heating before forming need to be controlled to assure that the resulting product retains the desired fine grain size and consequent good toughness, and to assure that excessive plate surface oxidation does not occur.

Cold working operations used in manufacturing boiler components are rolling, forging, bending and swaging. Detailed information about these processes and their effects on materials can be found elsewhere. (See References 6 and 7.)

Cold rolling of plate to make shells for drums is limited only by the capacity and diameter of available rolling equipment. This process is most often applied to carbon steel, and any postforming heat treatment performed is usually combined with postweld heat treatment of the completed drum. In some low pressure applications, tube to header or tube to drum connections may be made by roll expanding the tube into an internally grooved socket in the shell. The strength of the connection depends on the mechanical interference between the roll expanded tube, which generally deforms plastically, and the hole in the shell, which mostly deforms elastically.

Cold forging of boiler components is usually limited to final size forming of shells. Threaded fasteners used in boilers may have been cold headed or may have had their threads cold rolled. Effects of such forming operations are normally mitigated by heat treatments required by the fastener specification, but occasionally this heat treatment does not eliminate microstructural differences between the cold formed portion and the remainder of the fastener. This is particularly true of austenitic stainless steel or nickel alloy bolts, which do not transform during heat treatment. These bolts may be susceptible to cracking at the interface between the cold formed head and the shank in certain aqueous environments.

Cold bending is performed on many configurations of tubes and pipes for boilers. Boiler designers consider the effects of this process on the geometry and properties of the finished product.

With some exceptions, ferritic alloy tubes and pipe are usually not heat treated after bending. However, austenitic stainless steels and nickel alloys used in high pressure boilers are often exposed to temperatures at which the strain energy of the cold bending is sufficient to cause polygonization and recrystallization to a fine grain size during service. The service temperature is insufficient to produce grain growth and the fine grain size material has lower high temperature (creep) strength. To prevent this from happening, cold bends in these alloys are given a high temperature (solution) heat treatment to stabilize the coarse grain structure.

Welding

Joining of boiler pressure parts and of nonpressure parts to pressure parts is almost always accomplished by welding. This is particularly true of high temperature, high pressure boilers, whose service conditions are too severe for most mechanical joints (bolted flanges with gaskets) and brazed joints.

Welding is the joining of two or more pieces of metal by applying heat or pressure, or both, with or without the addition of filler metal, to produce a localized union through fusion or recrystallization across the interface.[8] There are many welding processes, but the most widely used for joining pressure parts is fusion welding with the addition of filler metal, using little or no pressure. Fig. 10 indicates the variety of processes.

Weld morphology Because of the heat distribution characteristics of the welding process, the weld joint is usually a chemically and mechanically heterogeneous composite consisting of up to six metallurgically distinct regions: a composite zone, the unmixed zone, the weld interface, the partially melted zone, the heat-affected zone (HAZ) and the unaffected base metal. These zones are shown in Fig. 11. The *composite zone* is the completely melted mixture of filler metal and melted base metal. The narrow region surrounding the composite zone is the *unmixed zone*, which is a boundary layer of melted base metal that solidifies before mixing in the composite zone. This layer is at the edges of the weld pool, with a composition essentially identical to the base metal. The composite zone and the unmixed zone together make up what is commonly referred to as the fusion zone. The third region is the *weld interface*, or the boundary between the

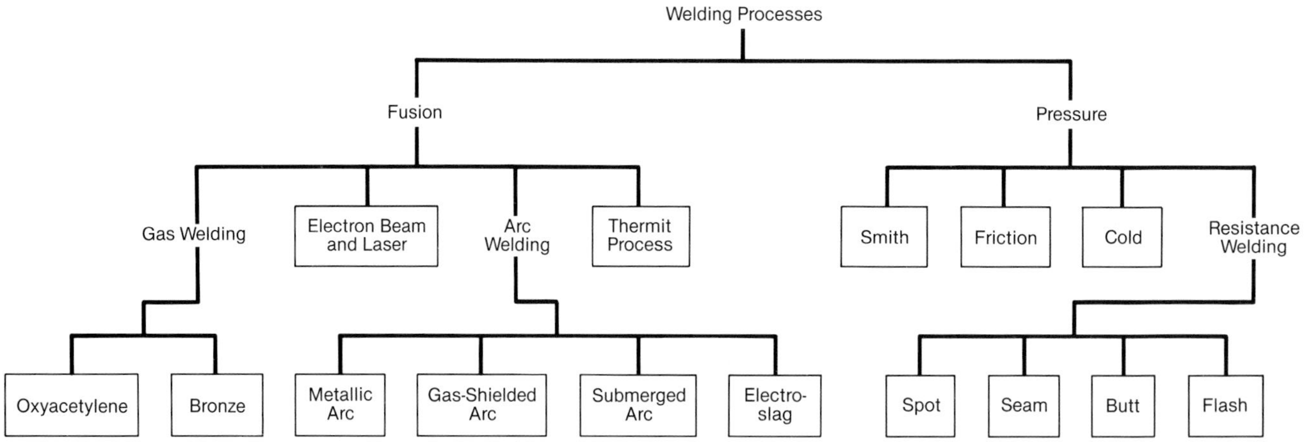

Fig. 10 Classification of welding processes.[4]

unmelted base metal on one side and the solidified weld metal on the other. The *partially melted zone* occurs in the base metal immediately adjacent to the weld interface, where some localized melting of lower melting temperature constituents, inclusions or impurities may have occurred. Liquation, for instance, of manganese sulfide inclusions can result in hot cracking or microfissuring. The *heat-affected zone* is that portion of the base material in the weld joint that has been subjected to peak temperatures high enough to produce solid state microstructural changes, but not high enough to cause melting. Finally, the last part of the workpiece that has not undergone a metallurgical change is the *unaffected base metal.*

Factors affecting weld quality

Ferrite content Austenitic stainless steel weld metals are susceptible to hot cracking or microfissuring as they cool from the solidus to about 1800F (980C). The microfissuring can be minimized by providing a small percentage of ferrite in the as-deposited welds.

Graphitization The shrinkage of the weld on freezing results in plastic deformation and high residual stresses in the weld joint. In carbon and carbon-molybdenum steels containing no stronger carbide-forming elements, the areas of localized strain adjacent to the heat-affected weld zones provide sites where the volume increase of the cementite to graphite decomposition can be more readily accommodated. At about 900F (482C), graphite nodules can precipitate on these planes of deformation. When samples of such materials are viewed in cross-section, the nodules appear to be arranged in rows or chains, and this condition has been termed *chain graphitization* (Fig. 6). The interfacial bond between the graphite and the ferrite matrix in such weldments is very low, much lower than that between ferrite and pearlite or ferrite and cementite. In the early 1950s, several failures of carbon-molybdenum main steam piping weldments occurred due to this phenomenon. The ruptures occurred with little warning because they were not preceded by swelling of the joints and, as a result, significant damage resulted. In consequence, the use of carbon-molybdenum main steam piping has been significantly restricted.

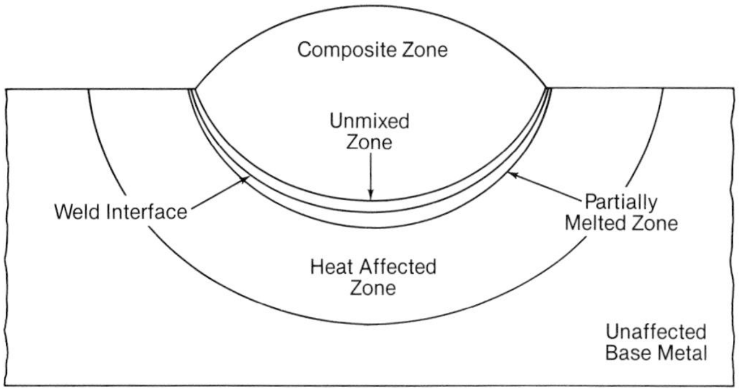

Fig. 11 Metallurgical zones developed in a typical weld (*courtesy of ASM*).[9]

Postweld heat treatment When cooling is complete, the welded joint contains residual stresses comparable to the yield strength of the base metal at its final temperature. The thermal relief of residual stresses by postweld heat treatment (PWHT) is accomplished by heating the welded structure to a temperature high enough to reduce the yield strength of the steel to a fraction of its magnitude at ambient temperature. Because the steel can no longer sustain the residual stress level, it undergoes plastic deformation until the stresses are reduced to the at-temperature yield strength. Fig. 12 shows the effect of stress relief on several steels. The temperature reached during the treatment has a far greater effect in relieving stresses than the length of time the weldment is held at temperature. The closer the temperature is to the critical or recrystallization temperature, the more effective it is in removing residual stresses, provided the proper heating and cooling cycles are used.[11]

Lamellar tearing Weld defect causes and inspection procedures are covered more extensively in References 9 and 11. However, one metallurgical effect of residual stresses should be mentioned in the context of boilers and pressure vessels: *lamellar tearing*. Lamellar tearing may result when an attachment is welded to a plate in the T-shaped orientation shown in Fig. 13, particularly if the plate contains shrinkage voids, inclusions, or other internal segregation parallel to the plate surface. In such an instance, the residual shrinkage stresses may be sufficient to open a tear or tears parallel to the plate surface to which the T-portion is welded.

Joining dissimilar metals It may be necessary to join austenitic and ferritic steels. Weld failures have occurred in these welds since the introduction of austenitic stainless steel superheater tubing materials. Nickel base filler metals have long been used to mitigate these problems, but these do not offer a permanent solution. Additional system stresses from component location, system expansion and bending can increase the potential for such failures.

Research is continuing toward the development of filler metals less likely to permit failures but none has become commercially available. The best alternative is to avoid dissimilar metal welds by using higher strength ferritic alloy materials, such as modified 9Cr-1Mo-V tubing and piping, when design conditions permit.

Materials

Almost all of the materials used in constructing boilers and pressure vessels are steels and the vast majority of components are made of *carbon steels*. Carbon steels are used for most types of pressure and nonpressure parts: drums, headers, piping, tubes, structural steel, flues and ducts, and lagging.

Carbon steels may be defined by the amount of carbon retained in the steel or by the steelmaking practice. These steels are commonly divided into four classes by carbon content: *low carbon*, 0.15% carbon maximum; *medium-low carbon*, between 0.15 and 0.23% C; *medium-high carbon*, between 0.23 and 0.44% C; and *high carbon*, more than 0.44% C. However, from a design viewpoint, high carbon steels are those over 0.35%, because these can not be used as welded pressure parts. Low carbon steels see extensive use as pressure parts, particularly in low pressure applications where strength is not a significant design issue. For most structural applications and the majority of pressure parts, medium carbon steels, with carbon contents between 0.20 and 0.35%, predominate.

Carbon steels are also referred to as killed, semi-killed, rimmed and capped, depending on how the carbon-oxygen reaction of the steel refining process was stopped. During the steelmaking process, oxygen, introduced to refine the steel, combines with carbon to form a gas. If the oxygen introduced is not removed or combined prior to or during casting by the addition of Si, Al, or some other deoxidizing agent, the gaseous products continue to evolve during solidification of the metal in the mold. The amount of gas evolved during solidification determines the type of steel and the amount of carbon left in the steel. If no gas is evolved and the liquid lies quietly in the mold, it is known as killed steel. With increasing degrees of gas evolution, the products are known as semi-killed and rimmed steels. Virtually all steels used in boilers today are fully killed.

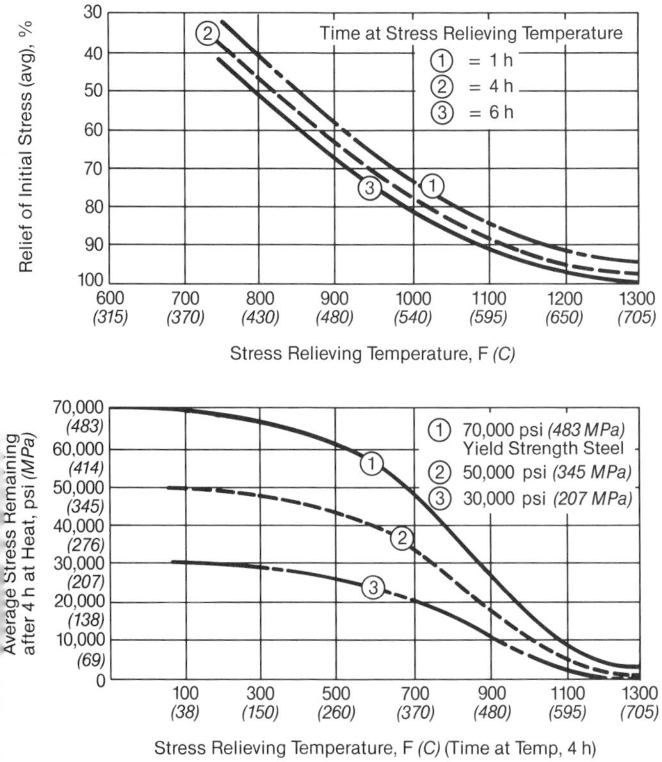

Fig. 12 Effect of temperature and time on stress relief in carbon steel (upper graph) and steels with varying as-welded strengths (*courtesy of AWS*).[10]

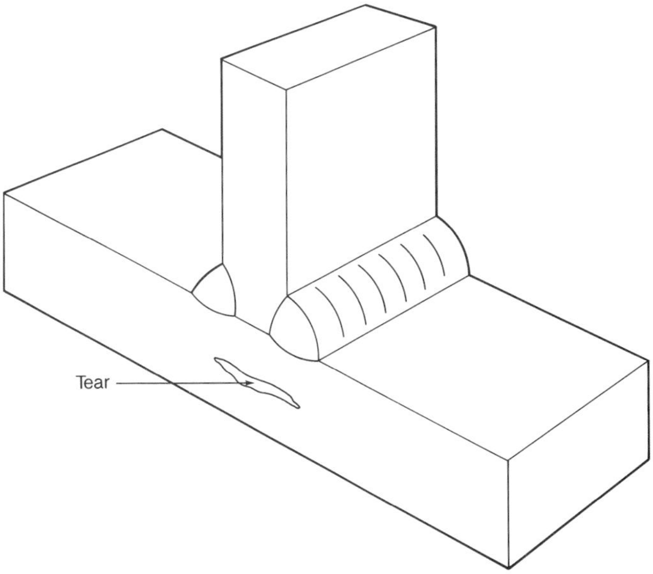

Fig. 13 Lamellar tearing.

Microalloyed steels are carbon steels to which small amounts (typically less than 1%) of alloying elements have been added to achieve higher strength. Common additions are vanadium and boron. Such steels are seldom used in pressure part applications, but they are gaining acceptance as structural steels.

Residual elements are present in steels in small amounts and are elements other than those deliberately added as alloying or killing agents during the steelmaking process. Their source is the scrap or pig iron used in the furnace charge. Cu, Ni, Cr, V and B are typical examples of residuals often found in carbon steels. S and P, also considered to be residual elements, usually are reported in chemical analyses of steels, and their concentrations are limited by specification because they degrade ductility. The residual elements S, P, Sb and tin (Sn) are also important contributors to temper embrittlement in steels.

Historically, residual elements other than S and P were neither limited nor reported. This practice is changing, however, and several residuals have established limits.

Low and medium alloy steels are the next most important category of steels used in boilers. These are characterized by Cr contents less than 11.5% and lesser amounts of other elements. The most common alloy combinations in this group encountered in boilers are: C-1/2Mo, 1/2Cr-1/2Mo, 1Cr-1/2Mo, 1-1/4Cr-1/2Mo-Si, 2-1/4Cr-1Mo, and 9Cr-1Mo-V-Ti-B. Other less common alloys in this group are 3Cr-1Mo, 5Cr-1/2Mo and 9Cr-1Mo.

Because of the exceptional strength-enhancing capability of Mo in carbon steel, it is not surprising that C-1/2Mo steel has many applications for pressure parts, particularly in the temperature range of about 700 to 975F (371 to 524C). C-Mo steels, however, are particularly prone to graphitization at temperatures above about 875F (468C). Inside the boiler, where graphitization failures do not present a safety hazard, C-Mo tubing has many uses up to 975F (524C), its oxidation limit. Because Al content promotes graphitization, C-Mo steel is usually Si-killed and it has a coarse grain structure as a consequence. Therefore, C-Mo components are somewhat prone to brittle failures at low temperatures. This is not a problem in service, because the design application range of this alloy is at high temperature.

The oxidation resistance of low alloy steels increases with Cr content. The first common alloy in the Cr-Mo family is 1/2Cr-1/2Mo. This steel was developed in response to the graphitization failures of C-Mo piping. It was found that the addition of about 0.25% Cr was sufficient to make the alloy immune to graphitization. Furthermore, 1/2Cr-1/2Mo has essentially the same strength as C-Mo and has therefore displaced it in many applications. Because the application of 1/2Cr-1/2Mo is virtually unique to the boiler industry, it is less readily available in certain sizes and product forms.

The next alloys in this series are the nearly identical 1Cr-1/2Mo and 1-1/4Cr-1/2Mo-Si; the Si-containing version is slightly more oxidation resistant. However, extensive analyses of the databases indicate that the 1Cr-1/2Mo version is stronger over the temperature range of 800 to 1050F (427 to 566C). As a result, this alloy is rapidly displacing 1-1/4Cr-1/2Mo-Si in most applications in this temperature regime.

Absent the addition of other alloying elements, the 2-1/4Cr-1Mo composition is the optimum alloy for high temperature strength. Where the need for strength at temperatures between 975 and 1115F (524 and 602C) is the dominant design requirement, 2-1/4Cr-1Mo is the industry workhorse alloy. The 3Cr through 9Cr alloys are less strong, but they have application where improved oxidation resistance is desired and lower strength can be tolerated. The increasing air-hardenability of these alloys with increasing Cr content makes fabrication processes more complex and their use is somewhat more costly as a result.

Mn-Mo and Mn-Mo-Ni alloys have limited use in fossil-fueled boilers. Their slightly higher strength compared to carbon steels promotes their application in very large components, where the strength to weight ratio is an important consideration. Their generally superior toughness has made them a popular choice for nuclear pressure vessels.

The *heat treatable low alloy steels*, typified by the AISI-SAE 4340 grade (nominally 0.40C-0.80Cr-1.8Ni-0.25Mo), are used for low temperature structural applications in boilers. The instability of their microstructures and therefore of their strength, with long exposure at elevated temperatures, has eliminated them from consideration for boiler pressure parts.

Higher Cr-Mo alloys Because of their tendency toward embrittlement, the martensitic 9Cr-1Mo and 12Cr-1Mo steels have not been widely used for pressure vessel and piping applications in North America prior to the 1980s. However in the early 1970s, the United States (U.S.) Department of Energy sponsored research to develop a 9Cr-1Mo steel with improved strength, toughness and weldability[12] for tubing in steam generators for liquid metal fast breeder nuclear reactors.[13] The alloy is 9Cr-1Mo-V-Ti-B, commonly called 9V or Grade 91. It has exceptional strength, toughness and stability at temperatures up to 1200F (649C). Because it is nearly twice as strong as 2-1/4Cr-1Mo at 1000F (538C), it is displacing that alloy in high pressure header applications. The resultant thinner vessels have significantly reduced thermal stresses and associated creep-fatigue failures compared to 2-1/4Cr-1Mo and 1-1/4Cr-1/2Mo-Si headers.[14,15] Because 9V is stronger than austenitic stainless steel up to about 1125F (607C), it is also displacing that alloy class in high pressure tubing applications. It has the added advantage of being a ferritic alloy, eliminating the need for many dissimilar metal welds between pressure parts. The Grade 91 gains its strength, toughness, and stability from its alloy additions and features the fully bainitic microstructure resulting from careful normalizing and tempering.

While not popular in North America because of the care necessary in handling very air-hardenable alloys during fabrication, the 12Cr-Mo and 12Cr-Mo-V alloys have had wide use in the European boiler industry. In addition, the experience being gained with 9V may eventually enhance the acceptance of the 12Cr group.

Austenitic stainless steel Every attempt is made to minimize the use of stainless steels in boilers because of their high cost, but the combination of strength and corrosion resistance they provide makes them the favored choices in certain applications. They are virtually the only choices for service above 1100F (593C). At lower temperatures, down to about 1050F (566C), they often displace the Cr-Mo ferritic steels, where the lower pressure drop afforded by the thinner stainless steel component wall is important. Wider acceptance of 9V may eliminate use within this range.

The common alloys of stainless steels used in boiler pressure parts are 18Cr-8Ni, 18Cr-8Ni-Ti, 18Cr-8Ni-Cb, 16Cr-12Ni-2Mo, 25Cr-12Ni, 25Cr-20Ni and 20Cr-30Ni. The last alloy in this group is technically a nonferrous alloy, because it has less than 50% Fe when its other minor alloying constituents are considered. However, because it is so similar to the other austenitic stainless steels, it may be considered one of them. These alloys are commonly designated the 300 series: 304, 321, 347, 316, 309 and 310 stainless steels. The 20Cr-30Ni alloy is commonly known as Alloy 800. Because the strength of these materials at high temperature is dependent on a moderate carbon content and usually on a coarse grain size, materials with those qualities are often specified for high temperature service. They carry the added designation of the letter H, e.g., 304H or 800H.

Of these alloys, 304H is the most commonly used. It provides an excellent balance of strength, oxidation resistance and corrosion resistance at the lowest cost of any alloy in this group.

All of the 300 series alloys are susceptible to sigma-phase formation after long exposure at temperatures of 1050 to 1700F (566 to 972C). Those with some initial ferrite, such as 309, can form the sigma phase earlier, but all eventually do so. This phase formation decreases toughness and ductility but has no effect on strength or corrosion resistance. It has been a problem in heavy-section piping components made of 316 stainless steels, but it is not a design consideration for smaller (tubing) components. Specifying materials with improved resistance to sigma phase formation adds cost without a parallel increase in reliability.

The 321 type is not as strong as the others in this series. While it is a stabilized grade and has important low temperature applications, the stability of the titanium carbide makes it extremely difficult to heat treat type 321 in one thermal treatment and obtain a resulting structure that is both coarse grained, for high temperature creep strength, and has stabilized carbides for sensitization resistance. It is possible to apply a lower temperature stabilizing heat treatment, at about 1300F (704C), following the solution treatment to achieve a stabilized condition and good creep strength. The stability of the columbium (niobium) carbides in type 347 is better, and this grade can be heat treated to obtain creep strength and sensitization resistance. This 18Cr-8Ni-Cb alloy is widely used at high temperatures because of its superior creep strength.

The Mo content of the 316 type increases its pitting resistance at lower temperatures. While this alloy has good creep strength, it is not often used because of its higher cost.

All of these austenitic stainless steel alloys require a high temperature heat treatment after cold or warm forming if they are to be used at high temperatures [above 1000F (538C)]. Otherwise, the internal strain energy of the cold work would eventually lead to recrystallization and a fine grain size with poor creep strength.[16] Also, these alloys are susceptible to stress corrosion cracking in certain aqueous environments. The 300 series alloys are particularly sensitive to the presence of halide ions. As a result, their use in water-wetted service is usually prohibited. The stress corrosion cracking experience with Alloy 800 has been mixed and, while this grade is permitted in water-wetted service, it is not common practice.

Types 309, 25Cr-12Ni, and 310, 25Cr-20Ni, have virtually identical strengths and corrosion resistance. They are not as strong as 304 or 347 but are more oxidation resistant. The high Ni alloys, 310 and Alloy 800, are somewhat more affected by sulfidation attack. They have been used as nonpressure fluidized-bed boiler components designed to remove particulate from hot gas streams.

Most of these alloys are available in a multiplicity of minor variations: H grades, with 0.04 to 0.10% C and a required high temperature anneal or coarse grain size (or both) for creep strength; L grades, with 0.035% maximum C for sensitization resistance; N grades, with 0.010% minimum N added for strength; LN grades, with 0.035% maximum C and 0.010% minimum N for sensitization resistance and strength; and straight (no suffix) grades, with 0.08% maximum C.

Ferritic stainless steels contain at least 10% Cr and have a ferrite-plus-carbide structure. *Martensitic stainless steels* are ferritic in the annealed condition but are martensitic after rapid cooling from above the critical temperature. They usually contain less than 14% Cr.[17] *Precipitation hardened stainless steels* are more highly alloyed and are strengthened by precipitation of a finely dispersed phase from a supersaturated solution on cooling. None of these steels are used for pressure parts or load carrying components in boilers because, at the high temperatures at which their oxidation resistance is useful, they are subject to a variety of embrittling, phase precipitation reactions, including 885F (474C) embrittlement and sigma phase formation. They are used as studs for holding refractories and heat absorbing projections and as thermal shields. These alloys are also difficult to weld without cracking.

Several *duplex alloys*, with mixed austenitic-ferritic structures, have been developed. They are useful in corrosive lower temperature applications such as those found in wet desulfurization equipment used as boiler flue gas scrubbers.

Bimetallic materials Weld cladding of one alloy with another has been available for many years. A more recent development has been the proliferation of *bimetallic* components, such as tubes and plate containing a load carrying alloy for their major constituent covered with a layer (usually external) of a corrosion resistant alloy. The first bimetallic tubes to see wide use in boilers were made from Alloy 800H clad with a 50Cr-50Ni alloy (Alloy 671) for coal ash corrosion resistance. The combination in widest use today is carbon steel clad with 304L, used in pulp and paper process recovery (PR) boilers. One of the latest to be developed is carbon steel or 1/2Cr-1/2Mo clad with Alloy 825 (42Ni-21.5Cr-5Mo-2.3Cu) used in refuse-fired boilers.[18] Other combinations that have been used are 1/2Cr-1/2Mo and 2-1/4Cr-1Mo clad with 309.

Cast irons Cast irons and steels (containing more than 2% or less than 2% C, respectively) have long had wide acceptance as wear resistant and structural components in boilers. Cast steels are also used for boiler pressure parts. The three types of cast iron used in boilers are white, gray and ductile iron.

White iron White cast iron is so known because of the silvery luster of its fracture surface. In this alloy, the carbon is present in combined form as the iron carbide cementite (Fe_3C). This carbide is chiefly responsible for the hardness, brittleness and poor machinability of white cast iron. Chilled iron differs from white cast iron only in its method of manufacture and it behaves similarly. This type of iron is cast against metal blocks, or chills, that cause rapid cooling at the adjacent areas, promoting the formation of cementite. Consequently, a white or mottled structure, which is characterized by high resistance to wear and abrasion, is obtained. Elverite alloys, a series of white iron, Ni-enriched cast materials developed by Babcock & Wilcox (B&W) for use in pulverizers and other wear resistant parts, have long been noted for their uniformity and high quality.

VAM 20®, a more recent development, is a 20% Cr white iron with a carbide-in-martensite matrix, very high hardness and good toughness (compared to other white irons). The hardness and wear resistance of VAM 20® are superior to those of the Elverites and similar alloys. It is always used in the heat treated condition, which accounts for its good toughness and uniformity. VAM 20® is used in grinding elements of coal pulverizers.

Malleable cast iron is white cast iron that has been heat treated to change its combined carbon (cementite) into free, or temper carbon (nodules of graphite). The iron becomes malleable because, in this condition, the carbon no longer forms planes of weakness.

Gray iron Gray cast iron is by far the most widely used cast metal. In this alloy, the carbon is predominantly in the free state in the form of graphite flakes, which form a multitude of notches and discontinuities in the iron matrix. The fracture appearance of this iron is gray because the graphite flakes are exposed. Gray iron's strength depends on the size of the graphite crystals and the amount of cementite formed with the graphite. The strength of the iron increases as the graphite crystal size decreases and the amount of cementite increases. Gray cast iron is easily machinable because the graphite carbon acts as a lubricant for the cutting tool; it also provides discontinuities that break the chips as they are formed. Modern gray iron having a wide range of tensile strength, from 20,000 to 90,000 psi (138 to 621 MPa), can be made by suitable alloying with Ni, Cr, Mo, V and Cu.

Ductile iron Another member of the cast iron family is *ductile cast iron*. It is a high carbon, Mg-treated ferrous product containing graphite in the form of spheroids or impacted particles. Ductile cast iron is similar to gray cast iron in melting point, fluidity and machinability, but it possesses superior mechanical properties. This alloy is especially suited for pressure castings. By special procedures (casting against a chill), it is possible to obtain a carbide-containing abrasion resistant surface with an interior of good ductility.

Cast iron was used extensively in early steam boilers for tubes and headers. This material is no longer used in the pressure parts of modern power boilers but is used in related equipment such as stoker parts and the grinding rings of coal pulverizers.

Cast steel Cast steels are used for many support and alignment applications in boilers, and for some pressure parts having complex shapes. The alloys range from carbon steel and 2-1/4Cr-1Mo to 25Cr-12Ni and 50Cr-50Ni.

Ceramics and refractory materials Ceramics and refractory materials are used primarily for their insulating and erosion resisting properties. While many early furnace designs featured brick furnace walls, these have mostly been replaced by steel membrane panels. (See Chapter 22.) However, in many applications, these walls may still have a rammed, troweled or cast refractory protection applied. Refractory linings are still important features of some furnaces, particularly those exposed to molten slag. In Cyclone furnaces (see Chapter 14) and other wet-bottom boilers, gunned and troweled alumina, and silicon carbide refractory products are generally used. Chromium-containing refractories are no longer used.

Cera-VAM® is a high density alumina ceramic used as an erosion liner in coal-air pipeline elbows, coal pulverizer internals, and pulverizer swing valves to reduce erosion and the associated maintenance costs. (See Chapter 12.) Structural ceramics have also been introduced as hot gas filters. These filters remove particulates from the flue gas of fluid-bed boilers before the gas enters the high temperature gas turbine of combined cycle plants. (See Chapter 29.)

Coatings Many types of coatings are applied to boiler metal parts. In addition to the cast, gunned and troweled types mentioned above, thinner carbide-containing, metallic matrix coatings are sprayed onto surfaces in fluid-bed boilers exposed to high velocity particulate erosion. Metallic coatings are also sprayed on boiler parts exposed to erosion and corrosion wastage by the *flame spraying, plasma spraying* and *high velocity oxy-fuel* processes. These are shop- and field-applied maintenance processes that protect and repair components that experience wastage. Proper surface preparation and process control must be exercised to ensure that these coatings adhere, have the proper density, and achieve the recommended thickness on all surfaces.

Chromizing In the mid 1970s, B&W pioneered the use of chemical vapor deposition (CVD) coatings for boiler components. *Chromizing*, a process previously applied to aircraft jet engine components, is applied to large surfaces on the interior of tubing and piping. The purpose of this process is to develop a high Cr-containing surface that is resistant to oxidation and subsequent exfoliation. High temperature steam carrying pressure parts suffer from oxidation on their internal surfaces. When the oxide layer becomes thick enough, it spalls off the surface and the particles are carried to the steam turbine, where the resulting erosion damage causes loss of efficiency and creates a risk of mechanical damage. Perfect coverage of tube ID surfaces is not necessary to reduce this condition. If 95% of the susceptible tube surface is chromized, a twenty-fold reduction in exfoliate particles will result.

In CVD processes, such as chromizing, the surfaces to be coated are usually covered with or embedded in a mixture containing powdered metal of the coating element, e.g., Cr, a halide salt, and a refractory powder, often alumina. When the parts and the mixture are heated to a sufficiently high temperature, the salt decomposes and the metal powder reacts with the halide ion to form a gas, e.g., $CrCl_2$ or $CrBr_2$. At the surface of the part being coated, an exchange reaction takes place. An Fe atom replaces the Cr in the gas and the Cr atom is deposited on the surface. The process is conducted at sufficient time and temperature to permit the Cr to diffuse into the base material. At the chromizing temperature, 2-1/4Cr-1Mo, for example, is fully austenitic. However, as Cr atoms are deposited on the surface, the Cr increases the stability of the ferrite phase. As a result, the diffusion front advances into the matrix concurrently with the phase transformation front. This results in a diffusion zone with a nearly constant Cr content. (See Fig. 14.) Typical depths of this zone range from 0.002 to 0.025 in. (0.051 to 0.64 mm). The diffusion layer on a 2-1/4Cr-1Mo substrate has a minimum Cr content of 13%.

Chromizing, though first developed to reduce solid particle erosion of turbines, is now being applied to external surfaces of boiler pressure parts to reduce or prevent corrosion and corrosion-fatigue damage. In these applications, near perfect continuity and integrity of the coating is required. A thicker coating is necessary to resist the more hostile external environments.

Aluminizing Aluminizing, a similar CVD process, has been used for many years to protect components in petrochemical process pressure vessels. However alumina, as silica, is soluble in high temperature, high pressure steam and it can be carried to the turbine, where pressure and temperature drops cause it to precipitate on the turbine components; this is undesirable.

Fused coatings Tungsten carbide/chromium carbide fused metallic coatings are also used for erosion protec-

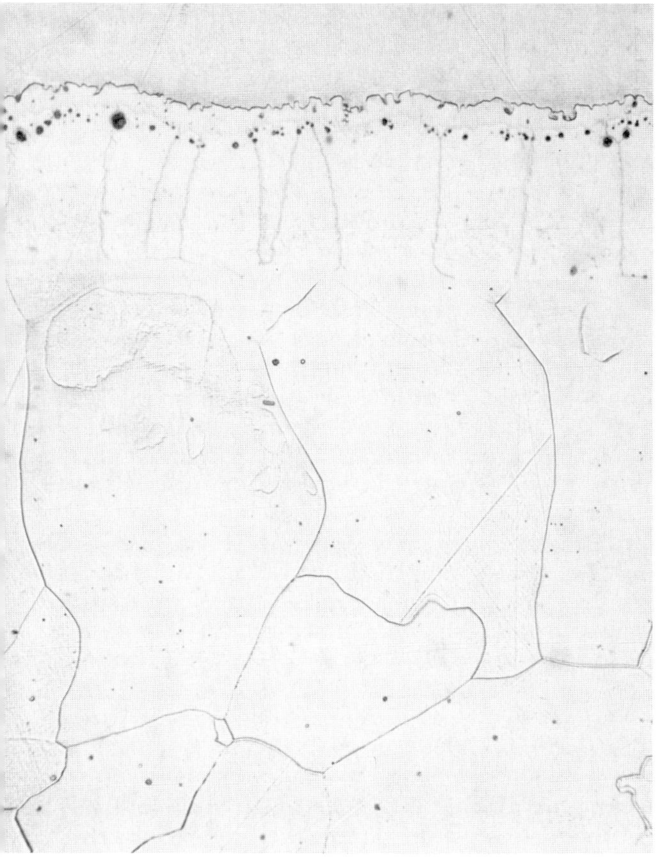

Fig. 14 Chromized 2-1/4Cr-1Mo at 400 x.

tion of tube membrane panels, for example, in basic oxygen furnace steelmaking furnace hoods. Fused coatings differ from sprayed coatings in being higher density and achieving better bonds due to the brazing-type action of the application process.

Galvanizing More mundane coatings, such as galvanizing, painting and organic rust prevention coatings, are also used on boiler components.

Galvanizing, a zinc coating usually applied by dipping in molten metal or by electroplating, is usually used on structural components external to the boiler, when erection is near a seashore or a petrochemical complex.[19] Galvanized components must be kept out of high temperature areas to avoid structural damage due to zinc grain boundary embrittlement.

Mechanical properties

Low temperature properties

Steels of different properties are used in boilers, each selected for one or more specific purposes. Each steel must have properties for both manufacturing and satisfactory service life. Each particular type, or grade, of steel must be consistent in its properties, and tests are normally run on each lot to demonstrate that the desired properties have been achieved.

Specifications standardizing all the conditions relating to test specimens, methods and test frequency have been formulated by the American Society of Testing and Materials (ASTM) and other authorities.

Tensile test

In the tensile test, a gradually applied unidirectional pull determines the maximum load that a material can sustain before breaking. The relationship between the stress (load per unit area) and the corresponding strain (change of length as a percent of the original length) in the test piece is illustrated in the stress-strain diagrams of Figs. 15 and 16. The metal begins to stretch as soon as the load is applied and, for some range of increasing load, the strain is proportional to the stress. This is the elastic region of the stress-strain curve, in which the material very closely follows Hooke's Law: strain, ε, is proportional to stress, σ. The proportionality constant may be considered as a spring constant and is called Young's modulus, E. Young's modulus is a true material property, characteristic of each alloy. Young's modulus for steel is approximately 30×10^6 psi at room temperature.

If the stress is released at any point in this region, the test specimen will return to very nearly its initial dimensions. However, if the stress is increased beyond a certain point, the metal will no longer behave elastically; it will have a permanent (plastic) elongation, and the linear relationship between stress and strain ceases. This value is known as the proportional limit of the material and, in this discussion, may be considered practically the same as the elastic limit, which may be defined as the maximum stress that can be developed just before permanent elongation occurs.

When a material has a well defined point at which it continues to elongate without further increase in load, this point is called the yield point. Many steels do not have a yield point and even in those that do, neither it nor the proportional or elastic limits can be determined with accuracy. By convention, therefore, engineers have adopted an arbitrary but readily measurable concept: the *yield strength* of a metal. This is defined as the stress at which the strain reaches 0.2% of the gauge length of the test specimen. This is illustrated in Fig. 16. (Other values, 0.1% or 0.5% are occasionally used, but 0.2% is most common.)

If the loading is continued after yielding begins, a test specimen of a ductile material with homogeneous composition and uniform cross-section will be elongated uniformly over its length, with a corresponding reduction in area.

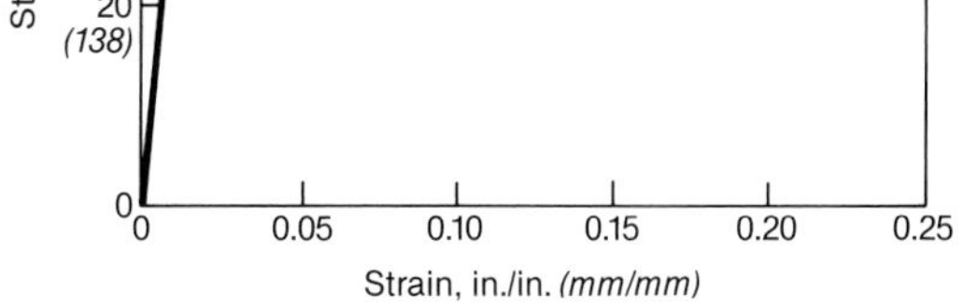

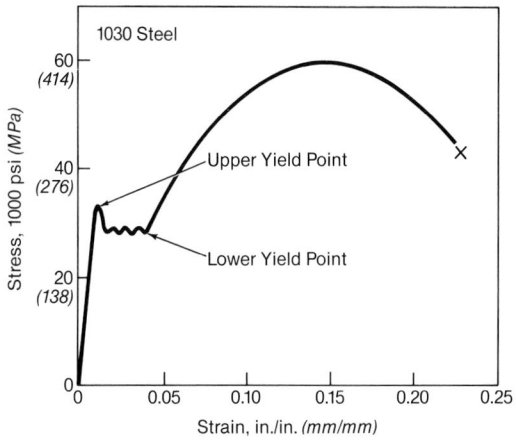

Fig. 15 Engineering stress-strain curve for 1030 carbon steel (*courtesy of Wiley*).[20]

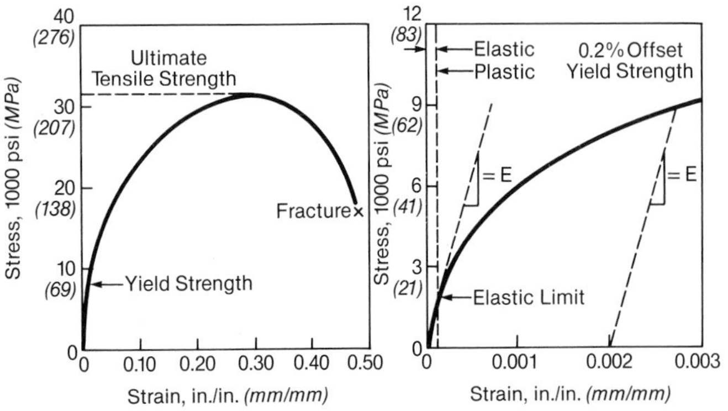

Fig. 16 Engineering stress-strain diagram for polycrystalline copper. Left, complete diagram. Right, elastic region and initial plastic region showing 0.2% offset yield strength.[20]

Eventually, a constriction or necking may occur. In some materials, localized necking may not occur, but the cross-section may reduce more or less uniformly along the full gauge length to the instant of rupture. In all ductile materials, however, an appreciable increase in elongation occurs in the reduced area of the specimen. The more ductile the steel, the greater is the elongation before rupture. The maximum applied load required to pull the specimen apart, divided by the area of the original cross-section, is known as the *ultimate tensile strength*. Brittle materials do not exhibit yielding or plastic deformation, and their yield point and ultimate tensile strength are nearly coincident.

The ductility of the metal is determined by measuring the increase in length (total elongation) and the final area at the plane of rupture after the specimen has broken, and is expressed as percent elongation or percent reduction of area.

Hardness test

Hardness may be defined as resistance to indentation under static or dynamic loads and also as resistance to scratching, abrasion, cutting or drilling. To the metallurgist, hardness is important as an indicator of the effect of heat treatment, fabrication processes, or service exposure. Hardness values are roughly indicative of the ultimate tensile strength of steels. Hardness tests are also used as easy acceptance tests and to explore local variations in properties.

Hardness is usually determined by using specially designed and standardized machines: Rockwell, Brinell, Vickers (diamond pyramid), or Tukon. These all measure resistance to indentation under static loads. The pressure is applied using a fixed load and for a specified time, and the indentation is measured either with a microscope or automatically. It is expressed as a hardness number, by reference to tables. Hardness can also be determined by a scleroscope test, in which the loss in kinetic energy of a falling metal weight, absorbed by indentation upon impact of the metal being tested, is indicated by the height of the rebound.

Toughness tests

Toughness is a property that represents the ability of a material to absorb local stresses by plastic deformation and thereby redistribute the stresses over a larger volume of material, before the material fails locally. It is therefore dependent on the rate of application of the load and the degree of concentration of the local stresses. In most steels, it is also temperature dependent, increasing with increasing temperature (although not linearly). Toughness tests are of two types, relative and absolute.

Notched bar impact tests are an example of the relative type. The most common is the Charpy test, in which a simple horizontal beam, supported at both ends, is struck in the center, opposite a V-shaped notch, by a single blow of a swinging pendulum. A Charpy specimen is illustrated in Fig. 17a. The energy absorbed by the breaking specimen can be read directly on a calibrated scale and is expressed in ft lb units. The specimen is also examined to determine how much it has spread laterally and how much of its fracture surface deformed in shear versus cleavage. The toughness is expressed in units of absorbed energy (ft lb), mils (thousandths of an inch) lateral expansion and percent shear. The values are characteristic not only of the material and temperature, but also of the specimen size. Therefore, comparison between materials and tests have meaning only when specimen geometries and other test conditions are identical. Specimens are inexpensive and the test is easy to do. Often, vessel designers are interested in the variation of toughness with temperature. Fig. 18 illustrates the variation in toughness with temperature of 22 heats of a fine grained carbon steel, SA-299, as determined by Charpy testing. This material displays a gradual transition from higher to lower toughness.

Another toughness test, and one that provides a more sharply defined transition, is the drop-weight test. The specimen for this test is shown in Fig. 17b. A known weight is dropped from a fixed height and impacts the specimen. This is a pass or fail test and is performed on a series of specimens at varying temperatures, selected to bracket the break versus no-break temperature within 10F (6C). If the impact causes a crack to propagate to either edge of the specimen from the crack-starter notch in the brittle weld

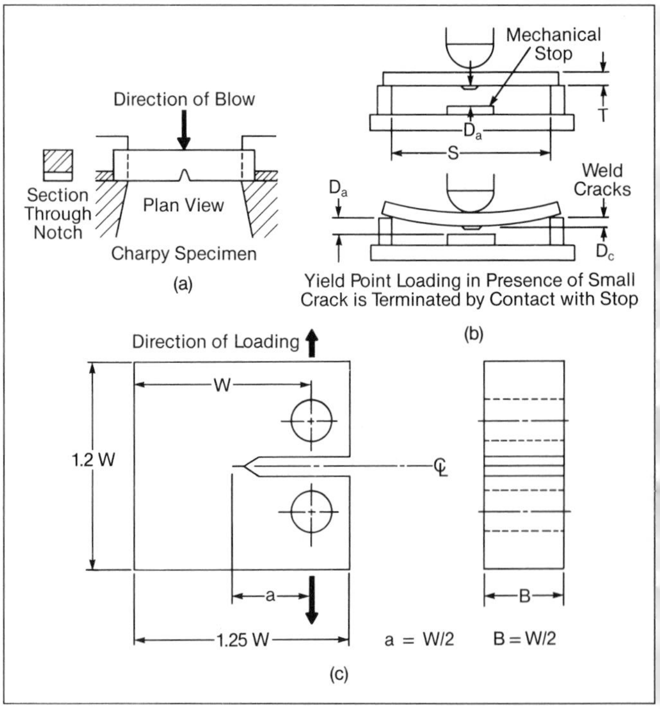

Fig. 17 (a) Charpy specimen, (b) nil ductility transition temperature (drop weight) test specimen, (c) compact tension specimen (*courtesy of Prentice-Hall*).[21]

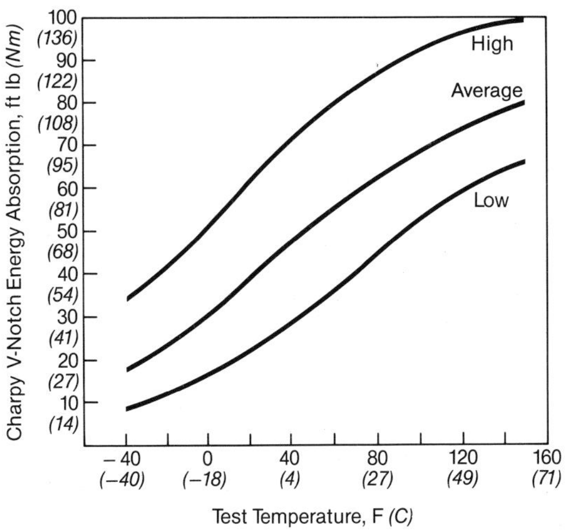

Fig. 18 Charpy V-notch impact energy versus test temperature for fine grained SA-299 plate material.

bead deposited on the face of the specimen, the specimen is considered to have broken at that temperature. The lowest temperature at which a specimen fails determines the nil-ductility transition temperature (NDTT). Fig. 19 shows a histogram of NDTTs from 20 heats of fine grained SA-299.

Fracture toughness tests measure true characteristics of a given metal. They are more complex and specimens are more costly. However, they produce values that can be used in analytical stress calculations to determine critical flaw sizes above which flaws or cracks may propagate with little or no increase in load. A typical fracture toughness specimen is shown in Fig. 17c. Variations of fracture toughness tests involve testing under cyclic rather than monotonically increasing load (fatigue crack growth testing) and testing in various environments to determine crack growth rates as a function of concurrent corrosion processes. The same specimen is used to determine fatigue crack growth behavior. Fig. 20 illustrates the difference in crack growth rate in air and in a salt solution for 4340 steel tempered to two strength levels.

Formability tests

Several different types of deformation tests are used to determine the potential behavior of a material in fabrication. These include bending, flattening, flaring and cupping tests. They furnish visual evidence of the capability of the material to withstand various forming operations.

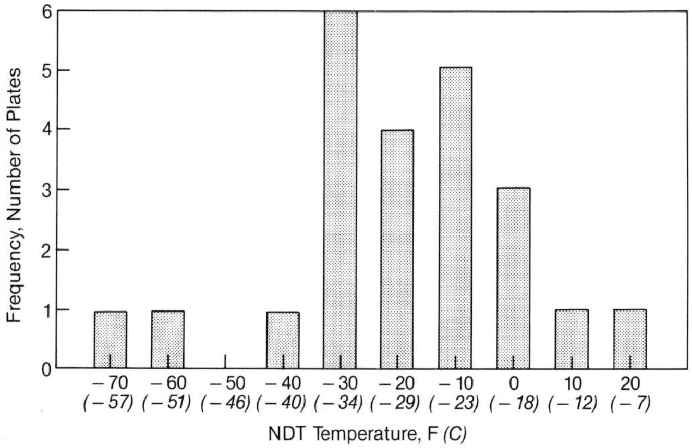

Fig. 19 Drop weight nil ductility transition temperature (NDTT) frequency distribution for 20 heats of fine grained SA-299 plate material.

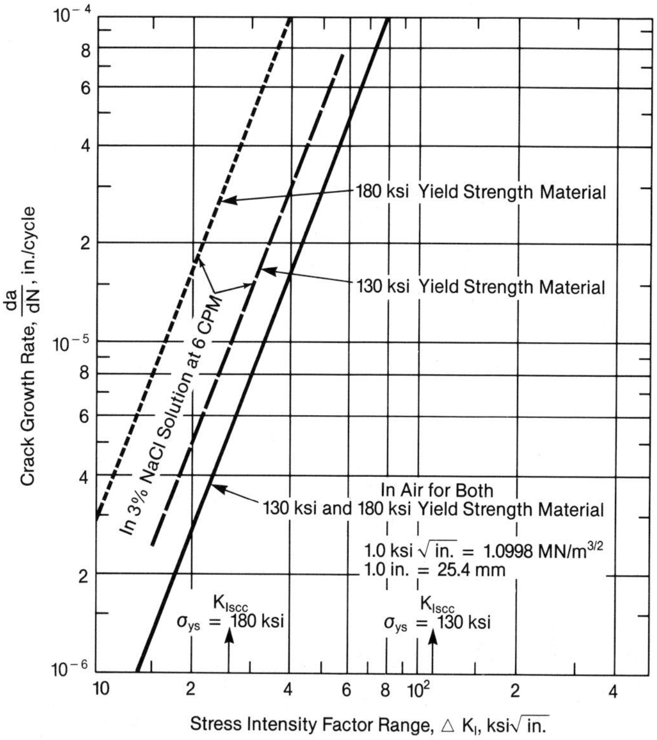

Fig. 20 Corrosion fatigue crack growth rates for 4340 steel.[21]

They are only a rough guide and are no substitute for full scale testing on production machinery.

High temperature properties

Tensile or yield strength data determined at ambient temperatures can not be used as a guide to the mechanical properties of metals at higher temperatures. Even though such tests are made at the higher temperatures, the data are inadequate for designing equipment for long term service at these temperatures. This is true because, at elevated temperatures, continued application of load produces a very slow continuous deformation, which can be significant and measurable over a period of time and may eventually lead to fracture, depending on the stress and temperatures involved. This slow deformation (creep) occurs for temperatures exceeding about 700F (371C) for ferritic steels and about 1000F (538C) for austenitic steels. The maximum allowable working stresses for ferrous materials in power boilers, set by the American Society of Mechanical Engineers (ASME), are based partially on long time creep-rupture tests.

The ASME Boiler and Pressure Vessel Code, Section I, *Power Boilers*, has established the maximum allowable stress values for pressure parts to be no higher than the lowest of:

1. 25% of the minimum specified tensile strength,
2. (1.1) ÷ 4 of the tensile strength at temperature,
3. 67% of the specified minimum yield strength at room temperature,
4. 67% of the yield strength at temperature, for ferritic steels; or 90% of the yield strength at temperature of austenitic steels and nickel base alloys,
5. a conservative average of the stress to give a creep rate of 0.01% in 1000 hours (1% in 100,000 hours), or
6. 67% of the average or 80% of the minimum stress to produce rupture in 100,000 hours.

Furthermore, the allowable stress at a higher temperature can not exceed that at a lower temperature, so no advantage is taken of strain aging behavior. The allowable stress is therefore the lower bound envelope of all these criteria. The tensile and yield strengths at temperature have a particular meaning in Code usage.

Tensile strength

Although the design of high temperature equipment generally requires use of creep and creep-rupture test data, the short time tensile test does indicate the strength properties of metals up to the creep range of the material. This test also provides information on ductility characteristics helpful in fabrication.

The ultimate strength of plain carbon steel and a number of alloy steels, as determined by short time tensile tests over a temperature range of 100F (38C) to 1300 to 1500F (704 to 816C), is shown in Fig. 21. In general, the results of these tests indicate that strength decreases with increase in temperature, although there is a region for the austenitic alloys between 400 and 900F (204 and 482C) where strength is fairly constant. An exception to the general rule is the increase in strength over that at room temperature of carbon and many low alloy steels (with corresponding decrease in ductility) over the temperature range of 100 to 600F (38 to 316C). As the temperature is increased beyond 600 to 750F (316 to 399C), the strength of the carbon and most of the low alloy steels falls off from that at room temperature with a corresponding increase in ductility.

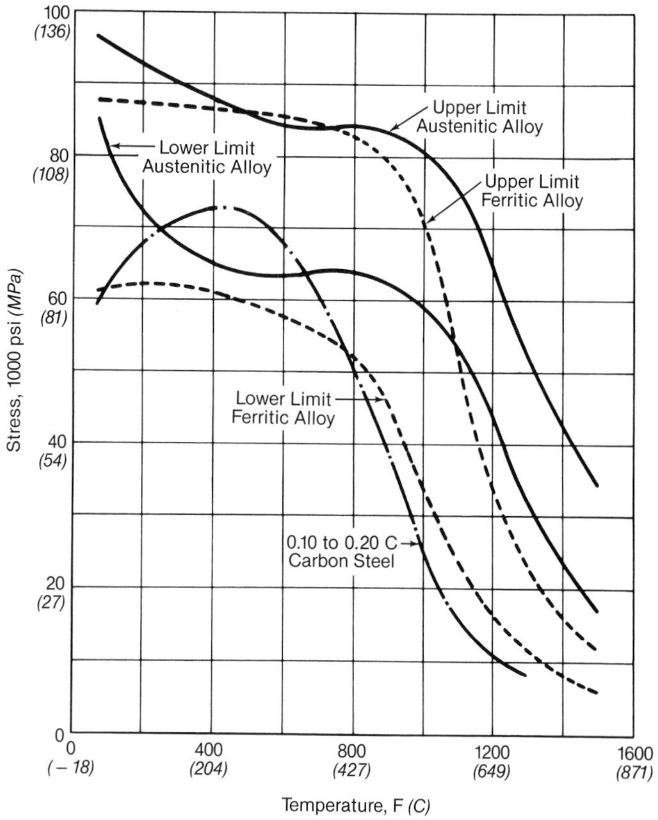

Fig. 21 Tensile strength of various steels at temperatures to 1500F (816C).

Creep and creep-rupture test

It has long been known that certain nonmetallic materials, such as glass, undergo slow and continuous deformation with time when subjected to stress. The concept of creep in metallic materials, however, did not attract serious attention until the early 1920s. Results of several investigations at that time demonstrated that rupture of a metallic material could occur when it is subjected to a stress at elevated temperatures for a sufficiently long time, even though the load applied is considerably lower than that necessary to cause rupture in the short time tensile test at the same temperature.

The earliest investigations of creep in the U.S. were sponsored by B&W in 1926. Many steels now used successfully in power generating units and in the petroleum refining and chemical industries were tested and proved in the course of these investigations, using the best equipment available at the time.

The creep-rupture test is used to determine both the rate of deformation and the time to rupture at a given temperature. The test piece, maintained at constant temperature, is subjected to a fixed static tensile load. The deformation of the test sample is measured during the test and the time to rupture is determined. The duration of the test may range from 1000 to 10,000 h, or even longer. A diagrammatic plot of the observed length of the specimen against elapsed time is often of the form illustrated in Fig. 22.

The curve representing classical creep is divided into three stages. It begins after the initial extension (0-A), which is simply the measure of deformation of the specimen caused by the loading. The magnitude of this initial extension depends on test conditions, varying with load and temperature and normally increasing with increases in temperature and load. The first stage of creep (A-B), referred to as primary creep, is characterized by a decreasing rate of deformation during the period. The second stage (B-C), referred to as secondary creep, is usually characterized by extremely small variations in rate of deformation; this period is essentially one of constant rate of creep. The third stage (C-D), referred to as tertiary creep, is characterized by an accelerating rate of deformation leading to fracture. Some alloys, however, display a very limited (or no) secondary creep and spend most of their test life in tertiary creep.

To simplify the practical application of creep data it is customary to establish two values of stress (for a material at a temperature) that will produce two corresponding rates of creep (elongation): 1.0% per 10,000 h and 100,000 h, respectively.

For any specified temperature, several creep-rupture tests must be run under different loads. The creep rate during the period of secondary creep is determined from these curves and is plotted against the stress. When these data are plotted on logarithmic scales, the points for each specimen often lie on a line with a slight curvature. The minimum creep rate for any stress level can be obtained from this graph, and the curve can also be extrapolated to obtain creep rates for stresses beyond those for which data are obtained. Fig. 23 presents such creep-rate curves for 2-1/4Cr-1Mo steel at 1000, 1100 and 1200F (538, 593 and 649C). The shape of the creep curve depends on the chemical composition and microstructure of the metal as well as the applied load and test temperature.

Creep-rupture strength is the stress (initial load divided by initial area) at which rupture occurs in some specified time, in an air atmosphere, in the temperature range in which creep takes place. The time for rupture at any temperature is a function of the applied load. A logarithmic-scale plot of stress versus time for fracture of specimens generally takes the form of the curves shown for 2-1/4Cr-1Mo steel in Fig. 24.

In general, rapid rates of elongation indicate a transgranular (ductile) fracture and slow rates of elongation indicate an intergranular (brittle) fracture. As a rule, surface oxidation is present when the fracture is transgranular, while visible intercrystalline oxidation may or may not be present when the fracture is intergranular. Because of the discontinuities produced by the presence of intercrystalline oxides, the time to rupture at a given temperature-load relationship may be appreciably reduced. In Fig. 24, the slope of the data at 1200F (649C) is steeper than those for lower temperatures. This is to be expected, because 1200F (649C) is above the usual temperature limit for maximum resistance to oxidation of 2-1/4Cr-1Mo. Therefore, excessive scaling occurs in the long time rupture tests conducted at 1200F (649C).

A complete creep-rupture test program for a given steel actually consists of a series of tests at constant temperature with each specimen loaded at a different level. Because tests are not normally conducted for more than 10,000 h, the values for rupture times longer than this are determined by extrapolation. The ASME Boiler and Pressure Vessel Code Committee uses several methods of extrapolation, depending on the behavior of the particular alloy for which design values are being established and on the extent and quality of the database that is available. Several informative discussions on these methods may be found in ASME publications.

Material applications in boilers

ASME specifications and allowable stresses

The ASME Boiler and Pressure Vessel Code Subcommittee on Materials is responsible for identifying and approving material specifications for those metals deemed

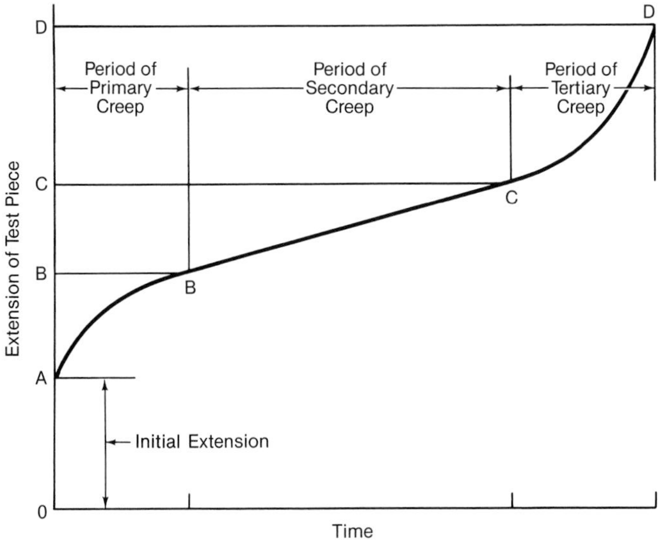

Fig. 22 Classic (diagrammatic) creep test at constant load and temperature.

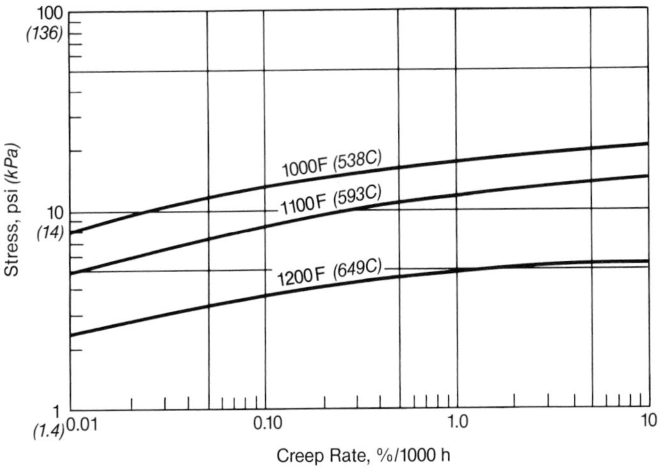

Fig. 23 Creep rate curves for 2-1/4Cr-1Mo steel.

suitable for boiler and pressure vessel construction and for developing the allowable design values for metals as a function of temperature. Most industrial and all utility boilers are designed to Section I of the Code, *Power Boilers*, which lists those material specifications approved for boiler construction. The specifications themselves are listed in Section II, Part A, *Ferrous Materials*, and Section II, Part B, *Non-Ferrous Materials*. The design values are listed in Section II, Part D, *Properties*. (Section II, Part C, *Specifications for Welding Rods, Electrodes, and Filler Metals*, contains approved welding materials.)

For many years, there were relatively few changes in either specifications or design values. Over the last five or ten years, however, many new alloys have been introduced and much new data have become available. The restructuring of the North American steel industry and the globalization of sources of supply and markets are partly responsible for this rapid rate of change. As a result, detailed tables of design values in a text such as this become obsolete much more rapidly than was once the case. A few examples of current allowable stresses are presented below for illustrative purposes. However, the reader is encouraged to consult Section II for an exposure to material specifications and the latest design values.

Pressure part applications

The metal product forms used in boiler pressure parts are tubes and pipe (often used interchangeably), plate, forgings and castings. Tubular products compose the

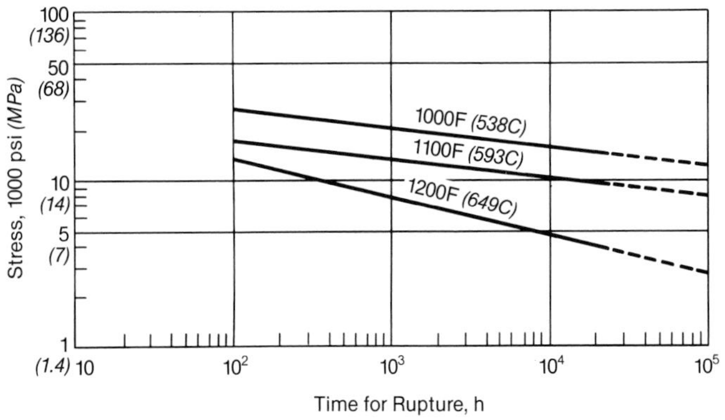

Fig. 24 Typical creep rupture curves for 2-1/4Cr-1Mo steel.

greatest part of the weight. The matrix in Table 1 shows the common pressure part material specifications used in fossil-fired boilers today, their minimum specified properties, recommended maximum use temperatures, and their applications. This list is not meant to be all inclusive, as there are many other specifications permitted by Section I and several of them see occasional use. Neither is it meant to be exclusive, as several of these specifications are used occasionally for components not checked in Table 1. Finally, the recommended maximum use temperatures represent one or more of a variety of limits. The temperature listed may be the highest for which stresses are listed in Section I, the oxidation limit for long term service, a temperature at which graphitization may be expected, or current commercial practice.

Boiler, furnace waterwall, convection pass enclosures and economizers

Boiler, furnace waterwall, and convection pass enclosure surfaces are generally made of carbon steel, C-Mo, and 1/2Cr-1/2Mo seamless or electric-resistance-welded (ERW) tubes. There is still some resistance to the use of ERW tubing in boilers, because of failures at the weld seam when the product was introduced more than two decades ago. That prejudice is unfounded today, as the improvements in starting strip quality, manufacturing procedures and inspection techniques have made today's product equal to seamless. ERW tubing is the standard in critical applications in pulp and paper process recovery (PR) boilers, where water leaks entail the risk of disastrous smelt-water explosions.

Lower carbon grades and 1/2Cr-1/2Mo alloy are used in high heat input regions to avoid the risk of graphitization in this region where tube metal temperatures may be subject to more fluctuation and uncertainty. Higher carbon grades and C-Mo are used in furnace floors, upper furnace walls, convection pass enclosures, and economizers.

Superheaters and reheaters

The highest metal temperatures of pressure parts in the steam generating unit occur in the superheater and reheater. Consequently, these tubes are made of material having superior high temperature properties and resistance to oxidation. Carbon steel is a suitable and economical material to about 850 to 950F (454 to 510C) metal temperature, depending on pressure. Above this range, alloy and stainless steels are required because of the low oxidation resistance and the low allowable stresses of carbon steel. Usually two or more alloys are used in the construction of the superheater. The lower alloys, such as carbon and C-Mo steels, are used toward the inlet section, while the low and intermediate alloy Cr-Mo steels are used toward the outlet, where the steam and metal temperatures are increasing. (See Chapter 18.)

Stainless steel tubes have been required in the hottest sections of the superheater. However, stainless steels are being replaced in many applications by 9Cr-1Mo-V. This high strength ferritic steel was developed initially by Oak Ridge National Laboratories for fast breeder nuclear reactor components. However, it has found many applications in fossil fuel-fired boilers because of its high strength and excellent toughness. Because it is ferritic, its use in place of stainless steel eliminates dissimilar metal weld failures.

Selection factors

Many factors influence material selection in a superheater. These include performance factors (heat transfer surface area required, final steam temperature, total mass flow through the tubes, and flow balancing among circuits); mechanical factors (internal pressure, design temperature, support systems and relative thermal expansion stresses); environmental factors (resistance to steam oxidation and out of service pitting corrosion on the ID and oxidation, fuel ash corrosion, and erosion on the OD); and manufacturing process and equipment limitations and considerations, such as cost of tube to tube butt welds.

Cost

Table 2 shows the allowable stresses for a series of seamless tubes that might be used in a typical utility boiler superheater. The last column in this table is as important a factor in determining which materials to use in construction of the superheater as are the allowable stresses or the fuel ash corrosion behavior of the alloys; it is the price per pound of the alloys as determined from vendor quotations for a utility boiler proposal. Because the prices reflect the lowest prices quoted from about six sources, including one each from France and Japan, it is clear that the foreign currency exchange rate can also be an important factor in material selection.

Using the allowable stresses in Table 2, minimum tube wall thicknesses can be calculated using the Code formula:

$$t = \frac{PD}{2S + P} + 0.005D$$

where t is the thickness in inches, P is the design pressure in psi, D is the tube OD in inches and S is the allowable stress in psi. (See also Chapter 45.) Using the formula:

$$\text{weight / ft} = 10.69 \times (D - 1.11t) \times (1.11t)$$

where 1.11 is a factor taking into account the over-thickness tolerance normally found in seamless tubes, the weight per foot of tubes can be found. Table 3 shows the minimum wall thicknesses and prices per foot of superheater surface for seven materials, using the price per pound in Table 2, for a tube OD of 2.5 in. (63.5 mm) and a design pressure of 2975 psi (205 bar). These results show that only three or four of these materials would be selected for the superheater: SA-210C; SA-213 T12, and either or both SA-213 T22 and SA-213 T91. The change from SA-210C to SA-213 T12 would occur between 750 and 800F (399 and 427C). The temperature at which the design would change from SA-213 T12 to SA-213 T22 would depend on mass flow requirements for the steam flowing through the superheater. This would most likely occur between 1000 and 1050F (538 and 566C). If so, the change from SA-213 T22 to SA-213 T91 would probably occur between 1050 and 1100F (566 and 593C). Because this might result in only a short length of SA-213 T22, the design might eliminate one weld and this material and switch directly from SA-213 T12 to SA-213 T91 at about 1050F (566C).

Fig. 25 shows allowable stresses plotted against temperature for several tube materials in their principal use ranges. One design consideration that can be addressed

Table 1 — Boiler Materials and Typical Applications (English Units)

Specification	Nominal Composition	Product Form	Min Tensile, ksi	Min Yield, ksi	High Heat Input Furn Walls	Other Furn Walls and Enclosures	SH RH Econ	Unheated Conn Pipe <10.75 in. OD	Headers and Pipe >10.75 in. OD	Drums	Recomm Max Use Temp, F	Notes
SA-178A	C-Steel	ERW tube	(47.0)	(26.0)	X	X	X				950	1,2
SA-192	C-Steel	Seamless tube	(47.0)	(26.0)	X	X	X	X			950	1
SA-178C	C-Steel	ERW tube	60.0	37.0		X	X				950	2
SA-210A1	C-Steel	Seamless tube	60.0	37.0	X	X	X	X			950	
SA-106B	C-Steel	Seamless pipe	60.0	35.0				X	X		950	3
SA-178D	C-Steel	ERW tube	70.0	40.0	X	X	X				950	2
SA-210C	C-Steel	Seamless tube	70.0	40.0		X	X	X			950	
SA-106C	C-Steel	Seamless pipe	70.0	40.0				X	X		950	3
SA-216WCB	C-Steel	Casting	70.0	36.0		X	X	X	X		950	
SA-105	C-Steel	Forging	70.0	36.0		X	X	X	X		950	3
SA-181-70	C-Steel	Forging	70.0	36.0		X	X	X	X		950	3
SA-266Cl2	C-Steel	Forging	70.0	36.0					X		800	
SA-516-70	C-Steel	Plate	70.0	38.0					X	X	800	
SA-266Cl3	C-Steel	Forging	75.0	37.5					X		800	
SA-299	C-Steel	Plate	75.0	40.0						X	800	
SA-250T1a	C-Mo	ERW tube	60.0	32.0		X	X				975	4,5
SA-209T1a	C-Mo	Seamless tube	60.0	32.0		X	X	X			975	4
SA-335P1	C-Mo	Seamless pipe	55.0	30.0				X			875	
SA-250T2	1/2Cr-1/2Mo	ERW tube	60.0	30.0	X		X		X		1025	6,7
SA-213T2	1/2Cr-1/2Mo	Seamless tube	60.0	30.0	X		X				1025	6
SA-250T12	1Cr-1/2Mo	ERW tube	60.0	32.0			X				1050	5,7
SA-213T12	1Cr-1/2Mo	Seamless tube	60.0	32.0			X				1050	8
SA-335P12	1/2Cr-1/2Mo	Seamless pipe	60.0	32.0					X		1050	8
SA-250T11	1-1/4Cr-1/2Mo-Si	ERW tube	60.0	30.0			X				1050	5
SA-213T11	1-1/4Cr-1/2Mo-Si	Seamless tube	60.0	30.0			X				1050	
SA-335P11	1-1/4Cr-1/2Mo-Si	Seamless pipe	60.0	30.0				X	X		1050	
SA-217WC6	1-1/4Cr-1/2Mo	Casting	70.0	40.0		X	X	X	X		1100	
SA-250T22	2-1/4Cr-1Mo	ERW tube	60.0	30.0			X				1115	5,7
SA-213T22	2-1/4Cr-1Mo	Seamless tube	60.0	30.0			X				1115	
SA-335P22	2-1/4Cr-1Mo	Seamless pipe	60.0	30.0				X	X		1100	
SA-217WC9	2-1/4Cr-1Mo	Casting	70.0	40.0			X	X	X		1115	
SA-182F22Cl1	2-1/4Cr-1Mo	Forging	60.0	30.0			X		X		1115	
SA-336F22Cl1	2-1/4Cr-1Mo	Forging	60.0	30.0					X		1100	
SA-213T91	9Cr-1Mo-V	Seamless tube	85.0	60.0			X				1200	
SA-335P91	9Cr-1Mo-V	Seamless pipe	85.0	60.0				X	X		1200	
SA-182F91	9Cr-1Mo-V	Forging	85.0	60.0			X				1200	
SA-336F91	9Cr-1Mo-V	Forging	85.0	60.0					X		1200	
SA-213TP304H	18Cr-8Ni	Seamless tube	75.0	30.0			X				1400	
SA-213TP347H	18Cr-10Ni-Cb	Seamless tube	75.0	30.0			X				1400	
SA-213TP310H	25Cr-20Ni	Seamless tube	75.0	30.0			X				1500	
SB-407-800H	Ni-Cr-Fe	Seamless tube	65.0	25.0			X				1500	
SB-423-825	Ni-Fe-Cr-Mo-Cu	Seamless tube	85.0	35.0			X				1000	

Notes:
1. Values in parentheses are not required minimums, but are expected minimums.
2. Requires special inspection if used at 100% efficiency above 850F.
3. Limited to 800F maximum for piping 10.75 in. OD and larger and outside the boiler setting.
4. Limited to 875F maximum for applications outside the boiler setting.
5. Requires special inspection if used at 100% efficiency.
6. Maximum OD temperature is 1025F. Maximum mean metal temperature for Code calculations is 1000F.
7. Requires use of a Code Case now. Will not later.
8. 32 ksi minimum yield requires use of Code Case 2070, which is being incorporated into the Code.

by families of such curves is the sensitivity to temperature excursions of the final design. Materials with a low rate of decreasing strength with increasing temperature are more tolerant of occasional upsets in operation that result in temperatures exceeding those used to establish the design.

Fuel ash corrosion considerations might dictate the use of higher alloys at lower temperatures. This is common in PR and refuse-fired boilers with very corrosive flue gas and ash. For example, alloy 825, SB-407-825 (42Ni-21.5Cr-3Mo-2.25Cu-0.9Ti-bal Fe) is used in the highly corrosive regions of refuse boiler superheaters, even at temperatures below 1000F (538C). In extreme cases, bimetallic tubes, with a core of a Code material for pressure retention and a cladding of a corrosion resistant alloys are used for both furnace wall and superheater application. Some common combinations are SA-210A1/304L, SA-210A1/Alloy 825, and SB-407-800H/50Cr-50Ni.

Headers and piping

Specifications for most of the commonly used pipe materials are listed in Table 1. As these components are usually not in the gas stream and are unheated, the major design factor, other than strength at temperature, is steam oxidation resistance. Carbon steels are not used above 800F (427C) outside the boiler setting, and C-Mo is limited to applications of small sizes [less than 10.75 in. (273 mm) OD] and below 875F (468C) to avoid graphitization.

The new alloy, 9Cr-1Mo-V, is beginning to replace 2-1/4Cr-1Mo for superheater outlet headers. This ma-

Table 2

Allowable Stresses,* ksi and Price/lb for Several Seamless Tube Materials**

For Metal Temperatures Not Exceeding, F

Material	700	750	800	850	900	950	1000	1050	1100	1150	1200	U.S. $/lb
SA-210C	16.6	14.8	12.0	7.8	5.0	3.0	1.5	—	—	—	—	0.5630
SA-209 T1a	15.0	14.8	14.4	14.0	13.6	8.2	4.8	—	—	—	—	0.6380
SA-213 T2	15.0	14.8	14.4	14.0	13.7	9.2	5.9	—	—	—	—	0.7045
SA-213 T12	14.8	14.6	14.3	14.0	13.6	11.3	7.2	4.5	2.8	1.8	1.1	0.6109
SA-213 T11	15.0	14.8	14.4	14.0	13.6	9.3	6.3	4.2	2.8	1.9	1.2	0.6109
SA-213 T22	15.0	15.0	15.0	14.4	13.6	10.8	8.0	5.7	3.8	2.4	1.4	0.9043
SA-213 T91	20.0	19.4	18.7	17.8	16.7	15.5	14.3	12.9	10.3	7.0	4.3	1.7800
SA-213 TP304H	15.9	15.6	15.2	14.9	14.7	14.4	13.8	12.2	9.8	7.7	6.1	3.3617

Source:* Table PG-23.1, Section I of the ASME Boiler and Pressure Vessel Code, 1989 Edition containing Addenda through 1991.

** Actual estimates obtained in late 1991.

terial is not operating in the creep range even at the 1000 to 1050F (538 to 566C) design temperatures of most such components. This factor and its very high strength lead to thinner components which are much less susceptible to the creep-fatigue failures observed in older 1-1/4Cr-1/2Mo-Si and 2-1/4Cr-1Mo headers. The use of forged outlet nozzles in place of welded nozzles has also reduced the potential for failure of these large piping connections.

Drums

Carbon steel plate is the primary material used in drums. SA-299, a 75,000 psi (517.1 MPa) tensile strength material, ordered to fine grain melting practice for improved toughness, is used for heavy section drums, those more than about 4 in. (101.6 mm) in thickness. SA-516 Gr 70, a fine grained 70,000 psi (482.7 MPa) tensile strength steel, is used for applications below this thickness, down to 1.5 in. (38.1 mm) thick shells. SA-515 Gr 70, a coarse grain melting practice steel, is used for thinner shells. Only in rare cases, where crane lifting capacity or long distance shipping costs are important considerations, are higher strength steels used.

Heat resistant alloys for nonpressure parts

High alloy heat resistant materials must be used for certain boiler parts that are exposed to high temperature and can not be water or steam cooled. These parts are made from alloys of the oxidation resistant, relatively high strength Cr-Ni-Fe type, many of them cast to shape as baffles, supports and hanger fittings. Oil burner impellers, sootblower clamps and hangers are also made of such heat resisting alloy steels.

Deterioration of these parts may occur through conversion of the surface layers to oxides, sulfides and sulfates, and in this condition they are referred to as burnt or oxidized. Experience indicates that 25Cr-12Ni and 25Cr-20Ni steels give reasonably good service life, depending on the location of the part in the flue gas stream and on the characteristics of the fuel. Temperatures to which these metal parts are exposed may range from 1000

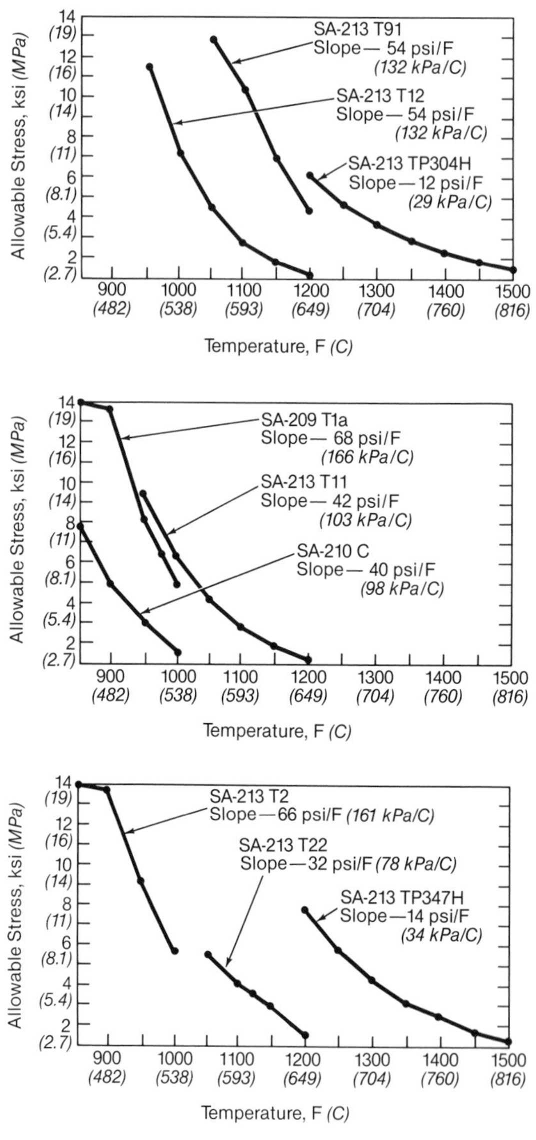

Fig. 25 Sensitivity of allowable stress to change in temperature.

Table 3

Minimum Wall Versus Temperature and Cost/ft Versus Temperature for Several Superheater Candidate Materials for 2.5 in. OD Tube and 2975 psi Design Pressure

Minimum Wall Thickness, in.*

Temperature Material	700	750	800	850	900	950	1000	1050	1100	1150
SA-210C	0.218	0.241	0.288	0.413	0.586	0.841	—	—	—	—
SA-209 T1a	0.238	0.241	0.247	0.253	0.259	0.396	0.604	—	—	—
SA-213 T2	0.238	0.241	0.247	0.253	0.257	0.360	0.516	0.922	—	—
SA-213 T12	0.238	0.238	0.241	0.250	0.267	0.310	0.472	0.678	—	—
SA-213 T22	0.238	0.238	0.238	0.247	0.267	0.310	0.413	0.523	0.666	—
SA-213 T91	0.185	0.191	0.197	0.205	0.217	0.231	0.248	0.271	0.328	0.451
SA-213 TP 304H	0.226	0.230	0.235	0.239	0.242	0.247	0.256	0.284	0.342	0.417

*—Indicates material not permitted.

Cost/ft for an OD = 2.5 in. and a Design Pressure = 2975 psi
Price/ft, U.S. $, at One Point in Time*

Temperature Material	700	750	800	850	900	950	1000	1050	1100	1150
SA-210C	3.29	3.59	4.19	5.63	7.23	8.79	—	—	—	—
SA-209 T1a	4.03	4.07	4.15	4.24	4.33	6.18	8.36	—	—	—
SA-213 T2	4.44	4.49	4.58	4.68	4.76	6.32	8.30	11.37	—	—
SA-213 T12	3.85	3.85	3.89	4.02	4.27	4.84	6.76	8.58	—	—
SA-213 T22	5.71	5.71	5.71	5.88	6.32	7.17	9.04	10.76	12.57	—
SA-213 T91	—	—	—	—	—	—	11.64	12.58	14.78	19.03
SA-213 TP 304H	—	—	—	—	—	—	—	—	24.73	33.85

*—Indicates material not permitted, or so thick as to be unrealistic, or so expensive as to not be of interest at lower temperatures.

to 2800F (538 to 1538C). Welding of such austenitic castings to ferritic alloy tubes must be avoided to minimize dissimilar metal weld failures. Special patented nonwelded constructions are used where such combinations are required.

Life may be shortened if these steels are exposed to flue gases from fuel oil containing vanadium compounds. Sulfur compounds formed from combustion of high sulfur fuels are also detrimental and act to reduce life. These may react in the presence of V and cause greatly accelerated rates of attack, especially when the temperature of the metal part exceeds 1200F (649C). Combinations of Na, S and V compounds are reported to melt at as low as 1050F (566C). Such deposits are extremely corrosive when molten because of their slagging action. In these circumstances, 50Cr-50Ni or 60Cr-40Ni castings are used to resist corrosion.

Environmental equipment applications

Flue gas desulfurization (FGD) systems are used to reduce SO_2 emissions from coal-fired boilers. Scrubbing SO_2 from the flue gas is accomplished by saturating the flue gas with a highly reactive reagent slurry in a countercurrent flow absorber tower. (See Chapter 35.)

Tower wet/dry interface

The tower inlet is classified as a wet/dry zone, because it is exposed to both the incoming dry, hot gas and the reagent slurry sprays. The tower inlet is exposed to the most severe corrosion area in the absorber; very acidic conditions with high chloride and fluoride levels can exist at this wet/dry interface.

Nickel base alloy metals, rather than nonmetallic linings, have generally been the materials of choice at this interface due to the high temperatures present. Materials such as Alloy C-276 (2.5Co max-15.5Cr-16Mo-3.75W-5.5Fe-Ni bal) and Alloy C-22 (2.5Co max-4Fe-21Cr-13.5Mo-0.3V-3W-Ni bal) have provided excellent service in this area.

Absorber spray-wetted zone

This area of the absorber tower is unique in that it is exposed to both corrosive and abrasive conditions. Slurry spray nozzles within the tower are oriented to guarantee total cross-sectional area coverage, thereby eliminating any possibility of untreated flue gas channeling through the sprays.

The Mo-containing austenitic stainless steels have seen considerable use in absorber tower fabrication. However, with the growing movement to minimal discharge closed loop systems, scrubber liquor chloride levels are exceeding the range in which stainless steels are effective. As a result, carbon steel lined with chlorobutyl elastomer or fiberglass reinforced plastic, duplex stainless steels such as Alloy 255, or high nickel alloys such as C-276 and C-22 are becoming the only viable material options. Because solid plate duplex stainless steels and high nickel alloys are so expensive, the wallpapering concept, in which 0.0625 in. (1.5875 mm) thick alloy sheets are welded onto the carbon steel shell plate, is gaining acceptance with utility customers who prefer to avoid relining.

Absorber recirculation tank

In the B&W absorber tower design, the recirculation tank is integral to the tower structure. Therefore, material options for the recirculation tank are quite similar to those for the spray-wetted zone of the tower.

While molybdenum-containing stainless or alloy steels could provide excellent service in this application, depending on system chemistry, the lowest initial capital cost will result from using a fiberglass reinforced plastic lining. Use of chlorobutyl rubber lining is the next best cost alternative. The principal drawback of such linings is their limited life, generally 10 to 15 years. Both lining systems are also susceptible to mechanical damage, requiring the installation of a brick overlay on floor surfaces.

Internal spray headers

This piping is unique in that it must have abrasion-resistant internal and external surfaces. Depending upon system chemistry and customer preference, stainless or alloy steels can be used but are expensive. Elastomer lined and coated carbon steel piping has also been used, but it is expensive and the extent of substrate corrosion is difficult to determine. With the advancements made in resin rich, high grade fiberglass reinforced plastic (FRP) pipe, this is now the industry standard for this service.

Moisture separators

In early FGD systems, moisture separators fabricated from stainless steel were used due to temperature excursion concerns. As tower spray reliability and temperature stability improved, a switch to plastic moisture separators was made for better corrosion resistance. Today, moisture separators made from fiberglass reinforced thermosetting or thermoplastic resins are common.

Absorber moisture separator and outlet zone

This zone is exposed to a different environment from what has previously been discussed. The abrasive, high chloride recycle slurry no longer predominates in this zone. Instead, blend water for moisture separator washing combines with flue gas residual SO_2 to form sulfurous acid (H_2SO_3).

Because abrasion resistance is no longer a material selection factor for this area, the use of an elastomer lining is unnecessary. A flake glass reinforced plastic lining over carbon steel is satisfactory.

Flue downstream of bypass

On some FGD units, a flue gas bypass around the absorber tower is provided. This bypass typically ties back into the tower's outlet flue work upstream of the stack. The outlet flue work at this tie-in is exposed to the highly corrosive environment of scrubbed and hot bypassed gas.

Due to the high temperatures [300F (149C)] at this section of flue, there are only a very limited number of plastic linings suitable for this service. Better success has been achieved with Alloy C-276 and Alloy C-22 wallpaper lining, or borosilicate foamed glass block in the application.

References

1. Cullity, B.D., *Elements of X-Ray Diffraction*, p. 32, Addison-Wesley, Reading, Massachusetts, 1956.

2. Darken, L.S., *The Physical Chemistry of Metallic Solutions and Intermetallic Compounds*, Her Majesty's Stationery Office, London, 1958.

3. Swalin, R.A., *Thermodynamics of Solids*, Wiley, New York, 1962.

4. Higgins, R.A., *Properties of Engineering Materials*, Hodder and Staughton, London, 1979.

5. Bain, E.C., and Paxton, H.W., *Alloying Elements in Steel*, American Society for Metals, Metals Park, Ohio, 1961.

6. McGannon, H.E., ed., *The Making, Shaping and Treating of Steel*, Ninth Edition, United States Steel, Pittsburgh, 1970.

7. Lankford, W.T., Jr., *et al.*, *The Making, Shaping and Treating of Steel*, Tenth Edition, Association of Iron and Steel Engineers, Pittsburgh, 1985.

8. Long, C.J., and DeLong, W.T., "The ferrite content of austenitic stainless steel weld metal," *Welding Journal*, Research Supplement, pp. 281S-297S, Vol. 52 (7), 1973.

9. Boyer, H.E., and Gall, T.L., eds., *Metals Handbook, Desk Edition*, American Society for Metals, Metals Park, Ohio.

10. Weisman, C., ed., *Welding Handbook*, Seventh Edition, American Welding Society, p. 272, Vol. 1, Miami, Florida, 1981.

11. *Ibid.*, p. 229.

12. Sikka, V.K., *et al.*, "Modified 9Cr-1Mo steel - an improved alloy for steam generator application," *Ferritic Steels for High Temperature Applications*, Proceedings of ASM International Conference on Production, Fabrication, Properties and Application of Ferritic Steels for High Temperature Service, pp. 65-84, Warren, Pennsylvania, October 6-8, 1981, Khare, A.K., ed., American Society for Metals, Metals Park, Ohio, 1963.

13. Swindeman, R.W., and Gold, M., "Developments in ferrous alloy technology for high temperature service," Widera, G.E.O., ed., *Trans. ASME, J. Pressure Vessel Technology*, p. 135, American Society of Mechanical Engineers, New York, May 1991.

14. Rudd, A.H., and Tanzosh, J.M., "Developments applicable to improved coal-fired power plants," presented at the First EPRI International Conference on Improved Coal-Fired Power Plants, Palo Alto, California, November 19-21, 1986.

15. Viswanathan, R., *et al.*, "Ligament cracking and the use of modified 9Cr-1Mo alloy steel (P91) for boiler headers," presented at the 1990 ASME Pressure Vessels and Piping Conference, Nashville, Tennessee, June 17-21, 1990, Prager, M., and Cantzlereds, C., *New Alloys for Pressure Vessels and Piping*, pp. 97-104, American Society of Mechanical Engineers, New York,1990.

16. Gold, M., *et al.*, "The effect of varying degrees of cold work on the stress-rupture properties of type 304 stainless steels," *Trans. ASME, J. Eng. Materials and Technology*, October 1975.

17. Benjamin, D., *et al.*, "Properties and selection: stainless steels, tool materials and special purpose metals," *Metals Handbook,* Ninth Edition, Vol. 3, American Society for Metals, Metals Park, Ohio, p. 17, 1980.

18. Barna, J.L., *et al.*, "Furnace wall corrosion in refuse-fired boilers," presented to ASME Twelfth Biennial National Waste Processing Conference, Denver, Colorado, June 1-4, 1986.

19. Morro, H., III, "Zinc," *Metals Handbook, Desk Edition*, pp. 11-1 to 11-3, Boyer, H.E., and Gall, T.L., eds., American Society for Metals, Metals Park, Ohio.

20. Hayden, H.W., *et al.*, *The Structure and Properties of Materials*, Vol. III, Mechanical Behavior, Wiley, New York, 1965.

21. Barson, J.M., and Rolfe, S.T., Fracture and Fatigue Control in Structures, 2nd ed., Prentice-Hall, Englewood Cliffs, New Jersey, 1987.

A large steam drum is prepared for lifting within power plant structural steel.

Chapter 7
Structural Analysis and Design

Equipment used in the power, chemical, petroleum and cryogenic fields often includes large steel vessels. These vessels may require tons of structural steel for their support. Steam generating equipment, for example, comprises pressure parts ranging from small diameter tubing to vessels weighing more than 1000 t (907 t_m). A large fossil fuel boiler may extend 300 ft (91.4 m) above the ground, requiring a steel support structure comparable to a 30 story building. To assure reliability, a thorough design analysis of pressure parts and their supporting structural components is required.

Pressure vessel design and analysis

Steam generating units require pressure vessel components that operate at internal pressures of up to 4000 psi (275.8 bar) and at steam temperatures up to 1050F (566C). Maximum reliability can be assured only with a thorough stress analysis of the components. Therefore, considerable attention is given to the design and stress analysis of steam drums, superheater headers, heat exchangers, pressurizers and nuclear reactors. In designing these vessels, the basic approach is to account for all unknown factors such as local yielding and stress redistribution, variability in material properties, inexact knowledge of loadings, and inexact stress evaluations by using allowable working stresses that include appropriate factors of safety.

The analysis and design of complex pressure vessels and components such as the reactor closure head, shown in Fig. 1, and the fossil boiler steam drum, shown in Fig. 2, requires sophisticated principles and methods. Mathematical equations based on the theory of elasticity are applied to regions of discontinuities, nozzle openings and supports. Advanced computerized structural mechanics methods, such as the finite element method, are used to determine complex vessel stresses.

In the United States (U.S.), pressure vessel construction codes adopted by state, federal and municipal authorities establish safety requirements for vessel construction. The most widely used is the American Society of Mechanical Engineers (ASME) *Boiler and Pressure Vessel Code*. Key sections include Sections I, *Rules for Construction of Power Boilers*; III, *Rules for Construction of Nuclear Power Plant Components*; and VIII, *Rules for Construction of Pressure Vessels*. A further introduction to the ASME Code is presented in Appendix 2.

Stress significance

Stress is defined as the internal force between two adjacent elements of a body, divided by the area over which it is applied. The main significance of a stress is its magnitude; however, the nature of the applied load and the resulting stress distribution are also important. The designer must consider whether the loading is mechanical or thermal, whether it is steady-state or transient, and whether the stress pattern is uniform.

Stress distribution depends on the material properties. For example, yielding or strain readjustment can cause redistribution of stresses.

Fig. 1 Head of nuclear reactor vessel.

Fig. 2 Fossil fuel boiler steam drum.

Steady-state conditions An excessive steady-state stress due to applied pressure results in vessel material distortion, progresses to leakage at fittings and ultimately causes failure in a ductile vessel. To prevent this type of failure a safety factor is applied to the material properties. The two predominant properties considered are yield strength, which establishes the pressure at which gross distortion occurs, and tensile strength, which determines the vessel bursting pressure. ASME Codes establish pressure vessel design safety factors based on the sophistication of quality assurance, manufacturing control, and design analysis techniques.

Transient conditions When the applied stresses are repetitive, such as those occurring during testing and transient operation, they significantly contribute to establishing the fatigue life of the vessel. The designer must consider transient conditions causing fatigue stresses in addition to those caused by steady-state forces.

Although vessels must have nozzles, supports and flanges in order to be useful, these features often embody abrupt changes in cross-section. These changes can introduce irregularities in the overall stress pattern called local or peak stresses. Other construction details can also promote stress concentrations which, in turn, affect the vessel's fatigue life.

Strength theories

Several material strength theories are used to determine when failure will occur under the action of multi-axial stresses on the basis of data obtained from uni-axial tension or compression tests. The three most commonly applied theories which are used to establish elastic design stress limits are the maximum (principal) stress theory, the maximum shear stress theory, and the distortion energy theory.

Maximum stress theory The maximum stress theory considers failure to occur when one of the three principal stresses reaches the material yield point in tension:

$$\sigma = \sigma_{y.p.} \qquad \textbf{(1)}$$

This theory is the simplest to apply and, with an adequate safety factor, it results in safe, reliable pressure vessel designs. This is the theory of strength used in the ASME Code, Section I, Section VIII Division 1, and Section III Division 1 (design by formula Subsections NC-3300, ND-3300 and NE-3300).

Maximum shear stress theory The maximum shear stress theory, also known as the Tresca theory,[1] considers failure to occur when the maximum shear stress in an element reaches the maximum shear stress at the yield strength of the material in tension. Noting that the maximum shear stress (τ), is equal to half the difference of the maximum and minimum principal stresses, and that the maximum shear stress in a tension test specimen is half the axial principal stress, the condition for yielding becomes:

$$\tau = \frac{\sigma_{max} - \sigma_{min}}{2} = \frac{\sigma_{y.p.}}{2} \qquad \textbf{(2)}$$

$$2\tau = \sigma_{max} - \sigma_{min} = \sigma_{y.p.}$$

The value 2τ is called the shear stress intensity. The maximum shear stress theory predicts ductile material yielding more accurately than the maximum stress theory. This is the theory of strength used in the ASME Code, Section VIII Division 2, and Section III Division 1, Subsection NB, and design by analysis, Subsections NC-3200 and NE-3200.

Distortion energy theory The distortion energy theory (also known as the Mises criterion[1]) considers yielding to occur when the distortion energy at a point in a stressed element is equal to the distortion energy in a uni-axial test specimen at the point it begins to yield. While the distortion energy theory is the most accurate, it is cumbersome to use and is not routinely applied in pressure vessel design codes.

Design criteria[1]

To determine the allowable stresses in a pressure vessel, one must consider the nature of the loading and the vessel response to the loading. Stress interpretation determines the required stress analyses and the allowable stress magnitudes. Current design codes establish the criteria for safe design and operation of pressure vessels.

Stress classifications Stresses in pressure vessels have three major classifications: primary, secondary and peak.

Primary stresses (P) are developed by mechanical loads which can cause gross failure of the pressure vessel. These stresses are further divided into general primary membrane (P_M), local primary membrane (P_L) and primary bending (P_B) stresses. A primary stress is not self-limiting, i.e., if the material yields or is deformed, the stress is not reduced. A good example of this type of stress is that produced by internal pressure such as in a steam drum. When it exceeds the vessel material yield strength, gross distortion appears and failure may occur.

Secondary stresses (Q), due to mechanical loads or differential thermal expansion, are developed by the constraint of adjacent components. They are self-limiting and are usually confined to local areas of the vessel. Local yielding or minor distortion can reduce secondary stresses. Although they do not affect the static bursting strength of a vessel, secondary stresses must be considered in establishing its fatigue life.

Peak stresses (F) are concentrated in highly localized areas at abrupt geometry changes. Although no appreciable vessel deformations are associated with them, peak stresses are particularly important in evaluating the fatigue life of a vessel.

Code design/analysis requirements Allowable stress limits and design analysis requirements vary with pressure vessel design codes.

According to ASME Code, Section I, the minimum vessel wall thickness is determined by evaluating the general primary membrane stress. This stress, limited to the allowable material tension stress S, is calculated at the vessel design temperature. The Section I regulations have been established to ensure that secondary and peak stresses are minimized; a detailed analysis of these stresses is normally not required.

The design criteria of ASME Code, Sections VIII Division 1, and Section III Division 1 (design by formula Subsections NC-3300, ND-3300 and NE-3300), are similar to those of Section I. However, they require cylindrical shell thickness calculations in the circumferential and longitudinal directions. The minimum required pressure vessel wall thickness is set by the maximum stress in either direction. Section III Division 1 and Subsections NC-3300 and ND-3300 permit the combination of primary membrane and primary bending stresses to be up to 1.5 S at design temperature. Section VIII Division 1 permits the combination of primary membrane and primary bending stresses to be 1.5 S at temperatures where tensile or yield strength sets the allowable stress S, and 1.25 S at temperatures where creep and rupture set the allowable stress.

ASME Code, Section VIII Division 2 provides formulas and rules for common configurations of shells and formed heads. It also requires detailed stress analysis of complex geometries with unusual or cyclic loading conditions. The calculated stress intensities are assigned to specific categories. The allowable stress intensity of each category is based on a multiplier of the Code allowable stress intensity value. The Code allowable stress intensity, S_m, is based on the material yield strength, S_y, or tensile strength, S_u.

Stress Intensity Category	Allowable Value	Basis for Allowable Value at k = 1.0 (Lesser Value)
General primary membrane (P_M)	kS_m	2/3 S_y or 1/3 S_u
Local primary membrane (P_L)	1-1/2 kS_m	S_y or 1/2 S_u
Primary membrane plus primary bending ($P_M + P_B$)	1-1/2 kS_m	S_y or 1/2 S_u
Primary plus secondary ($P_M + P_B + Q$)	3 S_m	2 S_y or S_u

The factor k varies with the type of loading, as listed below:

k	Loading
1.0	sustained
1.2	sustained and transient
1.25	hydrostatic test
1.5	pneumatic test

The design criteria for ASME Code, Section III Division 1, Subsection NB and design by analysis Subsections NC-3200 and NE-3200 are similar to those for Section VIII Division 2 except there is less use of design formulas, curves, and tables, and greater use of design by analysis in Section III. The categories of stresses and stress intensity limits are the same in both sections.

Stress analysis methods

Stress analysis of pressure vessels can be performed by analytical or experimental methods. An analytical method, involving a rigorous mathematical solution based on the theory of elasticity and plasticity, is the most direct and inexpensive approach when the problem is adaptable to such a solution. When the problem is too complex for this method, approximate analytical structural mechanics methods, such as finite element analysis, are applied. If the problem is beyond analytical solutions, experimental methods must be used.

Mathematical formulas[2] Pressure vessels are commonly spheres, cylinders, ellipsoids, tori or composites of these. When the wall thickness is small compared to other dimensions, vessels are referred to as membrane shells. Stresses acting over the thickness of the vessel wall and tangential to its surface can be represented by mathematical formulas for the common shell forms.

Pressure stresses are classified as primary membrane stresses since they remain as long as the pressure is applied to the vessel. The basic equation for the longitudinal stress σ_1 and hoop stress σ_2 in a vessel of thickness h, longitudinal radius r_1, and circumferential radius r_2, which is subject to a pressure P, shown in Fig. 3 is:

$$\frac{\sigma_1}{r_1} + \frac{\sigma_2}{r_2} = \frac{P}{h} \quad (3)$$

From this equation, and by equating the total pressure load with the longitudinal forces acting on a transverse section of the vessel, the stresses in the commonly used shells of revolution can be found.

1. Cylindrical vessel — in this case, $r_1 = \infty$, $r_2 = r$, and

$$\sigma_1 = \frac{Pr}{2h} \quad (4)$$

$$\sigma_2 = \frac{Pr}{h} \quad (5)$$

2. Spherical vessel — in this case, $r_1 = r_2 = r$, and

$$\sigma_1 = \frac{Pr}{2h} \quad (6)$$

$$\sigma_2 = \frac{Pr}{2h} \quad (7)$$

3. Conical vessel — in this case, $r_1 = \infty$, $r_2 = r/\cos\alpha$ where α is half the cone apex angle, and

$$\sigma_1 = \frac{Pr}{2h\cos\alpha} \quad (8)$$

$$\sigma_2 = \frac{Pr}{h\cos\alpha} \quad (9)$$

4. Ellipsoidal vessel — in this case (Fig. 4), the instanta-

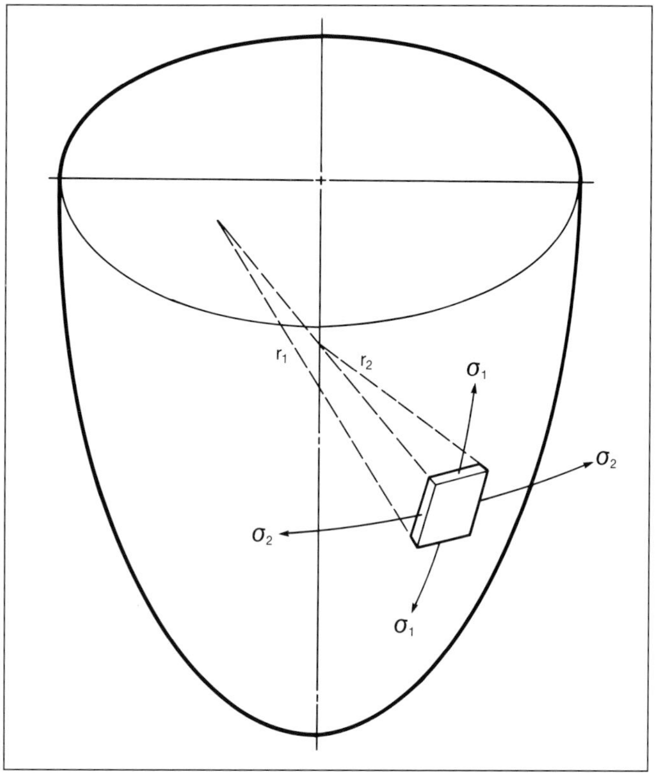

Fig. 3 Membrane stress in vessels (*courtesy of Van Nostrand Reinhold*).[2]

neous radius of curvature varies with each position on the ellipsoid, whose major axis is *a* and minor axis is *b,* and the stresses are given by:

$$\sigma_1 = \frac{Pr_2}{2h} \tag{10}$$

$$\sigma_2 = \frac{P}{h}\left(r_2 - \frac{r_2^2}{2r_1}\right) \tag{11}$$

At the equator, the longitudinal stress is the same as the longitudinal stress in a cylinder, namely:

$$\sigma_1 = \frac{Pa}{2h} \tag{12}$$

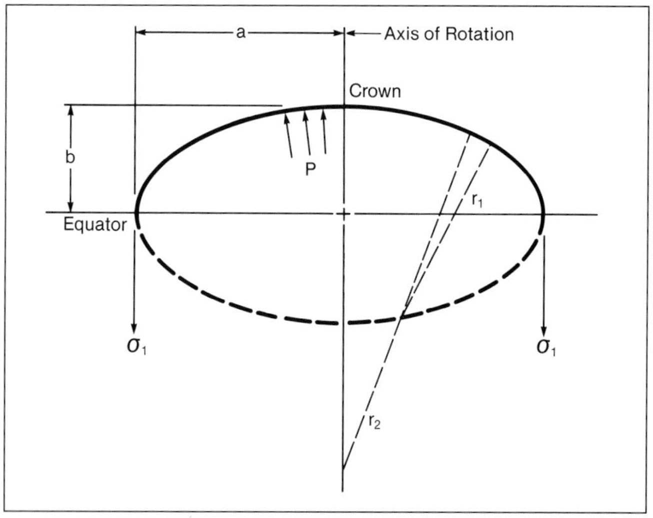

Fig. 4 Stress in an ellipsoid.[2]

and the hoop stress is:

$$\sigma_2 = \frac{Pa}{h}\left(1 - \frac{a^2}{2b^2}\right) \tag{13}$$

When the ratio of major to minor axis is 2:1, the hoop stress is the same as that in a cylinder of the same mating diameter, but the stress is compressive rather than tensile. The hoop stress rises rapidly when the ratio of major to minor axis exceeds 2:1 and, because this stress is compressive, buckling instability becomes a major concern. For this reason, ratios greater than 2:1 are seldom used.

5. Torus — in this case (Fig. 5), R_o is the radius of the bend centerline, θ is the angular hoop location from this centerline and:

$$\sigma_1 = \frac{Pr}{2h} \tag{14}$$

$$\sigma_2 = \frac{Pr}{2h}\left(\frac{2R_o + r\sin\theta}{R_o + r\sin\theta}\right) \tag{15}$$

The longitudinal stress remains uniform around the circumference and is the same as that for a straight cylinder. The hoop stress, however, varies for different points in the torus cross-section. At the bend centerline, it is the same as that in a straight cylinder. At the outside of the bend, it is less than this and is at its minimum. At the inside of the bend, or crotch, the value is at its maximum. Hoop stresses are dependent on the sharpness of the bend and are inversely proportional to bend radii. In pipe bending operations, the material thins at the outside and becomes thicker at the crotch of the bend. This is an offsetting factor for the higher hoop stresses that form with smaller bend radii.

Thermal stresses result from restricting a member that is attempting to expand or contract due to a temperature change, ΔT. They are classified as secondary stresses because they are self-limiting. If the material is restricted in only one direction, the stress developed is:

$$\sigma = \pm E\alpha\Delta T \tag{16}$$

where E is the modulus of elasticity and α is the coefficient of thermal expansion. If the member is restricted from expanding or contracting in two directions, as is the case in pressure vessels, the resulting stress is:

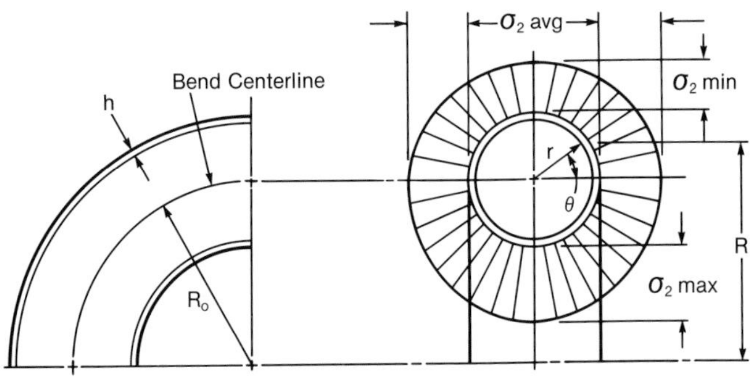

Fig. 5 Hoop stress variation in a bend.[2]

$$\sigma = \pm \frac{E\alpha\Delta T}{1-\mu} \quad \textbf{(17)}$$

where μ is Poisson's ratio.

These thermal stress equations consider full restraint, and therefore are the maximum that can be created. When the temperature varies within a member, the natural growth of one fiber is influenced by the differential growth of adjacent fibers. As a result, fibers at high temperatures are compressed and those at lower temperatures are stretched. The general equations for radial (σ_r), tangential (σ_t), and axial (σ_z) thermal stresses in a cylindrical vessel subject to a radial thermal gradient are:

$$\sigma_r = \frac{\alpha E}{(1-\mu)r^2}\left[\frac{r^2-a^2}{b^2-a^2}\int_a^b Trdr - \int_a^r Trdr\right] \quad \textbf{(18)}$$

$$\sigma_t = \frac{\alpha E}{(1-\mu)r^2}\left[\frac{r^2+a^2}{b^2-a^2}\int_a^b Trdr + \int_a^r Trdr - Tr^2\right] \quad \textbf{(19)}$$

$$\sigma_z = \frac{\alpha E}{(1-\mu)}\left[\frac{2}{b^2-a^2}\int_a^b Trdr - T\right] \quad \textbf{(20)}$$

where

E = modulus of elasticity
μ = Poisson's ratio
r = radius at any location
a = inside radius
b = outside radius
T = temperature

For a cylindrical vessel in which heat is flowing radially through the walls under steady-state conditions, the maximum thermal stresses are:

ASME Code calculations

In most U.S. states and Canadian provinces laws have been established requiring that boilers and pressure vessels comply with the rules for the design and construction of boilers and pressure vessels in the ASME Code. The complexity of these rules and the amount of analysis required are related to the factors of safety which are applied to the material properties used to establish the allowable stresses. When the stress analysis is simplified, the factor of safety is larger. When the stress analysis is more complex, the factory of safety is smaller. For conditions when material tensile strength establishes the allowable stress, ASME Code, Section IV, *Rules for Construction of Heating Boilers*, requires only a simple thickness calculation with a safety factor on tensile strength of five. ASME Code, Section I, *Rules for Construction of Power Boilers* and Section VIII, Division 1, *Rules for Construction of Pressure Vessels*, require a more complex analysis with additional items to be considered. However, the factor of safety on tensile strength is reduced to four. Section III, *Rules for Construction of Nuclear Components* and Section VIII, Division 2, *Rules for Construction of Pressure Vessels* require extensive analyses which are required to be certified by a registered professional engineer. In return, the factor of safety on tensile strength is reduced even further to three.

When the wall thickness is small compared to the diameter, membrane formulas (Equations 4 and 5) may be used with adequate accuracy. However, when the wall thickness is large relative to the vessel diameter, usually to accommodate higher internal design pressure, the membrane formulas are modified for ASME Code applications. Basically the minimum wall thickness of a cylindrical shell is initially set by solving the circumferential or hoop stress equation assuming there are no additional loadings other than internal pressure. Other loadings may then be considered to determine if the initial minimum required wall thickness has to be increased to keep calculated stresses below allowable stress values.

As an example, consider a Section VIII, Division 1, pressure vessel with no unreinforced openings and no additional loadings other than an internal design pressure of 1200 psi at 500F. The inside diameter is 10 in. and the material is SA-516, Grade 70 carbon steel. There is no corrosion allowance required by this application and the butt weld joints are 100% radiographed. What is the minimum required wall thickness needed? The equation for setting the minimum required wall thickness in Section VIII, Division 1, of the Code (paragraph UG-27(c)(1), 1992 Edition) is:

$$t = \frac{PR}{SE - 0.6P}$$

where

t = minimum required wall thickness, in.
P = internal design pressure, psi
R = inside radius, in.
S = allowable stress at design temperature, psi (Section VIII, Division 1, Subsection C) = 17,500 psi
E = lower of weld joint efficiency or ligament efficiency (fully radiographed with manual penetrations) = 1.0

For the pressure vessel described above:

P = 1200 psi
R = 5 in.
S = 17,500 psi
E = 1.0

$$t = \frac{(1200)(5)}{(17{,}500 \times 1.0) - (0.6 \times 1200)} = 0.358 \text{ in.}$$

Using commercial sizes, this plate thickness probably would be ordered at 0.375 in.

If Equation 5 for simple hoop stress (see Figure below) is used alone to calculate the plate thickness using the specified minimum tensile strength of SA-516, Grade 70 of 70,000 psi, the thickness h would be evaluated to be:

$$h = \frac{1200\ (5)}{70{,}000} = 0.0857$$

Therefore, the factor of safety (FS) based on tensile strength is:

$$FS = \frac{0.358}{0.0857} = 4.2$$

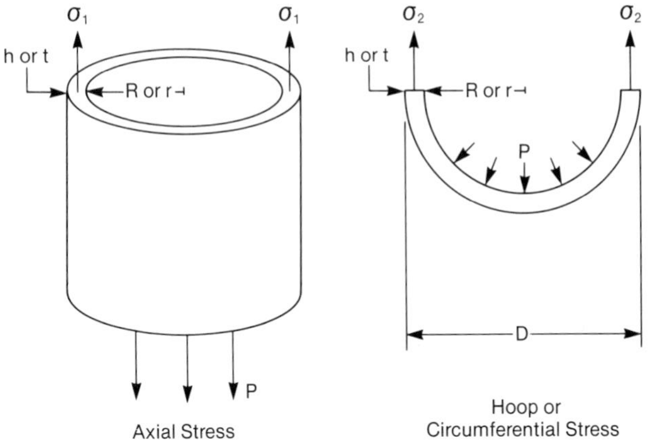

$$\sigma_{ta}(inside) = \frac{\alpha E T_a}{2(1-\mu)\ell n\left(\frac{b}{a}\right)}\left[1 - \frac{2b^2}{b^2 - a^2}\ell n\left(\frac{b}{a}\right)\right] \quad \textbf{(21)}$$

$$\sigma_{tb}(outside) = \frac{\alpha E T_a}{2(1-\mu)\ell n\left(\frac{b}{a}\right)}\left[1 - \frac{2a^2}{b^2 - a^2}\ell n\left(\frac{b}{a}\right)\right] \quad \textbf{(22)}$$

For relatively thin tubes and $T_a > T_b$, this can be simplified to:

$$\sigma_{ta} = \frac{-\alpha E \Delta T}{2(1-\mu)} \quad \textbf{(23)}$$

$$\sigma_{tb} = \frac{\alpha E \Delta T}{2(1-\mu)} \quad \textbf{(24)}$$

To summarize, the maximum thermal stress for a thin cylinder with a logarithmic wall temperature gradient is one half the thermal stress of an element restrained in two directions and subjected to a temperature change ΔT (Equation 17). For a radial thermal gradient of different shape, the thermal stress can be represented by:

$$\sigma = K\frac{E\alpha\Delta T}{1-\mu} \quad \textbf{(25)}$$

where K ranges between 0.5 and 1.0.

Alternating stresses resulting from cyclic pressure vessel operation may lead to fatigue cracks at high stress concentrations. Fatigue life is evaluated by comparing the alternating stress amplitude with design fatigue curves (allowable stress versus number of cycles or σ-N curves) experimentally established for the material at temperature. A typical σ-N design curve for carbon steel is shown in Fig. 6 and can be expressed by the equation:

$$\sigma_a = \frac{E}{4\sqrt{N}}\ell n\left(\frac{100}{100 - d_a}\right) + .01(TS)d_a \quad \textbf{(26)}$$

where

σ_a = allowable alternating stress amplitude
E = modulus of elasticity at temperature
N = number of cycles
d_a = percent reduction in area
TS = tensile strength at temperature

The two controlling parameters are tensile strength and reduction in area. Tensile strength is controlling in the high cycle fatigue region, while reduction in area is controlling in low cycle fatigue. The usual division between low and high cycle fatigue is 10^5 cycles. Pressure vessels often fall into the low cycle fatigue category, thereby demonstrating the importance of the material's ability to deform in the plastic range without fracturing. Lower strength materials, with their greater ductility, have better low cycle fatigue resistance than do higher strength materials.

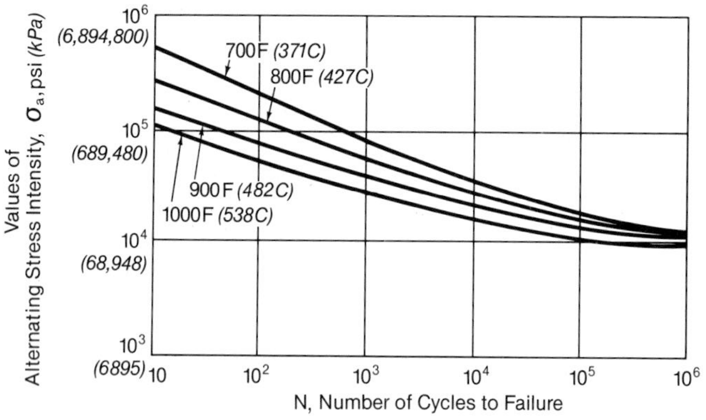

Fig. 6 Design fatigue curve.

Practical operating service conditions subject many vessels to the random occurrence of a number of stress cycles at different magnitudes. One method of appraising the damage from repetitive stresses to a vessel is the criterion that the cumulative damage from fatigue will occur when the summation of the increments of damage at the various stress levels exceeds unity. That is:

$$\sum \frac{n}{N} = 1 \quad \textbf{(27)}$$

where n = number of cycles at stress σ, and N = number of cycles to failure at the same stress σ. The ratio n/N is called the cycle damage ratio since it represents the fraction of the total life which is expended by the cycles that occur at a particular stress value. The value N is determined from σ-N curves for the material. If the sum of these cycle ratios is less than unity, the vessel is considered safe. This is particularly important in designing an economic and safe structure which experiences only a relatively few cycles at a high stress level and the major number at a relatively low stress level.

Discontinuity analysis method At geometrical discontinuities in axisymmetric structures, such as the intersection of a hemispherical shell element and a cylindrical shell element (Fig. 7a), the magnitude and characteristic of the stress are considerably different than those in elements remote from the discontinuity. A linear elastic analysis method is used to evaluate these local stresses.

Discontinuity stresses that occur in pressure vessels, particularly axisymmetric vessels, are determined by a discontinuity analysis method. A discontinuity stress results from displacement and rotation compatibilities at the intersection of two elements. The forces and moments at the intersection (Fig. 7c) are redundant or self-

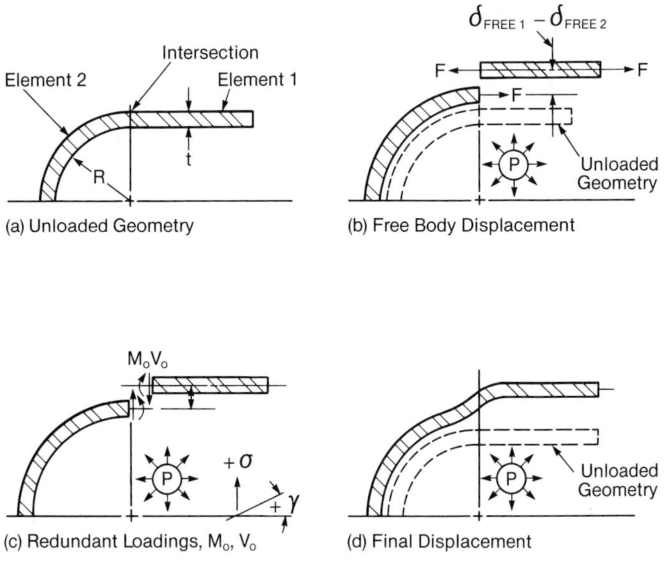

Fig. 7 Discontinuity analysis.

limiting loads because they are not required for static equilibrium. They develop solely to ensure compatibility at the intersection. As a consequence, a discontinuity stress can not cause failure in ductile materials in one load application even if the maximum stress exceeds the material yield strength. Such stresses must be considered in cyclic load applications or in special cases where materials can not safely redistribute stresses. The ASME Code refers to discontinuity stresses as secondary stresses. The application to the shell of revolution shown in Fig. 7 outlines the major steps involved in the method used to determine discontinuity stresses.

Under internal pressure, a sphere radially expands approximately one half that of a cylindrical shell (Fig. 7b). The difference in free body displacement results in redundant loadings at the intersection if Elements (1) and (2) are joined (Fig. 7c). The final displacement and rotation of the cylindrical shell are equal to the free body displacement plus the displacements due to the redundant shear force V_o and redundant bending moment M_o (Fig. 7d).

The direction of the redundant loading is unknown and must be assumed. A consistent sign convention must be followed. In addition, the direction of loading on the two elements must be set up consistently because Element (1) reacts Element (2) loading and vice versa. If M_o or V_o as calculated is negative, the correct direction is opposite to that assumed.

In equation form then, for Element (1):

$$\delta_{\text{FINAL 1}} = \delta_{\text{FREE 1}} - \beta_{\delta V1} V_o + \beta_{\delta M1} M_o \tag{28}$$

$$\gamma_{\text{FINAL 1}} = \gamma_{\text{FREE 1}} + \beta_{\gamma V1} V_o - \beta_{\gamma M1} M_o \tag{29}$$

Similarly for Element (2):

$$\delta_{\text{FINAL 2}} = \delta_{\text{FREE 2}} + \beta_{\delta V2} V_o + \beta_{\delta M2} M_o \tag{30}$$

$$\gamma_{\text{FINAL 2}} = \gamma_{\text{FREE 2}} + \beta_{\gamma V2} V_o + \beta_{\gamma M2} M_o \tag{31}$$

where

$$\delta_{\text{FREE 1}} = \frac{PR^2}{Et}\left(1 - \frac{\mu}{2}\right) \tag{32}$$

$$\delta_{\text{FREE 2}} = \frac{PR^2}{2Et}(1-\mu) \tag{33}$$

$$\gamma_{\text{FREE 1}} = \gamma_{\text{FREE 2}} = 0 \text{ in this case}$$

The constants β are the deflections or rotations due to loading per unit of perimeter, and are referred to as influence coefficients. These constants can be determined for a variety of geometries, including rings and thin shells of revolution, using standard handbook solutions. For example:

$\beta_{\delta V1}$ = radial displacement of Element (1) due to unit shear load
$\beta_{\delta M1}$ = radial displacement of Element (1) due to unit moment load
$\beta_{\gamma V1}$ = rotation of Element (1) due to unit shear load
$\beta_{\gamma M1}$ = rotation of Element (1) due to unit moment load

Because $\delta_{\text{FINAL 1}} = \delta_{\text{FINAL 2}}$ and $\gamma_{\text{FINAL 1}} = \gamma_{\text{FINAL 2}}$ from compatibility requirements, Equations 28 through 31 can be reduced to two equations for two unknowns, V_o and M_o, which are solved simultaneously. Note that the number of equations reduces to the number of redundant loadings and that the force F can be determined by static equilibrium requirements.

Once V_o and M_o have been calculated, handbook solutions can be applied to determine the resulting membrane and bending stresses. The discontinuity stress must then be added to the free body stress to obtain the total stress at the intersection.

Although the example demonstrates internal pressure loading, the same method applies to determining thermally induced discontinuity stress. For more complicated geometries involving four or more unknown redundant loadings, commercially available computer programs should be considered for solution.

Finite element analysis When the geometry of a component or vessel is too complex for classical formulas or closed form solutions, finite element analysis (FEA) can often provide the required results. FEA is a powerful numerical technique that can evaluate structural deformations and stresses, heat flows and temperatures, and dynamic responses of a structure. Because FEA is usually more economical than experimental stress analysis, scale modeling, or other numerical methods, it has become the dominant sophisticated stress analysis method.

During product development, FEA is used to predict performance of a new product or concept before building an expensive prototype. For example, a design idea to protect the inside of a burner could be analyzed to find out if it will have adequate cooling and fatigue life. FEA is also used to investigate field problems.

To apply FEA, the structure is modeled as an assembly of discrete building blocks called elements. The elements can be linear (one dimensional truss or beam), plane (representing two dimensional behavior), or solid (three dimensional bricks). Elements are connected at their boundaries by nodes as illustrated in Fig. 8.

Except for analyses using truss or beam elements, the accuracy of FEA is dependent on the mesh density. This refers to the number of nodes per modeled volume. As mesh density increases, the result accuracy also increases. Alternatively, in *p-method* analysis, the mesh density remains constant while increased accuracy is attained through mathematical changes to the solution process.

A computer solution is essential because of the numerous calculations involved. A medium sized FEA may require the simultaneous solution of thousands of equations, but taking merely seconds of computer time. FEA is one of the most demanding computer applications.

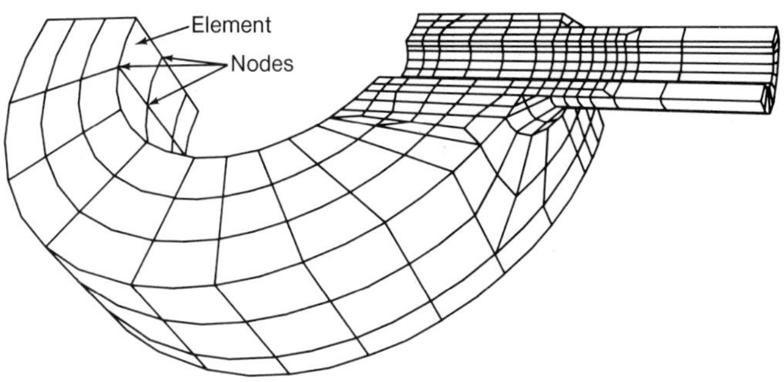

Fig. 8 Finite element model composed of brick elements.

FEA theory is illustrated by considering a simple structural analysis with applied loads and specified node displacements. The mathematical theory is essentially as follows.

For each element, a stiffness matrix satisfying the following relationship is found:

$$[k]\{d\} = \{r\} \quad \textbf{(34)}$$

where

$[k]$ = an element stiffness matrix. It is square and defines the element stiffness in each direction (degree of freedom)
$\{d\}$ = a column of nodal displacements for one element
$\{r\}$ = a column of nodal loads for one element

The determination of $[k]$ can be very complex and its theory is not outlined here. Modeling the whole structure requires that:

$$[K]\{D\} = \{R\} \quad \textbf{(35)}$$

where

$[K]$ = the structure stiffness matrix; each member of $[K]$ is an assembly of the individual stiffness contributions surrounding a given node
$\{D\}$ = the column of nodal displacements for the structure
$\{R\}$ = the column of nodal loads on the structure

In general, neither $\{D\}$ nor $\{R\}$ is completely known. Therefore, Equation 35 must be partitioned (rearranged) to separate known and unknown quantities. Equation 35 then becomes:

$$\left[\begin{array}{c|c} K_{11} & K_{12} \\ \hline K_{21} & K_{22} \end{array}\right]\left\{\begin{array}{c} D_s \\ \hline D_o \end{array}\right\} = \left\{\begin{array}{c} R_o \\ \hline R_s \end{array}\right\} \quad \textbf{(36)}$$

where

D_s are unknown displacements
D_o are known displacements
R_s are unknown loads
R_o are known loads

Equation 36 represents the two following equations:

$$[K_{11}]\{D_s\} + [K_{12}]\{D_o\} = \{R_o\} \quad \textbf{(37)}$$

$$[K_{21}]\{D_s\} + [K_{22}]\{D_o\} = \{R_s\} \quad \textbf{(38)}$$

Equation 37 can be solved for D_s and Equation 38 can then be solved for R_s.

Using the calculated displacements $\{D\}$, $\{d\}$ can be found for each element and the stress can be calculated by:

$$\{\sigma\} = [E][B]\{d\} \quad \textbf{(39)}$$

where

$\{\sigma\}$ are element stresses
$[E]$ and $[B]$ relate stresses to strains and strains to displacements respectively

FEA theory may also be used to determine temperatures throughout complex geometric components. (See also Chapter 4.) Considering conduction alone, the governing relationship for thermal analysis is:

$$[C]\{\dot{T}\} + [K]\{T\} = \{Q\} \quad \textbf{(40)}$$

where

$[C]$ = the system heat capacity matrix
$\{\dot{T}\}$ = the column of rate of change of nodal temperatures
$[K]$ = the system thermal conductivity matrix
$\{T\}$ = the column of nodal temperatures
$\{Q\}$ = the column of nodal rates of heat transfer

In many respects, the solution for thermal analysis is similar to that of the structural analysis. One important difference, however, is that the thermal solution is iterative and nonlinear. Three aspects of a thermal analysis require an iterative solution.

First, thermal material properties are temperature dependent. Because they are primary unknowns, temperature assumptions must be made to establish the initial material properties. Each node is first given an assumed temperature. The first thermal distribution is then obtained, and the calculated temperatures are used in a second iteration. Convergence is attained when the calculated temperature distributions from two successive iterations are nearly the same.

Second, when convective heat transfer is accounted for, the heat transfer at a fluid boundary is dependent on the surface temperature. Again, because temperatures are primary unknowns, the solution must be iterative.

Third, in a transient analysis, the input parameters, including boundary conditions, may change with time, and the analysis must be broken into discrete steps. Within each time step, the input parameters are held constant. For this reason, transient thermal analysis is sometimes termed quasi-static.

FEA applied to dynamic problems is based upon the differential equation of motion:

$$[M]\{\ddot{D}\} + [C]\{\dot{D}\} + [K]\{D\} = \{R\} \quad \textbf{(41)}$$

where

$[M]$ = the structure lumped mass matrix
$[C]$ = the structure damping matrix
$[K]$ = the structure stiffness matrix
$\{R\}$ = the column of nodal forcing functions

$\{D\}$, $\{\dot{D}\}$, and $\{\ddot{D}\}$ are columns of nodal displacements, velocities, and accelerations, respectively.

Variations on Equation 41 can be used to solve for the natural frequencies, mode shapes, and responses due to a forcing function (periodic or nonperiodic), or to do a dynamic seismic analysis.

Limitations of FEA involve computer and human resources. The user must have substantial experience and, among other abilities, he must be skilled in selecting element types and in geometry modeling.

In FEA, result accuracy increases with the number of nodes and elements. However, computer time is expensive and handling the mass of data can be cumbersome.

In most finite element analyses, large scale yielding (plastic strain) and deformations (including buckling instability), and creep are not accounted for; the material is considered to be linear elastic. In a linear structural analysis, the response (stress, strain, etc.) is proportional to the load. For example, if the applied load is doubled, the stress response would also double. For nonlinear analysis, FEA can also be beneficial. However, its cost and difficulty are several times that of linear analysis.

Although most FEA software has well developed three dimensional capabilities, some pressure vessel analyses are imprecise due to a lack of acceptance criteria.

Computer software consists of commercially available and proprietary FEA programs. This software can be categorized into three groups: 1) preprocessors, 2) finite element solvers, and 3) postprocessors.

A preprocessor builds a model geometry and applies boundary conditions, then verifies and optimizes the model. The output of a finite element solver consists of displacements, stresses, temperatures, or dynamic response data.

Postprocessors manipulate the output from the finite element solver for comparison to acceptance criteria or to make contour map plots.

The Babcock & Wilcox (B&W) Finite Element System (FES) is a comprehensive software system that promotes efficiency by automating the entire analysis process. It builds and organizes analyses, managing large volumes of data created by many users. A menu guides the user through analysis steps automating many recurring chores.

Application of FEA Because classical formulas and shell analysis solutions are limited to simple shapes, FEA fills a technical void and is applied in response to ASME Code requirements. A large portion of B&W's FEA supports pressure vessel design. Stresses can be calculated near nozzles and other abrupt geometry changes. In addition, temperature changes and the resulting thermal stresses can be predicted using FEA.

The raw output from a finite element solver can not be directly applied to the ASME Code criteria. The stresses must first be classified as membrane, bending, or peak (Fig. 9). B&W pioneered the classification of finite element stresses and these procedures are now used throughout the industry. An extension of this effort, the interpretation of three dimensional FEA results, is much more complex.

Piping flexibility, for example, is an ideal FEA application. In addition, structural steel designers rely on FEA to analyze complex frame systems that support steam generation equipment.

Finite element analysis is often used for preliminary review of new product designs. For example, Fig. 10 shows the deflected shape of two economizer fin configurations modeled using FEA.

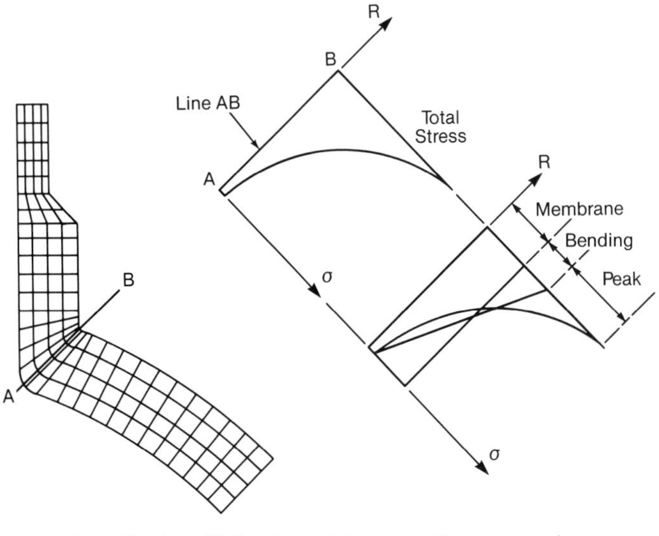

Fig. 9 Classification of finite element stress results on a vessel cross section for comparison to code criteria.[7]

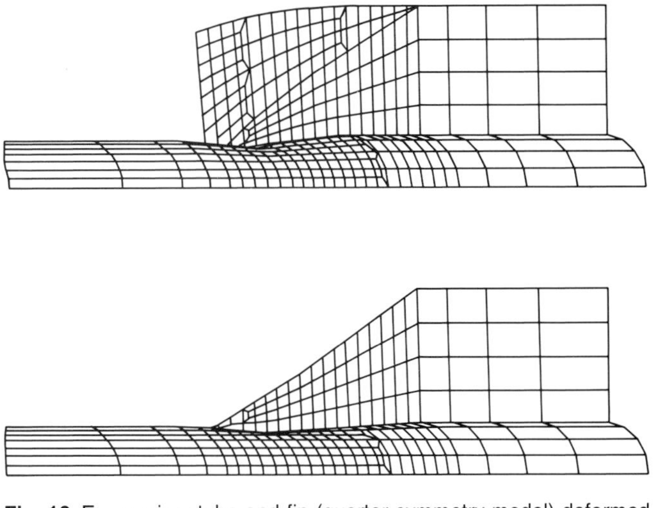

Fig. 10 Economizer tube and fin (quarter symmetry model) deformed shape plots before (upper) and after (lower) design modifications.

Fracture mechanics methods

Fracture mechanics provides analysis methods to account for the presence of flaws such as voids or cracks. This is in contrast to the stress analysis methods discussed above in which the structure was considered to be free of those kinds of defects. Flaws may be found by nondestructive examination (NDE) or they may be hypothesized prior to fabrication. Fracture mechanics is particularly useful to design or evaluate components fabricated using materials that are more sensitive to flaws. Additionally, it is well suited to the prediction of the remaining life of components under cyclic fatigue and high temperature creep conditions.

During component design, the flaw size is hypothesized. Allowable design stresses can be determined knowing the lower bound material toughness from accepted design procedures in conjunction with a factor of safety.

Fracture mechanics can be used to evaluate the integrity of a flawed existing structure. The defect, usually found by NDE, is idealized according to accepted ASME practices. An analysis uses design or calculated stresses based on real or hypothesized loads, and material properties are found from testing a specimen of similar material. Determining allowable flaw sizes strongly relies on accurate material properties and the best estimates of structural stresses. Appropriate safety factors are then added to the calculations.

During inspection of power plant components, minor cracks or flaws may be discovered. However, the flaws may propagate by creep or fatigue and become significant. The remaining life of components can not be accurately predicted from stress/cycles to failure (σ/N) curves alone. These predictions become possible using fracture mechanics.

Linear elastic fracture mechanics The basic concept of linear elastic fracture mechanics (LEFM) was originally developed to quantitatively evaluate sudden structural failure. LEFM, based on an analysis of the stresses near a sharp crack, assumes elastic behavior throughout the structure. The stress distribution near the crack tip depends on a single quantity, termed the stress intensity factor, K_I. LEFM assumes that unstable propagation of existing flaws occurs when the stress intensity factor becomes critical; this critical value is the fracture toughness of the material K_{IC}.

The theory of linear elastic fracture mechanics, LEFM, is based on the assumption that, at fracture, stress σ and defect size a are related to the fracture toughness K_{IC}, as follows:

$$K_I = C\sigma\sqrt{\pi a} \quad \textbf{(42a)}$$

and

$$K_I \geq K_{IC} \text{ at failure} \quad \textbf{(42b)}$$

The critical material property, K_{IC}, is compared to the stress intensity factor of the cracked structure, K_I, to identify failure potential. K_I is not to be confused with the stress intensity used in ASME design codes for unflawed structures. The term C, accounting for the geometry of the crack and structure, is a function of the crack size and relevant structure dimensions such as width or thickness.

C is exactly 1.0 for an infinitely wide center cracked panel with a through-wall crack length 2a, loaded in tension by a uniform remote stress σ. The factor C varies for other crack geometries illustrated in Fig. 11. Defects in a structure due to manufacture, in-service environment, or in-service cyclic fatigue are usually assumed to be flat, sharp, planar discontinuities where the planar area is normal to the applied stress.

ASME Code procedures for fracture mechanics design/analysis are presently given in Sections III and XI which are used for component thicknesses of at least 4 in. (102 mm) for ferritic materials with yield strengths less than 50,000 psi (3447.5 bar) and for simple geometries and stress distributions. The basic concepts of the Code may be extended to other ferritic materials (including clad ferritic materials) and more complex geometries; however, it does not apply to austenitic or high nickel alloys. These procedures provide methods for designing against brittle fractures in structures and for evaluating the significance of flaws found during in-service inspections.

The ASME Code, Section III uses the principles of linear elastic fracture mechanics to determine allowable loadings in ferritic pressure vessels with an assumed defect.

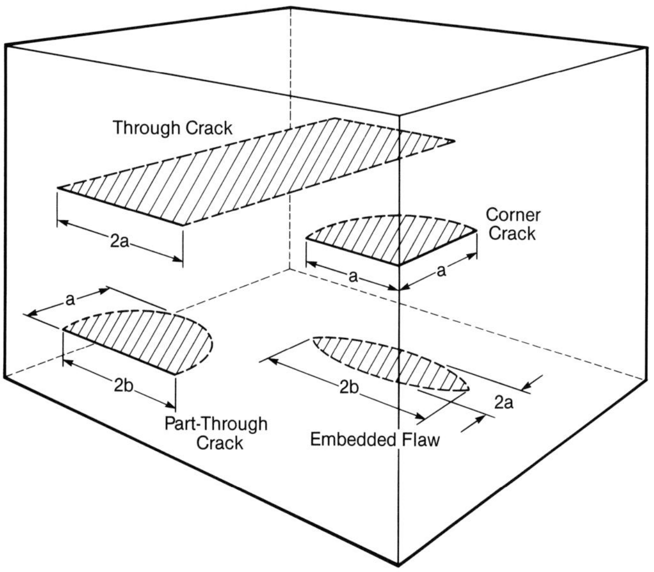

Fig. 11 Types of cracks.

The stress intensity factors (K_I) are calculated separately for membrane, bending, and thermal gradient stresses. They are further subdivided into primary and secondary stresses before summing and comparison to the allowable toughness, K_{IR}. K_{IR} is the reference critical stress intensity factor (toughness). It accounts for temperature and irradiation embrittlement effects on toughness. A safety factor of 2 is applied to the primary stress components and a factor of 1 is applied to the secondary components.

To determine an operating pressure that is below the brittle fracture point, the following approach is used:

1. A maximum flaw size is assumed. This is a semi-elliptical surface flaw one fourth the pressure vessel wall thickness in depth and 1.5 times the thickness in length.
2. Knowing the specific material's nil ductility temperature, and the design temperature K_{IR} can be found from the Code.
3. The stress intensity factor is determined based on the membrane and bending stresses, and the appropriate correction factors. Additional determinants include the wall thickness and normal stress to yield strength ratio of the material.
4. The calculated stress intensity is compared to K_{IR}.

The ASME Code, Section XI provides a procedure to evaluate flaw indications found during in-service inspection of nuclear reactor coolant systems. If an indication is smaller than certain limits set by Section XI, it is considered acceptable without further analysis. If the indication is larger than these limits, Section XI provides information that enables the following procedure for further evaluation:

1. Determine the size, location and orientation of the flaw by NDE.
2. Determine the applied stresses at the flaw location (calculated without the flaw present) for all normal (including upset), emergency and faulted conditions.
3. Calculate the stress intensity factors for each of the loading conditions.
4. Determine the necessary material properties, including the effects of irradiation. A reference temperature shift procedure is used to normalize the lower bound toughness versus temperature curves. These curves are based on crack arrest and static initiation values from fracture toughness tests. The temperature shift procedure accounts for heat to heat variation in material toughness properties.
5. Using the procedures above, as well as a procedure for calculating cumulative fatigue crack growth, three critical flaw parameters are determined:

   a_f = maximum size to which the detected flaw can grow during the remaining service of the component

   a_{crit} = maximum critical size of the detected flaw under normal conditions

   a_{init} = maximum critical size for nonarresting growth initiation of the observed flaw under emergency and faulted conditions
6. Using these critical flaw parameters, determine if the detected flaw meets the following conditions for continued operation:

$$a_f < 0.1\, a_{crit}$$
$$a_f < 0.5\, a_{init} \quad \textbf{(43)}$$

Elastic-plastic fracture mechanics (EPFM) LEFM provides a one parameter failure criterion in terms of the crack tip stress intensity factor (K_I), but is limited to analyses where the plastic region surrounding the crack tip is small compared to the overall component dimensions. As the material becomes more ductile and the structural response becomes nonlinear, the LEFM approach loses its accuracy and eventually becomes invalid.

A direct extension of LEFM to EPFM is possible by using a parameter to characterize the crack tip region that is not dependent on the crack tip stress. This parameter, the path independent J-integral, can characterize LEFM, EPFM, and fully plastic fracture mechanics. It is capable of characterizing crack initiation, growth, and instability. The J-integral is a measure of the potential energy rate of change for nonlinear elastic structures containing defects.

The J-integral can be calculated from stresses around a crack tip using nonlinear finite element analysis. An alternate approach is to use previously calculated deformation plasticity solutions in terms of the J-integral from the Electric Power Research Institute (EPRI) *Elastic-Plastic Fracture Analysis Handbook*.[8]

The onset of crack growth is predicted when:

$$J_I \geq J_{IC} \tag{44}$$

The material property J_{IC} is obtained using American Society for Testing and Materials (ASTM) test E813-89, and J_I is the calculated structural response.

Stable crack growth occurs when:

$$J_I(a, P) = J_R(\Delta a)$$

and

$$a = a_o + \Delta a \tag{45}$$

where

a = current crack size
P = applied remote load
$J_R(\Delta a)$ = material crack growth resistance (ASTM test standard E1152-87)
Δa = change in crack size
a_o = initial crack size

For crack instability, an additional criterion is:

$$\partial J / \partial a \geq \partial J_R / \partial a \tag{46}$$

Failure assessment diagrams Failure assessment diagrams are tools for the determination of safety margins, prediction of failure or plastic instability and leak-before-break analysis of flawed structures. These diagrams recognize both brittle fracture and net section collapse mechanisms. The failure diagram (see Fig. 12) is a safety/failure plane defined by the stress intensity factor/toughness ratio (K_r) as the ordinate and the applied stress/net section plastic collapse stress ratio (S_r) as the abscissa. For a fixed applied stress and defect size, the coordinates K_r, S_r are readily calculable. If the assessment point denoted by these coordinates lies inside the failure assessment curve, no crack growth can occur. If the assessment point lies outside the curve, unstable crack growth is predicted. The distance of the assessment point from the failure assessment curve is a measure of failure potential of the flawed structure.

In a leak-before-break analysis, a through-wall crack is postulated. If the resulting assessment point lies inside the failure assessment curve, the crack will leak before an unstable crack growth occurs.

The deformation plasticity failure assessment diagram (DPFAD)[3] is a specific variation of a failure assessment diagram. DPFAD follows the British PD 6493 R-6[4] format, and incorporates EPFM deformation plasticity J-integral solutions. The DPFAD curve is determined by normalizing the deformation plasticity J-integral response of the flawed structure by its elastic response. The square root of this ratio is denoted by K_r. The S_r coordinate is the ratio of the applied stress to the net section plastic collapse stress. Various computer programs are available which automate this process for application purposes.

Subcritical crack growth Subcritical crack growth refers to crack propagation due to cyclic fatigue, stress corrosion cracking, creep crack growth or a combination of the three. Stress corrosion cracking and creep crack growth are time based while fatigue crack growth is based on the number of stress cycles.

Fatigue crack growth Metal fatigue, although studied for more than 100 years, continues to plague structures subjected to cyclic stresses. The traditional approach to prevent fatigue failures is to base the allowable fatigue stresses on test results of carefully made laboratory specimens or representative structural components. These results are usually presented in cyclic stress versus cycles to failure, or σ/N, curves.

The significant events of metal fatigue are crack initiation and subsequent growth until the net section yields or until the stress intensity factor of the structure exceeds the material resistance to fracture. Traditional analysis assumes that a structure is initially crack free. However, a structure can have cracks that originate during fabrication or during operation. Therefore, fatigue crack growth calculations are required to predict the service life of a structure.

Fatigue crack growth calculations can 1) determine the service life of a flawed structure that (during its lifetime) undergoes significant in-service cyclic loading, or 2) determine the initial flaw size that can be tolerated prior to or during a specified operating period of the structure.

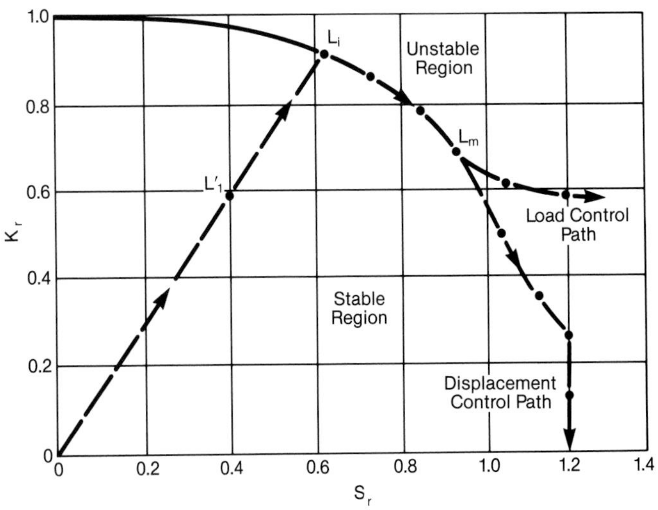

Fig. 12 Deformation plasticity failure assessment diagram in terms of stable crack growth.

The most useful way of presenting fatigue crack growth rates is to consider them as a function of the stress intensity difference, ΔK, which is the difference between the maximum and minimum stress intensity factors.

To calculate fatigue crack growth, an experimentally determined curve such as Fig. 13 is used. The vertical axis, da/dN, is the crack growth per cycle. ASME Code, Section XI contains similar growth rate curves for pressure vessel steels.

Creep crack growth Predicting the remaining life of fossil power plant components from creep rupture data alone is not reliable. Cracks can develop at critical locations and these cracks can then propagate by creep crack growth.

At temperatures above 800F (427C), creep crack growth can cause structural components to fail. Operating temperatures for selected fossil power plant components range from 900 to 1100F (482 to 593C). At these temperatures, creep deformation and crack growth become dependent on strain rate and time exposure. Macroscopic crack growth in a creeping material occurs by nucleation and joining of microcavities in the highly strained region ahead of the crack tip. In time dependent fracture mechanics (TDFM), the energy release rate (power) parameter C_t correlates[5] creep crack growth through the relationship:

$$da/dt = bC_t^q \quad \textbf{(47)}$$

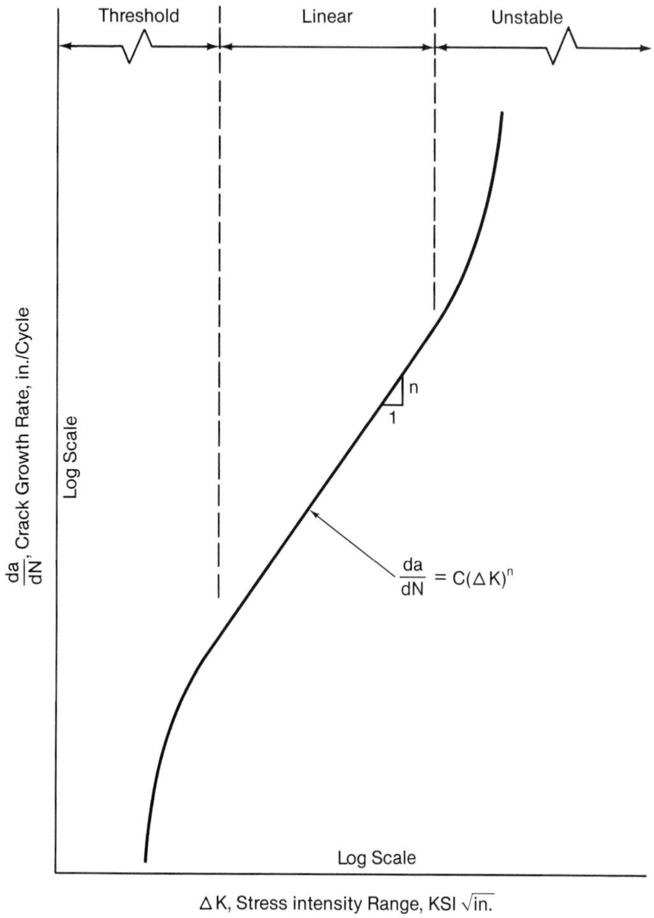

Fig.13 Relationship between da/dN and ΔK as plotted on logarithmic coordinates.

By using the energy rate definition, C_t can be determined experimentally from test specimens. The constants b and q are determined by a curve fit technique. Under steady-state creep where the crack tip stresses no longer change with time, the crack growth can be characterized solely by the path independent energy rate line integral C^*, analogous to the J-integral.

C^* and C_t can both be interpreted as the difference in energy rates (power) between two bodies with incrementally differing crack lengths. Furthermore, C^* characterizes the strength of the crack tip stress singularity in the same manner as the J-integral characterizes the elastic-plastic stress singularity.

The fully plastic deformation solutions from the EPRI *Elastic-Plastic Fracture Handbook* can then be used to estimate the creep crack tip steady-state parameter, C^*.

Significant data support C_t as a parameter for correlating creep crack growth behavior represented by Equation 47. An approximate expression[6] for C_t is as follows:

$$C_t = C^* \left[\left(t_T / t \right)^{\frac{n-3}{n-1}} + 1 \right] \quad \textbf{(48)}$$

where t_T is the transition time given by:

$$t_T = \frac{\left(1 - \vartheta^2\right) K_I^2}{(n+1) E C^*} \quad \textbf{(49)}$$

and n is the secondary creep rate exponent.

For continuous operation, Equation 48 is integrated over the time covering crack growth from the initial flaw size to the final flaw size. The limiting final flaw size is chosen based on fracture toughness or instability considerations, possibly governed by cold startup conditions. For this calculation, fracture toughness data such as K_{IC}, J_{IC} or the J_R curve would be used in a failure assessment diagram approach to determine the limiting final flaw size.

Construction features

All pressure vessels require construction features such as fluid inlets and outlets, access openings, and structural attachments at support locations. These shell areas must have adequate reinforcement and gradual geometric transitions which limit local stresses to acceptable levels.

Openings Openings are the most prevalent construction features on a vessel. They can become areas of weakness and may lead to unacceptable local distortion, known as bell mouthing, when the vessel is pressurized. Such distortions are associated with high local membrane stresses around the opening. Analytical studies have shown that these high stresses are confined to a distance of approximately one hole diameter, d, along the shell from the axis of the opening and are limited to a distance of $0.37\,(dt_{nozzle})^{1/2}$ normal to the shell.

Reinforcement to reduce the membrane stress near an opening can be provided by increasing the vessel wall thickness. An alternate, more economical stress reduction method is to thicken the vessel locally around the nozzle axis of symmetry. The reinforcing material must be within the area of high local stress to be effective.

The ASME Code provides guidelines for reinforcing openings. The reinforcement must meet requirements for the amount and distribution of the added material. A relatively small opening [approximately $d<0.2\,(Rt_s)^{1/2}$ where R is mean radius of shell and t_s is thickness of shell] remote from other locally stressed areas does not require reinforcement.

Larger openings are normally reinforced as illustrated in Figs. 14a and 14b. It is important to avoid excessive reinforcement that may result in high secondary stresses. Fig. 14c shows an opening with over reinforcement and Fig. 14a shows one with well proportioned reinforcement. Fig. 14b also shows a balanced design that minimizes secondary stresses at the nozzle/shell juncture. Designs a and b, combined with generous radii r, are most suitable for cyclic load applications.

The *ligament efficiency* method is also used to compensate for metal removed at shell openings. This method considers the load carrying ability of an area between two points in relation to the load carrying ability of the remaining ligament when the two points become the centers of two openings. The ASME Code guidelines used in this method only apply to cylindrical pressure vessels where the circumferential stress is twice the longitudinal stress. In determining the thickness of such vessels, the allowable stress in the thickness calculation is multiplied by the ligament efficiency.

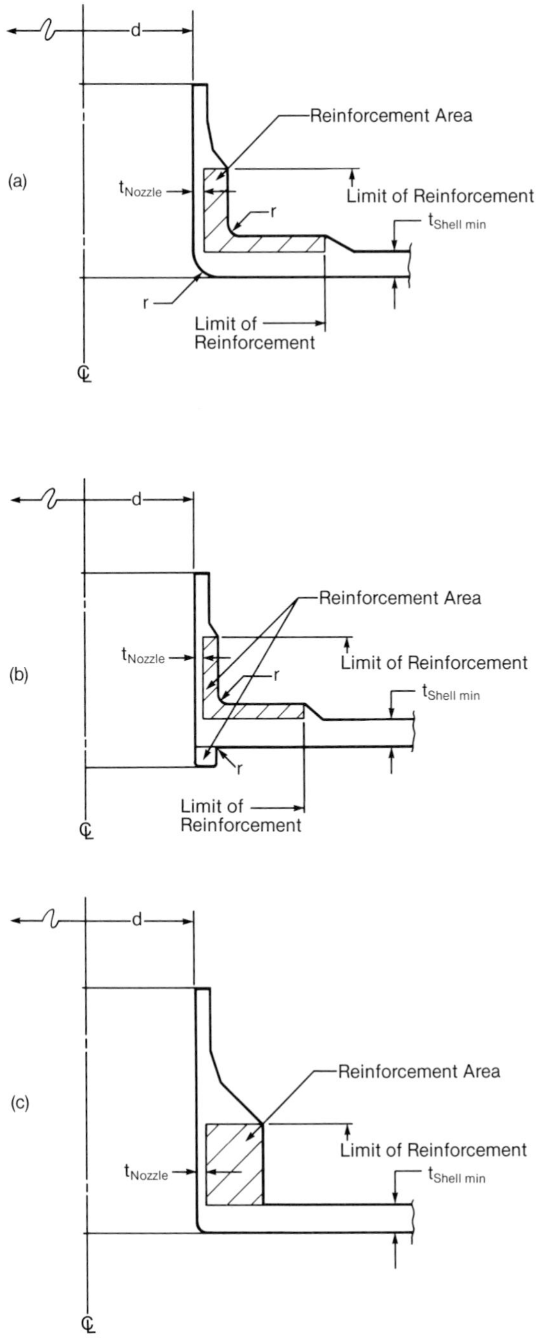

Fig. 14 Nozzle opening reinforcements.

Nozzle and attachment loadings When external loadings are applied to nozzles or attachment components, local stresses are generated in the shell. Several types of loading may be applied, such as sustained, transient and thermal expansion flexibility loadings. The local membrane stresses produced by such loadings must be limited to avoid unacceptable distortion due to a single load application. The combination of local membrane and bending stresses must also be limited to avoid incremental distortion under cyclic loading. Finally, to prevent cyclic load fatigue failures, the nozzle or attachment should include gradual transitions which minimize stress concentrations.

Pressure vessels may require local thickening at nozzles and attachments to avoid yielding or incremental distortion due to the combined effects of external loading, internal pressure, and thermal loading. Simple procedures to determine such reinforcement are not available, however FEA methods can be used. The Welding Research Council (WRC) Bulletin No. 107 also provides a procedure for determining local stresses adjacent to nozzles and rectangular attachments on cylindrical and spherical shells.

The external loadings considered by the WRC are longitudinal moment, transverse moment, torsional moment, and axial force. Stresses at various inside and outside shell surfaces are obtained by combining the stresses from the various applied loads. These external load stresses are then combined with internal pressure stresses and compared with allowable stress limits.

Use of the WRC procedure is restricted by limitations on shell and attachment parameters; however, experimental and theoretical work continues in this area.

Structural support components

Pressure vessels are normally supported by saddles, cylindrical support skirts, hanger lugs and brackets, ring girders, or integral support legs. A vessel has concentrated loads imposed on its shell where these supports are located. Therefore, it is important that the support arrangements minimize local stresses in the vessel. In addition, the components must provide support for the specified loading conditions and withstand corresponding temperature requirements.

Design criteria

Structural elements that provide support, stiffening, and/or stabilization of pressure vessels or components may be directly attached by welding or bolting. They can also be indirectly attached by clips, pins, or clamps, or may be completely unattached thereby transferring load through surface bearing and friction.

Loading conditions In general, loads applied to structural components are categorized as dead, live, or transient loads. Dead loads are due to the force of gravity on the equipment and supports. Live loads vary in magnitude and are applied to produce the maximum design conditions. Transient loads are time dependent and are expected to occur randomly for the life of the structural components. Specific loadings that are considered in designing a pressure component support include:

1. weight of the component and its contents during operating and test conditions, including loads due to static and dynamic head and fluid flow,
2. weight of the support components,
3. superimposed static and thermal loads induced by the supported components,
4. environmental loads such as wind and snow,
5. dynamic loads including those caused by earthquake, vibration, or rapid pressure change,
6. loads from piping thermal expansion,
7. loads from expansion or contraction due to pressure, and
8. loads due to anchor settlement.

Code design/analysis requirements Code requirements for designing pressure part structural supports vary. The ASME Code, Section I, only covers pressure part attaching lugs, hangers or brackets. These must be properly fitted and must be made of weldable and comparable quality material. Only the weld attaching the structural member to the pressure part is considered within the scope of Section I. Prudent design of all other support hardware is the manufacturer's responsibility.

The ASME Code, Section VIII, Division 1, does not contain design requirements for vessel supports; however, suggested rules of good practice are presented. These rules primarily address support details which prevent excessive local shell stresses at the attachments. For example, horizontal pressure vessel support saddles are recommended to support at least one third of the shell circumference. Rules for the saddle design are not covered. However, the Code refers the designer to the *Manual of Steel Construction*, published by the American Institute of Steel Construction (AISC). This reference details the allowable stress design (ASD) method for structural steel building designs. When adjustments are made for elevated temperatures, this specification can be used for designing pressure vessel support components. Similarly, Section VIII, Division 2, does not contain design methods for vessel support components. However, materials for structural attachments welded to pressure components and details of permissible attachment welds are covered.

Section III of the ASME Code contains rules for the material, design, fabrication, examination, and installation of certain pressure component and piping supports. The supports are placed within three categories:

1. plate and shell type supports, such as vessel skirts and saddles, which are fabricated from plate and shell elements,
2. linear supports which include axially loaded struts, beams and columns, subjected to bending, and trusses, frames, rings, arches and cables, and
3. standard supports (catalog items) such as constant and variable type spring hangers, shock arresters, sway braces, vibration dampers, clevises, etc.

The design procedures for each of these support types are:

1. design by analysis including methods based on the maximum shear stress and maximum stress theories,
2. experimental stress analysis, and
3. load rating by testing full size prototypes.

The analysis required for each type of support depends on the class of the pressure component being supported.

Typical support design considerations

Design by analysis involves determining the stresses in the structural components and their connections by accepted analysis methods. Unless specified in an applicable code, choosing the analysis method is the designer's prerogative. Linear elastic analysis (covered in depth here), using the maximum stress or maximum shear stress theory, is commonly applied to plate, shell type and linear type supports. As an alternate, the method of limit (plastic) analysis can be used for framed linear structures when appropriate load adjustment factors are applied.

Plate and shell type supports Cylindrical shell skirts are commonly used to support vertical pressure vessels. They are attached to the vessel with a minimum offset in order to reduce local bending stresses at the vessel skirt junction. This construction also permits radial pressure and thermal growth of the supported vessel through bending of the skirt. The length of the support is chosen to permit this bending to occur safely. See Fig. 15 for typical shell type support skirt details.

In designing the skirt, the magnitudes of the loads that must be supported are determined. These normally include the vessel weight, the contents of the vessel, the imposed loads of any equipment supported from the vessel, and loads from piping or other attachments. Next a skirt height is set and the forces and moments at the skirt base, due to the loads applied, are determined. Treating the cylindrical shell as a beam, the axial stress in the skirt is then determined from:

$$\sigma = \frac{-P_v}{A} \pm \frac{Mc}{I} \quad \textbf{(50)}$$

where

σ = axial stress in skirt
P_v = total vertical design load
A = cross-sectional area

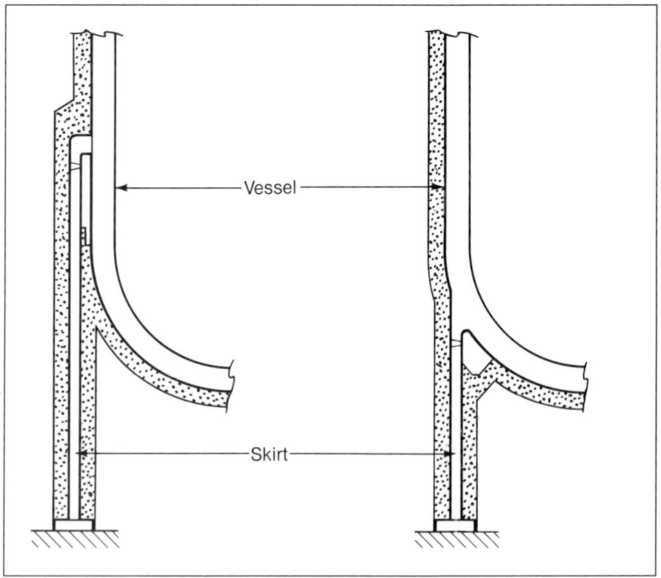

Fig. 15 Support skirt details.[2]

M = moment at base due to design loads
c = radial distance from centerline of skirt
I = moment of inertia

For thin shells ($R/t > 10$), the equation for the axial stress becomes:

$$\sigma = \frac{-P_v}{2\pi Rt} \pm \frac{M}{\pi R^2 t} \qquad (51)$$

where

R = mean radius of skirt
t = thickness of skirt

Because the compressive stress is larger than the tensile stress, it usually controls the skirt design. Using the maximum stress theory for this example, the skirt thickness is obtained by:

$$t = \frac{P_v}{2\pi RF_A} + \frac{M}{\pi R^2 F_A} \qquad (52)$$

where

F_A = allowable axial compressive stress

The designer must also consider stresses caused by transient loadings such as wind or earthquakes. Finally, skirt connections at the vessel and support base must be checked for local primary and secondary bending stresses. The consideration of overall stress levels provides the most accurate design.

Local thermal bending stresses often occur because of a temperature difference between the skirt and support base. The magnitudes of these bending stresses are dependent upon the severity of this axial thermal gradient; steeper gradients promote higher stresses. To minimize these stresses, the thermal gradient at the junction can be reduced by full penetration welds at the skirt to shell junction, which permit maximum conduction heat flow through the metal at that point, and by selective use of insulation in the crotch region to permit heat flow by convection and radiation. Depending on the complexity of the attachment detail, the discontinuity stress analysis or the linear elastic finite element method is used to solve for the thermal bending stresses.

Linear type supports Utility fossil fuel-fired steam generators contain many linear components that support and reinforce the boiler pressure parts. For example, the furnace enclosure walls, which are constructed of welded membraned tube panels, must be reinforced by external structural members (buckstays) to resist furnace gas pressure as well as wind and seismic forces. (See Chapter 22.) Similarly, vestibules, such as the burner equipment enclosure (windbox), require internal systems to support the enclosure and its contents as well as to reinforce the furnace walls. The design of these structural systems is based on linear elastic methods using maximum stress theory allowable limits.

The buckstay system is typically comprised of horizontally oriented beams or trusses which are attached to the outside of the furnace membraned vertical tube walls. As shown in Fig. 16, the buckstay ends are connected to tie bars that link them to opposing wall buckstays thereby forming a self-equilibrating structural system. The furnace enclosure walls are continuously welded at the corners creating a water-cooled, orthotropic plate, rectangular pressure vessel. The strength of the walls in the horizontal direction is considerably less than in the vertical direction, therefore the buckstay system members are horizontally oriented.

The buckstay spacing is based on the ability of the enclosure walls to resist the following loads:

1. internal tube design pressure (P),
2. axial dead loads (DL),
3. sustained furnace gas pressure (PL_S),
4. transient furnace gas pressure (PL_T),
5. wind loads (WL), and
6. seismic loads (EQ).

The buckstay elevations are initially established based on wall stress checks and on the location of necessary equipment such as sootblowers, burners, access doors, and ob-

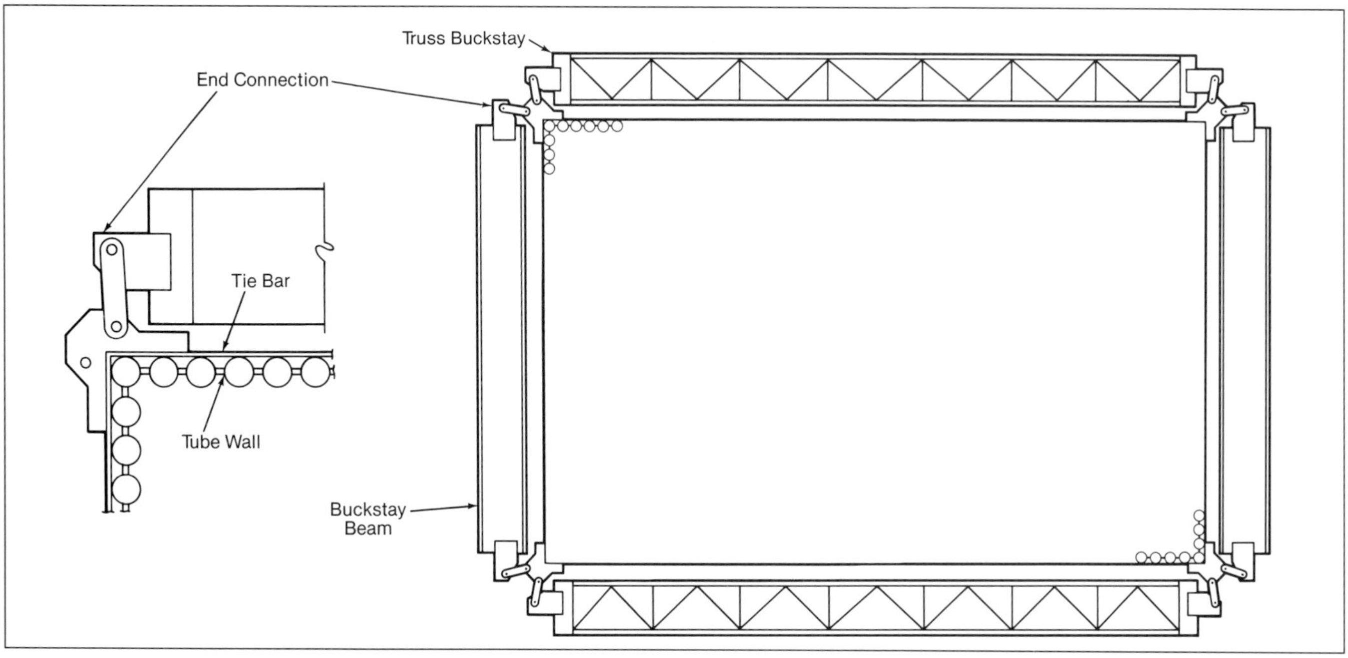

Fig. 16 Typical buckstay elevation, plan view.

Large coal-fired utility power plant under construction.

servation ports. These established buckstay elevations are considered as horizontal supports for the continuous vertical tube wall. The wall is then analyzed for the following load combinations using a linear elastic analysis method:

1. $DL + PL_s + P$,
2. $DL + PL_s + WL + P$,
3. $DL + PL_s + EQ + P$, and
4. $DL + PL_T + P$.

Buckstay spacings are varied to assure that the wall stresses are within allowable design limits. Additionally, their locations are designed to make full use of the structural capability of the membraned walls.

The buckstay system members, their end connections, and the wall attachments are designed for the maximum loads obtained from the wall analysis. They are designed as pinned end bending members according to the latest AISC ASD specification. This specification is modified for use at elevated temperatures and uses safety factors consistent with ASME Code, Sections I and VIII. The most important design considerations for the buckstay system include:

1. stabilization of the outboard beam flanges or truss chords to prevent lateral buckling when subjected to compression stress,
2. the development of buckstay to tie bar end connections and buckstay to wall attachments that provide load transfer but allow differential expansion between connected elements, and
3. providing adequate buckstay spacing and stiffness to prevent resonance due to low frequency gas side pressure pulsations common in fossil fuel-fired boilers.

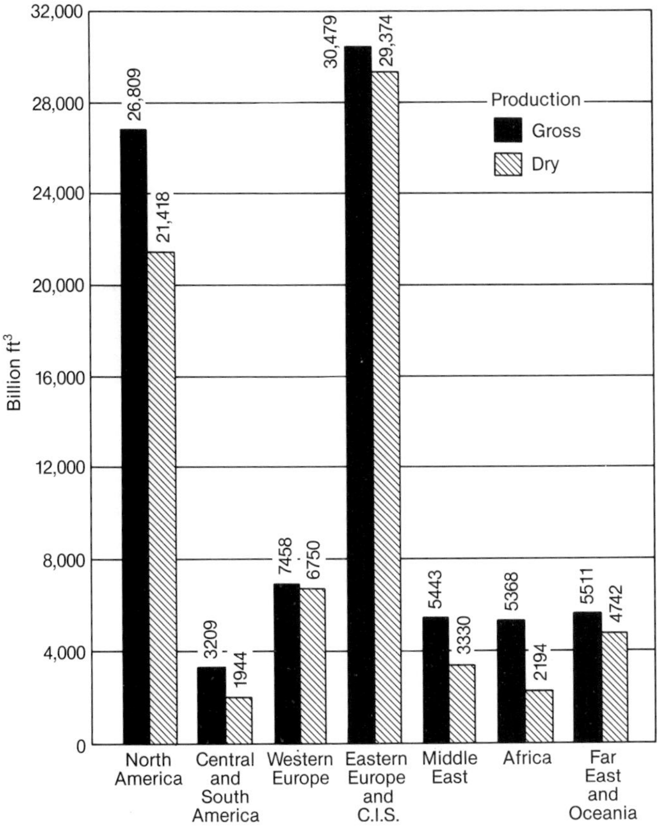

Fig. 17 1988 World natural gas production.

hydrocarbons (5 to 10% C). Gas containing H_2S is *sour* gas; conversely, *sweet* gas contains little or no H_2S.

Of all chemical fuels, natural gas is considered to be the most desirable for steam generation. It is piped directly to the consumer, eliminating the need for storage. It is substantially free of ash and mixes easily with air, providing complete combustion without smoke. Although the total hydrogen content of natural gas is high, its free hydrogen content is low. Because of this, natural gas burns less easily than some manufactured gases with high free hydrogen content.

Table 14
U.S. Natural Gas Consumption (Trillion ft³)

Year	Residential	Commercial	Industrial	Elec. Util.	Transportation	Total
1980	4.75	2.61	8.20	3.68	0.63	19.88
1981	4.55	2.52	8.06	3.64	0.64	19.40
1982	4.63	2.61	6.94	3.23	0.60	18.00
1983	4.38	2.43	6.62	2.91	0.49	16.83
1984	4.56	2.52	7.23	3.11	0.53	17.95
1985	4.43	2.43	6.87	3.04	0.50	17.28
1986	4.31	2.32	6.50	2.60	0.49	16.22
1987	4.31	2.43	7.10	2.84	0.52	17.21
1988	4.63	2.67	7.48	2.64	0.61	18.03
1989P	4.84	2.73	8.02	2.77	0.59	18.95

P = Preliminary data.

Note: Total may not equal sum of components due to independent rounding. Source: Energy Information Administration, *Annual Energy Review*, 1989.

The high hydrogen content of natural gas compared to that of oil or coal results in more water vapor being produced in the combustion gases. A correspondingly lower efficiency of the steam generating equipment results. (See Chapter 9.) This can readily be taken into account when designing the equipment.

Properties of natural gas

Analyses of natural gas from several U.S. fields are given in Table 15.

Other fuels

While coal, oil and gas are the dominant fuel sources, other carbonaceous fuels being used for boiler applications include petroleum byproducts; wood, its byproducts and wastes from wood processing industries; and certain types of vegetation, particularly bagasse and municipal solid waste.

Coke from petroleum

The heavy residuals from petroleum cracking processes are presently used to produce a higher yield of lighter hydrocarbons and a solid residue suitable for fuel. Characteristics of these residues vary widely and depend on the process used. Solid fuels from oil include delayed coke, fluid coke and petroleum pitch. Some selected analyses are given in Table 16.

The delayed coking process uses residual oil that is heated and pumped to a reactor. Coke is deposited in the reactor as a solid mass and is subsequently stripped, mechanically or hydraulically, in the form of lumps and granular material. Some cokes are easy to pulverize and burn while others are difficult.

Table 15
Selected Samples of Natural Gas from U.S. Fields

Sample No.	1	2	3	4	5
Source:	Pa.	S.C.	Ohio	La.	Ok.
Analyses:					
Constituents, % by vol					
H_2, Hydrogen	--	--	1.82	--	--
CH_4, Methane	83.40	84.00	93.33	90.00	84.10
C_2H_4, Ethylene	--	0.25	--	--	--
C_2H_6, Ethane	15.80	14.80	--	5.00	6.70
CO, Carbon monoxide	--	--	0.45	--	--
CO_2, Carbon dioxide	--	0.70	0.22	--	0.80
N_2, Nitrogen	0.80	0.50	3.40	5.00	8.40
O_2, Oxygen	--	--	0.35	--	--
H_2S, Hydrogen sulfide	--	--	0.18	--	--
Ultimate, % by wt					
S, Sulfur	--	--	0.34	--	--
H_2, Hydrogen	23.53	23.30	23.20	22.68	20.85
C, Carbon	75.25	74.72	69.12	69.26	64.84
N_2, Nitrogen	1.22	0.76	5.76	8.06	12.90
O_2, Oxygen	--	1.22	1.58	--	1.41
Specific gravity (rel to air)	0.636	0.636	0.567	0.600	0.630
HHV					
Btu/ft³ at 60F and 30 in. Hg	1,129	1,116	964	1,022	974
(kJ/m³ at 16C and 102 kPa)	(42,065)	(41,581)	(35,918)	(38,079)	(36,290)
Btu/lb(kJ/kg) of fuel	23,170 (53,893)	22,904 (53,275)	22,077 (51,351)	21,824 (50,763)	20,160 (46,892)

Table 16
Selected Analyses of Solid Fuels Derived from Oil

Analyses (dry basis) % by wt	Delayed Coke		Fluid Coke	
Proximate:				
VM	10.8	9.0	6.0	6.7
FC	88.5	90.0	93.7	93.2
Ash	0.7	0.1	0.3	0.1
Ultimate:				
Sulfur	9.9	1.5	4.7	5.7
Heating value,				
Btu/lb	14,700	15,700	14,160	14,290
(kJ/kg)	(34,192)	(36,518)	(32,936)	(33,239)

Fluid coke is produced by spraying hot residual feed onto externally heated seed coke in a fluidized bed. The fluid coke is removed as small particles, which are built up in layers. This coke can be pulverized and burned, or it can be burned in a Cyclone furnace or in a fluidized bed. All three types of firing require supplemental fuel to aid ignition.

The petroleum pitch process is an alternate to the coking process and yields fuels of various characteristics. Melting points vary considerably, and the physical properties vary from soft and gummy to hard and friable. The low melting point pitches may be heated and burned like heavy oil, while those with higher melting points may be pulverized or crushed and burned.

Oil emulsions

As previously discussed, coal-water slurries have potential as fuel oil substitutes in many combustion systems. In recent years applications for these fuels have ranged from utility boilers to diesel engines. With the discovery of large heavy hydrocarbon and bitumen reserves in Venezuela, considerable effort has been devoted to developing these sources as commercial fuels. This has led to the development of a bitumen oil emulsion.

Oil emulsions are liquid fuels composed of micron size oil droplets dispersed in water. Droplet coalescence is prevented by adding a small amount of a proprietary chemical. The fuel is characterized by relatively high levels of sulfur, asphaltenes and metals. The heating value, ash content and viscosity of the emulsion are similar to residual fuel oil. Emulsion handling and combustion performance are also similar to those of residual fuel oil. One of the concerns with burning these bitumen emulsions is the formation of corrosive vanadium compounds. This occurs by oxidation of the metallic compounds in the fuel.

Wood

Selected analyses and heating values of wood and wood ash are given in Table 17. Wood is composed primarily of carbohydrates. Consequently, it has a relatively low heating value compared with bituminous coal and oil.

Wood bark may pick up impurities during transportation. It is common practice to drag the rough logs to central loading points and sand is often picked up. Where the logs are immersed in salt water the bark can absorb the salt. Combustion temperatures from burning dry bark may be high enough for these impurities to cause fluxing of refractory furnace walls and fouling of boiler heating surfaces, unless sufficient furnace cooling surface is provided. Sand passing through the boiler banks can cause erosion of the tubes, particularly if the flue gas sand loading is increased by returning collected material to the furnace. Such collectors may be required with some bark burning equipment to reduce the stack discharge of incompletely burned bark.

Wood or bark with a moisture content of 50% or less burns quite well; however, as the moisture increases above this amount, combustion becomes more difficult. With a moisture content above 65%, a large part of the heat is required to evaporate the inherent moisture and little remains for steam generation. Burning this wet bark becomes a means of disposal rather than a source of energy.

Hogged wood and bark are very bulky and require relatively large handling and storage equipment. Uninterrupted flow from bunkers or bins through chutes is difficult to maintain.

Table 17
Analyses of Wood and Wood Ash

Wood analyses (dry basis), % by wt	Pine Bark	Oak Bark	Spruce Bark*	Redwood Bark*
Proximate analysis, %				
Volatile matter	72.9	76.0	69.6	72.6
Fixed carbon	24.2	18.7	26.6	27.0
Ash	2.9	5.3	3.8	0.4
Ultimate analysis, %				
Hydrogen	5.6	5.4	5.7	5.1
Carbon	53.4	49.7	51.8	51.9
Sulfur	0.1	0.1	0.1	0.1
Nitrogen	0.1	0.2	0.2	0.1
Oxygen	37.9	39.3	38.4	42.4
Ash	2.9	5.3	3.8	0.4
Heating value, Btu/lb	9030	8370	8740	8350
(kJ/kg)	(21,004)	(19,469)	(20,329)	(19,442)
Ash analysis, % by wt				
SiO_2	39.0	11.1	32.0	14.3
Fe_2O_3	3.0	3.3	6.4	3.5
TiO_2	0.2	0.1	0.8	0.3
Al_2O_3	14.0	0.1	11.0	4.0
Mn_3O_4	Trace	Trace	1.5	0.1
CaO	25.5	64.5	25.3	6.0
MgO	6.5	1.2	4.1	6.6
Na_2O	1.3	8.9	8.0	18.0
K_2O	6.0	0.2	2.4	10.6
SO_3	0.3	2.0	2.1	7.4
Cl	Trace	Trace	Trace	18.4
Ash fusibility temp, F				
Reducing				
Initial deformation	2180	2690		
Softening	2240	2720		
Fluid	2310	2740		
Oxidizing				
Initial deformation	2210	2680		
Softening	2280	2730		
Fluid	2350	2750		

* Salt water stored.

Wood wastes There are several industries using wood as a raw material where combustible byproducts or wastes are available as fuels. The most important of these are the pulp and turpentine industries. The nature and methods of utilization of the combustible byproducts from the pulp industry are discussed in Chapter 26.

The residue remaining after the steam distillation of coniferous woods for the production of turpentine is usable as a fuel. Some of the more easily burned constituents are removed in the distillation process with the result that the residue is somewhat more difficult to burn. Other than this, fuel properties are much the same as those of the raw wood and the problems involved in utilization are similar.

Bagasse

Mills grinding sugar cane commonly use bagasse for steam production. Bagasse is the dry pulp remaining after the juice has been extracted from sugar cane. The mills normally operate 24 hours per day during the grinding season. The supply of bagasse will easily meet the plant steam demands in mills where the sugar is not refined. Consequently, where there is no other market for the bagasse, no particular effort is made to burn it efficiently, and burning equipment is provided that will burn the bagasse as-received from the grinders. In refining plants, supplemental fuels are required to provide the increased steam demands. Greater efforts to obtain higher efficiency are justified in these plants. A selected analysis of bagasse is given in Table 8.

Other vegetation wastes

Food and related industries produce numerous vegetable wastes that are usable as fuels. They include such materials as grain hulls, the residue from the production of furfural from corn cobs and grain hulls, coffee grounds from the production of instant coffee, and tobacco stems. Fuels of this type are available in such small quantities that they are relatively insignificant in total energy production.

Table 18
Analyses of MSW and RDF Compared to Bituminous Coal

Constituent	Analyses, % by wt MSW	RDF	Bituminous Coal
Carbon	27.9	36.1	72.8
Hydrogen	3.7	5.1	4.8
Oxygen	20.7	31.6	6.2
Nitrogen	0.2	0.8	1.5
Sulfur	0.1	0.1	2.2
Chlorine	0.1	0.1	0
Water	31.3	20.2	3.5
Ash	16.0	6.0	9.0
HHV (wet), Btu/lb	5,100	6,200	13,000
(kJ/kg)	(11,836)	(14,421)	(30,238)

Municipal solid waste

One of the fastest growing energy sources in the U.S., Europe and Japan is municipal solid waste (MSW), or refuse. MSW is the combined residential and commercial waste generated in a given municipality. Formerly landfilled, MSW now fires numerous waste-to-energy boilers. It is burned as-received, called mass burning, or processed using size reduction and material recovery techniques to produce refuse-derived fuel (RDF).

Table 18 shows a typical analysis of raw refuse and RDF compared to bituminous coal. The relatively low calorific value and high heterogeneous nature of MSW provide a challenge to the combustion system design engineer. The design of MSW handling and combustion systems is discussed in Chapter 27.

References

1. *World Energy Resources 1990-91*, Oxford University Press, Oxford, U.K., 1990.

2. *1989 Survey of Energy Resources*, World Energy Conference, London, 1989.

3. "Estimate of U.S. coal reserves by coal type," Report DOE/EIA-0529, U.S. Energy Information Administration, Washington, D.C., 1989.

4. "Cost and quality of fuels for electric utility plants 1989," Report DOE/EIA-0191, U.S. Energy Information Administration, Washington, D.C., 1990.

5. "Gaseous fuels; coal and coke," Vol. 05.05, *Annual Book of ASTM Standards*, American Society for Testing and Materials, Philadelphia, 1991.

6. Vecci, S.J., Wagoner, C.L., and Olson, G.B., "Fuel and ash characterization and its effect on the design of industrial boilers," *Proceedings of the American Power Conference*, Vol. 40, pp. 850-864, 1978.

7. Wagoner, C.L., and Duzy, A. F., "Burning profiles for solid fuels," Technical Paper 67-WA-FU-4, American Society of Mechanical Engineers, New York, 1967.

8. Sala, D.L., Babu, S.P., and Bair, W.J., "Mild gasification of coal: Potential opportunities for value-added uses," *Coal: Targets of Opportunity Workshop; Proceedings*, Report CONF-880770-2, U.S. Department of Energy, Washington, D.C., pp. III.143 to III.161, September 1988.

9. Farthing, G.A., *et al.*, "Properties and performance characteristics of coal-water fuels," presented at the American Chemical Society National Meeting, Seattle, Washington, March 20-23, 1983.

450 MW midwest power station firing pulverized subbituminous coal.

Chapter 9
Principles of Combustion

A boiler requires a source of heat at a sufficient temperature to produce steam. Fossil fuel is generally burned directly in the boiler furnace to provide this heat although waste energy from another process may also be used.

Combustion is defined as the rapid chemical combination of oxygen with the combustible elements of a fuel. There are just three combustible elements of significance in most fossil fuels: carbon, hydrogen and sulfur. Sulfur, usually of minor significance as a heat source, can be a major contributor to corrosion and pollution problems. (See Chapters 20 and 32.)

The objective of good combustion is to release all of the energy in the fuel while minimizing losses from combustion imperfections and excess air. The combination of the combustible fuel elements and compounds in the fuel with all the oxygen requires *temperatures* high enough to ignite the constituents, mixing or *turbulence* to provide intimate oxygen-fuel contact, and sufficient *time* to complete the process, sometimes referred to as the three Ts of combustion.

Table 1 lists the chemical elements and compounds found in fuels generally used in commercial steam generation.

Concept of the mole

The mass of a substance in pounds equal to its molecular weight is called a pound-mole (lb-mole) of the substance. The molecular weight is the sum of the atomic masses of a substance's constituent atoms. For example, pure elemental carbon (C) has an atomic mass and molecular weight of 12 and therefore a lb-mole is equal to 12. In the case of carbon dioxide (CO_2), carbon still has an atomic mass of 12 and oxygen has an atomic mass of 16 giving CO_2 a molecular weight and a lb-mole equal to $(1 \times 12) + (2 \times 16)$ or 44. In SI, a similar system is based upon the molecular weight in kilograms expressed as kg-mole or kmole. In the United States (U.S.) power industry it is common practice to replace lb-mole with mole.

> For clarity, this chapter is provided in English units only. Appendix 1 provides a comprehensive list of conversion factors. Selected factors of particular interest here include: Btu/lb × 2.326 = kJ/kg; 5/9 (F-32) = C; lb × 0.4536 = kg. Selected SI constants include: universal gas constant = 8.3145 kJ/kmole K, one kmole at 0C and 1.01 bar = 22.4 m^3.

In the case of a gas, the volume occupied by one mole is called the molar volume. The volume of one mole of an ideal gas (a good approximation in most combustion calculations) is a constant regardless of its composition for a given temperature and pressure. Therefore, one lb-mole or mole of oxygen (O_2) at 32 lb and one mole of CO_2 at 44 lb will occupy the same volume equal to 394 ft^3 at 80F and 14.7 psi. The volume occupied by one mole of a gas can be corrected to other pressures and temperatures by the ideal gas law.

Because substances combine on a molar basis during combustion but are usually measured in units of mass (pounds), the lb-mole and molar volume are important tools in combustion calculations.

Fundamental laws

Combustion calculations are based on several fundamental physical laws.

Conservation of matter

This law states that matter can not be destroyed or created. There must be a mass balance between the sum of the components entering a process and the sum of those leaving: X pounds of fuel combined with Y pounds of air always results in X + Y pounds of products (see Note below).

Conservation of energy

This law states that energy can not be destroyed or created. The sum of the energies (potential, kinetic, thermal, chemical and electrical) entering a process must equal the sum of those leaving, although the proportions of each may change. In combustion, chemical energy is converted into thermal energy (see Note below).

Ideal gas law

This law states that the volume of an ideal gas is directly proportional to its absolute temperature and inversely proportional to its absolute pressure. The pro-

---

Note: While the laws of conservation of matter and energy are not rigorous from a nuclear physics standpoint (see Chapter 37), they are quite adequate for engineering combustion calculations. When a pound of a typical coal is burned releasing 13,500 Btu, the equivalent quantity of mass converted to energy amounts to only 3.5×10^{-10} lb.

Table 1 — Combustion Constants — Reference 1

No. Substance	Formula	Molecular Weight[a]	Density,[b] lb per ft³	Specific Volume[b] ft³ per lb	Specific Gravity[b] (air=1)	Heat of Combustion[c]: Btu per ft³ Gross	Btu per ft³ Net[d]	Btu per lb Gross	Btu per lb Net[d]	ft³ per ft³ of Combustible: Required for Combustion O_2	N_{2a}	Air	Flue Products CO_2	H_2O	N_{2a}	lb per lb of Combustible: Required for Combustion O_2	N_{2a}	Air	Flue Gas Products CO_2	H_2O	N_{2a}	Theor air lb/10,000 Btu
1 Carbon	C	12.0110	—	—	—	—	—	14,093	14,093	1.0	3.773	4.773	1.0	—	3.773	2.664	8.846	11.510	3.664	—	8.846	8.167
2 Hydrogen	H_2	2.0159	0.0053	187.970	0.0695	325.0	274.6	61,095	51,625	0.5	1.887	2.387	—	1.0	1.887	7.936	26.353	34.290	—	8.937	26.353	5.613
3 Oxygen	O_2	31.9988	0.0846	11.819	1.1053	—	—	—	—	—	—	—	—	—	—	—	—	—	—	—	—	—
4 Nitrogen	N_2	28.0135	0.0744	13.443	0.9717	—	—	—	—	—	—	—	—	—	—	—	—	—	—	—	—	—
4 Nitrogen (atm.)	N_{2a}	28.1610	0.0748	13.372	0.9769	—	—	—	—	—	—	—	—	—	—	—	—	—	—	—	—	—
5 Carbon Monoxide	CO	28.0104	0.0740	13.506	0.9672	321.9	321.9	4,347	4,347	0.5	1.887	2.387	1.0	—	1.887	0.571	1.897	2.468	1.571	—	1.897	5.677
6 Carbon Dioxide	CO_2	44.0098	0.1170	8.547	1.5284	—	—	—	—	—	—	—	—	—	—	—	—	—	—	—	—	—
Parafin series C_nH_{2n+2}																						
7 Methane	CH_4	16.0428	0.0424	23.574	0.5541	1013	912	23,875	21,495	2.0	7.547	9.547	1.0	2.0	7.547	3.989	13.246	17.235	2.743	2.246	13.246	7.219
8 Ethane	C_2H_6	30.0697	0.0803	12.455	1.0488	1792	1639	22,323	20,418	3.5	13.206	16.706	2.0	3.0	13.206	3.724	12.367	16.092	2.927	1.797	12.367	7.209
9 Propane	C_3H_8	44.0966	0.1196[e]	8.361[e]	1.5624	2592	2385	21,669	19,937	5.0	18.866	23.866	3.0	4.0	18.866	3.628	12.047	15.676	2.994	1.634	12.047	7.234
10 n-Butane	C_4H_{10}	58.1235	0.1582[e]	6.321[e]	2.0666	3373	3113	21,321	19,679	6.5	24.526	31.026	4.0	5.0	24.526	3.578	11.882	15.460	3.029	1.550	11.882	7.251
11 Isobutane	C_4H_{10}	58.1235	0.1582[e]	6.321[e]	2.0666	3365	3105	21,271	19,629	6.5	24.526	31.026	4.0	5.0	24.526	3.578	11.882	15.460	3.029	1.550	11.882	7.268
12 n-Pentane	C_5H_{12}	72.1504	0.1904[e]	5.252[e]	2.4872	4017	3714	21,095	19,507	8.0	30.186	38.186	5.0	6.0	30.186	3.548	11.781	15.329	3.050	1.498	11.781	7.267
13 Isopentane	C_5H_{12}	72.1504	0.1904[e]	5.252[e]	2.4872	4007	3705	21,047	19,459	8.0	30.186	38.186	5.0	6.0	30.186	3.548	11.781	15.329	3.050	1.498	11.781	7.283
14 Neopentane	C_5H_{12}	72.1504	0.1904[e]	5.252[e]	2.4872	3994	3692	20,978	19,390	8.0	30.186	38.186	5.0	6.0	30.186	3.548	11.781	15.329	3.050	1.498	11.781	7.307
15 n-Hexane	C_6H_{14}	86.1773	0.2274[e]	4.398[e]	2.9702	4767	4415	20,966	19,415	9.5	35.846	45.346	6.0	7.0	35.846	3.527	11.713	15.240	3.064	1.463	11.713	7.269
Olefin series C_nH_{2n}																						
16 Ethylene	C_2H_4	28.0538	0.0746	13.412	0.9740	1613	1512	21,636	20,275	3.0	11.320	14.320	2.0	2.0	11.320	3.422	11.362	14.784	3.138	1.284	11.362	6.833
17 Propylene	C_3H_6	42.0807	0.1110[e]	9.009	1.4500	2336	2185	21,048	19,687	4.5	16.980	21.480	3.0	3.0	16.980	3.422	11.362	14.784	3.138	1.284	11.362	7.024
18 n-Butene (Butylene)	C_4H_8	56.1076	0.1480[e]	6.757[e]	1.9333[e]	3086	2885	20,854	19,493	6.0	22.640	28.640	4.0	4.0	22.640	3.422	11.362	14.784	3.138	1.284	11.362	7.089
19 Isobutene	C_4H_8	56.1076	0.1480[e]	6.757[e]	1.9333	3069	2868	20,737	19,376	6.0	22.640	28.640	4.0	4.0	22.640	3.422	11.362	14.784	3.138	1.284	11.362	7.129
20 n-Pentene	C_5H_{10}	70.1345	0.1852[e]	5.400[e]	2.4191	3837	3585	20,720	19,359	7.5	28.300	35.800	5.0	5.0	28.300	3.422	11.362	14.784	3.138	1.284	11.362	7.135
Aromatic series C_nH_{2n-6}																						
21 Benzene	C_6H_6	78.1137	0.2060[e]	4.854[e]	2.6912[e]	3746	3595	18,184	17,451	7.5	28.300	35.800	6.0	3.0	28.300	3.072	10.201	13.274	3.380	0.692	10.201	7.300
22 Toluene	C_7H_8	92.1406	0.2431[e]	4.114[e]	3.1753[e]	4497	4296	18,501	17,672	9.0	33.959	42.959	7.0	4.0	33.959	3.125	10.378	13.504	3.343	0.782	10.378	7.299
23 Xlyene	C_8H_{10}	106.1675	0.2803	3.568[e]	3.6612[e]	5222	4970	18,633	17,734	10.5	39.619	50.119	8.0	5.0	39.619	3.164	10.508	13.673	3.316	0.848	10.508	7.338
Miscellaneous																						
24 Acetylene	C_2H_2	26.0379	0.0697	14.345	0.9106	1499	1448	21,502	20,769	2.5	9.433	11.933	2.0	1.0	9.433	3.072	10.201	13.274	3.380	0.692	10.201	6.173
25 Naphthalene	$C_{10}H_8$	128.1736	0.3384[e]	2.955[e]	4.4206[e]	5855	5654	17,303	16,707	12.0	45.279	57.279	10.0	4.0	45.279	2.995	9.947	12.943	3.434	0.562	9.947	7.480
26 Methyl alcohol	CH_3OH	32.0422	0.0846[e]	11.820	1.1052	868	767	10,258	9,066	1.5	5.660	7.160	1.0	2.0	5.660	1.498	4.974	6.472	1.373	1.124	4.974	6.309
27 Ethyl alcohol	C_2H_5OH	46.0691	0.1216[e]	8.224[e]	1.5884[e]	1600	1449	13,161	11,918	3.0	11.320	14.320	2.0	3.0	11.320	2.084	6.919	9.003	1.911	1.173	6.919	6.841
28 Ammonia	NH_3	17.0306	0.0456[e]	21.930[e]	0.5957[e]	441	364	9,667	7,986	0.75	2.830	3.580	—	1.5	3.330	1.409	4.679	6.088	—	1.587	5.502	6.298
29 Sulfur	S	32.0660	—	—	—	—	—	3,980	3,980	1.0	3.773	4.773	SO_2 1.0	—	3.773	1.000	3.320	4.320	SO_2 1.998	—	3.320	10.854
30 Hydrogen sulfide	H_2S	34.0819	0.0911	10.978[e]	1.1899[e]	646	595	7,097	6,537	1.5	5.660	7.160	SO_2 1.0	1.0	5.660	1.410	4.682	6.093	SO_2 1.880	0.529	4.682	8.585
31 Sulfur dioxide	SO_2	64.0648	0.1733	5.770	2.2640	—	—	—	—	—	—	—	—	—	—	—	—	—	—	—	—	—
32 Water vapor	H_2O	18.0153	0.0476	21.017	0.6215	—	—	—	—	—	—	—	—	—	—	—	—	—	—	—	—	—
33 Air	—	28.9660	0.0766	13.063	1.000	—	—	—	—	—	—	—	—	—	—	—	—	—	—	—	—	—

All gas volumes corrected to 60F and 30 in. Hg dry.

[a] 1987 Atomic Weights: C=12.011, H=1.00794, 0=15.9994, N=14.0067, S=32.066.

[b] Densities calculated from values given in grams per liter at 0C and 760 mm in International Critical Tables, allowing for known deviations from gas laws. Where no densities were available, the volume of the mole was taken as 22.415 liters. The ideal values for specific gravity may be found in Reference 2.

[c] For gases saturated with water at 60F, 1.74% of the Btu value must be deducted. Rossini, F.D. and others[3]

[d] Correction from gross to net heating value determined by deducting 1059.7 Btu/lb of water in products of combustion. ASME Steam Tables, 1983[4]

[e] Either the density or the coefficient of expansion has been assumed. Some of the materials cannot exist as gases at 60F and 30 in. Hg, in which case the values are theoretical ones. Under the actual concentrations in which these materials are present, their partial pressure is low enough to keep them as gases.

portionality constant is the same for one mole of any ideal gas, so this law may be expressed as:

$$v_M = \frac{\boldsymbol{R}T}{P} \qquad (1)$$

where:

v_M = volume, ft³/mole
P = absolute pressure, lb/ft²
T = absolute temperature, R = F + 460
$\boldsymbol{R}$ = universal gas constant, 1545 ft lb/mole R

Most gases involved in combustion calculations can be approximated as ideal gases.

Law of combining weights

This law states that all substances combine in accordance with simple, definite weight relationships. These relationships are exactly proportional to the molecular weights of the constituents. For example, carbon (molecular weight = 12) combines with oxygen (molecular weight = 32) to form carbon dioxide (molecular weight = 44) so that 12 lb of C and 32 lb of O_2 unite to form 44 lb of CO_2. (See *Application of Fundamental Laws* below.)

Avogadro's law

Avogadro determined that equal volumes of different gases at the same pressure and temperature contain the same number of molecules. From the concept of the mole, a pound mole of any substance contains a mass equal to the molecular weight of the substance. Therefore, the ratio of mole weight to molecular weight is a constant and a mole of any chemically pure substance contains the same number of molecules. Because a mole of any ideal gas occupies the same volume at a given pressure and temperature (ideal gas law), equal volumes of different gases at the same pressure and temperature contain the same number of molecules.

Dalton's law

This law states that the total pressure of a mixture of gases is the sum of the partial pressures which would be exerted by each of the constituents if each gas were to occupy alone the same volume as the mixture. Consider equal volumes V of three gases (a, b and c), all at the same temperature T but at different pressures (P_a, P_b and P_c). When all three gases are placed in the space of the same volume V, then the resulting pressure P is equal to $P_a + P_b + P_c$. Each gas in a mixture fills the entire volume and exerts a pressure independent of the other gases.

Amagat's law

Amagat determined that the total volume occupied by a mixture of gases is equal to the sum of the volumes which would be occupied by each of the constituents when at the same pressure and temperature as the mixture. This law is related to Dalton's law, but it considers the additive effects of volume instead of pressure. If all three gases are at pressure P and temperature T but at volumes V_a, V_b and V_c, then, when combined so that T and P are unchanged, the volume of the mixture V equals $V_a + V_b + V_c$.

Application of fundamental laws

Table 2 summarizes the molecular and weight relationships between fuel and oxygen for constituents commonly involved in combustion. The heat of combustion for each constituent is also tabulated. Most of the weight and volume relationships in combustion calculations can be determined by using the information presented in Table 2 and the seven fundamental laws.

The combustion process for C and H_2 can be expressed as follows:

C	+	O_2	=	CO_2
1 molecule	+	1 molecule	→	1 molecule
1 mole	+	1 mole	=	1 mole
(See Note below)	+	1 ft³	→	1 ft³
12 lb	+	32 lb	=	44 lb

$2H_2$	+	O_2	=	$2H_2O$
2 molecules	+	1 molecule	→	2 molecules
2 moles	+	1 mole	=	2 moles
2 ft³	+	1 ft³	→	2 ft³
4 lb	+	32 lb	=	36 lb

Note: When 1 ft³ of oxygen (O_2) combines with carbon (C) it forms 1 ft³ of carbon dioxide (CO_2). If carbon were an ideal gas instead of a solid, 1 ft³ of carbon would be required.

It is important to note that there is a mass or weight balance according to the law of combining weights but there is not necessarily a molecular or volume balance.

Table 2
Common Chemical Reactions of Combustion

Combustible	Reaction	Moles	Mass or weight, lb	Heat of Combustion (High) Btu/lb of Fuel
Carbon (to CO)	$2C + O_2 = 2CO$	2 + 1 = 2	24 + 32 = 56	3,950
Carbon (to CO_2)	$C + O_2 = CO_2$	1 + 1 = 1	12 + 32 = 44	14,093
Carbon monoxide	$2CO + O_2 = 2CO_2$	2 + 1 = 2	56 + 32 = 88	4,347
Hydrogen	$2H_2 + O_2 = 2H_2O$	2 + 1 = 2	4 + 32 = 36	61,095
Sulfur (to SO_2)	$S + O_2 = SO_2$	1 + 1 = 1	32 + 32 = 64	3,980
Methane	$CH_4 + 2O_2 = CO_2 + 2H_2O$	1 + 2 = 1 + 2	16 + 64 = 80	23,875
Acetylene	$2C_2H_2 + 5O_2 = 4CO_2 + 2H_2O$	2 + 5 = 4 + 2	52 + 160 = 212	21,502
Ethylene	$C_2H_4 + 3O_2 = 2CO_2 + 2H_2O$	1 + 3 = 2 + 2	28 + 96 = 124	21,636
Ethane	$2C_2H_6 + 7O_2 = 4CO_2 + 6H_2O$	2 + 7 = 4 + 6	60 + 224 = 284	22,323
Hydrogen sulfide	$2H_2S + 3O_2 = 2SO_2 + 2H_2O$	2 + 3 = 2 + 2	68 + 96 = 164	7,097

Molar evaluation of combustion

Gaseous fuel

Molar calculations have a simple and direct application to gaseous fuels, where the analyses are usually reported on a percent by volume basis. Consider the following fuel analysis:

Fuel Gas Analysis, % by Volume	
CH_4	85.3
C_2H_6	12.6
CO_2	0.1
N_2	1.7
O_2	0.3
Total	100.0

The mole fraction of a component in a mixture is the number of moles of that component divided by the total number of moles of all components in the mixture. Because a mole of every ideal gas occupies the same volume, by Avogadro's Law, the mole fraction of a component in a mixture of ideal gases equals the volume fraction of that component.

$$\frac{\text{Moles of component}}{\text{Total moles}} = \frac{\text{Volume of component}}{\text{Volume of mixture}} \quad \textbf{(2)}$$

This is a valuable concept because the volumetric analysis of a gaseous mixture automatically gives the mole fractions of the components.

Accordingly, the previous fuel analysis may be expressed as 85.3 moles of CH_4 per 100 moles of fuel, 12.6 moles of C_2H_6 per 100 moles of fuel, etc.

The elemental breakdown of each constituent may also be expressed in moles per 100 moles of fuel as follows:

C in CH_4	=	85.3 × 1	=	85.3 moles
C in C_2H_6	=	12.6 × 2	=	25.2 moles
C in CO_2	=	0.1 × 1	=	0.1 moles
Total C per 100 moles fuel			=	110.6 moles
H_2 in CH_4	=	85.3 × 2	=	170.6 moles
H_2 in C_2H_6	=	12.6 × 3	=	37.8 moles
Total H_2 per 100 moles fuel			=	208.4 moles
O_2 in CO_2	=	0.1 × 1	=	0.1 moles
O_2 as O_2	=	0.3 × 1	=	0.3 moles
Total O_2 per 100 moles fuel			=	0.4 moles
Total N_2 per 100 moles fuel			=	1.7 moles

The oxygen/air requirements and products of combustion can now be calculated for each constituent on an elemental basis. These requirements can also be calculated directly using Table 1. Converting the gaseous constituents to an elemental basis has two advantages. It provides a better understanding of the combustion process and provides a means for determining the elemental fuel analysis on a mass basis. This is boiler industry standard practice and is convenient for determining a composite fuel analysis when gaseous and solid/liquid fuels are fired in combination.

The following tabulation demonstrates the conversion of the gaseous fuel constituents on a moles/100 moles gas basis to a lb/100 lb gas (percent mass) basis.

Constituent	Moles/ 100 Moles		Mol Wt lb/ Mole		lb/ 100 Moles		lb/ 100 lb
C	110.6	×	12.011	=	1328.4	/1808.9 × 100 =	73.5
H_2	208.4	×	2.016	=	420.1	/1808.9 × 100 =	23.2
O_2	0.4	×	31.999	=	12.8	/1808.9 × 100 =	0.7
N_2	1.7	×	28.013	=	47.6	/1808.9 × 100 =	2.6
Total					1808.9		100.0

Solid/liquid fuel

The ultimate analysis of solid and liquid fuels is determined on a percent mass basis. The mass analysis is converted to a molar basis by dividing the mass fraction of each elemental constituent by its molecular weight.

$$\frac{\dfrac{\text{lb Constituent}}{\text{100 lb Fuel}}}{\dfrac{\text{lb Constituent}}{\text{Mole constituent}}} = \frac{\text{Mole constituent}}{\text{100 lb Fuel}} \quad \textbf{(3)}$$

The calculation is illustrated in Table 3.

The products of combustion and moles of oxygen required for each combustible constituent are shown. Note that when a fuel contains oxygen, the amount of theoretical O_2/air required for combustion is reduced (as designated by the brackets).

Composition of air

So far, combustion has been considered only as a process involving fuel and oxygen. For normal combustion and steam generator applications, the source of oxygen is air. Atmospheric air is composed of oxygen, nitrogen and other minor gases. The calculations and derivation of constants which follow in this text are based upon a U.S. standard atmosphere[5] composed of 0.20947 O_2, 0.78086 N_2, 0.00934 argon (Ar) and 0.00033 CO_2 moles per mole of dry air, which has an average molecular weight of 28.966. To simplify the calculations, N_2 includes argon and other trace elements; it is referred to as atmospheric nitrogen (N_{2a}) having an equivalent molecular weight of 28.161. (See Table 4.)

Air normally contains some moisture. As standard practice, the American Boiler Manufacturers Association (ABMA) considers moisture content to be 0.013 lb water /lb dry air, which corresponds to approximately 60% relative humidity at 80F. For combustion calculations on a molar basis, multiply the mass basis moisture by 1.608 (molecular weight of air divided by molecular weight of water). Therefore, 0.013 lb water/lb dry air becomes 0.0209 moles water/mole dry air.

Table 3
Calculation of Combustion Products and Theoretical Oxygen Requirements — Molar Basis

Fuel constituent (1)	% by wt (2)	Molecular weight (3)		Moles/100 lb fuel (2 ÷ 3) (4)	Combustion product (5)	Moles theoretical O_2 required (6)
C	72.0	12.011	=	5.995	CO_2	5.995
H_2	4.4	2.016	=	2.183	H_2O	1.091*
S	1.6	32.066	=	0.050	SO_2	0.050
O_2	3.6	31.999	=	0.113		(0.113)
N_2	1.4	28.013	=	0.050	N_2	0.000
H_2O	8.0	18.015	=	0.444	H_2O	0.000
Ash	9.0					
Total	100.0			8.835		7.023

* Column 6 is based upon moles of oxygen as O_2 needed for combustion. Therefore, the moles of H_2O need to be divided by 2 to obtain equivalent moles of O_2.

Table 4
Air Composition

	Composition of Dry Air % by vol	Composition of Dry Air % by wt
Oxygen, O_2	20.95	23.14
Atmospheric nitrogen, N_{2a}	79.05	76.86

The moisture content in air is normally determined from wet and dry bulb temperatures or from relative humidity using a psychrometric chart, as shown in Fig. 1. Air moisture may also be calculated from:

$$MFWA = 0.622 \times \frac{P_v}{(P_b - P_v)} \qquad \textbf{(4)}$$

where

$MFWA$ = moisture content in air, lb/lb dry air
P_b = barometric pressure, psi
P_v = partial pressure of water vapor in air, psi = $0.01\,(RH)\,(P_{vd})$, psi
P_{vd} = saturation pressure of water vapor at dry bulb temperature, psi
RH = relative humidity, %

P_v may also be calculated from Carrier's equation:

$$P_v = P_{vw} - \frac{(P_b - P_{vw})(T_d - T_w)}{2830 - (1.44 T_w)} \qquad \textbf{(5)}$$

where

T_d = dry bulb temperature, F
T_w = wet bulb temperature, F
P_{vw} = saturation pressure of water vapor at wet bulb temperature, psi

The following constants, with values from Table 4, are frequently used in combustion calculations:

moles air/mole O_2 or ft^3 air/ft^3 O_2	=	$\frac{100}{20.95}$	=	4.77
moles N_{2a}/mole O_2	=	$\frac{79.05}{20.95}$	=	3.77
lb air (dry)/lb O_2	=	$\frac{100}{23.14}$	=	4.32
lb N_{2a}/lb O_2	=	$\frac{76.86}{23.14}$	=	3.32

The calculations in Table 2 can be converted to combustion with air rather than oxygen by adding 3.77 moles of N_{2a}/mole of O_2 to the left and right side of each equation. For example, the combustion of carbon monoxide (CO) in air becomes:

$$2\,CO + O_2 + 3.77\,N_{2a} = 2CO_2 + 3.77\,N_{2a}$$

or for methane, CH_4:

$$CH_4 + 2O_2 + 2(3.77)\,N_{2a} = CO_2 + 2H_2O + 7.54\,N_{2a}$$

Theoretical air requirement

Theoretical air is the minimum air required for complete combustion of the fuel, i.e., the oxidation of carbon to CO_2, hydrogen to water vapor (H_2O) and sulfur to sulfur dioxide. In the combustion process, small amounts of sulfur trioxide (SO_3), nitrogen oxides (NO_x), unburned hydrocarbons and other minor species may be formed. While these may be of concern as pollutants, their impact is negligible with regard to the quantity of air and combustion products and, therefore, are not normally considered in these calculations.

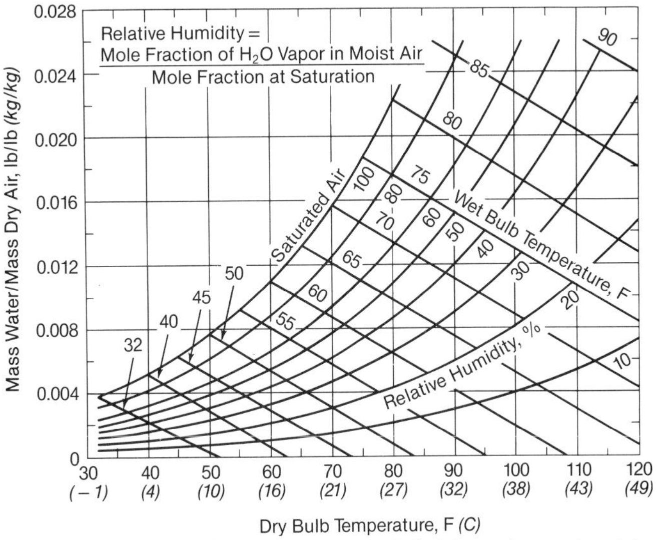

Fig. 1 Psychrometric chart — water content of air for various wet and dry bulb temperatures.

In practice, it is necessary to use more than the theoretical amount of air to assure complete combustion of the fuel. For the example shown in Table 3, consider completing the combustion calculations on a molar basis using 20% excess air. These calculations are summarized in Table 5.

Now consider the portion of the combustion products attributable to the air. The oxygen in the theoretical air is already accounted for in the products of combustion: CO_2, H_2O (from the combustion of hydrogen) and SO_2. That leaves N_{2a} in the theoretical air, N_{2a} in the excess air, O_2 in the excess air and H_2O in air (as calculated in Table 5) as the products in the combustion gas attributable to the wet combustion air. These constituents are in addition to the combustion products from fuel shown in Table 3.

Products of combustion — mass/mass fuel basis

Table 6 shows a tabulation of the flue gas products and combustion air on a molar (or volumetric) basis and the conversion to a mass basis (wet and dry). The products of combustion calculated on a molar basis in Tables 3 and 5 are itemized in column A. The moisture (H_2O) sources are separated from the dry products for convenience of calculating the flue gas composition on a wet and dry basis.

The water products shown in column A are from the combustion of hydrogen in the fuel, from moisture in the fuel and from moisture in the air. The N_{2a} is the sum of nitrogen in the theoretical air plus the nitrogen in the excess air. The N_{2a} is tabulated separately from the elemental nitrogen in the fuel to differentiate the molecular weight of the two. In practice, the nitrogen in the fuel is normally small with respect to the N_{2a} and can be included with the nitrogen in air. For manufactured gases that are formed

Table 5
Calculation of Wet Air Requirements for Combustion — Molar Basis

Line No.	Description	Source	Quantity (mole/ 100 lb fuel)
1	Theoretical combustion O_2	From Table 3	7.023
2	Molar fraction O_2 in dry air	O_2 Vol fraction from Table 4	0.2095
3	Theoretical dry combustion air	Line 1/Line 2	33.523
4	Excess air at 20%	Line 3 × 0.20	6.705
5	Total dry combustion air	Line 3 + Line 4	40.228
6	Molar fraction of H_2O in dry air	*	0.0209
7	H_2O in total dry air	Line 5 × Line 6	0.841
8	Molar fraction of N_{2a} in dry air	N_{2a} Vol fraction, from Table 4	0.7905
9	N_{2a} in theoretical dry air	Line 3 × Line 8	26.500
10	N_{2a} in dry excess air	Line 4 × Line 8	5.300
11	O_2 in dry excess air	Line 2 × Line 4	1.405

*Standard combustion air: 80F, 60% relative humidity; 0.013 lb H_2O/lb dry air; 0.0209 moles H_2O/mole dry air.

when combustible products oxidize with air (blast furnace gas, for example), the nitrogen in the fuel is predominately atmospheric nitrogen.

Flue gas products are normally measured on a volumetric basis. If the sample includes water products, it is measured on a *wet basis*, typical of in situ analyzers. Conversely, if water products are excluded, measurements are done on a *dry basis*, which is typical of extractive gas sample systems. (See *Flue Gas Analysis*.) Note that the flue gas products are summed on a dry and wet basis to facilitate calculation of the flue gas constituents on a dry and wet percent by volume basis in columns B and C. The molecular weight of each constituent is given in column D. Finally, the mass of each constituent on a lb/100 lb fuel basis is the product of the moles/100 lb fuel and the molecular weight.

The calculation of the mass of air on a lb/100 lb fuel basis, shown at the bottom of Table 6, follows the same principles as the flue gas calculations.

For most engineering calculations, it is common U.S. practice to work with air and flue gas (combustion products) on a mass basis. It is usually more convenient to calculate these products on a mass basis directly as discussed later. The mole method described above is the fundamental basis for understanding and calculating the chemical reactions. It is also the basis for deriving certain equations that are presented later. For those who prefer using the mole method, Table 7 presents this method in a convenient calculation format.

Alternate units — Btu method

It is customary within the U.S. boiler industry to use units of mass rather than moles for expressing the quantity of air and flue gas. This is especially true for heat transfer calculations, where the quantity of the working fluid (usually steam or water) is expressed on a mass basis and the enthalpy of the hot and cold fluids is traditionally expressed on a Btu/lb basis. Therefore, if the combustion calculations are performed on the mole basis, it is customary to convert the results to lb/100 lb.

Items that are expressed on a unit of fuel basis (mole/ 100 lb fuel, mass/mass fuel, etc.) can be normalized by using an input from fuel basis. For example, knowing that a coal has 10% ash only partly defines the fuel. For a 10,000 Btu/lb fuel, there are 10 lb ash per million Btu input, but for a 5,000 Btu/lb fuel there would be 20 lb ash per million Btu input. Considering that fuel input for a given boiler load does not vary significantly with heating value, a boiler firing the lower heating value fuel would encounter approximately twice the amount of ash.

The mass per unit input concept is valuable when determining the impact of different fuels on combustion calculations. This method is particularly helpful in theoretical air calculations. Referring to Table 8, in the first column, theoretical air has been tabulated for various fuels on a mass per mass of fuel basis. The resulting values have little significance when comparing the various fuels. However, when the theoretical air is converted to a mass per unit heat input from fuel basis, the theoretical air varies little between fuels. The common units are lb/10,000 Btu, abbreviated as lb/10KB. The fuel labeled

Table 6
Calculation of Flue Gas and Air Quantities — Mass Basis

		Flue Gas or Combustion Product				
	Constituent	A (From Tables 3 and 5) Moles/100 lb Fuel	B (A/A6) % Vol dry	C (A/A11) % Vol wet	D (From Table 1) Molecular weight	E (A × D) lb/100 lb Fuel
1	CO_2	5.995	15.25	14.02	44.010	263.8
2	SO_2	0.050	0.13	0.115	64.065	3.2
3	N_2 (fuel)	0.050	0.13	0.115	28.013	1.4
4	N_{2a} (air)	31.800 (26.500 + 5.300)	80.92	74.35	28.161	895.5
5	O_2	1.405	3.57	3.29	31.999	45.0
6	Total, dry combustion products	39.300	100.00			1208.9
7	H_2O combustion	2.183				
8	H_2O fuel	+0.444				
9	H_2O air	+0.841				
10	Total H_2O	3.468		8.11	18.015	62.5
11	Total, wet combustion products	42.768		100.00		1271.4
Air						
12	Dry air	40.228			28.966	1165.3
13	H_2O	0.843			18.015	15.2
14	Total wet air					1180.5

Table 7
Combustion Calculations - Molar Basis

	INPUTS (see also lightly shaded blocks)			FUEL - *Bituminous coal, Virginia*	
1	Excess air: at burner/at boiler/econ, %	*20/20*	4	Fuel input, 1,000,000 Btu/h	*330.0*
2	Moisture in air, lb/lb dry air	*0.013*	5	Unburned carbon loss, % efficiency	*0.40*
3	Fuel heating value, Btu/lb	*14,100*	6	Unburned carbon (UBC), [5]x[3]/14,500	*0.39*

COMBUSTION PRODUCTS CALCULATIONS

7	Ultimate Analysis, % Mass			8 Molecular Weight lb/mole	9 Moles /100 lb Fuel [7] / [8]	10 Moles O_2 /Mole Fuel Constituent	11 Moles Theo. O_2/100 lb Fuel [9] x [10]	12 Combustion Product
	Fuel Constituent	As-Fired	Carbon Burned (CB)					
A	C	*80.31*	*80.31*					
B	UBC [6]		*0.39*					
C	CB [A] – [B]		*79.92*	12.011	*6.654*	1.0	*6.654*	CO_2
D	S	*1.54*		32.066	*0.048*	1.0	*0.048*	SO_2
E	H_2	*4.47*		2.016	*2.217*	0.5	*1.109*	H_2O
F	H_2O	*2.90*		18.015	*0.161*			H_2O
G	N_2	*1.38*		28.013	*0.049*			N_2 (fuel)
H	O_2	*2.85*		31.999	*0.089*	– 1.0	*– 0.089*	
I	Ash	*6.55*						
K	Total	*100.00*			*9.218*		*7.722*	

	AIR CONSTITUENTS, Moles/100 lb Fuel		At Burner	At Blr/Econ
13	O_2-excess	[11K] x [1] / 100	*1.544*	*1.544*
14	O_2-total	[13] + [11K]	*9.266*	*9.266*
15	N_{2a}-air	[14] x 3.77	*34.933*	*34.933*
16	Air (dry)	[14] + [15]	*44.199*	*44.199*
17	H_2O-air	[16] x [2] x 1.608	*0.924*	*0.924*
18	Air (wet)	[16] + [17]	*45.123*	*45.123*

	FLUE GAS CONSTITUENTS		19 Moles /100 lb Fuel	20 Vol % Dry 100 x [19] / [19G]	21 Vol % Wet 100 x [19] / [19H]	22 Molecular Weight lb/mole	23 Flue Gas lb/100 lb Fuel [19] x [22]
A	CO_2	[9C]	*6.654*	*15.39*	*14.30*	44.010	*292.8*
B	SO_2	[9D]	*0.048*	*0.11*	*0.10*	64.065	*3.1*
C	O_2	[13]	*1.544*	*3.57*	*3.32*	31.999	*49.4*
D	N_2 (fuel)	[9G]	*0.049*	*0.11*	*0.11*	28.013	*1.4*
E	N_{2a} (air)	[15]	*34.933*	*80.82*	*75.07*	28.161	*983.7*
F	H_2O	[9E] + [9F] + [17]	*3.302*		*7.10*	18.015	*59.5*
G	Total dry	Sum [A] through [E]	*43.228*	*100.00*			*1330.4*
H	Total wet	Sum [A] through [F]	*46.530*		*100.00*		*1389.9*

	KEY PERFORMANCE PARAMETERS		At Burner	At Blr/Econ
24	Molecular weight wet flue gas, lb/mole	[23H] / [19H]		*29.871*
25	H_2O in wet gas, % by wt	100 x [23F] / [23H]		*4.28*
26	Dry gas weight, lb/10,000 Btu	100 x [23G] / [3]		*9.435*
27	Wet gas weight, lb/10,000 Btu	100 x [23H] / [3]		*9.857*
28	Wet gas weight, 1000 lb/h	[27] x [4] / 10		*325.3*
29	Air flow (wet), lb/100 lb fuel	[16] x 28.966 + [17] x 18.015	*1296.9*	
30	Air flow (wet), lb/10,000 Btu	100 x [29] / [3]	*9.198*	
31	Air flow (wet), 1000 lb/h	[30] x [4] / 10	*303.5*	

Table 8
Theoretical Air Required for Various Fuels

Fuel	Theoretical Air, lb/lb Fuel	HHV Btu/lb	Theoretical Air Typical lb/ 10^4 Btu	Theoretical Air Range lb/ 10^4 Btu
Bituminous coal (VM* >30%)	9.07	12,000	7.56	7.35 to 7.75
Subbituminous coal (VM* >30%)	6.05	8,000	7.56	7.35 to 7.75
Oil	13.69	18,400	7.46	7.35 to 7.55
Natural gas	15.74	21,800	7.22	7.15 to 7.35
Wood	3.94	5,831	6.75	6.60 to 6.90
MSW* and RDF*	4.13	5,500	7.50	7.20 to 7.80
Carbon	11.50	14,093	8.16	--
Hydrogen	34.28	61,100	5.61	--

*VM = volatile matter, moisture and ash free basis
MSW = municipal solid waste
RDF = refuse-derived fuel

MSW/RFD refers to municipal solid waste and refuse derived fuel. Note that the theoretical air is in the same range as that for fossil fuels on a heat input basis. Carbon and hydrogen, the principal combustible fuel elements, are shown for reference. Note that the coals listed in the table are limited to those with a volatile matter (moisture and ash free) greater than 30%. As volatile matter decreases, the carbon content increases and requires more excess air. To check the expected theoretical air for low volatile coals, refer to Fig. 2. The theoretical air of all coals should fall within plus or minus 0.2 lb/ 10,000 Btu of this curve. Table 9 provides the fuel analysis and theoretical air requirements for a typical fuel oil and natural gas.

Heat of combustion

In a boiler furnace (where no mechanical work is done), the heat energy evolved from combining combustible elements with oxygen depends on the ultimate products of combustion; it does not rely on any intermediate combinations that may occur.

For example, 1 lb of carbon reacts with oxygen to produce about 14,100 Btu of heat (refer to Table 2.) The reaction may occur in one step to form CO_2 or, under certain conditions, it may take two steps. In the multi-step process, CO is first formed, producing only 3960 Btu per pound of carbon. In the second step, the CO joins with additional oxygen to form CO_2, releasing 10,140 Btu per pound of carbon. The total heat produced is again 14,100 Btu per pound of carbon.

Measurement of heat of combustion

In boiler practice, a fuel's *heat of combustion* is the amount of energy, expressed in Btu, generated by the complete combustion, or oxidation, of a unit weight of fuel. *Calorific value, fuel Btu value* and *heating value* are terms also used.

The amount of heat generated by complete combustion is a constant for a given combination of combustible elements and compounds. It is not affected by the manner in which the combustion takes place, provided it is complete.

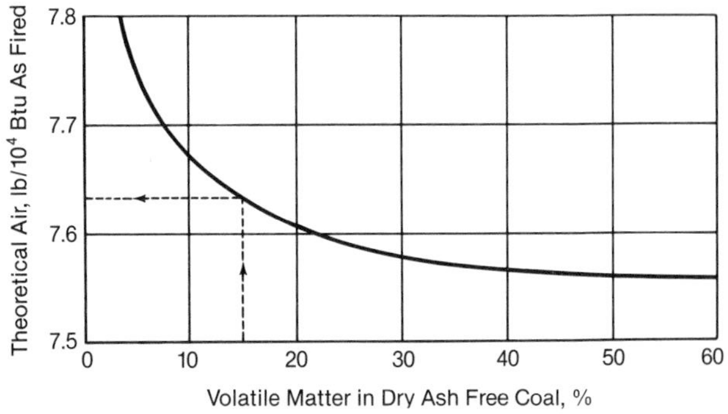

Fig. 2 Theoretical air in lb/10,000 Btu heating value of coal with a range of volatile matter.

A fuel's heat of combustion is usually determined by direct calorimeter measurement of the heat evolved. Combustion products within a calorimeter are cooled to the initial temperature, and the heat absorbed by the cooling medium is measured to determine the higher, or gross, heat of combustion (typically referred to as the higher heating value, or HHV).

For all solid and most liquid fuels, the bomb type calorimeter is the industry standard measurement device. In these units, combustible substances are burned in a constant volume of oxygen. When they are properly operated, combustion is complete and all of the heat generated is absorbed and measured. Heat from external sources can be excluded or proper corrections can be applied.

For gaseous fuels of 900 to 1200 Btu/ft^3, continuous or constant flow type calorimeters are industry standards. The principle of operation is the same as for the bomb calorimeter; however, the heat content is determined at constant pressure rather than at constant volume.

For most fuels, the difference between the constant pressure and constant volume heating values is small and is usually neglected. Because fuel is burned under

Table 9
Fuel Analysis and Theoretical Air for Typical Oil and Gas Fuels

Heavy Fuel Oil, % by wt		Natural Gas, % by vol	
S	1.16	CH_4	85.3
H_2	10.33	C_2H_6	12.6
C	87.87	CO_2	0.1
N_2	0.14	N_2	1.7
O_2	0.50	O_2	0.3
		Sp Gr	0.626
		Btu/ft^3, as-fired	1090
Btu/lb, as-fired	18,400	Btu/lb, as-fired	22,379

Theoretical Air, Fuel and Moisture

Theoretical air, lb/10,000 Btu	7.437	Theoretical air, lb/10,000 Btu	7.206
Fuel, lb/ 10,000 Btu	0.543	Fuel, lb/ 10,000 Btu	0.440
Moisture, lb/ 10,000 Btu	0.502	Moisture, lb/ 10,000 Btu	0.912

essentially constant pressure conditions, the constant pressure result is the technically correct value. See *Higher and Lower Heating Values*.

Gas chromatography is also commonly used to determine the composition of gaseous fuels. When the composition of a gas mixture is known, its heat of combustion may be determined as follows:

$$hc_{mix} = v_a hc_a + v_b hc_b + \ldots + v_x hc_x \quad \textbf{(6)}$$

where

hc_{mix} = heat of combustion of the mixture
v_x = volume fraction of each component
hc_x = heat of combustion of each component

For an accurate heating value of solid and liquid fuels, a laboratory heating value analysis is required. Numerous empirical methods have been published for estimating the heating value of coal based on the proximate or ultimate analyses. (See Chapter 8.) One of the most frequently used correlations is Dulong's formula which gives reasonably accurate results for bituminous coals (within 2 to 3%). It is often used as a routine check of calorimeter-determined values.

$$HHV = 14{,}544\,C + 62{,}028\,[H_2 - (O_2/8)] + 4050\,S \quad \textbf{(7)}$$

where

HHV = higher heating value, Btu/lb
C = mass fraction carbon
H_2 = mass fraction hydrogen
O_2 = mass fraction oxygen
S = mass fraction sulfur

A far superior method for checking whether the heating value is reasonable in relation to the ultimate analysis is to determine the theoretical air on a mass per Btu basis. (See *Alternate Units — Btu Method*.) Table 8 indicates the range of theoretical air values. The equation for theoretical air can be rearranged to calculate the higher heating value, *HHV*, where the median range for theoretical air for the fuel from Table 8, m_{air}, is used:

$$HHV = 100 \times \frac{11.51\,C + 34.29\,H_2 + 4.32\,S - 4.32\,O_2}{m_{air}} \quad \textbf{(8)}$$

where

HHV = higher heating value, Btu/lb
C = mass percent carbon, %
H_2 = mass percent hydrogen, %
S = mass percent sulfur, %
O_2 = mass percent oxygen, %
m_{air} = theoretical air, lb/10,000 Btu

Higher and lower heating values

Water vapor is a product of combustion for all fuels that contain hydrogen. The heat content of a fuel depends on whether this vapor remains in the vapor state or is condensed to liquid. In the bomb calorimeter, the products of combustion are cooled to the initial temperature and all of the water vapor formed during combustion is condensed to liquid. This gives the *HHV* or gross calorific value (defined earlier) of the fuel, and the heat of vaporization of water is included in the reported value.

For the lower heating value (*LHV*) or net calorific value (net heat of combustion at constant pressure), all products of combustion including water are assumed to remain in the gaseous state, and the water heat of vaporization is not available.

While the high, or gross, heat of combustion can be accurately determined by established [American Society of Testing and Materials (ASTM)] procedures, direct determination of the low heat of combustion is difficult. Therefore, it is usually calculated using the following formula:

$$LHV = HHV - 10.30\,(H_2 \times 8.94) \quad \textbf{(9)}$$

where:

LHV = lower heating or net calorific value, Btu/lb
HHV = higher heating or gross calorific value, Btu/lb
H_2 = mass percent hydrogen in the fuel, %

This calculation contains a correction for the difference between a constant volume and a constant pressure process and a deduction for the water vaporization in the combustion products. At 68F, the total deduction is 1030 Btu/lb of water, including 1055 Btu/lb for the enthalpy of water vaporization.

Ignition temperatures

Ignition temperatures of combustible substances vary greatly, as indicated in Table 10. This table lists minimum temperatures and temperature ranges in air for fuels and for the combustible constituents of fuels commonly used in the commercial generation of heat. Many factors influence ignition temperature, so any tabulation can be used only as a guide. Pressure, velocity, enclosure configuration, catalytic materials, air-fuel mixture uniformity and ignition source are examples of the variables. Ignition temperature usually decreases with rising pressure and increases with increasing air moisture content.

The ignition temperatures of coal gases vary considerably and are appreciably higher than those of the fixed

Table 10
Ignition Temperatures of Fuels in Air (Approximate Values or Ranges at Atmospheric Pressure)

Combustible	Formula	Temperature, F
Sulfur	S	470
Charcoal	C	650
Fixed carbon (bituminous coal)	C	765
Fixed carbon (semi-anthracite)	C	870
Fixed carbon (anthracite)	C	840 to 1115
Acetylene	C_2H_2	580 to 825
Ethane	C_2H_6	880 to 1165
Ethylene	C_2H_4	900 to 1020
Hydrogen	H_2	1065 to 1095
Methane	CH_4	1170 to 1380
Carbon monoxide	CO	1130 to 1215
Kerosene	--	490 to 560
Gasoline	--	500 to 800

carbon in the coal. However, the ignition temperature of coal may be considered as the ignition temperature of its fixed carbon content, because the gaseous constituents are usually distilled off, but not ignited, before this temperature is attained.

Adiabatic flame temperature

The *adiabatic flame temperature* is the maximum theoretical temperature that can be reached by the products of combustion of a specific fuel and air (or oxygen) combination, assuming no loss of heat to the surroundings and no dissociation. The fuel's heat of combustion is the major factor in the flame temperature, but increasing the temperature of the air or the fuel also raises the flame temperature. This adiabatic temperature is a maximum with zero excess air (only enough air chemically required to combine with the fuel). Excess air is not involved in the combustion process; it only acts as a dilutant and reduces the average temperature of the products of combustion.

The adiabatic temperature is determined from the adiabatic enthalpy of the flue gas:

$$H_g = \frac{HHV - \text{Latent heat } H_2O + \text{Sensible heat in air}}{\text{Wet gas weight}} \qquad \textbf{(10)}$$

where

H_g = adiabatic enthalpy, Btu/lb

Knowing the moisture content and enthalpy of the products of combustion, the theoretical flame or gas temperature can be obtained from Fig. 3 (see pages 12 and 13).

The adiabatic temperature is a fictitiously high value that can not exist. Actual flame temperatures are lower for two main reasons:

1. Combustion is not instantaneous. Some heat is lost to the surroundings as combustion takes place. Faster combustion reduces heat loss. However, if combustion is slow enough, the gases may be cooled sufficiently and incomplete combustion may occur, i.e., some of the fuel may remain unburned.
2. At temperatures above 3000F, some of the CO_2 and H_2O in the flue gases dissociates, absorbing heat in the process. At 3500F, about 10% of the CO_2 in a typical flue gas dissociates to CO and O_2. Heat absorption occurs at 4345 Btu/lb of CO formed, and about 3% of the H_2O dissociates to H_2 and O_2, with a heat absorption of 61,100 Btu/lb of H_2 formed. As the gas cools, the dissociated CO and H_2 recombine with the O_2 and liberate the heat absorbed in dissociation, so the heat is not lost. However, the overall effect is to lower the maximum actual flame temperature.

The term *heat available* (Btu/h) is used throughout this text to define the heat available to the furnace. This term is analogous to the energy term in the adiabatic sensible heat equation above except that one half of the radiation heat loss and manufacturer's margin are not considered available to the furnace.

Commercial combustion application issues

In addition to the theoretical combustion evaluation methodologies addressed above, several application issues are very important in accurate combustion calculations of commercial applications. These include the impact of injection of SO_2 sorbents and other chemicals into the combustion process, solid ash or residue, unburned carbon and excess air.

Sorbents and other chemical additives

In some combustion systems, chemical compounds are added to the gas side of the steam generator to reduce emissions. For example, limestone is used universally in fluidized-bed steam generators to reduce SO_2 emissions. (See Chapter 16.)

Limestone impacts the combustion and efficiency calculations by: 1) altering the mass of flue gas by reducing SO_2 and increasing CO_2 levels, 2) increasing the mass of solid waste material (ash residue), 3) increasing the air required in forming SO_3 to produce calcium sulfate, $CaSO_4$, 4) absorbing energy (heat) from the fuel to calcine the calcium and magnesium carbonates, and 5) adding energy to the system in the sulfation reaction ($SO_2 + {}^1/_2\ O_2 + CaO \rightarrow CaSO_4$). The impact of sorbent/limestone is shown as a correction to the normal combustion calculations presented later.

The limestone constituents that are required in the combustion and efficiency calculations are:

Reactive constituents:
  Calcium carbonate ($CaCO_3$)
  Magnesium carbonate ($MgCO_3$)
Water
Inerts

Some processes may use sorbents derived from limestone. These sorbents contain reactive constituents such as calcium hydroxide [$Ca(OH)_2$] and magnesium hydroxide [$Mg(OH)_2$].

For design purposes, the amount of sorbent is determined from the design calcium to sulfur molar ratio, *MOFCAS*. The sorbent to fuel ratio, *MFSBF*, is a convenient equation that converts sorbent products to a mass of fuel or input from fuel basis.

$$MFSBF = \frac{MOFCAS \times S}{MOPCA \times 32.066} \qquad \textbf{(11)}$$

and

$$MOPCA = \left(\frac{CaCO_3}{100.089} + \frac{Ca(OH)_2}{74.096} \right) \qquad \textbf{(12)}$$

where

MFSBF = mass ratio of sorbent to fuel, lb/lb
MOPCA = calcium in sorbent molar basis, moles/100 lb sorbent
S = mass percent sulfur in fuel, %
$CaCO_3$ = mass percent calcium carbonate in sorbent, %
$Ca(OH)_2$ = mass percent calcium hydroxide in sorbent, %

When calcium carbonate and magnesium carbonate are heated, they release CO_2, which adds to the flue gas products. This is referred to as *calcination*. Magnesium carbonate calcines readily; however, at the operating temperatures typical of atmospheric pressure fluidized beds, not all of the calcium carbonate is calcined. For design purposes, 90% calcination is appropriate for atmospheric fluidized-bed combustion. On an operating unit, the mass fraction of calcination can be determined by measuring the CO_2 in the ash residue and by assuming it exists as $CaCO_3$. The quantity of CO_2 added to the flue gas may be calculated from:

$$MQGSB = 44.01 \times MOGSB \times \frac{100}{HHV} \quad \textbf{(13)}$$

and

$$MOGSB = MFSBF \left(\frac{MFCL \times \mathrm{CaCO_3}}{100.089} + \frac{\mathrm{MgCO_3}}{58.320} \right) \quad \textbf{(14)}$$

where

$MQGSB$ = incremental CO_2 from sorbent, lb/10,000 Btu
$MOGSB$ = moles CO_2 from sorbent, moles/100 lb sorbent
$MFCL$ = fraction of available $CaCO_3$ calcined, lb/lb
$CaCO_3$ = mass percent calcium carbonate in sorbent, %
$MgCO_3$ = mass percent magnesium carbonate in sorbent, %

The water added to the flue gas, *MQWSB*, includes the free water and water evaporated due to dehydration of calcium and magnesium hydroxide products.

$$MQWSB = 18.015 \times MOWSB \times \frac{100}{HHV} \quad \textbf{(15)}$$

and

$$MOWSB = \mathrm{H_2O} + \left(\frac{\mathrm{Ca(OH)_2}}{74.096} + \frac{\mathrm{Mg(OH)_2}}{84.321} \right) \quad \textbf{(16)}$$

where

$MQWSB$ = water added to flue gas from sorbent, lb/10,000 Btu
$MOWSB$ = moles of water from sorbent, moles/100 lb sorbent
HHV = higher heating value, Btu/lb fuel
H_2O = free water from sorbent, moles/100 lb sorbent
$Ca(OH)_2$ = mass percent calcium hydroxide in sorbent, %
$Mg(OH)_2$ = mass percent magnesium hydroxide in sorbent, %

Spent sorbent refers to the solid products remaining due to the use of limestone. Spent sorbent is the sum of the inerts in the limestone, the mass of the reactive constituents after calcination ($CaCO_3$, CaO and MgO), and the SO_3 formed in the sulfation reaction.

$$MQSSB = MFSSB \times \frac{10,000}{HHV} \quad \textbf{(17)}$$

and

$$MFSSB = MFSBF - (0.4401 \times MOGSB) - (0.18015 \times MOWSB) + (250 \times \mathrm{S} \times MFSC) \quad \textbf{(18)}$$

where

$MQSSB$ = solids added to flue gas, lb/10,000 Btu
$MFSSB$ = solids added to flue gas, lb/lb fuel
HHV = higher heating value, Btu/lb fuel
$MFSBF$ = mass ratio of sorbent to fuel, lb/lb
$MOGSB$ = moles CO_2 from sorbent, moles/100 lb sorbent
$MOWSB$ = moles H_2O from sorbent, moles/100 lb sorbent
S = mass percent sulfur in fuel, %
$MFSC$ = mass fraction of sulfur in fuel captured, lb/lb

The combustion and efficiency values related to limestone (sorbent) are calculated separately in this text; they are treated as a supplement to the basic calculations. (See *Combustion and Efficiency Calculations*.)

Residue versus refuse

The term *residue* is used within this text to refer to the solid waste products that leave the steam generator envelope. This replaces the term refuse which is now used to refer to municipal solid waste fuels and their derivatives.

Unburned carbon

In commercial solid fuel applications, it is not always practical to completely burn the fuel. Some of the fuel may appear as unburned carbon in the residue or CO in the flue gas, although the hydrogen in the fuel is usually completely consumed. The capital and operating (energy) costs incurred to burn this residual fuel are usually far greater than the energy lost. In addition, the evolution of combustion equipment to reduce NO_x emissions has resulted in some tradeoffs with increases in unburned carbon and CO.

Unburned carbon impacts the combustion calculations and represents an efficiency loss. Therefore, when present, unburned carbon must be measured. The preferred procedure is to determine total carbon in the boiler flyash and bottom ash in accordance with ASTM Method D 3178 Instrumental Method. If carbonates are present, as in fluidized-bed boilers using limestone, the total carbon includes the carbon in the carbonates. Therefore, the CO_2 in the flyash/bottom ash must be determined in accordance with ASTM Method D 1756. The total carbon mass is corrected to unburned carbon mass by subtracting the carbon in the CO_2 ($CO_2 \times 12.01 / 44.01$).

The unburned carbon determined by these ASTM methods is on the basis of percent carbon in the flyash/bottom ash (lb carbon/100 lb residue). Unburned carbon may also be calculated on a lb/100 lb fuel basis (percent unburned carbon) from the following equation:

$$MPCU = MPCR \times MFR \quad \textbf{(19)}$$

and

$$MFR = \frac{AF + (100 \times MFSSB)}{(100 - MPCR)} \quad \textbf{(20)}$$

where

$MPCU$ = unburned carbon, lb/100 lb fuel
$MPCR$ = measured unburned carbon in residue mass, %
MFR = mass fraction residue of fuel, lb/lb fuel
AF = mass percent ash from fuel, %
$MFSSB$ = mass of spent sorbent, lb/lb fuel

Excess air

For commercial applications, more than theoretical air is needed to assure complete combustion. This *excess air* is needed because the air and fuel mixing is not perfect. Because the excess air that is not used for combustion leaves the unit at stack temperature, the amount of excess air should be minimized. The energy required to heat

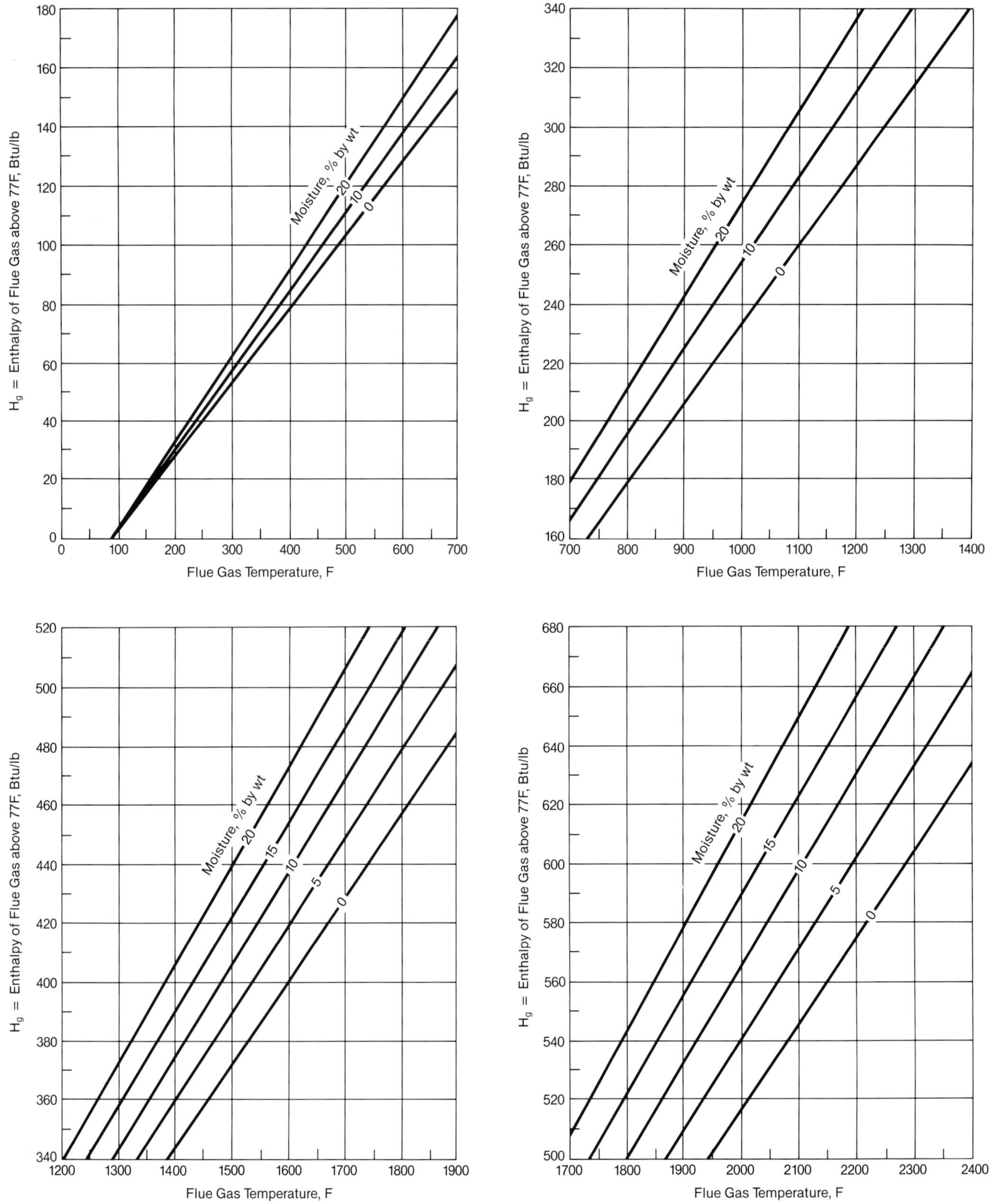

Fig. 3 Enthalpy of flue gas above 77F at 30 in. Hg.

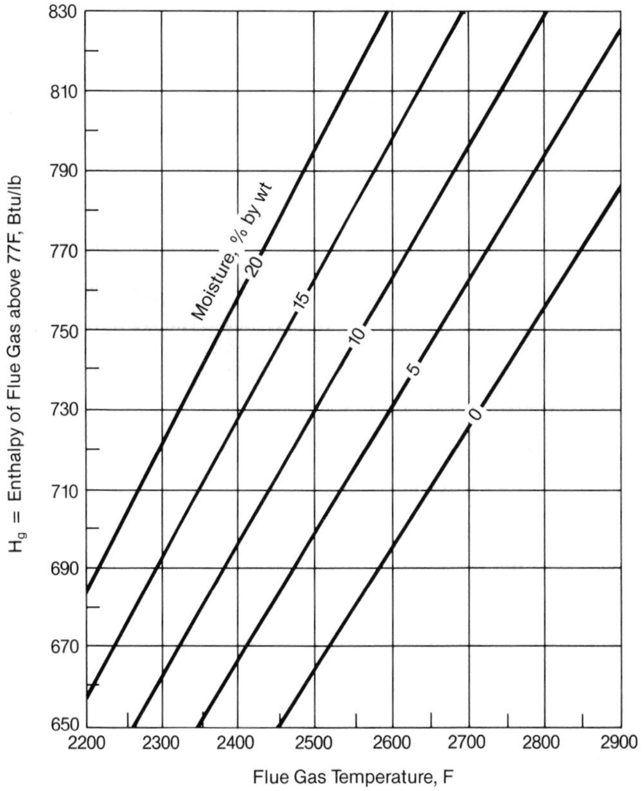
830
810
790
770
750
730
710
690
670
650
2200
2300
2400
2500
2600
2700
2800
2900
Moisture, % by wt
20
15
10
5
0
Hg = Enthalpy of Flue Gas above 77F, Btu/lb
Flue Gas Temperature, F

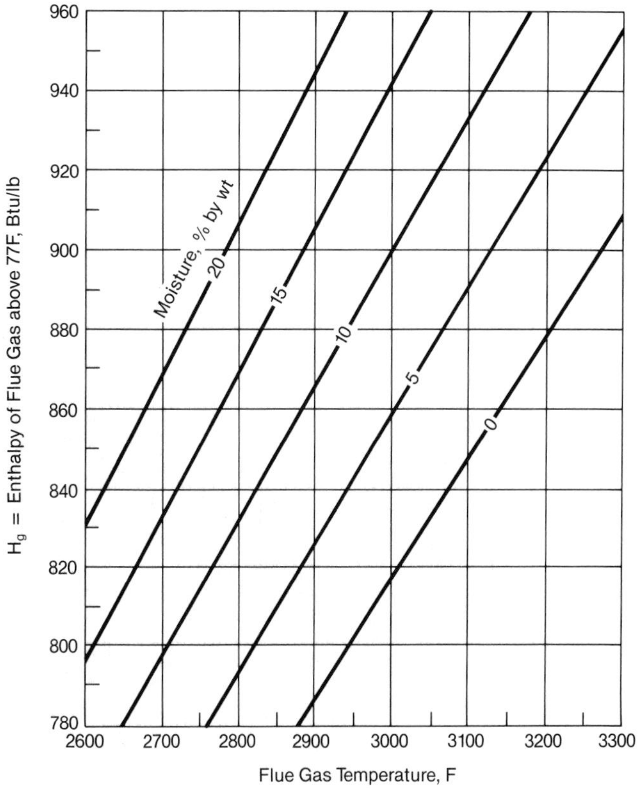
960
940
920
900
880
860
840
820
800
780
2600
2700
2800
2900
3000
3100
3200
3300
Moisture, % by wt
20
15
10
5
0
Hg = Enthalpy of Flue Gas above 77F, Btu/lb
Flue Gas Temperature, F

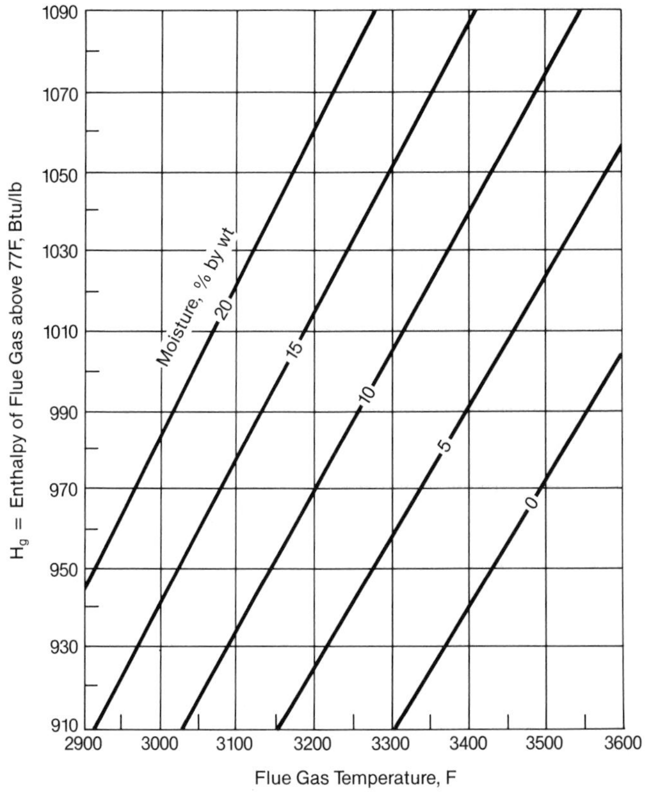
1090
1070
1050
1030
1010
990
970
950
930
910
2900
3000
3100
3200
3300
3400
3500
3600
Moisture, % by wt
20
15
10
5
0
Hg = Enthalpy of Flue Gas above 77F, Btu/lb
Flue Gas Temperature, F

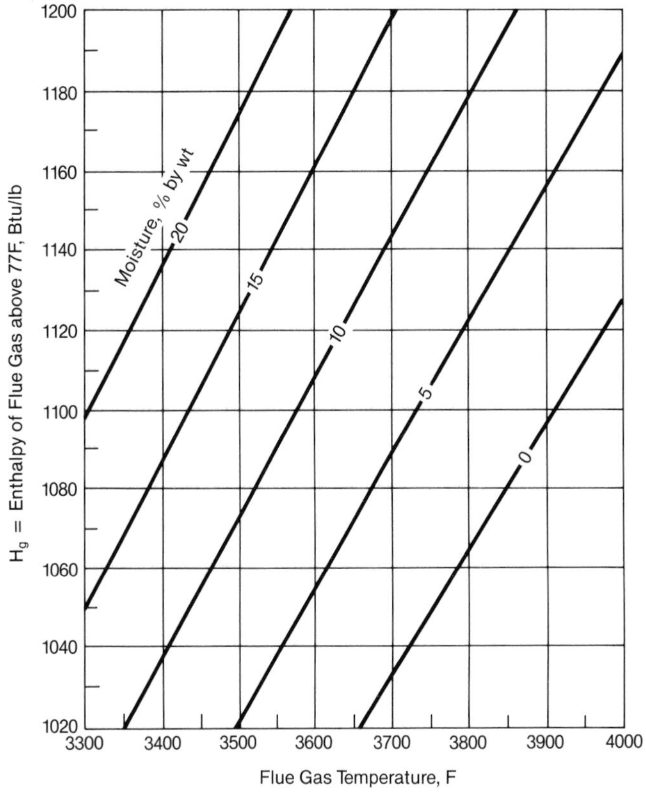
1200
1180
1160
1140
1120
1100
1080
1060
1040
1020
3300
3400
3500
3600
3700
3800
3900
4000
Moisture, % by wt
20
15
10
5
0
Hg = Enthalpy of Flue Gas above 77F, Btu/lb
Flue Gas Temperature, F

Table 11
Typical Excess Air Requirements at Fuel Burning Equipment

Fuel	Type of Furnace or Burners	Excess Air % by wt
Pulverized coal	Completely water-cooled furnace — wet or dry ash removal	15 to 20
	Partially water-cooled furnace	15 to 40
Crushed coal	Cyclone furnace — pressure or suction	13 to 20
	Fluidized-bed combustion	15 to 20
Coal	Spreader stoker	25 to 35
	Water-cooled vibrating grate stoker	25 to 35
	Chain grate and traveling grate	25 to 35
	Underfeed stoker	25 to 40
Fuel oil	Register type burners	3 to 15
Natural, coke oven and refinery gas	Register type burners	3 to 15
Blast furnace gas	Register type burners	15 to 30
Wood/bark	Traveling grate, water-cooled vibrating grate	20 to 25
	Fluidized-bed combustion	5 to 15
Refuse-derived fuels (RDF)	Completely water-cooled furnace — traveling grate	40 to 60
Municipal solid waste (MSW)	Water-cooled/refractory covered furnace reciprocating grate	80 to 100
	Rotary kiln	60 to 100
Bagasse	All furnaces	25 to 35
Black liquor	Recovery furnaces for Kraft and soda pulping processes	15 to 20

this air from ambient to stack temperature usually serves no purpose and is lost heat. Typical values of excess air required at the burning equipment are shown in Table 11 for various fuels and methods of firing. When substoichiometric firing is used in the combustion zone, i.e., less than the theoretical air is used, the values shown would apply to the furnace zone where the final air is admitted to complete combustion. The amount of excess air at the exit of the steam generator equipment (where it is usually monitored) must be greater than the air required at the burning equipment to account for setting infiltration on balanced draft units (or seal air on pressure-fired units). On modern units with membrane construction, this is usually only 1 or 2% excess air at full load. On older units, however, setting infiltration can be significant, and operating with low air at the steam generator exit can result in insufficient air at the burners. This can cause poor combustion performance.

Combustion and efficiency calculations

The combustion calculations are the starting point for all design and performance determinations for boilers and their related component parts. They establish the quantities of the constituents involved in the combustion process chemistry (air, flue gas, residue and sorbent), the efficiency of the combustion process and the quantity of heat released.[6]

The units used for the combustion and efficiency calculations are lb/10,000 Btu. The acronym MQxx also refers here to constituents on a mass per 10,000 Btu basis. For gaseous fuels, the volumetric analysis is converted to an elemental mass basis, as described in *Combustion of Gaseous Fuel*.

Combustion air — theoretical air

The combustion air is the total air required for the burning equipment; it is the theoretical air plus the excess air. Theoretical air is the minimum air required for complete conversion of the carbon, hydrogen and sulfur in the fuel to standard products of combustion. For some fuels and/or combustion processes, all of the carbon is not converted. In addition, when limestone or other additives are used, some of the sulfur is not converted to sulfur dioxide. However, additional air is required for the conversion of sulfur dioxide to sulfur trioxide in the sulfation reaction ($CaO + SO_2 + {}^1/_2\, O_2 \rightarrow CaSO_4$). Because the actual air required is the desired calculation result, the theoretical air is corrected for unburned carbon and sulfation reactions.

$$MQTHAC = THAC \times \frac{100}{HHV} \tag{21}$$

and

$$THAC = (11.51 \times CB) + (34.29 \times H_2) + \{4.32 \times S \times [1 + (0.5\, MFSC)]\} - (4.32 \times O_2) \tag{22}$$

where

$MQTHAC$	=	theoretical air, lb/10,000 Btu
$THAC$	=	theoretical air, lb/100 lb fuel
HHV	=	higher heating value, Btu/lb
CB	=	mass percent carbon burned
	=	percent carbon in fuel −*UBC*, %
H_2	=	mass percent hydrogen in fuel, %
S	=	mass percent sulfur in fuel, %
$MFSC$	=	mass fraction sulfur captured by furnace sorbent, lb/lb sulfur
O_2	=	mass percent oxygen in fuel, %
UBC	=	unburned carbon percent from fuel, %

For test purposes, the unburned carbon is measured. For design calculations, the unburned carbon may be calculated from the estimated unburned carbon loss, *UBCL*:

$$UBC = UBCL \times \frac{HHV}{14{,}500} \tag{23}$$

MFSC is the sulfur capture ratio or mass of sulfur retained per mass sulfur available from the fuel. It is zero unless a sorbent, e.g., limestone, is used in the furnace to reduce SO_2 emissions. See *Flue Gas Analysis* to determine *MFSC* for test conditions.

The mass of dry air, *MQDA*, water in air, *MQWA*, and wet air, *MQA*, are calculated from the following equations:

$$MQDA = MQTHAC \times \left(1 + \frac{PXA}{100}\right) \tag{24}$$

$$MQWA = MA \times MQDA \tag{25}$$

$$MQA = MQDA + MQWA = MQDA \times (1 + MA) \quad \textbf{(26)}$$

where

$MQDA$ = mass dry air, lb/10,000 Btu
$MQTHAC$ = theoretical air, lb/10,000 Btu
PXA = percent excess air, %
$MQWA$ = mass of moisture in air, lb/10,000 Btu
MA = moisture in air, lb/lb dry air
MQA = mass of wet air, lb/10,000 Btu

Flue gas

The total gaseous products of combustion are referred to as *wet flue gas*. Solid products or residue are excluded. The wet flue gas flow rate is used for heat transfer calculations and design of auxiliary equipment. The total gaseous products excluding moisture are referred to as *dry gas*; this parameter is used in the efficiency calculations and determination of flue gas enthalpy.

The wet flue gas is the sum of the wet gas from fuel (fuel less ash, unburned carbon and sulfur captured), combustion air, moisture in the combustion air, additional moisture such as atomizing steam and, if sorbent is used, carbon dioxide and moisture from sorbent. Dry flue gas is determined by subtracting the summation of the moisture terms from the wet flue gas.

Wet gas from fuel is the mass of fuel less the ash in the fuel, less the percent unburned carbon and, when sorbent is used to reduce SO_2 emissions, less the sulfur captured:

$$MQGF = [100 - AF - UBC - (MFSC \times \mathrm{S})] \times \frac{100}{HHV} \quad \textbf{(27)}$$

where

$MQGF$ = wet gas from fuel, lb/10,000 Btu
AF = mass percent ash in fuel, %
UBC = unburned carbon as mass percent in fuel, %
$MFSC$ = mass fraction of sulfur captured, lb/lb sulfur
S = mass percent sulfur in fuel, %
HHV = higher heating value, Btu/lb

Water from fuel is the sum of the water in the fuel, H_2O and the water produced from the combustion of hydrogen in the fuel, H_2:

$$MQWFF = [(8.94 \times \mathrm{H_2}) + \mathrm{H_2O}] \times \frac{100}{HHV} \quad \textbf{(28)}$$

where

$MQWFF$ = water from fuel, lb/10,000 Btu
H_2 = mass percent hydrogen in fuel, %
H_2O = mass percent moisture in fuel, %

Refer to *Sorbents and Other Chemical Additives* for calculating gas from sorbent (CO_2), *MQGSB*, and water from sorbent, *MQWSB*. The total wet gas weight, *MQG*, is then the sum of the dry air, water in air, wet gas from fuel and, when applicable, additional water, gas from sorbent (CO_2), and water from sorbent:

$$MQG = MQDA + MQWA + MQGF + MQWAD + MQGSB + MQWSB \quad \textbf{(29)}$$

where

MQG = total wet gas weight, lb/10,000 Btu
$MQDA$ = mass dry air, lb/10,000 Btu
$MQWA$ = moisture in dry air, lb/10,000 Btu
$MQGF$ = wet gas from the fuel, lb/10,000 Btu
$MQWAD$ = additional water such as atomizing steam, lb/10,000 Btu
$MQGSB$ = gas from the sorbent, lb/10,000 Btu
$MQWSB$ = water from the sorbent, lb/10,000 Btu

The total moisture in the flue gas, *MQWG*, is the sum of the water from fuel, water in air and, if applicable, additional water and water from sorbent.

$$MQWG = MQWFF + MQWA + MQWAD + MQWSB \quad \textbf{(30)}$$

Dry flue gas, *MQDG* in lb/10,000 Btu, is the difference between the wet flue gas and moisture in the flue gas:

$$MQDG = MQG - MQWG \quad \textbf{(31)}$$

The percent moisture in flue gas is a parameter required to determine its heat content or enthalpy (see *Air and Flue Gas Enthalpy*) and is calculated as follows:

$$MPWG = 100 \times \frac{MQWG}{MQG}, \% \quad \textbf{(32)}$$

For most fuels, the mass of solids, or residue, in the flue gas is insignificant and can be ignored. Even when the quantity is significant, solids do not materially impact the volume flow rate of flue gas. However, solids add to the heat content, or enthalpy, of flue gas and should be accounted for when the ash content of the fuel is greater than 0.15 lb/10KB or when sorbent is used.

The mass of residue from fuel, *MQRF* in lb/10,000 Btu, is calculated from the following equation:

$$MQRF = (AF + UBC) \times \frac{100}{HHV} \quad \textbf{(33)}$$

where

$MQRF$ = residue from fuel, lb/10,000 Btu
AF = mass percent (%) ash in fuel
UBC = unburned carbon as mass percent (%) in fuel

The mass percent of solids or residue in the flue gas is then:

$$MPRG = 100 \times \frac{MQRF + MQSSB}{MQG} \quad \textbf{(34)}$$

where

$MPRG$ = mass percent solids or residue in flue gas, %
$MQSSB$ = spent sorbent, lb/10,000 Btu
MQG = mass of gaseous combustion products excluding solids, lb/10,000 Btu

Efficiency

Efficiency, the ratio of energy output to input, is usually expressed as a percentage. The output term for a steam generator is the heat absorbed by the working fluid to produce useful energy external to the steam generator envelope. The energy input term is the maximum energy available when the fuel is completely burned, i.e., the mass flow rate (*MRF*) of fuel multiplied by the higher heating value of the fuel. This is conventionally expressed as:

$$\eta_f = 100 \times \frac{\text{Output}}{\text{Input fuel}} = 100 \times \frac{\text{Output}}{MRF \times HHV}, \% \quad \textbf{(35)}$$

and is commonly referred to as fuel or steam generator efficiency.

According to the law of conservation of energy, for steady-state conditions, the energy balance on the steam generator envelope can be expressed as:[7]

$$QRF = QRO + QHB, \text{ Btu/h} \quad \textbf{(36)}$$

where QRF is the input from fuel, Btu/h, QRO is the steam generator output, Btu/h, and QHB is the energy required by heat balance for closure, Btu/h. The heat balance energy associated with the streams entering the steam generator envelope and the energy added from auxiliary equipment power are commonly referred to as heat credits, QRB (Btu/h). The heat balance energy associated with streams leaving the steam generator and the heat lost to the environment are commonly referred to as heat losses, QRL (Btu/h). This steam generator energy balance may be written as:

$$QRF = QRO + QHB = QRO + QRB - QRL, \text{Btu/h} \quad \textbf{(37)}$$

and the efficiency may be expressed as:

$$\eta_f = 100 \times \frac{QRO}{QRO + QRL - QRB}, \ \% \quad \textbf{(38)}$$

When losses and credits are expressed as a function of percent input from fuel, QPL and QPB, the efficiency may be calculated from:

$$\eta_f = 100 - QPL + QPB, \ \% \quad \textbf{(39)}$$

Most losses and credits are conveniently calculated on a percent input from fuel basis. However, some losses are more conveniently calculated on a Btu/h basis. The following expression for efficiency allows the use of mixed units; some of the losses/credits are calculated on a percent basis and some on a Btu/h basis.

$$\eta_f = (100 - QPL + QPB) \times \left(\frac{QRO}{QRO + QRL - QRB} \right), \ \% \quad \textbf{(40)}$$

For a rigorous discussion of losses and credits, refer to the American Society of Mechanical Engineers (ASME) Performance Test Code, PTC 4, for steam generators.

The general form for calculating losses using the mass per unit of heat input basis to express the percent heat loss for individual constituents is:

$$QPL_k = \frac{MQ_k \times MCP_k \times (TO_k - TR)}{100} = \frac{MQ_k \times (HO_k - HR_k)}{100} \quad \textbf{(41)}$$

where

QPL_k = heat loss, Btu/10,000 Btu
MQ_k = mass flow, lb/10,000 Btu
MCP_k = mean specific heat between TO_k and TR, Btu/lb F
TO_k = outlet temperature, F
TR = reference temperature, F
HO_k = outlet enthalpy, Btu/lb
HR_k = reference enthalpy, Btu/lb

The reference temperature is usually taken as the temperature of the air leaving the forced draft fans or that entering an air to gas heat exchanger. When air preheater coils are the principal source of combustion air heat or when these coils are supplied by steam from the steam generator, the temperature of the air entering the preheater coils should be taken as the reference temperature. An arbitrary fixed value, such as 77F or a specified ambient temperature, may also be used as the reference temperature. An advantage of using the inlet air temperature is that it eliminates the need to calculate credits for entering air and moisture in air.

The general form for calculating credits (QPB_k) using the mass per unit of input basis to express the quantity of individual constituents is:

$$QPB_k = \frac{MQ_k \times MCP_k \times (TI_k - TR)}{100} = \frac{MQ_k \times (HI_k - HR_k)}{100}, \% \quad \textbf{(42)}$$

where

TI_k = inlet temperature, F
HI_k = inlet enthalpy, Btu/lb

and other terms were defined in Equation 41.

The terms used to calculate losses and credits that are a function of fuel input have been discussed previously. The other losses and credits are described below.

Radiation (and convection) loss

This is the heat lost to the atmosphere through the boiler casing surfaces, including flues and ducts, between the first and the last heat trap (commonly the boiler exit or air heater exit). It is a function of the difference between the average casing surface temperature and average ambient temperature and average surface velocity. The U.S. industry standards for calculating heat loss use a temperature differential of 50F and a surface velocity of 100 ft/min. For convenience, the American Boiler Manufacturers Association (ABMA) standard radiation loss chart, shown in Chapter 22, may be used for an approximation. This curve expresses the radiation loss on a percent of gross heat input basis as a function of steam generator output (percent gross heat input may be interpreted as heat input from fuel for most applications). This curve is generally conservative for the modern steam generator and the trend is toward calculating radiation/convection heat loss based on the flat projected area of the unit using standard or specified conditions and standard ASME Performance Test Code heat transfer coefficients.

Unburned carbon loss

For design of a unit, this is normally estimated based on historical data and/or combustion models. For an efficiency test, this item is calculated from measured unburned carbon in the residue. (See *Unburned Carbon.*)

Unaccounted for losses and manufacturers' margins

When designing a unit, usually only the major energy losses can be specifically accounted for (calculated based upon design data). The other minor losses are estimated or based on historical data. In addition, the manufacturer normally adds a margin, or safety factor, to the losses to account for unexpected performance deviations. Typical design values for these margins are 1.0% of heat input for gas, oil and coals with good combustion characteristics and slagging/fouling properties to 1.5% of heat input or higher for fuels with poor combustion characteristics and poor slagging/fouling characteristics. The manufacturer's margin is not accounted for in measuring actual efficiencies because it is applied as a safety factor to expected performance. In the evaluation of actual unit efficiency, the unaccounted for losses become unmeasured losses and should be estimated based on field data available.

Enthalpy

Enthalpy of air and gas

Enthalpy, H, in Btu/lb is an indication of the relative energy level of a material at a specific temperature and pressure. It is used in thermal efficiency, heat loss, heat balance and heat transfer calculations (see Chapter 2). Extensive tabulated and graphical data are available such as the ASME Steam Tables[4] summarized in Chapter 2. Except for steam and water at high pressure, the pressure effect on enthalpy is negligible for engineering purposes.

Enthalpies of most gases used in combustion calculations can be curve-fitted by the simple second order equation:

$$H = aT^2 + bT + c \quad \textbf{(43)}$$

where

H = enthalpy in Btu/lb
T = temperature in degrees, F

To determine the enthalpy of most gases used in combustion calculations at a temperature, T, Equation 43 can be used with the coefficients summarized in Table 12. Reference 8 is the source for the properties and the curve fits are in accordance with Reference 9. The curve fits are within plus or minus 0.2 Btu/lb for enthalpies less than 40 Btu/lb and within plus or minus 0.5% for larger values. If the enthalpy of a fluid is known, the temperature in degrees F can be evaluated from the quadratic equation:

$$T = \frac{-b + \sqrt{b^2 - 4a\,(c - H)}}{2a} \quad \textbf{(44)}$$

For mixtures of gases, such as dry air and water vapor or flue gas and water vapor, Equation 43 coefficient, a, b and c can be determined by a simple mass average:

$$n_{mix} = \sum x_i n_i \quad \textbf{(45)}$$

where

x_i = mass fraction of constituent i
n_i = coefficient a, b or c for constituent i
n_{mix} = equivalent coefficient a, b or c of the mixture

For convenience, Table 12 lists coefficients for a number of gas mixtures including standard wet air with 0.013 lb H_2O per lb dry air. In addition, Figs. 3 and 4 provide graphical representations of flue gas and standard air enthalpy.

Table 12 — Enthalpy Coefficients for Equation 43

Coefficient	a	b	c
Dry air (a)			
0 to 500	8.299003E-06	0.2383802	-18.43552
500 to 1500	1.474577E-05	0.2332470	-17.48061
1500 to 2500	8.137865E-06	0.2526050	-31.64983
2500 to 4000	4.164187E-06	0.2726073	-56.82009
Wet air (b)			
0 to 500	8.577272E-06	0.2409682	-18.63678
500 to 1500	1.514376E-05	0.2357032	-17.64590
1500 to 2500	8.539973E-06	0.2551066	-31.89248
2500 to 4000	4.420080E-06	0.2758523	-58.00740
Water vapor			
0 to 500	2.998261E-05	0.4400434	-34.11883
500 to 1500	4.575975E-05	0.4246434	-30.36311
1500 to 2500	3.947132E-05	0.4475365	-50.55380
2500 to 4000	2.413208E-05	0.5252888	-149.06430
Dry gas (c)			
0 to 500	1.682949E-05	0.2327271	-18.03014
500 to 1500	1.725460E-05	0.2336275	-18.58662
1500 to 2500	8.957486E-06	0.2578250	-36.21436
2500 to 4000	4.123110E-06	0.2821454	-66.80051
TEG gas (d)			
0 to 500	1.692527E-05	0.2326889	-18.02865
500 to 1500	1.726157E-05	0.2336052	-18.57088
1500 to 2500	8.955249E-06	0.2578280	-36.21585
2500 to 4000	4.123110E-06	0.2821454	-66.80848
Ash/SiO_2			
0 to 500	7.735829E-05	0.1702036	-13.36106
500 to 1500	2.408712E-05	0.2358873	-32.88512
1500 to 2500	1.394202E-05	0.2324186	-4.85559
2500 to 4000	1.084199E-05	0.2460190	-19.48141
N_{2a}- Atmospheric nitrogen (e)			
0 to 500	5.484935E-06	0.2450592	-18.93320
500 to 1500	1.496168E-05	0.2362762	-16.91089
1500 to 2500	8.654128E-06	0.2552508	-31.18079
2500 to 4000	3.953408E-06	0.2789019	-60.92904
O_2 - Oxygen			
0 to 500	1.764672E-05	0.2162331	-16.78533
500 to 1500	1.403084E-05	0.2232213	-19.37546
1500 to 2500	6.424422E-06	0.2438557	-33.21262
2500 to 4000	4.864890E-06	0.2517422	-43.18179
CO_2 - Carbon dioxide			
0 to 500	5.544506E-05	0.1943114	-15.23170
500 to 1500	2.560224E-05	0.2270060	-24.11829
1500 to 2500	1.045045E-05	0.2695022	-53.77107
2500 to 4000	4.595554E-06	0.2989397	-90.77172
SO_2 - Sulfur dioxide			
0 to 500	3.420275E-05	0.1439724	-11.25959
500 to 1500	1.366242E-05	0.1672132	-17.74491
1500 to 2500	4.470094E-06	0.1923931	-34.83202
2500 to 4000	2.012353E-06	0.2047152	-50.27639
CO - Carbon monoxide			
0 to 500	5.544506E-05	0.1943114	-15.23170
500 to 1500	2.559673E-05	0.2269866	-24.10722
1500 to 2500	1.044809E-05	0.2695040	-53.79888
2500 to 4000	4.630355E-06	0.2987122	-90.45853

Notes:
(a) Dry air composed of 20.947% O_2, 78.086% N_2, 0.934% Ar and 0.033% CO_2 by volume.
(b) Wet air contains 0.013 lb H_2O/lb dry air.
(c) Dry gas composed of 3.5% O_2, 15.4% CO_2, 0.1% SO_2 and 81.1% N_{2a} by volume.
(d) Turbine exhaust gas (TEG) composed of 16.97% O_2, 2.97% CO_2 and 80.06% N_{2a} by volume (No. 2 oil with 400% excess air).
(e) N_{2a} composed of the atomic nitrogen, Ar and CO_2 in standard air.

Source: JANAF Thermochemical Tables, 2nd Ed., NSRDS-NBS 37, 1971. Curve fits developed from NASA SP-273, 1971 correlations.

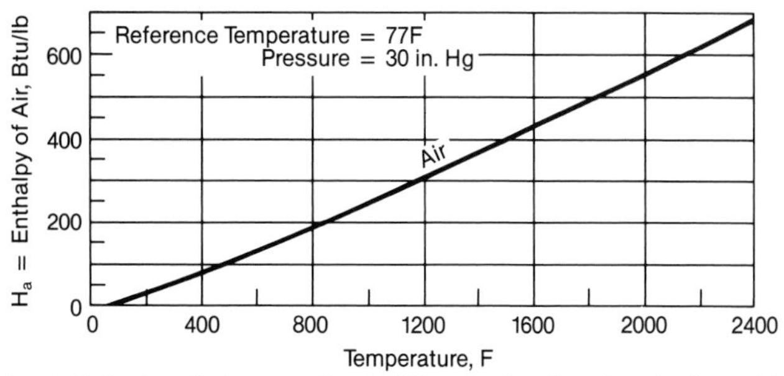

Fig. 4 Enthalpy of air assuming 0.987 mass fraction dry air plus 0.013 mass fraction of water vapor.

Another method of evaluating the change in specific enthalpy of a substance between conditions 1 and 2 is to consider the specific heat and temperature difference:

$$H_2 - H_1 = c_p (T_2 - T_1) \quad \textbf{(46)}$$

where
H = enthalpy, Btu/lb
c_p = specific heat at constant pressure, Btu/lb F
T = temperature, F

Enthalpy of solids and fuels

Enthalpy of coal, limestone and oil can be evaluated from the following relationships:

Coal:[10]

$$H = [(1 - W_F)(0.217 + 0.00248\, VM) + W_F](T - 77) \quad \textbf{(47)}$$

Limestone:

$$H = [(1 - W_F)\, H_{LS} + W_F](T - 77) \quad \textbf{(48)}$$

and

$$H_{LS} = (0.179\, T) + (0.1128 \times 10^{-3}\, T^2) - 14.45 \quad \textbf{(49)}$$

Oil:[11]

$$H = C_1 + C_2\,(API) + C_3 T + C_4\,(API)\, T + [C_5 + C_6\,(API)]\, T^2 \quad \textbf{(50)}$$

and

$$API = (141.5 - 131.5\, SPGR)/SPGR \quad \textbf{(51)}$$

where

H = enthalpy of coal, limestone or oil at T, Btu/lb
H_{LS} = enthalpy of dry limestone, Btu/lb
W_F = mass fraction free moisture in coal or limestone
VM = volatile matter on a moisture and ash free basis, %
T = temperature, F
API = degrees API
$SPGR$ = specific gravity, dimensionless
= density in lb/ft^3 divided by 62.4 at 60F
C_1 = −30.016
C_2 = −0.11426
C_3 = 0.373
C_4 = 0.143×10^{-2}
C_5 = 0.2184×10^{-3}
C_6 = 7.0×10^{-7}

Measurement of excess air

One of the most critical operating parameters for attaining good combustion is excess air. Too little air can be a source of excessive unburned combustibles and can be a safety hazard. Too much excess air increases stack gas losses.

Flue gas analysis

The major constituents in flue gas are CO_2, O_2, N_2 and H_2O. Excess air is determined by measuring the O_2 and CO_2 contents of the flue gas. Before proceeding with measuring techniques, consider the form of the sample. A flue gas sample may be obtained on a wet or dry basis. When a sample is extracted from the gas stream, the water vapor normally condenses and the sample is considered to be on a dry basis. The sample is usually drawn through water near ambient temperature to ensure that it is dry. The major constituents of a dry sample do not include the water vapor in the flue gas. When the gas is measured with an in situ analyzer or when precautions are taken to keep the moisture in the sample from condensing, the sample is on a wet basis.

The amount of O_2 in the flue gas is significant in defining the status of the combustion process. Its presence always means that more oxygen (excess air) is being introduced than is being used. Assuming complete combustion, low values of O_2 reflect moderate excess air and normal heat losses to the stack, while higher values of O_2 mean needlessly higher stack losses.

The quantity of excess O_2 is very significant since it is a nearly exact indication of excess air. Fig. 5 is a dry flue gas volumetric combustion chart that is universally used in field testing; it relates O_2, CO_2 and N_{2a} (by difference). For complete uniform combustion of a specific fuel, all points should lie along a straight line drawn through the pivot point. This line is referred to as the combustion line. The combustion line ranges for various fuels are indicated. Lines indicating constant excess air have been superimposed on the volumetric combustion chart. Note that excess air is essentially constant for a given O_2 level over a wide range of fuels. The O_2 is an equally constant indication of excess air when the gas is sampled on a wet or in situ basis because the calculated excess air result is insensitive to variations in moisture for specific types/sources of fuel.

The current industry standard for boiler operation is continuous monitoring of O_2 in the flue gas with in situ analyzers that measure oxygen on a wet basis.

For testing, the preferred instrument is an electronic oxygen analyzer. The Orsat unit, which measures (CO_2 + SO_2) and O_2 on a dry volumetric basis, remains a trusted standard for verifying the performance of electronic equipment. The Orsat uses chemicals to absorb the (CO_2 + SO_2) and O_2, and the amount of each is determined by the reduction in volume from the original flue gas sample. When an Orsat is used, the dry flue gas volumetric combustion chart should be used to plot the results. Valid results for any test with a consistent fuel should fall on a single combustion line (plus or minus 0.2 points of O_2/CO_2 is a reasonable tolerance). The Orsat has several disadvantages. It lacks the accuracy of more refined devices, an experienced operator is required, there are a limited number of readings available in a test, and the results do not lend themselves to electronic recording. Electronic CO_2 analyzers may be used in addition to oxygen analyzers to relate the O_2/CO_2 results to the fuel line on the volumetric combustion chart. When CO_2 is measured, by Orsat or a separate electronic analyzer, it is best to calculate excess air based on the O_2 result due to the insensitivity of excess air versus O_2 results in the fuel analysis.

Depending upon whether O_2 is measured or excess air known, the corresponding excess air, O_2, CO_2 and SO_2 can

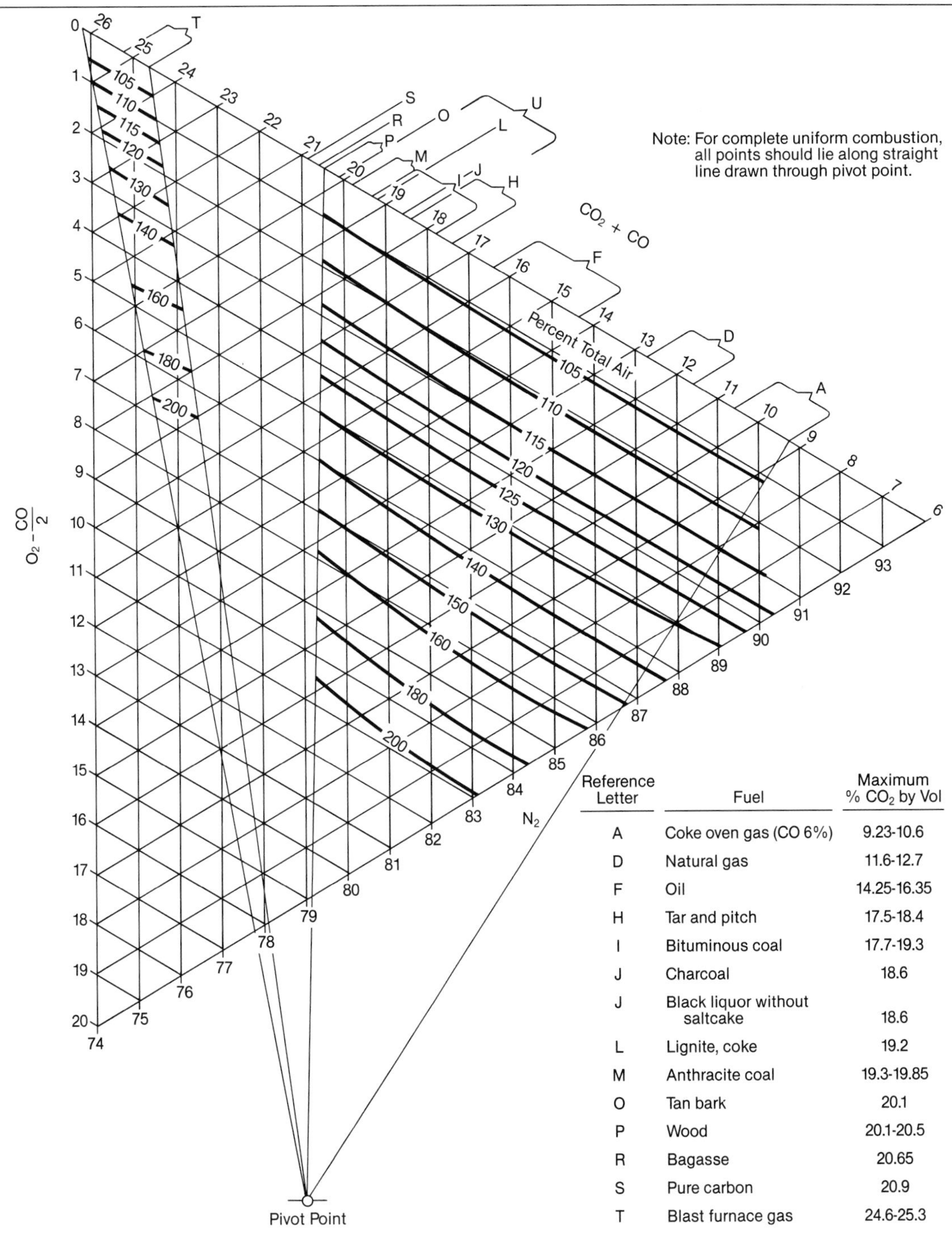

Note: For complete uniform combustion, all points should lie along straight line drawn through pivot point.
CO2 + CO
Percent Total Air
O2 - CO/2
N2
Pivot Point

Reference Letter	Fuel	Maximum % CO_2 by Vol
A	Coke oven gas (CO 6%)	9.23-10.6
D	Natural gas	11.6-12.7
F	Oil	14.25-16.35
H	Tar and pitch	17.5-18.4
I	Bituminous coal	17.7-19.3
J	Charcoal	18.6
J	Black liquor without saltcake	18.6
L	Lignite, coke	19.2
M	Anthracite coal	19.3-19.85
O	Tan bark	20.1
P	Wood	20.1-20.5
R	Bagasse	20.65
S	Pure carbon	20.9
T	Blast furnace gas	24.6-25.3
U	MSW/RDF	18.0-20.0

Fig. 5 Dry flue gas volumetric combustion chart.

be calculated using procedures provided in ASME Performance Test Code 4, *Steam Generators*.[6] The calculations are summarized in Table 15 at the end of this chapter in the *Combustion Calculations — Examples* section.

Flue gas sampling

To ensure a representative average gas sample, samples from a number of equal area points should be taken. Reference the Environmental Protection Agency (EPA) Method 1 standards and ASME Performance Test Code PTC 19.10. For normal performance monitoring, equal areas of approximately 9 ft^2 (0.8 m^2) up to 24 points per flue are adequate.

For continuous monitoring, the number of sampling points is an economic consideration. Strategies for locating permanent monitoring probes should include point by point testing with different burner combinations. As a guideline, four probes per flue located at quarter points have been used successfully on large pulverized coal-fired installations.

Testing heterogeneous fuels

When evaluating the performance of a steam generator firing a heterogeneous fuel such as municipal solid waste (MSW) (see Chapter 27), it is generally not possible to obtain a representative fuel sample. Waste fuel composition may vary widely between samples and is usually not repeatable.

For boiler design, an ultimate analysis for an average fuel and a range of the most significant components, such as moisture and ash, are used. Therefore, the design calculations are the same as those for homogeneous fuels.

When firing a heterogeneous fuel, the current industry practice used to evaluate average fuel properties and determine boiler efficiency is to test using the boiler as a calorimeter (BAC). The BAC method features the same principles for determining efficiency as those used when the fuel analysis is known. The significant difference is that the mass/volume flow rate of flue gas and moisture in the flue gas are measured directly rather than being calculated based upon the measured fuel analysis and O_2 in the flue gas.

The additional measurements that are required for the BAC test method versus conventional test methods are flue gas flow, moisture in flue gas, O_2 and CO_2 in the flue gas, and residue mass flow rates from the major extraction points.

BAC calculation method

This section describes how to calculate excess air, dry gas weight and water from fuel (water evaporated). The results are on a mass per unit of time basis and losses and credits, therefore, are calculated as Btu/h. Refer to the basic efficiency equations for application. (See Equations 37 and 38.)

The wet gas weight and water in the wet gas are measured. The dry gas weight is then calculated as the difference of the two.

The composition of flue gas is determined by measuring O_2 and CO_2. N_{2a} is determined by difference from 100%. The nitrogen in flue gas is considered to be atmospheric with a molecular weight of 28.161 lb/mole. Because waste fuel combustors operate at high levels of excess air and the nitrogen in the fuel is small, this nitrogen can be ignored.

The moisture in the flue gas may be from vapor or liquid sources. Vapor sources include moisture in the air and atomizing steam. Water sources are moisture in the fuel, moisture formed by combustion of H_2, water from ash quenching systems and fuel pit water spray. The moisture in air and that from other vaporous sources must be measured, so the sensible heat efficiency loss may be differentiated from the water evaporated loss. The water evaporated is the total moisture in the flue gas less the vaporous sources. The water evaporation loss is calculated in the same manner as the water from fuel loss and is analogous to the total water from fuel loss if miscellaneous water sources are accounted for.

The total dry air flow at the point of flue gas measurement is calculated from the nitrogen in the flue gas. Excess air is determined from the measured O_2 and theoretical air is calculated by difference from the total air flow. The percent excess air is calculated from the excess air and theoretical air weight flow rates.

Combustion calculations — examples

The detailed steps in the solution of combustion problems are best illustrated by examples. The examples in this section are presented through calculation forms which are a convenient method for organizing the calculations in a logical sequence. The input required to complete the forms is located at the top of the form. An elemental fuel analysis on a mass basis is used for all of the examples. For gaseous fuels, the analysis on a volume basis must be converted to an elemental mass basis as described in *Combustion of Gaseous Fuels* above. The calculations required are shown as a combination of item numbers (enclosed in brackets) and constants.

Mole method

The mole method is the fundamental basis for all combustion calculations. It is the source for the constants used in other more simplified methods. The only constants the user needs are the molecular weights of the fuel and air constituents. The reader should understand the mole method before proceeding with the Btu method.

Table 7 is an example of the combustion calculations for a bituminous coal on a molar basis. Items 1 through 6 are the required input. If the unburned carbon is known (Item 6), the unburned carbon loss (Item 5) is calculated. Provision is made for entering the excess air to the burners and excess air leaving the boiler if the user desires to account for setting infiltration (Item 1). For this example, the excess air to the burners is assumed to be the same as that leaving the boiler. An intermediate step in the calculations on a molar basis is the volumetric flue gas analysis (Items 20 and 21). Air and gas mass flow rates are shown on a lb/10,000 Btu basis as well as a 1000 lb/h basis.

Btu method

Once the reader understands the principles of the combustion calculations on a mole basis, the Btu method is the preferred method for general combustion calculations. The calculations provided in Table 13A are more comprehensive than the simple calculation of air and gas weights shown in Table 7. Provision is made for handling the impact of sorbent on the combustion calculations, the calculation of efficiency and finally heat available to the furnace.

The inputs to Table 13A are similar to those used in Table 7. The same fuel analysis and excess air are used and the calculated input from fuel is very nearly the same. These inputs are also the same as those used in the example performance problem in Chapter 21. Items 1 through 19 are the inputs and initial calculations required for the combustion calculations. For the efficiency calculations, Items 44 through 46 must be provided. If sorbent is used, Table 14, *Combustion Calculations — Sorbent*, must be completed first. (See Items 11 through 14 and 46). Because the entering air temperature and fuel temperature are the same as the reference temperature selected (80F), the efficiency credits are zero. The total fuel heat is calculated from the efficiency, Item 53, and steam generator output, Item 10. Flue gas and air flow rates are calculated from the fuel input and the results of the combustion gas calculations.

Table 13A shows the calculation results for a typical Eastern coal. A similar set of calculations can be made for a typical Western subbituminous coal which has an increased moisture content (30% by weight) and reduced

LHV (8360 Btu/lb). For the same boiler rating, the results for these coals can be compared to oil and gas on a lb per 10,000 Btu basis:

	Eastern Bitum.	Western Subbit.	No. 6 Oil	Natural Gas
Theoretical air	7.572	7.542	7.437	7.220
Dry air	9.086	9.050	8.924	8.664
Dry gas weight	9.442	9.463	8.965	8.194
Wet gas weight	9.864	10.303	9.583	9.236
H_2O in gas	0.422	0.840	0.618	1.042
Efficiency, %	86.91	82.10	85.30	80.79

The theoretical air, dry air and resulting dry gas weight are approximately the same for each coal. The wet gas weight and H_2O in gas are higher for the subbituminous coal due to the higher moisture content. Referring to the efficiency calculations and losses, the efficiency is lower for the subbituminous coal essentially due to the higher moisture content, not the lower heating value. The higher air weight required is primarily due to the lower efficiency, while the higher gas weight is due to the higher moisture in the fuel and the lower efficiency.

Table 13B is the same example as shown in Table 13A except that it is assumed that a limestone sorbent is used in a fluidized bed at a calcium to sulfur molar ratio of 2.5. A sulfur capture of 90% is expected. A higher unburned carbon loss is used, typical of this combustion process. It is necessary to complete the calculations shown in Table 14 to develop input for this Table. The net losses due to sorbent, Item 46 in Table 13B, are not overly significant. Therefore, the difference in efficiency from the example in Table 13A is primarily due to the difference in the assumed unburned carbon loss.

When testing a boiler, the excess air required for the combustion calculations is determined from measured O_2 in the flue gas. Table 15A, *Excess Air Calculations from Measured O_2*, demonstrates the calculation of excess air from O_2 on a wet basis. The fuel analysis and unburned carbon are the same as in Tables 7 and 13A. These Tables can also be used to determine the volumetric composition of wet or dry flue gas when excess air is known (Items 25 through 32). These values can be compared to the flue gas composition calculated on a molar basis, Table 7.

Table 15B is an example of calculating excess air from O_2 when a sorbent is used. All of the sulfur in the fuel will not be converted to sulfur dioxide. Therefore, the sulfur capture must first be determined from Table 16, *Sulfur Capture Based on Gas Analysis*. The example presented in Tables 13B and 14 is used as the basis for this example. The flue gas composition in Tables 13A and 13B can be compared to assess the impact of adding the sorbent.

When units firing municipal solid waste or refuse-derived fuels are tested, it is not practical to determine the ultimate analysis of the fuel. Table 17, *Combustion Calculations — Measured Gas Weight*, shows the combustion calculations for determining dry gas weight, water evaporated and excess air using measured gas weight.

References

1. *Gas Engineers Handbook*, The Industrial Press, New York, 1965.

2. Mason, D.McA., and Eakin, B.E., "Proposed standard method for calculating heating value and specific gravity from gas composition," American Gas Association Proceedings, CEP-61-11, 1961.

3. Rossini, F.D., *et al.*, "Selected values of physical and thermodynamic properties of hydrocarbons and related compounds," Research Project 44, American Petroleum Institute, Pittsburgh, 1953.

4. Meyer, C.A., *et al.*, *ASME Steam Tables*, 5th ed., American Society of Mechanical Engineers, New York, 1983.

5. *CRC Handbook of Chemistry and Physics*, 71st ed., CRC Press, Boca Raton, Florida, 1990.

6. Gerhart, P.M., Heil, T.C., and Phillips, J.T., "Steam generator performance calculation strategies for ASME PTC 4," Technical Paper 91-JPGC-PTC-1, American Society of Mechanical Engineers, New York, October 1991.

7. Entwistle, J., Heil, T.C., and Hoffman, G.E., "Steam Generation Efficiency Revisited," Technical Paper 88-JPGC/PTC-3, American Society of Mechanical Engineers, New York, September 1988.

8. *JANAF Thermochemical Tables*, 2nd ed., Publication NSRDS-NBS 37, United States National Bureau of Standards (now National Institute of Standards and Technology), Washington, D.C., 1971.

9. NASA Publication SP-273, 1971.

10. Elliot, M.A., ed., *Chemistry of Coal Utilization*, 2nd suppl. vol., Wiley, New York, 1981.

11. Dunstan, A.E., *The Science of Petroleum*, Oxford University Press, Oxford, U.K., 1938.

Table 13A
Combustion Calculations - Btu Method

	INPUT CONDITIONS - BY TEST OR SPECIFICATION		
1	Excess air: at burner/leaving boiler/econ, % by weight		*20/20*
2	Entering air temperature, F		*80*
3	Reference temperature, F		*80*
4	Fuel temperature, F		*80*
5	Air temperature leaving air heater, F		*350*
6	Flue gas temperature leaving (excluding leakage), F		*390*
7	Moisture in air, lb/lb dry air		*0.013*
8	Additional moisture, lb/100 lb fuel		*0*
9	Residue leaving boiler/economizer, % Total		*85*
10	Output, 1,000,000 Btu/h		*285.6*
	Corrections for sorbent (from Table 14 if used)		
11	Additional theoretical air, lb/10,000 Btu	Table 14, Item [21]	*0*
12	CO_2 from sorbent, lb/10,000 Btu	Table 14, Item [19]	*0*
13	H_2O from sorbent, lb/10,000 Btu	Table 14, Item [20]	*0*
14	Spent sorbent, lb/10,000 Btu	Table 14, Item [24]	*0*

FUEL - *Bituminous coal, Virginia*

	15 Ultimate Analysis		16 Theo Air, lb/100 lb fuel		17 H_2O, lb/100 lb fuel	
	Constituent	% by weight	K1	[15] x K1	K2	[15] x K2
A	C	*80.31*	11.51	*924.4*		
B	S	*1.54*	4.32	*6.7*		
C	H_2	*4.47*	34.29	*153.3*	8.94	*39.96*
D	H_2O	*2.90*			1.00	*2.90*
E	N_2	*1.38*				
F	O_2	*2.85*	– 4.32	*– 12.3*		
G	Ash	*6.55*				
H	Total	*100.00*	Air	*1072.1*	H_2O	*42.86*

18	Higher heating value (HHV), Btu/lb fuel		*14,100*
19	Unburned carbon loss, % fuel input		*0.40*
20	Theoretical air, lb/10,000 Btu	[16H] x 100 / [18]	*7.604*
21	Unburned carbon, % of fuel	[19] x [18] / 14,500	*0.39*

COMBUSTION GAS CALCULATIONS, Quantity / 10,000 Btu Fuel Input

			A At Burners	B Infiltration	C Leaving Furnace		D Leaving Blr/Econ	
22	Theoretical air (corrected), lb / 10,000 Btu	[20] – [21] x 1151 / [18] + [11]						*7.572*
23	Residue from fuel, lb/10,000 Btu	([15G] + [21]) x 100 / [18]						*0.049*
24	Total residue, lb/10,000 Btu	[23] + [14]						*0.049*
25	Excess air, % by weight		*20.0*	*0.0*		*20.0*		*20.0*
26	Dry air, lb/10,000 Btu	(1 + [25] / 100) x [22]				*9.086*		*9.086*
27	H_2O from air, lb/10,000 Btu	[26] x [7]			*0.118*	*0.118*	*0.118*	*0.118*
28	Additional moisture, lb/10,000 Btu	[8] x 100 / [18]			*0.000*	*0.000*	*0.000*	*0.000*
29	H_2O from fuel, lb/10,000 Btu	[17H] x 100 / [18]			*0.304*		*0.304*	
30	Wet gas from fuel, lb/10,000 Btu	(100 – [15G] – [21]) x 100 / [18]				*0.660*		*0.660*
31	CO_2 from sorbent, lb/10,000 Btu	[12]				*0.000*		*0.000*
32	H_2O from sorbent, lb/10,000 Btu	[13]			*0.000*	*0.000*	*0.000*	*0.000*
33	Total wet gas, lb/10,000 Btu	Summation [26] through [32]				*9.864*		*9.864*
34	Water in wet gas, lb/10,000 Btu	Summation [27] + [28] + [29] + [32]			*0.422*	*0.422*	*0.422*	*0.422*
35	Dry gas, lb/10,000 Btu	[33] – (34]				*9.442*		*9.442*
36	H_2O in gas, % by weight	[100] x [34] / [33]				*4.28*		*4.28*
37	Residue, % by weight	[9] x [24] / [33]				*0.42*		*0.42*

EFFICIENCY CALCULATIONS, % Input from Fuel

	Losses			
38	Dry gas, %	0.0024 x [35D] x ([6] – [3])		*7.02*
39	Water from fuel, as-fired %: Enthalpy of steam at 1 psi, T = [6]	H_1 = (3.958E–5 x T + 0.4329) x T + 1062.2	*1237.1*	
40	Water from fuel, as-fired %: Enthalpy of water at T = [3]	H_2 = [3] – 32	*48.0*	
41	Water from fuel, as-fired %	[29] x ([39] – [40]) / 100		*3.61*
42	Moisture in air, %	0.0045 x [27D] x ([6] – [3])		*0.16*
43	Unburned carbon, %	[19] or [21] x 14,500 / [18]		*0.40*
44	Radiation and convection, %	ABMA curve, Chapter 22		*0.40*
45	Unaccounted for and manufacturers margin, %			*1.50*
46	Sorbent net losses, % if sorbent is used	From Table 14 Item [41]		*0.00*
47	Summation of losses, %	Summation [38] through [46]		*13.09*
	Credits			
48	Heat in dry air, %	0.0024 x [26D] x ([2] – [3])		*0.00*
49	Heat in moisture in air, %	0.0045 x [27D] x ([2] – [3])		*0.00*
50	Sensible heat in fuel, %	(H at T[4] – H at T [3]) x 100 / [18]	*0.0*	*0.00*
51	Other, %			*0.00*
52	Summation of credits, %	Summation [48] through [51]		*0.00*
53	**Efficiency, %**	100 – [47] + [52]		*86.91*

	KEY PERFORMANCE PARAMETERS		Leaving Furnace	Leaving Blr/Econ
54	Input from fuel, 1,000,000 Btu/h	100 x [10] / [53]		*328.6*
55	Fuel rate, 1000 lb/h	1000 x [54] / [18]		*23.3*
56	Wet gas weight, 1000 lb/h	[54] x [33] / 10	*324.1*	*324.1*
57	Air to burners (wet), lb/10,000 Btu	(1 + [7]) x (1 + [25A] / 100) x [22]	*9.205*	
58	Air to burners (wet), 1000 lb/h	[54] x [57] / 10	*302.5*	
59	Heat available, 1,000,000 Btu/h	[54] x {([18] – 10.30 x [17H]) / [18] – 0.005		
	H_a = *66.0* Btu/lb	x ([44] + [45]) + H_a at T[5] x [57] / 10,000}	*335.2*	
60	Heat available/lb wet gas, Btu/lb	1000 x [59] / [56]	*1034.2*	
61	Adiabatic flame temperature, F	From Fig. 3 at H = [60], % H_2O = [36]	*3560*	

Table 13B
Combustion Calculations - Btu Method (with Sorbent)

	INPUT CONDITIONS - BY TEST OR SPECIFICATION		
1	Excess air: at burner/leaving boiler/econ, % by weight		*18/20*
2	Entering air temperature, F		*80*
3	Reference temperature, F		*80*
4	Fuel temperature, F		*80*
5	Air temperature leaving air heater, F		*350*
6	Flue gas temperature leaving (excluding leakage), F		*390*
7	Moisture in air, lb/lb dry air		*0.013*
8	Additional moisture, lb/100 lb fuel		*0*
9	Residue leaving boiler/economizer, % Total		*90*
10	Output, 1,000,000 Btu/h		*285.6*
	Corrections for sorbent (from Table 14 if used)		
11	Additional theoretical air, lb/10,000 Btu	Table 14, Item [21]	*0.0212*
12	CO_2 from sorbent, lb/10,000 Btu	Table 14, Item [19]	*0.0362*
13	H_2O from sorbent, lb/10,000 Btu	Table 14, Item [20]	*0.0015*
14	Spent sorbent, lb/10,000 Btu	Table 14, Item [24]	*0.0819*

FUEL - *Bituminous coal, Virginia; with sorbent*

	15 Ultimate Analysis		16 Theo Air, lb/100 lb fuel		17 H_2O, lb/100 lb fuel	
	Constituent	% by weight	K1	[15] x K1	K2	[15] x K2
A	C	*80.31*	11.51	*924.4*		
B	S	*1.54*	4.32	*6.7*		
C	H_2	*4.47*	34.29	*153.3*	8.94	*39.96*
D	H_2O	*2.90*			1.00	*2.90*
E	N_2	*1.38*				
F	O_2	*2.85*	− 4.32	*− 12.3*		
G	Ash	*6.55*				
H	Total	*100.00*	Air	*1072.1*	H_2O	*42.86*

18	Higher heating value (HHV), Btu/lb fuel		*14,100*
19	Unburned carbon loss, % fuel input		*2.50*
20	Theoretical air, lb/10,000 Btu	[16H] x 100 / [18]	*7.604*
21	Unburned carbon, % of fuel	[19] x [18] / 14,500	*2.43*

COMBUSTION GAS CALCULATIONS, Quantity / 10,000 Btu Fuel Input

22	Theoretical air (corrected), lb/10,000 Btu	[20] − [21] x 1151 / [18] + [11]	*7.427*
23	Residue from fuel, lb/10,000 Btu	([15G] + [21]) x 100 / [18]	*0.064*
24	Total residue, lb/10,000 Btu	[23] + [14]	*0.146*

			A At Burners	B Infiltration	C Leaving Furnace		D Leaving Blr/Econ	
25	Excess air, % by weight		*18.0*	*1.0*		*19.0*		*20.0*
26	Dry air, lb/10,000 Btu	(1 + [25] / 100) x [22]				*8.838*		*8.912*
27	H_2O from air, lb/10,000 Btu	[26] x [7]			*0.115*	*0.115*	*0.116*	*0.116*
28	Additional moisture, lb/10,000 Btu	[8] x 100 / [18]			*0.000*	*0.000*	*0.000*	*0.000*
29	H_2O from fuel, lb/10,000 Btu	[17H] x 100 / [18]			*0.304*		*0.304*	
30	Wet gas from fuel, lb/10,000 Btu	(100 − [15G] − [21]) x 100 / [18]				*0.646*		*0.646*
31	CO_2 from sorbent, lb/10,000 Btu	[12]				*0.036*		*0.036*
32	H_2O from sorbent, lb/10,000 Btu	[13]			*0.002*	*0.002*	*0.002*	*0.002*
33	Total wet gas, lb/10,000 Btu	Summation [26] through [32]				*9.637*		*9.712*
34	Water in wet gas, lb/10,000 Btu	Summation [27] + [28] + [29] + [32]			*0.421*	*0.421*	*0.422*	*0.422*
35	Dry gas, lb/10,000 Btu	[33] − [34]				*9.216*		*9.290*
36	H_2O in gas, % by weight	[100] x [34] / [33]				*4.37*		*4.35*
37	Residue, % by weight	[9] x [24] / [33]				*1.36*		*1.35*

EFFICIENCY CALCULATION, % Input from Fuel

	Losses				
38	Dry gas, %		0.0024 x [35D] x ([6] − [3])		*6.91*
39	Water from fuel, as-fired %	Enthalpy of steam at 1 psi, T = [6]	H_1 = (3.958E−5 x T + 0.4329) x T + 1062.2	*1237.1*	
40		Enthalpy of water at T = [3]	H_2 = [3] − 32	*48.0*	
41			[29] x ([39] − [40]) / 100		*3.61*
42	Moisture in air, %		0.0045 x [27D] x ([6] − [3])		*0.16*
43	Unburned carbon, %		[19] or [21] x 14,500 / [18]		*2.50*
44	Radiation and convection, %		ABMA curve, Chapter 22		*0.40*
45	Unaccounted for and manufacturers margin, %				*1.50*
46	Sorbent net losses, % if sorbent is used		From Table 14 Item [41]		*0.12*
47	Summation of losses, %		Summation [38] through [46]		*15.20*
	Credits				
48	Heat in dry air, %		0.0024 x [26D] x ([2] − [3])		*0.00*
49	Heat in moisture in air, %		0.0045 x [27D] x ([2] − [3])		*0.00*
50	Sensible heat in fuel, %		(H at T[4] − H at T [3]) x 100 / [18]	*0.0*	*0.00*
51	Other, %				*0.00*
52	Summation of credits, %		Summation [48] through [51]		*0.00*
53	**Efficiency, %**		100 − [47] + [52]		*84.80*

	KEY PERFORMANCE PARAMETERS		Leaving Furnace	Leaving Blr/Econ
54	Input from fuel, 1,000,000 Btu/h	100 x [10] / [53]		*336.8*
55	Fuel rate, 1000 lb/h	1000 x [54] / [18]		*23.9*
56	Wet gas weight, 1000 lb/h	[54] x [33] / 10	*324.6*	*327.1*
57	Air to burners (wet), lb/10,000 Btu	(1 + [7]) x (1 + [25A] / 100) x [22]	*8.878*	
58	Air to burners (wet), 1000 lb/h	[54] x [57] / 10	*299.0*	
59	Heat available, 1,000,000 Btu/h	[54] x {([18] − 10.30 x [17H]) / [18] − 0.005		
	H_a = *66.0* Btu/lb	x ([44] + [45]) + H_a at T[5] x [57] / 10,000}	*342.8*	
60	Heat available/lb wet gas, Btu/lb	1000 x [59] / [56]	*1056.1*	
61	Adiabatic flame temperature, F	From Fig. 3 at H = [60], % H_2O = [36]	*3624*	

Table 14
Combustion Calculations - Sorbent

	INPUTS (see also lightly shaded blocks)			FUEL-*Bituminous coal, Virginia*	
1	Sulfur in fuel, % by weight	*1.54*	6	Sulfur capture, lb/lb sulfur	*0.90*
2	Ash in fuel, % by weight	*6.55*	7	Reference temperature, F	*80.0*
3	HHV of fuel, Btu/lb	*14,100*	8	Exit gas temperature (excluding leakage), F	*390.0*
4	Unburned carbon loss, % fuel input	*2.5*	9	Sorbent temperature, F	*80.0*
5	Calcium to sulfur molar ratio	*2.5*			

SORBENT PRODUCTS

		[10] Chemical Analysis % Mass	[11] Molecular Weight lb/mole	[12] Ca mole/100 lb sorb [10]/[11]	[13] Calcination Fraction	[14] Molecular Weight lb/mole	[15] CO_2 lb/100 lb sorb [10]x[13]x[14]/[11]	[16] H_2O lb/100 lb sorb [10]x[13]x[14]/[11]
A	$CaCO_3$	*89.80*	100.089	*0.897*	*0.90*	44.010	*35.529*	
B	$MgCO_3$	*5.00*	84.321		1.00	44.010	*2.610*	
C	$Ca(OH)_2$	*0.00*	74.096	*0.000*	1.00	18.015		*0.000*
D	$Mg(OH)_2$	*0.00*	58.328		1.00	18.015		*0.000*
E	H_2O	*1.60*	18.015		1.00	18.015		*1.600*
F	Inert	*3.60*						
G	Total Ca, mole/100 lb sorbent			*0.897*		Total	*38.139*	*1.600*

SORBENT/GAS CALCULATIONS, lb/10,000 Btu Except as Noted

17	Sorbent, lb/lb fuel	[1] x [5] / [12G] / 32.066	*0.1339*
18	Sorbent, lb/10,000 Btu	10,000 x [17] / [3]	*0.0950*
19	CO_2 from sorbent, lb/10,000 Btu	[15G] x [18] / 100	*0.0362*
20	H_2O from sorbent, lb/10,000 Btu	[16G] x [18] / 100	*0.0015*
21	Additional theoretical air, lb/10,000 Btu	216 x [1] x [6] / [3]	*0.0212*
22	SO_2 reduction, lb/10,000 Btu	200 x [1] x [6] / [3]	*0.0197*
23	SO_3 formed, lb/10,000 Btu	0.2314 x [21] + [22]	*0.0246*
24	Spent sorbent, lb/10,000 Btu	[18] − [19] − [20] + [23]	*0.0819*
25	Unburned carbon, lb/10,000 Btu	[4] x 100 / 14,500	*0.0172*
26	Residue from fuel, lb/10,000 Btu	[2] x 100 / [3] + [25]	*0.0637*
27	Total residue, lb/10,000 Btu	[24] + [26]	*0.1456*

LOSSES DUE TO SORBENT, % Input from Fuel

28	H_2O from sorbent, %	H of steam at 1 psi, T = [8]	H_1 = (3.958E − 5 x [8] + 0.4329) x [8] + 1062.2	*1237.1*	
29		H of water	H_2 = [9] − 32	*48.0*	
30		0.01 x [20] x ([28] − [29])			*0.018*
31	Sensible heat sorbent (dry), %	[18] x (1.0 − [10E]/100) x (H at T = [9] − H at T = [7])/100			
		H of limestone (dry) = (0.1128E−3 x T + 0.179) x T − 14.45			*0.000*

Calcination/Dehydration, %

32	$CaCO_3$, %	[10A] x [13A] x [18] x 766 / 10,000	*0.588*	
33	$MgCO_3$, %	[10B] x 1.0 x [18] x 652 / 10,000	*0.031*	
34	$Ca(OH)_2$, %	[10C] x 1.0 x [18] x 636 / 10,000	*0.000*	
35	$Mg(OH)_2$, %	[10D] x 1.0 x [18] x 625 / 10,000	*0.000*	
36	Heat gain due to sulfation, %	[6] x [1] x 6733 / [3]	*0.662*	
37	Total of losses due to chemical reactions, %	[32] + [33] + [34] + [35] − [36]		*− 0.043*

Sensible Heat Of Residue Loss, %

	Location	[38] Temp Residue, F	[39] Mass Flow Rate, % Total	x	[27] lb/10,000 Btu	x (	H at T = [38] Btu/lb	−	H at T = [7] Btu/lb	) / 10,000 =	Loss %
A	Bed drain	*1500*	*10*	x	*0.1456*	x (	*376.3*	−	*0.5*	) / 10,000 =	*0.055*
B	Economizer	*600*	*10*	x	*0.1456*	x (	*116.2*	−	*0.5*	) / 10,000 =	*0.017*
C	Flyash	*390*	*80*	x	*0.1456*	x (	*64.3*	−	*0.5*	) / 10,000 =	*0.074*
	H Residue = ((− 2.843E − 8 x T + 1.09E − 4) x T + 0.16) x T − 12.95									40 Total	*0.146*
41	Summation losses due to sorbent, %				[30] − [31] + [37] + [40]						*0.121*

Table 15A
Excess Air Calculations from Measured O_2
Bituminous coal, Virginia-O_2 on wet basis

	INPUTS (see also lightly shaded blocks)			SORBENT DATA (if applicable)	
1	Moisture in air, lb/lb dry air	0.013	6	CO_2 from sorbent, moles/100 lb fuel, Table 16	0
2	Additional moisture, lb/100 lb fuel	0.00	7	H_2O from sorbent, moles/100 lb fuel, Table 16	0
3	HHV fuel, Btu/lb	14,100	8	Sulfur capture, lb/lb sulfur fuel, Table 16	0
4	Unburned carbon loss, % fuel input	0.40			
5	Unburned carbon (UBC), [3] x [4] / 14,500	0.39			

COMBUSTION PRODUCTS

9	Ultimate Analysis, % Mass			10 Theoretical Air lb/100 lb Fuel		11 Dry Products from Fuel mole/100 lb Fuel		12 Wet Products from Fuel mole/100 lb Fuel	
	Fuel Constituent	As-Fired	Carbon Burned (CB)	K1	[9] x K1	K2	[9] / K2	K3	[9] / K3
A	C	80.31	80.31						
B	UBC [5]		0.39						
C	CB [A]−[B]		79.92	11.51	919.9	12.011	6.654		
D	S	1.54		4.32	6.7	32.066	0.048		
E	H_2	4.47		34.29	153.3			2.016	2.217
F	H_2O	2.90						18.015	0.161
G	N_2	1.38				28.013	0.049		
H	O_2	2.85		− 4.32	− 12.3				
I	Ash	6.55							
K	Total	100.00			1067.6		6.751		2.378

13	Dry products of combustion, mole/100 lb fuel	[11K] − [11D] x [8] + [6]	6.751
14	Wet products of combustion, mole/100 lb fuel	[12K] + [13] + [7]	9.129
15	Theoretical air (corrected), mole/100 lb fuel	([10K] + [8] x [9D] x 2.16) / 28.966	36.857

EXCESS AIR WHEN O_2 KNOWN

16	O_2, % volume (input)					3.32
17	O_2 measurement basis	0 = Dry 1 = Wet	1	Dry	Wet	
18	Moisture in air, mole/mole dry air			0.0	[1] x 1.608	0.021
19	Dry/wet products of combustion, mole/100 lb fuel			[13]	[14]	9.129
20	Additional moisture, mole/100 lb fuel			0.0	[2] / 18.016	0.000
21	Intermediate calculation, step 1	[15] x (0.7905 + [18])				29.909
22	Intermediate calculation, step 2	[19] + [20] + [21]				39.038
23	Intermediate calculation, step 3	20.95 − [16] x (1 + [18])				17.560
24	Excess air, % by weight	100 x [16] x [22] / [15] / [23]				20.0

O_2, CO_2, SO_2 WHEN EXCESS AIR KNOWN

25	Excess air, % by weight				20.0
26	Dry gas, mole/100 lb fuel	[13] + [15] x (0.7905 + [25]/100)			43.258
27	Wet gas, mole/100 lb fuel	[14] + [15] x (0.7905 + [18] + (1 + [18]) x [25]/100) + [20]			46.565
			Dry	Wet	
28	O_2, % by volume	[25] x [15] x 0.2095 / ([26] or [27])	[26]	[27]	3.32
29	CO_2, % by volume	100 x ([11C] + [6]) / ([26] or [27])	[26]	[27]	14.29
30	SO_2, % by volume	100 x (1 – [8]) x [11D] / ([26] or [27])	[26]	[27]	0.1031
31	H_2O, % by volume	H_2O = 0.0 if dry or 100 x ([27] − [26]) / [27]	NA	[27]	7.10

Table 15B
Excess Air Calculations from Measured O_2

Bituminous coal, Virginia: with sorbent–O_2 on wet basis

	INPUTS (see also lightly shaded blocks)			SORBENT DATA (if applicable)	
1	Moisture in air, lb/lb dry air	*0.013*	6	CO_2 from sorbent, moles/100 lb fuel, Table 16	*0.116*
2	Additional moisture, lb/100 lb fuel	*0.00*	7	H_2O from sorbent, moles/100 lb fuel, Table 16	*0.012*
3	HHV fuel, Btu/lb	*14,100*	8	Sulfur capture, lb/lb sulfur fuel, Table 16	*0.90*
4	Unburned carbon loss, % fuel input	*2.50*			
5	Unburned carbon (UBC), [3] x [4] / 14,500	*2.43*			

COMBUSTION PRODUCTS

9	Ultimate Analysis, % Mass			10 Theoretical Air lb/100 lb Fuel		11 Dry Products from Fuel mole/100 lb Fuel		12 Wet Products from Fuel mole/100 lb Fuel	
	Fuel Constituent	As-Fired	Carbon Burned (CB)	K1	[9] x K1	K2	[9] / K2	K3	[9] / K3
A	C	*80.31*	*80.31*						
B	UBC [5]		*2.43*						
C	CB [A]–[B]		*77.88*	11.51	*896.4*	12.011	*6.484*		
D	S	*1.54*		4.32	*6.7*	32.066	*0.048*		
E	H_2	*4.47*		34.29	*153.3*			2.016	*2.217*
F	H_2O	*2.90*						18.015	*0.161*
G	N_2	*1.38*				28.013	*0.049*		
H	O_2	*2.85*		– 4.32	*– 12.3*				
I	Ash	*6.55*							
K	Total	*100.00*			*1044.1*		*6.581*		*2.378*

13	Dry products of combustion, mole/100 lb fuel	[11K] – [11D] x [8] + [6]	*6.654*
14	Wet products of combustion, mole/100 lb fuel	[12K] + [13] + [7]	*9.044*
15	Theoretical air (corrected), mole/100 lb fuel	([10K] + [8] x [9D] x 2.16) / 28.966	*36.149*

EXCESS AIR WHEN O_2 KNOWN

16	O_2, % volume (input)				*3.31*
17	O_2 measurement basis	0 = Dry 1 = Wet	*1* Dry	Wet	
18	Moisture in air, mole/mole dry air		0.0	[1] x 1.608	*0.021*
19	Dry/wet products of combustion, mole/100 lb fuel		[13]	[14]	*9.044*
20	Additional moisture, mole/100 lb fuel		0.0	[2] / 18.016	*0.000*
21	Intermediate calculation, step 1	[15] x (0.7905 + [18])			*29.335*
22	Intermediate calculation, step 2	[19] + [20] + [21]			*38.379*
23	Intermediate calculation, step 3	20.95 – [16] x (1 + [18])			*17.570*
24	Excess air, % by weight	100 x [16] x [22] / [15] / [23]			*20.0*

O_2, CO_2, SO_2 WHEN EXCESS AIR KNOWN

25	Excess air, % by weight				*20.0*
26	Dry gas, mole/100 lb fuel	[13] + [15] x (0.7905 + [25]/100)			*42.460*
27	Wet gas, mole/100 lb fuel	[14] + [15] x (0.7905 + [18] + (1 + [18]) x [25]/100) + [20]			*45.761*
			Dry	Wet	
28	O_2, % by volume	[25] x [15] x 0.2095 / ([26] or [27])	[26]	[27]	*3.31*
29	CO_2, % by volume	100 x ([11C] + [6]) / ([26] or [27])	[26]	[27]	*14.42*
30	SO_2, % by volume	100 x (1 – [8]) x [11D] / ([26] or [27])	[26]	[27]	*0.0105*
31	H_2O, % by volume	H_2O = 0.0 if dry or 100 x ([27] – [26]) / [27]	NA	[27]	*7.21*

Table 16
Sulfur Capture Based on Gas Analysis
Bituminous coal, Virginia

	INPUTS						
1	SO_2, ppm	105	/ 10,000 = %	0.0105	2	O_2 Flue gas at location SO_2 measured, %	3.31
	Data from Table 15, Excess Air Calculations from Measured O_2						
3	Moisture in air, lb/lb dry air			0.013	7	Theoretical air, lb/100 lb fuel	1044.1
4	Additional moisture, lb/100 lb fuel			0	8	Dry products of combustion, mole/100 lb fuel	6.581
5	Sulfur in fuel, % by weight			1.54	9	Wet products of combustion, mole/100 lb fuel	2.378
6	HHV fuel, Btu/lb fuel			14,100			
	Data from Table 14, Combustion Calculations-Sorbent						
10	CO_2 from sorbent, lb/100 lb sorbent			38.139	12	Sorbent, lb sorbent/lb fuel	0.134
11	H_2O from sorbent, lb/100 lb sorbent			1.600			

	CALCULATIONS, Moles/100 lb Fuel Except As Noted				
	SO_2 / O_2 Measurement basis	0 = Dry 1 = Wet 1	Dry	Wet	
13	Moisture in air, mole/mole dry air		0.0	[3] x 1.608	0.0209
14	Additional moisture		0.0	[4] / 18.015	0.000
15	Products of combustion from fuel		[8]	[8] + [9]	8.959
16	H_2O from sorbent	[11] x [12] / 18.015	0.0	Calculate	0.012
17	CO_2 from sorbent	[10] x [12] / 44.01			0.116
18	Intermediate calculation, step 1	(0.7905 + [13]) x [7] / 28.966			29.247
19	Intermediate calculation, step 2	Summation [14] through [18]			38.334
20	Intermediate calculation, step 3	1.0 − (1.0 + [13]) x [2] / 20.95			0.8387
21	Intermediate calculation, step 4	(0.7905 + [13]) x 2.387 − 1.0			0.9368
22	Intermediate calculation, step 5	[1] x [19] x 32.066 / [5] / [20]			9.993
23	Intermediate calculation, step 6	[21] x [1] / [20]			0.0117
24	Sulfur capture, lb/lb sulfur	(100 − [22]) / (100 + [23])			0.90
25	SO_2 released, lb/1,000,000 Btu	20,000 x (1.0 − [24]) x [5] / [6]			0.22

Table 17 Combustion Calculations-Measured Gas Weight

	INPUTS			
		A Wet Analysis (not req'd)		B Dry Analysis
1	O_2, % volume	9.28	Measured dry or 100 / (100 − [3A]) x [1A]	10.55
2	CO_2, % volume	8.56	Measured dry or 100 / (100 − [3A]) x [2A]	9.73
3	H_2O, % volume	12.00		
4	Mass flow wet gas, 1000 lb/h			539.2
5	Moisture in wet gas, lb/lb wet gas			0.0754
6	Moisture in air, lb/lb dry air			0.0130
7	Additional moisture (sources other than fuel and air), 1000 lb/h			0
	CALCULATIONS			
8	Water in wet gas, 1000 lb/h		[4] x [5]	40.7
9	Dry gas weight, 1000 lb/h		[4] − [8]	498.5
10	N_{2a} in dry gas, % dry volume		100 − [1B] − [2B]	79.72
11	Molecular weight of dry gas, lb/mole		0.32 x [1B] + 0.4401 x [2B] + 0.28161 x [10]	30.11
12	Dry gas, 1000 moles/h		[9] / [11]	16.56
13	Dry air weight, 1000 lb/h		0.28161 x [10] x [12] / 0.7685	483.8
14	Water in dry air, 1000 lb/h		[13] x [6]	6.3
15	Water evaporated, 1000 lb/h		[8] − [7] − [14]	34.4
16	Excess air, 1000 lb/h		[1B] x [9] x 0.32 / 0.2315 / [11]	241.4
17	Theoretical air, 1000 lb/h		[13] − [16]	242.4
18	Excess air, % by weight		100 x [16] / [17]	99.6

Fig. 1 Crude oil and natural gas supply begins with facilities like this 62-well offshore platform in the Gulf of Mexico.

ited for conservation reasons and the cost of the pipeline to provide the peak rate would be prohibitive. Therefore, to meet fluctuations in demand, it is usually necessary to provide local storage or to supplement the supply with manufactured gas for brief periods.

Above ground methods of storage include: 1) large water seal tanks, 2) in-pipe holders laid parallel to commercial gas lines, and 3) using the trunk transmission line as a reservoir by building up the line pressure. In consumer areas where depleted or partially depleted gas and oil wells are available, underground storage of gas pumped back into these wells provides, at minimum cost, the large storage volume required to meet seasonal variations in demand. In liquid form, natural gas can be stored in insulated steel tanks or absorbed in a granular substance, released by passing warm gas over the grains.

Fuel properties

Natural gas is comprised primarily of methane and ethane. Physical properties of practical importance to boiler applications include ultimate analysis, heating value, specific gravity, sulfur content and flammability.

In general, natural gas suitable for combustion with conventional utility and industrial burner designs must contain at least:

70% methane (CH_4)
or 70% propane (C_3H_8)
or 25% hydrogen (H_2) by volume

Oil and gas combustion — system design

The burner is the principal equipment component for the combustion of oil and natural gas (Fig. 2). In utility and industrial steam generating units, the burner admits fuel and air to the furnace in a manner that ensures safe and efficient combustion while realizing the full capability of the boiler. Burner design determines mixing characteristics of the fuel and air, fuel particle size and distribution and size and shape of the flame envelope.

The means of transporting, measuring and regulating fuel and air to the furnace, together with the burners, ignitors and flame safety equipment comprises the overall combustion system. The following factors must be considered when designing the combustion system and when establishing overall performance requirements:

Fig. 2 Typical oil and gas utility boiler burner front.

1. the rate of feed of the fuel and mand on the boiler over a pred
2. the types of fuel to be fired inc ents and characteristic prop
3. the efficiency of the combus unburned combustibles and
4. imposed limitations on emis
5. physical size and complexity to establish the most efficien
6. hardware design and materia tion equipment to ensure reli for long firing periods, and
7. safety standards and procedu ers and boiler, including starti and variations in fuel.

The combustion system must flexibility of operation, includir tions in fuel type, fuel firing r burners in and out of service. Co direct to ensure rapid response

Combustion air is typically by forced draft fans. To improv bustion efficiency and further combustion air is normally prel of 400 to 600F (204 to 316C) b downstream of the fans. The far livering adequate quantities of tion at a pressure sufficient to ov air preheaters, burners, control ing duct work. The total combus to theoretically burn all the fue sary for complete combustion. (

The fuel delivery system mus pressure and flow to the burners in accordance with applicable Proper distribution of fuel to t burner applications, is critical eration of the combustion system be designed for allowable velocit sure requirements and pressure

Performance requirement

Excess air

Excess air is the air supplied f ing of idle burners in excess of tha for complete oxidation of the fuel required to compensate for impe livery system that results in mal tion air to the burners. Excess air for imperfect mixing of the air an full load, with all burners in serv for gas and oil firing, expressed a cal air, is typically in the range upon fuel type and the requirem system. Operation at excess air ues is possible if combustion effic rate. Combustion efficiency is me bon monoxide, unburned combus particulate matter and stack op design of the burners and the air air can be held to a minimum, th sible heat loss to the stack.

Chapter 10
Oil and Gas Utilization

Before the industrial revolution, distilled petroleum products were used primarily as a source of illumination. Today, petroleum finds its primary importance as an energy source and greatly influences the world's economy. In the modern world the production, distribution and consumption of oil and natural gas dictate international relations and often govern matters of foreign policy. The following deals with the use of petroleum products and natural gas as energy sources for steam generation.

Fuel oil

Preparation

Petroleum or crude oil is the source of various fuel oils used for steam generation (Fig. 1, facing page). Most petroleum is refined to some extent before use although small amounts are burned without processing. Originally, refining petroleum was simply the process of separating the lighter compounds, higher in hydrogen, from the heavier compounds by fractional distillation. This yielded impure forms of kerosene, gasoline, lubricating oils and fuel oils. Through the development of refining techniques, such as thermal cracking and reforming, catalytic reforming, polymerization, isomerization and hydrogenation, petroleum is now regarded as a raw material source of hydrogen and carbon elements that can be combined as required to meet a variety of needs.

In addition to hydrocarbons, crude oil contains compounds of sulfur, oxygen and nitrogen and traces of vanadium, nickel, arsenic and chlorine. Processes are used during petroleum refinement to remove impurities, particularly compounds of sulfur. Purification processes for petroleum products include sulfuric acid treatment, sweetening, mercaptan extraction, clay treatment, hydrogen treatment and the use of molecular sieves.

The refining of crude oil yields a number of products having many different applications. Those used as fuel include gasoline, distillate fuel, residual fuel oil, jet fuels, still gas, liquefied gases, kerosene and petroleum coke. Products for other applications include lubricants and waxes, asphalt, road oil and petrochemical feedstock.

Fuel oils for steam generation consist primarily of residues from the distillation of crude oil. As refinery methods improve, the quality of residual oil available for utility and industrial steam generation is deteriorating. High sulfur fuels containing heavy components create problems during combustion that range from high particulate and sulfur oxide emissions to higher maintenance costs due to the corrosive constituents in the flue gas.

Transportation, storage and handling

The high heating value per unit of volume of oil, its varied applications and its liquid form have fostered a worldwide system of distribution. The use of supertankers for the transportation of crude oil has significantly reduced transportation costs and has allowed refineries to be located near centers of consumption rather than adjacent to the oil fields. Large supertankers, up to 250,000 t (227,000 t_m), are capable of transporting nearly 2,000,000 bbl (318,000 m^3) of crude oil at a time to deepwater ports.

Tanker and barge shipments on coastal and inland waterways are by far the cheapest method of transporting the various grades of oil. With the depletion of oil fields in the Eastern United States (U.S.), crude oil trunk lines were developed in the early 1900s to transport oil from points west of the Mississippi River to the east coast refineries. Today, more than 170,000 mi (274,000 km) of pipeline, including small feeder lines, are used for the transportation of oil within the U.S. Much smaller quantities of oil are shipped overland by rail and truck because of the higher cost of haulage.

Fuel oil systems require either underground or surface storage tanks. Oil is usually stored in cylindrical shaped steel tanks to eliminate evaporation loss and to protect it from lightning. Loss in storage of the relatively nonvolatile heavy fuel oils is negligible. Lighter products, such as gasoline, may volatilize sufficiently in warm weather to cause appreciable loss. In this instance, storage tanks with floating roofs are used to eliminate the air space above the fuel where vapors can accumulate. The National Fire Protection Association (NFPA) has prepared a standard set of rules for the storage and handling of oils. These rules serve as the basis for many local ordinances and form a practical guide for the safe transportation and handling of fuel products.

Extensive piping and valving and suitable pumping and heating equipment are necessary for the transportation and handling of fuel oil. Storage tanks, piping and heaters for heavy oils must be cleaned periodically because of fouling or sludge accumulation.

Fuel properties

Safe and efficient transportation, handling and combustion of fuel oil requires a knowledge of fuel characteristics. Principal physical properties of fuel oils, important to boiler applications, are summarized below:

Viscosity The viscosity of an oil is the measure of its resistance to internal movement, or flow. Viscosity is important because of its effect on the rate at which oil flows through pipelines and on the degree of atomization obtained by oil firing equipment.

Ultimate analysis An ultimate analysis is used to determine theoretical air requirements for combustion of the fuel and also to identify potential environmental emission characteristics.

Heating value The heating value of a liquid fuel is the energy produced by the complete combustion of one unit of fuel [Btu/lb (J/kg)]. Heating value can be reported either as the gross or higher heating value (HHV) or the net or lower heating value (LHV). To determine HHV, it is assumed that any water vapor formed during combustion is condensed and cooled to the initial temperature. The heat of vaporization of the water formed is included in the HHV. For LHV, it is assumed that the water vapor does not condense. The higher heating value determines the quantity of fuel necessary to achieve a specified heat input.

Specific gravity Specific gravity is the ratio of the density of oil to the density of water. It is important because fuel is purchased by volume, in gallons (l) or barrels (m^3). The most widely used fuel oil gravity scale is degree API devised by the American Petroleum Institute; its use is recommended by the U.S. Bureau of Standards and the U.S. Bureau of Mines. The scale is based on the following formula:

$$\text{degrees API} = \frac{141.5}{\text{sp gr at 60/60F (16/16C)}} - 131.5$$

[sp gr at 60/60F (16/16C) means when both oil and water are at 60F (16C)]

Flash and fire point Flash point is the lowest temperature at which a volatile oil will give off explosive or ignitable vapors. It is important in determining oil handling and storage requirements. The fire point is the temperature to which a liquid must be heated to produce vapors sufficient for continuous burning when ignited by an external flame.

Pour point The pour point is the temperature at which a liquid fuel will first flow under standardized conditions.

Distillation Distillation determines the quantity and number of fractions which make up the liquid fuel.

Water and sediment Water and sediment are a measure of the contaminates in a liquid fuel. The sediment normally consists of calcium, sodium, magnesium and iron compounds. Impurities in the fuel provide an indication of the potential for plugging of fuel handling and combustion equipment.

Carbon residue Residue that remains after a liquid fuel is heated in the absence of air is termed carbon residue. The tests commonly used to determine carbon residue are the Conradson Carbon Test and the Ramsbottom Carbon Test. Carbon residue gives an indication of the coking tendency of a particular fuel.

Asphaltene content Asphaltenes are long chain, high molecular weight hydrocarbon compounds. The asphaltene content of a petroleum product is the percentage by weight of wax free material insoluble in n-hep-

tane but soluble in hot l
high temperatures and
completely. Higher aspl
potential to produce pa

Burning profile Bur
which a sample of fuel l
as temperature is incre
profile is a characterist
under standard conditi
absolute kinetic and the
ate combustion charact
tive basis to determine
necessary for complete

Natural gas

Preparation

Natural gas, found ir
solved in the oil or as a
associated gas. Natural
that contain no oil and i

Natural gas, directly
to produce commercial
natural gas undergoes a
which is distilled to prod
lized gasoline. Propane
bottle gas. They are distr
der pressure. When the
boils, producing a gaseou

Natural gas may conta
fur compounds to be tro
removed at the source. N
amounts of hydrogen sul
gas, can be treated by a
Sweetening removes hyd
dioxide. Additional treat
mercaptan by soda fixat
chain hydrocarbons.

Where natural gas is u
manufactured gas, it is sc
heating value in line with
ral gas may also be mixe
gas to increase the heatin

Transportation, storage a

Pipelines are an economi
ral gas in its gaseous form.
tion of natural gas in areas t
in an extensive system of lo
gas can also be transported
der pressure producing liqu

The distribution of natu
tical limitations because of
portation. High pressures,
kPa), are necessary for eco
over long distances. Compr
specified intervals to boost
the line.

In general, it is not practi
ral gas to accommodate the
in consumer demand. For
tance pipelines should ope
The rate of withdrawal fron

Operation at partial load requires additional excess air. When operating with all burners in service at reduced load, lower air velocity at the burner results in reduced mixing efficiency of the fuel and air. Increasing the excess air improves combustion turbulence and maintains overall combustion efficiency. Additional excess air and improved burner mixing also compensates for lower furnace temperature during partial load operation. In some instances, boiler performance dictates the use of higher than normal excess air at reduced loads to maintain steam temperature or to minimize cold end corrosion.

Additional excess air is also necessary when operating with burners out of service. Sufficient cooling air must be provided to idle burners to prevent overheat damage. Permanent thermocouples installed on select burners measure metal temperature and establish efficient excess air necessary to maintain burner temperatures below the maximum use limits of the steel. Excess air for burner cooling varies with the percentage of burners out of service.

Stability and turndown

Proper burner and combustion system design will permit stable operation of the burners over a wide operating range. A stable burner, best determined through visual observation, is one where the flame front remains relatively stationary and the root of the flame is securely anchored to the burner fuel element. To ensure stable combustion, the burner must be designed to prevent blowoff or flashback of the flame for varying rates of fuel and air flow.

It is often desirable to operate over a wide boiler load range without taking burners out of service. This reduces partial load excess air requirements to cool idle burners. The burners must therefore be capable of operating in a turned down condition. Burner turndown is defined as the ratio of full load fuel input to partial load input while still maintaining stable combustion. Limitations in burner turndown are generally dictated by fuel characteristics, fuel and air velocity, full load to partial load fuel pressures and adequacy of the flame safety system. Automated and reliable flame safety supervision, with proper safeguards, must be available to achieve high burner turndown ratios.

With gas firing, a turndown ratio of 10:1 is not uncommon. Natural gas is easily burned and relatively easy to control. Residual oil, on the other hand, is more difficult to burn. Combustion characteristics are highly sensitive to particle size distribution, excess air and burner turbulence. A typical turndown ratio for oil is in the order of 5:1, depending upon fuel characteristics, flexibility of the delivery system and atomization technique.

Burner pulsation

Burner pulsation is a phenomenon frequently associated with natural gas firing and, to a lesser degree, with oil firing. Pulsation is thought to occur when fuel rich or oxygen rich pockets of gas suddenly ignite within the flame envelope. The resultant pulsating burner flame is often accompanied by a noise referred to as combustion rumble. Combustion rumble may transmit frequencies that coincide with the natural frequency of the furnace enclosure resulting in apparent boiler vibration. In some instances, these vibrations may become alarmingly violent.

Boiler vibration on large furnaces can sometimes be attributed to a single burner. Minor air flow adjustment to a given burner, or removing select burners from service, may suddenly start or stop pulsation. Pulsation problems can be corrected through changes to burner hardware that affect mixing patterns of the fuel and air. Changes to the burner throat profile to correct anomalies in burner aerodynamics or changes to the fuel element discharge ports have successfully eliminated pulsation.

Historical operating data has enabled the development of empirical curves that are useful in designing burners to avoid pulsation. These curves relate the potential for burner pulsation to the ratio of burner fuel to air velocity. Together with careful consideration of furnace geometry, burner firing patterns and burner aerodynamics, problems with burner pulsation are becoming less common.

Combustion efficiency

Many factors influence combustion efficiency including excess air, burner mixing, fuel properties, furnace thermal environment, residence time and particle size and distribution. Complete combustion occurs when all combustible elements and compounds of the fuel are entirely oxidized. In utility and industrial boilers, the goal is to achieve the highest degree of combustion efficiency with the lowest possible excess air. As thermal efficiency decreases with increasing quantities of excess air, combustion performance is measured in terms of the efficiency loss due to incomplete combustion together with the efficiency loss due to sensible heat in the stack gases.

From the standpoint of optimum combustion efficiency the following factors are critical to proper design:

1. careful distribution and control of fuel and air to the burners,
2. burner and fuel element design that provides thorough mixing of fuel and air and promotes rapid, turbulent combustion, and
3. proper burner arrangement and furnace geometry to provide sufficient residence time to complete chemical reactions in a thermal environment conducive to stable and self-sustained combustion.

With oil firing, depending upon fuel properties, heat loss due to unburned combustibles (UBC) may be in the order of 0.2%. In most cases, however, unburned combustible loss when firing oil or natural gas is virtually negligible. Combustion efficiency with these fuels is usually measured in terms of carbon monoxide (CO) emissions, particulate emissions and stack opacity. Generally, CO levels in the range of 50 to 150 ppm are considered satisfactory.

Emission control techniques

Ever increasing concern over atmospheric pollutants is changing the focus of boiler and combustion system design. The combustion of fossil fuels produces emissions that have been attributed to the formation of acid rain, smog, changes to the ozone layer and the so called *greenhouse* effect. To mitigate these problems, federal and local regulations are currently in place that limit oxides of nitrogen, oxides of sulfur, particulate matter and stack opacity. While emission limits vary depending upon state and local regulations, the trend is toward more stringent control. (See also Chapter 32.)

Many combustion control techniques have emerged to reduce fossil fuel emissions. These techniques generally focus on the reduction of nitrogen oxides (NO_x), as changes to the combustion process can greatly influence NO_x formation and destruction.

Oxides of nitrogen

Nitrogen oxides in the form of NO and NO_2 are formed during combustion by two primary mechanisms: thermal NO_x and fuel NO_x. Thermal NO_x results from the dissociation and oxidation of nitrogen in the combustion air. The rate and degree of thermal NO_x formation is dependent upon oxygen availability during the combustion process and is exponentially dependent upon combustion temperature. Thermal NO_x reactions occur rapidly at combustion temperatures in excess of 2800F (1538C). Thermal NO_x is the primary source of NO_x formation from natural gas and distillate oils because these fuels are generally low or devoid of nitrogen. Fuel NO_x, on the other hand, results from oxidation of nitrogen organically bound in the fuel. Fuel bound nitrogen in the form of volatile compounds is intimately tied to the fuel hydrocarbon chains. For this reason, the formation of fuel NO_x is linked to both fuel nitrogen content and fuel volatility. Inhibiting oxygen availability during the early stages of combustion, during fuel devolatilization, is the most effective means of controlling fuel NO_x formation.

Numerous combustion process NO_x control techniques are commonly used. These vary in effectiveness and cost. In all cases, control methods are aimed at reducing either thermal NO_x, fuel NO_x, or a combination of both (Fig. 3).

Low excess air Low excess air (LEA) effectively reduces NO_x emissions with little, if any, capital expenditure. LEA is a desirable method of increasing thermal efficiency and has the added benefit of inhibiting thermal NO_x. If burner stability and combustion efficiency are maintained at acceptable levels, lowering the excess air may reduce NO_x by as much as a 10 to 20%. The success of this method depends largely upon fuel properties and the ability to carefully control fuel and air distribution to the burners. Operation may require more sophisticated methods of measuring and regulating fuel and air flow to the burners and modifications to the air delivery system to ensure equal distribution of combustion air to all burners.

Burners out of service Essentially a crude form of two-stage combustion, burners out of service (BOOS) is a simple and direct method of reducing NO_x emissions. When removing burners from service in multiple burner applications, active burner inputs are typically increased to maintain load. Without changing total air flow, increased fuel input to the active burners results in a fuel rich mixture, effectively limiting oxygen availability and thereby limiting both fuel and thermal NO_x formation. Air control registers on the out of service burners remain open, ostensibly serving as overfire air ports.

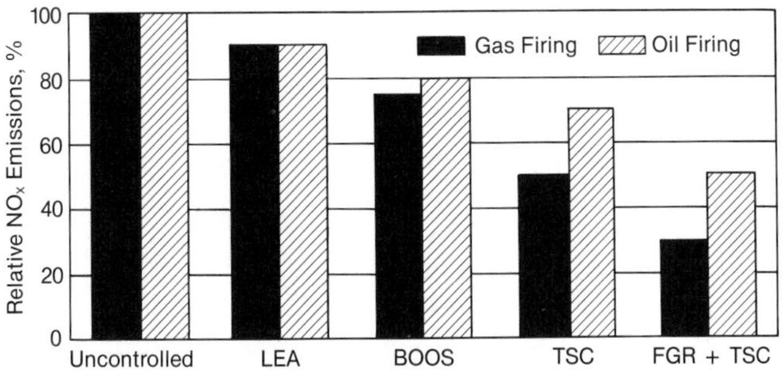

Fig. 3 Approximate NO_x emission reductions for oil and gas burners using various control techniques. [LEA (low excess air); BOOS (burners out of service); TSC (two-stage combustion); FGR (flue gas recirculation)].

While a fairly significant NO_x reduction is possible with this method, lower NO_x is frequently accompanied by higher levels of CO in the flue gas and boiler back-end oxygen (O_2) imbalances. With oil firing, an increase in particulate emissions and increased stack opacity are likely. Through trial and error, some patterns of burners out of service may prove more successful than others. A limiting factor is the ability of existing burners to handle the increased input necessary to maintain full load operation. Short of derating the unit, changes to fuel element sizes may be required.

Two–stage combustion Two-stage combustion is a relatively long standing and accepted method of achieving significant NO_x reduction. Combustion air is directed to the burner zone in quantities less than that required to theoretically burn the fuel, with the remainder of the air introduced through overfire air ports. By diverting combustion air away from the burners, oxygen concentration in the lower furnace is reduced, thereby limiting the oxidation of chemically bound nitrogen in the fuel. By introducing the total combustion air over a larger portion of the furnace, peak flame temperatures are also lowered.

Appropriate design of a two-stage combustion system can reduce NO_x emissions by as much as 50% and simultaneously maintain acceptable combustion performance. The following factors must be considered in the overall design of the system.

1. *Burner zone stoichiometry* The fraction of theoretical air directed to the burners is predetermined to allow proper sizing of the burners and overfire air ports. Normally a burner zone stoichiometry in the range of 0.85 to 0.90 will result in desired levels of NO_x reduction without notable adverse effects on combustion stability and turndown.
2. *Overfire air port design* Overfire air ports must be designed for thorough mixing of air and combustion gases in the second stage of combustion. Ports must have the flexibility to regulate flow and air penetration to promote mixing both near the furnace walls and toward the center of the furnace. Mixing efficiency must be maintained over the anticipated boiler load range and the range in burner zone stoichiometries.
3. *Burner design* Burners must be able to operate at lower air flow rates and velocities without detriment to combustion stability. In a two-stage combustion system, burner zone stoichiometry is typically increased with decreasing load to ensure that burner air velocities are maintained above minimum limits. This further ensures positive windbox to furnace differential pressures at reduced loads.
4. *Overfire air port location* Sufficient residence time from the burner zone to the overfire air ports and from the ports to the furnace exit is critical to proper system design. Overfire air ports must be located to optimize NO_x reduction and combustion efficiency and to limit change to furnace exit gas temperatures.

5. *Furnace geometry* Furnace geometry influences burner arrangement and flame patterns, residence time and thermal environment during the first and second stages of combustion. Liberal furnace sizing is generally favorable for lower NO_x as combustion temperatures are lower and residence times are increased.
6. *Air flow control* Ideally, overfire air ports are housed in a dedicated windbox compartment. In this manner, air to the NO_x ports can be metered and controlled separately from air to the burners. This permits operation at desired stoichiometric levels in the lower furnace and allows for compensation to the flow split as a result of air flow adjustments to individual burners or NO_x ports.

Additional flexibility in controlling burner fuel and air flow characteristics is required to optimize combustion under a two-stage system. Consequently, improved burner designs are emerging to address the demand for tighter control of air flow and fuel firing patterns to individual burners.

In the reducing gas of the lower furnace, sulfur in the fuel forms hydrogen sulfide (H_2S) rather than sulfur dioxide (SO_2) and sulfur trioxide (SO_3). The corrosiveness of reducing gas and the potential for increased corrosion of lower furnace wall tubes is highly dependent upon H_2S concentration. Two-stage combustion is therefore not recommended when firing high sulfur residual fuel oils.

Flue gas recirculation Flue gas recirculation (FGR) to the burners is instrumental in reducing NO_x emissions when the contribution of fuel nitrogen to total NO_x formation is small. For this reason, the use of gas recirculation is generally limited to the combustion of natural gas and fuel oils. By introducing flue gas from the economizer outlet into the combustion air stream, burner peak flame temperatures are lowered and NO_x emissions are significantly reduced. (See Fig. 4.)

Air foils are commonly used to mix recirculated flue gas with the combustion air. Flue gas is introduced in the sides of the secondary air measuring foils and exits through slots downstream of the air measurement taps. This method ensures thorough mixing of flue gas and combustion air before reaching the burners and does not affect the air flow metering capability of the foils.

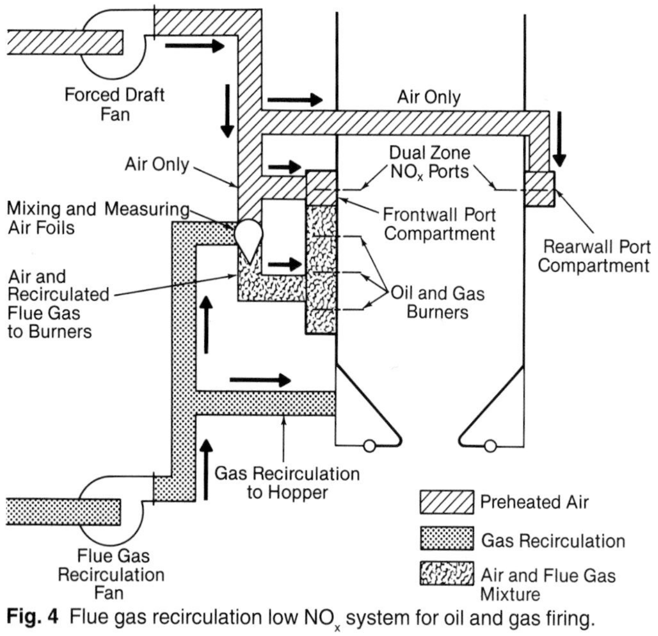

Fig. 4 Flue gas recirculation low NO_x system for oil and gas firing.

In general, increasing the rate of flue gas recirculation to the burners results in an increasingly significant NO_x reduction. Target NO_x emission levels and limitations on equipment size and boiler components dictate the practical limit of recirculated flue gas for NO_x control. Other limiting factors include burner stability and oxygen concentration of the combustion air. Oxygen content must be maintained at or above 17% on a dry basis for safe and reliable operation of the combustion equipment.

The expense of a flue gas recirculation system can be significant. Gas recirculation (GR) fans are required for the desired flow quantities at static pressures capable of overcoming losses through the flues, ducts, mixing devices and the burners themselves. Additional controls and instruments are also necessary to regulate GR flow to the windbox at desired levels over the load range. In retrofit applications, significant cost is associated with routing of flues and ducts to permit mixing of the flue gas with combustion air. Also, the accompanying increase in furnace gas weight at full load operation may require modifications to convection pass surfaces or dictate changes to standard operating procedures.

From an operational standpoint, the introduction of flue gas recirculation as a retrofit NO_x control technique must, in virtually all cases, be accompanied by the installation of overfire air ports. Oil and gas burners, initially designed without future consideration to FGR, are not properly sized to accommodate the increase in burner mass flow as a result of recirculated flue gas. The quantity of flue gas necessary to significantly reduce NO_x emissions will, in all likelihood, result in burner throat velocities that exceed standard design practices. This, in turn, may cause burner instability, prohibitive burner differentials and in the case of gas firing, undesirable pulsation. Therefore, the installation of overfire air ports in conjunction with FGR serves two useful purposes, 1) lower NO_x emissions through two-stage combustion, and 2) a decrease in mass flow of air to the burners to accommodate the increased burden of recirculated flue gas.

When employing flue gas recirculation in combination with overfire air, it is desirable to house the overfire air ports in a dedicated windbox compartment separate from the burners. In this manner, it is possible to introduce recirculated flue gas to the burners only. This permits more efficient use of the GR fans and overall system design as only that portion of flue gas introduced through the burners is considered effective in controlling NO_x emissions.

Reburning Reburning is a relatively new in-furnace NO_x control technique. By effectively staging both fuel and combustion air, NO_x emissions as low as 40 to 60 ppm (corrected to 3% O_2) are possible when firing residual oil. Even lower NO_x emissions are possible when firing natural gas. Heat input is spread over a larger portion of the furnace, with combustion air carefully regulated to various zones to achieve optimum NO_x reduction (Fig. 5).

In reburning, the lower furnace or main burner zone provides the major portion of the total heat input to the furnace. Similar to two-stage combustion, air less than that theoretically required to burn the fuel is introduced into this zone. Combustion gases from the main burner zone then pass through a second combustion zone termed the reburning zone. Here, burners provide the remaining heat input to the furnace to achieve full load operation but at a significantly lower stoichiometry. By injecting reburn fuel above the main burner zone, a NO_x reducing region is produced in the furnace where hydrocarbon radicals from the partially oxidized reburn fuel strip oxygen from the NO molecules, leaving elemental nitrogen to ultimately form molecular nitrogen (N_2). Overfire air ports are installed above the reburning zone where the remainder of air is introduced to complete combustion in an environment both chemically and thermally nonconducive to NO_x formation.

Application of this technology must consider a number of variables. System parameters requiring definition include: fuel split between the main combustion zone and the reburn zone, stoichiometry to the main burners, stoichiometry to the reburn burners, overall stoichiometry in the reburn zone of the furnace, residence time in the reburn zone and residence time required above the overfire air ports to complete combustion. An optimum range of values has been defined for each of these parameters through laboratory tests and field application and is largely dependent upon the type of fuel being fired.

Implementation of reburning technology adds considerable complexity to operation and maintenance of the overall combustion system. Initial costs may be prohibitive, particularly for retrofits. From an economic standpoint, the potential benefits and technical merit of the reburning process must be commensurate with long term goals for NO_x abatement.

Oxides of sulfur

The sulfur content of fuel oils can range anywhere from a fraction of a percent for lighter oils to 3.5% for some residual oils. During the combustion process, all sulfur contained in the fuel is converted to either sulfur dioxide, SO_2, or sulfur trioxide, SO_3 (SO_x emissions). The control of SO_x emissions is of increasing environmental concern and sulfur compounds in the flue gas can also cause boiler and downstream equipment corrosion problems.

SO_3 will form sulfuric acid when cooled in the presence of water vapor. In addition to corrosion problems, it can produce emissions of acid smut and visible plume opacity from the stack. Emissions of SO_3 are best controlled through low excess air operation and can also be reduced through using magnesium based fuel additives.

Techniques to control sulfur oxides during the combustion process have been investigated in laboratory and pilot scale tests with varying degrees of success. At present, however, the most effective and commercially accepted method, short of firing low sulfur fuels, is to install flue gas cleanup equipment (Chapter 35).

Particulate matter

Particulate matter in the form of soot or coke is a byproduct of the combustion process resulting from carryover of inert mineral matter in the fuel and from incomplete combustion. Primarily a concern on oil-fired units, particulate matter becomes apparent when fuel oil droplets undergo a form of fractional distillation during combustion, leaving relatively large carbonaceous particles known as cenospheres.

Cenospheres are porous, hollow particles of carbon that are virtually unaffected by further combustion in a conventional furnace environment. Cenospheres can also absorb sulfur oxides in the gaseous phase and thereby further contribute to the formation of acid smut.

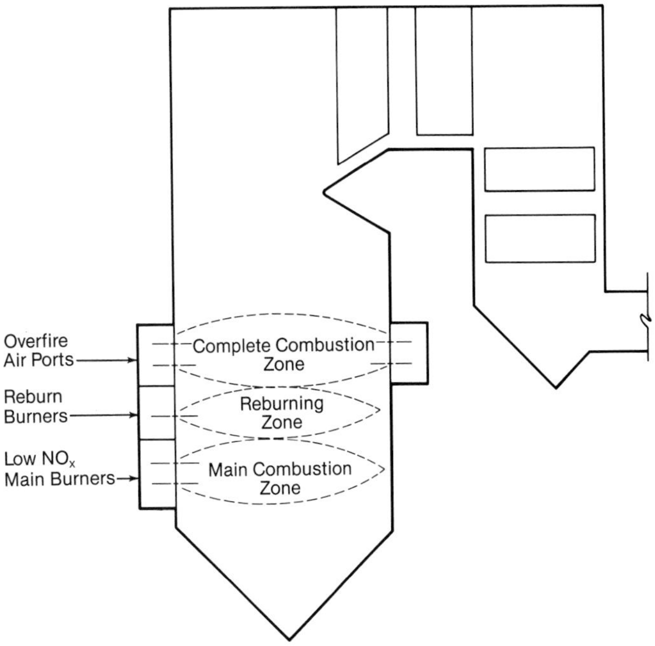

Fig. 5 Boiler side view showing reburn principle and combustion zones for a utility boiler.

Particulate matter from the combustion of fuel oil is, in large part, a function of the fuel properties. Ash content of the fuel oil plays a significant role in forming submicron particulate emissions. These ultrafine particulate emissions are potentially more dangerous to the environment than larger particles as they tend to stay suspended in the atmosphere. The fuel oil property most closely linked to forming cenospheres during the combustion process is asphaltene concentration. Asphaltenes are high molecular weight hydrocarbons that do not vaporize when heated. The combustion of fuel oils high in asphaltene content produces a greater quantity of large and intermediate size particulate emissions. Carbon residue, commonly determined by the Conradson Carbon Test, is also a means of evaluating the tendency to form particulate matter during combustion.

Control of particulate emissions is best achieved through proper atomization of the fuel oil and careful design of the burners and combustion control system to ensure thorough and complete mixing of fuel and air. Liberal residence time in a high temperature environment is favorable for complete combustion of fuel and low particulate emissions, although not conducive to low NO_x emissions. Oil emulsification also has a favorable impact on reducing particulate emissions. Oil emulsified with water further breaks up individual oil droplets when the water vaporizes during combustion. In addition, many fuel additives are now available to promote carbon burnout. These additives are primarily transition metals such as iron, manganese, cobalt and nickel that act as catalysts for the further oxidation of carbon particles.

Opacity

A visible plume emanating from the stack is undesirable from both a regulatory and public relations standpoint. Stack opacity is controlled in much the same manner as particulate emissions. Careful selection of fuels and complete combustion are keys to minimizing plume visibility.

Dark plumes are generally the result of incomplete combustion and can be controlled by careful attention to the combustion process and through transition metal based additives. White plumes are frequently the result of sulfuric acid in the flue gas and can be controlled by low excess air operation or alkaline based fuel additives that neutralize the acid.

The simultaneous control of all criteria pollutants poses a significant challenge to the burner and boiler designer. Techniques, or operating conditions, effective in reducing one form of atmospheric contaminant are frequently detrimental to controlling others. For this reason, hardware design and combustion controls are becoming increasingly complex. Modern boilers must have the added flexibility necessary to optimize thermal efficiency and combustion performance in conjunction with sound environmental practices.

Burner selection and design

As environmental concerns continue to dictate boiler and combustion system design, higher standards of performance are imposed on the fuel burning equipment. Control techniques to reduce NO_x emissions are in direct conflict with proven methods of good combustion performance, e.g., time, temperature and turbulence.

Potential increases in carbon monoxide emissions, particulate emissions and stack opacity as a result of low NO_x operation suggests that the burners must be capable of continued reliable mechanical operation and of providing the flexibility necessary to optimize combustion under a variety of operating conditions.

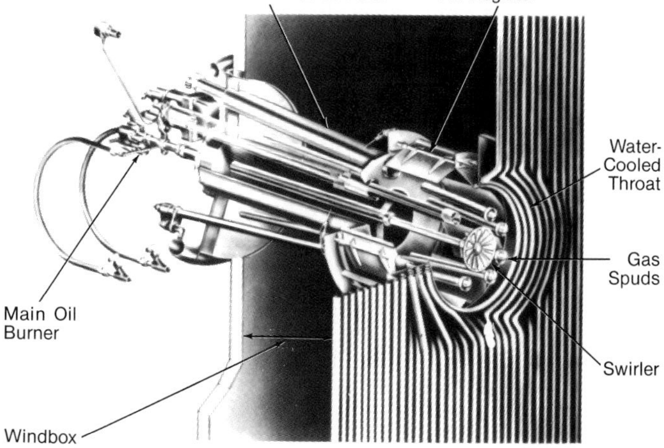

Fig. 6 Circular register burner with water-cooled throat for oil and gas firing.

Circular burner

Shown in Fig. 6, the circular type burner has long been the standard design for oil and gas firing applications. The tangentially disposed doors of the circular burner air register provide the turbulence necessary to mix the fuel and air and produce short compact flames. This burner typically operates with high secondary air velocities providing rapid, turbulent combustion for high combustion efficiency. Fuel is introduced to the burner in a

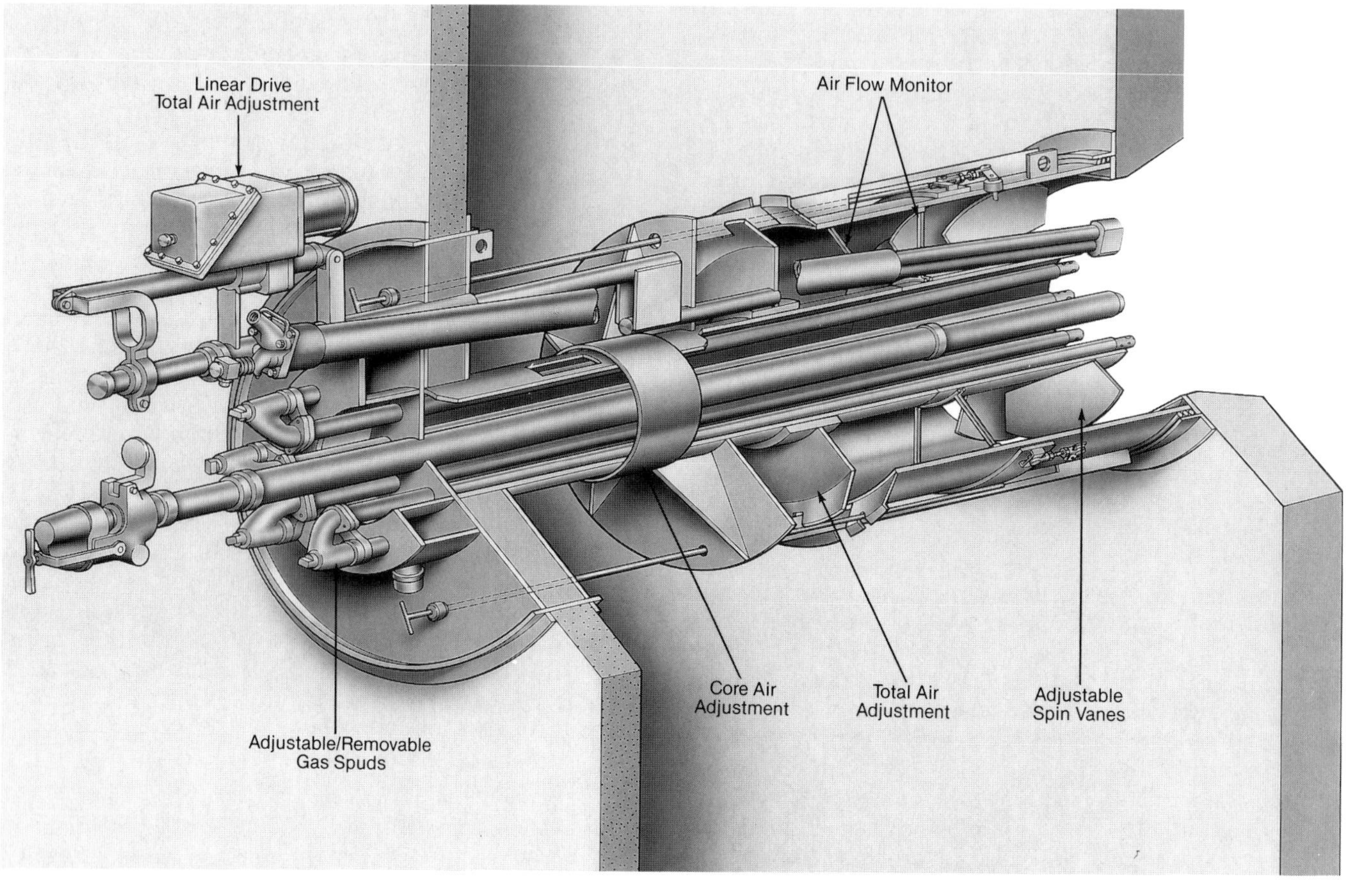

Fig. 7 S-type burner for oil and gas.

fairly dense mixture in the center. The direction and velocity of the air and dispersion of the fuel result in complete and thorough mixing of fuel and air.

S-type burner

The S-type oil and gas combination burner as shown in Fig. 7 was developed to replace the circular burner. To meet the demand for added flexibility and improved control of combustion air flow, the S-type burner incorporates several design features not available with the circular register. This burner has two air zones: the inner or core zone and the outer secondary air zone. When firing natural gas or oil, combustion air is introduced to the core zone through slots located around the periphery of the inner sleeve. An oil impeller or swirler is unnecessary with this burner; control of the flow entering the core zone ensures stable ignition. The core zone houses the main gas fuel elements and the main fuel oil atomizer.

The majority of combustion air enters the burner through the outer air zone. Axially disposed spin vanes are located in the outer sleeve to impart swirl to the combustion air and a sliding disk separately controls total air to the burner independent of swirl. The burner is equipped with an air measuring device upstream of the spin vanes. This device provides a relative indication of air flow to each burner and, on multiple burner applications, permits balancing of air flow from burner to burner. The outer zone also houses the ignitor and flame detection equipment.

Because it can measure air flow to individual burners and regulate total air flow independent of swirl, the S-type burner provides the added flexibility needed when employing combustion control techniques aimed at reducing NO_x emissions. This burner is ideally suited as an upgrade to existing circular burners as burner throat pressure part modifications can be avoided.

DRB-XCL™ type burner

The DRB-XCL™ oil and gas burner was developed specifically for NO_x reduction. The burner incorporates internal air and fuel staging into its design. Shown in Fig. 8, the oil atomizer and gas elements are centrally located in the burner to limit air and fuel interaction at the root of the flame. The fuel elements are housed in a single, central flame stabilizer that improves flame stability and turndown while separating the fuel elements from the combustion air. By controlling the rate of combustion and apparent stoichiometry, peak NO_x formation is reduced.

Combustion air is regulated by dual air zones with multistage swirl vanes. Adjustable swirl vanes are located in the inner air sleeve to optimize combustion for stability and NO_x reduction. The outer air sleeve houses a set of fixed spin vanes to improve peripheral air flow distribution and a set of adjustable spin vanes for air swirl control. Similar to the S burner, total air flow is controlled independent of swirl by an adjustable sliding disk. The DRB-XCL™ is also equipped with an air measurement device upstream of the spin vanes to permit air balancing from burner to burner.

The DRB-XCL™ burner is capable of low NO_x emissions. Design combustion air velocities with DRB-XCL™ are much lower than older vintage burner designs. This suggests that burner throat pressure part modifications will be necessary in retrofit applications.

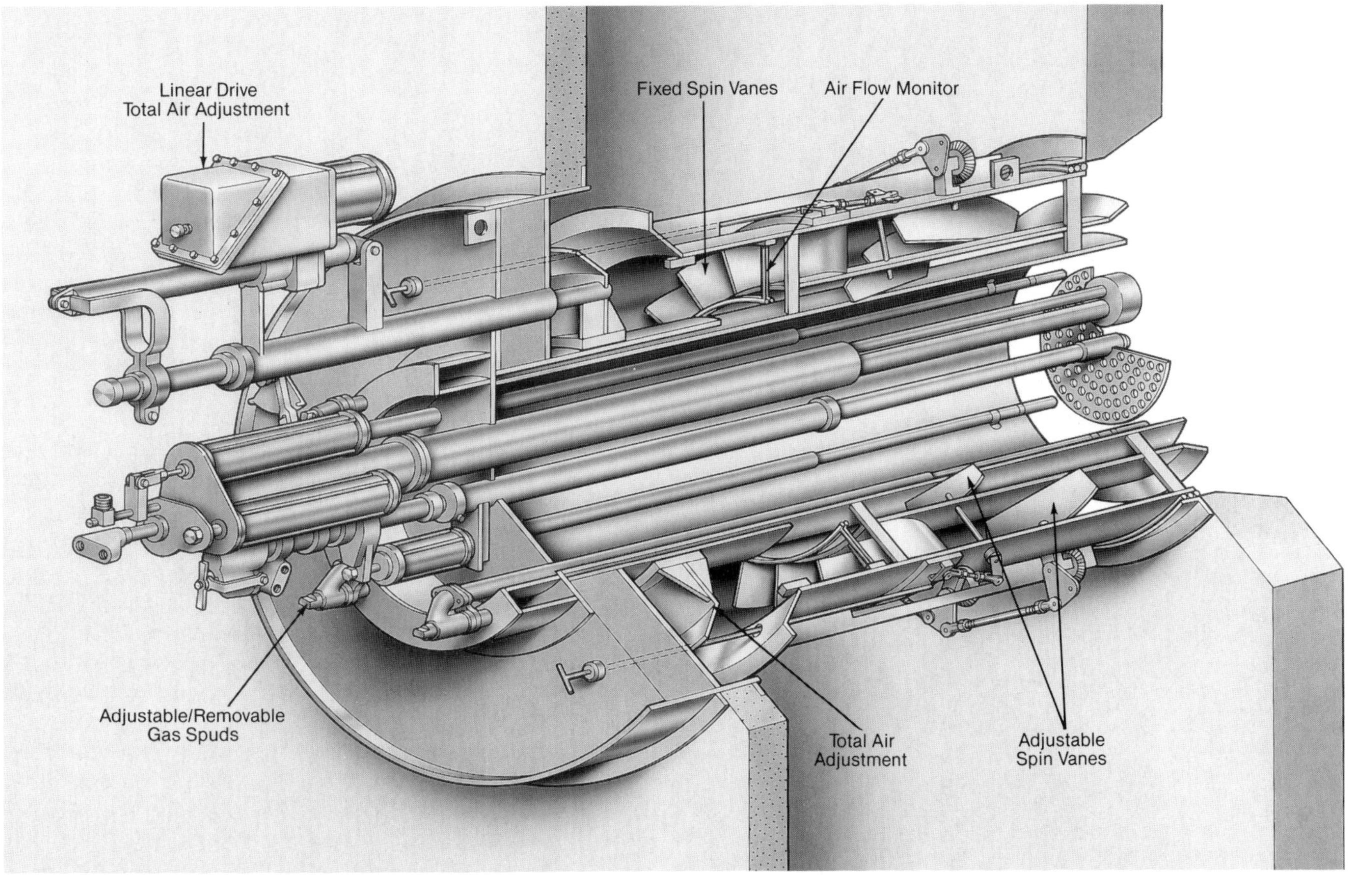

Fig. 8 DRB-XCL™ type burner for reduced emissions.

Fuel oil equipment

Oil for combustion must be atomized into the furnace as a fine mist and dispersed into the combustion air stream. Proper atomization is the key to efficient combustion and reduced particulate emissions. Atomization quality is measured in terms of droplet size and droplet size distribution. High quality atomization occurs when oil droplets are small, producing high surface to volume ratios and thereby exposing more surface to the combustion air. A convenient means of expressing and comparing atomization quality produced by various atomizer designs is the Sauter mean diameter (D_{sm}). This is the ratio of the mean volume of the oil droplets over the mean surface area, expressed in microns. The lower the Sauter mean diameter, the better the atomization.

Much attention has been given to atomization techniques for oil firing. Shown in Fig. 9, Babcock & Wilcox's (B&W) atomization facility at its Alliance Research Center can quantitatively and qualitatively characterize atomization through state-of-the-art laser diagnostic techniques. This facility has been used to characterize numerous atomizer designs.

For proper atomization, oil, of grades heavier than No. 2, must be heated by means of steam or electric heaters to reduce their viscosity to between 100 and 150 SSU (Seconds Saybolt Universal). In heating fuel oils, caution must be observed to ensure that temperatures are not raised to the point where vapor lock may occur. Vapor lock results when volatile fractions of the fuel separate in the fuel supply system causing flow interruptions and subsequent loss of ignition. Fuel supplied to the atomizers must also be free of acid, grit and fibrous or other foreign matter likely to clog or damage hardware components.

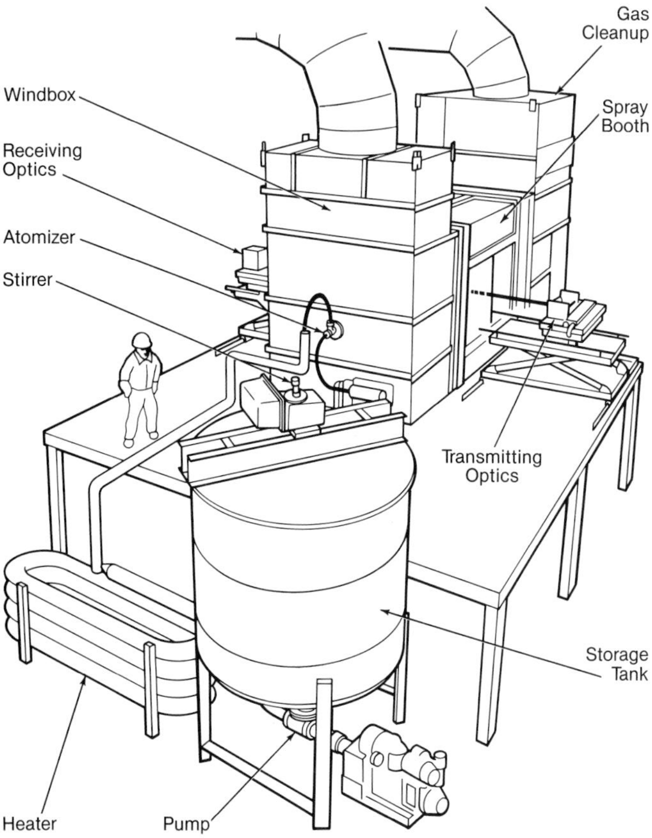

Fig. 9 Diagram of atomization facility operated by Babcock & Wilcox at the Alliance Research Center in Ohio.

Fuel oil is atomized by either mechanical or dual fluid atomizers that use steam or air as the medium. The choice of whether to use mechanical or steam atomization is determined by the boiler design and operating requirements. In general, steam assisted atomizers produce a higher quality spray and are more appropriate for low NO_x applications or where particulate emissions and stack opacity are of primary concern. At some installations, however, the heat balance is such that steam can not be used economically for oil atomization. Another factor to consider is the necessity to conserve boiler feedwater. In these instances, mechanical atomization may be more appropriate.

Mechanical atomizers With mechanical atomizers, the pressure of the fuel itself provides the energy necessary for atomization. These atomizers require relatively high pressure oil for proper performance. Three conventional types are in common use: the Uniflow, the Return Flow and the Steam Mechanical atomizer.

The Uniflow atomizer is used in small and medium sized stationary power plants, as well as in naval and merchant marine boilers. This atomizer is simple to operate and made with as few parts as possible. Fuel is introduced into ports that discharge tangentially into a whirl chamber. The fuel is spun out of the whirl chamber and passes through an orifice into the combustion chamber as a well atomized conical spray. Required oil pressure at the atomizer is above 300 psig (2.07 MPa gauge) for heat inputs of 70 to 80 x 10^6 Btu/h (20.5 to 23.4 MW_t).

The Return Flow atomizer (Fig. 10) is used on either stationary or marine boilers that require wide capacity ranges. This atomizer is designed to minimize, or eliminate entirely, the need for changing sprayer plates or the number of burners in service during normal operation. A wide range of operation is possible by maintaining a high flow through the sprayer plate slots even at reduced firing rates. At reduced firing rates, because the quantity of oil supplied is greater than the required firing rate, the excess oil is returned to a low pressure point in the piping system. The required oil pressure at the atomizer must be either 600 or 1000 psig (4.14 or 6.90 MPa gauge) depending upon fuel, capacity and load range requirements. Maximum input is in the order of 200 x 10^6 Btu/h (58.6 MW_t).

The Steam Mechanical atomizer combines the features of steam and mechanical atomization permitting operation over wide ranges, including low loads with cold furnaces. At high loads it can be operated as a mechanical atomizer. At reduced loads mechanical atomization is augmented by the use of steam. Required oil pressure is 200 to 300 psig (1.38 to 2.07 MPa gauge) depending upon capacity requirements. Steam pressure at the atomizer must be 10 to 15 psig (0.07 to 0.10 MPa gauge) above oil pressure not to exceed 125 psig (0.86 MPa gauge). Maximum capacity of this atomizer is in the order of 80 to 90 x 10^6 Btu/h (23.4 to 26.4 MW_t).

Residual fuel oils contain heavy residues that may condense on cold surfaces creating potentially hazardous conditions. To avoid this, a certain number of steam atomizers should be provided for satisfactory operation when boiling and drying out new units and for continued operation at very low capacities.

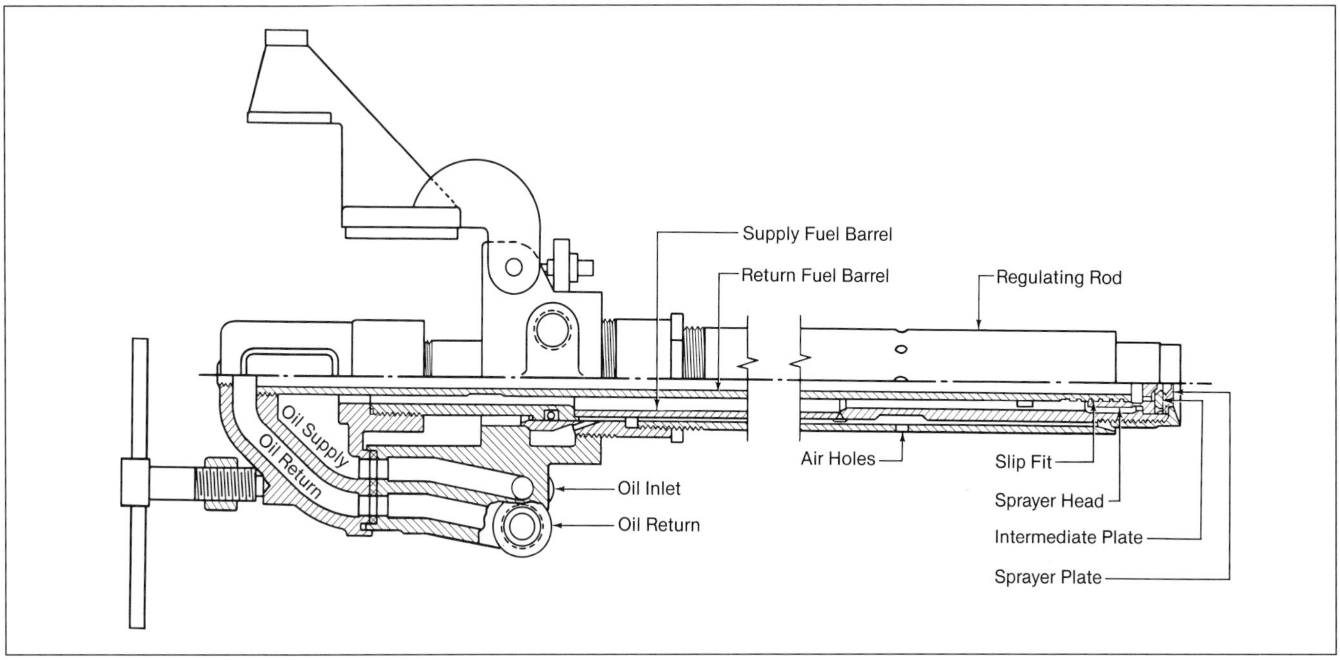

Fig. 10 Mechanical Return Flow oil atomizer assembly.

Steam atomizers Due to better operating and safety characteristics, steam assisted atomization is preferred. Steam atomization generally produces a finer spray. The steam-fuel emulsion produced in a dual fluid atomizer reduces oil droplet size when released into the furnace through rapid expansion of the steam. Dry saturated steam at the specified pressure must be used for fuel oil atomization to avoid the potential for burner pulsation. If necessary, moisture free compressed air can be substituted.

Several steam atomizer designs are available in sizes up to 300 x 10^6 Btu/h (87.9 MW_t) or about 16,500 lb/h (2.08 kg/s) of fuel. Required oil pressure is much lower than for mechanical atomizers. Steam and oil pressure requirements are dependent upon the specific atomizer design. Maximum oil pressure can be as much as 300 psig (2.07 MPa gauge) with steam pressures as high as 150 psig (1.03 MPa gauge). The four most common steam atomizer designs are the Y-jet, Racer, T-jet and I-jet. (See Fig. 11.) Each design is characterized by different operating ranges, steam consumption or atomization quality.

The Y-jet is designed for a wide firing range without changing the number of burners in service or the size of the sprayer plate. It can be used on any type of boiler with either steam or air atomization. Fuel oil and atomizing medium flow through separate channels to the atomizer sprayer plate assembly where they mix immediately before discharging into the furnace. Required oil pressure at the atomizer for maximum capacity ranges from 65 to 90 psig (0.45 to 0.62 MPa gauge). The Y-jet is a constant differential atomizer design requiring steam pressure to be maintained at 40 psig (0.28 MPa gauge) over the oil pressure throughout the normal operating range. Steam consumption with this design is in the order of 0.1 lb steam/lb oil.

The Racer atomizer is a refinement of the Y-jet. It was developed primarily for use where high burner turndown and low atomizing steam consumption are required. The name Racer is the class of merchant ship where this atomizer was first used. Required design oil pressure for maximum capacity is 300 psig (2.07 MPa gauge). Steam pressure is held constant throughout the load range at 150 psig (1.03 MPa gauge). Atomizing steam consumption at full load is approximately 0.02 lb steam/lb oil.

The T-jet and I-jet atomizers also allow a wide range of operation without the need for excessive oil pressure.

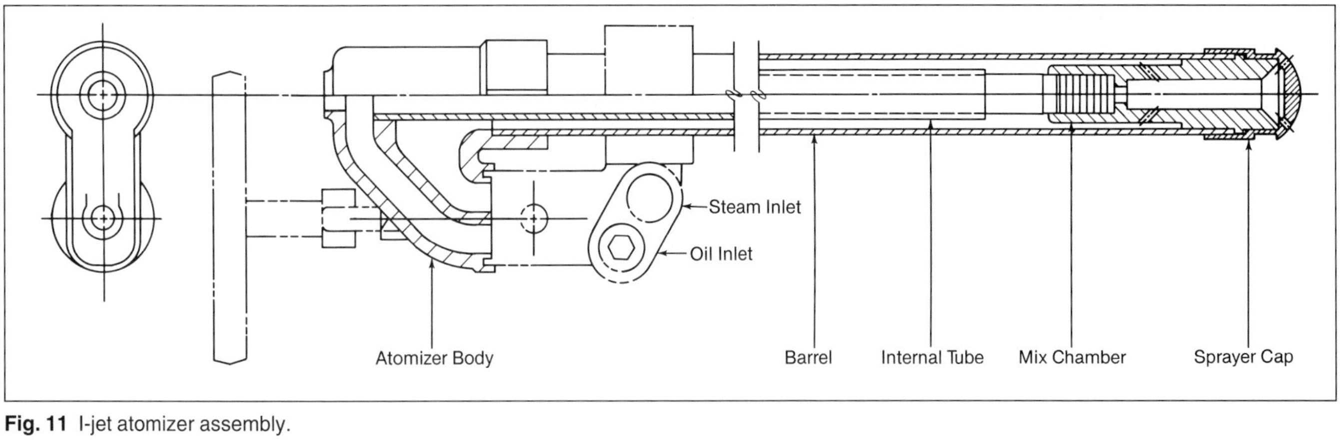

Fig. 11 I-jet atomizer assembly.

These atomizer designs are unique in that the steam and oil are mixed in a chamber prior to discharging through the sprayer plate (Fig. 11). Oil pressure may range from 90 to 110 psig (0.62 to 0.76 MPa gauge) depending upon fuel and capacity requirements. Of the constant differential variety, the I-jet and T-jet require a steam pressure of approximately 20 to 40 psig (0.14 to 0.28 MPa gauge) above oil pressure. Steam pressures may be adjusted to obtain optimum combustion performance. Steam consumption rates vary with these atomizer designs depending upon actual operating pressures. Consumption rates may be as high as 0.2 lb steam/lb oil. In general, higher steam consumption rates result in improved atomization quality.

A disadvantage of the steam atomizer is its consumption of steam. For a large unit, this can amount to a sizable quantity of steam and consequent heat loss to the stack. When the boiler supplies a substantial amount of steam for a process where condensate recovery is small, the additional makeup for the steam atomizer is inconsequential. However, in a large utility boiler where turbine losses are low and there is little makeup, the use of atomizing steam can have a significant effect.

Natural gas equipment

The three most common types of gas elements available for the combustion of commercial high Btu natural gas are the variable mix multi-spud, hemispherical multi-spud and the radial spud. The variable mix multi-spud gas element was developed for use with the circular type burner. The goal was to improve ignition stability when flue gas recirculation or two-stage combustion are implemented for NO_x control. This design uses a manifold outside the furnace, with a number of individual gas elements, or spuds, projecting through the windbox into the throat. The following is a list of the distinctive features of this gas element:

1. individual gas spuds are removable with the boiler in service to enable cleaning or redrilling of the gas nozzles as required,
2. individual spuds can be rotated to orient the discharge holes for optimum firing conditions with the burner in service,
3. the location of the spud tip, with respect to the burner throat, can be varied to a limited degree for optimum firing conditions, and
4. individual gas spud flame retainers provide added stability and are more conducive to lower NO_x emissions.

This gas element design adds needed flexibility for optimizing combustion under a variety of operating conditions. With proper selection of control equipment, the operator can change from one fuel to another without a drop in load or boiler pressure. Simultaneous firing of natural gas and oil in the same burner is acceptable on burners equipped with the multi-spud arrangement.

The hemispherical gas element (Fig. 12) was developed for use with the low NO_x DRB-XCL™ type burner and S-type burner. With the exception of the flame retainers, the hemispherical gas element offers the same features as the variable mix element design. This gas element, however, can achieve lower NO_x emissions by virtue of the tip profile, drilling pattern and location within the burner.

The radial spud design consists of a manifold located inside the windbox with individual gas elements projecting into the burner throat. Unlike the variable mix and hemispherical design, the radial spud is not conducive to low NO_x emissions. Use of the radial spud is reserved for applications where coal and gas are fired in the same burner. Radial spuds are less obtrusive to combustion air flow through the burner which eliminates undesirable flow characteristics when firing pulverized coal.

The maximum practical limit for gas input per burner is in the order of 200 x 10^6 Btu/h (58.6 MW_t). Frequently, physical arrangement of the gas hardware is a limiting factor in fuel input capability. With all gas element designs, the spud tip drilling is determined by adhering to empirical curves aimed at eliminating the potential for burner pulsation. Allowable gas discharge velocity criteria are established and the resulting required manifold pressure is determined. Gas pressure at the manifold at full burner capacity is generally in the range of 8 to 12 psig (60 to 80 kPa).

In many respects, natural gas is an ideal fuel since it requires no preparation for rapid and intimate mixing with the combustion air. However, this characteristic of easy ignition under most operating conditions has, in some cases, led to operator carelessness and damaging explosions.

To ensure safe operation, gas flames must remain anchored to the gas element discharge ports throughout the full range of allowable gas pressures and air flow conditions. Ideally, stable ignition should be possible at minimum load with full load air flow through the burner and at full load with as much as 25% excess air. With this latitude in air flow it is not likely that ignition can be lost, even momentarily, during upset conditions.

Byproduct gases

Many industrial applications utilize coke oven gas, blast furnace gas, refinery gas or other byproduct gases to produce steam. (See Chapter 31.)

Steel mill blast furnaces generate a byproduct gas containing about 25% carbon monoxide by volume. This fuel can be burned to produce steam for mill heating and power applications. Many mills also have their own coke producing plant, another source of byproduct fuel. Coke oven gas is an excellent fuel that burns readily because of its high free hydrogen content. With these gases, the heat

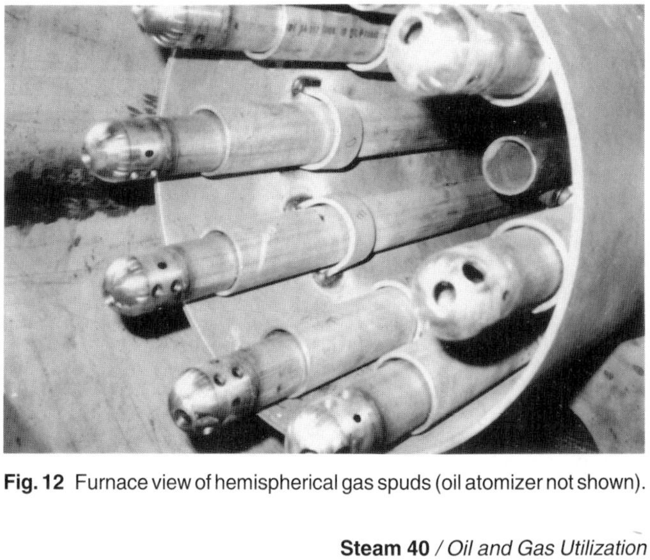

Fig. 12 Furnace view of hemispherical gas spuds (oil atomizer not shown).

release per unit volume of fuel may be different from that of natural gas. Therefore, gas elements must be designed to accommodate the particular characteristics of the gas to be burned.

Burners have been designed specifically to fire these byproduct fuels. To overcome the problem of impurities with blast furnace gases, burners are equipped with large scrolls at the entrance to the gas nozzle. Much of the foreign matter contained in the gas is deposited on the relatively cool walls of the scroll inlet. Cleanout doors are provided to permit removing these accumulated deposits. The nozzle itself has a large area so that any deposits that may accumulate do not affect burner capacity due to plugging or bridging. Coke oven gas nozzles are also constructed for ease of firing and maintenance. Dirt deposits, although not as severe as with blast furnace gases, are easily removed through cleanout doors and ports.

In the petroleum industry, catalytic cracking units produce byproduct carbon monoxide gas. CO boilers have been developed to reclaim the thermal energy present in this gas. CO boilers are equipped with supplementary oil and gas burners to provide steam for starting the catalytic cracking units. These burners are necessary to raise the temperature of the CO gases to the ignition point and assure complete burning of combustibles in the CO gas stream. Supplementary fuel burners and CO nozzles are radially positioned to impart a cyclonic motion to the gases in the cylindrical furnace. This ensures thorough mixing and promotes rapid, complete combustion.

Cold start and low load operation

Cold boiler startup requires firing at low heat input for long periods of time to avoid expansion difficulties and possible overheat of superheaters or reheaters. Low pressure [200 psig (1.38 MPa gauge)] boilers without superheaters may need only about an hour of low load operation. Larger high pressure units, however, may need four to six hours for startup. During startup, combustion efficiency is typically poor, especially with residual oils, due largely to low furnace and combustion air temperatures.

Low load operation may also result in poor distribution of air to the burners in a common windbox. Low air flow and accompanying low windbox to furnace pressure differentials may result in significant stratification of air in the windbox and adversely affect combustion performance. Stack effect between the bottom and top rows of burners may further alter the quantity of air reaching individual burners. At extremely low loads, firing oil or gas, the fuel piping itself can also introduce poor distribution and accordingly affect burner operation. For a given supply pressure, some burners may have adequate fuel while others do not receive enough fuel for continuous operation.

The low ignition temperature and clean burning characteristics of natural gas make it an ideal fuel for startup and low load operation. However, extreme care must be taken to avoid a momentary interruption in the flame. Large quantities of water are generated by the combustion of gaseous fuels with a high percentage of hydrogen. For units equipped with regenerative type air heaters, the air heater should not be placed in service until the flue gas temperature to the air heater has reached 400F (204C). This prevents condensation of water from the flue gas on the air heater surface and subsequent transport of this water to the burners by the combustion air. Water in the combustion air may cause loss of ignition.

Fuel oil is potentially more hazardous than gas for startup and low load operation. Although ignition stability is generally not a problem, the low temperature furnace environment may result in the accumulation of soot and carbon particles on air heater and economizer surfaces and may also create a visible plume. If deposits accumulate over an extended period of time they pose a fire hazard. To reduce the incidence of fires, some operators use sootblowers that blow steam continuously on the surfaces of regenerative type air heaters as they rotate in the flue gas stream.

When oil must be used for startup, the following firing methods are listed in order of preference:

1. Use of dual fluid atomizers and light fuel oil, No. 1 or 2, with steam or compressed air as the atomizing medium. This will provide a clean stack and limit the deposit of carbonaceous residue on boiler back-end surfaces.
2. Use of dual fluid atomizers with steam atomization and fuel oil no heavier than No. 6, heated as required for reduced viscosity and proper atomization.
3. Use of dual fluid atomizers with compressed air as the atomizing medium and fuel oil no heavier than No. 6, heated as required.

Startup with oil using mechanical atomizers is not recommended due to the inherent lower atomization quality. Stationary boilers equipped with mechanical atomizers for oil firing at loads between 20 and 100% are generally started up with natural gas or by one of the methods previously described.

Lighters and pilots

Lighting a burner requires an independent source of ignition. Lighters range from hand inserted torches to automatic spark ignited components that are an integral part of the burner. For safe boiler startup, each burner is lighted individually using either natural gas or atomized light fuel oil. Fig. 13 shows the CFS retractable gas lighter. Lighter capacity is tailored to meet requirements of the burner and flame safety system with a maximum input of approximately 25 x 10^6 Btu/h (7.3 MW_t). The flame of the lighter is applied to the burner main fuel stream until ignition is self-sustaining. In addition to lighting the main burner, lighters can also be used to stabilize combustion of difficult to burn fuels and, on a limited basis, to warm up the furnace.

A minimum combustion air temperature of 70F (21C) is necessary for lighter stability. When the lighter is to be operated for extended periods during boiler startup, steam or water coil air heaters should be used to raise the secondary air temperature to at least 150F (66C) to assure complete combustion and acceptable stack appearance.

As a minimum, the lighter assembly consists of a fuel nozzle, spark ignition source and an energy source to produce the spark. Lighters can be either stationary (Fig. 14) or equipped with retracting mechanisms for protection from furnace radiation (Fig. 13). Programmable lighter controls are available that automatically sequence

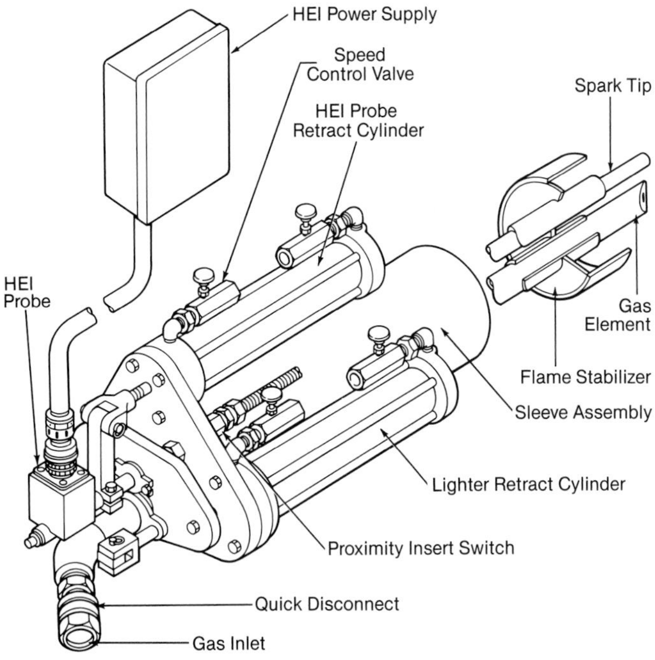

Fig. 13 CFS retractable gas lighter with high energy spark probe.

all lighter functions including on/off control of the fuel, atomizing and purge mediums, lighter and spark probe insertion and retraction and adjustable time intervals for trial ignition, condensate purge and atomizer purge.

In some applications, lighter or pilots must operate continuously. This is particularly true in the use of a byproduct fuel, such as gas from a chemical process. In most installations such fuels are piped directly to the boiler house without an intervening accumulator. The quantity and quality of fuel supplied to the boiler are subject to malfunction in the chemical process. Supply pressure may vary beyond the range of main burner stability, or the combustible content of the gas may change. Such variations require continuous operation of a pilot. It may also be necessary to provide supplementary fuel to each burner continuously to keep the process gas burning during abnormal conditions in the process.

Safety precautions

The NFPA has prepared a standard set of operating guides, procedures and recommended interlocks and trips for the safe and reliable operation of gas and oil combustion processes. Recommendations of this standard along with governing local codes and ordinances are followed in the design of gas and oil burner systems, burner interlock and trip systems and burner sequence control systems.

Five rules of prime importance in safe and reliable operation of gas and oil fired combustion systems, whether employing a manual or automatic control system, are described below.

1. Never allow oil or gas to accumulate anywhere, other than in a tank or lines that form part of the fuel delivery system. The slightest odor of gas must be cause for alarm. Steps should be taken immediately to ventilate the area thoroughly and locate the source of the leak.

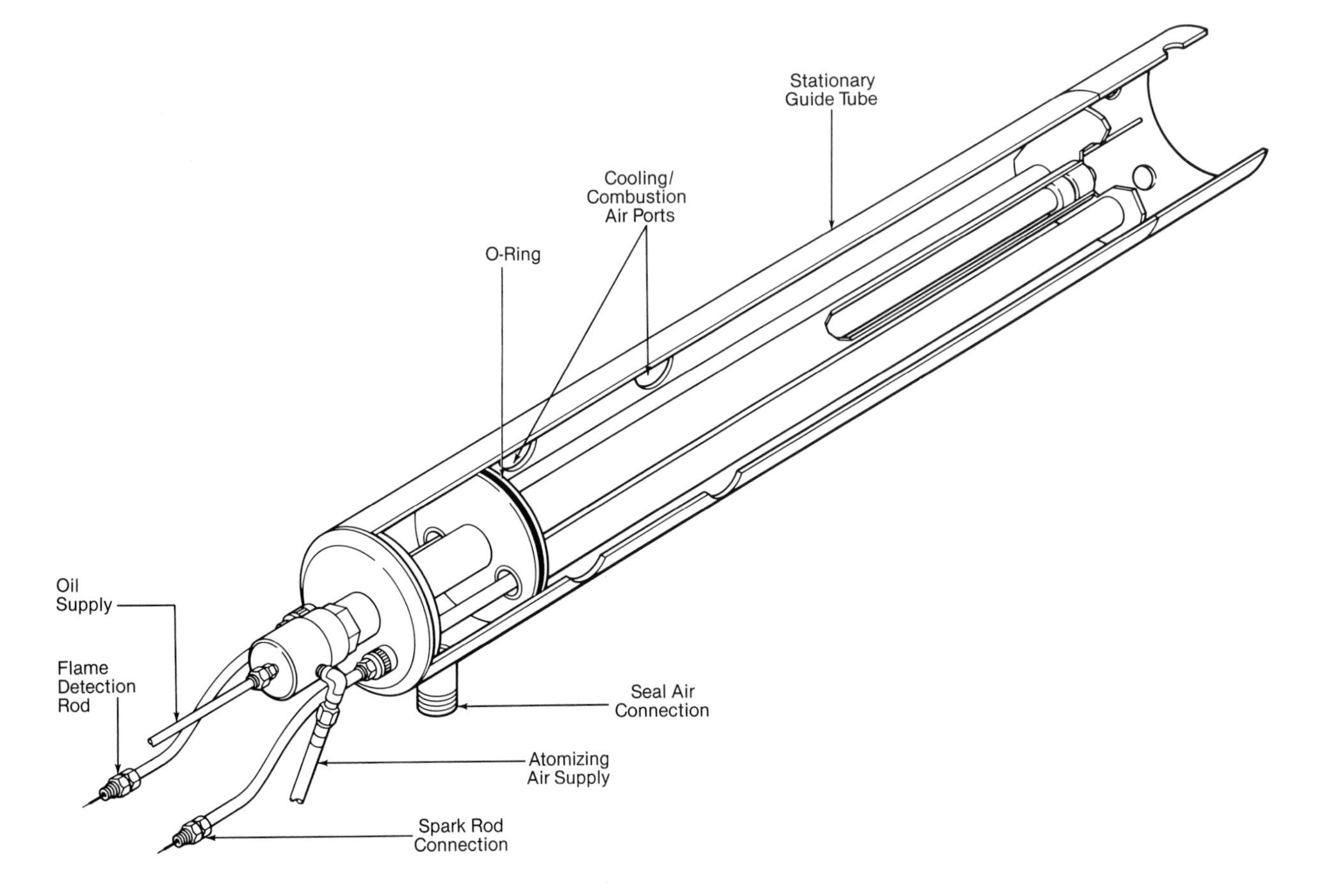

Fig. 14 FPS stationary air atomized oil lighter.

2. A minimum purge rate air flow not less than 25% of full load volumetric air flow must be maintained during all stages of boiler operation. This includes pre-purging the setting and lighting of the ignitors and burners until the firing rate air requirement exceeds the purge rate air flow. During startup, shutdown, or low load operation, not less than 25% of full load volumetric air flow must be maintained as a continuous purge of the furnace and setting.
3. A spark producing device or lighted torch must be in operation before introducing any fuel into the furnace. The ignition source must be properly placed with respect to the burner and must continuously provide a flame or spark of adequate size until combustion of the main burner is self-sustaining.
4. A positive air flow through the burners into the furnace and up the stack must be maintained at all times.
5. Adequate fuel pressure for proper burner operation must be maintained at all times. In the case of oil firing, fuel pressure and temperature must be maintained for proper atomization. In dual fluid atomizer applications, adequate steam or air pressure must be available at the atomizer.

Equipment requirements, sequence of operation, interlock systems and alarm systems are equally important, if not more so, when implementing combustion control techniques to reduce NO_x emissions.

In observance of the recommended rules of operation, an automatic control system should include the following:

1. purge interlocks requiring a specified minimum air flow for a specific time period to purge the setting before the fuel trip valve can be opened,
2. flame detectors on each burner, connected to an alarm and interlocked to shut off the burner fuel valve upon loss of flame,
3. closed position limit switches for burner shutoff valves, requiring that individual burner shutoff valves be closed to permit opening the fuel trip valve,
4. shutoff of fuel on failure of the forced or induced draft fan,
5. shutoff of fuel in the event of low fuel pressure (and low steam or air pressure to oil atomizers),
6. shutoff of fuel in the event of low oil temperature, and
7. shutoff of fuel in gas fired units in the event of excessive fuel gas pressure.

Any of the commonly used fuels can be burned with safety when using the proper equipment and operating skill. Hazards are introduced when, through carelessness or misoperation of equipment, the fuel is no longer burned in a safe manner. While a malfunction should be corrected promptly, panic must be avoided. Investigation of explosions of boiler furnaces equipped with good recording apparatus reveal that conditions leading to the explosion had, in most cases, existed for a considerable period of time. Sufficient time was available for someone to have taken unhurried corrective action before the accident. The concern that malfunctions of safety equipment cause frequent, unnecessary, unit trips is not consistent with the facts.

Utility power plant coal storage and handling facilities.

Chapter 11
Solid Fuel Processing and Handling

Coal remains the dominant worldwide source of energy for steam generation. However, additional solid fuels such as wood byproducts and municipal wastes are also gaining widespread use. The large scale continuous supply of such solid fuels for cost effective and reliable steam power generation requires the effective integration of recovery (e.g., mining), preparation, transportation and storage technologies. The relationships between these are illustrated in Fig. 1. While each fuel offers unique challenges, a discussion of the processing and handling of the dominant fuel source — coal — identifies many of the common issues and considerations for all solid fuels. Selected additional topics about the special aspects of some other solid fuels are covered in Chapters 26 through 28.

Mining is the first step in producing coal. Raw coal can be treated to remove impurities and to provide a more uniform feed to the boiler. The resulting reduction in ash and sulfur can significantly improve overall boiler performance and reduce pollution emissions. The transportation of coal to the plant may represent a major portion of the plant's total fuel cost although mine mouth generating stations can minimize these transportation costs. Storage and handling of large coal quantities at the plant site require careful planning to prevent service interruptions.

Coal mining

As discussed in Chapter 8, electric utility coal consumption for power generation dominates the market for coal produced in the United States (U.S.) and the rest of the world. U.S. coal production reached more than 1 billion tons per year in 1990 with 77% being used for electric utility power generation. Corresponding worldwide production in 1990 was approximately 5.2 billion tons.

In the U.S., coal production is split between surface mining (59%) and underground mining (41%). Because of the geologic locations of coal deposits, or *seams*, and cost considerations, surface mining dominates coal production west of the Mississippi River and underground mining dominates in the East. The emergence of large, high volume surface mines producing low sulfur coal in the West has resulted in a continuing shift of coal production from the Eastern U.S. to the West.

Surface mining

Coal may be recovered from relatively shallow seams by removing the overlying earth, or *overburden,* to expose the coal seam. Typically, topsoil is first removed and stored for later use in reclamation of the site. The remaining overburden is drilled and blasted to loosen the rock for removal with a dragline or excavating shovel. The overburden is then methodically stripped away and stored for restoration of the land to the original contour following removal of the coal. A dragline is commonly used to expose the coal seam (Fig. 2). The coal may then be removed using a bulldozer and front-end loader or a shovel. A mobile crusher and screen may be set up in the mine for initial sizing of the raw coal. The coal is then loaded into haulage trucks for delivery to a cleaning facility or steam generating plant.

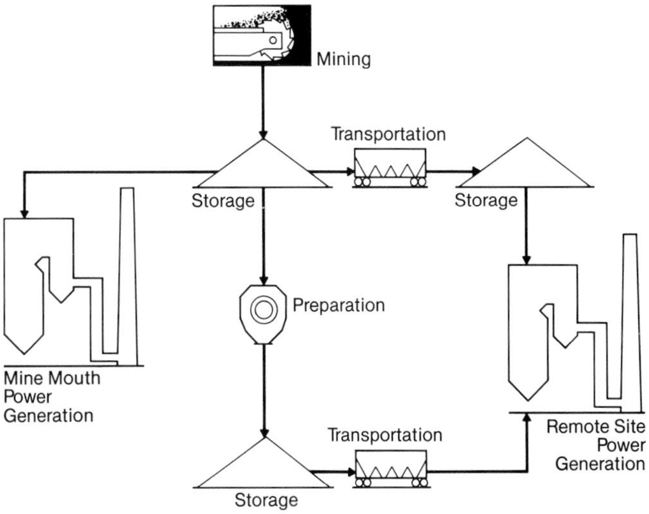

Fig. 1 Fuel supply chains for coal-fired power generation.

The *stripping ratio* is defined as the unit amount of overburden which must be removed to access a unit amount of coal. In general, surface mines in the Western U.S. have lower stripping ratios than mines in the East.

Strict environmental regulations limit the amount of land surface area which may be exposed at any one time, control water runoff and establish land reclamation procedures.

Underground mining

Approximately 68% of the total U.S. coal reserves are accessible only by underground mining. (See Chapter 8.) Underground coal mine production is dominated by the use of continuous mining machines (Fig. 3) which presently account for approximately 70% of the total underground production.

Fig. 2 Large coal dragline in operation (*courtesy of National Coal Association*).

Room and pillar mining Most U.S. coal is produced using an underground technique known as room and pillar mining. A series of headings or parallel entries are cut into the coal seam to provide passages for the mining machinery, uncontaminated ventilation air and conveying equipment. These headings are typically 18 to 20 ft (5.5 to 6.1 m) wide and may be several miles (km) long. Rooms are driven off of the main headings to the property limits of the mine. Typically, only about 50% of the coal is removed when the rooms are first mined. The remaining coal is left in place to support the roof. Roof bolts are installed where the coal has been removed for additional roof support. At the completion of the development cycle, the remaining coal may be removed by retreat mining, and the unsupported roof is allowed to collapse.

Longwall mining In longwall mining, shearers or plows are pulled back and forth across a panel of coal to break it loose from the seam. The coal falls onto a flight conveyor and is transported to the main haulage line. Because essentially all of the coal is removed, artificial roof supports, known as shields or chocks, are used to cover the plow and conveyor. The plow, conveyor and roof supports are advanced using hydraulic jacks as the coal is removed from the mining face. The unsupported roof is allowed to cave in behind the supports. In 1990, there were 96 longwall sections operating in the U.S. The average panel width exceeded 700 ft (213.4 m) and panels as wide as 1000 ft (304.8 m) were in operation. Longwall panel lengths may reach 10,000 ft (3048 m) or more.

Fig. 3 Continuous mining machine in operation (*courtesy of National Coal Association*).

Longwall mining can be a very high capacity production system provided that the geologic conditions are suitable and the longwall has been carefully integrated with the existing panel development scheme and coal haulage infrastructure. Production capacities exceeding 6000 t of raw coal per shift have been reported. The annual production capacity for a longwall is a function of the seam height, panel width and length, and the time required to move the equipment between operating panels.

Raw coal size reduction and classification

Sizing requirements

Size reduction operations at steam power plants are usually confined to crushing and pulverizing although it is sometimes more economical to purchase pre-crushed coal for smaller plants, especially stoker-fired units. Screening at the plant is generally not required, except to remove large impurities and trash using simple grids (or grizzlies) and rotary breakers. Techniques for determining coal particle size distribution are discussed in Chapter 8 while the American Society for Testing and Materials (ASTM) Standard D 431 details the commonly accepted screen size designations.

Coal particle size degradation, which occurs in transportation and handling, must be considered when establishing supply size specifications. This may be critical where the maximum quantity of coal fines is set by firing equipment limitations.

For stoker-fired installations, it is customary to specify purchased coal sized to suit the stoker (see Chapter 15), so that no additional sizing is required at the plant [typically 1.5 in. x 0 (38.1 mm x 0)].

For pulverized coal-fired boilers, a maximum delivered top size is usually specified with no limitation on the percentage of fines, so that the delivered coal is suitable for crushing and pulverizing in the available equipment. The coal is crushed to reduce particle size and then ground to a very fine size in the pulverizer. (See Chapter 12.)

Crushers alone may be used to provide the relatively coarser sizes required for Cyclone furnaces. (See Chapter 14.) A properly sized crusher efficiently reduces particle size while producing minimal fines.

Size reduction equipment selection

Size reduction equipment is generally characterized by the maximum acceptable feed size and the desired product top size. The reduction ratio is defined as follows:

$$\frac{p_{feed}}{p_{product}} \qquad (1)$$

where

p_{feed} = feed particle size in which 80% of the particles pass a given screen size or mesh

$p_{product}$ = product particle size in which 80% of the particles pass a given screen size or mesh

Once-through crushing devices, which discharge the fines without significant re-crushing, are used to minimize the production of fines. Rotary breakers and roll crushers are commonly used to reduce the coal top size without producing a significant amount of fines.

The *rotary breaker*, illustrated in Fig. 4, reduces the coal to a predetermined maximum size and rejects larger refuse, mine timbers, trash and some tramp metal. It consists of a large cylinder of steel screen plates which rotates at approximately 20 rpm. The size of the screen openings determines the top size of the coal. The coal fed at one end of the cylinder is picked up by lifting shelves and is carried up until the angle of the shelf permits the coal to drop onto the screen plate. It shatters and is discharged through the screen openings. The harder rock does not break as readily; it travels along the screen and is rejected at the discharge end. Wood, large rocks and other trash which do not pass through the screen are separated from the coal. Rotary breakers may be installed at the mine, preparation plant, or steam generating plant.

The elements of a *single-roll crusher* are illustrated in Fig. 5. This crusher consists of a single toothed roll which forces the coal against a plate to produce the crushing action. The maximum product particle size is determined by the gap between the roll and the plate. To prevent jamming by large impurities such as tramp metal, the roll is permitted to rise or the plate can swing away, allowing the impurities to pass through. This is an old type of crusher which is commonly used for reducing run-of-mine bituminous coal to a maximum product of 1.25 to 6 in. (31.8 to 152 mm). The abrasive action between the coal and the plate produces some fines; however, they are discharged with minimal re-breakage.

In a *double-roll crusher* the coal is forced between two counter-rotating toothed rolls (Fig. 6). The mating faces of both rolls move in a downward direction, pulling the coal through the crusher. The size of the roll teeth and the spacing between rolls determine the product top size. One of the rolls may be spring loaded to provide a means for passing large, hard impurities. Double-roll crushers are used for reducing run-of-mine coal to smaller sizes at preparation and steam generating plants.

In *retention crushers* such as *hammer mills* and *ring crushers*, coal is retained in the breakage zone until it is sufficiently fine to pass through a screen to the discharge. The re-breakage action produces considerable fines, consequently these mills are not used in applications where fines are objectionable. They are often used to reduce run-of-mine coal to an acceptable size for feed to a stoker or pulverizer, e.g., 0.75 in. x 0 (19 mm x 0).

A *hammer mill* is depicted in Fig. 7. In this mill, the coal is broken by impact with the hammers, which are mounted on a central shaft and permitted to swing freely as the shaft is rotated. The coal is fed at the top of the mill and is forced

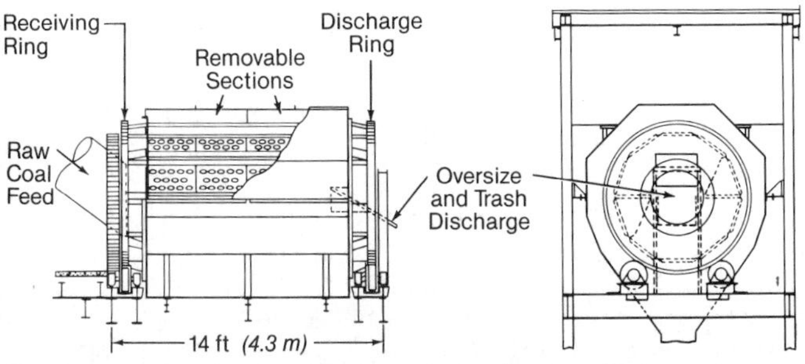

Fig. 4 Rotary breaker for use at the mine and power plant.

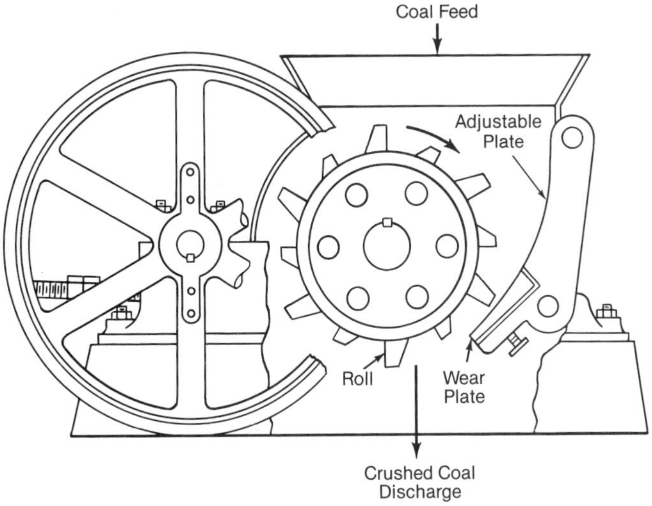

Fig. 5 Single-roll crusher — diagrammatic section.

down and outward to the grate bars as it is struck by the hammers. The spacing of the bars determines the maximum size of the finished product. The coal remains in the mill, and breakage continues until the particles are fine enough to pass through the grate. A trap is usually provided for collection and removal of tramp metal.

Screen selection

Screening is usually performed at the mine or preparation plant to remove unwanted material and size coal for various uses prior to shipment to the steam generating plant. The run-of-mine coal is usually passed over a grid of steel bars, or grizzly, to remove mine timbers and trash. The raw coal may be separated into various size fractions to meet contract specifications or for further processing in a preparation plant by passing the coal through various screens. Common screen types include gravity bar screens, trommels or revolving screens, shaker screens and vibrating screens.

A *gravity bar screen* consists of a number of sloped parallel bars. The gaps between bars, the slope, and the length of the bars determine the separating size. The bars

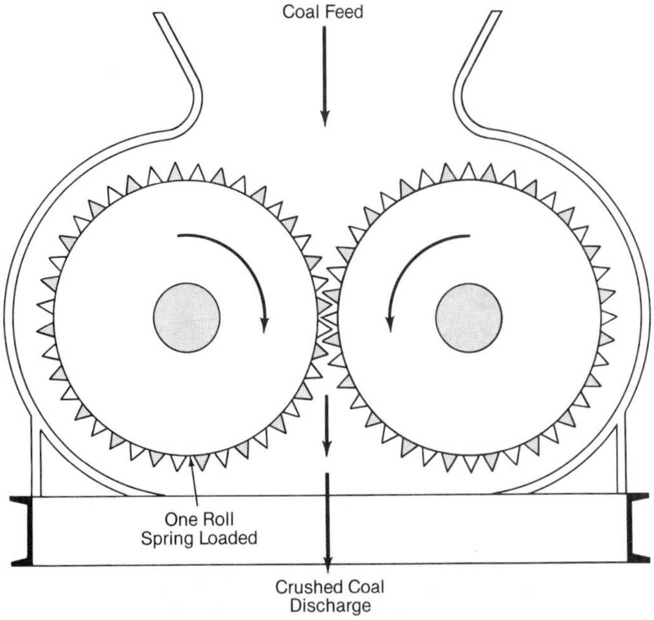

Fig. 6 Double-roll crusher — diagrammatic section.

have tapered cross sections; the gaps between the bars are smaller on the top side than on the bottom. This design reduces plugging.

A *revolving screen* consists of a slowly rotating cylinder with a slight downward slope parallel to the axis of coal flow. The cylinder is comprised of a perforated plate or a wire cloth, and the size of the openings determines the separating size. Because of the repeated tumbling as the coal travels along the cylinder, considerable breakage can occur. For this reason, revolving screens are not used for sizes larger than about 3 in. (76.2 mm). Because only a small portion of the screen surface is covered with coal, the capacity per area of screen surface is low.

A *shaker screen* consists of a woven wire mesh mounted in a rectangular frame which is oscillated back and forth. This screen may be horizontal or sloped slightly downward from the feed end to the discharge end. If the screen is horizontal, it is given a differential motion to help move the coal along its surface.

Bituminous coal generally fractures into roughly cubicle shapes, while commonly associated slate and shale impurities fracture to form relatively thin slabs. This shape difference enables the impurities to be separated with a *slotted shaker*.

Vibrating screens are similar to shaker screens except that an electric vibrator is used to apply a high frequency, low magnitude vibration to the screen. The screen surface is sloped downward from the feed to the discharge end. The vibration helps to keep the mesh openings clear of wedged particles and helps to stratify the coal so that fine particles come in contact with the screen surface. For screening fine, wet coal, water sprays are used to wash fine particles through the coal bed and the screen surface. Vibrating screens are the most widely used types for sizing and preliminary dewatering.

It is common practice to separate the fines from the coarse coal to improve the efficiency of subsequent cleaning and dewatering processes. The fines may be discarded, cleaned separately, or bypassed around the cleaning process and then blended back into the coarse clean coal product.

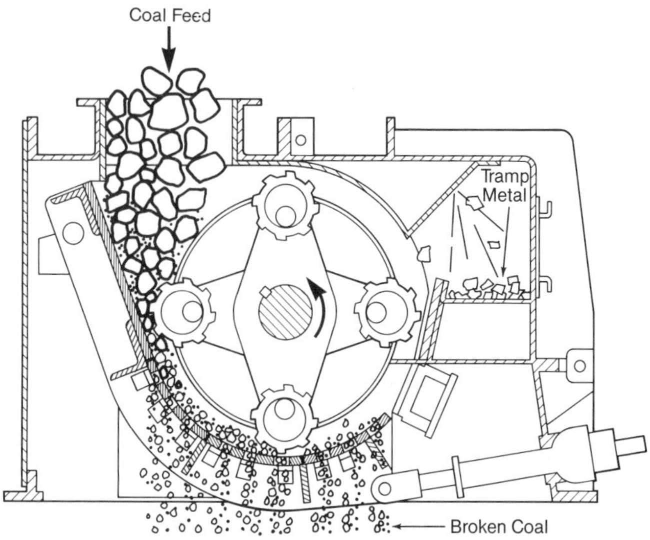

Fig. 7 Ring hammer mill crusher — diagrammatic section (*courtesy of Pennsylvania Crusher Corporation*).

Coal cleaning and preparation

The demand for coal cleaning has increased in response to environmental regulations restricting sulfur dioxide (SO_2) emissions from coal-fired boilers. The demand is also due to a gradual reduction in run-of-mine coal quality as higher quality seams are depleted and continuous mining machines are used to increase production. Approximately 70% of coal mined for electric utility use is cleaned in some way.

Coal cleaning and preparation cover a broad range of intensity, from a combination of initial size reduction, screening to remove foreign material, and sizing discussed previously, to more extensive processing to remove additional ash, sulfur and moisture more intimately associated with the coal.

The potential benefits of coal cleaning must be balanced against the associated costs. The major costs to consider, in addition to the cleaning plant capital and operating costs, include the value of the coal lost to the refuse product through process related inefficiencies and the cost of disposing of the refuse product. Generally, the quantity of coal lost increases with the degree of desired ash and sulfur reduction. An economic optimum level of ash and sulfur reduction can be established by balancing shipping and postcombustion cleanup costs against precombustion coal cleaning costs.

Coal characterization

Coal is a heterogeneous mixture of organic and inorganic materials as described in detail in Chapter 8. Coal properties vary widely between seams and within a given seam at different elevations and locations. The impurities associated with coal can generally be classified as inherent or extraneous. *Inherent impurities* such as organic sulfur can not be separated from the coal by physical processes. *Extraneous impurities* can be partly segregated from the coal and removed by physical coal cleaning processes. The extent to which these impurities can be economically removed is determined by the degree of material dissemination throughout the coal matrix, the degree of liberation possible at the selected processing particle size distribution, and physical limitations of the processing equipment.

Mineral matter associated with the raw coal forms ash when the coal is burned. Ash forming mineral matter may also be classified as inherent or extraneous. Inherent mineral matter consists of chemical elements from plant material organically combined with coal during its formation. This mineral matter generally accounts for less than 2% of the total ash. Extraneous mineral matter consists of material which was introduced into the deposit during or after the coalification process, or is extracted with the coal in the mining process.

Sulfur is always present in coal and forms SO_2 when the coal is burned. If the sulfur is not removed before combustion, the SO_2 that forms is exhausted through the stack or removed by postcombustion flue gas treatment, discussed in Chapters 32 and 35. Sulfur is generally present in coal in three forms: pyritic, organic or sulfate.

Pyritic sulfur refers to sulfur combined with iron in the minerals pyrite (FeS_2) or marcasite. Pyrite may be present as lenses, bands, balls or as finely disseminated particles. *Organic sulfur* is chemically combined with molecules in the coal structure. *Sulfate sulfur* is present as calcium or iron sulfates. The sulfate sulfur content of coal is generally less than 0.1%.

The total sulfur in U.S. coals can vary from a few tenths of a percent to more than 8% by weight. The pyritic portion may vary from 10 to 80% of the total sulfur and is usually less than 2% of the coal by weight (Table 1).

The larger pyritic sulfur particles can generally be removed by physical cleaning, but finely disseminated pyritic sulfur and organic sulfur can not. Advanced physical and chemical cleaning technologies under development to remove these sulfur forms have not yet proven economical.

Moisture can also be considered an impurity because it reduces the heating value of raw coal. Inherent moisture varies with coal rank, increasing from 1 to 2% in anthracite to 45% or more in lignite. Surface moisture can generally be removed by mechanical or thermal dewatering. This drying requires an energy expense at the cleaning plant or the steam generating plant (in pre-drying or during combustion). Drying before shipment reduces transportation costs on a per-Btu basis. When predrying is used, atmospheric oxidation will tend to be increased for low rank coals because of the exposure of additional oxidation sites in the particles.

The distribution of ash and sulfur in a coal sample can be characterized by performing a *washability analysis*. This analysis consists of separating the raw coal into relatively narrow size fractions and then dividing each fraction into several specific gravity fractions. The coal in each size/specific gravity fraction is then analyzed for ash, sulfur and heating value content. The hardness and distribution of the impurities relative to the coal determines if the impurities are concentrated in the larger or smaller size fractions. Relatively soft impurities are generally found in the finer size fractions. In general, the lowest specific gravity fractions have the lowest ash content, as indicated in Table 2.

The information generated by these *float/sink* characterization tests can be used to predict the degree of ash and sulfur reduction possible using various specific gravity based cleaning technologies discussed below. In general, the more material that is present near the desired specific gravity of separation, the more difficult it is to make an efficient separation.

Table 1
Distribution of Sulfur Forms in Various Coals (%)

Mine Location County, State	Coal Seam	Total Sulfur	Pyritic Sulfur	Organic Sulfur
Henry, MO	Bevier	8.20	6.39	1.22
Henry, MO	Tebo	5.40	3.61	1.80
Muhlenburg, KY	Kentucky #11	5.20	3.20	2.00
Coshocton, OH	Ohio #6	4.69	2.63	2.06
Clay, IN	Indiana #3	3.92	2.13	1.79
Clearfield, PA	Upper Freeport	3.56	2.82	0.74
Franklin, IL	Illinois #6	2.52	1.50	1.02
Meigs, OH	Ohio #8A	2.51	1.61	0.86
Boone, WV	Eagle	2.48	1.47	1.01
Walker, AL	Pratt	1.62	0.81	0.81
Washington, PA	Pittsburgh	1.13	0.35	0.78
Mercer, ND	Lignite	1.00	0.38	0.62
McDowell, WV	Pocahontas #3	0.55	0.08	0.46
Pike, KY	Freeburn	0.46	0.13	0.33
Kittitas, WA	Big Dirty	0.40	0.09	0.31

Table 2
Typical Ash Contents of Various Bituminous Coal Specific Gravity Fractions

Specific Gravity Fraction	Ash Content % by wt
1.3 to 1.4	1 to 5
1.4 to 1.5	5 to 10
1.5 to 1.6	10 to 35
1.6 to 1.8	35 to 60
1.8 to 1.9	60 to 75
Above 1.9	75 to 90

Coal cleaning and preparation operations

The initial steps in the coal cleaning process include removal of trash, crushing the run-of-mine coal and screening for size segregation. These preliminary operations and associated hardware were discussed previously. The following operations are then used to produce and dewater a reduced ash and sulfur product. Fig. 8 provides a general layout of coal cleaning unit operations.

Gravity concentration Concentration by specific gravity and the subsequent separation into multiple products is the most common means of mechanical coal cleaning. Concentration is achieved because heavier particles settle farther and faster than lighter particles of the same size in a fluid medium. Coal and impurities may be segregated by their inherent differences in specific gravity, as indicated in Table 3.

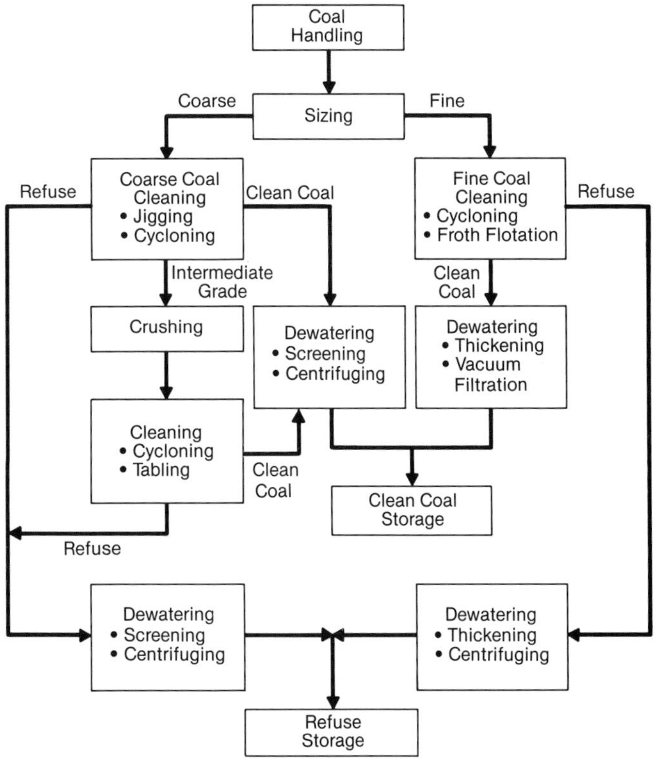

Fig. 8 General layout of coal cleaning operations.

Table 3
Typical Specific Gravities of Coal and Related Impurities

Material	Specific Gravity
Bituminous coal	1.10 to 1.35
Bone coal	1.35 to 1.70
Carbonaceous shale	1.60 to 2.20
Shale	2.00 to 2.60
Clay	1.80 to 2.20
Pyrite	4.80 to 5.20

The fluid separating medium may consist of a suspension of the raw coal in water or air, a mixture of sand and water, a slurry of finely ground magnetite or an organic liquid with an intermediate specific gravity. Aqueous slurries of raw coal and magnetite are currently the most common separating media.

If the effective separating specific gravity of the media is 1.5, particles with a lower specific gravity are concentrated in the clean coal product and heavier particles are in the reject or refuse product. Several factors prevent ideal separation in practice.

Gravity separation processes concentrate particles by mass. The mass of a particle is determined by its specific gravity and particle size. Raw coal consists of particles representing a continuous distribution of specific gravities and sizes. It is quite possible for a larger, less dense particle to behave similarly to a smaller particle with a higher specific gravity. For example, a relatively smaller pyrite particle may settle at a similar rate as a larger coal particle. The existence of *equal settling* particles can lead to separating process inefficiency. Fine pyrite in the clean coal product and coarse coal in the refuse are commonly referred to as *misplaced material*. The amount of misplaced material is determined by the quantity and distribution of the raw coal impurities, the specific gravity of separation, and the physical separation efficiency of the segregated material.

A significant amount of material with a specific gravity close to the desired specific gravity of separation results in a more inefficient separation. If the amount of *near gravity material* exceeds approximately 15 to 20% of the total raw coal, efficient gravity separation is difficult.

The most common wet gravity concentration techniques include jigging, tabling and dense media processes. Each technique offers technical and economic advantages.

Jigging In a coal jig, a pulsating current of water is pushed upward in a regular, periodic cycle through a bed of raw coal supported on a screen plate. This upward or pulsion stroke of the cycle causes the bed to expand into a suspension of individual coal and refuse particles. The particles are free to move and generally separate by specific gravity and size, with the lighter and smaller pieces of coal moving to the upper region of the expanded bed. In the downward or suction stroke of the cycle, the bed collapses, and the separation is enhanced as the larger and heavier pieces of rock settle faster than the coal. The pulsion/suction cycle is repeated continuously. The separated layers are split at the discharge end of the jig to form a clean coal and a refuse product. The bed depth at which the cut is made determines the effective specific gravity of separation.

The upward water pulsation can be induced by using a diaphragm or by the controlled release of compressed air in an adjacent compartment. Operation of a Baum jig is illustrated in Fig. 9. This type of jig may be used to process a wide feed size range. Typically, the specific gravity of separation ranges from 1.4 to 1.8. The separation efficiency may be enhanced by pre-screening the feed to remove the fines for separate processing.

Tabling A concentrating, pitched table is mounted so that it may be oscillated at a variable frequency and amplitude. A slurry of coal and water is continuously fed to the top of the table and is washed across it by the oncoming feed. Diagonal bars, or *riffles*, are spaced perpendicular to the flow of particles. The coal-water mixture and oscillating motion of the table create a *hindered settling* environment in which the lower gravity particles rise to the surface. Higher specific gravity particles are caught behind the riffles and transported to the edge of the table, away from the clean coal discharge.

Tables are generally used to treat 0.375 in. x 0 (9.53 mm x 0) coal. Three or four tables may be stacked vertically to increase throughput while minimizing plant floor space requirements.

Dense media separation In dense or heavy media separation processes, the raw coal is immersed in a fluid with a specific gravity between that of the coal and the refuse. The specific gravity differences cause the coal and refuse to migrate to opposite regions in the separation vessel. In coal preparation, the heavy media fluid is usually an aqueous suspension of fine magnetite in water.

Flotation Coal and refuse separation by *froth flotation* is accomplished by exploiting differences in coal and mineral matter surface properties rather than specific gravities. Air bubbles are passed through a suspension of coal and mineral matter in water, which is agitated to prevent particles from settling out. Air bubbles preferentially attach to the coal surfaces which are generally more hydrophobic, or difficult to wet. The coal then rises to the surface where it is concentrated in a froth on top of the water. The mineral matter remains dispersed (Fig. 10). Chemical reagents, referred to as collectors and frothers, are added to enhance the selective attachment of the air bubbles to the coal and to permit a stable froth to form.

Flotation is generally used for cleaning coal finer than 48 mesh (300 microns). The efficiency of the process can be enhanced by carefully selecting the type and quantity of reagents, fine grinding to generate discrete coal and refuse particles, and generating fine air bubbles.

Dry processing Dry coal preparation processes account for a small percentage of the total coal cleaned in the U.S. In general, pneumatic processing is only applied to coal less than 0.5 in. (12.7 mm) in size with low surface moisture.

Dewatering Dewatering is a key step in the preparation of coal. Reducing the fuel's moisture content increases its heating value per unit weight. Because coal shipping charges are based on tonnage shipped, a reduction in moisture content results in lower shipping costs per unit heating value.

Coarse coal, greater than 0.375 in. (9.53 mm) particle size, can be sufficiently dewatered using vibrating screens. Intermediate size coal, 0.375 in. (9.53 mm) by approximately 28 mesh (600 microns), is normally dewatered on vibrating screens followed by centrifuges.

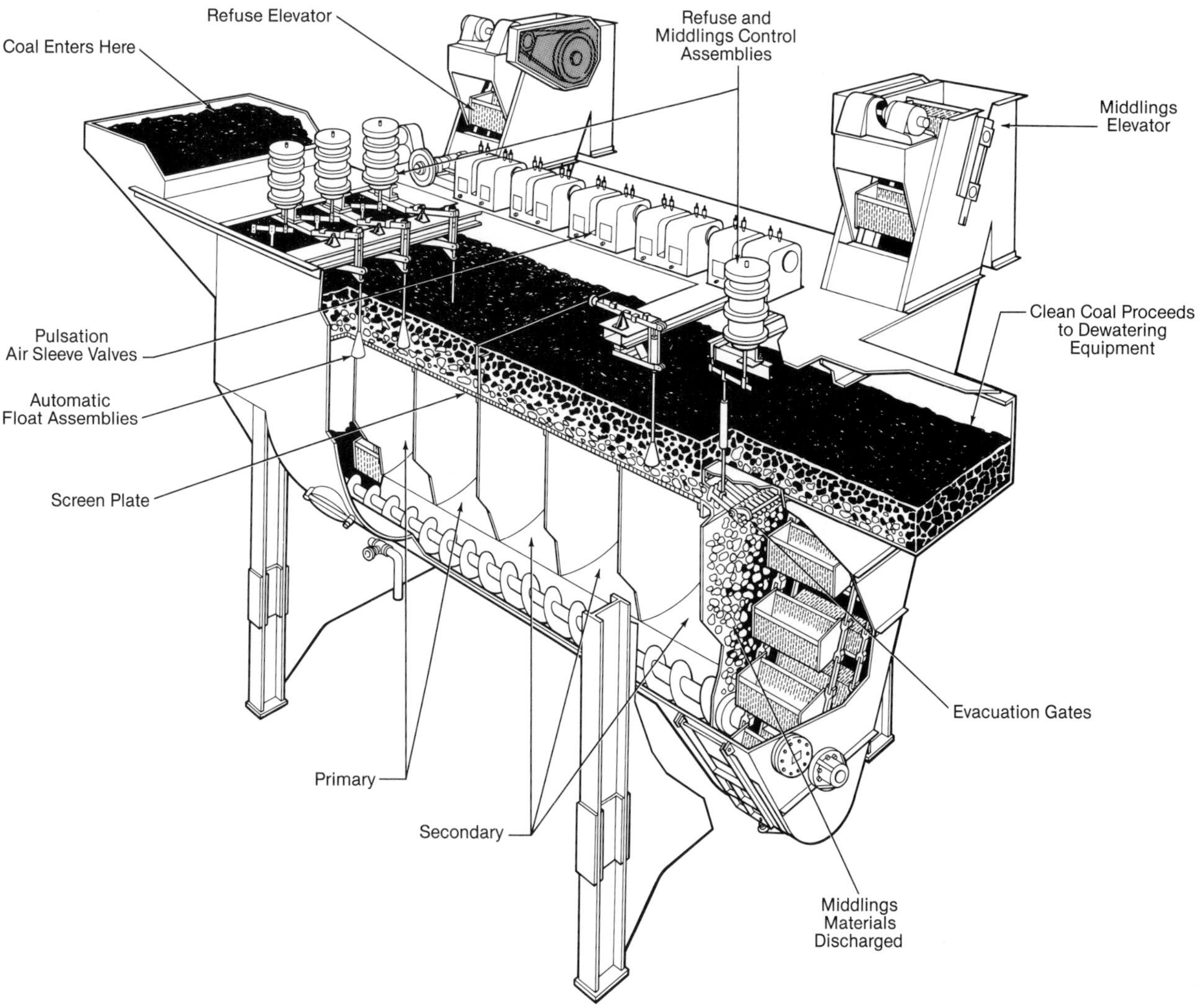

Fig. 9 Baum jig for coal preparation *(courtesy of McNally Pittsburg Mfg. Corp.).*

Fine coal dewatering often involves the use of a thickener to increase the solids content of the feed to a vacuum drum, vacuum disc filter or high gravity centrifuge. The filter cake may be mixed with the coarser size fractions to produce a composite product satisfying the specifications. Fine coal dewatering also serves to clarify the water for reuse in the coal preparation plant. Fines must be separated from the recycled water to maximize the efficiency of the separation processes.

Thermal dewatering may be necessary to meet product moisture specifications when the raw coal is cleaned at a fine size to maximize ash and sulfur rejection. The various types of thermal dryers include rotary, cascade, reciprocating screen, suspension and fluidized-bed dryers. Cyclones or bag filters are used to prevent fine dust emissions from the dryer. The collected fine coal may be recycled to support dryer operation. Thermal drying represents an economic tradeoff of reduced product moisture content versus heat required to fire the dryer.

Impact on steam generator system operations

The principal benefit of coal cleaning is the reduction in ash and sulfur content. Reduced ash content results in lower shipping costs and reduced storage handling requirements at the plant on a cost per unit heating value basis. Boiler heat transfer effectiveness may increase as a result of reduced ash deposition on tube surfaces. A reduction in sulfur content leads directly to reduced SO_2 emissions. Lower sulfur feed coal may preclude the need for or reduce the performance requirements of postcombustion SO_2 emission control systems. A reduction in sulfur content may also reduce spontaneous combustion during storage and

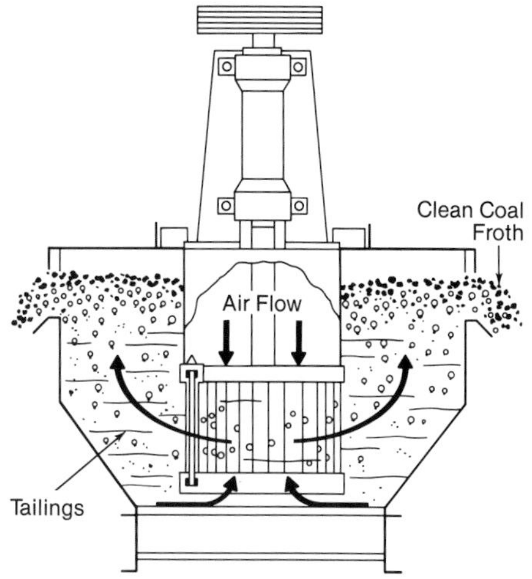

Fig. 10 Flotation cell.

corrosion in coal handling and storage equipment. Reduced ash content can result in reduced maintenance through removal of abrasive pyrite and quartz from the coal. Reduction of clay in the coal can improve handling but this may be offset by the effects of higher fines content and higher surface moisture on cleaned coal.

Coal transportation

The means of transportation and the shipping distance significantly influence the total fuel cost, reliability of supply, and fuel uniformity at the power plant. In some cases where Western U.S. coal is shipped over an extended distance, freight costs may represent 75 to 80% of the total delivered fuel cost. At the other extreme, transportation costs may be negligible for mine mouth generating stations. In transit, the coal's handling characteristics may be changed by freezing, increased moisture content or size degradation. When open rail car or barge transport is used, the moisture content of the delivered coal depends on the weather conditions in transit, the initial moisture level and the particle size distribution. Size degradation during shipping is dependent on the coal friability (ease of crumbling) and the techniques and number of transfers. As previously stated, for pulverized coal applications, size degradation is generally not a concern.

Coal is primarily shipped by rail, barge, truck and conveyor. The volume and distribution of coal transported by various means is summarized in Table 4. Combinations of these methods are often used to obtain the lowest delivery cost. Available transportation infrastructure, haulage distance, required flexibility, capital cost and operating cost are important factors in selection of a system for delivering coal to the power plant.

In general, barge transport represents the lowest unit cost per ton per mile followed by rail, truck and conveyor in terms of increasing cost. Combinations of these four transportation systems may be used to move coal to loading docks for overseas shipment. The major coal export ports in the U.S. are Hampton Roads, Virginia; Baltimore, Maryland and Philadelphia, Pennsylvania.

Transportation systems are generally designed to minimize intermediate storage of coal to control inventory costs, reduce insurance costs and minimize the effects of changes which can reduce the commercial value of coal. Potentially harmful changes include a reduction of heating value, particle size degradation, and loss due to self-ignition or wind and water erosion.

Rail

Approximately 57% of the coal delivered to power plants in the U.S. is shipped by rail. About half of this total is shipped in unit trains. Unit trains are dedicated rail shipments of coal normally consisting of 100 or more cars with a total of 10,000 t of coal or more. Bottom dump rail cars (100 t capacity) are typically used. The high capacity rail cars are generally not uncoupled from the time they are loaded at the mine until they arrive at the plant. In 1989 coal accounted for 40% of the rail industry's total freight tonnage and 22% of the revenues.

Rail transport provides for the movement of large quantities of coal over distances ranging from 10 to 1500 mi (16 to 2414 km).[1] Dedicated service between one mine and the steam generating plant simplifies management of coal deliveries.

The advantages of rail transport are somewhat offset by the restricted rail access. Generally only one rail line is available to transport coal from a mine or to a specific steam generating plant. The installation of dedicated rail lines must be factored in to the cost of the coal handling and storage system. Rail spurs to a specific mine location are useful only for the life of the mining activity. Transit time is typically on the order of 4 to 20 days.[1] The rail car unloading system and intermediate storage facilities must be designed to quickly process the cars to avoid demurrage (delay) charges at the plant.

Table 4
Distribution of Coal Transportation Methods (1988)[2]

	10^6 t/yr	10^6 t_m/yr	% of Total
Rail	550	499	57.4
Barge	155	141	16.2
Truck	128	116	13.3
Conveyor/pipeline	126	114	13.1

Barge

Barge transport of coal is the most cost effective alternative to rail. Approximately 16% of all coal shipped to steam generating plants in the U.S. is delivered by barge as illustrated in Fig. 11. Coal is the second largest single barge commodity and coal traffic accounts for a large fraction (23% in 1985) of annual barge tonnage.

Two standard sizes of open top coal barges are commonly applied to coal transport. The 1000 t (907 t_m) capacity Pittsburgh standard barge measures 175 ft (53 m) long x 26 ft (8 m) wide with a draft of 9 ft (2.7 m). The jumbo barge has a nominal capacity of 1500 t (1361 t_m) and is 195 ft x 35 ft (59 x 11 m) with a 12 ft (3.6 m) draft.[2] A single tow, or group of 20 barges, can carry 20,000 to 30,000 t (18,144 to 27,216 t_m).

The major waterways for coal traffic in the U.S. are the Ohio, Mississippi and Black Warrior-Tombigbee Rivers.[3] The quantity of coal shipped in a single tow or string of barges is determined by the lock requirements of the river system being navigated. For example, on the Ohio River system, a tow of three barges wide by five barges long is commonly used because of the River's lock requirements.[4]

Fig. 11 Typical barge transport for large quantities of coal.

However, on the relatively unobstructed lower Mississippi, tows of 30 barges are not uncommon.

There is a significant degree of competition and transport prices are generally stable. Some cost differences between upstream and downstream travel are common.

Barge transportation of coal to steam generating plants is constrained by the location and characteristics of the available river systems. Close proximity to waterways for direct loading and unloading is needed for efficient barge transportation. Barge delivery must be supported by truck, rail or belt conveyor trans-loading at the mine location or the steam generating plant. The natural river network is not always the most direct route and may result in increased delivery time. The channel width and seasonal variability in water level are natural limitations for barge traffic. River lock sizes and condition of repair may restrict the maximum permitted tow size. Delays due to deteriorating locks and congestion may be significant on some river systems. In some areas lock repair costs are recovered through a surcharge on tonnage shipped through the lock. Barges are not self-unloading. The capital investment required for the unloading facilities may restrict barge deliveries to plants using more than 50,000 t/yr.[1]

Truck

For power plants located near mines, trucks loaded at the mine deliver coal directly to the power plant storage site. Trucking accounts for about 13% of the total tonnage of coal delivered to steam generating plants. Usually, a small receiving stockpile for truck deliveries is set up separate from the main storage pile to permit isolation until it is determined that the coal satisfies the required quality specifications.

Trucking also pays a key role in both rail and barge transport. Approximately 70% of the coal transported by rail or barge is first trucked to the loading dock or involves truck transfer at some point.[5] Highway trucks typically carry 15 to 30 t (14 to 27 t_m) of coal over distances ranging up to 70 mi (113 km). Off-road vehicles can handle 100 to 200 t (91 to 181 t_m) over a range of 5 to 20 mi (8 to 32 km) at mine mouth generating stations.

Trucking is the most flexible mode of coal transportation. It is relatively easy to adjust to changes in demand to meet the generating plants' variable supply requirements. The short haulage distances, and therefore short delivery times, can be used to minimize storage requirements at the generating plant. Trucks are simple to unload and a minimum of on-site handling and distribution is needed. Use of the existing highway infrastructure provides for flexible delivery routes and reduces travel restrictions associated with rail and river transport. Trucks are very efficient for short haulage distances and for smaller generating plants. Trucking is the least capital intensive mode of transporting coal and a high degree of competition exists.

Truck transportation is characterized by a high operating cost per ton mile relative to barge or rail transport. Practical haulage distances are usually limited to 50 mi (80 km). State and local transportation regulations often limit loads to 25 t (23 t_m) or less. A large generating plant would require a significant amount of truck traffic and congestion at the delivery site may be severe. Truck deliveries require the highest degree of monitoring at the plant. Frequently, every truck must be weighed.

Continuous transport

Coal may be transported from the mine to the generating plant by continuous belt conveyors or slurry pipelines. In 1988 continuous transport systems accounted for approximately 13% of the total coal deliveries. Belt conveyors are normally limited to lengths of 5 to 15 mi (8 to 24 km). The coal delivery rate is a function of the belt width, operating speed and the number of transfer points. Only one major coal slurry pipeline is in operation. The 273 mi (439 km) long Black Mesa pipeline runs from a mine in Arizona to a generating plant in Nevada. The coal transport rate is determined by the pipe diameter, slurry velocity and solids loading.[6]

Continuous systems can move large amounts of coal cost-effectively over short distances. Often, continuous systems can be used where the terrain limits the use of other modes of transport. Social and environmental impacts are minimal.

The application of continuous transportation systems is limited by the proximity to the generating plant, a low degree of operating flexibility due to the fixed carrying capacity, the inflexibility of the loading and discharge locations, high capital cost and a relatively high energy consumption per ton mile of coal delivered. Pipeline builders must overcome significant opposition in obtaining rights of way and water resource allocation. The added costs associated with dewatering the coal at the generating plant must also be considered.

Coal handling and storage at the power plant

Bulk storage of coal at the power plant is necessary to provide an assured continuous supply of fuel. The tonnage of coal stored at the site is generally proportional to the size of the boiler. A 100 MW plant burns approximately 950 t/d (862 t_m/d), while a 1300 MW plant requires approximately 12,000 t/d (10,887 t_m/d). For some public utilities, a minimum of 60 to 90 day supply must be stored at the plant by law. However, stored coal represents substantial working capital and requires land which may be otherwise productive. Economic considerations are a key factor in determining when to purchase coal and how much coal to store at the plant. Additional considerations, such as the changes to coal characteristics due to weathering, restrict the maximum amount of coal stored on site.

For smaller, industrial boiler applications, bin or silo storage may be preferred over stockpile storage. The advantages of bin storage include shelter from weather and ease of reclamation. Prefabricated bins with capacities up to 94,500 ft^3 (2674 m^3) holding approximately 2400 t (2177 t_m) are commercially available.[7]

The complexity of the coal storage and handling operations increases in proportion to the size of the steam generating plant. Efficient techniques have been developed for large and small plants. The components of a sophisticated coal storage and handling system for a large, 1000 MW electric generating plant are illustrated in Fig. 12. Coal is delivered in self-unloading, bottom dump railcars and is transferred to a large stockpile. An automatic reclaim system recovers coal from the stockpile for crushing and distribution to in-plant storage silos. The system is automated and a two man crew can handle 7000 t/d (6350 t_m/d) of coal. All the equipment from the reclaim feeders to the in-plant silos and boiler is unmanned and controlled by the central control room operator.

The storage and handling operations of a utility boiler are depicted in the chapter frontispiece.

Raw coal handling

An extensive array of equipment is available for unloading coal at the plant site and distributing it to stockpile and bin storage locations. Equipment selection is generally based on the method of coal delivery to the plant, the boiler type, and the required coal capacity. For small plants, portable conveyors may be used to unload rail cars, to reclaim coal from yard storage piles and to fill bunkers. Larger plants require dedicated, installed handling facilities to meet the demand for a continuous fuel supply. However, even relatively small plants may benefit from the improved plant appearance, cleanliness and reduced coal handling labor requirements associated with mechanical handling systems.

The coal handling system components are determined by the design and requirements of the boiler. A crusher is normally integrated into the system to generate a uniform top size coal feed to the pulverizer or boiler. The system normally includes a magnetic separator to remove misplaced mining tools and roof support bolts which could damage the pulverizer. Crushing and tramp metal removal needs are generally less critical for stoker-fired boilers than for pulverized coal units.

The coal handling system capacity is determined by the boiler's rate of coal use, the frequency of coal deliver-

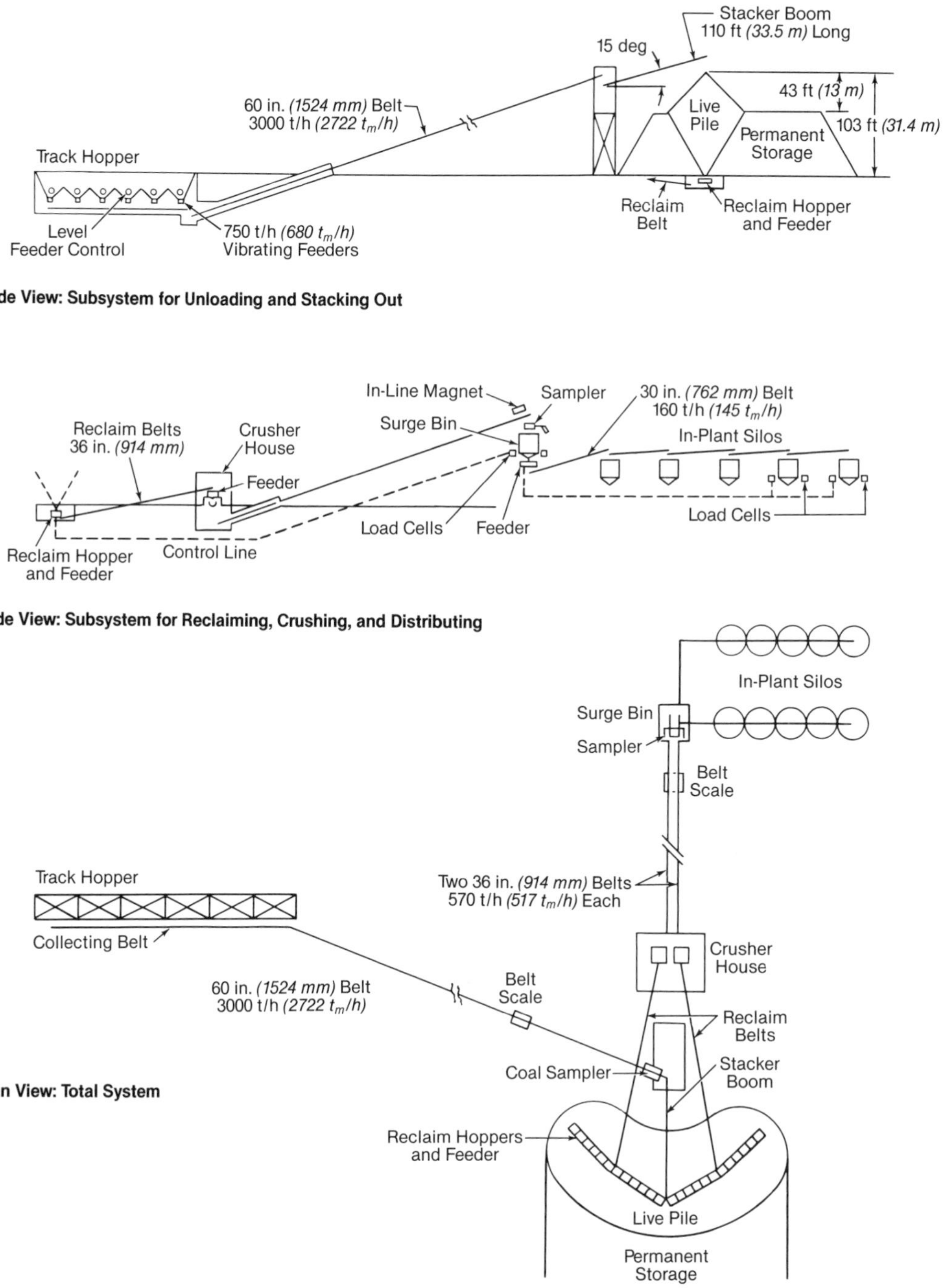

Fig. 12 Typical coal handling system and subsystems for a 1000 MW coal-fired power generation plant.

ies to the plant, and the time allowed for unloading. In most large plants, only four to six hours per day are dedicated to unloading coal deliveries.

Rail car unloading

In automatic rotary car dumping systems, the rail cars are hydraulically or mechanically clamped in a cradle, and the cradle is rotated so the coal falls into a hopper below the tracks. The dumping is completed without uncoupling the cars. The rotary dump system advantages of short cycle time and high capacity are offset by a relatively high capital cost.

With bottom dump cars, unloading is relatively simple when the coal is dry and free flowing. However, high surface moisture can cause the coal to hang up in the car and, in cold weather, freeze into a solid mass. In hot, dry weather, high winds can create severe dust clouds at the unloading station unless special precautions are taken. The coal supplier frequently sprays the coal with oil or an anti-freezing chemical such as ethylene glycol as the car is loaded to settle the fines and to ease handling in freezing weather. The treatment does not appreciably affect combustion or cause problems in the pulverizers. There is also some evidence that the treatment may reduce hangups in bunkers and chutes.

A rail car unloading system which includes a crusher and magnetic separator is illustrated in Fig. 13. A screw conveyor is used to distribute coal along the length of the bunker. The capacity of the bucket elevator generally limits this system to relatively small plants.

A rail car unloading and handling system for a large plant using at least 3000 t/d (2722 t_m/d) of coal is shown in Fig. 14. Coal from the car dump hopper is fed to a rotary breaker, in which the coal breaks into smaller pieces as it is tumbled and passes through a screen shell. The broken coal is then conveyed to the storage bunker, where a tripper belt conveyor distributes it over the length of the bunker.

Barge unloading

The simplest barge unloader consists of a clamshell bucket mounted on a fixed tower. The barge is positioned under the bucket and is moved as necessary to allow emptying. With this type of unloader, the effective grab capacity of the buckets is only 40 to 50% of the nominal bucket capacity. A shore mounted bucket wheel or elevator unloader can increase the efficiency and capacity of this unloading operation. Modern ocean-going vessels are often equipped with a bucket wheel for self-unloading.

Truck unloading

Trucks may dump coal through a grid into a storage hopper. This grid separates large pieces of wood and other trash from the coal. At some plants, the trucks are directed to a temporary storage area, where coal from various mines can be blended prior to crushing or feeding to the boiler.

An effective truck delivery and coal handling system for a small to medium size (30 to 300 t/d) stoker coal-fired boiler is illustrated in Fig. 15. The elevator and storage bunker are located outside and no provisions are made for crushing or tramp metal removal. Transfer chutes should be angled at least 60 deg from horizontal to minimize coal hangups.

Stockpile storage

Careful consideration should be given to storage pile location. The site must be conveniently accessible by barge, rail or truck. Frequently, provisions must be made for more than one method of coal delivery. The site should be free of underground power lines. Other underground utilities that would not be accessible after the storage pile is constructed must also be avoided. A thorough evaluation and environmental survey of the proposed site topography should include analysis of the soil characteristics, bedrock structure, local drainage patterns and the potential for flooding. Climatic data, such as precipitation records and prevailing wind patterns, should also be evaluated. Protection from tidal action or salt water

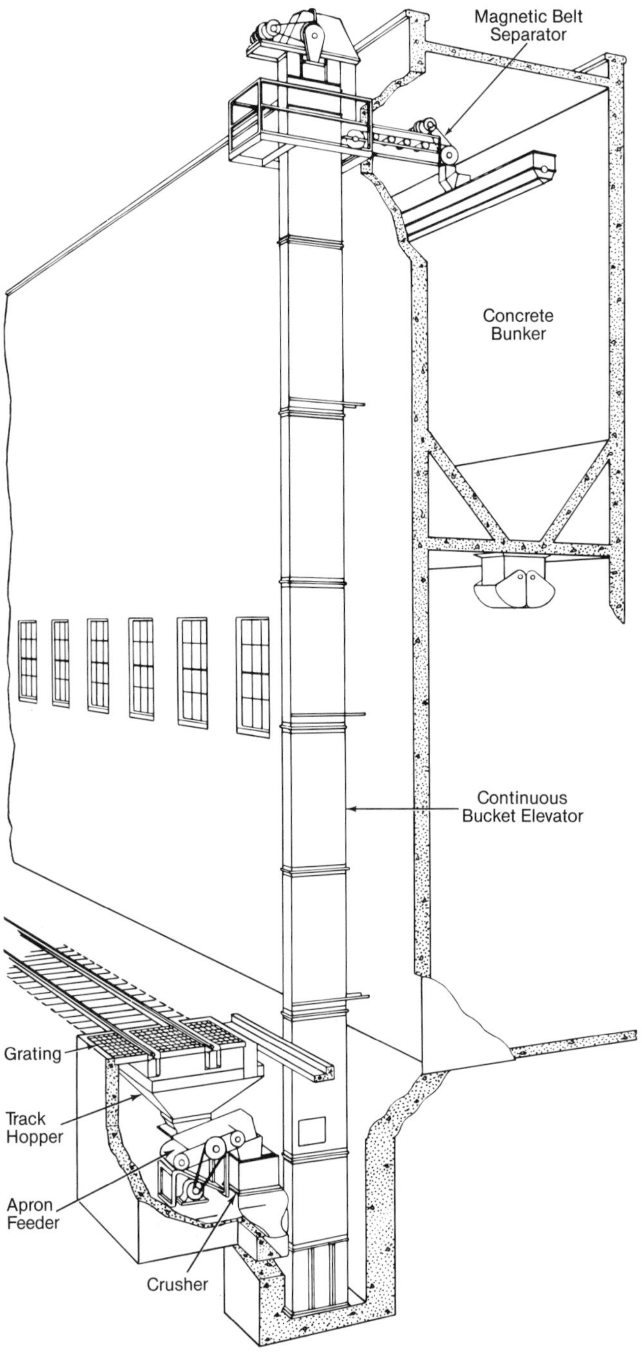

Fig. 13 Rail car dumping system for a small power plant.

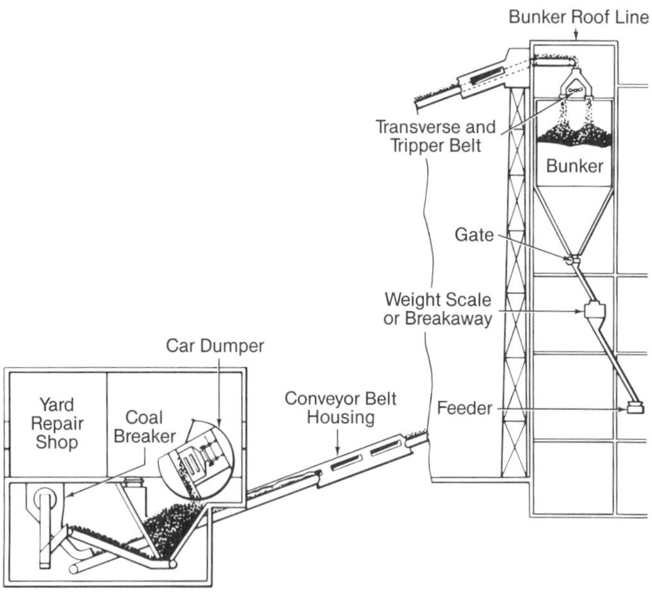

Fig. 14 Rail car unloading and coal handling system for a large power plant.

spray may be needed in coastal areas. The potential effects of water runoff and dust emissions from the pile must be considered. Site preparation includes removing foreign material, grading for drainage, compacting the soil and providing for collection of site drainage.

The shape of a stockpile is generally dependent on the type of equipment used for pile construction and for reclaiming coal from the pile. Conical piles are generally associated with a fixed stacker while a radial stacker generates a kidney shaped pile. A rail mounted traveling stacker can be used to form a rectangular pile. Regardless of the shape of the pile, the sides should have a shallow slope.

Bituminous coal, subbituminous coal and lignite should be stockpiled in multiple horizontal layers. To reduce the potential for spontaneous combustion, coal piles are frequently compacted to minimize air channels. These channels can function as chimneys which promote increased air flow through the pile as the coal heats. For bituminous coal, an initial layer, 1 to 2 ft (0.3 to 0.6 m) thick, is spread and thoroughly packed to eliminate air spaces. A thinner layer is required for subbituminous coal

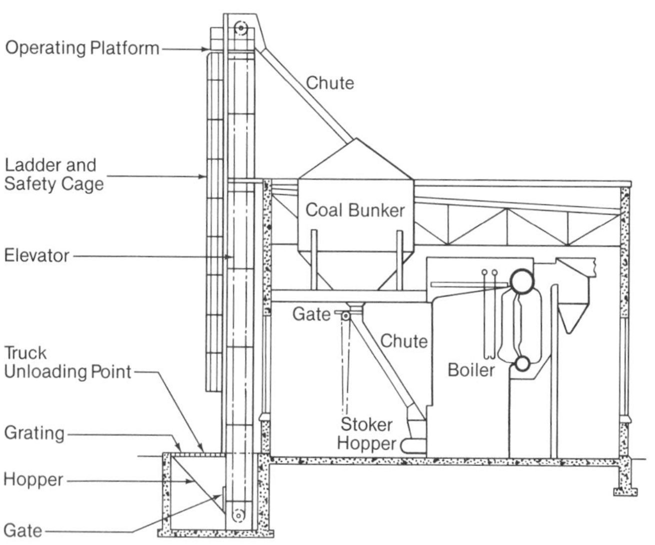

Fig. 15 Coal handling equipment for truck delivery.

and lignite to assure good compaction. Care should be taken to avoid coal pile size segregation by blending coal during pile preparation.

For long term storage (see Fig. 16), the top of the pile may be slightly crowned to permit even rain runoff. All exposed sides and the top may be covered with a 1 ft (0.3 m) thick compacted layer of fines and then capped with a 1 ft (0.3 m) layer of screened lump coal. It is not practical to seal subbituminous and lignite piles with coarse coal because the coarse coal would weather and break apart to a smaller size in a short period of time. At smaller industrial plants, where heavy equipment for compaction can not be justified, a light coating of diesel oil can help to seal off the outer surface of the pile.

The quantity of coal stored in a stockpile can be estimated using geometry and some assumptions about the characteristics of coal. The volume of the pile can be estimated based on its shape. Approximate values for the material's bulk density and its angle of repose are required to complete tonnage calculations. Both of these parameters are particle size dependent. For typical utility storage pile applications, a loose coal bulk density of 50 lb/ft^3 (801 kg/m^3) and a 40 deg angle of repose may be used. For well compacted piles, a bulk density of 65 to 72 lb/ft^3 (1041 to 1153 kg/m^3) is more appropriate.

Storage pile inspection and maintenance Visual inspections for hot spots should be made daily. In wet weather, a hot area can be identified by the lighter color of the surface coal dried by escaping heat. On cold or humid days, streams of water vapor and the odor of burning coal are

Fig. 16 Long term coal storage — typical example of thorough packing with minimum size segregaton. This pile contains about 200,000 t of coal.

signs of heating or air flowing through the pile. Hot spots may also be located by probing the pile with a metal rod. If the portion of the rod in contact with the coal is too hot to be held as it is withdrawn, the coal temperature is dangerously high.

It is also important to rotate areas of the storage pile from long to short term storage on a planned schedule. This minimizes harmful degradation of the coal.

Bulk storage reclaim and transfer The specific system for reclaiming the stored coal depends upon plant size and economic tradeoffs between operating and capital costs. For higher capacity systems [150 to 4000 t/h (136 to 3629 t_m/h)], underground conveyors (Fig. 12) and bucketwheel reclaimers are used. As discussed earlier and shown in Fig. 12, coal is transferred to the in-plant silos or bunkers by inclined conveyors and, where needed, bucket elevators. Appropriate coal sizing and monitoring steps can be incorporated: crushing, magnetic tramp iron removal, screening, weighing and sampling. Bucket elevators (see Figs. 13, 15 and 17) are primarily used at small installations because of their limited capacity.

Silo storage

Fig. 17 illustrates a common silo design with an approximate storage capacity of 600 t (544 t_m). An internal shelf permits maximum use of the storage space. As coal is used from above the shelf, reserve storage can be reclaimed from the lower part of the silo. Silo storage requires less building volume and structural steel for a given capacity than bunker storage.

Bunker storage

Coal bunkers provide intermediate, short term storage ahead of the pulverizers or other combustion zone coal feed equipment. The complexity of bunker and chute design has increased as the size of individual boilers and consequently the rate of coal consumption have increased.

When the bunkers and transfer chutes have been properly selected and sized, the condition of the coal becomes the dominant factor in determining the effectiveness of the fuel supply system. Steady flow of fine coal can be difficult to maintain when the surface moisture is 5 to 10%. *Rat holes* or *pipes* can form in the bunker, resulting in intermittent flow or complete flow stoppage.

To assure a continuous flow of coal from the bunker, the design and construction must be integrated with the upstream bulk storage and handling techniques and the anticipated coal properties at the bunker inlet.

Bunker design Once constructed, coal bunkers are an integral part of the boiler house structure and are usually not modified significantly over the life of the plant. The required capacity, construction material, shape and location are important considerations in designing a bunker.

The capacity of a bunker or series of bunkers should normally be sufficient to provide 30 hours of fuel supply at maximum boiler operating load. The storage capacity must also be adequate to cover weekend or holiday periods during which coal handling labor or equipment may not be available. Storage bunkers provide flexibility to optimize plant operating labor requirements and to permit maintenance and repair of coal handling equipment independent of steam generator operation.

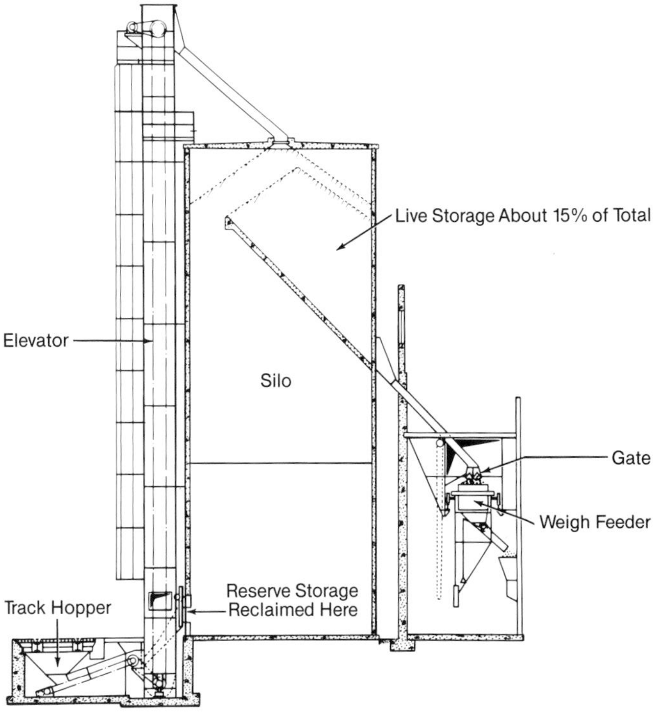

Fig. 17 Silo system for live and reserve storage.

Coal bunkers have been built of tile, reinforced concrete, carbon steel plate, stainless steel clad plate, steel plate lined with acid resistant concrete and steel plate lined with rubber. When unlined carbon steel plate is used, a periodic inspection program is required to monitor corrosion and to assure the safety of the structure.

The final shape of a bunker usually represents a compromise between space limitations and the optimum design for coal flow. Dead pockets, in which coal flow is restricted, should be avoided because of possible spontaneous combustion. Several common bunker shapes are shown in Fig. 18. The silo bunker is less susceptible to rat holing and hangups than other shapes. This design is equally suitable for small and large plants.

The bunker should be located so that coal flow to the pulverizer or furnace is nearly vertical. Generally, bunkers are located in the upper levels of the plant to permit gravity flow to pulverizer or furnace feeders underneath. Storage bunkers should be as far as possible from furnace exit gas flues, hot air ducts, steam pipes, or other external sources of heat which could contribute to spontaneous coal ignition. It may be necessary to insulate the bunker and provide ventilation to reduce heat transfer from nearby steam lines or breachings. Bunkers should also be designed to provide a means of emptying the coal for an extended forced outage.

Chute design Modeling of coal flow in transfer chutes has identified several design features which can minimize flow interruptions. Chutes should generally be circular, short and as steep as possible. Reductions in cross-sectional area and sudden changes in direction should be avoided. When two streams merge, a minimum angle of convergence should be used. Finally, except in pressurized systems discussed below, a *breakaway* (a vented sudden enlargement in diameter) should be used when significant changes in angle can not be avoided in order to relieve lateral pressure and pipe friction.

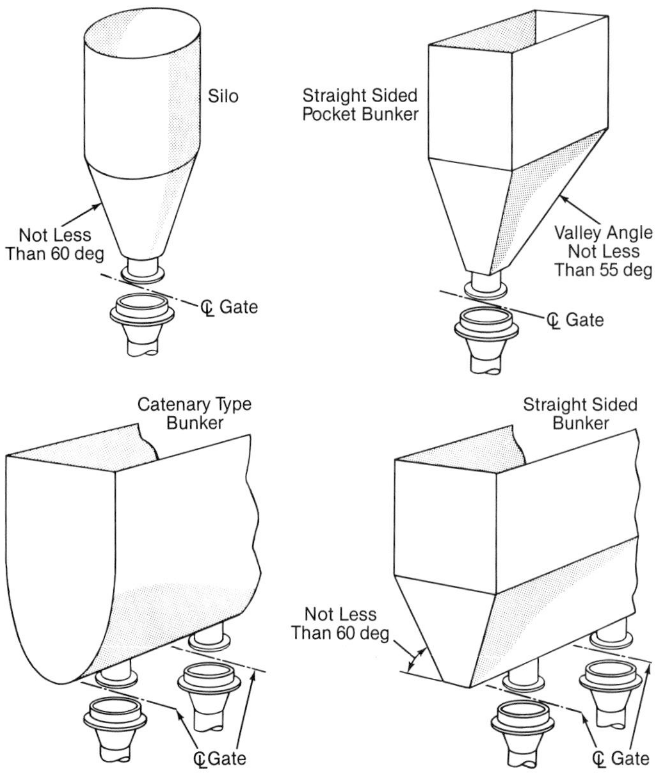

Fig. 18 Four commonly used shapes in coal bunker design.

In pressurized pulverizer feed applications, special consideration must be given to chute design because the coal inside the chute also serves as the seal to minimize air loss from the pressurized pulverizer to the bunker. A minimum required height (*seal height*) and a sealed chute/feeder system are required. Breakaways can not be used. A typical bunker to pulverizer system is shown in Fig. 19. Vertical, constant-diameter chutes connect the bunker to the feeder and the feeder to the pulverizer. Appropriate couplings and valves complete the system.

Feeder design

Feeders are used to control coal flow from the storage bunker at a uniform rate. Feeder selection should be based on an analysis of the material properties (maximum particle size, particle size distribution, bulk density, moisture content and abrasiveness), the desired flow rate and the degree of flow control required. A variety of feeder designs have been used for coal-fired applications with increasing sophistication over time as more accurate control of coal flow has become necessary. Feeders for modern pulverized coal applications can generally be classified as *volumetric* or *gravimetric*.

Volumetric feeders, as the name implies, are designed to provide a controlled volume rate of coal to the pulverizer. Typical examples include drag, table, pocket, apron and belt feeders. Belt feeders, perhaps the most accurate type, have a *level bar* to maintain the flow of coal at a constant height and width while the belt speed sets the velocity of the coal through the opening. As with all volumetric designs, however, the belt feeder does not compensate for changes in coal bulk density. This results in variations in the energy input to the pulverizer and ultimately to the burners.

Gravimetric feeders (see Fig. 20) compensate for variations in bulk density due to moisture, coal size and other factors. They provide a more precise weight flow rate of coal to the pulverizer and therefore more accurate heat input to the burners and boiler. Even variations in coal moisture have a larger relative impact on coal bulk density than on heating value. Therefore, modern gravimetric feeders offer an accurate, commercially accepted technology to control fuel and heat input to the burners and boiler. This can be a very significant issue where more accurate control of fuel/air ratios is needed to: 1) minimize the formation of nitrogen oxides (NO_x), 2) control furnace slagging, and 3) maximize boiler thermal efficiency by reducing excess air levels. (See Chapter 13.)

In the most common gravimetric feeder system, coal is carried on a belt over a *load cell* which monitors the coal weight on the belt. The feedback signal is used to maintain the weight flow by either: 1) adjusting the height of a leveling bar to control the cross-sectional coal flow area while the belt speed remains constant, or 2) fine tuning the belt speed while the cross-sectional area remains constant. The overall set point flow rate is adjusted by varying the base belt speed.

Coal blending

When coals from two or more sources fuel a single boiler, effective coal blending or mixing is required to provide a uniform feed to the boiler. The use of multiple coals can be driven by economics, coal sulfur content to

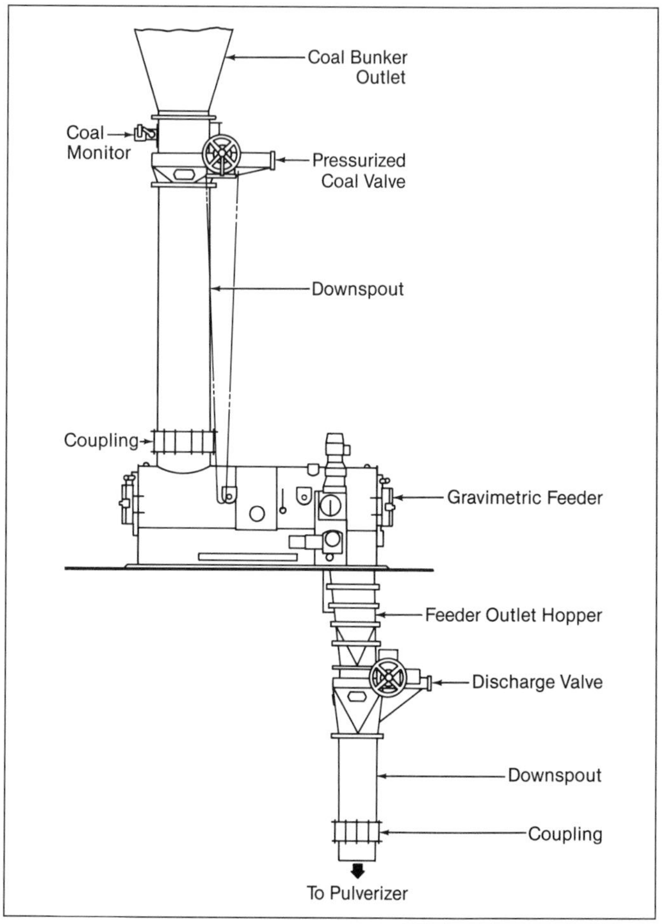

Fig. 19 Arrangement of bunker discharge to pulverizer showing typical feed system.

meet emission requirements and/or the effects of different coals on boiler operation. The goal of effective blending is to provide a coal supply with reasonably uniform properties which meet the blend specification typically including sulfur content, heating value, moisture content and grindability.

Coal blending may occur at a remote location or at the steam generating plant. Off-site blending eliminates the need for separate coal storage and additional fuel blending facilities. Steam plant on-site blending may be accomplished through a variety of techniques. It may be sufficient to provide separate stockpiles for each coal source and use front-end loaders to transfer the appropriate quantities to a common pile or hopper for blending prior to crushing. Coal may also be reclaimed from the various stockpiles using a boom or bridge-type reclaimer. Coal from the various sources may be stored in separate bins with a feeder from each bin used to meter the desired quantities onto a common transfer belt. On-site blending provides more flexibility in coal sourcing and in adjusting to actual on-site coal variations.

Particular care must be maintained to ensure proper and complete blending. Significant variations in the blended coal can have a major impact on operation of the pulverizers, burners, sootblowers and postcombustion cleanup equipment. If uniform blending does not occur, pulverizer performance can deteriorate (see Chapter 12), the boiler may experience excessive slagging and fouling, and electrostatic precipitator particulate collection efficiency may decline, among others.

Resolution of common coal handling problems

Dust suppression

Water, oil and calcium chloride ($CaCl_2$) are common agents used to suppress dust emissions at coal handling and transportation transfer points. A water or oil mist may be sprayed onto a stream of coal falling from a chute or loading boom. The oil reduces dust emissions by causing the dust to adhere to larger pieces of coal and by forming agglomerates which are less easily airborne. Use of $CaCl_2$ should be limited because of its potentially harmful side effects on boiler operation. (See Chapter 20.)

Oxidation

Coal constituents begin to oxidize when exposed to air. This oxidation may be considered as a very slow, low temperature combustion process, because the end products, carbon dioxide (CO_2), carbon monoxide, water and heat, are the same as those from furnace coal combustion. Furnace combustion of coal may be viewed as a very rapid oxidation process. Although there is evidence that bacterial action causes coal heating, the heating primarily occurs through a chemical reaction process. If spontaneous combustion is to be avoided, heat from the oxidation should be minimized by retarding oxidation or removing the generated heat.

Coal oxidation is primarily a surface action. Finer coal particles have more surface area for a given volume and, therefore, oxidize more rapidly. Freshly crushed coal also has a high oxidation rate. Coal's oxygen absorption rate at constant temperature decreases with time. Once a safe storage pile has been established, the rate of oxidation has been slowed considerably. Coal should be kept in dead storage undisturbed until it is to be used.

Fig. 20 Typical gravimetric feeder.

The rate of oxidation also increases with moisture content. High moisture, western coals are particularly susceptible to self-heating.

Frozen coal

The difficulties associated with handling frozen coal may be avoided by thermally or mechanically drying the fines following coal preparation or by spraying the coal with an oil or anti-freezing solution mist. The cost of this oil spray treatment can be reduced by using waste oil. A heavy coat of oil may also be sprayed on the rail car hoppers to prevent the coal from freezing to the sides. The use of salt or $CaCl_2$ may result in accelerated ash deposition or boiler heating surface corrosion and therefore is generally not recommended.

Permanent installations for thawing frozen coal in rail cars include steam heated thawing sheds, oil-fired thawing pits, and radiant electric thawing systems. Steam heated systems are reliable and efficient, but are relatively expensive. Oil-fired systems, which prevent direct flame impingement on the cars, provide reliable operation and rapid thawing. Electric thawing systems are used at many plants that handle unit train coal shipments.

Coal pile fires

A primary concern in coal storage is the potential for spontaneous combustion in the pile as a result of self-heating properties.

A coal pile fire may be handled in several ways depending on its extent. The hot region should be isolated from the remainder of the pile. This may be accomplished by trenching and sealing the sides and top of the hot area with an airtight coating of road tar or asphalt. Caution should be used in working the hot area with heavy equipment as subsurface coal combustion can affect the stability and load bearing characteristics of the pile. Water should not be used unless it is necessary to control flames. Pouring water on a smoldering pile induces more pronounced channeling and promotes greater air flow through the pile. The use of asphalt or road tar for air-tight sealing of the entire pile is not recommended.

Bunker flow problems

Hangups of fine coal are liable to begin when surface moisture reaches 5 to 10% by weight. Improved bulk storage techniques and careful coal reclaiming are the best preventive measures for reducing bunker flow problems.

Fine, wet coal feeding from a bottom outlet bunker tends to rat hole or pipe all the way to the top surface. When this occurs, coal flow to the feeders is intermittent and may be completely interrupted.

Bunkers may be equipped with ports located near the outlet. These ports permit the use of air lances for restoring flow. Air lances may also be effectively used from above. Small boring machines can be mounted above the bunker to loosen coal jams at the outlet. Service companies can be contracted to remove flow obstructions using boring tools. Air blasters or air cannons have been successfully used to promote flow of coal in bunker hoppers. If there is any possibility of fire, these devices must be charged with nitrogen or carbon dioxide to prevent triggering a dust explosion.

Castable polyurethane can be used for lining bunkers and chutes to improve abrasion resistance and reduce flow resistance. Bunker hoppers or silo cones are frequently lined with stainless steel sheet. The sheet is typically specified as cold-finished and all welds are ground and polished to an equivalent finish. This smooth and corrosion-resistant surface resists the formation of deposits which provide anchors for arches or stagnant coal masses.

Bunker fires

A fire in a coal bunker is a serious danger to personnel and equipment and must be dealt with promptly. The coal feed to the bunker should be stopped. An attempt should be made to smother the fire while quickly discharging the coal. Continuity and uniformity of the hot or burning coal discharge from the bunker is especially important; interruption of coal flow aggravates the danger. The bunker should be emptied completely; no fresh coal should be added until the bunker has cooled and the cause of the fire determined.

The fire may be smothered using steam or CO_2. CO_2 settles through the coal and displaces oxygen from the fire zone because it is heavier than air. Permanent piping connections to the bottom of the bunker may be made to supply CO_2 on demand. The CO_2 should fill the bunker, displace the air and smother the fire.

It is highly desirable to completely extinguish the fire before emptying the bunker. This is rarely possible because of boiler load demands and the difficulty of eliminating air flow to the fire. However, the use of steam or CO_2 to smother the fire can minimize the danger.

Bunker flow problems which result in dead zones may contribute to fires. Thermocouples installed in the bunker can monitor the temperature of the stored coal. The coal feed to the bunkers may also be monitored to prevent loading the bunker with hot coal.

Additional remarks on dealing with bunker fires in pulverized coal plants are provided in Chapter 12.

Environmental concerns

Water which percolates through the coal storage pile can become a source of acidic drainage which may contaminate local streams. Runoff water must be isolated by directing the drainage to a holding pond where the pH may be adjusted. (See also Chapter 32.)

Airborne dust from stockpiles creates a public nuisance, has potentially harmful effects on surrounding vegetation and may violate regulated dust emission standards. The potential for dust emissions is determined by the surface layer coal characteristics (particle size distribution, moisture), stockpile design (exposed area, height), and local climatic conditions (wind velocity, rainfall).[8]

Alternate solid fuel handling

Economic and environmental concerns have led to increasing steam generation from solid fuels derived from residential, commercial and industrial byproducts and wastes. Key among these are municipal solid waste (MSW), wood and biomass as discussed in Chapters 27 and 28. The properties of these solid fuels require storage, handling and separation considerations different from those applied to coal.

MSW can either be burned with little pre-combustion processing (mass burn) or as a refuse-derived fuel (RDF). As-received refuse for mass-burn units is delivered to the tipping area and stored in an open concrete storage pit. The refuse pit is usually enclosed and kept under a slightly negative pressure to control odors and dust emissions. The tipping bay is designed to facilitate traffic flow based upon the frequency of deliveries and the size of the delivery trucks or trailers. The pit is usually equipped with a water spray system to suppress fires which may arise in part due to heat generated from decomposition of the refuse. An overhead crane is used to mix the raw MSW in the storage pit, to remove bulky items and to transfer material to the boiler feed charging hoppers. Large objects and potentially explosive containers are located and removed prior to combustion. (See Chapter 27.) A full capacity spare crane is recommended. Storage capacity is typically three to five days in order to accommodate weekends, holidays and other periods when refuse delivery may not be available. Longer term storage of refuse is not normally recommended.

MSW may be processed to yield a higher Btu, lower ash RDF. The degree of processing required is determined by economics and by the fuel properties necessary for efficient boiler operation. MSW is usually delivered to an enclosed receiving floor. Front-end loaders can be used to spread the refuse, remove oversized and potentially dangerous items and feed the MSW to the RDF processing system as needed. RDF processing includes an integrated system of conveying, size reduction, separation, ferrous metal recovery, sizing and other equipment discussed in depth in Chapter 27. MSW may be processed into RDF at the power plant site or at a remote location. The selection is based upon a number of economic factors, but operation of the RDF processing system at the boiler site typically will enhance availability to support uninterrupted steam generation.

Wood waste generally consists of bark, sawdust, saw mill shavings and lumber rejects. Material is generally shipped by truck to the steam generator site near the source. Material can be dumped directly on the storage pile or an unloading facility can be used. The unloading and handling equipment must be designed to handle extremely dusty conditions with a very abrasive material. Wood products can be stored in large outdoor piles or inside bins or silos. Wood is not typically stored in piles for more than six months or in silos or bins for more than three to five days. This fuel is typically screened to remove oversized material for further size reduction. Oversized material is reduced by a shredding machine, or *hog*, and either returned to storage or sent directly to the combustor. Mechanical belt conveyors are the most popular method of transporting the fuel on site, although pneumatic systems can be effective with a finely ground, clean fuel such as sawdust. Tramp iron is usually removed by a magnetic separator. While most modern wood-fired boilers can burn materials with a moisture content of up to 65% as-received (see Chapter 28), pre-drying may be required. Mechanical hydraulic presses and hot gas drying, or both, are used.

Economics

The selection of the fuel source, degree of cleaning, and transportation system are closely tied to providing the lowest plant fuel cost. The selections must not be made in isolation, but in concert with evaluating the impact of the specific fuel on the boiler and bypass system operation. For example, use of a new less expensive fuel may result in significant deterioration in boiler performance and availability due to more severe slagging and fouling tendencies of the flyash. (See Chapter 20.) The relative contributions of coal cleaning, transportation and base fuel price vary widely.

References

1. Bessett, R.D., "Coal transportation," appearing in *Handbook of Energy Systems Engineering,* L.C. Wilbur, ed., Wiley, New York, 1985.

2. Schwieger, R.G., "Coal handling," appearing in *Standard Handbook of Powerplant Engineering*, T.C. Elliot, ed., McGraw-Hill, New York, 1989.

3. *Facts About Coal, 1989*, National Coal Association, Washington, D.C., 1989.

4. Mahr, D., "Coal transportation and handling," *Power Engineering*, pp. 38-43, November 1985.

5. Scott, R.H., "Competitive coal transportation," presented at the conference Coal Marketing Days: Is the Turnaround at Hand?, Pittsburgh, 1983.

6. Edgar, T.F., *Coal Processing and Pollution Control*, Gulf, Tulsa, Oklahoma, 1983.

7. Yu, A.T., "Transfer and storage," appearing in *Society of Mining Engineers Mining Engineering Handbook*, A.B. Cummins, and I.A. Given, eds., Vol. 2, Section 18.6, Society of Mining Engineers, New York, 1973.

8. Smitham, J.B., and Nicol, S.K., "Physico-chemical principles controlling the emission of dust from coal stockpiles," *Powder Technology*, Vol. 64, No. 3, pp. 259-270, February 1991.

Bibliography

Pfleider, E.P., ed., *Surface Mining*, The American Institute of Mining, Metallurgical, and Petroleum Engineers, New York, 1968.

Leonard, J.W., ed., *Coal Preparation*, The American Institute of Mining, Metallurgical, and Petroleum Engineers, New York, 1979.

Lotz, C.W., *Notes on the Cleaning of Bituminous Coal*, West Virginia University, Charleston, West Virginia, 1960.

Given, I.A., ed., *Society of Mining Engineers Mining Engineering Handbook*, The American Institute of Mining, Metallurgical, and Petroleum Engineers, New York, 1973.

Coal pulverizers at a modern power station.

Chapter 12
Coal Pulverization

The development and growth of coal pulverization closely parallels the development of pulverized coal-firing technology. Early systems used ball-and-tube mills to grind coal and holding bins to temporarily store the coal before firing. Evolution of the technology to eliminate the bins and direct fire the coal pneumatically transported from the pulverizers required more responsive and reliable grinding equipment. Vertical air-swept pulverizers met this need.

The first Babcock & Wilcox (B&W) vertical air-swept pulverizing mills were the E type design introduced in 1929 at Commonwealth Edison's Powerton Station in the United States (U.S.). Today, B&W offers a broad line of proven MPS pulverizer mills (see Fig. 1) to meet utility needs and EL type mills to meet lower load industrial requirements. Both use rolling elements on rotating tables to finely grind the coal which is swept from the mill by air for pneumatic transport directly to the burners.

Reliable coal pulverizer performance is essential for sustained full load operation of modern coal-fired electric generating stations. Also, an effective pulverizer must be capable of handling a wide variety of coals and accommodating load swings. The MPS pulverizer, through conser-

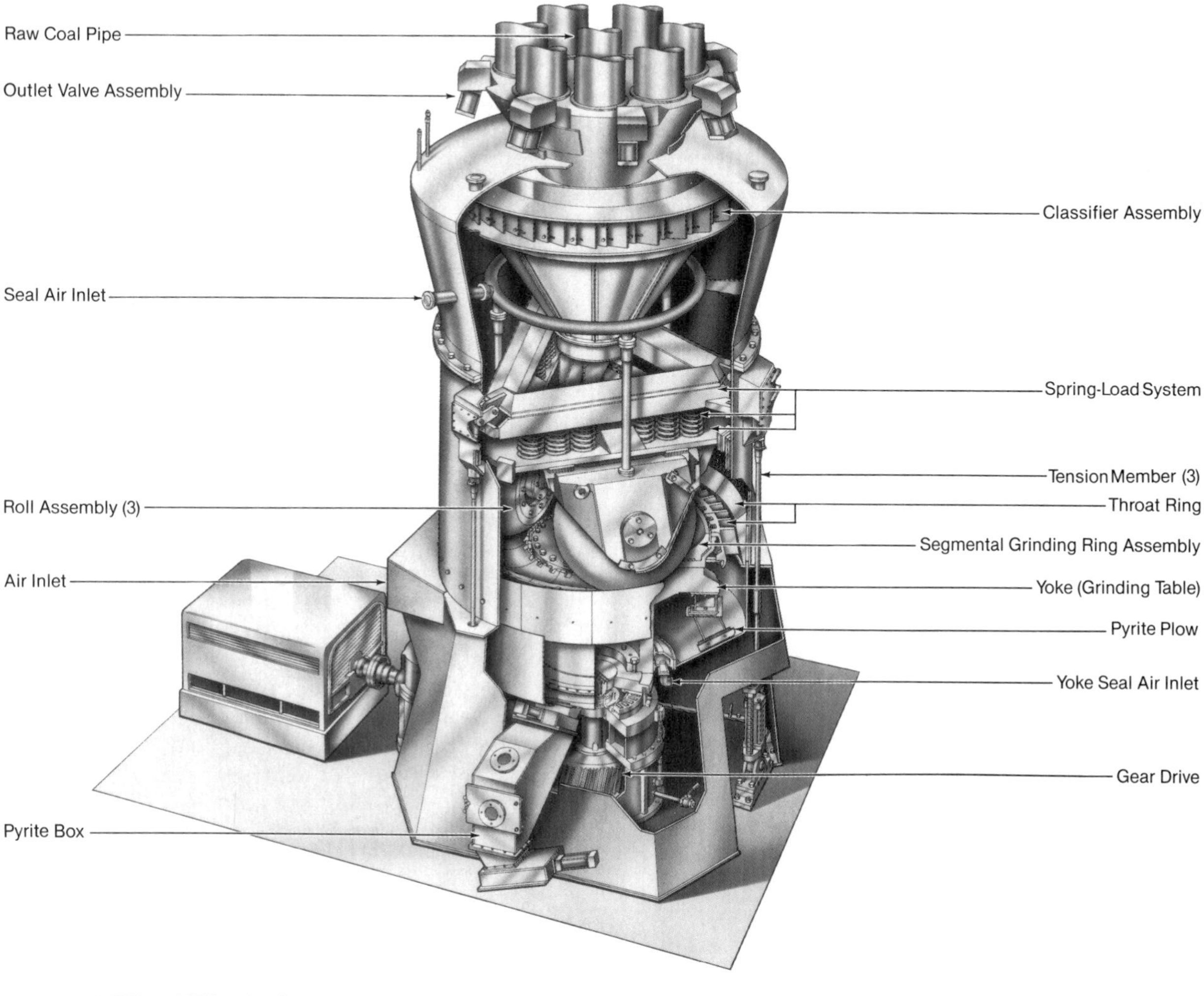

Fig. 1 Babcock & Wilcox MPS pulverizer.

vative design, reserve capacity and long grinding element life, has set the standard for high availability, reliability and low maintenance, contributing to stable boiler performance.

A key difference between much of the boiler system and the pulverizer is that the pulverizer is sized and operated as a mass flow machine while the boiler is a thermal driven machine. Therefore, the heating value of the fuel plays a key role in integrating these two components.

Vertical air-swept pulverizers

Principles of operation

The elements of a rolling action grinding mechanism are shown in Fig. 2. The roller passes over a layer of granular material, compressing it against a moving table. The movement of the roller causes motion between particles, while the roller pressure creates compressive loads between particles. Motion under applied pressure within the particle layer causes attrition (particle breakup by friction) which is the dominant size reduction mechanism. The compressed granular layer has a cushioning influence which reduces grinding effectiveness but also reduces the rate of roller wear dramatically. When working surfaces in a grinding zone are close together, near the dimensions of single product particles, wear is increased by three body contact (roller, particle and table). Wear rates can be as much as much as 100 times those found in normal pulverizer field experience. Wear from the three body contact has also been observed in operating mills when significant amounts of quartz bearing rock are present in sizes equal to or greater than the grinding layer thickness.

As grinding proceeds, fine particles are removed from the process to prevent excessive grinding, power consumption and wear. Fig. 3 presents a simplified MPS vertical pulverizer, showing the essential elements of a vertical air-swept design. A table is turned from below and rollers, called *tires*, rotate against the table. Raw coal is fed into the mill from above and passes between the rollers and the rotating table. Each passage of the particles under the rollers reduces the size of the coal. The combined effects of centrifugal force and displacement of the coal layer by the rollers spills partly ground coal off the outside edge of the table. An upward flow of air fluidizes and entrains this coal.

The point where air is introduced is often called the air port ring, nozzle ring or throat. Rising air flow, mixed

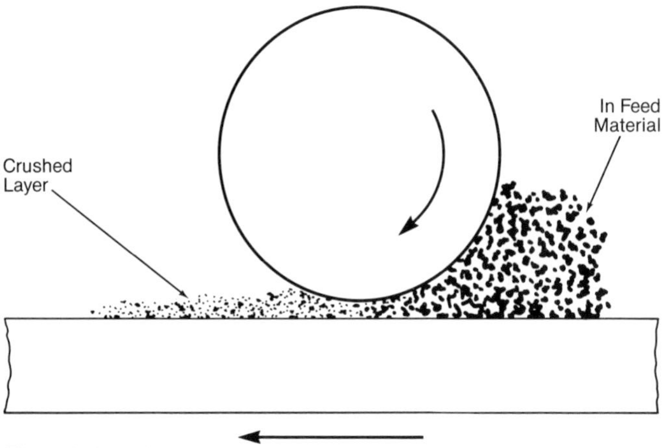

Fig. 2 Roller mill grinding mechanism.

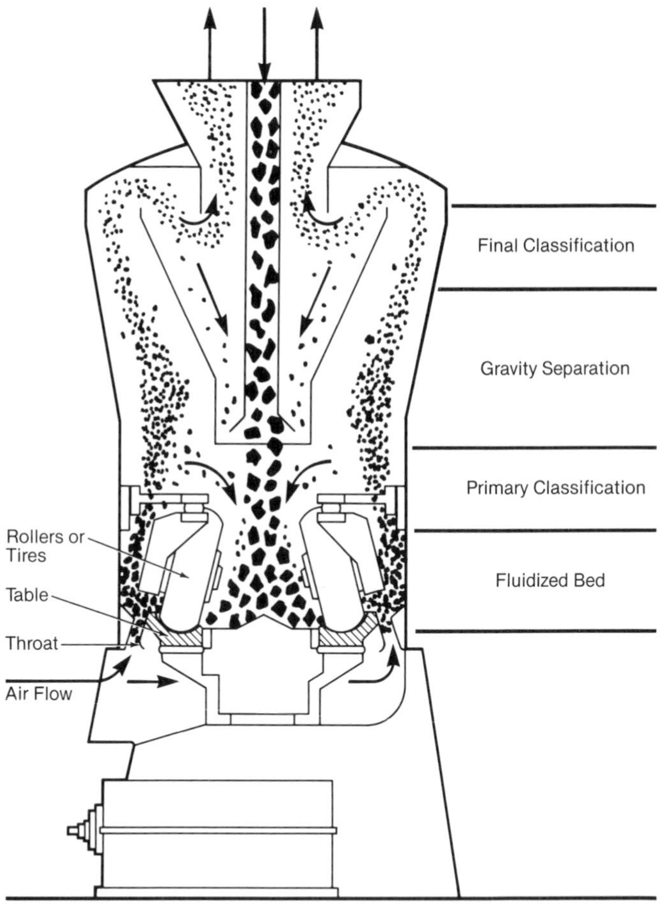

Fig. 3 Pulverizer coal recirculation.

with the coal particles, creates a fluidized particle bed just above the throat. The air velocity is low enough so that it entrains only the smaller particles and percolates with them through the bed. The air-solids flow leaving the bed forms the initial stage of size separation or classification. The preheated air stream also dries the coal to enhance the combustion process.

Vertical pulverizers are effective drying devices. Coals with moisture content up to 40% have been successfully handled in vertical mills. Higher moisture levels are possible, but the primary air temperature needed would require special structural materials and would increase the chance of pulverizer fires. A practical moisture limit is 40%, by weight, requiring air temperatures up to 750F (399C).

As the air-solids mixture flows upward, the flow area increases and velocity decreases returning larger particles directly to the grinding zone. The final stage of size separation is provided by the classifier located at the top of the pulverizer. This device is a centrifugal separator. The coal-air mixture flows through openings angled to impart spin and induce centrifugal force. The coarser particles impact the perimeter, come out of suspension and fall back into the grinding zone. The finer particles remain suspended in the air mixture and exit to the fuel conduits.

Pulverizer control

There are two input streams into an air-swept pulverizer, air and coal. Both must be controlled for satisfactory operation. Many older methods of coal flow control are still used successfully. Either volumetric or gravimetric belt feeders are the currently preferred method of coal flow regulation. Accurate measurement of coal flow has allowed parallel control of air and coal (Fig. 4).

Pulverizer design requirements

Many different pulverizer designs have been applied for coal firing. Successful designs have met certain fundamental goals and requirements:

1. optimum fineness for design coals over the entire pulverizer operating range,
2. rapid response to load changes,
3. stable and safe operation over the entire load range,
4. continuous service over long operating periods,
5. acceptable maintenance requirements, particularly grinding elements, over the pulverizer life,
6. ability to handle variations in coal properties,
7. ease of maintenance (minimum number of moving parts and adequate access), and
8. minimum building volume.

Pulverizer designs

Mill development efforts by B&W were, for many years, based on variations of ball-and-ring (ball-and-race) grinding elements. In the late 1920s, B&W introduced the type E pulverizer. During the early 1950s, a redesign was installed and tested at Lancaster, South Carolina. This design, which uses two vertical axis horizontal rings, was designated E (Lancaster) or EL (Fig. 5). The lower ring rotates while the upper ring is stationary and is spring loaded to create grinding pressure. A set of balls is placed between the rings. The force from the upper ring pushes the balls against the coal layer on the lower ring.

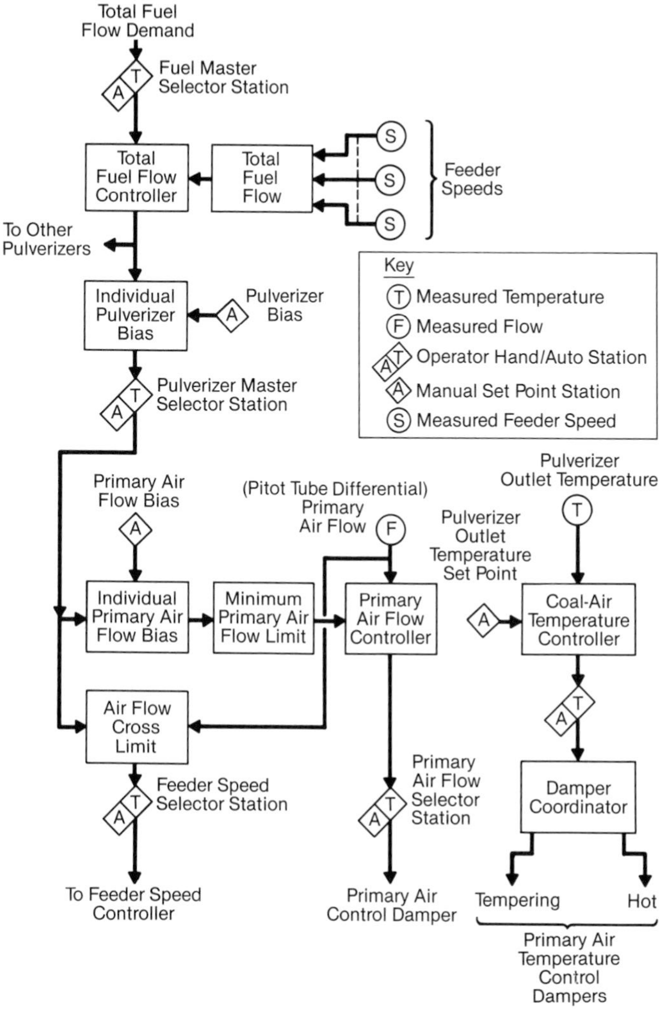

Fig. 4 Schematic of parallel cross limited pulverizer control system.

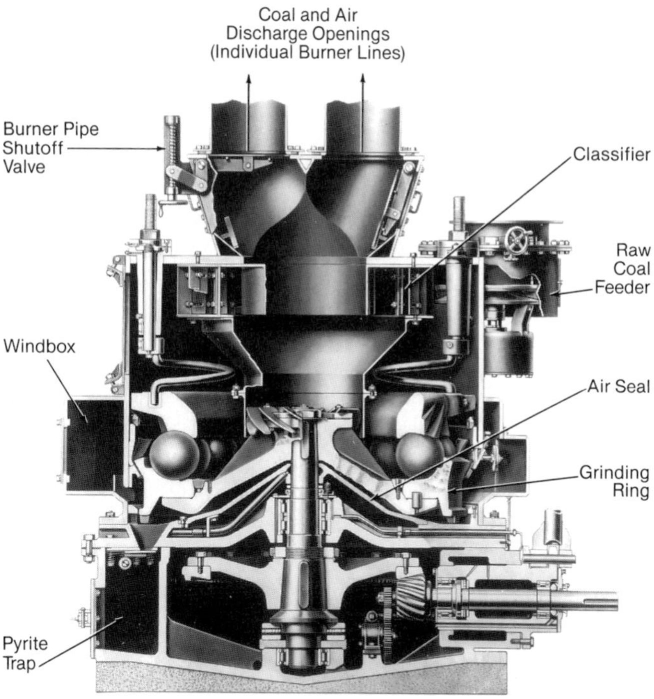

Fig. 5 B&W type EL ball-and-race mill.

Type E and EL mills have size designations which indicate the diameter, in inches, of the midline of the grinding track. Type E mills have been built as large as the E-70 with a capacity of 17 t/h (15 t_m/h). Type EL mills have been built as large as the EL-76 with a maximum capacity of 20 t/h (18 t_m/h). Approximately 1600 E and EL mills have been installed, with more than 1000 still in service. The major wear parts of E and EL pulverizers are the two rings and the balls which are made of abrasion resistant alloys and are easily replaced.

In 1970, B&W introduced the type MPS pulverizer to the U.S. market. A current MPS pulverizer design is shown in Fig. 1. The MPS pulverizer is an air-swept, roller type vertical pulverizer; however, it differs in significant ways from other roller mills. The rollers (tires) of the MPS are supported and loaded by a unique system which loads the three rollers simultaneously. The common loading system allows independent radial movement of each roller. This, in turn, allows continuous realignment with the grinding track as rollers wear. The rollers can also accommodate large foreign objects such as tramp iron or rocks which inadvertently enter the grinding zone. The MPS can maintain design performance with as much as 40% of its tires' weight lost due to wear.

MPS mills operate at speeds which produce centrifugal force at the grinding track midline of about 0.8 times the force of gravity. This very low speed contributes to low vibration levels and the ability to handle large foreign objects.

This mill also introduced a nonintegral gear drive to coal pulverizer design. This feature is now popular with most mill designers in the U.S. These drives allow the most complex and difficult to repair component to be removed in case of failure and exchanged with a standby unit. Repair can then be undertaken using a variety of options while the mill is back in service.

The major replaceable wear parts of MPS pulverizers include the tires and ring segments plus minor items such as wear guide plates, ceramic linings, and parts for the throat or air port ring. Wear life of all parts in contact with the coal is dependent upon the coal abrasiveness, and ranges from 8000 to more than 100,000 hours.

The B&W MPS mill size designations are also based upon the grinding track centerline diameter, in inches. Units have been furnished ranging from the MPS-56 to the MPS-118 with capacities ranging from 17 to 105 t/h (15 to 95 t_m/h) respectively. More than 1000 MPS mills have been sold with the MPS-89 being the most popular. There are more than 700 MPS-89 units currently in use.

Table 1 lists important features and characteristics of MPS and EL mills for comparison.

Horizontal air-swept pulverizers

High speed pulverizers

Horizontal pulverizers are categorized as either high speed or low speed. One high speed design which has been used in the U.S. serves the same applications as the vertical types. This machine operates at about 600 rpm and grinds by both impact and attrition. Coal enters at the impact section for initial size reduction then passes between moving and stationary parts for final size reduction (Fig. 6). At the final stage, an exhauster fan provides the air flow for drying and transport. There is no classifier and the design relies upon the action of the rotor and stator to achieve size control in a single pass of the coal. The coal passes through quickly so there is little storage in the mill. This pulverizer is limited to coals with moisture content of 20% or less because of the short residence time for drying.

Fan/beater mills are used outside North America (Fig. 7). These are similar in some respects to the attrition mills described above. These mills are used for grinding and

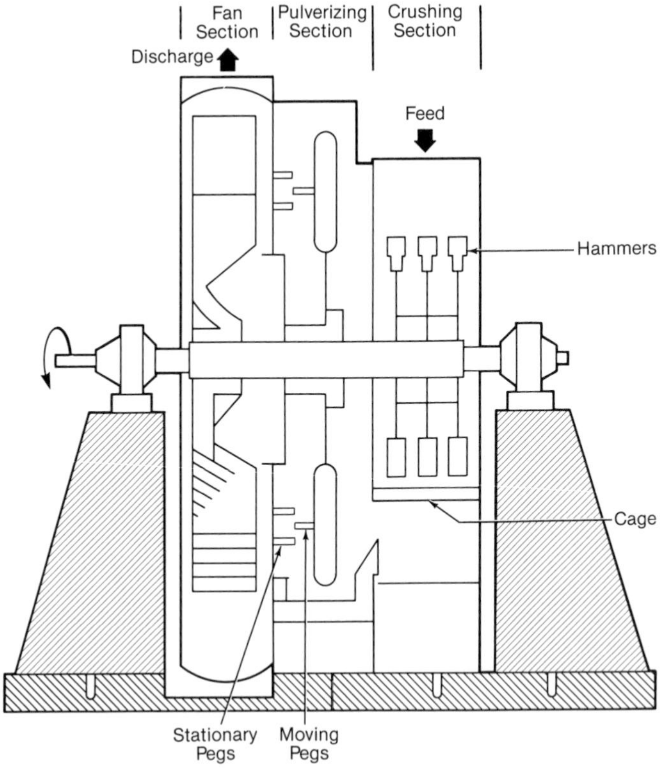

Fig. 6 Horizontal high speed pulverizer.

Table 1
Characteristics of B&W Type EL and MPS Pulverizers

	Type EL Ball-and-Race	Type MPS Roll-and-Race
Size range	EL-17 to EL-76	MPS-56 to MPS-118
Capacity, t/h (t_m/h)	1.5 to 20 (1.4 to 18)	17 to 105 (15 to 95)
Motor size, hp (kW)	25 to 300 (18 to 224)	200 to 1250 (149 to 933)
Speed	Medium	Slow
Table, rpm	231 to 90	32 to 21
Operates under	Pressure	Pressure
Classifier	Internal, centrifugal	Internal, centrifugal
Classification adjustment	Internal	Internal (standard) External (special)
Drying limit	40% H_2O or 700F (371C) Primary air temperature	40% H_2O or 750F (399C) Primary air temperature
Moisture load correction	None up to temperature limit	Load correction above 4% surface moisture
Maximum exit temperature limit	250F (121C)	210F (99C)
Effect of wear on performance	None if fill-in balls added	Power increase up to 15% at fully worn condition
Air-coal control system	Mill level with table feeders, parallel control with belt feeders	Parallel coal and air flow control
Air/coal weight ratio	1.75:1 at full load	1.75:1 at full load
Internal inventory	Medium, 2 to 3 min of output	High, 5 to 6 min of output
Load response	>10%/min	>10%/min
Specific power, kWh/t (kWh/t_m)	Low, 14 (15) including primary air fan	Low, 14 (15) including primary air fan
Noise level	Above 90 dBA	Above 90 dBA, 85 dBA attenuated
Vibration	Moderate	Low

Note: Capacities are those for bituminous coal with a 50 HGI and 70% fineness passing through 200 mesh (74 microns). Specific power is that at full load. Externally driven variable speed rotating classifiers have been applied to EL and MPS pulverizers.

drying the fossil fuels referred to as brown coals. These are very low grade lignite type coals with high ash and moisture contents and low heating values. Moisture content of brown coal often exceeds 50% and combined ash and moisture contents may exceed 60%.

Beater mills grind by impact and attrition with very rapid drying. They achieve drying despite short residence time by replacing primary air with extremely high temperature gas extracted from the upper furnace at temperatures of about 1900F (1038C). Gas from a cooler location such as the air heater outlet may be mixed with the hot gas for mill outlet temperature control. Coal is mixed with the gas stream ahead of the mill for initial drying to reduce the inlet temperature. The final stage is the fan section which maintains a negative pressure in the mill at all times. Beater mills operate at variable speed to control grinding performance at varying coal feed rates. One and two shaft machines are used trading simplicity in the former for more flexibility in the latter. Because of the low coal quality, beater mills must handle huge amounts of coal and are made in very large sizes. There are very few coal deposits in North America which would require this grinding/drying technology. However deposits in Germany, Eastern Europe, Turkey and Australia make brown coal a very important fuel for power generation throughout these regions.

Low speed pulverizers

The oldest pulverizer design still in frequent use is the ball-and-tube mill. This is a horizontal cylinder, partly filled with small diameter balls (Fig. 8). The cylinder is lined with wear resistant material contoured to enhance the action of the tumbling balls and the balls fill 25 to 30% of the cylinder volume. The rotational speed is 80% of that at which centrifugal force would overcome gravity and cause the balls to cling to the shell wall. Grinding is caused by the tumbling action which traps coal particles between balls as they impact.

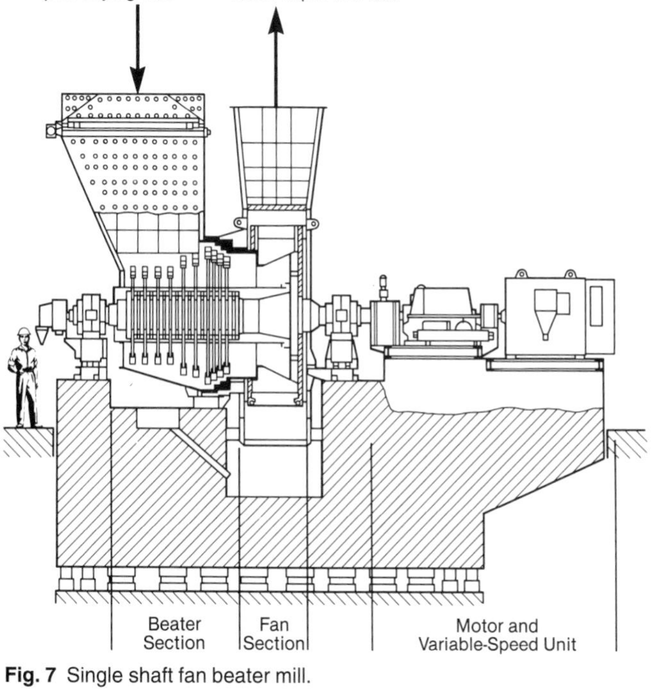

Fig. 7 Single shaft fan beater mill.

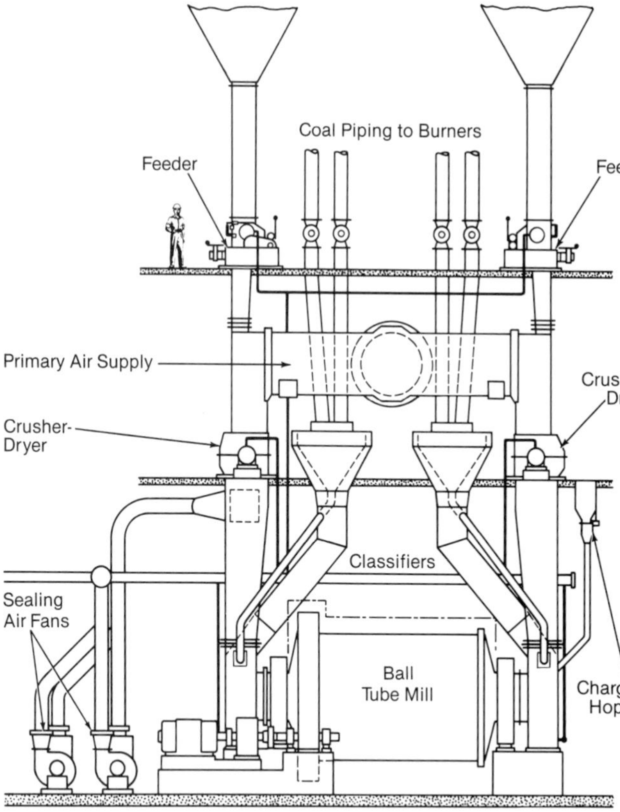

Fig. 8 Typical pressurized ball-and-tube coal pulverizing system.

Ball-and-tube mills may be either single or double ended. In the former, air and coal enter through one end and exit the opposite. Double ended mills are fed coal and air at each end and ground-dried coal is extracted from each end. In both types, classifiers are external to the mill and oversize material is injected back to the mill with the raw feed. Ball-and-tube mills do not develop the fluidized bed which is characteristic of vertical mills and the poor mixing of air and coal limits the drying capability. When coals with moisture above 20% must be ground in ball-and-tube mills, auxiliary equipment, usually crusher dryers, must be used.

Ball-and-tube mills have largely been supplanted by vertical air-swept pulverizers for new boilers. They typically require larger building volume and higher specific power consumption than the vertical air-swept pulverizers. They are also more difficult to control and have higher metal wear rates. They are, however, well suited for grinding extremely abrasive, low moisture, and difficult materials such as petroleum coke. Their long coal residence time makes them effective for fine grinding.

Application engineering

The arrangement of coal-fired system components must be determined according to economic factors as well as the attributes of the coal. The performance in terms of product fineness, mill outlet temperature, and air-coal ratio must all be determined as part of overall combustion system design.

Pulverizer systems

Pulverizers are part of larger systems, normally classified as either direct-fired or storage. In direct firing, coal leaving each mill goes directly to the combustion process. The air, evaporated moisture, and the thermal energy which entered the mill, along with the ground coal, all become part of the combustion process. Storage systems separate the ground coal from the air, evaporated moisture and the thermal energy prior to the combustion process. Stored ground coal is then injected with new transport air to the combustion process. Bin storage systems are seldom used in steam generation today, but are still used with special technologies such as coal gasification and blast furnace coal injection. Of the 1000 or so MPS pulverizers in service in the U.S. more than 99% are used in direct-firied systems.

The essential elements of a direct-fired system are:

1. a raw coal feeder that regulates the coal flow from a silo or bunker to the pulverizer,
2. a heat source that preheats the primary air for coal drying,
3. a pulverizer (primary air) fan that is typically located ahead of the mill (pressurized mill) as a blower, or after the mill (suction mill) as an exhauster.
4. a pulverizer, configured as either a pressurized or suction unit,
5. piping that directs the coal and primary air from the pulverizer to the burners,
6. burners which mix the coal and balance of combustion air, and
7. controls and regulating devices.

These components can be arranged in several ways based on project economics. With pressurized pulverizers, the choice must be made between hot primary air fans with a dedicated fan for each mill, or cold fans located ahead of a dedicated air heater and a hot air supply system with lateral branches to the individual mills. Hot fan systems have a lower capital cost because a dedicated primary air heater is not required. Cold fan systems have lower operating costs which, on larger systems, may offset the higher initial cost. Figs. 9 and 10 show these systems.

The terminology for air-swept pulverizers refers to the air introduced for drying and transport as primary air. Control of primary air is of vital importance to proper pulverizer system operation. For direct-fired or storage systems and for hot fan and cold fan systems, common control elements are found. Primary air must be controlled for flow rate and pulverizer outlet temperature. This control is achieved by three interrelated dampers.

Two of these, hot and cold air dampers, regulate air temperature to the mill and these dampers are usually linked so that as one opens, the other closes. The third damper is independent and controls air volume. Some manufacturers use only two dampers, but lack of stability or slow load change response can offset the cost advantages.

Because direct-fired pulverizers are closely linked to the firing system, system engineering must coordinate the design performance of the mills and burners. A set of curves can be used to relate important operating characteristics of volume flow, velocities at critical locations, and system pressure losses through the load range of the boiler. The curves consider numbers of mills in service and the output range of the individual mills. An example set of curves is shown in Fig. 11. Study of these curves provides information on many aspects of pulverizer and system operation. The lower curve, labeled A, shows boiler steam flow versus coal output per pulverizer. The individual lines show the number of mills in operation. In this example, a full boiler load of 2.5 x 10^6 lb/h (315 kg/s) steam flow can be reached with five mills in service at about 89,000 lb/h (11.2 kg/s) each. The maximum load also can be reached with four mills at about 111,000 lb/h (14.0 kg/s) each. The maximum steam flow with three mills is just over 2.1 x 10^6 lb/h (265 kg/s). The minimum load line represents a typical limit of turndown based on coal properties, ignition stability and the onset of mechanical vibration. This

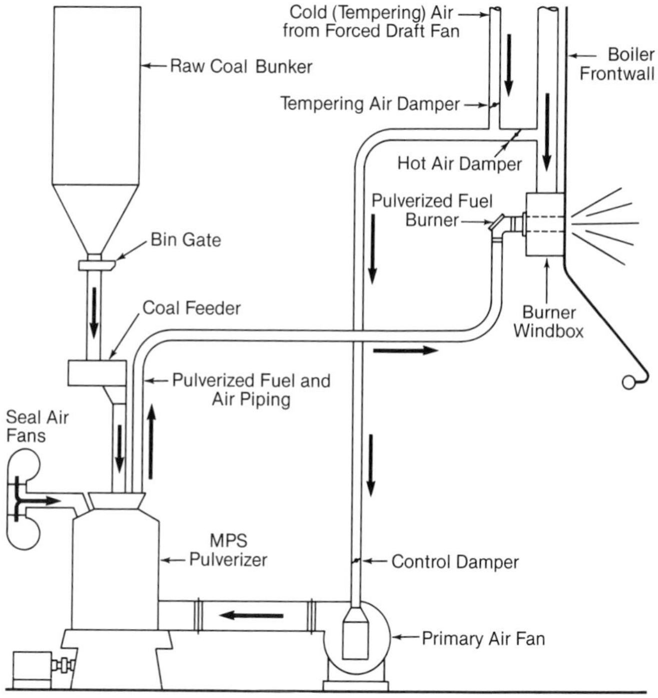

Fig. 9 Direct-fired, hot fan system for pulverized coal.

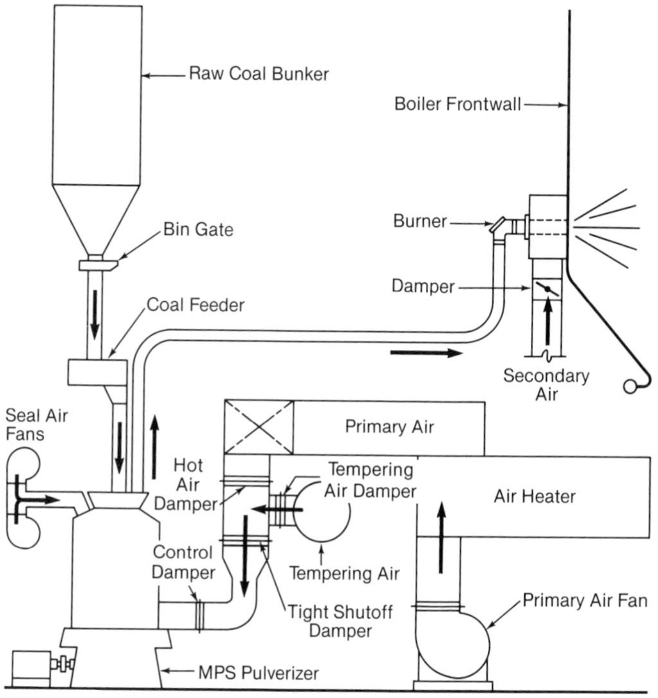

Fig. 10 Direct-fired, cold fan,fuel-air system for pulverized coal.

limit varies and the value shown of 30,000 lb/h (3.8 kg/s) of coal represents a relatively high turndown ratio.

Curve B shows primary air flow at mill exit conditions. Maximum flow is 55,800 ft^3/min (26.34 m^3/s) for an MPS-89N mill, corresponding to 124,000 lb/h (15.6 kg/s) coal flow. The minimum equipment design flow is approximately 55% of this or 30,700 ft^3/min (14.49 m^3/s). However, the exact minimum flow may need adjustment to permit stable burner operation. This is illustrated by 37,000 ft^3/min (17.46 m^3/s) minimum shown in Fig. 11, curve B, plus the minimum in curves D and E.

Curve C is the air/fuel ratio expressed in ft^3 of air per lb of coal. This ratio is critical to stable ignition at low loads and is influenced by coal rank and fineness. If the air flow indicated by curve B allows the air-fuel mixture to fall below the stability limit for the fuel being used, this limit will set the minimum coal flow.

Curve D is an internal velocity in the pulverizer. The velocity in the throat must be at least 7000 ft/min (35.56 m/s) to prevent spillage of coal into the mill windbox or air plenum. This velocity is calculated at mill inlet conditions. With high moisture coals, the required high drying temperature results in high specific volume and velocity is seldom a problem at low load. With very dry coal, it may be necessary to raise the minimum air flow or reduce the throat area and raise the velocity. However, throat restrictions can cause excessive pressure loss at high loads.

Curve E shows primary air velocity in the burner pipes versus pulverizer coal flow. This curve is plotted by dividing air volume at mill exit conditions by the total flow

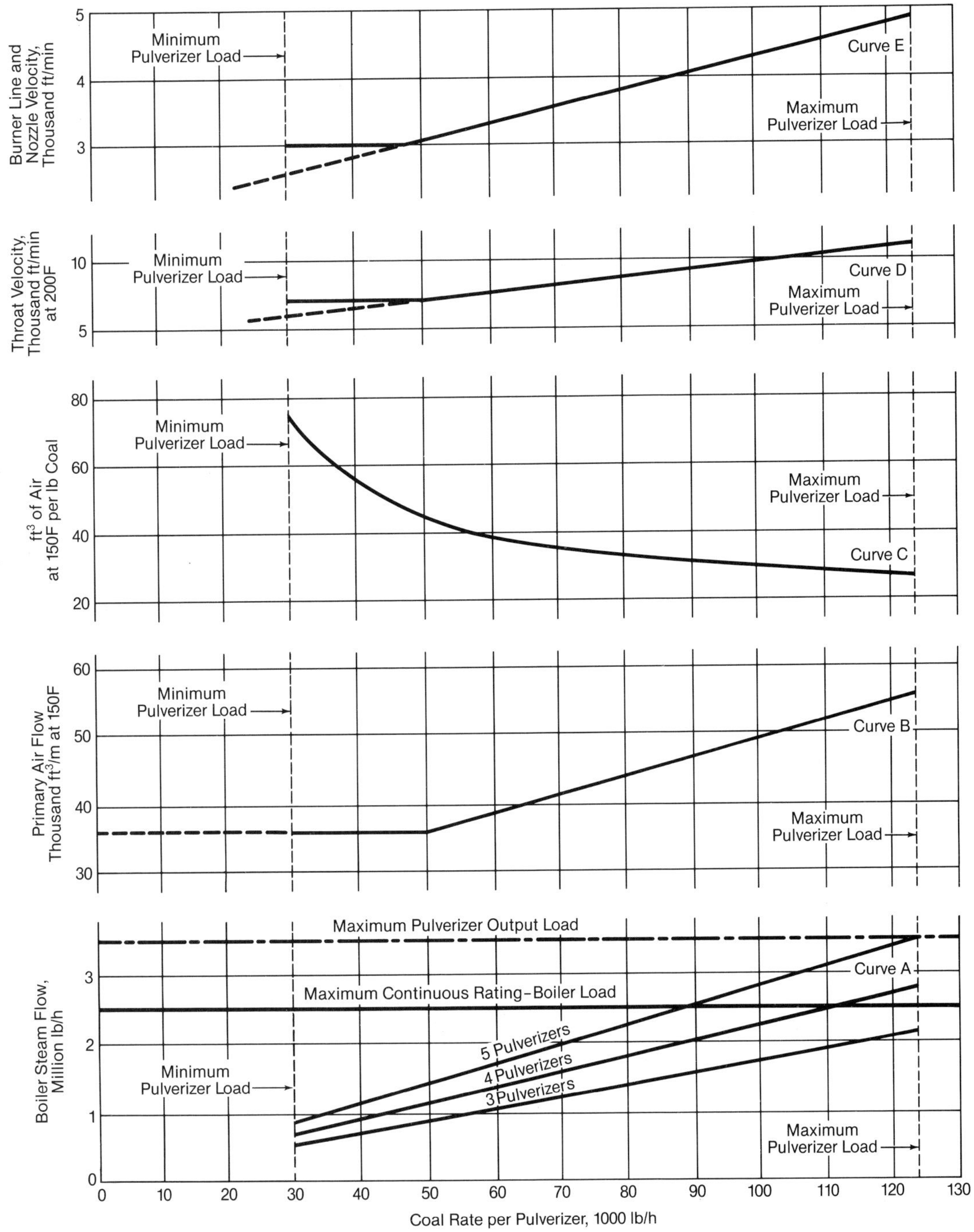

Fig. 11 MPS-89 pulverizer-burner coordination curves.

area of the pipes connecting the mill to the burners. The minimum velocity allowed is 3000 ft/min (15.2 m/s), at which the pulverized coal can be kept entrained in the primary air stream. The minimum, or saltation, velocity is particularly important in long horizontal spans of pipe. This velocity limit is influenced by air density and viscosity, particle loading and particle size. For the relatively narrow range of applications in burner pipes, the 3000 ft/min (15.2 m/s) limit is adjusted only for density and particle loading (air/coal ratio). Fig. 12 is a correction curve for minimum velocity.

An important pulverizer performance requirement is particle size leaving the mill. There is a mixture of particle sizes with any pulverizer because of the statistical distribution of particle sizes produced by the pulverizing process. The most frequently referenced size fraction for coal combustion is that portion smaller than a 200 U.S. standard sieve opening (mesh), or 74 microns. The portion passing this sieve size has been used as a gauge of pulverizer capacity as well as for assuring good combustion and good carbon burnout. The required particle size (fineness) is determined by the combination of coal combustion characteristics and the combustion chamber. Table 2 shows requirements used for various firing systems according to coal rank.

For bituminous coals burned in water-cooled enclosures, the required fineness is 70% passing through 200 mesh (74 microns) or better. The capacity rating of pulverizers in the U.S. is based also on 70% fineness. This value for good combustion characteristics has come under close scrutiny with the widespread use of low NO_x burners and the need for lower flame temperatures. Reduced flame temperature increases the required residence time in the combustion chamber. Coarse particles, those larger than 100 mesh (150 microns), contribute to high unburned carbon loss at traditional particle size distributions. The proven means to reduce the amount of plus 100 mesh (150 microns) product is to reduce the overall product size by increasing the amount passing through 200 mesh (74 microns) to 80%. Increasing the fineness from 70 to 80% causes a 20% pulverizer capacity reduction with stationary classifiers.

Outlet temperature

Coal with low volatile content may require higher air-coal temperatures to assure stable combustion, especially

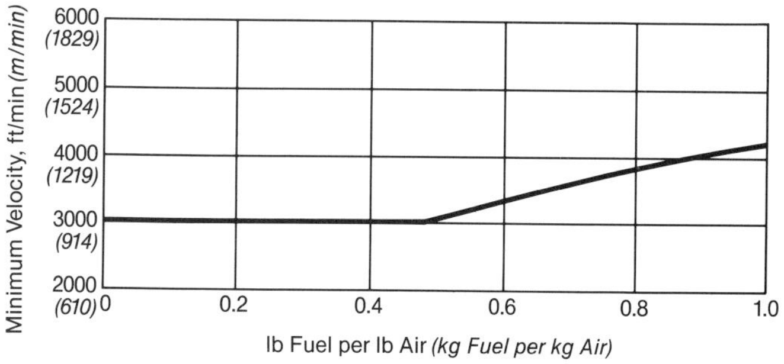

Fig. 12 Correction curve for minimum velocity.

at lower burner inputs or low furnace loads. The usual pulverizer exit temperature is 150F (66C). Higher temperatures, up to 210F (99C) for coal, may be used if needed. Higher temperatures lead to lubrication problems in the grinding roller bearings. In addition, high outlet temperatures require high inlet temperatures which increase the risk of pulverizer fires. Some pulverizer applications, notably bin storage systems, will require higher temperatures to assure complete drying and prevent handling problems with fine coal. Usually, 180F (82C) is adequate to assure drying of coal having raw coal moisture up to 10%. For direct firing, the outlet temperature requirements are determined by volatile content and the need for stable combustion. Table 3 lists mill outlet temperatures for various coal types.

Effects of coal properties

Grindability When determining pulverizer size, the most important physical coal characteristic to consider is *grindability*. This characteristic is indicative of the ease with which coal can be ground; higher grindabilities indicate coals which are easier to pulverize. The test procedure to determine the commonly used Hardgrove Grindability Index (HGI) is described in The American Society for Testing and Materials (ASTM) D 409. The operative principle with this procedure is the application of a fixed, predetermined amount of grinding effort or work on a prepared and sized sample. The apparatus shown in Fig. 13 is used to determine grindability. By rotating exactly 60 revolutions, this miniature pulverizer does a fixed amount of work on each sample. The amount of new, fine material produced is a measure of the ease of grinding. The index value of grindability used by B&W is HGI = 50, i.e., at HGI = 50, the capacity correction factor is 1.0. The index is open ended with no upper limit on the HGI scale.

Table 2
Typical Pulverized Coal Fineness Requirements — Percent Passing 200 U.S. Standard Sieve

	High Rank * Fixed Carbon,%			Low Rank (Fixed Carbon < 69%) Heating Value, Btu/lb**		
Furnace or process:	97.9 to 86.0	85.9 to 78.0	77.9 to 69.0	Above 13,000	13,000 to 11,000	Below 11,000
Marine boiler	—	85	80	80	75	—
Water-cooled	80	75	70	70	65	60
Cement kiln	90	85	80	80	80	—
Blast furnace	N/A	N/A	N/A	80	80	N/A

* ASTM classification
** Btu/lb x 2.326 = kJ/kg

Table 3
Typical Pulverizer Outlet Temperature

Fuel Type	Volatile Content, %*	Exit Temperature F (C)**
Lignite and subbituminous	—	125 to 140 (52 to 60)
High volatile bituminous	30	150 (66)
Low volatile bituminous	14 to 22	150 to 180 (66 to 82)
Anthracite, coal waste	14	200 to 210 (93 to 99)
Petroleum coke	0 to 8	200 to 250 (93 to 121)

* Volatile content is on a dry, mineral-matter-free basis.
** The capacity of pulverizers is adversely affected with exit temperatures below 125F (52C) when grinding high moisture lignites.

Grindability is not strictly a matter of hardness. Some materials, fibrous in nature, are not hard but are very difficult to grind. Sticky or plastic materials can also defy grinding.

The procedure described in ASTM D 409 was originally developed as a purely empirical method but provides a valid capacity calculation for vertical pulverizers when grinding bituminous coal. When lower rank coals are tested in the Hardgrove apparatus, the correlation between laboratory tests and field operating equipment is often quite poor. There is a strong but unpredictable effect of sample moisture on test results. A more accurate index of grindability for low rank coals is determined in a laboratory sized MPS-32 pulverizer. This equipment has all the features of a field installation and can emulate field conditions. The capacity of the mill is about 800 lb/h (0.10 kg/s) and a coal capacity test requires a 1000 lb (453.6 kg) sample. Fig. 14 is an arrangement diagram of this test apparatus. Results are interpreted as apparent grindability and the values are valid for capacity calculations.

Wear properties Wear in coal pulverizers results from the combined effects of abrasion and erosion. The mechanism which dominates in the wear of a particular machine component depends upon the designed function of the component and on the properties of the coal being ground. Abrasiveness and erosiveness are the important properties to be considered when evaluating the influence of a candidate coal on expected maintenance cost. Unfortunately, these two important properties are inherently difficult to measure, especially with a material as variable as coal.

Abrasiveness One indication of the likely pulverizer wear part life is abrasiveness. It is primarily related to the quantity and size distribution of quartz and pyrite found in the as-received coal, especially for particles larger than 100 mesh (150 microns). While standard and accepted procedures are available to determine the quantity of these two minerals in a sample (ASTM D 2492 and ASTM D 2799 for pyrite and quartz respectively), accepted procedures to establish quartz and pyrite particle sizes are not available.

Unfortunately, direct small-scale laboratory abrasiveness testing and correlation to field wear conditions are not satisfactory because of operating differences between small laboratory equipment and power plant pulverizers. However, satisfactory abrasiveness measurements have been made using the MPS-32 laboratory pulverizer with 3000 lb (1362 kg) samples. Although expensive, wear life testing with this procedure can typically predict field performance with an error of 10% or less. While other abrasiveness tests, such as the Yancey-Geer Price apparatus, are used to provide some insight into relative wear rates, they are of very limited value in predicting actual field wear rates.

Erosiveness Erosion is not the same as abrasion. It can be defined as the progressive removal of material from a target on which a fluid borne stream of solids impinges. For a given combination of particles and target material, wear rate increases with velocity, with particle size, and

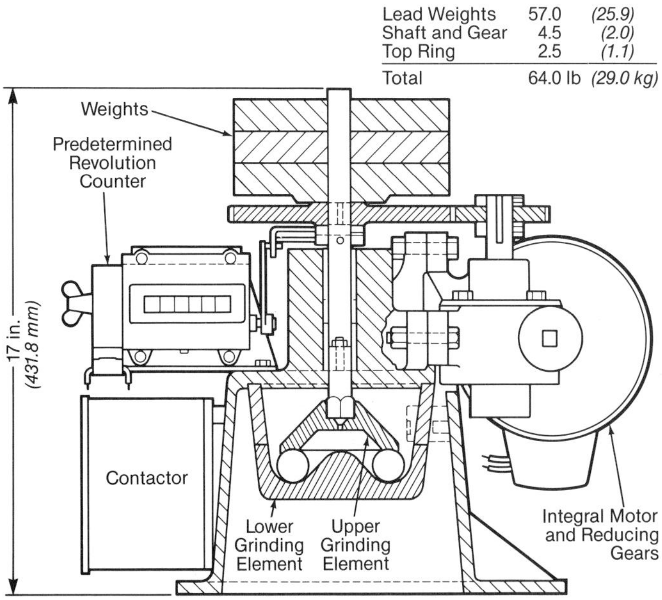

Fig. 13 Hardgrove grindability testing machine.

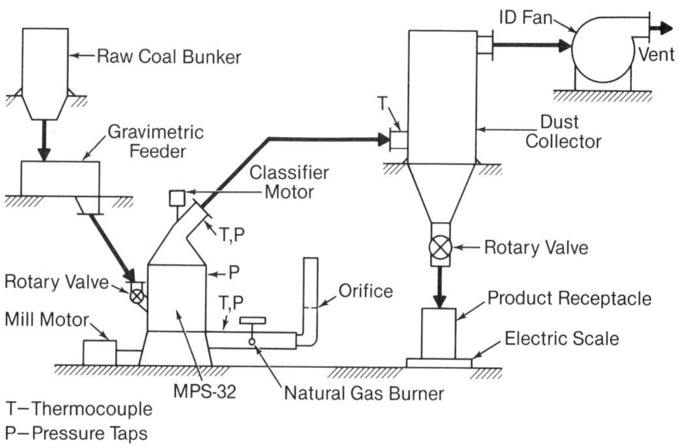

Fig. 14 B&W's MPS-32 test apparatus.

with the solids content of the stream. For ductile target materials, the wear rate increases up to an impingement angle of approximately 35 to 45 deg, at which point wear begins to decline.[1] For brittle materials, the rate of erosion increases up to the angle of approximately 70 deg, where rebounding material interferes with the impinging stream and measurement becomes erratic.

Moisture Moisture content in the fuel is a key design parameter. This inherent or fuel bound moisture can strongly influence grindability for subbituminous and lignite coals (Fig. 15). In addition, the surface moisture plus inherent moisture strongly influence the amount of drying needed and therefore strongly influence pulverizer air inlet temperature. Coal moisture is highly variable and depends more on coal type than on the amount of water introduced after mining. Moisture may be inherent, that which is present in the geological deposit, or may be surface moisture which is water introduced during handling, transport, processing or storage, or may be an artificial value called equilibrium moisture. This is the stable value reached after thermally dried coal reabsorbs atmospheric moisture. Coals used in the U.S. range from inherent moisture levels of 2% for Appalachian bituminous to near 40% for lignites. Brown coals may range up to 70%. Moisture in coal is determined according to the procedures in ASTM D 3302.

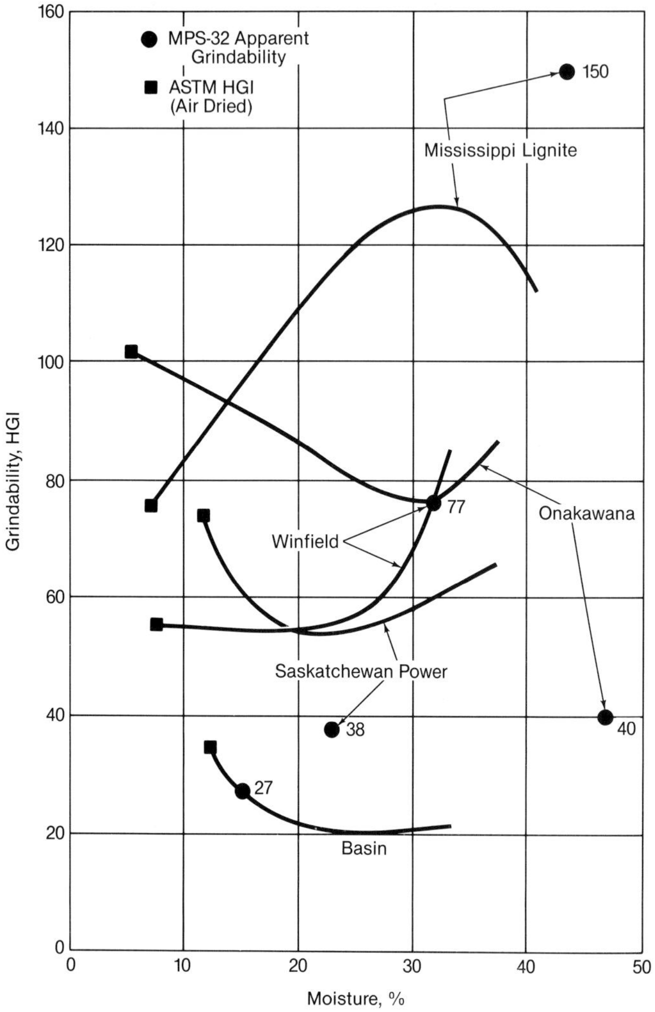

Fig. 15 Hardgrove Grindability Index versus fuel moisture.

Pulverizer size selection

The ultimate task in pulverizing application is the selection of the size and number of mills for the proposed project. The total boiler heat input and coal flow requirements are established from the combustion calculations (see Chapter 9) and the specified boiler steam flow requirements. The coal flow rate is then divided by the capacity correction factor to establish the equivalent required pulverizer capacity. The correction factor not only includes the fineness and grindability factors shown in Fig. 16, but also the appropriate fuel moisture correction. Most roller mills, including the MPS, require a moisture correction for capacity loss while the EL mills require no correction.

The number of pulverizers is then evaluated by dividing the equivalent required capacity by the unit pulverizer rated capacity based upon 70% passing through a 200 mesh (74 microns) screen and a grindability of 50. The unit pulverizer rated capacity is based upon pulverizer size, type and manufacturer. The tradeoff between fewer larger mills and more smaller mills is based upon balancing total capital cost with the proposed operating requirements such as boiler turndown. Frequently an extra pulverizer is specified to permit the boiler to operate at full load while one mill is out of service for maintenance.

Pulverizer selection and sizing are complicated by the frequent need to consider various coal sources and emission requirements. Computer programs can analyze pulverizer performance quickly, even for a large number of candidate coals, to learn which will govern mill selection. If the coal setting the pulverizer size is greatly different from the intended primary use coal, the result may be oversized mills and limits on turndown on the primary coal. In such a case, it may be necessary to reconsider whether this coal should remain on the list of possible fuels.

Performance testing

The most rigorous performance tests are done to assure compliance with the purchase contract. This acceptance testing will usually be guided by the procedures in The American Society of Mechanical Engineers (ASME) Performance Test Code, PTC 4.2 which references other applicable codes and standards. Testing may also be done as part of an overall boiler test, or simply to learn whether maintenance or adjustments are needed.

Fineness testing

Among the more daunting tasks in performance testing is the collection of valid fineness samples. The method of sampling is well described in PTC 4.2 and ASTM D 197. While this procedure has inherent weaknesses, it is the basis for measuring one of the fundamental performance parameters to verify pulverizer capacity. Periodically, alternative sampling procedures or new hardware are proposed to overcome the weaknesses in the ASTM method and apparatus. One such device is the ISO probe, also called the Rotorprobe. This well conceived device can sweep around circles that represent equal concentric areas of the pulverized coal pipe cross section. It is possible that the ISO probe may collect a more representative sample than the ASTM methods. However, the ASTM probe is currently the accepted standard on which capacity correction factors for fineness are based and whatever

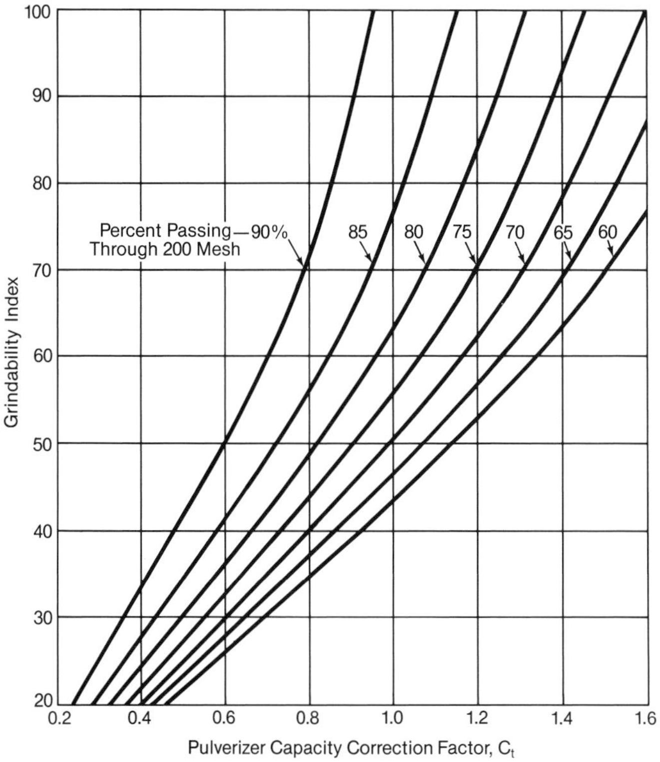

Fig. 16 Capacity correction factor.

errors are inherent in its application are included in its basis for use. It is important that new methods do not introduce new errors or controversies. They should produce results at least as consistent as the ASTM methods.

When tests are run to determine the condition of the pulverizer system or to evaluate a proposed new coal, comparative values between successive tests may be adequate. In such cases, alternatives, such as the ISO probe, may serve well if they provide acceptable repeatability and are not sensitive to variables such as velocity or temperature. Use of panel board instruments, if they can be read with sufficient precision, will serve as well as calibrated instruments for periodic testing.

Evaluating test results

The approach to evaluation depends on the purpose of the tests. If the parties have agreed upon test protocols as prescribed by PTC 4.2, results evaluation is simply a calculation process. When results indicate that the performance requirements have not been met but that the shortfall is slight, reference to the ASTM standards, D 409 and D 197, may show that the shortfall could lie within the limits of laboratory analysis repeatability. The protocols should provide for resolution of small deviations in the test results.

The more usual case of testing to evaluate the equipment or its systems requires a different approach. It is a mistake to try to read excessive precision into test results. It is more important to discover trends as guides to corrective action or the frequency of retesting. In such testing, it is useful to develop checklists or troubleshooting guides to note what each abnormality indicates. For example, acceptable fineness but excessive pressure differential probably indicates a need to adjust the pulverizer load springs. Acceptable fine fraction fineness but poor coarse fraction fineness indicates erosion damage and internal short circuiting of some air-coal flow. Also, it is common to discover that recalibration of primary measuring devices is needed on a regular schedule.

When equipment tests are first undertaken, it is important to collect all data which could possibly prove useful. As test experience grows, extraneous data can be removed. It is usually only a short time before standard data sheets evolve and testing becomes a routine task for an individual plant.

Operations

Power consumption

The main costs of coal pulverizer operation are capital, power consumption and maintenance. The capital cost is usually an annual levelized charge including capital, taxes and several other factors. Power and maintenance are often expressed as costs per ton of coal ground, and are influenced by operating plant practices.

When comparing pulverizer designs, it is common to combine primary air fan and pulverizer power consumption. The power consumption for B&W EL or MPS pulverizers and their respective fans is about 14 kWh/t (15 kWh/t_m) with coal at 50 HGI and 70% fineness, operating at rated output and with new grinding parts. These conditions seldom exist for any length of time during operation. Most modern installations include extra capacity for future coal variation. For most mills, there will be either a decline in capacity or an increase in power consumed as wear progresses. Therefore, because of the combined effect of pulverizer oversizing and wear, a realistic power value might be 20 kWh/t (22 kWh/t_m) for a well sized mill on a base loaded boiler or 22 kWh/t (24 kWh/t_m) on a load following unit.

There are various ways of calculating the cost of auxiliary power. The lowest cost which can be calculated is the cost of coal used to generate the power. This is a heat rate cost and, for a hypothetical boiler using coal with a 13,000 Btu/lb (30,238 kJ/kg) heating value and a thermal heat rate of 10,000 Btu/kWh (34.1% efficiency), 22 kWh/t (24 kWh/t_m) is about 0.85% of the fuel energy input to the boiler. As heating value declines, more power is spent grinding and moving material which does not produce thermal input, and the percentage represented by 22 kWh/t (24 kWh/t_m) increases.

Power can be expected to increase as the grinding parts wear. The depth of the grinding track may play a part in the power increase as well. There is little information to show whether power increases uniformly with time or if it takes place mostly at the end of wear life.

Maintenance

Costs for material and labor depend on wear life which varies with the abrasiveness of the coal. Wear life on MPS mills grinding U.S. coals ranges from under 10,000 to more than 100,000 hours, and normally between 25,000 and 60,000 hours. Therefore, average annual maintenance costs can vary widely. Costs will also vary by mill size, where part costs are proportionately higher for smaller MPS mills. For MPS-67 mills, taking all variables into account, maintenance will cost about 60% of the power costs cited above at a wear life of 25,000 hours and as little as 20% at 60,000 hours. Respective values for MPS-89 are 40% and 10%. For a very rough estimate, maintenance costs will be about one half the power costs for small mills and about one third for larger mills.

Maintenance and power costs are also influenced by operating practices. Experienced operators may notice that, for example, an increase in pulverizer pressure differential indicates a need for spring readjustment. There may be cost effective decisions between maintenance and operations regarding rebuild frequency to minimize the costly power increases near the end of wear life.

Ceramics

The use of ceramic tiles for erosion resistance has reduced repair costs. Erosion resistance of ceramic, 97% alumina, is at least eight times greater than an equivalent thickness of carbon steel. Ceramic lined panels are used in classifier cones, in the housing above the throat, and in the discharge turret in MPS mills. Ceramic lining is used extensively in burner pipe elbows and for wear shields on MPS roller brackets. When ceramic tiles are used, they must be carefully fitted to avoid undercutting and loss of the lining (Fig. 17). Ceramic tiles are not suitable for all forms of erosion protection. High angles of impingement are destructive to brittle materials.

Record keeping

Records are useful for pulverizer maintenance. Critical dimensions as recommended by the manufacturer should be taken and recorded during internal inspections. Suspicious wear data should be recorded and given special attention at the next inspection. When lubricants are drained, the amount should be noted and recorded and a sample obtained for analysis. Many owners are now including periodic vibration measurements in their data collection and record keeping. Such data may indicate impending bearing or gear failure.

Fires and explosions

It must always be remembered that a coal pulverizer grinds fuel to a form suitable for good combustion. Combustion can and does occur at unplanned locations with costly results. Pulverizer fires are serious and should be treated with preplanned emergency procedures. The surprising speed with which a fire can develop requires prompt action.

Pulverizer fires can develop in the so-called high temperature areas of mills or in the coal rich low temperature areas. Usually, there should be no coal in the areas upstream of the throat — the high temperature areas. Excessive spillage, because of insufficient air flow or throat wear, can cause coal accumulation in the plenum or windbox of the pulverizer. If this coal is not rejected promptly to the pyrite removal system, it will be ignited by the high temperature primary air. Coal will also enter this area when a mill is tripped or shut down in an abnormally fast mode. Coal in the windbox may burn out without damage and, indeed, this form of fire may go undiscovered. Fires in the cooler, fuel rich zones may develop more slowly, but these fires are fed by enormous amounts of coal and can quickly destroy the internal pulverizer parts.

Fig. 17 Ceramic tile lining.

The most common fire indicator is mill outlet temperature. Temperature indicators are slow in response because they are sheltered by erosion protection, but they are inherently reliable and their indications should be taken seriously.

Control room operators can gain useful information about potential hazards by following consistent operating practices, especially on multiple mill installations. If, for example, all mills are at the same coal feed rate, and if primary air control is in the automatic mode, all mills should have the same air/coal ratio. Under these conditions, a significant temperature difference from inlet to outlet in one mill may indicate a fire and should be investigated immediately. Of course, there are other possible causes of inlet temperature differences such as faulty feeder calibration, faulty air flow calibration, or coal moisture variation between the respective mills. The significant point is that there is important information to be drawn from careful observation of the vital signs of pulverizer system performance. Use of all vital signs as leading indicators of potential hazards is important to safe operations.

Severe fire damage has also been caused by feeding burning coal from silos or bunkers. All pulverizer manufacturers prohibit this practice. Silo fires must be dealt with separately by emptying the fire through special diverter chutes or by extinguishing agents applied to the coal surface.

In mill trips, the sudden loss of air flow causes the collapse of the fluid bed above the throat; the coal then falls into the windbox and contacts the hot metal surfaces. Low rank coals can begin smoldering within a few minutes at temperatures of 450F (232C) or more, and, upon agitation, dust clouds can be created leading to explosions. The agitation may be from mechanical action of the pyrite plows upon restart, or from the start of primary air flow before restart. The danger of explosion is minimized by removing the coal under rigidly controlled conditions. It is first necessary to assure personnel safety by evacuating all workers from places which could be affected by an explosion if the enclosures of the mill, ducts or burner pipes are breached. The mill is isolated from all air flow and an inert internal atmosphere is created by injecting inert gas or vapor. Brief operation of the mill, while isolated, rejects the coal to the pyrite removal system. The inert atmosphere prevents any smoldering coal from igniting the dust cloud raised internally by mill operation.

When a mill explosion occurs it is nearly always during changes in operating status, that is, during startup and shutdown.[2] During startup, the air-coal mixture in the mill passes from extremely fuel lean to fuel rich. There is a transitional mixture which is ideal for a dust explosion if a sufficiently strong ignition source exists. One recurring source of ignition is smoldering fires which develop in the residual coal left in the hot zone of a mill following a trip. The coal removal procedure just described was devised to avoid or safely remove this ignition source.

Undiscovered fires can cause dust explosions during otherwise normal shutdowns. These are not common, but experience has led many operators to actuate audible and visible alarms before routine starts and stops. This is a recommended safety precaution for all coal pulverizer operators.

In general, operators have reported a declining rate of incidents since 1985 and those with formal event investigation procedures have reported the greatest decline. The clear message from recent industry surveys is that experience and constant review of safety and emergency procedures will pay dividends in preventing injury and equipment damage from pulverizer fires and explosions. Proper safety systems and procedures are reducing the frequency and damage.

Controls and interlocks

The most widely followed standard for design and application of coal pulverizer systems is the National Fire Protection Association Standard NFPA-8503 (formerly NFPA 85F). This sets the minimum standards listed below for safety interlocks to be used with coal pulverizers:

1. failure of primary air flow trips the pulverizer system,
2. failure of the pulverizer trips the coal feeder and primary air flow,
3. closing of all burner pipe valves trips the pulverizer, the feeder, and the primary air flow, and
4. primary air flow below the manufacturer's minimum trips the pulverizer.

Other interlocks not mandated by NFPA but required by B&W are:

1. loss of the lube oil pump, or low pressure, trips the pulverizer (MPS mill only), and
2. flame safety systems trip the pulverizer if minimum flame detector indications are not met.

Additional indications of malfunctions which may require manually tripping a mill include:

1. excessively high or rapidly rising mill exit temperature probably indicates a fire,
2. inability to maintain an exit temperature of at least 125F (52C) may lead to loss of grinding capacity,
3. loss of seal air pressure or flow will eventually lead to bearing failure, and
4. high lube oil temperature indicates a loss of cooling water flow which may lead to gear drive damage.

The faults indicated by numbers 3 and 4 above are not strictly safety related. They are, however, examples of designers' choices which avoid unnecessary mill trips and the special safety procedures which must be followed to restart tripped mills.

Special safety systems

The procedure for removing coal after a pulverizer trip requires a separate control system. This system must bypass the control logic which prevents operation of the pulverizer drive motor without adequate primary air flow. To prevent misuse of this bypass feature, the coal removal logic (referred to as inert and clear logic) is enabled by actuation of the mill trip logic. A schematic logic diagram is shown in Fig. 18.

Another special system has been developed to deal with a phenomenon which has been observed in ducts after trips. A fine layer of dust is often seen in the clean air portion of the primary air system. This is thought to be due to a turbulent dust cloud which persists after collapse of the fluid bed. An array of water fog nozzles under the throat will intercept this cloud if actuated with

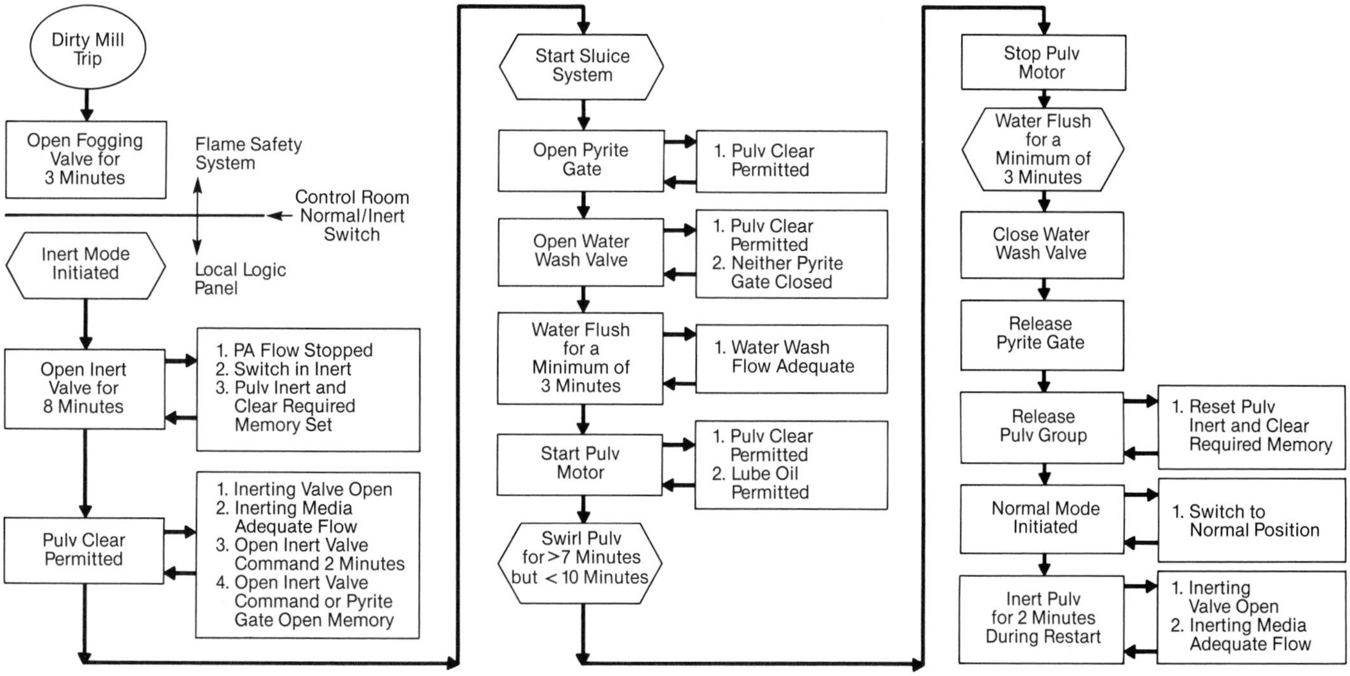

Fig. 18 Safety system logic diagram.

the mill trip signal. Applied along with the automatic spray system is hardware for tangential wash nozzles to enhance the coal removal action of the pyrite plows during the inert and clear cycle. The wash nozzles are actuated manually at the appropriate time in the cycle.

Advanced designs

Rotating classifiers

As noted under *Application Engineering*, product fineness is under review after being generally standardized at 70% passing through 200 mesh (74 microns) for direct firing of bituminous coal. Raising fineness to 80% will meet most early 1990s emission limits, but more demanding requirements are expected. Fineness values in excess of 95% passing through 200 mesh (74 microns) can be achieved with externally driven rotating classifiers. This is a recent development for coal pulverizers, emerging since the mid 1970s, although rotating classifiers have been applied to large vertical rock grinding mills for many years. The design developed for EL pulverizers is shown in Fig. 19. Interestingly, as shown in Fig. 20, this classifier is more effective at reducing percentages of coarse particles than finer particles. The value shown of 95% passing through 200 mesh (74 microns) may be near the practical limit for large vertical pulverizers. Other configurations for classifier rotors have been developed including squirrel-cage, various vane types and two-stage using a stationary first stage. Each of these represents the designer's solution to problems with retrofit, with particle size distribution, with drive power consumption or with overall pulverizer performance. Classifiers developed for MPS pulverizers have taken forms similar to that shown in Fig. 19 as well as two-stage and squirrel-cage designs.

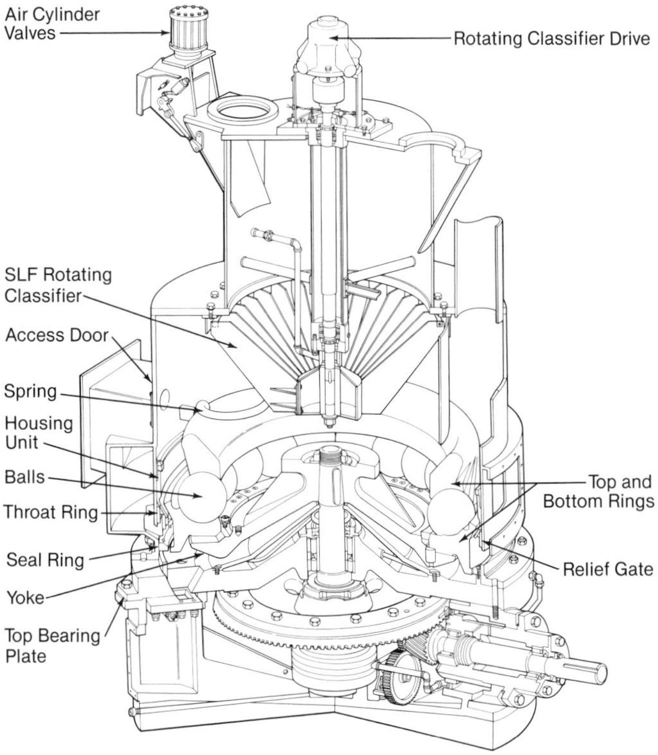

Fig. 19 Rotating classifier for EL mill.

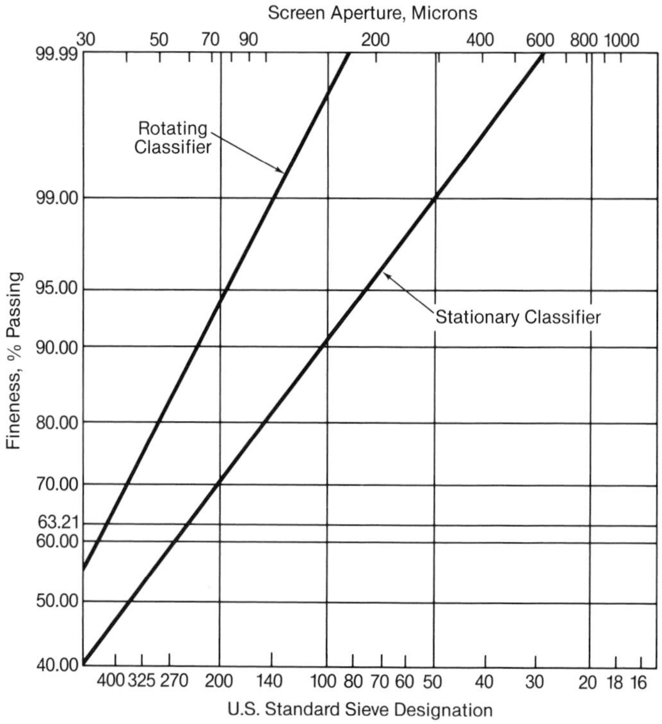

Fig. 20 Particle size distribution.

Product improvements

The 1980s was a period of refinement and product improvement for vertical pulverizers. One major improvement for MPS pulverizers was the introduction of rotating throat rings which improved performance and reduced erosive wear costs. A more subtle improvement was variable spring loading systems. It has been normal practice to operate vertical mill loading systems at constant preload. If, however, the spring preload is caused to follow coal feed rate, as illustrated in Fig. 21, the benefits will be reduced power consumption at low loads, ability to operate at lower output without vibration, and, in some instances, reduced pressure loss at high load by increasing preload above the previous fixed values.

Fine grinding

Vertical roller mills may have a practical limit of about 95% passing through 200 mesh (74 microns) for product fineness. As the product becomes finer, the recirculated material in the grinding chamber also becomes finer and the compressed layer becomes more fluid and less stable.

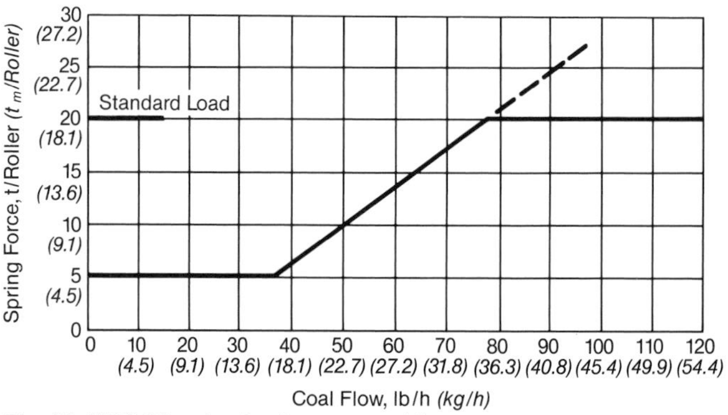

Fig. 21 MPS-89 spring load versus coal flow.

For very fine grinding other machines are needed. One of these is the ball-and-tube mill, offering the long residence time needed for an extremely fine product. In addition, ball-and-tube mills are built in the large sizes needed for direct firing.

It is not clear that the high capacity of ball-and-tube mill will be needed to meet foreseen fine grinding applications. An interesting application of fine grinding is to provide fuel for coal-fired lighters to reduce oil consumption for startup and low load flame stabilization. Smaller capacity mills could be used for this application.

Mills for fine grinding have been used to grind pigments, food and pharmaceuticals. These are small mills, generally 5 t/h (4.5 t_m/h) or less. Grinding principles include fluid energy, stirred ball, and high speed attrition. Fluid energy mills cause solids bearing fluid streams to impinge; particle impacts cause the size reduction. Stirred ball mills, usually wet grinders, use a charge of small balls in a cylindrical vessel which are agitated by a rotor, grinding the solids flowing through the charge.

High speed attrition mills grind by a combination of first stage impact and final stage attrition. It is these mills which have begun to attract some attention for preparation of oil replacement lighter fuel. Fig. 22 illustrates the grinding chamber of a high speed attrition mill. The rotor is a series of discs, each with impact bars, and each forming the driver for a grinding stage. The grinding stages are separated by diaphragms and coal must move inward against centrifugal force to pass from one stage to the next. Coal captured in each stage is said to provide an effective wearing surface to protect the mill shell from rapid wear. These mills are made in capacities up to about 5 t/h (4.5 t_m/h) and may be once-through or be equipped with external centrifugal classifiers for size control. Manufacturers of these mills refer to the product as micronized coal. When the term was adopted in the early 1980s, micronized coal referred to coal which was 100% finer than a 325 mesh (44 microns). The output of vertical high speed attrition mills is about 98% passing through 200 mesh (74 microns), 86% passing through 325 mesh (44 microns). However, this coal is reported to be fine enough to be easily ignited without the need for preheated primary or combustion air. These mills are high in power consumption and motor power is said to be sufficient to provide the necessary energy for drying most coals. The economic attractiveness of finely ground coal as a stabilizing and startup fuel depends on oil prices.

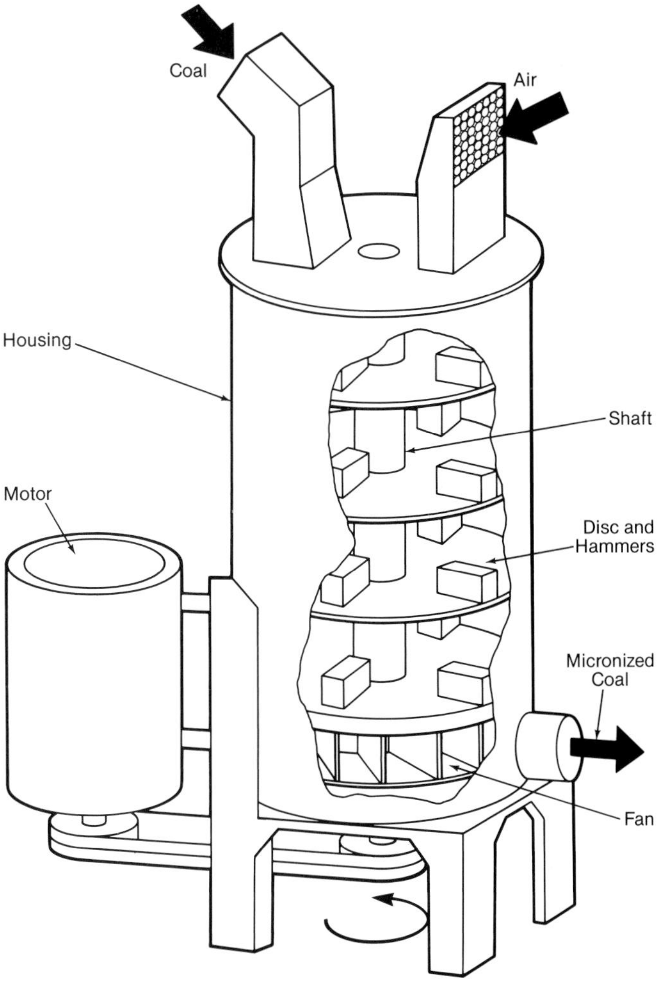

Fig. 22 High speed attrition mill.

References

1. Johnson, T.D., *et al.*, Central Electricity Generating Board, England, "Pulverized fuel system erosion," presented at EPRI Conference on Coal Pulverizers, pp. 18-22, 1985.

2. Zalosh, R.G., "Review of coal pulverizer fire and explosion incidents," ASTM STP 958, Cashdollar, K.L., and Hertzberg, M., eds., American Society for Testing and Materials, Philadelphia, p. 194, 1987.

Bibliography

"Coal — Sampling of pulverized coal conveyed by gases in direct fired coal systems," Draft International Standard ISO/DIS 9931, International Standards Organization, 1989.

Donais, R.T., Tyler, A.L., and Bakker, W.T., "The effect of quartz and pyrite on abrasive wear in coal pulverizers," *Proceedings of the International Conference on Advances in Material for Fossil Power Plants*, American Society for Metals, Metals Park, Ohio, p. 643, 1987.

Piepho, R.R., and Dougan, D.R., "Grindability measurements on low rank fuels," *Proceedings, Coal Technology '81*, Vol. 3, pp. 111-121, 1981.

Wiley, A.C., *et al.*, "Micronized coal for boiler upgrade/retrofit," Proceedings of Power-Gen '90, PennWell, Houston, pp. 1155-1170, 1990.

These three 550 MW B&W boilers fire pulverized subbituminous coal.

Chapter 13
Burners and Combustion Systems for Pulverized Coal

Coal is an abundant and low cost fuel for boilers. It can be burned in a number of ways depending upon the characteristics of the coal and the particular boiler application. Cyclone, stoker, and fluidized bed firing methods are used for a variety of applications as discussed in Chapters 14 through 16. However, pulverized coal (PC) firing — burning coal as a fine powder suspension in an open furnace — is the dominant method in use today. PC firing has made possible the large, efficient utility boilers used as the base load capacity in many utilities worldwide. Modern PC burners provide high combustion efficiency and low emissions through full integration with the entire boiler design.

Pulverized coal firing differs from the other coal combustion technologies primarily through the much smaller particle size used and the resulting high combustion rates. The combustion rate of coal as a solid fuel is, to a large extent, controlled by the total particle surface area. By pulverizing coal to a nominal 50 micron diameter or smaller (see Chapter 12), the coal can be completely burned in approximately one to two seconds. This approaches the rate for oil and gas. In contrast, the other technologies discussed in subsequent chapters use crushed coal of various sizes and provide substantially longer combustion zone residence times (up to 60 seconds or longer).

Pulverized coal was first used in the 1800s as a cost effective fuel for cement kilns.[1] The ash content enhanced the properties of the cement and the low cost resulted in the rapid displacement of oil and gas as a fuel. However, early pulverizing equipment was not highly reliable and, as a result, an indirect bin system was developed to temporarily store the pulverized coal prior to combustion, providing a buffer between pulverization and combustion steps. Use in the steel industry followed closely. In this application, the importance of coal drying, particle size control and uniform coal feed were recognized.

Success for boiler PC firing applications required modification of the boiler furnace geometry to effectively use the PC technology. In the early 1900s, fireboxes and crowded tube banks had to be expanded to accommodate PC firing. By the late 1920s, the first water-cooled furnaces were used with advanced pulverized coal-fired systems, and burner design advanced to improve flame stability, provide better mixing, increase flame temperatures and improve combustion efficiency. Coupled with improvements in pulverizer design and reliability, improved PC burners permitted the use of direct firing with coal transported directly from the pulverizer to the burner. (See Fig. 1.)

The exponential increase in the United States (U.S.) electric power generation and the increase in boiler size from 100 MW to as large as 1300 MW could not have been achieved economically without the development and refinement of pulverized coal firing. Today, nearly all types of coal from anthracite to lignite can be burned through pulverized firing. Combustion efficiencies of most coals approach those of oil and gas. Current research is focusing on continuing reduction in emissions of nitrogen oxides (NO_x) for environmental protection without sacrificing boiler performance or availability. (See also Chapter 34.)

Fig. 1 Early pulverized coal-fired radiant boiler.

Combustion

The manner in which pulverized coal burns depends on its rank and properties as well as the furnace conditions. As a coal particle enters the furnace (see Fig. 2), its surface temperature increases due to radiative and convective heat transfer from furnace gases and other burning particles. As particle temperature increases, the moisture is vaporized and volatile matter is released. This volatile matter, which ignites and burns almost immediately, further raises the temperature of the char particle, which is primarily composed of carbon and mineral matter. The char particle is then consumed at high temperature leaving the ash content and a small amount of unburned carbon. The volatile matter, fixed carbon (char precursor), moisture and ash content of the fuel are identified on a percentage basis as part of the proximate analysis discussed in Chapter 8.

Volatile matter content

Volatile matter is critical for maintaining flame stability and accelerating char burnout. Coals with minimal volatile matter, such as anthracites and low volatile bituminous, are more difficult to ignite and require specially designed combustion systems. The amount of volatile matter evolved from a coal particle depends on coal composition, the temperature to which it is exposed, and the time of this exposure. The American Society for Testing and Materials (ASTM) Method D 3175 stipulates a temperature of 950 ± 20C for seven minutes for volatile matter content determination.[2] Raising the temperature would increase volatile yield with other factors held constant. Coals with higher volatile matter content also benefit from more effective NO_x control by combustion methods. Ignition is influenced by the quality and the quantity of volatile matter. Volatile matter from bituminous and higher rank coals is rich in hydrocarbons and high in heating value. Volatile matter from lower rank coals includes larger quantities of carbon monoxide and moisture (from thermal decomposition) and consequently has a lower heating value. Volatile matter from higher rank coals can provide twice the heating value per unit weight as that from low grade coals.

Char particles

The speed of the char particle combustion depends on several factors including particle size, porosity, thermal environment, and oxygen partial pressure. Char reactions often begin as the coal particle is heated and devolatilizes, but they continue long after devolatilization is complete. Devolatilization is mostly completed after 0.01 seconds but char-based reactions continue for one to two seconds. The char particle retains a fraction of the hydrocarbons. Small particles, with 10 to 20 micron diameters, benefit from high surface to mass ratios and heat up rapidly, while coarse particles heat more slowly. Many coals go through plastic deformation and swell by 10 to 15% when heated. These changes can significantly impact the porosity of the coal particle.

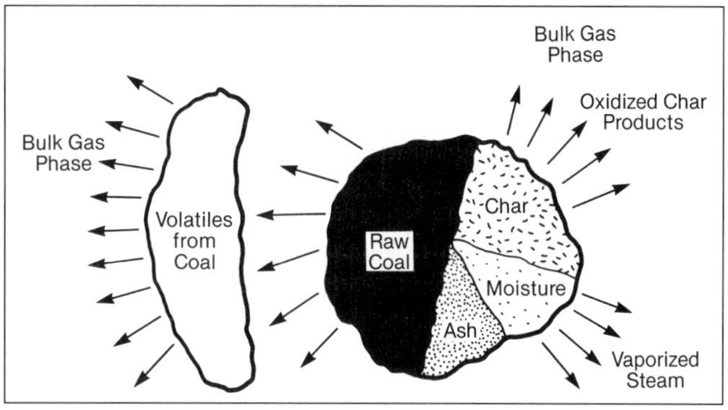

Fig. 2 Coal particle combustion.

Char oxidation requires oxygen to reach the carbon in the particle and the carbon surface area is primarily within the particle interior structure. Char combustion generally begins at relatively low particle temperatures. Reaction rates are primarily dependent upon local temperature as well as oxygen diffusion and char reactivity. For larger particles, the solid mass is reduced as carbon monoxide (CO) and carbon dioxide (CO_2) form, but particle volume is maintained. Coarse particles, more than 100 micron diameter, burn out slowly as a result of their lower surface to mass ratios. Longer burnout times cause these larger char particles to continue reacting downstream where the flame temperature has moderated.

Rapid heat transfer to and combustion of smaller particles lead to higher particle temperatures. Reaction rates increase exponentially with temperature, and oxygen (O_2) diffusion into the particle becomes the controlling parameter. Particle diameter and density change in the process. At higher particle temperatures, char reactions are so fast that oxygen is consumed before it can penetrate the particle surface. The particle shrinks as the outer portions are consumed, and transport of oxygen from the surroundings to the particle is the factor governing combustion rate.

Effect of moisture content

The moisture content of the coal also influences combustion behavior. Direct pulverized coal-fired systems convey all of the moisture to the burners. This moisture presents a burden to coal ignition; the water must be vaporized and superheated as the particles devolatilize. Further energy is absorbed at elevated temperatures as the water molecules dissociate.

Moisture content increases as rank decreases as discussed in Chapter 8. 15% moisture is common in high volatile bituminous coals, 30% is seen in subbituminous, and more than 40% is common in some lignites. Moisture contents in excess of 40% exceed the ignition capability of conventional PC-fired systems. Alternate systems are then required to boost drying during fuel preparation and/or divert a portion of the evaporated moisture from the burners. Char burnout is impaired by moisture which depresses the flame temperature. This is compensated for in part by the generally higher inherent reactivities and porosities of the higher moisture coals.

Effect of mineral matter content

The mineral matter, or resulting ash, of the coal is inert and dilutes the coal's heating value. Consequently, more fuel by weight is required as ash content increases in order to reach the furnace net heat input.

The ash absorbs heat and interferes with radiative heat transfer to coal particles, inhibiting the combustion process noticeably with high ash coals. The forms of mineral matter determine the slagging tendencies of the ash. These impact combustion by influencing the burner/furnace arrangement and the resulting flame temperatures and residence time. The additional effects of ash on overall boiler design are discussed in Chapter 20.

Pulverized coal combustion system

Effect of coal type

Conventional PC Pulverized coal firing is adaptable to most types of coal. Its versatility has made it the most prevalent method of coal firing for power generation worldwide. The majority of coals are PC-fired in conventional combustion systems as shown in Fig. 3. In these systems, the burners are located in the lower portion of the furnace, usually on one or two walls. Primary air typically at 130 to 200F (54 to 93C) conveys pulverized coal directly to the burners at a rate set by the combustion controls based on steam generation requirements. Secondary air is supplied by the forced draft fans and is typically preheated to about 600F (316C). All or most of the secondary air is supplied to the windboxes enclosing the burners. A portion of the secondary air may be diverted from the burners to NO_x ports (discussed later) in order to control the formation of NO_x. The secondary air supplied to the burners is mixed with the pulverized coal in the throat of the burner. This permits the coal to ignite and burn.

The combustion process continues as the gases and unburned fuel move away from the burner and up the furnace shaft. Final burnout of the char depends on the coal properties, particle fineness, excess air, air-fuel mixing, and thermal environment. The products of combustion eventually leave the furnace and enter the convection pass after being cooled sufficiently to minimize convection surface fouling.

Low volatile coals While the majority of bituminous, subbituminous, and lignite PC-fired units are arranged in this manner, alternate designs are required to accommodate other coals. Coals with low volatile matter content, particularly anthracites, are difficult to ignite. Their low levels of rapid burning volatiles limit this critical source of ignition energy. In addition, their advanced coalification (see Chapter 8) has reduced char reactivity, resulting in elevated char ignition and burnout temperatures. These factors require hotter furnaces to sustain combustion. A downshot firing system (see Fig. 4) accomplishes this in combination with an enlarged, refractory-lined furnace. The downshot arrangement causes the flame to travel down and then turn upward to leave the furnace. The char reactions, which build in intensity some distance from the burners, return their heat near the burners as the gases return and leave the furnace. This heat supplies energy to ignite the fuel introduced at the burners. The refractory lining in the furnace inhibits heat transfer and elevates gas temperatures to sustain combustion. The enlarged lower furnace also provides increased residence time to accommodate the slower burning char. In addition, very high coal fineness is used to accelerate char reactions and reduce unburned carbon loss. Growth in unit size and NO_x emission regulations have limited downshot fired units in the U.S. although they are applied internationally where low volatile coals are a fuel source.

High moisture coals The other major category of nonconventional PC-fired units is that used for lignites with very high moisture content. Moisture content in these coals typically ranges from 50 to 70% by weight, which is well beyond the level that air-swept pulverizers can adequately dry. (See Chapter 12.) Instead, hot gases, removed from the upper portion of the furnace, are used in combination with beater mills to dry and prepare the fuel. These hot gases, near 1832F (1000C), are more suitable for coal drying than preheated air, with a practical upper limit of 752F (400C). Furthermore, the low oxygen content of the flue gas provides a relatively inert atmosphere in the grinding equipment, lessening the threat of fire or explosion as the hot gas mixes with these reactive coals. The fuel is conveyed from the mills to the burners located on the walls, and combustion proceeds as secondary air mixes with the fuel in the furnace. Units of this design are prevalent in parts of Europe and Australia where high moisture brown coals are used.

Combustion system and boiler integration

The most fundamental factors that determine the boiler design are the steam production requirements and the coal to be fired. The thermal cycle defines the heat absorption requirements of the boiler which supplies the rated steam flow at design temperature and pressure. Gas side parameters, based on the design coal, are used to estimate boiler efficiency. Heat input requirements from the coal can then be determined, setting the coal firing rate at maximum load. The number and size of pulverizers are then selected. Frequently, the pulverizer

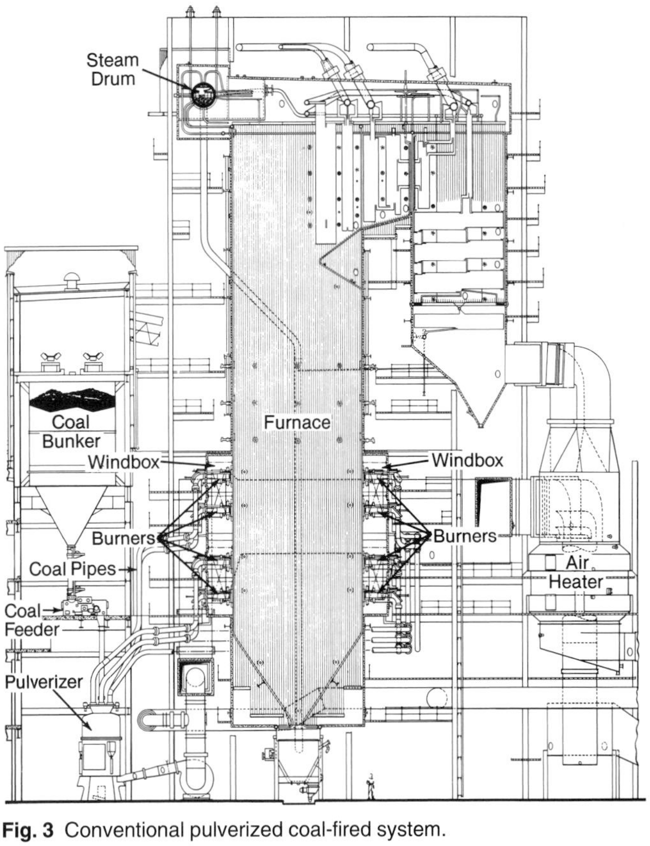

Fig. 3 Conventional pulverized coal-fired system.

selection is based on meeting maximum requirements with one mill out of operation. This permits maintenance on one pulverizer without limiting boiler load.

The size and configuration of the furnace are designed to accommodate the combustion and slagging/fouling characteristics of the coal as discussed in Chapter 20. NO_x emission control factors are also incorporated into the layouts of modern combustion systems. The number of burners and the heat input per burner are selected to minimize flame impingement on the furnace walls. Multiple burners at moderate inputs, in an opposed-fired arrangement, are favored to provide uniform heat distribution across the furnace. This reduces thermal gradients and helps prevent localized slagging, while improving NO_x emissions control and combustion efficiency. The burner design is based on fuel parameters and NO_x emission control requirements in order to provide complete combustion and minimum pollutant emissions.

Theoretical and excess air quantities are determined for the required coal input rate from combustion air calculations as discussed in Chapter 9. Fans, ductwork, air heaters, windboxes, and burners are sized to satisfy flow, pressure loss, air preheat and velocity requirements for the unit. If needed to meet emission requirements, separate staged combustion can be employed in the furnace through air injection after the burner zone (NO_x ports).

Burner fundamentals

Burners are the central element of effective combustion system designs which incorporate fuel preparation, air-fuel distribution, furnace design and combustion control. The burners provide for the introduction and mixing of the fuel and combustion air as shown in Fig. 5.

Control The quantity of pulverized coal and primary air supplied to the burners is set by the coal feed rate and primary air control dampers. Restrictors are usually included in the piping from the pulverizer to its burners to more evenly distribute coal among several burners. Secondary air flow is controlled by dampers at the forced draft fans to maintain excess air for good combustion and proper boiler operation. The distribution of secondary air among burners is accomplished by duct/windbox dampers and burner adjustments.

Flame stability A burner introduces the primary air and pulverized coal to the secondary air in a manner that establishes a stable flame. This involves producing a flame front close to the burner over the range of operating conditions. An ignitor or lighter is required to initiate combustion as fuel is first introduced to the burner and to sustain combustion when flames would otherwise be unstable. The burner normally sustains a stable flame by using heat from coal combustion to ignite the incoming pulverized coal. A flame safety system, included with modern burners, electronically scans the flame to verify stabilization and triggers corrective action if the flame becomes unstable.

Air-fuel mixing The burner also promotes local air-fuel mixing to completely burn the fuel. The overall mixing is a combination of burner-induced local mixing and the resultant more global furnace-induced mixing. Furnace mixing results from expansion of the air-fuel mixture in the furnace due to rapid temperature increase during combustion, from flow variations introduced by the burners, and from the size and shape of the furnace enclosure.

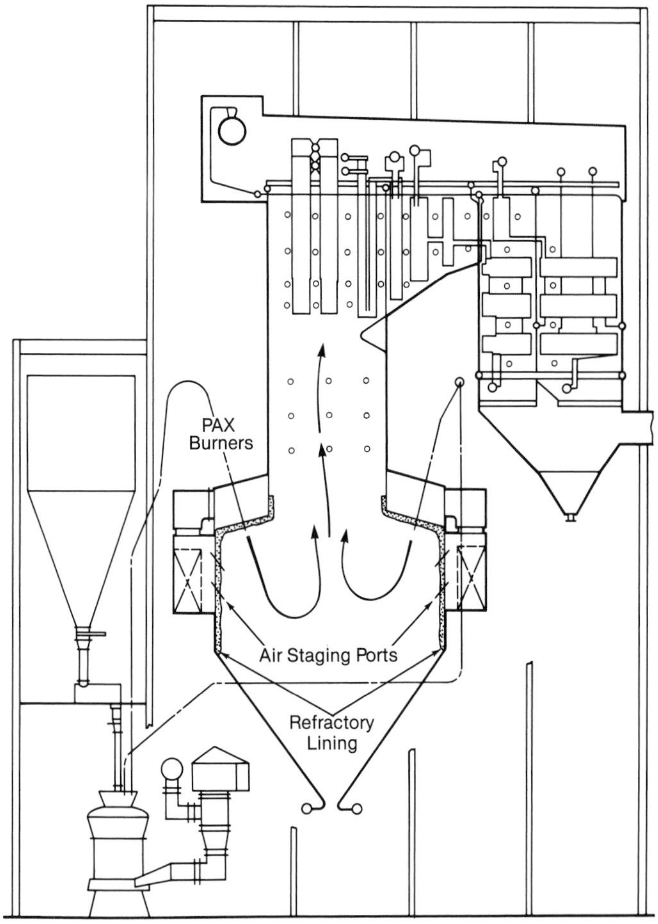

Fig. 4 Downshot-fired unit with refractory-lined furnace for low rank coals and PAX (primary air exchange) burners.

The amount of air-fuel mixing produced by the burner varies considerably with burner type. The simplest burners inject the fuel and air in parallel or concentric streams without the benefit of other burner-induced mixing. Entrainment of adjacent flow streams occurs as the jets develop and due to jet expansion from combustion. Tangentially-fired furnaces operate in this manner, as do some roof-fired designs. Flame length and combustion performance greatly depend on the effectiveness of furnace mixing. Mixing effectiveness decreases with boiler load as air-fuel flow rates drop. This can result in poor combustion performance at reduced boiler loads.

Modern wall-fired burners provide further controlled air-fuel mixing. The amount varies with the firing rate of the burners. Air-fuel mixing is achieved at reduced boiler loads by operating with fewer burners in service and by operating those burners at higher firing rates. Burner mixing can be induced by using the primary air/pulverized coal (PA/PC) stream, by using the secondary air, or by a combination of the two. Looking first at the PA/PC system, the most frequently used burner mixing devices are deflectors, bluff bodies, and swirl generators. Deflectors are frequently installed near the exit of the burner nozzle to cause the PA/PC stream to disperse into the secondary air. These deflectors, or impellers, also reduce axial momentum of the fuel jet, reducing flame length. Impellers may also induce radially-pitched swirl of the fuel jet to further accelerate mixing. Bluff bodies are sometimes used in or adjacent to the burner nozzle exit. Flow locally accelerates around the upstream side of the bluff body and recirculates on the downstream side. The recirculation promotes mixing. The bluff body can also be used to increase residence time for a portion of the fuel near the burner, thereby improving flame stability.

Secondary air swirl is the most common way to induce air-fuel mixing in circular throat burners. Swirl generators are used upstream of the burner throat to impart rotating motion to the secondary air. This air leaves the burner throat with tangential, radial and axial velocity components. Radial and axial pressure gradients form in the flow field downstream of the throat, with the lowest pressure being near the center of the throat.[3] As swirl is increased the pressure gradients increase. This causes the flow to reverse and travel along the axis of the flame toward the low pressure zone. A recirculating flow pattern is then generated near the burner.

Primary air and coal flow rates The primary air requirements are determined by the pulverizer, and the burner piping must be accommodated by the burners. Most pulverizers require 40 to 70% of their full load primary air requirements at their minimum output level. (See Chapter 12.) In addition, the PA/PC mixture traveling to the burners must be transported at a minimum of 3,000 ft/min (15 m/s). This velocity serves to prevent the coal particles from dropping out of suspension in horizontal runs of coal pipe. Minimum primary air flow is the greater of the minimum PA flow required for the pulverizer or the minimum required to satisfy burner line velocity limits.

The primary air and coal mixture conveyed to the burners reaches a maximum velocity and solids loading at full pulverizer load and follows the pulverizer output as mill loading is reduced. As the burner nozzle velocity increases, the ignition point gradually moves farther from the burner. At some point, continued increases in nozzle velocity lead to *blowoff* of the flame, a potentially hazardous condition where coal ignition and flame stability are lost. High burner nozzle velocities also result in accelerated erosion of burner hardware. The weight ratio of coal to primary air typically reaches a peak of 0.4 to 0.65 at full load and a minimum of 0.15 to 0.3 for minimum pulverizer load. Solid fuel loading reduction eventually leads to flame instability as the flame temperature drops.

Temperature The temperature of the PA/PC mixture supplied to the burners is based on the raw coal properties. A minimum of 130F (54C) is required with high moisture, reactive coals. A value of 150F (66C) is typical for high volatile bituminous coal. A maximum of 200F (93C), a result of pulverizer mechanical constraints, is used with low volatile coals to improve ignition.

Unburned carbon loss A small portion of the coal is not completely burned in the boiler, typically reducing efficiency by less than 1%. This unburned carbon loss is predicted based on the furnace-combustion system configuration, thermal environment in the furnace, coal reactivity, coal fineness and excess air levels. Unburned carbon loss is included in the final determination of total boiler efficiency. (See Chapters 9 and 21.)

Performance requirements

Pulverized coal-fired equipment should meet the following performance conditions:

1. The coal and air feed rates must comply with the load demand over a predetermined operating range. For modern applications with high volatile bituminous or subbituminous coal, flames should be stable without the use of lighters from about 30% to full load. The minimum load depends on the coal, burner design, and pulverizer load; operation below this load is performed in combination with lighters in service.

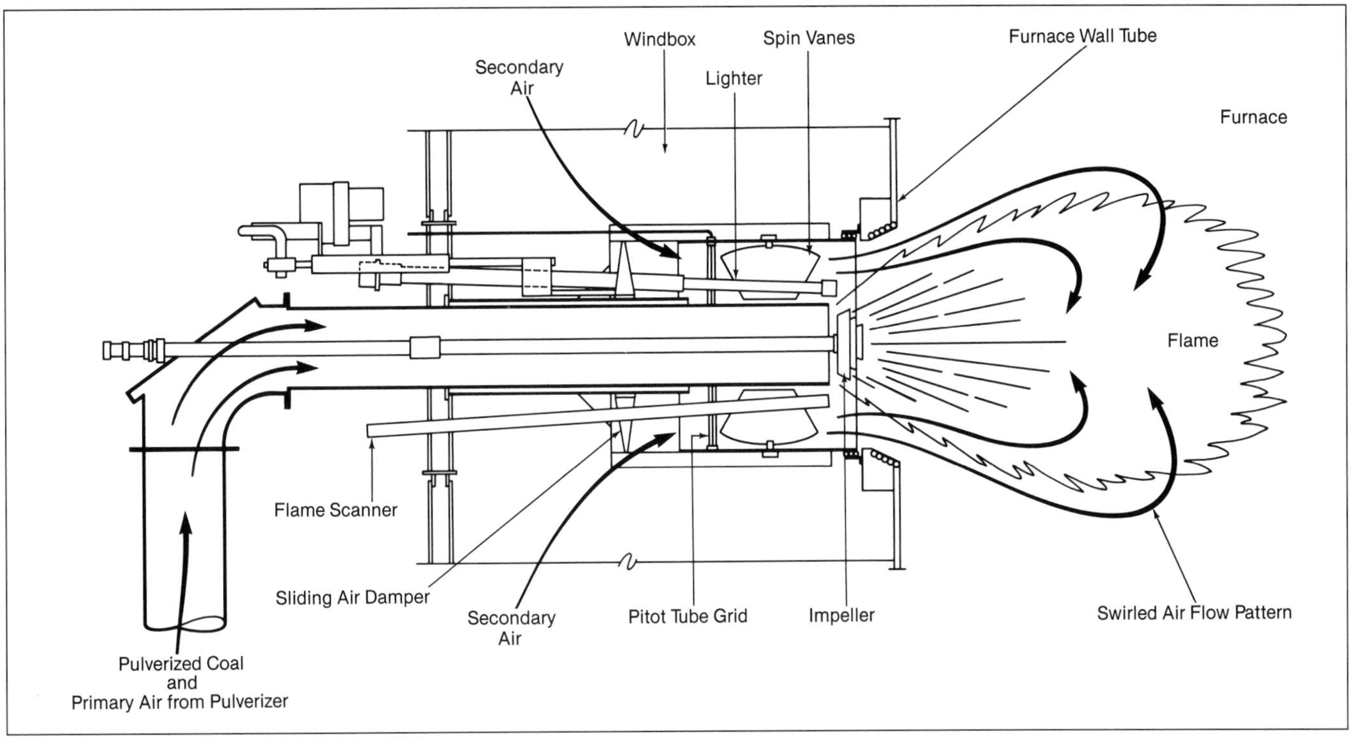

Fig. 5 S-type burner and components.

2. Unburned combustible loss (UCL) should reduce efficiency by less than 1%. UCL less than 0.2% should be expected with reactive subbituminous or lignite coals. UCL increases with less reactive bituminous coals and in boilers with less than 400,000 lb/h (50.4 kg/s) steam capacity.
3. The burner should not require continual adjustment to maintain performance. The unit should be designed to avoid the formation of localized slag deposits that may interfere with burner performance or damage the boiler.
4. Only minor maintenance should be necessary during the annual outage. To avoid high temperature damage, alloy steel should be used for burner parts exposed to furnace radiative heat transfer. Burner parts subject to PC erosion may require more frequent repair or replacement.
5. Safety must be paramount under all operating conditions. Automated flame safety and combustion control systems are recommended and required in many cases.

Conventional PC burners

Prior to 1971 in the U.S., the primary focus of combustion system development was to permit the design of compact, cost effective boilers. As a result, the burner systems developed focused on maximizing heat input per unit volume with small furnace volumes, rapid mixing burners, and very high flame temperatures. An unintended side effect was the production of high levels of NO_x. Many boilers with this technology remain in operation today. Burners used on such boilers include the conventional circular burner, the cell burner, and a more mechanically reliable version — the S-type burner.

Fig. 6 Circular register pulverized coal burner with water-cooled throat.

Conventional circular burner

The circular burner (Fig. 6) was one of the earliest forms of swirl-stabilized PC-fired burners. Due to its success, this burner has been used for six decades firing a variety of coals in many boiler sizes. The circular burner may still be supplied for conventional use in selected cases. The burner is composed of a central nozzle to which PA/PC is supplied. The nozzle is equipped with an impeller at the tip to disperse the coal into the secondary air. Secondary air is admitted to the burner through a register. The register consists of interlinked doors arranged in a circular pattern between two plates. The doors are closed to cooling position when the burner is out of service, are partially open for lightoff and are more fully open for normal operation.

Opening the register doors allows more air into the burner but reduces swirl. Flame shape can also be adjusted using the registers. However, air distribution among burners is also impacted by register position, and high combustion efficiency depends on uniform air-fuel distribution.

The circular burner can fire natural gas, oil or pulverized coal. Simultaneous firing of more than one fuel is not recommended due to the potential for overfiring the burners and the resulting combustion problems. Variations of the circular burner are in operation at inputs from 50 to 300 × 10^6 Btu/h (15 to 88 MW). Moderate input units provide more uniform heat distribution to the combustion zone, improving operation.

Cell burner

The cell burner combines two or three circular burners into a vertically stacked assembly that operates as a single unit (Fig. 7). In the 1960s and 1970s the cell burner was applied to numerous utility boilers which had compact burner zones. While highly efficient, the cell burners produced high levels of NO_x emissions.

S-type burner

The S-type burner shares the functional attributes of the circular burner in an improved configuration. (See Fig. 5.) The burner nozzle is the same as that in the circular burner. However, secondary air flow and swirl are separately controlled. Secondary air quantity is controlled by a sliding disk as it moves closer to or farther from the burner barrel. Secondary air swirl is provided by adjustable spin vanes positioned in the burner barrel. An air-measuring pitot tube grid is installed in the barrel ahead of the spin vanes. Secondary air can be measured for each burner to distribute the flow uniformly among burners using the sliding disks or other means. Swirl control for flame shaping is controlled separately by spin vanes. The S-type burner provides higher combustion efficiency and mechanical reliability than the circular unit and is a direct plug-in replacement.

Low NO_x combustion systems

NO_x formation

NO_x is an unintended byproduct from the combustion of fossil fuels and its emissions are regulated in the U.S. and many other parts of the world. (See Chapter 32.)

While a number of options are available to control and reduce NO_x emissions from boilers, as discussed in Chapter 34, the most cost effective means is the use of low NO_x combustion technology either alone or in combination with other techniques. The effectiveness of the pulverized coal NO_x control technology, however, depends upon the fuel characteristics and the overall system design. For pulverized coal wall-fired units, NO_x emissions from conventional combustion systems discussed above typically range from 0.8 to 1.6 lb/10^6 Btu (984 to 1968 mg/Nm^3: see Note below). Low NO_x PC combustion systems are capable of reducing NO_x to 0.2 to 0.7 lb/10^6 Btu (246 to 861 mg/Nm^3).

NO_x control by combustion

More than 75% of the NO_x formed during conventional PC firing is fuel NO_x; the remainder is primarily thermal NO_x. Consequently, the most effective combustion countermeasures are those limiting fuel NO_x formation. Fuel NO_x is formed by oxidation of fuel-bound nitrogen during devolatilization and char burnout. Coal typically contains 0.5 to 2.0% nitrogen bound in its organic matter. High oxygen availability and high flame temperatures during devolatilization encourage the conversion of volatile-released nitrogen to NO_x. Nitrogen retained in the char has a lower conversion efficiency to NO_x, primarily due to lower oxygen availability during char burnout. Reactive coals with high volatile matter and low fixed carbon/volatile matter (FC/VM) ratios have tended to be the most amenable to NO_x control by combustion modification. Coals with higher nitrogen content tend to produce higher NO_x emissions.

The most effective means of reducing fuel-based NO_x formation is to reduce oxygen (air) availability during the critical step of devolatilization. Additional air can then be added later in the process to complete char reactions and maintain high combustion efficiency.

Oxygen availability can be reduced during devolatilization in two ways. One method is to remove a portion of the combustion air from the burners and introduce it elsewhere in the furnace. This method is referred to as *air staging*. PC combustion systems typically operate with 15 to 20% excess air at maximum firing rate. (See Chapter 9.) In some staging situations, reducing the air flow by 10% at the burners is sufficient for a given level of NO_x emissions control. This permits the burners to continue to operate in an excess air condition. Reducing the air flow to less than stoichiometric (theoretically required for complete combustion air flow) conditions can further reduce NO_x emissions. However, further reductions in burner stoichiometry below 70% of the theoretical air requirements can cause NO_x to increase again. This is due largely to the significant fraction of combustion which is deferred and the high air flows and mixing rates necessary to subsequently complete this combustion. Burner operation below theoretical air requirements can also increase unburned carbon loss and slagging in the combustion zone with bituminous coals while increasing the risk of furnace corrosion. Unburned carbon loss

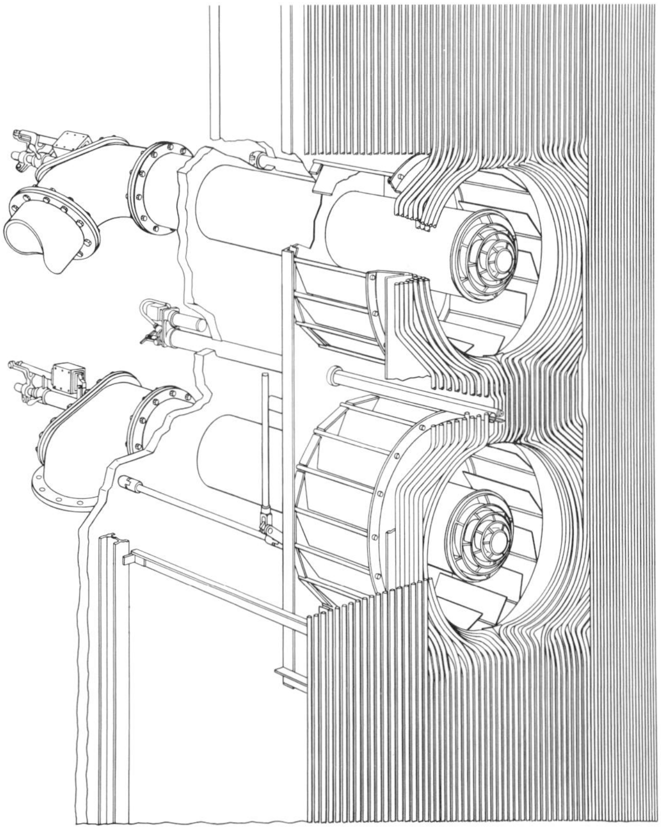

Fig. 7 Conventional cell burner.

can be controlled by increasing coal fineness and by careful management of air-fuel distribution among burners and air staging ports. Slagging can be minimized by proper arrangement of burners and by burner design. Tube corrosion damage by hydrogen sulfide (H_2S) attack is a concern when firing high sulfur coals under substoichiometric conditions. H_2S is formed during the combustion of sulfur bearing coals under oxygen deficient or reducing conditions. The resulting corrosion potential is dependent upon the H_2S concentration, tube material, tube temperature and operating conditions. This corrosion potential can be reduced by switching to low sulfur coal or by coating the burner zone tubes with a corrosion resistant material.

A second method of reducing oxygen availability during coal devolatilization is by burner design. The burner can be designed to supply all the combustion air but to limit its rate of introduction to the flame. Only a fraction of the air is permitted to mix with the coal during devolatilization. The remaining air is then mixed downstream in the flame to complete combustion. However, overall mixing is reduced and the flame envelope is larger compared to rapid mixing conventional burners. NO_x emission reduction using low NO_x burners ranges from 30 to 60% compared to uncontrolled levels. These burners can be used in combination with air staging through NO_x ports (discussed later) to satisfy emission limits. However, from a cost and performance perspective, the use of NO_x ports should be minimized where possible. Advanced low NO_x burners can frequently meet emission control requirements without the use of NO_x ports.

Another method of reducing NO_x emissions is referred to as *reburning* or *fuel staging*. This control technique

Note: The International Energy Agency conversion has been adopted for NO_x emission rates: 1230 mg/Nm^3 equals 1 lb/10^6 Btu for dry flue gas, 6% O_2, 350 Nm^3/GJ for coal.

destroys NO_x after it has formed. Fuel staging involves introducing the fuel into the furnace in steps. The bulk of the fuel is burned in the furnace at near stoichiometric conditions. The balance of the fuel with a limited amount of air is then injected to create a reducing zone part way through the combustion process. The reducing conditions form hydrocarbon radicals which strip the oxygen from previously formed NO_x, thereby reducing overall NO_x emissions. The balance of the air necessary to complete the combustion is then added. Some fuel-staged systems involve separate burners or fuel injectors followed by NO_x ports which supply the remaining combustion air. Some advanced low NO_x burners embody fuel staging while supplying all the fuel to the burners. A portion of the fuel is introduced in a manner to generate hydrocarbon radicals; these are mixed into the flame later to reduce previously formed NO_x. These burners can achieve NO_x reductions of 50 to 70% from baseline uncontrolled levels.

Dual register burner

B&W began experimenting with techniques to reduce NO_x from PC-fired burners in the 1950s. A slot burner, which incorporated controlled air-fuel mixing, was first developed.[4] Air introduced through the burner throat mixed rapidly in the flame while air diverted to the external slots mixed gradually. While effective, the design with slots external to the burner throat was difficult to apply to furnaces. In 1972, B&W developed the dual register burner (DRB) for firing pulverized coal. The DRB (Fig. 8) provided two air zones, each controlled by a separate register, around an axially positioned coal nozzle. A portion of the secondary air was admitted through the inner air zone to aid ignition and flame stability. The remainder passed through the outer air zone and was mixed downstream in the flame. Recirculation induced by the swirled outer air also improved flame stability. The burner nozzle was equipped with a venturi to disperse the primary air-coal mixture. The mixture was then injected into the throat without deflection. The results were a stable flame with a fuel-rich core and gradually completed mixing downstream. The first unit retrofit was completed in 1973 and achieved 50% NO_x reduction relative to prior circular burners. This retrofit required no modifications to boiler pressure parts, fans, or pulverizers, and NO_x ports were not used. The U.S. government granted a patent to B&W for the DRB in January 1974. Numerous variations have been used by B&W since its introduction.

The DRB is often used in combination with enlarged furnace combustion zones to reduce thermal NO_x and with a compartmented windbox to better control air distribution (Fig. 8). This design has the burners coupled with each pulverizer situated in an individual windbox compartment. Secondary air is metered and controlled separately for each compartment. This approach corrects much of the flow imbalance among burners that was experienced in open windbox designs. High combustion efficiency with minimal burner zone slagging results.

The DRB typically reduces NO_x 50 to 60% from uncontrolled levels. This efficiency enables its use on new boilers without NO_x ports or other air diversion systems. Utility boiler NO_x emissions with the DRB range from 0.27 to 0.70 lb/10^6 Btu (332 to 861 mg/Nm^3).

DRB-XCL™ burner

The dual register XCL burner is an advanced version of the original DRB. The DRB-XCL™ is reconfigured mechanically as shown in Fig. 9. A DRB-XCL™ burner being prepared for shipment is shown in Fig. 10. Air flow to the burner is regulated by a sliding disk similar to that used on the S-type burner. An impact/suction pitot tube grid is located in the burner barrel to measure secondary air flow. With this information, the secondary air can be uniformly distributed among all burners by adjusting the sliding disks or other means. Balanced air and

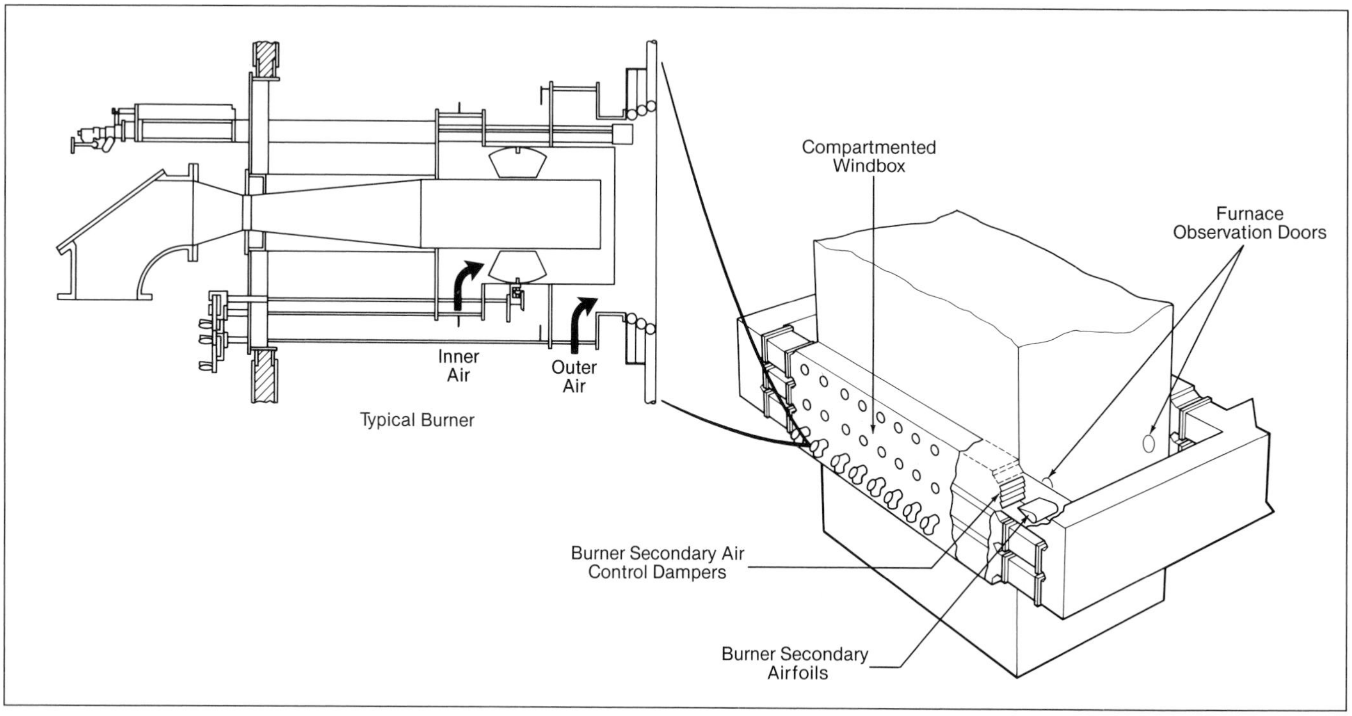

Fig. 8 Dual register burner and compartmented windbox.

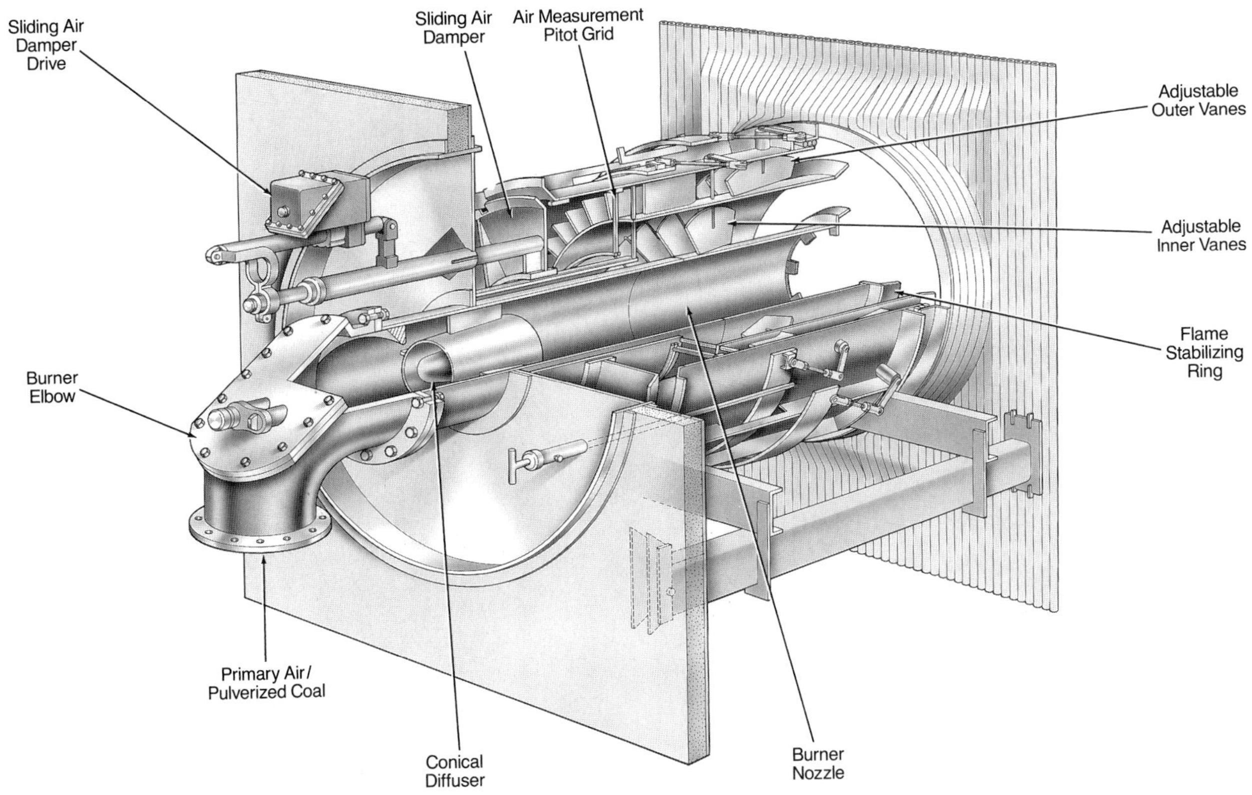

Fig. 9 DRB-XCL™ low NO_x burner for pulverized coal firing.

fuel distribution among all burners is critical to combustion efficiency, particularly with low NO_x systems. Downstream of the pitot tube grid are the inner and outer air zones. Adjustable vanes in the inner zone stabilize ignition at the burner nozzle tip. The outer zone, the main air path, is equipped with two stages of vanes. The upstream set is fixed and improves peripheral air distribution within the burner. The downstream set is adjustable and provides proper mixing of this secondary air into the flame. The burner nozzle is equipped with a conical diffuser and flame stabilizing ring. These combine to improve flame stability while further reducing NO_x by incorporating fuel staging technology. The DRB-XCL™ burner reduces NO_x 50 to 70% from uncontrolled levels without using NO_x ports. This level of NO_x reduction is possible through the creation of the flame combustion zones as shown in Fig. 11. The burner is specifically designed to generate rapid heating and high temperatures in the fuel rich flame core. This causes more of the coal to burn as volatile matter and releases a larger portion of the fuel nitrogen early in the combustion process, leaving less in the char. By limiting available oxygen in the flame core, NO_x formation is minimized. At the same time, reducing species are generated from the volatile materials. These propagate into the flame to aid in reducing NO_x formed in the later zones of the flame. Char oxidation occurs last at reduced temperature and oxygen concentrations, thereby limiting NO_x formation during char burnout. NO_x ports can be added when further NO_x reduction is required. In retrofit applications, the DRB-XCL™ burner can be equipped with modified coal impellers to shape the flame to existing furnaces.

The air and fuel staging technology of the DRB-XCL™ burner is also well suited for firing fuel oil or natural gas and for multi-fuel applications with pulverized coal. Chapter 10 discusses the oil and gas performance.

Low NO_x Cell™ burner

The cell burners discussed earlier produced high NO_x emissions, typically 1.0 to 1.8 lb/10^6 Btu (1230 to 2214 mg/Nm^3). The unique arrangement of the cell burner (Fig. 7) was incompatible with direct replacement by standard

Fig. 10 DRB-XCL™ burner being prepared for shipment.

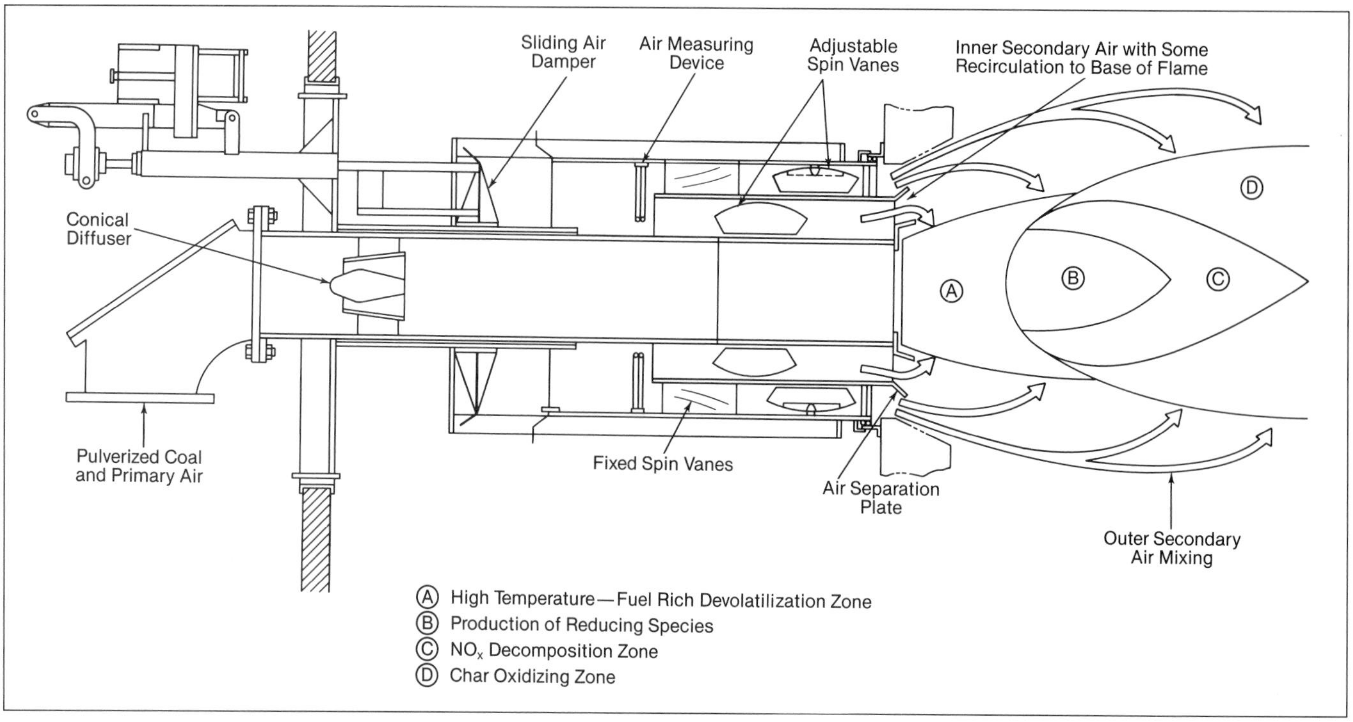

Fig. 11 DRB-XCL™ low NO_x combustion zones.

low NO_x burners. The close spacing of the burner throats prevented direct installation of larger, multi-zone low NO_x burners. One option is to rearrange the burners, increasing space between them and installing the DRB-XCL™ low NO_x burner. This solution is effective but costly. Costs increase due to the need for new boiler tube wall panels, coal piping, burners, ignition system plus windbox alterations. Another alternative is the Low NO_x Cell™ burner (LNCB™). (See Fig. 12.) This burner is designed to fit into existing cell burner locations and reduce NO_x. The LNCB™ was conceived by B&W and developed in cooperation with the Electric Power Research Institute. In essence, all of the coal is supplied to the lower burner throat with a portion of the secondary air. The balance of the secondary air is supplied to the upper burner throat. The upper throat serves as an integral NO_x port for each burner location. The coal piping is altered to supply all the coal to the lower throat, usually by joining the pipes in a Y connection. A larger burner nozzle is installed in the lower throat and can be equipped with an impeller to shape the flame. The lower throat burner assembly is essentially a conventional S-type burner with an oversized nozzle. The upper throat assembly is equipped with a sliding disk and pitot tube grid to set secondary air flow. It is also equipped with vanes to control mixing of this air into the flame. NO_x levels can be reduced by 50% with the LNCB™ while maintaining high combustion efficiency. Retrofit costs are reduced because wall panel and windbox alterations are avoided and because coal piping alterations are reduced.

NO_x ports

Removing a portion of the secondary air from PC-fired burners effectively reduces NO_x emissions as discussed above. The air is diverted to ports which introduce it later in the combustion process. These ports may be located close to the burners, as in the LNCB™, or at some distance from all of the burners. In the majority of applications, the ports are placed above the burner zone in furnaces arranged for gases to travel upward and out. Such ports are sometimes referred to as overfire air (OFA) ports. In some applications, ports are placed beneath or within the burner zone.

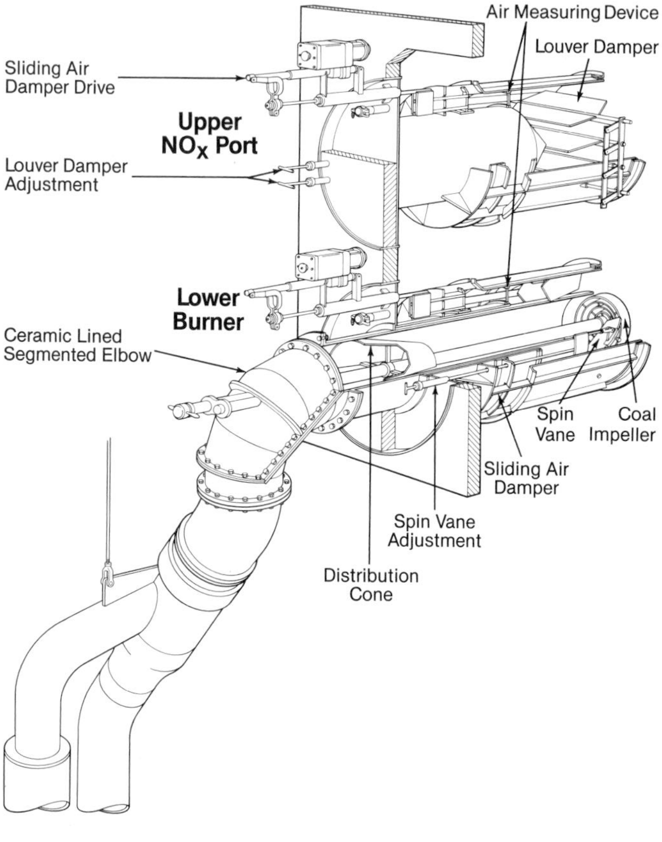

Fig. 12 Low NO_x Cell™ burner.

B&W pioneered the development and application of NO_x ports and air staging as NO_x emission controls. This work, which began in the 1950s, was initially directed at NO_x reduction for oil- and gas-fired boilers in California. R.M. Hardgrove of B&W filed a patent for a *Method for Burning Fuel* on June 18, 1959, disclosing key aspects of air-staging influences on NO_x emissions. The patent was granted in 1962, and NO_x ports have been used subsequently by B&W.

Despite the proven effectiveness of NO_x ports, they are not always used with pulverized coal. This is due to the potential for increased slagging and corrosion in the furnace and the loss of combustion efficiency. The H_2S corrosion potential is a concern with higher sulfur coals, particularly those exceeding 2 lb/10^6 Btu (860 g/GJ) sulfur content. High and severe slagging bituminous coals are not suitable for NO_x ports due to the potential for wet or plastic slag forming in the furnace. This slag can block burner throats or the furnace hopper and can be corrosive to boiler tubes. Protective coatings of some stainless steels or aluminum can greatly reduce the corrosion potential. However, substoichiometric operation with high sulfur coals is not recommended. Low NO_x burners alone, or those with slight staging to maintain a stoichiometry of at least 1.05 at the burners, are better solutions with such coals.

To maintain high combustion efficiency, NO_x port jet size and velocity must be sufficient to provide penetration across the furnace and mixing with crossflow gases. However, high velocity jets can miss combustion product gases flowing between the ports. B&W developed the Dual Zone NO_x Port (Fig. 13) to address this problem. The inner zone of the port provides a jet to penetrate across the furnace. The outer zone has vanes to deflect air and entrain nearby gases. The ports have adjustable disks and vanes to permit tuning during commissioning. Each port is also equipped with pitot tubes to measure flow and to allow uniform air distribution among ports. Systems intended to operate at low burner zone stoichiometries, e.g., 0.70 to 0.85, benefit from two levels of NO_x ports. This better controls introduction of oxygen in the latter stages of combustion and minimizes associated NO_x formation. NO_x port system performance benefits from the inclusion of equipment to measure and regulate flow to both the burners and the ports through the boiler load range.

An emerging technology which facilitates this integration is numerical modeling. (See Chapters 3 and 4.) Computer programs provide detailed information about mixing effectiveness throughout the furnace. These programs can be used to optimize NO_x port size and placement, burner swirl orientation, and furnace geometry. Major improvements in mixing can be achieved by this parametric evaluation;[5] however, models must be validated with test data.

Burners for difficult coals

A few coals have proven to be considerably more difficult to burn than others, including unreactive coals and reactive coals which would burn readily except for an overabundance of moisture and/or ash. B&W has developed burners specifically for use with these problem coals. The selection of burner type is based on an empirical ignition factor derived from full scale experience with difficult coals. This factor relates variations in fuel ratio (FC/VM), moisture content, ash content, and coal heat-

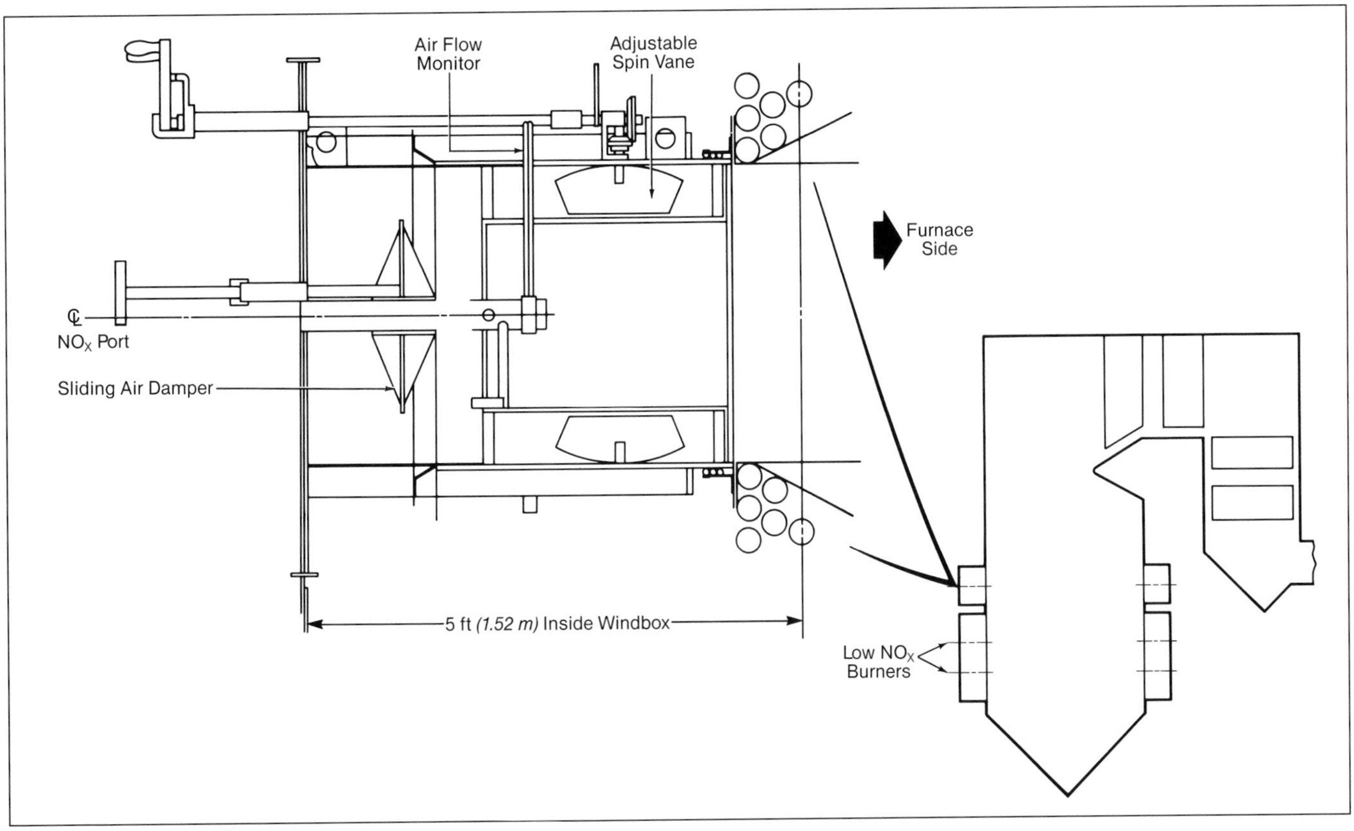

Fig. 13 Dual Zone NO_x Port.

ing value to ignition behavior. Burner choices for these applications include conventional and low NO_x burners discussed above, the enhanced ignition (EI) burner, and the primary air exchange (PAX) burner.

Enhanced ignition burner

The enhanced ignition (EI) burner resembles the DRB-XCL™ burner. However, the EI burner uses a larger coal nozzle to reduce the primary air and coal mixture velocity as it enters the furnace. The conical diffuser produces a fuel-rich ring and a fuel-lean core; at low velocity, this combination is readily ignited. The dual air zones are adjusted relative to the DRB-XCL™ burner to increase recirculation of hot gases along the fuel jet. The EI burner is well suited for firing high moisture (more than 35%) lignites and low volatile bituminous coal.

Primary air exchange burner

The primary air exchange (PAX) burner is designed to fire anthracite coals. Some anthracites can be wall-fired conventionally with the PAX burner, while those lowest in volatile matter must be fired in a downshot arrangement (Fig. 4). The register design for the PAX burner is similar to that of the EI burner. However, in the case of downshot firing, air staging ports direct air away from the burner to raise flame temperature.

The anthracite fineness is increased to improve combustion efficiency. In addition, mill exit temperature is increased to the maximum for the pulverizer, often limited to 200F (93C) for mechanical reasons. Tests have shown the benefit of higher temperatures on ignition performance. The PAX burner raises the primary air and pulverized coal (PA/PC) temperature by a patented design which separates half of the primary air from the PA/PC mixture and vents it into the furnace. This primary air is replaced by secondary air near 700F (371C), significantly raising fuel temperatures as it enters the furnace. The PAX nozzle tip is also sized for low velocity to further assist ignition. Even with these techniques, good combustion is dependent upon high furnace temperatures. Low load operation still requires lighters for flame stabilization.

Auxiliary equipment

Oil/gas firing equipment

In some cases, PC-fired furnaces are required to burn fuel oil or natural gas up to full load firing rates. These fuels can be used when the PC system is not available for use early in the life of the unit, due to an interruption in coal supply, or as a NO_x control strategy. Some operators avoid installing a spare pulverizer by firing oil or gas when a mill is out of service.

To fire oil, an atomizer is installed axially in the burner nozzle. Erosion protection for the atomizer is recommended. A source of air is needed when firing oil or gas to purge the nozzle and improve combustion. This air system must be sealed off prior to PC firing. Steam-assisted atomizers are recommended for best performance.

Gas elements of several designs can be used in PC-fired burners. The burner gas manifold can be inside or outside of the windbox. Installation in the windbox clears the burner front of considerable hardware but prevents on-line adjustment or repair of gas elements. Gas elements are equipped with flame stabilizing cans to protect the root of the flame at each spud. Retrofitting multiple gas elements into PC-fired burners can disrupt air flow in the throat and impair performance. Compatibility of the gas elements with the PC burner design is important. Retrofit of gas or oil elements increases the forced draft fan load in many cases. Primary air is eliminated which results in higher quantities of secondary air compared to PC. As a result, fan flow and static margins need to be considered.

See Chapter 10 for more information on oil and gas firing equipment.

Lighters

A lighter, or ignitor, is required to initiate combustion as pulverized coal is first introduced to the burner, as the burners are being normally shut down, and as otherwise required for flame stability. Lighters typically use an electrically generated spark to ignite a lighter fuel, which is usually natural gas or No. 2 fuel oil. Most modern units have automated controls to activate, operate, purge, and shut down the lighters. Lighter systems are described in Chapter 10.

Flame safety system

Modern PC-fired boilers are equipped with a flame safety systems (FSS). The FSS uses scanners at each burner to electronically monitor flame conditions. Flame scanners are used to evaluate the lighter and main flames for intensity and frequency. The lack of satisfactory flame signals causes the FSS to automatically take corrective action, which is to shut down the burner and its associated group. This prevents unburned fuel from entering the furnace and significantly reduces the risk of an explosion. See Chapter 41 for additional information on the FSS.

Safety and operation

Uncontrolled ignition of pulverized coal may result in an explosion which can severely damage equipment and injure personnel. Explosions result from ignition of an accumulated combustible mixture within the furnace or associated boiler passes, ducts, and fans.[6,7] The magnitude and intensity of the explosion depend on the quantity of accumulated combustibles and the air-fuel mixture at ignition. Explosions result from improper operation, design, or malfunction. The National Fire Protection Association (NFPA) publishes standards in the following areas: *Pulverized Fuel Systems* (NFPA 85F) and *Prevention of Furnace Explosions in Pulverized Coal-Fired Multiple Burner Boiler-Furnaces* (NFPA 85E). These are excellent sources of information concerning design and equipment issues, operation and control of PC-fired equipment.

Another key source of information is the operating instructions. Operators of PC-fired equipment should expect the instructions to provide specific information concerning the purpose, design, calibration and adjustment, startup, operation and shutdown of the various equipment. The instructions also provide maintenance and troubleshooting information.

Other PC applications

Pulverized coal in the metals and cement industries

The application of pulverized coal firing to copper- and nickel-ore smelting and refining has been standard practice for many years. With the use of pulverized coal, high purity metal can be obtained because the furnace atmosphere and temperature can be easily controlled. Pulverized coal may be favored over other fuels for several reasons: it may be less expensive, it offers a high rate of smelting and refining, and it readily oxidizes sulfur.

Copper reverberatory furnaces for smelting and refining are fitted with waste heat boilers for steam generation. These boilers supply a substantial portion of power for auxiliary equipment and also provide the means for cooling the gases leaving the furnace.

In producing cement, the fuel cost is a major expense item. Except in locations where the cost favors oil or gas, pulverized coal is widely used in the industry. Direct firing, with a single pulverizer delivering coal to a single burner, is common practice in the majority of U.S. cement plants. The waste heat air from the clinker cooler, taken from the top of the kiln hood through a dust collector, usually serves as the preheated air for drying the coal in the pulverizer.

Pulverized coal has recently been introduced in the steel industry where it is injected into the blast furnace tuyères. The pulverized coal replaces a similar weight of higher priced coke. Theoretically, 40% of the coke could be replaced by pulverized coal.

References

1. Orning, A.A., "The combustion of pulverized coal," Chapter 34, appearing in *Chemistry of Coal Utilization*, H.H. Lowry, ed., Wiley, New York, pp. 1522-1567, 1945.

2. "Standard test method for volatile matter in the analysis sample of coal and coke," D3175-89a, *Annual Book of ASTM Standards*, Vol. 5.05, American Society for Testing and Materials, Philadelphia, pp. 329-331, 1991.

3. Beer, J.M., and Chigier, N.A., *Combustion Aerodynamics*, Chapter 5, Krieger Publishing, Malabar, Florida, 1983.

4. Brackett, C.E., and Barsin, J.A., "The dual register pulverized coal burner," presented at the Electric Power Research Institute NO_x Control Technology Seminar, San Francisco, February 1976.

5. Fiveland, W.A., Latham, C.E., and LaRue, A.D., "Combustion system optimization by advanced modeling technology," presented at the International Joint Power Generation Conference, Boston, October 1990.

6. *NFPA 85C — Prevention of Furnace Explosions / Implosions in Multiple Burner Boiler-Furnaces*, National Fire Protection Association, Quincy, Massachusetts, 1991.

7. *NFPA 85F — Installation and Operation of Pulverized Fuel Systems*, National Fire Protection Association, Quincy, Massachusetts, 1991.

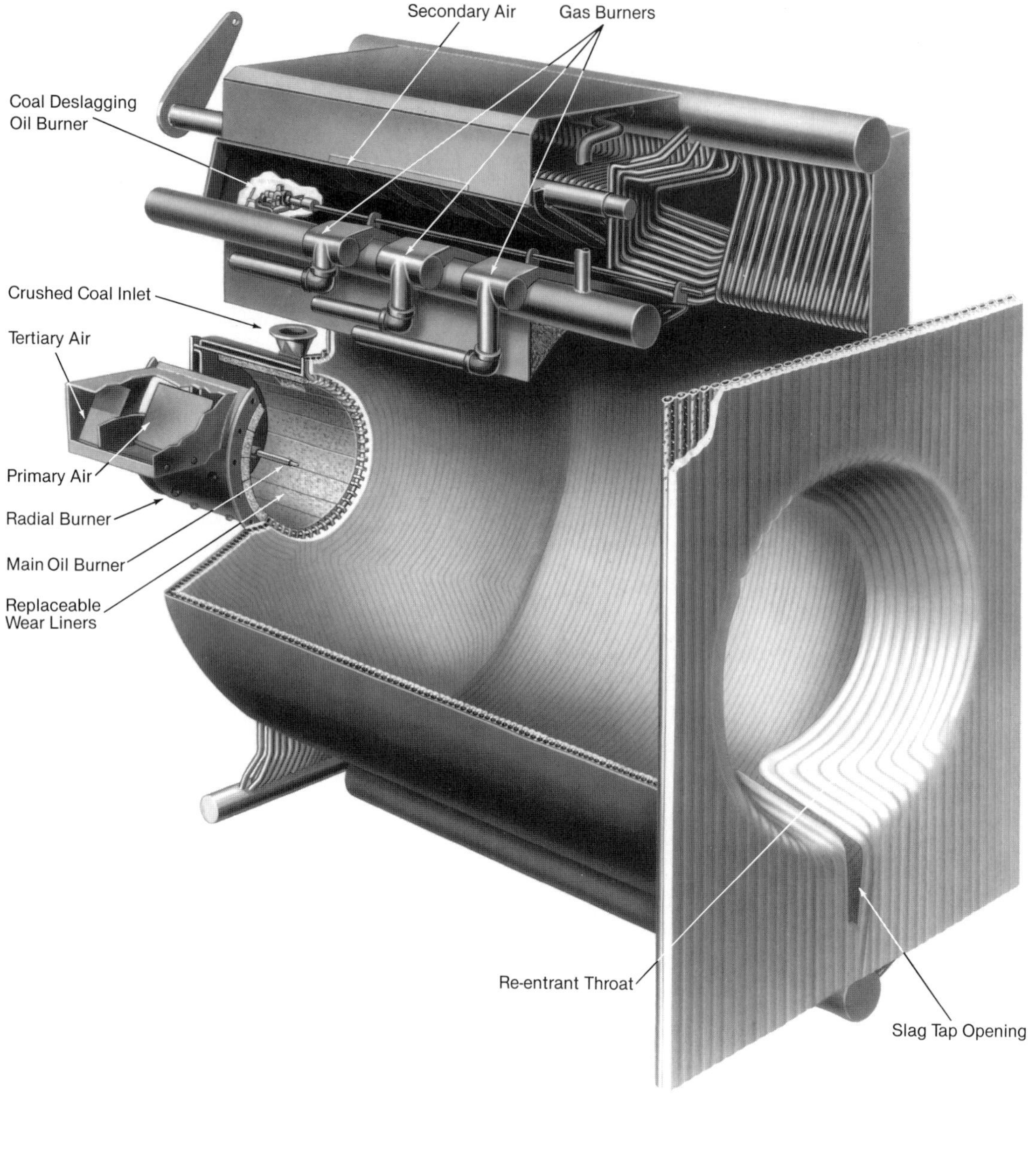
Secondary Air
Gas Burners
Coal Deslagging
Oil Burner
Crushed Coal Inlet
Tertiary Air
Primary Air
Radial Burner
Main Oil Burner
Replaceable
Wear Liners
Re-entrant Throat
Slag Tap Opening

Cyclone furnace.

Chapter 14
Cyclones

Babcock & Wilcox (B&W) developed the Cyclone furnace concept in the 1940s to burn coal grades that are not well suited for pulverized coal (PC) combustion. The ash from these coals has a low melting (fusion) temperature and would enter the superheaters of PC units in a molten state, creating severe slagging.

The Cyclone furnace was originally designed to take advantage of:

1. lower fuel preparation capital and operating costs (crushers only),
2. a smaller furnace, and
3. less flyash and convection pass fouling (15 to 30% of the fuel ash enters the convection pass instead of 80% for PC firing).

Cyclone furnace arrangement

As shown in Fig. 1 and the facing page figure, the Cyclone furnace consists of a horizontal cylindrical barrel 6 to 10 ft (1.8 to 3.0 m) in diameter, attached to the side of the boiler furnace. The Cyclone barrel is of water-cooled tangent tube construction. Inside the Cyclone barrel short pin studs are welded to the outside surface of the tubes in a very dense pattern. The studs are covered with a refractory lining material. This insulation maintains the Cyclone at a high enough temperature to permit adequate slag tapping from the bottom of the unit and significantly reduces the potential for corrosion.

Crushed coal and some air (primary and tertiary) enter the front of the Cyclone through specially designed burners in the frontwall of the Cyclone. In the main Cyclone barrel a swirling motion is created by the tangential addition of the secondary air in the upper Cyclone barrel wall. A unique combustion pattern and circulating gas flow structure result (discussed below). The products of combustion eventually leave the Cyclone furnace through the re-entrant throat. A molten slag layer develops and coats the inside surface of the Cyclone barrel. The slag drains to the bottom of the Cyclone and is discharged through the slag tap.

Principles of operation

To understand the Cyclone concept, the basics of solid fuel combustion must first be considered, particularly as they relate to PC firing. During coal combustion in a boiler furnace, the volatiles burn without difficulty; however, combustion of the fuel carbon char particles requires special measures that ensure a continuing supply of oxygen to unburned carbon particles. A thorough mixing of coal particles and air must occur with sufficient turbulence to remove combustion products and to provide fresh air at the particle surface.

Pulverized firing achieves these requirements by reducing the coal to a very fine powder, 70% of which passes through a 200 mesh (74 micron) screen. This powder then

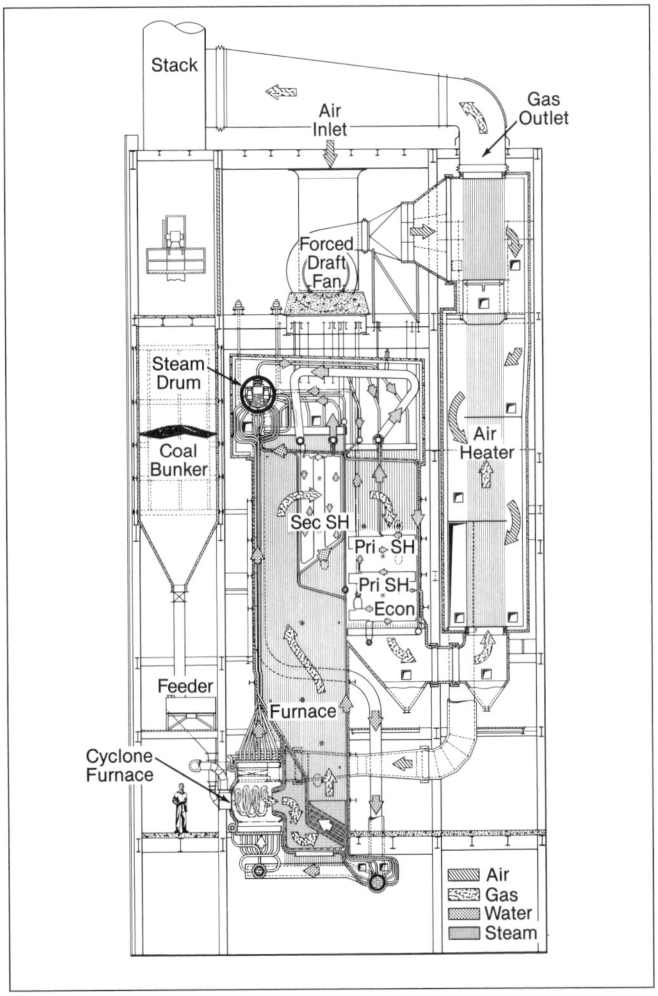

Fig. 1 Cyclone furnace boiler showing air, flue gas, water and steam flows.

mixes with the turbulent combustion air. After this initial phase, the small particles are carried in the air stream with much less mixing. Without the continued rapid mixing action the coal particulate combustion must be completed through diffusion of the oxygen and combustion products around the particle. The relatively large PC furnace provides sufficient residence time for oxygen to penetrate the combustion products blanket around the particles as well as cooling of the ash to minimize convection pass fouling.

The Cyclone furnace, on the other hand, fires relatively large crushed coal particles of which approximately 95% pass through a 4 mesh screen (fuel dependent). Fuel particulates of this size are too large to burn completely in suspension and would pass through the PC-fired boiler without burning all of the carbon. Therefore, the large particles must be retained in place with the air passing over the particle (air scrubbing) for complete combustion to occur. The Cyclone furnace accomplishes this by forming a molten sticky slag layer which captures and holds the heavier particles. While the large particles are trapped in the slag layer, the volatiles and fine coal particles burn in suspension providing the intense radiant heat required for slag layer combustion. Ideally, all of the large coal particles become trapped in the molten slag where they complete carbon burnout leaving behind ash to replenish the slag layer.

With most of the combustion occurring in the confines of one or more Cyclone furnaces, the main boiler furnace can be relatively small compared to a pulverized coal furnace.

Cyclone combustion

In the Cyclone, fuel is fired under intense heat input to maximize combustion efficiency. Injecting the main combustion air tangentially creates a swirling motion which throws the large coal particles against the Cyclone inside surface where they are trapped in the slag layer and burn to completion. The hot gases then exit through the Cyclone core and depart through the re-entrant throat into the main boiler furnace.

Air distribution

The cyclonic gas flow does not follow a simple corkscrew path from entry to exit. (See Fig. 2.) The main combustion air (*secondary air*) enters the Cyclone furnace tangentially at high velocity and passes along the Cyclone periphery, also known as the *recirculation zone*. In this zone, the air-gas flow along the chamber is away from the exit opening and toward the burner, pulled along by the vortex vacuum created near the burner end of the Cyclone. Closer to the center of the Cyclone, the general gas flow is toward the exit (re-entrant throat). Eventually, the hot combustion gases flow from the recirculation zone to the Cyclone core, known as the *vortex*. Once in the vortex, the gases and particulate are immediately pulled out of the Cyclone and into the main furnace. The fuel residence time in the recirculation zone is affected directly by the velocity of the secondary air and inversely by the size of the re-entrant throat opening.

The balance of the combustion air is admitted as primary and tertiary air as discussed below. The air injection locations are identified in Fig. 2 and later in Fig. 5.

The *primary* air enters the burner tangentially in the same rotational direction as the secondary air, carrying

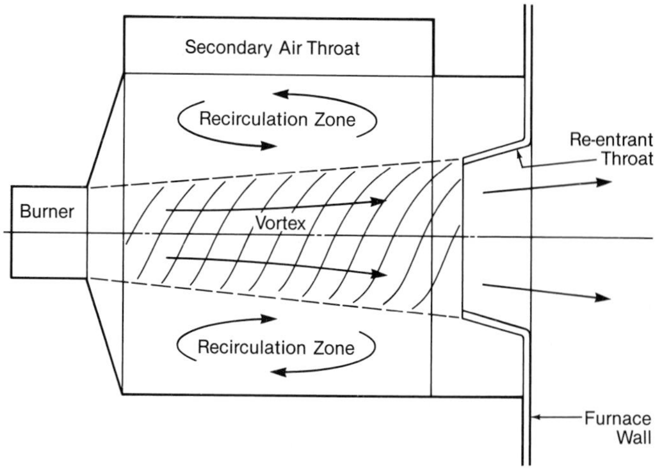

Fig. 2 Cyclone furnace gas recirculation pattern.

the coal into the Cyclone. This primary air controls the coal distribution within the main Cyclone chamber. For critical coals, the primary air must be minimized to avoid throwing the raw coal too deep or near the vortex, but must not be set so low as to create a fuel buildup near or in the burner.

The *tertiary* air enters the center of the burner along the Cyclone axis, directly into the Cyclone vortex. It controls the vortex vacuum and consequently determines the position of the main combustion zone, the primary source of radiant heat. An increase in tertiary air reduces the vortex vacuum at the burner, allowing the main combustion zone to move deeper into the Cyclone toward the re-entrant throat and main boiler furnace.

Heat rates

Within the Cyclone the fuel burns at a heat release rate of 450,000 to 800,000 Btu/h ft^3 (4.66 to 8.28 MW$_t$/m^3) developing gas temperatures of more than 3000F (1649C). Heat absorption rates by the water-cooled walls are relatively low because the unit has a relatively small surface protected by a refractory coating — 40,000 to 80,000 Btu/h ft^3 (414 to 828 kW/m^3). The high heat release and low heat absorption rates combine to ensure the very high temperatures needed to complete the combustion and maintain the slag layer in a molten state. The high temperature and high heat release rates also produce high levels of nitrogen oxides (NO_x) which are characteristic of bituminous firing Cyclones. These combustion characteristics also permit occasional overfiring which does not necessarily cause immediate thermal damage to the Cyclone. The Cyclone furnace traps and burns only as much coal as it can handle. The excess passes into the main furnace and boiler back-end as unburned carbon carryover. During overfiring, long term corrosion from iron sulfide attack and associated tube deterioration could occur if the Cyclone is operated under a reducing or oxygen deficient atmosphere. In addition, overfiring may cause problems in the main furnace and convection pass.

Slag layer

The intense radiant heat and high temperatures melt the ash into a liquid slag coating which covers the entire Cyclone interior surface except for the area immediately in front of the secondary air opening. The unit's refractory lining further assists this molten condition by limiting heat absorption to the water-cooled walls. The slag coating is kept hot and fluid by combustion and flows constantly from the Cyclone into the main furnace where it drains through a floor tap opening into a water-filled tank. The slag layer must be constantly replenished by the ash from incoming coal. For this reason bituminous coals fired in Cyclones need a minimum of 6% ash (dry basis); subbituminous coals must contain at least 4% ash (dry basis).

Optimal combustion conditions are achieved when excess air is maintained at about 2-1/2% oxygen (O_2). Cyclone operating temperatures tend to decline at higher and lower levels. However, in actual operating practice, most large Cyclone equipped boilers are fired with higher outlet O_2 levels in order to protect the Cyclones from a possible reducing atmosphere. The primary indicator of Cyclone combustion temperature is the slag temperature at the slag tap. Slag temperature is a function of radiant heat input to the Cyclone. The degree of radiant heat input indicates the portion of the combustion occurring within the Cyclone. Relative slag temperatures can be monitored with an optical or ultraviolet pyrometer from the furnace side ports or by sighting through an inactive Cyclone on double wall-fired units. The traditional method of aiming a pyrometer through the Cyclone front observation port has serious limitations.

Under ideal combustion conditions, the Cyclone can capture approximately 70% of the original fuel ash as slag and drain it to the furnace for disposal in the slag tank. (See Fig. 3.) Smaller boilers can retain more of the ash by capturing additional quantities on the walls and screen tubes. Other details of the Cyclone furnace and boiler arrangements are discussed under *Design Features*.

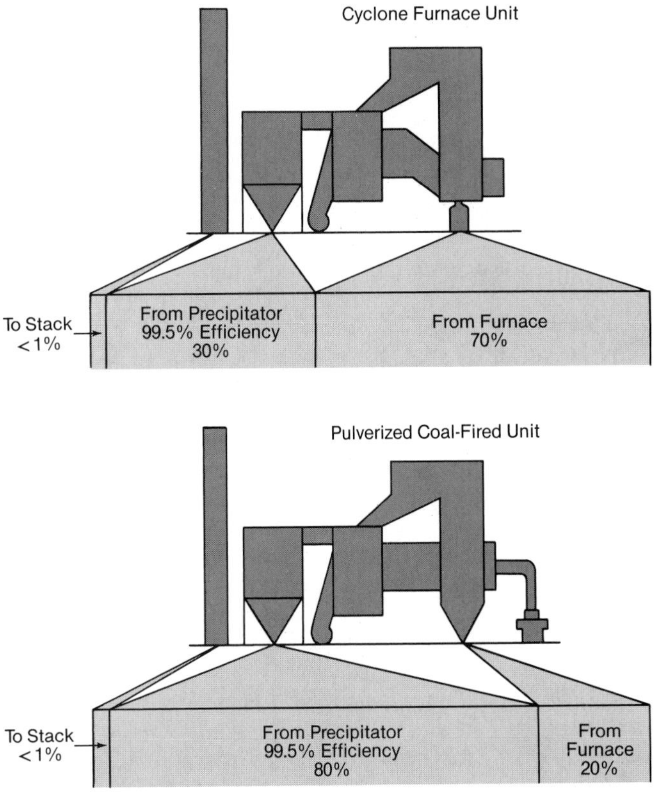

Fig. 3 Comparison of ash distribution from a large Cyclone furnace unit and typical pulverized coal unit.

Suitable fuels

The Cyclone furnace can handle a wide range of coals from low volatile bituminous to lignite depending on the fuel preparation and delivery system. Cyclones have also successfully co-fired solid waste fuels such as wood chips, bark, coal chars, refuse-derived fuel (RDF), petroleum coke and tire-derived fuel (TDF). Fuel oils or gases (natural and other) can be burned in Cyclone furnaces as the primary, contingency or startup fuel.

On a continuous basis the Cyclone must fire either a solid fuel or a liquid/gaseous fuel. Long term firing of oil or gas should only be pursued in a bare tube Cyclone because these fuels can destroy the slag and refractory lining. The resulting debris would severely erode the tubing, leading to eventual pressure part failure. Cyclones switching from coal to oil or gas for an extended period should have all refractory removed, the slag tap direction reversed and all air ducts cleaned. The pin studding (discussed earlier) can remain as it has a minimal effect on heat absorption. When switching back to coal, the refractory must be reinstalled and the slag tap must be set in the correct direction.

Fuel criteria

Coals and co-fired fuels (solid or otherwise) must be evaluated against several criteria. Volatiles must be higher than 15% on a dry basis to ensure stable combustion within the Cyclone. The ash content must be at least 6% for bituminous coals or 4% for subbituminous coals, but can not exceed 25% on a dry basis. With low ash coals the proper slag coating can not be developed, nor can it be maintained. Thin, low viscosity slag coatings do not protect the refractory. To compensate for this condition the slag can be thickened by increasing excess air and reducing the Cyclone combustion temperature if other operating conditions permit.

The moisture content for standard bituminous coal-firing Cyclones should not exceed 20%. Higher moisture content requires better crushing and higher secondary air temperatures to dry the fuel for proper combustion. Particularly moist fuels may require the pre-dry system used on lignite-firing Cyclone-equipped boilers.

Another important coal ash property is its tendency to create iron and initiate a corrosive iron sulfide attack on the refractory and Cyclone barrel. This can be evaluated by comparing the total amount of sulfur to the iron/calcium and iron/magnesium ratios. (See Fig. 4.) Frozen iron puddles found on the Cyclone floor during outages provide an early indication of iron sulfide attack.

Slag viscosity factor — T_{250} value The most important evaluation consideration for Cyclone coals is the slag viscosity characteristic or T_{250} value. This ultimately separates coals into categories suitable and unsuitable for Cyclone operation. The T_{250} value denotes the temperature at which the coal slag has a viscosity of 250 centipoise. At this viscosity the slag flows on a horizontal surface. A slag with a higher viscosity would fail to flow steadily from the Cyclone and main furnace and would also be too stiff to trap the unburned coal particles for proper combustion. The currently recommended maximum T_{250} value for all Cyclone bituminous coals is 2450F (1343C). With the advent of in-

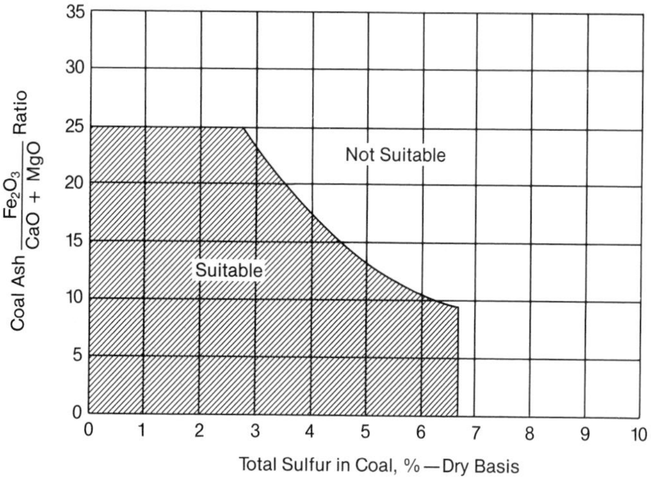

Fig. 4 Coal suitability for Cyclone furnaces based on tendency to form iron and iron sulfide.

Table 1
Summary of Cyclone Furnace Coal Suitability

Ash	Bituminous — minimum	> 6% dry basis
	Subbituminous — minimum	> 4% dry basis
	Maximum	< 25% dry basis
Moisture	Bin firing system (Bit.)	< 20% as-fired
	Direct firing system (Subb.)	< 30% as-fired
	Pre-dry firing system	< 42% as-fired
Volatile		> 15% dry basis
T_{250}	Bituminous	< 2450F (1343C)
	Subbituminous coal	< 2300F (1260C)
Coal type		Bituminous
		Subbituminous
		Lignite
Coal ash iron/sulfur tendencies		See Fig. 4

creasing low sulfur subbituminous coal use in the 1990s, B&W has established a T_{250} limit of 2300F (1260C) for subbituminous coals fired in standard Cyclones. This value compensates for the effect of increased coal moisture on Cyclone combustion temperatures.

The preferred method of establishing the T_{250} value is by experimental measurements on actual fuel ash samples. However, due to the large database accumulated from past testing, the T_{250} value can also be estimated based upon calculations using the coal ash analysis. (See Chapter 8.)

For some coals the T_{250} value of the slag can be lowered by altering its ash base to acid (B/A) constituent ratio (discussed in Chapter 20) through the addition of an agent such as limestone (fluxing). Any such fluxing should be done on the coal conveyor belt prior to the crusher to ensure blending uniformity. B&W does not generally recommend this approach due to the potential of plugging the main boiler furnace and forcing an outage.

If a Cyclone unit must burn high fusion temperature coal, the best approach is to adjust the T_{250} value by blending in a different coal; however, this usually represents higher coal handling costs. Both blending and fluxing create concerns with maintaining acceptable consistency.

Co-fired fuels must be evaluated on an individual basis with emphasis placed on the resulting ash content as a percentage of the total heat input.

The general characteristics of coals suitable for Cyclone furnace firing are summarized in Table 1.

Cyclone coal burner types

In the United States (U.S.), Cyclone furnaces have been equipped with three different types of coal burners — scroll, vortex and radial, as shown in Fig. 5. All types inject coal from the front end of the Cyclone and impart a swirl to the crushed coal in the same rotation as the secondary (main) combustion air. The *scroll* burner which was used in the first Cyclones combines the primary air and coal at the feeder outlet and injects the mixture in a cone shaped distribution pattern. Tertiary air, admitted at the center of the burner, controls the position of the main flame within the Cyclone. However, this injection pattern resulted in a high concentration of recirculating coal particles around the burner outlet, leading to heavy burner wear-block erosion. The premixing of primary air and coal required the addition of a rotary seal to protect the feeder from backflow of hot

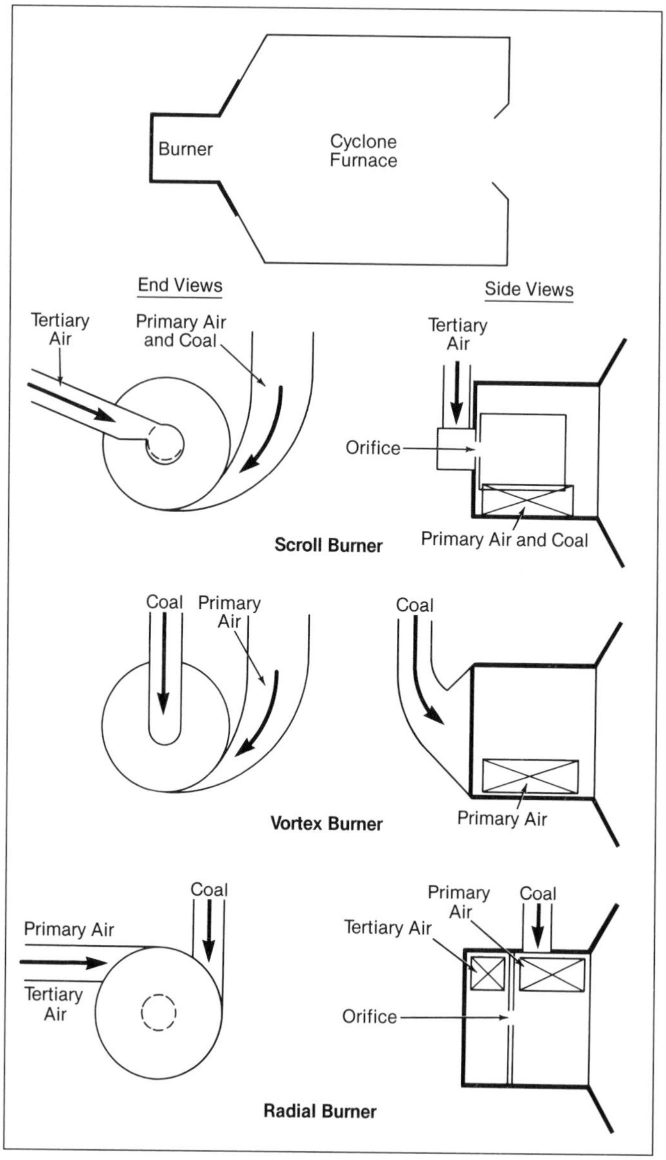

Fig. 5 Three Cyclone coal burners.

combustion air. Later scroll burner applications eliminated the rotary valve by using a higher head of coal at the feeder inlet to prevent hot air backflow into the feeder and coal bunker.

The *vortex* burner eliminated the rotary seal by feeding coal and primary air to the burner separately. The primary air entered the Cyclone burner tangentially as with the scroll design, but the coal poured in at the burner center. This configuration eliminated the tertiary air orifice; however, adjusting feeder seal air had the same effect on the Cyclone vortex zone. The vortex burner produced a similar cone shaped injection pattern as the scroll model. Advanced ceramic wear blocks and a new burner door design have significantly reduced the erosion problem.

The development of the *radial* burner in the 1960s eliminated the severe wear-block erosion. Like the vortex burner, the radial version did not combine coal and primary air until both had entered the burner chamber. However, the coal entered the radial burner tangentially in the same rotation as the primary air. The coal particulate formed a long rope (concentrated stream of coal) as it swept across the burner wear blocks and entered the Cyclone. This approach greatly reduced the concentration of coal recirculating around the burner, effectively reducing wear-block erosion. Tertiary air was admitted again using the same axial entry location as with the scroll burner.

The radial burner remains the modern standard for bituminous coal-fired Cyclones. The scroll burner is utilized on modern lignite-fired Cyclone units because the pre-dry system requires mixing of the primary air and coal during coal preparation, prior to the burner inlet. (See later discussion.) As the scroll, vortex and radial type burners have slightly different coal inlet locations, switching from one burner to another requires a change in coal feeder length.

Design features

Boiler furnace

Commercial Cyclone furnaces have been built in sizes ranging from 6 to 10 ft (1.8 to 3 m) in diameter, with a maximum heat input ranging from 160 to 425 x 10^6 Btu/h (47 to 124.6 MW_t), respectively.

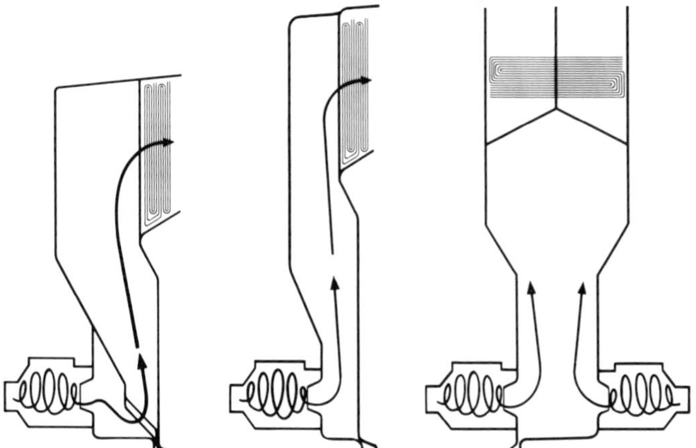

Fig. 6 Firing arrangements used for Cyclone furnaces.

As illustrated in Fig. 6, the Cyclone furnace has been used on three boiler arrangements — single wall firing with screen, open furnace single wall firing and double wall firing. Full boiler drawings are provided in Figs. 1, 7 and 8. Boiler and Cyclone furnace arrangements have ranged from one 6 ft (1.8 m) Cyclone in a single wall unit to twenty-three 10 ft (3 m) Cyclones on an 1100 MW double wall unit. In all cases, the main furnace is relatively small to maintain furnace temperature over the furnace floor slag taps and to promote slag flow on the furnace walls. The lower furnace chamber also contains a protective refractory lining held in place by thousands of pin studs which are welded to the tubes. Generally, the Cyclone boiler can be operated continuously down to about 50% of total capacity. Below this point the slag freezes on the furnace floor and plugs the floor taps. Many units have been equipped with larger floor slag taps.

Coal preparation

The Cyclone equipped boilers feature three fuel delivery systems — bin, direct-fired and direct-fired with predrying configurations. (See Fig. 9.) The bin system, the

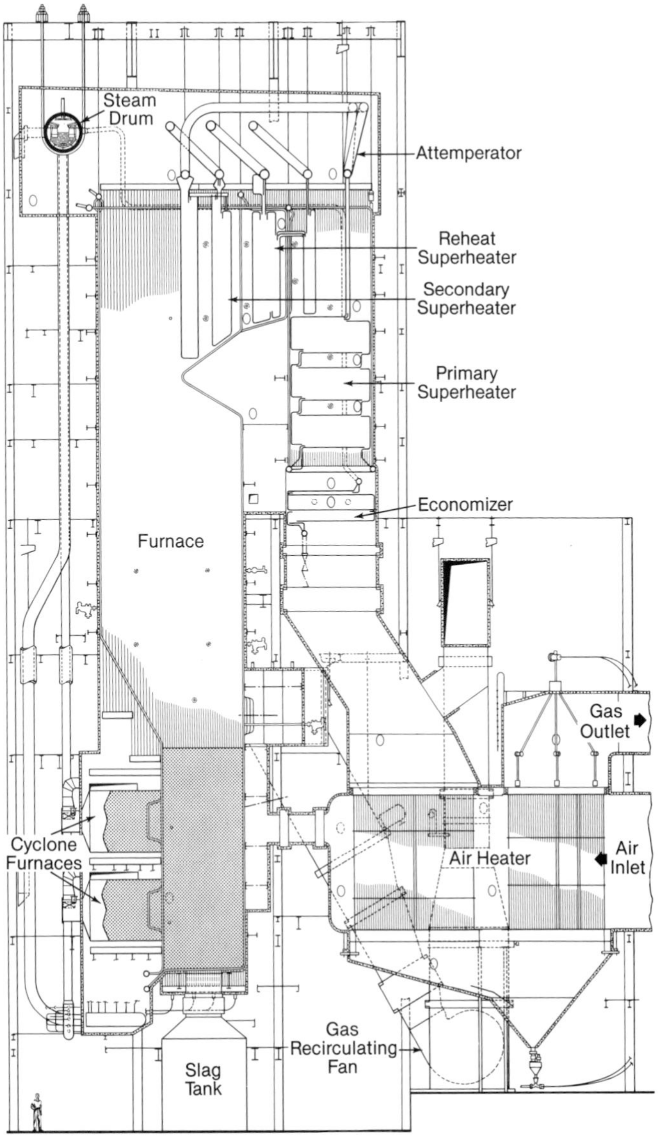

Fig. 7 Radiant boiler with Cyclone furnaces (one wall) and bin system for coal preparation and feeding.

simplest, least expensive and most common, uses a pair of large crushers in a central tower to prepare the coal for overhead storage bunkers (bins). Because the crushed coal has a relatively large particle size, the storage hazards associated with pulverized coal do not exist. The only requirement is adequate venting to remove freshly released combustible gases from the crushed coal. With the bin system, a short crusher outage does not interrupt boiler operations. An existing bin system can be upgraded to fire high moisture coals by converting the coal piping into a simple pre-dry system (see Fig. 9).

The direct-fired system uses a smaller separate crusher, sometimes called a coal conditioner, between the coal feeder and burner on each individual Cyclone furnace. These crushers are swept by hot air which removes moisture from the freshly crushed coal. This produces an advantage by improving crusher performance and fuel ignition when firing coals with a moisture content up to 30%. The direct-fired system has also been used where the existing plant layout could not accommodate the bin system.

The direct-fired with pre-dry system represents the Cyclone fuel preparation system for firing lignite and other high moisture coals. In addition to the hot air-swept individual crushers, this system can also include crusher classifiers and mechanical cyclone moisture separators. The classifiers increase the coal fineness which results in more moisture extraction and better combustion. The mechanical cyclone separators remove the moisture and fines from the coal-air mixture and vent the mixture directly to the boiler furnace through the gas recirculation plenum area. The exclusion of moisture from the main coal increases Cyclone combustion temperatures and subsequently en-

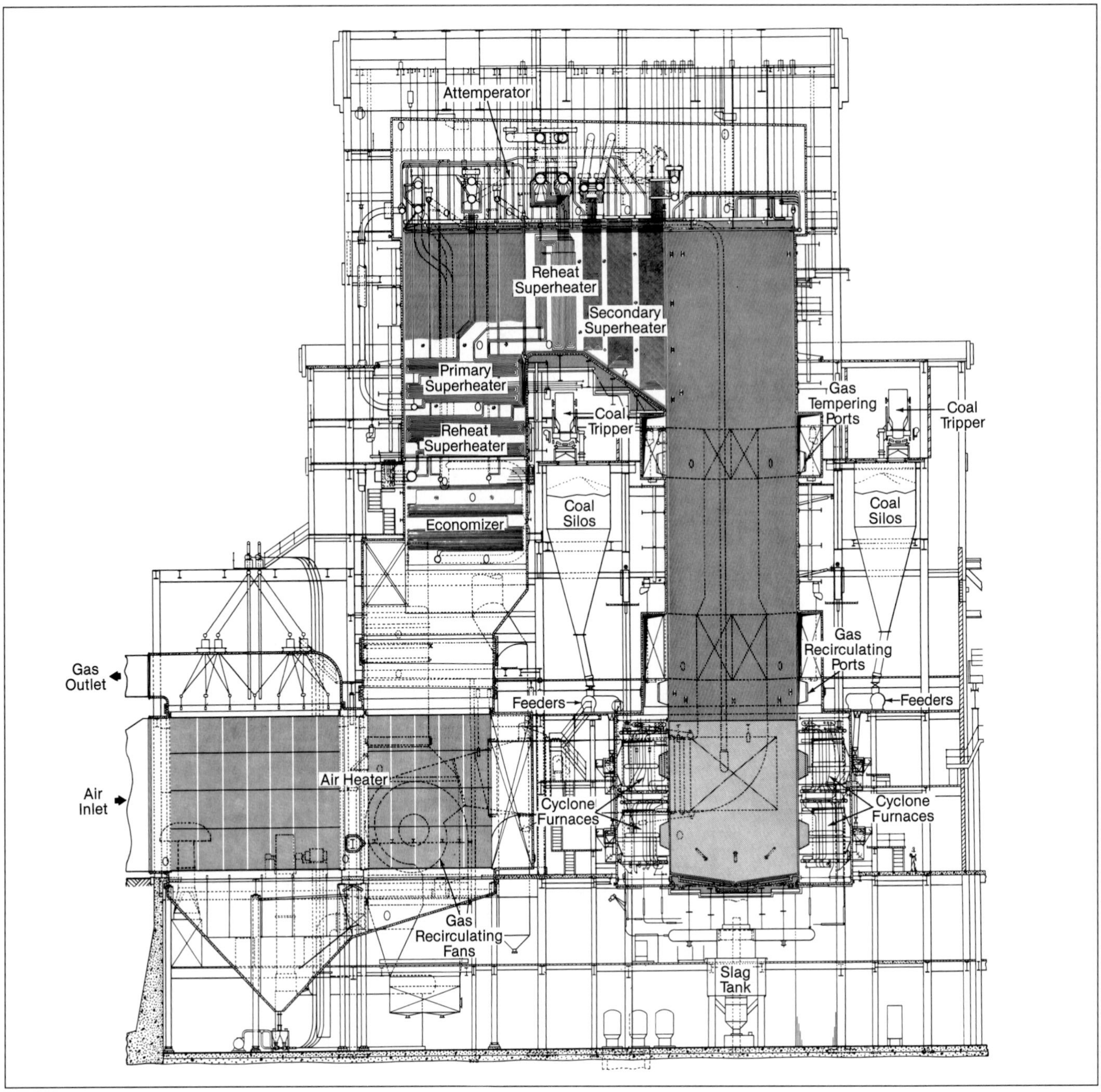

Fig. 8 Universal-Pressure boiler with opposed wall Cyclone furnaces and bin system for coal preparation and feeding.

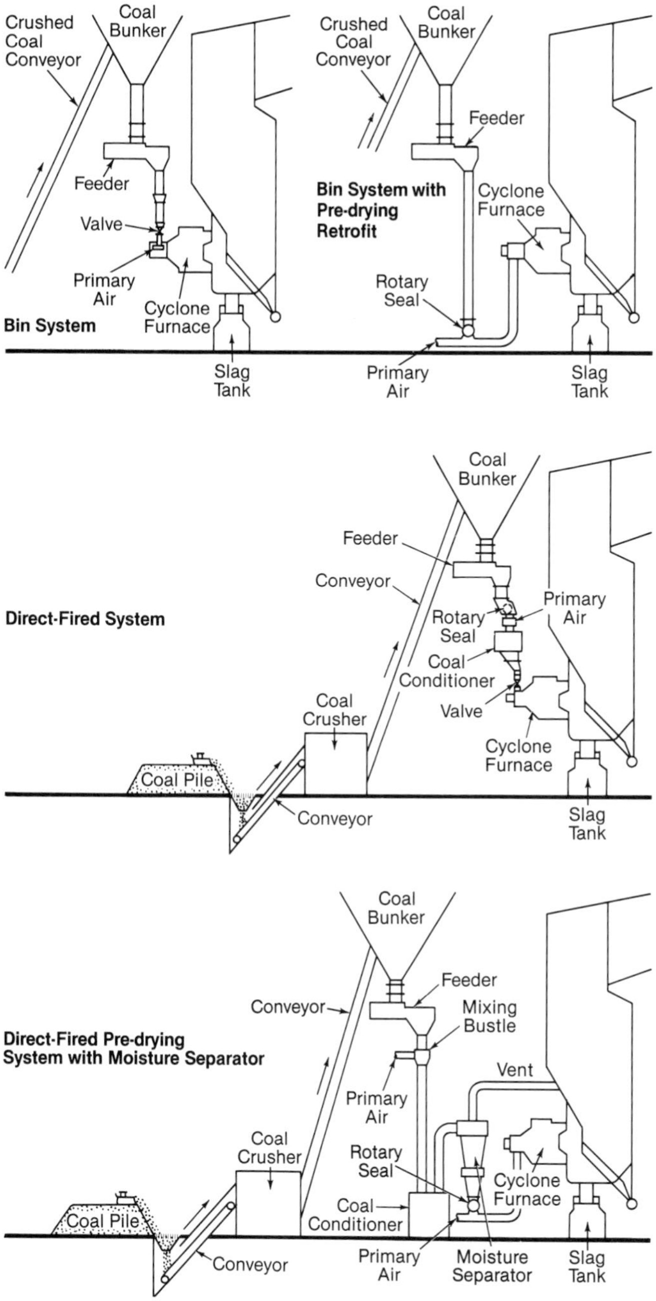

Fig. 9 Bin, direct-fired and pre-drying bypass systems for coal preparation and feeding to a Cyclone furnace.

hances slag tapping. With moisture separators, this system, designed for lignites with moisture higher than 36%, handles poor quality, low sulfur coals as easily as standard bin system Cyclones fire high grade bituminous coal.

The preferred coal size distribution for various coal grades is shown in Fig. 10. Although a Cyclone can only handle a limited amount of pulverized coal, it remains very difficult to crush coal to a fineness that would impair Cyclone combustion. Crushers simply can not approach the fineness capability of a pulverizer. With high moisture, high ash fusion or other difficult-firing coals, every attempt should be made to produce the highest percentage of coal fines.

Coal feeders

A number of different feeders have been used with Cyclones since their inception. Early units had small

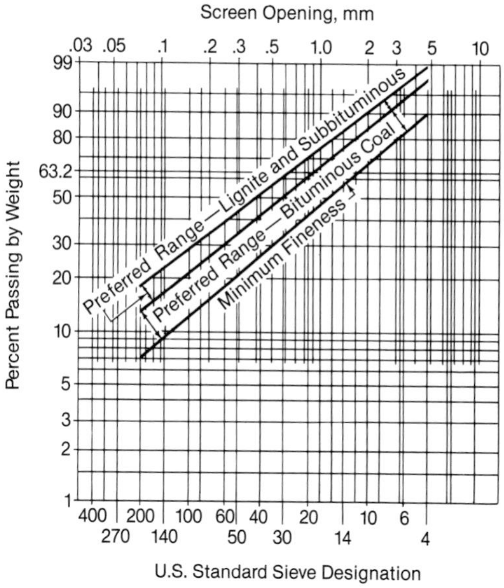

Fig. 10 Crushed coal sizing requirements for Cyclone furnaces.

table type feeders, followed by drag types, then volumetric belt types and now gravimetric belt types, some of which have been upgraded with microprocessor controls. With the introduction of each model the control and continuity of the coal flow improved.

The preferred feeder type today is the gravimetric (weighing) type with a microprocessor control upgrade. Accurate fuel-air control is needed to optimize Cyclone performance — requiring a modern gravimetric feeder.

Slag handling equipment

Cyclone boilers feature a batch removal system for disposing of slag after it is discharged from the main furnace. (See Fig. 11.) The molten slag continuously flows through the furnace floor tap and into a tank beneath the floor where it is quenched and solidified in a water bath. At intervals, the accumulated slag is broken down by a clinker grinder and removed in batches. However, floating slag can create gas buildup at the top of the tank. Similarly, a

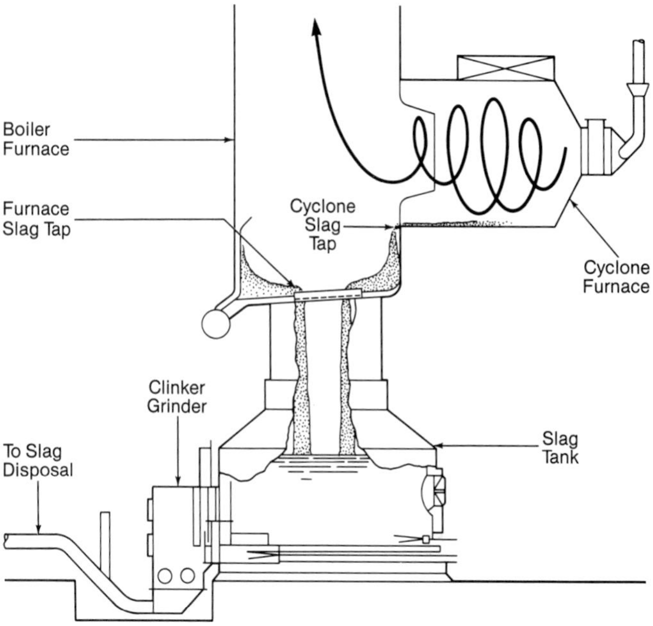

Fig. 11 Batch-removal slag handling system for Cyclone furnace boiler.

plugged or partially closed furnace floor tap can result in carbon monoxide accumulation. Both potentially explosive situations can generally be prevented by installing and properly maintaining a flue gas vent line from the top of the tank to the boiler economizer outlet.

Oil and gas burners

Although coal is the primary Cyclone fuel, oil and gas have been used as startup, auxiliary and main fuels using the combustion system shown in Fig. 12. Due to their high combustion-air pressure drop, Cyclones can not compete with other burner types if gas or oil were the sole primary fuel.

Originally, the oil burner could either be a simple transverse pipe across the secondary air inlet or a single oil gun through the coal burner door. The transverse pipe burner or *roof burner*, has a single row of radial holes directed into the Cyclone tangent. This simple design has proven to be an excellent combustor. However, injecting the oil directly into the Cyclone furnace recirculation zone resulted in an extremely high heat release rate and the bare tubes would experience long term thermal damage. The oil roof burner continues to be used as an emergency Cyclone furnace deslagger for units firing coals with high T_{250} values. It is also used to deslag a Cyclone prior to a boiler outage.

The alternate design uses an oil burner located at the Cyclone centerline firing directly into the Cyclone furnace vortex. This design created boiler *rumbling* problems in some boilers at the upper load range because of the low level of atomization produced by early mechanical atomizers. Large [up to 450 x 10^6 Btu/h (32 MW_t)] air/steam atomizers have been used to successfully reduce boiler rumbling.

The Cyclone gas burner is located in the secondary air throat, injecting gas through three flat nozzles. Although it also injects fuel directly into the recirculation zone, the lower radiant heating of natural gas produces a lower heat absorption rate within the Cyclone furnace. Therefore, long term thermal damage has not been observed with a gas burner. Natural gas assists in deslagging plugged Cyclones, but its low radiant heat transfer rate to the Cyclone does not heat slag deposits as well as oil firing.

Cyclone units firing coal may switch to oil or gas for a short period without serious damage. However, for long term gas and oil operation, the Cyclone should be stripped of all slag and refractory to prevent erosive scrubbing of the tubes as previously discussed.

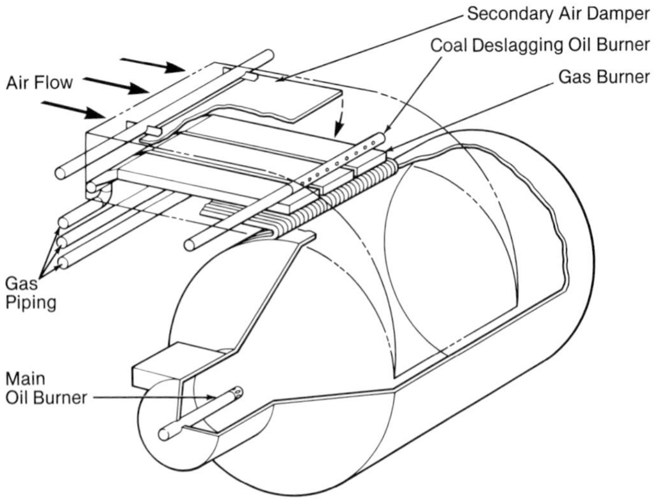

Fig. 12 Arrangement of gas and oil burner in Cyclone furnace.

Waste burning

Due to their high heat release rate, the Cyclone furnaces have always been suited for firing a number of waste fuels with coal. A nominal blend ratio has been typically in the range of 5 to 20% heat input from the co-fired waste. However, waste or refuse fuel should first be thoroughly tested on the prospective Cyclone equipped boiler to determine these limits.

Refuse-derived fuel (RDF) RDF consists of the lighter, more easily burned materials from municipal waste. The ideal point of injection is the secondary air throat to maximize fuel particle time in the Cyclone. A scaled up insulated version of the gas burner nozzle provides the best injector. However, the resistance welded flat studs found in original Cyclones may not withstand the associated particulate erosion.

Tire-derived fuel (TDF) Along with stokers and wet slagging PC units, Cyclones have been viewed as one of the best proposed methods for disposing of scrap automobile tires. Tire rubber provides an excellent low sulfur fuel, but the material's elasticity prevents it from being crushed or pulverized like coal. It generally appears that a 1/2 to 1 in. (13 to 25 mm) sizing is close to optimum. Smaller grinds can be provided by the waste rubber industry, but the cost greatly exceeds that of coal. The TDF should fire easily when mixed with the coal. Although the steel content from bead and radial ply tires does not affect the iron sulfide factor noticeably, the bead wiring may be difficult to melt.

Petroleum coke A number of Cyclone operators have successfully blended petroleum coke with their coal. However, the lack of volatiles in petroleum coke can delay fuel burnout, increasing furnace exit gas temperature and resulting in superheater slagging.

Wood chips Wood chips, sawdust, bark and other similar solid waste products have long been co-fired in Cyclones. Moisture content and sizing of the prospective waste fuel determine the blend ratios.

Combustion controls

The fuel/air ratio at each Cyclone remains the critical item in the combustion control system. When the air flow becomes too low a reducing atmosphere develops. This imbalance leads to a corrosive iron sulfide attack on the refractory and pin studding, eventually resulting in pressure part failure. The balancing of the fuel/air ratio is relatively easy to accomplish and monitor on smaller boilers with only one, two or three Cyclones. Individual ducts supply the total combustion airflow to each Cyclone furnace. On larger units, multiple Cyclones are housed within common windboxes for the supply of secondary air. This arrangement is not well adapted to accurate monitoring of air to individual Cyclone furnaces. As a result, larger Cyclone equipped boilers are typically operated at higher excess air levels than smaller units to reduce the likelihood of forming a reducing atmosphere in one or more Cyclones due to flow imbalances. Modern measurement and control techniques are improving the accuracy of air flow measurement. When retrofitted with accurate gravimetric coal feeds for each Cyclone, such devices permit tighter individual Cyclone furnace control.

Operation

Lighters

Cyclones have been equipped with oil or gas lighters, the former being the most common option. Both lighter types are located at the Cyclone front in the secondary air throat outlet. Original Cyclones used a retractable 17 x 10^6 Btu/h (5 MW_t) oil lighter with a mechanically atomizing nozzle. A cleaner, more powerful model, the MPO Lighter, has been adapted to units where startup operations remain critical. The MPO model produces up to 25 x 10^6 Btu/h (7.3 MW_t), provides cleaner operation due to its air/steam atomizing nozzle and features a separately retractable ignitor.

The gas lighter design remains simple and effective. It has no moving parts and can be uprated from an original output of 17 to 25 x 10^6 Btu/h (5 to 7.3 MW_t).

Sectional secondary air control dampers

Cyclone furnaces can be equipped with sectional dampers across the width of the secondary air inlet. This arrangement provides an additional level of control for the Cyclone furnace combustion. Biasing these dampers permits adjustment to the Cyclone combustion pattern and increases fuel retention time. The effectiveness of this additional control technique is higher when high moisture, low sulfur fuels are fired.

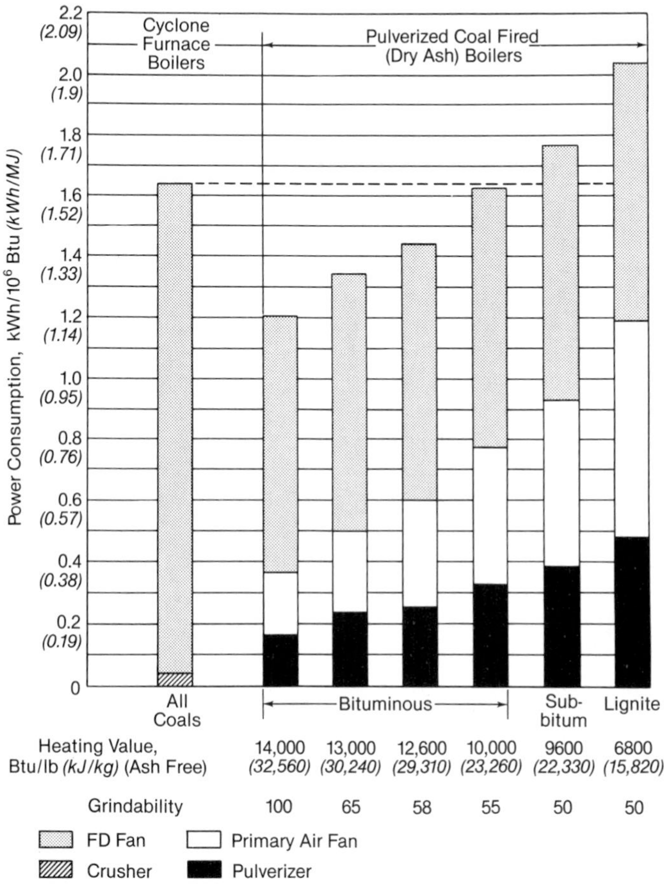

Fig. 13 Auxiliary power requirements for typical high capacity, pressure-fired Cyclone furnace and pulverized coal units.

Low load operations

Typically Cyclone furnaces can not operate below half load without the slag freezing. In addition, the main boiler furnace floor typically stops tapping slag below half load. In either case the slag taps tend to plug solid. An individual Cyclone can continue to operate and should eventually start retapping. However, a boiler with a solidly plugged floor tap must be shut down for manual slag removal.

Power requirements

Cyclone furnace boilers have significantly different power requirements than pulverized coal units. Fuel preparation power consumption for Cyclone boilers is very low because the coal is only crushed, not pulverized, and the primary air fan is not used. However, the forced draft fan power usage is substantially higher due to the relatively high Cyclone furnace pressure drop [typically 20 to 40 in. wg (5 to 10 kPa)]. This is illustrated in Fig. 13. The difference between Cyclone and PC firing is dependent upon the fuel type and heating value. PC firing has an advantage for high heat content, high grindability, bituminous coal firing applications. Cyclone firing has an advantage for low heat content and lower rank fuels such as subbituminous, lignite and brown coals which are also harder to pulverize. In the case of the high heat content bituminous coals, the lower operating costs of PC firing must be balanced against the lower capital costs of Cyclone based firing systems.

Maintenance

Corrosion and erosion within the Cyclone are the two most critical maintenance items. The Cyclone's wet slagging environment produces a potentially corrosive iron sulfide attack on the pressure part tubing.

Erosion is also a problem in an area opposite the secondary air throat where a protective slag coating can not form. The coal particulate wears away the edges of the protective flat studding and can cut a channel between the tangent tubes.

Tubing — pin studs and refractory

In areas coated by molten slag, the tubes are protected by a refractory layer held in place by pin studs (Fig. 14). In addition to retaining the refractory, the pin studs cool the

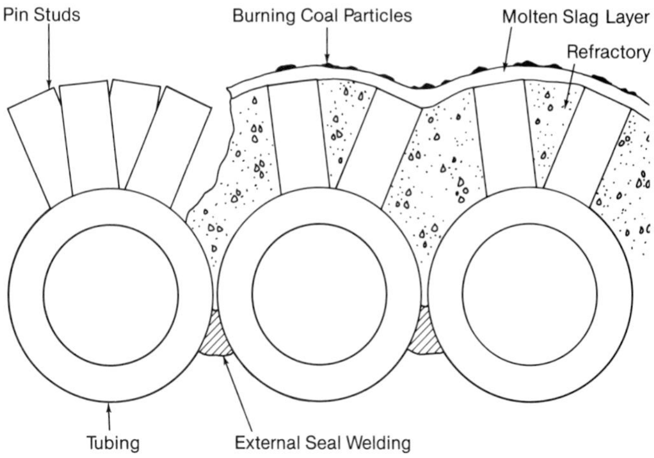

Fig. 14 Cyclone furnace stud and refractory section.

refractory surface in contact with the corrosive slag and retard the corrosive chemical action. The pin studs protect the refractory and the refractory in turn protects the pin studs. Over time, experience has demonstrated that tighter, or more dense, pin stud spacing improves refractory performance. The denser pin stud patterns have provided very good performance in their ability to protect refractory and resist corrosion (Figs. 15a and b).

Tubing — flat studs

Cyclone tubes in the erosive zone opposite the secondary air throat were originally protected from fuel particle erosion by rectangular steel stock welded to the tubes. This design was found to have a number of disadvantages from a maintenance perspective.

An advanced substitute design has been developed and installed by B&W to reduce maintenance. The new flat staggered stud design, using a hand applied fillet weld, offers the following advantages:

1. more precise stud manufacturing and closer spacing,
2. minimum potential for channeling and accelerated wear between studs,
3. excellent heat transfer which reduces metal temperature and erosion rates, and
4. thicker stud sizes to extend life.

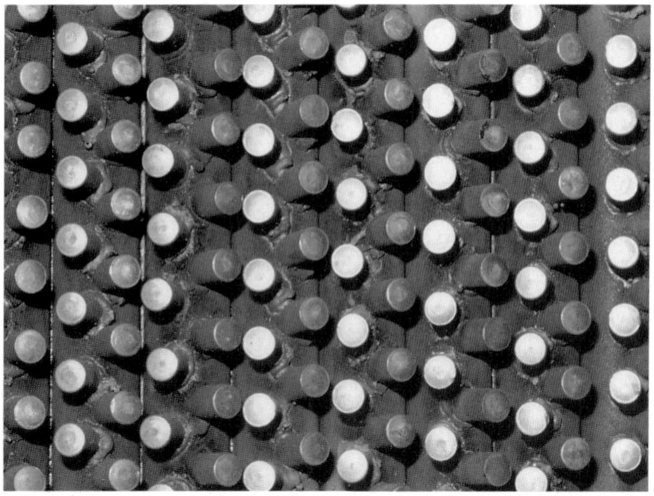

Fig. 15a Recently developed high density Cyclone pin studding.

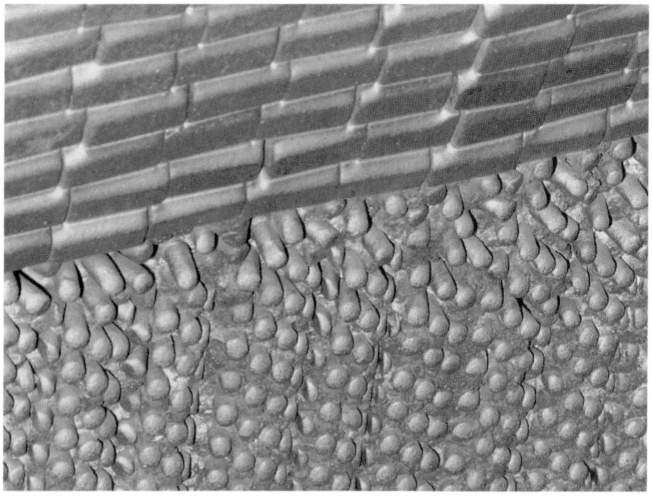

Fig. 15b Modern *super dense* pin stud pattern after five years without maintenance.

Recent field experience with this new design indicates that this technology will last substantially longer than existing techniques, perhaps as long as the life of the Cyclone furnace itself. (See Fig. 16.)

Metallization

To enhance tube life, the plasma arc flame spraying of alloy metals onto the tube surface has been used. Despite a number of experiments with different metallization powders the results generally remain inconclusive. Some applications using expensive coatings resulted in a nickel sulfide attack which consumed the coating in the same manner as iron sulfide corrodes studs and tube surfaces. Furthermore, the pin studding does not lend itself to consistent spray applications. Some coatings prohibit installing replacement pin studs unless the coating and any interacting tube surface have been removed.

Refractory

Field experience has demonstrated that corrosive slag in any form should be kept away from the tubes by a refractory coating. Experience on operating units has proven that the most durable refractories are ram-type high density formulations. The specific refractory selection may be contingent upon the specific plant fuel.

Coal crusher

Cyclone coal crushers have generally remained unchanged over the years. With the increased use of difficult to fire subbituminous low sulfur coals and the need to crush the fuel as fine as possible, a change in maintenance practice is recommended. Cages should be adjusted and mills should be reversed more frequently than with standard Cyclone furnace coals. In addition, the hammers should be discarded at their half life to maintain adequate striking mass.

Additional advancements

Additional retrofit design features continue to be developed to improve the performance and reduce maintenance costs of Cyclone boilers. Seal welding design, wearblock development, stainless steel cooling water jackets for burners, among others, are available for applications to meet site specific needs.

Air pollution control

Flyash

To meet modern particulate emission standards, a Cyclone equipped boiler precipitator must be about the same size as that for a pulverized coal unit. However, the production of more slag and less flyash can offer a significant disposal benefit. Furnace slag is much easier to dispose of than flyash in most cases. The constituent minerals are tightly bound in Cyclone slag minimizing any leachate problems when landfilled. Cyclone slag physical properties have led to its reuse for road bed fill and sandblasting material among other applications.

A number of Cyclones have used flyash re-injection at the burner to burn off any remaining carbon and to convert more of the flyash to slag for easier disposal. Due to the higher wear experienced by the burner and Cyclone,

Fig. 16 State-of-the-art flat studs after five years of service.

the use of flyash re-injection has been minimal. In light of improvements in burner wear blocks and protective flat studding, flyash re-injection remains an option to minimize ash disposal concerns.

Sulfur dioxide (SO_2) reduction

As with other combustion systems, SO_2 emissions from Cyclone units are a function of the sulfur content in the fuel. The easiest method for reducing SO_2 emissions is frequently to switch to lower sulfur coal. In the case of Cyclone furnace boilers, additional care is required in such a change because Cyclone furnaces are more sensitive to the ash composition, ash quantity, heating value and especially moisture. If high moisture subbituminous or lignite fuels are to be substituted for low moisture bituminous coal, it may be necessary to install the direct-fired pre-dry system. This system has been successfully used for lignite coal-fired Cyclone furnaces.

Provided that the fuel selected meets the T_{250} value discussed earlier, a standard Cyclone equipped boiler can potentially be modified to handle low sulfur, moist subbituminous or lignite coals through a series of hardware and operational changes. Key changes focus on preventing coal fineness from deteriorating, maintaining Cyclone furnace temperatures as high as possible, continuing an aggressive maintenance program and making provision for rapid deslagging. The actions required are unit specific and are usually established following an engineering study and test fuel burn.

Coal washing systems may be installed to remove pyritic sulfur. Although washing makes the coal more difficult to burn as it raises the T_{250} value, SO_2 emissions from combustion of washed coal are reduced.

Nitrogen oxides (NO_x) reduction

Historically, Cyclone equipped boilers have produced relatively high levels of NO_x emissions, typically ranging from 0.8 to 1.9 lb of NO_x as NO_2 per million Btu (1201 to 2850 mg/Nm^3) input, corrected to 3% O_2. Selective catalytic reduction (SCR) equipment, discussed in Chapter 34, can be applied to Cyclone equipped boilers. Special combustion modifications may also be applied.

Two-stage combustion (see Chapter 13) can reduce NO_x levels up to 50% as in the case of pulverized coal units. However, the reducing conditions in the Cyclone furnace have the strong probability of resulting in unacceptable corrosion.

Low NO_x Cyclone reburn technology has been specifically developed to reduce NO_x emissions levels from Cyclone equipped boilers while permitting successful Cyclone furnace operation. The low NO_x Cyclone reburn concept is illustrated in Fig. 17. In this system, the Cyclone furnace is operated under fully oxidizing conditions, but at a reduced load — typically 65 to 85% of full load air and fuel flow. The balance of the fuel is injected directly into the main boiler furnace with minimum air for fuel transport to create a reburning zone. In this zone, the reburn fuel creates an oxygen deficient or reducing zone where the NO_x created in the Cyclone furnace is reduced or decomposed into molecular nitrogen through a series of complex interactions with free hydrocarbon radicals. Overfire air ports located above the reburn zone permit injection of the balance of the air to produce a final stoichiometry of 1.15 to 1.20 and complete the fuel combustion. The use of coal, oil or gas as the reburn fuel depends upon an economic evaluation balancing the higher capital cost of a coal based system against the fuel cost differential for oil or gas.

Applications

The Cyclone boilers gained wide acceptance due to their ability to burn a substantial reserve of coals deemed unsuitable for pulverized coal firing. The addition of low NO_x Cyclone reburn technology and potentially postcombustion deNO_x technology should permit the retention of the Cyclone boiler as a major source of steam power.

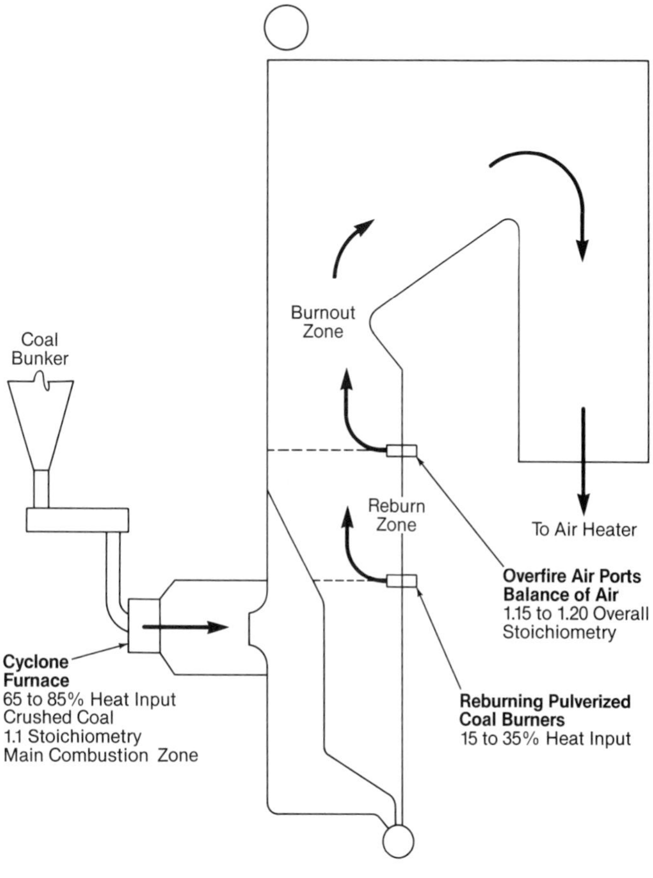

Fig. 17 Cyclone furnace reburn NO_x control system using pulverized coal as the reburn fuel.

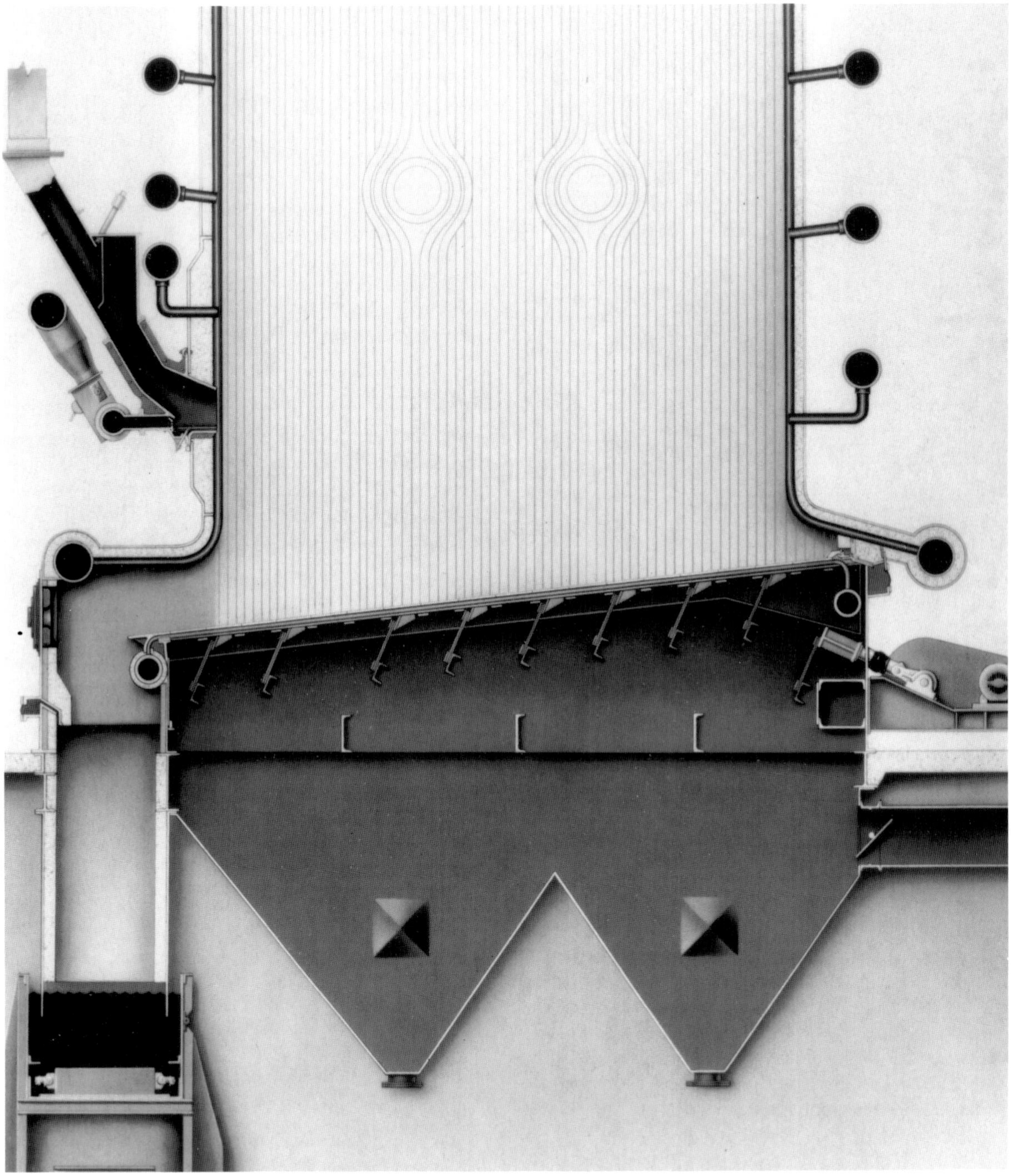

Vibrating grate stoker.

Chapter 15
Stokers

Many technologies have evolved to convert a wide variety of fuels to alternate forms of energy. One important technology is the mechanical stoker. All stokers are designed to feed fuel onto a grate where it burns with air passing up through it. The stoker is located within the furnace and is designed to remove the ash residue after combustion. Evolving from the hand-fired boiler era, mechanical stoker/grate system designs are available to burn a wide range of fuels for industrial, small utility and cogeneration applications. Fuels range from all forms of coal to wood wastes, bagasses, rice hulls, peach pits, almond shells, vineyard trimmings and coffee grounds as well as residential and commercial refuse. Modern mechanical stoker firing systems are composed of:

1. a stoker or fuel admission system,
2. a stationary or moving grate assembly to support the burning mass of fuel and admit most of the combustion air to the fuel,
3. an overfire air system to complete combustion and limit atmospheric pollutant emissions, and
4. an ash or residual discharge system.

These components are then integrated into the overall furnace design to optimize combustion and heat recovery while minimizing unburned fuel, atmospheric emissions and cost.

A successful installation requires selecting the correct type and size of stoker for the fuel being used and for the load conditions and capacity being served. There are two general types of systems, *underfeed* and *overfeed*. Underfeed stokers supply both the fuel and air from under the grate while overfeed stokers supply fuel from above the grate and air from below the grate. The overfeed stokers are further divided into two types, *mass feed* and *spreader*. In the mass feed stoker, fuel is continuously fed to one end of the grate surface and travels horizontally across the grate as it burns. The residual ash is discharged from the opposite end. Combustion air is introduced from below the grate and moves up through the burning bed of fuel. In the spreader stoker, combustion air is again introduced primarily from below the grate but the fuel is thrown or spread uniformly across the grate area. The finer fraction of the fuel burns in suspension as it falls against the upward moving air flow. The remaining heavier fraction of the fuel burns on the grate surface with any residual ash removed from the discharge end of the grate.

There is little demand in today's market for the underfeed and small mass overfeed coal-fired units because of cost and environmental considerations. This market has been replaced with shop assembled oil- and gas-fired units and to some extent by overfeed spreader-stoker systems.

The stoker/grate systems are provided in many mechanical configurations depending upon the manufacturer. Table 1 summarizes several variations of basic stoker designs by type, fuel, heat release rate and approximate largest capacity available.

For a given boiler steam capacity, the typical fuel burning rates in Table 1 generally determine the plan area of the grate and furnace in which it is installed. Practical considerations limit stoker size and, consequently, the maximum steam generation rates. For coal firing, this maximum is about 350,000 lb/h (44.1 kg/s); for wood or biomass firing it is about 700,000 lb/h (88.2 kg/s).

Almost any coal can be burned on some type of stoker. Many other solid fuels such as refuse, wood, bark, coffee grounds, rice hulls and orchard prunings can be burned alone on a grate or in combination with fuels such as pulverized coal, oil or natural gas.

The spreader stoker, in combination with the various grate types, is most commonly used for a steaming capacity range from 75,000 to 700,000 lb/h (9.5 to 88.2 kg/s). It responds rapidly to changes in steam demand, has good turndown capability and can use a wide variety of fuels. It is not, however, suitable for low volatile fuels such as anthracite and petroleum coke because of carbon burnout problems.

Underfeed stokers

There are two general types of underfeed stokers, the horizontal feed side ash discharge type shown in Fig. 1 and the gravity feed rear ash discharge type shown in Fig. 2.

In the side ash discharge type, coal is fed from a hopper to a central trough, called a *retort*, by a screw or a ram pusher. In the larger units, a ram assisted by pusher blocks or a sliding retort bottom (fuel distributors) moves the fuel upward and into the retort. As the coal moves upward and over the retort edges and spreads out over the active grate area, it is exposed to air and radiant heat. Drying occurs and distillation of volatiles begins. As the coal moves to the sides and/or rear, the distillation is completed, leaving coke

Table 1
Stoker/Grate System Overview

Stoker Type	Grate Type	Fuel	Typical Release Rate 1000 Btu/h ft^2 (MW_t/m^2)		Steam Capacity 1000 lb/h (kg/s)	
Underfeed:						
Single retort	--	Coal	425	(1.34)	25	(3.15)
Double retort	--	Coal	425	(1.34)	30	(3.78)
Multiple retort	--	Coal	600	(1.89)	500	(63.0)
Overfeed:						
Mass	Vibrating: water-cooled	Coal	400	(1.26)	125	(15.8)
	Traveling chain	Coal	500	(1.58)	80	(10.1)
	Reciprocating	Refuse	300	(0.95)	350	(44.1)
Spreader	Vibrating:					
	air-cooled	Coal	650	(2.05)	150	(18.9)
		Wood	1100	(3.47)	700	(88.2)
	water-cooled	Wood	1100	(3.47)	700	(88.2)
	Traveling	Coal	750	(2.37)	390	(49.1)
		Wood	1100	(3.47)	550	(69.3)
		RDF*	750	(2.37)	400	(50.4)

* Refuse-derived fuel

which is burned out near the edges or end of the grate. High pressure overfire air is used to produce high turbulence and reduce smoke.

Burning coal in this fashion increases the probability of clinkering (producing large agglomerates of ash slag) or matting (layers of ash slag). To reduce this tendency alternate fixed and moving grate sections are applied to the underfeed stoker design to agitate the fuel. Coal characteristics are critical to underfeed stoker performance. Table 2 outlines coal specifications for stationary and moving grates, although underfeed stokers have burned coals outside of these guidelines. To burn these coals, some deviation from the normal maximum grate release rate may be required. A reduction in the percentage of fines helps to keep the feed bed porous and extends the range of coals with a higher coking index.

With suitable coal, single and double retort units are generally limited to 25,000 to 30,000 lb/h (3.2 to 3.8 kg/s) steam flow. Typical grate release rates are 425,000 Btu/h ft^2 (1.34 MW_t/m^2) with water-cooled walls and 300,000 Btu/h ft^2 (0.95 MW_t/m^2) with refractory walls. Capacities up to 500,000 lb/h (63 kg/s) steam are possible with multiple retort rear ash discharge units similar to that shown in Fig. 2. Grate release rates up to 600,000 Btu/h ft^2 (1.89 MW_t/m^2) are practical with a 20 to 25 deg (0.35 to 0.44 rad) grate inclination from the horizontal.

Mass feed stokers

Two types of mass feed stokers are used for coal firing, the water-cooled vibrating grate and the moving (chain and traveling) grate stokers. Another mass feed stoker system used for municipal refuse is discussed in Chapter 27.

Mass feed stokers are characterized by the gravity feed of fuel onto an adjustable grate that controls fuel bed height. The method of firing involves a fuel bed that moves along a grate with air being admitted under the grate perpendicular to the fuel flow. As it enters the furnace, the layer of coal is heated by furnace radiation to drive off volatiles and to promote ignition. The coal continues to burn as it is conveyed along the depth of the furnace. The fuel bed decreases in thickness until all the fuel has burned and cool ash discharges into a pit. With this method of fuel entry, the undergrate air must be sectionalized along the length of the grate, because the quantity of air required for ignition, burning and burnout are different and must be regulated. This method of fuel entry and combustion inherently produces low ash carryover. However, it is more sensitive to variations in fuel characteristics that affect ignition without a larger ignition arch. Mass feed stokers require nonsegregating coal feed hoppers. Without them, the fines migrate to the sides and severe clinkering along the sidewalls can occur.

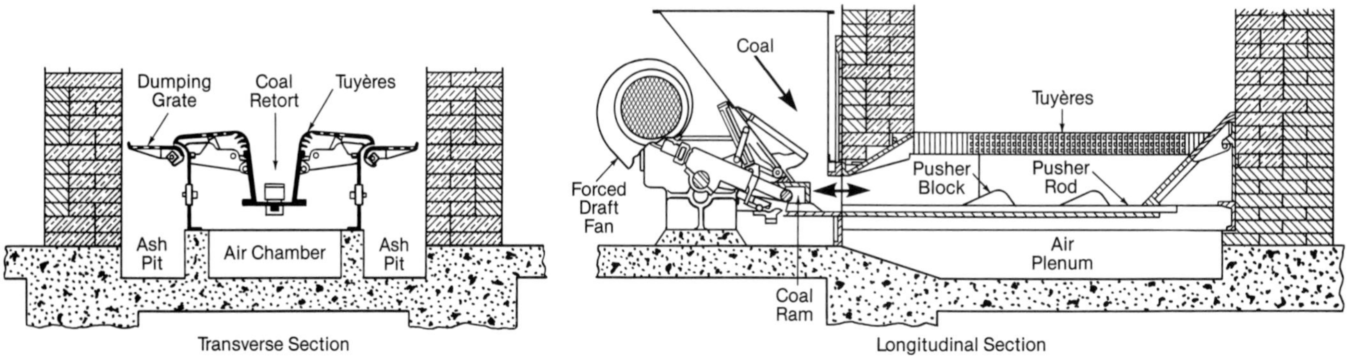

Fig. 1 Single retort underfire stoker with horizontal feed, side ash discharge.

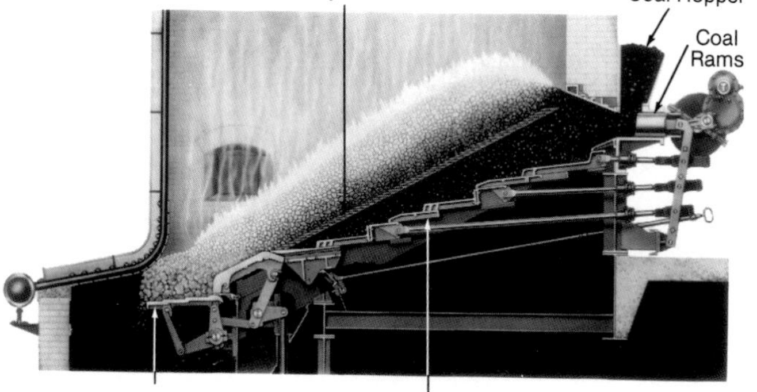

Fig. 2 Multiple retort gravity fed underfeed stoker with rear ash discharge.

A water-cooled vibrating grate stoker is shown in Fig. 3. The vibrating grate consists of tuyère grate surface mounted on and in intimate contact with a grid of water tubes. The grate is connected to the boiler circulation system for cooling. The entire structure is supported by a number of flexing plates that allow the grid and its grate to move freely in a vibrating action that conveys the coal from the feed hopper to the ash discharge. Vibration of the grates is intermittent and is adjustable to convey fuel into the furnace and control ash bed thickness and discharge as needed. A rear arch extends over approximately the rear third of the grate as shown in Fig. 3. It assists burnout and directs the higher excess air gases forward to mix with the rich volatile gases from the ignition zone. A short front arch is adequate for most coals. If ignition is inadequate due to low volatile fuel, refractory can be added to the short arch to increase radiation and assist ignition.

High pressure air, up to 30 in. wg (7.5 kPa), is injected through the front arch to promote turbulence and combustion. Water cooling of the grates makes this stoker more flexible with gaseous and liquid fuels, because a shift to either does not require special grate protection other than the normal bed of ash left from coal firing. Burning rates of these stokers vary with different fuels; in general, the grate release rate should not exceed 400,000 Btu/h ft^2 (1.26 MW$_t$ /m^2). Due to the limited number of moving parts, this stoker/grate system typically requires little maintenance.

Chain and traveling grate stokers, shown in Fig. 4, are similar to each other. Both form an endless belt arrangement that passes over drive and idler sprockets or return bends. They both convey coal from the hopper through the furnace to the ash discharge and return under the grate. In the chain grate design, the chain continues around to the return side. The traveling grate unit differs in that it uses a grate bar to provide better control of sifting (fine ash falling through the grate) when firing anthracite. The moving chain stoker requires more maintenance than the water-cooled vibrating grate unit.

Chain and traveling grate stokers can burn a wide range of solid fuels — peat, lignite, subbituminous, bituminous, anthracite coals and coke breeze. Typical coal characteristic ranges are provided in Table 3. Generally, these stokers use furnace arches (front and/or rear, not shown in Fig. 4) to improve combustion by reradiating heat to the fuel bed. When burning low volatile anthracite or coke breeze, rear arches direct the incandescent fuel particles and combustion gases toward the front of the stoker, where they assist ignition of the incoming fuel.

Burning rates on chain and traveling grate stokers vary with different fuels. The lower ash (8 to 12%) and lower moisture (10%) fuels permit rates to 500,000 Btu/h ft^2 (1.58 MW$_t$/m^2); higher moisture (20%) and higher ash (20%) fuels would limit the rate to 425,000 Btu/h ft^2 (1.34 MW$_t$ /m^2). For low volatile anthracite, the rate should not exceed 350,000 Btu/h ft^2 (1.10 MW$_t$ /m^2).

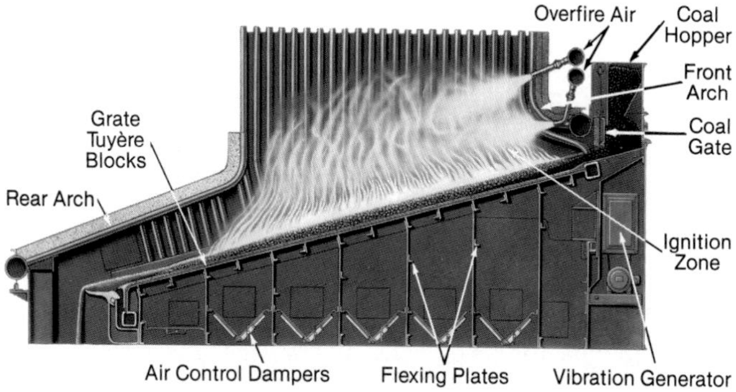

Fig. 3 Water-cooled vibrating grate stoker.

Table 2
Typical Undergrate Feed Stoker Coal Characteristics

	Stationary Grate	Moving Grate
Moisture	0 to 10%	0 to 10%
Volatile matter	30 to 40%	30 to 40%
Fixed carbon	40 to 50%	40 to 50%
Ash	5 to 10%	5 to 10%
Btu/lb (kJ/kg), as-fired	12,500 (29,075) min	12,500 (29,075) min
Free swelling index	5 max	7 max
Ash softening temperature*	2500F (1371C)**	2500F (1371C)**
Coal size, in. (mm)	1 (25.4) top size x 0.25 (6.4) max 20% through 0.25 (6.4) with round screen	Equal portions: –0.25, 0.25 to 0.5, 0.5 to 1.0 (–6.4, 6.4 to 12.7, 12.7 to 25.4)

* The *ash softening temperature* here is the temperature at which the height of a molten globule is equal to half its width under reducing atmosphere conditions.

** Below 2500F (1371C) the moving grate is derated linearly to 70% of its rated capacity at 2300F (1260C) ash fusion temperature. Stationary grates are derated linearly to 70% at 2100F (1149C) ash fusion temperature and use steam for tempering below about 2400F (1316C) fusion temperature.

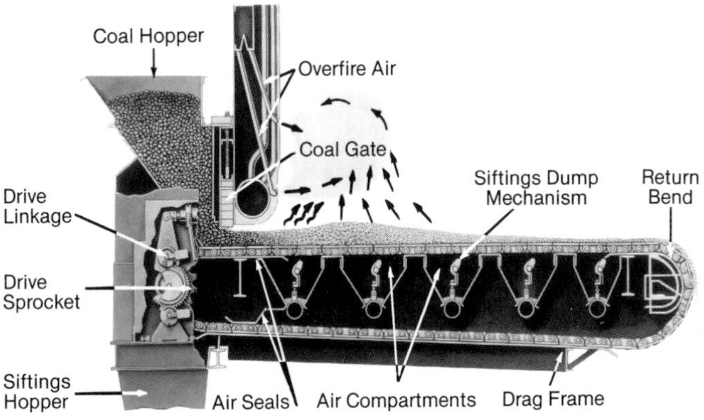

Fig. 4 Chain grate stoker (*courtesy of Detroit Stoker Company*).

Spreader stokers

In the spreader stoker, the fuel is uniformly thrown into the furnace across the grate area. Fines ignite and burn in suspension; the coarser fuel particles fall to the grate and combust on a thin, fast burning bed. Because the fuel is evenly distributed across the active grate area, the air is uniformly distributed under and through the grate. The undergrate air plenums may or may not be compartmented depending on the grate type and application. A portion of the total combustion air is admitted through ports above the grate as overfire air. The modern spreader stoker is the most versatile and the most commonly used stoker system.

Spreader coal firing

Fig. 5 shows a modern boiler equipped with a traveling grate spreader stoker and designed to fire an eastern bituminous coal. A straight, water-cooled membrane furnace wall construction minimizes refractory. For the typical stoker coal-fired application, ignition or combustion arches are not used. The installation consists of:

1. state-of-the-art feeder-distributor units which distribute fuel uniformly over the grate,
2. specifically designed air metering grates,
3. dust collection and reinjection equipment,

Table 3
Typical Mass Feed Stoker Coal Characteristics

	Water-Cooled Vibragrate	Chain/Traveling Grate
Moisture	0 to 10%	0 to 20%
Volatile matter	30 to 40%	30 to 40%
Fixed carbon	40 to 50%	Remainder
Ash	5 to 10%	6 to 20%
Btu/lb (kJ/kg), as-fired	12,500 (29,075) min	10,500 (24,423) min
Free swelling index	—	5 max
Ash softening temperature*	2300F (1260C)	2100F (1149C)
Iron oxide, % in ash	20% max	20% max
Coal size, in. (mm)	1 to 0.75 (25.4 to 19) x 0 max 40% through 0.25 (6.4) screen	1 to 0 (25.4 to 0) top size max 60% through 0.25 (6.4) screen

* See Table 2 for definition.

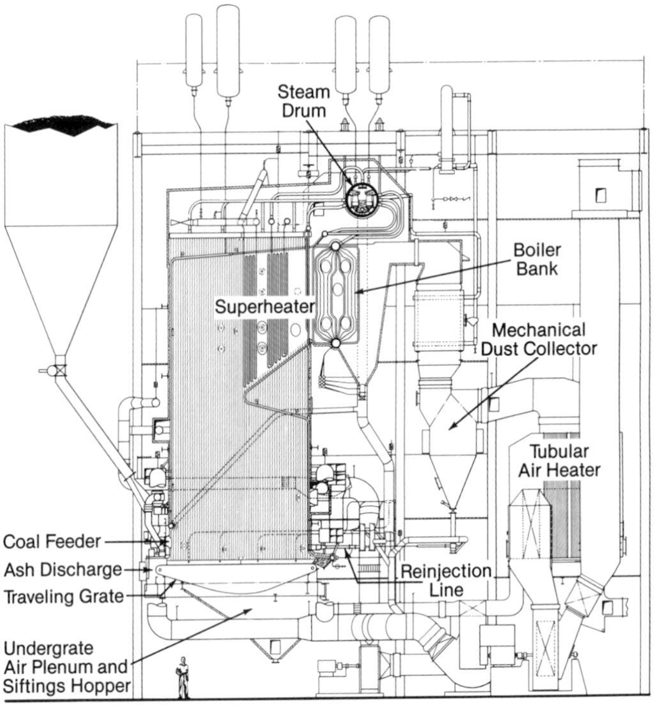

Fig. 5 Typical spreader-stoker coal-fired boiler: 290,000 lb/h (36.5 kg/s) steam.

4. a combustion air system including forced draft fans for undergrate and overgrate air, and
5. combustion controls to coordinate fuel and air supply with steam demand.

Spreader feeders

Spreader feeders have the capability to uniformly feed coal into a device that can propel it along the depth of a grate in an evenly distributed pattern. Many designs have been used successfully over the years. The coal feed mechanisms include gravity, reciprocating plates and metering chain conveyors. The mechanisms that propel the coal into the furnace include steam and air injection as well as underthrow and overthrow rotors. Steam or air assist can be used with the rotor systems. Fig. 6 shows a feeder-distributor with overthrow rotor. The metering chain moves coal from a small hopper to fall on an over-

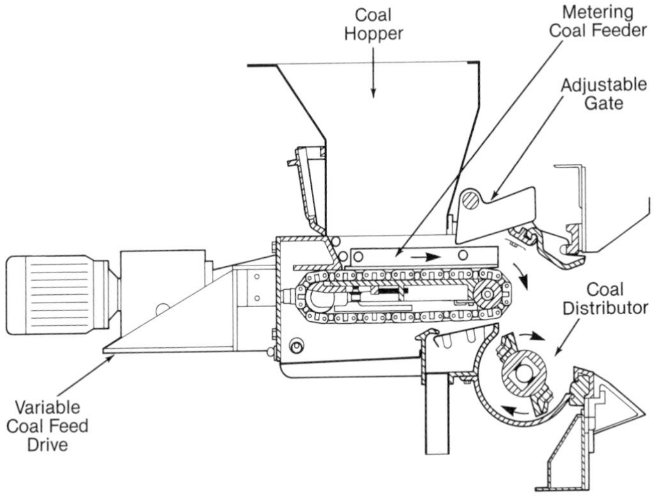

Fig. 6 Detroit Stoker chain-type coal feeder Model OT (*courtesy of Detroit Stoker Company*).

throw rotor. The rotor is equipped with curved blades to uniformly distribute the coal over the grate area. Although spreader designs vary, the overthrow design has been the most common.

With increasingly stringent emission regulations, stoker manufacturers have been studying design variations that reduce the formation of nitrogen oxides (NO_x) and carbon monoxide (CO). Fig. 7 depicts a state-of-the-art mechanical-pneumatic feeder-distributor with underthrow rotor. In this device, the coal is fed from a hopper by a metering chain into an underthrow rotor with air assist. This has improved the feed and distribution capabilities, and 10 to 15% reductions in NO_x have been recorded.

Grates for spreader stokers

As with mass stokers, there has been a wide range of grates used in spreader-stoker coal firing. Stationary and dumping-type units are no longer used. Traveling and/or vibrating air-cooled grates are the most common. The grate shown in Fig. 5 is a high resistance air metering grate. The resistance air metering concept eliminates the need for undergrate air plenum compartmentation for good air distribution and control. Moving adjustable air seals are provided at the front and rear. The grates are bottom supported and therefore require an expansion joint at the interface to the water-cooled furnace. The grate is a continuous moving chain. It consists of a series of chains to which are attached the grate bars, which contain the air metering holes. The grate bars are contoured to interface with adjacent bars, minimizing air leakage between bars. The chains and grate bars are supported by and slide on grate rails. The sliding interface between the grate bar and rail also serves as a seal to prevent excessive air from bypassing the air admission nozzles. The grate travels from rear to front, or towards the fuel feed end. This permits the optimum fuel distribution pattern and the maximum residence time for burnout of larger fuel particles. The ash is continuously conveyed and spills off the end of the grate into a hopper.

Although the traveling grate is a durable and proven design, it has many moving parts and is subject to wear. To minimize wear, the speed of the grate should generally be kept below 40 ft/h (12.2 m/h). This may limit its use in some high ash fuel applications or it may require limiting the grate plan area release rate and input per unit width of stoker.

The drive of the grate is typically through the front sprocket and shaft to keep the top of the chain in tension; mechanical and hydraulic drive systems are available. Some systems use a ratchet concept, which results in loading and unloading the chain at each stroke. This leads to higher grate component wear than does a continuous and uniformly torqued drive system.

Other types of traveling grates are similar to the chain grate shown in Fig. 4. Many of these consist of a chain with several links to form an endless belt. Air admission is through the gaps in the links. These types of traveling grates typically have compartmented undergrate air plenums to control air distribution.

Air-cooled vibrating grates, similar to those shown in Fig. 3, are also used in spreader-stoker applications. Except for a smaller size limitation, the application to spreader coal firing is similar to that of the traveling grate.

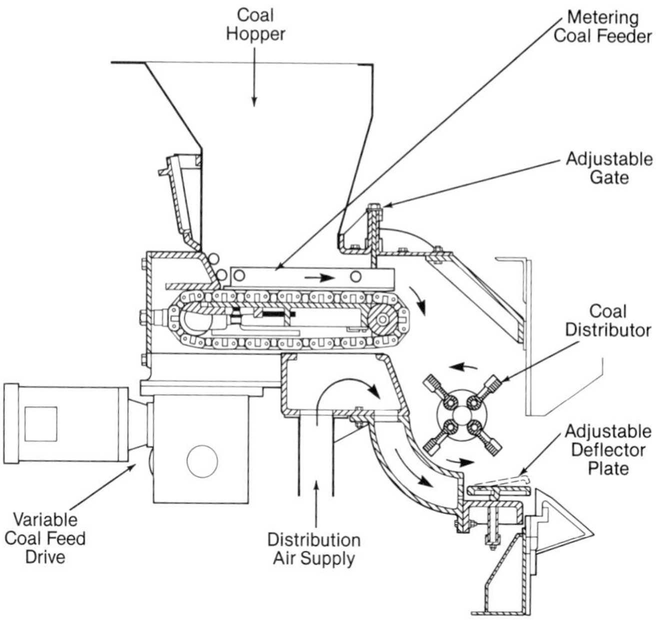

Fig. 7 Detroit Stoker chain-type coal feeder Model UT (*courtesy of Detroit Stoker Company*).

Carbon reinjection systems

The high degree of suspension burning results in greater carryover of partially combusted fuel particles. To achieve the highest efficiency, these particles are captured in a dust collector and returned to the furnace for complete combustion. The system on the unit in Fig. 5 is pneumatic. The carbon particles and some ash are collected in a mechanical dust collector and routed to a pickup box, where air is injected to convey the carbon back to the furnace. Multiple injection ports are located across the width of the unit to uniformly mix the unburned carbon into the combustion zone for enhanced burnout. Reinjection of significant ash is undesirable because it can contribute to boiler surface erosion and grate clinkering. Because a mechanical dust collector is more effective in collecting larger particles and because the majority of the unburned carbon is the larger size fraction of the total carryover, the bias towards collecting the unburned particles can be effected by limiting the design efficiency of the mechanical collector. Carbon reinjection improves coal-fired boiler efficiency by 2 to 4%.

Combustion air system

In spreader stoker-fired units, there is typically 25% excess air at the furnace exit at the design full load input. This air is split between the undergrate air, overfire air and stoker distribution air. Due to the high degree of suspension burning, air is injected over the fuel bed for mixing to assist the fuel burnout and to minimize smoking. This dictates that 15 to 20% of the total air be used as overfire air. This air is injected at pressures of 15 to 30 in. wg (3.7 to 7.5 kPa) through a series of small nozzles arranged along the frontwalls and rearwalls. (See Fig. 8.)

Spreader-stoker firing, with the air split between undergrate and overfire, is a form of staged combustion and is effective in controlling NO_x. By deeper staging, further reductions are possible. To take advantage of this characteristic, the modern unit shown in Fig. 5 is designed for a total excess air of 25%. The total air flow is

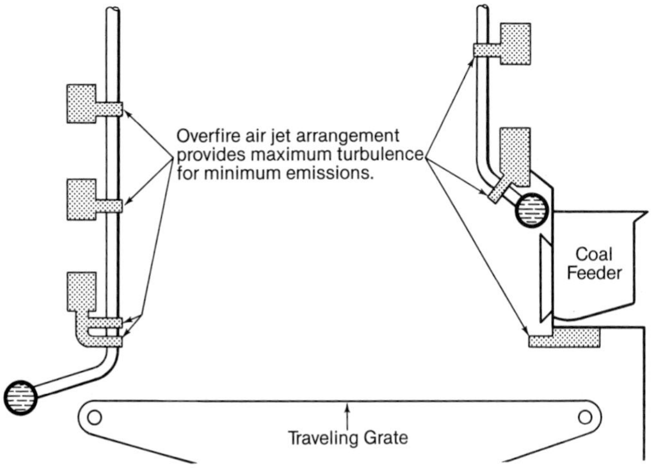

Fig. 8 Stoker overfire air system.

split 65% undergrate and 35% overfire. The 35% includes any air to the coal feeders. An additional layer of overfire air nozzles is installed in the frontwalls and rearwalls above the coal feeder. (See Fig. 8.) This permits deeper staging and delays adding the last of the combustion air until the hot fuel bed gases have radiated some heat to the waterwalls and are at a lower temperature. The overfire air is admitted through nozzles designed for high penetration and mixing. The overfire air system is designed for a maximum static pressure of 30 in. wg (7.5 kPa).

Spreader-stoker coal characteristics

As noted earlier, the spreader method of feeding and combusting coal is versatile. It can operate satisfactorily on the full range of coals from lignite to bituminous. Fuels having less than 18% volatile matter are not generally suitable. Table 4 summarizes the range of coal properties for spreader-stoker application.

Bituminous coals burn readily on a traveling grate without preheat. However, an air heater may be required in the unit design for improved efficiency. In these instances the design air temperature should be limited to below 350F (177C). The use of preheated air may limit the selection of fuels to the lower iron, high fusion coals to prevent undesirable grate-fired bed slagging and agglomerating. The use of preheated air at 350 to 400F (177 to 204C) is necessary for the higher moisture subbituminous coals and lignites.

Table 4
Typical Spreader Stoker Coal Characteristics

Moisture	25% max*
Volatile matter	18% min
Fixed carbon	65% max
Ash	15% max
Free swelling index	Not applicable
Ash softening temperature**	2000F (1093C) min
Coal sizing, in. (mm)	1.25 (31.8) max top size; 0.75 (19.1) min top size with max of 40% through 0.25 (6.4) screen

* Higher moisture may require preheated combustion air.
** See Table 2 for definition.

Higher ash coals can also be satisfactorily burned. However, to keep grate speeds reasonable, it may be necessary to lower the grate heat release rate and/or reduce the input per unit frontal width.

Coal size segregation can present a problem on any stoker, but the spreader is more tolerant because the feeder rates can be adjusted; much of the fines burn in suspension and not on the bed. Proper selection of the coal feed equipment can significantly reduce the tendency for coal size segregation.

Selection of a spreader stoker

Spreader stoker-fired traveling grates for coal are designed with release rates up to 750,000 Btu/h ft^2 (2.37 MW_t /m^2). The air-cooled vibrating grate is limited to 650,000 Btu/h ft^2 (2.05 MW_t /m^2). The length of the grate is limited to that over which the coal can be distributed. For vibrating grates, the length limit is 18 ft (5.5 m); for the traveling grate, it is 21 ft (6.4 m). The width of the grate is the main variable in providing sufficient total grate area. Sufficient width is also required to install enough feeders and to keep the heat input per foot of width below about 13.5 x 10^6 Btu/h ft (13.0 MW_t /m). A grate heat release rate of 750,000 Btu/ft^2 h (2.4 MW/m^2) or less also applies with full reinjection from a mechanical dust collector. Higher inputs increase slagging potential and cause excessive fuel entrainment and carryover.

Most stoker vendors supply traveling grates up to about 17 ft (5.2 m) wide with a single driveshaft. For higher capacities, two opposite hand grates provide the desired width. With a practical grate width limit of 34 ft (10.4 m) and a length limit of about 21 ft (6.4 m), the largest practical grate area is about 700 ft^2 (65 m^2). This equates to an upper steaming capacity limit of about 390,000 lb/h (49.1 kg/s) steam when firing coal.

Ash removal

When properly sized and operated, the ashes discharging from a spreader stoker-fired unit are relatively cool and clinker free. Ash discharges into a hopper and may be removed by a conventional ash transport system, generally without the need for clinker grinders.

Spreader-stoker firing of bark, wood and other biomass fuels

Fig. 9 shows a modern boiler equipped with a traveling grate spreader stoker designed to fire bark and wood residue. The furnace features a controlled combustion zone, membrane wall construction. The installation consists of state-of-the-art air-swept spouts in widths and numbers required to uniformly distribute fuel over the air metering grates. It features forced draft fans for undergrate and overfire air, a mechanical dust collector and combustion controls to coordinate fuel and air supply with steam demand.

Bark distributor feeders

Fig. 10 shows a modern air-swept spout. Bark is fed from a metering bin through a chute into the inlet of the spout. High pressure 20 in. wg (4.4 kPa) distribution air is introduced through an annulus across the width of the air-swept spout. In combination with the momentum of

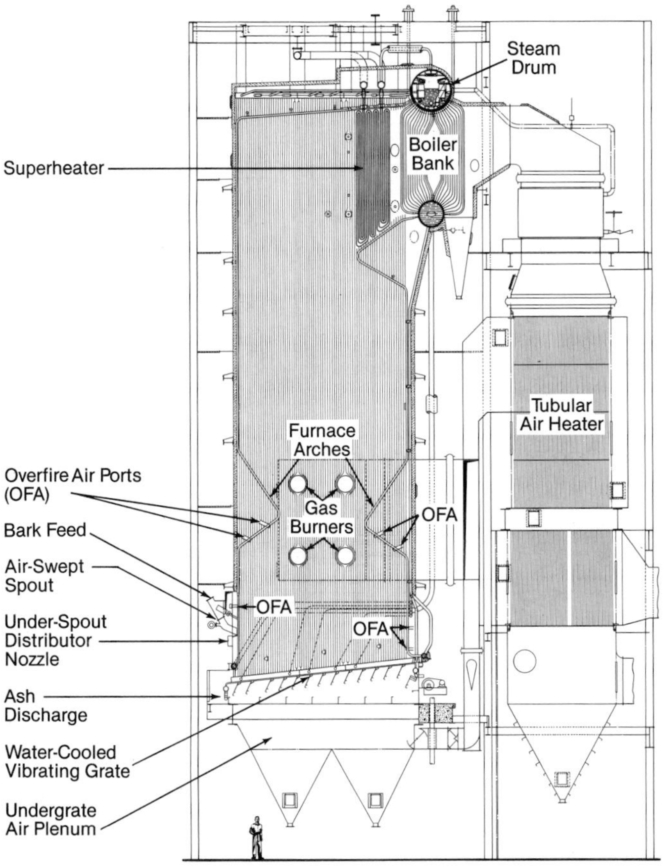

Fig. 9 Typical wood-fired stoker boiler with Controlled Combustion Zone (CCZ™) furnace.

the falling bark, the air propels the bark into and across the grate depth. The distribution air is pulsed by rotating dampers or by a preprogrammed distributor to spread the bark uniformly across the grate depth. The discharge of the air-swept spout is shaped to produce the optimum trajectory of the fuel particles. Some vendors provide an adjustable plate to control the discharge and subsequent distribution; some also supply mechanical bark feeders having features similar to those depicted earlier for stoker coal firing. Sufficient spouts are installed across the width of the unit to feed the necessary quantity of bark and to control the side to side distribution and grate coverage.

Of equal importance is the system that feeds the fuel to the air-swept spouts. Bark is conveyed across a series of smaller metering bins in a quantity greater than that needed for combustion; the excess is returned to storage. The individual metering bins are equipped with screw feeders that convey the bark in accordance with fuel demand. They are capable of speed biasing across the width. This system minimizes fuel shortages at the boiler front, uses smaller bins, has a reduced tendency to have the fuel hang up or bridge in the bins and hoppers, and provides consistent fuel size distribution to the bins and across the width of the unit. Other types of feed systems, such as large live-bottom bins, are successful when applied with the proper fuel characteristics. Some bark and other biomass fuels are stringy and can agglomerate and segregate. These characteristics must be considered when designing feed systems for the air-swept spout.

Grates for bark (biomass firing)

A traveling grate for bark firing is very similar to that used for spreader-stoker coal firing. The bark is fed over the fuel bed and distributed uniformly across the grate area. This design is essentially an air-cooled grate; therefore it is important to retain a layer of ash on the grate to shield the grate bars from furnace radiation. When firing low ash barks it is common to operate the grates intermittently, so an inventory of ash can build and be retained. In addition, high alloy grate bars having higher resistance to thermal degradation can be used.

Bark is a high volatile fuel that, in combination with the fines, has a high degree of suspension burning. High volatile and low ash characteristics permit sizing traveling grates with heat release rates to 1,100,000 Btu/h ft^2 (3.47 MW$_t$ /m^2). The width limitation of dual stokers is also about 34 ft (10.4 m) and the depth is mechanically limited to about the equivalent furnace depth of 20 ft (6.1 m). At these size limits, steam flow capabilities are limited to about 550,000 lb/h (69.3 kg/s).

The ash in bark, predominately silica, is very abrasive. This, in combination with high temperature grate bar exposure, results in high maintenance. As a result, some vendors have recently reintroduced the water- or air-cooled vibrating grate stoker. A water-cooled version is shown in Fig. 11. The combustion concept is no different than that for traveling grates. A complement of grate bars with air admission holes is attached to and supported by a water-cooled tubular grid. A heat conduct-

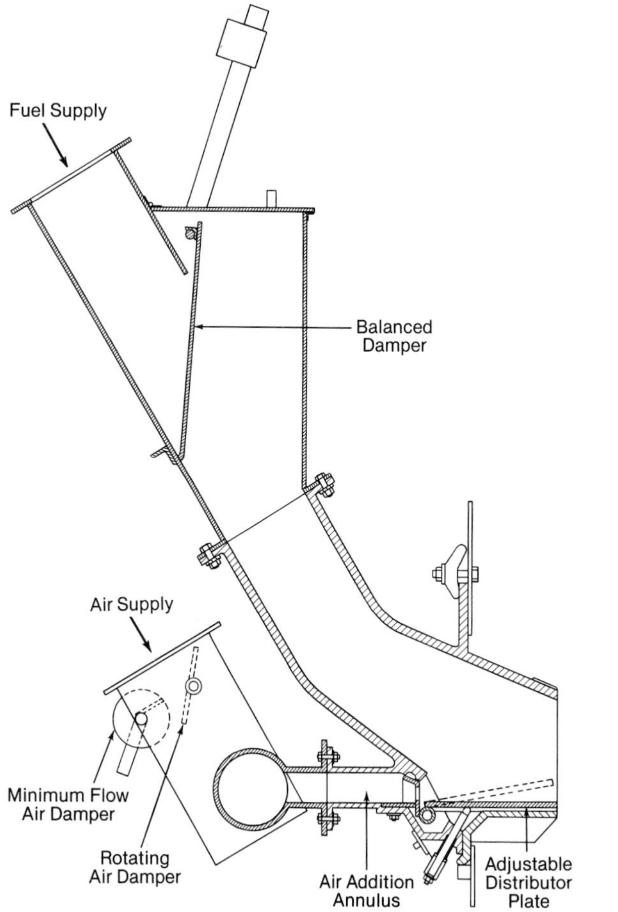

Fig. 10 Wood and biomass air-swept fuel distributor (*courtesy of Detroit Stoker Company*).

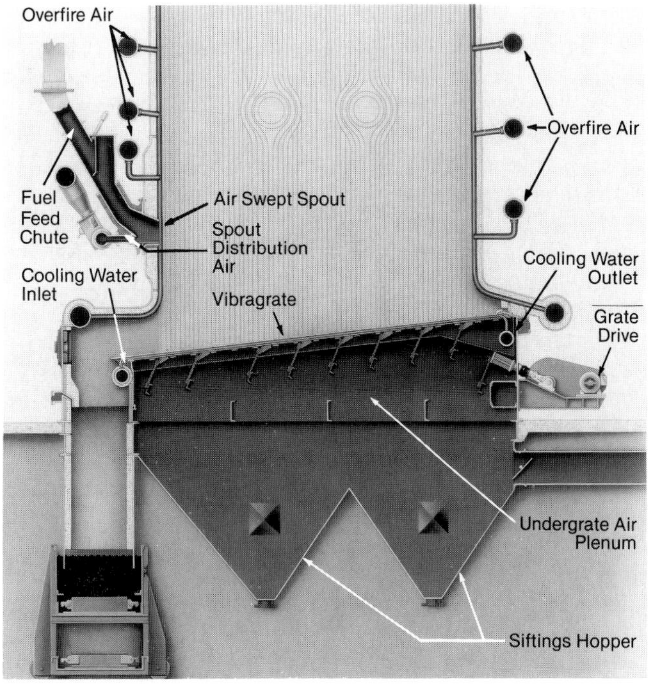

Fig. 11 Water-cooled vibrating grate stoker (*courtesy of Detroit Stoker Company*).

ing cement is used to enhance grate cooling. The water-cooled grid is then supported by a complement of flexible straps. To convey ash, a back and forth motion is imparted to the grid which, supported on the flexible straps, imparts a looping motion to the grate. This motion, at predetermined intervals, is sufficient to convey and discharge the ash to the hopper. A variant of the water-cooled grate is the air-cooled type. In the air-cooled grate, the water tube grid is replaced by a mechanical grid to support the grate bars; components are now cooled by the flow of undergrate air. The advantages of the water and air-cooled grates are their simplicity, minimal moving parts and lower maintenance. Furthermore, the mechanical limitations of the traveling grate are not present and grate depths are permitted to the limit of fuel throw and distribution of 26 ft (7.9 m). When sized to grate release rate limits of 1,100,000 Btu/h ft^2 (3.47 MW_t /m^2), steam flow capacities of 600,000 to 700,000 lb/h (75.6 to 88.2 kg/s) are available on bark and wood residue firing.

Combustion air system

The excess air for bark, wood and most biomass fuels is set at 25% for stoker firing. Because biomass fuels are typically highly volatile on a dry basis, heterogeneous in size and burn in suspension, combustion air systems are designed to provide more overfire air than that used for coal. Modern designs permit undergrate and overfire quantities of 40 to 60% of total air respectively. Overfire air system design and arrangement in combination with furnace geometry play an important role in completely burning the fuels. Fig. 9 shows a furnace design that promotes recirculation and mixing at and below the lower furnace bustle. Multiple elevations of large overfire air nozzles permit high energy jets of air to penetrate and mix, enhancing combustion. The system is designed to permit flexibility in distributing the air within the overfire air system as fuel characteristics vary.

Fig. 12 shows the recommended sizing distribution for wood or bark; however, this distribution is rarely achieved. As the size varies, the degree of grate or suspension burning varies. A portion of the fuel that ignites in suspension may be too large to complete burning; it therefore goes out with the flue gas as unburned carbon. Many factors such as fuel sizing, grate release rates, moisture content and reinjection affect the unburned carbon loss. In addition, furnaces sized to a maximum liberation rate of about 18,000 Btu/h ft^3 (186 kW_t/m^3) can control unburned carbon losses to 1 to 3%. Further reductions can be effected with flyash reinjection in certain biomass-fired units. Due to the high silica content and abrasiveness of wood and bark flyash, reinjection systems are not frequently used with these fuels unless sand classifiers are installed, because of high maintenance costs. As an example, the unit shown in Fig. 9 does not feature a flyash reinjection system. Because the carbon content is typically larger than that of the flyash, a mechanical dust collector is used to collect the larger flyash fraction and reduce the loading and the carbon content and tendency for fires in the flue gas cleanup equipment.

Grate sizing

For wood and bark firing with moisture contents at or below 50%, it is common practice to select a grate area that results in a design heat release rate of 1,100,000 Btu/h ft^2 (3.47 MW_t /m^2). At 35% moisture, the release rate may approach 1,250,000 Btu/h ft^2 (3.94 MW_t /m^2). At moisture levels higher than 55%, the fuel becomes more difficult to ignite and burn. Traveling grate stokers are mechanically limited to about 20 ft (6.1 m) of equivalent furnace depth and about 34 ft (10.4 m) in width. Grates are generally longer than they are wide because this design is the least expensive. For a water-cooled grate, the depth is limited by the stoker's ability to distribute the bark over the grate depth. This distance is about 26 ft (7.9 m). A reasonable width limit for this grate is also 34 ft (10.4 m).

When firing wood and bark, it is best to preheat the combustion air. With seasonal variation in fuel moisture and

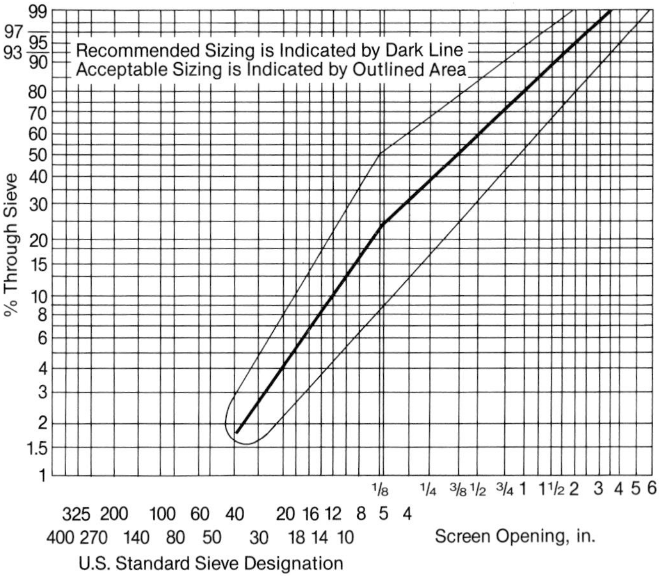

Fig. 12 Recommended sizing for wood or bark spreader stoker firing (*courtesy of Detroit Stoker Company*).

the growing interest in cofiring wood and bark with sludges, the air temperature should be at least 550F (288C). This is generally the temperature limit for the air-cooled traveling grate with ductile iron bars. The water-cooled vibrating grate can withstand air temperatures to 650F (343C). Hot air is preferred for both undergrate and overfire air.

There are many other biomass fuels, such as straw, rice hulls, bagasse, peach pits, coffee grounds and demolition debris, that may be fired on spreader-type stokers. Most of the criteria for sizing and using a spreader stoker are similar to those for wood.

Ash removal

The ash content in bark and wood is relatively small; it is only a few percent and is predominantly silica. It is very abrasive and, for this reason, slow speeds and abrasion resistant materials should be used. Bottom ash typically drops into a refractory lined hopper from which it is intermittently removed. Submerged chain conveyors are also suitable. Due to the carbon content and the tendency for fires, the flyash from the hopper and dust collecting equipment is removed continuously. Wet sluice or inert gas (flue gas) systems are generally used to convey the flyash to the storage silos. Wet systems are preferred due to fire and dusting resistance.

Emissions

The installation of any new steam generator or a major rebuild or upgrade to an existing one requires environmental permitting. Therefore, the ability to predict and control the various emissions is important. Add-on pollution control equipment is generally required to meet the increasingly stringent regulated levels. However, it is also important to control or minimize the source emissions where possible to reduce the cost of this add-on equipment.

Table 5 lists typical uncontrolled emission values for spreader-stoker firing of various coals and wood/bark. These values will vary with fuel composition and equipment selection.

NO_x is formed from the oxidation of the nitrogen compounds in the combustion air and in the fuel. With stoker firing it is believed that most of the NO_x is derived from fuel-bound nitrogen (fuel NO_x); the contribution due to oxidation of the nitrogen in the air (thermal NO_x) is small due to relatively low furnace temperatures. NO_x emissions can be effectively controlled by staging combustion, inherent in spreader-stoker firing, and by controlling excess air levels. For both coal and wood/bark stoker firing, the excess air level in low NO_x stoker systems is about 25%. To control NO_x to the lower end of the range shown in Table 5, designers are now using deeper staging, i.e., lower undergrate and higher overfire air flows. For spreader firing, feeders are also designed to improve fuel distribution and combustion on the grates. Other factors that reduce NO_x formation include minimizing the quantity of fines in the fuel and using ambient temperature combustion air.

Many waste fuels such as straw and other nonwood fibers have high fuel nitrogen content. Commercial wastes such as demolition wood are dry and will combust at higher temperatures, producing higher NO_x. It is therefore important that a thorough investigation of the fuel be made to assess its impact on emissions.

For most stoker-fired units burning coal or biomass which contains sulfur, sulfur dioxide (SO_2) will be present in the flue gas. Therefore, for prediction purposes and for sizing of SO_2 control equipment it is assumed that all the fuel sulfur becomes SO_2.

Carbon monoxide (CO) and volatile organic compound emissions are generally a function of the efficiency of the combustion process and the quantity and control of fines and excess air. CO will tend to increase as NO_x is reduced.

Table 5
Typical Uncontrolled Emissions for Spreader-Stoker Firing

Fuel	NO_x (as NO_2) lb/10^6 Btu	CO lb/10^6 Btu	Unburned Carbon Loss (% of Heat Input) With Reinjection	Unburned Carbon Loss (% of Heat Input) Without Reinjection
Bituminous	0.35 to 0.5	0.05 to 0.30	0.5 to 2.0	3 to 6
Subbituminous	0.3 to 0.5	0.05 to 0.30	0.5 to 1.5	3 to 5
Lignite	0.3 to 0.5	0.10 to 0.30	0.5 to 1.5	3 to 5
Wood/bark	0.2 to 0.35	0.20 to 0.35	0.5 to 1.5	2 to 5

Approximate conversion: 1 lb/10^6 Btu = 1230 mg/Nm^3 (dry flue gas 6% excess O_2, 350 Nm^3/GJ).

Circulating fluidized-bed power plant by B&W.

Chapter 16
Atmospheric Pressure Fluidized-Bed Boilers

The fluidized-bed process

Fluidized-bed combustion (FBC) technology has distinct advantages for burning solid fuels and recovering energy to produce steam. The process features a mixture of particles suspended in an upwardly flowing gas stream, the combination of which exhibits fluid-like properties. Combustion takes place in the bed with high heat transfer to the furnace and low combustion temperatures. Key benefits of this process are fuel flexibility and reduced emissions. FBC is technically accepted for industrial sized units and is emerging as an accepted technology for electric power generation.

The following example will help visualize the fluidized-bed process. Fig. 1a shows a container with an air supply plenum at the bottom, a distributor plate that promotes even air flow through the bed, and an upper chamber filled with sand or other granular material.

If a small quantity of air flows through the distributor plate into the sand, it will pass through the voids of an immobile mass of sand. For low velocities, the air does not exert much force on the sand particles and they remain in place. This condition is called a fixed bed and is shown in Fig. 1b.

By increasing the flow rate, the air exerts greater forces on the sand and thereby reduces the contact forces between the sand particles caused by gravity. By increasing the air flow further, a point is reached where the drag forces on the particles counterbalance the gravity forces and sand particles become suspended in the upward flowing air stream. The point where the bed starts to behave as a fluid is called the *minimum fluidization condition*. As shown in Fig. 1c, the increase in bed volume is insignificant when compared with the nonfluidized case.

As the air flow increases further, the bed becomes less uniform, bubbles of air start to form and the bed becomes violent. This is called a *bubbling* fluidized bed which is shown in Fig. 1d. The volume occupied by the air-solids mixture increases substantially. For this case, there is an easily seen bed level and a distinct transition between the bed and the space above.

By increasing the air flow further, the bubbles become larger and begin to coalesce, forming large voids in the bed. The solids are present as interconnected groups of high solids concentrations. This condition is called a *turbulent* fluidized bed.

If the solids are caught, separated from the air and returned to the bed they will circulate around a loop. This type of system is defined as a *circulating* fluidized bed (CFB) and is shown in Fig. 1e. Unlike the bubbling bed, there is no distinct transition between the dense bed in the bottom of the container and the dilute zone above. The solids concentration gradually decreases between these two zones.

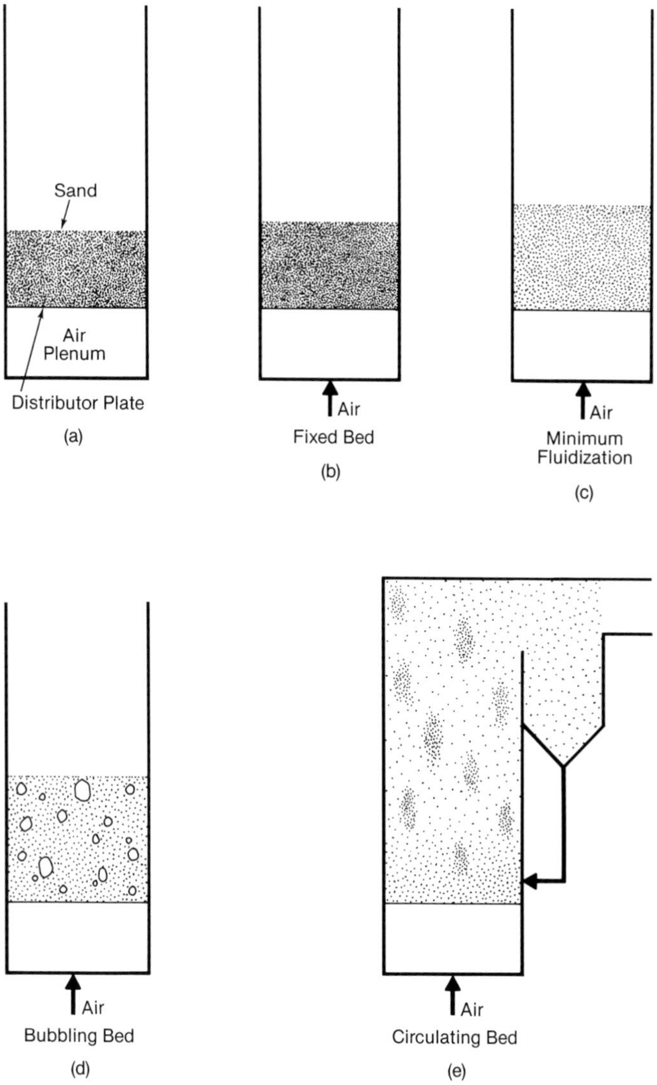

Fig. 1 Typical fluidized-bed conditions.

The weight of solids recirculated from the outlet back to the bed zone can be hundreds of times the weight of air flowing through the system, and the quantity of solids in the container is proportional to the amount of sand recycled from the collector. As a result, the pressure differential will increase to the value required to support the solids in the container.

The pressure differential between the top and the bottom of the container changes with air flow, as shown in Fig. 2. At low air flow rates, the pressure differential increases with flow until the minimum fluidization velocity is reached. At this point, the sand is supported by the air and the pressure differential is determined only by the mass of sand in the bed. The pressure differential is independent of further increases in air flow until the air velocity becomes high enough to convey the sand out of the container. At higher air flows, the pressure differential starts to decrease as mass is lost from the system.

Of the fluidization conditions just described, only bubbling and circulating beds are currently used by the power industry to generate steam.

Background and evolution

One of the earliest recorded applications of a fluidized bed used coal as the feed stock. In the mid 1920s, the fluidized-bed coal gasification process, invented by Fritz Winkler, was used commercially to generate gas from coal. (See also Chapter 17.) This gas was then used for fuel or as a feed stock for chemical syntheses. While there are a few Winkler gasifiers operating today, industry has found it both easier and less expensive to use natural gas or oil as fuels and for synthesis gas production.

During the late 1930s and early 1940s, a large research and development effort was directed to fluidized beds. This work identified the advantages of a fluidized bed as a solids-gas contacting device and led to development of the fluid catalytic cracker to produce gasoline and other petroleum based products. Today, fluidized beds are used worldwide for a variety of processes in many industries.

In the early 1960s, national attention turned to reducing sulfur dioxide (SO_2) and nitrogen oxides (NO_x) emissions from power plants. The fluidized-bed combustion process offered the potential for reducing these emissions. The effort to develop the coal-fired fluidized-bed boiler began.

In the early 1970s, Babcock & Wilcox (B&W) conducted an extensive study to evaluate the application of atmospheric pressure fluidized-bed combustion to large boilers. In 1977, B&W, in cooperation with the Electric Power Research Institute (EPRI), constructed and placed into operation a 6 × 6 ft (1.8 × 1.8 m) bubbling-bed test unit at the B&W Research Center in Alliance, Ohio. Test results from this facility have contributed significantly to the advancement of bubbling fluidized-bed boiler technology. Recent applications of large, coal-fired bubbling beds have included retrofits to existing boilers.

In the early 1980s, the market for coal-fired fluidized-bed boilers moved toward circulating beds. To meet this changing market, B&W used its knowledge of bubbling-bed technology and the technical concepts of Studsvik AB of Sweden to advance into the circulating-bed field. A schematic of a pilot-scale circulating fluidized bed built at B&W's research center is shown in Fig. 3.

Comparison with other combustion methods

The fluidized-bed combustion process, as with other firing methods, provides a means for mixing fuel with air to convert the chemical heat contained in the fuel into recoverable, sensible heat. Normally, fluidized-bed combustors are used to burn solid fuels.

In a pulverized coal-fired furnace, the combustion process consists of oxidizing fine (70% less than 200 mesh), widely dispersed fuel particles suspended in air and combustion gases. (See Chapter 13.) The volume around the burners is the hottest zone in the furnace with temperatures reaching 3000 to 3500F (1649 to 1927C). Also, the residence time of the particles in the furnace is close to the flue gas residence time.

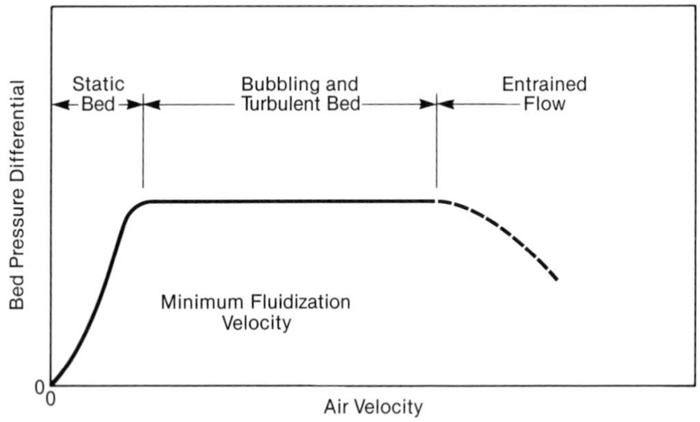

Fig. 2 Effect of velocity on bed pressure drop.

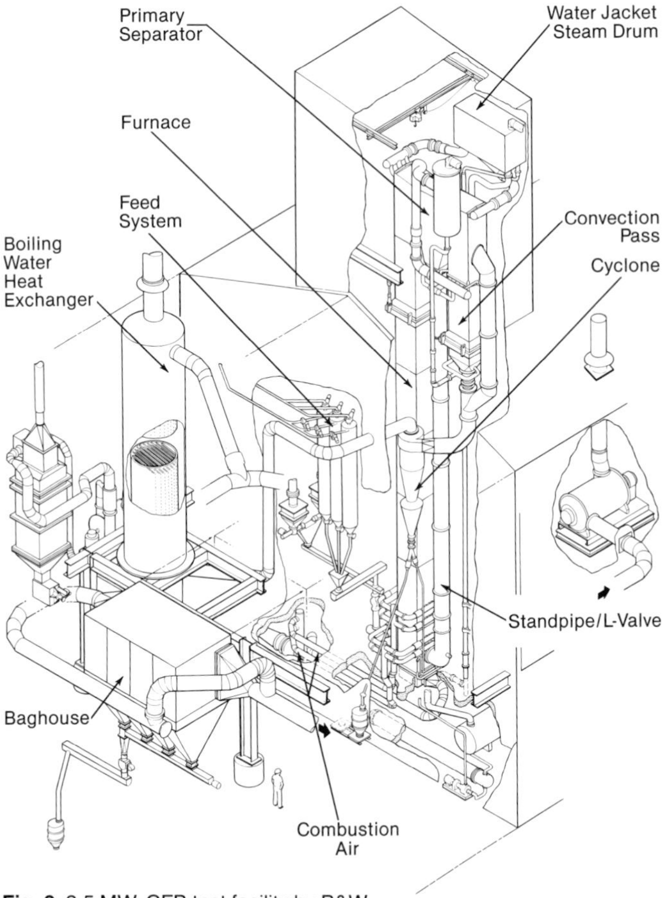

Fig. 3 2.5 MW_t CFB test facility by B&W.

Stoker firing uses considerably larger fuel particles than pulverized coal firing. Fuel sizing is typically 1 to 1.25 in. (25.4 to 31.8 mm) top size for bituminous coal. Most of the fuel is burned as an immobile mass on some type of moving grate, with the air and combustion gas passing through the fixed bed of fuel. (See Chapter 15.) Temperatures in the fuel bed exceed 3000F (1649C) and the fuel residence time is determined by the grate speed.

The fluidized-bed combustion process falls in between pulverized coal and stoker firing with respect to the size of the fuel feed. Coal is typically crushed to less than 0.25 in. (6.4 mm). Depending on coal properties larger coal, 1.25 in. (31.8 mm), or smaller coal, 0.125 in. (3.18 mm), is used. Fuel is fed into the lower portion of a fluidized-bed boiler furnace. The bed has a density of approximately 45 lb/ft^3 (721 kg/m^3) for a bubbling bed and 35 lb/ft^3 (561 kg/m^3) for a circulating bed. The solids are maintained at a temperature of 1500 to 1600F (816 to 871C) in an upwardly moving stream of air and combustion gas.

When fuel is introduced into the bed it is quickly heated above its ignition temperature, ignites and becomes part of the burning mass. The flow of the air and fuel to the dense bed is controlled so that the desired amount of heat is released to the furnace on a continuous basis. Typically, the fuel is burned with 20% excess air. Due to the long fuel residence time and high intensity of the mass transfer process, the fuel can be efficiently burned in the fluidized-bed combustor at temperatures considerably lower than in conventional combustion processes.

The fuel particles remain in the dense bed until they are entrained by combustion gas or removed with the bed drain solids. As the fuel particles burn, their size falls below a given value where the terminal and gas velocities are equal, which allows them to be entrained. Therefore, the residence time is determined by the initial fuel particle size and by the reduction of the initial size resulting from combustion and attrition.

In bubbling fluidized beds, combustion occurs mostly in the bed due to lower gas velocity and coarser fuel feed size. The residence time of the fine fuel particles carried out of the bed with the combustion gas is, in many cases, increased by collecting and recycling the particles to the furnace.

In circulating beds, more particles are blown from the bed (elutriated) than for a bubbling bed. The particles are then collected by a particle separator and recirculated to the furnace. The residence time of the particles is determined by the collection efficiency of the particle separator and the solids circulation rate. As a result of the recirculation process, the effective particle residence time greatly exceeds the gas residence time.

The concentration of fuel in the dense bed is normally quite low. For a reactive fuel such as wood, it is difficult to find a measurable amount of carbon in the bed. Normally, the carbon content in a bed burning bituminous coal is less than 1%. The remaining portion of the bed is made up of fuel ash, lime and calcium sulfate when a sorbent is used for sulfur capture, and sand or other inert material when a sorbent is not used.

Overall carbon conversion efficiencies are near 100% for wood and highly reactive fuels, greater than 98% for bituminous coals, and slightly lower for less reactive fuels and waste coals.

Fluidized-bed combustion advantages

The primary driving force for the development of fluidized-bed combustors in the United States (U.S.) is reduced SO_2 and NO_x emissions. By implementing this technology, it is possible to burn high sulfur coals and achieve low SO_2 emission levels without the need for additional back-end sulfur removal equipment. As the technology developed, it also became apparent that the process could burn low grade fuels that are difficult or impractical to burn with other methods.

Designers of coal burning equipment have known for many years that coal properties vary depending on the fuel type and source and that the firing equipment must be compatible with the specific fuel. While fluidized-bed combustors are more tolerant of these variations, the same basic rules of combustion apply, so fuel type, chemical composition and heating value must be considered on each design.

Fluidized-bed boilers are designed so that the bed operating temperature is between 1500 and 1600F (816 and 871C); the ability to operate at this low temperature results in several operating advantages.

Reduced emissions — SO_2 and NO_x

Because of lower operating temperatures, it is possible to use an inexpensive material, such as limestone or dolomite, as a sorbent to remove SO_2 from flue gas. When limestone or dolomite is added to the bed, a reaction occurs in the furnace between the resulting calcium oxide (CaO) and the SO_2 in the gas. SO_2 emissions can be reduced by 90% or more depending on the sulfur content of the fuel and the amount of sorbent added.

Nitrogen and oxygen will react at high temperatures, above 2700F (1482C), to form nitric oxide. The rate of reaction decreases rapidly as the temperature is reduced from this value. With an operating bed temperature between 1500 and 1600F (816 and 871C), the amount of NO_x formed in the fluidized bed is less than in conventional units which operate at higher temperatures. For some bubbling-bed and all CFB boilers, additional suppression of NO_x formation is achieved by air staging. Fluidized beds can operate in this manner with less impact on combustion efficiency than in pulverized coal furnaces. With the addition of postcombustion reduction techniques, even lower NO_x emissions can be achieved.

Fuel flexibility

Fuel ash properties In addition to reduced emissions, the lower combustion temperatures permit burning high fouling and slagging fuels at temperatures below their ash fusion temperature. As a result, many of the boiler operating problems associated with these fuels are greatly reduced. However, care is still required as high alkali metal concentrations in the bed may cause sintering and in-bed tube bundle fouling. In addition, excessive amounts of above-bed burning may cause a significant increase in the furnace exit gas temperature and cause deposits to form in the superheater. For these reasons, bed temperatures around 1500F (816C) are normally chosen for fuels whose ash is high in alkali metals.

Low Btu fuels The fluidized-bed combustion process can also burn fuels with very low heating values. This capability results from the rapid heating of the fuel particles by the large mass of hot bed material and the long residence time that the fuel spends in the bed, both of which offset the effects of lower combustion temperatures. When burning high moisture fuels it is also necessary to consider the additional gas weight, resulting from water vapor, in the design of the convection pass and other related components.

Fuel preparation For high ash coals, the fluidized-bed boiler offers an advantage in fuel preparation over pulverized coal systems. These fuels require greater installed pulverizer capacity and the pulverizers generally require frequent maintenance. The crushed fuel, less than 0.25 in. (6.4 mm), required for a fluidized-bed boiler is easier and less costly to prepare.

A fluidized-bed boiler can be designed to burn a wider range of fuels than alternate firing methods. However, once the boiler is designed for a specified set of fuels there are limitations to the degree of deviation from the specified values that can be fired before design limits are exceeded. A CFB boiler has greater fuel flexibility than a bubbling bed.

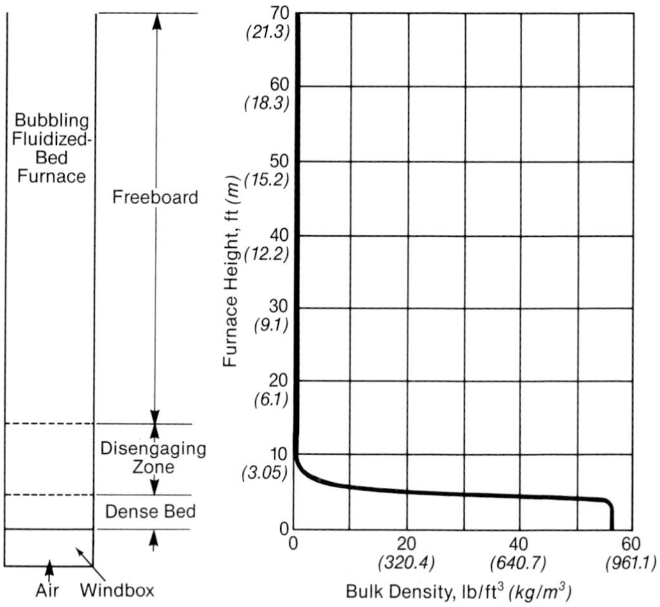

Fig. 5 Typical atmospheric pressure bubbling-bed furnace density profile.

Atmospheric pressure fluidized-bed boilers

Bubbling bed

Fig. 4 shows the main features of a bubbling fluidized-bed boiler. The bed itself is usually 4 ft (1.2 m) deep in its expanded or fluidized condition. Fig. 5 shows a typical furnace bulk density profile curve. The sharp drop in density indicates the top of the bed.

Normally, the heat transfer surface, which is in the form of a tube bundle, is placed in the bed to achieve the desired heat balance and bed operating temperature. For fuels with low heating values the amount of surface can be minimal or absent. In all cases, the bed temperature is uniform, plus or minus 25F (14C), as a result of the vigorous mixing of gas and solids.

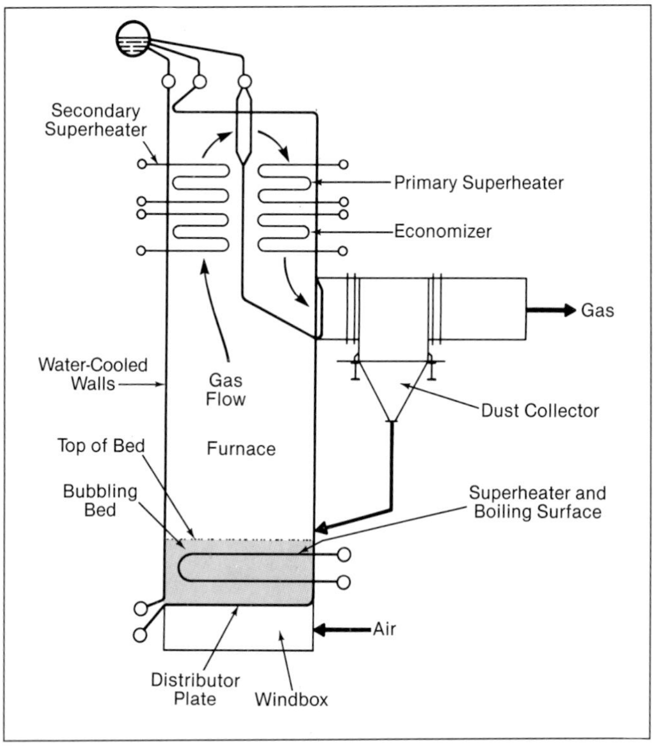

Fig. 4 Typical bubbling fluidized-bed boiler schematic.

Coal-fired bubbling-bed boilers normally incorporate a recycle system that separates the solids leaving the economizer from the gas and recycles them to the bed. This maximizes combustion efficiency and sulfur capture. Normally, the amount of solids recycled is limited to about 25% of the combustion gas weight. For highly reactive fuels this recycle system can be omitted.

Bubbling beds that burn coal usually operate in the range of 8 to 10 ft/s (2.4 to 3 m/s) superficial flue gas velocity at maximum load. The bed material size is 30 mesh (590 microns) and coarser, with a mean size of about 18 to 16 mesh (1000 to 1200 microns).

Circulating bed

Fig. 6 shows the main features of a circulating fluidized-bed boiler and Fig. 7 shows the furnace density profile. The dense bed does not contain any in-bed tube bundle heating surface. The furnace enclosure and internal division wall type surfaces provide the required heat removal. This is possible because of the large quantity of solids that are recycled internally and externally around the furnace. Because the mass flow rate of recycled solids is many times the mass flow rate of the combustion gas, furnace temperatures remain uniform. Also, the heat transferred to the furnace walls is adequate to provide the heat absorption required to maintain the target bed temperature of 1500 to 1600F (816 to 871C). B&W circulating fluidized-bed boilers usually operate at about 20 ft/s (6.1 m/s) superficial flue gas velocity at full load. The size of the solids in the furnace is usually smaller than 30 mesh (590 microns), with the mean particle size in the 150 to 200 micron range.

Emissions

U.S. federal and state governments have imposed limits on emissions from most large boilers and combustion processes. These emissions limits vary by region and gov-

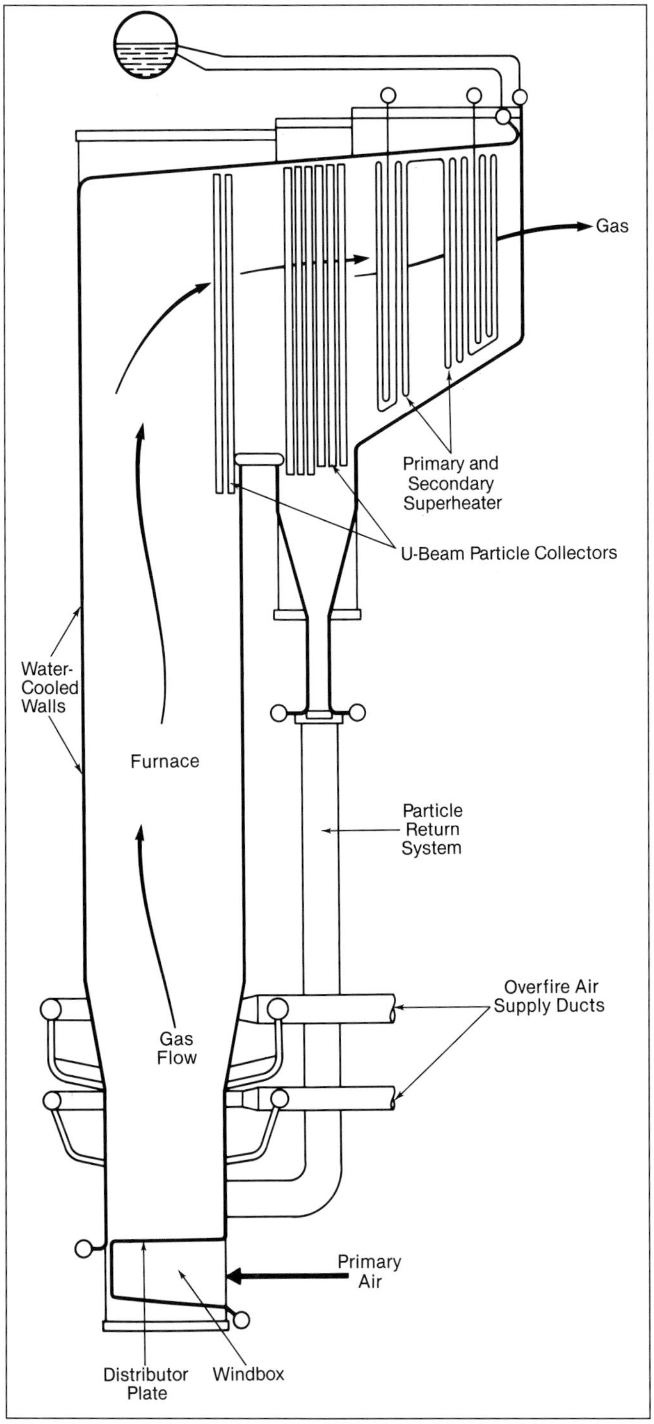

Fig. 6 Typical circulating-bed boiler schematic.

ernment, but the compounds and materials controlled are generally the same. These are SO_2, NO_x, carbon monoxide (CO), hydrocarbons and particulate matter. (See Chapter 32.) Fluidized-bed boilers are designed, primarily, to burn solid fuel while controlling many of these emissions.

Sulfur dioxide

When sulfur bearing fuels burn, most of the sulfur is oxidized to SO_2 which becomes a component of the flue gas. When limestone is added to the bed it undergoes a transformation called *calcination* and then reacts with the SO_2 in the flue gas to form calcium sulfate ($CaSO_4$). The calcining reaction is endothermic and is described by:

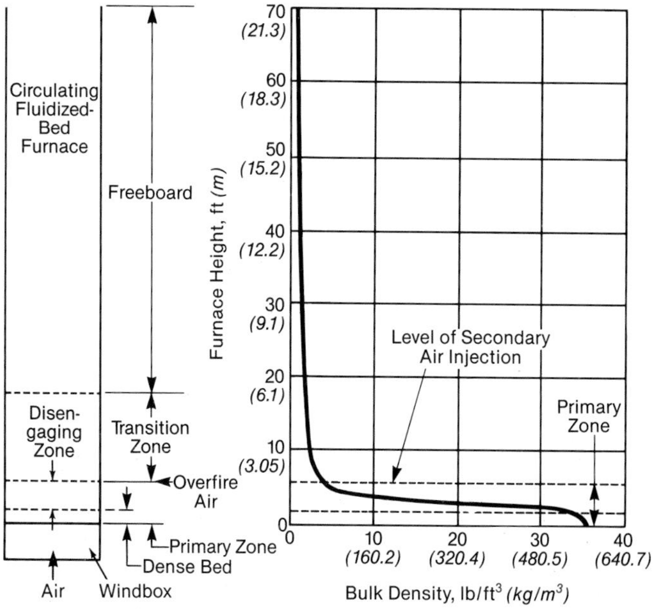

Fig. 7 Typical atmospheric pressure circulating-bed furnace density profile.

$$CaCO_3\ (s) + 766\ Btu/lb\ (of\ CaCO_3) \rightarrow CaO(s) + CO_2\ (g)$$

Once formed, solid CaO (lime) reacts with gaseous SO_2 and oxygen exothermically to form $CaSO_4$ according to the following reaction:

$$SO_2\ (g) + {}^1\!/_2\ O_2\ (g) + CaO\ (s)$$
$$\rightarrow CaSO_4\ (s) + 6733\ Btu/lb\ (of\ S)$$

$CaSO_4$ is chemically stable at fluidized-bed operating temperatures and is removed from the system as a solid for disposal.

Early coal-fired fluidized-bed combustion work was carried out on a once-through principle. In this case, the coal and limestone were fed to the combustor, reacted and passed out of the system. Combustion efficiency and sulfur capture fell short of desired values. To overcome this, a portion of the solids leaving the furnace (flyash, $CaSO_4$, carbon and lime) is separated from the gas by a dust collector located between the economizer and the air heater and is recycled to the furnace for further reaction. Fig. 8 shows the effect of solids recycle on sulfur capture for a bubbling bed. Normally, recycle rates are limited to a maximum of 2.5 times the fuel feed rate which is a result of practical considerations of equipment size and arrangement.

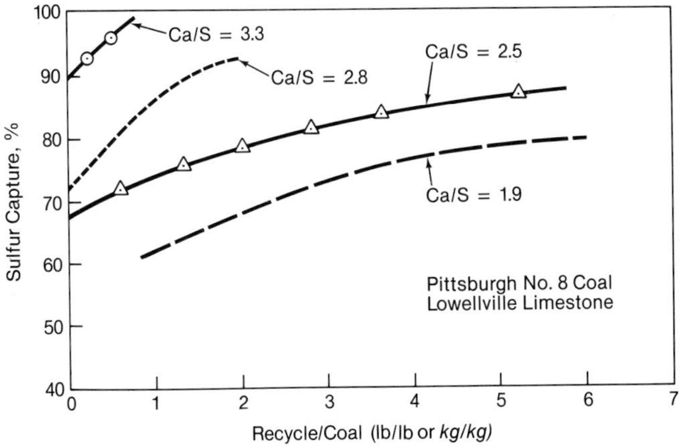

Fig. 8 Sulfur capture as a function of recycle ratio.

In a CFB, a primary solids collector is located immediately after the furnace and is designed to recycle all of the captured solids. As a result, sulfur capture and combustion efficiency are improved over bubbling beds. Some designs also include a secondary collector after the convection pass which further enhances sulfur capture and combustion efficiency by increasing the concentration of fine particles in the furnace. These improvements result from longer solids-gas contact times and higher specific surface of the fine particles in contact with the gas.

SO_2 reductions of 90% are typically achieved in a circulating bed with calcium to sulfur (Ca/S) mole ratios of 2 to 2.5, depending on the sulfur content of the fuel and the reactivity of the limestone. Slightly higher ratios are required in bubbling beds. The lower the sulfur concentration in the fuel the greater the calcium to sulfur mole ratio must be for a given SO_2 removal level.

For removal requirements greater than 90%, the amount of limestone needed increases rapidly and alternative SO_2 removal methods, such as conventional pulverized coal-fired boilers with scrubbers, may become the economic choice. (See Chapter 35.)

Nitrogen oxides

The nitrogen oxides that are present in the flue gas come from two sources, the oxidation of nitrogen compounds in the fuel and the reaction between nitrogen and oxygen in the combustion air. It is common to refer to the nitrogen oxides in the flue gas as NO_x, where the subscript x implies the presence of several different nitrogen/oxygen compounds. The NO_x formed by oxidizing fuel nitrogen compounds is referred to as *fuel* NO_x and the NO_x formed from nitrogen and oxygen in the combustion air is called *thermal* NO_x, because it is a product of a high temperature process above 2700F (1482C).

As a result of the low temperatures at which a fluidized bed operates, thermal NO_x makes only a minor contribution to overall emissions. A fluidized-bed boiler can also suppress the amount of fuel NO_x formed. This is accomplished by putting less than the theoretical amount of combustion air through the distributor plate and adding the remainder of the combustion air to the furnace above the dense bed. As a result, some of the fuel nitrogen compounds decompose into molecular nitrogen rather than forming NO_x. This process is referred to as *staged combustion* and is also used with other firing methods to reduce NO_x emissions.

Staged combustion is used for bubbling beds that do not contain in-bed surface and for circulating beds. However, it is not used for designs where cooling tube bundles are submerged in the bed. The reason is corrosion. By burning the fuel with less than theoretical air (substoichiometrically) the combustion gas contains many reduced chemical species that cause rapid metal loss from the furnace cooling tubes. When staged combustion is used, the furnace tube walls in the reducing zone are protected with a thin layer of refractory.

The combination of low temperatures and staged combustion permits fluidized-bed boilers to operate with significantly lower NO_x emissions. Typical values for NO_x emissions are within the 100 to 200 ppmdv (parts per million dry volume) range for a CFB boiler burning coal.

Carbon monoxide and hydrocarbons

When designing a boiler, it is necessary to maximize combustion efficiency by minimizing unburned carbon and the quantity of CO and hydrocarbons in the flue gas. This is done by choosing the proper number of fuel feed points, by proper design of the overfire air system, if used, and by providing sufficient furnace residence time for mixing and complete combustion (burnout).

Typical flue gas concentrations are less than 200 ppmdv for CO and 20 ppmdv for hydrocarbons in a CFB boiler burning coal.

Particulates

The ash contained in solid fuel is released during the combustion process. Some of this ash remains in the fluidized bed and is discharged by the bed material removal or *drain* system. This ash is normally larger than 140 mesh (105 microns) and is easy to handle and transport. The remaining ash leaves the boiler in the flue gas. This material is typically less than 325 mesh (44 microns) and requires a high efficiency collection device. Normally, a fabric filter is used with atmospheric pressure fluidized-bed boilers because it is less sensitive to the ash properties, such as size, concentration and resistivity, than electrostatic precipitators (ESP). (See Chapter 33 for a discussion of particulate collection.)

Fluidized-bed boiler furnace design

Numerous factors and parameters affect fluidized-bed furnace design. Some are specified by the boiler owner/operator. Many are selected by the designer and can only be derived from empirical data. Some are interrelated, making the design task more challenging.

The following list is typical of the type of information used when establishing the design of a fluidized-bed boiler.

Owner specified

1. unit capacity — steam flow requirements,
2. fuel — type, ash and moisture content, size, reactivity, chemical analyses, attrition characteristics, and fouling and sintering properties,
3. limestone — type, reactivity, size and attrition characteristics,
4. sulfur capture requirements,
5. NO_x emission limits, and
6. load turndown range.

Designer specified

1. type of fuel feed system — in-bed, under-bed or above-bed,
2. number and location of fuel feed points,
3. fuel combustion efficiency,
4. number and location of sorbent injection points,
5. primary/secondary air split when used and the location of the overfire air nozzles,
6. bed operating temperature,
7. bed operating velocity,
8. size of bed particles,
9. amount of solids that leaves through the bed drain and the amount that leaves with combustion gas to the final solids collector, and
10. amount, temperature and location of solids recycled to the furnace from particle separators and bed drain classifiers.

The following discusses some of the major areas involved with the functional design of fluidized-bed furnaces. The design of the convection pass surface is similar to a conventional boiler design process as discussed in Chapter 18.

Importance of bed material

Combustion and inventory requirements For a fluidized bed to operate properly it must be continuously supplied with a sufficient quantity of particles of the proper size distribution. If the particles are too coarse, the bed will defluidize and become fixed. If the particles are too fine, they will blow out of the furnace making it impossible to maintain an adequate bed inventory. There is a range of bed particle size which is needed to maintain a stable fluidized-bed process. The supply and retention of these particles must be controlled to provide the required inventory.

In the case of a bubbling bed, it is easy to visualize that a properly fluidized bed must be present to receive, suspend, mix and burn the fuel particles. If the bed density is too lean or if the bed is too shallow, an accumulation of fuel particles is possible in a localized area of the bed. This may lead to high temperature zones where the fuel and ash will turn into a sintered mass. At the other extreme, if too many large particles are fed into the bed it will defluidize with a similar result.

In circulating beds, even though the solids inventory is distributed over the furnace height, a dense bed is still required in the lower furnace to support and mix the fuel during combustion to avoid the problems just mentioned.

Normally, when coal is burned, most of the ash is liberated from the burning pieces as fine particles. These are so fine that they quickly blow out of the bed and are carried out of the furnace by the flue gas.

In bubbling beds, this material does not contribute to bed inventory. In circulating beds, this material is caught and returned to the furnace where it becomes part of the circulating mass. However, the fines do not make a significant contribution to the dense bed itself.

Because of the wide variation in fuel ash properties, the ash is not usually depended upon to form a stable bed, and a second inert material of the desired size distribution such as sand is added to the system. When SO_2 capture is required, limestone replaces the sand to achieve in-bed SO_2 capture.

Ideally, the size of the sorbent fed to the boiler would simply be that required to form a stable bed. However, in the process of heating, calcining and sulfating, the sorbent's size and physical properties change. The final size of the sorbent-derived bed material, after these chemical and physical changes, can not, in many cases, be reliably predicted. In addition, some limestones are softer than others and wear faster. For these reasons, limestone characteristics and some trial and error testing during initial operation are needed to establish the proper limestone feed size and makeup flow rate.

The above discussion assumes a typical bituminous coal. In the case of low ash fuels like wood, sand is used for the bed. While sand properties are important, sand does not break down as rapidly as limestone, its makeup rates are lower and the size of the bed material is more predictable.

High ash waste coals require special consideration. Typically, these fuels contain large amounts of ash not tied to organic matter. This ash frequently consists of rock that has been generated by coal beneficiation and is called *tramp ash*. Because tramp ash does not break down into fine particles, it forms a large percentage of the bed material. For this reason, the size of the fuel feed must be chosen carefully so that the ash complements the bed material rather than causing fluidization problems. There are fluidized-bed boilers in operation where the size and consistency of the fuel ash will form a stable bed without the need for additional bed material makeup.

Bed drain classification systems are often used in circulating-bed units in addition to proper fuel and/or sorbent sizing to control the size and inventory of the bed material. These systems help to remove oversized noncirculating material while maintaining the required inventory of circulating particles.

Particle characterization and measurement In fluidized-bed boiler furnaces, particle motion is affected by the combined influence of gravitational and aerodynamic forces, as well as the impact with other particles and the walls of the boiler.

For an individual particle, the three most important characteristics are its size, density and shape. Particles of the same size but with different densities or with the same density but varying sizes do not act the same. The particle's shape, from spherical to flat, determines how it will react to the forces present in the furnace.

From an analytical standpoint, the ideal particle is a homogenous sphere and the ideal mixture consists of many homogenous spheres of equal diameter. In practice, mixtures that consist of a variety of particles with different sizes, densities and shapes are encountered.

Fluidized-bed heat transfer and pressure drop calculations assume that the mixtures of particles can be characterized by a mean particle diameter, a mean particle density and a mixture bulk density. The formulas for calculating mean diameters of mixtures assume spherical particles, and a sphericity correction term is applied to the calculated mean diameter to approximate the influence of the nonspherical nature of the actual material. Mixtures containing a significant quantity of particles that could be described as rods or flakes are difficult to characterize.

The equations that are used to calculate mean diameters are based on the weight fractions of incremental size groups. The first step of the determination is sieving a representative sample of the mixture through a stack of screens having a successively finer mesh and weighing the amount of solids retained on each screen. For calculation purposes, the average size of the particles is assumed to be the average of the mesh opening of the screen the particles were retained on and the next larger screen through which they all passed.

Two characteristic diameters are used in fluid-bed work, the weight mean diameter and the Sauter mean diameter. The Sauter mean diameter is also called the volume-surface mean diameter or the harmonic mean diameter. The particle size is typically given in microns. The weight mean diameter is calculated from the weight fraction of particles in each size cut as follows:

$$D_{WM} = D_1 X_1 + D_2 X_2 + \ldots + D_N X_N \tag{1}$$

which can be rewritten as:

$$D_{WM} = \sum_{1}^{N} D_N X_N \quad \textbf{(2)}$$

where

D_{WM} = weight mean particle diameter, microns
D_1 to D_N = average diameter of first to last size cut, microns
X_1 to X_N = weight fraction of first to last size cut

The Sauter mean diameter for a mixture of particles is calculated from the ratio of the average volume to the average surface area. This diameter is used when predicting the hydrodynamic performance of particle mixtures. This is calculated from the weight fractions of particle mixtures using the following equation:

$$D_{VS} = \frac{X_1 + X_2 + \ldots + X_N}{\frac{X_1}{D_1} + \frac{X_2}{D_2} + \ldots + \frac{X_N}{D_N}} = \frac{1}{\sum_{1}^{N} \frac{X_N}{D_N}} \quad \textbf{(3)}$$

where

D_{VS} = Sauter mean particle diameter, microns

When all the particles have the same diameter, the weight mean and the Sauter mean diameters are equal. For mixtures with a fairly narrow size distribution of particles, the weight mean and Sauter mean diameters will be similar with the Sauter diameter being smaller. For particle mixtures with a wide range of diameters, the Sauter mean diameter will be considerably smaller than the weight mean diameter.

Pressure loss

Pressure loss in a fluidized-bed boiler furnace is of great importance because it defines the total amount of solids inventory in the furnace, which is a major variable that influences heat transfer. Because the concentration of solids and the pressure profile in a fluidized-bed furnace are closely related, the determination of the pressure loss is a primary task when establishing furnace performance.

Bubbling bed In the case of bubbling-bed boilers, pressure loss is of special interest only in the bed itself. For the remainder of the boiler, the pressure loss is calculated by conventional boiler furnace fluid flow equations. (See Chapter 21.) Fig. 5 shows the furnace density profile and identifies the different zones used in establishing pressure drop and heat transfer.

The following equation is used to calculate the dense bed pressure drop in a bubbling bed.

$$\Delta P = (C)(1-e)(\rho_s - \rho_g)(L) \quad \textbf{(4)}$$

where

ΔP = pressure loss
C = units conversion constant
e = bed void fraction
L = bed height
ρ_s = particle density
ρ_g = gas density at bed conditions

e is primarily a function of particle size, particle density, bed gas velocity and gas viscosity.

Various methods are used to predict bed voidage, including those proposed by Leva,[1] Babu, *et al.*,[2] and Staub and Canada.[3]

Circulating bed The density profile of a CFB boiler furnace is more complex than in a bubbling bed. It is normal practice to establish a dense bed, bubbling or turbulent, in the bottom of the furnace. This is achieved by staging the admission of air to the furnace and supplying between 50 and 70% of the total air flow to the distributor plate. This reduces the gas velocity in the primary zone below the overfire air ports and allows the maintenance of the dense bed with comparatively low solids recirculation rates. Injection of overfire air converts the solids-gas flow regime to a circulating bed above the overfire air ports. The upward flow of solids decreases with increased furnace height, resulting in a reduction of local furnace density. Fig. 7 shows the density profile and the location of the various density zones in the furnace.

Pressure loss in a CFB furnace conforms to the basic equation below:

$$\Delta P = (C)(\rho_b)(L) \quad \textbf{(5)}$$

where

ΔP = pressure loss
C = units conversion constant
ρ_b = average bulk density in the furnace or furnace section associated with L
L = height of the furnace or furnace part of interest

To use the equation above, a density profile as shown in Fig. 7 is developed. This curve is a function of many variables and has been derived from empirical data. The important variables are:

D_p = average particle size above the dense bed zone
D_{DB} = average particle size in the dense bed zone
V = nominal gas velocity
T = nominal furnace temperature
W_S = external solids flux, lb/h ft^2 (kg/s m^2)
ρ_s = particle density
ϕ = particle shape factor
D_e = furnace equivalent diameter

Because the mixture bulk density in a furnace varies significantly with height, the furnace is divided into a number of zones (see Fig. 7 for definitions): dense bed, disengaging, transition and freeboard, wherein an average density value is calculated for each based on experimental data. The pressure loss equation, above, is then applied to each zone and summed to get the total furnace pressure loss.

Heat transfer

In conventional furnaces, the combustion gas carries a portion of the fuel ash with it as it goes through and out of the furnace. In general, this fuel ash represents less than 10 lb (4.54 kg) of inert solids per 1000 lb (454 kg) of gas. Also, the heat transfer from the gas to the furnace enclosure walls is predominately by radiation.

In a circulating fluidized-bed furnace, the amount of solids in the gas leaving the furnace may exceed 5000 lb (2268 kg) of solids per 1000 lb (454 kg) of gas. As a result of this high solids content, additional heat transfer mechanisms must be considered in the design. Heat transfer to the in-bed tubes of a bubbling bed and to the walls of a circulating bed includes solids and gas convection and solids and gas radiation in decreasing order of importance. In a conventional boiler furnace, gas radiation is the most important and solids convection is the least important to overall heat transfer.

The influence of the high solids concentration is significant. For equal temperatures, the heat transfer coefficients in a fluidized-bed boiler furnace are considerably higher than those in a conventional furnace. However, because the temperatures in the fluidized bed are between 1500 and 1600F (816 and 871C) the overall heat fluxes in the two systems are similar. Typical values for overall heat transfer coefficients in a fluidized-bed boiler furnace vary between 15 and 60 Btu/h ft^2 F (85 and 341 W/m^2 K).

Bubbling-bed heat transfer For heat transfer purposes, a bubbling-bed boiler is divided into three zones: bubbling bed or dense bed, disengaging, and upper furnace or freeboard. (See Fig. 5.)

Dense-bed heat transfer to tube banks The equation for the overall heat transfer coefficient for any tube is:

$$U_o = \frac{1}{\dfrac{1}{h_c + h_r} + R_m + R_{ft}} \quad \textbf{(6)}$$

where

U_o = overall heat transfer coefficient, Btu/h ft^2 F (W/m^2 K)
h_c = convection heat transfer coefficient for the tube bank, Btu/h ft^2 F (W/m^2 K)
h_r = radiation heat transfer coefficient for the tube bank and walls, Btu/h ft^2 F (W/m^2 K)
R_m = metal wall resistance, h ft^2 F/Btu (m^2 K/W)
R_{ft} = tube fluid film resistance, h ft^2 F/Btu (m^2 K/W)

The convection heat transfer coefficient h_c is given by Equation 7. Two single tube equations that are used are shown as follows. Equation 8 is a modified Vreedenberg form[4] and applies primarily to beds with particles less than 800 micron average. Equation 9, a Glicksman-Decker[5] type, applies well when the average particle size in the bed exceeds 800 microns.

$$h_c = (h_{st})(FAB) \quad \textbf{(7)}$$

$$h_{st} = 900\,(1-e)\left(\frac{k}{d_t}\right)\left[\left(\frac{G d_t \rho_s}{\rho_g \mu}\right)\left(\frac{\mu^2}{D_p{}^3 \rho_s{}^2 g}\right)\right]^{0.326} (\mathrm{Pr})^{0.3} \quad \textbf{(8)}$$

for $D_p < 800$ microns

$$h_{st} = \frac{k\,(1-e)}{D_p}\left[C_1 + (C_2)\left(\frac{3600\, D_p\, \rho_g C_p\, V}{k}\right)\right] \quad \textbf{(9)}$$

for $D_p > 800$ microns

where

h_{st} = convection heat transfer coefficient for a single tube, Btu/h ft^2 F
e = bed voidage, dimensionless
k = gas thermal conductivity, Btu/h ft F
d_t = tube outside diameter, ft
G = mass velocity or flux of the gas, lb/s ft^2
ρ_s = particle density, lb/ft^3
μ = gas viscosity, lb/ft s
ρ_g = gas density, lb/ft^3
D_p = average particle diameter, ft
g = acceleration constant, 32.2 ft/s^2
Pr = Prandtl number, dimensionless
C_1 = experimental constant, dimensionless
C_2 = experimental constant, dimensionless
C_p = gas specific heat, Btu/lb F
V = nominal bed gas velocity, ft/s

To convert the single tube heat transfer coefficients to those suitable for tube banks, the equations below are applied:

$$FAB = \left[1 - \left(\frac{D_o}{S_n}\right)\left(\frac{2D_o + S_p}{D_o + S_p}\right)\right]^{0.25} \quad \textbf{(10)}$$

where

FAB = bank arrangement factor (staggered arrangement only), dimensionless
D_o = tube outside diameter, in. (mm)
S_n = tube spacing normal to flow, in. (mm)
S_p = tube spacing parallel to flow, in. (mm)

Other variables are as defined previously. The equation for FAB is as derived by Gel'perin, *et al.*[6]

For the radiation heat transfer component, h_r, the following equation may be used:

$$h_r = (\sigma)(\epsilon)\left[(T_b)^4 - (T_w)^4\right]/(T_b - T_w) \quad \textbf{(11)}$$

where

ϵ = average overall emissivity, dimensionless
σ = 0.1713×10^{-8} Btu/h ft^2 R^4
T_b = absolute bed gas temperature, R
T_w = absolute wall temperature, R

The average overall emissivity in bubbling beds will be about 0.8 depending on wall emissivity and particle size. Typically, the overall heat transfer coefficient for an in-bed tube bundle is between 40 and 60 Btu/h ft^2 F (227 and 341 W/m^2 K).

Dense-bed heat transfer to wall Many equations for the vertical wall convection heat transfer coefficient have been proposed. An example is one proposed by Mickley,[7] which takes the following form:

$$h_{cw} = (C_3)\left[\frac{3600\,(\rho_s)(1-e)(\rho_g)(V)}{D_p^3}\right]^{0.263} \quad \textbf{(12)}$$

where C_3 is an experimental constant and the other variables are as previously defined. If the walls in the bubbling bed zone are coated with refractory, that resistance must be added to the other resistances in the equation for U_o. Refractory will also impact the calculation of h_r, but will not affect h_{cw} significantly.

Disengaging zone heat transfer to tube banks During periods of low bed level operation, the topmost tubes of the in-bed tube banks will be uncovered. While this portion of the bank is actually in the disengaging zone (called the splash zone in its lower portion), it is convenient to handle the heat transfer calculation as a special part of the bed. The solids content of the gas stream is much less than in the bed and decreases almost exponentially with height. As a result, the heat transfer rate to the exposed portion of the bank drops off rapidly. The heat transfer rates for this surface are based on experimental results.

As an example, Tang, *et al.*[8] developed the following empirical equation:

$$\frac{h}{h_m} = \exp\left(-\left(\frac{10 + H_L}{25.8}\right)^{2.2}\right) \qquad \textbf{(13)}$$

where

h = outside film heat transfer coefficient for an unsubmerged tube
h_m = outside film heat transfer coefficient for a fully submerged tube
H_L = unsubmerged tube height above the bed level, in.

Disengaging zone heat transfer to walls For the vertical walls in this zone, the convection rate conforms to an equation of the type shown below:

$$h_{cw} = C_4 \,[1 - C_5(1-e)] + h_{cg} \qquad \textbf{(14)}$$

where C_4 and C_5 are experimental constants, e is the dense bed voidage, and h_{cg} is given by the following equation:

$$h_{cg} = 0.023\frac{k}{D_e}\left(\frac{3600\, C_p\mu}{k}\right)^{0.3}\left(\frac{D_e G}{\mu\, e}\right)^{0.8} \qquad \textbf{(15)}$$

where

k = gas thermal conductivity, Btu/h ft F
D_e = equivalent diameter of vessel, ft
C_p = gas specific heat, Btu/lb F
μ = gas viscosity, lb/ft s
G = mass velocity or flux of the gas, lb/s ft^2

Note that the voidage, e, is for the dense bed and not the disengaging zone.

As a general rule, the height of this zone may be assumed to extend for a residence time of one second.

Upper furnace heat transfer (freeboard) This portion is handled similarly to conventional boilers. The main difference is the emissivity of the solids-gas mixture. The solids content of the gas will be quite high compared to conventional boilers that burn high ash coals and may alter the radiating properties.

Circulating-bed heat transfer Circulating fluidized-bed boilers do not incorporate tube bank surface and rely entirely on heat absorption of the containment walls and internal partitions, such as division walls and wingwalls. The heat transfer in CFB boiler furnaces is handled by breaking the furnace into two distinct regions. One comprises the dense bed and the other the remainder of the furnace.

Dense-bed heat transfer Dense bed heat transfer is similar to that described for the vertical walls of bubbling-bed boilers. There is some difference because the flow regime will usually be turbulent instead of bubbling.

Disengaging zone and upper furnace heat transfer The zone just above the dense bed, but below the point where secondary air is added to produce a circulating fluid bed, is defined as the disengaging zone. The upper furnace includes the transition zone and the freeboard as indicated in Fig. 7. Bed-to-wall heat transfer in these zones can be predicted by considering three parallel processes: particle convection, gas convection and radiation.

In particle convection, heat is removed from particles at the cooling wall surfaces by conduction. The energy loss is replenished by the material and energy exchange with the upwardly flowing central core of solids and combustion gases.

Gas convection is the dominant mode of heat transfer over the fractions of the surface not in contact with particles. This gas component is of minor consequence where the solids content is high. Even in the upper portions of a furnace, where solids content is relatively low, gas convection is usually small compared to the radiative component.

Radiant heat transfer occurs in a manner similar to that in conventional furnaces. Procedures are available to account for the combined emissivities and scattering effects of solids mixed with gases, as long as the void fraction is near or above 0.8. From a practical standpoint, this applies to the furnace above the dense bed.

The overall effective emissivity is a function of the pertinent radiative properties of the gases and solids found in a given mixture, as well as that of the heat absorbing surface. Typically, overall emissivities are about 0.5.

Various equations have been proposed to predict particle convection heat transfer coefficients. Some are complex and include many parameters. Two parameters have the most influence, particle size and mixture bulk density. Reliable convection equations have been developed considering only these two variables.

At full load conditions, where the solids circulation rate is high, a CFB furnace operates at approximately isothermal conditions from top to bottom. The overall heat transfer, in this case, is determined from the vertical bulk density distribution and a proper mean particle size. When the solids circulation rate is reduced, as occurs at low loads, the furnace becomes less isothermal, and more complex procedures are needed to calculate furnace heat absorption. One such procedure involves dividing the furnace into a larger number of vertically arranged zones. This permits the variation in temperature and bulk density to be properly handled in each zone wherein the variation is small.

The equations that follow are typical of those that predict furnace heat transfer:

$$U_o = \frac{1}{\dfrac{1}{h_c + h_r} + R_{ref} + R_m + R_{ft}} \qquad \textbf{(16)}$$

where

$h_c = h_{cp} + h_{cg}$
$h_{cp} = C_6\,(\rho_b)^m/(D_p)^n$
$h_r = (\sigma)\,(\epsilon)\,[(T_g)^4 - (T_w)^4]\,/\,(T_g - T_w)$

and where

T_g = absolute local bulk gas temperature, R
h_{cp} = particle heat transfer coefficient, Btu/h ft^2 F
C_6, m, n = experimentally derived constants

The value of h_{cg} is obtained with Equation 15 and all other variables are as previously defined.

Published literature provides a great deal of theoretical information, laboratory experimental data and test results from operating furnaces. Typical of the latter category is the paper of Kobro and Brereton.[9] Figs. 9 and 10 are from that paper. They show measured overall heat transfer rates for a range of bulk densities for two specific average particle sizes. Similar curves are made from commercial boiler field test data and are used to help make empirical corrections to the basic equations.

Heat and material balance

One of the first steps in designing a boiler is to calculate the overall heat and material balances. This establishes the flow quantities, compositions and temperatures of all streams entering and leaving the system. The boiler system is then subdivided and balances are calculated for the pieces. This process is repeated until all systems are defined.

The sensible heat of the solids, as a fraction of gas sensible heat, is quite large in fluidized-bed boilers. It represents the major difference between the heat and material balance for a fluidized-bed boiler and a conventional boiler. (See Chapter 9.) The information that follows describes the many flow streams that are considered when preparing the heat and material balance for a fluidized-bed boiler.

Material balance When limestone is used as a sorbent, the following chemical reactions which affect the solids balance occur in the furnace:

$$CaCO_3 \rightarrow CaO + CO_2$$
$$MgCO_3 \rightarrow MgO + CO_2$$
$$CaO + SO_2 + 1/2\ O_2 \rightarrow CaSO_4$$

There is a net solids weight loss during calcination as CO_2 in the carbonate is driven off endothermically and a solids weight gain as a result of the exothermic sulfation reaction. The Ca/S ratio selected, the amount of sulfur reacted and the ultimate analysis of the limestone will determine the net influence on boiler performance.

For the purpose of illustration, the material balance for an atmospheric pressure CFB boiler is shown in Fig. 11.

From an overall material balance standpoint, the solids that enter the furnace leave as relatively coarse material through the bed drain or as fine material collected after the convection pass. There is a significant amount of internal solids recirculation that takes place within the system, and these streams and their effects must be considered. Also, as discussed in the heat transfer section, the distribution of the solids in the furnace influences overall furnace thermal performance.

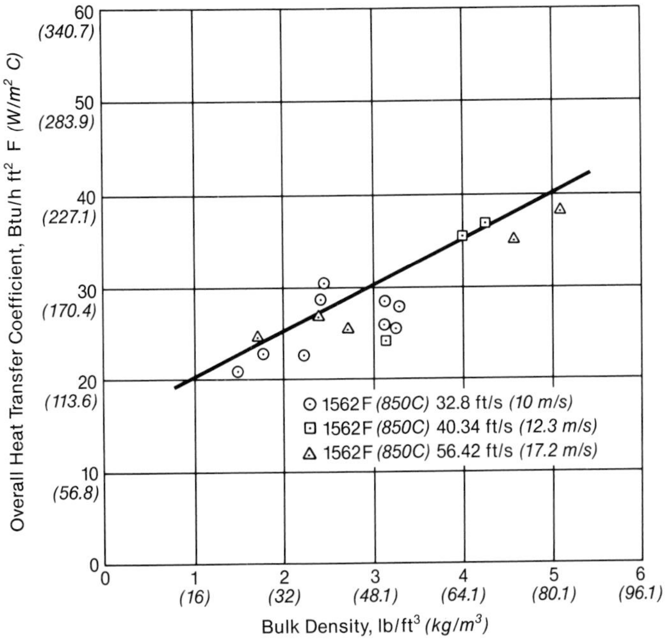

Fig. 9 Heat transfer coefficient versus density in a circulating fluidized bed, 250 micron mean diameter sand.

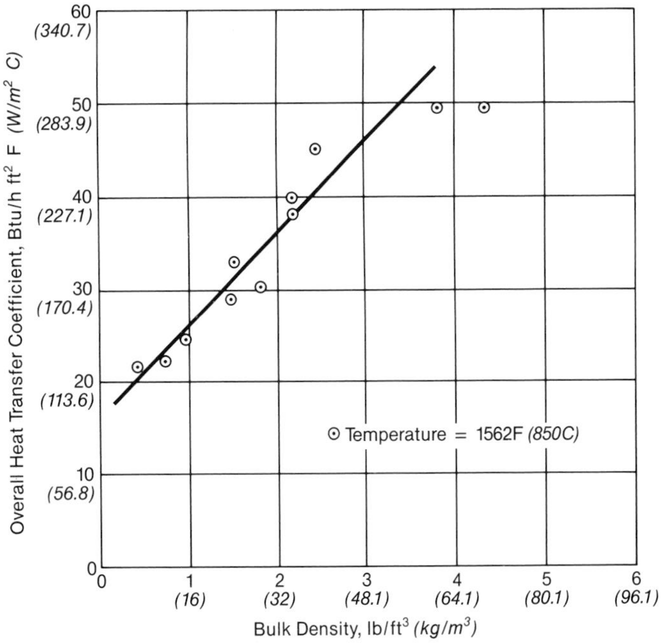

Fig. 10 Heat transfer coefficient versus density in a circulating fluidized bed, 170 micron mean diameter sand.

The fine particles that leave the bed and circulate are called *elutriated solids*. These solids ultimately leave the system after the convection pass with the gas leaving the dust collector or as purged from the multi-clone, if required. Changes in this fraction only change the split between the quantity of solids that leave through the bed drain and the quantity that leave the back end. In the discussions that follow, the sorbent input is considered to have already undergone calcination and sulfation reactions, and the fuel has completely combusted, leaving only ash. These flows should not be confused with the nonreacted sorbent and fuel fed into the boiler.

Definitions of the terms shown in Fig. 11 and the equations which are used to define the mass balance of a B&W CFB boiler are listed in Table 1.

The elutriated fractions Ea, Es and Ei depend on the specific fuel ash, sorbent and inert material in the system. Numerical values are determined from laboratory testing or from operating boilers. N_{UB} and N_{MC} depend on the mechanical arrangement of the collectors plus the size distribution and physical properties of solids circulating (ash, sorbent and inert) in the system. The values are determined empirically. The values of N_{UB}, N_{MC} and E for each input component and for the total solids are interdependent as shown in the following material balance equation.

$$ESF(1-N_{UB})(1-N_{MC}) + MCP = E(ISF) \qquad \textbf{(17)}$$

The external mass flux, W_S, is a variable controlled by the designer and is cited under the section on heat balance.

The outlines that follow describe the steps carried out when establishing the material balances for fluidized-bed boilers. For CFB boilers:

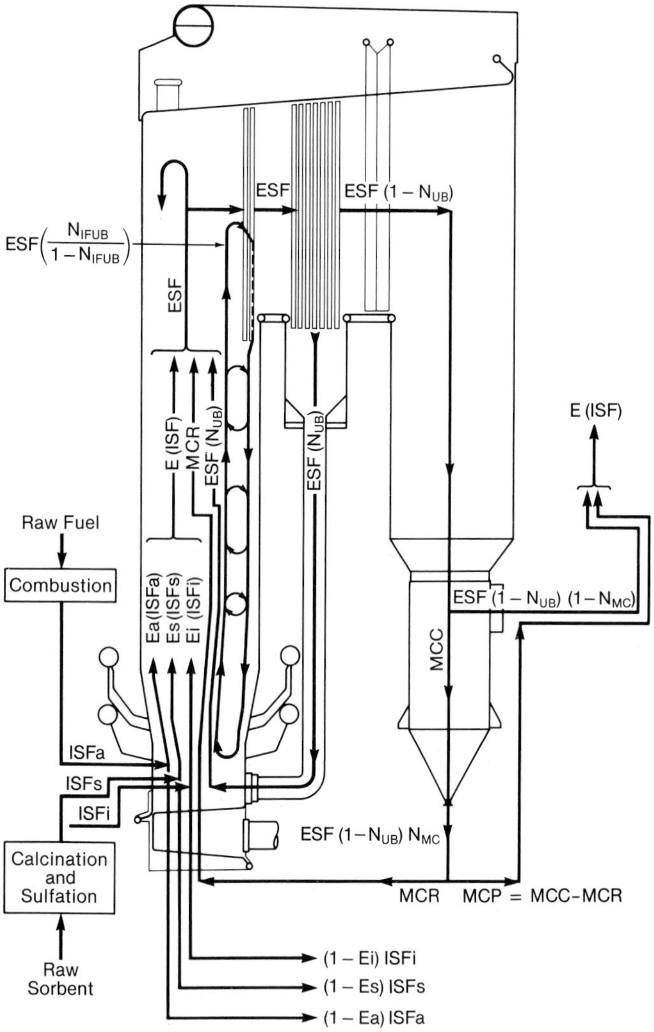

Fig. 11 Circulating fluidized-bed boiler material balance schematic.

1. ISF is determined from the required fuel and limestone inputs. Normally, the limestone input is proportional to the fuel input and is set to provide the required sulfur retention.
2. E is selected based on empirical data.
3. ESF is selected based on furnace design considerations.
4. N_{MC} and N_{UB} values are selected based on empirical data. It should be verified that the selected efficiency values are tied by Equation 17. All other solids flow, external and internal, can be determined from the information above.
5. BDF = (1-E) ISF
6. LVF = ESF (N_{UB})
7. CPSF = ESF (1-N_{UB})
8. BHC = ESF (1 - N_{UB}) (1 - N_{MC})
9. MCC = ESF (1-N_{UB}) N_{MC}
10. MCP = E(ISF) - BHC = MCC - MCR
11. MCR = ESF (1 - N_{UB}) - E(ISF) = CPSF - E(ISF)

Note that the baghouse loss is considered negligible. With properly selected E, N_{MC} and N_{UB} values, the CPSF will not exceed its maximum value corresponding to the allowable solids loading in the convection pass. In most cases, a circulating fluidized-bed boiler can be designed so the multi-clone purge (MCP) is zero. For bubbling-bed boilers:

1. ISF is as above
2. E is selected based on empirical data.
3. BDF = (1-E) ISF
4. Capacity of multi-clone recycle system (MCR_c) is selected based on empirical data in proportion to fuel input.
5. N_{MC} is selected based on empirical data and desired sorbent utilization and carbon conversion.
6. CPSF = ISF (E) + MCR_c
7. MCC = CPSF (N_{MC})
8. BHC = CPSF (1-N_{MC}) (assumes baghouse loss is negligible)
9. MCP = MCC - MCR_c
10. MCR = MCR_C
11. If MCP < 0, then set MCP = 0 and MCR = MCC.
12. BHC = E(ISF)
13. $MCR = \dfrac{E(ISF)\,N_{MC}}{1-N_{MC}}$
14. $CPSF = \dfrac{ISF\ (E)}{1-N_{MC}}$

Heat balance In furnaces fired by pulverized coal burners, Cyclones and stokers, little heat is removed from the combustion gases in the zone of maximum heat release. As a result, the flue gas reaches a high temperature before it is subsequently cooled by the furnace enclosure surface.

In fluidized-bed combustion, heat is removed from the zone of maximum heat release at a much higher rate. As a result, the maximum temperature of the flue gas is limited to a predetermined level.

Table 1
Mass Balance Terms

	Input solids flow (ISF), lb/h (kg/s)	
ISF	=	ISFa + ISFs + ISFi
ISFa	=	Fuel ash and unburned carbon
ISFs	=	Postcalcination and postsulfation sorbent
ISFi	=	Inert bed material (SiO_2 and H_2O)
	Elutriated fractions leaving the boiler, dimensionless	
Ea	=	Ash
Es	=	Sorbent
Ei	=	Inert material
E	=	[Ea(ISFa) + Es(ISFs) + Ei(ISFi)]/ISF
N_{IFUB}	=	In-furnace U-beam efficiency
	Additional terms	
BDF	=	Bed drain flow, lb/h (kg/s)
BHC	=	Baghouse catch flow, lb/h (kg/s)
CPSF	=	Convection pass solids flow, lb/h (kg/s)
ESF	=	External solids flow, lb/h (kg/s)
LVF	=	L-valve flow, lb/h (kg/s)
MCC	=	Multi-clone catch flow, lb/h (kg/s)
MCP	=	Multi-clone recycle purge flow, lb/h (kg/s)
MCR	=	Multi-clone recycle flow, lb/h (kg/s)
MCR_c	=	Multi-clone recycle system capacity, lb/h (kg/s)
N_{IFUB}	=	In-furnace U-beam efficiency
N_{MC}	=	Multi-clone dust collector efficiency
N_{UB}	=	Hot particle collector efficiency
W_S	=	External mass flux based on plan area of the upper furnace shaft (ESF/A), lb/h ft^2 (kg/m^2 s)

In bubbling-bed systems, the heat removal process is accomplished by having the cooling surface immersed in the active bed of hot solids and burning fuel.

In circulating beds, the large quantity of solids circulating in the system removes heat from the active combustion zone and transfers it to the heating surface throughout the furnace.

The ability to change the amount of heat removed from the combustion process to achieve the desired bed temperature provides flexibility when designing a fluidized-bed boiler. In a bubbling bed, the heat balance around the bed itself determines the bed temperature. The excess heat, which is the difference between the quantity of heat entering the bed with the fuel plus air and the amount of heat leaving the bed in the combustion products and through radiation to the rest of the furnace, must be absorbed by the in-bed tubes and the enclosure surrounding the bed. If the heating value of the fuel decreases, less in-bed surface is required to achieve the same bed temperature. If the heating value of the fuel increases, more in-bed surface is required. At the lower extreme of fuel heating value, no in-bed surface is used and the enclosure is often covered with refractory to minimize heat loss. This latter example is typical when high moisture sludges are burned.

In the CFB designs, all furnace heat transfer surfaces must be considered when making a heat balance. The amount of material circulating from the furnace to the primary collector and back to the furnace determines the inventory or average density of solids in the furnace. The heat transfer coefficient in the furnace is proportional to bulk density. Therefore, in the case of CFB boilers, the furnace heat absorption depends on the total furnace surface and the external recycle rate (which is a direct function of W_S, the external mass flux).

The furnace heat balance is also influenced by the way a particular fuel burns. When fuel is introduced into a fluidized bed, the majority of it burns in the dense bed. However, some of the fuel burns above the dense bed and the heat release pattern of the above-bed burning must be considered when calculating the furnace exit gas temperature. The actual split between in-bed and above-bed burning depends on fuel properties such as type, volatility, size and feed system. Because of the large amount of recirculated solids, the heat release pattern in a CFB furnace does not have a strong influence on the temperature distribution. For this reason, CFB furnaces are more tolerant of fuel changes than bubbling beds.

The heat balance around the convection surface is calculated similarly to conventional boilers. However, it is necessary to include the effects of the solids in the gas because they can have a substantial influence on the convection pass heat balance.

Fluidized-bed boiler arrangements

Boiler subsystems

The unique features discussed include the distributor plate, the bubble caps and the overfire air system. Also discussed are some of the features that are common to all water-cooled boilers designed by B&W.

Distributor plate and bubble caps The distributor plate is located at the bottom of the furnace and separates the windbox from the furnace. The distributor plate is fitted with bubble caps to provide a uniform distribution of combustion air to the entire furnace cross-section over the boiler load range. (See Fig. 12.) For this reason, distributor plates designed by B&W have a pressure drop of about 16 in. (406 mm) of water column across the bubble caps at full load and a minimum drop of 4 in. (102 mm) of water column at minimum load. Typically, 50 to 70% of the combustion air flows through the distributor plate of a CFB boiler at full load. A bubbling fluidized-bed boiler distributor plate is designed for 85 to 100% of the combustion air when staged combustion is not used.

The distributor plate must form an air-tight seal, other than the bubble caps between the furnace and the windbox and must support the weight of a slumped bed and resist the uplift generated from the air pressure drop across it during operation. In general, most distributor plates used in boilers are constructed from water-cooled membraned tubes.

In addition to the items discussed above, the bubble caps are designed to split the air into small streams for good flow distribution, avoid the formation of large bubbles in the bed, minimize erosion and prevent back sifting of bed solids into the windbox.

Overfire air system The overfire, or secondary, air system is part of the larger combustion and emission control system. As previously described under *Emissions*, all CFB boilers and some bubbling fluidized-bed boilers, when no tube bundle is present, use a staged combustion process to control the emission of NO_x. Overfire air is sometimes used in bubbling-bed boilers to improve free-

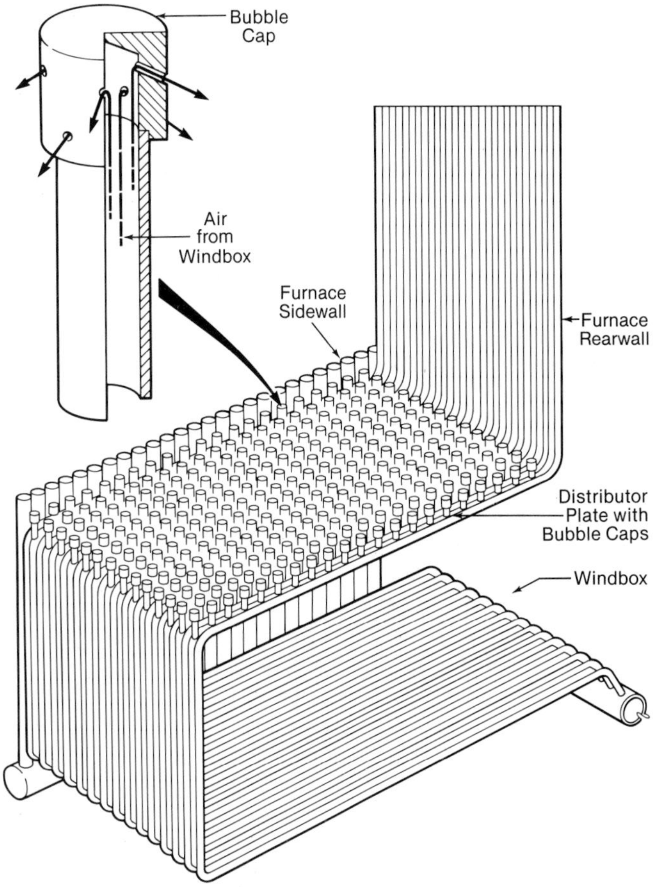

Fig. 12 Furnace distributor plate and bubble caps.

board mixing and burning. This technique is used when fuel is fed over-bed or when there are excessive fuel fines.

With overfire air, the combustion air that does not flow through the bubble caps is injected into the combustion gases shortly after initial coal ignition to complete combustion. The overfire air system must provide adequate penetration and thorough mixing of the overfire air with the combustion gases to achieve complete fuel burnout and minimize the amount of CO discharged to the atmosphere.

Overfire air penetration depends on the size of the overfire nozzles and the air and gas velocities and densities. In circulating beds, the solids density at the point of injection must also be considered.

Boiler enclosure The furnace and convection pass enclosures are conventional. (See Chapter 22.) They are constructed of water-cooled membraned tubes welded together to form a gas-tight enclosure. This enclosure is also used to support the fluidized bed, windbox, superheater and other components because the boilers are usually top supported.

Auxiliary equipment

Fuel feed systems The fuel feed system has had a greater impact on the evolution of fluidized-bed boilers than any other auxiliary or support system. As a result, three classifications of fuel feed systems have been developed and are in use today. These are the under-bed, over-bed and in-bed systems.

The under-bed feed system is usually applied to bubbling-bed boilers burning bituminous coal. An under-bed feed system is essentially a pneumatic transport system that moves the coal from a storage silo to the bed.

When burning bituminous coal and less reactive fuels, it is necessary to spread the coal as evenly as possible throughout the bed for good fuel-air mixing. To do this, fuel feed pipes are spaced on about 4 ft (1.2 m) centers throughout the bed.

When compared with the over-bed feed system described below, a well designed under-bed feed system can provide between 2 and 4% better combustion efficiency with bituminous coals.

On the negative side, an under-bed feed system is complicated. The fuel, typically, must be crushed to less than 0.25 in. (6.4 mm); the fuel must be dried to less than about 6% moisture or the transport pipes will plug; transport piping erosion can be a problem; the equipment needed to pressurize the coal for transport is a high maintenance item; and costs are generally higher than for an over-bed feed system.

Over-bed feed systems are also used on bubbling beds. They are used for reactive fuels and bituminous coals where the simplicity of the system can justify any reduction in carbon conversion.

The feeders are located above the bed where the furnace gas pressure is slightly below atmospheric. This location simplifies feeding since the coal stream does not have to be pressurized. An over-bed system uses the same fuel feed equipment as a spreader stoker. (See Chapter 15.)

While over-bed feed systems are simpler than under-bed systems, they still have requirements and limitations. The coal for an over-bed feed system is crushed to a top size of 1.25 in. (31.8 mm), and the amount of fine coal (less than 30 mesh) is limited to prevent excessive burning in the freeboard zone immediately above the bed. Also, the amount of fines in the fuel must be relatively consistent.

Fig. 13 compares the combustion efficiency for under-bed and over-bed coal feed systems for two coals, Sarpy Creek and Kentucky No. 9. Sarpy Creek is a subbituminous coal and Kentucky No. 9 is a bituminous coal. It can be seen from Fig. 13 that the Sarpy Creek coal is much more reactive than the Kentucky coal. Also, under-bed feed results in higher combustion efficiency than over-bed. Recycling solids collected in a multi-clone to the furnace also increases combustion efficiency.

In-bed feed systems are used for CFB boilers and some sludge-burning bubbling beds. Typically, some form of a feed screw or air assisted chute is used. The fuel is injected through an enclosure wall just above the distributor plate. Because the furnace pressure at this point can be as high as 50 in. (1270 mm) of water gauge, the feed system must seal against this pressure to allow the fuel to be introduced to the furnace. The pressure seal is accomplished by the use of a rotary seal or a head of coal between the coal silo and a belt feeder.

A CFB with its higher fluidizing velocity is a better fuel mixing process than a bubbling bed with in-bed tubes. As a result, fewer feed points are required.

Sorbent feed systems As previously discussed under *Emissions*, limestone or dolomite is added to a fluidized-bed boiler to capture SO_2. In B&W designs, the sorbent is usually fed into the dense bed area of the furnace. While good limestone distribution is desired to reduce sorbent requirements, it is not nearly as critical as distribution in the case of under-bed coal feed.

The design of the sorbent feed system must consider where the material will be injected into the furnace, the furnace gas pressure at that point and the method of addition. In the case of under-bed and in-bed feed, the sorbent can be mixed and injected with the fuel. Sorbent can not be fed properly with a spreader due to its fineness. Sorbent has been blown into the furnace through separate pneumatic feed points and by gravity from a storage silo.

Bed ash removal systems When an ash bearing fuel is burned in a fluidized-bed boiler, the ash is liberated from the coal in the boiler furnace. Also, sorbent or inerts are fed to the boiler. Therefore, means must be provided to remove the solids from the system to prevent accumula-

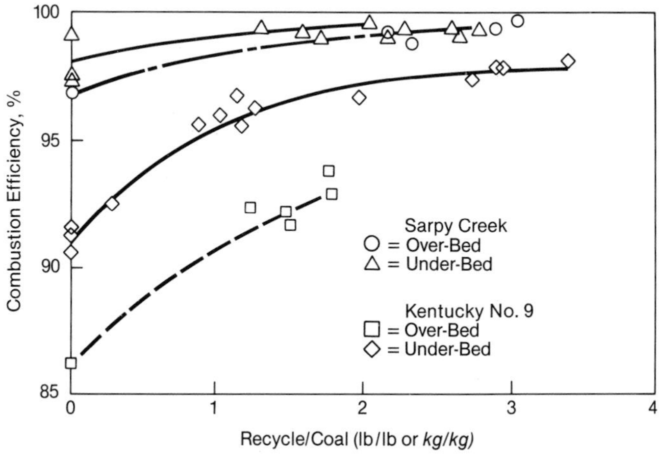

Fig. 13 Comparison of under-bed and over-bed coal feed.

tion. There are two major places in a fluidized-bed boiler system where solids are removed, the bed drain and the baghouse or electrostatic precipitator.

Normally, the design of the bed ash removal system includes a large safety margin because of the uncertainty as to where the solids will exit the system and to handle changes in fuels. Operating results have shown that the bed drain flow can range from 0 to more than 50% of the total solids output.

In the case of fuels with a high alkali content in the ash, it may be necessary to drain the bed material from the boiler at a rate higher than the solids buildup rate to prevent a concentration buildup of the alkali in the bed.

Under these conditions, it is also necessary to make up any excessive bed material lost from the system. Typically, when the concentration of alkali exceeds 5 to 6% of the bed weight, the probability of forming agglomerates increases significantly.

A second case that requires greater than normal bed drain flow rates is created when the fuel contains a high percentage of large rocks and ash, greater than 0.5 in. (12.7 mm). There is a natural tendency for these large pieces of noncombustible material to accumulate in the bed and cause defluidization. To avoid this, it is necessary to move the oversize material to the bed drain for removal. This is done by maintaining the bed drain flow rate above a minimum experimentally-determined value. It is necessary to make up any excess bed material lost in the rock removal process.

Sootblowers Chapter 23 discusses the application of sootblowers to various types of firing systems. Because the fuel in a fluidized-bed boiler is burned at temperatures below the ash softening point, the flyash never reaches the plastic state. As a result, it forms a dry powder that is easy to remove from the heating surfaces.

There are fluidized-bed boilers in operation that have no sootblowers. As long as the heat transfer surface is properly spaced and the dusting that does occur is accounted for in the design of the unit, the boiler will perform satisfactorily, achieving the desired values of superheat temperature and boiler efficiency. However, when fuels with high sodium and/or potassium ash contents (low ash fusion temperatures) are fired, especially with an over-bed feed system, there is a higher likelihood for convection pass fouling and, therefore, sootblowers may be required.

Typical fluidized-bed boiler designs

At the time of this writing, the majority of applications for bubbling fluidized beds had been directed at installations of less than 200,000 lb/h (25.2 kg/s) of steam. Most of the new, large capacity, coal-fired fluidized-bed boilers have been of the circulating type. However, several large boilers have been retrofitted with bubbling beds because the bubbling bed was physically more compatible with the changes, as compared with circulating beds. In some of these cases, bubbling beds were used to reduce SO_2 emissions. In other cases, fuel-related operating problems have provided the incentive for change.

Another area where bubbling beds, both new and retrofitted, are being used is for burning sewage sludge and pulp mill and recycled de-inking paper sludge.

The design of a boiler and its auxiliaries is a complex task. Because other chapters describe the conventional aspects of boiler design, the following descriptions are limited to unique fluidized-bed boiler features.

Bubbling-bed retrofit During the winter of 1986, the moving grates were removed from Unit No. 2 at Montana-Dakota Utilities' R. M. Heskett station and replaced with a bubbling fluidized bed. (See Fig. 14.) This boiler was originally designed to generate 650,000 lb/h (81.9 kg/s) steam at 1300 psig (89.6 bar gauge) and 950F (510C) superheat burning Beulah lignite. However, high sodium content in the fuel ash caused severe furnace slagging and superheater fouling. Before the bubbling-bed modification, long term output was limited to about 50 MW of the rated capacity of 72 MW. The fluidized bed was installed to reduce furnace operating temperatures, to avoid slagging and fouling problems, and to achieve full power operation.

The new fluidized bed measured 40 × 26 ft (12.2 × 7.9 m) and was fitted under the boiler with only minor pressure part changes to the furnace division walls. The distributor plate and enclosure walls are water-cooled. In this unit, both superheat and boiling surface are placed in the bed to meet steam generation and turbine superheat temperature requirements and to limit the bed operating temperature to 1500F (816C). The superficial bed velocity is 12 ft/s (3.7 m/s) and the bed is 54 in. (1372 mm) deep in the expanded condition. Eight windbox compartments are used for unit turndown and startup. Because Beulah lignite is a highly reactive fuel, no flyash recycle was installed. Additionally, because of the low sulfur and high alkali content of this fuel, sand is used as the bed material.

The boiler was restarted in May 1987. This unit is now producing 80 MW and avoiding fuel related slagging and fouling problems. However, operators must avoid high bed and furnace temperatures because of the fuel's fouling tendencies.

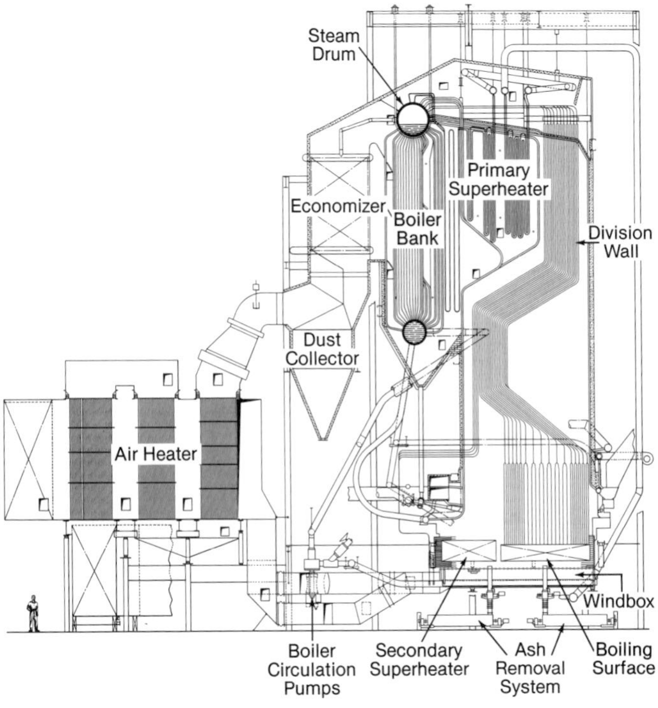

Fig. 14 80 MW bubbling fluidized-bed retrofit.

Circulating fluidized-bed boiler The B&W circulating fluidized-bed boiler design is based on a completely water-cooled setting. This feature provides a modern gas-tight enclosure suitable for operating with a positive pressure in the furnace. It has no high temperature refractory lined flues in the vicinity of the primary particle collector and therefore requires minimal building space and reduces furnace refractory maintenance. This construction is possible due to the use of an impingement-type primary solids separator (U-beams) which can easily be integrated into the boiler enclosure.

Fig. 15 shows the side view of a 55 MW CFB boiler designed to burn low volatile bituminous coal. The unit produces 465,000 lb/h (58.6 kg/s) steam at 1550 psig (107 bar gauge) and 955F (513C). The furnace is 30 ft (9.1 m) wide, 15 ft (4.6 m) deep, 85 ft (25.9 m) high and contains full height, water-cooled division walls and steam-cooled wingwalls in the upper furnace.

Fuel and sorbent are fed to the bed through the lower furnace frontwall. The ash and spent sorbent are removed through drain pipes in the floor. The solids collected by the U-beams and multi-clone are returned to the lower furnace through the rearwall.

The primary air enters the furnace through the distributor plate and secondary air is injected at elevations approximately 6 and 12 ft (1.8 and 3.7 m) above the distributor plate.

The entire lower furnace, up to 22 ft (6.7 m) above the distributor plate, is covered by a thin layer of highly conductive refractory held to the water tubes by pin studs. Refractory is used in the lower furnace to protect the tubes from corrosion and erosion. The remaining portion of the furnace enclosure consists of bare tubes.

The B&W CFB boiler design uses an impact separator to collect and recycle solids to the furnace. The primary solids separation system consists of staggered rows of U-shaped channel members, or U-beams, suspended from the boiler roof. (See insert, Fig. 15.) Material strik-

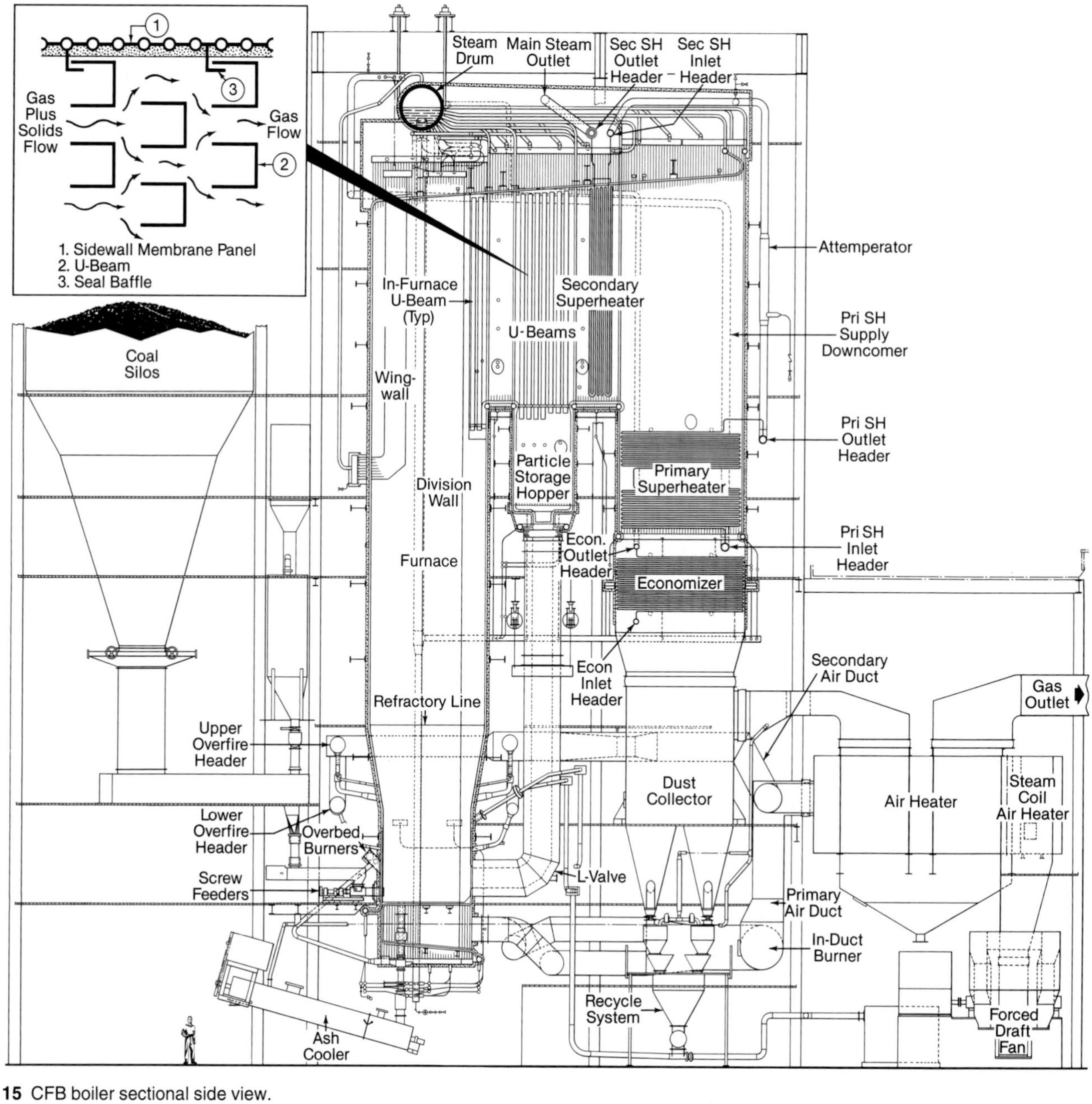

Fig. 15 CFB boiler sectional side view.

ing the U-beams is separated from the gas, flows down the U-channel and discharges from the bottom. The most recent designs use two stages of U-beam collectors which result in better overall collection efficiency than was provided in the first generation of boilers. The first stage is located in the upper furnace and returns the solids directly to the lower furnace. The second stage is located after the furnace and above a particle storage hopper. Bed material collected by the second stage U-beams is recycled to the lower furnace by flow-controlling L-valves.

The L-valve is a nonmechanical device for returning solids to the furnace (Fig. 16). Solids collected in the particle storage hopper flow to the standpipes and serve as a source of inventory for the furnace. Solids flow is induced by injecting a small amount of aeration into the L-valve. With this arrangement, hundreds of thousands of pounds (kg) of solids per hour can be circulated with air flows on the order of 10 ACFM (4.7×10^{-3} m^3/s).

Startup and operation

Startup

Fluidized-bed boilers are brought into operation by first establishing air flow through the unit, heating the bed material to a temperature that is above the fuel's autoignition temperature, and then introducing the fuel into the hot solids and air. The fact that the bed solids are above the fuel's autoignition temperature ensures that combustion is initiated safely.

The heat for raising the bed material to the desired temperature is normally supplied by a burner in the air supply duct that is capable of heating the air and solids to the required temperature, a burner located above the bed that fires into the bed, or a combination of both methods.

The above-bed type of burner is generally used for circulating beds and bubbling beds that do not have in-bed surface. The duct type of burner is used for bubbling beds that contain in-bed cooling surface. Both the above-bed and duct burners must have complete flame safety systems which include independent flame detection and burner safety trip control circuits.

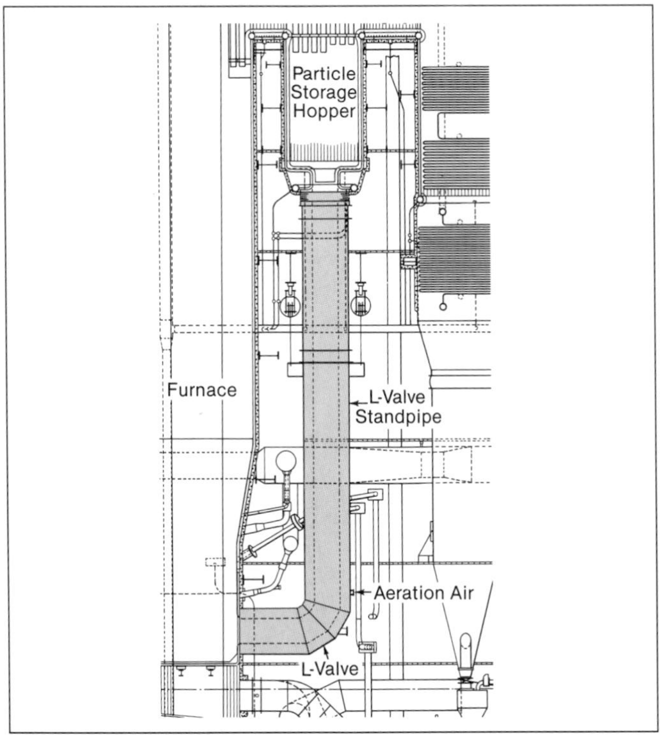

Fig. 16 Particle recirculation for a CFB.

Different fuels have varying ignition temperatures. The following table lists several fuels and the minimum bed temperature that must be attained before the fuel can be fed into the bed.

Fuel	Minimum Bed Temperature, F (C)
Bituminous coal	900 to 950 (482 to 510)
Lignite	900 (482)
Anthracite	1000 to 1050 (538 to 566)
Wet wood	1200 to 1250 (649 to 677)
Oil	1400 (760)
Natural gas	1400 (760)

The actual bed temperature for a particular fuel must be determined by demonstration. Once the minimum safe bed temperature is established, it is only necessary to measure bed temperature to make sure it is above the minimum value whenever feeding fuel to the bed. As fuel is introduced into the bed, it ignites, releases its energy and increases the bed temperature above the minimum value. Further increases in fuel feed will result in higher bed temperatures. This process is continued until the desired boiler load and bed temperature are achieved.

System control

Many of the operating and control features of fluidized-bed and conventional boilers are the same. There are two areas where fluidized-bed and conventional boilers vary significantly.

One is the need for the fluidized-bed boiler control system to monitor and control the transport of large quantities of solids. The other is the need to control the primary and secondary air ratio to achieve minimum emissions and unburned carbon loss. With the ability to control solids and air flows, the unique fluidized-bed parameters of bed temperature and bed inventory can be controlled.

Bed temperature control

Bubbling bed The temperature of a fluidized bed is controlled by manipulating the heat removal process. In a bubbling-bed boiler, the bed itself is limited to the lowermost 4 ft (1.2 m) of the furnace.

Because the heat absorbing surface is located within the bed and solids recycle has only a minor influence on bed temperature, the heat balance around the bed establishes the bed operating temperature. As the load (fuel flow) is changed, it is necessary to change the bed heat balance to achieve the desired bed operating temperature. The heat balance can be altered by the following methods.

For the case of a load reduction, bed depth can be reduced by removing inventory from the system and/or reducing the velocity of the combustion gas through the bed. By decreasing the bed depth, less heat transfer surface will be submerged in the solids. As a result, the amount of heat removed decreases and the bed temperature is maintained at the desired value.

There are limits to the degree of turndown achievable by reducing the bed level and a point is reached where a portion of the bed must be removed from service, or *slumped*. Under these conditions, little heat is removed from the bed in the slumped area. This control requirement dictates that the windbox be designed so that portions of the bed can be isolated and removed from service.

Bed temperatures can also be controlled by excess air and air preheat. By increasing the excess air that flows through the bed, the amount of heat required to heat the air increases and the temperature of the bed decreases. By decreasing the air preheat temperature, the temperature of the bed can also be made to decrease.

Circulating bed In CFB boilers, the solids inventory is distributed more uniformly throughout the furnace than in a bubbling bed. (See Figs. 5 and 7.) The combination of this condition and the large amount of solids that recirculate at high boiler loads results in uniform furnace gas and solids temperatures. In the section on heat transfer, it was pointed out that the heat transfer coefficient in a CFB furnace is greatly dependent on the solids inventory in the furnace. Therefore, by changing furnace inventory, the heat transfer from the gas and solids to the cooling surface can be varied. The net result of this process is the ability to control furnace bed temperature. Therefore, as load changes or as fuel varies, it is possible to increase or decrease furnace temperatures by controlling furnace inventory. In some CFB designs, furnace temperature is moderated with an external heat exchanger. As in bubbling beds, the bed temperature may also be changed by varying excess air and air preheat temperature.

Bed inventory control

The ability to maintain the required bed inventory, as described above, is highly dependent on the properties of the solids that make up the inventory as well as the performance characteristics of the solids collection and recycle system. Furnace inventory is made up of sorbent and/or inert bed material and fuel ash, but each must be properly controlled.

For the system to be under control, the quantity of solids in the furnace must equal or exceed the amount required to support stable combustion and for bed temperature control. If too many solids leave the boiler with the flue gas, furnace inventory will decrease, combustion may become unacceptable and furnace temperatures will increase. This condition must be corrected by modifying the particle size distribution of the solids fed to the boiler or by providing some kind of inventory control system.

Bubbling bed Bubbling-bed inventory is dependent upon the balance between feed solids and the bed drain purge flow, plus the solids that are elutriated and lost from the system through back-end collectors. The amount of fine material leaving the boiler with the flue gas (elutriated fraction) depends on particle size and fluidizing velocity. When solids are recycled from a dust collector, the flow of solids in the freeboard increases significantly. However, bed inventory, which consists mostly of nonentrainable particles, is not affected.

From a practical operating and control standpoint, the minimum solids feed flow rate must provide enough material to maintain the bed drain flow required to remove oversized particles plus make up the back-end loss. With solids input exceeding the minimum value, the rate of bed drain purge is adjusted to maintain the desired bed inventory. In practice, bed inventory is controlled by measuring the pressure drop across the bed and maintaining it at a value corresponding to the desired inventory.

Circulating bed In a circulating fluidized bed, the amount of solids leaving the furnace with the flue gas and the bed inventory are interdependent. Solids leaving the furnace must be collected and returned to the furnace to maintain bed inventory. Overall inventory control is determined by the solids input-output balance similar to a bubbling bed and is controlled in the same way.

In B&W's circulating fluidized-bed system, the inventory consists of solids in the furnace plus those in the particle storage hopper and L-valve standpipes. The inventory distribution between the furnace and storage is established by the external recycle rate which is controlled by nonmechanical L-valves. For a constant total inventory, a reduction in L-valve flow rate transfers inventory from the furnace to the particle storage hopper and L-valve standpipes and vice versa.

Because the solids flow rate circulating through the L-valves is many times larger than the solids input rate, inventory can be exchanged between the furnace and particle storage hopper at the rate of 10 to 20% per minute. This feature allows rapid control response to changes in fuel flow or boiler load.

The inventory split between the dense bed and the upper furnace is dependent upon the particle size distribution of the solids in the furnace and gas velocities in the primary zone and upper furnace. To provide sufficient inventory in the upper furnace and satisfy heat transfer requirements, the total system inventory is controlled by the bed drain purge rate. On the other hand, the bed drain purge rate should be sufficient to prevent accumulation of coarse material in the dense bed.

If the bed drain purge rate required for coarse material

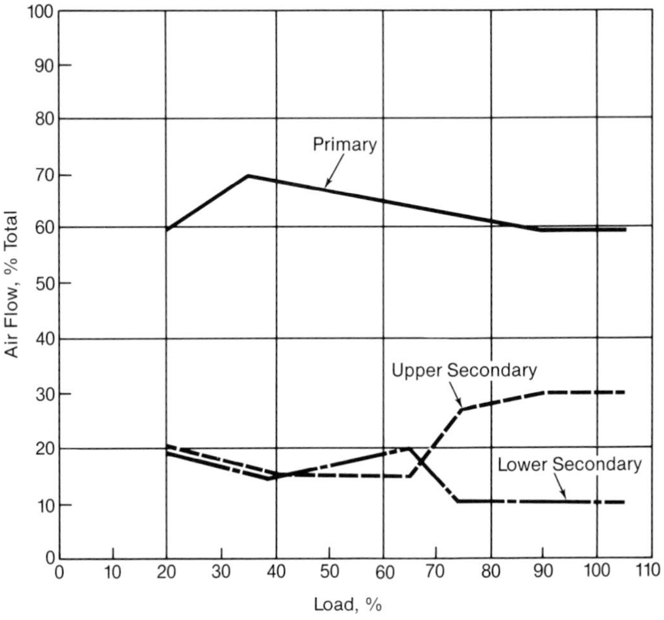

Fig. 17 Air flow distribution.

removal exceeds that needed for total inventory control, a depletion of upper furnace inventory will result. In this case, means must be provided to bring the system back into balance. Furnace inventory size distribution can be corrected by changing the particle size distribution of the input solids or by classifying and recirculating a portion of the bed drain flow. Like the bubbling bed, furnace pressure drop measurements provide the required control signals.

Overfire air control

The split between primary air and overfire air is established, primarily, to optimize fuel burnout and CO and NO_x emissions. Fig. 17 shows a typical curve of air flow versus load for a bituminous coal-fired boiler. As a result of variations among fuels, final air splits are established as a result of unit operation and control system tuning for optimum combustion and furnace inventory distribution.

Summary

By traditional power industry development standards, fluidized-bed boilers represent an emerging technology. However, during the past 20 years the efforts have brought the technology from concept to commercial status. Activities are continuing to improve existing designs and extend unit size to larger capacities.

In the quest to reduce electric power costs and emissions, pressurized fluidized-bed boiler plants are being built and operated. (See Chapter 29.) In addition, investigators are re-examining fluidized-bed gasification to determine if improved power generating systems can be designed. (See Chapter 17.)

Today thousands of fluidized-bed units are in operation and are providing industry with a technique capable of burning a wide variety of fuels in an environmentally improved manner.

References

1. Leva, M., *Canadian Journal of Chemical Engineering*, Vol. 35, pp. 71-76, August 1957.

2. Babu, S.P., Shah, B., and Talwalkar, A., AIChE Symposium Series No. 176, Vol. 74, pp. 176-186, 1978.

3. Staub, F.W., and Canada, G.S., *Fluidization*, Cambridge University Press, pp. 339-344, 1978.

4. Vreedenberg, H.A., *Chemical Engineering Science*, Vol. 9, pp. 52-60, 1958.

5. Glicksman, L.R., and Decker, N.A., Proceedings of the 6th International Fluidized Bed Combustion Conference, Atlanta, Georgia, pp. 1152 to 1158, 1980.

6. Gel'perin, N.I., Ainshtein, V.G., and Korotyanskaya, L.A., *International Chemical Engineering*, Vol. 9, No. 1, pp. 137 to 142, January 1969.

7. Mickley, H.S., and Trilling, C.A., *Industrial Engineering Chemistry*, Vol. 41, No. 6, June 1949.

8. Tang, J.T., *et al.*, AIChE Spring National Meeting, Houston, Texas, March 27-31, 1983.

9. Kobro, H., and Brereton, C., Circulating Fluidized-Bed Technology Proceedings of the First International Conference on Circulating Fluidized Beds, Halifax, Nova Scotia, Canada, pp. 263 to 272, November 1985.

Portion of an entrained-flow gasifier under fabrication at B&W.

Chapter 17
Coal Gasification

Definition of coal gasification

Coal gasification is a process that converts coal from a solid to a gaseous fuel through partial oxidation. Once the fuel is in the gaseous state, undesirable substances, such as sulfur compounds and coal ash, may be removed from the gas by established techniques. The net result is a clean, transportable gaseous energy source.

When coal is burned, its potential chemical energy is released in the form of heat. Oxygen from the air combines with the carbon and hydrogen in the coal to produce gaseous carbon dioxide (CO_2) and water plus thermal energy. Under normal conditions, when there is an abundance of air (oxygen), all of the chemical energy in the coal is converted into heat and the process is called combustion.

However, if the available oxygen is reduced, less heat is released from the coal and new gaseous reaction products appear. These byproducts, including hydrogen, carbon monoxide (CO) and methane (CH_4), also contain potential chemical energy. If the objective is to maximize the chemical energy in the gas byproducts, it would appear logical to continue decreasing the available oxygen. However, a point is reached where an increasing percentage of the coal is no longer converted to gas, leaving behind unreacted carbon, and the process becomes inefficient. When the oxygen supply is controlled such that both heat and a new gaseous fuel are produced as the coal is consumed, the process is called *gasification*.

Current state of development

Coal is a major energy source for electric power generation worldwide. (See Chapter 8.) The continued popularity of coal depends partly on our ability to control its undesirable combustion byproducts. In this respect, gasification-based systems are favored.

Coal gasification processes which supply synthesis gas for chemical production were commercialized in the 1950s. Improved second and third generation gasifier designs have been developed and large synthesis gas commercial units were built in the 1980s.

Coal gasification for electrical power generation was demonstrated in the United States (U.S.) at the Cool Water power station in California (92 MW) and the Plaquemine power station in Louisiana (155 MW). The first fully integrated gasification combined cycle (IGCC) plant is the 250 MW unit at Buggenum in the Netherlands. The plant is oxygen-blown, using a cold acid gas removal process typical of the petroleum industry.

Chemistry of gasification processes

Gasification reactions

In contrast to combustion reactions which are achieved with a controlled excess of oxygen, gasification consists of incomplete combustion with an oxygen deficiency. Carbon monoxide and hydrogen, which are combustible gases, are the most common products of combustion. Only a fraction of the carbon in the coal is completely oxidized to CO_2. The heat released by partial combustion provides most of the energy necessary to break chemical bonds in the coal and raise the products to reaction temperature.

The chemistry of coal gasification is complex; only a few of the important reactions are discussed here. The following are the major reactions involved in the gasification process:

Exothermic reactions — releasing heat

Carbon combustion:	$C + O_2 = CO_2$	**(1)**
	$C + \frac{1}{2} O_2 = CO$	**(2)**
Water-gas shift:	$CO + H_2O = CO_2 + H_2$	**(3)**
Methanation:	$CO + 3H_2 = CH_4 + H_2O$	**(4)**
	$C + 2H_2 = CH_4$	**(5)**

Endothermic reactions — absorbing heat

Boudouard reaction:	$C + CO_2 = 2CO$	**(6)**
Steam-carbon reaction:	$C + H_2O = CO + H_2$	**(7)**
Hydrogen liberation:	$2H \text{ (in coal)} = H_2 \text{ (gas)}$	**(8)**

The methanation reactions are important in lower temperature systems and are favored by high pressures. The other reactions are more prominent in high temperature gasification systems. The combination of these reactions is autothermic, i.e., sufficient heat is initially released by the carbon combustion reactions to provide energy for the endothermic reactions.

Sulfur in coal is converted primarily to hydrogen sulfide (H_2S) and a small amount of carbonyl sulfide (COS) is also formed. High temperatures and low pressures favor coal nitrogen conversion to N_2, while the opposite conditions favor some ammonia (NH_3) formation with small amounts of HCN also being formed. In lower temperature processes [< 1200F (< 649C)], tars, oils and phenols are not destroyed and therefore exit with the raw gas.

Reaction stages

There are two main reaction stages in the gasification process: devolatilization and char gasification. The devolatilization stage is one of transition; coal becomes coal char as the temperature rises. Weaker chemical bonds are broken and tars, oils, phenols and hydrocarbon gases are formed. Char, or fixed carbon, which remains after devolatilization is gasified by reaction with oxygen, steam, carbon dioxide and hydrogen. The gases react with each other to produce the final mixture.

The type of gasification process has a strong bearing on the devolatilization products. In fixed- or moving-bed gasifiers (see below), these products exit the gasifier with the product gas due to the low temperatures and lack of oxygen. In fluidized-bed and entrained-flow processes, uniform high temperatures cause hydrocarbon cracking, or breaking down the more complex molecules to simpler ones. Also, oxygen is available to react with the devolatilization products, producing hydrogen, CO and CO_2. These reactions are the most complete in an entrained-flow process.

The basic objective of gasification is to convert coal into a combustible gas containing the maximum remaining heating value. If the gas is used for chemical synthesis rather than fuel for combustion, its composition must be adjusted for the stoichiometry of the synthesized product. Hydrogen for synthesis gas production must be provided by the steam-carbon reaction or by the water-gas shift. The influence of the end product may affect the choice of gasification process. The water-gas shift reaction can be promoted by supplying extra steam in the fuel bed or in a separate catalytic conversion stage downstream. This produces increased hydrogen. Where hydrogen is the only product required, CO_2, an inevitable product of splitting water by reaction with carbon, must be removed.

In the case of synthetic natural gas (SNG) production with low temperature, high pressure moving-bed processes, much of the methane can be produced in the fuel bed by carbon hydrogenation (Reaction 5). However, all of the methane can not be produced by direct hydrogenation. Some must be produced indirectly in a separate catalytic stage, e.g., by Reaction 4.

Classification of gasification processes

There are many coal gasification processes. However, they can be separated into three major reactor types:[1,2]

1. moving-bed, or countercurrent,
2. fluidized-bed, or back-mixed, and
3. entrained-flow, or plug-flow.

Fig. 1 shows the three generic coal gasification reactors and their temperature profiles. The locations of coal, steam and oxidant inputs and synthetic gas and ash outputs are also indicated. Table 1 summarizes the important characteristics of each gasifier. Gasifiers may be characterized by the bottom ash as dry or slagging. If the temperature is maintained below the ash fusion temperature, the ash is removed dry. The ash can also be made to melt and be run off as a liquid slag.

Moving bed

The moving-bed gasifier (Fig. 1a) is also referred to as a fixed-bed. In this unit, a column or bed of crushed coal is

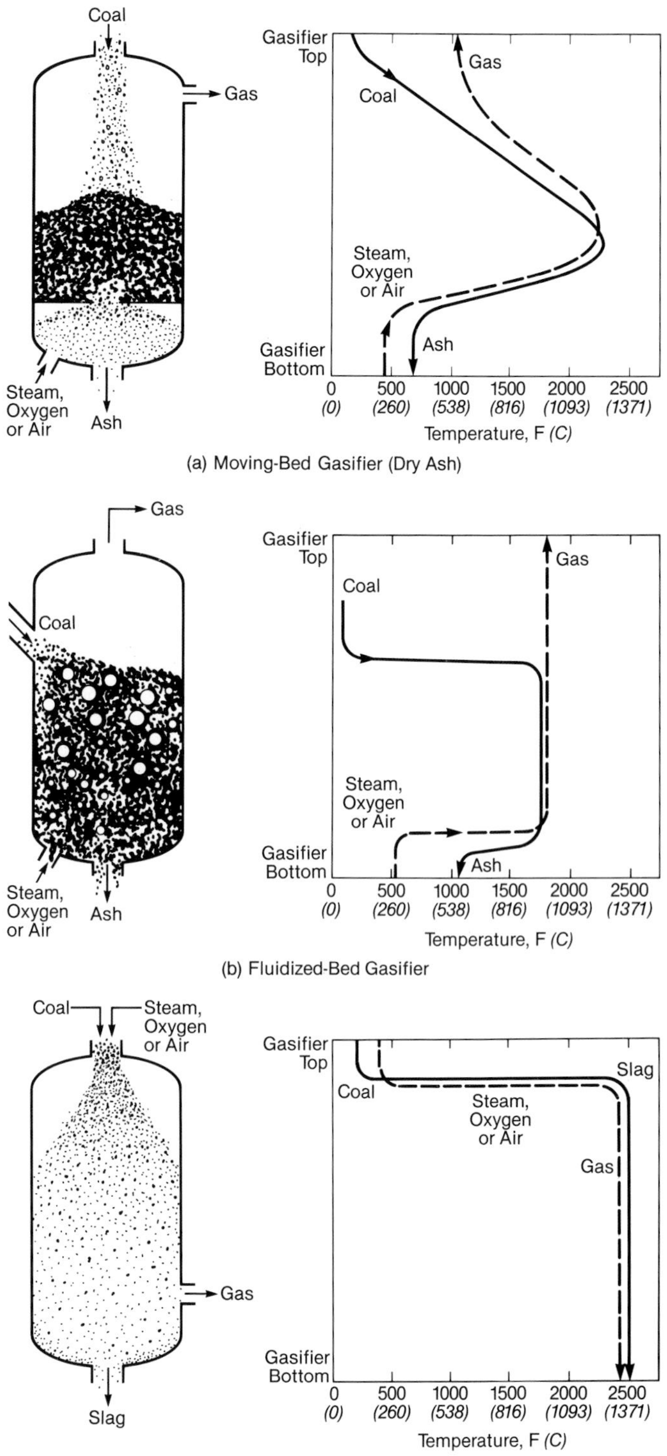

Fig. 1 Typical reactors for coal gasification (*courtesy of the Electric Power Research Institute*).[1,2]

supported by a grate and the process involves a series of countercurrent reactions. At the top, the coal is heated and dried while cooling the product gas. The coal is further heated and devolatilized as it descends through the carbonization zone. Below this area, the devolatilized coal is gasified by reaction with steam and CO_2. The highest temperatures are reached in the combustion zone near the bottom of the gasifier. The char-steam reaction together with the presence of excess steam keeps the temperature in the combustion zone below the ash slagging temperature.

Table 1
Gasifier Characteristics[1]

Moving Bed

Ash Conditions	Dry Ash	Slagging
Feed coal characteristics:		
Size	Coarse [<2 in. (<51 mm)]	Coarse [<2 in. (<51 mm)]
Acceptability of fines	Limited	Better than dry ash
Acceptability of caking coal	Yes (with modifications)	Yes (with modifications)
Preferred coal rank	Low	High
Operating characteristics:		
Exit gas temperature	Low [800 to 1200F (427 to 649C)]	Low [800 to 1200F (427 to 649C)]
Oxidant requirement	Low	Low
Steam requirement	High	Low
Key distinguishing characteristics	Hydrocarbon liquids in the raw gas	
Key technical issue	Utilization of fines and hydrocarbon liquids	

Fluidized Bed

Ash Conditions	Dry Ash	Agglomerating
Feed coal characteristics:		
Size	Crushed [<0.25 in. (<6.4 mm)]	Crushed [<0.25 in. (<6.4 mm)]
Acceptability of fines	Good	Better
Acceptability of caking coal	Possibly	Yes
Preferred coal rank	Low	Any
Operating characteristics:		
Exit gas temperature	Moderate [1700 to 1900F (927 to 1038C)]	Moderate [1700 to 1900F (927 to 1038C)]
Oxidant requirement	Moderate	Moderate
Steam requirement	Moderate	Moderate
Key distinguishing characteristics	Large char recycle	
Key technical issue	Carbon conversion	

Entrained Flow

Ash Conditions	Slagging
Feed coal characteristics:	
Size	Pulverized [<100 mesh (<149 microns)]
Acceptability of fines	Unlimited
Acceptability of caking coal	Yes
Preferred coal rank	Any
Operating characteristics:	
Exit gas temperature	High [>2300F (>1260C)]
Oxidant requirement	High
Steam requirement	Low
Key distinguishing characteristics	Large amount of sensible heat energy in the hot raw gas
Key technical issue	Raw gas cooling

The distinguishing characteristics of a moving-bed gasifier are:

1. produces hydrocarbon liquids such as tars and oils,
2. has limited ability to handle fine particles,
3. requires special steps to handle caking coals,
4. produces relatively high gas methane content, and
5. has low oxidant requirements.

The main differences among moving-bed gasifiers are the ash conditions (dry or slagging) and design provisions for handling fines, caking coals and hydrocarbon liquids.

Fluidized bed

The fluidized-bed gasifier in Fig. 1b is a back-mixed reactor because feed coal particles are mixed with the particles undergoing gasification. While it consists of a discrete bed, the crushed coal particles are kept in constant motion by the upward gas flow. The fluidized bed is maintained below the ash fusion temperature to avoid clinkers which could cause defluidization. A clinker is a large solid mass of coal ash agglomerated by ash slagging. Char particles entrained with the hot, raw gas are recovered and recycled to the gasifier.

The distinguishing characteristics of a fluidized-bed gasifier are:

1. has large char recycle,
2. requires special steps to obtain high carbon conversion of high rank coals,
3. requires special steps to handle caking coals,
4. features uniform and moderate temperature, and
5. requires moderate oxygen and steam inputs.

The main differences among fluidized-bed gasifiers are the ash conditions (dry or agglomerated) and design provisions for char recycle. An agglomerated fluidized bed contains a hot zone where the ash particles are caused to cluster into small pellets prior to removal. Agglomerated ash operation enables the gasification of high rank coals. Dry ash fluidized-bed gasifiers operate most efficiently on low rank coals.

Entrained flow

The entrained-flow gasifier shown in Fig. 1c is a plug-flow reactor (not back-mixed). Fine, pulverized coal particles co-currently react with steam and oxidant with very short residence time. An entrained-flow gasifier has the following characteristics:

1. can gasify all coals regardless of rank, caking characteristics, or amount of fines,
2. requires substantial heat recovery due to the large amount of sensible heat in the raw gas,
3. is a high temperature slagging operation,
4. has a large oxidant requirement, and
5. requires special steps to avoid molten slag carryover to downstream heat recovery surfaces.

The main differences among entrained-flow gasifiers are the coal feed system (slurry or dense phase) and the design configurations for raw gas cooling and sensible heat recovery. This unit is also called a *suspension* gasifier because it consists of a two-phase system of finely divided solids dispersed in a gas.

Molten iron bath gasification

Another generic type of gasification introduced as an adjunct to steel convertor technology is molten iron bath gasification. This process is a modification of the basic oxygen furnace which is commonly used in steel production. Although not fully developed, its main application would be combined steel production and power generation.

The current major development in molten bath gasification is the German Klockner process.[3] This process has been piloted in a steelmaking convertor that was re-equipped as a gasifier. Ground coal [250 t/d (227 t_m/d)] and oxygen are injected into the molten bath through bottom nozzles and medium Btu gas, composed of CO and H_2, is produced. A claimed advantage of this process is sulfur removal in the slag. Lime or limestone added to the coal combines with the coal ash to form a basic liquid slag. Furthermore, sulfur reacts with the calcium to form calcium sulfide which is stable in the slag. This process is similar to entrained-flow gasification in that the product gas and slag leave at the reactor temperature of about 2700F (1482C).

Practical applications

Moving bed

The first full scale moving-bed coal gasification plant was constructed in 1936. Since then, 164 Lurgi gasifiers have been built by a number of companies. All of these gasifiers except one have been oxygen-blown units. The SASOL plant in South Africa uses 97 units to produce gasoline from coal. In addition, the Great Plains Synthetic Natural Gas Project at Beulah, North Dakota has 14.

The configuration of the moving-bed dry ash gasifier is shown in Fig. 2. It is a high pressure cylindrical unit, operating at 350 to 450 psig (2410 to 3100 kPa). The main gasifier shell is surrounded by a cooling or water jacket. Sized coal enters the top through a lock hopper and moves down the bed under the control of a rotating grate. Temperature in the combustion zone near the bottom is about 2000F (1093C), whereas the gas leaving the drying and devolatilization zone near the top is approximately 1000F (538C). This moving-bed process is capacity limited. Almost all of these units have a diameter of 13.1 ft (4 m) and a nominal dry gas capacity rating of 33,333 SCFM (14.9 Nm^3/s). This is equivalent to about 650 t/d (590 t_m/d) of MAF (moisture- and ash-free) coal. Lurgi and SASOL are operating a 16.4

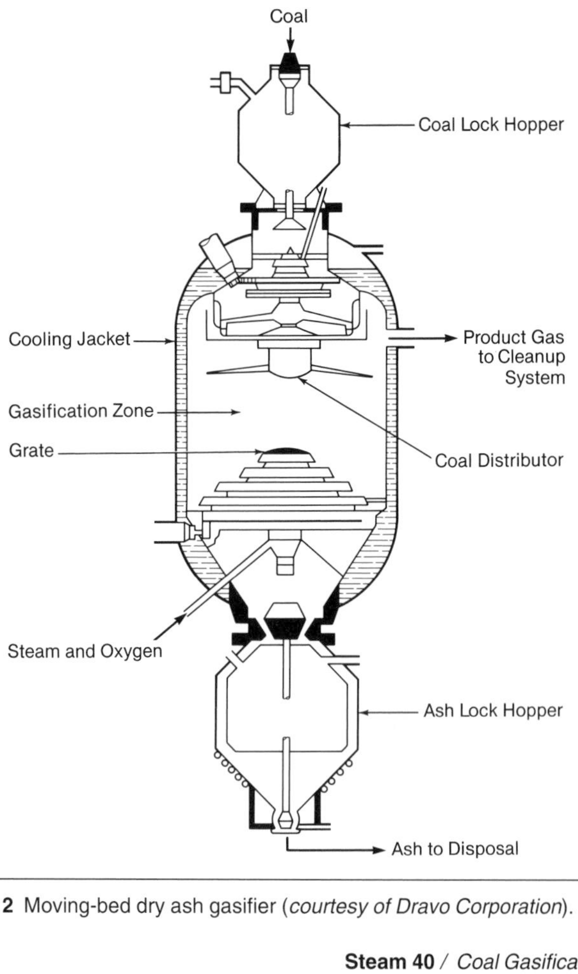

Fig. 2 Moving-bed dry ash gasifier (*courtesy of Dravo Corporation*).

ft (5 m) diameter Mark V gasifier having a capacity of 1000 t/d (907 t_m/d) and generating about 50,000 SCFM (22.35 Nm^3/s). The cold gas efficiency, defined as the ratio of sulfur-free gas to coal heat content, is 80% and this efficiency increases to 89% if the hydrocarbon liquids are included. Cold gas efficiency is expressed as [higher heating value (HHV) of sulfur-free gas at 60F/HHV of feed coal] × 100.

Fluidized bed

The fluidized-bed gasifier was the first commercial coal gasification unit and was the initial application of fluidized-bed technology. The first of these gasifiers was put into operation in Leuna, Germany in 1926. Since then, about 70 *Winkler* gasifiers have been constructed worldwide. However, these units were superseded by entrained-flow gasifiers and by pressurized moving-bed units. Only three of these gasifiers at two plants remain in operation today.[4] Low capacity and high operating costs have limited further use of the conventional Winkler gasifier.

Entrained flow

The entrained-flow technology began as an atmospheric pressure process for producing synthesis or fuel gas from solid or liquid carbonaceous fuels. The original bench-scale development of this gasifier type was done by Dr. Friedrich Totzek in the late 1930s while working for Heinrich Koppers GmbH of Essen, Germany. The process was jointly piloted in the U.S. by H. Koppers GmbH; Koppers Company, Inc. of Pittsburgh; and the U.S. Bureau of Mines. The first commercial entrained-flow gasification plant was built in France in 1949. Since then, 39 units have been built in various countries with most still in operation. The primary application is hydrogen production for ammonia synthesis.

Fig. 3 shows the entrained-flow atmospheric pressure gasifier. The coal is pulverized and fed by screw conveyors through opposing burners into a horizontal, elliptically-shaped gasifier. The fuel is oxidized, producing a flame zone temperature of about 3500F (1927C). Heat losses and endothermic reactions reduce the gas temperature to about 2700F (1482C). The hot product gas is further cooled by direct water quenching to about 1700F (927C) to solidify entrained liquid slag particles before they enter the heat recovery (waste heat) boiler.

The cold gas efficiency of the original Koppers-Totzek oxygen-blown gasifier is 67%. Although this appears to be low, an additional portion of the coal's energy is converted into heat which is recovered by the cooling jacket and heat recovery boiler. Combining these energy sources yields an overall efficiency of 85%. Later improvements in modern entrained-flow gasification systems can increase the total efficiency into the high 90s. Most gasifiers use a two headed design featuring two opposing burners. These have a total capacity of 10,583 SCFM (4.73 Nm^3/s), which is equivalent to about 230 t/d (209 t_m/d) of MAF coal. Two plants with four headed designs have about twice this capacity each.

Babcock & Wilcox gasifier development

Babcock & Wilcox (B&W) gasifier development dates back to 1951. Several pilot plants were built in-house and a full scale commercial gasifier was built to produce synthesis gas. All of B&W's gasification processes are the entrained-flow type.

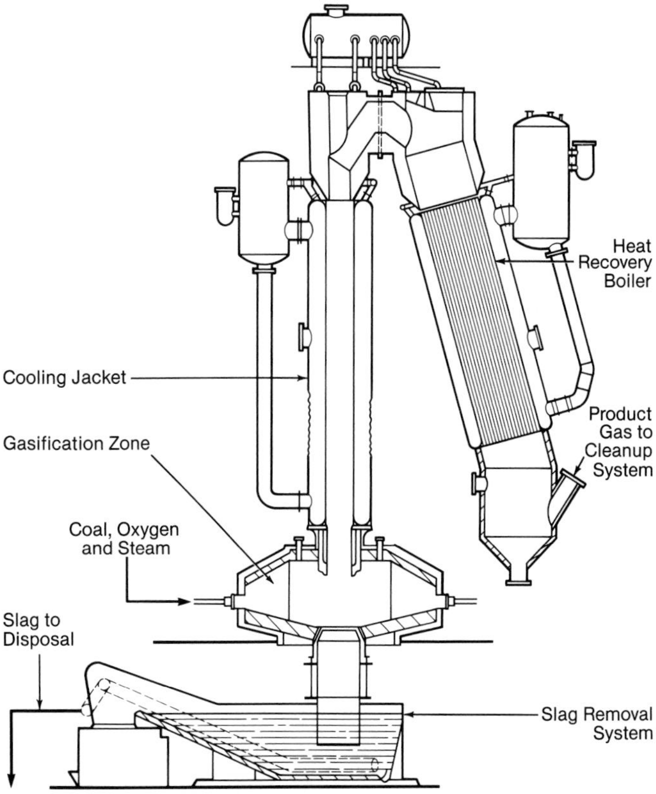

Fig. 3 Entrained-flow gasifier (*courtesy of Dravo Corporation*).

Early B&W gasifiers

The first gasifier designed and constructed by B&W (Fig. 4) was put into operation in 1951 at the U.S. Bureau of Mines station at Morgantown, West Virginia (currently the U.S. Department of Energy Morgantown Energy Technology Center). Operating at atmospheric pressure, this unit was oxygen-blown and was capable of gasifying 500 lb/h (0.063 kg/s) of coal, producing about 330,000 ft^3/d (147.3 Nm^3/h) of synthesis gas. It was refractory lined and was comprised of a primary and secondary reaction zone. The primary zone produced temperatures exceeding 3000F (1649C) and the secondary zone operated at about 2200F (1204C). The unit was operated successfully for more than 1200 hours.

B&W also assisted the U.S. Bureau of Mines in its early pressurized gasification studies. One unit designed by B&W featured a down-fired axial coal feed [500 lb/h (0.063 kg/s)] with tangential steam and oxygen feeds. This unit was designed for operation at 450 psig (3100 kPa). It was operated during the 1950s and 1960s and was occasionally used in the 1980s.

A reduced-scale gasifier was constructed at Belle, West Virginia in 1951.[5] This unit was rated at 3000 lb/h (0.38 kg/s) of coal and produced 1389 SCFM (0.62 Nm^3/s) of synthesis gas. When installed, it was the largest pulverized coal gasifier in the U.S. The unit operated for more than 5000 hours and established the design basis for a full scale gasifier at the same plant.

In 1955, a full scale gasifier began operation at Belle, West Virginia. The unit was 15 ft (4.6 m) in diameter and 88 ft (26.8 m) tall. It was designed to gasify 17 t/h (15.4 t_m/h) of

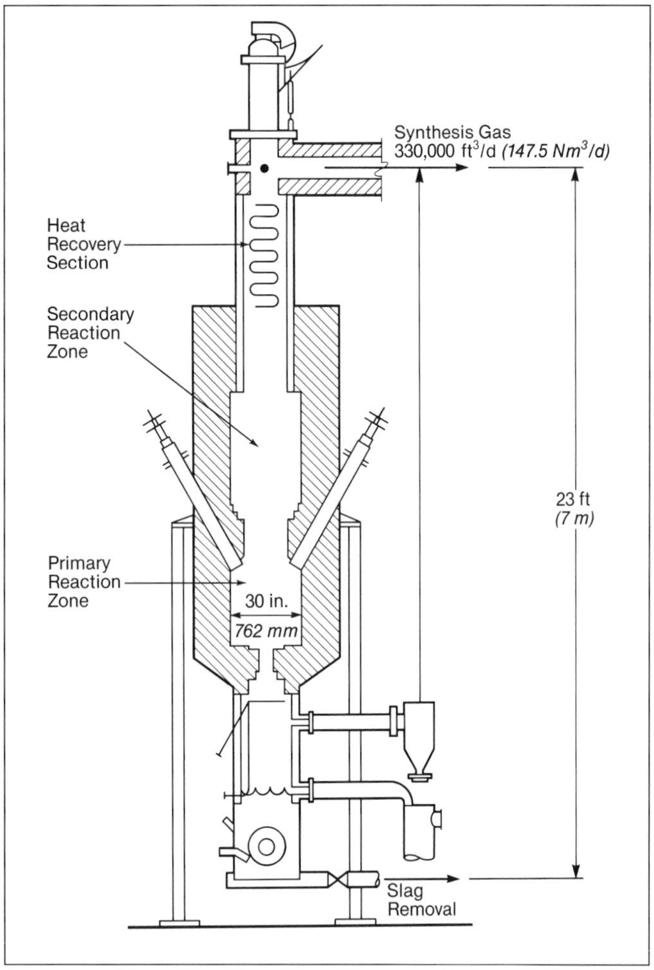

Fig. 4 Early B&W oxygen-blown gasifier.

coal, producing 17,360 SCFM (7.75 Nm^3/s) of carbon monoxide and hydrogen. The gasification zone was refractory lined and the floor and walls were water cooled. The gasifier operated for more than a year, when it was shut down because low cost natural gas became available.

Other B&W gasification activities

In the early 1960s, another pilot unit was constructed by B&W at its research facility in Alliance, Ohio (Fig. 5). This air-blown unit was used in a cooperative program with General Electric to study combined gas turbine/steam turbine cycles. It consisted of a 3 t/h (2.7 t_m/h) gasifier with char recycle, mechanical gas cleanup equipment and a gas-cooled cyclone combustor. This facility was operated for three years beginning in 1960, and horizontal and vertical slag-tap, vortex-fired gasifiers were studied.

B&W also built a 5 t/h (4.5 t_m/h), 1500 psig (10,340 kPa) gasifier for the Bi-Gas pilot plant at Homer City, Pennsylvania (Figs. 6 and 7). This plant, which opened in 1976, demonstrated the production of synthetic natural gas from coal. The project was sponsored by the Office of Coal Research, the U.S. Department of Interior and the American Gas Association.

B&W entrained-flow gasifier characteristics

B&W has emphasized entrained-flow gasification systems featuring pulverized coal feed. The firing and operation of an entrained-flow gasifier are similar to that of a pulverized coal-fired boiler.[6,7] As a result, there are inherent advantages in using pulverized coal as a fuel for entrained-flow gasifiers. A pressurized 850 to 1000 t/d (771 to 907 t_m/d) gasifier design with a waste heat recovery boiler is shown in Fig. 8.

The gas-tight membrane wall enclosure provides an annular space which separates the gasification reactor and the pressure vessel. Therefore, the pressure vessel, at a relatively low temperature, is not in contact with the corrosive raw gas. In the lower portion of the unit, the gasification zone, the tubes are covered with refractory to maintain high temperatures necessary to keep the slag fluid. In the upper portion of the gasifier, the freeboard zone, the tubes are exposed to provide maximum cooling. Cyclone wet bottom boilers (see Chapter 14) use the studded refractory-coated surfaces. The durability of this type of construction is well proven. This gasifier, featuring a 2 in. (51 mm) thick vessel wall, is designed for relatively low pressure operation [50 psig (345 kPa)]. Temperature in the combustion zone is about 3400F (1871C) and the gas leaving the unit is at 1800F (982C).

The advantages and disadvantages of the dry feed, entrained-flow gasifier are generally summarized below. Special features include:

1. Insensitive to coal characteristics — Unlike moving-bed or fluidized-bed gasifiers, it can accept caking coals, fines and any rank coal. With dense-phase coal feed, as contrasted to slurried coal, it can accommodate readily the wide range of coal moisture contents.
2. Ease of handling and mixing permits large unit sizes — Basically, as in pulverized coal-fired boilers, unit size can be increased by adding burners. The higher throughput velocity allows a high quantity of coal per hour to be treated in a given size gasifier.
3. Rapid control response — This is due to its low solids inventory compared to other gasifiers.

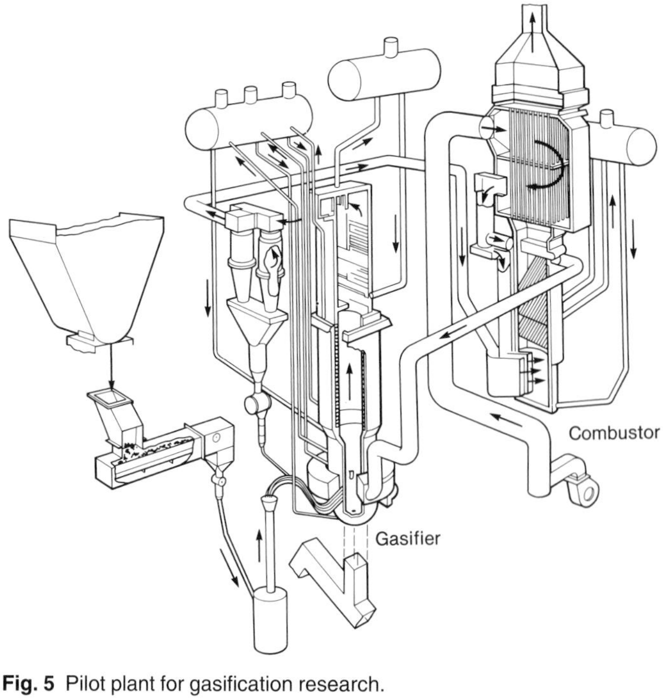

Fig. 5 Pilot plant for gasification research.

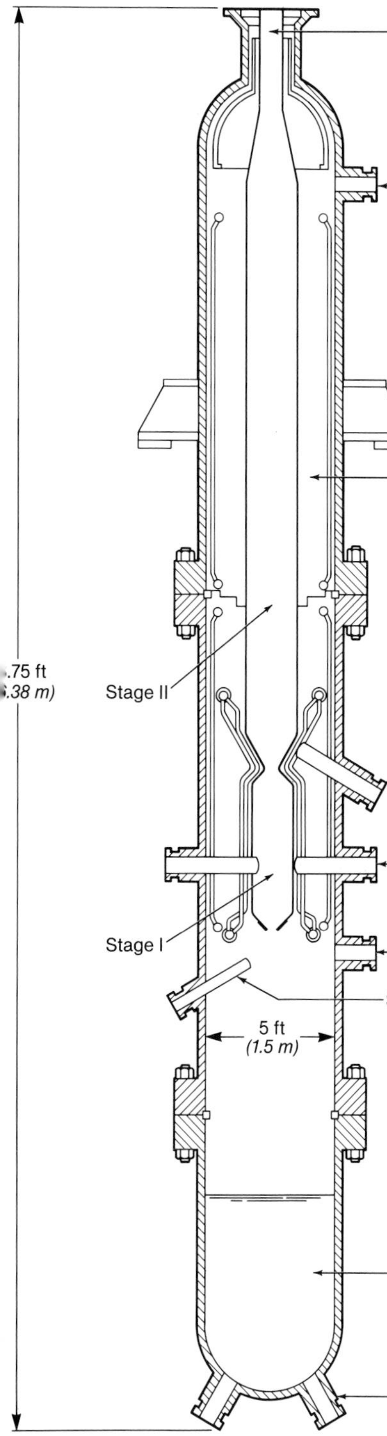

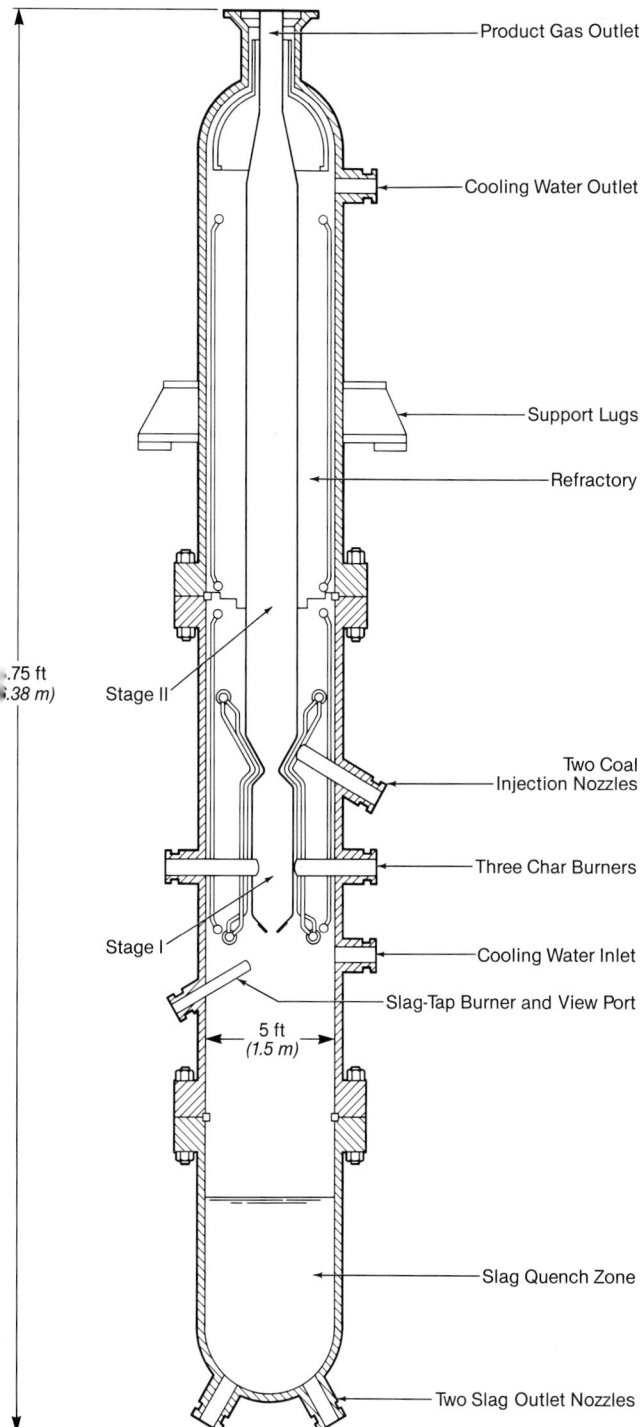

Fig. 6 Bi-Gas gasifier by B&W.

4. Dense granular slag is easily disposed — Similar material from slag-tap boilers has been used as road fill. It does not create a dust or water pollution problem.
5. High yield of synthesis gas ($CO + H_2$) — Conversely, the yield of CO_2 and H_2O is low. For example, the dry-bottom moving-bed product gas can contain 60% (by volume) H_2O and 10% CO_2 as contrasted to about 2% each of these components in a dry feed entrained gasifier.
6. Generates no hydrocarbon liquids — Tars, phenols and oils are absent. Special handling problems associated with production of such chemicals and special treatment requirements for scrubber water are avoided. Reinjection of unreacted char is considerably simpler than such tar-like streams.
7. High reliability (no moving parts in furnace) — The requirement for mechanical devices, such as a rotating grate, is eliminated because the pulverized coal is dispersed through the gasifier and rapidly heated through the plastic temperature range without coming into contact with adjacent coal particles.
8. Cooled membrane wall reduces maintenance — The durability of the studded tube/refractory-coated gasifier cooling enclosure has been proven in Cyclone furnace and slag-tap boilers. This contrasts to the higher maintenance requirements of noncooled, refractory-lined gasifiers.

Fig. 7 Bi-Gas gasifier prepared for shipment.

The disadvantages are:

1. May not be economical for very small sizes — With its high gasification capacity for a given size vessel, fewer gasifiers are needed to supply the required gas. Scale down would require a detailed evaluation of the scaling factors for the pressure vessel, up front coal preparation plant (pulverizers) and downstream heat recovery equipment.
2. Generates negligible methane (if methane is desired) — Methane production is generally favored in a lower temperature, high pressure, moving-bed type gasifier.
3. Control of coal rate is more complex — Maintenance of uniform heat release and slagging conditions requires precise control of coal flows. Dense-phase pulverized coal feed systems require inerting for safety. Coal feed and char recycle are simplified with coal slurries and higher pressures are possible, but there are associated penalties in carbon burnout and flexibility of coal feed sourcing.
4. With some coals turndown may be limited — Turndown to approximately 25% of capacity should generally be possible, but certain high ash-fusion temperature coals may constrain proper slagging at low loads.
5. Relatively high oxygen and waste heat recovery duty — These characteristics are associated with the high temperature of the entrained-flow process. Heat recovery costs exceed those of the fluidized or moving bed. Because the gasifier walls and heat recovery boiler absorb about 85% of the energy released as sensible heat (25% of the energy in the coal), it has been estimated that the total recovered and chemical energy efficiency of the gasifier can exceed 96% for most coals.

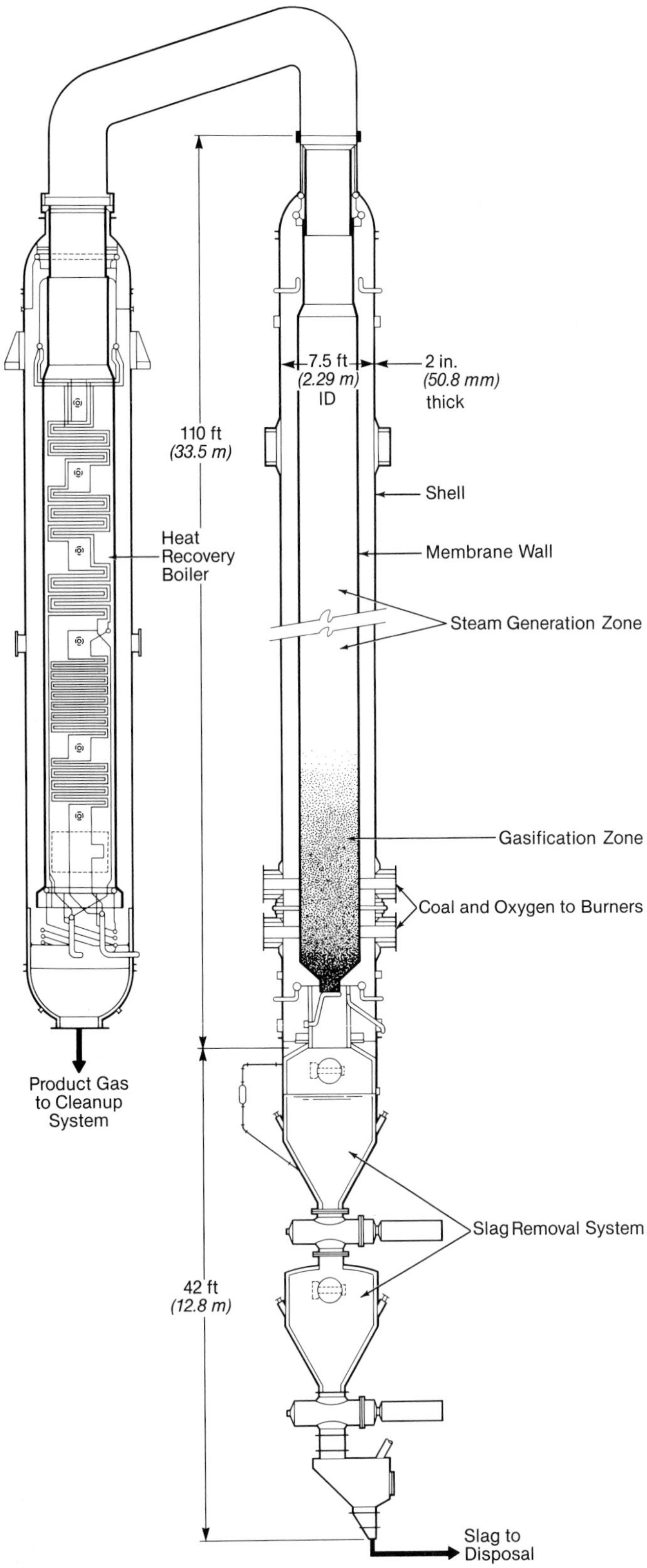

Fig. 8 Typical 1000 t/d entrained-flow design.

Current development

A new generation of coal gasifiers is in service making synthesis gas for chemical production. Entrained-flow gasifiers have been installed worldwide in several large chemical feedstock plants [800 to 1650 t/d (726 to 1497 t_m/d)]. Other sizable synthesis gas plants are the high temperature fluidized-bed [580 t/d (526 t_m/d)] and other entrained-flow [790 t/d (717 t_m/d)] installations in Germany.

The most significant recent developments of coal gasification for power generation are found in entrained-flow, oxygen-blown gasifiers. Both dry and slurry coal feed systems are used.

A 1000 t/d (907 t_m/d) gasifier was used in the 92 MW Cool Water, California coal gasification combined cycle (CGCC) demonstration plant. This plant was operated by Southern California Edison from 1984 to 1989. Another gasifier, piloted at 250 t/d (227 t_m/d), has been scaled up to 2000 t/d (1814 t_m/d) in the 250 MW integrated coal gasification combined cycle (IGCC) plant at Buggenum in The Netherlands. Another gasifier has been operated at 1600 t/d (1451 t_m/d) in a 155 MW IGCC plant in Plaquemine, Louisiana.

Texaco gasifier

The Texaco entrained-flow coal gasification process is shown in Fig. 9. The downflow gasifier is fed with a coal-water slurry of 60 to 65% solids by weight and oxygen. It operates at up to 900 psi (6205 kPa) and is refractory lined. The raw gas leaves the unit at 2300 to 2700F (1260 to 1482C) and is separated from the slag. The syngas is cooled by a radiant boiler, followed by a convection boiler, which generates 1600 psi (11,031 kPa) saturated steam. These boilers are called *syngas coolers*.

There are several options for gas cooling. One is to use the radiant and convection coolers (as shown in Fig. 9) which results in the maximum efficiency. A second involves replacing the radiant cooler with a direct water syngas quencher and eliminating the convection cooler, which minimizes cost. A third option is to use the radiant cooler only, which provides partial recovery of the syngas heat and is intermediate in efficiency and cost.

The cold gas efficiency for the process shown in Fig. 9 is 77%. If the energy of the steam produced is added to that of the fuel gas, the unit efficiency increases to 95%. Overall carbon conversion of 96.9% and 97.8%, respectively, is obtained with Illinois No. 6 and Pittsburgh No. 8 coals.[1,8] Carbon conversion is defined as the percent of carbon in the coal converted to gases or liquid (tar) products.

Shell gasifier

The Shell gasification process is shown in Fig. 10. The coal is pulverized, dried and fed to lock hoppers for pressurization. An operating pressure of 350 psig (2410 kPa) is lower than that of the Texaco unit. The membraned walls are water cooled. Burners are opposed and in a reactor configuration similar to that of the Koppers-Totzek design. The raw gas leaves the unit at 2500 to 3000F (1371 to 1649C) and most of the coal ash exits through the slag-tap in molten form. The syngas contains a small quantity of unburned carbon and a significant fraction of molten ash. To keep the ash particles from sticking together, the hot exiting gas is quenched with cold recycle gas. Further cooling takes place in the syngas cooler, consisting of radiant and convective sections. Some steam superheating is achieved with the hot, raw syngas.

The Shell gasifier cold gas efficiency is at least 80%. The combined chemical and thermal energy recovery is at least 97%. Feed coal drying reduces the net efficiency for low rank

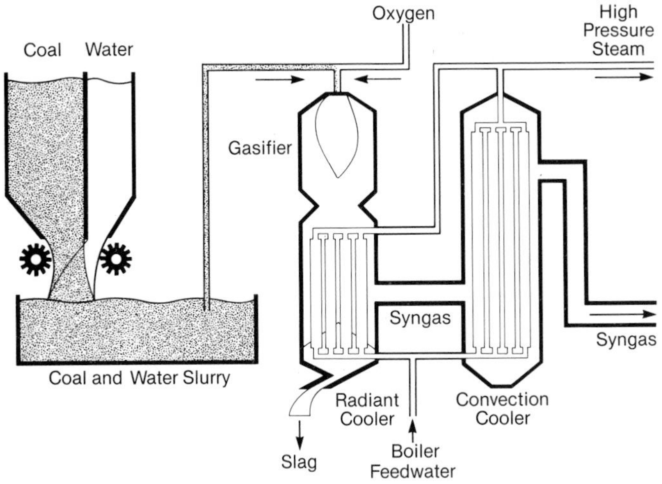

Fig. 9 Typical entrained-flow process (*courtesy Texaco Development Corp.*).

coals. The efficiency is also reduced when direct coal combustion is not used for drying. The effect of coal drying could lower the efficiency to approximately 94%. Finally, the net efficiency is about 84% when the energy required for oxygen production is also included.[1,9] Better than 99% carbon burnout has been obtained for most coals.[10]

Dow gasifier

The Dow gasifier is a two-stage, slurry feed, entrained-flow, slagging gasifier unit. Coals are slurried with water to produce a solids loading of 50 to 55% by weight. About 75% of the slurry is gasified with oxygen in the first stage. The hot gas leaving this stage at about 2600F (1427C) is used to gasify the remaining 25% of the coal in the second stage. Both stages are refractory lined and uncooled. The first stage reactor is similar to the Koppers-Totzek unit with two opposing burners and slag removal. This arrangement assures high carbon conversion and optimal slag removal.

The direct injection of coal slurry at the second stage entrance quenches the hot gas and gasifies the additional coal, providing an exit gas temperature of about 1900F (1038C). This eliminates the need for a radiant boiler and requires reduced heat recovery downstream. There is no requirement for quench recycle gas or the compression energy associated with recycling large volumes of cooled gas. Dow's unit includes a fire tube boiler followed by a steam superheater and economizer to recover heat from the raw product gas.[11]

BGC/Lurgi slagging gasifier

Lurgi has worked with the British Gas Corporation (BGC) in developing the BGC/Lurgi slagger. This gasifier, shown in Fig. 11, is very similar to the conventional dry ash Lurgi system. The key difference is that the BGC/Lurgi unit slags the coal ash. The BGC/Lurgi slagger partly alleviates the coal acceptability problems associated with the dry ash Lurgi by permitting the use of higher rank, low ash coals. Tars and oils are re-injected, providing a cold gas efficiency of 88%. The value rises to

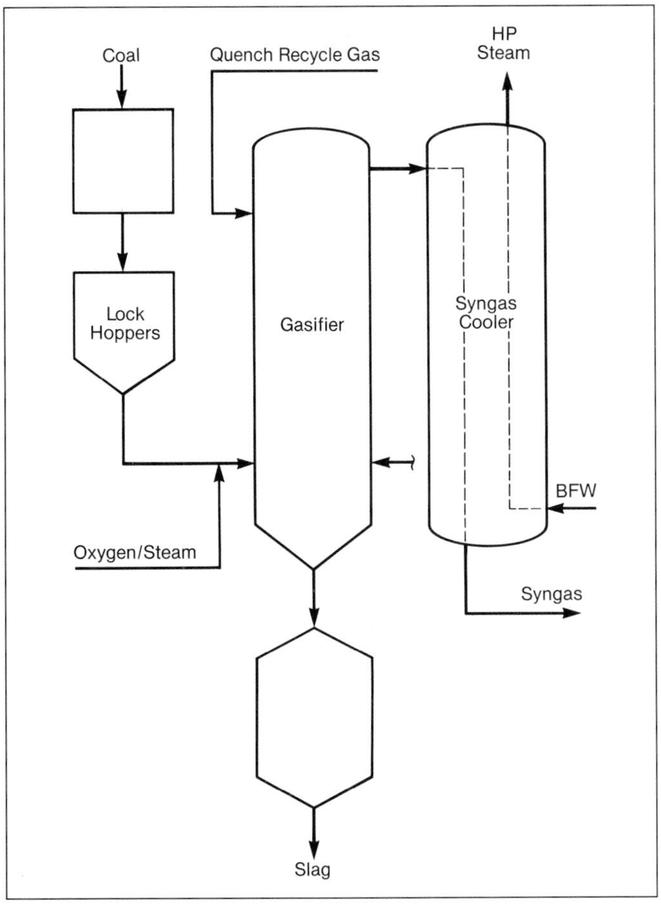

Fig. 10 Typical process with syngas cooler (*courtesy of Synfuels Business Development, Div. of Shell Oil Company*).

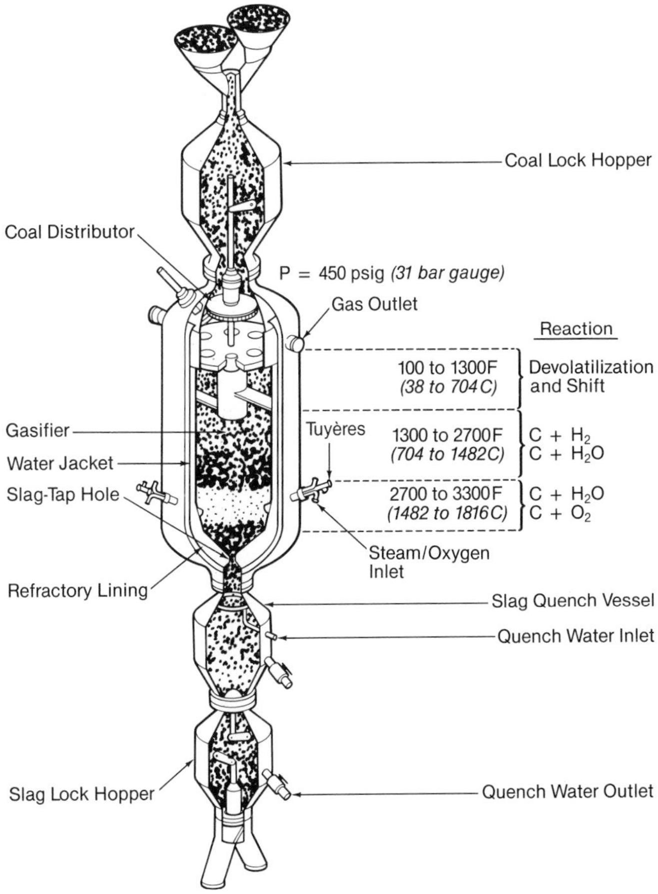

Fig. 11 Slagging gasifier (*courtesy of British Gas PLC*).

about 90% when considering additional hydrocarbon liquids.[1] A substantial advantage of slagging the coal is that the steam requirement is only about 15% of that for the conventional Lurgi when gasifying bituminous coal.

High temperature Winkler gasifier

The High Temperature Winkler (HTW) process is an extension of the old Winkler fluidized-bed technology. The HTW process is being developed by Rheinbraun (Rheinische Braunkohlenwerke AG) in Cologne, Germany in cooperation with its engineering partner, Uhde GmbH, Dortmund. The dry ash gasification of reactive German brown coal along with peat and wood is being researched. Recycling of fines entrained with the raw gas results in better carbon conversion. Furthermore, by operating at an increased pressure of 130 psig (896 kPa), the synthesis gas production rate for a given gasifier diameter has more than doubled. A 580 t/d (526 t_m/d) HTW plant has been operating since 1986 at Huerth (Berrenrath brown coal plant), West Germany, producing 22,917 SCFM (10.24 Nm^3/s) of synthesis gas from brown coal.

The HTW unit relies on the use of non-caking coals to achieve high carbon conversion. Approximately 95% conversion can be reached with very reactive German brown coals. The bed operates at temperatures of 1400 to 1500F (760 to 816C). Cold gas efficiency is about 82% for the oxygen-blown HTW; the efficiency increases to 85% if the net steam generation (deducting energy for coal drying and steam fed to gasifier) is added.[1] Note that there is a high energy requirement for drying the high moisture, low rank coal.

Developmental issues

The major application of coal gasification in the last several decades has been for chemical feedstock production. Although numerous units have been piloted, gasifiers for electric power generation are still in the developmental stage. To date, there have been only two coal gasification combined cycle plant demonstrations: 92 MW at Cool Water, California and 155 MW at Plaquemine, Louisiana.

The reliability of gasifier hardware impacts overall system reliability. Much of the IGCC plant equipment represents relatively mature technology. However, the IGCC system is more complex than a conventional steam plant and overall plant integration has a major bearing on performance and operability. The reliability of the gasification system is the least proven of all IGCC components. Advanced gas turbines being developed to enhance plant efficiency must also be tested over time.

In addition to the gasifier itself, system hardware components include: 1) syngas coolers, 2) waterwalls or refractory liners, 3) burners, 4) slag or dry ash removal systems, 5) fuel feed systems, 6) particulate collectors, and 7) char/unreacted byproduct recycle systems.

Syngas cooler materials

Syngas cooler material failures have accounted for a substantial percentage of downtime at the IGCC demonstration plants. Material selection in syngas coolers requires special attention.[12,13]

There are major differences between syngas coolers and coal-fired boilers:

1. The highly reducing and corrosive raw syngas presents a different environment for material qualification. Protective oxide scales which usually form on low alloy boiler steels do not readily form in syngas coolers, where H_2S and HCl are present.
2. Because of the high operating pressure, generally 300 to 600 psi (2070 to 4140 kPa), the syngas dew point is generally much higher than that in atmospheric pressure boilers. This is especially true when the steam content of the syngas is high, as with a coal-water slurry feed system. Precautions are required to prevent operation below the acid dew point and to avoid aqueous corrosion during downtime.
3. Pressurized operation, requiring a vessel enclosure for the exchangers, constrains the design layout. This requires special attention to assure accessibility for inspection and repair.

Low alloy steels are not suitable for syngas coolers because of excessive corrosion rates caused by sulfidation and chlorination. Sulfidation is the major corrosion mechanism in syngas coolers. For low alloy steels, the sulfidation rate is dependent on H_2S partial pressure and the rate is higher with high sulfur coals. Iron sulfide (FeS) and mixed sulfide-oxide scales form; these are generally less protective than oxide scales.

Because low alloy steels produce excessive corrosion rates, stainless steels, particularly SS310, are used in syngas coolers. Corrosion rates must be less than about 4 mils/yr (0.1 mm/yr) to obtain a service life of 25 years.

The effects of syngas chlorides are not clearly defined. Sulfidation rates in syngas with 300 to 500 ppm HCl become variable and are generally higher, presumably because of $FeCl_2$ formation and spalling during thermal cycling. Chlorides greatly accelerate aqueous corrosion during downtime if condensation occurs. The presence of HCl in syngas increases its dew point 54 to 90F (30 to 50C).

Preliminary material selection guidelines have been formulated by the Electric Power Research Institute (EPRI) based on currently available laboratory and field data.[12,13] These are presented here. EPRI's preliminary syngas cooler material recommendations are aimed at achieving a 25 year service life. Actual service life will depend on the severity of the conditions to which the individual material components are exposed, which in turn are dependent on the specific design, fuel used and operating conditions.

Economizer service Low alloy steels are usually adequate for this application. However, economizers must be operated above the syngas dew point to avoid corrosion. For coals with high sulfur content, a chromized coating may be required to obtain a 25 year service life.

Evaporator service Current EPRI material recommendations for syngas coolers used in evaporator service [1500 to 2500 psi (103.4 to 172.4 bar)] [600 to 660F (316 to 349C)] are:

Coal	Good Downtime Corrosion Control	Reduced Downtime Corrosion Control
Low S (<1%), Low Cl (<0.05%)	Low alloy steel (T-11) with chromized coating or with SS304 cladding.	Modified Cr-V coating on T-11 steel or cladding with SS311.
High S (>1.5%), High Cl (>0.1%)	Low alloy steel (T-11) with SS310 cladding.	Cladding containing 25 to 30% Cr and 3 to 4% Mo on T-11 steel.

Superheater service There is little field experience from which to select materials for superheater service with raw syngas. In laboratory studies, SS310 and other alloys with at least 25% chromium appear to be adequate up to about 1000F (538C). However, limited plant data indicate that these alloys develop a sulfur-rich scale which may not be protective. Vanadium additions decrease the sulfur content and scale thickness and increase its Cr_2O_3 content, thereby forming a more protective scale. Vanadium-modified alloys need further development and are not commercially available.

Cleaning the products of gasification

Raw coal gas exiting the gasifier must be treated to remove impurities. The acceptable level of purity depends on the final use of the gas. For example, the gas produced for power generation would typically be fired in a gas turbine as part of a combined cycle plant. For this application, the gas must be cleaned to avoid fouling and corrosion of the turbine and to comply with emissions requirements. When the coal gas is used for chemical synthesis, additional refinement is necessary.

The gasification process, when combined with selected gas cleanup processes, emits less pollutant than a conventional coal combustion system. Emissions of sulfur dioxide (SO_2), nitrogen oxides (NO_x) and particulates can be reduced with an oxygen-blown gasifier coupled with conventional purification processes.

Product gas composition

Raw gas composition is highly dependent on the gasifier process, operating conditions and coal used. Gas from an oxygen-blown unit consists primarily of CO, H_2, CO_2, CH_4 and H_2O. N_2 is also a major constituent when air is used as the oxidant. Fuel sulfur is predominantly converted to hydrogen sulfide (H_2S) and some carbonyl sulfide (COS), while fuel nitrogen compounds are mainly converted to ammonia (NH_3) and hydrogen cyanide (HCN). Many other minor impurities are produced, depending on their presence in the raw fuel. These include chlorides, vaporous alkali species and heavy metals. Heavy hydrocarbons are also in the raw gas from lower temperature gasifier processes.

Coal gas has traditionally been classified by heating value, expressed in Btu per dry standard cubic foot [Btu/DSCF (MJ/Nm^3)].[14] Air-blown gasification produces a low Btu gas due to nitrogen dilution. This gas ranges from 90 to 200 Btu/DSCF (3.5 to 7.9 MJ/Nm^3). Oxygen-blown units produce a medium Btu gas at the lower end of the 270 to 600 Btu/DSCF (10.6 to 23.6 MJ/Nm^3) range. Post-gasifier shift and methanation reactions are required to produce heating values at the higher end of the medium Btu range and ultimately to produce high Btu gas of approximately 1000 Btu/DSCF (39.3 MJ/Nm^3). High Btu gas is pipeline quality and is therefore referred to as synthetic natural gas (SNG).

Raw gas composition as a function of gasifier type and oxidant is listed in Table 2. Heating values vary widely between product gas from air- and oxygen-blown moving-bed gasifiers, which can influence the combustion characteristics of the gas. Raw gas from a moving-bed gasifier contains significant quantities of ammonia and heavy condensable organics such as tars, oils and phenols. It also has a higher heating value than gas from entrained-flow processes due to the increased methane content. The two advanced entrained-flow gasifiers produce negligible quantities of organics due to their high temperature operation. The one process is slurry fed, producing a product gas lower in CO and higher in H_2, CO_2 and H_2O compared to the dry feed process. The data presented in Table 2 illustrates some contrasts between different gasification conditions.

Many fuel and operating variables such as fuel composition, operating pressure, and steam and oxygen inputs influence the final gas composition. Fig. 12 illustrates the effect of operating pressure on methane produced in a Lurgi gasifier for different fuel types.[14]

Gas purification

Impurities in the raw gas must be removed to produce a commercial-quality product. The processes applied to purify coal gas for power production can be classified as conventional or developmental, depending on whether the gasifier is oxygen- or air-blown. The raw gas (often called *acid gas* because of its H_2S and CO_2 content) from oxygen-blown gasifiers is first cooled and then treated for acid removal. Such cold removal processes have been developed for cleaning product gas used for chemical synthesis and for pipeline natural gas.

Air-blown gasifiers are generally limited to producing gas for power generation because the inherent nitrogen in the product gas is not suitable for chemical synthesis. The low Btu fuel gas from an air-blown unit may have to be fired hot to support stable combustion in a high efficiency gas turbine and to avoid the large heat loss and efficiency penalties associated with cooling the gas.

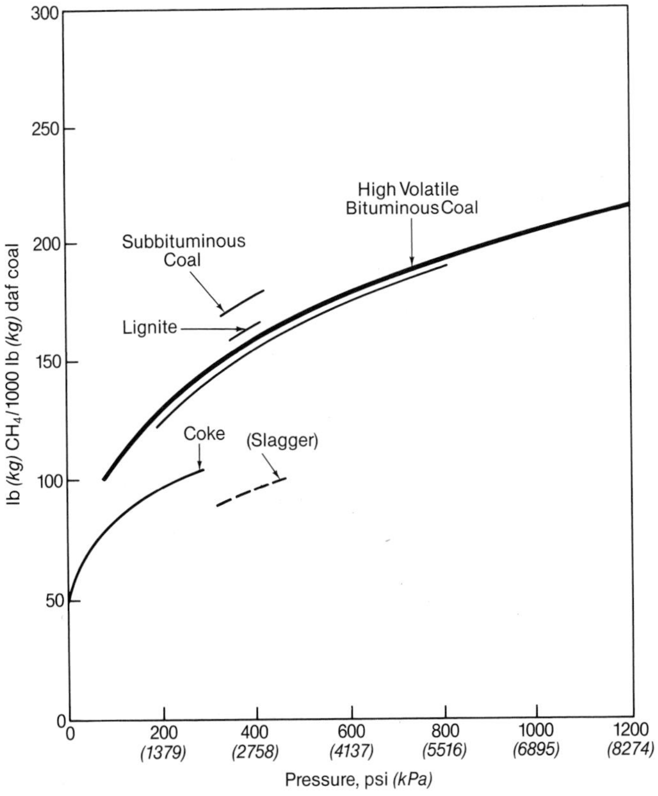

Fig. 12 Methane formation with pressure.[14]

Table 2
Raw Gas Composition for Selected Gasifier Types

Gasifier Type	Moving Bed, Dry	Moving Bed, Slagging	Fluidized Bed	Entrained Flow (Slurry)	Entrained Flow (Dry Feed)
Oxidant:	Air	Oxygen	Oxygen	Oxygen	Oxygen
Fuel type:	Subbituminous	Bituminous	Lignite[Note 3]	Bituminous	Bituminous[Note 3]
Fuel analysis, % by wt					
C	41.1	61.2	56.9	61.2	66.1
H	4.6	4.7	3.8	4.7	5.0
N	0.8	1.1	0.8	1.1	1.2
O	20.5	8.8	15.9	8.8	9.5
S	0.6	3.4	1.0	3.4	3.7
Ash	16.1	8.8	9.6	8.8	9.5
Moisture	16.3	12.0	12.0	12.0	5.0
HHV, Btu/lb	11,258	11,235	9,914	11,235	12,128
(kJ/kg)	(26,189)	(26,136)	(23,063)	(26,136)	(28,213)
Fuel feed method	Dry	Dry	Dry	Slurry (66.5 wt % solids)	Dry
Operating pressure, psi	295	465	145	615	365
(kPa)	(2,034)	(3,206)	(1,000)	(4,240)	(2,516)
Raw gas composition, % by vol					
CO	17.4	46.0	48.2	41.0	60.3
H_2	23.3	26.4	30.6	29.8	30.0
CO_2	14.8	2.9	8.2	10.2	1.6
H_2O	--[Note 1]	16.3	9.1	17.1	2.0
N_2	38.5	2.8	0.7[Note 4]	0.8[Note 4]	4.7[Note 4]
CH_4 + CnHm	5.8	4.2	2.8	0.3	--
H_2S + COS	0.2	1.1	0.4	1.1	1.3
NH_3+ HCN	--[Note 2]	0.3	--[Note 2]	0.2	0.1
HHV, Btu/dscf	196	333	309	278	297
(MJ/Nm^3)	(7.7)	(13.1)	(12.2)	(10.9)	(11.7)
References	2	1	1	1	1

Notes:
1. Dry analysis
2. Not reported
3. Dried fuel
4. Includes argon

Conventional processes The primary purpose of acid removal is to remove sulfur compounds from the fuel gas. Initial treatment of the raw gas is required to remove undesirable materials that would contaminate the solvents used for acid removal.

Initial treatment The initial treatment of gas from moving-bed gasifiers requires many steps to remove the tars and oils, as shown in Fig. 13.[15] An initial quench and cooling step removes heavy tars and particulate for recycling to the gasifier. Final cooling steps condense and remove lighter oils and water.

The initial treatment of higher temperature raw gases from fluid bed and entrained-flow gasifiers is much simpler. The gas is cooled by quenching or by passing it through a heat exchanger; this is followed by particulate removal. Dry particulate removal devices, such as cyclones or ceramic filters, may be used to provide a product that can be recycled to improve carbon conversion and to minimize waste water. A final wet particulate removal system, such as a venturi scrubber, ensures that virtually no particulate passes to the acid removal system. Final downstream particulate emissions of less than 0.01 lb/10^6 Btu (4.3 g/GJ) have been reported.[16]

Water washing, along with further cooling of the raw gas, also removes trace metals and condensable/soluble impurities such as NH_3, HCN, chlorides and alkali compounds.

Acid gas removal The sulfur compounds in the raw gas from a gasifier are more efficiently removed than is SO_2 from conventional coal-fired boiler flue gas. For example, a wet scrubber on a coal-fired steam plant must process up to 150 times more volume than a comparable acid removal system downstream of a 500 psig (3450 kPa) gasifier, due to differences in pressure and nitrogen content.[1] Solvent-based acid removal processes have been available for more than 60 years. Developments to improve these processes are ongoing to reduce energy consumption, optimize plant operations and meet more stringent environmental regulations.[17]

Acid removal processes are distinguished by the solvents used, which can be chemical, physical, or a combination of both. Chemical solvent processes use various amines or

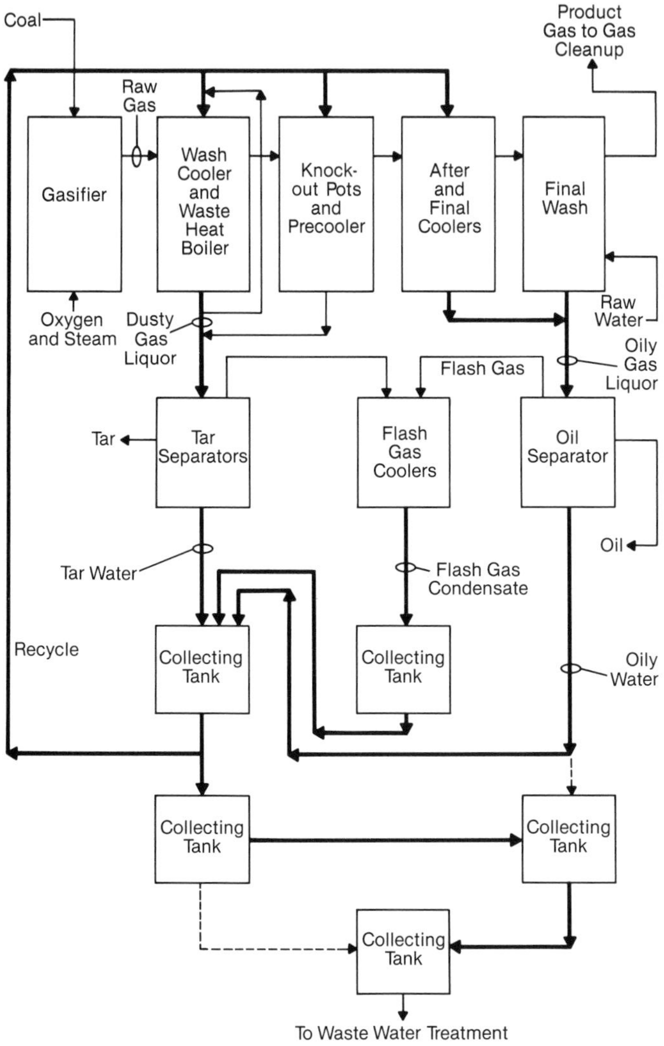

Fig. 13 Moving-bed demonstration plant process schematic (*courtesy of British Gas PLC*).[15]

alkali carbonates to react with the acid gas. In an amine process, lean solvent [typically at 80 to 120F (27 to 49C)] is passed countercurrently to the gas in an absorption tower. This solvent combines with H_2S, COS and CO_2 to form a rich solvent. The rich solvent is regenerated by applying heat and reducing pressure, producing concentrated acid gas for further treatment. The solvent selectivity of H_2S and COS over CO_2 and the solvent resistance to degradation by certain gas constituents can be important factors in choosing a removal process.

Physical solvents place the acid gas in solution, generally with increased solubility at decreasing solvent temperature. For example, the Rectisol process uses a methanol solvent at temperatures well below 0F (-18C). Regeneration is conceptually similar to chemical processes and is usually less energy intensive.[17]

The concentrated acid gas released during regeneration is sent to a sulfur recovery system, such as a Claus plant, which converts most sulfur compounds to saleable elemental sulfur. The Claus process is the most widely used process for the recovery of sulfur from the sour gases separated from natural gas and petroleum refinery and chemical plant streams. It has been used as well on the acid gas streams from gasification processes. The typical Claus process can recover 95 to 97% of the sulfur in an acid gas stream. Tail gas from a Claus plant containing the remaining 3 to 5% of the sulfur has become a regulated emission source, requiring further modification of the process or the addition of a tail gas treating process. The Claus plant when combined with a SCOT (Shell Claus Off-gas Treatment) unit can provide reduction of H_2S to well below 10 ppmv or 99+%.[18] Fig. 14 depicts a typical gas cleanup system for an entrained-flow oxygen-blown gasifier.

There are many factors, including economics, that influence the selection of a particular acid removal process. For example, eliminating the off-gas treatment reduces cost with the penalty of increased sulfur emissions. Fuel sulfur content and the subsequent concentration of H_2S in the raw gas can also affect process selection. The final selection must be based on a comparative analysis of allowable emissions, capital and operating costs. Application of these commercial processes produces a coal gas that, when burned, results in sulfur and particulate emissions far below current federal New Source Performance Standards in the U.S. (See Chapter 32.)

NO_x emissions There are no emissions from the gasification system itself (other than the small amount from the sulfur recovery plant) because a product gas is being created for power generation or chemical production. For power generation, the medium Btu gas from an oxygen-blown gasifier is favored for gas turbine fuel in a combined cycle. In this application, NO_x emissions can be extremely low, depending on the gas turbine used. Because most of the nitrogen from the coal is scrubbed from the raw gas during initial gas cleaning steps, most of the NO_x exiting the turbine combustor is thermally created from diatomic nitrogen. By diluting the clean gas with water, saturation steam, or nitrogen injection prior to combustion, NO_x emissions less than 0.10 lb/10^6 Btu (43 g/GJ) can be achieved.[16] NO_x values lower than 0.05 lb/10^6 Btu (21.5 g/GJ) without steam injection have been reported on an experimental basis.[19]

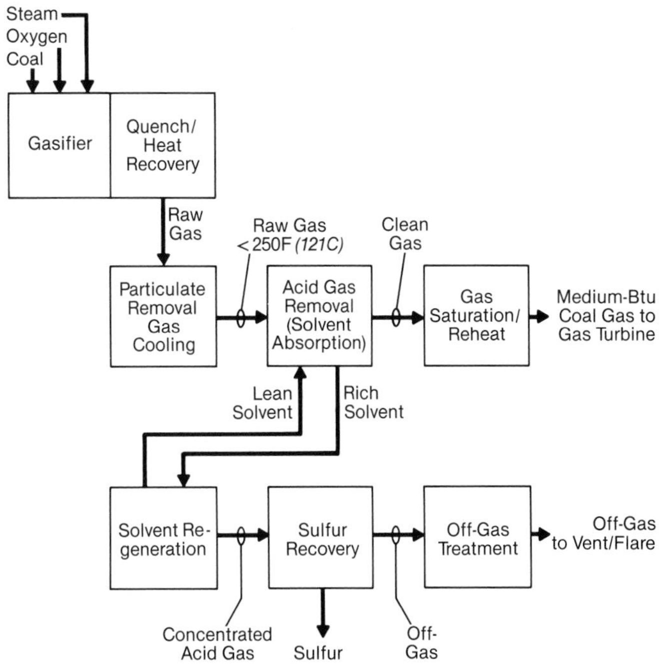

Fig. 14 Schematic of a typical gas cleanup system for an entrained-flow oxygen-blown gasifier.

Developmental processes Pressurized air-blown gasifiers are being evaluated for application to combined cycle power plants. Eliminating the oxygen plant may reduce costs and improve plant efficiency. One of the major challenges in designing air-blown systems is to develop reliable and cost effective hot gas cleanup systems. By cleaning gas hot, it can be fed directly from the gasifier to a gas turbine to maximize efficiency. The U.S. Department of Energy and others are sponsoring significant efforts to develop commercial particulate and sulfur removal systems which operate at gas temperatures greater than 1000F (538C).[20,21]

Material limitations in hot cleanup systems permit a maximum operating range of 1200 to 1400F (649 to 760C). As a result, they are more suitable for fluid-bed and moving-bed gasifiers. The sorbents being considered operate and regenerate dry which minimizes waste water.

Particulate removal Rigid, porous ceramic barrier filters are appropriate for hot particulate removal. (See also Chapter 29.) The ceramic material must be porous enough to allow gas penetration. However, it must capture fine particulate and permit solids removal and pressure drop recovery. It must also be resistant to chemical attack from the high temperature reducing environment. Different filter arrangements are being developed to permit scale-up while maintaining cleanability and maintainability. These devices have been tested on a small scale with encouraging results, but long term, high temperature exposure in a gasification environment is still needed.

Granular bed filters are also being developed. In these systems, dirty gas passes through a moving bed of solids, which traps the particulate. The bed material is removed, cleaned and returned to the process.

Hot desulfurization The U.S. Department of Energy has been funding research on hot gas desulfurization since the 1970s. A mixed metal oxide sorbent called zinc ferrite ($ZnFe_2O_4$) has emerged as a prime candidate for regenerable systems. Small scale testing has shown this material can remove gas phase sulfur species to levels below 10 ppmv and still be regenerated.[20]

In this process, zinc ferrite pellets are housed in reactor vessels through which the sulfur-laden acid gas passes. Some systems are fixed beds in parallel, with hot valves diverting gas between beds as they absorb and regenerate. Other systems use a moving-bed concept, as shown in Fig. 15. The sorbent is regenerated by passing a controlled flow of air through the pellet bed; the air reacts exothermically to produce a concentrated SO_2 off gas which must be further treated. Fluidized-bed reactors are also being developed. These have the advantage of more precise temperature control and have the potential for heat extraction during regeneration. Keeping the temperature below 1200F (649C) is critical to prevent potentially hazardous zinc vaporization and sintering.[21] Other sorbents such as zinc titanate ($ZnTiO_3$) are being developed to increase the operating temperature to as high as 1400F (760C). These materials would reduce zinc vaporization and increase the strength and longevity of the sorbent. Sorbent life is a critical operating consideration due to its high cost.

Zinc ferrite has been tested in bulk removal and polishing modes. Bulk removal uses the sorbent to perform the total sulfur removal duty. In the polishing mode, the sorbent is a secondary means of sulfur removal; primary removal is achieved within the gasifier. Fluid bed gasifiers using a throwaway sorbent such as limestone or dolomite can capture some sulfur. Entrained-flow units have been fed with iron oxide or dolomite to capture sulfur in the slag.[22]

Many development issues must be resolved before hot gas cleanup systems can be offered commercially. In addition to particulate and sulfur removal issues, the effects of raw gas chlorides, alkali species and other trace volatile elements on the performance of downstream equipment must be addressed. Ammonia in the gas will not be condensed and may result in high NO_x emissions from a gas turbine combustor. The challenge for these developmental processes is to achieve the excellent environmental performance of conventional processes while being reliable, efficient and economical.

Water effluents

The gasification systems discussed use water for many functions, including cooling hot raw gas, quenching molten slag, making coal slurry, saturating clean gas and acting as a gasifying reactant in the form of steam. In addition, water produced in the form of raw gas moisture is condensed during cooling processes. The amount of waste water created and the degree of treatment required vary greatly between the different gasification processes. Much of the water is often recycled within the system, but continuous blowdown is required to limit the accumulation of corrosive salts and other impurities washed from the gas stream.

Contaminants such as ammonia, sulfide, cyanide and formate are typically present in all gasifier waste water. Their quantities are related to sulfur and nitrogen content of the coal and to gasifier operating temperature. Also, the concentration of organic contaminants and phenolic compounds is high in moving-bed gasifier waste water. However, these contaminants are virtually nonexistent in today's entrained-flow units.[15,23,24] Suspended fine particulates are present in the waste water from wet scrubbers.

Waste water from coal gasifiers can be treated to achieve acceptable emission levels. The treatment process varies depending on the contaminants present and

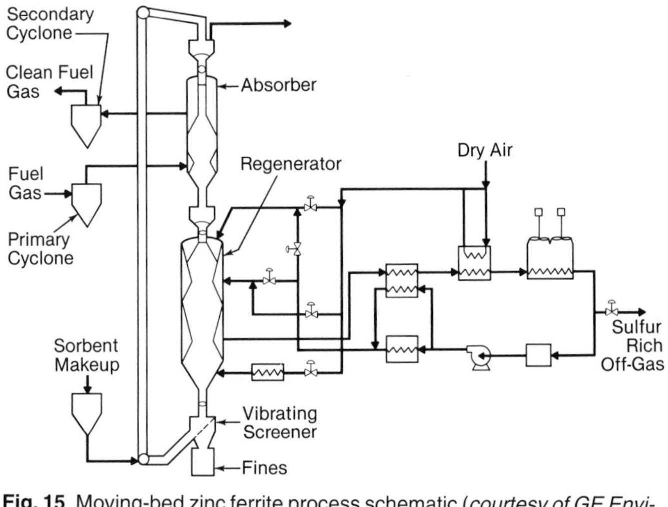

Fig. 15 Moving-bed zinc ferrite process schematic (*courtesy of GE Environmental Services, Inc.*).

the discharge requirements. Typical treatments include steam stripping, flocculation and clarification of particulates, biological oxidation of inorganics and organics, and chemical addition.

Federal effluent discharge standards for coal gasification plants have not yet been developed. Significantly, coal gasifiers have operated worldwide with adequate control of effluents.[1] Some gasifier suppliers have added water treatment to their process scope of supply and have done extensive testing to demonstrate system performance.[23,24]

Solid discharges

The primary source of solid discharge from a gasifier is the rejected coal ash from the bottom of the unit. Solids entrained in the product gas can also become waste if they are not recycled after being captured. In some arrangements, the entrained solids are discharged as clarifier sludge from the waste water treatment system. The disposition of these solids depends largely on the gasifier process and the coal properties.

A major environmental advantage of entrained-flow gasifiers compared to conventional coal-fired boilers is the inert nature and reduced volume of solid wastes. The high gasifying temperature converts most of the fuel ash to molten slag tapped from the bottom of the gasifier. After water quenching, the slag becomes a vitrified solid that is very resistant to leaching.[16,25] Ash recycling has been investigated, primarily as road bed material and light weight aggregate. In the acid removal system, minor solid wastes are generated in the form of spent catalysts from the sulfur recovery and off gas treatment processes. Also, the sulfur recovery plant produces elemental sulfur, which can be sold.

The bottom ash from slagging moving-bed gasifiers has tested as nonhazardous.[26] However, ash from a nonagglomerating fluidized-bed gasifier is not molten and can contain reactive solids. Typically, a calcium-based sorbent is added to the bed for sulfur capture. Some calcium sulfide can be present, which would require external oxidation to stabilize the calcium to sulfate form.[21]

Coal gasification for power generation

Synthesis gas from a coal gasifier has many potential applications. Further processing of the CO- and H_2-laden gas into ammonia, methanol, acetic anhydride, gasoline and other byproducts through additional chemical conversion steps is being done commercially around the world. However, these applications are frequently costly compared to petroleum and natural gas based chemical production.

Synthesis gas can also be used for power generation with the nearest term application being combustion in a gas turbine. By recovering waste heat from the gasifier and gas turbine exhaust for steam production, a gas turbine (Brayton) cycle and steam turbine (Rankine) cycle can be efficiently combined into an *Integrated Gasification Combined Cycle* (IGCC). Although technically viable, this approach to power generation from coal has not been applied commercially because conventional combustion systems have been less costly and less complicated. However, further development and increasingly stringent environmental regulations have renewed interest in IGCC.

IGCC power plant

Production of power from gasifying coal and combusting the fuel gas in a combined cycle power plant requires a high degree of component integration, as depicted in Fig. 16. In simple terms, the hot, combusted fuel gas expands in a gas turbine to drive the air compressor and generator, with some of the compressed air used to gasify the coal.

The gas turbine exhaust, hotter than 1000F (538C) with modern machines, passes through a heat recovery steam generator (HRSG) to produce superheated steam to drive a steam turbine-generator. (See Chapter 31.) The gasification process releases significant thermal energy which must also be recovered into the steam cycle to achieve high overall plant efficiency.

The design of an IGCC system is quite complex, with many factors to be considered to achieve the right balance of capital cost, plant efficiency, operability and environmental protection for a given application. For example, the choice of gasifier type affects the amount of fuel gas heat recovery duty.

An oxygen-blown entrained-flow gasifier with its high operating temperature requires more raw gas cooling in efficiently designed cycles. These coolers must perform in a harsh gas environment and are a critical component in plant reliability.

Integrating this large quantity of lower level steam energy (limited by metal temperatures) into the steam cycle tends to complicate plant control and operation. In contrast, cooling the raw gas by quenching eliminates the capital cost and complexity of heat recovery, but at a substantial efficiency penalty (as much as 10% less efficient than with total heat recovery).

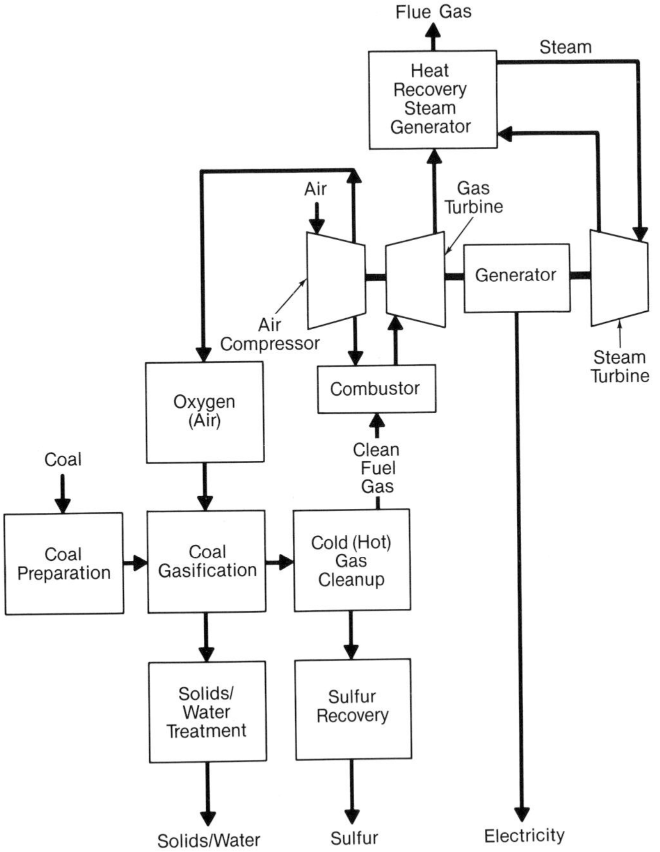

Fig. 16 Integrated coal gasification combined cycle schematic.

Oxygen plant integration is another design option. Fig. 16 shows the turbine-driven air compressor directly feeding the oxygen plant, which produces oxygen more efficiently by reducing external compression requirements. However, this places more demand on turbine controls due to the large system volume between the compressor and turbine.

Other technical factors, such as fuel characteristics, influence process design. A high inherent moisture fuel such as lignite may not be a good candidate for a coal water slurry fed gasifier due to reduced efficiency. Gas turbines have varied abilities concerning fuel gas minimum heating value, emissions performance and compressor range for handling the larger mismatch between air flow and gas flow when firing relatively high volume, low Btu coal gas. Ambient temperature effects and utility load demand characteristics add to the list of design considerations.

Plant size is centered around the gas turbine. A single gas turbine/steam turbine installation can produce more than 250 MW, with about 60% of the power from the gas turbine. Larger plants require multiple gas turbines, with economies of scale realized in maintaining a single steam turbine and in larger balance-of-plant systems such as fuel handling, electrical and controls, and water treatment. Phased construction of an IGCC plant by incrementally installing a natural gas-fired turbine (simple cycle), then steam turbine and HRSG (combined cycle) and finally a gasification system (IGCC) allows the owner flexibility in meeting changing power demands from peaking to base loading and accommodating a relative rise in fuel costs of natural gas versus coal. Although simple in concept, modifications to the gas turbine and steam side heat balance to switch from natural gas to coal gas can be significant.

The myriad of technologies available and the external design factors combine to make IGCC plant design a complex, site-specific process. Concurrently, this flexibility offers opportunity to produce a highly optimized system for each application.

Benefits of IGCC

The major force behind development and implementation of IGCC is the demand for a cleaner environment. Conventional coal-fired plants continue to add environmental controls to meet ever-tightening emissions requirements. (See Chapters 32 through 36.) An IGCC plant has the potential to inherently achieve very low emissions. Oxygen-blown gasifiers with a cold-gas cleanup system are particularly well suited to deliver low emissions of SO_2, NO_x, solid wastes and air toxins.

As mentioned above, sulfur removal efficiencies in excess of 99% are possible with an IGCC plant equipped with a Claus plant and tail-gas cleanup system. Wet scrubbers in a conventional coal plant can practically be designed for similar performance through use of more expensive enhanced sorbents and/or more auxiliary power for increased sorbent to gas interaction.

With fuel nitrogen compounds removed in the cold gas cleanup system, NO_x emissions from an IGCC system are strictly determined by gas turbine combustor performance.

NO_x values less than 0.05 lb/10^6 Btu (21.5 g/GJ) are feasible with state-of-the-art combustors, depending on fuel gas fired heating value (moisture and/or nitrogen can be added) and turbine inlet temperature, both of which affect peak flame temperature and thermal NO_x generation. By comparison, conventional coal plants can only reach such values through low NO_x combustion techniques (see Chapter 13) in combination with post-combustion NO_x controls (see Chapter 34).

Solid byproducts from an IGCC plant with an entrained-flow gasifier consist primarily of nonreactive quenched slag from the gasifier and elemental sulfur. If the sulfur is sold as a marketable commodity, the disposal stream is reduced to the fuel ash. Depending on fuel sulfur content, a conventional coal plant may produce up to twice as much solid waste due to sulfur dioxide (SO_2) byproducts. (See Chapters 1 and 32.) However, this may be partially offset by the production of saleable gypsum in conventional scrubber systems or other byproducts from regenerable scrubber systems.

The environmental impacts of heavy metals in fuel ash, emissions of air toxins and extremely fine particulate matter are now being scrutinized. IGCC systems also offer advantages in these areas.

High cycle efficiency of an IGCC system is another benefit relative to a conventional steam plant. With heightened awareness worldwide of CO_2 emissions, efficiency is becoming as much an environmental issue as an economic factor related to fuel costs.

Theoretical IGCC cycle efficiency is typically in the range of 38 to 43% depending on capital cost tradeoffs, degree of plant integration, fuel type and other specifics. Higher efficiencies are possible as gas turbines with higher firing temperatures are developed. A large state-of-the-art Rankine cycle steam plant operating at advanced supercritical steam conditions has an efficiency of approximately 38%, while subcritical steam cycles have an efficiency of about 35%.

Air-blown IGCC systems with hot gas cleanup are being developed as an alternative to eliminate oxygen separation and the chemical refinery-like cold gas cleanup process. This reduces the plant design to components and processes more familiar to the power industry. Lower capital costs and higher efficiencies than oxygen-blown systems are targeted. The environmental performance of hot gas cleanup is critical, with further developments required to achieve low NO_x emissions in the gas turbine combustor with fuel nitrogen compounds present, to remove particulate at temperatures greater than 1000F (538C) and to demonstrate reasonable performance and life.

Ultimately, the cost of producing power with an IGCC system, advanced Rankine cycle steam plants and other competing systems, such as pressurized fluidized bed power systems, will determine the extent to which each system will be used in the future.

Advanced gasifier applications

Development work is proceeding on advanced power cycles that incorporate a coal gasifier. Studies have been made of coproduction of chemicals (such as methanol) and power from a gasification plant to increase plant use as power demand varies. Fuel cells are also being developed that would electrochemically convert highly purified coal gas directly into electrical energy at high efficiency (greater than 50%) with extremely low emissions. Significant materials and scale-up issues must be resolved before commercialization of fuel cell power systems.

References

1. *Coal Gasification Systems: A Guide to Status, Applications, and Economics*, EPRI Report AP-3109, Synthetic Fuels Associates Inc., June 1983.

2. Elliot, M.A., ed., *Chemistry of Coal Utilization*, Second Supplementary Volume, Chapters 23, 24, 25, Wiley, New York, 1981.

3. "SFA quarterly real project list," *The SFA Quarterly Report*, SFA Pacific, Inc., Mountain View, California, December 1987.

4. Huebler, J., and Janka, J.C., "Fuels, synthetic," *Encyclopedia of Chemical Technology*, Third Edition, Volume 11, Wiley-Interscience, New York, 1983.

5. Grossman, P.R., and Curtis, R.W., "Pulverized-coal-fired gasifier for production of carbon monoxide and hydrogen," Transactions of the ASME, Paper No. 53-A-49, July 1953.

6. Probert, P.B., "Industrial fuel gas from coal," American Ceramic Society, Washington, D.C., May 5-8, 1975.

7. James, D.E., and Peterson, M.W., "Suspension type gasifiers," AIChE, Los Angeles, CA, November 16-20, 1975.

8. "Cool Water coal gasification program: Fifth progress report," EPRI Report AP-5931, October 1988.

9. "Dry feed entrained-flow coal gasification processes: Technology status," *The SFA Quarterly Report*, SFA Pacific, Inc., Mountain View, California, December 1988.

10. Krewinghaus, A.B., and Richards, P.C., "Coal flexibility of the Shell coal gasification process," Pittsburgh Coal Conference, Pittsburgh, September 25-29, 1989.

11. "Dow coal gasification project and technology analysis," *The SFA Quarterly Report*, SFA Pacific, Inc., Mountain View, California, April 1987.

12. Bakker, W.T., "Materials for coal gasification, an EPRI perspective," EPRI Conference on Gasification Power Plants, Palo Alto, California, October 18, 1990.

13. Bakker, W.T., "Materials for syngas coolers in integrated gasification-combined cycle power plants," EPRI Technical Brief RP2048, 1988.

14. Meyers, R.A., ed., Part 3, *Handbook of Synfuels Technology*, McGraw-Hill, New York, 1984.

15. Peterson, D.L., *et al.*, "Treatment of gasification wastewater: Treatability test results and development of a process design manual," Ninth Annual EPRI Conference on Coal Gasification Power Plants, Palo Alto, California, October 17-19, 1990.

16. Rib, D.M., "Cool Water environmental performance utilizing four coal feedstocks," Eighth Annual EPRI Coal Gasification Contractor's Conference, Palo Alto, California, October 19-20, 1988.

17. "Acid gas removal," *The SFA Quarterly Report*, SFA Pacific, Inc., Mountain View, California, April 1988.

18. "The Claus sulfur recovery process," *The SFA Quarterly Report*, SFA Pacific, Inc., Mountain View, California, July 1990.

19. Allen, R.P., Battista, R.A., and Ekstrom, T.C., "Characteristics of an advanced gas turbine with coal derived fuel gases," Ninth Annual EPRI Conference on Coal Gasification Power Plants, Palo Alto, California, October 17-19, 1990.

20. Bajura, R.A., and Bechtel, T.F., "Update on DOE's IGCC programs," Ninth Annual EPRI Conference on Coal Gasification Power Plants, Palo Alto, California, October 17-19, 1990.

21. Notestein, J.E., "Update on Department of Energy hot gas cleanup programs," Eighth Annual EPRI Conference on Coal Gasification, Palo Alto, California, October 19-20, 1988.

22. Robin, A.M., "Integration and testing of hot desulfurization and entrained-flow gasification for power generation systems," Tenth Annual Gasification and Gas Stream Cleanup Systems Contractors Review Meeting, Morgantown, West Virginia, August 28-30, 1990.

23. Baker, D.C., *et al.*, "Environmental characterization of the Shell coal gasification process, II, aqueous effluent," Sixth Annual International Pittsburgh Coal Conference, Pittsburgh, September 1989.

24. Klock, B.V., Vuong, D.C., and Webster, G.H., "Texaco coal gasification wastewater treatment pilot plant studies," Ninth Annual EPRI Conference on Coal Gasification Power Plants, Palo Alto, California, October 17-19, 1990.

25. Perry, R.T., *et al.*, "Environmental characterization of the Shell coal gasification process, III, solid by-products," Seventh Annual Pittsburgh Coal Conference, Pittsburgh, September 11-13, 1990.

26. Ebbins, J.R., and Ruhl, E., "The BGL gasifier: Recent environmental results," Eighth Annual EPRI Conference on Coal Gasification, Palo Alto, California, October 19-20, 1988.

Bibliography

Handbook of Gasifiers and Gas Treatment Systems, USERDA Report FE-1772-11, Dravo Corporation, February 1976.

May, M.P., "Suspension gasifier offers economics of scale," *Iron and Steel Engineer*, pp. 46-52, November 1977.

Superheater installation in a coal-fired utility boiler.

Chapter 18
Boilers, Superheaters and Reheaters

In a modern steam generator, various components are arranged to efficiently absorb heat from the products of combustion and provide steam at the rated temperature, pressure and capacity. These components include the boiler, superheater, reheater, economizer and air heater. They are supplemented by systems for steam-water separation (see Chapter 5) and the control of steam outlet temperature. The entire boiler system can be divided into two general sections, the furnace and convection pass. The furnace provides a large open volume with water-cooled enclosure walls where combustion takes place and the combustion products are cooled to an appropriate furnace exit gas temperature (FEGT). The convection pass contains tube bundles which consist of the superheater, reheater, boiler bank and economizer. The convection pass is usually followed by the air heater. The boiler or evaporation-circulation system usually includes the furnace enclosure walls, steam drum and steam-water separation equipment, boiler bank tube bundle, and associated connecting piping (downcomers, supplies and risers as discussed in Chapter 1). This chapter focuses on the boiler, superheater and reheater components plus systems for steam temperature control, steam bypass and unit start-up. Chapter 19 discusses economizers and air heaters.

Boilers

Boiler surface is defined as the tubes, drums and shells which are part of the steam-water circulation system and which are in contact with the hot gases. Although the term *boiler* now frequently refers to the overall steam generating system, the term *boiler surface* excludes the economizer, superheater, reheater or any component other than the steam-water circulation system itself.

While boilers can be broadly classified as shell, fire tube and water tube types, as discussed in the introduction to *Steam*, modern high capacity boilers are of the water tube type. In the water tube boiler, the water and steam flow inside the tubes and the hot gases flow over the outside surfaces. The boiler circulation system is constructed of tubes, headers and drums joined in such a way that water flow is provided to generate steam while cooling all parts. The water tube construction allows greater boiler capacity and higher pressure than shell or fire tube designs. Also, the water tube boiler offers greater versatility in arrangement; this permits the most efficient use of the furnace, superheater, reheater and other heat recovery components.

Boiler configurations

Modern high capacity boilers come in a variety of designs, sizes and configurations to suit a broad range of applications. Sizes range from 1000 to 10,000,000 lb/h (0.13 to 1260 kg/s) and pressures range from one atmosphere to above the critical pressure.

The boiler configuration is largely determined by the combustion system, fuel, ash characteristics, operating pressure and total capacity. The diversity in configurations is illustrated in Figs. 1 through 5.

Typical industrial and small power boilers The Integral Furnace boiler shown in Fig. 1 is an oil- and gas- fired, low pressure two drum package boiler. In small capacities it can be entirely shop assembled and shipped to the site. (See Fig. 1, Chapter 1.) Because it burns a clean fuel, provision has not been made for flyash collection or surface cleaning, and a small furnace volume can be used. A boiler bank, closely spaced tubes between a steam drum and lower drum, provides the heat transfer surface necessary for the rated steaming capacity. Fig. 2 illustrates a two drum Stirling® power boiler (SPB) with a controlled combustion zone (CCZ™) furnace specially designed for effective firing of high moisture wood and biomass. (See also Chapter 28.) A spreader-stoker firing system is supplied and a boiler bank is provided for sufficient steam generating surface.

Additional unique designs include fluidized-bed boilers, process recovery boilers and waste-to-energy boilers discussed in Chapters 16, 26 and 27.

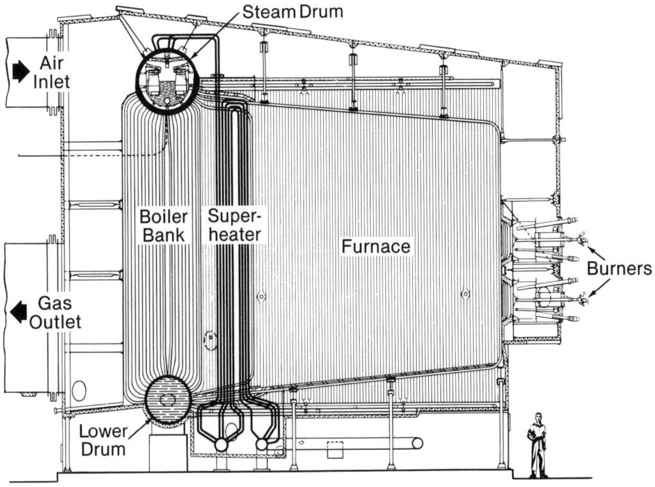

Fig. 1 Integral Furnace industrial boiler for oil and gas firing.

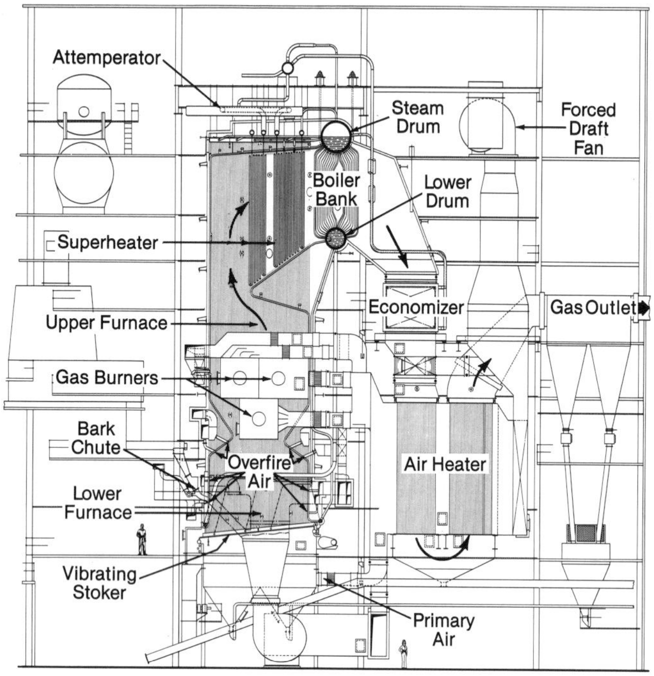

Fig. 2 Two drum Stirling® power boiler — CCZ™ stoker furnace configuration for bark firing.

Large utility boiler designs Figs. 3 and 4 illustrate two variations of the Babcock & Wilcox (B&W) Radiant boiler (RB) for natural circulation drum type steam generating systems. Fig. 5 illustrates one version of the B&W Universal Pressure (UP) boiler designed for once-through flow at supercritical or subcritical pressures. As discussed in Chapter 24, all of these units feature gas-tight, fully water-cooled furnace enclosure walls and floors made of all-welded membrane panel construction. Each design normally includes a single reheat section, although the supercritical once-through boiler has also been supplied as a double reheat unit. Fig. 3 shows an RB unit for coal firing. This is the Carolina-type design (RBC) with downflow convection backpass which minimizes overall steam generator height. Provision is made for sootblower surface cleaning and for flyash collection. A Tower-type (RBT) configuration is also available which features fully drainable convection pass surfaces and a minimum plan area. (See Chapter 24.) Fig. 4 is a Radiant boiler of the El Paso (RBE) configuration for oil and gas firing. This unit is very compact because of the relatively clean fuels being used. The compact design minimizes the boiler footprint (plan area) and support steel. Selected provision may be made for cleaning equipment.

Fig. 5 is a 1300 MW pulverized coal unit for supercritical pressures using vertical furnace wall tubes. As discussed in Chapter 24, the once-through circuitry design eliminates the need for a steam drum. Multiple fluid passes through the furnace minimize the imbalances in steam temperature around the furnace periphery which could result from nonuniform heat input. Other UP designs are available for subcritical pressure operation with spiral circuitry furnaces. (See Fig. 11 of Chapter 24.) Inclined tubes wrapping around the furnace make multiple fluid passes through the furnace enclosure unnecessary.

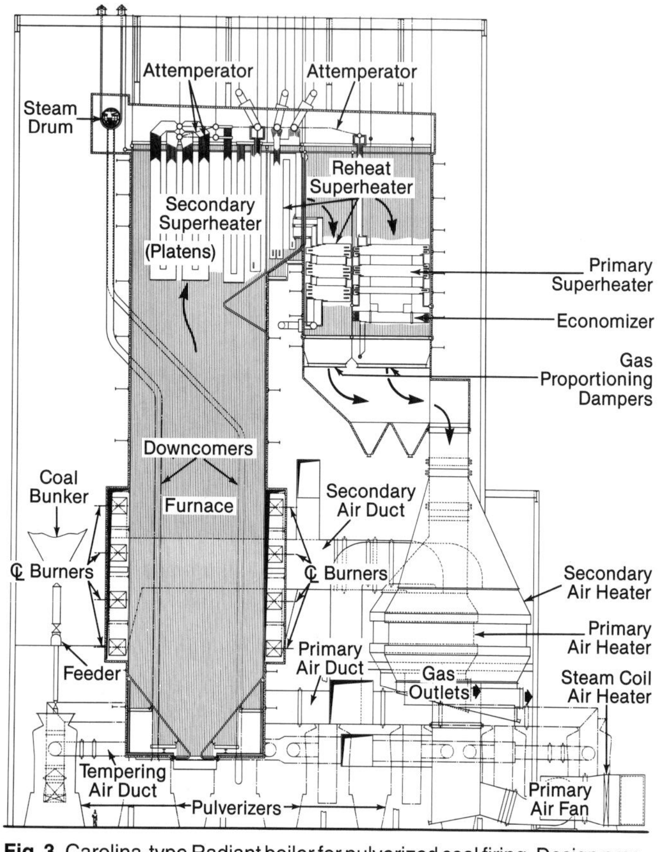

Fig. 3 Carolina-type Radiant boiler for pulverized coal firing. Design pressure 2950 psig (203.4 bar gauge); primary and reheat steam temperatures 1005F (541C); capacity 4,900,000 lb steam/h (617 kg steam/s).

Boiler design

Regardless of the size or configuration, modern boiler design remains driven by four key factors: 1) efficiency (boiler and cycle), 2) reliability, 3) capital and operating cost and 4) environmental protection. These factors, combined with specific applications, produce the diversity of designs presented above and discussed at length in Chap-

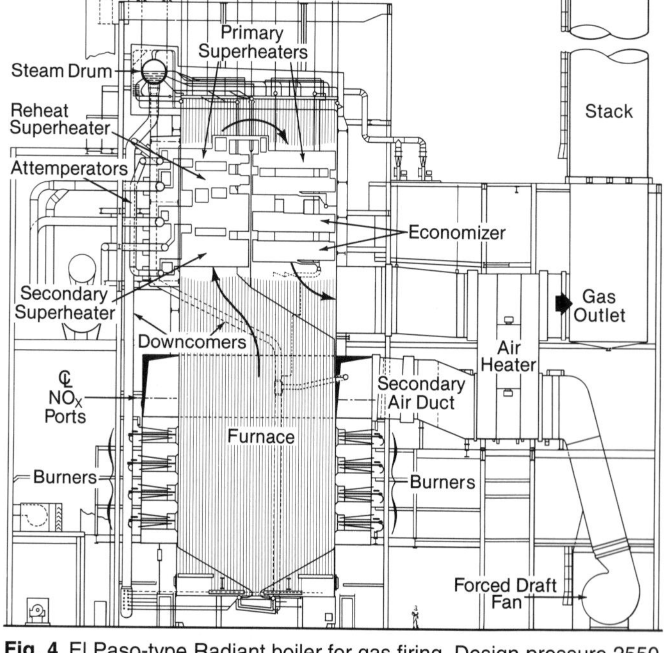

Fig. 4 El Paso-type Radiant boiler for gas firing. Design pressure 2550 psig (175.8 bar gauge); primary and reheat superheat temperatures 955F (513C); capacity 3,825,000 lb steam/h (482 kg steam/s).

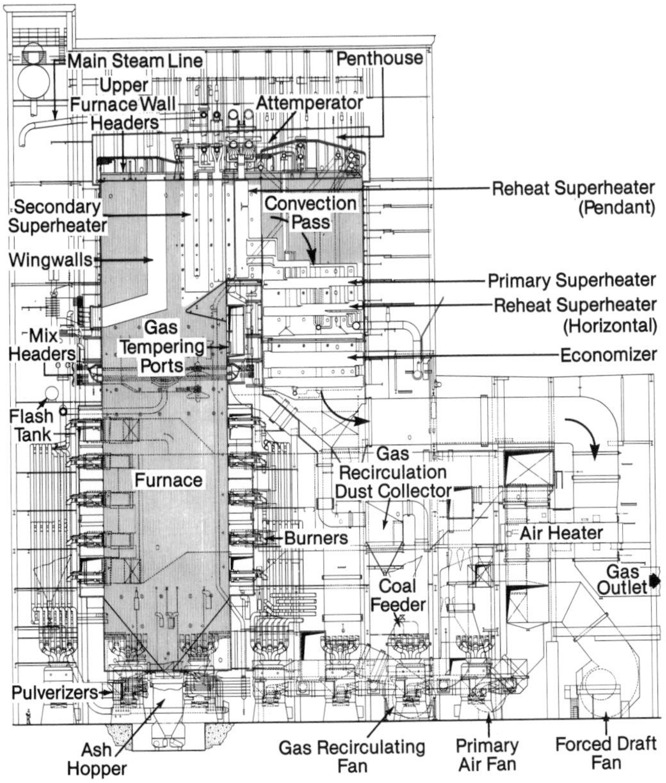

Fig. 5 Universal Pressure boiler for pulverized coal firing. Superheater outlet pressure 3845 psig (265.1 bar gauge); primary and reheat steam temperatures 1010F (543C); capacity 9,775,000 lb steam/h (1232 kg steam/s).

Table 1
Use-Derived Specifications for Boiler Design

Specified Parameter	Comments
Steam use	Flow rates, pressures, temperatures — for utility boilers, the particular power cycle and turbine heat balance.
Fuel type and analysis	Combustion characteristics, fouling and slagging characteristics, ash analysis, etc.
Feedwater supply	Water source, analysis and economizer inlet temperatures.
Pressure drop limits	Gas side and steam side.
Government regulations	Including emission control requirements.
Site-specific factors	Geographical and seasonal characteristics.
Steam generator use	Base load, cycling, etc.
Customer preferences	Specific design guidelines such as flow conditions, equipment preferences and steam generator efficiency.

ters 24 through 31. However, all of these units share a number of fundamental elements upon which the site- and application-specific design is based.

The boiler evaluation begins by identifying the overall application requirements specified in Table 1. These are generally selected in an iterative process balancing initial capital cost, operating costs (especially fuel), steam process needs and operating experience. The selection of these parameters can dramatically impact cost and thermal efficiency. These are addressed further in Chapters 1 and 37.

From a boiler evaluation perspective, the temperature-enthalpy diagram shown in Fig. 6 (for a typical high pressure, single reheat unit) provides important design information about the unit configuration. In this example, the relative heat absorption for water preheating, evaporation and superheating are 30%, 32% and 38%, respectively. Reheating the steam increases the total heat absorption by approximately 20%. For cycles at supercritical operating pressures, a second stage of reheat may be added. For process applications, only the preheating and evaporation steps may be required.

Boilers can be designed for subcritical or supercritical pressure operation. At subcritical pressures, the furnace enclosure is cooled by constant-temperature boiling water, and the flow circuits must be designed to accommodate the two-phase steam-water flow and boiling phenomena addressed in Chapter 5. At supercritical pressures, the water acts as a single-phase fluid with a continuous increase in temperature as it passes through the boiler. These designs require special consideration to avoid excessive unbalances in metal temperatures due to variations in heat pickup in different flow circuits. In addition, special heat transfer phenomena must also be addressed. (See Chapter 5.)

Two basic fluid circulation systems are used — natural circulation and once-through. In natural circulation systems which operate at subcritical pressures, water is only partially evaporated in the boiler circuits producing a steam-water mixture at the tube outlets. Steam-water separation equipment is provided to separate the steam and water, supply saturated (dry) steam to the superheater and recirculate water back to the boiler circuits. Natural circulation is the result of the density difference between the hot and cold legs of the loop. Indus-

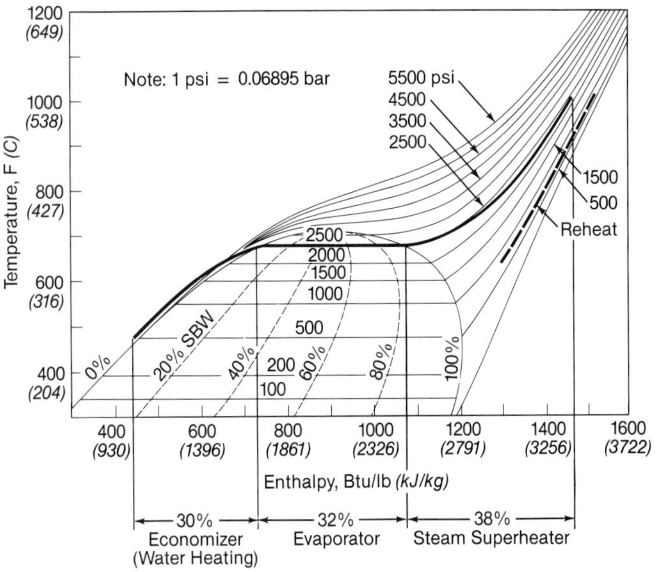

Fig. 6 Temperature-enthalpy diagram for subcritical pressure boiler absorption — one reheat section.

try accepted water chemistry limits are typically less stringent, especially at low pressures, and steam-water pressure drop inside the tubes is less of an issue. In B&W UP once-through designs, the steam drum and internal steam separation equipment are eliminated and a separate startup system added. UP boilers have been designed for both subcritical and supercritical operation. At supercritical pressures, the system can increase overall power cycle efficiency but at a higher initial capital cost. More precise operation is needed (see Chapter 43) and more stringent water treatment is required. (See Chapter 42.) Hybrid recirculating and once-through systems are presented in Chapter 5.

Design criteria Within the preceding framework, the important items which must be accomplished in boiler design are the following:

1. Define the energy input based upon the steam flow requirements, feedwater temperature, and an assumed or specified boiler thermal efficiency.
2. Evaluate the energy absorption needed in the boiler and other heat transfer components.
3. Perform combustion calculations to establish fuel, air and gas flow requirements. (See Chapter 9.)
4. Determine the size and shape of the furnace, considering the location and space requirements of the burners or other combustion system, and incorporating sufficient furnace volume for complete combustion and low emissions. Provision must be made for handling the ash contained in the fuel and cooling the flue gas so that the furnace exit gas temperature (FEGT) meets design requirements.
5. Determine the placement and configuration of convection heating surfaces. The superheater and reheater, when provided, must be placed where the gas temperature is high enough to produce effective heat transfer, yet not so high as to cause excessive tube temperatures or ash fouling. All convection surfaces must be designed to minimize the impact of slag or ash buildup and permit surface cleaning without erosion of the pressure parts.
6. Provide sufficient saturated boiler surface to generate the remainder of the steam not generated in the furnace walls. This can be accomplished with or without an economizer.
7. Design pressure parts in accordance with applicable codes using approved materials.
8. Provide a gas-tight boiler setting or enclosure around the furnace, boiler, superheater, reheater and economizer.
9. Design pressure part supports and the setting for expansion and local conditions, including wind and earthquake loading.

Fuel selection and specification are particularly important. Boiler systems are designed for specific fuels and frequently will encounter combustion, slagging, fouling or ash handling problems if a fuel with characteristics other than those originally specified is fired. All potential boiler fuels must be assessed to determine the most demanding fuel.

Procedures for optimizing steam pressure and temperature, and for evaluating the value of specific auxiliary equipment in a given application are outlined in Chapter 37.

As discussed in Chapter 9, the boiler or combustion efficiency is usually evaluated as 100 minus the sum of the heat losses expressed as a percentage. Chapter 9 provides procedures for calculating the corresponding fuel, air and gas flow rates.

Enclosure surface design

The furnace of a large pulverized coal-, oil- or gas-fired boiler is essentially a large enclosed volume where fuel combustion and cooling of the combustion products take place prior to their entry into the convection pass tube bundles. Excessive gas temperatures entering these tube banks could lead to elevated metal temperatures or unacceptable fouling and slagging. Heat transfer to the furnace enclosure walls is basically controlled by radiation. The walls can be cooled by either boiling water (subcritical pressure) or high velocity supercritical pressure water.

The convection pass enclosure contains the horizontal and vertical downflow gas passes, where most of the superheater, reheater and economizer surfaces are located. (See Fig. 3.) These enclosure surfaces can be water or steam cooled.

The furnace enclosures and convection pass enclosures are usually made of water-cooled tubes in an all-welded membrane construction. These enclosures have also been made from tangent tube construction or closely spaced tubes with an exterior gas-tight seal. For a membrane construction, the tube wall and membrane surfaces are exposed on the furnace side to the combustion process while insulation and lagging (sheet metal) on the outside protect the boiler, minimize heat loss and protect operating personnel. (See Chapter 22.)

Furnace size versus cycle requirements Besides providing the volume necessary for complete combustion and a means to cool the gas to an acceptable FEGT, the furnace enclosure also provides much of the steam generating surface in a boiler. These roles may not perfectly match one another. In coal-fired units, the minimum furnace volume is usually set to provide the fuel ash specific FEGT. Frequently this results in too much evaporator surface in high pressure boilers and too little surface in low pressure units to meet the thermodynamic requirements for the desired steam exit temperature.

Fig. 7 illustrates the effect of the steam cycle and the operating pressure and temperature on the relative energy absorption between the boiler/economizer and superheater/reheater. As the pressure and temperature increase, the total unit absorption for a given power production progressively declines because of increased cycle efficiency. The boiler and economizer absorption represents the relative amount of heat added to the entering feedwater to produce saturated steam or reach the critical point in a supercritical pressure UP boiler. As the operating pressure increases, the amount of heat required to produce saturated steam declines. Conversely, the amount of heat required for the superheater and reheater increases. The change in required boiler/economizer absorption may not seem significant. However, a 1% shift in absorption is equivalent to approximately 10F (6C) of superheat or reheat temperature.

On a drum unit, the furnace and water-cooled convection pass enclosure walls are the boiler surface. On low pressure units, the amount of heat absorbed in the furnace is usually not adequate to produce all the saturated steam required, and a boiler bank is installed after the superheater. (See Figs. 1 and 2.) On a high pressure unit, the heat absorbed by the furnace and economizer is adequate to produce all the saturated steam required. As the furnace size increases, the economizer can be made smaller to produce the same amount of steam. It can be envisioned that as the furnace is made larger, some point will be reached where an economizer will not be required. However, as the furnace is enlarged to reduce the FEGT, too much steam would be produced, leaving insufficient energy in the flue gas to meet design superheat and reheat temperatures. This situation requires special design features to meet all of the design goals.

Furnace design criteria The furnace is basically a large open volume enclosed by water-cooled walls for combustion. Its shape and volume are established by the selection of fuel and combustion system. For wall firing using circular burners (see Chapter 13), minimum clearances between individual burners, between burners and walls, as well as between burners and the furnace floor are established based upon physical clearances and functional criteria for complete combustion. These clearances prevent fuel stream and flame interaction, assure complete combustion, avoid unacceptable flame impingement on the walls (which could lead to tube overheating or excessive deposits) and minimize the formation of nitrogen oxides (NO_x). The maximum fuel input rate, number of burners and the associated clearances establish: 1) the furnace cross-sectional area, 2) the height of the combustion zone, 3) the height of the overfire air injection zone (if used), and 4) the distance between the burner zone and the furnace floor. Where the fuel is burned on stokers, the furnace cross-sectional area is established by a specified heat release rate per unit bed area. (See Chapter 15.)

As discussed in detail in Chapter 13, combustion system design and its impact on furnace volume and shape have become complex and critical as emissions limits have been reduced. Not only are low NO_x burners such as the B&W DRB-XCL™ used to reduce NO_x emissions, but other techniques, such as in-furnace staging with overfire air (NO_x ports), fuel reburning and reactant injection, can also be considered for NO_x reduction. In addition, some techniques for furnace sorbent injection to reduce sulfur dioxide (SO_2) emissions have also been developed. (See Chapter 35.) Each technique can have an impact on the furnace size and configuration.

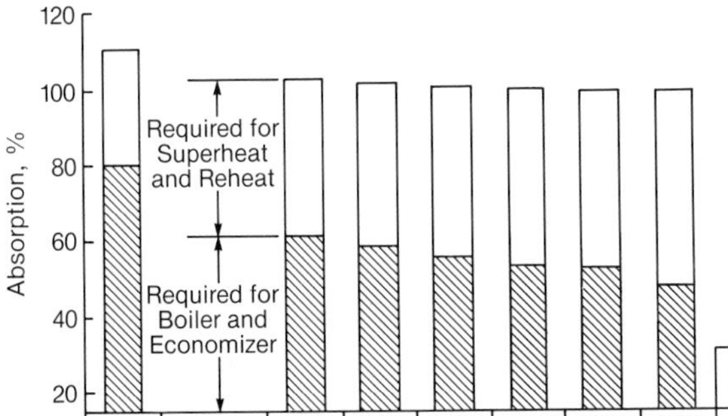

1250 (86.2)		1800 (124.1)	1800 (124.1)	2100 (144.8)	2400 (165.5)	2520 (173.8)	3500 (241.3)	psig (bar gauge)
950 (510)		950 (510)	1000 (538)	1000 (538)	1000 (538)	1000 (538)	1000 (538)	SH deg F (C)
—		950 (510)	1000 (538)	1000 (538)	1000 (538)	1000 (538)	1000 (538)	RH deg F (C)

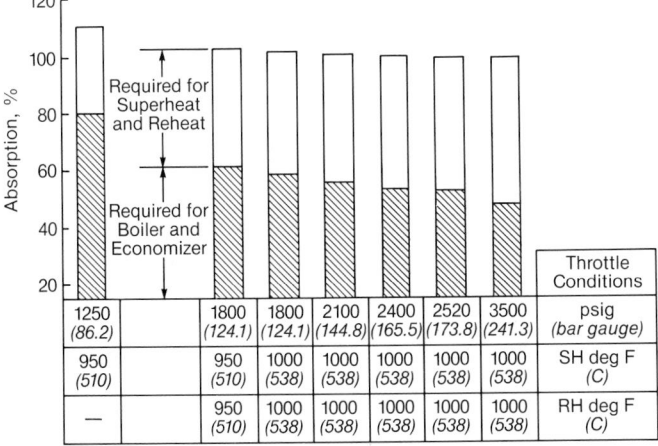

Throttle Conditions

Fig. 7 Relative heat absorption by operating pressure and steam temperature.

The overall furnace height is established by several criteria. For clean fuels, such as natural gas, the furnace volume and height are generally set to cool the combustion products to an FEGT which will avoid superheater tube overheating. For fuels such as coal and some oils which contain significant levels of ash, the furnace volume and height are established to cool the products of combustion to an FEGT which will prevent excessive fouling of the convection surfaces. The relationships between furnace volume and FEGT are explored in Chapter 4. The specific effects of ash on furnace design are discussed below and in Chapter 20. The furnace height must also be set to provide at least the minimum time to complete combustion and to meet minimum clearance requirements from the burners and NO_x ports to the arch and convective surface.

Ash effects In the case of coal and to a lesser extent with oil, an extremely important consideration is ash in the fuel. If this ash is not properly considered in the unit design and operation, it can deposit on the furnace walls, sloping surfaces and throughout the convection pass tube banks. Ash not only reduces the heat absorbed by the unit, but it increases draft loss, erodes pressure parts and eventually can result in unit outages for cleaning and repairs. (See Chapter 20.)

In coal-fired furnaces the ash problems are the most severe. There are two general approaches to ash handling: the dry ash or dry-bottom furnace and the slag-tap or wet-bottom furnace.

In the dry ash furnace, which is particularly applicable to coals with high ash fusion temperatures, a hopper bottom (Figs. 3 and 5) and sufficient cooling surface are provided so that the ash impinging on furnace walls or the hopper bottom is solid and dry; it can be removed essentially as dry particles. When pulverized coal is burned in a dry ash furnace, about 80% of the ash is carried through the convection banks. The chemistry of the ash can have a dramatic impact on the furnace volume necessary for satisfactory dry-bottom unit operation. This is illustrated in Fig. 8 which compares the furnace volume required for a nominal 500 MW boiler burning a low slagging bituminous or subbituminous coal to that for a high slagging lignite coal. The relationships defining the furnace size requirements are discussed in detail in Chapter 20.

With many coals having low ash fusion temperatures, it is difficult to use a dry-bottom furnace because the slag is molten or sticky; it clings and builds up on the furnace walls and hopper bottom. The slag-tap or wet-bottom furnace has been developed to handle these coals. The most successful form of the slag-tap furnace is that used with Cyclone furnace firing. (See Figs. 7 and 8, Chapter 14.) The furnace comprises a two-stage arrangement. In the lower part of the furnace, sufficient gas temperature is maintained so that the slag drops onto the floor in liquid form. Here, a pool of liquid slag is maintained and tapped into a slag tank containing water. In the upper part of the furnace, the gases are cooled below the ash fusion point, so ash carried over into the convection banks is dry and does not cause excessive fouling. Because of high NO_x emissions, slag-tap designs are used infrequently on new boilers.

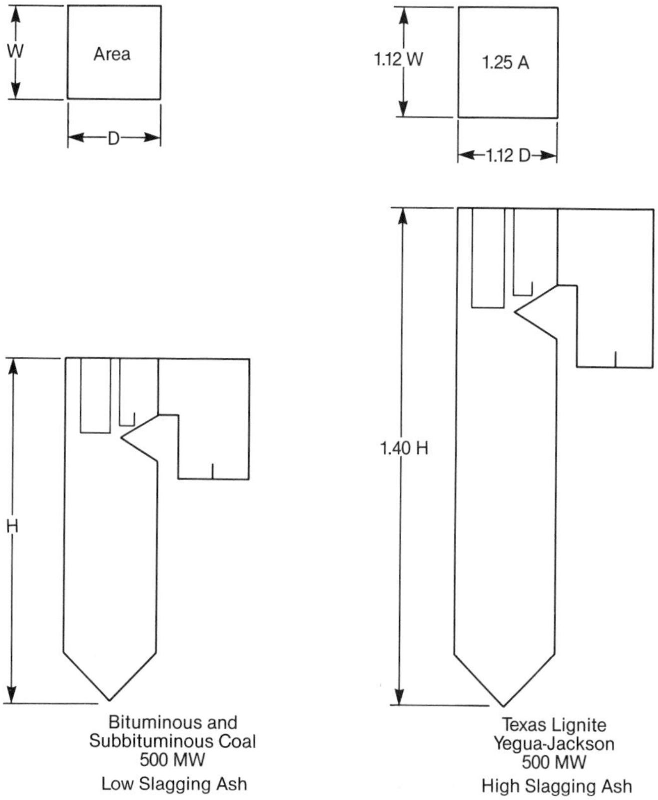

Fig. 8 Boiler size comparison for alternate coal types.

Water-cooled walls Most boiler furnaces have all water-cooled membrane walls. This reduces maintenance on the furnace walls and reduces the temperature of the gas entering the convection bank to the point where slag deposits and superheater corrosion can be controlled by sootblowing.

Furnace wall tubes are spaced on close centers to obtain maximum heat absorption while maintaining tube and membrane temperatures and thermal stresses within limits. The membrane panels (see Fig. 9) are composed of tube rows spaced on centers wider than a tube diameter and joined by a membrane bar securely welded to the adjacent tubes. This results in a continuous wall surface of rugged, pressure-tight construction capable of transferring the required heat from the furnace gas to the steam-water mixture in the tubes. The width and length of individual panels are suitable for economical manufacture and assembly, with bottom and top headers shop-attached prior to shipment for field assembly. Size limitations are frequently governed by shipping clearances and erection constraints. Membrane construction with refractory lining is used in the lower furnace walls of Cyclone-fired, refuse and fluidized-bed units.

Convection boiler surface

Some designs include boiler tubes as the first few rows of tubes in the convection bank. The tubes are spaced to provide gas lanes wide enough to prevent ash and slag pluggage and to facilitate cleaning for dirty fuels. These widely spaced boiler tubes, known as the *slag screen* or *boiler screen*, receive heat by radiation from the furnace and by radiation and convection from the combustion gases passing through them. Another option is the use of water- or steam-cooled wingwalls in the upper furnace. This provides additional radiant boiler surface in the furnace while allowing the furnace size to be optimized.

In the larger high pressure units, superheater surface generally forms the furnace outlet plane. The gas temperature entering the superheater must be high enough to give the desired superheat temperature with a reasonable amount of heating surface and the use of economical materials. The arrangements in Figs. 1 through 5 illustrate various configurations of superheater surface at the furnace outlet. Also, to optimize superheater design, widely spaced steam-cooled platens or wingwalls may be incorpo-

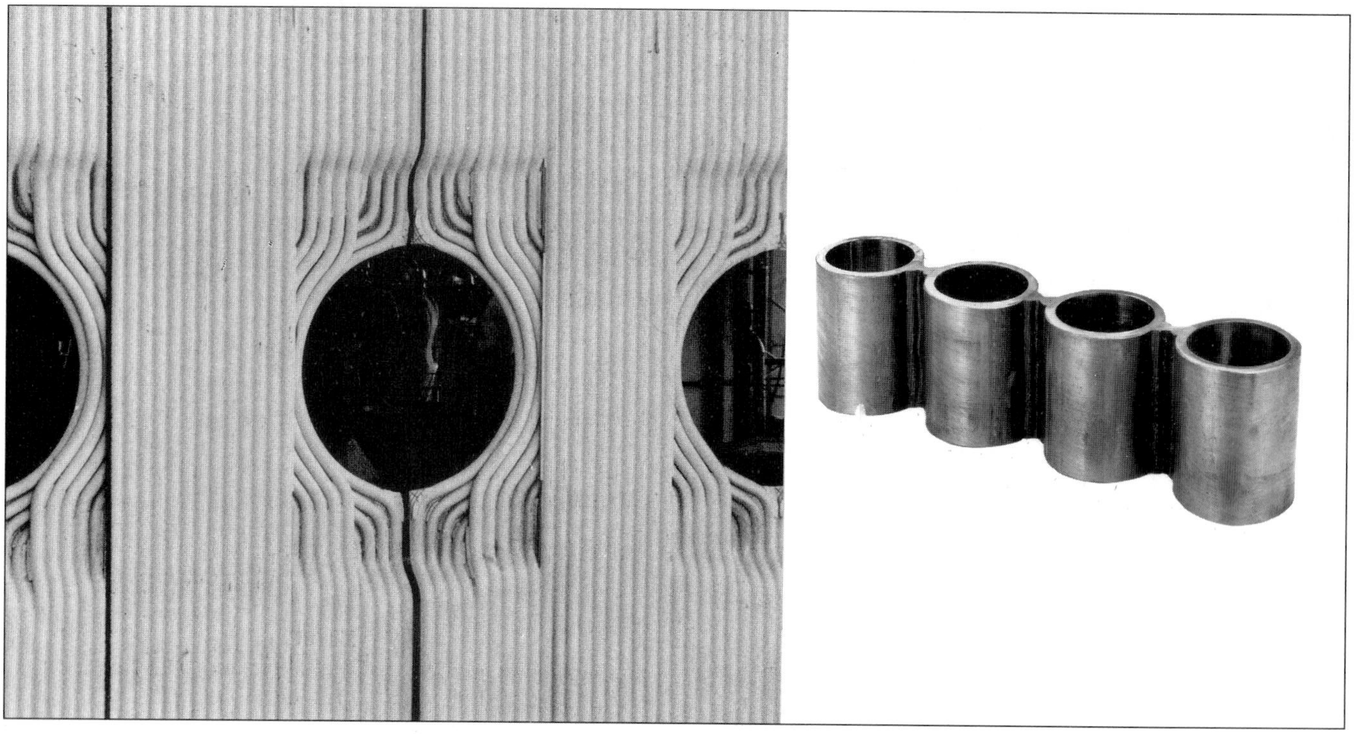

Fig. 9 Membrane wall construction at burner openings.

rated in the upper furnace as shown in Figs. 3 and 5. Fig. 10 shows a plan arrangement of convection surface and the change in average gas temperature.

Design of boiler surface after the superheater depends on the type of unit (industrial or utility), desired gas temperature drop and acceptable gas side resistance (draft loss) through the boiler surface. Typical arrangements of boiler and superheater surface for various types of boilers are illustrated in the preceding section on boilers. In designing convection heating surfaces, the objective is to establish the proper combination of several parameters to provide the desired gas temperature drop with allowable flue gas resistance. These parameters are tube diameter and length, tube spacing, number and orientation of tubes, and gas baffling.

In the gas side design of convection surfaces, the quantity of heating surface (ft^2 or m^2) needed for a given load or duty (Btu/h or W) is generally inversely related to gas side flow resistance or pressure loss. Design changes which increase resistance, such as tightening tube spacing perpendicular to flow, result in higher heat transfer rates (Btu/h ft^2 or W/m^2). This in turn reduces the amount of heating surface needed to carry the desired total thermal load. An optimal gas mass flux and design result from balancing the capital cost of the heating surface against the operating cost of fan power needed to overcome the resistance.

For a given gas flow rate, a considerably higher gas film heat transfer coefficient, heat absorption and draft loss result when the gases flow at right angles to the tubes (crossflow) compared to flow parallel to the tubes (long flow). Gas turns between tube banks generally add draft loss and maldistribution with little benefit to heat absorption. Turns leaving the upstream bank and entering the downstream bank of tubes should therefore be designed for minimum resistance and optimum distribution.

Determination of the amount of radiation and convection boiler surface required for a specified heat transfer is illustrated in Chapter 21.

Numerical analysis

When one dimensional analysis and experimental data are insufficient for design alone, numerical computer modeling can be used to determine local heat transfer and flue gas conditions in boilers, superheaters and reheaters. These numerical tools are complex multi-dimensional computer codes which solve conservation equations for mass, momentum, energy and species. (See Chapters 3 and 4.)

A numerical model of the superheater and/or reheater regions generally includes the boiler enclosure, burners and any other geometric aspects which may affect the flow. Downstream and upstream components are sometimes added to provide the best boundary conditions to the area of interest. In applying these computer codes, model size and complexity must be balanced against the associated computer calculation time required. Required numerical inputs include:

1. geometric data — boiler and heat exchanger surface configurations,
2. performance data — heat exchanger pressure drops and heat absorptions,
3. operating data — stoichiometry and furnace exit gas temperature,

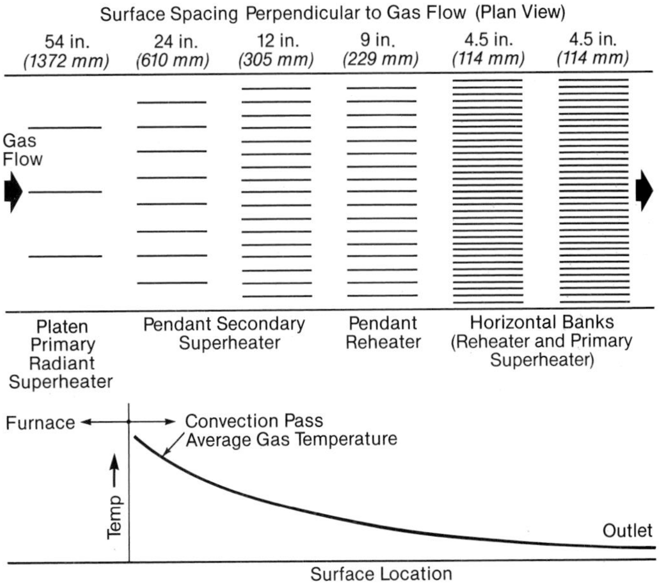

Fig. 10 Schematic plan arrangement of convection surface and change in average gas temperature.

4. burner setup — vane angles and turbulence intensity, and
5. fuel properties — fuel and flue gas quantities, composition and coal fineness, if applicable.

Use of operating data along with predictions based on company standards improves the accuracy of the model by tuning it to actual conditions. This is a necessary step because actual local conditions, e.g., ash thickness and property variation, are generally not known. Often, published data and sound engineering judgment are relied upon to obtain otherwise unavailable information. A numerical model can then be run once the boundary and inlet conditions are calculated and fluid properties are known.

After the model has been successfully run, the results can be postprocessed to determine the dominant physical characteristics. Further postprocessing can be used to obtain derived quantities such as slagging resistance or mixing effectiveness.

Numerical analyses fall into three general categories: 1) verification of proposed designs, 2) evaluation of design modifications, and 3) investigation of localized problems. Figs. 11, 12 and 13 provide sample numerical computer model results for illustration purposes. Figs. 11a and 11b are typical streamline plots containing curves that are tangent to the gas flow in a proposed boiler design. The design was modified as shown because the amount of heat transfer surface being bypassed by the gas flow lowered the expected heat absorption. Such evaluation of design modifications can be performed prior to fabrication to reduce long term costs and improve unit performance.

A second numerical model application includes an analysis to determine the optimum NO_x port arrangement in a boiler system. Such an analysis can use mixing effectiveness to rank the individual arrangements. Fig. 12 contains plots of fuel-air stoichiometry at an elevation 15 ft (4.6 m) above the NO_x ports before (a) and after (b) NO_x port adjustment. The more uniform distribution with improved mixing shown in (b) results in less NO_x formation.

Numerical modeling is also well suited to investigate boiler problems. The model takes into account the actual boiler geometry and performance characteristics, i.e., heat

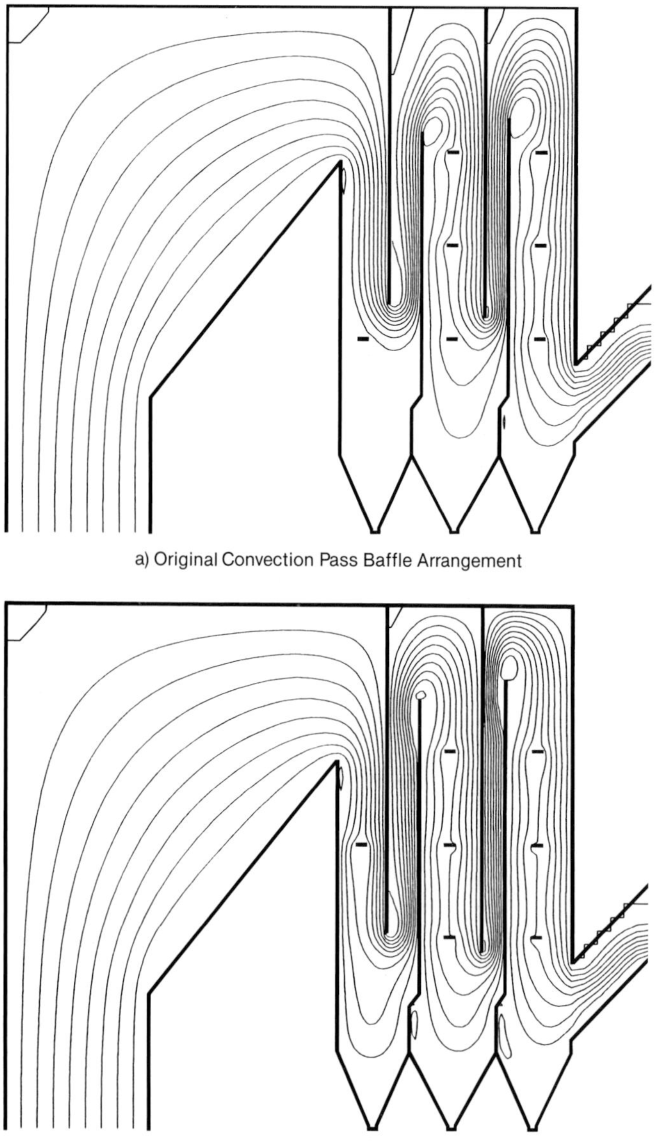

a) Original Convection Pass Baffle Arrangement

b) Improved Convection Pass Baffle Arrangement

Fig. 11 Numerical modeling results — flow streamlines.

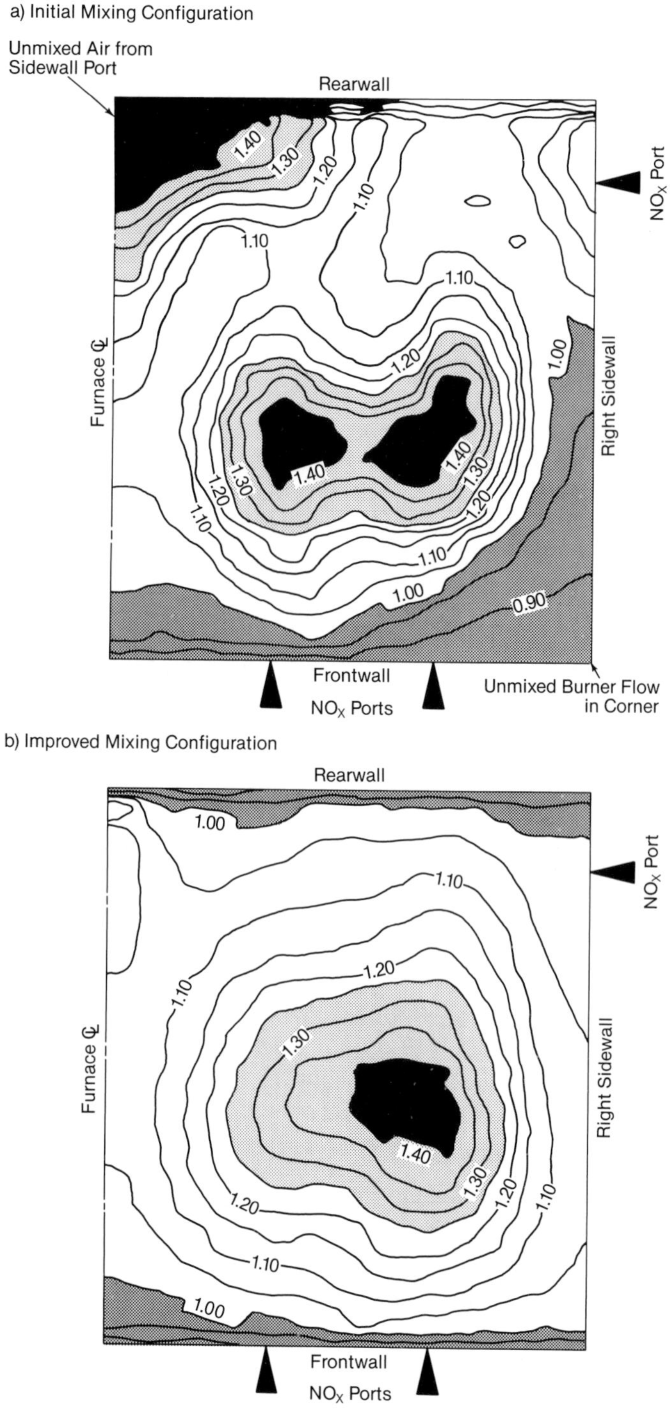

Fig. 12 Numerical modeling results — local combustion fuel-air stoichiometry.

absorption, temperature and pressure drop. The model also provides a more detailed and clearer performance picture than can be attained using analytical, field or laboratory methods. As an example, a unit was experiencing excessive slag accumulation on the lower leading edge of the secondary superheater inlet bank. Numerical modeling was used to determine the cause of the slagging and to evaluate several design modifications to reduce it. High velocities and associated high ash loadings were found to occur on the lower surfaces with a bias toward the sidewalls. Temperature contour plots for the as-built nominal case and for a larger furnace arch case (shown in Fig. 13) indicated that increasing the size of the arch moved the hotter furnace gases away from the secondary superheater problem area, virtually eliminating the hot ash impaction. This reduction of about 100F (56C) in the maximum temperature entering the secondary superheater enhanced the unit's resistance to slagging.

Accurate modeling of boilers and boiler components depends heavily upon the availability of information to determine the proper inlet conditions, boundary conditions and fluid properties. Careful qualification of the model is required and is augmented by known global characteristics such as pressure drop and heat absorption. Correct application of accurate information to develop numerical models will improve designs and determine the cause of localized problems.

Design of pressure parts

Boilers have achieved today's level of safety and reliability through the use of sound materials and safe practices for determining acceptable stresses in drums, headers, tubes and other pressure parts. Boilers must be designed to applicable codes. Stationary boilers in the United States

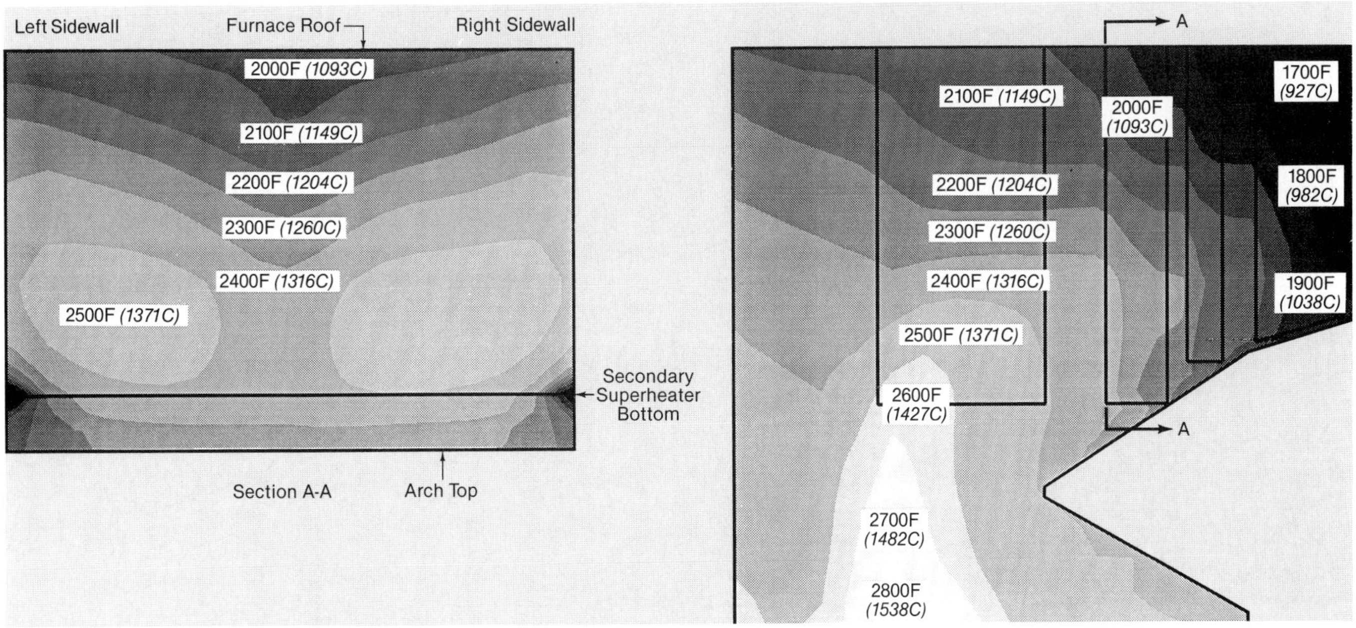

Fig. 13 Numerical modeling results — flue gas temperatures at furnace exit.

(U.S.) are designed to the American Society of Mechanical Engineers (ASME) Boiler and Pressure Vessel Code. (See Chapter 7 and Appendix 2.) The allowable design stress depends on the maximum temperature to which the part is subjected and, therefore, it is important that pressure part design temperatures are known and not exceeded in operation. Drum boiler enclosure material temperatures are a function of upset spot heat flux, design pressure, metal conductivity and the saturation temperature corresponding to the maximum boiler operating pressure. These parameters are also used to determine each tube outside diameter and thickness. For boiler tubes, temperatures are maintained at known levels by providing a sufficient flow of water to prevent the occurrence of CHF, or critical heat flux phenomena. (See Chapter 5.) An adequate saturated water velocity must exist for each tube and particular attention must be given to high heat flux zones and sloped tubes with heat on top.

Because steam drums have thick walls, it is necessary to limit the heat flow through them to avoid excessive thermal gradients during startup, shutdown and normal operation. This is particularly important where the drum is exposed to flue gas. Where the drum is penetrated by a number of tube holes, the flow of water through these holes serves to cool the drum wall. Where the heat input through a drum would be too high because of high gas temperature or velocity, insulation may be provided on the outside of the drum or the drum relocated out of the heat.

In a drum-type boiler, steam separation equipment is provided to maintain steam moisture and solids at acceptable levels. (See Chapter 5.) In once-through boilers, all moisture is evaporated in the tubes, so that boiling and superheating occur sequentially. In boilers of this type, steam purity depends on maintaining adequate feedwater purity. (See Chapter 42.)

Boiler safety valves constitute very important protection items. (See Chapter 23.) The ASME Code stipulates that the boiler design pressure must not be less than the high-set safety valve relief pressure. As a practical matter, to avoid unnecessary losses and maintenance from frequent popping of the safety valves, the first valve should be set to relieve at not less than the boiler operating pressure plus 5%. The operating pressure in the steam drum, in turn, depends on the pressure required at the point of use and the intervening pressure drop. As an example, where the steam is used in a turbine, the boiler operating pressure is determined by adding the turbine throttle pressure and the pressure drop through the steam piping, nonreturn valve, superheater and drum internals at maximum unit steam flow.

Boiler enclosure

The methods used to provide a tight boiler setting and a tight enclosure around the superheater, reheater, economizer and air heater are described in Chapter 22.

Boiler supports

Furnace wall tubes are usually supported by the headers to which they are attached, and generating bank tubes and screens are supported by the drum or headers to which they are connected. As discussed in Chapter 7, the following considerations for proper support design are important:

1. The tubes must be arranged and aligned so that they are not subjected to excessive bending-moment stresses in supporting the weight of the tubes, headers, drums, attachments and fluid within. When the unit is bottom supported, the tubes must satisfy column buckling requirements.
2. The holding strength of the tube seats must not be exceeded.
3. Provision must be made to accommodate the expansion of the pressure parts. For a top supported unit, the hanger rods which tie the pressure parts to structural steel must be designed to swing at the proper angle. They must be long enough to withstand the movement without excessive stresses in the rods or the pressure parts. Bottom supported boilers should be anchored only at one point, guided along one line and allowed to expand freely in all other directions. To reduce the frictional forces and resultant stresses in the pressure parts, roller saddles or mountings are desirable for bottom supported heavy loads.

Superheaters and reheaters

Advantages of superheat and reheat

When saturated steam is used in a turbine, the work done results in a loss of energy by the steam and subsequent condensation of a portion of the steam, even though there is a drop in pressure. The amount of work that can be done by the turbine is limited by the amount of moisture that it can handle without excessive turbine blade wear. This is normally between 10 and 15% moisture. It is possible to increase the amount of work done with moisture separation between turbine stages, but this is economical only in special cases. Even with moisture separation, the total energy that can be transformed to work in the turbine is small compared to the amount of heat required to raise the water from feedwater temperature to saturation and then evaporate it. Therefore, moisture content constitutes a basic limitation in turbine design.

Because a turbine generally transforms the energy of superheat into work without forming moisture, this energy is essentially all recoverable in the turbine. This is illustrated in the temperature-entropy diagram of the ideal Rankine cycle (shown in Fig. 6, Chapter 2). While this is not always entirely correct, the Rankine cycle diagrams in Chapter 2 indicate that this is essentially true in practical cycles.

The foregoing discussion is not specifically applicable at steam pressures at or above the critical pressure. In fact, the term *superheat* is not truly accurate in defining the temperature of the working fluid in this region. However, even at pressures exceeding 3208 psi (221 bar), heat added at temperatures above 705F (374C) is essentially all recoverable in a turbine.

The benefit of superheat is indicated by the reduction of cycle heat rate when steam temperatures entering the turbine are raised. For example, in a simple calculation of a 2400 psig (165.5 bar gauge) ideal Rankine cycle with a single reheat stage, an increase in superheat temperature from 900 to 1100F (482 to 593C) reduces the gross heat rate from approximately 7550 to 7200 Btu/kWh. This is more than a 4.5% efficiency improvement attributable to the superheat.

Superheater types

Two basic types of superheaters are available depending upon the mode of heat transfer from the flue gas. The original type was the *convection superheater*, for gas temperatures where the portion of heat transfer by radiation from the flue gas is small. With a unit of this design, the steam temperature leaving the superheater increases with boiler output because of the decreasing percentage of unit heat input that is absorbed in the furnace. This results in more heat available for superheater absorption. Because convection heat transfer rates are almost a direct function of gas flow rate and therefore boiler output, the total absorption in the superheater per pound of steam, and therefore steam temperature, increase with boiler output. (See Fig. 14.) This effect is increasingly pronounced the farther the superheater is located from the furnace and the lower the gas temperature entering the superheater.

A *radiant superheater* receives energy primarily by thermal radiation from the furnace with little energy from convective heat transfer. It usually takes the form of widely spaced [24 in. (609.6 mm) or larger side spacing] steam-

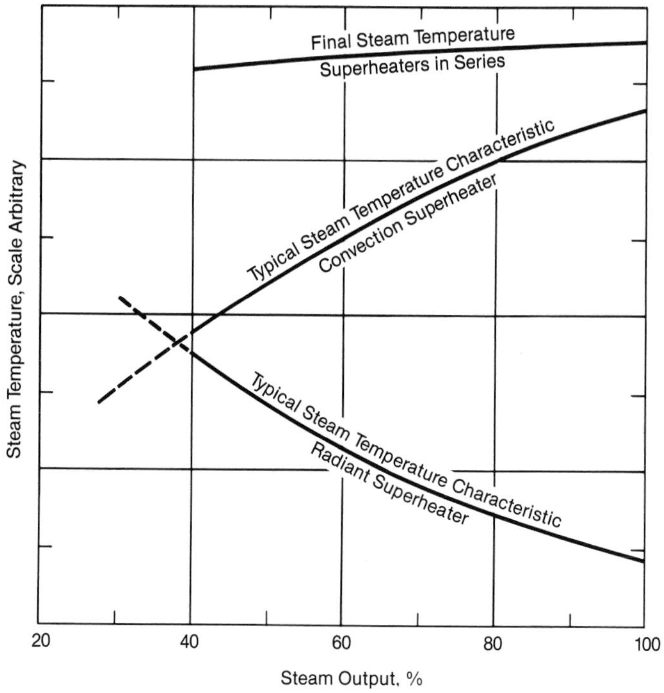

Fig. 14 A substantially uniform final steam temperature over a range of output can be attained by a series arrangement of radiant and convection superheater components.

cooled wingwalls or pendant superheat platens located in the furnace. It is sometimes incorporated into the furnace enclosure curtain walls. Because the heat absorption by furnace surfaces does not increase as rapidly as boiler output, the radiant superheater outlet temperature declines with an increasing boiler output, as shown in Fig. 14.

In certain cases the two opposite sloping curves have been coordinated by the series combination of radiant and convection superheaters to give a flat superheat curve over a wide load range, as indicated in Fig. 14. A separately-fired superheater can also be used to produce a flat superheat curve.

The design of radiant and convective superheaters requires extra care to avoid steam and flue gas distribution differences which could lead to tube overheating. Superheaters generally have steam mass fluxes of 100,000 to 1,000,000 lb/h ft^2 (136 to 1356 kg/m^2 s) or higher. These are set to provide adequate tube cooling while meeting allowable pressure drop limits. The mass flux selected depends upon the steam pressure and temperature as well as superheater thermal duty. In addition, the higher pressure loss associated with higher velocities improves the steam side flow distribution.

The fundamental considerations governing superheater design also apply to reheater design. However, the pressure drop in reheaters is critical because the gain in heat rate with the reheat cycle can be nullified by too much pressure loss through the reheater system. Therefore, steam mass fluxes are generally somewhat lower in the reheater.

Tube sizes

Bare cylindrical tubes of 1.75 to 2.75 in. (44.5 to 69.9 mm) outside diameter are typical in current superheaters and reheaters. Steam pressure drop is higher and alignment is more difficult with the smaller diameters, while larger diameters result in higher pressure stresses.

Recent designs have called for greater spans between supports for horizontal superheater tubes and for wider tube spacing or fewer tubes per row to avoid slag accumulation. The 2.5 in. (63.5 mm) tube has met these new conditions with minimum sacrifice of the smaller tube advantages; 2.75 or 3 in. (69.9 or 76.2 mm) tubes are used to advantage in some cases. When steam temperatures increase, the allowable stresses may force a return to the smaller diameter, thinner-walled tubes.

Bare tubes are used almost exclusively in superheaters. Extended surface on superheater tubes in the form of fins, rings or studs makes gas side cleaning difficult, and the added thickness can increase metal temperature and thermal stress beyond tolerable limits.

Relationships in superheater design

Effective superheater design must consider several parameters including:

1. the steam temperature specified,
2. the range of boiler load over which steam temperature is to be controlled,
3. the superheater surface required to give this steam temperature,
4. the gas temperature zone in which the surface is to be located,
5. the type of steel, alloy or other material best suited for the surface and supports,
6. the rate of steam flow through the tubes (mass flux or velocity), which is limited by the permissible steam pressure drop but which, in turn, exerts a dominant control over tube metal temperatures,
7. the arrangement of surface to meet the characteristics of the anticipated fuels, with particular reference to the spacing of the tubes to prevent accumulations of ash and slag or to provide for easy removal of these formations in their early stages, and
8. the physical design and type of superheater as a structure or component.

A change in any of these items may require a counterbalancing change in some or all of the other items.

The steam temperature desired in advanced power station design is typically the maximum for which the superheater designer can produce an economical component. Economics in this case requires the assessment of two interrelated costs — initial investment and the subsequent cost of upkeep to minimize operating problems, outages and replacements. The steam temperature desired is, therefore, based upon an iterative evaluation of variations in factors 3 through 7, past operating experience and the requirements of the particular project. Operating experience in recent years has resulted in the use of 1000F (538C) or 1050F (566C) steam temperatures for the superheat and reheat in nearly all new large utility boilers.

After the steam temperature is specified, the amount of surface necessary to provide this superheat must be established. This is dependent on parameters 5 through 8. Because there is no single correlation, this quantity must be determined by trial and error, locating the superheater in a zone of gas temperature that satisfies the design criteria. In standard boilers, the zone is fairly well established by unit arrangement and by the space designated for superheater surface.

After the amount of surface is established for the optimum location and tube spacing, the steam mass flux or velocity, steam pressure drop, and superheater tube metal temperatures are calculated considering material options and thickness requirements. The material determination is then made based on economic and functional optimization of the tubes, headers and other components. It may be necessary to compare several arrangements to obtain an optimum combination that:

1. requires an alloy of less cost,
2. gives a more reasonable steam pressure drop without jeopardizing the tube temperatures,
3. gives a higher steam mass flux or velocity to lower tube temperatures,
4. gives the tube spacings that minimize ash accumulations with various types of fuel,
5. permits closer spacing of the tubes, thereby making a more economical arrangement for a favorable fuel supply,
6. gives an arrangement of tubes which reduces the draft loss for an installation where this parameter evaluation is crucial, and
7. permits the superheater surface to be located in a zone of higher gas temperature, with a subsequent saving in surface.

It is possible to achieve a practical design with optimum economic and operational characteristics and with all criteria reasonably satisfied, but a large measure of experience and the application of sound physical principles are required for best results. The calculation methods for superheater performance are given in Chapter 21.

Relationships in reheater design

Superheaters and reheaters are similar in design, but the reheater is limited by the permissible steam pressure drop. Steam mass fluxes or velocities in reheater tubes should be sufficient to keep the difference between the bulk steam and metal surface temperature below 150F (83C). Ordinarily this is done with a pressure drop through the reheater tubes that is 4 to 5% of reheater inlet pressure. This allows another 4 to 5% pressure drop for the reheat piping and valves without exceeding the usual 8 to 10% total allowable pressure loss. The pressure drop allocated for piping is usually distributed with one-third to the cold (inlet) reheat piping and two-thirds to the hot (outlet) reheat piping.

Tube metals

Oxidation resistance, allowable stress and economics determine the materials used for superheater and reheater tubes. The use of carbon steel is maximized. However, carefully selected alloy steels are used where required. Additional information on these materials including maximum metal temperatures are given in Chapter 6.

Variations in basic superheater heat transfer surface arrangements are shown in Fig. 15. These variations in surface arrangement permit the economic tradeoff between material unit costs and differences in the quantity of surface required due to thermal-hydraulic considerations. The main purpose of the counterflow versus parallel flow comparisons is to show basic relationships between heat transfer surface requirements and the corresponding tube metal temperatures and requirements.

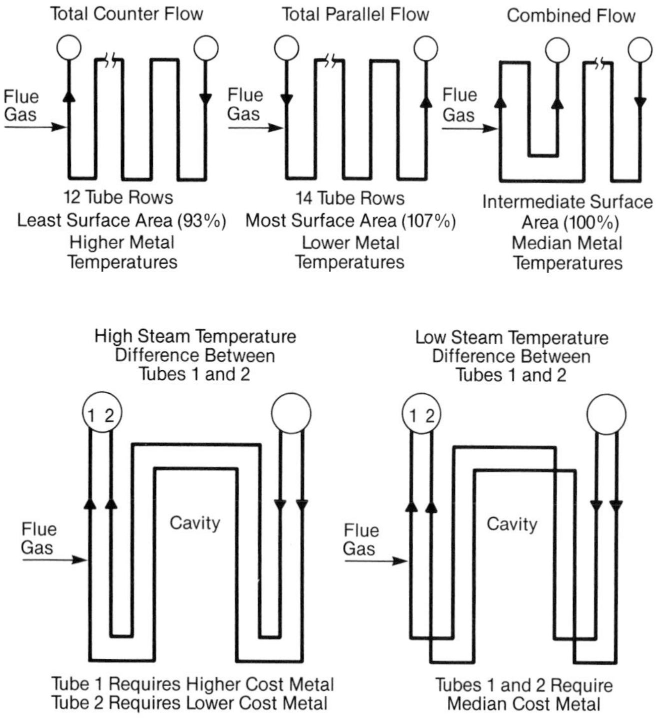

Fig. 15 Typical superheater heat transfer surface arrangements.

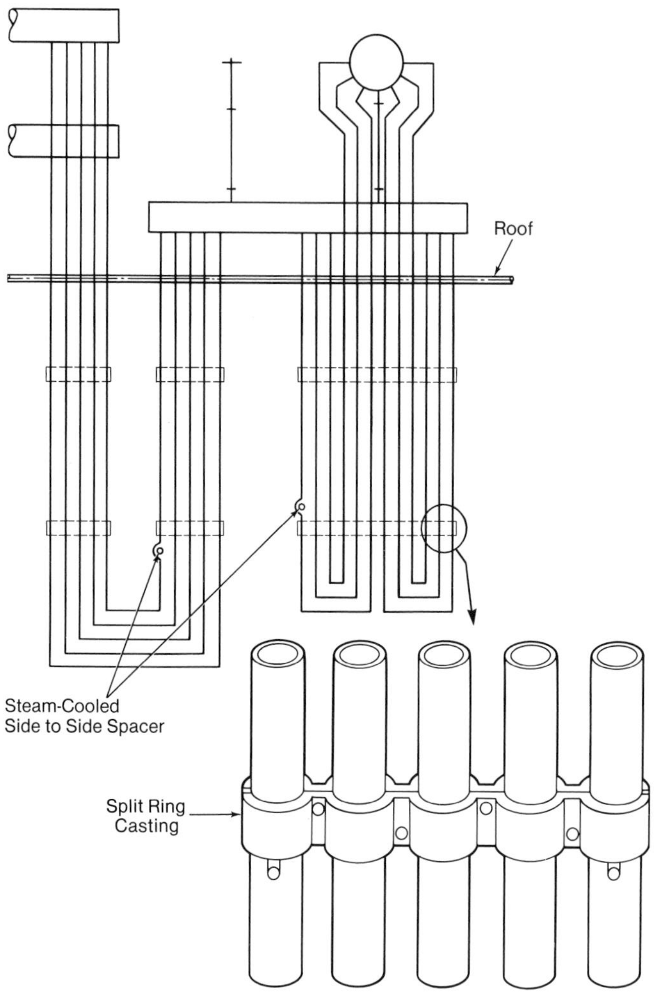

Fig. 16a Pendant superheater section with split ring casting supports.

Supports for superheaters and reheaters

Because superheaters and reheaters are located in zones of relatively high gas temperature, it is preferable to have the major support loads carried by the tubes themselves. For pendant superheaters, the major support points are located outside of the gas stream, with the pendant loops supporting themselves in simple tension. Figs. 16a and 16b illustrate standard support arrangements for a pendant superheater outlet section with major section supports above the roof line. Where adequate side spacing is available and the ash cleaning is not abrasive, steam-cooled wraparound guides are used. Where abrasive ash cleaning is anticipated (coal firing), high chromium-nickel alloy ring-type guides are used. In addition, in the higher gas temperature zones, steam-cooled side to side ties are used to maintain side spacings. For closer side spaced elements, steam-cooled wraparounds are not practical and mechanical ties, such as D-links, are used to maintain alignment as shown in Fig. 17. In this case the clear backspacing between tubes and the size of attachments have been kept to a minimum. This serves to reduce the thermal stresses imposed on the tube wall. Fig. 17 also shows a typical arrangement of a pendant reheater section and illustrates the support of a separated bank by a special loop of reheater, which permits all major supports to be kept above the roof and out of the gas stream.

In horizontal superheaters, the support load is usually transferred to boiler or steam-cooled enclosure tubes or economizer stringer tubes by lugs, one welded to the support tubes and the other to the superheater tubes as indicated in Fig. 18. These lugs are made of carbon steel through high chromium-nickel alloy. Depending upon design considerations, they must slide on one another to provide relative movement between the boiler tubes and the superheater tubes. Saddle-type supports, also shown in Fig. 18, provide for relative movement between adjacent tubes of the superheater.

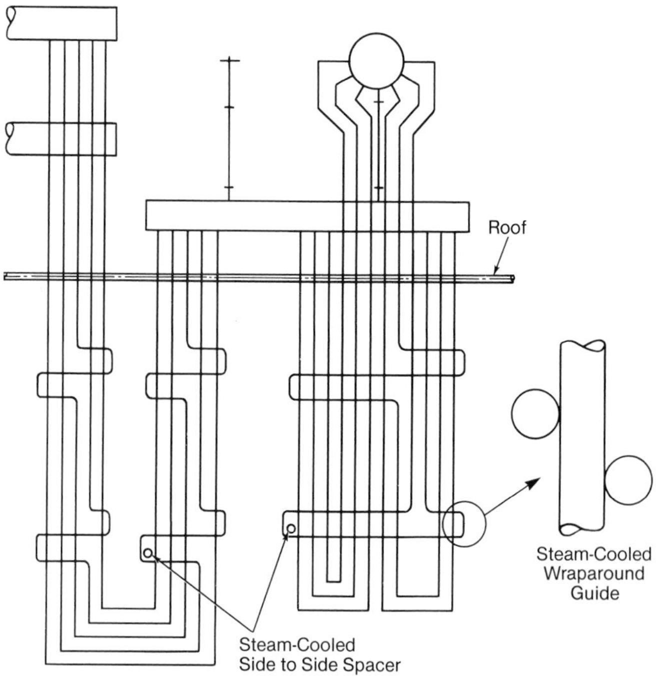

Fig. 16b Pendant superheater section with steam-cooled wraparound guide.

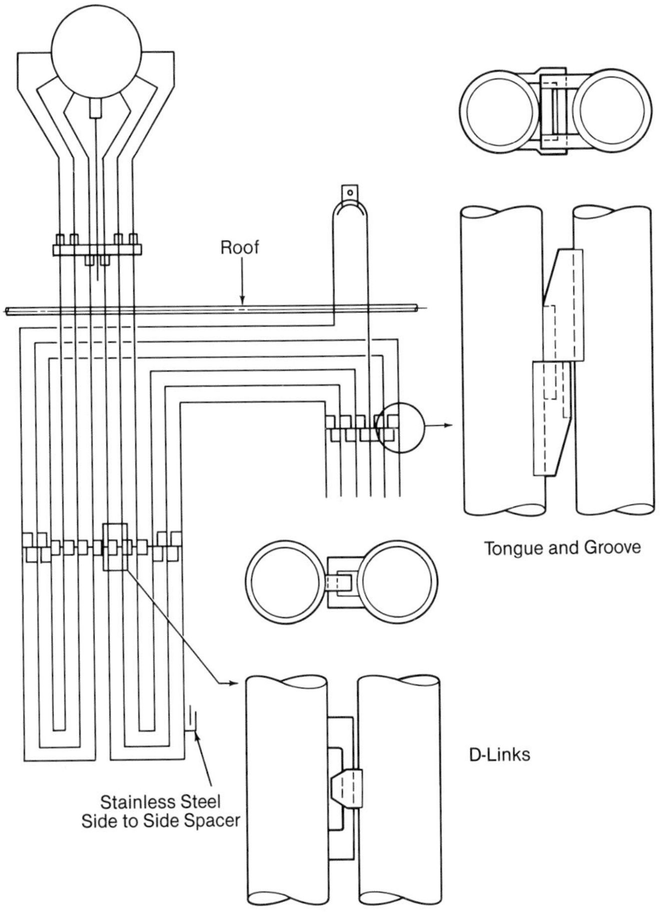

Fig. 17 Pendant reheater section with supports.

As units grow in size, the span of the superheater tubes may become so great that it is impossible to end-support these tubes. Most of the larger units use stringer tubes, generally hung from the economizer outlet, to support the superheater tubes. Tube spacing within the section is maintained by saddle-type supports.

Internal cleaning

The internal cleaning of superheater and/or reheater surfaces is not normally required; however, under certain circumstances these surfaces have been chemically cleaned. This is discussed in depth in Chapter 42. During startup, steam line blowing is used to remove residual scale, oils and residual debris.

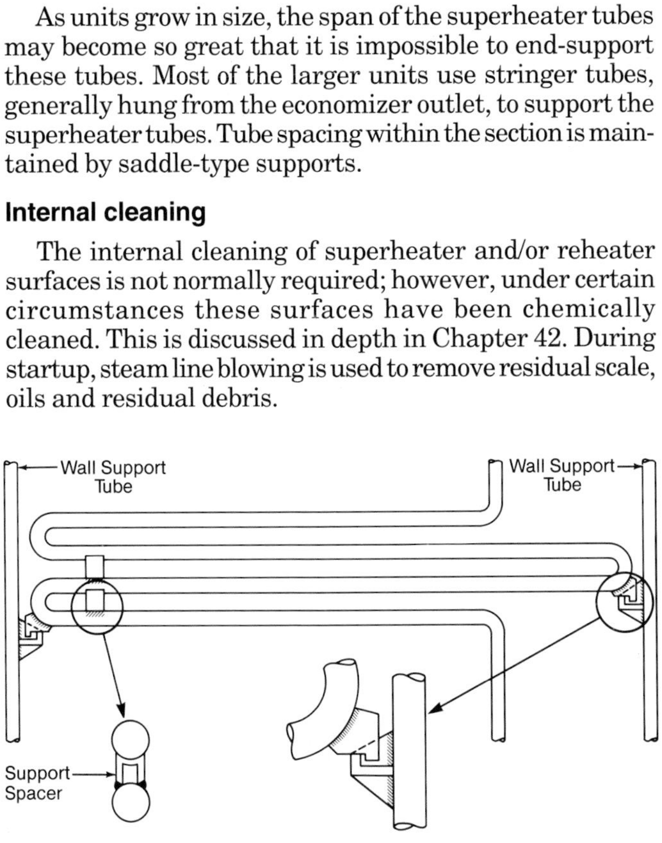

Fig. 18 Horizontal superheater section end supports on walls.

External cleaning and surface spacing

Units purchased today are designed for continuous operation, in some cases for 18 to 24 months between outages, so gas side cleanability is critical. Usually one year between outages is considered acceptable. To enhance cleanability, the superheater sections (platens) of modern utility boilers are spaced according to the gas temperature and the fuel fired. Fig. 10 illustrates the spacing for pulverized coal-fired units. The backspacing in the direction of gas flow is usually set at 0.50 to 0.75 in. (12.7 to 19.1 mm) clear space between tubes in the high temperature zones, with increased spacing allowable in the horizontal surface zones of less than 1500F (816C) entering gas temperature. These spacings are empirical; they are based on tube fouling and erosion experience and on manufacturing requirements. Surface arrangement and cleanability are discussed in Chapter 23 under *Sootblowers*.

Steam temperature adjustment and control

Improvement in the heat rate of modern boiler units and turbines results in large part from the high cycle efficiency possible with high steam temperatures. The importance of regulating steam temperatures within narrow limits is illustrated by the fact that a change of 35 to 40F (19 to 22C) corresponds to a change of about 1% in heat rate at pressures above 1800 psi (124.1 bar).

Other important reasons for accurate steam temperature regulation are to prevent failures due to excessive metal temperatures in the superheater, reheater or turbine; to prevent thermal expansion from dangerously reducing turbine clearances; and to avoid erosion from excessive moisture in the last stages of the turbine.

The control of temperature fluctuations from variables of operation, such as slag or ash accumulation, is important. However, superheater and reheat steam temperatures in steam generation are mainly affected by changes in steam output.

With drum-type boilers, steam output and pressure are maintained constant by firing rate, while the resulting superheat and reheat steam temperatures depend on basic design factors, such as total surface quantity and the ratio of convection to radiant heat absorbing surface. Steam temperatures are also affected by other important operating variables such as excess air, feedwater temperature, changes in fuel that affect burning characteristics and ash deposits on the heating surfaces, and the specific burner combination in service. In the Universal Pressure once-through boiler, which has a variable steam-water transition zone, steam output, pressure and temperature are controlled by the coordination of the firing rate and the boiler feedwater flow rate, leaving reheat steam temperature as a dependent variable. (See Chapter 41.) Standard performance practice for steam generating equipment usually permits a tolerance of ± 10F (6C) in a specific steam outlet temperature.

Definitions of terms

Adjustment is a change in the arrangement of heating surface which affects steam temperature but can not be used to vary steam temperature during operation. This could be the addition or deletion of component surface or boiler surface ahead of the superheater and/or reheater.

Control is the regulation of steam temperature during operation without changing the arrangement of surface. Examples are operation of attemperators, gas proportioning dampers and gas recirculation fans.

The *attemperator* is an apparatus for reducing and controlling the temperature of a superheated fluid passing through it. This is accomplished by spraying high purity water into an interconnecting steam pipe usually between superheater stages or upstream of a reheater inlet.

Effect of operating variables

Many operating variables affect steam temperatures in drum-type units. To maintain constant steam temperature, means must be provided to compensate for the effect of these variables.

Load As load increases, the quantity and temperature of the combustion gases increase. In a convection superheater (see *Superheater Types* above), steam temperature increases with load, the rate of increase being less the closer the superheater surface is to the furnace. In a radiant superheater (see *Superheater Types* above), steam temperature decreases as load increases. Normally a proportioned combination of radiant and convection superheater surface is installed in series in a steam generating unit to maintain substantially constant steam temperature over the control range of the unit. (See Fig. 14.)

Excess air For a change in the amount of excess air entering the burner zone there is a corresponding change in the quantity of gas flowing over the convection superheater; therefore an increase in excess air generally raises the steam temperature.

Feedwater temperature An increase in feedwater temperature causes a reduction in superheat because, for a given steam flow, less fuel is fired, less gas flows over the superheater and the gas temperatures are lower.

Heating surface cleanliness Removal of ash deposits from heat absorbing surfaces ahead of the superheater reduces the temperature of the gas entering the superheater and, subsequently, the steam temperature. Removal of deposits from the superheater surface increases superheater absorption and raises steam temperature.

Use of saturated steam If saturated steam from the boiler is used for sootblowers or auxiliaries, such as pump or fan drives, an increased firing rate is required to maintain constant main steam output; this raises the steam temperature.

Blowdown The effect of blowdown is similar to the use of saturated steam but is to a lesser degree because of the low enthalpy of water as compared to steam.

Burner operation The distribution of heat input among burners at different locations or a change in burner adjustment usually has an effect on steam temperature due to changes in furnace heat absorption rate.

Fuel Variations in steam temperature may result from changing the type of fuel burned or from day to day changes in the characteristics of a given fuel.

Adjustment

A power generating unit represents a large capital investment, and means should be provided at a reasonable cost for adjusting steam temperature to meet changing conditions. For instance, if a long term fuel change will have a considerable effect on steam temperature, it is good engineering practice to design for compensating physical alterations of the equipment.

Adjustment for regulating steam temperature is required when the operating conditions depart from the conditions on which the design is based. To provide the ultimate adjustment to meet such variations in operation, the design should accommodate anticipated changes at minimum expense.

The basic method of adjustment for regulating steam temperature is the addition or removal of superheater and/or reheater surface. A good design will provide an economical way of doing so.

On certain unit designs, adjustment is also possible by a reduction or increase in the amount of saturated surface ahead of the superheater surface. Such alterations modify the gas temperature at the inlet to these superheaters. If saturated surface is removed to increase steam temperature, this type of adjustment is relatively simple and, in general, costs less than the addition of superheater and/or reheater surface. However, the addition of saturated surface to decrease steam temperature can be difficult and expensive, or even impractical.

Limited use of a refractory coating on selected areas of water-cooled furnace surfaces is also permissible to increase gas temperatures entering the superheater surfaces. This can have a favorable effect on combustion and carbon loss, but refractory should not be added in areas where undesirable ash would deposit. This may also add to maintenance costs.

One of the simplest, least expensive and most effective means of adjustment in regulating steam temperature is to change the mass velocity of the gas flowing over the superheater elements by baffle modifications, if the unit design permits. Several standardized boilers, especially in the smaller sizes, have an adjustable baffle suitable for steam temperature regulation. This feature permits varying the steam temperature control range as much as 20%. The limit of variation is the effect on draft loss and efficiency. A 10F (6C) increase in the temperature of the gas at the boiler outlet reduces efficiency by about 0.25%.

Control

Control is necessary to regulate steam temperature within required limits in order to correct fluctuations caused by operating variables — particularly boiler load. Besides boiler load variation, ash deposition on heat transfer surfaces is the most frequent cause of steam temperature fluctuations. This condition can usually be corrected by changes in the sequence or frequency of sootblower operation. Selective operation of furnace wall blowers or implementation of unit load reductions to induce slag shedding from the furnace walls can reduce gas temperatures entering the superheater surfaces.

The time in which a turbine may be brought to full load is established by its manufacturer in accordance with a safe steam temperature-time curve. Because the temperature of the steam is directly related to the degree of expansion of the turbine elements and consequently the maintenance of safe clearances, this temperature must be regulated within permissible limits by an accurate control device.

The removal of feedwater heaters or pulverizers from service can impact steam temperature and require steam temperature control.

Among the means of control for regulating steam temperature are: attemperation, gas proportioning dampers, gas recirculation, excess air, burner selection, movable burners, divided furnace with differential firing, and separately-fired superheaters. With attemperation, steam temperature is regulated by diluting high temperature steam with low temperature water or by removing heat from the steam. By comparison, the other methods of control are based on varying the amount of heat absorbed by the steam superheating surfaces.

Attemperation Attemperators may be classified as two types — direct contact and surface. The direct contact design is exemplified by the spray type, where the steam and the cooling medium (water and saturated steam) are mixed. In the surface design, which includes the shell type and the drum type, the steam is isolated from the cooling medium by the heat exchanger surface. Surface-type attemperators are rarely used on current utility boiler designs, while spray attemperators are generally used on all units which have attemperation requirements.

The superheater attemperator may be located in one of two places: at some intermediate point between two sections of the superheater or at the superheater outlet. The ideal location for the superheater attemperator for *process control* would be at the superheater outlet. Control would be direct and there would be no time lag. However, problems with this location include that: 1) water may carry over into the turbine, and 2) the spray does not protect the superheater metals from overheating. The superheater attemperator, located between superheater stages, addresses these concerns and is therefore generally preferred. The steam temperature leaving the superheater does not exceed the maximum temperature desired. In addition, steam from the elements of the first stage superheater is so thoroughly mixed that it enters the second stage superheater at a uniform temperature. For reheat superheater applications, the attemperator is located at the reheater inlet.

The superheater spray attemperator, illustrated in Fig. 19, has proven most satisfactory for regulating steam temperature. High purity water is introduced into the superheated steam line through a spray nozzle at the throat of a venturi section within the line. Because of the spray action at the nozzle and the high velocity of the steam passing through the venturi throat, the water vaporizes, mixes with and cools the superheated steam. An important construction feature is the continuation of the venturi section into a thermal sleeve downstream from the spray nozzle; this protects the high temperature piping from thermal shock. This shock could result from nonevaporated water droplets striking the hot surface of the piping. The reheat attemperator is similar but does not have a venturi section.

The spray attemperator provides a quick acting and sensitive control for regulating steam temperature. It is important that the spray water be of highest purity, because solids entrained in the water enter the steam and may cause troublesome deposits on superheater tubes, piping or turbine blades. High pressure heater drains are a source of extremely pure water but require a separate high pressure corrosion resistant pump if used for attemperator supply. Normally, boiler feedwater is satisfactory, provided condenser leakage and makeup do not introduce too much contamination. The total solids concentration in the spray water should not exceed 2.5 ppm and the spray water should not add more than 40 ppb solids to the steam flow.

Three attemperator arrangements are possible depending on the boiler performance requirements:

1. *Single-stage attemperator* A single attemperator may be installed in each of the connecting pipes between two stages of superheat.
2. *Tandem attemperator* A single-stage attemperator with two spray water nozzles may be installed in series in the connecting pipe between two stages of superheat. This arrangement is used where the spray quantity exceeds the capacity of a single nozzle or where the required turndown can not be achieved with a single spray water nozzle. The usual application requires a spray control valve for each spray nozzle. The operation is sequential with the control valve for the downstream spray nozzle opening first and closing last.
3. *Two-stage attemperator* Two single-stage attemperators are used. The first unit is located in the connecting pipes between the first and second stages of superheat and the second is between the second and third stages of superheat. A spray control valve is required for each stage of spray. The first stage spray attemperator is used first, with the maximum spray flow based on a minimum allowable difference between the temperature of the steam leaving the attemperator and the saturation temperature. The second stage attemperator is used after the flow limit is reached on the first stage unit.

In most instances, the pressure loss from the boiler feed pump through the feedwater heaters, piping and boiler to the attemperator location in the superheater results in sufficient boiler feed pump discharge pressure to provide the required pressure differential in the attemperator system. In certain cases, however, the feed pump discharge pressure is not sufficient due to the low pressure loss in the boiler and feedwater system or due to the high pressure differential the spray systems require to suit particular boiler characteristics. In these cases, a booster stage in the boiler feed pump is desirable to raise the spray water to the required pressure. Other options are a separate spray water booster pump or an additional feed line valve to increase boiler side resistance at the required loads.

Over the past years there has been an increasing frequency of problems in industrial steam power cycles, involving deposits in the superheater and on turbine blading, which has resulted in failures. Many of these problems are due to impurities in the attemperator spray water. To as-

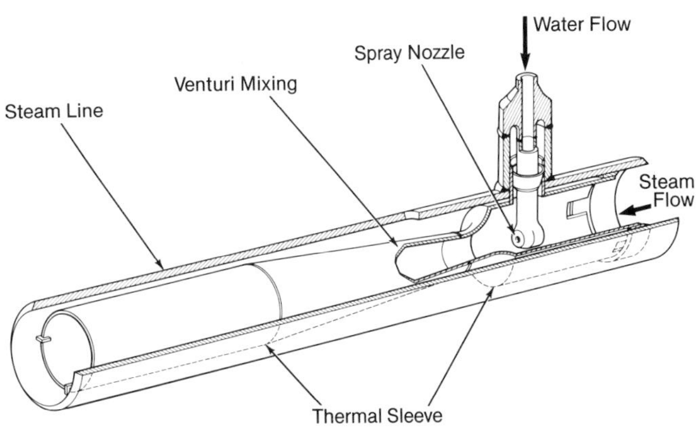

Fig. 19 Spray attemperator showing thermal sleeve.

sure high quality spray water, additional cleanup equipment can be installed on the condensate return and feedwater makeup.

For low pressure units with feedwater purity less than that required for spray attemperation, the B&W condenser attemperator system is an economical and reliable system used to produce high quality spray water. (See Chapter 25.)

Steam attemperation may be used on drum and once-through boilers as described in the *Bypass and Startup* section of this chapter.

UP boiler attemperator applications UP boilers are supplied without superheater spray attemperators. However, they can be supplied when specified to reduce variations in superheater outlet temperatures during transients. Spray attemperators are installed between the primary superheater outlet and the secondary superheater inlet, upstream of the high pressure superheater stop valve. Spray water is supplied from the UP boiler economizer inlet. Spray attemperation corrects main steam temperature deviations only on a temporary basis and must not be the primary means of steam temperature control. Under steady-state conditions, steam temperature is determined by the ratio of firing rate to feedwater flow. Special rules apply to surface and boiler design for spray attemperators supplied with a UP boiler.

Gas proportioning dampers As shown in Fig. 3, the horizontal convection tube banks in the back end of a boiler can be divided into two or more separate gas passes separated by a baffle wall. The use of dampers in these gas passes then permits proportioning of the gas over the heat transfer surfaces and the control of reheat and superheat temperatures.

Design considerations for such systems include the following:

1. Dampers must be placed in a cool gas zone to assure maximum reliability (typically downstream of all boiler heat transfer surfaces).
2. Draft loss through the unit could increase for some designs, particularly with alternate fuels, so this parameter must be optimized.
3. Control system design and tuning are critical because damper control response is slower than with spray attemperators. Therefore, spray attemperators are used for transient control.
4. Under maximum bias conditions, the gas temperatures at the dampers and the heat transfer surfaces nearest the dampers will be at their highest. These temperatures then set the metal design requirements.

Gas proportioning dampers are combined with spray attemperation for overall optimal steam temperature control systems. Spray attemperators provide for short term transient temperature control. The gas proportioning dampers provide longer term control and adjustment between superheat and reheat temperatures with a minimum impact on overall unit efficiency.

Gas recirculation Another method of controlling superheat or reheat is gas recirculation. As the name implies, gas from the boiler, economizer or air heater outlet is reintroduced to the furnace by fans and flues. For the sake of clarity, recirculated gas introduced in the immediate vicinity of the initial burning zone of the furnace and used for steam temperature control is referred to as *gas recirculation* and recirculated gas introduced near the furnace outlet and used for control of gas temperature is referred to as *gas tempering*. Fig. 20 shows an application of gas recirculation through the hopper bottom and gas tempering in the upper furnace of a Radiant boiler. In most instances the gas is obtained from the economizer outlet. The recirculated gas must be introduced into the furnace in a manner that avoids interference with the fuel combustion. The amount of recirculated gas is expressed as a percentage of the gas that remains downstream of its withdrawal point.

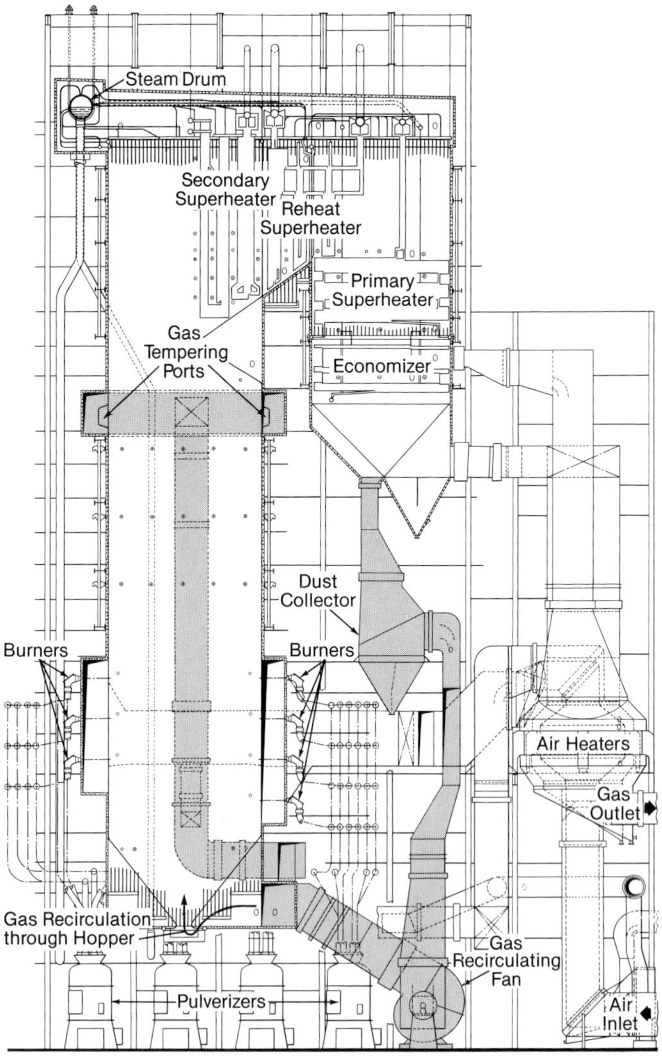

Fig. 20 Radiant boiler with gas tempering for gas temperature control and gas recirculation for control of furnace absorption and reheat temperature.

While recirculated gas may be used for several purposes, its basic function is to alter the heat absorption pattern within a steam generating unit. Recirculated gas has the special advantage of providing heat absorption adjustment that may be used as a design factor in initial surface arrangement and as a method of controlling the heat absorption pattern under varying operating conditions.

An important feature of recirculated gas is that its use changes only the pattern of heat absorption through a boiler; it has a negligible effect on the total boiler heat absorption and the weight of the gas sent up the stack. The thermal effect of recirculated gas depends on the amount of gas recirculated, the location of gas introduction and the furnace heat release rate.

Fig. 21 shows the variation in heat absorption with gas recirculation into the hopper. Introduction of gas at this

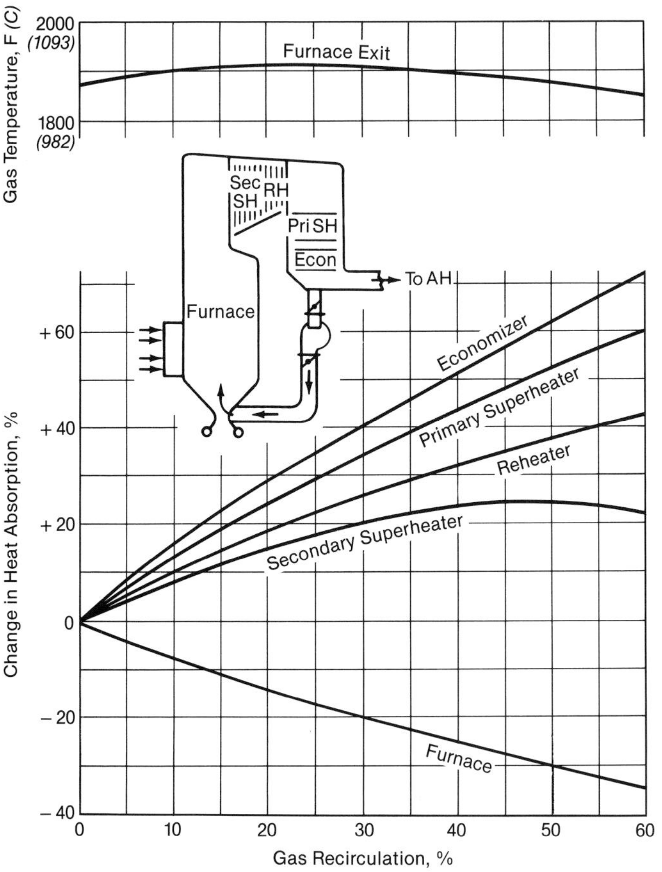

Fig. 21 Effect of gas recirculation on heat absorption pattern at a constant firing rate.

location produces a marked reduction in furnace absorption and increases the absorption of the convection section. Furnace heat absorption is primarily a function of the gas temperatures and gas temperature patterns throughout the furnace, because the heat is mainly transferred by radiation. Therefore, the introduction of gas recirculation into the furnace hopper reduces furnace absorption by altering the gas temperature pattern.

The major portion of the heat absorbed in the superheater, reheater and economizer is transferred by convection, which depends on gas temperature and gas mass flow rate. Both parameters are affected by gas recirculation. Therefore, when the mass velocity of gas flowing through a convection bank is increased by gas recirculation, the amount of heat transferred may increase, decrease or remain unchanged depending on changes in the relationship between the temperature and weight of the gas entering the bank. Fig. 21 illustrates a condition in which the gas temperature entering the secondary superheater (FEGT) is relatively unchanged by gas recirculation. Increasing the amount of gas recirculation, therefore, increases the heat absorption in the secondary superheater. The heat absorption in the reheater, primary superheater and economizer is also increased, with the greatest increase occurring at the cold end of the unit. This is a typical example of the variation in convection pass heat absorption pattern by gas recirculation.

While gas recirculation into the hopper always reduces furnace heat absorption, its effect on the FEGT depends mainly on furnace rating. This exit gas temperature may increase, decrease or, as shown in Fig. 21, be essentially unchanged by gas recirculation. In general, recirculated gas introduced in the hopper decreases the FEGT of a unit operating at high furnace loading and increases this gas temperature at low loading.

Fig. 22 illustrates the effect of introducing tempering gas at a point near the furnace exit. Because the portion of the furnace in which the bulk of heat absorption occurs is unaffected by the recirculated gas, the furnace heat absorption is decreased only slightly. There is, however, a large decrease in FEGT caused by dilution of the hot combustion gases with cooler recirculated gas.

In the case of tempering gas introduced near the furnace outlet, the reduction in FEGT is usually sufficient to overbalance the effect of gas weight increase, and the heat absorption in the secondary superheater is decreased. The effect of gas tempering on the primary superheater and economizer follows the pattern that was shown in Fig. 21, with the greatest change in heat absorption again occurring at the cold end. Because of the location of the reheater in Fig. 22, its absorption remains constant regardless of the percentage of gas tempering.

Figs. 21 and 22 illustrate the effect of introducing recirculated gas into the hopper or at a point adjacent to the furnace exit. Introduction of gas at intermediate points results in heat absorptions and gas temperatures between those shown. To show the effect of recirculated gas only, Figs. 21 and 22 have been based on constant firing rates.

Excess air Boiler operators have long known that the steam outlet temperature of a convection superheater on a drum type or separator type unit can be increased at fractional loads by decreasing the furnace heat absorption through an increase in the amount of excess combustion

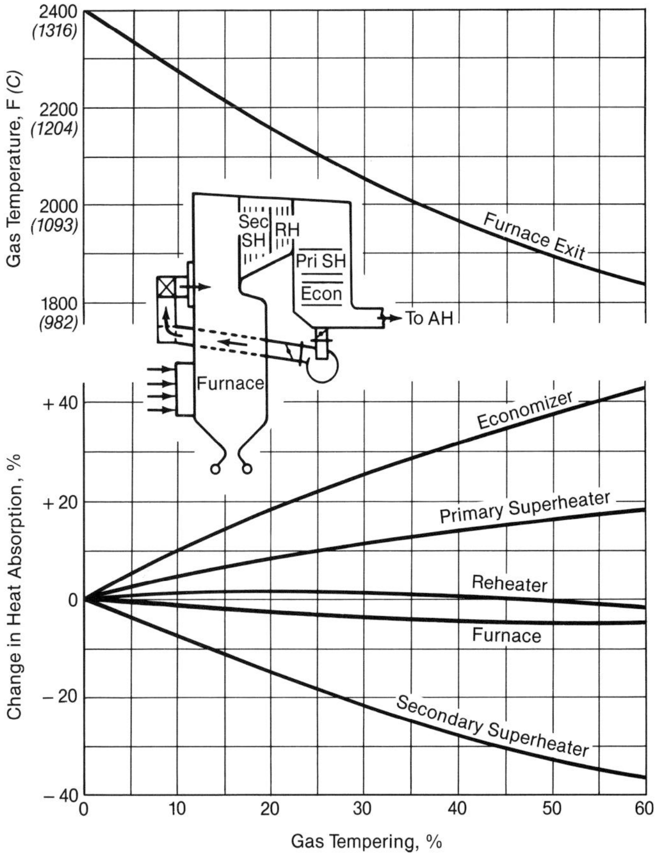

Fig. 22 Effect of gas tempering on heat absorption pattern at a constant firing rate.

air. The resulting greater weight of gas sent to the stack increases the stack loss. However, the drop in boiler efficiency can be offset by the increase in turbine efficiency.

Burner selection It is often possible to regulate steam temperature by selective burner operation. Higher steam temperatures may be obtained at less than full load by operating only the burners giving the highest furnace outlet temperature. When steam temperature reduction is required, firing may be shifted to the lower burners. This method of control can be improved by distributing the burners over an extended height of the burner wall or by installing a special burner near the furnace outlet.

Movable burners Regulation of steam temperature by changing the furnace absorption pattern can also be effected by using movable burners to raise or lower the main combustion zone in the furnace. Tilting burners are used for this purpose.

Differentially-fired divided furnaces In some divided furnaces, the superheater receives heat from one section of the furnace only, while the other section of the furnace generates only saturated steam or may include a reheater. Steam temperature is regulated by changing the proportion of fuel input between the two furnaces. This arrangement, similar in principle to the separately-fired superheater, was formerly widely used in marine practice. The differentially-fired divided furnace method of steam temperature control is no longer used today.

Separately-fired superheaters A superheater that is completely separate from the steam generating unit and independently fired may serve one or several saturated steam boilers. This arrangement is not generally economical for power generation, where a large quantity of high temperature steam is needed.

Reheat steam temperature

The need for regulating the temperature of reheated steam and the methods of adjustment and control to do so are, in general, the same as for superheated steam. However, the designer does not have the same freedom of action as when only a superheater is used. For instance, in a drum-type unit, the removal of boiler heating surface ahead of the superheater to increase superheat temperature or the removal of superheat surface to reduce superheat steam temperature results in increased reheat steam temperature, which may be undesirable. Furthermore, to reduce the gas temperature below slagging limits for convection tube banks and to give the desired steam temperature, such a large proportion of the total input must be absorbed in the furnace and in the superheater and reheater that there is no boiler surface ahead of the superheater available for adjustment purposes. Some boilers have furnace division walls or water-cooled wingwalls, which may provide adjustment surface if furnace exit gas temperature permits.

Most of the control methods described for regulating main steam temperature also affect the reheat temperature if a reheater is incorporated.

Bypass and startup systems

High pressure drum and once-through Universal Pressure boilers must respond to the change in required operating modes of large fossil fueled generating plants. This requires rapid, frequent and reliable unit startups and load changes to meet demand with economical electrical production. During startup or low load (less than 20%) conditions, regular water attemperator steam temperature control systems are usually ineffective as the outlet steam temperature tends to follow the flue gas temperature because the gas flow is substantially higher than the steam flow. However, provision must still be made to match the differing flow, pressure and temperature needs of the steam turbine and boiler. To address these special requirements, three bypass and startup systems have been developed for drum and once-through UP boilers.

Drum boiler bypass system

The B&W Radiant boiler bypass system minimizes startup time, controls shutdowns in anticipation of restarts, provides control of steam temperature to match turbine metal temperature, and allows dual pressure operation of boiler and turbine for better load response. These features reduce stresses in the turbine for improved turbine availability and reduced maintenance costs. The drum bypass system consists of a control system and carefully engineered steam valves and piping as shown in Fig. 23. Probes monitor gas temperatures at the superheater and reheater outlet tubes to permit control of firing rates and gas temperatures during startup.

Operational benefits

Decreased unit startup time The bypass system substantially reduces the time required for a cold startup because it controls the temperature differences of the saturated boiler surface, superheater surface and turbine.

This is accomplished by providing direct control of the steam temperature by mixing saturated steam with superheat and reheat outlet steam as shown in Fig. 23. This arrangement provides the desired steam temperature for the turbine without restricting the startup firing rate of the boiler.

Rapid load changing The system has a set of superheater stop and bypass control valves, which allow dual pressure operation with the throttle pressure controlled separately from drum pressure. This control system permits constant pressure operation of the major boiler components and variable pressure operation of the turbine during load changes. Dual pressure operation minimizes thermal stresses in the boiler and turbine. A dual pressure shutdown keeps the boiler near full pressure and the turbine metal near maximum temperature in preparation for a quick restart. In addition, it allows more rapid load changes than full variable boiler pressure operation.

A superheater bypass diverts excess steam from the boiler to the condenser, thereby separating firing rate from drum pressure control during shutdown and startup. This feature is used during shutdown and next morning restart to keep the turbine and steam piping near full temperature for quick starts and minimum stresses.

Operation The flexible bypass system can conform to most operating changes. For example, it can adopt to a turbine malfunction by adjusting the steam temperature and flow through any load range from synchronization to load points. Although the bypass system enhances unit operation, the boiler can be operated as a conventional nonbypass unit at any time. The following valves (see Fig. 23), with their associated functions, provide the required system flexibility:

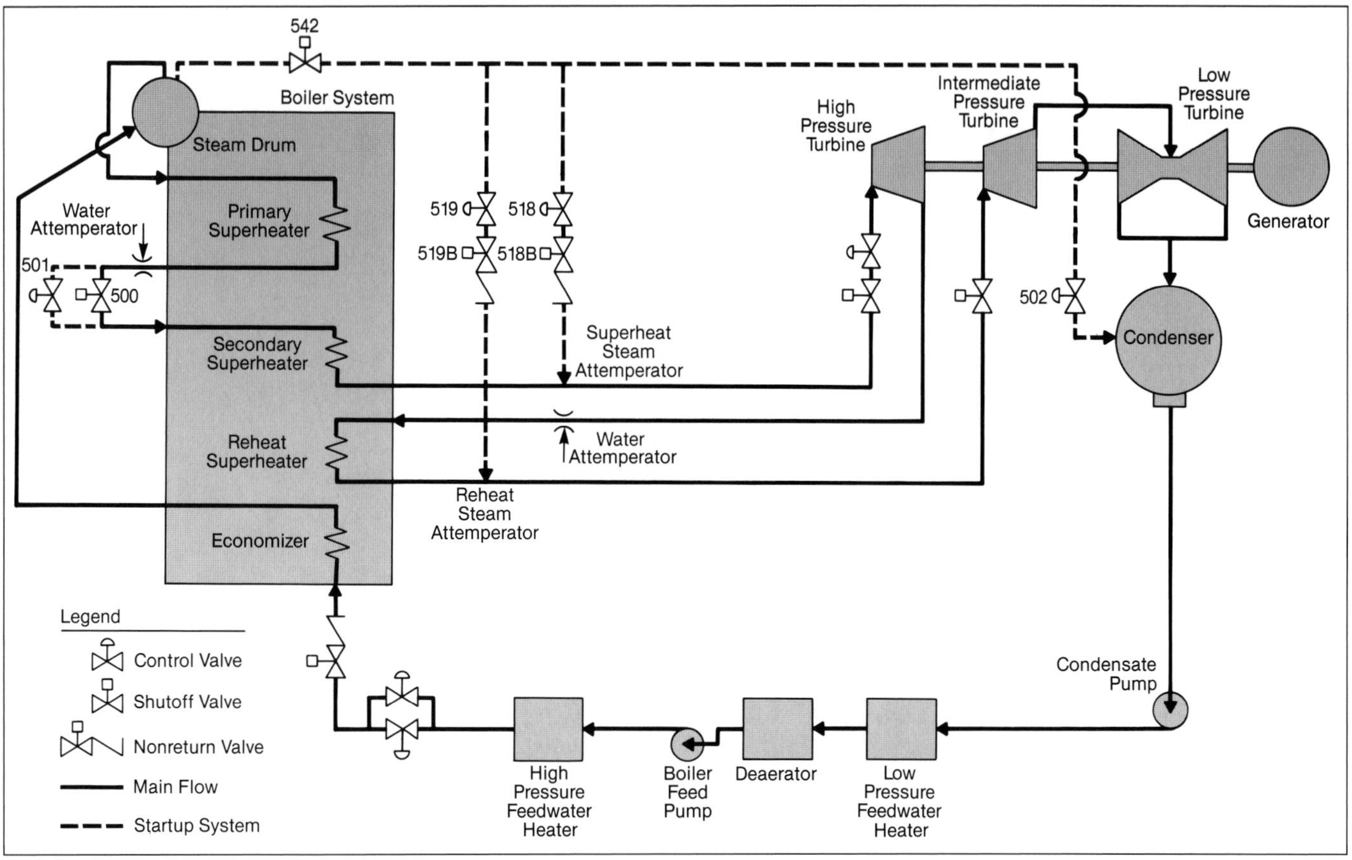

Fig. 23 Drum boiler bypass system schematic.

Valve No.	Function
500	*Secondary superheater stop valve* Separates the secondary superheater from the rest of the steam generator. Closed when using bypass system. Usually designed to open above 70% load.
501	*Secondary superheater stop valve bypass valve* Controls steam pressure leaving the secondary superheater below drum pressure for startup, variable pressure operation and steam attemperation. Design capacity is usually 70% of boiler design flow.
502	*Primary superheater bypass to condenser control valve* Permits firing for control of gas temperature entering the superheater to achieve high steam temperature during startup. Used primarily during hot starts, hot restarts and shutdowns. Closed at loads above 20%.
518	*Main steam outlet attemperator control valve* Reduces main steam temperature to match first-stage turbine metal temperature during cold starts and starts following weekend outages. Closed at loads above approximately 20%. The 518B shutoff valve provides isolation when the 518 valve is closed.
519	*Reheat outlet steam attemperator control valve* Reduces reheat steam temperature to match reheat bowl temperature during cold starts and postweekend starts. Closed at loads above approximately 20%. The 519B shutoff valve provides isolation when the 519 valve is closed.
542	*Primary superheater bypass system shutoff valve* Used to isolate bypass system during normal operation. Closed when the 518, 519 and 502 valves are closed.

Design considerations

Drum pressure and superheater outlet pressure control On restarts following overnight shutdowns, the gas temperature leaving the furnace has to be kept high. This maintains high main steam and reheat temperatures to match high turbine metal temperatures. The turbine must be rolled at a low pressure to prevent a large adiabatic throttling temperature drop when steam is admitted.

The saturation surface can be isolated from the secondary superheater surface by means of the secondary superheater stop valve (500) and bypass valve (501). The boiler can then be fired at the desired rate to raise steam temperatures or drum pressure while maintaining low pressures in the secondary superheater and entering the high pressure turbine. If the drum pressure increases too rapidly or reaches its limit, the primary superheater bypass valve (502) relieves pressure from the drum to the condenser, avoiding condensate waste. A better turbine metal temperature match is obtained with maximum heat input to the boiler and the firing limit dictated by metal protection of the convection pass surface.

Reheat steam temperature control Control of reheat steam temperature is maintained by the steam attemperator valve (519) during startups and shutdowns. Before steam is taken to the turbine, the reheater is without flow. The reheat metal absorbs heat from the flue gas and eventually reaches the flue gas temperature, which can approach

1000F (538C). When steam is first admitted to the turbine and passes through the reheater, the reheater outlet steam temperature rises very rapidly to the gas temperature level, resulting in a poor match with reheat bowl temperatures in the intermediate pressure turbine for cold starts and starts after a weekend shutdown. Water attemperators are not effective at low loads where reheater steam temperatures are controlled by flue gas temperatures in the reheater banks.

Reheat steam attemperation with saturated steam from the drum limits the rise of reheat outlet steam temperature when steam is first admitted to the turbine and offers positive reheat steam temperature control up to about 20% of full load.

Main steam temperature control The main steam attemperator valve (518) controls main steam temperature. Water attemperators are not effective during low boiler loads, where superheater steam temperatures are determined by the flue gas temperatures in the superheater banks. Therefore, the main steam temperature is controlled by a steam attemperator valve, using saturated steam from the drum to lower temperatures.

Design conditions

Cold start A cold start is defined as a unit startup from no drum pressure or from ambient furnace gas temperature. Turbine metal temperatures are less than 300F (149C) and prewarming is required.

Warm start A warm start is defined as a unit startup after a two day shutdown, such as a weekend shutdown. Turbine metal temperatures are at least 300F (149C). Remaining drum pressure can be as high as 500 psig (34.5 bar gauge).

Hot start A hot start is defined as a unit startup following a six to eight hour outage. The boiler is closed up to retain the maximum internal energy. Boiler drum pressure is quite high and vacuum will have normally been maintained. Turbine metal temperatures of 900 to 925F (482 to 496C) may exist following a controlled shutdown and a decay in turbine metal temperatures of 100F (56C) may occur during the shutdown. Therefore, minimum steam temperatures of 800 to 825F (427 to 441C) are needed during the restart. The steam temperature is usually less than the temperature limit for gas entering the superheater during a no-flow condition.

Hot restart If the unit is tripped but ready for a restart very quickly, turbine metal temperatures may be 1000F (538C) and drum pressure may be near the pressure existing at the time of the trip. To achieve the high gas temperatures needed, it may be necessary to continue unit firing and dump excess steam to the condenser.

Controlled shutdown When shutting the unit down, the bypass system may be used to facilitate the operation that follows the shutdown. If maintenance on the unit is scheduled, the turbine metal temperature and boiler pressure should be kept as low as possible while still carrying load. The throttle pressure is 501 valve controlled and necessary sprays to reduce steam and metal temperatures per the cooldown curves are controlled by the 518 and 519 valves. After the unit is tripped, the fans may remain on to cool the boiler.

Variable pressure operation Under most circumstances, the unit heat rate can be increased at partial loads if boiler pressure is reduced with load. This mode of operation is normally referred to as *variable pressure operation*.

The secondary superheater stop valve (500) and the stop valve bypass valve (501), used with the B&W bypass system, permit operating with the drum and primary superheater pressures at constant levels while the secondary superheater outlet pressure is varied with load. Operating in this mode maintains turbine steam temperatures at the design level over a greater load range than is possible under constant throttle pressure operation.

Maintaining drum and primary superheater pressure relatively constant at reduced loads permits rapid load pickup. The savings in pump power, which could be achieved by permitting drum pressure to vary with load, are eliminated when the rapid load pickup feature is used.

Startup without bypass system The unit can be started and operated without use of the bypass system. Fully opening the 500 valve allows starting up the unit like a conventional drum boiler.

UP boiler startup system

A key requirement of UP startup and bypass systems is the need for minimum design circulation flows in high heat absorbing circuits for cooling before the unit can be fired. Additional important features include providing a turbine bypass until steam pressure and temperatures are matched, reducing the bypass flow pressure and temperature before condenser and auxiliary equipment steam admission, recovering heat during startup, providing clean water for full startup, accelerating the startup processes, and providing greater unit flexibility through dual pressure boiler operation. UP boiler plants, while originally designed for base load operation, have been adapted to load cycling, including daily operation on the bypass system.

Two different UP systems are available — one for constant furnace pressure operation over the load range and a second for variable furnace pressure operation over the load range.

Constant furnace pressure startup system For constant furnace pressure the startup system has the steam separator (*flash tank*) located in a bypass that can be isolated from the boiler during normal operation. The general arrangement is shown in Fig. 24.

The boiler feed pump supplies the minimum required flow of feedwater during startup and low load operation to protect the furnace circuitry. Included in the startup circuitry are the economizer, furnace convection pass enclosure and primary superheater.

The fluid leaving the primary superheater (at full pressure) bypasses the secondary superheater through the pressure reducing valve (207) to the flash tank, where the steam-water mixture is separated during startup.

Water level in the flash tank is controlled by drain valves 230 and 241, with the 230 valve controlling the flow to the deaerator for maximum heat recovery. Excess water (above the capability of the deaerator) is discharged to the condenser through valve 241. If the drains are not within water quality limits (see Chapter 42), all of the flow is through valve 241 to the condenser and polishing system.

The 242 block valve remains closed until a level is established in the flash tank to assure that water does not enter the steam lines. Once a level is established, the 242 valve is opened so the deaerator steam line from the flash tank can be used to hold pressure in the deaerator (controlled by the 231 valve). This permits returning all of the drains to the condenser through the 241 valve during a hot cleanup without using an auxiliary steam source for maintaining deaerator pressure, and also serves to recover the heat in the flash tank steam during cleanup.

The low pressure steam line, the low pressure superheater nonreturn valve (205), and the downstream high pressure steam line connect the flash tank to the secondary superheater inlet downstream of the high pressure superheater block valve (200) and stop/control valve (401). After a water level is established in the flash tank, 205 valves normally open at 300 psig (20.7 bar gauge) and dry steam flows to the secondary superheater for warming steam lines. The turbine bypass valve (210) normally opens at 300 psig (20.7 bar gauge) main steam pressure to assist with warming and boiling out the superheater during the initial stage of startup. When sufficient steam is available, the turbine is rolled and placed on line.

Steam separated in the flash tank in excess of that required is relieved through the 240 valve to the condenser. This valve also acts as an overpressure relief valve to avoid tripping spring loaded safety valves on the flash tank. The 240 valve has an adjustable set point, which can be set to hold various flash tank pressures at particular load points during startup.

The entire bypass system is sized to handle minimum required flow during startup and to permit operating at minimum load on the flash tank.

The transition from operation on the flash tank to once-through flow is made at minimum load. As the steam entering and leaving the flash tank at this time is dry and superheated, the transition from flow through the bypass to once-through flow is accomplished, with minimum fluctuation in steam temperature, by opening the 200 block valve and the 401 combination stop/control valve and by closing the 207 and 205 bypass valves.

Variable throttle pressure Above minimum load the unit can be operated at constant or variable throttle pressure, with the 401 valve controlling the throttle pressure from minimum to 100% load, while maintaining full pressure in the upstream circuits.

The variable throttle pressure feature permits operating the unit with the throttle valves essentially wide open. This eliminates turbine metal temperature changes resulting from valve throttling and permits rapid load changes without being limited by turbine heating or cooling rates. Shutdown with variable throttle pressure maintains high temperatures in the turbine metals and permits rapid hot restarting.

Steam temperature control The means for controlling main steam and reheat temperatures at normal operating loads are not effective during startup or at very low loads.

The startup system shown in Fig. 24 includes provision for steam attemperation from the flash tank to the main and reheat steam outlet headers for precise control of steam conditions during startup to meet the turbine metal temperature requirements.

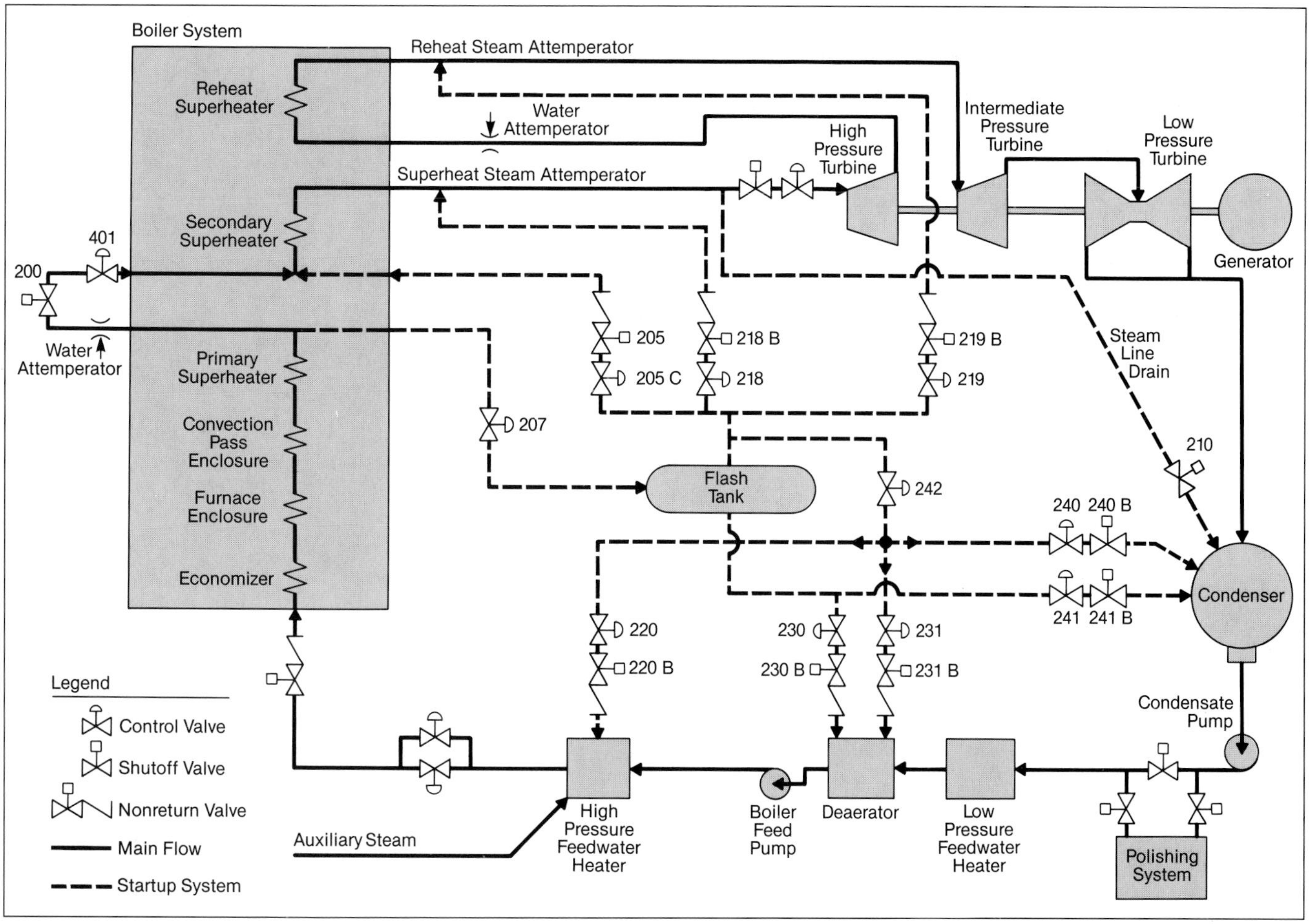

Fig. 24 Universal Pressure boiler startup system — constant pressure furnace operation.

The superheater outlet steam attemperator valve (218) is used at loads less than 20% to introduce saturated steam from the flash tank to the superheater outlet header. Initial rolling of the turbine, for a cold start, may be done with saturated steam passing from the flash tank through the 218 valve. This steam may be mixed with a limited quantity of steam passing through the 205 valves and the secondary superheater to control the high pressure turbine inlet temperature down to about 550F (288C). The 205C control valve achieves the necessary pressure drop between the flash tank and the secondary outlet header for attemperation.

The reheat outlet steam attemperator valve (219) is used at loads below about 20% of full load to introduce flash tank steam to the reheat outlet header. The ratio of flow through the attemperator valve to that through the high pressure turbine is limited mainly for turbine and turbine control considerations.

Overpressure relief The bypass system is also used to relieve excessive pressure in the boiler during a load trip. This is accomplished by the use of the 207 valve, which sends excess steam to the flash tank.

Variable furnace pressure startup system For units capable of operating with variable furnace pressure over the load range, the startup system has the steam separator located in the main flow path upstream of the primary superheater. The general arrangement is shown in Fig. 25.

The boiler feed pump supplies the minimum required flow of feedwater during startup and low load operation to protect the furnace circuits. Included in the startup circuitry are the economizer, the furnace enclosure, the wingwall pass (on units where the wingwalls are upstream of the primary superheater), and the steam separator.

The steam-water mixture is separated in the vertical steam separator during startup and low load operation. The unit is started with low furnace pressures to obtain the maximum amount of steam early in the startup; the 330 and 341 valves control the drains from the separator. The 330 valve also controls the flow to the deaerator for maximum heat recovery.

The 342 block valve remains closed until there is a steam-water mixture entering the separator, and it is closed any time a high level is indicated in the separator to prevent water from entering the superheater and reheater through the steam attemperators. Once the 342 valve is opened, the deaerator steam line holds pressure in the deaerator (controlled by the 331 valve). This permits returning all of the drains to the condenser through the 341 valve during a hot cleanup without using an auxiliary steam source for maintaining deaerator pressure; it also serves to recover the heat in the separated steam during cleanup.

During the initial stage of startup, the steam flow from the superheater is through the main steam line drains and 310 valve to warm the steam lines.

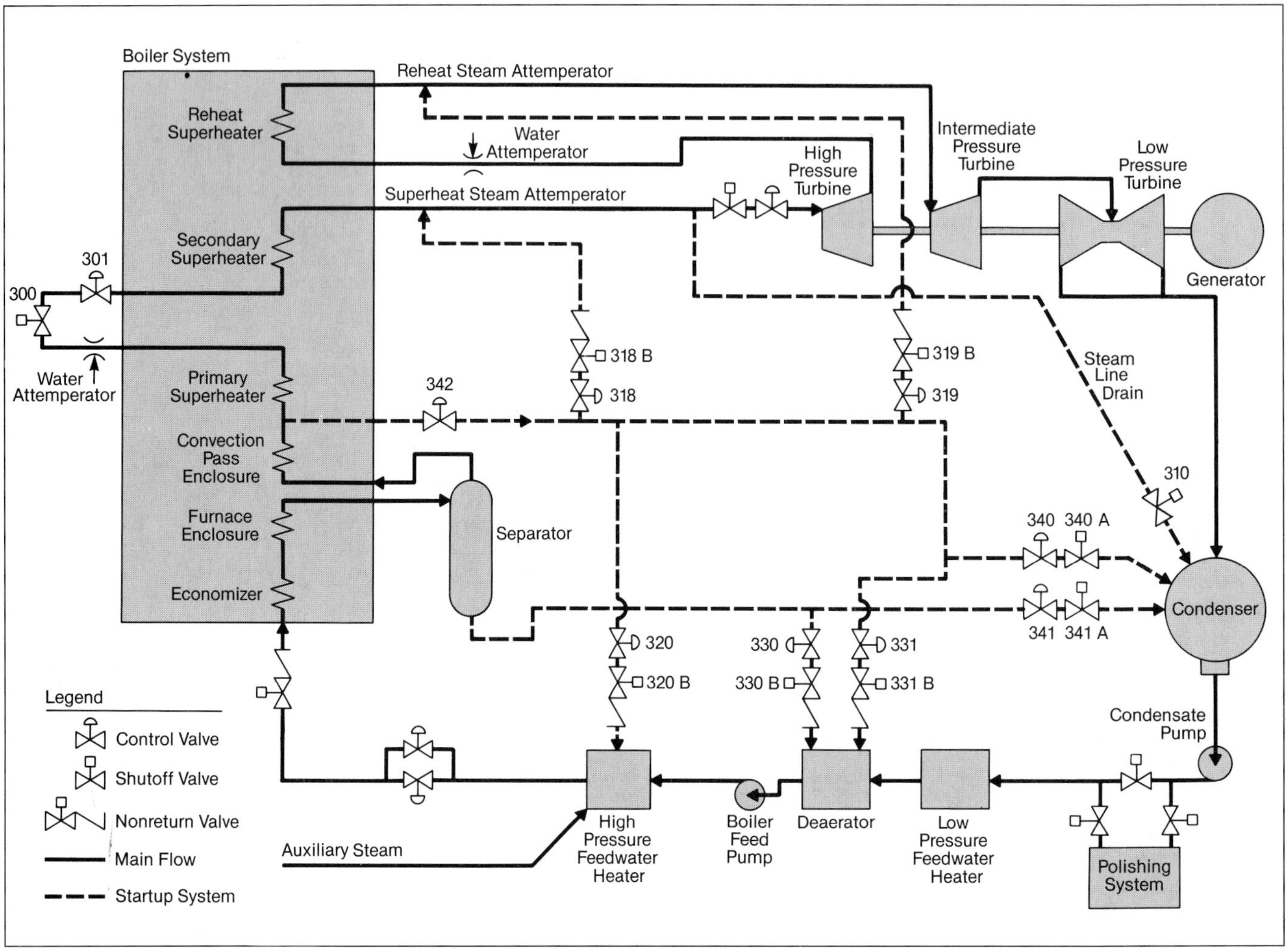

Fig. 25 Universal Pressure boiler startup system — variable pressure furnace operation.

As the enthalpy of the fluid entering the separator increases, the drains diminish until there is dry steam entering the separator. The drain valves are closed at this time, the deaerator is being controlled with turbine extraction steam, and the unit is on once-through operation with all the flow going to the turbine.

Division valves (dual pressure capability) On hot starts, including starts following overnight shutdowns, the gas temperature leaving the furnace has to be kept high to maintain high main steam and reheat temperatures. This generally results in an excessively rapid rise in throttle pressure, which is undesirable because of the resulting large throttling temperature drop when admitting steam to the turbine.

By means of the superheater combination stop and control valve (301), the boiler surface can be isolated from the secondary superheater surface. The overfiring required to raise and maintain steam temperatures can be allowed to raise saturation temperature or boiler pressure while maintaining the desirable low pressures in the secondary superheater and entering the high pressure turbine. Dual pressure operation also permits more rapid rates of load change. As the boiler temperature and pressure can be maintained at the higher level, the unit also responds more quickly to a change in load demand.

Superheater bypass to condenser When reaching the maximum desired boiler pressure or when starting up with the superheater stop valve open, the superheater bypass to the condenser valve (340) provides a means to control boiler pressure during hot start conditions. During the transient loading period for the unit following a hot restart, the superheater bypass valve (340) may be opened to permit higher firing rates and therefore sustain raised steam temperature until the boiler control load is reached.

The superheater bypass to the condenser is used to relieve excessive pressure in the boiler during a load trip. It can also be used as an overpressure relief valve to supplement and avoid popping spring loaded safety valves.

Steam temperature control The startup system shown in Fig. 25 includes provision for attemperation with saturated steam. This steam is taken from downstream of the separator and injected into the main and reheat steam outlet headers for precise control of steam conditions during startup in order to meet the turbine metal temperature requirements.

The superheater outlet steam attemperator valve (318) is used in a manner similar to the 218 valve discussed above under *Constant Furnace Pressure, Steam Temperature Control.*

The division valve (301) is used to achieve the necessary pressure drop between the separator and the secondary outlet header for attemperation.

The reheat outlet steam attemperator valve (319) is used at loads below 20% of full load to introduce separator steam to the reheat outlet header. The ratio of flow through the attemperator valve (319) to the flow through the high pressure turbine is limited due to turbine and turbine control considerations.

Permissible rates of load change

The permissible rates of load change, based on the allowable rates of temperature change in the boiler components for various modes of operation, are shown in Fig. 26.

The longer time required for load change with variable pressure operation is due to the greater change in furnace enclosure temperature and the need to restrict the rate of temperature change to avoid tearing casing attachments from the enclosure walls. With dual pressure operation, the furnace enclosure remains at constant pressure and practically constant temperature; however, the secondary superheater inlet header goes through a large temperature change with variable pressure operation, and the time required for load change is based on the limiting rate of temperature change for the outlet header.

If the entire unit is operated at constant pressure, there is virtually no boiler imposed limit on the rate of load change except for the rate of change possible from the firing equipment. With a large load change the expected rate would be a maximum of about 5% of full load per minute. The rate of load change is also restricted by the turbine metals which go through a large temperature swing due to the throttling at low loads and are limited by the design number of cycles.

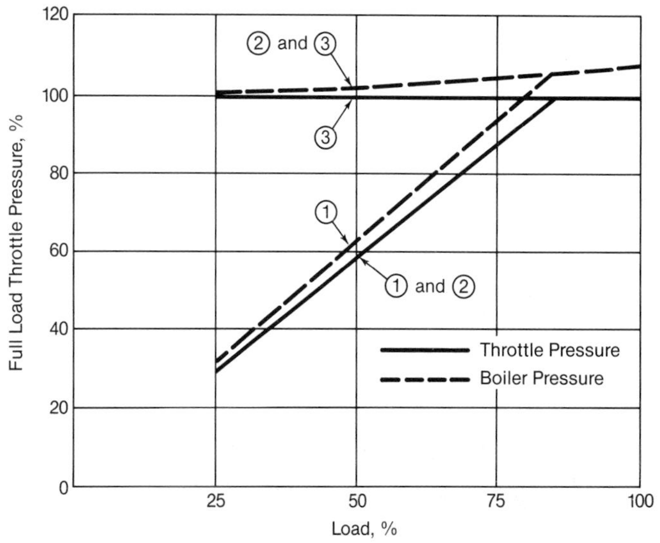

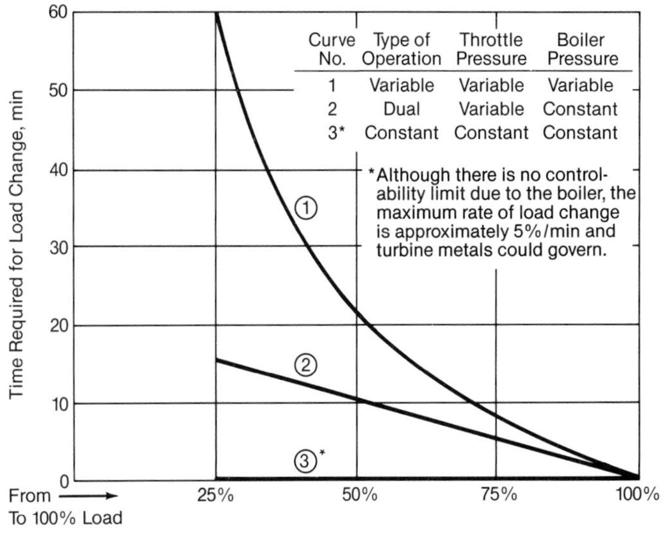

Fig. 26 Permissible rates of load change for three boiler operating configurations.

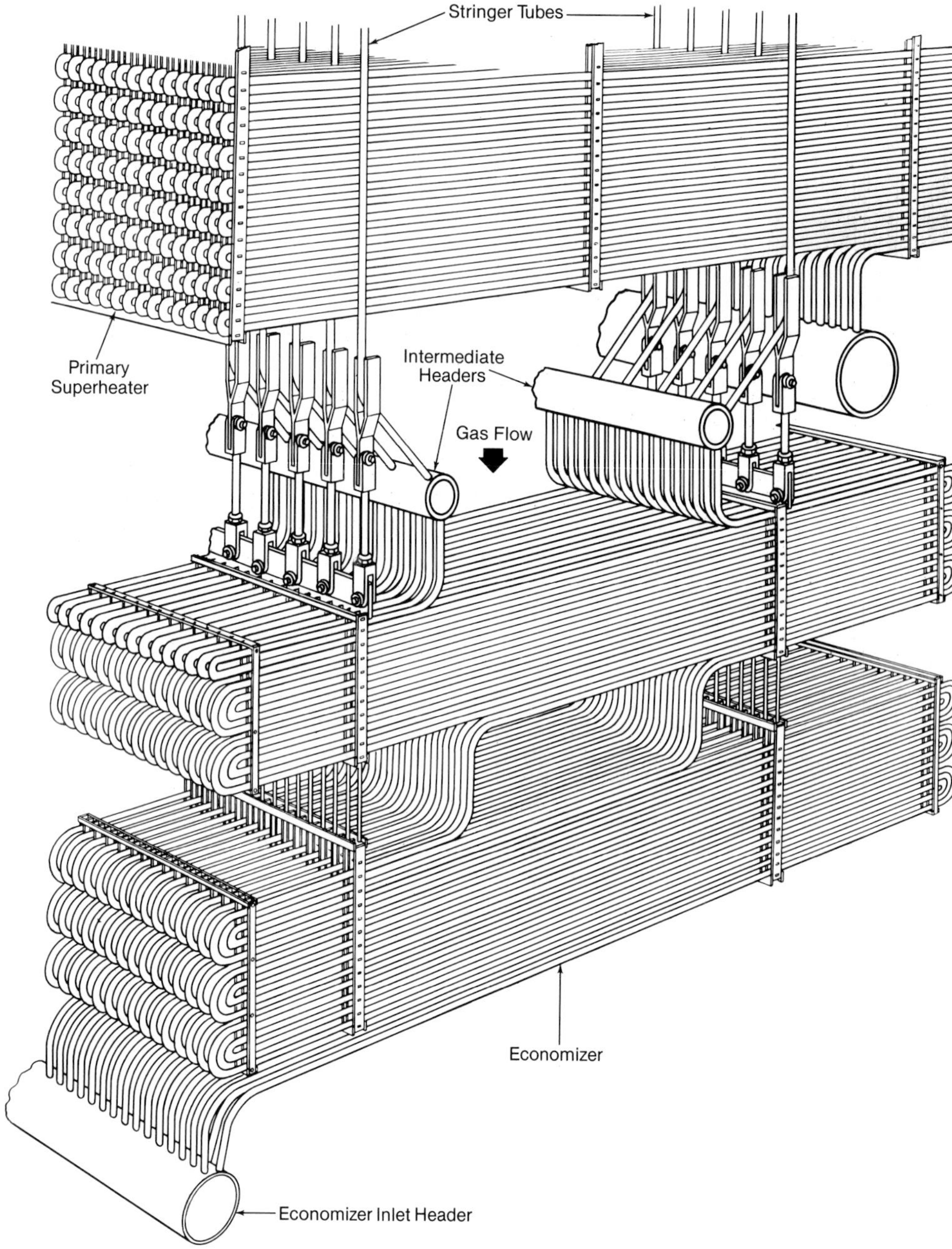

Typical utility boiler economizer.

Chapter 19
Economizers and Air Heaters

Economizers and air heaters perform a key function in providing high overall boiler thermal efficiency by recovering the low level, i.e., low temperature, energy from the flue gas before it is exhausted to the atmosphere. For each 40F (22C) that the flue gas is cooled by an economizer or air heater, the overall boiler efficiency increases by approximately 1% (Fig. 1). Economizers recover the energy by heating the boiler feedwater while air heaters heat the combustion air. Air heating also enhances the combustion of many fuels and is critical for pulverized coal firing for drying the coal and ensuring stable ignition.

In comparison to the furnace waterwalls, superheater and reheater, economizers and air heaters require a large amount of heat transfer surface per unit of heat recovered. This is because of the relatively small difference between the temperature of the flue gas and the temperature of either the feedwater or the combustion air. Use and arrangement of the economizer and/or air heater depend upon the particular fuel, application, boiler operating pressure, power cycle and overall minimum cost configuration.

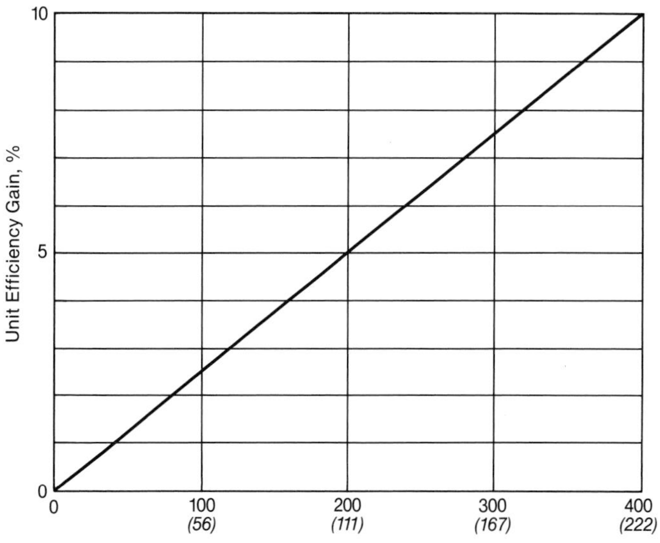

Fig. 1 Approximate unit efficiency increase due to an economizer and air heater.

Economizers

Economizers are basically tubular heat transfer surfaces used to preheat boiler feedwater before it enters the drum (recirculating units) or furnace surfaces (once-through units). The term economizer comes from early use of such heat exchangers to reduce operating costs or economize on fuel by recovering extra energy from the flue gas. Economizers also reduce the potential of thermal shock and strong water temperature fluctuations as the feedwater enters the boiler drum or waterwalls. Fig. 2 shows an economizer location on a coal-fired boiler. The economizer is typically the last water-cooled heat transfer surface upstream of the air heater. (See facing page.)

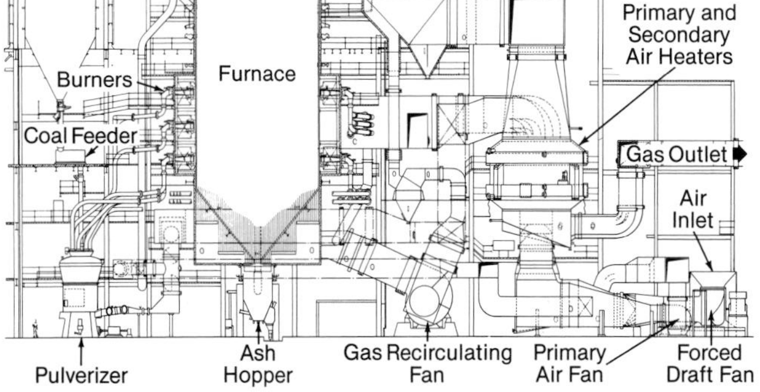

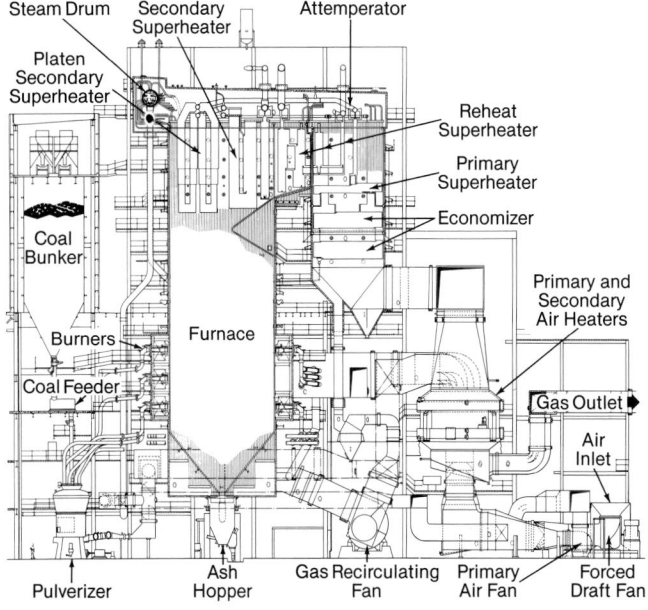

Fig. 2 Economizer and air heater locations in a typical coal-fired boiler.

Economizer surface types

Bare tube

The most common and reliable economizer design is the bare tube, in-line, crossflow type. (See Fig. 3a.) When coal is fired, the flyash creates a high fouling and erosive environment. The bare tube, in-line arrangement minimizes the likelihood of erosion and trapping the ash as compared to a staggered arrangement shown in Fig. 3b.

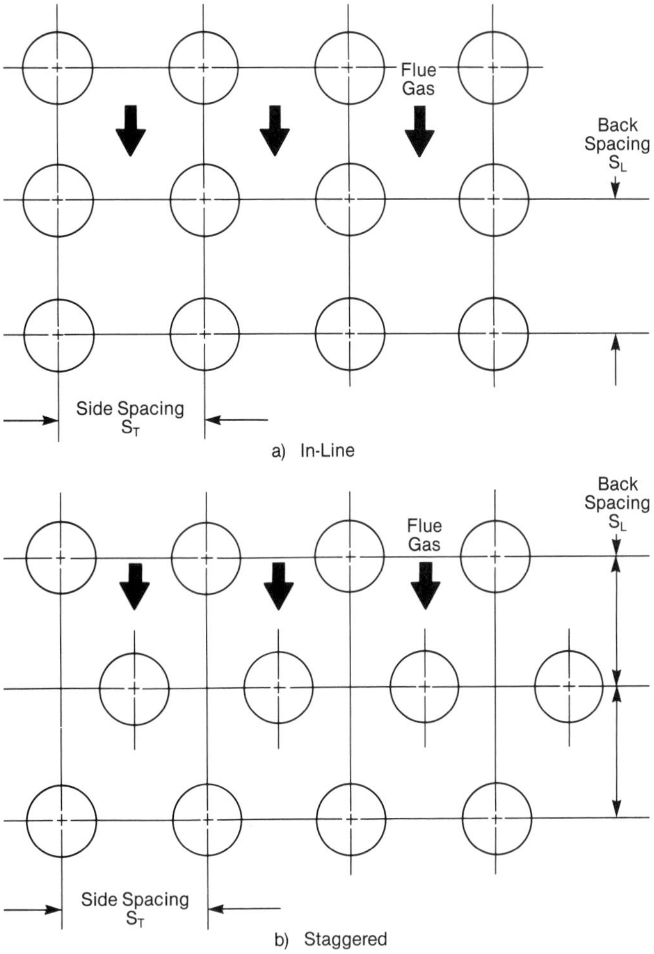

Fig. 3 Bare tube economizer arrangements.

It is also the easiest geometry to be kept clean by sootblowers. However, these benefits must be evaluated against the possible larger weight, volume and cost of this arrangement.

Extended surfaces

To reduce capital costs, most boiler manufacturers have built economizers with a variety of fin types to enhance the controlling gas side heat transfer rate. Fins are inexpensive nonpressure parts which can reduce the overall size and cost of an economizer. However, successful application is very sensitive to the flue gas environment. Surface cleanability is a key concern. In selected boilers, such as Cyclone furnace units (see Chapter 14), extended surface economizers are not recommended because of the coarser flyash characteristics.

Stud fins Stud fins have worked reasonably well in gas-fired boilers. However, stud finned economizers can have higher pressure loss than a comparable unit with helically-finned tubes. Studded fins have performed poorly in coal-fired boilers because of high erosion, loss of heat transfer, increased pressure loss and plugging resulting from flyash deposits.

Longitudinal fins Longitudinally-finned tubes in staggered crossflow arrangements, shown in Fig. 4, have also not performed well over long operating periods. Excessive plugging and erosion in coal-fired boilers have resulted in the replacement of many of these economizers. In oil- and gas-fired boilers, cracks have occurred at the

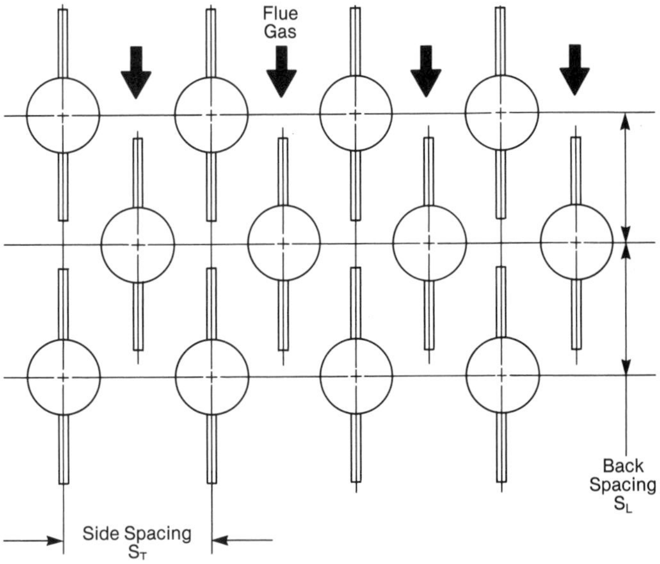

Fig. 4 Longitudinal fins, staggered tube arrangement. (Fin width exaggerated for clarity.)

points where the fins terminate. These cracks have propagated into the tube wall and caused tube failures in some applications.

Helical fins Helically-finned tubes (Fig. 5) have been successfully applied to some coal-, oil- and gas-fired units. The fins can be tightly spaced in the case of gas firing due to the absence of coal flyash or oil ash. Four fins per inch (1 fin per 6.4 mm), a fin thickness of 0.06 to 0.075 in. (1.5 to 1.9 mm) and a height of 0.75 in. (19.1 mm) are typical. For 2 in. (51 mm) outside diameter tubes, these fins provide ten times the effective area of bare tubes per unit tube length. If heavy fuel oil or coal is fired, a wider fin spacing must be used and adequate measures taken to keep the heating surface as clean as possible. Economizers in units fired with heavy fuel oil can be designed with helical fins, spaced at 0.5 in. (13 mm) intervals. Smaller fin spacings promote plugging with oil ash, while greater spacings reduce the amount of heating surface per unit length. Sootblowers are required and the maximum bank height should not exceed 4 to 5 ft (1.2 to 1.5 m) to assure reasonable cleanability of the heating surface. An in-line arrangement also facilitates cleaning and provides a lower gas side resistance.

Rectangular fins The square or rectangular fins, arranged perpendicular to the tube axis on in-line tubes as shown in Fig. 6, have had some success in retrofits. The fin spacing typically varies between 0.5 and 1 in. (13 and 25 mm) and the fins are usually 0.125 in. (3.18 mm) thick. There is a vertical slot down the middle because the two halves of the fin are welded to either side of the tube. Most designs are for gas velocities below 50 ft/s (15.2 m/s).

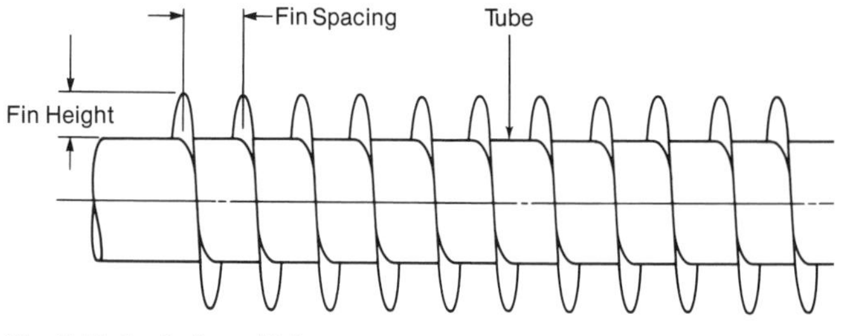

Fig. 5 Helically-finned tube.

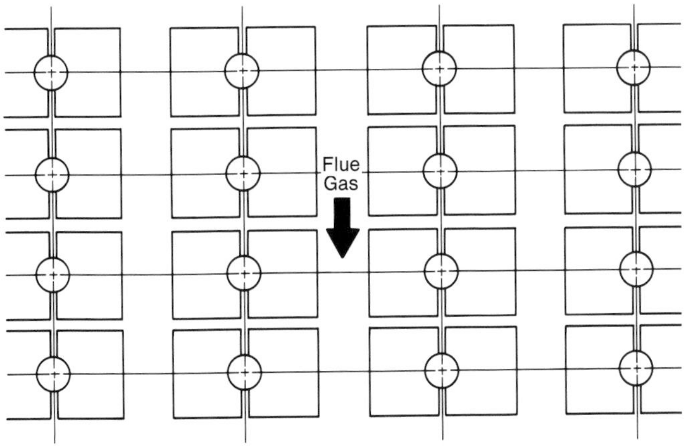

Fig. 6 Rectangular fins, in-line tube arrangement.

Baffles The tube ends should be fully baffled (Fig. 7) to minimize flue gas bypass in finned bundles. Such bypass flow can reduce heat transfer, produce excessive casing temperatures and with coal firing, can lead to tube bend erosion because of very high gas velocities. Baffling is also used with bare tube bundles but is not as important as for finned tube bundles.

Velocity limits

The ultimate goal of economizer design is to achieve the necessary heat transfer at minimum cost. A key design criterion for economizers is the maximum allowable gas velocity (defined at the minimum cross-sectional free flow area in the tube bundle). Higher velocities provide better heat transfer and reduce capital cost. For clean burning fuels, such as gas and low ash oil, velocities are typically set by the maximum economical pressure loss. For high ash oil and coal, gas side velocities are limited by the erosion potential of the flyash. This erosion potential is primarily determined by the percentage of Al_2O_3 and SiO_2 in the ash, the total ash in the fuel and the gas maximum velocity. Experience dictates acceptable velocities. Fig. 8 provides sample base velocity limits as a function of ash characteristics.

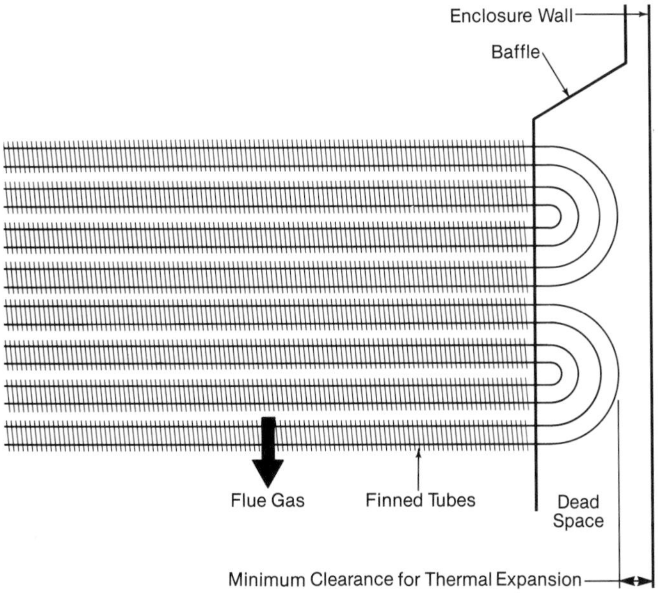

Fig. 7 Baffling of bare tube return bends for finned tube bundles.

Further criteria may also be needed. For example, a 5 ft/s (1.5 m/s) reduction in the base velocity limit is recommended when firing coals with less than 20% volatile matter. In other cases, such as Cyclone boilers, high flue gas velocities can be used because much less flyash is carried into the convection pass, as much of the ash (> 50%) is collected in the bottom of the boiler as slag. Particles that carry over are also less erosive. (See Chapter 14.)

For a given tube arrangement and boiler load, the gas velocity depends on the specific volume of flue gas which falls as the flue gas is cooled in the economizer. To maintain the gas velocity, it can be economical to decrease the free flow gas side cross-section by selecting a larger tube size in the lower bank of a multiple bank design. This achieves better heat transfer and reduces the total heating surface.

Other types of economizers

Fig. 9 depicts an industrial boiler with a long flow economizer, often used in chemical recovery boilers. Such heating surfaces consist of vertical, longitudinally-finned (membraned) tubes through which the feedwater flows upward. The gas flows downward in pure counterflow, outside the tubes and fins. While the heat transfer is less efficient than crossflow banks of tubes, there is minimal gas side resistance and fouling products are removed through hoppers at the bottom of the enclosure.

Steaming economizers

Steaming economizers are defined as meeting the following enthalpy relationships:

$$H_2 - H_1 \geq 2/3 \left(H_f - H_1\right) \tag{1}$$

where

H_2 = enthalpy of fluid leaving economizer (to drum)
H_1 = enthalpy of fluid (water) entering economizer
H_f = enthalpy of saturated water at economizer outlet pressure

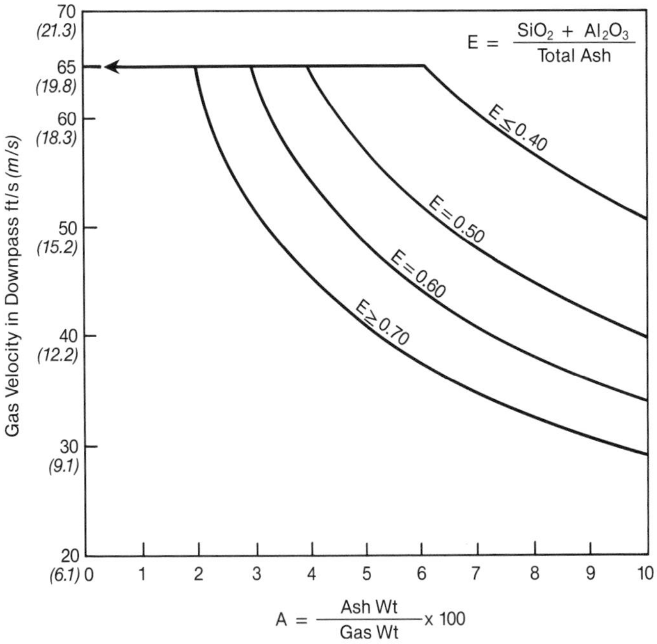

Fig. 8 Base maximum allowable velocity for pulverized coal-fired boiler economizers.

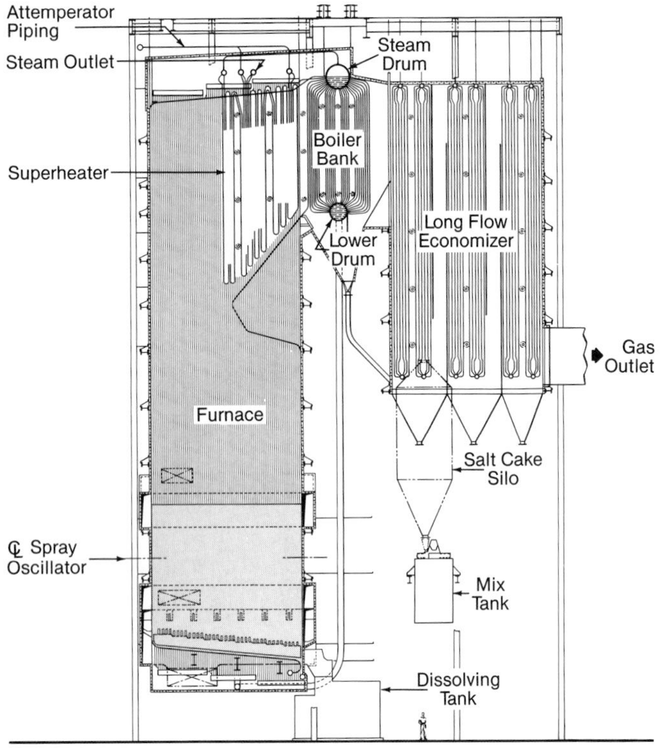

Fig. 9 Long flow economizer for a chemical recovery boiler.

These economizers can be economical in certain boilers. They require careful design and must be oriented so the water flows upward and the outlet is below the level of the drum. This avoids water hammer and excessive flow instabilities. The enthalpy equation accounts for possible steaming due to flow imbalances and differences in individual circuit heat absorptions.

High pressure, drum type units are usually sensitive to feedwater temperatures close to saturation. To enhance circulation, feedwater temperature to the drum should normally be at least 50F (28C) below saturation temperature.

Performance

Heat transfer

Bare tubes The equations discussed in Chapter 4 can be used to evaluate the quantity of surface for an economizer. For the economizer shown in Fig. 2 with the upflow of water and downflow of gas and nonsteaming conditions, the bundle can be treated as an ideal counterflow heat exchanger with the following characteristics:

1. bundle log mean temperature difference correction factor = 1.0,
2. heat absorbed by the tube wall enclosures and heat radiated into the tube banks from various cavities can generally be neglected,
3. all of the energy lost by the flue gas is absorbed by the water, i.e., no casing heat loss,
4. the water side heat transfer coefficient is typically in the range of 2000 Btu/h ft^2 F (11,357 W/m^2 K) and has only a small overall impact on the economizer performance, and
5. the effect of gas side ash deposition can be accounted for by a cleanliness factor based upon experience.

In general the heat transfer rate is primarily limited by the gas side heat transfer for in-line bare tube bundles. In this case, the overall heat transfer coefficient (flue gas to feedwater) used in the heat exchanger calculation can be approximated by the following relationship:

$$U = 0.98\ (h_c + h_r)\ k_f \qquad \textbf{(2)}$$

where

U = overall heat transfer coefficient, Btu/h ft^2 F (W/m^2 K)
h_c = gas side heat transfer coefficient for a bare tube bundle, Btu/h ft^2 F (W/m^2 K) (Chapter 4, Equations 54 and 55)
h_r = inter-tube radiation heat transfer coefficient, Btu/h ft^2 F (W/m^2 K) $\approx$ 1.0 for coal firing
k_f = surface effectiveness factor = 0.7 for coal, 0.8 for oil and 1.0 for gas

Finned tubes Heat transfer performance of finned tube economizers can be evaluated in a similar fashion except that appropriate relationships for extended surfaces should be used. In addition, because the gas side heat transfer has been enhanced, the water side heat transfer coefficient and tube wall thermal resistance are more significant and must be included in the evaluation. (See Chapter 4.) As a general guideline, the overall heat transfer coefficient can be approximated by the following relationship for most types of economizer fins:

$$U = 0.95\ (h_g k_f) \qquad \textbf{(3)}$$

where h_g is the gas side heat transfer coefficient for the heat transfer across finned tube bundles evaluated with the procedures defined in Chapter 4. A calculated example is provided in Chapter 21 for a bare tube economizer tube bundle.

Gas side resistance

The gas side pressure loss across the economizer tube bank can be evaluated using the crossflow correlations presented in Chapter 3. The pressure loss should be adjusted for the number of tube rows using the correction factors provided. The gas side resistance across the in-line finned tube banks is approximately 1.5 times the resistance of the underlying bare tubes.

Water side pressure drop

The water side pressure loss can be evaluated using the procedures in Chapter 3 where the total pressure loss ΔP_T is calculated:

$$\Delta P_T = \Delta P_f + \Delta P_l + \Delta P_z \qquad \textbf{(4)}$$

where

ΔP_f	= friction pressure loss	Ch. 3, Eq. 42
ΔP_l	= sum of the local losses (entrance, bends and exits)	Ch. 3, Eq. 47
ΔP_z	= static head loss	Ch. 5, Eq. 10

The design pressure is then evaluated by the sum of the drum design pressure and the total pressure loss ΔP_T rounded up to the nearest 25 psig (1.7 bar gauge).

If the calculated water side pressure drop is excessive, the number of parallel flow paths must be increased. If the gas velocity can be increased, the water side pressure drop can also be reduced by increasing the tube size, usually in increments of 0.125 in. (3 mm). As indicated in Chapter 3, the dynamic pressure drop is inversely proportional to the fifth power of the inside tube diameter. This is significant and may be advantageous for retrofit applications. Sometimes, a material upgrade (from SA-210A1 to SA-210C, for example) can permit use of the optimum tube wall thickness.

Economizer support systems

Economizers are located within tube wall enclosures or within casing walls, depending on gas temperatures. In general, casing enclosures are used at or below 850F (454C) and inexpensive carbon steel can be used. If a casing enclosure is used, it must not support the economizer. However, tube wall enclosures may be used as supports.

The number of support points is determined by analyzing the allowable deflection in the tubes and tube assemblies. Deflection is important for tube drainability. Figs. 10 through 12 show typical support arrangements for bare tube economizers.

Wall or end supports are usually chosen for relatively short spans and require bridge castings or individual lugs welded or attached to the tube wall enclosures. (See Fig. 10.) Another possibility exists if enclosure wall (usually primary superheater circuitry) headers are present above the economizer (for example, Fig. 11).

Quarter point stringer supports are used for spans exceeding the limits for end supports (Fig. 12). The stringers are mechanically connected to the economizer sections, which are held up by ladder type supports. The supports exposed to hot inlet gases may be made of stainless steel, while lower grade material is normally used to support the lower bank which is exposed to reduced gas temperatures. In Babcock & Wilcox (B&W) designs, stringer tubes also usually support other horizontal convection surfaces above the economizer. (See frontispiece.) Bottom support is sometimes used if the gas temperature leaving the lowest economizer bank is low enough.

Bank size

The bank size is limited by the following constraints:

1. type of fuel,
2. fabrication limits,
3. sootblower range,
4. maximum shipping dimensions,
5. construction considerations, especially for retrofits, and
6. maintenance.

Bank depths greater than 6 ft (1.82 m) are rare in new boilers, while larger banks can be tolerated in retrofits.

Access requirements

Cavities around the banks are needed for field welding, tube leg maintenance and sootblower clearance. A sufficient number of access doors must be arranged in the enclosure walls to access these cavities. Cavity access can be provided from the outside through individual doors or from the inside through special openings across stringers or collector frames. The minimum cavity height should be 2 ft (0.6 m) of crawl space.

Headers

B&W economizer header designs are based on American Society of Mechanical Engineers (ASME) Code requirements. Inlet headers are frequently located inside the gas stream and may receive feedwater through one or both ends. Regardless of design, it is necessary to properly seal the inlet pipe where it penetrates the enclosure by using brackets and flexible seals. The seal becomes especially important in pressurized (forced draft) units. Other important considerations are tube leg flexibility, differential expansion and potential gas temperature imbalances and upsets.

The outlet headers, not to be confused with intermediate headers seen in larger boilers and stringer supports, receive the heated feedwater and convey it to the drum

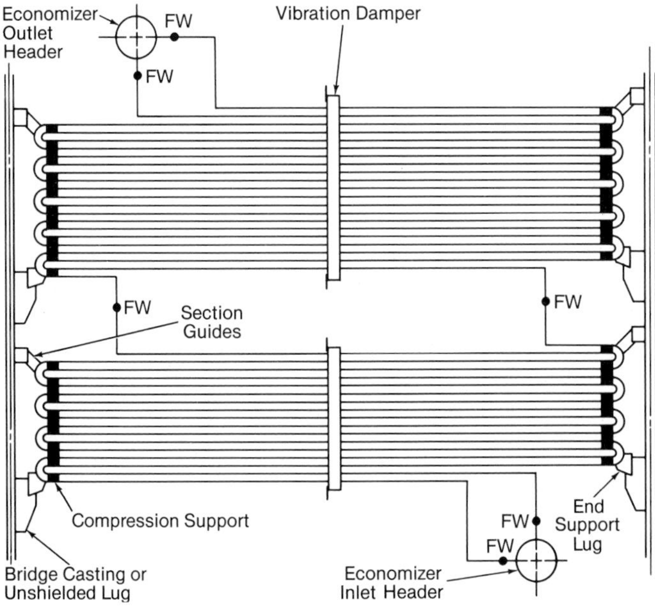

Fig. 10 Economizer supports — sample waterwall support arrangement. (FW = field weld.)

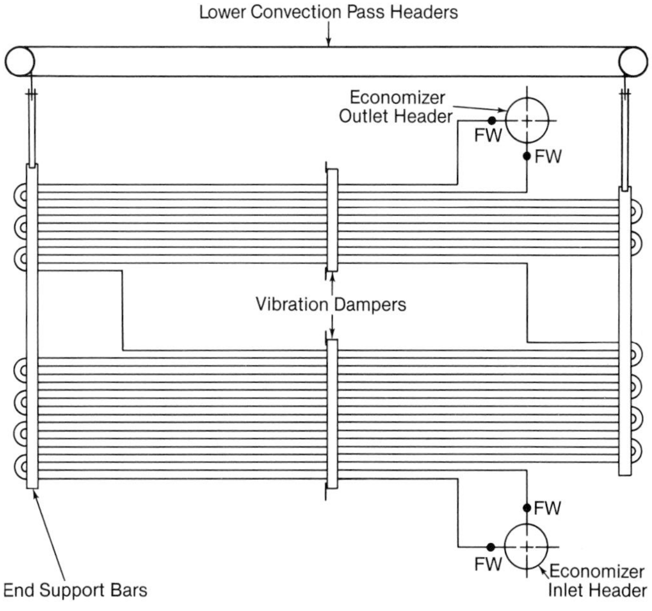

Fig. 11 Economizer supports — sample lower waterwall header arrangement. (FW = field weld.)

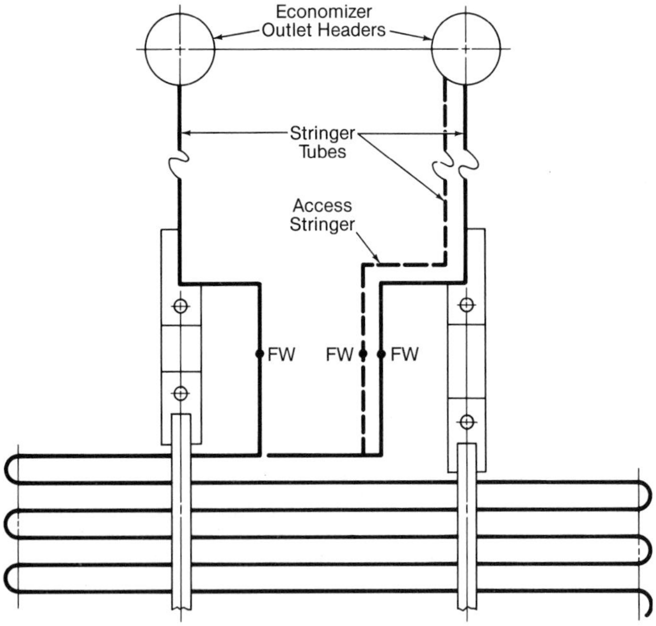

Fig. 12 Economizer supports — sample stringer support arrangement.

or, in the case of once-through boilers, to the downcomer supplying the furnace circuitry. Inlet and outlet headers must be large enough to assure reasonable water flow distribution in the economizer banks. Flow velocities are typically less than 20 ft/s (6 m/s).

Vibration ties

Vibration ties or tube guides are required on some end-supported tube sections. These ties may be needed if the natural frequencies within the boiler load range are in or near resonance with the vortex shedding frequency.

Stringer tubes are also subject to vibration. This vibration is magnified by long unsupported stringer tube lengths near the large cavity below the convection pass roof.

Tube geometry, materials and code requirements

Economizer tube diameters typically range between 1.75 and 2.5 in. (44.5 and 63.5 mm). Tubes outside this range are sometimes used in retrofits. Smaller tubes are normally used in once-through, supercritical boilers where water side pressure drop is less of a consideration. In these units, tube wall thicknesses are minimized.

The ASME Code requires that the design temperature for internal boiler pressure parts is at least 700F (371C). The calculated mean tube wall temperature in economizers seldom reaches this temperature. It usually lies 10 to 20F (6 to 12C) above the fluid temperature, which seldom exceeds 650F (343C) along any economizer circuit.

The minimum tube wall thickness is determined in accordance with procedures outlined in Chapter 7.

In coal-fired boilers, the side spacing is usually determined by the maximum allowable gas velocity and gas side resistance, which are functions of a given tube size. If fins are used, the side and back tube spacings should permit the fin tips to be at least 0.5 in. (13 mm) apart. For bare tubes, a minimum clear spacing of 0.75 in. (19 mm) is desirable.

The minimum back (vertical) spacing of the tubes should be no less than 1.25 times the tube outside diameter. Smaller ratios can reduce heat transfer by as much as 30%. Ratios larger than 1.25 have relatively little effect on heat transfer but increase the gas side resistance and bank depth.

Air heaters

Air heaters are used in most steam generating plants to heat the combustion air and enhance the combustion process. Most frequently, the flue gas is the source of energy and the air heater serves as a heat trap to collect and use waste heat from the flue gas stream. This can increase the overall boiler efficiency by 5 to 10%. Air heaters can also use extraction steam or other sources of energy depending upon the particular application. These units are usually employed to control air and gas temperatures by preheating air entering the main gas-air heaters.

Air heaters are typically located directly behind the boiler, as depicted in Fig. 2, where they receive hot flue gas from the economizer and cold combustion air from the forced draft fan. The hot air produced by air heaters enhances combustion of all fuels and is needed for drying and transporting the fuel in pulverized coal-fired units.

Classification of air heaters

Air heaters are classified according to their principle of operation as recuperative or regenerative.

Recuperative

In a recuperative heat exchanger, heat is transferred continuously through stationary solid heat transfer surfaces which separate the hot flow stream from the cold flow stream. The most common heat transfer surfaces are tubes and parallel plates. Recuperative heat exchangers function with little cross-contamination, or leakage, between streams.

Tubular air heaters In a typical tubular air heater, energy is transferred from the hot flue gas flowing inside many thin walled tubes to the cold combustion air flowing outside the tubes. The unit consists of a nest of straight tubes that are roll expanded or welded into tubesheets and enclosed in a steel casing. The casing serves as the enclosure for the air or gas passing outside of the tubes and has both air and gas inlet and outlet openings. In the vertical type (Fig. 13), tubes are supported from either the upper or lower tubesheet while the other (floating) tubesheet is free to move as tubes expand within the casing. An expansion joint between the floating tubesheet and casing provides an air/gas seal. Intermediate baffle plates parallel to the tubesheets are frequently used to separate the flow paths and eliminate tube damaging flow induced vibration.

Carbon steel or low alloy corrosion resistant tube materials are used in the tubes which range from 1.5 to 4 in. (38 to 100 mm) in diameter and have wall thicknesses of 18 to 11 gauge [0.049 to 0.120 in. (1.24 to 3.05 mm)]. Larger diameter, heavier gauge tubes are used when the potential for tube plugging and corrosion exists. Tube arrangement may be in-line or staggered with the latter being more thermally efficient.

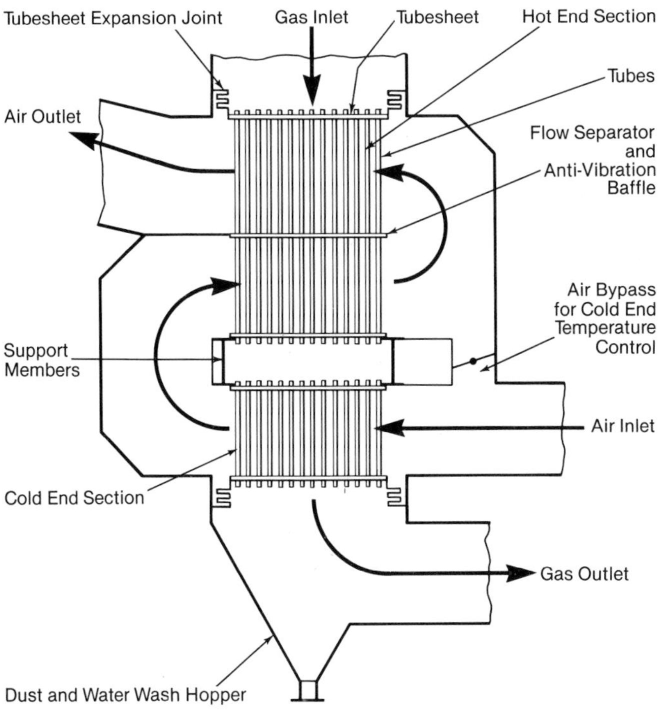

Fig. 13 Vertical type tubular air heater.

The most common flow arrangement is counterflow with gas passing vertically through the tubes and air passing horizontally in one or more passes outside the tubes. A variety of single and multiple gas and air path arrangements are used to accommodate plant layouts. Designs frequently include provisions for cold air bypass or hot air recirculation to control cold end corrosion and ash fouling. Modern tubular air heaters are shop assembled into large, transportable modules. Several arrangements are shown in Fig. 14.

Cast iron air heaters Cast iron tubular air heaters are heavy, large and durable. Their use is mainly limited to the petrochemical industry, but some are used on electric utility units. Cast iron is used because of its superior corrosion resistance. Rectangular, longitudinally split tubes are assembled from two cast iron plates and individual tubes are assembled into air heater sections. Air heaters are usually arranged for a single gas pass and multiple air passes with air flow inside the tubes. Heat transfer is maximized by fins cast into inside and outside tube surfaces.

Plate air heaters This type of air heater transfers heat from hot gas flowing on one side of a plate to cold air flowing on the opposite side, usually in crossflow. Heaters consist of stacks of parallel plates. Sealing between air and gas streams at plate edges is accomplished by weld-

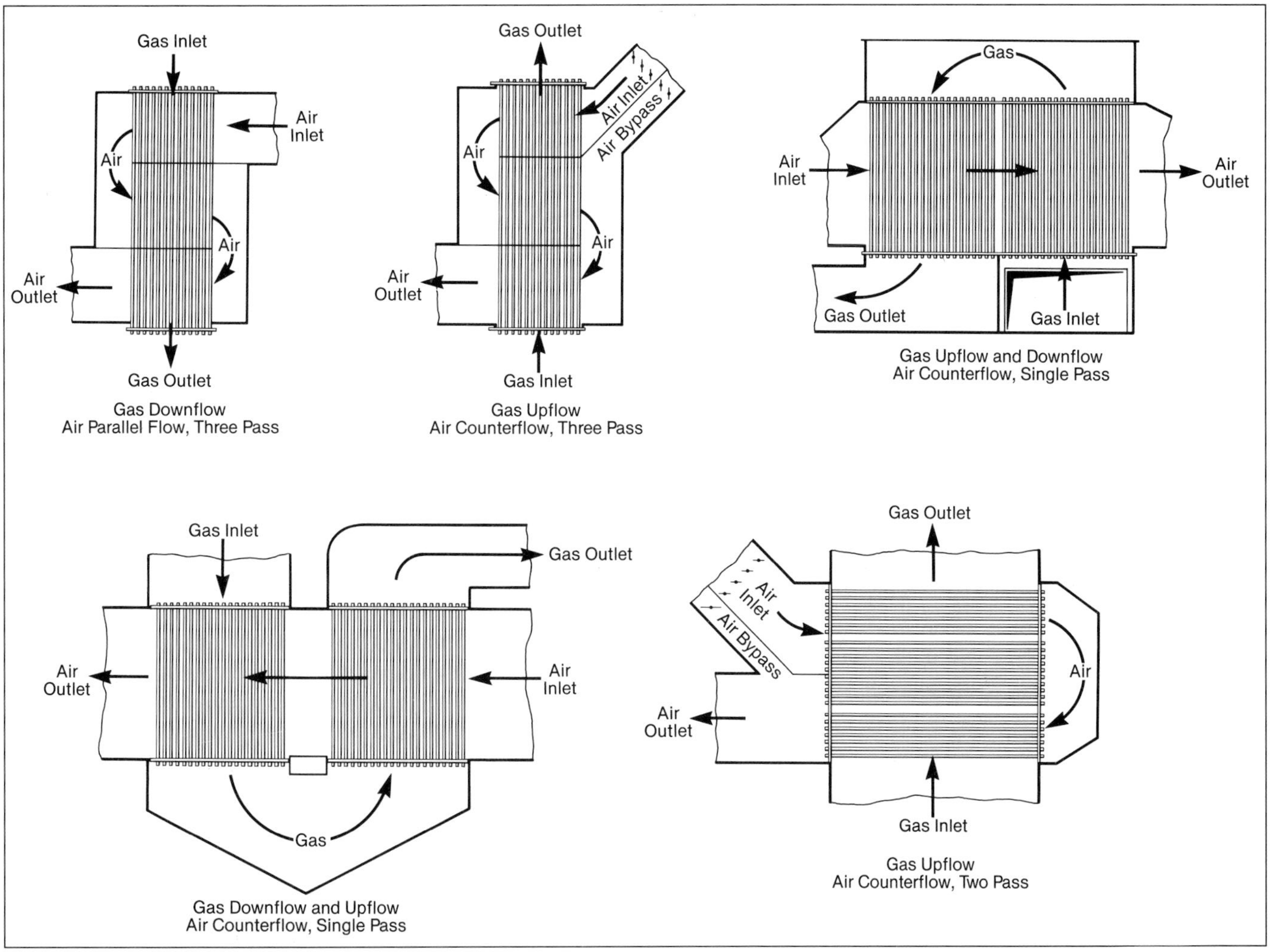

Fig. 14 Various tubular air heater arrangements.

ing or by a combination of gaskets, springs and external compression of the plate stack. Plate materials and spacing can be varied to accommodate operating requirements and fuel types.

Steel plate air heaters were some of the earliest types used, but their use declined due to plate to plate sealing problems. However, recent sealing developments have prompted increased use in industrial and small utility applications. Plate modules may be combined to make different size air heaters with a variety of flow path arrangements. A single gas pass, two air pass plate air heater is shown in Fig. 15. Modern plate units are somewhat smaller than tubular units for a given capacity and exhibit minimal air to gas leakage.

Steam coil air heaters Steam coil and water coil recuperative air heaters are widely used in utility steam generating plants to preheat combustion air. Air preheating reduces the corrosion and plugging potential in the cold end of the main air heater. Occasionally, they serve as the only source of preheated combustion air. These heaters consist of banks of small diameter, externally finned tubes arranged horizontally or vertically in air ducts between the combustion air fan and main air heater. Combustion air, passing in crossflow outside the tubes, is heated by turbine extraction steam or feedwater flowing inside the tubes. Ethylene glycol is sometimes used as the hot fluid to prevent out of service freezing damage.

Heat pipe A heat pipe is a simple, highly efficient device for transporting thermal energy. The basic thermosyphon type heat pipe used in steam generation unit air heaters consists of an evacuated sealed pipe which has been partially filled with a heat transfer fluid (Fig. 16). The evaporator end of the pipe is exposed to a heat source (hot flue gas) and the other end, the con-

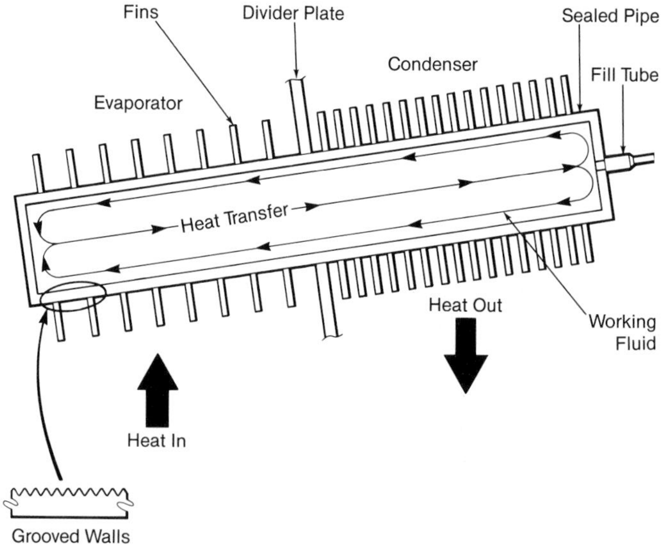

Fig. 16 Heat pipe schematic.

denser, is placed in a heat sink (cold combustion air). Heat absorbed from the flue gas evaporates the fluid which travels to the combustion air end where heat is released as the fluid condenses. The condensed fluid returns by capillary action and gravity to the evaporator end. Fluid circulation within the tube is continuous as long as there is a temperature difference between evaporator and condenser ends; fluid temperature is nearly isothermal at approximately the average of air and gas temperatures. Large quantities of heat are transferred as the heat of condensation and vaporization are used. Heat pipes operate with the evaporator end lower than the condenser end; they are inclined from the horizontal. Internal surfaces of heat pipes are roughened or grooved to assist fluid circulation and external surfaces are usually finned to increase heat transfer surface area.

Heat pipe air heaters consist of bundles of parallel heat pipe tubes. About half of the tube length is exposed to flue gas flow and the remaining length is exposed to air flow. A central divider plate separates air and gas and supports the tube bundle. Heat pipe bundles or modules can be combined and enclosed in casing to build air heaters capable of accepting a variety of flow configurations.

Tube bundle configuration is usually in-line for dirty gas applications such as coal and heavy oil and staggered for natural gas and light oil. Typically, 2 in. (51 mm) diameter carbon steel tubes up to 40 ft (12.2 m) long are used. They feature three steel fins per in. (1 per 8.5 mm) on the gas side and up to ten fins per in. (1 per 2.5 mm) on the air side. Corrosion resistant alloy materials can increase cold end corrosion life.

Heat pipe air heaters are smaller than tubular air heaters and air to gas leakage is minimal as in other types of recuperative air heaters. Due to the isothermal behavior of each tube, these units can operate at a lower gas outlet temperature for a specific minimum metal temperature (MMT) compared to a tubular or regenerative air heater. This can permit operation at higher boiler efficiency and can reduce the potential for air heater cold end corrosion.

Of utmost importance is the heat transfer fluid's long term compatibility with tube wall material. Incompatibility can result in internal corrosion which produces noncon-

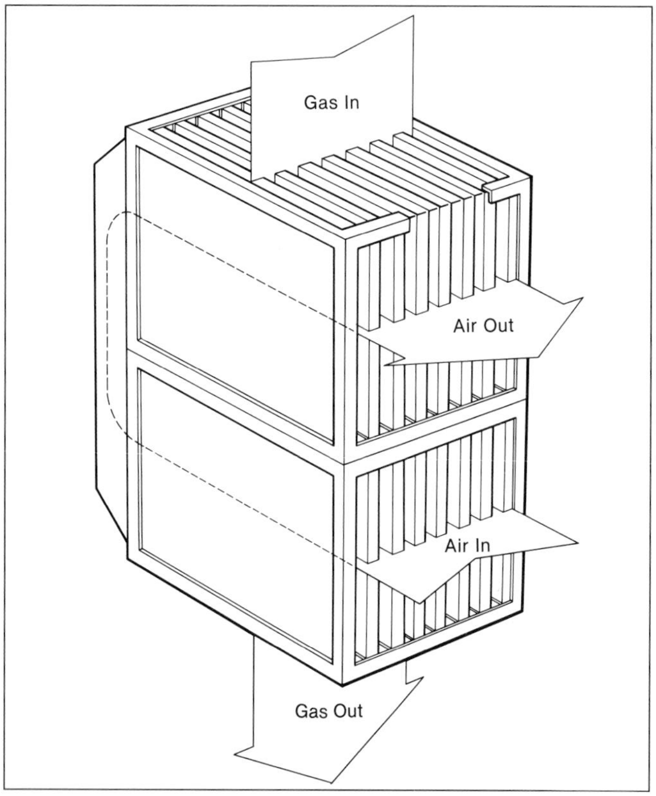

Fig. 15 Single gas pass, two air pass plate air heater.

densable gases, reduces heat transfer rates and jeopardizes tube pressure integrity. Hydrocarbons and water based fluids, commonly used in carbon steel tubes, are subject to temperature limits of 400 to 800F (204 to 427C). Water based fluids can not be allowed to freeze. High alloy steel or nonferrous tube materials may be used to extend temperature limits or to allow more fluid choices, but their expense precludes use in large heaters.

Heat pipe air heaters have been used successfully in the petrochemical industry. A limited number have also been applied to electric utility steam generation units. More widespread adoption is expected when long term reliability is proven.

Regenerative

In a regenerative air heater, heat is transferred indirectly as a heat storage medium is alternately exposed to hot and cold flow streams. A variety of materials can be used as the medium and periodic exposure to hot and cold flow streams can be accomplished by rotary or valve switching devices. In steam generating plants, tightly packed bundles of corrugated steel plates serve as the storage medium. In these units either the steel plates, or surface elements, rotate through air and gas streams or rotating ducts direct air and gas streams through stationary surface elements.

Regenerative air heaters are relatively small and are the most widely used type for combustion air preheating in electric utility steam generating plants. Their most notable operating feature is that a small but significant amount of air leaks into the gas stream due to the rotary operation.

Ljungström The most prevalent regenerative air heater is the Ljungström type (Fig. 17), which features a cylindrical shell plus a rotor which is packed with bundles of heating surface elements and is rotated through counterflowing air and gas streams. The rotor is enclosed by a stationary housing which has ducts at both ends. Air flows through one half of the rotor and gas flows through the other half. Metallic leaf-type seals minimize air to gas leakage and flow bypass around the rotor. Bearings in upper and lower beam assemblies support and guide the rotor at the central shaft. A rotor speed of one to three rpm is provided by a motor driven pinion engaging a rotor encircling pinrack. Both vertical and horizontal shaft designs are used to accommodate various plant air and gas flow schemes. The vertical shaft design is more common.

Rothemühle The Rothemühle-type regenerative air heater uses stationary surface elements and rotating ducts (Fig. 18). The surface elements are supported and contained within a stationary cylindrical shell called the stator. On both sides of the stator, a double wing symmetrical hood rotates synchronously on a common vertical shaft. The central shaft is supported by bearings within the stator and the hoods are driven slowly by a pinion which engages a pinrack encircling the lower hood. Stationary housings surround the hoods. Heat is transferred as flow streams are directed through the heating surface in counterflow fashion, one flow stream inside the hoods and the other outside. Either air or gas may pass through the hoods. However, air is more common because it requires less fan power. Special spring mounted sealing systems employing cast iron seals are used at hood rotating to stationary interfaces to minimize air to gas leakage.

Several features distinguish the Rothemühle from the Ljungström heater. Its relatively low rotating weight (20% of total weight) contributes to high reliability. The stationary stator permits air heater loads to be distributed equally to a number of surrounding points, permitting transfer of significant duct loads through the air

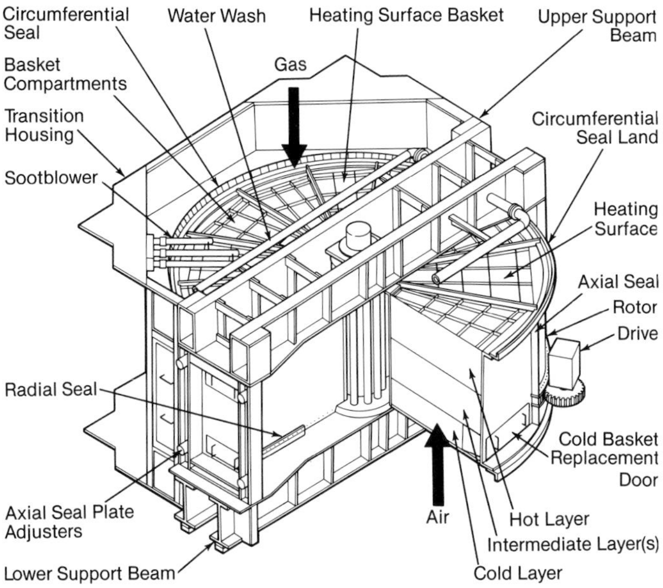

Fig. 17 Ljungström-type air heater.

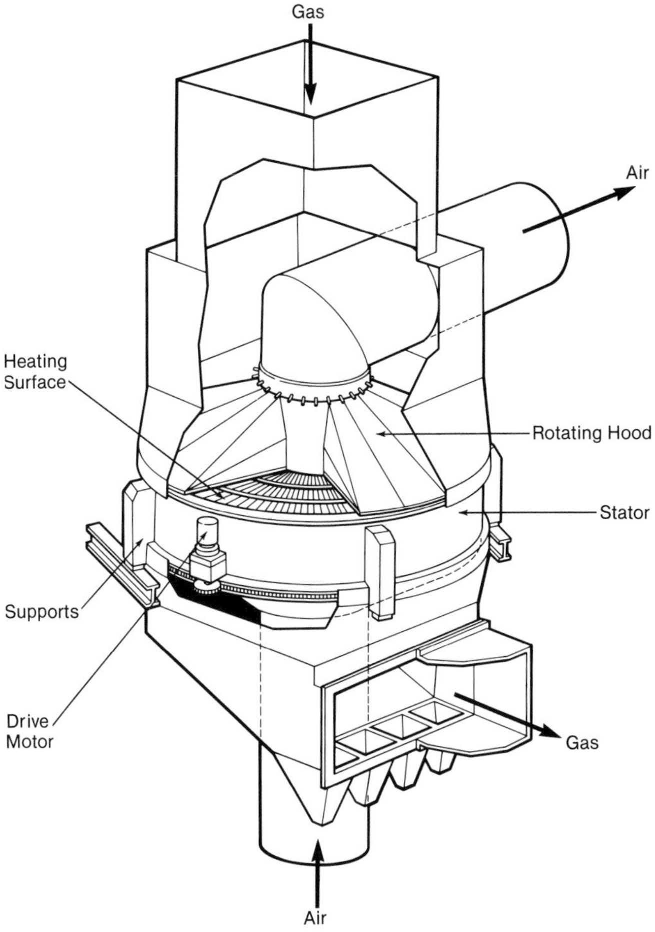

Fig. 18 Rothemühle-type air heater.

heater into structural steel. The spring mounted hood sealing system, which adapts to the curvature of the stator during operation, allows hot starts without overloading the drive motor. A simple, permanent early warning fire detection system can be embedded in the stator.

Regenerative heating surface Regenerative air heater surface elements are a compact arrangement of two specially formed metal plates. Each element pair consists of a combination of flat, corrugated or undulated plate profiles. The roll formed corrugations and undulations serve to separate the plates to maintain flow paths, increase heating surface area and maximize heat transfer by creating flow turbulence. The steel plates, 26 to 18 gauge thick, are typically spaced 0.2 to 0.4 in. (5 to 10 mm) apart. Closely spaced, highly profiled element pairs exhibit a high heat transfer rate, pressure drop and fouling potential while widely spaced element combinations, where one plate is flat, exhibit a low heat transfer rate, low pressure drop and reduced fouling potential. The combination of plate profile, material and thickness is selected for maximum heat transfer, minimum pressure drop, good cleanability and high corrosion resistance.

Surface elements are stacked and bundled into self-contained baskets and are installed into air heater rotors and stators in two or more layers. The surface layer at the air inlet side, designated the cold layer, is distinguished from other layers by design. Cold layers, which are subject to corrosion and ash fouling, are typically 12 in. (300 mm) deep for economical replacement. Heavy gauge, open profile elements are used for corrosion resistance and cleanability. Practically all cold layer elements are low alloy corrosion resistant steel or, when high corrosion potential exists, porcelain enamel coated steel. Hot and intermediate surface layers are more compact than cold layers and use thinner plates. Figs. 17 and 19 illustrate several heating surface element profiles and air heater surface arrangements.

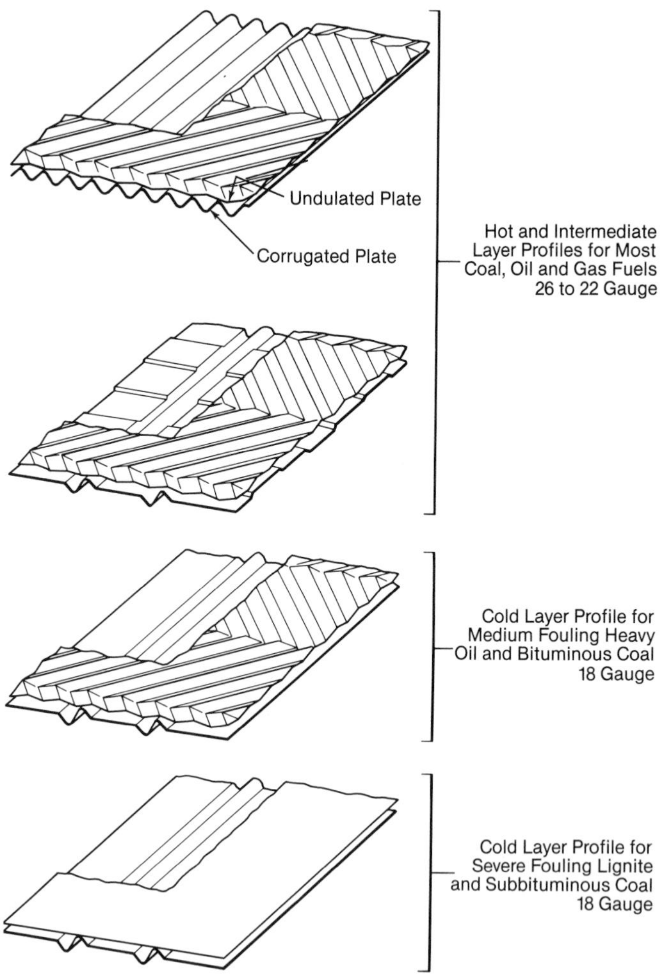

Fig. 19 Regenerative air heater surface element profiles.

Advantages and disadvantages

Many subtle differences exist between air heater designs within a particular type. However, there are some general advantages and disadvantages associated with each type which are listed in Table 1. Note that the recuperative heat pipe air heater is listed separately.

Performance and testing

Air heaters are designed to meet performance requirements in three areas: thermal, leakage and pressure drop. Low performance in any area increases boiler operating costs and may cause unit load curtailment.

Thermal performance

The thermal performance and surface area (A) of a recuperative air heater can be evaluated by:

$$A = Q / (U \quad LMTD \quad F) \qquad \textbf{(5)}$$

where Q is the total thermal load [Btu/h (W)], U is the overall heat transfer coefficient, $LMTD$ is the log mean temperature difference and F is the corresponding geometry correction factor. The performance, U, $LMTD$ and F can be evaluated using the correlations and methodology presented in Chapter 4. The overall U should include convection and radiation components as well as the appropriate gas and air side fouling factors. U typically ranges from 3 to 10 Btu/h ft^2 F (17 to 57 W/m^2 K).

Performance verification Thermal performance is measured by comparing the test gas outlet temperature to its design value. The true outlet temperature is obtained by correcting the measured temperature for air heater leakage and deviations from design conditions.

The ASME Performance Test Code, Section 4.3 (PTC 4.3), provides the following equation which is based on an air heater mass flow heat balance and assumes that the source of all leakage is from the entering air:

Table 1
Advantages and Disadvantages of Air Heater Types

Type	Advantage	Disadvantage
Recuperative	Low leakage No moving parts	Large and heavy Difficult to replace surface
Heat pipe	Low leakage High minimum metal temperatures No moving parts	Difficult to clean Temperature restrictions
Regenerative	Compact Easy to replace surface	Leakage High maintenance Fire hazard

$$T_2 = T_{2m} + \left(\frac{\% \, lkg}{100}\right)\frac{c_{pa}}{c_{pg}}(T_{2m} - T_1) \qquad \textbf{(6)}$$

where

T_2 = air heater gas outlet temperature corrected for leakage, F (C)
T_{2m} = measured gas temperature leaving air heater, F (C)
$\% \, lkg$ = percent air leakage with respect to inlet gas flow
c_{pa}, c_{pg} = specific heat of air and gas respectively, Btu/lb F (J/kg C)
T_1 = air inlet temperature, F (C)

The measured gas outlet temperature must also be corrected for deviations in various operating parameters such as mass flow rates and operating temperatures in order to accurately assess performance. Suppliers and the ASME Performance Test Codes provide various correction curves and factors for this purpose.

Leakage

Air flow passing from the air side to the gas side is called leakage. It is quantified in pounds per hour (kg/s) but is frequently expressed as a percentage of the gas inlet flow. Leakage is undesirable primarily because it represents fan power wasted in conveying air which bypasses the boiler combustion zone. Leakage can also reduce an air heater's thermal performance.

All air heaters leak. Recuperative units may begin operation with essentially zero leakage, but leakage occurs as time and thermal cycles accumulate. With regular maintenance, leakage can be kept below 3%.

Air heater leakage is inherent with the rotary regenerative design. There are two types of leakage, gap and carryover. Gap leakage occurs as higher pressure air passes to the lower pressure gas side through gaps between rotating and stationary parts. Its rate is given by the following general expression:

$$w_l = KA \ (2g_c \ \Delta P \rho)^{1/2} \qquad \textbf{(7)}$$

where

w_l = leakage flow rate, lb/h (kg/s)
K = discharge coefficient, dimensionless (generally 0.4 to 1.0)
A = flow area, ft^2 (m^2)
g_c = 32.17 lbm ft/lbf $s^2 \times (3600 \text{ s/h})^2$
= 4.17×10^8 lbm ft/lbf h^2 (1 kg m/N s^2)
ΔP = pressure differential across gap, lb/ft^2 (kg/m^2)
ρ = density of leaking air, lb/ft^3 (kg/m^3)

Carryover leakage is the air carried into the gas stream from each rotor (stator) heating surface compartment as the surface passes from the air stream to the gas stream. This leakage is directly proportional to the void volume of the rotor and the rotation speed.

Regenerative air heater design leakage ranges from 5 to 15% but increases over time as seals wear. During recent years effective automatic sealing systems, which nearly eliminate leakage rise due to seal wear, have been applied. These systems monitor and adjust rotating to stationary seals on-line.

Another source of air to gas flow, which appears as air heater leakage, is outside air infiltration into lower pressure gas streams. Infiltration may occur at casing cracks or holes, flue expansion joints and access doors or gaskets. This sometimes neglected source can be significant and difficult to detect if leaks occur under lagging and insulation.

Air heater leakage can be obtained directly as the difference between air or gas side inlet and outlet flows based on velocity measurements. However, because velocity measurements are difficult to obtain accurately in large duct cross-sections, air heater leakage is more accurately based on calculated gas weights using gas analysis, boiler efficiency and fuel analysis data. (See Chapter 9.) Approximate air heater leakage can be determined by the following formula based on gas inlet and outlet analysis (dry basis).

$$\% \text{ Leakage} = \frac{\% \ O_2 \text{ Leaving} - \% \ O_2 \text{ Entering}}{21 - \% \ O_2 \text{ Leaving}} \times 90 \qquad \textbf{(8)}$$

Test air heater leakage should be corrected for deviations from design cold end air to gas differential pressure and inlet air temperature before comparison to design leakage.

Pressure drop

In recuperative air heaters, gas or air side pressure drop arises from frictional resistance to flow, inlet and exit shock losses and losses in return bends between flow passes. In regenerative air heaters, the main cause is heating surface frictional flow resistance. In both cases, pressure drop is proportional to the square of the mass flow rate. Typical values at full load flows are 2 to 7 in. wg (0.5 to 1.7 kPa).

Air and gas side pressure drop values are the differences between terminal inlet and outlet static gauge pressures. Correction of measured pressure drops for deviations from design flows and temperatures is necessary before comparison to design values.

Operational concerns

There are several operating conditions and maintenance concerns common to most air heaters. These include corrosion, plugging and cleaning, erosion and fires. Air heaters used with high ash and/or high sulfur content fuels require more attention and maintenance than those firing clean fuels such as natural gas.

Corrosion

Air heaters used on units firing sulfur bearing fuels are subject to cold end corrosion of heating elements and nearby structures. In a boiler, a portion of the sulfur dioxide (SO_2) produced is converted to sulfur trioxide (SO_3) which combines with moisture to form sulfuric acid vapor. This vapor condenses on surfaces at temperatures below its dew point of 250 to 300F (120 to 150C). Because normal air heater cold end metal temperatures are frequently as low as 200F (93C), acid dew point corrosion potential exists. The obvious solution would be to operate at metal temperatures above the acid dew point but this results in unacceptable overall boiler heat losses. Most air heaters are designed to operate at MMTs somewhat below the acid dew point, where the efficiency gained more than balances the additional maintenance costs. B&W recommends limiting MMTs to the values in Figs. 20 and 21 when burning sulfur bearing fuels.

When fuel sulfur levels are high, or ambient temperatures or operating loads are low, MMTs may be unacceptably low. These situations dictate the use of active or passive cold end corrosion control methods. Active systems used to raise MMT include: 1) steam- or water-coil air heaters to preheat inlet air, 2) cold air bypass, in which a portion of the inlet air is ducted around the air heater, and 3) hot air recirculation, in which a portion of the hot outlet air is ducted to combustion air fan inlets.

Passive corrosion control methods incorporated in air heater design include: 1) thicker cold end materials, such as 11 or 14 gauge (3 or 2 mm) tubes and 18 gauge (1 mm) regenerative surface elements, 2) low or high alloy cold end surface materials which have at least twice the corrosion life of carbon steel, 3) nonmetallic coating, such as porcelain enamel, teflon, or epoxies on cold elements, 4) nonmetallic cold end surface materials such as extruded ceramic in regeneratives and borosilicate glass tubes in tubulars, and 5) tubular air heater cold end tube arrangements which maximize MMT by providing higher gas flow velocities.

Plugging and cleaning

Plugging is the fouling and eventual closing of heat transfer flow passages by gas-entrained ash and corrosion products. It can occur at the air heater hot end but is most common at the cold end where ash particles adhere to acid moistened surfaces. Plugging increases air heater pressure drop and can limit unit load when fan capacity is reached at less than full load. Air heater deposits are controlled and removed by sootblowing, cold end temperature control, surface design, off-line cleaning and furnace additives depending upon the particular application.

Erosion

Heat transfer surfaces and other air heater parts can suffer erosion damage through impact of high velocity, gas-entrained ash particles. Erosion usually occurs near gas inlets where velocities are highest. However, areas near seals in regenerative air heaters can also be damaged as ash is accelerated through seal gaps. The undesirable effects of erosion are structural weakening, loss of heat transfer surface area and perforation of components which can cause air to gas or infiltration leakage. Erosion rate is a function of velocity, gas stream ash loading, physical nature of ash particles and angle of particle impact. It is controlled by reducing velocities, removing erosive elements from the gas stream, or using sacrificial material.

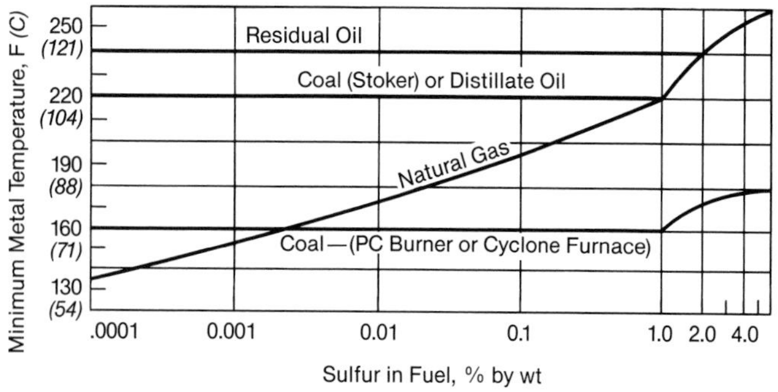

Fig. 20 Recuperative air heater cold end MMT limits when burning sulfur bearing fuels.

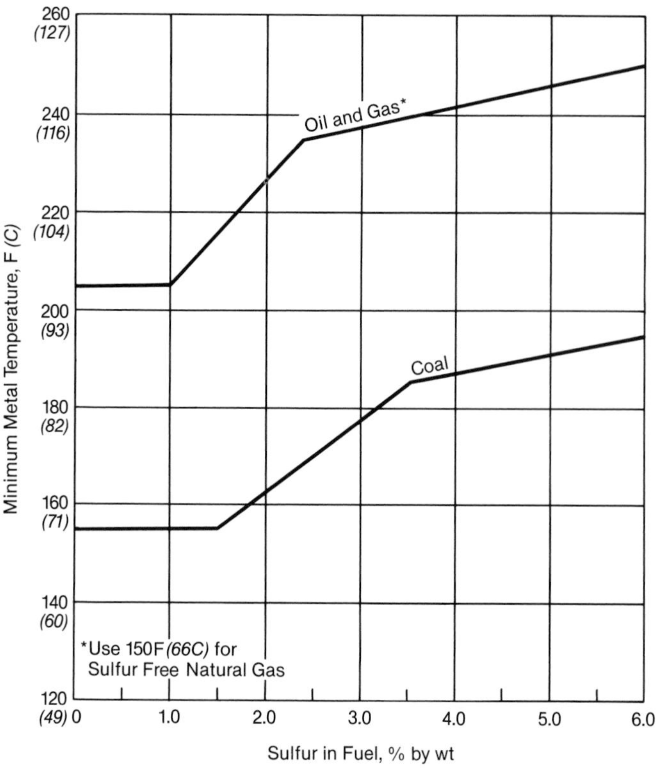

Fig. 21 Regenerative air heater cold end MMT limits when burning sulfur bearing fuels.

In the design stage, air heaters used with fuels containing highly erosive ash can be sized to limit gas inlet velocities to 50 ft/s (15 m/s). Inlet flues can also be designed to evenly distribute gas over the air heater inlet to eliminate local high velocity areas. Dust collectors, or strategically located screens and hoppers, may be used ahead of air heaters to remove some of the ash. In existing problem air heaters, flow distribution baffles may be installed to eliminate local high velocities, sacrificial materials such as abrasion resistant steel or ceramics may be placed over critical areas, or parts can be replaced with thicker materials for longer life.

Erosion in tubular air heaters frequently occurs within about 1 ft (0.3 m) of the gas inlet end due to turbulence as gas enters the tubes. Replaceable sacrificial sleeves may be installed in tube ends or egg crate-type flow straightening grids can be installed at tubular air heater inlets to reduce erosion.

Fires

Air heater fires are rare but do occur, particularly in regenerative units. They may be severe enough to completely destroy an air heater and are detected by thermocouples as well as special early warning systems. Fires usually start near the cold end, which can be fouled with unburned combustible materials. Most fires occur during startup as unburned fuel oil deposited on ash fouled heating surfaces is ignited. Leaking bearing lubrication equipment and heavy accumulations of flyash are also fire hazards. Fires can be avoided by maintaining a clean air heater and proper tuning of boiler firing equipment. Frequent sootblowing during startup and just before shutdown is a strongly recommended fire prevention practice.

Utility applications

Gas to air recuperative and regenerative air heaters are usually used in utility units, primarily to enhance unit efficiency. Small increments of increased efficiency in large units amount to substantial fuel savings. Utility units generally use multiple air heaters for plant arrangement convenience, type of firing and maximum unit availability.

Pulverized coal-fired units require two streams of hot combustion air, i.e., primary air supplied at high pressure to pulverizers and secondary air supplied at lower pressure directly to burners. Two basic air flow systems are used, hot primary air and cold primary air. Each system uses air heaters. In the hot primary air scheme, used for smaller units, about one third of the combustion air heated in a secondary heater is ducted to hot primary air fans, where it is boosted in pressure and passed to the pulverizers; the remaining two thirds is ducted to the burners. The cold primary air system uses separate air heaters supplied by separate primary and secondary (forced draft) fans. In some units, the primary and secondary air are heated in a single regenerative unit.

If separate regenerative primary and secondary air heaters are used, the primary air heaters, which operate at high air to gas pressure differentials, exhibit twice as much leakage as the secondary units. For this reason, low leakage recuperative air heaters may be used for primary air heating and regeneratives may be used for secondary air heating.

Recuperative and regenerative air heaters are used on oil- and gas-fired units. In general, regardless of fuel type, larger units use regenerative air heaters because of their smaller size and lower initial cost. However, for air to gas pressure differentials above 40 in. wg (10 kPa), in fluidized bed applications for example, recuperative air heaters are usually preferred.

Industrial applications

Industrial units fire a variety of fuels such as wood, municipal refuse, sewage sludge and industrial waste gases as well as coal, oil and natural gas. As a result, many air heater types are used. In the small units, tubular, plate, heat pipe and cast iron heaters are widely used. Fuels fired on stoker grates, such as bituminous coal, wood and refuse, do not require high air temperatures, therefore water- or steam-coil air heaters can be used.

Environmental heat exchanger application

For environmental reasons, emission of certain fossil-fired combustion products may be limited by law. (See Chapter 32.) Systems developed to limit emissions of two objectionable flue gas constituents, NO_x and SO_2, may require the use of specially modified heat exchangers.

NO_x removal

Noncombustion NO_x reduction systems introduce ammonia into flue gas streams. The ammonia (NH_3) reacts with NO_x thermally at high temperatures and in the presence of a catalyst at lower temperatures to form molecular nitrogen (N_2) and water. NH_3 also reacts with some of the SO_3 in the flue gas to form ammonia-sulfur compounds which condense at temperatures below 530F (277C). Because air heaters or gas coolers are usually downstream of NO_x reduction equipment, the heat exchanger surface is subject to rapid fouling and increased corrosion potential, particularly in the 510 to 340F (266 to 171C) range. Regenerative heat exchangers used in these situations are designed with special features to minimize plugging and corrosion. These features include a minimum number of surface layers to minimize plugging between layers; heavy gauge, low alloy, corrosion resistant or enamel coated surface material for long corrosion life; open profile heating surface design for ease of cleaning; and hot and cold end sootblowers. Heat exchangers with these features can operate reliably without off-line water washing for a year or more. As discussed in Chapter 34, some regenerative air heaters have been modified to simultaneously serve as selective catalytic NO_x reduction systems.

SO_2 reduction

When sulfur emission reduction is required, flue gas desulfurization (FGD) systems are frequently used. These systems remove SO_2 from the flue gas by reaction with injected compounds such as limestone. In most cases, the scrubbed flue gas exits the FGD system at a saturation temperature of 120 to 130F (49 to 54C) before entering the stack. In cases where acid dew point corrosion of flues and stack liners is a concern or increased gas buoyancy is needed to improve stack plume dispersal, gas exiting the FGD system is reheated to 180F (82C) or higher. Regenerative and recuperative heat exchangers, similar to those used for air heating, are used for this application.

Bibliography

Dubbel's Taschenbuch für den Maschinenbau, 11th Edition, 1958.

McAdams, W.H., *Heat Transmission*, 3rd ed., McGraw-Hill, New York, 1954.

Ledinegg, M., *Dampferzeugung*, Dampfkessel, Feuerungen Springer Verlag-Wien, 1952.

Mayer, E.H., and McCarver, G.M., *Economizer Technology for Utility and Industrial Boilers, Course for the Center for Professional Advancement*, E. Brunswick, New Jersey, 1991.

Two coal-fired boilers: one 685 MW pulverized coal and one 844 MW Cyclone.

Chapter 20
Fuel Ash Effects on Boiler Design and Operation

The effective utilization of fossil fuels for power generation depends to a great extent on the capability of the steam generating equipment to accommodate the inert residuals of combustion, commonly known as *ash*. The quantity and characteristics of the ash inherent to a particular fuel are major concerns to both the designer and the operator of the equipment.

With few exceptions, most commercial fuels contain sufficient ash to warrant specific design and operating considerations. The following focuses on these design and operating considerations, primarily as they relate to pulverized coal firing. Fuel ash characteristics relating to petroleum fuels are also discussed.

Ash dilutes the heating value of fuel, placing additional burdens on fuel storage, handling and preparation equipment. Extensive facilities are also needed to collect, remove and dispose of the ash. These material handling requirements represent significant costs in terms of equipment and real estate which are directly proportional to the amount of ash in the fuel. In the case of coal, ash quantities can be substantial. Consider, for example, a 650 MW utility steam generator firing a coal with a heating value of 10,000 Btu/lb (23,250 kJ/kg) containing 10% ash by weight. The unit would burn approximately 300 tons per hour (272 t_m/h) of coal, generating more than 700 tons per day (635 t_m/d) of ash.

In pulverized coal-fired boilers, most of this ash is carried out of the furnace by the gaseous products of combustion (flue gas). Abrasive ash particles suspended in the gas stream can cause erosion problems on convection pass heating surfaces. However, the most significant ash related problem is deposition. During the combustion process, the mineral matter that forms ash is released from the coal at temperatures in the range of 3000F (1649C), well above the melting temperature of most mineral matter compounds. Ash can be released in a molten fluid or sticky plastic state. A portion of the ash, which is not cooled quickly to a dry solid state, impacts on and adheres to the furnace walls and other heating surfaces. Because such large total quantities of ash are involved, even a small fraction of the total can seriously interfere with boiler operation. Accumulation of ash deposits on furnace walls impedes heat transfer, delaying cooling of the flue gas, and increasing the gas temperature leaving the furnace. Elevated temperatures at the furnace exit raise steam temperature and can extend deposition problems to pendant superheaters and other heat absorbing surfaces in the convection pass. In extreme cases, uncontrolled ash deposits can develop to the point where flow passages in tube banks are blocked, impeding gas flow and ultimately requiring unit shutdown for manual removal. Large deposits in the upper furnace or radiant superheater can become dislodged and fall, damaging pressure parts in the lower furnace. Under certain conditions, ash deposits can also cause fireside corrosion on tube surfaces.

Minimizing the potential for these ash related problems is a primary goal of both the designers and operators of coal-fired boilers. The extent to which coal ash characteristics affect boiler design is illustrated in Fig. 1 which compares the relative size of a gas-fired and coal-fired boiler. Both are sized for the same steam generating capacity and similar steam conditions. While the combustion characteristics of coal play a role in sizing the furnace, the deposition and erosion potential of the ash are the primary design considerations driving the overall size and arrangement.

Understanding coal ash characteristics is not only important to boiler designers, but also to operators interested in evaluating alternate coal sources for existing units. Many such studies are driven by increasingly restrictive emissions regulations.

The variability of ash behavior is one of the biggest problems. Although boilers are often designed to burn a wide range of coals satisfactorily, no unit can perform equally well with all types of coal. Design optimization with respect to performance, reliability, availability and cost requires that ash related problems be carefully identified and considered.

Ash content of coal

The ash content of coal varies over a wide range. This variation occurs not only in coals from different geographical areas or from different seams in the same region, but also from different parts of the same mine. These variations result primarily from the wide range of conditions that introduced foreign material during or following the formation of the coal. (See Chapter 8.) Ash content can also be influenced by extraneous mineral matter introduced during the mining operation. Before being sold, some commercial coals are cleaned or washed to remove a portion of what would be labeled ash in the laboratory. However, the

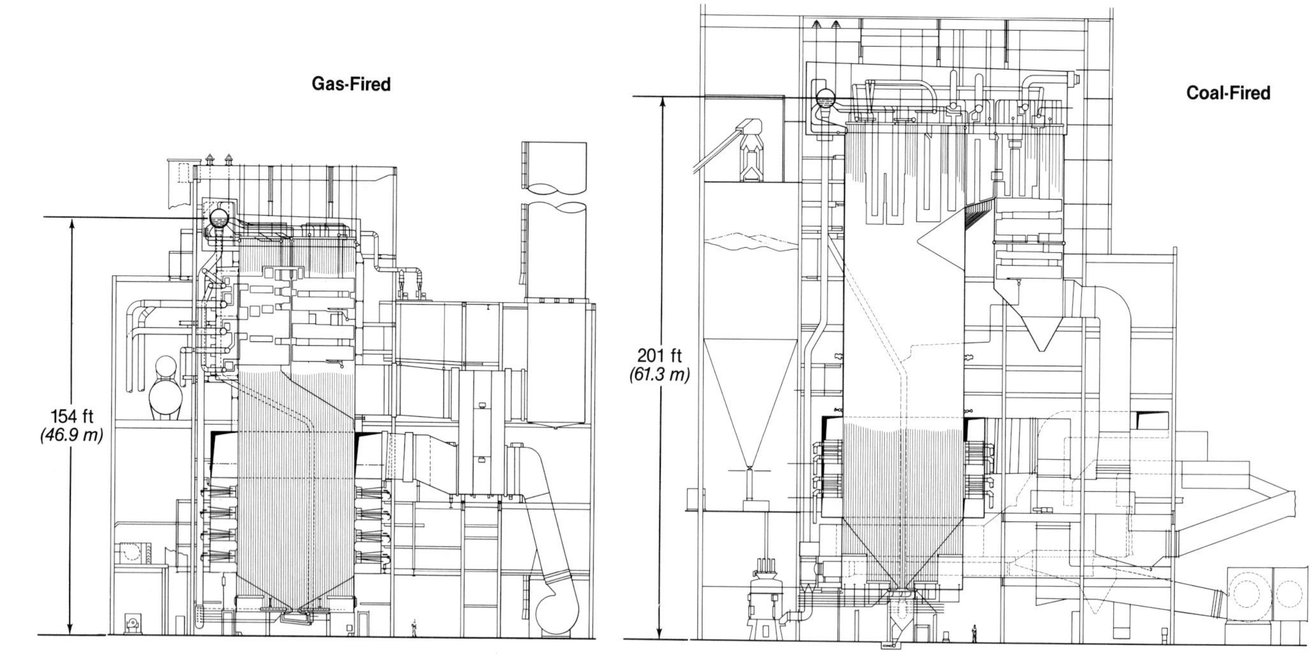

Fig. 1 Size comparison of gas-fired and coal-fired utility boilers.

ash content of significance to the user is the content at the point of use. The values noted below are on that basis.

Most of the coal used for power generation in the United States (U.S.) has an ash content between 6 and 20%. Low values of 3 to 4% in bituminous coals are rare and these coals find other commercial uses, particularly in the metallurgical field. On the other hand, some coals may have ash contents as high as 40%. Many high ash fuels can be successfully burned in utility (electric power generation) boilers. Their use has increased in areas where they offer an economic advantage.

Evaluation of ash content on a weight percentage basis alone does not take into account the heat input associated with the coal, which is also related to moisture content. It is common, for design and fuel evaluation purposes, to consider ash content on the basis of weight per unit of heat input, generally expressed as pounds of ash per million Btu. This factor is calculated as follows:

$$\frac{\text{Ash (\% by weight)}}{\text{HHV (Btu/lb)}} \times 10^4 = \text{lb ash}/10^6 \text{ Btu}$$

or **(1)**

$$\frac{\text{Ash (\% by weight)}}{\text{HHV (kJ/kg)}} \times 10^3 = \text{kg ash/MJ}$$

where HHV is the higher heating value of the fuel.

The relevance of this factor is illustrated in Table 1, which provides proximate analyses for three selected coals. Each coal has a moderate ash content of 9 to 10% by weight. However, on a heat input basis, ash quantities vary significantly. The lignite in this example would introduce almost three times as much ash compared to the high volatile bituminous coal at an equivalent heat input.

Furnace design for ash removal

Historically, two distinctly different types of furnace design were used to handle the ash from coal firing in large utility boilers. These are commonly referred to as the *dry ash* or *dry-bottom* furnace and the *slag-tap* or *wet-bottom* furnace.

All modern pulverized coal-fired boilers use the dry-bottom arrangement. The coal-fired boiler in Fig. 1 is typical of this design. In a dry-bottom unit most of the ash, typically 70 to 80%, is entrained in the flue gas and carried out of the furnace. This portion of the ash is commonly known as *flyash*. Some of the flyash is collected in hoppers arranged under the economizer and air heaters, where coarse particles drop out of suspension when gas flow direction changes. The finer ash particles remain in suspension and are carried out of the unit for collection by particulate control equipment. (See Chapter 33.) The remaining 20 to 30% of the ash that settles in the furnace, or is dislodged from the furnace walls, is collected in a hopper formed by the frontwall and rearwall tube panels at the bottom of the furnace. This *bottom ash* is discharged through a 3 to 4 ft (0.9 to 1.2 m) wide opening that spans the entire width of the hopper.

Slag-tap furnaces were originally developed to resolve ash deposition and removal problems when firing coals with low ash fusion temperatures in dry-bottom furnaces. These units are intentionally designed to maintain ash in a fluid state in the lower furnace. Molten ash is col-

Table 1
Proximate Analyses of Three Selected Coals — Ash Content as Weight Per Unit of Heat Input

Rank	High Volatile Bituminous	Subbituminous	Lignite
Moisture, %	3.1	23.8	45.9
Volatile matter, %	42.2	36.9	22.7
Fixed carbon, %	45.4	29.5	21.8
Ash, %	9.4	9.8	9.6
Heating value, Btu/lb	12,770	8,683	4,469
lb Ash/10^6 Btu	7.4	11.3	21.5

lected on the furnace walls and other surfaces in the lower furnace and drained continuously to openings called *slag-taps* in the furnace floor. Water tanks positioned beneath the slag-taps solidify the liquid ash for disposal.

Slag-tap furnaces have been used with both pulverized coal and Cyclone furnace firing systems. (See Chapter 14.) Application is limited to coals having ash viscosity characteristics which would ensure that ash fluidity could be maintained over a reasonable load range. Much of the coal ash research conducted by Babcock & Wilcox (B&W) concerning the viscosity-temperature relationship of coal ash was initially directed at defining coal ash suitability limits for wet-bottom and Cyclone furnace applications. A minimum coal ash content was also specified to ensure sufficient ash quantities to maintain the required slag coating. One benefit of wet-bottom firing was a significant reduction in flyash quantity. In pulverized coal wet-bottom applications, as much as 50% of the total ash was collected in the furnace. Units equipped with Cyclone furnaces could retain up to 80% of the ash in the furnace.

The application of slag-tap units for pulverized coal firing began to decline in the late 1940s, primarily due to design improvements in dry-bottom units that minimized ash deposition problems. Slag-tap units equipped with Cyclone furnaces continued to be applied until the early 1970s when the federal Clean Air Act mandated control of nitrogen oxides (NO_x) emissions. The high furnace temperatures required for wet-bottom operation were highly conducive to NO_x formation.

Ash deposition

Regardless of the firing method, when coal is burned, a relatively small portion of the ash will cause deposition problems. Ash passing through the boiler is subject to various chemical reactions and physical forces which lead to deposition on heat absorbing surface. The process of deposition and the structure of deposits are variable due to a number of factors. Particle composition, particle size and shape, particle and surface temperatures, gas velocity, flow pattern and other factors influence the extent and nature of ash deposition.

Due primarily to the differences in deposition mechanisms involved, two general types of high temperature ash deposition have been defined as *slagging* and *fouling*.

Slagging is the formation of molten, partially fused or resolidified deposits on furnace walls and other surfaces exposed to radiant heat. Slagging can also extend into convective surface if gas temperatures are not sufficiently reduced.

Most ash particles melt or soften at combustion temperatures. The time-temperature history or cooling rate of the particle determines its physical state (solid, plastic or liquid) at a given location in the furnace. Generally, in order to adhere to a clean surface and form a deposit, the particle must have a viscosity low enough to wet the surface.

Slag deposits seldom form on clean tube surfaces. A conditioning period is required before significant deposition occurs. Assuming there is no direct flame impingement, as ash particles approach a clean tube, most tend to be resolidified due to the relatively lower temperature at the tube surface. The particles fracture on impact and partially disperse back into the gas stream. Over a period of time, however, a base deposit begins to form on the tube. The base deposit may be initiated by the settling of fine ash particles or the gradual accumulation of particles with very low melting point constituents. As the base deposit thickens, the temperature at its outside face increases significantly above the tube surface temperature. Eventually, the melting point of more of the ash constituents is exceeded and the deposit surface becomes molten. The process then becomes self-accelerating with the plastic slag trapping essentially all of the impinging ash particles. Ultimately, the deposit thickness reaches an equilibrium state as the slag begins to flow, or the deposit becomes so heavy that it falls away from the tubes. Depending on the strength and physical characteristics of the deposit, steam or air sootblowers (see Chapter 23) may be able to control or remove most of the deposit. However, the base deposit generally remains attached to the tube, allowing subsequent deposits to accumulate much more rapidly.

Fouling is defined as the formation of high temperature bonded deposits on convection heat absorbing surfaces, such as superheaters and reheaters, that are not exposed to radiant heat. In general, fouling is caused by the vaporization of volatile inorganic elements in the coal during combustion. As heat is absorbed and temperatures are lowered in the convective section of the boiler, compounds formed by these elements condense on ash particles and heating surface, forming a glue which initiates deposition.

Areas in which slagging and fouling can occur are shown in Fig. 2. Figs. 3 and 4 show heavily slagged and fouled surfaces. The characteristics of coal ash and their influence on slagging and fouling are discussed in the following sections.

Characteristics of coal ash

Sources of coal ash

Mineral matter is always present in coal and forms ash when the coal is burned. This mineral matter is usually classified as either inherent or extraneous. (See Chapter 8.) Inherent mineral matter is organically combined with the coal. This portion came from the chemical elements existing in the vegetation from which the coal was formed and from elements chemically bonded to the coal during its formation. Extraneous mineral matter is material that is foreign to the organic structure of the coal. This includes airborne and waterborne material that settled into the coal deposit during or after formation. It usually consists of mineral forms associated with clay, slate, shale, sandstone or limestone and includes pieces ranging from microscopic size to thick layers. Other extraneous material may be introduced through the mining process.

Mineralogical composition

There are no standardized methods that are used routinely for determining the specific mineral constituents of coal. Mineralogical analysis requires the use of a low temperature ashing technique to separate the mineral matter

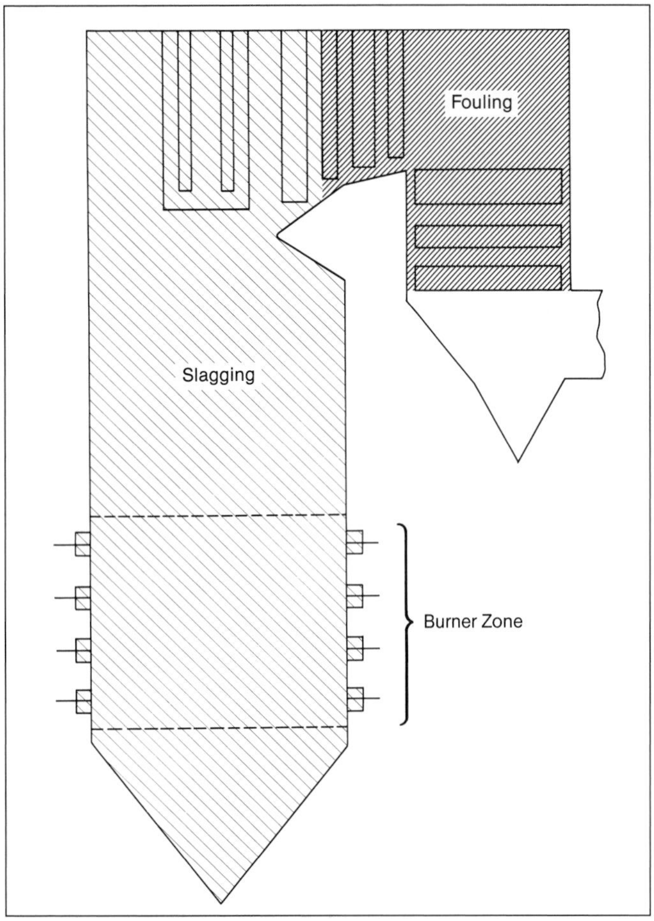

Fig. 2 Deposition zones in a coal-fired boiler.

from the organic portion of the coal. Standard high temperature ashing procedures would significantly alter the mineral forms. However, a number of researchers, using a variety of low temperature ashing methods and sophisticated analytical techniques, have identified an enormous variety of mineral species in coal, encompassing the entire spectrum of major mineral forms found in the earth's crust. Most of these minerals fall into one of several groups: clay minerals (aluminosilicates), sulfides/sulfates, carbonates, chlorides, silica/silicates and oxides. Some of the more common minerals in these groups are shown in Table 2.

Chemical composition

Because both quantitative and qualitative evaluation of mineral matter forms are extremely difficult, relatively simple chemical analyses are commonly used to determine the percentages of the major elements in the ash. Elemental ash analysis is performed on a coal ash sample produced in accordance with the American Society for Testing and Materials (ASTM) D 3174 ashing procedure. Pulverized coal is burned in a furnace with an oxidizing atmosphere at 1290 to 1370F (699 to 743C). The elements present in the ash are quantitatively measured using a combination of emission spectroscopy and flame photometry and are reported as weight percents of their oxides. Coal ash is consistently found to be composed mainly of silicon, aluminum, iron and calcium with smaller amounts of magnesium, titanium, sodium and potassium. The elemental analysis also identifies phosphorus as P_2O_5 and sulfur as sulfur trioxide (SO_3). Phosphorus is usually present in very small quantities and is sometimes omitted. Sulfur is reported as SO_3 because it is normally present as the sulfate form of one of the metals.

Percentages of the individual elements vary over a wide range for different coals; however, characteristic differences are evident between the older, high rank coals common in the Eastern U.S. and the younger, low rank Western coals. Bituminous coals typically have higher levels of silica, aluminum and iron, while the lower rank subbituminous coals and lignites generally have higher levels of the alkaline earth metals, calcium and magnesium, and the alkali metal sodium. These trends are evident in the ash analyses shown in Table 3.

Although the ash constituents are reported as oxides, they actually occur in the ash predominately as a mixture of silicates, oxides and sulfates, with smaller quantities of other compounds. The silicates originate mainly from quartz and the clay minerals which contribute silicon, aluminum, sodium and much of the potassium. A principal source of iron oxide is pyrite (FeS_2) which is oxidized to form Fe_2O_3 and sulfur oxides. Part of the organic and pyritic sulfur that is oxidized combines with calcium and magnesium to form sulfates. Calcium and magnesium oxides result from the loss of carbon dioxide from carbonate minerals such as calcite ($CaCO_3$) and dolomite [$(Ca,Mg)CO_3$)]. In low rank coals, a major por-

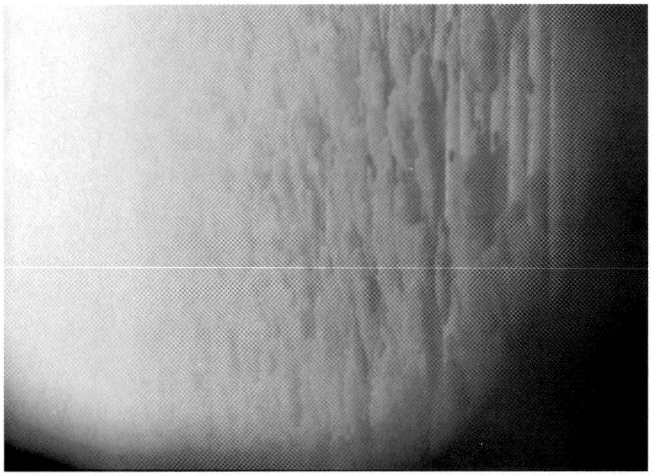

Fig. 3 Heavily slagged surface.

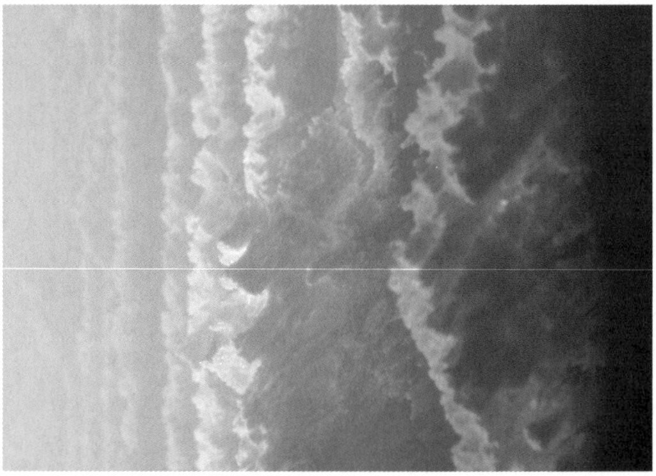

Fig. 4 Heavily fouled surface.

tion of the sodium, calcium and magnesium oxides can originate from organically bound elements in the coal.

Laboratory ash is prepared from a coal sample in a controlled atmosphere at controlled temperatures in order to provide a reproducible and uniform ash. The actual ashing process during combustion in a pulverized coal-fired furnace is a much more complex process. In a boiler furnace, pulverized coal is burned in suspension as discrete particles. If all of the mineral matter were evenly distributed through the coal, the composition of each resulting ash particle would be the same as the bulk ash composition determined by the analysis of ASTM ash. A coal with no extraneous mineral matter might approach this hypothetical case, because organically combined inherent material would be expected to be evenly distributed. In reality, however, all coals contain non-uniformly distributed extraneous mineral matter in some of the wide variety of mineral forms shown on Table 2. When the coal is pulverized, some of the particles will be mostly coal with only inherent mineral matter, some will be pure mineral matter and others will be combinations of both. Because the coal particles are burned discretely in suspension, the composition of an individual ash particle will depend on the specific mineral form or forms that were included in the coal particle. As a result, individual particle composition can vary significantly from the bulk ash composition.

During combustion, ash particles are exposed to temperatures as high as 3000F (1649C) and a variety of heating and cooling rates. The atmosphere in the burner zone can range from highly oxidizing to highly reducing. Depending on the composition of the particular particle, mineral forms in the ash can react with each other, with the organic and inorganic constituents of the coal, and with gaseous elements, such as sulfur dioxide (SO_2), in the flue gas. The compounds that are ultimately formed by these interactions are the materials that cause deposition problems. The compounds can have a wide variety of melting temperatures and viscosity-temperature characteristics. Some compounds combine to form eutectic mixtures that have melting temperatures lower than either of the original compounds. Particles that melt at lower temperatures and stay sticky long enough to reach a furnace wall become slag deposits. Volatile compounds that vaporize in the furnace tend to condense on and foul cooler convective heating surfaces.

Table 2
Common Minerals Found in Coal

Clay minerals:	
Montmorillonite	$Al_2Si_4O_{10}(OH)_2 \cdot H_2O$
Illite	$KAl_2(AlSi_3O_{10})(OH_2)$
Kaolinite	$Al_4Si_4O_{10}(OH)_8$
Sulfide minerals:	
Pyrite	FeS_2
Marcasite	FeS_2
Sulfate minerals:	
Gypsum	$CaSO_4 \cdot 2H_2O$
Anhydrite	$CaSO_4$
Jarosite	$(Na,K)Fe_3(SO_4)_2(OH)_6$
Carbonate minerals:	
Calcite	$CaCO_3$
Dolomite	$(Ca,Mg)CO_3$
Siderite	$FeCO_3$
Ankerite	$(Ca,Fe,Mg)CO_3$
Chloride minerals:	
Halite	NaCl
Sylvite	KCl
Silicate minerals:	
Quartz	SiO_2
Albite	$NaAlSi_3O_8$
Orthoclase	$KAlSi_3O_8$
Oxide minerals:	
Hematite	Fe_2O_3
Magnetite	Fe_3O_4
Rutile	TiO_2

Elemental ash analyses do not directly identify the compounds that cause deposition, or directly identify the mechanisms of deposit formation. Despite these limitations, no other data pertaining to coal ash composition are as widely available as the chemical analyses of ASTM ash. A large part of the coal ash research that has been conducted over the last sixty years has been directed at correlating analysis data and other characteristics of ASTM ash to observed ash behavior both in full scale boilers and test facilities that closely simulate full scale conditions. Evaluation methods have been developed based on these correlations to characterize ash behavior and predict deposition potential.

Ash fusibility

The measurement of ash fusibility temperatures is by far the most widely used method for predicting ash behavior at elevated temperatures. The preferred procedure in the U.S. is outlined in ASTM Standard D 1857, *Fusibility of Coal and Coke Ash*. An ash sample is prepared by burning coal under oxidizing conditions at temperatures of 1470 to 1650F (799 to 899C). The ash is pressed in a mold to form a triangular pyramid (cone) 0.75 in. (19 mm) in height with a 0.25 in. (6.35 mm) triangular base. The cone is heated in a furnace at a controlled rate to provide a temperature increase of 15F (8C) per minute. The atmosphere in the furnace is regulated to provide either oxidizing or reducing conditions. As the sample is heated, the temperatures at which the cone fuses and deforms to specific shapes, as shown in Fig. 5, are recorded. Four deformation temperatures are reported as follows:

1. *Initial deformation temperature* (IT or ID) — the temperature at which the tip of the pyramid begins to fuse or show signs of deformation.
2. *Softening temperature* (ST) — the temperature at which the sample has deformed to a spherical shape where the height of the cone is equal to the width at the base (H = W). The softening temperature is commonly referred to as the fusion temperature.
3. *Hemispherical temperature* (HT) — the temperature at which the cone has fused down to a hemispherical lump and the height equals one half the width of the base (H = 1/2 W).
4. *Fluid temperature* (FT) — the temperature at which the ash cone has melted to a nearly flat layer with a maximum height of 0.0625 in. (1.59 mm).

Table 3
Ash Content and Ash Fusion Temperatures of Some U.S. Coals and Lignite

Rank:	Low Volatile Bituminous	High Volatile Bituminous				Sub-bituminous	Lignite
Seam	Pocahontas No. 3	No. 9	No.6	Pittsburgh		Antelope	
Location	West Virginia	Ohio	Illinois	West Virginia	Utah	Wyoming	Texas
Ash, dry basis, %	12.3	14.1	17.4	10.9	17.1	6.6	12.8
Sulfur, dry basis, %	0.7	3.3	4.2	3.5	0.8	0.4	1.1
Analysis of ash, % by wt							
SiO_2	60.0	47.3	47.5	37.6	61.1	28.6	41.8
Al_2O_3	30.0	23.0	17.9	20.1	21.6	11.7	13.6
TiO_2	1.6	1.0	0.8	0.8	1.1	0.9	1.5
Fe_2O_3	4.0	22.8	20.1	29.3	4.6	6.9	6.6
CaO	0.6	1.3	5.8	4.3	4.6	27.4	17.6
MgO	0.6	0.9	1.0	1.3	1.0	4.5	2.5
Na_2O	0.5	0.3	0.4	0.8	1.0	2.7	0.6
K_2O	1.5	2.0	1.8	1.6	1.2	0.5	0.1
SO_3	1.1	1.2	4.6	4.0	2.9	14.2	14.6
P_2O_5	0.1	0.2	0.1	0.2	0.4	2.3	0.1
Ash fusibility							
Initial deformation temp, F							
Reducing	2900 +	2030	2000	2030	2180	2280	1975
Oxidizing	2900 +	2420	2300	2265	2240	2275	2070
Softening temp, F							
Reducing		2450	2160	2175	2215	2290	2130
Oxidizing		2605	2430	2385	2300	2285	2190
Hemispherical temp, F							
Reducing		2480	2180	2225	2245	2295	2150
Oxidizing		2620	2450	2450	2325	2290	2210
Fluid temp, F							
Reducing		2620	2320	2370	2330	2315	2240
Oxidizing		2670	2610	2540	2410	2300	2290

The determination of ash fusion temperatures is strictly an empirical procedure, developed in standardized form, which can be duplicated with some degree of accuracy. Strict observance of test conditions is required to assure reproducible results. ASTM specified tolerances on reproducibility of the individual temperature measurements range from 100 to 150F (56 to 83C) when the test is performed by different operators and apparatus.

An earlier version of the ASTM D 1857 procedure specified the use of only a reducing atmosphere and had loosely defined criteria for identifying the softening and fluid points. When the atmosphere is not specified, it is generally assumed to be reducing. Reported softening temperatures are assumed to be the ST (H = W) point unless otherwise specified. Methods for determining fusibility of ash used by other countries are similar to the ASTM procedure but results may vary considerably due to differences in procedures or the definition of terms.

The gradual deformation of the ash cone is generally considered to result from differences in melting characteristics of the various ash constituents. As the temperature of the sample is increased, compounds with the lowest melting temperatures begin to melt, causing the initial deformation. As the temperature continues to increase, more of the compounds melt and the degree of deformation proceeds to the softening and hemispherical stages. The process continues until the temperature is higher than the melting point of most of the ash constituents and the fluid stage is reached.

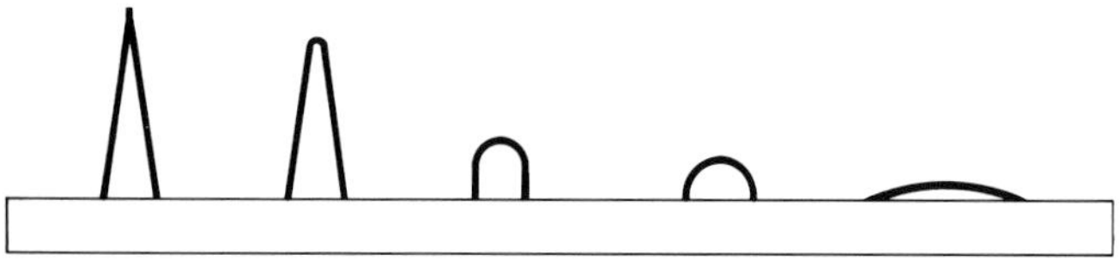

Fig. 5 Specific shapes as ash fuses and deforms with temperature.

Fusibility testing was originally developed to evaluate the clinkering (agglomerating) tendency of coal ash produced by combustion on a grate. In several respects, the test method is a somewhat better simulation of stoker firing than suspension burning of pulverized coal. During the fusion test, at a heating rate of 15F (8C) per minute, the transition from the IT to the FT stage may take up to two hours or more for a high fusion ash. Rather than slow heating and gradual melting of the ash, the process in a pulverized coal furnace is essentially reversed. Ash particles are rapidly heated, and then cooled at a relatively slow rate, as they pass through the furnace. During combustion, coal particles are heated almost instantaneously to temperatures ranging up to 3000F (1649C). As heat is removed from the flue gas, the ash is cooled over a period of less than two seconds to temperatures around 1900 to 2200F (1038 to 1204C) at the furnace exit.

In practical terms, for dry-bottom furnaces, fusion temperatures provide an indication of the temperature range over which portions of the ash will be in a molten fluid or semi-molten, plastic state. High fusion temperatures indicate that ash released in the furnace will cool quickly to a nonsticky state resulting in minimal potential for slagging. Conversely, low fusion temperatures indicate that ash will remain in a molten or plastic state longer, exposing more of the furnace surface to potential deposition.

When temperatures in the furnace are below the measured initial deformation temperature, the majority of the ash particles are expected to be in a dry solid state. In this form, particles impacting on heating surface will bounce off and be re-entrained in the gas stream, or, at worst, settle on the surface as a dusty deposit which can be readily removed by sootblowers. At temperatures above the IT, the ash becomes increasingly more plastic in nature and impacting particles have a greater potential to stick to heating surface.

Fusibility temperatures also provide an indication of deposit characteristics as they relate to control and cleanability. When the temperature at a deposit surface is at or above the fluid temperature of the ash, slag will tend to flow or drip from the surface. While fluid slag can not be controlled with sootblowers, the deposits tend to be self-limiting in thickness and do not interfere significantly with heat transfer effectiveness. However, if the deposit surface temperature is in the plastic range, between the initial deformation and hemispherical temperatures, the slag will be too viscous to flow and will continue to build in thickness. Wide IT to HT differentials can result in deposits that build quickly to large proportions and are difficult to control, because sootblowers can be ineffective in penetrating the plastic shell that forms on the deposit surface.

In practice, very high and very low fusion values are relatively easy to interpret as being troublesome or non-troublesome with respect to slagging. Unfortunately, however, most coals fall in an intermediate range where evaluations can be much more difficult. Fusion temperatures have their most valid significance when used on a comparative basis against corresponding data from other fuels of known full scale performance. Even comparisons can be misleading, however, when differences in data are within the range of reproducibility of the test. Actual ash viscosity measurements (described later) provide a much more accurate and less subjective definition of the viscosity/temperature relationship and are considered by B&W to provide a better assessment of slagging potential.

Influence of ash elements

Ash classification

Coal ash is classified into two categories based on its chemical composition. *Lignitic* ash is defined as having more (CaO + MgO) than Fe_2O_3. *Bituminous* ash is defined as having more Fe_2O_3 than the sum of CaO and MgO. Bituminous ash is generally characteristic of higher rank coals from the Eastern U.S. Lower rank Western coals typically have lignitic ash. As a result, bituminous ash is sometimes referred to as Eastern ash and lignitic ash is sometimes referred to as Western ash. However, ash classification is not specific to ASTM rank or geographical origin. In rare cases, lignites and subbituminous coals can have bituminous ash and bituminous coals can have lignitic ash. For example, the Utah coal shown in Table 3 is classified a bituminous, but has lignitic ash.

Effect of iron

Iron has a dominating influence on the slagging characteristics of coals with bituminous type ash. As shown in Table 2, iron can be present in coal in several mineral forms. These include pyrite (FeS_2), siderite ($FeCO_3$), hematite (Fe_2O_3), magnetite (Fe_3O_4) and ankerite [(Ca,Fe,Mg)CO_3]. Pyrite is the major form of iron in most Eastern coals. In areas of the furnace where there is sufficient oxygen, pyrite is converted to Fe_2O_3 and SO_2. If the local atmosphere is reducing, however, pyrrhotite (FeS) is formed along with the lesser-oxidized iron forms such as FeO and metallic iron, Fe. The reduced forms have significantly lower melting temperatures than the oxidized forms. When completely oxidized to Fe_2O_3, iron tends to raise all four values of ash fusion temperatures: initial deformation, softening, hemispherical and fluid. In the lesser oxidized form (FeO) it tends to lower all of these values. The effect of iron in each of these forms is indicated in Fig. 6, plotted for a large number of ash samples from U.S. coals. The data show that as the amount of iron in the ash increases, there is a greater difference in ash fusibility between oxidizing and reducing conditions.

These effects may be negligible with coal ash containing small amounts of iron. Coals with lignitic ash generally have small amounts of iron and the ash fusion temperatures are affected very little by the state of iron oxidation. In fact, lignitic ash containing high levels of calcium and magnesium may have ash fusion temperatures that are lower on an oxidizing basis than a reducing basis. The ash analysis and fusion temperatures shown for the subbituminous coal in Table 3 illustrate this effect.

Base to acid ratio

The constituents of coal ash can be classified as either basic or acidic. The basic constituents are iron, the alkaline earth metals calcium and magnesium, and the alkali metals sodium and potassium. Acidic constituents are silicon, aluminum and titanium. Bases and acids tend to combine to form compounds with lower melting temperatures. Experience has shown that the relative proportions of basic and acidic constituents provide an indication of the melting behavior and viscosity characteristics of coal ash.

The elemental analysis is used to calculate the percent base, percent acid and the base to acid ratio as follows:

$$\text{Percent base} = \frac{(Fe_2O_3 + CaO + MgO + Na_2O + K_2O)\,100}{SiO_2 + Al_2O_3 + TiO_2 + Fe_2O_3 + CaO + MgO + Na_2O + K_2O} \quad (2)$$

$$\text{Percent acid} = \frac{(SiO_2 + Al_2O_3 + TiO_2)\,100}{SiO_2 + Al_2O_3 + TiO_2 + Fe_2O_3 + CaO + MgO + Na_2O + K_2O} \quad (3)$$

$$\text{Base / acid ratio} = \frac{Fe_2O_3 + CaO + MgO + Na_2O + K_2O}{SiO_2 + Al_2O_3 + TiO_2} \quad (4)$$

The range of base to acid ratio extends from approximately 0.1 for highly acidic ash to 9 for ash that is high in base content.

Ash that is either highly acidic or highly basic generally has high ash fusion and melting temperatures. However, the presence of basic constituents in an acidic ash tends to flux or reduce the melting temperature and vis-

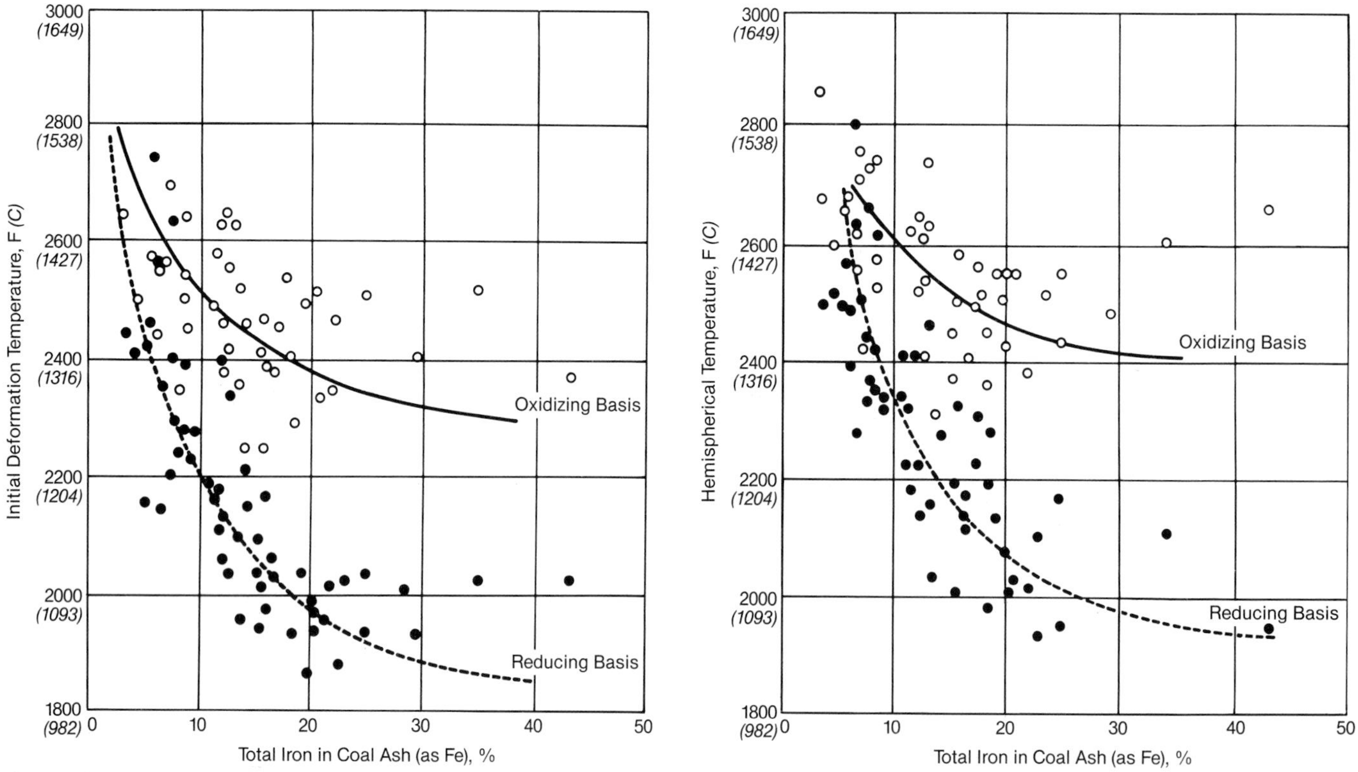

Fig. 6 Influence of iron on coal ash fusion temperatures.

cosity of the mixture. Conversely, the melting temperature and viscosity of a basic ash are reduced by relative proportions of acidic constituents. When the percent base and percent acid are nearly equal, fusion temperatures and ash viscosity tend to be reduced to minimum levels. The general trend is shown in Fig. 7. Minimum fusion temperatures typically occur at approximately 40 to 45% base which equates to base to acid ratios in the range of 0.7 to 0.8. Ratios in the range of 0.5 to 1.2 are generally considered to indicate high slagging potential.

The base to acid ratio considers all of the basic and acidic constituents to have equal effects on ash melting characteristics. However, research has shown that the various acids and bases have different fluxing strengths which must also be considered.

Studies conducted by B&W on the relationship of ash composition to ash viscosity have provided additional factors which improve the simple base to acid relationship. Ash viscosity is an important criterion for determining the suitability of a coal ash for use in a slag-tap furnace. Experience has shown that slag will flow readily at or below a viscosity of 250 poise. The temperature at which this viscosity occurs is called the T_{250} temperature of the ash. The preferred maximum T_{250} for wet-bottom applications is 2450F (1343C). Trends in T_{250} temperatures have been shown to correlate with ash fusion temperatures. Low T_{250} temperatures indicate low fusion temperatures and increased slagging potential.

Ash viscosity can be measured directly in a high temperature viscometer. B&W's facility is described later in this chapter. Because viscosity measurements require a considerable amount of coal ash that may not be readily available and are costly and time consuming, methods were developed to determine viscosity from chemical analysis of the coal ash. Based on a large number of direct viscosity measurements of bituminous and lignitic ash samples, T_{250} temperatures were related to ash composition as shown in Figs. 8 and 9. Fig. 8 is for bituminous ash and lignitic ash with an acidic content above 60%. At base to acid ratios less than 0.3 the silicon (SiO_2)/aluminum (Al_2O_3) ratio is taken into account. Silicon and aluminum are both acidic constituents; however, higher percentages of silicon tend to raise the T_{250} and the melting temperature.

Fig. 9 is for lignitic ash with an acidic content less than 60%. T_{250} is a function of both the percent base and the dolomite percentage which is defined as:

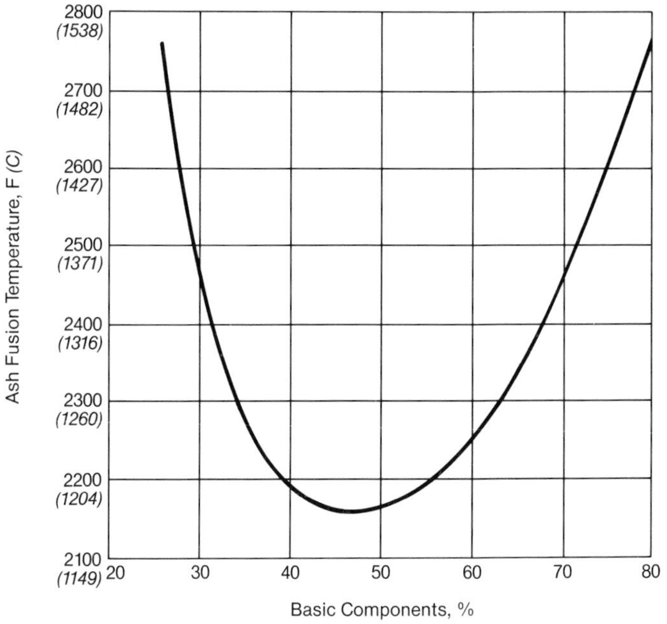

Fig. 7 Fusion temperatures and viscosities versus acidic constituents.

$$\text{Dolomite percentage} = \frac{(CaO + MgO)100}{Fe_2O_3 + CaO + MgO + Na_2O + K_2O} \quad \textbf{(5)}$$

At a given percent base, higher dolomite percentages increase the T_{250} temperature, indicating that calcium and magnesium tend to raise ash viscosity and fusion temperature. Increasing amounts of the other base constituents (iron, sodium and potassium) tend to lower the T_{250} temperature.

Taken together, these trends indicate higher melting temperatures and higher viscosities at a given temperature for ash that is predominately composed of either silicon and aluminum or calcium and magnesium. Lower melting temperatures result from intermediate mixtures of these elements. However, in all combinations, iron, sodium and potassium act to flux the ash and increase the slagging potential.

As previously noted, the fluxing strength of iron is related to its state of oxidation. Metallic iron (Fe) and ferrous iron (FeO) are stronger fluxes than Fe_2O_3 and tend to reduce fusion temperatures and slag viscosity at a given temperature. The degree of iron oxidation is normally expressed as the ferric percentage where:

$$\text{Ferric percentage} = \frac{Fe_2O_3 \times 100}{Fe_2O_3 + 1.11\,FeO + 1.43\,Fe} \quad \textbf{(6)}$$

The effect of ferric percentage on slag viscosity for a typical bituminous ash is shown in Fig. 10. Note that the T_{250} temperature can vary over a wide range depending on the degree of iron oxidation. Experience has shown that slag from boiler furnaces operating under normal conditions with 15 to 20% excess air has a ferric percentage of approximately 20%. The curves in Fig. 8 are based on this value.

Influence of alkalies on fouling

The alkali metals, sodium and potassium, have long been associated with the fouling tendencies of coal ash. Volatile forms of these elements are vaporized in the furnace at combustion temperatures. Subsequent reactions with sulfur in the flue gas and other elements in the ash form compounds that contribute to the formation of bonded deposits on convection heating surface.

Research conducted by B&W dating back to the 1950s identified a relationship between the total alkali content in bituminous coals and fouling potential. The specific laboratory procedure developed to establish this relationship, called the sintering strength test, is described in detail later in this chapter. Basically, the test involves measuring the compressive strength of flyash pellets heated in air for a period of time at temperatures of 1500 to 1800F (816 to 982C). The application of this method, combined with observations of fouling conditions in operating boilers, showed that high fouling coals produced flyash with high sintered strength. Conversely, low strength flyash was associated with low fouling coals. Correlation of standard ASTM ash analysis data with the sintering test results indicated a significant relationship (Fig. 11) between total alkali content (Na_2O and K_2O, expressed as equivalent total Na_2O) and flyash sintered strength. These correlations formed the basis for the first *fouling index* for bituminous coals which used the total alkali content in the coal to predict fouling potential.

Because ASTM ash produced in the laboratory could not be expected to represent the physical and chemical properties of flyash produced by full scale combustion, sintering strength testing required actual flyash samples

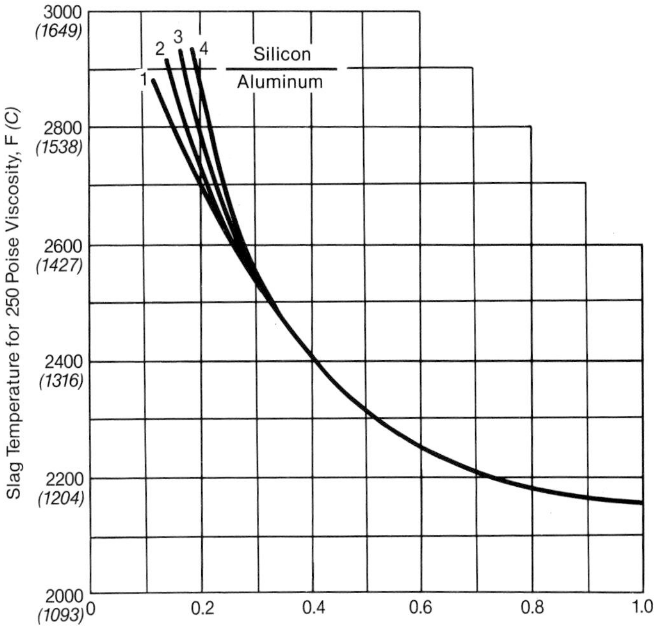

Fig. 8 Plot of temperature for 250 poise viscosity versus base to acid ratio — based on ferric percentage of 20.

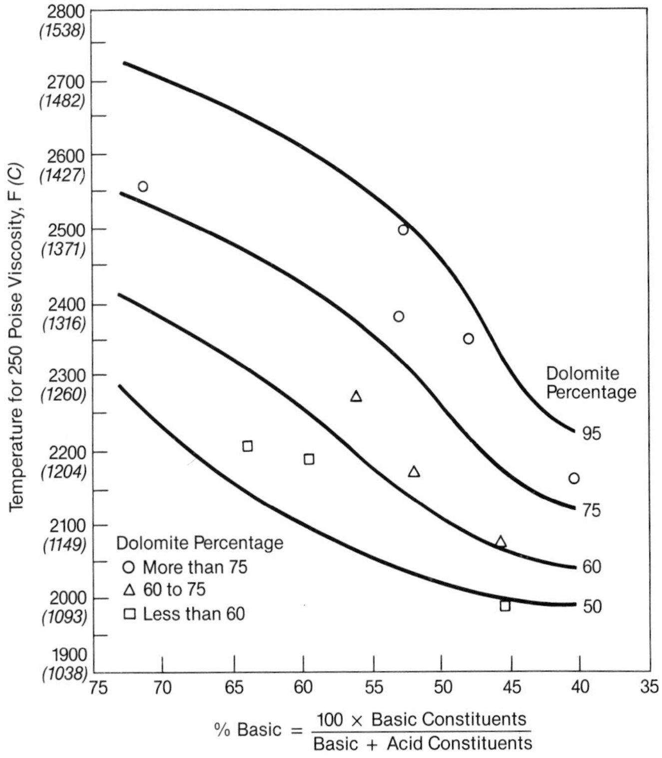

Fig. 9 Basic content and dolomite percentage of ash versus temperature for 250 poise viscosity.

aspirated from the flue gas in operating boilers. This meant full scale tests under steady-state conditions with a consistent coal supply, which became increasingly more difficult as unit size increased. In order to improve the efficiency and accuracy of obtaining data, a small laboratory ashing furnace (LAF) was constructed to burn pulverized coal at controlled conditions similar to those in a commercial boiler.

Subsequent tests on flyash produced in the LAF from a wide variety of bituminous coals demonstrated that sodium was the most important single factor affecting ash fouling. Potassium, which had been included in the previous alkali fouling indices, was found to make no significant contribution to sintering strength. Additionally, it was found that water soluble sodium, which was related to the more readily vaporized forms of sodium, had a major effect on sintered strength. This result was obtained by washing coals with hot condensate in the laboratory to remove the water soluble sodium. The washed coals were ashed in the LAF and sintered at various temperatures. Results for a high fouling Illinois coal are shown in Table 4. Water washing decreased the sodium content in the ash by approximately 70%, while the potassium content, which was initially higher than the sodium content, decreased by only 4%. Removing the soluble sodium resulted in a reduction in sintering strength at 1700F (927C) from 17,300 psi (119.3 MPa) for the raw coal to 550 psi (3.8 MPa) for the washed coal. Because the coal had a high chlorine content, it was concluded that most of the volatile sodium was probably in the form of NaCl. The insoluble potassium was likely associated with clay minerals or feldspar which would not readily decompose and vaporize during combustion.

The relationship of sintering strength to the percentage of soluble sodium in the ash was also found to be a function of the base to acid ratio, as shown in Fig. 12.

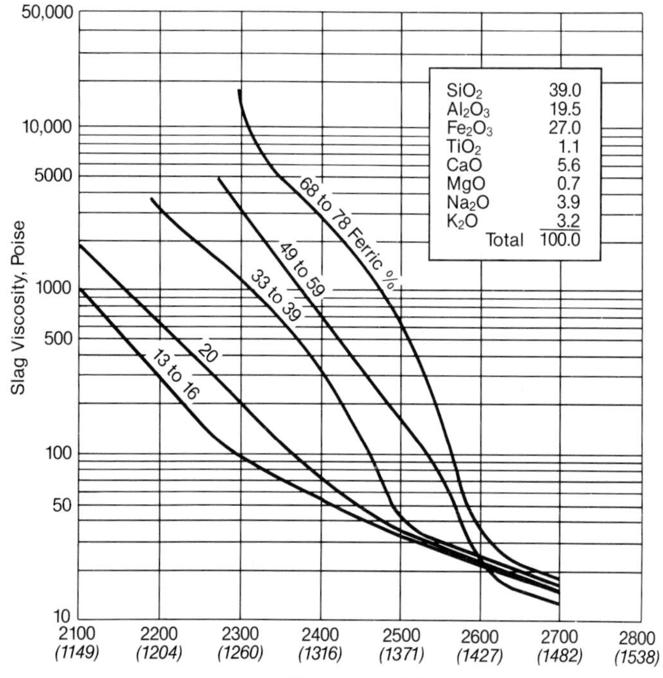

Fig. 10 Viscosity-temperature plots of a typical slag showing effect of ferric percentage.

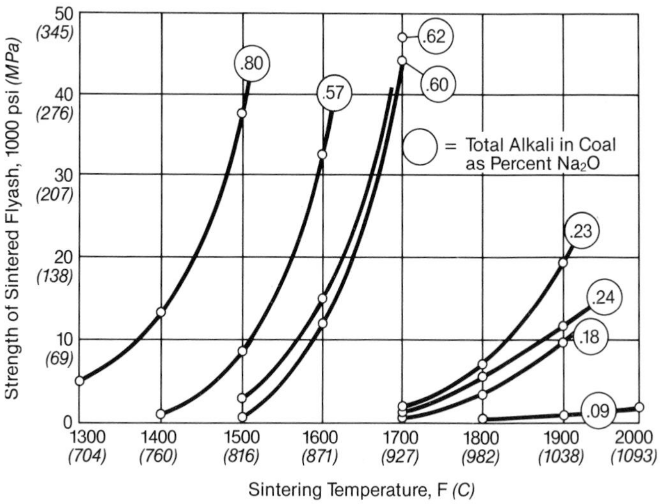

Fig. 11 Effect of alkali content in coal.

The combination of high sodium and high base to acid ratios resulted in the highest sintering strengths. Low ratios and sodium contents resulted in reduced flyash strength at the same sintering temperature. Similar trends were noted for variations in sintering strength as a function of base to acid ratio and total Na_2O in the ash. Statistical evaluations of these relationships were used to develop the fouling index currently used for coals with bituminous ash.

Similar tests on the sintering characteristics of lignitic ash indicated that the sintering criteria associated with fouling for bituminous ash did not apply to lignitic ash with high alkaline (CaO, MgO) contents. However, sintering strength was found to be directly proportional to the total sodium content in the ash shown in Fig. 13. Full scale and pilot scale tests conducted by the U.S. Bureau of Mines at the Grand Forks Coal Research Laboratory in North Dakota also established a correlation between fouling rate and sodium content for coals with lignitic ash. As shown in Fig. 14, deposition rates were found to increase sharply as the Na_2O content increased up to approximately 6% and then level off at higher percentages of sodium.

As previously noted, in low rank coals, a major portion of the alkali and alkaline earth metals can be organically bound in the coal. Because they are intimately mixed

Table 4
Effect of Soluble Sodium on Sintered Strength

Ash Analysis	Raw Coal	Washed Coal
SiO_2	45.0	49.8
Al_2O_3	18.0	20.9
Fe_2O_3	21.0	22.9
TiO_2	0.8	1.0
CaO	8.8	1.6
MgO	0.9	1.0
Na_2O	1.6	0.5
K_2O	2.4	2.3
Ash sintered strength, psi	17,300	550
(MPa)	(119.3)	(3.8)

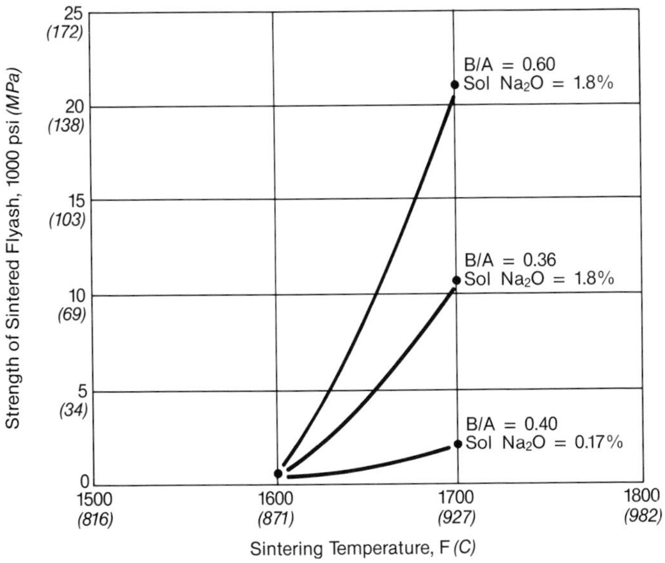

Fig. 12 Bituminous (Eastern) ash fouling effect of base to acid ratio and soluble sodium on sintered strength.

with the coal, it is believed that alkalies in this form are readily vaporized during combustion and play a dominating role in fouling. The organically associated elements occur in the form of cations chemically bonded to the organic structure of the coal. Ion exchange techniques have been developed to remove the cations from the coal for measurement. The method employed by B&W uses an ammonium acetate solution to provide a source of NH_4^+ ions which extract the ion-exchangeable cations. Ion exchange data for a high fouling North Dakota lignite and a severe fouling Montana subbituminous coal

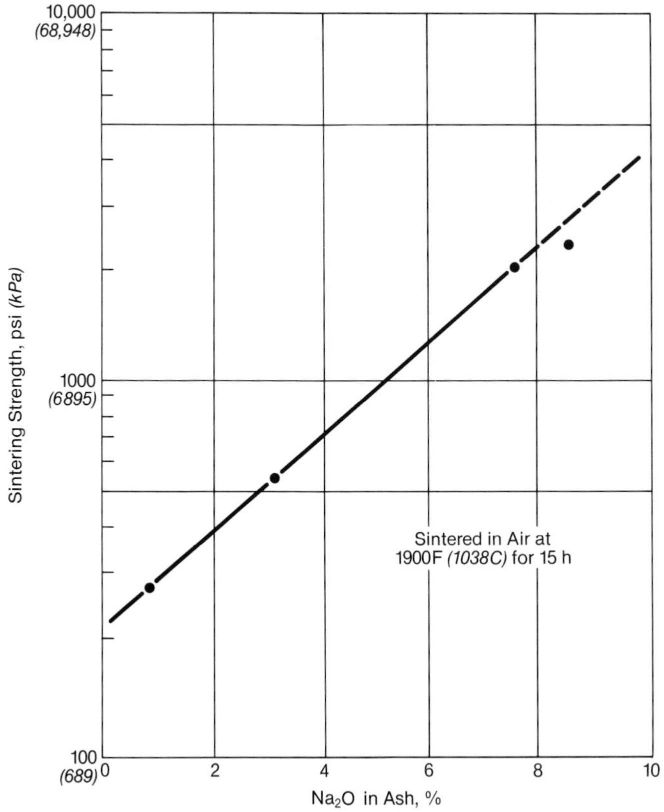

Fig. 13 Effect of Na_2O on sintering strength (North Dakota lignite ash).

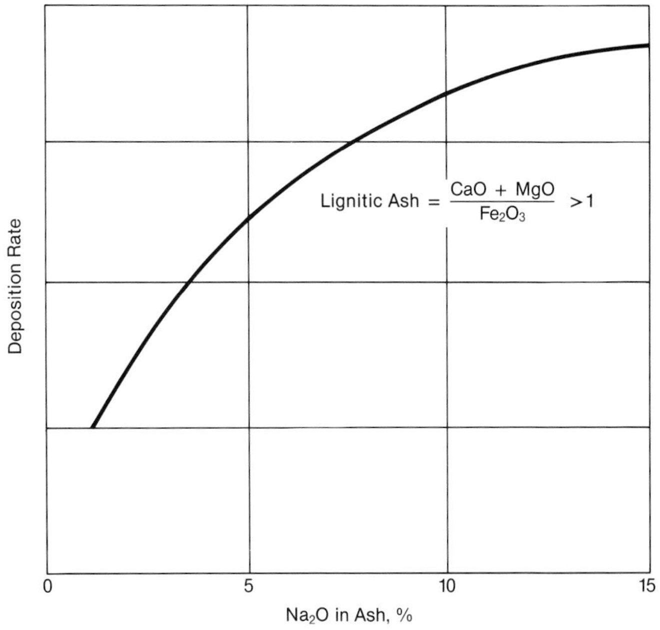

Fig. 14 Effect of Na_2O on deposition rate.

are shown in Table 5. The data show that essentially all of the sodium in both coals is organically bound. In the lignite, the ion exchangeable sodium actually exceeded the total sodium measured in ASTM ash. The difference most likely results from a loss of sodium due to vaporization during the high temperature ashing procedure. The relatively low percentages of ion-exchangeable K_2O indicate that most of the potassium exists in stable mineral forms.

Viscosity-temperature relationship of coal ash

The characteristics of slag deposits which form on furnace walls and other radiant surface are a function of deposit temperature and deposit composition. Deposit composition, in turn, is a function of the local atmosphere, particularly for ash with a significant iron content. Relationships between these factors determine the physical state of the deposit, which can range from a dry solid to plastic or even a viscous liquid if temperatures are suf-

Table 5
Ion Exchange Data — High and Severe Fouling Coals

	Source:	North Dakota	Montana
	Rank:	Lignite	Subbituminous
Ash, dry basis, %		11.2	5.4
Total alkali, dry coal basis, %	Na_2O	4.25	6.74
	K_2O	0.37	0.65
Ion exchangeable alkali, dry coal basis, %	Na_2O	4.52	6.37
	K_2O	0.10	0.13
Relative ion exchange alkali, %	Na_2O	106%	95%
	K_2O	27%	20%

ficiently high. Dry deposits are usually not troublesome; they tend to be loosely bonded to the tube surface and relatively easy to remove by sootblowing. If deposits are allowed to build in thickness, the temperature increases and the surface of the deposit can become semi-molten or plastic. The plastic slag traps other transient ash particles and continues to build more and more rapidly as the surface temperature continues to increase. Ultimately, the deposit reaches an equilibrium state as the slag begins to flow.

Field experience has shown that plastic slag tends to form large deposits that are highly resistant to removal by conventional ash cleaning equipment. This observation led to an extensive study of the relationship between ash viscosity and potential slagging tendency. Viscosity measurements that had previously been used to determine flow characteristics for wet-bottom furnace applications were extended to higher viscosity ranges to define the temperature range where a given ash would exhibit plastic characteristics.

As liquid ash is cooled, the logarithm of its viscosity increases linearly with decreasing temperature as shown in Fig. 15. At some point, the progression deviates from the linear relationship, and viscosity begins to increase more rapidly as the temperature continues to decrease. This transition into the plastic region is caused by the selective separation of solid material from the liquid, resulting from crystallization of the higher melting point constituents of the ash. The temperature at which this deviation takes place is called the temperature of critical viscosity (T_{cv}). T_{cv} varies depending on ash composition but normally occurs in a range between 100 and 500 poise. The end of the plastic region is the point of solidification, or freeze point, of the slag. The freeze point typically occurs at a viscosity of approximately 10,000 poise. For convenience in comparing the viscosity-temperature relationship of various ashes, the viscosity range of 250 to 10,000 poise has been defined as the plastic region.

The temperature at which the plastic region begins and the range of temperature over which the ash is plastic provide an indication of the slagging tendency. The lower the temperature within this range and the wider the range, the greater the potential for slagging. Viscosity-temperature curves for a high slagging Illinois coal and a low slagging east Kentucky coal, shown in Fig. 16, illustrate this effect. The plastic range for the Illinois coal begins at a relatively low temperature and extends over a wide temperature range. In contrast, the east Kentucky coal has a very narrow plastic range which begins at a much higher temperature. In comparison to the Kentucky coal, the Illinois coal ash would be expected to cool quickly below the temperature where the ash is plastic, exposing much less of the furnace to potential deposition.

As previously noted, the iron content of coal ash and its degree of oxidation have a significant influence on the viscosity of the ash. This effect is illustrated in Fig. 17 which shows the viscosity-temperature relationship for the high iron Illinois coal under both oxidizing and reducing conditions. Under reducing conditions, the viscosity at a given temperature is significantly lower and the ash remains plastic over a much wider temperature range.

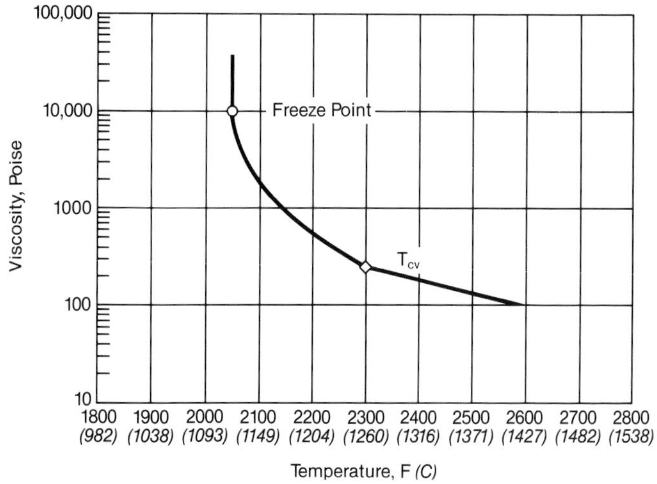

Fig. 15 Viscosity increase with decreasing temperature.

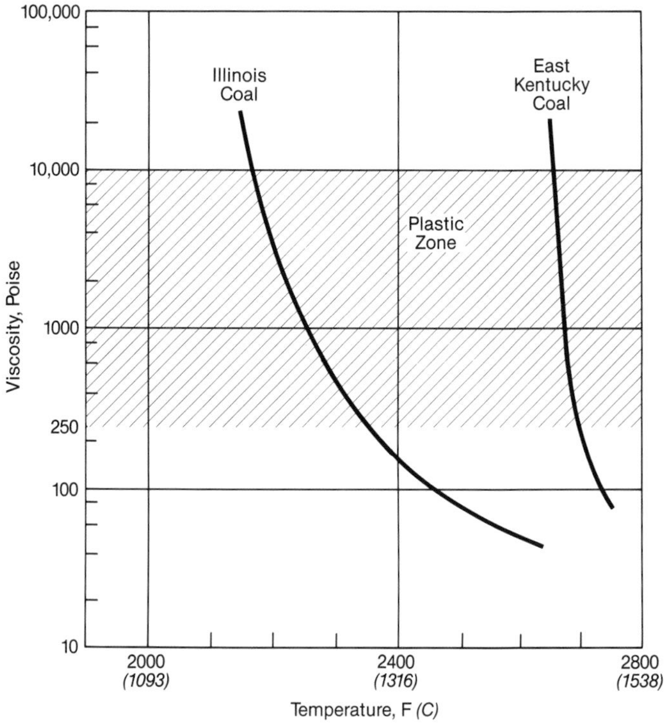

Fig. 16 Ash viscosity comparison for a high slagging and low slagging coal (oxidizing atmosphere).

Ash characterization methods

Several slagging and fouling indices have been developed by B&W to provide criteria for various aspects of boiler design. Slagging indices establish design criteria for the furnace and other radiant surface while fouling indices establish design criteria for convective surface. Deposition characteristics are generally classified into four categories: low, medium, high and severe.

For the most part, the indices described below are based on readily available ASTM ash analysis and fusibility data. In actual practice, when evaluating coals, designers take into account full scale experience on similar fuels and results of nonroutine testing which can, in some cases, modify the classification. These indices can also be used on a comparative basis to rank coals with respect to their slagging and fouling potential when evaluating a new coal supply for an existing unit.

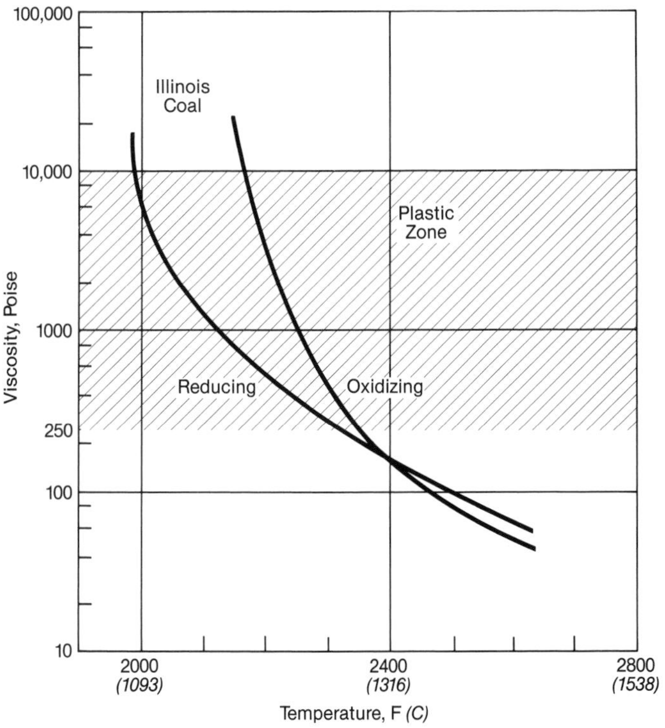

Fig. 17 Ash viscosity comparison — oxidizing and reducing conditions.

Ash classification

Because the characteristics of bituminous and lignitic ash vary significantly, the first step in calculating slagging and fouling indices is the determination of ash type. In accordance with the criteria previously described, ash is classified as bituminous when:

$$Fe_2O_3 > CaO + MgO \tag{7}$$

Ash is classified as lignitic when:

$$Fe_2O_3 < CaO + MgO \tag{8}$$

Slagging index — bituminous ash (R_s) Calculation of the slagging index (R_s) for bituminous ash takes into account the base to acid ratio and the weight percent, on a dry basis, of the sulfur in the coal. The base to acid ratio indicates the tendency of the ash to form compounds with low melting temperatures. The sulfur content provides an indication of the amount of iron that is present as pyrite. The calculation is as follows:

$$R_s = \frac{B}{A} \times S \tag{9}$$

where

B = $CaO + MgO + Fe_2O_3 + Na_2O + K_2O$
A = $SiO_2 + Al_2O_3 + TiO_2$
S = weight % sulfur, on a dry coal basis

Classification of slagging potential using R_s is as follows:

$R_s < 0.6$ = low
$0.6 < R_s < 2.0$ = medium
$2.0 < R_s < 2.6$ = high
$2.6 < R_s$ = severe

Slagging index — lignitic ash (R_s*) The slagging index for lignitic ash (R_s*) is based on ASTM ash fusibility temperatures. As previously noted, fusibility temperatures indicate the temperature range where plastic slag is likely to exist. The index is a weighted average of the maximum hemispherical temperature (HT) and the minimum initial deformation temperature (IT) as follows:

$$R_s^* = \frac{(\text{Max HT}) + 4\ (\text{Min IT})}{5} \tag{10}$$

where

Max HT = higher of the reducing or oxidizing hemispherical softening temperatures, F
Min IT = lower of the reducing or oxidizing initial deformation temperatures, F

Classification of slagging potential using R_s* is as follows:

$2450 < R_s^*$ = low
$2250 < R_s^* < 2450$ = medium
$2100 < R_s^* < 2250$ = high
$R_s^* < 2100$ = severe

Slagging index — viscosity (R_{vs}) As previously noted, B&W's most accurate method for predicting slagging potential is based on the viscosity-temperature relationship of the coal ash. This index (R_{vs}) is applicable to both bituminous and lignitic ash coals; however, measured ash viscosities are required.

$$R_{vs} = \frac{(T_{250\ oxid}) - (T_{10,000\ red})}{97.5\ (fs)} \tag{11}$$

where

$T_{250\ oxid}$ = temperature, F, corresponding to a viscosity of 250 poise in an oxidizing atmosphere
$T_{10,000\ red}$ = temperature, F, corresponding to a viscosity of 10,000 poise in a reducing atmosphere

and fs is a correlation factor based on the average of the oxidizing and reducing temperatures (T_{fs}) corresponding to a viscosity of 2000 poise. Values for fs as a function of T_{fs} are provided in Fig. 18.

Classification of slagging potential using R_{vs} is as follows:

$R_{vs} < 0.5$ = low
$0.5 < R_{vs} < 1.0$ = medium
$1.0 < R_{vs} < 2.0$ = high
$2.0 < R_{vs}$ = severe

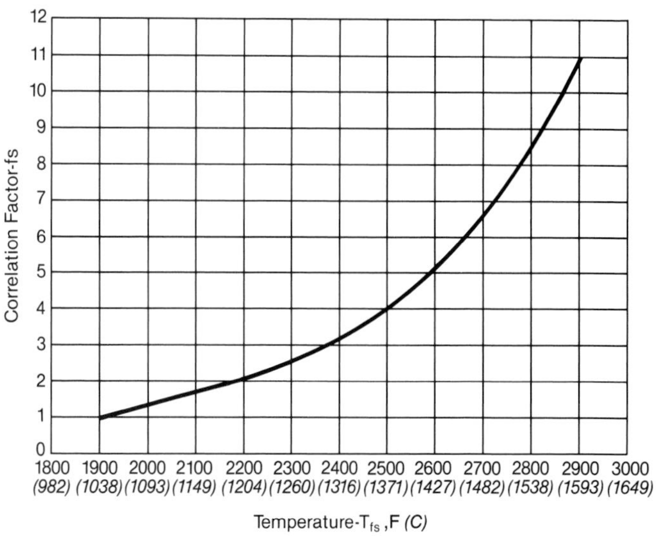

Fig. 18 Slagging index correction factor fs.

Fouling index — bituminous ash (R_f) The fouling index for bituminous ash is derived from sintering strength characteristics using the sodium content of the coal ash and the base to acid ratio as follows:

$$R_f = \frac{B}{A} \times Na_2O \qquad (12)$$

where

B = $CaO + MgO + Fe_2O_3 + Na_2O + K_2O$
A = $SiO_2 + Al_2O_3 + TiO_2$
Na_2O = weight % from analysis of coal ash

Classification of fouling potential using R_f is as follows:

$R_f < 0.2$ = low
$0.2 < R_f < 0.5$ = medium
$0.5 < R_f < 1.0$ = high
$1.0 < R_f$ = severe

Fouling index — lignitic ash The fouling classification for lignitic ash coals is based on the sodium content in the ash as follows:

When $CaO + MgO + Fe_2O_3 > 20\%$ by weight of coal ash

$Na_2O < 3$ = low to medium
$3 < Na_2O < 6$ = high
$Na_2O > 6$ = severe

When $CaO + MgO + Fe_2O_3 < 20\%$ by weight of coal ash

$Na_2O < 1.2$ = low to medium
$1.2 < Na_2O < 3$ = high
$Na_2O > 3$ = severe

Coal ash effects on boiler design

Furnace design

The key to a successful overall gas side design is proper sizing and arrangement of the furnace. As a first priority, the furnace must be designed to minimize slagging and to provide effective control of slag where and when it does form.

Ash deposition in the furnace can cause a number of problems. Slag deposits reduce furnace heat absorption and raise gas temperature levels at the furnace exit. This, in turn, can cause slagging and can aggravate fouling in the convection banks where ash deposits become increasingly more difficult to control as gas temperatures increase. The shift in heat absorption from the furnace to the superheater and reheater results in increased attemperator spray flow for control of steam temperatures, reducing cycle efficiency. Slag buildup at the top of a tall furnace is dangerous. Large deposits can become dislodged and fall, causing failures of furnace hopper tubes and loss of availability. Excessive slagging in the lower furnace can interfere with ash removal.

Experience has shown that several interrelated furnace design parameters are critical for slagging control. These parameters focus on keeping ash particles in suspension and away from furnace surfaces, distributing heat evenly to avoid high localized temperatures, and removing enough heat to achieve temperatures at the furnace exit that will minimize deposition on convection surface.

In the context of gas side design, the furnace basically serves two functions. It must provide sufficient volume to completely burn the fuel, and provide sufficient heat transfer surface to cool the flue gas and ash particles to a temperature suitable for admission to the convection surface. In general, for a coal-fired unit, it is the second criterion that determines the minimum furnace size.

The slagging classification of the coal establishes the upper limit on furnace exit gas temperature (FEGT) required to minimize the potential for slagging both in the radiant superheater and the close-spaced convection surface. As described in Chapter 21, furnace exit gas temperature is a function of furnace heat release rate. Limiting the FEGT, therefore, limits the heat release rate, resulting in lower average temperatures in the furnace. FEGT limits and corresponding heat release rates have been established by experience for different types of coal. In general, units using coals with low or medium slagging tendencies can have higher heat release rates and higher FEGTs. Units firing coals with high or severe slagging potential require lower heat release rates and lower FEGTs.

Ideally, the furnace would be an open box, sized with sufficient wall surface to cool the furnace gas and ash particles to the desired temperature before they reached any superheater surface. However, thermodynamic considerations in modern high pressure and high temperature cycles require that a significant portion of the total heat absorption be accomplished in the superheater and reheater. This requirement places a practical limit on the amount of furnace wall surface which, in a drum boiler, is dedicated to generating saturated steam. In order to achieve the required FEGT it becomes necessary to replace water-cooled furnace wall surface with steam-cooled superheater surface. These surfaces are generally in the form of widely spaced platens located in the upper radiant zone of the furnace. Because platen surface is located in a relatively high gas temperature zone and subject to ash particle impaction, the side spacing must be sufficient to limit the potential for bridging and provide a degree of self-cleaning. Typical side spacing between platen sections is 4 to 5 ft (1.2 to 1.5 m). When platen superheater surface is used, the slagging classification of the coal establishes the upper limit on platen inlet gas temperature, in addition to limiting the FEGT.

An alternate method of controlling furnace exit gas temperature that has been widely used is gas tempering by flue gas recirculation. In this method, relatively cool gas from the economizer outlet is mixed with hot furnace gas near the furnace exit. Gas tempering offers a number of advantages. The FEGT can be limited with less furnace surface while the increased gas weight improves the thermal head for heat transfer, reducing the surface requirements in the convection pass. Proper introduction of the tempering flue gas provides a flat temperature profile at the furnace exit, reducing the possibility of localized slagging and fouling. Once the choice of gas recirculation is made, the system can also be used to control reheat steam temperature at partial loads. For this purpose, flue gas from the economizer outlet is introduced into the furnace through the furnace hopper opening. The cool gas reduces furnace heat absorption and makes more heat available to the reheater which offsets its natural characteristic of decreasing outlet steam temperature at partial loads.

The major disadvantages of gas recirculation are fan maintenance and power requirements. Fan erosion can be minimized to some extent by proper design and operation of a mechanical dust collector ahead of the fan. Extracting the recirculated gas after a hot precipitator offers the best potential for a relatively clean recirculated gas source.

In addition to having sufficient volume and heating surface, the furnace also must be correctly proportioned with respect to width, depth and height to minimize slagging. A significant design parameter in this regard is heat input to the furnace per square foot of furnace plan area at the burners. Maximum limits on plan area heat release rate are a function of the slagging potential of the coal. Limits typically range from 1.5 to 1.8×10^6 Btu/h ft^2 (4.7 to 5.7 MW_t/m^2) for severe slagging and low slagging coals respectively.

The furnace must also be designed to limit the potential for ash particle impaction on furnace surfaces. Ample clearance must be provided between the burners and furnace walls as well as the furnace hopper and arch. These critical dimensions have been established by operating experience and keyed to the slagging classification of the coal.

The slagging classification also determines the locations, quantity and spacing of furnace wall blowers and long retractable sootblowers in the pendant radiant surface. (See Chapter 23.) These allow control of the deposition that inevitably occurs and are essential for maintaining furnace surface effectiveness and furnace exit gas temperature within the range provided for in the design. Some degree of slagging may be permitted above the burner zone but only to the extent that it can be controlled by selective operation of the wall blowers. The control of these deposits can help maintain steam temperature at reduced loads. Slag deposits on furnace walls must be avoided below and between burners, however, where they can not be controlled by sootblowers.

Effect of slagging potential on furnace sizing

Referring to Fig. 19, three large utility boilers are shown sized for 660 MW at maximum continuous load. The boilers are assumed to have the same width for purposes of illustration, with the boiler setting height and furnace depth varied to accommodate the slagging characteristics of the different fuels. Boiler (a) is designed to fire a bituminous coal having a low to medium slagging potential. The slightly larger boiler (b) is designed to fire a subbituminous coal classified as having a high slagging potential. The difference in size can be attributed primarily to the difference in slagging potential. The furnace (b) depth has been increased to control slagging by reducing the input per plan area. The input and gas weight are higher for the subbituminous coal due to its higher moisture content and resulting lower boiler efficiency. This increases the required furnace surface and the furnace exit area in order to maintain acceptable gas velocities entering the convection pass. Comparing boiler (c), firing a severe slagging lignite, to boiler (b), the furnace depth has again been increased due to the increased slagging potential. The furnace surface has also been increased to reduce the gas temperature leaving the furnace. The size differential of the three units is quantified in Table 6. This table shows the proportionate differences or increases using boiler (a) as a base. Boiler (a) is assigned a size factor of 1.0 for the various parameters shown.

Convection pass design

The key to successfully preparing a design that will control convection pass fouling reverts back to a furnace design that will maintain the furnace exit gas temperature at predicted levels. Temperature excursions at the furnace exit result in corresponding higher temperature levels throughout the convection pass which can cause

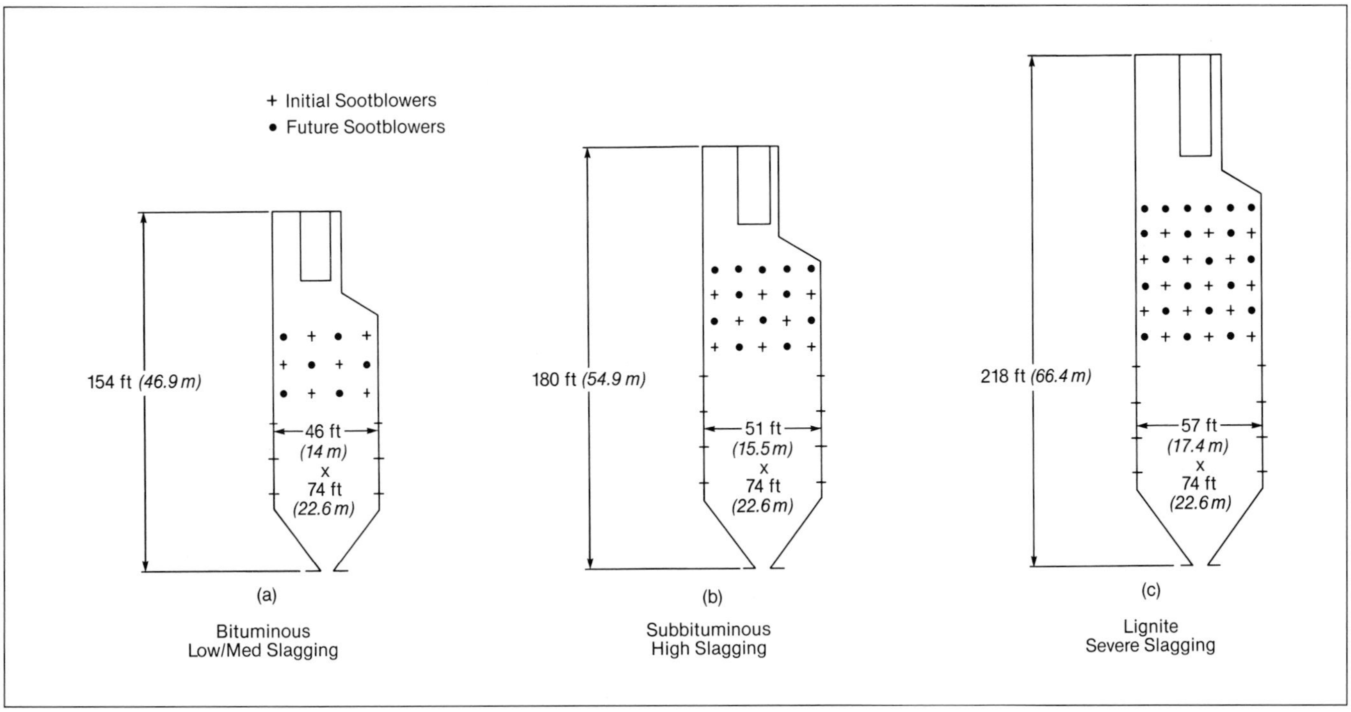

Fig. 19 Influence of slagging potential on furnace size. (See Table 6.)

Table 6
Boiler Size Versus Slagging Classification

	Boiler		
	(a)	(b)	(c)
Coal Rank	Bituminous	Subbituminous	Lignite
Slagging	Low/Med	High	Severe
Furnace plan area	1.0	1.11	1.24
Furnace surface	1.0	1.18	1.50
No. of furnace wall blowers	30	36	70

deposition problems even with coals which normally would be considered to have a low or moderate fouling tendency.

In general, convective heating surface, both pendant and horizontal, is arranged to minimize the potential for bridging and obstruction of the gas lanes between adjacent sections. The minimum clear side spacing (measured perpendicular to the gas flow) between sections in a bank varies as a function of the average flue gas temperature entering the bank. The widest spacing is required in the superheater banks which are in close proximity to the furnace exit, where the gas temperature and fouling potential are high. As the flue gas temperature is reduced, the side spacing in succeeding banks can also be reduced. The specific side space dimensions at a given temperature entering the bank depend on the fouling classification of the coal. Severe fouling coals require the widest spacing. Adequate side spacing must be maintained even in low temperature horizontal banks such as economizers. While these surfaces are not normally subject to bonded deposits, sufficient clear space must be maintained between sections to ensure that accumulations of ash dislodged from upstream surfaces will not bridge and plug the gas lanes. (See Chapter 19.)

Bank depths (measured parallel to the direction of gas flow) are established as a function of fouling potential, clear side spacing and the temperature entering the bank. Cavities between the banks provide locations for long retractable sootblowers. At high gas temperatures, shallow bank depths are required to ensure adequate sootblower effectiveness. Sootblower jet penetration increases as temperatures are reduced and bank depths can be increased incrementally in cooler areas.

Flyash erosion

The metal loss on convection pass tubes due to flyash erosion is proportional to the total ash quantity passing through the boiler and is an exponential function of flue gas velocity. While with a given fuel there is no control of the ash quantity, erosion problems can be eased by reducing flue gas velocities. Velocity limits are determined based on the ash quantity on a pounds per million Btu basis and the relative proportion of abrasive constituents in the ash. Typical limits range from 65 ft/s (19.8 m/s) for relatively nonabrasive low ash coals to 45 ft/s (13.7 m/s) or less for coals with high ash quantities and/or abrasive ash.

Effect of operating variables

Although the predominant factors affecting deposition are ash characteristics and boiler design, operating variables can also have a significant impact on slagging and fouling.

In general, operating variables associated with combustion optimization (see Chapter 13) tend to reduce the potential for deposition problems. These variables include air distribution, fuel distribution, coal fineness and excess air.

Air and fuel imbalances can result in high excess air at some burners while others operate with less than theoretical air. This, in turn, results in localized reducing conditions in the burner zone which can aggravate slagging, especially with coals having high iron content. High coal/air ratios can also delay combustion and upset heat distribution, resulting in elevated temperatures in the upper furnace and at the furnace exit. Long burnout times also increase the potential for burning particles to contact furnace walls and other heat transfer surfaces.

Secondary air imbalances can be minimized by adjusting burner registers to provide a flat O_2 profile at the economizer outlet. Care must be exercised to avoid burner adjustments that cause flame impingement on furnace walls. On the fuel side, burner line resistances should be balanced to maintain uniform coal flow to each burner. Coal feeders should be calibrated and adjusted to provide uniform coal flow to each pulverizer.

Low pulverizer fineness (see Chapter 12) can also cause problems associated with delayed combustion. Coarse particles require longer residence times for burnout and can cause slagging in the lower furnace.

Excess air has a tempering effect on average temperatures within the furnace and on furnace exit temperature. Excess air also reduces the potential for localized reducing conditions in the furnace when it is introduced through the burners. Air infiltration into the furnace or convection pass is far less beneficial and should be corrected or taken into account when establishing excess air requirements. While there is an associated efficiency loss, raising excess air above normal design levels is usually an effective tool for controlling deposition problems. In some cases, high excess air may also upset superheater/reheater absorption and steam temperatures.

Sootblowers (see Chapter 23) are the primary means of dealing directly with furnace wall slagging and convection pass fouling. The most important fundamental requirement is to use this equipment in a preventive, rather than corrective, manner. Sootblowers are most effective in controlling dry, loosely bonded deposits which typically occur in the early stages of deposition. If furnace slag is allowed to accumulate to the point that it becomes plastic or wet, or if convection pass deposits are allowed to build and sinter for long periods of time, removal becomes much more difficult. Sootblower sequencing requirements must be established by initial operating experience and updated when required as the unit seasons or when fuel characteristics change. Boiler diagnostic systems, which are discussed in the following section, can assist in optimizing sootblower operation.

The least desirable operating technique for controlling deposition problems is load reduction. The most severe situations may require a permanent derate. However, in many marginal situations, temporary load reductions during off peak periods may provide sufficient cooling to shed slag and allow sootblowers to regain effectiveness.

Application of boiler diagnostic systems

Awareness of slagging and fouling conditions is a critical key to achieving reliability and availability on a coal-fired utility boiler. However, boiler surface cleanliness has been, traditionally, one of the most difficult operating variables to quantify. Typical indications of surface fouling appear to the operator indirectly in the form of steam temperatures, spray attemperation flows and draft losses. In some cases, experienced operators who are familiar with the operating characteristics of a unit can make judgments on slagging and fouling conditions based on operating conditions, but these secondary indications can be misleading. For example, the furnace can be slagged, causing undesirably high gas temperatures entering the convection surface. However, the steam temperatures and spray attemperation may be normal if the convection surfaces are also fouled.

Another indication of surface cleanliness is draft loss. By watching draft loss across a bank, an alert operator can determine that sootblowing is probably required. Usually, however, by the time a change in draft loss is detected across widely spaced pendant sections, the banks are already bridged and it may be too late for removal by the sootblowers.

Visual observation is frequently used to further quantify cleanliness conditions. In many instances, however, access is limited and subjective evaluations can leave considerable room for error.

Computer based boiler performance monitoring systems have been developed to provide a direct and quantitative assessment of furnace and convective surface cleanliness. B&W's TotalScope® Performance Management System is derived from the computer models developed for boiler design. Measurements of temperatures, pressures, flows, and gas analysis data are used to perform heat transfer analysis in the furnace and convective section on a bank by bank basis. With a quantitative indication of surface cleanliness, potential slagging and fouling problems can be recognized early in their development and selective sootblowing can be directed at the specific problem area. Sootblower sequencing can be optimized based on actual cleaning requirements rather than on fixed time cycles which can waste blowing medium, increase cycle time and cause erosion by blowing clean tubes.

An overview of the cleanliness of the unit is presented graphically in a form similar to that found in Fig. 20. Cleanliness values for each section of surface are compared for the four most recent time intervals (right hand bar of each component being the most recent). By presenting the relative cleanliness in such a manner, not only can the current values of cleanliness be observed but the trends and their rates of change can also be seen. All levels of cleanliness are referenced to normal for each boiler as determined by full load range testing.

To optimize sootblowing, the degree of cleanliness must be related to sootblowing requirements. Fig. 21 is a typical example of surface cleanliness plotted versus time for a single section. In general, surface effectiveness or cleanliness decreases with time between sootblowing cycles. The rate of decay of the surface effectiveness is dependent upon the ash characteristics and gas temperature of the zone. In some cases, the surface effectiveness may stabilize at some level, or may continue to decrease if sootblowers are not operated. To establish sootblowing requirements, three ranges of cleanliness are considered — normal, medium fouled and severely fouled.

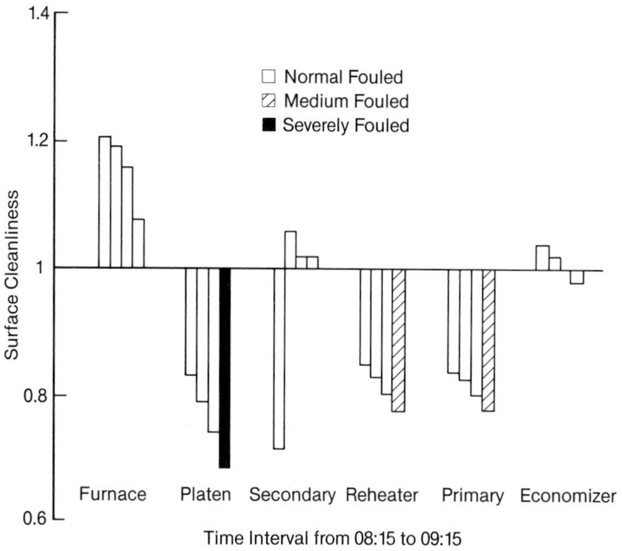

Fig. 20 Unit cleanliness as relayed to the operator.

When cleanliness is in the normal range, sootblowers are operated or placed on hold, as required, to optimize superheater and reheater absorption as well as boiler exit gas temperature. If cleanliness conditions drop below the normal range into the medium fouled range, sootblowing is shifted from the normal pattern to the problem bank. There is no immediate concern about ash sintering or plugging in the range. In extreme situations, surface cleanliness may fall below the medium fouled range into the severely fouled range. At the time a given bank enters this range, concentrated sootblowing is immediately directed to the problem section. If immediate action is not taken, there is a danger of ash sintering and plugging. The concentrated blowing consists of continuously cycling a small subset of blowers (determined by pretesting to have the greatest short term impact on deposit removal) until the problem is rectified. The subsets will significantly reduce cycle time.

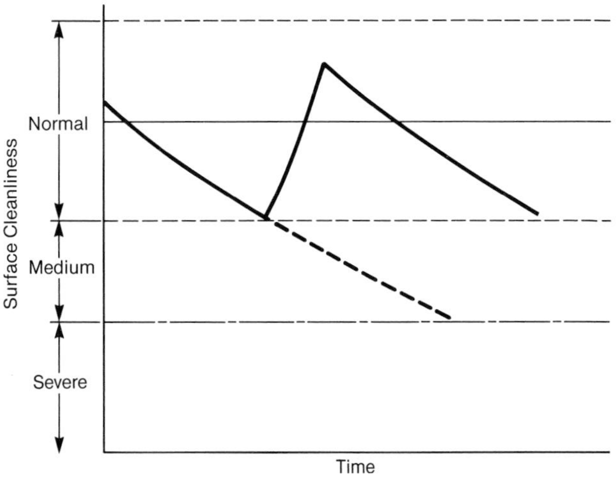

Fig. 21 Ranges of fouling with time.

The computer generated sootblowing priorities can be displayed for operator assistance and conceivably used for direct sootblower control. In the case of direct control, the computer would initiate or stop specific groups of blowers to optimize unit performance.

Nonroutine ash evaluation methods

The following describes the laboratory equipment and test procedures, referenced earlier, that are used to supplement the standard ASTM coal ash characterization methods.

Laboratory ashing furnace

As noted, the ASTM ashing procedure does not duplicate the ashing process that actually occurs in a boiler. The laboratory ashing furnace (LAF) provides a means to obtain flyash and deposit samples that are comparable to those obtained from full scale installations operating under similar conditions.

The LAF, shown in Fig. 22, is designed to fire pulverized coal at rates typically between 5 and 10 lb/h (2.3 and 4.5 kg/h). The facility consists of a fuel feed system, pulverized coal burner and a refractory lined chamber. The combustion chamber is surrounded by an electrically heated guard furnace which controls the rate of heat removal from the chamber to simulate full scale furnace temperatures. The firing rate is established to approximate full scale furnace residence time. A deposition section located at the furnace exit contains air- or water-cooled probes. The surface temperature of the probes can be adjusted to simulate furnace and superheater tube operating temperatures. The probes are instrumented to allow measurement of metal temperatures, cooling fluid flow rates, and cooling fluid inlet and outlet temperatures. These data permit calculation of the total heat flux from the flue gas through the deposit and into the probe. The deposition section is also fitted with sootblowers to evaluate the effectiveness of ash removal equipment. Fig. 23 shows ash particles impacting a simulated superheater tube during a deposition test.

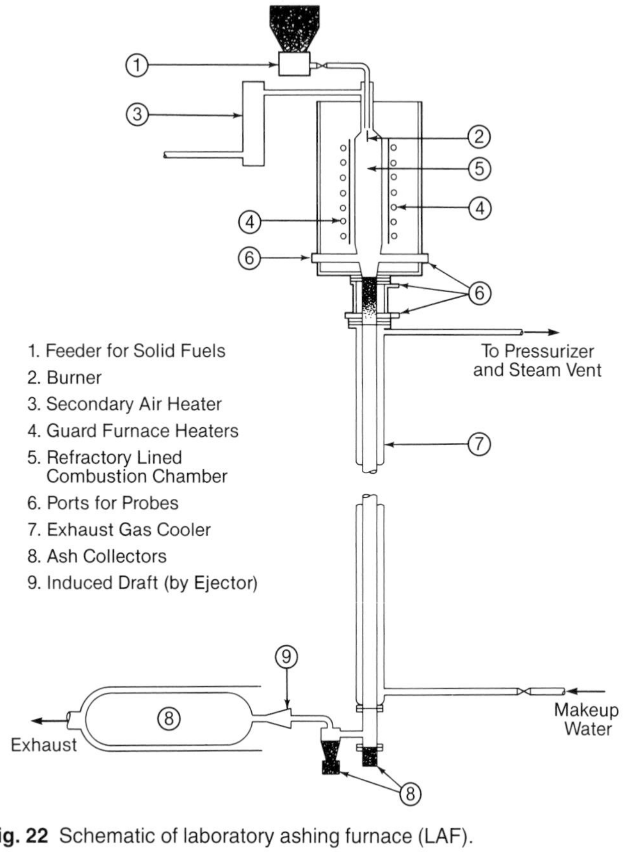

Fig. 22 Schematic of laboratory ashing furnace (LAF).

Measurement of ash viscosity

Viscosity of coal ash is measured in a high temperature rotating-bob viscometer (Fig. 24). The ash under study is contained in a cylindrical platinum-rhodium crucible, and a cylindrical bob is rotated in the liquid at a constant speed through a calibrated suspension wire. The torque or amount of twist produced in the suspension wire is proportional to the viscosity. The amount of twist is measured and recorded as the interval between impulses from light beams reflected from mirrors attached to the ends of the wire. The suspension wires are calibrated against viscosity standard oils obtained from the Bureau of Standards.

The electrically heated furnace is of the Globar tube type with temperature regulation provided through a controlling type potentiometer actuated by a thermocouple located in the furnace adjacent to the sample crucible. A thermocouple imbedded in the ash crucible support indicates sample temperature. Provision is made for controlling the atmosphere within the furnace. Ash is introduced into the crucible at an elevated temperature [2600 to 2800F (1427 to 1538C)] and held at that temperature until it becomes uniformly fluid. The temperature is then decreased in predetermined steps and the viscosity of the ash is measured at each temperature.

Ash sintering strength

The sintering strength test is performed on a flyash sample prepared in the LAF under a standard set of fir-

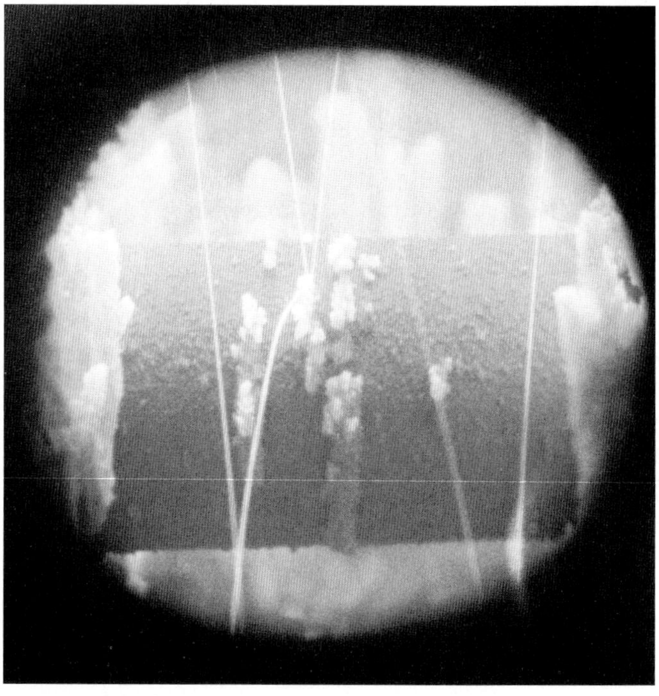

Fig. 23 Deposit formation on simulated superheater tube.

ing conditions established by the work of Attig and Barnhart (B&W). The flyash is passed through a 60 mesh (U.S. standard) (250 micron) screen to remove any particles of slag and then ignited to constant weight at 900F (482C) to remove any carbon that might be present. The ignited ash is then reduced to a minus 100 mesh size and at least 24 cylindrical specimens [0.6 in. (15.2 mm) diameter by 0.85 in. (21.6 mm) long] are formed in a hand press at a pressure of 150 psi (1034 kPa). At least six specimens are heated in air, usually at each of four temperature levels [1500, 1600, 1700 and 1800F (816, 871, 927 and 982C)] for 15 hours.

After the specimens have cooled slowly in the furnace, they are removed, measured and then crushed in a standard metallurgical testing machine. The sintered or compression strength is then computed from the applied force and the cross-sectional area of the sintered specimen. The average strength of six specimens is used as the strength of the sintered flyash at a particular sintering temperature.

Measurement of ion exchangeable cations in coal

Twenty grams of an air-dried minus 60 mesh coal sample are mixed with 100 ml of 1 N ammonium acetate in a 300 ml three-neck round bottom flask. A thermometer is inserted into the slurry. The slurry is stirred constantly and heated to 60 +/- 5C. The coal slurry sample is refluxed for 18 hours. The sample is filtered through a cellulosic filter media with 0.45 μ average pore size and washed twice with 25 ml of 1 N ammonium acetate solution.

The above procedure is repeated on the filtered coal except that the time is shortened to three hours. The combined filtrates are acidified by adding 2% by volume of glacial acetic acid and stored for inductive coupled plasma atomic emission spectrometric (ICPAES) analysis of Na, K, Ca and Mg.

Coal ash corrosion

Serious external wastage or corrosion of high temperature superheater and reheater tubes was first encountered in coal-fired boilers in 1955. Tube failures resulting from excessive thinning of the tube walls, as shown in Fig. 25, occurred almost simultaneously in the reheater of a dry ash furnace boiler and the secondary superheater of a slag-tap furnace unit. Corrosion was confined to the outlet tube sections of the reheater and the secondary superheater, which were made from chrome ferritic and stainless steel alloys, respectively.

Significantly, these boilers were among the first to be designed for 1050F (566C) main and reheat steam temperatures; also, both units burned high sulfur, high alkali coals from Central and Southern Illinois, which were causing chronic ash fouling problems at the time.

Early investigations showed that corrosion was found on tube surfaces beneath bulky layers of ash and slag. When dry, the complex sulfates were relatively innocuous; but when semi-molten [1100 to 1350F (593 to 732C)], they corroded most of the alloy steels that might be used in superheater construction, as well as other normally corrosion resistant materials.

At first, it appeared that coal ash corrosion might be confined to boilers burning high alkali coals, but complex sulfate corrosion was soon found on superheaters

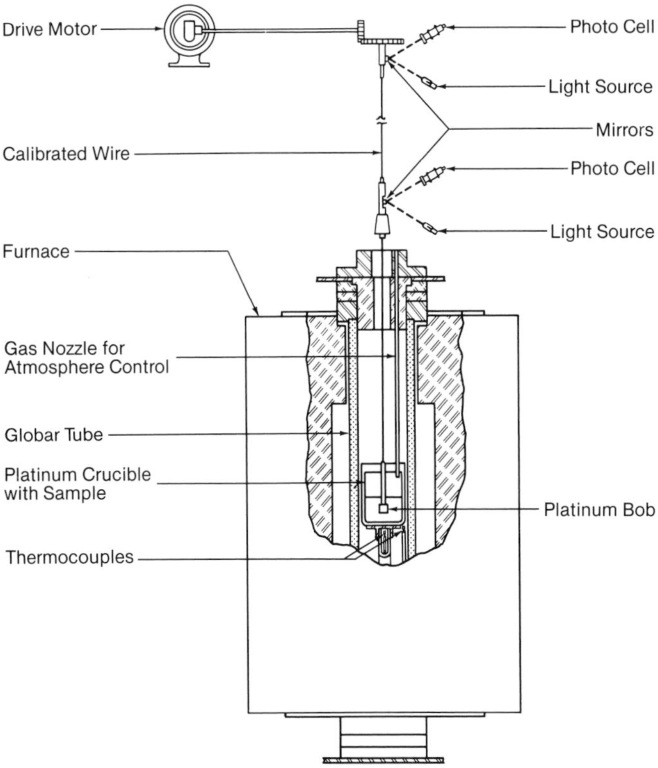

Fig. 24 Section through furnace of high temperature viscometer.

and reheaters of several boilers burning low to medium alkali coals. Where there was no corrosion, the complex sulfates were either absent or the tube metal temperatures were moderate [less than 1100F (593C)]. The general conclusions drawn from this survey of corrosion were:

1. All bituminous coals contain enough sulfur and alkali metals to produce corrosive ash deposits on superheaters and reheaters, and those containing more than 3.5% sulfur and 0.25% chlorine may be particularly troublesome.
2. Experience has shown that corrosion rate is affected by both tube metal temperature and gas temperature. Fig. 26, which is used as a guide in design, indicates stable and corrosive zones of fuel ash corrosion as a function of gas and metal temperatures.

Based on this information, B&W modified the design of its boilers to greatly reduce the corrosion of superheaters and reheaters. These modifications included changes in furnace geometry, burner configuration, superheater arrangement and the use of gas tempering, all of which reduced metal and gas temperatures and reduced temperature imbalances. Experience from these installations has shown that it is possible to operate boilers with main and reheat steam temperatures up to 1050F (566C) with little, if any, corrosion from most coals.

Meanwhile, there was a gradual return to the 1000F (538C) steam conditions for new plants, due primarily to economic factors and secondarily to coal ash corrosion. This temperature level permits the use of lower cost alloys in the boiler, steam piping and turbine, with substantial savings in investment costs; it also provides a greater margin of safety to avoid corrosion. Steam temperatures will therefore probably remain on the current 1000F (538C) plateau until economics dictate the use of high temperature alloys

Fig. 25 Typical corroded 18Cr-8Ni tube from secondary superheater.

and until methods are developed for avoiding corrosion at higher steam temperatures. (See Chapter 18.)

General characteristics of corrosion

External corrosion of superheaters and reheaters is concentrated on the upstream side of the tube, as shown in Fig. 27. The greatest metal loss usually occurs on the 10 and 2 o'clock sectors of the tubes, and it tapers off to little or none on the back side of the tubes. The corroded surface of the tube is highly sculptured by a shallow macropitting type of attack. The amount of corrosion, as measured by reduction in tube wall thickness, varies considerably along the length of the tube, depending on local conditions, i.e., the position of the tube in the bank or platen, the proximity of sootblowers, the composition of ash deposits and, most importantly, the gas and metal temperatures.

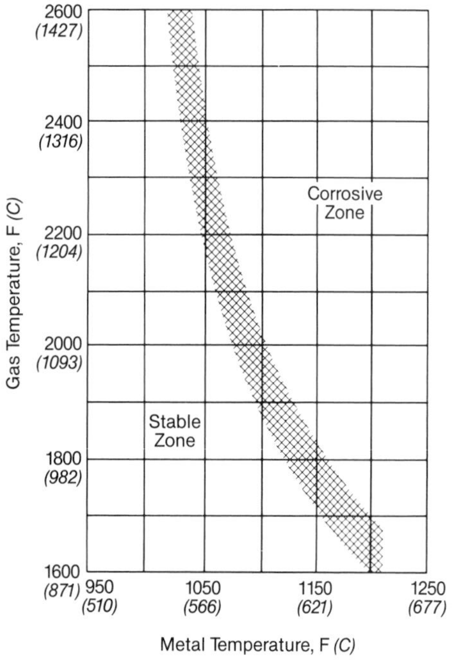

Fig. 26 Coal ash corrosion — stable and corrosive zones.

The corrosion rate is a nonlinear function of metal temperature (Fig. 28). The corrosion of both chrome ferritic and 18Cr-8Ni stainless steels increases sharply above a temperature of 1150F (621C), passes through a broad maximum between 1250 and 1350F (677 and 732C) and then decreases rapidly at still higher temperatures.

The highest corrosion rates are generally found on the outlet tubes of radiant superheater or reheater platens opposite retractable sootblowers. Values ranging from 50 to 250 mils/yr (1.27 to 6.35 mm/yr) have been observed on 18Cr-8Ni stainless steel tubes under these adverse conditions. When similar high temperature surfaces [1100 to 1175F (593 to 635C)] are arranged in convection tube banks so they are shielded from direct furnace radiation and sootblower action, corrosion rates are much lower, ranging between 5 and 20 mils/yr (0.13 to 0.51 mm/yr).

Corrosive ash deposits

Corrosion is rarely found on superheater or reheater tubes having only dusty deposits. It is nearly always associated with sintered or slag type deposits that are strongly bonded to the tubes. Such deposits consist of at least three distinct layers. The outer layer, shown diagrammatically in Fig. 29, constitutes the bulk of the deposit and has an elemental composition similar to that of flyash. Though often hard and brittle, this layer is a porous structure through which gases may diffuse. Innocuous by itself, it plays an important part in the formation of an intermediate layer that contains the corrosive agents.

The intermediate layer, frequently called the white layer, is a white to yellow colored material which varies in thickness from 0.0313 to 0.25 in. (0.794 to 6.35 mm). It usually has a chalky texture where corrosion is mild or nonexistent, but is fused and semi-glossy where corrosion is severe. In the latter condition this layer is difficult to remove as it is so firmly bonded to the corroded surface beneath.

Upon heating in the air, the intermediate layer melts around 1000F (538C) and slowly discolors and hardens into a hard mass resembling rust. Chemical analyses of this layer show that it contains higher concentrations of potassium, sodium and sulfur than does the parent coal ash. A large part of this deposit is water soluble and the water soluble fraction is always acidic. The identification of compounds making up the intermediate layer is difficult because its constituents are not well crystallized. The normal sulfates are conspicuously absent and the complex alkali sulfates are detected irregularly. The most common compounds found are $Na_3Fe(SO_4)_3$ and $KAl(SO_4)_2$, although other complex sulfates are thought to be present.

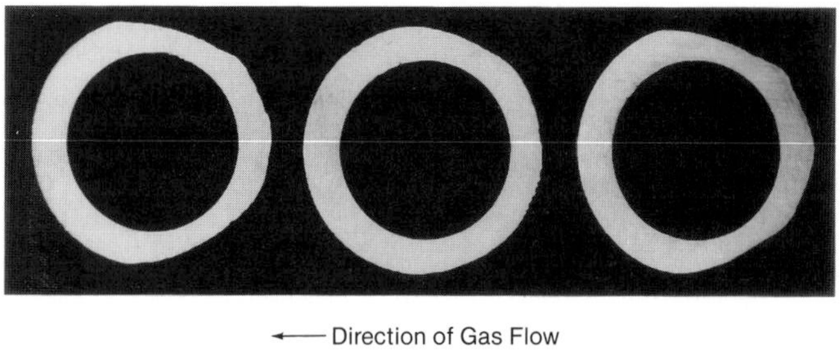

Fig. 27 Transverse sections of corroded tubes from secondary superheater platens.

Laboratory studies have shown that complex alkali sulfates, when molten, rapidly corrode most, if not all, superheater alloys. Corrosion begins between 1000 and 1150F (538 and 621C), depending on the relative amounts of complex sodium and potassium sulfates present and whether these are predominantly iron or aluminum base compounds. Corrosion usually begins at the lower temperature where the sodium-iron-sulfate system is the major part of the intermediate layer, but corrosion is more severe and persists into a higher temperature range when the potassium-aluminum-sulfate system dominates.

If the intermediate layer is carefully removed, a black, glassy inner layer is revealed, which appears to have replaced the normally protective oxide on the tube. This layer is composed primarily of corrosion products, i.e., oxides, sulfides, and sulfates of iron and other alloying constituents in the tube metal. It seldom exceeds 0.063 in. (1.59 mm) thickness on corroded 18Cr-8Ni stainless steel tubes, probably because of its strong tendency to spall when the tubes cool. The layer containing corrosion products from chrome ferritic alloys often reaches 0.125 in. (3.18 mm) thickness and exhibits little tendency to spall as the tube cools.

Corrosion mechanisms

The elements in coal ash corrosion (sodium, potassium, aluminum, sulfur and iron) are derived from the mineral matter in coal. The minerals supplying these elements include shales, clays and pyrites which are commonly found in all coals.

During the combustion of coal, these minerals are exposed to high temperatures and strongly reducing effects of carbon for very short periods of time. Although comparatively stable, the mineral matter undergoes rapid decomposition under these conditions. Some of the alkalies are released or volatilized as relatively simple compounds, which have dew points in the 1000 to 1300F (538 to 704C) range. Furthermore, the pyrite is oxidized, releasing SO_2 with the formation of a small amount of SO_3, leaving a residue of iron oxide (Fe_2O_3).

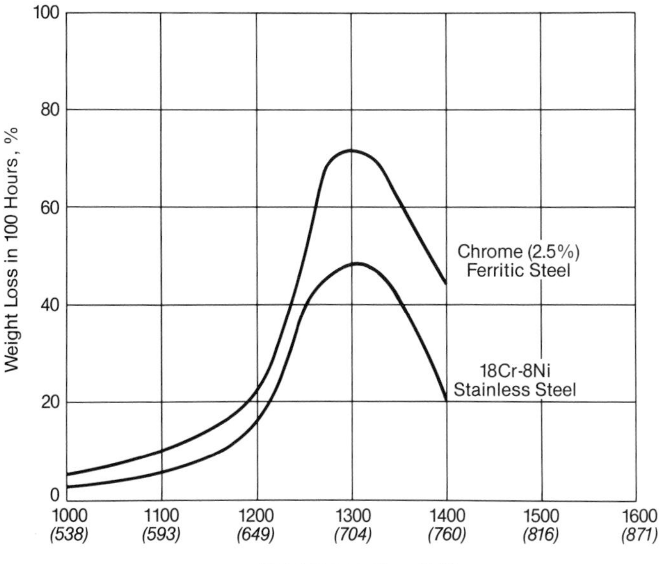

Fig. 28 Effect of temperature on corrosion rate.

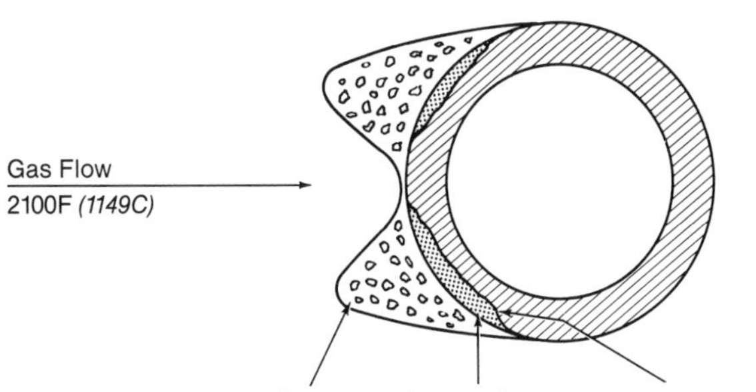

	Outer Layer % by wt	Intermediate Layer % by wt	Inner Layer % by wt
SiO_2	23.5	23.3	7.6
Al_2O_3	14.0	11.5	1.7
Fe_2O_3	36.0	11.0	70.5
TiO_2	0.9	<0.1	<0.1
CaO	1.3	<0.1	<0.1
MgO	1.3	1.1	<0.1
Na_2O	0.3	1.7	0.15
K_2O	2.9	13.5	1.3
NiO	<0.1	<0.1	0.3
Cr_2O_3	<0.1	<0.1	7.0
SO_3	7.3	27.5	10.0
Cl	0.02	<0.01	<0.01
Water Soluble, %	9.0	45.4	9.0
pH	3.0	2.2	4.3
Excess SO_3, %	0.5	11.2	11.8

Fig. 29 Analyses of typical ash deposit from 18Cr-8Ni superheater tube.

By far the largest portion of the mineral matter or its derived species react to form the glassy particulates of flyash. The flyash and volatile species in the flue gases tend to deposit on the tube surfaces in a selective manner and subsequent reactions between these materials occur over long periods of time.

In the formation of corrosive deposits, flyash first deposits on the superheater and reheater tubes. Slowly, over a period of weeks, the alkalis and the sulfur oxides diffuse through the layer of flyash toward the tube surface. In the lower temperature zone of the ash deposit, chemical reactions between the alkalis, the sulfur oxides, and the iron and aluminum components of the flyash result in the formation of the complex alkali sulfates as follows:

$$3K_2SO_4 + Fe_2O_3 + 3SO_3 \rightarrow 2K_3Fe(SO_4)_3$$

and **(13)**

$$K_2SO_4 + Al_2O_3 + 3SO_3 \rightarrow 2KAl(SO_4)_2$$

Similar reactions occur with sodium sulfate (Na_2SO_4), although the complex sodium sulfates are less apt to form at high temperatures because of their lower stability.

Work at B&W's Research Center has shown that SO_3 concentrations in ash deposits must be very high (1000 to 1500 ppm) compared to the level in the flue gas (10 to 25 ppm) in order to form the complex alkali sulfates in the intermediate layer. Therefore the bulk of the SO_3 must come from the catalytic oxidation of SO_2 in the outer layer of the deposit.

When the SO_3 produced in the outer deposit exceeds the partial pressure of SO_3 necessary for stability, the complex sulfates form through the above reactions. When the opposite is true, the complex sulfates begin to decompose according to the reverse of these reactions until a new equilibrium is reached. Because the formation of SO_3

is temperature dependent, the reversibility of these reactions is also temperature dependent. As shown in Fig. 28, the corrosion rate increases with temperature, passes through a maximum between 1250 and 1350F (677 and 732C), and then falls to a comparatively low level at higher temperatures.

The temperature range of this rapid liquid-phase attack is bracketed by: 1) the melting temperature of the mixture of complex alkali sulfates present, and 2) their thermal stability limits. The extreme width of this temperature band is approximately 400F (222C); corrosion due to the complex alkali sulfates may range from as low as 1000F (538C) to a maximum of 1400F (760C), depending on the species present in the intermediate layer.

Corrective measures

Various methods of combatting corrosion of superheater and reheater tubes have been used or suggested, including the following:

1. the use of stainless steel shields to protect the most vulnerable tubes,
2. coal selectivity,
3. improvement of combustion conditions, i.e., providing proper coal fineness, fast ignition, good mixing and proper excess air,
4. the use of more corrosion resistant alloys and ceramic coatings on the most vulnerable superheater and reheater tubes. (See Chapter 18.)

Fuel oil ash

The ash content of residual fuel oil seldom exceeds 0.2%, an exceedingly small amount compared to that in coal. Nevertheless, even this small quantity of ash is capable of causing severe problems of deposition and corrosion in boilers. Of the many elements that may appear in oil ash deposits, the most important are vanadium, sodium and sulfur. Compounds of these elements are found in almost every deposit in boilers fired by residual fuel oil and often constitute the major portion of these deposits.

Origin of ash

As with coal, some of the ash forming constituents in the crude oil had their origin in animal and vegetable matter from which the oil was derived. The remainder is extraneous material, resulting from contact of the crude oil with rock structures and salt brines, or is picked up during refining processes, storage and transportation. (See Chapter 8.)

In general, the ash content increases with increasing asphaltic constituents in which the sulfur acts largely as a bridge between aromatic rings. Elemental sulfur and hydrogen sulfide have been identified in crude oil, and simpler sulfur compounds are found in the distillates of crude oil including thio-esters, disulfides, thiophenes and mercaptans.

Vanadium, iron, sodium, nickel and calcium in the fuel oil were probably derived from the rock strata but some elements, such as vanadium, nickel, zinc and copper, probably came from organic matter from which the petroleum was derived. Vanadium and nickel especially are known to be present in organo-metallic compounds known as porphyrins which are characteristic of certain forms of animal life. Table 7 indicates the amounts of vanadium, nickel and sodium present in residual fuel oils from various crudes.

Table 7
Vanadium, Nickel and Sodium Content of Residual Fuel Oils (ppm by wt)

Source of Crude Oil	Vanadium	Nickel	Sodium
Africa:			
1	5.5	5	22
2	1	5	--
Middle East:			
3	7	--	1
4	173	51	--
5	47	10	8
U.S.:			
6	13	--	350
7	6	2.5	120
8	11	--	84
Venezuela:			
9	--	6	480
10	57	13	72
11	380	60	70
12	113	21	49
13	93	--	38

Crude oil as such is not normally used as a fuel but is further processed to yield a wide range of more valuable products. For example, in a modern U.S. refinery the average product yield, as a percentage of total throughput, is:

Gasoline	44.4
Lube oil	16.4
Jet fuel	6.2
Kerosene	2.9
Distillates	22.5
Residual fuel	7.6

Virtually all metallic compounds and a large part of the sulfur compounds are concentrated in the distillation residue, as illustrated for sulfur in Table 8. Where low sulfur residual fuel oils are required, they are obtained by blending with suitable stocks, including both heavy distillates and distillation from low sulfur crudes. This procedure is also used occasionally if a residual fuel oil must meet specifications such as vanadium or ash content.

Table 8
Sulfur Content in Fractions of Kuwait Crude Oil

Fraction	Distillation Range, F (C)	Total Sulfur % by wt
Crude oil	---	2.55
Gasoline	124 to 253 (51 to 123C)	0.05
Light naphtha	257 to 300 (125 to 149C)	0.05
Heavy naphtha	307 to 387 (153 to 197C)	0.11
Kerosene	405 to 460 (207 to 238C)	0.45
Light gas oil	477 to 516 (247 to 269C)	0.85
Heavy gas oil	538 to 583 (281 to 306C)	1.15
Residual oil	588 to 928 (309 to 498C)	3.70

Source: Hixon, F.E., Shell Refining and Marketing Co., Ltd., *Chemistry and Industry*, p. 333, March 26, 1955.

Release of ash during combustion

Residual fuel oil is preheated and atomized to provide enough reactive surface so that it will burn completely within the boiler furnace. (See Chapter 10.) The atomized fuel oil burns in two stages. In the first stage the volatile portion burns and leaves a porous coke residue and in the second stage the coke residue burns. In general, the rate of combustion of the coke residue is inversely proportional to the square of its diameter, which in turn is related to the droplet diameter. Therefore, small fuel droplets give rise to coke residues that burn very rapidly, and the ash forming constituents are exposed to the highest temperatures in the flame envelope. The ash forming constituents in the larger coke residues from the larger fuel droplets are heated more slowly, partly in association with carbon. Release of the ash from these residues is determined by the rate of oxidation of the carbon.

During combustion, the organic vanadium compounds in the residual fuel oil thermally decompose and oxidize in the gas stream to V_2O_3, V_2O_4 and finally V_2O_5. Although complete oxidation may not occur and there may be some dissociation, a large part of the vanadium originally present in the oil exists as vapor phase V_2O_5 in the flue gas. The sodium, usually present as chloride in the oil, vaporizes and reacts with sulfur oxides either in the gas stream or after deposition on tube surfaces.

Subsequently, reactions take place between the vanadium and sodium compounds, with the formation of complex vanadates having melting points lower than those of the parent compounds; for example:

$$Na_2SO_4 + V_2O_5 \rightarrow 2NaVO_3 + SO_3\uparrow$$

Melting points: 1625F 1275F 1165F **(14)**

Excess sodium or vanadium in the ash deposit, above that necessary for the formation of the sodium vanadates (or vanadylvanadates), may be present as Na_2SO_4 and V_2O_5, respectively.

The sulfur in residual fuel oil is progressively released during combustion and is promptly oxidized to SO_2. A small amount of SO_2 is further oxidized to SO_3 by a small amount of atomic oxygen present in the hottest part of the flame. Also, catalytic oxidation of SO_2 to SO_3 may occur as the flue gases pass over vanadium rich ash deposits on high temperature superheater tubes and refractories. (See Chapter 35.)

Oil slag formation and deposits

The deposition of oil ash constituents on the furnace walls and superheater surfaces can be a serious problem. This deposition, coupled with corrosion of superheater and reheater tubes by deposits, was largely responsible for the break in the trend towards higher steam temperatures that occurred in the early 1960s.

Practically all boiler installations are now designed for steam temperatures in the 1000 to 1015F (538 to 546C) range to minimize those problems and to avoid the higher capital costs of the more expensive alloys required in the tubes, steam piping and turbine for 1050 to 1100F (566 to 593C) steam conditions.

There are many factors affecting oil ash deposition on boiler heat absorbing surfaces. These factors may be grouped into the following interrelated categories: characteristics of the fuel oil, design of the boiler and operation of the boiler.

Characteristics of fuel oil ash

Sodium, sulfur and vanadium are the most significant elements in the fuel oil because they can form complex compounds having low melting temperatures, 480 to 1250F (249 to 677C), as shown in Table 9. Such temperatures fall within the range of tube metal temperatures generally encountered in furnace and superheater tube banks of many oil-fired boilers. However, because of its complex chemical composition, fuel oil ash seldom has a single sharp melting point, but rather softens and melts over a wide temperature range.

An ash particle that is in a sticky, semi-molten state at the tube surface temperature may adhere to the tube if it is brought into contact by the gas flow over the tube. Even a dry ash particle may adhere due to mutual attraction or surface roughness. Such an initial deposit layer will be at a higher temperature than that of the tube surface because of its relatively low thermal conductivity. This increased temperature promotes the formation of adherent deposits. Therefore, fouling will continue until the deposit surface temperature reaches a level at which all of the ash in the gas stream is in a molten state, so that the surface is merely washed by the liquid without freezing and continued buildup.

In experimental furnaces, it has been found that the initial rate of ash buildup was greatest when the sodium-vanadium ratio in the fuel oil was 1:6, but an equilibrium thickness of deposit [0.125 to 0.25 in. (3.175 to 6.35 mm)] was reached in approximately 100 hours of operation. When the fuel oil contained more refractory constituents,

Table 9
Melting Points of Some Oil Ash Constituents

Compound	Melting Point, F	(C)
Aluminum oxide, Al_2O_3	3720	(2049)
Aluminum sulfate, $Al_2(SO_4)_3$	1420*	(771)
Calcium oxide, CaO	4662	(2572)
Calcium sulfate, $CaSO_4$	2640	(1449)
Ferric oxide, Fe_2O_3	2850	(1566)
Ferric sulfate, $Fe_2(SO_4)_3$	895*	(479)
Nickel oxide, NiO	3795	(2091)
Nickel sulfate, $NiSO_4$	1545*	(841)
Silicon dioxide, SiO_2	3130	(1721)
Sodium sulfate, Na_2SO_4	1625	(885)
Sodium bisulfate, $NaHSO_4$	480*	(249)
Sodium pyrosulfate, $Na_2S_2O_7$	750*	(399)
Sodium ferric sulfate, $Na_3Fe(SO_4)_3$	1000	(538)
Vanadium trioxide, V_2O_3	3580	(1971)
Vanadium tetroxide, V_2O_4	3580	(1971)
Vanadium pentoxide, V_2O_5	1275	(691)
Sodium metavanadate,		
$Na_2O{\cdot}V_2O_5(NaVO_3)$	1165	(629)
Sodium pyrovanadate, $2Na_2O{\cdot}V_2O_5$	1185	(641)
Sodium orthovanadate, $3Na_2O{\cdot}V_2O_5$	1560	(849)
Sodium vanadylvanadates,		
$Na_2O{\cdot}V_2O_4{\cdot}V_2O_5$	1160	(627)
$5Na_2O{\cdot}V_2O_4{\cdot}11V_2O_5$	995	(535)

* Decomposes at a temperature around the melting point.

such as silica, alumina and iron oxide, in addition to sodium and vanadium, an equilibrium condition was not reached and the tube banks ultimately plugged with ash deposits. However, these ash deposits were less dense, i.e., more friable, than the glassy slags encountered with a 1:6 sodium-vanadium fuel oil. Both the rate of ash buildup and the ultimate thickness of the deposits are also influenced by physical factors such as the velocity and temperature of the flue gases and particularly the tube metal temperature.

In predicting the behavior of a residual oil insofar as slagging and tube bank fouling are concerned, several fuel variables are considered including: 1) ash content, 2) ash analysis, particularly the sodium and vanadium levels and the concentration of major constituents, 3) melting and freezing temperatures of the ash, and 4) the total sulfur content of the oil. Applying this information in boiler design is largely a matter of experience.

Boiler design

Generally speaking, progressive fouling of furnaces and superheaters should not occur as long as the ash characteristics are not severe compared to the tube metal temperatures. If such trouble is encountered, the solution can usually be found in improving combustion conditions in the furnace and/or modifying the sootblowing procedures.

Studies on both laboratory and field installations have shown that the rate of ash deposition is a function of the velocity and temperature of the flue gases and the concentration of oil ash constituents in the flue gases. The geometry of the furnace and the spacing of tubes in the convection banks are selected in the design of a boiler to minimize the rate of deposition. It is common practice to use in-line tube arrangements with progressively wider lateral spacings for tubes located in higher gas temperature zones. This makes bridging of ash deposits between tubes less likely and facilitates cleaning of tube banks by the sootblowers.

Boiler operation

Poor atomization of the fuel oil results in longer flames and frequently increases the rate of slag buildup on furnace walls which, in turn, makes it more difficult to keep the convection sections of the boiler clean. Completing combustion before the gases pass over the first row of tubes is especially important. Relatively large carbonaceous particles have a far greater tendency to impinge on the tubes than do the smaller ash particles. If these larger particles are in a sticky state, they will adhere to the tubes where oxidation will proceed at a slow rate with consequent formation of ash. Fouling from this cause is difficult to detect by inspection during boiler outages because the carbonaceous material has usually disappeared completely. It can generally be detected during operation because flames are usually long and smoky, and sparklers may be carried along in the flue gases.

Regular and thorough sootblowing can have a decisive effect on superheater and reheater fouling. (See Chapter 23.) To be fully effective, however, sootblowing cycles should be frequent enough so that ash deposits can not build to a thickness where their surfaces become semi-molten and difficult to remove. In instances of extreme slagging, it is sometimes necessary to relocate sootblowers, to install additional sootblowers to control deposition in a critical zone, or to use additives.

The boiler load cycle can also have a significant effect on the severity of slagging and superheater fouling. A unit that is base loaded for long periods is more apt to have fouling problems on a borderline fuel oil than a unit that takes daily swings in load. In the latter instance, the furnace generally remains cleaner due to periodic shedding of slag, with the result that the gas temperatures through the superheaters are appreciably lower. This eases the burden on the sootblowers and substantially controls ash deposit formation in the superheater-reheater tube banks. Overloading the boiler, even for an hour or two a day, should be avoided, especially if excess air has to be lowered to the point where some of the burners are starved of air. The furnace is apt to become slagged and ash deposition can creep into the superheater and reheater tube banks.

Oil ash corrosion

High temperature corrosion

The sodium-vanadium complexes, usually found in oil ash deposits, are corrosive when molten. A measurable corrosion rate can be observed over a wide range of metal and gas temperatures, depending on the amount and composition of the oil ash deposit. Fig. 30 shows the combined gas and metal temperature effects on corrosion for a specific fuel oil composition of 150 ppm vanadium, 70 ppm sodium and 2.5% sulfur. As the vanadium concentration of the fuel oil varies, the amount of corrosion,

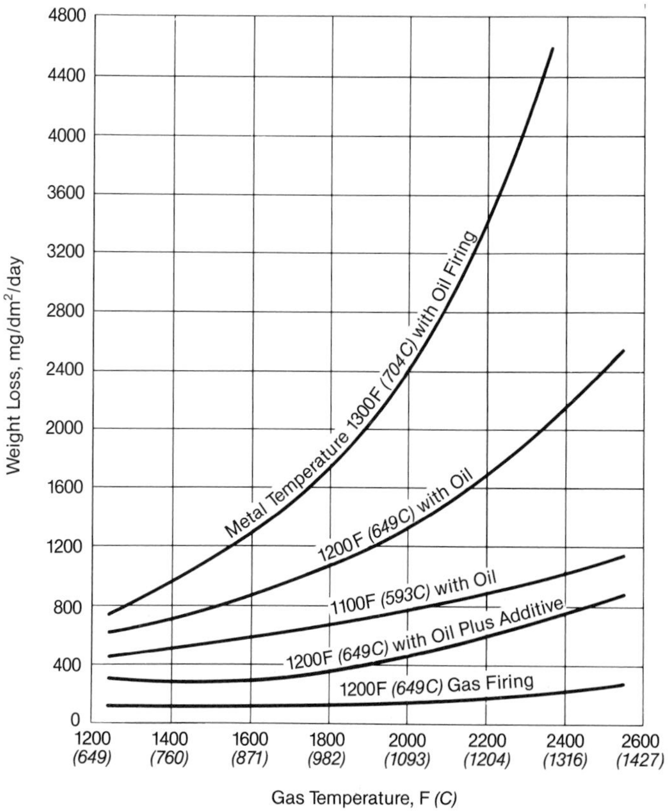

Fig. 30 Effect of gas and metal temperatures on corrosion of 304, 316 and 321 alloys in a unit fired with oil containing 150 ppm vanadium, 70 ppm sodium and 2.5% sulfur. Test duration 100 hours.

compared to a 150 ppm vanadium fuel, will increase or decrease according to the curve shown in Fig. 31.

The effect of the sodium level in the fuel oil is not as clear because combustion conditions and the chloride content of the fuel oil may be controlling. The sodium content does, however, definitely affect the minimum metal temperature at which corrosion will be significant.

At the present time there appears to be no alloy that is immune to oil ash corrosion. In general, the higher the chromium content of the alloy, the more resistant it is to attack. This is the main reason for the use of 18Cr-8Ni alloys for high temperature superheater tubes. High chromium contents, greater than 30%, give added corrosion resistance but at the expense of physical properties; 25Cr-20Ni has been used as a tube cladding but even this alloy has not provided complete protection. The presence of nickel in high temperature alloys is needed for strength and high nickel high chromium alloys may be fairly resistant to oil ash attack under oxidizing conditions. The higher material cost must be justified by longer life, which is not always predictable.

Low temperature corrosion

In oil-fired boilers the problem of low temperature corrosion resulting from the formation and condensation of sulfuric acid from the flue gases is similar to that previously described for coal firing.

Oil-fired boilers are more susceptible to low temperature corrosion than are most coal-fired units for two reasons: 1) the vanadium in the oil ash deposits is a good catalyst for the conversion of SO_2 to SO_3, and 2) there is a smaller quantity of ash in the flue gases. Ash particles in the flue gas react with and reduce the amount of SO_3 vapor in the gas. Because oil has considerably less ash than coal, significant differences would be expected. Furthermore, coal ash is more basic than oil ash and tends to neutralize any acid deposited; oil ash generally lacks this capability.

Under certain conditions, oil-fired boilers may emit acidic particulates from their stacks that stain or etch painted surfaces in the vicinity of the plant. The acidic deposits, or smuts, are generally caused by metallic surfaces (air heaters, flues and stacks) operating well below the acid dew point of the flue gases or by soot which has absorbed sulfuric acid vapor in its passage through the boiler.

Methods of control

The methods of control that have been used or proposed to control fouling and corrosion in oil-fired boilers are summarized in Table 10, but in every instance economics govern their applicability. There is no doubt that reducing the amount of ash and sulfur entering the furnace is the surest means of control and that minimizing the effects of the ash constituents, once they have deposited on the tubes, is the least reliable. Because the severity of fouling and corrosion depends not only on the fuel oil characteristics but also on boiler design and operating variables, a generalized solution to these problems can not be prescribed.

Fuel oil supply

Although fuel selection and blending are practiced to some extent in the U.S., the common purpose is to provide safe and reliable handling and storage at the power plant rather than to avoid fouling difficulties. Because the threshold limits of sodium, sulfur and vanadium are not accurately defined for either fouling or corrosion, use of these means of control can not be fully exploited.

Processes are available for both the desulfurization and de-ashing of fuel oils. Water washing of residual fuel oil has been successfully applied to a few marine-type boilers, but it is doubtful that it will be widely used because only sodium and sediment, mainly rust and sand, are removed by the process. Use of low sulfur, low ash crudes and desulfurized fuel oil is expected to increase.

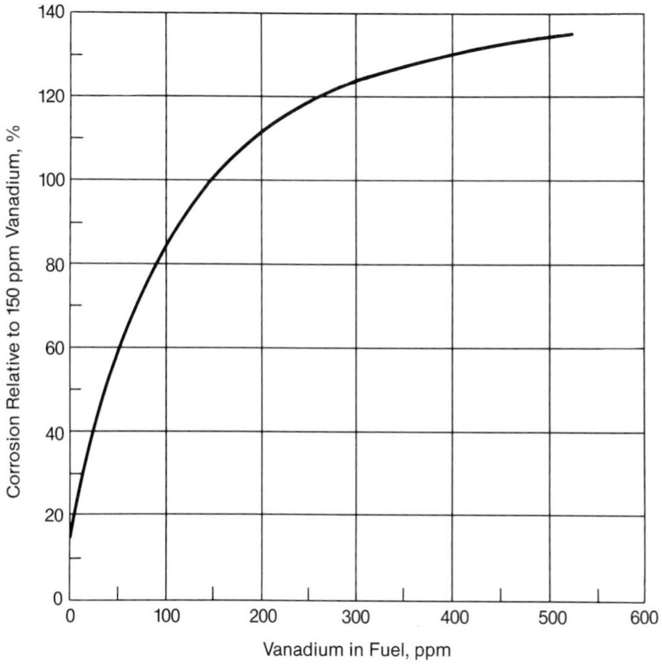

Fig. 31 Effect of vanadium concentration on oil ash corrosion.

Table 10
Classification of Methods for Controlling Fouling and Corrosion in Oil-Fired Boilers

	Fuel Oil Supply
Reduce amount of fuel ash constituents to the furnace	Selection Blending Purification
	Design
Minimize amounts of fuel ash constituents reaching heat transfer surfaces	Furnace geometry Tube bank arrangement Metal temperature Gas temperature Sootblower arrangement
	Operation
Minimize effects of bonding and corrosive compounds in ash deposits	Load cycle Sootblowing schedule Combustion — excess air Additives Water washing

Fuel oil additives

An approach that is effective where the fuel oil ash is most troublesome involves adding, to the fuel or furnace, small amounts of materials that change the character of the ash sufficiently to permit its removal by steam or air sootblowers or air lances.

Additives are effective in reducing the problems associated with superheater fouling, high temperature ash corrosion and low temperature sulfuric acid corrosion. Most effective are alumina, dolomite and magnesia. Kaolin is also a source of alumina. Analyses of typical superheater deposits from a troublesome fuel oil, before and after treating it with alumina or dolomite, are shown in three bar graphs on the left of Fig. 32. The results for a different oil treated with magnesia are shown in the bar graph at the right.

The reduction of fouling and high temperature corrosion is accomplished basically by producing a high melting point ash deposit that is powdery or friable and easily removed by sootblowers or lances. When the ash is dry, corrosion is considerably reduced.

Low temperature sulfuric acid corrosion is reduced by the formation of refractory sulfates by reaction with the SO_3 gas in the flue gas stream. By removing the SO_3 gas, the dew point of the flue gases is sufficiently reduced to protect the metal surfaces. The sulfate compounds formed are relatively dry and easily removed by the normal cleaning equipment.

In general, the amount of additive used should be about equal to the ash content of the fuel oil. In some instances, slightly different proportions may be required for best results, especially for a high temperature corrosion reduction, in which it is generally accepted that the additive should be used in weight ratios of 2:1 or 3:1, based on the vanadium content of the oil.

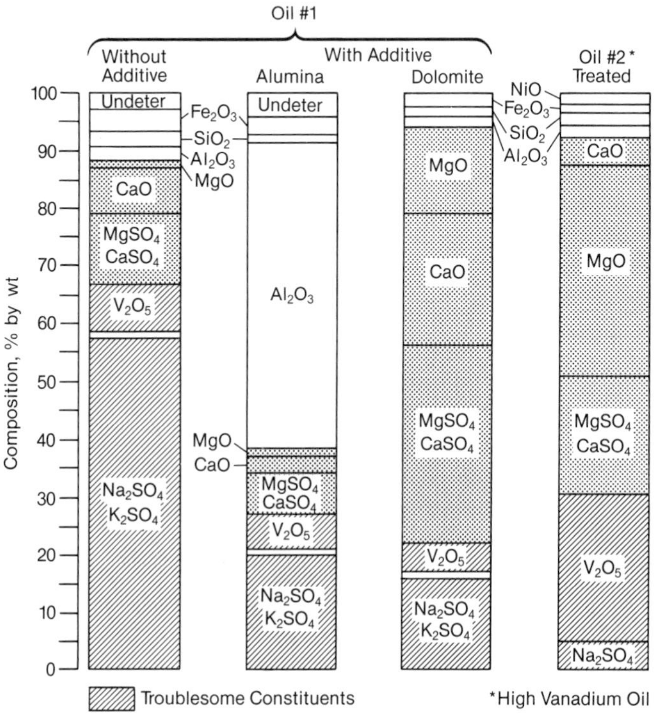

Fig. 32 Effect of fuel oil additives on composition of oil ash deposit.

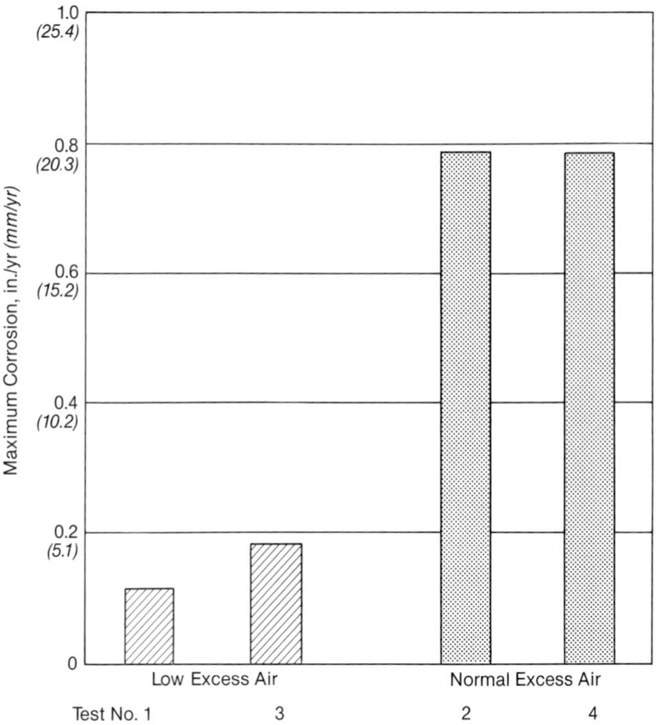

Fig. 33 Effect of low excess air combustion on high temperature oil ash corrosion.

Several methods have been successfully used to introduce the additive materials into the furnace. The one in general use consists of metering a controlled amount of an additive oil slurry in the burner supply line. The additive material should be pulverized to 100% through a 325 mesh (44 micron) screen for good dispersion and minimum atomizer wear.

For a boiler fired by a high pressure return flow oil system, it has been found advantageous to introduce the additive powders by blowing them into the furnace at the desired locations. The powder has to be reduced in size to 100% through a 325 mesh (44 micron) screen for good dispersion.

The choice of a particular additive depends on its availability and cost to the individual plant and the method of application chosen. For example, alumina causes greater sprayer plate wear than the other materials when used in an oil slurry.

The quantity of deposit formed is, of course, an important consideration for each unit from the aspect of cleaning. A comparison of the amounts of deposit formed with different additives shows that dolomite produces the greatest quantity because of its sulfating ability, magnesia is intermediate, and alumina and kaolin form the least. However, when adequate cleaning facilities are available, the deposits are easily removed and the quantities formed should not be a problem.

Excess air control

As mentioned previously the problems encountered in the combustion of residual fuels — high temperature deposits (fouling), high temperature corrosion and low temperature sulfuric acid corrosion — all arise from the presence of vanadium and sulfur in their highest states of

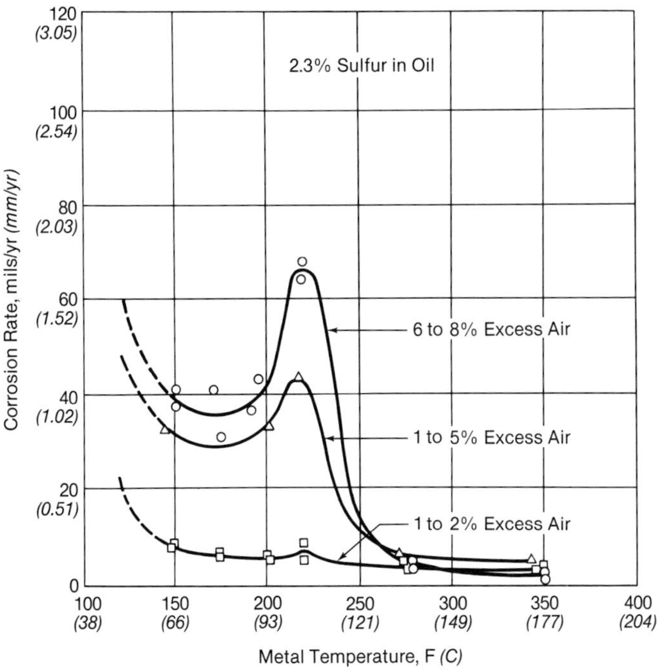

Fig. 34 Effect of excess air on low temperature corrosion of carbon steel.

oxidation. By reducing the excess air from 7% to 1 or 2%, it is possible to avoid the formation of fully oxidized vanadium and sulfur compounds and, thereby, reduce boiler fouling and corrosion problems.

In a series of tests on an experimental boiler, it was found that the maximum corrosion rate of type 304 stainless steel superheater alloy held at 1250F (677C) in 2100F (1149C) flue gas was reduced more than 75% (Fig. 33) when the excess air was reduced from an average of 7% to a level of 1 to 2%. Moreover, the ash deposits that formed on the superheater bank were soft and powdery, in contrast to hard, dense deposits that adhered tenaciously to the tubes when the excess air was around 7%. Also, the rate of ash buildup was only half as great. Operation at the 1 to 2% excess air level practically eliminated low temperature corrosion of carbon steel at all metal temperatures above the dew point of the flue gases (Fig. 34). However, much of the beneficial effects of low excess air combustion are lost if the excess air at the burner fluctuates even for short periods of time to a level of about 5%. Carbon loss values for low excess air were approximately 0.5%, which is generally acceptable for electric utility and industrial practice.

A number of large industrial boilers both in the U.S. and in Europe have been operating with low excess air for several years. As a result, the benefits of reducing low temperature corrosion are well established. However, the benefits on high temperature slagging and corrosion are not wholly conclusive. In any event, great care must be exercised to distribute the air and fuel oil equally to the burners, and combustion conditions must be continuously monitored to assure that combustion of the fuel is complete before the combustion gases enter the convection tube banks.

Midwest power station burning bituminous coal.

Chapter 21
Performance Calculations

The evaluation of boiler performance involves many complex factors as indicated in preceding chapters. Only a few of these factors are subject to precise analysis; many others are the result of data taken from operating units. Ash in the fuel has perhaps the most dramatic impact on boiler performance as discussed in Chapter 20. In spite of the large number of variables, boilers are designed, built and operated in conformance with design specifications.

A well designed and operated boiler completes combustion within the furnace. Gas temperatures leaving the furnace can be predicted by the methods presented in Chapter 4. Beyond the furnace, heat transfer surface arrangements represent a balance of temperature difference, space, pressure drop and draft losses. The final selection of these surface arrangements represents a compromise on the designer's part in meeting performance requirements while controlling ash deposition, corrosion and erosion.

This chapter introduces the basic principles of boiler performance calculations. It illustrates the use of heat transfer, thermodynamics and fluid mechanics to determine heat and material balances for given boiler heat transfer components through practical application. These principles, as well as fundamental relationships, experimental data, operational experience and designer knowledge, are being incorporated into advanced numerical computational models for boiler evaluation. As discussed in Chapters 3, 4, 18 and elsewhere in *Steam*, these models are becoming increasingly useful in the design of boilers.

For the fossil fuel-fired steam generators being considered, the hotter heat transfer medium consists of the products of combustion, or flue gases. The cooler medium is superheated steam, steam-water mixtures at saturation, water or air depending upon the heat transfer component under consideration. Heat transfer surfaces can be categorized into one of four cases according to relative hot and cold medium flow direction and temperature as shown in Fig. 1. Typically, boiler banks or screens are Case I, superheaters and reheaters are either Case II or III, economizers are Case II or III, and air heaters are generally Case IV.

To avoid the confusion of dual units, this chapter is provided in English units only. See Appendix 1 for SI conversion factors.

Performance calculations are typically used to establish one of three parameters: temperature, heat transfer surface area, or surface cleanliness. As in most thermal analysis problems, the evaluation of boiler performance is an iterative process. To evaluate flue gas and

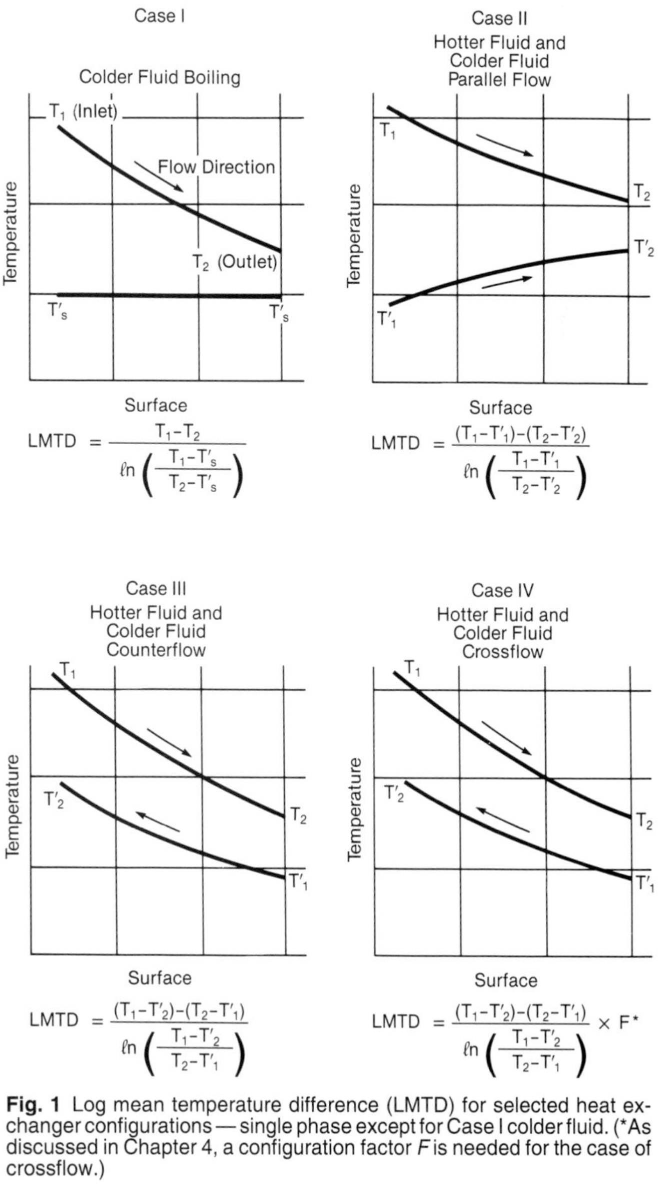

Fig. 1 Log mean temperature difference (LMTD) for selected heat exchanger configurations — single phase except for Case I colder fluid. (*As discussed in Chapter 4, a configuration factor *F* is needed for the case of crossflow.)

steam temperatures for a known boiler design arrangement, the surface area and surface cleanliness are normally known while the temperatures are assumed. The outlet temperature calculation updates subsequent iterations until convergence between assumed and calculated temperatures is achieved.

Heat transfer surface area or sizing can be determined for given fluid temperatures and surface cleanliness by assuming an initial surface arrangement and then confirming the desired thermal performance by calculation. High calculated outlet gas temperatures indicate the need for additional surface whereas low calculated outlet gas temperatures indicate the need to remove surface. Surface area is adjusted until calculated and specified temperatures converge.

Finally, for a given boiler configuration, measured temperatures can be used to assess surface cleanliness. An initially assumed cleanliness is used with the measured temperatures and known surface to verify the temperature data. The cleanliness factors are varied until temperature convergence is achieved.

When calculating the thermal performance of heat transfer equipment, initial temperatures are selected in part based upon past experience. These are used to establish the thermophysical properties, calculate mean temperature differences and solve the problem. The resulting final temperatures are compared to the original assumptions. Generally, if an error of 5F or less is observed between iteration steps, the solution is considered complete. Otherwise, the new temperature is used to repeat the analysis.

To illustrate performance calculations, a small boiler (Fig. 2) with a simplified furnace and heating surface arrangement will serve as an example. The objective is to evaluate the outlet flue gas temperatures for the given surface area and cleanliness. The procedures developed here are fundamental and as such, can be used on a wide range of boiler applications.

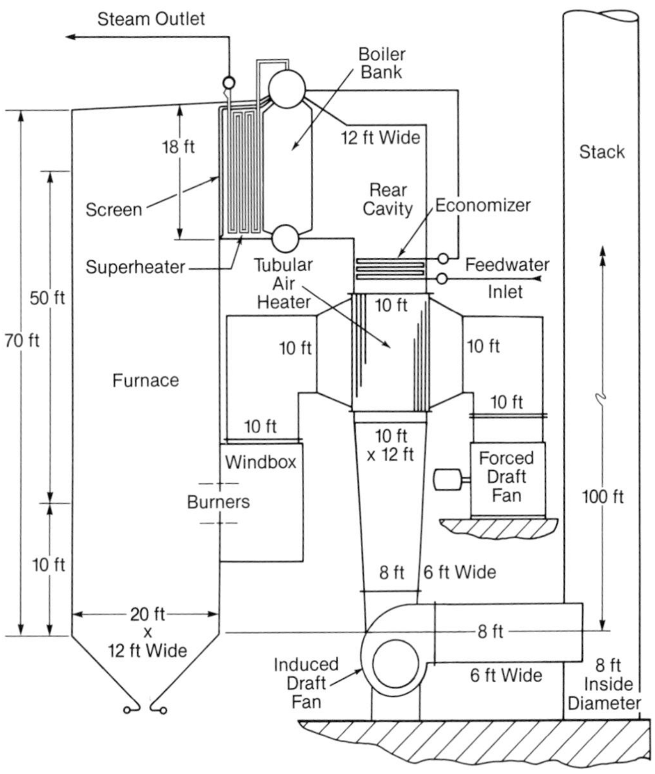

Fig. 2 Example coal-fired industrial boiler, sectional side view.

Operating conditions

Boiler performance specifications are defined by the customer. These specifications normally include: steam output conditions — pressure, temperature and flow; feedwater conditions; fuel and ash analysis; load range; capability; and efficiency. The final boiler design efficiently meets the specifications with a minimum of surface, materials and flow losses.

The specific analysis procedure is started at slightly different steps depending upon whether an existing piece of equipment is being analyzed or a new plant is being designed. For an existing installation, boiler performance calculations begin by establishing the geometry of all heat transfer equipment and by defining the required operating conditions. Heat and material balances, including combustion calculations for the steam generator as a whole, are then determined. These calculations provide the information required to analyze each heat recovery component. The performance of each component is often intimately tied to other components within the boiler. As discussed in Chapter 1, the calculation process normally follows the direction of gas flow from furnace to stack and that is the approach used in the following example. Predictions of unit performance are only complete when the performance of all devices within the boiler envelope agrees with the overall unit heat and material balances.

If a new plant is being analyzed, the process begins with the heat and material balances to establish air, fuel and flue gas handling requirements. The process then continues with the physical sizing of equipment and components as discussed above.

Because air resistance and draft loss calculations, including stack effects, are dependent on established air and gas temperature profiles through the boiler and accessories, they are determined after any thermal analysis. Pressure drop calculations are included in each component evaluation. Flue and duct resistance calculations and the evaluation of stack effects are the final steps in the performance calculations.

For the example considered here, dimensional parameters are contained in Table 1, while Table 2 contains the specified operating conditions. The remainder of this chapter focuses on predicting performance for the hypothetical boiler.

Heat and material balances

Heat and material balances begin with combustion calculations. For this example, combustion calculations are determined by the *Btu method* as presented in Chapter 9. For the fuel analysis and losses specified in Table 2, the combustion calculations are summarized in Table 3.

The unit is expected to produce 250,000 lb/h superheated steam at 450 psig and 650F with feedwater conditions of 470 psig and 220F entering the economizer. The energy flowing out with the steam is calculated from the energy balance using the information summarized in Table 2:

Table 1
Physical Arrangement – Furnace

Construction: 2.5 in. outside diameter (OD) tubes on 3 in. centers with membrane construction. (See Chapter 22 for description.)
Width: 12 ft — Volume: 18,000 ft^3 to superheater entrance plane
Depth: 20 ft — Surface: 5,050 ft^2 flat projected area, not including superheater exit plane

Physical Arrangement — Components

Parameter	Units	Screen	Superheater (Note 1)	Boiler Bank	Economizer	Air Heater
Tube OD	in.	2.5	2.5	2.5	2	2
Backspacing (centerline)	in.	6	3.25	4	3	2.5
Sidespacing (centerline)	in.	6	6	4	3	3.5
Rows deep		2	12	28	10	53
Rows wide		23	23	35	47	41
Tube length	ft	18	18	17 (Note 2)	10	16
Heating surface (Note 3)	ft^2	542	3250	10,900	2450	18,205
Free flow area (Note 4)	ft^2					
Gas		130	130	80	42	39.2
Air						82.7

Notes:
1. The superheater is a counterflow configuration; however, the steam flows in two parallel paths from steam drum to superheater outlet header. This is also called a two-flow arrangement.
2. Boiler bank tubes vary in length; the value listed represents an average length.
3. Heating surface is the external surface area of the tubes exposed to the flue gases except for the air heater where the flue gas flows through rather than over the tubes.
4. Free flow area is the minimum clear area between tubes, perpendicular to the direction of gas or air flow except for air heaters. The gas side free flow area for air heaters is the area defined by the inside diameter (ID) of the air heater tubes.

$$\text{Output} = \dot{m}\,(H_2 - H_1) = 285.6 \times 10^6 \text{ Btu/h} \qquad \textbf{(1)}$$

where

$\dot{m}$ = steam flow rate = 250,000 lb/h
H_2 = steam outlet energy per unit mass (enthalpy) = 1331.5 Btu/lb
H_1 = water inlet energy per unit mass (enthalpy) = 189.2 Btu/lb

The combustion evaluation summarized in Table 3 also establishes the three flow rates which guide much of the subsequent equipment design — heat input, gas weight and air weight. The heat input as fuel is calculated by dividing the required steam flow energy output by the boiler efficiency:

$$\text{Input} = \text{Output/Efficiency} = 328.6 \times 10^6 \text{ Btu/h} \qquad \textbf{(2)}$$

where

Output = 285.6×10^6 Btu/h
Efficiency = 86.9%

The products of combustion or flue gas weight flowing through the boiler can be determined from the fuel heat input and the wet gas weight (including moisture) established in the combustion calculations:

$$\text{Gas weight} = \text{Input} \times \text{Wet gas weight} = 324,100 \text{ lb/h} \qquad \textbf{(3)}$$

where

Input = 328.6×10^6 Btu/h
Wet gas weight = 9.864 lb/10^4 Btu

Using the assumed excess air of 20% for pulverized coal firing from Table 3, combustion air to the burners can be calculated:

$$\text{Air mass flow rate} = \text{Input} \times \text{Moisture correction} \times \text{Dry air mass} = 302,500 \text{ lb/h} \qquad \textbf{(4)}$$

where

Input = 328.6×10^6 Btu/h
Moisture correction = 1.013 lb moist air/lb dry air
Dry air mass = 9.086 lb/10^4 Btu

The dry air mass includes excess air, therefore air mass flow rate represents total air to the furnace. Other information determined in the combustion calculations (Table 3) are heat available from the fuel (1034 Btu/lb) and moisture in the flue gas (4.28%). These values are noted for future reference.

This completes the heat and material balances and combustion calculations for the defined boiler envelope. As shown in Fig. 3, all pressures, temperatures and flows crossing the unit boundaries are established and calculations can now proceed on each component. A more detailed analysis would also account for items such as air infiltration, continuous boiler water discharge (blowdown), saturated steam extractions and steam reheaters if used in the unit.

Component performance calculations

Furnace

Furnace exit gas temperatures must be determined to design downstream heat transfer components. Through testing and correlation of gas temperature data, furnace exit gas temperature has been found to have a relation-

Table 2
Operating Conditions
Fuel: Bituminous Coal, Virginia
Analysis As-Fired

Ultimate, % by wt		Proximate, % by wt	
C	80.31	Moisture	2.90
H_2	4.47	Volatiles	22.05
S	1.54	Fixed carbon	68.50
O_2	2.85	Ash	6.55
N_2	1.38		100.00
H_2O	2.90		
Ash	6.55		
	100.00		

Higher heating value (HHV), as-fired: 14,100 Btu/lb

Excess air	20.0% by wt
Unburned carbon loss	0.4% by wt
Unaccounted loss	1.5% by wt
ABMA radiation loss (see Chapter 22)	0.40% by wt
Furnace exit gas temperature	2000F
Superheater outlet:	
Steam flow	250,000 lb/h
Steam temperature	650 F
Steam pressure	450 psig
Steam enthalpy	1331.5 Btu/lb
Economizer inlet:	
Water flow	250,000 lb/h
Water temperature	220 F
Water pressure	470 psig
Water enthalpy	189.2 Btu/lb
Air heater:	
Air temperature entering	80 F
Barometric pressure	30 in. Hg
Gas temperature leaving	390 F

ship to the heat input of the fuel and to the effectiveness of the furnace walls. In Chapter 4, furnace exit gas temperature curves for various fuels are approximated. This extended family of curves represents the accumulation of extensive field experience and analytical evaluation. They are dependent on fuel and furnace geometry.

The heat which can be absorbed by the furnace was determined from the combustion calculations to be 1034 Btu/lb of flue gas. The layout of the furnace (Table 1) provides a flat projected surface of 5050 ft^2. For 2.5 in. OD furnace tubes on 3 in. centers, Fig. 31 in Chapter 4 indicates an effectiveness factor of 1.0. The heat release rate to the furnace is then:

$$\text{Heat release rate} = \frac{\text{Heat available} \times \text{Gas mass flow rate}}{\text{Flat projected area} \times \text{Effectiveness factor}} = 66.4 \times 10^3 \text{ Btu/h ft}^2 \qquad (5)$$

where

Heat available = 1034 Btu/lb
Gas mass flow rate = 324.1×10^3 lb/h
Flat projected area = 5050 ft^2
Effectiveness factor = 1.0

Furnace exit gas temperature is 2000F from Chapter 4, Fig. 32. Sufficient information is now known to begin analysis of the convection pass, i.e., screen, superheater, boiler bank, economizer and air heater.

The performance of the convection pass, however, is a function of the cleanliness, or effectiveness, of the heat transfer surfaces. The values of heat transfer coefficient, as described in Chapter 4, apply to heat transfer surfaces free from ash and slag deposits. For calculation purposes, the effect of ash or other deposits on the heat transfer surfaces can be accounted for by a cleanliness factor:

$$\text{Cleanliness factor} = \frac{\text{Actual heat transfer rate}}{\text{Clean surface heat transfer rate}} \qquad (6)$$

Therefore, a clean new surface would have a cleanliness factor of 1.0. In certain coal- or refuse-fired boilers where slag and deposits are difficult to remove, a factor of less than 1.0 may be required. The cleanliness factor for each heat transfer surface and flue gas composition is multiplied times the product of the overall heat transfer coefficient, U, the area, A, and the temperature difference, ΔT, to establish the overall heat transfer rate. The calculations that follow use a cleanliness factor of 1.0 and the cleanliness factor has been omitted for simplicity.

Screen

Heat transfer

In this boiler design, the gases leaving the furnace first pass across the screen tubes. In this case, the tubes contain boiling water. These tubes control the amount of furnace radiation reaching the superheater surface. The heat transfer relationship for this surface is as follows:

$$q = U\,A\,(\text{LMTD}) = \dot{m}_g\, c_p\, \Delta T_g = \dot{m}_g\, c_p\, (T_1 - T_2) \qquad (7)$$

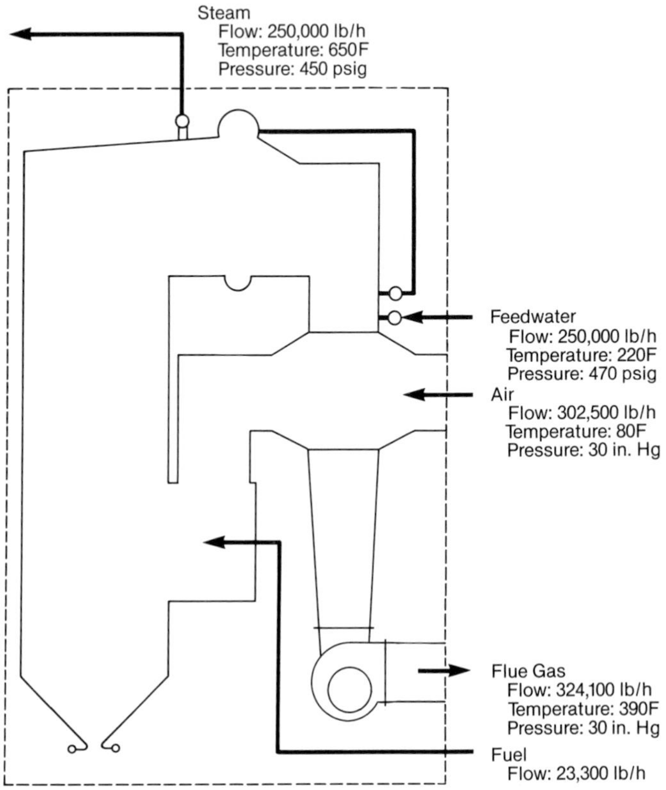

Fig. 3 Example boiler fuel, air, gas, water and steam flow streams.

Table 3
Combustion Calculations - Btu Method

	INPUT CONDITIONS - BY TEST OR SPECIFICATION	
1	Excess air: at burner/leaving boiler/econ, % by weight	20/20
2	Entering air temperature, F	80
3	Reference temperature, F	80
4	Fuel temperature, F	80
5	Air temperature leaving air heater, F	350
6	Flue gas temperature leaving (excluding leakage), F	390
7	Moisture in air, lb/lb dry air	0.013
8	Additional moisture, lb/100 lb fuel	0
9	Residue leaving boiler/economizer, % Total	85
10	Output, 1,000,000 Btu/h	285.6
	Corrections for sorbent (from Table 14, Chapter 9 if used)	
11	Additional theoretical air, lb/10,000 Btu Table 14, Item [21]	0
12	CO_2 from sorbent, lb/10,000 Btu Table 14, Item [19]	0
13	H_2O from sorbent, lb/10,000 Btu Table 14, Item [20]	0
14	Spent sorbent, lb/10,000 Btu Table 14, Item [24]	0

FUEL - *Bituminous coal, Virginia*

	15 Ultimate Analysis		16 Theo Air, lb/100 lb fuel		17 H_2O, lb/100 lb fuel	
	Constituent	% by weight	K1	[15] x K1	K2	[15] x K2
A	C	80.31	11.51	924.4		
B	S	1.54	4.32	6.7		
C	H_2	4.47	34.29	153.3	8.94	39.96
D	H_2O	2.90			1.00	2.90
E	N_2	1.38				
F	O_2	2.85	− 4.32	− 12.3		
G	Ash	6.55				
H	Total	100.00	Air	1072.1	H_2O	42.86

18	Higher heating value (HHV), Btu/lb fuel		14,100
19	Unburned carbon loss, % fuel input		0.40
20	Theoretical air, lb/10,000 Btu	[16H] x 100 / [18]	7.604
21	Unburned carbon, % of fuel	[19] x [18] / 14,500	0.39

COMBUSTION GAS CALCULATIONS, Quantity / 10,000 Btu Fuel Input

			A At Burners	B Infiltration	C Leaving Furnace		D Leaving Blr/Econ	
22	Theoretical air (corrected), lb/10,000 Btu	[20] − [21] x 1151 / [18] + [11]						7.572
23	Residue from fuel, lb/10,000 Btu	([15G] + [21]) x 100 / [18]						0.049
24	Total residue, lb/10,000 Btu	[23] + [14]						0.049
25	Excess air, % by weight		20.0	0.0		20.0		20.0
26	Dry air, lb/10,000 Btu	(1 + [25] / 100) x [22]				9.086		9.086
27	H_2O from air, lb/10,000 Btu	[26] x [7]			0.118	0.118	0.118	0.118
28	Additional moisture, lb/10,000 Btu	[8] x 100 / [18]			0.000	0.000	0.000	0.000
29	H_2O from fuel, lb/10,000 Btu	[17H] x 100 / [18]			0.304		0.304	
30	Wet gas from fuel, lb/10,000 Btu	(100 − [15G] − [21]) x 100 / [18]				0.660		0.660
31	CO_2 from sorbent, lb/10,000 Btu	[12]				0.000		0.000
32	H_2O from sorbent, lb/10,000 Btu	[13]			0.000	0.000	0.000	0.000
33	Total wet gas, lb/10,000 Btu	Summation [26] through [32]				9.864		9.864
34	Water in wet gas, lb/10,000 Btu	Summation [27] + [28] + [29] + [32]			0.422	0.422	0.422	0.422
35	Dry gas, lb/10,000 Btu	[33] − [34]				9.442		9.442
36	H_2O in gas, % by weight	[100] x [34] / [33]				4.28		4.28
37	Residue, % by weight	[9] x [24] / [33]				0.42		0.42

EFFICIENCY CALCULATIONS, % Input from Fuel

	Losses				
38	Dry gas, %		0.0024 x [35D] x ([6] − [3])		7.02
39	Water from fuel, as-fired %	Enthalpy of steam at 1 psi, T = [6]	H_1 = (3.958E−5 x T + 0.4329) x T + 1062.2	1237.1	
40		Enthalpy of water at T = [3]	H_2 = [3] − 32	48.0	
41			[29] x ([39] − [40]) / 100		3.61
42	Moisture in air, %		0.0045 x [27D] x ([6] − [3])		0.16
43	Unburned carbon, %		[19] or [21] x 14,500 / [18]		0.40
44	Radiation and convection, %		ABMA curve, Chapter 22		0.40
45	Unaccounted for and manufacturers margin, %				1.50
46	Sorbent net losses, % if sorbent is used		From Table 14 Item [41, Chapter 9]		0.00
47	Summation of losses, %		Summation [38] through [46]		13.09
	Credits				
48	Heat in dry air, %		0.0024 x [26D] x ([2] − [3])		0.00
49	Heat in moisture in air, %		0.0045 x [27D] x ([2] − [3])		0.00
50	Sensible heat in fuel, %		(H at T[4] − H at T [3]) x 100 / [18]	0.0	0.00
51	Other, %				0.00
52	Summation of credits, %		Summation [48] through [51]		0.00
53	**Efficiency, %**		100 − [47] + [52]		86.91

	KEY PERFORMANCE PARAMETERS		Leaving Furnace	Leaving Blr/Econ
54	Input from fuel, 1,000,000 Btu/h	100 x [10] / [53]		328.6
55	Fuel rate, 1000 lb/h	1000 x [54] / [18]		23.3
56	Wet gas weight, 1000 lb/h	[54] x [33] / 10	324.1	324.1
57	Air to burners (wet), lb/10,000 Btu	(1 + [7]) x (1 + [25A] / 100) x [22]	9.205	
58	Air to burners (wet), 1000 lb/h	[54] x [57] / 10	302.5	
59	Heat available, 1,000,000 Btu/h	[54] x {([18] − 10.30 x [17H]) / [18] − 0.005		
	H_a = 66.0 Btu/lb	x ([44] + [45]) + H_a at T[5] x [57] / 10,000}	335.2	
60	Heat available/lb wet gas, Btu/lb	1000 x [59] / [56]	1034.2	
61	Adiabatic flame temperature, F	From Fig. 3, Chap. 9 at H = [60], % H_2O = [36]	3560	

where

q = heat transfer rate, Btu/h
U = h_g = combined heat transfer coefficient or overall heat transfer coefficient, Btu/h ft^2 F where the boiling water film and tube wall resistance are assumed negligible
h_g = $h_{rg} + h_{cg}$ = overall gas side heat transfer coefficient, Btu/h ft^2 F
h_{rg} = radiation heat transfer coefficient (gas side), Btu/h ft^2 F
h_{cg} = convection heat transfer coefficient (gas side), Btu/h ft^2 F
A = total surface area, ft^2
LMTD = log mean temperature difference, gas and saturated water (T_s'), F
$\dot{m}_g$ = gas mass flow rate, lb/h
c_p = mean specific heat of gas, Btu/lb F
ΔT_g = $T_1 - T_2$
T_1 = gas temperature entering tube bank, F
T_2 = gas temperature leaving tube bank, F
T_s' = saturation temperature of boiler water, F

Using the nomenclature listed above, the log mean temperature difference for Case I of Fig. 1 is defined by:

$$\text{LMTD} = \frac{(T_1 - T_2)}{\ell n \dfrac{T_1 - T_s'}{T_2 - T_s'}} \quad \textbf{(8)}$$

Heat is transferred to the screen by direct furnace radiation, intertube radiation and convection. Furnace radiation to the screen is considered first. Radiation from the furnace to the screen is calculated from relationships developed in Chapter 4 using an effectiveness of 0.2 which approximates flue gas and screen surface emissivities. The temperature differential ($T_1^4 - T_2^4$) is defined by the furnace exit gas temperature and the screen tube temperature. Due to the large difference between gas and screen tube temperatures and the exponents in the formulation, it is sufficiently accurate to assume that the screen tube temperature is equal to the saturation temperature.

$$q'' = \sigma F_e (T_1^4 - T_2^4) = 12{,}280 \text{ Btu/h ft}^2 \quad \textbf{(9)}$$

where

q'' = heat flux, Btu/h ft^2
σ = Stefan-Boltzmann constant
= 1.71×10^{-9} Btu/h ft^2 R^4
F_e = effectiveness factor = 0.2
T_1 = furnace exit gas temperature = 2000F = 2460R
T_2 = saturation temperature = 462F = 922R

The entrance to the screen is 18 ft high and 12 ft wide for a flat projected area of 216 ft^2. The heat transferred to the screen by furnace radiation is 216 ft^2 $\times$ 12,280 Btu/h ft^2 or 2.65×10^6 Btu/h.

Based on the configuration of the screen (two rows deep on 6 in. side- and back-spacing), some of the radiant heat is absorbed by the screen and the remainder is absorbed by the superheater. From Chapter 4, Fig. 31, Curve 1, an effectiveness factor of 0.55 (i.e., 55% of the radiant energy entering any row is absorbed) is used to determine radiant screen absorption. On a row by row basis, the radiation from the furnace is distributed as follows:

Furnace radiation to first screen row:	2.65×10^6 Btu/h
First screen row absorption:	
$0.55 \times 2.65 \times 10^6$ Btu/h =	$\underline{1.46} \times 10^6$ Btu/h
Furnace radiation to second screen row:	1.19×10^6 Btu/h
Second screen row absorption:	
$0.55 \times 1.19 \times 10^6$ Btu/h =	$\underline{0.65} \times 10^6$ Btu/h
Furnace radiation to superheater:	0.54×10^6 Btu/h

Furnace radiation does not affect the flue gas temperature drop across the screen; however, the furnace radiation absorbed by the screen is taken into account when determining screen steam generation rate. Furnace radiation passing through the screen will be absorbed by the superheater. The evaluation of furnace radiation on superheater performance will be addressed in the superheater section.

The heat transfer to the screen tubes is by convection and intertube radiation, and is calculated from the formulas provided below. Calculations start by assuming a gas temperature leaving the screen which is later verified. In this case, 1920F is assumed. Log mean temperature difference:

$$\text{LMTD} = \frac{(T_1 - T_2)}{\ell n \dfrac{T_1 - T_s'}{T_2 - T_s'}} = 1498\text{F} \quad \textbf{(10)}$$

where

T_1 = furnace exit gas temperature = 2000F
T_2 = gas temperature leaving screen = 1920F
T_s' = saturation temperature = 462F

Gas mass flux:

$$G_g = \dot{m}_g / A_g = 2493 \text{ lb/h ft}^2 \quad \textbf{(11)}$$

where

$\dot{m}_g$ = gas mass flow rate = 324,100 lb/h — Table 3
A_g = minimum gas free flow area = 130 ft^2 — Table 1

Gas film temperature:

$$T_f = T_s' + (\text{LMTD}/2) = 1211\text{F} \quad \textbf{(12)}$$

where

T_s' = 462F
LMTD = 1498F

Reynolds number:

$$\text{Re} = K_{\text{Re}} G_g = 5734 \quad \textbf{(13)}$$

where

K_{Re} = gas properties factor = 2.3 h ft^2/lb — Fig. 4
G_g = 2493 lb/h ft^2

Gas film convection heat transfer coefficient from Equation 55, Chapter 4:

$$h_{cg} = h_c' F_{pp} F_a F_d = 7.65 \text{ Btu/h ft}^2 \text{ F} \quad \textbf{(14)}$$

where

h_c' = basic convection crossflow geometry and velocity factor
= 62.5 Btu/h ft^2 F — Ch. 4, Fig. 17
F_{pp} = physical properties factor = 0.133 — Ch. 4, Fig. 18
F_a = arrangement factor = 0.92 — Ch. 4, Fig. 20
F_d = heat transfer depth factor = 1.0 — Ch. 4, Fig. 21

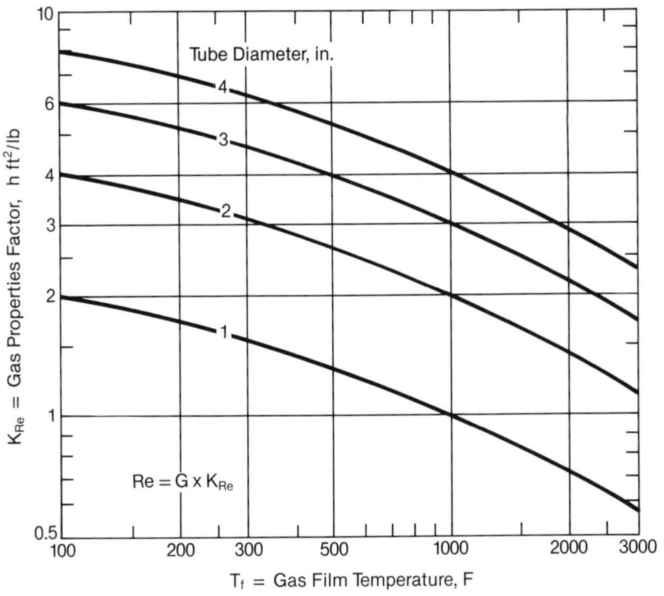

Fig. 4 Determination of Reynolds number, Re.

In this calculation, F_d is 1.0 because the furnace flue gases turn prior to entering the screen.

The total radiation absorption in the screen is comprised of direct furnace radiation and intertube radiation. Direct furnace radiation affects screen steam generation rate, but does not affect the gas temperature leaving the screen. The intertube radiation, however, is a direct function of gas temperature leaving the bank. Also, direct furnace radiation is proportional to the planar area it crosses while intertube radiation is proportional to the total bank heating surface.

To determine intertube radiation, the radiation heat transfer coefficient must be adjusted to eliminate direct furnace radiation through the use of an effectiveness factor based on areas.

$$F_s = \frac{\text{Effective surface}}{\text{Total surface}} = \frac{A - A_p}{A} = 0.681 \quad \textbf{(15)}$$

where

A = total bank heating surface = 542 ft² — Table 1

A_p = planar area of the bank credited with radiation absorption. In this example, 80% of the direct furnace radiation was absorbed by the screen
$= 0.8 \times 12 \times 18 = 172.8$ ft²

Because this calculation subtracts the effect of direct furnace radiation, it will be added back in when the total screen absorption is finally determined.

Gas side radiation heat transfer coefficient adjusting for effective surface:

$$h_{rg} = h'_r \, K \, F_s = 2.21 \text{ Btu/h ft}^2 \text{ F} \quad \textbf{(16)}$$

where

h'_r	= 8.1 Btu/h ft² F	Fig. 5
p_r	= partial pressure = 0.19 atm	Fig. 6
L	= mean radiating length = 1.33 ft	Fig. 7
K	= fuel factor = 0.4	Fig. 8
F_s	= 0.681	

Combined heat transfer coefficient:

$$h_g = h_{cg} + h_{rg} = 9.86 \text{ Btu/h ft}^2 \text{ F} \quad \textbf{(17)}$$

where

h_{cg} = 7.65 Btu/h ft² F
h_{rg} = 2.21 Btu/h ft² F

Overall heat transfer rate:

$$q = U \, A \text{ (LMTD)} = 8.0 \times 10^6 \text{ Btu/h} \quad \textbf{(18)}$$

where

U = h_g = 9.86 Btu/h ft² F
A = 542 ft²
LMTD = 1498F

To verify the flue gas exit temperature assumption of T_2 = 1920F for the screen tubes, the gas temperature leaving the screen can be calculated by an energy balance between the energy absorbed by the screen tubes (excluding furnace radiation) and the energy lost by the flue gas.

$$T_2 = T_1 - [q / \dot{m}_g \, c_p)] = 1920\text{F} \quad \textbf{(19)}$$

where

T_1	= 2000F	
q	= 8.0×10^6 Btu/h	
$\dot{m}_g$	= 324,100 lb/h	Eq. 3
c_p	= 0.31 Btu/lb F	Fig. 9

Solving the heat balance for T_2 indicates agreement with the earlier assumption of 1920F, and no recalculation is

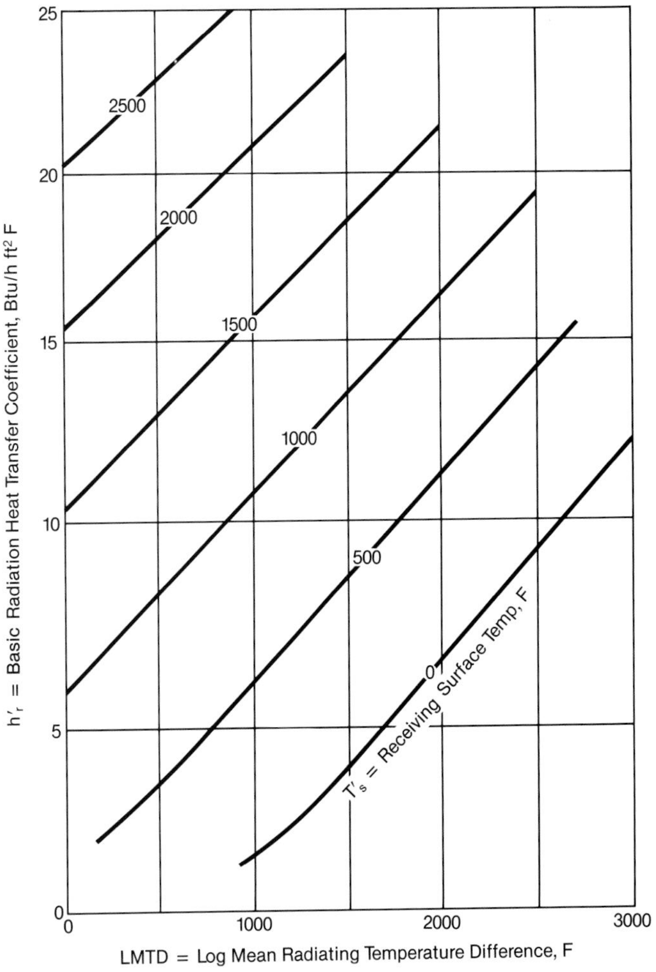

Fig. 5 Basic radiation heat transfer coefficient, h'_r, for Equations 16, 36, and 48.

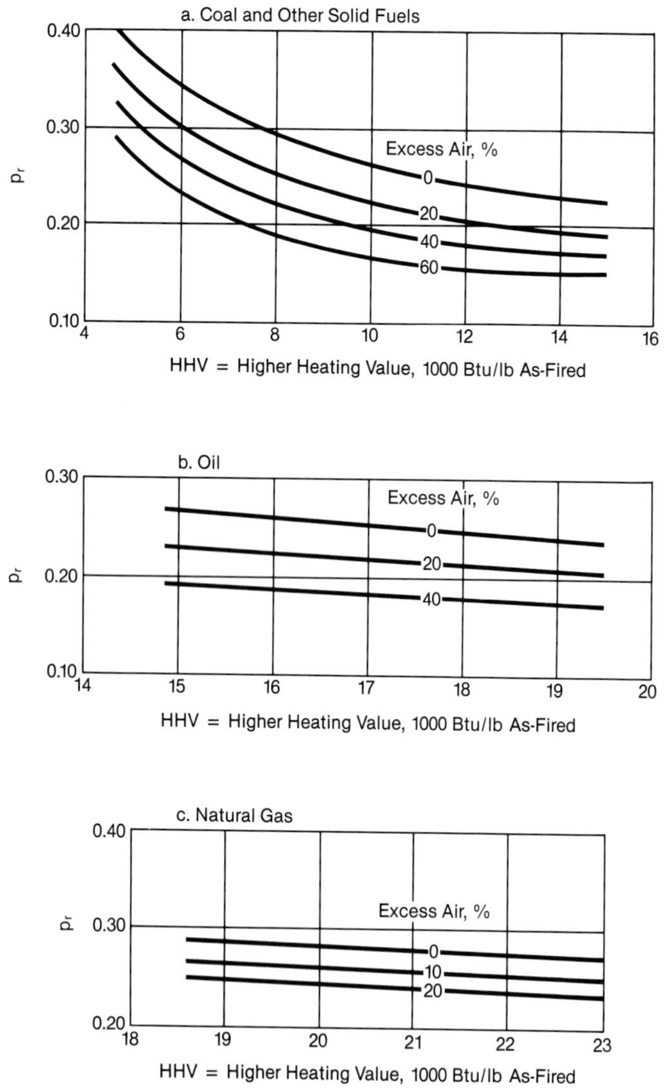

Fig. 6 Partial pressure, p_r, of principal radiating constituents ($CO_2 + H_2O$) of combustion gases in atmosphere for various fuels, heat values and excess airs.

required. Otherwise, T_2 from Equation 19 would be used in Equation 10 and the calculation repeated.

To complete the screen heat transfer analysis, the total screen absorption is determined. Screen absorption is the sum of the convection and intertube radiation heat transfer rates (8.0×10^6 Btu/h) and direct furnace radiation rate (2.11×10^6 Btu/h) for a total screen absorption of 10.11×10^6 Btu/h.

Draft loss

Screen gas side draft loss is calculated from Equations 47 and 51 of Chapter 3:

$$\Delta P = (f\ N\ F_d)\left(\frac{30}{B}\right)\left(\frac{T+460}{1.73\times 10^5}\right)\left(\frac{G}{10^3}\right)^2$$

$$= 0.04 \text{ in. wg} \qquad \textbf{(20)}$$

where

f	= friction factor = 0.24	Ch. 3, Fig. 16
N	= number of tube rows = 2	Table 1
F_d	= tube bundle correction factor = 1.12	Ch. 3, Fig. 15
B	= barometric pressure = 30 in. Hg	Table 2

T = 0.95 $(T_1 + T_2)/2$ = 1862F (Note 1 below)
G = 2493 lb/h ft^2 Eq. 11

Pressure drop

The screen tubes are part of the furnace circuitry and tube side pressure drop calculations for this surface are included in a circulation analysis. Refer to Chapter 5 for circulation information.

Superheater

Heat transfer

The governing heat transfer equations for superheater surfaces are:

$$q = U\ A\ (\text{LMTD}) \qquad \textbf{(21)}$$

$$\text{LMTD} = \frac{(T_1 - T_2') - (T_2 - T_1')}{\ell n \dfrac{(T_1 - T_2')}{(T_2 - T_1')}} \qquad \textbf{(22)}$$

Fig. 1, Case III

$$q = \dot{m}_g\ c_p\ \Delta T_g = \dot{m}_g\ c_p\ (T_1 - T_2) \qquad \textbf{(23)}$$

$$q = \dot{m}_s\ \Delta H \qquad \textbf{(24)}$$

where

q	= heat transfer rate, Btu/h
U	= $(h_g\, h_s)/(h_g + h_s)$ = combined heat transfer coefficient, Btu/h ft^2 F (assuming negligible wall resistance)
h_g	= $h_{rg} + h_{cg}$ = overall gas side heat transfer coefficient, Btu/h ft^2 F
h_{rg}	= radiation heat transfer coefficient (gas side), Btu/h ft^2 F
h_{cg}	= convection heat transfer coefficient (gas side), Btu/h ft^2 F
h_s	= convection heat transfer coefficient (steam side), Btu/h ft^2 F

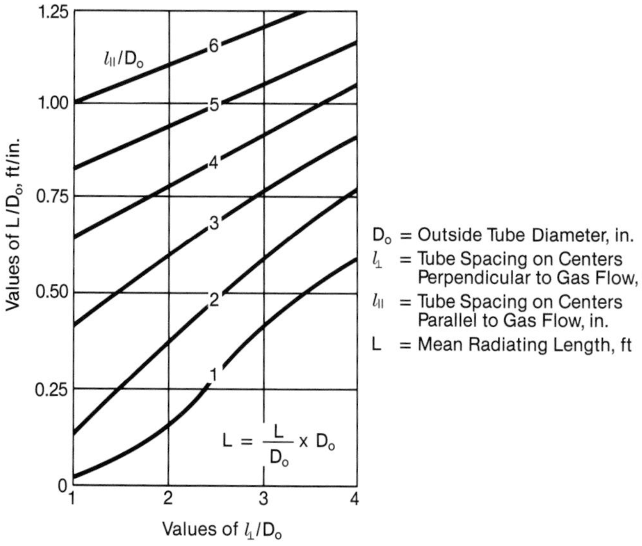

Fig. 7 Mean radiating length, L, for various tube diameters and arrangements or pitches — in-line tubes.

Note 1: Under most conditions, this equation is a good approximation of effective flue gas temperature.

A = total surface area, ft^2
LMTD = counterflow log mean temperature difference, gas and steam, F
T_1 = gas temperature entering superheater, F
T_2 = gas temperature leaving superheater, F
T_1' = steam temperature entering superheater, F
T_2' = steam temperature leaving superheater, F
$\dot{m}_g$ = mass flow of gas, lb/h
c_p = mean specific heat of gas, Btu/lb F
ΔT_g = $T_1 - T_2$, gas temperature differential, F
$\dot{m}_s$ = mass flow of steam, lb/h
ΔH = steam enthalpy difference, Btu/lb

Superheater steam side design conditions are:

Outlet: $T_2' = 650F$, $P_2' = 450$ psig, $H_2' =$ 1331.5 Btu/lb
Inlet: $T_1' = 462F$, $P_1' = 460$ psig, $H_1' =$ 1204.8 Btu/lb
$\Delta H =$ 126.7 Btu/lb

The outlet conditions are specified in Table 2 while the inlet conditions are assumed to be saturated steam at the drum pressure. Drum pressure is determined by superheater pressure drop and is assumed here based upon experience and verified later.

The heat transfer rate to the superheater is calculated as follows:

$$q = \dot{m}_s \ \Delta H = 31.68 \times 10^6 \text{ Btu/h} \qquad \textbf{(25)}$$

where

$\dot{m}_s$ = 250,000 lb/h — Table 2
ΔH = 126.7 Btu/lb

Previous calculations determined that the superheater will receive 0.54×10^6 Btu/h furnace radiation. Therefore, the heat transferred by convection and intertube radiation is:

$$q_{ci} = q - q_r = 31.14 \times 10^6 \text{ Btu/h} \qquad \textbf{(26)}$$

where

q = 31.68×10^6 Btu/h
q_r = 0.54×10^6 Btu/h

By rearranging Equation 23, the gas temperature leaving the superheater can be determined:

$$T_2 = T_1 - [q_{ci} / \dot{m}_g \ c_p)] = 1605\text{F} \qquad \textbf{(27)}$$

where

q_{ci} = 31.14×10^6 Btu/h

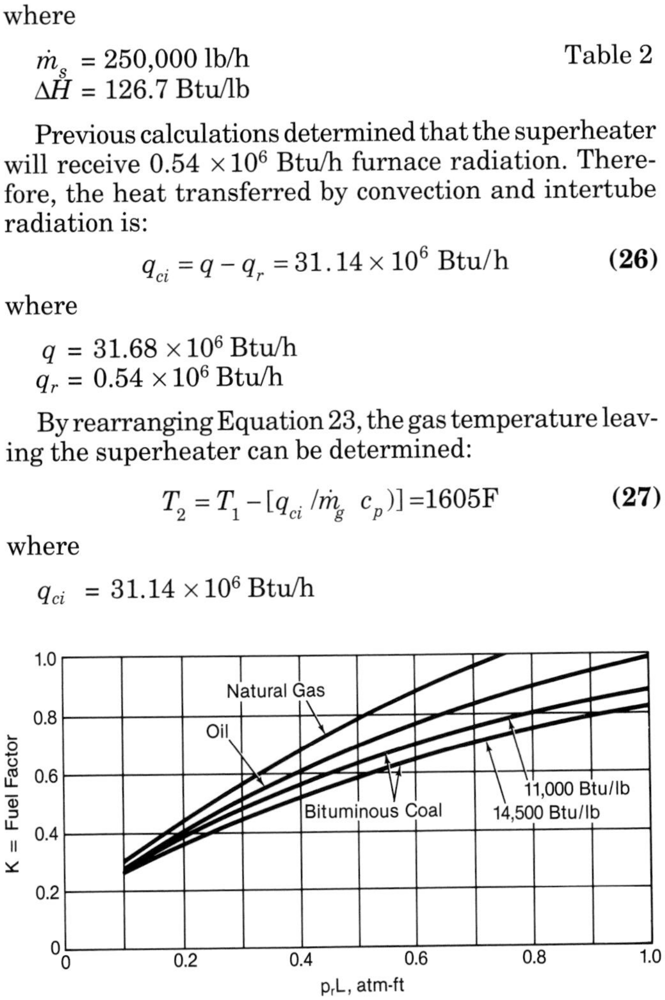

Fig. 8 Effect of fuel, partial pressure (H_2O and CO_2) and mean radiating length on radiation heat transfer coefficient.

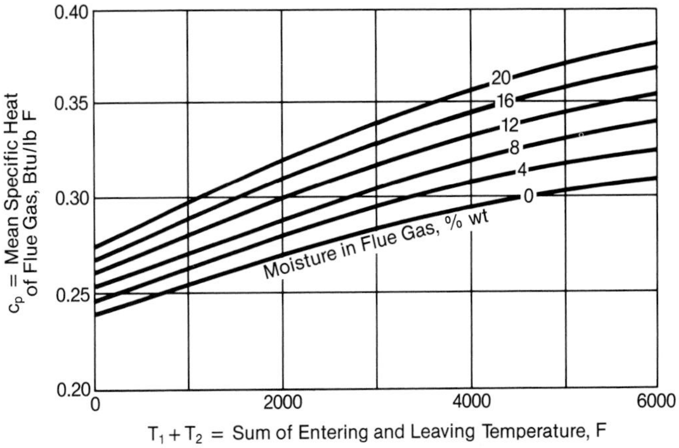

Fig. 9 Approximate mean specific heat, c_p, of flue gases.

T_1 = gas temperature entering superheater
= gas temperature leaving screen
= 1920F
$\dot{m}_g$ = 324,100 lb/h
c_p = 0.305 Btu/lb F — Fig. 9

Superheater log mean temperature difference is:

$$\text{LMTD} = \frac{(T_1 - T_2') - (T_2 - T_1')}{\ell n \dfrac{(T_1 - T_2')}{(T_2 - T_1')}} = 1205\text{F} \qquad \textbf{(28)}$$

where

T_1 = 1920F
T_2 = 1605F
T_1' = 462F
T_2' = 650F

The average gas side film temperature is:

$$T_f = [(T_1' + T_2')/2] + [(\text{LMTD})/2] = 1159\text{F} \qquad \textbf{(29)}$$

where

T_1' = 462F
T_2' = 650F
LMTD = 1205F

Superheater tube material and thickness are selected according to the American Society of Mechanical Engineers (ASME) Code and manufacturing capabilities which are discussed in Chapters 6 and 7. For this example, a 2.5 in. OD carbon steel seamless tube with 0.165 in. wall thickness has been selected. Thickness is set by tube bending limitations and is normally greater than that required by Code. Allowing for manufacturing tolerances (+15% for pressure tubing), the average inside diameter of the tube is calculated to be 2.12 in. The flow area corresponding to this diameter is 3.53 in.^{2}. The total steam flow area is 1.13 ft^2 (2 flow × 23 rows × 3.53 in.^{2} × ft^2/144 in.^{2}).

Steam mass flux:

$$G_s = \dot{m}_s / A_s = 221{,}200 \text{ lb/h ft}^2 \qquad \textbf{(30)}$$

where

$\dot{m}_s$ = 250,000 lb/h
A_s = 1.13 ft^2

Steam Reynolds number:

$$Re = \frac{G_s D_e}{\mu} = 814{,}000 \quad \textbf{(31)}$$

where

D_e = 2.12 in. = 0.1767 ft
μ = steam absolute viscosity
= 0.048 lb/h ft — Ch. 3, Fig. 5
G_s = 221,200 lb/h ft^2

Gas mass flux:

$$G_g = \dot{m}_g / A_g = 2493 \text{ lb/h ft}^2 \quad \textbf{(32)}$$

where

$\dot{m}_g$ = 324,100 lb/h
A_g = 130 ft^2 — Table 1

Gas side Reynolds number:

$$Re = K_{Re}\, G_g = 5734 \quad \textbf{(33)}$$

where

K_{Re} = 2.3 h ft^2/lb — Fig. 4
G_g = 2493 lb/h ft^2

Gas film convection heat transfer coefficient from Equation 55, Chapter 4:

$$h_{cg} = h'_c F_{pp} F_a F_d = 6.10 \text{ Btu/h ft}^2 \text{ F} \quad \textbf{(34)}$$

where

h'_c = 62.6 Btu/h ft^2 F — Ch. 4, Fig. 17
F_{pp} = 0.13 — Ch. 4, Fig. 18
F_a = 0.75 — Ch. 4, Fig. 20
F_d = 1.0 — Ch. 4, Fig. 21

To obtain the gas side radiation heat transfer coefficient (h_{rg}), a factor, F_s, must be included to account for the furnace radiation absorbed in the superheater. In the screen calculations, it was shown that 80% of the furnace radiation was absorbed in the screen while 20% passed through the screen and was absorbed by the superheater. Similar to the screen calculations, superheater intertube radiation will be determined by eliminating direct furnace radiation from the radiation heat transfer coefficient through the use of an effectiveness factor:

$$F_s = \frac{A - A_p}{A} = 0.987 \quad \textbf{(35)}$$

where

A = 3250 ft^2
A_p = 0.2 (12 × 18) = 43.2 ft^2 — Fig. 2

Gas side radiation heat transfer coefficient:

$$h_{rg} = h'_r K F_s = 2.20 \text{ Btu/h ft}^2 \text{ F} \quad \textbf{(36)}$$

where

h'_r = 7.2 Btu/h ft^2 F — Fig. 5
p_r = 0.19 atm — Fig. 6
L = 0.83 ft — Fig. 7
K = 0.31 — Fig. 8
F_s = 0.987

In a superheater, the resistance to heat transfer through the steam film inside the tubes can not be assumed to be negligible as was done with the screen.

Steam film convection heat transfer coefficient from Equation 52, Chapter 4 corrected to the OD surface area:

$$h_s = h'_l F_{pp} F_T D_i / D_o = 183 \text{ Btu/h ft}^2 \text{ F} \quad \textbf{(37)}$$

where

h'_l = 615 Btu/h ft^2 F — Ch. 4, Fig. 12
F_{pp} = 0.35 — Ch. 4, Fig. 15
F_T = temperature factor = 1.0 — Ch. 4, Fig. 16
D_i = 2.12 in.
D_o = 2.50 in.

Overall heat transfer coefficient:

$$U = \frac{h_g h_s}{h_g + h_s} = \frac{\left(h_{rg} + h_{cg}\right) h_s}{h_{rg} + h_{cg} + h_s} = 7.94 \text{ Btu/h ft}^2 \text{ F} \quad \textbf{(38)}$$

where

h_{rg} = 2.20 Btu/h ft^2 F
h_{cg} = 6.10 Btu/h ft^2 F
h_s = 183 Btu/h ft^2 F

Overall heat transfer rate:

$$q = U\, A\, (\text{LMTD}) = 31.10 \times 10^6 \text{ Btu/h} \quad \textbf{(39)}$$

where

U = 7.94 Btu/h ft^2 F
A = 3250 ft^2 — Table 1
LMTD = 1205F

Because this agrees with 31.14 × 10^6 Btu/h from Equation 26, no iteration is required. If these heat transfer rate calculations do not agree, then one of the terminal temperatures must be re-estimated and the calculations repeated until agreement is achieved.

Draft loss

Superheater gas side draft loss is determined by combining Equations 47 and 51, Chapter 3.

$$\Delta P = (f\, N\, F_d)\left(\frac{30}{B}\right)\left(\frac{T + 460}{1.73 \times 10^5}\right)\left(\frac{G}{10^3}\right)^2$$
$$= 0.09 \text{ in. wg} \quad \textbf{(40)}$$

where

f = 0.1 — Ch. 3, Fig. 16
N = 12 — Table 1
F_d = 1.0 — Ch. 3, Fig. 15
B = 30 in. Hg
T = 0.95 (T_1 + T_2)/2 = 1674F
G = 2493 lb/h ft^2

Steam pressure drop

Superheater steam side pressure drop is the sum of friction or straight-flow losses, entrance and exit losses, and bend losses. Equations 42 and 46 in Chapter 3 can be combined as follows:

$$\Delta P = \Delta P_f + \Delta P_{e+e} + \Delta P_b$$

$$= \left[\frac{f\,L}{D_i} + \frac{1.5}{12} + \frac{N_b}{12}\right]\left[v\left(\frac{G}{10^5}\right)^2\right] = 7.0 \text{ psi} \quad \textbf{(41)}$$

where

ΔP_f = frictional pressure drop (fL/D_i)

ΔP_{e+e} = entrance (1/12) and exit (0.5/12) pressure drop
ΔP_b = bend loss (N_b/12)
G = steam mass flux
= 221,200 lb/h ft^2
f = 0.013 Ch. 3, Fig. 1
L = length of one continuous superheater tube from steam drum to superheater outlet header = 140 ft
D_i = 2.12 in.
v = average steam specific volume
= 1.16 ft^3/lb
N_b = bend loss factor = 2.94

The superheater is a two-flow design. Steam side pressure drop will be determined for the steam path with the highest bend loss factors which, in this example, is the path with three short radius 180 degree bends. A composite bend loss factor for this path is determined as follows:

N_b: Three 180 degree bends, R/D = 0.77, 3 × 0.64	=	1.92	Ch. 3, Fig. 9
Two 180 degree bends, R/D = 2.3, 2 × 0.28	=	0.56	Ch. 3, Fig. 9
Two 90 degree bends, R/D = 2.3, 2 × 0.23	=	0.46	Ch. 3, Fig. 9
Composite N_b	=	2.94	

Drum pressure and saturation temperature can now be determined and verified. The steam drum pressure is equal to the outlet steam pressure plus the superheater pressure loss calculated above, plus an assumed pressure loss across the secondary steam separation equipment located in the drum outlet:

Superheater outlet pressure	450 psig	Table 2
Superheater pressure drop	+ 7 psi	
Steam separation equipment	+ 3 psi	(assumed)
Drum pressure	460 psig	
Sat. temperature (at 460 psig)	462 F	

This saturation pressure is in agreement with that originally assumed.

Boiler bank

Heat transfer

The function of the boiler bank, like the screen tubes, is to boil water, and the governing heat transfer equations defined for the screen are also applicable. Heat is transferred by convection, intertube radiation, and radiation from the rear cavity. In this example there are no cavities in the screen or superheater; however, in many applications cavities exist to accommodate sootblowers. Whenever cavities surround a bank of tubes, whether front or rear, the impact of the cavities on heat transfer must be considered (see discussion on cavity heat transfer, page 21-12).

Calculations begin by assuming the gas temperature leaving the boiler bank to be 819F based upon prior experience. The log mean temperature difference is calculated from Fig. 1, Case I:

$$\text{LMTD} = \frac{(T_1 - T_2)}{\ell n \dfrac{T_1 - T_s'}{T_2 - T_s'}} = 675\text{F} \qquad \textbf{(42)}$$

where

T_1 = gas temperature entering the boiler bank
= gas temperature leaving the superheater
= 1605F
T_2 = assumed bank exit temperature
= 819F
T_s' = saturation temperature
= 462F

Gas mass flux:

$$G_g = \dot{m}_g / A_g = 4051 \text{ lb/h ft}^2 \qquad \textbf{(43)}$$

where

$\dot{m}_g$ = 324,100 lb/h
A_g = 80 ft^2 Table 1

Gas film temperature:

$$T_f = T_s' + (\text{LMTD}/2) = 800\text{F} \qquad \textbf{(44)}$$

where

T_s' = 462F
LMTD = 675F

Reynolds number:

$$\text{Re} = K_{\text{Re}}\, G_g = 10{,}530 \qquad \textbf{(45)}$$

where

K_{Re} = 2.6 h ft^2/lb Fig. 4
G_g = 4051 lb/h ft^2

Gas film heat transfer coefficient:

$$h_{cg} = h_c' \, F_{pp} \, F_a \, F_d = 9.06 \text{ Btu/h ft}^2 \text{ F} \qquad \textbf{(46)}$$

where

h_c' = 83.9 Btu/h ft^2 F Ch. 4, Fig. 17
F_{pp} = 0.12 Ch. 4, Fig. 18
F_a = 0.90 Ch. 4, Fig. 20
F_d = 1.0 Ch. 4, Fig. 21

For the screen it was determined that 80% of the direct furnace radiation is absorbed in a two row bank. It can be shown that all rear cavity radiation is absorbed in the boiler bank because it is 28 rows deep. To calculate intertube radiation, an effectiveness factor must again be determined:

$$F_s = \frac{A - A_p}{A} = 0.98 \qquad \textbf{(47)}$$

where

A = 10,900 ft^2 Table 1
A_p at 100% = 12 × 18 = 216 ft^2 Fig. 2

Gas side radiation heat transfer coefficient:

$$h_{rg} = h_r' \, K \, F_s = 0.98 \text{ Btu/h ft}^2 \text{ F} \qquad \textbf{(48)}$$

where

h_r' = 4.0 Btu/h ft^2 F Fig. 5
p_r = 0.19 Fig. 6
L = 0.5 ft Fig. 7
K = 0.25 Fig. 8
F_s = 0.98

Combined heat transfer coefficient:

$$h_g = h_{cg} + h_{rg} = 10.04 \text{ Btu/h ft}^2 \text{ F} \qquad \textbf{(49)}$$

where

h_{cg} = 9.06 Btu/h ft^2 F
h_{rg} = 0.98 Btu/h ft^2 F

Overall heat transfer rate assuming negligible wall and boiling resistances:

$$q = U\ A\ (\text{LMTD}) = 73.9 \times 10^6 \text{ Btu/h} \quad \textbf{(50)}$$

where

U	$= h_g = 10.04$ Btu/h ft^2 F	
A	$= 10{,}900$ ft^2	Table 1
LMTD	= 675F	

Checking the gas temperature leaving the boiler bank:

$$T_2 = T_1 - [q/(\dot{m}_g\ c_p)] = 819\text{F} \quad \textbf{(51)}$$

where

T_1	= 1605F	
q	$= 73.9 \times 10^6$ Btu/h	
$\dot{m}_g$	= 324,100 lb/h	
c_p	= 0.29 Btu/lb F	Fig. 9

This agrees with the original outlet temperature assumption and therefore no recalculation is required. Otherwise, T_2 would be used in Equation 42 and the calculations repeated.

Draft loss

The governing equation for the gas side draft loss in the boiler bank is:

$$\Delta P = (f\ N\ F_d)\left(\frac{30}{B}\right)\left(\frac{T+460}{1.73\times10^5}\right)\left(\frac{G}{10^3}\right)^2 = 1.41 \text{ in. wg} \quad \textbf{(52)}$$

where

f	= 0.33	Ch. 3, Fig. 16
N	= 28	Table 1
F_d	= 1.0	Ch. 3, Fig. 15
B	= 30 in. Hg	Table 2
T	$= 0.95\ (T_1 + T_2)/2 = 1151$F	
G	= 4051 lb/h ft^2	

Pressure drop

As with the screen tubes, the boiler bank is integral to the furnace circuitry. Pressure loss to the steam-water flow is discussed in Chapter 5.

Cavity: boiler bank to economizer

Heat transfer

Heat is transferred from each cavity to the cooler banks which form its boundaries. Cavity radiation is most significant at higher gas temperatures. In this example, cavity radiation has little impact on the overall results, but is included to illustrate the procedure for evaluating other configurations.

Assume the gas temperature leaving the cavity, entering the economizer to be 815F and the water temperature leaving the economizer to be 286F. Both assumptions are verified later. Consider cavity radiation to the boiler bank first.

Mean temperature difference:

$$\text{LMTD} = 0.5\,(T_1 + T_2) - T_s' = 355\text{F} \quad \textbf{(53)}$$

where

T_1 = gas temperature entering the cavity
= gas temperature leaving the boiler bank
= 819F
T_2 = 815F (assumed)
T_s' = saturation temperature = 462F

Mean radiating length:

$$L = 3.4\frac{V_L}{A} = 7.1 \text{ ft} \quad \textbf{(54)}$$

where

V_L	= volume of the cavity = 12 ft × 18 ft × 10 ft	
	= 2160 ft^3	Fig. 2
A	= area = 2 (12 × 18 + 12 × 10 + 10 × 18)	
	= 1032 ft^2	Fig. 2

Gas side radiation heat transfer coefficient:

$$h_{rg} = h_r'\ K = 2.57 \quad \textbf{(55)}$$

where

h_r'	= 2.7 Btu/h ft^2 F	Fig. 5
p_r	= 0.19 atm	Fig. 6
L	= 7.1 ft	
K	= 0.95	Fig. 8

Because radiation is the only significant mode of heat transfer in the cavity, the overall heat transfer coefficient $U = h_{rg}$. Then overall heat transfer rate to the boiler bank is then:

$$q = U\ A\ (\text{LMTD}) = 197{,}100 \text{ Btu/h} \quad \textbf{(56)}$$

where

U	$= h_{rg} = 2.57$ Btu/h ft^2 F	
A	= 12 × 18 = 216 ft^2	Fig. 2
LMTD	= 355F	

Radiation to the economizer follows the same logic.

$$\text{LMTD} = 0.5\,(T_1 + T_2) - T_2' = 531\text{F} \quad \textbf{(57)}$$

where

T_1 = gas temperature entering the cavity
= 819F
T_2 = gas temperature leaving the cavity
= 815F
T_2' = assumed economizer water outlet temperature
= 286F

$$h_{rg} = h_r'\ K = 1.90 \text{ Btu/h ft}^2 \text{ F} \quad \textbf{(58)}$$

where

h_r'	= 2.0 Btu/h ft^2 F	Fig. 5
p_r	= 0.19 atm	Fig. 6
L	= 7.1 ft	
K	= 0.95	Fig. 8

$$q = U\ A\ (\text{LMTD}) = 121{,}100 \text{ Btu/h} \quad \textbf{(59)}$$

where

U	$= h_{rg} = 2.45$ Btu/h ft^2 F	
A	= 12 × 10 = 120 ft^2	Fig. 2
LMTD	= 531F	

Total heat transfer is the sum of the rates to the boiler bank and economizer, or 318,200 Btu/h. Checking the gas temperature leaving the cavity:

$$T_2 = T_1 - [q/(\dot{m}_g\ c_p)] = 815\text{F} \qquad \textbf{(60)}$$

where

T_1 = 819F
q = 318,200 Btu/h
$\dot{m}_g$ = 324,100 lb/h
c_p = 0.28 Btu/lb F — Fig. 9

This is the same as the assumed value and further iteration is not needed for the cavity loss. Otherwise, T_2 from Equation 60 would be used in Equation 53 and the calculations repeated. Verification of the economizer water outlet temperature is shown in the next section.

The total boiler bank absorption can now be determined. Absorption due to convection and intertube radiation is 73.9×10^6 Btu/h (Equation 50) while that from cavity radiation is 0.23×10^6 Btu/h (Equation 56) for a total boiler bank absorption of 74.1×10^6 Btu/h.

Economizer

Heat transfer

Economizer heat transfer follows the same formulations established for the superheater:

$$q = U\ A\ (\text{LMTD}) \qquad \textbf{(61)}$$

$$q = \dot{m}_g\ c_p\ \Delta T_g = \dot{m}_g\ c_p\ (T_2 - T_1) \qquad \textbf{(62)}$$

$$q = \dot{m}\ \Delta H \qquad \textbf{(63)}$$

where the terms are generally defined after Equations 21 to 24 except the tube side fluid is water instead of steam.

Economizer outlet water temperature was previously assumed to be 286F. It was also established that 121,100 Btu/h is transferred from the cavity preceding the economizer. The total heat transfer rate to the economizer is calculated to be:

$$q = \dot{m}\ \Delta H = 16.73 \times 10^6 \text{ Btu/h} \qquad \textbf{(64)}$$

where

$\dot{m}$ = 250,000 lb/h — Table 2
T_2' = 286F
H_2 = 256.1 Btu/lb
T_1' = 220F — Table 2
H_1 = 189.2 Btu/lb — Table 2
$\Delta H = H_2 - H_1$ = 66.9 Btu/lb

The heat transfer by convection and intertube radiation is:

$$q_{ci} = q - q_r = 16.61 \times 10^6 \text{ Btu/h} \qquad \textbf{(65)}$$

where

q = total heat transfer rate = 16.73×10^6 Btu/h
q_r = cavity radiation = 121,100 Btu/h

As can be seen from this result, cavity thermal radiation at low temperatures is not usually significant.

By rearranging Equation 62, the gas temperature leaving the economizer can be determined:

$$T_2 = T_1 - [q_{ci}/(\dot{m}_g\ c_p)] = 624\text{F} \qquad \textbf{(66)}$$

where

q_{ci} = 16.61×10^6 Btu/h
$\dot{m}_g$ = 324,100 lb/h — Table 3
c_p = 0.268 Btu/lb F — Fig. 9
T_1 = 815F

From Fig. 1, Case III, for counterflow heat exchanger, the log mean temperature difference is:

$$\text{LMTD} = \frac{(T_1 - T_2') - (T_2 - T_1')}{\ell n \dfrac{(T_1 - T_2')}{(T_2 - T_1')}} = 464\text{F} \qquad \textbf{(67)}$$

where

T_1 = economizer inlet gas temperature = 815F
T_2 = economizer outlet gas temperature = 624F
T_1' = economizer water inlet temperature = 220F — Table 2
T_2' = economizer water outlet temperature = 286F

Average gas film temperature:

$$T_f = 0.5\,(T_1' + T_2') + 0.5\,(\text{LMTD}) = 485\text{F} \qquad \textbf{(68)}$$

where

T_1' = 220F
T_2' = 286F
LMTD = 464F

Gas mass flux:

$$G_g = \dot{m}_g / A_g = 7717 \text{ lb/h ft}^2 \qquad \textbf{(69)}$$

where

$\dot{m}_g$ = 324,100 lb/h — Table 3
A_g = 42 ft^2 — Table 1

Reynolds number:

$$\text{Re} = K_{\text{Re}}\ G_g = 20{,}800 \qquad \textbf{(70)}$$

where

K_{Re} = 2.7 h ft^2/lb — Fig. 4
G_g = 7717 lb/h ft^2

Gas film heat transfer coefficient:

$$h_{cg} = h_c'\ F_{pp}\ F_a\ F_d = 14.28 \text{ Btu/h ft}^2 \text{ F} \qquad \textbf{(71)}$$

where

h_c' = 136 Btu/h ft^2 F — Ch. 4, Fig. 17
F_{pp} = 0.105 — Ch. 4, Fig. 18
F_a = 1.0 — Ch. 4, Fig. 20
F_d = 1.0 — Ch. 4, Fig. 21

As for the preceding components, an effectiveness factor based on the total economizer surface area is determined:

$$F_s = \frac{A - A_p}{A} = 0.951 \qquad \textbf{(72)}$$

where

A = 2460 ft^2
A_p at 100% = (12 ft) (10 ft) = 120 ft^2 — Fig. 2

Gas side radiation heat transfer coefficient:

$$h_{rg} = h'_r\ K\ F_s = 0.32 \text{ Btu/h ft}^2 \text{ F} \qquad \textbf{(73)}$$

where

h'_r = 1.6 Btu/h ft^2 F — Fig. 5
p_r = 0.19 — Fig. 6
L = 0.30 ft — Fig. 7
K = 0.21 — Fig. 8
F_s = 0.951

The water film and tube wall resistances are negligible so the total heat transfer coefficient is:

$$U = h_g = h_{cg} + h_{rg} = 14.60 \text{ Btu/h ft}^2 \text{ F} \qquad \textbf{(74)}$$

where

h_{cg} = 14.28 Btu/h ft^2 F
h_{rg} = 0.32 Btu/h ft^2 F

Overall heat transfer rate:

$$q = U\ A\ (\text{LMTD}) = 16.60 \times 10^6 \text{ Btu/h} \qquad \textbf{(75)}$$

where

U = 14.60 Btu/h ft^2 F
A = 2450 ft^2 — Table 1
LMTD = 464F

After adding cavity radiation (121,100 Btu/h), the total heat transfer rate to the economizer is 16.72 × 10^6 Btu/h.

Verifying the water outlet temperature assumption, the outlet enthalpy is evaluated from:

$$H_2 = H_1 + q/\dot{m} = 256.1 \text{ Btu/lb or } T'_2 = 286\text{F} \qquad \textbf{(76)}$$

where

q = 16.72 × 10^6 Btu/h
$\dot{m}$ = 250,000 lb/h
H_1 = 189.2 Btu/lb

From steam tables at an enthalpy of 256.1 Btu/lb, the water temperature is verified to be 286F. Therefore, the total economizer absorption is 16.72 × 10^6 Btu/h.

Draft loss

Economizer gas side draft loss is calculated from Equations 47 and 51, Chapter 3.

$$\Delta P = (f\ N\ F_d)\left(\frac{30}{B}\right)\left(\frac{T+460}{1.73\times10^5}\right)\left(\frac{G}{10^3}\right)^2 = 1.38 \text{ in. wg} \qquad \textbf{(77)}$$

where

f = 0.35 — Ch. 3, Fig. 16
N = 10 — Table 1
F_d = 1.0 — Ch. 3, Fig. 15
B = 30 in. H_2O
T_1 = 815F
T_2 = 624F
T = 0.95 $(T_1 + T_2)/2$ = 684F
G = 7717 lb/h ft^2

Water side pressure drop

In this example, the economizer tubes are 2 in. OD with 0.148 in. wall thickness. Considering the manufacturing tolerance of plus 15% in the pressure tubing wall thickness, the tube inside diameter is 1.66 in. Water flow area is 2.16 in.^{2}/tube and the total flow area is 0.705 ft^2 (2.16 in.^{2}/tube × 47 tubes × 1.0 ft^2/144 in.^{2}). Water mass flux is:

$$G = \dot{m}/A = 355{,}000 \text{ lb/h ft}^2 \qquad \textbf{(78)}$$

where

$\dot{m}$ = 250,000 lb/h
A = 0.705 ft^2

Water side Reynolds number:

$$\text{Re} = \frac{GD_i}{\mu} = 149{,}000 \qquad \textbf{(79)}$$

where

G = 355,000 lb/h ft^2
D_i = 1.66 in. = 0.138 ft
μ = 0.33 lb/h ft — Ch. 3, Fig. 3

Economizer pressure drop is the sum of friction losses, entrance and exit losses, and bend losses.

$$\Delta P = \left[\frac{f\ L}{D_i} + \frac{1.5}{12} + \frac{N_b}{12}\right]\left[v\left(\frac{G}{10^5}\right)^2\right] = 0.34 \text{ psi} \qquad \textbf{(80)}$$

where

f = 0.017 — Ch. 3, Fig. 1
L = 105 ft
D_i = 1.66 in.
N_b = nine 180 degree 0.90 R/D bends = 9 × 0.55 = 4.95 — Ch. 3, Fig. 9
v = specific volume = 0.017 ft^3/lb — Ch. 2, Table 3
G = 355,000 lb/h ft^2

The total pressure drop (economizer inlet to the steam drum) must include the static head of water for that elevation difference (25 ft) plus fittings and friction losses. Assuming the feedwater piping is large enough that piping and fitting losses are negligible ($\Delta P_{piping} = 0$), then the static head is calculated by:

$$\Delta P_{static} = \frac{\Delta Z}{144\,v} = 10.2 \text{ psi} \qquad \textbf{(81)}$$

where

ΔZ = elevation between drum centerline and economizer inlet header = 25 ft
v = 0.017 ft^3/lb — Ch. 2, Table 3

Total pressure drop from economizer inlet to drum is:

$$\Delta P = \Delta P_{econ} + \Delta P_{static} + \Delta P_{piping} = 10.5 \text{ psi} \qquad \textbf{(82)}$$

where

ΔP_{econ} = 0.34 psi
ΔP_{static} = 10.2 psi
ΔP_{piping} = 0 (assumed)

Air heater

Heat transfer

The air heater is the last heat transfer component before the stack in this example. In the overall heat and material balances, the air heater exit gas temperature is assumed to be 390F. The air heater, when sized properly, will have sufficient surface to provide the required air temperature to the fuel equipment (burners, pulverizers, etc.) and to lower the gas temperature to that assumed in the combustion calculations. For the air heater, the heat transfer rate is determined as follows:

$$q = \dot{m}_g \, c_p (T_1 - T_2) = 20.1 \times 10^6 \text{ Btu/h} \quad \textbf{(83)}$$

where

$\dot{m}_g$	= 324,100 lb/h	Table 3
c_p	= 0.265 Btu/lb F	Fig. 9
T_1	= gas temperature entering the air heater = gas temperature leaving the economizer = 624F	
T_2	= assumed air heater exit temperature = 390F	

For the air side, the temperature rise is:

$$T'_2 = T'_1 + [q / \dot{m}_a \, c_p)] = 351\text{F} \quad \textbf{(84)}$$

where

T'_1	= 80F	Table 2
q	= 20.1×10^6 Btu/h	
$\dot{m}_a$	= 302,500 lb/h	Table 3
c_p	= 0.245 Btu/lb F	Fig. 10

The tubular air heater in this example is a crossflow design. The log mean temperature difference is determined from Fig. 1, Case IV. This requires not only the two inlet and two outlet temperatures, but also a crossflow correction factor, F:

$$\text{LMTD} = \frac{(T_1 - T'_2) - (T_2 - T'_1)}{\ell n \dfrac{(T_1 - T'_2)}{(T_2 - T'_1)}} \times F = 262\text{F} \quad \textbf{(85)}$$

where

T_1	= 624F	
T_2	= 390F	
T'_1	= 80F	
T'_2	= 351F	
F	= crossflow correction factor = 0.90	Ch. 4, Fig. 24

In an air heater, gas and air film heat transfer coefficients are approximately equal. For this example, film temperatures are approximated by the following calculations.

Gas: $$T_f = 0.5\,(T_1 + T_2) - 0.25\,(\text{LMTD}) = 442\text{F} \quad \textbf{(86)}$$

where

T_1 = 624F
T_2 = 390F
LMTD = 262F

Air: $$T_f = 0.5\,(T'_1 + T'_2) + 0.25\,(\text{LMTD}) = 281\text{F} \quad \textbf{(87)}$$

where

T'_1 = 80F
T'_2 = 351F
LMTD = 262F

The gas flows through 2173 tubes (53 rows of 41 tubes per row). The air heater tubes are electric resistance welded (ERW), 2 in. OD with a 0.083 in. wall thickness. Allowing for manufacturing tolerances (9% for nonpressure tubing), the average tube inside diameter is 1.819 in. Gas flow area is 2.60 in.^{2}/tube for a total of 39.2 ft^2. The gas mass flux is:

$$G_g = \dot{m}_g / A_g = 8268 \text{ lb/h ft}^2 \quad \textbf{(88)}$$

where

$\dot{m}_g$	= 324,100 lb/h	Table 3
A_g	= 39.2 ft^2	

Reynolds number:

$$\text{Re} = K_{\text{Re}} \, G_g = 21{,}500 \quad \textbf{(89)}$$

where

K_{Re}	= 2.60 h ft^2/lb	Fig. 4
G_g	= 8268 lb/h ft^2	

Gas film heat transfer coefficient is the sum of the convection heat transfer coefficient from longitudinal gas flow inside the air heater tubes and a small gaseous radiation component from within the tube. The gas convection heat transfer coefficient h_{cg} is calculated from Equation 52, Chapter 4.

$$h_{cg} = h'_l \, F_{pp} \, F_T \, D_i / D_o = 8.52 \text{ Btu/h ft}^2 \text{ F} \quad \textbf{(90)}$$

where

h'_l	= 44.8 Btu/h ft^2 F	Ch. 4, Fig. 12
F_{pp}	= 0.190	Ch. 4, Fig. 13
F_T	= 1.1	Ch. 4, Fig. 16
D_i	= 1.819 in.	
D_o	= 2.0 in.	Table 1

Gas side radiation heat transfer coefficient:

$$h_{rg} = h'_r \, K = 0.18 \text{ Btu/h ft}^2 \text{ F} \quad \textbf{(91)}$$

where

h'_r	= 1.1 Btu/h ft^2 F	Fig. 5
p_r	= 0.19	Fig. 6
L	= 1.819/12 = 0.15 ft	Fig. 7
K	= 0.16	Fig. 8

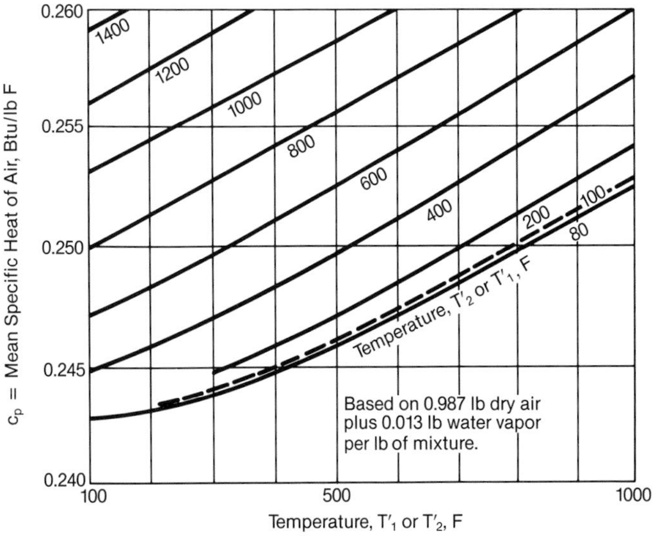

Fig. 10 Mean specific heat, c_p, of air at one atmosphere.

From Table 1, the air side free flow area is 82.7 ft², hence the air mass flux is calculated to be:

$$G_a = \dot{m}_a / A_a = 3658 \text{ lb/h ft}^2 \qquad \textbf{(92)}$$

where

$\dot{m}_a$ = 302,500 lb/h — Table 3
A_a = 82.7 ft²

Reynolds number:

$$\text{Re} = K_{\text{Re}}\, G_a = 11{,}700 \qquad \textbf{(93)}$$

where

K_{Re} = 3.2 — Fig. 4
G_a = 3658 lb/h ft²

The crossflow convection heat transfer coefficient for air is obtained from Equation 55, Chapter 4.

$$h_{ca} = h'_c\, F_{pp}\, F_a\, F_d = 8.06 \text{ Btu/h ft}^2 \text{ F} \qquad \textbf{(94)}$$

where

h'_c = 86.1 Btu/h ft² F — Ch. 4, Fig. 17
F_{pp} = 0.104 — Ch. 4, Fig. 19
F_a = 0.90 — Ch. 4, Fig. 20
F_d = 1.0 — Ch. 4, Fig. 21

Assuming negligible wall resistance, the overall heat transfer coefficient is:

$$U = \frac{\left(h_{cg} + h_{rg}\right) h_{ca}}{h_{cg} + h_{rg} + h_{ca}} = 4.18 \text{ Btu/h ft}^2 \qquad \textbf{(95)}$$

where

h_{cg} = 8.52 Btu/h ft² F
h_{rg} = 0.18 Btu/h ft² F
h_{ca} = 8.16 Btu/h ft² F

The total heat transfer rate for the air heater is:

$$q = U\, A\ (\text{LMTD}) = 19.94 \times 10^6 \text{ Btu/h} \qquad \textbf{(96)}$$

where

U = 4.18 Btu/h ft² F
A = 18,205 ft² — Table 1
LMTD = 262F

Air heater exit gas temperature is calculated to be:

$$T_2 = T_1 - [q / \dot{m}_g\, c_p)] = 390\text{F} \qquad \textbf{(97)}$$

where

T_1 = 624F
q = 19.94 × 10⁶ Btu/h
$\dot{m}_g$ = 324,100 lb/h
c_p = 0.263 Btu/lb F — Fig. 9

which is in agreement with the temperature assumed. Otherwise, T_2 would be substituted into Equation 85 and the calculations repeated.

Draft loss (gas inside tubes)

Air heater draft loss is comprised of friction (ΔP_f) plus entrance and exit losses (ΔP_{e+e}). Modifying and combining Equations 42 and 51 of Chapter 3 for average gas temperatures, draft loss can be expressed as follows:

$$\Delta P = \Delta P_f + \Delta P_{e+e} = \left(12 \frac{f\,L}{D_i} + 1.5\right)\left(\frac{30}{B}\right) \times \left(\frac{T + 460}{1.73 \times 10^5}\right)\left(\frac{G_g}{10^3}\right)^2 = 1.52 \text{ in. wg} \qquad \textbf{(98)}$$

where

f = 0.025 — Ch. 3, Fig. 1
L = 16 ft — Table 1
D_i = 1.819 in.
T_1 = 624F
T_2 = 390F
T = $(T_1 + 2T_2)/3$ = 468F (Note 2 below)
G_g = 8268 lb/h ft²
B = 30 in. Hg

Air resistance (air crossflow over tubes)

The draft loss due to air flow across the air heater tubes is calculated from Equations 47 and 51, Chapter 3.

$$\Delta P = (f\ N\ F_d)\left(\frac{30}{B}\right)\left(\frac{T + 460}{1.73 \times 10^5}\right)\left(\frac{G_a}{10^3}\right)^2 = 0.54 \text{ in. wg} \qquad \textbf{(99)}$$

where

f = 0.20 — Ch. 3, Fig. 16
N = 53 — Table 1
F_d = 1.0 — Ch. 3, Fig. 15
B = 30 in. Hg
T'_1 = 80F
T'_2 = 351F
T = 0.95 (80 + 351)/2 = 205F (Note 3 below)
G_a = 3658 lb/h ft²

Boiler thermal performance summary

Thermal performance of the example boiler is shown in Fig. 11. This illustrates how the relative absorption is distributed within the unit. It also shows the relationship between temperature difference and the calculated absorption rates. The furnace and boiler bank are shown to be the components in which most of the heat transfer takes place. The economizer and air heater are used to reduce gas temperature to the stack and to reduce the sensible heat loss in the flue gas, thereby increasing boiler efficiency.

Flues, ducts and stack

Performance calculations are not yet complete. Component draft loss and air resistance calculations have been made, but air and gas side performance still requires the evaluation of flues, ducts and stack effects. Once these are established, the designer can evaluate forced draft and induced draft fan conditions.

Note 2: An approximation used for mean gas temperature.

Note 3: Under most conditions, this equation is a good approximation for effective flue gas temperature.

Air side loss — forced draft fan outlet to furnace

Air resistance calculations for forced draft fan outlet to windbox inlet are considered first. Static pressure at the windbox for this example is set at 5 in. wg. Windbox pressure is normally a function of the burner or fuel equipment design and is specified to assure proper operation. Starting at the windbox and working toward the forced draft fan:

Air mass flux:

$$G_a = \dot{m}_a / A_a = 2520 \text{ lb/h ft}^2 \qquad \textbf{(100)}$$

where

$\dot{m}_a$ = 302,500 lb/h — Table 3
A_a = 10 × 12 = 120 ft^2 — Fig. 2

Reynolds number:

$$\text{Re} = \frac{G_a D_e}{\mu} = 4.7 \times 10^5 \qquad \textbf{(101)}$$

where

G_a = 2520 lb/h ft^2
D_e = hydraulic diameter = 4 × area/perimeter
= 4 × 120/44 = 10.9 ft
μ = 0.058 lb/h ft — Ch. 3, Fig. 4

The air resistance from windbox inlet to air heater outlet is:

$$\Delta P = \left(\frac{f\,L}{D_e} + N\right)\left(\frac{30}{B}\right)\left(\frac{T+460}{1.73\times 10^5}\right)\left(\frac{G_a}{10^3}\right)^2 = 0.04 \text{ in. wg} \qquad \textbf{(102)}$$

where

f = 0.013 — Ch. 3, Fig. 1
L = 25 ft
D_e = 10.9 ft
B = 30 in. Hg
T = 351F
G_a = 2520 lb/h ft^2
N = $N_{bend} + N_{expansion}$ = 1.35
N_{bend} = 1.3 — Ch. 3, Fig. 11
$N_{expansion}$ = 0.05 — Ch. 3, Fig. 7

The frictional component of this equation (fL/D_e) is negligible. The draft loss from the air heater inlet to the forced draft fan transition outlet, neglecting friction, is:

$$\Delta P = N\left(\frac{30}{B}\right)\left(\frac{T+460}{1.73\times 10^5}\right)\left(\frac{G_a}{10^3}\right)^2 = 0.03 \text{ in. wg} \qquad \textbf{(103)}$$

where

N = $N_{contraction} + N_{bend}$ = 1.48
$N_{contraction}$ = 0.18 — Ch. 3, Fig. 6
N_{bend} = 1.3 — Ch. 3, Fig. 11
B = 30 in. Hg
T = T'_2 for the air heater = 80F — Table 2
G_a = 2520 lb/h ft^2

The net static pressure at the forced (FD) draft fan outlet transition is then:

$$\Delta P_{total} = P_{windbox} + \Delta P_{windbox\ to\ air\ heater} + \Delta P_{air\ heater\ to\ FD\ fan} + \Delta P_{air\ heater} = 5.61 \text{ in. wg} \qquad \textbf{(104)}$$

where

$P_{windbox}$ = 5.0 in. wg (set by burners)
$\Delta P_{windbox\ to\ air\ heater}$ = 0.04 in. wg
$\Delta P_{air\ heater\ to\ FD\ fan}$ = 0.03 in. wg
$\Delta P_{air\ heater}$ = 0.54 in. wg

The static pressure at the furnace exit in balanced draft boilers is controlled to be slightly negative; a value of – 0.1 in. wg is used in this example. Furnace stack effect is determined from methods developed in Chapter 23, and is calculated as follows:

$$\Delta P_{SE} = SE \times Z = -0.58 \text{ in. wg} \qquad \textbf{(105)}$$

where

Z = centerline of furnace exit to centerline of windbox = – 50 ft — Fig. 2
SE = stack effect = 0.0116 in. wg/ft — Ch. 23, Eq. 3, and Table 3
T_1 = adiabatic temperature = 3560F — Table 3
T_2 = furnace exit gas temperature = 2000F

By controlling the furnace outlet to – 0.1 in. wg, the net static pressure in the furnace at the burner elevation is approximately – 0.68 in. wg.

From screen, superheater, and boiler bank component calculations, gas side draft losses were determined to be 0.04, 0.09 and 1.41 in. wg, respectively. The net static pressure at the boiler bank outlet is:

$$P_{boiler\ bank\ outlet} = P_{furnace} - \Delta P_{screen} - \Delta P_{superheater} - \Delta P_{boiler\ bank} = -1.64 \text{ in. wg} \qquad \textbf{(106)}$$

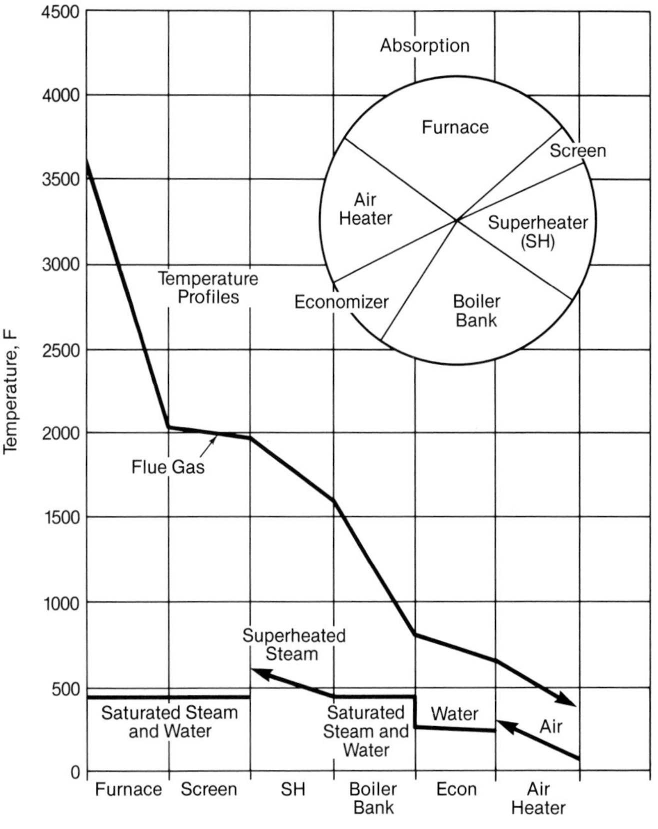

Fig. 11 Example boiler thermal performance summary.

where

$P_{furnace}$ = – 0.10 in. wg
ΔP_{screen} = 0.04 in. wg
$\Delta P_{superheater}$ = 0.09 in. wg
$\Delta P_{boiler\ bank}$ = 1.41 in. wg

The calculations for the boiler bank outlet flue to the economizer inlet are handled the same as those for the air resistance calculations.

Gas mass flux:

$$G_g = \dot{m}_g / A_g = 2700 \text{ lb/h ft}^2 \quad \textbf{(107)}$$

where

$\dot{m}_g$ = 324,100 lb/h
A_g = 120 ft^2 — Fig. 2

Reynolds number:

$$\text{Re} = \frac{G_g D_e}{\mu} = 373{,}000 \quad \textbf{(108)}$$

where

G_g = 2700 lb/h ft^2
D_e = hydraulic diameter = 4 × area/perimeter = 4 × 120/44 = 10.9 ft
μ = 0.079 lb/h ft — Ch. 3, Fig. 4

Draft loss:

$$\Delta P = \left(\frac{f\,L}{D_e} + N\right)\left(\frac{30}{B}\right)\left(\frac{T+460}{1.73\times 10^5}\right)\left(\frac{G_g}{10^3}\right)^2 = 0.08 \text{ in. wg} \quad \textbf{(109)}$$

where

f = 0.013 — Ch. 3, Fig. 1
L = 15 ft
D_e = 10.9 ft
N = 1.38 — Ch. 3, Fig. 11
T = 1/2 (819 + 815) = 817F
G_g = 2700 lb/h ft^2
B = 30

From the economizer component calculations, draft loss was calculated to be 1.38 in. wg. Stack effect from the boiler bank outlet to economizer outlet is:

$$\Delta P_{SE} = SE \times Z = 0.11 \text{ in. wg} \quad \textbf{(110)}$$

where

SE = 0.0074 in. wg/ft — Ch. 23, Eq. 3, Table 3
Z = 15 ft

The net static pressure at the economizer outlet is then calculated:

$$P_{economizer\ outlet} = P_{boiler\ bank\ outlet} - \Delta P_{boiler\ bank\ to\ economizer} - \Delta P_{economizer} + \Delta P_{SE} = -2.99 \text{ in. wg} \quad \textbf{(111)}$$

where

$P_{boiler\ bank\ outlet}$ = –1.64 in. wg — Eq. 106
$\Delta P_{boiler\ bank\ to\ economizer}$ = 0.08 in. wg — Eq. 109
$\Delta P_{economizer}$ = 1.38 in. wg — Eq. 77
ΔP_{SE} = 0.11 in. wg — Eq. 110

The flue gas resistance from the economizer outlet to the air heater is due to friction only. Previous calculations showed that the frictional loss component of the draft loss equation ($f\,L/D_e$), is negligible. Air heater gas side draft loss was calculated to be 1.52 in. wg. Referring to Fig. 2, the flue cross-section from the air heater outlet to the induced fan inlet decreases from 120 ft^2 to 48 ft^2. Again, frictional losses are negligible.

The mass flux is:

$$G_g = \dot{m}_g / A_g = 6750 \text{ lb/h ft}^2 \quad \textbf{(112)}$$

where

$\dot{m}_g$ = 324,100 lb/h — Table 3
A_g = 6 ft × 8 ft = 48 ft^2 — Fig. 2

Draft loss:

$$\Delta P = N\left(\frac{30}{B}\right)\left(\frac{T+460}{1.73\times 10^5}\right)\left(\frac{G_g}{10^3}\right)^2 = 0.08 \text{ in. wg} \quad \textbf{(113)}$$

where

N = 0.34 — Ch. 3, Fig. 6
B = 30 in. Hg
T = T_2 for the air heater = 390F
G_g = 6750 lb/h ft^2

The stack effect from the economizer outlet to the induced draft (ID) fan inlet is determined to be:

$$\Delta P_{SE} = SE \times Z = 0.30 \text{ in. wg} \quad \textbf{(114)}$$

where

SE = 0.0059 in. wg/ft — Ch. 23, Eq. 3
Z = 50 ft — Fig. 2

The net static pressure at the induced draft fan inlet is calculated:

$$P_{ID\ fan\ inlet} = P_{economizer\ outlet} - \Delta P_{air\ heater} - \Delta P_{air\ heater\ to\ ID\ fan\ inlet} + \Delta P_{SE} = -4.29 \text{ in. wg} \quad \textbf{(115)}$$

where

$P_{economizer\ outlet}$ = – 2.99 in. wg
$\Delta P_{air\ heater}$ = 1.52 in. wg
$\Delta P_{air\ heater\ to\ ID\ fan\ inlet}$ = 0.08 in. wg
ΔP_{SE} = 0.30 in. wg

A straight flue runs from the induced draft fan outlet to the stack breaching. Friction again is negligible; however, an expansion loss at the stack breaching is included:

$$\Delta P = N\left(\frac{30}{B}\right)\left(\frac{T+460}{1.73\times 10^5}\right)\left(\frac{G_g}{10^3}\right)^2 = 0.22 \text{ in. wg} \quad \textbf{(116)}$$

where

N = 1.0 — Ch. 3, Fig. 7
B = 30 in. Hg
T = 390F
G_g = 6750 lb/h ft^2

For the stack, two components must be determined: stack draft and stack resistance. For standard air with 0.013 lb wg/lb dry air (v_a = 13.70 ft^3/lb at 80F and 30 in. Hg) and a typical flue gas (v_g = 13.23 ft^3/lb at 80F and 30 in. Hg), Equation 3 of Chapter 23 and the ideal gas law (Equation 11a, Chapter 3) can be combined to calculate the stack draft, ΔP_{SD}:

$$\Delta P_{SD} = 7.84\, Z\, (0.00179 - 1/T)\,(B/30)$$
$$= 0.45 \text{ in. wg} \qquad \textbf{(117)}$$

where

D_i = stack diameter = 8 ft		Fig. 2
Z = stack height = 100 ft		Fig. 2
T_1 = stack inlet gas temperature = 390F		
T_2 = stack exit gas temperature = 340F		Ch. 23, Fig. 19
T = 0.5 (390 + 340) = 365F = 825R		
B = 30 in. Hg		

The stack resistance is calculated from Equation 7, Chapter 23.

Gas mass flux:

$$G_g = \dot{m}_g / A_g = 6443 \text{ lb/h ft}^2 \qquad \textbf{(118)}$$

where

$\dot{m}_g$ = 324,100 lb/h
A_g = $\pi D_i^2/4$ = 50.3 ft^2 — Table 3

Reynolds number:

$$\text{Re} = \frac{G_g D_i}{\mu} = 8.6 \times 10^5 \qquad \textbf{(119)}$$

where

G_g = 6443 lb/h ft^2
D_i = 8 ft
μ = 0.06 lb/h ft — Ch. 3, Fig. 4

Stack resistance:

$$\Delta P_{SR} = \frac{2.76}{B}\frac{T}{D_i^4}\left(\frac{\dot{m}_g}{10^5}\right)^2\left(\frac{f\,L}{D} + N_e\right) = 0.22 \text{ in. wg} \qquad \textbf{(120)}$$

where

B = 30 in. wg
T = 0.5 (390 + 340) = 365F = 825R
D_i = 8 ft
$\dot{m}_g$ = 324,100 lb/h — Table 3
f = 0.012 — Ch. 3, Fig. 1
L = 100 ft — Fig. 2
N_e = stack exist loss = 1.0

The net static pressure at the induced draft fan outlet is calculated:

$$P_{\text{ID fan outlet}} = \Delta P_{SD} - \Delta P_{\text{ID fan outlet to breaching}} - \Delta P_{SR} = 0.01 \text{ in. wg} \qquad \textbf{(121)}$$

where

ΔP_{SD} = 0.45 in. wg
$\Delta P_{\text{ID fan outlet to breaching}}$ = 0.22 in. wg
ΔP_{SR} = 0.22 in. wg

The net operating conditions for the fans are summarized in Table 4.

Fan purchasing specifications add test block factors to each net condition to accommodate deviations from design. See Chapter 23 for further discussion.

Table 4
Fan Operating Conditions

		Fans	
Net design conditions	Units	Forced draft	Induced draft
Flow	lb/h	302,500	324,100
Static pressure rise	in. wg	5.61	4.30
Inlet temperature	F	80	390

Summary

This chapter is intended to give the reader a realistic yet basic overview of boiler design performance calculations. Although there are many variables in these calculations, the designer must pay special attention to the slagging and fouling characteristics of the fuel ash. These factors are particularly detrimental because they reduce thermal performance and increase draft losses. The best analysis of boiler performance is one that considers all variables.

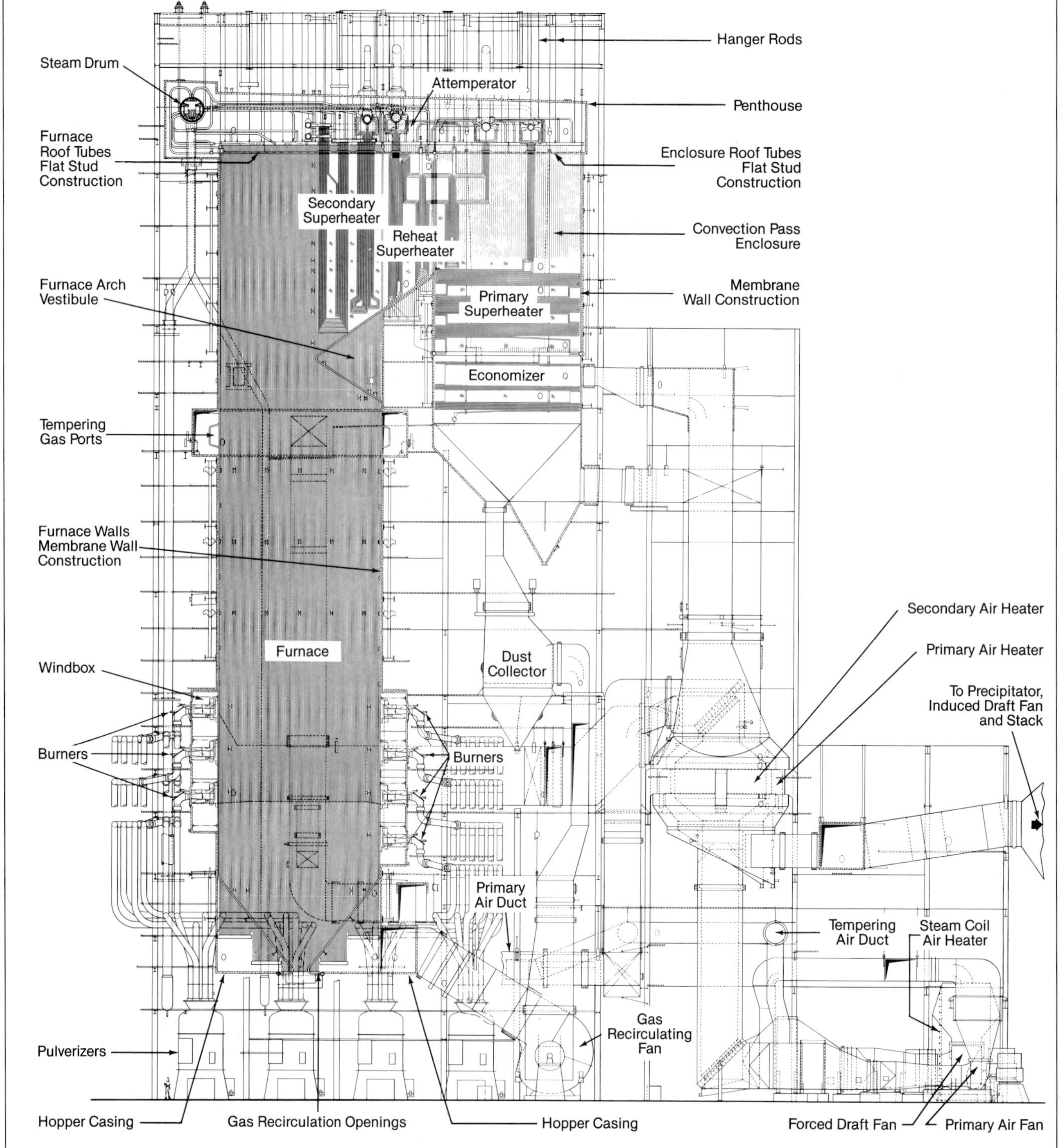

Typical coal-fired utility boiler setting and enclosure.

Chapter 22
Boiler Enclosures, Casing and Insulation

Boiler setting

The term *boiler setting* originally applied to the brick walls enclosing the furnace and heat transfer surfaces of the boiler. Today, boiler setting comprises all the water-cooled walls, casing, insulation, outer covering and reinforcement steel that form the outside envelope of the boiler and furnace enclosure. The term *enclosure* may refer to either the entire setting or to a part of it.

As larger capacity steam generating units were demanded, boiler settings underwent a long evolution from uncooled brick surfaces to today's water-cooled walls. Water-cooled walls began as widely spaced tubes covered with insulating block and progressed to tangent tubes covered with refractory. These tubes were backed with refractory and exposed directly to the combustion process. They gradually evolved to the present day construction of membrane tubes.

Design requirements

The boiler settings must safely contain high temperature pressurized gases and air. Leakage, heat loss and maintenance must be reduced to acceptable values. The following factors must be considered in the setting design:

1. Enclosures must withstand the effects of temperatures up to 3500F (1927C).
2. The effects of ash and slag, or molten ash, must be considered because:
   a. destructive chemical reactions between slag and metal or refractory can occur under certain conditions,
   b. accumulation of ash on the waterwalls can significantly reduce heat absorption,
   c. ash accumulations can fall causing injury to personnel or damage to the boiler, and
   d. high velocity ash particles can erode the pressure parts and refractory.
3. Provisions must be made for the thermal expansion of the enclosure and for differential expansion of attached components.
4. The buckstay system must accommodate the effects of thermal expansion, temperature and pressure stresses, as well as wind and earthquake loading appropriate to the plant site.
5. The effect of explosions and implosions must be considered to lessen the probability of injury to personnel and damage to equipment.
6. Vibrations caused by combustion pulsations and the flow characteristics of gas and air must be limited to acceptable values.
7. Insulation of the enclosures should limit the heat loss to an economical minimum.
8. Neither the exterior surface temperature nor the ambient air temperature should cause discomfort or hazard to operating personnel.
9. Enclosures must be gas-tight to minimize leakage into or out of the setting.
10. Settings of outdoor and indoor units that require periodic washdown must be weatherproof.
11. Settings must be designed for economic fabrication, erection and service life.
12. Serviceability, including access for inspection and maintenance, is essential.
13. Good appearance, in conjunction with cost and maintenance requirements, is desirable.

Tube wall enclosures

In today's units, water- or steam-cooled tubes, or both, are used as the basic structure of the enclosure in high temperature areas of the setting. Important types of water-cooled enclosures are membrane tubes, membrane tubes with refractory lining, flat stud tubes and tangent tubes. The facing page illustrates present day construction for tube wall enclosures.

Membrane tubes

Fig. 1 illustrates a typical furnace wall with membrane construction. These walls are water-cooled and constructed of bare tubes joined by thin membrane bars. The walls are gas-tight and do not require an exterior casing to contain the products of combustion. Insulation is placed on the outside of the wall and sheet metal or lagging is installed over the insulation to protect it.

Membrane tubes with refractory lining

There are several locations in selected types of boilers that require refractory lining on the furnace side of the tubes to protect the tubes from either erosion or corrosion from the products of combustion. Some of the most common applications are:

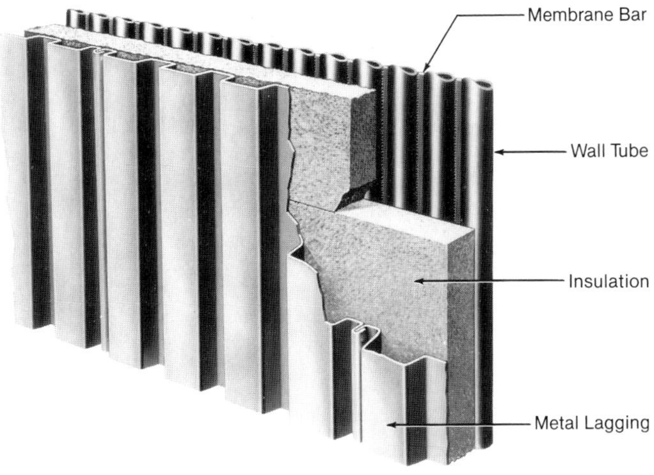

Fig. 1 Membrane wall construction.

1. Cyclone-fired units: lower furnace and cyclone burner walls. (See Chapter 14.)
2. Circulating fluidized-bed boilers: lower furnace. (See Chapter 16.)
3. Refuse boilers: lower furnace. (See Chapter 27.)
4. Pulverized coal-fired boilers: burner throats. (See Chapter 24.)

Cylindrical pin studs, welded on the hot side of the tubes at close intervals, hold the refractory in place (Fig. 2). Lining the wall with refractory can also increase furnace temperature by reducing heat absorption where this is desired. The increase in temperature helps to maintain the coal, peat, or lignite ash in a liquid state, thereby preventing large ash buildup and allowing better removal of slag. These issues are discussed in more depth in Chapter 20. However, because of maintenance problems it is usually desirable to avoid refractory where technically acceptable.

Flat stud tube walls

These walls consist of tubes with small, flat bar studs welded at the sides (Fig. 3). These walls are typically backed by one of two construction methods which are usually found in the convection pass enclosure.

In the current method, the flat studded tubes are backed with refractory covered with a welded inner hot casing that is insulated and covered with metal lagging for protection. The casing is supported from channel tie bars welded to the tubes at each buckstay row. The walls are reinforced with buckstays and the inner casing is reinforced with stiffeners. Stiffener spacing and size are set by the design pressure of the walls between buckstays. This system provides a better gas-tight enclosure than the former method.

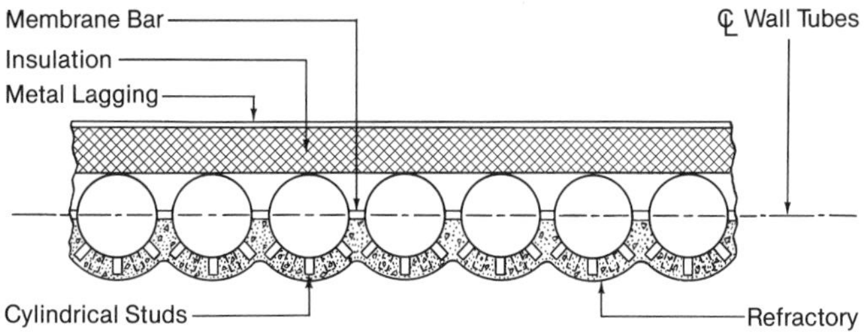

Fig. 2 Fully studded membrane walls.

In former practice, the flat studded tubes are backed with refractory, followed by a dense insulation and an outer cold casing. The casing is supported from the buckstays with expansion folds at the attachments. These folds minimize stresses in the casing caused by differential expansions between the hot tube wall and cold casing. This method is now obsolete, but found on many old boilers that are still in service.

While the construction of the casings described in the preceding paragraphs applies to areas of horizontal buckstay reinforcement, some industrial boiler designs require a vertical casing with a tie bar that is welded vertically to a bar between the tubes.

Tangent tube walls

These walls are constructed of bare tubes placed next to each other with a typical gap of 0.03125 in. (0.7937 mm). The refractory backup, casing and insulation system design has also been used with tangent tube walls. These walls are typically found in the furnace area of older boiler designs (Fig. 4).

Flat stud and tangent tube wall upgrades

In recent years, two methods have been used to provide a better enclosure seal on units with either inner (hot) or outer (cold) casing as the gas seal. In one method, on boiler enclosure areas with tangent tubes, a round bar is seal welded between each tube for the full length (Fig. 5). In the other method, where boiler enclosure areas have widely spaced tubes with flat studs, a flat bar is seal welded between the tubes just behind the flat studs for the full length (Fig. 6).

These methods have been effective on many boilers, providing an improved gas seal with considerably less maintenance and longer life than the casing seal they replaced. Their biggest drawback is high installation costs because the entire boiler must be stripped of its existing casing and insulation, then a new insulation and lagging system must be installed.

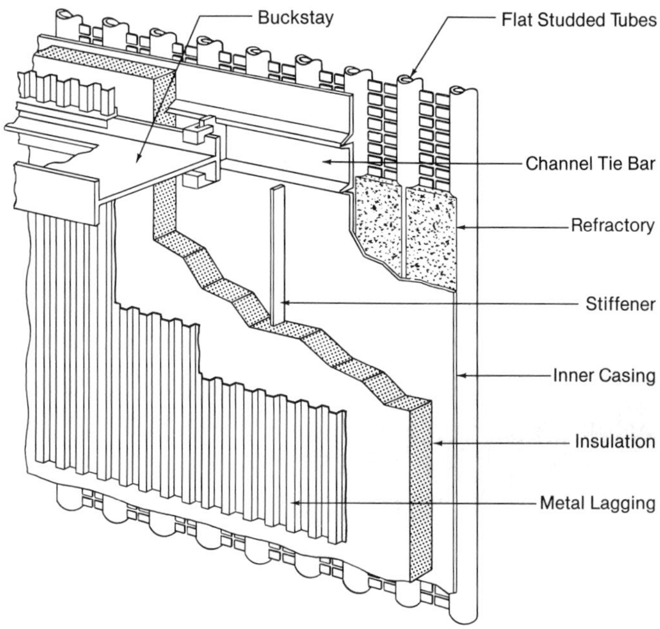

Fig. 3 Flat stud tube wall construction with inner casing shown.

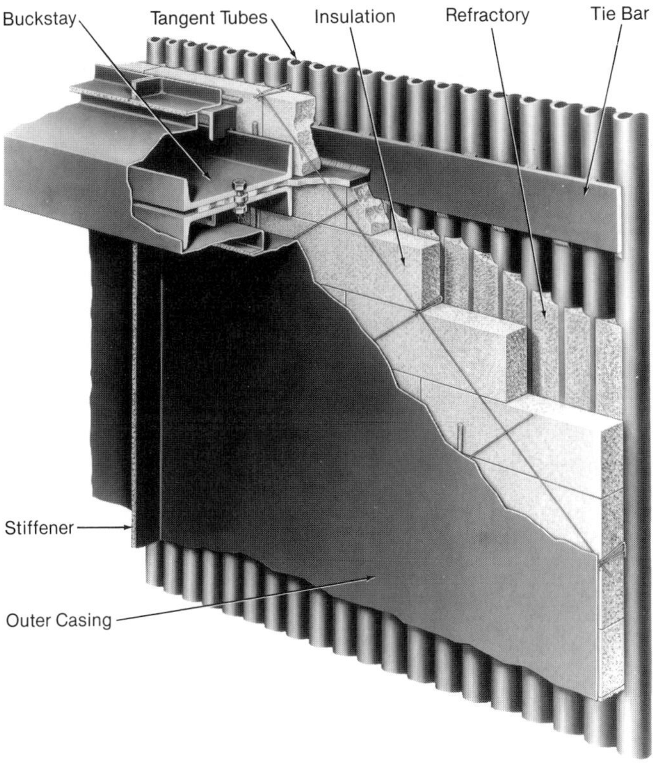

Fig. 4 Tangent tube wall construction with outer casing shown.

Casing enclosures

The casing is the sheet or plate attached to pressure parts for supporting, insulating, or forming a gas-tight enclosure.

A boiler unit contains many cased enclosures that are not water-cooled. These enclosures must be designed to withstand relatively high temperatures while having external walls that minimize heat loss and protect operating personnel.

Casings are constructed of sheet or plate reinforced with stiffeners to withstand the design pressures and temperatures. When the casing is directly attached to the furnace walls, expansion elements are added to allow for differential thermal expansion of the tubes and casing. The frontispiece shows typical enclosures including the hopper casing, windbox, tempering gas plenum and penthouse casing.

Hopper

Hopper enclosures are used in various areas of the boiler setting including the economizer hopper, furnace hopper enclosure and the wash hopper for dry bottom units.

The enclosure provided by the hopper casing may also serve as a plenum for the recirculating gas which leaves the economizer hopper through ports and enters the furnace through openings between tubes in the furnace hopper.

Windbox

The windbox is a reinforced, metal-cased enclosure that attaches to the furnace wall and houses the burners and distributes the combustion air. It may be located on one furnace wall or on all furnace walls using a wraparound configuration. The attachments to the furnace walls must be gas-tight and permit differential thermal expansion between the tubes and casing.

For large capacity boilers, the windbox is compartmented and placed only on the front and rear furnace walls. The windboxes are compartmented for better combustion air control.

Tempering gas plenum

This enclosure provides for the distribution and injection of gas which is used to temper the furnace gases and control the ash fouling of heating surfaces. It is constructed similarly to the windbox, but is normally protected on the inside by stainless steel shields opposite the gas ports.

Penthouse

The penthouse casing forms the enclosure for all miscellaneous pressure parts located above the furnace and convection pass roofs. It is a series of reinforced flat plate panels welded together and to the top perimeter of the furnace pressure parts. Various seals are used at the penetrations through the penthouse walls, roof and roof tubes. Some examples are cylindrical bellows or flexible cans sealing the suspension hangers, large fold (pagoda) seals around the steam piping and refractory or casing seals around heating surface tube penetrations through the roof tubes. On many utility and some industrial boilers a gas-tight roof casing is used on top of the roof tubes as the primary gas seal. The penthouses may or may not be designed as pressure-tight enclosures with seal air. It depends upon whether the boiler is a pressure fired or a balanced draft unit and whether the roof seals are seal welded gas-tight or are the refractory type.

Design considerations

Resistance to ash and slag

Ash has a tendency to shed from a water- or steam-cooled metal surface, particularly when its temperature is well below the ash softening point. Wall type sootblowers remove ash in high temperature areas where it tends to adhere to the walls. (See Chapter 23.)

Extensive areas of exposed refractory must be avoided because large accumulations of ash could fall into the furnace damaging equipment or causing injury to personnel. Also, crotches in the tube wall should be designed to prevent ash and slag accumulation.

Pressure part erosion is reduced to acceptable levels by limiting the gas velocity through the unit. However, local high velocities can still occur in areas where gas

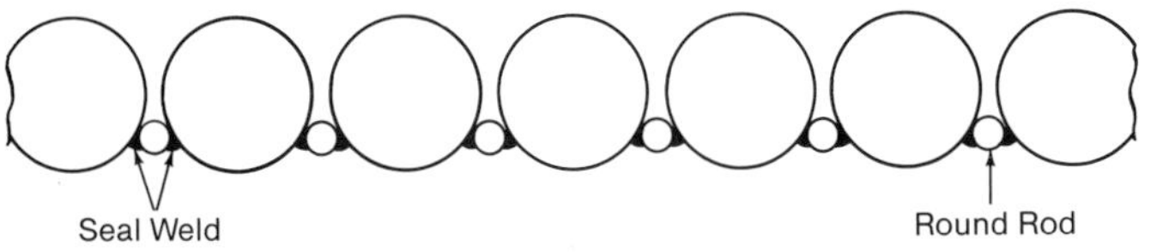

Fig. 5 Tangent tubes with closure rods.

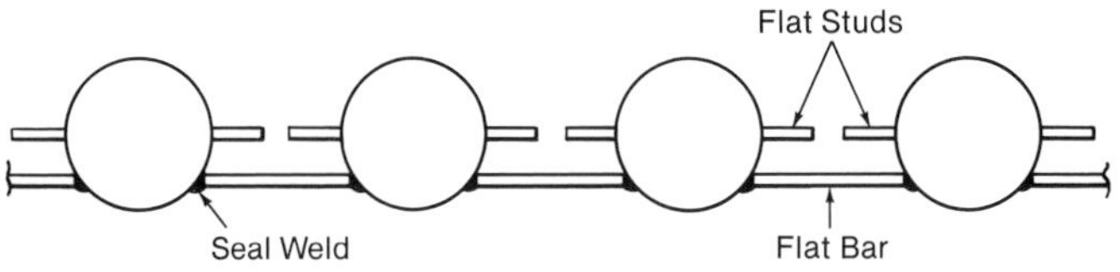

Fig. 6 Widely spaced tubes with flat studs and closure bars.

bypasses baffles or heating surfaces. These high velocity lanes are best eliminated by proper design, baffle installation and routine maintenance.

Some unit designs such as process recovery, refuse and circulating fluidized-bed boilers require large areas of refractory on tube surfaces. In these units, the refractory system is designed to eliminate the effects of corrosion or erosion, or both, on pressure parts while minimizing the reduction in heat absorption by pressure parts.

Expansion

With the inner cased unit, Fig. 3, temperature differentials can occur between the casing and the tubes during startup. Expansion of the wall in the horizontal direction is governed by the temperature of the tie channel. Because the casing and the channels are at the same temperature, they can be welded together. Vertical expansion differences are accommodated by flexing of the casing flanges at the top and bottom of each casing section.

With a bottom supported unit, such as the PFI integral furnace boiler (Fig. 7) which is designed for pressure firing, the structure is fixed at a point at one end of the lower drum. Clearances, seals and supports are designed for known expansions in all directions from the fixed point.

With a top supported unit (see frontispiece), the expansion occurs downward from one elevation. Unless one of the walls is fixed to the building steel, the expansion will occur outward from the center of the unit.

Flues and ducts, piping, ash tanks and burner lines must be designed with expansion joints or seals to accommodate movement. Flexible metal bellows or elastomer belts are used in flues and ducts while metal hoses and sliding or toggled gasketed couplings are used in piping. Water seals are generally used between ash hoppers or slag tanks and the associated furnace. With large units, the expansion may be as great as 12 in. (305 mm) between adjacent parts yet the joints must remain pressure-tight.

Support

The support of boilers is discussed in Chapter 39. It is generally more economical to support the smaller units from the bottom and the larger units from the top. In either case, the boiler setting is formed by the waterwalls when these are available.

For bottom supported units, the enclosure is usually supported from a common foundation with the boiler as shown in Fig. 7. For top supported units, cased enclosures are supported from the pressure parts with the exception of the penthouse enclosure which is supported directly from the structural steel by the hanger rods.

Explosions

In the design of settings, the effect of possible explosions must be considered to eliminate the possibilities of personnel injury and serious equipment damage. A better understanding of the technical problems and the development of adequate design and operating codes have eliminated most explosions. On units with fluid or fluidized fuels, care must be taken to avoid puffs that can occur from improper fuel and air mixture during startup. (See Chapter 10.)

The enclosure is normally designed to withstand common puffs and minor explosions. In the event of a major

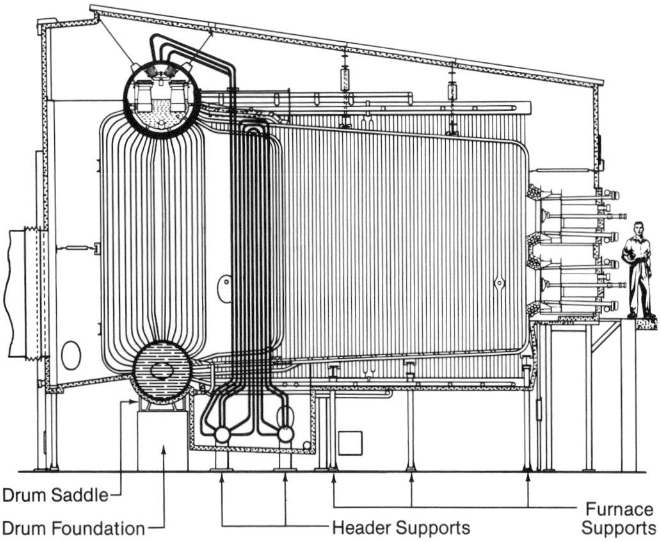

Fig. 7 Bottom supported unit.

furnace explosion, the design should provide for the failure of studs, stud attachments and welds rather than failure of tube walls. This minimizes the release of large quantities of steam or furnace gases.

Explosion doors were once used on small furnaces to relieve excessive internal furnace pressure. These doors are no longer used because the rapid internal pressure increase from a fuel explosion is not significantly relieved by opening one or more doors. Explosion doors may also be more of a hazard than a safety margin because, in the event of a puff, they may discharge hot gases that would otherwise be completely contained within the setting.

The forces from furnace puffs and normal operating negative or positive furnace pressures are contained by bars and channels welded to tubes to form continuous bands around the setting. Beams (*buckstays*) are attached to the tie bars with slip connections to keep the walls from bowing inward or outward.

Because the buckstays are outside of the insulation, special corner connections are required that allow the walls to expand (Fig. 8). Forces generated by explosions concentrate at the corner connections. These connections must be tight during startup when the walls have not fully expanded and during the normal operating fully expanded position.

The tube span between the buckstays acts as a beam to resist the internal furnace pressure. The larger the tube diameter and the heavier the tube wall, the farther

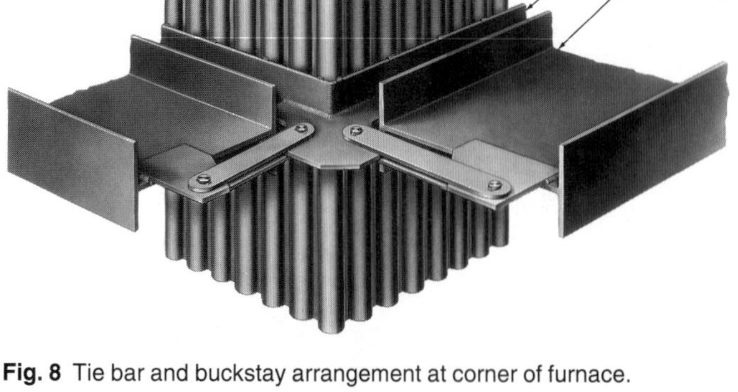

Fig. 8 Tie bar and buckstay arrangement at corner of furnace.

apart the buckstays may be spaced. The size of the buckstay beam is determined by the permissible deflection and the positive or negative pressure loading.

Implosions

Implosions are usually caused by the improper operation of dampers on units with high static pressure induced draft fans, or an extremely rapid decay of furnace pressure due to sudden loss of fuel supply. The rules for determining minimum continuous and transient design pressures for the furnace enclosures can be found in the National Fire Protection Association (NFPA) 85G Standard. In addition, induced draft fan controls are specified to minimize possible operating or control errors and to reduce the degree of furnace draft excursion following a fuel trip.

Vibration

Excessive vibration can cause failures of the tubes, insulation, casing and supports. These vibrations can be produced by external rotating equipment, such as turbines and fans, with vibration transmitted through the building's steel, piping, flues and ducts; furnace pulsations from the uneven combustion of the fuel; or turbulence in the flowing streams of air or gas in flues, ducts and tube banks.

The walls, flues and ducts are designed to limit vibration during normal operating conditions. For the walls, the section modulus of elasticity of the buckstays is usually selected to limit wall deflection. Flues, ducts and casings are similarly stiffened by bars or structural shapes. This stiffening is particularly necessary in sections of flues and ducts where the flow is highly turbulent, as in the fan discharge connecting piece. Every effort should be made to eliminate the sources of severe vibration, such as unbalanced rotating equipment, poor combustion and highly turbulent or unbalanced air or gas flow.

Heat loss

Heat loss from a boiler setting is reduced by insulation, usually as an integral part of the boiler enclosure. (See Figs. 1 through 4.) There is an economical balance between the value of the heat loss and the cost of the insulation and installation.

For most steam generating units, the insulation system is designed to provide both safety for personnel and minimal heat loss. In addition, indoor units require ventilation for both operator comfort and room air change.

The materials used most frequently for heat insulation are listed below.

Mineral wool

This material is comprised of molten slag, glass or rock, blown into fibers by steam or an air jet or spun by high speed wheels.

Mineral wool base block Mineral wool fibers and clay, molded under heat and pressure, are used to insulate membrane tube walls and boiler casings up to temperatures of 850, 1200 or 1900F (454, 649 or 1038C) depending upon the grade.

Mineral wool blanket Mineral wool fibers, compressed into blanket form and held in shape between hexagonal wire mesh or expanded metal lath, are used on all types of enclosures with external metal lagging or casing and for piping inside cased enclosures. The temperature limit is normally 1200F (649C).

Calcium silicate block

Reacted hydrous calcium silicate block is used on enclosures and piping, generally below 1200F (649C).

High temperature plastic

Insulating cement made of mineral wool fibers processed into nodules and dry mixed with clay forms a tough, fibrous monolithic insulation in final dried condition. Drying shrinkage is as much as 40%, but there is a tendency to crack upon drying. This material is used principally on irregularly shaped valves and fittings and to fill gaps between block insulation. Insulating cement is available in grades useable up to 1900F (1038C).

Ceramic fiber

High purity ceramic fibers, with melting points above 3000F (1649C), are occasionally used for tube enclosure seals where resiliency or high temperature insulation is required.

Heat loss calculations

Calculations of the heat flow through a composite wall are discussed in Chapter 4. Thermal conductivities of a wide range of commercial refractory and insulating materials, at the temperatures for which they are suitable, are given in Fig. 9. Combined heat losses (radiation plus convection) per square foot (square meter) of outer wall surface are given in Fig. 10 for various ambient air velocities and surface air temperature differences. The American Boiler

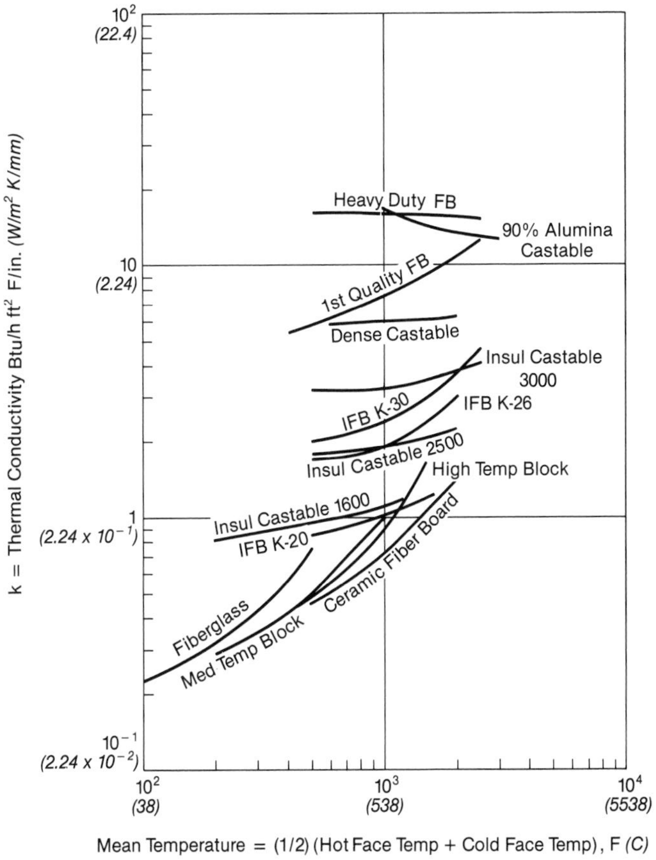

Fig. 9 Thermal conductivity of various refractory materials.

Manufacturers Association (ABMA) Radiation Loss Chart provides a quick approximation for radiation loss, expressed as a percentage of gross heat input (Fig. 11).

Ventilation, surface temperature, conditions

To maintain satisfactory working conditions for personnel around a boiler, a cold face temperature of 130 to 150F (54 to 66C) is usually considered satisfactory. Heat losses, corresponding to these surface temperatures, range from 90 to 130 Btu/h ft^2 (284 to 410.1 W/m^2), which can be readily absorbed by the air circulation generally provided in present day boiler rooms.

Insulating a boiler to reduce heat loss to a value that can readily be absorbed by the total volume of room air does not in itself assure comfortable working conditions. Proper air circulation around all parts of the boiler is also necessary to prevent the accumulation of heat in the areas frequented by the operating personnel. This can be helped by using grating rather than solid floors, by ample aisle space between adjacent boilers, by the location of fans to assist the circulation of air around the boiler and by installing ventilating equipment to assure adequate air change.

Good ventilation does not greatly increase overall heat loss. Air velocity affects the surface heat transfer coefficient. This can be verified by data from Fig. 10. However, surface conductance is only a small part of the total resistance to heat flow. For example, an increase in air velocity from 1 to 10 ft/s (0.3 to 3 m/s), for the conditions given in Fig. 12, will increase the heat loss rate through the wall by only 2%.

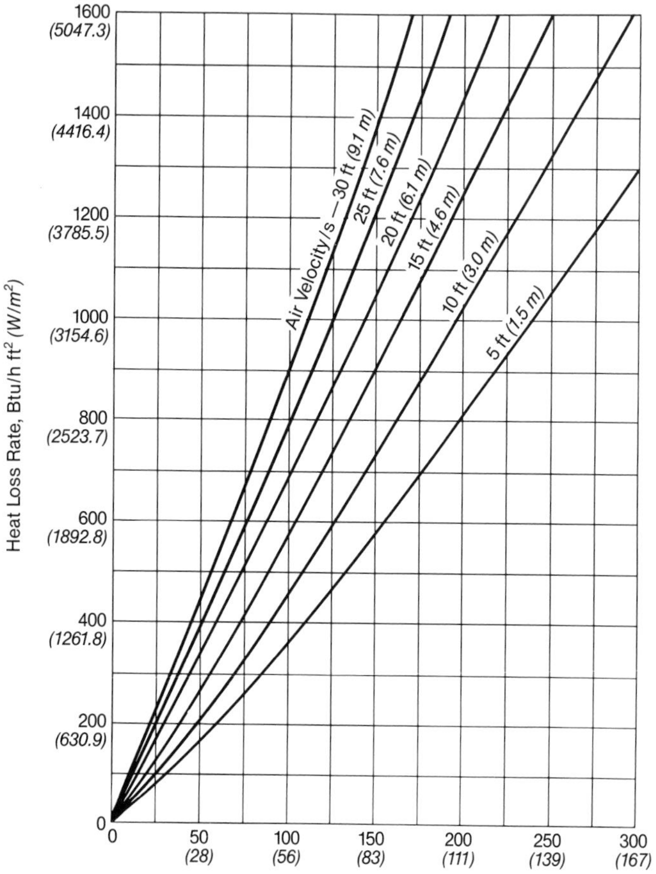

Fig. 10 Heat loss from wall surfaces (radiation + convection). *(Source — ASTM Standards, Part 13,* 1969.*)*

Unlike heat loss, outer surface temperature is considerably affected by the surrounding conditions. In the situation shown in Fig. 13, where two walls of similar temperatures are close together, the radiant heat transfer from either wall is negligible. The natural circulation of air through such a cavity is inadequate to cool the walls to a temperature suitable for personnel working in the vicinity. From Fig. 14 it can be seen that a considerable change in surface film resistance will cause an appreciable change in lagging or surface temperature while not affecting the heat loss through the wall to any extent.

Increased insulation thickness would not significantly reduce the surface temperature in the cavity shown in Fig. 13. Cavities should therefore be avoided in areas where operators work. Ventilating ducts to reduce the air temperature in such a cavity can be installed if necessary.

Leakage

Continuing efforts have been made over the years to reduce air infiltration into boiler settings. Such leakage increases gas flow and the heat rejected to the stack, thereby lowering boiler efficiency and increasing the amount of induced draft fan power. (See Chapter 9.)

Corrosion

One of the advantages of membrane walls over cased walls is that they eliminate flue gas corrosion on the cold face of the enclosure walls. Most flue gases contain sulfur; therefore, metal parts of the setting must either be kept above the dew point of the gases or out of contact with the gases. (See Chapter 20.) The dew point generally ranges between 150 and 250F (66 and 121C) and is dependent on the fuel, its sulfur content and the firing method.

Flues carrying low temperature spent gases should be insulated on the outside to inhibit corrosion. This is particularly necessary on outdoor units. Water-cooled doors and slag tap coils require water temperatures above 150F (66C) to keep the cooling coils above the dew point of the gases.

When casing is located outside of insulation or refractory, it is still subject to the action of the flue gases. When this type of casing is subjected to a temperature below the dew point, an asphalt mastic or other type of coating is needed to protect it from corrosion on the inside. This problem requires special attention in the design of outdoor installations where temperatures may, at times, be below flue gas dew point temperatures.

With the use of externally insulated casing, corrosion problems are greatly reduced because the flue gases are completely contained by a metal skin, which is well above the dew point temperature. However, even with the inner casing, seals and expansion joints must be insulated properly to avoid cold spots and consequent corrosion.

Resistance to weather

Outdoor boiler installations are possible in mild climates. While the initial cost of the plant is reduced, maintenance of the boiler and auxiliary equipment must be considered. Severe weather can extend outage time and increase maintenance expenses. These units must

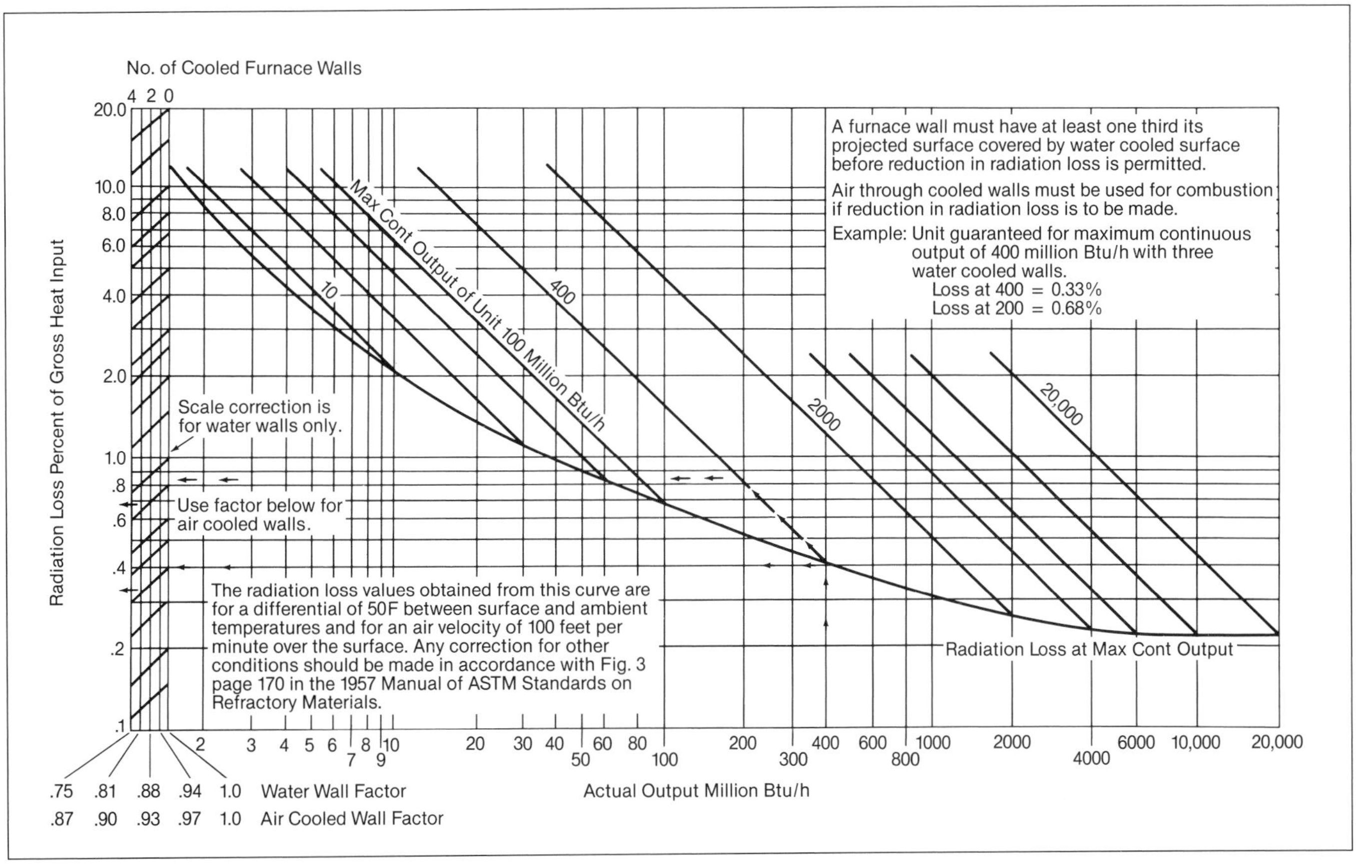

Fig. 11 Radiation loss in percent of gross heat input (American Boiler Manufacturers Association).

also have sufficient reinforcement to withstand the pressure and suction forces of the wind.

Lagging is an outer covering over a wall used for protecting the insulation from water or mechanical damage. It is relatively simple to make a metal lagged unit rainproof. Joints and flange connections are overlapped and flashings are used around openings. Welding of joints or the use of mastic compounds is necessary in areas which are difficult to seal.

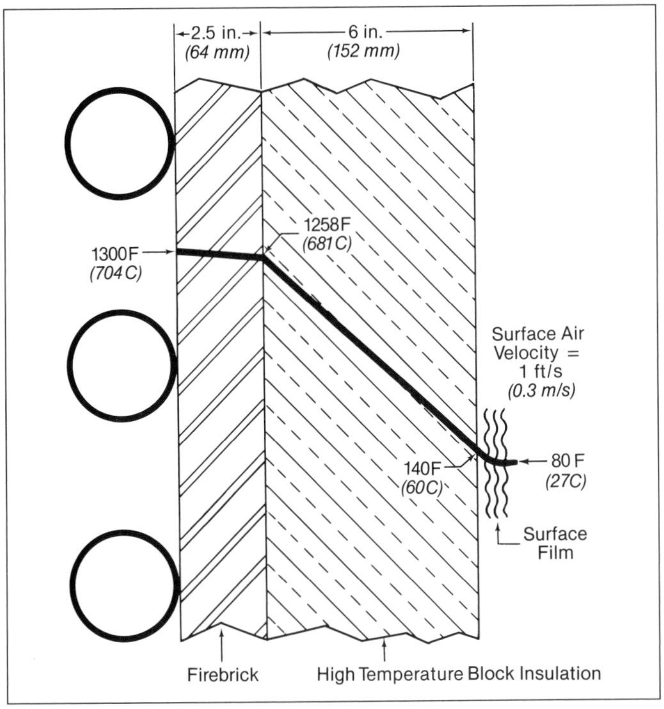

Fig. 12 Temperature gradients through tube and brick wall.

Sloping roofs are required and are particularly important on aluminum lagging where pockets of water would eventually stain the surface. Direct contact between aluminum and steel must be avoided to prevent galvanic corrosion of the aluminum in the presence of moisture. Copper lines or roof flashings should be designed so that water runoff does not wet the aluminum.

Weather hoods should be used to keep rain, snow and ice from contacting outdoor safety valves. Nozzle and valve necks must be insulated and protected with sheet metal or outer waterproof covering. Outdoor control lines containing air or flue gas, drain and sampling lines and intermittently operated steam and water lines should be insulated and protected by electric resistance heating wires. Steam pipe tracer lines may also be used in some cases. Dry air should be supplied for control lines, and sootblowers. Steam and water lines outside the setting must be completely drainable.

Fabrication and assembly

The setting must be designed for economical fabrication and assembly. This requires integration of all shop and field methods and practices. Small units can be completely shop assembled. For larger units the trend has been toward shop subassembly of large components.

Shipping clearances usually limit the size of shop-assembled wall panels to 16 ft (4.9 m) in width or 105 ft (32 m) in length. These size criteria can not be used simulta-

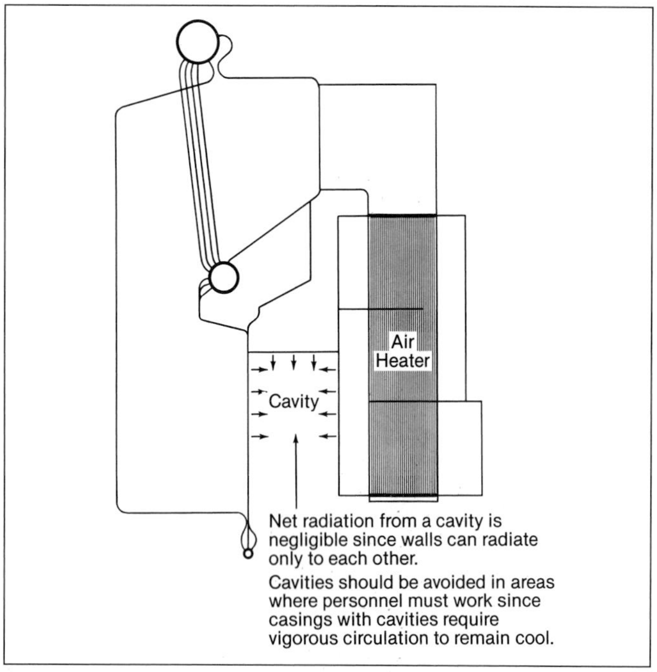

Fig. 13 Cavities tend to raise wall surface temperatures.

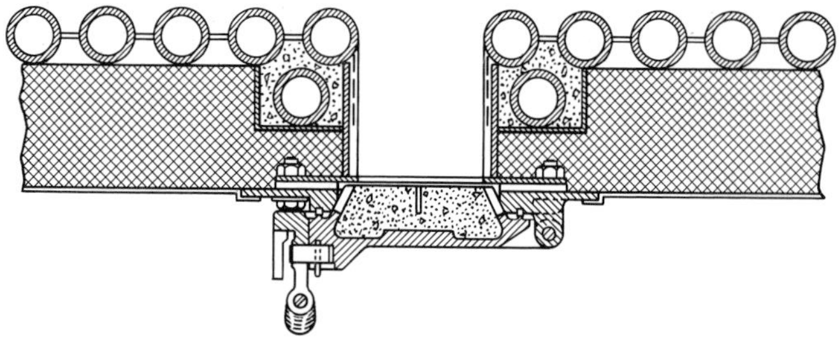

Fig. 15 Inspection door for balanced draft furnace.

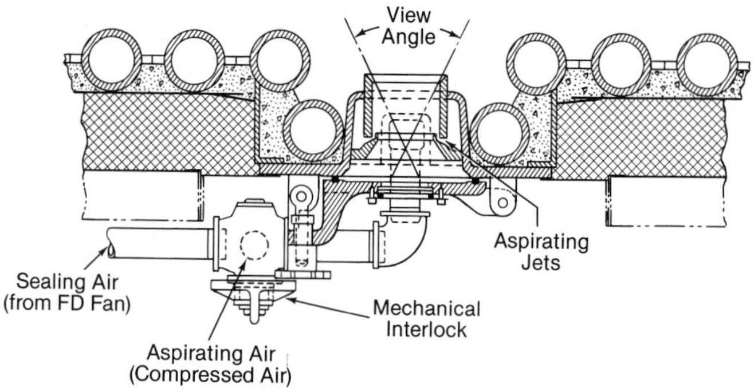

Fig. 16 Inspection door for pressurized furnace.

neously. That is, as the panels are made longer, the maximum width will be less and as the width is increased the maximum length decreases. Shop assembly of components permits better quality control of the more complicated parts.

Casing enclosures, tube connections to headers, tie bars, doors and other attachments are normally of welded construction. New and improved materials and attachment methods reduce the manhour requirements for insulating boilers and installing metal lagging.

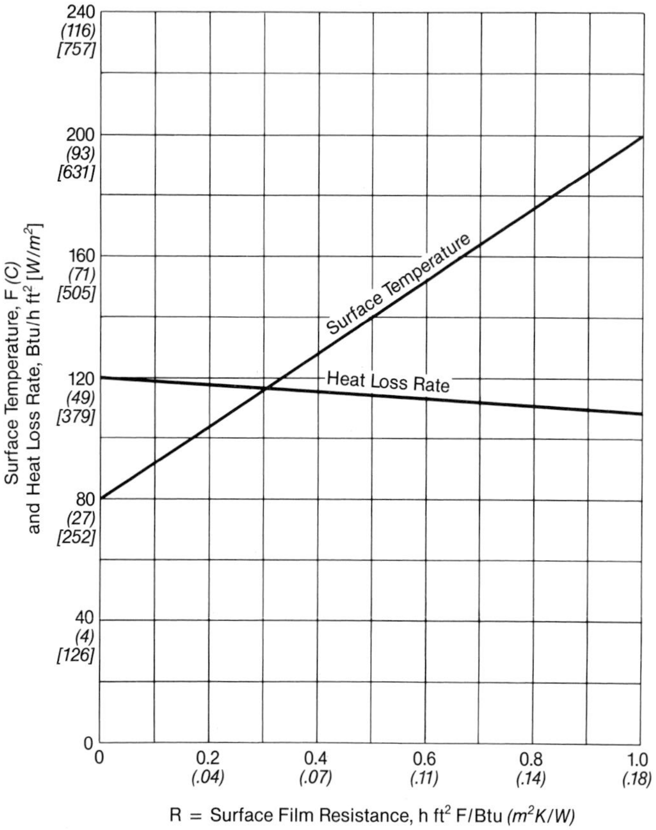

Fig. 14 Effect of surface film resistance on surface temperature and heat loss rate.

Serviceability

Many setting design details must be resolved to simplify operation and maintenance. Working areas around the unit should have adequate lighting and comfortable temperatures. Clearances for servicing and removing parts should be provided. Access through the setting is necessary for inspection of boiler internals. Suitable platforms for access doors, sootblowers, instruments and controls are essential.

Inspection doors allow observation of combustion conditions and the cleanliness of heat absorbing surfaces. Figs. 15 and 16 illustrate inspection doors for balanced draft and pressurized settings. Safety is provided by two types of interlocks which assure that compressed air is properly aspirating the aperture before the door is opened. A feature of this door is that the aspirating jet does not restrict the comparatively wide view angle.

The tube bends that form openings must have the smallest possible radius. The length of the stud plate closures around the opening is minimized so that the plates can be adequately cooled by welded contact to the tubes, thereby preventing burn back and subsequent overheat of seals.

Appearance

The setting should present a good appearance and be designed so that it can be retained indefinitely with a minimum of housekeeping. The outer surface should be easy to clean. Equipment handling flue gas, coal, ash or oil should be designed to minimize leakage.

Light gauge metal lagging is generally used as an outer covering. This is particularly true for outdoor units where it is relatively simple to make the metal lagging water-tight.

Many types of covering are found in older installations including plastic insulation, insulating cement, canvas and welded steel casing.

Figs. 1, 2 and 3 show metal lagging. Light gauge galvanized steel, or clad aluminum sheets are commonly used. Galvanized steel is generally less expensive than aluminum, but for outdoor units it may be necessary to paint the galvanized steel after weathering, unless the climate is dry. The clad aluminum may be preferable because it only requires painting under more severe conditions.

Wall sootblower inspection.

Chapter 23
Boiler Auxiliaries

A variety of components beyond the systems already discussed are needed in order for the modern steam generating system to function effectively and efficiently. While space prevents an in depth review of all items, several deserve special attention. Sootblowers are used to clean the gas side of heat transfer surfaces permitting the boiler to operate at peak efficiency. Safety and relief valves are critical to assure the continued safe operation of the boiler. Ash handling removes and conditions the ash or combustion refuse for ultimate disposal. Proper air and gas flow required for good combustion are provided and controlled by the dampers, stack and fans. Finally, a specialty condenser is used to provide high purity water for attemperator spray in industrial boilers where such water is not typically available.

Sootblowers

Sootblowers are mechanical devices used for on-line cleaning of gas side boiler ash and slag deposits on a periodic basis. They direct a cleaning medium through nozzles against the soot or ash accumulated on the heat transfer surfaces of boilers to remove the deposits and maintain the heat transfer efficiency. They also prevent plugging of the gas passes. Chapters 18 and 20 provide information regarding boiler types and the effect of slagging and fouling characteristics of various ash laden fossil fuels on design.

The type of sootblower used in an application varies with the location in the boiler, the cleaning coverage required and the severity of the deposit accumulation. Sootblowers basically consist of: 1) a tube element or lance which is inserted into the boiler and carries the cleaning medium, 2) nozzles in the tip of the lance to accelerate and direct the cleaning medium, 3) a mechanical system to insert or rotate the lance, and 4) a control system. (See Fig. 1.)

Cleaning media

The cleaning media used in sootblowers may be saturated steam, superheated steam, compressed air or water. Combinations of water with other media, such as steam or air, have also proven effective.

In most cases, superheated steam has become the preferred cleaning medium because field experience indicates that erosion of tube surfaces can occur because of the moisture contained in saturated steam. Also, superheated steam, when compared with saturated steam on a pound for pound basis, has a greater cleaning potential due partially to the higher sonic velocity through the sootblower nozzles. This increase in sonic velocity more than offsets the loss of jet energy due to the lower media density.

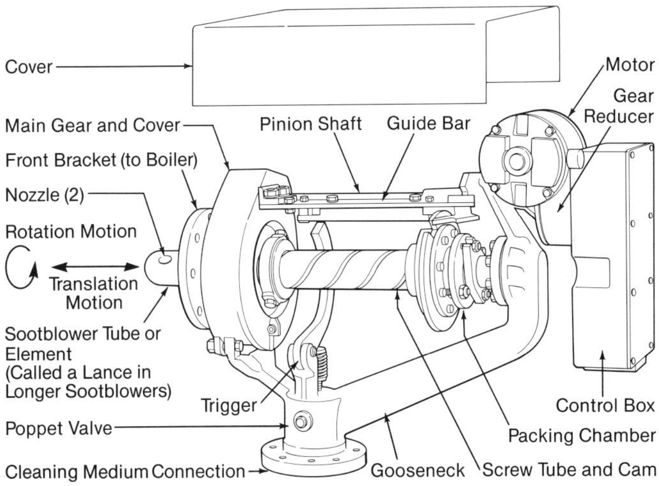

Fig. 1 Model IR wall blower.

On larger boilers, compressed air is often used as the cleaning medium. The source may be either high pressure 350 to 500 psig (24.1 to 34.5 bar gauge) reciprocating compressors or high flow rate centrifugal compressors discharging at pressures in the range of 150 to 225 psig (10.3 to 15.5 bar gauge).

Normal nozzle pressures for sootblowers will vary from 70 to 350 psig (4.8 to 24.1 bar gauge) for steam and from 60 to 220 psig (4.1 to 15.2 bar gauge) for air, depending upon the ash deposit being removed and the type of sootblower. Line pressure is controlled by valves and pressure reducing stations (PRV) in the blower piping.

Steam for sootblowing may be taken from intermediate superheater headers, cold reheat inlet headers, secondary superheater outlet headers or hot reheat outlet headers. Each source has advantages and disadvantages, often complicated by the boiler operating mode, i.e., constant, hybrid or variable pressure.

Water, at sootblower inlet pressures of 150 to 300 psig (10.3 to 20.7 bar gauge), may also be used as a cleaning medium either alone or in combination with steam or air.

Water may also be injected into sootblowers to cool retractable sootblower tubes or lances exposed to high temperature gas zones.

The choice between the cleaning mediums of air and steam is usually based upon an economic analysis of operating costs and technical issues.

Reliability, investment costs and expected annual operating costs are affected by the following major differences:

1. Steam sootblowers must be designed to permit warm-up of system piping, drainage or removal of condensate in the piping, and protection from freezing, corrosion and erosion. Availability of makeup water also must be considered.
2. Steam sootblowers are expected to require more maintenance than air sootblowers. However, this cost may be offset by the maintenance cost of air compressors.
3. Increasing the capacity of a steam system is usually easier to accomplish because the steam supply from the boiler is normally limited only by the pressure reducing valves.
4. Air systems demand a higher flow rate to cool long retractable blowers. This is caused by the better heat transfer characteristics of steam.

For certain high temperature ranges in which the deposit is plastic, or where the deposit strongly sinters to the tube (some Western low sulfur coals), neither air nor steam is effective. In these cases, water is required as a cleaning medium to remove the deposit.

Sootblower terminology

Peak impact pressure

Peak impact pressure (PIP) is a measure of the energy delivered to the deposit at a given distance from the sootblower nozzle and is a key application parameter. It has been found to be a function of nozzle size and configuration as well as fluid pressure, temperature and quality.

Cleaning radii and bank penetration

Cleaning radii and bank penetration refer to the effective cleaning distance for an unobstructed cleaning jet and for a cleaning jet through a tube bank respectively. They are functions of sootblower position, surface geometry, PIP, ash characteristics and deposit temperature.

Jet progression velocity

Jet progression velocity is the linear velocity [ft/s (m/s)] of the sootblowing medium traveling across the surface as the sootblower head moves. It is a function of the nozzle rotational speed, translational speed and distance from the surface.

Sootblower types

Fixed position blowers

The G9B fixed position sootblower, Fig. 2, is a nonretractable sootblower, either rotating or nonrotating, used to remove dusty or lightly sintered ash from tube banks or duct systems. The fixed element is generally more economical to install and operate than the retractable sootblower, but can only be used where lower gas temperature areas permit and where high mass energy from large nozzles is not required. Nozzle sizes range from 0.25 to 0.375 in. (6.35 to 9.53 mm) with a typical value of 0.3125 in. (8 mm).

Fig. 2 Electric motor driven rotary blower.

Short retractable furnace wall blower

A short travel retractable-type unit (IR blower) used principally for cleaning furnace water wall tubes is illustrated in Fig. 1. The normal nozzle position, when the blower is fully extended, is approximately 1.5 in. (38 mm) from the face of the tubes. The cleaning radius is a function of nozzle pressure, the type of nozzle used, the nature of the deposit and the surface to which the deposit is adhering. In general, the normal effective cleaning area of an IR blower is an oval shape with a vertical axis of 12 ft (3.66 m) and a horizontal axis of 10 ft (3.05 m). These data are based on either 150 psig (10.3 bar gauge) of air or 200 psig (13.8 bar gauge) of saturated steam.

The rotating arc of the sootblower may be cut to reduce the blowing rotation from 360 deg (6.3 rad) to a lesser angle to reduce blowing medium erosion of adjacent side wall tubes. Nozzle sizes can vary, but are typically 1 in. (25.4 mm).

Long retractable blower

The model IK 500 Series blower is a long retractable type of cleaning device. Fig. 3 shows an IK severe duty long retractable blower used on chemical recovery boilers. A set of cleaning media nozzles at the end of a lance tube extends into a boiler cavity to clean tube banks. These blowers have a travel range of 2 ft (0.6 m) to a maximum of 56 ft (17.1 m). Long retractable blowers are applied in either vertical or horizontal cavities and typically use either air or steam as a blowing media. In some cases, special nozzles are employed to use water as a cleaning medium where deposits are especially difficult to remove.

The *translation speed* is the speed at which the lance tube travels into the boiler. Translation speeds may be 35 in./min (0.89 m/min) up to as high as 200 in./min (5.1 m/min). The helix pattern due to the translation and rotational speed is normally 4, 5, 6 or 8 in. (102, 127, 152

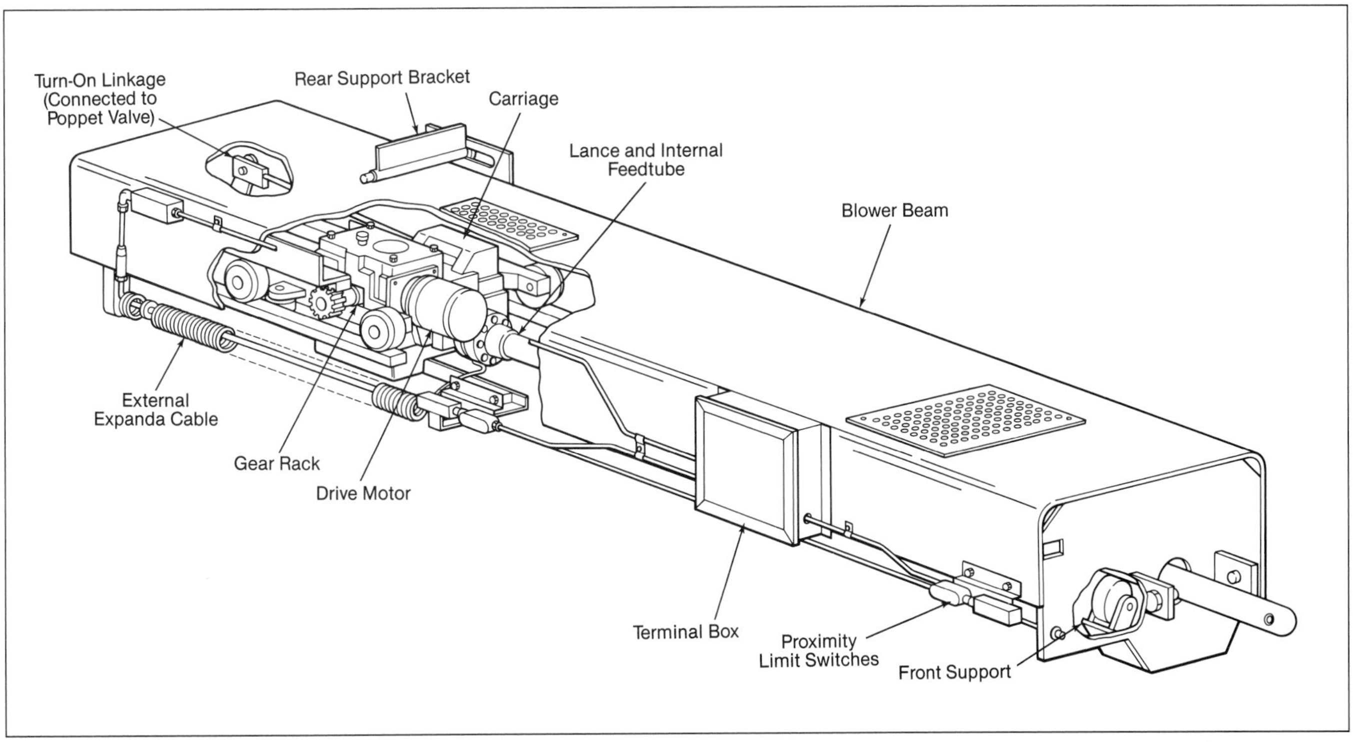

Fig. 3 Model IK-SD sootblower, overall view.

or 203 mm) depending upon design details and the length of the extended lance tube. For travel distances up to 45 ft (13.7 m), the lance indexes when it reverses to clean a larger area (Fig. 4). On some designs, this fixed helix pattern may be shifted slightly after each blower operation to cause the cleaning media jet to impact at a different position during each blowing cycle.

The pressures used in a long retractable sootblower are a function of blowing media, nozzle size, the area of application and the fuel deposit characteristics. In general, the pressures for air would range from 60 to 180 psig (4.1 to 12.4 bar gauge) while the pressures for steam may vary from 70 to 350 psig (4.8 to 24.1 bar gauge). Steam pressure ranges are higher than air pressure ranges because steam is normally used for the most difficult to clean deposits resulting from the combustion of lignites, subbituminous coal and pulping liquors. Flow rates will vary depending upon nozzle size, gas zone temperature and the type of ash being cleaned. When the lances are exposed to high gas zone temperatures additional flow is required to cool the lance. Deposit strength can vary within the convection pass. Experiments are currently underway to evaluate variable translation speed and variable pressure as tools to modify cleaning capability and blowing medium consumption as a function of deposit tenacity.

The typical nozzle size for air is 0.625 in. (15.9 mm) and for steam 0.875 in. (22.2 mm) or 1 in. (25.4 mm).

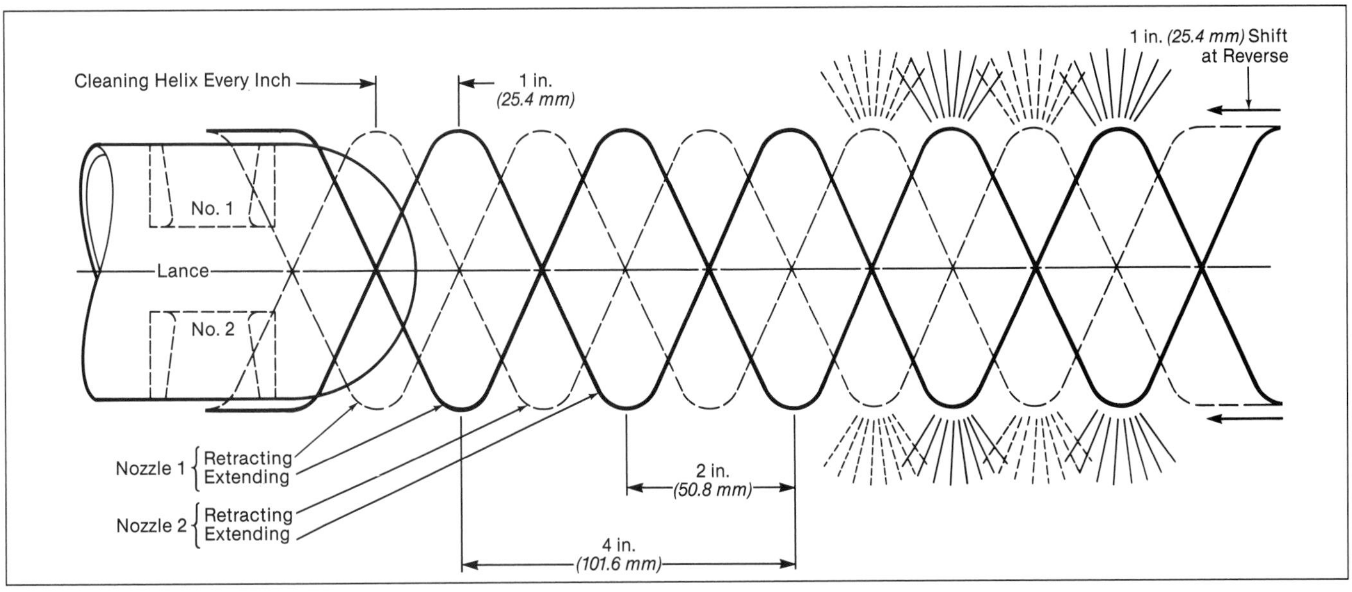

Fig. 4 Nozzle cleaning pattern.

IK water lance blower

The IK-4M-WL variable speed water lance is used on furnace wall surfaces where slag deposits can not be removed by conventional air or steam wall blowers. Examples are on installations firing lignites or subbituminous coals having highly reflective ash. The travel and rotational speeds vary during the cleaning cycle to obtain a uniform dwell time (jet progression velocity), to maintain optimum use of water for cleaning and to limit thermal shock. Water is normally used only during forward (insert) travel to avoid thermal shock on clean tubes during retracting.

Blowers for air heater cleaning

Tubular air heaters are cleaned using a special nonrotating, long travel blower mechanism with an internally supported, multi-nozzled element similar to the nozzles used in fixed position, rotating sootblowers. These types of blowers are identified as IK straight line sootblowers or rake-type sootblowers.

The element is normally located at the gas outlet end of the air heater with the nozzles positioned to blow into the open tube ends. This tee element moves back and forth across the face of the tube sheet and may consist of one, two or three parallel elements supplied by the moving lance. The number of nozzles per element is determined by the arrangement of air heater tubes, but the nozzle area of the elements must not exceed the capacity of the lance tube to provide blowing medium.

The blowing medium may be either compressed air or steam. When steam is used superheated steam is preferred to reduce moisture entering the cold end of the air heater tubes. Nozzle pressure of 125 psig (8.6 bar gauge) with a source enthalpy of 1377 Btu/lb (3203 kJ/kg) is recommended for steam. For higher nozzle pressures, higher source enthalpy must be provided. The blowing medium does not penetrate the entire tube length but keeps the colder section of tubes open. Out-of-service water washing is frequently necessary to completely clean the entire tube length.

The regenerative-type air heater is cleaned by a special nozzled, internally supported traveling lance. This mechanism is driven slowly across the basket surface by a nonrotating, long travel blower mechanism so that the entire surface is cleaned as the air heater baskets rotate. Either air or steam may be used as the blowing medium. Blowing medium must be able to achieve a maximum nozzle pressure of 200 psig (13.8 bar gauge), although the initial operating pressure will be set somewhat lower. The initial setting is a function of cold end basket material and construction. Because pluggage and corrosion of the baskets is possible if metal temperatures drop below the dew point, the steam source should have at least 300F (167C) of superheat. This usually requires a separate steam source for air heater cleaning.

Special designs

A variety of sootblowing equipment has been developed for specialty applications. These include oscillating sootblowers where the nozzles rotate in a limited arc to concentrate jet energy in a selective area, duct blowers to remove ash and dust found in flues or air ducts, fan cleaners, vertical mounted retractable blowers to permit use in difficult to access locations, and blowers for flue gas desulfurization system gas reheaters. (See Chapter 35.) Additional designs and configurations are developed as special cleaning applications arise.

Application to various boiler types

Pulverized coal-fired boilers

Fig. 5 is a sectional side view of a pulverized coal-fired utility boiler showing typical sootblower locations.

Slag and ash deposits on the furnace walls are removed by short retracting IR wall blowers, the locations determined by the coal being fired. For a bituminous-type coal, the vertical spacing of the rows of blowers would be between 10 and 14 ft (3.0 and 4.3 m) with side to side spacing of 8 to 14 ft (2.4 to 4.3 m). For a severe slagging lignite coal, the vertical spacing of the rows would be closer to 10 ft (3 m), while the side to side spacing would be 7 to 10 ft (2.1 to 3.0 m).

For some lignite and subbituminous coals having severe slagging or reflective ash deposits, IK water lance sootblowers might be installed in place of IR wall blowers. Water lances have proven to be an effective cleaning method in this application.

The pendant superheater and reheater sections, as well as the convection section of horizontal tubes, are cleaned by long retractable sootblowers. Vertical spacing depends upon the gas temperature and the coal ash characteristics. Spacing may be 12 to 14 ft (3.7 to 4.3 m). In some cases this spacing may be as low as 8 ft (2.4 m) with some locations being designated for future sootblower additions.

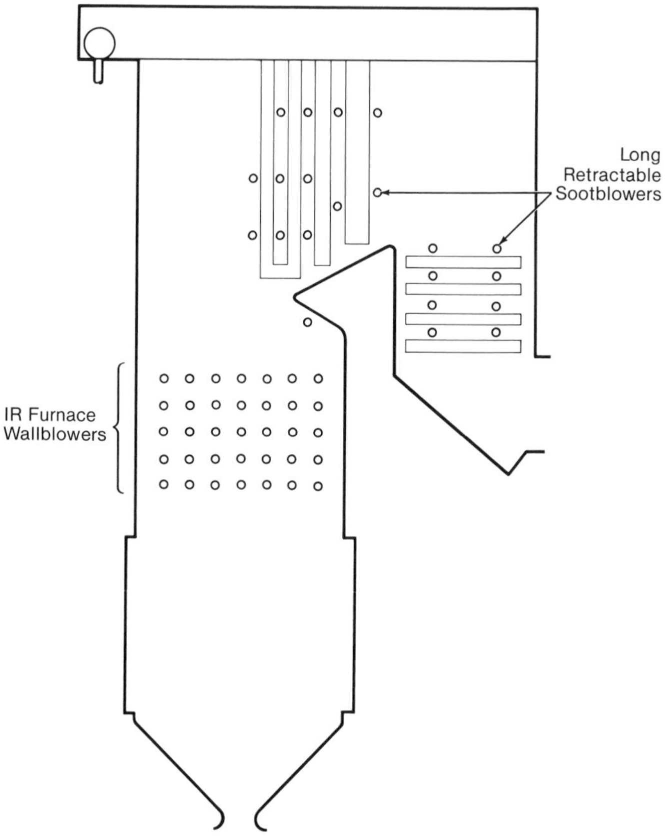

Fig. 5 Typical sootblower locations in a large coal-fired boiler.

Oil-fired boilers

Boilers fired only by oil do not use IR wall blowers because oil ash has a very low melting temperature and the furnace areas of the boiler are normally running with molten ash.

Long retractable IK blowers are not used in gas temperature zones above 1750F (954C) unless a chemical additive is used to raise the melting temperature of the ash. These additives (aluminum oxide, calcium magnesium carbonate or magnesium oxide) must be used for oil fuels having a high percentage of vanadium content.

The cleaning radius and bank penetration of the sootblower jet for a boiler firing only oil are approximately the same as for boilers firing high fouling coals.

Chemical recovery boilers

IR wall blowers have not been used on chemical recovery boilers as they are ineffective in removing low melting temperature ash deposits. Retractable steam blowing sootblowers are used extensively in the superheater, boiler bank and economizer sections. (See Chapter 26.)

Waste-to-energy refuse boilers

As discussed in Chapter 27, refuse boilers may be of two basic types, mass burn or refuse derived fuel. Steam sootblowers are used to keep the furnace and convection gas passes open and prevent plugging and excessive draft loss. The frequency of operation must be carefully monitored. The surface can be too clean and tube corrosion caused by flue gas constituents (chlorides) and/or erosion caused by flyash has been a problem.

Long retractable sootblowers are used to remove ash deposits from the superheater and boiler bank, while either long retractable blowers or fixed position rotary blowers may be used in the economizer. Retractable blowers are spaced vertically at approximately 8 to 10 ft (2.4 to 3.0 m) and have a cleaning flow rate comparable to that used with bituminous coal firing.

Mechanical rapping has been used in either, or both, the superheater and economizer areas either in conjunction with or in lieu of sootblowers.

Fluidized-bed boilers

Fluidized-bed boiler ash deposits on tube surfaces are normally light and easy to remove. However, there are cases where the buildup on horizontal tube surfaces will reduce heat transfer and promote plugging of the gas passages. In these cases, either long retractable steam blowing or fixed position rotary blowers are necessary. (See Chapter 16.)

Others

Application standards exist for bagasse (a byproduct of sugar cane production), wood and bark, waste heat, coke oven gas, spreader stokers, underfeed stokers, coke breeze, blast furnace gas and copper reverberatory units. Although these boilers use conventional equipment, the blowing medium application standards differ.

Controls

A reliable and flexible sootblower control system is essential to obtain the ultimate performance of the equipment. The control system operates the sootblowers either in a fixed sequence or in an operator selected variable sequence.

Sootblower control systems must contain many features and interlocks to protect the mechanical equipment and to ensure successful system operation. Among these are:

1. blowing medium source pressure alarms and interlocks to prevent equipment operation without proper medium,
2. high/low blowing medium flow alarms to warn of equipment malfunction,
3. blower-type interlocks to prevent over instantaneous demand of blowing medium source,
4. long retractable blowing medium failure interlocks to protect long lances,
5. motor overload protection and alarm to indicate malfunction,
6. motor stall alarms,
7. elapsed time alarm to ensure proper equipment operation, and
8. protection in the event of a boiler trip.

In addition to operator programmable systems, intelligent sootblower automation systems have been developed in which the operation and sequencing of selective blowers is based on measured and calculated surface cleanliness conditions of the individual heat transfer surfaces in the boiler. Advanced optical based systems are also being developed which remotely monitor surface cleanliness. Infrared imaging techniques to measure furnace wall emissivity have been developed. Emissivity probes connected to a Cleaning Advisor™ provide continuous clear images, otherwise not visible, of the interior surface of the boiler, trends of heat transfer surface cleanliness, and guidance to the operator as to where and when to clean. Such automated systems permit the most cost effective use of cleaning medium and energy.

Safety and relief valves

The most critical valve on a boiler is the safety valve. Its purpose is to limit the internal boiler pressure to a point below its safe operating level. To accomplish this goal, one or more safety valves must be installed in an approved manner on the boiler pressure parts so that they can not be isolated from the steam space. The valves must be set to activate at approved set point pressures (discussed below) and then close when the pressure drops to some level below the set point. When open, the set of safety valves must be capable of carrying all of the steam which the boiler is capable of generating without exceeding the specified pressure rise.

The American Society of Mechanical Engineers (ASME) Boiler and Pressure Vessel Code, Section I, outlines the minimum requirements for safety and safety relief valves applying to new stationary water tube power boilers. The Code also covers requirements for safety and safety relief valves for other applications beyond the scope of this text. By Code definition, a *safety valve* is used for gas or vapor service, a *relief valve* is used primarily for liquid service, and a *safety relief valve* may be suitable for use as either a safety or a relief valve.[1]

Fig. 6 shows a typical Code approved spring loaded safety valve for steam service. The valve must be installed

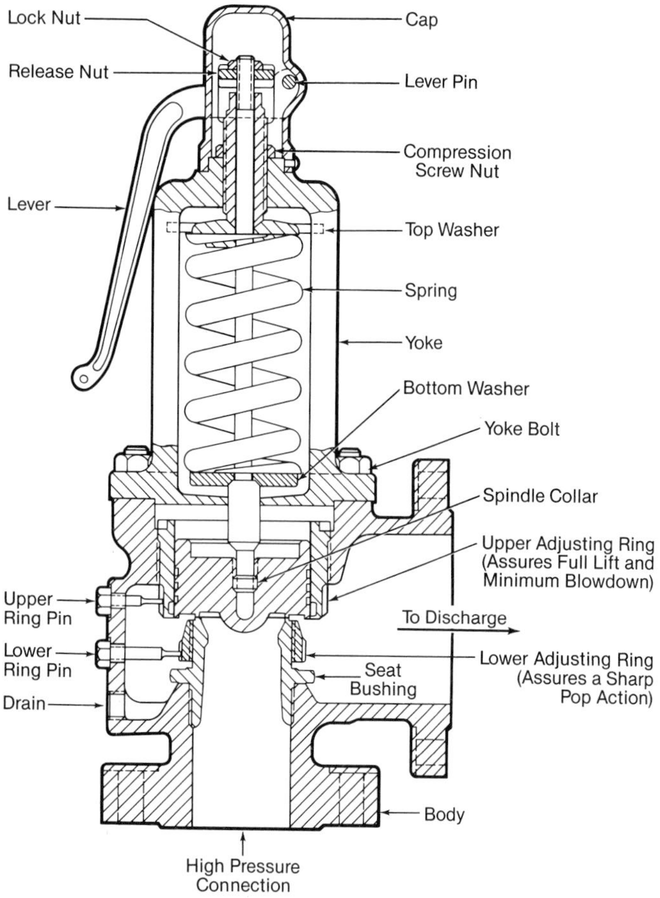

Fig. 6 Spring loaded safety valve (*courtesy of Dresser Industries, Inc.*).

independently and close to the pressure part without intervening valves or valves on the discharge side.[1] The inlet nozzle opening must not be less than the area of the valve inlet and unnecessary pipe fittings must not be installed. These valves are designed for large initial opening at the upstream static pressure set point and for maximum discharge capacity at 3% above set point pressure.[1]

Fig. 7 shows a typical power actuated type safety valve that may be used in some Code approved applications. Power actuated valves are fully opened at the set point pressure by a controller with a source of power such as air, electricity, hydraulic fluid or steam.

Fig. 8 shows a typical spring loaded relief valve for liquid service designed for a small initial opening at the upstream static pressure set point, followed by further opening with pressure increases above set point pressure to prevent additional pressure rise.

Because of variations in power boiler designs, Code interpretations are sometimes necessary. It is also necessary to comply with local ordinances. The customer's approval must be obtained for all safety valve settings.

For drum boilers with superheaters, Babcock & Wilcox (B&W) prefers to follow the Code allowed procedure of setting the safety valves so that the superheater valve(s) lift first at all loads, thereby maintaining a flow of steam through the superheater(s) to provide a measure of overheat protection. This method permits the lowest design pressure for piping and valves downstream of the superheater and is required on hand controlled units, stoker or other fuel bed fired units and brick set units. Another method may be used for all other units that permits the drum safety valves to lift first. This method could cause a reduced flow condition to occur in the superheater while the boiler is still at a high heat input level. As a result some superheater materials can exceed temperature limits.

The required valve relieving capacities for waste heat boiler applications are determined by the manufacturer. Auxiliary firing shall be considered in the selection of safety or safety relief valves. The Code required relieving capacity shall be based on the maximum boiler output capabilities by waste heat recovery, auxiliary firing or the combination of waste heat recovery with auxiliary firing.[1]

Additional Code requirements are applicable for modified existing boilers or new boilers installed in parallel with old boilers, or boilers operated at an initial low pressure, but designed for future high pressure (increases or decreases in operating pressure). Additional requirements, including mounting, operation, mechanical, material, inspection and testing of safety and safety relief valves are specified in the Code.

Code requirements for once-through boilers

For safety valve requirements on once-through boilers, the Code allows a choice of using the rules for a drum boiler or special rules for once-through boilers. Power operated valves may be used as Code required valves to account for 10 to 30% of the total required relief valve capacity. If the power relief valve discharges to an intermediate pressure (not atmospheric), the valve does not have to be capacity certified, but it must be marked with the design capacity at the specified relieving conditions.

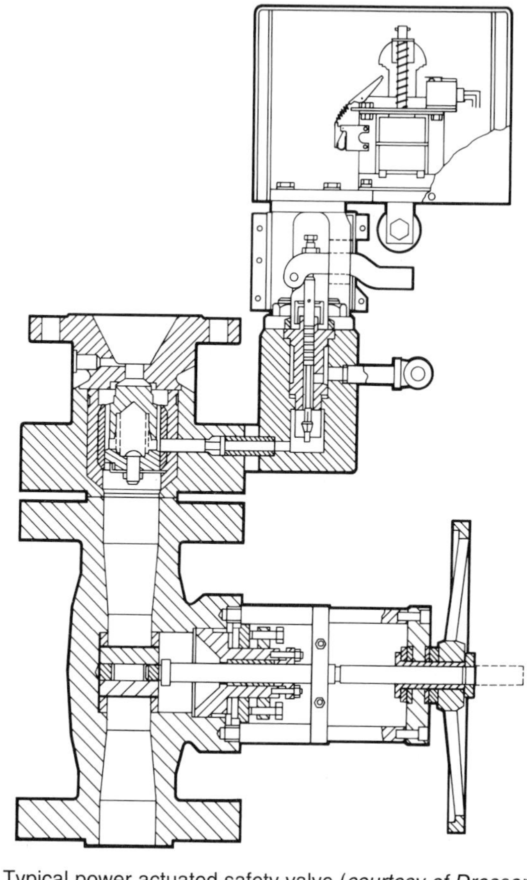

Fig. 7 Typical power actuated safety valve (*courtesy of Dresser Industries, Inc.*).

The power relief valve shall be in direct communication with the boiler and its controls must be part of the plant's essential service network, including required pressure recording instruments. A special isolating stop valve may be installed for power relief valve maintenance provided redundant relieving capacity is installed. Provided all the Code requirements are met or exceeded, the remaining required relief capacity is met with spring loaded valves set at 17% above master stamping pressure. The superheater division valves are part of the power relief valve system so that credit may be taken for the spring loaded superheater outlet valve(s) relieving capacity as part of the total required relieving capacity. The superheater valve(s) capacity must be a minimum of 10% of the maximum boiler steaming rate or equal to 6 lb/h relieving capacity per each ft^2 of superheater heating surfaces (0.008 kg/s per m^2). The blowdown of the spring loaded valves shall not be less than 8% nor more than 10% of set pressure. (See Code, Section I.)[5]

Ash handling systems

Ash and residue from the combustion of solid fuels are discharged at four different points in the boiler flue gas stream as shown in Fig. 9: furnace bottom, boiler back pass, air heater (flyash) and particulate collection device (flyash). A small amount of the ash in the fuel is also rejected as pyrite from the pulverizing mills. (See Chapter 12.) Particle size, density, velocity and physical arrangement in the flue gas path dictate at which point in the flue gas path the ash particles are no longer carried along in the flue gas. Typically, the bottom ash from pulverized coal and stoker-fired boilers is water quenched as it leaves the furnace. This ash, ranging in temperatures up to 2400F (1316C), is cooled to temperatures suitable for the ash handling equipment. The boiler back pass ash and flyash discharges are at low enough temperatures that further cooling is not required. These ash streams are typically conveyed to silos for storage prior to disposal.

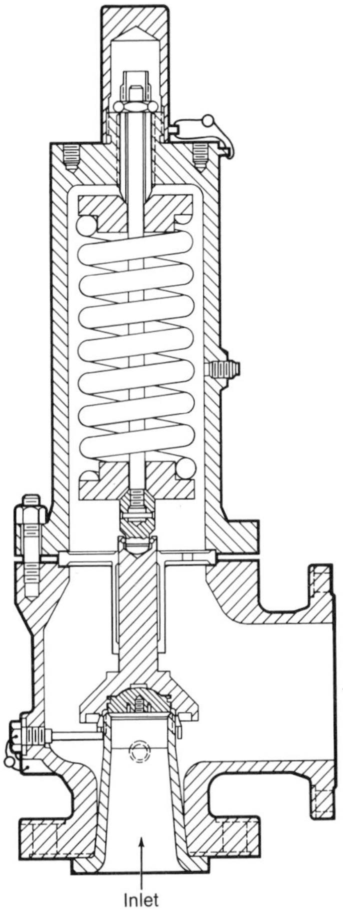

Fig. 8 Spring loaded pressure relief valve (*courtesy of Crosby Valve & Gage Company*).

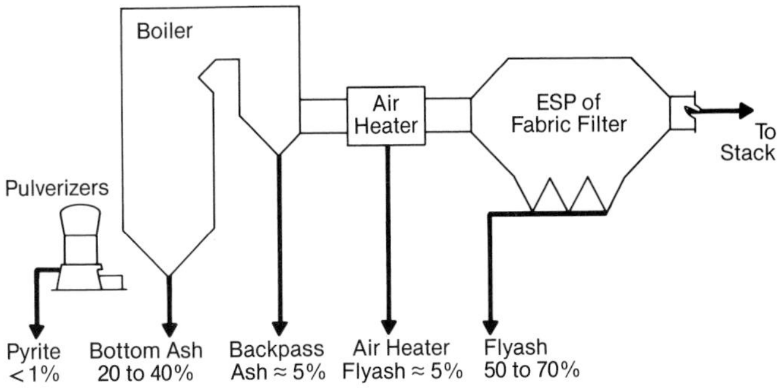

Fig. 9 Typical pulverized coal-fired boiler coal ash discharge locations and approximate amounts.

The systems that handle and transport the ash and flyash must be designed for the characteristics of each ash stream. These include the particle size, density and temperature, as well as the chemical composition and surface characteristics of the ash particles. The ash handling systems for each of these streams are usually sized to handle more than 100% of the expected ash produced by the design fuel. This provides capacity to handle surges or accommodate firing of higher ash content fuels.

Bottom ash handling systems

Bottom ash handling systems can be categorized into either intermittent or continuous systems. The water impounded hopper system is of the intermittent type which cycles off and on. During the off mode ash is stored for up to several hours before removal. The submerged chain conveyor is a continuous type which removes ash as it is discharged from the boiler.

Water impounded hopper system

Water impounded hopper systems can be used on dry-bottom boilers. Variations of this system are used on pulverized coal or Cyclone-fired furnaces and on various stoker-fired units. They consist of one or more water filled hoppers located under the furnace throat opening as shown in Fig. 10. The elevation of the furnace throat opening is usually at least 25 ft (7.6 m) above ground to allow space for the hopper. The bottom ash drops into the water filled hopper where the water absorbs the impact and quenches the ash. A water seal trough is located at the top of the hopper with seal plates attached to the bottom of the furnace to maintain the gas seal.

When the hopper reaches its storage capacity the ash is removed. The ash passes through a clinker grinder located at the discharge of each hopper. This grinder will crush the larger ash and slag pieces to an acceptable size for processing and downstream removal.

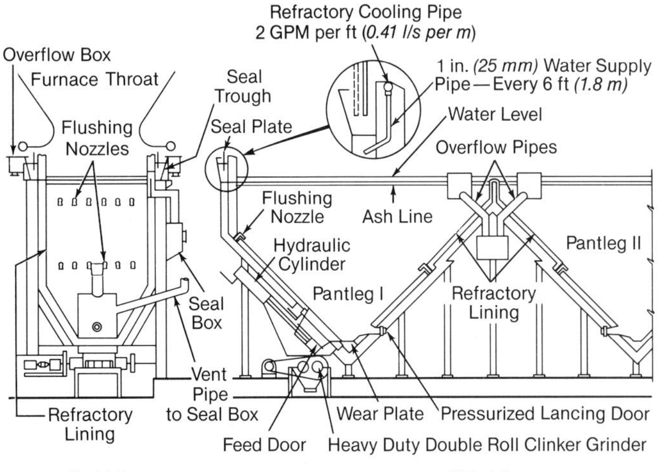

Fig. 10 Water impounded hopper (*courtesy of Allen-Sherman-Hoff*).

Sluice water passes through a jet pump to draw the crushed ash and water mixture from each hopper. The efficiency of a jet pump is low, requiring several pounds of water to convey one pound of ash. This ash-water mixture is routed through abrasion resistant pipes to a disposal area away from the boiler unit. Another method uses abrasion resistant suction pumps to draw the ash-water mixture from the hoppers.

Disposal of the ash-water mixture varies from unit to unit. It can be discharged into large ash ponds. The water in the ash pond can be used as the sluice water source for the transport of the ash from the water impounded hoppers. When ash ponds are not feasible or a dewatered ash product is required, the ash-water mixture can be discharged into large dewatering systems. The water from the settling tanks provides the water for the high pressure sluice water pumps.

Other variations of the water impounded hopper systems, such as dry-ash storage with water introduced only during the conveying cycle, are also available.

Submerged chain conveyor system

The submerged chain conveyor (SCC) (Figs. 11 and 12) can be applied to the same boiler unit designs as water impounded hopper systems. The SCC is a mechanical conveyor with a water-filled upper trough and a dry lower trough. One end of the conveyor has an incline rising out of the water. Seal plates attached to the furnace bottom are immersed in the upper water-filled trough to maintain the gas seal.

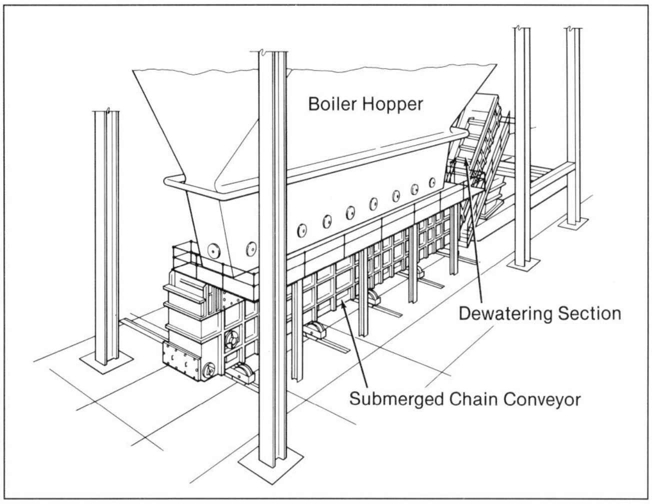

Fig. 11 Submerged chain conveyor (*courtesy of TLT-Babcock, Inc.*).

Ash from the furnace bottom drops into the upper, water-filled trough. Flights, which are pulled slowly by two chains in the SCC, convey the ash. The incline section of the conveyor dewaters the ash. The moisture content of the dewatered ash is approximately 30% (by weight), but this varies with the application and the consistency of the ash. Makeup water must be added to the SCC to compensate for the moisture content of the dewatered ash and water lost by evaporation. The conveying chain speed is slow to minimize chain and flight wear and to maximize ash dewatering.

The dewatered ash from the SCC can be handled in various ways, depending upon the specific plant requirements. If space is available at the discharge of the SCC, the ash can be loaded into a truck for further transport. If this space is not available, belt or mechanical conveyors can be used to transfer the ash away from the SCC to a convenient dumping point.

The water in the SCC conveying trough should be maintained at 140F (60C) to prevent high temperature scaling. Depending on the application, cooling water may be required to maintain the proper temperature. If more cooling water is required than makeup water, the SCC will overflow; this overflow must be accounted for in the design of the plant. The overflow water can be transferred to an ash pond if available. However, many plants take advantage of the SCC's inherent design to minimize water losses and use a closed loop cooling system to maintain a proper trough water temperature.

Some types of ash are very porous. When large, hot pieces of these ashes drop into water, water is drawn into them and turns to steam so quickly that they explode. An SCC specifically designed to inhibit ash explosions by gradually immersing the ash in water is available.

The SCC requires approximately 12 to 15 ft (3.7 to 4.6 m) under the furnace throat opening on larger utility boiler applications; smaller boilers require less distance. This allows the boiler to be located closer to the ground and reduces the cost of structural steel, coal piping, flues, ducts and other components.

A variation of the SCC is the platform conveyor which has a continuous band of ash carrying pans, instead of flights, between the two chains. Therefore, the platform conveyor transports a continuous bed of ash. This pro-

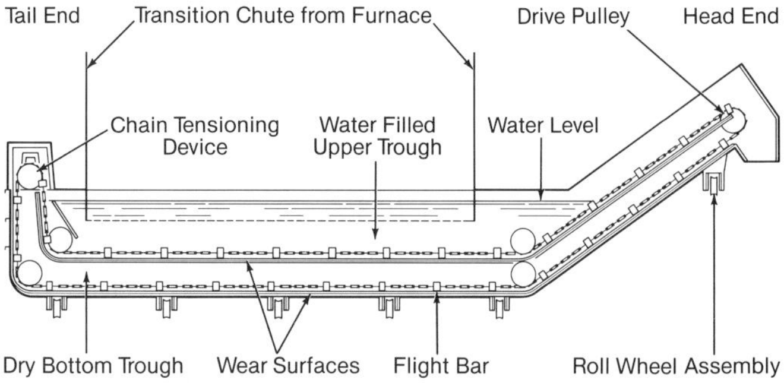

Fig. 12 Submerged bottom ash chain conveyor, longitudinal cross-section (*courtesy of TLT-Babcock, Inc.*).

vides the ability to convey large amounts of ash or ash residue from municipal solid waste (MSW) units.

Plunger ash extractor

The ash discharged from the stoker on mass burn MSW units consists of ash and slag similar to that derived from other combusted fuels, except that the MSW fuel has a high ash content and includes oversize bulky wastes (OBW). This system is discussed in more detail in Chapter 27.

Pulverizer pyrites system

Pyrites are the tramp iron and rocks mined with coal which are separated from the coal in the pulverizers and can be removed manually, hydraulically, or pneumatically. In the hydraulic system, sluice water passing through a jet pump draws the pyrites from a tank. The pyrites and conveying water are transported through piping to the disposal area. Due to the abrasive nature of the pyrites, abrasion resistant piping is used. In the pneumatic system, motive air from either a vacuum or pressure system conveys the pyrites to the disposal area.

The pyrites can be conveyed to an intermediate dewatering bin, ash pond or bottom ash collection system for disposal.

Boiler back pass ash systems

Because the boiler back pass ashes are neither fine like flyash nor coarse like bottom ash, many variations for handling them exist. Some plants extend the flyash handling system to collect and transport these ashes, while others transport them to the bottom ash system. Others keep them separate from either the bottom ash or the flyash systems. Back pass systems use hydraulic, pneumatic or mechanical conveying means.

Dry flyash handling systems

Dry flyash handling is primarily accomplished by pneumatic or mechanical conveying or a combination of the two. This flyash is accumulated in hoppers and discharged into the ash conveying system. The boiler back pass ash and flyash are accumulated in the hoppers and discharged into the respective ash conveying system.

Pneumatic conveying

Pneumatic conveying uses an air flow stream through a piping system as the conveying medium for the ash. It can be one of two types, a pressure system or a vacuum system. In the pressure system, the conveying piping system is arranged with a blower at the beginning of the conveying line. The piping system operates at a positive pressure. It is routed beneath each ash disposal hopper and includes an air seal system at every ash feed point. The ash is conveyed to a storage bin or silo which includes a discharge filter for the fine ash particulates from the discharged conveying air, plus the air displaced when filling the silo (Fig. 13).

In the vacuum system, the conveying air producer is at the discharge, or silo, end . This creates a negative pressure in the conveying piping. A blower or a steam or water eductor can be used as the motive air producer. Because the conveying lines are at a negative pressure, a single valve is used for isolation and feed control at each ash pickup point. The ash conveying line enters a primary collector, then a secondary collector. These separate the ash from the conveying air and discharge it into the silo (Fig. 14).

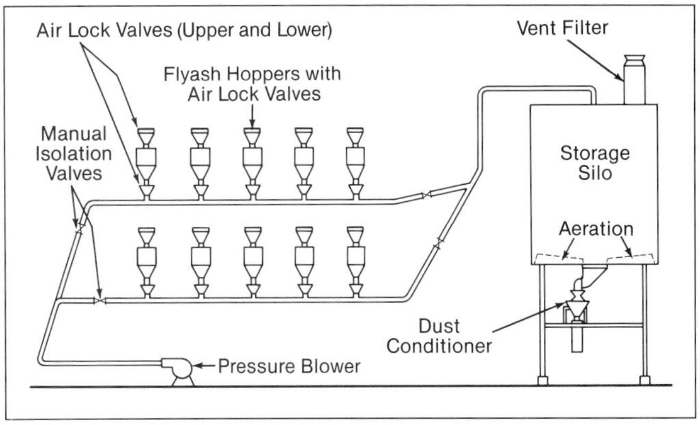

Fig. 13 Pressure flyash system (*courtesy of Allen-Sherman-Hoff*).

When selecting a pneumatic system, the design should specify the system capacity for added capacity to accommodate surges and to provide some down time to perform minor equipment maintenance. Another consideration in specifying the ash system is redundancy of conveying air blowers to assure maximum system availability. Because the ash conveyed is an abrasive material, the conveying piping is normally abrasion resistant.

The pneumatic systems discussed above are of the dilute phase, which means a high conveying air/solids ratio. A variation of the pressure system is the dense phase transport or conveying system. In this system design, a much lower air/solids ratio is used. The ash from the hopper is discharged into a dense phase conveying pot. The pot is pressurized with high pressure conveying air and the ash is conveyed as a dense slug. The dense phase concept is typically limited to small particle size material such as flyash. Larger particle size material can be dense phase conveyed, but is limited to shorter conveying distances.

Mechanical conveying

Mechanical conveying of the dry ash streams can be accomplished using a combination of drag chain conveyors, en masse conveyors, screw conveyors or bucket el-

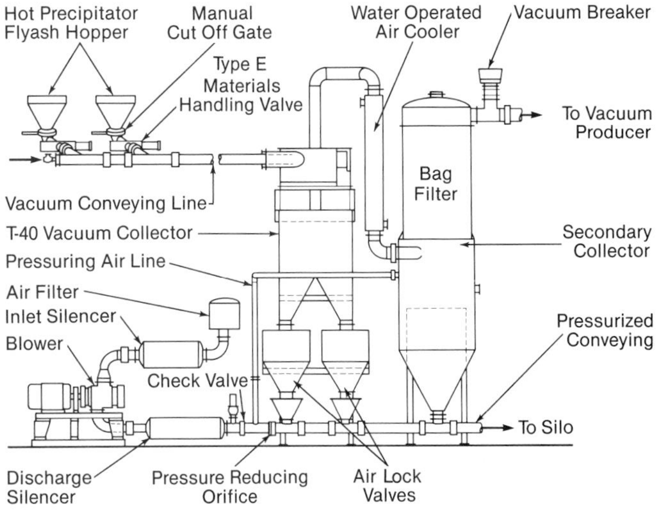

Fig. 14 Vacuum-pressure flyash system (*courtesy of Allen-Sherman-Hoff*).

evators. Due to the high temperatures and dusty nature of the ash streams, belt conveyors are not applicable. See Chapter 11 for more information on mechanical material handling devices.

Drag chain conveyors are very similar in design to the SCC used for bottom ash handling and move the ash horizontally or on an incline. The chain and flights are totally enclosed in a dust-tight housing with ash feed through openings in the top of the housing. Air lock devices, such as rotary seal feeders or tipping valves, are used to seal the internal positive or negative pressures of the boiler gas stream. The volume of ash conveyed is normally determined by the height of the flight bars. En masse conveyors are similar in design but move an entire pile of material which totally fills the conveyor cross-section.

It is highly recommended to limit the use of screw conveyors in ash handling to conveyor lengths that will not require intermediate hanger bearings due to the highly abrasive nature of ash and the resultant wear. The alternative is to select screw conveyors of sufficient diameter keeping the level of fill of ash low enough so that bearing exposure is minimized. This oversizing typically makes screw conveyors economically unattractive in this application.

Bucket elevators (see Chapter 11) provide mechanical means of conveying vertically. Because of the fluid characteristics of the ash, the incline angle of other conveyors is normally limited to 30 deg (0.52 rad). En masse conveyors can attain much steeper incline angles, sometimes approaching 90 deg (1.56 rad). A combination of the various conveyor types may be necessary in order to move the ash in the most functionally appropriate, cost effective manner.

Mechanical versus pneumatic transport considerations

Mechanical conveyors have very low drive power requirements compared to the blower drive requirements of a pneumatic system. Mechanical systems normally operate continuously, yet the cumulative power consumption on a 24 hour basis is less than the intermittently operating pneumatic system. However, pneumatic systems offer much more flexibility in design due to the easier routing of piping versus conveyors.

A combination of mechanical conveying to a small intermediate storage hopper with pneumatic conveying from the intermediate hopper to a storage silo provides the benefits of continuous ash removal with the flexibility and simplicity of a pneumatic system.

The mechanical handling of ash is advantageous for boilers firing fuels that result in significant levels of unburned combustibles in the various ash streams such as wood-fired or refuse-fired boilers. These hot unburned combustibles can ignite when exposed to the air stream of a pneumatic conveying system and result in explosions or continued burning in the system and storage silo. Mechanical conveying systems more readily accommodate large ash particles which can result from unit operation excursions. A pneumatic conveying system is prone to plugging from oversized ash particles. An ash stream which may be high in moisture content, such as from a flue gas scrubber, can plug a pneumatic system during moisture level excursions.

Ash storage and unloading

The dry ash is normally conveyed to a silo for storage prior to discharge into ash transport trucks or rail cars. Ash loading can be accomplished by loading the material in its dry, dusty state or by moisture conditioning. Dry loading uses closed transport vehicles such as bulk tank trucks. The material is totally contained during the loading to minimize dust emissions.

The conditioning system is more commonly used. Ash is discharged from the silo into a mechanical mixer (rotary drum or pug mill) with water to produce a dust free ash containing 15 to 35% moisture by weight. Particular care must be exercised with high calcium content flyash to avoid adding too much water which could result in the ash setting up into a cement type product.

Air and flue gas dampers

Dampers are used to control flow and temperature of air and flue gas. They can also be used to isolate equipment in the air or flue gas streams when such equipment is out of service or requires maintenance.

The Air Movement and Control Association in its Publication 850-84, *Flue Gas Dampers*, defines both the isolation and control functions for the selection of dampers.[2]

Isolation dampers may be the nominal shutoff or zero leakage type. Shutoff dampers are used in applications where some limited leakage is tolerable. All types of dampers, with appropriate blade seals, can be used. Zero leakage dampers are designed to prevent any flow media leakage. This is accomplished by overpressurizing the blade seal periphery with seal air, which leaks back into the system. Guillotine-type dampers are best suited for this service. A pair of dampers with an overpressurized air block between them can also be used.

Control dampers are capable of providing a variable restriction to flow and may be of several varieties.[2]

Balancing dampers are used to balance flow in two or more ducts. *Preset position* dampers are normally open or normally closed dampers which move to an adjustable preset position on a signal. *Modulating* dampers are designed to assume any position between fully open and fully closed in response to a varied signal (either pneumatic or electric). A positioner with feedback indication is normally required.[2]

Dampers may also be classified by shape or configuration. Damper shape classifications that are commonly used in the fossil fuel-fired steam generation industry are *louver*, *round* and *guillotine*.

Louver dampers

A louver damper, as shown in Fig. 15., is characterized by one or more blades that mount into bearings located in a rigid frame. One blade shaft end extends far enough beyond the frame so that a drive can be mounted for damper operation. Blade shape is determined by the amount of pressure drop that can be tolerated across the open damper. A flat plate is the simplest shape but offers the greatest pressure drop. Air foil blades have lower pressure losses. If uniform downstream flow distribution is required, opposed blade rotation is used; otherwise, parallel blade rotation may be used. (See Fig. 16.)

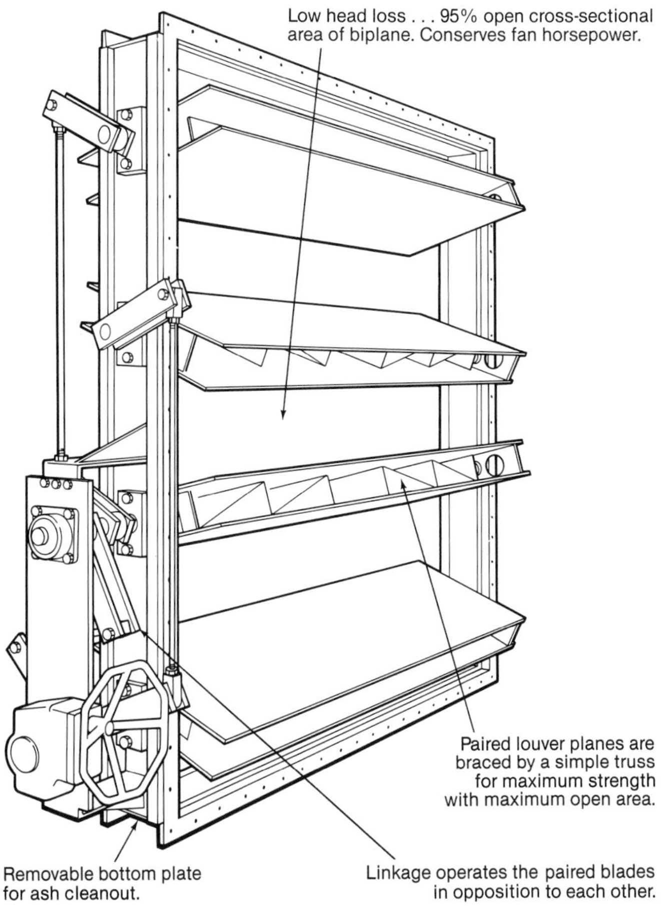
Low head loss . . . 95% open cross-sectional area of biplane. Conserves fan horsepower.
Paired louver planes are braced by a simple truss for maximum strength with maximum open area.
Removable bottom plate for ash cleanout.
Linkage operates the paired blades in opposition to each other.

Fig. 15 Louver damper (*courtesy of Damper Design, Inc.*).

If the louver damper is to be used for isolation purposes, seal strips are mounted on the blades to minimize leakage around the closed blades. Seal strips are not required in applications where louver dampers are used for flow control.

Depending on the application, louver dampers may have internal or external bearings. Internal bearings are machined castings with a self-cleaning feature. They do not require lubrication or penetration of the frame with the associated sealing maintenance. External bearings are the self-aligning and self-lubricating sleeve type, which require a mounting block and packing box with provision for seal air.

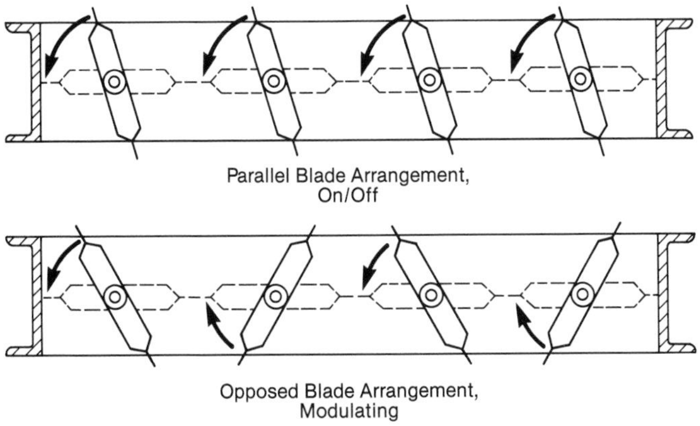
Parallel Blade Arrangement, On/Off
Opposed Blade Arrangement, Modulating

Fig. 16 Louver damper blade arrangement options (*courtesy of Air Movement and Control Association, Inc.*).

Round dampers

Round dampers, as shown in Fig. 17, can be used for control or shutoff service. Due to higher velocity limits, they are relatively smaller than louver dampers and have a smaller seal edge to area ratio which makes them more efficient for shutoff applications.

Guillotine dampers

Guillotine dampers have an external frame and drive system that can insert and withdraw the blade which acts as a blanking plate in the full cross-section of the duct. This minimizes the sealing edge required for any given duct size. The blade periphery is surrounded by, and forced between, flexible metal sealing strips. On zero leakage dampers, such as shown in Fig. 18, the blade edge extends through the frame into a bustle that is pressurized by a seal air blower. Seals for this type of damper are normally designed to eliminate reverse flexing. The blade is actuated by a chain and sprocket, or screw-type mechanism, connected to a drive which is mounted on the frame. Guillotines are typically large and used for isolation service, preferably in horizontal ducts. It is not uncommon for blade thickness to reach 0.75 in. (19 mm), and blade thicknesses exceeding 1 in. (25.4 mm) have been used.

Stacks and draft

An adequate flow of air and combustion gases is required for the complete and effective combustion of fossil and chemical fuels. Flow is created and sustained by stacks and fans. Either the stack alone, or a combination of stack and fans, produces the required pressure differential for the flow.

Draft is the difference between atmospheric pressure and the static pressure of combustion gases in a furnace, gas passage, flue or stack. The flow of gases through the boiler can be achieved by four methods of creating draft, referred to as forced draft, induced draft, balanced draft and natural draft.

Forced draft boilers operate with the air and combustion products maintained above atmospheric pressure.

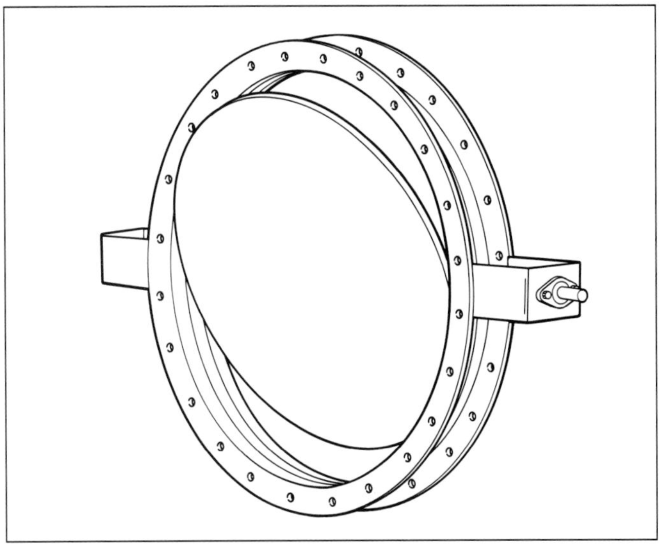

Fig. 17 Wedge seat round damper (*courtesy of Damper Design, Inc.*).

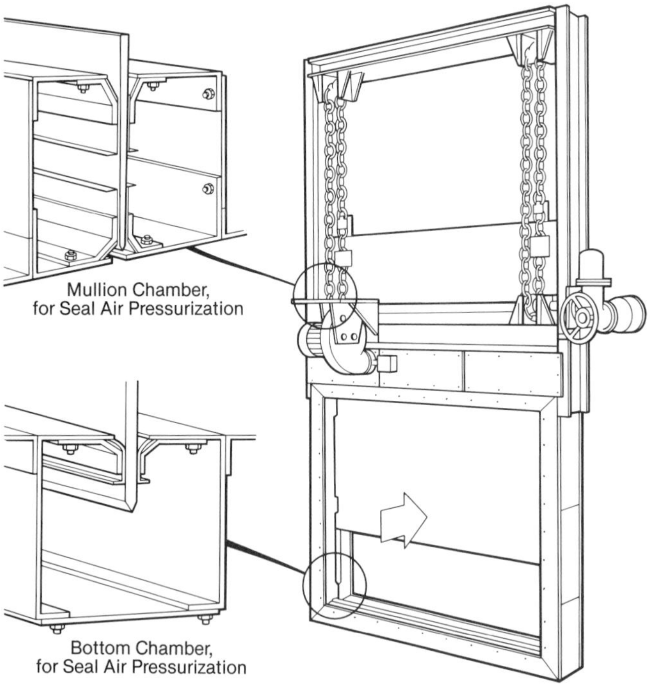

Fig. 18 Guillotine damper (*courtesy of Effox, Inc.*).

Fans at the inlet to the boiler system provide sufficient pressure to force the air and flue gas through the system. Any openings in the boiler settings, such as opened doors, allow air or flue gas to escape unless the opening is also pressurized.

Induced draft boilers operate with air and gas static pressure below atmospheric. The static pressure is progressively lower as the gas travels from the air inlet to the induced draft fan. The required flow through the boiler can be achieved by the stack alone when the system pressure loss is low or when the stack is tall; this is called *natural draft*. For most modern boilers, a fan at the boiler system outlet is needed to draw flow through the boiler. Unlike forced draft units, air from the boiler surroundings enters through (infiltrates) any openings in the boiler setting.

Balanced draft boilers have a forced draft air fan at the system inlet and an induced draft fan near the system outlet. The static pressure is above atmospheric at the forced draft fan outlet and decreases to atmospheric pressure at some point within the system (typically the lower furnace). The static pressure is subatmospheric and progressively decreases as the gas travels from the balance point to the induced draft fan. This scheme reduces both flue gas pressure and the tendency of hot gases to escape. There are also power savings for this method because forced draft air fans require smaller volumetric flow rates and therefore less energy for a given mass flow. Most modern boilers are balanced draft for these reasons.

Draft loss is the reduction in static pressure of a gas caused by friction and other nonrecoverable pressure losses associated with the gas flow under real conditions. As discussed in Chapter 3, static pressure is related to the total pressure at a location by the addition of the velocity or dynamic pressure.

$$P_{total} = P_s + P_v = P_s + \frac{V^2}{2g_c v} \quad (1)$$

where

P_{total} = total pressure, lb/ft² (N/m²)
P_s = static pressure, lb/ft² (N/m²)
P_v = velocity pressure, lb/ft² (N/m²) = $V^2/2g_c v$
V = average gas velocity
v = specific volume of the gas
g_c = conversion constant
= 32.17 lbm ft/lbf s² = 1 kgm/Ns²

Stack effect

Stack effect, or chimney action, is the difference in pressure caused by the difference in elevation between two locations in vertical ducts or passages conveying heated gases at zero gas flow. It is caused by the difference in density between air and heated gases. The stack effect is independent of gas flow and can not be measured with draft gauges. Draft gauges combine stack effect and flow losses. The intensity and distribution of this pressure difference depends on the height, the arrangement of ducts and the average gas temperature in the duct and ambient air temperature.

Based upon this definition, the overall stack effect in its most general form is defined as:

$$\Delta P_{SE} = \frac{g}{g_c} Z\left(\rho_a - \rho_g\right) = \frac{g}{g_c} Z\left(\frac{1}{v_a} - \frac{1}{v_g}\right) \quad (2)$$

where

ΔP_{SE} = stack draft effect driving pressure, lb/ft² (N/m²)
g = acceleration of gravity = 32.17 ft/s² (9.8 m/s²)
g_c = 32.17 lbm ft/lbf s² (1 kgm/Ns²)
Z = elevation between points 1 and 2
ρ_a = density of air at ambient conditions, lb/ft³ (kg/m³)
ρ_g = average density of flue gas, lb/ft³ (kg/m³)
v_a = specific volume of air at ambient conditions, ft³/lb (m³/kg)
v_g = average specific volume of flue gas, ft³/lb (m³/kg)

The customary English units used for draft calculations are inches of water for draft pressure loss and feet for stack height or elevation. Using this system of units, the incremental stack effect (inches of pressure loss per foot of stack height) can be evaluated from:

$$SE = \left(\frac{1}{v_a} - \frac{1}{v_g}\right)\left(\frac{1}{5.2}\right) \quad (3)$$

where

SE = stack effect, in./ft

For convenience, Table 1 provides the specific volume of air and flue gas at one atmosphere and 1000R (556K). Assuming air and flue gas can be treated as ideal gases, the ideal gas law in Chapter 3 permits calculation of the specific volume at other conditions:

$$v = v_b\left(\frac{T_f}{T_R}\right)\left(\frac{B_R}{B}\right) \quad (4)$$

where

v = specific volume at T_f and B
T_f = average fluid temperature, R (K)
= T_f(F) + 460 (T_f(C) + 256)
B = barometric pressure (see Table 2)
T_R = 1000R (556K)
B_R = 30 in. Hg (7.5 kPa)

Table 1
Sample Specific Volumes at 1000R (556K) and One Atmosphere

Gas	v_b (ft^3/lb)	v_b (m^3/kg)
Dry air	25.2	1.57
Combustion air (0.013 lb water/lb dry air)	25.4	1.58
Flue gas (3% water by wt)	24.3	1.5
Flue gas (5% water by wt)	24.7	1.54
Flue gas (10% water by wt)	25.7	1.60

Table 3 provides a reference set of values for SE at one atmosphere. The total theoretical draft effect of a stack or duct at a given elevation above sea level can be calculated from Equation 1 or from:

$$\text{Stack draft} = Z\ (SE)\left(\frac{B_{\text{elevation}}}{B_{\text{sea level}}}\right) \qquad \textbf{(5)}$$

where

Z = stack height, ft (m)
SE = stack effect, in./ft (Table 3 or Equation 2)
$B_{\text{sea level}}$ = barometric pressure at sea level (Table 2)
$B_{\text{elevation}}$ = barometric pressure at elevation (Table 2)

The average gas temperature in these calculations is assumed to be the arithmetic average temperature entering and leaving the stack or duct section. For gases flowing through an actual stack there is some heat loss to the ambient air through the stack structure. There is also some infiltration of cold air. The total loss in temperature in a stack depends upon the type of stack, stack diameter, stack height, gas velocity and a number of variables influencing the outside stack surface temperature. Fig. 19 indicates an approximate stack exit temperature relative to height, diameter and inlet gas temperature.

Sample stack effect calculation

Fig. 20 illustrates the procedure used in calculating stack effect. The stack effect can either assist or resist the gas flow through the unit. The three gas passages are at different temperatures and the example is at sea level. For illustrative purposes, assume atmospheric pressure, i.e., (draft = 0) at point D.

Table 2
Barometric Pressure, B — Effect of Altitude

Ft Above Sea Level	Pressure in. Hg	Pressure kPa	Ft Above Sea Level	Pressure in. Hg	Pressure kPa
0	29.92	760	6000	23.98	609
1000	28.86	733	7000	23.09	586
2000	27.82	707	8000	22.22	564
3000	26.82	681	9000	21.39	543
4000	25.84	656	10,000	20.58	523
5000	24.90	632	15,000	16.89	429

Values from Publication 99, Air Moving and Conditioning Association, Inc., 1967.

Stack effect always assists up flowing gas and resists down flowing gas. Plus signs are assigned to up flows and minus signs to down flows. Using values from Table 3 for an ambient air temperature of 80F, the stack effect in inches of water for each passage is:

Stack effect C to D $= +(110 \times 0.0030) = +0.33$ in. H_2O
Stack effect B to C $= -(100 \times 0.0086) = -0.86$ in. H_2O
Stack effect A to B $= +\ (50 \times 0.0100) = +0.50$ in. H_2O

If draft gauges are placed with one end open to the atmosphere at locations A, B, C and D of Fig. 20 the theoretical zero flow draft readings are:

Draft at D = 0 in. H_2O
Draft at C = draft at D minus stack effect C to D
$= 0 - (+0.33) = -0.33$ in. H_2O
Draft at B = draft at C minus stack effect C to B
$= -0.33 - (-0.86) = +0.53$ in. H_2O
Draft at A = draft at B minus stack effect A to B
$= +0.53 - (+0.50) = +0.03$ in. H_2O

Note that because the calculation of stack effect in this example is opposite to the gas flow, stack effects are subtracted in calculating static pressures or drafts. If the stack effect is in the direction of gas flow, stack effects should be added.

The net stack effect from A to D in Fig. 20 is the sum of all three stack effects and is -0.03 in. For this reason, fans or stack height must be selected not only to provide the necessary draft to overcome flow loss through the unit, but also to allow for the net stack effect of the system.

In some boiler settings, gases leak from the upper portions when the unit is operating at very low loads or when it is taken out of service. The leakage can occur even though the outlet flue may show a substantial negative draft. The preceding example illustrates this condition with a suction or negative pressure at the bottom of the uptake flue C-D and positive pressures at both points A and B in Fig. 20.

The chimney or stack

Early boilers operated with natural draft caused by the stack effect alone. This is also true for many of the smaller modern units. However, for large units equipped

Table 3
Reference Set of Stack Effect (SE) Values
in. of H_2O/ft of Stack Height — English Units Only

Reference Conditions:
Air — 0.013 lb H_2O/lb dry air: 13.7 ft^3/lb, 80F, 30 in. Hg
Gas — 0.04 lb H_2O/lb dry gas: 13.23 ft^3/lb, 80F, 30 in. Hg
Barometric pressure 30 in. Hg

Avg. Temp in Flue or Stack, T_g	Ambient Air Temperature, T_a, F			
F	40	60	80	100
250	0.0041	0.0035	0.0030	0.0025
500	0.0070	0.0064	0.0059	0.0054
1000	0.0098	0.0092	0.0087	0.0082
1500	0.0112	0.0106	0.0100	0.0095
2000	0.0120	0.0114	0.0108	0.0103
2500	0.0125	0.0119	0.0114	0.0109

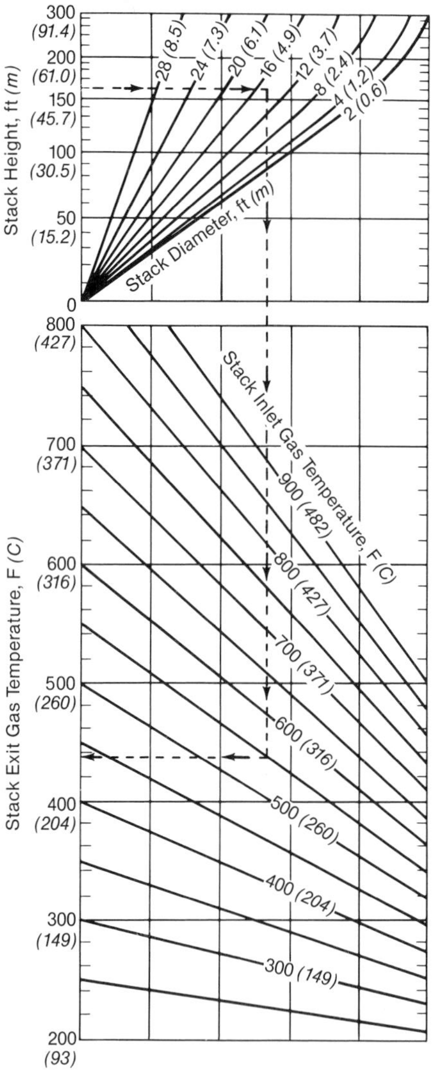

Fig. 19 Approximate relationship between stack exit gas temperature and stack dimensions.

with superheaters, economizers and especially air heaters, it is not practical or economical to operate the entire unit from stack induced draft alone. These units require fans to supplement the stack induced draft. The entire unit might be pressurized by a forced draft fan or the unit might use both induced and forced draft fans for balanced draft operation. The combination of only an induced draft fan and stack is not commonly used.

The required height and diameter of stacks for natural draft units depend upon:

1. draft loss through the boiler from the point of balanced draft to the stack entrance,
2. average temperature of the gases passing up the stack and the temperature of the surrounding air,
3. required gas flow from the stack, and
4. barometric pressure.

No single formula satisfactorily covers all of the factors involved in determining stack height and diameter. The most important points to consider are: 1) temperature of the surrounding atmosphere and temperature of the gases entering the stack, 2) drop in temperature of the gases within the stack because of heat loss to the atmosphere and air infiltration, and 3) stack draft losses associated with gas flow rate (due to fluid friction within the stack and the kinetic energy of gases leaving the stack).

Stack flow loss

The net stack draft, or available induced draft at the stack entrance, is the difference between the theoretical draft calculated by Equation 1 or Equations 2, 3 and 4 and the pressure loss due to gas flow through the stack.

From Equation 41 in Chapter 3 for friction loss, plus one velocity head exit loss ($G^2 v/2g_c$):

Stack flow loss = ΔP_l = Friction loss + Exit loss

$$\Delta P_l = f \frac{L}{D}\frac{G^2}{2g_c} v + \frac{G^2}{2g_c} V \qquad (6)$$

where

ΔP_l = stack pressure loss, lb/ft^2 (N/m^2)
f = friction factor, dimensionless (Fig. 1, Chapter 3: ≈ 0.014 to 0.017)
L = length of stack = Z, ft (m)
D = stack diameter, ft (m)
G = mass flux = $\dot{m}$/A, = lb/s ft^2 (kg/m^2 s)
$\dot{m}$ = mass flow rate, lb/s (kg/s)
A = stack cross-sectional area, ft^2 (m^2)
v = specific volume at average temperature, ft^3/lb (m^3/kg)
g_c = 32.17 lbm ft/lbf s^2 (1 kgm/Ns^2)

For English units this reduces to:

Stack flow loss =

$$\Delta P_l = \frac{2.76}{B}\frac{T_g}{D_i^4}\left(\frac{\dot{m}}{10^5}\right)^2\left(\frac{fL}{D_i}+1\right) \qquad (7)$$

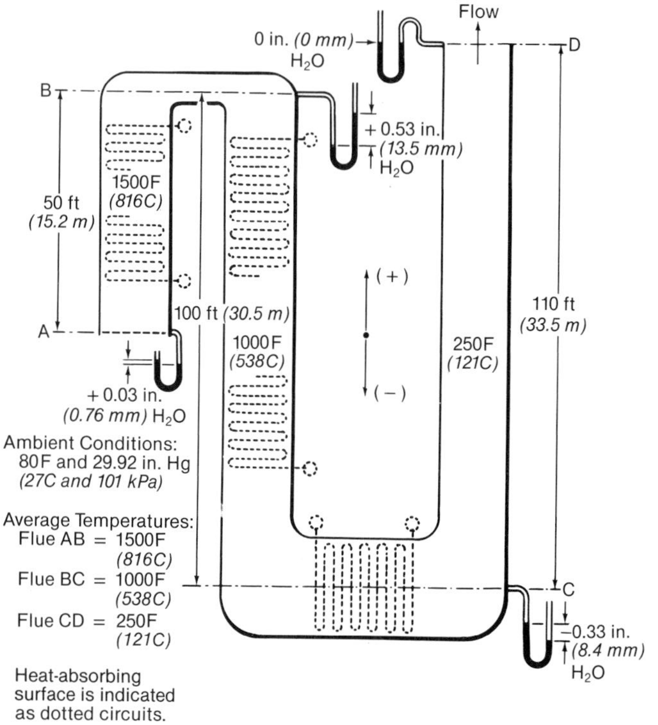

Fig. 20 Diagram illustrating stack effect, or chimney action, in three vertical gas passes arranged in series.

where

P_l = stack flow loss, in. of water
D_i = internal stack diameter, ft
L = stack height above gas entrance, ft
f = friction factor from Fig. 1, Chapter 3, dimensionless
T_g = average absolute gas temperature, R
B = barometric pressure, in. Hg

Stack flow losses for natural draft units are typically less than 5% of the theoretical stack draft. Also, that part of the loss due to unrecoverable kinetic energy of flow (exit loss) is from three to seven times greater than the friction loss, depending on stack height and diameter.

Sample stack size selection

Tentative stack diameter and height for a given draft requirement can be calculated using English units with Figs. 19, 21 and 22 and an assumed stack exit gas temperature. Adjustments to these values are then made as required, by verification of the assumed stack exit temperature, a flow loss check and altitude correction, if necessary. The following example illustrates this sizing procedure:

Unit Specifications:	
Fuel	Pulverized coal
Steam generated, lb/h	360,000
Stack gas flow, lb/h	450,000
Stack inlet gas temp., F	550
Required stack draft (from point of balanced draft to stack gas entrance), in. H_2O	1.0
Plant altitude	Sea level
Initial Assumption:	
Stack exit gas temperature, F	450

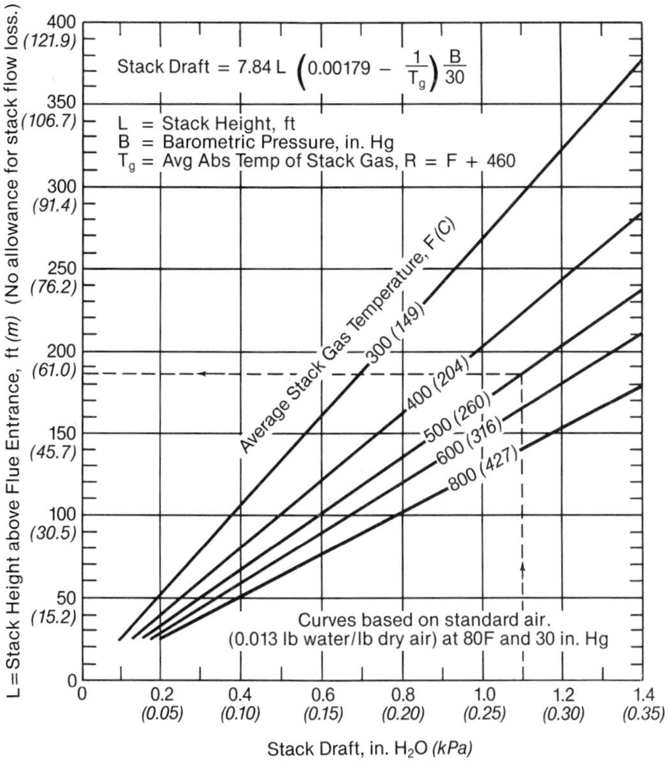

Fig. 21 Stack height required for a range of stack drafts and average stack gas temperatures.

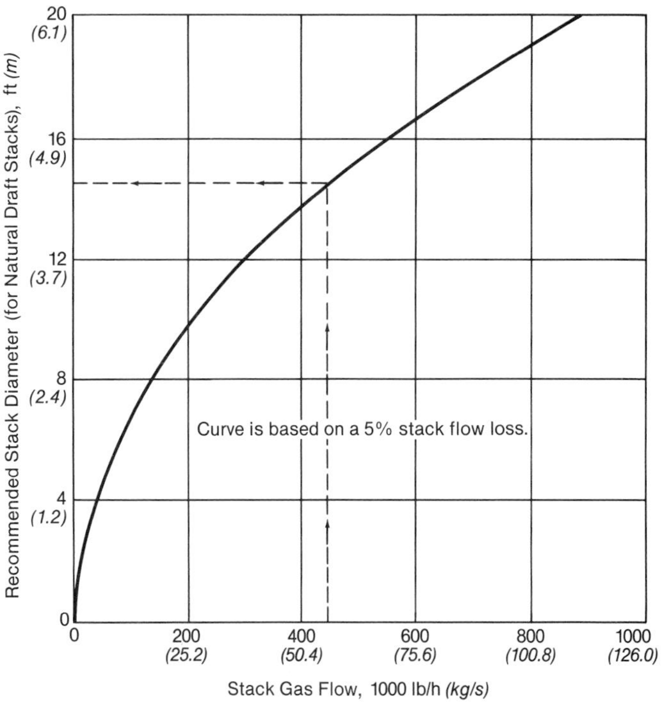

Fig. 22 Recommended stack diameter for a range of gas flows.

If the stack gas flow is not specified the following approximate ratios may be used:

Type of Firing	Gas Weight/Steam Flow Ratio
Oil or gas	1.15
Pulverized coal	1.25
Stoker	1.50

The stack diameter to the nearest 6 in. increment from Fig. 22 for 450,000 lb/h stack gas flow is 14 ft 6 in. For the required stack draft of 1.0 in. (increased to 1.1 in. for safety) and an average stack gas temperature of 500F, based on the specified inlet temperature of 550F and assumed exit temperature of 450F, Fig. 21 gives an approximate height of 187 ft. A check of the assumed stack exit temperature is obtained from Fig. 19, with the tentative height of 187 ft, diameter of 14 ft 6 in. and inlet temperature of 550F. This result is 430F, or an average stack temperature of 490F and draft of 1.1 in. H_2O. Fig. 21 is again used to establish a stack height neglecting stack flow losses. This height is 190 ft.

Assuming a stack flow loss of 5%, the final required stack height is 200 ft (190/0.95). This represents the active height of the stack. The height of any inactive section from foundation to stack entrance must also be included.

The stack flow loss is checked using the above values for diameter, height, average gas temperature and gas flow in Equation 6. A check of available net draft, using Equation 2, indicates that the 1.0 in. draft requirement is amply covered.

If the plant is not located at sea level, the draft requirement of the unit must be increased by multiplying the draft by the altitude factor 30/B and the theoretical stack draft decreased by multiplying the theoretical draft by B/30, where B is the normal barometric pressure, inches of mercury, at the boiler site (Table 2).

External factors affecting stack height

The stack also functions to disperse combustion gases. Increasing stack height enlarges the area of dispersion. In narrow valleys or locations where there is a concentration of industry, it may be necessary to provide increased stack height.

Some power plants located near airports are prohibited from using stacks high enough to provide adequate dispersion. In such cases the stack may be necked down at the top to increase the discharge velocity, simulating the effect of a higher stack. However, necking down the stack adds an appreciable amount of flow resistance which can only be accommodated by a mechanical draft system.

Stack design

After the correct stack height and diameter are established, there are economic and structural factors to consider in designing the stack. Stack material selection is influenced by material and erection costs, stack height, means of support, i.e., whether the stack is supported from a steel structure or a foundation, and erosive and corrosive constituents in the flue gas. After selecting the material, the stack is checked for structural adequacy, making both a static and a dynamic analysis of the loads.

Stack operation and maintenance

All connections to the stack should be air-tight and sealed with dampers when not in use. Cold air leaks into the stack during operation, reducing the average gas temperature and the stack effect. Leaks also increase the gas flow and erosion potential in the stack.

A stack is subjected to the erosive action of particulate, acid corrosion from sulfur products and weathering. Erosion is most common at the stack entrance, throats, necked down sections and locations where the direction or velocity of gas changes. Abrasion resistant materials or erosion shields at these locations can reduce stack maintenance.

Fans

A fan moves a quantity of air or gas by adding sufficient energy to the stream to initiate motion and overcome resistance to flow. The fan consists of a bladed rotor, or impeller, which does the actual work and usually a housing to collect and direct the air or gas discharged by the impeller. The power required depends upon the volume of air or gas moved per unit of time, the pressure difference across the fan, and the efficiency of the fan and its drives.

Power

Power may be expressed as shaft horsepower, input horsepower to motor terminals if motor driven, or theoretical horsepower which is computed by thermodynamic methods.

Fan power consumption can be expressed as:

$$\text{Power} = k\frac{\Delta P\dot{v}}{\eta_f C} \quad \mathbf{(8)}$$

where

Power = shaft power input, hp (kW)
ΔP = pressure rise across fan, in. wg (kPa)
$\dot{v}$ = inlet volume flow rate, ft^3/min (m^3/s)
η_f = fan mechanical efficiency, 100% = 100
k = compressibility factor, dimensionless (see Table 4)
C = constant of 6354 (1.00 for SI)

Approximate ranges of fan efficiencies and compressibility factors for use in Equation 8 are provided in Table 4. The term fan efficiency can be misleading because there are a number of ways it can be defined. Fan efficiency can be calculated across the fan rotor only, across the fan housing (inlet to outlet) with no allowance for efficiency losses caused by inlet or outlet duct configuration, or across the housing with losses induced by inlet and outlet ducting included. The fan vendor can usually recommend the best duct arrangement at the fan inlet and outlet to minimize these losses. To select the proper fan motor, shaft input power must be calculated using the efficiency that accounts for all of the losses associated with the fan type, including losses caused by inlet and outlet duct arrangement.

Another method used to calculate power consumption involves the concept of adiabatic head. If total pressure rise is known, adiabatic head can be calculated using the following formula:

$$Hd = \frac{(k)(\Delta P)(C)}{\rho} \quad \mathbf{(9)}$$

where

Hd = developed adiabatic head of gas column, ft (m)
k = compressibility factor
ΔP = total pressure increase, in. wg (kPa)
ρ = actual density, lb/ft^3 (kg/m^3)
C = constant of 5.20 (1.00 for SI)

Using the adiabatic head concept as defined by Equation 9, fan shaft input power can be calculated as:

$$\text{Power} = \frac{(Q)(Hd)(C)}{\eta_f} \quad \mathbf{(10)}$$

where

Q = gas flow, lb/h (kg/h)
C = constant of 0.505×10^{-6} (2.724×10^{-6} for SI)

Table 4
Mechanical Efficiency — Approximate Ranges (η_f)

Centrifugal fan	
Paddle blade	45 to 60%
Foward curved blade	45 to 60%
Backward curved blades	75 to 85%
Radial tipped blades	60 to 70%
Air Foil	80 to 90%
Axial flow fan	85 to 90%

Approximate compressibility factors (air)

ΔP/P	0	0.03	0.06	0.09	0.12	0.15	0.18
k	1	0.99	0.98	0.97	0.96	0.95	0.94

Fan performance

Stacks seldom provide sufficient natural draft to cover the requirements of modern boiler units. These higher draft loss systems require the use of mechanical draft equipment and a wide variety of fan designs and types is available to meet this need.

There are essentially two different kinds of fans:

1. The centrifugal fan in which gas or air accelerates radially outward in a rotor from heel to tip of blades, discharging into a surrounding scroll casing. (Figs. 23 and 24.)
2. The axial flow fan in which the fluid is accelerated parallel to the fan axis. (See Fig. 25.)

Fan performance is best expressed in graphic form as fan curves (Fig. 26) which provide static pressure (head), shaft horsepower and static efficiency as functions of capacity or volumetric flow rate. Because fan operation for a given capacity must match single values of head and horsepower on the characteristic curves, a balance between fan static pressure and system resistance is required.

Varying the operating speed (rpm) to yield a family of curves, as shown in Fig. 27, will change the numerical performance values of the curve characteristics. However, the shape of the curves remains substantially unaltered. Changes in operation of fans can generally be predicted from the *Laws of Fan Performance*:

1. *Fan speed variation* (for constant fan size, density and system resistance)
   a) Capacity [ft^3/min (m^3/min)] varies directly with speed.
   b) Pressure varies as the square of the speed.
   c) Power varies as the cube of the speed.
2. *Fan size variation* (geometrically similar fans, constant pressure, density and rating)
   a) Capacity varies as the square of wheel diameter.
   b) Power varies as the square of wheel diameter.
   c) Rpm varies inversely as wheel diameter.
3. *Gas density variation* (constant size and speed plus constant system resistance or point of rating).
   a) Capacity remains constant.
   b) Pressure varies directly as gas density.
   c) Power varies directly as gas density.

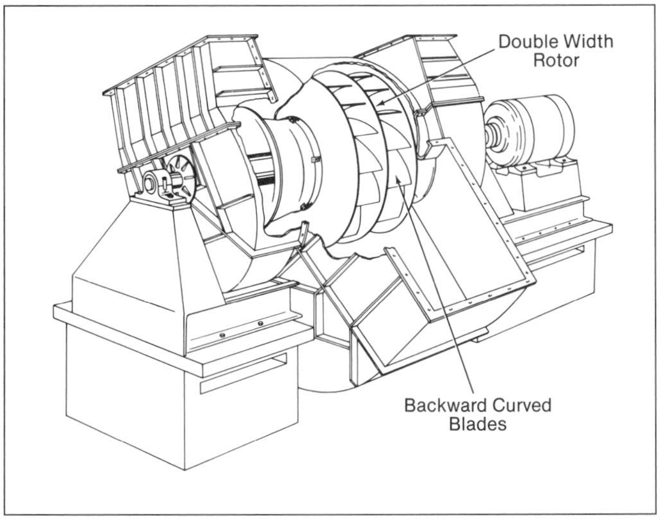

Fig. 23 Double width, double inlet centrifugal fan with backward curved blades.

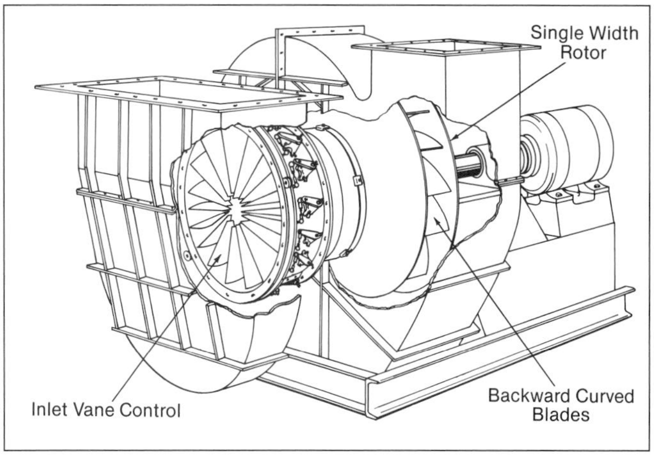

Fig. 24 Single width, single inlet centrifugal fan with backward curved blades and inlet vane control.

Geometrically similar fans have similar operating characteristics. Therefore, the performance of one fan can be predicted by knowing how a smaller or larger fan operates. The two main performance factors (speed and head) are linked in the concept of specific speed and specific diameter.

Specific speed is the rpm at which a fan would operate if reduced proportionately in size so that it delivers 1 ft^3/min of air at standard conditions, against a 1 in. wg static pressure.

Specific diameter is fan diameter required to deliver 1 ft^3/min standard air against a 1 in. wg static pressure at a given specific speed. From the fan laws we get these equations:

$$\text{Specific speed } (N) = \frac{\text{rpm } (\text{ft}^3/\text{min})^{\frac{1}{2}}}{(SP)^{\frac{3}{4}}} \qquad \textbf{(11)}$$

$$\text{Specific diameter } (D) = \frac{D(SP)^{\frac{1}{4}}}{(\text{ft}^3/\text{min})^{\frac{1}{2}}} \qquad \textbf{(12)}$$

where flow in ft^3/min is at standard conditions, *SP* is static pressure (in. wg) and *D* is fan diameter (in.).

Because there is only one value of specific speed at the point of maximum efficiency for any fan design, that value serves to identify the particular design. The same is true for specific size. If either specific speed or specific size can be established from the requirements of an application, only those designs with corresponding identifying values need to be considered.

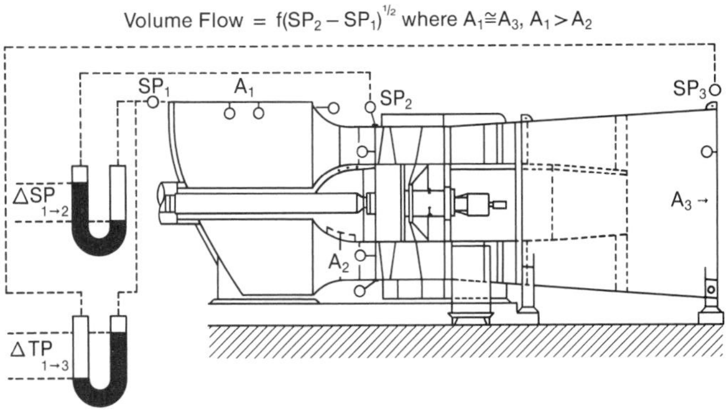

Fig. 25 Flow and head determination of axial flow fan.

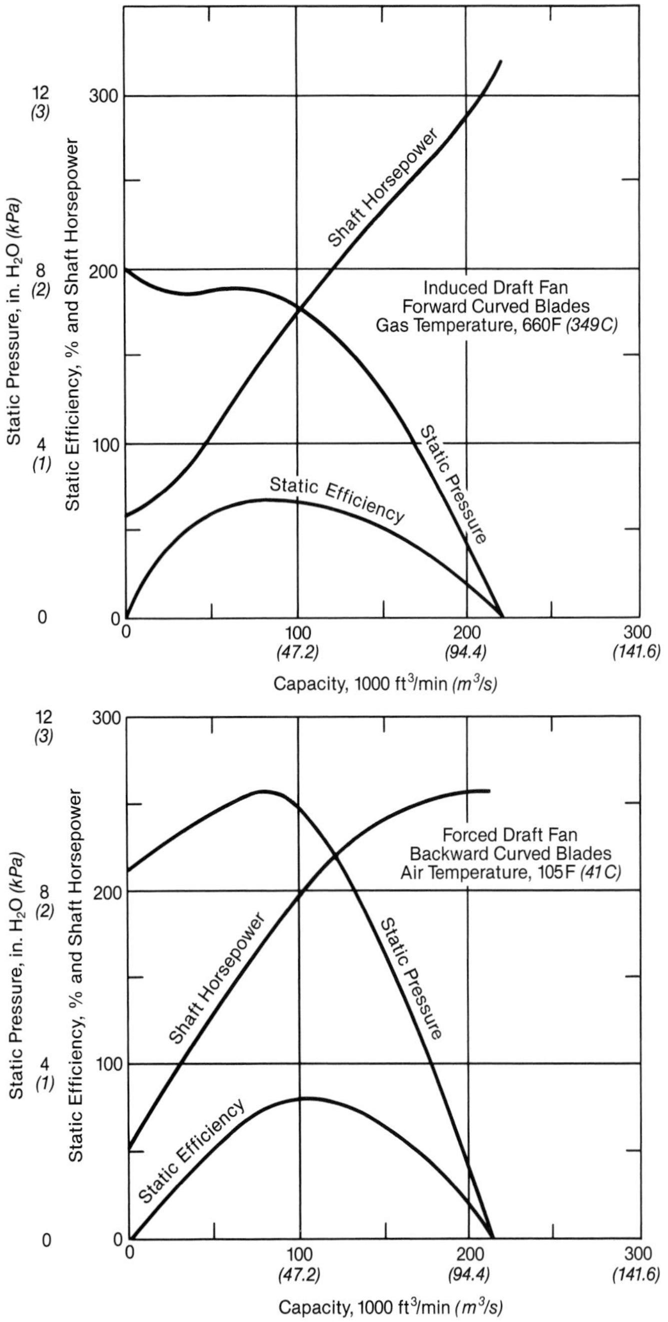

Fig. 26 Characteristic curves for two types of centrifugal fans operating at 5500 ft (1676 m) elevation and 965 rpm.

Aerodynamic characteristics

Fig. 28 shows a fan selection diagram used to determine the most economical fan for a given set of performance conditions.

The performance requirements for a 500 MW coal-fired boiler using pairs of induced, forced and primary fans are superimposed on the fan type selection diagram.

When specific speed increases, the diameter ratio of a centrifugal fan increases toward unity. As unity is approached the design becomes less practical in the high specific speed range because the diameter ratios decrease.

System resistance is plotted in Figs. 27 and 29 along with the fan characteristic of static pressures at various speeds. If the fan operates at constant speed, any flow rate less than that shown at the intersection of the system resistance and specified rpm curves must be obtained by adding resistance through throttling the excess fan head. This represents a waste in power that can be avoided by using a variable speed drive.

Plotting characteristics on a percentage basis shows some of the many variations available in different fan designs. Fig. 30 is such a plot with 100% rated capacity selected at the point of maximum efficiency. Some fans give steep head characteristics, while others give flat head characteristics. Some horsepower characteristics are concave upward, others are concave downward. The latter have the advantage of being self-limiting, so that there is little danger of burning out the driver or little need for oversizing the motor.

Centrifugal fans have a steep head characteristic and are attractive for high static, low flow rate applications. Fans with steep head characteristics also require high tip speeds for the specified head. Therefore, centrifugal fans are usually equipped with wear liners when there is heavy dust loading in the gas. In contrast, axial flow fans are more suitable for use where static pressure requirements tend to decrease as flow requirements de-

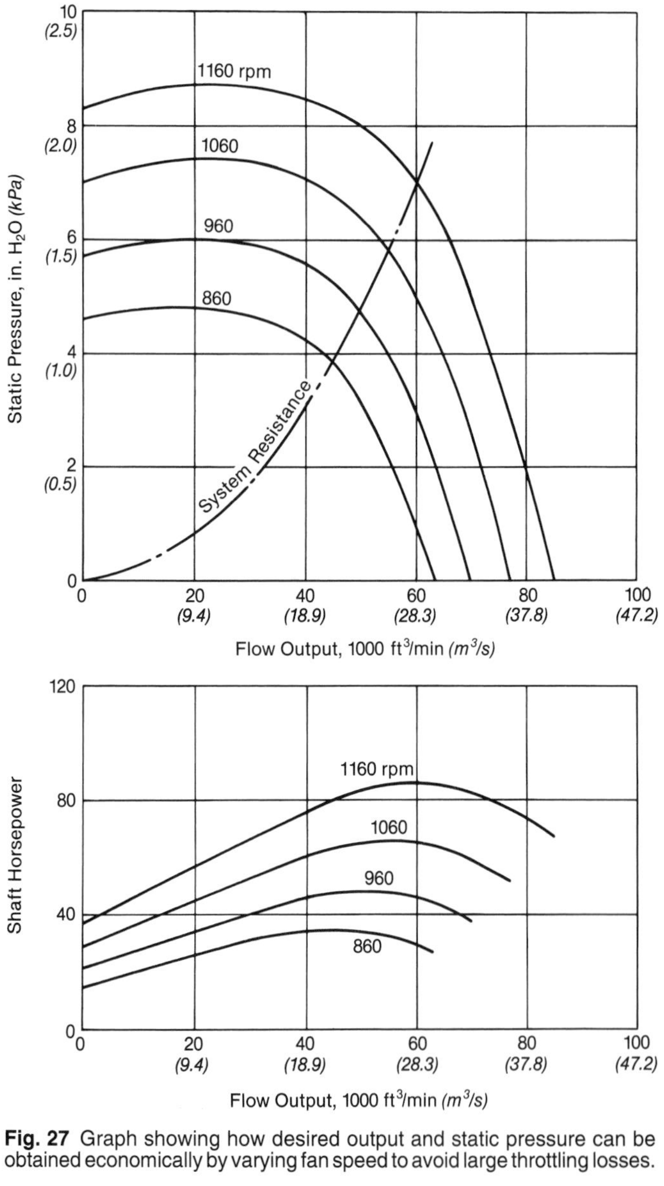

Fig. 27 Graph showing how desired output and static pressure can be obtained economically by varying fan speed to avoid large throttling losses.

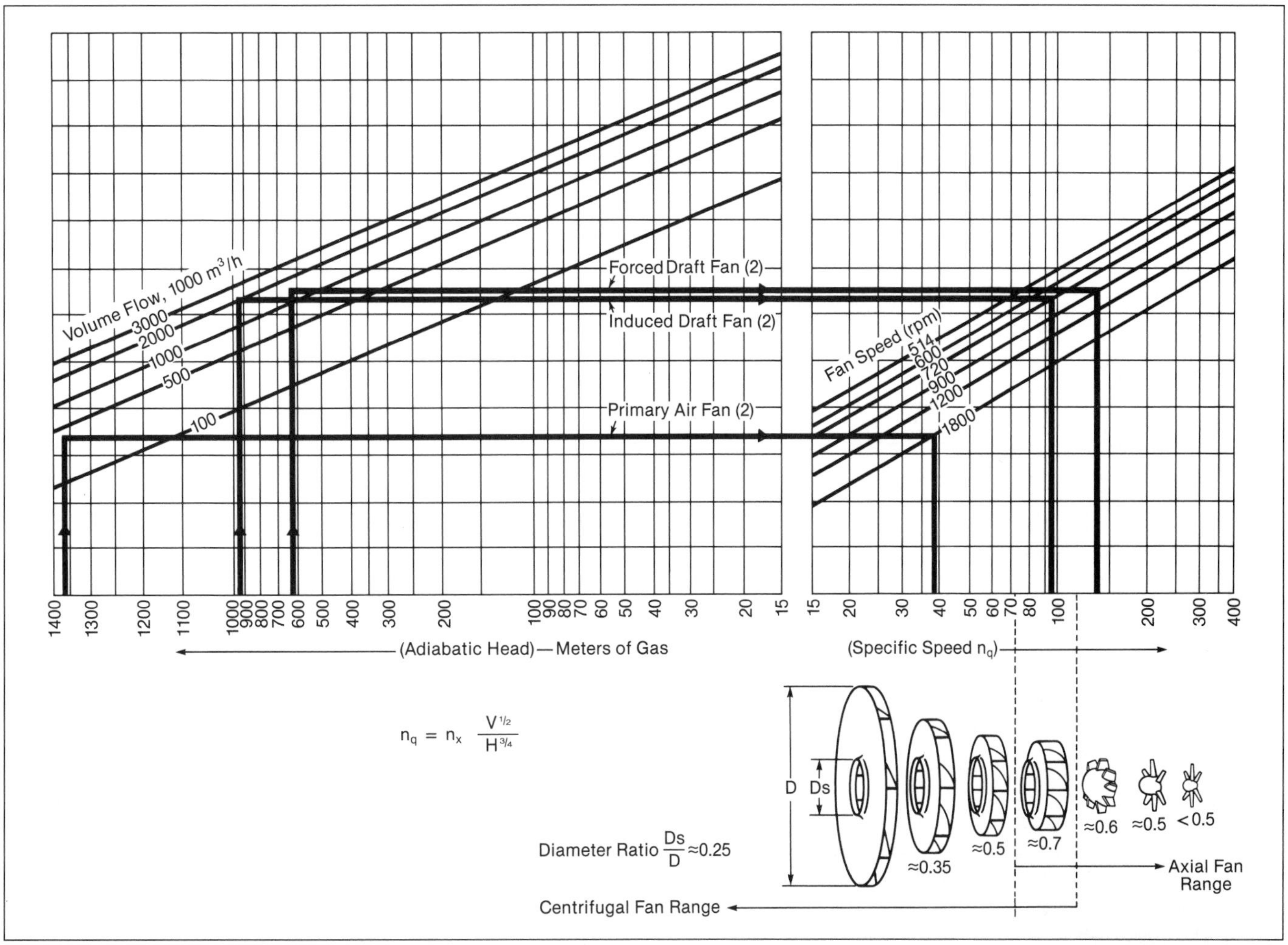

Fig. 28 Fan type selection diagram.

crease. Application of axial flow fans to systems where pressure requirements may suddenly increase at reduced flow conditions must be carefully reviewed in order to make certain that the change in static requirements does not cause the fan to stall.

Fan testing

It is difficult to obtain consistent data from a field test of fans installed in flue and duct systems because it is seldom possible to eliminate flow disturbances from such things as bends, change in flow area and dampers. Structural arrangements at the fan entrance and discharge also materially affect field performance results. The consistent way to verify fan performance is on a test stand. (See Fig. 31.)

Control of fan output — centrifugal fan

Very few applications permit fans to operate continuously at the same pressure and volume discharge rate. Therefore, to meet requirements of the system, some means of varying the fan output are required such as damper control and variable speed control.

Damper control introduces sufficient variable resistance in the system to alter the fan output as required. The advantages are:

1. lowest initial capital cost of all control types,
2. ease of operation or adaptation to automatic control,
3. least expensive type of fan drive; a constant speed induction type A-C motor may be used, and
4. continuous rather than a step type of control making this method effective throughout the entire range of fan operation.

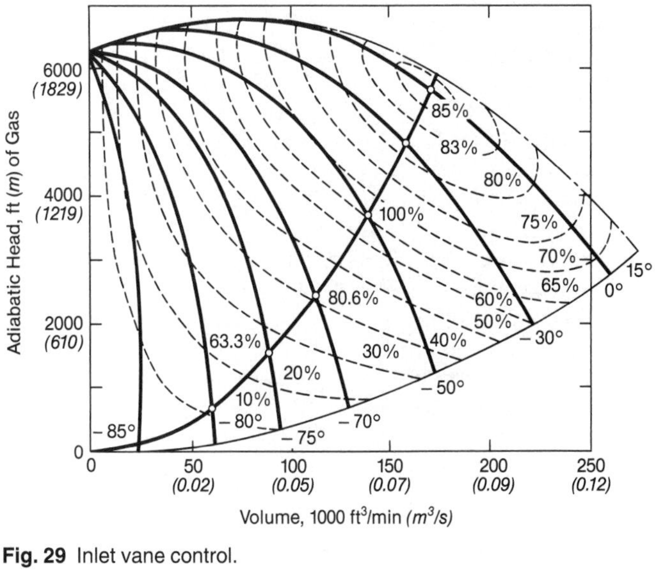

Fig. 29 Inlet vane control.

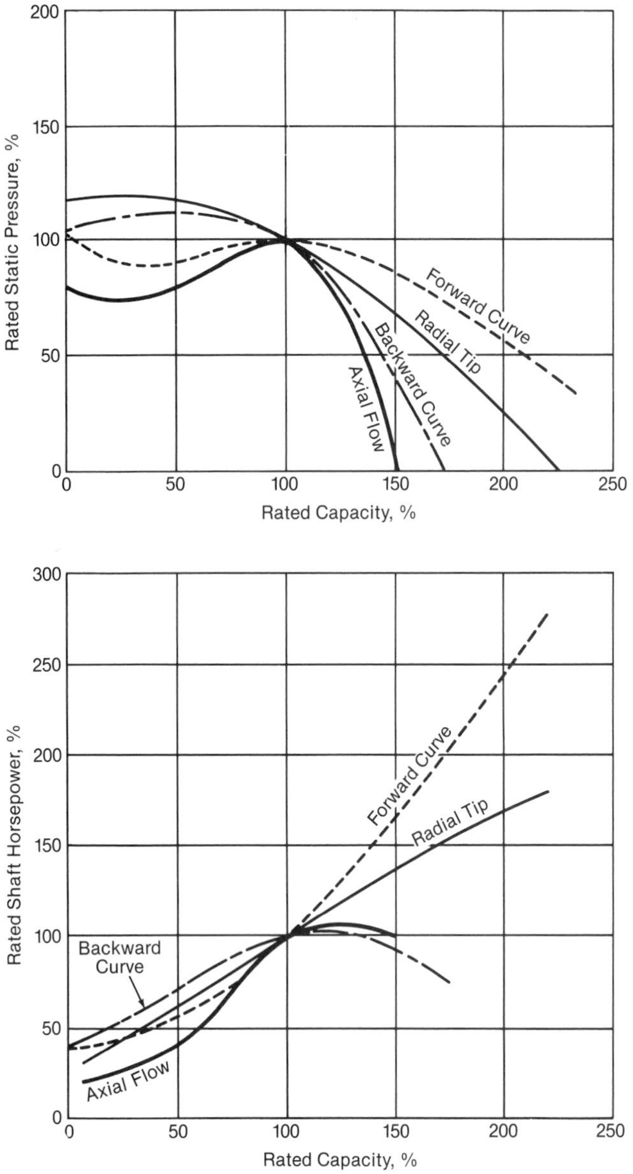

Fig. 30 Selected centrifugal and axial flow fan characteristic curves.

However, damper control causes wasted power because of the excess pressure energy which must be dissipated by throttling. The most economical control of centrifugal fans is accomplished with inlet vanes which are designed for use with dirty air as well as clean air.

Operating experience on forced draft, primary air and induced draft fans has proven that the inlet vane control is reliable and reduces operating cost. It also controls stability, controls accuracy and minimizes hysteresis. Inlet vane control (Fig. 29) regulates air flow entering the fan and requires less horsepower at fractional loads than outlet damper control. The inlet vanes give the air a varying degree of spin in the direction of wheel rotation enabling the fan to produce the required head at proportionally lower power. Although vane control offers considerable savings in efficiency over damper control at any reduced load, it is most effective for moderate changes close to full load operation. The initial cost is more than for damper control but less than for variable speed control.

Variable speed motors are attractive because they reduce shaft speed (energy) at reduced flow rates. However, variable speed motors require a higher initial cost which may not be offset by lower power requirements. Speed control also results in some loss in motor efficiency because no variable speed driver works as efficiently throughout the entire fan load range as a direct connected constant speed A-C motor. The loss in efficiency depends upon the type of speed variation.

A number of commonly used variable speed arrangements are hydraulic coupling, variable speed D-C motor and variable speed steam turbine.

Fan drives

Electric motors are normally used for fan drives because they are less expensive and more efficient than other drives. For fans of more than a few horsepower, squirrel cage induction motors predominate. This type of motor is relatively inexpensive, reliable and highly efficient over a wide load range. It is frequently used in large sizes with a magnetic or hydraulic coupling for variable speed installations.

For some variable speed installations, particularly in the smaller sizes, wound rotor (slip ring) induction motors are used. If a D-C motor is required, the compound type is usually selected. The steam turbine drive costs more than a squirrel cage motor, but is less expensive than any of the variable speed electric motor arrangements in sizes more than 50 hp (37 kW). A steam turbine may be more economical than the electric motor drive in plants where exhaust steam is needed for process, or on large utility units using the exhaust steam for feedwater heating.

Fan capacity margins

To make sure that the fans will not limit a boiler's performance, margins of safety are added to the calculated or net fan requirements to arrive at a satisfactory test block specification. These margins are intended to cover conditions encountered in operation that can not be specifically evaluated. For example, variation in fuel ash characteris-

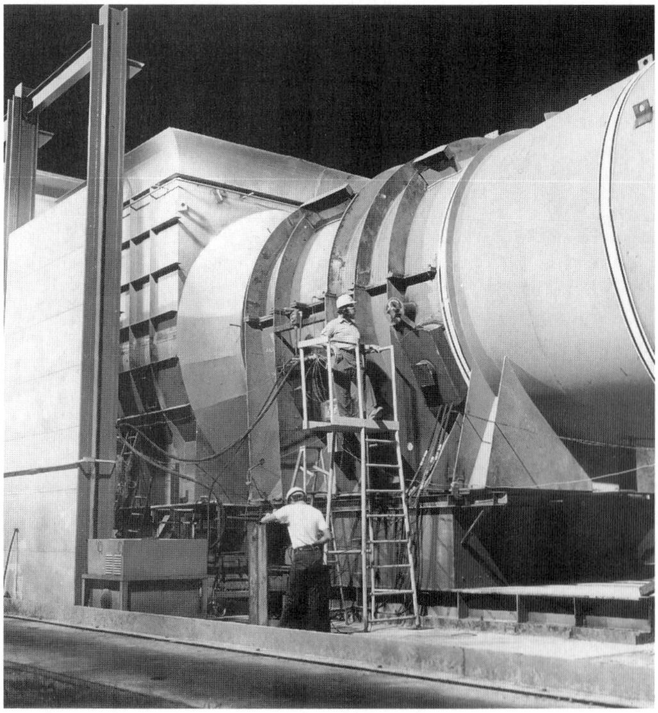

Fig. 31 Full scale testing of variable pitch axial flow fan.

tics or unusual operating conditions slag or foul heating surfaces. The unit then requires additional draft. Air heater leakage can increase to higher than expected levels because of incorrect seal adjustment or seal wear. Stoker-fired boilers, burning improperly sized coal, may require more than normal pressure to force air through the fuel bed. A need for rapid load increase or a short emergency overload often calls for over capacity of the fans. The customary margins to allow for such conditions include: 1) 15 to 20% increase in the net weight flow of air or gas, 2) 20 to 30% increase in the net head, and 3) 25F (14C) increase in the air or gas temperature at the fan inlet.

Forced draft fan

Boilers operating with both forced and induced draft use the forced draft fan to push air through the combustion air supply system into the furnace. (See Fig. 32.) The fan must have a discharge pressure high enough to equal the total resistance of air ducts, air heater, burners or fuel bed and any other resistance between the fan discharge and the furnace. This makes the furnace the point of balanced draft or zero pressure. Volume output of the forced draft fan must equal the total quantity of air required for combustion plus air heater leakage. In many installations, greater reliability is obtained by dividing the total fan capacity between two fans operating in parallel. If one fan is out of service, the other usually can carry 60% or more of full boiler load, depending on how the fans are sized.

To establish the required characteristics of the forced draft fan, the system resistance from fan to furnace is calculated for the actual weight of air required for combustion plus the expected leakage from the air side of the

Fig. 32 Forced draft centrifugal fan for a 364 MW outdoor unit.

air heater. It is design practice to base all calculations on 80F (27C) air temperature entering the fan. The results are then adjusted to test block specifications by the margin factors previously discussed.

Forced draft fan selection should consider the following general requirements:

Reliability Boilers must operate continuously for long periods (up to 18 months in some instances) without shutdown for repairs or maintenance. Therefore, the fan must have a rugged rotor and housing and conservatively loaded bearings. The fan must also be well balanced and the blades shaped so that they will not collect dirt and disturb this balance.

Efficiency High efficiency over a wide range of output is necessary because boilers operate under varying load conditions.

Stability Fan pressure should vary uniformly with volumetric flow rate over the capacity range. This facilitates boiler control and assures minimum disturbance of air flow when minor adjustments to the fuel burning equipment change the system resistance. When two or more fans operate in parallel, the pressure output curves should have characteristics similar to the radial tip or backward curved blade fans in order to share the load equally near the shutoff point.

Overloading It is desirable for motor driven fans to have self-limiting horsepower characteristics, so that the driving motor can not overload. This means that the horsepower should reach a peak and then drop off near the full load fan output (Fig. 27).

Induced draft fan

Units designed to operate with balanced furnace draft or without a forced draft fan require induced draft to move the gaseous products of combustion.

The gas weight used to calculate net induced draft requirements is the weight of combustion product gas at maximum boiler load, plus any air leakage into the boiler setting from the surroundings and from the air side to the gas side of the air heater. Net gas temperatures are based on the calculated unit performance at maximum load. Induced draft fan test block specifications of gas weight, negative static pressure and gas temperature are obtained by adjusting from net values by margins similar to those used for forced draft fans.

An induced draft fan has the same basic requirements as a forced draft fan except that it handles higher temperature gas which may contain erosive ash. Excessive maintenance from erosion can be avoided by protecting casing and blades with replaceable wear strips. Because of their lower resistance to erosion, air foil blades should be treated with caution when considering an induced draft application. Air foil blades are very susceptible to dust erosion and, if hollow, they can fill with dust and cause rotor imbalance should the blade surface wear through. Bearings, usually water-cooled, have radiation shields on the shaft between the rotor and bearings to avoid overheating.

Gas recirculation fans

As discussed in Chapter 18, gas recirculation fans are used variously for controlling steam temperature, furnace heat absorption and slagging of heating surfaces. They are generally located at the economizer outlet to extract gas and inject it into the furnace at locations dependent on the intended function. These multiple purposes are also an important consideration in properly sizing and specifying gas recirculation fans. Selection may be dictated by the high static pressure required for tempering furnace temperatures at full load on the boiler unit, or by the high volume requirement at partial loads for steam temperature control.

Even though gas recirculation fans have the same basic requirements as induced draft fans, there are additional factors to be considered. The gas recirculation fan operates at higher gas temperatures, so intermittent service may cause thermal shock or unbalance. When the fan is not in service, tight shutoff dampers and sealing air must be provided to prevent the back flow of hot furnace gas and a turning gear is often used on large fans to rotate the rotor slowly to avoid distortion.

Primary air fans

Primary air fans on coal-fired boilers supply pulverizers with the air needed to dry the coal and transport it to the boiler. Cold primary air fans should be designed for duty similar to forced draft fans. Primary air fans may be located before the air heater (cold primary air system) or downstream of the air heater (hot primary air system). The cold primary air system has the advantage of working with a smaller volumetric flow rate for a given mass flow rate. This method will pressurize the air side of the air heater and encourage leakage to the gas side. The hot primary air system avoids primary air heater leakage, but requires a higher fan design temperature and larger volumetric flow rate.

Fan maintenance

Fans require frequent inspection to detect and correct irregularities that might cause trouble. However, they should also have long periods of continuous operation compared with other power plant equipment. This can be assured by proper lubrication and cooling of fan shafts, couplings and bearings.

A fan should be properly balanced, both statically and dynamically, to assure smooth and lasting service. This balance should be checked after each maintenance shutdown by running the fan at full speed, first with no air flow and second with full air flow.

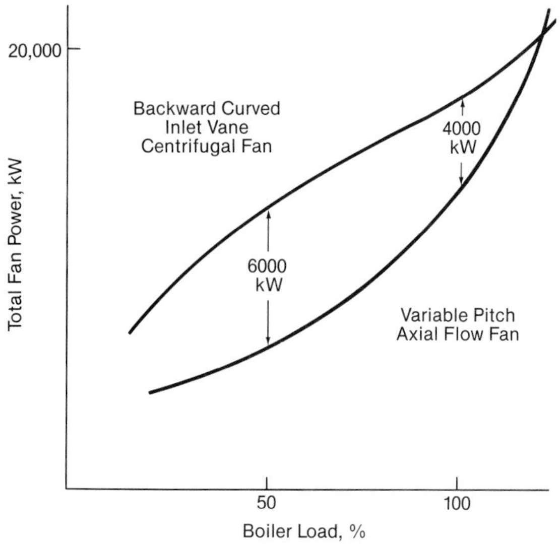

Fig. 33 Power savings.

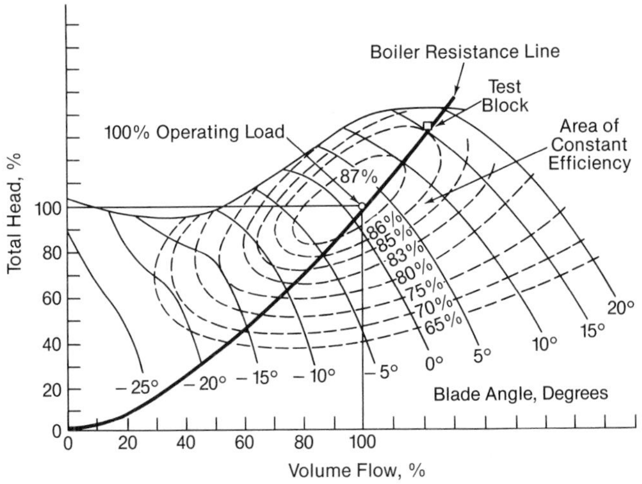

Fig. 34 Performance field for variable pitch axial flow fan.

Fans handling gases with entrained abrasive dust particles are subject to erosion. Abrasive resistant materials and liners can be used to reduce such wear. In some cases, beads of weld metal are applied to build up eroded surfaces.

Axial flow fans

One way to reduce auxiliary power requirements is to install variable pitch axial flow fans in fossil power generating systems. Fig. 33 compares the power consumption of the primary air, forced draft and induced draft fans for a typical 500 MW coal-fired unit using variable pitch axial flow fans with the same unit using backward curved, inlet vane controlled centrifugal fans. At 100% unit load, auxiliary power savings using a variable pitch axial flow fan will be 4000 kW, or about 7% of the total auxiliary power consumption.

Performance and control characteristics

Fig. 34 shows the characteristic performance field of a variable pitch axial flow fan. Several major benefits observed from this figure for axial flow fans include:

1. The areas of constant efficiency run parallel to the boiler resistance line resulting in high efficiency over a wide boiler load range.
2. There is also a large control range above, as well as below, the area of maximum efficiency permitting the fan to be designed for net boiler conditions, while the test block point remains within the control range.
3. The lines of constant blade angle are actually individual fan curves for a given blade setting. Because the curves are very steep, a change in resistance produces very little volume change.
4. As the blade angle is adjusted from minimum to maximum position the flow change is nearly linear, as shown in Fig. 35.

These last two characteristics provide stable fan and boiler control.

Parallel operation

Variable pitch axial flow fans can be operated in parallel provided care is taken to avoid operating either fan in the stall area discussed below. With two fans in op-

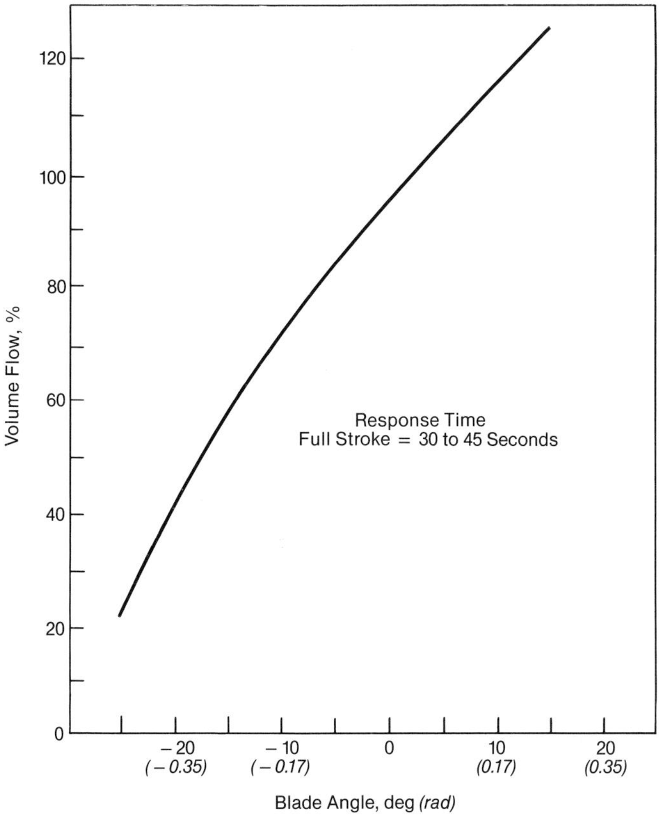

Fig. 35 Variable pitch blade control characteristics.

eration the resistance line for one fan is influenced by the other fan as well as by the boiler conditions. Two fans together will develop the pressure required to overcome the boiler resistance but individual volume flows need not be equal. However, to obtain the most efficient fan operation and to avoid operation in a range close to the stall line, it is best to keep both fans operating in parallel in their design condition.

The suggested control logic for starting, stopping and supervising the operation of variable pitch axial flow forced draft and induced draft fans is very similar to that of centrifugal fans. An additional requirement of the logic is to prevent damage to the fan while maintaining an open flow path through the boiler to prevent furnace pressure excursions.

Stall characteristics

Axial flow fans have a unique characteristic called *stall*. Stall is the aerodynamic phenomenon which occurs when a fan operates beyond its performance limits and flow separation occurs around the blade. If this happens the fan becomes unstable and no longer operates on its normal performance curve. Extended operation in the stall region should be avoided. Unpredictable flow vibrations occur which can lead to rotating blade damage.

The curves in Fig. 36, marked A, are the normal fan performance curves for a constant blade angle. Each blade angle curve has an individual stall point, identified as S on the diagram. The curve C connects all the stall points S and is generally referred to as the stall line.

The dashed curves B are the characteristic stall curves for three different blade angles. The curves show the path that the fan will follow when operating in a stalled condition.

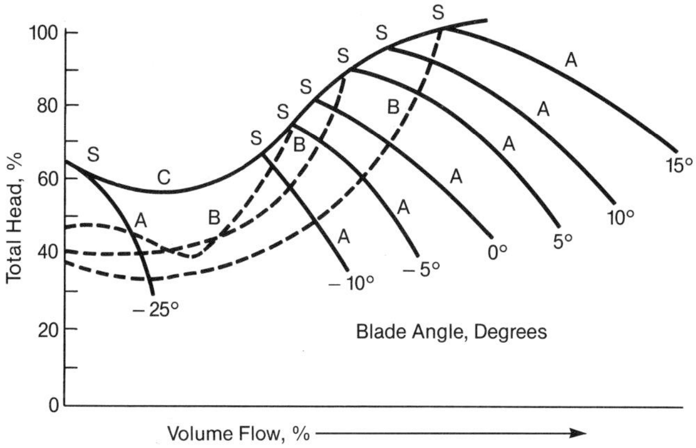

Fig. 36 Actual stall curves.

Fig. 37 explains the stall phenomenon in relation to the fan and boiler system. If the normal boiler system resistance (curve B) increases for any reason (for instance, a furnace pressure excursion caused by a main fuel trip) the normal operating point X will change to meet a new higher system resistance (curve B_1) by traveling along the fan performance curve A. If the operating point arrives at point S, the fan will stall. Because of the relationship between the fan performance curve D in the stall area and the upset system resistance (curve B_1), a new operating point X_1 will be found where the system resistance (curve B_1) and the stall curve B intersect. When the system resistance is reduced to curve B, the fan will recover from the stall and return to its normal performance curve A.

In the case of an upset as described above, the blade angle can be reduced until the fan regains stability. The fan will be stable when the new performance curve A_1 provides a stall point S, which is higher than the system resistance (curve B_1).

Stall prevention

When axial fans are sized properly and the system resistance line is parabolic in shape the probability of experiencing stall is low. The possibility of a stall increases if a fan is oversized in regard to volume capacity, if the system resistance increases significantly, or if the fans are operated improperly.

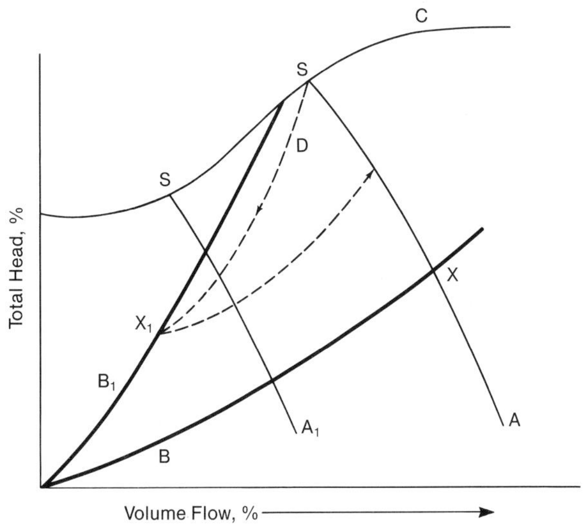

Fig. 37 Stall as related to the boiler.

The degree of stall protection desired in the control system depends on individual owner philosophy. A visual indication of head and volume (or blade angle) in the control room is a minimum requirement for satisfactory operation.

Fan arrangement

Axial flow fans designed for today's large fossil fueled steam generating systems are compact and relatively light. By concentrating the load carrying parts of the rotor on a small radius, a low Wk^2 is obtained. The low weight, unbalanced forces and inertia of the axial flow fan permit either horizontal or vertical arrangement on steel construction using a dampening frame. This provides greater arrangement flexibility. The normal ground level installation can be accomplished with reduced concrete foundation requirements.

A horizontal fan arrangement is generally the economic preference for forced draft and primary air fans. Induced draft fans can be arranged horizontally, or vertically inside the stack should available plant space be at a premium.

Acoustic noise

Fans produce two distinct types of noise:

1. *Single tone noise* is generated when the concentrated fluid flow channels leaving the rotating blades pass a stationary object (straightener vanes or nose). The distance from the rotating blades to stationary objects affects the sound intensity, with the blade passing frequency and its first harmonic being most dominant.
2. Broad band noise is produced by high velocity fluid rushing through the fan housing and, as the name implies, covers a wide frequency range.

From the outside, the apparent source of both types of fan noise is the fan housing. The sound travels out of the inlet box opening, through the discharge ductwork and through the casing of the fan. All three areas must be acoustically analyzed and individually treated to achieve an acceptable installation.

Inlet sound levels from forced draft and primary air fans are most commonly controlled by the installation of an absorption-type silencer. Fan casing noise can usually be effectively controlled by the use of mineral wool insulation and acoustic lagging. Fan discharge noise, however, requires a more detailed evaluation to determine the most cost effective method of control. For forced draft and primary air fans, insulating the outlet ducts or installing an absorption-type discharge silencer can be effective. For induced draft fans, installation of thermal insulation and lagging on the outlet flues will generally be sufficient. For situations where stack outlet noise must be reduced, an absorption-type discharge silencer will not work on coal-fired units as the sound absorbing panels will become plugged with flyash. In this instance a resonant-type discharge silencer with a self-cleaning design must be used.

Induced draft fan blade wear

Fan blade wear for induced draft fans is a function of dust loading (including excursions due to particulate collection equipment malfunction), particle hardness, particle size distribution, the relative blade particle velocity, the blade material and the angle of attack. The fan designer has little control over the first three and must therefore look to other areas to address blade wear issues. Wear is inherently more uniform over variable pitch axial flow fan blades. By selecting a two-stage fan at a lower speed (rpm) instead of faster one-stage fan, the tip velocity can be reduced. Abrasion resistant material coatings can be added to the blade wear areas. The blades can be made easily removable to permit fast replacement.

Condensing attemperator system

Superheater spray water attemperation is the primary method of steam temperature control on most boilers. This spray water is introduced between stages of the superheater or, in some cases, immediately downstream of the superheater outlet. If feedwater is used as the source for this attemperation, it must have very low solids content in order to avoid introducing potentially damaging deposits in the superheater or in the turbine.

Many industrial plants do not have the water treatment facilities capable of meeting the low solids feedwater requirements for spray attemperation. If feedwater quality does not meet the solids criteria established for spray water, a condensing attemperator system can be used to provide a low solids spray water source.

A schematic diagram of a typical condensing attemperation system is shown in Fig. 38. The system typically consists of condenser, spray water flow control valve, spray water attemperator and occasionally, a condensate storage tank and a condensate pump. The basic design philosophy of this system is that saturated steam from the boiler steam drum is routed to the shell side of the vertical condenser, where it is condensed and subcooled. This condensate then flows through the spray water piping and control valve to an interstage attemperator or to a terminal attemperator.

Most condensing attemperation systems are driven by the pressure differential between the drum and the point in the steam path where the spray water is introduced. Condensers are normally located above the steam drum in order to increase the hydrostatic head available to the

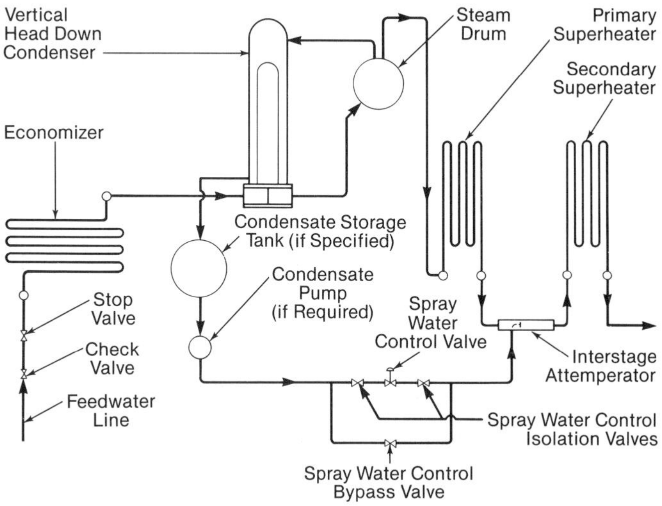

Fig. 38 Schematic arrangement, vertical condenser system.

attemperator. If the available pressure differential is not sufficient to overcome the system resistance developed at the design spray water flow, then a condensate pump must be used. In order to avoid the need for a condensate pump, system resistance losses are usually minimized by using a low pressure drop attemperator if system spray characteristics over the boiler operating range permit its use.

Feedwater is the cooling medium that flows through the tube side of the condenser. Fig. 38 shows the feedwater flowing from the economizer outlet to the condenser and from the condenser to the drum. Alternate locations may be upstream of the economizer inlet or an intermediate header in a multi-bank economizer.

The spray water condenser is a vertical, head down shell and tube heat exchanger arrangement (Fig. 39). The feedwater tube bundle is of the inverted U-tube type construction. Feedwater enters and leaves the tube bundle at the bottom of the inverted U-tubes. The cylindrical shell enclosure features a drum steam inlet near the top of the shell and a condensate outlet close to the bottom of the shell.

During normal operation, the condenser system will experience cyclic operation. At times there will be no spray flow demand and the condensate level on the shell side of the condenser will rise to the top of the tube bundle. This is a desirable operating characteristic of the inverted U-tube arrangement. It prevents steam pockets from being trapped at the top of the condenser shell which could cause water hammer and subsequent cracking of the shell.

The spray water condenser must be designed to withstand significant temperature transients and gradients. At times the condenser will be almost flooded with water and the whole assembly will be close to feedwater temperature. A sudden requirement for spray water can drop this water level quickly, exposing significant lengths of shell and internals to saturated steam temperature. In some cases this can mean a sudden rise in shell temperature of as much as 300F (167C).

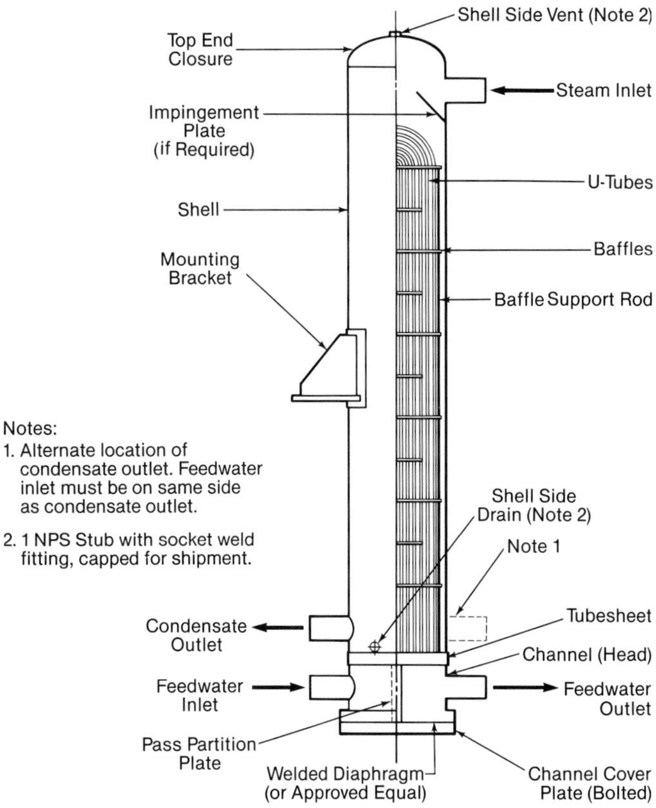

Fig. 39 General arrangement, vertical head down condenser.

References

1. ASME Boiler and Pressure Vessel Code, Section I.

2. Air Movement and Control Association, Publication 850-84, Arlington Heights, Illinois.

Section III
Applications of Steam

The eight chapters in this section illustrate how the subsystems described in Section II are combined to produce modern steam generating systems for specific applications. An expanded number of steam generating unit and system designs for various applications are described and illustrated.

The section begins with a discussion of large fossil fuel-fired equipment used to generate electric power. Both large and small industrial units, as well as those for small electric power applications, are then described in Chapter 25. The next six chapters address specialized equipment for specific applications. Unique designs for steam producing systems are used in pulp and paper mills, waste-to-energy plants and biomass-fired units. Pressurized fluidized-bed combustion, an advanced system for electric utility applications, is covered in Chapter 29, while unique features for marine applications are discussed in Chapter 30. A discussion of combined cycles, waste heat recovery and other steam systems concludes this section.

This facility features three 580 MW Carolina-type B&W boilers firing subbituminous coal.

Chapter 24
Fossil Fuel Boilers for Electric Power

Most of the electric power generated in the United States (U.S.) is produced in steam plants using fossil fuels and high speed turbines. These plants deliver a kilowatt hour of electricity for each 8500 to 9500 Btu (8968 to 10,023 kJ) supplied from the fuel, for a net thermal efficiency of 36 to 40%. They use steam driven turbine-generators of up to 1300 MW capacity with boilers generating from one million to ten million pounds steam per hour. A typical coal-fired facility is shown on the facing page.

Modern fossil fuel steam plants use cycles with nominal steam pressures from 1800 to 3500 psi (124 to 241 bar) at the turbine throttle and with superheater and reheater steam temperatures from 950F (510C) to more than 1000F (538C) delivered to the turbine throttle. Reheat plant cycles are used because of increased station thermal efficiency and reduced plant heat rates.

Most power plants in the U.S. and around the world are owned and operated by: 1) investor owned electric companies, 2) federal, state or local governments, or 3) financed companies.

These owners, whether public or private, have been generally known as *utilities*. During the 1980s, there was a trend away from traditional utility companies adding new electric generation plants. Instead, a new form of company, referred to as a cogenerator, independent power producer (IPP) or nonutility generator (NUG), emerged as a source of new electric power generation. It is anticipated that this new type of power company, which will build, own and operate a significant portion of the new power plants, will follow the traditional approaches to selecting power generation equipment.

Selection of steam generating equipment

The owner has several technologies to choose from based upon fuel availability, emissions requirements, reliability and project timing. One of the common choices for modern electric power generation is the high pressure, high temperature, fossil fuel-fired boiler.

Each new electric generating unit must satisfy the user's specific needs in the most economical manner. Achieving this requires close cooperation between the designer and the owner's engineering staff or consultants. The designer, owner and engineering group must identify those equipment features and characteristics that will reliably produce low cost electricity. The primary costs of electricity include: 1) capital equipment, 2) financing charges, 3) fuel, and 4) operation and maintenance. The owner, prior to issuing equipment specifications, reviews and surveys all cost factors. (See Chapter 37.)

The capital cost survey must include all direct costs such as the boiler, steam turbine and electric generator, emissions control equipment, condenser, feedwater heaters and pumps, fuel handling facilities, buildings and real estate. In addition, finance charges, including interest rates, loan periods, source of funds and tax considerations must be added. Fuel costs need to be evaluated based on the initial costs of the fuels, plant capacity variations expected during the life of the plant and forecasts of fuel cost changes during plant lifetime. The operation and maintenance costs should be estimated based on other current plants with similar equipment, fuels and operating characteristics. Operating and maintenance costs are heavily affected by manpower requirements, and consideration should be given to the availability of skilled personnel as well as to the cost of retaining the skilled staff during the plant lifetime.

Plant efficiency, fuel use and capital cost are critically related. Higher plant thermal efficiency obviously reduces annual fuel costs; however, fuel savings are partially offset by the associated higher capital costs. Therefore, selection of the desired plant efficiency requires consideration of the classical economic tradeoffs between capital and operating costs.

Other important criteria are the location of the electric generating plant with respect to fuel supply and the areas where electricity is used. In some cases, it is more economical to transport electricity than fuel. Some large steam generating stations have been built at the coal mine mouth to generate electricity which is then used several hundred miles away. If the user is a member of a broader grid of interconnected utility companies, the future requirements of other system members may also be an important factor.

In power plant planning, considerable time and effort are required to establish accurate basic plant assessment data with comprehensive consideration of engineering factors and plans for future expansion or changes. The accuracy of these data is critical if the experience and craftsmanship of the boiler manufacturer and other suppliers are to fully benefit the plant designer and owner. The owner should, at the outset, decide who is to prepare these data. If the owner lacks personnel with the necessary qualifications, the services of consulting engineers should be used. A thorough discussion with the boiler manufacturer will provide many details to help the user make correct decisions.

Before the boiler and other major equipment items can be selected, the basis of operation and arrangement of the entire steam plant must be planned. Ultimately the available data must be translated into the form of equipment specifications so that the manufacturers of various components can provide apparatus in accordance with the user's requirements. After equipment selection, construction drawings must be prepared for the foundations, building, piping and walkways. The construction work must be coordinated utilizing modern schedule and control techniques for effective management and completion of erection.

Boiler designer's requirements

The most important factors to the modern boiler designer are the steam conditions, fuel and environmental constraints. Steam conditions identify the amount of steam required and the temperature and pressure of the primary and reheat steam. Of equal importance is the type and chemical composition of the fuel to be used and the mandated emissions requirements for the specified plant site.

The boiler designer needs a complete set of data pertinent to steam generation to enable him to produce the most economical steam generating equipment and satisfy the needs of the user.

The requirements and conditions that form the basis for the designer's equipment selection can be outlined as follows:

1. Fuels — sources presently available or planned for future use; proximate, ultimate and ash analyses of each fuel as well as costs and future trends.
2. Steam requirements —
   a. Pressure and temperature at each point of use, including the outlet of steam generating units, as well as allowable variations in temperature.
   b. Rate of heat delivery (or steam flow) to each point of use, to boiler house auxiliaries and feedwater heating, to blowdown and to the outlet of the steam generating unit. Variations (minimum, average and maximum) and predictable future requirements are also important.
3. Boiler feedwater — source, chemical analysis and temperature of the feedwater entering the steam generating unit.
4. Environmental requirements — site limitations; permissible levels for nitrogen oxides (NO_x), sulfur dioxide (SO_2), particulate and other atmospheric emissions; solid waste specifications; and all other regulatory and commercial requirements.
5. Space and geographical considerations — space limitations, relation of new equipment to existing boiler equipment, earthquake and wind resistance requirements, elevation above sea level, foundation conditions, climate and accessibility for service and construction.
6. Auxiliary power — medium used and evaluated cost of energy.
7. Operating personnel — experience level of operating and maintenance personnel as well as the cost of labor.
8. Guarantees.
9. Evaluation basis — for unit efficiency, auxiliary power required, building volume and various fixed charges.

With this information, the boiler designer is able to analyze the user's specific needs. As a result, an economical, reliable steam generator can be built and initial costs can be balanced with long term savings.

Design practice

The boiler designer usually works with standardized, pre-engineered components. Because the detailed engineering of these components has been completed, shop fabrication is expedited, and operating capability and reliability have been proven. Examples of such pre-engineered components include burners, pulverizers, furnace sections, steam drums and selected pressure parts.

There has been little standardization of complete boiler designs for utility applications primarily because of each user's unique requirements. These variations are primarily due to differences in the types and range of fuels to be fired and the user's plans for operating the steam generating unit within the system. Such variations require changes in detail and overall arrangement of boiler components. When combined with ever-changing costs of financing, fuel, materials and labor, these factors have made full unit standardization impractical.

At the start of a project, the designer reviews the available database and past experience to determine which large shop-fabricated components, called *modules*, may be applied to the design. Reviewing shipping sizes, traffic methods and routing, and the general economic tradeoff of greater shop assembly versus reduced field erection time are standard project practices. Modularization is now common practice but needs to be assessed on each project to determine the minimum cost.

Fuels

The basic types of fuels used for U.S. electricity generation in 1990 are shown in Fig. 1. The dominant fuel in modern U.S. central stations is coal, either bituminous, subbituminous or lignite. A similar picture emerges if worldwide electric power production is considered. (See Fig. 2.) While natural gas or fuel oil may be the fuel of choice for selected future fossil fuel power plants, coal is

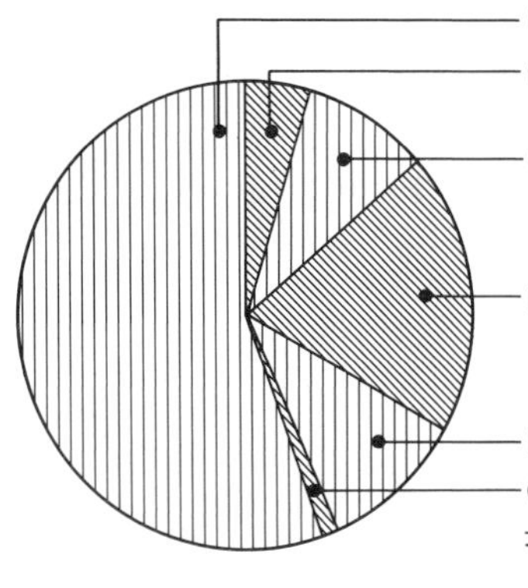

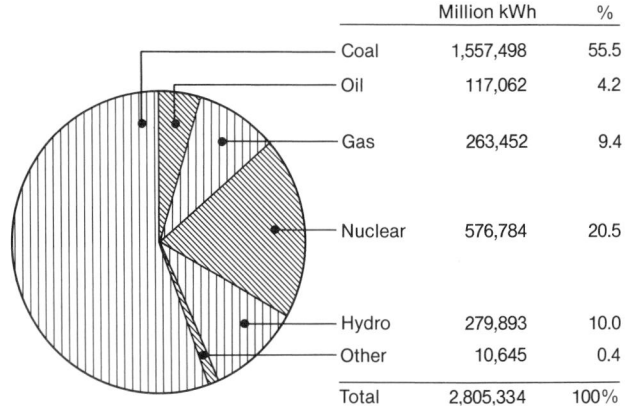

	Million kWh	%
Coal	1,557,498	55.5
Oil	117,062	4.2
Gas	263,452	9.4
Nuclear	576,784	20.5
Hydro	279,893	10.0
Other	10,645	0.4
Total	2,805,334	100%

Fig. 1 U.S. net electricity generation by energy source during 1990 — 2,805.3 billion kWh.

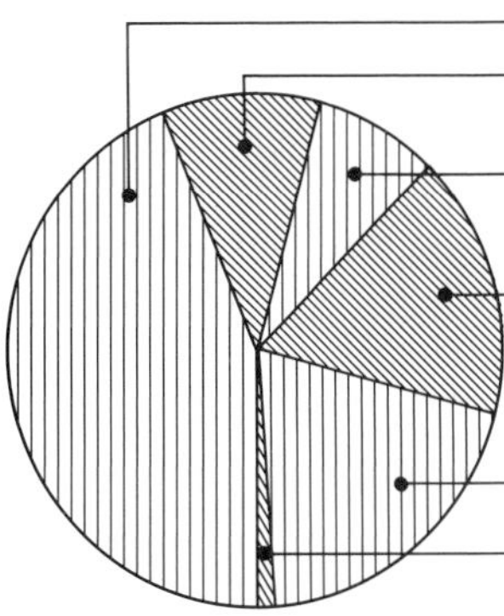

Fig. 2 Worldwide net electricity generation by energy source during 1988 — 10,537.6 billion kWh.

expected to continue its dominant role in supplying energy to new, base loaded utility power station boilers. With this in mind, the following material focuses primarily on coal-fired utility boilers but includes selected comments and discussion of other fossil fuels. General discussion of steam production as part of combined cycle and waste heat recovery applications is also provided in Chapter 31. Specialized designs for unique, smaller power production facilities are discussed in Chapter 25.

Coal Coal is the most abundant fuel in the U.S. and many parts of the world. However, of the major fossil fuels, it is also the most complicated and troublesome to burn. Its use involves unloading, storage and handling facilities; preparation before firing using crushers and pulverizers; ash disposal equipment; emissions control equipment and sootblowers. In the U.S., Europe, Japan, and increasingly in other parts of the world, there are also the societal and political difficulties in locating, siting and permitting of a coal-fired power plant producing airborne, liquid and solid emissions. There is a wide variation in the properties of coal and its ash. As a result, while the design of the steam producing unit must provide optimum performance when firing the intended coals, it must also accommodate reasonable alternate coals if necessary.

The coals selected for a prospective installation should be tabulated to indicate chemical analysis, heating value and grindability. In addition, each coal's ash analysis and ash temperature profile should be listed. The ash composition has a marked effect on a coal's slagging and fouling characteristics; the analysis must be provided to the boiler supplier for proper selection of unit geometry, features and surfaces. (See Chapter 20.)

Steam requirements

Modern turbine-generators are designed for nominal steam pressures from 1800 to 3500 psi (124 to 241 bar) at the turbine throttle with superheat and reheat temperatures from 950F (510C) to more than 1000F (538C).

The steam temperature specified at the superheater outlet is considered to be equal to that required at the turbine plus 5F (3C) to compensate for the loss in temperature in the steam lines. For optimum performance and turbine maintenance, the steam temperature should remain constant over a broad load range. Means for controlling superheat temperatures are required in utility practice. Steam temperature variation and control are discussed in Chapters 18 and 41.

The steam pressure at the superheater outlet is normally specified by the plant designer after evaluating the economics of pressure drop versus pipe sizing.

It has been customary stationary boiler practice in the U.S. to hold the main steam pressure constant for all loads, on the premise that this condition satisfies all pressure and quantity requirements of the steam-using equipment (constant pressure). This is particularly the case where the unit is base loaded and significant cycling operation is not anticipated. In addition, constant pressure operation permits more rapid load transients while minimizing transient stresses in the boiler. However, constant pressure operation assesses some heat rate penalty at part load conditions and results in some additional stresses in the turbines during rapid load changes.

Outside the U.S., mainly in Europe, the steam pressure varies with boiler load (variable pressure). This is more energy efficient at reduced load conditions and permits better temperature control to the turbine under load transients. However, the boiler may be less responsive and additional stress is placed on the unit under transient conditions due to the saturated water temperature change.

A hybrid design is also available with the unit pressure control divided between the primary and secondary superheaters (split pressure). The boiler can be maintained at a constant pressure thereby minimizing stress and maximizing response to load changes. The temperature of the steam entering the turbine can be separately maintained, minimizing stress and maximizing load response of the steam turbine. All three control approaches plus their implications are addressed in Chapters 18 and 41.

Central station units use the reheat cycle because of the increased station thermal efficiency compared to the nonreheat cycle. The reheater, located within the boiler enclosure, receives steam exhausted from an intermediate pressure stage of the turbine. The steam temperature is raised and the steam is returned to the low pressure section of the turbine.

The double reheat cycle, where steam is returned to the boiler twice for reheating and then delivered to low pressure sections of the turbine, results in even higher station thermal efficiency than the single reheat cycle. However, the gain in plant efficiency must be weighed against the cost of additional piping systems and turbine and boiler equipment used to accommodate the flow of low pressure steam.

Steam flow requirement

Steam producing equipment must be of sufficient capacity, range of output and responsiveness to ensure prompt response to the turbine steam demands. The demand may be steady, as in base loaded systems, or it may fluctuate widely and rapidly, as in cycling units. The steam flow requirements should, therefore, be accurately established for peak flow, maximum continuous flow (usual steady maximum flow), minimum flow and rate of change in flow. The peak load establishes the capacity of the steam producing equipment and all its auxiliaries.

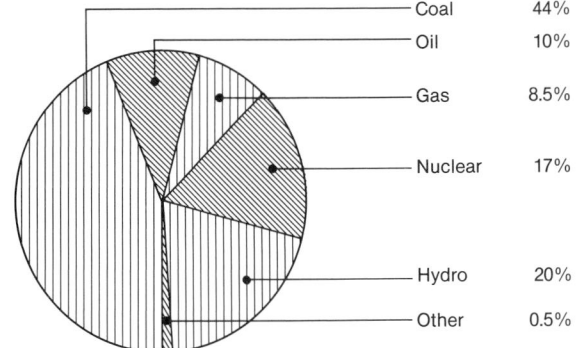

The two principal methods of obtaining peak ratings in utility cycles are by operating at 5% overpressure at the turbine throttle valve and by removing one or more feedwater heaters from service. The method used is dependent upon the steam turbine design parameters and limitations. For the condition of 5% overpressure at the superheater outlet, the steam generating equipment is designed to provide a steam flow 5% greater than the normal maximum continuous flow. Approximately 5% additional output from the turbine is obtained by passing this increased steam flow at higher pressure through the turbine.

The alternative method of obtaining peak ratings from the cycle is by removing feedwater heaters from service. Some of the steam normally supplied to feedwater heaters from intermediate points on the turbine is continued through the turbine, thereby increasing its output. This method results in higher than normal heat rates, i.e., lower efficiencies, while overpressure operation reduces heat rate.

The range from minimum to maximum output is an important factor in selecting the firing equipment. To maintain ignition and optimal combustion, this range must also be considered in designing the furnace.

The required rate of change in steam flow (load) may affect the entire design of the steam producing equipment. Within a specified load range, good response is easily obtained with well designed firing equipment for pulverized coal, oil and gas. A multiplicity of burners (and pulverizers, where used) widens the available range, but if there is a frequent change in load, remote manual on/off operation of burners and pulverizers may be objectionable. It is, therefore, important that the designer, manufacturer and operator understand what can be done by full automatic firing and what is required beyond that.

In selecting equipment to produce steam at minimum total cost, it is also necessary to establish the probable capacity factor — the ratio of the average output to the rated output. This capacity factor serves as a basis for establishing the value of incremental increases in the efficiency of steam generation. For example, the initial cost necessary to achieve high efficiency in a plant operating continuously at high output may not be justified for a plant operating as a cycling unit.

Boiler feedwater

Raw water from surface or subsurface sources invariably contains in solution some degree of troublesome scale-forming materials, free oxygen and sometimes acids. Because good water conditioning is essential in the operation of any steam cycle, these impurities must be removed.

Dissolved oxygen will attack steel and the rate of attack increases sharply with a rise in temperature. High chemical concentrations in the boiler water and feedwater cause furnace tube deposition and allow solids carryover into the superheater and turbine, resulting in tube failures and turbine blade deposition or erosion.

As steam plant operating pressures have increased, the water treatment system has become more critical. This has led to the installation of more complete and refined water treatment facilities. (See Chapter 42).

The temperature of the feedwater entering an economizer should be high enough to prevent condensation and acid attack on the gas side of the tubes. Dew point and rate of corrosion vary with the sulfur content of the fuel and with the type of firing equipment.

Environmental considerations

Protection of the environment is an integral part of the boiler and power system design process. As outlined in Chapters 1 and 32, boilers emit varying levels of NO_x, SO_2 and particulate into the air as well as discharging ash. Government regulations tightly control these emissions in most parts of the world. The specific local regulations will govern the extent of back-end cleanup equipment required and many of the boiler design parameters. A new unit burning coal in the U.S. today might look like the unit shown in Fig. 3. The boiler is designed with low NO_x burners, an enlarged furnace, and overfire air or NO_x ports to minimize NO_x emissions. (See Chapters 13 and 34.) An electrostatic precipitator or fabric filter collects particulate (Chapter 33), while a wet flue gas desulfurization system or scrubber removes most of the SO_2 (Chapter 35). Where required, supplemental NO_x control equipment, such as a selective catalytic reduction (SCR) system discussed in Chapter 34, may also be added. Careful review and economic evaluation of all local regulations and permitting requirements are needed to provide the required environmental protection in the most cost effective manner.

Space and geographical considerations

Space and geographical considerations are extremely site specific. In some cases, a new boiler is added to an existing plant to maximize the use of existing equipment and structures. The most extreme case involves *repowering* an existing plant, where the old boiler is removed and a newer unit is installed in the same space. In many cases, a modern high efficiency system can double the steam output of older units.

The boiler and its support structures must be designed to accommodate expected earthquake loads, wind loadings, and local structural laws and regulations. The site elevation above sea level significantly affects the design and sizing of ducts, fans and the stack. Expected power consumption will also vary. Finally, boilers can be designed for indoor or outdoor installation, depending upon the prevailing climate.

Power for driving auxiliaries

Modern practice in central stations generally calls for electric motor drives for rotating auxiliaries such as pumps, fans, pulverizers and crushers. The convenience and cost of the electrical drive substantiate this preference. Where variable speed may be desirable or where the steam generating unit is large, a steam turbine is frequently chosen for an auxiliary drive, such as for feed pumps or forced draft fans. The combined rating of all auxiliaries in a modern plant may amount to 15% of the main unit rating.

Guarantees

In power plant design the designer relies on sound engineering fundamentals. However, it is common practice to obtain performance guarantees from component suppliers to validate the expected performance of the unit.

The boiler manufacturer is usually requested to provide the following guarantees, depending on the arrangement, type of fuel and type of boiler.

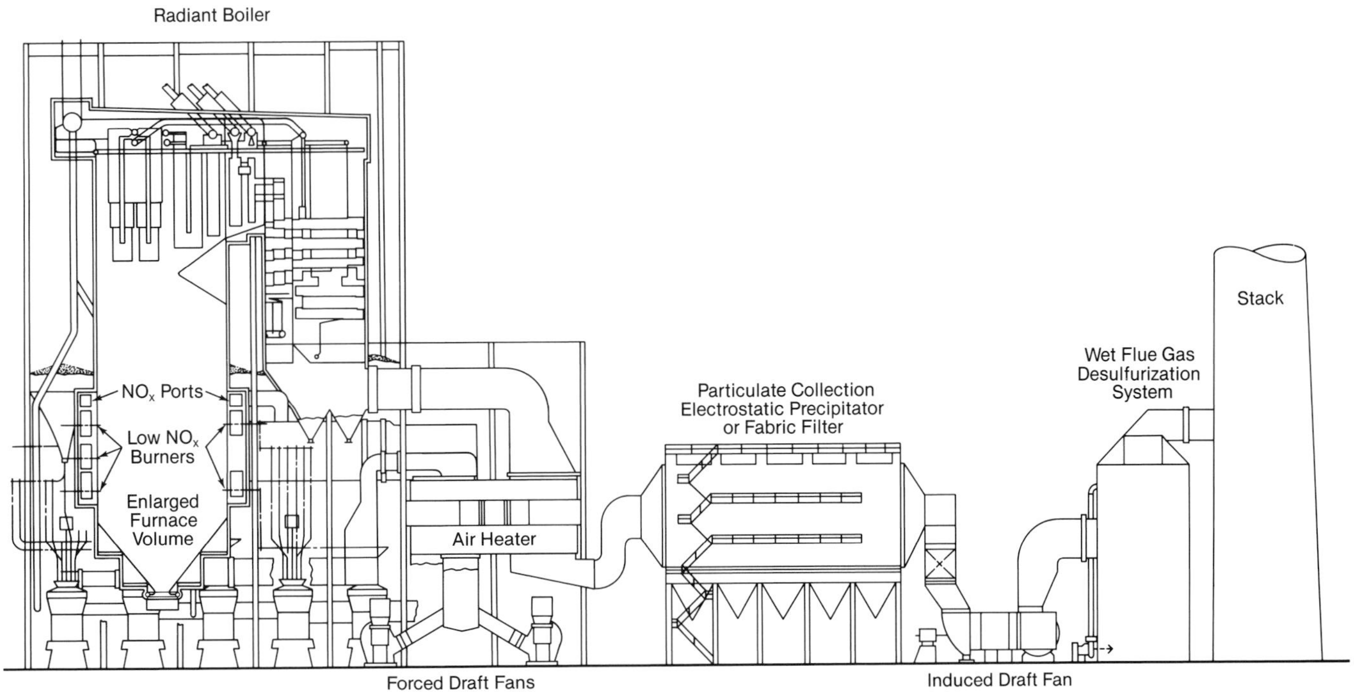

Fig. 3 Integrated pulverized coal-fired boiler power system with emissions control.

1. For a given load point and fuel:
   a. efficiency,
   b. superheater steam temperature,
   c. reheater steam temperature,
   d. pressure drop,
   e. total air and gas resistance, and
   f. solids in steam.
2. Unit capacity.
3. Superheater and reheater control ranges.
4. Pulverizer (where applicable):
   a. fineness, and
   b. power requirements.
5. NO_x emissions.

Pulverized coal firing

The size of large pulverized coal-fired boilers and turbine-generators has leveled off at 1300 MW. The equipment can be designed to burn practically any bituminous coal, subbituminous coal or lignite commercially available. Anthracite can be successfully burned in pulverized form, but the special attention and additional expenses associated with plants designed for this fuel preclude its use in most modern units.

The overall aspects of pulverized coal firing, as applied to boiler units, are as follows:

1. is suitable for almost any coal mined throughout the world,
2. is economically suitable for a very wide range of boiler capacities,
3. provides wide flexibility in operation and high thermal efficiency,
4. must have proper coal preparation and handling equipment, including moisture removal,
5. must have proper means of handling the ash refuse, and
6. must have controls for atmospheric emissions arising from elements in the coal and from the combustion process.

Babcock & Wilcox (B&W) pulverized coal-fired boiler types for electric power

Prior chapters provide information on steam generation and power plant design and include fundamentals that are applied by designers and manufacturers worldwide. In particular, Chapter 18 discusses general boiler design, while Chapters 12 and 13 address the preparation and combustion of pulverized coal. The following portion of this chapter will depart from that approach and instead concentrate on the B&W design philosophy as it applies to large reheat steam generators for electric power generation.

B&W has a broad base of experience which can be applied to most customer needs. B&W's basic design philosophy focuses on the key operating issues of availability and reliability, and on incorporating technology advances into each new unit design. Examples of the company's strengths include rapid startup and load shedding features, extended control ranges for superheater and reheater outlet temperatures, and experience in burning a wide variety of coals.

Radiant boiler

The Radiant boiler (RB) is so named because the heat absorption of a saturated surface is largely by radiant energy transfer. Its components are pre-engineered with enough flexibility to adapt the design to various fuels and a broad range of steam conditions.

In modern units, the furnace is a completely water-cooled, balanced draft design and usually features dry ash removal. Superheater and reheater surfaces are of the vertical pendant and/or horizontal designs. Superheat temperature control is accomplished by an attemperator. Reheat temperature control is by gas proportioning dampers or excess air. Economizers and air heaters are required to obtain an acceptable efficiency. The size of the unit dictates field assembly, although many component parts are shop-assembled.

Range in capacity, steam output — about 300,000 lb/h (37.8 kg/s) to a maximum that may exceed 7,000,000 lb/h (882 kg/s).

Pressure — subcritical, usually 1800 to 2400 psi (124 to 165 bar) throttle pressure with 5% overpressure capability.

Superheater and reheater outlet temperatures — as required, usually 1000F (538C).

Radiant boiler for pulverized coal — Carolina

The steam generating unit shown in Fig. 4 is a balanced draft-fired Carolina-type Radiant boiler (RBC). It is arranged with a water-cooled dry-bottom furnace, and superheater, reheater, economizer and air heater components. The unit is designed to burn coal, usually pulverized to a fineness of which at least 70% passes through a 200 mesh (75 micron) screen. The horizontal convection pass and vertical pendant heat transfer surfaces provide the benefits of: 1) high temperature supports outside of the flue gas stream, 2) minimal motion between the boiler roof penetrations, and 3) more control over pendant section spacing in the design. The lower furnace height compared to other pulverized coal designs also reduces structural steel and erection costs.

Fuel flow Raw coal is discharged from the feeders to the pulverizers. The pulverized coal is transported by the primary air to the burners through a system of pressurized fuel and air piping.

Air and gas flow Air from the forced draft fans is heated in the air heaters, then routed to the windbox where it is

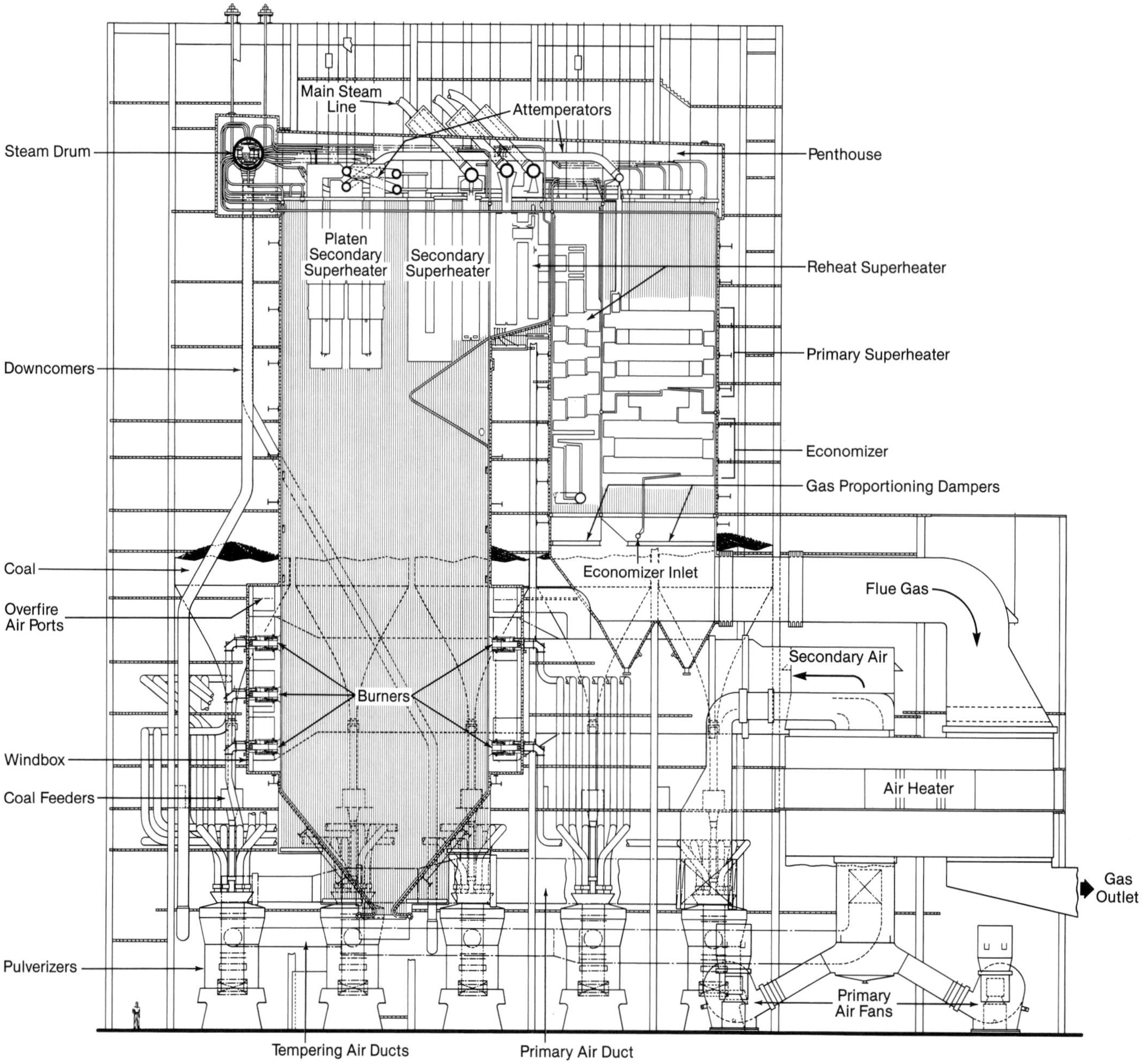

Fig. 4 455 MW Carolina-type Radiant (RBC) boiler for pulverized coal firing.

distributed to the burners as secondary air. In the arrangement shown in Fig. 4, high pressure fans provide air from the atmosphere to a separate section of the air heater known as the primary section. A portion of the air from the primary fans is passed unheated around the primary air heater as tempering primary air. Controlled quantities of preheated and tempering primary air are mixed before entering each pulverizer to obtain the desired pulverizer fuel-air mixture outlet temperature. The primary air is used for drying and transporting fuel from the pulverizer through the burners to the furnace.

Hot gas from the furnace passes successively across the finishing banks of the superheater and reheater. Before exiting the boiler, the gas stream is divided into two parallel paths — one gas stream passing over a portion of the reheater and the other stream passing over a portion of the superheater. Proportioning the gas flow between these two paths as unit load changes provides an economical tool for reheat steam temperature control. The flow quantities are adjusted by a set of dampers at the boiler exit. A controlled amount of gas passes over a portion of the reheater to obtain the reheater steam temperature set point and the remaining gas travels a parallel path across the superheater heating surface. Attemperators are provided in the reheater and superheater systems. The reheater attemperator is used during transient loads and upset conditions, with reheater spray quantities held to a minimum to maximize cycle efficiency. Superheater spray is used to maintain main steam temperatures. This arrangement of convection surface and damper system provides extended capability to obtain design steam temperatures in the superheater and reheater over a broad load range.

The gases leaving the superheater and reheater sections of the convection pass cross the economizer, pass to the air heater(s) and then travel to the appropriate environmental control equipment.

Water and steam flow Feedwater enters the bottom header of the economizer. The water passes upward through the economizer tube bank into support tubing that is located between tube rows of the primary superheater. The heated feedwater is collected in outlet headers at the top of the unit, then routed into the steam drum. As shown in Fig. 5, natural circulation provides water flow from the steam drum down through large diameter downcomer pipes to multiple supply distributor tubes or *supplies* connecting to the individual lower furnace headers. The fluid rises as it is heated in the furnace tubes and passes through riser tubes back into the steam drum.

The mixture entering the steam drum is separated into steam and water flows by cyclone separators. These provide essentially steam-free water to the downcomer inlets. The steam is further purified by passing it through the primary and secondary steam scrubbers as discussed in Chapter 5.

As illustrated in Fig. 6, steam from the steam drum passes through multiple connections to a header supplying the furnace roof and convection pass rearwall tubes, which directly connect to the primary superheater inlet header. Steam leaving the drum is also routed to a series of headers, which provide cooling steam for the convection pass sidewalls, before being directed to the primary superheater inlet header.

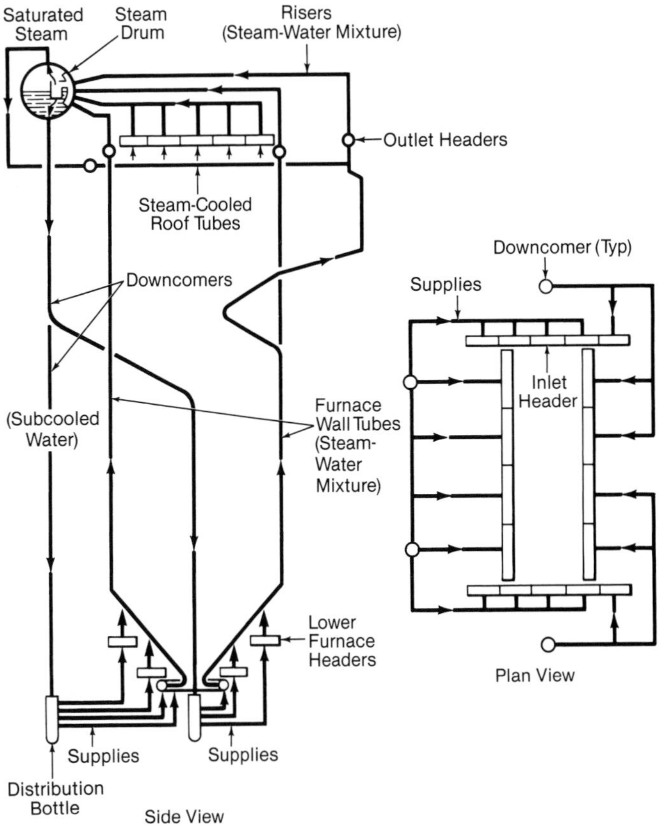

Fig. 5 Typical furnace circulation diagram for a Radiant boiler (RB).

As shown in Fig. 7, the steam rises through the primary superheater, discharges to its outlet header and flows through connecting piping equipped with a spray attemperator. The partially superheated steam then enters the secondary superheater and flows through the various superheater sections to its outlet headers and discharge pipes, which terminate at points outside of the unit penthouse. The superheated steam is directed to the high pressure section of the steam turbine. After partial expansion (see Chapter 2), the low pressure steam is returned to the boiler for reheating.

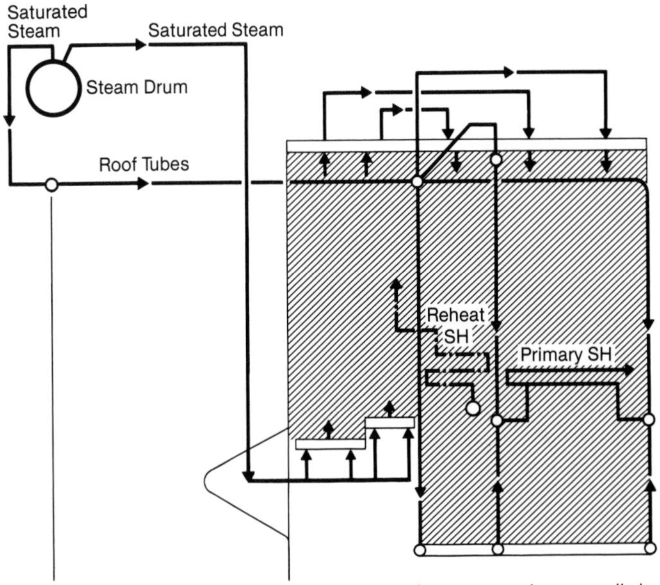

Fig. 6 Typical steam-cooled roof and convection pass enclosure wall circuits for a Radiant boiler (RB).

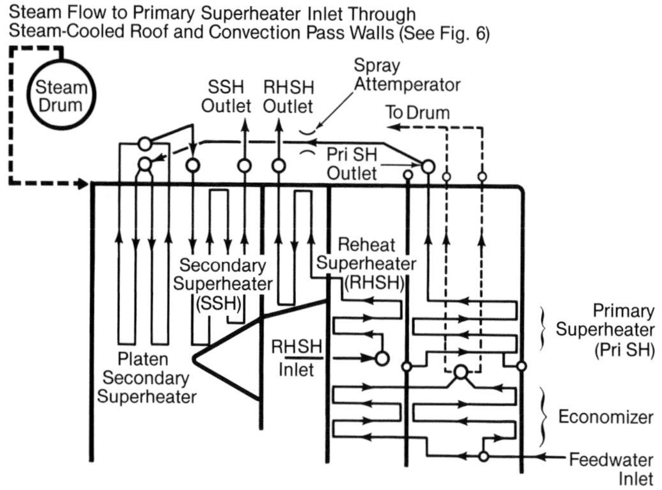

Fig. 7 Superheater, reheater and economizer circuits for a Radiant boiler.

The low pressure steam is reintroduced to the boiler at the reheater inlet header (RHSH inlet) and flows through the reheater tube bank to the reheater outlet header (RHSH outlet). Reheated steam is then routed to the intermediate pressure and then low pressure sections of the steam turbine-generator set.

Radiant boiler for pulverized coal — Tower

An alternate configuration of the Radiant boiler is the Tower design (RBT) shown in Figs. 8 and 9. A special feature of this design is that all surfaces are drainable. This feature was initially designed for service in Northern latitudes to provide freeze protection. The Tower design may be equipped with gas recirculation and tempering for extended superheater and reheater temperature control, and it has been used in more temperate climates where the economic balances of limited shop labor and extensive field labor dictate its selection. Additional advantages of the Tower design include uniform gas flow entering all convection sections, minimum tube bend erosion potential because of no convection pass turns, and removal of most of the ash and slag through the main furnace hopper.

Tower boiler components are also pre-engineered and provide sufficient flexibility to adapt the design to various fuels and a broad range of steam conditions.

General characteristics of the design are water-cooled furnace, balanced draft, natural circulation, top supported and drainable. The draining feature is accomplished by placing all steam- and water-contained surfaces in the horizontal position so that they may be completely drained when the unit is taken out of service.

Fuel flow Similar to the RBC, the RBT is arranged so that raw coal is discharged from the feeders to the pulverizers. Primary air is used to transport the pulverized coal from the pulverizers to the burners through a system of pressurized fuel and air piping.

Air and gas flow Air is introduced to the RBT in the same pattern as the RBC. The geometric differences of the two units create varying paths for gas flow. The hot gases from the furnace pass vertically up and over the superheater and reheater banks located in the upper section of the boiler. The flue gas is then directed down to the economizer located outside of the boiler setting, finally passing to the air heater system and then to the environmental control equipment for cleanup.

Water and steam flow Feedwater is introduced to the boiler system through the economizer where it is heated and discharged through a header/piping system to the steam drum. With this natural circulation system, the furnace circuitry follows the same general flow patterns as the RBC described above.

Saturated steam from the steam drum is directed through a series of pipes down to the outlet of the furnace, where a small bank of primary superheater tubes (primary superheater 1) is placed with its outlet tubing forming the internal steam-cooled support structure for the other banks of surface located farther up in the gas stream. (See Fig. 8.)

Steam from the support tubing is collected and directed to the finishing banks of primary superheater 2. The primary and secondary superheaters are connected with piping that contains a spray attemperator for superheater steam temperature control. Steam enters the secondary superheater and flows through the tube sections to the outlet header, which discharges to a terminal located outside of the unit. The superheated steam is directed to the high pressure section of the steam turbine. After partial expansion, low pressure steam is then returned to the boiler for reheating.

Steam returned to the boiler is introduced at the reheater inlet header. The steam flows through the reheater tube bank and is then routed to the outlet header and steam piping connection before being sent to the intermediate and low pressure sections of the steam turbine-generator.

Universal Pressure boiler

The Universal Pressure (UP) boiler, sometimes referred to as the once-through boiler, can be designed for all commercial temperatures and pressures, subcritical

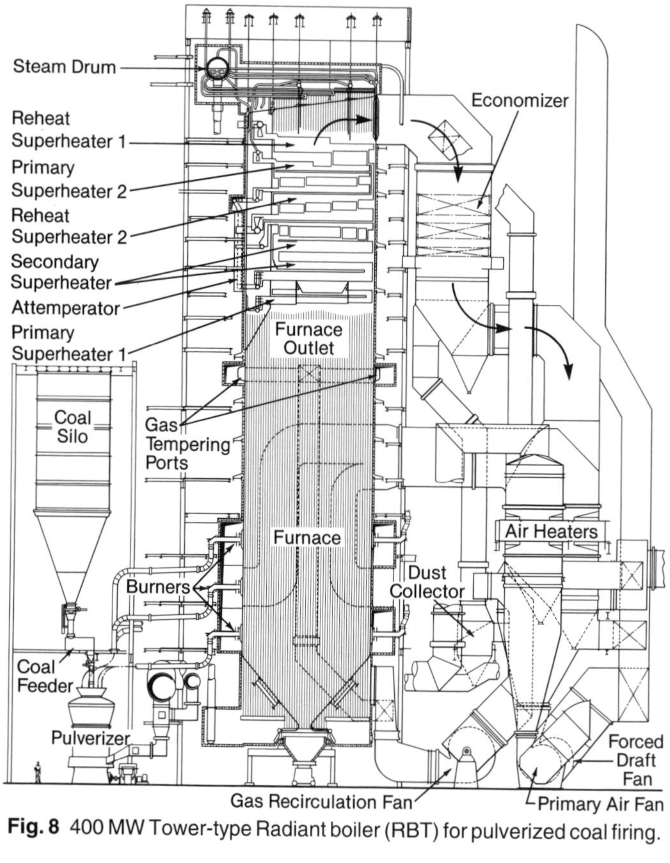

Fig. 8 400 MW Tower-type Radiant boiler (RBT) for pulverized coal firing.

Fig. 9 This plant in South East Asia now features four (two shown) 400 MW Tower units designed for indigenous coal and oil.

and supercritical. It is most economically attractive in the larger sizes at the supercritical range.

Range in capacity, steam output — about 300,000 lb/h (38 kg/s) to more than 10,000,000 lb/h (1260 kg/s).

Operating pressure — subcritical, usually at 2400 psi (166 bar) throttle pressure, and supercritical usually at 3500 psi (241 bar) throttle pressure with 5% overpressure.

Superheater steam temperatures — as required, usually 1000F (538C). Constant main steam temperature can be maintained to minimum load.

The principle of operation is that of the once-through or Benson cycle. The water, pumped into the unit as a subcooled liquid, passes sequentially through all the pressure part heating surfaces where it is converted to superheated steam as it absorbs heat; it leaves as steam at the desired temperature. There is no recirculation of water within the unit and, for this reason, a conventional drum is not required to separate water from steam.

The furnace is completely fluid-cooled and is usually designed for balanced draft operation. Heat transfer surface for single or two-stage reheat may be incorporated in the design. (See Chapter 2.)

Firing rate, feedwater flow, superheater division valves and turbine throttle valves are coordinated to control steam flow and pressure. Superheater steam temperature is controlled by coordinating firing and pumping rate. Reheater steam temperature is controlled by gas proportioning dampers, gas recirculation, excess air and attemperation, separately or in combination.

The UP boiler is designed to maintain a minimum flow inside the furnace circuits to prevent furnace tube overheating during all operating conditions. This flow must be established before startup of the boiler. A bypass system, integral with the boiler, turbine, condensate and feedwater system, is provided. This system assures that the minimum design flow is maintained through pressure parts that are exposed to high temperature combustion gases during the startup operations and at other times when the required minimum flow exceeds the turbine steam demand. (See Chapter 18).

Universal Pressure boiler for pulverized coal

The steam generating unit shown in Fig. 10 is a balanced draft B&W Universal Pressure coal-fired boiler (UPC), comprising a water-cooled dry-bottom furnace and superheater, reheater, economizer and air heater components. The unit is designed to fire coal usually pulverized to a fineness of at least 70% through a 200 mesh (75 micron) screen. The B&W UPC unit is particularly suited for high intermediate or base load duty.

Fuel flow Raw coal is discharged from the feeders to the pulverizers, which can be located at the front or sides of the unit. Pulverized coal is transported by the primary air to the burners through a system of pressurized fuel and air piping.

Air and gas flow The air flow arrangement and routing are similar to that used on the RBC previously described.

As noted in Fig. 10, hot gases leaving the furnace pass successively across the fluid-cooled surface at the top of the furnace (wingwalls), the secondary superheater and the pendant reheater, which are located in the convection pass out of the high radiant heat transfer zone of the furnace. The gas turns downward (convection pass) and crosses the horizontal primary superheater, horizontal reheater and economizer before passing to the air heaters.

Water and steam flow Feedwater enters the bottom header of the economizer and passes upward through the economizer to the outlet header. It is then piped to the lower furnace area from which multiple connecting pipes (supplies) are routed to the lower furnace headers.

From the furnace wall headers the fluid is then passed upward through the vertical furnace tubes. A set of headers is provided partway through the furnace to mix the fluid from all of the furnace tubes. This equalizes the temperature of the fluid before it completes the furnace tube pass and exits to the upper furnace wall headers. The nonuniform fluid temperature arises from local variations of absorption throughout the furnace and must be equalized to prevent structural problems in the furnace walls and outlet headers.

After discharging to the upper furnace wall headers, all of the fluid is piped to the front roof header, then through the roof tubes to the rear roof headers where mixing again takes place. It is then passed through a pipe distribution system to the convection pass enclosure wall lower headers. The fluid flows up through the wall tubes and the superheater screen. From here, pipes then convey the fluid to a common header and then to the primary superheater inlet header.

The fluid is collected and partially mixed before entering the primary superheater, then partially mixed again as it flows from the primary to the secondary superheater. Side to side crossover connections between primary and secondary superheaters reduce temperature

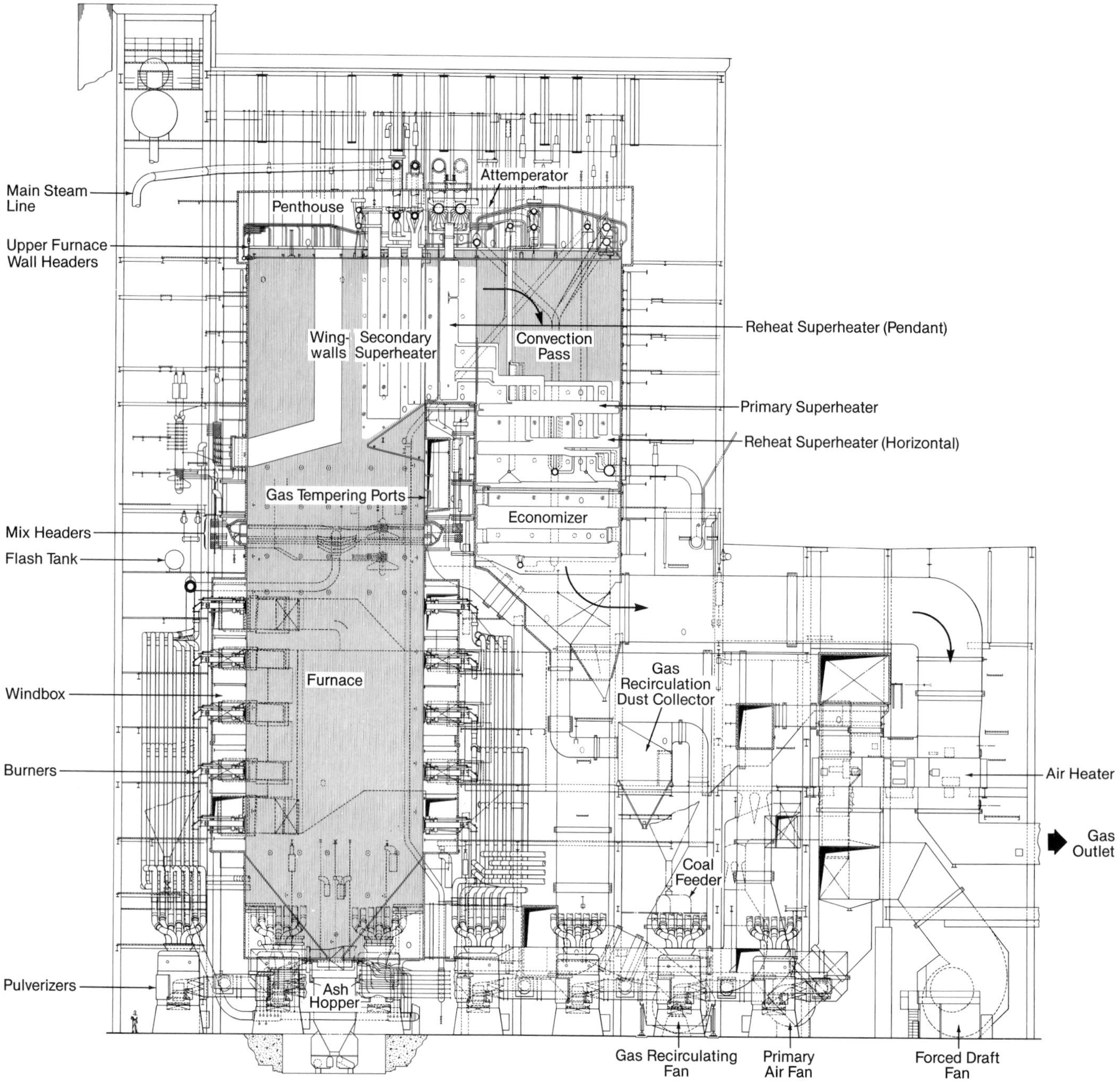

Fig. 10 1300 MW Universal Pressure (UPC) boiler for pulverized coal firing.

imbalances. The superheated steam is then partially expanded through the high pressure turbine.

The low pressure steam from the high pressure turbine outlet is then introduced to the horizontal reheater inlet header. It then flows through the tubes in the horizontal and pendant reheater sections to the outlet headers before passing to the intermediate and low pressure turbines. In some cases, a second reheat stage is added if economically viable.

As discussed in Chapter 18, tempering can be used for gas temperature control, while gas recirculation or excess air are used to control the furnace absorption.

Spiral Wound UP boiler for pulverized coal

The B&W Spiral Wound Universal Pressure (SWUP) boiler, another type of UPC unit, may be used where utility service loads require rapid startup and load-following capability. This boiler type is in all aspects similar to the UPC boiler, except for the configuration of the furnace circuitry. The SWUP geometry, using once-through technology, includes the tubes which wrap around the furnace box as they gain elevation and gradually absorb heat. (See Fig. 11.) Therefore, the water introduced from the economizer piping is heated at essentially the same rate to the same temperature, minimizing thermal upsets which restrain rapid load change. The capability of this design to operate at variable pressure is further enhanced by using a startup and bypass system specifically designed for rapid load change. (See Chapter 18.)

This design is used in Europe and Japan, where units are extensively cycled. B&W has supplied the only units of this type in the U.S. This design is particularly attractive for smaller, once-through boiler designs where furnace cooling, especially at low loads, may be a more critical design issue. When combined with appropriate startup and recirculation systems, the SWUP boiler is capable of full variable pressure operation over a wide load range (15 to 100% load). The spiral furnace design reduces the number of tubes in the furnace, which increases the mass flux or velocity in each tube and provides adequate cooling under all load conditions. At the same time, the high per-tube flow rate also increases the pressure drop and associated pumping power compared to designs with vertical tubes. In addition, the inclined furnace tubes can not support the furnace walls alone as in vertical tube designs. Therefore a supplemental support system is required.

Oil- and gas-fired utility boilers

The use of oil and gas as fuels for new utility boilers has declined except for selected locations and conditions. This is due in part to the price of these premium fuels, their availability and government regulations. At the same time, advancements in gas turbine and gas turbine combined cycle systems (Chapter 31) have made the use of oil and gas in these systems more cost effective.

However, in selected cases, the proposed boiler location, fuel availability, local price or existing facility constraints may make oil or gas the preferred fuel. In addition, low sulfur oil- and natural gas-fired new utility boilers have the distinct advantage of low environmental emissions; NO_x, SO_2 and particulate emissions are minimal in these units.

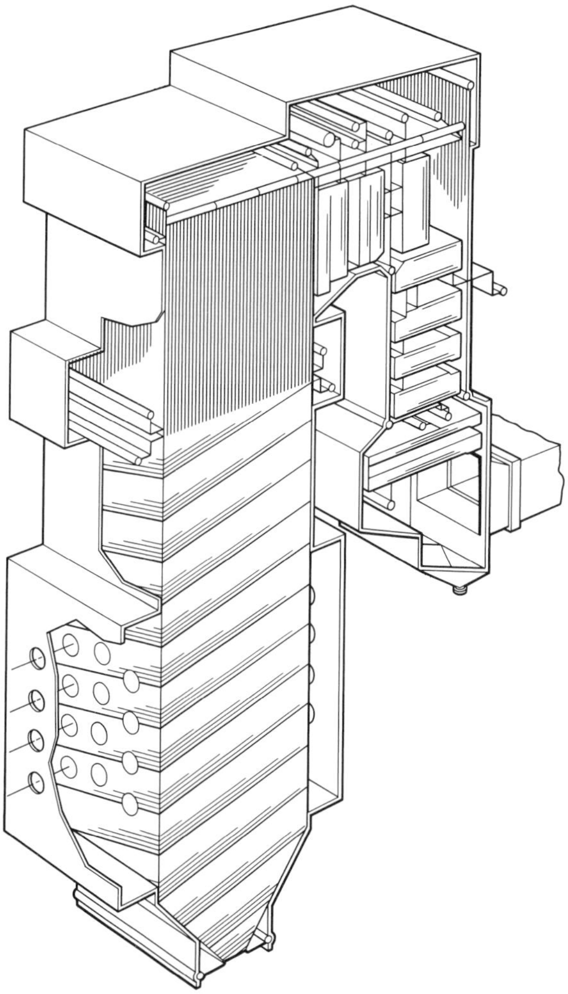

Fig. 11 Schematic of a Spiral Wound Universal Pressure (SWUP) boiler showing tube furnace waterwall arrangement.

Boilers designed to burn these fuels can take unique advantage of the relatively clean burning fuel characteristics as compared to coal and other solid fuels.

Range in capacity, steam output — about 300,000 lb/h (37.8 kg/s) to a maximum that may exceed 7,000,000 lb/h (882 kg/s).

Pressure — subcritical, usually 1800 to 2400 psi (124 to 165 bar) throttle pressure with 5% overpressure capability.

Superheater and reheater outlet temperatures — as required, usually 1000F (538C).

Fuels

Natural gas Natural gas (Chapter 10) has the fewest design restrictions of the major fuels because it is clean and easy to burn. If only natural gas is burned, fuel storage facilities, ash hoppers, ash pits and ash handling equipment are unnecessary. Sootblowers can be omitted and dust collectors are not needed. Control of heat input to the boiler furnace is simplified. Heating surfaces can be arranged for optimum heat transfer and draft loss without consideration of ash deposits and erosion. The total enclosure volume is at a minimum and the adaptability for outdoor service is increased.

Fuel oil Fuel oil (Chapter 10) has many of the desirable features of natural gas, including ease of handling and elimination of ash hoppers and ash pits. However, it requires storage, heating and pumping facilities. Oils with high sulfur and vanadium contents can cause troublesome deposits on surfaces throughout the unit. (See Chapter 20.) These deposits can be minimized by arranging the heating surfaces for optimum cleaning by sootblowing equipment. Provision should be made for water washing the furnace and all convection surfaces when the unit is shut down for maintenance. Air heater protection devices (a steam coil or hot water coil) should be installed to prevent gas condensation and acid attack on the gas side of the air heater surfaces. (See Chapter 19.)

B&W Radiant boiler for natural gas and oil

The steam generator shown in Figs. 12 and 13 is an El Paso-type Radiant boiler (RBE) unit. It is arranged with a water-cooled hopper-bottom furnace and superheater, reheater, economizer and air heater components. The unit is designed to use natural gas and oil separately or in combination. The general characteristics of the Radiant boiler are discussed in *Pulverized Coal Firing*. The unique aspects of the El Paso design are the elimination of the pendant convection pass and the inclusion of the upflow and downflow convection passes within the footprint occupied by the boiler furnace. The flow rates and pressure ranges for the coal-fired Radiant boiler are also applicable here.

Air and gas flow Air from the forced draft fan is heated in the air heater and is distributed to the burner windbox. Hot gas from the furnace passes successively over the horizontal secondary superheater and reheater sections and one bank of the primary superheater. The gas then turns, flows downward and crosses the remainder of the primary superheater and economizer sections. It then travels out of the water-cooled enclosure to the air heater.

Water and steam flow Feedwater enters the bottom header of the economizer. The water flows upward through the economizer and discharges through the outlet header into piping that directs it to the steam drum. By means of natural circulation, water flows downward through downcomer pipes, then through supply distributor tubes to the lower furnace headers. It then rises through the furnace tubes (which enclose the convection area) and flows to the upper headers and through connecting riser tubes into the steam drum.

The steam-water mixture in the steam drum is passed through cyclone separators, which provide essentially steam-free water for the downcomers. The steam is further purified by passing it through the primary and secondary steam scrubbers within the drum.

Steam from the drum passes through multiple connections at the rear of the convection enclosure into the primary superheater inlet header.

The steam rises through part of the primary superheater and, by means of connecting tubes, is directed into

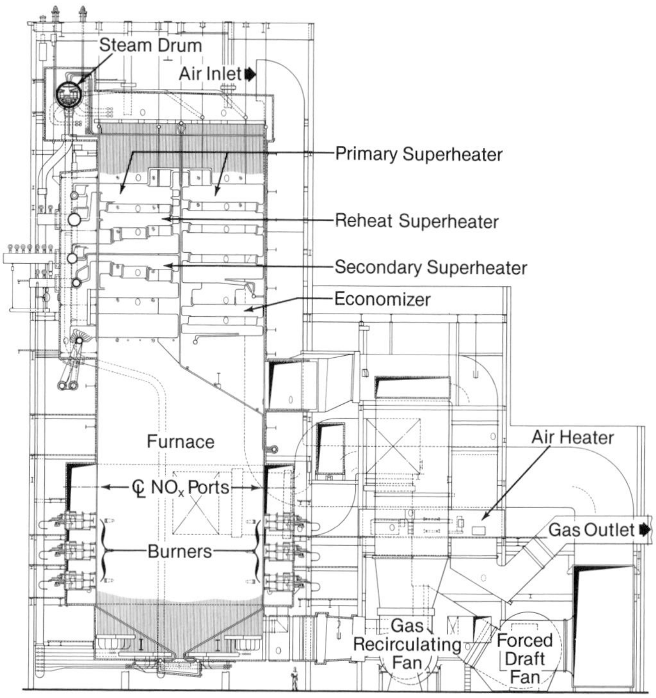

Fig. 12 El Paso-type Radiant (RBE) boiler for oil and gas firing.

Fig. 13 This facility includes a 125 MW oil- and gas-fired El Paso-type unit (left), as well as a 350 MW Carolina-type boiler that burns coal, oil and refuse-derived fuel.

the remainder of the primary superheater in the first gas pass. The flow discharges through the primary superheater outlet header into connecting piping equipped with a spray attemperator. The steam enters the secondary superheater inlet header and flows through the secondary superheater sections to the outlet header and to a discharge pipe, which terminates outside the casing at the front of the unit.

Low pressure steam, introduced to the reheater inlet header, flows through the reheater sections and out through the reheater outlet header to a connection terminating outside the casing, also at the front of the unit.

Type PFI Integral Furnace boiler installed in an industrial plant.

Chapter 25
Fossil Fuel Boilers for Industry and Small Power

Most manufacturing industries require steam for a variety of uses. Basic plant heating and air conditioning, prime movers such as turbine drives for blowers and compressors, drying, constant temperature reaction processes, large presses, soaking pits, water heating, cooking and cleaning are all examples of how steam is used.

Industrial steam can also be used to generate electricity in a cogeneration mode which uses a conventional steam turbine for electric power generation and low pressure extraction steam for the process. The electricity is then used by the plant or sold to a local electric utility company. As an alternate cogenerating system, a gas turbine can be used for power generation with a heat recovery steam generator for steam.

Many thousands of boilers are installed in industrial and municipal plants, providing lower pressure and temperature steam than utility boilers dedicated to large, central station electric power generation. In an industrial plant, the dependability of steam generating equipment is critical. Most often the industrial operation has a single steam plant with one or more boilers. If the steam flow is interrupted, production can be seriously impacted. Accordingly industrial boilers must be very reliable because plant productivity relies so heavily on their availability. Loss of a boiler for a short time can stop production for days if, for example, materials cool and solidify in process lines. For this reason, some industries prefer multiple smaller units.

The principles governing the selection of boilers and related equipment are discussed in Chapter 37. Proper equipment selection can be accomplished only in the framework of a sound technical and cost evaluation. This requires a working knowledge and understanding of the performance of the different components of the steam generating unit under various conditions, including the significance of the many different arrangements of heat absorbing surfaces, the characteristics of available fuels, combustion methods and ash handling. The owner must also establish the present and future steam conditions and requirements. All pertinent environmental regulations must also be considered. A brief summary of boiler specifications is provided in Table 1.

Table 1
Typical Industrial Boiler Specification Factors

1. Steam pressure
2. Steam temperature and control range
3. Steam flow
   - Peak
   - Minimum
   - Load patterns
4. Feedwater temperature and quality
5. Standby capacity and number of units
6. Fuels and their properties
7. Ash properties
8. Firing method preferences
9. Environmental emission limitations — sulfur dioxide (SO_2), nitrogen oxides (NO_x), particulate, other compounds
10. Site space and access limitations
11. Auxiliaries
12. Operator requirements
13. Evaluation basis

Industrial boiler design

Industrial boilers generally have different design characteristics than utility boilers. These are most apparent in steam pressures and temperatures as well as the fuel burning equipment.

Industrial units are built in a wide range of sizes, pressures and temperatures — from 2 psig (1.2 bar gauge) and 218F (103C) saturated steam for heating to 1800 psig (125 bar gauge) and 1000F (538C) for plant power production.

In addition, industrial units often supply steam for more than one application. For some applications steam demand may be cyclic or fluctuating, thereby complicating unit operation and control of the equipment.

Babcock & Wilcox (B&W) industrial boilers, such as the unit shown in Fig. 1, generally rely upon natural circulation for steam-water circulation. Exceptions include bubbling fluidized-bed retrofit units and steam flood boilers.

Most utility boilers are designed to burn pulverized or crushed coal, oil, gas, or a combination of oil or gas with a solid fuel. Industrial boilers can be designed for the above fuels as well as coarsely crushed coal for stoker firing.

Many industrial processes generate byproducts which can serve as boiler fuels, significantly contributing to the plant operating efficiency and effectively reducing product cost. Examples of these are gas products from the steel industry (blast furnace and coke oven gas), products from the petroleum industry [carbon monoxide (CO), refinery gas, petroleum coke], products from agriculture (sugar mill bagasse, peanut hulls, coffee grounds), waste from the pulp and paper industry (wood, bark, process chemicals, sludge) and municipal solid waste. Steam generation and fuel handling for some of these fuels and applications have become quite specialized. The following chapters are devoted to these units:

Chapter 26 — Chemical and Heat Recovery in the Paper Industry
Chapter 27 — Waste-to-Energy Installations
Chapter 28 — Wood and Biomass Installations
Chapter 30 — Marine Installations
Chapter 31 — Combined Cycles, Waste Heat Recovery and Other Steam Systems.

One of the distinguishable features of most industrial boilers is a large saturated water *boiler bank* between the steam drum and lower drum. (See Figs. 1 and 4.) The boiler bank serves the purpose of preheating the inlet feedwater to the saturation temperature and then evaporating the water while cooling the flue gas to a cost effective exit temperature.

In lower pressure boilers, insufficient heating surface is available in the furnace enclosure to absorb all of the energy needed to accomplish this function. Therefore, a boiler bank located downstream of the furnace and superheater (if included) in the flue gas stream provides the rest of the necessary heat transfer surface.

As shown in Fig. 2, as pressure increases, the amount of heat absorption to evaporate the water declines rapidly, and absorption for superheating and water preheating increases. In some modern very high pressure industrial units, a smaller boiler module separate from the steam drum provides the same function as the traditional boiler bank (see Fig. 5) but at a lower cost.

A separate economizer and/or air heater can be used downstream of the boiler bank to further reduce the flue gas exit temperature to an economic value.

Steam requirements

To assure prompt fulfillment of all steam demands, i.e., the delivery of heat to all points of use at the required rates, it is necessary to select steam producing equipment of sufficient capacity, range of output and responsiveness. The demand may be steady, as in most space heating systems, or it may fluctuate widely and rapidly, as in a heavy forging plant. Many steam heated processes, as in the initial heating of a liquid batch, require high peak flows of short duration. Rapidly changing rates of steam flow characterize requirements to produce the electrical power to drive a steel rolling mill. The steam flow requirements should, therefore, be accurately established for a number of conditions to ensure that the boiler system selected will meet all of its demand conditions, i.e., peak flow, maximum continuous flow (usual steady maximum flow), minimum flow, and rate of change in flow.

The peak load will, of course, establish the top capacity for the steam producing equipment and all of its auxiliaries. For widely fluctuating loads it is advisable to establish the 15 minute peak. In most systems, peaks of shorter duration can be met by the *storage* of heat inherent in the steam generating equipment.

Steam for process and heating The pressure of saturated steam (no superheat) used for process heating is such that the corresponding condensing steam tempera-

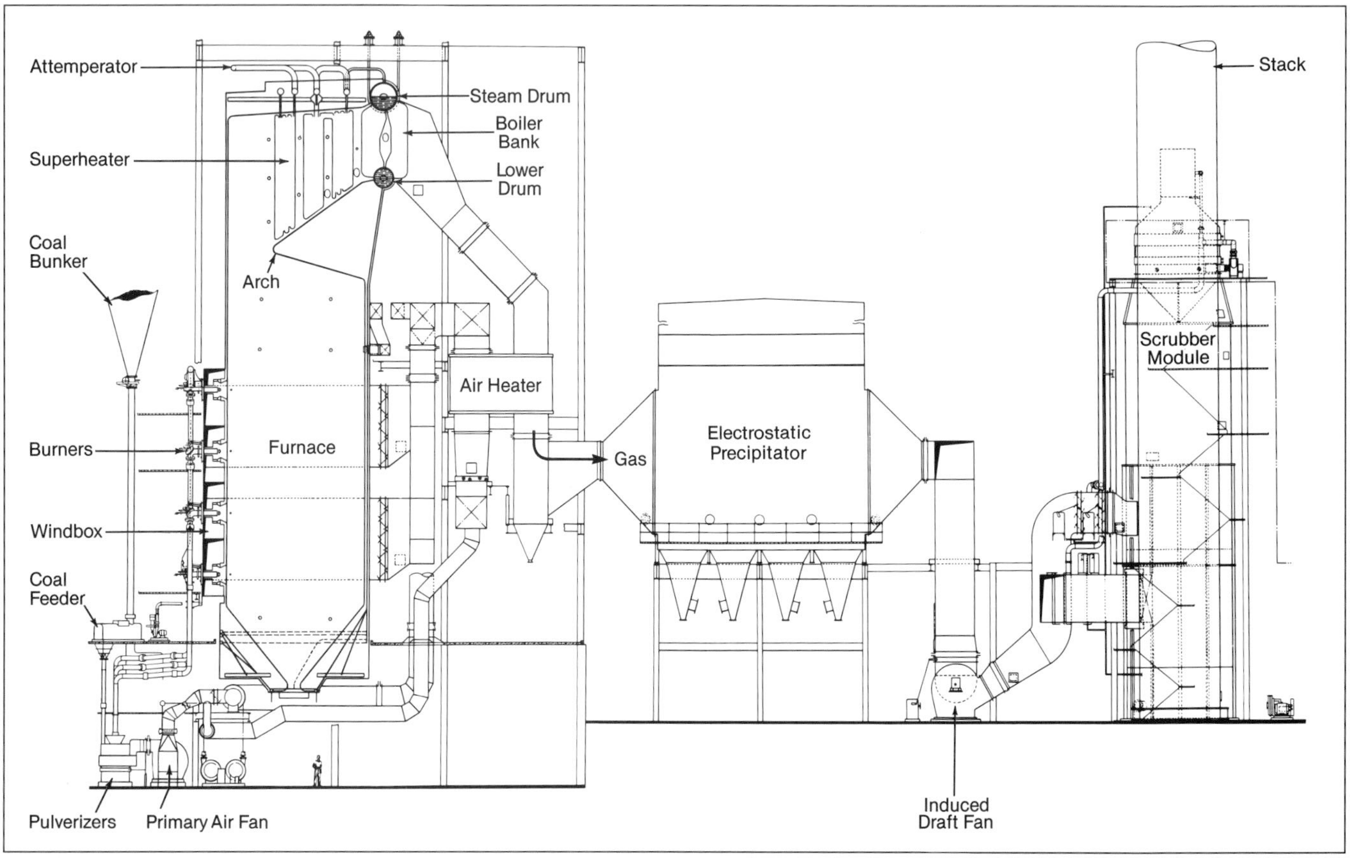

Fig. 1 Two drum Stirling® power boiler system for pulverized coal with environmental control equipment.

ture is somewhat above the required temperature of the materials to be heated. Generally superheat is of no value for this kind of service and is often undesirable because of its interference with temperature control. Reclaiming or devulcanizing rubber, where the rubber in a caustic solution is heated to 400F (204C) by condensing saturated steam at 250 psig (17.2 bar gauge) and 407F (208C) in the jacket of the devulcanizer, is a typical example of process heating with steam.

Pressures of saturated steam for comfort heating of buildings range from 2 psig (0.14 bar gauge) to as high as 80 psig (5.5 bar gauge) in the case of space heaters. It is seldom economical to distribute steam through long lines at pressures below 150 psig (10.3 bar gauge) because of piping costs. Furthermore, the usual requirements for steam within the boiler house for sootblowers, feed pumps and other auxiliaries make it desirable to operate boilers at a minimum of 125 psig (8.6 bar gauge). Consequently few steam plants of any size are operated below this pressure. If the pressure required at points of use is lower, it is common practice to use pressure reducing stations at or near these locations.

The pressure required at the outlet of steam producing equipment for process heating service usually ranges from 125 to 250 psig (8.6 to 17.2 bar gauge) and superheat is usually not required. For this service, boiler manufacturers have generally standardized on a pressure of 250 psig (17.2 bar gauge) for small water tube boilers.

It is customary American stationary boiler practice to hold the main steam line pressure practically constant for all loads, on the premise that this condition satisfies all pressure and quantity requirements of the steam using equipment. Automatic combustion control apparatuses are accordingly designed to function on this basis.

Combination heating and power service, or cogeneration

Many manufacturing operations, such as in paper and textile mills, in the production of chemicals and in processing rubber, require mechanical or electrical power as well as steam for process heating. For such applications, studies are made of the relative merits and costs of: 1) a plant where the power is purchased and steam is generated to supply the heating requirements only, and 2) a plant where steam and power are generated in the same system. A sound appraisal of the relative merits of the two alternatives requires a knowledge of the steam and power requirements, ability to correlate these requirements, economic studies and good judgement. The following general summary may be of assistance.

1. The basic economic advantage in generating steam and power in the same system arises from the use of a much larger portion of the heat supplied in the fuel. When generating electricity alone, as much as 60% of the heat supplied in the fuel is lost to the condensing system, even in a modern central station. (See Chapter 2.)
2. Despite this fundamental thermodynamic advantage, it is frequently more economical to purchase power when it is available at reasonable rates from a dependable source, except where:
   a. waste fuels and waste heat, such as bagasse, blast furnace gas, sawdust or hogged wood, and hot gases are available at low cost from the plant process, and

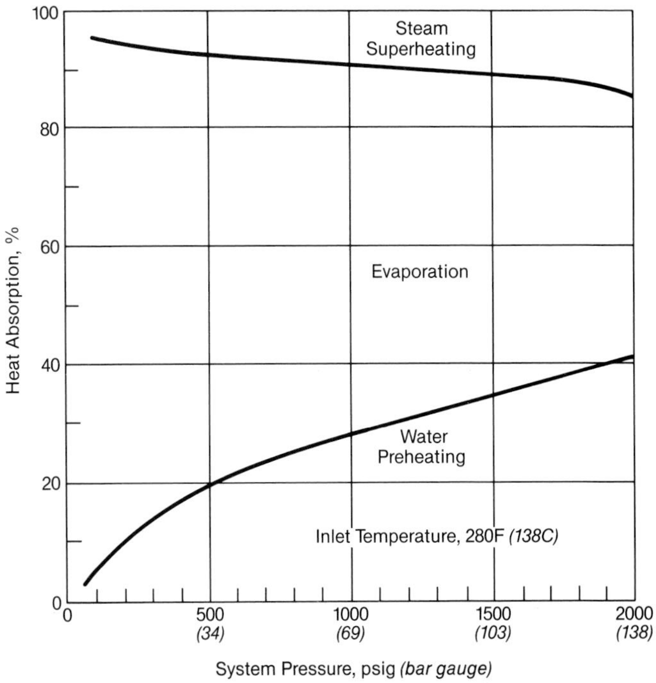

Fig. 2 Effect of system pressure on evaporation in industrial boilers — 100F (56C) constant superheat.

   b. the steam heating and power demands are reasonably parallel and relatively large, i.e., 50,000 lb/h (6.3 kg/s) of steam or more.
3. Two approaches are used for cogeneration. Where natural gas is inexpensive and available on-site, a gas turbine can be used to generate power, with the waste heat in the turbine exhaust gas used to produce steam in a heat recovery steam generator (HRSG). (See Chapter 31.) Where a waste fuel such as petroleum coke or other fossil fuel such as coal is the economic fuel of choice, a steam turbine topping cycle is used. Here high pressure, high temperature steam is produced in the boiler system and is first passed through a steam turbine to generate power. The exhaust steam is then conditioned (brought to appropriate temperature and pressure) and then sent to the process.
4. Variations in process heat and power demands usually do not coincide. To compensate for the differences, a variety of options can be used depending upon the economic evaluation:
   a. for a gas turbine based system
      1.) If steam demand typically exceeds power demand, auxiliary burners are supplied with the HRSG.
      2.) If electrical load demand typically exceeds the steam demand requirements, the remainder of the power may be purchased outside.
   b. for a steam based power system
      1.) If steam demand typically exceeds power demand, this turbine exhaust steam flow can be supplemented with boiler auxiliary firing, and then the additional steam can be passed through a pressure reducing and desuperheating system.
      2.) If power demand is typically higher, either an extraction condensing turbine system can be used or the incremental power can be purchased outside.

If the process requirements for steam and power are reasonably parallel and steady, cogeneration including capital, operating and maintenance costs can be beneficial. Where discontinuous service, low capacity factor or significantly different steam and power requirements exist, electrical supply from the local power grid and on-site steam generation is frequently more cost effective.

Power generation Except for small isolated installations, the high speed turbine is the prime mover of choice for steam power generation because of its efficiency, compactness and low cost. Continued improvement in reliability, reduction in cost and availability of packaged systems have made on site power generation more popular. Where natural gas is available and cost effective, simple gas turbines, especially package units, have tended to dominate on site power production needs. Where a waste fuel (or low cost coal) is available a boiler and steam turbine system can prove to be the most economic system to supply on site power. The selection of steam pressure and temperature for such systems depends upon an economic evaluation along the guidelines presented in Chapter 37. Steam temperature control using one or more of the methods outlined in Chapter 18 is usually provided when electrical output exceeds 25 MW. Steam temperature control is also very important where variations in the flow of the fuel and the fuel quality would otherwise lead to wide swings in power output.

Power for driving auxiliaries

Power is required to drive feed pumps, fans, stokers and pulverizers, while additional energy is needed for feedwater heating in all steam plants. It is common practice in industrial process heating plants to use sufficient steam driven auxiliaries to provide enough exhaust steam for the feedwater heater and to use motor drives for the other auxiliaries. There are instances, however, where the demand for low pressure exhaust steam is great enough that all the auxiliaries can be economically steam driven, and thereby the expense of purchased power can be avoided. Sometimes both motor and turbine drives are provided for each auxiliary — particularly appropriate for cold startup.

Boiler feedwater

Boiler life is directly related to water quality as discussed in Chapter 42. The lack of proper water quality is frequently the cause of failure and lack of availability in industrial boilers. As pressures and temperatures have increased, so has the required level of water quality. However, the need for water quality is often underestimated and the result is expensive downtime for pressure part replacement. Higher water quality translates to higher steam quality, which is needed to protect superheaters and turbines.

The use of a spray water condenser (see Chapter 23) was a step taken to produce higher quality spray water to minimize the steam contamination that is often prevalent when using feedwater for spray attemperation. A sample condenser installation is shown in Fig. 3.

The maximum allowable boiler-water concentration (total solids in boiler water) in relation to the pressure at the outlet of a steam generating unit (applicable to the boiler types in the following descriptive outlines) is given in Table 2 as compiled by the American Boiler Manufacturers Association (ABMA).

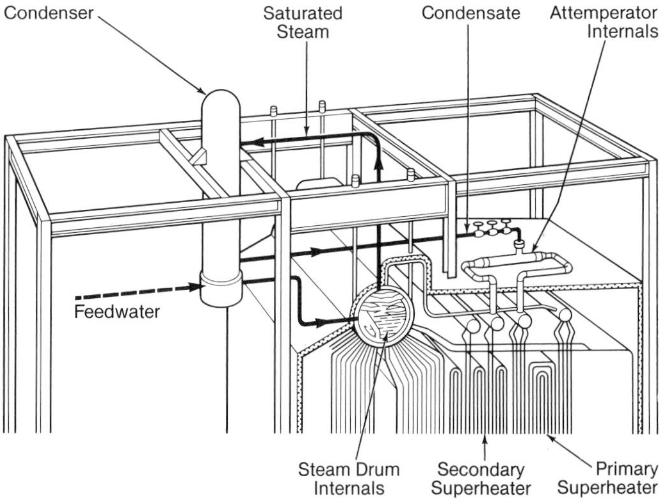

Fig. 3 B&W condenser-attemperator system supplies pure attemperator spray water by condensing steam from the steam drum with incoming boiler feedwater.

In summary, many present day industrial boilers require better water quality and the lack of this can be very expensive.

Other design requirements

The other factors affecting industrial boiler evaluation and selection as outlined in Table 1 are addressed in depth in Chapter 37. Key among these are fuel and ash identification and evaluation. Chapters 8 and 20 address the characterization of fuels and the effect of the non-combustible residue or ash on boiler design. Chapters 10 through 16 discuss the various fuel preparation and combustion systems which may be used. The designs of the boiler, superheater, economizer and air heater systems are covered in Chapters 18 and 19 while auxiliaries are addressed in Chapter 23.

Environmental control

Atmospheric emissions from industrial boilers have come under increasingly stringent regulation from federal, state and local jurisdictions. (See Chapter 32.) Primary controlled pollutants include SO_2, NO_x and particulate. Additional pollutants are controlled in special applications

Table 2
Limits for Solids Content of Boiler Water in Drum Boilers, ppm

Pressure at Outlet of Steam Generating Unit, psi	Total Solids	Total Alkalinity	Suspended Solids
0 to 300	3500	700	300
301 to 450	3000	600	250
451 to 600	2500	500	150
601 to 750	2000	400	100
751 to 900	1500	300	60
901 to 1000	1250	250	40
1001 to 1500	1000	200	20
1501 to 2000	750	150	10

such as waste-to-energy plants discussed in Chapter 27. A typical large industrial power system with environmental controls is shown in Fig. 1. This Stirling® power boiler (SPB), discussed below, is fitted with low NO_x burners to limit NO_x formation, an electrostatic precipitator to limit flyash emissions to less than 0.2%, and a wet limestone flue gas desulfurization (FGD) system to remove SO_2.

SO_2 control in industrial plants can take three general forms: burning low sulfur fuel, removal of SO_2 in the combustion process through the application of fluidized-bed technology, or postcombustion SO_2 removal through furnace sorbent injection, either dry or wet FGD. These technologies are discussed in Chapter 35. When burning lower sulfur fuels, dry FGD systems using lime as a reagent, combined with fabric filters, have been used.

NO_x control is typically centered on limiting NO_x formation during the combustion process. The technology applied is closely tied to the combustion system selected — low NO_x burners for pulverized coal, oil and gas units; overfire air systems for stokers; and low temperature combustion in fluidized-bed combustors. These are discussed in Chapters 10, 13, 15 and 16. If needed, postcombustion selective catalytic reduction (SCR) and selective noncatalytic reduction (SNCR) $deNO_x$ systems are available. (See Chapter 34.)

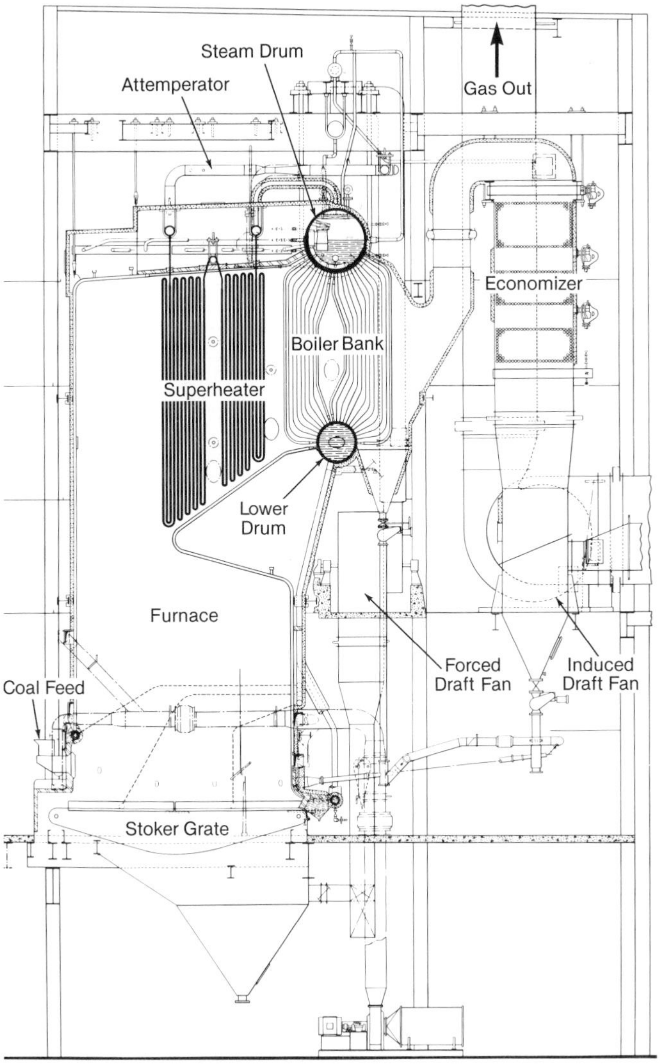

Fig. 4 Two drum Stirling® boiler for spreader-stoker firing.

Finally, postcombustion particulate control is required for burning virtually all fuels other than clean gas or selected fuel oils. Electrostatic precipitators (ESP) or fabric filters (baghouses) meet these requirements. (See Chapter 33.)

B&W boiler types for industrial applications

Brief descriptions of the boiler types that follow are intended to serve as an introduction to the types of steam generating units that meet the wide range of fuel and performance requirements of the industrial sector.

Stirling® power boiler

Description The SPB is a top supported, two drum, single gas pass unit. (See Figs. 1 and 4.) In some cases it is cost effective to replace the two drum design with a single drum and a smaller shop-assembled boiler module. (See Fig. 5.)

The furnace is completely water cooled using membrane wall construction [normally 3 in. (76.2 mm) tubes on 4 in. (101.6 mm) centers] and is satisfactory for either pressurized or balanced draft operation. Shop assembly of wall panels is maximized to facilitate field erection. Wall panels always come with the headers attached, regardless of the number of shipping pieces.

Cyclone steam-water separators, discussed in Chapter 5, along with primary and secondary steam scrubbers, are included in the steam drum to provide the high quality dry steam needed for present day superheater and turbine designs.

Furnaces include a nose arch which serves to direct gas flow over the superheater section and to shield the superheater from high temperature furnace radiation.

SPBs are capable of firing solid, liquid or gaseous fuels. There are several furnace configurations to complement the type of fuel fired, as shown in Fig. 6. A hopper bottom furnace is used for pulverized coal firing; a flat floor for gas or oil firing; and an open bottom to receive a stoker for stoker coal, wood, bagasse, biomass, refuse-derived fuel (RDF) and as-received municipal solid waste (MSW).

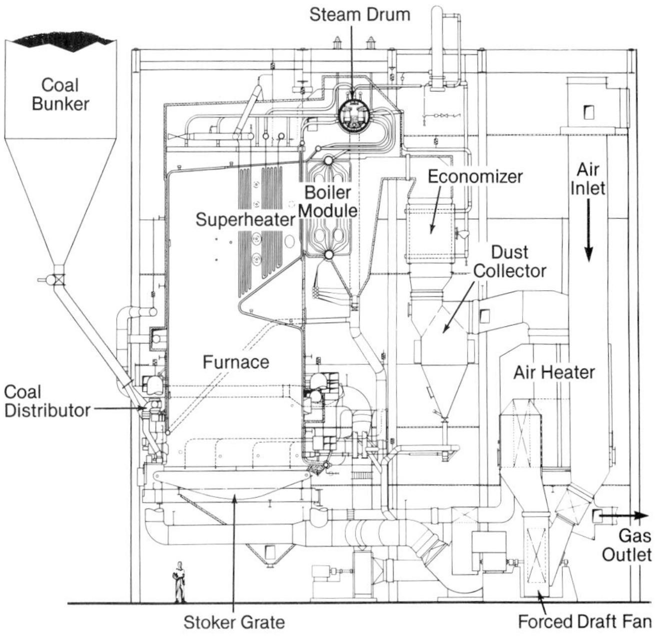

Fig. 5 Single drum Stirling® boiler for spreader-stoker firing.

Furnaces for fuels having a significant amount of fines and/or high moisture such as wood, biomass, bagasse and RDF are also arranged with front and rear furnace arches. These arches help define the combustion zone and allow better location of overfire air nozzles, which are desired for solid fuels with high levels of fines. This dual arch design, developed by B&W, is referred to as a controlled combustion zone (CCZ ™) furnace.

The SPB is equipped with an economizer and/or air heater to provide for economical heat recovery. For many fuels, air heating is important for combustion. Pulverized coal requires hot air to dry the fuel while hot air is required to promote combustion of moist fuels such as wood, bagasse and biomass.

Because of the design features, the SPB is the preferred industrial boiler for many applications. There are some other designs that serve special fuels, capacities, pressures and temperatures, that make them a good alternative under special conditions.

Pre-engineered While the SPB is custom designed to meet specific steam and fuel conditions, the design is done within a framework of pre-engineered components to minimize engineering costs and delivery time.

The furnace width and depth are pre-engineered in 1

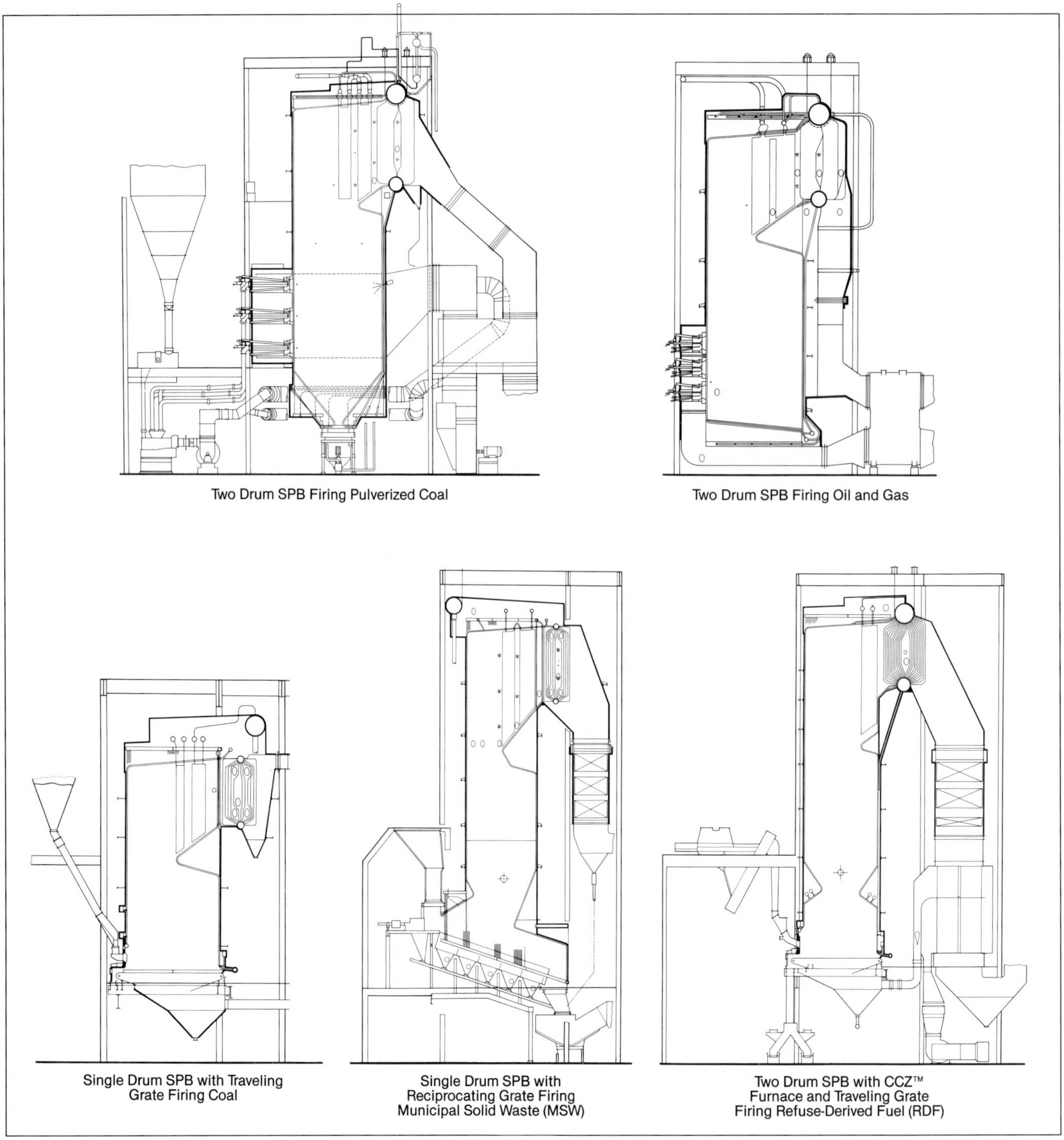

Fig. 6 Typical Stirling® power boiler furnace configurations.

ft (0.3 m) increments so that all of the closures at the corners are established. Drum centerlines, in 2 ft (0.6 m) increments between 16 and 32 ft (4.9 and 9.8 m), have been pre-engineered to locate all of the access doors, sootblower openings, buckstays and platforms. Combinations of steam and lower drum sizes are designed so that all of the bend angles for drum entry are established.

These pre-engineered increments allow flexibility in the design to satisfy the job-specific requirements of furnace exit gas temperature, burner clearances, residence time, grate size, gas velocity, convection spacing, etc.

SPB design range

Capacity	
pulverized coal, oil, gas	150,000 to 1,200,000 lb/h (18.9 to 151.2 kg/s)
stoker coal	150,000 to 400,000 lb/h (18.9 to 50.4 kg/s)
stoker wood, bagasse, and biomass	180,000 to 600,000 lb/h (22.7 to 75.6 kg/s)
Steam pressure to	1800 psig (124 bar gauge) design
Steam temperature to	1000F (538C)

Towerpak® boiler

Description The Towerpak® is a version of the SPB designed for lower capacities that are often required by smaller industrial plants. (See Fig. 7.) It incorporates many of the features of the SPB including membrane walls, cyclone steam-water separators, and furnace wall arches for wood or biomass.

Towerpak® boilers are two drum, bottom supported units. For the smaller sizes, they can be shipped in a single unit or in modules for ease of field assembly. Larger units follow the SPB format of maximum subassembly of wall panels, again for ease of field assembly. This unit is a preferred design at low steam capacity for hard to burn solid fuels such as wood, biomass and stoker coal.

Pre-engineered Like the SPB, these units are custom designed to meet specific conditions for each application, but within the framework of pre-engineered components.

Towerpak® design range

Capacity	40,000 to 180,000 lb/h (5.0 to 18.9 kg/s)
Steam pressure to	1100 psig (76 bar gauge) design
Steam temperature to	900F (482C)

PFI boiler

Description The PFI boiler is a two drum, bottom supported, multiple gas pass unit designed specifically to burn liquid and/or gaseous fuels. (See Fig. 8.) Due to the large furnace enclosure, the PFI is an excellent choice for harder to burn byproduct fuels such as blast furnace gas and refinery catalytic cracker CO gas.

The furnace is completely water cooled using membrane wall construction [2.5 in. (63.5 mm) tubes on 3 in. (76.2 mm) centers] and is satisfactory for either pressurized or balanced draft firing.

The PFI was developed for maximum shop assembly of components. For example, the furnace is shipped in as few as ten membrane wall panels with headers attached at the top and bottom. The panels include two for each side wall and two each for roof, front wall and floor. The burner throats are integral with the front wall panel. The unit is bottom supported on simple concrete piers and incorporates a drainable superheater. A unique plenum encloses the upper front wall, roof and rear of the unit. This plenum serves as an integral air duct to the windbox and, with a division plate in the rear, a flue gas outlet. Either an air heater or economizer is used to provide an economical exit gas temperature and heat recovery.

This design features a gas pass the full length of the boiler bank. Flue gas flows horizontally and parallel to the drums through the bank. A gas baffle is used to direct the gas across the tubes in multiple passes to maximize heat transfer.

An inverted loop, drainable superheater is located behind a screen at the furnace outlet to protect it from direct furnace radiation. This superheater location gives

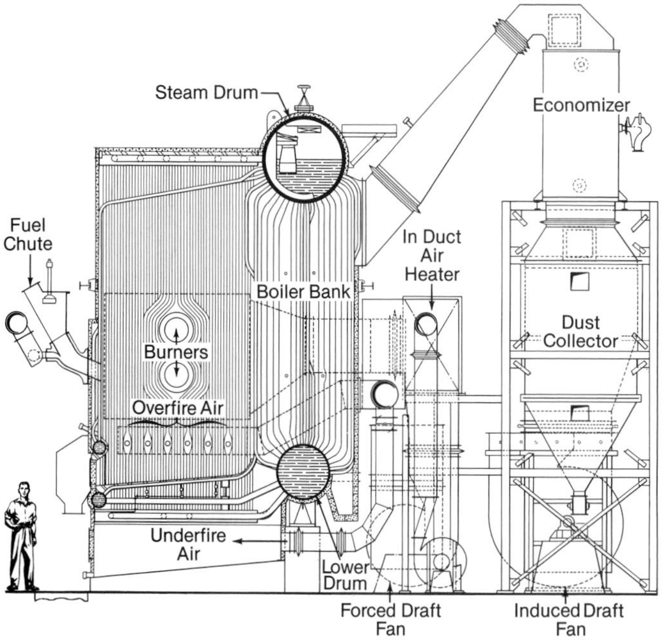

Fig. 7 Towerpak® boiler.

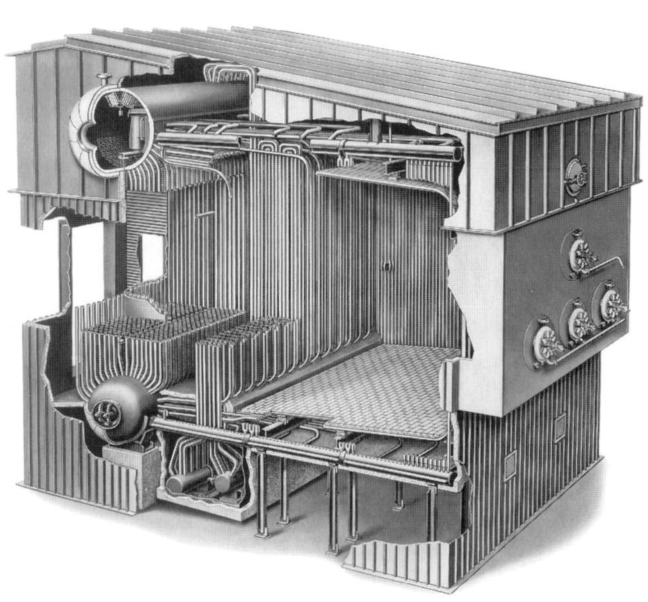

Fig. 8 Type PFI Integral Furnace boiler.

a semi-radiant heat transfer characteristic that produces a relatively flat temperature curve across the load range. This temperature profile minimizes attemperation for steam temperature control.

Cyclone steam separators with primary and secondary scrubbers are included in the steam drum to produce the high quality steam needed for present day superheaters and turbines.

Pre-engineered The PFI unit is totally pre-engineered in a number of frame sizes to satisfy the capacity range of this boiler design. Several superheater arrangements are also pre-engineered for each frame size.

Units come in three different drum centerlines. Each centerline has a specific furnace depth and three or four furnace widths.

PFI design range

Capacity	100,000 to 500,000 lb/h (12.6 to 63.0 kg/s)
Steam pressure to	1150 psig (79 bar gauge) design
Steam temperature to	950F (510C)

PFT boiler

Description The PFT boiler was developed as an extension of the PFI design to accommodate the development of higher pressure and temperature turbine cycles. This unit incorporates many of the PFI features: two drum, bottom supported, modularized furnace membrane walls [3 in. (76.2 mm) tubes on 4 in. (101.6 mm) centers], drum cyclones and a drainable superheater. (See Fig. 9.)

Some of the differences include an alternate pendant superheater and gas flow path the full width of the boiler bank with flow in vertical directions.

PFT units are particularly well suited to burn high ash liquid fuels, blast furnace gas and CO gas because cavities provide space for retractable sootblowers for cleaning requirements.

Pre-engineered The PFT, like the PFI, is totally pre-engineered in a number of frame sizes. The unit is designed with two drum centerlines and several furnace depths and widths to satisfy the capacity range of this design.

PFT design range

Capacity	300,000 to 800,000 lb/h (37.8 to 100.8 kg/s)

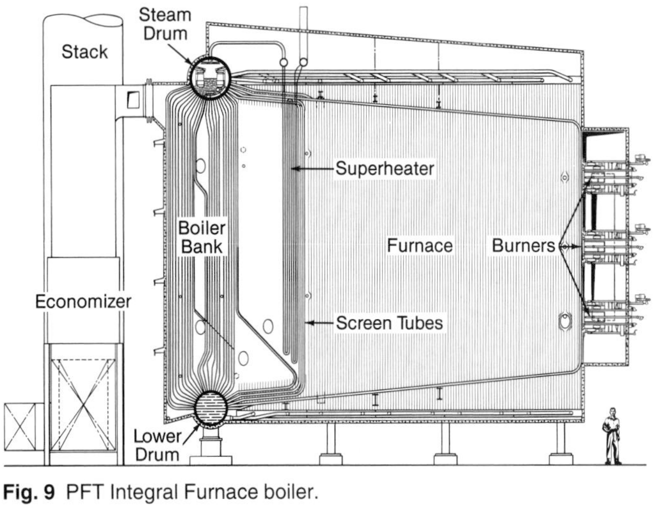

Fig. 9 PFT Integral Furnace boiler.

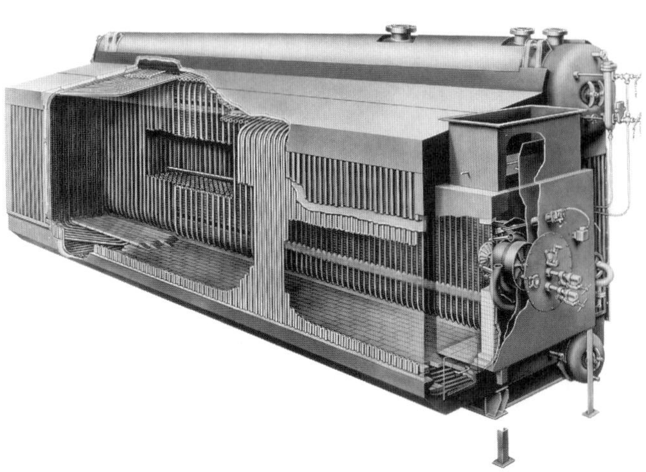

Fig. 10 Type FM Integral Furnace boiler — membrane wall construction.

Steam pressure to	1800 psig (124 bar gauge) design
Steam temperature to	1000F (538C)

FM boiler

Description The FM is a shop-assembled (package), two drum, bottom supported boiler. (See Fig. 10.) Package boilers, by most definitions, can be shipped by rail or truck. This is a D boiler design which has the furnace on one side and boiler bank on the other separated by a baffle wall. The unit is fired parallel to the drums toward the rear wall where the gas turns 180 degrees and flows frontward to the gas outlet. Many units are arranged with an economizer or air heater for fuel efficiency.

Gas-tight furnaces are used because these units are pressure-fired (forced draft fan only). For the smaller size units (FM9 through FM101), which operate at relatively low furnace pressures, a studded furnace construction and inner casing are used. Larger units (FM103 through FM120) which operate at higher furnace pressures have a membrane furnace.

With today's advanced burner designs and steam separation technology, B&W offers rail shippable designs to 200,000 lb/h (25.2 kg/s). These units are ideal for small scale process, heating or power needs. (See Fig. 11.)

Pre-engineered The FM is totally pre-engineered in frame sizes to satisfy the capacity range of this package

Fig. 11 Two shop-assembled FM package boilers.

design. Series available are FM9, FM10, FM103, FM106, FM117 and FM120. Pre-engineered set furnace depths are the only variables in this series.

FM design range

Capacity	10,000 to 200,000 lb/h (1.3 to 25.2 kg/s)
Steam pressure, units to 30,000 lb/h (37.8 kg/s)	525 psig (36 bar gauge) design
Steam pressure, units above 30,000 lb/h (37.8 kg/s)	1050 psig (72 bar gauge) design
Steam temperature to	750F (399C) on oil 825F (441C) on natural gas

High Capacity FM boilers

Description The High Capacity FM (HCFM) and FM300 boilers are extensions of the FM D-type boiler design. (See Fig. 12.) High capacity boilers can be dock or field assembled. Shipping dimensions of these units require that they be shipped by barge or ocean vessel. The FM300 is a modular design that is rail shippable.

The design is for oil and gas firing and uses membrane furnace walls for pressurized operation. Multiple burners are used for the increased capacity.

Pre-engineered Like the FM, these boilers are completely pre-engineered and set furnace depths are the only variable.

HCFM and FM300 design range

Capacity	
HCFM	200,000 to 350,000 lb/h (25.2 to 44.1 kg/s)
FM300	150,000 to 300,000 lb/h (18.9 to 37.8 kg/s)
Steam pressure to	1050 psig (72 bar gauge) design
Steam temperature to	825F (441C)

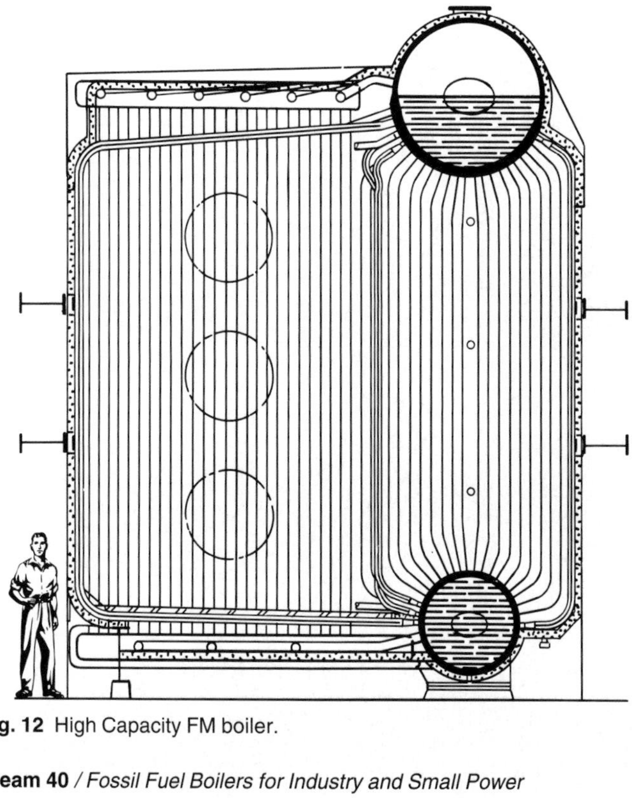

Fig. 12 High Capacity FM boiler.

PFM boiler

Description Another higher capacity and design pressure D-type boiler, the PFM, is also designed for dock or field assembly and must be shipped by barge or ocean vessel. (See Fig. 13.) If shipped assembled, the size of these units requires special handling at the installation site. This design is also for gas and oil firing and uses a membrane furnace for pressure operation. Again, multiple burners are used for the increased capacity.

Pre-engineered Like the FM and HCFM, this boiler series, which includes PFM 140, 180, 220, 250 and 280, is completely pre-engineered and set furnace depths are the only variable.

PFM design range

Capacity	200,000 to 600,000 lb/h (25.2 to 75.6 kg/s)
Steam pressure to	1800 psig (124 bar gauge) design
Steam temperature to	900F (482C)

Circulating fluidized-bed boiler

Description Fluidized-bed boilers feature a unique concept of burning fuel in a bed of particles to control the combustion process and, when required, control SO_2 and NO_x emissions. Two options are offered — the circulating fluidized-bed boiler (CFB) and the bubbling fluidized-bed boiler (BFB). (See Figs. 14 and 15.) The CFB has been used for many new boiler applications while the BFB has proven particularly attractive in retrofit applications.

Fig. 13 Type PFM Integral Furnace very high capacity package boiler being prepared for shipment.

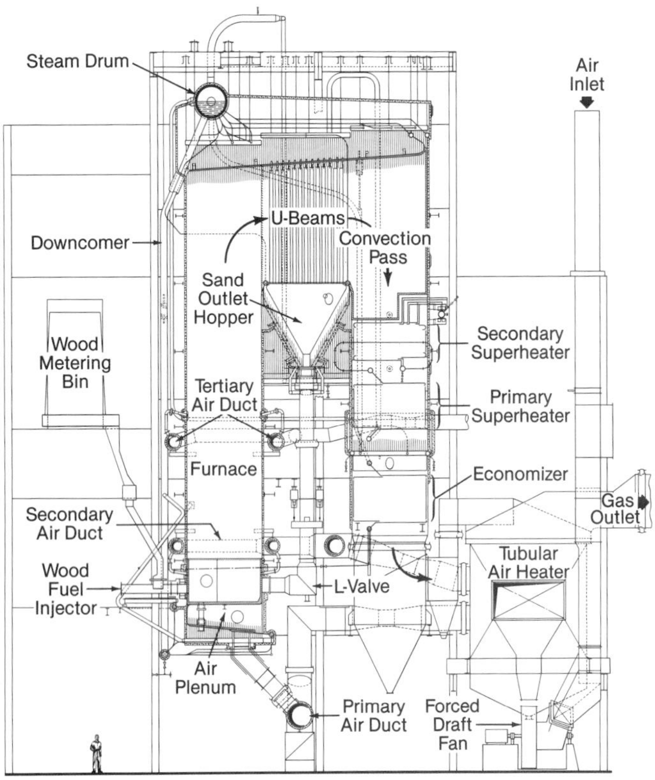

Fig. 14 Wood-fired atmospheric pressure circulating fluidized-bed boiler.

Both fluidized-bed technologies are discussed in depth in Chapter 16.

The CFB is a top supported boiler. (See Fig. 14.) One or two drums are used depending on the need for a generating bank to absorb heat. Fuel is admitted to the lower part of the furnace by screws, chain feeders or air-swept spouts, depending on the fuel.

When SO_2 removal is required, the bed medium is limestone and when SO_2 removal is not an issue, sand is used. Compared to the fuel quantity present in the unit, the circulating bed material is many times greater. Total solids in the flue gas passing upwards through the furnace are a function of how much heat must be absorbed by the waterwalls.

Varying the bed density maintains the desired constant temperature necessary for maximum SO_2 removal [about 1550F (843C)]. Solids laden flue gas exits the furnace to the U-beam particle separators. Solids (about 98% collected) drop into a hopper and are recirculated to the furnace. The recirculation rate is controlled by L-valves to provide the flow necessary to maintain the required bed temperature and density. Flue gas exiting the U-beams proceeds over convection surfaces, similar to other boiler designs.

The CFB has been selected for application with high sulfur fuels (petroleum coke, coal, sludge and oil pitches) and for wood and other biomass fuels. The CFB, because it operates at a much lower combustion temperature, inherently generates about one half the NO_x as the other solid fuel-fired industrial boilers previously described.

The CFB is an alternative to the pulverized coal or stoker coal-fired SPB which frequently must be equipped with a wet or dry scrubber (for SO_2 removal) and ammonia injection, catalytic or noncatalytic reduction (NO_x removal) equipment. The choice of technologies requires in depth evaluation of a number of factors including required amount of emissions removal, fuel cost, reagent cost and capital cost.

Pre-engineered The CFB is custom designed to meet each specific application, but like the SPB, it is designed within a framework of pre-engineered components to minimize engineering costs and delivery time.

CFB design range	
Capacity to	700,000 lb/h (88.2 kg/s) with expected growth to 1,000,000 lb/h (126 kg/s) or greater
Steam pressure to	1850 psig (128 bar gauge) design
Steam temperature to	1000F (538C)

Bubbling fluidized-bed boiler

Description The bubbling fluidized-bed boiler (BFB) (Fig. 15) is similar to the CFB in some elements. The unit is top supported, can use one or two drum designs, and can burn a wide variety of fuels cleanly and efficiently. It differs from the CFB in that the air velocity is kept low enough that the bed material (except for elutriated fines) is held in the bottom of the unit — the solids do not circulate through the rest of the furnace enclosure. This feature makes it particularly attractive in retrofit applications where the bottom of an existing furnace can be removed and replaced with a BFB without major modifications to the balance of the furnace, convection pass enclosures and heat transfer surfaces. Such conversions have been effective in regaining boiler capacity lost because of a fuel change or a change in ash characteristics which are not compatible with the original boiler furnace design. Such retrofits also provide one option to reduce SO_2 and NO_x emissions from industrial and small utility boilers.

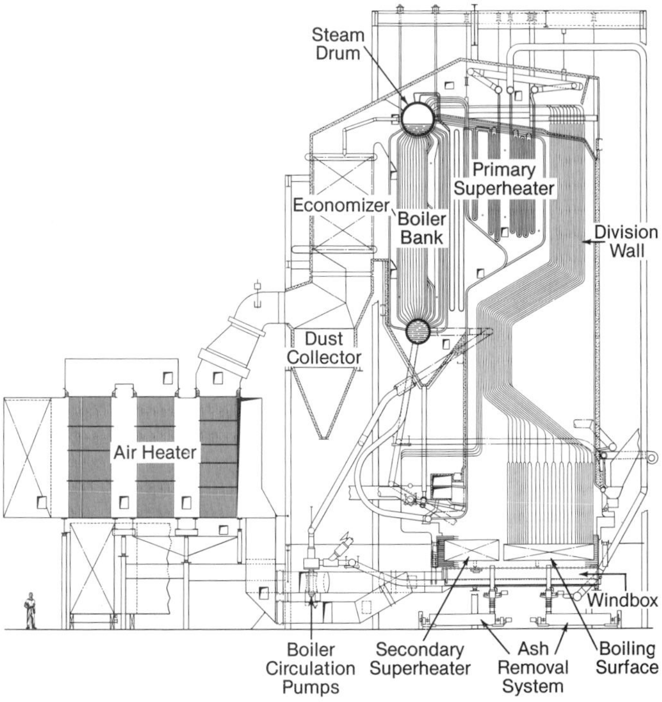

Fig. 15 Atmospheric pressure bubbling fluidized-bed boiler retrofit.

In new boiler applications the BFB is particularly well suited for high moisture waste fuels such as sewage sludge and the various sludges produced in pulp and paper mills and recycle paper plants.

Pre-engineered The BFB is custom designed to meet each specific application, but like the CFB, this is done within a framework of pre-engineered components.

Enhanced Oil Recovery boiler

Description The Enhanced Oil Recovery (EOR) boiler unit (Fig. 16) was developed to meet a single market need, as the name implies. High pressure, wet steam (approximately 80% quality) is produced by the boiler and then injected into strata containing heavy oils. The steam enhances the recovery of oil by heating the heavy oil which reduces its viscosity and thereby aids in moving the oil to the producing wells.

A once-through steam water circuitry is used. Feedwater flows continuously in a single tube circuit through the economizer section to the furnace section where the water is boiled to 80% (by weight) steam quality level.

By maintaining wide flame clearances from the furnace walls and low heat releases, relatively poor feedwater (100,000 ppm solids) can be tolerated. This permits minimal feedwater treatment and allows the water that is separated from the recovered oil to be recycled to the boiler with minimum cleanup.

Control of the process is achieved by pumping the required amount of feedwater at the specified pressure [up to 2500 psig (172 bar gauge)] into the economizer and by regulating burner firing rate to maintain measured outlet steam quality.

Fig. 16 Enhanced Oil Recovery boiler on location.

Pre-engineered and design range Units come in pre-engineered sizes from 5 to 50 × 10^6 Btu/h (1.5 to 14.7 MW_t) output. The units are shop-assembled with units through 40 × 10^6 Btu/h (11.7 MW_t) trailer mounted in one piece for shipment. Larger sizes are shop-assembled in several sections for final field assembly.

EOR design range (oil and gas)

Capacity to	48,000 lb/h (6.0 kg/s)
Steam pressure to	2500 psig (172 bar gauge) design

This pulp and paper facility was a turnkey project for Babcock & Wilcox.

Chapter 26
Chemical and Heat Recovery in the Paper Industry

In the United States (U.S.), the pulp and paper industry is the fourth largest industrial consumer of energy, and the third largest in energy purchases. The industry is the leading cogenerator of electric power with a 1985 capacity of 7000 MW.[1]

Approximately one half of the steam and power consumed by this industry is generated from fuels that are byproducts of the pulping process. The main source of self-generated fuel is the spent pulping liquor, followed by wood and bark. The energy required to produce pulp and paper products has been significantly reduced. Tremendous progress has also been made in reducing air emissions. Process improvements allowed U.S. pulp and paper manufacturers to reduce energy consumption to just 26.7 x 10^6 Btu/t (31 x 10^6 MJ/t_m) of production in 1989, compared to 33.0 x 10^6 Btu/t (38.4 x 10^6 MJ/t_m) in 1972, a decrease of about 19%.[2]

Pulp and paper mill electric power requirements have increased disproportionately to process steam requirements. This factor, coupled with steadily rising fuel costs, has led to the greater cycle efficiencies afforded by higher steam pressures and temperatures in paper mill boilers. The increased value of steam has produced a demand for more reliable and efficient heat and chemical recovery boilers.

The heat value of the spent pulping liquor solids is a reliable fuel source for producing steam for power generation and process use. A large portion of the steam required for the pulp mills is produced in highly specialized heat and chemical recovery boilers. The balance of the steam demand is supplied by boilers designed to burn coal, oil, gas and biomass.

Major pulping processes

The U.S. and Canada are the leading producers of pulp. North America produced 53% of the world's sulfate pulp in 1989, and consumed 42.6% of the total pulp. The U.S. alone consumed 33% of the total pulp production in 1989, resulting in the highest per capita consumption of paper and board in the world. (See Table 1.)

By the end of 1989 the U.S. had installed 29% of the world's capacity for paper and board production and 32.3% of the pulping capacity. This U.S. capacity is installed in 544 paper and board mills and 215 pulp mills, of which about one half produce bleached pulp. Total 1989 pulp production in the U.S. was divided among four principal processes: 80.3% sulfate, 2.6% sulfite, 7.2% semichemical and 9.9% mechanical pulping. These processes produced 54,950,000 t (49,850,000 t_m) of pulp in 1989.[3]

Table 1
Paper and Board 1989 Consumption — lb (kg)/person

United States	670.0	(303.9)
Canada	490.7	(222.6)
Japan	490.7	(222.6)
Federal Republic of Germany	451.3	(204.7)
United Kingdom	371.5	(168.5)
Taiwan	347.9	(157.8)
Mexico	69.0	(31.3)
Brazil	61.9	(28.1)
People's Republic of China	27.8	(12.6)
Indonesia	11.0	(5.0)
India	6.4	(2.9)

The dominant North America pulping process is the sulfate process, deriving its name from the use of sodium sulfate (Na_2SO_4) as makeup chemical. The paper produced from this process was originally so strong in comparison with alternative processes that it was given the name *kraft*, which is the Swedish and German translation for strong. Kraft is an alkaline pulping process, as is the soda process which derives its name from the use of sodium carbonate, Na_2CO_3 (soda ash), as makeup chemical. The soda process has limited use in the U.S. and is more prominent in countries pulping nonwood fiber. Recovery of chemicals and the production of steam from waste liquor are well established in the kraft and soda processes. The soda process accounts for less than 1% of alkaline pulp production and its importance is now largely historic.

Kraft pulping and recovery process

Kraft process

The kraft process flow diagram (Fig. 1) shows the typical relationship of the recovery boiler to the overall pulp and paper mill.[4] The kraft process starts with feeding wood chips, or alternatively a nonwood fibrous material, to the digester. Chips are cooked under pressure in a steam heated aqueous solution of sodium hydroxide (NaOH) and sodium sulfide (Na_2S) known as *white liquor* or cooking liquor. Cooking can take place in continuous or batch digesters.

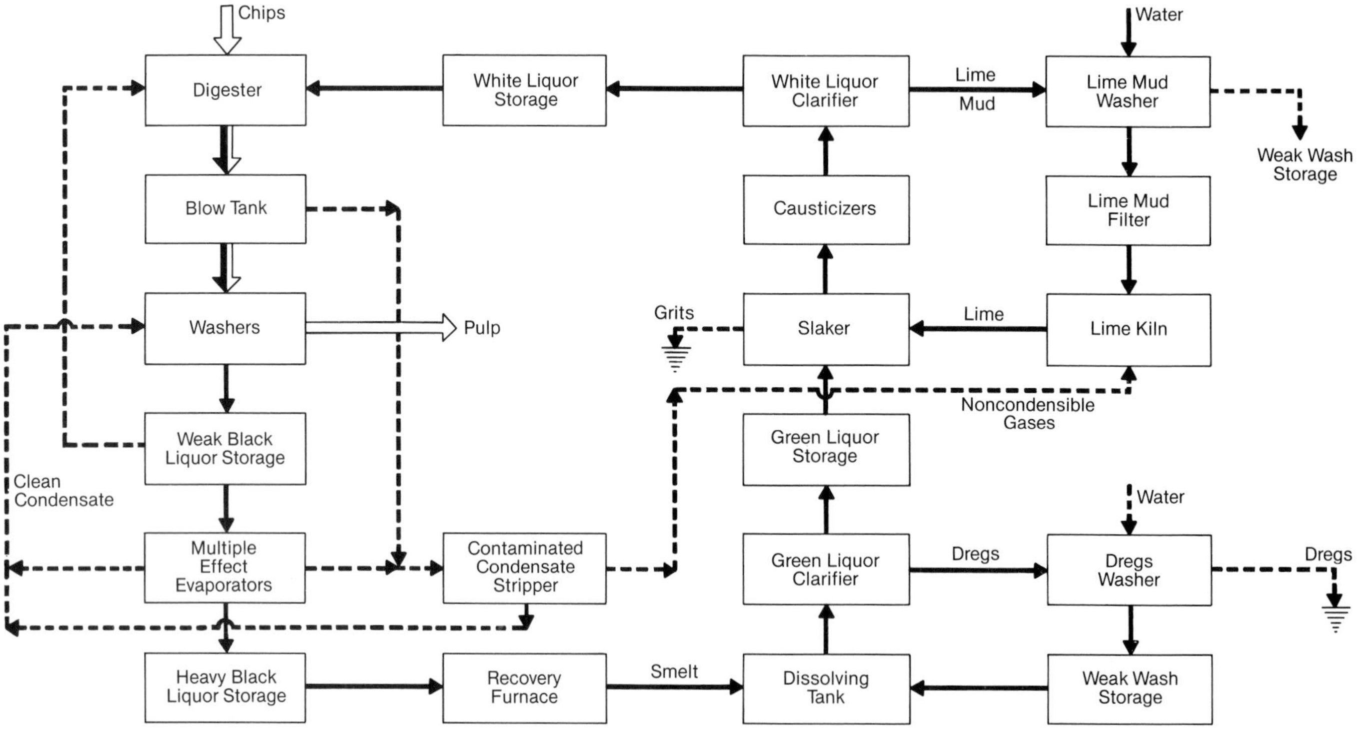

Fig. 1 Kraft process diagram.

After cooking, pulp is separated from the residual liquor in a process known as brown stock washing. The most common method features a countercurrent series of vacuum drum washers which displace the liquor with minimum dilution. Following washing, the pulp is screened and cleaned to remove knots and shives and to produce fiber for use in the final pulp and paper products.

The *black liquor* rinsed from the pulp in the washers is an aqueous solution containing wood lignins, organic material and inorganic compounds oxidized in the cooking process. Typically, the combined organic and inorganic mixture is present at a 13 to 17% concentration of solids in weak black liquor. The kraft cycle processes this black liquor through a series of operations, including evaporation, combustion of organic materials, reduction of the spent inorganic compounds and reconstitution of the white liquor. The physical and chemical changes in the unit operations are shown in Fig. 2.[5]

The unique recovery boiler furnace was developed for combusting the black liquor organic material while reducing the oxidized inorganic material in a pile, or bed, supported by the furnace floor. The molten inorganic chemicals or *smelt* in the bed are discharged to a tank and dissolved to form *green liquor*. Green liquor active chemicals are Na_2CO_3 and Na_2S.

Green liquor contains unburned carbon and inorganic impurities from the smelt, mostly calcium and iron compounds, and this insoluble material, or *dregs*, must be removed through clarification. This operation is basically settling of sediment and decantation of clear green liquor that can be pumped to the slaker. The dregs are pumped out of the clarifier as a concentrated slurry. Normal operation is to water wash the dregs before landfill disposal. The water wash liquid containing the recovered sodium chemical is known as *weak wash*. The sodium chemicals are recovered by using the weak wash to dissolve the smelt in the dissolving tank.

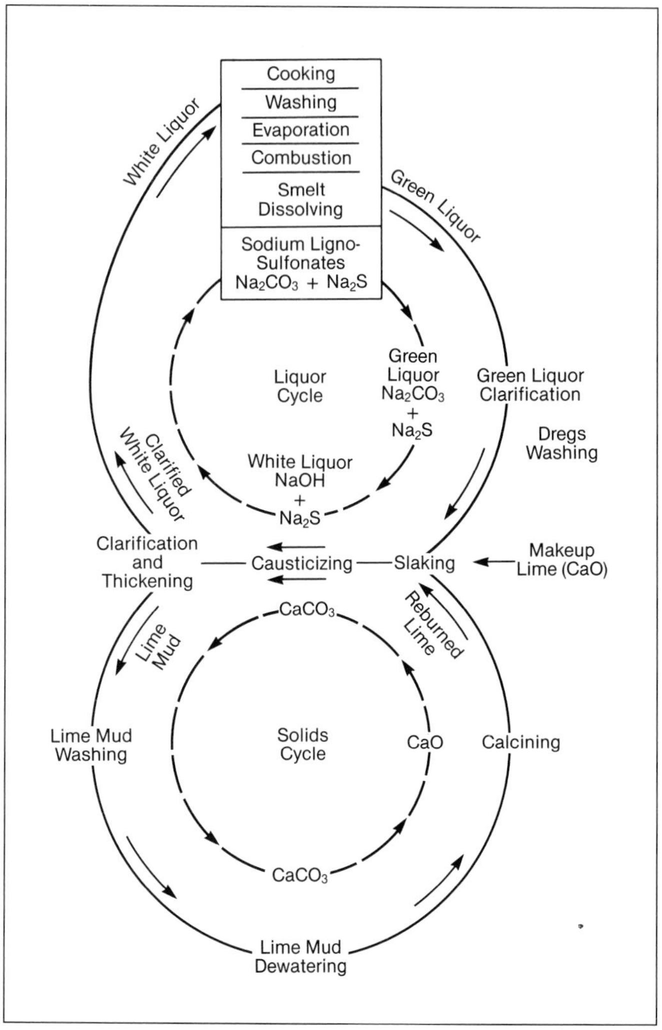

Fig. 2 Kraft process cycle.

Clarified green liquor and lime (CaO) are continuously fed to a slaker where high temperature and agitation promote rapid slaking of the CaO into calcium hydroxide ($Ca(OH)_2$). The liquor from the slaker flows to a series of agitated tanks that allow the relatively slow causticizing reaction to be carried to completion. The function of the causticizing plant is to convert sodium carbonate into active NaOH. The calcium carbonate ($CaCO_3$) formed in the conversion reaction precipitates in the causticizing operation to form a suspended *lime mud.*

The causticizing product must be clarified to remove the $CaCO_3$ precipitate and produce a clear white liquor for cooking. This clarification is carried out by either settling and decanting in a manner similar to green liquor clarification, or by using pressure filters. In pressure filtration, the white liquor is filtered through a medium to provide a separation of clear white liquor from the lime mud. The lime mud is then washed to remove sodium chemicals that can lead to increased kiln emissions and clinkering, and further filtered to obtain the desired consistency for feed to the kiln.

The lime kiln calcines the washed lime mud feed into reburned lime. Calcination is the chemical breakdown with heat of the $CaCO_3$ into active lime and carbon dioxide (CO_2). The calcined lime is then slaked as previously described.

The reactions occurring in the solids cycle operations are as follows:

Slaking: $CaO + water = Ca(OH)_2 + heat$
Causticizing: $Ca(OH)_2 + Na_2CO_3 = CaCO_3 + 2NaOH$
Calcination: $CaCO_3 + heat = CaO + CO_2$

The combination of these process steps is referred to as *recausticization.*

In parallel with the reduction of sulfur compounds to form smelt, energy is released in the recovery furnace as the black liquor organic compounds are combusted. This combustion energy is used in the process recovery boiler to produce steam from feedwater. The steam can be introduced to a turbine generator to supply a large portion of the energy demand of the pulp and paper mill. Steam extracted from the turbine at low pressure is used for process requirements such as cooking wood chips, evaporation, recovery furnace air heating and drying the pulp or paper products.

Rated capacity of a recovery unit

The capacity of a pulp mill is based on the daily tons of pulp produced. The primary objectives of a recovery boiler are to reclaim chemicals for reuse and to generate steam by burning the black liquor residue. Accordingly, the capacity of the recovery boiler should be based on its ability to burn or process the dry solids contained in the recovered liquor. Because the proper measure of recovery boiler capacity is the heat input to the furnace, Babcock & Wilcox (B&W) has established a 24 hour heat input unit of 19,800,000 Btu (20,890 MJ). This unit, known as a B&W-Btu ton, corresponds to the heat input from 3000 lb (1361 kg) of solids (approximately equivalent to one ton of pulp produced) having a heating value of 6600 Btu/lb (15,352 kJ/kg) of solids. These were averages for the typical black liquor solids generated from a ton of kraft pulp production when the unit was originally defined. The black liquor solids produced from modern operations generally are characterized by considerable variation in the quantity and heating value of the solids per ton of pulp product. However, the B&W-Btu ton still provides a measure of recovery boiler rating.

The nominal size of a B&W kraft recovery boiler can be determined by application of a simple formula, as follows:

$$\text{Nominal size} = \frac{A \times B \times C}{19{,}800{,}000}, \text{ B\&W - Btu tons} \quad \textbf{(1)}$$

where

A = dry solids recovered, lb/t of pulp
B = pulp output of mill, t/24 h
C = heating value of dry solids, Btu/lb

and 19,800,000 is the product of 3000 lb/t and 6600 Btu/lb.

Process flows through the recovery boiler

The kraft process recovery boiler is similar in many respects to a conventional fossil fuel-fired boiler. The concentrated black liquor fuel is introduced into the furnace along with combustion air. Inside the furnace, the residual water is evaporated and the organic material is combusted. The inorganic portion of the black liquor solids is recovered as sodium compounds. Most of the sulfur is in the reduced form of Na_2S and most of the remaining sodium is Na_2CO_3. The requirement to recover sulfur in a reduced state is the most unique aspect of recovery boiler design. Fig. 3 illustrates a typical modern recovery boiler.

Combustion air is introduced into the furnace at staged elevations — primary, secondary and tertiary. One fourth to one half of the air enters at the primary level near the furnace floor. The balance is staged at the secondary and tertiary levels. Heavy black liquor is fed to the furnace through multiple burners between the secondary and tertiary air levels.

The gases generated by the black liquor combustion rise out of the furnace and flow across convection heat transfer surface. Superheater surface is arranged at the entrance to the convection pass, followed by steam generating surface and finally the economizer. In designs featuring direct contact evaporators, the flue gas may flow from the boiler bank to the evaporator with no economizer surface provided, or a relatively small economizer may be required.

Feedwater enters the recovery boiler at the bottom of the first pass economizer. Heated water from the second pass economizer is discharged into the steam drum. From the drum, saturated water is routed through pipe downcomers to lower furnace enclosure wall headers and the boiler bank. From these steam generating circuits the steam-water mixture is returned by natural circulation to the steam drum where the mixture is separated. From the drum, steam-free water is again returned to the furnace and boiler bank circuits, and water-free steam is directed to the superheater. After flowing through the superheater sections, the steam leaves the recovery boiler and is typically piped to a turbine-generator.

Boiler thermal performance

The thermal efficiency of a recovery boiler is defined as the ratio of energy output to energy input. The boiler output is a measure of the energy transferred to the feed-

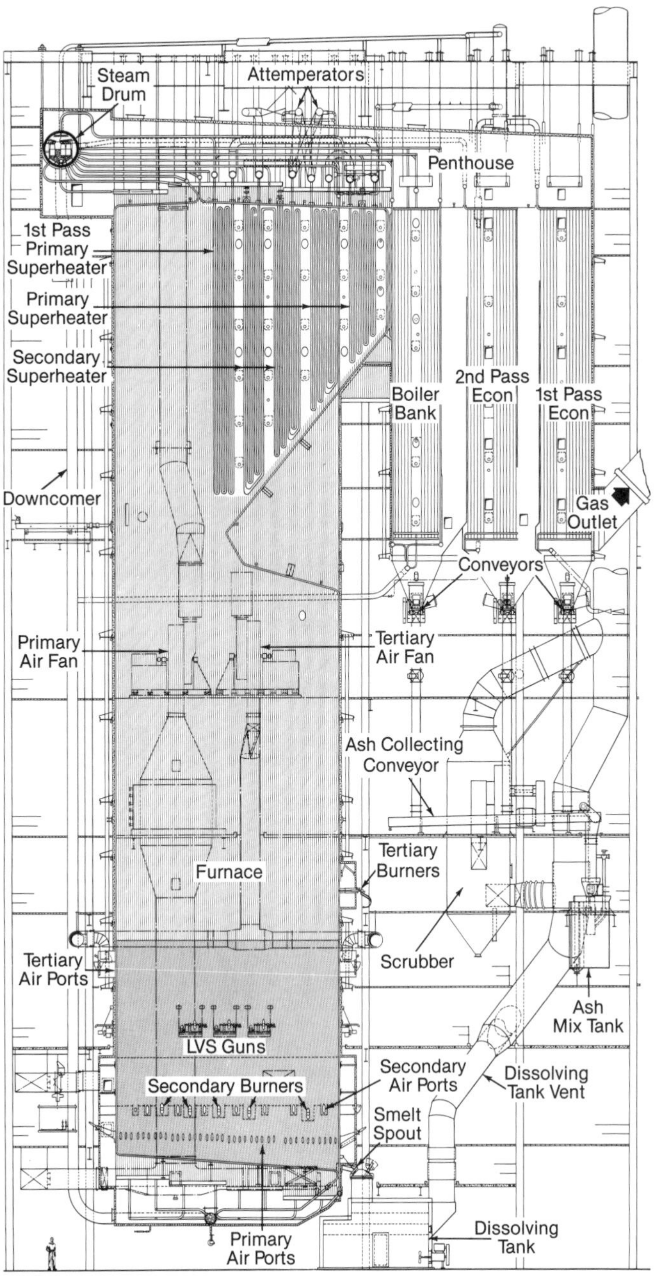

Fig. 3 Typical modern recovery boiler.

water in generating steam and can be expressed as:

$$\text{Output} = (H_s - H_{fw})m, \text{Btu/h (J/s)} \qquad \textbf{(2)}$$

where

H_s = enthalpy of steam leaving superheater, Btu/lb (J/kg)
H_{fw} = enthalpy of entering feedwater, Btu/lb (J/kg)
m = steam or water flow rate, lb/h (kg/s)

Boiler water is frequently withdrawn from the steam drum as blowdown to maintain steam purity. Steam may also be withdrawn prior to the final superheater stage for use in sootblowing. In these instances, the output expression must be corrected to account for the energy leaving the boiler prior to the superheater outlet.

The portion of the input energy available to generate steam can be determined by calculating a steady-state heat and material balance around the boiler. Because steady-state output must equal the input less energy losses, boiler efficiency can also be expressed as:

$$\text{Boiler efficiency} = \frac{\text{Output}}{\text{Input}} = \frac{\text{Input} - \text{Losses}}{\text{Input}} \qquad \textbf{(3)}$$

Fig. 4 illustrates the major streams crossing the heat and material balance boundaries. The total heat input can be calculated by summing the chemical and thermal energy contained in the streams entering the boundary. The total losses are then calculated by summing the heat losses due to endothermic reactions occurring within the boiler and the thermal energy losses of the exiting streams.

In practice, it is not feasible to precisely measure all streams entering and leaving the system boundaries. An unaccounted for heat loss and a manufacturer's margin are added to the total losses to correct the calculated efficiency for these limitations.

The gross heating value or chemical energy of black liquor is determined by combusting a black liquor sample with an excess of oxidant, under pressure, in a bomb calorimeter. Under these laboratory conditions, the combustion products predominantly exist as CO_2, water, Na_2CO_3, Na_2SO_4 and sodium chloride (NaCl). A key process in black liquor combustion is the reclamation of sodium compounds in a reduced state. The reduction reactions occurring in the recovery furnace result in different combustion products than those resulting from the bomb calorimeter procedure. These endothermic reactions account for a portion of the black liquor heating value that is not available in the recovery furnace to generate steam. To accurately determine recovery boiler efficiency, the bomb calorimeter gross heating value must be corrected for the heats of reaction of these different combustion products.

The heat of reaction correction is the difference be-

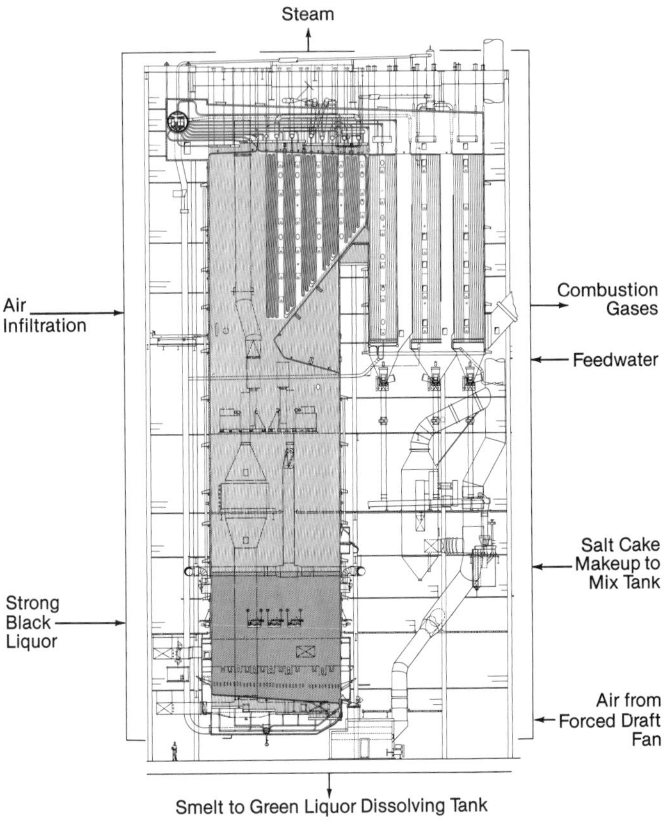

Fig. 4 Streams entering and leaving a recovery boiler.

tween the standard heat of formation of the bomb products and the heat of formation of the furnace products. Application of the heats of formation to determine a reaction correction is illustrated for kraft liquor in Fig. 5.

Step 1 This is the gross heating value of the black liquor sample determined in the bomb calorimeter.

Step 2 From the quantitative analysis of the fully oxidized bomb calorimeter compounds, the heat required to convert these products to their elemental state can be calculated from the standard heats of formation of the compounds from their elements.

Step 3 Similarly, from the quantity of each chemical compound present in the furnace combustion products, the heat of formation for the actual furnace products can be calculated.

The difference between Step 2 and Step 3 is the heat of reaction correction.

Sulfur dioxide (SO_2) and Na_2S are the most significant recovery furnace combustion products that differ from those formed under bomb calorimeter conditions. The heat of reaction correction for Na_2S is calculated as follows:

$$Na_2SO_4 = 2Na + S + 2O_2 \text{ (Step 2)}$$

$$2Na + S = Na_2S \text{ (Step 3)} \qquad \textbf{(4)}$$

The calculation is simplified by combining Steps 2 and 3 and using standard heats of formation:

$$Na_2SO_4 = Na_2S + 2O_2 \qquad \textbf{(5)}$$

ΔH_f^0 (Na_2S)	=	89.2 kcal/gmole
ΔH_f^0 (O_2)	=	0
ΔH_f^0 (Na_2SO_4)	=	- 330.9
Heat of reaction correction	=	- 241.7 kcal/gmole
	=	- 5550 Btu/lb Na_2S (-12,909 kJ/kg)

Similarly, the heat of reaction correction for sulfur dioxide can be determined from standard heats of formation from the bomb calorimeter combustion products:

$$Na_2SO_4 + CO_2 = SO_2 + Na_2CO_3 + \tfrac{1}{2}\, O_2 \qquad \textbf{(6)}$$

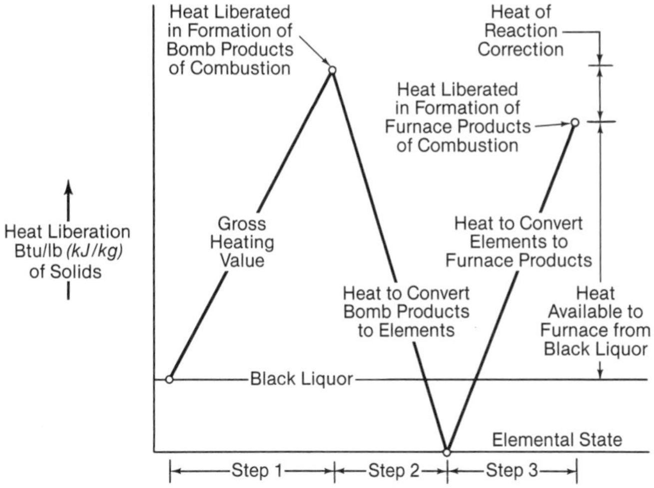

Fig. 5 Determination of black liquor heat of reaction correction.

ΔH_f^0 (SO_2)	=	71.0 kcal/gmole
ΔH_f^0 (Na_2CO_3)	=	270.3
ΔH_f^0 (O_2)	=	0
ΔH_f^0 (Na_2SO_4)	=	- 330.9
ΔH_f^0 (CO_2)	=	- 94.1
Heat of reaction correction	=	- 83.7 kcal/gmole
	=	- 2360 Btu/lb SO_2 (- 5489 kJ/kg)

In actual furnace operations, there is a variety of partially reduced, partially oxidized combustion products. However, accounting only for the presence of SO_2 and Na_2S in correcting the bomb calorimeter gross heating value closely approximates black liquor combustion in a recovery furnace.

Salt cake makeup and other additives to the black liquor are treated in a manner similar to the heat of reaction correction in calculating recovery boiler efficiency. The heat of formation or gross heating value of the Na_2SO_4 salt cake is accounted for as a contribution to the total system energy input. The subsequent reduction of Na_2SO_4 to Na_2S and O_2 is then taken as a heat loss.

The black liquor elemental analysis and gross heating value are used to determine the chemical and thermal performance of the recovery boiler. A typical black liquor analysis is presented in Table 2.

Table 3 lists the inputs and losses for a recovery unit firing 250,000 lb/h (32 kg/s) dry solids at 70% black liquor concentration based on the composition and heating value given in Table 2. Industry practice is to express the various heat losses as a percentage of the total heat input, also shown in Table 3. The system boundaries for the heat and material balance are shown diagrammatically in Fig. 6.

The black liquor gross heating value is the predominant energy input to the recovery boiler system. The balance of the input is the sum of the sensible heats contributed by those process streams entering the boiler above a base reference temperature. The black liquor is typically preheated to 220 to 260F (104 to 127C) prior to firing. A portion of the combustion air is also generally preheated to promote stable furnace conditions.

The heat of reaction correction is expressed as a heat loss due to the endothermic reduction reactions in calculating recovery boiler efficiency. To determine this heat loss, the fraction of sodium and sulfur converted to Na_2S, Na_2SO_4 and SO_2 must be calculated from the chemical

Table 2
Black Liquor Analysis

	Dry solids, % by wt
Sodium (Na)	18.20
Sulfur (S)	3.70
Hydrogen (H_2)	3.50
Carbon (C)	37.20
Oxygen (O_2)	35.10
Inerts	0.30
Potassium (K)	1.40
Chlorine (Cl)	0.60
Total solids	100.00

Solids gross heating value = 6300 Btu/lb (14,654 kJ/kg)

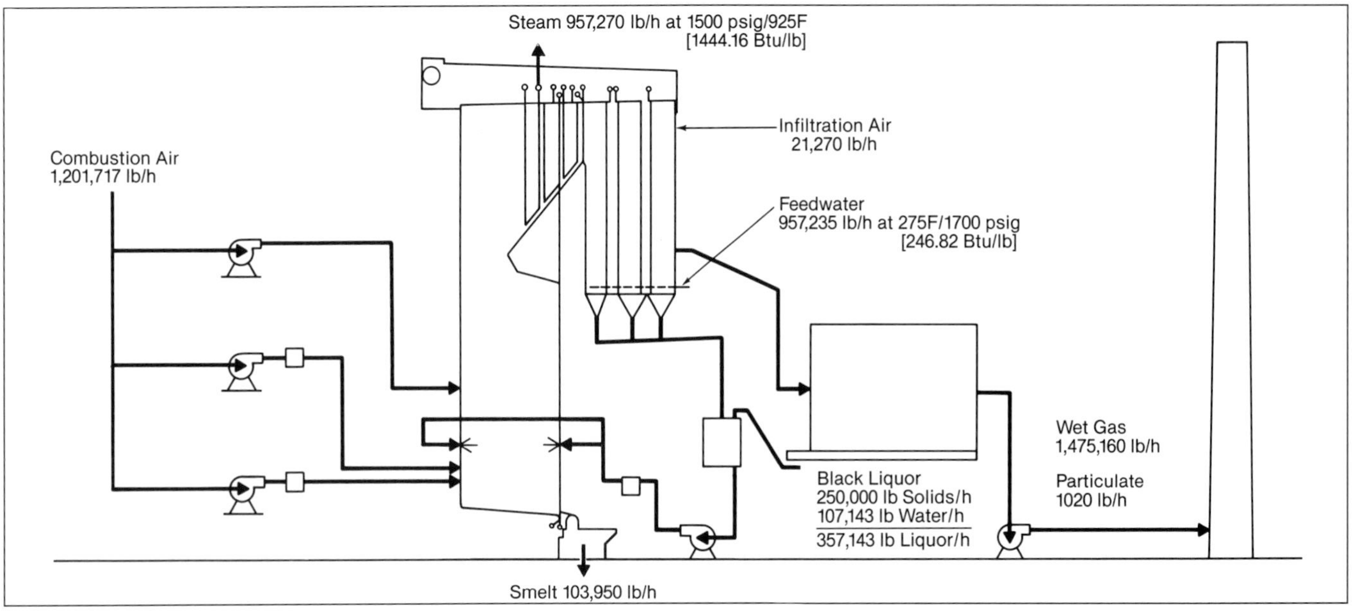

Fig. 6 Plant heat balance diagram. (See also Table 3.)

analysis of the smelt and flue gas leaving the recovery boiler. For the example presented in Table 3, 0.084 lb of Na_2S is formed for each pound of black liquor solids entering the recovery boiler. The heat of reaction correction or heat loss associated with the formation of Na_2S is calculated as follows:

$$\frac{0.084 \text{ lb } Na_2S}{\text{lb solids}} \times \frac{5550 \text{ Btu}}{\text{lb } Na_2S} \times \frac{250,000 \text{ lb solids}}{\text{h}} =$$

$$116.6 \times 10^6 \text{ Btu/h}$$

In addition, 0.0016 lb of SO_2 is formed in the reduction of Na_2SO_4 to Na_2CO_3:

$$\frac{0.0016 \text{ lb } SO_2}{\text{lb solids}} \times \frac{2360 \text{ Btu}}{\text{lb } SO_2} \times \frac{250,000 \text{ lb solids}}{\text{h}} =$$

$$0.9 \times 10^6 \text{ Btu/h}$$

The heat loss due to the reduction reactions is the sum of these heat of reaction corrections:

$$\text{Reduction reaction heat loss} = 116.6 \times 10^6 + 0.9 \times 10^6 = 117.5 \times 10^6 \text{ Btu/h}$$

In addition to the heat of reaction correction and the heat loss attributed to reducing salt cake makeup, energy is lost from the boiler in the form of sensible heat. Heat is also lost through water vaporization and through the molten smelt. Typically, smelt leaving the recovery furnace represents 532 Btu/lb (1237 kJ/kg) of heat consumed to melt the smelt and raise its temperature to a nominal 1550F (843C). The balance of the heat losses are determined in a manner similar to those for conventional power boilers. (See Chapter 21.)

The contribution to boiler efficiency offered by the remaining streams crossing the recovery unit heat and material balance is primarily established by their sensible heat content at the temperature at which they cross the system boundary. The minimum temperature of the flue gas leaving the boiler is selected to minimize corrosion. The heat transfer surface arrangement and ther-

Table 3
Material and Energy Balances for a Recovery Boiler Firing 250,000 lb/h Dry Solids at 70% Liquor Concentration

Material balance:			
Entering combustion air	=	1,201,717 lb/h	
Entering infiltration air	=	21,270 lb/h	
Entering black liquor	=	357,143 lb/h	
Total in		1,580,130 lb/h	
Smelt leaving	=	103,950 lb/h	
Wet gas leaving	=	1,475,160 lb/h	
Particulate leaving	=	1,020 lb/h	
Total out		1,580,130 lb/h	
Energy balance:		10^6 Btu/h	% Total
Chemical heat in liquor	=	1575.00	94.40
Sensible heat in liquor	=	41.06	2.46
Sensible heat in air	=	52.35	3.14
Input	=	1668.41	100.00
Sensible heat in dry gas	=	89.59	5.37
Moisture from air	=	2.17	0.13
Moisture from hydrogen	=	92.93	5.57
Moisture from liquor	=	126.47	7.58
Reduction reactions	=	117.46	7.04
Heat in smelt	=	55.22	3.31
Radiation	=	5.00	0.30
Unaccounted for and manufacturer's margin	=	33.37	2.00
Losses	=	522.21	31.30

Boiler efficiency =

$$\frac{\text{Input} - \text{Losses}}{\text{Input}} = \frac{1668.41 - 522.21}{1668.41} = 68.70\%$$

Output = Efficiency × Input =

$$\frac{68.70}{100} \times 1668.41 \times 10^6 = 1146.2 \times 10^6 \text{ Btu/h}$$

Steam flow =

$$\frac{\text{Output}}{H_s - H_{fw}} = \frac{1146.2 \times 10^6}{(1444.16 - 246.82)} = 957,270 \text{ lb/h}$$

modynamic considerations then dictate the economic limit for the flue gas exit temperature, typically 350 to 400F (177 to 204C).

The gross heating value of a given black liquor sample is strongly influenced by its carbon content. As this content increases, the heating value typically increases, as illustrated in Fig. 7. An increased liquor heating value also generally corresponds to an increased hydrogen content, with a corresponding decrease in inorganic sodium and sulfur contents. These factors result in an increased quantity of theoretical air (see Chapter 9) required to combust the black liquor. The overall trends can be summarized as follows:

1. Carbon (and hydrogen) content increases with increasing heating value.
2. Inorganic sodium and sulfur contents decrease with increasing heating value.
3. Theoretical air increases with increasing carbon and hydrogen contents.
4. Theoretical air increases with increasing heating value.

These trends can be used as quick checks on laboratory results for a given black liquor sample chemical analysis and gross heating value.

Black liquor as a fuel

Black liquor

Black liquor is a complex mixture of inorganic and organic solids partially dissolved in an aqueous solution. Heavy or strong black liquor introduced to the recovery furnace ranges from 60 to 80% solids by weight. The organic fraction of the solids is principally derived from the hemicellulose and the lignin removed from the cellulose strands of the wood chips. The solids inorganic fraction is primarily Na_2CO_3, sodium hydrosulfide (NaHS) and oxidized sulfur compounds. Black liquor also contains various chemical elements which enter the process with the wood, as impurities in makeup limestone and salt cake, and as contaminants in makeup water. These elements include potassium, chlorine, aluminum, iron, silicon, manganese, magnesium and phosphorous. The waste stream from a chlorine dioxide generator in a bleached mill can also contribute NaCl. Potassium and chlorine directly impact the recovery boiler design and operation if they are present in the black liquor in sufficient quantities.

Black liquor is sprayed into the furnace as coarse droplets which fall to the floor in a dry and partially combusted state to form a char bed. The mounded bed consists of a matrix of carbon and inorganic sodium chemicals rising 3 to 6 ft (1 to 2 m). The black liquor droplets sprayed into the furnace must be large enough to minimize droplet entrainment in the rising combustion gases, yet small enough so that they fall to the bed nearly dry. Wet liquor droplets reaching the bed can quench the burning char and cause a bed blackout, or result in high sulfur emissions.

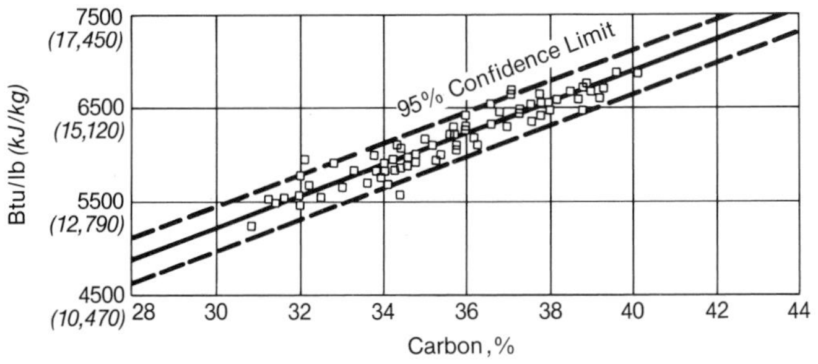

Fig. 7 Black liquor high heating value as a function of carbon content in dry solids.

The design of the recovery furnace must promote combustion of the black liquor in parallel with the efficient reduction of sodium compounds. The absolute reduction efficiency is determined by the degree to which sulfur is present in the smelt in a reduced state, such as Na_2S and NaHS.

$$\text{Reduction efficiency} = \frac{Na_2S + NaHS}{\text{Total sodium sulfur compounds}} \times 100, Na_2O \quad (7)$$

A common industry simplification is as follows:

$$\text{Reduction efficiency} = \frac{Na_2S}{Na_2S + Na_2SO_4} \times 100, Na_2O \quad (8)$$

The industry practice is to express the compounds in the equations as the equivalent weight of Na_2O.

Emissions

The black liquor combustion process is never theoretically complete. This results in small concentrations of unburned combustibles, typically carbon monoxide (CO), organic and sulfur compounds, and hydrogen sulfide (H_2S), being discharged to the atmosphere. The volatile organic compounds, or VOC, are generally expressed in terms of equivalent methane (CH_4) and are sometimes more specifically referred to as nonmethane volatile organic compounds (NMVOC). H_2S and sulfur-bearing organic compounds such as mercaptans are grouped together as total reduced sulfur (TRS). Trace amounts of SO_2 also exist in addition to TRS-bound sulfur. As in most combustion processes, nitrogen oxides (NO_x) are present and are expressed in terms of equivalent nitrogen oxide (NO_2). Black liquor combustion also creates particulate matter.

The modern recovery boiler achieves effective NO_x control by staged air combustion, control of excess air, and a uniform distribution of the black liquor through multiple burners. A recovery furnace inherently produces lower NO_x emissions compared to fossil fuel boilers.

SO_2 emissions are a function of the sulfidity of the smelt. Data from a series of B&W tests on one boiler show the variation of SO_2 with sulfidity (Fig. 8). An environmental benefit of the increased black liquor concentration fired in recovery boilers is a reduction in SO_2 emissions.

Concentrations of TRS in the combustion gases leaving a modern boiler are readily controlled below 5 ppm, as the H_2S and volatile organic sulfide compounds are oxidized in the high temperature furnace. VOC emissions can be controlled by proper furnace design and operation. A hot furnace and thorough mixing of combustion air with the generated volatiles are essential in minimizing VOC, TRS and CO emissions. Particulate is removed from the combustion gases in a high efficiency electrostatic precipitator.

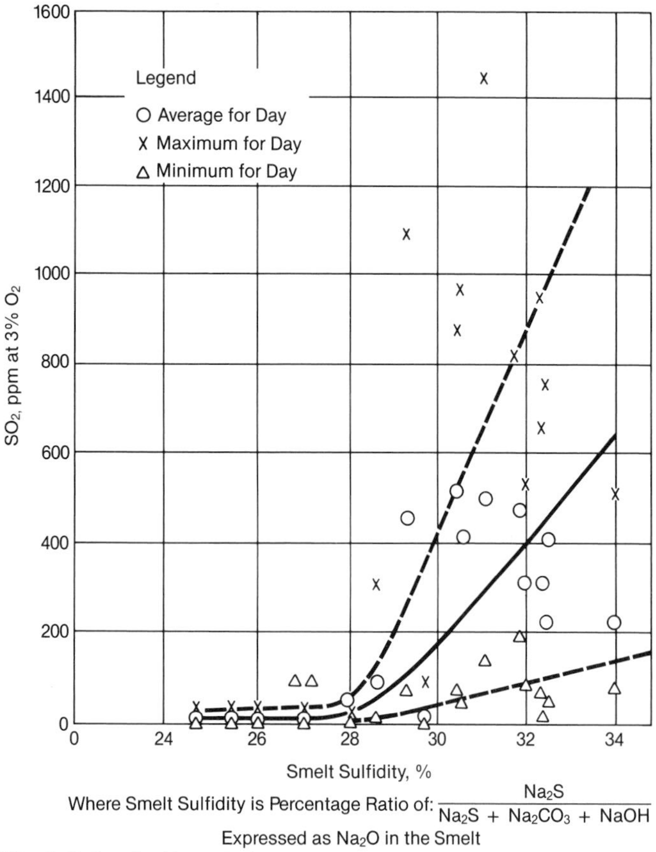

Fig. 8 Sulfur dioxide emissions.

Ash

The characteristics of ash from black liquor combustion impact the design of the process recovery boiler. Approximately 45% by weight of the dry, as-fired solids is inorganic ash. The majority of these inorganics are removed from the furnace as Na_2S and Na_2CO_3 in the molten smelt. A significant amount of ash is present as particulate entrained in the existing flue gases. Generally, about 8% by weight of the entering black liquor solids leaves the furnace as ash.

Ash is generally categorized as *fume* or *carryover*. Carryover consists of char particles and black liquor droplets that are swept away from the char bed and liquor spray by the upward flue gas flow. Entrainment occurs when small particles caught in the furnace gases are not of sufficient size, shape or density to fall back into the furnace. Entrainment causes combustion of black liquor in the upper furnace which affects temperature and ash deposit properties. Entrainment of smelt and char materials is a major cause of convection surface plugging.

Once entrained, the black liquor carryover droplet follows the gas flow. When complete particle burnout occurs, the entrained droplet can settle out of the gas flow as a smelt bead. Otherwise, the partially combusted particles form sparklers which deposit on tubes as char, then continue to burn and yield a smelt deposit. At low loads with lower furnace gas flow rates, entrained droplets have time to burn out, and only small smelt droplets show up as carryover. As load is increased, larger drops can be entrained by the correspondingly increased gas flow. The particles can then include small smelt droplets and large char particles. Carryover can be controlled by furnace size and by proper design and operation of the firing and combustion air systems.

Fume consists of volatile sodium compounds and potassium compounds rising into the convection sections of recovery boilers. These volatiles condense into submicron particles that deposit onto the superheater, boiler bank and economizer surfaces. Fume particles in kraft recovery boilers are usually 0.25 to 1.0 microns in diameter and consist primarily of Na_2SO_4 and a much lower content of Na_2CO_3. Fume also contains potassium and chloride salts.

The much larger carryover particles, typically 5 to 100 microns, are easily distinguishable from the submicron fume particles. Fume and carryover ash are also different in their chemical analyses. Carryover is similar in composition to the smelt. Fume is mostly Na_2SO_4 and is enriched in potassium and chloride relative to their concentration in the smelt.

Fume can contribute to deposit formation and plugging in the convection heat transfer sections of the boiler, particularly if allowed to sinter and harden. Fume particles are also the predominant source of particulate emissions from recovery boiler stacks.

Furnace temperature controls the fuming rate. A rate just sufficient to capture the sulfur released during combustion should be established. This minimizes the dust load to the precipitator and the SO_2 to the stack.

As potassium (K) has a higher vapor pressure than sodium (Na), the fume contains a higher ratio of K/Na than that found in smelt or black liquor. This is referred to as potassium enrichment. Chlorides are also found at higher concentrations in fume. Potassium and chlorides can contribute to severe plugging in the recovery boiler convection surfaces.

Ash enrichment by chloride reduces the ash melting point. This characteristic sticky temperature has been defined as the temperature where 20% of an ash deposit sample is molten. The presence of K in combination with chlorides further reduces the sticky temperature of deposits. The recovery boiler should be designed to reduce the gas temperature entering the boiler bank to below the ash sticky temperature to avoid bank plugging. Decreasing the Cl level in the black liquor can also decrease the plugging tendency of the resulting ash.

Recovery boiler design evolution

The kraft recovery process evolved in Danzing, Germany some 25 years after the soda process was developed in the United Kingdom in 1853. In 1907, the kraft recovery process was introduced in North America. From its inception, a variety of furnace types competed for a successful commercial design, including rotary and stationary furnaces. During the late 1920s and early 1930s, significant design developments were achieved by G.H. Tomlinson, working in conjunction with B&W engineers.

The first Tomlinson recovery boiler was supplied by B&W Canada in 1929 at the Canada Paper Company's Windsor Mills, Quebec plant (Fig. 9). This black liquor recovery boiler had refractory furnace walls that proved costly to maintain. The steam generated with the refractory furnace was also much less than that theoretically possible. Tomlinson decided that the black liquor recov-

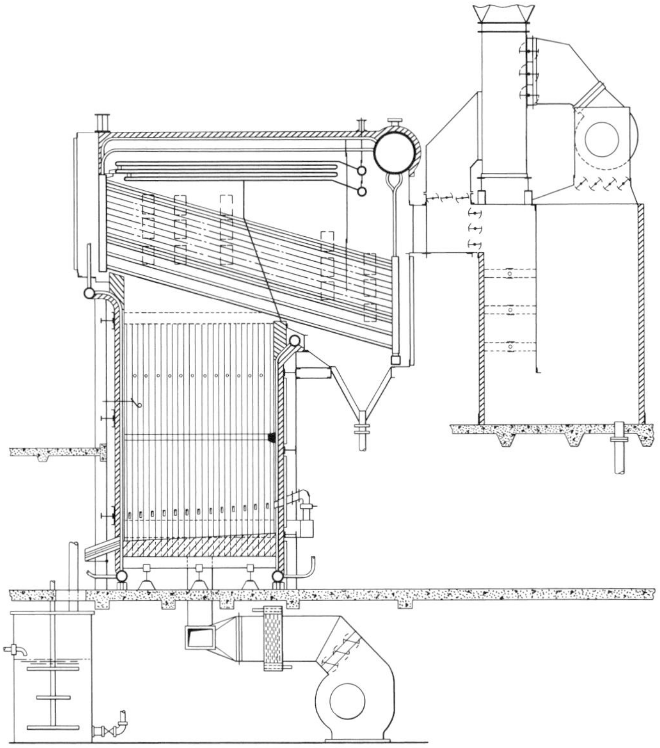

Fig. 9 First Tomlinson recovery boiler.

ery furnace should be completely water-cooled, with tube sections forming an integral part of the furnace. This new concept boiler, designed in cooperation with B&W, was installed at Windsor Mills in 1934. The water-cooled design was a complete success, and the boiler operated until 1988. The first Tomlinson recovery boilers in the U.S. were two 90 B&W-Btu t/day units sold to the Southern Kraft Corporation in Panama City, Florida in 1935.

The Tomlinson design evolved with a technique of spraying black liquor onto the furnace walls. The liquor is dehydrated in flight and on the furnace walls, where pyrolysis begins with the release of volatile combustibles and organically bound sodium and sulfur. As the liquor mass builds on the furnace walls, its weight eventually causes it to break off and fall to the hearth. There, pyrolysis is completed and the char is burned, providing the heat and carbon required in the reduction reaction.

By the end of World War II, the recovery boiler design (Fig. 10) had evolved to the general two-drum arrangement that represented B&W's standard product until the mid 1980s. Retractable sootblowers using steam as a medium eliminated hand lancing in the 1940s; this significant development made large recovery boiler designs practical.

Wall construction

By 1946, wall construction had evolved from tube and refractory designs to a completely water-cooled furnace enclosure, using flat plate studs to close the space between tubes and to minimize smelt corrosion and the resultant smelt leaks. The flat stud design was superseded in 1963 with membrane tube construction where the gas-tight seal is along the plane of the wall rather than formed by casing behind the wall.

The 1963 furnace wall construction had 3 in. (76 mm) OD tubes on 4 in. (102 mm) centers. The advantages of this construction included less air infiltration, reduced refractory maintenance and a completely gas-tight unit. The design used cylindrical pin studs for corrosion protection of the tubes in the reducing zone of the lower furnace. The pin studs held solidified smelt, forming a barrier to the corrosive furnace environment. The current construction calls for 64 half inch (13 mm) diameter studs per linear foot (.3 m) of tubing.

The lower furnace design continued to evolve in the 1980s from the traditional pin stud arrangement to the use of composite or bimetallic tubes. The composite tubes are comprised of an outer protective layer of AISI 304L stainless steel and an inner core layer of standard American Society of Testing and Materials (ASTM) A 210 Grade A1 carbon steel. The composite tube inner and outer components are metallurgically bonded.

The outer layer of austenitic stainless steel, which is also used to cover the furnace side of the carbon steel membrane bar, protects the core carbon steel material from furnace corrosion.

Increased industry emphasis on high pressure and temperature operation, along with the higher availability demanded of large boilers and the trend of many mills being dependent upon a single recovery boiler, have re-

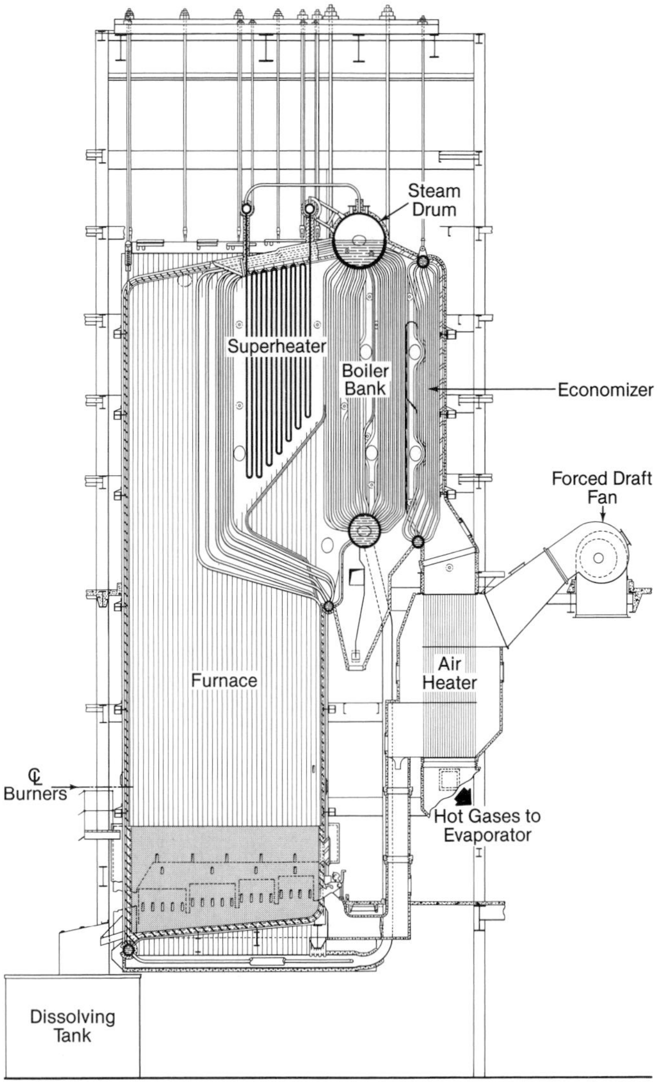

Fig. 10 General two-drum arrangement of the 1940s.

quired the decreased maintenance afforded by modern composite tubes. The single-drum boiler designed in 1987 featured readily available 2.5 in. (64 mm) OD composite tubes with 0.5 in. (13 mm) wide membrane bars. Fig. 11 chronicles the evolution of furnace wall construction with the decreasing width between tube seal bars.

Evolution of the modern, single-drum design

The 1980s saw an increase in the pulp and paper industry's need for high pressure and temperature steam generation from the recovery boiler. This trend was paralleled by the demand for large, conservatively sized furnaces and general acceptance of the single-drum, all welded boiler design. B&W commissioned its first modern single-drum boiler in 1989 at Gaylord Container Corporation in Bogalusa, Louisiana.

In the two-drum design arrangement, the drums are exposed to combustion gases which limit the drum length that can be effectively supported. This maximum length established the design capacity of the two-drum arrangement at about 5 x 10^6 lb (2.3 x 10^6 kg) of solids per day. In the single-drum arrangement, the steam drum is moved out of the gas flow path, thereby removing this limitation and allowing recovery boilers designed to process daily solids rates of 8 x 10^6 lb (3.6 x 10^6 kg).

Design for low odor

The air pollution legislation of the mid 1960s forced major changes in recovery boiler design. To reduce malodorous emissions, the direct contact evaporator was replaced by additional multiple effect evaporator capacity to obtain the optimum liquor concentration. Economizer surface was added for flue gas cooling that was previously accomplished in the direct contact evaporator.

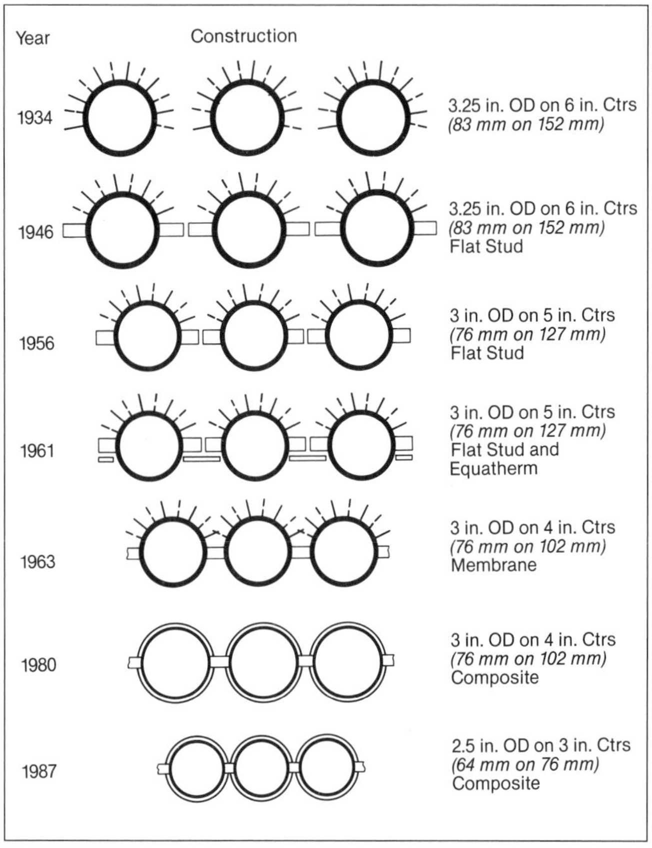

Fig. 11 Evolution of wall construction.

Superheater design

B&W's first recovery boiler designed for elevated pressure and temperature was placed in operation in 1957 at Continental Can Company (now Stone Container) in Hodge, Louisiana; it generated steam at 1250 psig (86.2 bar gauge) and 900F (482C).

B&W's high steam temperature design philosophy is reflected in the arrangement of superheater surface. The inlet primary superheater bank is placed following the furnace cavity, with steam flowing through the bank parallel to the gas flow. This results in the coolest available steam flowing through the superheater tubes exposed to the hottest gas temperatures and the radiant heat from the furnace. This arrangement minimizes the metal temperature of the superheater tubes.

Ash buildup and superheater surface plugging have frequently limited availability of the recovery boiler. To avoid these conditions, in 1968 B&W established a 12 in. (305 mm) side spacing for the entire superheater bank, abandoning the conventional 5 and 6 in. (127 and 152 mm) spacing in the secondary superheater construction. This increased the clear side spacing in the superheater from 3.5 to 9.5 in. (89 to 241 mm).

Combustion air system

The first recovery boiler to introduce air at three furnace levels — primary, secondary and tertiary — was built by B&W in the late 1940s. In 1956, a concentrated development effort was successful in providing fuller utilization of tertiary air, which had been largely ineffective in earlier designs. Today's Advanced Air Management system is a result of extensive laboratory scale and computer flow modeling, theoretical consideration for the air penetration across the furnace at the secondary and tertiary air levels, and testing of different air system configurations on operating boilers. Several features resulting from this development program include the use of variable velocity control dampers on the secondary and tertiary air ports to better regulate air penetration across the furnace area, an interlaced port arrangement to provide better gas mixing within the furnace, and optimization of port location.

Design considerations for B&W recovery boiler

Furnace design

The design of the recovery boiler and its associated equipment systems must first consider efficient black liquor combustion.

The black liquor solids concentration determines whether the droplets are dried in suspension or deposited on the furnace walls. Suspension drying is used for solids concentrations above about 68%. At lower levels, additional droplet drying time is required. This is achieved by spraying the liquor on the walls between the secondary and tertiary air port levels where it is dehydrated prior to falling onto the hearth char bed.

Primary air enters the furnace around the perimeter of the hearth bed. The controlled reducing atmosphere at the hearth burns the char at the bed surface to effect maximum reduction of Na_2SO_4 to Na_2S in the smelt. The remaining air is admitted at the secondary and tertiary air zones. The total air admitted through the primary and secondary air ports is approximately the stoichiometric requirement for black liquor combustion. High pressure air entering through large secondary ports penetrates across the furnace to assure mixing with the volatile gases rising from the char bed. Combustion at the secondary air level achieves a maximum furnace temperature zone below the liquor spray for drying the liquor. Secondary air also limits the height of the char bed by providing air for combustion across the bed surface.

Further turbulence and mixing are created by admission of tertiary air, which assures complete combustion of unburned gases rising from the secondary zone and of volatiles escaping from the sprayed liquor. Tertiary air mass penetration also provides a uniform temperature and velocity profile of combustion gases entering the convection surface.

The recovery boiler furnace must also be designed for efficient removal of the molten inorganic chemicals as smelt. Finally, the combustion gases and particulate carryover must be adequately cooled in the furnace to minimize deposition on convection surfaces.

The first step in the design of a new recovery boiler is the selection of furnace plan area, defined as the furnace width times depth. The plan area is generally set to achieve a black liquor solids heat input of 800,000 to 850,000 Btu/h ft^2 (2.52 to 2.68 MW_t/m^2). As the solids heating value decreases, a larger furnace plan area is desirable. However, an oversized plan area can lead to local cold spots on the smelt bed, which in turn limit reduction efficiency. Cold spots can also lead to unstable furnace blackouts. A large plan area further constrains load turndown with stable combustion. An undersized plan area generally leads to increased fume and particle carryover.

Once the plan area is established, the width and depth dimensions are selected. Maintaining a depth to width ratio between 1.0 and 1.15 allows an effective arrangement of the combustion air ports and generally permits an economical arrangement of convection pass heat transfer surfaces. In smaller recovery furnaces designed for high steam temperature, a higher aspect ratio permits increasing the depth to accommodate the large superheater surface. Furnace height is then determined by the radiant furnace heat transfer surface required to cool the combustion gases below 1700F (927C).

The surface of the floor and wall tubes in the lower furnace must be protected against the corrosiveness of the smelt and partially combusted gases. The most widely accepted approach today is to build the furnace and floor of composite tubes. This tube construction should extend to 3 ft (0.9 m) above the tertiary air ports. Above this elevation, carbon steel tube and membrane construction is adequate.

Wide closure plates attached to tubes bent to form air ports and other openings can result in potentially high localized stress areas. Wide closure plates suffer from corrosion, burn back and cracking. This adversely affects the air flow area of the port opening and has the potential for closure plate cracking that propagates into the furnace wall tubes. The modern furnace openings are designed without closure plates to minimize this potential (Fig. 12).

Structural attachments must also minimize tube stresses. In high stress areas, a plate stamping is welded to the tubes, and the structural member is attached to this plate. The windbox attachment to the tube wall, shown in Fig. 12, uses plate stampings which are shop-attached.

Fig. 12 Burner and secondary air port windbox attachments.

Upper furnace and arch arrangement

Combustion is completed in the tertiary zone. The water-cooled furnace walls and volume above this zone provide the necessary surface and retention time to cool the gas to temperatures where sootblowers can effectively remove the chemical ash from convection surfaces.

The furnace arch, or *nose*, serves several important functions. The arch shields the superheater from the radiant heat of the furnace. The high temperature steam loops of the superheater are completely protected. Penetration of the arch into the furnace uniformly distributes the gas entering the superheater. An eddy above and behind the arch tip causes the gas to recirculate in the superheater tube bank, with a reverse gas flow between the superheater and the upper arch face preventing hot gas from bypassing the superheater surfaces. The angle of the arch is set to minimize the repose of deposited ash on its surface.

Furnace screen

In some recovery boilers, the superheater surface is insufficient to adequately cool the combustion gases before they enter the boiler bank. This is common in boilers designed for low steam temperature. A furnace screen can be used to absorb the additional heat, thereby maintaining an acceptable temperature of the gas entering the boiler bank section. However, it is preferable to avoid a furnace screen, thereby eliminating an additional generating circuit. The required heat absorption can often be accomplished through added furnace surface (height) and/or an oversized superheater.

When a furnace screen is required, the screen tubes are designed with wide side spacing to reduce pluggage potential. The horizontal section of the screen originates inside the furnace arch to limit its length. Tubes in the horizontal section of the screen are joined to provide structural integrity.

Convection surface

After leaving the furnace, the flue gases pass across the steam-cooled superheater banks to the long flow boiler bank and finally to the economizer sections. (See

Fig. 3.) As the gas is cooled, entrained ash becomes less sticky and adheres less to the tube surfaces. As a result, it is possible to space the tubes in the convection banks progressively closer together. The closer spacing results in higher gas velocities and improved convection heat transfer rates, which in turn permit a more economical design as less heat transfer surface area is required.

Superheater

The superheater surfaces are exposed to the highest gas temperatures and, consequently, are arranged on a 12 in. (305 mm) side spacing. This results in very low gas velocities.

In arranging superheater surface, it is desirable to maintain low tube temperatures. Lower temperatures reduce the potential for high temperature corrosion and allow the use of less expensive low alloy steel. Temperatures are reduced by establishing a high steam flow through each tube, by arranging the coolest steam to flow through the superheater tubes exposed to the hottest gas temperatures, and by locating the majority of the superheater tube banks behind the furnace arch tip, shielded from furnace radiation.

From the drum, saturated steam enters the front tube row in the first or primary inlet superheater bank and flows through successive tube loops in parallel with the flue gas flow. (See Fig. 3.) The secondary superheater is located in the cooler gas region behind the inlet primary bank. Steam flow in this secondary superheater is generally opposite to the gas flow. Typically, a third bank of counterflow primary superheater surface follows in the direction of gas flow behind the secondary bank, which results in reasonable gas to steam approach temperatures throughout the superheater section. Careful selection of superheater tube materials permits final steam temperatures up to 950F (510C) with this surface arrangement.

The superheater banks are top supported with the tube elements expanding downward. The tubes are interconnected with flexible support ties which allow independent tube expansion. Tube movement is critical to effective sootblowing. However, lack of tube restraints within a bank can lead to failure at the bank's top supports.

Boiler bank

Today's kraft recovery boiler generally incorporates a single steam drum, with a *long flow* boiler bank arranged downstream of the superheater. In passing across the superheater and rear wall screen tubes, the flue gas should be cooled below the ash sticky temperature prior to entering the boiler bank. For ashes with extremely low sticky temperatures, particular attention must be given to sootblower locations.

The boiler bank is constructed of shop-assembled tube sections arranged as modules inside a water-cooled enclosure (Fig. 13). Tubes in each section are connected to headers, with water entering the lower header and the steam-water mixture exiting the upper headers. As the flue gas enters the bank, it turns downward and flows parallel to the tube length, providing easy cleanability. Within the bank, a central cavity accommodates fully retractable sootblowers. The cavity permits personnel access for visual tube inspection adjacent to sootblower lance entry. Ash deposits dislodged during sootblowing are collected in a trough hopper connected to the bank enclosure. Tube section inlet headers are widely spaced and vertically staggered to facilitate ash dropping into the hopper.

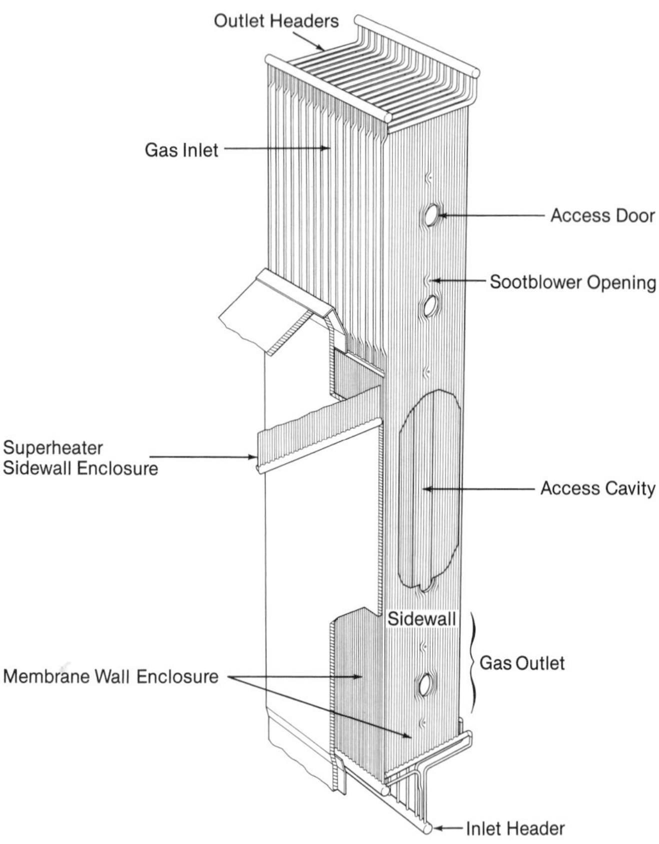

Fig. 13 Boiler bank isometric.

Impact-type particle deposition on the boiler bank tubes is less likely to occur with the gas long flow orientation. As a result, the allowable gas velocity in the downflow portion of the bank can be increased.

To improve heat transfer, longitudinal fins are welded to the front and back of each tube. Fins are tapered at the ends and welded to the tube on both sides. The welds are terminated by wrapping around the end of the fin. This combination of welding technique and tapered ends assures minimal stress concentration at fin termination for fins as large as 2.5 in. (62.5 mm).

Economizer

The boiler bank surface area is typically set to achieve a nominal exit gas temperature of about 800F (427C). This temperature maintains a reasonable differential with the saturated steam temperature [610F (321C) for a 1650 psig (113.8 bar gauge) drum pressure] and allows the use of carbon steel casing to enclose the downstream economizer banks. The modular economizer has vertical finned tubes arranged in multiple sections with upward water flow and downward long flow of gas (Fig. 14). The common arrangement features two banks. The flue gas enters at the upper end and discharges at the lower end of each bank. Gas flows down the length of the bank to provide good cleanability. As in the boiler bank, a central cavity dimensioned for personnel access accommodates fully retractable sootblowers. Trough hoppers are attached to the economizer casing to collect dislodged ash deposits.

The economizer surface area is set to achieve a final gas outlet temperature approximately 100F (56C) higher

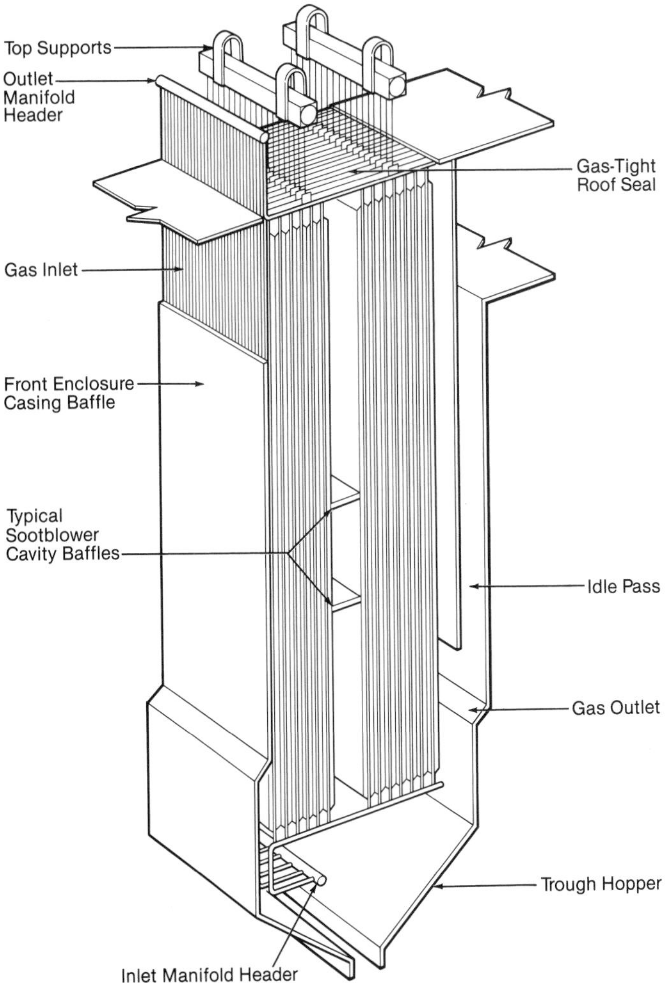

Fig. 14 Economizer isometric.

than the feedwater temperature. Although it is possible to achieve an exit gas temperature closer to that of the feedwater, the decreased temperature differential results in substantially increased surface requirements for small improvements in end temperature. In addition to this thermodynamic limitation, concern for cold end corrosion generally establishes a minimum gas exit temperature around 350F (177C). The minimum recommended temperature of the feedwater entering the economizer is 275F (135C) for corrosion protection of the tube surface. With special considerations, the feedwater entering the economizer can be designed for as low as 250F (121C).

Emergency shutdown system

An emergency shutdown procedure for black liquor recovery boilers has been adopted by the Black Liquor Recovery Boiler Advisory Committee in the U.S. An immediate emergency shutdown must be performed whenever water enters the furnace and can not be stopped immediately, or when there is evidence of a leak in the furnace setting pressure parts. The boiler must be drained as rapidly as possible to a level 8 ft (2.4 m) above the mid point of the furnace floor.

An auxiliary fuel explosion can occur when an accumulated combustible mixture is ignited within the confined spaces of the furnace and/or the associated boiler passes, duct work and fans which convey the combustion gases to the stack. A furnace explosion will result from ignition of this accumulation if the quantity of the combustion mixture and the proportion of air to fuel are within the explosive limit of the fuel involved. The magnitude and intensity of the explosion will depend upon both the quantity of accumulated combustibles and the proportion of air in the mixture at the moment of ignition.

Contacting molten smelt with water can also result in a very powerful explosion. The mechanism for a smelt-water explosion is keyed to the contact of water with hot liquid smelt. Rapid water vaporization causes the propagation of a physical detonation or shock wave.

In the design and operation of black liquor recovery boilers, every effort is made to exclude water from the furnace, including water entrainment in combustion air. For example, furnace attachment details are designed to prevent external tube loads, which can lead to stress assisted corrosion.

Recovery boiler auxiliary systems

Black liquor evaporation

The high black liquor solids concentration required for efficient burning is achieved by evaporating water from the weak black liquor. Large amounts of water can be economically evaporated by multiple effect evaporation. A multiple effect evaporator, illustrated in Fig. 15, consists of a series of evaporator bodies, or effects, operating at different pressures.[6] Vapor from one body becomes the steam supply to the next, operating at a lower pressure. As a general rule, each pound of water evaporated from the weak liquor results in one additional pound of high pressure steam generation.

Modern evaporator systems integrate a concentrator into the flow sequence to achieve the final liquor concentration. Several manufacturers use forced circulation evaporator bodies designed specifically to control heat transfer surface scaling that can develop from soluble and insoluble compounds in the black liquor. These concentrators operate on the principle of controlling precipitation and crystallization of the supersaturated liquor constituents. Evaporation and concentration of the liquor results in certain salts and inorganic compounds exceeding their solubility limits and precipitating to produce a sludge. Recycling the small sludge particles results in the precipitating compounds depositing on the relatively high surface area of the particles instead of the tube walls, thereby minimizing heat transfer surface scaling. High velocity and pressure of recirculation liquor in the tubes suppress boiling in the heat exchanger in contrast to a film type concentrator, where boiling must occur in the tubes. Boiling in the tubes at high concentrations can bake the insoluble compounds on the tube surfaces.

High solids evaporation is not possible with liquor from some fiber sources, particularly fibrous plants. These species are generally associated with high viscosity and silica contents that can cause severe evaporator surface fouling above about 45% solids. Commonly, these installations use direct contact evaporation. In the direct contact evaporator, liquor and flue gas are brought together and mass transfer of liquor water vapor to the gas occurs across the liquor-gas interface. While there is a gas temperature decrease in the evaporator, the total enthalpy of the exiting gas and evaporated water is nearly the same as the enthalpy of the

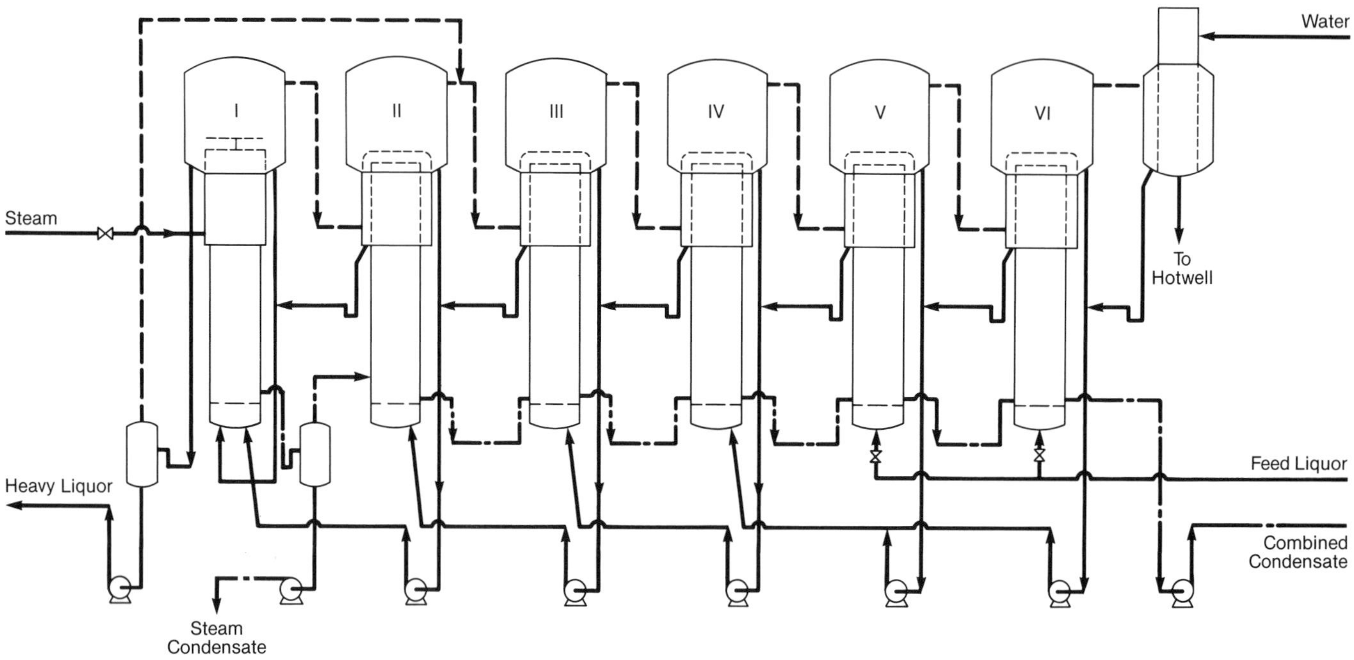

Fig. 15 Multiple effect evaporator.

gas and liquor entering the evaporator. Any difference may be accounted for by radiation loss from the evaporator, air leakage into the evaporator, and the sensible heat given up or absorbed by the liquor.

In the direct contact evaporator, adequate liquor surface must be provided for the heat and mass transfer. The gas contact acidifies the liquor by absorbing CO_2 and SO_2, which decreases the solubility of the dissolved solids and requires continuous agitation. The acidification also results in the release of malodorous compounds into the flue gas.

There are two types of direct contact evaporators used in the recovery unit, cyclone and cascade. The cyclone evaporator (Fig. 16) is a vertical, cylindrical vessel with the flue gas admitted through a tangential inlet near the conical bottom. The gas flows in a whirling helical path to the cylinder's top and leaves through a concentric re-entrant outlet. Black liquor is sprayed across the gas inlet to obtain contact with the gas. The liquor droplets mix intimately with the high velocity gas and are centrifugally forced to the cylinder wall. Recirculated liquor flowing down the cylinder wall carries the droplets and any dust or fumes from the gas to the conical bottom, out through the drain, and into an integral sump tank. Sufficient liquor from the sump tank is recirculated to the nozzles at the top of the evaporator to keep the interior wall wet, preventing ash accumulation or localized drying.

In the cascade evaporator, horizontally spaced tubular elements are supported between two circular side plates to form a wheel that is partially submerged in a liquor pool contained in the lower evaporator housing. The wetted tubes are slowly rotated into the gas stream. As the tubes rise above the liquor bath, the surface coated with black liquor contacts the gas stream flowing through the wheel.

Black liquor oxidation

When a direct contact evaporator is used, odor can be reduced by oxidation of sulfur compounds in the liquor before introduction to the evaporator. The oxidation stabilizes the sulfide compounds to preclude their reaction with flue gas in the evaporator and the consequent release of mercaptan compounds. Oxidation can effectively reduce, but does not eliminate, discharge of malodorous gas compounds. The direct contact evaporator is the prime source of odor.

Odor is generated in direct contact evaporators when the hot combustion gases strip hydrogen sulfide gas from the black liquor:

$$2NaHS + CO_2 + H_2O = Na_2CO_3 + 2H_2S \qquad \textbf{(9)}$$

Oxidation stabilizes the black liquor sulfur by converting it to thiosulfate:

$$2NaHS + 2O_2 = Na_2S_2O_3 + H_2O \qquad \textbf{(10)}$$

Black liquor oxidation involves high capital and operating costs; the oxidation step also robs the liquor of heating value. Most modern recovery facilities incorporate multiple effect evaporators which eliminate the need for a direct contact evaporator.

Black liquor system

Recovery boilers are operated primarily to recover pulping chemicals. This objective is best realized by maintaining steady-state operation. Recovery boilers are base loaded at a selected black liquor feed flow or heat input, in contrast to power boiler applications, where fuel flow is varied in response to demand for steam generation.

The black liquor solids concentration can vary with the rate that recirculated ash sheds from heat transfer surfaces, the rate it is collected in the precipitator, and the rate it is returned to the black liquor stream system. Considerable fluctuations can occur, depending on which surface is being cleaned by sootblowers, the frequency of sootblower operation, and the rapping sequence of precipitator collection surfaces. The black liquor system design should provide uniform dispersion of the ash into the liquor to minimize the fluctuation of solids at the burner.

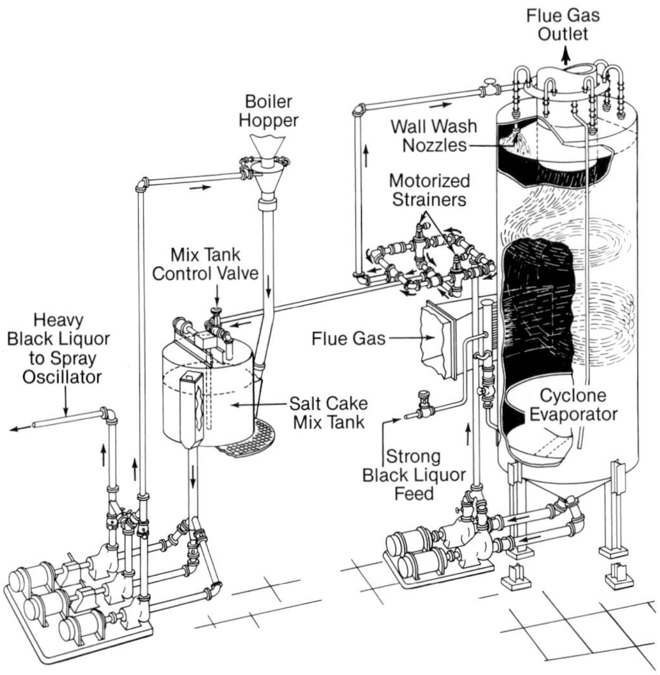

Fig. 16 Cyclone evaporator.

This is increasingly important as the liquor solids concentration is increased into the regime where burners are maintained in a fixed position and liquor drying is in-flight (not on the walls). A fluctuation in ash flow would change the solids concentration by several percent, significantly impacting the characteristics of the liquor.

A heater is used to adjust the black liquor temperature to that required for optimum combustion and minimum liquor droplet entrainment in the gas stream. The black liquor is typically heated in a tube-and-shell heat exchanger, using low pressure steam on the shell side and black liquor on the tube side. There are two general categories of heater designs that can be operated with minimum scaling of the heat transfer surface. The first uses a conventional heat exchanger with once-through flow of liquor at high velocity in the tubes. The tubes have polished surfaces to inhibit scale formation. This heat exchanger is satisfactory for long periods of operation without cleaning.

The second approach is to recirculate liquor through a standard heat exchanger with stainless steel tube surface to maintain high velocities that inhibit scale formation. This approach also permits long periods of operation without cleaning.

As the liquor solids concentration increases above 70%, storage can become a problem. Moreover, it becomes increasingly difficult to blend recycled ash into the highly concentrated liquor. Recycled ash can be returned to an intermediate concentration liquor stream (of about 65% concentration) prior to final evaporation in a concentrator. In this type of arrangement, liquor is routed to the recovery furnace from an evaporator system product flash tank. The as-fired liquor temperature is established by controlling the flash tank operating pressure.

There are two designs of liquor burners, the oscillator and the limited vertical sweep (LVS). Oscillators are used with liquors concentrated to 68% solids or less and can be used in combination with LVS burners in small furnaces for higher concentrations. Above 68% solids concentration, where drying can be completed in-flight, LVS burners are used. Both types of burners utilize a nozzle splash plate to produce a sheet spray of coarse droplets.

The oscillator sprays the black liquor on the furnace walls, where it is dehydrated and falls to the char bed. The oscillator burners, located in the center of the furnace wall between the secondary and tertiary air ports, are continuously rotated and oscillated, spraying liquor in a figure eight pattern to cover a wide band of the walls above the hearth.

In LVS burners (Fig. 17), black liquor is sprayed into the furnace for in-flight drying and devolatization of the combustible gas stream rising from the char bed. The objective of the LVS burner is to minimize the liquor on the wall. The LVS gun is normally used in a fixed position, but can sweep vertically to burn low solids liquor or those with poor burning characteristics.

The temperature and pressure of atomized liquor directly impact recovery furnace operations. Lower temperature and pressure generally create a larger particle or droplet of atomized liquor. This minimizes the entrainment of liquor in the combustion gases passing to the heat absorbing surfaces. Where wall drying is carried out, large liquor droplets maximize the liquor sprayed on the wall and minimize in-flight drying. For oscillator firing, liquor at approximately 230F (110C) and 30 psig (2.1 bar gauge) generally provides the most satisfactory operation.

As the liquor sprayed on the walls builds, it eventually falls to the char hearth. The majority of the char falling from the wall is deposited in front of the primary air ports, requiring 40 to 50% of the primary air to be introduced through the primary ports.

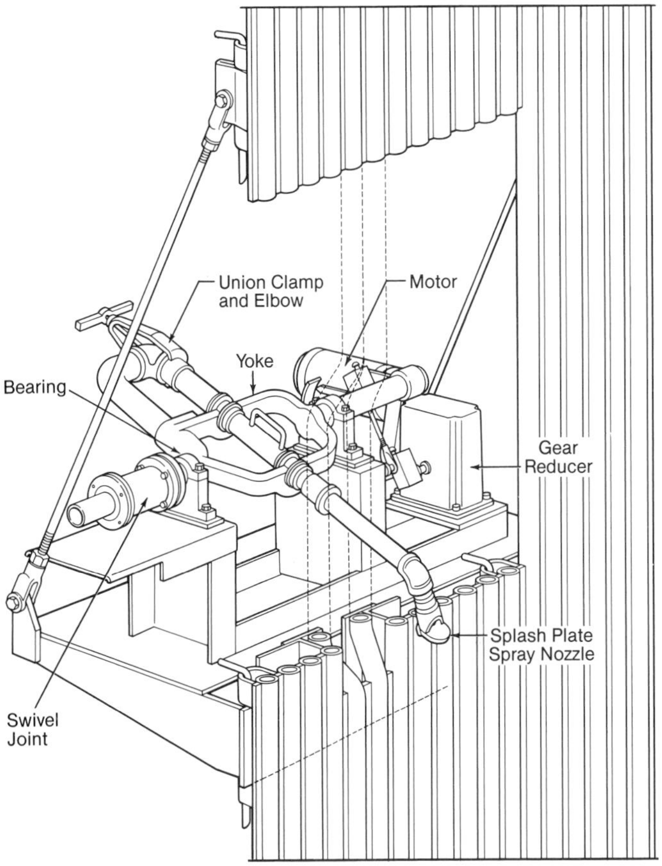

Fig. 17 Limited vertical sweep burner.

In-flight drying deposits a minimum of char in the primary air zone around the periphery of the unit, as all the drying and a majority of the devolatization are in-flight over the furnace area. Consequently, less primary air flow is required to keep the char from in front of the primary air ports. Because the higher liquor solids concentration dictated by in-flight drying translates into less water evaporated in the furnace, a greater fraction of the combustion heat released is available to maintain bed temperatures. Less primary air is therefore required when burning liquor in-flight in contrast to the lower concentration liquors encountered in oscillator burner applications. Typically, 30 to 40% of the total combustion air is staged at the primary level with the in-flight drying promoted by LVS liquor burners.

Ash system

The sodium compounds entrained in the flue gas originate from fume generation and liquor droplet carryover from the lower furnace. The resultant ash drops out of the flue gas stream and is collected in trough hoppers located below the boiler and economizer modules. The electrostatic precipitator removes nearly all the remaining ash.

The majority of the entrained ash is Na_2SO_4, and is commonly referred to as salt cake ash, or simply, salt cake. The ash collected in the trough hoppers and precipitator must be returned to the black liquor to recover its significant sodium and sulfur contents. This is accomplished by mixing the ash into the liquor in a specially designed tank. The salt cake ash is transferred to the mix tank through a wet ash sluice system or a drag chain ash conveyor system.

With a dry ash system (Fig. 18), mechanical drag chain conveyors are bolted to the bottom of the boiler and economizer trough hoppers, which extend across the full unit width. The drag chain conveyors are equipped with heat treated, high alloy forged link chains to support and convey the flights. The conveyors discharge through rotary seal valves into a collection conveyor. The collection conveyor discharges to the mix tank where the ash is uniformly mixed with the liquor.

Drag chain conveyors are also provided across the floor of the electrostatic precipitator, beneath the collecting surfaces. The conveyors discharge to mechanical combining conveyors, which in turn discharge the salt cake ash through rotary valves into the mix tank. Frequently, the dry ash system is arranged with two mix tanks, one serving the boiler and economizer hopper ash system and one dedicated to the precipitator.

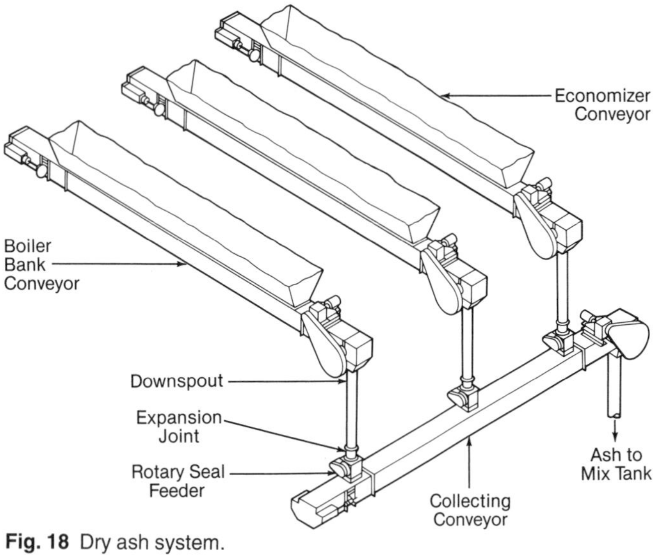

Fig. 18 Dry ash system.

In a wet ash removal system, black liquor is flowed through the hoppers to sluice the collected ash. The sluice discharges directly from the hoppers through large pipes to the mix tank. However, these pipes can become plugged with ash and overflow liquor which create safety and cleanliness problems. This has led to wider acceptance of the dry ash mechanical conveyor system. However, the mechanical conveyors generally require more maintenance than the wet ash sluice designs.

The mix tank (Fig. 19) includes a mechanically scraped screen to assure that all material which passes to the fuel pumps is small enough to readily pass through the burner nozzles. Scraping is provided by flights on a low horsepower, slow moving agitator. The tank is also designed to receive the makeup salt cake feed for the mill. Typically, the makeup is stored in a day tank located above the mix tank and is introduced at a regulated rate through a screw feeder.

Combustion air system

B&W's Advanced Air Management system provides combustion air at three elevations of the furnace: primary, secondary and tertiary (Fig. 20). The use of three levels allows performance optimization of the respective furnace zones — the lower furnace reducing zone, the intermediate liquor drying zone, and the upper furnace burnout or combustion completion zone.

The primary air flow quantity must be sufficient to produce stable combustion and to provide the hot reducing zone for the molten smelt. Increasing primary air flow beyond that required to achieve these objectives increases the amount of Na_2S re-oxidized to Na_2SO_4. The balance of the air at or less than the stoichiometric requirement is introduced above the char bed at the secondary level to control the rate at which the liquor dries and the volatiles combust, and to minimize the formation of NO_x. The additional air required to complete combustion is introduced at the tertiary level.

The air system may be arranged as a single-fan, two-fan (one primary and one secondary/tertiary) or three-fan system. On larger systems, additional provisions are made to bias the air flow at each level. Side to side biasing is provided on the primary and secondary air, as is front to rear biasing for the tertiary air. This gives additional flexibility to the operator.

The primary air ports are arranged on all four furnace walls about 3 ft (0.9 m) above the floor. Air is introduced at a low velocity and 3 to 4 in. wg (0.75 to 1.0 kPa) static pressure which prevents it from penetrating the bed. The air lifts the carbon char in front of the port back onto the bed and maintains ignition.

Approximately one half of the total air is admitted at the secondary zone. The air is introduced at the pressure and velocity needed to penetrate the furnace 4 to 6 ft (1.2 to 1.8 m) above the primary air ports. A velocity damper is used on each port.

Proper secondary air mixing with the gases rising from the char bed results in volatile combustion generating the heat required for in-flight liquor drying. This secondary

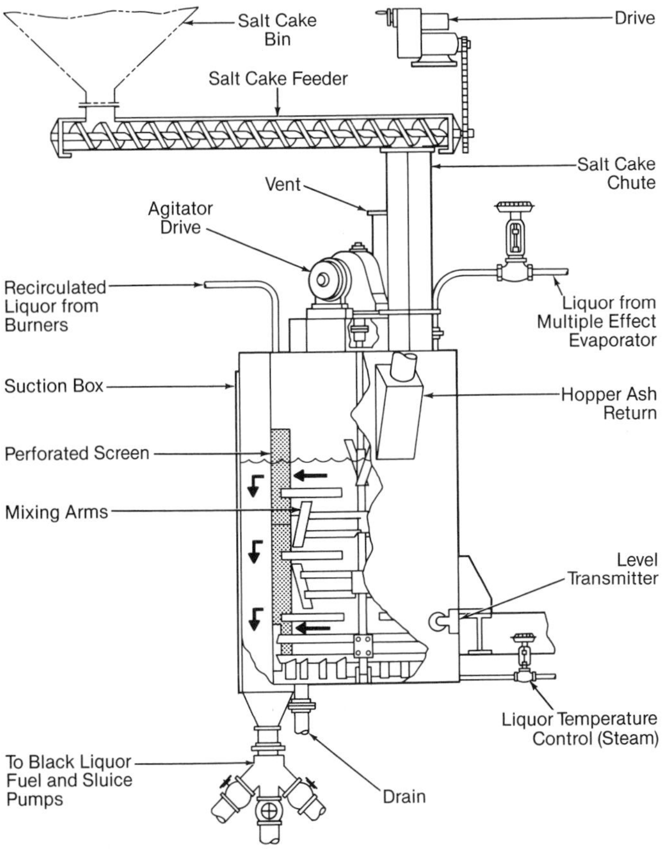

Fig. 19 Salt cake mix tank.

zone typically exhibits the highest temperatures in the recovery furnace. The quantity of secondary air is dictated by the amount of burning required to dry liquor and control bed height, and decreases as the black liquor solids concentration increases.

The balance of the air is admitted at ambient temperature through the tertiary air ports located above the liquor guns. A velocity damper is again used on each port. The tertiary flow increases proportionately as the solids increase.

The secondary air ports are normally arranged on the longest furnace wall, generally the sidewalls, as the furnace is deeper than it is wide. In contrast, the tertiary air ports are arranged on the front and rear walls.

Buildup of smelt and char that restricts the air port openings can cause a performance deterioration. A reduction in port area affects air pressure at the port or air flow through the port, depending on which is being controlled. When air flow is controlled, the effect of plugged ports is to increase air pressure and push the bed farther from the wall. If pressure is controlled, the effect is a resultant decrease in air flow. This results in less effective burning of the primary zone with a decrease in furnace temperature and increased emissions.

Inadequate manual air port rodding, generally performed every two hours, will limit effective recovery boiler operation. Automatic port rodders provide continuous cleaning to maintain a constant flow area. The automatic port rodding system stabilizes lower furnace combustion for maximum thermal efficiency and is vital in achieving low emissions.

Secondary and tertiary air ports also require periodic rodding, although the plugging that occurs at these ports is generally less severe than at the primary air ports. In the secondary and tertiary ports, the rodding equipment can be integrated with the velocity control damper. This is necessary to provide synchronization of the damper and rodder drives.

To further enhance stability in the lower furnace, the primary and secondary air is preheated in a steam coil air heater. Low pressure steam, normally 50 to 60 psig (3.4 to 4.5 bar gauge) is used for preheating. Additional preheating is provided by steam at 150 to 165 psig (10.3 to 11.4 bar gauge), typically achieving a 300F (149C) combustion air temperature. When burning liquors with a low heating value or from a nonwood fiber, air should be preheated to about 400F (204C).

Flue gas system

Combustion gas exiting the economizer is routed through flues to the electrostatic precipitator and induced draft fan before discharge through a stack to the atmosphere.

The induced draft fan speed controls the pressure inside the furnace. The fan is normally located after the precipitator, allowing the fan to operate in the cleaner gas.

A two-chamber precipitator is normally used. Each chamber is equipped with isolation gates or dampers and a dedicated induced draft fan discharging to a common stack. Flue gas exiting the economizer is divided into two flues, which route the gases to the precipitator's two chambers. Each chamber typically has sufficient capacity to operate the recovery boiler at 70% load, corresponding to a stable black liquor firing mode. When designing the pre-

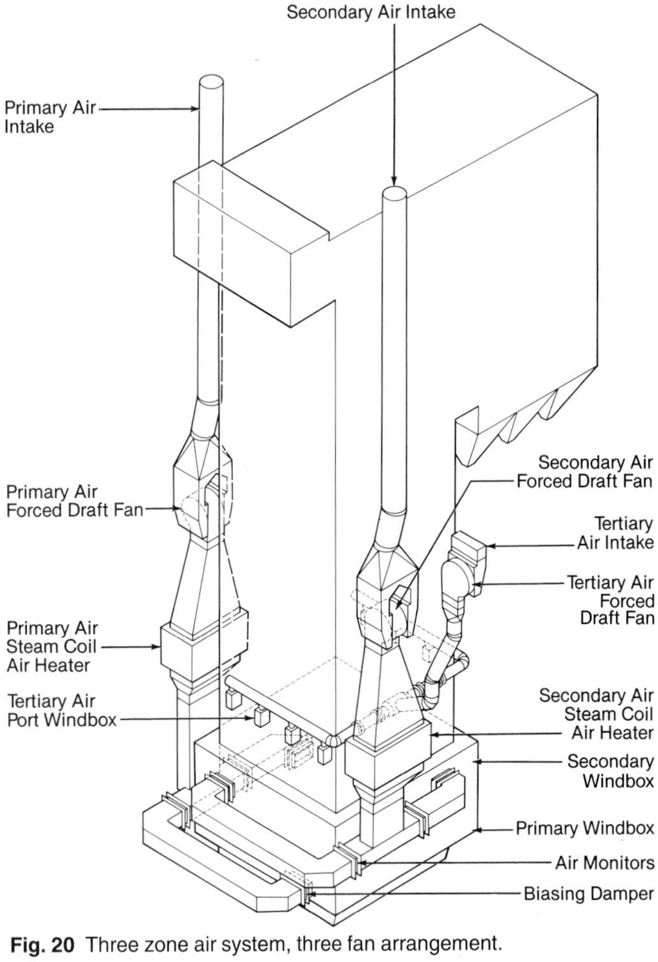

Fig. 20 Three zone air system, three fan arrangement.

cipitator chamber, the expected gas flow rate should include that of the sootblower steam and the increased excess air under which the boiler would operate at a reduced rating. The maximum permissible particulate discharge rate then establishes the electrostatic precipitator size.

Cleaning system

Ash entrainment in furnace gases is affected by gas velocity, air distribution and liquor properties. The design of all recovery unit heating surfaces should include sootblowers using steam as the cleaning medium. Gas temperatures must be calculated to make certain that velocities and tube spacings are compatible with the sootblowers.

High levels of chloride and potassium in the black liquor may require greater cleaning frequency. As unit overload is increased, increased entrainment of ash and sublimated sodium compounds also leads to more frequent water washing. In addition to excess quantities of flue gas ash, velocities and temperatures throughout the unit are increased, making ash deposits more difficult to remove.

Auxiliary fuel system

The primary objective of the recovery boiler is to process black liquor. However, the unit can fire auxiliary fuel, usually natural gas and/or fuel oil. This fuel is fired through specially designed burners, arranged at the secondary air level. These burners raise steam pressure during startup, sustain ignition while building a char bed, stabilize the furnace during upset conditions, carry load while operating as a power boiler, and burn out the char bed when shutting down.

In some installations, the recovery boiler must be able to generate full load steam flow and temperature on auxiliary fuel. These applications typically arise in mills that cogenerate electricity and/or have limited power boiler steam generating capability. Increased auxiliary fuel capacity can be accommodated by adding auxiliary burners above the secondary air level.

Upper level burners allow combination black liquor and auxiliary fuel firing with minimal interference with lower furnace operations. This is not possible when operating the secondary level auxiliary burners. Also, the upper level burners can provide higher steam temperatures at lower loads during startup, which can be important in mills operating high steam temperature turbine-generators.

Smelt spout system

Smelt exits the furnace through specially designed openings at the low point of the floor and is conveyed to the dissolving tank in a sloped water-cooled trough, or smelt spout (Fig. 21). The spout is bolted to the furnace wall mounting box, and the machined face of the spout is positioned against the furnace wall tubes surrounding the opening. The trough is a V-shape to correspond to the V-shaped bottom of the wall opening. The spouts are constructed of a double wall carbon steel trough, with a continuous flow of cooling water passing between the inner and outer walls. Given the explosive nature of smelt-water reactions and the extreme temperature of the molten smelt, the spout must receive adequate cooling water. A dedicated cooling system with built in redundancy and multiple sources of backup water assures system reliability (Fig. 22). The cooling water is treated in a dedicated, closed cycle to minimize scale-forming contaminants.

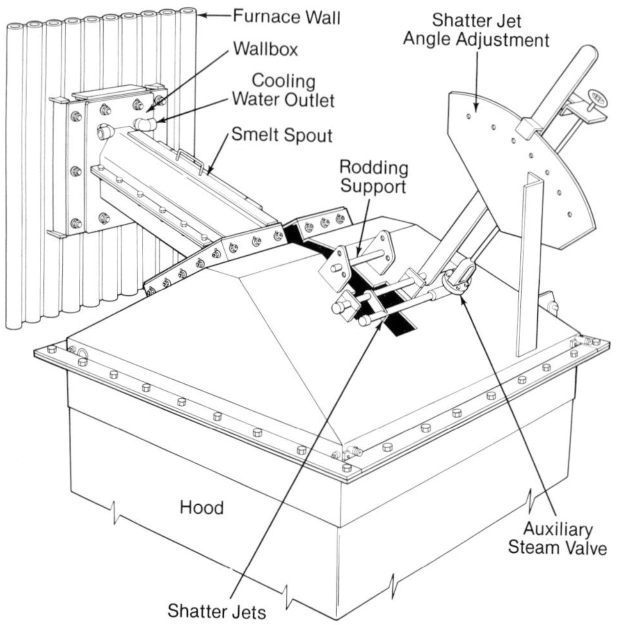

Fig. 21 Integral smelt spout hood with shatter jet assembly.

Green liquor system

The smelt is dissolved in green liquor in an agitated tank. Green liquor is withdrawn from the tank at a controlled density and the volume replaced with weak wash. The entering smelt stream must be finely dispersed, or shattered, to control the smelt-water reaction by rapid dissolution of the smelt into the green liquor. Excessively large smelt particles entering the tank can lead to explosions. The outer surface of a large smelt particle cools quickly as it contacts the green liquor, forming an outer shell around its hot core. As the shrinking forces build, the particle shell explodes, exposing the hot core to water and resulting in the sudden release of steam. To eliminate large smelt particles from reaching the dissolving

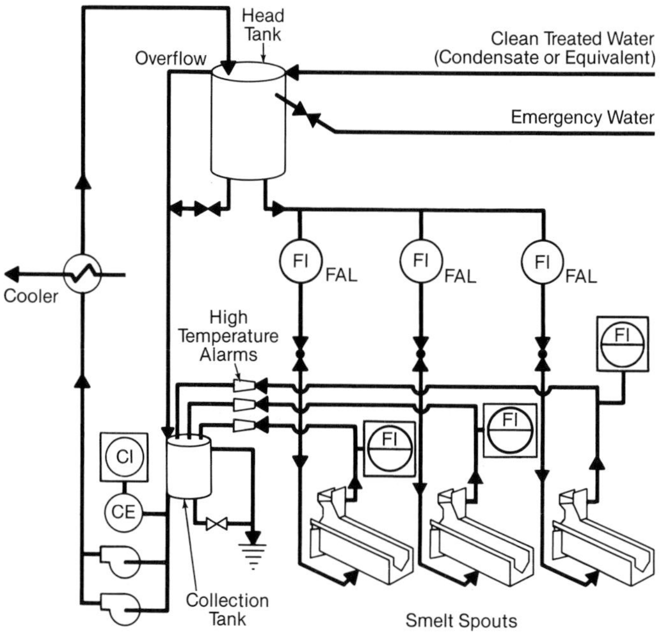

Fig. 22 Smelt spout cooling system.

tank contents, steam shatter jets are located above the smelt stream to disperse the flow into the tank.

The smelt spout discharge is enclosed by a hood (Fig. 21). The steam shatter jets are mounted on the top of the hood, with the nozzle angle adjusted to direct the jet at the smelt stream cascading off the spout end. The periphery of the hood is equipped with wash headers to flush the hood walls and prevent smelt buildup. An opening in the hood permits operator lancing for manually dislodging solidified smelt in the spout trough or opening.

The dissolving tank is of heavy construction and is equipped for agitation. One or more agitators, either side or top entering, are generally used. A stainless steel band is commonly used in the carbon steel tank at the liquor level for corrosion protection, while the floor is protected with steel grating or poured refractory.

As the smelt is cooled and dissolved into the green liquor, large quantities of steam are released. The steam vapors are pulled through an oversized atmospheric vent on the tank to quickly relieve pressure in the event of a surge or explosion. Smelt particles and green liquor droplets are entrained in the steam-air mixture vented from the tank. These vented gases also contain H_2S that must be removed prior to discharge to the atmosphere. Typically, a vent stack scrubber (Fig. 23) is used to reduce H_2S emissions and to trap entrained particulates. Weak wash is used as the scrubbing medium to take advantage of its residual NaOH, which absorbs the malodorous gases.

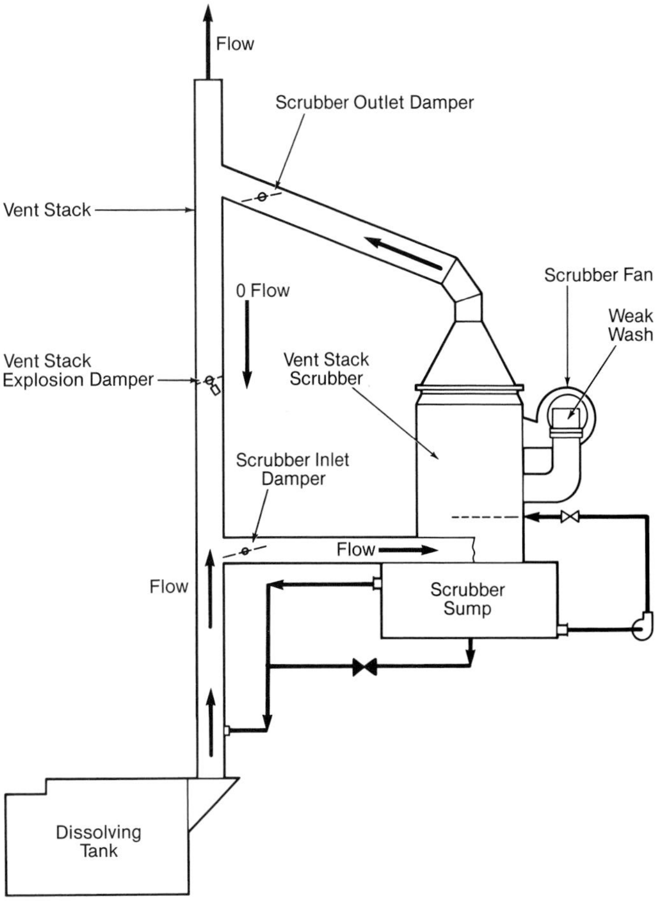

Fig. 23 Vent stack scrubber.

Soda process recovery boiler

With few exceptions, the requirements and features of the kraft recovery boiler apply to the soda recovery boiler. Because sulfur is not present in the soda process, there is no sulfur reduction in the recovery furnace. Na_2CO_3, or soda ash, is added to the recovered green liquor. The ash collected from the boiler hoppers and the electrostatic precipitator is in the form of Na_2CO_3, and can be added directly to the green liquor in the dissolving tank. Consequently, a salt cake mix tank is not required.

In the furnace, the soda liquor does not form a suitably reactive char for burning in a bed. It is finely atomized and sprayed into the furnace by multiple steam-atomizing soda liquor burners. The fine spray dehydrates in flight and combustion takes place largely in suspension in an oxidizing atmosphere. Combustion air is admitted through primary and secondary ports around the furnace periphery, with the hottest part of the furnace just above the hearth. The Na_2CO_3 collects in molten form on the hearth and discharges through the smelt spouts.

The Na_2CO_3 smelt has a higher melting point than the kraft process smelt. This makes the soda smelt more difficult to tap from the furnace and shatter. Auxiliary fuel burners can be located low in the sidewalls, close to the spout wall, to keep the smelt hot for easier tapping.

Bleached chemi-thermomechanical pulp (BCTMP) mills produce a low heating value effluent high in sodium content. The effluent from a BCTMP process can be evaporated to recover water for recycling. A recovery boiler designed for incinerating BCTMP effluent is illustrated in Fig. 24. The concentrated effluent is burned in an oxidizing combustion zone and forms ash from the inorganic matter. Natural gas is burned continuously to

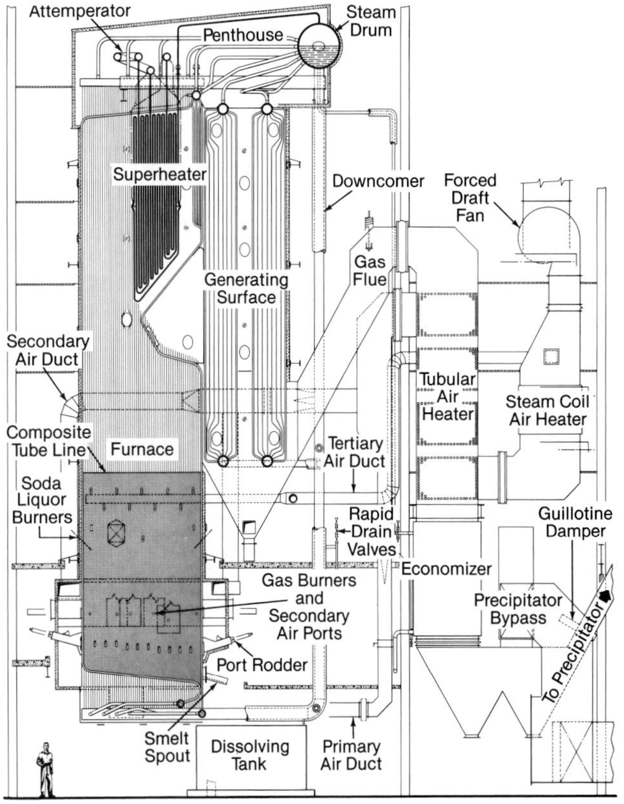

Fig. 24 Effluent recovery boiler.

augment the low heating value of the liquor. Auxiliary fuel quantity is minimized by providing 600F (316C) combustion air to the furnace. The inorganic ash is tapped as molten smelt from the furnace and chemicals are recovered for reuse in the mill.

Recovery unit design for nonwood fiber liquor

Many countries do not have adequate forest resources for pulp production. As a result, alternative fiber sources are used, such as bamboo, sugar cane bagasse, reeds and straw. The U.S. Department of Agriculture has also developed kenaf as a viable fiber source. Black liquor from these fibrous materials requires special consideration in recovery system design. The liquor is generally characterized by high viscosity and a high silica content.

The high viscosity limits the level at which liquor can be concentrated, generally well below that achieved with wood fiber liquor. A further limit to multiple effect evaporator operation is the formation of insoluble scale. In many cases, high silica content limits the liquor product to 40 to 45% solids concentration. A cascade evaporator is commonly used to further concentrate the liquor. Some installations have successfully used cyclone evaporators.

Increased combustion stability can be accomplished by providing the maximum air temperature achievable within the mill system. The hot air compensates for low liquor concentrations and allows flexibility in adjusting the firing conditions.

Sulfite process

Pulp produced by the sulfite processes is divided into two broad categories, semi-chemical and chemical. Semi-chemical pulp requires mechanical fiberizing of the wood chips after cooking. Chemical pulp is largely manufactured by the acid sulfite and bisulfite processes, which differ from the alkaline processes in that an acid liquor is used to cook the wood chips. The acid sulfite process is characterized by an initial cooking liquor pH of 1 to 2. The bisulfite process operates with a pH of 2 to 6. Cooking liquor used in a neutral sulfite process for manufacture of semi-chemical pulps has an initial pH of 6 to 10. Spent sulfite liquor, separated from chemical and semi-chemical pulp and containing the residual cooking chemicals and dissolved constituents of the wood, is evaporated and burned, and the chemicals are recovered in a system particular to each base.

Sulfite process pulp mills normally use one of four basic chemicals for digestion of wood chips to derive a large spectrum of pulp products: sodium, calcium, magnesium or ammonium. The principal differences in the sulfite waste liquor of the four bases are the physical properties of the base and the products of combustion.

Sodium

Incineration of sodium liquor yields a mixture of compounds requiring a relatively complex secondary chemical reprocessing system. This system reconstitutes the base in a form suitable for reuse in the pulping cycle. The sodium-base liquor may be burned alone or in combination with black liquor in a kraft recovery unit. The base chemical is recovered as a smelt of Na_2S with some Na_2CO_3 which would be a suitable makeup chemical for a nearby kraft mill. The combination of sulfite liquor and kraft liquor, referred to as cross recovery, is common to pulp mill operations; it has the advantage of providing chemical recovery without a complex secondary recovery operation. The proportion of sulfite liquor in cross recovery is generally the equivalent of the sodium makeup requirement of the kraft mill cycle.

Calcium

Calcium-base liquor is concentrated and burned in specially designed furnaces. The furnace and boiler design applicable to magnesium-base liquor combustion can also be applied to calcium. When calcium-base spent liquor is burned, the calcium is present as calcium oxide and sulfate entrained in the flue gas as finely divided flyash. This ash is separated from the combustion gases in electrostatic precipitators for disposal. The large content of calcium sulfate makes it generally uneconomical to reuse the ash in acid preparation. The process is largely historic today due to its unacceptable effluents.

Magnesium

With the magnesium base, a simple system is available for recovery of heat and total chemicals. This is due to the chemical and physical properties of the base. The spent liquor is burned at elevated temperatures in a controlled oxidizing atmosphere, and the base is recovered in the form of an active magnesium oxide (MgO) ash. The oxide can be readily recombined in a simple secondary system with the SO_2 produced in combustion, thereby reconstituting the cooking acid for pulping.

Magnesium-base pulping and recovery process

Industrial interest in the improved pulp from a variety of wood species stimulated the development of pulping techniques using a magnesium base. The two major pulping procedures, by which a variety of pulps can be produced, are magnesium acid sulfite and bisulfite, or Magnefite™. The basic magnesium recovery system is appropriate for each of the pulping processes.

In the pulping process, the pulp and spent sulfite liquor are first discharged into a blow tank from which the digester contents, diluted with additional weak liquor, are pumped to stage washers. The first stage washer filtrate, having 13 to 15% solids, enters the weak red liquor storage tank. This liquor is concentrated in the multiple effect evaporator and transferred to the strong liquor tank at the concentration required for boiler operation. The as-fired solids concentration may vary between 60 and 65%. The liquor is fired at about 230F (110C), and can be preheated by an indirect steam heater.

The heavy liquor is fired with steam-atomizing burners located in opposite furnace walls. The combustion products of the liquor's sulfur and magnesium are discharged from the furnace in the gas stream as sulfur dioxide and solid particles of MgO ash. Gases are cooled in passing over heat transfer surface to generate steam for process and power. The major portion of the MgO is removed from the gas stream in a mechanical collector or electrostatic precipitator and is then slaked to magnesium hydroxide ($Mg(OH)_2$). The SO_2 is recovered by reaction with the $Mg(OH)_2$ to produce a magnesium bisulfite acid in an absorption system. This acid is passed through a fortification or bisulfiting system and is fortified with makeup SO_2. The finished cooking acid is filtered and placed in storage for reuse in the digester.

Burning magnesium-base liquor

Fig. 25 illustrates a high capacity, low maintenance magnesium-base liquor design developed to eliminate the earlier firebrick furnace and provide auxiliary fuel capacity in the red liquor burners. The recovery unit incorporated a water-cooled furnace, where the combustion zone absorption is minimized by applying refractory over studded tubes which, in turn, cool the refractory.

Small liquor particles are injected into the furnace through steam atomizers similar to those used to burn oil. (See Chapter 10.) The opposed burners provide a highly turbulent condition in a combustion chamber maintained at elevated temperatures by the refractory lining. An oxidizing atmosphere is maintained by introducing a controlled quantity of air slightly in excess of that required for theoretical combustion.

Combustion must be complete to produce a MgO ash that is essentially carbon free. The ash has a carbon content of less than 0.1% by weight when produced in a properly operated furnace having an exit gas temperature in excess of 2400F (1316C). Combustion air at about 700F (371C) is supplied through the burner windbox.

Sulfur dioxide absorption

In a secondary recovery system, the flue gas leaving the furnace contains roughly 1% SO_2 by volume. Absorption is carried out by contacting the combustion gases with recirculated acid containing magnesium sulfite absorbent. A constant level of magnesium sulfite is maintained by feeding $Mg(OH)_2$ slurry to the recirculated acid sprayed into the gas stream. The magnesium sulfite contacting the gas stream absorbs SO_2 to form magnesium bisulfite acid.

Ammonium

Ammonium-base liquor is the ideal fuel for producing a low ash combustion product. Burning can be accomplished in a simple recovery boiler. The ammonia decomposes on burning to nitrogen and hydrogen, the latter oxidizing to water vapor and thereby destroying the base. Concentrated ammonium sulfite liquor is burned in a water-cooled furnace enclosure faced with refractory similar to that shown in Fig. 25. The combustion products, including recoverable SO_2, are discharged from the furnace and cooled in passing over convection heat transfer surface to generate steam for power and process. SO_2 produced in combustion can be absorbed in a secondary system to yield cooking acid for pulping. A small quantity of ash results from noncombustible solids and is separated in a mechanical collector. The SO_2 reacts in an absorption system with an anhydrous or aqueous ammonium makeup chemical to produce ammonium bisulfite acid. In a neutral sulfite semi-chemical plant, the absorption system produces cooking liquor consisting essentially of ammonium sulfite.

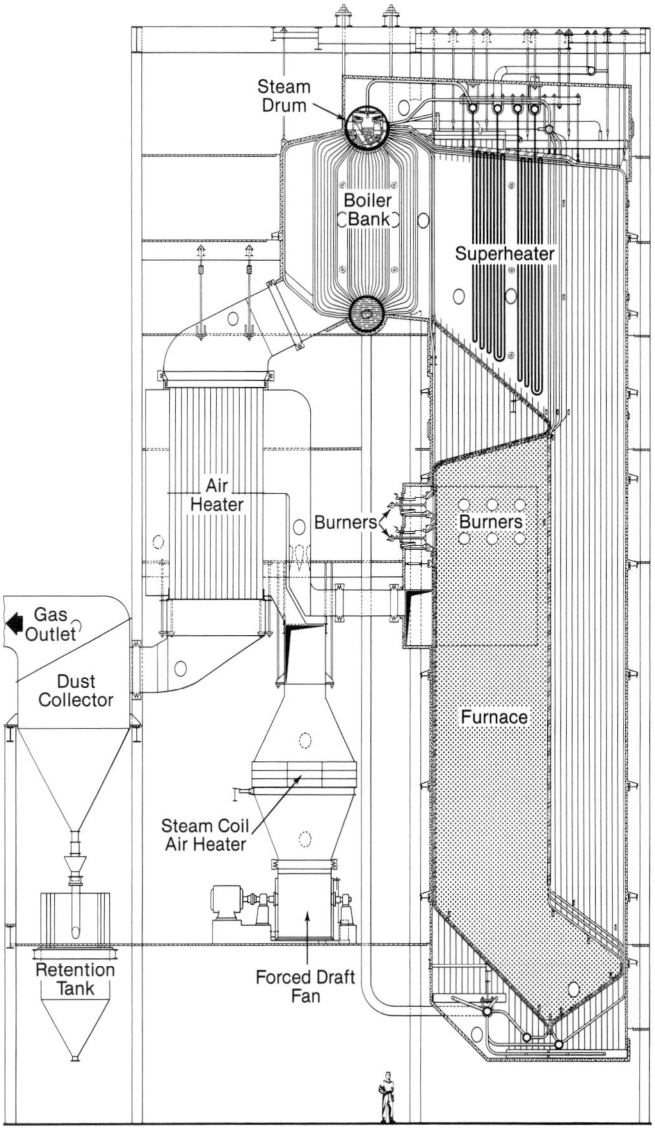

Fig. 25 Red liquor recovery boiler.

References

1. Elaahi, A., and Lowitt, H.E., *The Pulp and Paper Industry: An Energy Perspective*, DOE/RL/01830-T57, April 1988.

2. *Pulp and Paper 1990 North American Factbook*, Miller Freeman.

3. *1991 International Fact and Price Book*, Pulp and Paper International, Miller Freeman.

4. Adapted from Hough, G.W., *Chemical Recovery in the Alkaline Pulping Processes*, TAPPI, p. 197, 1985.

5. Adapted from Smook, G.A., *Handbook for Pulp and Paper Technologists*, Joint Textbook Committee of the Paper Industry TAPPI/CPPA, p. 69, 1986.

6. Adapted from Whitney, R.P., *Chemical Recovery in Alkaline Pulping Processes*, TAPPI, Monograph Series No. 32, p. 40, 1968.

This facility features three 750 ton per day mass-fired units by B&W.

Chapter 27
Waste-to-Energy Installations

The disposal of garbage is a problem that has been with us since civilization began. At various times throughout history, composting, animal feed, landfill and incineration have all been popular disposal methods. Today, refuse disposal methods are determined by cost and the effect on our environment.

The most common means of refuse disposal is still landfilling. Even in the late 1970s, nearly all of the refuse generated in North America was landfilled. Incineration with no heat recovery was a popular option that became economically unacceptable with the advent of environmentally responsible air pollution regulations and inexpensive landfill alternatives.

In Europe and Japan, where new landfill sites were less available, incineration continued as a viable option and those plants became the predecessors of today's refuse-to-energy plants. Heat recovery was added in the form of waste heat boilers which were originally hot water boilers and later low pressure and temperature steam boilers. These incinerators with waste heat boilers then evolved into waterwall boilers with integral stokers.

Refuse-fired boiler design parameters and operating characteristics are strongly affected by the components of the refuse, which change with time. The components also vary greatly by location. In North America, typical municipal solid waste (MSW) is high in paper and plastics content (Fig. 1) and typically has a lesser moisture content and greater heating value than that found worldwide. In a less industrialized country the refuse tends to have a greater moisture content and lesser heating value. Table 1 shows representative refuse analyses ranging from 3000 to 6000 Btu/lb (6978 to 13,956 kJ/kg) higher heating value (HHV) basis.

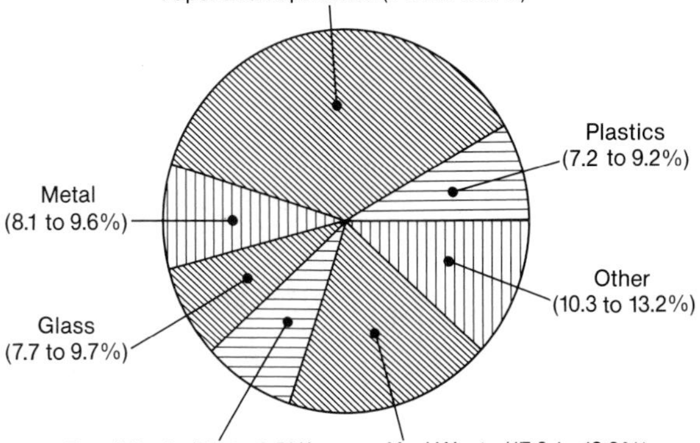

Fig. 1 U.S. municipal solid waste generation — 1990.

In the United States (U.S.) and North America, the refuse characteristics have changed dramatically in a short period of time. With more and more convenience foods, plastics, packaging, containers, and less food scraps due to home garbage disposals, the average refuse heating value has increased and the moisture content has decreased (Table 2). As more recycling programs are implemented, the analysis will continue to change. As glass, aluminum and other metals are recycled the refuse

Table 1
Range Of As-Received Refuse Fuel Analysis

	Weight Percent As-Received						
HHV, Btu/lb (kJ/kg)	3,000 (6,978)	3,500 (8,141)	4,000 (9,304)	4,500 (10,467)	5,000 (11,630)	5,500 (12,793)	6,000 (13,956)
Carbon	16.88	19.69	22.50	25.32	28.13	30.94	33.76
Hydrogen	2.33	2.72	3.10	3.49	3.88	4.27	4.66
Oxygen	12.36	14.42	16.49	18.55	20.62	22.68	24.75
Nitrogen	0.22	0.26	0.30	0.34	0.38	0.42	0.46
Sulfur	0.15	0.18	0.21	0.24	0.27	0.30	0.33
Chlorine	0.34	0.38	0.42	0.46	0.50	0.54	0.58
Moisture	35.72	32.35	29.98	28.60	25.22	22.85	21.46
Ash	32.00	30.00	27.00	23.00	21.00	18.00	14.00
Total	100.00	100.00	100.00	100.00	100.00	100.00	100.00

Table 2
U.S. Refuse Trends

Increasing heating value per ton of refuse
1960 — 4,200 Btu/lb (9,769 kJ/kg)
1980 — 4,500 Btu/lb (10,467 kJ/kg)
2000* — 5,200 Btu/lb (12,095 kJ/kg)

More paper and paperboard
33% in 1970
41% in 2000

More plastics
2.7% in 1970
9.8% in 2000

Less food wastes
11.5% in 1970
6.8% in 2000

*Estimate

heating value will increase; as paper and plastics are recycled the heating value will decrease.

In 1990 approximately 490,000 t (444,521 t_m) per day of MSW were generated in the U.S. About 13% of that total was recycled, 15% sent to refuse-to-energy facilities, and the balance landfilled. Early in the 21st century this is expected to reach 40% for recycling and 50% for refuse-to-energy facilities. The growth of refuse-to-energy facilities in the U.S. accelerated in the 1980s (Fig. 2) due to the growing disposal costs for landfills and a government-created market for the sale of electric power.

As old landfills closed, new landfills became more difficult and costly to open and tended to be located farther from the source of the refuse, increasing transportation costs. Concerns about ground water contamination resulted in more expensive landfill designs with several containment layers and leachate monitoring and control systems. The passage of the Public Utility Regulatory Policies Act of 1978 (PURPA) required public utilities to purchase the electric power generated by refuse-to-energy plants. This created a revenue flow that helped offset the inherent high capital cost of these plants. These market forces resulted in a proliferation of refuse-to-energy facilities in the northeast U.S. where the costs of landfill and other disposal options were the highest, and selectively throughout North America in response to local environmental or economic factors.

Refuse disposal is a major problem worldwide and there is no single solution. An environmentally sound refuse disposal program includes generating less refuse, recycling components that can be economically reused, combustion of the balance of the refuse (including the efficient generation of electric power), and the landfill of the resulting ash.

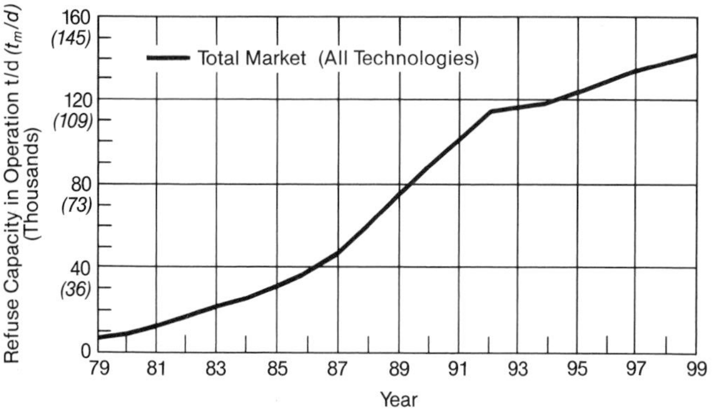

Fig. 2 U.S. refuse-to-energy market.

Refuse combustion alternatives

Two main techniques are used for burning municipal refuse, distinguished by the degree of fuel preparation. The first technique, known as mass burning, uses the refuse in its as-received, unprepared state (Fig. 3). Only large or noncombustible items such as tree stumps, discarded appliances, and other bulky items are removed. Refuse collection vehicles dump the refuse directly into storage pits. Overhead cranes equipped with grapples move the refuse from the pit to the stoker charging hopper. Hydraulic rams move the refuse onto the stoker grates. The combustible portion of the refuse is burned off and the noncombustible portion passes through and drops into the ash pit for reclamation or disposal.

The second burning technique uses prepared refuse, or refuse-derived fuel (RDF), where the as-received refuse is first separated, classified, and reclaimed in various ways to yield salable or otherwise recyclable products (Fig. 4). The remaining material is then moved to the boiler feeders and fed through multiple feeders onto a traveling grate stoker. The RDF is burned, part in suspension and part on a stoker. More finely shredded RDF can also be fired in suspension to supplement conventional fuels in large boilers used for power generation.

Corrosion

Combustion products from municipal refuse are very corrosive. The components that are present in coal, oil and other fuels that contribute to corrosion, as well as to high slagging and high fouling, are all present in refuse (Table 3). Corrosion in refuse-fired boilers is usually caused by the chlorides which deposit on the furnace, superheater and boiler tubes. Several modes of chloride corrosion may occur:

1. corrosion by hydrochlorides (HCl) in the combustion gas,
2. corrosion by NaCl and KCl deposits on tube surfaces,
3. corrosion by low melting point metal chlorides (mainly $ZnCl_2$ and $PbCl_2$), and
4. out of service corrosion by wet salts on the tube surface.

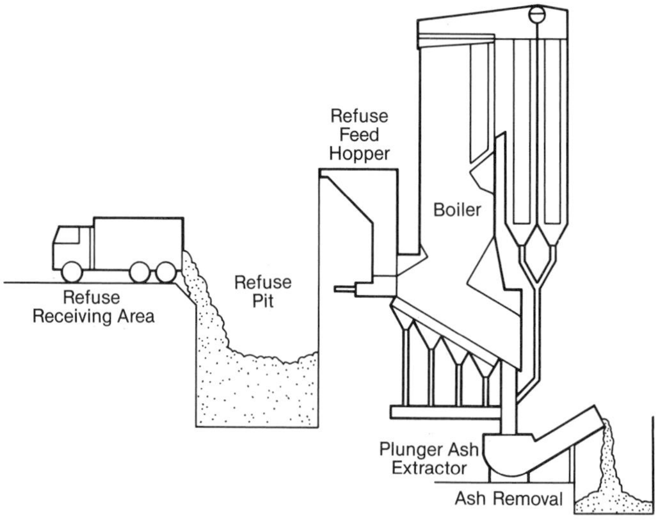

Fig. 3 Mass burning schematic.

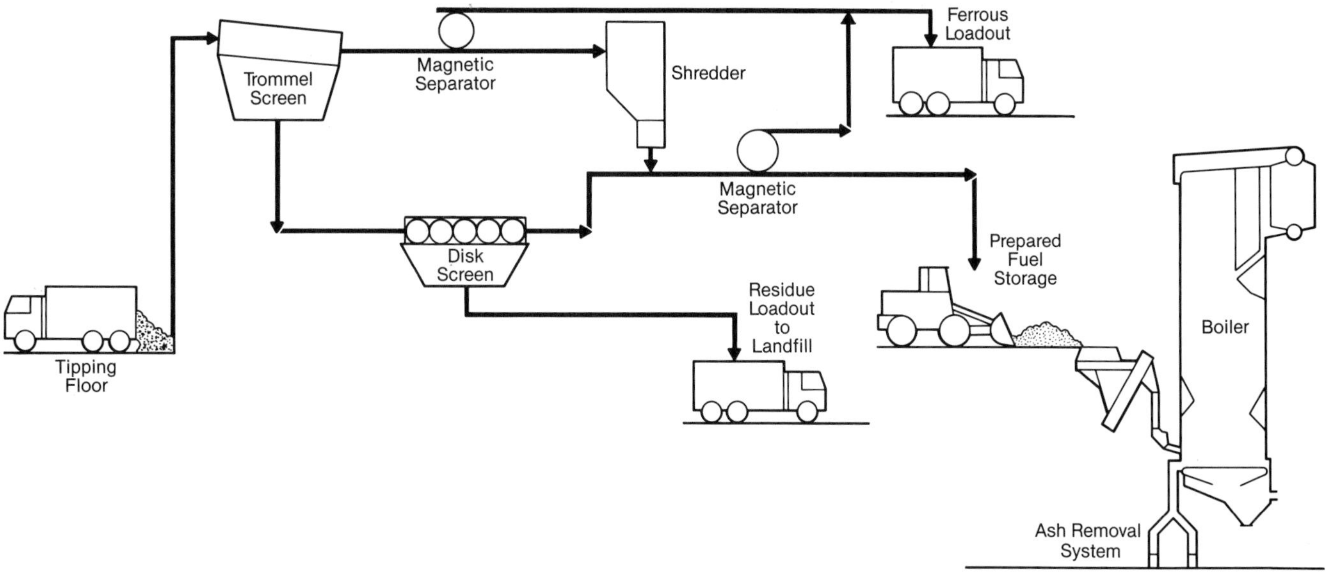

Fig. 4 RDF burning schematic.

The rate of tube metal loss due to corrosion is temperature dependent with high metal temperatures correlating with high rates of metal loss (Fig. 5). Refuse boilers operating at higher steam pressures have higher temperature saturated water in the furnace tubes and, therefore, these furnace tubes have higher metal temperature. Superheater tube metal temperatures are directly related to the steam temperature inside the tubes. In both cases, it is the temperature of the water or steam inside the tube that largely controls the tube metal temperature, rather than the temperature of the flue gas outside of the tube.

Furnace-side corrosion can be aggravated by poor water chemistry control. If water-side deposits are permitted to form, tube wall metal temperatures will rise and furnace corrosion will be accelerated. Standards for feedwater and boiler water quality are based on boiler operating pressures. These standards are no more stringent for refuse-fired units than for other fuels. However, the maintenance of feedwater and boiler water quality, and adherence to those standards, is more critical on refuse-fired boilers due to the highly corrosive nature of the fuel.

Lower furnace corrosion

The lower furnace environment of both mass-fired and RDF-fired units is constantly changing between an oxidizing atmosphere (an excess of O_2 beyond that needed for combustion) and a reducing atmosphere (a deficiency of O_2 below that needed for combustion) which can rapidly accelerate corrosion. Therefore, some form of corrosion protection is needed. Typically, the area of protection will encompass all four walls up to 30 ft (9.1 m) above the grate where there is reasonable assurance that oxidation zones are predominant.

Table 3
Corrosive Constituents in Fuels

Coal	Oil	Refuse	
Sodium	Sodium	Sodium	Chloride
Sulfur	Sulfur	Sulfur	Lead
Potassium	Vanadium	Potassium	Zinc
		Vanadium	

Mass-fired units

Virtually all mass-fired refuse boilers incorporate some type of pin stud and silicon carbide (SiC) refractory to protect the membraned lower furnace walls (Fig. 6). The quality and physical characteristics of the silicon carbide refractory must be maintained through proper application and curing. Lack of control during installation will result in spalling, deterioration and increased maintenance. The refractory material should have high thermal conductivity rates to minimize reducing the effectiveness of the water-cooled surface it is protecting. However, such characteristics may reduce its resistance to erosion as experienced along the grate line due to the scrubbing action of the refuse fuel and ash as it moves along the grate to the ash discharge. Increased erosion-resistant SiC materials are available for these zones. They do, however, have lower thermal conductivities.

A better alternative near the grate line is the use of armour blocks or refractory blocks, rigidly attached to the furnace walls. These blocks extend up the full height of the charging hopper opening which is about 4 ft (1.2 m)

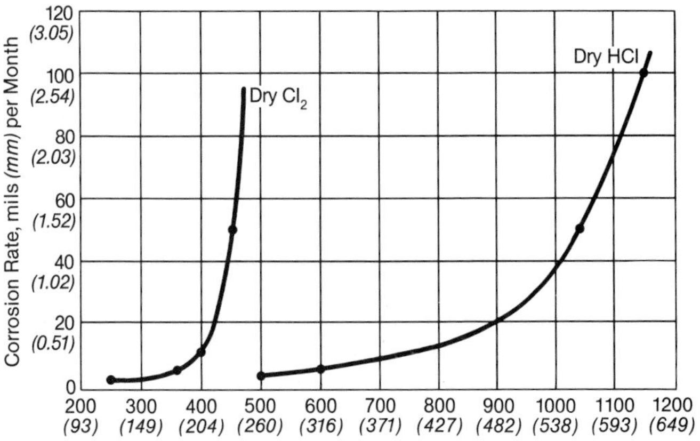

Fig. 5 Corrosion of carbon steel in chlorine and hydrogen chloride.

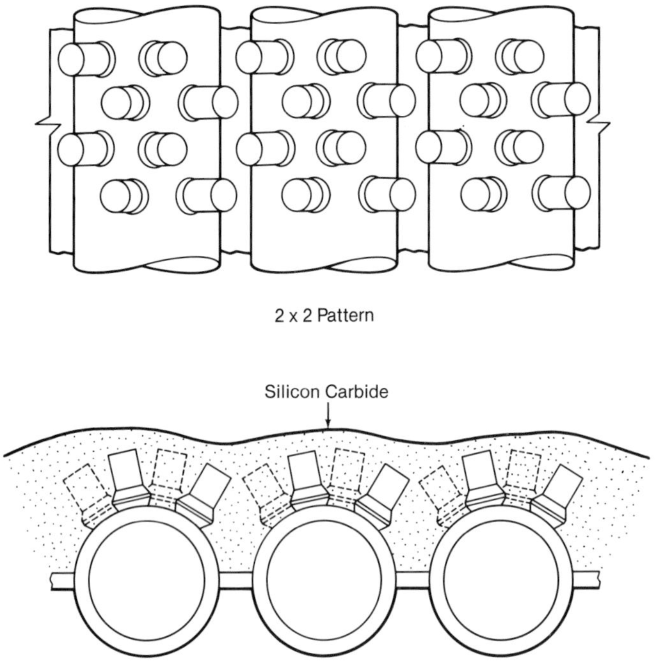

Fig. 6 Lower furnace studs and refractory, mass-fired unit.

high at the front of the furnace and tapers off to about 1 ft (0.3 m) high at the ash discharge end (Fig. 7). These blocks are designed for easy replacement as they wear.

The pin stud pattern, pin stud length and pin stud diameter must be carefully chosen for its ability to hold the refractory in place and to maximize the heat transfer through the stud to the furnace wall tubes. This, in turn, serves two purposes. One is to provide maximum cooling to keep as low a refractory temperature as possible. Maintaining a low refractory surface temperature has a dramatic effect on refractory life, furnace wall fouling and maintenance costs. Secondly, with more heat removed in this lower furnace area, less heating surface is required in the upper furnace to achieve the desired flue gas temperature leaving the furnace.

RDF-fired units

Prior to the late 1980s, RDF boilers were installed with bare carbon steel tubes in the lower furnace and no corrosion protection. It was thought that with the more even combustion with a processed fuel, corrosion would not be a concern in the lower furnace. Early units, operating at low steam pressure and temperature, did not experience corrosion problems. However, as higher pressure and temperature units went into operation, corrosion increased and lower furnace protection was needed.

The same pin stud and refractory design used on mass-fired units has also been tried on RDF units. This solved one problem but created another. Inherent in the RDF combustion process is a high degree of suspension firing and high flame temperatures in the lower furnace. When pin studs and refractory are applied, the lower furnace tubes are insulated, resulting in less heat transfer and hotter flue gas temperatures in the lower furnace. This, in turn, can result in significant slagging on the refractory wall surface. Pin stud and refractory was tried at two RDF facilities. However, increased furnace slagging resulted, and eventually the pin studs and refractory were removed.

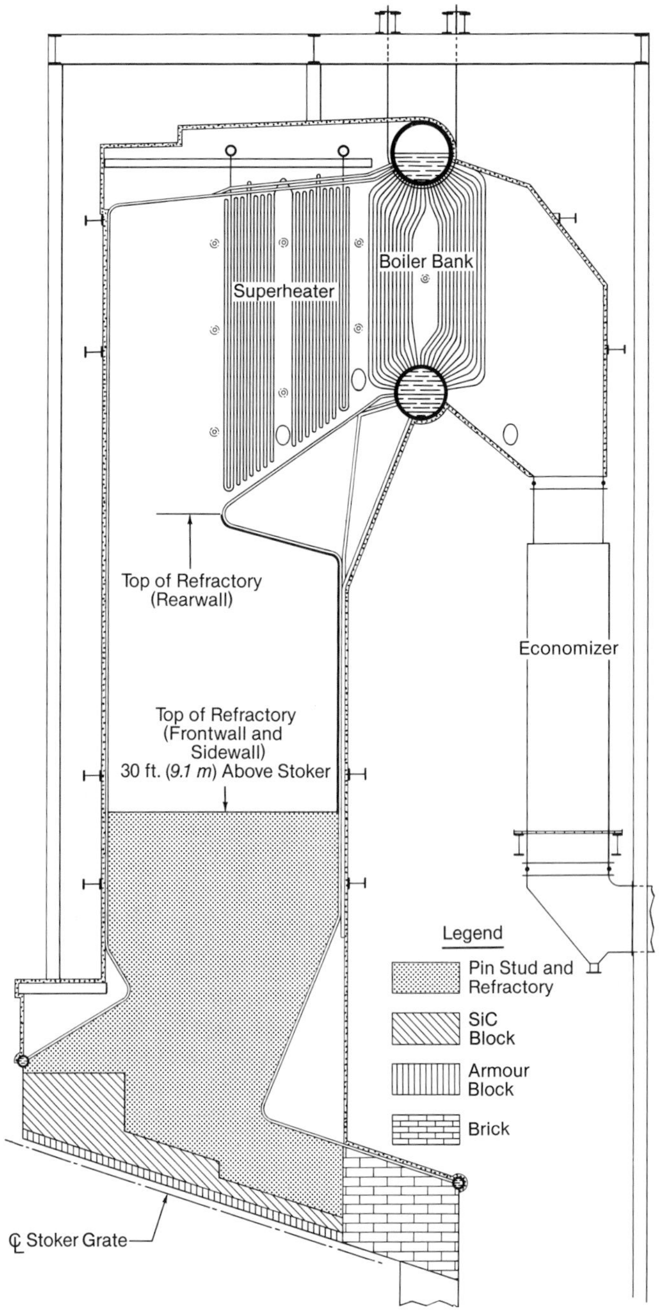

Fig. 7 Refractory type and location, mass-fired unit.

What was needed was a material that was resistant to the chloride corrosion found in refuse boilers while not insulating the lower furnace tubes. Babcock & Wilcox (B&W) pioneered the use of Inconel material as a solution to this lower furnace corrosion problem. In 1986, following rapid corrosion of the bare carbon steel tubes, the lower furnace of the Lawrence, Massachusetts unit was covered with a weld overlay of Inconel material. This overlay proved to be effective in minimizing corrosion in the lower furnace. Based on this early experience the industry followed B&W's lead and Inconel weld overlay was field applied to the lower furnace of a number of operating boilers.

For the RDF boilers supplied as part of the refuse-to-energy plant in Palm Beach County, Florida, the decision was made to add the Inconel protection prior to manufacture. A bimetallic tube construction was used consisting of a carbon steel inner tube co-extruded with an Inconel outer tube (Fig. 8). This allows for a more uniform protective coating than does a weld overlay. These boilers went into operation in 1989 and the Inconel bimetallic tubes have experienced no corrosion in their initial years of operation. Today the bimetallic tube is the industry standard for lower furnace corrosion protection in RDF-fired boilers.

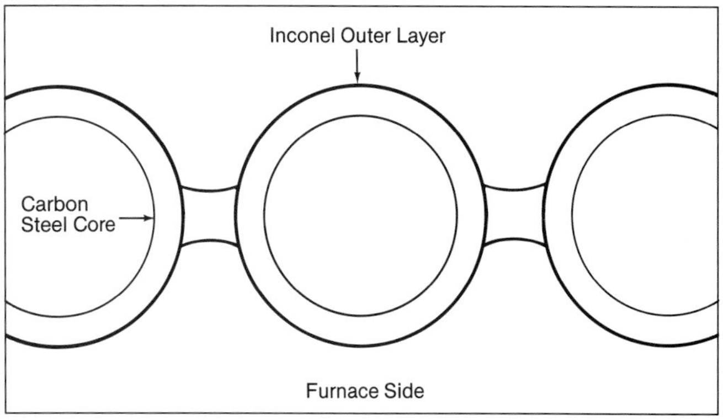

Fig. 8 Bimetallic tube.

Mass burning

Mass burning is the most common refuse combustion technology worldwide. There are more than 1600 mass burn units in operation throughout the world, predominantly in Europe and Japan. When the market for refuse-to-energy facilities expanded rapidly in the U.S. in the early 1980s, many of these refuse plants adopted this well proven mass burning technology (Fig. 9). However, some major differences in the U.S. application resulted in some operational problems. U.S. applications tend to require large units (Fig. 10) to accommodate larger regional facilities instead of the small local and community plants typical of European and Japanese applications. U.S. plants were designed to operate at significantly higher operating pressures and temperatures to take advantage of the economics of production and sale of electric power while typical non-U.S. applications produced hot water and low pressure steam for heating applications. U.S. refuse fuel typically has a higher heating value and lower moisture content. Finally, U.S. units began to be installed at a time when environmental concerns were increasing.

The net result of these characteristics of the market resulted in many early U.S. refuse units experiencing operating problems related to:

1. high rates of slagging in the furnace,
2. higher gas temperature leaving the furnace resulting in overheating of superheaters and excessive fouling in the convection section,
3. tube failures from accelerated corrosion that were metal temperature related, and
4. concerns about the creation of dioxins (Polychlorinated Dibenzoparadioxin, PCDD) and furans (Polychlorinated Dibenzofurans, PCDF) during the combustion process

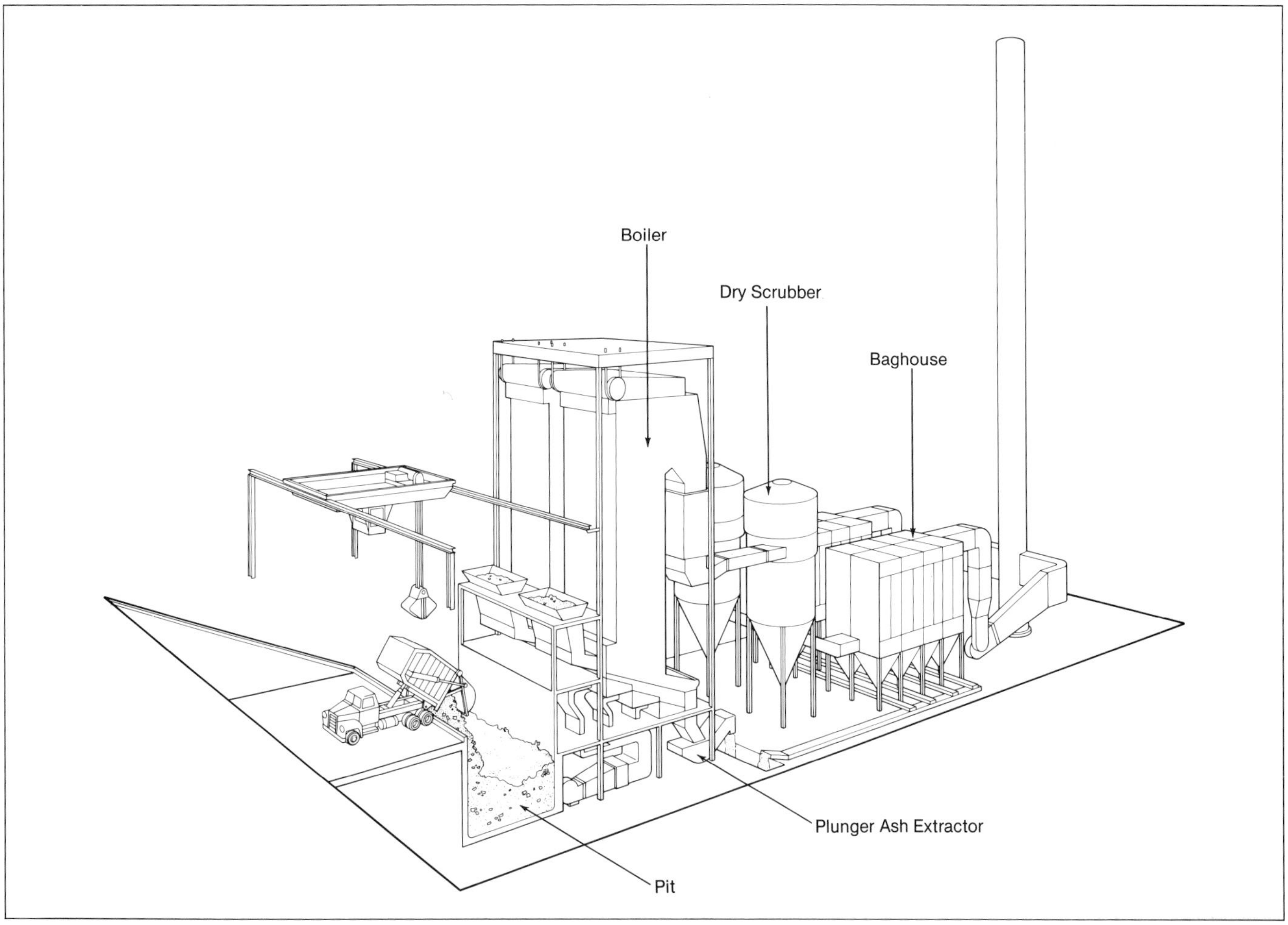

Fig. 9 Typical mass burning refuse-to-energy system.

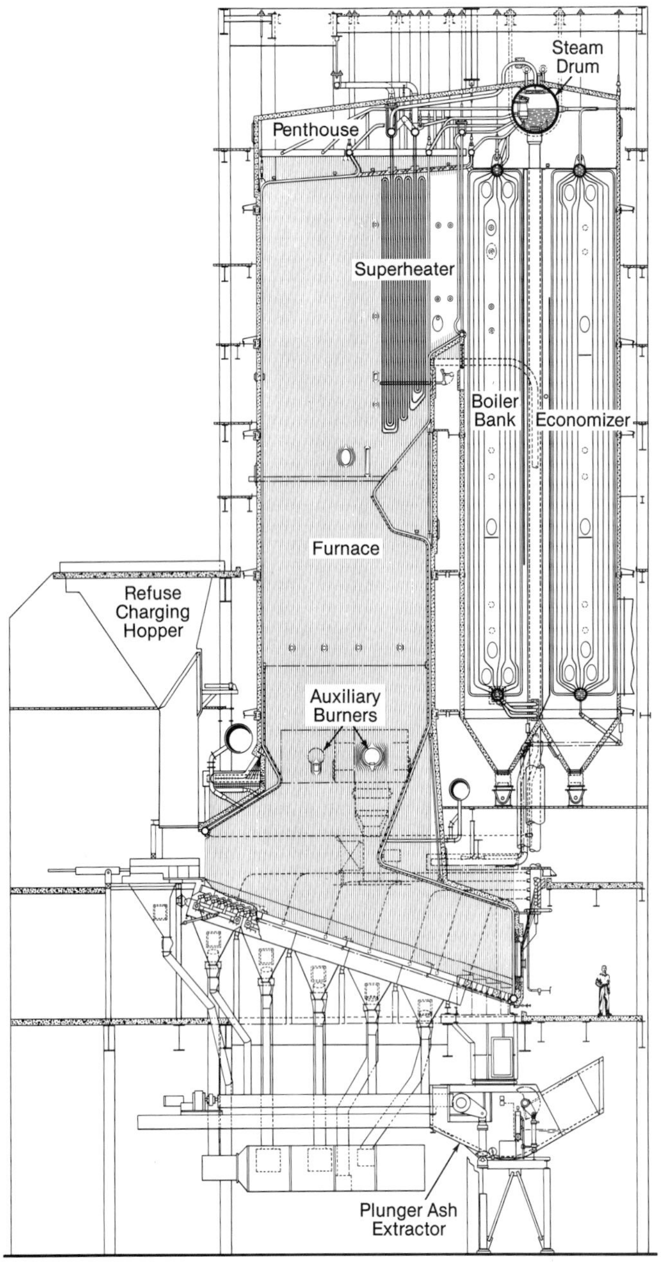

Fig. 10 Typical mass burning unit.

that were related to less than optimum combustion systems, particularly less than optimum turbulence and mixing of fuel and air in the lower furnace.

Boiler plant sizing

A refuse plant must be sized to handle the physical amount of refuse that is delivered to that plant, regardless of the refuse heating value. A refuse boiler, on the other hand, is a heat input device and must be sized for the maximum heat input expected. When designing a refuse boiler you need to know the design tons per day of refuse to be combusted and the typical range of heating values that is expected for the refuse in that location.

The boiler is typically designed for the maximum ton per day input at the maximum refuse heating value. A 1000 t/d (907 t_m/d) refuse plant is actually not the same size plant in all locations. A plant in the Northeast U.S. would typically be designed to handle refuse at heating values as high as 5500 Btu/lb (12,793 kJ/kg), or 458 x 10^6 Btu/h (134 MW_t) total heat input to the boilers. At the other extreme, a plant in a less industrialized country would be designed to handle refuse with a heating value in the range of 3500 Btu/lb (8140 kJ/kg), or 292 x 10^6 Btu/h (85.6 MW_t) total heat input. Both are 1000 t/d (907 t_m/d) plants, but one has refuse boilers that have a 50% larger capacity.

For many of the early U.S. plants, good data were not available on the true range of heating values of the refuse. Refuse boilers were sized for typical heating values of 4500 Btu/lb (10,467 kJ/kg). When the actual heating values were found to be as high as 5200 to 5500 Btu/lb (12,095 to 12,793 kJ/kg), the boilers were actually undersized and could not process the available refuse on a ton per day basis.

Stoker capacity

A refuse stoker has both a heat input limit and a ton per day of refuse limit. If a typical 1000 t/d (907 t_m/d) refuse plant has two 500 t/d (454 t_m/d) boilers, and the design refuse heating value is 5000 Btu/lb (11,630 kJ/kg), each boiler would have a maximum heat input limit of 208.3 x 10^6 Btu/h (61.1 MW_t). If the actual refuse heating value is above 5000 Btu/lb (11,630 kJ/kg), the maximum heat input limit can not be exceeded and therefore the unit's t/d capacity would be reduced below 500 t/d (454 t_m/d). On the other hand, if the actual refuse heating value is below 5000 Btu/lb (11,630 kJ/kg), then the unit could actually process more than 500 t/d (454 t_m/d) of refuse, up to the maximum ton per day limit of that stoker.

The ton per day limit is usually set by a refuse capacity per unit of width, a limit for optimum fuel feed and distribution, or a weight per square foot (m^2), a structural limit. These limits are in the range of 30 t/d (27 t_m/d) per front foot (0.3 m) of width and 65 lb/h ft^2 (2.74 kg/h m^2) of grate area. The grate surface area is set by a grate release rate generally in the range of 300,000 to 350,000 Btu/h ft^2 (946,350 to 1,104,080 W/m^2), but may be lower for low heating value, high moisture fuels. The stoker width and depth are also related to the specific fuel.

A high heating value, low moisture fuel would require a wider, less deep stoker because the fuel will tend to burn more rapidly. A low heating value, high moisture fuel would require a narrow, deeper stoker because more residence time on the stoker is usually needed. The combination of all these criteria will set the maximum ton per day rating of the stoker.

There is also a minimum load that can be effectively handled on a given stoker. This load is also set by both a ton per day limit (minimum fuel inventory on the grate) and a heat input limit (minimum heat input for good combustion).

All of these limits can be incorporated into a capacity diagram, which provides the operator with the boundary limitations around a family of heating value curves. Fig. 11 is such a diagram for a typical 500 t/d (454 t_m/d) boiler burning 5000 Btu/lb (11,630 kJ/kg) refuse.

Stoker design

The combustion of MSW requires a rugged, reliable stoker to successfully convey and burn unsorted refuse. Most stokers use some variation of a reciprocating grate action, with either forward moving or reverse acting grate movement. Some arrangement of moving and stationary grates is used to move the refuse through the furnace and allow time for complete combustion.

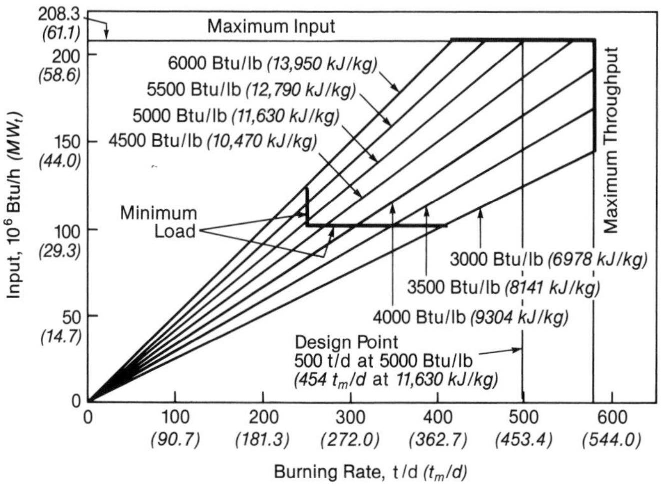

Fig. 11 Stoker capacity diagram.

The stoker illustrated in Fig. 12 is typical of the forward moving reciprocating grate stoker. This grate is designed with alternate moving and stationary rows of grates in a stairstep construction with a downward slope to help move the refuse through the furnace. Each row of grates overlaps the row beneath it and the alternate rows are supported from a moving frame driven by hydraulic cylinders (Fig. 13). These grates move the refuse over the stationary grates, where it is picked up by the next row of moving grates and moved through the furnace. The action of these reciprocating grates rolls and mixes the refuse, constantly exposing new material to the high temperatures in the bed and allowing the combustion air to contact all the burning refuse.

For low heating value, high moisture refuse, drop off steps are often incorporated into the stoker design. The steps are located at the end of each grate module resulting in one to three steps depending on the overall stoker length. These steps promote a tumbling and rolling action as the burning refuse falls off the step. This type of design was used on a number of the early refuse units in the U.S. but was found to be unnecessary with the higher heating value and lower moisture refuse. In fact, it can be a detriment as the tumbling can also result in excursions of high carbon monoxide (CO) emissions.

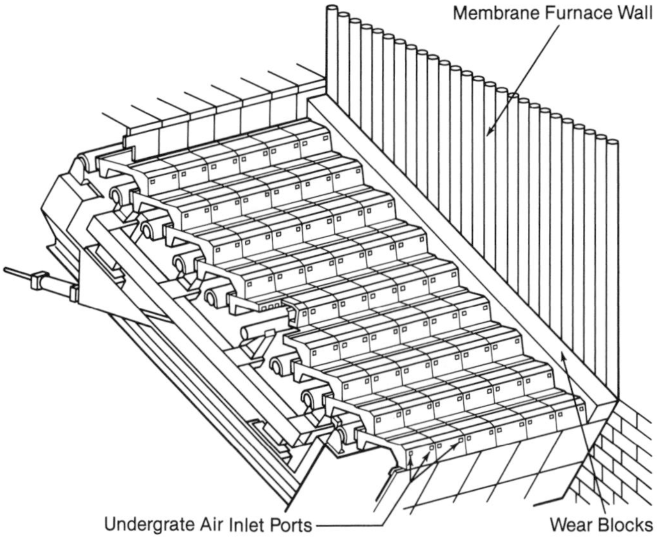

Fig. 12 Forward moving reciprocating grate stoker *(courtesy von Roll Inc.)*.

The grates are usually constructed in a series of standard modules with independent drives and air plenums. This allows the individual grate modules to be factory- assembled to limit field construction time and provide complete duplication of parts for easy maintenance and repair. This modular construction allows any size stoker to be constructed from a small number of standard modules. A typical stoker is usually from two to four modules in length and one to four modules in width. This method of construction also allows for complete zoned undergrate air control to the individual burning areas of the grate and provides complete freedom in the operating speed of the individual grate modules to provide the required feed rate along the grate for complete burnout of the fuel. This ability to control the undergrate air in multiple air zones along the width and depth of the stoker is an important factor in minimizing CO and nitrogen oxides (NO_x) emissions.

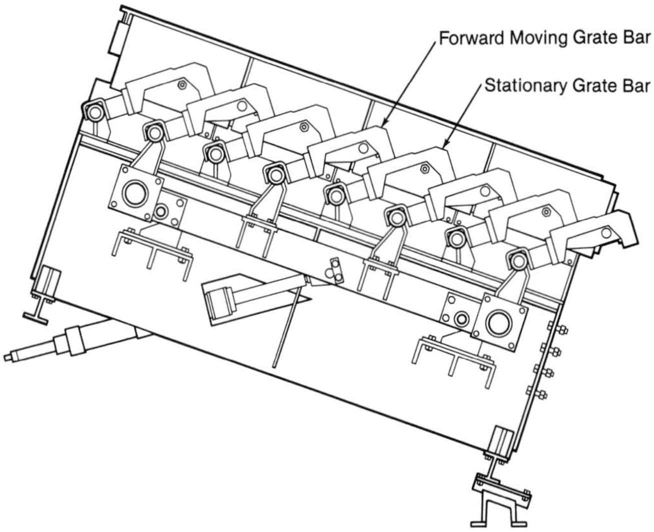

Fig. 13 Reciprocating grate, longitudinal section *(courtesy von Roll Inc.)*.

Fuel handling

The MSW delivered to the refuse plant is generally dumped directly into the storage pit. This large pit also provides a place to mix the fuel. This mixing is done using the crane and grapple to move and restack the refuse as it is dumped into the pit. This produces a fuel, in both composition and heating value, as consistent as possible for the boilers. This is an essential job for the crane operator and any time not required to feed the furnaces is used to mix the fuel. It is not uncommon for the crane operator to mix four grapple loads back into the pit for every one load that goes to the charging hopper.

Fuel feed system

Controlled feed of the fuel is necessary for good combustion to minimize CO and NO_x emissions and to maintain constant steam output. At the bottom of the charging hopper feed chute, a hydraulic ram pushes the fuel into the furnace and onto the stoker grates at a controlled rate. On larger units, multiple charging rams are used across the width of the unit to provide a continuous fuel feed with optimum side to side distribution. The hydraulic rams stroke forward slowly and then retract quickly to provide the positive continuous fuel feed. These rams are simple to control with feed rate adjustments made by either the speed of travel or the number of strokes per hour.

Combustion air system

The primary combustion air, or undergrate air, is fed to the individual air plenums beneath each grate module. A control damper at the entrance to each air plenum controls the undergrate air to each section of the grate (Fig. 14). The grate surface is designed to meter the primary combustion air to the burning refuse uniformly over the entire grate area. This is accomplished by providing small air ports or tuyères in the surface of the individual grate bars. These air ports provide openings equal to approximately 3% of the grate area which results in sufficient pressure drop of air resistance across the grate to assure good distribution of the air flow through the grate, regardless of the depth of refuse on the grate. Undergrate air systems are generally designed for 70% of the total air to be undergrate air with expected normal operation of 60%.

Because refuse contains a high percentage of volatiles, a large portion of the total combustion air should enter the furnace as secondary, or overfire, air through the furnace walls. These secondary air ports are located only in the front and rear furnace walls so that the air flow parallels the normal flow pattern through the unit. Older design units generally provided 25 to 30% of the total air as overfire air. With today's emphasis on better combustion and lower emissions, the overfire air systems are designed for 50% of the total air to be overfire air with expected normal operation at 40%.

The basic function of the overfire air is to provide the quantity of air and the turbulence necessary to mix the furnace gases with the combustion air and to provide the oxygen necessary for complete combustion of the volatiles in the lower furnace. Excess air in the furnace is usually maintained in the range of 80 to 100% and complete combustion is demonstrated by a CO value in the furnace of 100 ppm or less.

To aid in the combustion of wet fuels during extended periods of rainy weather, the air system includes steam coil air heaters designed to provide air temperatures in the range of 300 to 350F (149 to 177C) to help dry these fuels

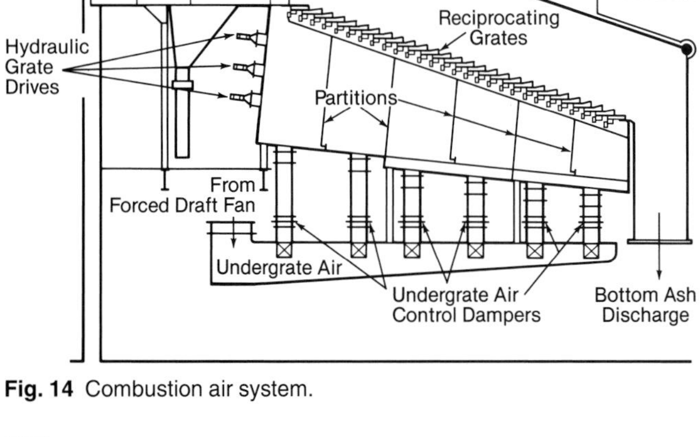

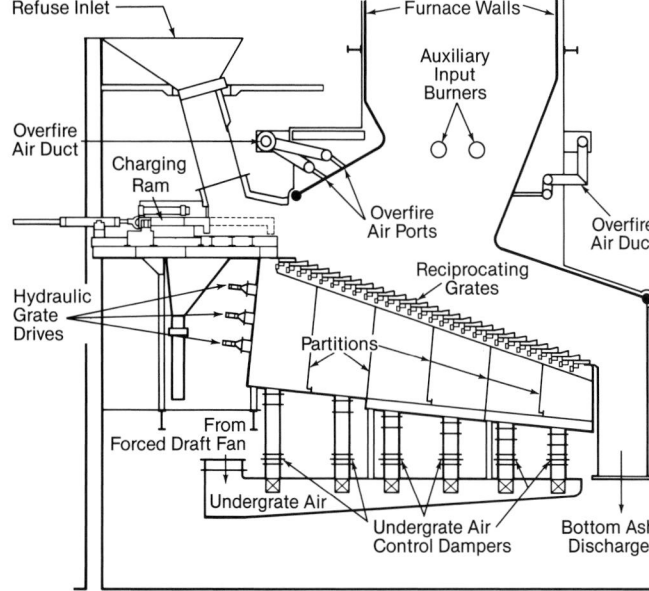

Fig. 14 Combustion air system.

and maintain furnace temperature. These steam coil air heaters are commonly used only for the undergrate air because this is the air flow which directly aids in drying wet fuel. These air heaters must be conservatively designed with fin spacing not exceeding 4 to 5 fins/in. (1 fin/6.4 to 5.1 mm). Most plants take combustion air from the storage pit area to help minimize odors. This air is normally contaminated with dust and lint which could plug the steam coil. Some type of cleaning arrangement or filters must therefore be included to keep the steam coil clean.

When high moisture fuels are encountered, the first action by the operator is to use the steam coil air heater to provide hot air. However, for very high moisture fuels it may also be necessary to use the auxiliary fuel burners to stabilize combustion in the furnace. These cases are the exception, and in normal operation neither the auxiliary burners nor the steam coil air heater are needed for good combustion.

Ash handling systems

When refuse is burned, the ash takes the form of either light ash, called flyash, or coarse ash, which comes off the stoker. The flyash is entrained in the gas stream until it is removed in the particulate collection device or falls out into the boiler, economizer, or air heater hoppers. The stoker ash consists of ash from the fuel, slag deposits on the grate, ash from the furnace walls and, in some designs, ash from the superheater. The stoker ash is discharged through the stoker discharge chute and from the stoker siftings hoppers.

Plunger ash extractor

The ash from the stoker discharge on mass-fired units may contain large pieces of noncombustible material, in addition to the normal ash from combustion. Ash consistency can vary from fine particles to large and heavy noncombustible objects in the fuel. The ash from the stoker discharge chute falls into a water bath in the plunger ash extractor (Fig. 15) that quenches the ash and controls dusting. After the ash is quenched, a slow moving, hydraulically operated ram cycles forward and back to push and squeeze the accumulated ash up an inclined dewatering section to the discharge of the extractor. The ram cycle continues at a slow speed to push the ash out of the extractor. The dewatered ash has a moisture content of 15 to 20% as a result of the squeezing process on the incline. The lower moisture content can have an economic advantage as the cost of landfilling ash is based on total weight, which includes the weight of water in the ash.

To keep the ash system simple and to minimize costs, ash from the extractor can discharge directly into a truck or bin for final disposal. To move the ash away from the vicinity of the stoker discharge, vibrating and belt-type conveyors are used. A short vibrating conveyor is placed at the discharge of the ash extractor; its metal trough can take the impact of the oversized, noncombustible material falling from the extractor. The vibrating conveyor then transfers the ash to a belt conveyor, minimizing the wear on the belt conveyor that would occur if the extractor discharged directly onto it.

Double gate ash hopper

Double gate ash hoppers can be used (Fig. 16) in lieu of an ash extractor. The hopper has gates on the top and

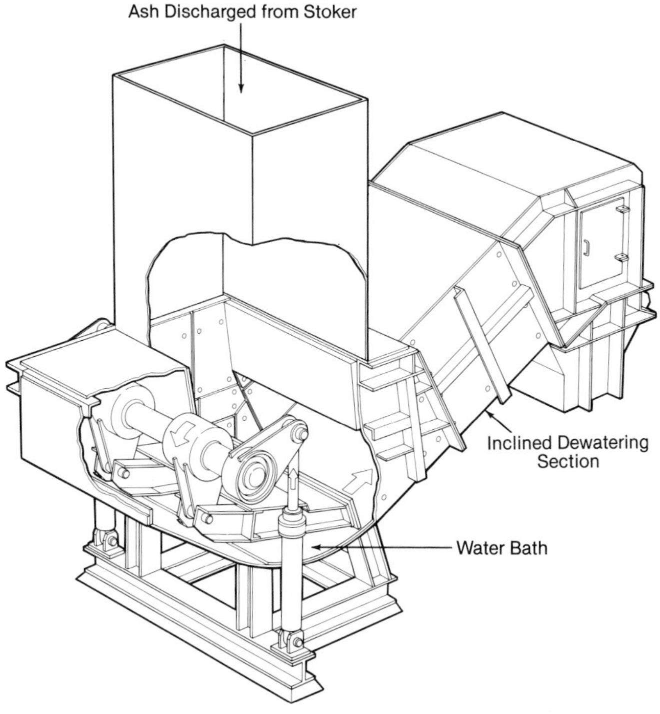

Fig. 15 Plunger ash extractor.

the bottom to control the flow of ash entering and discharging from the hopper. Water spray nozzles help quench the ash in the hopper. The ash can be discharged directly into a truck or onto a conveyor for disposal. This system is simple with low capital cost.

Scrubber, precipitator and baghouse flyash

The flyash collected in the scrubber, precipitator or baghouse hoppers can be handled by dry mechanical screw or chain-type conveyors. These units operate continuously to minimize hopper pluggage problems and they discharge onto a collecting conveyor, which is usually a dry chain type. Because the mechanical conveyors are dust-tight, but not designed to be gas-tight, separate sealing devices, such as rotary seals or double flop valves, are used. The collecting conveyor will collect the flyash discharged by all the conveyors under the rows of hoppers, and move it to a single collection point for ultimate disposal.

RDF firing

RDF technology was developed in North America as an alternative to the mass burning method. Initially, RDF was used as a supplementary fuel for large, usually coal-fired, utility boilers. For this application the RDF was finely processed and sized to 1.5 in. (38.1 mm) maximum size. The resulting RDF was nearly all light plastics and paper.

For supplemental firing, B&W guidelines call for a maximum RDF input of 20% on a heat input basis and no RDF input until the boiler is operating above 50% load. In most cases the RDF is blown into the furnace sidewalls at the pulverized coal burner elevation through an RDF burner with a fuel distribution impeller. Most of the RDF burns in suspension in the high heat input zone of the pulverized coal fire. However, some of the heavier fuel fraction falls out in the lower furnace. Dump grate stokers located in the neck of the ash hopper allow more complete burnout of these heavier pieces before they are discharged into the ash system (Fig. 17). RDF has been successfully co-fired in B&W boilers at Lakeland, Florida, Ames, Iowa, and Madison Gas & Electric in Wisconsin.

RDF has also been successfully co-fired in B&W Cyclone furnaces where the finely processed and sized RDF is injected into the Cyclone secondary air stream moving tangentially inside of the Cyclone barrel. (See Chapter 14.) This method of RDF combustion is used at the Baltimore Gas & Electric Company's Crane station in Maryland.

Dedicated RDF-fired boilers

From this supplemental fuel experience, RDF then became the main fuel for boilers specifically designed to generate full load steam flow when burning RDF (Fig. 18). In some cases where steam flow was required even when refuse was not available, the boiler was designed so that it could also reach full load on wood, coal or natural gas. More commonly there would only be auxiliary gas or oil burners for startup and shutdown.

The first boilers in the world to fire RDF as a dedicated fuel were B&W units which began operation in 1972 in Hamilton, Ontario, Canada. The first of such boilers in the U.S. went into operation in 1979 in Akron, Ohio. The boiler design was highly influenced by the proven technology of wood-fired boilers with respect to their fuel feed system, stoker design, furnace sizing and overfire air system. The transfer of this technology from wood firing to RDF firing was successful in many areas, but in other areas design adjustments were needed to accommodate the unique aspects of RDF.

The operating experience from the first generation designs at Hamilton, Akron and other plants led to second generation designs with improved RDF processing systems, fuel feed systems and boiler design. Specific improvements included the first fuel feeder designed specifically for RDF and the use of alloy weld overlay in the lower furnace for corrosion protection.

The third generation of facilities is essentially today's state-of-the-art design (Fig. 19). This boiler design has a unique lower furnace arch arrangement and an enhanced

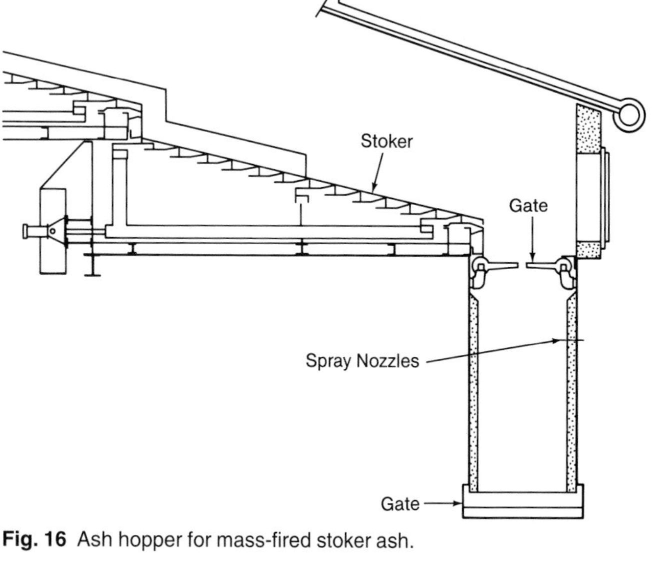

Fig. 16 Ash hopper for mass-fired stoker ash.

overfire air system to significantly improve combustion efficiency. These third generation designs also incorporate dramatically improved fuel processing systems.

RDF preparation systems

The first generation RDF processing systems were crunch and burn systems. The incoming refuse first went to a hammermill type shredder that produced an RDF with 6 x 6 in. (152 x 152 mm) top size. Ferrous metal was removed by magnetic separators. There was no other material separation and many undesirable components entered the boiler. Shredded particles of glass were embedded in wood and paper resulting in a very abrasive fuel entering the boiler in suspension. Also, the RDF was generally stored in a hopper or bin. RDF is compactible and in nearly every case, significant problems were encountered getting the RDF out of the storage bins.

Second generation RDF processing systems recognized and corrected some of the problems. The shredder for final fuel sizing was moved to the back of the processing system and some type of rough sizing shredder was used as the first piece of equipment in the system. This reduced, but did not eliminate, the problem of abrasive particles embedding in the fuel. Some size separation equipment was introduced, generally removing the small size fraction which is less than 1.5 in. (38.1 mm) composed mostly of broken glass, ceramics and dirt, which was sent to landfill. RDF was stored on the floor rather than in bins or hoppers and was moved by front-end loaders to conveyor belts. This greatly improved the reliability of fuel flow to the boiler.

In third generation RDF processing systems (Fig. 20) the first piece of equipment became a flail mill or similar equipment whose main function was to break open the garbage bags. The refuse was still size separated using a trommel or disk screen with the minus 1.5 in. (38.1 mm) size destined for landfill. Generally, a device such as an air density separator was added to remove the light fraction (paper, plastics, etc.) from this stream to achieve maximum heat recovery. Where it was economically attractive, aluminum separation was added to the plus 1.5 in. (38.1 mm.) minus 6 in. (152 mm) stream.

RDF yield

The ash content of the RDF is directly related to the yield of the processing system or the percentage of RDF produced from a given quantity of MSW. A 70% yield means that 70 t of RDF is produced for every 100 t of incoming MSW. In a processing system with a lower yield, the portion of the

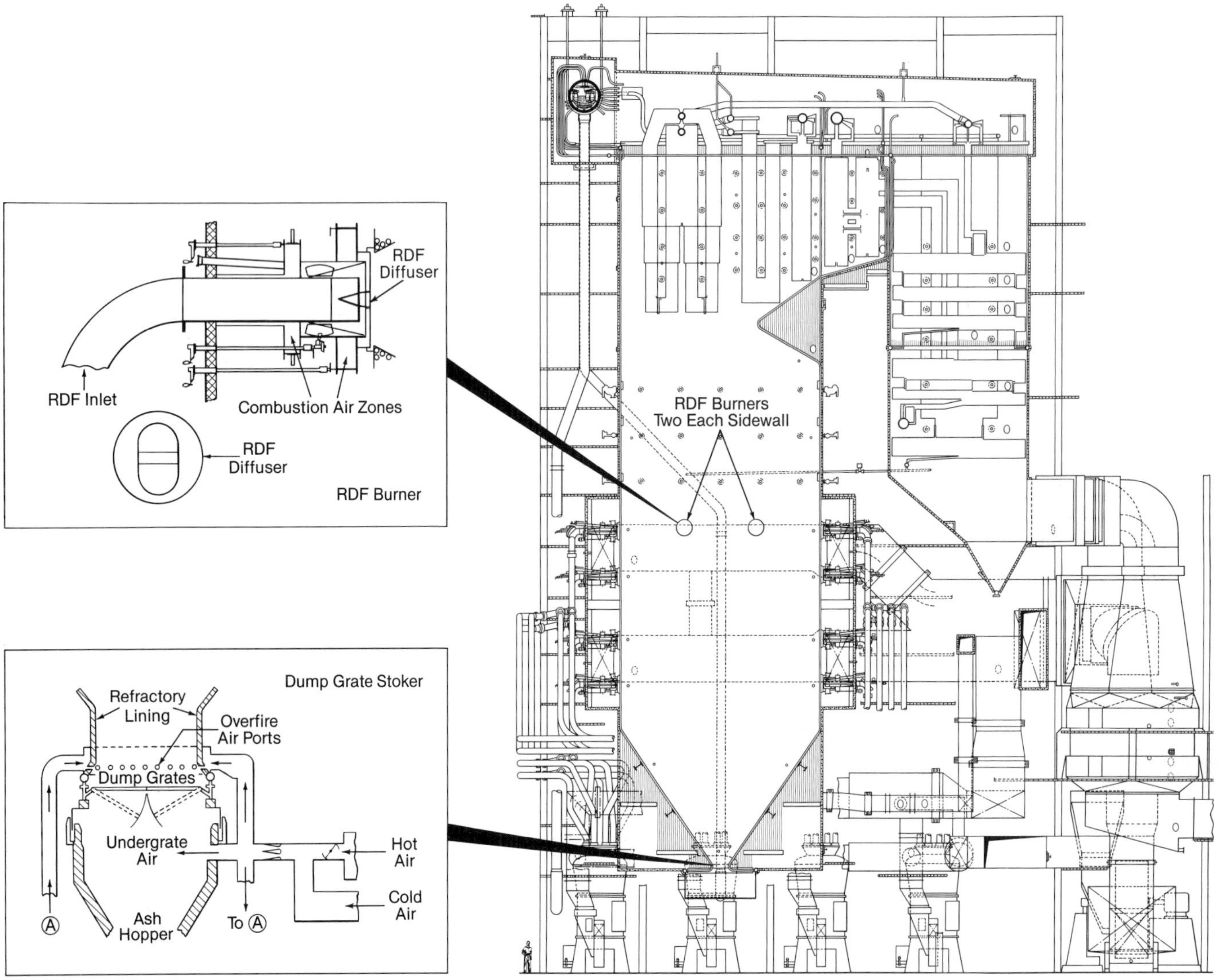

Fig. 17 Typical B&W RB type utility boiler firing RDF as a supplementary fuel.

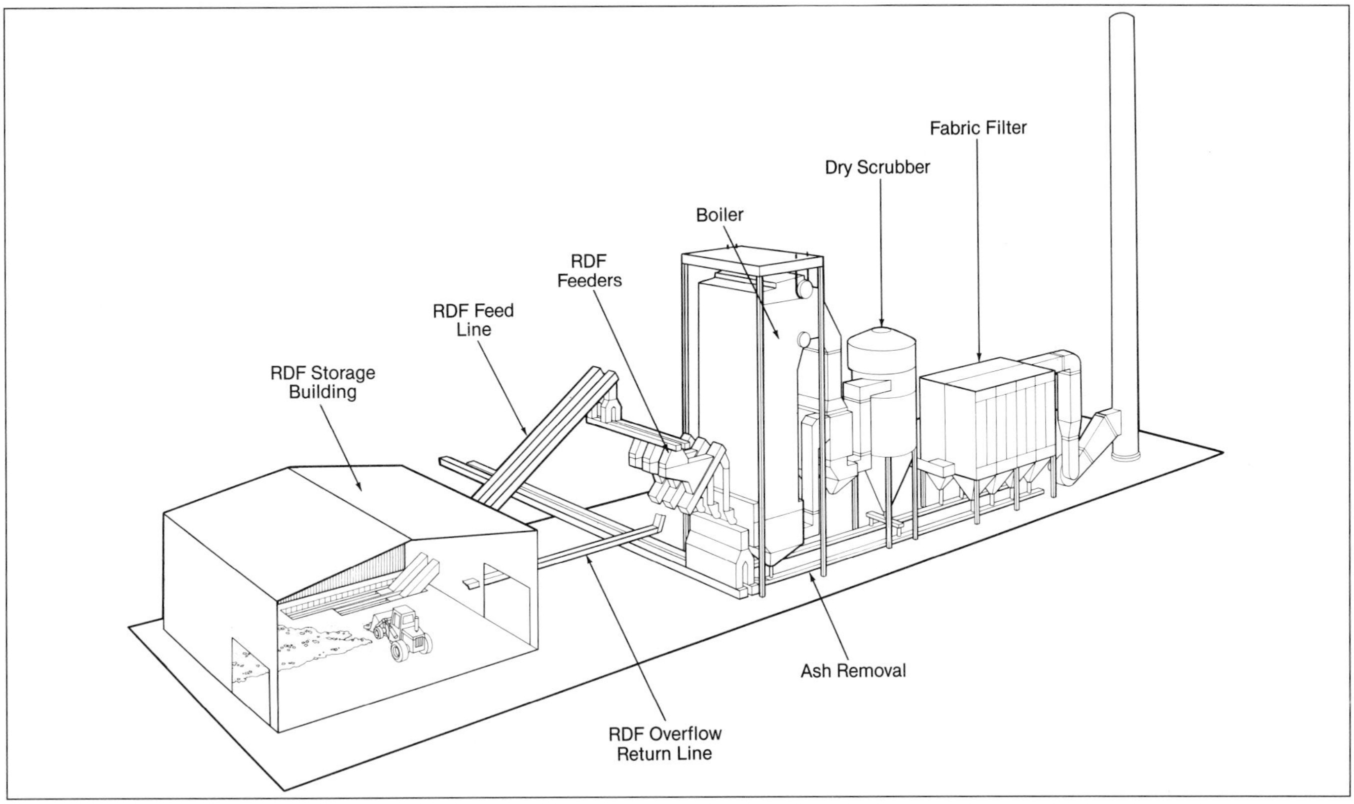

Fig. 18 Typical RDF refuse-to-energy system.

MSW that is rejected is generally high in ash and inerts content and, therefore, the resulting RDF is low in ash content. As the RDF processing system is designed to obtain a higher yield, more of the ash is carried over into the RDF fuel fraction and the ash content is increased.

The RDF heating value is inversely related to the yield; the higher the yield, the lower the heating value. In a high yield system, most of the rejects are ferrous metals and inerts (glass, ceramics, dirt) which have no heating value. While some fuel is also rejected, the quantity is small resulting in a higher net heating value for the RDF. A yield of about 93% represents a crunch and burn type system in which only the ferrous metal is removed. In such a system, the RDF heating value is only marginally higher than the heating value of the incoming MSW.

A typical MSW might have a composition comparable to the reference waste shown in Table 4. The majority of the waste is combustible materials, which have ash contents ranging from approximately 4% for wood to 12% for glossy magazine paper. The glass fraction, yard waste and mixed combustibles may also contain varying quantities of sand, grit and dirt. The predicted composition of the RDF will vary depending on the type of processing system and the resulting yield. Table 5 shows how the ash content in the fuel and the heating value of the fuel will vary as the RDF yield varies with different processing systems. Two cases are considered, one which assumes no front end recycling (curbside recycling or separate recycling facility), and a second case which assumes that such a system is in place in the community.

RDF quality

RDF used to supplement pulverized coal in utility boilers should be low in ash; have minimum ferrous metal, aluminum, and other nonferrous metal; and be small enough in particle size to be fed pneumatically to the boiler. The processing system for such a fuel would generally be a very low yield system, between 40 and 60%.

RDF for dedicated traveling grate stoker boilers should

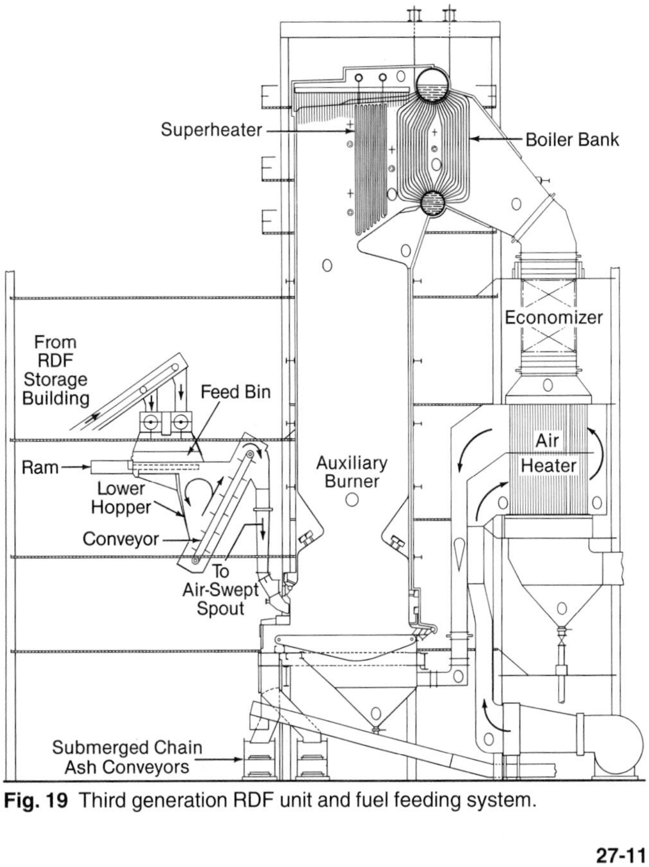

Fig. 19 Third generation RDF unit and fuel feeding system.

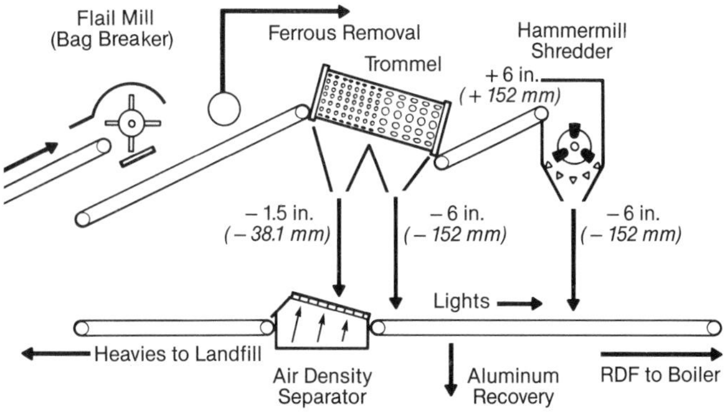

Fig. 20 RDF processing system.

be low in ash, consistent with a high RDF yield from the MSW; as free as possible of ferrous metal, aluminum and other nonferrous metals; and of a particle size distribution that is considerably larger than the particle size of RDF for use in Cyclone or pulverized coal boilers. The processing system for such a fuel will be a higher yield, around 70 to 85%.

RDF produced in a crunch and burn system, in which solid waste is shredded and only the ferrous metal removed, has a yield of about 93%; an inherently high ash content; and contains 100% of the aluminum, other non-ferrous metals, glass, stones and ceramics in the original MSW. While this is a high yield system which is desirable for a dedicated traveling grate stoker boiler, it also contains large quantities of aluminum, glass, and other inerts which results in higher wear on the stoker and lower furnace. Another result is a more conservatively designed stoker and furnace as well as a larger ash handling system.

RDF processing systems

An optimum RDF processing system for a dedicated boiler application achieves the highest yield with the highest heat recovery, while removing ferrous, aluminum and glass before entering the boiler. Such a system (Fig. 21) includes the following:

In-feed conveyors From the tipping floor, the solid waste is fed by front-end loaders to steel pan apron conveyors which feed the flail mill in-feed conveyors.

Initial size reduction The flail mill tears open the plastic garbage bags, coarsely shreds the refuse, and also breaks glass bottles to a size of approximately 1.5 in. (38.1 mm) or less.

Ferrous metal recovery Ferrous metal is extracted from the coarsely shredded MSW in each line by a single-stage overhead magnet. Recovered ferrous metal is moved to a ferrous air classifier where tramp materials such as paper, plastics or textiles are removed, thereby providing a clean ferrous product. A ferrous recovery of 90% is possible.

Size classification and final size reduction After ferrous removal, shredded waste is fed into a rotating trommel screen, a size separating device about 10 ft (3.0 m) in diameter by 60 ft (18.3 m) long. The trommel performs the following functions:

1. removes glass, sand, grit and nonferrous metal less than 1.5 in. (38.1 mm) in size, and
2. removes the minus 6 in. (152 mm), plus 1.5 in. (38.1 mm) fraction which is the proper fuel size without additional processing and contains the bulk of the aluminum cans.

The trommel oversize material, plus 6 in. (152 mm), is then shredded in a horizontal secondary shredder. Because the secondary shredder is a major consumer of energy and has high hammer maintenance costs, the RDF process is specifically designed to reduce the secondary shredder's load by shredding only those combustibles too large for the boiler. Particle size is controlled with a disk screen which recycles oversize material back to the secondary shredder.

Separation of glass, stones, grit and dirt Trommel undersize material, minus 1.5 in. (38.1 mm), passes over an air density separator (ADS) designed to remove dense particles from less dense materials through vibration and air sweeping. This device can efficiently remove glass, stones, grit and dirt, as well as nonferrous metals. The light fraction, which can range from approximately 50 to 90% of the ADS feed, consists essentially of combustibles with high fuel value which are recovered and blended into the main fuel stream.

Aluminum can recovery To optimize aluminum can recovery, an air classifier is provided for the plus 1.5 in. (38.1 mm), minus 6 in. (152 mm) undersize fraction. The air classifier removes the light organic portion of the stream, allowing aluminum cans to be more visible for hand pickers. The air classifier heavy fraction drops onto a conveyor moving at approximately 2.5 ft/s (0.76 m/s) with numerous hand-picking stations on either side of the belt. Cans

Table 4
Typical Reference Refuse

	Reference	
Component Analysis	MSW (% by wt)	RDF (% by wt)
Corrugated board	5.53	—
Newspapers	17.39	—
Magazines	3.49	—
Other paper	19.72	—
Plastics	7.34	—
Rubber, leather	1.97	—
Wood	0.84	—
Textiles	3.11	—
Yard waste	1.12	—
Food waste	3.76	—
Mixed combustibles	17.75	—
Ferrous	5.50	—
Aluminum	0.50	—
Other nonferrous	0.32	—
Glass	11.66	—
Total	100.00	
Ultimate Analysis		
Carbon	26.65	31.00
Hydrogen	3.61	4.17
Sulfur	0.17	0.19
	(max. 0.30)	(max. 0.36)
Nitrogen	0.46	0.49
Oxygen	19.61	22.72
Chlorine	0.55	0.66
	(max. 1.00)	(max. 1.20)
Water	25.30	27.14
Ash	23.65	13.63
Total	100.00	100.00
Heating value	4,720 Btu/lb (10,979 kJ/kg)	5,500 Btu/lb (12,793 kJ/kg)
Fuel value recovery, % MSW	96	
Mass yield, % RDF/MSW	83	

Table 5
RDF Yield Versus Ash Content and Fuel Heating Value

Mode	RDF Yield %	Ash	Btu/lb (kJ/kg)
Without front-end recycling:			
Mass burn	100%	23.64	4,814 (11,197)
Crunch and burn	93%	19.87	5,146 (11,970)
RDF	83	11.72	5,641 (13,121)
	to 70%	to 8.87	to 5,834 (13,570)
With front-end recycling:			
Mass burn	100%	19.58	5,513 (12,823)
Crunch and burn	93%	17.16	5,898 (13,714)
RDF	85	9.91	6,328 (14,719)
	to 71%	to 6.59	to 6,491 (15,098)

go into hoppers and, by conveyor, to a can flattener. A pneumatic conveyor then transfers the flattened cans into a trailer. An eddy current separator, for the removal of aluminum cans, can replace hand picking if the expected amount of cans is high enough to justify the additional capital cost. Aluminum recovery of 60% is possible with hand picking or the eddy current separator.

Oversized bulky waste (OBW) The OBW shredder is generally a horizontal hammer mill used to shred ferrous metal recovered by the RDF processing lines and preseparated oversized material which includes white goods such as refrigerators and washing machines, furniture and tree limbs. The ferrous metal is magnetically recovered and given a final cleaning by an air scrubber to remove tramp materials. The nonferrous material is integrated into the RDF stream.

Tire shredding line If there is a sufficient supply of tires, a separate tire shredding line can be included. A shear shredder, used specifically for shredding tires, can shred 500 passenger car tires per hour. The shredder includes a rotary screen classifier (trommel) for returning shredded tire chips above 2 in. (51 mm) back to the shredder. A tire chip 2 x 2 in. (51 x 51 mm) or less is the final product which is then blended in with the RDF stream.

RDF storage building RDF from each processing line is conveyed to an RDF storage building. From there, it is either fed directly to the boiler or fed directly to a shuttle conveyor and storage pile. When RDF feed is direct to the boiler, excess RDF from the boiler feed system is returned to the RDF storage building. RDF not being fed directly to the boiler is retrieved from the storage pile by a front-end loader and loaded onto inclined conveyors which transport the RDF to the boiler feed system.

Fuel feed system: metering feeders

A successful RDF metering feeder must meet the following design criteria:

1. controlled metering of fuel to meet heat input demand,
2. homogenization of material to produce even density,
3. liberal access to deal with oversized material problems,
4. maintainability, in place, and
5. fire detection and suppression devices.

A reliable RDF metering feeder (see Fig. 19) is a key feature of the second generation RDF boiler design. One feeder is used for each air-swept fuel distributor spout. Each feeder has an upper feed bin which is kept full at all times by an over-running conveyor to ensure a continuous fuel supply. The fuel in this hopper is transferred to a lower hopper by a hydraulic ram. The ram feed from the upper hopper is controlled by level control switches in the lower hopper. The RDF is fluffed into a uniform density by a variable speed inclined pan conveyor which sets up a churning motion in the lower hopper. The pan conveyor delivers a constant volume of RDF per flight which is carried up the pan conveyor and deposited into the air-swept spout. The rate at which the fuel is deposited into the spout is based on fuel demand.

Air-swept distributor spouts

Air-swept fuel spouts, used extensively in the pulp and paper industry, proved to be equally effective for RDF firing. (See Chapter 26.) Lateral fuel distribution by multiple spouts across the width of the furnace delivers fuel evenly over the grate. Longitudinal distribution is accomplished by continuously varying the pressure of the air sweeping the spout floor. A major feature of this design is its simplicity.

Traveling grate stoker

To date, only traveling grates have been used for spreader-stoker firing of RDF. These grates move from the rear of the furnace to the front, into the direction of

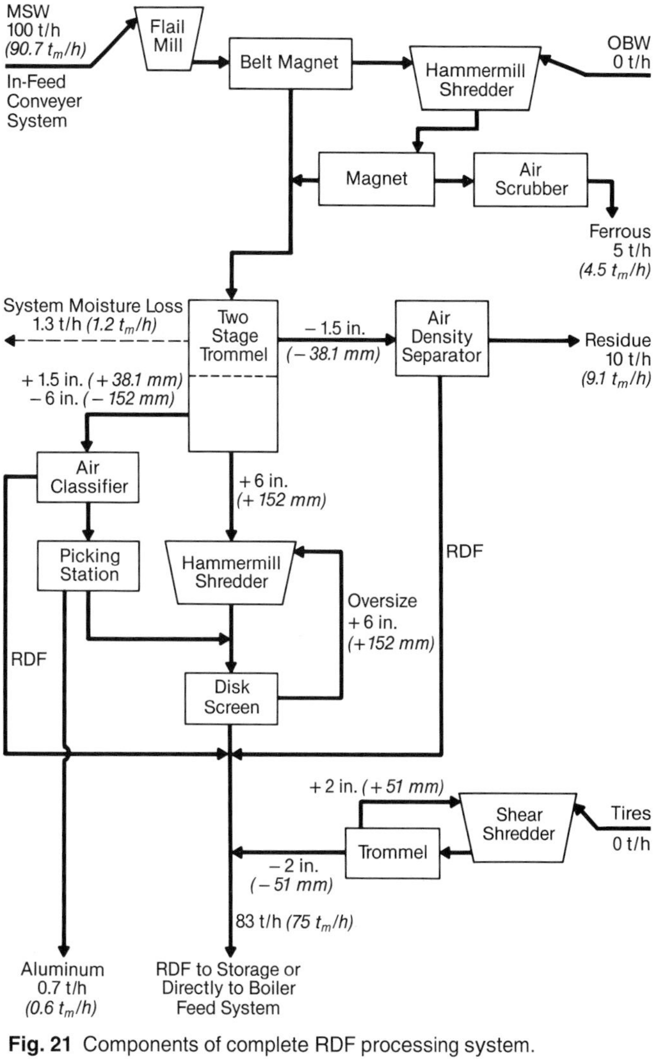

Fig. 21 Components of complete RDF processing system.

fuel distribution. A single undergrate air plenum is used. There is a wealth of experience worldwide with traveling grate stokers burning a myriad of waste and hard to burn fuels. The parameters for unit design shown in Table 6 were developed from this experience and the uniqueness of the RDF. (See also Chapter 15.)

On mass-fired stokers, a large volume of fuel at the front slowly burns down to a small volume of ash at the back. For an RDF stoker the key is to maintain an even 8 to 10 in. (203 to 254 mm) bed over the entire stoker area. Grate problems are usually due to a shallow ash bed. Operator tendency, when confronted with poor metering and/or fuel distribution, is to run the grates faster. While this technique can minimize bed upset, it will shorten grate life due to higher wear rates and the overheating of the grate bars. With the recommended ash bed thickness, tramp material is minimized, grate temperatures are lowered, wear is reduced and grate life is increased. To achieve this optimum ash bed requires controlled metering of the RDF and proper distribution of the fuel to the grates, as previously described.

A second problem is the accumulation of melted aluminum. The best solution is total removal of the aluminum from the fuel stream. If this is not practical, experience has shown that maintaining proper ash bed thickness will cause the aluminum to solidify in the ash bed rather than on the grate.

Fluidized-bed combustion, both circulating- and bubbling-bed designs, has been considered for RDF. For the high furnace pressure circulating beds, feeding the compressible RDF is a concern. For the bubbling beds, the feed system would be the same as for a stoker unit. In either case, RDF can be properly combusted. For coal firing, the inherent advantage of a fluid bed is the in-bed SO_2 capture. In-bed HCl capture, however, can not be achieved at the required removal efficiencies and therefore a back end scrubber is still required. A fluid bed boiler can achieve slightly lower NO_x emission levels and as emission requirements become more stringent, fluid bed combustion may become a viable alternative for the combustion of RDF.

Lower furnace design configuration

The lower furnace designs of early RDF boilers were largely based upon technology used for wood-fired boilers. This included modest overfire air (OFA) systems with multiple small diameter nozzles designed for 25 to 30% of the total air supply, straight wall furnaces and carbon re-injection systems. The result was less than desired combustion performance due to inadequate turbulent mixing in the furnace. Today's RDF units are designed with fewer, large diameter OFA nozzles designed for 50% of the total air supply with nominal operation at 40%.

In addition, B&W adapted its proven controlled combustion zone (CCZ™) lower furnace design for RDF firing. Developed originally in the early 1970s for very high moisture wood firing, this design consists of twin arches in the lower furnace with the overfire air nozzles directed down into the lower furnace from the arches (Fig. 22). The CCZ™ design for RDF applications without a carbon re-injection system has achieved lower unburned carbon loss than earlier designs which required the use of carbon re-injection systems.

The current B&W state-of-the-art RDF boiler system is exemplified by the Palm Beach County, Florida refuse facility which began operation in 1989. (See Fig. 23.) This design is currently used for all new B&W RDF boilers.

Furnace exit gas temperature

Gas temperatures leaving the furnaces of first and second generation designs were higher than anticipated. There were not enough data on RDF firing to accurately predict the relationship between furnace surface area and furnace exit gas temperature. Compounding this problem was a continual increase in the heating value of the RDF due to changes in the composition of the raw refuse and the development of more efficient processing equipment. To achieve the desired furnace exit gas temperatures, the size of the third generation furnace has increased significantly. The furnace width and depth are set by the size of the stoker, therefore the furnace height has increased to achieve the required furnace exit gas temperatures.

Table 6
Stoker Design Criteria (English units)

Parameters	RDF	Wood
Grate heat release, 10^6 Btu/h ft^2	0.750	1.100
Input per ft of grate width, 10^6 Btu/h	15.5	29
Fuel per in. of distributor width, lb/h	450	1,000
Feeding width as % of grate width	45 to 50	45 to 50
Grate speed, ft/h	25	N/A

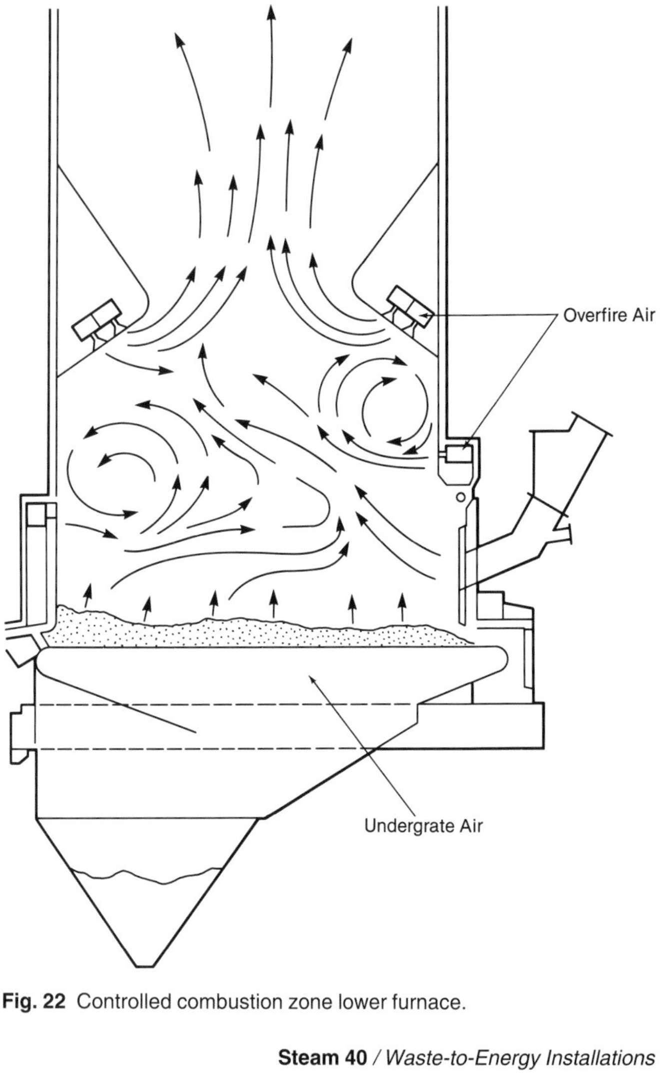

Fig. 22 Controlled combustion zone lower furnace.

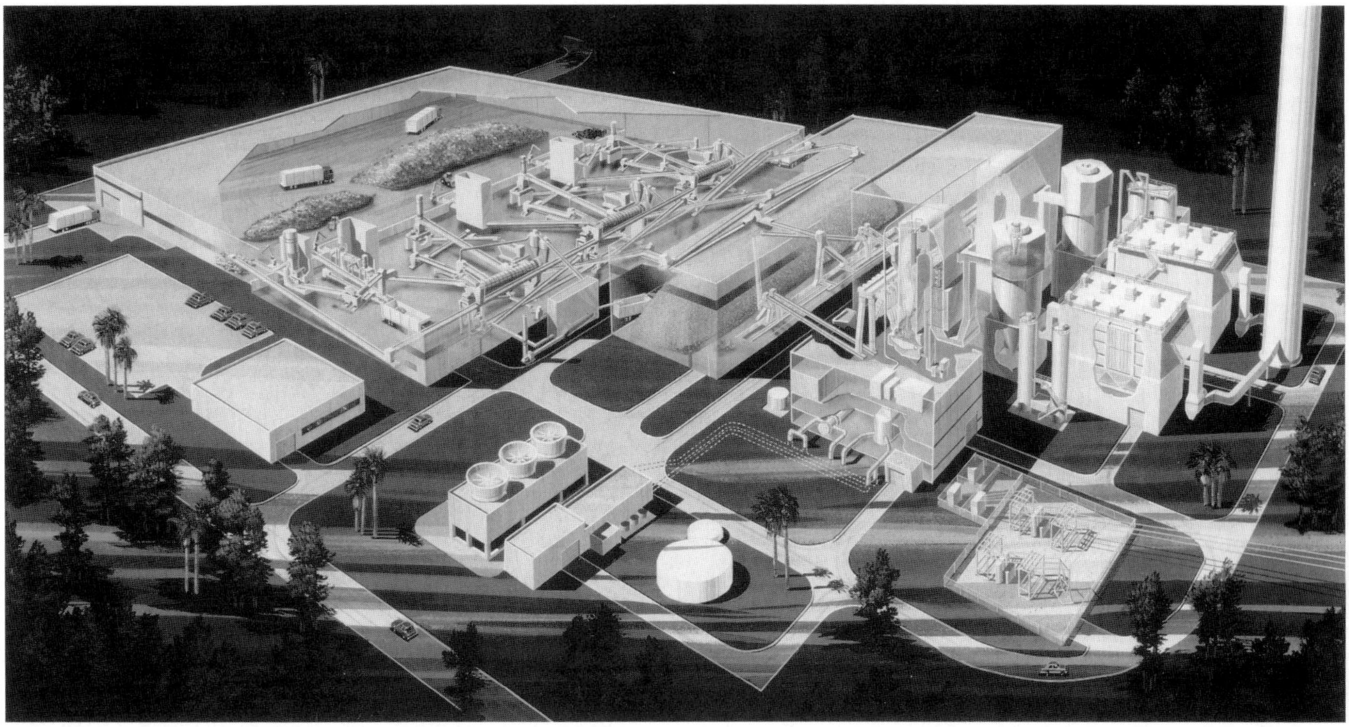

Fig. 23 West Palm Beach waste-to-energy facility.

Ash handling systems

Much of the noncombustible material in the RDF system is removed before it is fed to the boiler. Although systems vary, there is generally some effort made to remove ferrous metals and aluminum, both of which can be troublesome once they reach the stoker grates. Non-combustibles and most of the ash from combustion collect on the traveling grate stoker and discharge off the front into a submerged chain conveyor system.

The submerged chain conveyor (Fig. 24) is a mechanical conveyor that consists of a water filled trough and a dry return trough, with two endless chain strands with flights connected between the strands. The return trough can be either above or below the water filled trough. Ash from the stoker discharge chute drops into the water filled trough. The water absorbs the impact of any larger ash pieces, quenches the ash, and provides a gas-tight seal with the stoker discharge chute.

The chains are usually driven by a variable speed drive to handle varying ash rates. The ash residue is conveyed from the bottom of a water filled trough up an incline section where the ash dewaters and discharges directly into a truck, a storage bin, or onto another type conveyor for final disposal and transport. Because this conveyor uses a dragging action to convey the ash, it is not used on mass-fired units where it can have problems in dragging the large noncombustible items up the incline section.

Fine ash from the boiler siftings hopper, flyash from the boiler, economizer and air heater hoppers, and flyash from the scrubber, baghouse or precipitator hoppers are all handled the same way as previously discussed for mass-fired boilers.

Retrofits to RDF

Most dedicated RDF boilers are new installations. However, it is possible to retrofit existing boilers to become dedicated RDF boilers. To be candidates, the existing boilers must be conservative designs for solid fuels, such as wood or coal. Typically, these are older units which are underutilized or used as standby units. These plants are often located near large metropolitan areas, a source of large quantities of refuse. The conversion of such an older power plant could represent a cost effective solution to that community's refuse disposal problems.

B&W has converted several such boilers from coal-fired to dedicated RDF. Each retrofit is unique in that each of the coal-fired boilers was of different design and originally supplied by different manufacturers. Each retrofit was also the same in that all were designed to the same standards as new RDF units.

The principal modification involves enlarging the furnace to obtain the proper furnace volume for combustion. Although coal-fired boilers have conservatively sized furnaces, refuse firing requires even larger volumes. This is achieved by removing the existing stoker and lower furnace and installing new membrane furnace wall panel extensions (Fig. 25). The new lower furnace is protected from corrosion using either Inconel weld overlay or Inconel bimetallic tubes.

Other pressure part modifications could include:

1. converting the superheater to a counterflow design while adding the proper metals for corrosion protection,

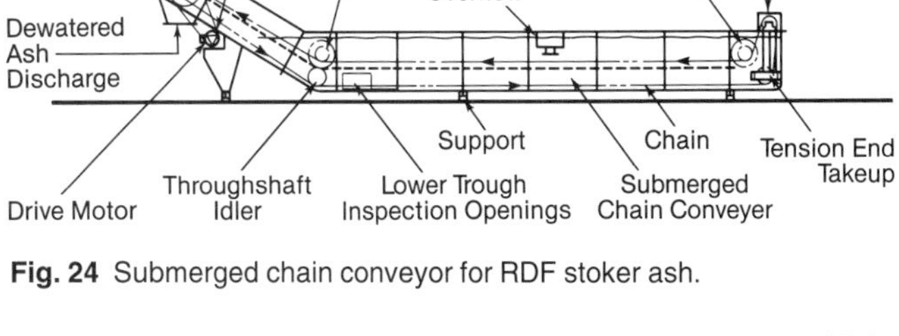

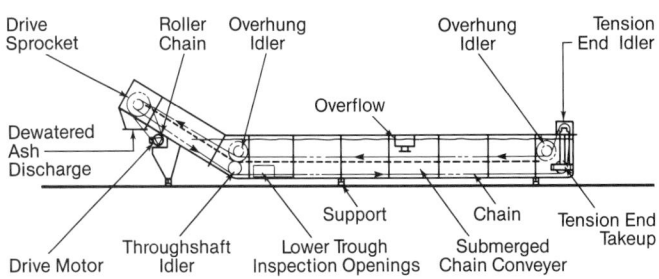

Fig. 24 Submerged chain conveyor for RDF stoker ash.

2. modifying the boiler, economizer and air heater surface for the proper heating surface distribution and to meet refuse standards for velocities, tube spacing, etc., and
3. possibly adding screen surface to lower the flue gas temperature entering the superheater.

In some cases, the coal stoker can be reused for RDF firing; in other cases it must be replaced. In either case the grate release rate, and other design criteria, must be set to the same design standards as new refuse boilers. Properly executed, the retrofit of an existing boiler to RDF firing will result in an RDF boiler as conservative as a new RDF boiler and capable of operating equally well.

Superheater

Superheater design is critical in both mass burn and RDF-fired refuse boilers because of the highly corrosive nature of the products of combustion. This is compounded in the U.S. by the desire for the highest possible steam temperature and pressure to maximize income from power production sales while still disposing of refuse in an environmentally safe manner. B&W has pioneered the 900 psig (62.1 bar gauge), 830F (443C) high pressure, high temperature steam cycle for refuse boiler application. This 50% increase in pressure and 80F (44C) increase in temperature over the more conventional refuse boiler designs has resulted in a significant improvement in cycle efficiency.

To accomplish this improvement, the superheater must be specifically designed for corrosion protection. Superheater corrosion is a function of many variables including flue gas temperature, flue gas velocity, tube spacing, tube metal temperature, tube metallurgy and ash cleaning equipment. Even a lower steam temperature superheater designed for one or a few of these criteria can experience rapid corrosion. For example, specifying only a low furnace exit gas temperature will not assure long superheater life.

Of these criteria, tube metal temperature and tube material are most critical. To obtain satisfactory refuse boiler superheater performance, two key design features are needed:

1. a parallel flow superheater design as shown in Fig. 26 where the coolest steam conditions are exposed to the hottest gas temperatures and the hottest steam temperatures are matched with the coolest gas temperatures. The result is a design with the lowest maximum superheater metal temperatures.
2. use of Incoloy tube material in the highest tube metal temperature sections of the superheater. Carbon steel is still used in the superheater sections with lower superheater metal temperatures.

B&W refuse boilers were the first to verify successful commercial operation at the 900 psig (62.1 bar gauge), 830F (443C) steam cycle with the first three units going into operation at the Westchester County, New York refuse-to-energy facility in 1984. As of 1992, there were 19 B&W refuse boilers in operation at these steam conditions with more than 75 cumulative years of operating experience. There has been no significant corrosion in any of these superheaters.

B&W is currently extending this leadership position with the first refuse boiler design to use the 1500 psig (103.4 bar gauge) and 930F (499C) steam cycle. This design is based on laboratory corrosion research and full scale superheater test sections installed in operating

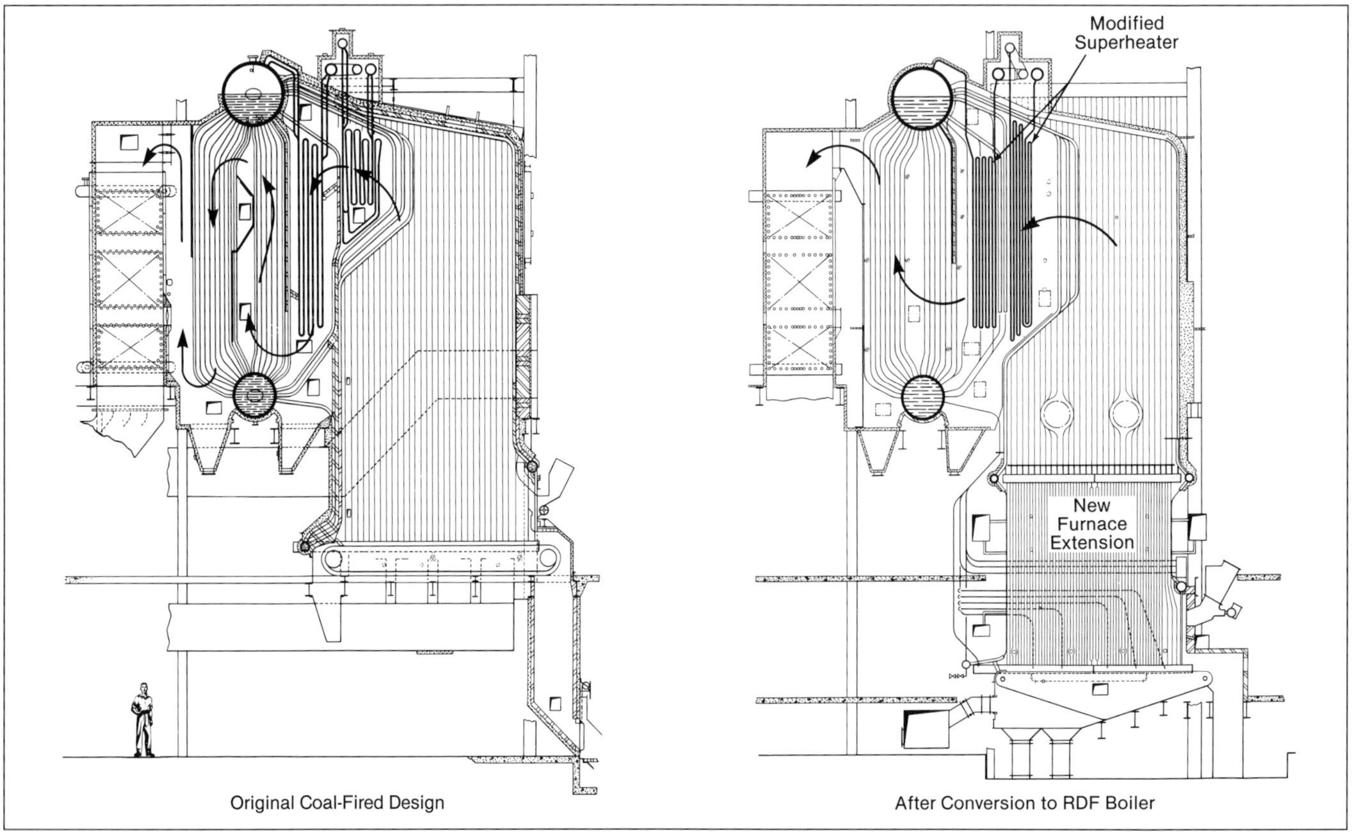

Fig. 25 Coal-fired unit converted to RDF.

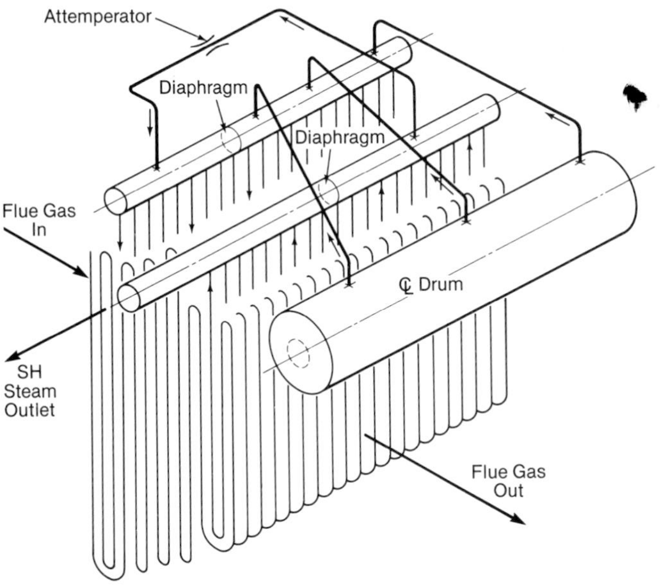

Fig. 26 Parallel flow superheater (SH).

refuse boilers. In one test, two full scale superheater sections, composed of a variety of tube metallurgies, were installed in an operating 900 psig (62.1 bar gauge), 830F (443C) design refuse boiler with steam flow through the sections controlled to simulate 950F (510C) operation. One section was removed after one year of operation and the second after 26 months. This test provided the basis for tube metallurgy selection for the 1500 psig (103.4 bar gauge), 930F (499C) design.

In addition to corrosion concerns, the superheater must be designed to minimize fouling and the potential for erosion due to excessively high flue gas velocities. Maximum design velocity is 30 ft/s (9.1 m/s), but in practice it is usually in the 10 to 15 ft/s (3 to 4.6 m/s) range. Minimum superheater side spacing is 6 in. (152 mm).

Boiler design

The lower furnace design, refuse stokers and refuse feed systems are markedly different for mass-fired and RDF boilers. However, the design requirements for the upper furnace, generating surface and economizer are the same. This is also true for auxiliary equipment such as burners and ash cleaning equipment.

Upper furnace design

The upper furnace must be sized to provide adequate heat transfer surface to reduce the flue gas temperature entering the superheater to an acceptable level. This helps minimize fouling in the superheater and maintain low superheater tube metal temperatures to minimize corrosion. A certain amount of furnace volume is required for complete burnout of the fuel in the furnace and minimum CO emissions. The required volume should be measured from the point where all the combustion air has entered the furnace (the highest level of overfire air ports) to the point where the flue gas enters the first convective heating surface (at the tip of the furnace arch at the bottom of the superheater). Measured in this manner, the required furnace volume per unit of heat input is the same for both mass-fired and RDF boilers.

The furnace must also contain sufficient heating surface to lower the flue gas temperature to help reduce fouling in the first convection section, superheater or boiler bank. These limits are 1600F (871C) entering the superheater and 1400F (760C) entering the boiler bank. As a general rule, the furnace size is set by volumetric requirements in smaller capacity boilers and by maximum gas temperature limits in larger capacity boilers.

Boiler generating bank

Refuse boilers in operation use both the one-drum and two-drum design. In the two-drum design there is both a steam drum (upper drum) and a lower drum, interconnected by the boiler generating bank tubes.

In the one-drum design the steam drum is located outside of the flue gas stream; there is no lower drum. The steam generating bank tubes are shop-assembled modules. These modules may be of either the vertical longflow (Fig. 27) or a vertical crossflow design. Minimum side spacing in the two-drum design and for the generating bank modules used with the one-drum design is 5 in. (127 mm). Maximum design flue gas velocity is set at 30 ft/s (9.1 m/s).

Economizer

The economizers can be either vertical longflow or horizontal crossflow. Economizer side spacing should be no less than 4 in. (102 mm) with a maximum flue gas velocity of 45 ft/s (13.7 m/s).

Air heater

Air heaters may be used for two reasons: 1) to supply preheated air to help dry and ignite the refuse on the stoker, and/or 2) to increase thermal efficiency where high feedwater temperatures preclude designing to lower exit gas temperatures with economizers. RDF-fired units have typically used air heaters to preheat the combustion air to the 300 to 350F (149 to 177C) range. Both tubular and regenerative air heaters have been used successfully. Due to air leakage into the air heater and the potential for fouling, regenerative types have been limited to the outlet side of hot electrostatic precipitators where the flue gases are relatively clean.

When either tubular or regenerative air heaters are used, the design and arrangement should minimize the potential for low-end temperature corrosion. To some extent, the surface arrangement in a tubular air heater will maintain adequate protection. Steam coil air heaters are required at the air inlet, on either type, to preheat the incoming ambient air and maintain temperature above acid dew points. (See also Chapter 19.)

Ash cleaning equipment

To maintain the effectiveness of all convective heating surfaces and to prevent pluggage of gas passages, it is necessary to remove ash and slag deposits from external tube surfaces. Steam or air sootblowers are most commonly used. Saturated steam is preferred for its higher density and better cleaning ability. One disadvantage of sootblowing is that localized erosion and corrosion can occur in areas swept too clean by the blowing medium. This problem can be addressed by installing tube shields on all tubes adjacent to each sootblower for localized protection.

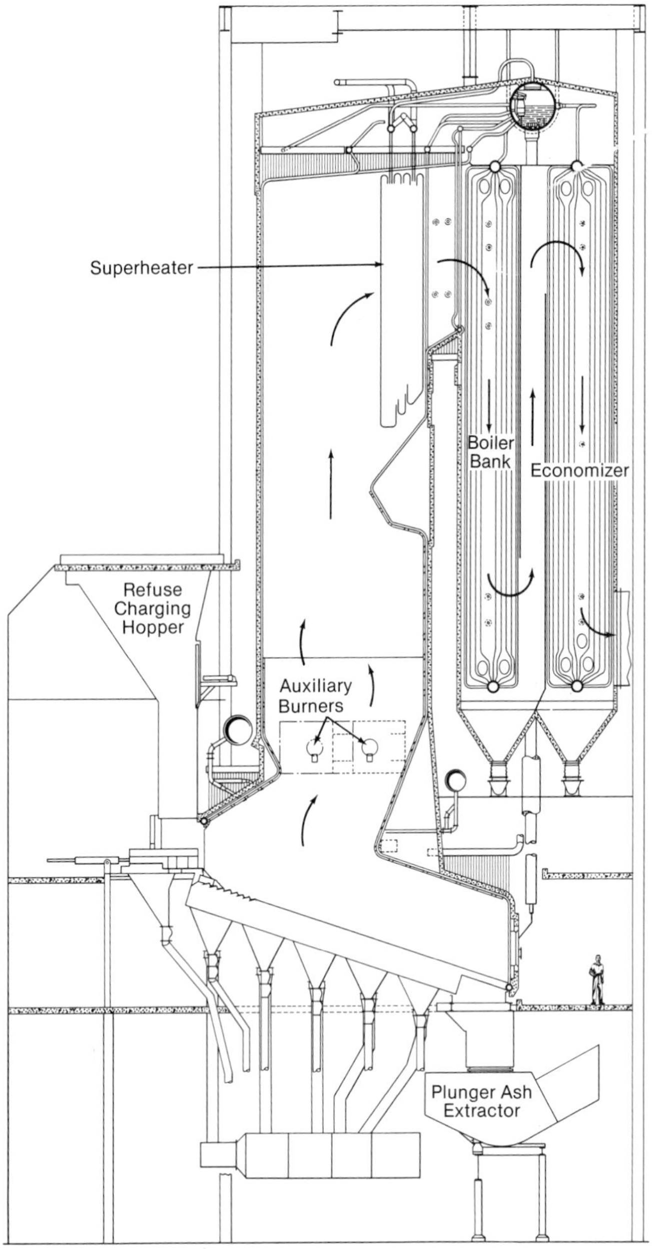

Fig. 27 Typical refuse unit with vertical longflow economizer.

A mechanical rapping system (Fig. 28) can be used to complement the sootblowers. In this system, a number of anvils strike designated pins to impart an acceleration through the superheater tube assembly. The purpose is to remove the bulk of the ash while leaving a light layer of ash on the tubes for corrosion protection. Mechanical rapping systems will not eliminate the need for sootblowers, but will reduce the number of sootblower cleaning cycles required.

Auxiliary input burners

Auxiliary fuel burners are used to maintain furnace temperature during startup, shutdown, and upset conditions since operation at low furnace temperatures could result in the incomplete destruction of volatile organic compounds. In most cases, the auxiliary fuel (oil or gas) burners are designed for only 25 to 30% of the boiler's maximum heat input.

When not in service, the typical gas- or oil-fired burner

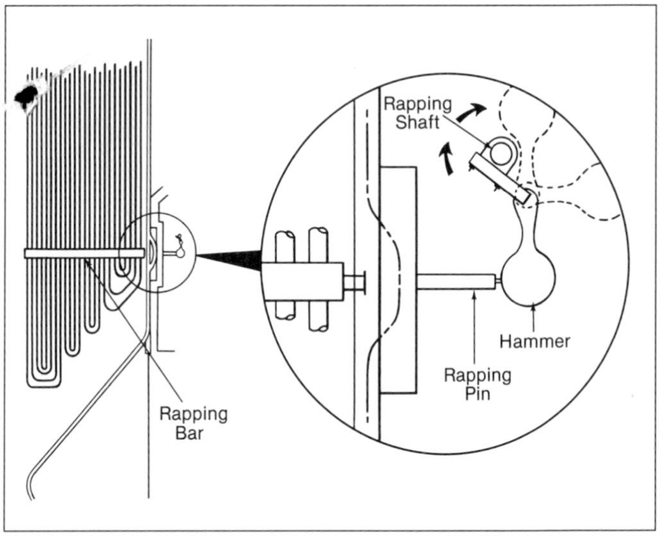

Fig. 28 B&W mechanical rapping system for cleaning superheaters.

requires some amount of air flow through the idle burner for protection against overheating. Because this air leakage represents an efficiency loss, and because these burners are used infrequently, a special design auxiliary input burner (AIB) is used for refuse boilers. The AIB is designed with a retractable burner element which is inserted toward the furnace when in use, and retracted when out of service. There is also a movable refractory block which provides protection against furnace radiation when the burner is out of service. With the burner in service, this refractory block is retracted to one side and the burner is inserted through an opening in the refractory block (Fig. 29).

Upper furnace maintenance platforms

Because refuse is a high fouling fuel, it is necessary to have good access to the convection sections. Maintenance platforms (Fig. 30) are often used to allow access to the superheater area for inspection and maintenance. Either retractable or light weight aluminum support beams are inserted into the furnace from access doors in the front wall to the superheater arch, where the beams are locked in place. Corrugated decking material is then inserted into the furnace through special sidewall access doors, and arranged on top of the support beams. This system provides both a platform for working in the superheater and upper furnace and provides some protection to those working in the lower furnace.

Air pollution control equipment

Various boiler fuels have specific components unique to that fuel. Some of these components, such as sulfur, create specific air pollution emissions that require unique boiler designs or specific air pollution control equipment. These fuels, such as high sulfur coal, are homogenous. This means the fuel will be the same in the future as it is today, and will be the same from one day to the next.

Refuse is a nonhomogenous fuel. It not only changes over the long term, but can change from day to day. Nearly every component of a fuel that can result in an unwanted air pollutant is present in refuse. However, in the early 1980s when the population of refuse-fired boil-

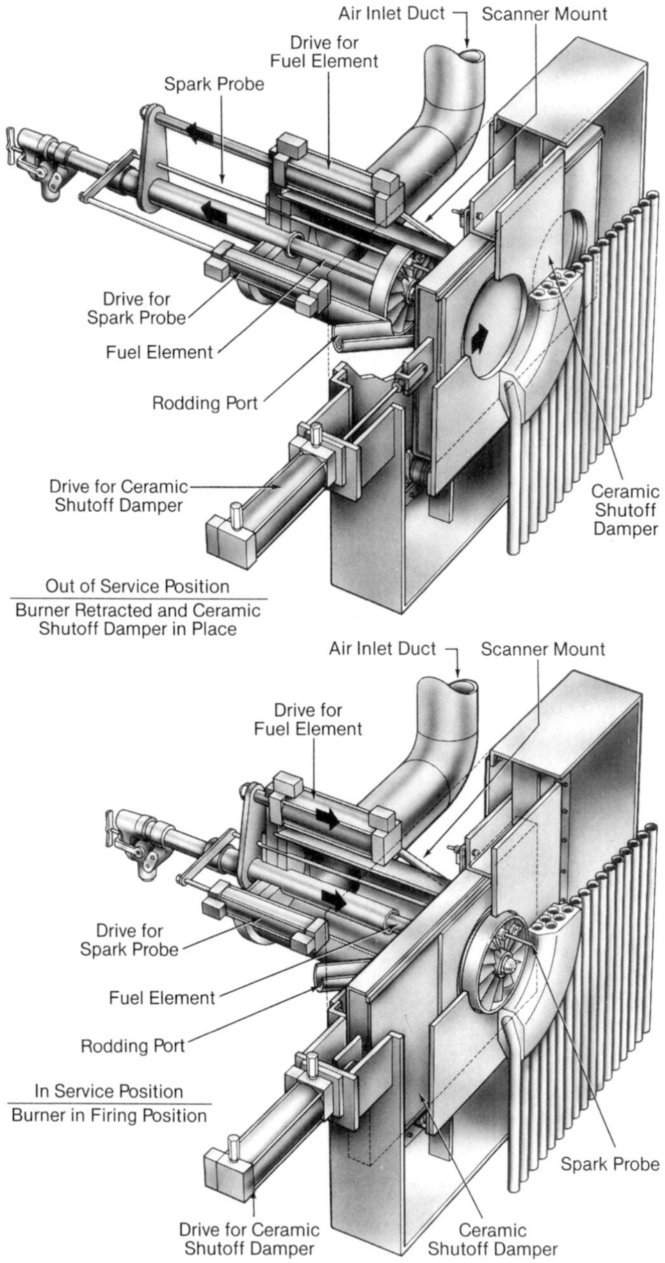

Fig. 29 Auxiliary input burner (out of service and in service positions).

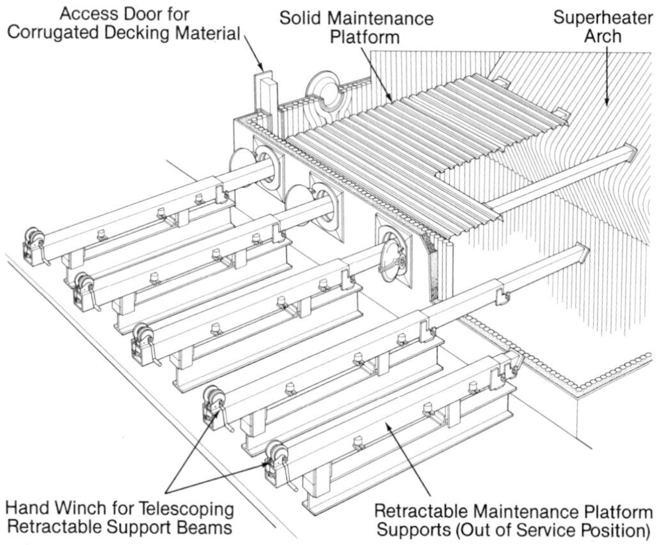

Fig. 30 Upper furnace maintenance platform.

Table 7
Permissible Stack Emissions (1990)

Pollutant	Emission Concentration
NO_x	200 to 350 ppmdv*
CO	20 to 100 ppmdv
VOC	< 10 ppmdv
SO_2	< 35 ppmdv**
HCl	< 20 ppmdv***
Particulate	< 0.01 gr/DSCF
PCDD/PCDF	< 10 ng/Nm^3****

Stack emission levels are test data values for units equipped with a dry scrubber and baghouse/precipitator, and may not represent values achievable in all normal daily operations. All stack emission concentrations are corrected to a 7% O_2 reference basis.

* NO_x emissions without add-on NO_x control technology.

** Typical SO_2 emissions with back-end control efficiency in the 70 to 90% range.

*** Typical HCl emissions with back-end control efficiency in the 90 to 98% range.

**** Total dioxin and furan emissions, including all tetra through octa homologies.

ers began to rapidly grow in the U.S., the only emission requirements were on particulates, NO_x and SO_2. Refuse boilers, due to their relatively cool burning systems and the generally low level of fuel bound nitrogen, are low NO_x generators. There are also very low levels of sulfur in refuse. Therefore, early boilers were generally equipped only with an electrostatic precipitator (ESP) for particulate control. As more boilers went into operation and further air emissions data were obtained, additional emission requirements were applied. Initially hydrochlorides were targeted for control. Soon the various state air pollution agencies set regulations for the control of dioxins and furans as well as a long list of heavy metals. Dry scrubbers, used for years to control SO_2 emissions from coal-fired units, were found to be equally effective in controlling HCl emissions from refuse units. These same dry scrubbers were also found to be very effective in controlling dioxin, furan and heavy metal emissions.

With the initial use of dry scrubbers, there was a split in the preferred particulate collection system between the ESP and baghouse. ESPs were used in earlier applications due to their more extensive history of proven performance. However, it has been fairly well documented that the layer of ash and lime that collects on the bags themselves allows improved sorbent utilization for the removal of SO_2 and HCl. This allows better capture of pollutants for the same lime slurry rates, or the same level of pollutant capture at slightly reduced lime slurry rates. Today, the preferred system for nearly all refuse boilers is the dry scrubber/baghouse combination.

In the late 1980s, lower level NO_x emissions were also being required. In-furnace ammonia or urea injection systems were installed on several refuse boilers and were shown to achieve NO_x reduction efficiencies in the 40% range. This NO_x control technology, termed selective noncatalytic reduction (SNCR), was quickly accepted by the regulatory agencies as best available control technology (BACT). In certain geographical areas, which are nonattainment areas for NO_x emissions, selective catalytic reduction (SCR) systems are also under evaluation for even greater NO_x control. The SCR systems have been demonstrated to achieve up to 90% NO_x reduction on fossil fuel boilers. However, the catalyst itself is fairly easily poisoned, and therefore rendered less effective, by a multitude of substances, all of which are found in refuse to various degrees. At this time it is not clear what the long term life of the SCR catalyst would be on a refuse-fired boiler. (See Chapters 32 through 35.)

During this same time period the emission requirements for CO have also been driven to lower levels. These requirements have been met by a combination of:

1. better overfire air system,
2. more control of undergrate air (more compartments with individual air control),
3. better combustion control systems,
4. larger furnace volumes, and
5. operator training.

The air emissions from refuse-fired boilers are as tightly regulated as those from any combustion system. However, the technology exists today to meet these requirements of the 1990s (Table 7) and will be available to meet future requirements.

Bibliography

Barna, J.L., Blue, J.D., and Daniel, P.L., "Furnace wall corrosion in refuse fired boilers," Proceedings of the ASME National Waste Processing Conference, Denver, Colorado, pp. 221-228, June 1-4, 1986.

Barsin, J.A., *et al.*, "Initial operating results of coal-fired steam generators converted to 100% refuse derived fuel," Proceedings of the American Power Conference, Chicago, April 18-20, 1988.

Barsin, J.A., "Initial operational results of a coal-fired steam generator converted to 100% RDF fuel," Proceedings of the ASME Joint Power Conference, Miami, Florida, October1987.

Bielawski, G.T., *et al.*, "Resource recovery in Palm Beach," Proceedings of the International Conference on Municipal Waste Combustion, Hollywood, Florida, April 11-14, 1989.

Blue, J.D., *et al.*, "Waste Fuels: Their Preparation, Handling and Firing," Section 3, Chapter 6, Standard Handbook of Powerplant Engineering, McGraw-Hill, pp. 3.117 to 3.151, February 1989.

Blue, J.D., Gibbs, D.R., and Hepp, M.P., "Design and operating experience with high temperature and high pressure refuse-fired boilers," Proceedings of the ASME National Waste Processing Conference, Philadelphia, pp. 93-99, May 1-4, 1988.

Blue, J.D., and Strempek, J.R., "Considerations for the design of refuse-fired water wall incinerators," Proceedings of the Conference on Energy From Municipal Waste, Washington, D.C., October 24-25, 1985.

Boman, E., *et al.*, "The greater Bridgeport regional resource recovery project: A waste-to-energy reincarnation," Proceedings of the ASME National Waste Processing Conference, Long Beach, California, pp. 345-351, June 3-6, 1990.

Burnham, D.E., *et al.*, "The evolution of RDF boiler designs through state-of-the art system at Palm Beach," Proceedings of the EPRI 1989 Conference on Municipal Solid Waste as a Utility Fuel, Springfield, Massachusetts, pp. 3-45 to 3-61, October 10-12, 1989.

Campobenedetto, E.J., *et al.*, "City of Lakeland to generate power and burn municipal waste," Proceedings of The American Power Conference, Chicago, April 26-28, 1982.

Chambliss, C.W., Hestle, J.T., and McNertney, R.M., "Operating experience is reflected in Nashville's new refuse boiler," Proceedings of the Joint ASME/IEEE Power Generation Conference, Portland, Oregon, October 20-23, 1986.

Clunie, J.F., *et al.*, "The importance of proper loading of refuse fired boilers," Proceedings of the ASME National Waste Processing Conference, Orlando, Florida, pp. 169-177, June 3-6, 1984.

Cramblitte, D.I., *et al.*, "Akron recycle system is on-line," Proceedings of The ASME Waste Processing Conference, Washington, D.C., pp. 463-474, May 11-14, 1980.

Gibbs, D.R., and Kreidler, L.A., "From Hamilton to Palm Beach: The evolution of dedicated RDF plants," Proceedings of the Processed Fuels and Material Recovery from Municipal Solid Waste Symposium, Washington, D.C., Section 7, December 1-2, 1986.

Gibbs, D.R., and Kreidler, L.A., "What RDF has evolved into," pp. 251-262, *Waste Age*, April 1989.

Gibbs, D.R., and Hepp, M.P., "Mass-burn plant achieves high availability, low emissions," *Power*, April 1990.

Gibbs, D.R., "Refuse to energy," pp. 14, 15 and 32, *Pennsylvanian*, June 1990.

Gibbs, D.R., "Examine utility options in refuse-to-energy business," pp. 53-55, *Electrical World*, August 1990.

Hanson, W.C., and Scheatzle, J.E., "The design and start-up of the Elk River's conversion to RDF firing," Proceedings of the 1989 EPRI Conference on Municipal Solid Waste as a Utility Fuel, Springfield, Massachusetts, pp. 3.3 to 3.12, October 10-12, 1989.

Hepp, M.P., Nethercutt, R.M., and Wood, J.F., "Considerations in the design of high temperature and pressure refuse fired power boilers," Proceedings of the ASME National Waste Processing Conference, Denver, Colorado, pp. 113-118, June 1-4, 1986.

Herrmann, R.H., "Improvements in the quality of RDF," Proceedings of the EPRI 1989 Conference on Municipal Solid Waste as a Utility Fuel, Springfield, Massachusetts, pp. 3-13 to 3-24, October 10-12, 1989.

Herrmann, R.H., "Palm Beach County waste processing facility," Proceedings of the GRCDA Resource Recovery Symposium '90, West Palm Beach, Florida, January 16-18, 1990.

Herrmann, R.H., "RDF-fired systems," Proceedings of Waste-To-Energy '87: Exploring the Total Market, Washington, D.C., September 21-22, 1987.

Jackson, B.W., Scheatzle, J.E., and Taylor, D.A., "Design to convert a three boiler power plant to RDF firing," Proceedings of the Industrial Power Conference, Houston, Texas, October 1988.

Johnson, N.H., and Reschly, D.C., "MSW and RDF: An examination of the combustion process," Proceedings of the ASME Joint ASME/IEEE Power Generation Conference, Portland, Oregon, October 20-23, 1986.

Larsen, P.S., "Continuously monitoring furnace temperature in refuse boilers using acoustic pyrometry," Proceedings of the Waste Technology 1986 Conference, Chicago, October 20-22, 1986.

Rochford, R.S., and Witkowski, S.J., "Considerations in the design of a shredded municipal refuse burning and heat recovery system," Proceedings of the ASME Waste Processing Conference, Chicago, May 7-10, 1978.

Schueler, P.H., "The case for dedicated refuse conversion of utility boilers," Proceedings of the Power-Gen '88 Conference, Orlando, Florida, December 6-8, 1988.

Strach, L., "Operating experience of refractory linings in mass-refuse fired waterwall boilers," Proceedings of the ASME National Waste Processing Conference, Philadelphia, pp. 1-8, May 1-4, 1988.

Scavuzzo, S.A., Strempek, J.R., and Strach, L., "The determination of the thermal operating characteristics in the furnace of a refuse-fired power boiler," Proceedings of the ASME Waste Processing Conference, Long Beach, California, pp. 397-404, June 3-6, 1990.

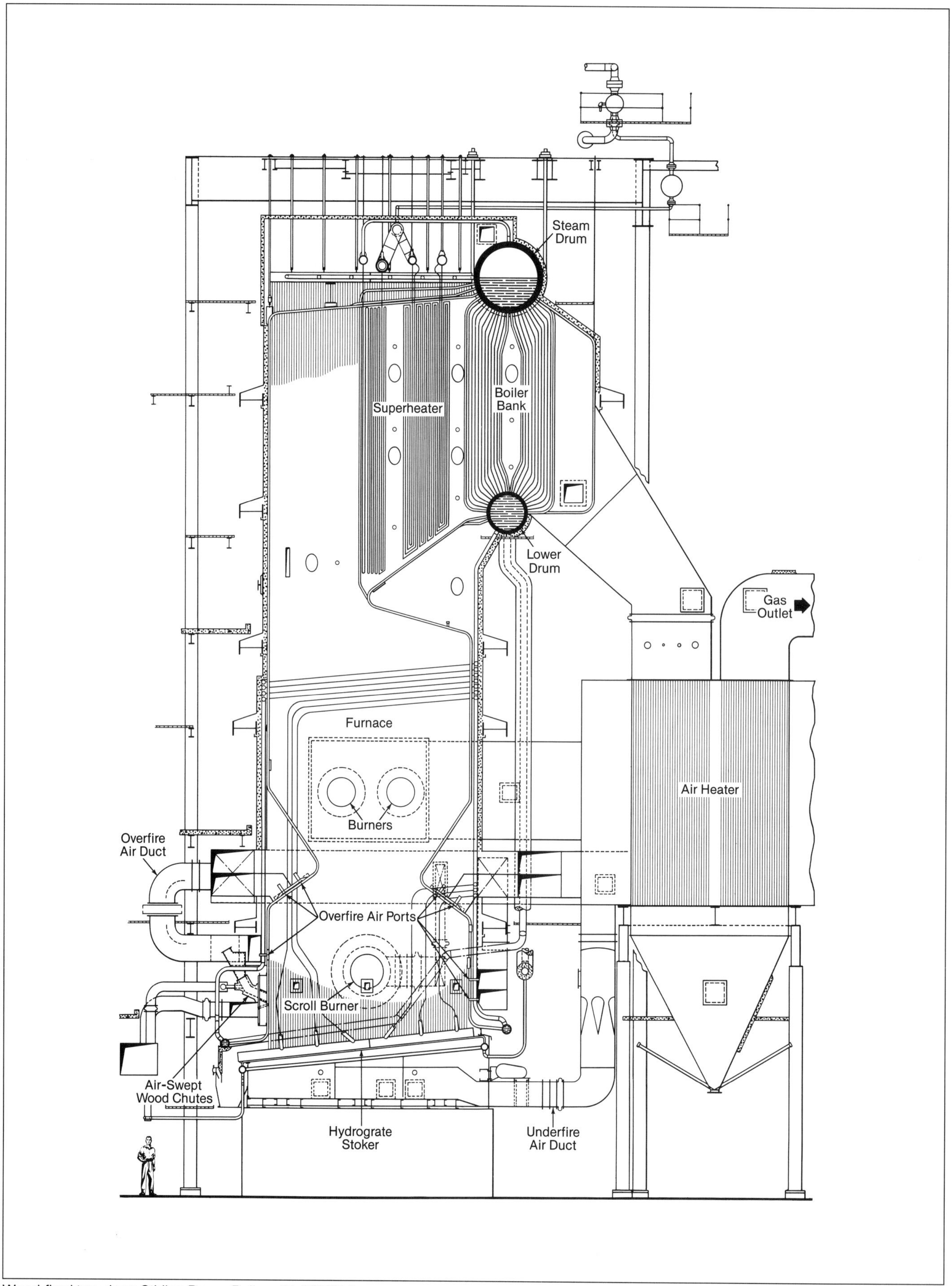

Wood-fired two-drum Stirling Power Boiler with CCZ™ furnace.

Chapter 28
Wood and Biomass Installations

The category of wood and biomass covers a very wide range of material that can be used as a source of chemical energy. Wood encompasses a number of sources, such as bark, wood sticks, sawdust, sander dust, over- and under-sized wood chips rejected from the pulping process, whole tree chips and scrap pallets. Biomass is anything that is, or recently was, alive such as vine clippings, leaves, grasses, bamboo and sugar cane (called bagasse once the sugar has been extracted), coffee grounds and rice hulls from the food processing industry. All can be used as fuel sources to generate steam. Municipal waste contains a great deal of biomass.

While wood and other biomass fuels were some of the first materials used as energy sources, their use declined around the turn of the century when more consistent and easily transportable fossil fuels such as coal, oil and gas became available. Nevertheless, there have always been particular applications where wood and other biomass have been the preferred fuels. There has therefore been a steady if somewhat slow progress in the development of equipment for firing these fuels. Many factors have led to the more recent rapid development in this area including the rising cost of some fossil fuels, technology development to allow better use of industrial byproducts, and the trend toward cogeneration in many industries, producing both electric power and process steam.

Equipment for chemical and heat recovery in the paper industry is discussed in Chapter 26. Equipment for municipal solid waste is discussed in Chapter 27.

Steam supply to process

Pulp and paper

The pulp and paper industry is the major user of biomass fuels because wastes such as bark, sawdust, shavings, lumber rejects and clarifier sludge are byproducts of the pulping, paper making and lumber manufacturing processes. There is a large amount of energy available in these products making them useful energy sources.

The production of pulp and paper requires vast quantities of mechanical energy for grinding, chipping, cooking and refining. In order to produce a saleable product, the pulp must be dried, using heated air or heated surfaces such as paper machine dryer rolls. These energy needs are met using steam in a variety of equipment (steam engines or turbines, steam coil air heaters, dryer rolls, and indirect heaters) and by direct steam injection.

These requirements, coupled with the availability of the waste products, make boilers fired by wood and wood waste a logical choice for the pulp and paper industry. Most of the developments and improvements in equipment for these boilers have been driven by this industry's needs.

Food processing

Another industry which has energy needs that are provided by steam is food processing. Mechanical preparation, cooking, drying and canning all require a source of energy. Many foods leave behind waste byproducts rich in cellulose or other organic (hydrocarbon) material; instant coffee production generates coffee grounds, sugar making leaves bagasse from sugar cane, coconut preparation discards husks, rice has its hulls removed before packaging and many types of nuts are sold roasted with their shells removed.

Many producers have now installed boilers which burn such biomass material, usually based very closely on equipment originally designed for the pulp and paper industry. These boilers produce steam which is then used as an energy source for the plant.

Steam supply for power production

Cogeneration in industry

Most heating requirements can be met with saturated steam at 150 psig (10.3 bar gauge) or less. Cogeneration, or the simultaneous production of electrical (or mechanical) energy and heat energy to a process, has very high thermal efficiency, as waste heat from electricity production is absorbed into a usable process rather than being rejected into the atmosphere.

In a cogeneration facility, relatively high pressure superheated steam passes through a steam turbine, or steam engine, in which energy is extracted. The exhaust steam is then used as a heat source in a process. The conversion efficiency, or the amount of heat absorbed in the steam turbine plus the heat absorbed in the process, versus the amount initially contained in the steam, approaches 100%. The overall thermal efficiency of a cogeneration process is closely approximated by the thermal efficiency of the boiler alone.

As well as providing high energy efficiency, there is a second and often more compelling reason for industries, especially the pulp and paper industry, to practice cogeneration. These industrial plants are often located far from economical and reliable sources of electricity and conventional fossil fuels in order to be close to their source of fiber, the forests. Cogeneration facilities equipped with wood-fired power boilers are therefore often justified.

Biomass-fired utilities

Changing economic conditions and environmental regulations have made utility plants fired by biomass fuels a practical source of electrical energy, in spite of their relatively high capital costs. These changes have resulted in the emergence of the independent power producer (IPP) who builds a biomass-fired plant whose sole purpose is to generate revenue by producing electric power for sale to a utility. Sometimes the plant is built with a condensing turbine and other times it is built beside (or within) a plant which can use exhaust steam. Economic conditions favoring such facilities include the unpredictable and sometimes very high cost of conventional fossil fuels, the relatively low cost of wood waste and other biomass fuels, and the high cost to transport and dispose of waste biomass by landfill.

Occasionally an installation is justified on the high cost of emissions control equipment. Most wood and wood wastes generate lower nitrogen oxides (NO_x) and sulfur dioxide (SO_2) emissions than fossil fuels.

In one installation in California, it was found that the amount of NO_x produced by gathering and incinerating grape vine clippings and other wastes could be substantially reduced by burning the collected material in a wood-fired utility installation in a controlled manner. This produced environmental benefits while generating electricity which would otherwise be produced by burning conventional fossil fuels.

Fuels

Constituents

Wood and most biomass fuels are composed predominantly of cellulose and moisture. The high proportion of moisture is significant because it acts as a heat sink during the combustion process. The latent heat of evaporation (H_{fg}) depresses the flame temperature, contributing to the difficulty of efficiently burning biomass fuels.

Cellulose, as well as containing the chemical energy released during combustion, contributes fuel bound oxygen. This oxygen decreases the theoretical air required for combustion and therefore the amount of nitrogen included in the products of combustion.

Most natural biomass fuels contain little ash. However, some byproducts, such as de-inking sludges, do contain a great deal of ash, in some cases up to 50% ash on a dry basis. De-inking sludges are particularly difficult to burn because they usually have high moisture and ash contents and low fuel bound oxygen content.

Burning wood and biomass

The following general guidelines for wood and biomass combustion have been developed from experience:

1. Stable combustion can be maintained in most water-cooled furnaces at fuel moisture contents as high as 65% by weight, as received.
2. The use of preheated combustion air reduces the time required for fuel drying prior to ignition and is essential to spreader-stoker combustion systems. The design air temperature will typically vary directly with fuel moisture content.
3. A high proportion of the combustible content of wood and biomass fuels burns in the form of volatile compounds. A large portion of the combustion air requirement is therefore added above the fuel, as overfire air.
4. Solid chars produced in the initial stages of combustion of these fuels are of a very low density. Conservative selection of furnace size is used to reduce gas velocity and keep char entrainment at an acceptable level. Typical furnace selection criteria include a grate heat release rate of 1 x 10^6 Btu/h ft^2 (3.15 MW_t/m^2) of grate surface area, a furnace liberation rate of 17,000 Btu/h ft^3 (176 kW/m^3) of furnace volume and an upward gas velocity of 20 ft/s (6.1 m/s). This results in furnace residence times of approximately three seconds for larger units to enhance particulate burnout and minimize emissions.

Burning in combination with traditional fuels

Biomass can be burned on a traveling grate with stoker coal. The biomass is introduced to the furnace through a separate conveying system and either a separate windswept spout below the coal feeder or through a combination feeder. (See Chapter 15.)

Biomass can be burned with pulverized coal, oil or natural gas, using dedicated burners for the latter. In this case, the grate is selected for wood firing.

When burning biomass with substantial quantities of stoker or pulverized coal, the amount of ash from the coal is greater than that from biomass and therefore the design parameters for the coal (slagging and fouling index) will govern the design.

When burning biomass with heavy oil having high sulfur and vanadium content, the ash which forms on the convective surfaces can be tenacious and once removed can be very abrasive. It is preferred to design for low flue gas temperatures and low flue gas velocities, regardless of the specified contaminants in the heavy oil and in the biomass, as both flue gas parameters can vary widely over the typical range of boiler operating conditions. (See Chapter 10.)

Sludge burning

As mentioned above, paper mill sludges, especially de-inking sludges, are difficult fuels to burn. Their high moisture and ash content and low fuel bound oxygen may limit their allowable proportion of the total heat input to the furnace.

Specific limits will vary with the sludge composition and combustion system. Higher sludge inputs for a given system can normally be achieved by combined firing with better quality fuels.

Wood waste mixed with sludge can prove difficult for the fuel handling system. Frequently the sludge segregates from the other fuels, at transfer points on belt conveyors for example, and can be fed preferentially to one feeder.

In spreader-stoker applications, the wet, dense sludge can pile up in one place on the grate very quickly. Therefore, when a portion of the waste fuel is sludge, the boiler operator must continually inspect the grate to determine whether adjustments are required.

Combustion systems

There have been many methods developed to burn wood and other biomass fuels. The best known and successful methods use the following equipment.

Dutch oven

A dutch oven is a refractory-walled cell connected to a conventional boiler setting. (See Fig. 1.) Water cooling is sometimes provided to protect and extend the life of the refractory walls. Wood waste is introduced through an opening in the roof of the dutch oven and burns in a pile on its floor. Overfire air is introduced around the periphery of the cell through rows of holes or nozzles in the refractory walls.

The principal advantage of the dutch oven is that only a small portion of the energy released in combustion is absorbed, because of its high percentage of refractory surface. Therefore it is able to burn high moisture fuels (up to 60% moisture). In addition, as the fuel is pile burned, there is a high thermal inventory in the cell that makes the unit less sensitive to interruptions in the fuel supply.

In spite of being able to burn wood with high moisture content the dutch oven has distinct disadvantages when compared to more modern methods. The unit operates best when at a steady load and when burning a consistent fuel. It does not respond quickly to load demand.

The refractory is subject to damage from the following sources: spalling and erosion caused by rocks or tramp metal introduced with the fuel, rapid cooling from contact with very wet fuels and overheating when very dry wood is fired.

The dutch oven cell must be shut down regularly to allow manual removal (rake out) of the ashes which have accumulated. During this period, either load must be drastically reduced, as done with multiple cell units, or auxiliary fuel must be fired, as is required for units equipped with a single dutch oven.

During the raking out period, many operators run the furnace with the highest possible negative draft in order to reduce the possibility of hot gas, flames and hot ash being blown out towards the cleaning personnel. Operation with high furnace draft draws large quantities of tramp air into the dutch oven. This large air volume causes any auxiliary fuel to be burned at low thermal efficiency due to the high excess air. It can result in high flue gas velocities through the convective surfaces of the unit which can promote erosion, especially when quantities of char, ash and clinkers are drawn from the dutch oven into the boiler with the air.

Pinhole grate

The pinhole grate (Fig. 2) is a water-cooled grate formed by cast iron grate blocks, sometimes referred to as Bailey blocks, that are clamped and bonded to the spaced floor tubes of a water-cooled furnace. The grate blocks have venturi type air holes to admit undergrate air to the fuel on the grate. This grate is used in conjunction with either mechanical fuel distributors or windswept fuel spouts. Both produce a semi-suspension mode of burning, wherein the finer portion of the fuel is burned in suspension and the heavier fraction accumulates and burns on the grate. The ash and foreign material which stay on the grate are removed by raking. Typically, 75% of the air for combustion of wood is introduced as overfire air through nozzles in the lower furnace walls and 25% is undergrate air.

This combustion system can follow minor load swings by varying both the fuel flow and air flow and is suitable for biomass fuels containing up to approximately 55% moisture content. Little refractory is used and the maintenance requirements are low.

The main disadvantage is that the grate requires manual raking of the ashes. Therefore, biomass firing must be stopped on a regular basis. Manual raking also limits the depth of the furnace and therefore the steam capacity that can be economically built. Mechanical raking machines have been developed to increase the allowable furnace depth and to speed the raking process, while providing a certain degree of protection to the operator from hot gas, flames and hot ash. These machines have

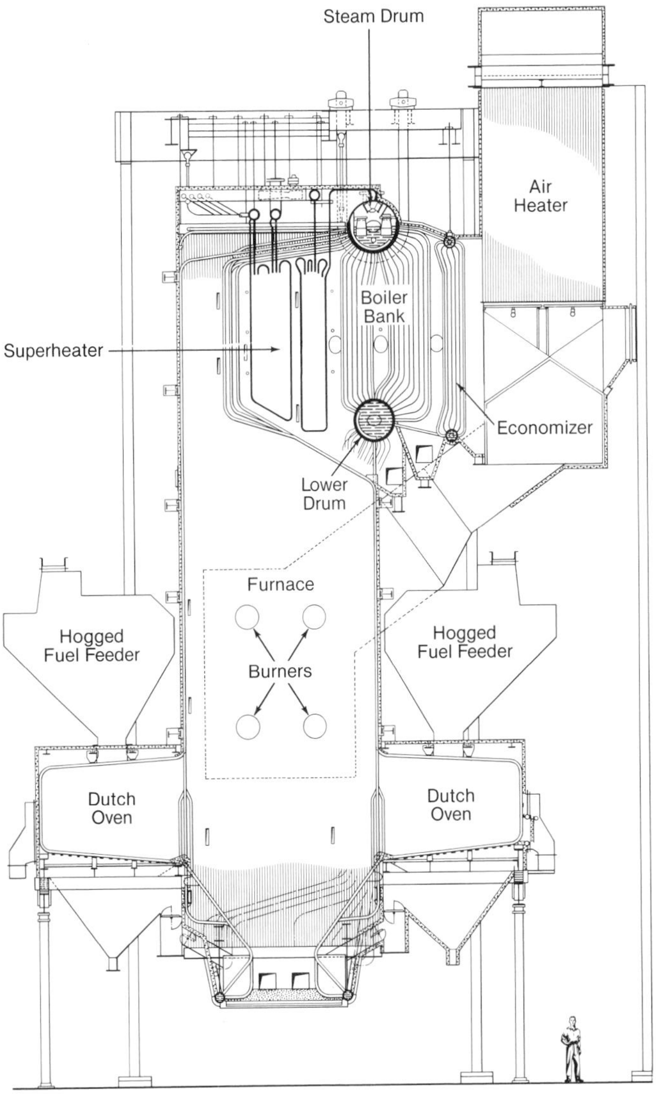

Fig. 1 Dutch oven furnace.

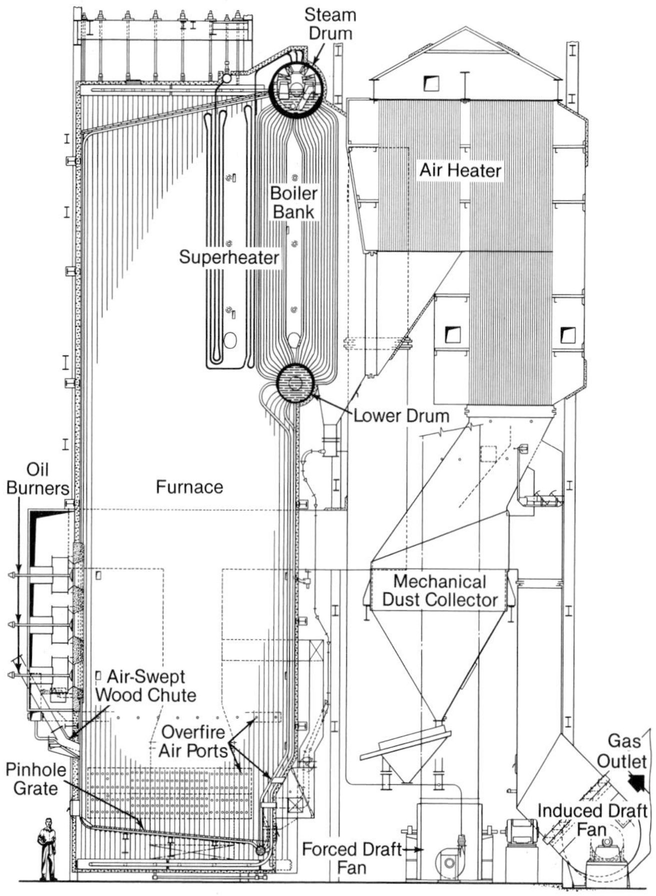

Fig. 2 Wood-fired boiler with pinhole grate.

proven to be unreliable, both in terms of their raking performance and their protection function.

Traveling grate

The traveling grate was introduced as an improvement over the pinhole grate. (See Fig. 3.) It is a moving grate that allows continuous automatic ash discharge and consists of cast iron grate bars attached to chains that are driven by a slow moving sprocket drive system. The grate bars have holes in them to admit undergrate air that is also used to cool the grate bar castings. This cooling function requires that 60 to 85% of the combustion air be undergrate and only 15 to 40% of the air be introduced as overfire. Rows of nozzles in the furnace frontwall and rearwall are used for overfire air. Fuel spreaders and burning mode are identical to those used with the pinhole grate. The main advantages of this grate are its ability to follow load swings and the automatic ash discharge that permits continuous operation on biomass fuel.

The traveling grate was originally developed for spreader-stoker firing of bituminous and subbituminous coals. With coal, the quantity of overfire air required for efficient combustion can be as low as 15% of the total air. The quantity of ash in coal is also much higher than in wood. It is therefore possible to develop a relatively large bed of ash on the grate in order to protect the grate from high temperatures and also to help distribute the undergrate air flow. Lower moisture coal can often be burned without air preheating and rarely requires air temperatures in excess of 350F (177C).

The traveling grate, while an improvement over the pinhole grate, must be considered a compromise design for the burning of biomass because the use of preheated air and the usually low ash content of biomass reduce the cooling available to the grate. This grate has many moving parts that are subjected to the furnace heat, resulting in high maintenance costs.

Vibrating grate

The vibrating grate allows intermittent, automatic ash discharge. The grate consists of cast iron grate bars attached to a frame that vibrates on an intermittent basis, controlled by an adjustable timer. There are two major types which have been used for biomass fuels, one air-cooled and the other water-cooled.

The water-cooled vibrating grate (Fig. 4) is used in conjunction with the semi-suspension firing mode. Because the grate is water-cooled, high temperature undergrate air, up to 650F (343C), can be used along with very high percentages of overfire air and a relatively thin fuel bed. A key advantage of the vibrating grate is the low number of parts moving, highly stressed, or in sliding contact. This results in reduced maintenance requirements. The vibration is intermittent and is at a maximum of five cycles per second for a short duration of about five seconds every five minutes. The vibration time and the dwell time can be adjusted to suit the fuel characteristics. The vibrating action can help

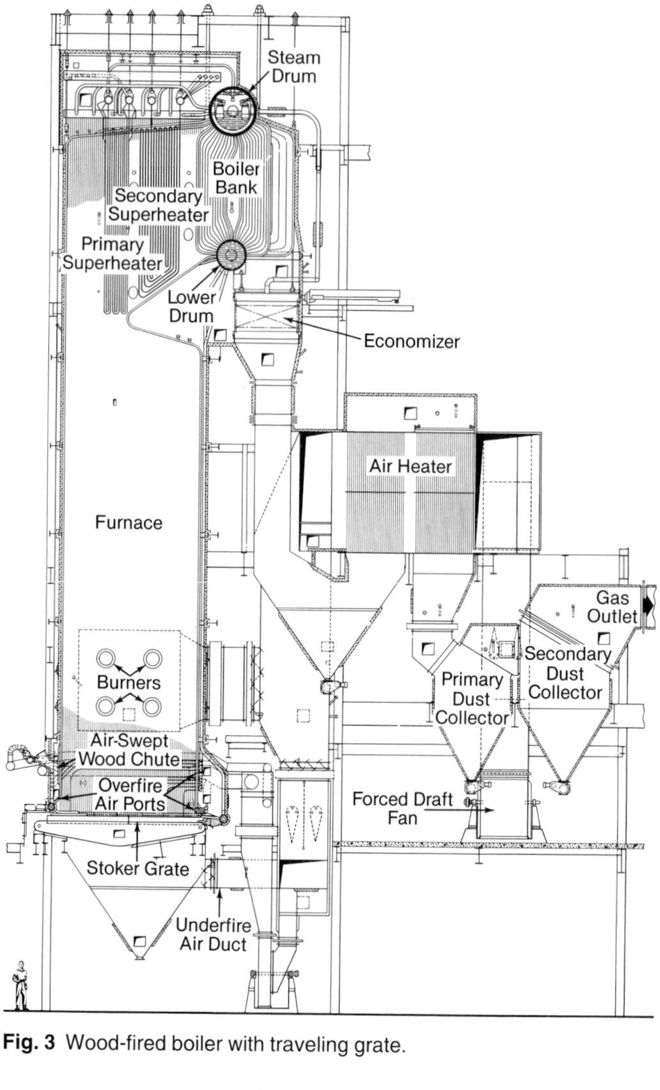

Fig. 3 Wood-fired boiler with traveling grate.

improve fuel distribution on the grate by causing fuel piles on the grate to collapse. This style of biomass combustion system is being used very successfully at a number of installations for a wide range of wood fuel moisture contents.

The air-cooled vibrating grate can be used for applications similar to those suitable for the water-cooled vibrating grate, but the maximum allowable undergrate air temperature is 550F (288C), even with the use of stainless steel components. Because it can be installed with a horizontal grate surface, the air-cooled vibrating grate can be a very effective replacement for a traveling grate, when the traveling grate has been deemed unacceptable due to high maintenance or repair costs.

Controlled Combustion Zone furnace

The Controlled Combustion Zone (CCZ™) furnace was developed by Babcock & Wilcox (B&W) in the late 1970s specifically for biomass combustion. (See chapter frontispiece.)

The design uses arches in the front and rear walls of the furnace to create a lower furnace zone in which combustion of the biomass can be confined. Overfire air from nozzles located within and below the arches penetrates and burns off volatile fuel released from the bed, as well as solid particles entrained in the upward flow.

During initial development, a 0.25 scale air flow modeling study was conducted to optimize the location of the arches and the configuration of overfire nozzles with regard to the following goals: high turbulence levels in all sections of the lower furnace, uniform flow in the upper furnace, thorough mixing of air and fuel, and minimum scouring of the grate surface.

Concurrent with this initial development work, the first CCZ™ boiler was built for Crown Zellerbach Corporation (now Fletcher Challenge) at their Elk Falls mill in British Columbia, Canada. This boiler is rated at 400,000 lb/h (50.4 kg/s) steam flow at 650 psi (44.8 bar) and 750F (399C) from wood or oil and has been in service since 1980. By 1992, there were 24 B&W CCZ™ boiler installations in four countries.

Computer modeling of furnace flow and combustion patterns was first applied to the CCZ™ design in the mid 1980s. The codes, originally developed by B&W for utility boiler analysis, are run on a personal computer and have greatly reduced the time required to evaluate specific furnace settings and air systems. Computer modeling has verified the benefits of the furnace configuration in providing superior mixing of overfire air in the lower furnace. In the same regard, it has also demonstrated the superiority of medium pressure [20 in. wg (4.98 kPa)] large diameter [4 to 6 in. (101.6 to 152.4 mm)] overfire air nozzles over high pressure, small diameter nozzles.

The CCZ™ furnace is used most often with the water-cooled vibrating grate to provide a most reliable and efficient means to burn biomass fuels. It is particularly effective in burning high moisture fuels (60%) without auxiliary fuel firing.

Dryers and pulverizers

If the biomass fuel available is particularly high in moisture, or if the capacity of an existing installation is to be increased, it is sometimes more economical to dry the fuel with boiler flue gas before firing it in the boiler furnace, rather than pressing the fuel to remove moisture or modifying the boiler for increased capacity. Dryers of the rotating drum type and cascading fuel type are available from several manufacturers. The dried fuel is then fired using one of the previously mentioned combustion systems.

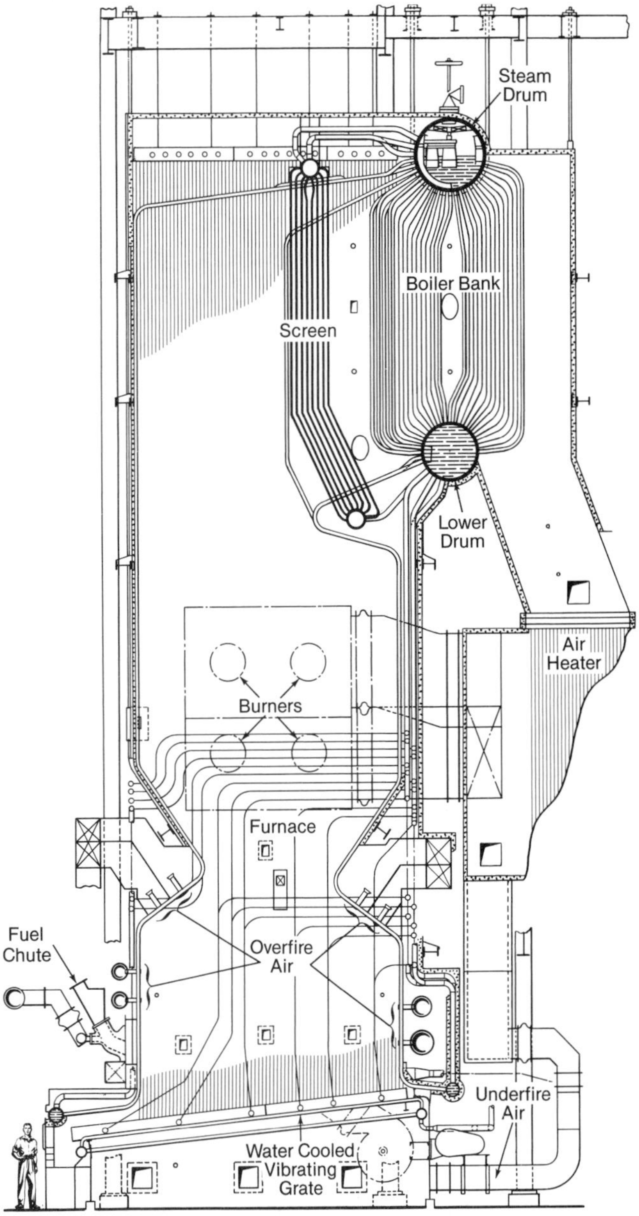

Fig. 4 Water-cooled vibrating grate unit with CCZ™ furnace.

Another means of burning biomass is by pulverizing/drying. This can be accomplished by mixing the biomass with hot gases removed from the boiler exit, pulverizing it in a fan/beater mill and returning the mixture to the furnace through a burner. Milled peat is fired in several installations in the Nordic countries using this process. The milled peat, except for larger pieces, does not undergo much size reduction. Most other types of pulverizing systems have proven to require very high maintenance and to have poor availability and significant power consumption in biomass applications.

Fluidized bed

Fluidized-bed combustion has been successfully applied to a range of wood waste fuels and offers a number of features which may be advantageous in specific applications.

Only 2 to 3% of the bed is carbon; the remainder is comprised of inert material (sand). This inert material provides a large inventory of heat in the furnace, thereby dampening the effect of brief fluctuations in fuel heating value on steam generation.

Fluid beds typically operate at 1400 to 1600F (760 to 871C), a range considerably lower than combustion temperatures for spreader stoker units [2200F (1204C)]. The lower temperatures produce less NO_x and would be expected to provide the most benefit on high nitrogen wood and biomass fuels.

SO_2 emissions from wood waste and biomass firing are generally considered insignificant, but where sulfur contamination of the fuel stream is a problem, limestone can be added to the fluid bed to achieve a high degree of sulfur capture. Fuels that are typically sulfur contaminated include construction waste and some paper mill sludges.

The type of fluid bed selected will be a function of the as-fired calorific value of the wood waste or biomass fuel. Bubbling bed technology is generally selected for fuels of lower calorific value. For fuels of higher calorific value, the circulating fluid bed is most suitable.

Designs for fluidized-bed combustion are discussed in Chapters 16 and 25 and a biomass-fired circulating fluidized-bed unit is shown in Chapter 18.

Boiler component design for wood and biomass burning

Grate

The grate forms the furnace floor and provides a surface on which the larger fuel particles burn. The grate may be air-cooled or water-cooled and stationary or arranged for automatic, continuous ash removal. Most grates consist of some form of cast iron or cast alloy grate bars.

Fuel distributor

The two most common devices for introducing fuel into the furnace for semi-suspension firing are mechanical distributors and wind-swept spouts. They are both designed to distribute fuel as evenly as possible over the grate surface.

The mechanical distributor uses a rotating paddle wheel to distribute the fuel. The speed of the wheel is varied to suit the fuel characteristics. In some installations, a continually varying speed is used to ensure good fuel distribution.

The wind-swept spout uses high pressure air, which is continuously varied by a rotary damper, to distribute the fuel. Adjustments to the supply air pressure are made by the operator as the characteristics of the fuel change. The trajectory of the fuel leaving the spout is altered by a ramp at the bottom of the spout.

Burners

Burners are sometimes used to burn all or a portion of the biomass fuel. The burners are somewhat similar in design to a pulverized coal burner. (See Chapter 13.) Fuels which can be fired in a burner include sander dust, sawdust of less than 35% moisture content and the fine material collected from a fuel dryer. Due to the possibility of inconsistent fuel flow and quality, a continuously operated auxiliary fuel pilot flame is recommended.

Furnace

A properly designed furnace has two main functions. The first is to provide a volume in which the fuel can be burned completely. The second is to absorb sufficient heat to cool the flue gas to a temperature at which the entrained flyash will not foul the convective surfaces. This must be accomplished while matching the dimensions of the grate and while providing sufficient clearance dimensions from auxiliary burners to prevent flame impingement on furnace walls.

Modern boilers are typically of membrane wall construction, but in certain limited cases such as installations in developing countries, reverting to a tube and tile type of furnace construction may be appropriate. In such circumstances, the frequent maintenance requirement is outweighed by the reduction in first cost and the ease of operation due to the higher temperatures in the furnace.

Superheater

The sizing of the superheater on a wood-fired boiler can be complicated by several factors. For a given fuel being fired, the setting of the surface depends on the final steam temperature and on the control range required. The side and back spacing are selected to minimize fouling and erosion potential. (See Chapters 18 and 20.)

However, a wood-fired boiler rarely burns a consistent fuel. Moisture content and analysis change, which affect the steam to flue gas ratio, and a variety of auxiliary fuels, e.g., oil, gas or coal, may be available. Therefore, when designing a superheater, the full range of operating conditions must be fully understood.

Constituents in the ash can affect superheater design. For example, high levels of chloride are often found in bark from logs floated in sea water and can necessitate the use of high alloy materials such as SS310 to minimize the corrosion rate of the superheater tubes in the high temperature zones.

Boiler bank

Due to the relatively high ratio of flue gas flow to steam flow and the relatively low pressures and temperatures at which most wood-fired boilers are operated, a large amount of saturated surface is required. Furthermore, due to relatively low adiabatic flame temperatures, the amount of heat absorption in the furnace is usually low compared to fuels such as oil or gas. Therefore, a large portion of the total heating surface in a wood-fired boiler is usually provided as boiler bank.

In some cases, the amount of furnace surface is augmented by water screens in front of superheaters in order to lower flue gas temperatures entering the superheater and to protect if from radiation from the active burning areas in the furnace.

The amount of boiler bank surface is usually very substantial. This surface can be arranged as cross flow surface as normally found in a two-drum Stirling boiler (chapter frontispiece), or may be arranged for longitudinal flow as found in a smaller bottom supported Towerpak boiler. (See Fig. 5.)

Because wood fuels frequently contain sand or other mineral matter in addition to the ash, flue gas velocities in the convection pass or boiler bank must be kept low, typically below 60 ft/s (18.3 m/s).

Economizer

In most cases, when an economizer is required to reduce the back end temperature to a specified level, it is located between the boiler bank and the tubular air heater. The economizer is designed to reduce the flue gas temperature to that required at the air heater gas inlet.

Occasionally, however, this usual order of equipment is reversed. For example, it may be necessary to provide a relatively high gas exit temperature as part of an installation which incorporates a fuel dryer. The dryer needs the hot flue gas to remove moisture from the fuel. The same installation may require a low gas exit temperature during periods when the dryer is out of service for maintenance or repair. The low temperature is desired in order to obtain the best thermal efficiency. To provide hot air during both operating conditions, the economizer must be used as the final heat trap. A bypass, preferably on the gas side rather than the water side, allows the temperature to be controlled.

In the above example, it should be remembered that the gas exit temperature would be lower when burning dry fuel from the dryer and would be higher when burning high moisture fuel with the dryer out of service, if no control method was provided.

There are no special mechanical design considerations for economizers on wood-fired units, other than to limit the flue gas velocity. In virtually all cases, a continuous bare tube economizer is used. (See also Chapter 19.)

Air heater

Due to the requirement to provide hot air to burn all but the driest of wood fuels, wood-fired boilers are usually equipped with air heaters. Because of the ash, sand and char in the flue gas, a recuperative type is normally selected, usually tubular. (See Chapter 19.) It is B&W practice to provide the tubes on square or rectangular pitch, rather than on triangular pitch, to lower air side resistance and allow easier maintenance.

The common arrangement is for the flue gas to pass through the tubes and the air to flow around the outside of the tubes. A two gas pass design is favored for economic reasons. The low flue gas velocity in the hopper and the 180 deg change in direction also promote the separation of large, heavier particles of char and sand. The tubular air heater, when arranged for two gas passes, can act as a low efficiency (approximately 50%) mechanical dust collector.

Experience has shown the use of 2.5 in. (63.5 mm) OD tubes to usually be the most cost efficient. Available space limitations may require the use of smaller 2 in. (50.8 mm) OD tubes. In cases with a history of plugging by contaminants in the flue gas, it may be necessary to use 3 in. (76.2 mm) OD tubes.

Auxiliary equipment

Fans Wood-fired boilers require *forced draft* (FD), *induced draft* (ID) and usually *overfire air* (OFA) fans. (See Chapter 23.)

Forced draft fans require no special design considerations other than determining the required capacity and static pressure. The design may be determined by wood firing alone, wood firing in combination with auxiliary fuel, or auxiliary fuel alone, depending on the quantity and

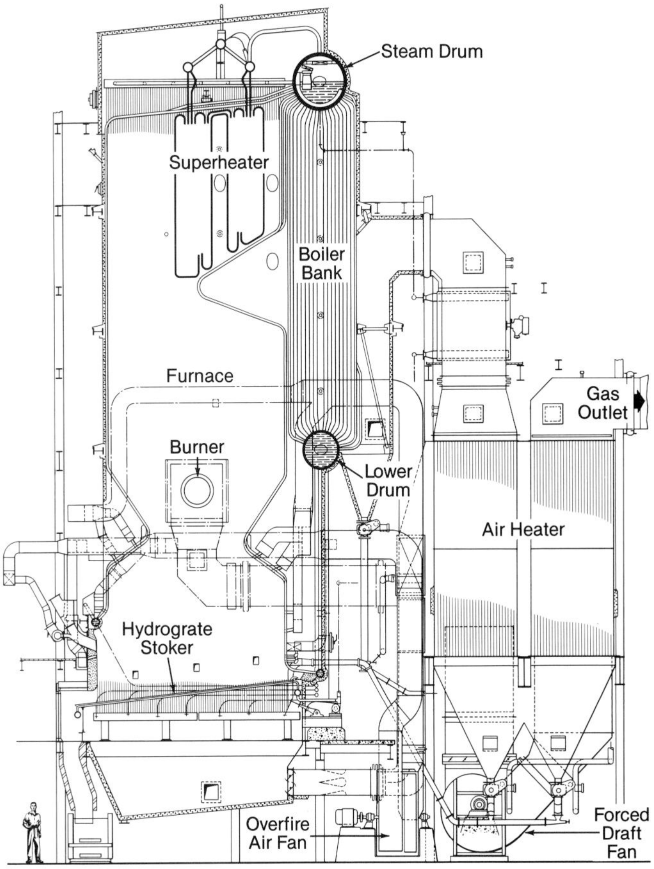

Fig. 5 Longitudinal flow boiler bank.

pressure of air required for each of these conditions. Normal test block margins for FD fans are sufficient. Usually these fans are controlled to maintain a constant pressure at the tubular air heater air outlet plenum.

Induced draft fans must be designed to take into account the abrasiveness of the flue gas, the quantity of flue gas to be handled, the draft losses to be overcome and the temperature of the flue gas. The abrasiveness of the flue gas depends on the type and efficiency of the dust collection equipment installed and whether the fan is located before or after this equipment. The ID fans are used to control the furnace pressure to a set point, usually at -0.1 to -0.5 in. wg (-0.025 to -0.12 kPa).

The quantity of gas and the draft loss used to specify the ID fan must take into account not only the expected operation of the boiler but also possible wide variations in heating value, theoretical air and moisture content of the fuel.

Overfire air fans are often exposed to severe service, as typically the air to be brought to 30 in. wg (7.5 kPa) has already been heated in the tubular air heater to as high as 650F (343C) before reaching the fan inlet. When specifying an overfire air fan, it is particularly important to specify the maximum air temperature that it must be capable of handling.

Sootblowers Because biomass-fired boilers are susceptible to ash and carbon carryover, the convective heat transfer surfaces must be designed to accommodate sootblowers. Retractable sootblowers must be used for the superheater and high temperature boiler bank surfaces. Rotary sootblowers can be used for the low temperature boiler bank areas and the economizer, but retractable sootblowers are preferred, provided there is sufficient space. Traveling rake type sootblowers are typically used above the tubular air heater tubesheets on the flue gas side. (See Chapter 23.)

These sootblowers use steam, either saturated or superheated, to clean the gas lanes between the tubes. Blowing pressures are typically 150 to 250 psi (10.3 to 17.2 bar) and the blowing sequence is usually initiated once a working shift. The high pressure jet of steam from the sootblower nozzle cleans the tube surfaces with about a 5 ft (1.52 m) radius around the sootblower lance.

The ash is typically nonsticky and is relatively easy to remove by sootblowing. If allowed to accumulate, the ash can plug gas passages, cause flow unbalances, affect boiler circulation and heat transfer, and ultimately necessitate a forced outage of the boiler. If unburned carbon particles are allowed to accumulate, they can create the potential for a fire, particularly in the back-end equipment.

Fuel handling systems Biomass fuel systems can be quite complicated and maintenance intensive due to the varying characteristics of the fuel. Typically, the fuel is continuously conveyed from storage to small surge bins at each fuel distributor which are kept full, with any overfeed returned to storage. The surge bins are equipped with variable speed screw feeders or chain feeders to control the rate of biomass fuel fed into the furnace.

These variable speed feeders must be capable of operation over a turndown range of four to one on automatic control. They must also be able to operate at very low speeds during startup conditions in order to build a fuel bed on the furnace grate. The feeder drives must be of sufficient horsepower to allow the feeder to be started when the surge bin is full of fuel.

Upstream of the surge bins, it is common to have a large live bottom storage bin with four to eight hours of biomass fuel inventory. This is to avoid interruptions in the fuel feed to the boiler when there are problems with the outside fuel handling equipment. (See Chapter 11.)

Ash handling Ash handling systems on biomass-fired boilers can be divided into two main areas, bottom ash and flyash.

Bottom ash is the ash that is raked or conveyed off the grate, plus the ash that falls through the grate bar holes into the undergrate hopper (called a riddlings or siftings hopper). Bottom ash consists mainly of sand and stones. The ash at the grate discharge is typically collected using a submerged drag chain conveyor with a dewatering incline at the discharge end. The siftings can be collected with drag chain or screw conveyors.

Flyash is the fine ash and unburned carbon that is collected from all the boiler bank, economizer, air heater and emission control equipment hoppers. The ash handling equipment can be drag chains, screw conveyors or wet sluicing systems. Because the flyash contains a high percentage of hot carbon, it is important that rotary seal valves be used at each hopper discharge to prevent air infiltration that could create a fire in the hopper. For the same reason, all ash conveyors should be sealed. (See Chapter 23.)

In some instances, the flyash from the boiler bank and air heater hoppers is reinjected into the furnace to lower the unburned carbon loss and to reduce the quantity of material that must be disposed of. However, the high maintenance requirements of these systems have limited their use.

Air systems

Air systems can be categorized as underfire or undergrate air and overfire air.

Undergrate air is typically low pressure [3 in. wg (0.75 kPa)] and depending on the type of grate used, can be anywhere between 25 to 75% of the total air required for combustion. The purpose of undergrate air is to help dry the fuel, to promote release of the volatiles, to provide the oxygen necessary for the combustion of the devolatilized char resting on the grate and, in the case of an air-cooled grate, to cool the grate bars. The pinhole grate and vibrating grate can be provided with multiple undergrate air compartments with separate dampers for the operator to bias the undergrate air to the area of the furnace where the fuel is concentrated. Traveling grates can be provided with only one compartment per drive section.

Overfire air system capacities are varied and can range from 25 to 75% of the total air. Varying nozzle sizes and air pressures are used to obtain adequate penetration of the air into the rising stream of volatiles from the grate. Typically, modern overfire air nozzles are 3 to 6 in. (76.2 to 152.4 mm) in diameter and use air pressures up to 20 in. wg (4.98 kPa). Levels of nozzles are controlled independently such that the overfire air can be varied with load and fuel characteristics. Where very high temperature overfire air is used and it constitutes more than 40% of the total air flow, it is usually economical and energy efficient to provide a high pressure FD fan, rather than a low pressure FD fan plus a large high pressure OFA fan.

Emission control equipment

Dust collector Mechanical dust collectors are used after the last heat trap on the boiler to collect the larger size flyash particulate, sometimes as protection for the ID fan. They typically consist of multi-cyclone tubes enclosed in a casing structure. The tubes consist of outer inlet tubes with spin vanes and inner tubes used without recovery vanes. The dust collector efficiency is in the range of 65 to 75% at an optimum draft loss of 2.5 to 3.0 in. wg (0.62 to 0.75 kPa). Due to the abrasive nature of the flyash, the outer collection tubes and cones are made of high hardness (450 Brinell) abrasion resistant material.

Precipitator Electrostatic precipitators are typically used after the mechanical collector to reduce the particulate concentration in the flue gas to meet environmental requirements. Due to the high carbon content in the flyash, it is important to reduce the fire potential in the precipitator. It is necessary to ensure no tramp air enters the precipitator and that the flyash is continuously removed from the hoppers. Hopper level detectors and temperature detectors alert the operator.

Installations can be equipped with fire fighting apparatus such as steam inerting. Other suppliers recommend de-energizing the precipitator if a predetermined oxygen content in the flue gas is exceeded.

Fabric filter or baghouse Due to the fire potential, baghouse collectors have been rarely used for biomass fuels to date. The advent of high temperature metallic bag materials may change this in the future.

Wet scrubbers Wet scrubbers have also been used to control particulate emissions on biomass-fired boilers. Their main disadvantages are the high flue gas pressure drop which increases the ID fan horsepower requirements and the high water consumption. Also, there is the need for a wet ash collection system and a water separation and clarification system. Wet scrubbers have given way to electrostatic precipitators as the preferred means of final flue gas cleanup, provided there is no need for a scrubber to reduce SO_2 emissions from auxiliary fuel.

Wet scrubbers with numerous small spray nozzles in a chamber can be used where low pressure drop and low water consumption are required. This is particularly well suited to retrofit applications where the scrubber can replace the mechanical dust collector for improved collection efficiency.

Environmental impact

Particulate emissions

About 80 to 95% of the total ash residue produced by a bark- and wood-fired spreader stoker is in the form of gas borne particulate. This particulate is composed of a number of materials, including ash, sand contaminants introduced during fuel handling, unburned char from the furnace and salt fume (usually present only where logs are sea water flumed).

The ash content of wood and bark fuels is low (0.2 to 5.3%, dry basis). Therefore, if fuel contaminants are not appreciable, the particulate will usually contain high percentages of unburned char.

The particulate loading in the flue gas exiting the boiler is influenced by factors related to both combustion and aerodynamics. Combustion related factors affect particulates by determining the degree of burnout for the entrained char. They include plan and volumetric heat release rate, two design parameters which affect furnace temperature, residence time and consequently, char burnout.

The importance of aerodynamic factors is based on the fact that bark- and wood-fired spreader-stoker units are designed to operate with some degree of suspension burning. Variables which would tend to increase the ratio of furnace velocity to mean char particle size would therefore tend to increase particulate loading.

Some of these factors include fuel moisture and fines content, boiler plan area, air staging and excess air level.

For wood and bark fuels not containing appreciable quantities of sand, particulate loading at the air heater exit of a modern spreader-stoker unit would typically be in the range of 1 to 3 grains/DSCF (2.4 to 7.2 g/Nm^3).

Nitrogen oxides

NO_x emissions from wood and bark firing are low compared with those from traditional fossil fuels. Combustion temperatures in wood firing are sufficiently low, and little thermal NO_x is formed from the nitrogen in the combustion air. Corresponding NO_x emissions are therefore predominantly a function of the fuel nitrogen content. (See also Chapter 34.)

Conversion of fuel bound nitrogen to NO_x is dependent on a number of operating conditions including excess air, air staging, heat release rate and fuel moisture content. Empirical studies have also found NO_x to vary inversely with fuel moisture content, although the magnitude of this correlation is less significant.

All factors considered, NO_x emissions while stoker firing most wood and bark vary between 0.2 and 0.4 lb/10^6 Btu (0.09 and 0.17 g/MJ) heat input, expressed as nitrogen dioxide. Fuel contaminants which may introduce nitrogen compounds (glues and chemicals, for example) should receive special consideration.

Sulfur dioxide

Wood and bark typically contain 0.0 to 0.1% elemental sulfur on a dry basis. During the combustion process some of this sulfur can be converted to flue gas SO_2 but the conversion ratio is typically low (10 to 30%). Because the quantities of both wood sulfur and flue gas SO_2 are near the low end detection limit of corresponding analytical instruments, correlation of the two is not practical.

Typically, SO_2 emissions for stoker-fired wood and bark fuels do not exceed 0.03 lb/10^6 Btu (0.01 g/MJ) heat input. Special consideration should be given to fuels where sulfur bearing contaminants may be present.

Carbon monoxide

Of all emissions commonly associated with wood and bark firing, carbon monoxide (CO) is usually the most variable. As a gaseous product of incomplete combustion, CO is dependent on time, temperature and turbulence considerations.

At normal excess air levels, consistency of both fuel heating value and fuel distribution are considered the most important determinants of CO emissions. Typically, test data showing the highest standard deviation in CO correspond to the highest mean CO.

Conditions of high excess air, low excess air, high fuel moisture and reduced load (<70% of the maximum continuous rating) have all been demonstrated to increase flue gas CO concentration.

Modern spreader-stoker units firing wood and bark only, and operating at steady-state, typically emit CO in the range of 0.1 to 0.5 lb/10^6 Btu (0.04 to 0.22 g/MJ) heat input.

Volatile organic compounds

Volatile organic compounds (VOCs) are also gaseous products of incomplete combustion. As such, the emission of VOCs during wood firing is influenced by the same factors affecting CO.

Typically, VOC emissions while stoker firing wood and bark fuels do not exceed 0.05 lb/10^6 Btu (0.02 g/MJ) heat input, expressed as methane.

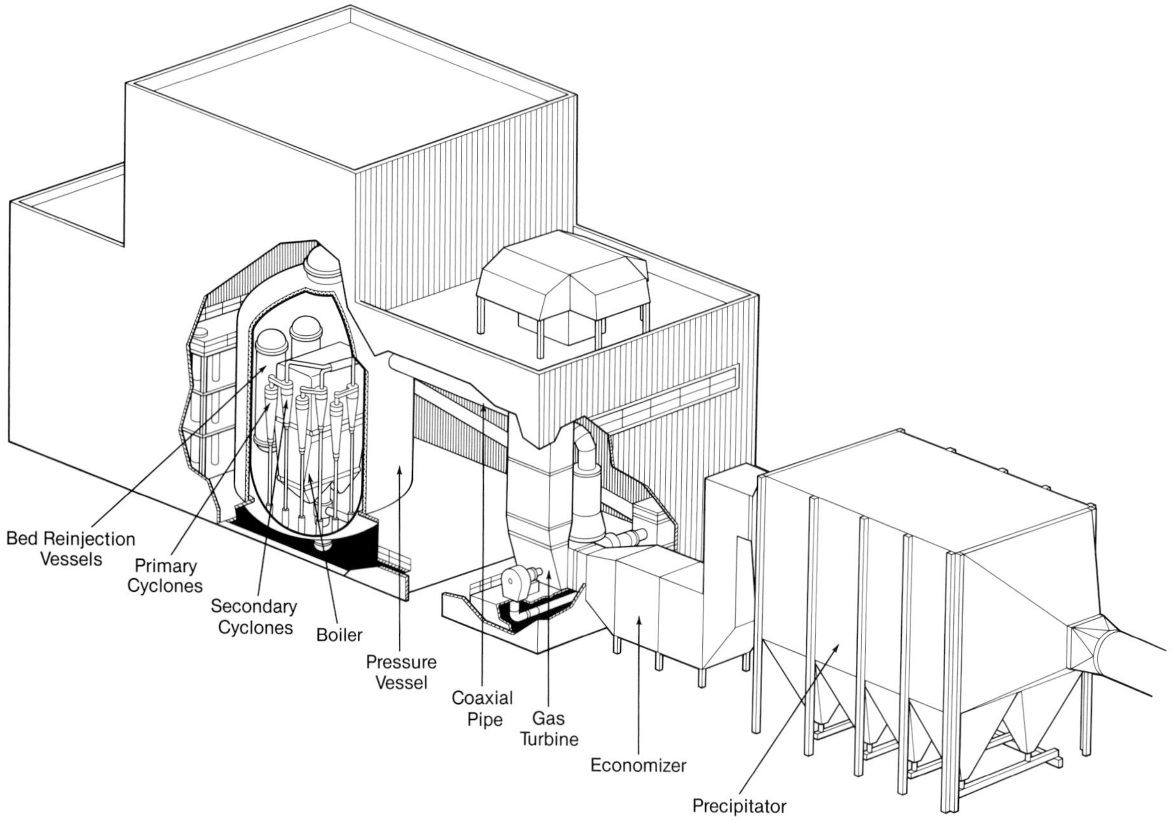

Pressurized fluidized-bed combustion power plant schematic.

Chapter 29
Pressurized Fluidized-Bed Combustion

In the 1970s and 1980s, great attention was given to the fluidized-bed combustion of coal to reduce emissions. Pressurized fluidized-bed combustion (PFBC) is an outgrowth of this technology, operating a fluidized bed at elevated pressures and combining fluidized-bed combustion with both a gas turbine and a steam turbine to produce a combined cycle.

Chapter 16 presents the basic chemical processes and considerations involved in coal-fired fluidized-bed boilers operating under atmospheric pressure conditions. The chemical processes in these systems are essentially the same as in the PFBC process. Pressurizing the combustion process permits further reduction of sulfur emissions, improved combustion efficiency and, most significantly, produces hot pressurized exhaust gases which can be used to power a gas turbine. The gas turbine can drive a compressor to supply the pressurized process air and, in a combined cycle configuration, produce electrical power in a generator.

In an atmospheric fluidized-bed power plant, the boiler is the key component. It receives its combustion air from fans and extracts as much heat as practical in the steam cycle before exhausting the combustion gases to the stack. In a pressurized fluidized-bed plant the steam and gas cycles are synergistic. Coal is combusted in the bed and the resulting gases are maintained at high temperature to power the gas turbine. Simultaneously, the boiler absorbs heat from the bed to drive the steam turbine.

General history of PFBC

The concept of burning coal in a fluidized bed found early application in the Winkler gasification process invented by Fritz Winkler in the mid 1920s.

The combustion of coal in a fluidized bed did not receive significant attention until the 1960s when public interest in reducing emissions of oxides of sulfur and nitrogen from pulverized coal (PC) fired power plants increased. The lower combustion temperature in a fluidized bed and the ability to use a sulfur sorbent as the bed material offered a potentially less expensive and more efficient method of using coal than conventional PC systems.

With increasing desires to improve plant heat rates while reducing emissions, interest in operating the fluid bed process under pressurized conditions escalated in the early 1970s. Bench scale testing in the United States (U.S.), the United Kingdom, Sweden, Germany and other countries led to larger pilot scale units in the mid 1970s.

Babcock & Wilcox's involvement in PFBC

In the late 1970s, Babcock & Wilcox (B&W) began to develop PFBC boiler designs. Because of B&W's strengths in boiler and pressure vessel design and fabrication, most of the efforts focused on these components. Initial work centered on air-cooled cycles, but the need for high temperature materials made these early designs impractical.

In the mid 1980s, B&W submitted a proposal through ASEA Babcock, a partnership between B&W and Risinge, Inc., for a portion of the PFBC equipment to repower Ohio Power (a subsidiary of American Electric Power) Tidd plant with a 70 MW PFBC unit. The project was selected by the U.S. Department of Energy and the Ohio Coal Development Office for funding under their Clean Coal Technology programs.

PFBC experience in the early 1980s identified erosion/corrosion of the in-bed tube bundles and gas turbine blades, as well as component reliability/serviceability, as the key uncertainties of this emerging technology. The Tidd PFBC demonstration project was intended to address these issues. To further understand the behavior of fuels and sorbents under pressurized conditions, B&W built a small reactor at its Alliance Research Center in Ohio to perform coal and sorbent characterization testing.

Also in the 1980s, advances in hot gas cleaning and desires to improve performance even further led to several studies of advanced plant cycles. Advanced cycles incorporate coal gasification and gas turbine topping into a PFBC configuration and are expected to increase net plant efficiencies to percentages approaching the mid 40s.

The PFBC process

General principles

As in atmospheric pressure designs, two types of pressurized fluid beds have been applied to power generation: bubbling and circulating. Due to the elevated pressure and air density, deeper beds and lower fluidizing velocities are possible. Bubbling-bed designs have been developed and operated while circulating designs are in the pilot stage.

In addition to powering a gas turbine and a steam turbine, pressurized bubbling beds operating at 174 to 232 psi (1200 to 1600 kPa) are typically more than twice as deep as equivalent atmospheric fluid beds with one third to one fourth of the fluidizing velocity. These differences result in several advantages such as improved cycle efficiency, reduced emissions, improved combustion, reduced boiler size, reduced tube erosion and modularity.

Advantages of pressurized fluidized-bed combustion

Improved cycle efficiency (lower heat rate) The major advantage of pressurizing the fluidized-bed combustion process is that plant efficiency can be significantly improved by combining the Rankine steam cycle with the Brayton gas cycle. Because the combustion process occurs at pressures of 174 psi (1200 kPa) or higher, the hot pressurized gases from the boiler can power a gas turbine while steam is generated to power a steam turbine. One limitation is that unit capacity is determined by the ability of the gas turbine to supply combustion air. Therefore, the sizes of suitable gas turbines available determine the PFBC unit size.

For first generation PFBC combined cycles, efficiencies approaching 40% and heat rates of about 8500 Btu/kWh can be achieved compared with about 9600 Btu/kWh (36% efficiency) for atmospheric fluid beds. Second generation advanced combined cycles are expected to attain efficiencies more than 45% and heat rates as low as about 7500 Btu/kWh.

Reduced emissions and improved combustion Increasing the process pressure results in several advantages beyond permitting combined cycle operation. The increased pressure and corresponding air/gas density allow much lower superficial fluidizing velocities, about 3 ft/s (0.9 m/s), which reduce erosion risk for immersed tubes. The higher available pressure drop also permits much deeper beds. For a bubbling PFBC, full load bed depths can range from 12 to 15 ft (3.6 to 4.6 m), depending on pressure. The combined effect of lower velocity and deeper bed results in greatly increased in-bed gas residence time which reduces emissions of sulfur oxides (SO_x) and improves combustion efficiency. In fact, the deeper bed allows 50% of the total residence time to be in the bed where it is more effective compared with 10 to 15% in the shallower atmospheric bubbling fluid beds.

Reduced boiler size Because air mass flow ($\dot{m}$) is equal to the product of its density (ρ), superficial velocity (V) and bed plan area (A), $\dot{m} = \rho VA$, the high air/gas density results in a much smaller required bed plan area. For the same air mass flow, a bubbling PFBC at 174 psi (1200 kPa) with a superficial velocity of 3 ft/s (0.9 m/s) will require 28% of the bed plan area of an atmospheric bubbling fluid bed and about 56% of that required for an atmospheric circulating fluid bed. The lower velocity also significantly reduces the total height required for the bed and freeboard since the furnace height is the product of the velocity and the total residence time desired. The result of these factors on the PFBC boiler size is illustrated in Fig. 1. The outline of the combustor vessel is also shown for comparison.

Reduced tube wastage In-bed tube wastage is a function of several parameters affecting erosion and corrosion. Tube temperature, tube material, tube bundle geometry, gas velocities, bed material properties and fuel characteristics can have adverse effects on in-bed tubes and supports. The most significant effect of pressurized operation on tube wastage is a result of the reduced gas velocities which are about one third of those experienced in atmospheric fluidized beds. Although most of the causes of in-bed wastage are the same for atmospheric or pressurized fluid beds, the substantially lower gas side velocities and more compact tube bundle geometries lead to significantly reduced erosion potential.

Notes:
1. Furnace shaft size is bed plan area and height.
2. Based on:

	AFB	CFB	PFB
Pressure	1 atm *(100 kPa)*	1 atm *(100 kPa)*	12 atm *(1200 kPa)*
Velocity	10 ft/s *(3 m/s)*	20 ft/s *(6 m/s)*	3 ft/s *(0.9 m/s)*
Total Residence Time	6 s	4.5 s	6 s

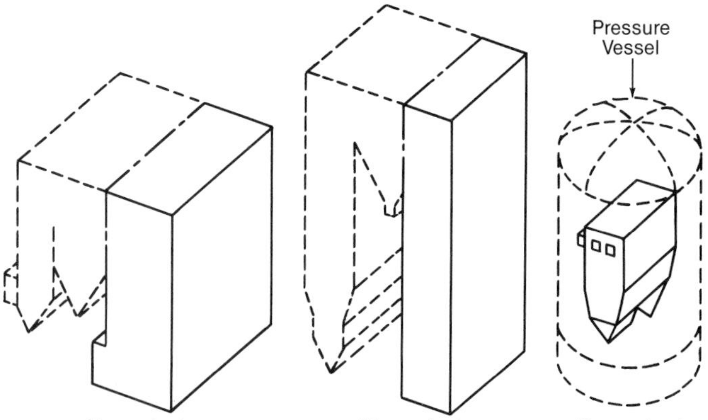

Fig. 1 Effect of process pressure on reducing boiler size.

Test facilities show that higher temperature tubes, such as those in superheaters, tend to form an erosion resistant oxide layer which nearly eliminates erosion of these surfaces. Lower temperature evaporative surfaces, however, remain susceptible to erosion. Various methods including coatings, fins and other surface treatments are used to give reasonable tube life. Bed material particle size, hardness and fuel characteristics can also contribute to in-bed wastage. Larger particle sizes and harder sorbents tend to increase erosion. Some fuel constituents can cause corrosion of metals and destruction of the protective oxide layer on hotter surfaces.

Modularity Because the gas turbine compressor capability sets the requirements of the boiler and major components, PFBC lends itself to a high degree of standardization. Once a gas turbine is selected, the air flow, pressure and resulting boiler size are set. Variations between plants are then a result of fuel and sorbent handling requirements and steam cycle conditions.

The range of PFBC design sizes is set by the compatible gas turbine sizes. Because the boiler size is greatly reduced by the process pressure and is enclosed within a combustor vessel, a high degree of modularity is possible. The degree of modularity, however, depends largely upon the plant location.

When a plant is located beside a navigable waterway, large modules are possible. As an example, for the Tidd PFBC demonstration project, modularity began with the boiler and ended with the shipment of the nearly completed combustor vessel, with internals (Fig. 2), to the site.

Fig. 2 PFBC pressure vessel being prepared for shipment.

The entire 1200 t (1088 t_m) module was loaded on a barge at Mt. Vernon and shipped more than 750 miles (1207 km) to the Tidd plant located on the Ohio River near Steubenville, Ohio. Upon arrival at the site, the entire module was moved off the barge by sliding it onto a specially designed steel frame. The tracks were lubricated with ship launching grease, and hydraulic jacks pushed it up about a 3% grade for approximately 900 ft (274 m) onto its final foundation. Fig. 3 shows the unloading at the site.

A larger 350 MW plant located near a navigable waterway would also allow the boiler to be modularized. However, it would have to be shipped in sections and considerably more site assembly would be necessary. The vessel sections, boiler modules and other components would be shipped to the site in large modules weighing as much as 1000 t (907 t_m).

PFBC plant cycles

Three basic cycle configurations have been proposed for PFBC: the turbocharged, combined and advanced combined cycles.

Turbocharged cycle The turbocharged cycle was initially developed as an intermediate step toward a combined cycle. Due to the high gas temperatures of the combined cycle, above 1580F (860C), conventional gas cleaning methods could not be used and the state-of-the-art in hot gas cleaning was not sufficiently developed. To reduce the temperature of the gases leaving the boiler for conventional gas cleaning, heat transfer surfaces are

Fig. 3 Pressure vessel arrival at Tidd site.

placed above the bed. This increases steam production capability but reduces power to the gas turbine. Under these conditions, the gas side is designed to achieve self-sustaining conditions for the gas turbine with essentially all of the electrical generation coming from the steam cycle. This cycle incorporates more conventional technology at the expense of efficiency. Fig. 4 shows a typical turbocharged cycle.

Combined cycle The best combination of the gas and steam cycles is the combined cycle. Unlike the turbocharged configuration, there are no convective heat transfer surfaces above the fluid bed to cool the hot pressurized gases. The gas leaving the boiler is cleaned and sent directly into the gas turbine. This produces gas turbine inlet conditions in the range of 174 to 232 psi (1.2 to 1.6 MPa) at about 1550F (843C). Process pressures as high as 290 psi (2.0 MPa) have been proposed. This produces a power split between the gas and steam cycles which results in about 80% of the electrical generation from the steam turbine and 20% from the gas turbine. The power to the gas turbine is sufficient to drive both a compressor to provide the preheated, pressurized air to the fluidized-bed boiler for the combustion process and an electrical generator. Fig. 5 shows a typical combined cycle configuration.

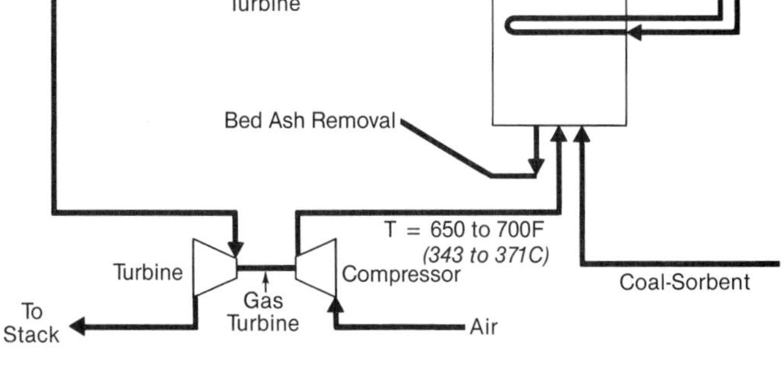

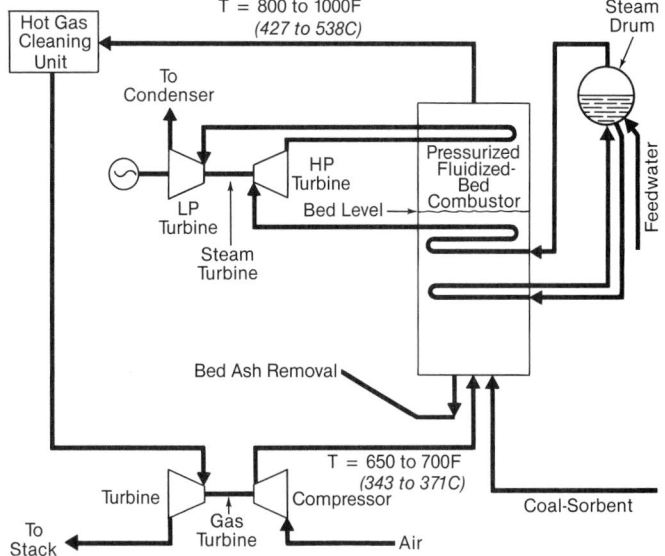

Fig. 4 The turbocharged PFBC cycle.

Advanced combined cycle To prevent ash slagging, minimize the formation of thermal NO_x, maximize the sulfur capture reactions and avoid formation of alkalies in the gas stream, the bed operates at about 1580F (860C). This produces a gas turbine entry temperature around 1525F (830C). To further increase the contribution of the gas turbine to the plant efficiency, the turbine entry temperature must be increased.

Conventional gas turbines operate at up to 2000F (1093C) inlet temperatures with new designs in the range of 2300F (1260C). By increasing the gas temperature from the PFBC, higher outputs can be produced. Because coal is the primary fuel for the PFBC boiler, the most popular method of producing fuel for a topping combustor is by some form of partial gasification where the resulting char is fed back into the fluidized bed (see Chapter 17). Application of this method requires advanced hot gas cleaning systems to remove particulate from the fluidized bed gas stream and from the gas produced by the partial gasifier before sending the gases to the turbine. Fig. 6 illustrates the essential configuration of this second generation advanced PFBC cycle.

Coupled with the increased gas turbine entry temperature, advanced supercritical steam cycles with steam conditions above 4000 psi (300 bar), 1100F (593C) superheat and 1100F (593C) reheat will further improve plant output. Advanced emissions reduction methods for NO_x and SO_x may achieve levels below 0.1 lb NO_x as NO_2 per 10^6 Btu (123 mg/Nm^3 at 6% O_2) and more than 98% sulfur capture. However, successful application of these technologies at acceptable costs has not yet been demonstrated.

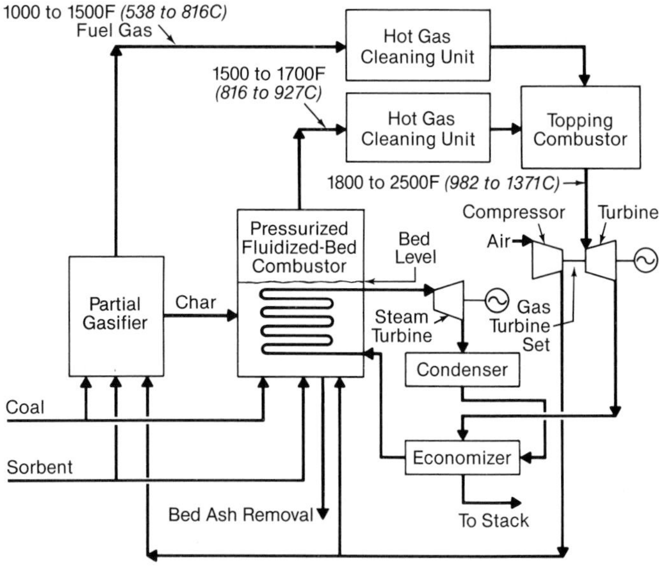

Fig. 6 The advanced PFBC cycle.

The combined cycle design

Fluidized bed process

Though pressurized fluidized beds behave basically the same way as atmospheric pressure fluidized beds, the elevated pressure does affect fluidization, heat transfer and combustion efficiency.

Fluidization Good fluidization is important to achieve in bubbling beds to avoid excessive elutriation while main-

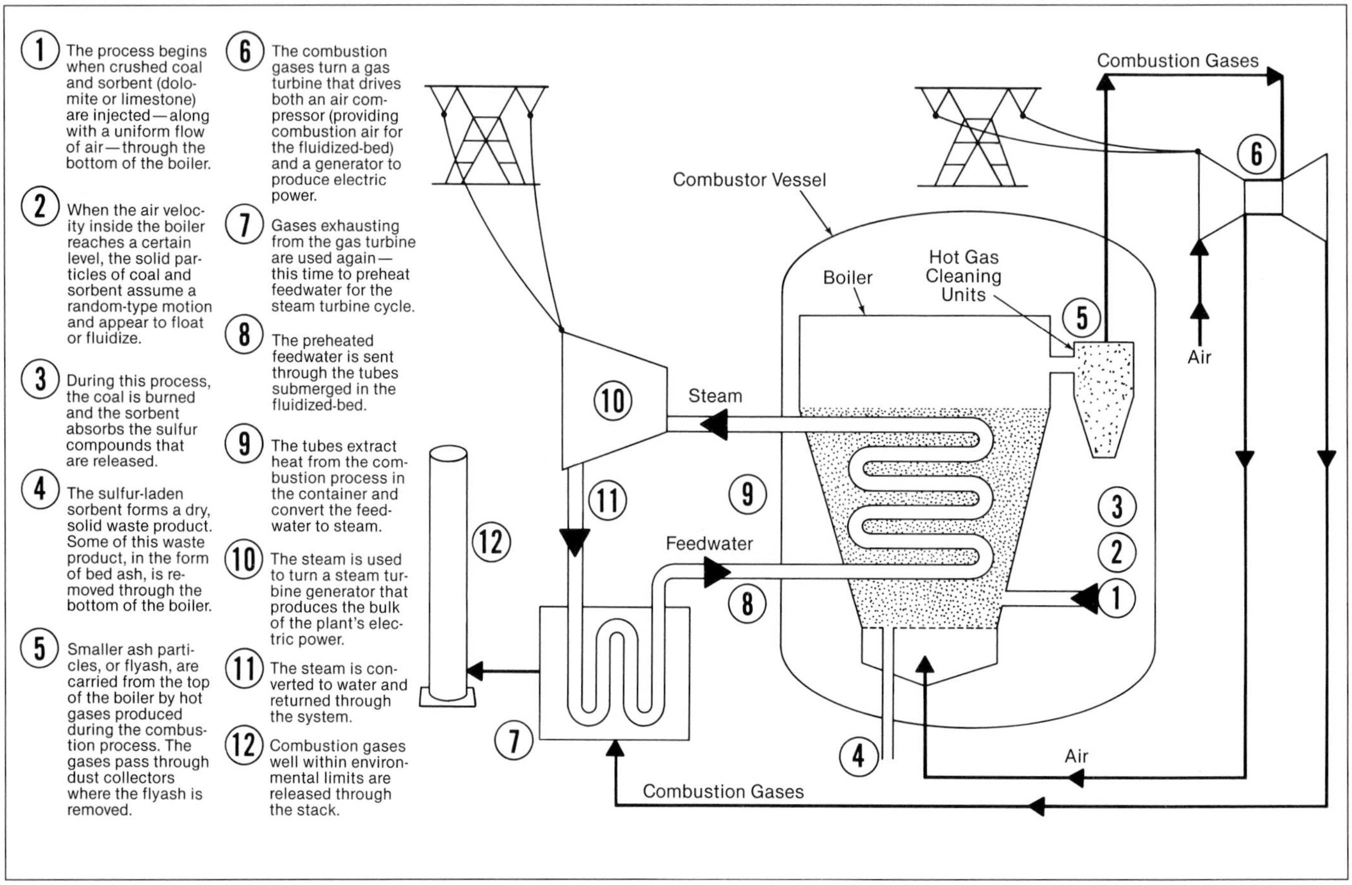

Fig. 5 Combined cycle PFBC.

taining the particles in suspension with adequate mixing to keep the heavier particles from settling to the bottom of the bed. Good fluidization falls between the velocity at which the particles become fluidized, called the *minimum fluidizing velocity,* and the velocity at which a particle leaves the bed and is elutriated, called the *terminal velocity*.

Fluidization is affected by particle size and shape, temperature and process pressure. It has been shown that increasing this pressure will decrease both the minimum and terminal velocities as much as 30% for particle sizes in the range of 1500 microns but has negligible effect for particles in the range of 200 microns. Bed temperature has a similar effect, but it is much less significant.

Heat transfer As in atmospheric fluidized bubbling beds, three zones of heat transfer must be considered: in-bed, splash zone and above-bed. Predicting the heat transfer in these zones is the single most important factor in determining part load performance.

Heat transfer to immersed tubes is by both gas and particle convection and radiation. The convective portion is the most significant and has been correlated by the equation:

$$h = 13.5\, k\, Ar^{0.12}\, D^{-0.5} + 0.46\, k\, Ar^{0.15}\, D^{-1}\, \varepsilon^{-0.7} \quad (1)$$

where

h = convective heat transfer coefficient, $W/m^2 K$
k = mean gas conductivity in film, W/m K
Ar = Archimedes number = $gD^3\rho_g(\rho_p\text{-}\rho_g)/\mu^2$
D = surface mean particle diameter, m
ε = mean voidage within tube bank
g = gravitational constant, m/s^2
ρ_g = mean gas density in film, kg/m^3
ρ_p = mean particle density, kg/m^3
μ = mean gas viscosity in film, kg/m s

Increasing temperature, pressure or both will increase the convective heat transfer coefficient. However, pressure has the much greater effect. For a given bed temperature, increasing the pressure from about 85 to 230 psi (590 to 1600 kPa) will increase the convective coefficient by as much as 28%. When a radiation coefficient is added to the convective coefficient calculated from this correlation, the result is the total in-bed heat transfer coefficient.

The magnitude of the radiation coefficient is typically about 75% of the convective coefficient. The radiation coefficient is significantly affected by the bed and tube metal temperatures. A 400F (222C) change in bed temperature will change the radiation heat transfer coefficient by as much as 50%.

Heat transfer in the splash zone varies between the total in-bed coefficient and the above-bed coefficient in the freeboard. The in-bed coefficient is generally about four times the above-bed coefficient. Though the splash zone is not large, it must be considered to obtain an accurate prediction of heat absorption as load is reduced.

Above-bed heat transfer in a bubbling-bed boiler is determined in basically the same manner as in conventional pulverized coal fired boilers. However, the gas-solids mixture in the freeboard, which is high even when compared to a conventional boiler burning a high ash coal, does affect radiation emissivity and convective heat transfer. In combined cycle PFBC, very little heat transfer occurs above the bed due to the intent to maintain a high gas temperature for the gas turbine. To minimize above-bed heat transfer, the freeboard is internally insulated.

Combustion efficiency Combustion efficiency is a function of excess air, fluidizing velocity, bed height and absolute bed temperature. These parameters have been correlated by:[2]

$$(1-E) = K((1+X)^{-3} t_g^{\,1.69} ((1300 - T_b)/T_b)^2) \quad (2)$$

where

E = the fractional combustion efficiency
K = a parameter that mainly depends on fuel characteristics ranging from 12 for high volatile fuels to 27 for low volatile fuels
X = the fractional excess air
t_g = the in-bed gas residence time, s
T_b = the absolute bed temperature, K

As this correlation shows, increasing in-bed gas residence time also improves combustion efficiency. Increasing process pressure reduces fluidizing velocity, increases allowable design bed height and corresponding in-bed gas residence time. Therefore, combustion efficiency in PFBC is improved over bubbling atmospheric fluidized beds.

Emissions

Nitrogen oxides The principle factors affecting emissions of nitrogen oxides (NO_x) from a fluidized bed are the nitrogen contained in the coal and the excess air level. The effect of these variables on NO_x (based on 3% oxygen at typical bed temperatures) has been correlated by:[2]

$$NO_x = 20.5\, N(O_2 + 0.5) \text{ in ppmv} \quad (3)$$

or

$$NO_x = 0.028\, N(O_2 + 0.5) \text{ in lb}/10^6 \text{Btu} \quad (4)$$

where

N = nitrogen content in the coal, by weight dry as-fired
O_2 = oxygen content of the dry gas, mole %

As can be seen, both the nitrogen content in the coal and the excess air affect NO_x production. However, due to the low combustion temperature, thermal NO_x production is minimal. Pressure has a negligible effect.

Sulfur capture Sulfur capture is a function of the calcium to sulfur molar ratio (Ca/S), in-bed gas residence time, particle size, bed temperature and sorbent reactivity. These parameters have been correlated by:[2]

$$R = 100\,(1 - \exp(-mC)) \quad (5)$$

and

$$m = A''\,(t_g/d)^{1/2} \exp(-n/T_b) \quad (6)$$

where

R = the sulfur capture, %
C = the calcium to sulfur molar ratio (Ca/S)
t_g = the in-bed gas residence time, s
d = a particle size parameter (surface mean diameter of sorbent), mm
n = a constant related to the apparent activation energy for the sulfur capture reactions, K
T_b = the absolute bed temperature, K
A'' = the sorbent reactivity index, $(mm/s)^{1/2}$

For given bed operating conditions and sorbent, the in-bed gas residence time is the key parameter. Again, as process pressure increases, gas velocities decrease and design bed depths increase resulting in four to six times the in-bed gas residence time available in atmospheric bubbling fluidized beds, improving sulfur capture appreciably.

Steam-water cycle

A boiler forced circulation schematic is shown in Fig. 7. In the steam-water cycle, water from the condensate polishing system is warmed in the feedwater heaters and forced by the boiler feed pumps into the economizer or the boiler inlet if a heat recovery steam generator is used after the gas turbine. From the economizer, the subcooled water enters the boiler. After passing through the bed and freeboard enclosures, it enters the in-bed evaporator surface where boiling occurs. Fluid is conveyed to the steam-water separator. The evaporator outlet fluid condition is a two-phase saturated mixture below 40% load and is slightly superheated at or above 40% load. At lower loads, the water from the separator is recirculated through the boiling surface with pump assistance and at higher loads the recirculation ceases. The steam from the separator then enters the in-bed primary superheater. From the primary superheater outlet, spray attemperation is used to control final steam temperature leaving the in-bed secondary superheater to the turbine. Main steam pressure is controlled by the steam turbine throttle valves.

When reheat is required, the surface must be placed in the bed in a manner which produces the desired heat absorption characteristic with load while minimizing pressure drop. Because reheat flow does not begin until the steam turbine is rolled, the reheat surface is in the upper portion of the bed to prevent excessive metal temperatures and minimize the effects of the slumped bed condition. Reheat temperature is controlled by attemperation at the inlet.

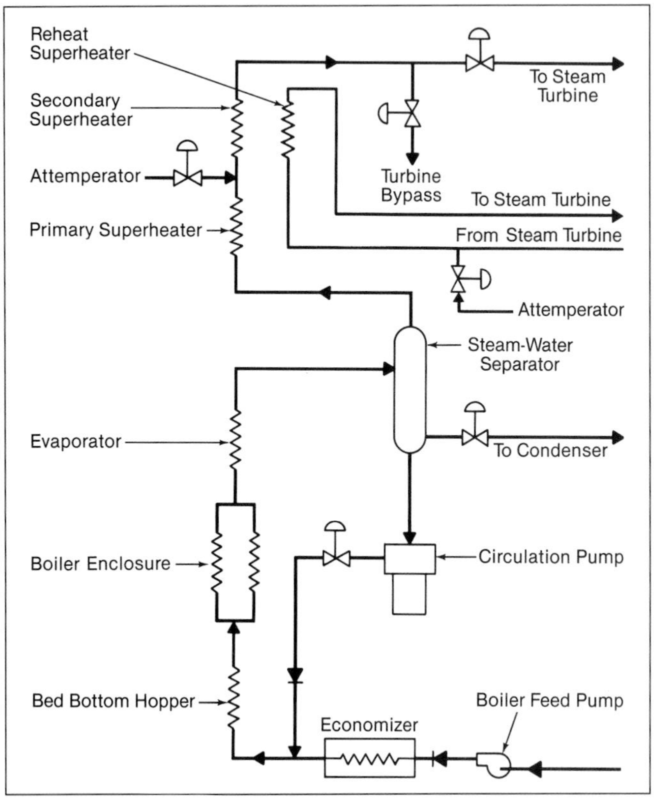

Fig. 7 Typical steam-water forced circulation system schematic.

During startup and in the event of a steam turbine trip, a turbine bypass system to the condenser disposes of the steam while controlling the boiler pressure/temperature decay and conserving as much of the treated water as practical. In the event of loss of plant power or the boiler feed pumps during operation, a backup feedwater system is also needed to maintain water flow to the boiler circuits exposed to the heat contained in the slumped bed.

Air-gas cycle

In the air-gas cycle at full load, ambient air enters the gas turbine, passes through the low pressure compressor and is cooled by an intercooler. The intercooler controls the final temperature of the air sent to the combustor vessel. After the intercooler, the air is further pressurized to the design pressure somewhere between 170 and 230 psi (1200 and 1600 kPa) and heated to about 600F (316C) in the high pressure compressor. The hot compressed air is then directed through the outer annulus of a coaxial pipe into the combustor vessel. The air from the top of the pressure vessel proceeds downward through ash coolers, which recover heat from the ash to the air prior to entering the air distribution system. From the air distributor, the air enters the fluidized bed where the combustion takes place.

The gases from the bed pass through the boiler freeboard and into the hot gas cleaning equipment. From the hot gas cleaning units, the cleaned hot gases flow through the inner annulus of the coaxial pipe to the gas turbine. The pressurized hot gases are expanded through the high pressure and low pressure turbines providing the energy to drive the compressors and produce about 20% of the plant electrical output. After the gas turbine, the gases flow through an economizer, or heat recovery unit, where they are further cooled. If conventional cyclones are used for hot gas cleaning, the gases then pass through a precipitator or fabric filter and are finally discharged to the stack.

Energy losses

Because of the integration of several components and the high gas temperature leaving the boiler, conventional boiler efficiency is a misleading indication of the efficiency of the PFBC system. To adequately understand the energy and mass balance, the entire contents of the combustor vessel and the gas turbine should be considered. In addition, the bed energy balance is of interest because the calcination and sulfation processes affect the in-bed energy.

The bed energy balance considers entering solids, fuel and sorbent, the calcination and sulfation heats of reaction and unburned carbon losses. Both the entering solids and the net result of the calcination and sulfation reactions add heat to the stream; therefore, the loss is attributed to unburned carbon. Assuming the unburned carbon loss of 1%, bed efficiency is generally more than 99%.

If the combustor vessel is considered the thermodynamic boundary, losses include the bed ash, ash from the hot gas cleaning units, unburned carbon and radiation to the plant from the outer surface of the vessel. Considering these losses, efficiencies more than 98% are typical.

When the entire PFBC system is considered, the additional losses are to the stack and external coolers for nonworking fluids. Efficiencies of the PFBC system are in the 80 to 85% range. Considering the fact that 20% of the plant's electrical power generation has already been produced by the gas turbine, 80 to 85% compares well with conventional boiler efficiencies of 85 to 89%.

PFBC systems

Fig. 8 shows the PFBC system schematic for the Tidd PFBC demonstration project. The PFBC system consists of a gas turbine, a boiler and associated systems, hot gas cleaning systems, a load control system, fuel preparation and feeding systems, sorbent preparation and feeding systems and ash removal systems. Many of these systems are either partially or totally contained in a pressure vessel.

Major components

The gas turbine compressor capability sets the air flow, pressure and temperature which in turn determine the quantity of fuel which can be fired at the specified excess air level, usually about 20%. The fuel is burned in a fluidized bed contained within the boiler enclosure. The heat transfer surfaces are located in the bed and are completely immersed at full load. In the combined cycle configuration, this maintains the temperature of the gas sent to the gas turbine near the bed temperature while removing heat to control bed temperature. For turbocharged boilers, convective surfaces are located above the bed to reduce the exit gas temperature range to between 800 and 1000F (427 and 538C) before entering the gas turbine.

The boiler is the central component in the PFBC system because it must be designed to operate compatibly with both the steam and gas cycles throughout the load range. As the firing rate is reduced, the heat absorbed by the steam cycle should be reduced proportionally and the bed level should be maintained as high as practical to maximize in-bed gas residence time while minimizing the cooling of the gas. In the combined cycle system, the gas temperature must be maintained as high as possible. These constraints can only be satisfied by careful placement of the tubes within the bed. In-bed tube placement is further complicated by fluid dynamic considerations to provide even cooling of the bed and good mixing of the bed material while minimizing erosion potential. Therefore, both the enclosure design and the design of the in-bed tubing must meet several unique criteria, as well as conventional steam side and fluid bed considerations.

Steam generator

Boiler type A PFBC boiler must produce steam at the desired conditions and provide sufficient controllability to accommodate the steam turbine and the electrical grid. Simultaneously, it must provide the exhaust gas conditions corresponding to the power required by the gas turbine while containing and regulating the fluid bed combustion process throughout the load range. It must also be designed to withstand the abnormal conditions resulting from equipment trips and plant power loss.

The method of steam-water circulation for a PFBC boiler must be carefully considered. Natural circulation, pump assisted drum-type and once-through methods have been developed, each having advantages and disadvantages. Natural and pump assisted circulation have advantages at low loads while the once-through design is best at high loads. Therefore, the ideal boiler for PFBC would incorporate some aspects of each.

The once-through circulation design has the advantage of converting heating surface from evaporating to superheating duty. This can be used to balance the process requirements with load turndown (Fig. 9a). The surface

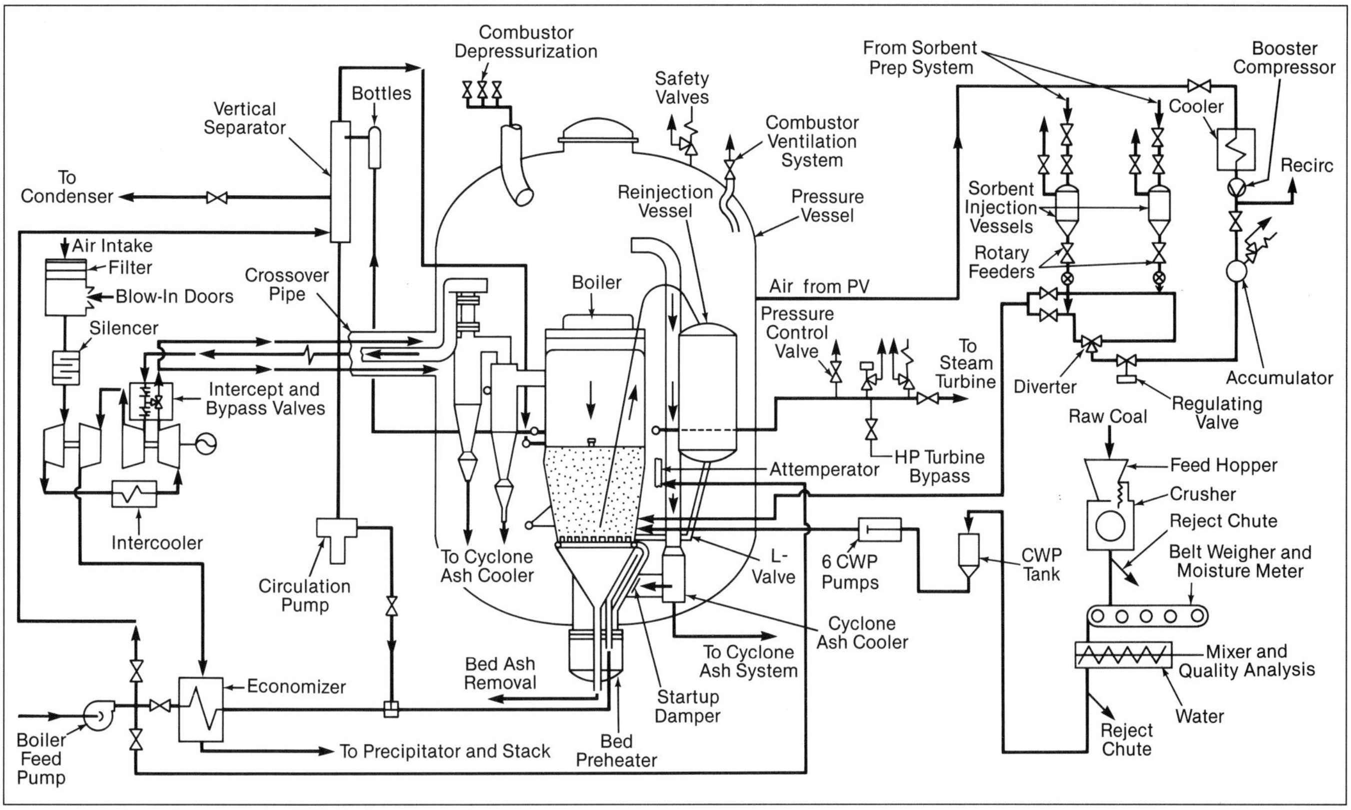

Fig. 8 PFBC system schematic (Tidd project).

can be placed so that as bed level is reduced for turndown, the conversion of surface to superheating duty increases the tube metal temperatures and less heat is removed from the gas before it enters the gas turbine. Because these tubes are not in the bed where heat transfer rates are high, the metal temperatures are still not excessive. Further, the simple operation of only the required boiler feed pump improves efficiency, maintenance, reliability and cost.

Since PFBC boilers are significantly more compact than conventional boilers, it is difficult to achieve the drum elevation necessary to provide adequate pumping head needed for natural circulation (Fig. 9b). Vertical tube bundles needed to produce adequate mass velocities are more difficult to arrange within the bed height at the various loads. If horizontal tube bundles are used, these circuits would normally be pump assisted (Fig. 9c). Pump assistance can also reduce the required drum elevation.

In a combined cycle configuration, when the bed height is reduced and tube surface is exposed above the bed level, the in-bed water-cooled surface in a natural circulation or pump assisted boiler will decrease the gas temperature more than the surface of a once-through design.

B&W's design is a once-through boiler with a vertical steam separator and circulation pump for low load operation. This design operates like a drum boiler at low loads, below about 40%, where a steam-water mixture is leaving the in-bed evaporator surface. The vertical steam separator removes the water from the steam, which flows to the primary and secondary superheaters. The water is circulated back through the enclosure and in-bed evaporator surface at a relatively high circulation rate. Once the load is raised to achieve slightly superheated conditions at the evaporator outlet, the vertical separator can be bypassed and the circulation pump shut down. The boiler will now operate in a true once-through mode. This design is suitable for both subcritical and supercritical steam cycles where pressure would be raised above the critical pressure using the turbine throttle valve.

Boiler enclosure B&W's boiler enclosure (Fig. 10) is top-supported from rods while a conventional buckstay system supports the walls against the relatively high pressure difference between the air and gas sides. Conventional, pulverized coal-fired boilers generally operate in a balanced draft mode but their enclosures are designed for a 30 to 40 in. wg (7 to 10 kPa) differential. Atmospheric fluid bed boilers may have as high as 50 in. wg (13 kPa) pressure difference across the enclosure walls. However, the pressure difference across the enclosure wall for PFBC is 7 psi (50 kPa) or more at full load. This significantly higher pressure difference requires closer spaced buckstays. However, since the enclosure is compact, the buckstay spans are greatly reduced resulting in the ability to use conventional methods for supporting the walls.

The higher potential pressure differences possible due to a tube leak or too rapid depressurization of the combustor vessel affect the design of buckstays and supports. A system is also required to protect the enclosure by opening a flow path between the freeboard to the combustor vessel if the pressure exceeds the prescribed limit.

The enclosure consists of three distinct sections: the boiler bottom, the bed enclosure and the freeboard enclosure, each having unique functional and mechanical design considerations. Unlike conventional boilers, the enclosure is not externally insulated since the air temperature inside the combustor vessel nearly matches the enclosure temperature at higher loads. The freeboard is internally insulated and lined to maintain boiler exit gas temperature.

The *boiler bottom* supports the air distribution system and collects the bed ash, which is cooled by combustion air and conveyed out of the combustor vessel. It is constructed of membraned waterwalls arranged into inverted pyramidal hoppers which fill with bed ash during operation. An air inlet duct, which contains the air preheating system for startup, conveys the air to the windbox during normal operation.

The *bed enclosure*, where the combustion process and the majority of the heat transfer to the steam cycle take place, contains the fluidized bed and the in-bed heat transfer surfaces. Its membraned waterwalls are tapered

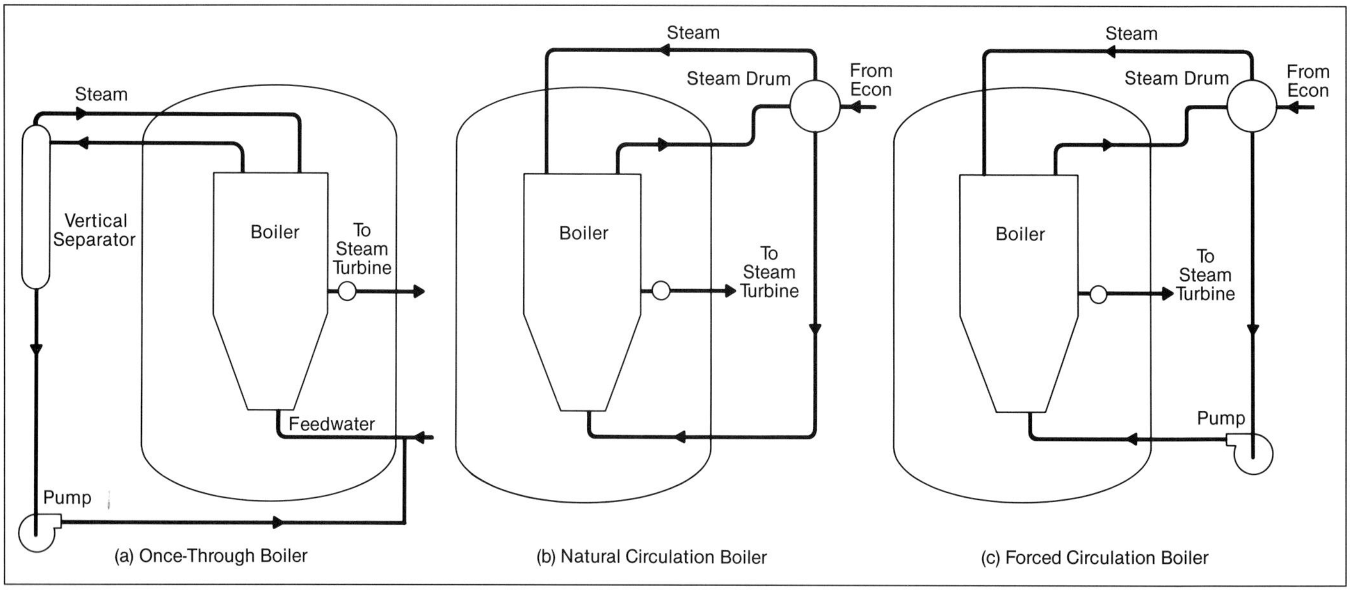

Fig. 9 PFBC boiler circulation systems.

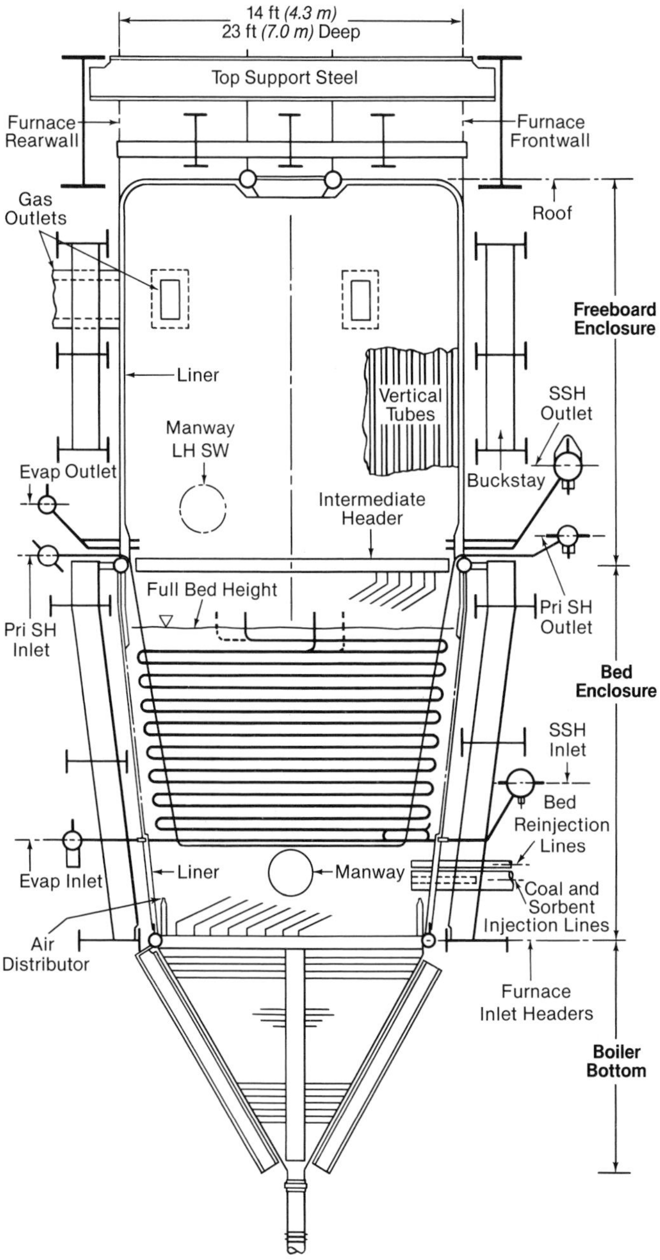

Fig. 10 Boiler for the Tidd project.

to enhance the process by increasing the plan area as bed level increases. This allows somewhat higher bed levels at lower loads and improves material flow. The lower enclosure is spiral wound resulting in fewer circuits and high mass flows per tube required to withstand the high heat transfer rates in the fluidized bed without requiring internal refractory or insulation. Though no internal insulation is used in the zone containing the tube bundle, a short section below it is internally insulated to minimize startup bed preheating requirements. This lower spiral wound portion of the enclosure is supported vertically by straps and vertically oriented buckstays. The loads from the vertical buckstays are collected in horizontally oriented members which are connected at the boiler corners forming a self-equilibrating system.

The lower, spiral wound enclosure steam-water circuits empty into an intermediate header which distributes the flow to the upper enclosure walls. This intermediate header also acts as the main device supporting the tube bundle loads from the primary superheater support tubes and transferring them to the enclosure walls.

The *freeboard enclosure* is also of membraned waterwall construction with vertically or spirally oriented tubes. It is internally insulated and lined to minimize heat absorption above the bed while maintaining the gas temperature as close to the bed temperature as practical at full load. The freeboard height is set based on desired gas residence time and tube bundle maintenance considerations. To allow tube bundle maintenance, sufficient space is provided in the freeboard to lift the tube bundles for repair, and a removable panel or door is located in the roof. The buckstay system supporting these walls is of conventional design using horizontal members.

Tube bundle The in-bed tube bundle is supported from the bed enclosure. It consists of evaporator plus the primary, secondary, and reheat superheater surfaces. A water attemperator is placed in the piping between the primary and secondary and reheat superheaters. The tubing and supports are arranged to achieve the spacing and density desired to promote good mixing in the bed and minimize erosion potential. The tube bundle is designed to resist the dynamic forces of the bed material and absorb the correct amount of heat from the bed at a given bed level while providing a boiler exit temperature which is compatible with the gas turbine at that load.

The tube bundle consists of several platens. Each platen is composed of two or more serpentine tube circuits of evaporator or superheater supported from a U-shaped primary superheater tube. This design allows for support of the platens by the primary surface through the enclosure wall just above the full load bed height. Support tubes in the freeboard are undesirable because they complicate tube bundle maintenance, reduce the gas temperature and potentially increase gas side pressure drop.

The platens are designed to require the minimum number of tube cuts for lifting into the boiler freeboard or for removal from the combustor vessel for servicing.

Backup feedwater system In the event of a gas turbine trip and loss of feedwater, a large mass of slumped bed material at high temperature surrounds the tube bundle. The heat must be removed to prevent failure of low alloy components designed for low temperatures. Heat rejection is provided by the once-through bypass system. The injection tank storage capacity is used for reserve feedwater requirements. Even in conventional pump assisted circulation boilers, the heat rejection from the water and pressure parts is limited by the convection rate obtained from air cooling of the furnace. For the PFBC, the circulation of water through external coolers is carried out to provide protection of the metals during trip conditions for a reasonable period of time. Makeup water to replace vented steam and water shrinkage on cool down must be provided until the bed temperature is safe for the tube metals.

Air distributor Two different design principles for air distribution in fluidized beds are presently in use. One is the windbox design where the boiler bottom is an integral floor, usually water-cooled, from which the air nozzles protrude. The other is the open bottom with sparge ducts distributing the air to fluidizing nozzles.

The water-cooled, bubble cap floor design is better able to endure the severe temperature changes imposed by the startup preheater and is lower cost since sparge ducts are constructed of stainless steel. Both designs provide the possibility of cooling the bed ash as it leaves the boiler. Cooling the bed ash before it leaves the boiler bottom also means that the penetrations in the combustor vessel wall for the bed ash ducts are not subject to high temperatures and can be a simpler design.

Gas turbine

The gas turbine compressor replaces the fans and air heater (heat of compression) in a conventional plant to supply the hot, pressurized air for combustion and to fluidize the bed. However, unlike fans which are independent of the combustion process, the gas turbine is powered by the exhaust gases from the boiler. The energy remaining after the air is compressed is used to generate electricity. Because the gas turbine is driven by the hot pressurized gases from the boiler and simultaneously supplies the combustion air to the boiler and generates electricity, certain characteristics are desirable to efficiently accommodate the subsequent unconventional conditions it must meet.

A gas turbine ideally suited for PFBC throughout the load range would accept the relatively low inlet gas temperatures of about 1550F (843C) associated with the fluid bed process (assuming no topping cycle) and would not be significantly affected by changes in ambient air conditions. It would provide a volumetric flow characteristic which would permit nearly constant fluidizing velocity, excess air ratio and velocity into the hot gas cleaning equipment (important for cyclones). It would also balance the opposing desires for a low air flow to the boiler, a high air flow to the gas turbine at low load, and withstand some particulate loading in the gases without significant damage.

Gas turbines are presently available with a single shaft and multiple shafts. The single shaft machine is capable of constant speed for both the turbine and compressor while multiple shaft machines allow the high pressure turbine stages to be coupled with the generator and sometimes a portion of the compressor rotating at constant speed. The low pressure stages drive either all or a portion of the compressor and are free to run at variable speeds. When the compressor is divided into a constant speed high pressure portion and a variable speed low pressure portion, an intercooler can be added to control the air temperature leaving the compressor and improve the high pressure compressor efficiency.

The turbine installed at Tidd and those being designed for scaleup to 350 MW have been modified to suit the PFBC process (Fig. 11). These turbines result in PFBC plants which produce 80 and 350 MW respectively. They are both in-line two shaft machines. On one shaft, the variable speed, low pressure turbine is coupled to the low pressure compressor. On the other shaft, the high pressure turbine drives both the compressor to which it is mechanically coupled and the electric generator. There is also an intercooler between the low and high pressure compressors. The turbine internal design is also more rugged to minimize the effects of the small amount of fine particulate passing through the turbine. The two shaft design has significant operating advantages. The free spinning, low pressure turbine can accommodate the reduced gas temperature and resulting reduction in air flow as load is reduced while maintaining constant speed

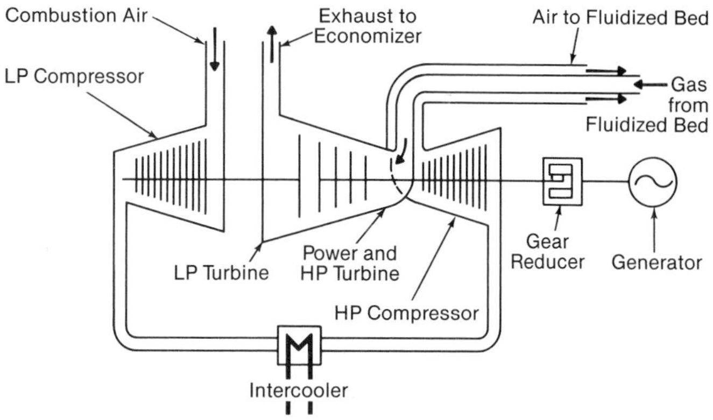

Fig. 11 Gas turbine for PFBC.

at the generator. The intercooler between the low and high pressure compressors controls the air temperature to the combustor vessel allowing use of conventional materials. The generator can be used as a motor to start up the turbine, which provides air to fire the oil preheater and ignite coal. Once firing begins, the gas temperature entering the turbine begins to increase and power requirements for the motor decrease. As the load continues to be raised, the turbine will become self-sustaining (zero power required or produced). Passing through that point, the motor will become a generator and begin to produce power up to full load.

Auxiliary equipment

PFBC requires several auxiliary systems. The most significant are: combustor vessel, fuel preparation and feed, sorbent feed, ash removal, bed material re-injection for load control, hot gas cleaning, economizer and the balance of plant equipment.

The three primary components in a PFBC plant are the gas turbine, the steam turbine and the boiler. Though the other systems support one of the primary components, they must be carefully designed since their energy losses can significantly impact plant efficiency and they are often major maintenance concerns.

Combustor vessel For the process to take place under pressure, the fluidized bed must be contained in a pressure boundary. In addition, it is expedient to maintain several of the other systems at the high pressure. By locating equipment inside the pressure boundary, it can be designed for the much lower air to gas side pressure differences. Any heat losses from the equipment are then recovered to the combustion air. Individual components can be enclosed in separate pressure vessels or the major components can be grouped together within a single large combustor vessel. Though the individual vessel approach lends itself to several smaller shippable modules, the single large vessel approach can also be shipped as a module for the smaller PFBC sizes while the larger vessels, though shipped in relatively large sections, must be assembled on site.

For the Tidd PFBC demonstration project, the boiler, re-injection system, cyclones and cyclone ash coolers were arranged within a 44 ft (13.4 m) diameter, 68 ft (20.7 m) high cylindrical combustor vessel. The vessel is 2.875 in. (73 mm) thick carbon steel with a total weight of 1340 t (1215 t_m) including most of the contents and is barge transportable.

The vessel for the 350 MW plant, which contains a similar scope of equipment, is about 64 ft (19.5 m) in diameter and 150 ft (46 m) high. It is made of about 3.75 in. (95.3 mm) thick carbon steel.

Fuel feed systems The fuel preparation and feed system must convey coal from the coal yard into the boiler in a safe manner while overcoming the high process pressure. The most common injection methods are pneumatic and coal-water mixtures. The pneumatic methods require a means of inerting the fuel and a source of boosted air for conveying. Erosion may be a problem. Coal-water mixtures are inherently inert and slippery. Assuming the mixture can be pumped and the moisture content is less than about 30% by weight, the moisture will have little impact on efficiency since about one third of the energy used for vaporization can be recovered in the gas turbine. Coal-water mixtures have also been shown to burn more evenly, therefore producing more uniform bed temperatures and requiring significantly fewer feed points than pneumatic systems.

The optimum system design depends on the ash and sulfur content of the coal. For fuels with low ash content, coal-water mixtures are advantageous. The quantity of water needed for fuels with high ash content can adversely affect efficiency and pneumatic feed becomes more attractive.

A coal-water mixture using 25% water, by weight, has been called a coal-water paste (CWP). The Tidd plant uses CWP for fuel feed. A crusher is used to reduce the raw 0.75 in. (19.1 mm) top size coal to the proper combination of large and fine material to produce a pumpable paste. From the crusher, coal and water are brought together in a controlled manner into a mixer. When sufficiently mixed, the CWP is stored in a continuously agitated surge hopper before being fed into the boiler by hydraulic dual piston fuel injection pumps, similar to those used in the concrete industry. Each pump feeds its own nozzle and compressed air is injected near its tip to properly break up the paste stream as it enters the bed.

Sorbent feed systems Because sorbents are not combustible, the greatest challenge in system design is overcoming the process pressure. Generally, pneumatic feed systems have been used. However, in some cases where the coal ash and sulfur content is low and relatively low sorbent feed rates are required, it has been combined with the coal in a CWP to simplify the feed systems.

In a pneumatic feed system, sorbent is crushed to 0.125 in. (3.18 mm) top size, dried and fed into a lockhopper system. Two types of pneumatic conveying systems have been used for fluidized-bed boilers, dense phase and dilute phase. The most common is the dilute phase. Dense phase systems require conveying air pressures which are significantly higher than the process pressure but use less air flow. Much lower velocities also minimize erosion concerns. Dilute phase systems require lower pressures relative to the bed pressure but a significant air flow.

Two types of dilute phase systems are generally used, intermittent and continuous feed systems. In intermittent feed systems, the lockhoppers are sequenced so that as one fills, the other is pressurized and is feeding material into the boiler through a transport piping system. During the switch from one vessel to the other, there is a brief interruption in sorbent flow to the bed. In continuous feed systems, the main feed vessel is maintained at process pressure and a smaller vessel, usually located directly above the main feed vessel, is used to fill, pressurize, empty into the main feed vessel, depressurize and refill. Since there are no interruptions in sorbent flow to the bed, the continuous feed system results in the least upsets to the bed process and is simpler to maintain. Air for the sorbent transport system is taken from the combustor vessel and boosted by a compressor.

Bed re-injection system Load is controlled by changing bed level through a bed material re-injection system. This system pneumatically conveys material from the boiler during load reductions, stores it in vessels located inside the combustor vessel and re-injects it into the boiler through an L-valve when an increase in load is desired.

Gas cleaning system In PFBC, the dust-laden hot gases leaving the boiler must be cleaned before entering the gas turbine. Though several methods of hot gas cleaning are in the experimental stage, high efficiency cyclones have been successfully tested. Though the cyclones will clean the gas sufficiently for the gas turbine, further cleaning by an electrostatic precipitator or a fabric filter is necessary to meet New Source Performance Standards (NSPS). Though cyclones may be adequate for first generation PFBC plants, they alone will be inadequate for the advanced plant designs. If any coal ash, which remains in the PFBC flue gases, is exposed to temperatures above the fusion temperature, problems could result in the gas turbine. If a reliable method could be developed to clean the hot gases efficiently, it would not only eliminate the precipitator or fabric filter in first generation PFBC plants, but would also pave the way for higher gas turbine inlet temperatures in advanced PFBC plants. Consequently, hot gas cleaning has become an important technology to PFBC.

Several hot gas cleaning technologies are being developed in parallel with PFBC. These include candle and ceramic tube filters as well as granular bed and crossflow designs. These designs are being developed and tested at several locations. Many hours of testing have been conducted on candle filter designs but most designs have had failures associated with thermal cycling and the tensile stresses induced into the candles during the back pulse cleaning (Fig. 12). Granular bed filters have proven to be quite large and ceramic crossflow designs have experienced failures in the elements.

Because of PFBC, B&W has been interested in hot gas cleaning. As a result, B&W obtained a license to design and supply the ceramic tube filter design developed by Asahi Glass Co. of Japan. This filter is somewhat similar to the ceramic candle filter design except the clean gases are on the outside instead of the inside of the tubes (Fig. 13). This means that the stresses induced in the tubes by back pulse cleaning are in the stronger, compressive direction.

Since all high temperature gas filtration systems are still developmental, one or more of these designs may prove applicable to PFBC.

Ash removal systems Two ash removal systems are used in PFBC systems, one for the ash from the gas cleaning system and one for bed ash. The ash from the gas cleaning system is continuously removed and conveyed from the combustor vessel to an atmospheric silo. To overcome the difficulty of handling the 1580F (860C) ash and the need to reduce it to atmospheric pressure, it is usu-

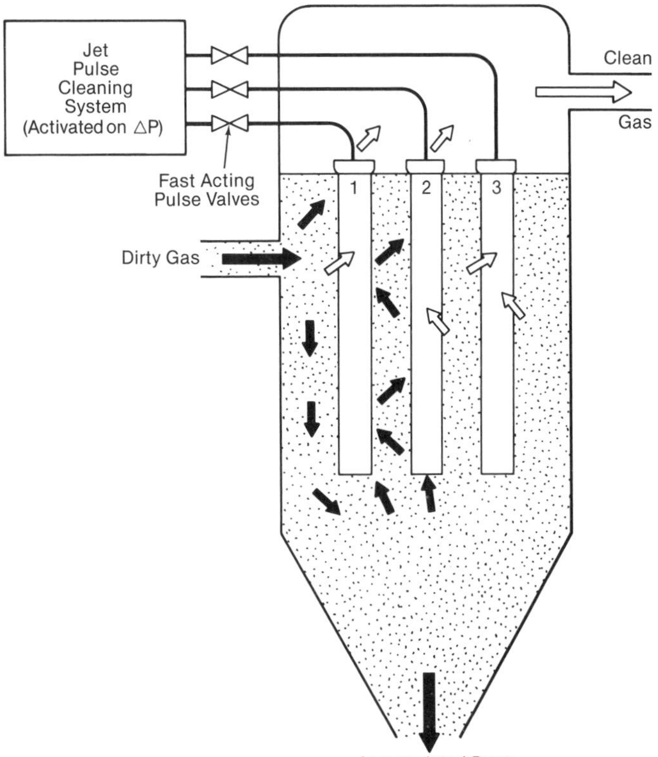

Fig. 12 Typical candle filter for high temperature gas filtration.

ally cooled and depressurized to an acceptable level for handling before leaving the combustor vessel and being transported to storage silos.

The bed ash, collected and cooled in the boiler bottom and removed at a rate coordinated with fuel feed rates to maintain bed level, is generally metered from the boiler bottom into a lockhopper system. It is either reduced to atmospheric pressure and transported by conveyors, or it is transported by a dense phase pneumatic system to a silo.

PFBC operation

General control principles

Due to the close coupling between the boiler and gas turbine, control of a combined cycle PFBC plant is somewhat more complicated than a conventional PC-fired plant. The controls must regulate the fluid bed process while maintaining compatibility with the steam turbine and the gas turbine.

The process is controlled, basically, by changing bed level. As bed level is increased, load increases. The introduction of additional bed material reduces bed temperature. Because the firing rate is regulated based on bed temperature, fuel feed will increase to maintain bed temperature. At lower loads, sufficient excess air is present to support the increased combustion. As bed level is increased at essentially constant bed temperature, more of the heat transfer surfaces are submerged. The increase in heat absorption calls for additional feedwater flow and steam production increases. The increase in water and steam flow also tends to decrease bed temperature, but the firing rate will compensate accordingly.

As the steam flow and power production from the steam turbine increase, so does the power to the gas turbine. This is because the amount of heat transfer surface above the bed decreases and the gas temperature leaving the boiler increases. This additional power to the gas turbine causes the turbine, which produces a constant air volume flow, to increase the air pressure and corresponding air mass flow to the fluidized bed. In addition to increasing power to the compressor, power also increases to the high pressure turbine coupled with the motor/generator. The result is either a reduction of motor power requirements, up to about 25% load, or electrical generation between 25 and 100% load. This process continues until full load conditions are achieved. Full load is set by the maximum air flow the gas turbine can produce and the minimum excess air allowed.

Startup

To start the process, a shallow bed is fluidized and heated by the oil-fired preheater. During this stage, a portion of the air flow is temporarily diverted through the preheater where fuel oil is burned at high excess air ratios. The hot gases are directed into the startup bed through the air distribution system. When the bed material has reached the required ignition temperature, CWP is injected and the preheater is ramped off.

During startup, the gas turbine is driven by the generator which is used as an electric motor to produce the required air flow. As load increases and power to the turbine increases, a decreasing amount of power is used from the

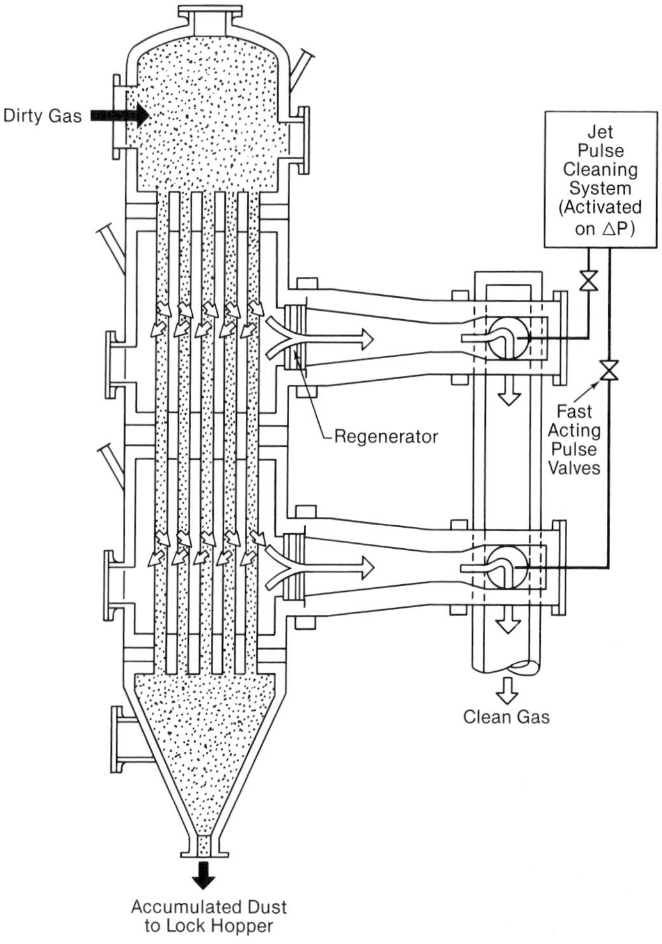

Fig. 13 Advanced ceramic tube filter for high temperature gas filtration.

electrical grid until the turbine becomes self sustaining. Above that point, the gas turbine generates electrical power.

For a once-through boiler, the vertical separator and circulating pump are used up to about 40% load. Above 40% load, the boiler becomes once-through and the pump is shut down. Feedwater flow is established at 5 to 10% of full load flow and the pump recirculates water through the boiler enclosure and evaporator to maintain minimum cooling flow. Until the steaming rate reaches about 40% flow, the separator drains will remove excess water to the condenser. While the boiler circulating pump is operating, feedwater flow is regulated to maintain separator level. Once it is shut down, feedwater flow is regulated to control steam temperature; a minimum water level is maintained to permit a pump restart if necessary.

Until sufficient steam flow, temperature and pressure are available to roll the steam turbine, the pressure ramp is controlled by the bypass system. Once the turbine is rolled, the turbine throttle valve regulates pressure.

Normal operation and load changing

Due to the unique synergistic relationship between the boiler and the gas turbine, the steam cycle will normally be operated in a turbine-following mode. Load changing of up to about 2% per minute (4% per minute can be provided) is accomplished by changing the bed level. This will expose more or less of the in-bed surface to the lower convective heat transfer rates above the bed. This will simultaneously reduce heat input to the steam cycle and reduce the gas temperature to the gas turbine. The reduction in power to the gas turbine results in a corresponding reduction in air flow to the combustor. The firing rate is adjusted to maintain the bed temperature at a constant 1580F (860C) at a specified excess air ratio. Feedwater flow is regulated to protect the heat transfer surface, absorb the proper amount of heat from the bed and, along with attemperation, control steam temperature. Steam pressure is maintained by modulating the steam turbine throttle valve.

Shutdown

Shutdown is accomplished by reducing the bed to the minimum level, stopping the fuel and allowing the gas turbine to coast down. This cools the remaining bed material to prevent combustion. At the appropriate time, the intercept/bypass valve is closed and the gas turbine is tripped.

Abnormal conditions

In the event of a gas turbine trip, the combustor will be immediately isolated from the gas turbine by an intercept/bypass valve at the turbine inlet. The loss of air flow will cause the hot bed to slump. Sufficient feedwater flow will be maintained to protect the in-bed surface. In addition, special systems have been designed to depressurize the combustor vessel, prevent generation of combustible gases, cool and remove the bed material in preparation for a restart.

A steam turbine trip will result in bypassing the steam flow to the condenser while venting the remainder to the atmosphere using a pressure control valve to regulate steam pressure. Heat input is rapidly reduced and the unit is shut down in the normal manner.

A station blackout is accommodated similarly to a gas turbine trip with all essential service valves and equipment on standby power. If the feedwater pumps are not steam driven, or if they are not on standby power, a special backup feedwater system is needed to supply the necessary water flow to protect the heat transfer surface at compatible conditions. This system is also needed in the event of a loss of the main feedwater pumps.

References

1. Clark, R.K., *et al.*, "Comparative performance of the Grimethorpe PFBC combustor when fed with dry coal and coal water mixtures," 10th International Fluidized Bed Combustion Conference, San Francisco, April 30-May 3, 1989.

2. Hoy, H.R., *et al.*, "Part-load operation of pressurized fluidized bed combustors: a review," 9th International Fluidized Bed Combustion Conference, Boston, May 3-7, 1987.

Bibliography

Boersma, D., *et al.*, "NO_x emissions from pressurized fluidized bed combustion," 10th International Fluidized Bed Combustion Conference, San Francisco, April 30-May 3, 1989.

Geldart, D., *Gas Fluidization Technology*, Wiley, New York, 1986.

Kinsinger, F.L., "Pressurized fluidized bed combustion — utility boiler in a pressure vessel," *National Board Bulletin*, National Board of Boiler and Pressure Vessel Inspectors, Vol. 45, No. 8, April 1989.

Kinsinger, F.L., and McDonald, D.K., "Combined cycle using PFBC," American Power Conference, Chicago, 1988.

McDonald, D.K., and Weitzel, P. S., "The PFBC boiler for AEP's Tidd Plant; design for performance and operation," 1988 EPRI Seminar on Fluidized Bed Combustion Technology for Utility Applications, Palo Alto, California, May 3-5, 1988.

Minchener, A.J., *et al.*, "NO_x studies on the PFBC facility at the British Coal Research Establishment," 10th International Fluidized Bed Combustion Conference, San Francisco, April 30-May 3, 1989.

Mudd, M., "Status of AEP's Tidd PFBC demonstration plant," 10th International Fluidized Bed Combustion Conference, San Francisco, April 30-May 3, 1989.

Podolski, W.F., *et al.*, *Pressurized Fluidized Bed Combustion Technology*, Noyes Data, Park Ridge, New Jersey, 1983.

Stringer, J., EPRI, "Current information on metal wastage in fluidized bed combustors," 9th International Fluidized Bed Combustion Conference, Boston, May 3-7, 1987.

Weitzel, P.S., *et al.*, "The use of advanced ceramics for hot gas clean up," 1990 AIChE Annual Meeting, Chicago, November 1990.

Naval steam power then and now: U.S.S. *Missouri* (top) and U.S.S. *John F. Kennedy*.

Chapter 30
Marine Applications

The ingenuity of many inventors and engineers has been devoted to the application of fossil fuel-fired steam boilers and engines to ship propulsion. The need for higher power, lower weight and smaller volume designs for ships drove many of the boiler system advances which ultimately appeared in stationary boilers designed and used today.

The primary propulsion systems in present day ships have, however, progressed to diesel engines, gas turbines and even nuclear power systems in naval applications. (See Chapter 40.) In spite of this, there remain requirements aboard new ships to produce steam for a variety of needs. In addition, a large existing population of modern ships, such as the Lash Class container ships (Fig. 1), continues to use steam boilers as their primary source of power. Finally, increasing fuel costs have led to investigations of burning coal-water mixtures, coal-oil mixtures and other fuels in steam based systems for potential future use.

Babcock & Wilcox (B&W) marine boilers have established a reputation for dependability and efficiency. The history of these boilers reflects sound principles of design and fabrication. Since the first installation in the S.S. *Monroe* in 1875, B&W boilers have been installed in more than 4,000 ships of the United States (U.S.) Navy and the Merchant Marine.

During World War I, B&W boilers were installed in 50 mine sweepers, 100 destroyers, and 500 Shipping Board vessels. During World War II, B&W boilers were furnished to the Navy for all but two battleships (such as the U.S.S. *Missouri*, shown in the chapter frontispiece), all the cruisers, all aircraft carriers (see U.S.S. *John F.*

Fig. 1 S.S. *President Tyler*, Lash Class container ship (*courtesy of American President Companies, Ltd.*).

Kennedy in the chapter frontispiece), 90% of the destroyers, 33% of the destroyer escorts and numerous miscellaneous small craft.

General design considerations

Steam generating equipment for marine service is designed in accordance with the principles and considerations that apply to stationary units, with modifications to meet specific requirements for operation at sea. The boiler must fit within a minimum of engine room space, yet be accessible for operation, inspection and maintenance. Although light in weight, it must be sufficiently rugged to operate dependably under adverse sea conditions to absorb vibration and forces resulting from rolling and pitching in heavy seas or shocks which may result from accidental causes such as groundings and collisions. Special steam drum design requirements must be met to accommodate *roll* of the ship from side to side, *pitch* variations from bow up to bow down, permanent *list* to either side and permanent *trim* to either bow or stern. In the case of naval vessels, the shock effects of the detonation of explosives must also be considered. Operation over a wide load range, with operating characteristics compatible with a high degree of automation, is also required. Finally, the factors used in both the thermal and structural design must be conservative so that continuous operation over extended periods of time will be provided with minimum maintenance. These factors combined with the use of several different fuels have led to a variety of B&W marine boiler designs.

Integral-Furnace naval boiler

The Integral-Furnace naval type boiler (Fig. 2) is fitted with welded high strength, low weight stainless or low-alloy steel casings. Furnace roof, side and rear walls are water cooled by 2 in. (50.8 mm) outside diameter (OD) close-spaced tubes. Front (burner) wall and floor are usually refractory, but some recent boilers have water-cooled frontwalls. Steam drum diameters range from 54 to 60 in. (1.37 to 1.52 m) and water drums vary from 27 to 36 in. (0.69 to 0.91 m). Two to four rows of 2 in. (50.8 mm) OD tubes form the superheater screen. The superheater has 1 or 1.25 in. (25.4 or 31.8 mm) OD tubes arranged in a maximum of 8 rows deep with provision for complete drainability. The boiler bank is usually inclined and has from 18 to 23 rows of 1 in. (25.4 mm) OD tubes. For maximum efficiency, the boiler is fitted with a stud-tube economizer and, in some cases, steam air heaters.

Range in size, steam output:
  To 350,000 lb/h (44.1 kg/s) in no fixed increments
Operating pressure:
  Up to 1200 psig (82.7 bar gauge)
Steam temperature:
  Saturation to 1000F (538C)
Fuel:
  Navy special fuel oil (residual)
  Multi-purpose fuel oil
Operational control:
  Manual to complete automatic, combustion and feedwater regulation.
Draft loss at maximum output:
  Up to about 75 in. wg (18.7 kPa) total through all components
Dimensions outside setting:
  Smallest — 14 ft 2 in. high × 15 ft 1 in. wide × 10 ft 6 in. deep (4.32 m × 4.60 m × 3.20 m)
  Largest — 21 ft 5 in. high × 18 ft 11 in. wide × 16 ft deep (6.53 m × 5.77 m × 4.88 m).

Indicated field of application

The primary use of these boilers is for combat or auxiliary naval vessels or special installations.

General comments

The nature of the construction and the rating at which these boilers operate limit their use to naval or high speed naval auxiliary vessels requiring maximum power in minimum space. They are designed for maximum efficiency at cruising speed, and some attainable efficiency is sacrificed to develop high power to weight and power to boiler volume ratios. The construction includes lightweight tubes, high-tensile drum plates, and other special features to minimize weight.

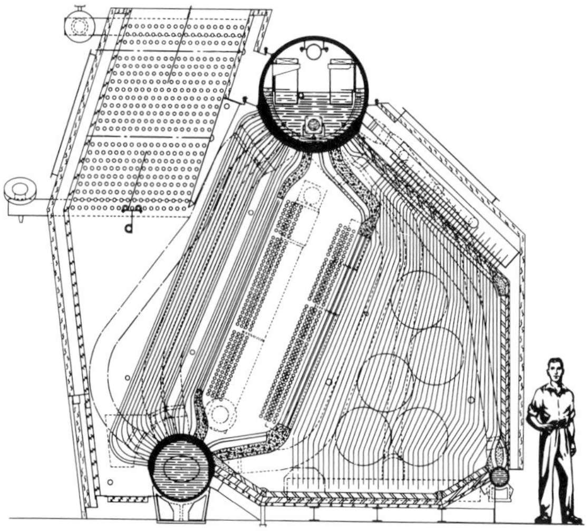

Fig. 2 Two-drum boiler, naval type.

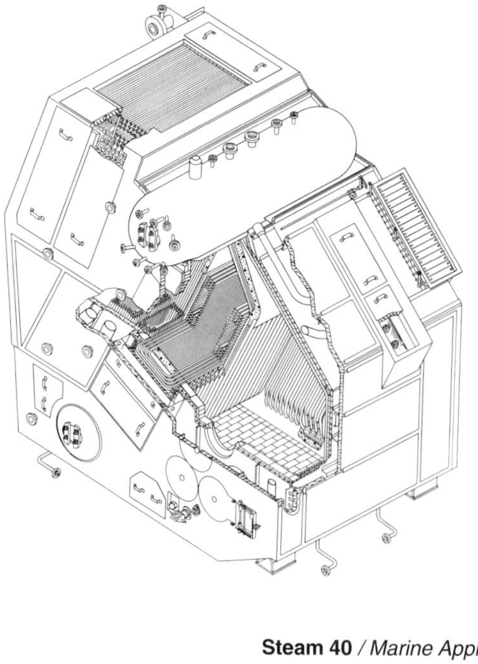

Integral-Furnace merchant boiler

The Integral-Furnace merchant boiler became the marine industry standard after World War II. The boilers were originally designed as tangent tube (Figs. 3 and 4) and refractory construction (water-cooled furnace) and more recently membrane furnace construction (water-cooled welded wall furnace). Both types have a single gas pass with a 2 row, 2 in. (50.8 mm) OD screen before the superheater, steam drum diameters ranging from 54 to 72 in. (1.37 to 1.83 m) and water drums from 30 to 36 in. (0.76 to 0.91 m).

The tangent tube furnace is water cooled by closely spaced 2 in. (50.8 mm) OD tubes on the side, roof, front and rearwalls. Recent designs have sloped, bare furnace floor tubes with consequent reduction in exposed refractory. In designs with refractory floors, the floor tubes are usually buried in the floor to reduce refractory temperature and increase refractory life.

The membrane furnace is water cooled by welded wall construction consisting of 2.75 in. (69.9 mm) OD tubes on 3-9/16 in. (90.5 mm) centers. This applies to the furnace floor, front, rear and sidewalls. The floor is covered with firebrick to protect the tubes from overheating.

In either style boiler, an inclined or vertical boiler tube bank, composed of 2 to 4 rows of 2 in. (50.8 mm) OD screen tubes and 17 to 24 rows of 1.25 in. (31.8 mm) OD generating tubes, may be used. The superheater consists of either horizontal or vertical 1.25 in. (31.8 mm) OD tubes and has one or two access cavities to facilitate water washing, cleaning, inspection, and maintenance of the superheater and boiler bank. A single cavity is normally provided when the fouling characteristics of Bunker C oil are average, and two cavities are used when fouling characteristics are more severe. The boiler is generally equipped with steam atomizing burners, retractable sootblowers, and an air heater or an economizer.

Range in size, steam output:
  To 400,000 lb/h (50.4 kg/s) in no fixed increments
Operating pressure:
  100 to 1200 psig (6.9 to 82.7 bar gauge)
Steam temperature:
  Saturation to 1000F (538C)
Fuel:
  Oil, (light fractions to heavy residuals) and liquefied natural gas (LNG)
Operational control:
  Manual or completely automatic, including feedwater flow, combustion and steam temperature controls
Draft loss at maximum output:
  From 15 to 30 in. wg (3.7 to 7.5 kPa) total through all components
Dimensions outside setting:
  Smallest, 7 ft 8 in. high × 5 ft 5 in. wide × 5 ft 6 in. deep (2.34 m × 1.65 m × 1.68 m)
  Largest, 41 ft 6 in. high × 34 ft 3 in. wide × 24 ft 1 in. deep (12.65 m × 10.44 m × 7.34 m)

Indicated field of application

These designs are primarily used for propulsion power, although they are also used without a superheater for auxiliary or heating service in the smaller sizes where space is at a premium, a compact lightweight design is needed, and feedwater is of good quality.

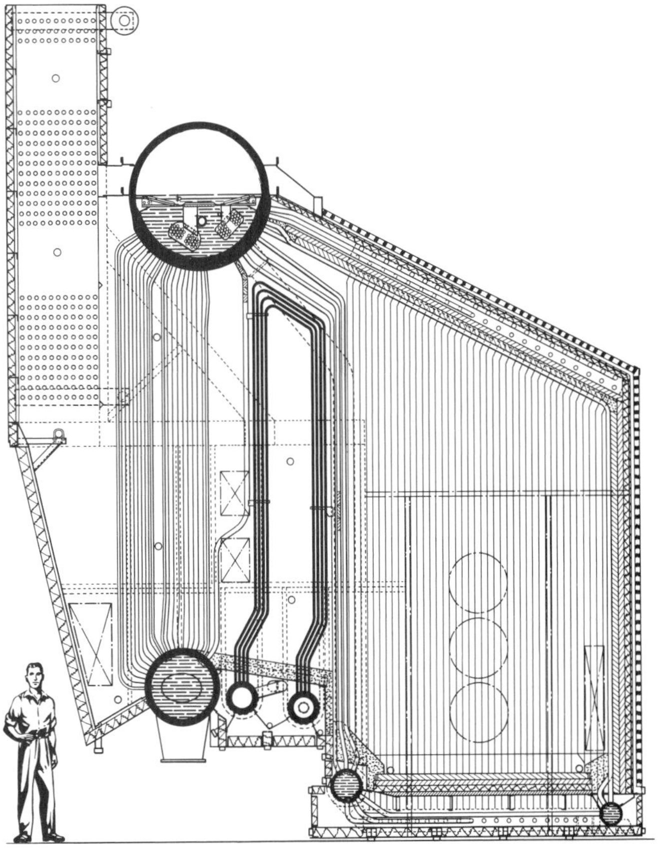

Fig. 3 Two-drum Integral-Furnace boiler with vertical superheater.

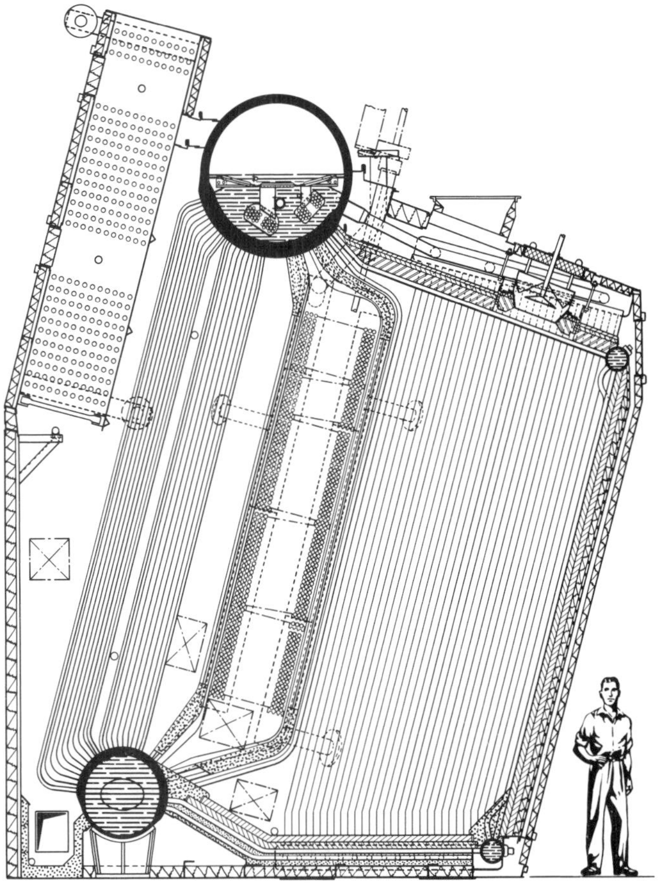

Fig. 4 Two-drum Integral-Furnace boiler with horizontal superheater.

General comments

Designs of this type are suitable for vessels in which large power plants must be installed in a minimum of space, and where weight saving is a vital consideration. Many possible variations in configuration permit application under limited space conditions. Mass action retractable sootblowers are recommended for the superheater zones.

Reheat boilers

Although reheat boilers have not been extensively used in the marine industry, there have been several applications. In considering a reheat boiler for shipboard applications, one design feature must be considered — protecting the reheater tubes when the main propulsion turbine is in astern operation or stopped. B&W has designed the boiler with damper and bypass systems as well as with extensive screen tube protection systems in front of the reheater.

The marine reheat boiler is of the two-drum divided inclined boiler bank type, capable of burning fuel oil or natural gas. Other than the reheater, reheat screen, baffle wall and damper system, the reheat boiler is similar to the tangent tube boiler.

Coal-fired marine boilers

Up until the late 1970s the last new coal-fired marine boilers put into service were for vessels on the Great Lakes. These units were built in the late 1950s and were for relatively low pressure and temperature conditions. The boiler designs were either the sectional header type or the two-drum type. This two-drum boiler has the same physical appearance as many of the newer oil-fired two-drum boilers. However, coal firing requires a much lower furnace heat release rate than oil firing for proper combustion. This results in less steam generating capacity for the same shipboard space availability.

As with a land-based unit, boiler design begins with the stoker. The maximum stoker grate release rate for good operation and for prevention of grate overloading is 750,000 Btu/h ft^2 (2.37 MW_t/m^2) of grate surface. Another requirement for good stoker operation, particularly when flyash and unburned carbon are re-injected into the furnace, is a limit of input to the stoker of 13×10^6 Btu/h ft (12.5 MW_t/m) of stoker width. Exceeding this limit can cause poor fuel distribution on the grate, resulting in uneven burning.

The furnace volume should be set so that the furnace liberation does not exceed 30,000 Btu/h ft^3 (0.31 MW_t/m^3). Adherence to this value will ensure sufficient residence time in the furnace to properly burn the fuel and minimize slagging and fouling.

As illustrated by Fig. 5, this coal-fired marine boiler has a totally different appearance than that of the customary oil-fired marine boiler. The furnace extends considerably below the centerline of the lower drum. This dimension, commonly called the setting height of the furnace, is the dimension from the top of the stoker to the centerline of the water drum. Generally, on units of this capacity, setting heights range from 16 to 20 ft (4.8 to 6.01 m).

This boiler has been specially designed to make full power by either burning 100% coal or 100% oil. When burning coal and with a grate release rate of 750,000 Btu/h ft^2 (2.37 MW_t/m^2), the approximate turndown ratio of the stoker will be 3:1. However, with the oil burners, the turndown ratio can be as high as 16:1. Therefore, if required during maneuvering modes of operation, the boiler can be readily fired by oil and have the full flexibility and response time needed for this condition. The design operating conditions are 875 psig (60.3 bar gauge) and 900F (482C) with steam capacity ranging from 60,000 to 150,000 lb/h (7.6 to 18.9 kg/s).

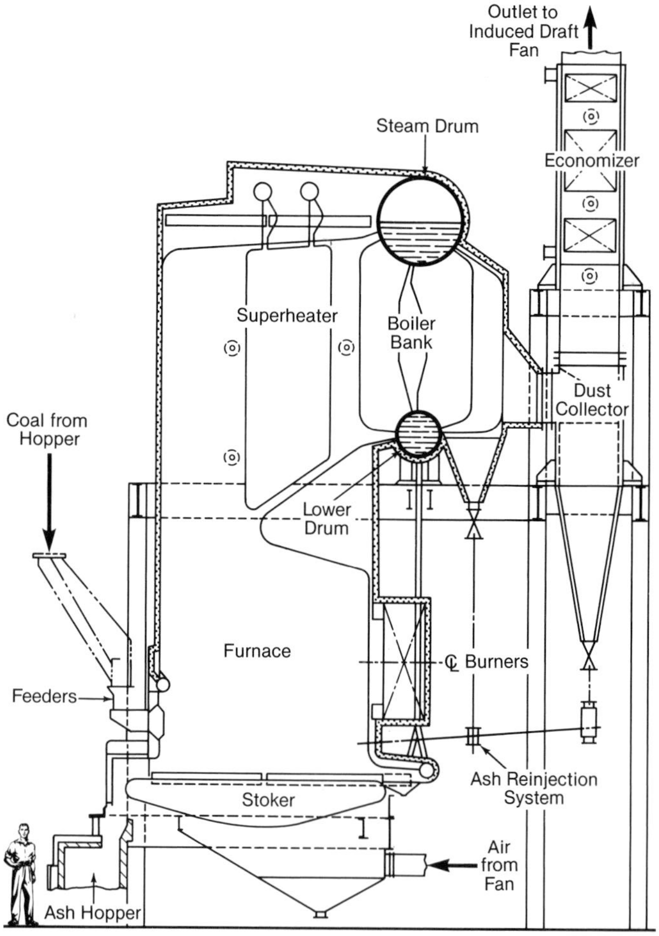

Fig. 5 Coal-fired stoker marine boiler.

Additional features of this design address other impacts of coal and its associated ash loading. Retractable sootblowers, hoppers, dust collection equipment and flyash re-injection are all included.

Auxiliary package marine boilers

As the marine industry has shifted its emphasis from main steam propulsion systems to diesel propulsion, the need developed for a simple, low cost, self-contained auxiliary boiler.

B&W developed an integral furnace, two-drum D-type package boiler (FMB) that could operate as a stand alone unit or in conjunction with a diesel or gas turbine exhaust waste heat boiler.

The boiler in Fig. 6 represents the marine package design conforming to the requirements of U.S. Coast Guard (USCG) and American Bureau of Shipping (ABS) Rules. The unit is shop-assembled, including furnace and firing equipment. It is not intended for ship propulsion, but as a means of supplying auxiliary steam.

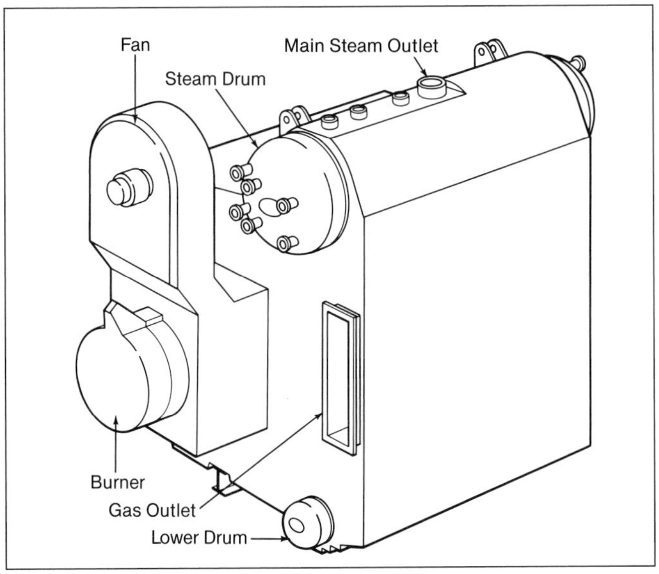

Fig. 6 Auxiliary package marine boiler.

The boiler incorporates a compact vertical bank of 2 in. (50.8 mm) OD tubes arranged between and terminating in an upper and lower drum. The tube ends are expanded into tube seats which are grooved to obtain maximum tightness.

A 54 in. (1.37 m) nominal diameter steam drum is required to promote water level stability, even during wide variations in load, and helps to ensure dry, saturated steam. The requirements for this drum include satisfactory operation under the following conditions:

30 degree roll to each side for a 15 second period,
10 degree pitch from bow up to bow down for a 6 second period,
15 degree permanent list to either side, and
5 degree permanent trim to either the bow or stern.

Under any of these adverse conditions, the downcomers and boiler bank generating tubes will not uncover.

The furnace and boiler are covered by a 20 gauge (0.9 mm) nonpressurized galvanized steel ribbed lagging except in areas such as the steam drum joints and bent tube portions of the walls, which are covered by a 12 gauge (2.7 mm) carbon steel casing.

The boiler furnace unit is arranged for upward expansion and is mounted on a welded structural base frame, ready for positioning on the ship's foundations. The unit has a furnace with a refractory covered flat floor, and a boiler bank. The furnace sidewall, roof, floor and rearwall are completely water cooled and form an integral part of the boiler circuitry. One burner is used in the refractory frontwall.

Water-cooled furnace wall tubes, comprising the sidewall, roof and floor, are of membrane construction. The rear furnace wall is 2 in. (50.8 mm) OD flat-studded tubes on 4 in. (101.6 mm) centers. A 10 gauge (3.4 mm) steel inner casing backs the flat-studded wall. The furnace wall tubes receive their flow from the lower drum and discharge the steam-water mixture into the steam drum. The 2.75 in. (69.9 mm) OD furnace, roof and floor tubes are a part of the furnace sidewall circuitry. The furnace frontwall (burner wall) is of refractory anchored construction and is backed by insulation and a gas-tight casing.

These marine auxiliary package boilers are designed to provide unattended operation with controls that are fully automatic after the boiler is on line.

Range in size, steam output:
To 100,000 lb/h (12.6 kg/s) in no fixed increments
Operating pressure:
225 psig (15.5 bar gauge)
Steam temperature:
Saturation only
Fuel:
Oil, light fractions to heavy residuals
Dimensions outside setting:
Overall width is 12 ft 6.25 in. (3.82 m), overall height (base to face of steam outlet flange) is 16 ft 9.75 in. (5.12 m) with drum centers at 12 ft 10 in. (3.91 m). These dimensions are constant for all boiler sizes. The length is variable to establish the various boiler sizes and ranges from about 12 to 18 ft (3.66 to 5.49 m).

Waste heat marine boilers

With the use of diesel or gas turbine propulsion systems, such as aboard the LaJolla Class ship, shown in Fig. 7, it is often advantageous to incorporate a waste heat (exhaust gas) boiler system which, in addition to an auxiliary boiler, supplies saturated steam and improves overall plant efficiency.

The waste heat boiler system is composed of a forced circulation, spiral wound, finned tube or studded tube boiler (Fig. 8); auxiliary steam drum; and control equipment.

The waste heat boiler unit is arranged for upward expansion and is ready for positioning on the ship's foundation. It is designed to operate under either a wet or dry condition and can serve as a muffler or sound reducer whether filled with water or in a dry state.

The horizontal waste heat boiler uses the B&W continuous stud tube extended surface or a spiral wound, finned type surface. The elements including the inlet and outlet headers are located inside a welded gas-tight casing, arranged for bottom support. This boiler consists of 1.5 in. (38.1 mm) OD steel tubes and studs or fins.

The boiler elements, headers, casing, integral support steel and framing angles are shop-assembled. The tubes are supported by tube support plates. The unit is single cased and of welded construction. The casing provides a

Fig. 7 M.V. *Potomac Trader*, LaJolla Class vessel with diesel propulsion plus waste heat and auxiliary steam boilers (*courtesy of Penn-Attransco Corp.*).

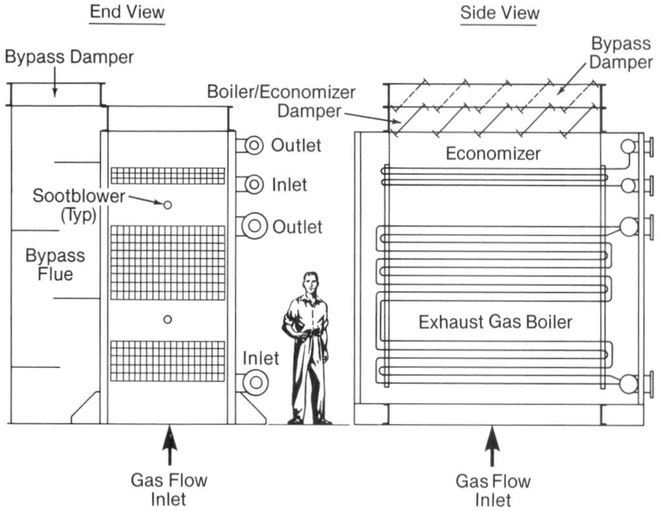

Fig. 8 Waste heat marine boiler.

gas-tight seal and is fitted with access panels for inspection purposes. Support of the exhaust gas boiler is on the ship's structural members.

A 36 or 42 in. (0.91 or 1.07 m) nominal diameter auxiliary steam drum is provided to meet anticipated list and trip operating conditions as well as to promote water level stability. The effect of shrink and swell is controlled by forced circulation with a minimum 4:1 circulation ratio, cyclone steam separators and the large diameter steam drum.

One mode of operation has an auxiliary boiler operating in conjunction with the waste heat boiler system. (See Fig. 9.) The main diesel engine exhaust gas is directed to the waste heat boiler. The gas then enters the boiler and crossflows over the in-line extended surface to the waste heat boiler exit where it is then sent to the stack.

The auxiliary boiler feed pump supplies condensate to the auxiliary boiler steam drum. The circulating pumps take suction from the auxiliary boiler lower drum and feed the waste heat boiler. Flow is upward through the waste heat boiler to its outlet where the resulting steam-water mixture is conducted through piping to the auxiliary boiler's steam drum. All steam for ship services is drawn through the auxiliary boiler's cyclone steam separators so that the auxiliary boiler acts as the steam drum of the exhaust gas boiler. The auxiliary boiler will fire automatically to maintain the steam pressure within an adjustable band.

The alternate mode of operation for this system, with the auxiliary boiler, has the boiler feed pump supply condensate to the auxiliary steam drum. The circulating pumps take suction from the auxiliary steam drum and feed the waste heat boiler. Flow is upward through the boiler to its outlet where the resulting steam-water mixture is conducted through piping to the auxiliary steam drum. The auxiliary boiler may be isolated and unfired in this condition for maintenance or to accommodate steam demand. It may also be fired automatically to maintain the steam pressure within an adjustable band.

Marine boiler design

The techniques used in the design of marine boilers are similar to those applied to stationary units (Chapters 18, 19 and 20). However, the design of marine boilers is also subject to constraints and requirements that are specific to marine plants. These include space conditions, fireroom configuration, list and trim, pitch and roll, and ship maneuverability. In addition to the codes and standards used for stationary boilers (ASME, ANSI, etc.), other codes apply. Examples include the United States Coast Guard Marine Engineering Regulations, subchapter F and other subchapters; American Bureau of Shipping (ABS) Rules; and Lloyd's Register of Shipping Rules.

Design procedures and guidelines for marine boilers are generally set in the Society of Naval Architects and Marine Engineers (SNAME) Technical Bulletin No. 3-32 (*Furnace Performance Criteria for Gas, Oil and Coal Fired Boilers*, 1981) and No. 3-11 (*Marine Steam Plant Heat Balance Practices*).

Fuel characteristics can have considerable bearing on the design. Regardless of whether oil or coal is used, the chemical analysis and ash characteristics are particularly important and should be specified so that heat transfer surface can be arranged to minimize corrosion and deposition from ash and slag.

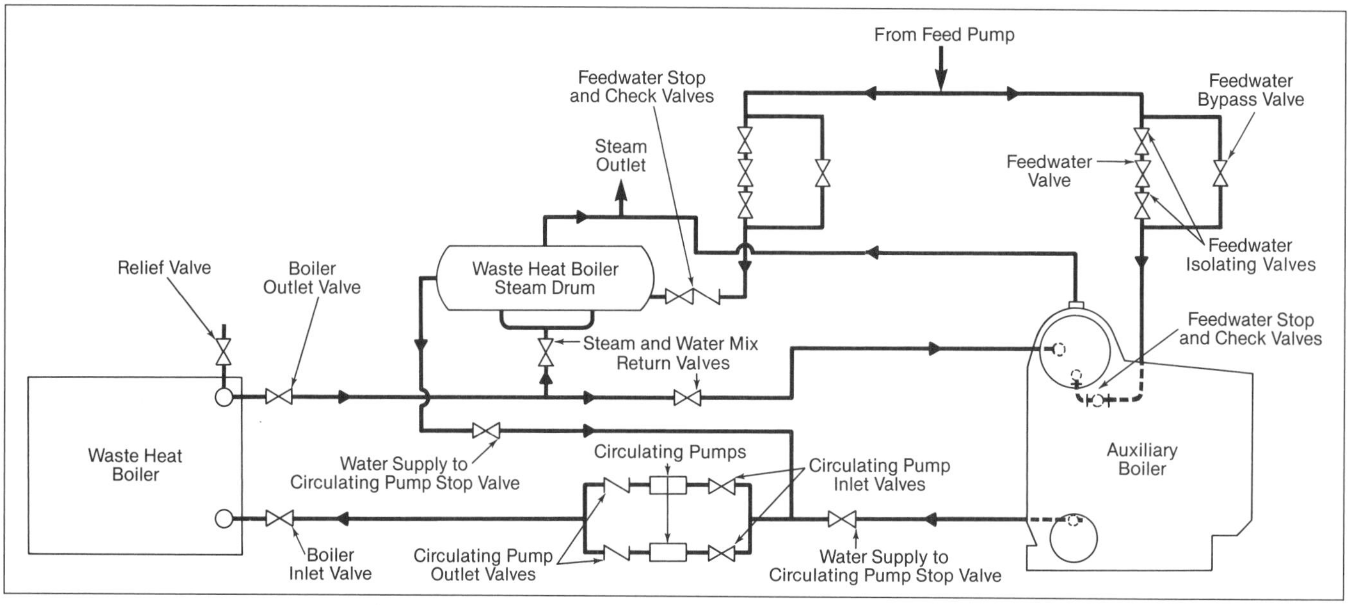

Fig. 9 Shipboard steam system combining auxiliary boiler and waste heat boiler.

Boiler design for a particular application starts by establishing the dimensions and arrangement of the furnace. Within this volume the desired fuel burning equipment can be installed and the fuel burned efficiently, without exceeding allowable furnace heat absorption rates. This is a major concern where small furnaces are needed to fit into restricted shipboard engine room space such as in the Seabee Class ships shown in Fig. 10.

The furnace arrangement and dimensions generally establish the overall size of the boiler bank. Therefore, the first step in designing a new boiler is to prepare a preliminary layout of a suitable furnace. To do this, the amount of fuel to be fired is calculated based on the required steam flow, pressure and temperature as well as on feedwater temperature and efficiency. It is then possible to estimate the type and number of burners to be used, the allowable pressure drop through the burners for good combustion conditions, the size and shape of the furnace to properly accommodate the burners for good combustion, and the allowable furnace heat absorption rate.

Fig. 10 Seabee Class container ship.

Oil burners

The type and number of oil burners selected depend on the fuel rate and the allowable air resistance. The number is usually not less than two, so that at least one burner operates at all times, even when changing sprayer plates or cleaning atomizers of the idle burners. Many types of oil burners are available. The burners described in Chapter 10 are for stationary boilers and are for burning natural gas and oil, as well as other gaseous type fuels or mixtures.

Marine burners are generally designed solely for burning oil. Since World War II three types or sizes of registers were applied to most merchant marine and naval vessels. They were the Iowa, Saratoga and Progress types and covered oil flow rates up to 7,000 lb/h (0.9 kg/s) or 130×10^6 Btu/h (38.1 MW_t) input per register with excess air of 10 to 20%.

In the 1970s with the pursuit of more efficient steam propulsion systems, B&W developed a low excess air burner (5 to 10%) to be used on merchant ships. These burners were designed to burn natural gas or oil up to 130×10^6 Btu/h (38.1 MW_t) input. The burners are available in throat sizes from 15 to 30 in. (381 to 762 mm) (Fig. 11) and have pressure drops of 8 in. wg (2 kPa) at normal operating conditions.

Prior to 1970, most steamships used a mechanical atomizer. However, B&W developed and introduced the wide range *Racer* steam atomizers or sprayer plate. This steam atomizing sprayer plate provided coverage over the full range of operation. The properly sized and selected sprayer plate will cover the range from minimum to the designed maximum overload firing rate, as long as atomizing steam is available. This steam atomizer and burner are capable of operating at extremely low firing rates without loss of good combustion or ignition as long as proper fuel oil temperature and proper air flow (air/oil ratio) are maintained.

When maneuvering, or at any time during port operation or normal operation at sea, it should not be necessary to secure burners unless the steam demand is less than that produced with all burners in service at the minimum fuel oil pressure. During normal operation all burners should be kept in use for all rates of operation.

For all fuel oil pressures the atomizing steam pressure should be constant and should be maintained at 135 psig (9.3 bar gauge) at the burner. Due to the pressure drop between the atomizing steam header and the burner, this is equivalent to a steam pressure of approximately 150 psig (10.3 bar gauge) at the atomizing steam header.

When burning heavy residual fuel oil, certain minimum clearances are required around the burners to prevent flame impingement and carbon deposits. To assure complete combustion of the fuel, the furnace depth is usually limited to a minimum of 6 ft (1.83 m), although both diesel and bunker oils have been burned successfully in furnaces 5 ft (1.52 m) or less in depth.

Furnace heat absorption

In the marine field, the heat release rate per unit furnace volume is frequently used for comparing boilers, without regard to similarity of design. This ratio is not an important design criterion as it provides only an approximation of the time required for the products of combustion to pass through the furnace. However, it may be used to indicate the operating range for which the firing equipment is to maintain satisfactory combustion conditions. The availability of suitable oil burners has permitted the installation (in naval vessels) of high capacity lightweight boiler units with heat released at a rate of 200,000 Btu/h ft^3 (2.07 MW_t/m^3) of furnace volume at cruising conditions and more than 1,000,000 Btu/h ft^3 (10.3 MW_t/m^3) at the maximum evaporative condition with satisfactory results. For merchant ships, arbitrary limits of 75,000 to 90,000 Btu/h ft^3 (0.77 to 0.93 MW_t/m^3) at the normal rate of operation are commonly specified.

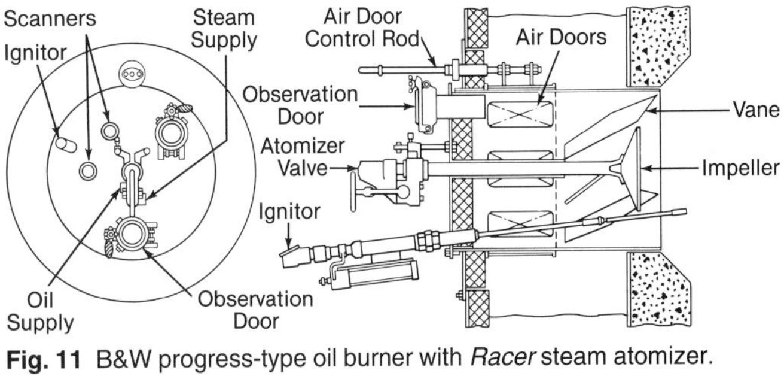

Fig. 11 B&W progress-type oil burner with *Racer* steam atomizer.

A more meaningful criterion of furnace design is the heat absorbed by the cold surfaces of the furnace, expressed as the heat absorbed per square feet of radiant heat-absorbing surface as discussed in Chapter 18. The considerations described also apply to marine units, with particular emphasis on the determination of the furnace exit gas temperature to assure satisfactory superheater design and to establish the heat absorption by the furnace wall tubes for adequate circulation margins.

Marine boilers operate with higher heat input rates per square foot of boiler and superheater surfaces than stationary boilers designed for the same steam outputs. They have less heating surface and smaller overall dimensions. Flame temperatures more closely approach the adiabatic temperature, and furnace exit gas temperatures are considerably higher than those encountered in stationary boilers. Fig. 12a indicates the effect of excess air on furnace temperature.

The effectiveness of the water-cooled surface in the furnace is determined by applying the factors shown in Chapter 4 to the flat projected areas of the furnace walls and tube banks facing the furnace. Bare tangent tubes or membrane tubes are used wherever possible in furnace waterwalls. Where tubes must be spaced to facilitate replacement, it is preferable to limit the width of the exposed refractory areas between tubes to 1 in. (25.4 mm) or less to avoid wastage resulting from sodium compounds in the oil ash. The effect of wall tube spacing, or the ratio of water-cooled surface to refractory surface, on furnace heat absorption is shown in Fig. 12b. The heat absorption rate in the furnace increases with increased firing rate. However, furnace absorption as a percentage of the total boiler absorption decreases with increased firing rate as indicated in Fig. 12c.

Furnace tube temperatures

In all boiler tubes, and particularly in tubes exposed to the high heat absorption rates of marine furnaces, adequate circulation must be provided to avoid critical heat flux (CHF) or departure from nucleate boiling (DNB) as discussed in Chapter 5. This is generally accomplished by empirical data and methods based on tests and operating experience. Tube wall thickness is usually set close to the minimum required for the design pressure, because weight is a primary consideration and excess thickness increases external tube temperature. The heat transfer coefficient across the boiling water film in the furnace steam generating tubes can be as high as 20,000 Btu/h ft^2 F (114 kW/m^2 K). However, in estimating tube temperature, a conservative value may be used, e.g., 2000 Btu/h ft^2 F (11.4 kW/m^2 K), to obtain a somewhat higher estimated tube temperature, resulting in lower allowable stress in the tube metal and therefore a somewhat more conservative tube thickness.

Scale deposits on the water side of boiler tubes are a long recognized cause of tube failure. These deposits can be particularly serious in the furnace tubes of marine units because of the high heat absorption rates. As an example, a calcium sulfate scale deposit on the inside of a tube with a thickness of only 0.024 in. (0.61 mm) and a thermal conductivity of 0.83 Btu/h ft F (1.44 W/m K) results in a 362F (201C) temperature differential across the scale. For a boiler operating at 665 psig (45.9 bar gauge), this scale would increase the tube outside metal temperature to 1004F (540C). This temperature exceeds the oxidation limit for steel, and oxidation and ultimate tube burnout will likely occur, even though boiler circulation may be adequate. Scale must be avoided by proper water conditioning. (See Chapter 42).

Boiler tube banks

The boiler bank, composed of multiple rows of tubes in which steam is generated, generally consists of a screen ahead of the superheater and a convection bank behind it. Both sections generate saturated steam but they are considered separately during the design stage. The screen tubes, by virtue of their location, absorb heat at a considerably higher rate than the main bank tubes. Conse-

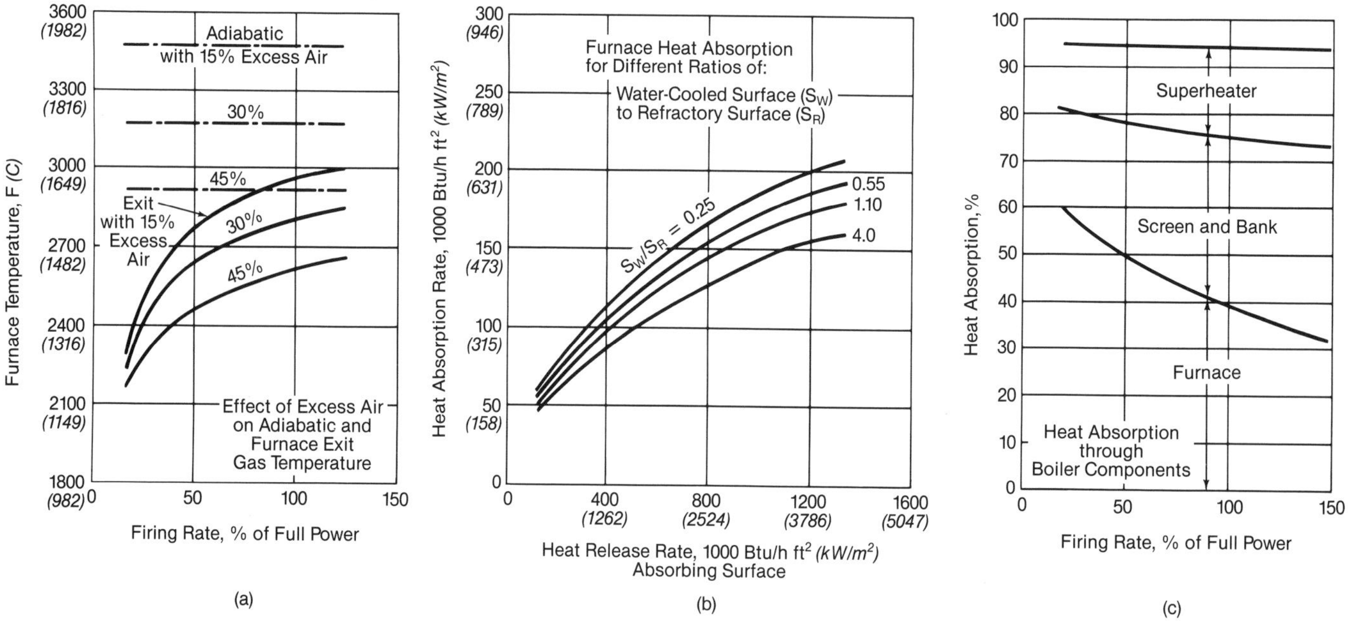

Fig. 12 General effect of excess air (a), ratio of heat-absorbing to refractory surface (b), and firing rate (c) on furnace heat absorption and temperature, based on 18,500 Btu/lb (43.0 MJ/kg) fuel oil.

quently, to assure an adequate flow of water, the furnace screen tubes should be larger in diameter. For this reason, marine boilers are designed with two sizes of steam generating tubes. The diameters of screen tubes are generally 2 in. (50.8 mm) while those of the other generating tubes are 1, 1.25 or 2 in. (25.4, 31.8 or 50.8 mm), swaged to 1.5 in. (38.1 mm) at the drums. Circulation and the amount of heating surface required to obtain the desired gas temperature leaving the tube bank are the major factors in determining the tube size and the number of tube rows to be installed, although resistance to gas flow is also a factor.

In the design of the superheater screen, consideration must be given to the effect of tube spacing and the number of rows, the desired superheater outlet temperature and the maximum allowable superheater tube metal temperatures. The screen must be designed so that the gas temperature entering the superheater and the radiant heat penetration from the furnace will provide the desired steam temperature with a superheater of reasonable size and arrangement. Fig. 13 shows the effect of radiant heat penetration on the performance of the superheater over the designed load range of the boiler. The number of rows in the boiler bank is usually established by analyzing the economic advantages of economizers and air heaters as compared with boiler surface. The proper proportions of boiler surface and additional heat-absorbing surface beyond the boiler bank are based on the efficiency desired.

Superheaters

Superheaters of the convection type are generally used in marine boilers. While some radiant superheaters have been used, the difficulty of providing adequate cooling during fast startups and under maneuvering conditions has severely limited their application.

In the convection superheater there is usually enough absorption by radiant heat penetration of the screen to give a flatter steam temperature characteristic than would be obtainable by convection alone. The characteristic steam temperature curves of radiant and convection superheaters are shown in Chapter 18.

With a properly designed screen between the convection superheater and the furnace, it is possible to maintain a high temperature differential between the gas and the steam. This minimizes the amount of heating surface necessary to obtain the desired steam temperature and reduces the size and weight of the superheater. Superheater heating surface usually consists of U-shaped tubes connected to headers at each end, although continuous tube superheaters are used in special designs.

Either a vertical or a horizontal arrangement may be used in drum-type boilers. The horizontal arrangement can be vented, drained, and readily cleaned by mechanical or chemical means. However, complete draining and venting is possible only with the vessel on an even keel. Vertical superheaters have inverted loops and, although drainable at all times, they can not be vented. Typical superheater arrangements for a two-drum boiler are shown in Figs. 3 and 4.

Superheater surface is arranged in loops of in-line tubes (usually one to four loops), and the most commonly used tube size is 1.25 in. (31.8 mm) OD. Because the tubes are relatively short, adequate strength is available in the usual range of thicknesses to maintain proper alignment with a minimum of supports. The number of steam passes is selected to provide sufficient pressure drop to obtain proper steam distribution and to assure satisfactory tube metal temperatures.

Good practice requires that maldistribution of gas and steam be kept to a minimum to maintain proper superheater tube temperatures. However, with various fuels and operating demands it is not possible to maintain the conditions necessary for perfect distribution. To provide a satisfactory margin when calculating steam and tube temperatures, the average gas side heat transfer rate is increased and the steam side rate is reduced.

For steam temperatures below 850F (454C), tubes are usually expanded into headers to minimize leakage and reduce maintenance costs. Above 850F (454C) rolled and seal welded joints, or joints of the stub welded type, are necessary for satisfactory service.

In addition to being supported by the headers, superheater tubes may also be supported at one or more points by alloy castings dovetailed into brackets welded to water-cooled support tubes. With these designs the superheater tubes can be replaced without removing the supports.

Superheater fouling and high temperature corrosion

The use of higher steam temperatures has resulted in rapid fouling of superheater surfaces and corrosion and wastage of superheater supports. The heavy ash deposits come from burning oils with 0.05 to 0.20% or more ash by weight. The most significant constituents of the ash are vanadium, sodium and sulfur. Experience indicates that heavy deposits occur with a combination of high gas and heat-receiving surface temperatures. The heaviest ash accumulations occur on the furnace side of a superheater, where the gas temperatures are relatively

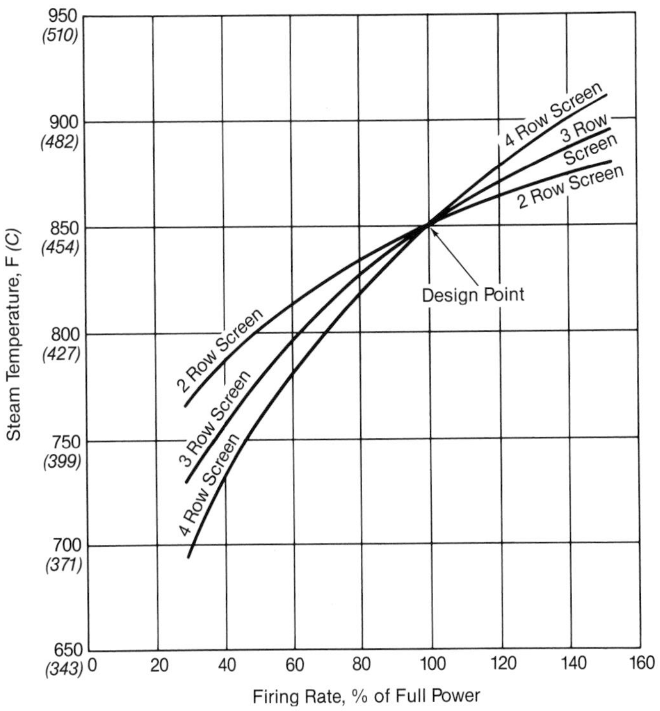

Fig. 13 Typical curves showing the effect of superheater location and firing rate on superheat.

high, while less ash forms at the back side, where the gas temperatures are much lower. Because of the potential of high corrosion rates, superheater metal temperatures should not exceed 1100F (593C). For further discussion of high temperature corrosion in oil-fired boilers, see *Oil-Ash Corrosion* in Chapter 20.

Steam temperature control

In the marine power plant, just as in the stationary power plant, accurate control of steam temperature is required. In most cases steam temperature controls are used to assure that the allowable metal temperatures of the main steam piping and the main turbine are not exceeded. Close control of the maximum steam temperature permits the use of less expensive alloys in the superheater and piping. Several methods are used for superheat control, as described in Chapter 18. One or more methods may be used to alter the characteristic steam temperature curve of a particular unit design.

One method used in marine boilers is the dual or two-furnace boiler design. As shown in Fig. 14, the furnace is divided into two sections separated by a superheater and screen tube section. The flue gas flows successively through the superheat furnace (left hand furnace), the superheater heat transfer surfaces, the second furnace (where fuel and air are burned) and, finally, the boiler bank before being exhausted to the stack. By varying the oil firing rate in the superheater furnace, the quantity and temperature of the gases flowing across the superheater are controlled to obtain the desired steam temperature. At the same time, the firing rate in the saturated furnace, which contains no superheater surface, is adjusted to hold the desired steam pressure.

In most modern two-drum boilers, steam temperature is controlled, where required, through the use of a drum-type surface attemperator or *desuperheater*. With this type of control, all of the flue gas passes over the superheater, but a portion of the steam is passed through the attemperator, which is a steam-to-water heat exchanger submerged in boiler water in the steam or water drum. Fig. 15 indicates two piping arrangements for control attemperators. In a submerged desuperheater where the inlet steam temperature exceeds 850F (454C), 16-1 Cr-Ni alloy should be considered to protect against methane gas corrosion.

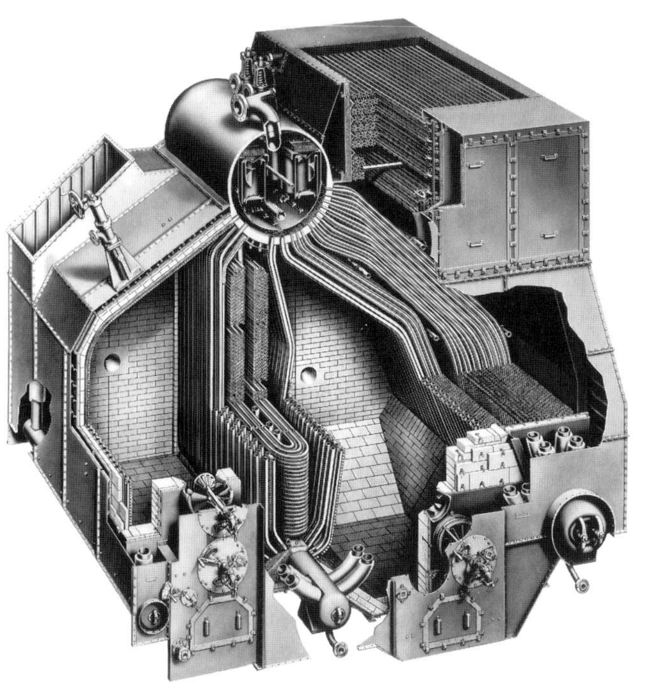

Fig. 14 Twin furnace design used for marine boiler superheat temperature control.

Auxiliary desuperheaters

Steam for auxiliary use may be required at a lower temperature than that delivered by the superheater to the main engine. To satisfy these demands, auxiliary desuperheaters are used, often in conjunction with pressure reducing stations. These desuperheaters may be inside the drum as discussed above or separate external spray types.

Reheaters

The use of steam reheat in a marine plant is more attractive as the speed and power requirements of the ship increase. Although reheaters increase the complexity of the boiler design and the first cost of the plant, fuel savings in high horsepower, high utilization plants can be significant. Design considerations are similar to those for superheaters but must be augmented by the requirements to protect the reheater from overheating during periods when there is no steam flow during maneuvering, running astern or at a stop bell.

Economizer and air heater cycles

Economizers or air heaters, and in some instances both, are required if a high boiler efficiency is to be obtained. The temperature of flue gas leaving the boiler bank at full power (the design rate) is a function of the saturation temperature corresponding to the drum pressure at which the unit is operating. Space, weight and economic considerations usually result in a boiler bank sized to reduce the exit gas temperature to within 50 to 100F (28 to 56C) of the saturation temperature. Gas temperatures in the range of 550 to 650F (288 to 343C) leaving the boiler bank are typical of merchant ship boilers. To reduce this temperature sufficiently in order to obtain acceptable efficiency, economizers and/or air heaters may be added. The choice of which to use depends on the design of the power plant and the desired performance characteristics of the unit.

Where the design includes a deaerating feedwater heater and a single stage of feedwater heating to supply water at 240 to 280F (116 to 138C), an economizer can be used, either alone or in conjunction with a steam air heater, to provide reasonably high boiler efficiency. In this range of feed temperature, an economizer can be economically designed to reduce the products of combustion to within 30F (17C) of the inlet water temperature at the normal rating. Therefore, with 280F (138C) feedwater, a boiler efficiency of 88.6% can be attained with an economizer alone and about 88.8% if a steam air heater is added.

When the feedwater temperature to the economizer is higher, the efficiency is limited, because the temperature of the gas leaving the economizer can not be lower

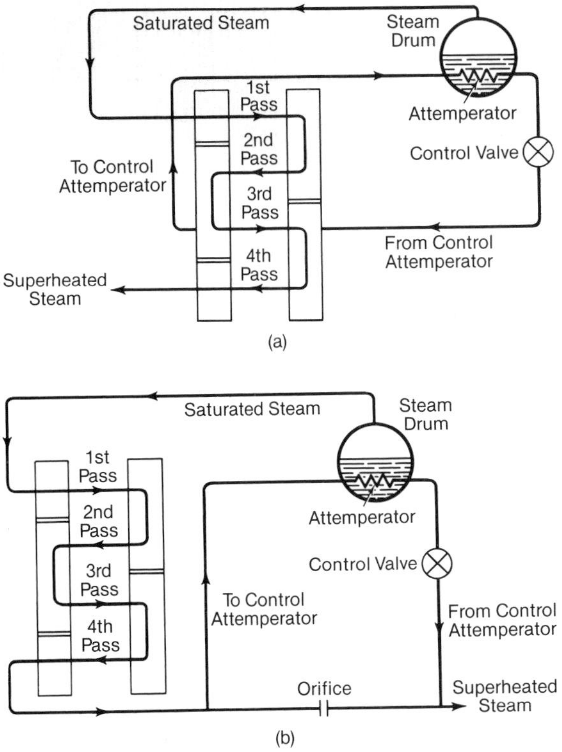

Fig. 15 Internal attemperators — (a) inter-pass, (b) after-pass.

than the inlet feedwater temperature. Consequently, when additional stages of regenerative feedwater heating are used, the inlet feedwater to the economizer may be at a temperature of 300 to 450F (149 to 232C), and it may not be economical to use an economizer unless it is followed by an air heater. Cycle efficiency is increased approximately 1% for each 100F (56C) rise in feedwater temperature through regenerative feedwater heating.

An air heater is particularly attractive because it contributes additional efficiency even when the maximum practical amount of regenerative feedwater heating is used. At normal operating rates and an inlet air temperature of 100F (38C), exit gas temperatures of 300 to 320F (149 to 160C) are readily obtained, corresponding to 88.5 to 88.0% efficiency. Air temperatures leaving air heaters are normally in the range of 300 to 450F (149 to 232C).

As an alternate to the standard economizer/air heater cycle, a dual economizer steam air heater cycle (DESAH), using a primary and secondary economizer, can be used. This arrangement will improve the cycle efficiency and reduce the fuel rate by utilizing turbine bleed steam to a high pressure feedwater heater. The actual cycle improvement will depend on the turbine characteristics with the additional bleed for the high pressure heater. The overall boiler efficiency will remain essentially unchanged over that with the regular economizer cycle; however, the fuel oil consumed will decrease because of additional heat added to the feedwater.

The cycle would use a feed temperature leaving the deaerator feed heater of about 285 to 300F (141 to 149C), presently being used on single economizer, steam air heater cycles. A feedwater temperature lower than this could result in economizer corrosion and plugging problems. If higher feedwater temperatures are desired in order to more efficiently use turbine bleeds, it should be pointed out that the boiler efficiency will correspondingly decrease. This is due to the higher gas temperatures occurring at the secondary economizer outlet because a terminal end temperature difference of 25F (14C) is required.

The feedwater would pass through the secondary economizer and be heated to about 335F (168C). It would then pass through a high pressure feedwater heater, which would raise the feedwater temperature about 50F (28C) before returning the water to the primary economizer and the steam drum (Fig. 16).

This cycle shows a reduction in all purpose fuel use over that associated with the single economizer steam air heater cycle while being only slightly more complicated. Equipment costs rise somewhat due to the addition of a high pressure feedwater heater and its control. In addition, in order to realize the same exit gas temperature, the economizer heating surface would have to be increased 1.5 times that used with the single economizer cycle.

Economizers Two general types of economizers, the bare tube and extended surface types, are used in marine service. Both are nonsteaming and are almost always arranged for counterflow of water and flue gas to obtain the best possible heat transfer characteristics. (See Chapter 19.) The bare tube type is used where high feedwater temperature makes the application of both an economizer and air heater desirable. These economizers

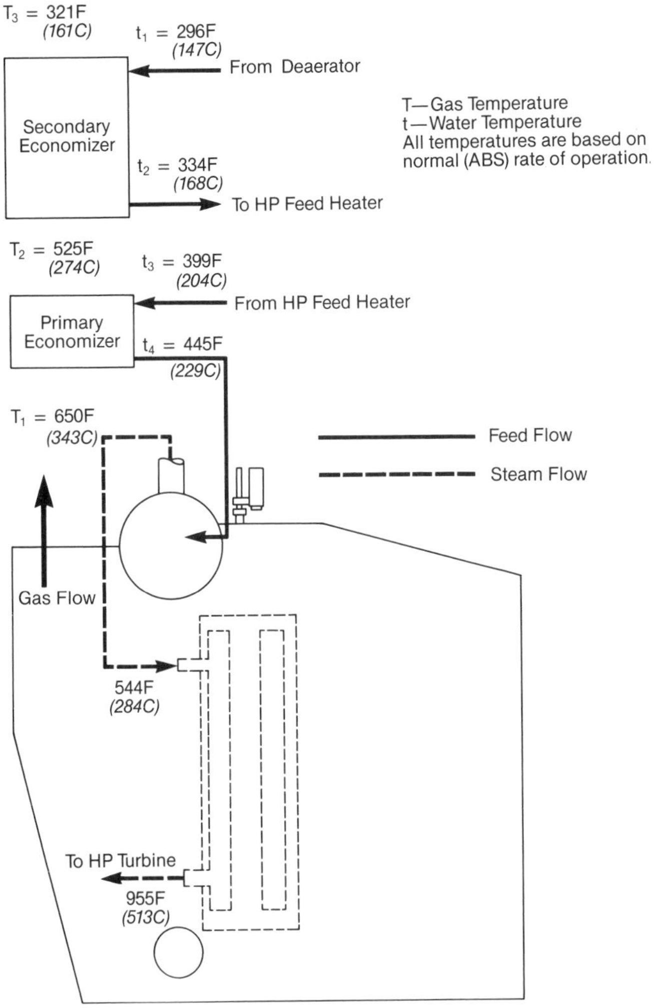

Fig. 16 Dual economizer steam air heater (DESAH) cycle schematic.

are designed to reduce the gas temperature about 100F (56C), with the remainder of the required temperature drop obtained with an air heater. All are drainable and ventable.

The second and more common type uses extended surface to reduce the size of the economizer. There are many types of extended surface, including cast iron or aluminum gill rings, spiral fins and small metal studs welded to the tubes. B&W has chosen the latter for its marine economizers due to the shape and size of the stud and because the method of attachment eliminates soot-collecting crevices where corrosion usually begins.

Extended surface economizers are usually constructed of 1.5 in. (38.1 mm) OD tubes formed into continuous loops. This construction requires only two headers. Tube-to-header joints are welded, or expanded and seal welded, to eliminate seat leakage. In the most recent designs, handhole fittings, except those required for inspection, are eliminated by externally welding the tubes in sockets or to stubs on the headers. A typical arrangement of an extended surface economizer with studded tubes is shown in Fig. 17.

When used alone, economizers are generally designed to reduce gas temperatures by 200 to 300F (111 to 167C), with a corresponding increase in the water temperature of 70 to 100F (39 to 56C). If the inlet water temperature is 280F (138C), the exit water temperature normally will range from 350 to 365F (177 to 185C), far enough below saturation temperature at a pressure of 600 psig (41.4 bar gauge) to prevent steaming.

In designing an economizer, the possibility of sulfuric acid corrosion of the economizer tubes must be considered. (See Chapter 19). The tube metal temperature of bare tube economizers is essentially the same as that of the water within the tube. This is also true for the tube metal of extended surface elements, but the tip temperature of the studs is considerably higher.

In order to minimize gas side corrosion of carbon steel economizer tubes, B&W recommends that feed temperature be kept above 246F (119C) when burning Navy Special Fuel Oil (NSFO) and above 270F (132C) when burning Bunker C oil. Where these temperatures are not obtainable, the use of a corrosion resistant alloy tube is advised. The use of cast iron is not recommended as the corrosion resistance of cast iron is less than that of standard carbon steel, and any improvement in tube life with cast iron cladding is only due to its greater mass.

Fig. 17 Extended surface stud tube economizer.

A bypass line is usually provided for operation of the boiler with the economizer out of service. During such operation, it is necessary to fire more fuel to maintain the required evaporative rating because of the decreased efficiency. This increases the steam temperature and requires either more attemperation or a reduction in rating to prevent overheating of superheater tubes or turbine. Normally there is no danger of metal oxidation in the economizer during bypass operation, because the gas temperatures entering the economizer are usually less than 850F (454C).

When economizers are included in the design, deaerating feedwater heaters should be installed to remove all traces of oxygen from the feedwater and to prevent internal corrosion of tubes and headers. (See Chapter 19).

Air heaters (gas to air) Tubular and regenerative air heaters (see Chapter 19) have both been used for marine boiler applications. However, the tubular air heaters have become virtually obsolete due primarily to their extremely large space requirements and their relatively poor metal temperature characteristics.

Air heater corrosion can be reduced or practically eliminated by several means. For a given exit gas temperature, metal temperatures are somewhat higher with a regenerative air heater than with a recuperative type. Therefore, for a given satisfactory metal temperature, a lower exit gas temperature and a higher unit efficiency can be obtained with a tubular type air heater. With a regenerative air heater, air temperatures of 550F (288C) or higher can be produced with a resulting exit gas temperature of 235F (113C) or less. This in turn can result in a boiler efficiency of about 90.3% when using low excess air values.

As with the tubular air heater, corrosion can be a problem when the regenerative air heater is designed for low exit gas temperature and the corresponding high air temperature. This can result in an average cold-end metal temperature well below the acid dew point of the combustion gas. Corrosion can be reduced by the use of porcelain-enameled cold-end surface.

Another alternative available when using a rotary regenerative air heater is to place a small economizer in the gas stream and/or a steam air heater on the air side ahead of the regenerative air heater. Without the economizer, the regenerative air heater becomes large due to the relatively low heat transfer rates inherent in a gas-to-air heat exchanger. Often space and weight limitations, along with the increased cost associated with a large air preheater, dictate a reduction in its size. The addition of an economizer, to reduce the gas temperature entering the air heater by 100 to 125F (56 to 69C), reduces the size of the air heater although the air temperature to the furnace is correspondingly reduced. This can frequently provide a better arrangement and at a lower overall cost.

Boiler casing

In marine service the walls which form the boiler and furnace enclosure, together with those surrounding the economizer and air heater, are normally made gas-tight by using metal casing. Therefore, the term *casing* is generally used for the walls surrounding a marine unit.

To provide a comfortable fireroom, an outer casing temperature of 130F (54C) is usually specified. In the double case arrangement used for tangent tube boiler construction the space between the inner and outer casing is pressurized with air at or near the forced draft fan discharge pressure, to prevent leakage of combustion gases into the fireroom and to help maintain the desired outer casing temperature. With welded wall construction, the boiler is, for the most part, single cased and is pressure sealed to prevent leakage of combustion gases into the fireroom.

Circulation and steam separation

Satisfactory circulation characteristics and efficient steam separation are of prime importance in the successful operation of any boiler. These factors take on increased importance in a marine boiler rolling and pitching in a seaway or undergoing rapid load changes. In analyzing circulation, the procedures outlined in Chapter 5 are applied to the design at the maximum anticipated rate of evaporation.

Generally, steam drum diameters range from 54 to 72 in. (1.37 to 1.83 m) for merchant vessels and from 54 to 60 in. (1.37 to 1.52 m) for naval units. The drum must accommodate the desired number of tube rows and provide sufficient space for steam separation, feedwater introduction, distribution and treatment, and water level fluctuations caused by sudden load variations.

Marine steam drum baffling for lower rated boilers is usually simple in construction and arrangement. A V-type baffle using a triple layer of perforated steel plates is used in moderately rated drum boilers. With relatively small clearances between the plates it is installed (Fig. 17) just below the normal water level to break up steam and water jets issuing from the riser circuits. In higher rated two-drum boilers the duplex compartment baffle is more effective. In all present day B&W marine boiler designs, horizontal steam separators and scrubbers are used (Fig. 18). Cyclone separators and scrubbers are used to ensure that dry steam is sent to the superheater where moisture and dissolved boiler water containment carryover would not be acceptable. The principles of these cyclone separators are similar to those of the vertical cyclones used in stationary units. (See Chapter 5.)

Drum boilers are generally rated conservatively and do not require external downcomers. Adequate circulation can be maintained by the rear rows of boiler bank tubes as downcomers, where the gas temperature is 850F (454C) or less. At low loads, the first several rows of tubes act as risers, and the remaining tubes serve as downcomers. As the firing rate increases, the high temperature gas zone moves deeper into the tube bank, the number of tubes acting as risers increases, and the number of downcomers declines. Excessive firing rates and rates beyond those contemplated in the design can reduce the number of downcomers below the minimum requirements. For highly rated units, it is necessary to install unheated downcomers to supplement the number of boiler bank tubes acting as downcomers. Circulation is benefited by interposing a convection superheater in the boiler bank, because the heat absorbed by the superheater reduces the temperature of the gas flowing over the boiler tubes beyond it and more boiler tubes consequently act as downcomers over a wide operating range. In vertical superheater units, external downcomers are also required because of physical arrangement constraints.

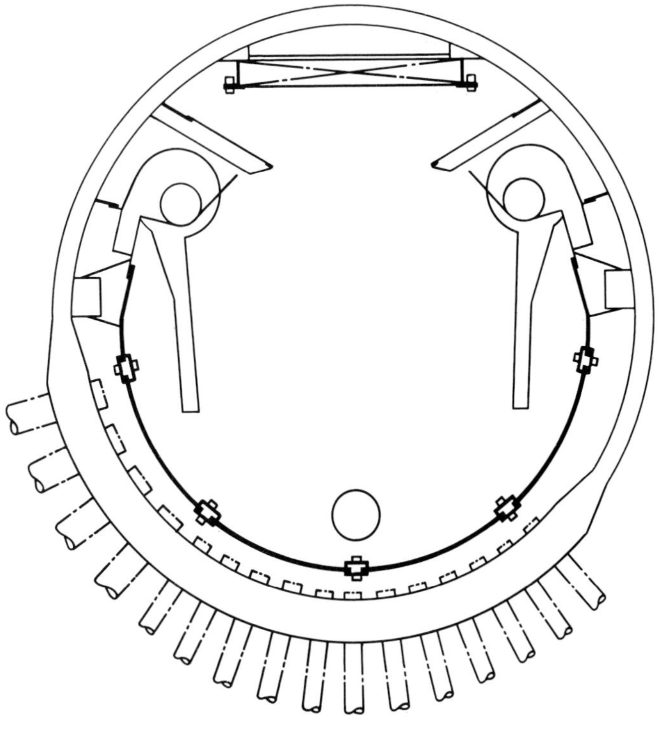

Fig. 18 Arrangement of horizontal cyclone separators in a marine boiler drum.

In most horizontal superheater, drum boilers (Fig. 4), furnace waterwall supply headers receive water from the lower drum through supply tubes located below the furnace floor. These tubes are spaced along the length of the waterwall headers to assure even distribution of water to the high duty furnace tubes and provide cooling for the furnace floor refractory. Naval and other high rated designs have external downcomer tubes from the steam drum to the lower wall header because the supply of water from the boiler bank alone would be inadequate. In vertical superheater drum boilers, (Fig. 3) furnace waterwall supply headers receive water from the steam or lower drum through external downcomers or feeder tubes.

Drum boilers should not be located with the drum longitudinal axis athwart ship. In such a design, rolling motion may stop circulation in some riser tubes located in the sides of the boiler and can result in tube overheating and failures. While this difficulty is alleviated by locating the drums fore and aft, ship rolling can still affect boiler circulation. During rolling, some of the rear boiler tubes in the drum may become uncovered and the overall downcomer water flow to the lower drum reduced. However, marine boilers operating at conservative load ratings are usually able to accommodate such conditions.

Heat recovery steam generator for combined cycle installation.

Chapter 31
Combined Cycles, Waste Heat Recovery and Other Steam Systems

The idea of using waste energy for increased steam generation in industry has been around for many years. The progressive increase in fuel costs, the need to capture heat from various industrial processes and the increasingly stringent environmental regulations have created the need for using waste heat to its fullest potential.

In the power industry, the waste heat from one power system such as a gas turbine can serve as the heat source for a steam turbine cycle. Such *combined cycles* can push overall electrical power cycle efficiency to nearly 50%. Overall energy use can substantially exceed even this when electrical generation is combined with process steam use.

Industries such as steel making, oil refining, pulp and paper, and food processing have used many unique steam generating systems to get the most out of their waste heat. These systems allow reduced consumption of traditional fuels, recovery of waste heat for safety and economy, and elimination of process byproducts. More recent developments for waste energy recovery include a system to destroy hazardous organic elements in wastes that contain sufficient heat content to sustain combustion.

A variety of other systems which use more unconventional sources of energy are also under development and demonstration. Solar and geothermal energy sources are now tapped to produce steam. Development efforts are continuing in advanced systems to combine direct electricity generation with the more conventional steam based Rankine cycle. All of these systems have created the need for specialized designs and applications of steam generating equipment.

Combined cycles and cogeneration

In its broadest terms, a *combined cycle* plant consists of the integration of two or more thermodynamic power cycles to more fully and efficiently convert input energy into work or power. With the advancements in reliability and availability of gas turbines, the term combined cycle plant today usually refers to a system composed of a gas turbine, heat recovery steam generator and a steam turbine. Thermodynamically, this implies the joining of a high temperature Brayton gas turbine cycle with a moderate and low temperature Rankine cycle — the waste heat from the Brayton cycle exhaust serves as the heat input to the Rankine cycle. The challenge in such systems is the degree of integration needed to maximize efficiency at an economic cost. Where the heat recovery steam generator supplies at least part of the steam for a process, the application can be referred to as *cogeneration*. Chapter 2 discusses the thermodynamic processes and benefits of both applications.

While the following brief discussion focuses on the combination of a gas turbine with a heat recovery steam generator and steam turbine, other advanced combined cycle systems are also available or under development. Some of these include pressurized fluidized-bed combustion (PFBC) discussed in Chapter 29, the integrated gasification combined cycle (IGCC) discussed in Chapter 17, and the magnetohydrodynamics (MHD) system discussed later in this chapter. Other combinations for repowering conventional boilers with gas turbines are reviewed in Chapter 57.

Simple combined cycle system

As shown schematically in Figs. 1a and 1b, the simple combined cycle system can consist of a single gas turbine-generator, heat recovery steam generator (HRSG), single steam turbine-generator, condenser and auxiliary systems. In addition, if the environmental regulations require, a selective catalytic reduction (SCR) system to control emissions of nitrogen oxides (NO_x) can be directly integrated within the steam generator. (See Chapter 34.) This is particularly attractive because the SCR catalyst can be positioned in an optimal temperature window within the HRSG. The gas temperature leaving the gas turbine is usually in the range of 950 to 1050F (510 to 566C) while the optimal SCR catalyst temperature is 675 to 840F (357 to 449C).

A variety of more complex configurations are possible. A key improvement in the steam cycle efficiency can be obtained by adding multiple separate pressure circuits to the HRSG to supply low pressure steam for deaeration and feedwater heating. This replaces the steam extraction regenerative feedwater heating used in conventional steam power cycles.

Commercial combined cycle systems

Actual configurations are typically more complex because of application requirements and the degree of integration. The gas turbine-generators, steam turbine-generators, and HRSGs are commercially available in a number of specific sizes and arrangements. Frequently multiple gas turbines with HRSGs may feed a single steam turbine system. A gas bypass stack and silencer

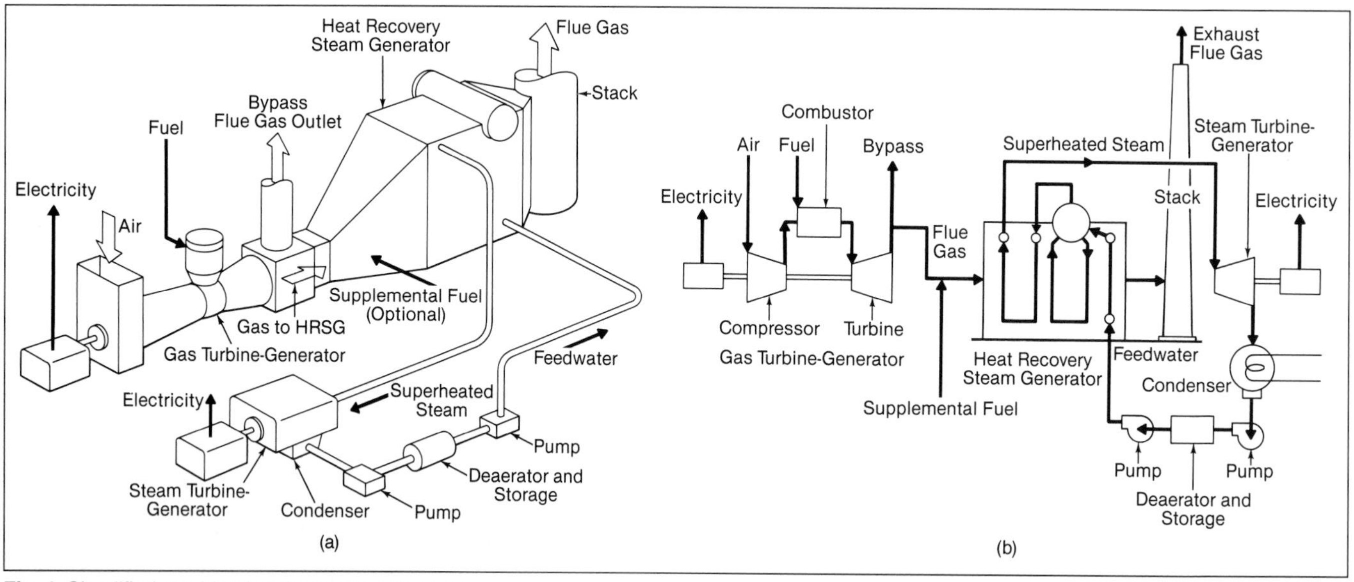

Fig. 1 Simplified combined cycle system schematics.

are frequently installed downstream of the gas turbine so that it can be operated independently of the steam cycle. With the high levels of oxygen remaining in the gas turbine exhaust, supplemental firing systems can be installed upstream of the HRSG. This permits greater operating flexibility, improved steam temperature control, and higher overall power capacity. As discussed below, the HRSG can be designed with one to four separate operating pressure circuits to optimize heat recovery and cycle efficiency. In selected cases further cycle efficiency gains are possible with the addition of steam reheat.

A range of cycle efficiencies is possible depending upon the complexity of the system and components. Sample overall electrical power generation cycle efficiencies for a system using a gas turbine with a 2200F (1204C) turbine inlet temperature are provided in Table 1. These are based upon the fuel higher heating value.

The environmental emissions from combined cycle systems are generally low. If natural gas is fired, sulfur dioxide (SO_2) and particulate emissions will be negligible. With current gas turbine combustor designs, NO_x emissions from the gas turbine will be low, ranging generally between 10 and 70 ppm. The final NO_x emissions then depend upon the supplemental firing system used (if any) and the potential incorporation of an SCR NO_x control system.

Table 1
Sample Cycle Efficiencies and Heat Rates

System	Efficiency (%)	Heat Rate (Btu/kWh)
Simple gas turbine	32	10,700
Gas turbine plus unfired single pressure steam cycle	42	8,200
Advanced gas turbine plus unfired multiple pressure steam system	48	7,100
Gas turbine plus dual pressure steam system plus process steam use (cogeneration)	61	—

Note: All values are calculated using the higher heating value (HHV) of the fuel. Use of the lower heating value (LHV) of the fuel would increase the efficiencies listed.

Beyond thermal efficiency gains and low environmental emissions, potential benefits of a gas turbine combined cycle plant include:

Schedule — Depending upon equipment size and complexity, delivery and construction of the gas turbine can take about one year. The steam turbine cycle system delivery and erection can frequently be completed in an additional year. (See Fig. 2.)

Flexibility — The gas turbine can be used alone for rapid startup, peaking service. The HRSG boiler system can usually be brought from a cold start to full load steam generation in about 60 minutes.

Capital — The system capital cost is typically low as a result of smaller standardized components, modular construction, prompt delivery, rapid erection, and minimum support system costs.

In the system selection, these benefits must be weighed against the usually higher cost of the cleaner premium

Fig. 2 Rapid installation of modular components minimizes schedules.

fuels needed for the gas turbine, potential maintenance and availability issues, and load requirements.

Cogeneration

While gas turbine combined cycle systems focus on the production of electricity, the gas turbine-generator and heat recovery steam generator portion of the system can also be adapted to supplying process steam as well as electrical power. In such systems, the waste heat from the gas turbine can be used to produce steam for space heating or process heat input. Through utilization of the waste heat, the total energy utilization can approach 80% as compared to the 40 to 50% in the best gas turbine combined cycle system without process steam use.

Heat recovery steam generator

The heat recovery steam generator (HRSG) is sometimes referred to as a waste heat recovery boiler (WHRB) or turbine exhaust gas (TEG) boiler. The latter designation indicates the main application for these units today — waste heat recovery and steam generation from gas turbine exhaust gas. The HRSG is a key element in a combined cycle plant affecting initial capital cost, operating cost and overall cycle efficiency.

HRSGs are flexible in design depending upon the specific application. The gas flow through the unit can either be horizontal or vertical with a tradeoff being made between the cost of floor space for the horizontal flow unit and structural steel requirements of the vertical flow unit. The horizontal flow case is the most common. The HRSG may be designed for operation with multiple separate pressure steam-water loops to meet application requirements and maximize heat recovery. Circulation may be forced or natural (see Chapter 5) with most horizontal gas flow units using natural circulation. HRSGs may be unfired, i.e., only use the sensible heat of the gas as supplied, or may include supplemental fuel firing to raise the gas temperature to reduce heat transfer surface requirements, increase steam production, control superheat steam temperature, or meet process steam temperature requirements. Finally, HRSGs may be designed to incorporate an SCR catalyst as discussed in Chapter 34.

Babcock & Wilcox (B&W) supplies a full range of HRSGs suitable for use with gas turbine sizes, from a 1 MW unit up to the current state-of-the-art 220 MW machines. (See Figs. 3, 4, and 5.)

B&W heat recovery steam generators are modular, natural circulation designs, applicable to the unique re-

Fig. 4 Large HRSG for enhanced oil recovery/cogeneration.

quirements of a variety of combined cycle systems. Depending on the cycle configuration, the HRSG may consist of one to four separate boiler circuits (high pressure, two types of intermediate pressure and low pressure) within the same casing. The high pressure boiler is used for power generation [up to 1005F (541C) superheat]. The intermediate pressure boilers can be used for power generation, steam injection for NO_x control (water or steam injected into the gas turbine combustor to limit NO_x formation) and/or process steam. The low pressure boiler is normally used for feedwater heating and/or deaeration.

These boilers are designed to handle large gas flows with minimal pressure drop resulting in greater gas turbine net electrical output. Special consideration is also given to the configuration of the interconnecting flue gas ductwork, transitions, and diverter valves to minimize pressure losses caused by changes in flow patterns or excessively high velocities. Heat losses through the boiler casing and ductwork are minimized through the use of an insulation system.

In the natural circulation design, the vertical tubes provide the pumping head necessary to develop sufficient circulation ratios for stable circulation. This design also produces a more rapid response during load transients, common with combined cycle systems. Circulation pumps are eliminated.

Because most heat recovery requirements are too large to permit shop assembly of an entire HRSG, B&W units are designed for maximum shop assembly of pressure parts. The basic unit of construction, the *section*, consists

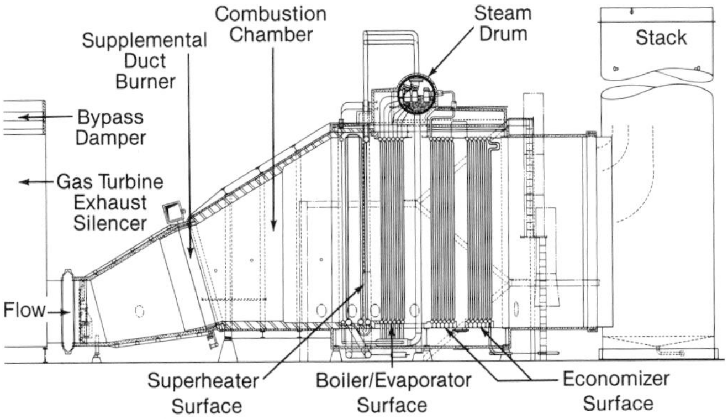

Fig. 3 Heat recovery steam generator (HRSG) arrangement.

Fig. 5 HRSG small scale installation.

of staggered rows of helically finned tubes welded into the upper and lower headers. These sections are then assembled into *modules* for shipment. Some of the smaller HRSG units can be entirely shop-assembled and skid mounted.

To accommodate rapid startup, the design includes provisions for differential expansion between pressure parts and nonpressure parts. Expansion joints are used where a gas-tight seal is required.

Typical parameter ranges for B&W HRSG units are summarized in Table 2.

Special HRSG designs are also available for enhanced oil recovery applications. The gas turbine exhaust gas is used in the HRSG to generate wet steam (approximately 80% quality) at pressures up to 2500 psig (172 bar gauge). The steam is injected into wells to enhance heavy oil recovery. A unique feature is that relatively dirty (up to 10,000 ppm dissolved solids) feedwater is used in a once-through steam generator design. See Chapter 25 for a discussion of direct-fired steam generation for this application.

Technical considerations The HRSG is basically a counterflow heat exchanger composed of a series of superheater, boiler (or evaporator), and economizer sections positioned from gas inlet to gas outlet to maximize heat recovery and supply the rated steam flow at proper temperature and pressure. To provide the most economical design, it is necessary to evaluate five key parameters:

1. allowable back-pressure,
2. steam pressure and temperature,
3. pinch point,
4. superheater and economizer approach temperatures, and
5. stack outlet temperature.

These parameters are evaluated and set based upon experience and economic considerations. Back-pressure is significantly influenced by the HRSG cross-sectional flow area. Higher back-pressures reduce HRSG cost but also reduce gas turbine efficiency. Back-pressures are typically 10 to 15 in. wg (2.5 to 3.7 kPa) in most units.

The pinch point temperature and approach temperatures have a significant impact on overall unit size. They are illustrated in Fig. 6 for a single pressure HRSG. Small pinch point and superheater approach temperatures result in larger heat transfer surfaces and higher capital costs, while the economizer approach temperature is typically set to avoid economizer steaming at the design point. Experience has generally indicated that the following ranges provide economical and technically satisfactory designs although lower values may be appropriate in specific applications:

Pinch point = ΔT_P = 20 to 50F (11 to 28C)
Superheater approach = ΔT_{SH} = 40 to 60F (22 to 33C)
Economizer approach = ΔT_E = 10 to 30F (6 to 17C)

The economizer inlet water temperature must be set to minimize acid dew point corrosion if sulfur is present in the flue gas. The minimum feedwater temperature is typically set at 240F (116C) if sulfur is present in the fuel. As with the feedwater temperature, the minimum flue gas exit or stack temperature needs to be controlled to avoid corrosion due to acid condensation. Typical values are identified in Chapter 19.

The steam pressure and temperature are selected to provide an economical design. In general, higher steam pressures increase system efficiency but can limit total heat recovery from the flue gas in single pressure HRSGs due to the higher saturation temperature. To overcome this problem, multiple pressure HRSGs are offered. One to four separate pressure sections may be used. The superheater, boiler and economizer sections at each pressure are arranged to reduce overall cost and increase heat recovery.

Steaming in the economizer is inevitable at off-design points, and HRSG economizers must incorporate features to prevent problems. Such features include: 1) up-flow in the final section before the drum, 2) a recirculation line to be used during startup to minimize steam generation while there is no feedwater flow, and 3) the feedwater passing through steam-water separation equipment in the drum.

Table 2
HRSG Parameters

Turbine application:	
Gas turbine sizes	1 MW to 220 MW
Gas flow	25,000 to 5,000,000 lb/h (0.32 to 630 kg/s)
Gas turbine outlet temperature	≤ 1200F (≤ 649C)
Supplemental firing temperature	≤ 1650F (≤ 899C)
Steam flow:	15,000 to 600,000 lb/h (1.9 to 76 kg/s)
Operating pressures:	
High	≥ 400 psig (28 bar gauge)
Intermediate	50 to 400 psig (3.4 to 28 bar gauge)
Low	15 to 50 psig (1.03 to 3.4 bar gauge)
Steam temperature	up to 1005F (541C)
Supplemental fuels	#2 oil, natural gas

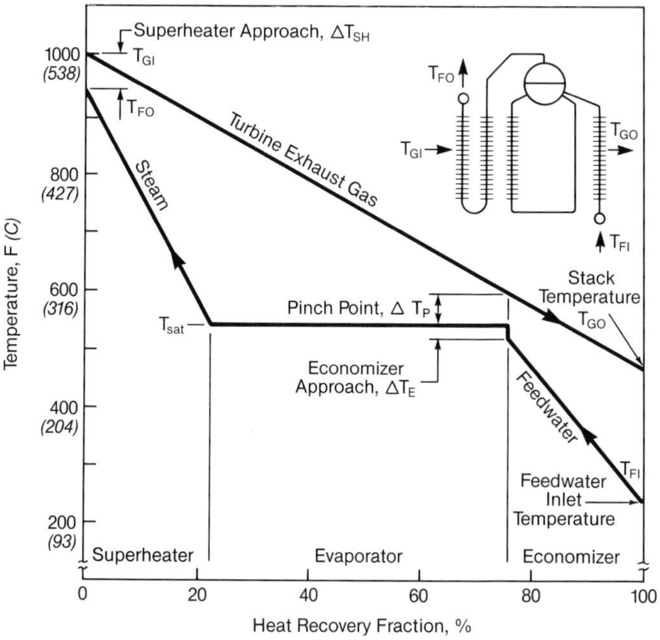

Fig. 6 Temperature profile in single pressure HRSG.

Steam systems based on waste heat

As fuel costs increase, use of the heat in the exhaust gases from industrial process furnaces becomes more important. Properly designed boiler equipment absorbs the heat in what was formerly waste gas, and can often generate all the steam required for an industrial process. Where the waste gases carry some of the noncombustible process material in suspension, suitable hoppers will collect a portion of the material, and the cooled gases leaving the boiler may be passed through dust collectors to recover the remaining particulate. Many types of boilers are necessary to meet the wide range of requirements in this field. Boiler design depends on the chemical nature of the gases and their temperature, pressure, quantity and dust loading.

Heat transfer from waste gases

The rate of heat transfer from the gas to the boiler water depends on the temperature and thermophysical properties of the gases, velocity and direction of flow over the absorbing surfaces, and the surface cleanliness, as discussed in Chapter 4. Temperatures of many process gases are relatively low as shown in Table 3.

To obtain the proper velocity of the gases over the surfaces, a sufficient pressure difference or draft must be provided, either by a stack or a fan. The draft must overcome the pressure losses caused by the flow of gases through the unit, with adequate allowance for normal heating surface fouling.

The radiation heat transfer component is low and the tendency is to design many waste heat boilers for higher gas velocities than prevail on fuel-fired units. However, high velocities with dust-laden gases can cause tube abrasion, particularly where there are changes in gas flow direction. Therefore, process-specific velocity limits must be met.

Diagrams A and B, Fig. 7, show the approximate convection heating surface required for usual conditions in waste heat boilers.

A water-cooled furnace is a feature of many waste heat boiler applications. This open furnace cools the gases to the temperature necessary to prevent slagging in the following tightly spaced convection surfaces. The approximate amount of surface required for such furnace applications is given in Diagram C of Fig. 7.

Table 3
Temperature of Waste Heat Gases

Source of Gas	Temperature F	Temperature C
Ammonia oxidation process	1350 to 1475	732 to 802
Annealing furnace	1100 to 2000	593 to 1093
Cement kiln (dry process)	1150 to 1500	621 to 816
Cement kiln (wet process)	800 to 1100	427 to 593
Copper reverberatory furnace	2000 to 2500	1093 to 1371
Diesel engine exhaust	1000 to 1200	538 to 649
Forge and billet heating furnaces	1700 to 2200	927 to 1204
Open hearth steel furnace, air blown	1000 to 1300	538 to 704
Open hearth steel furnace, oxygen blown	1300 to 2100	704 to 1149
Basic oxygen furnace	3000 to 3500	1649 to 1927
Petroleum refinery	1000 to 1400	538 to 760
Sulfur ore processing	1600 to 1900	871 to 1038

Application factors

The design of a boiler for a particular application depends on many factors, most of which vary from process to process and even within an industry. The cost of equipment, auxiliary power and maintenance must be compared with the expected savings. The boiler design depends somewhat on the cost of power at the plant. A smaller unit with closely spaced tubes will cost less but will require more fan power, because of high draft loss. A larger unit will be more expensive but have a lower draft loss. Other important factors are the available space, locations for the proper flue connections, the corrosive nature of the gases, the effect of dust loading on

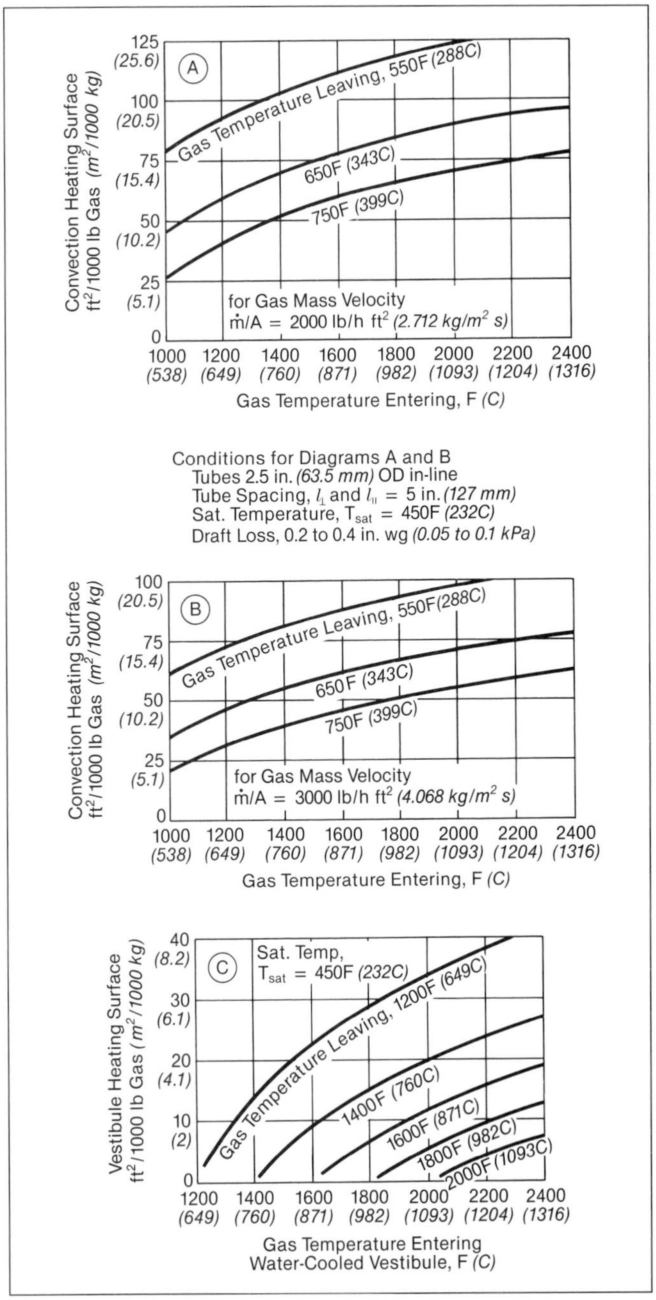

Fig. 7 Approximate surface required in convection tube bank and vestibule for various entering and leaving waste gas temperatures.

erosion, and the process operating pressure conditions, i.e., pressurized or induced draft.

If the gases carry dust, attention must be given to tube spacing and to removal of dust dislodged from the heating surfaces. The tubes must be spaced reasonably close for good heat transfer yet far enough apart to prevent bridging of deposits or excessive pressure loss. Often the boiler is arranged for wide tube spacing where the gases are hottest, with spacing reduced where the gases are cooler to maintain gas velocities and good heat transfer.

Sometimes the particulate or dust carried into the boiler from the process can be removed by mechanical cleaners or sootblowers (see Chapter 23). In other cases, deposits from processes may require periodic manual cleaning with high pressure air, steam or water to keep the boiler passes open. In either case, suitable hoppers should be provided to collect the deposit material removed from the tubes.

Gases from oil- or gas-fired process furnaces are relatively clean and, therefore, can be used in units with 1.0 in. (25.4 mm) clear spacing between tubes with little likelihood of deposit bridging and plugging.

Basic oxygen furnace/oxygen converter hood

Unlike older steelmaking furnaces, the basic oxygen furnace (BOF), Fig. 8, is blown with pure oxygen through a retractable water-cooled lance mounted vertically above the furnace. After each charge of molten iron, scrap steel and fluxing material are loaded into the furnace, the oxygen lance is lowered into position above the charged material and a blowing period of 15 to 20 minutes begins.

During the blowing time, the oxygen starts a chemical reaction which brings the charge up to temperature by burning out the silicon and phosphorous impurities and reducing the carbon content, as required, for high grade steel. Large amounts of carbon monoxide gas, at 3000 to 3500F (1649 to 1927C), are released in this conversion process. The gas is collected in a water-cooled hood and burned with air introduced at the mouth of the hood. The products of combustion are then cooled by adding excess air, injecting spray water, or water cooling of the hood. Combinations of these cooling methods may be used.

Fig. 8 Basic oxygen furnace.

The high temperature of the gases discharged from the BOF and their high carbon monoxide content (about 70% by volume) make them ideal for burning in the water-cooled hood located above the BOF. While there are basic similarities to usual boiler service, there are also significant differences, particularly the carryover of iron bearing slag from the BOF and the short, intermittent operating periods. The criteria established for good hood design should include the following:

1. Adequate structural strength: This is a rugged type of service where equipment is handled roughly, and the oxygen converter hood should be built accordingly.
2. Smooth hood surface: The hood surface that is in contact with the furnace gases must be smooth to more easily shed the skulls or slag that are heavy with iron.
3. Minimum of crevices: There should be a minimum of crevices, cracks, sharp corners and openings in the fore part of the hood which would permit anchorage of the slag.
4. Uniform water cooling: Positive and uniform water cooling of all surfaces exposed to the furnace gases is important. There should be a minimum of temperature differences between any and all water circuits with no eddy currents or uncooled corners.
5. Hood waterwalls: The hood waterwalls should be cooled with treated and deaerated water to prevent internal deposits of hard scale or oxygen corrosion. Good quality water is necessary because of the high rate of heat absorption.
6. Water circulation: Water circulation should be maintained at a high rate through the entire cycle. Recirculation makes possible high flow rates without excessive makeup.
7. Hood cooling system: The hood cooling system should be suitable for pressurizing to permit the generation of steam or high temperature water in keeping with industrial boiler practice.

Membrane wall construction for oxygen furnace hoods meets these design criteria (Fig. 9). The membrane wall can be formed into a variety of hood configurations depending on plant layout. The hood may be a long flue type used to transport the gases to an evaporation or quench chamber, or it may be a bonnet type that collects the gases and immediately discharges them into a *spark box* where the temperature is sufficiently reduced with spray water for use in the gas cleanup system.

The hood with water-cooled membrane walls may be applied to the oxygen converter process in one of the following ways:

1. It may be operated as a boiler at pressures from 100 to 1500 psi (6.9 to 103.4 bar) to generate steam for general plant use.
2. It may generate steam which is condensed in a closed system with the heat dissipated by an air-cooled heat exchanger.
3. Water in a closed system may be heated in the membrane walls of the hood and the heat dissipated by an air-cooled heat exchanger.

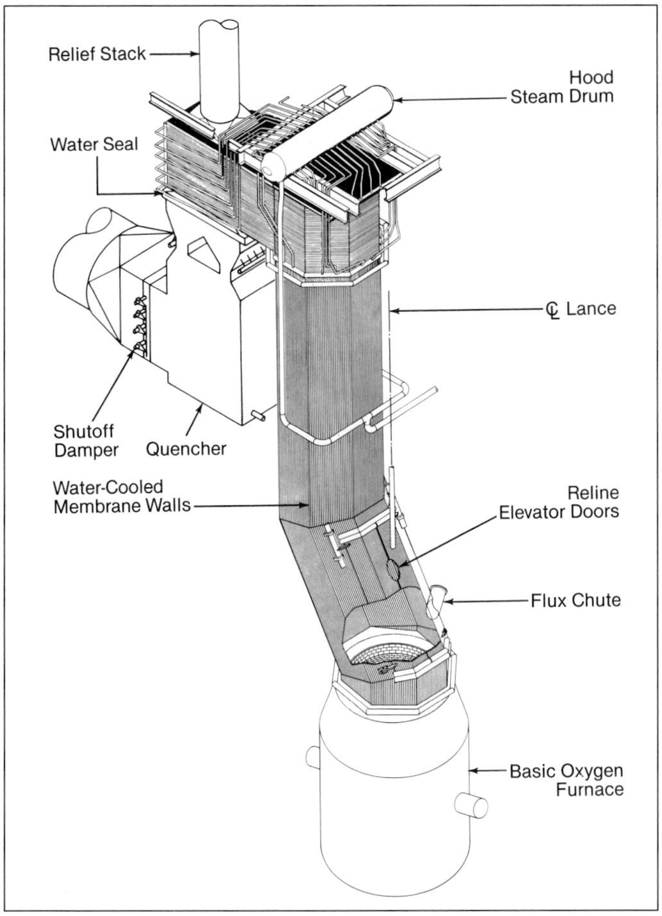

Fig. 9 Oxygen converter hood arrangement with wet scrubber.

Steam generator hood The oxygen converter hood, when equipped with a steam drum, boiler circulation pumps, boiler mountings and controls, is an effective steam generator during the oxygen blowing portion of the converter cycle. Steam generation is limited to the time of the oxygen blowing period, which is intermittent or cyclical in operation because of the steelmaking process. The rate of steam generation varies from zero to a maximum and back to zero during this period which is normally about 20 minutes for a converter cycle of 40 to 45 minutes. This cyclic operation, coupled with the outage time required for relining the converter vessel every few weeks, limits the steam production of a single hood to 12 to 15% of the life of the lining.

The steam, during its period of generation, may be discharged directly into the plant steam system. The cyclic type of operation, combined with short period high rates of generation, imposes widely fluctuating load swings within the steam system. The effect of load swings can be reduced by operating the hood boiler unit at a higher pressure and discharging the steam into an accumulator. Heat is stored in water at saturation temperature by the rising accumulator pressure. When the steam production rate in the hood boiler drops, the heat stored in the accumulator is released to produce steam at the lower plant pressure.

Steam-pressurized hood closed circuit system There are BOF plants which can not utilize the potential steam production of hood boilers. However, if use can be anticipated, the units can be arranged for closed circuit operation as illustrated in Fig. 10. The closed circuit system assures an

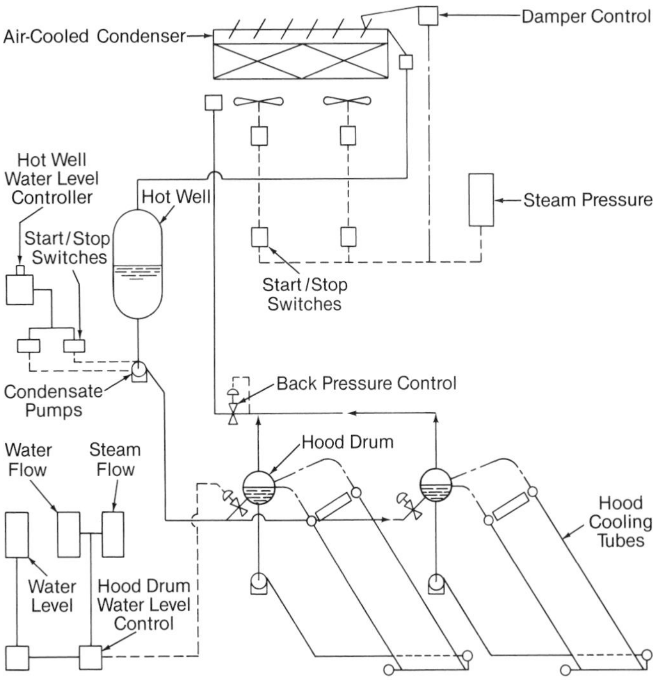

Fig. 10 Closed-circuit steam generating hood with air-cooled condenser.

ample supply of good quality boiler water without the need for an elaborate treatment plant. A portion of the heat absorbed during the blowing period raises system pressure to 250 to 450 psi (17 to 31 bar). Any excess heat is discharged to the atmosphere through the air-cooled condenser operating under system pressure. The condensate collected is then returned to the hot well and from there to the hood drum, completing the cycle.

The air-cooled condenser of the closed pressurized circuit is small because of the high temperature difference, about 350F (194C), between condensing steam and cooling air. The power required for dissipating the heat is small when compared to the pump power for an equivalent system using cooling water as the condensing medium. The power required for water circulation is also small. Makeup is limited to the wastage that occurs at pump packings and valve stems, plus blowdown.

This closed circuit system can be altered to supply plant steam by tapping into the steam line from the hood. Steam can be taken from the hood drum, thereby reducing the load on the air-cooled condenser.

Steam-pressurized, high temperature water, closed circuit system Some steel mills do not find it advantageous to recover heat absorbed by the hoods. For these plants, the steam-pressurized, high temperature water, closed circuit system is preferred. This installation is simpler to control and is less costly than the equivalent closed circuit steam generating system.

The steam-pressurized, high temperature water system serves the same purpose as the closed circuit steam system. The only real difference is that saturation temperature water is generated in the hood and discharged to the system's steam-pressurized expansion tank. The saturation temperature water from the expansion tank is pumped through the air-cooled heat exchanger to lower its temperature and to return it to the hood completing the circuit. With this system, the high temperature water is pressurized to 250 to 450 psi (17 to 31 bar) by controlling the air flow over the heat exchanger.

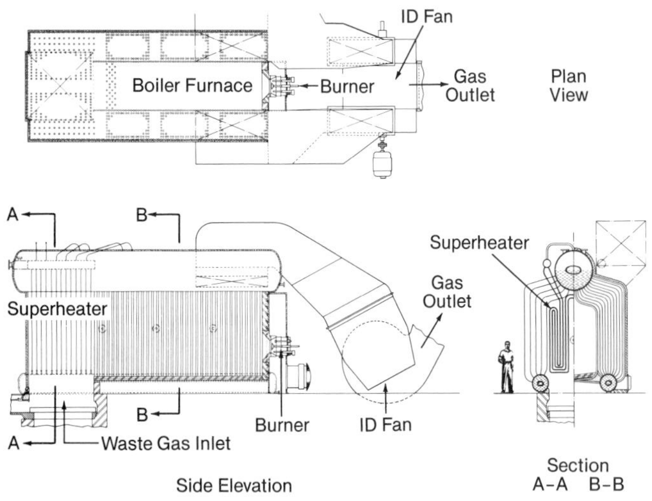

Fig. 11 Waste heat boiler for oxygen blown open hearth furnace.

Waste heat boilers for open hearths

The open hearth process for producing steel has been in use since the 1950s. This process produces a highly dust-laden waste gas with gas temperatures as high as 2100F (1150C). Waste heat boilers (see Fig. 11) are used to recover this waste heat.

For many installations, it is desirable to maintain the steam flow, during furnace charging and maintenance periods, by firing an auxiliary fuel, which requires a boiler furnace for its combustion. Therefore, the waste heat boiler for oxygen blown open hearths is a versatile unit that considers the space availability, amount of waste gas, steam capacity, cleanability and supplementary fuel firing. The steam capacity for a single unit serving a single open hearth may be as high as 150,000 lb/h (18.9 kg/s).

Waste heat boilers for special conditions

Other types of waste heat boilers are used to recover the heat from waste gases or from industrial process fluids; these boilers are often designed for special conditions of space, temperature, pressure and draft. When all of the requirements are known, waste heat recovery boilers can be designed to accommodate almost any set of conditions.

The progressive increase in the cost of fuel has fostered technical progress in the utilization of waste energy, including specialized designs and applications of boilers. Many industries have specific problems with waste product recovery. For example, the oil industry has a tremendous source of energy in the gas discharged from the catalytic regenerators; steel mills have their blast furnace gas; the sugar industry has its cane refuse; and the wood and pulp industry has sawdust, hogged fuel, bark and process liquors as wastes. All these waste products or residuals can be burned as fuels to generate steam. In addition, there are industries or processes that have large quantities of high temperature gases from which the sensible heat may be extracted for steam generation. Such gases are produced in copper reverberatory furnaces; annealing, forge and billet heating furnaces; and fired kilns of many types.

Three-drum waste heat boiler An example of a simple three-drum waste heat boiler, specially designed to operate with dust-laden gases, is shown in Fig. 12. This type of unit is particularly well suited for use with high solids content waste gases from cement kilns. Maximum precipitation of solids is assured by the horizontal flow of the gases through the vertical tube banks and by the effective arrangement of baffles. With this tube arrangement, hand lancing is possible from both sides of the unit. Every space in the full width of the boiler can be lanced. Hand lancing can also be done through the roof and above the lower drums, making all absorbing surfaces accessible with a short hand lance.

With high solids content gases, it is often possible to reduce the amount of hand lancing by using long retractable sootblowers. These are located at one or sometimes two elevations along the depth of tube banks at gaps formed by eliminating a single row of tubes.

To maintain optimum heat transfer conditions without changing the direction of gas flow, the tubes in the rear sections of the boiler are more closely spaced than those at the inlet.

Circulation in this boiler is simple, with the boiler tubes in the hot gas end acting as risers and the tubes in the cooler gas zones acting as downcomers or supply tubes. The boiler has a relatively long drum in which steam separation takes place without the use of baffles. The steam is collected in a dry pipe located in the quiet, cool gas end of the drum. Feedwater is thoroughly mixed with the boiler water rising into the steam drum.

Expansion and contraction of the drums and tubes have no effect on the steel casing, firebrick, or insulation. The most common source of brickwork trouble is therefore eliminated and air infiltration is reduced to a minimum. All pressure parts rest on supports located below the lower drums.

The location of the superheater can be altered for superheat temperature requirements. To increase heat absorption further, an economizer can be installed in the flue leaving the boiler. The economizer is arranged for downward flow of gases to aid in the collection of solids.

Solids collected in the hoppers under the boiler and economizer can be readily removed while the boiler is in service. In a single boiler behind a cement kiln, from 20 to 40 t (18.14 to 36.3 t_m) of cement dust may be recovered each day from these hoppers.

Steam producing systems with unique fuels

CO boilers

In the petroleum industry, the operation of a fluid catalytic cracking (FCC) unit, depending on the FCC arrangement, produces gases rich in carbon monoxide (CO). To reclaim the thermal energy in these gases, the fluid catalytic cracking unit can be designed to include a CO boiler to generate steam.

For refineries generating large quantities of CO, field-erected boilers such as the Integral Furnace boiler are used. (See Chapter 25.) There are also many smaller refineries that have cracking units in the general range of 12,000 bbl (1908 m^3) per day or less which produce from 75,000 to 175,000 lb/h (9.5 to 22.1 kg/s) of CO gas. CO boilers for this capacity are often small enough to be shop-assembled. A shop-assembled boiler modified for CO firing is shown in Fig. 13. CO gas is admitted through ports in the sidewalls and frontwall to promote mixing and rapid combustion. The

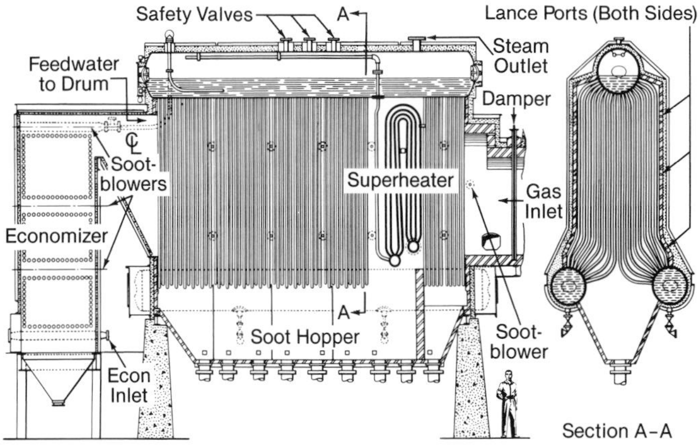

Fig. 12 Three-drum waste heat boiler with lance ports and sootblowers.

burners, for firing supplementary fuel, are located in a refractory frontwall and fire into a horizontal furnace.

The maximum steam requirements of the cracking unit may occur at normal, full load operation or during startup of the cracking unit, depending on the plant steam cycle. The supply of CO is normally not sufficient to generate the maximum amount of steam required; supplementary fuel is then needed.

Supplementary fuel is also required to raise the temperature of the CO gases to the ignition point and to assure complete burning of the combustibles in the CO gas stream. The following design criteria have been established:

1. The basic firing rate should produce a temperature of 1800F (982C) in the furnace to provide safe and stable combustion of the fuels.
2. Air is supplied by the forced draft fan to provide 2% oxygen leaving the unit when burning CO gases and supplementary fuel.
3. Supplementary firing equipment is provided which is capable of raising the temperature of the CO gases to 1450F (788C), the temperature needed for ignition of the combustibles.

Because of possible variations in the combustible and oxygen content of the CO gases, the sensible heat of the CO gases and the amount of supplementary firing, it is impractical to set up a fuel-air relationship. Consequently, it is necessary to directly determine the amount of excess oxygen leaving the unit. This may be determined intermittently by an Orsat oxygen analyzer, or continuously by an oxygen analyzer or combination oxygen-combustible recorder.

Water seal tanks are installed upstream of the CO boiler to act as shutoff valves in the large gas lines so that the CO gases from the catalyst regenerator may be passed through the boiler or sent directly to the stack. This permits independent operation of the CO boiler without interfering with the operation of the regenerator. Water seal tanks are preferred to mechanical shutoff dampers because of the high gas temperature, the size of the CO ducts and the need for leakproof construction.

The operation of the CO boiler is coordinated with that of the catalytic cracker. Normally, the boiler will be required to supply steam for the operation of the catalytic cracking unit and will be started using supplementary fuel. The CO boiler should always be started using only the supplementary fuel burners and bypassing the gases from the regenerator to the atmosphere. CO gases should not be introduced into the boiler until it is brought up to temperature because these gases usually are at or below 1000F (538C) and, consequently, tend to cool the furnace. They ignite quite readily and burn with a nonluminous flame. As the CO is introduced to the boiler, it is necessary to reduce the supplementary fuel and the combustion air. This readjustment in the air requirement is determined from the oxygen recorder reading.

Because there are only slight variations in the operation of the catalytic cracking unit, the CO boiler is normally base loaded. It handles all the gases from the regenerator regardless of the carbon dioxide (CO_2) to CO ratio. A change in this ratio merely affects the quantity of supplementary fuel necessary to maintain the required furnace temperature of 1800F (982C). This temperature provides a reasonable operating margin for possible variation in the operation of the regenerator or the boiler. Stable operation can be maintained at a furnace temperature as low as 1500F (816C), but the margin above the ignition temperature of the CO gas is considerably reduced.

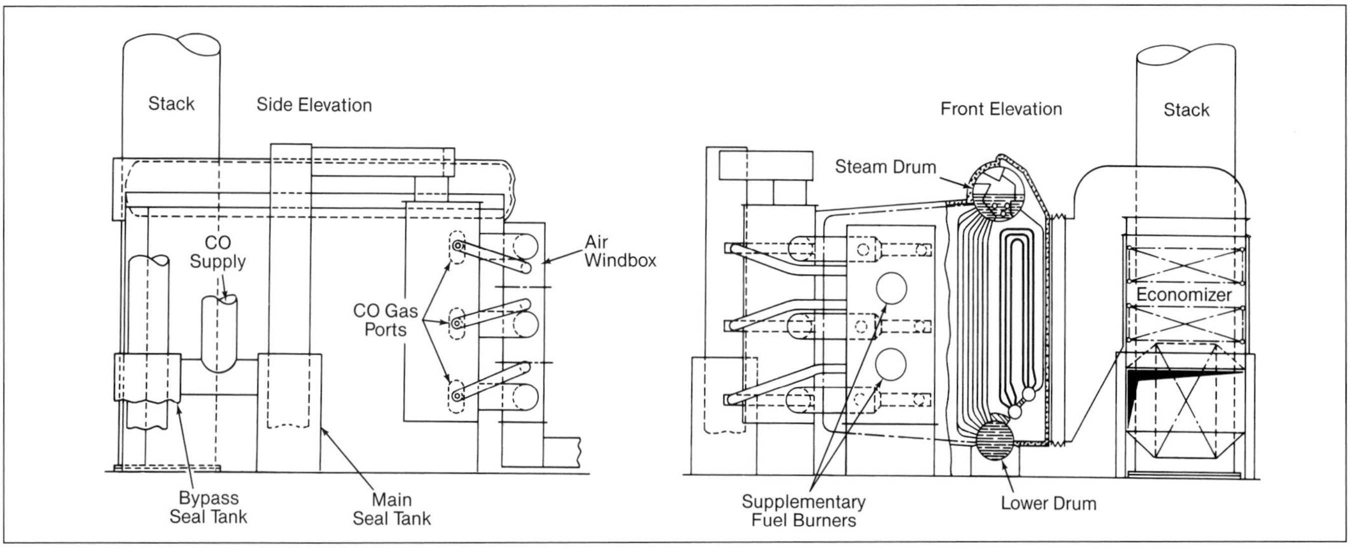

Fig. 13 Shop-assembled CO boiler.

The economics of the CO boiler depend on the amount of available heat in the regenerator exhaust compared with an equivalent amount of heat from an alternate fuel. The heat from the CO gases is calculated by taking the sensible heat above an assumed boiler stack temperature plus all the heat from the combustibles. The additional steam generated in the CO boiler by the supplementary fuel is comparable with the steam generated in a conventional power boiler. Normally, the supplementary fuel requirement will account for one fourth to one third of the output when the temperature of the entering CO gas is maintained at 1000F (538C).

Changes in FCC catalysts and in process conditions have permitted reductions in the CO content of the gases leaving the unit. These changes also result in the gas temperature to the CO boiler increasing from the 1000F (538C) level to as high as 1450F (788C). Many existing CO boilers have been upgraded to handle the new conditions. Changes have included elimination of combustion zone refractory, use of membraned water walls and resizing of heat transfer surface. (See Fig. 14.) New heat recovery boilers on FCC units are designed for these new process conditions.

Boilers burning blast furnace and coke oven gases

Steel mill processes generate blast furnace gas and coke oven gas (COG). Blast furnace gas contains about 25% CO by volume and is heavily dust-laden. COG has a high free hydrogen content and, therefore, burns readily. Blast furnace gas and COG are used as fuel for steam generation in steel mills. Typically COG is used for continuous pilots and main fuel on units which burn blast furnace gas. Blast furnace gas is often cleaned by washing and electrostatic precipitation before entering the burners of the boilers.

A safety consideration when using blast furnace gas is handling the upset when the furnace undergoes a *slip*. The charge of iron, coke and chemicals in a blast furnace occasionally arches over. The collapse of this arch is known as a furnace slip. A slip causes a momentary surge in gas pressure throughout the system which can extinguish the flames from the burners. Consequently, provision for immediate re-ignition is made in the design of blast furnace gas-fired boilers to prevent serious explosions of furnace gas. Modern boiler units use little or no refractory in order to minimize maintenance, and continuous burning pilots are used to re-establish ignition after a slip.

Boilers burning RCRA-defined hazardous wastes

In 1976, the United States (U.S.) Congress passed the original Resource Conservation and Recovery Act (RCRA). Under this act, materials are reviewed based upon certain criteria: ignitability, reactivity, toxicity and corrosivity. If a material exceeds the requirements for any of these criteria, the material is a RCRA hazardous waste. Disposal of an official hazardous waste is subject to RCRA regulations. These regulations contain many

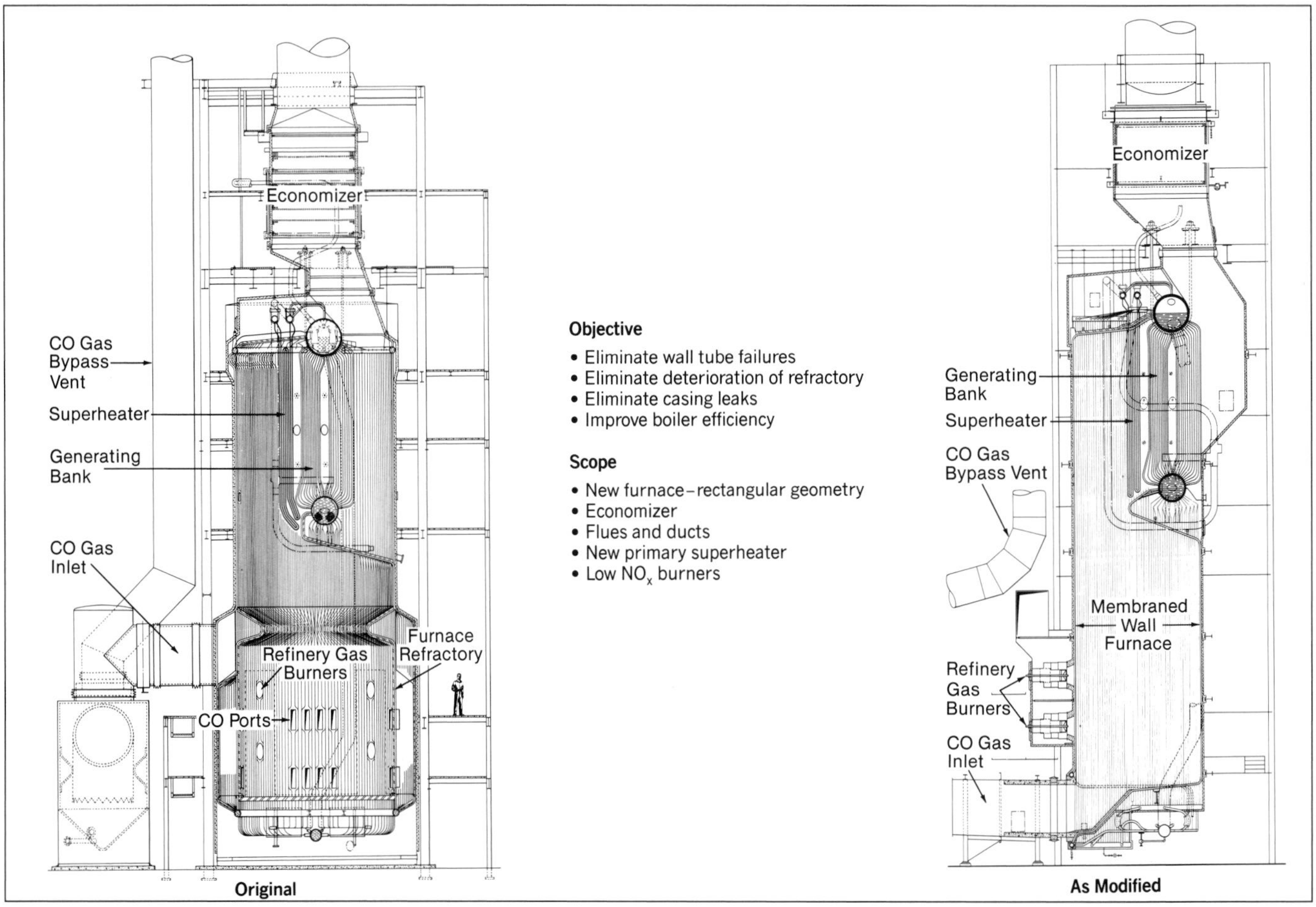

Fig. 14 Circular CO boiler modernization.

requirements including a minimum destruction and removal efficiency (DRE) when hazardous organic wastes are burned. The DRE requirement is presently 99.99% for most principal organic hazardous constituents (POHCs). Exceptions include polychlorinated biphenyls (PCBs) and dioxins, which must be destroyed to 99.9999% because of greater concern about the health effects. DRE is defined as:

$$DRE = [(W_{in} - W_{out}) / W_{in}] \times 100)$$

where

W_{in} = mass feed rates of the POHCs in the waste stream feeding the combustion unit
W_{out} = mass emission rate of the POHCs present in the exhaust emission prior to the release to atmosphere

Many official RCRA hazardous wastes are suitable boiler fuels and many liquid wastes are disposed of in boilers, serving the dual purpose of destroying the POHCs and generating steam for plant requirements.

The design of steam boilers to meet RCRA emission regulations requires attention to both the combustion process and the emissions issues.

Combustion must provide the required DRE of the hazardous constituents, in addition to traditional combustion considerations. The specific waste fuel dictates the boiler and burner design. In many instances, furnace refractory in the combustion chamber, hot combustion air, limited individual burner inputs and conservatively sized furnaces are the approaches to ensuring the 99.99% DRE.

Emissions requirements include the traditional controls of NO_x, SO_2 and particulate. In addition, control of heavy metal emissions and hydrochloric acid (HCl) are required. The level of emissions permitted depends on federal and local air quality control agency requirements. (See Chapter 32.)

Systems with unique energy sources

Geothermal systems for producing steam

Geothermal power uses the heat of the earth's interior to generate steam for low temperature, low pressure turbine-generators.

Geothermal power potential is limited to selected locations where suitable high temperature subterranean rock formations are present and economically accessible through drilling technology. Projects have been developed in several countries around the world including Iceland, Indonesia, Italy, the Philippines and the U.S.

Two technologies are frequently used to recover this energy. The preferred method involves production by obtaining dry steam directly from the earth. Fig. 15 provides a simplified schematic. However, the more common method requires the use of a flash evaporator. The hot, higher pressure water is drawn from the earth and passed through a flash evaporator vessel. In the evaporator, the pressure of the hot water is reduced and a portion of the hot water is flashed to steam which is then sent to a low pressure turbine. The remaining well water is returned to re-injection wells. This process is presented in Fig. 16.

Key parameters in geothermal power development are the purity and chemical composition of the hot feedwater which is often quite brackish. Plants utilizing hot water containing totally dissolved solids as high as 300,000 ppm have been installed. Poor water quality requires the use of expensive alloy steam turbines and flash vessels which increase plant cost and, in some situations, result in reduced plant availability.

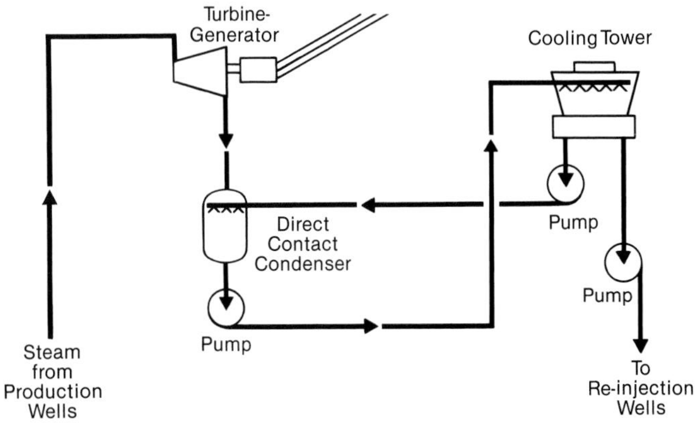

Fig. 15 Schematic of a dry steam, geothermal power system.

Solar power generation system

Generation of electric power with solar energy has achieved operational reliability. These plants include three key subsystems: a solar energy collection system, a solar receiver and the steam turbine-generator. Fig. 17 shows a solar plant conceptual schematic.

Advantages of solar energy include its inherent environmental benefits and its use of a cost free energy source. These must be balanced against the relatively high capital equipment requirements and only periodic availability of the energy source.

Solar energy production is dependent on the daily and seasonal variations in the sun's energy transmitting patterns. Variations in cloud cover cause random and unplanned system upsets which must be automatically compensated for. Fig. 18 shows seasonal and daily variations in absorbed thermal power. Fig. 19 illustrates the variations in absorbed thermal power due to cloud cover taken at short time intervals over the desert at Daggett, California, where one would expect limited cloud cover.[1] To ensure reliable power generation, solar plants usually incorporate backup fossil-fired boilers, or energy storage within the system.

Conceptual cycle Fig. 20 shows a conceptual cycle which uses thermal storage instead of fossil-fired equipment to provide backup when energy production falls be-

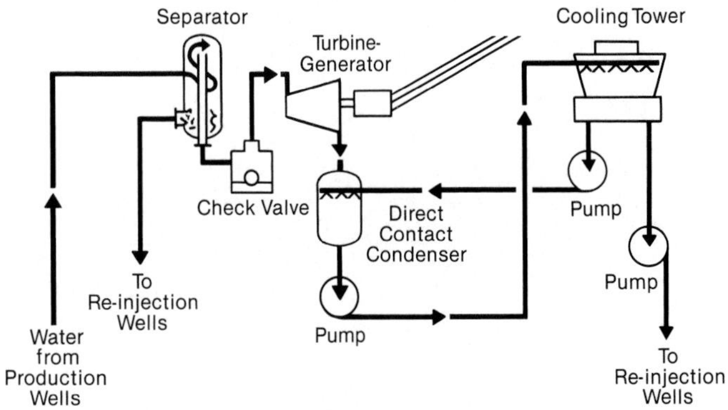

Fig. 16 Schematic of a single flash, geothermal power system.

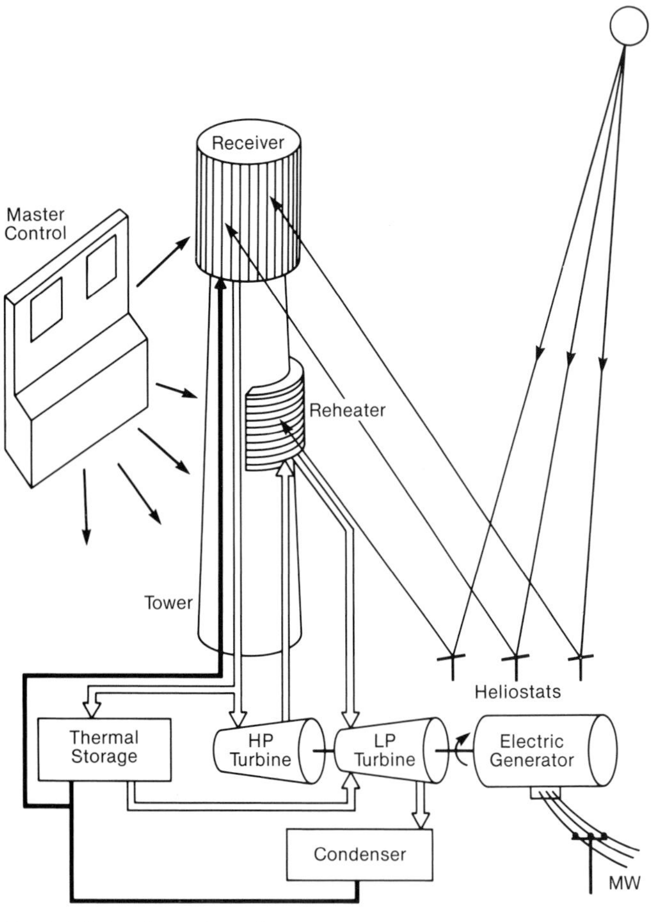

Fig. 17 Solar plant schematic.

low demand. This cycle shows two stages of thermal storage consisting of oil and salt.

MHD system

Magnetohydrodynamic or MHD electric generation occurs when hot, partially ionized combustion gases (plasma) are expanded through a magnetic field. Electrodes in the collection channel (see Fig. 21) pick up energy from the moving gas. The electrical-conducting gas

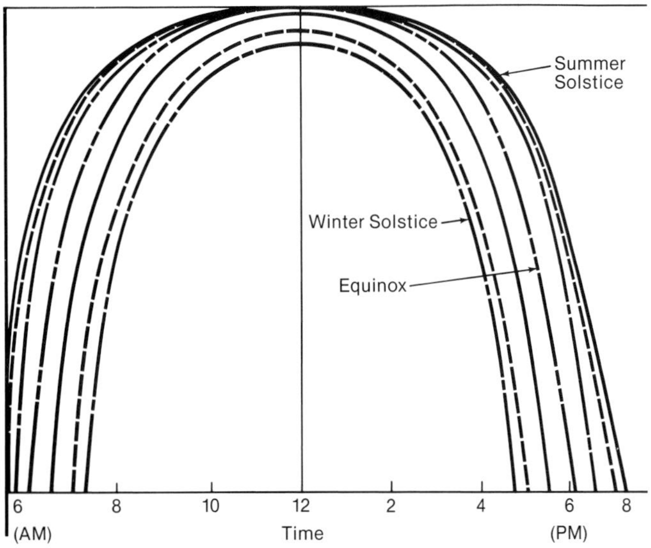

Fig. 18 Seasonal-diurnal variations in absorbed thermal power.

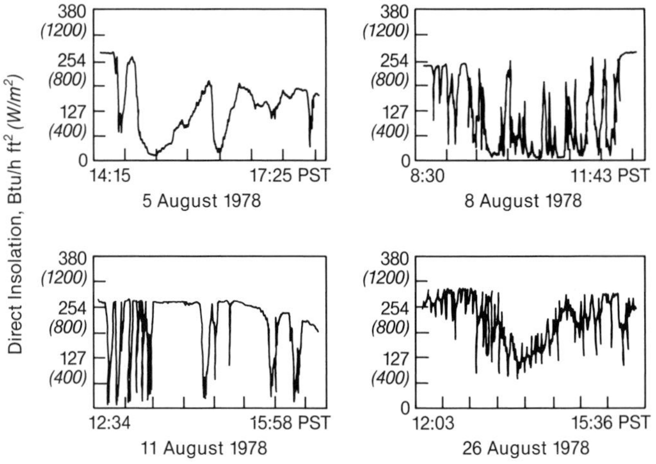

Fig. 19 Direct insolation, Daggett, California.[1]

passing through the magnetic field creates a voltage potential similar to moving an electrical wire through a magnetic field. This creates a direct current. The direct current is conditioned and inverted to alternating current feeding conventional electric power distribution systems.

The hot gas is produced in a coal combustor at temperatures approaching 5000F (2760C). Even at high gas temperatures it is necessary to increase the available gas ionization by seeding the gas with an easily ionized material. Potassium compounds are preferred. The spent seed compounds are treated and recycled for economic and environmental reasons.

Cycle efficiency Selected research and development efforts on MHD continue because MHD generation offers potential cycle efficiencies of nearly 60% versus conventional cycle efficiencies of 35 to 38%.

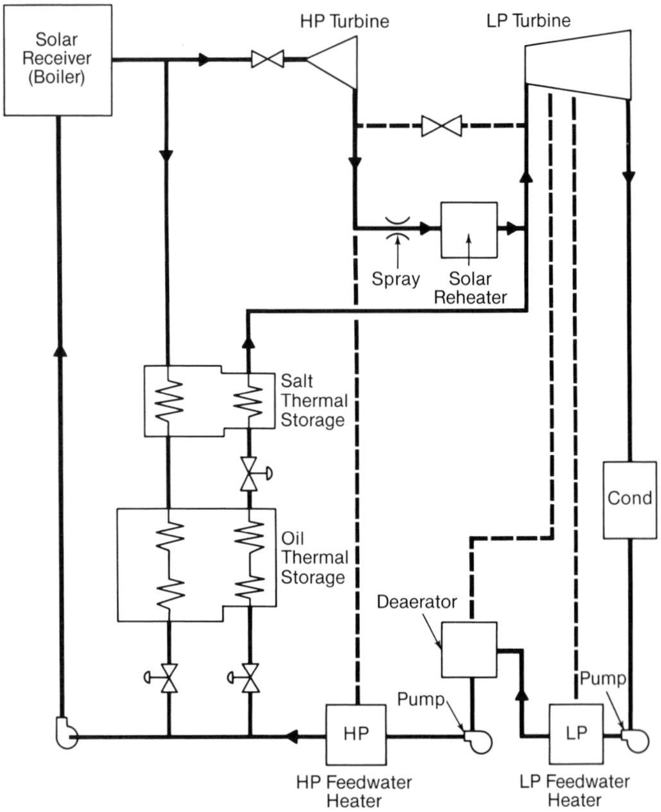

Fig. 20 Flow diagram featuring solar power system with reheater.

MHD systems potentially address cost and environmental concerns by providing significant cycle efficiency improvements. Many elements of the MHD system require further development before commercialization.

MHD cycle The MHD system consists of the high temperature coal combustor with seeding capability producing high temperature plasma which enters the magnetic field through a nozzle. The gases expand through the magnetic field and then enter a high temperature ceramic air heater. The high gas temperature required for the plasma makes it necessary for the combustion air/oxygen to have a temperature of about 3000F (1649C). Downstream of the air heater, the cooled gases enter the steam bottoming portion of the plant cycle. The bottoming portion of the plant consists of a conventional steam generator capable of generating high temperature and pressure steam which powers the steam turbine-generator.

Studies based upon the cycle shown in Fig. 21 have conceptualized an advanced design 1000 MW output MHD steam plant. Advanced concepts used in this study are 3100F (1704C) direct-fired ceramic air preheaters, a high efficiency motor driven axial compressor, high pressure (1.467 kPa) combustion with low heat loss, superconducting magnets, low heat loss ceramic channel electrodes, high electrical stress design, moderate pressure [800 psi (1.4 MPa)] channel cooling and a high performance diffuser. The advanced concept MHD topping cycle, combined with an advanced steam bottoming cycle results in a plant having a net efficiency of 60.4% on a higher heating value basis. The combustion air preheat for the MHD combustor would be 3100F (1704C) with a combustor pressure of 210 psi (14.5 bar). The bottoming cycle main steam throttle pressure is 5000 psi (345 bar); superheater outlet steam temperature is 1200F (649C); and reheat outlet temperatures are 1050 to 1150F (566 to 621C).[2] The environmental impact of this cycle is projected to be significantly better in the areas of SO_2, NO_x, CO_2, particulates, solid wastes, cooling heat rejection and total water consumption than a conventional steam plant producing the same amount of electricity. However, only further development will determine if such improvements are economical.

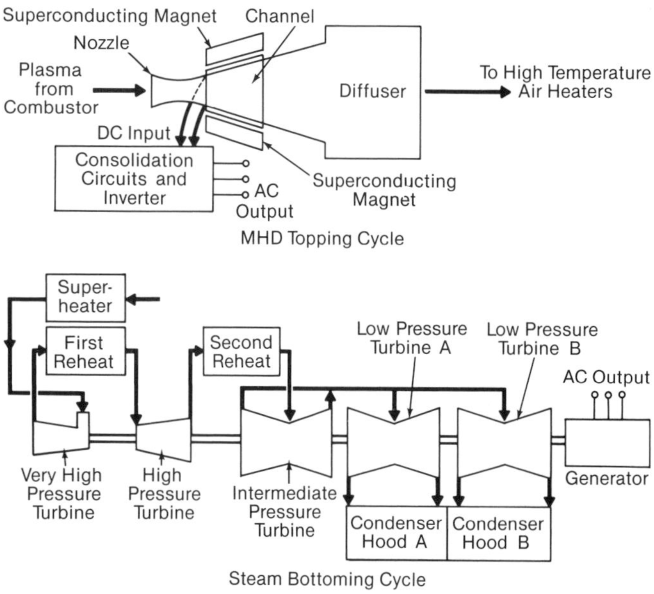

Fig. 21 MHD plant power systems.

References

1. Durrant, O. W., "Solar power generation systems," presented at the Second Conference on Air Quality Management in the Electric Power Industry, Austin, Texas, January 22-25, 1980.

2. "MHD Technical Support Services," U.S. DOE PON Number DE-AC22-87PC79338, Subcontract No. 08-8244-0104.

Section IV
Environmental Protection

Environmental protection and the control of solid, liquid and gaseous effluents or emissions are key elements in the design of all steam generating systems. The emissions from combustion systems are tightly regulated by local and federal governments, and specific rules and requirements are constantly changing. At present the most significant of these emissions are sulfur dioxide (SO_2), oxides of nitrogen (NO_x) and fine airborne particulate. All of these require specialized equipment for control.

This section begins with an overview of current regulatory requirements. Even though the regulatory environment is changing rapidly and specific requirements treated here will soon be dated, a general framework is provided. Following this overview, Chapters 33, 34 and 35 discuss specific equipment to control atmospheric emissions of particulate, NO_x and SO_2 respectively. The NO_x discussion focuses on post-combustion technologies; combustion-related control options are addressed in Chapter 10 and Chapters 13 through 17.

Finally, a key element in a successful emissions control program is measurement and monitoring. Chapter 36 addresses a variety of issues and outlines a number of technologies for flue gas monitoring. These systems not only monitor regulatory compliance but also permit optimal operation of environmental protection equipment.

Integrated environmental control system on a 450 MW coal-fired power plant.

Chapter 32
Environmental Considerations

Since the early 1960s, there has been an increasing worldwide awareness that industrial growth and energy production from fossil fuels are accompanied by the release of potentially harmful pollutants into the environment. Studies to characterize emissions, sources and effects of various pollutants on human health and the environment have led to increasingly stringent legislation to control air emissions, waterway discharges and solids disposal.

Comparable concern for environmental quality has been manifest worldwide. Since the 1970s, countries of the Organization for Economic Cooperation and Development have reduced sulfur dioxide (SO_2) and nitrogen oxides (NO_x) emissions from power plants in relation to energy consumption. In at least the foreseeable future, emission trends are expected to continue downward due to a combination of factors: change in fuel mix to less polluting fuels, use of advanced technologies, and new and more strict regulations. In Japan, the reductions in SO_2 emissions were particularly pronounced due to strong environmental measures taken in the 1970s. In the United States (U.S.), use of flue gas desulfurization systems and low sulfur coal have resulted in an approximate 30% decrease in utility SO_2 emissions between 1970 and 1989, while electricity produced from coal increased approximately 120%.

Environmental control is primarily driven by government legislation and the resulting regulations at the local, national and international levels. These have evolved out of a public consensus that the real costs of environmental protection are worth the tangible and intangible benefits now and in the future. To address this growing awareness, the design philosophy of energy conversion systems such as steam generators has evolved from providing the lowest cost energy into providing low cost energy with an acceptable impact on the environment. Air pollution control with emphasis on particulate, NO_x and SO_2 emissions is perhaps the most significant environmental concern for fired systems and is the subject of Chapters 33, 34 and 35, respectively. However, minimizing aqueous discharges and safely disposing of solid byproducts are also key issues for modern power systems.

Sources of plant emissions and discharges

Fig. 1 identifies most of the significant waste streams from a modern coal-fired power plant. Typical discharge rates for the primary emissions from a new 500 MW coal-fired drum boiler, with and without control equipment, are summarized in Table 1.

Atmospheric emissions arise primarily from the byproducts of the combustion process [SO_2, NO_x, particulate flyash, volatile organic compounds (VOC) and some trace quantities of other materials] and are exhausted from the stack. A second source of particulate is fugitive dust from coal piles and related fuel handling equipment. This is especially significant for highly dusting Western U.S. subbituminous coals. Some low temperature devolatilization of the coal can also emit other organic compounds. A final source of air emissions is the cooling tower and the associated thermal rise plume which contains heat and some trace materials along with the water vapor.

Solid wastes arise primarily from collection of the coal ash from the bottom of the boiler, economizer and air heater hoppers, as well as from the electrostatic precipitators and fabric filters. Pyrite collected in the pulverizers (see Chapter 12) is usually also included. Most of the ash is transported to an ash settling pond where it settles out and is prepared for either landfill or other use. The chemical composition and characteristics of various ashes are discussed in Chapter 20.

The second major source of solids is the byproduct from the flue gas desulfurization (FGD) scrubbing process. Most frequently, this is a mixture containing calcium sulfate or calcium sulfite. After dewatering, processing and treatment, the byproduct may be sold as gypsum or landfilled. Additional sources of solids include the sludge from cooling tower basins, wastes from the water treatment system and wastes from the periodic boiler chemical cleaning.

Aqueous discharges arise from a number of sources. These include once-through cooling water (if used), cooling tower blowdown (if used), sluice water from the ash handling system (via the settling pond), FGD waste water (frequently minimal), coal pile runoff from rainfall, boiler chemical cleaning solutions, gas side water washing waste solutions, as well as a variety of low volume wastes including ion exchange regeneration solutions, evaporator blowdown (if used), boiler blowdown and power plant floor drains. Many of these streams are chemically characterized in Chapter 42. Additional discussions of these systems as well as the controlling regulations are provided in References 1 and 2.

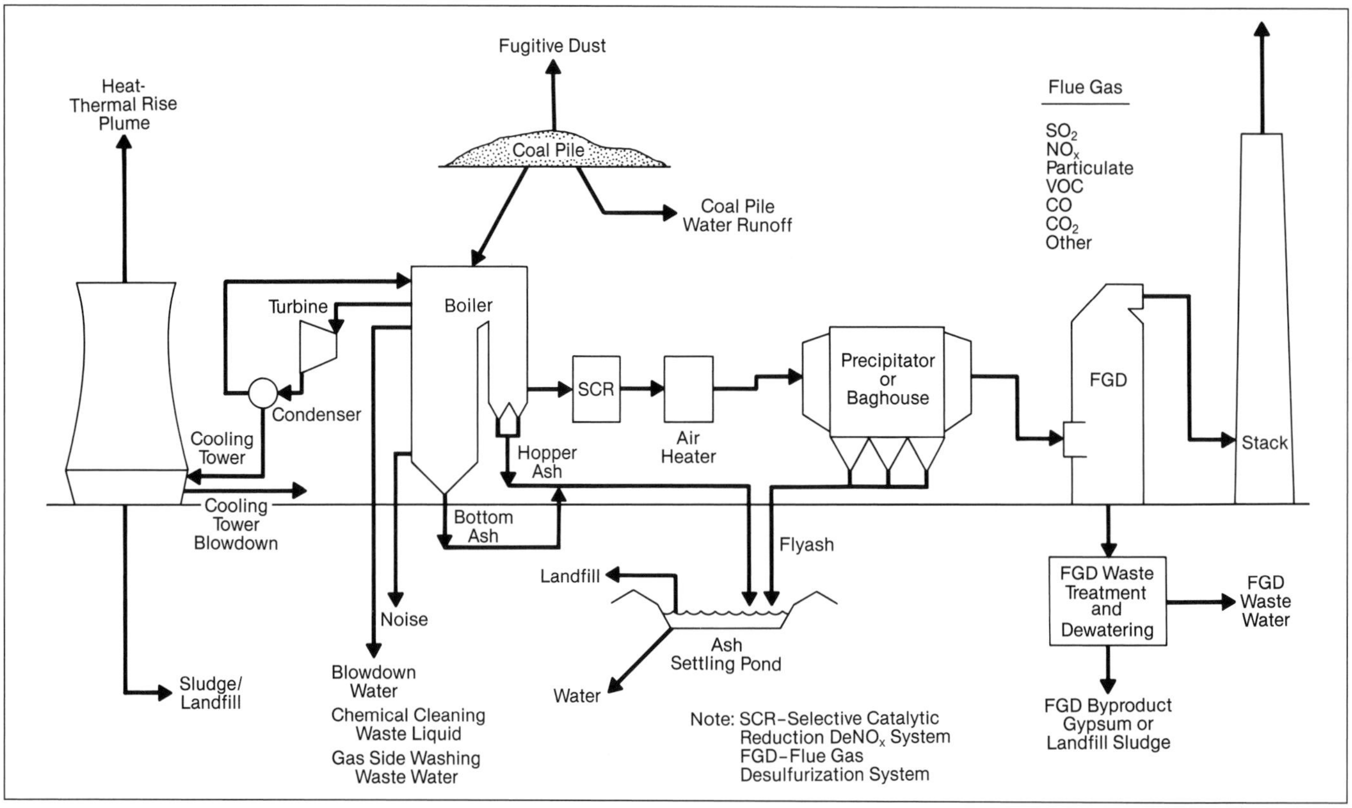

Fig. 1 Typical power plant effluents and emissions.

Air pollution control

U.S. legislation — Clean Air Act

The 1990 Amendments to the Clean Air Act (CAA), signed into law on November 15, 1990, have significantly expanded the original Act. The Amendments represent a significant increase in the U.S. commitment to an improvement in environmental quality. These Amendments build on the framework of the CAA which was originally enacted in 1963 and periodically amended since that time.

The Clean Air Act serves as the national basis for maintaining and improving the air quality through a series of minimum national requirements in a number of areas. The primary objective of the Act is *to protect and enhance the quality of the nation's air resources so as to promote the public health and welfare and the productive capacity of its population.*[3] The legislation generally provides for the U.S. Environmental Protection Agency (EPA) to set the minimum national requirements and to provide regulations, guidelines and guidance to the state and local regulatory agencies for implementation. The state and local government agencies are responsible to develop and implement plans to at least meet the federal requirements. However, they may also adopt more stringent regulations. The Act as amended prior to 1990 addressed the following areas of potential interest to boiler operators.

National Ambient Air Quality Standards These define acceptable air quality levels to protect public health and thereby set objectives for improvement. National Ambient Air Quality Standards have been defined for the six *Criteria Pollutants*: sulfur dioxide (SO_2), nitrogen dioxide (NO_2), carbon monoxide (CO), ozone (O_3), particulate matter and lead.

Table 1
Typical 500 MW Coal-Fired Steam Generator Emissions and Byproducts
[2.5% sulfur, 16% ash, 12,360 Btu/lb (28,749 kJ/kg)]

Emission	Typical Control Equipment	Discharge Rate — t/h (t_m/h) Uncontrolled		Controlled	
SO_x as SO_2	Wet limestone scrubber	9.3	(8.4)	0.9	(0.8)
NO_x as NO_2	Low NO_x burners	2.9	(2.6)	0.7	(0.7)
CO_2	Not applicable	485	(440)	485	(440)
Flyash to air*	Electrostatic precipitator or baghouse	22.9	(20.8)	0.05	(0.04)
Water stream thermal discharge	Natural draft cooling tower	2.8×10^9 Btu/h	(821 MW_t)	~0	
Ash to landfill*	Controlled landfill	9.1	(8.3)	32	(29)
Scrubber sludge: gypsum plus water	Controlled landfill or wallboard quality gypsum	0		25	(27.7)

* As flyash emissions to the air decline, ash shipped to landfills increases.

State Implementation Plans (SIP) These plans are submitted by states to identify how the national standards will be implemented. They divide each state into a number of identifiable areas known as air quality control regions, in which at least the minimum National Ambient Air Quality Standards must be met and control requirements imposed as needed to achieve compliance.

New Source Performance Standards (NSPS) For more than 60 types of new and modified sources, these provide that best demonstrated available control technology must be used to satisfy requirements as set forth in Section III, Title 40 of the U.S. Code of Federal Regulations (CFR).[4]

Air toxics The Act authorized the setting of standards to limit human exposure to air toxics, which include any potentially hazardous non-criteria air pollutants. The standards are risk based and require an ample margin of safety to prevent any adverse effects (essentially zero risk).

New source review This activity is intended to prevent significant environmental deterioration from industrial growth in the form of new plants or expansion of existing plants.

Enforcement The Act provides the EPA and state agencies with the authority to bring civil and criminal actions against violators of the statutes.

Evolution of U.S. clean air legislation and key elements

The Clean Air Act has evolved from the first legislation passed in 1963 to establish authority for setting air quality standards through the Clean Air Act of 1970, which was more comprehensive and provided greater statutory strength for implementing regulatory efforts in the 1970s and 1980s. During the Middle East Oil Embargo of October 1973, the U.S. Congress, in an attempt to ensure an adequate energy supply, modified some air quality requirements through the Energy Supply and Environmental Coordination Act. In 1977 the Clean Air Act was revised to address prevention of significant deterioration of the environment and defined requirements for new sources in areas where National Ambient Air Quality Standards had not been attained.

Elements of the pre-1990 Amendments which have had the most impact on utility and industrial boilers include National Ambient Air Quality Standards and the New Source Performance Standards. National Ambient Air Quality Standards promulgated by the EPA established minimum numerical criteria for the country. Two types of standards have been established, primary standards aimed at prevention of adverse impact on human health and secondary standards to prevent damage to the environment. While ambient air quality standards have been set for the six Criteria Pollutants [CO, SO_2, NO_2, ozone (O_3), particulates and lead], there has been intense controversy in establishing what constitutes safe concentrations, accounting for the combined effects of exposure to multiple pollutants and the limited number of pollutants covered.

Before the 1977 CAA Amendments, FGD systems installed on utility boilers were designed to limit SO_2 emissions to not more than 1.2 lb per million Btu heat input. Burning low sulfur coal with less than 0.7% sulfur required no scrubbing. Coal with 3% sulfur required only 75 to 80% SO_2 removal efficiency.

The Amendments of 1977 required new utility power stations to reduce SO_2 emissions by 90% for high sulfur coals, by 70% for low sulfur coals and by a sliding scale for intermediate sulfur coals. (See Fig. 2.) A key element of the legislation and prior amendments was that a large population of existing operating utility boilers was not required to lower emissions, i.e., grandfathered, unless they were significantly modified or upgraded.

The Amendments of 1990 enhanced the previous legislation in a number of areas which have an impact on both existing and new fossil fuel-fired utility generation capacity.

Nonattainment areas (those areas not achieving National Ambient Air Quality Standards) were given new classifications and deadlines to achieve attainment under Title I of the Act. These deadlines were set on the basis of present pollution levels. This portion of the legislation may have a dramatic but indirect impact on boiler system NO_x emissions in selected nonattainment areas because NO_x is one precursor to ozone formation.

Air toxics were defined in Title III to include a list of 190 hazardous air pollutants, and the EPA was directed to promulgate control standards for sources of these emissions. While air toxics were addressed in the Act of 1970, standards had only been promulgated for seven substances by 1990. The EPA has been directed to impose tight controls through a two-phase strategy. The first phase is technology based and would require application of Maximum Achievable Control Technology (MACT) to selected categories of emission sources which emit more than minimum or threshold quantities of the listed pollutants. The second phase is risk based and may require certain facilities to apply additional controls beyond MACT to reduce remaining residual risk. Utility boilers are initially exempted from this section until the EPA studies the emissions of the air toxics from utility plants and their impact.

Acid rain (discussed below) was addressed under Title IV of the 1990 Amendments in the form of new requirements applying mainly to coal-fired power plants.

Title IV, Acid Deposition Control The 1990 Amendments establish a new market based control methodology which focuses on the control of SO_2 and NO_x emissions from fossil-fired power plants. Annual U.S. emissions of SO_2 are to be reduced by 10 million tons and annual U.S. emissions of NO_x as NO_2 are to be reduced by 2 million tons by the year 2000. This reduction is planned to take place in two phases, with the first beginning in 1995. The market based system requires utilities to use an Allowance (a permit to

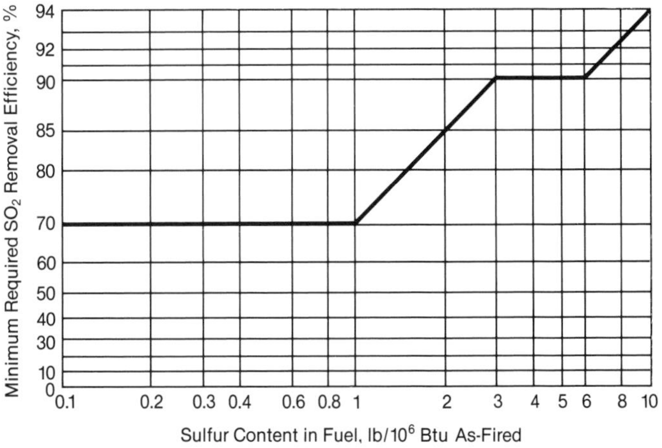

Fig. 2 Effective New Source Performance Standards for coal-fired utility boilers prior to November 15, 1990.

emit one ton of SO_2 on or after a certain year) for each ton of SO_2 emitted after a certain date. Each existing utility boiler site receives a certain number of Allowances per year. The plant owner then has the option to either: 1) reduce emissions through fuel switching, the application of control technology or other options, or 2) acquire additional Allowances from other plants to cover the actual SO_2 emissions. Any unused Allowances may be sold or kept (*banked*) for future use. The total number of new Allowances issued each year is fixed and new power plants will not receive an Allowance allocation. This effectively caps total utility SO_2 emissions.

The NO_x control section of the acid rain legislation is part of the more traditional command and control approach where specific units or types of equipment will be allowed only certain levels of NO_x emission rates based upon fuel heat input (pounds of NO_x as NO_2 per million Btu input) although averaging will be permitted between plants.

As in prior environmental legislation, local or state environmental regulatory agencies may impose stricter regulations on individual plants, areas or locations.

New Source Performance Standards These regulations establish maximum emission rates for selected pollutants from new boilers. The NSPS were conceived to be a technology — forcing approach to provide an incentive to develop new control technology. Again, these were national maximum limits which could be tightened by the local or state implementing agencies. An example of NSPS requirements for utility boilers with a capacity greater than 250 x 10^6 Btu/h firing coal, oil or gas is provided in Table 2. Fig. 2 provides an overview of the SO_2 reduction and control requirements for coal-fired units. The details for specific applications and other fuels and technologies may be found in Title 40, Part 60, Subparts Ca, D, Da, Db, Dc, E, Ea and BB of the Code of Federal Regulations.[4] The 1990 Amendments to the Act direct the EPA to revise the NSPS.

In December 1987, the EPA issued similar guidelines for smaller industrial units which were updated in December 1989. Selected elements of these rules are summarized in Table 3. Refer to Title 40, Part 60, Subpart Db of the CFR for site specific details and rules.[4]

The NSPS apply not only to new boilers but also to boilers which undergo substantial modifications. However, the extent of the modifications and the change in pollutant emissions which trigger NSPS are currently under review.

Prevention of significant deterioration PSD relates to avoiding a reduction in air quality from industrial growth in areas that already meet air quality standards. In these PSD areas, new major sources and major modifications to existing sources are required to use Best Available Control Technology (BACT). Air quality analyses are also performed to show that the air quality standards will not be exceeded.

Nonattainment Areas that do not meet the air quality standards for a given pollutant are referred to as nonattainment areas for that pollutant. Different rules apply for new sources in these areas. New sources must apply state-of-the-art (SOA) technology and must offset the remaining emissions after application of SOA technology by reducing emissions in existing sources.

In applying the SOA technology, the Lowest Achievable Emission Rate (LAER) is to be attained. This means that the emission rate must meet the most stringent limit in any SIP or the most stringent limitation achieved by plants in the same industrial category, whichever is more stringent.

Table 2
Selected Summary of Federal NSPS for Electric Utility Steam Generators >250 million Btu/h Commencing Construction After September 18, 1978 (Notes 1 and 2)

Fuel	Pollutant	Max. Emissions Rate (lb/10^6 Btu) (Notes 3 and 6)	Req'd. Reduction in Potential Emissions, % (Note 3)
Coal	SO_2	1.2 or	90 or
		0.6	70
	NO_x as NO_2:		
	Subbituminous	0.5	(Note 4)
	Bituminous	0.6	
	Particulate	0.03 (Note 5)	99
Oil	SO_2	0.8 or	90 or
		0.2	0
	NO_x as NO_2	0.3	(Note 4)
	Particulate	0.03 (Note 5)	70
Gas	SO_2	0.8 or	90 or
		0.2	0
	NO_x as NO_2	0.2	(Note 4)
	Particulate	0.03 (Note 5)	—

Notes:
1. Source: 40CFR60 Subpart Da (6/11/79).
2. For reference only: see source for details.
3. *Maximum Emissions Rate* and *Req'd. Reduction in Potential Emissions* must both be met.
4. Selected % reductions are also identified for NO_x but meeting the Maximum Emissions Rate stated constitutes compliance with the % reduction requirements.
5. Separate opacity limit of 20% may be controlling.
6. Approximate SI conversions:[5] mg/Nm^3 per lb/10^6 Btu = 1230 for coal at 6% O_2, 1540 for oil at 3% O_2 and 1590 for gas at 3% O_2.

Best Available Control Technology Classification of a control technology as BACT is determined on a case by case basis by the EPA. It is then up to the owner of the new plant or plant expansion to justify deviations from BACT based on its inapplicability to the proposed plant.

Increments of air quality In addition to the application of BACT, the resulting emissions from the source must meet the available increment of air quality. All areas in the U.S. which meet the National Ambient Air Quality Standards are divided into Class I areas (that are to be kept in a pristine condition) and Class II areas (where industrial growth will be allowed). For each class an allowable degradation increment is determined. The air quality increment for a given area is set so that the resulting air quality after degradation from a new source is cleaner than what is required to satisfy the ambient air quality standard. As might be expected, Class I areas allow almost no degradation. Air quality increments have been set for SO_2, NO_x and particulates.

International regulations — air pollution control

The passage of the Clean Air Act in the U.S. in 1963 marked the first enactment of air pollution control legislation by a major industrial nation. Since that time, air pollution control regulations have become more widespread in industrial and developing nations, particularly in Ja-

Table 3
Selected Summary of Federal NSPS for Industrial Steam Generators >100 million Btu/h Commencing Construction, Modification or Reconstruction After June 19, 1984 (Notes 1 and 2)

Fuel	Pollutant: Technology	Max. Emissions Rate (lb/10^6 Btu) (Notes 3 and 6)	Req'd. Reduction in Potential Emissions, % (Note 3)
Coal	SO_2: All	1.2	90 (Note 3)
	NO_x as NO_2:		
	Spreader-stoker	0.6	—
	Mass-feed stoker	0.5	—
	Pulverized coal	0.7	—
	Fluidized bed	0.6	—
	Particulate	0.05 (Note 5)	—
Oil	SO_2	0.8 or	90 or
(Resid.)		0.5	0 (Note 3)
	NO_x as NO_2	0.4/0.3 (Note 4)	—
	Particulate	0.10 (Note 5)	—
Gas	SO_2	—	—
	NO_x as NO_2	0.2/0.1 (Note 4)	—
	Particulate	0.10 (Note 5)	

Notes:
1. Source: 40CFR60, Subpart Db (12/18/89).
2. For reference only: see source for details.
3. *Maximum Emissions Rate* and *Req'd. Reduction in Potential Emissions* must both be met.
4. Higher rate for heat release rates >70,000 Btu/h ft^3.
5. Separate opacity limit of 20% may be controlling.
6. Approximate SI conversions: (see Table 2).

pan, Canada and Western Europe. As in the U.S., steam generating plants have been one focus of these regulatory measures, and the primary emissions of concern from combustion processes are SO_2, NO_x and particulates. CO_2 and air toxics have received some recent attention. The detailed regulations continue to evolve rapidly and are quite country specific. However, two trends are widespread:

1. allowable emission limits for controlled pollutants will continue to decline with time, and
2. a wider array of species will be considered for control.

Without attempting to be comprehensive, the following items provide a brief overview of worldwide SO_2 and NO_x regulatory efforts. While these items will probably be quickly out of date, they provide a general indication of the range and application of control measures.

Control approaches One or more of the following of measures have typically been adopted to control emissions:

1. *Emission standards* These limit the mass of SO_2 or NO_x emitted by volume, by heat input, or by unit of time (hourly, daily, annually).
2. *Percent removal requirements* These specify the portion of the uncontrolled emissions which must be removed from the flue gas.
3. *Fuel requirements* Primarily aimed at SO_2 control, these either limit the type of fuel which can be burned or the fuel sulfur content.
4. *Technology requirements* These typically indicate the type of control technology specifically required or indicate the use of the best available control technology or reasonably available control technology at the time of installation. These requirements depend in many cases on some level of economic feasibility.

The most widely used control approach is emission standards, although this is usually combined with one or more of the other approaches. Emissions from new plants are usually more tightly controlled than emissions from existing capacity. Occasionally, older plants are not controlled, although this is changing.

Local, regional or national control of emission standards depends upon the country. In the U.S. and the Netherlands, federal standards provide minimum requirements which local authorities may tighten or apply to a broader range of applications. In Canada and Australia, federal governments can only provide emission control guidelines and local or regional governments set plant limits.

SO_2 control Based upon a recent compilation by the International Energy Agency,[5] Table 4 lists selected SO_2 emission standards for new coal-fired boilers (New Source Performance Standards).

NO_x control As with SO_2 control, NO_x emissions are most frequently regulated through emission standards of mass per unit flue gas volume or per unit heat input. Most specify NO_x emissions as NO_2. Table 5 from Reference 6 provides a summary of selected current and pending NO_x control regulations.

Evolving requirements Korea, Hong Kong, the People's Republic of China, and the countries of Eastern Europe, as well as an array of developing nations, are expected to implement SO_2 and other air pollution control strategies in the near future.[5,6]

Kinds of pollutants, sources and impacts

Air pollutants are contaminants in the atmosphere which, because of their quantity or characteristics, have deleterious effects on human health and/or the environment. The sources of these pollutants are classified as stationary, mobile or fugitive. Stationary sources generally include large individual point sources of emissions such as electric utility power plants and industrial furnaces where emissions are discharged through a stack. Mobile sources are those associated with transportation activities. Fugitive emissions generally include discharges to the atmosphere from pumps, valves, seals and other process points not vented through a stack. They also include emissions from area sources such as coal piles, landfills, ponds and lagoons. They most often consist of particulates and occur in industry related activities in which the emissions are not collected.

The focus of this chapter is stationary emission sources, particularly fired utility and industrial boiler systems. Key pollutants from these sources are SO_2, NO_x, CO and particulate matter. Another class of emissions is called air toxics. These are potentially hazardous pollutants which generally occur in only trace quantities in the effluents from fired processes. However, they are undergoing more intense examination because of their potential health effects.

During the 1980s, concern increased about the potential impact of carbon dioxide (CO_2) emissions from man-made sources on the global climate. CO_2 is one of several so-called *greenhouse* gases which are said to contribute to a potential global warming phenomenon. It is emitted from a variety of naturally occurring and man-made

Table 4
Selected International SO_2 Emission Limits for New Coal-Fired Plants (Note 1)

Country	Plant Type	Plant Size (Note 6) MW$_t$	10^6 Btu/h	Emissions Standard mg/Nm^3	lb/10^6 Btu	% Removal
Austria	Boilers	10 to 50	34 to 171	400	0.33	—
	Lignite-fired	>50	>171	400	0.33	—
	Hard coal	>50	>171	200	0.16	—
Belgium		50 to 100	171 to 341	2000	1.6	—
		100 to 300	341 to 1024	1200	1.0	—
		>300	>1024	400 (250 Note 2)	0.33 (0.2)	—
Canada	Utilities	All	All	740 (Note 3)	0.6	—
Denmark	Utilities	>50	>171	860 (Note 3)	0.7	—
Finland		50 to 150	>171 to 512	660	0.54	—
		>150	>512	400	0.33	—
Germany	Conventional	1 to 100	3 to 341	2000	1.6	—
	Boilers	100 to 300	341 to 1024	2000	1.6	60
		>300	>1024	400	0.33	85
	FBC boilers	>1	>3	400	0.33	75
Italy	Utilities	>100	>341	400	0.33	—
Japan	All plants	All	All	Plant specific (Note 4)		
Netherlands		50 to 300	171 to 1024	700	0.57	—
		>300	>1024	400	0.33	85
Spain	Utilities	Hard coal	—	2400	2.0	—
		Brown coal	—	9000	7.3	—
	Industry	Hard coal	—	2400	2.0	—
		Brown coal	—	6000	4.9	—
Sweden	Coal-fired	>0.5	>1.7	290	0.24	—
Taiwan		All	—	3150/4000 (Note 5)	2.56/3.25	—
UK		>700	>2388	—	—	90
USA	Utility	>73	>250	1480	1.2	70 to 90
	Industry	>29	>100	1480	1.2	90

Notes:
1. Source: IEA Coal Research, October 1989.
2. Lower limit applies from 1995.
3. Guidelines.
4. Set on a plant by plant basis according to nationally defined formulae.
5. Lower limit applies to imported coal.
6. SI conversion: 8.14×10^{-4} lb/10^6 Btu per mg/Nm^3, 3.412×10^6 Btu/MWh, 350 m^3 flue gas/GJ input and 30 GJ/tonne.

sources which include the combustion of all fossil and hydrocarbon based fuels. Obviously, improving the power cycle efficiency (more power from less fuel) and the use of fuels with less carbon content are potential methods to address CO_2 emissions from any combustion source. However, there remains disagreement in the scientific discussion of the impact of CO_2 emissions as well as the methods and level of control. Therefore, further discussion of this issue awaits scientific and public clarification and consensus.

Sulfur oxides This category of pollutants includes mainly SO_2 with small quantities of sulfur trioxide (SO_3). The main source of sulfur oxides is from the combustion of coal, with lesser amounts from other fuels such as residual oil. Based on 1985 National Acid Precipitation Assessment Program (NAPAP) data,[7,8] the utility and industrial sectors (smelters, iron and steel mills, refineries) are the largest emitters of sulfur oxides (see Table 6). In the presence of particulate matter, sulfur oxides have been related to irritation of the human respiratory system, reduced visibility, materials corrosion and varying effects on vegetation. The reaction of sulfur oxides with moisture in the atmosphere has been identified as contributing to acid rain.

Nitrogen oxides This category includes numerous species comprised of nitrogen and oxygen, although nitric oxide (NO) and nitrogen dioxide (NO_2) are the most significant in terms of quantity released to the atmosphere. NO is the primary nitrogen compound formed in high temperature combustion processes where nitrogen present in the fuel and/or combustion air combines with oxygen. The quantity of NO_x formed during combustion depends on the quantity of nitrogen and oxygen available, the temperature, the level of mixing and the time for reaction. Control of these parameters has formed the basis for a number of control strategies involving combustion process control and burner design. Based on 1985 NAPAP data, utilities account for 33% of NO_x emitted in the U.S., with the transportation sector emitting 43%. Of the total utility NO_x emissions, 89% comes from coal-fired boilers. The most deleterious effects come from NO_2 which can form from the reaction of NO and oxygen. NO_2 also absorbs the full visible spectrum and can reduce visibility. NO_x has been associated with respiratory disorders, corrosion and degradation of materials, and damage to vegetation. NO_x has also been identified as a precursor to ozone and smog formation.

Carbon monoxide This colorless, odorless gas is formed from incomplete combustion of carbonaceous fuels. CO

Table 5
Selected International NO_x Emission Limits for Coal-Fired Plants (Note 1)

	New Plants		Existing Plants	
	mg NO_2/Nm^3	lb/10^6 Btu	mg NO_2/Nm^3	lb/10^6 Btu
Austria	200 to 400	0.16 to 0.33	200 to 400	0.16 to 0.33
Belgium	200 to 800	0.16 to 0.65	—	—
Denmark	650 (Note 2)	0.53	(Note 2)	—
European Community (EC)	650 to 1300	0.53 to 1.06	—	—
Finland	200 to 400	0.16 to 0.33	400 to 620	0.33 to 0.50
Germany	200 to 500	0.16 to 0.41	200 to 1300	0.16 to 1.06
Italy	200 to 650	0.16 to 0.53	200 to 650	0.16 to 0.53
Japan	410 to 510	0.33 to 0.41	620 to 720	0.50 to 0.60
Netherlands	400 to 800	0.33 to 0.65	1100	0.90
Sweden	140	0.11	140 to 560	0.11 to 0.46
Switzerland	200 to 500	0.16 to 0.41	200 to 500	0.16 to 0.41
Taiwan	600 to 850	0.49 to 0.69	600 to 850	0.49 to 0.69
UK	650	0.53	—	—
USA	615 to 980	0.50 to 0.80	553 to 614	0.45 to 0.50 (Note 3)

Notes:
1. Source: IEA Coal Research Report, IEA/CR-30, December 1990.
2. In addition to *bubble principle* for utilities.
3. Dry-bottom wall-fired and tangential-fired only; other limits pending.
4. SI conversion: 8.14×10^{-4} lb/10^6 Btu per mg/Nm^3, 350 m^3 flue gas/GJ input and 30 GJ/tonne.

emissions from properly designed and operated utility boilers are a relatively small percentage of total U.S. combustion source CO emissions, most of which come from the internal combustion engine in the transportation sector. The primary environmental significance of CO is its effect on human and animal health. It is absorbed by the lungs and reduces the oxygen carrying capacity of the blood. Depending on the concentration and exposure time, it can cause impaired motor skills and physiological stress.

Particulate matter Solid and liquid matter of organic or inorganic composition which is suspended in flue gas or the atmosphere is generally referred to as particulate. Particle sizes from combustion sources are in the 1 to 100 µm range, although particles smaller than 1 µm can occur through condensation processes. Among the effects of particulate emissions are impaired visibility, soiling of surrounding areas, aggravation of adverse effects of SO_2, and human respiratory problems.

Table 6
1985 U.S. Anthropogenic Emissions
NAPAP Data (Version 2)[7,8]

Category	Emissions (10^3 tons) SO_2	NO_x	VOC	TSP
Utility combustion	16,055	6,662	40	570
Industrial combustion	2,679	3,198	97	304
Other combustion	613	790	1,862	1,171
Industrial processes	2,931	926	3,715	1,099
Transportation	864	8,835	8,800	4,195
Other	4	130	7,558	1,044
Total	23,146	20,541	22,072	8,383

Notes: SO_2 — Sulfur dioxide
NO_x — Nitrogen oxides as NO_2
VOC — Volatile organic compounds
TSP — Total suspended particulate

VOC Volatile organic compounds, or more commonly VOC, represent a wide range of organic substances. These compounds consist of molecules containing carbon and hydrogen and include aromatics, olefins and paraffins. A major source is the refining and use of petroleum products. Also included among VOCs are compounds derived from primary hydrocarbons including aldehydes, ketones and halogenated hydrocarbons. The major source of these compounds is the transportation and the commercial/residential combustion sectors. VOCs are environmentally significant because of their role in the formation of photochemical smog through photochemical reactions with NO_x. Control of VOCs has been the primary means of addressing areas of ozone nonattainment. Smog arising from VOC emissions can cause respiratory problems, eye irritation, damage to vegetation and reduced visibility.

Toxic and hazardous materials This is a large category of air pollutants which could have hazardous effects. The EPA had only promulgated standards for arsenic, asbestos, benzene, beryllium, mercury, radionuclides and vinyl chlorides before the passage of the 1990 Amendments to the Clean Air Act. This Act, under Title III, established a list of 190 toxic pollutants for which emissions are to be regulated. The list includes a wide range of simple and complex industrial organic chemicals and a small number of inorganics, particularly heavy metals. The EPA has identified hundreds of categories of air toxics sources, among which are utility boilers, pulp and paper plants, and municipal solid waste incinerators.

Air pollution control technologies

The strategies for control of all emissions from a utility or industrial boiler are formulated by considering design fuels, kind and extent of emission reduction mandated, and economic factors such as boiler design, location, new or existing equipment, age and remaining life.

SO_2 control strategies and technologies SO_2 emissions from coal-fired boilers can be reduced using precombustion techniques, combustion modifications and postcombustion methods.

Precombustion These techniques include the use of oil or gas in new units or the use of cleaned (beneficiated) coal or fuel switching in existing units. By using gas, sulfur emissions can be reduced to almost zero while the use of low sulfur oil will minimize SO_2 emissions. While the low sulfur content of oil and gas is advantageous, the price volatility and availability of these fuels make them less attractive except where local circumstance dictates. Switching to oil and gas in existing boilers requires attention be given to receiving equipment, storage facilities, combustion equipment including safety systems, boiler design, and unit postcombustion FGD performance. In the case of new systems, oil or gas firing can significantly reduce steam system capital costs. Even switching from one coal to another low sulfur coal can have a dramatic impact on fuel handling, combustion and particulate collection equipment. These effects are explored in more detail in Chapters 20 and 46.

Combustion modifications These techniques are primarily used to reduce NO_x emissions but can also be used to control SO_2 emissions in fluidized-bed combustion where limestone is used as the bed material. The limestone can absorb more than 90% of the sulfur released during the combustion process. (See Chapters 16 and 29.)

Sorbent injection technologies Sorbent injection, while not involving modification of the combustion process, is applied in temperature regions ranging from those just outside the combustion zone in the upper furnace to those at the economizer and duct work following the air heater. Sorbent injection involves adding an alkali compound to the coal combustion gases for eventual reaction with SO_2. Typical calcium sorbents include limestone [calcium carbonate ($CaCO_3$)], lime (CaO), hydrated lime [$Ca(OH)_2$] and modifications of these compounds with special additives. Sodium based compounds are also used. The manner in which injected sorbents react with sulfur oxides and the efficiency of the processes depend on the temperature of injection, sorbent type, sorbent surface area, and molar ratio of sorbent to sulfur. These processes are discussed in Chapter 35.

Wet and dry scrubbing technology Worldwide, wet and dry scrubbing systems are the most commonly used technologies in the coal-fired electric utility industry. 1989 data indicate that in the U.S. approximately 20% of the coal-fired utility capacity used wet scrubbing for SO_2 emission control. Both wet and dry scrubbing use slurries of sorbent and water to react with SO_2 in flue gas, producing wet and dry waste products, respectively.

In the wet scrubbing process, a sorbent slurry consisting of water mixed with lime, limestone, magnesium promoted lime or sodium carbonate (Na_2CO_3) is contacted with flue gas in a reactor vessel. Wet scrubbing is a highly efficient (> 90% at calcium/sulfur molar ratios close to 1.0), well established technology which can produce usable byproducts.

Dry scrubbing involves spraying an aqueous sorbent slurry into a reactor vessel so that the slurry droplets dry as they contact the hot flue gas [~300F (~149C)]. The SO_2 reaction occurs during the drying process and results in a dry particulate containing reaction products and unreacted sorbent entrained in the flue gas, along with flyash. These materials are captured downstream in the particulate control equipment.

Dry scrubbing is a well established technology with considerable operational flexibility, but fouling of downstream duct work is possible. The waste residue is dry.

NO_x control technologies NO_x emissions from fossil fuel-fired industrial and utility boilers arise from the nitrogen compounds in the fuel and molecular nitrogen in the air supplied for combustion. Conversion of molecular and fuel nitrogen into NO_x is promoted by high temperatures and high volumetric heat release rates found in boilers. The main strategies for reducing NO_x emissions take two forms: 1) modification of the combustion process to control fuel and air mixing and reduce flame temperatures, and 2) postcombustion treatment of the flue gas to remove NO_x. (See Chapters 10, 13, 15 and 34.)

Combustion modification This approach to NO_x reduction can include the use of low NO_x burners, combustion staging, gas recirculation or reburning technology.

Low NO_x burners slow and control the rate of fuel and air mixing, thereby reducing oxygen availability in the ignition and main combustion zones. Low NO_x burners can reduce NO_x emissions by 50% or more, depending upon the initial conditions, are relatively low cost and are applicable to new plants as well as retrofits.

Staged combustion uses low excess air levels in the primary combustion zone with the remaining (overfire) air added higher in the furnace to complete combustion. Significant NO_x reductions are possible with staged combustion, although reducing zones and potential for corrosion and slagging exist.

Flue gas recirculation reduces oxygen concentration and combustion temperatures by recirculating some of the flue gas to the furnace without increasing total net gas mass flow. Large NO_x reductions are possible with oil and gas firing while moderate reductions are possible with coal firing. Modifications to the boiler in the form of ducting and an efficiency penalty due to power requirements of the recirculation fans can make the cost of this option higher than some of the other in-furnace NO_x control methods.

Reburning is a technology used to reduce NO_x emissions from Cyclone furnaces (Chapter 14) and other selected applications. In reburning, 75 to 80% of the furnace fuel input is burned in the Cyclone furnace with minimum excess air. The remaining fuel (gas, oil or coal) is added to the furnace above the primary combustion zone. This secondary combustion zone is operated substoichiometrically to generate hydrocarbon radicals which reduce NO formed in the Cyclone to N_2. The combustion process is then completed by adding the balance of the combustion air through overfire air ports in a final burnout zone in the top of the furnace.

Postcombustion The two main postcombustion techniques for NO_x control are selective noncatalytic reduction (SNCR) and selective catalytic reduction (SCR). In SNCR, ammonia or other compounds such as urea (which thermally decomposes to produce ammonia) are injected downstream of the combustion zone in a temperature region of 1400 to 2000F (760 to 1093C). If injected at the optimum temperature, NO_x is removed from the flue gas through reaction with ammonia. SCR is being used worldwide where high NO_x removal efficiencies are required in gas-, oil- or coal-fired industrial and utility boilers. SCR systems remove NO_x from flue gases by reaction with ammonia in the presence of a catalyst.

Particulate control technologies Particulate emissions from boilers arise from the noncombustible, ash forming mineral matter in the fuel which is released during the combustion process and is carried by the flue gas to the stack. Another source of particulate is the incomplete combustion of the fuel which results in unburned carbon particles. A brief description of the principal options for particulate emissions control in industrial and utility boilers follows while Chapter 33 provides an in depth discussion.

Coal cleaning Historically, physical coal cleaning has been applied to reduce mineral matter, increase energy content and provide a more uniform boiler feed. Although reduction in flue gas particulate loading is one of the potential benefits, coal cleaning has been driven by the many other boiler performance benefits related to improved boiler maintenance and availability and, more recently, the reduction in SO_2 emissions.

Mechanical collectors These are generally cyclone collectors and have been widely used on small boilers when less stringent particulate emission limits applied. Cyclones are low cost, simple, compact and rugged devices. However, conventional cyclones are limited to collection efficiencies of about 90% and are poor at collecting the smallest particles. Improvements in small particle collection are accompanied by high pressure drops.

Fabric filters These filters, also commonly referred to as baghouses, are available in a number of designs (reverse air, shake/deflate and pulse jet), each having advantages and disadvantages in various applications. Applications include industrial and utility power plants firing coal or solid wastes, plants using sorbent injection and dry scrubbing FGD, and fluidized-bed combustors. Collection efficiency can be expected to be at least 99.8% or greater. Fabric filters have the potential for enhancing SO_2 capture in installations downstream of sorbent injection and dry scrubbing systems as discussed in Chapter 35.

Electrostatic precipitators ESPs are available in a broad range of sizes for utility and industrial applications. Collecting efficiency can be expected to be 99.8% or greater of the inlet dust loading. ESPs are considered to be less sensitive to plant upsets than fabric filters because their materials are not as sensitive to maximum temperatures. They also have a very low pressure drop. Power usage of ESPs and fabric filters tend to be similar because the high fan power needed to overcome the higher fabric filter pressure drop is approximately equal to the power consumed in the ESP transformer rectifier sets. ESP performance is sensitive to flyash loading, ash resistivity and coal sulfur content. Lower sulfur concentrations in the flue gas can lead to lower ESP collection efficiency.

Water pollution control

U.S. legislation — Clean Water Act

The objective of the Clean Water Act (CWA) is to *restore, maintain and enhance the chemical, physical and biological integrity of the nation's waters.*[9] The Act does this by setting standards for the quality of bodies of water and limitations on effluents from industrial and municipal activity. The Federal Water Pollution Control Act of 1972 and the amendments included in the Clean Water Act of 1977 authorized the EPA to control toxic pollutants discharged into waterways from point sources. These included pipes, channels and ditches. The 1972 legislation identified 129 priority pollutants which are considered hazardous wastes and which must be limited in discharges into domestic waters. During the 1980s, regulatory activities broadened to include nonpoint sources of toxic pollutants and water quality in lakes and ground water, in addition to previously regulated water quality in rivers and streams.

Requirements under the Act Industries must achieve discharge limits based on *best practicable*, followed by *best available,* treatment technologies as determined by the EPA. Industrial facilities saw an increase in compliance from 36 to 78% between 1972 and 1982. The positive response to the CWA requirements has also had a negative result. Treatment of waste water from point sources results in large quantities of sludge which, in many cases, contain toxic hazardous wastes. The wastes are generally landfilled and have resulted in numerous and large disposal sites throughout the U.S. The effects of complying with both the CWA and the CAA have shifted the burden of ultimate disposal to land based facilities.

Power plant discharge sources

In 1982 the EPA promulgated effluent limits, pretreatment standards and new source standards for 21 major industrial categories which included steam electrical power generation. Key limits in this category for utility power plants are given in Table 7. The EPA surveyed industrial discharges to determine for each discharge which of the 129 priority pollutants were present in sufficient concentration to require national regulation. Most of the organic compound priority pollutants were excluded from the utility plant discharges because of their absence or low concentration. The EPA did not initially set limits for four types of utility plant aqueous wastes, reserving these for future rule making. These are nonchemical metal cleaning wastes, FGD waste water, runoff from materials storage and construction areas (excluding coal piles), and thermal discharges. The following describes the principal aqueous discharge streams from utility power plants.

Once-through cooling water Water from rivers, lakes or oceans is used to absorb heat from the steam condenser. The cooling water exiting the steam condenser is at an elevated temperature and can be returned to the source or pumped to a cooling tower for evaporative cooling before being returned to the steam condenser. In the former case, the cooling water contains significant concentrations of only one principal regulated pollutant, total residual chlorine (TRC), which arises out of chlorine addition for condenser fouling control. The duration of each chlorination event is limited. The concerns over TRC discharge include toxicity to living organisms and the generation of halogenated hydrocarbons.

Cooling tower blowdown When the heated cooling water from the main steam condenser is cooled in an evaporative cooling tower, a buildup in dissolved solids and suspended matter occurs. Most of this buildup is removed from the system by cooling tower blowdown. Some of the suspended matter can settle out in the cooling tower basin and is removed at infrequent intervals. All of the dissolved solids and the remaining suspended solids are removed largely by cooling tower blowdown, although a small amount is discharged by drift in cooling tower exhaust. Blowdown flow is adjusted to keep the concentration of dissolved and suspended solids below the limits required to control condenser tube fouling and corrosion. Blowdown can vary from 3 to 65% of the makeup flow, depending on whether fresh water or salt water is used and whether a portion of the recirculating stream is continuously treated for deposit forming substances. Sources of chemical pollutants in blowdown include chlorine and proprietary organic chemicals for control of biofouling, corrosion inhibitors (consisting of chromate, zinc, polyphosphates, etc.), chemicals for scale control and products of corrosion. Although some of these maintenance chemicals appear on the 126 priority pollutants list, none are permitted to be present in significant levels in cooling tower blowdown (except for chromium and zinc, which are separately regulated).

Ash handling water waste Ash produced from the combustion of fuel, whether oil or coal, is collected at different points in the combustion process. Flyash is the finer size ash collected by particulate collection systems and bottom ash is removed from hoppers at the furnace bottom. Additional hoppers at intermediate points also accumulate ash. In many cases, ash is moved from these points with sluice water, which then goes to a settling pond and can typically contain 5% of suspended solids by weight.

The ash settling pond overflow can contain dissolved and suspended solids, the quantities of which will depend on the source of the ash, the type of combustion process and the point from which it is extracted from the combustion process. In general, oil ash can contain oxides and salts of vanadium, nickel and iron, carbon, organometallic compounds and magnesium compounds when a magnesium oxide additive is used for corrosion control. Oil ash is more soluble than coal ash and settles more slowly because of its particle size. Coal ash, because of the significantly greater quantity of mineral matter in coal compared to oil, is produced in much greater quantities than oil ash (300,000 t/y versus 2000 t/y for a typical 1000 MW utility plant). Bottom ash from coal combustion tends to be a mixture of vitreous metal oxides and silica, is low in solubility and tends to settle quickly because of its coarse size compared to oil ash. Coal flyash can have solubility of several percent and, because of its fineness and the presence of low density hollow cenospheres, tends to settle slowly. Coal ash contains, in addition to the eight or nine major elemental constituents, a large number of trace elements that can appear in pond overflow and which may need to be treated by best practicable technology.

Coal pile runoff Open storage of large quantities of coal is required for an uninterrupted fuel supply to utility plants [on the average, 800 to 2400 yd^3 (611 to 1834 m^3) per megawatt of rated capacity is kept on hand]. The water and oxygen from the air react with the minerals in the coal to produce a leachate contaminated with ferrous sulfate and sulfuric acid. The low pH from the acid accelerates dissolution of many of the metals present in the coal minerals.

FGD blowdown In wet FGD systems, a portion of the absorber slurry, which is sprayed into the flue gas stream to remove SO_2, is removed from the absorber tank for dewatering. In the dewatering process, the solid reaction products are separated from the liquor. The liquor is recycled to the absorber tank where additional sorbent is added. Recycling of the liquor can result in chloride buildup which, in turn, can cause increased sulfate scaling by upsetting the sulfite/sulfate balance. This buildup can be controlled by the loss of liquor retained in the dewatered sludge or by a blowdown. An aqueous blowdown discharge would typically be a saturated solution of calcium sulfate, calcium sulfite and sodium chloride. Also, depending on flyash carryover, traces of metal ions could also be present. In setting effluent limitations in 1982, the EPA reserved regulating FGD aqueous discharge to a future date.

Table 7
1982 Adopted Aqueous Discharge Limits for Steam Power Generating Systems (Note 4)

Source and Pollutant (Note 1)	Effluent Limits, mg/l BAT (Note 2) Maximum	Average
All discharges		
pH	6 to 9	6 to 9
PCBs	No discharge	No discharge
Low volume waste (Note 5), bottom ash and flyash transport water (Note 3):		
TSS	100	30
OG	20	15
Chemical metal cleaning wastes:		
TSS	100	30
OG	20	15
Copper	1.0	—
Iron	1.0	—
Once-through cooling water, total residual chlorine (TRC)	0.2	—
Cooling tower blowdown:		
Free available chlorine	0.5	0.2
Zinc	1.0	1.0
Chromium	0.2	0.2
Other 124 priority pollutants	No detectable amount	
Coal pile runoff:		
TSS (1980)	50	—

Notes:
1. Nomenclature: TSS - total suspended solids; OG - oil and grease; BAT - best available technology.
2. 30 day rolling daily average (Average); maximum any one day (Maximum).
3. Best conventional technology basis.
4. Adapted from Reference 1.
5. Low volume wastes include ion exchange, water treatment, evaporator blowdown, boiler blowdown, lab and floor drains, plus FGD waste water.

Metal cleaning wastes These aqueous wastes can arise from either *chemical* or *nonchemical* cleaning of metal heat transfer surfaces in the boiler.

Chemical metal cleaning uses chemical solvents for water-side cleaning of boiler system components to remove corrosion products. Cleaning intervals are measured in years for large utility boilers, and produce three to four boiler volumes [20,000 to 100,000 gal (75,708 to 378,540 l)] of waste water per cleaning. The composition of the waste solvents depends on the construction material of the feedwater system, but largely consists of iron with lesser amounts of copper, nickel, zinc, chromium, calcium and magnesium. The disposal method for the spent solvent depends on the type of chemical cleaning solvents used. When hydrochloric acid based solvents are used, spent solvent is treated on-site by neutralization and is discharged subject to the effluent limits in Table 7 or more stringent water quality standards. With approval from appropriate regulating bodies, organic based solvent wastes are often incinerated in other operating boilers at the site. The metals in the chemical cleaning wastes are retained with the normal boiler ash.

Nonchemical cleaning is used to remove fireside deposits by means of high pressure jets of water. The waste water can contain the same metals and pollutants contained in the ash deposits being removed. Because the deposit composition varies with position in the boiler, the wash water composition will depend on the location of the area being cleaned. These waste waters may be classified as either low volume wastes or metal cleaning wastes and are treated according to the corresponding effluent limits.

Low volume wastes These include discharges from ion exchange water treatment, evaporator blowdown, boiler blowdown, cooling tower basin cleaning, laboratory and floor drains, and drains and losses from recirculating house service water systems. FGD blowdown is also included until the EPA develops specific regulations for this stream. By EPA definition, low volume wastes are those from all sources taken collectively as if they were from one source. Excluded are those wastes for which specific effluent limits are established.

Water pollution control technologies

The technologies for waste water treatment used to meet limits for discharge include clarification and filtration. (See also Chapter 42.)

Clarification This is used to settle out larger suspended particles and condition smaller colloidal particles to make them settle and allow filtration for removal. A pond, reservoir or tank is used to allow larger particles to settle in a matter of hours. The finer particles overflow and are made to settle more quickly by the addition of chemical agents, coagulants and polymers that cause agglomeration to sizes large enough to settle out of suspension.

Filtration This uses a porous barrier across flowing liquid to remove suspended materials. Filtration can be used to supplement clarification and permits reducing suspended solids to the parts per billion level. Filter types include sand filters which are generally slow and do not handle fine clay well. Preconditioning with coagulants can improve filtration rates. Dual media filters improve on sand filters by superimposing a coarse, granular material over the fine bed. This allows more of the filter bed to be used, reduces head loss, and provides higher flow rates and longer operating cycles before cleaning.

As required, and with approvals from appropriate regulating bodies, final waste stream pH is controlled by combining various plant streams to provide a neutral pH product. Where needed, acid or alkali addition can be used to achieve the final pH. Other treatments are also available to address other criteria pollutants where concentration warrants.

In selected cases, zero discharge water management is provided which does not return any waste water to water sources. Effectively all water brought into the plant is evaporated through cooling towers, ponds or stack. Residual solids are then sent for disposal.

Solid waste disposal

The Resource Conservation and Recovery Act

The rapid growth of industrial activity and use of consumer goods by our society have resulted in an explosive growth in solid wastes. One of the first attempts to protect health and the environment, reduce waste, conserve natural resources and control hazardous waste production was the Solid Waste Disposal Act of 1965. This was followed in 1976 by the Resource Conservation and Recovery Act (RCRA) which amended the original Act to ensure proper future solid waste management.

The RCRA requires improved solid waste management practices, defines what constitutes hazardous waste and greatly expanded provisions for hazardous waste management. Three notable parts of the Act with far reaching implications are Subtitles C, D and I. Revisions made in 1984 were significant in expanding the regulatory scope of the RCRA through the Hazardous and Solid Waste Amendments.

The definition of a hazardous waste is very broad. It is based on the premise that a material is hazardous when its quantity, concentration or physical/chemical/infectious characteristics *(1) cause, or significantly contribute to, an increase in mortality, or an increase in serious irreversible or incapacitating reversible illness, or (2) pose a substantial present or potential hazard to human health or the environment when improperly treated, stored, transported or disposed of, or otherwise managed.*[10]

Under Subtitle C, hazardous waste can include virtually any type of waste (solid, semisolid, liquid and gaseous) which fits one or more of the following criteria:

1. possesses the characteristics of ignitability, corrosivity, reactivity and toxicity as determined by standard analytical tests and procedures (commonly referred to as a characteristic hazardous waste).
2. is listed as a specific hazardous waste (commonly referred to as a listed hazardous waste).
3. is a mixture of wastes which contains a listed hazardous waste.
4. has not been excluded from RCRA regulation as a hazardous waste.
5. is produced as a byproduct from the treatment of any hazardous waste.

Listed hazardous wastes fall into three categories based on their sources:

1. *Nonspecific source wastes* These may be derived from various industrial processes and are viewed as generic in nature, such as degreasing solvents and electroplating waste water sludges.
2. *Specific source waste* This type of waste is specifically named for and identified in terms of the industrial process from which it is produced, i.e., K047, pink/red water from TNT operations; K050, heat exchanger bundle cleaning sludge from the petroleum refining industry.
3. *Commercial chemical products* These consist of specific chemicals such as organic and inorganic compounds, pesticides and often products which are no longer of use and are identified for disposal.

In the enactment of the RCRA, there were a number of legislative exclusions of waste types which were produced in very large quantities but did not present a danger to health or the environment, or that were covered by other environmental regulations. Some of the exclusions are of significance to the power industry and the management of municipal solid wastes. Among the excluded wastes were household wastes, agricultural wastes and wastes from municipal resource recovery plants. In 1980 additional temporary exclusions were made to include wastes from fossil fuel combustion, mining operations, and oil and gas explorations. These exclusions are being examined by the EPA for possible regulation under Subtitle C.

The general trend, whether through federal or state and local action, is toward tighter control and restrictions. Depending on the state and local jurisdiction, high volume waste streams from power plants such as scrubber sludge, flyash and bottom ash are subject to different and highly variable disposal requirements. In some cases, stabilization of the solids is required. In addition, landfills are beginning to use leachate collection systems with single or double linings and extensive monitoring wells.

Kinds of solid wastes

The principal solid waste streams in coal- and oil-fired utility boilers include the following:

1. *Bottom ash* is that portion of fuel ash which falls to the bottom of the furnace or from the stoker discharge. In coal-fired Cyclone furnace boilers, the bottom ash consists of slag which drops from the bottom of the furnace into a slag tank for solidification.
2. *Flyash* is the finer ash material which is borne by the flue gas from the furnace to the back end of the boiler; it drops out in the economizer and air heater hoppers or is collected by particulate control equipment.
3. *Pyrite* is iron sulfide, an impurity which is separated from coal in the pulverizer and which is combined with bottom ash for disposal.

FGD waste characteristics depend on the particular technology used:

1. *Wet scrubbing (calcium based system)* Using natural oxidation produces a wet sludge containing a mixture of calcium sulfite and calcium sulfate reaction products, trace amounts of flyash and unreacted limestone. In a forced oxidation system, the principal difference in the waste is that the reaction product is almost totally in the form of calcium sulfate or gypsum, which is more easily dewatered to a filter cake for landfill or other use.
2. *Dry scrubbing* Waste is dry and contains calcium sulfite, calcium sulfate, flyash and unreacted sorbent (hydrated lime).
3. *Dry lime injection* Waste is dry and contains calcium sulfate, flyash and a large proportion of calcium oxide (CaO).

Solid waste treatment methods

In order to dispose of waste materials from wet collection systems, treatment methods are applied to ultimately produce a solid. These methods include dewatering, stabilization and fixation and are designed to achieve waste volume reduction, stability and better handling, or liquid recovery for reuse.

Dewatering This is used to physically separate water from solids to increase solids content of the product and recover water for further treatment and potential reuse.

A settling pond is the simplest method for dewatering, is not sensitive to inlet solids content, requires low maintenance and is highly reliable. Ponds are often used for ash or limestone scrubber slurries. Sizing provides low flow velocity so that solids can settle undisturbed by gravity. Settling ponds are unpopular with regulatory agencies, require substantial acreage and must be shut down for solids removal.

Thickeners are large cylindrical tanks with a center column that drives radial rakes extending from the bottom of the shaft. The rakes carry plows to stir material on the bottom which slopes toward the center. (See Chapter 35 and 42.) The plows push settled material toward the underflow discharge. Thickeners rely on gravity to separate high specific gravity solids and are often applied to dewatering wet scrubber slurry. Although thickeners are complicated and have high capital and maintenance costs, they have high throughput rates and require less land area than settling ponds.

Cyclone collectors or hydroclones are sometimes used in place of thickeners to remove solids from slurries by centrifugal and liquid shear effects. Cyclones separate and collect particles down to a particular size, with finer particles staying with the liquid overflow. Cyclones do not separate material less than 5 μm effectively and are not efficient with slurries containing more than 15% solids. They are low in cost, have low space requirements and produce low solids content in the liquid fraction and a high liquid content in the solids fraction.

Vacuum filters, either of the drum or belt type, are generally used for second stage dewatering of wet scrubber slurries. They take little space and produce a high solids content product, up to 65% for FGD slurry and 75% for ash slurry. However, vacuum filters are higher in cost and maintenance and are mechanically complex.

Stabilization This increases solids content of waste by adding dry solids such as flyash. Stabilization is applied to impart greater physical stability to the waste, making it easier to place in the landfill and making it less susceptible to future problems. Stabilization and fixation are generally applied to scrubber wastes as the final treatment step after dewatering. Bottom ash and flyash, because of their granular nature, generally dewater easily and do not require stabilization for disposal. For stabilization, a dry solid such as soil or flyash is mixed with the waste slurry and spreads the water in the waste over a larger mass of solids. Also, there is improvement in particle size distribution which leads to closer packing, lower permeability and lower combined volume. Stabilization can be reversible and if the waste is rewetted it may fluidize and fail structurally.

Fixation This involves the addition of an agent such as lime to produce a chemical reaction to bind free water and produce a dry product. Fixation includes a number of processes. Mixing suitable proportions of scrubber slurry with alkaline flyash containing sufficient CaO produces a chemical reaction which results in a material with compressive strength comparable to concrete and with very low permeability. Both characteristics contribute to ease of placement and minimal leaching problems.

When the flyash does not have sufficient alkalinity, lime may be added with the flyash scrubber slurry to produce the cementitious reaction. Four percent addition of lime has produced material with the necessary physical properties for disposal or use. The cured material is suitable for structural fill, providing a site which can be used for building construction after completion of the landfill. Comparable fixation reactions with scrubber sludge have been obtained with additions of 5 to 10% of blast furnace slag or Portland cement.

Disposal and utilization methods and requirements

Ultimate disposition of utility plant wastes (ashes and FGD residues) is by disposal (in landfills or impoundments) or by utilization. Where disposal is used, the waste stream is analyzed, and the site is permitted and approved by the appropriate regulatory agencies.

Disposal methods These can be either wet or dry, depending on the physical condition of the material. Wet disposal requires construction of a pond which may be below or above grade with impermeable barriers or dikes provided. Below grade construction may be considered and depends on suitable geology and hydrology at the site. With wet disposal, the waste is placed in slurry or liquid. After placement and settling, the liquid which has separated is collected, treated and either released or recycled. Dry disposal can use a simple method of landfill construction in which the waste is placed and compacted to form an artificial hill. The trend is toward dry disposal because of smaller volumes, more options for site or material reclamation, and the developing interest in dry scrubbing. Baghouse particle collection systems would also encourage the use of dry disposal.

Utilization methods These become more attractive as waste management costs increase. Bottom ash, flyash and boiler slag are used in applications where they can be substituted for sand or gravel. The characteristics of boiler slag and bottom ash also make these materials useful for blasting grit, roofing granules and controlled fills. Flyash, because of its chemistry and physical properties, is applicable in the manufacture of Portland cement and concrete mixes. The value of these materials is so low that the cost of transportation severely limits their use to applications close to the producing power plant.

FGD byproduct use is potentially in the areas of agriculture, metals recovery, sulfur recovery and gypsum. Agricultural use is limited to waste which contains lime; this can be substituted for agricultural lime. However, trace elements from flyash contamination could have an unacceptable impact and make wide use doubtful. Use for metals recovery is limited by undeveloped technology and questionable cost effectiveness. Use for sulfur recovery is limited by incomplete technology development, low market price of sulfur and large amounts of residual byproduct after sulfur extraction.

Only FGD waste from forced oxidation wet scrubbing systems, which consists of almost pure gypsum, has seen some commercial use for drywall production.

References

1. Rice, J.K., "Legislation and pollution sources," appearing in *Standard Handbook of Power Plant Engineering*, T.C. Elliot, ed., McGraw-Hill, New York, 1989.

2. Corbitt, R.A., ed., *Standard Handbook of Environmental Engineering*, McGraw-Hill, New York, 1990.

3. Clean Air Act, 42 USCA S7401 *et seq*., Sec. 101 (b) (1).

4. Title 40, Code of Federal Regulations, Part 60, Subparts Ca, D, Da, Db, Dc, E, Ea, and BB, U.S Government Printing Office, Washington, D.C., July 1, 1991.

5. Vernon, J.L., "Market impacts of sulphur control: The consequences for coal," Report IEA/CR18, International Energy Agency/Coal Research, London, October 1989.

6. Hjalmarsson, A.-K., and Soud, H. N., "Systems for controlling NO_x from coal combustion," Report IEACR/30, International Energy Agency/Coal Research, London, December 1990.

7. Zimmerman, D., *et al*., "Anthropogenic emissions data for the 1985 NAPAP inventory — final report," U.S. Govt. Report No. EPA-600/7-88-022, November 1988.

8. Saeger, M., *et al*., "The 1985 NAPAP emissions inventory (version 2): Development of annual data and modelers tape," U.S. Govt. Report, EPA-600/7-89-012a, November 1989.

9. Clean Water Act, 33 USCA S1251 *et seq*., Sec. 101 (a).

10. Resource Conservation and Recovery Act, 42 USCA S9601 *et seq*., Sec. 1004 (5).

Large utility electrostatic precipitator installation.

Chapter 33
Particulate Control

As steam is generated by the combustion of most fossil fuels, the flue gas carries particulate matter or ash from the furnace. Except for natural gas, all other fossil fuels contain some quantity of ash or noncombustibles which form the majority of the particulate. Unburned carbon also appears as particulate. Combustion in most steam generation applications using nonfossil fuel also produces particulate. Control is needed to collect this material and to limit its release to the atmosphere.

All coals contain some amount of ash. Content varies depending on the type of coal, location, depth of mine and mining method. In the United States (U.S.), eastern bituminous coals typically contain 5 to 15% ash while the western subbituminous coal ash content may range from 5 to 30% ash by weight. Texas lignites also contain up to 30% ash. Mining methods on thin seams of coal may also contribute to higher ash quantities.

When coal is burned in conventional boilers, a portion of the ash drops out of the bottom of the furnace while the remainder of the ash is carried out of the furnace in the flue gas. It is this remaining ash (*flyash*) that must be collected before exhausting the flue gas to the atmosphere.

Different combustion methods contribute different proportions of the total coal ash content to the flue gas. With pulverized coal firing, 70 to 90% of the ash is carried out of the boiler with the flue gas. A stoker-fired unit will emit about 40% of its ash in the flue gas along with some amount of unburned carbon. With cyclone firing, only 15 to 30% of the ash is normally carried by the flue gas. On circulating fluidized-bed boilers, all of the ash, along with the fluidized-bed material, is carried by the flue gas. Therefore, the selection and design of particulate control equipment are closely tied to the type of firing system.

An American Society for Testing and Materials (ASTM) spectrographic analysis test of a coal sample reveals the major ash components and determines total ash content. Ash constituents are typically reported in the oxide form and include silicon dioxide, titanium dioxide, iron oxide, aluminum dioxide, calcium oxide, magnesium oxide, sodium oxide, potassium oxide, sulfur trioxide and diphosphorous pentoxide. Trace quantities of many more elements are also found in ash. The proportion of the major constituents varies significantly between coal type and mine location. The analysis and composition of flyash are discussed in greater detail in Chapters 8 and 20.

Other significant coal ash properties are particle size distribution and shape, both of which are dependent on the type of firing method. Stoker-fired units generally produce the largest particles. Pulverized coal-fired boilers produce smaller, spherical shaped particles of 7 to 12 microns (Fig. 1). Particles from cyclone-fired units, also mostly spherical, are smaller yet. Fluidized-bed units produce a wide range of particles that are generally less spherical and are shaped more like crystals. Knowledge of ash properties is important in the selection of the correct particulate control equipment.

Regulation of particulate emissions

Particulate control equipment was first used by utilities in the 1920s and before that time in some industrial applications.[1] Prior to 1971, however, controls were installed mostly on a best effort basis. In 1971, the first Environmental Protection Agency (EPA) performance standard limited outlet particulate emissions to 0.1 lb/10^6 Btu (123 mg/Nm^3 at 6% O_2) heat input and stack opacity to 20% for those units larger than 250 x 10^6 Btu/h (73.3 MW_t) heat input. *Opacity*, measured by a transmissometer, is the portion of light which is scattered or absorbed by particulate as the source of light passes across a flue gas stream. Therefore, both the amount and appearance of the stack emissions are regulated. Since 1979 the EPA New Source Performance Standards (NSPS) for particulate control

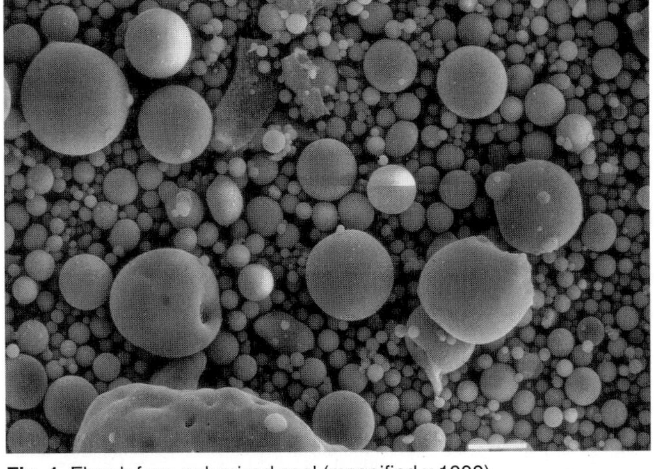

Fig. 1 Flyash from pulverized coal (magnified x 1000).

permit a maximum of 0.03 lb/10^6 Btu (36.9 mg/Nm^3 at 6% O_2) heat input for these units. A 20% opacity is still permissible.

Federal and state EPA regulations set the primary guidelines for particulate emissions. In addition, many local regulatory bodies have established stricter regulations than those set by the EPA. There are separate emissions standards for a variety of fired processes including steam generators firing coal, oil, refuse and biomass. Currently there are three major classification levels for steam generating units: one for units greater than 250 x 10^6 Btu/h (73.3 MW_t), one for the 100 to 250 x 10^6 Btu/h (29.3 to 73.3 MW_t) units and a third for those units less than 100 x 10^6 Btu/h (29.3 MW_t) heat input.[2, 3] Finally, if a new plant is in a nonattainment area, the permissible particulate emissions and opacity may be significantly reduced from nominal control levels. (See Chapter 32.)

Particulate control equipment

Particulate emissions from the combustion process are collected by particulate control equipment (Fig. 2). This equipment must remove the particulate from the flue gas, keep the particulate from re-entering the gas and discharge the collected material. There are several major types of equipment available including electrostatic precipitators, fabric filters, mechanical collectors and venturi scrubbers. Each of these uses a different collection process with different factors affecting the collection performance.

Electrostatic precipitators

An electrostatic precipitator (ESP) electrically charges the ash particles in the flue gas to collect and remove them. The unit is comprised of a series of parallel vertical plates through which the flue gas passes. Centered between the plates are charging electrodes which provide the electric field. Fig. 3 is a plan view of a typical ESP section which indicates the process arrangement.

Charging

The collecting plates are typically electrically grounded and are the positive electrode components. The discharge electrodes in the flue gas stream are connected to a high voltage power source, typically 55 to 75 kV DC, with a negative polarity. An electric field is established between the discharge electrodes and the collecting surface. As the flue gas passes through the electric field, the particulate takes on a negative charge which, depending on particle size, is accomplished by field charging or diffusion.

Collection

The negatively charged particles are attracted toward the grounded collection plates and migrate across the gas flow. Some particles are difficult to charge, requiring a strong electric field. Other particles are charged easily and driven toward the plates but also may lose the charge easily requiring recharging and recollection. Gas velocity between the plates is also an important factor in the collection process since lower velocities permit more time for the charged particles to move to the collection plates and reduce the likelihood of re-entrainment. In addition, a series of plates and discharge electrodes are necessary to maximize overall particulate collection by increasing the opportunities of the individual ash particles to be charged and collected.

The ash particles form an ash layer as they accumulate on the collection plates. The particles remain on the collection surface due to the forces from the electric field as well as the molecular and mechanical cohesive forces between particles. These forces also tend to make the particles agglomerate or cling together.

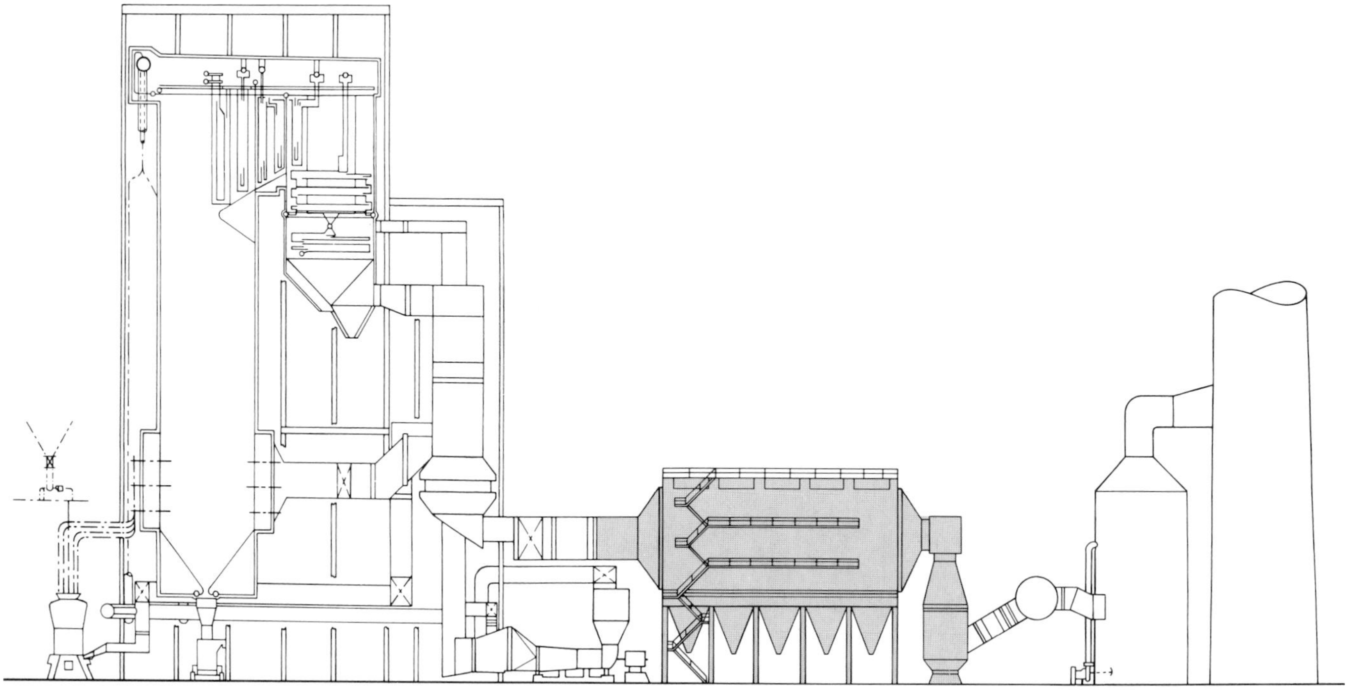

Fig. 2 Particulate control equipment — plant side view.

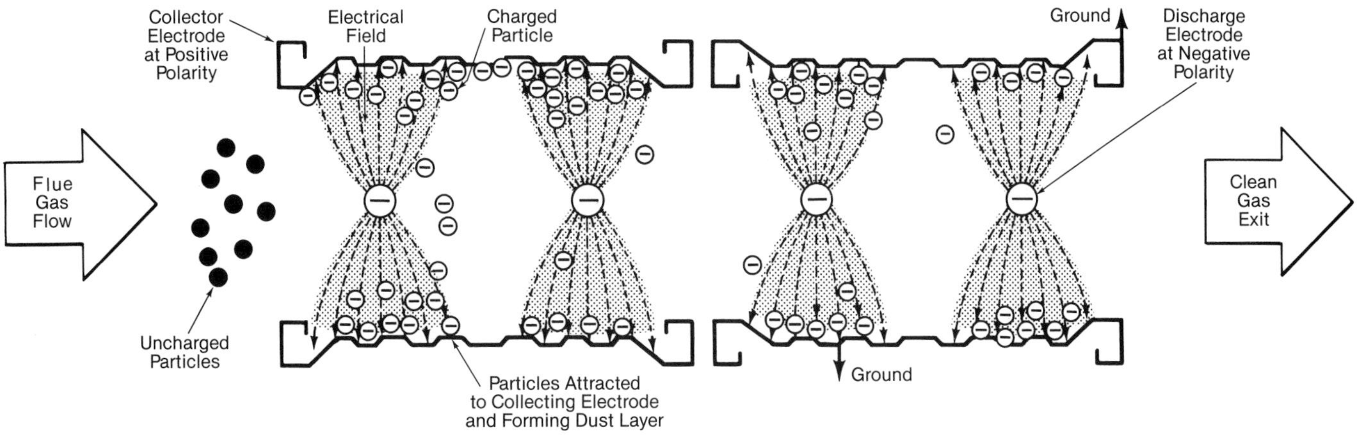

Fig. 3 Particle charging and collection within an ESP.

Rapping

The ash layer must be periodically removed. The most common removal method is rapping which consists of suddenly striking the collection surface; this rapping force dislodges the ash. Because particulate tends to agglomerate, the ash layer is removed in sheets. This sheeting is important to prevent the re-entrainment of individual particles into the flue gas stream, requiring recharging and collection downstream.

While most of the particles are driven to the collection surface, some positively charged particles attach to the discharge electrodes. A separate rapping system is therefore used to clean these electrodes to maintain the maximum charging forces.

Ash removal

The dislodged particulate falls from the collection surface into hoppers. Once the particulate has reached the hopper it is important to ensure that it remains there until the hopper is emptied. (See Chapter 23 for hopper ash removal methods and equipment.)

Precipitator sizing factors

An ESP is sized to meet a required performance or particulate collection efficiency. The sizing procedure determines the amount of collection surface to meet the specified performance. An equation which relates the collection efficiency (E) to the unit size, the particle charging and the collection surface is the Deutsch-Anderson equation:[1]

$$E = 100\left(\frac{\text{Inlet dust loading} - \text{Outlet dust loading}}{\text{Inlet dust loading}}\right) \quad (1)$$

$$1 - E = e^{\frac{-wA}{V}} \quad (2)$$

or

$$A = \left[\ell n\left(\frac{1}{1-E}\right)\right]\frac{V}{w} \quad (3)$$

where

E = ESP removal efficiency, %
w = migration velocity, ft/min (m/s)
A = collection surface area, ft^2 (m^2)
V = gas flow, ft^3/min (m^3/s)

Migration velocity is the theoretical average velocity at which the charged particles travel toward the collection surface. This velocity is dependent upon how easily the particulate is charged, and the value is normally selected by empirical means based on experience. The factors which affect migration velocity are the fuel and ash characteristics, the operating conditions and the effects of gas flow distribution.

These factors also have an effect on the ability of the particulate to accept a charge. A commonly used indication of this effect is resistivity, measured in ohm-cm. Fig. 4 illustrates typical resistivity curves for two fuel ashes. High resistivity ashes result in low migration velocities and large collection surface areas while average resistivity ashes result in moderately sized surface areas.

Fuel and ash characteristics The fuel and ash constituents which reduce resistivity or which are favorable to ash collection in an ESP include moisture, sulfur, sodium and potassium. Applications with sufficient quantities of these components usually result in moderately sized precipitators. Constituents which hamper ash collection and increase outlet emissions include calcium and magnesium. High percentage concentrations of these items without offsetting quantities of the favorable constituents result in poorer collection and larger precipitators.

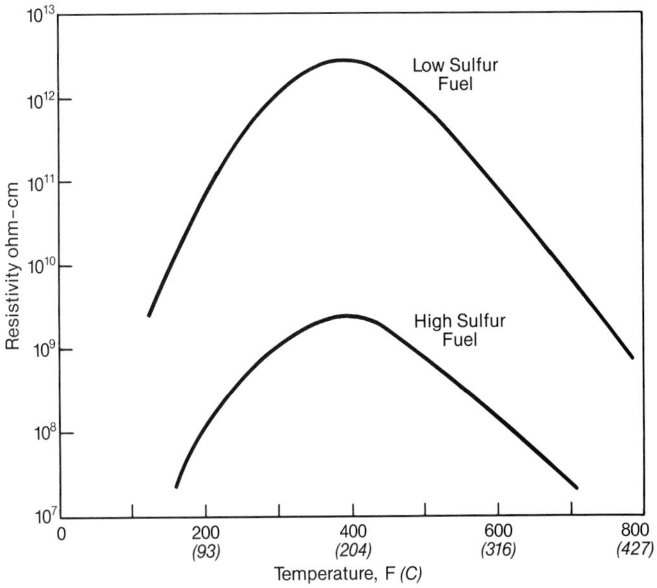

Fig. 4 Typical ash resistivity.

The fuel and ash constituents and their relative quantities must be reviewed in the sizing process to determine the overall effect on migration velocity. The migration velocity/resistivity can then be altered to some extent by controlling the content of the critical constituents.

Operating conditions The design operating conditions from the boiler or process also affect precipitator sizing and performance. As indicated on the resistivity curve, gas temperature has a direct effect on resistivity and on the gas volume passing through the ESP. Gas flow from the boiler also has an effect on sizing as indicated by the Deutsch-Anderson equation. There is an optimum gas velocity range within an ESP for maximum performance which must be considered as part of the design selection.

In addition to the quantity of particulate sent to the precipitator, particle size also affects ESP design and performance. A particle size distribution versus collection efficiency curve (Fig. 5) indicates that an ESP is less efficient for smaller particles (less than 2 microns) than for larger ones. Therefore, ESP applications with a high percentage of particles less than 2 microns will require more collection surface and/or lower gas velocities.

Flow distribution Maximum ESP efficiency is achieved when the gas flow is distributed evenly across the unit cross section. Uniform flow is assumed in the ESP sizing calculations and should be verified during the design stage by using a flow model. These models should include the precipitator as well as the inlet and outlet flues. Flow uniformity is typically achieved by installing distribution devices in the flue transition sections immediately upstream and downstream of the ESP. Hopper design must also prevent high velocity areas to avoid flyash re-entrainment. The industry standard for flow distribution and modeling is the Industrial Gas Cleaning Institute EP-7.[5]

Precipitator components

All ESPs have several components in common (Fig. 6) although there is some shape and size variation between units.

Discharge electrodes As described in the section on charging, the discharge electrodes, connected to the high voltage power source, are located in the gas stream and

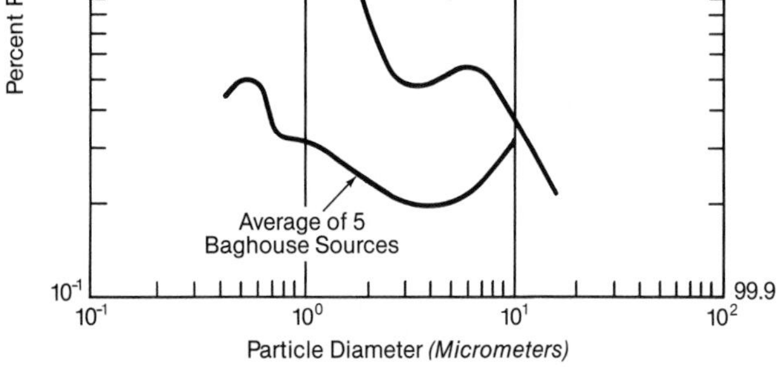

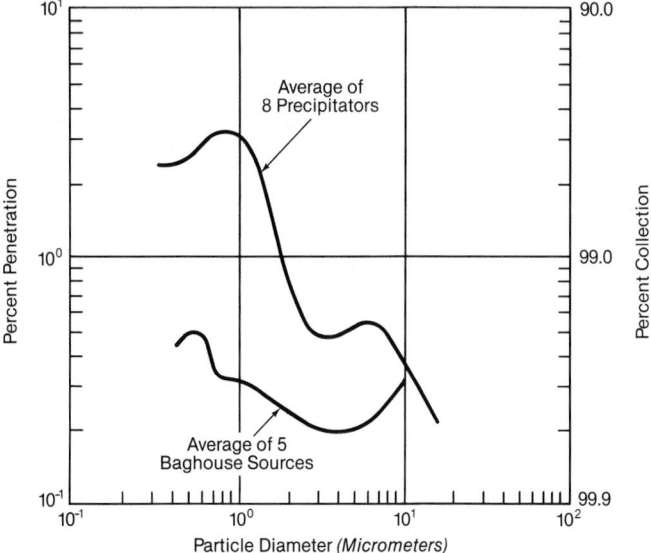

Fig. 5 Summary of fine particulate collection (adapted from Reference 4).

serve as the source of the corona discharge. These electrodes are the central components of the discharge system which is electrically isolated from the grounded portions of the ESP. An electrical gap of 6 to 8 in. (152 to 203 mm), depending on plate spacing, must be maintained throughout the ESP between the discharge and the grounded components.

Discharge electrodes are found in several shapes. Common types include the rigid frame, rigid electrode and weighted wire. The rigid frame, shown in Fig. 7, consists of strips of electrode supported between sections of frame tubing. Each frame is attached to a structural carrier, both front and rear. This assembly is supported by insulators forming a four point suspension system. The rigid electrode consists of a member with proprietary shape which is top supported and hangs the full height of the precipitator. The typical rigid electrode top support is also a frame hanging from insulators. The lower ends of the rigid electrode have a guide bar and side to side spacers. The third type of electrode, or weighted wire design, consists of a round discharge electrode (wire) supported at the top and held straight and in tension with a weight at the bottom. The upper frame is supported from insulators and there is a lower steadying frame to guide and space the electrodes.

For highest equipment reliability, either the rigid frame or rigid electrode is the most common configuration. Discharge electrode failure in the form of broken wires has been a recurring problem with the weighted wire electrodes, particularly with lengths of 30 ft (9.1 m) or more, which results in performance deterioration.

Collection surface As previously stated, the collection surface area in the Deutsch-Anderson equation is the total plate area required for particulate collection. Shown in Fig. 6, the collection surface typically consists of a series of roll formed collector plates assembled into a curtain and supported from the top. The curtains are spaced in rows across the width of the precipitator, typically on 12 or 16 in. (305 or 406 mm) centers. In depth, the curtains are arranged into fields. For calculating surface area, the curtain assembly is treated as a plane and includes both sides of the plates. The rolled plates can be up to 50 ft (15.2 m) in length with a shop straightness tolerance of 0.5 in. (12 mm). Collector curtains may also be large flat plates with stiffener bars added to maintain straightness. For optimum performance with a uniform electric field and with no electrical arcing (high current spark), the alignment of collection surface and electrodes must be maintained within tight tolerances.

Rapping systems As shown in Fig. 6, the most effective method of cleaning the collector curtains is to rap each one separately and in the direction of gas flow. This method assures that each curtain receives a rapping force. The rapping system shown is a tumbling hammer type, where the hammer assemblies are mounted on a shaft extending across the ESP in a staggered arrangement. The shaft is turned slowly by an external drive controlled by timers for rapping frequency and optimum cleaning. Hammer size is selected to match the application and size of the collector curtain. Although not as effective for many applications, top rapping of the collection surface is also popular. Typically, more than one collector curtain is rapped at a time with this method, and the rapping force is in the downward direction on the top edge of the curtains. Both a drop rod and magnetic impulse are the drive mechanisms used.

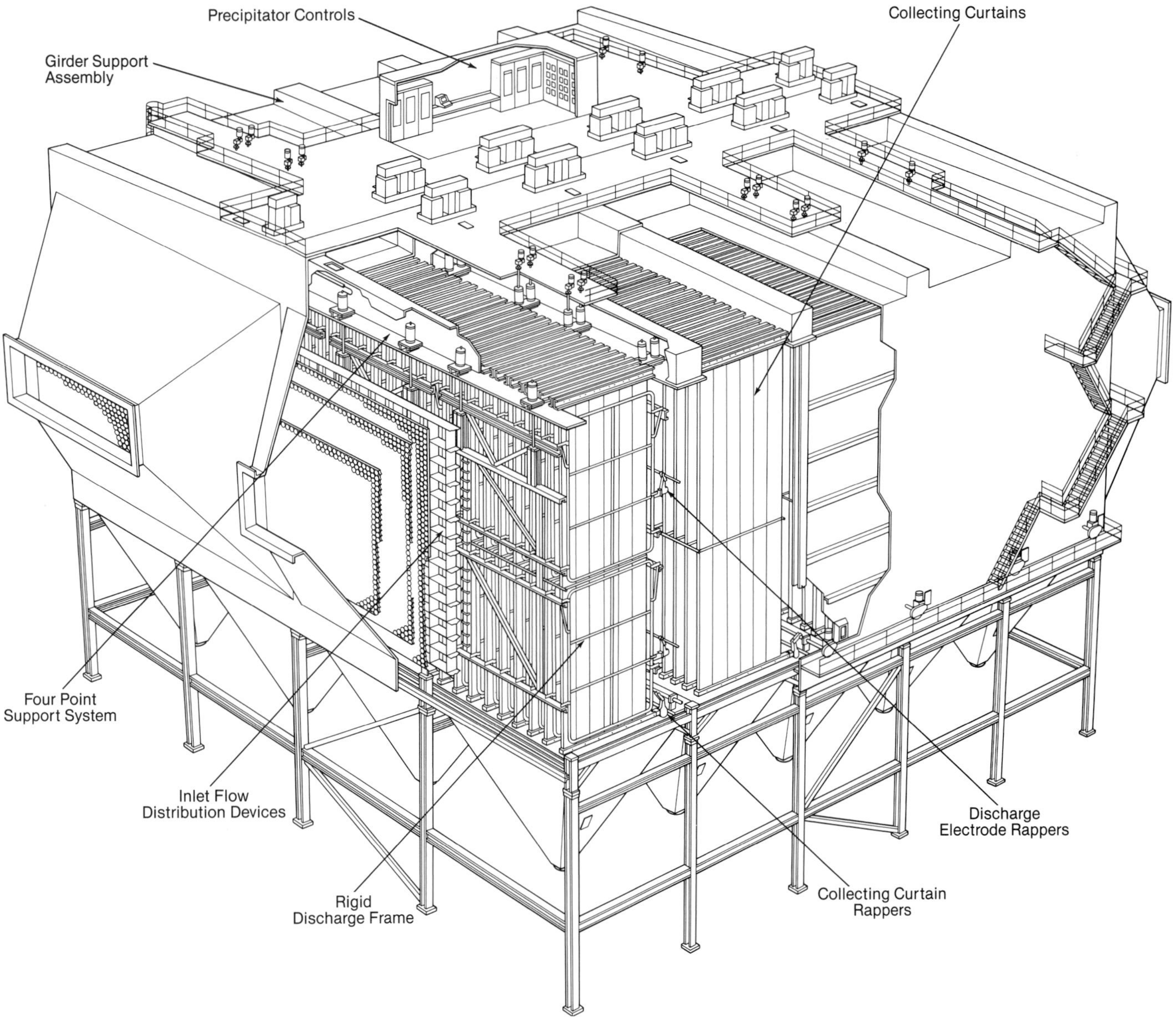

Fig. 6 B&W/Rothemuhle rigid frame electrostatic precipitator.

Due to the difficulty of cleaning high resistivity flyash from the collection surface, considerable tests have been performed to ensure that adequate rapping forces are transmitted across the entire collection surface. A minimum acceleration of 100 g's applied at the farthest point from impact has been established as an industry standard.

Typically, rapping of the rigid frame discharge electrodes is accomplished using a tumbling hammer system as shown in Fig. 7. The hammer assemblies are mounted in a staggered arrangement on a shaft across the width of an electrical section. Note that a smaller hammer than that used for the collector system is required for proper cleaning of the discharge electrodes. An external drive unit mounted on the precipitator roof is used to slowly turn the rapper shaft and, because it is attached directly to the carrier frame, the drive shaft must also be electrically isolated with an insulator. As with the collector system, top rapping of the discharge electrodes is another method of cleaning.

A rapping system is sometimes used on the flow distribution devices at the precipitator inlet. On those applications where the particulate is sticky or tends to accumulate, a rapping system is needed to keep the surfaces clean and to allow the distribution devices to provide uniform flow to the precipitator inlet at an acceptable pressure loss.

Enclosure As shown in Fig. 6, the structure forming the sides and roof of an ESP is a gas-tight metal cased enclosure. The structure rests on a lower grid, which serves as a base and is free to move as needed to accommodate thermal expansion. All of the collecting plates and the discharge electrode system are top supported from the plate girder assemblies. The entire enclosure is covered with insulation and lagging. Access doors in the casing and adequately sized walkways between the fields assist in maintenance access for the internals.

Materials for the precipitator enclosure and internals are normally carbon steel, ASTM A-36 or equivalent, because gas constituents are noncorrosive at normal operating gas and casing temperatures. Projects with special conditions may warrant an upgrade in some component materials.

Hoppers Metal pyramid shaped hoppers are supported from the lower grid and are made of externally stiffened

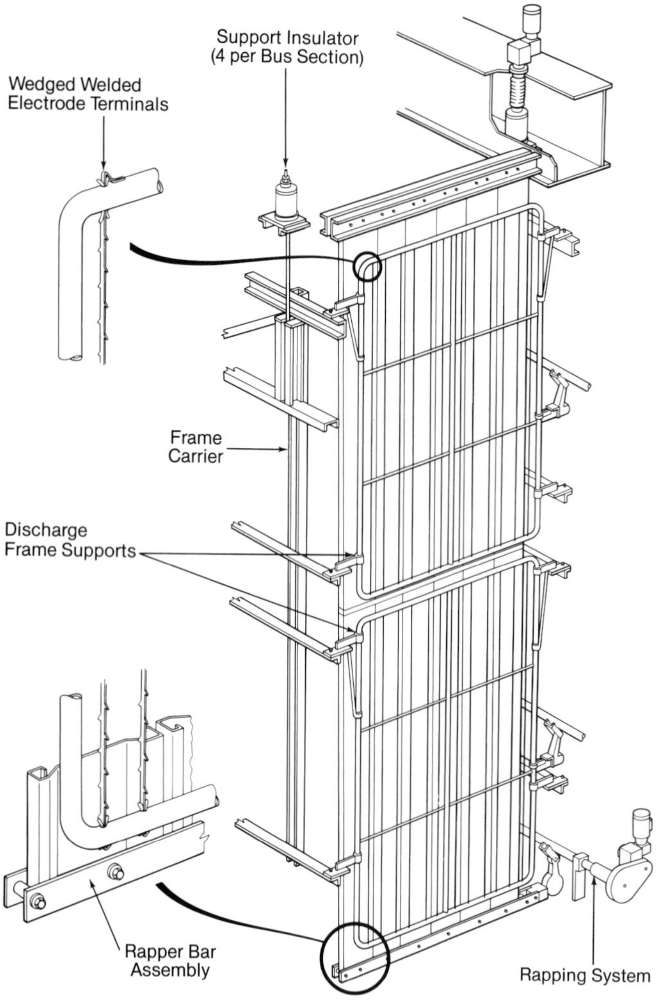

Fig. 7 Rigid frame discharge electrode and rapping system for an ESP.

casing. The hoppers provide the lower portion of the overall enclosure and complete the gas seal. Their sides are designed with an inclination angle of at least 60 deg from horizontal. Hoppers are generally designed as particulate collection devices which can store ash for short periods of time when the ash removal system is out of service.

Because many ash removal systems are noncontinuous, the following items are normally supplied with the precipitator hoppers to ensure good particulate removal: hopper heaters, electromagnetic vibrators, poke holes, anvil bars and level detectors. Hot air fluidizing systems are also sometimes supplied to assist in ash removal.

Power supplies and controls A transformer rectifier (TR) set along with a controller supply the high voltage power to the discharge electrode system. Several TR sets are normally needed to power a precipitator. With this combination of electrical components, the single-phase 480 V AC line voltage is regulated in the controller and then transformed into a nominal 55,000 to 75,000 V before being rectified to a negative DC output for the discharge system. Electrically, a precipitator most closely resembles a capacitive load. Due to the capacitive load and the nature of the precipitator internals, the TR set must be designed to handle the current surges caused by arcs between the discharge electrodes and the grounded collection surface. A current-limiting reactor in series with the TR set primary also helps to temporarily limit the current surges.

Traditionally, a voltage controller tries to maximize the voltage input to the precipitator. To achieve this input and when operating as designed, the controller must periodically raise the voltage to the point of sparking between the discharge electrode and the collection surface. The controller must then also detect the sparks and reduce the voltage to avoid an arc.

Today's microprocessor electronics with quick response times, interface advantages and programming capabilities provide many functions to optimize particulate collection.

Applications and performance

Utility Because coal is the most common fuel for steam generation, collection of the coal ash particles is the greatest use of a particulate collector. The electrostatic precipitator has been the most commonly used collector. To meet the particulate control regulations for utility units and considering the resulting high collection efficiency, special attention must be given to details of precipitator sizing, rapping, flow distribution and gas bypass around the collector plates. The result will then be a collector which can be confidently and consistently designed and operated to meet the outlet emissions requirements. Operating collection efficiencies which exceed 99.9% are common on the medium and higher ash coals with outlet emissions levels of 0.01 to 0.03 lb/10^6 Btu (12.3 to 36.9 mg/Nm^3 at 6% O_2) heat input common on all coals.

Industrial Other common noncoal-fired industrial units where ESPs are successfully being applied include municipal refuse incinerators and wood-, bark- and oil-fired boilers. For these, the resistivity of the ash in the flue gas is typically lower than coal flyash so an ESP of modest size will easily collect the particulate. The moisture content in the refuse, wood and bark is the major contributor to the low resistivity. The carbon content of the residue, ash and unburned combustibles also contributes to low resistivity.

Pulp and paper In the pulp and paper industry, precipitators are used on power boilers and recovery boilers. The power boiler particulate emissions requirements are the same as those for the industrial units using the same fuels. For the recovery boilers, precipitators are used to collect the residual salt cake in the flue gas. Refer to Chapter 26 for further information on the recovery boiler processes and the reuse of the collected material.

A recovery boiler is a unique application for a precipitator due to the small particulate size and the tendency for the ash to stick together. The resistivity of the particulate is low so it is collected easily in the ESP. Because the particulate is so small, gas bypass around collector plates and re-entrainment of rapped particulates in the flue gas are more of a design concern. Re-entrainment is minimized by lower gas velocities. Precipitator collection efficiencies are 99.7 to 99.8% to meet the 20% opacity and the local emissions requirements. Due to the characteristics of the salt cake particulates, a drag chain conveyor across a precipitator floor, rather than a normal hopper, is used for salt cake removal. In addition, casing corrosion is a more significant concern and as a result more insulation is required to reduce casing heat loss. Finally, in order to improve system reliability, two precipitator chambers are commonly used, each capable of handling 70% of the gas flow and each equipped with separate isolation capabilities.

Precipitators have also been applied in the steel industry to collect and recover the fine dust given off by some processes.

Performance enhancement

A change in fuel, a boiler upgrade, a change in regulation, or performance deterioration may call for a precipitator performance enhancement. Enhancement techniques include additional collection surface, gas conditioning, improved flow distribution, control upgrades and internals replacement. Gas conditioning alters resistivity by adding sulfur trioxide (SO_3), ammonia, moisture, or sodium compounds while the other modifications involve only mechanical hardware changes.

After identifying the causes of current or anticipated performance deterioration, the equipment is surveyed to determine the need for replacement or upgrade. Additional collection surface, in series or in parallel with existing surface, may be needed to meet improved particulate collection needs. Gas conditioning may be used to offset some collection surface deficiency or to enhance the performance of a marginal precipitator. Large dust accumulations near the precipitator entrance, flow patterns on the collection surface and a velocity traverse across the precipitator face indicate possible flow maldistribution. In addition, TR set controllers made before 1975 can potentially benefit from an upgrade to improve performance. Finally, a detailed internal inspection will determine a possible need for replacement of collection surface and discharge electrodes and the need to upgrade the rapping system. A combination of enhancement techniques may be needed.

Wet electrostatic precipitators

The collection of acid mists consisting of fine particulate has been accomplished with wet ESPs in some industrial processes. These units differ from the dry, or conventional, ESPs in materials; however, the collection mechanism is basically the same. Typical operation is at the flue gas moisture dew point temperature and particulate loading is low compared to normal coal-fired applications. To withstand the corrosive atmosphere, the construction materials are critical. Typical materials for the wet ESP components contacting the flue gas include:

1. Discharge system — Alloy 276
2. Collection surface — Alloy 276, lead lined, plastic
3. Enclosure — lead lined, acid brick, plastic or plastic coated

Instead of a rapping system, water spray or a water film removes the collected particulate. Most applications are small industrial units; on larger units a modularized concept is used because the configuration and shapes of the components are nonstandard compared to the dry ESP. Collection efficiencies of 99% have been reported with wet ESPs when precipitator sections or modules are placed in series.

Fabric filters

A fabric filter, or baghouse, collects the dry particulate matter as the cooled flue gas passes through the filter material. The fabric filter is comprised of a multiple compartment enclosure (Fig. 8) with each compartment containing up to several thousand long, vertically supported, small diameter fabric bags. The gas passes through the porous bag material which separates the particulate from the flue gas.

Operating fundamentals

With the typical coal-fired boiler, the particle laden flue gas leaves the boiler and air heater and enters the filter inlet plenum which in turn distributes the gas to each of the compartments for cleaning. An outlet plenum collects the cleaned flue gas from each compartment and directs it toward the induced draft fan and the stack. Inlet and outlet dampers then allow isolation of each compartment for bag cleaning and maintenance. Each compartment has a hopper for inlet gas flow as well as for particulate collection and removal by conventional equipment, as discussed further in Chapter 23. The individual bags are closed at one end and connected to a tubesheet at the other end to permit the gas to pass through the bag assembly. The layer of dust accumulating on the bag is usually referred to as the *dustcake*.

Collection of the particulate on the bag fabric is the heart of the control process. The major forces causing this collection include impingement by either direct contact or impaction and dustcake sieving. Minor forces which assist in the collection are diffusion, electrostatic forces, London-Van der Waal's forces and gravity.[6] The dustcake is formed by the accumulation of particulate on the bags over an operating period. Once formed, the dustcake and not the filter bag material provides most of the filtration. Although impingement collection is most effective on the larger particles and the sieving process collects all particle sizes, a dustcake must form to maximize overall collection.

As the dustcake builds and the flue gas pressure drop across the fabric filter increases, the bags must be cleaned. This occurs after a predetermined operating period or when the pressure drop reaches a set point. Each compartment is then sequentially cleaned to remove the excess dustcake and to reduce the pressure drop. A residual dust coating is preferred to enhance further collection.

Two fabric filter design parameters are air/cloth (A/C) ratio and drag. A/C ratio is the gas volumetric flow rate divided by the exposed bag surface area. Industry standards, along with operating experience, establish the design A/C ratios. These values are typically stated with one compartment out of service for cleaning (net condition). The pressure drop includes the drop across the bags, the dustcake and the attachment of the bag to the tubesheet. The calculation of drag for each compartment is useful in evaluating performance.

Types of fabric filters

Bag cleaning methods distinguish the types of fabric filters, with the three most common types being reverse air, shake deflate and pulse jet. The cleaning method also determines the relative size by the A/C ratio and the filtering side of the bag. Both the reverse air and the shake deflate are inside-the-bag filters with gas flow from inside the bag to outside; the pulse jet is an outside-the-bag filter with the flow from outside to inside (Fig. 9). Note that the tubesheet on the inside-bag filtering is located below the bags and for pulse jet the tubesheet is above the bags.

A reverse air, more correctly termed reverse gas, fil-

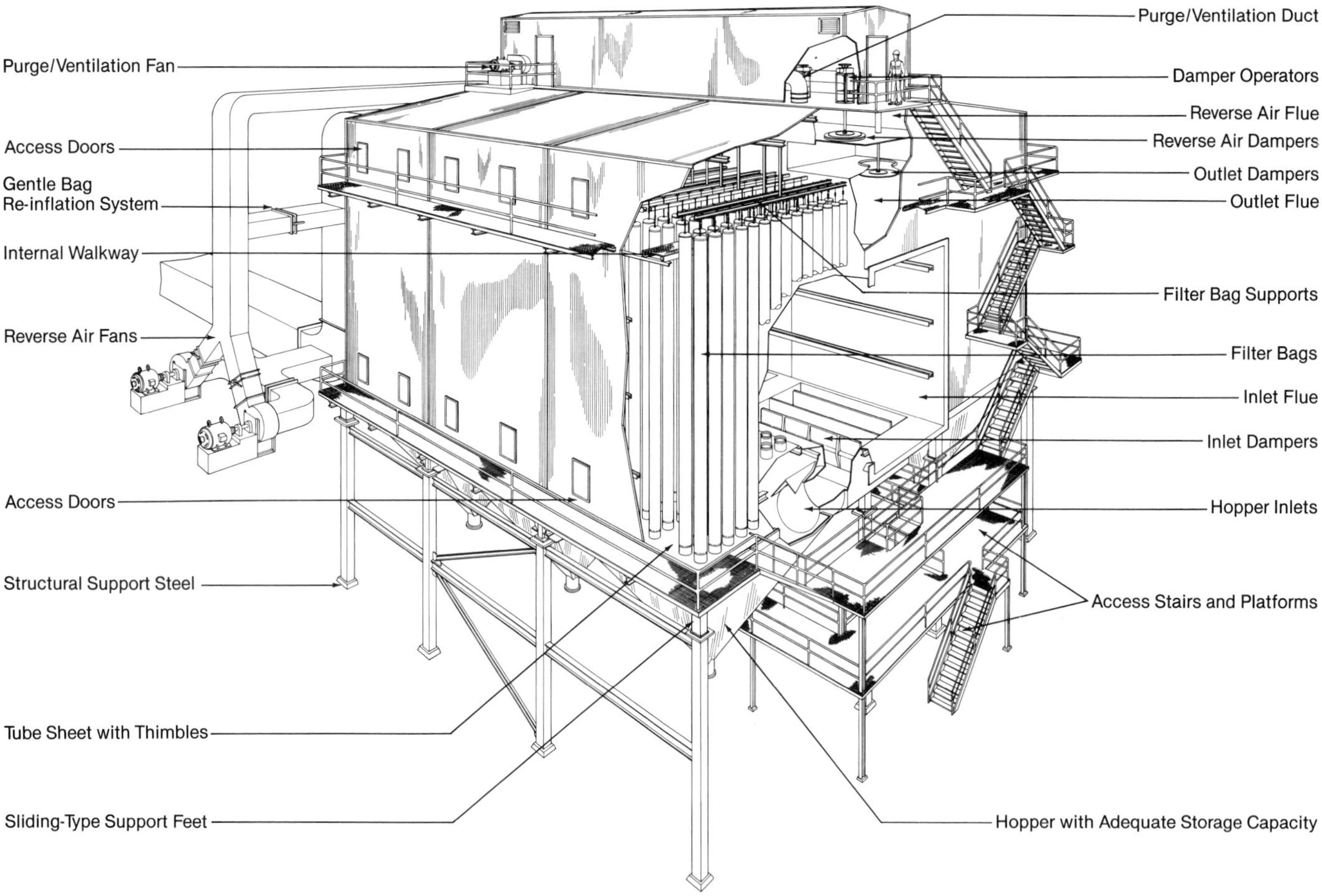

Fig. 8 Fabric filter or baghouse.

ter reverses the flow of clean gas from the outlet plenum back into the bag compartment to collapse the bags in an isolated compartment and dislodge the dustcake. This is a gentle cleaning motion. Once the dislodged particulate falls to the hopper, the bags are gently re-inflated before full gas flow is allowed for filtering. This system requires a reverse gas fan to supply the cleaning gas flow along with additional dampers for flow control. This type of filter system has been used in most large utility power plant fabric filters in the U.S. Experience with this type of fabric filter on some coal flyash applications has demonstrated that reverse gas cleaning alone does not provide an acceptable operating pressure drop. Therefore, some units have added sonic horns to each compartment to assist in the cleaning.

Shake deflate filters are similar to reverse air units in that the cleaning occurs in an isolated compartment and a small amount of cleaned flue gas is used to slightly deflate the bags. In addition, a mechanical motion is used to shake the bags and dislodge the accumulated dustcake.

Pulse jet technology is a more rigorous cleaning method and can be used when the compartment is either isolated or in service. A pulse of compressed air is directed into the bag from the open top which causes a shock wave to travel down its length dislodging the dustcake from the outside surface of the bag. A unique aspect of the pulse jet system is the use of a wire cage in each bag to keep it from collapsing during normal filtration. The bag hangs from the tubesheet. A series of parallel pulse jet pipes are located above the bags with each pipe row having a solenoid valve. This permits the bags to be pulsed clean one row at a time. The initial experience with pulse jet filters has been on industrial units, with limited use on utility units to date; however, use of pulse jet units in utility boilers is increasing.

Bag materials and supports Substantial research and development on bags and their materials has taken place to lengthen their life and to select bags for various applications. The flexing action during cleaning is the major factor affecting bag life. Bag blinding, which occurs when small particulate becomes trapped in the fabric interstices, limits bag life by causing excessive pressure drop in the flue gas. Finishes on the bag surface are also used to make some bags more acid resistant and to improve cleaning.

The most common bag material in coal-fired utility units with reverse fabric filters is woven fiberglass. Typical bag size is 12 in. (305 mm) diameter with a length of 30 to 36 ft (9.1 to 11.0 m). Bag life of three to five years is common. The shake deflate filters also use mostly fiberglass bags. On both of these units the fiberglass bag is fastened at the bottom to a thimble in the tubesheet. At the top, a metal cap is sown into the bag and the bag has a spring loaded support for the reverse air filters. The upper operating temperature limit is 500F (260C) for most fiberglass bags.

In addition to fiberglass, the industrial size filters use synthetic materials with trade names like Nomex®, Ryton®, Gortex® and Hyglass®. Advantages of the synthetic materials include better abrasion resistance and

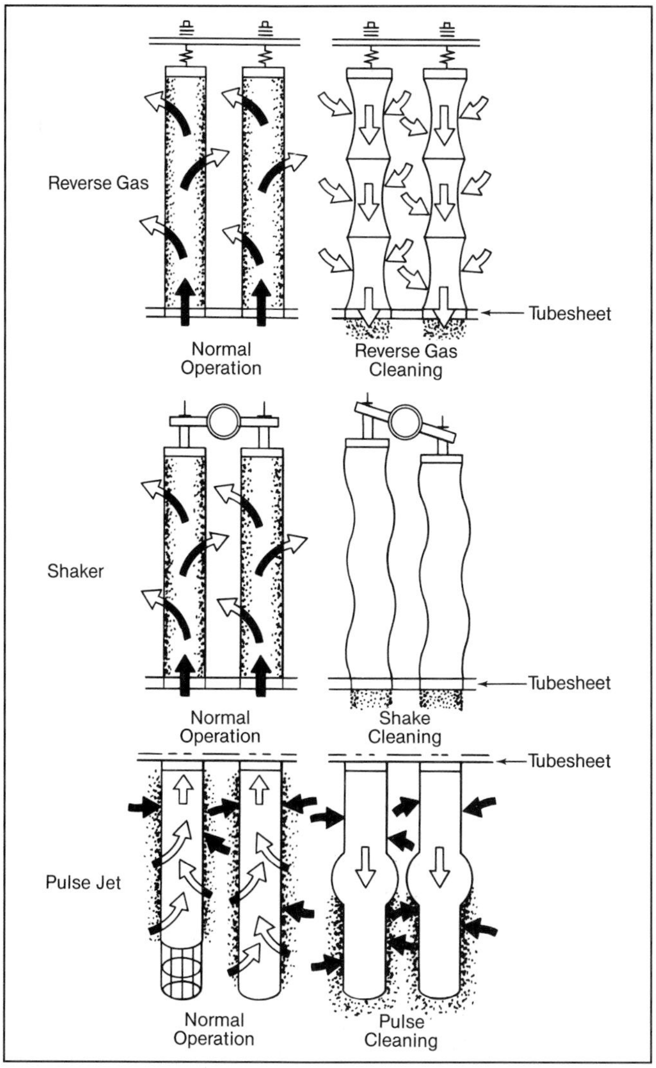

Fig. 9 Fabric filter types.

resistance to acid attack. Disadvantages include higher cost and limited temperature capabilities. For the pulse jet filters, the typical bag size is 5 or 6 in. (130 or 150 mm) diameter with a length of 10 to 20 ft (3 to 6 m).

Enclosure The fabric filter is a metal encased structure with individual bag compartments. The inlet and outlet plenums are typically located between two rows of compartments to provide short inlet and outlet flue connections. This enclosure rests on a support steel structure. For reverse air units, interior access is required at both the lower tubesheet and bag support elevations. In a pulse jet filter, access is required above the tubesheet for bag cage removal. This is provided by large roof access doors or by a top plenum and a side manway. Typical enclosure materials are carbon steel ASTM A-36 or equivalent under normal coal-fired boiler conditions. The entire enclosure is covered with insulation and lagging to keep metal temperatures high and to minimize corrosion potential.

Hoppers Each filter compartment has a hopper to collect the dislodged particulate and to channel its flow to the ash removal system. Most filters also use the hopper as part of the flue gas inlet to each compartment. Therefore, the hopper is designed with steep sides for ash removal along with considerations for proper gas flow distribution. Hopper heaters, level detectors, poke holes and an access door are common hopper features.

Performance and applications

The first utility fabric filter in the U.S. was installed with Babcock & Wilcox's participation on an oil-fired Southern California boiler in the mid 1960s.[7] After a few years of limited activity, several utilities had installed fabric filters by the late 1970s. Interest in these systems continues to grow due to the high particulate removal efficiency. They are the preferred collector for some applications. Well designed filters routinely achieve greater than 99.9% particulate removal, meeting all U.S. EPA and local regulations.

Besides the utility coal-fired applications, fabric filters are used on circulating fluidized-bed boilers, industrial pulverized and stoker coal units, refuse-fired units in combination with a dry flue gas scrubber, and in the steel industry. A unique advantage with fabric filters in this application is that all of the flue gas passes through the dustcake as it is cleaned; when the dustcake has high alkalinity, it can be used to remove other flue gas constituents and acid gases, such as sulfur dioxide (SO_2). (See Chapter 35.)

Mechanical collectors

Mechanical dust collectors, often called cyclones or multiclones, have been used extensively to separate large particles from a flue gas stream. The cyclonic flow of gas within the collector and the centrifugal force on the particulate drive the particulate out of the flue gas (Fig. 10). Hoppers below the cyclones collect the particulate and feed an ash removal system. The mechanical collector is most effective on particles larger than 10 microns. For smaller particles, the collection efficiency drops considerably below 90%.

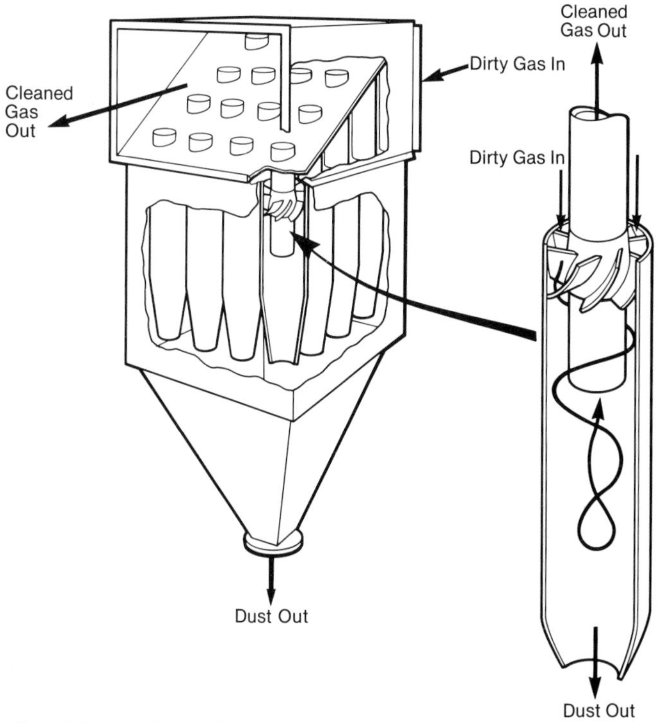

Fig. 10 Mechanical collector.

Mechanical collectors were adequate when the emissions regulations were less stringent and when popular firing techniques produced larger particles. These collectors were frequently used for re-injection to improve unit efficiency on stoker firing of coal and biomass. With stricter emissions regulations, mechanical collectors can no longer be used as the primary control device. However, with the onset of fluidized-bed boilers, there has been a resurgence of mechanical collectors for recirculating the bed material. A high efficiency collector is then used in series with the mechanical one to meet particulate emissions requirements. (See Chapters 16 and 29 for more information on fluidized-bed combustion.)

Wet scrubbers

A wet scrubber can be used to collect particulate from a flue gas stream with the intimate contact between a gas stream and the scrubber liquid. The venturi-type wet scrubber (Fig. 11) is used to transfer the suspended particulate from the gas to the liquid. Collection efficiency, dust particle size and gas pressure drop are closely related in the operation of a wet scrubber. The required operating pressure drop varies inversely with the dust particle size for a given collection efficiency; or, for a given dust particle size, collection efficiency increases as operating pressure drop increases.

Due to the excessive pressure drop and the stringent particulate regulations, wet scrubbers are used infrequently as the primary collection device. However, on most coal-fired applications, wet scrubbers are required in series with a high efficiency collector for control of acid gas emissions, so the extra particulate removal is an added benefit.

Other collection devices

Other more specialized particulate control devices include ELECTROSCRUBBER® filters and ceramic tube filters. The ELECTROSCRUBBER® filter combines the technologies of granular filtration and electrostatic collection and uses electrostatic forces on the particulate as the flue gas passes through a recirculating bed of gravel. Charged particles attach to the gravel which is cleaned as the lower bed gravel is recirculated to the top of the bed. The collected dust is sent to disposal and the cleaned gas stream flows to the stack. This collection device is popular on wood-fired units because it aids fire prevention by its handling of glowing embers.

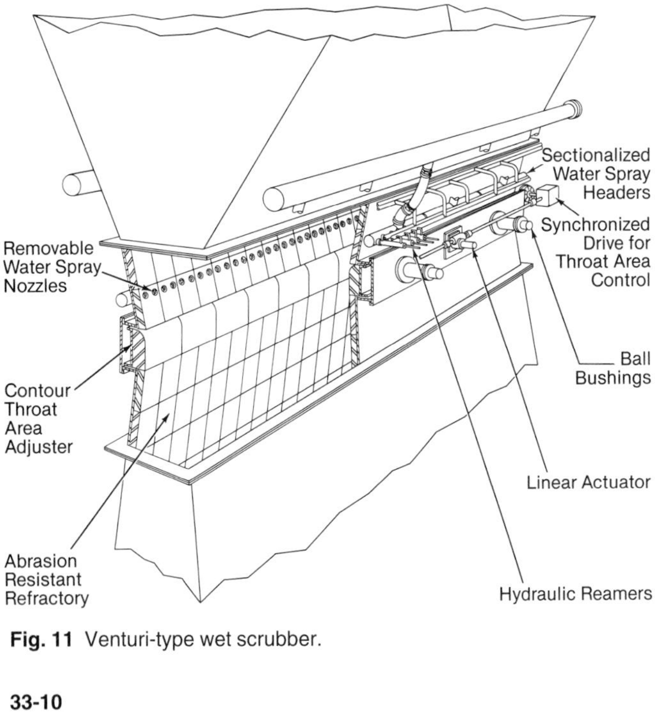

Fig. 11 Venturi-type wet scrubber.

Ceramic tube filters are being developed for high temperature and pressure applications such as coal gasifiers and pressurized fluidized-beds. (See Chapters 17 and 29.)

Equipment selection

Major evaluation factors to consider when selecting particulate control equipment include emissions requirements, boiler operating conditions with resulting particulate quantity and sizing, allowable pressure drop/power consumption, combined pollution control requirements, capital cost, operating cost and maintenance cost. For new units which must meet the stringent federal, state and local regulations, the selection is reduced to a comparison of electrostatic precipitators and fabric filters because these are the only high efficiency, high reliability choices. For retrofits on operating units, the performance of existing control equipment as well as unique flue gas conditions may require specialized equipment.

The advantages of a well designed ESP are high total collection efficiency, high reliability, low flue gas pressure loss, resistance to moisture and temperature upsets, and low maintenance. Advantages of a fabric filter include high collection efficiency throughout the particle size range, high reliability, resistance to flow upsets, little impact of ash chemical constituents on performance, and good dustcake characteristics for combination with dry acid gas removal equipment. A comparison of overall and particle size collection efficiencies for precipitators and fabric filters is shown in Fig. 5. An application where small particulate dominates would favor a fabric filter for maximum control as long as bag blinding does not occur.

For those applications where an ESP or fabric filter is technically acceptable and high collection efficiencies are required, some general guidelines on capital costs are: 1) on small units, a pulse jet fabric filter is more economical, 2) on large units with medium or high sulfur coal, an ESP is economical, and 3) on low sulfur coal-fired large units, a reverse air fabric filter may be more economical. However, when operating and maintenance costs are also considered, the lowest capital cost may not provide the lowest overall cost. Therefore, it is important to perform a detailed engineering study to quantify all of the variables for a specific site to obtain a true assessment of the real cost.

International particulate control

Particulate control regulations and control equipment usage have been emphasized in Western Europe, Scandinavia, South Africa, Australia, Japan and Canada, as well as the U.S. The traditional high efficiency collection device in all countries has been the ESP for many applications; however, the fabric filter has gained acceptance in Australia and the U.S. for coal-fired units. On new refuse-fired units, the combination of spray dryer and fabric filter is popular in Scandinavia, Western Europe and the U.S.

Control challenges for the future

Particulate control will be required for steam generation combustion equipment as long as ash bearing fuels are used. New combustion methods may change the ash characteristics, permitting less costly or more efficient control equipment. Furthermore, new material developments, increased use of electronics and overall equipment size reduction are expected to continue.

References

1. White, H.J., *Industrial Electrostatic Precipitation*, Addison-Wesley, Reading, Massachusetts, 1963.

2. United States Code of Federal Regulations, Title 40, Part 60, Subparts D, Da, Db and Dc, July 1988.

3. Federal Register, Environmental Protection Agency, 40 CFR Part 60, Vol. 55, No. 177, September 12, 1990.

4. Lane, W.R., and Khosla, A., "Comparison of baghouse and electrostatic precipitator fine particulate, trace element and total emissions," ASME-IEEE Joint Power Conference, Indianapolis, Indiana, September 27, 1983.

5. *Gas Flow Model Studies*, Publication EP-7, Industrial Gas Cleaning Institute, January 1987.

6. Bustard, C.J., *et al.*, *Fabric Filters for the Electric Utility Industry*, Vol. 1, Report CS-5161, Electric Power Research Institute, Palo Alto, California, 1988.

7. Bagwell, F.A., Cox, L.F., and Pirsh, E.A., *Design and Operating Experience With a Filterhouse Installed on an Oil-Fired Boiler*, Air Pollution Control Association, June 1968.

Bibliography

Bickelhaupt, R.E., "A technique for predicting flyash resistivity," Report EPA-600/7-79-204, U.S. Environmental Protection Agency, Research Triangle Park, North Carolina, July 1979.

Billings, C.E., *et al.*, *Handbook of Fabric Filter Technology*, GCA Corporation, Bedford, Massachusetts, December 1970.

Faulkner, M.G., and DuBard, J.L., "A mathematical model of electrostatic precipitation (Rev. 3)," Report EPA-600/7-84-069, U.S. Environmental Protection Agency, Research Triangle Park, North Carolina, July 1984.

Katz, J., *The Art of Electrostatic Precipitator*, Precipitator Technology, Inc., Munhall, Pennsylvania, 1981.

Selective catalytic reduction (SCR) NO_x emission control system retrofit on a coal-fired boiler.

Chapter 34
Nitrogen Oxides Control

Nitrogen oxides (NO_x) are one of the primary pollutants emitted during combustion processes, with transportation systems accounting for the largest contribution in the United States. (See Fig. 1.) Along with sulfur oxides (SO_x) and particulate matter, NO_x emissions have been identified as contributors to acid rain and ozone formation, visibility degradation and human health concerns. As a result, NO_x emissions from most combustion sources are regulated and require some level of control as discussed in Chapter 32. A number of approaches can be used to control NO_x emissions.

NO_x formation mechanisms

NO_x refers to the cumulative emissions of nitric oxide (NO), nitrogen dioxide (NO_2) and trace quantities of other species generated during combustion. Combustion of any fossil fuel generates some level of NO_x due to high temperatures and the availability of oxygen and nitrogen from both the air and fuel.

NO_x emissions from fired processes are typically 90 to 95% NO with the balance being NO_2. However, once the flue gas leaves the stack, the bulk of the NO is eventually oxidized in the atmosphere to NO_2. It is the NO_2 in the flue gas which creates the brownish plume often seen in a power plant stack discharge. Once in the atmosphere, the NO_2 is involved in a series of reactions which form secondary pollutants. The NO_2 can react with sunlight and hydrocarbon radicals to produce photochemical (urban) smog and acid rain constituents.

There are two common mechanisms of NO_x formation, thermal NO_x and fuel NO_x.

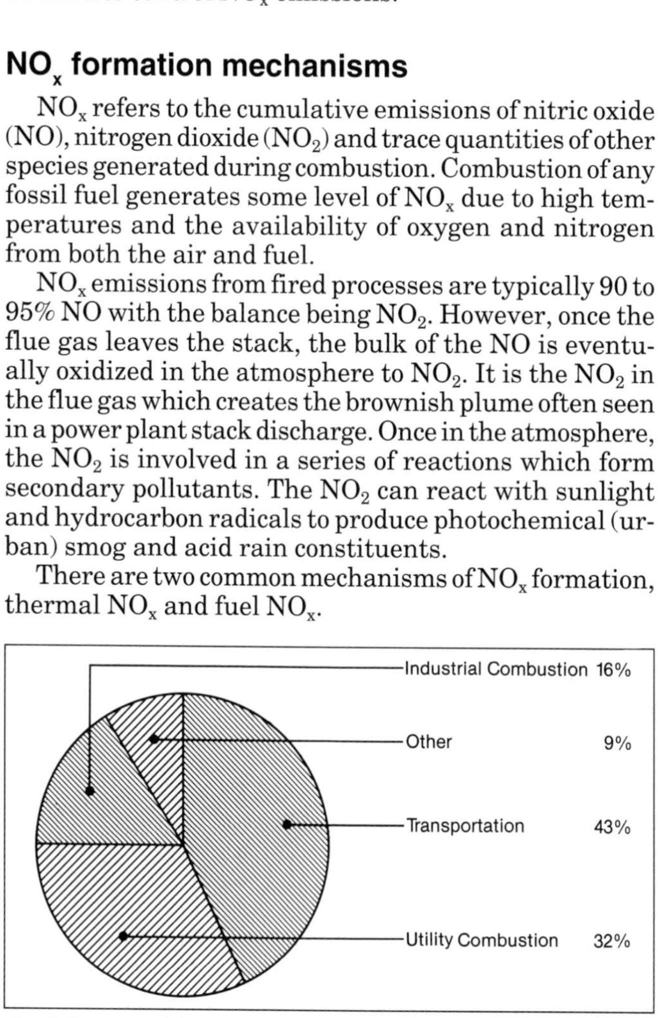

Fig. 1 Sources of United States NO_x emissions — 1985.

Thermal NO_x

Thermal NO_x refers to the NO_x formed through high temperature oxidation of the nitrogen found in the combustion air. The formation rate is a strong function of temperature as well as the residence time at temperature. Significant levels of NO_x are usually formed above 2200F (1204C) under oxidizing conditions with exponential increases as the temperature is increased. At these high temperatures, molecular nitrogen (N_2) and oxygen (O_2) in the combustion air dissociate into their atomic states and participate in a series of reactions. One product of these reactions is NO. The three principal reactions proposed for this process are:[1,2]

$$N_2 + O \rightarrow NO + N \quad \textbf{(1)}$$

$$N + O_2 \rightarrow NO + O \quad \textbf{(2)}$$

$$N + OH \rightarrow NO + H \quad \textbf{(3)}$$

The traditional factors leading to complete combustion (high temperatures, long residence time and high turbulence or mixing) all tend to increase the rate of thermal NO_x formation. Therefore, some compromise between effective combustion and controlled NO_x formation is needed.

Thermal NO_x formation is typically controlled by reducing the peak and average flame temperatures. This can be accomplished through a number of combustion system changes. Controlled mixing burners can be used to reduce the turbulence in the near burner region of the flame and to slow the combustion process. This typically reduces the flame temperature by removing additional energy from the flame before the highest temperature is reached. A second approach is staged combustion where only part of the combustion air is initially added to burn the fuel. The fuel is only partially oxidized and then cooled before the remaining air is added separately to complete the combustion process. A third alternative is to mix some of the flue gas with the combustion air at the burner, referred to as flue gas recirculation. This increases the gas weight which must be heated by the chemical energy in the fuel, thereby reducing the flame temperature.

These technologies have been used effectively with gas, oil and coal firing to reduce NO_x formation. Specific use of each technology or a combination of technologies depends upon the costs, fuel and regulatory requirements. The specific applications of these different combustion system modifications are presented in more detail in Chapters 10 and 13.

For fuels which do not contain significant amounts of chemically bound nitrogen, such as natural gas, thermal NO_x is the primary overall contributor to NO_x emissions. In these cases, the technologies described above are particularly effective in NO_x emission control.

Fuel NO_x

The major source of NO_x emissions from nitrogen bearing fuels such as coal and oil is the conversion of fuel bound nitrogen to NO_x during combustion. Laboratory studies indicate that fuel NO_x contributes approximately 50% of the total uncontrolled emissions when firing residual oil, and more than 80% when firing coal. Nitrogen found in fuels such as coal and residual oils is typically bound to the fuel as part of organic compounds. During combustion, the nitrogen is released as a free radical to ultimately form NO or N_2. Although it is a major factor in NO_x emissions, only 20 to 30% of the fuel bound nitrogen is converted to NO.

The majority of NO_x formation from fuel bound nitrogen occurs through a series of reactions which are not fully understood. It does appear, however, that this conversion occurs by two separate paths. These paths, primarily for coal, are represented in Fig 2.

The first path involves the oxidation of volatile nitrogen species during the initial phase of combustion. During the release, and prior to the oxidation of the volatile compounds, nitrogen reacts to form several intermediate compounds in the fuel rich flame regions. These intermediate compounds are then oxidized to NO or reduced to N_2 in the postcombustion zone. The formation of either NO or N_2 is strongly dependent on the local fuel/air stoichiometric ratio. It is estimated that this volatile release mechanism accounts for 60 to 90% of the fuel NO_x contribution.

The second reaction path involves the release of nitrogen radicals during combustion of the char fraction of the fuel. These reactions occur much more slowly than the reactions involving the volatile species.

Conversion of fuel bound nitrogen to NO_x is strongly dependent on the fuel/air stoichiometry but is relatively independent of variations in combustion zone temperature. Therefore, this conversion can be controlled by reducing oxygen availability during the initial stages of combustion. With this reduction, a major portion of the nitrogen released during devolatilization reduces to N_2. This is because hydrocarbon radicals, also released during devolatilization, compete with the nitrogen for the available free oxygen. Techniques such as controlled fuel-air mixing and staged combustion provide a significant reduction in NO_x emissions by controlling stoichiometry in the initial devolatilization zone.

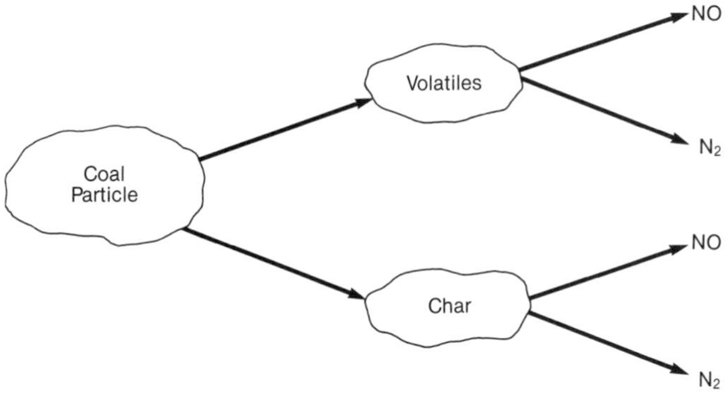

Fig. 2 NO_x formation from nitrogen contained in coal during combustion.

A portion of the NO_x that is formed by oxidation of the fuel bound nitrogen under fuel rich conditions is referred to as *prompt NO_x*, identified through the research of C.P. Fenimore.[3] The name is derived from its formation very early during the combustion process. It was originally identified as the difference between observed and predicted thermal NO emissions.

Prompt NO_x occurs through the formation of intermediate hydrogen cyanide (HCN) species and the reaction between molecular nitrogen and hydrocarbon compounds. This reaction is then followed by the oxidation of HCN to NO.

Although prompt NO_x formation normally has a weak temperature dependence, this dependence can become strong under fuel rich conditions. Modern burners are designed to reduce peak flame temperatures by controlling the rate of fuel and air mixing. Combustion is initiated under fuel rich conditions and this fuel rich zone is where prompt NO_x is formed. Prompt NO_x can contribute from near zero to more than 100 ppmv of NO.

Health and environmental effects of NO_x

Once released into the atmosphere, NO_x becomes a contributor to ozone/photochemical smog, acid rain, particulate matter and potential airborne carcinogens. Each of these pollutants has a significant effect on human health and the environment. However, as discussed in Chapter 32, automobiles and other vehicles are the major source of NO_x with utility power plants contributing less than one third of the total United States (U.S.) emissions.

Ozone/photochemical smog

Naturally occurring ozone, found in the upper atmosphere, is necessary for man's existence. However, manmade ozone, formed in the lower atmosphere, is considered a pollutant. This man-made ozone, formed by a reaction of hydrocarbons and NO_x in the presence of sunlight, is a major component of urban photochemical smog.[4]

Another major component of photochemical smog is NO_2. Once emitted, the NO contained in the flue gas is oxidized to NO_2, which is a yellow-brown gas and is also an oxidant. Urban smog is also made up of a variety of other components.

Acid rain

NO_x and sulfur dioxide (SO_2) contribute to the formation of acid rain,[4] which includes a dilute solution of sulfuric and nitric acids with small amounts of carbonic and other organic acids. The NO_x and SO_2 react with water vapor to form acidic compounds; these acids account for more than 90% of acid rain. Nitric acid is formed through the reaction:

$$NO_2 + OH \rightarrow HNO_3 \quad (4)$$

Most acid rain control has concentrated on SO_2 contributions instead of those due to NO_x because it is estimated that NO_x contributes less than one third of the acid rain generated.

Particulate matter

NO_x can also contribute to ambient particulate matter. In the atmosphere NO_x reacts with other airborne chemicals to create nitrates. NO_x also promotes the transformation of SO_2 into sulfate particulate compounds.

Controlling NO_x emissions in steam generating equipment

An understanding of the mechanisms and chemical reactions that produce NO_x emissions has enabled engineers and scientists to develop techniques for reducing emissions of this pollutant. Because NO_x is formed during combustion, research has centered on controlling NO_x at the source. The application of techniques that control fuel-air mixing rates and peak combustion temperatures has been effective in reducing these emissions.

NO_x emissions may be controlled through careful fuel selection and selective fuel switching. Once the fuel is chosen, NO_x emissions are minimized by using low NO_x combustion technology and postcombustion techniques. Typically, the fuel is selected by overall economic considerations. Combustion technology based NO_x control is then typically the lowest overall cost approach provided that it will meet local and federal emissions requirements. Where further control is needed, postcombustion NO_x control is added.

Precombustion fuel effects

Unlike sulfur and particulate constituents in coal, nitrogen species contained in the fuel can not be easily reduced or removed. Therefore, the most common precombustion option for reducing NO_x levels is to switch to a fuel with inherently lower nitrogen content. For boilers capable of firing multiple fuels with minimal impact on the steam cycle, this may be the most cost effective solution.

Coal combustion generally produces the highest NO_x emissions. Oil combustion generates less NO_x while gas firing produces even less. When firing oil fuels, a reduction in fuel nitrogen results in reduced NO_x emissions. However, there does not appear to be a similar correlation between coal nitrogen content and NO_x generation. Other factors in coal chemistry including volatile species, oxygen and moisture content appear to dominate the formation of NO_x during coal combustion. Therefore, reducing coal nitrogen content may not provide a corresponding NO_x reduction.

Combustion techniques

NO_x formation is promoted by rapid fuel-air mixing. This produces high peak flame temperatures and excess available oxygen which, in turn, promotes NO_x emissions. Combustion system developments responsible for reducing NO_x formation include low NO_x burners, such as the coal-fired burner shown in Fig. 3, staged burning techniques, and flue gas recirculation (FGR). The specific NO_x reduction mechanisms include controlling the rate of fuel-air mixing, reducing oxygen availability in the initial combustion zone, and reducing peak flame temperatures. Modern systems can emit less than one third of the NO_x produced by older units.

Additional information regarding the details of the combustion process, combustion techniques and hardware used to control NO_x can be found in Chapters 10, 13, 15, 16, 17 and 29. Information on uncontrolled and controlled NO_x emissions is also presented.

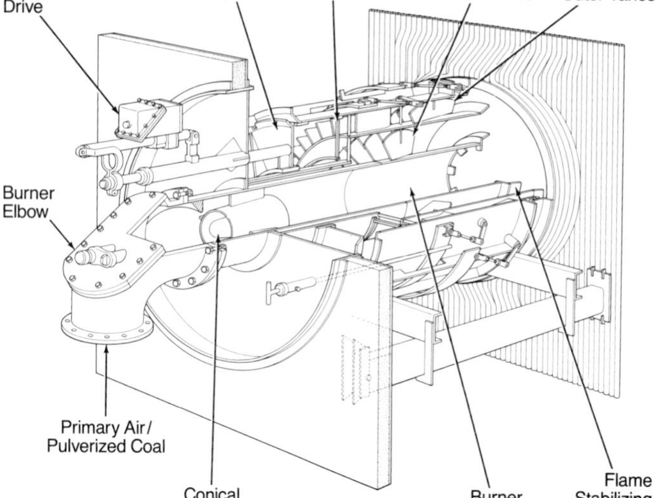

Fig. 3 DRB-XCL™ low NO_x burner for coal firing.

Postcombustion techniques

Many localities require lower NO_x emissions than economically possible by combustion modifications alone. To achieve further reductions, additional techniques are applied downstream of the combustion zone. These postcombustion control systems are referred to as selective noncatalytic reduction (SNCR) and selective catalytic reduction (SCR). In either technology, NO_x is reduced to nitrogen (N_2) and water (H_2O) through a series of reactions with a chemical agent injected into the flue gas.

The most common chemical agents used in commercial applications are ammonia and urea for SNCR and ammonia for SCR systems. Most ammonia-based systems have used anhydrous ammonia (NH_3) as the reducing agent. However, due to the hazards of storing and handling NH_3, many systems use aqueous ammonia at a 25 to 28% concentration. Urea [$(NH_2)_2CO$] can be stored as a solid or mixed with water and stored in solution. An additional byproduct of the urea injection is carbon dioxide (CO_2).

Selective noncatalytic reduction technology

There are currently two basic selective noncatalytic reduction (SNCR) processes available. An ammonia-based system has been developed by Exxon and is called Thermal DeNO$_x$®. A urea-based technology has been developed under the sponsorship of the Electric Power Research Institute (EPRI). Although there are distinct differences to each technology, the overall processes are similar.

Both SNCR technologies inject a reducing agent into NO_x laden flue gas within a specific temperature zone or window. In addition, it is important to properly mix the chemical agent with the flue gas. Finally, the mixture must have adequate residence time for the reduction reactions to occur. The chemical reactions for both processes can be represented by:

Ammonia $$4NO + 4NH_3 + O_2 \rightarrow 4N_2 + 6H_2O \quad (5)$$

Urea $2NO + (NH_2)_2CO + 1/2\ O_2 \rightarrow 2N_2 + 2H_2O + CO_2$ **(6)**

The acceptable temperature range for either reaction is 1400 to 2000F (760 to 1093C), although temperatures above 1700F (927C) are preferred. Below 1600F (871C), chemical enhancers, such as hydrogen, are needed to assist the reactions. As the temperature increases within this range, the ammonia or urea may react with available oxygen to form NO_x. This reaction becomes significant at temperatures in excess of 2000F (1093C) and may become dominant as temperatures approach 2200F (1204C).

The application of SNCR technology to a municipal solid waste boiler is shown in Fig. 4. Multiple injection levels are used to maintain NO_x reduction efficiencies at acceptable levels as boiler load changes. Based on operating experience, four or more injection levels may be required for larger units. Multiple levels are required because the flue gas temperature profile changes with boiler load and the injection point for the ammonia or urea must be adjusted accordingly.

The SNCR system consists of storage and handling equipment for the ammonia or urea, equipment for mixing the chemical with the carrier (compressed air, steam or water), and the injection equipment. The key component, the injection system, consists of nozzles generally located at various elevations on the furnace walls to match the expected flue gas operating temperature. The number and location of the nozzles are established by the supplier and are based on obtaining good reagent distribution within the flue gas. Fig. 5 provides a simplified flow diagram for an ammonia-based SNCR system.

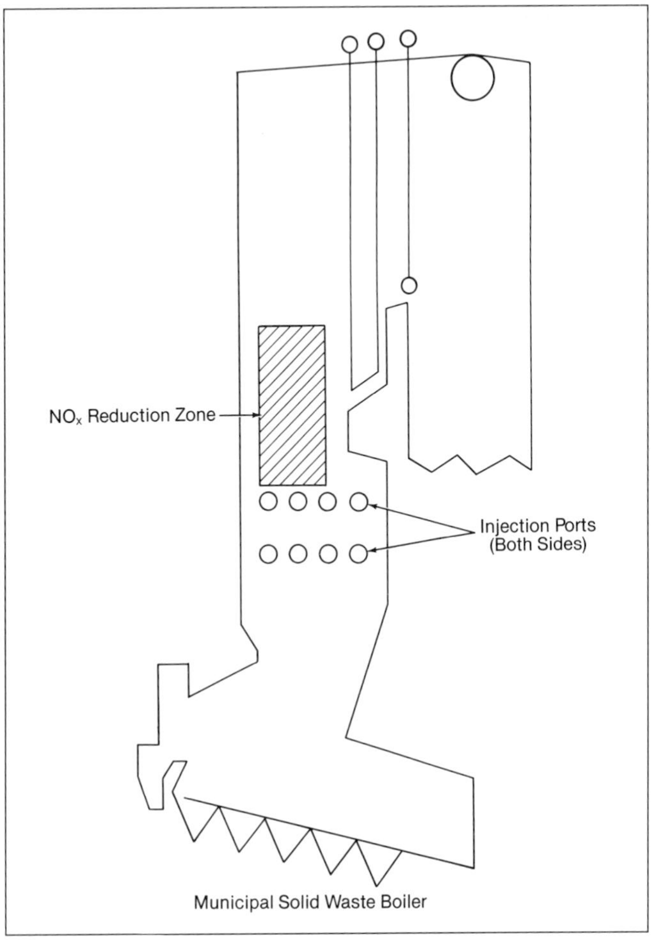

Fig. 4 Typical SNCR application.

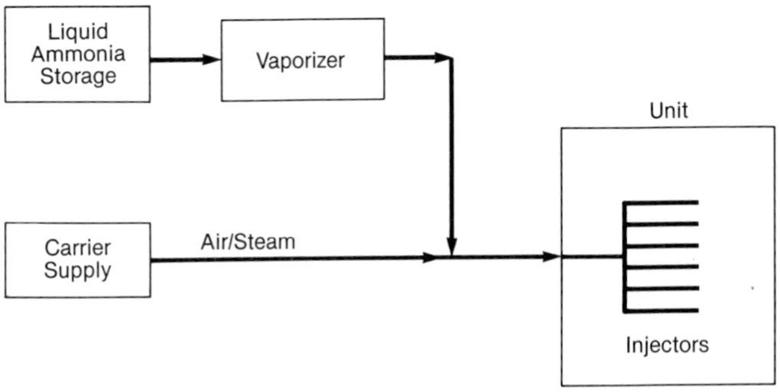

Fig. 5 SNCR deNO_x system.

One major difference between the ammonia and urea based processes is that the ammonia is usually injected into the gas stream in a gaseous state, whereas the urea is injected as an aqueous solution. The urea technology requires a longer residence time for reactions due to the time required to vaporize the liquid droplets once they are in the gas stream. However, as aqueous ammonia systems become more frequently specified, ammonia may also be injected as a liquid with additional residence time required for vaporization.

When injecting ammonia or urea it is important to control the excess unreacted ammonia. As flue gas temperatures are reduced, the excess ammonia can react with other combustion species, primarily sulfur trioxide (SO_3), to form ammonium salts. The major ammonium products formed are ammonium sulfate [$(NH_4)_2SO_4$] and ammonium bisulfate (NH_4HSO_4). Ammonium sulfate is a dry, fine particulate (1 to 3 microns in diameter) that may contribute to plume formation. The ammonium bisulfate is a highly acidic and sticky compound which, when deposited on downstream equipment such as air heaters, contributes to significant fouling and corrosion.

While NO_x reduction levels of 70% are possible under carefully controlled conditions, 30 to 50% reductions in NO_x emissions are more typically used in practice to maintain acceptable levels of reagent consumption and unreacted ammonia carryover.

Most of the current applications for these technologies have been on smaller municipal solid waste or biomass fueled boilers, where the appropriate temperature range is located in the upper furnace, as shown in Fig 4. (See also Chapter 27.) In large utility units, the proper temperature range occurs in the convection pass cavities, making application, especially on retrofits, more challenging. For these applications, overall control may be limited to the 20 to 40% range. New boiler applications may require special design of the boiler convection pass with a dedicated open cavity in an optimal temperature window for the reactions to occur (Fig. 6).

Selective catalytic reduction technology

Fundamental design considerations

Selective catalytic reduction (SCR) systems catalytically reduce flue gas NO_x to N_2 and H_2O using ammonia in a chemical reduction. This technology is the most effective method of reducing NO_x emissions especially where high removal efficiencies (70 to 90%) are required.

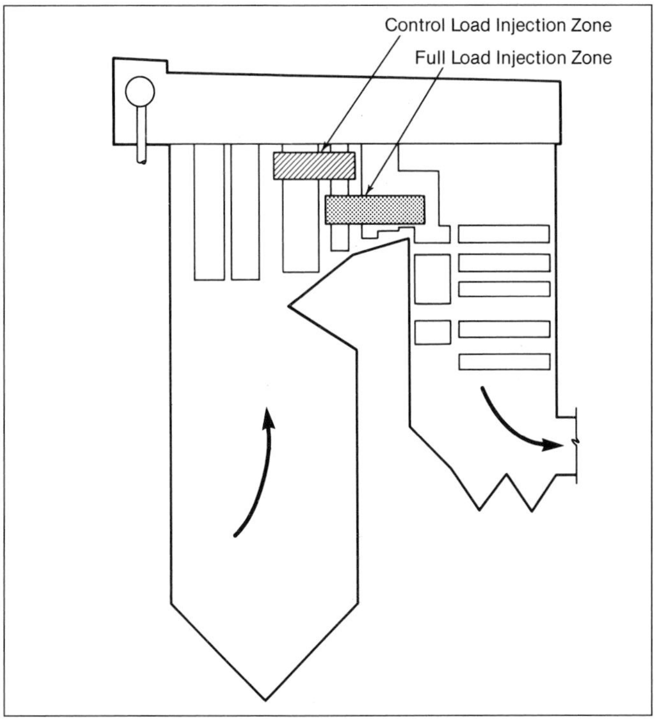

Fig. 6 Coal-fired utility boiler with SNCR temperature window location at full and part load.

Initial development of SCR technology began in Japan around 1963. Early laboratory tests included the evaluation of many catalyst formulations and their life expectancies, as well as process design optimizations. Development tests were followed by several catalytic pilot plant installations during the early 1970s; the first commercial installations occurred in 1978. As mandated by environmental regulations, SCR systems have been installed on utility boilers in Japan, Germany and other European countries. SCR technology has also been applied to U.S. installations primarily for gas turbines, combined cycle systems, and small industrial boilers and process heaters. Legislation passed in areas of California requires the use of this technology on utility installations. Environmental permitting considerations will likely require this technology on many new large installations during the 1990s. Typical examples of new and retrofit applications are shown in Figs. 7 through 9. There are more than 400 SCR systems throughout the world, covering a variety of fuels including natural gas, oil, coal and wood.

The NO_x reduction reactions take place as the flue gas passes through the catalyst chamber. Before entering the catalyst, ammonia is injected into and mixed with the flue gas as shown in Fig. 10. Once the mixture enters the catalyst, the NO_x reactions with NH_3 can be represented as follows:

$$4NO + 4NH_3 + O_2 \rightarrow 4N_2 + 6H_2O \quad \textbf{(7)}$$

$$2NO_2 + 4NH_3 + O_2 \rightarrow 3N_2 + 6H_2O \quad \textbf{(8)}$$

As with SNCR systems, the SCR reactions take place within an optimal temperature range. A variety of catalysts are available. Most can operate within a range of 450 to 840F (232 to 449C), but optimum performance occurs between 675 and 840F (357 and 449C). The minimum temperature varies and is based on fuel, flue gas specifications and catalyst formulation. In addition, this minimum temperature tends to increase with flue gas sulfur dioxide content. This results in a smaller operating range as sulfur content increases in order to eliminate the formation of ammonium sulfate salts in the catalyst bed. Above the recommended temperature range, a number of catalyst materials tend to become less effective.

Catalyst formulation is the key to SCR system performance. Although catalyst formulations are proprietary, material typically falls into one of three categories: base metal, zeolite and precious metal.

Most of the operating experience to date has been with base metal catalysts. These catalysts use titanium oxide with small amounts of vanadium, molybdenum, tungsten or a combination of several other active chemical agents. The base metal catalysts are selective and operate in the specified temperature range. The major drawback of the base metal catalyst is its potential to oxidize SO_2 to SO_3; the degree of oxidation varies based on catalyst chemical formulation. The quantities of SO_3 which are formed can react with the ammonia carryover to form the ammonium sulfate salts as previously discussed. Proper system design and catalyst formulation can minimize this problem.

Zeolite catalysts are relatively new to NO_x emission control. These catalysts are aluminosilicate materials which function similarly to base metal catalysts. One potential advantage of zeolite is its higher operating temperature of 970F (521C). This catalyst may also oxidize SO_2 to SO_3 and must be carefully matched to the flue gas conditions.

Precious metal catalysts are generally manufactured from platinum and rhodium. As with the zeolite units, there is little operating experience with these catalysts. They also require careful consideration of flue gas constituents and operating temperatures. While effective in reducing NO_x, these catalysts can also act as oxidizing catalysts, con-

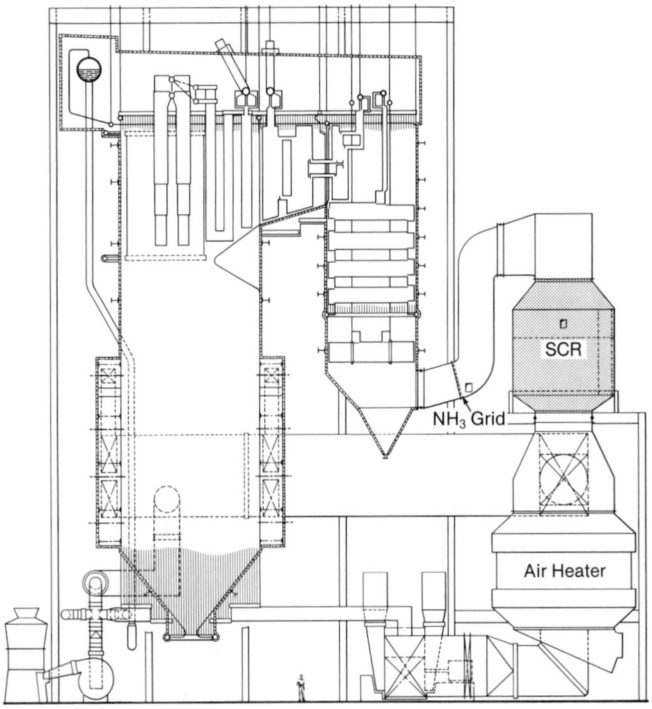

Fig. 7 New utility boiler with SCR.

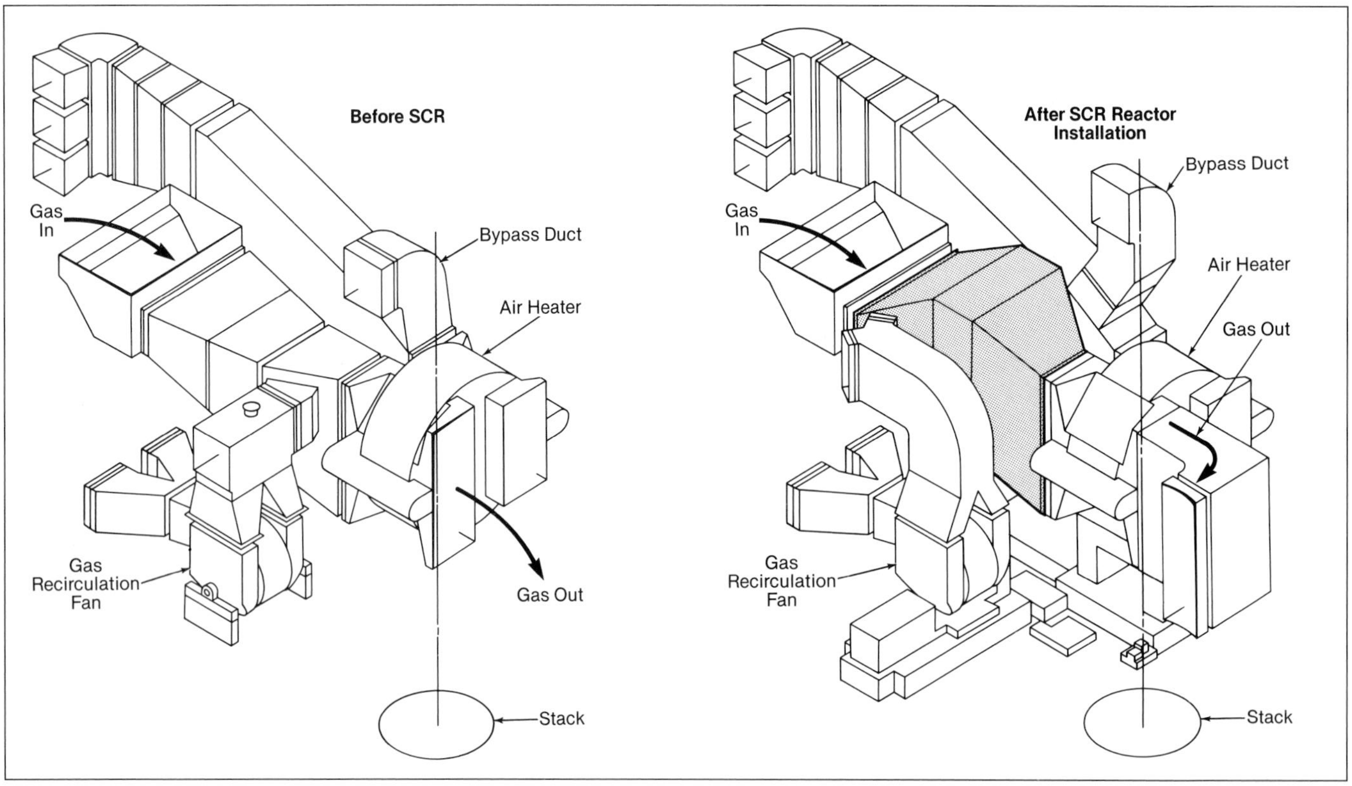

Fig. 8 Retrofit utility boiler application (half of installation shown).

verting CO to CO_2 under proper temperature conditions. However, SO_2 oxidation to SO_3 and high material costs often make precious metal catalysts less attractive.

Early technology used pelletized catalysts in a packed bed configuration. This material was effective for NO_x reduction but could be difficult to handle and frequently added significant pressure drop to the system. As a result, catalyst manufacturing technology evolved to larger, uniform catalyst blocks. Most modern SCR systems use a block type catalyst which is manufactured in the parallel plate or honeycomb configurations shown in Figs. 11 and 12, respectively. For ease of handling and installation, these blocks are fabricated into larger modules.

Each catalyst configuration has its advantages. The plate type unit offers less pressure drop and is less susceptible to plugging and erosion when particulate-laden flue gas is treated in the SCR reactor. The honeycomb configuration often requires less reactor volume for a given overall surface area.

The catalyst is housed in a reactor which is strategically located within the system. This location permits catalyst exposure to proper SCR reaction temperatures. The reactor design includes a sealing system to prevent flue gas bypassing and to provide internal support for the catalyst. The reactor configuration can be vertical or horizontal depending on the fuel used, space available, and upstream and downstream equipment arrangement. Uniform flow distribution is required for maximum performance. Fig. 13 shows a vertical SCR reactor design and Fig. 14 shows a coal-fired boiler installation.

Ammonia and stoichiometry considerations

A diagram of an ammonia control and supply system is shown in Fig. 15. Ammonia can be anhydrous or aqueous, although most commercial experience has been with anhydrous NH_3. However, due to safety concerns, aqueous NH_3 is required for selected SCR system installations.

Ammonia is transported to the site and stored in specially designed tanks. A controlled amount of ammonia

Fig. 9 HRSG with SCR.

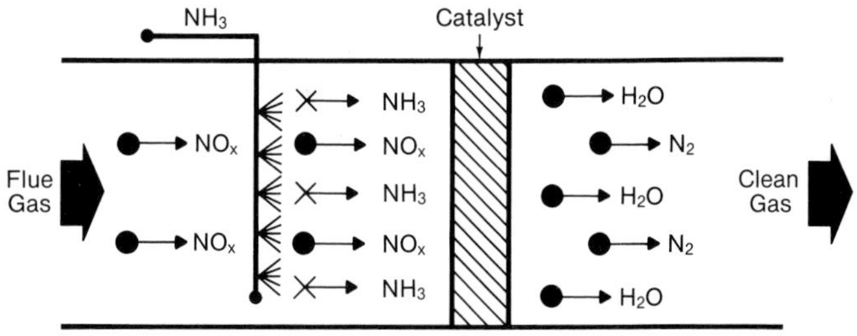

Fig. 10 Principles of NO_x removal process for SCR.

Fig. 11 Plate-type catalyst block.

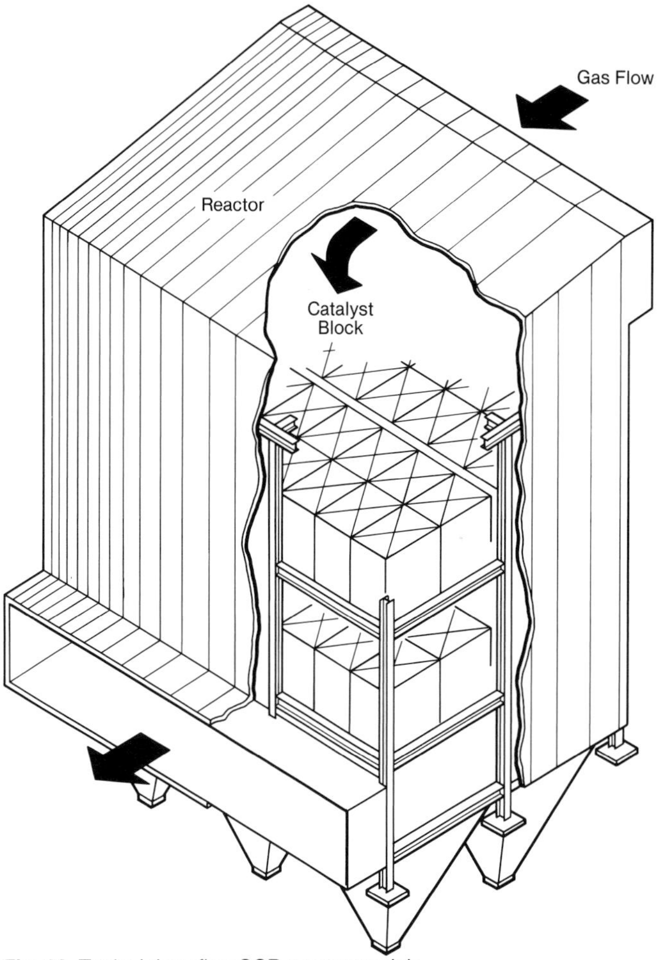

Fig. 13 Typical downflow SCR reactor module.

is then fed to a vaporizer/mix chamber where it is mixed with air or steam at about a 1:20 volume ratio.

The mixture is then introduced into the flue gas through an injection grid system. The unit-specific grid is designed to provide an even distribution of ammonia throughout the flue gas. The grid configuration is based on the duct size and the distance from the injection grid to the inlet of the catalyst bed. Longer distances require fewer injectors for adequate mixing.

The fundamental process control provides ammonia flow at a constant NH_3/NO_x mole ratio. The product of the inlet NO_x concentration and the boiler flue gas flow yields a NO_x flow signal. Ammonia flow control is then established by multiplying the NO_x flow signal by the NH_3/NO_x mole ratio set point. (See Fig. 15.)

In general, the stoichiometry of NO_x reduction is a 1:1 mole ratio of NH_3 to NO_x. Based on this stoichiometry, for example, a theoretical mole ratio of 0.80 is required for 80% NO_x removal. However, the actual mole ratio required is slightly higher to account for unreacted ammonia carryover from the reactor (NH_3 slip). To account for reaction lag time during abrupt boiler load changes, an ammonia-rich ratio is provided for increasing loads. Similarly, a lean mixture is provided for decreasing loads.

Related topics

A continuous emissions monitoring (CEM) system is frequently required to monitor all atmospheric pollutants. A detailed discussion of these systems is found in Chapter 36. Data generated from the CEM system can

Fig. 12 Honeycomb catalyst geometry.

Fig. 14 SCR installation on a coal-fired boiler.

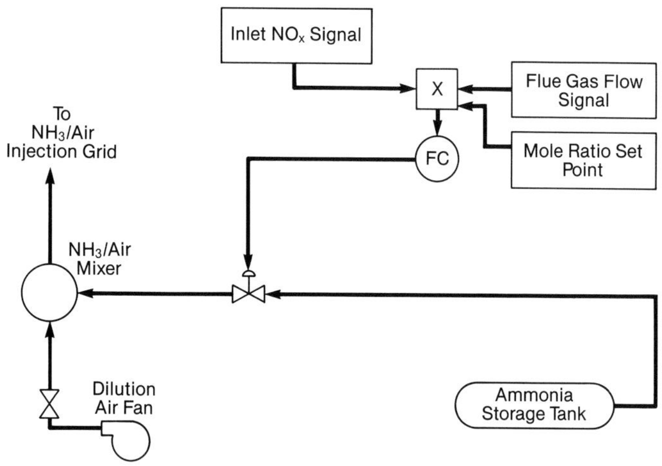

Fig. 15 NH_3/dilution air supply and control system.

be used to control the ammonia flow while achieving the required NO_x emissions level.

The design of each SCR system is unique. The major items to be considered include space constraints, location of existing equipment (retrofit projects), temperature requirements, fuel and cost. Fig. 16 shows several potential options of how the SCR system can fit into the boiler system. For new boilers, or when space is available on a retrofit, the preferred configuration is System (a), which provides the optimum temperature profile. The configuration shown in System (b) was originally developed as a retrofit due to space limitations. Unfortunately, its capital and operating costs are substantially higher than those for System (a) due to the required additional heat exchangers and duct burner (external heat source).

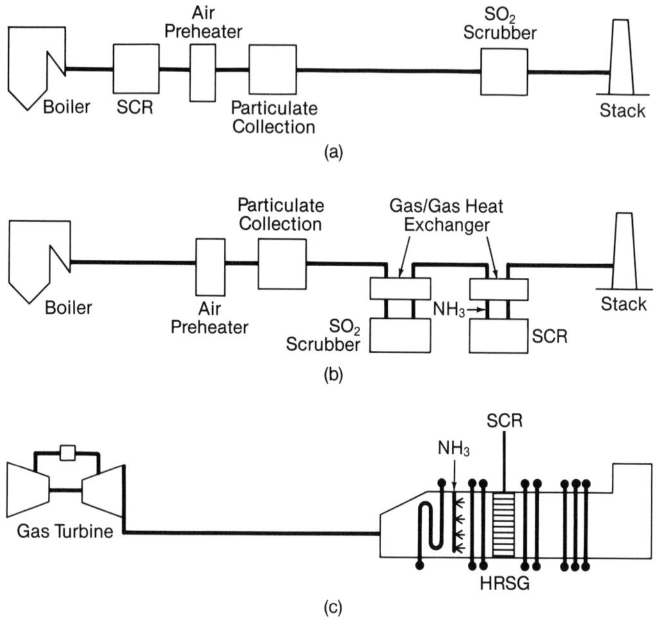

Fig. 16 Examples of SCR deNO$_x$ system configuations: (a) preferred SCR location, (b) possible retrofit configuration, and (c) combined cycle heat recovery steam generator (HRSG) location.

The application of SCR technology to combined cycle systems with heat recovery steam generators (HRSG) is shown as System (c). In this application, the catalyst placement within the HRSG is based on gas temperature requirements.

Selective catalytic reduction must be considered when low NO_x combustion technology provides insufficient NO_x reduction to meet local requirements. The SCR technology has proven to be reliable and is capable of up to 90% NO_x reductions and higher in some selected cases.

Future technology

Selective catalytic and noncatalytic processes have been widely accepted since the mid 1970s. SCR advancements have centered on catalyst developments for improved resistance to erosion, resistance to chemical deactivation, lower SO_2 to SO_3 conversion, and an expanded temperature performance range. Catalyst materials such as zeolite and precious metals are recent innovations.

Most SNCR process experimentation has considered alternatives to ammonia injection. Urea is the likely alternative. Two other alternatives receiving attention include cyanuric acid and ammonium sulfate. In addition, the use of methanol extenders (injected with the NO_x reagent) is being studied. The goal is to extend the useful NO_x reduction temperature range.

A special case of injection technology involves organoamines. For example, methylamine reacts with HNO_2, yielding methanol and nitrogen. This is a low temperature gas phase reaction as opposed to the current high temperature reactions. However, the gas phase concentration of HNO_2 is limited by the oxidation rate of NO to NO_2. This technology, under development, has not been demonstrated on a commercial scale.

A variety of advanced postcombustion NO_x control systems are at various stages of development and demonstration. Many of these are combined NO_x and SO_2 control systems. While the SNCR and SCR systems discussed here basically feature dry reduction of NO_x by ammonia or urea, advanced systems under development offer a variety of options, e.g., wet (aqueous) or dry absorption by solids, absorption plus oxidation by a liquid, and absorption plus reduction by a liquid, among others. However, only the SCR and SNCR systems have achieved widespread commercialization to date. Further development and demonstration of the other systems will be required before significant acceptance is likely and attractive economics are achieved.

References

1. Zeldovich, J.B., "Oxidation of nitrogen in combustion and explosion," *Academic des Sciences de l'URSS-Comptes Rendus (Doklady)*, Vol. 51, No. 3, pp. 217-220, January 30, 1946.

2. Sarofim, A.F., and Pohl, J.H., "Kinetics of nitric oxide formation in premixed laminar flames," presented at the Fourteenth Symposium on Combustion, Pennsylvania State University, University Park, Pennsylvania, August 20-25, 1972.

3. Miller, J.A., and Bowman, T., "Mechanism and modeling of nitrogen chemistry in combustion," Fall Meeting of the Western States Section/The Combustion Institute, Dana Point, California, October 17-18, 1988.

4. "Why NO_x? The costs, consequences and control of nitrogen oxides in the human and natural environment," American Lung Association Report, April 1989.

Bibliography

Hurst, B.E., and White, C.M., "Thermal deNO_x: A commercial selective noncatalytic NO_x reduction process for waste-to-energy applications," proceedings of the ASME 12th Biennial National Waste Processing Conference, Denver, Colorado, June 2, 1986.

Lim, K.J., *et al.*, "Environmental assessment of utility boiler combustion modification NO_x controls," U.S. Environmental Protection Agency Report, Contract No. 68-02-2160, April 1980.

Perhac, R.M., "Environmental effects of nitrogen oxides," appearing in *1989 Symposium on Stationary Combustion Nitrogen Oxide Control*, Report GS 6423-Vol. 1, Electric Power Research Institute, Palo Alto, California, July 1989.

Pohl, J.H., and Sarofim, A.F., "Devolatilization and oxidation of coal nitrogen," 16th Symposium on Combustion, Massachusetts Institute of Technology, Cambridge, Massachusetts, pp. 491-501, August 15, 1976.

Sourcebook NO_x Control Technology Data, prepared for Control Technology Center U.S. Environmental Protection Agency, October 1990.

1600 MW capacity wet SO_2 flue gas desulfurization retrofit.

Chapter 35
Sulfur Dioxide Control

Sulfur appears in the life cycle of most plants and animals. Most sulfur emitted to the atmosphere originates in the form of hydrogen sulfide from the decay of organic matter. These emissions slowly oxidize to sulfur dioxide (SO_2). Under atmospheric conditions, SO_2 is a reactive, acrid gas which can be rapidly assimilated back to the environment. However, the combustion of fossil fuels, in which large quantities of SO_2 are emitted to relatively small portions of the atmosphere, can stress portions of the ecosystem in the path of these emissions.

Man is responsible for the majority of the SO_2 emitted to the atmosphere. Annual worldwide emissions are estimated to be 180 to 200 million tons, nearly half of which are from industrial sources. The two principal industrial sources are fossil fuel combustion and metallurgical ore refining.

When gaseous (g) SO_2 combines with liquid (ℓ) water, it forms a dilute aqueous (aq) solution of sulfurous acid (H_2SO_3). Sulfurous acid can easily oxidize in the atmosphere to form sulfuric acid (H_2SO_4). Dilute sulfuric acid is a major constituent of *acid rain*. (Nitric acid is the other major acidic constituent in acid rain.) The respective reactions are written:

$$SO_2(g) + H_2O(\ell) \rightarrow H_2SO_3(aq) \quad (1)$$

$$O_2(g) + 2H_2SO_3(aq) \rightarrow 2H_2SO_4(aq) \quad (2)$$

The SO_2 can also oxidize in the atmosphere to produce gaseous sulfur trioxide (SO_3). Having a high affinity for water vapor, the sulfur trioxide reactions are written:

$$2SO_2(g) + O_2(g) \rightarrow 2SO_3(g) \quad (3)$$

$$SO_3(g) + H_2O(g) \rightarrow H_2SO_4(\ell) \quad (4)$$

While Equations 1 and 2 describe the mechanism by which SO_2 is converted to sulfuric acid in acid rain, Equations 3 and 4 characterize dry deposition of acidified dust particles and aerosols.

The pH scale is the method used to quantify the acidity of acid rain. The scale is defined by:

$$pH = -\log_{10}[H^+] \quad (5)$$

where $[H^+]$ is the concentration of hydrogen ions in solution. A pH below 7.0 is considered acidic, while a value above 7.0 is alkaline. The pH of pure distilled water is 7.0. If rain water contained no sulfuric or nitric acid, its pH would be approximately 5.7 due to absorption of carbon dioxide (CO_2) from the atmosphere. The contributions of man-made SO_2 and nitrogen oxides (NO_x) further reduce the pH of rain water. No uniformly accepted definition exists as to what pH constitutes acid rain. Some authorities believe that a pH of about 4.6 is sufficient to cause sustained damage to lakes and forests in the Northeastern portion of North America and in the Black Forest region of Europe. Rains with pH values of 4.2 have been observed.

SO_2 emissions regulations

Legislative action has been responsible for most industrial SO_2 controls. Major regulations include the Clean Air Amendments of 1970, 1977 and 1990 in the United States (U.S.), the Stationary Emissions Standards of 1970 in Japan, and the 1983 SO_2 Emissions Regulations of the Federal Republic of Germany.

Since the mid 1980s, SO_2 emissions regulations have been proposed in many other industrialized nations. Most members of the European Economic Community are regulated and Canadian laws are similar to those in the U.S. A more detailed discussion of regulation is included in Chapter 32.

SO_2 control

A variety of SO_2 control processes and technologies are in use and others are in various stages of development. Commercialized processes include wet, semi-dry (slurry spray with drying) and completely dry processes. Of these, the wet SO_2 scrubber has been the dominant worldwide technology for the control of SO_2 from utility power plants. As such, wet flue gas desulfurization (FGD) systems will be the primary focus of this chapter with other systems, especially dry FGD systems, also discussed.

From 1970 to 1988, SO_2 emissions in the U.S. diminished from 28.4 to 20.7 million tons per year. In that same time span SO_2 emissions by electric utilities were reduced from 15.8 to 13.6 million tons per year. In 1970 the average sulfur content of coal consumed by these utilities was about 2.3%. The use of coal by the utilities has grown in this time period from 320 million tons in 1970 to 758 million tons in 1988. If no SO_2 emission controls had been implemented in 1970, it is estimated that SO_2 emissions from U.S. electric utilities would have grown to about 34.7 million tons per year by 1988.

The utilities primarily used two strategies for control, switching to low sulfur coal and installing scrubbers. By 1988, the average sulfur content of fuels burned was 1.34% for coal and 1.09% for oil. With fuel switching alone, SO_2 emissions for electric utilities would have been 21.25 million tons in 1988; however, scrubbers, primarily wet, removed 7.65 million tons of SO_2 in 1988. In all, fuel switching accounts for about 64% of the SO_2 control and scrubbers have accounted for 36% of utility SO_2 reductions. (Fuel switching technical considerations are discussed in Chapter 46.)

Wet scrubbers have been used by many utilities since 1970 to meet SO_2 emission regulations. 68,000 MW of FGD capacity were installed in the U.S. between 1970 and 1990.[1] In Japan, 32,000 MW of capacity were added during this period. 48,000 MW of capacity were installed between 1979 and 1990 in Germany. The rest of the world accounts for approximately 4,000 MW. (See Fig. 1).

Wet scrubbers offer the following advantages:

1. As an alternative to fuel switching, the utility can use its normal source of fuel supply.
2. As an alternative to and in contrast to other traditional and emerging flue gas desulfurization methods, wet scrubbers provide high SO_2 removal efficiency and high reagent utilization.

Of the 180,000 MW of worldwide FGD capacity, 85% are wet scrubbers with the balance being predominantly dry scrubbers. Wet scrubbing processes are often categorized by reagent. Table 1 lists several scrubber processes and, of these, the limestone process has been the most widely applied. Within each category several process variations have been developed and commercialized. These are broadly outlined in Table 1 and are discussed later in this chapter. The primary advantages of limestone are its wide availability and cost effectiveness; roughly 5 to 6% of the earth's crust consists of calcium and magnesium carbonates and silicate. Limestone, consisting mostly of calcium carbonate ($CaCO_3$), is easily mined, transported and stored. Its storage and conveying at the plant site are similar to coal handling. (See Chapter 11.)

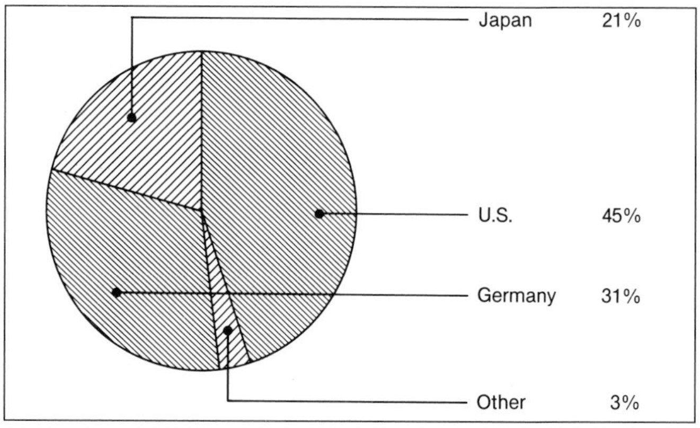

Fig. 1 Wet FGD systems installed throughout the world, on a capacity basis — through 1990.

Wet scrubbers

Design

The first FGD wet scrubbers installed in the U.S. were combined particulate collectors and SO_2 absorbers. However, the energy requirements for the venturi scrubbers used for particulate collection proved to be excessive. High efficiency dust collectors, usually electrostatic precipitators, soon replaced venturi scrubbers. (See Chapter 35.) Fig. 2 displays the typical arrangement.

The dust collector is placed upstream of the wet scrubber. In this arrangement, the induced draft fan is between the dust collector and the scrubber, permitting the fan to operate in a particulate-free, dry flue gas. In Germany, where nearly all FGD systems were retrofits to existing boilers, some induced draft fans did not have sufficient capacity. In some instances, a booster fan is placed downstream of the scrubber to handle the additional scrubber draft. These fans, incorporating a wash system to prevent excessive solid deposition from the scrubber, generally have rubber-lined housings and high alloy rotors.

The maximum size of a wet scrubber depends upon several design factors. Scrubbers which treat more than 6×10^6 lb/h (756 kg/s) of flue gas have been built and no fundamental engineering limits to wet scrubber size have been identified. Prior to 1990, U.S. environmental regulations encouraged utilities to purchase FGD systems which included multiple module designs with spare capacity. Current regulations now provide a framework within which single module scrubber systems can form a part of the compliance strategy. In Germany and Japan, utilities have favored large single modules for each boiler; in some instances, one scrubber treats the flue gas from two or more boilers.

The most popular wet scrubber design is the *spray tower* depicted in Fig. 3. The tower is designed so that, at maximum load, the average superficial gas velocity does not exceed the design gas velocity. For most spray towers, the average gas velocity varies from about 8 to 13 ft/s (2.4 to 4 m/s) based upon scrubber outlet conditions. A typical design velocity for a limestone wet scrubber is about 10 ft/s (3.1 m/s).

Because the flue gas enters the absorber from the side, gas flow nonuniformity in the tower is a potential prob-

Table 1
Wet Scrubber Processes

Limestone —
- with no oxidation inhibition (*natural oxidation*)
- with inhibited oxidation
- with in situ forced oxidation
- with ex situ forced oxidation
- with soluble organic or inorganic buffers

Lime —
- with inorganic buffers (such as magnesium oxide)
- without buffers

Dual alkali —
- sodium carbonate/calcium hydroxide
- sodium carbonate/calcium carbonate

Soda ash —
- with regeneration by steam stripping
- without regeneration

Magnesium oxide with thermal regeneration

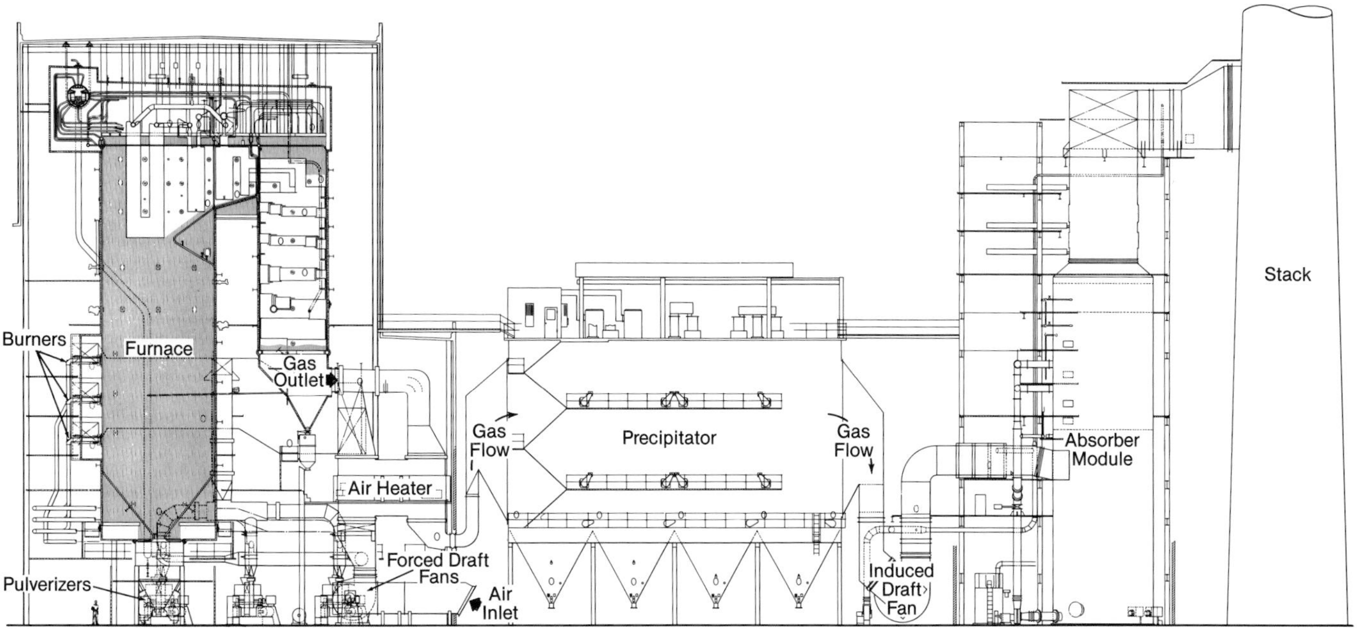

Fig. 2 Typical emission control systems for a coal-fired utility boiler.

lem. This nonuniformity reduces overall SO_2 removal performance and aggravates mist eliminator carryover. The absorber design depicted in Fig. 3 incorporates a sieve or perforated plate tray which reduces flue gas flow maldistribution. The pressure drop across the tray is usually between 1 to 3 in. wg (0.2 to 0.7 kPa). Towers with multiple trays have also been built. The design of the tower is influenced by the reagent (lime or limestone, for example), the desired SO_2 removal level, the tradeoff between fan power and recirculation slurry pump power, and several other factors.

The dominant design variable for all FGD wet scrubbers is the ratio of slurry flow to gas flow in the tower, known as L/G. In the U.S., this term is expressed as gallons per thousand cubic feet of flue gas evaluated at scrubber outlet conditions. In Japan and Germany, the L/G units are l/m^3. A wide range of L/G values have been used on wet scrubbers. Soda scrubbers operate at as low as 10 gal/1000 ft^3 (1.34 l/m^3) and some limestone scrubbers operate at up to 150 gal/1000 ft^3 (20 l/m^3).

Spray nozzles are used in wet scrubbers to control the mixing of slurry with flue gas. The operating pressures typically vary between about 5 and 20 psi (34 and 138 kPa). Spray nozzles without internal obstructions are favored to minimize plugging by tramp debris. Although plugging could be minimized by using a minimum number of large spray nozzles, flow maldistribution would most likely occur. Therefore, several smaller nozzles are usually preferred. Ceramic nozzles are commonly used.

The large tank at the bottom of the tower is called the *reaction tank* or the *recirculation tank*. The volume of this tank permits several chemical and physical processes to approach completion. Some of these rate processes will be described in more detail later.

Flue gas/scrubber considerations

Flue gas enters the side of the scrubber module at a temperature of 250 to 350F (121 to 177C) and is evaporatively cooled to its adiabatic saturation temperature by a slurry spray. The scrubber inlet must be designed to prevent deposition of slurry solids at the wet-dry interface. Because the inlet flue is at the flue gas temperature [for example 300F (149C)] and the shell of the scrubber is at the saturation temperature [typically 125F (52C)], there is a point where the surface temperature abruptly changes. Deposits are most likely to form at this point. Deposition is minimized by a combination of features which prevent periodic slurry splashing on the hot, dry side of the wet-dry line.

Flue gas passes vertically upward through the scrubber. In the unit illustrated in Fig. 3, the gas flow is uniformly distributed by a specially designed perforated plate or sieve tray. This tray serves as a gas-liquid contacting device. Gas-liquid contact is enhanced by a froth of slurry which forms on the tray.

Above the tray, flue gases pass through several spray levels where additional gas-liquid contact is achieved. Each spray level consists of a set of headers and spray nozzles as illustrated in Fig. 3. The spray nozzles produce a relatively coarse spray with mass median droplet diameters of about 2000 to 2500 microns. This suspension of droplets is in countercurrent contact with the flue gas for about one to three seconds. A majority of the SO_2 absorption occurs during this short contact time. The spray zone in combination with the froth on the tray is referred to as the gas-liquid contact zone of the wet scrubber.

A disengagement height is provided above the top spray zone before the flue gases reach the mist eliminators. The purpose is to allow disengagement and return of the largest slurry droplets to the spray zone. For a scrubber operating at an average gas velocity of 10 ft/s (3 m/s), droplets larger than about 600 microns may have sufficient time and mass to fall back to the spray zone.

The mist eliminator design in most wet scrubbers uses *chevrons* to remove additional moisture from the flue gas (see Fig. 3). Chevrons are closely spaced corrugated plates which collect slurry deposits by impaction. They efficiently collect droplets larger than about 20 microns in diameter.

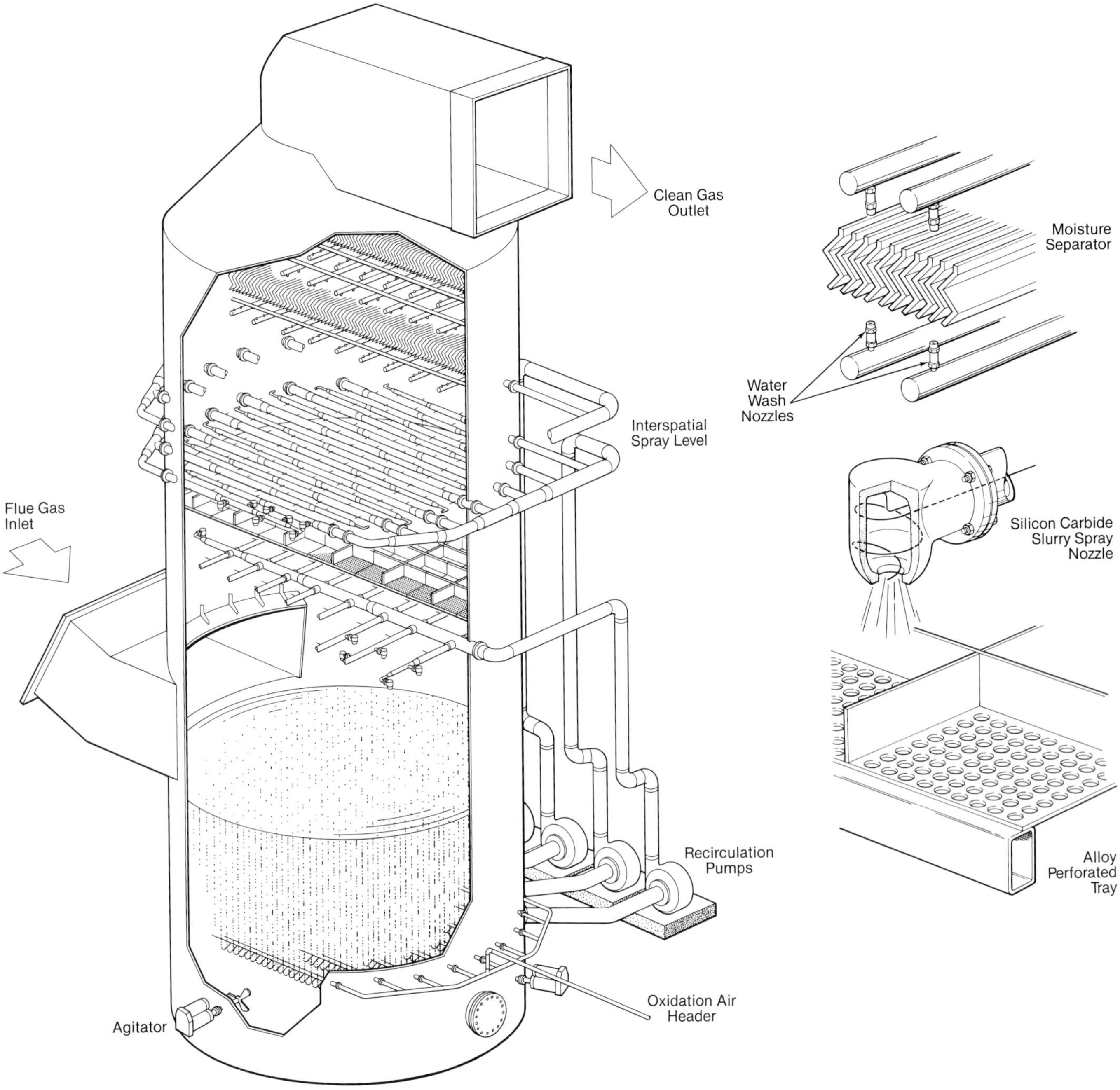

Fig. 3 Wet flue gas desulfurization scrubber module.

The design of the flue from the exit of the wet scrubber to the stack is an important facet of the system design. The potential for severe corrosion and deposition in these flues is well documented. This potential for severe corrosion arises from a combination of facts. First, the flue gas leaves the mist eliminator saturated with water vapor. Second, some carryover of slurry droplets smaller than 20 microns is inevitable. These droplets will usually be slightly acidic and may contain high concentrations of dissolved chlorides. The flue gases will contain some residual SO_2 and ample oxygen to oxidize some of the SO_2 to SO_3. Because the flue gas is saturated with water vapor, surface condensation is inevitable. This condensate can become severely acidic (pH less than 1.0), and calcium salts can deposit on the walls.

Two approaches are used to minimize these effects, flue gas reheat and flue/stack lining. The former option involves reheating the flue gas so that no more droplets remain. Reheating the flue gas that is leaving the scrubber has been accomplished by various systems:

1. steam coil heaters,
2. mixing with some hot flue gas which is bypassed around the scrubber,
3. mixing with hot air,
4. mixing with hot gases generated by combustion of a clean fuel, and
5. regenerative heat exchangers which transfer heat from the hot flue gas inlet to the cooler flue gas outlet.

Several problems are associated with each of the reheat methods. Deposition and corrosion occur in the heat exchangers. Reheating with bypass gas reduces the overall FGD system effectiveness. Finally, the evaporation of droplets from the scrubber concentrates the corrosive constituents in the slurry. As a result, operation without flue gas reheat, i.e., with a *wet stack*, has become popular in the U.S. Under these conditions, the flues from the scrubber to the stack are lined with corrosion resistant materials, and the stack is lined with acid resistant brick or other suitable material. A drainage system is also included to accommodate condensation of water vapor.

Wet scrubber — limestone and lime FGD processes

All of the limestone and lime processes listed in Table 1 are classified as nonregenerable. This means that the reagent is consumed by the process and must therefore be continually replenished. A generic diagram of nonregenerative processes is illustrated in Fig. 4. Each consists of four process steps: reagent preparation, SO_2 absorption, slurry dewatering and final disposal. Within each of these process steps many variations exist. Essentially all wet FGD installations have some unique aspects. A description of some of the more frequently used approaches to each process step is presented below followed by examples of two specific processes.

Reagent preparation

Limestone In North America, most limestone wet FGD systems feature on-site wet grinding for slurry preparation. In most cases, the system of choice is a closed loop ball mill. A typical ball mill circuit is shown in Fig. 5. The energy required to achieve a given grind size is estimated by the Bond relationship:

$$W = \frac{10\,W_i}{\sqrt{D_P}} - \frac{10\,W_i}{\sqrt{D_F}} \qquad \textbf{(6)}$$

where

W = power, kWh/t of product
W_i = Bond work index, kWh $(\text{micron})^{1/2}$/t
D_P = diameter in microns for which 80% of product is finer
D_F = diameter in microns for which 80% of feed is finer

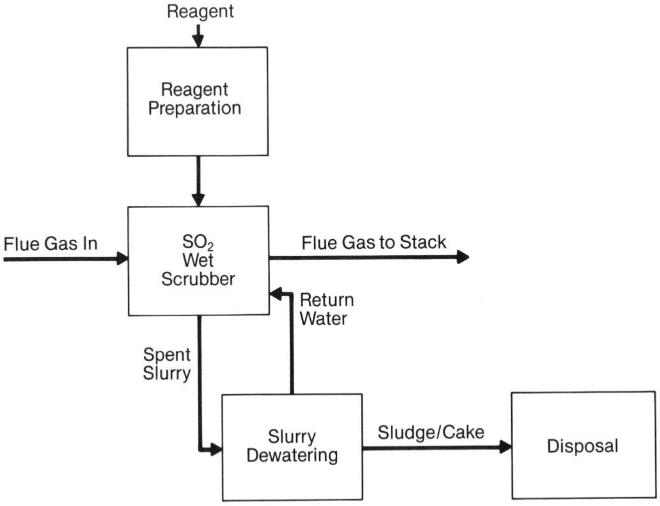

Fig. 4 Wet FGD system diagram.

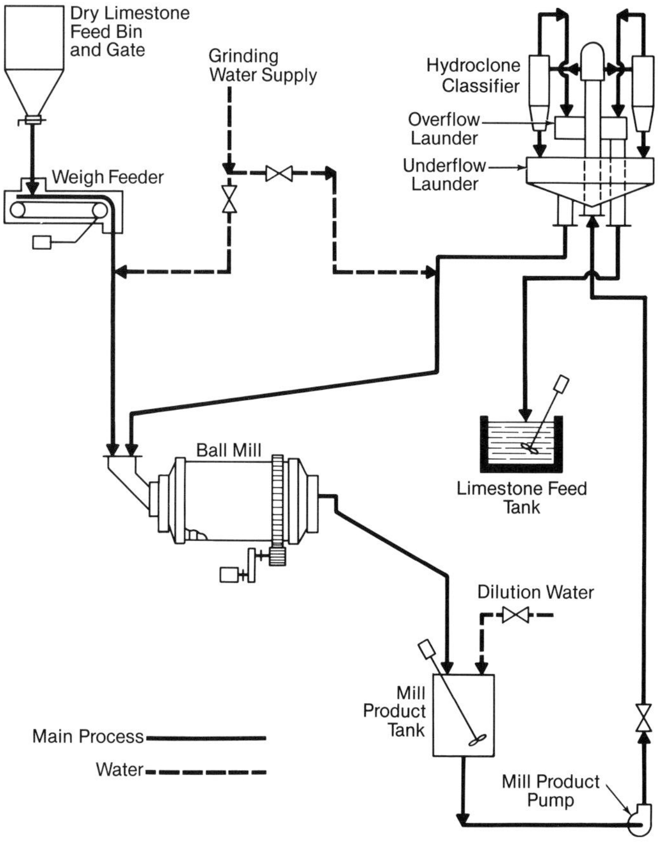

Fig. 5 Limestone reagent preparation system.

The Bond work index for calcitic limestones typically ranges from about 8 to 15 kWh $(\text{micron})^{1/2}$/t.

For on-site grinding systems, the limestone is typically received with a diameter of about 1 in. (25 mm) or less and is fed through a weigh belt feeder to the ball mill. Either fresh or recycled water is added at the feed chute in proportion to the feed rate of the stone. The output from the mill overflows to the mill product tank where it is pumped to a set of hydroclones. These hydroclones separate the fines from the coarse fraction. The coarse fraction (*underflow*) is returned to the ball mill and the fine material is sent to the limestone feed tank. The water balance is maintained to provide 25 to 35% suspended limestone solids in the feed tank.

The limestone grind is usually expressed as a percent passing a certain sieve size. The coarsest limestone grind used in wet FGD systems is one where about 70% passes through a 200 mesh (75 micron) sieve. Fine limestone grinding produces a product where about 95% of the limestone is finer than a 325 mesh (44 micron) sieve. Fine grinding requires a larger ball mill system and higher operating expense. The finer material provides higher limestone utilization, better reactivity with SO_2, and higher SO_2 removal for a given stoichiometry, while permitting a smaller reaction tank for some systems. Stoichiometry is defined as the molar ratio of the reactant, $CaCO_3$ for limestone systems, to the SO_2. It is common practice in the U.S. to define the stoichiometry by using the quantity of SO_2 removed for wet scrubbing systems and by the inlet quantity of SO_2 for other removal systems.

In Germany, off-site grinding is preferred. Limestone is dry ground at a central location and transported by truck to the FGD system. At the site, the pulverized limestone is pneumatically conveyed to a storage bunker. It is then fed to water-filled slurry preparation tanks. This system requires less space than on-site grinding facilities and reduces some of the FGD operating burden.

Lime Lime based FGD systems are common in the U.S. Three types of slaking systems are used: the detention slaker, the paste slaker and the ball mill slaker. The detention slaker is the simplest but produces the poorest quality slaked lime. The paste slaker, while producing high quality lime, is the most complex system and grit separation can be troublesome. The ball mill slaker produces an intermediate slaked lime quality but requires no postslaking separation of the grit and slaked lime.

A typical slurry preparation system is shown in Fig. 6. This system consists of storage for pebble lime, a weigh belt feeder, slaker, grit removal system, dilution system and slaked lime storage tank. Lime storage is usually maintained at 20 to 25% suspended solids.

Slurry dewatering and disposal

The disposal of reaction products from lime/limestone wet scrubbers has taken many forms. Several approaches, some unique to U.S. units, are used.

Ponding of the spent slurry without dewatering is the simplest disposal method. However, limited site availability and the cost of pond management have restricted its use. The combination of primary dewatering, secondary dewatering and landfilling is the most common approach used in the U.S. Primary dewatering is accomplished using thickeners or hydroclones. A thickener (Fig. 7) is capable of concentrating slurries with particle sizes as small as 5 microns to between 20 and 30% suspended solids. Lime/limestone systems with natural oxidation produce fine crystals and therefore require thickeners.

If the slurry is limited to particles larger than about 15 microns, hydroclones can be used for primary dewatering. These devices, much smaller and simpler than thickeners, are preferred where the particle size distribution is suitable. The gypsum crystals produced by lime/limestone systems with forced oxidation are usually suitable for hydrocyclones.

Secondary dewatering of FGD wastes is accomplished using a number of methods: vacuum drum filters, vacuum belt filters, solid bowl centrifuges and vertical basket centrifuges. Vacuum drum filters are the most common U.S. application, accounting for about 80% of the installations. Germany uses vacuum belt filters on more than 90% of its units.

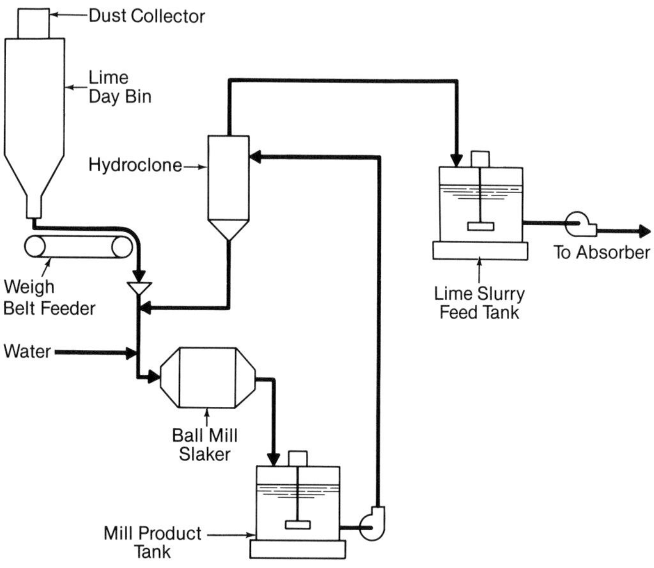

Fig. 6 Lime reagent preparation system.

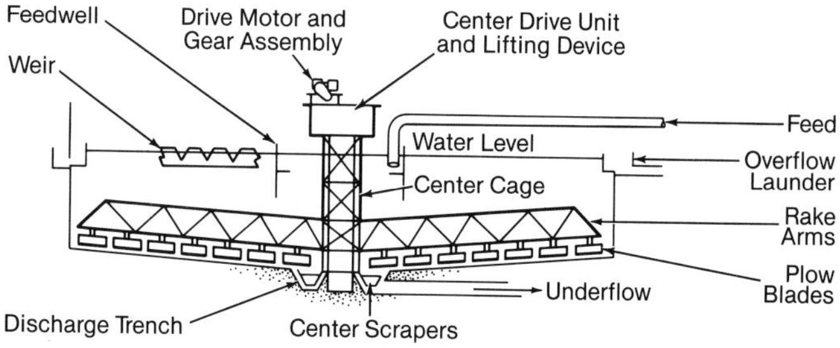

Fig. 7 Thickener system for primary dewatering of FGD scrubber effluent.

When landfilling is used for sulfite sludge disposal, additional treatment is required for sludge management at the landfill. *Sludge fixation* is used to increase the compressive strength of the sludge. Sulfite sludge is mixed with flyash and lime to yield a material suitable for landfilling. If gypsum is to be landfilled it needs no special treatment after dewatering.

Overall process flow diagrams and mass balance

All lime/limestone FGD wet scrubbers require fresh water for mist eliminator wash. This fresh water is usually combined with recycle water from the clarifier, thickener and/or vacuum filter to provide the total wash flow. Fresh water is also used for pump seals, reagent preparation and filter cake washing. Water losses from the system include evaporative losses, moisture in the waste solids, waters of crystallization and, in some cases, a controlled water purge. Careful management is required to assure adequate water supplies.

A typical limestone based wet scrubbing process using in situ forced oxidation is depicted in Fig. 8. In this example, the reagent preparation system includes a closed circuit ball mill system producing a slurry of ground limestone to a fineness of 95% minus 325 mesh (44 microns). This means that 95% by weight of the ground limestone will pass through a screen designated as a 325 mesh (44 micron) sieve. This slurry contains 30% by weight of suspended solids. The water used in the milling system is recycled from the spent slurry dewatering system. The feed slurry is pumped to the absorber reaction tank at a controlled rate to maintain the pH of the slurry in this tank at a set point of 5.5. Air is also pumped to this reaction tank and distributed by spargers located near the bottom. The oxygen in the air reacts with any sulfite present in the slurry to produce gypsum (calcium sulfate dihydrate).

The slurry is pumped from the reaction tank to the spray headers shown in Fig. 3. The slurry is sprayed countercurrently into the flue gas where it absorbs the SO_2. The slurry falls to the perforated plate tray where additional SO_2 is absorbed into the froth created by the interaction of the flue gas and slurry on the tray. The slurry then drains back to the reaction tank.

A small fraction of the slurry being pumped to the spray nozzles is diverted to the dewatering system. The spent slurry typically contains about 15% suspended solids. In this example, a hydroclone is used to concentrate the slurry. The underflow from the hydroclone is

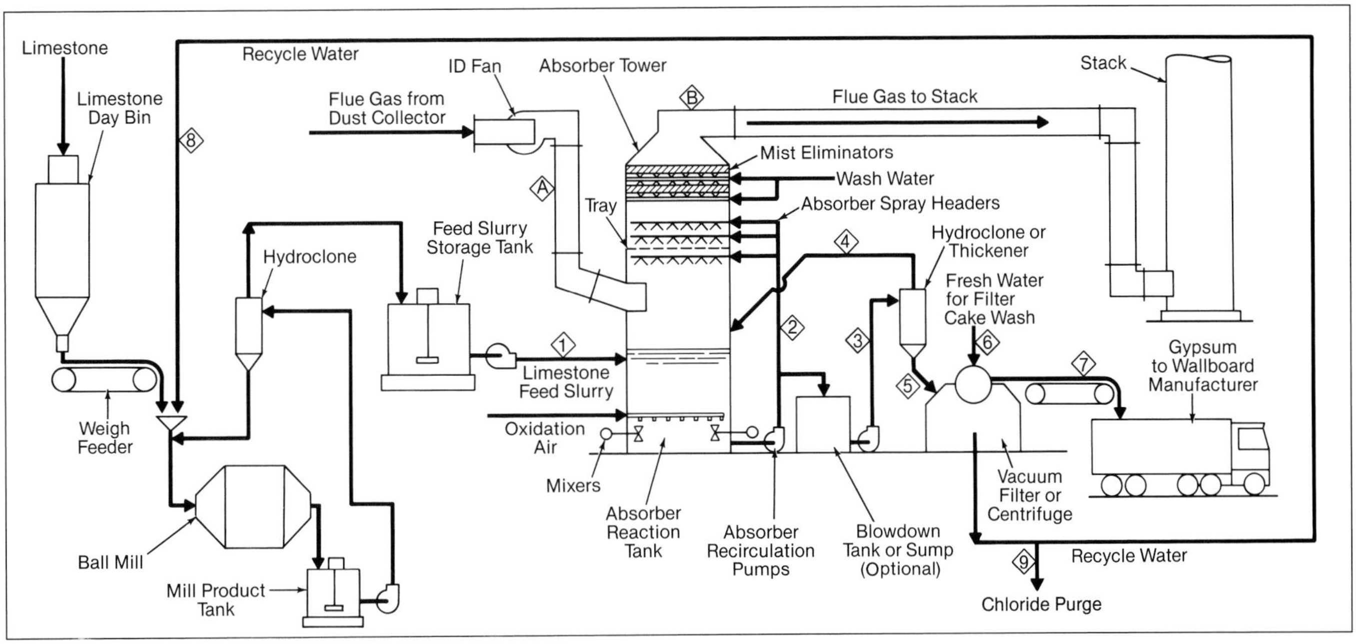

Fig. 8 Wet scrubber FGD system flow diagram (see Table 2 for material balance).

concentrated to 25% solids. The overflow containing 4% suspended solids is sent back to the absorber.

The underflow from the hydroclone is directed to a vacuum filter where the filtered solids are washed with fresh water and concentrated to form a filter cake containing about 10% free moisture. The cake is then sent by truck to a wallboard manufacturer. A mass balance for this example is presented in Table 2. A summary of the major power requirements for this limestone to gypsum process is shown in Table 3.

Fig. 9 illustrates a typical flow sheet for an FGD process using a lime produced from a dolomitic limestone. This lime contains about 5% magnesium oxide (MgO), 90% calcium oxide and 5% inerts. Within the process, the MgO is converted to $MgSO_3$, where, because of its relatively high solubility (approximately 10 g/l), the sulfite acts as a buffer to increase the SO_2 capacity of the slurry. The overall reaction is:

$$Ca(OH)_2(s) + SO_2(g) \rightarrow CaSO_3 \cdot {}^1\!/\!_2\, H_2O(s) + {}^1\!/\!_2\, H_2O(\ell) \quad (7)$$

In this instance, a closed circuit ball mill system is used to produce a slaked lime:

Table 2
Mass Balance for The Limestone Forced Oxidation System Shown in Fig. 8

Gas side	Inlet	Exit
Stream number	A	B
Flow rate, ACFM	626,724	500,000
Total mass flow rate, lb/h	1,905,400	1,982,091
Mass flow rate of H_2O, lb/h	76,466	157,700
Mass flow rate of SO_2, lb/h	8,565	600
Mass flow rate of HCl, lb/h	344	0
Static pressure, in. wg	6	1
Temperature, F	330	123

180 MW boiler burning coal with 3.2% sulfur
Limestone stoichiometry = 1.03
L/G = 130 gal/1000 ACF
SO_2 efficiency = 93%

Liquid side	Feed Slurry	Recirc	To Hydroclone	Hydroclone Overflow	Hydroclone Underflow	Filter Wash Water	Filter Cake	Recycle Water	Chloride Purge
Stream number	1	2	3	4	5	6	7	8	9
Flow rate, GPM	59	65,260	319	236	83	26	22	51	36
Total flow rate, lb/h	38,819	37,275,000	180,196	125,136	55,060	13,215	24,471	25,732	18,072
Flow rate of sus. solids, lb/h	13,586	5,607,000	27,029	5,005	22,024	0	22,024	0	0
% suspended solids	35	15	15	4	40	0	90	0	0
Chloride conc., ppm in liquid	25,000	25,000	25,000	25,000	25,000	50	100	18,514	18,514
pH	7.8	5.7	5.7	5.7	5.7	7	7	5.7	5.7
Temperature, F	100	123	123	123	123	70	70	70	70

Table 3
Typical Power Requirements for Limestone Scrubber with Forced Oxidation

Absorber System	Avg. Power (kW)	% Power
Oxidation air blower	375	
Absorber recirc. pump #1	312	
Absorber recirc. pump #2	367	
Absorber recirc. pump #3	380	
Absorber recirc. tank agitators	60	
Mist eliminator wash water pump	19	
Misc. pumps and agitators	24	
Subtotal	1537	60.3%
Dewatering Area	**Avg. Power (kW)**	**% Power**
Vacuum pump for filter	55	
Filter wash water tank heater	16	
Reclaim water pump	14	
Hydroclone overflow pump	15	
Filter feed tank agitator	7	
Clarifier overflow sump pump	6	
Misc. pumps and agitators	13	
Subtotal	126	4.9%
Reagent Preparation	**Avg. Power (kW)**	**% Power**
Ball mill driver	220	
Mill product tank pump	5	
Limestone feed tank agitator	25	
Misc. pumps and agitators	6	
Subtotal	256	10.0%
Other Systems	**Avg. Power (kW)**	**% Power**
General-instrument air	50	2.0%
Differential induced draft fan power	580	22.8%
Subtotal	630	
Total	2549	

Notes:
180 MW boiler burning coal with 2.5% sulfur
Heating value of coal = 12,767 Btu/lb
Absorber L/G = 130 gal/1000 ACF
Total pressure drop = 5 in. wg
Parasitic power = 2.549 MW/180 MW
= 1.42%

$$CaO + H_2O \rightarrow Ca(OH)_2 + Heat \quad (8)$$

$$MgO + H_2O \rightarrow Mg(OH)_2 \quad (9)$$

The absorber towers used for this process are smaller than their limestone counterparts because the L/G required to achieve a comparable level of SO_2 absorption is typically only about 20% of L/G for the limestone units.

The size distribution of the particles in the spent slurry is much smaller than the gypsum particles produced in the limestone forced oxidation process described above. A hydroclone would be incapable of achieving adequate concentration. Therefore, a thickener as illustrated in Fig. 7 is used. By comparison to a hydroclone, a thickener has the advantage of producing a clear overflow. The clear overflow means that this water can be recycled to the process and used for washing mist eliminators. The space requirement for a thickener is many times greater than that required by a hydroclone, however. The underflow from the thickener is directed to a vacuum filter where the unwashed cake is filtered to 45 to 55% solids. This cake is then mixed with lime and flyash to form an aggregate suitable for landfilling.

Wet scrubber chemistry

SO_2 absorption in a wet scrubber and its subsequent reaction with alkaline earth materials such as limestone is an elementary acid-base reaction. However, the chemical processes involved are complex. SO_2 is a relatively insoluble gas in water. Calcium carbonate ($CaCO_3$) has a low solubility in water. The principal reaction products are calcium sulfite hemihydrate ($CaSO_3 \cdot {}^1/_2 H_2O$) and calcium sulfate dihydrate ($CaSO_4 \cdot 2\,H_2O$), or gypsum. Both of these salts also have low solubilities.

In a limestone system with inhibited oxidation, the following reaction model can be used to describe the process using the chemical species in Table 4.

In the gas-liquid contact zone:

$$SO_2(g) \leftrightarrow SO_2(aq)$$
$$\textit{Dissolving gaseous } SO_2 \quad \textbf{(10)}$$

$$SO_2(aq) + H_2O \leftrightarrow HSO_3^- + H^+$$
$$\textit{Hydrolysis of } SO_2 \quad \textbf{(11)}$$

$$CaCO_3(s) + H^+ \leftrightarrow Ca^{++} + HCO_3^-$$
$$\textit{Dissolution of limestone} \quad \textbf{(12)}$$

$$HCO_3^- + H^+ \leftrightarrow CO_2(aq) + H_2O$$
$$\textit{Acid – base neutralization} \quad \textbf{(13)}$$

$$CO_2(aq) \leftrightarrow CO_2(g)$$
$$CO_2 \textit{ stripping} \quad \textbf{(14)}$$

Table 4
Chemical Species Found In Scrubber Reaction Model

SO_2	sulfur dioxide
H_2O	water
HSO_3^-	bisulfite ion
H^+	hydrogen ion
$CaCO_3$	calcium carbonate (main limestone constituent)
Ca^{++}	calcium ion
CO_2	carbon dioxide
HCO_3^-	bicarbonate ion
$CaSO_3 \cdot 1/2\ H_2O$	calcium sulfite hemihydrate
$CaSO_4 \cdot 2\ H_2O$	calcium sulfate dihydrate (gypsum)
(ℓ)	denotes liquid phase
(g)	denotes gaseous phase
(aq)	denotes dissolved specie in water
(s)	denotes solid phase

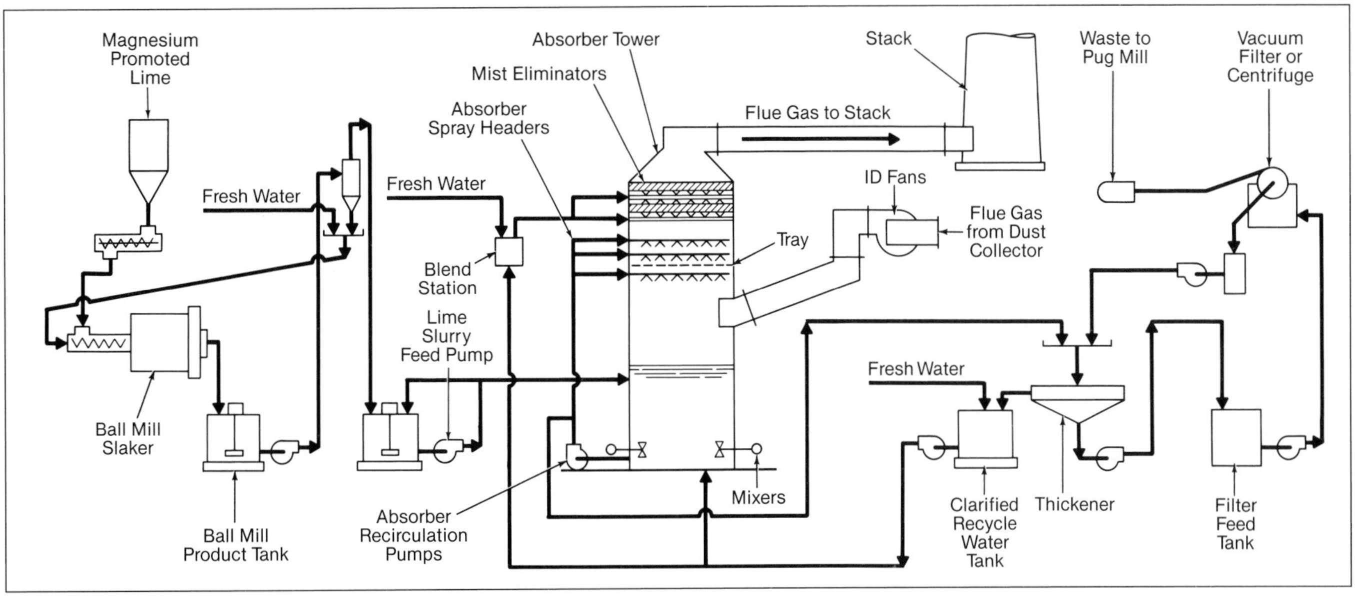

Fig. 9 Wet lime FGD system flow schematic.

$$CaSO_3 \cdot {}^1\!/_2\, H_2O + H^+ \leftrightarrow Ca^{++} + HSO_3^- + {}^1\!/_2\, H_2O$$

Dissolution of calcium sulfite **(15)**

In the reaction tank:

$$CaCO_3(s) + H^+ \leftrightarrow Ca^{++} + HCO_3^-$$

Dissolution of limestone **(16)**

$$HCO_3^- + H^+ \leftrightarrow CO_2(aq) + H_2O$$

Acid – base neutralization **(17)**

$$CO_2(aq) \leftrightarrow CO_2(g)$$

CO_2 *stripping* **(18)**

$$Ca^{++} + HSO_3^- + {}^1\!/_2\, H_2O \leftrightarrow CaSO_3 \cdot {}^1\!/_2\, H_2O + H^+$$

Precipitation of calcium sulfite **(19)**

Reaction 10 expresses the mass transfer rate of SO_2 from the gas phase to the liquid or aqueous phase. Its mass transfer rate can be expressed by:

$$\frac{d(Gy)}{dV} = k_g a(y - y^*) \qquad \textbf{(20a)}$$

or

$$N_g = \int \frac{dy}{y - y^*} = \int \frac{k_g a}{G} dV \qquad \textbf{(20b)}$$

where

G = molar gas flow rate, moles/s
y = mole fraction of SO_2 in flue gas
k_g = gas film mass transfer coefficient, moles/m^{2}s
a = interfacial surface area, m^2/m^3
y^* = equilibrium SO_2 concentration at the gas-liquid interface
V = volume of the gas-liquid regime, m^3
N_g = number of gas phase transfer units, dimensionless

Although k_g can be approximated, the interfacial surface area can not. The gas phase mass transfer rate must be determined experimentally. This involves operating the scrubber under conditions in which $y^* \rightarrow 0$. Equation 20 can then be integrated to:

$$N_g = -\ell n(1 - E) = k_g a\, V / G \qquad \textbf{(21)}$$

where

N_g = overall number of gas phase transfer units, dimensionless
E = overall SO_2 fractional efficiency

Many factors determine the number of gas phase transfer units (N_g). These include the slurry spray rate, the droplet size and spacial distributions, the gas phase residence time (height of spray zone), the liquid residence time, wall effects and the gas flow distribution.

In a limestone based wet scrubber, the rate-limiting reactions in the gas-liquid contact zone are believed to be Reactions 12 and/or 15. The reaction rate for limestone dissolution can be expressed by:

$$\frac{d[CaCO_3]}{dt} = k_c([H^+] - [H^+]_{eq})S_{pc}[CaCO_3] \qquad \textbf{(22)}$$

where

$[CaCO_3]$ = calcium carbonate concentration in the slurry, moles/l
$[H^+]$ = hydrogen ion concentration, moles/l
$[H^+]_{eq}$ = equilibrium (H^+) at the limestone surface, moles/l
S_{pc} = specific surface area of limestone in slurry

Fig. 10 is a plot of limestone dissolution rates at various pH values and CO_2 partial pressures. The dissolution rate of calcium sulfite can be expressed in a form similar to Equation 22. The calcium sulfite will dissolve only if the H^+ concentration exceeds its equilibrium value in the gas-liquid contact zone.

The equilibrium pH for calcium sulfite is approximately 6.3 at a CO_2 partial pressure of 0.12 atm (12 kPa). Therefore, in order for Reaction 15 to proceed to the right, the pH in the gas-liquid contact zone must be below the

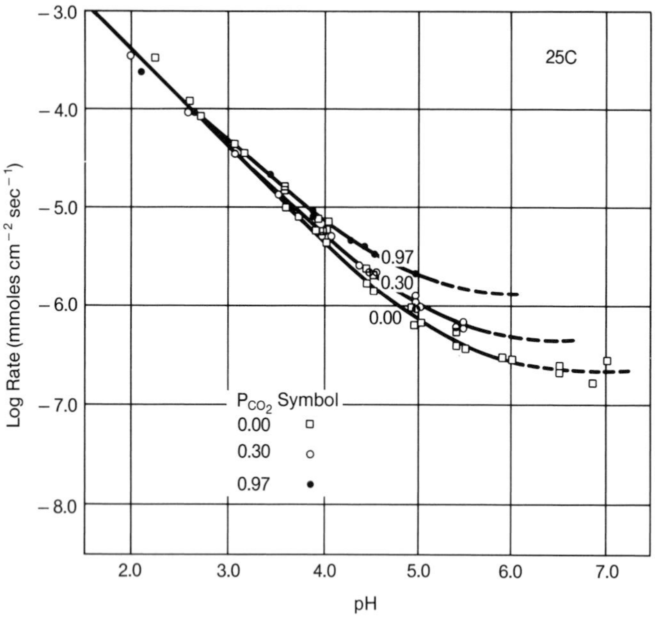

Fig. 10 Limestone dissolution rates at various pH values and CO_2 partial pressures (atm).[2]

equilibrium pH. If the pH were to stay above 6.3, then Reaction 15 would proceed to the left. This is undesirable because scaling due to sulfite precipitation in the scrubber is a likely consequence.

The reaction tank permits Reactions 16 through 19 to approach completion. In a limestone scrubber, limestone is added directly to the reaction tank. The pH of the slurry returning from the gas-liquid contact zone to the reaction tank can be as low as 3.5. The pH in the reaction tank is usually 5.2 to 6.2. Therefore, the overall reaction in the reaction tank is:

$$CaCO_3(s) + H^+ + HSO_3^- + \tfrac{1}{2}\ H_2O \rightarrow CaSO_3 \cdot \tfrac{1}{2} H_2O(s) + CO_2(g) \qquad \textbf{(23)}$$

Some oxidation of sulfite to sulfate inevitably occurs in the gas-liquid contact zone of the scrubber:

$$HSO_3^- + \tfrac{1}{2}\ O_2 \rightarrow SO_4^= + H^+ \qquad \textbf{(24)}$$

This is called *natural oxidation* to distinguish it from *forced oxidation* in which air is sparged through the slurry. The extent of natural oxidation is typically 15 to 30%, although it may be as high as 50% or higher.

The sulfate combines with dissolved calcium and water to form calcium sulfate dihydrate, or gypsum. The reaction proceeds as follows:

$$Ca^{++} + SO_4^= + 2H_2O \leftrightarrow CaSO_4 \cdot 2H_2O \qquad \textbf{(25)}$$

The rate of gypsum crystallization can be expressed by:

$$\frac{d[CaSO_4 \cdot 2H_2O]}{dt} = k(R-1)Sp_g[CaSO_4 \cdot 2H_2O] \qquad \textbf{(26)}$$

where

$R = A_{Ca^{++}} A_{SO_4^=} / K_{Sp}$
$A_{Ca^{++}}$ = activity of Ca^{++} ion
$A_{SO_4^=}$ = activity of $SO_4^=$ ion
K_{sp} = solubility product of gypsum
Sp_g = specific surface area of gypsum

R is a measure of the level of supersaturation. If R is greater than 1, the solution is supersaturated in gypsum. If R is less than 1, the solution is subsaturated in gypsum.

The reaction tank serves a second important function in lime/limestone scrubber systems. The tank is sized to provide sufficient time for the dissolved gypsum to crystallize and precipitate. Typically, the reaction tank is designed for six to ten minutes of residence time based upon the recirculation rate.

Gypsum scaling might be more of a problem in wet scrubbers that use lime or limestone and operate in the natural oxidation mode if it were not for a phenomenon known as *co-precipitation*. The principal reaction product in these wet scrubbers is $CaSO_3 \cdot \tfrac{1}{2} H_2O$. Co-precipitation occurs when the sulfate ion ($SO_4^=$) is substituted for the sulfite ion ($SO_3^=$) in the calcium sulfite crystal lattice. This has the effect of removing sulfate from solution thereby reducing the gypsum supersaturation.

The reaction model for lime based wet scrubbers is similar to that for limestone. Two principal differences exist. First, calcium hydroxide is much more soluble than limestone, approaching 1 g/l at 77F (25C). Secondly, solid $Ca(OH)_2$ has a much higher specific surface area. For example, ground limestone has a specific surface area of about 0.2 to 0.8 m^2/g. By contrast, slaked lime typically varies from about 5 to 15 m^2/g. The equilibrium pH of $Ca(OH)_2$ at 77F (25C) is 12.4 compared to a value of 7.8 for $CaCO_3$. In many respects, lime behaves like a highly reactive limestone. The following lime reactions could be substituted for Reactions 12 and 13.

$$Ca(OH)_2(s) + H^+ \leftrightarrow CaOH^+ + H_2O \qquad \textbf{(27)}$$

$$CaOH^+ + H^+ \leftrightarrow Ca^{++} + H_2O \qquad \textbf{(28)}$$

Performance enhancing additives

The steady-state hydrogen ion concentration of the slurry in the gas-liquid contact zone is determined by the balance between the rate of H^+ generation in Reaction 11 and the rate of H^+ consumption by Reactions 12, 13 and 15. As the hydrogen ion concentration increases, i.e., as the pH drops, the equilibrium SO_2 vapor pressure increases. This equilibrium relationship can be expressed by:

$$y^* = \frac{C' \times 10^3 [H^+]}{k_1(k_2 + [H^+])} \qquad \textbf{(29)}$$

where

k_1 = the equilibrium constant for Reaction 10, moles/l atm
k_2 = the equilibrium constant for Reaction 11, moles/l
C' = total concentration of dissolved SO_2, millimoles/l
= $[SO_2]_{aq} + [HSO_3^-]$
$[H^+]$ = steady-state hydrogen ion concentration
y^* = SO_2 vapor pressure expressed as ppm (assuming barometric pressure = 1 atm)

At 122F (50C):

k_1 = 0.4643 moles/l atm
k_2 = 7.162×10^{-3} moles/l

As the SO_2 vapor pressure rises, the rate of SO_2 absorption diminishes and approaches zero as $y^* \rightarrow y$. (See Equation 20.) If a simple means existed to reduce the hydrogen ion concentration in the gas-liquid contact zone, then the SO_2 vapor pressure could be controlled and the SO_2 absorption rate would be maximized. A buffer is a chemical specie which performs this function. A class of weak organic acids has been found suitable for use in limestone scrubbers to control pH and improve overall SO_2 removal. These buffers can be described by:

$$H^+ + A^- \leftrightarrow AH \qquad \textbf{(30)}$$

where AH is the generalized acid group. When the pH falls in the gas-liquid contact zone, Reaction 30 is driven to the right. Some organic acids which buffer in this range include adipic, formic and succinic acid. Because these are water soluble, the pH buffering is nearly instantaneous compared to the lime or limestone dissolution. Because the SO_2 vapor pressure, i.e., the equilibrium SO_2 concentration of the gas-liquid interface, is proportional to the hydrogen ion concentration, buffers minimize the rise in SO_2 vapor pressure. The buffer concentration required to achieve a given absorption depends upon SO_2 concentration and L/G ratio (the ratio of slurry flow to gas flow). Typically, the concentration of these additives which is required to achieve adequate control ranges from 3 to 30 millimoles/l.

A second additive used in wet FGD systems is magnesium oxide, which reacts with SO_2 to form magnesium sulfite. Because magnesium sulfite is highly soluble, the sulfite ion is the primary reactant in the gas-liquid interface as follows:

$$SO_3^= + H^+ \leftrightarrow HSO_3^- \qquad \textbf{(31)}$$

Sodium carbonate can also be added to the lime/limestone system for a similar benefit.

The total concentration of dissolved alkaline species such as $CO_3^=$, HCO_3^-, $SO_3^=$ and OH^- in the slurry is referred to as *dissolved alkalinity*. If the dissolved alkalinity is sufficiently high, the scrubber may become gas-phase diffusion controlled. This is illustrated in Fig. 11.[2] Under these conditions the rate of SO_2 absorption is dependent only upon the amount of interfacial surface area, i.e., the spray droplet surface area plus the tray froth surface area.

Additives for scale control

Although many scrubber additives provide performance improvements, others are used specifically for scale control. Scaling, a potential problem in all lime/limestone scrubbers, is primarily due to calcium sulfate ($CaSO_4 \cdot 2H_2O$: gypsum) precipitation onto surfaces within the gas-liquid contact zone. The large reaction tanks used in these scrubbers alleviate this condition but do not eliminate it. Sodium thiosulfate ($Na_2S_2O_3$) is a recognized oxidation inhibitor which, when added to a limestone slurry, reduces the level of natural oxidation. The effectiveness of thiosulfate is proportional to the product of its molar concentration, the calcium ion concentration, and the moles of SO_2 absorbed.

An economic alternative to using $Na_2S_2O_3$ is to add colloidal sulfur to the lime/limestone slurry. The sulfur reacts with the bisulfite in solution as follows:

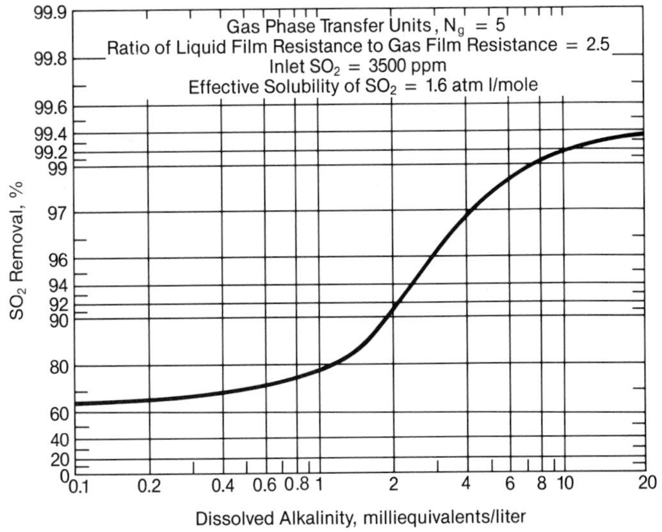

Fig. 11 Influence of dissolved alkalinity on SO_2 performance (1 milliequivalent = 50 ppm $CaCO_3$).

$$S(s) + HSO_3^- \leftrightarrow S_2O_3^= + H^+ \qquad \textbf{(32)}$$

The concentration of sulfur required ranges from 0.004 to 0.006 moles/l.

In addition to reducing scale formation in the scrubber, inhibiting sulfite oxidation permits easier removal of free water from the waste product (dewatering). Dewatering is simplified because the calcium sulfite crystal grows larger in the absence of sulfate ions.

Dry scrubbers

Dry scrubbing is the principal alternative to wet scrubbing for SO_2 control on utility boilers. Since 1980, 7200 MW of dry scrubbers have been installed at U.S. electric utilities. Dry scrubbing is also a popular choice for smaller industrial applications and for combined HCl and SO_2 control on waste-to-energy units. In the U.S., dry scrubbers have mainly been applied to units burning low sulfur fuels. Of the 43 U.S. utility dry scrubber installations, the majority lie west of the Mississippi River.

The advantages of dry scrubbing over wet scrubbing include:

1. less costly construction materials,
2. dry waste products,
3. fewer unit operations, and
4. simplicity of operation.

Dry scrubbing is sometimes referred to as spray absorption, spray drying or semi-wet scrubbing. It involves spraying a highly atomized slurry or aqueous solution of an alkaline reagent into the hot flue gas to absorb the SO_2. Fig. 12 depicts a utility size dry scrubber installation coupled with a baghouse. Unlike a wet scrubber installation, the dry scrubber is positioned before the dust collector. Flue gases leaving the air heater at a temperature of 250 to 350F (121 to 177C) enter the dry scrubber through an array of Turbo-Diffusers® as shown in Fig. 13 for a horizontal flow design. Vertical flow modules such as that shown in Fig. 14 are also supplied.

The quantity of water in the atomized spray is limited so that it completely evaporates in suspension. SO_2 absorp-

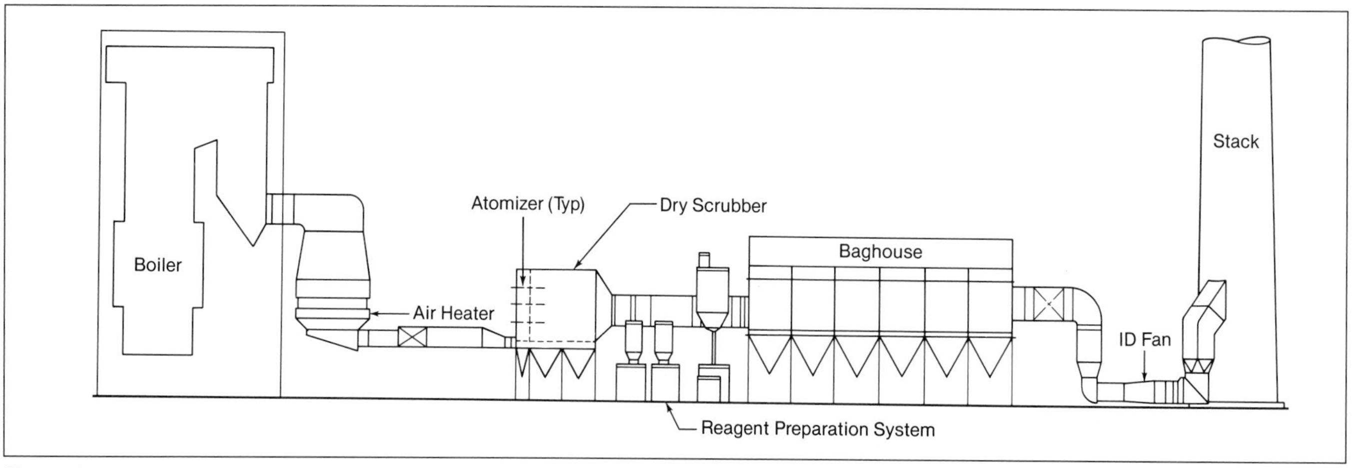

Fig. 12 Dry scrubber emissions control system configuration.

tion takes place primarily while the spray is evaporating and the flue gas is adiabatically cooled by the spray. The difference between the temperature of flue gas leaving the dry scrubber and the adiabatic saturation temperature is known as the *approach temperature*. Reagent stoichiometry and approach temperature are the two primary variables which dictate the scrubber's SO_2 removal efficiency.

The flue gas temperature leaving the dry scrubber may be too low for proper operation of the dust collector. In those instances the gases may be heated before entering the dust collector. The dust collector can be either an electrostatic precipitator (ESP) or a baghouse. The ESP is more forgiving of temperature variation but the baghouse has the advantage of being a better SO_2-lime reactor.

The predominant reagent used in dry scrubbers is slaked lime. The system shown in Fig. 15 consists of storage facilities for pebble lime (CaO), a ball mill slaking system, a system for mixing slaked lime with recycle material from the dust collector, the dry scrubber and the dust collector. Atomization of the slurry is accomplished by pneumatic or rotary atomization. A typical pneumatic atomizer is shown in Fig. 13.

Dry scrubber chemistry

SO_2 absorption in a dry scrubber is similar to that attained by wet scrubbing. The majority of the reactions take place in the aqueous phase; the SO_2 and the alkaline constituents dissolve into the liquid phase where ionic reactions produce relatively insoluble products. The reaction path can be described as follows:

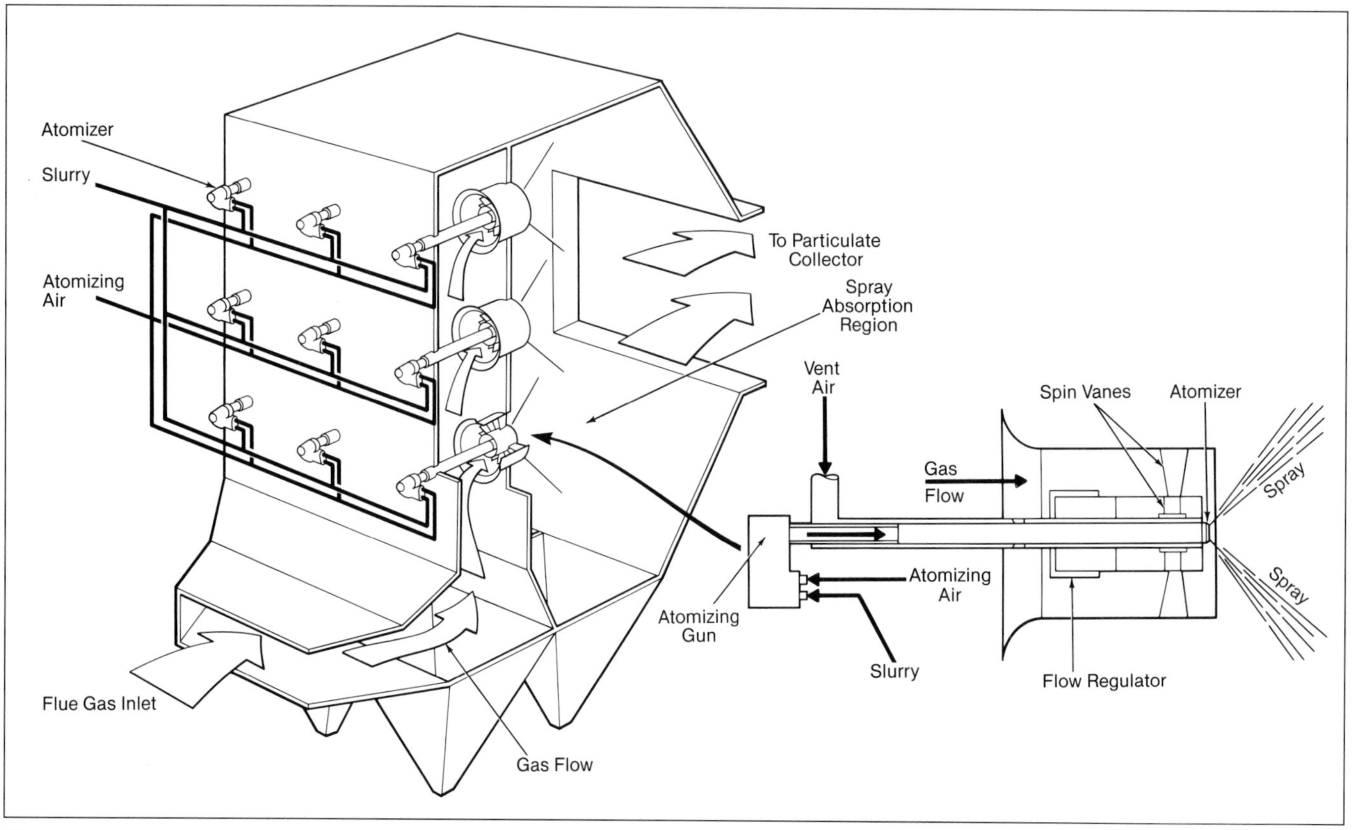

Fig. 13 Dry scrubber reactor module — horizontal flow configuration showing Turbo-Diffuser® insert.

$$SO_2(g) \leftrightarrow SO_2(aq)$$
Absorption **(33)**

$$Ca(OH)_2(s) \rightarrow Ca^{++} + 2OH^-$$
Dissolution **(34)**

$$SO_2(aq) + H_2O \leftrightarrow HSO_3^- + H^+$$
Hydrolysis **(35)**

$$SO_2(aq) + OH^- \leftrightarrow HSO_3^- \qquad (36)$$

$$OH^- + H^+ \leftrightarrow H_2O \qquad (37)$$

$$HSO_3^- + OH^- \leftrightarrow SO_3^= + H_2O$$
Neutralization **(38)**

$$Ca^{++} + SO_3^= + \tfrac{1}{2}\ H_2O \rightarrow CaSO_3 \cdot \tfrac{1}{2}\ H_2O\,(s)$$
Precipitation **(39)**

These dry scrubber reactions usually take place in the pH range of 10 to 12.5. The reaction scheme for wet scrubbers depicted by Equations 10 to 19 occurs below a pH of 7.0.

Dry scrubbers are normally sized for a certain gas-phase residence time, which depends on the design approach temperature and the degree of atomization. For example, if the approach temperature is 25F (14C) and the atomization system is designed for a Sauter mean diameter of 50 microns, the reactor residence time would be designed for about ten seconds.

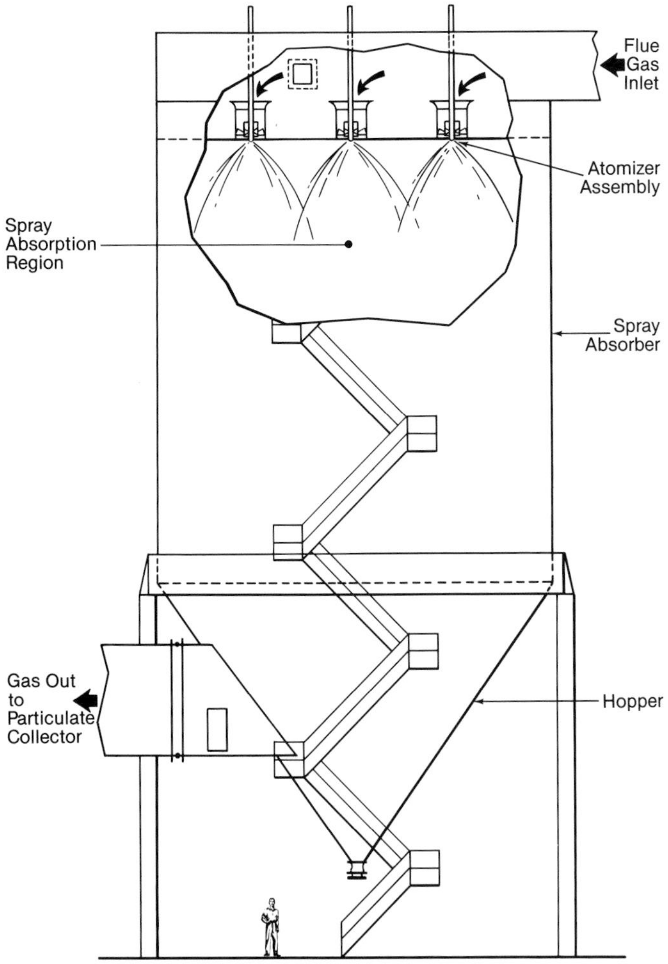

Fig. 14 Dry scrubber reactor module — vertical flow configuration.

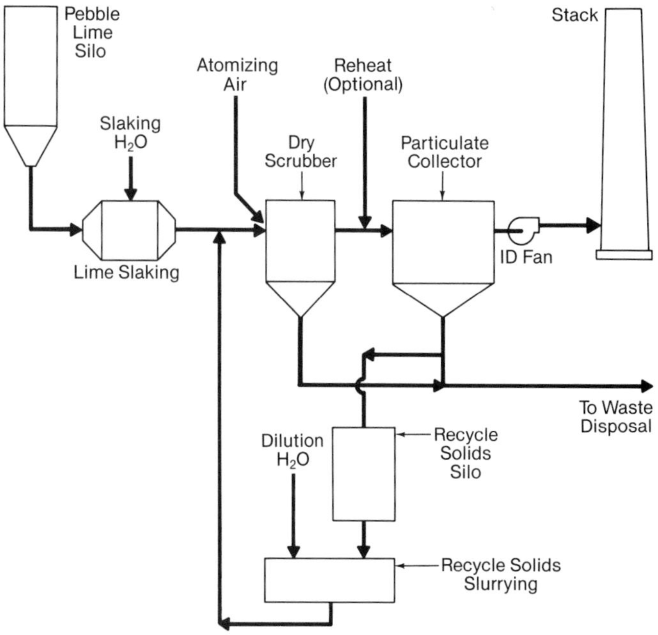

Fig. 15 Dry FGD scrubber system schematic.

The application of dry scrubbers to large electric utility boilers (see Fig. 16) is generally limited to those burning low sulfur coals. This is due primarily to the higher reagent costs for dry scrubbing. However, for smaller utility and industrial boiler applications, the simplicity and lower capital costs of dry scrubbing make it an attractive alternative.

The primary class of additives used in dry scrubbers is deliquescent salts which, when added to lime, reduce the drying rate and increase the equilibrium moisture content. SO_2 removals above 95% are possible with these additives. Of these salts, calcium chloride is the most popular. Ammonia injection upstream of a dry scrubber also increases SO_2 performance significantly.

Other commercialized technologies

A number of additional approaches and technologies are also available to control SO_2 emissions from fired processes. These range from reducing sulfur in the fuel to special combustion systems to postcombustion technology. Sulfur removal using coal cleaning processes is discussed in Chapter 11. Fluidized beds, illustrating SO_2 control during combustion, are discussed in Chapters 16 and 29. Finally, sulfur removal during coal gasification is covered in Chapter 17.

Other examples of commercialized FGD technology include:

1. furnace sorbent injection,
2. nacholite/trona injection, and
3. activated carbon.

Furnace sorbent injection has developed during the past 20 years. The technology involves the pneumatic in-

Fig. 16 470 MW utility boiler with dry scrubber and baghouse.

jection of limestone, dolomite or hydrated lime at a gas temperature of 2000 to 2300F (1093 to 1260C). Normally, the injection point is near the nose of the boiler. (See Fig. 17.) Using hydrated lime, 50 to 60% SO_2 capture is achievable with a calcium/sulfur ratio of 2.

Trona and nacholite are naturally occurring forms of sodium carbonate and bicarbonates which react with SO_2 at air heater exit temperatures. A relatively simple injection system for powdered reagent is placed between the air heater and baghouse. SO_2 reactions take place in the flue ahead of the baghouse and on the surface of the bag material. However, sodium carbonates have been observed to catalyze the oxidation of nitric oxide (NO) to nitrogen dioxide (NO_2). Whereas nitric oxide is a colorless gas, NO_2 is a brown gas which creates a visible stack plume.

Limited applications of activated carbon moving beds for SO_2 removal are seen in Germany and Japan. These beds have the added feature of low temperature SCR catalysis for the selective reduction of NO with ammonia.

Although a major worldwide effort has been expended to develop these and other alternatives, more than 97% of all SO_2 emissions control is currently accomplished by conventional wet and dry scrubbing.

Developments and trends

In addition to the commercialized FGD technologies, many new FGD processes are under development at the pilot plant scale. Some of these are at the demonstration stage on operating boilers.

The uncertainty as to which new FGD processes will make it successfully to the marketplace is great. However, it is certain that the successful processes will be those which meet the new regulations in the most economic manner. Processes which minimize capital expenses will be preferred on older power plants nearing retirement and those that are the least costly to build and operate will be preferred on newer power plants.

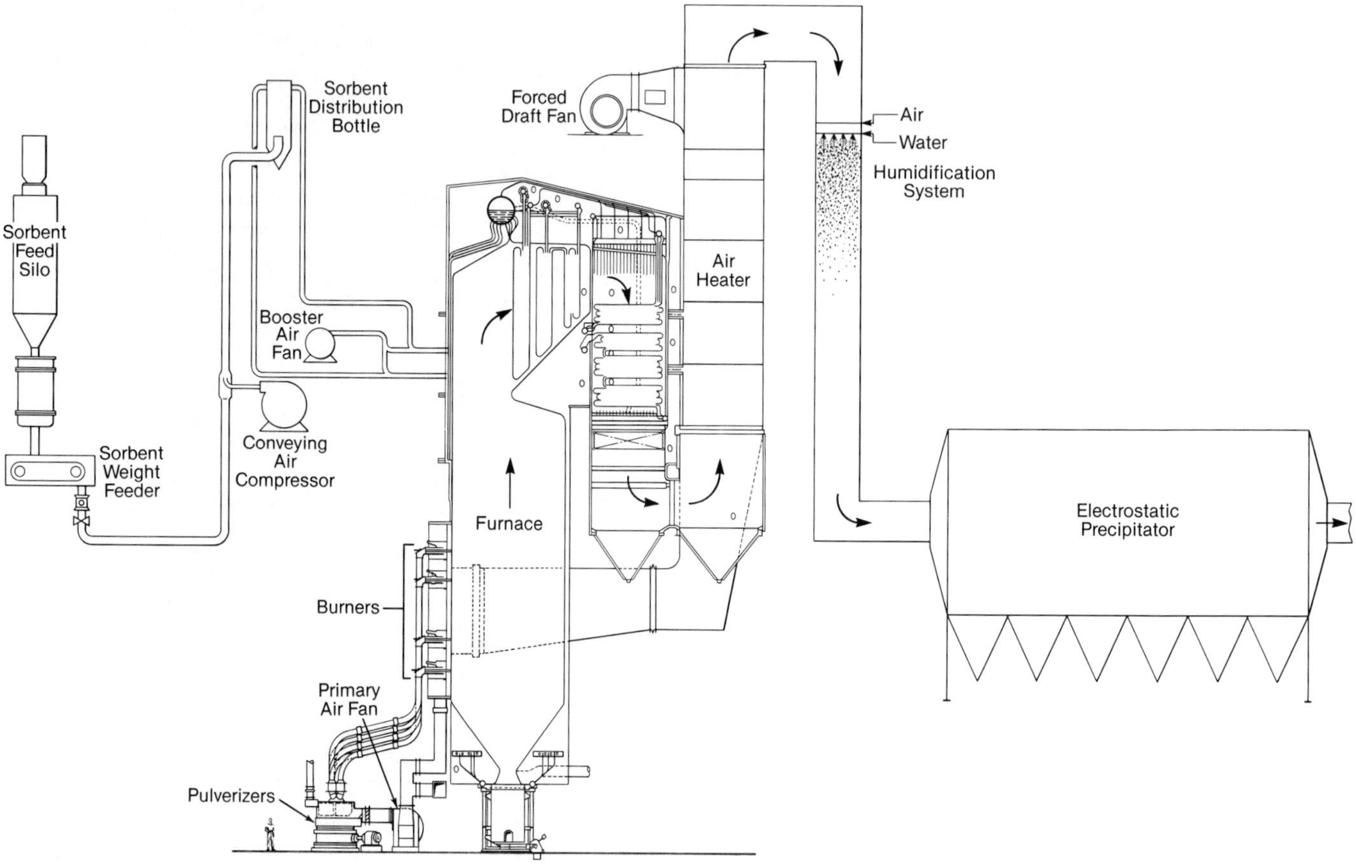

Fig. 17 Furnace sorbent injection system for SO_2 control — includes humidifier system to enhance particulate collection and SO_2 removal efficiency.

References

1. *Annual Energy Review 1990*, Report DOE/IRA 0381 (90), U.S. Energy Information Agency, Washington, D.C., 1991.

2. Plummer, L.N., Wigley, T.M.L., and Parkhurst, D.L., "The kinetics of calcite dissolution in CO_2-water systems at 5C to 60C and 0.0 to 1.0 atm CO_2," *American Journal of Science*, Vol. 278, pp. 179-216, 1978.

Flue gas desulfurization system installation in the Western U.S.

Chapter 36
Environmental Measurement

Environment related measurements at modern power plants are primarily performed to ensure compliance with government regulations. However, measurements are also necessary to monitor and optimize process performance, assess guarantees, calibrate continuous emission monitoring systems (CEMS) and verify design information. As discussed in Chapter 32, fossil fuel power plants emit air, water and solid pollutants. This chapter focuses on measurements for monitoring air pollution systems and, in particular, equipment for controlling sulfur dioxide (SO_2), nitrogen oxides (NO_x) and particulate emissions. Special emphasis is placed on measurements associated with wet SO_2 scrubbing systems which involve perhaps the most complex process.

Postcombustion emissions can be controlled by the use of one or more of the following systems: 1) particulate — electrostatic precipitators or baghouses, 2) NO_x — selective catalytic reduction (SCR) deNO_x systems or selective noncatalytic reduction (SNCR) deNO_x systems, and 3) SO_2 — wet or dry flue gas desulfurization (FGD) scrubbing systems or other forms of sorbent injection. These systems are discussed in Chapters 33, 34 and 35. In each case, the flow rate and composition must be measured for the associated gas, liquid and solid streams. In addition, the overall stack emissions are monitored and recorded.

Instrumentation

Primary measurements taken on environmental equipment include pH, liquid level, solids level, solids concentration (density), continuous emission monitoring for regulated pollutants, static and differential pressure, temperature and flow of liquid, flue gas, air and steam.

Chapter 49 discusses the measurement of pressure, temperature and fluid flow.

pH measurement

The acidic or alkaline nature of a liquid stream is monitored by measuring pH or the potential of hydrogen which is expressed as the negative log of the hydrogen ion concentration ($-\log_{10}[H^+]$). This is the most important parameter in controlling wet FGD systems. pH measurement is also important in water softening, acid neutralization, and other auxiliary processes. The alkalinity, or lack of acidity, is generally proportional to the reagent use in the FGD system. Solutions with a pH less than 7 are considered to be acidic; those having a pH greater than 7 are alkaline.

pH measurements are used to control the feed of fresh reagent slurry to the wet FGD system. (See Chapter 35.) A bleed stream of recirculation slurry is passed through one or two pH sensors. The outputs are typically combined with signals for boiler load, SO_2 level, and fresh reagent density and flow to establish the flow of fresh reagent to the FGD absorber module. An increasing pH reduces the fresh reagent feed flow and a decreasing pH increases the flow.

pH is measured by means of electrodes and a voltmeter. The electrode (Fig. 1) is an electrochemical device comparable to a battery whose voltage changes with pH. In combination electrodes, one of the half cells is the measure-

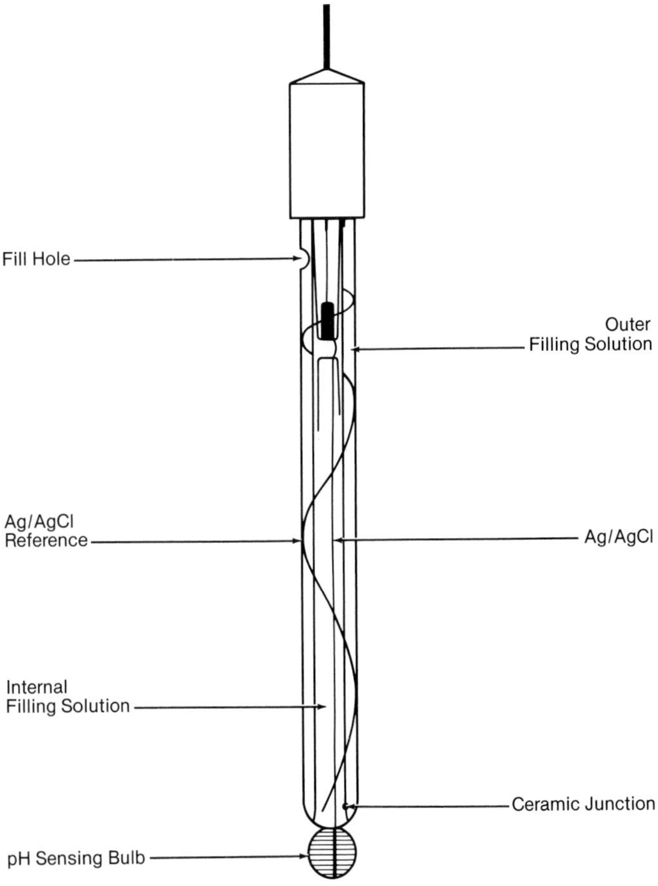

Fig. 1 Combination pH electrode (laboratory version).

ment electrode or pH sensing bulb and the other is the reference electrode or reference junction. Because the measurement electrode reacts to the pH of the solution, its potential is variable though repeatable. Theoretical electrode output varies from +414 mV at a pH of 0 to -414 mV at a pH of 14. The actual output is sensitive to the reference solution concentration at 77F (25C).

The reference electrode completes the electrical circuit and produces a comparative voltage, so changes in the overall sensor are only functions of the measuring electrode. The reference electrode is comprised of a potassium chloride (KCl) salt solution with a controlled amount of silver chloride (AgCl) dissolved in it. An Ag/AgCl electrode is placed into this electrolyte. The reference potential is a function of the KCl and AgCl concentrations. For example, a 1.0 M (molar) KCl electrolyte produces an offset (E_x), or voltage difference from theoretical, of -8 mV, while 3.3 M KCl produces a -45 mV offset (E_x). The difference, which is consistent across the entire range, is compensated for by standardization or zero adjustment.

The pH sensor is also affected by temperature. The slope (mV per pH unit) changes 1 mV per 5C (see Fig. 2). Temperature is compensated for by slope or temperature adjustment or by automatic temperature compensation.

The pH electrode used in wet FGD processes is subject to fouling and aging. This is compounded by the slurry's abrasive or scaling nature which can cause plugging of the liquid junction. This junction is the point where the KCl reference solution meets the solution being measured. The liquid junction can be a small hole or may be made of fiber, wood, porous ceramic, teflon or kynar. It holds the electrolyte in the probe and enables the electrolyte to form a salt bridge with the measured solution.

There are three types of pH sensors: dip, flow-through and insertion. The dip sensor is inserted into the tank and can be removed for maintenance and calibration. The flow-through sensor is placed in a flowing stream of process solution. The insertion sensor is a variation of the flow-through sensor, but does not require sample lines. It is inserted through packing and a valve into the flowing stream. To maintain or calibrate the insertion sensor, it is retracted through the valve, which is then closed to isolate the process.

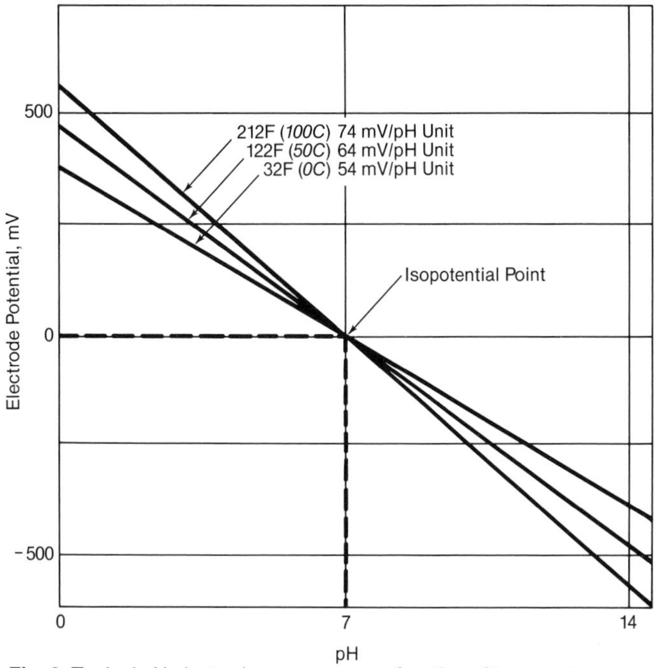

Fig. 2 Typical pH electrode response as a function of temperature.

All sensor types have advantages and disadvantages. The dip sensor is easy to maintain but prone to leakage from scrubber vessels that operate at positive pressure. Sample lines for flow-through sensors can plug. Both flow-through and insertion types are subject to high rates of erosion. All types are subject to scaling, although ultrasonic cleaning devices, which are designed to remove brittle, insoluble coatings, have been used with some success. Ultrasonic cleaners work best if used intermittently, but they can cause probe breakage.

The pH meter is a voltmeter that measures the electrode potential, converts the potential at a given temperature into pH terms, and corrects for nonideal behavior. pH meter operation depends upon proper standardization and calibration. This is best done with stable pH solutions known as buffers. Four of the best buffers, measured at 77F (25C), are pH 4.01, 6.87, 9.18 and 12.45 because they are traceable to the National Bureau of Standards. Two common buffers, pH 7 and 10, are unstable and should be avoided.

Standardization (offset or buffer adjustment) is best done at the isopotential point (pH 7). Standardization adjusts the offset so the meter reads 0.0 mV at pH 7.0. Calibration, or slope adjustment, compensates for alterations in an electrode's response to pH changes. To perform the calibration, the buffer should span the pH of the measured solution. For example, if the expected process pH is 5.4, then a buffer near pH 4 is preferred. The standardization/calibration procedure is summarized below:

1. standardize using a pH 6.87 buffer,
2 calibrate with a buffer that spans the process pH using a different adjustment than that used for standardization (calibration or slope), and
3. use the correct temperature setting, use an automatic temperature compensator or use the same temperature for the two buffers and the process.

Liquid level

Liquid level measurement is needed in any liquid storage tank which is not of the overflow design to control flow out of the tank. Examples in a wet FGD system (Chapter 35) include the tanks for absorber slurry recirculation, reagent storage, thickener underflow and quench recycle. Common liquid level measurement devices are listed in Table 1.

Differential pressure level transmitters are the most common for absorber recirculation tank level control because of their proven reliability. However, they are not the most accurate because of errors due to pressure above the liquid level or due to changes in slurry density.

Other liquid level devices have varying degrees of accuracy due to the tank conditions. For example, most tanks used in FGD systems contain slurry and are therefore agitated, causing waves. The liquid surface level in the absorber tower recirculation tank is also disrupted by: 1) the very high spray flow rates impinging on the surface, 2) foaming due to gas released during the reactions or entrained by the spray, and 3) oxidation air sparged into the tank. Sensors immersed in the slurry also tend to scale. Some level devices use stilling wells to reduce the effect of agitation and to shield the level probe from spray.

Table 1
Common Liquid Level Measurement Devices for Environmental Control Systems

Sensor Type	Theory of Operation	Environmental Control System Applications	Comments
Differential pressure	Measures differential static pressure with sensors or pressure taps mounted on the wall of the vessel. The sensor contacts the fluid.	Slurry tank level. This technique is often used with a ram or a flush system.	Reliable, subject to plugging in slurry application. Two types are used — flange mounted and extended diaphragm.
Ultrasonic	Emits ultrasonic pulse to the liquid surface. The transit time of the return pulse reflected from surface is converted to the liquid level.	Open tanks containing most fluids or slurries.	Sensors do not contact the liquid. Some problems result from waves, dust, vapor, foam, etc. Background noise can often be filtered electronically.
Electrical capacitance	Detects level by measuring the capacitance between the probe and the tank wall.	High/low level alarm. Discrete level capacitance probes can detect liquid surface.	Some problems with solids buildup or scale.
Float	Float moves up and down with level. The float is on an arm inserted into the vessel or within a housing.	Clean water, mill product tanks (nonscaling): high/low level alarm or constant tank volume.	Magnetic float types are available but are not considered good for scaling applications.
Radio frequency (phase tracking)	Detects an electrical signal which passes through a sensor and is reflected back at both the liquid surface and the bottom of the sensor. The signal phase shift provides a continuous indication of level.	Liquid and solids level.	Relatively new, promising technology. The electric field is much larger than the conductors.
Optical	Detects changes in light intensity.	Sludge level in thickeners.	For sludge level, the change in light intensity is measured at a fixed distance from the surface.

Solids level

Many types of sensors used for measuring liquid level are also used for solids level measurement. However, solids level measurement presents some unique problems due to the often dusty environments, the uneven level caused by bridging or normal angles of repose, or the changes in density or properties due to compaction or aeration. For environmental equipment, solids level measurement is used mainly for storage silos (ash, lime, limestone, gypsum) but is also used for hoppers on precipitators or dry scrubbers. Types of silo measurement devices include electrical capacitance, ultrasonic, and radio frequency (phase tracking) sensors. In addition, load cells or strain gauges, which measure the weight change of the silo as it fills or empties, are common. Another type of solids level device is a cable measurement system in which a weight or float is lowered from the top of the bin or silo. When the float reaches the solids surface, loss of cable tension is detected. The device then measures the distance traveled by counting electronic pulses as the float retracts. The measured level is therefore the silo maximum level less the distance traveled by the probe. This device should not be used when the silo is being filled.

A typical device used to determine high hopper level is a nuclear transmitter. When the ash level rises above the sensor, the strength of the signal from the nuclear source to the receiver is decreased. This signal is converted to a level indication.

Another high/low solids level device is the paddle wheel, which includes a motor that turns a paddle. When material is present, rotation of the paddle stops. The increased torque of the motor as it attempts to turn the stopped paddle triggers a microswitch alarm.

Finally, vibrating probes are sometimes used for single point solids level detection. The vibration of a piezoelectric transducer-driven probe inserted into the vessel is dampened by contact with material. This is detected as a voltage change, which is electronically amplified to yield a level indication.

Solids concentration (density)

The most commonly used devices for density measurement are nuclear absorption meters (see Fig. 3.) These devices measure the absorption of gamma rays from a radioactive source, and the degree of absorption is proportional to the solids content. Nuclear density meters do not contact the slurry; instead, they are installed on the outside of a slurry pipe. Some disadvantages of nuclear density meters are:

1. The signal is not linear with solids content unless a linearizer is used. If accurate chemical analysis of the slurry is available, the device can be precalibrated.
2. A nuclear license is required in the United States (U.S.) and in most other countries.
3. The device does not distinguish between suspended and dissolved solids.
4. Errors can result due to the buildup of solids or scale in the pipe.

Nuclear density meters are typically low maintenance

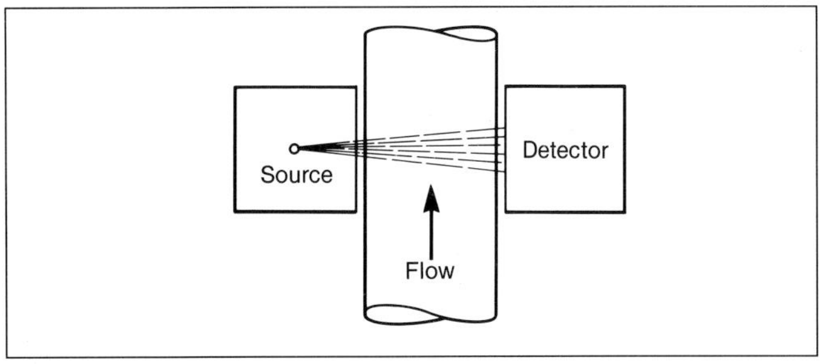

Fig. 3 Nuclear absorption meter for density measurement.

devices. The inaccuracies mentioned above are routinely verified by manual sampling and measurement.

Other devices used for density measurement include differential pressure and ultrasonic devices and vibrating reed instruments. The reed instruments electronically convert the slurry dampening effect on electrically-driven coil vibrations to a density measurement.

Pressure, flow and temperature

Pressure, pressure differential, flow and temperature measurements are common in environmental system applications. They indicate process performance, energy consumption, operational problems and ability to meet design or operating requirements. Theory, instrumentation and application for these devices are discussed in detail in Chapter 40. Selected uses in environmental systems are discussed in Chapters 33 through 35. Comments with particular importance in environmental systems include:

1. Slurries tend to be highly erosive and potentially corrosive, and provision must be made for appropriate environments.
2. The presence of solids in liquid or gas flows can lead to plugging of measuring devices. Purging systems are frequently included to ensure long term operation.
3. Temperature monitoring can be particularly important for freeze protection, crystallization prevention, thermal gradient prevention (thickeners and clarifiers), corrosion prevention (precipitators, fabric filters, dry FGD systems and flues/ducts) and general process control. Elastomeric linings and fiberglass or plastic components must be protected from high temperatures. Unlined stacks must be protected from low temperatures.

Continuous emission monitoring systems

Early power plant emission measurements focused on controlling combustion. The Orsat gas analyzer, which sampled oxygen (O_2), carbon monoxide (CO) and carbon dioxide (CO_2) flue gas concentrations, was used extensively. However, this analyzer is not a continuous monitor.

Largely due to increasing regulation, power plants have been forced to institute continuous emission measurements. In addition, continuous monitoring devices permit process control of the emissions control equipment.

During the 1980s, CEMS were used primarily to measure particulate (opacity), SO_2, NO_x and CO_2. By the early 1990s, CEMS to monitor hydrogen chloride (HCl), CO, ammonia and volatile organic compounds (VOC) were introduced for commercial combustion applications.

The common methods for continuously measuring flue gas constituents are described in Table 2 and are further detailed in Table 3. All of the analyzers are similar in that they use a measuring cell and a reference cell. They are set to zero with air or a calibration standard, the span is set with a calibration standard, and a sample is taken. The result is obtained by comparing the voltages detected across the sample, the zero standard and span standard.

There are three types of analyzers — extraction, dilution extraction and in situ.

In the extraction method (Fig. 4) a sample of flue gas is withdrawn from the process stream and directed to an external analyzer for the constituent concentration determination. This type of analyzer is typically used where there are short distances between the sample point and the analyzer. The probe and sample line are often heated to prevent condensation. Because the gas sample is typically drawn by vacuum to the external analyzer, sample line leaks can compromise the integrity of the sample.

Table 2
CEM Technologies

Technology	Operating Characteristics
Infrared radiation (IR)	An infrared beam passes through a measurement filter and is absorbed by the constitutent gas. A light detector creates a signal which is used to monitor concentrations.
Ultraviolet absorption (UV)	A split beam, with optical filters, phototubes and amplifiers, measures the difference in light beam absorption between the reference and sample.
Chemiluminescence	Ozone (O_3) is injected into the sample to react with NO_x, generating light that is measured by a photocell.
Flame ionization detection	Hydrocarbons are ionized with strong light. The signals are received by the flame ionization detector.
Transmissometer	Light is passed through the stack where it is reflected by a mirror on the opposite side. The quantity of light returning is proportional to particulate matter and aerosols in the flue gas. (See Fig. 5.)
Electrochemical cells	The voltage measured when an O_2 sample is injected into a solution with a strong base is compared to a reference voltage.
Chromatography	A sample, zero and calibration gas are eluted through a column. The output as measured by flame photometric or thermal conductivity detectors is compared.

Table 3
Continuous Emission Monitoring
(Reference Table 2 for Technologies)

Constituent	CEM Technology
Particulate matter (opacity)	Transmissometer, beta ray absorption
Sulfur dioxide (SO_2)	UV absorption, IR pulsed fluorescence
Nitrogen oxides (NO_x)	Chemiluminescence, UV spectroscopy, IR
Hydrogen chloride (HCl)	IR with gas filter
Carbon monoxide (CO)	IR
Carbon dioxide (CO_2)	IR
Oxygen (O_2)	Electrochemical cell
Volatile organic compounds (VOC)	Flame ionization detection
Other organic air toxics	Chromatography
Ammonia (NH_3)	Same as NO_x*

* NH_3 is converted to NO_x in one of two split streams. Both streams are analyzed for NO_x. NH_3 is determined as the difference.

The dilution-extraction method uses a carrier/dilution media, typically instrument air, which is supplied to the extraction probe in the flue gas process stream. Problems associated with dirty, wet and corrosive flue gases are drastically reduced by the 50 to 200 dilution ratio provided by the sample probe. Precise dilution is assured by accurately controlling the dilution air pressure which yields a constant air flow. The air flow creates a vacuum on the downstream side of a critical sample orifice. The gas sample rate, limited by sonic flow through the orifice, is therefore held constant. Furthermore, because the sample lines from the probe to the analyzer are under positive pressure, minor sample line leakage does not affect the integrity of the sample concentration as it would in a vacuum system. This method has gained wide acceptance in utility and other systems where longer sample lines are required.

In situ analyzers, such as that shown in Fig. 5, provide measurement of flue gas concentrations directly in the process stream. Several types of detection devices are available, and all are based on absorption spectroscopy. Since gas stream particulates can reduce the level of light transmission, ceramic filters are used to exclude these particles. A single in situ probe can measure several gas species and opacity. Problems with in situ analyzers generally occur in the optic or electronic components. Because a separate analyzer is required for each sample point, installations which require multiple sample points frequently find the extractive systems more cost effective.

Continuous analyzers generally require flue gas conditioning. Sample conditioning includes filtration to remove particulate (see Fig. 6), chilling to remove water, heating to maintain process temperature, and dilution (used with extractive monitor and in situ probe).

The need for conditioning coupled with zero and span calibration make extractive systems complicated, costly and vulnerable to plugging, corrosion and leaking. These

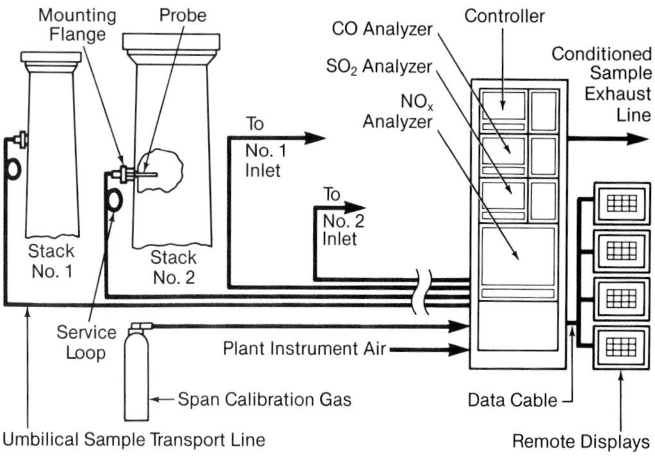

Fig. 4 Extractive stack/inlet CEM system.

systems are also slow to respond due to the time lapse between sampling and analysis. In situ systems use conditioning but do not rely on long sample lines. These systems have faster response, fewer components and better availability than extractive systems. However, they are prone to interference and fouling and are less accessible for service. Extractive systems can be expanded by adding sample points as shown in Fig. 4. It is important to consider the location and ambient conditions in which CEMS equipment is placed. Many of the problems associated with CEMS can be traced to dirty conditions, excessive heat or cold, vibration, humidity, etc.

Monitoring process performance

Environmental control systems (ECS) involve physical and chemical processes. Chemical methods are used to maintain desired removal or emission levels, avoid operating problems, improve reagent utilization, improve process efficiency, troubleshoot equipment, and monitor process changes and catalyst life.

In an FGD system, the quality of the reagent and the recirculating slurry stream is monitored. Analysis is more critical for processes which produce a usable byproduct such as gypsum. In this case, the analysis extends into product purity and feed stream impurities. The quality of the reagent is normally measured extensively when

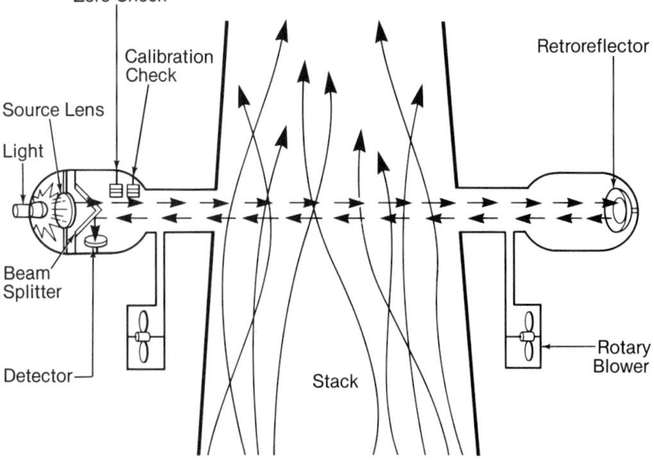

Fig. 5 In situ transmissometer system.

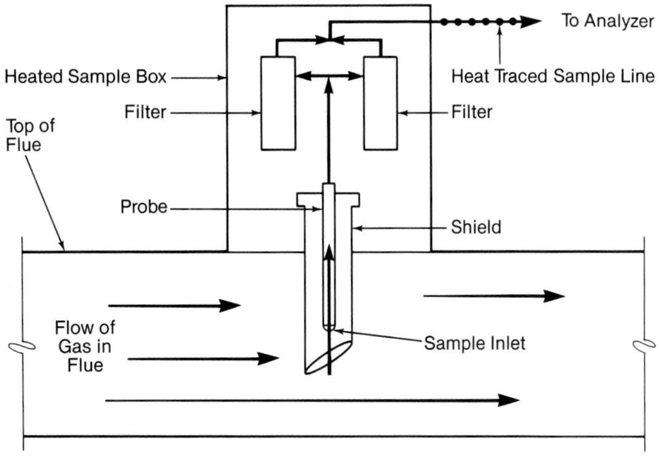

Fig. 6 Extractive sample probe with dust conditioning.

the reagent source is selected. Later it is evaluated if the source is changed or process problems occur.

Tables 4 through 7 list the important analyses for common wet scrubber FGD systems. Additional analyses are required for makeup water streams, sodium-based FGD systems, dual loop systems, regenerative systems, ex situ oxidation processes, water treatment, and sulfuric acid or SO_2 recovery processes. However, the major species and methods would be similar to those presented in Tables 4 through 7; only the concentrations would vary. Other procedures used for design or troubleshooting include relative saturation, which is determined by seed crystal procedures, and ion balance (balance of anions and cations), which is a check of the analysis accuracy.

Performance testing

As with any system, performance tests are conducted on environmental control systems to determine if the units are operating properly. A typical new environmental control system might include low NO_x burners, wet FGD scrubber modules, gypsum processing equipment, and an electrostatic precipitator. An SCR deNO_x system could also be added. The ten most common parameters measured to determine performance for such a system include:

1. system pressure drop,
2. particulate removal efficiency,
3. stack opacity,
4. NO_x emission concentration,
5. SO_2 removal efficiency,
6. makeup water requirements,
7. reagent consumption,
8. gypsum purity,
9. mist eliminator carryover, and
10. electric power consumption.

Other performance parameters that would be measured depend on the type of system. Performance testing is usually extensive. It is common for a new ECS to have two performance tests. The first test is scheduled soon after installation and the final test is performed after one year of operation.

Sample location selection

The first step in the performance test is often done before the ECS is built — selecting the sample locations. Outlet sampling is not difficult if it is done in the stack. However, the inlet and intermediate samples are more difficult to obtain. In Fig. 7, the system inlet at the far left is a section of flue followed by an immediate upward bend. The precipitator outlet is at the induced draft (ID) fan inlet. Fig. 7 also shows test ports in the downward flue section after the absorber tower and at the stack entrance. Test port locations are selected to give the best flow profile within the plant layout limitations.

The number and location of traverse points are selected by Environmental Protection Agency (EPA) Method 1. Application of Method 1 is illustrated in Fig. 8. The distance downstream from the flow disturbance,

Table 4
Lime Quality Analysis

Parameter	Suggested Method*	Suggested Frequency	Use of Data
Slaking rate	ASTM Method C-110-76a	Duplicate analysis on all candidate limes	Temperature rise rate is related to reactivity and lime quality.
Grit	Acid dissolution	Duplicate analysis on all candidate limes	High grit levels indicate poor lime quality.
Weight loss on ignition	Gravimetric	Duplicate analysis on all candidate limes	Weight loss is an indirect measure of lime quality via carbonate and hydroxide content.
Available lime index	Rapid sugar test	Duplicate analysis on all candidate limes	Lime quality
Calcium/ magnesium	Atomic absorption (AA) EDTA titration or Ion Chromatograph (IC)	Duplicate analysis on on all candidate limes then as needed to check reagent quality	Analysis is required for processes producing a usable byproduct or high magnesium lime systems.
Lime slurry solids content, % by wt	Gravimetric	Daily	Check density meter calibration and indicate slaker performance.

* Many methods have been standardized through the American Society of Testing and Materials, Philadelphia, Pennsylvania.

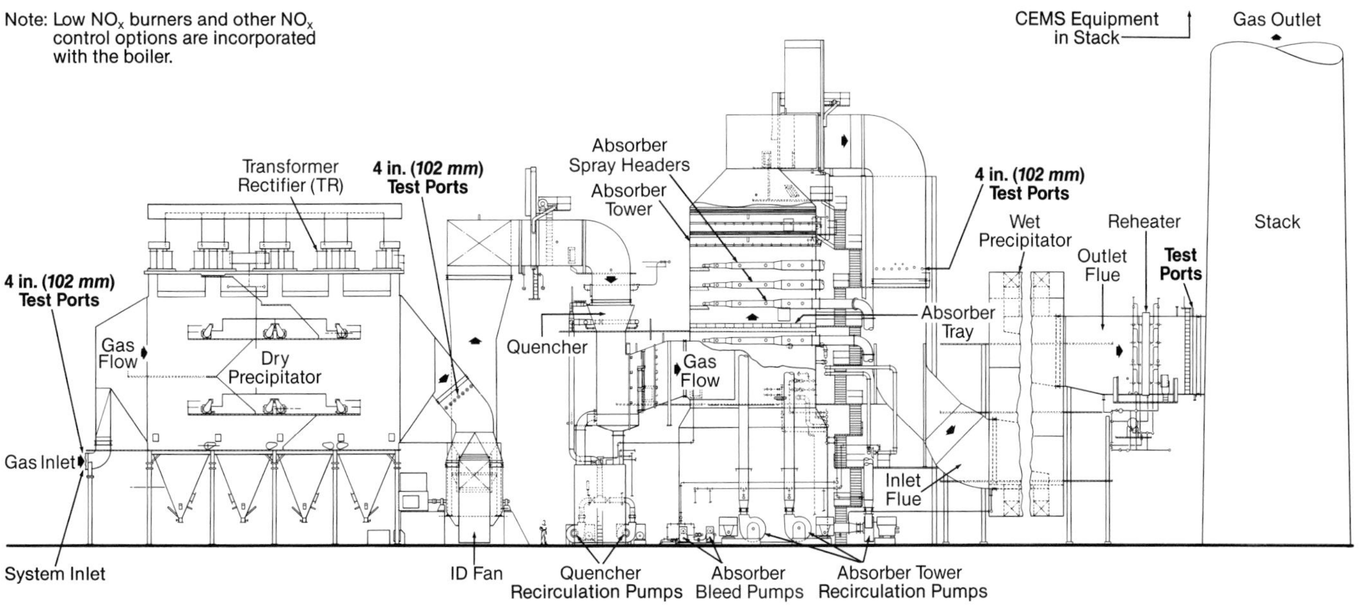

Fig. 7 Typical test port locations indicating compromises made in measurement due to equipment arrangement.

which in this case is the stack inlet, is 138 ft (42.1 m). This is more than eight diameters (138/16 = 8.6). The upstream distance to the stack exit is 5 diameters. Therefore, 12 traverse points are required by Method 1. For rectangular flues, Method 1 is applied by first determining an effective or equivalent diameter which is calculated as follows:

$$DC = \frac{2LW}{(L+W)} \tag{1}$$

where

DC = equivalent diameter
L = length
W = width

Then a matrix layout is used for sample location selection. The cross-section is divided into as many equal rectangular areas as there are traverse points, and the sample is taken at the centroid of each area.

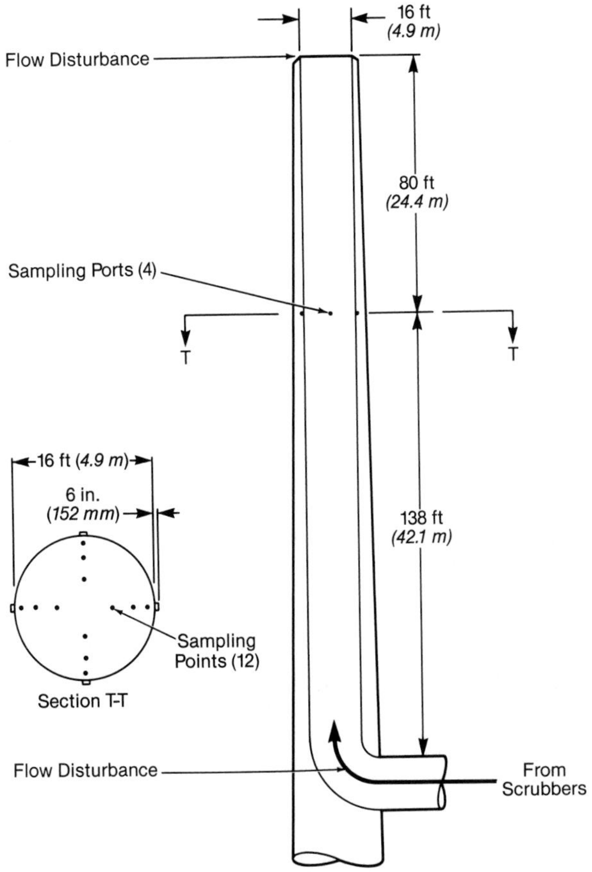

Fig. 8 Stack dimensions and sampling port locations for EPA Method 1.

Gas side measurements

For a performance test of an ECS, a combination of EPA Methods 5 and 8 is the most often used procedure to simultaneously measure particulate, SO_2, and SO_3 or acid mist (H_2SO_4) levels. The Method 5 sampling train, shown in Fig. 9, is used with Method 8 impingers shown in this figure. The data obtained from this traverse also represent Methods 1 through 4. Results include gas velocity by S-type pitot tube (Fig. 10) and CO_2, O_2, moisture, SO_2, particulate, and SO_3 or H_2SO_4 levels. The following calculations are used:

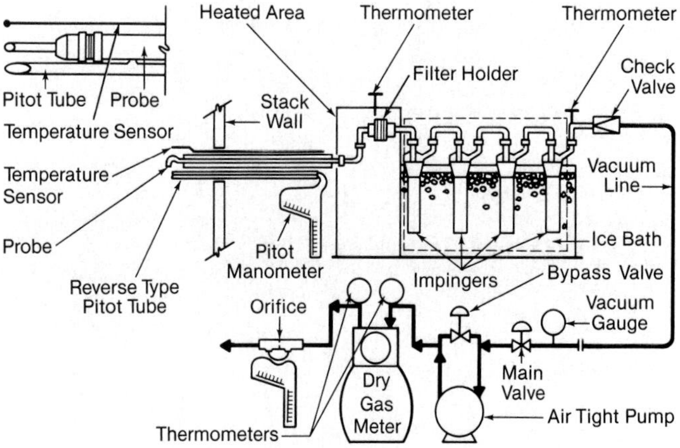

Fig. 9 EPA Method 5 particulate and SO_2 emissions sampling train.

Determination of moisture content in stack gas

$$B_w = \frac{(MWC)\ (1.339)}{[(MWC)\ (1.339)] + \frac{(DGV)\ (P_m)}{T_m}\ (499.9)} \qquad \textbf{(2)}$$

where

B_w = moisture fraction
MWC = impinger weight gain, g
DGV = sampled dry gas volume at meter conditions, ft^3
P_m = pressure at meter, in. Hg
T_m = temperature at meter, R
1.339 = constant for converting grams of H_2O to liters of H_2O vapor at standard conditions, l/g
499.9 = conversion of cubic feet at meter conditions to liters at standard conditions, l R/ft^3/in. Hg

Determination of flue gas molecular weight — dry basis

$$M_d = [(44)(\%\ CO_2\ dry) + (32)(\%\ O_2\ dry) + (28)(\%\ N_2\ dry)] \times 0.01 \qquad \textbf{(3)}$$

where

M_d = molecular weight of gas on dry basis, lb/lb-mole
44 = molecular weight of CO_2, lb/lb-mole
32 = molecular weight of O_2, lb/lb-mole
28 = molecular weight of N_2, lb/lb-mole

Determination of flue gas molecular weight — wet basis

$$M_w = M_d(1 - B_w) + 18B_w \qquad \textbf{(4)}$$

where

M_w = molecular weight of gas on wet basis, lb/lb-mole
M_d = molecular weight of gas on dry basis, lb/lb-mole
B_w = moisture fraction, lb-mole H_2O/lb-mole wet gas
18 = molecular weight of H_2O, lb/lb-mole

Determination of gas humidity

$$H = \frac{18B_w}{M_d(1 - B_w)} \qquad \textbf{(5)}$$

where

H = gas humidity, lb H_2O/lb dry gas
B_w = moisture fraction, lb-mole H_2O/lb-mole wet gas
M_d = molecular weight of dry gas, lb/lb-mole
18 = molecular weight of H_2O, lb/lb-mole

Determination of stack gas velocity

$$V_s = 85.48\ C_p \left(\frac{T_s\ \Delta P}{P_s\ M_w} \right)^{1/2} \qquad \textbf{(6)}$$

where

85.48 = pitot tube constant in the following units:

$$\frac{ft}{s} \left(\frac{(lb / lb\text{-}mole)\ (in.\ Hg)}{(R)\ (in.\ H_2O)} \right)^{1/2}$$

V_s = velocity of the flue gas, ft/s
ΔP = average differential pressure measured by S-type pitot tube, in. H_2O
T_s = average gas temperature in duct, R
P_s = absolute duct pressure, in. Hg

Table 5
Limestone Quality Analysis

Parameter	Suggested Method*	Suggested Frequency	Use of Data
Reactivity	pH stat test followed by particle size measurement	Duplicate analysis of candidate limestones	Compared to reference to determine acceptability.
Grindability (Bond Work Index)	Laboratory ball mill test	Duplicate analysis of candidate limestones	Limited by specification of wet milling system design.
Particle size of limestone as delivered	Sieve method	Duplicate analysis of candidate limestones then monitor monthly	Limited by specification of wet milling system design.
Inerts	Acid dissolution	Duplicate analysis of candidate limestones then monitor weekly for gypsum systems	Quality indicator; important for gypsum systems.
Carbonate	CO_2 evolution, $Ba(OH)_2$ absorption, alkalinity titration	Duplicate analysis of candidate limestones then monitor weekly for gypsum systems	Check reagent quality; important for gypsum systems.
Calcium and magnesium	AA, IR, EDTA titration	Duplicate analysis of candidate limestones then monitor weekly for gypsum systems	Check reagent quality.
Limestone slurry solids content, % by wt	Gravimetric	Daily	Check density meter and indicate problems in milling circuit.
Limestone slurry particle size	Wet sieve	Daily	Troubleshooting problems in milling circuit or process such as low utilization or poor gypsum quality.

* Many methods have been standardized through the American Society of Testing and Materials, Philadelphia, Pennsylvania.

Table 6
Recirculation Slurry/Blowdown Analysis

Parameter	Suggested Method*	Suggested Frequency	Use of Data
pH	2-buffer method calibration or grab sample	Once per shift by grab sample, daily by buffer method	Troubleshoot pH meter and maintain process efficiency.
Slurry suspended solids content	Gravimetric	Weekly	Calibrate density meter and maintain process efficiency.
Cations: Ca^{++} and Mg^{++} primary; K^{+} and Na^{+} secondary	IC, AA or wet chemistry methods	Daily for Ca^{++} and Mg^{++} in calcium-based systems	To calculate stoichiometry and monitor process chemistry.
Anions: $SO_3^{=}$, $CO_3^{=}$, $SO_4^{=}$ and Cl^{-} primary; F^{-} and NO_3^{-} secondary	IC or wet chemistry methods	Daily for $SO_3^{=}$, $CO_3^{=}$, $SO_4^{=}$, Cl^{-}	To calculate stoichiometry, measure chloride concentration, and monitor process performance and gypsum production.
Dissolved metals: Fe, Mn, Al, etc.	AA	Infrequently	Process troubleshooting.
Additives: adipic acid, dibasic acid, formate, thiosulfate, scale inhibitors, etc.	IC or wet chemistry methods	Daily	Determine or correct dosage.
TDS (total dissolved solids)	Gravimetric	Weekly	Calibrate density meter for suspended solids; important for mag-lime, high chloride or suspended solids.
Dissolved alkalinity	Acid-base titration	Weekly	Measure alkali loss in filter cake of mag-lime process; important for mag-lime or sodium systems.
Settling tests	Graduated cylinder	Weekly	Predict thickener performance and assist process troubleshooting.
Filter leaf test	Filter and weight	During startup	Predict filter cake properties.
Particle size distribution	Coulter counter	During startup	Determine settling or dewatering characteristics; important for gypsum systems.

* Many methods have been standardized through the American Society of Testing and Materials, Philadelphia, Pennsylvania.

M_w = wet gas molecular weight, lb/lb-mole
C_p = S-type pitot tube correction factor (normally 0.84)

Volumetric flow rate — actual conditions

$$Q_{ac} = 60\, V_s A \qquad (7)$$

where

Q_{ac} = actual volumetric flow rate, ACFM
V_s = velocity of the flue gas, ft/s
A = cross-sectional area of the duct, ft^2
60 = conversion factor, s/min

Volumetric flow rate — dry standard conditions

$$Q_{sd} = \frac{528\ Q_{ac} P_s \ (1 - B_w)}{29.92\ T_s} \qquad (8)$$

where

Q_{sd} = dry volumetric flow rate at standard conditions, DSCFM
Q_{ac} = actual volumetric flow rate, ACFM
B_w = moisture fraction, lb-mole H_2O/lb-mole wet gas
528 = standard temperature, R
T_s = average gas temperature in duct, R
29.92 = standard pressure, in. Hg
P_s = absolute duct pressure, in. Hg

Total particulate and H_2SO_4 — grains/ACF

$$C_a = \frac{(0.001)\ (15.43)\ (528)\ (1 - B_w)\ P_s M_n}{29.92\ V_d\ T_s} \qquad (9)$$

where

C_a = concentration of specie in the flue gas at duct conditions, grains/ACF
M_n = mass of specie collected, mg
V_d = dry gas volume sampled at standard conditions, DSCF
T_s = average gas temperature in duct, R
P_s = absolute duct pressure, in. Hg
B_w = moisture fraction of gas, lb-mole H_2O/lb-mole wet gas
0.001 = conversion of milligrams to grams, g/mg
15.43 = conversion of grams to grains, grains/g
528 = standard temperature, R
29.92 = standard pressure, in. Hg

Table 7
Dewatering System Analysis

Parameter	Suggested Method*	Suggested Frequency	Use of Data
Thickener or hydroclone underflow solids, %	Gravimetric	Daily	Troubleshoot process and calibrated lime density meters
Clarified water, water analysis (anions, cations, suspended solids)	See Table 6	Weekly	Monitor thickener performance
Supernatent, filtrant solids content, % by wt	Gravimetric	Daily	Monitor filter cloth status
Gypsum Quality Analysis			
Ca^{++}, Mg^{++}, Na^{+}, K^{+}, Fe_2O_3, R_2O_3, SiO_2	AA	Weekly	Product purity and properties
Cl^-, $SO_4^=$	IC	Daily	Product purity and properties
$SO_3^=$, $CaSO_3 \cdot ½\ H_2O$	Iodimetric titration	Daily	Product purity and properties
Moisture content	Gravimetric	Daily	Important for salable gypsum and monitoring dewatering system performance.
Surface area	BET analysis (sub-sieve sizer)	Startup or process change	N_2 gas is condensed on surface and absorbed; N_2 corresponds to surface area.
Combined water	Heat to 482F (250C)	Weekly	Gypsum purity
Mean particle size	Coulter counter, Sedigraph plus sieve analysis	Startup or process change	Product purity and properties
Aspect ratio	Scanning electron microscope	Startup or process change as required by end users	Product purity and properties
Total water soluble salts	Sum ($Na^+ + K^+ + Mg^{++}$)	As needed	Product purity and properties
Flyash	Determined as acid insoluble	Daily	Determined as acid insolubles along with silica (SiO_2) and other impurities.
pH	Electrode	Daily	Product purity and properties

* Many methods have been standardized through the American Society of Testing and Materials, Philadelphia, Pennsylvania.

Total particulate, H_2SO_4, condensible total particulate and condensible H_2SO_4

$$C_s = \frac{(0.001)\ (15.43)\ M_n}{V_d} \tag{10}$$

where

C_s = concentration of specie in the flue gas at dry conditions, grains/DSCF
M_n = mass of specie collected, mg
V_d = dry gas volume sampled at standard conditions, DSCF
0.001 = conversion of milligrams to grams, g/mg
15.43 = conversion of grams to grains, grains/g

Total particulate, H_2SO_4, NO_x and SO_2 — lb/h

$$C_m = C_s (1.428 \times 10^{-4})(60)(Q_{sd}) \tag{11}$$

where

C_m = mass flow rate of specie in the flue gas, lb/h
C_s = concentration of specie in the flue gas at dry standard conditions, grains/DSCF
Q_{sd} = dry volumetric flow rate at standard conditions, DSCFM
60 = conversion of minutes to hours, min/h
1.428×10^{-4} = conversion of grains to pounds, lb/grain

Total particulate, H_2SO_4, NO_x and SO_2

$$E = \frac{20.9\ C_m F_d}{60\ Q_{sd}\ (20.9 - \%\,O_{2ds})} \tag{12}$$

where

E = emission rate on a dry basis at 0% O_2, lb/10^6 Btu
C_m = mass flow rate of specie in the flue gas, lb/h
Q_{sd} = dry volumetric gas flow rate at standard conditions, DSCFM
% O_{2ds} = O_2 in the gas on a dry basis, %
F_d = dry F-factor at 0% O_2, determined from fuel analysis, DSCF/10^6 Btu (see Equation 17)
20.9 = O_2 in air, used with % O_2 in the gas to correct to 0% O_2, %
60 = conversion of minutes to hours, min/h

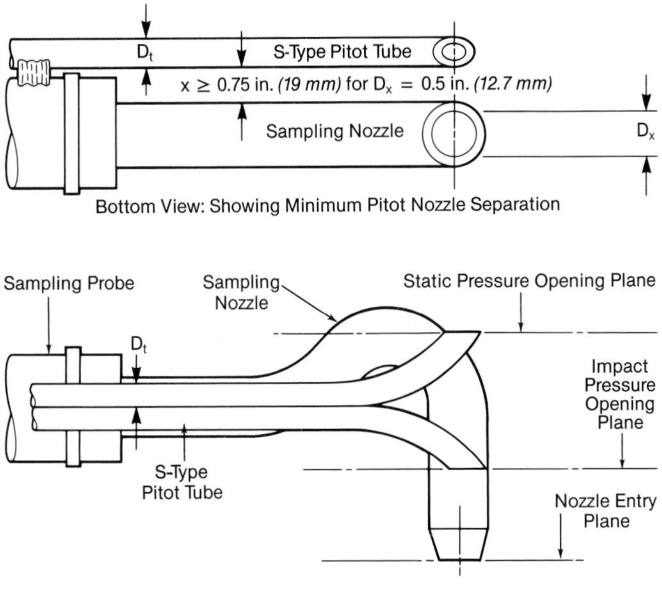

Fig. 10 S-type pitot tube with sampling probe.

SO_2 — parts per million (ppm) wet

$$C_{ppm,w} = \frac{(8.48\times10^{-4})\ (10^6)\ V_s C_i\ (1-B_w)}{V_d} \qquad \textbf{(13)}$$

where

$C_{ppm,w}$ = gas SO_2 concentration on a wet basis, ppm wet
V_s = total volume of H_2O_2 impinger solution
C_i = SO_2 concentration in the impinger solution, mg/liter
V_d = gas volume sampled at dry standard conditions, DSCF
B_w = gas moisture fraction, lb-mole H_2O/lb-mole wet gas
8.48×10^{-4} = conversion of millimoles to standard cubic feet, SCF/mM
10^6 = conversion from fraction to ppm

SO_2 — ppm dry

$$C_{ppm,d} = \frac{20.9\ C_{ppm,w}}{(1-B_w)\ (20.9-\%\,O_{2ds})} \qquad \textbf{(14)}$$

where

$C_{ppm,d}$ = gas specie (SO_2) concentration on a dry basis, corrected to 0% O_2, ppm dry
$C_{ppm,w}$ = gas specie (SO_2) concentration on a wet basis, ppm wet
B_w = moisture fraction of the gas, lb-mole H_2O/lb-mole wet gas
$\%\,O_{2ds}$ = O_2 in the gas on a dry basis, %
20.9 = O_2 in air, used with % O_2 in the gas to correct to 0% O_2, %

NO_x — ppm dry and ppm wet

NO_x is evaluated directly by monitoring systems as parts per million, dry basis. It can be converted to ppm wet if needed by solving Equation 14 for $C_{ppm,w}$.

Collection/removal efficiency — total particulate, H_2SO_4 and SO_2

$$C_{eff} = \left[(C_{in} - C_{out}) / C_{in}\right] \times 100 \qquad \textbf{(15)}$$

where

C_{eff} = collection/removal efficiency, %
C_{in} = concentration of matter at inlet, lb/10^6 Btu
C_{out} = concentration of matter at stack, lb/10^6 Btu

Percent isokinetic

$$\%\,I = (V_n \,/\, V_s) \times 100 \qquad \textbf{(16)}$$

where

$$V_n = \frac{V_m T_s P_m}{60\ t A_n T_m\ P_s (1-B_w)}$$

and where

$\%\,I$ = % isokinetic
V_s = average flue gas velocity in duct during sampling, ft/s
V_n = average sample gas velocity through the nozzle during sampling, ft/s
V_m = dry gas volume sampled at meter conditions, ft^3
t = elapsed sampling time, min
A_n = cross-sectional area of nozzles, ft^2
B_w = gas moisture fraction, lb-mole H_2O/lb-mole wet gas
T_s = flue gas temperature in the duct, R
T_m = average gas temperature at meter, R
P_m = absolute pressure at the meter, in. Hg
P_s = absolute pressure in the duct, in. Hg
60 = conversion from minutes to seconds, min/s

F-factor

The F-factor used in making emissions rate calculations was developed from chemical and combustion analyses of the fuel. This factor is the ratio of the theoretical volume of dry gases at 0% excess air (O_2) given off by complete combustion of a known amount of fuel, to the gross caloric value of the burned fuel. The data generated from the ultimate analysis of the fuel are used to calculate the F-factor as follows:

$$F_d = (10^6)\,\frac{3.64(\mathrm{H}) + 1.53(\mathrm{C}) + 0.57(\mathrm{S}) + 0.14(\mathrm{N}) - 0.46(\mathrm{O})}{HHV} \qquad \textbf{(17)}$$

where

F_d = dry F-factor at 0% O_2, DSCF/10^6 Btu
HHV = higher heating value, Btu/lb
H = hydrogen in the fuel, % by wt
C = carbon in the fuel, % by wt
S = sulfur in the fuel, % by wt
N = nitrogen in the fuel, % by wt
O = oxygen in the fuel, % by wt

F-factors are constant for a given fuel category (within ± 3%). Average values are:

Fuel	DSCF/10^6 Btu
Bituminous coal	9820
Oil	9220
Gas	8740
Wood bark	9640
Wood chips	9280

Particulate and SO_2 removal efficiencies

The SO_2 and particulate removal efficiencies are determined by inlet and outlet sampling using Equations 12 and 15.

Stack opacity

Stack opacity is determined by the installed transmissometer or by EPA Method 9.

NO_x concentration

NO_x concentration is determined by EPA Method 7. This measurement is taken in the stack on a composite sample. NO_x concentration may be adjusted to design conditions from actual test conditions.

Makeup water requirements

Makeup water requirements are measured at the appropriate terminal point with a certified flow meter. Adjustments from test conditions may be necessary due to variations in gas volume or temperature or other process deviations from design conditions.

Reagent consumption

There are two common ways to measure reagent consumption. The first method is to measure the solid reagent flow (for lime/limestone systems) using a calibrated weight belt feeder. The measurement is time averaged. This mass flow method is cumbersome because the initial and final storage tank volumes and densities must also be measured. This method is best done over a period of at least 24 hours. The second and preferred method is chemical analysis. The reagent consumption is calculated from the stoichiometric ratio, limestone (or reagent) analysis, and gas side measurements by the following equation:

$$LS = \frac{100\ (SR)\ (SO_2)}{64\ (CaCO_3)} \tag{18}$$

where

LS	=	limestone consumption, lb/h
SR	=	stoichiometric ratio (see Chapter 35)
SO_2	=	SO_2 removed, lb/h
64	=	molecular weight of SO_2, lb/lb-mole
100	=	molecular weight of $CaCO_3$, lb/lb-mole
$CaCO_3$	=	available $CaCO_3$ in limestone (wt fraction: lb available $CaCO_3$/lb limestone)

Reagent consumption may be adjusted for deviations from SO_2 removal rate and mass flow (or concentration) design conditions.

Gypsum purity

A composite sample of gypsum, taken over the test period, is analyzed as outlined in Table 7. Adjustments may be necessary for flyash contamination, chloride concentration, and inerts if the inlet levels to the FGD system substantially deviate from design.

Other test methods

There are numerous procedures for measuring ECS parameters. The example in this chapter is a formal performance test; other methods could be used. For example, SO_2 can be measured with a CEMS or by other temporary instrument methods. Particulate level is measured by EPA Method 5 or 17 or alternate methods such as the probe shown in Fig. 11.

The future of environmental measurement

Advances in environmental measurement techniques will certainly continue. Technologies such as fiber optics, video imaging and Fourier transform infrared analysis (FTIR), which have been used successfully in the laboratory, are expected to move into the plant. Measurement will be extended to include other air and water pollutants, and air toxics monitoring will become increasingly important.

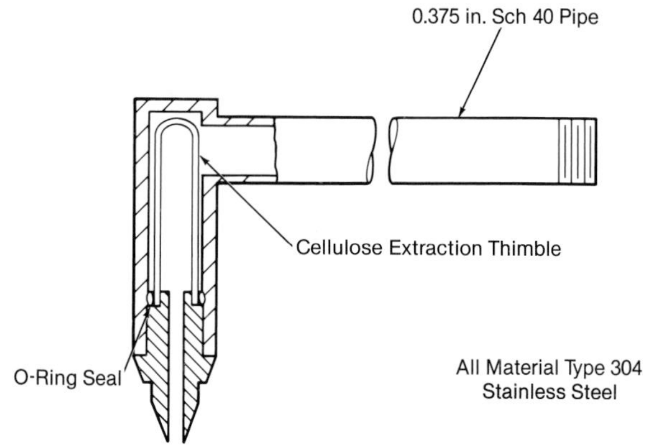

Fig. 11 Probe detail, isokinetic dust sampling.

Bibliography

ASME Performance Test Codes: PTC 1 on General Instructions, PTC 2 on Definitions and Values, and PTC 3.2 on Solid Fuels.

ASME PTC Supplements on Instruments and Apparatus: PTC 19.2 on Pressure Measurement, PTC 19.3 on Temperature Measurement, PTC 19.6 on Electrical Measurements (IEEE Standard 120), and PTC 19.10 on Flue and Exhaust Gas Analyses.

United States Code of Federal Regulations, CFR 40, parts 53 to 80.

Continuous Emission Monitoring Guidelines: Update, Report CS-5998, Electric Power Research Institute, Palo Alto, California, September 1988.

FGD Chemistry and Analytical Methods Handbook, Report CS-3612, Vols. 1 and 2, Electric Power Research Institute, Palo Alto, California, July 1984.

Handbook of Electrode Technology, Orion Research, Cambridge, Massachusetts, 1982.

Liptak, B.G., ed., *Instrument Engineers Handbook*, Chilton, Radnor, Pennsylvania, 1982.

Performance Test Code Committee No. 40, Flue Gas Desulfurization Units, Draft VII, The American Society of Mechanical Engineers, New York, October 1989.

Elliot, T.C., "Monitoring pollution," *Power*, pp. 51-56, September 1986.

Elliot, T.C., "Level monitoring," *Power*, pp. 41-58, September 1990.

Gilbert, G., "Three ways to measure flow in power plant ducts," *Power*, pp. 63-72, June 1987.

Kiser, J.V.L., "Continuous emissions monitoring: a primer," *Waste Age*, pp. 64-68, May 1988.

Kiser, J.V.L., "More on continuous emissions monitoring," *Waste Age*, pp. 119-122, June 1988.

FGD Manual, The McIlvaine Company, Northbrook, Illinois, August 1989.

Robertson, R.L., "Continuous emission monitoring and quality assurance requirements for new power plants," appearing in *Continuous Emission Monitoring Systems for Power Plants — The State-of-the-Art*, The American Society of Mechanical Engineers, pp. 13-15, New York, 1988.

Rothstein, F., and Fisher, J.E., "pH measurement: the meter," *American Laboratory*, September 1985.

Spriggs, D., "Calibration and troubleshooting of pH loops," presented to American Water Works Association Ohio Section Northeast District Meeting, Painesville, Ohio, July 20, 1989.

Section V
Specification, Manufacturing and Construction

This section begins with an in-depth discussion of the specification, evaluation and procurement process for large capital expense items. This includes discussions about project scope, terms and conditions, and general bid evaluation. Also discussed are power system economics, and procedures for the evaluation of equipment characteristics in terms of justifiable expenditures. Worked examples are included for both utility and industrial units.

This is followed by a discussion of the manufacturing processes for fossil fuel-fired equipment. Welding and metal removal techniques as well as fabrication of the various components and component parts are covered as well as examination and quality control. The section ends with a discussion of various construction techniques, manpower requirements, on-site considerations, safety issues, and post-construction testing prior to unit startup.

Chapter 37
Equipment Specification, Economics and Evaluation

Preparation of accurate specifications and the in depth evaluation of proposed product offerings are the core of an effective procurement process. Specifications are a critical link between the buyer and supplier to promote a complete understanding of the buyer's needs. Their preparation requires the balancing of specific product needs with a maximum amount of flexibility, allowing the suppliers to offer options based upon their experience and knowledge.

The evaluation process is equally important. Most project proposals from suppliers include a number of options and exceptions, and the impact of these on the overall capital cost evaluation must be clearly understood. The significance of the initial capital investment must be weighed against short and long term unit operating and maintenance expenses. Only then can the buyer determine the least cost option for generating the new steam supply needs.

Specifications

Specifications for steam generating equipment are derived from a detailed evaluation of the steam requirements and available fuel options. For a utility power plant, the need is identified from the forecast of both peak and average long term electrical demand (load) of the system. The overall power and steam generation plan must also recognize unit retirements within the system, spinning reserve, reserve for scheduled and unscheduled outages, and current generating equipment mix. The specific power plant type selected is based upon an evaluation of the amount and type of demand, assessment of the current boiler types and sizes within the system, long term fuel availability and cost, condensing water supply, equipment types available, existing facilities and locations, environmental impact and, in many cases, public opinion. Extensive feasibility studies are undertaken to provide such data.[1,2]

The needs for industrial steam production are usually tied to requirements for process steam or heating. The timing of equipment addition, the steam flow rate and the frequency of operation are governed by the application. In addition, with the recent growth of cogeneration and independent power production, steam supply needs have been increasingly influenced by the demands of electrical power production. In all cases, the steam flow requirements will determine the type, size and number of steam generating units required. Particular attention should be given to the variety, cost and availability of fuel sources including process byproducts, wastes and waste heat.

Based upon the detailed evaluation of the need, fuel sources, and available steam generating options, the specification for new or replacement steam generating equipment must include at least the following information:

1. steam flow characteristics,
2. steam temperature and pressure conditions,
3. feedwater supply,
4. fuel options and characteristics including ash properties,
5. boiler efficiency requirements,
6. performance guarantees, and
7. environmental regulations.

The specification should also include the complete scope of supply including auxiliaries, commercial terms and conditions, project schedule, site access restrictions and evaluation criteria to be used in assessing design and operating tradeoffs. Beyond these items, it is usually appropriate to only specify material and workmanship quality requirements. Potential suppliers can then use their experience and knowledge to propose the most dependable and cost effective design.

Performance

Steam flow and number of units Primary performance specifications include the steam flow requirements. For utility and other units which produce steam for electrical power generation, these usually include:

1. maximum continuous rating (MCR),
2. minimum steam flow rate,
3. type of service (base load, cycling or peaking), and
4. expected cycle of operation.

In power station practice, typically one boiler is matched to an individual turbine-generator set with sizes ranging from 25 to 1300 MW. Depending on the load requirements, one or more units are purchased. In general, economies of scale dictate the purchase of the fewest number of units possible to meet overall load and system requirements.

For industrial units, specified steam flow requirements include:

1. maximum steam flow capacity,
2. firm capacity (maximum plant output with the largest boiler out of service),
3. turndown (ratio of maximum to minimum load), and
4. expected steam load pattern variations (daily, weekly or seasonal).

In general, industrial unit operators frequently prefer several smaller boilers rather than a single large unit giving them more operating flexibility. The number of units is established by assessing the load pattern, load duration, availability requirements, maintenance periods, types of boilers in the system and possible expansion plans. Minimum standby requirements are set by the minimum required steam flow plus an allowance for routine maintenance and forced outages. Where the load is seasonal and steam is not a critical process requirement, standby may not be required. Where steam is required throughout the year, some level of economical excess capacity will be needed (Table 1).

The number of units purchased is then dependent upon the total cost with the cost per unit steam flow rising as smaller boilers are purchased. Thermal efficiency and likely load pattern are also important. A large unit will be less efficient when run at partial load conditions for extended periods of time.

Steam temperature and pressure The boiler is typically designed for the maximum steam pressure and temperature conditions specified, plus an appropriate design allowance. If a turbine-generator set is specified, then the turbine cycle selected will determine the pressure and temperature relationships. The turbine specifications will typically establish acceptable deviation limits in outlet steam temperature as well as the maximum rate of change. Constant pressure operation or variable pressure operation, each with advantages and disadvantages, may be selected to meet operator needs. (See Chapter 24.) When the application calls for superheat and reheat steam conditions, the load range and load pattern will help determine the steam temperature control ranges. The benefits of higher steam temperatures at lower loads must be compared to the higher costs of providing an extended range. The evaluation penalties and allowable variations in steam temperature will also affect the type of temperature control system included. Allowance must also be made for pressure and heat loss from piping between the boiler outlet and the steam use location.

In industrial applications where there may be some flexibility in the selection of steam pressure and temperature, interaction with the steam system supplier may permit selection of a more economical set of operating conditions.

Table 1
Boiler Capacity with Standby Units

Total Number of Units	Number of Standby Units	Per Unit Capacity, %	Total Capacity, %
2	1	100	200
3	1	50	150
4	1	33	133
5	1	25	125

Feedwater The temperature and chemical analysis of the feedwater must be specified. The minimum required feedwater temperature helps set the boiler firing rate and the water chemistry will affect water treatment needs. Recommended water chemistry specifications are provided in Chapter 42. For the smallest units, a deaerator treats the feedwater before it enters the boiler. The resulting feedwater temperature is typically 220 to 240F (104 to 116C). Oxygen in the feedwater must be minimized to prevent excessive boiler corrosion.

In larger high temperature, high pressure boilers, closed feedwater heaters are used to improve cycle efficiency as discussed in Chapter 2. The feedwater temperature versus load curve must be provided with the boiler specification.

For large industrial boilers operating in process plants or involved in cogeneration, properly treated desuperheater spray water is not available when the extraction steam is not recovered from the condenser. Under these circumstances, a *condenser-attemperator system* may be used. In this system, drum steam is condensed by incoming boiler feedwater and is injected between superheater stages. Steam pressure drop through the primary superheater is often sufficient to provide the pressure differential needed for injection without using booster pumps. (See Chapter 23.)

Fuel The fuel selected has a major impact on steam generating equipment design and capital cost because of the type of applicable combustion equipment, fouling characteristics of any residual ash, and heat transfer properties of the flue gas (including moisture content). Gas and light oil are clean burning fuels which result in capital costs up to 50% less than coal-fired units with comparable steam flow capacities. Refuse, biomass and most byproduct or waste fuels require dramatically different combustion systems than pulverized coal, oil or gas and can also result in different boiler corrosion characteristics. Boilers with provision for multi-fuel firing must represent a compromise between various fuels particularly for steam temperature control. They therefore may be more expensive than boilers designed to burn a single fuel.

After selecting the fuel or fuels, complete fuel analyses should be provided with the specification. For each fuel to be fired, ultimate analysis, proximate analysis, heating value and ash analysis should be provided. The analyses should specify the appropriate basis — moisture and ash free, moisture free or as-fired. Additional information needed includes as-received solid fuel sizing, liquid fuel viscosities and liquid/gaseous fuel pressures. The effect of the fuel and ash on unit design is covered in detail in Chapter 20.

Boiler system efficiency Maximizing boiler operating efficiency minimizes fuel usage and reduces costs. To achieve the highest efficiency, *heat traps*, which extract the last heat from the flue gas, are used. Heat traps include air heaters and economizers. (See Chapter 19.) The temperature of the flue gas exiting the final trap should be low to maximize boiler efficiency. However, this temperature should normally exceed the dew point of the gas to prevent corrosion in the precipitator, baghouse and stack as well as heat trap. The temperature must also be high enough to promote proper operation of dry sulfur dioxide (SO_2) scrubbers if used. If a single heat trap is required, an air heater is usually chosen when firing pulverized coal, wood or biomass to aid in combustion. For other fuels, an economizer is frequently selected because of lower cost, possible increased efficiency and reduced nitrogen oxides (NO_x) levels compared to an air heater installation (increased air temperatures can promote NO_x formation in the furnace).

If the boiler supplier is to design the heat traps effectively, the specification must include evaluation factors such as fuel cost and capitalization period.

Performance guarantees Specific guarantees usually include boiler efficiency and MCR capacity. Auxiliary power consumption, draft loss and air resistance are usually guaranteed at MCR. Steam purity and stack emissions are specified at all loads. Superheater and reheater temperature control ranges are guaranteed. Finally, boiler NO_x emissions have come under tighter control and are guaranteed by the boiler supplier.

Environmental limits Meeting appropriate environmental limits will be essential. To determine if the emissions are governed by federal regulations or if local laws are more restrictive, Babcock & Wilcox recommends joint discussions between the buyer, equipment suppliers and regulating agencies for a clear understanding of permissible emissions limits. Key pollutants include NO_x, SO_2 and particulate. Other airborne emission requirements are highly site specific. NO_x control can significantly affect boiler design. SO_2 control will affect fuel selection and postcombustion equipment. Particulate control for oil and solid fuels requires an electrostatic precipitator or baghouse. (See Chapters 32 through 35.)

Scope

Choices The buyer may specify whether or not bids should be for a complete *turnkey* project, in which all components of the steam generating system are within the supplier's scope. Another approach requiring more of the buyer's resources is the *system island* scope.

For a turnkey project, the supplier assumes full performance liability for the completed facility. This liability is outlined in guarantees, often secured by liquidated damages. Other turnkey options include operation and maintenance or equity participation by the supplier.

In the system island approach, the buyer releases several smaller scopes to equipment suppliers. These may include the boiler island, turbine island and design engineering services. As opposed to the turnkey approach, the overall risk is assumed by the buyer.

Sample scope A basic boiler scope for a pulverized coal-fired unit would include gravimetric feeders, pulverizers, burners, windbox, furnace, superheater, reheater, boiler bank (low pressure units), air heater, economizer, flues and ducts, support steel, access platforms, trim valves and mountings, as well as fans and drives. A scope extension may include the environmental control equipment — precipitator or baghouse and wet or dry SO_2 scrubbers. Burner management controls could also be added. A clearly defined scope provides the most consistent and useful bids.

Options Some options can improve operation of the unit. Spare pulverizers, burners, fans and drives are options to consider. Purchasing these options depends on intended unit service, load/capacity factors, and criticality of operation. For example on smaller units, instead of buying a spare pulverizer, gas can temporarily be used to replace coal when a pulverizer is out of service in order to maintain the steam flow requirement. This would substitute an operating cost for the capital cost.

General specifications

Components Scope components, such as pulverizers, burners and feeders, may be further defined in the specification. While clear performance expectations should be outlined, excessive detail may add unnecessary cost. It is often advisable to rely on the supplier's expertise in these areas. Examples of supplier defined items include metal surface preparation, painting, welding and certain manufacturing techniques. However, applicable codes must be clearly defined by the buyer.

Schedule The specification should accurately state schedule requirements using supplier inputs on proposal, engineering, fabrication, erection and startup timing. Allowing sufficient proposal preparation time is particularly important. The project initiation date is determined by subtracting the appropriate number of weeks from the necessary on-line date.

Site conditions The specification should clearly describe the site conditions including dimensioned plot plans. Rail, truck and personnel access to the site should be indicated. Sufficient laydown area adjacent to the site generally reduces handling costs. However, provision can be made for off-site fabrication and extended modularization if space is limited. It is important to state whether an existing facility will remain on-line during the construction. Site visits and supplier cost saving suggestions should be encouraged.

Personnel It is appropriate to discuss the supplier's project management and construction teams. Information from these discussions indicates the supplier's capabilities and provides additional data for comparing bids.

Commercial terms and conditions

Payment terms Matching payment terms to a supplier's cash flow needs minimizes project cost by minimizing working capital requirements. *Milestone payments* would provide a payment upon receipt of order, with additional payments made following drawing submittal, major material orders and completion of major component fabrication. *Progress payments* are made at given date intervals following project initiation.

It is customary for the buyer to retain a small portion of the contract price (usually 5%) until performance guarantees are demonstrated. However, a bank letter of credit can be obtained by the buyer to reduce the supplier's retainer cost. This guarantees the buyer access to funds if problems arise in the future while not limiting working capital for the supplier before completion of the contract.

Major clauses Additional major clauses which can be part of the contract include indemnity, warranty, liability limitation, liquidated damages, cancellation protection and insurance. In many instances these are prepared for specific proposals to meet special supplier and buyer needs. In other cases, the United States (U.S.) Uniform Commercial Code or other laws, rules or regulations provide protection for the buyer and supplier in the absence of specific disclaimers.

Evaluation criteria The design of steam generating units involves complex tradeoffs between hardware options and operating costs. For the supplier to provide the best possible offering to meet the buyer's needs, it is important to include the general criteria and factors to be used by the buyer to evaluate competing proposals. Equipment preferences, pressure drop limitations, auxiliary steam costs, power costs, demand charges and evaluation penalties are among the factors which should be covered.

General bid evaluation

Once the specification has been prepared and proposals submitted, the bids are evaluated. A consulting engineer may be used to help with the evaluation. Although the best bid is usually price competitive, it is not always the lowest. A thorough review must consider whether the supplier has correctly bid the intended scope, and it should evaluate any exceptions to the specification. Bid evaluation includes a review of the scope, operating ease, maintenance and operating costs, service, design and construction features, hardware, schedule, backup, experience, terms and price.

Scope

Scope of supply typically differs between suppliers to the same specification. It is important to verify what items are explicitly covered in each bid. Inconsistencies between bids can be resolved by discussions with the suppliers and by assessing a cost for discrepancies in the bid evaluation.

Operating ease

Design differences will appear among the product offerings. Seemingly minor items, if acceptable, can significantly reduce costs. It is appropriate to have boiler operators evaluate certain designs; their experience is valuable.

Maintenance costs

Designs that require minimal maintenance provide ongoing cost savings. Areas that previously demanded high maintenance, such as refractory and brickwork, have generally been eliminated from modern boiler designs. Current gas-tight water-cooled furnace construction has replaced the tube and tile design. However, some corner seals, hangers, supports and spacers may still be high maintenance items if not properly designed. Pulverizer designs can vary widely resulting in significantly different annual maintenance costs. These factors can be incorporated into the final evaluation discussed below.

Operating costs

Annual boiler operating costs can approach the initial unit investment; therefore, fuel and auxiliary power costs should be quantitatively evaluated. Based on an assumed fuel cost, discount rate and the predicted unit life, present values can be assigned to calculate annual fuel and power costs. Another critical operating cost is wet or dry scrubber reagent usage.

Service organization

The quality of a supplier's service department is important. An experienced staff can troubleshoot problems quickly and minimize costly downtime. Service factors to consider include the proximity of the supplier's office or representative, number of available personnel, training of the personnel and portion of the scope they are expected to handle.

Design and construction features

A thorough bid evaluation must review the boiler's design and construction features and determine if certain technologies favor one supplier. Although some designs may present a low initial cost, accompanying higher maintenance requirements may quickly offset these savings.

The buyer should review the steam drum internals to determine if there is a provision for steam/water separation, and should check the design drum water level. If the level is low, load swing capabilities are reduced. Other areas for review include, but are not limited to:

1. review the low NO_x burners to determine if they have been effective on a similar installation,
2. check the pulverizer design regarding maintenance requirements and degradation between overhauls, and
3. review the superheater supports. Closely spaced tubes are more reliably supported but are less efficient due to reduced heat transfer capabilities.

Schedule

Bid schedules should be compared to one another and to the specification. The evaluation should include a comparison of drawing submittal dates, fabrication schedule, shipping times and erection span.

Project management

A supplier's project manager will greatly influence project performance. His central role focuses on providing the specified product on the contract schedule within the project cost constraints. Additional duties include, but are not limited to, managing subsuppliers and vendors, responding to buyer changes, providing liaison between various project team members and monitoring project compliance. This critical function requires experienced personnel knowledgeable in executing similar projects.

Support services

It is important to determine the level of service that will be provided after the sale and project completion, and to review the supplier's organization to determine if quality engineering consultation services are available. Backup areas include feedwater analysis, fuel selection, laboratory analyses, equipment upgrades, capacity increase capabilities, operating optimization, life extension and condition assessment.

Experience

Although relatively new suppliers may be successful, demonstrated successes in similar boiler projects best indicate a quality supplier. The buyer should contact organizations that have purchased similar units; visits to these operational plant sites can provide important insights.

Terms and price

Special payment terms are expensive and often unavailable. In addition, the lowest cost is obtained when the buyer and supplier share a reasonable amount of risk. Finally, note that performance guarantees rely on a supplier's ability to back them up. This ability should be evaluated before awarding a contract.

Partnering

A relatively new approach to the buyer/seller relationship is the formation of partnering alliances. This procurement concept is consistent with principles developed by today's leading quality experts.

The concept is based on selecting partners for designing and procuring major components. Partners are selected based on a comprehensive analysis of resources, quality programs and management commitment. Once chosen, the partner provides design and fabrication inputs from the early stages of the project.

A partnering approach has been successful in several projects. Design expertise is gained early in the project, and up front costs, such as specification preparation and bid solicitation and evaluation, are reduced or eliminated.

Power system economics

The relative economics of the options must be compared as part of the evaluation of alternate power systems or different product offerings. Unfortunately, this is not a straightforward, simple task. Large projects involve a complex series of time dependent cash flows which usually include an initial near term investment in plant and equipment, and a long term, uneven series of operating costs, maintenance costs and revenues from the sale of power or end product. This is further complicated by the time value of money and the effects of inflation. Money today is generally worth more than the same amount of money a year from now because of: 1) the potential interest which could be earned during the next year, and 2) the erosion in the future buying power due to inflation. The following provides a very brief overview of several methodologies used to place the total evaluated costs of different projects on the same financial basis so they can be directly compared. More comprehensive discussions of such evaluations are provided in References 3 to 6. This discussion is broken down into the following subsections:

constant versus current cost basis,
definitions,
system cost evaluation,
evaluation parameters,
methods for evaluating operating costs,
other considerations, and
examples.

Constant versus current cost basis

Analyses are conducted on either a current or constant cost basis. The current cost basis presents costs as they would be expected to appear on a company balance sheet in the future while the constant cost basis does not include inflation.

The selection of the constant or current cost approach usually does not change the economic choice between similar projects. For nearer term projects, such as plant upgrades or the addition of small components, current costs are frequently used because of near term impact. For longer term projects such as new plants, constant costs are frequently used as at least one of the evaluation tools so that the effect of inflation does not distort real cost trends. In either case, it is important to use the same basis for evaluating all alternatives. The examples provided below are evaluated on a constant cost basis.

Definitions

The following parameters provide the basis for the economic evaluation:

Discount rate (k) This is the cost of money for a power system owner and includes the weighted cost of capital for each class of debt and equity. Establishing the appropriate cost of capital or discount rate is increasingly complex. References 4 and 7 provide detailed discussion of the sources of funds and the evaluation of their costs. A simplified discount rate (k_c) estimate can be obtained from the following formula which takes into account that interest payments on debt are generally tax deductible expenses:

$$k_c = (FD)\, k_d (1 - TR) + (FE)\, k_e \tag{1}$$

where

FD = fraction of capital consisting of debt
FE = fraction of capital consisting of equity
k_d = annual debt interest rate
k_e = annual stockholder return
TR = fractional income tax rate

The annual debt interest rate and stockholder return reflect the anticipated impact of inflation and are used in current cost basis analyses. If a constant cost based analysis is desired, the *current cost* discount rate (k_c) can be related to the *constant cost* discount rate (k_f) through the escalation or inflation rate (e) in the following equation:

$$k_f = [(1 + k_c)/(1 + e)] - 1 \tag{2}$$

The escalation rate may have to be treated as a time dependent variable if it is expected to change significantly over the period of analysis.

Fixed charges Also referred to as carrying charges, these are the fixed annual costs associated with the initial capital investment and are generally incurred whether or not the plant is operated. The components include the return on investment, capital recovery (or depreciation), property taxes and insurance as well as federal and state taxes. Over the life of a new plant the annual revenue requirement declines because of income tax effects and depreciation of the initial capital investment. However, a weighted average or levelized value is frequently used for evaluations. To obtain a levelized value the actual annual fixed charges are discounted to the present and then levelized using the formulas discussed below. Fixed charge rates vary from 12 to 21% of the initial total capital investment.

Interest during construction This refers to the cost of money during construction prior to initial plant operation. It is not included if all project costs are incurred in a single year.[3]

Capacity factor This is the total power produced in a given period (typically one year) divided by the product of the net unit capacity and the given time period. It accounts for part load operation and outage times. The capacity factor varies over the plant life, typically climbing to a maximum in the first few years, remaining relatively constant for a period of several years and then falling as the unit begins to age.

Load factor This is the fraction of the full load or MCR at which the steam generating system is normally expected to operate for extended periods of time.

Net plant heat rate (NPHR) This is the total fuel heat input expressed in Btu divided by the net busbar power leaving the power plant expressed in kWh. Plant total energy efficiency (E) can be obtained from the following relationship:

$$E = \frac{3412 \text{ Btu/kWh}}{NPHR} \times 100\% \qquad \textbf{(3)}$$

The *NPHR* typically varies with plant load.

Demand and auxiliary power charges These are the charges placed against the internal use of energy for auxiliary equipment. Separate rates are established by electricity and steam usage. The size of the charges depends upon the cost of fuel and other operating costs.

System cost evaluation

In the economic evaluation of steam generating systems, it is important to include all appropriate costs and expenses. Table 2 provides a list of typical capital, variable operating and maintenance costs which could be associated with a steam generating system installation. Capital costs basically include all expenditures for the purchase, installation and startup of the steam generating system and its auxiliaries. Care must be taken to ensure that different product offerings contain the same scope of work. To obtain the total plant investment, interest during construction, initial inventory items (such as fuel) and contingency are added. Variable operating costs include delivered fuel cost, other consumable material purchases, power and steam demand charges, and waste disposal or byproduct use.

Fixed operating and maintenance costs cover operating expenses which are likely to be incurred whether or not the plant is in operation. These change from year to year and generally climb as the plant ages. They also tend to vary with the unit load pattern, plant size and primary fuel.

In the financial evaluations, selected penalties or operating costs can be associated with superheater, reheat superheater and economizer pressure drops (steam-water side and gas side). Depending upon the buyer's perspective, these can be explicitly listed as separate operating costs or they may be indirectly accounted for as part of the general electrical and steam power requirements listed above. Reheat superheater steam side pressure loss is of particular interest because of its strong connection with thermal cycle efficiency.

Table 2
Typical Steam Generating System Cost Items

Capital Costs
Boiler (nominal scope including burner system and air heater)
Environmental control equipment as specified
Fans and drives
Appropriate flues and ducts
Coal crushers
Pulverizers
Pyrite removal (pulverized coal units)
Dust collectors
Ash handling
Steam coil air heaters
Steam piping
Injection water piping
Startup pump and drive
Feedwater heaters and piping (if used)
Feedwater treatment
Instrumentation
Controls
Foundations and electrical connections
Structural steel
Building
Platforms
Freight
Sales taxes
Site preparation
Erection/insulation/painting
Startup
Performance tests
Miscellaneous engineering and supervision services

Variable Operating Costs
Primary fuel cost
Electrical and steam power demand:
Fan power
Pulverizer power
Recirculating pump power
Boiler feedpump power
Crusher power
Makeup water and chemicals
Waste water treatment
Steam demand:
Sootblowers
Flue gas reheat
Air heating
Compressed air demand
Ash removal and disposal
Auxiliary fuel for low load stability
Other consumables (ammonia, limestone, etc.)

Fixed Operating and Maintenance Costs
Plant operators
Maintenance labor
Maintenance material
Chemical cleaning
Replacement material
Supervision
Other overhead items

Evaluation parameters

Present value With an economic life stretching from three to six decades, power plants experience operating expenses over extended periods of time. It is necessary to account for the time value of money in evaluating projects which have differing initial capital investments and future expenses over several years. This is usually accomplished by using a *present value* (*PV*) calculation for each annual cash flow. The *PV* of a *future value* or payment (*FV*) in year n is defined as:

$$PV = FV(1+k)^{-n} = FV(PVF) \quad (4)$$

where

$PVF = (1+k)^{-n}$ = present value factor
k = interest or discount rate
n = number of years or periods at rate k

An alternate use of the *PVF* is to evaluate the future value of a present value:

$$FV = PV/PVF \quad (5)$$

Where a series of equal future annual payments, A, will be made for n years, the *PV* is defined as:

$$PV = A\left(\frac{(1+k)^n - 1}{k\ (1+k)^n}\right) = A\ (SPVF) \quad (6)$$

where

SPVF = uniform annual *series present value factor*

$$= \frac{(1+k)^n - 1}{k\ (1+k)^n}$$

As with the single payment present value factor, *SPVF* can be used to convert a *PV* into a series of future equal payments for n years:

$$A = PV/SPVF \quad (7)$$

For illustration purposes, Table 3 provides values of *PVF* and *SPVF* for a discount rate of 6% from 1 to 30 years.

Levelization In evaluating various power system options, it is frequently desirable to compare weighted average annual values of future costs such as fuel instead of the lumped sum present values. Such a parameter would be a levelized cost evaluated using the following formula:

$$\text{Levelized value} = \frac{\sum_{j=1}^{n} (FV_j)(PVF_j)}{SPVF} \quad (8)$$

where

PVF_j = present value factor for year j and rate k
FV_j = future annual payment in year j
SPVF = uniform annual series present value factor in year n for discount rate k
j = individual years up to n years
k = annual discount rate

For the special case where the annual payment escalates or is inflated from an initial value of S_1 with a constant rate, e, the future value of the annual payment in year j is:

$$S_j = (1+e)^j S_1 \quad (9)$$

The resulting levelization equation then becomes:

Levelized value of S =

$$S_1\left(\frac{1-[(1+e)/(1+k)]^n}{k-e}\right) / \, SPVF \quad (10)$$

Methods for evaluating operating costs

There are several approaches to the evaluation of operating costs. The revenue requirements method and the capitalized cost method are reviewed here and examples are given. While each method used involves a large number of inherent assumptions, these are the dominant methods used to evaluate steam supply equipment. See References 3 to 7 for more discussion.

Revenue requirements method This method determines the minimum gross receipts which are required to make each alternative pay its way. The alternative that requires the lowest receipt of revenues is the economic choice as it will permit profitable operation with the smallest revenues. This method takes into account the present value of all future expenditures and costs.

The capital (first cost) expenditure which is justified in order to put two alternatives on an equal basis is less than the difference in future revenue requirements. The future revenue requirements are converted to current dollars by multiplying each future expenditure by the PVF for each year in which the expenditure will occur. By summing the present value of the future revenue requirements, a total present value is determined. To convert this sum into a justifiable capital expense, it must be levelized, or converted to an equivalent equal annual cost. This is done by dividing the total present value by the uniform annual series present value factor (SPVF) (which is 13.765 for a 30 year period and 6% interest rate, used in the examples below). A tabulation of present value factors (PVF) for 6% interest rate is shown in Table 3. Present value factors for other interest rates and evaluation periods can be evaluated from Equations 4 and 6 above. After converting the total into an equal annual cost, the justifiable capital expenditure is determined by dividing the annual cost by the fixed charge rate.

Capitalized cost method This method assumes that the equipment being operated has an infinite lifetime and that the costs of operating the unit will continue indefinitely. It is a simpler evaluation procedure to use than the revenue requirements method described above, but is not as exact or flexible in accounting for future changes in such items as fuel and maintenance costs which will change during the life of the unit.

Table 3
6% Compound Interest Factors

Year	Single Payment Present Value Factor (PVF)	Uniform Annual Series Present Value Factor (SPVF)	Year	Single Payment Present Value Factor (PVF)	Uniform Annual Series Present Value Factor (SPVF)
1	0.9434	0.943	16	0.3936	10.106
2	0.8900	1.833	17	0.3714	10.477
3	0.8396	2.673	18	0.3503	10.828
4	0.7921	3.465	19	0.3305	11.158
5	0.7473	4.212	20	0.3118	11.470
6	0.7050	4.917	21	0.2942	11.764
7	0.6651	5.582	22	0.2775	12.042
8	0.6274	6.210	23	0.2618	12.303
9	0.5919	6.802	24	0.2470	12.550
10	0.5584	7.360	25	0.2330	12.783
11	0.5268	7.887	26	0.2198	13.003
12	0.4970	8.384	27	0.2074	13.211
13	0.4688	8.853	28	0.1956	13.406
14	0.4423	9.295	29	0.1846	13.591
15	0.4173	9.712	30	0.1741	13.765

The amount calculated using this method is the justifiable capital expense which can be spent to purchase the additional equipment and still give an operating advantage.

For example, if it is found that additional air heater surface will achieve higher unit efficiency, i.e., fuel savings over the lifetime of the unit, the additional air heater surface is justified if its capitalized cost is less than the cost of the fuel which can be saved over the remaining unit life.

Other considerations

Risk All new projects involve some level of risk or uncertainty. Short term projects using established technology typically are considered low risk while long term projects using new, untried technology are generally considered higher risk. A number of approaches are used to account for the risk or uncertainty during project evaluations. One approach is to establish special guarantee provisions in the specifications and contract. This will typically increase the price from the suppliers since they assume more of the risk. A second approach is to increase the buyer's internal contingency account in the project cost evaluation (see Reference 3). A third approach involves making adjustments to the discount rate to accommodate the level of risk or uncertainty with higher rates applied to riskier projects (see Reference 7). One or more of these options, as well as other approaches, may be used in the ranking of project options depending upon the buyer's internal needs.

Unequal periods When comparing project alternatives of equal life, the various evaluation techniques generally provide similar rankings. However, when project lives are significantly different, additional care must be exercised in performing the analysis and interpreting the results. As an example, the costs of additional future equipment replacement may need to be considered when short term and long term repairs to a particular equipment problem are considered. In addition, when interpreting the evaluation results, each financial model uses different assumptions about capital re-investment which could result in different project rankings.

Examples

Four examples follow which illustrate the use of the evaluation methodologies. The first two examples compare the use of the revenue requirement method and the capitalization method in evaluating the choice between a subcritical pressure 2400 psi utility steam cycle and a supercritical pressure 3500 psi utility steam cycle assuming that a new power plant is needed. Example 3 provides a comparison between two industrial boiler bids where one bidder has a higher price but higher boiler efficiency. Example 4 evaluates the lifetime auxiliary power difference between two industrial boiler bids. All of these examples are evaluated on a constant U.S. dollar basis, i.e., excluding inflation.

Example 1
Revenue Requirements Method Evaluation of 2400 psi and 3500 psi Steam Cycles

A) Problem: Determine the economic choice between a 2400 psi cycle and a 3500 psi cycle with 1000F superheat (SH) and reheat superheat (RH) temperatures using a constant cost basis.

B) Given:

	Unit 1	Unit 2
1. Throttle pressure, psi	2,400	3,500
2. Temperature, SH/RH,F	1,000/1,000	1,000/1,000
3. Plant size, kW	600,000	600,000
4. Net plant heat rate (NPHR), Btu/kWh: 100% Load	9,000	8,860
80% Load	9,000	8,860
60% Load	9,180	9,040
35% Load	10,080	9,920
5. Fuel cost, $/million Btu	$2.00	$2.00
6. Evaluation period, yrs	30	30
7. Interest rate or discount rate, %	6	6
8. Maintenance costs	Same	Same
9. Availability	Same	Same
10. Operating costs (excluding fuel)	Same	Same
11. Fixed charge rate, % (Interest, taxes, insurance, debt return)	14.0	14.0

12. Plant capacity factors (Units 1 and 2)

Years	1 to 10	11 to 20	21 to 30
100% Load	44%	23%	6%
80% Load	28%	23%	23%
60% Load	13%	34%	34%
35% Load	6%	11%	23%
0% Load	9%	9%	14%
	100%	100%	100%
Average capacity factor for period	76.30%	65.65%	52.85%
Average capacity factor for lifetime		64.93%	

C) Solution:

1. Cost of fuel consumed per year — 2400 psi unit
   (h/yr) × capacity factor × kW load × Btu/kWh = Btu/yr
   (Btu/yr) × ($/Btu) = $/yr

(a) Years 1 to 10 — Btu/yr

$$8{,}760 \times 0.44 \times 1.00 \times 600{,}000 \times 9{,}000 = 20{,}814 \times 10^9$$
$$8{,}760 \times 0.28 \times 0.80 \times 600{,}000 \times 9{,}000 = 10{,}596 \times 10^9$$
$$8{,}760 \times 0.13 \times 0.60 \times 600{,}000 \times 9{,}180 = 3{,}764 \times 10^9$$
$$8{,}760 \times 0.06 \times 0.35 \times 600{,}000 \times 10{,}080 = \underline{1{,}113} \times 10^9$$
$$36{,}287 \times 10^9$$

$$36{,}287 \times 10^9 \times \$2.00/10^6 = \$72{,}574{,}000/\text{yr}$$

(b) Years 11 to 20

$$8{,}760 \times 0.23 \times 1.00 \times 600{,}000 \times 9{,}000 = 10{,}880 \times 10^9$$
$$8{,}760 \times 0.23 \times 0.80 \times 600{,}000 \times 9{,}000 = 8{,}704 \times 10^9$$
$$8{,}760 \times 0.34 \times 0.60 \times 600{,}000 \times 9{,}180 = 9{,}843 \times 10^9$$
$$8{,}760 \times 0.11 \times 0.35 \times 600{,}000 \times 10{,}080 = \underline{2{,}040} \times 10^9$$
$$31{,}467 \times 10^9$$

$$31{,}467 \times 10^9 \times \$2.00/10^6 = \$62{,}934{,}000/\text{yr}$$

(c) Years 21 to 30

$$8{,}760 \times 0.06 \times 1.00 \times 600{,}000 \times 9{,}000 = 2{,}838 \times 10^9$$
$$8{,}760 \times 0.23 \times 0.80 \times 600{,}000 \times 9{,}000 = 8{,}704 \times 10^9$$
$$8{,}760 \times 0.34 \times 0.60 \times 600{,}000 \times 9{,}180 = 9{,}843 \times 10^9$$
$$8{,}760 \times 0.23 \times 0.35 \times 600{,}000 \times 10{,}080 = \underline{4{,}265} \times 10^9$$
$$25{,}650 \times 10^9$$

$$25{,}650 \times 10^9 \times \$2.00/10^6 = \$51{,}300{,}000/\text{yr}$$

2. Present value of fuel costs — 2400 psi unit
   Fuel costs for period × PVF for period
   = PV of fuel costs for period

(a) Years 1 to 10
$72,574,000/yr × 7.36 = $534,145,000
(b) Years 11 to 20
$62,934,000/yr × (11.47 – 7.36) = $258,659,000
(c) Years 21 to 30
$51,300,000/yr × (13.76 – 11.47) = $117,477,000
$910,281,000

3. Cost of fuel consumed per year — 3500 psi unit
(h/yr) × capacity factor × kW load × Btu/kWh = Btu/yr
(Btu/yr) × ($/Btu) = $/yr

(a) Years 1 to 10 — Btu/yr

$8,760 \times 0.44 \times 1.00 \times 600,000 \times 8,860 = 20,490 \times 10^9$
$8,760 \times 0.28 \times 0.80 \times 600,000 \times 8,860 = 10,431 \times 10^9$
$8,760 \times 0.13 \times 0.60 \times 600,000 \times 9,040 = 3,706 \times 10^9$
$8,760 \times 0.06 \times 0.35 \times 600,000 \times 9,920 = 1,095 \times 10^9$
$35,722 \times 10^9$

$35,722 \times 10^9 \times \$2.00/10^6 = \$71,444,000/yr$

(b) Years 11 to 20

$8,760 \times 0.23 \times 1.00 \times 600,000 \times 8,860 = 10,711 \times 10^9$
$8,760 \times 0.23 \times 0.80 \times 600,000 \times 8,860 = 8,569 \times 10^9$
$8,760 \times 0.34 \times 0.60 \times 600,000 \times 9,040 = 9,693 \times 10^9$
$8,760 \times 0.11 \times 0.35 \times 600,000 \times 9,920 = 2,007 \times 10^9$
$30,980 \times 10^9$

$30,980 \times 10^9 \times \$2.00/10^6 = \$61,960,000/yr$

(c) Years 21 to 30

$8,760 \times 0.06 \times 1.00 \times 600,000 \times 8,860 = 2,794 \times 10^9$
$8,760 \times 0.23 \times 0.80 \times 600,000 \times 8,860 = 8,569 \times 10^9$
$8,760 \times 0.34 \times 0.60 \times 600,000 \times 9,040 = 9,693 \times 10^9$
$8,760 \times 0.23 \times 0.35 \times 600,000 \times 9,920 = 4,197 \times 10^9$
$25,253 \times 10^9$

$25,253 \times 10^9 \times \$2.00/10^6 = \$50,506,000/yr$

4. PV of fuel costs — 3500 psi unit

(a) Years 1 to 10
$71,444,000/yr × 7.36 = $525,828,000
(b) Years 11 to 20
$61,960,000/yr × (11.47 – 7.36) = $254,656,000
(c) Years 21 to 30
$50,506,000/yr × (13.76 – 11.47) = $115,659,000
$896,143,000

5. Saving in fuel costs — 3500 psi versus 2400 psi

PV of fuel costs	
2400 psi unit	$910,281,000
3500 psi unit	$896,143,000
PV of savings	$ 14,138,000

$14,138,000 / 600,000 kW = $23.56/kW

6. Justifiable capital expenditure — 3500 psi versus 2400 psi

Justifiable additional expenditure =

$$\frac{\text{PV of fuel cost saving}}{\text{Fixed charge rate} \times \text{PVF}}$$

PVF for 30 years at 6% interest = 13.76

Fixed charge rate = 14% = 0.14

Justifiable additional expenditure =

$$\frac{\$14,138,000}{0.14 \times 13.76} = \$7,339,000$$

Justifiable expenditure/kW =

$7,339,000 / 600,000 kW = $12.23/kW

D) Result:
1. In terms of today's worth, the 3500 psi cycle will save $14,138,000 or 23.56/kW in fuel costs over its life as compared with the 2400 psi cycle.
2. The saving will justify the expenditure of $7,339,000 or $12.23/kW more for the 3500 psi cycle than for the 2400 psi cycle all other factors being equal.

Essentially any auxiliary power usage, difference in boiler efficiency, or difference in net plant heat rate can be treated in the same manner as this example. This method basically converts the differences between designs into fuel costs. The fuel costs, which account for future load factor, cycle heat rates and future fuel costs, are converted into present day dollars using present value mathematics for the projected interest rate over the life of the unit. The justified capital expenditure can then be calculated.

Example 2
Capitalized Cost Method Evaluation of 2400 psi and 3500 psi Steam Cycles

A) Problem: Compare the economics of a 2400 psi cycle and a 3500 psi cycle using a constant cost basis.

B) Given: Same information as Example 1.

C) Solution:

1. Average NPHR — 2400 psi cycle — Btu/kWh

100% Load: (0.44 + 0.23 + 0.06) × 9,000 = 6,570
80% Load: (0.28 + 0.23 + 0.23) × 9,000 = 6,660
60% Load: (0.13 + 0.34 + 0.34) × 9,180 = 7,436
35% Load: (0.06 + 0.11 + 0.23) × 10,080 = 4,032
24,698

Average NPHR = 24,698/2.68* = 9,216

* This is the sum of all capacity factors in Item B12, Example 1, except the factors for zero load.

2. Average NPHR — 3500 psi cycle — Btu/kWh

100% Load: (0.44 + 0.23 + 0.06) × 8,860 = 6,468
80% Load: (0.28 + 0.23 + 0.23) × 8,860 = 6,556
60% Load: (0.13 + 0.34 + 0.34) × 9,040 = 7,322
35% Load: (0.06 + 0.11 + 0.23) × 9,920 = 3,968
24,314

Average NPHR = 24,314/2.68 = 9,072

3. Capitalized value of heat rate difference

Difference = 9,216 – 9,072 = 144 Btu/kWh

Average capacity factor for lifetime = 64.93% = 0.6493

Capitalized value =

$$\frac{\text{NPHR} \times \text{kWh generation / yr} \times \text{Fuel cost}}{\text{Fixed charge rate}}$$

kWh generation/yr = Capacity factor × Capacity × h/yr

Capitalized value =

$$\frac{144 \times 0.6493 \times 600{,}000 \times 8{,}760 \times \$2.00 / 10^6 \text{ Btu}}{0.14}$$

$$= \$7{,}020{,}000$$

$$\$7{,}020{,}000 / 600{,}000 \text{ kW} = \$11.70/\text{kW}$$

D) Result:
For these conditions the expenditure of up to $11.70/kW is economically justified for a 3500 psi unit versus a 2400 psi unit. (Note that this is quite close to the $12.23/kW answer reached by the revenue requirements method in Example 1.)

The capitalized cost method of evaluation can be adapted to evaluate auxiliary power, boiler efficiency, or other operating costs by converting them to fuel cost.

Example 3
Comparing Industrial Boiler Bids with Differing Boiler Efficiencies and Capital Cost

A) Problem: Determine which of two boiler bids provides the most economical system offering.

B) Given:
1. Boiler produces 600,000 lb/h superheated steam at 1550 psi and 955F (H_g = 1460 Btu/lb from Mollier diagram, Chapter 2)
2. Boiler feedwater is 400F (H_f = 375 Btu/lb, Chapter 2)
3. Fuel cost = $3.00/10⁶ Btu (gas)
4. Load factor (average annual fraction of MCR) = 0.9
5. Use factor (fraction of in service time) = 0.9
6. Discount rate = 10%
7. Constant dollar basis
8. Expected unit life = 30 years
9. SPVF = 9.43 (30 years; 10% discount rate)
10. Bid No. 1 guarantees 86.5% efficiency with a price of $10,400,000
11. Bid No. 2 guarantees 87.5% efficiency with a price of $11,200,000

C) Solution:
Equations:

Unit input = (steam flow rate) (ΔH) (load factor)/efficiency

Annual fuel cost = (per Btu cost) (annual unit output) (use factor)

Net PV of fuel savings = Δ fuel cost × SPVF

For bid No. 1:

Unit input = (600,000 lb/h) (1460-375 Btu/lb) (0.9) / 0.865

= 677.3×10^6 Btu/h

Annual fuel cost = ($\$3.00/10^6$ Btu) (677.3×10^6 Btu/h) × (8760 h/yr) (0.9) = $16,020,000/yr

For bid No. 2:

Unit input = (600,000 lb/h) (1460-375 Btu/lb) (0.9) / 0.875

= 669.6×10^6 Btu/h

Annual fuel cost = ($\$3.00/10^6$ Btu) (669.6×10^6 Btu/h) × (8760 h/yr) (0.9) = $15,837,000/yr

Difference between bids = $16,020,000 - $15,837,000 = $183,000/yr

Net PV = $183,000/yr × 9.43 = $1,725,700

Difference in price = $11,200,000 - $10,400,000 = $800,000

D) Result:
Because the net present value of the energy savings ($1,725,700) is greater than the price for the more efficient unit ($800,000), bid No. 2 provides the more economical offering.

Example 4
Evaluating the Cost Impact of Different Auxiliary Power Requirements of Two Industrial Boiler Bids

A) Problem: Determine the economic impact of differing auxiliary power requirements of two boiler bids.

B) Given:
1. Power cost = $0.025/kWh
2. Load factor = 0.9
3. Use factor = 0.9
4. Auxiliary power requirements

Component	Bid No. 1 kW at MCR	Bid No. 2 kW at MCR
Pulverizer	430	320
Coal feeder	3	3
Primary air fans	200	175
Forced draft fans	450	425
Auxiliary fans and blowers	30	30
Regenerative air heater	10	10
Total maximum auxiliary power	1,123	963

5. Assume the same discount rate, unit life, and SPVF as in Example 3.

C) Solution:
Total maximum auxiliary power corrected for capacity and load factors:

Bid No. 1 = (1,123 kW) (0.9) (0.9) = 910 kW
Bid No. 2 = (963 kW) (0.9) (0.9) = 780 kW

Annual power costs:
Bid No. 1 = (910 kW) ($0.025/kW) × (8760 h/yr) = $199,290
Bid No. 2 = (780 kW) ($0.025/kW) × (8760 h/yr) = $170,820
Difference = $ 28,470

PV of power cost differential:

Net PV = $28,470/yr × 9.43 = $268,472

D) Result:
The power system supplied as bid No. 1 will cost approximately $270,000 more to operate than the system supplied under bid No. 2 on a present value basis.

References

1. Aschner, F.S., *Planning Fundamentals of Thermal Power Plants*, Wiley, New York, 1978.

2. Li, K.W., and Priddy, A.P., *Power Plant System Design*, Wiley, New York, 1985.

3. *TAG Technical Assessment Guide: Vol. 1 Electrical Supply — 1989 (Revision 6)*, Report EPRI P-6587-L, Electric Power Research Institute, Palo Alto, California, September 1989.

4. Brigham, E.F., and Gapenski, L.C., *Financial Management: Theory and Practice*, 5th ed., Dryden, Hinsdale, Illinois, 1988.

5. Grant, E.L., and Ireson, W.G., *Principles of Engineering Economy*, 5th ed., Ronald Press, New York, 1970.

6. Leung, P., and Durning, R.F., "Power system economics: On selection of economic alternatives," *Journal of Engineering for Power*, Vol. 100, pp. 333-346, April 1978.

7. Copeland, T.E., and Weston, J.F., *Financial Theory and Corporate Policy*, 3rd ed., Addison-Wesley, Reading, Massachusetts, 1988.

Seal welding the tangent tubes of a Cyclone furnace.

Chapter 38
Manufacturing

The installation of a steam generating unit involves the design and manufacture of the components and final erection of the unit. The power generation industry has achieved unprecedented levels of quality and reliability in producing modern steam generating equipment. State-of-the-art design and shop processing techniques, as well as the use of advanced materials, are responsible for these gains.

Component manufacturing involves a variety of skills and unique manufacturing methods and requires facilities for virtually all welding, metal removal and metal forming processes. Manufacturing operations are continuously evolving to meet the increased quality and productivity demanded by today's markets. Of equal importance, quality control techniques are improving to ensure that each product meets or exceeds intended design goals. These manufacturing and quality evolutions have enabled the fabrication of large, complex vessels once thought to be unachievable.

Similarly, value, i.e., achieving essential functionality at low cost, is increasing in the mature fossil power equipment industry. Realizing the associated high quality levels often dictates focusing on relatively small details. It is, therefore, important for Manufacturing to participate early in design discussions. The resultant design-for-manufacturability improvements add value to the product. *Value analysis*, a structured approach that enhances the manufacturing process, is used by Babcock & Wilcox (B&W) to maintain its leadership position in the industry.

Once a design approach is determined, communication of the design information must be accurate and timely. Today's computers and communication technology provide unprecedented capability to handle these tasks.

Traditional techniques as well as recent innovations implemented by B&W to manufacture steam generating equipment are described in the following sections.

Welding

Welding and welding related processes are central to steam generator fabrication. These processes are used in manufacturing a large percentage of components produced by B&W. The company pioneered many of today's basic welding technologies.

A *weld* can basically be defined as a coalescence of one or more materials by the application of heat localized in the region of the intended joint. For most welding processes, the induced heat raises the temperature of the material(s) above the melting point to create a weld pool. The heat source is then removed, allowing the pool to solidify and creating a metallurgical bond between the joined surfaces. Depending upon the application, filler material may be added to the weld pool to fill a cavity between the parts being joined or to modify the properties of the resultant weld.

There are many welding processes from which to select. The choice is primarily based on the best method of generating and conveying the energy needed to create the weld. Electric arc welding is by far the most common group of processes used in fabricating fossil fueled steam generating equipment, although other processes also apply. The most widely used arc welding processes are gas tungsten arc welding (GTAW), gas metal arc welding (GMAW), submerged arc welding (SAW), shielded metal arc welding (SMAW) and plasma arc welding (PAW). Of the non-arc welding processes, electroslag welding (ESW), high frequency resistance welding (HFRW) and electron beam welding (EBW) are the most common. Table 1 summarizes the advantages and disadvantages of each process and provides examples of steam generator applications. The *Welding Handbook* and other sources provide specific information on the distinguishing features, attributes and limitations of each process.[1]

Selecting a welding procedure requires consideration of several factors. The composition of the materials and the anticipated service conditions dictate the appropriate filler material. Base material composition also affects the use of preheat and postweld heat treatment to control the resultant hardness of the material and to minimize the susceptibility to cracking. Also, the structural mass of the component and the orientation and design of the weld joints influence the welding processes selected and the process parameters used. The quality and productivity associated with each welding process must also be factored into the selection.

Welding automation

A given welding process can be implemented in a number of ways. Historically, the most common approach was manual welding, in which the operator moves the welding torch along the joint to be welded and makes the necessary adjustments to the welding equipment. However, automation is becoming more common because of its more consistent quality and often lower cost.

Table 1
Comparison of Welding Processes

Process	Advantages	Disadvantages	Applications
EBW	Very deep penetration Very high quality Little distortion Good mechanical properties	High capital expense Equipment difficult to maintain Not portable Very good fit-up needed Productivity and size of work piece limited by vacuum chamber pump-down time	
ESW	Joins very thick sections in one pass Inexpensive joint preparation High productivity	Poor mechanical properties	
GTAW	High quality Low capital expense	Shallow penetration High operator skill required Low productivity Arc radiation hazard	Tube-to-tube weld root pass Join superheater and economizer tube sections
GMAW	Moderate productivity Moderate quality	Shallow penetration Contact tip wear requires period stoppages Arc radiation hazard Readily automated	Panel wall repair and attachments Tube-to-tube weld fill pass
PAW	Deep penetration High productivity High quality	Complex process Difficult to maintain High capital expense Arc radiation hazard	Sootblower tube-to-tube welds
SAW	Very high productivity High quality Operator protected from arc radiation	Limited in welding position	Header fabrication
SMAW	Very low capital expense Very portable	Can not be automated High operator skill required Lower weld quality Low productivity	Repair Tube-to-header welds
HFRW	Very high productivity High quality	High capital expense Very limited in application	Tube manufacturing

One form of automation is *orbital* welding, used with the GTAW process to join superheater and economizer tube sections. In this process, the welding head is driven around the OD of the tube section to produce a weld. A computer coordinates the motion of the welding head, the weld filler wire feed rate and the process parameters to maintain high weld quality.

Robotic welding and cutting are also becoming more common. Robots are used in a number of welding applications including seal welding the tangent tubes in a 10 ft (3 m) diameter Cyclone furnace (see chapter frontispiece), fabricating coal pulverizer components, applying nonferrous overlay cladding to steam generator tubesheets and in nozzle bores and making complex tube-to-tubesheet welds inside 1.5 in. (38 mm) bores. (Fig. 1.) The flexibility of robots also promotes their use to plasma cut nozzle openings in shells and to contour burn nozzle mating surfaces. A number of robot types, including rectilinear, articulated arm, cylindrical coordinate and gantry style units, are used. One application uses a robot with a coordinated 100 t (91 t_m) positioner. In a second case, one robot is mobile on a track to perform various fabrication and inspection operations while another positions precision welding devices.

Fig. 1 Robotic welding is used extensively for modern boiler equipment.

Automated determination of a weld joint position has also been implemented. Electromechanical devices, called *probes*, are used to track the weld groove to ensure proper placement of the weld head during submerged arc welding. An alternative guidance method, which has been used with the GMAW process, is *through the arc* tracking. This method senses the welding arc's electrical characteristics to guide the weld head. A more recent development for GTAW, GMAW and PAW is laser based vision tracking, in which a noncontact sensor determines the torch position within the groove and adjusts the torch as necessary. This type of sensor also allows automated adaptive control of the welding process parameters by real time response to changes in weld joint topography.

Welding power supplies with built in microprocessors are also being used for *synergic* GMAW. In this method, the operator need only adjust one process parameter to effect a desired change; the other parameters are then automatically adjusted to maintain proper welding conditions.

Fig. 3 Horizontal boring mill.

Metal removal

Metal removal is performed either to prepare a component for another manufacturing operation, such as machining weld joint preparations, or to produce a finished shape. Traditional machining processes are considered to be for chip removal or abrasive removal. The chip removal operations used in steam generation equipment production include turning, milling, drilling, boring and broaching, while grinding, sanding and grit blasting are examples of abrasive operations. Nontraditional metal removal operations, such as electrodischarge machining (EDM) and arc gouging, are also used.

B&W's fabricating shops have a variety of manufacturing equipment ranging in size from small hand tools to machines with 100 t (91 t_m) programmable rotary tables. Larger pressure vessels require larger machine tools. For example, the vertical boring mill shown in Fig. 2 can accommodate a part 16 ft (4.9 m) in diameter and 10 ft (3.0 m) high, while the horizontal boring mill in Fig. 3 has a rotary table capable of supporting 100 t (91 t_m). B&W's newest addition has a work envelope of 41 ft (12.5 m) X-axis, 21 ft (6.4 m) Y-axis, 7 ft 10 in. (2.4 m) Z-axis and 5 ft 3 in. (1.6 m) W-axis.

Fig. 2 Vertical boring mill.

As in other manufacturing areas, the most apparent innovation in machining is computer control of the equipment. Part configurations, programmed off line, are transmitted by distributed numerical control (DNC) to many machines on the shop floor. (See Fig. 4.) More recent user friendly computer numerical control equipment also permits sequences to be programmed by the machine operator. Computer numerical control (CNC) machining centers with automatic tool changing are common. In addition, multi-spindle numerical control (NC) programmable drills are often used to prepare steam generator tube bundle support plates. (See Chapter 52.) These and other refinements permit some tools to be accurate within 0.00005 in. (0.00127 mm).

Fig. 4 Computer numerical control machining.

CNC thermal cutting machines, oxy-fuel and plasma, make use of down loaded computer generated nesting programs to maximize material use. The addition of a multi-station CNC punch, featuring an integral plasma burning head, to the metal forming inventory complements the thermal cutting tables.

Electrical discharge machining (EDM), a nontraditional method, removes metal by spark erosion between an electrode and the workpiece. In this process, intricate shapes, matching those of the solid carbon electrode, can be burned into the workpiece. The solid electrodes are being replaced in many cases by the CNC wire electrode machine.

Automation is increasing as computers often control entire shop sections. Material, automatically moved into storage buffers then to CNC machines, is scheduled and routed using computerized identification systems. Probes are used to check the machined dimensions.

In 1986 B&W implemented a major loose tube automation program at its West Point, Mississippi manufacturing plant. Electronically integrating the design, routing, scheduling and manufacturing process control, the plant reduced its typical order turnaround time from 17 to 3 days.

Forming

Forming is the permanent plastic deformation of a workpiece accomplished by applying mechanical forces through tooling to the workpiece. The primary objective of forming is the production of a desired shape.

Basic forming elements include the material to be formed, the tooling to change the shape, the lubrication between the tools and the workpiece and the machine used to hold the tooling and to apply the deformation forces.

In sheet forming operations, the forces are primarily applied in the plane of the sheet. The thickness is not constrained between opposing tools but may change due to in-plane strains. Sheet forming operations also include tube forming processes using an internal or external tool and sometimes both. Most sheet forming operations are performed cold.

B&W's principal forming application is plate and tube bending. Plate is bent to form large cylindrical shells, while tubes are bent to form flow loops. Two methods are used to bend plate: press die forming to make half shell courses (two seam welds) and roll bending to form full round shell courses (one seam weld). (See Chapter 52.)

Tube forming is done primarily on draw benders. The clamp die, bend die and pressure die are always used; the wiper die and mandrel are only required for tight radius bends with thin wall tubing. Forming dimensions are typically expressed relative to tube diameter. For example, a 1D bend has a centerline radius equal to the tube diameter. Tubes with wall thickness greater than 10% of the OD do not require a mandrel or wiper die for forming.

A variation of draw bending is *booster bending*. Here the tube is clamped ahead of the bend die and an axial force is applied to push the tube into the bend region. This shifts the neutral axis toward the outside of the bend to reduce thinning on the outside and increase thickening on the inside.

Fig. 5 Tube bending by computer numerical control.

The application of numerical control to bending, as shown in Fig. 5, is highly advantageous for tubes containing multiple bends of the same radius. The tube is fed into the bender, gripped by the axial transporter and moved to a reference location. All bends are then made automatically as programmed. The transporter rotates the tube for out-of-plane bends and feeds it axially to the next tangent point.

Fabrication of fossil fuel equipment

Drum fabrication

A large steam drum, complete and ready for shipment, is shown in Fig. 6. The drum consists of multiple cylindrical sections, or *courses,* and two hemispherical heads. Large stubs and nozzles along its length serve as flow

Fig. 6 Utility steam drum prepared for shipment.

connections for the steam generating and superheating surfaces of the boiler. The large nozzles along the bottom and on the hemispherical heads are connections for the downcomers, which carry water to the various generating system circuits.

Drum fabrication begins by pressing flat plate into half cylinders or by rolling the plate into full cylindrical shells. When short shell sections must be used because of material dimensions or heat treatment requirements, the plate is rolled into a cylinder. For large drums, the plate is pressed into half cylinders then welded longitudinally, forming a course. The drum length is achieved by circumferentially welding the required number of shell sections. Longitudinal and circumferential welds are made using the automatic submerged arc process.

Drum heads may be hot formed by pressing or spinning flat plate in suitable dies. The heads are attached to the completed cylinders by circumferential welds.

Automatic welding techniques are often used to join the various shell courses and heads. Fig. 7 shows a two wire submerged arc welding machine on a longitudinal seam. Filler wire is continuously fed to the weld area, which is completely covered by a granular flux to prevent oxidation and enhance metal properties. The electrode carriage, moving along the weld joint, is guided by a seam tracker. Depending on the plate material and thickness and the configuration of the weld, multiple passes may be required.

For submerged arc circumferential welds, the drum is rotated while the welding head and flux applying equipment are stationary. An electromechanical guidance device may also be used to keep the arc tracking the weld groove.

Prior to and throughout all welding operations, preheat may be applied to minimize stresses and metallurgical transformations. In some cases, heating is maintained after welding until the vessel has been stress relieved.

Engineering advances in designs and the use of high strength, high temperature alloys have required the development of new weld techniques, especially for stainless steels, Inconel and other high nickel alloys. The tungsten inert gas (TIG) process is often specified to attain full root penetration. These processes also assure a uniform inside surface contour that does not interfere with fluid flow.

Nozzles and stubs are machined from hot forged billets, pipe or tube. Larger connections may be integral with the heads, formed from rolled plate, or made from heavy wall pipe as shown along the bottom of the drum in Fig. 6. Attachments are manually made to the drum with the shielded metal arc process using coated electrodes, by automatic satellite welding with the submerged arc process, or by semi-automatic welding with the metal inert gas (MIG) process. Other techniques are being developed that will significantly contribute to welded joint quality and increase the deposition rate of weld metal. Narrow weld grooves in thick plate are used to reduce weld volume in the joint.

Every carbon steel drum in which the material thickness is greater than 0.75 in. (19.1 mm) must be subjected to a final postweld heat treatment. In some cases, several short cycle stress relief operations may be applied to reduce preheating. For applications requiring high impact strength, formed heavy plate is water quenched to cool it rapidly from 1600 to 600F (871 to 316C). This heat treatment is followed by a normal stress relief, which consists of heating to 1150F (621C), holding for one hour per inch of thickness and furnace cooling to 600F (316C).

All longitudinal and circumferential drum welds and the adjacent base material must be subjected to radiographic examination. For this purpose extensive x-ray equipment is available. This equipment ranges from machines rated at 400 kV to the 8 MeV linear accelerator shown in Fig. 8. Ultrasonic evaluations are used

Fig. 7 Welding longitudinal drum seam with double submerged arc machine.

Fig. 8 Linear accelerator (8 MeV) for inspection of thick wall welds.

where x-ray examinations are difficult or impractical. Surface and subsurface defects may be located using magnetic particle examination. Liquid penetrant examination also reveals surface flaws. Ultrasonic examination of all heavy plate is now standard procedure.

After all pressure parts have been welded, the vessel is ready for the finishing operations, which include the installation of attachments and internals. Fig. 9 shows an internal baffling arrangement that is partially installed in a large utility boiler steam drum prior to attaching the second hemispherical head. The balance of the structure is disassembled then re-assembled inside the drum for shipment.

Headers as terminal pressure vessels

Headers are used in steam boilers to connect two or more tubes to a circuit. Headers are generally less than 36 in. (914 mm) ID and can be fabricated from seamless tubing, pipe or hollow forgings. Access to their interiors is through handholes. Headers vary in length and can be up to 100 ft (30 m) long. Fig. 10 shows a superheater header nearing completion.

To fabricate a header, several pipes or hollow forgings are first joined by submerged arc welding. The ends of the header stock may be spun hot until closed, may have spherical or elliptical forged heads welded in place, or may be machined to accept a welded-in flat closure plate. With a spun closure, the closure center is machined away and replaced by a fusion welded steel plug, which assures a tight seal.

Completion of the header includes drilling and machining the tube and nozzle openings and fitting and welding the nozzles and stubs (Figs. 11 and 12). Other attachments are added by manual or semiautomatic welding followed by heat treating, cleaning and finishing steps. These operations are conducted following strict quality control procedures.

While most headers are round in cross-section, square

Fig. 10 Large superheater header nearing completion.

headers are sometimes used. In fabricating the square sections, the stock is formed at forging temperatures.

Tubes as heat transfer surfaces

Tubes, in a variety of arrangements, are the heat absorbing surfaces and flow circuits in today's steam generating units. Tubes range from 0.875 to 5.0 in. (22.23 to 127.0 mm) OD; they may be plain or fitted with studs for extended surface in some furnace wall panels. Fig. 13 shows an opening in a wall panel surrounded by studded tubes. Materials range from inexpensive carbon steel to chromium molybdenum alloys to composites with a metallurgically bonded stainless or Inconel sheath over carbon steel.

Although past practice was to ship and field assemble each wall panel tube, the tubes are now shop assembled into panels prior to shipping. A wall panel is a mem-

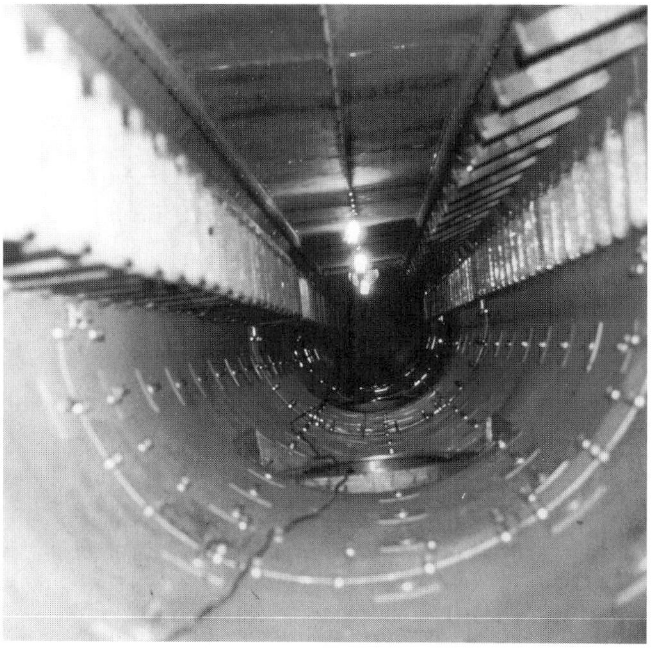

Fig. 9 Partially installed steam drum internals.

Fig. 11 Header drilling on vertical boring mill.

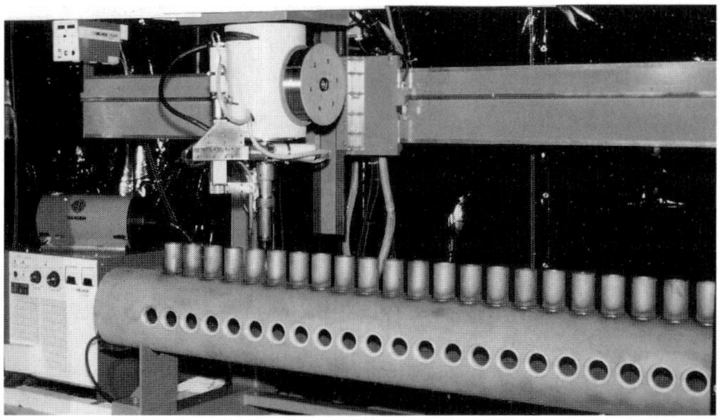

Fig. 12 Header tube stub installation.

Fig. 14 Wall panel membraning operation.

braned assembly of tubes. The assembly is begun by placing a metal strip between each row of tubes (Fig. 14). This strip is then welded to the tubes on automatic equipment, which can perform as many as 6 welds simultaneously. A gas-tight metal wall surrounding the enclosed volume is created. The membraned wall panel design eliminates the need for a pressure-tight casing and reduces field erection costs.

Walls are typically assembled in sections up to 10 ft (3.0 m) wide × 90 ft (27.4 m) long, although larger panels have been fabricated. The panels may be formed into various configurations by using one of two bending machines (Fig. 15). To further reduce field erection costs, some panels and sections with headers are preassembled into modules.

In place of wall panels, convection superheater surface uses continuous tube sections. A typical horizontal superheater section, shown in Fig. 16, is made from relatively short lengths of straight tube that are first bent, then welded together to give the proper configuration.

Close return bends are made by first hot upsetting a section of straight tube to increase the wall thickness at the bend location. Bending and hot sizing then form the tube to the desired dimension. The prepared *candy canes*, or *hairpin* bends, are joined by computer controlled orbital TIG welding to form a section. Upsetting and hot bending operations are being replaced by CNC booster bending. All other bends in a section are cold formed.

To complete a wall panel or section, supporting lugs, brackets and, in the case of panels, wall boxes are welded into position. Sections are hydrotested at 1.5 times design pressure, as required by Code. The sections are painted and often flow coated with an asphaltic material. Wall panels are painted white to enhance worker vision within the boiler during erection.

Fig. 13 Studded wall panel inspection.

Fig. 15 Wall panel bending.

Fig. 16 Superheater sections undergoing test.

Long flow economizers

A *long flow economizer* is an assembly of finned tubes and headers installed within a boiler. This unit absorbs much of the residual heat from the low temperature flue gases to improve boiler efficiency.

Value analysis has been beneficial in redesigning long flow economizers, reducing manufacturing costs, lead time and field erection costs. One economizer design change eliminates bending the tube ends before fitting them into the inlet and outlet headers. The ends may be swaged to a smaller diameter in an automated process where the OD and ID of the swaged section remain concentric. Fins, formed from levelled cold rolled flat stock and sheared to size, are automatically submerged arc welded on both sides to a shot blasted tube. After adding the fins, all welds are inspected. The tube ends are welded to inlet and outlet headers in jigs to form platens, which are furnace stress relieved. Finally the platens are assembled into an economizer module in a fixture. This fixture, with stiffeners added, is also used as a shipping frame for the unit. The frame is then used at the job site to assemble the economizer module into the boiler (Fig. 17).

Structural steel and casings

As much as 11,000 t (10,000 t_m) of structural steel may be required to support a modern utility boiler. Most of this steel is available in standard mill shapes; however, some members are fabricated from plate material into T- and H-shaped sections using submerged arc welding.

In modern units, water- or steam-cooled tubes constitute the basic structure of the boiler enclosure. Most boilers also contain nonwater-cooled, or cased, enclosures. (See Chapter 22.)

Fuel burning equipment

B&W manufactures a great variety of burners for many fuels including oil, gas, coal, refuse, black liquor, wood bark, combinations of fuels and special fuels. Because many duplicate burners are built, product standardization and volume manufacturing techniques have evolved. One example is the manufacture of Cyclone furnaces (Fig. 18).

Cyclone barrels, ranging from 5 to 10 ft (1.5 to 3.0 m) in diameter, are composed of many duplicate tubes bent to a semicircular shape and joined to form a cylinder. Jigs, fixtures and power tools are used for these repetitive operations.

Cyclone furnaces are completely shop assembled in erection frames that allow stress relieving the final product. This permits close quality control to assure that American Society of Mechanical Engineers (ASME) Code and B&W standards are met. Before shipment, a refractory coating is applied to the studded inside surfaces and the outside is covered with a protective casing. See Chapter 14 for further details on Cyclone furnaces.

The B&W coal pulverizer is another product that benefits from standardization, although some variations are required due to installation configurations. The EL pulverizer series for smaller boiler applications has been supplemented by the MPS series (Fig. 19). A large MPS pulverizer has the capacity to grind 105 t (95 t_m) of coal per hour. (See Chapter 12.)

The top, intermediate and lower pulverizer housings are fabricated from plate material. The grinding ring, cast wheels of the MPS and cast balls of the EL are cast from wear resistant alloys by the V-process.

The pulverizer gear box, a welded and machined fabrication, encloses precision bevel and helical gears, which are flooded with lubricating oil. Each gear box is tested to assure proper performance. The pulverizer gear box assembly is also tested before being shipped.

Tube manufacturing

Thousands of feet (meters) of tubes made from a variety of steels and alloys are required for a large steam gen-

Fig. 17 Long flow economizer prepared for shipment.

erating unit. Tubes are used for heat transfer and steam generating surfaces, including furnace walls and floors and in superheaters, reheaters, economizers and air heaters. Headers may also be classified as tubes, because their manufacturing process is similar to that of tubes. However, at 3 to 36 in. (76 to 914 mm) OD, they are generally larger than tubes.

Two types of tubes are used: seamless and welded. Each has applications based on user and Code requirements that define diameter, wall thickness and chemistry. (See Chapter 6.)

Seamless tubes Seamless tubes are pierced from solid rounds of steel, which are produced in electric arc furnaces. The process is carefully monitored to produce quality steel consistent with application requirements.

Tubes for boiler application are furnished hot finished or cold drawn, depending on size, tolerance and finish desired. Metallurgically, the two are similar; the differences are surface finish and permissible tolerances. To obtain a smooth, even surface and close tolerances, stainless tubing is generally finished by cold drawing only. This process also permits grain size control in final heat treatment, which is important in high temperature applications of austenitic stainless steels. The production of seamless tubes requires special tools and careful control of heating and manufacturing procedures.

Electric resistance welded (ERW) tubes The ERW tube making process begins with a coil of high quality, fully killed steel strip that has been shot blasted for surface cleanliness. The strip is uncoiled and levelled and its edges are skived to remove defects. Passing through sequential forming rolls, the tube edges are fused without filler metal by an electrical current of 450 kHz. An exclusive feature of the process is on-line electronic weld imaging, which determines weld integrity. This is typically joined at the welding station by ultrasonic testing.

ERW tubing is produced in sizes from 1.0 to 4.5 in. (25 to 114 mm) OD with walls from 0.095 to 0.650 in. (2.413 to 16.51 mm) in carbon, carbon-molybdenum and alloy steel grades. Tubes can be delivered as cold drawn and normalized.

Fig. 19 MPS pulverizer lifted for barge shipment.

Fig. 18 Cyclone furnace manufacturing bay.

Engineering-manufacturing interface

Communications between the key areas of manufacturing and engineering are continually improving. As in other manufacturing areas, computers are largely responsible for many of the improvements. Engineering drawings can be electronically transmitted to the manufacturing facilities and videophones are often used to resolve plant problems.

Value analysis, a logical planning tool, has also streamlined projects from the engineering analysis through the manufacturing steps. The reduction of component complexity and improvements in product quality and delivery are results of this technique.

Equally important as computerization and value analysis is the early participation of manufacturing personnel in the product design. Their years of experience provide objective analyses of potential design changes.

In summary, B&W's manufacturing emphasis continues to be on high quality, competitively priced boiler components. Proficient engineering design and follow up quality control assure that components meet all codes and customer specifications.

Quality assurance and quality control

The quality assurance group establishes planned and systematic practices to assure that all products conform to technical requirements. In addition, quality control personnel ensure that all products meet applicable codes and specifications.

The ASME Boiler and Pressure Vessel Code (see Appendix 2, Codes and Standards Overview) provides rules for the safe construction of boilers and other pressure vessels. In the United States, these rules are also usually part of state laws. The ASME Code is written by industry members, inspection agencies and users of boilers and pressure vessels. It strikes an acceptable balance between safety and practicality.

Section I of the Code, *Power Boilers*, requires a boiler manufacturer to have the fabrication procedures, as well as the design and material selection, monitored by a third party inspector. This authorized inspector is usually employed by an insurance agency and is licensed by the National Board of Boiler and Pressure Vessel Inspectors. In plants that produce a large volume of boiler components, this inspector may be a full time resident. He checks drawings, reviews material certifications, inspects welds, reviews radiographs, witnesses hydrostatic tests and monitors the general workmanship achieved by the fabricator. At the completion of a job, when he is satisfied that the parts comply with all aspects of the Code, he signs the ASME Code data sheet.

All welding performed on power boilers and pressure vessels must be done by Code qualified personnel following qualified procedures. The personnel must be trained to use the correct filler material, welding position and process and to apply any required preheat and postweld heat treatments. In addition, procedures must be qualified to assure consistent, high quality welds. To maintain employment of these qualified personnel and assure appropriate implementation of procedures, B&W operates welding schools and employs trained instructors at its manufacturing facilities.

Nondestructive examination

Nondestructive examination provides a quality control of base metals and weldments. There are four predominant methods that pertain to the pressure vessel industry: radiography, ultrasonics, magnetic particle and dye penetrant. The first two are used for volumetric examination and the latter two provide surface examination.

Radiography Radiography (x-ray) has been a primary weld quality control technique for many years. The first x-ray units were rated at 200 kV, but the largest is now an 8 MeV accelerator, shown in Fig. 8. These machines are supplemented by gamma ray equipment; this equipment uses the radioisotope iridium-192 for activity from 10 to 100 curies and cobalt-60 from 10 to 2500 curies.

The choice of using x-ray machines or radioisotopes is based on the mobility of the part to be examined and the availability of an x-ray facility. X-ray units can best be used in shops where fixed position equipment is available. Isotopes may be used to reduce the time needed to examine a circumferential weld seam. This examination is done by locating the isotope, usually cobalt-60, on the centerline of the unit, wrapping the outside diameter of the weld seam with x-ray film and, using remote control, exposing the film. For field use or when the component can not be readily moved, radioisotopes are preferred.

To assure correct exposure time, a small three hole penetrometer is placed on the object being x-rayed; this causes an image to form on the film. The penetrometer thickness and hole size are specified in applicable regulatory documents. When properly used, an image of the holes appears on the film and, therefore, flaws of equal or larger size are also expected to appear. The ASME Code specifies acceptance standards with respect to the size of porosity and slag inclusion defects. It also specifies those welds that must be radiographed in pressure vessels.

A recent development is real time x-ray, shown in Fig. 20, which is used to examine tube-to-tube welds. This technology permits viewing the weld seam on a video screen while the radiographic process is occurring. The results may also be recorded on video tape.

A significant advantage of radiography over other nondestructive methods is that a permanent film record is made of each inspection. These records are retained for at least five years or as required by a contract.

Ultrasonics (UT) UT examination has proven to be a valuable tool in evaluating welds, plate and forgings of pressure vessels. This examination can be applied to most sizes and configurations of materials.

In generating the sound for UT testing, piezoelectric materials are used. When pulsed by an electric current, these materials vibrate and generate a sound wave. In addition, when the materials receive sound wave vibrations, they also generate electronic impulses. These impulses may be viewed on an oscilloscope to locate a defect in the material or weld. Sound is generated in frequencies ranging from 200 kHz (1 Hz = 1 cycle/s) to 25 MHz but is usually in the range of 1 to 5 MHz.

Electrical pulses are directed from an oscillator to a transducer that is applied to the material being examined; a liquid couplant is used for good contact. The transducer generates high frequency sound that is directed into the material. If a flaw is encountered, some of the sound is reflected to the transducer, which in turn generates an electrical impulse. This impulse is then dis-

Fig. 20 Real time x-ray of tube welds.

played on an oscilloscope, showing the size and location of the defect. With proper manipulation of the transducer, the entire sample can be examined. Flaw acceptance standards are given in the ASME Code.

In addition to flaw detection, ultrasonic techniques are used to measure material thickness.

Magnetic particle examination (MT) Many fabrications within B&W rely on magnetic particle examination. Although limited to magnetic materials, the test may involve the dry method, using alternating current (AC) or direct current (DC), or it may consist of the wet method, which uses normal or ultraviolet light for examination. All procedures involve generating a magnetic field in the part being examined. A flaw on or near the surface becomes discernible by the accumulation of magnetic particles. The method used depends on the size, shape and surface condition of the material being tested.

Magnetic particle inspection provides an inexpensive method of determining the surface and near surface condition of magnetic materials. It can be used in the manufacturing facilities or in the field.

Liquid penetrant examination (PT) In contrast to magnetic particle, liquid penetrant examination can be used for magnetic or nonmagnetic materials. However the examination only reveals surface flaws. While procedures vary, all examinations begin by coating the surface with a red dye. The bulk of the dye is removed, leaving it only in any small openings in the surface. Although it may not be readily visible, the remaining dye, indicating any flaws, is then brought out by applying a white developer.

Fabricators also use liquid penetrant examination for welds or base metals of nonmagnetic material, such as stainless steel or Inconel. This examination can be performed during manufacturing or in the field. The equipment is inexpensive and portable and personnel can be easily trained for the procedure.

Quality control of fossil fuel equipment

A quality control system must be maintained throughout the manufacturing operation. The initial step is the inspection of incoming material to assure that it conforms to the purchase order requirements and the appropriate ASME Codes.

In addition to examinations stipulated by the Code, B&W performs other nondestructive tests as process quality control measures. Shop welded tubes in boiler wall panels are radiographed. A percentage of the stub-to-header welds is examined by the magnetic particle method. Radiography is used to examine a percentage of the butt welds in the superheater, reheater and economizer sections. These sections may also be hydrostatically tested.

Internal cleanliness of pressure parts is an important part of quality control. Tube IDs are cleaned with sponges and steel balls, while ODs are shot blasted and coated with a rust preventive.

In-process and final inspection, which consists of checking critical dimensions and verifying that all required nondestructive examinations have been completed, is part of the quality program. A final inspection of car loading is made to assure safe delivery of components to the job site.

Technologies

The development and use of state-of-the-art manufacturing technologies have continued to be important to B&W. The company maintains an entire group dedicated to manufacturing technology research and development.

Work has been conducted to improve all aspects of the welding operation, including planning, materials, process, new applications and automation. In the area of weld planning, computer based models have been generated to assist in weld procedure optimization. Work is also underway to reduce the cost of welding consumables by eliminating the weld wire drawing operations. When commercially available alloys are found to be lacking in a given operation, new materials, which can provide the required mechanical and corrosion properties, are developed.

Improved welding procedures are continuously being developed for the common processes as well as for the more exotic approaches, such as explosion welding, laser welding, electron beam welding, plasma MIG and variable polarity plasma arc welding. In the area of welding automation, B&W has advanced off-line programming and sensor technologies to improve the flexibility of automated systems. One outcome of B&W's research efforts is the development of a manufacturing process that allows complex components to be produced entirely of weld metal. The patented technique, known as *shape melting*, promises to reduce cost and lead time while improving the properties of components currently produced by casting or forging.

Dramatic changes in manufacturing are occurring with the advent of computer technologies. Newer machine tools are expected to provide greater accuracy in motion control, repeatability and in process measurement. Bends may be made in sequence rather than in a batch mode, thereby reducing part inventories. Heat treating is improving with computer control of furnace temperature and atmosphere. Inspection is being automated through the use of computer controlled coordinate measuring equipment and vision systems and real time radiography. Furthermore, manufacturing operations are being integrated with computers on a plant wide scale to facilitate material flow and to apply statistical process control tools.

Reference

1 *Welding Handbook*, Vols. 1 and 2, 8th edition, American Welding Society, 1987.

1300 MW coal-fired power plant under construction.

Chapter 39
Construction

Construction plays a major role in the success of an overall project. Improvements in quality, safety and cost, together with reduced construction schedules, are required as contractors and owners strive to minimize field operations.

During initial project development, the construction plan and requirements needed for material handling must be incorporated into the overall design. In addition, project schedule requirements will affect manpower and the erection methods that the contractor must use. These points, as well as safety and other site considerations, are important and must be evaluated early in a project's development cycle.

A number of boiler types are small enough to permit shipment completely assembled. However, larger industrial and central station boilers are shipped to the job sites in various stages of fabrication and subassembly. A central station boiler, with its associated heat transfer equipment and auxiliaries, may weigh more than 12,000 t (10,885 t_m), requiring the equivalent of 500 railroad cars to ship the material to the site over a period of several months. The modularized construction practices in use today result in some boiler components weighing 400 to 1000 t (363 to 907 t_m). In these cases, overland and barge shipment becomes a unique operation.

The field assembly of a high capacity steam generating unit, whether fossil fueled or nuclear, requires efficient, well engineered, well organized erection methods to permit installation of the equipment within a reasonable time and at minimum cost, without sacrifice of quality. Computerized drafting, modeling, scheduling and tracking have become the standard approach for complex or fast track projects in the construction industry. The cost of erecting a steam generating unit, such as that shown in Fig. 1, represents a sizable part of the total investment in the plant.

Sound field assembly is essential for correct functional performance. For this reason, the construction organization must apply techniques that will complement the designer's skill and the fabricator's craftsmanship. Components must be on the job site at the right time with the necessary tools, equipment, manpower and supervision for assembly. The construction organization must be experienced in estimating, planning, quality control, cost analysis, labor relations, tool and equipment design, technical services and finance. Direct liaison should be maintained with the engineering and manufacturing departments, making field assembly a continuation of shop fabrication.

Fig. 1 Construction of a central station boiler for a large electric generating facility.

Cost and schedule

Cost and time estimating are two of the most important functions of a construction organization. The process begins with a complete understanding of the project specifications and scope. A detailed list of specific activities or tasks on a unit basis (which can be individually estimated) is prepared covering the entire project. This is obtained from the customer or is developed by the estimator based upon customer specifications and drawings. Direct manhours for each craft labor category are then assigned to each activity based upon historical data and general experience. As an example, insulation is estimated based upon the number of hours per unit area [ft^2 (m^2)] times the insulated area of the proposed equipment. The total hours for each craft and for the total project are then calculated. The totals are frequently corrected for a variety of factors, e.g., local labor productivity, overtime and site factors. To the direct labor total, indirect labor such as supervisors, foremen and housekeeping, etc., are added to produce a total project estimate.

While detailed estimates are required for specific projects, total estimates for nominal scope jobs frequently fall into general ranges. For coal-fired utility boilers of nominal scope, erection manhours usually range from 1 to 1.75 manhours per kilowatt. Refuse-fired boilers range from 85 to 110 manhours per ton of refuse. In other cases, such as flue gas desulfurization systems, the estimates do not fit such ranges well because of the dramatically different equipment scopes possible for a given size unit.

The elapsed time required to complete a project is largely a function of manhours. There is a limit, however, for any given unit, to the number of workers that can be used during any given stage of erection. If more than an optimum number are employed, the lack of adequate work space results in diminishing returns and excessive costs. Fig. 2 shows the typical number of weeks required for various manhour expenditures for a single boiler unit and the maximum number of people that can be efficiently employed during peak operations.

Erection time span can also be affected by performing on a shift or overtime basis. With these alternatives, additional costs are incurred due to the inefficiency result-

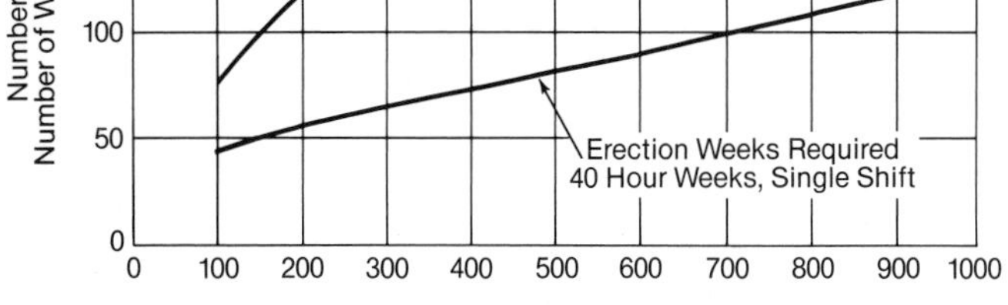

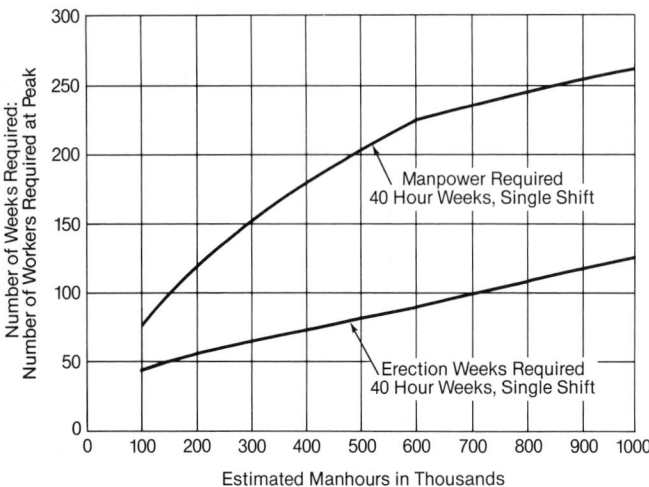

Fig. 2 Graph showing typical number of weeks and number of workers required for boiler erection related to total manhours.

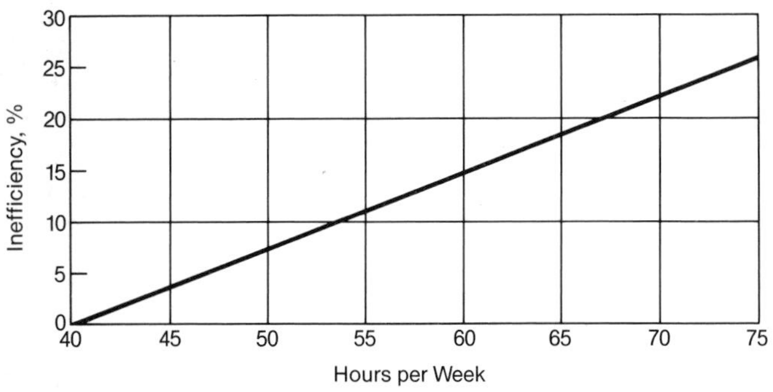

Fig. 3 Labor inefficiency factor for overtime.

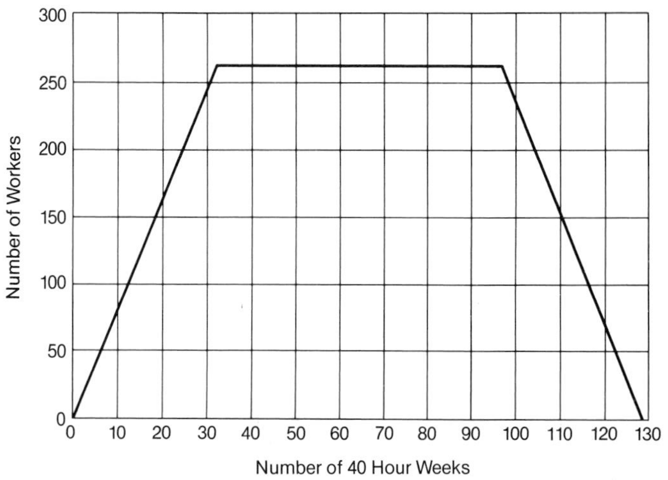

Fig. 4 Manpower graph for unit requiring 1,000,000 manhours for erection.

ing from working longer hours, hours not worked, fatigue or shift turnover inefficiencies. Many combinations can be worked to fit even the tightest schedule. (See Fig. 3.)

Detailed schedules are developed by sequencing all activities according to their corresponding relationships and then including elapsed times. Work activities can be organized into work breakdown structures which are useful in organizing and monitoring progress. Schedules are established using computer based scheduling systems. A broad range of detailed schedules and reports are derived from such systems that permit the overall coordination of labor by craft or specialty, erection equipment commitments, material and equipment delivery dates and other material and personnel resource requirements. Fig. 4 provides a sample overall manpower loading chart for a one million manhour project employing 263 people at the peak of construction.

Approaches to construction

There are three basic approaches that may be used in a construction project: 1) *knocked down*, the traditional method, 2) *field modularization*, and 3) *shop modularization*. The knocked down approach uses components sized for rail or truck shipping. Field modularization combines these traditional ship units into larger components, or modules, at the construction site for final installation. Off-site assembly, or shop modularization, combines the traditional ship units into subassemblies at the manufacturer's plant. While modularization may be cost effective, Babcock & Wilcox (B&W) has learned that the extent of effective modularization is site specific.

Modularization is an integrated approach to power plant construction. The plant materials are divided into subsets, or modules, which are assembled before being brought to the construction site. (See Fig. 5.)

Boiler components have been modularized for some time. Fig. 6 depicts furnace wall panels in which loose boiler tubing and membrane stock are welded to form the panels. A large modularized component, shown in Fig. 7, was fabricated at B&W's Mount Vernon, Indiana facility before shipment to the construction site.

A modularized project requires more engineering and design time than a conventional project. Planning must consider the water and land routes used for shipping the

Fig. 5 Modular components for an environmental control installation.

modules. Construction sites located on navigable waterways have the greatest potential for shipment of large modularized components. (See Fig. 8.)

Module unloading, site movements and erection procedures must be carefully considered. Special consideration must also be given to foundations, backfilling and compacting at the site. Other concurrent site activities must be evaluated when planning for modularization.

Construction technology

A successful construction project requires management expertise for the tasks and crafts involved. This effort requires: 1) a thorough knowledge of the scope of work to be performed, 2) a familiarity with the details and function of the products included within the scope, 3) the development and maintenance of a plan to complete the activities, and 4) construction tools, equipment, trained personnel, systems and methods to support the work plan.

Selection of erection method

Construction participation by all involved parties is required in planning any project to enhance constructability and identify intended erection methods. The erection method is influenced by the following factors.

Retrofit versus new plant installation The construction logistics associated with retrofitting, repairing or altering an existing unit are much different from those required for constructing a new unit. Logistics for new plant construction are further diversified, depending on whether the unit is to be built on an existing plant site, adjacent to existing units or on a new site. Retrofit operations are generally performed in a pre-outage/outage/post-outage mode.

In planning retrofit work, an erection method that permits performing many tasks concurrently is chosen. This optimizes equipment and manpower usage and minimizes outage times. When constructing on an existing plant site, the erection method must maintain a productive interface between construction activities and ongoing plant operations. Construction on a new site offers a contractor the greatest amount of freedom in selecting the most cost effective, efficient method for executing the work; however, potential obstructions include those areas of work occupied by other contractors.

Plant arrangement The overall plant arrangement plays a key role in establishing the sequence for erecting components and equipment. Plant arrangement defines the available access routes for material flow from the laydown and storage areas to the final installation points.

Scope The scope of work directly affects the construction plan and completion of the construction. Tools and equipment needed for one phase of the project may also be influenced by the requirements of other phases.

Parameters Product and component parameters, including size, weight and shipping configurations, dictate the type and size of material handling equipment and

Fig. 6 Raising furnace wall section into position. Buckstays and other parts were attached during ground assembly.

Fig. 7 Large modular component being shipped to site.

the need for special handling procedures, methods, jigs and fixtures. Material type, thickness and required erection tolerances are considered in establishing methods for component alignment, welding, postweld heat treatment, stress relieving and nondestructive examination.

Site conditions Existing site conditions and accessibility play major roles in determining the size, weight and configuration of the components to be shipped. Water accessibility represents the greatest opportunity for shipping large shop-assembled modules; otherwise, rail or truck shipments are required.

Shipping accessibility only partially controls the extent of material and equipment shipments to the job site. Upon arrival, the material must be off-loaded and transported to its storage area or final installation point. Material off-loading schemes depend upon the mode of shipment and the site conditions at the off-loading area. Transportation of the material is then limited by the width, overhead clearances and load limitations of existing roadways.

Contractor interface Interface with other contractors is an important part of a construction project. Once portions of the project have been completed by individual contractors, effective interfacing between contractors is required to achieve certain location tolerances and to complete the erection.

Schedule The project schedule, including customer specified start, finish and milestone dates, dictates the required erection sequence, manpower and timing for project completion. The quality and skills of the local labor force play a major role in the erection plan and overall sequencing of the work.

Because cost is an important criterion governing selection of the erection method, the economic advantages and disadvantages of the other selection criteria must also be evaluated. Costs of construction operations versus component shipping configurations are often arranged in a matrix. The shipping configuration reflecting the lowest total cost is considered to be the most effective approach and becomes the preferred choice for erection. Tabulated construction costs are coupled with corresponding costs for component engineering, shop fabrication and assembly, and shipment to the job site to identify the most effective overall approach for a specific project.

Extent and location of field welds Overall product quality and constructability are enhanced by minimizing the number of field welds; the required welds must be readily accessible to allow the use of state-of-the-art equipment and processes.

The scope of work to be performed and the component configuration influence the extent and location of field welds. The construction engineer must provide input regarding preferred field weld locations, suggested component configurations to suit these weld locations, and intended welding processes.

Construction engineering and rigging design

Background

B&W has historically taken an engineered approach to the execution of heavy lifts. Specialized tools, equipment, rigs, jigs and fixtures are provided for each particular project.

Modern construction includes modularized components that are erected on short timetables. To reduce construction and shipping costs the weight of support steel components has been reduced. As a result, the calculation of imposed erection loads is particularly critical.

In recent years, customer specifications have required the submittal of engineered rigging drawings and procedures prior to construction. These submittals often require the stamp of a licensed professional engineer.

Fig. 8 Barge shipment of shop-assembled unit.

Engineering principles

The role of the construction engineer involved with product erection is two fold. First, constructability is enhanced by the engineer's familiarity with and knowledge of the product's construction details and function. Second, the construction engineer is responsible for the engineering, design and specification of all systems, tools, equipment, devices and mechanisms required for erection. This dual role requires a command of engineering fundamentals and a knowledge of principles taken from civil, structural, mechanical and electrical engineering disciplines.

Engineered systems

Engineered systems required to support construction operations may generally be identified by:

1. material handling (discussed below),
2. rigs, jigs and fixtures,
3. temporary supports, shoring and reinforcement of existing structural components,
4. access and protection structures,
5. temporary structures and construction facilities,
6. civil/site work, and
7. specialized tooling, machining and equipment.

Material handling and setting considerations

Like any phase of a construction project, material handling operations must be planned and integrated into the total construction plan. The size, weight and shipping schedules of material and components to be received and installed must be known in order to plan for storage area needs and to schedule crews and equipment. Special handling provisions and special inspection and storage requirements must be known and accommodated. In addition, the plant structure must be designed to support the loads involved during construction and the clearances required to get equipment in place.

Material handling plans must reflect actual site conditions including rail and truck access, locations and condition of haul roads, and availability and location of laydown areas and warehouses. Although most material shipped to a construction site can be handled with conventional construction equipment, material handling plans must identify any special equipment or fixtures required to handle large or unusual components.

Equipment used for material handling operations may generally be grouped by function into two categories: 1) transporting, and 2) lifting.

Equipment for transporting material is limited to that intended for on-site application, and does not include trucks, railroad cars or barges listed for material delivery to the job site.

A wide range of equipment is available for material transport and lifting on the job site. Selection and use depend on the configuration and weight of the material to be handled and the relative distance between transport points and available site clearances. The varying techniques and types of equipment which can be utilized for material transport and lifting are shown in Table 1 (selected equipment is shown in Figs. 9 through 11).

While this listing is intended to provide an idea of the range of tools, methods and equipment available for material handling, it is not all inclusive. Various suppliers and manufacturers maintain and supply their own version of the generic equipment identified. The selection, specification and use of this equipment requires a thorough construction engineering effort from the planning phase of the project through the execution of the work.

Fig. 9 Lampson crawler transporter.

Depending on the equipment selected or available, various operations are involved to get the components to their final in-place position. These operations, represented in Fig. 12, may be accomplished by use of a single hydraulic crane or may require a sophisticated handling system custom designed for the application.

Serviceability

The serviceability of a material handling system is a measure of how well the system performs its intended function subject to conditions produced in its work environment.

Fig. 10 Straddle carrier.

Table 1
Material Transportation and Lifting

Material Handling System	Description	Capacity t (t_m)
Site Transport:		
Flatbed trailers	Bed dimension 8 × 40 ft *(2.4 × 12.2 m)* — deck height 60 in. *(1524 mm)* used to transport materials from storage to staging area.	20 *(18)*
Extendable trailers	Bed dimension up to 8 × 60 ft *(2.4 × 18.3m)* — deck height 60 in. *(1524 mm)* used to transport materials from storage to staging area.	15 *(14)*
Lowboy and dropdeck	Bed dimension up to 8 × 40 ft *(2.4 × 12.2 m)* — deck height of 24 in. *(610 mm)* used to transport materials from storage to staging area.	60 *(54)*
Crawler transporter	Specially designed mechanism for handling heavy loads; see Fig. 9, Lampson crawler transporter for an example of the Lampson design.	700 *(635)*
Straddle carrier	Mobile design to transport structural steel, piping and other assorted items; see Fig. 10, straddle carrier for an example of this design.	30 *(27)*
Rail	Track utilized to transport materials to installed location. Continuous track allows material installation directly from delivery car.	as designed
Roller and track	Steel machinery rollers located relative to component center of gravity handle the load. Rollers traverse the web of a channel welded to top flange of structural member below.	2000 *(1814)*
Plate and slide	Sliding steel plates. Coefficient of friction — 0.4 steel on steel, 0.09 greased steel on steel, 0.04 Teflon on steel. See Fig. 11, sliding plate transport for movement of 1200 t *(1089 t_m)* vessel.	as designed
Air bearings or air pallets	Utilizes film of air between flexible diaphragm and flat horizontal surface. Air flow 3 to 200 ft^3/min *(0.001 to 0.09 m^3/s)*. 1 lb *(4.5 N)* lateral force per 1000 lb *(454 kg)* vertical load.	75 *(68)*
High line	Taut cable guideway anchored between two points and fitted with inverted sheave and hook.	5 *(4.5)*
Lifting:		
Chain hoist	Chain operated geared hoist for manual load handling capability. Standard lift heights 8 to 12 ft *(2.4 to 3.7 m)*.	25 *(23)*
Hydraulic rough terrain cranes	Telescopic boom mounted on rubber tired self-propelled carrier.	90 *(82)*
Hydraulic truck cranes	Telescopic boom mounted on rubber tired indepedent carrier.	450 *(408)*
Lattice boom truck cranes	Lattice boom mounted on rubber tired independent carrier.	800 *(726)*
Lattice boom crawler cranes	Lattice boom mounted on self-propelled crawlers.	2200 *(1996)*
Fixed position crawler cranes	Lattice boom mounted on self-propelled crawlers and equipped with specifically designed attachments and counterweights.	750 *(680)*
Tower gantry cranes	Tower mounted lattice boom gantry for operation above work site.	230 *(209)*
Guy derrick	Boom mounted to a mast supported by wire rope guys. Attached to existing building steel with load lines operated from independent hoist. Swing angle 360 deg *(6.28 rad)*.	600 *(544)*
Chicago boom	Boom mounted to existing structure which acts as mast, and to which is attached boom topping lift and pivoting boom support bracket. Load lines operated from independent hoist. Swing angle from 180 to 270 deg *(3.14 to 4.71 rad)*.	function of support structure
Stiff leg derrick	Boom attached to mast supported by two rigid diagonal legs and horizontal sills. Horizontal angle between each leg and sill combination ranges from 60 to 90 deg *(1.05 to 1.57 rad)*; swing angle from 270 to 300 deg *(4.71 to 5.24 rad)*.	700 *(635)*
Monorail	High capacity load blocks suspended from trolleys which traverse monorail beams suspended from boiler support steel. Provides capability to lift and move loads within boiler cavity.	400 *(363)*
Jacking systems	Custom designed hydraulic or mechanical system for high capacity special lifts.	as specified

Fig. 11 Sliding steel plate transporter.

Imposed loads and forces

The weight of the component being handled is a major consideration in the design and selection of the material handling system; however, component weight is not the only load imposed. All systems operate dynamically while simultaneously generating and being acted on by forces that include steady-state and time-varying components. Forces are introduced by the operating environment, by the work being done, by the inertia of the masses put into motion and by the weights of the load and the system components.[1]

Imposed loads are generally categorized as gravity or nongravity. Furthermore, both loadings may be static or dynamic. While dynamic loading must be recognized in designing the material handling system, the magnitude of acceleration/deceleration loads is small in most cases. As a result, static loading generally governs the system design.

Static gravity loading conditions are categorized as lifted load, dead load, live load, friction or impact.

The *lifted load* is the calculated rigging weight of the component being handled. This calculation must account for manufacturing variability and normally includes a safety factor. Shapiro suggests multiplying the calculated weight by 1.03 for mass produced items and by 1.07 for singular devices.[1]

Dead load is defined as the weight of the material handling system components themselves independent of the component being lifted. In calculating this weight, the same considerations must be made with regard to potential dimensional variations in components. The suggested weight parameters act as multipliers as they relate to imposed loading or induced stress and as divisors as they provide stability or uplift resistance. The *imposed dead load* is the cumulative weight of all other components supported from the component being designed.

The *live load* is an allowance that accounts for the weight of construction personnel, scaffolding, ash (slag or dust accumulations) and snow or ice accumulations on the handled component. Because these loads are frequently unknown when the material handling system is designed, they are applied on the basis of a specified unit weight per surface area or a specified density per volume.

Friction loads are those associated with a body's resistance to movement. Friction takes on various forms depending on the system arrangement and load handling application.

The coefficients of static, rolling and sliding friction define friction loads. Table 1 lists some of these coefficients. Depending on the arrangement of load handling components, friction may become a significant part of the total design load.

Impact loads are associated with sudden shocks that occur during a load handling operation. An impact allowance, 25% of the handled load, is typically applied in the direction of load movement. Minimal impact loading is achieved by moving components at constant, slow speeds.

Nongravity loading conditions include wind loading, seismic loading and stabilization. Most applicable codes and specifications have reduced these dynamic conditions to static load equivalents. For the purpose of material handling system design, these conditions are therefore treated as such.

Wind loading, when imposed on material handling system components, is evaluated and applied in accordance with Uniform Building Code (UBC) or American National Standards Institute (ANSI) guidelines.

Seismic loading is not a major design consideration for most material handling systems. Nevertheless, seismic forces may be evaluated with UBC or ANSI specifications.

Stabilization forces are those forces required to maintain lateral stability of compression and flexural structural members. This force is arbitrarily considered to be 2% of the total force developed in the compression zone of the member being stabilized. In the case of compression members, the compression zone is the total cross-section; for flexural members, this zone is limited to the compression flange.

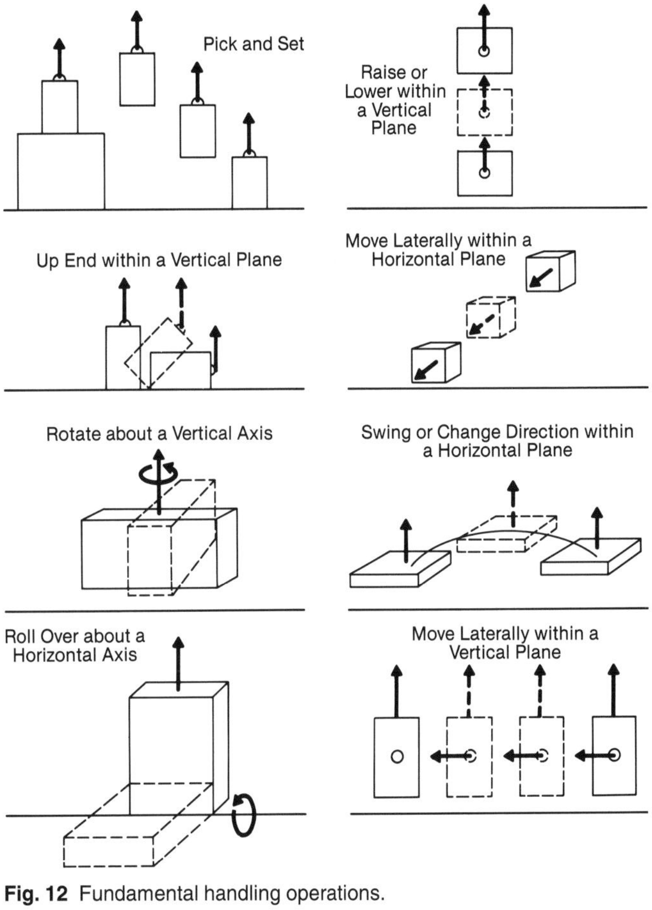

Fig. 12 Fundamental handling operations.

Load distribution

Throughout the load handling operation, the component being handled seeks a position of static equilibrium. A lifted load always vertically aligns its center of gravity with the centerline of the lifting hook. Prudent arrangement and design of the material handling system is based on a knowledge of the component's center of gravity as well as the magnitude of the resultant loads and forces acting on the system.

Depending on the handling operations involved, the relative position of a component's center of gravity with respect to its supports can vary with time. This variation produces a time-dependent variation in the magnitude of the support reactions, and thereby, the load imposed on the material handling system. Consider the steam drum shown in Fig. 13. Two separate sets of load blocks and rigging are used for raising the drum to its final position in the boiler support steel. Throughout the course of the raising operation, it may be necessary to incline the drum to clear local interferences with boiler support steel. As the drum is inclined at some angle θ from the horizontal, the load imposed on each set of blocks varies in accordance with the following expressions:

$$W_{L1} = {}^{1}\!/_{2}\ (1 + \frac{Y}{L}\ \tan\ \theta)\ W_{L} \quad (1)$$

$$W_{L2} = {}^{1}\!/_{2}\ (1 - \frac{Y}{L}\ \tan\ \theta)\ W_{L} \quad (2)$$

where

W_{L1} = portion of W_L supported by high end blocks
W_{L2} = portion of W_L supported by low end blocks
W_L = weight of component being raised
Y = 2 x (transverse distance between longitudinal centerline of drum and centerline of lifting lugs)
L = center to center spacing between lifting lugs

Per these expressions, when the drum is horizontal, i.e., $\theta = 0$ deg:

$$W_{L1} = W_{L2} = \frac{W_L}{2} \quad (3)$$

That is, the rigging weight of the drum is shared equally by each set of load blocks. However, as the angle θ is increased, the load transferred to the blocks on the upper end of the drum is increased. In this regard, each set of load blocks, rigging and drum lifting lugs are conservatively sized to accommodate the total rigging weight of the drum (Fig. 14).

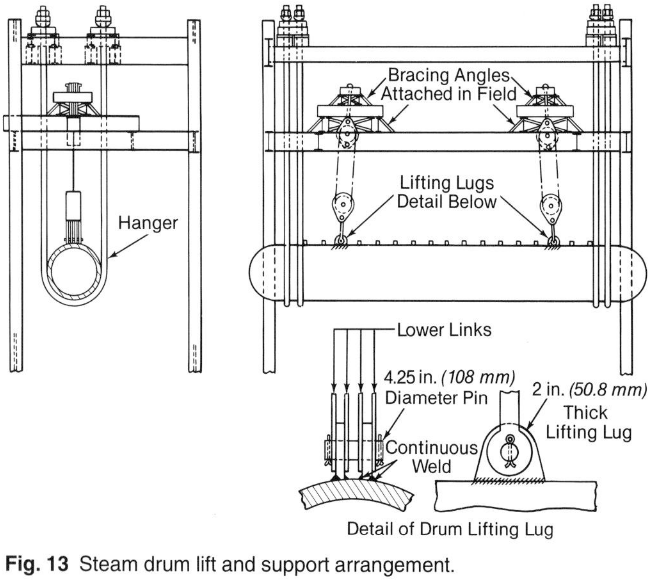

Fig. 13 Steam drum lift and support arrangement.

Fig. 14 Typical utility boiler steam drum lift.

Design loading

The material handling system design loading is established by identifying the maximum of several critical load combinations and then by multiplying this value by a safety factor of 1.10. Typical combinations include:

lifted load + dead load + live load + impact,
lifted load + dead load + live load + friction, and
lifted load + dead load + live load + impact + greatest of wind, seismic or stabilization load.

Friction and impact loading are mutually exclusive events, and therefore are never considered to be a viable combination.

Other site considerations — storage

Storage is also a critical issue. Storage areas must be laid out and material placed in storage to provide ready retrievability to support the construction effort. Generally, a numbered grid system is used in laydown yards and warehouses to identify specific storage locations. Fig. 15 shows a typical storage layout.

To support the planned construction activities, material must not only be on site and available, but must be in good condition. In addition to inspection of the material upon receipt, the material and the storage areas are inspected periodically to satisfy specific storage requirements and to ensure proper protection of the material.

With the amount of material involved in most construction projects, it is essential that an effective documentation system be used. A complete listing of items to be received is needed to ensure that a complete inventory is maintained. For each item, material receiving and storage records must effectively document receipt and disbursement status, condition of the material and current storage location.

Fig. 15 Site laydown area.

Erection procedure for a fossil fuel unit

The installation of a top supported steam generating unit is unique in that the heaviest components are at or near the top and most other major components are hung from steel rods. Field assembly of the boiler begins by positioning the major upper components in a manner that provides a design deflection after completing the boiler installation. While the following discussion is primarily concerned with top supported units, many of the basic procedures apply to all units.

Structural supports

The design of the supporting structural steel must consider the loading and stresses encountered during the installation of the boiler. The major construction-imposed loads must be evaluated. Because certain structural members may be temporarily omitted to permit access for large subassemblies, the associated additional structural loads must also be considered.

Before erection is started, all foundations must be completed and checked for specification compliance. Because most large units are top supported, as discussed in Chapter 24, erection of the structural supporting steel is the first step in building the unit. This is done by conventional methods, using cranes and derricks. Stairways and walkways are completed to the fullest possible extent to provide safe access to all parts of the unit.

Because many of the structural members govern the location of boiler and auxiliary components, supporting steel must be aligned with established building centerlines. After the structure has been aligned and plumbed, the members are rigidly connected with high tensile bolts or by welding before supporting any structural loads.

Assembly of pressure components

The construction schedule and plan include a timely sequence for installation of boiler components. The steam drum is usually the first major assembly placed in the structure. All large or heavy components, superheater modules, top headers, interconnecting pipes and wall tube panels are positioned, following the plan, in the boiler cavity while unrestricted space is available to access and hoist them.

Drums Drums for the largest boilers may weigh more than 400 t (363 t_m); they may be 100 ft (30.5 m) in length and may be located more than 200 ft (61 m) above the ground. To clear supporting steel, the drum may be inclined by as much as 60 deg (1.05 rad) during the lifting operation. Capacity and safety of the drum rigging are of primary importance in selecting the equipment and methods.

Fig. 14 shows a typical drum lifting arrangement using lugs that are shop-welded to the drum. Attachment pieces for the top of tackle blocks are located on temporary steel members above the boiler structure. These blocks should permit simple linkage connections at the top support and drum attachments. The capacity of the hoisting engine and the sizes of tackle blocks and wire rope are determined with ample safety factors.

When the drum is in its final position it is supported from the top structural steel by U-bolts that permit linear movement with temperature variations. Fig. 16 shows a typical drum and support arrangement. The drum is a major anchor point for other boiler components and, for this reason, accurate location is important. The structural members, platforms and stairways that were omitted for drum erection are then installed.

Downcomers, headers and large pipes Plant access consideration is important and the sequential plan should be carefully followed in the erection of downcomers, loose headers and large pipes. These components are shipped in the longest lengths possible to avoid costly field welds. Large downcomers are often placed in the boiler well while steel is being erected and are moved to their final location when convenient. Without proper integration into the planning sequence, these components must be shipped in shorter length or additional structural steel must be omitted to provide clearances during erection.

Fig. 16 Radiant boiler steam drum in position.

Tube walls To reduce cost and the time required for field erection a large percentage of furnace enclosure tubes are shop-assembled into membrane panels. B&W's preference is to furnish wide, long panels. However, as with any modularization approach, shipping and field conditions determine the panel sizes that can be provided. The wall panels are erected as-received from the supplier or, if conditions permit, the construction plan may include on-site assembly into larger wall sections. Panels may even be ground-assembled into complete walls with headers, casing, buckstays and doors attached. (See Fig. 6.)

Superheater, reheater and economizer These components can be furnished with loose headers and individual elements or in shop-assembled modules with or without headers. Again, the design features of the particular unit and project determine the degree of field or shop modularization. When large, heavy surface modules are to be erected, the construction plan normally includes early installation while access and hoisting clearance are available.

Tube connections Tubes are attached to drums or headers by welding, expanding or a combination of the two. Generally, tubes that withstand pressures greater than 1500 psi (103 bar) are expanded and seal welded (Fig. 17) or welded to shop-attached stubs. Shop-assembled tube to header connections, such as those in wall panels and superheater modules, are usually welded directly into the headers. (See Fig. 18.)

Design standards permit expanded tubes for low pressure boilers. Tube expanding, or rolling, is a process of cold working the end of a tube into contact with the metal of the drum or header tube hole or seat. The end of the tube is inserted into the hole and then plastically expanded by internal pressure; relieving of the pressure leaves the tube tightly seated in the drum or header.

A typical roller expander is shown in Fig. 19. This tool contains rolls set at a slight angle to the body of the expander causing the tapered mandrel to feed inward when it is turned. This is discussed in more detail in Chapter 44. For conditions of widely fluctuating temperatures and bending loads, the expanded joint must be seal welded or replaced by a shop-attached tube stub.

Field welding

The welds that are made in the erection of a steam generating unit can be divided into two classes — those for assembly of plate and structural materials and those for assembly of tubes, pipes and other pressure parts. Welding on plate and structural materials must be of acceptable quality and appearance and must satisfy sealing and/or strength requirements.

Welding that involves boiler pressure parts must conform with the requirements of the American Society of Mechanical Engineers (ASME) Boiler and Pressure Vessel Code, Section I. The Code prescribes qualification tests for procedures and operators as well as inspection requirements.

Various conditions make field welding different in many respects from welding in the shop (discussed in Chapter 38). A typical field welding environment includes dust, wind, variable temperature, rain, high scaffolding, locations remote from supply and maintenance facilities and a mobile labor force. Under these conditions, important factors in selecting welding processes and equipment include simplicity, reliability, portability, ease of maintenance and availability of spare parts and trained welders.

Manual and semiautomatic welding methods are therefore favored for field erection of standard boiler components. While there is a continuing search for and evaluation of reliable automated welding equipment for field erection and repair, it is seldom used. For example, efforts to use automatic orbital tungsten inert gas (TIG) welding machines for tube to tube welds have been limited due to difficulty in getting the precise joint fitup required, limited access for equipment and lack of specially trained operators.

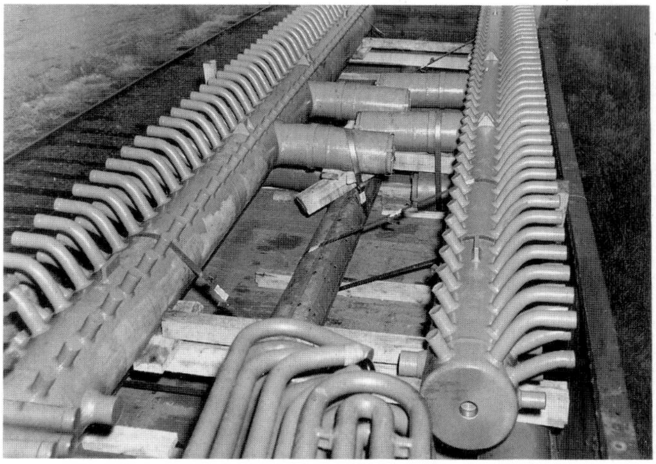

Fig. 18 Headers with shop-assembled stubs.

Fig. 17 Expanded and seal welded tube connections.

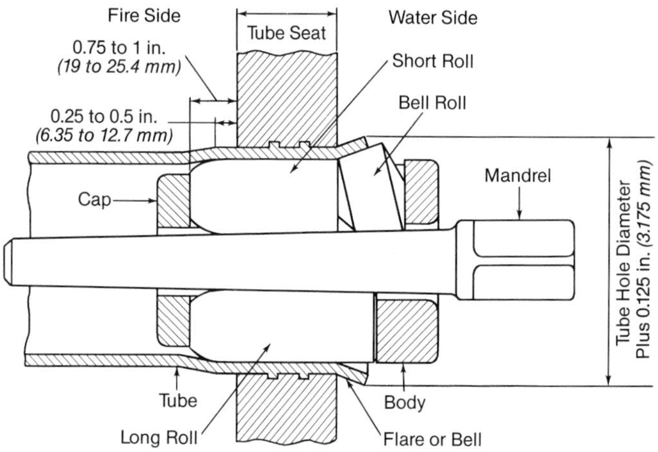

Fig. 19 Position of expander and mandrel after tube is expanded and flared.

Processes and equipment

Manual shielded metal arc welding (SMAW) and TIG welding are the most common processes used in the field. (See Chapter 38, Table 1.) These processes are versatile and can be used on all field weld materials utilizing equipment that is simple and easily maintained. In addition, most field welders are qualified in both processes.

For welding nonpressure parts or making nonpressure attachments to pressure parts, there is increasing use of semiautomatic welding. In the hands of a qualified welder, the gas metal arc (GMAW) or the flux core arc welding (FCAW) processes can make quality welds at significantly higher deposition rates compared to the SMAW process. The new, small diameter filler wires are effective in making butt and filler welds in all positions.

As with the TIG process, adequate precautions must be taken to protect the weld area from the elements, especially excessive wind. A flux core welding electrode that is self-shielded, i.e., does not require external gas for protection of the molten weld puddle, is used for erecting boiler casing and other structural components.

Applications

Pressure welds Welds joining dissimilar metals are made in the shop because they are often difficult to make and special qualification is required. Field butt welds in tubes and pipes are made using premachined weld joints that have been covered with a protective cap. Thorough inspection and cleaning are still required before fitup and welding to avoid defects.

Butt welds may be made with an open groove joint or with a backing ring that is left in place. When an open groove joint is used, the root pass is typically made with the TIG process because of the inherent greater control of the inside weld surface contour. The groove is filled using the manual SMAW process.

When the weld joint contains a backing ring, the entire weld is usually made with the SMAW process. This method is normally used for the original erection of riser and downcomer piping. Rings are never used for furnace walls. However, they may be used for other pressure boundary tube welds depending upon engineering or customer requirements or, if permitted, the preference of construction personnel.

Welding closely spaced furnace wall panel tubes or any tube with poor access to the weld groove is very difficult with the SMAW process, because the welder can not properly direct the coated electrode into the weld joint. This results in improper gas shielding leading to porosity (gas holes). Weld porosity can be reduced by using type E-7010 A1 coated electrodes rather than the low hydrogen type. Because joining furnace wall panels is time consuming and difficult, on-ground pre-assembly is maximized.

Seal welds Seal welds are used to make mechanical joints fluid-tight. Joint strength is developed by pipe threads, tubes rolled into drum holes containing grooves, or header master handholes fitted with specially designed plugs. The maximum size of seal welds made without postweld heat treatment is 0.375 in. (9.5 mm).

Special precautions are required to obtain sound seal welds; welding over pipe threads is not permitted. After seal welding, tubes should be lightly rerolled. Handhole plug seal welds should be thoroughly cleaned and preheated to avoid cracking.

Qualification of welding procedures and personnel

The requirements for qualification of welding are contained in the ASME Code, Section IX. The National Board Inspection Code (NBIC), published by the National Board of Boiler and Pressure Vessel Inspectors, references this Code for qualification of repairs and alterations to in-service boilers.

A welding procedure qualification test verifies that required mechanical properties are obtained when designated materials are welded by following a specific technique. B&W has established many procedure qualification records based upon the ASME Code, and additional qualification tests at the job site are occasionally required for special repair or alteration work.

A welder qualification test is required to verify that an individual has the skill to deposit sound weld metal in a given position while following an established procedure. Test welds on an inclined tube or pipe are usually made at each job site. The ASME has also approved an approach whereby several contractors have welders qualified at a central testing site.

Thermal treatment

Preheat Preheating the weld metal is the most effective way of preventing cracks in low alloy steel welds because it reduces the cooling rate of the metal. The required preheat temperature is specified in the field weld schedule. Preheating methods include the use of electric resistance blankets and oxy-gas torches. Special temperature-indicating crayons are used to assure the proper preheat temperature has been attained. (See also Chapter 38.)

Postweld heat treatment Requirements for postweld stress relief are also specified in the field weld schedule. Electric resistance heaters are commonly used for field postweld heat treatment and operate on conventional manual metal arc welding power supplies. These heaters offer the advantages of ease of application and temperature control.

Codes, inspections and examinations

The ASME Code, Section I, *Power Boilers*, contains requirements for the field assembly of power boilers and internal piping. ASME B31.1, *Power Piping*, contains requirements for the installation of boiler external piping, and applies by reference in ASME Code, Section I. Although most boilers built in the U.S. comply with the ASME Code, this Code has no legal status unless adopted by a state, municipality or other responsible body. The Code is administered by the National Board of Boiler and Pressure Vessel Inspectors.

The ASME Code requires a valid ASME Certification of Authorization to authorize field assembly of power boilers. To obtain this an ASME Certificate of a Quality Control System complying with the Code is required. A boiler installed in accordance with the Code must be inspected at certain stages by an authorized inspector. Inspections to be performed during the life of the project are usually determined by mutual agreement of the authorized inspector and the assembler at the beginning of the project.

The Code applies to new construction before boiler operation. Postoperation repair or modifications are governed by the NBIC. As with the ASME Code, the NBIC is mandatory only when adopted by a legally responsible body.

Table 2 summarizes the common nondestructive examination (NDE) methods used in the installation of boilers and boiler external piping. The type and extent of NDE are specified in the applicable ASME Code and must be considered during proposal planning and scheduling. For all construction, the applicable codes should be clearly specified in project specifications. The nondestructive technologies are discussed in more depth in Chapters 38 and 45.

Construction safety

A comprehensive safety program must be implemented at each construction site. The objective is to promote an awareness of safety among personnel and create a safe working environment to prevent injuries. B&W has developed a Safety Program Manual in support of a construction site safety program. The manual has been specifically designed for construction and includes the requirements of 29 CFR 1926 Safety and Health Regulations for Construction by the Occupational Safety and Health Administration (OSHA). The contents of the Safety Program Manual include, but are not limited to, the safety procedures outlined in Table 3.

Handhole and manhole fittings

Handholes and manholes are provided in boiler components for access to tubes that must be expanded or welded during erection and for subsequent inspection, maintenance, cleaning and repairs. Gasketed handhole fittings can be used in headers designed for pressures and temperatures up to 1200 psi (82.7 bar) and 850F (454C), respectively. (See Fig. 20.) For higher pressures and temperatures, a welded plug-type closure is used. (See Fig. 21.)

When gasketed fittings are installed, all surfaces must be clean and free of burrs. The studs, nuts and washers should be lubricated with a mixture of graphite and kerosene or a suitable commercial product. Handhole fittings should be initially tightened by applying a torque of approximately 300 ft lb (407 N m) and then retightened during hydrostatic testing, applying the same torque.

Welded-type handhole fittings are designed to enter the header through a master opening. The master handhole plug is of similar design but is shaped to enter the header through the hole that it closes. In each case the shoulder of the plug, which seats on a machined surface of the opening, provides the mechanical closure strength. The 0.375 in. (9.5 mm) throat seal weld of the plug on the outside of the header maintains joint fluid tightness. Headers with welded stubs for tube connections usually have one master handhole at each end for inspection.

Manhole openings in drums are circular or elliptical. Woven gaskets are used on manhole covers for pressures up to 500 psi (34.5 bar); metallic gaskets are used for higher pressures. Woven gaskets are coated with a mixture of graphite and kerosene; metal gaskets are installed dry. Manhole covers are usually attached to the drum with a hinge bar.

Although the covers are interchangeable, the outside yokes are fitted to the contour of the drum. The cover, yokes and drumhead are match-marked to ensure that all parts are replaced as originally fitted. Manhole studs should be tightened when installed and again under hydrostatic test pressure, with a torque of approximately 400 ft lb (542 N m).

Assembly of flanged joints

The general precautions taken when installing handhole and manhole fittings should be followed in the assembly of flanged joints. However, nothing is gained by tightening a flanged joint under pressure. Sufficient

Table 2
Application of Nondestructive Examination for Construction

	Visual	Magnetic Particle	Liquid Penetrant	Radiography[4]
Furnace wall tubes[1]	X			X
Superheater tubes[1]	X			X
Economizer tubes[1]	X			X
Supply tubes[1]	X			X
Riser tubes[1]	X			X
Downcomers[1]	X			X
Main steam piping[1]	X	X		X
Attachments[2]	X	X	X	
Boiler external piping[3]	X	X	X	X

Notes:
1. Includes NDE required by Section I of the ASME Code and typical customer requirements for the boiler.
2. Attachments are items such as lifting lugs and alignment lugs.
3. The application of the NDE method is dependent upon temperature, pressure and pipe size.
4. Where radiography is impractical, ultrasonic examination may be used.

Table 3
Selected B&W Safety Procedures for Construction

Posting requirements	Welding, thermal cutting and compressed gas cylinders
Medical facilities and treatment	Fire protection
Safety training	Steel erection
Hazard communication	Confined space entry
Accident and personal injury reporting	Lockout/tagout
Personal protective equipment	Excavation and trenching
Housekeeping	Radiation protection
Lifting equipment	Documentation and record-keeping
Hand tools	Project site safety inspections
Electrical	Material safety data sheets
Scaffolds and ladders	Fall protection
	Floor, wall openings and stairways

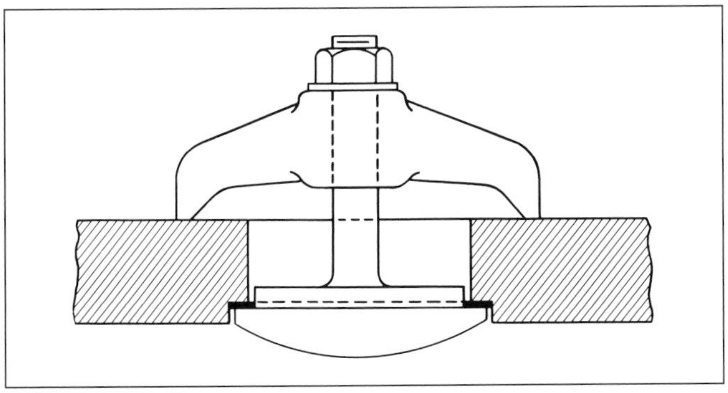

Fig. 20 Gasketed handhole fitting for moderate pressure and temperature.

and uniform tightening of flange studs is of primary importance. A feeler gauge can be used to check the alignment of mating surfaces and to obtain uniform pressure on the gasket.

Stress in the studs of a flanged joint must be sufficient to resist internal pressure and to exert a bearing load on the gasket. Flanged joints on B&W boilers are designed for a nominal stud stress of 30,000 psi (206,844 kPa). The stress in studs may be measured as they are tightened by checking the elongation. An increase of 0.001 in. per inch of stud length measured between centers of nuts produces 30,000 psi (206,844 kPa) stress. This increase may be measured with a dial indicator mounted in a frame.

If the surfaces of stud and nut threads are smooth and uniformly lubricated, a relationship between torque and stress can be determined experimentally.

Internal cleanliness of pressure parts

Manufacturing and shipping Many problems can develop during operations because of dirt or other foreign matter in the circulating system of a boiler. During the fabrication of drums, tubes, headers, pipes or other pressure parts, precautionary measures must be taken to assure internal cleanliness. For boilers which use welded stub construction, the ends of all tubes and pipe, as well as the stubs in the drums and headers, are closed with metal caps and sealed with plastic tape. This usually makes it unnecessary to clean these components in the field. Inevitably there are a few closure caps displaced during unloading and handling. In such instances, the construction crew must examine the components involved, perform any necessary cleaning and replace the caps as soon as possible.

Field assembly The pressure parts of boilers should be searched before handhole and manhole closure fittings are installed. Reasonably straight tubes between headers or drums may be probed by passing a ball through them. Bent tubes which will not clear themselves of foreign material by gravity may be probed by blowing a tight fitting sponge through them. The butt welds in small diameter tubes are examined radiographically for misalignment or obstructions that may cause flow restrictions. Headers are checked with a light and mirror immediately before the end handhole fittings are installed.

No foreign material may enter the pressure part system during assembly. Closures are left in place until removal for welding or positioning equipment. If shop-installed closures are removed to permit installation, temporary closures are placed over the openings to prevent debris from entering the equipment. Prior to flushing or final closing of equipment, a visual inspection is made and foreign objects are removed from the equipment.

Removal of tube sections during repair or alteration must be performed in a manner that prevents foreign material from entering the tubes during the cutting operation. Ends of existing tubes are plugged before machining new weld preparations.

Postconstruction testing

Hydrostatic testing

When all pressure part connections have been completed, the steam generating unit is tested at a pressure specified by the applicable code; this is usually 1.5 times the maximum allowable working pressure (MAWP). Before the test is applied, a final inspection is made of all welding. External connections, including all fittings, are completed within code requirements. Connections for flange safety valves are usually blanked. Welded safety valves are closed with an internal plug.

Hydrostatic testing water must be clean and warm enough to bring pressure part metal temperatures to at least 70F (21C). Nondrainable superheaters and reheaters are filled with demineralized water or condensate, if available. As a boiler is being filled, it is vented at every available connection to allow water to reach every circuit.

Hydrostatic pressure is applied after the boiler is filled. If a boiler feed or other high capacity pump is used, precautions must be taken to control the pressure at all times. While water is being pumped into the boiler, continuous inspection is made to detect any leakage. During hydrostatic testing, safety precautions must be observed, and nonessential personnel must be kept away from the test area.

The test pressure must be held long enough to allow minute leaks to be detected. Following this holding period, pressure is reduced to MAWP and retained for an inspection period. During final inspection, every area of the pressure system is viewed for leakage, and repaired if necessary.

After inspection the unit is drained and all nondrainable parts are protected from freezing, if necessary. The refractory, insulation and casing work is then completed in areas left open for inspection.

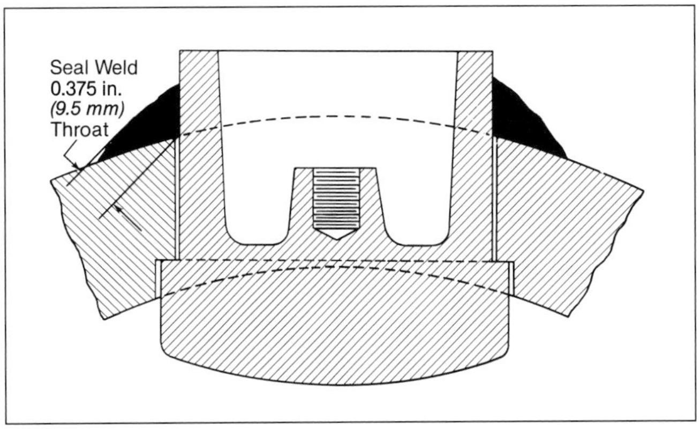

Fig. 21 Welded handhole fitting for higher pressure boilers.

Air testing

Gas-bearing components, such as the boiler, flues and ducts, are field tested for leakage and tightness. All boilers are tested by one of two procedures. Boilers designed to operate with a positive pressure, wherein furnace leaks would emit gases from the boiler setting, are tested following a pressure-fired procedure. Units designed for suction or balanced draft are tested following the balanced draft procedure.

The pressure-fired air test procedure is a pressure drop test. A test is considered satisfactory when the pressure is raised to 15 in. wg (3.74 kPa) maximum and does not drop more than 5 in. wg (1.25 kPa) in ten minutes. If pressure can not be maintained, leaks are sought and corrected and the test is rerun until satisfactory tightness is achieved. On large units, it is common to install temporary bulkheads in flue and duct runs for individual tests.

The balanced draft testing procedure is commonly performed by pressurizing the system with the forced draft fan while throttling the dampers at the downstream test terminal. With the system pressurized, all field welds are inspected by sight and sound to detect leaks.

Assembly of nonpressure parts and auxiliaries

While pressure parts are being assembled and tested, work is in progress on nonpressure parts and auxiliary equipment. These components, including air heaters, fuel equipment, fans, duct work, refractory, insulation and casing, require a large portion of the construction manpower. They must be scheduled with consideration for items that must be assembled in a prescribed sequence. For example, air heaters are finished prior to erection of flues and ducts. Pulverizers are located early enough for piping, drives and other auxiliaries to be completed.

Refractory materials

The furnace and other wall areas of modern fossil fuel boilers are made almost entirely of water-cooled tubes. The increased use of membrane walls has reduced the use of refractory in these areas. However, castable and plastic refractories may still be used to seal flat studded areas, wall penetrations and door and wall box seals. (See Chapter 22.)

Other than membrane wall construction, when tubes are tangent or flat studded, several types of plastic refractory materials are applied to the outside of tubes for insulation or sealing purposes. When an inner casing is to be applied directly against the tubes, the refractory serves principally as an inert filler material for the lanes between tubes. It has a binder which cements it to the tubes and it is troweled flush with the surface of the wall.

Smaller boilers have furnaces constructed of tubes on wide spaced centers backed with a layer of brick or tile. Brickwork of this type is supported by the pressure parts and held in place by studs. The brick is insulated and made air-tight by various combinations of plastic refractory, plastic or block insulation and casing.

High quality workmanship is mandatory in the application of refractory material. Construction details are clearly outlined on drawings and instructions which also designate the materials to be used. These materials must be applied to correct contour and thickness without voids or excessive cracking. Skilled mechanics and close supervision are essential.

Casing

In general, inner or outer boiler casing is used in areas which are not membraned. Outer casing is applied outside of insulating materials and has, for the most part, been replaced by sheet metal lagging. Inner casing is positioned inside of insulation and may be pressure or nonpressure casing.

The primary purpose of pressure inner casing is to prevent air and gas leakage from a steam generating unit. The inner casing is seal welded to prevent leakage from the boiler setting and also serves as a base for the application of insulation and lagging.

In most cases, modern boiler casings feature welded construction. For pressure casing, it is essential that the strength of the weld equal the strength of the plate, that the welds be free of cracks and pinholes and that they present a good appearance. For pressure casing, tightness is checked by a pressure drop air test. Nonpressure casings are given a thorough visual inspection.

Insulation

Selecting the type and thickness of insulation applied to boiler surfaces is an essential part of the design work. Specifications are written and drawings are made to indicate the applied materials and the method of attachment.

Typical standards specify insulation blocks or insulation blankets. Some plastic insulation is used for filling voids. The thickness of the insulation for all surfaces is selected to give a specific surface temperature, e.g., 130F (54C) with 80F (27C) ambient air temperature and 50 ft/s (15.2 m/s) surface velocity. The prime requisites for installing boiler insulation are that the material be tight, free of voids, well anchored and reinforced where necessary.

Ductwork with heavy stiffeners at frequent intervals presents a difficult insulating job. With the stiffeners outside the flue and duct plate, the insulation thickness is often increased over minimum requirements to assure adequate coverage of the stiffeners. If the stiffeners are placed in a manner that standard insulation coverage is too costly, the insulation may be humped over the stiffeners. As an alternative, inner lagging may be applied over the stiffeners and the minimum required insulation is placed over the lagging. When insulation is applied at an appreciable distance from a hot surface with a long vertical run, air may circulate in the void. This increases heat transfer into the insulation, which is undesirable. Where vertical runs exceed 10 ft (3 m), horizontal barriers are installed between the duct and/or tube wall and insulation.

Sheet metal lagging

Outdoor boilers require the use of a durable waterproof lagging. A light gauge metal lagging (galvanized steel or aluminum) or outer casing is used on walls, flues, ducts, air heaters, vestibules, windboxes, downcomers, steam/water lines and other exposed components. The metal covering does not seal against air infiltration but serves as a barrier to protect the insulation against water and physical damage. Its cost is reasonable and it is extensively used in indoor as well as outdoor installations.

Pre-operational inspection

A final inspection is made after erection work is completed and the unit is ready for operation. This is a combined responsibility of the erection, service and operating organizations.

Externally, all components of the unit are checked for expansion clearances. Obscure corners are examined for any construction blocking or bracing that might have been left in place. Points reviewed for expansion clearances are platforms and walkways constructed adjacent to external members that move with the unit.

Of special importance in fossil fuel units is the removal of combustible materials that may present an explosion hazard during initial firing. Internal cavities are checked to ensure that all debris has been removed. Tubes are given a final inspection for alignment, particularly those which might interfere with sootblower operations. Movable connections between tubes are examined for expansion clearances and to see that all attachments are properly anchored.

On most units, construction schedules do not permit completion of insulation before the unit is fired. It is important that this work be completed in areas that will be inaccessible or cause a safety hazard when the unit is in operation.

Reference

1. Shapiro, H., *Cranes & Derricks*, McGraw-Hill, New York, 1980.

Section VI
Operations

With proper design, manufacture and construction, modern steam generating systems are capable of operating efficiently for long periods of service. However, successful operation requires adherence to basic operating principles. These principles begin with the careful monitoring of operating conditions so that the unit functions within design limits. Chapter 40 describes the instrumentation for monitoring pressures, temperatures and flows - the key process parameters. (Specialized instrumentation for environmental equipment is covered in Section V, Chapter 36.) These operating parameters then serve as the inputs to the control system. The fundamentals of control theory and modern integrated control systems are reviewed in Chapter 41. These systems have advanced rapidly during the past two decades to provide greater operator knowledge and flexibility to optimize plant performance.

Successful long term operation of steam producing systems requires careful attention to water treatment and water chemistry control. Chapter 42 provides an expanded discussion of water treatment practices from startup through operation and chemical cleaning. Drum and once-through boilers have different requirements, and each boiler requires individual consideration.

General operating principles and guidelines outlined in Chapter 43 conclude this section. Each steam generating system is unique and requires specific operating guides. However, a number of general principles covering initial operations serve as a basis. Additional specialized procedures apply to Cyclone furnaces, once-through units, fluidized-bed systems and Kraft recovery boilers.

Virtually all areas of a power plant are continually monitored to ensure efficient and reliable daily operation.

Chapter 40
Pressure, Temperature, Quality and Flow Measurement

Instruments for measuring pressure, temperature, fluid flow, and the quality and purity of steam are essential in the operation of a power plant. The measurements obtained permit safe, economical and reliable operation of the steam generating equipment. Instrumentation may range from a simple indicating device to a complex automatic measuring device. The fundamental purpose of the instrument is to convert some physical property of the material being measured into useful information. The methods used to obtain the measurement depend on available technology, economics and the purpose for which the information is being obtained.

Test instrumentation, often portable, may be used to determine flow, pressures and temperatures required to satisfy the user and the boiler equipment supplier that the design operating conditions have been achieved. Requirements for these instruments are summarized in the American Society of Mechanical Engineers (ASME) Performance Test Codes. Many test instruments require skilled technical operators, careful handling and frequent calibration. Some instruments and measurement methods may not be suitable for long term continuous operation because of a harsh environment, material limitations, or excessive cost. Instruments used for continuous measurement may require compromises in accuracy for long term, dependable, reliable operation.

Pressure measurement

The pressure gauge is probably the earliest instrument used in boiler operation. Today, with complex control systems in operation and more than one hundred years after the first water tube safety boiler, a pressure gauge is still used to determine steam drum pressure. The Bourdon tube pressure gauge is shown in Fig. 1. Although improvements have been made in construction and accuracy, its basic principle of operation remains unchanged. A closed end oval tube in a semicircular shape straightens with internal pressure. The movement of the closed end is converted to an indication (needle position).

Pressure measuring instruments take various forms, depending on the magnitude of the pressure, the accuracy desired and the application.

Manometers are considered an accurate means of pressure or pressure differential measurement. These instruments, containing a wide variety of fluids depending on the pressure, are capable of high accuracy with careful use. The fluids used vary from those lighter than water for low pressures to mercury for relatively high pressures. Fig. 2 illustrates an inclined manometer used to measure low pressure differentials at low static pressure. Differential diaphragm gauges with a magnetic linkage are also used for low pressure measurement. Fig. 3 shows a high pressure mercury manometer. For greater precision in measuring small pressure differentials, such as the measurement of flow orifice differentials, hook gauges or micromanometers may be used.

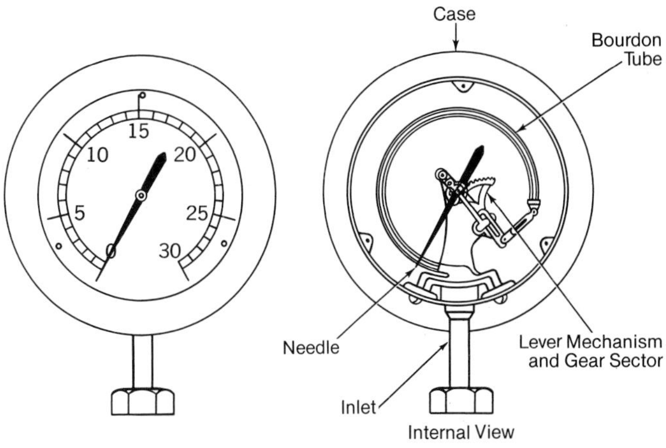

Fig. 1 Bourdon gauge.

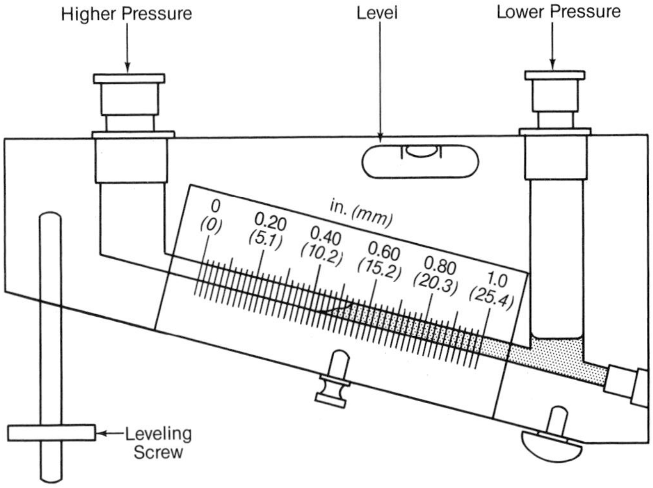

Fig. 2 Inclined differential manometer (*courtesy of Dwyer Instruments, Inc.*).

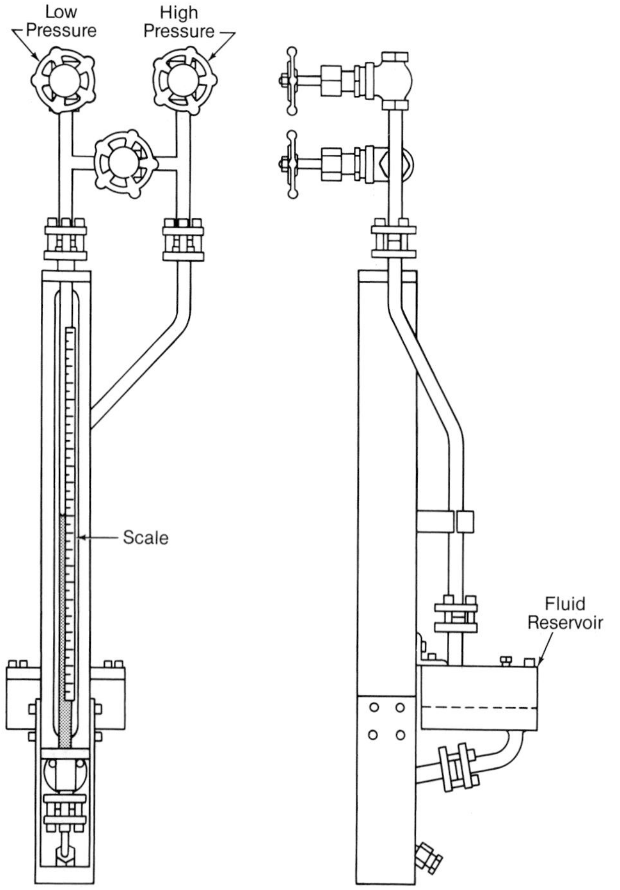

Fig. 3 High pressure mercury manometer (*courtesy of Meriam Instrument Company*).

Bourdon tube pressure gauges are available for measuring a wide range of static pressures in varying degrees of precision and accuracy. Pressure gauges used as operating guides need not be of high precision and normally have scale subdivisions of about 1% of the full scale range. For certain tests, such as hydrostatic testing of pressure parts and boiler efficiency, more precision is required. Gauges with 0.1% full scale subdivisions are used. For performance calculations, high precision temperature and pressure measurements are required to accurately determine steam and water enthalpies. Dead weight gauges or calibrated pressure transmitters are preferable to Bourdon gauges for high precision pressure measurement.

Diaphragm type gauges are used for measuring differential pressures. A slack diaphragm pressure gauge can be used to measure small differentials where total pressure does not exceed about 1.0 psig (6.9 kPa). For high static pressures, opposed bellows gauges (Fig. 4) read a wide range of differential pressures. They are suitable for measuring differential pressure across boiler circuits and can be used to measure differentials from 2 to 100 psi (0.14 to 6.9 bar) at static pressures up to 6000 psi (413.7 bar).

Incorporated into many pressure measurement devices is the capability of producing an output signal proportional to the measurement. The output signal may be transmitted to a central measurement or control system. Pneumatic transmitted signals are used in control systems, but the more common designs use electrical circuitry to sense the changing position of basic mechanical motion and to produce an electrical output signal to be transmitted. These are readily adaptable to computer based systems. Some transducers use a piezoelectric crystal which changes electrical resistance as the element is deformed under pressure. Other pressure transmitter designs use a diaphragm with strain gauges attached; minor changes in diaphragm strain are transferred to the strain gauges, which change resistance. Electrical circuits recognize the resistance change in the crystal or strain gauges and produce an indication of pressure. Advanced applications have used an optic glass fiber embedded in a metal diaphragm, with pressure measurement determined as a change in the light beam traversing the fiber. Pressure transducers incorporating electrical components have the advantage of fewer mechani-

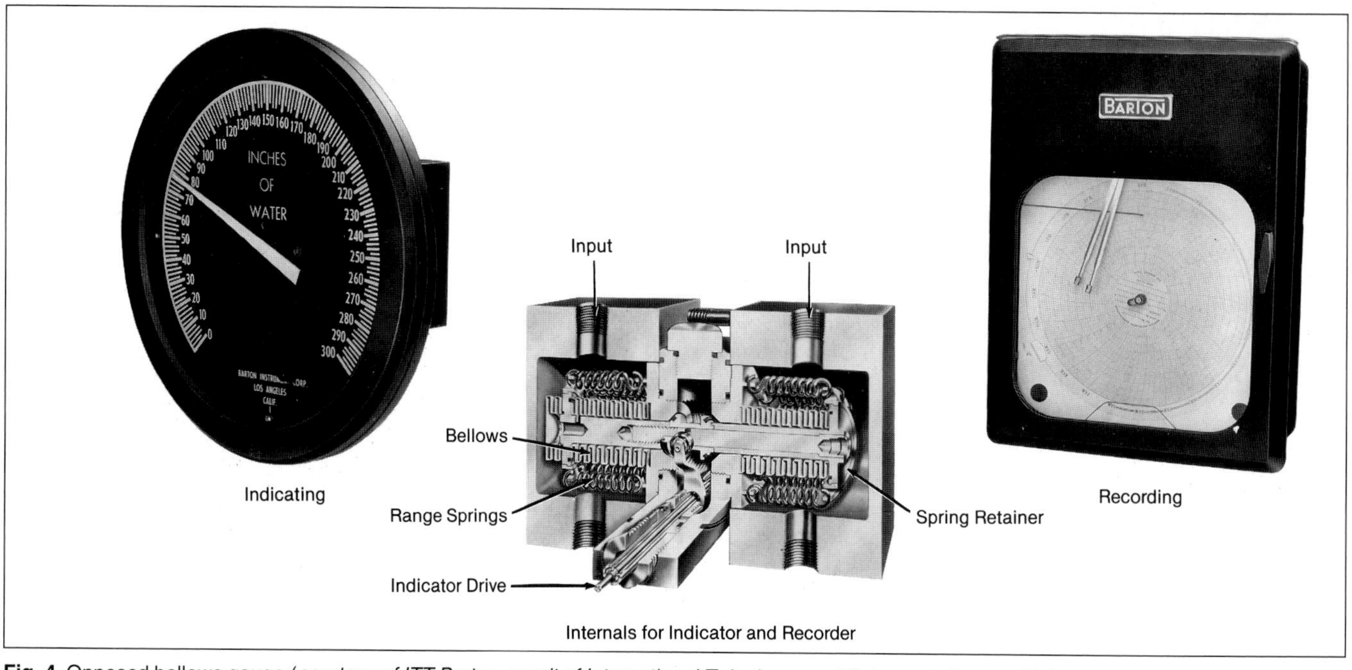

Fig. 4 Opposed bellows gauge (*courtesy of ITT Barton, a unit of International Telephone and Telegraph Corporation*).

cal parts for improved durability. However, this may restrict the installation location due to temperature limitations of the electrical components. Instruments configured to generate electric signals have been adapted to incorporate computing capability to assist in calibration. Pressure transmitters and transducers have been readily adapted for power plant computer data acquisition and control systems.

Instrument corrections

In recording and reporting pressure readings, corrections may be necessary for water leg and for conversion to absolute pressure by adding atmospheric pressure. Effective water leg is the added pressure imposed on the gauge by the leg of water standing above the gauge. Fig. 5 illustrates a water leg correction to a pressure gauge reading. The density of the water in the leg may change with ambient temperature and, for precise measurement, the water leg temperature should be monitored. On some gauges it may be possible to zero the gauge at zero operating pressure with the water leg completely filled to compensate for the water leg static head. Care must be taken in performing water leg adjustments as this may affect the instrument calibration over the operating range. Differential pressure measurements require a water leg correction for both sides of the measuring instrument. For some liquid level devices, the differential pressure measurements from two water legs of different elevations are used to provide the indication of liquid level. Differential pressure measurements across orifices, nozzles or pitot tubes to measure flow are described below.

Location of pressure measurement connections

The criteria for selecting a connection location to the pressure source for the measuring device are the same regardless of the magnitude of the pressure, the type of measuring device or the fluid being measured. Pressure connections, or *taps*, in piping, flues or ducts should be located to avoid impacts or eddies; this assures an accurate static pressure measurement. Under certain circumstances a through the wall pressure connection may not be representative. It may be necessary to use a probe incorporating a tip configuration that minimizes the effects of the flowing fluid to obtain a representative static pressure measurement.

Connecting lines to instruments should be as short and direct as possible and must be leak-free. For differential pressure readings, it is preferable to use a differential pressure measuring device rather than to calculate the difference between the readings on two instruments. Particular attention should be paid to the placement of the instrument, the potential for condensation in the impulse lines, and the accumulation of debris or noncondensable gases in the completed installation.

Temperature measurement

Today, the Fahrenheit and Celsius (formerly centigrade) scales are the most common and firmly established temperature scales.

The Fahrenheit scale is fixed at the freezing point (32F) and the boiling point (212F) of pure water at atmospheric pressure, with 180 equal degrees between the two points. The centigrade scale was originally based on the same freezing (0C) and boiling (100C) points of water at standard atmospheric pressure, using the convenient 100 degree interval of the decimal system.

In 1960, the General Conference of Weights and Measures changed the defining fixed point from the freezing point to the triple point (0.0C) of water. The triple point is the condition under which the three phases of matter (solid, liquid and vapor) coexist in equilibrium. This point is more easily and accurately reproduced than the freezing point. At standard atmospheric pressure the interval between the triple point for water and steam on the centigrade scale became 99.99C instead of 100C.

In 1990, temperature standards were further refined by the International Temperature Scale 1990 (ITS-90). These changes, however small, reflect the continued review of temperature measurement. As measured by the current standard, water at standard conditions boils at 211.953F (99.974C).

In thermometer calibrations, the fusion and vaporization of various pure substances have been carefully determined and are known as fixed points. Two additional scales in scientific work use the absolute zero of temperature, i.e., -273.16C or -459.69R. The absolute scale using Celsius degrees as the temperature interval, is called the Kelvin (K) scale; the absolute scale using Fahrenheit degrees is called the Rankine (R) scale.

Methods of measuring temperature

Heat affected properties of substances, such as thermal expansion, radiation and electrical effects, are incorporated into commercial temperature measuring instruments. These instruments vary in their precision depending on the property measured, the substance used and the design of the instrument. The care taken in selecting the correct instrument largely determines the accuracy of the results. It is the engineer's responsibil-

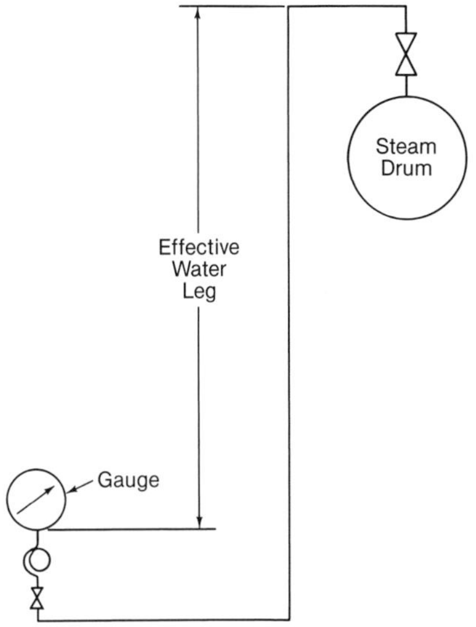

Fig. 5 Application of water leg correction to pressure gauge reading (Correction $\Delta P = g\rho L/g_c$).

ity to make certain that the application is correct and that the results are not affected by extraneous factors.

Changes of state

Fusion For a pure chemical element or compound such as mercury or water, fusion, or change of state from solid to liquid, occurs at a fixed temperature. The melting points of these materials are suitable fixed reference points for temperature scales.

The fusion of pyrometric cones is widely used in the ceramic industry as a method of measuring high temperatures in refractory heating furnaces. These cones, small pyramids about 2 in. (50.8 mm) high, are made of oxide and glass mixtures that soften and melt at established temperatures. The pyrometric cone is suitable for temperatures ranging from 1100 to 3600F (593 to 1982C). Its use in the power industry is generally limited to the laboratory. An adaptation of this method of temperature determination is used when observing ash and slag accumulations soften and flow in a coal-fired furnace or superheater.

Fusion pyrometers are also made in the form of crayons, paint and pellets. The crayons and paint, applied to a metal surface, produce a flat-finish mark. A change in the finish from flat to glossy indicates that the surface temperature has reached or exceeded a selected value. The pellets melt at specified temperatures when in contact with a hot surface and may be seen more easily than crayon marks.

Vaporization The vapor pressure of a liquid depends on its temperature. When the liquid is heated to its boiling temperature, the vapor pressure is equal to the total pressure above the liquid surface. The boiling points of various pure chemical elements or compounds at standard atmospheric pressure [29.92 in. Hg (101.33 kPa)] can be used as thermometric fixed points. If the liquid and vapor are confined, the increase in vapor pressure may be used to measure temperature using a calibrated pressure gauge. (See Fig. 6.)

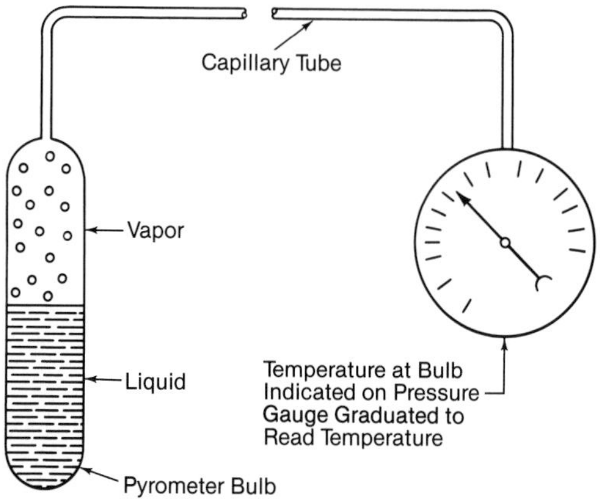

Fig. 6 Schematic assembly of vapor pressure thermometer.

Expansion properties

Most substances expand when heated and, for many materials, the amount of expansion is almost directly proportional to the change in temperature. This effect is the basis of various thermometer types using gases, solids or liquids.

Gases The expansion of gases follows the relationship, in English units:

$$Pv_M = RT \tag{1}$$

where

P = absolute pressure, lb/ft^2
v_M = volume, ft^3/mole of gas
R = universal gas constant, 1545 ft lb/mole R
T = absolute temperature, R = F + 460

At high pressures or near the condensation point, gases deviate considerably from this relationship. Under other conditions, however, the deviation is small.

Two types of gas thermometers are based on this relationship. In one, a constant gas volume is maintained and pressure changes are used to measure changes in temperature. Very accurate instruments of this type have been developed for laboratory work. The constant volume type thermometer is widely used commercially.

Nitrogen is commonly used for the gas-filled thermometer in industrial applications. It is suitable for a temperature range of -200 to 1000F (-129 to 538C). The construction is similar to the vapor pressure thermometer with nitrogen gas replacing the liquid and vapor. Expansion of the heated nitrogen in the bulb increases the pressure in the system and actuates a temperature indicator.

Liquids Liquid expansion properties may be incorporated into a thermometer by using a bulb and capillary to confine the liquid. In this thermometer, the bulb and capillary tube are completely filled with liquid and a calibrated pressure gauge is used to measure temperature. Mercury has been used for a temperature range of -40 to 1000F (-40 to 538C). Instrument readings can be in error if the capillary tubing is subjected to temperature changes.

The liquid-in-glass thermometer is a simple, direct reading, portable instrument. Low precision thermometers are inexpensive and instruments of moderate precision are available for laboratory use. This thermometer features a reservoir of liquid in a glass bulb that is connected directly to a glass capillary tube with graduated markings. Mercury, the most common liquid, is satisfactory from -40F (-40C), just above its freezing point, to about 600F (316C) if the capillary space above the mercury is evacuated. The upper temperature limit may be 900F (482C) or higher if this space is filled with pressurized nitrogen or carbon dioxide (CO_2).

The use of unprotected glass thermometers is restricted to laboratory or research applications. For more severe service there are various designs of industrial thermometers with the bulb and stem protected by a metal casing and usually arranged in a thermometer well. Response to rapid temperature changes is slower than with the unprotected laboratory type instrument.

Solids The expansion of solids when heated is recognized in thermometers using a bimetallic strip. Flat ribbons of two metals with different coefficients of thermal expansion are joined face to face by riveting or welding to form a bimetallic strip. When the strip is heated, the expansion is greater for one metal than for the other. A flat strip bends, or changes curvature if it was initially in a spiral form. Bimetallic strips are seldom used in power plant thermometers but are widely used in inexpensive thermometers, household thermostats and other temperature control and regulating equipment. They are particularly useful for automatic temperature compensation in other instrument mechanisms.

Radiation properties

All solid bodies emit radiation. The amount is very small at low temperatures and larger at high temperatures. The quantity of radiation may be calculated by the Stefan-Boltzman formula:

$$q / S = \sigma \varepsilon T^4 \qquad (2)$$

where

q = radiant energy per unit time, Btu/h
S = surface area, ft^2
σ = Stefan-Boltzman constant, 1.71×10^{-9} Btu/h ft^2 R^4
ε = emissivity of the surface, dimensionless (usually between 0.80 and 0.95 for boiler materials)
T = absolute temperature, R = F + 460

At low temperatures the radiation is primarily in the invisible infrared range. As the temperature rises, an increasing proportion of the radiation is in shorter wavelengths, becoming visible as a dull red glow at about 1000F (538C) and passing through yellow toward white at higher temperatures. The temperature of hot metals [above 1000F (538C)] can be estimated by color. For iron or steel, the approximate color scale is:

dark red	1000F (538C)
medium cherry red	1250F (677C)
orange	1650F (899C)
yellow	1850F (1010C)
white	2200F (1204C)

Within the visible range, two types of temperature measuring instruments, the optical pyrometer and the radiation pyrometer, sense the emitted radiation to indicate temperature.

Optical pyrometers The optical pyrometer visually compares the brightness of an object to a reference source of radiation. The typical internal calibration source is an electrically heated tungsten filament. A red filter may also be used to restrict this visual comparison to a particular wavelength. This instrument measures the temperature of surfaces with an emissivity of 1.0, which is equivalent to that of a blackbody. By definition, a blackbody absorbs all radiation incident upon it, reflecting and transmitting none. When calibrated and used properly, the pyrometer yields excellent results above 1500F (816C). Temperature measurement of the interior surface of a uniformly heated enclosure, such as a muffle furnace, is a typical application. When used to measure the temperature of an object outside a furnace, the optical pyrometer always reads low. The error is small [20F (11C)] for high emissivity bodies, such as steel ingots, and large [200 to 300F (111 to 167C)] for unoxidized liquid steel or iron surfaces.

The optical pyrometer is widely used for measuring objects in furnaces at steel mills and iron foundries. It is not applicable for measuring gas temperature because clean gases do not radiate in the visible range.

Radiation pyrometers In one type of radiation pyrometer, all radiation from the hot body, regardless of wavelength, is absorbed by the instrument. The heat absorption is measured by the temperature rise of a delicate thermocouple (TC) within the instrument. The thermocouple is calibrated to indicate the temperature of the hot surface at which the pyrometer is sighted, on the assumption that the surface emissivity equals 1.0. The hot surface must fill the instrument's entire field of view.

This type of radiation pyrometer has high sensitivity and precision over a wide range of temperatures. The instrument gives good results above 1000F (538C) when used to measure temperatures of high emissivity bodies, such as the interiors of uniformly heated enclosures. Because operation of the instrument does not require visual comparison, radiation pyrometers may be used as remotely operated temperature indicators. Errors in temperature measurements of hot bodies with emissivities of less than 1.0, especially if they are in the open, are extremely large.

Radiation pyrometers sensitive to selective wavelengths in the infrared band give good results when measuring temperatures of bodies or flames. One design uses a system of lenses, a lead sulfide cell and electronic circuitry to produce a measurement.

An advancement of the single band pyrometer is the two color pyrometer, which measures the intensities of two selected wave bands of the visible spectrum emitted by a heated object. This unit computes the total of the emitted energies and converts it into a temperature indication. As with the optical pyrometer, indications depend on sighting visible rays and therefore its minimum temperature measurement is 1000F (538C).

Infrared radiation is produced by all matter at temperatures above absolute zero and detectors sensitive to the infrared band may be used to measure objects at low temperatures where the radiation is not in the visible band. The development of metal-silicide detectors permits sensitivities in the 1 to 5 μm band. Metal-silicides are formed by the reaction of metals, such as platinum or palladium, with p-type silicon. When coupled with electronic circuitry for producing an image, the detectors are sophisticated devices for low temperature measurement.[1]

Infrared thermal imaging equipment is available for quantitative measurement or qualitative imaging. The instruments for making quantitative measurements and providing target temperature indications are called *thermographic imagers* or *imaging radiometers*. Qualitative imaging instruments are called *thermal viewers*. These viewers are mechanically or electronically scanned; the electronically scanned units are called *pyroelectric vidicons* or *pyrovidicons*.[2]

Infrared temperature measurement is applicable to surveys of surfaces, such as boiler casings and insulated steam pipes, for locating temperature variations. These instruments may also assist in determining electrical component temperatures in operating circuits. Radiation pyrometers are not capable of determining gas temperatures.

Optical properties

Optical fibers have been used in fixed (end) and distributed (average temperature along length) temperature measurement. Development is continuing in the application of this method to temperatures above 450F (232C).[3]

Electrical properties

Two classes of widely used temperature measuring instruments, the thermocouple and the electrical resistance thermometer, are based on the relation of temperature to the electrical properties of metals.

Because of its versatility, convenience and durability, the thermocouple is of particular importance in power plant temperature measurements. A separate section later in this chapter is devoted to the thermocouple and its applications.

Resistance thermometer The electrical resistance thermometer, used over a temperature range of -400 to 1800F (-240 to 982C), depends on the increase in electrical resistance of metals with increasing temperature; this relationship is nearly directly proportional. If the electrical resistance of a calibrated wire material is measured (by a Wheatstone bridge or other device), the temperature of the wire can be determined.

In the simplest form of the Wheatstone bridge system, shown in Fig. 7a, the reading would be the sum of the resistances of the calibrated wire and the leads connecting this wire to the Wheatstone bridge. However, this value would be subject to error from temperature changes in the leads. By using a more refined circuit shown in Fig. 7b, the resistance of the leads can be eliminated from the instrument reading. To localize the point of temperature measurement, the resistance wire may be made in the form of a small coil. From room temperature to 250F (121C), commercial instruments usually have resistance coils of nickel or copper. Platinum is used for higher temperatures and in many high precision laboratory instruments to cover a wide range of temperatures.

The electrical resistance thermometer can be used for the remote operations of indicating, recording or automatic controlling. If proper precautions are taken, it is stable and accurate, but it is less rugged and less versatile than a thermocouple.

Acoustic properties

Acoustic properties of gases vary with temperature and are used as the basis for temperature measuring instruments. The velocity of sound in a gas varies directly with the square root of its absolute temperature, as indicated in the following formula:

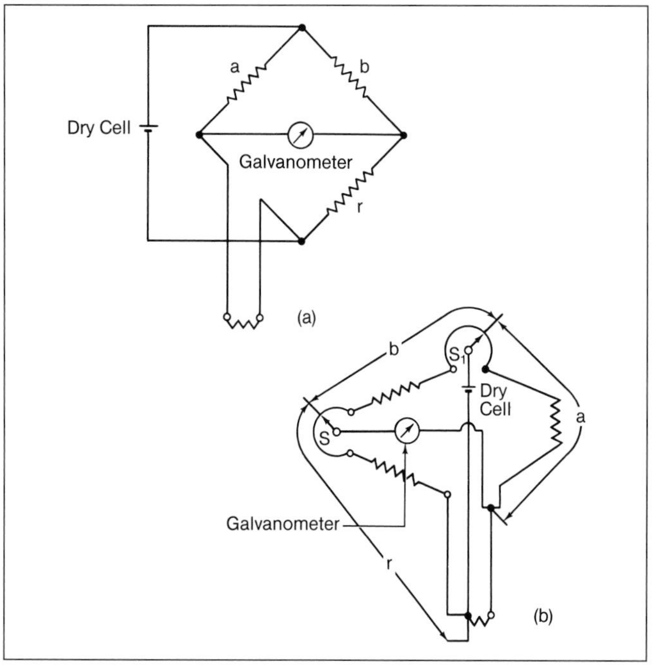

Fig. 7 Diagrams of electrical circuits for resistance thermometers (*courtesy of Leeds and Northrup Company*).

$$c = \sqrt{g_c kRT / M} \quad \textbf{(3)}$$

where

c = speed of sound, ft/s
g_c = proportionality constant, 32.17 lbm ft/lbf s^2
k = specific heat ratio, dimensionless
R = universal gas constant, 1545 ft lb/mole R
T = absolute temperature, R = F + 460
M = molecular weight, lb/mole

By measuring the time required for sound to traverse a specific distance, the average gas temperature can be calculated. This phenomenon has been commercially applied to measuring furnace gas temperatures. By developing a number of independent temperature paths a thermal contour may be developed.

Other properties

Other material properties that vary with temperature are also used as the basis for temperature measuring instruments. By flowing gas through two orifices in series, a method has been developed for measuring gas temperatures up to 2000F (1093C) commercially and up to 4500F (2482C) in the laboratory. Temperatures ranging from absolute zero to 5000F (2760C) have been accurately determined by using a laboratory instrument that measures electron motions. High temperature research has been stimulated by gas turbine and jet propulsion applications, and further development of new and improved temperature measuring instruments is expected.

The thermocouple

A thermocouple consists of two electrical conductors of dissimilar materials joined at the ends to form a circuit. If one of the junctions is maintained at a temperature higher than the other, an electromotive force (emf) is generated, producing current flow through the circuit, as shown in Fig. 8.

The magnitude of the net emf depends on the temperature difference between the two junctions and the materials used for the conductors. No unbalance, or net emf, is generated if the two junctions are at the same temperature or if the conductors are of the same material at different temperatures. If one junction of a calibrated thermocouple is maintained at a known temperature, the

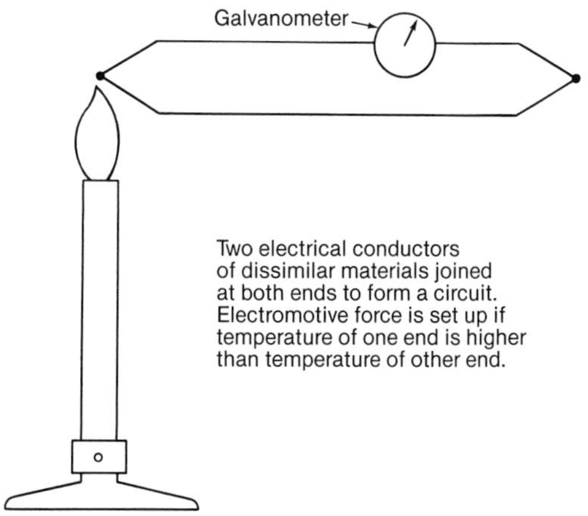

Galvanometer
Two electrical conductors of dissimilar materials joined at both ends to form a circuit. Electromotive force is set up if temperature of one end is higher than temperature of other end.

Fig. 8 Principle of the thermocouple illustrated.

temperature of the other junction can be determined by measuring the net emf produced; this emf is almost directly proportional to the temperature difference of the two junctions. The relationship between emf and the corresponding temperature difference has been established by laboratory tests throughout the temperature ranges for common thermocouple materials. These values are plotted in Fig. 9. The thermocouple is a low cost, versatile, durable, simple device providing fast response for accurate temperature measurement. It also allows convenient centralized reading.

In the potentiometer circuit, copper leads may connect the reference junction terminals to the measuring instrument without affecting the net emf of the thermocouple. If the instrument is at a uniform temperature, no emf is set up between the copper conductors and slide wire materials within the potentiometer. However, if temperature differences exist within the instrument, there will be disturbing emf values in the circuit and the readings will be affected.

Multiple circuits

If two or more thermocouples are connected in series, the net emf at the outside terminals is equal to the sum of the emf values developed by the individual thermocouples. When all of the individual hot and cold junctions are maintained at the same respective temperatures, as in the device known as the thermopile, this multiplied emf value makes it possible to detect and measure extremely small temperature variations.

Two or more thermocouples may be connected in parallel to obtain a single reading of average temperature. In this case, the resistance of each thermocouple must be the same. The emf across the terminals of such a circuit is the average of all the individual emf values and may be read on a potentiometer normally used for a single thermocouple.

Practical application and multiple circuits

The limitation of the potentiometer circuitry to measure small emf values has led to various circuits of multiple thermocouples connected in series or parallel. These circuits make it possible to measure extremely small temperature variations or average temperatures. Thermocouple circuit considerations of wire diameter and length, while less important for steady-state measurement, will affect thermocouple readings during dynamic temperature conditions. These applications require detailed circuit design investigation.

The application of thermocouples requires attention to the insulation of the lead wire to prevent erroneous circuits. Parasitic couples may be introduced by mechanical, soldered, or brazed connections if temperature variations exist within different metals of the connections.

Selection of materials The combinations of metals and alloys most frequently used for thermocouples are listed in Table 1 with their general characteristics and useful temperature ranges. Selection depends largely on the ability to withstand oxidation attack at the maximum expected service temperature. Durability depends on wire size, use or omission of protection tubes, and nature of the surrounding atmosphere.

All thermocouple materials deteriorate when exposed at the upper portion of their temperature range to air or flue gases and when in contact with other materials. Platinum in particular is affected by metallic oxides and by carbon and hydrocarbon gases when used at temperatures above 1000F (538C) and is subject to calibration drift. The upper temperature limits shown in Table 1 are well below the temperatures at which rapid deterioration occurs.

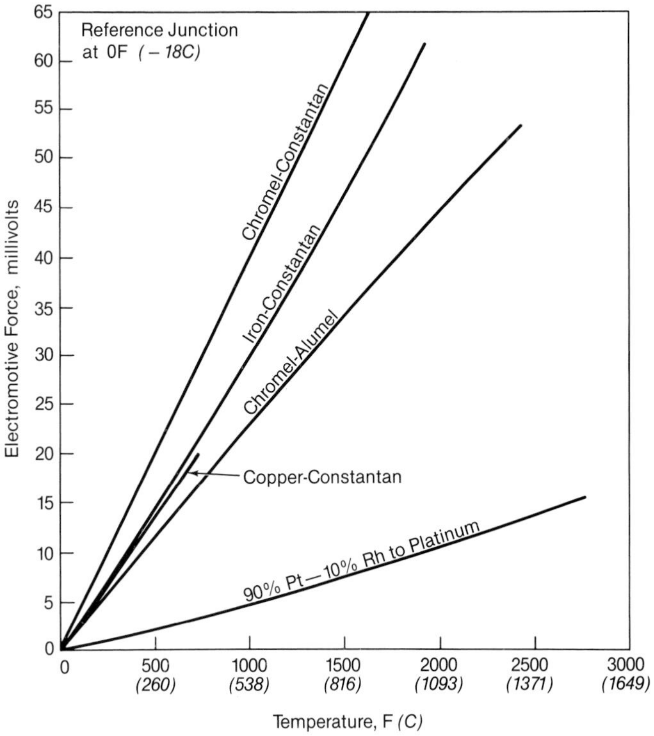

Fig. 9 Relation of temperature to electromotive force produced by several commonly used thermocouples.

For high temperature duty in a permanent installation or where destructive contact is likely, service life may be extended, at some sacrifice in response speed, by using closed-end protection tubes of alloy or ceramic material. The arrangement should permit removal of the thermocouple element from the protection tube for calibration or replacement. Omission of the protection tube may require frequent calibration and corrections for calibration drift. For short periods of service within the useful range of the thermocouple selected, the correction for calibration change is usually negligible.

Sheathed-type thermocouple Sheathed-type magnesium oxide insulated thermocouples have been in use for some time. The thermocouple wires are insulated with inert magnesium oxide insulation that protects the wires from the deteriorating effects of the environment. Sheaths can be made of stainless steel and other resistant materials, ensuring relatively long life and resistance to oxidizing, reducing, or otherwise corrosive atmospheres. Sheathed thermocouples are available as grounded or nongrounded types, as illustrated in Fig. 10. The grounded type has a more rapid response to temperature change but can not be used for connection in series or parallel, because it is grounded to the sheath. For these applications, the nongrounded type should be used. The grounded type is susceptible to separation or parting of the thermocouple wire where long leads at high temperatures are used, whereas the nongrounded type appears satisfactory for this service.

Sheathed thermocouples may be equipped with pads for welding to tubes, as illustrated in Fig. 11. Pad-type ther-

Table 1
Types of Thermocouples in General Use

Type of Thermocouple (Note 1)	Useful Temperature Range, F (C)		Maximum Temperature, F (C)	Millivolts at 500F (260C) (Note 2)	Magnetic Wire
(+) Copper to constantan (-)	-300 to 650	(-184 to 343)	1100 (593)	13.24	--
(+) Iron to constantan (-)	0 to 1400	(-18 to 760)	1800 (982)	15.01	Iron (+)
(+) Chromel to constantan (-)	-300 to 1600	(-184 to 871)	1800 (982)	17.94	--
(+) Chromel to alumel (-)	0 to 2300	(-18 to 1260)	2500 (1371)	11.24	Alumel (-)
(+) 90% Pt — 10% Rh to platinum (-)	900 to 2600	(482 to 1427)	3190 (1754) (Note 3)	2.048	--

Notes:
1. Nominal composition: constantan, 55% Cu, 45% Ni; chromel, 90% Ni, 10% Cr; alumel, 95% Ni, 5% Al, Si and Mn combined.
2. Reference junction 0F (-18.0C).
3. Melting point.

mocouples should not be used when the pad is exposed to temperatures that are markedly different from that of the body being measured, because this added metal surface can radiate or absorb heat. Typical applications of pad-type thermocouples are drum, superheater header, or tube surface temperature measurements. (See Fig. 12.)

Thermocouple and lead wire There are two classes of wire for thermocouples, the closely standardized and matched thermocouple wire and the less accurate compensating lead wire. For thermocouples of noble metal, extension leads of copper and copper-nickel alloy, which have an emf characteristic approximating the noble metal pair, are used to reduce cost. For thermocouples of base metal, the extension lead wire is the same composition as the thermocouple wires, but it is less expensive because the control in manufacture and calibration is less rigorous.

For accuracy, the matched thermocouple wire should be used at the hot junction and continued through the zone of greatest temperature gradient to a point close to room temperature, where a compensating lead wire may be spliced in for connection to the reference junction. The reference junction is usually located at the recorder or central observation point. Where a number of thermocouples are used, it may be economical to establish a zone box (see Fig. 13) from which copper conductors are extended to the measuring instrument.

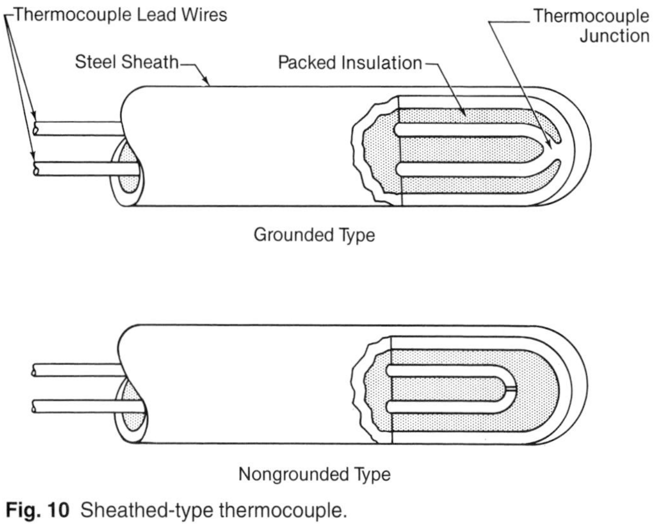

Fig. 10 Sheathed-type thermocouple.

Splicing of extension lead wires may be done by twisting the wire ends together, by using screw clamp connectors, or by fusion welding, soldering or brazing.

Care should be taken to maintain correct polarity by joining wires of the same composition. Polarity is usually identified by color code or tracers in the wire covering.

Hot junction Wire and pipe-type elements for hot junctions may be purchased or may be made from stock thermocouple wire, as shown in Fig. 14. These types are useful for direct immersion in a gas stream, for insertion into thermometer wells, or for thermal contact with solid surfaces.

Where the temperature of a metallic surface is to be measured, a versatile hot junction may be formed by peening the wires separately into holes drilled in the surface of the metal. Steps for making this type of junction are illustrated in Fig. 15.

The peened junction has the advantage of being completely mechanical, with a tight and remarkably strong attachment and minimum interference with the temperature of the object. The temperature indicated is essentially that of the metallic surface, which is the first point of contact with the conductors. The depth of the drilled hole has no temperature effect; it only provides mechanical strength. However, the junction contact points are subject to thermal conduction in the wires. Where this type of error is significant, precautions should be taken to prevent temperature gradients in the wires. Drilled holes for the wire ends should allow a snug fit before peening. For installations requiring a significant

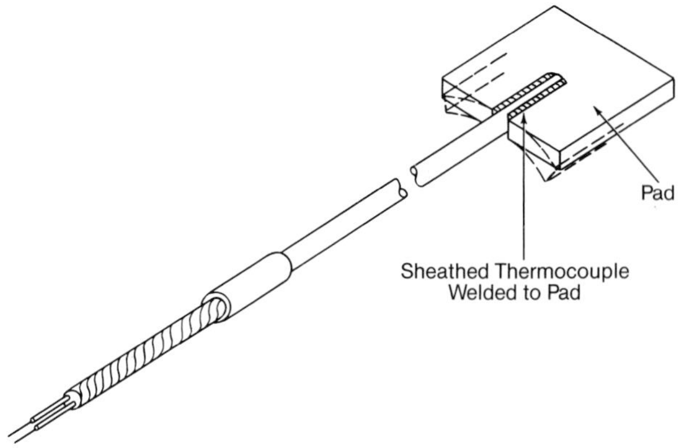

Fig. 11 Sheathed thermocouple with pad.

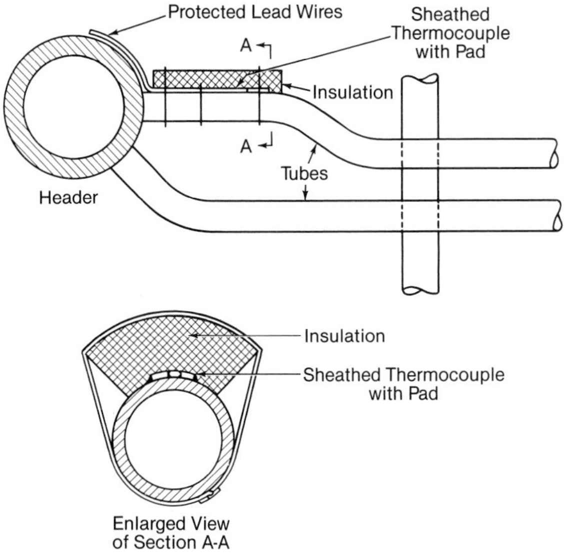

Fig. 12 Application of pad-type thermocouple.

number of thermocouples, there are portable spot welding machines available to join the ends of the wire directly to a steel surface.

Millivolt potentiometer

The temperature-emf relationships of standardized thermocouple wires, as established and published by the manufacturer, should be used to convert the potentiometer readings to equivalent temperature values. These tabulated values represent the net emf impressed in the potentiometer terminals when the reference junction is at 0F (-18C) and the measuring junction is at the temperature listed. The millivolt potentiometer may be used with any type of thermocouple. It is frequently used when several different thermocouple types are installed.

It is usually impractical to maintain the reference junction at 0F (-18C) while taking readings. Correction for any reference junction temperature can be made by adding to each observed emf reading the emf value corresponding to the reference junction temperature. The temperature of the measuring junction can then be determined. Most millivolt potentiometers are equipped with a compensator, which should be used to correct for the reference junction temperature. Direct millivolt readings then correspond to the actual thermocouple temperature.

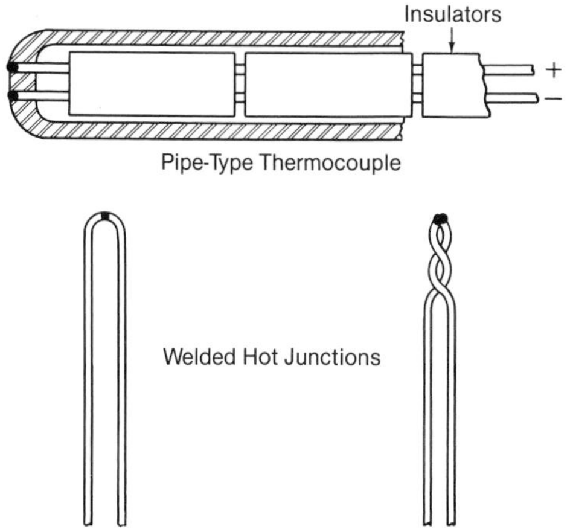

Fig. 14 Thermocouple elements and hot junctions.

Direct reading potentiometer

A potentiometer may be graduated to read the hot junction temperature directly instead of in millivolts. When calibrated for use with a specific type of thermocouple, the reference junction temperature is usually automatically compensated for by an internal resistor circuit. This circuit compensates for deviations from standard room temperature. The resistor circuit characteristics are specific to the type of thermocouple material for which the potentiometer is calibrated. Some potentiometers provide a selector switch to directly read different types of thermocouples.

Temperature recorders

In temperature recorders, a power operated mechanism automatically adjusts the slide wire of a potenti-

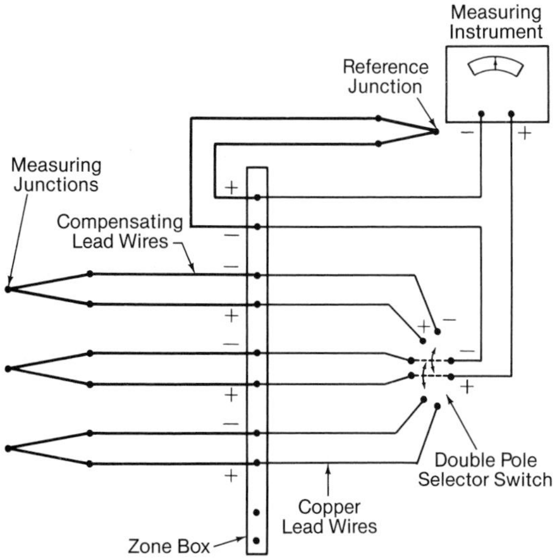

Fig. 13 Thermocouple and lead wire arrangement to measuring instrument through selector switch.

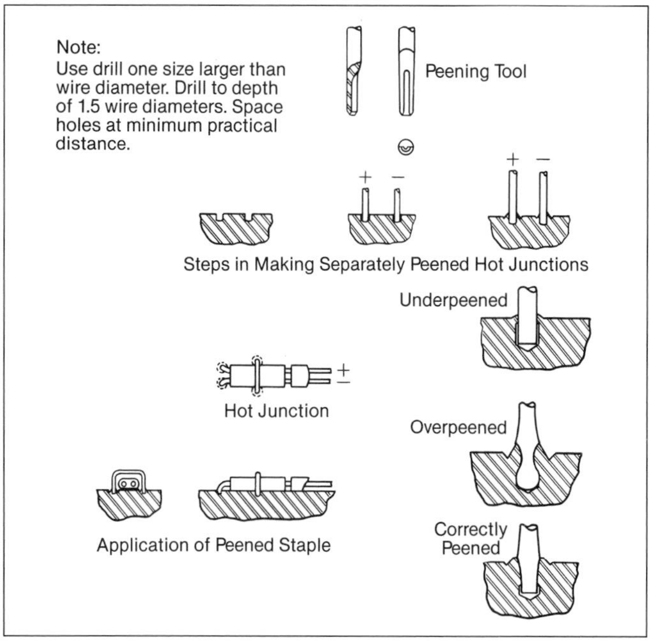

Fig. 15 Procedure for installing a peened-junction thermocouple to measure surface temperature of metal.

ometer circuit to balance the emf received from the thermocouple. The action of the slide wire is coordinated with the movement of a recording pen or print wheel that is drawn over, or impressed on, a moving temperature graduated chart.

As a result, an essentially continuous record may be traced for one thermocouple, or a rapid scanning of many thermocouples may be recorded.

Checking recorder calibration Compared with the manually operated potentiometer, all recorders are inherently less accurate because of wear and lost motion in the actuating mechanism, shrinkage, or printing inaccuracies of paper and chart scales.

Temperature of fluids inside pipes

The temperature of a fluid (liquid, gas or vapor) flowing under pressure through a pipe is usually measured by a glass thermometer, an electrical resistance thermometer, or a thermocouple. Each thermometer is inserted into a well, or thermowell, projecting into the fluid. The thermowell is preferred but a thermocouple properly attached to the outside of a pipe wall can provide good results. The thermowell provides an average temperature if temperature stratification exists in the fluid, because thermal conduction along the well equalizes the temperature.

A metal tube closed at one end, screwed into or welded to a pipe wall, and projecting into the fluid, serves as a thermowell (Fig. 16). It must have strength and rigidity to withstand hydrostatic pressure, resistance to bending and vibration caused by its resistance to fluid flow, and material compatibility for a welded attachment. To minimize heat conduction to the surroundings and to give rapid response to temperature changes, the well should be as small as possible. The portion outside the pipe should also be small. The material of the well must resist the erosive or corrosive action of the fluid. Because the temperature may be locally depressed by acceleration of a compressible fluid through a constriction in a pipe, the well should be carefully positioned, taking these factors into account. Care should also be observed when locating the well for measuring steam temperature downstream of spray water attemperator injection points, as inadequate mixing of the steam and spray water may result in fluid temperature measurement errors. Projecting parts of the thermowell and the pipe wall should be thoroughly insulated to prevent heat loss. Substantial error can result from air circulation within the thermowell and should be minimized by packing with insulating material.

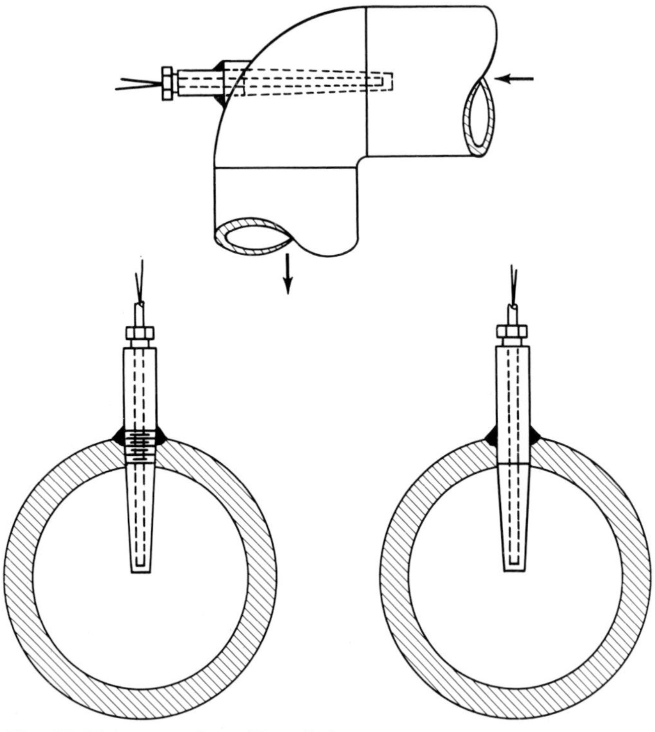

Fig. 16 Thermocouple well installation.

Thermowell designs vary. If the fluid is a liquid or a saturated vapor, good heat transfer is assured and a plain well is satisfactory. If the fluid is a gas or superheated vapor, a finned well is sometimes used. The method of attachment, a mechanical thread or welded attachment to the pipe wall, is optional provided safety code requirements are met (See Fig. 16).

Tube temperature measurement

It is frequently desirable to know the metal temperature of tubes in different classes of service. These classes include furnace wall or boiler bank tubes that are cooled by water and steam at saturation temperature, economizer tubes cooled by water below the temperature of saturation, and superheater and reheater tubes cooled by steam above saturation temperature. These temperature measurements may be used to determine the safety of pressure parts, the uniformity among tubes in parallel flow circuits, or the fluid temperature increase between inlet and outlet conditions. The peened-type hot junction, illustrated in Fig. 17, is satisfactory in many cases and is probably the simplest form of thermocouple application. Sheathed-type thermocouples with pads are generally used to permanently monitor temperatures of tubes not exposed to external heat. (See Fig. 11.) Properly applied insulation of sufficient thickness in the vicinity of the thermocouple attachment will reduce the effects of thermal conduction along the tube. The thermocouple wire wrapped around the tube will minimize the conduction along the thermocouple wire. These techniques are illustrated in Fig. 17. When properly installed the surface thermocouple may be used to measure both metal and fluid temperatures.

Furnace wall tubes

In measuring water-cooled furnace wall tube temperatures, special protection for the thermocouple and its lead wires must be provided because of the destructive high temperature furnace atmosphere and, in some cases, the accumulation and shedding of ash and slag deposits. If a simple peened thermocouple is used, the results may lack accuracy because of errors from conduction; in most cases, the service life would be short because of physical damage or wire deterioration from overheating. Cover plates welded to the tubes have been used to protect the wires, but these plates interfere with the normal heat transfer and cause local ash deposits to create abnormal conditions at the temperature measurement point.

The use of chord-drilled holes through which the thermocouple wires are laced (Fig. 18) was developed by Babcock & Wilcox (B&W); this is a satisfactory tempera-

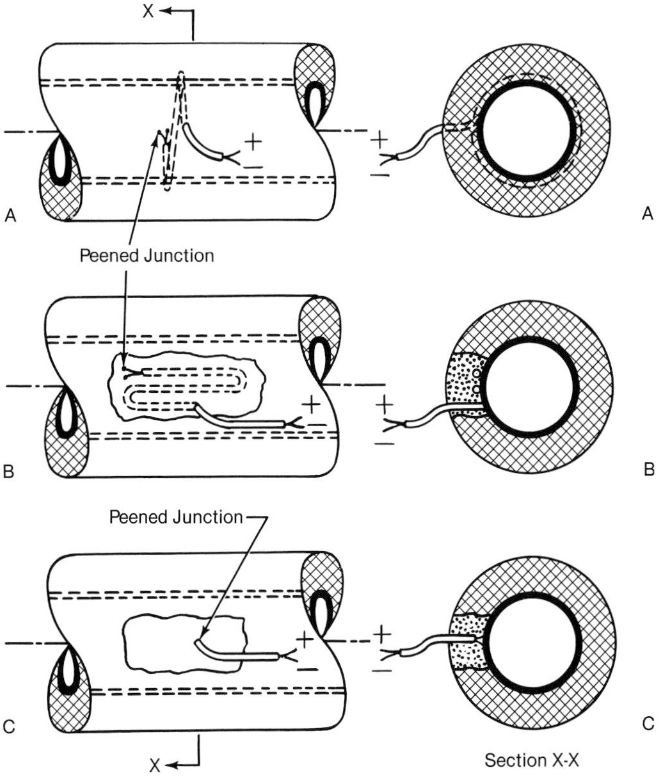

A—Good practice—Thermocouple wire wrapped around pipe underneath pipe insulation.

B—Good practice—Thermocouple wire laid along pipe wall before reinsulation.

C—Bad practice—Thermocouple wire led directly to outside.

Note: Any opening made in pipe insulation for access to pipe wall for installation of thermocouple must be carefully reinsulated with dry insulation firmly held in place.

Fig. 17 Thermocouple wires extending from hot injunction disposed on pipe wall before leading outside.

ture measurement method for furnace wall tubes. The tube surface is free of projections, the wires are protected, and the thermal conduction effect at the hot junction is minimized because the wires pass through an essentially isothermal zone before emerging into cooler surroundings at the rear of the tube. To minimize the effect on the heat flow pattern within the tube metal, the chord-drilled holes should be as small as possible. The effect of these holes on the strength of the tube is small. This effect is minimal in the critical direction of hoop stress and is readily tolerated in the direction of longitudinal stress. The use of chord-drilled holes in the thicker, high pressure wall tubes is a practical method of obtaining surface metal temperatures.

Gradient thermocouple

Measuring the temperature gradient through a tube wall is a means of determining the heat flow rate through the wall and of detecting the accumulation of certain types of internal deposits.

To obtain the best results, a large temperature gradient through the wall is desirable. However, ideal conditions may be compromised to permit an arrangement that provides measurable metal temperatures and is practical to fabricate. Fig. 18 shows a section through a tube which illustrates the drilling and installation of typical surface and depth thermocouples.

Calibration Sheathed thermocouples are suitable for this temperature gradient application. For greater accuracy the thermocouples may be laboratory calibrated. No procedure is available to accurately or economically calibrate the thermocouple tube assembly in the field. The most satisfactory method is as follows.

Following installation, obtain a series of temperature readings on all thermocouples at two or three different ratings on the boiler, from a low rate to maximum rate, while the tube is known to be internally clean; read all temperatures as simultaneously as possible. Several series of readings should be taken on the thermocouples at each rating. An average should be calculated for each thermocouple to correct for minor differences in temperatures resulting from heat input variations. This also compensates for the time lag between surface and depth thermocouples. By using the surface and depth dimensions, the equivalent lengths can be determined and the temperature gradients can be plotted. In practice, the heat flows through a tube wall in a path of decreasing sectional area and the temperature gradient through the metal is not linear with respect to thickness. The gradient, however, may be drawn as a straight line if temperatures are plotted against an equivalent length of flow path (l_e). This represents an equivalent flat plate having a uniform flow path with an equivalent but greater thickness. This is illustrated in Fig. 19.

The equivalent plate thickness is calculated in the following manner:

$$l_e = R \ \ell n \ (R/r) \qquad \textbf{(4)}$$

where

R = outside radius of tube, in.
r = inside radius of tube, in.
l_e = equivalent flat plate thickness, in.

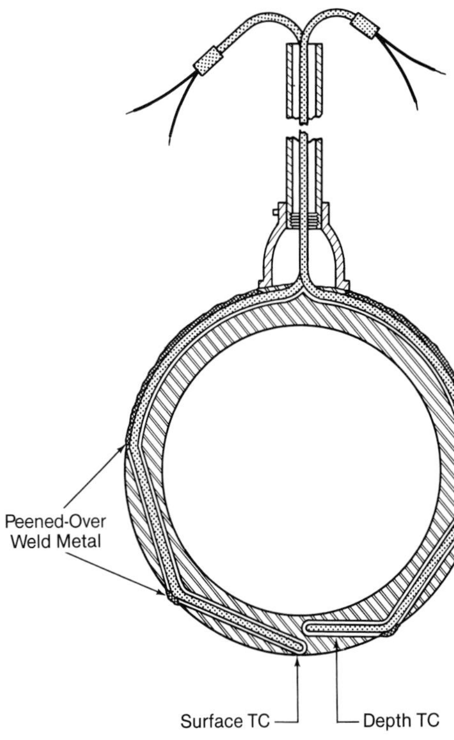

Fig. 18 Chordal thermocouple.

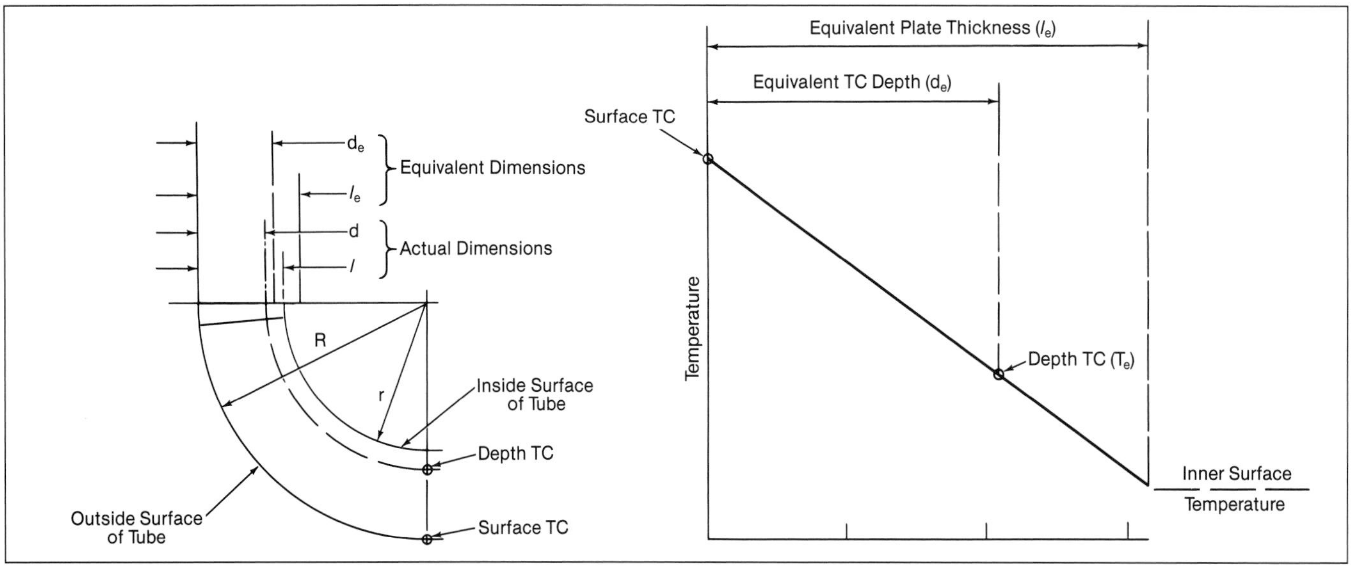

Fig. 19 Thermocouple installation.

In many cases, thermocouple error may represent a significant portion of the gradient being measured. Ideally, lines drawn through the surface and depth temperature (Fig. 20) should intersect the inner surface equivalent thickness line slightly above the fluid temperature as measured by a thermocouple peened on the outside surface of the tube opposite the furnace. The amount above the fluid temperature represents the temperature drop across the fluid film and this drop increases with increased heat input. It is difficult to obtain consistently drilled thermocouple holes that reflect the pattern shown in Fig. 20. If, at low inputs, the line drawn through the surface and depth temperature intersects the inner surface equivalent thickness line at fluid temperature or slightly above, it may be considered as satisfactory. If it falls below or well above this point, a correction or calibration may be made by ignoring the measured equivalent depth dimension, plotting a line through the outer and inner surface temperatures, then adjusting the equivalent depth to a value represented by the intersection of the adjusted gradient and depth thermocouple temperature (Fig. 21).

Heat flux measurement Once a series of temperature plots shows consistent results, heat inputs may be calculated:

$$q'' = \frac{k(T_1 - T_2)}{l_e} \tag{5}$$

where

q'' = heat flux, Btu/h ft^2
k = thermal conductivity, Btu/h ft F
T_1 = outside surface thermocouple temperature, F
T_2 = depth thermocouple temperature, F
l_e = equivalent plate thickness (surface to depth), ft (Fig. 19)

Scale detection If the temperature gradients are to be used as a scale detector, then thermocouples should be read periodically, plotted, and checked for a change in the temperature above saturation.

To assure a close comparison, the thermocouple temperatures should always be read with the calibrated equipment and under operating conditions close to those existing during the original setup. An accumulation of internal deposits in the thermocouple zone is marked by an increased temperature difference from inner surface to saturation.

When the average of the surface and depth temperatures approaches the limit for the tube metal, the boiler should be shut down and the inside tube surfaces inspected. Sufficient allowance should be made for tube to tube temperature variations.

Limitations of gradient thermocouples The gradient thermocouple method of measuring heat flow through tube

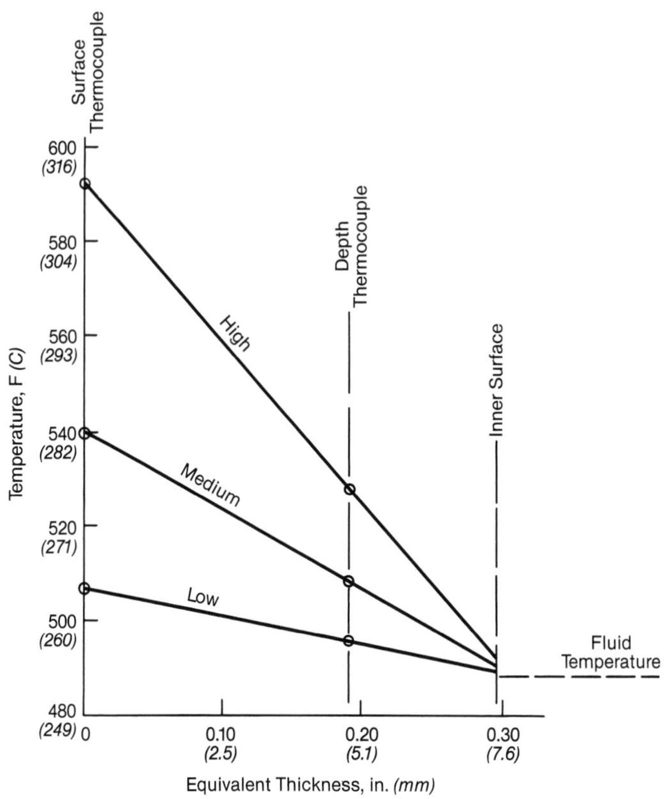

Fig. 20 Sample plot of temperature gradients for low, medium and high heat rates.

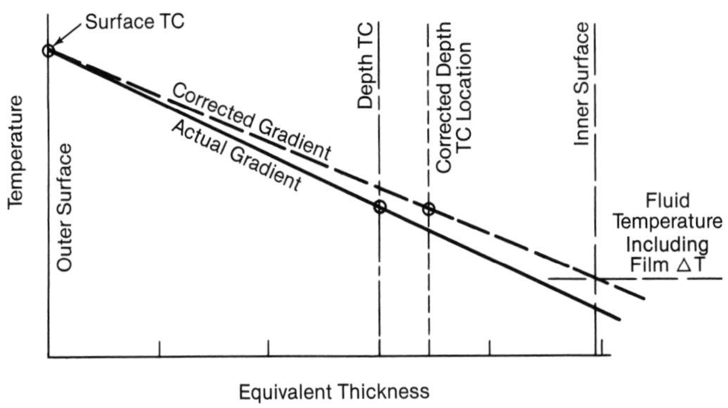

Fig. 21 Adjustment of temperature gradient.

walls is a guide rather than an absolute measure. Relatively small dimensions and small differential temperatures exist between surface thermocouple and depth thermocouple locations. Also, any small error in determining temperatures or in metal thickness between thermocouples usually represents a large percentage of this small difference.

The effectiveness of the gradient temperature plot as an internal scale detector is also dependent on the nature of the scale. Certain types of scale, such as carbonate or silica deposits, accumulate uniformly, while iron oxide deposits can accumulate irregularly. The location of nonuniform deposits is uncertain and the gradient thermocouple method is not reliable for detecting such accumulations.

For a quick determination of internal tube deposit changes, the difference between surface and depth temperature readings can be plotted against the difference between surface and fluid temperature readings. With a clean tube, points should fall along a straight line over a range of inputs. As deposits form, the difference between surface and fluid temperature increases for an unchanging surface and depth temperature difference. Fig. 22 shows a typical plot of such conditions.

Superheater and reheater applications

Chordal surface thermocouples can be used to measure metal temperatures of superheater or reheater tubes in the gas stream, using the principles previously described. However, special provision must be made to protect the wires between the point of measurement and the exit from the boiler setting. This can be done by containing the sheathed thermocouple in stainless steel tubing that is welded to the superheater or reheater tube. This maintains the sheath and protection tubing at a temperature which is approximately that of the superheater or reheater tube. Additional cooling of the stainless steel protection tube is frequently required.

Gas temperature measurement

Temperature measurement of combustion gas at different locations within the boiler is important to the boiler design engineer and the operating plant engineer. Accurate temperature measurements of gas entering and leaving the heat absorbing components can confirm design predictions and operating performance.

In all cases of gas temperature measurement, the temperature sensitive element approaches a temperature that is in equilibrium with the conditions of its environment. While it receives heat primarily by convection transfer from the hot gases in which it is immersed, it is also subject to heat exchange by radiation to and from the surrounding surfaces and by conduction through the instrument itself. If the temperature of the surrounding surfaces does not differ from that of the gas (gas flowing through an insulated duct), the temperature indicated by the instrument should accurately represent the gas temperature. If the temperature of the surrounding surfaces is higher or lower than that of the gas, the indicated temperature will be correspondingly higher or lower than the gas temperature.

The variation from the true gas temperature depends on the temperature and velocity of the gas, the temperature of the surroundings, the size of the temperature measuring element, and the construction of the element and its supports. To correct for the errors in temperature measurement caused by the surroundings, it is best to calibrate the instrument to a known and reliable source.

As an example, consider a 22 gauge (0.7112 mm) bare thermocouple. When it is used to measure the gas temperature in boiler, economizer, or air heater cavities with surrounding walls cooler than the gas, the error in the observed readings may be found from Fig. 23 (line for bare TC).

High velocity thermocouple

The design and operation of steam generating units depend on the evaluation of gas temperatures in the fur-

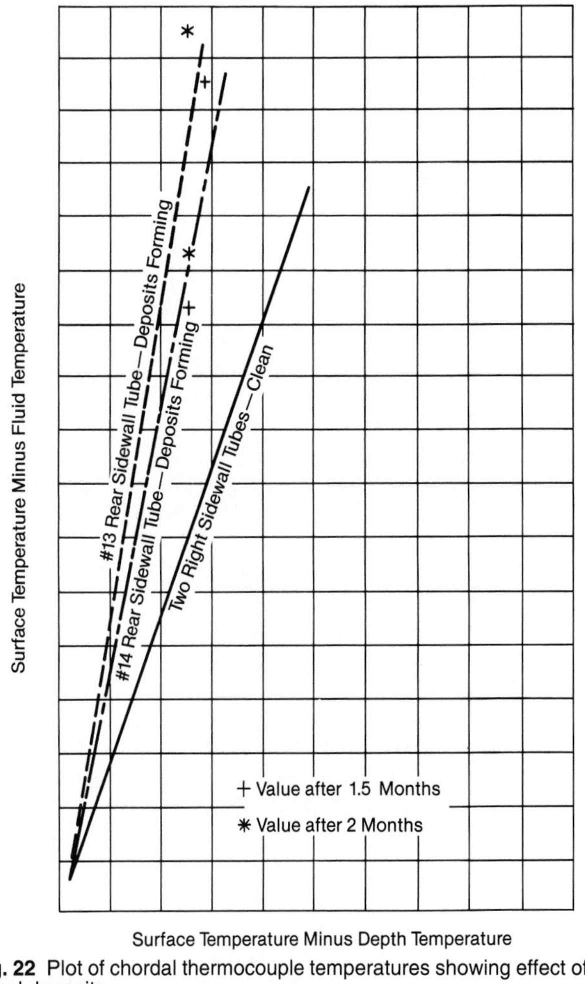

Fig. 22 Plot of chordal thermocouple temperatures showing effect of internal deposits.

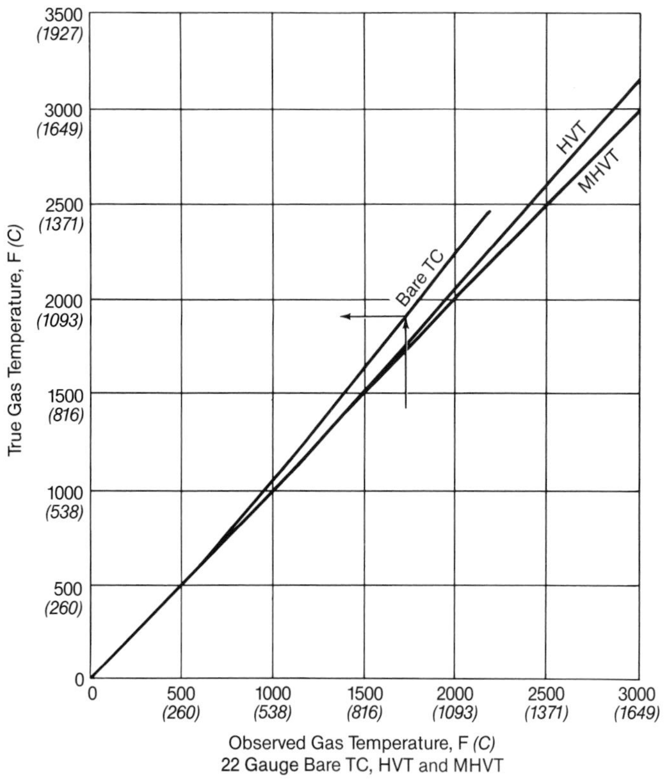

Fig. 23 General magnitude of error in observed readings when measuring temperature in boiler cavities with thermocouples.

nace and superheater sections of the equipment. Boiler design to achieve successful thermal performance must take into account the limitations imposed by the allowable metal temperatures of superheater tubes and by the fusing characteristics of ash and slag from the current or expected fuel. The overall complexity of combustion and heat transfer relationships prevents exact calculation and accurate measurement of the gas temperatures provides information to confirm calculation methods.

The optical pyrometer and radiation pyrometer are not designed to measure gas temperature. Excessive error is also encountered when gas temperatures in the furnace and superheater areas are measured with a bare thermocouple.

The high velocity and multiple-shield high velocity thermocouples (HVT and MHVT), developed to correct for radiation effects, are the best instruments available for measuring high gas temperatures in cooler surroundings or low gas temperatures in hotter surroundings. Cross-sections through single and multiple-shield high velocity thermocouples developed for use in boiler testing are shown in Fig. 24.

The surfaces (water-cooled walls, or superheater or boiler tube banks) surrounding the usual location of a gas traverse are cooler than the gases. Consequently, the readings from a bare unshielded thermocouple indicate lower temperatures than those obtained with an HVT. For the same reason, an HVT generally indicates lower values than an MHVT. A comparison between bare thermocouple, HVT and MHVT results in typical boiler furnaces and cavities is given in Fig. 23.

MHVT measurements closely approach true gas temperatures. In this design, the thermocouple junction is surrounded by multiple shields, all of which receive heat by convection induced by the high gas flow rate. In this manner, the heat transfer by radiation is so reduced that there is virtually no heat exchange between the junction and the innermost shield. Because of the small flow areas that rapidly become clogged by ash, use of the MHVT is limited to clean gas conditions. Where traverses are taken in dust- or slag-laden gases, it is usually necessary to use the HVT. The readings are corrected by comparison with results obtained under clean gas conditions from an MHVT.

For temperatures exceeding 2200F (1204C), noble metal thermocouples are required and it is important to protect the thermocouple from contamination by the gases or entrained ash. Various coverings, shown in Fig. 24, provide some protection for the wires, especially when fouling occurs from molten slag at temperatures above 2400F (1316C). When platinum thermocouples are used in gas above 2600F (1427C), appreciable calibration drift may occur even while taking measurements requiring only several minutes of exposure time. The thermocouple elements should be checked before and after use with corrections applied to the observed readings. When the error (ΔT) reaches 40 to 60F (22 to 33C), the contaminated end of the thermocouple should be removed and a new hot junction should be made using the sound portion of the wire.

Because heat transfer by radiation is proportional to the surface area, emissivity, and difference of the source and receiver absolute temperatures, the effects of radiation increase as the temperature difference between the thermocouple hot junction and the surrounding surfaces

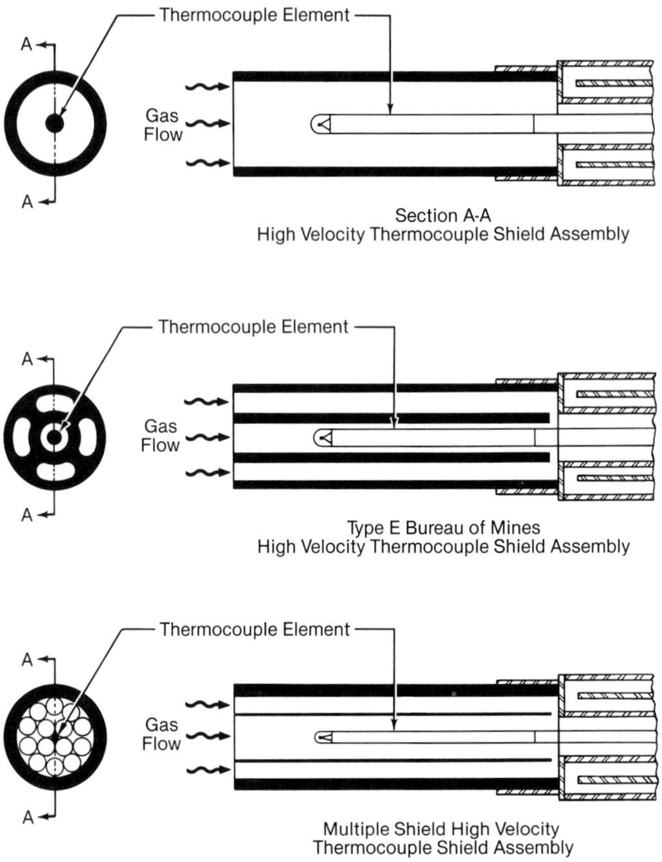

Fig. 24 Shield assemblies for high velocity thermocouple (HVT) and multiple-shield high velocity thermocouple (MHVT).

increases. Also, because heat transfer to the thermocouple by convection is proportional to the gas mass velocity and the temperature difference between gas and thermocouple, the junction temperature may be brought closer to the true temperature of the gas by increasing the rate of mass velocity and convection heat transfer at the thermocouple while shielding the junction from radiation.

A special high velocity thermocouple probe for measuring high gas temperatures in boilers is illustrated in Fig. 25. This portable assembly is primarily used for making test traverses in high duty zones by insertion through inspection doors or other test openings in the setting.

This thermocouple is supported by a water-cooled probe. The measuring junction is surrounded by a tubular porcelain radiation shield through which gas flow is induced at high velocity by an attached aspirator. The gas aspiration rate over the thermocouple can be checked by an orifice incorporated with the aspirator and connected to the probe by a flexible hose. The gas mass velocity (or mass flux) over the thermocouple junction should be at least 15,000 lb/h ft^2 (20.34 kg/m^2 s). Convection heat transfer to the junction and shield is simultaneous and both approach the temperature of the gas stream. Radiation transfer at the junction is diminished by the shield. Because the shield is exposed to the radiation effect of the surroundings, it may gain or lose heat and its temperature may be slightly different from the junction temperature.

With the increasing size of steam generators, handling long HVT probes has become more difficult but, to date, no acceptable point by point measurement of high temperature gases has been fully developed.

Measurement evaluations

Large boilers may have significant variation in actual measurements at a particular measurement plane. Where significant data variation exists, data reduction may require mathematical methods to weight-average individual measured values during the data reduction process.

Flue gas temperature measurements at the economizer outlet during performance tests are an example of a large number of data points which require reduction to a single representative temperature.

A weighted average temperature becomes time consuming and the need for such accuracy must be justified. An average of the individual temperature measurements is usually acceptable if a sufficient number of points are obtained. The average gas temperature may be approached by increasing the number of points or by instrument design to satisfy a particular requirement. For example, in a heat loss efficiency test, multiple measuring locations are used to indicate flue gas exit temperatures for determining dry gas loss. The number of points required by the ASME Performance Test Code permits efficiency accuracies within 0.05%.

Under certain circumstances, the maximum, rather than average, gas temperature may be required for equipment protection. A moveable probe with one or two thermocouples may be used to determine the location of this high temperature. For instance, during boiler pressurization and before steam is flowing through the superheater or reheater tubes, a bare thermocouple temporarily installed in the gas stream immediately before the tubes may be used to indicate the highest temperature to prevent tube overheating. These thermocouples are normally removed after steam flow is adequate for cooling. On units designed for remote operation, it is possible to remotely retract and insert these thermocouples using a sootblower carriage. The remotely insertable thermocouple is called a thermoprobe.

Insulation and casing temperature

Outer surfaces

The measurement method should be carefully selected to avoid significant errors in measuring uncased insulation surface temperature. Portable contact thermocouple instruments designed to be pressed against a surface are unsatisfactory on insulated surfaces because the instrument cools the surface at the point of contact, and the low rate of heat transfer through the insulating material prevents adequate heat flow from surrounding areas to the contact point.

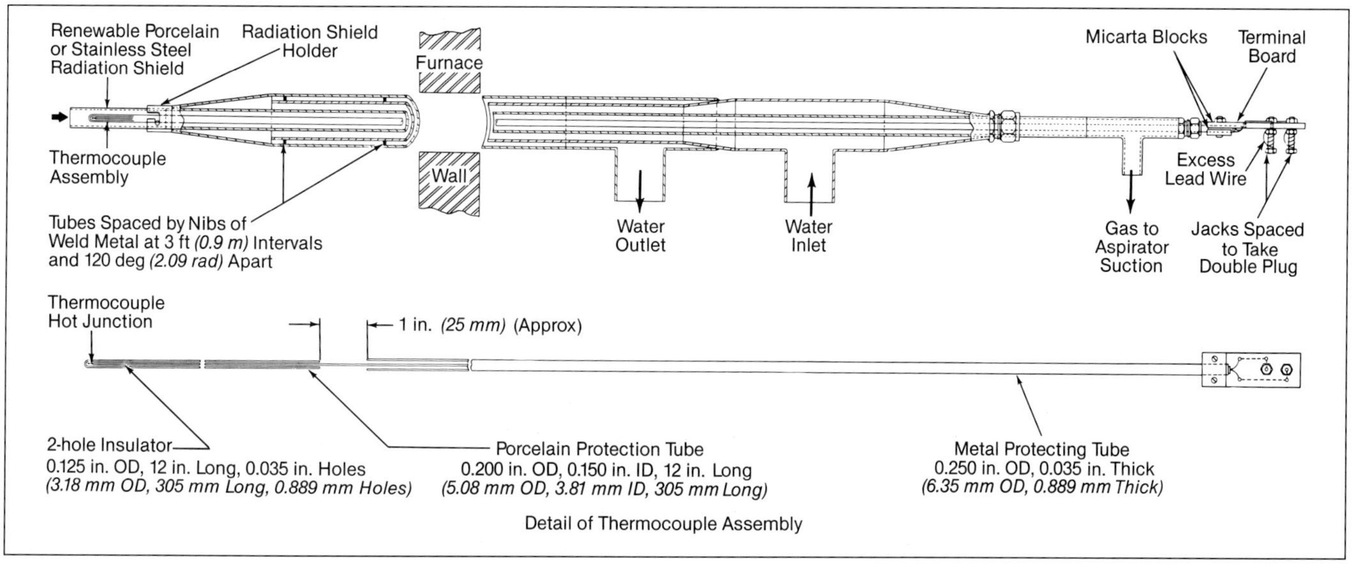

Fig. 25 Rugged water-cooled high velocity thermocouple (HVT) for determination of high gas temperatures.

Thermocouple attachment to the insulation surface must not appreciably alter the normal rate of heat transmission through the insulation and from the surface to the surroundings. Fine wires can be attached and maintained at the surface temperature more easily than heavy wires. If the insulation is plastic at the time of application, press the thermocouple junction and several feet of lead wire into the surface of the insulation; the thermocouple will adhere when the insulation hardens. If the insulation is hard and dry, the junction and lead wires may be cemented to the surface using a minimum of cement. Fastening the wires to the surface with staples introduces conduction errors. Covering the wires with tape changes the heat transfer characteristics of the surface and imposes an undesirable insulation layer between the wire and the ambient air.

Steel casings

The temperature of steel boiler casings may be accurately measured with portable contact thermocouple instruments, because lateral heat flow from adjoining metal areas quickly compensates for the small quantity of heat drawn by the instrument at the contact point. Thermocouple wires may be peened into or fused onto the metal surface to form the hot junction. Thermal contact between the lead wires and the surface should be maintained for several feet. The wire installation should minimize any disturbance of heat transfer from the surface to the surrounding air. Approximate surface temperatures may be conveniently measured with fusion paints or crayons.

Thermometers are sometimes fastened to metal surfaces with putty. This method gives an approximate surface temperature indication if the metal is massive and near ambient temperature. It is not recommended for boiler casing temperature measurement and is completely unsatisfactory for measuring insulation surface temperature.

Through-steel components, such as metal ribs imbedded in insulation and studs or door frames extending through insulation, cause considerable local upsets in surface temperatures and their influence may spread laterally along a metal casing. These effects must be considered in planning and interpreting surface temperature measurement.

Infrared cameras may be used to measure surface temperatures. These instruments can minimize the need for complex thermocouple installations, especially if large surface areas are to be measured. Improved equipment permits videotape recording for further review.

Measurement of steam quality and purity

The most common methods of field testing for steam quality or purity are:

1. sodium tracer (flame photometry),
2. electrical conductivity (for dissolved solids),
3. throttling calorimeter (for direct determination of steam quality), and
4. gravimetric (for total solids).

Each of these methods is described in ASME Performance Test Code 19.11, *Water and Steam in the Power Cycle*. The throttling calorimeter determines steam quality directly, whereas the other methods determine the steam solids content.

Most of the solids content of steam comes from the boiler water, largely in the carryover of water droplets. (See also Chapter 5.) It is customary to relate steam quality and solids content of steam by the following equation:

$$x = 100 - \left[\frac{\text{solids in steam} \times 100}{\text{solids in boiler water}}\right] \qquad \textbf{(6)}$$

where x equals % steam quality in percent by weight and solids are expressed as equivalent parts per million (ppm) (by weight) of steam or water.

By the use of Equation 6, steam quality can be determined if the solids content is known. This relationship is subject to error resulting from carryover of solids in solution in the steam or in vaporized form. This effect occurs principally with silica at pressures above 2000 psi (137.9 bar).

The four steam quality measurement methods may be summarized as follows:

1. The sodium tracer method, with continuous recording of total dissolved solids in steam, is used for the highest accuracy.
2. The conductivity method is useful within certain limitations but is less accurate.
3. The calorimeter method is not suitable for measuring extremely small quantities of carryover and at pressures exceeding 600 psi (41.4 bar).
4. Gravimetric analyses require large samples and do not detect carryover peaks.

Obtaining the steam sample

If the steam quality results are to be accurate, the instruments must be supplied with a representative steam sample. The method of obtaining a steam sample and the operation of the boiler during testing are fundamentally the same for each testing method.

The sampling nozzle design should be as recommended by ASME Performance Test Code 19.11. It should be located after a run of straight pipe equal to at least ten diameters. Locations in the order of preference are:

1. vertical pipe, downward flow,
2. vertical pipe, upward flow,
3. horizontal pipe, vertical insertion, and
4. horizontal pipe, horizontal insertion.

The nozzle should be installed in the saturated steam pipe in the plane of a preceding bend and in a position that ensures the sampling ports directly face the steam flow. On boilers with multiple superheater supply tubes, sampling nozzles should be located in tubes spaced across the width of the drum. These sampling points should be at no greater than 5 ft (1.52 m) intervals.

When a calorimeter is used the connection from the sampling nozzle to the calorimeter must be short and well insulated to minimize radiation losses. Connections must be steam-tight so the insulation remains dry.

When steam purity is tested by conductivity or sodium tracer methods, the tubing from the sampling nozzles to the condenser should be steel (preferably stainless), with an inside diameter not exceeding 0.25 in. (6.4 mm). It should be of minimal length to reduce the storage capacity of the line. Multiple connections can be run to a common line and then to the condenser. However, these connections should be valved so that each one can be sampled individually to isolate possible selective carryover. Cooling coils, or condensers, should be located close to the sampling nozzles to minimize settling of solids in the sample line.

Sodium tracer method

The sodium tracer technique permits measuring dissolved solids impurities (carryover) in steam condensate to as low as 0.001 ppm.

Sodium is present in all boiler water where chemical treatment is in the form of solids. The ratio of the total dissolved solids in the steam condensate to the total dissolved solids in the boiler water is proportional to the ratio of the sodium in the steam condensate to the sodium in the boiler water. By determining the sodium content of the steam condensate and boiler water and the total dissolved solids in the boiler water, the solids content of the steam and the percent moisture carryover may be calculated. The sodium content of steam condensate and water is usually determined by using a flame photometer. The effect of any change in boiler operating conditions on carryover is promptly indicated, facilitating problem analysis.

The operation of a flame photometer is illustrated in Fig. 26. The condensed steam sample is aspirated through a small tube in the burner into the oxygen-hydrogen flame. The flame, at 3000 to 3500F (1649 to 1927C), vaporizes the water and excites the sodium atoms, which emit a characteristic yellow light having a definite wavelength. The intensity of the emitted yellow light is a measure of the sodium in the sample. The intensity of the light is measured with a spectrophotometer equipped with a photomultiplier attachment.

The light from the flame is focused by the condensing mirror and is directed to the diagonal entrance mirror. The entrance mirror deflects the light through the entrance slit and into the monochromator to the plane mirror. Light striking the plane mirror is reflected to the fiery prism where it is dispersed into its component wavelengths. The desired light wavelength is obtained by rotating a wavelength selector which adjusts the position of the prism. The selected wavelength is directed back to the plane mirror where it is reflected through the adjustable exit slit and lens. The light impinges on the photomultiplier tube, causing a current gain which registers on the meter. The amount of sodium in the sample is obtained by comparing the emission from the water sample to emissions obtained from solutions of known sodium concentration.

Conductivity method

Electric conductivity can be used to determine steam purity in certain boilers. This method is applicable to units operating with significant solids concentration in the boiler water and with a total steam solids content in excess of 0.5 ppm.

The conductivity method is based on the fact that dissolved solids, whether acids, bases or salts, are completely

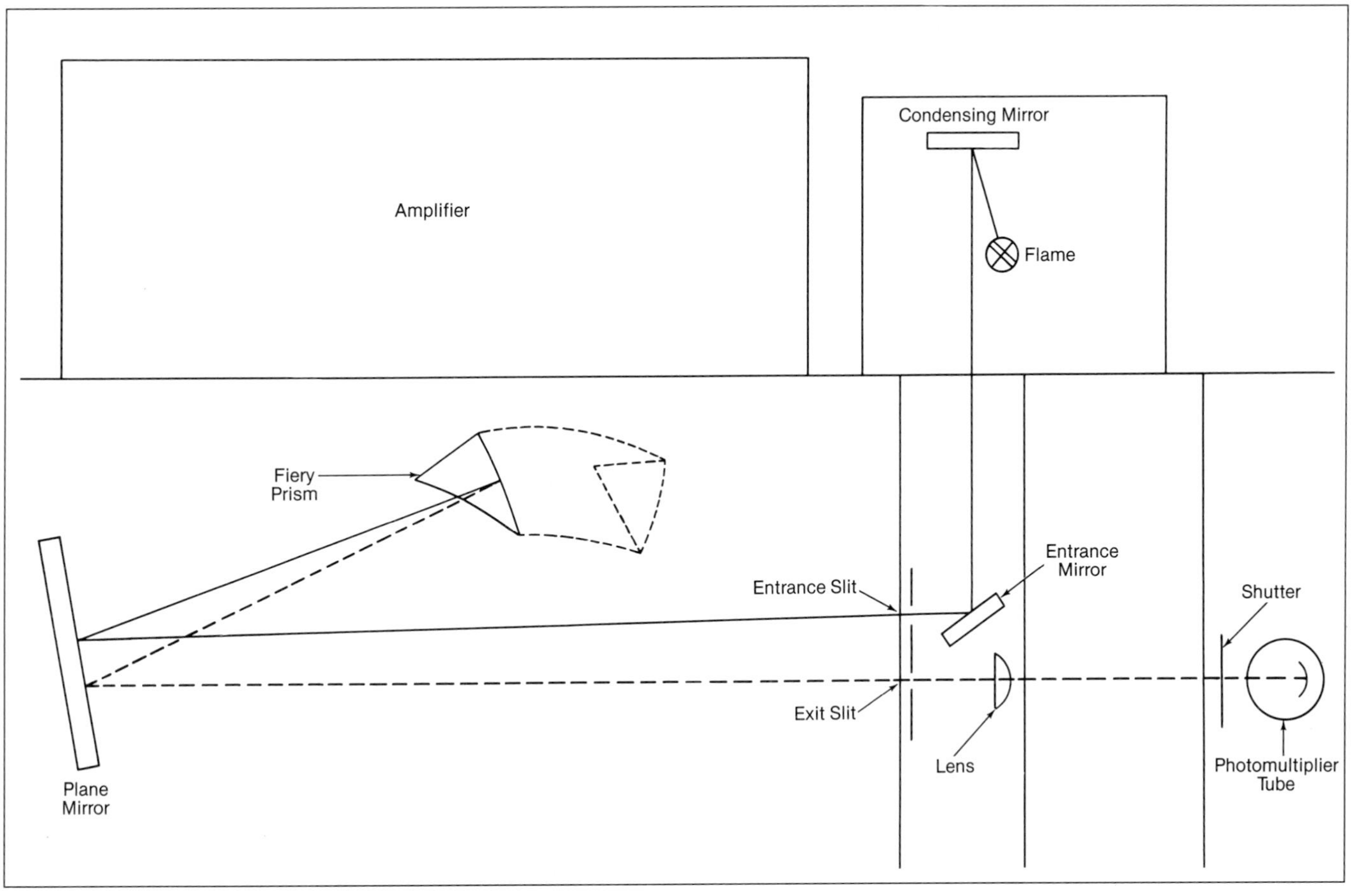

Fig. 26 Schematic arrangement of flame photometer.

ionized in dilute solution and, therefore, conduct electricity in proportion to the total solids dissolved. On the basis of solids normally present in boiler water, the solids content in parts per million equals the electrical conductivity of the sample in micromho per mm times 0.055.

The condensed sample should ideally be free from dissolved gases, especially ammonia (NH_3) and CO_2, which contribute nothing to the solids content of the steam but have a significant effect on conductivity.

Throttling calorimeter

When steam expands adiabatically without doing work, as through an orifice, the enthalpies of the high and low pressure steam are equal, provided there is no net change in steam velocity. Such an expansion is termed throttling. As can be seen from a Mollier chart (see Chapter 2), wet steam with an enthalpy exceeding 1150 Btu/lb (2675 kJ/kg) becomes superheated when throttled to atmospheric pressure. The temperature of the expanded steam determines the enthalpy, which may be used with the pressure of the wet steam to determine the percent moisture in the wet steam sample.

The foregoing principle is used in various forms of throttling calorimeters. Each incorporates a small orifice for expansion of the steam sample into an exhaust chamber, where the temperature of the expanded steam is measured at atmospheric pressure. Velocity changes in properly designed calorimeters are negligible. The variations in the different designs are chiefly in the means of shielding the unit from external temperature influences. At pressures below 600 psi (41.4 bar), the calorimeter shown in Fig. 27 is used. This is a small, low cost, simple, accurate instrument which gives good results for steam having appreciable quantities of moisture: up to 4.3% at 100 psi (6.9 bar), 5.6% at 200 psi (13.8 bar) and 7.0% at 400 psi (27.6 bar).

Fig. 27 shows the calorimeter installed and ready for use. The connection should be short; the connection and the calorimeter should be well insulated. It is essential that the calorimeter discharge be completely unobstructed, so no backpressure can build in the exhaust chamber. The orifices must be clear, full opening and of the correct diameter: 0.125 in. (3.175 mm) for pressures from atmospheric to 450 psi (31.0 bar) and 0.0625 in. (1.588 mm) from 451 to 600 psi (31.1 to 41.4 bar). The thermometer should be immersed in oil of suitably high flashpoint. The calorimeter is operated by fully opening the shutoff valve and letting the steam discharge through the unit to the atmosphere. The calorimeter must be thoroughly warmed up and in temperature equilibrium before normal temperature or test readings are taken.

In obtaining calorimeter readings, the temperature of the expanded and superheated steam is measured by a thermometer inserted in the thermometer well. Due to radiation from the calorimeter installation, thermometer corrections and orifice irregularities, the observed temperature is generally lower than the actual temperature. To determine a suitable correction, the as-installed temperature of the calorimeter should be determined. This can be done by taking readings when the boiler is known to be delivering dry saturated steam. This is the case when the boiler output is steady at about 20% of rated capacity with low water concentration and steady water level. The normal correction is obtained by subtracting the as-installed temperature from the theoretical temperature, read from the zero moisture curve of Fig. 28. The normal correction should not exceed 5F (2.8C). If it does, the calorimeter orifice is clogged, the insulation is faulty, or some other test feature is incorrect. When the unit is used to determine the quality of a wet steam sample, the percent moisture can be found from the curves of Fig. 28 using measured drum pressure and corrected calorimeter temperature. When properly installed, insulated and operated, the calorimeter can give accurate results for steam moisture contents as low as 0.25% for low pressure boilers. For pressures above 600 psi (41.4 bar) and for greater accuracy of steam purity measurement in the low ppm range, other methods of measurement should be used.

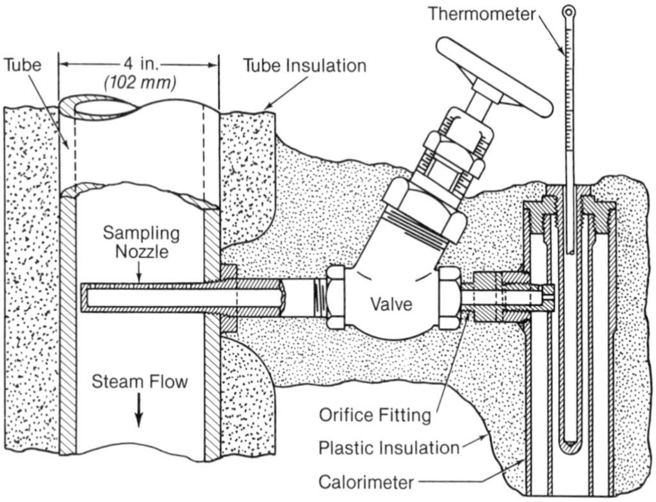

Fig. 27 Throttling calorimeter showing sampling tube installed in steam pipe.

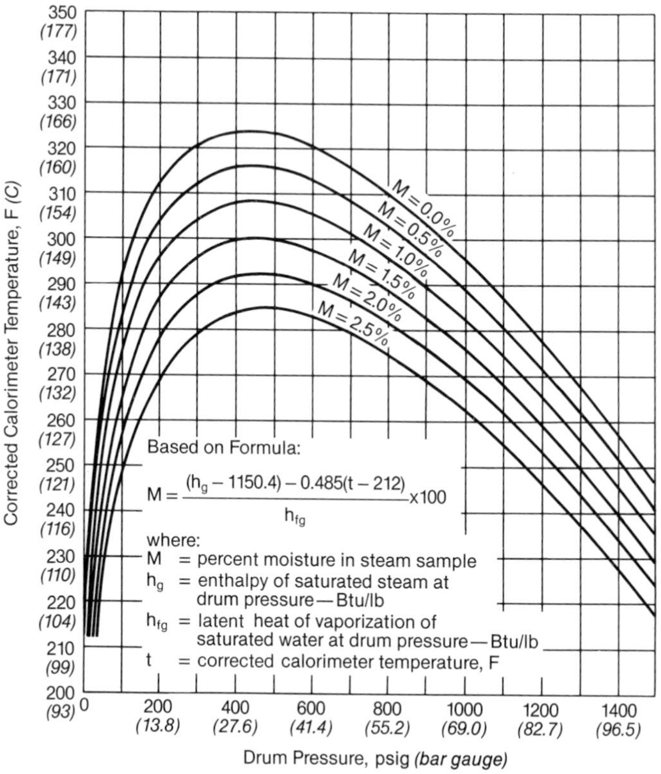

Fig. 28 Percent moisture in steam versus calorimeter temperature.

Gravimetric analysis

Gravimetric analysis is used to determine solids levels, particularly total solids, in a condensed steam sample. It consists of evaporating a known quantity of condensed steam to complete dryness and weighing the residual. This can be done on a batch basis or by a continuous sample system.

While a gravimetric analysis provides an accurate measure of total solids in water, its main disadvantage is that it requires a relatively large quantity of water obtained over an extended period of time. It therefore does not localize carryover peaks when they occur. Time lag in other methods of steam purity measurement is usually small enough to permit attributing peaks of carryover to periods of high water level, high boiler water solids, or other operating upsets.

Flow measurement

Fluid flow includes water, steam, air and gas flow. While there are many means of measuring flow, the basic methods for the accuracy required by the ASME Performance Test Code use the orifice, flow nozzle (flow tube), or venturi tubes as primary elements. The pressure drop or differential pressure created by these restrictions can be converted into a flow rate.

The flow of any fluid through an orifice, nozzle or venturi tube may be determined by the equation:

$$\dot{m} = C_q YA\sqrt{\frac{2g_c\rho_1(P_1 - P_2)}{1-\beta^4}} \tag{7}$$

where

$\dot{m}$ = flow rate, lb/s
C_q = coefficient of discharge, dimensionless, dependent on the device used, its dimensions and installation
Y = compressibility factor, dimensionless, equals 1.0 for most liquids and for gases where the pressure drop across the device is less than 20% of the initial pressure
A = cross-sectional area of throat, ft^2
g_c = proportionality constant, 32.17 lbm ft/lbf s^2
P_1 = upstream static pressure, lb/ft^2
P_2 = downstream static pressure, lb/ft^2
β = ratio of throat diameter to pipe diameter, dimensionless
ρ_1 = density at upstream temperature and pressure, lb/ft^3

Details of primary element sizing, fabrication and flow calculations can be obtained from the ASME publication *Fluid Meters, Their Theory and Application*.

Table 2 is extracted from the ASME Performance Test Code 19.5, Section 4, *Flow Measurement,* to show the advantages and disadvantages of the three primary element types. The throat tap nozzle has the highest accuracy of the primary flow measuring devices.

The primary elements listed should be fabricated of erosion and corrosion resistant materials. The orifice and flow nozzles are shown in Fig. 29; the venturi tube is shown in Fig. 30.

Even though these elements can be sized with considerable accuracy by calculation, they should be laboratory calibrated prior to precision testing. Calibration is usually by weighed water tests using scales. For commercial use, calculated flow rates or rates based on prototype testing or calibration are adequate.

Certain important factors of the primary element installation should be considered for accurate measurement:

1. location in the piping in relation to bends or changes in cross-section,
2. possible need for approach straightening vanes,

Table 2
Advantages and Disadvantages of Various Types of Primary Elements

Advantages	Disadvantages
Orifice	
1. Lowest cost	1. High nonrecoverable head loss
2. Easily installed and/or replaced	2. Suspended matter may build up at the inlet side of horizontally installed pipe unless eccentric or segmental types of orifices are used with the hole flush with the bottom of the pipe
3. Well established coefficient of discharge	3. Low capacity
4. Will not wiredraw or wear in service during test period	4. Requires pipeline flanges, unless of special construction
5. Sharp edge will not foul with scale or other suspended matter	
Flow Nozzle	
1. Can be used where no pipeline flanges exist	1. Higher cost than orifice
2. Costs less than venturi tubes and capable of handling same capacities	2. Same head loss as orifice for same capacity
	3. Inlet pressure connections and throat taps when used must be made very carefully
Venturi Tube	
1. Lowest head loss	1. Highest cost
2. Has integral pressure connections	2. Greatest weight and largest size for a given size line
3. Requires shortest length of straight pipe on inlet side	
4. Will not obstruct flow of suspended matter	
5. Can be used where no pipeline flanges exist	
6. Coefficient of discharge well established	

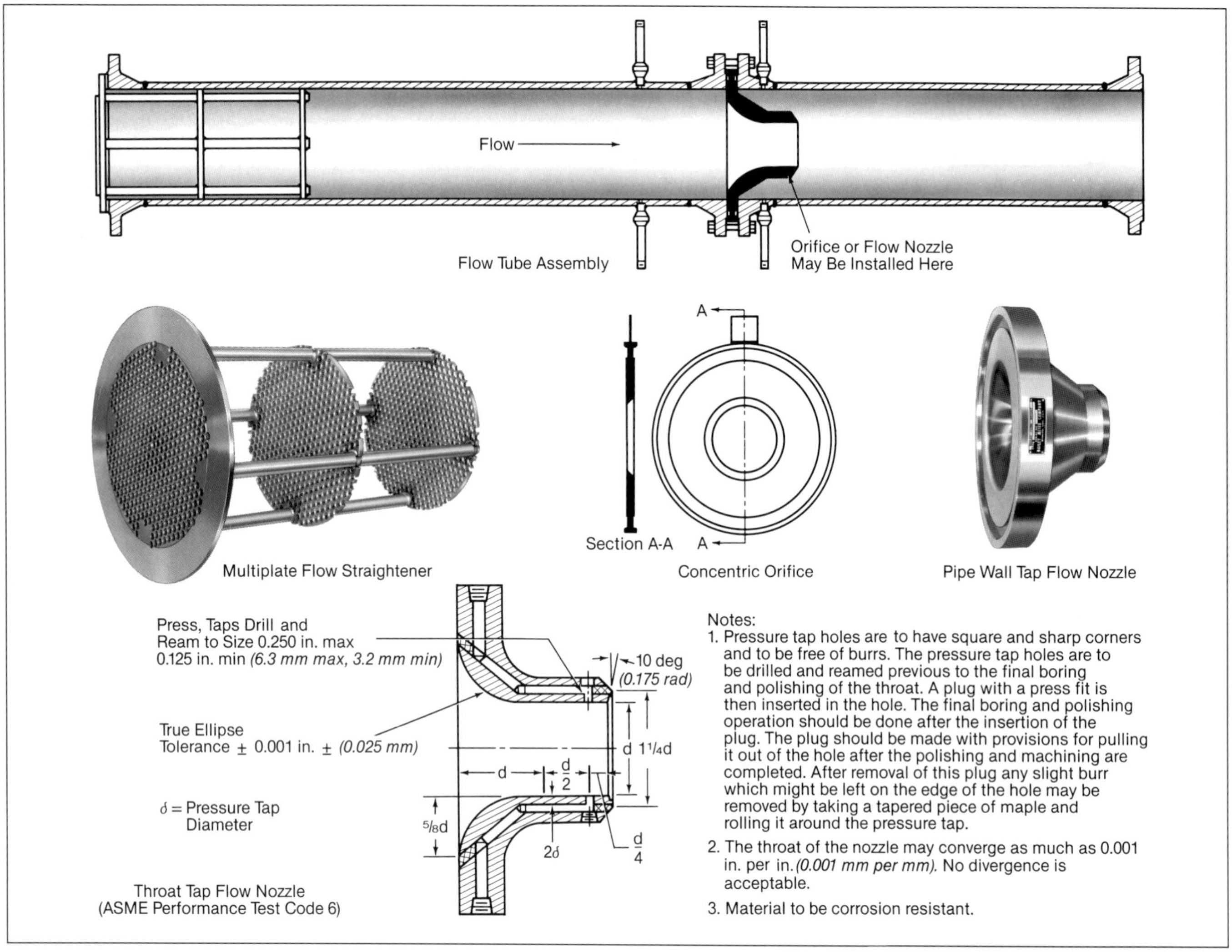

Fig. 29 Orifice and flow nozzles.

3. location and type of pressure taps,
4. dimensions and condition of surface or piping before and after the element,
5. position of the element relative to direction of fluid flow, and
6. type and arrangement of piping from primary element to differential pressure measuring instrument.

Details of these requirements for precision testing are also covered in the ASME publication *Fluid Meters, Their Theory and Application*. These devices produce a differential pressure across the element and a differential pressure gauge should be used for the flow measurement.

Flow measurement of combustion air or flue gas generally does not require a high degree of precision. Orifices, flow nozzles, or venturis are also used, but the rigid requirements of construction and location are usually not possible because of space limitations. As a substitute, a component of the steam generator, located in the flow path, can create a suitable pressure differential, which is used to obtain a flow measurement. For example, the draft loss across the gas side of an air heater may be used as an index of the flue gas flow. Because pressure differentials vary with a change in surface cleanliness, these are not highly reliable methods. The use of orifices, flow nozzles, venturi air foils, or impact suction tubes in areas free of entrained dust and at relatively stable temperatures prove more satisfactory. Fig. 31 illustrates two types of venturi sections used in air ducts for measuring combustion air flow. Fig. 32 illustrates an impact suction pitot tube arrangement for metering primary air flow to a pulverizer. For a dependable determination of air flow, the primary element should be calibrated at normal operating pressure and temperature. Test connections, or sampling points, for this purpose are located at a zone in the duct or pipe where good flow characteristics are obtainable, and the calibrating flow measurements are made using a hand held pitot tube or equivalent probe. The pitot tube, when inserted facing the air or gas flow stream, measures velocity pressure, that is, the difference between total and static pressure. The velocity pressure measurement can be converted into a velocity reading for English units only by:

$$V = 1097\sqrt{\frac{h_V}{\rho}} \qquad \textbf{(8)}$$

where

V = velocity, ft/min
h_V = velocity head as indicated by pitot tube differential, in. wg
ρ = density at the temperature of the sampling location, lb/ft^3

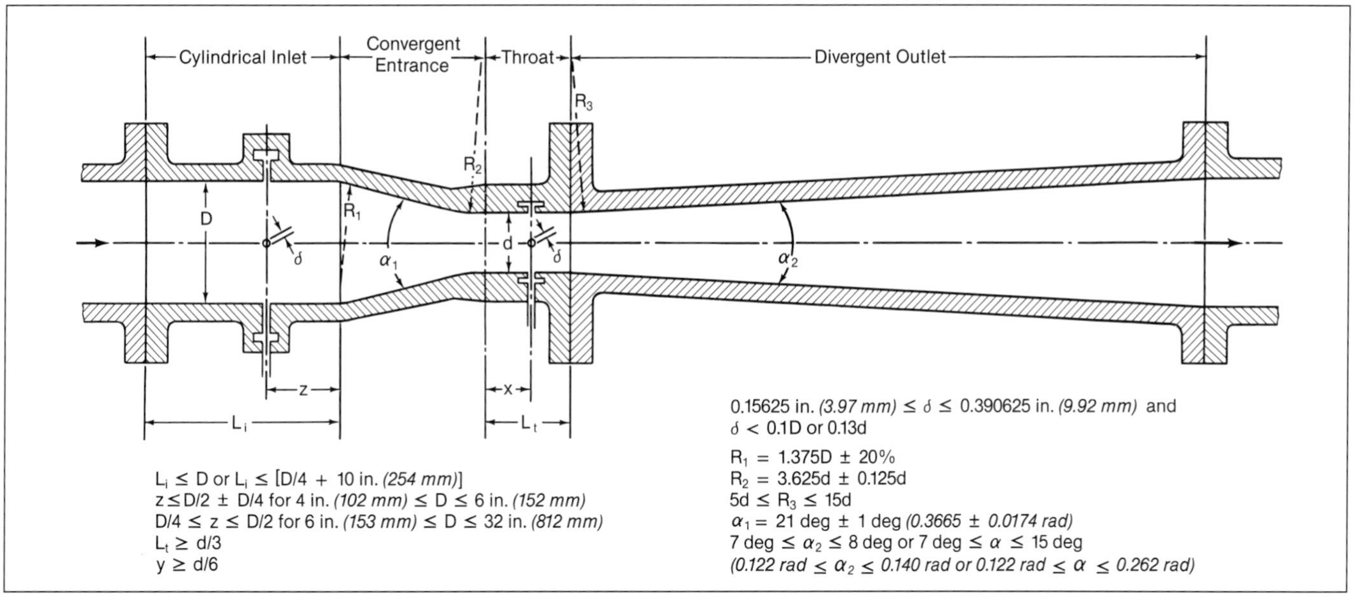

Fig. 30 Dimensional proportions of classical (Herschel) venturi tubes with a rough-cast, convergent inlet cone. (Source: ASME, *Fluid Meters, Their Theory and Application*, Sixth Edition, 1971.)

A typical pitot tube and measuring manometer are illustrated in Fig. 33. The pitot tube has a coefficient of 1.0, which is applied to velocity head readings, eliminating the need for corrections. Fig. 34 shows a method of traversing a round duct using a pitot tube.

An additional portable flow measuring device that can be used for calibration purposes is the Fechheimer probe. It also has a coefficient of 1.0 and incorporates a null balance feature that permits determining when the probe faces the direction of gas flow. This probe is illustrated in Fig. 35. In this schematic, the outer holes in the probe are each at exactly 39.25 deg from the centerline hole, placing them at a point of zero impact pressure and providing a reading of true static pressure. When the probe faces the gas flow, a manometer connected across the two outer holes shows a zero, or null, balance. Because the centerline hole receives full impact or total pressure when facing into the gas stream, a manometer connected across one outer hole and the centerline hole gives true velocity pressure, that is, impact pressure less static pressure.

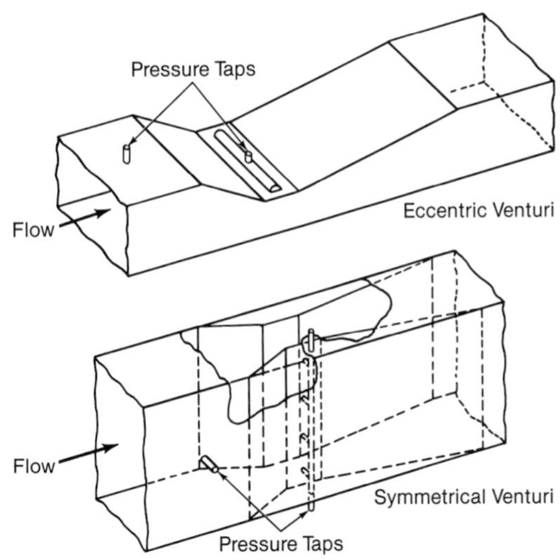

Fig. 31 Venturi sections for air ducts.

The impact suction or reversed type pitot tube, illustrated in Fig. 36, produces a differential pressure greater than one velocity head, which gives a magnified reading. This type of pitot tube measures flow in either direction. To use the reversed type pitot tube for flow measurement, calibration is required for the Reynolds number range as well as for the geometry of the flow channel and its orientation in the stream.

Permanently installed pitot tubes are used where relatively high pressure losses due to the primary measuring element are undesirable. Flow tubes are also used in these cases. These tubes differ from most other primary elements

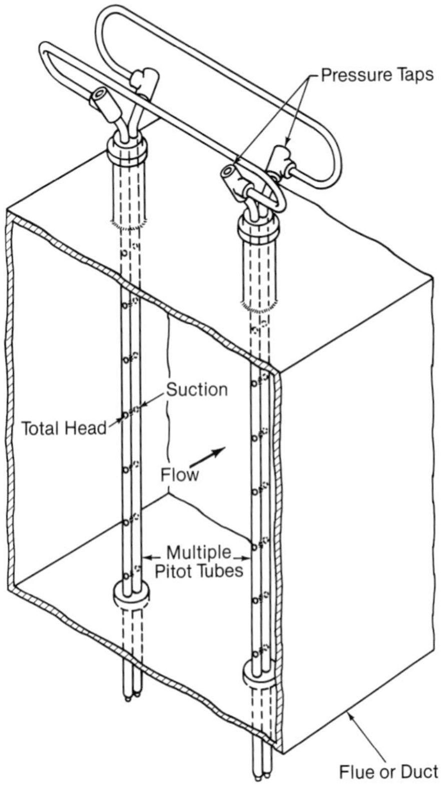

Fig. 32 Averaging pitot tubes.

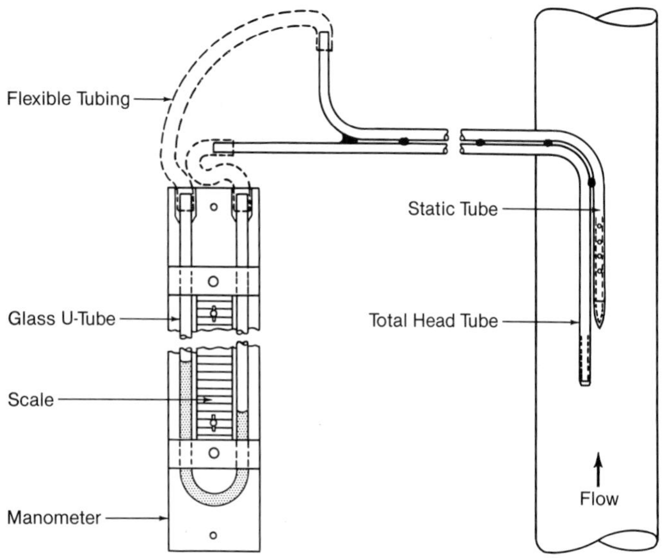

Fig. 33 Arrangement of pitot tube and manometer.

in that the flow channel cross-section varies symmetrically in the inlet and exit directions from the contoured throat section.

The throat section is equipped with two sets of pressure taps, one set pointing upstream and one downstream, similar to impact suction pitot tubes. Each set of taps is interconnected and the two sets in turn are connected to a differential pressure gauge. The pressure taps are located at cross-sections of equal area so that the differential developed is a result of impact and suction pressure differences only and is, therefore, a function of the velocity head alone.

Many other types of flow measuring devices exist. The pitot tube grid incorporates an array of pitot tubes in a flow path. The primary signal is a pressure differential. The array of pitot tubes measures a number of flows and minimizes measurement errors caused by flow unbalance and upstream conditions. The hot wire anemometer measures flow using a probe with a small wire at the tip. Flow over the wire removes heat, and flow may be measured when the hot wire is in equilibrium with electrical power input and heat lost. Doppler phenomena may be applied to flow measurement with laser or acoustic transmissions. The transmission signal and type of fluid make these methods very application specific.

Computer application to measurement

The application of computer technology to instrumentation provides the capability to further understand the operating characteristics of many interrelated power plant components and processes at a much faster rate than was previously possible. Measurements have progressed from readings recorded at local gauges, strip charts, or manually recorded test gauges to significant quantities of information acquired in seconds.

The instrument readings are not made directly at the instrument, but rather transmitted signals from the instrument are entered into a data acquisition system and the reading is indicated at a computer. The computer gathers, stores, interprets and displays the information. The display appears as an extension of the instrument. The information gathered may be further manipulated for process control or equipment performance computations.

Performance computation

Computer computation permits calculations in near real time for boiler efficiency and other plant performance indicators. These performance indicators provide rapid updates to operating information and provide a check on measurement errors. Measurements of temperature, pressure and flow that might be in error because of instrument calibration are identified rapidly.

Process control

A computer based control system receives its basic information from a data acquisition system which accumulates information from the plant instrumentation. Process control is achieved according to specific mathematical algorithms. The control system, updated by current data or operator interface, produces output signals to modulate components based on these algorithms.

Data acquisition system

A typical data acquisition system consists of computer hardware and related software. The basic hardware components are the computer, a signal multiplexer and the measuring instrument. The computer program (software) controls the hardware operation and data recording. In operation, the program causes the signal multiplexer to read a specific instrument analog output signal. The signal is converted from an analog to a digital signal, which is conveyed to the computer and stored in memory. The first

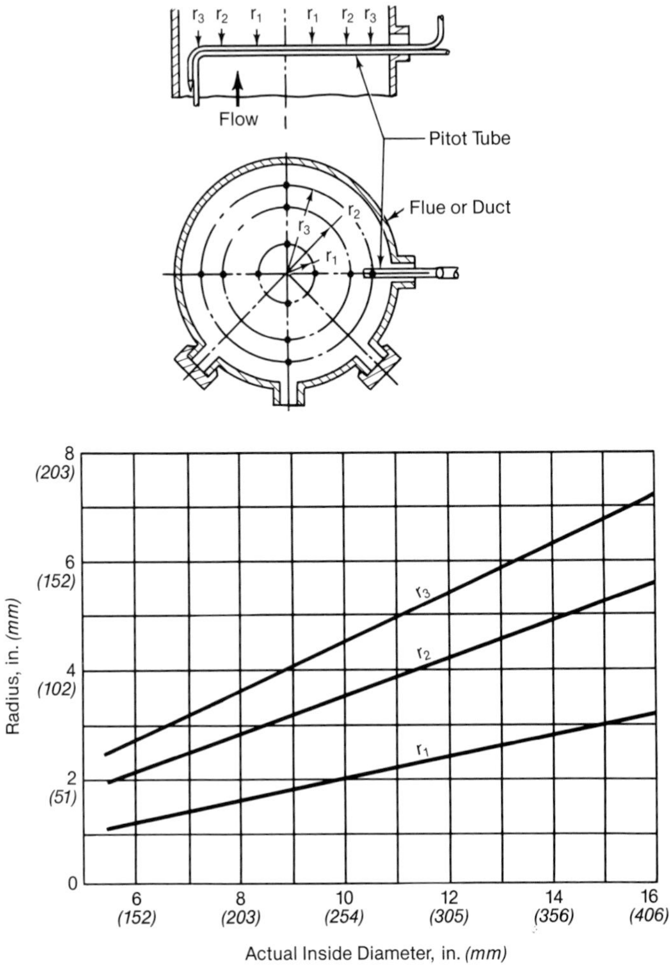

Fig. 34 Pitot tube traverse.

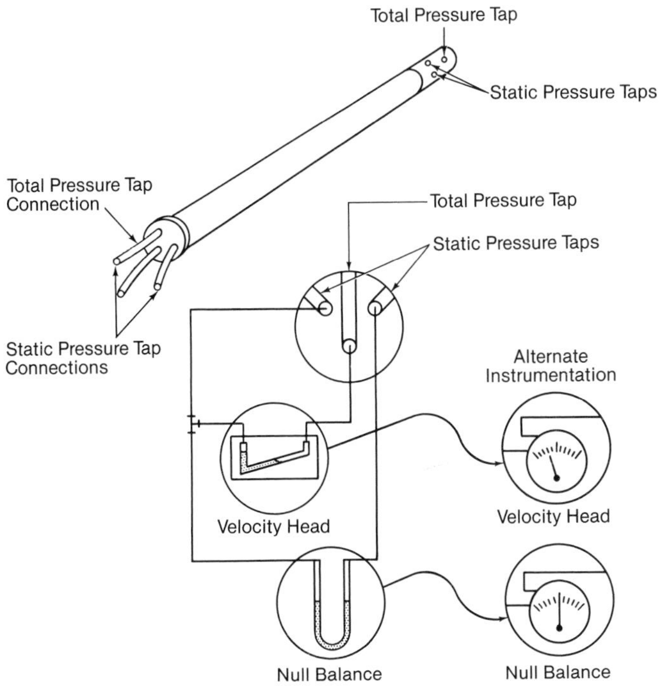

Fig. 35 Fechheimer probe.

processing of the data is the storage of the digitized signal in a data array. Data acquisition, incorporating many instruments, other computer outputs and manual inputs, requires that each of the signals be assigned a specific location in the array. The stored signal may be converted to engineering units and stored in another array. Furthermore, the data may be averaged over time and then stored. Performance calculations are generally performed using data converted to engineering units and averaged over time. Portable data acquisition systems are generally used for specific test purposes and may use high precision transmitters and thermocouples.

Signal multiplexer

The signal multiplexer in a data acquisition system provides an interface between the transmitted instrument signal and the computer. The typical multiplexer provides signal conditioning, analog to digital signal conversion and switching capability for multiple inputs. It produces a digital output signal for computer processing and assignment to a position in the data array. Significant engineering design is required to accommodate the variety of signals entering the multiplexer.

Data acquisition computer program

The data acquisition software may be program oriented or menu oriented but should readily permit changes to the quantity of signals, the rate at which the computer reads each signal and the computations performed with each signal. Data storage and retrieval are but two of many design considerations for computer program configurations.

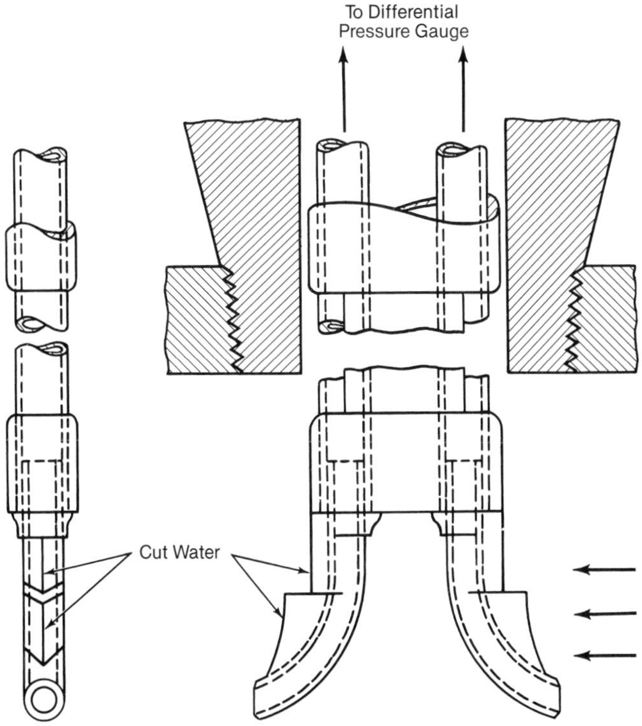

Fig. 36 Pitot tube, combined type with reversed static tube. (Source: ASME, *Fluid Meters, Their Theory and Application*, Sixth Edition, 1971.)

References

1. Tower, J.R., "Staring PtSi IR cameras: More diversity, more applications," *Photonics Spectra,* (USPS 448870), Laurin, Pittsfield, Massachusetts, February 1991.

2. Kaplan, H., "What's new in IR thermal imagers," *Photonics Spectra,* (USPS 448870), Laurin, Pittsfield, Massachusetts, February 1991.

3. Weiss, J., and Esselman, W., Electric Power Research Institute and R. Lee Foster-Miller Inc., "Assess fiberoptic sensors for key power plant measurements," *Power*, p. 55, October 1990.

Bibliography

1986 Digital Systems Techniques, Instruments and Apparatus, Supplement to ASME Performance Test Codes, ANSI/ASME PTC 19.22, American Society of Mechanical Engineers, New York.

Manual on the Use of Thermocouples in Temperature Measurement, ASTM Special Technical Publication 470B, American Society for Testing and Materials, Philadelphia, 1981.

1986 OMEGA Complete Temperature Measurement Handbook and Encyclopedia™, Omega Engineering, Inc., Stamford, 1985.

Temperature Measurement Designer's Guide, Thermo Electric Co. Inc., Saddle Brook, New Jersey, 1986.

Typical modern control console (*courtesy of Elsag Bailey, Inc.*).

Chapter 41
Controls for Fossil Fuel-Fired Steam Generating Plants

Instruments and controls are essential to all steam generating installations promoting safe, economic and reliable operation of the equipment. They range from the simplest manual devices to complete systems that automatically control the boiler and all associated equipment.

Boiler control systems, particularly those applied to boiler turbine-generator units, are of several types that perform varied control functions. In the past, the accepted practice has often been to identify these functions as separate and independent systems, often applied to the boiler by different suppliers. Today, a power plant control system is most often an integrated package with demand requirements applied simultaneously to the boiler, turbine and major auxiliary equipment. This minimizes the number of complex interactions between subsystems.

Various types of boiler control systems for fossil fuel boilers include:

1. boiler instrumentation,
2. combustion (fuel and air) control,
3. steam temperature control for superheater and reheater outlet,
4. drum level and feedwater flow control,
5. burner sequence control and management systems,
6. bypass and startup,
7. systems to integrate all of the above with the turbine and electric generator control,
8. data processing, event recording, trend recording and display,
9. performance calculation and analysis,
10. alarm annunciation system,
11. management information system, and
12. unit trip system.

An overall integrated control system for a fossil fuel-fired drum boiler is described later in this chapter.

Basic control theory

Boiler control means the regulation of the boiler outlet conditions of steam flow, pressure and temperature to their desired values. In control terminology, the boiler outlet steam conditions are called the outputs or controlled variables. The desired values of the outlet conditions are the set points or input demand signals. The quantities of fuel, air, and water are adjusted to obtain the desired outlet steam conditions and are called the manipulated or control variables.

Disruptive influences on the boiler, both internal and external, such as fuel heat content variations, unit load change or change in cycle efficiency, are called disturbance inputs. The controller or control system looks at the desired (set points) and actual (output variables) values of the outlet steam conditions and adjusts the amounts of fuel, air and water (manipulated variables) to make the outlet conditions match their desired values. The controller can be manual, with an operator making the adjustments, or automatic, with a pneumatic system, electronic analog computer, or a digital computer making the adjustments. A block diagram of a conceptual boiler control system is shown in Fig. 1.

While it is theoretically possible to operate a boiler with manual control, the operator must maintain a tedious, constant watch for a disturbance. Time is needed for the boiler to respond to a correction and this can lead to overcorrection with further upset to the boiler. An automatic controller, on the other hand, once properly tuned will make the proper adjustment quickly to minimize upsets and will control the system more accurately and reliably.

The various types of automatic control can be illustrated by Fig. 2, which represents the control of one out-

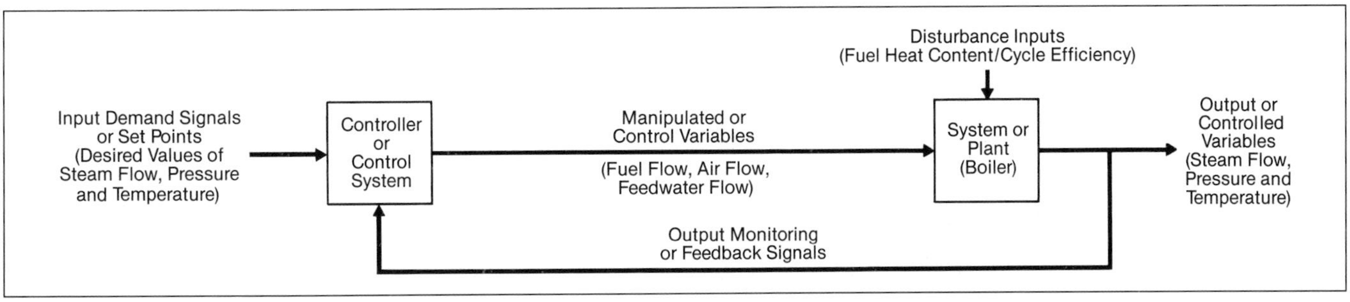

Fig. 1 Block diagram of a conceptual boiler control system.

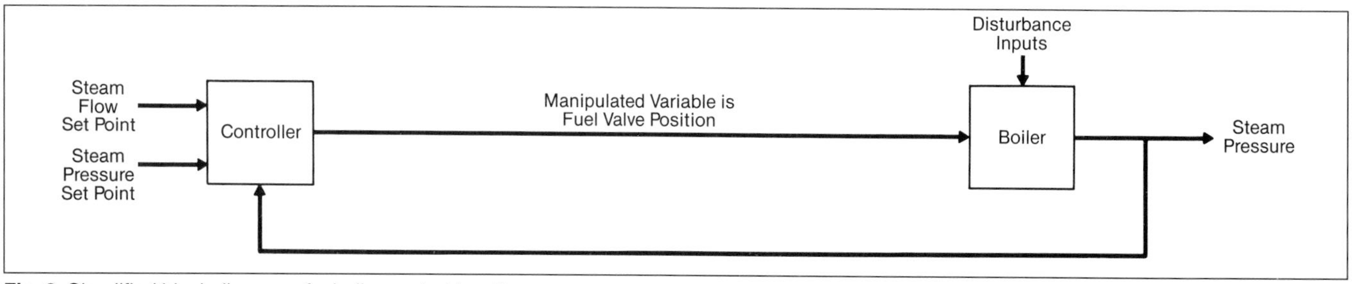

Fig. 2 Simplified block diagram of a boiler control function.

put variable, steam pressure, in a hypothetical boiler. The set points for steam pressure and steam flow will be used as reference values, and the manipulated variable is the fuel valve position. Disturbance inputs act on the boiler, the system to be controlled.

Open loop control

The simplest control mode is an open loop, feedforward, or nonfeedback control. The manipulated variables of fuel, air and water are adjusted only from the input demand signals without monitoring the outlet conditions or output variables. As an example, Fig. 3 illustrates an open loop control system used to accomplish the control function illustrated by Fig. 2. Fig. 3a is a block diagram representing the action to be taken, which is feedforward only. Fig. 3b is a calibration curve expressing fuel valve position as f (D), a function of steam flow demand. This curve, established by calibrating or programming the fuel valve position required to maintain the desired steam flow with a constant steam pressure, is entered into the controller. The response of this open loop control is fast and depends only on the accuracy of the calibration curve.

One difficulty with open loop control is that the calibration curve is accurate only as long as the boiler conditions are the same as when the calibration was set. Another problem is that the output changes as the load changes on the output signal (the fuel valve in Fig. 3). For example, a change in friction on the valve stem will cause a variation in valve position in response to a given input demand signal. This introduces an error into the performance of the open loop control. Such disadvantages normally outweigh the open loop advantages of stability and simplicity.

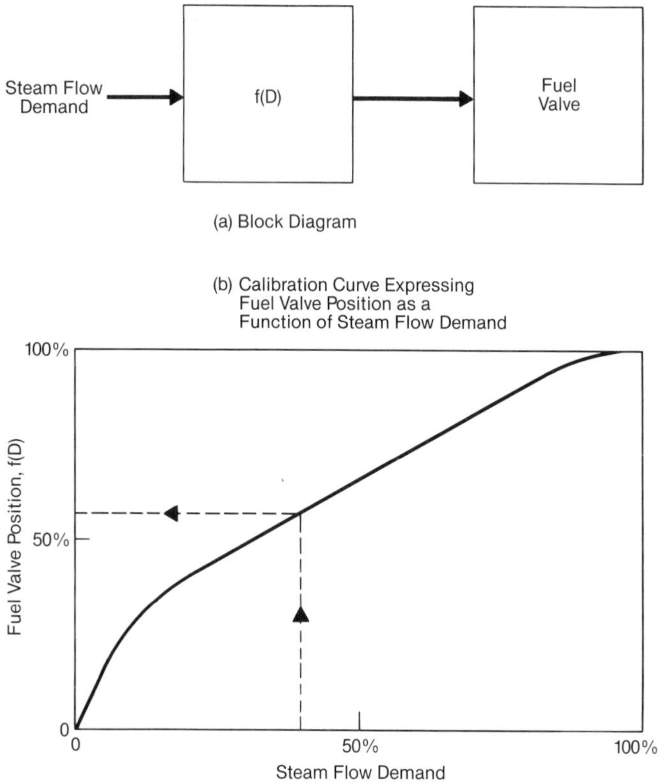

Fig. 3 Open loop mode using steam flow demand to vary fuel input.

Closed loop control

Usually the system requirements can not be met by an open loop control, and a closed loop or feedback control must be used. In closed loop control, the actual output of the system is measured and compared to the input demand signal with the difference between the two signals (the error signal) used to eliminate this difference.

The open loop system of Fig. 3 could be closed by having an operator observe the measured steam pressure and then manually adjust the fuel valve position to obtain the fuel flow necessary to maintain the desired steam pressure and flow. To avoid the disadvantages of manual control, an automatic controller can be provided in a closed loop or feedback system.

Proportional control

The simplest type of closed loop system is proportional control. The manipulated variable or controller output is proportional to the deviation of the controlled variable from its desired value. The deviation is called the error signal. Depending upon controller arrangement, the output sig-

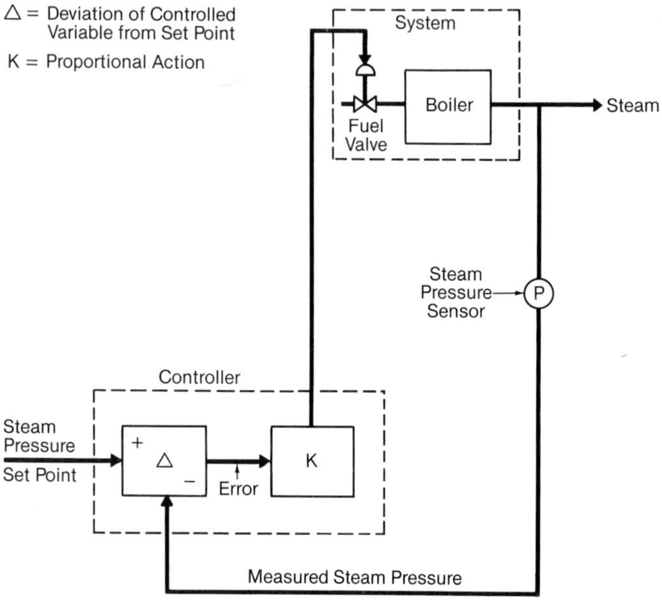

Fig. 4 Proportional control system.

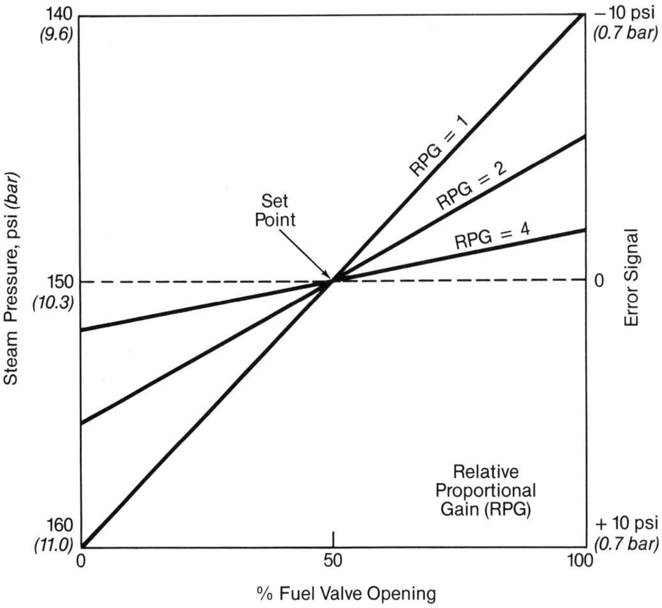

Fig. 5 Fuel valve opening for various pressure deviations in a proportional control system.

nal of a proportional control system will always be either directly or inversely proportional to the controlled variable.

Fig. 4 illustrates the application of proportional control to the control function of Fig. 2. Table 1 explains symbols in Fig. 4 and subsequent figures. In this system, for every deviation of steam pressure from its set point, the fuel valve will be moved to a specific position, as shown in Fig. 5, for three values of relative proportional gain. In Fig. 5, the relative proportional gain could be set so that a pressure error of 10 psi (0.7 bar) initiates a 50% change in position of the fuel valve which in the system calibration would be called a gain of 1.0. If the control were set to be two or four times as sensitive, the relative proportional gain would be said to be 2.0 or 4.0 respectively. In the latter case, a change of 2.5 psi (0.2 bar) in steam pressure initiates a 50% change in fuel valve opening.

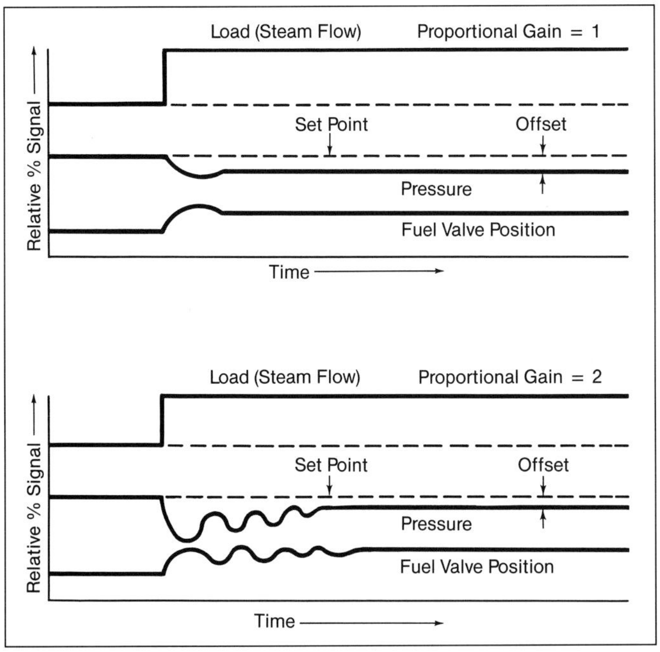

Fig. 6 Response of steam pressure to step increase load in a proportional control system.

Table 1
Control Symbols

O	— Transmitter
K	— Proportional action (gain)
∫	— Integral action
Σ	— Summing action
Δ	— Difference or subtracting action
<	— Low select auctioneer
>	— High select auctioneer
d/dt	— Derivative (rate)
Σ/n	— Averaging
T	— Transfer
±	— Bias action
f (x)	— Operating curves of valves, drives, etc.

The response of steam pressure to step increases in load on the boiler is shown in Fig. 6 for two values of proportional gain on the controller. A step increase in load or steam flow immediately decreases the steam pressure which, through the proportional controller, causes the fuel valve to open by a proportional amount. The additional heat input due to the increased fuel flow causes the steam pressure to return toward its set point, which in turn reduces the fuel valve opening.

There may be a certain amount of cycling or searching, depending on the proportional gain of the controller, before the steam pressure stabilizes. It is noted in Fig. 6 that the steam pressure does not stabilize at its set point, but is offset to a lower value. A characteristic of proportional only control is that an error or offset is necessary to provide a steady-state fuel valve opening which will support the desired load, except for a single load condition.

An increase in the relative proportional gain or sensitivity will reduce the final offset in the steam pressure as shown in Fig. 6. An increase in proportional gain may increase the time required for the system to reach a steady-state condition. This offset is always present with proportional control and, if the gain is increased even more to reduce the offset, undamped oscillations will result.

Proportional plus integral control

The offset may be eliminated by adding integral or reset control to the proportional control system (Fig. 7). The response of the boiler to a step increase in load, when using a proportional plus integral control system, is shown in Fig. 8. The steam pressure is returned to its set point without the offset. However, the system may be less stable as it takes a longer period of time for the steam pressure to stabilize with the offset completely eliminated.

Integral control, as its name implies, is based on the repetitive integration of the difference between the controlled variable and its set point over the time the deviation occurs. Integral control is also referred to as reset control because the band of proportional action is shifted or reset so that the manipulated variable operates about a new base point.

Proportional plus integral plus derivative control

The stability and response of the system can be improved further by adding a third mode of action called derivative or rate control. Derivative control is a function of the rate of controlled variable change from its set point, as

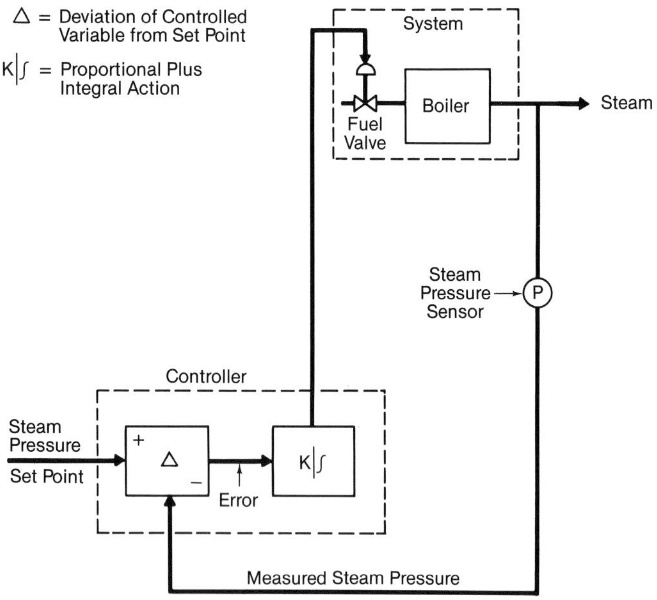

Fig. 7 Proportional plus integral control.

shown in Fig. 9. The addition of this is shown in Fig. 10. As soon as the step change is made, the pressure starts to drop and the proportional mode begins to open the fuel valve. The derivative mode will also open the fuel valve further, as a function of the rate at which the pressure is changing, anticipating where the valve should be positioned. When the rate of steam pressure change decreases, the derivative control has less effect and the proportional and integral modes do the final valve positioning.

Feedforward/feedback control

In a closed loop system, the controlled variable must deviate from its set point before the controller initiates a corrective action. The open loop system therefore has a faster response because it takes corrective action before the controlled variable starts to change. When the open loop or feedforward system is combined with the closed loop or feedback system, the result is a fast response system that can compensate for changes in the calibration curve. Fig. 11 represents a feedforward/feedback control system for the control function of Fig. 2.

As the step change in load occurs, the feedforward signal immediately positions the fuel valve to meet the calibration curve requirements. If this curve is exact, no error develops and the feedback loop has no work to do. If the calibration curve is in error, the feedback loop readjusts the fuel valve position to eliminate any pressure error which may develop due to shifts in the calibration. However, the

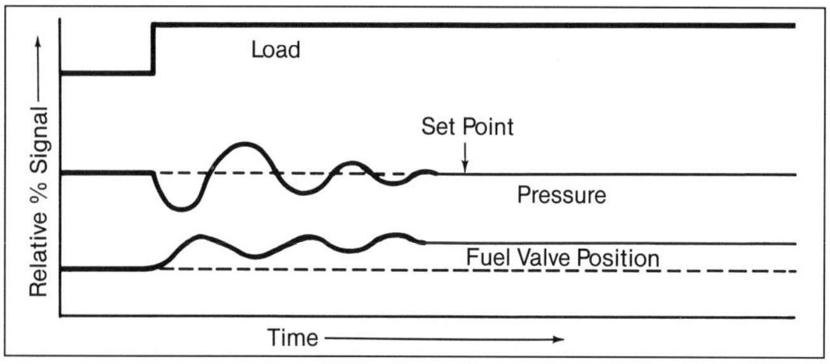

Fig. 8 Response of steam pressure to step increase of load in a proportional plus integral control system.

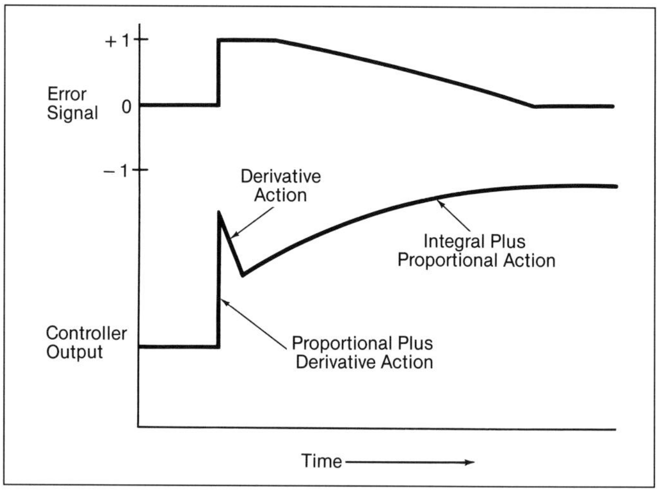

Fig. 9 Proportional plus integral plus derivative action.

response will be faster and the magnitude of the system upset will be smaller because the feedforward signal moves the valve near its final steady-state position leaving a smaller range of action for the feedback loop.

Control system design philosophy for a utility boiler

The effectiveness of a boiler turbine-generator unit, as a power producing system, depends on how well it can respond to demands. To achieve optimum response from the unit while maintaining stable pressure and temperature, the control system must coordinate the boiler and turbine. Because the boiler can not produce rapid changes in steam generation by changes in throttle pressure, the turbine valves open and close for initial load response. This initial change permits the new steam flow requirements to be met at the existing throttle pressure. The control system then acts to restore the proper amount of stored energy to the boiler by coordinating the boiler and turbine. This must be done without exceeding the operating limits of the boiler or turbine-generator.

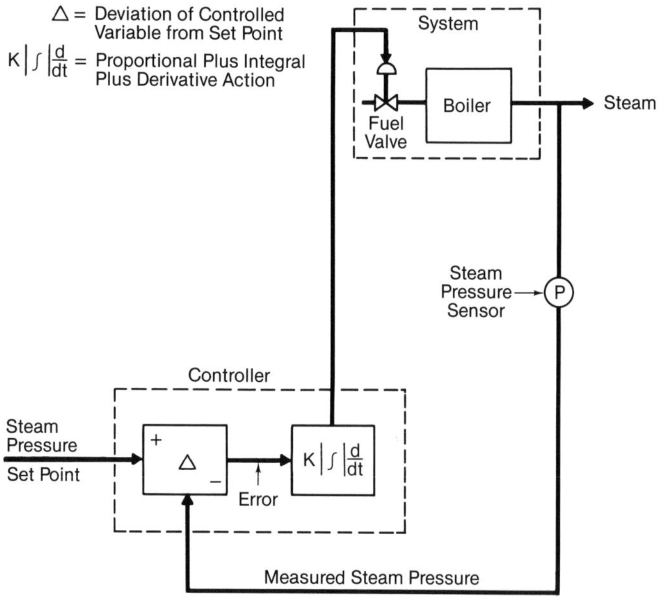

Fig. 10 Proportional plus integral plus derivative control.

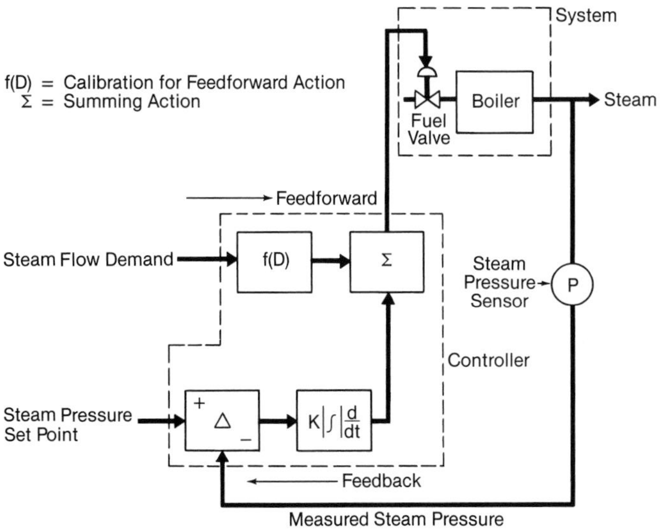

Fig. 11 Feedforward/feedback control.

Characteristics of different control modes

Boiler-following control This mode of operation is designed for boiler response to follow turbine response. Megawatt load control is the responsibility of the turbine-generator. The boiler is assigned secondary responsibility for throttle pressure control. The demand for a load change repositions the turbine control valves. Following a load change, the boiler control modifies the firing rate to reach the new load level and to restore throttle pressure to its normal operating value.

Load response with this type of system is rapid because the stored energy in the boiler provides the initial change in load. The fast load response is obtained at the expense of less stable throttle pressure control. A boiler-following control system is illustrated in Fig. 12.

Turbine-following control In this operating mode, turbine response follows boiler response. Megawatt load control is the responsibility of the boiler while the turbine-generator is assigned secondary responsibility for throttle pressure control. With increased load demand, the boiler control increases the firing rate which, in turn, raises throttle pressure. To maintain a constant throttle pressure, the turbine control valves open, increasing megawatt output. When a decrease in load is demanded, this process is reversed. Load response with this type of system is rather slow because the turbine-generator must wait for the boiler to change its energy output before repositioning the turbine control valves to change load. However, this mode of operation will provide minimal steam pressure and temperature fluctuation during load change. A conventional turbine-following control system is shown in Fig. 13.

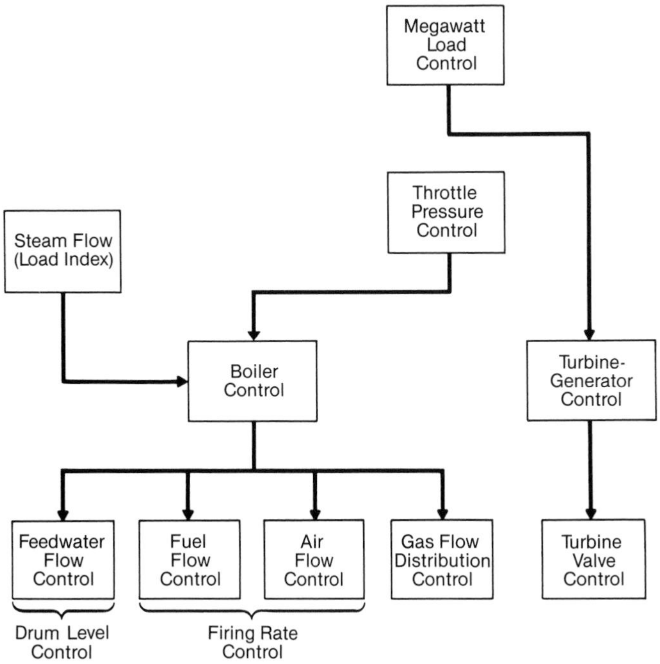

Fig. 12 Boiler-following control system.

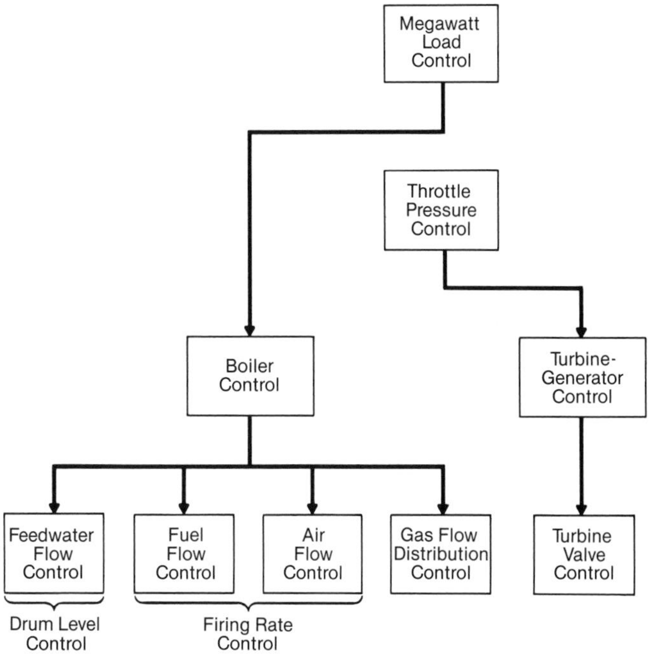

Fig. 13 Turbine-following control system.

Coordinated boiler-turbine control While both the above systems can provide satisfactory control, each has inherent disadvantages and neither fully exploits the capabilities of both the boiler and turbine-generator. The turbine-following and boiler-following systems may be combined into a coordinated control system giving the advantages of both systems and minimizing the disadvantages. The coordinated boiler turbine-generator control system is shown in Fig. 14. Megawatt load control and throttle pressure control are the responsibility of both the boiler and the turbine-generator. One type of an integrated system assigns the responsibility of throttle pressure control to the turbine-generator, taking advantage of the stability of a turbine-following system. In addition, the system uses the stored energy in the boiler, taking advantage of the fast load response of a boiler-following system.

Because the boiler is not capable of producing rapid changes in steam generation at constant pressure, the turbine is used to provide the initial load response. When a change in load is demanded, the throttle pressure set point is modified using megawatt error (the difference between the actual load and the load demand), and the turbine control valves respond to the change in set point to give the new load level quickly. As the boiler modifies the firing rate

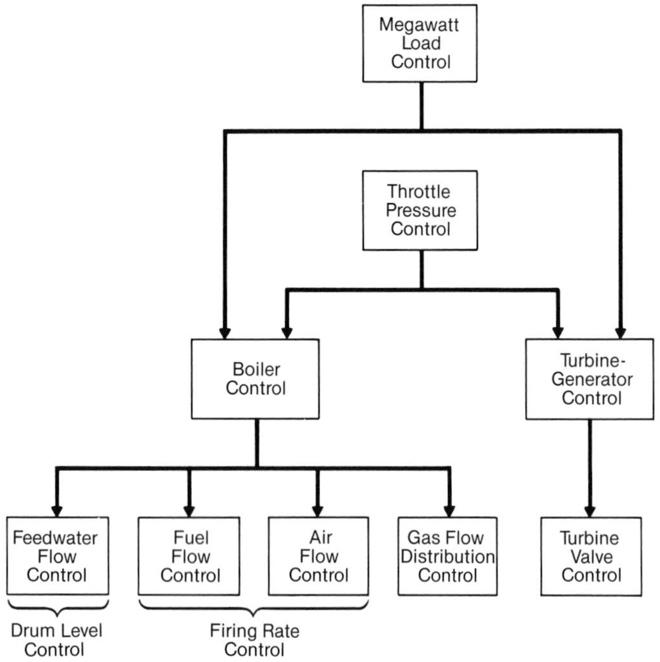

Fig. 14 Coordinated boiler turbine-generator control system.

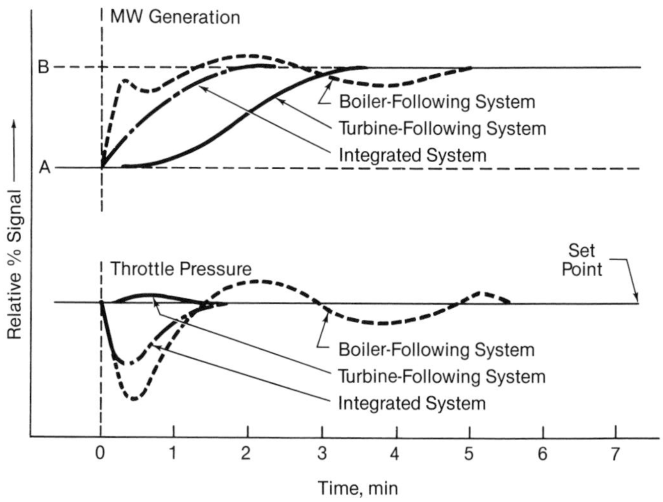

Fig. 15 Megawatt load change and throttle pressure deviation.

to reach the new load and restore throttle pressure, the throttle pressure set point returns to its normal value. The result is a fast and efficient production of electric power through proper coordination of the boiler and turbine-generator. The response is much faster than the turbine-following system, shown in Fig. 13. However, it is not as fast as the boiler-following system because the effect of megawatt error is limited to maintain a balance between boiler response and stability. Wider limits would provide more rapid response while narrower limits would provide more stability.

A comparison of response characteristics of various arrangements of a boiler turbine-generator control system for a large change in load demand is shown in Fig. 15.

Development of the integrated control system received added impetus when electric power generating companies introduced the wide area economic load dispatch system. This system requires precise load control. Each controlled unit must produce a specific level of megawatts and the area load control system adjusts the turbine control valves until this level is reached. Even relatively minor interactions between steam flow and throttle pressure can create a continuous interaction between the boiler, the turbine-generator, and the area load control. The conventional boiler-following system, which is subject to such interactions between steam flow and throttle pressure, does not lend itself to area load control because two highly responsive systems would be adversely interacting with each other.

Because wide area economic load dispatch systems are being increasingly used, greater emphasis is placed on controlling the boiler and turbine-generator together in an integrated system. This achieves automatic control over a wide operating load range.

Integrated boiler turbine-generator control The integrated control system, in its basic form, consists of ratio controls that monitor pairs of controlled inputs, including:

1. boiler energy input to generator energy output,
2. superheater spray water flow to feedwater flow,
3. fuel flow to feedwater flow,
4. fuel flow to air flow,
5. recirculated gas to air flow; this, in effect, is a ratio of reheater absorption to absorption in primary water and steam, and
6. fuel to primary air flow for pulverized coal-fired units.

Fig. 16 is a simplified diagram showing all system inputs controlled in a parallel relationship by ratio controls.

The integrated control system, shown in Fig. 17, coordi-

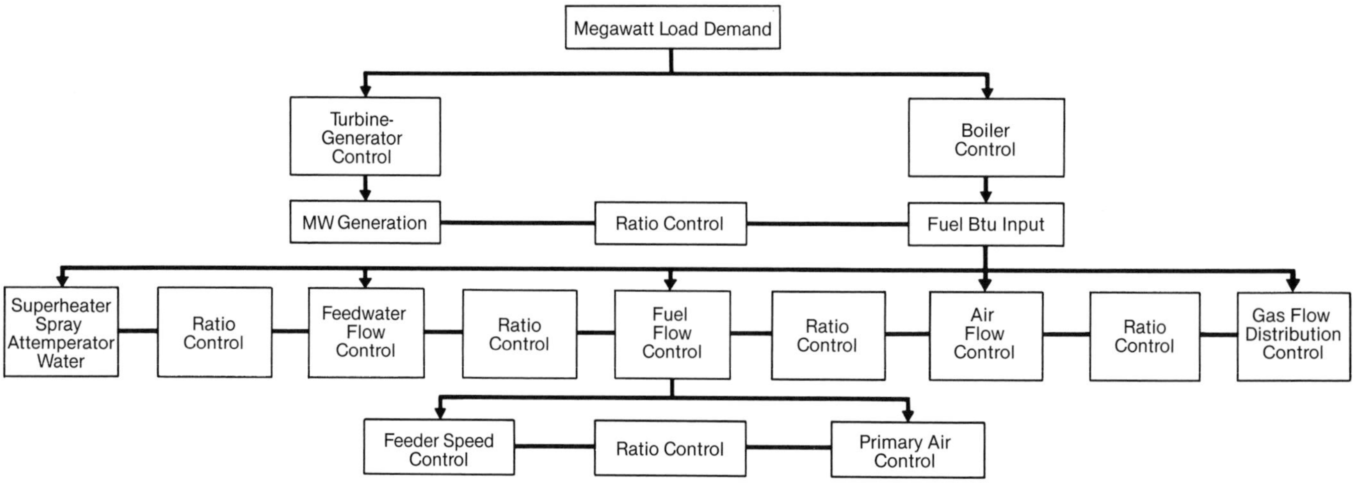

Fig. 16 Simplified diagram of ratio control.

nates the boiler and turbine-generator for fast and efficient response to load demand initiated by the automatic load dispatch system.

Variable pressure operation

A variable pressure boiler operates with the throttle pressure varying with load. In its purest form, the throttle valves on the turbine are left wide open and throttle pressure varies directly with load. There are recognized advantages of variable pressure operation. These include minimum change in turbine metal temperature due to the elimination of the adiabatic throttling effect of turbine valves. Improved cycle efficiency at reduced loads results from reduced boiler feed pump power and higher turbine inlet steam temperatures. However, there are disadvantages including sluggish response to load change demand because of little energy storage during operation. Typically, the load demand change response rate for variable pressure operation is approximately 50% of the response rate for constant pressure operation.

The Babcock & Wilcox (B&W) drum boiler bypass and startup system adds another dimension to variable pressure operation. The system, shown in Fig. 18, can generate and deliver steam to the turbine with a wide range of controlled throttle pressure and temperature.

This system combines the advantages of constant pressure operation of the boiler with variable pressure operation of the superheater and turbine. The throttle pressure is controlled by modulating the superheater division valves with the drum pressure held at or near full load pressure. The control strategy and operating procedure are equivalent to variable throttle pressure operation with load or steam flow controlled by the turbine control valves.

Cycling operation

Most of the fossil fuel-fired drum boilers being installed today will encounter cycling operation. Cycling involves taking a unit off-line on nights and weekends and reloading it on weekday mornings.

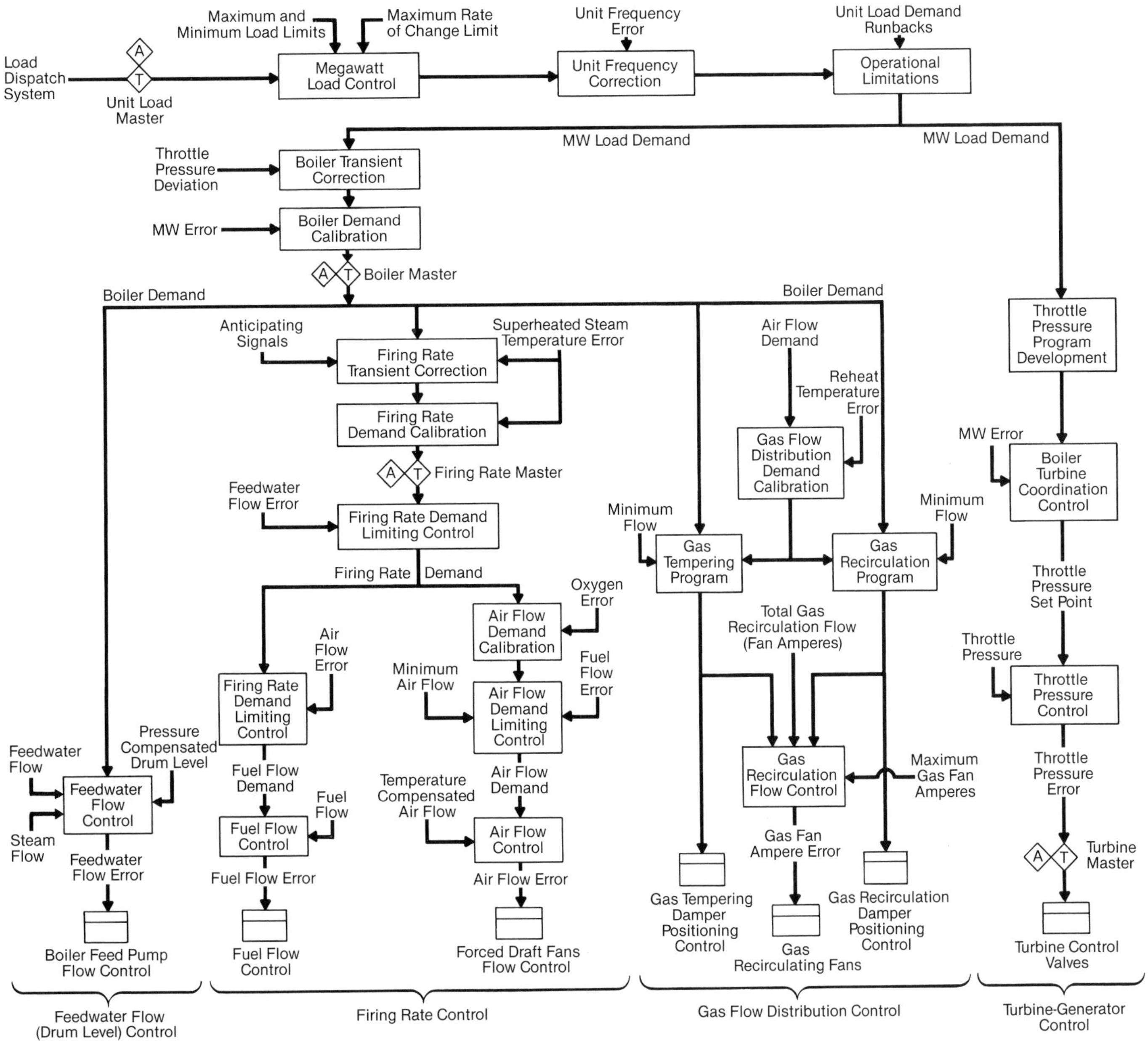

Fig. 17 Integrated boiler turbine-generator control system.

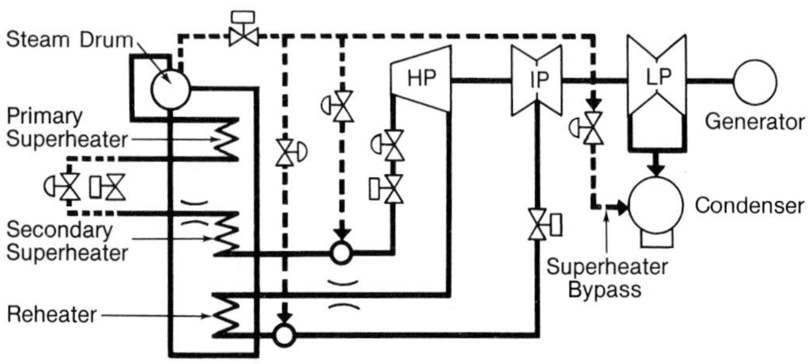

Fig. 18 Superheater bypass and startup system for drum boiler.

This analysis of cycling units assumes that the daily load cycle starts at full load, full temperature operation and passes successively through a load reduction, an off period and an idle period. This is followed by a restart and a reloading back to full load, full temperature. The basic difficulty is that steam temperatures must be controlled to limit cyclic thermal stresses on the turbine. The boiler lacks capability for temperature control during startup and low load. Therefore, the turbine requirements and the boiler capability are not compatible unless an interfacing system is provided. The startup system described above (Fig. 18) provides this interface.

A more aggressive integrated control strategy is needed to take full advantage of the boiler turbine system capabilities while recognizing the limitations of each. The principles applied to such a control system include:

1. The fuel heat content effectively released in the boiler furnace will establish and maintain the megawatt output from the unit on a steady-state basis. Therefore, the only valid boiler load index for fuel and air input is megawatt demand and/or generation.
2. The turbine valve control will control steam flow and megawatt generation but only on a short term transient basis.
3. The relationship between steam flow and megawatt generation is not dependent on the turbine valve position.

Feedwater control systems

Feedwater control regulates the flow of water to a drum type boiler to maintain the level in the drum within the desired limits. The control system will vary with the type and capacity of the boiler as well as with load characteristics.

For higher capacity boilers, and those operating at higher pressures, a pneumatic or electrically operated feedwater control system is used. Controls of these types use varying degrees of control action to suit the particular application. They are classified as one-, two- or three-element feedwater control systems.

In single element feedwater control (Fig. 19) the water in the drum is at the desired level when the signal from the level transmitter equals its set point. If an undesirable water level exists, the controller applies proportional plus integral action to the difference between the drum level and set point signals to change the position of the regulating valve. A hand-automatic station gives the operator complete control over the valve. A valve positioner can be included in the control valve assembly to match the valve characteristics to the individual requirements of the system.

Single element control will maintain a constant drum level for slow changes in load, steam pressure or feedwa-

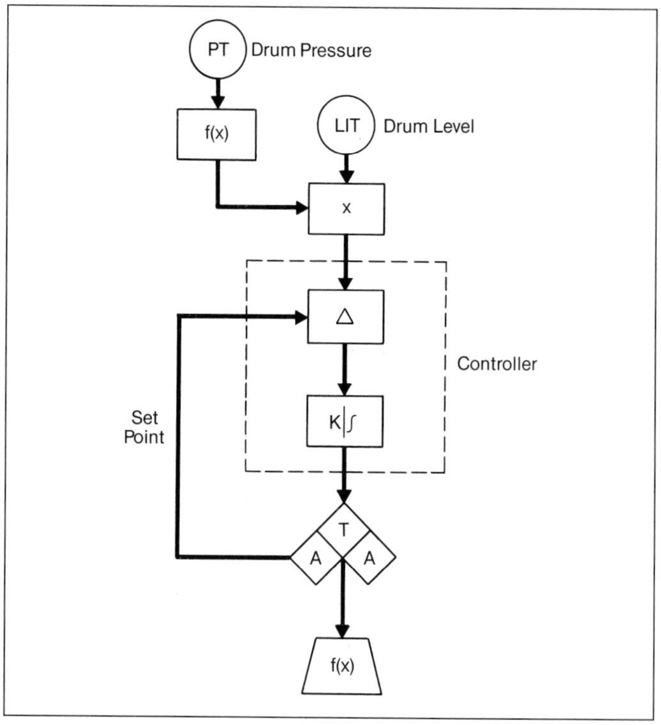

Fig. 19 Single element feedwater control.

ter pressure. However, because the control signal satisfies the requirements of drum level only, excessive *swell* or *shrink* effects will result in wider drum level variations and a longer time for restoring drum level to set points following a load change with single element control than with three-element control.

The most widely used feedwater control system, especially in the utility industry, is the three-element feedwater control.

Three-element control is a *cascaded-feedforward* control loop which maintains water flow input equal to feedwater demand. Drum level measurement keeps the level from changing due to flow meter errors, blowdown, or other causes.

The control applies proportional action to the error between the drum level signal and its set point. The sum of the drum level error signal and the steam flow signal is the feedwater demand signal. This is the output of the *summer*. The feedwater demand signal is compared with the water flow input and the difference is the combined output of the controller. Proportional plus integral action is added to provide a feedwater correction signal for valve regulation or pump speed control. In single boiler turbine units, the turbine first stage shell pressure can be used as a measure of steam flow. The details of this system are illustrated in Fig. 20, using the symbols given in Table 2.

Three-element feedwater control systems can be adjusted to restore a predetermined drum level at different loads. In boilers with severely fluctuating loads, the system can be modified to permit drum water level to vary with boiler load to compensate for swell and shrink effects. If the drum level is allowed to vary in this manner, a nearly constant inventory of water, as opposed to a constant level, is maintained.

During low load operation (0 to 20%), it is difficult to obtain steam and water flows accurately because flow transmitters are usually calibrated for high load operation.

It is advisable to transfer the feedwater control system to single element control. Drum level is the only variable used in the control scheme. Automatic transfer as a function of load between three-element and single element control is often provided.

Drum water level is one of the most important measurements for safe and reliable boiler operation. If the level is too high, water can flow into the superheater with droplets carried into the turbine. This will leave deposits in the superheater and turbine, and in the extreme, cause superheater tube failure. Consequences of low water level are even more severe. An insufficient head of water may cause a reduction in furnace circulation, causing tube overheating and failure. To provide a more reliable system B&W recommends the use of three independent drum level transmitters as inputs to an auctioneered remote level signal system. To provide an additional safeguard, the control logic for drum level should be independent of the boiler tripping logic. It is also important to install and calibrate drum level instrumentation with care. Instrumentation to be provided must be in accordance with The American Society of Mechanical Engineers (ASME) *Boiler and Pressure Vessel Code*. Minimum drum water level instrumentation used on utility size power boilers should be one water gauge glass and two indirect water level indicators with alarm and trip points clearly marked.

Furnace draft control

The induced draft fan control portion of the system maintains flue gas side furnace pressure at a desired set point and provides furnace overpressure and implosion protection.

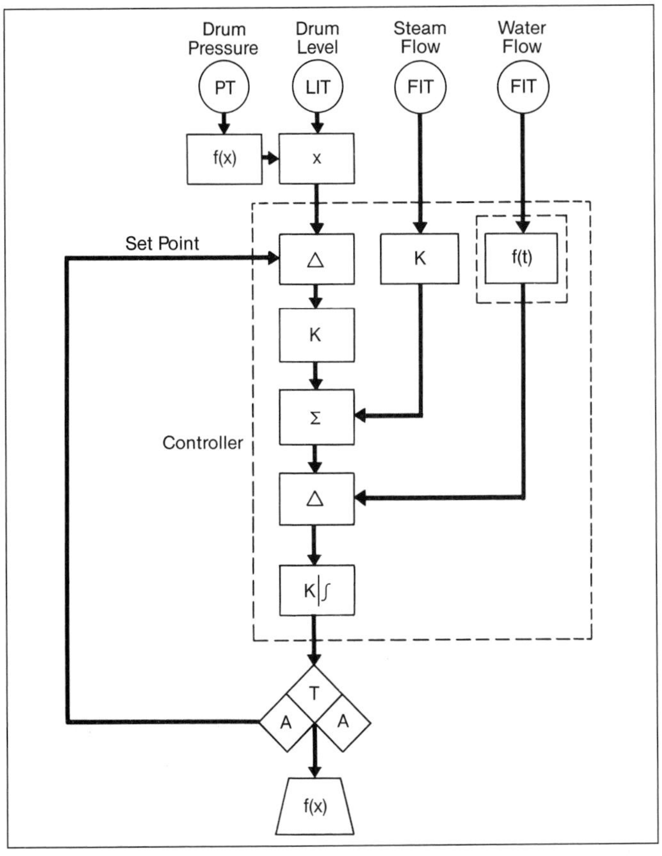

Fig. 20 Three-element cascade feedforward control loop for feedwater flow control.

Table 2
Feedwater Control Symbols

(FIT)	Flow indicating transmitter
(LIT)	Level indicating transmitter
◇A	Manual signal generator
f(x)	Final controlling function
A/T	Hand/automatic control station
△	Subtracting unit
K	Proportional controller
K ∫	Proportional plus integral controller
Σ	Summer
f(t)	Signal lag unit

Overpressurization may create a major disturbance in the furnace resulting in partial or complete loss of flame. If fuel flow is not stopped immediately, furnace explosion may result. Typical system protection is to initiate a unit load runback on high furnace pressure until the induced draft (ID) fans can maintain the desired furnace pressure. If the furnace pressure exceeds the high limit for a predetermined time period, a boiler trip is desirable.

Furnace implosion is the result of a combination of a unit trip (flame collapse) and an equipment malfunction resulting in a vacuum condition in the furnace. The control system design should recognize that a main fuel trip has occurred and take action to reduce the negative pressure excursion which occurs as the result of the rapid decrease in furnace gas temperature when the combustion process is stopped. The control system should drive the induced draft fan damper or inlet vanes towards a closed position. If the furnace pressure drops below the low limit, the normal control action should be overridden and the induced draft fan dampers or inlet vanes are to be driven toward the closed position. The furnace implosion protection elements are normally introduced downstream of the manual stations of the induced draft fans so that human action is not permitted. Fig. 21 illustrates the integration of the furnace draft control overpressure protection and furnace implosion protection systems.

Superheat and reheat temperature control

From the viewpoint of control system designers, there are inherent advantages to considering boiler outlet steam temperature control as an independent temperature control system. A greater understanding is gained if the temperature control systems are recognized as a system for distributing the boiler heat input between steam generation, steam superheating and steam reheating. Starting from that viewpoint, some of the inherent limitations of boilers to control temperatures and the difficulties in reaching design solutions can be identified. For the various methods of steam temperature control, refer to Chapter 18.

Combustion control systems (air and fuel flow control)

A combustion control system regulates the fuel and air input, or firing rate, to the furnace in response to a load index. The demand for firing rate is, therefore, a demand for energy input into the system to match a withdrawal of energy at some point in the cycle. For boiler operation and control systems, variations in the boiler outlet pressure are often used as an index of an unbalance between fuel-energy input and energy withdrawal in the output steam.

A great variety of combustion control systems have been developed over the years to fit the needs of particular applications. Load demands, operating philosophy, plant layout, and types of firing must be considered. The following examples show some of the systems for various types of fuel firing. The control symbols shown in these figures are listed in Tables 1 and 2.

Oil- and gas-fired boilers Fig. 22 illustrates a system for burning gas and oil, separately or together. The fuel and air flows are controlled from steam pressure through the boiler master with the fuel readjusted from fuel flow and air flow. The oil or gas header pressure may be used as an index of fuel flow and the windbox to furnace differential as an index of air flow on a per burner basis. Such indices are common for single burner boilers. For multiple burner boilers, a pitot tube or other flow measuring instrument is used to provide more accurate air flow measurement.

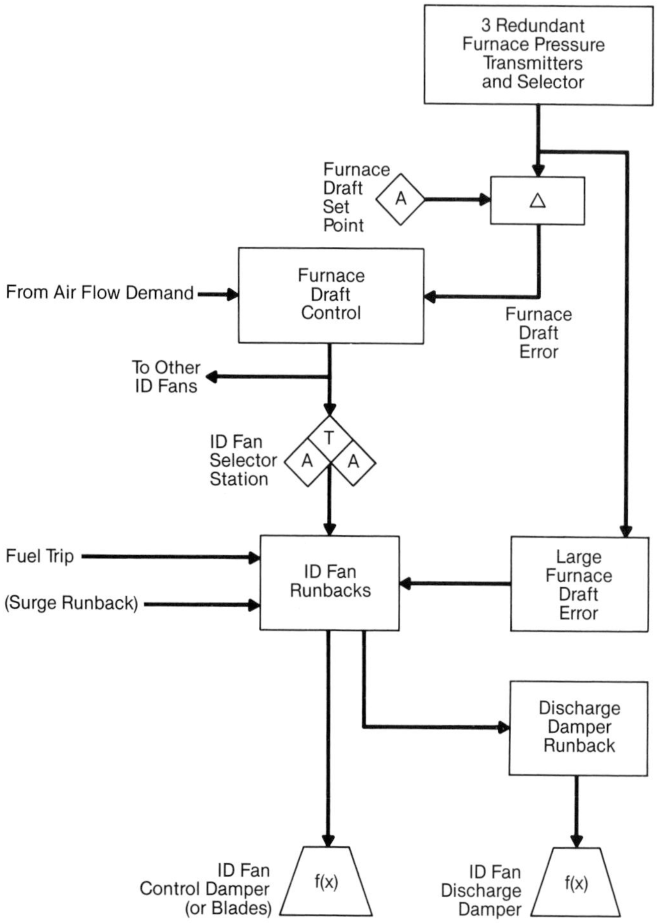

Fig. 21 Furnace draft control.

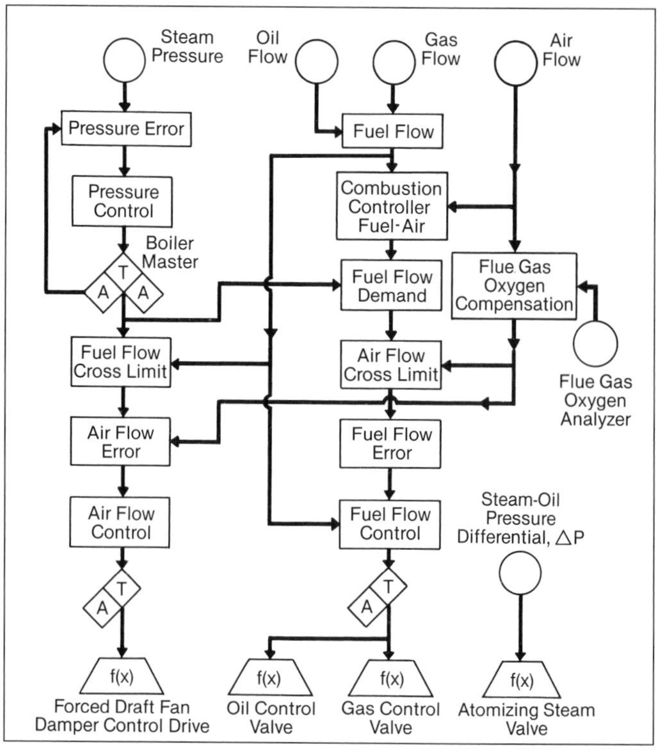

Fig. 22 Diagram of combustion control for a gas- and oil-fired boiler.

Pulverized coal-fired boilers Fig. 23 illustrates a more sophisticated combustion control that would be used on larger boilers having several pulverizers. Each pulverizer supplies a group of burners that are compartmented off from each other. Primary and secondary air is admitted and controlled on a per- pulverizer basis.

The boiler firing rate demand is compared to the total measured fuel flow (summation of all feeders in service delivering coal) to develop the demand to the pulverizer master. The pulverizer master demand signal is then applied in parallel to all operating pulverizers.

The individually biased pulverizer demand signal is then applied in parallel to meet demands for coal flow, primary air flow, and total pulverizer group air flow (primary plus secondary air flows). When an error develops between demanded and measured primary air flow or total air flow, proportional plus integral action will be applied, adjusting the primary or secondary air dampers to reduce the error to zero. A low primary air flow or total air flow cutback is applied in the individual pulverizer control. If either measured primary air flow or total air flow is low relative to coal rate demand, this condition is recognized in the low-select auctioneer in the coal feeder demand to reduce the demand to that equivalent to the measured air flows. A minimum pulverizer loading limit, a minimum primary air flow limit, and a minimum total air flow limit are applied to the respective demands to keep the pulverizers above their minimum safe operating load. (See Chapter 12.) A minimum limit maintains sufficient burner nozzle velocities at all times, and maintains primary air/fuel and total air/fuel ratios above prescribed levels.

High capacity vertical shaft pulverizers, such as the MPS pulverizer (Chapter 12), are being supplied on most large boilers today to achieve the required steam flow capacities with coal having low heating values. Pulverizer response can be improved by overshooting both primary air flow and feeder speed. This overfeed is obtained with present control systems by the use of kickers on primary air flow and coal flow developed from a change in pulverizer demand.

Cyclone-Fired boilers The Cyclone boiler (see Chapter 14) controls shown in Fig. 24 are similar to those described for pulverized coal-fired units. Although the Cyclone functions as an individual furnace, the principles of combustion control are the same.

On multiple Cyclone installations, the feeder drives are calibrated so that all feeders run at the same speed for the same master signal. The total air flow is controlled by the velocity damper in each Cyclone to maintain the proper air/fuel relationship. This air flow is automatically temperature compensated to provide the correct amount of air under all boiler loads. The total air flow to the Cyclone is controlled by the windbox to furnace dif-

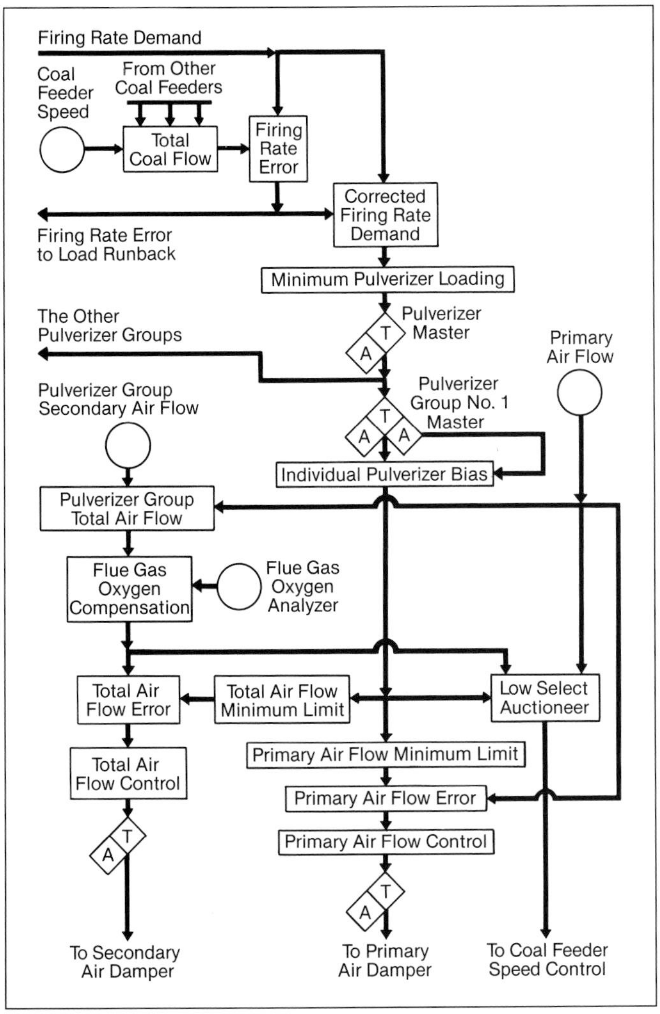

Fig. 23 Diagram of combustion control for a pulverized coal-fired boiler.

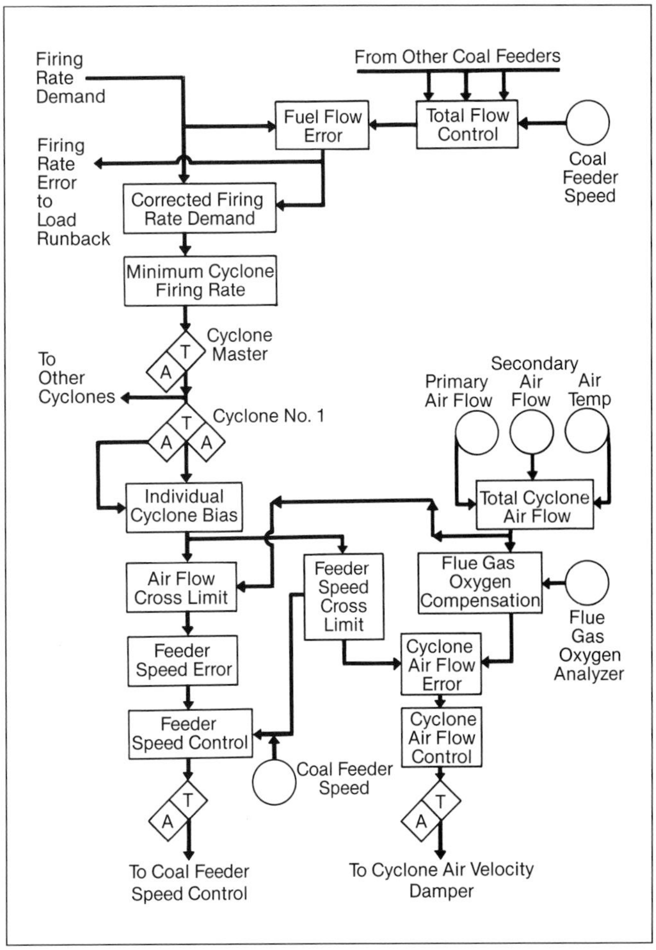

Fig. 24 Diagram of combustion control for a cyclone-fired boiler.

ferential pressure, which varies with load to increase or decrease the forced draft fan output.

Automatic compensation for the number of Cyclones in service has been incorporated, along with an oxygen analyzer. This gas analyzer helps the operator monitor excess air for optimum firing conditions.

Atmospheric fluidized-bed boilers The circulating fluidized bed thermal performance (load, power consumption, furnace temperature, combustion efficiency, gaseous emissions) is strongly dependent on the solids mass or inventory distribution within the system, solids size, and solids flow within and through the system. (See Chapter 16.) Specific examples of the special role of solids in a circulating fluidized-bed process include:

1. furnace heat absorption is controlled through changes in solids inventory,
2. fuel combustion is affected by the solids inventory distribution in the furnace,
3. solids size strongly affects furnace heat transfer and inventory distribution, U-beam and multiclone dust collector efficiencies, solids flow in the system, and solids losses,
4. carbon conversion efficiency and sorbent consumption depend on amount and distribution of solid effluents,
5. L-valve operation and U-beam collection efficiency depend on the solids level in the particle storage hopper, and
6. forced draft fan operating point, power consumption and air flow depend on the furnace inventory.

Different areas of circulating fluidized bed process control are interrelated, with the solids inventory control function playing a central role. The complexity of process control makes it difficult to analyze and predict system performance for multiple combinations of the process variables. Additional problems are being caused by continuous variation of material and energy inputs and widely different speeds of the system dynamic responses to these variations. Much is to be learned from operation of commercial and laboratory circulating fluidized-bed units to define more exactly the optimal control strategy for different operating conditions. (For information on pressurized fluidized-bed combustion see Chapter 29.)

Control system design philosophy for Universal Pressure once-through boilers

The use of the once-through boiler design in the United States (U.S.) stems from continuing efforts to reduce electric power generating costs. The once-through boiler can operate at full steam temperature over a wider load range, and can operate at supercritical pressures to increase cycle efficiency. The B&W Universal Pressure boiler is one type of a once-through boiler. It can be designed to operate at either subcritical or supercritical pressures.

The once-through concept can best be visualized as a single tube as discussed in Chapters 1 and 24. Feedwater is pumped into one end, heat is applied along the length of the tube, and steam flows out of the other end. The output is a function of the feedwater flow and the amount of heat supplied. The outlet fluid enthalpy, or heat content, depends only on the ratio of heat input to feedwater flow. A valve at the outlet provides a means of varying the pressure level. When the pressure level is maintained constant, the outlet steam temperature is also dependent only on the ratio of heat input to feedwater flow.

At steady-state conditions, feedwater flow equals steam flow. The pressure level will be influenced by the valve restriction at the outlet and by the density of the fluid throughout the system. Therefore, a change in heat input will influence both pressure and temperature. It is possible to vary the flow and pressure at the outlet by changing the amount of valve restriction, or at the inlet by changing both the feedwater flow and the heat input. It is important to note that a change in feedwater flow without a corresponding change in the heat input will result in a change in the outlet steam temperature. During transient conditions, other factions, such as the fluid and energy storage requirements which change with load, must be considered because they influence the feedwater flow and heat inputs.

Combustion guides

It is necessary to provide the operating personnel with additional information to allow manual or automatic proportioning of air and fuel.

The level of excess air is one index commonly used to determine the performance of the unit and to guide its operation. Excess air is the amount of air supplied above the theoretical air needed for combustion. (See Chapter 9.) It is always necessary to supply some excess air to assure complete combustion of the fuel. Any excess air not used constitutes a loss in boiler efficiency by exhausting excess heat up the stack.

Steam flow-air flow

The steam flow-air flow concept (Fig. 25) can give an immediate indication of fuel consumption efficiency. Because boiler efficiency at a specific load remains essentially constant for a given amount of excess air, fuel consumption can be determined from steam flow, which is established by standard flow measuring devices. Steam flow is used as an index of heat absorption. The air flow indication or index can be established by primary elements located on either the air side or the flue gas side of the unit.

Air flow measurement

From a control standpoint, the preferred location for air flow measurement is in the clean air duct between the air heater and the burners. Available air measuring devices are discussed in Chapter 40.

The alternate location for air flow measurement is on the flue gas side, with connections to measure pressure differences resulting from the flow of combustion products through a section of the main boiler or air heater. The products of combustion at a given flue gas temperature are proportional to the amount of air supplied; therefore, the flue gas flow can be an index of air flow.

The measurement of combustion air supplied to a boiler is made difficult by several factors:

1. The air measuring device is typically of the restrictive type, similar to the primary elements used in other flow measurements. The resulting unrestored head loss means added fan power.
2. On large and complicated units it is usually impossible to find one duct through which all of the combustion air passes. Metering equipment capable of accurately totaling several flows must then be used.

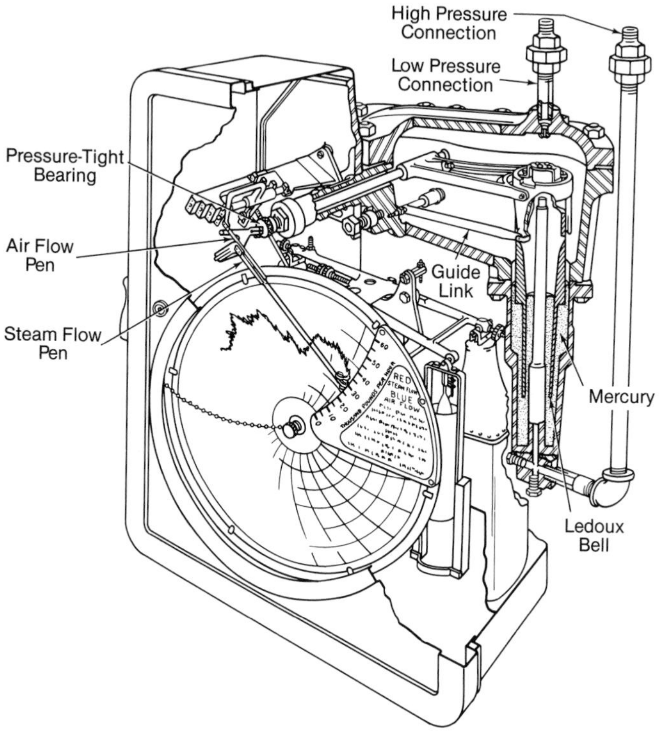

Fig. 25 A steam flow-air flow combustion guide for maintaining desired excess air. Fuel flow-air flow device is similar but may have a different flow measuring mechanism.

3. Air density must be considered in both the calibration and operation of any air flow meter. Either manual or automatic temperature compensation must be applied to the air flow indication for it to reflect the mass rate of air flow.

The steam flow-air flow meter is calibrated after the unit is placed in operation, and calibration is based on results from combustion tests. Proper calibration allows a visual presentation of the relative correctness of the fuel-air proportioning.

Fuel flow-air flow

Where it is practical to measure the rates of admission of fuel and air flow before combustion, a fuel flow-air flow ratio guide can be established. This may be direct control or backup for adjustment or setting up limits of fuel or air flow.

Fuel flow measurement

The measurement of gaseous fuels must consider the physical properties of the gas, particularly temperature, pressure and specific gravity. Where widely changing properties are encountered, automatic compensation should be used. A primary element installed in the main header will produce a differential pressure to actuate a recording meter. On less complex installations, particularly those with single burners, the pressure at the burner can be used to indicate flow.

With liquid fuels, various types of flow meters can be installed directly in the supply header to indicate and record fuel flow. As with gaseous fuels, oil-header pressure can be used to indicate flow.

The measurement of solid fuel flow can be accomplished through weight scales or feeder speeds. The use of mass flow, however, is more accurate. The feeder can be calibrated in units of fuel flow per feeder revolution to produce a continuous flow rate. Gravimetric feeders (see Chapter 11) provide controlled weight flow rates to the pulverizers.

The new concept of *megawatt generation-air flow* is now being applied in new boiler instrumentation and control systems. Megawatts generated is a direct index of heat input to the unit.

Gas analysis

Gas analyzers are used to determine correct fuel-air relationships. Representative flue gas samples are continuously analyzed either chemically or electronically to record the oxygen and combustibles present in the products of combustion. Because there is direct correlation between oxygen content in the flue gas and the quantity of excess air supplied to the combustion zone, the operator receives a continuous and direct reading of combustion efficiency. The indicated or recorded oxygen signal can be used in adjusting the total air flow.

Comparison of combustion guides

Each combustion guide has its particular merits; none is infallible. The advantages of one may be used to overcome the disadvantages of another.

The fuel flow-air flow ratio guide proportions fuel and air continuously during severe load swings. It is, therefore, popular on gas- and oil-fired units. It has the disadvantage of being in error for wide variations in fuel heating value. When the fuel heating value changes, the calibration of the correct fuel-air relationship also changes. Therefore, to maintain a given excess air, the proportions of air and fuel must also be changed. Often, the fuel flow-air flow ratio guide is used for coal-fired units having a variable heating value fuel. However, a feedback calibration of air flow based on oxygen measurement or a heat input index such as steam flow or megawatt generation is added.

On major load changes, the steam flow-air flow device is temporarily in error because of the overfiring and underfiring necessary for responsive load change. The fuel consumption is normally higher or lower than indicated by the steam flow load index. This guide is also affected by changing feedwater or steam temperatures, because a variation of either temperature demands more or less transfer of heat per unit of steam produced at the specified pressure and temperature. These errors are inherently minimized where megawatt generation is used as the index of heat absorption. It is necessary, when substantial changes may occur in steam or feedwater temperature, to provide manual or automatic compensation to the steam flow-air flow circulation.

A gas analyzer gives true excess air determinations. There is some delay because combustion must be completed before a correct sample can be obtained. If the sample is taken near the combustion zone, this delay is not usually significant. Considerable study may be needed to locate sample points that will give correct average excess air values. The dirty, hot, and corrosive conditions of the flue gases at these points make periodic maintenance necessary for continuous, dependable sampling. The gas analyzer is an accurate index for feedback control.

At least one of these combustion guides is needed for proper control of the combustion process. The flow index guides serve as a direct signal to the firing equipment and act to maintain a correct relationship of the combustion conditions.

Burner control systems

Burner control systems of various types are now applied to almost all boilers to prevent continued operation during hazardous furnace conditions, and to assist the operator in starting and stopping burners and fuel equipment.

The most important burner control function is to prevent furnace or pulverizer explosions which could threaten the safety of operating personnel and damage the boiler and auxiliary equipment. The control system must also prevent damage to burners and fuel equipment from maloperation while avoiding false trips of fuel equipment when a truly unsafe condition does not exist.

Other important factors in the design of the burner control system are the method and location to be used in the startup, shutdown, operation and control of the fuel equipment. These factors must be understood before purchase, design, or application of a burner control system.

The different categories of burner control are summarized below.

Manual control

With manual control (Fig. 26) it is necessary to operate the burner equipment at the burner platform. Checks on the existing conditions are dependent on observation

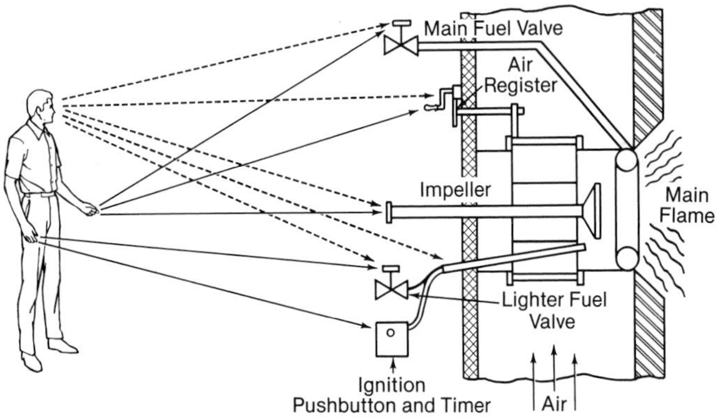

Fig. 26 Fuel burner manual control.

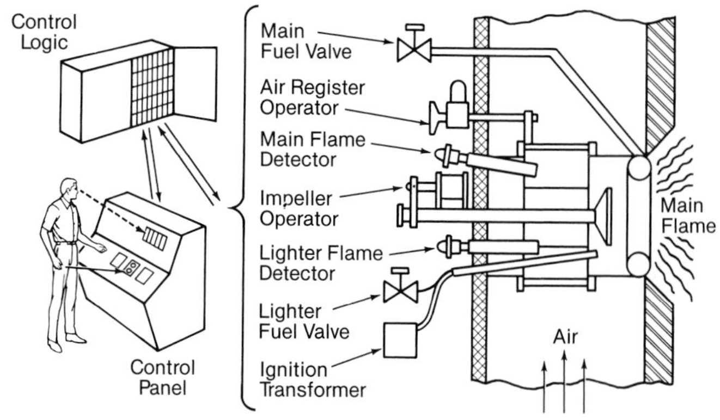

Fig. 28 Automatic sequence control.

and evaluation by the local operator. Good communication between the local operator and the control room is needed to coordinate the startup and shutdown of burner equipment. Manual supervision and control are of interest principally on older boilers, because some form of automatic control is provided with almost all new units.

Remote manual sequence control

Illustrated in Fig. 27, this system represents a major improvement over manual control, permitting remote manual operation using instrumentation systems and position switches in the control room for intelligence. This system requires various burner permissives (prequalifying conditions) and interlocks, and trips should also be provided to account for the position of fuel valves and air registers in addition to main flame detection. It is not recommended for remote operation where safety is dependent only on flame detection. With this system, the operator participates in the operation of the fuel equipment. He controls each sequence of the burner operating procedure from the control room and no steps are taken except by his command.

Automatic sequence control

The next logical step (Fig. 28) is to automate the sequence control to permit startup of burner equipment from a single pushbutton or switch control. Automation, then, replaces the operator in controlling the operating sequences. Because the operator initiates the demand for each fuel utilization unit, it is expected that he participates in, or at least monitors the operating sequence as indicated by signal lights and instrumentation signals, as the startup process proceeds to completion with the fuel utilization unit in service on automatic control. This category has been widely applied to gas burners, and also to oil and coal burning systems.

Fuel management

A degree of fuel system automation can be achieved that will permit fuel equipment to be placed in service without operator supervision (Fig. 29). A fuel management system can be applied that will recognize the level of fuel demand to the boiler. This system will know the operating range of the fuel equipment in service, will reach a decision concerning the need for starting up or shutting down the next increment of fuel equipment, and will select the next increment based on the firing pattern of burners in service. Such demands for the startup or shutdown of fuel preparation and burning equipment can be initiated by the management system without the immediate knowledge of the operator.

The degree of operating flexibility allowed with a burner control system is closely related to the degree of operator participation. A higher level of automation reduces the flexibility of the operator in handling situations where a piece of equipment fails to perform as expected. One method used to provide more operating flexibility is to allow operator participation at two or more levels. Increased flexibility is also obtained by the careful grouping of equipment so that a fault anywhere in the system will affect a limited amount of equipment. An example of this is the grouping of burners on a pulverizer so that the failure of a piece of hardware on a burner will only affect the failed burner, and not the pulverizer and its remaining burners.

Fig. 30 illustrates the provision for manual intervention in a fuel management type system. In this figure,

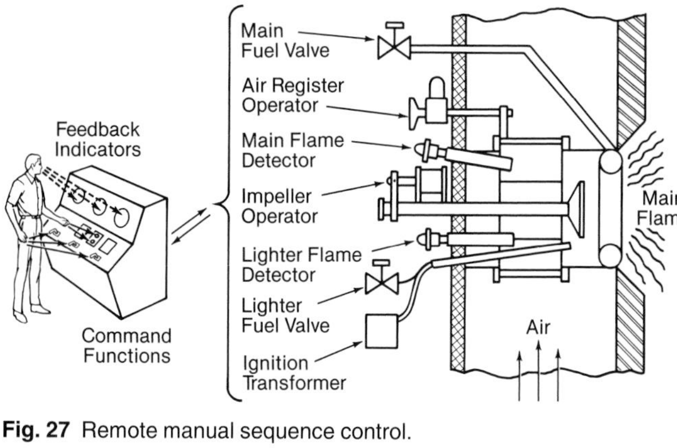

Fig. 27 Remote manual sequence control.

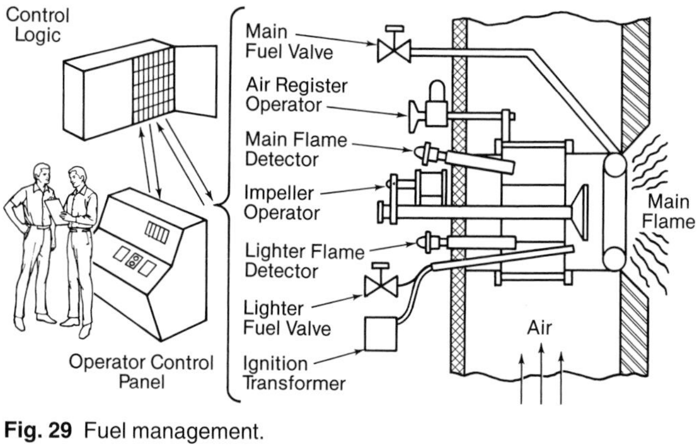

Fig. 29 Fuel management.

functions are detailed at four levels of a pulverizer burner control system. Manual intervention is provided at level 2 to permit remote manual operation with interlock protection features.

Flame detection

One of the most important items required for a burner control system is individual burner flame detection, regardless of fuel.

Ultraviolet (UV) flame detectors have been successfully applied to natural gas-fired burners due to the abundance of ultraviolet radiation produced by the combustion of the hydrogen in natural gas. Flicker or infrared detectors are better suited for coal- or oil-fired burners. A flicker detector uses the high frequency dynamic flickering of the primary combustion process. An infrared detector uses the infrared radiation produced in the primary combustion of the flame.

The proper location of the flame detectors on a burner is dependent on many factors, including burner control system design. It should, therefore, be determined by the boiler manufacturer in conjunction with the burner control manufacturer as an integral part of the burner control system design.

Flame monitoring devices in current use are designed for on/off operation based on the presence or absence of flame. Although the basic detector units provide a variable analog output signal which can be read on a meter, there is no direct claim to an established relationship between this relative output signal and the flame quality. Therefore, the analog signals should never be relied upon as flame quality indicators. They should be used only to provide helpful information for initial setup adjustments and continuous on-line observance.

Several power plants today are using closed circuit television for continuous furnace observation by the operator from the control room. Closed circuit TV has been successfully applied to all fuels and more applications are expected in the future.

Classification of lighters

The combustion of solid, liquid and gaseous fuels is hazardous because upset conditions may lead to an air-fuel mixture that could cause furnace explosion. These hazardous conditions are especially prevalent during startup and shutdown of the boiler. Therefore, it is important to have reliable instrumentation, well designed safety interlocks, and clear understanding of operating sequences by the operator. However, a reliable and noninterruptible ignition source is probably the single most important factor. The National Fire Protection Association (NFPA) provides guidelines on the application of igniters. The following is an extract from the NFPA 8500 Standards.

Class 1 (continuous igniter). This is an igniter applied

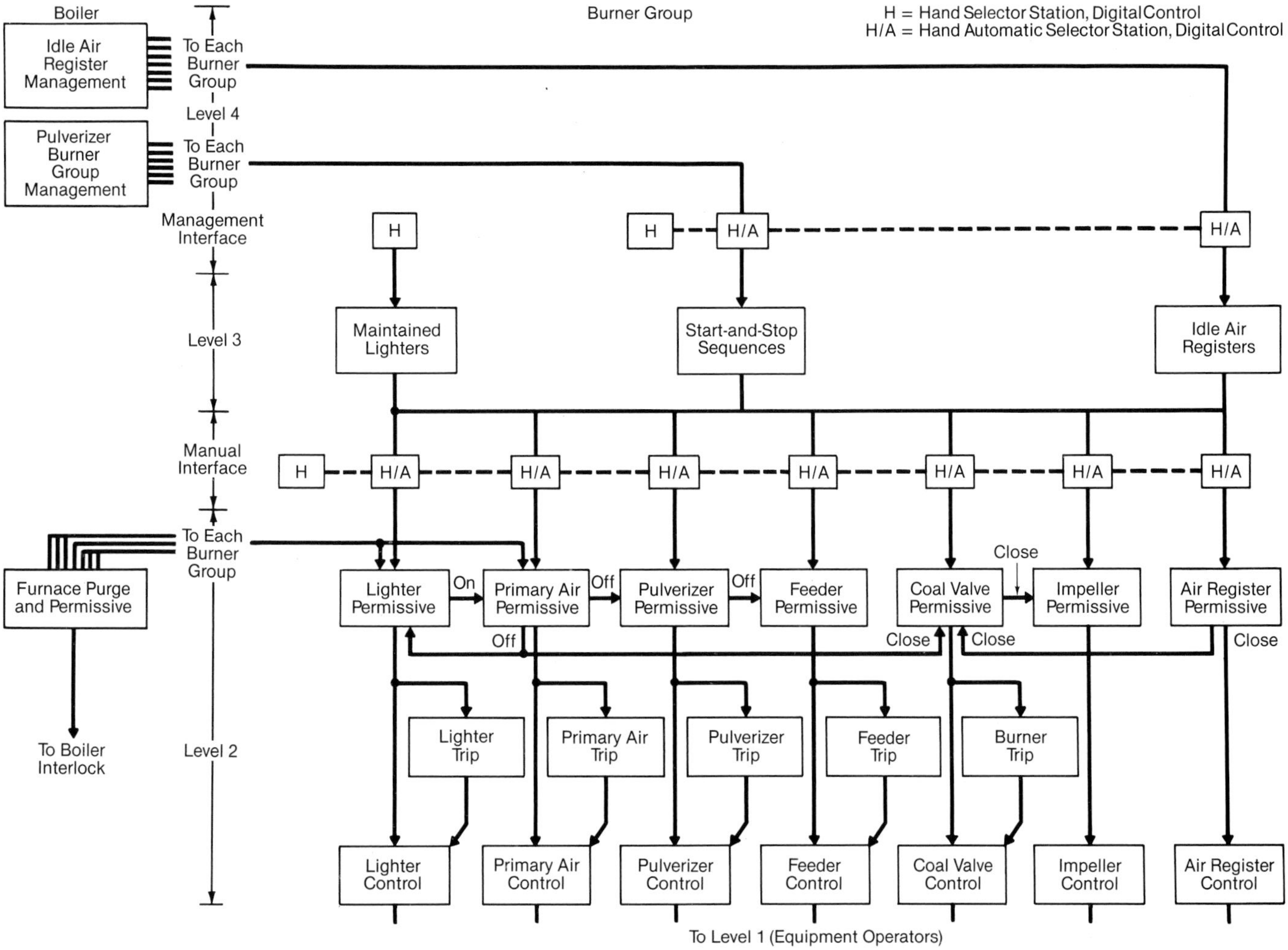

Fig. 30 Pulverizer-burner control system.

to ignite the fuel input through the burner and to support ignition under any burner light off or operating conditions. Its location and capacity are such that it will provide sufficient ignition energy, generally in excess of 10% of full load burner input, at its associated burner to raise any credible combination of burner inputs of both fuel and air above the minimum ignition temperature.

Class 2 (intermittent igniter). This is an igniter applied to ignite the fuel input through the burner under prescribed light off conditions. It is also used to support ignition under low load or certain adverse operating conditions. The range of capacity of such igniters is generally 4 to 10% of full load burner fuel input. It shall not be used to ignite the main fuel under uncontrolled or abnormal conditions. The burner shall be operated under controlled conditions to limit the potential for abnormal operation, as well as to limit the charge of fuel to the furnace in the event that ignition does not occur during light off. Class 2 igniters may be operated as Class 3 igniters.

Class 3 (interrupted igniter). This is a small igniter applied to gas and oil burners to ignite the fuel input to the burner under prescribed lightoff conditions. The capacity of such igniters generally does not exceed 4% of the full load burner fuel input. As a part of the burner light off procedure, the igniter is turned off when the timed trial for ignition of the main burner has expired. This is to ensure that the main flame is self-supporting, is stable, and is not dependent upon ignition support from the igniter. The use of such igniters to support ignition or to extend the burner control range shall be prohibited.

Class 3 Special (direct electric igniter). A special Class 3 high energy electrical igniter is capable of directly igniting the main burner fuel. This type of igniter shall not be used unless supervision of the individual main burner flame is provided.

Failure mode analysis

Because the burner control system is also known as the flame safety system it is implied that the safety of the boiler is supervised by this system. Therefore, the control system designer must consider carefully every aspect of the operation and equipment limitations. A great number of failure modes must be analyzed and addressed. Due to space limitation, the following are only a few of the issues:

1. The only safe way to implement the burner control system is de-energize to trip the equipment. As the description indicates, a circuit should be energized continuously. De-energizing of the circuit represents an abnormal condition. Therefore, the equipment must be removed from service. For an energize to trip system, in the event of specific power failure, the equipment can not be stopped.
2. Redundancy of control hardware may be necessary or desirable in many areas. However, too many backups may render the system too cumbersome for practical use.
3. System design must not prevent the operator from shutting down any equipment he desires. The system will function to complement the operator's action so that he or she will not generate a hazardous operating condition.

Recent developments in instrumentation and controls

Power plant automation

Many power plants today are essentially automatic when operating between one third and full load using the conventional systems discussed earlier in this chapter. To achieve complete automation of a plant, it is first necessary to add control subloops for the auxiliaries which were previously considered to be the exclusive responsibility of the operator. All of these subloops must then be integrated through unit management control to provide complete plant automation. Fig. 31 shows the arrangement of such a control system.

The basic concept of full automation is to provide the maximum amount of control and monitoring at the lowest possible level. The organization of the different levels, as shown in Fig. 32 for a fossil fuel plant, is summarized below:

Level 1— equipment hardware being controlled
Level 2— internal control of the individual equipment items or group of functionally identical items of equipment always operated in parallel

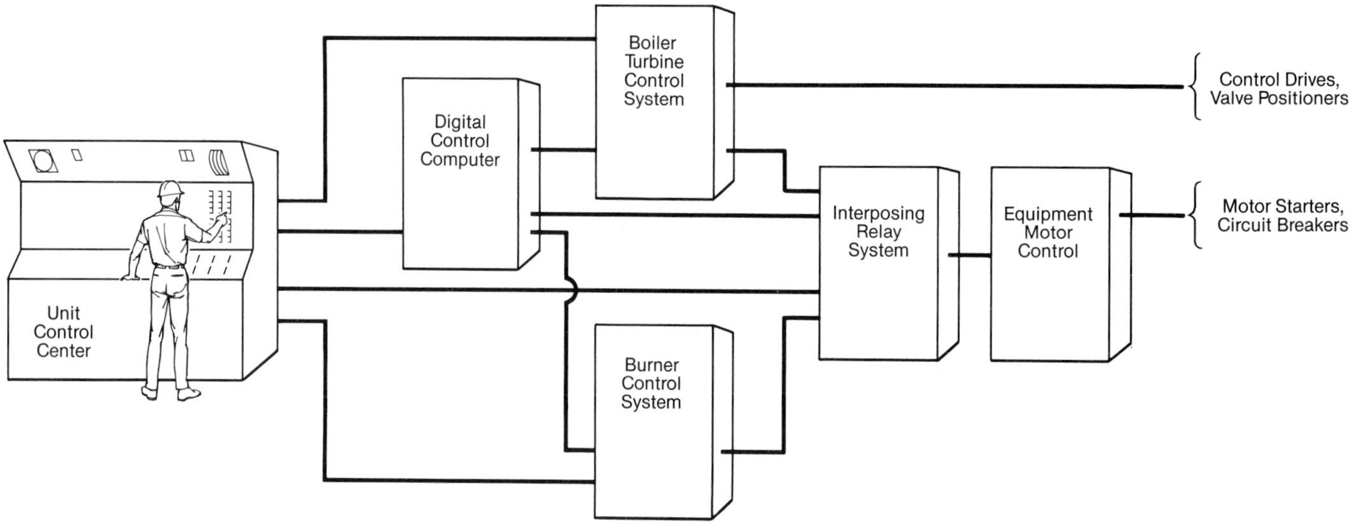

Fig. 31 Composite of automation components.

Level 3— coordination of the sequence of actions and interlocking with respect to more than one equipment group
Level 4— modulation and sequencing of major cycle components involving the entire unit
Level 5— generation of the major load demand signals to coordinate the operation of the overall plant

Advantages obtained with the hierarchical control concept include self-regulation of equipment groups. This allows the equipment group to protect itself through low level interlocking, which provides greater reliability and control decentralization and permits continued plant operation with some portions of the control system out of service.

Some of the advantages of overall unit automation are:

1. improved protection of personnel and equipment through more complete instrumentation, simplified information display, and more extensive and thorough supervisory control,
2. reduced outages, reduced maintenance, and longer equipment life through more uniform and complete control procedures of startup, on-line operation, and shutdown,
3. better plant efficiency through continuous and automatic adjustments to the controls to optimize plant operation, and
4. more efficient use of manpower during startup and on-line operation.

The trend of boiler automation is clear and its future scope can be predicted with considerable confidence. The achievement of some goals will be difficult. There are operating problems to be solved, traditions to be overcome, and new philosophies and approaches to be accepted. Reliable and adequate equipment is already available and its application, through the cooperation of manufacturers and users, will make the automatic power plant a reality. More work needs to be done to improve the peripheral equipment such as instrumentation systems, position indicators, equipment operators, and other sensors in order to couple the control system in a reliable manner to the equipment it is controlling.

Human factor engineering

Until recently, most of the control rooms have been designed with only one thing in mind, i.e., how to group the indicators, recorders, and operator stations within reach and sight of an operator who stands in front of the panel and operates the unit. With a state-of-the-art control room, the operators are surrounded by display screens and keyboards, and human factor engineering becomes very important. (See Fig. 33 and chapter frontispiece.) It will affect the physical well being of opera-

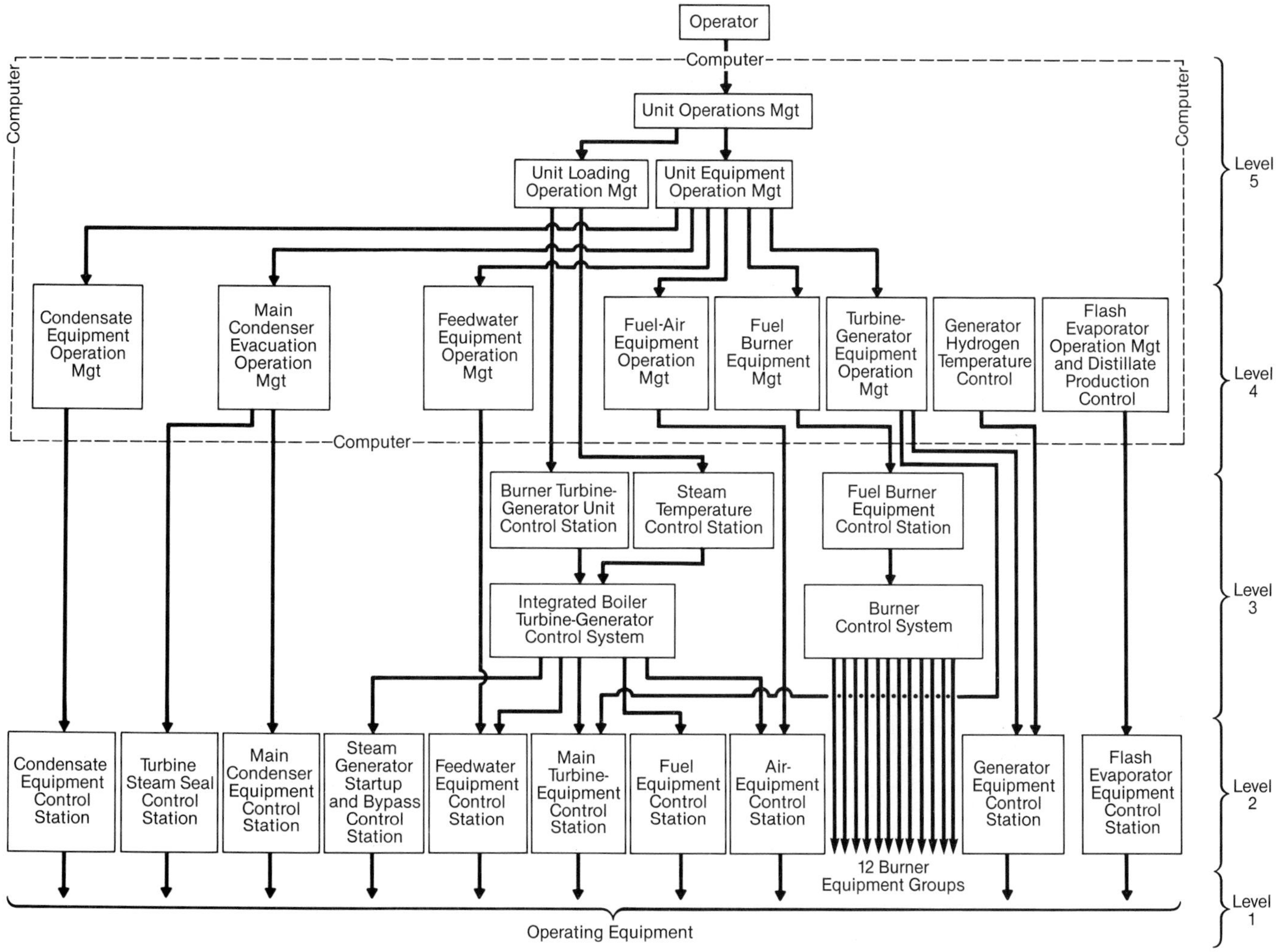

Fig. 32 Multi-level approach to power plant automation.

tors as well as his or her ability to operate the unit effectively. The following are just a few examples of why human factor engineering is important.

1. With distributed control systems, the amount of information available to the operator has increased drastically. The system has to be designed so that important information is presented and instantly understood by the operator. For example, an alarm list that is not recognized properly will cause a great deal of confusion. On the other hand, several levels of alarms using different color codes will help the operator make correct decisions.
2. To be able to read display screens for long periods of time, lighting in the control room must be designed to avoid eye strain of the operators.
3. Furnishings to accommodate display screens and keyboards to maximize operator alertness must be provided.

Smart transmitters

In the past a transmitter was used to measure pressure, temperature, and flow (differential pressure) from remote locations. A signal representing the calibrated value of that measurement is sent to the control system located in or near the central control room. The signal is then conditioned to provide a usable signal to the control system, indicator, or recorder. The smart transmitters available today can condition the signal locally — for example, square root extraction to convert differential pressure to flow and temperature compensation for flow measurements. The smart transmitter can also provide remote configuration, reranging, rezeroing, calibration and diagnostics features to provide direct access to the field device from the control room, virtually eliminating periodic maintenance visits to the instruments.

Installation and service requirements

It is important to make certain that instrumentation and control equipment are properly installed and serviced. Supervision of installation and the required calibration adjustments should be performed by qualified and experienced personnel. Adequate time to tune the controls should be provided during initial boiler operation.

Maintenance can be performed by trained plant personnel. The personnel who will be responsible for this maintenance should be available to observe and assist the control manufacturer's representatives in the installation and calibration of the control system. Most control manufacturers offer training programs to familiarize customer personnel with the control equipment. A planned program of preventive maintenance should be developed for the control system.

Current systems print out a written description of the condition which is causing the alarm. Because this operation is slow compared to the time available in an emergency, cathode ray tube displays are being used in some systems.

Another approach is to have the computer store and display analog values on demand. By storing the calibration curves for the individual transducers in the computer, the display can be made in engineering units such as psi (bar) or ft/s (m/s). Some high speed electronic displays al-

Fig. 33 Modern pulp and paper control room (*courtesy of Elsag Bailey, Inc.*).

low the readings to be updated on each data collection cycle after the point is selected.

Performance monitoring

The continuing growth in size and complexity of units has added importance to the frequent analysis of performance. Many operating companies are analyzing their performance evaluation programs and developing monitoring techniques that make possible the detection of malfunctions before substantial losses have occurred.

A number of performance evaluation procedures are practiced today. In some power companies, the performance of each unit is checked periodically by extensive heat rate tests. Techniques and procedures for performing such evaluations are covered under the ASME Performance Test Codes.

The computer is also being used to conduct on-line performance calculations periodically, hourly and daily in most systems. This information can be compared with expected performance data stored in the computer and any deviation recorded for future correction.

Future application of the performance computer will include its use as an on-line control where the performance calculations will be used to obtain optimum plant performance by automatically modifying set points and programs in the normal on-line control system.

The performance computer can also be used to record operating hours accumulated on pieces of equipment to signal when maintenance is due and to provide a record of the equipment operation.

To date, a few computer programs have been developed to scan several criteria, evaluate the data, and report to the operator an impending malfunction condition. As improved instrumentation systems and more knowledgeable techniques are developed, it is expected that greater use will be made of such programs.

Degassed conductivity analyzer for water quality analysis.

Chapter 42
Water Chemistry, Water Treatment and Corrosion

The goals of modern water technology are to maintain the integrity and performance of steam boiler components. Although proper water treatment practices alone can not assure meeting these goals, the lack of proper water treatment will most certainly assure that they will not be obtained.

Proper water treatment and chemistry control can result in a number of important benefits:

1. Minimal water side deposits of impurities and corrosion products. If permitted to grow uncontrollably, such deposits can lead to appreciable increases in tube metal temperatures and under-deposit concentration of boiler water impurities that may produce accelerated corrosion. Such deposits can also cause unacceptable pressure drop increases, especially in once-through boilers.
2. Limited corrosion rates so that boiler components meet their expected lives without premature failure or damage.
3. Prevention of excessive carryover of impurities with the steam. Such carryover may result in damage to superheaters, steam turbines or downstream process equipment.

Experience has shown that integration of water technology with boiler design can be very important. Mechanical designs which minimize cracks, crevices and high stress zones reduce the likelihood of accelerated corrosion attack. Coordination of material selection and water treatment practices can reduce operational problems and component failures. Proper pretreatment during startup can minimize initial internal boiler surface deposits.

Definitions

Some definitions of water technology terminology are important in the following discussions. *Condensate* refers to steam which has been condensed for return to the boiler. *Makeup* water is the treated raw water which is added to a system to replace steam and water lost to process requirements, blowdown, evaporation, sampling or venting. *Feedwater* is the total flow supplied to the boiler and is the sum of condensate and makeup. *Carryover* refers to the transport of impurities and water in the steam leaving the drum. *Sludge* describes the boiler water solids which settle out in headers, drums and boiler surfaces. Finally, *blowdown* is the stream of water which is bled from the boiler drum or steam supply system to control the concentration of total solids in the boiler water. Blowdown can be continuous or intermittent. *Condensate polishing* refers to in-line treatment of the returning condensate flow to reduce the concentration of soluble and insoluble contaminants. Finally, chemical and contaminant concentrations in boiler water are frequently so low that they are measured in *parts per million (ppm)* or *parts per billion (ppb)* on a weight basis.

Development and evolution

Water technology has undergone continuous development and evolution from the late 1800s to the present as steam supply demands, temperatures and pressures have increased. Hard scale and sludge led to failures and explosions in boilers with no water treatment in the late 1800s and early 1900s. Soda ash injection was initially used for internal boiler water treatment. Soda ash was almost completely superseded by sodium phosphates in the late 1920s and early 1930s. This was eventually replaced with the modern coordinated phosphate treatment which better balances acid and alkaline constituents. The advent of very high (supercritical) pressure once-through boilers accelerated the development of the all volatile treatment (AVT). This scheme avoids solid chemical additives so all of the additives depart the boiler with the steam. AVT is combined with condensate polishing systems to maintain water quality. All of these developments as well as many others have led to today's water treatment practice. Developments continue with the evolution of controlled oxygen treatment to reduce corrosion product transport and deposition. Chemical cleaning of boilers joined the water treatment arsenal and became prevalent after 1951. This technology permitted efficient removal of deposits in areas inaccessible to mechanical cleaning (most areas in modern power boilers).

These developments have largely been accomplished by boiler operators and water chemistry specialists who bear the ultimate responsibility for avoidance of deposition and corrosion problems. Many of the developments were achieved or coordinated by industry cooperatives under the auspices of associations such as the American Society of Mechanical Engineers (ASME), Edison Electric Institute (EEI), Electric Power Research Institute

(EPRI), Vereinigung der Grosskesselbetreiber (VGB), Central Electric Generating Board (CEGB), and the Central Research Institute of Electric Power Industry (CRIEPI). For example, the landmark paper ushering in the coordinated phosphate control was that of Whirl and Purcell of the Duquesne Light Company,[1] and a major research effort to address internal corrosion in high pressure boilers was sponsored and coordinated by ASME and EEI.[2] Oxygenated water treatment was largely pioneered by VGB,[3] and equilibrium phosphate treatment was pioneered by Ontario Hydro.[4] The balance of this chapter and References 5 through 7 provide an overview of current water technology practices.

Corrosion, deposits and control

Corrosion and deposits in boiler systems arise from a number of sources: preboiler component corrosion products, condenser cooling water contaminant in-leakage, makeup water impurities, pre-existing oxides found in new components, and corrosion products from the boiler materials themselves. Water technology practice addresses these issues through a number of activities:

1. raw makeup water treatment,
2. pre-operational cleaning and procedures,
3. internal water treatment for drum and once-through boilers,
4. condensate treatment,
5. chemistry and corrosion control, and
6. chemical cleaning.

These are the subjects of the following material. Waste water treatment, sampling, measuring techniques and corrosion mechanisms are also discussed. This chapter is considered a general overview with a fossil fuel-fired boiler focus. Issues of particular concern to nuclear operations are provided in Chapter 53.

Raw makeup water treatment

Pure water rarely exists; all natural waters contain varying amounts of dissolved and suspended matter. The type and amount of impurities vary with the source, such as lake, river, well or rain, and with the location of the source.

Rain water brings the atmospheric gases of oxygen, nitrogen and carbon dioxide into solution. As it percolates through the soil, the water dissolves and picks up minerals that are harmful to boiler operation. Surface waters frequently contain organic matter that must be removed before the water is used in a boiler. Solids appear as both suspended and dissolved solids in the raw water.

Suspended solids are those that do not dissolve in water and can be removed or separated by filtration. Examples of suspended forms are mud, silt, clay and some metallic oxides.

Dissolved solids are those which are in solution and can not be removed by filtration. The major dissolved materials in water are silica, iron, calcium, magnesium and sodium. Metallic constituents occur in various combinations with bicarbonate, carbonate, sulfate and chloride. In solution many of these impurities divide into their component parts called *ions*. The bicarbonate, carbonate, sulfate and chloride ions are negatively charged and are referred to as *anions*. The positively charged ions such as sodium and ammonium are called *cations*.

Scaling occurs when calcium or magnesium compounds in the water (*water hardness*) precipitate and adhere to boiler internal surfaces. These hardness compounds become less soluble as temperature increases, causing them to separate from solution. Scaling causes damage to heat transfer surfaces by decreasing the heat exchange capability. The result can be overheating of tubes, followed by failure and equipment damage.

External water treatment is required when the level of feedwater impurities can not be tolerated by the boiler system. The selection of raw water treatment equipment should consider the raw water composition, quantity of makeup required, boiler water composition, and boiler operating pressure.

The first step in water processing involves coagulation and filtration of the suspended material. Natural settling in still water removes relatively coarse suspended solids. The settling time depends on specific gravity, shape, and size of particles and currents within the settling basin. This process can be expedited by coagulation. Coagulation is the process by which finely divided materials are combined by the use of chemicals to produce large particles capable of rapid settling. Typical coagulation chemicals are alum and iron sulfate. The preliminary treatment involves chlorination of the water for the destruction of living organisms. Several manufacturers offer equipment that operates on a completely automated basis, as illustrated in Fig. 1.

Following coagulation and settling, the water is normally passed through filters. Activated charcoal filters may be necessary to remove the final traces of organics and excess chlorine. After removing the suspended materials, the scale forming constituents still remain and require further treatment for their removal. A number of processes are currently used to remove scale constituents from the water.

Sodium cycle softening

This process, often called *sodium zeolite softening*, uses resin materials (denoted by R) that can exchange sodium (Na) for the hardness constituents, calcium (Ca) and magnesium (Mg). The process continues until the sodium ions in the resin become depleted or, conversely, the resin can no longer absorb the calcium and magnesium. A typical reaction for exhaustion of sodium zeolite is as follows:

$$2R\text{-}Na + Ca^{++} + CO_3^{=} \rightarrow R_2Ca + 2Na^{+} + CO_3^{=} \qquad \mathbf{(1)}$$

where $CO_3^{=}$ is the carbonate ion.

The forward reaction is chemically favored. Regeneration is accomplished by passing a high concentration of salt solution (NaCl) through the resin. This increases the Na concentration on the right side of the equation. The reaction is then overwhelmed with sodium, forcing it to the left. This is known as the mass action effect.

Hot lime zeolite softening In this process, hydrated lime reacts with the bicarbonate alkalinity of the raw water. The precipitate, calcium carbonate, is filtered from the solution. To reduce silica, the natural magnesium of the raw supply can be precipitated or magnesium hydroxide

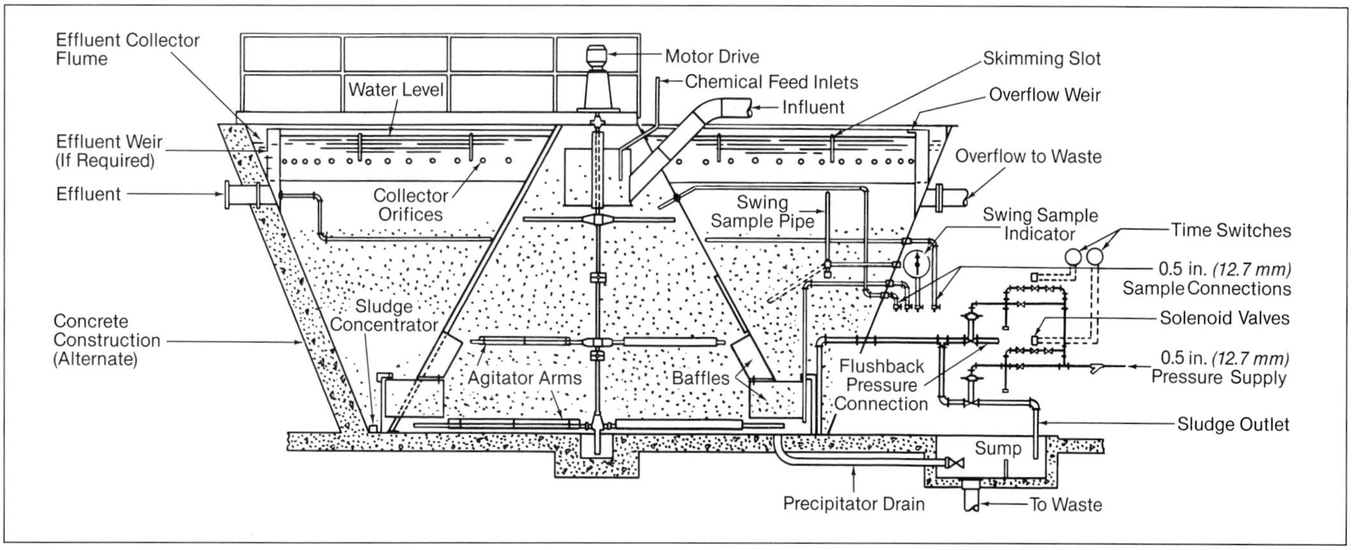

Fig. 1 Sludge contact softener. (*Source: Betz Handbook of Industrial Water Conditioning.*)

[$Mg(OH)_2$] can be added as a natural silica absorbent. These reactions are carried out in a tank that is just ahead of the zeolite softener tank. The effluent from this tank is filtered and then introduced into the zeolite softener. There is always some residual hardness leakage from the hot process softener that must be removed in the final zeolite process. This system is shown in Fig. 2.

Hot lime zeolite — split stream softening This process requires a second tank that has a zeolite resin in the hydrogen form in addition to the standard tank with the resin in the sodium form; the two tanks are operated in parallel. In the second tank, calcium and magnesium ions are replaced by hydrogen ions. The effluent from this tank is acidic and has a lower total solids content. The flow can be proportioned between the two tanks to produce an effluent with any desired alkalinity as well as to provide excellent hardness removal. As the two streams mix, part of the carbonate evolves as carbon dioxide which is vented in a decarbonator. When the hydrogen resin is exhausted, it is regenerated with acid. A typical layout of a split stream softener is illustrated in Fig. 3.

Demineralization

At boiler drum pressures above 1000 psi (69 bar), demineralization or evaporative distillation of makeup water is desirable. Treated makeup water that closely approaches theoretical chemical purity can be obtained by either process.

Evaporation This is a distillation process. A typical evaporator consists of a steel shell and tube heat exchanger and an external condenser. The tubes are submerged in makeup water. Steam supplied to the inside of the tubes gives up heat to the surrounding water. The heat causes the makeup water to evaporate, forming a vapor which leaves the evaporator. The vapor is recondensed, forming purified water.

Ion exchange This process is similar to zeolite softening. The cations in solution are replaced with hydrogen ions by the process described for the hydrogen zeolite system. In addition, the anions (bicarbonate, carbonate, sulfate and chloride) in solution are replaced by hydroxide ions through contact with an anion resin regenerated in the hydroxide form.

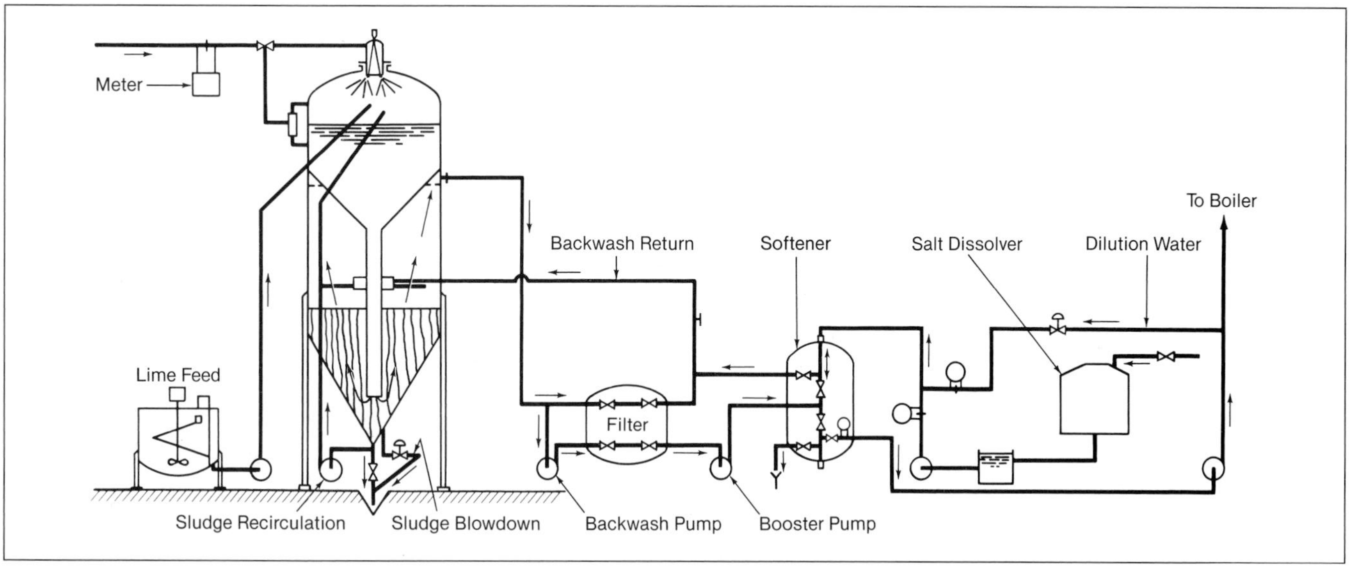

Fig. 2 Flow sheet of typical hot lime zeolite softening process. (*Source: Betz Handbook of Industrial Water Conditioning.*)

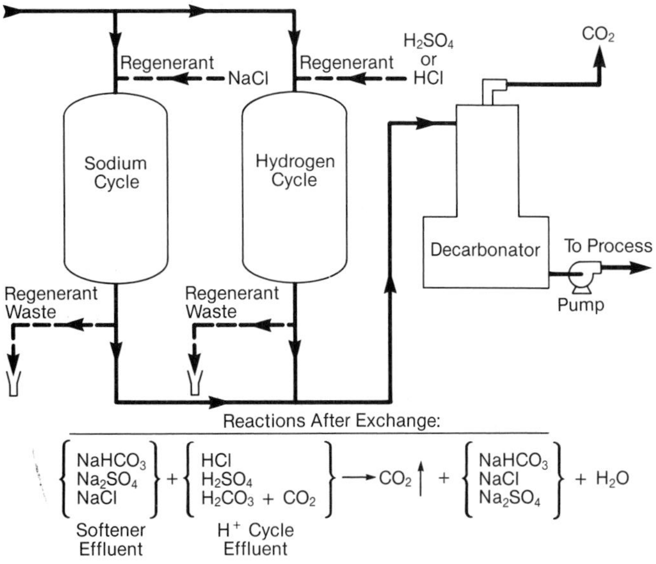

Fig. 3 Dealkalization by split stream.

The cation and anion resins can be located in separate containers or mixed in one vessel. For makeup water treatment two tanks are normally used in series in a cation-anion sequence. The anion resin is usually regenerated with a solution of sodium hydroxide. The cation resin is regenerated with hydrochloric or sulfuric acid. Some cation leakage always occurs in a cation exchanger, resulting in alkalinity leakage from the anion exchanger.

Another type of arrangement, known as a *mixed-bed demineralizer*, can be used to increase effluent water quality from that of a standard two bed system. The standard mixed bed is predominantly used for polishing boiler condensate but is also used to polish makeup water for boilers operating above 2000 psi (137.9 bar) and for all once-through boilers. In the mixed-bed demineralizer, cation and anion exchange take place virtually simultaneously. Regeneration is possible in a mixed bed because the two resins can be hydraulically separated. The cation resin is approximately twice as dense as the anion resin. Resins can be regenerated in place or sluiced to external tanks.

The effluent from demineralization is approximately neutral (pH near 7). With nearly all salts removed, minimal chemical control of the boiler water is required.

Other purification techniques

Several other processes are available to purify water although they are not extensively used to treat boiler makeup.

1. Osmosis is the process in which solvent flows through a semi-permeable membrane from a less concentrated solution to one with a higher salt concentration. This normal osmotic flow can be reversed (reverse osmosis) by applying hydraulic pressure to the more concentrated solution to produce high purity water. (See Fig. 4.)
2. Electrodialysis is a membrane process where an applied electrical charge draws impurity ions through permeable membranes to create high purity feedwater streams and low purity waste streams. (See Fig. 5.)
3. Ultrafiltration is another process that forces water through a filtering membrane by means of a pressure gradient, leaving large particles behind in a concentrated reject stream.

Pre-operational cleaning

In general, all boiler systems receive an alkaline boilout — hot circulation of an alkaline mixture with frequent blowdown and final draining of the unit. Many systems also receive a pre-operational chemical cleaning. Chemical cleaning of the superheater and reheater is not necessary for initial operation. However, these surfaces should be given a conventional steam blow — a period of high velocity steam flow which carries debris from the system. Chemical cleaning of superheater and reheat surfaces has been effective in reducing the number of steam blows to obtain clean surfaces.

Hydrostatic test water

After the pressure parts are assembled, a hydrostatic test at 1.5 times the design pressure is applied to all new boilers; the pressure is maintained for a sufficient time to detect leaks. Treated demineralized, deaerated water [defined as having a specific conductance of less than 1 μS(micro Siemens)/cm)] should be used for all hydrostatic testing whenever possible to prevent fouling and corrosion. The treated water also has an oxygen scavenger (such as hydrazine, N_2H_4) added and has enough ammonia (NH_3) to give a pH of 10 (approximately 10 mg/l as NH_3).

Temperature plays an important part in hydrostatic testing. First, the metal and water temperatures should be above the ambient air dew point temperature to prevent condensate formation on the parts being tested. Condensation interferes with the detection of small leaks. Second, the water temperature should remain below 120F (49C) so that close inspections can be made without injury to personnel. Finally the water and metal parts being tested should always be above 70F (21C) to avoid the danger of brittle fracture.

If the unit is drained, the water must be replaced by pressurized nitrogen to exclude air.

Alkaline boilout

All new boilers should be flushed and given an alkaline boilout to remove debris, oil, grease and paint. This boilout can be accomplished with a combination of trisodium phosphate (Na_3PO_4) and disodium phosphate (Na_2HPO_4), with a small amount of surfactant added to act as a wetting

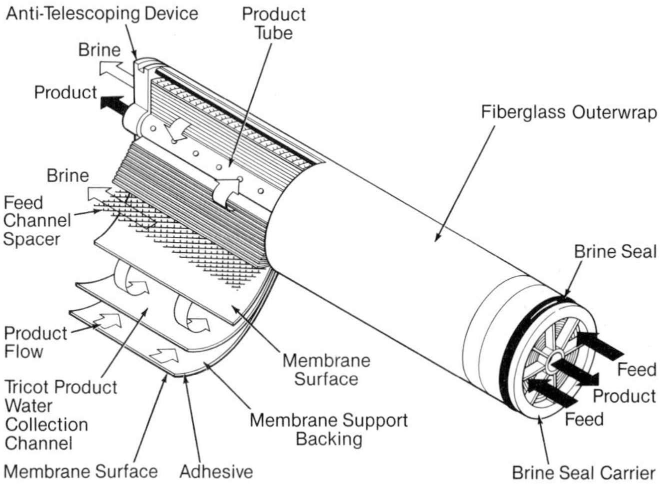

Fig. 4 Spiral wound reverse osmosis unit (*courtesy of E.I. Du Pont de Nemours & Co.*).

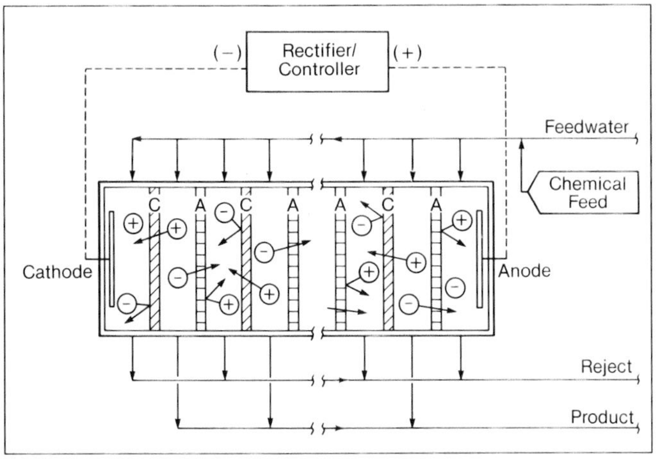

Fig. 5 Schematic of electrodialysis stack for water purification.[5]

agent. The use of caustic NaOH and/or soda ash (Na_2CO_3) is not recommended. If either is used, special precautions must be taken to protect the boiler components.

Chemical cleaning

After boilout and flushing are completed, corrosion products may remain in the feedwater system and boiler in the form of iron oxide and mill scale. It is recommended that chemical cleaning be delayed until full load operation has carried the loose scale and oxides from the feedwater system to the boiler. The exception would be units that incorporate a full flow condensate polishing system. In general, these units can be chemically cleaned immediately following the pre-operational boilout.

Different solvents and cleaning processes can be used for the pre-operational chemical cleaning. The three most frequently used are: 1) inhibited 5% hydrochloric acid with 0.25% ammonium bifluoride; 2) 2% hydroxyacetic/1% formic acids with 0.25% ammonium bifluoride and a corrosion inhibitor; and 3) inhibited ammonium salts of ethylenediaminetetraacetic acid (EDTA).

Steam line blowing

The steam line blow procedure depends on the design of the unit. Temporary piping to the atmosphere is required with all procedures. This piping must be anchored to resist the high nozzle reaction force created during blowing.

All normal startup precautions should be observed for steam line blowing. The unit should be filled with demineralized water. Sufficient feedwater pump capacity and condensate storage must be available to replace the water lost during the blowing period.

Numerous short blows are most effective. The color of the steam discharged to the atmosphere indicates the debris being removed from the piping. Coupons (targets) of polished steel attached to the end of the exhaust piping are typically used as final indicators.

Water treatment for drum boilers

Water quality limits

Direct boiler water treatment, usually referred to as *internal treatment*, is used to prevent deposit formations caused by hardness constituents and to provide pH control for corrosion prevention. Fig. 6 illustrates internal deposits resulting from poor water treatment. Fig. 7 shows a screen tube with low deposit loading from good water treatment. The permissible amounts of contaminants and treatment chemicals entering the boiler decrease with rising boiler pressures.

As boiler operating pressures increase, the boiler water chemistry may have a significant impact on the transport of contaminant solids to the superheaters and turbine. Such impurities in the boiler water will have a significant detrimental effect if too high. Solids are carried into the steam in one of two ways: 1) mechanical carryover of water droplets in the steam (usually limited to 0.025% by weight), and 2) vaporous carryover of constituents such as silica in the steam. Fig. 8 shows a representative relationship between dissolved boiler solids and average solids in steam at various operating pressures. The relationship depends on the boiler water chemistry. Fig. 9 indicates the reduction in boiler water silica content that must occur as operating pressures increase to limit silica carryover to less than 20 ppb, which normally passes through the superheater without deposition. The curve is valid for boiler water pH of 9.5. Babcock & Wilcox (B&W) recommends the feedwater limits listed in Table 1 and the maximum total boiler water solids identified in Fig. 10 for drum boilers. A total hardness limit of zero ppm means as close as possible to zero. The purchaser of the feedwater treatment system must evaluate the relative merits of increased capital costs versus increased frequency of pressure part chemical cleaning. In general, feedwater hardness in the range of 0.1 ppm should permit about three to five years between chemical cleanings. An organic limit of zero ppm means as close as possible to zero total organic carbon (TOC). Organic contamination of feedwater can lead to drum level instability, carryover, superheater tube failures and turbine problems. The degree to which any of these difficulties occur depends on the concentration and composition of the organic contaminant. TOC analyzers are commercially available with detection levels of at least 40 ppb. Further discussion is provided under the drum carryover and steam purity section below.

There are various methods of internal treatment used in natural circulation drum boilers: hydroxide, phosphate (conventional, coordinated, congruent and equilibrium),

Fig. 6 An example of internal deposits resulting from poor boiler water treatment. These deposits, besides hindering heat transfer, allowed boiler water salts to concentrate, causing corrosion.

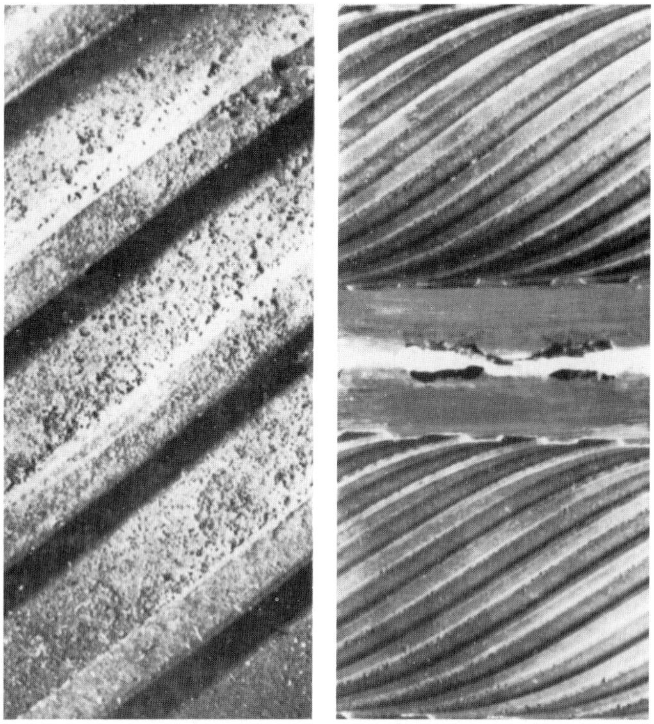

Fig. 7 Three years of operation resulted in light deposits because of good water treatment. The upper right figure is the heated side and the lower right figure is the unheated side.

volatile, chelant and polymer. The treatment method is generally dictated by the pressure range of the unit and the nature of the feedwater as well as other factors.

Hydroxide treatment This is used in very low pressure industrial boilers. In this treatment regime, a few ppm of NaOH are maintained in the boiler water to control boiler water pH in the range of about 10.0 to 11.5. Because NaOH is very soluble at operating temperatures, hideout (accumulation during normal operation) is not a problem, and operators find it easy to control within established limits. The alkalinity present provides a limited amount of neutralizing capacity for mineral acids which are generally formed from salts, such as magnesium and calcium chlorides, and sulfates which enter during periods of condenser leakage.

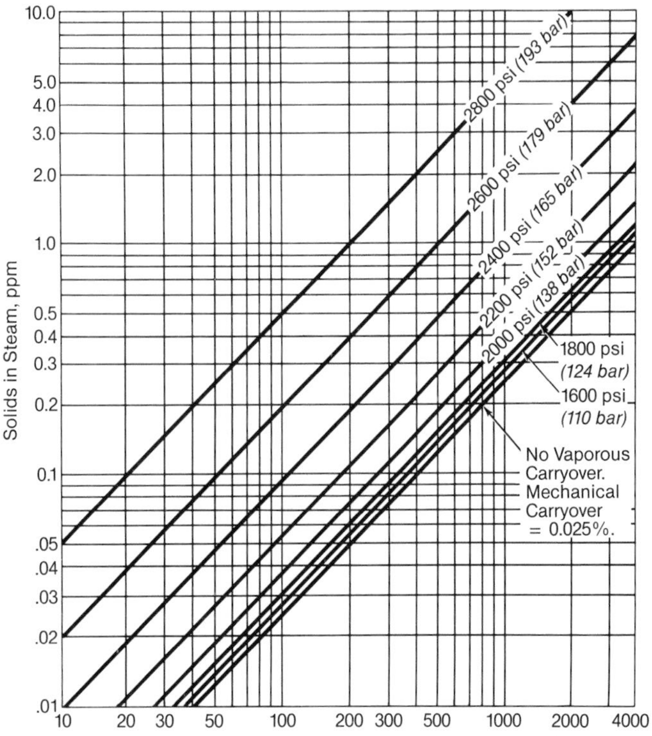

Fig. 8 Solids in steam versus dissolved solids in boiler water.

However, it has been well established that NaOH itself can be very corrosive when concentrated in local areas such as under porous water side deposits. The concentrated NaOH reacts with the tube's protective magnetite film and results in severe localized attack commonly referred to as *caustic gouging*. Many thousands of boiler tube failures have resulted from this mechanism over the years, although they usually occurred at much higher NaOH levels in the boiler water than are presently being maintained.

Conventional phosphate treatment This method, in which phosphate and caustic are added to the boiler water, is common for industrial boilers operating below 1000 psi (69 bar). The primary purpose of phosphate addition is to precipitate the hardness constituents. Calcium reacts with phosphate under the proper pH conditions to precipitate calcium phosphate as calcium hydroxyapatite [$Ca_{10}(PO_4)_6(OH)_2$]; this adheres less to boiler sur-

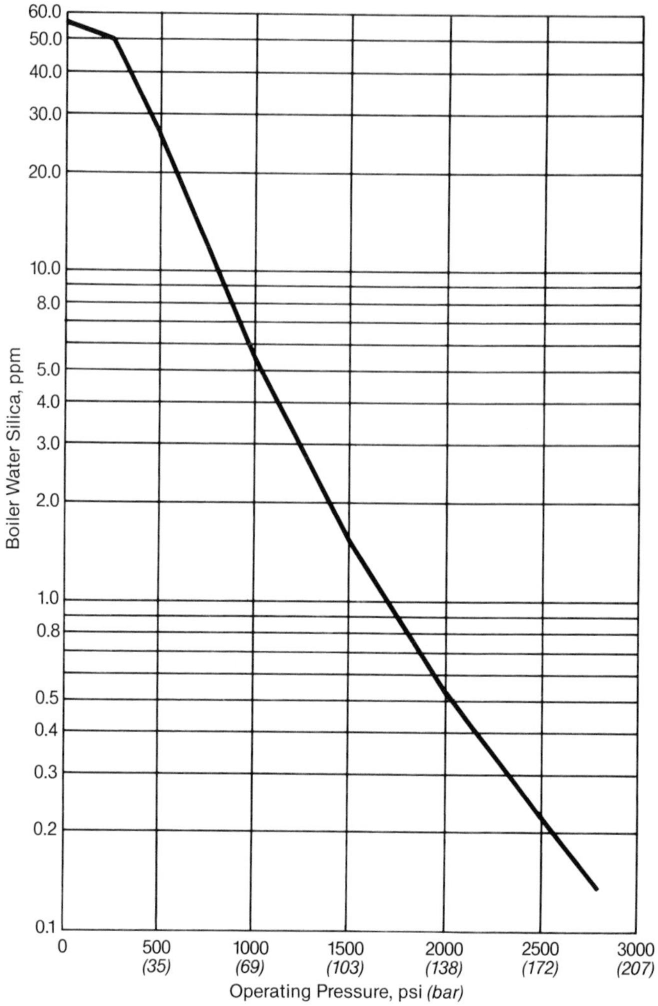

Fig. 9 Recommended maximum silica concentrations in boiler water at pH 9.5 (drum-type boilers).

Table 1
Recommended Limits in Boiler Feedwater

Drum Pressure	Below 600 psi	600 to 1000 psi	1000 to 2000 psi	Above 2000 psi
Total solids, ppm	---	---	0.15	0.05
Iron, ppm	0.1	0.05	0.01	0.01
Copper, ppm	0.05	0.03	0.005	0.002
Oxygen, ppm	0.007	0.007	0.007	0.007
pH	8.0 to 9.5	8.0 to 9.5	8.5 to 9.5	8.8 to 9.5*
Total hardness as ppm $CaCO_3$	0	0	0	0
TOC	0	0	0	0

* 8.8 to 9.2 with copper alloy feedwater heaters.
9.2 to 9.5 with steel feedwater heaters.

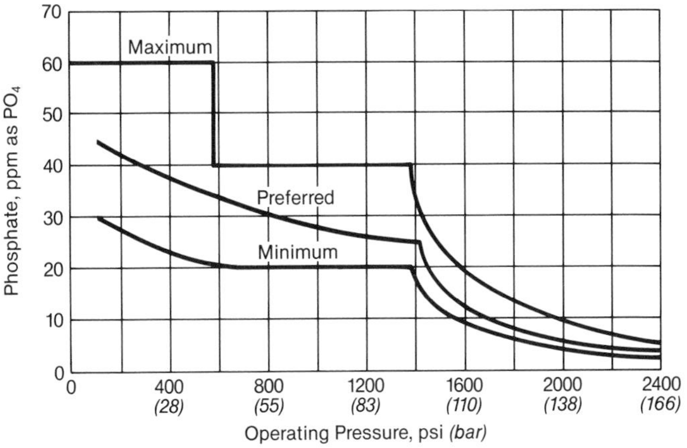

Fig. 11 Recommended phosphate concentration in boiler water at various boiler operating pressures (phosphate-hydroxide treatment).

faces than simple tricalcium phosphate [$Ca_3(PO_4)_2$] which is precipitated below 10.2 pH. Caustic reacts with magnesium to form magnesium hydroxide or brucite, $Mg(OH)_2$, which is formed in preference to magnesium phosphate above 10.5 pH as it is less adherent.

The recommended phosphate concentrations for various boiler operating pressures are shown in Fig. 11. At higher pressures, comparatively low phosphate residuals must be maintained to avoid appreciable phosphate hideout. The preferred alkalinity as a function of pressure is given in Fig. 12.

Coordinated and congruent phosphate treatment Coordinated phosphate pH treatment was designed to provide boiler water alkalinity with mixtures of disodium and trisodium phosphate to minimize free sodium hydroxide and its corrosion potential when concentrated under deposit or in crevice locations. The control parameters involved maintaining an Na to PO_4 ratio at or slightly below 3:1. Fig. 13 shows the relationship between phosphate concentration and the resulting pH when trisodium phosphate is dissolved in water. The precipitate from a concentrated solution of trisodium phosphate (Na_3PO_4) at elevated temperatures has a sodium to phosphate ratio of less than 3, and the supernatant liquid is correspondingly rich in sodium hydroxide (NaOH). The sodium hydroxide can destroy the magnetite protective film and boiler surfaces.

To assure that no free caustic is present, a boiler water phosphate concentration corresponding to a sodium to phosphate mole ratio of 2.6 (congruent phosphate treatment) is recommended above 1000 psi (69 bar), as shown in Fig. 13. Free hydroxide alkalinity is less critical in boilers operating below 1000 psi (69 bar). The shaded areas on Fig. 13 indicate the recommended operating range of PO_4 and the resulting pH for boiler pressures to 2000 psi (138 bar).

Caution should be used in calculating the weights needed to provide the proper sodium to phosphate mole ratios. Phosphates commonly occur in the form of $Na_3PO_4{\cdot}12H_2O$ and $Na_2HPO_4{\cdot}7H_2O$. A mixture of 65% $Na_3PO_4{\cdot}12H_2O$ and 35% $Na_2HPO_4{\cdot}7H_2O$ corresponds to an Na/PO_4 mole ratio of 2.6. If the pH is low, it may be corrected by increasing the ratio of trisodium to disodium phosphate. If monosodium phosphate or mostly disodium phosphate must routinely be added to avoid excessively alkaline boiler water, then there is excessive ingress or

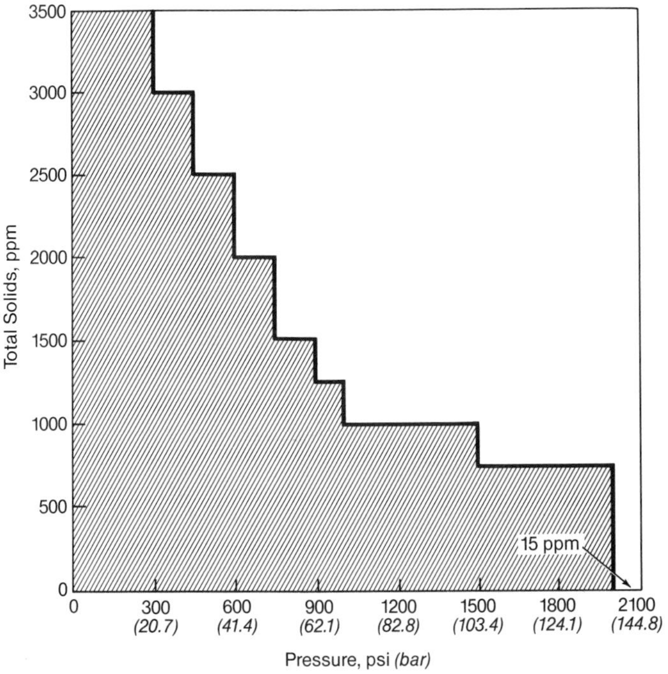

Fig. 10 Maximum total boiler water solids limits required to limit carryover. Above 2000 psi, total solids limits are set on a case by case basis depending on the desired carryover guarantee.

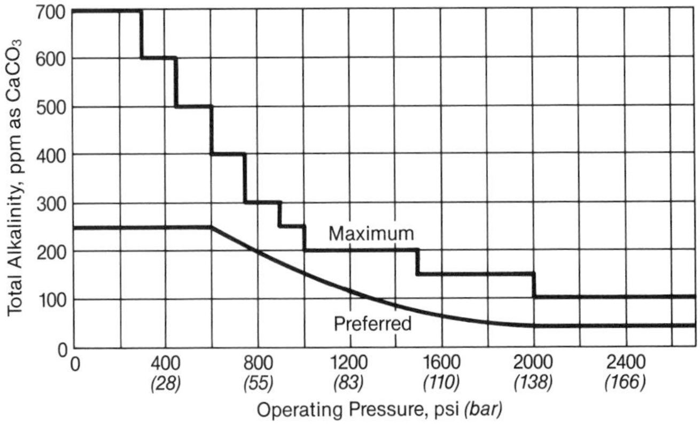

Fig. 12 Recommended alkalinity of boiler water at various boiler operating pressures (phosphate-hydroxide treatment).

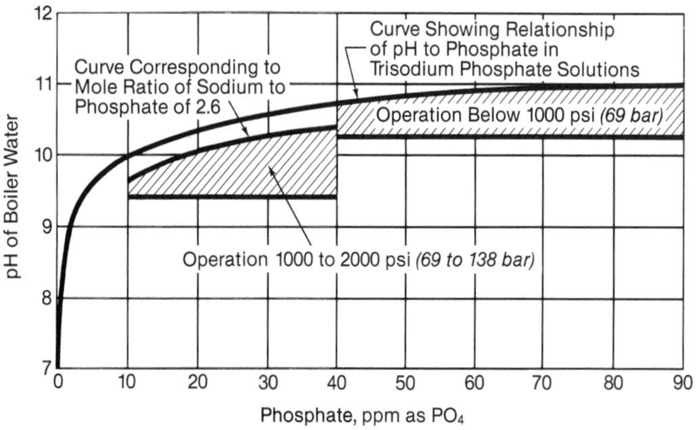

Fig. 13 Recommended phosphate content of boiler water for drum-type boilers using coordinated phosphate treatment below 2000 psi.

generation of alkaline constituents that should be addressed. Feeding of phosphate under hideout conditions generally aggravates pH control and hideout problems.

For drum pressures from 2000 to 2835 psi (138 to 196 bar), the boiler water ideally should contain 2 to 7 ppm phosphate with a corresponding pH of 9.0 to 10.0 to buffer against condenser leaks and other upsets. However, the amount of phosphate that can be carried without excessive hideout (greater than 5 ppm change in phosphate concentration between high and low load) varies among boiler units even at the same operating pressure. Generally, the higher the pressure, the lower the amount of phosphate that can be carried in the boiler water. In addition, cycling boilers and boilers with substantial water-side deposition must operate with reduced phosphate levels to avoid excessive hideout. Because each unit is unique, boiler operators must be sensitive to the phosphate limits for each specific boiler and to how the limits change with time. Operation at phosphate limits below 2 ppm is described in the following section on *Equilibrium Phosphate*. Unusual hideout behavior may be indicative of deposition problems that may require immediate attention.

Equilibrium phosphate At operating pressures above 2500 psig (172 bar gauge), the concentrations of total sodium phosphate chemicals which can be maintained in the boiler water may be limited to a few ppm or less (sometimes as low as 0.5 ppm). One factor contributing to this effect is the retrograde solubility of sodium phosphates, whereby their solubility limits decrease as saturation temperatures increase. Soluble chemicals such as sodium phosphates are further concentrated in the layer of boiling water immediately adjacent to the tube wall increasing the likelihood of exceeding the solubility limits. This concentration effect is greatly accentuated by the presence of porous deposits. The amount of phosphate that can be carried in the boiler water is also limited by adsorption of phosphate on and reaction with magnetite and other constituents of the porous water-side deposits.[4,8,9] This loss of phosphates from the boiler water has become known as *phosphate hideout*. It has long been a concern in the industry, and research is still being conducted in search of a better understanding of this phenomenon and how it is related to boiler deposits and corrosion. The phosphate concentration in the bulk boiler water which will produce hideout in a clean boiler varies from unit to unit, but it is believed to be governed primarily by saturation temperature, heat flux, steam quality, mass velocity, as well as the extent and nature of water-side boiler deposits.

Some operators of high pressure utility boilers have experienced considerable difficulty in controlling boiler water phosphate chemistry within the *congruent control* regime and its attendant concentration guidelines. It has become necessary to evaluate phosphate behavior in units on an individual basis to establish proper sodium phosphate addition and control parameters.

Equilibrium phosphate[4] involves determining the level of phosphate which is stable, i.e., no hideout, at the maximum operating pressure and not exceeding this phosphate level. Chemical injection practices, as well as the loss of phosphate from the boiler water, should be monitored by material balance calculations. Boiler blowdown to control the concentration of chemical additives to the boiler water should not be used except in emergencies. The concentration of additives should be controlled by the feed rate.

Volatile treatment B&W recommends this method of treatment be used only on units operating with condensate polishing. Normally it is used on units operating at or above 2000 psi (138 bar) drum pressure. In this method, no solid chemicals are added to the boiler or preboiler cycle. As a result, the volatile carryover of solids is eliminated and turbine deposits are avoided. Cycle pH is controlled between 9.0 and 9.5 with a volatile amine such as ammonia. Hydrazine, or a suitable substitute, is added as an oxygen scavenger. With volatile treatment, the feedwater must not contain hardness or condenser leak constituents.

Use of chelants This method of water treatment has become popular with industrial boiler operators. These organic agents react with the residual divalent ions such as calcium, magnesium and iron in the feedwater to form soluble complexes. The resultant complexes are removed through continuous blowdown. B&W recommends limiting the use of chelants to units below 1000 psi (69 bar).

Certain precautions are necessary in using this treatment. The agents do not chelate ferric iron. The combined presence of chelating agents and oxygen in the boiler or preboiler cycle must be avoided. Deaeration must be good during operation and the boiler must be protected from oxygen during off-line periods.

It is difficult to control chelant feed based on chelant residual in the boiler water; excess chelant attacks clean boiler surfaces. B&W therefore recommends that chelant feed be based on known quantities of hardness in the feedwater; the objective is to maintain a residual approximating 1 ppm of chelant. To protect the boiler from upsets resulting from heat exchanger leakage or makeup plant overrun, a phosphate residual should also be maintained in the boiler water. If sludge deposits accumulate, the chelant feed should be increased by 1 to 2 ppm. If the boiler is found to be exceptionally clean, the chelant feed should be decreased. A light gray deposit coating on internal boiler surfaces characterizes the ideal condition.

Polymer treatment Besides their use as dispersants, synthetic polymers (polyelectrolytes) have been used in all polymer treatment programs to control both hardness and iron oxide deposition. Their limitation is thermal stability and they are not normally used above 1200 psi (83 bar), but this may change as new polymer treatments are developed.

Drum carryover and steam purity

The trend toward higher pressures and temperatures in steam power plants imposes a severe demand on steam-water separation equipment. (See Chapter 5). Carryover may result from limited mechanical separation and from the vaporization of boiler water salts. Total carryover is the sum of the mechanical and vaporous components.

Mechanical carryover is the entrainment of small boiler water droplets in the separated steam. Because these droplets contain solids in the same concentration as the boiler water, the amount of impurities in steam contributed by mechanical carryover is the sum of the boiler water impurities concentration multiplied by the steam moisture content.

Boiler water foaming results in gross mechanical carryover. The common causes of foaming are excessive solids or alkalinity and the presence of organic matter such as oil. Maintaining dissolved solids at a level that prevents foaming requires continuous or periodic blowdown of the boiler. Fig. 10 gives recommended maximum total solids concentration which prevent excessive carryover at various operating pressures. It is advisable to run well below these limits.

High boiler water alkalinity increases carryover, particularly in the presence of suspended matter. This effect may be corrected by various methods depending on its cause. For example, if trisodium phosphate (Na_3PO_4) is being added to the boiler water, a less alkaline phosphate, such as disodium phosphate (Na_2HPO_4), reduces alkalinity. Also, it is preferable to eliminate external sources of excess alkalinity.

Water used in a spray attemperator should be of the highest quality. Solids entrained in the water enter the steam and can cause deposits on superheater tubes and turbine blades.

Carryover of volatile silica is generally a problem only at pressures above 1000 psi (69 bar). Refer to the silica limits of Fig. 9.

Vaporous carryover prevention is much more difficult than the correction of mechanical carryover. The only effective method is to reduce the boiler water solids content.

Measurement of steam quality and purity

The most common methods for testing steam quality and purity in the field are: 1) sodium tracer (selective ion electrode or flame photometry), 2) electrical conductivity (for dissolved solids), 3) throttling calorimeter (for direct determination of steam quality), and 4) gravimetric (for total solids).

Each of these methods is described in the ASME Performance Test Code 19.11, *Water and Steam in the Power Cycle*. The throttling calorimeter determines steam quality directly, whereas the other methods determine the steam solids content.

Most of the solids content in a low pressure boiler comes from boiler water droplet carryover. (See also Chapter 5). It is customary to relate steam quality and solids content by the formula:

$$x = 100 - \left[\frac{\text{Solids in steam} \times 100}{\text{Solids in boiler water}}\right] \qquad (2)$$

where x = % steam quality, and solids are expressed as equivalent ppm by weight of steam or water.

Obtaining the steam sample If steam quality or purity determinations are to be accurate, the measuring instruments must be supplied with a representative steam sample. The method of obtaining a steam sample and the boiler operating procedure during testing are fundamentally the same for all testing methods. (See also *Sampling* section.)

The sampling nozzle design should be as recommended by the ASME Performance Test Code 19.11. The nozzle should be located after a run of straight pipe equal to at least, and preferably greater than, ten diameters. Locations in the order of preference are: 1) vertical pipe, downward flow, 2) vertical pipe, upward flow, 3) horizontal pipe, vertical insertion, and 4) horizontal pipe, horizontal insertion.

The nozzle should be installed in the saturated steam pipe in the plane of a preceding bend and the sampling ports must directly face the steam flow.

On boilers with multiple superheater supply tubes, nozzles should be spaced across the width of the steam drum. Sampling intervals should be no greater than 5 ft (1.5 m).

When a calorimeter is used in low pressure boilers, the connection from the sampling nozzle to the calorimeter must be short and well insulated to minimize radiation losses; connections must be steam-tight.

When steam purity is tested by the conductivity or sodium tracer methods, the connection from the sampling nozzles to the condenser is preferably stainless steel, with an inside diameter not exceeding 0.25 in. (6.35 mm); the length should be minimal to reduce the storage capacity of the line. Multiple connections can be run to a common line and then to the condenser, but these should be individually valved.

Sodium tracer method The sodium tracer technique permits measuring dissolved solids steam impurities to values as low as 0.001 ppm.

Sodium is present in all boiler water where solid chemical treatment is used. The steam condensate dissolved solids to boiler water dissolved solids ratio is proportional to the steam condensate sodium to boiler water sodium ratio. By determining the sodium content of the steam condensate and boiler water and the total dissolved solids in the boiler water, the solids content of the steam and the percent moisture carryover may be calculated. A selective ion sodium monitor is typically used to determine the sodium contents of steam condensate and water. The sodium ion electrode is not subject to solution interference for color, turbidity, colloidal matter, oxidants or reductants.

A continuous flame photometer can also be used to measure sodium content. The condensed steam sample is aspirated through a small tube in the burner into the oxygen-hydrogen flame. The high temperature flame [3000 to 3500F (1649 to 1927C)] vaporizes the water and excites the sodium atoms which emit a characteristic yellow light. The intensity of the emitted light is a measure of the sodium in the sample.

Conductivity method Electrical conductivity can be used to determine steam purity in units with significant boiler water solids concentration and a total steam solids content in excess of 0.5 ppm. Dissolved solids, whether acids, bases or salts, are completely ionized in a dilute solution and therefore conduct electricity in proportion to the total solids dissolved. The steam solids content in ppm approximately equals the electrical conductivity of the sample in microSiemens per centimeter (μS/cm) times 0.55.

The condensed sample should ideally be free from dissolved gases, especially NH_3 and CO_2, which have a marked effect on conductivity. This can be accomplished by several analyzing systems, which are described in the ASME Performance Test Code 19.11. Additional procedures are described in American Society for Testing and Materials (ASTM) Standards D1066, *Standard Practice for Sampling Steam* and D4519, *Standard Test Method for On-Line Monitoring of Electrical Conductivity to Determine Anions and Carbon Dioxide in High Purity Water*.[10]

Throttling calorimeter method This method determines the amount of moisture in wet saturated steam by throttling a continuous sample of steam from its initial pressure to atmospheric pressure.

The determination of impurity content in steam is usually accomplished by analyzing a condensed steam sample for the concentration of one or more impurities, such as sodium, or for a physical property, such as electrical conductivity. However, in some cases, particularly for small industrial or commercial boilers, it is sufficient to measure the moisture content of the steam sample to check for excessive carryover of boiler water with the steam. Throttling calorimetry (see Chapter 40) is useful for this purpose. Throttling calorimetry is considered to be a generally obsolete art for all but a few applications. This apparatus gives good results for steam having appreciable quantities of moisture: up to 4.3% at 100 psi (6.9 bar), 5.6% at 200 psi (14 bar), and 7.0% at 400 psi (28 bar). For pressures greater than 600 psi (41 bar), the other measurement methods should be used.

In obtaining calorimeter readings, the temperature of the expanded and superheated steam is measured by a thermometer inserted in a well. Due to errors arising principally from radiation, the observed temperature is lower than theoretical and must be corrected. The ASME Performance Test Code 19.11 contains a detailed procedure for the throttling calorimeter.

Gravimetric analysis method This procedure consists of evaporating a known quantity of condensed steam to complete dryness and weighing the residual. This can be done on a batch or continuous sample basis. Its main disadvantage is that it requires a relatively large quantity of water obtained over an extended time period.

Water treatment for once-through Universal Pressure boilers

Satisfactory operation of the once-through boiler and associated turbine requires that the total feedwater solids be less than 0.05 ppm. Table 2 lists recommended limits for feedwater contaminants.

Water treatment chemicals must be volatile. All cycles should have condensate polishing systems; a schematic of a deep bed condensate treatment system is shown in Fig. 14. These systems perform as filters of suspended material and remove ionized particles. Ammonia is added to control pH. Fig. 15 indicates the amount of ammonia required to give a certain pH in the system. Hydrazine, or a suitable volatile substitute, is used for oxygen scavenging.

Iron pickup from preboiler components can be minimized by maintaining a feedwater pH of 9.3 to 9.5.

Ammonia is injected downstream of the condensate polishers. The oxygen scavenger is generally fed at the exit of the condensate polishing system and/or at the boiler feed pump suction. A total hardness and organics

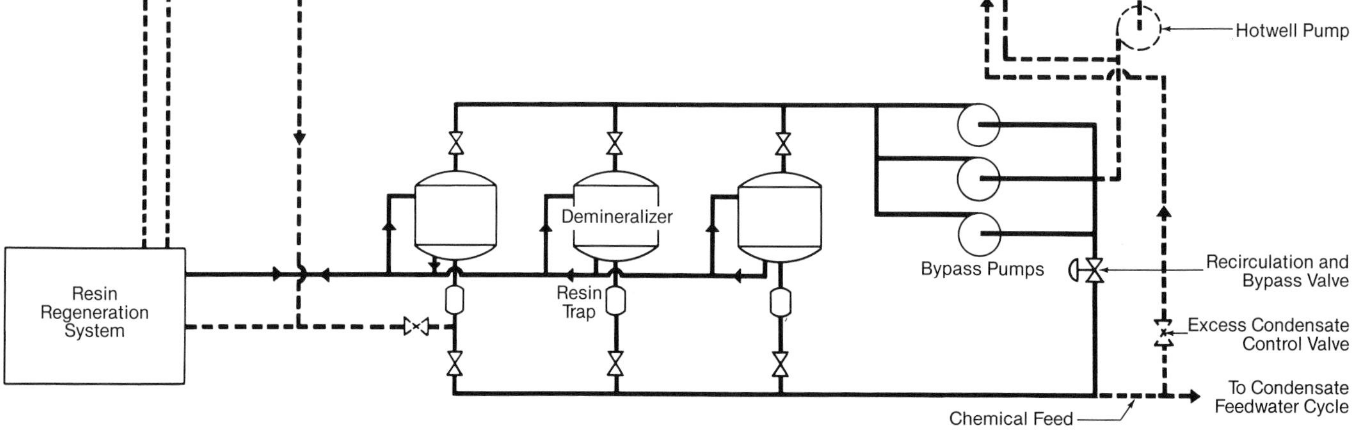

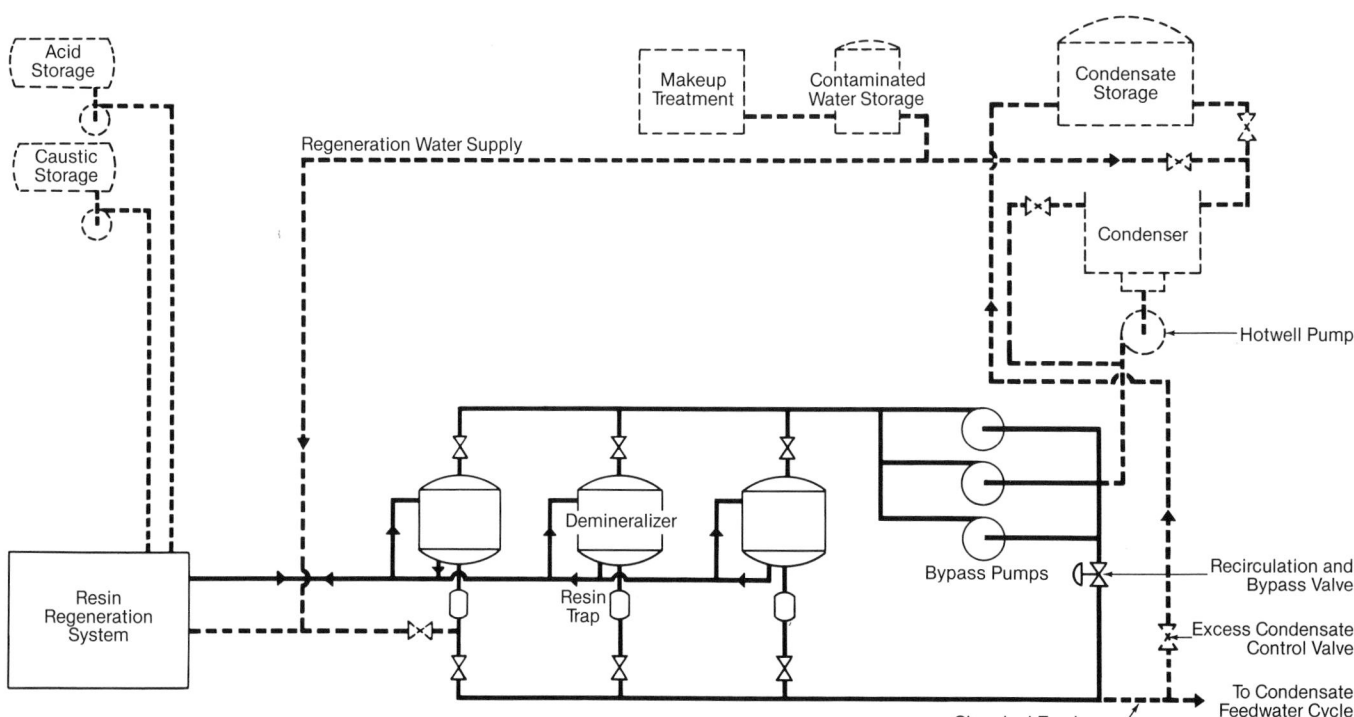

Fig. 14 Schematic of condensate polishing system with high quality makeup treatment (4-bed ion exchange or equivalent).

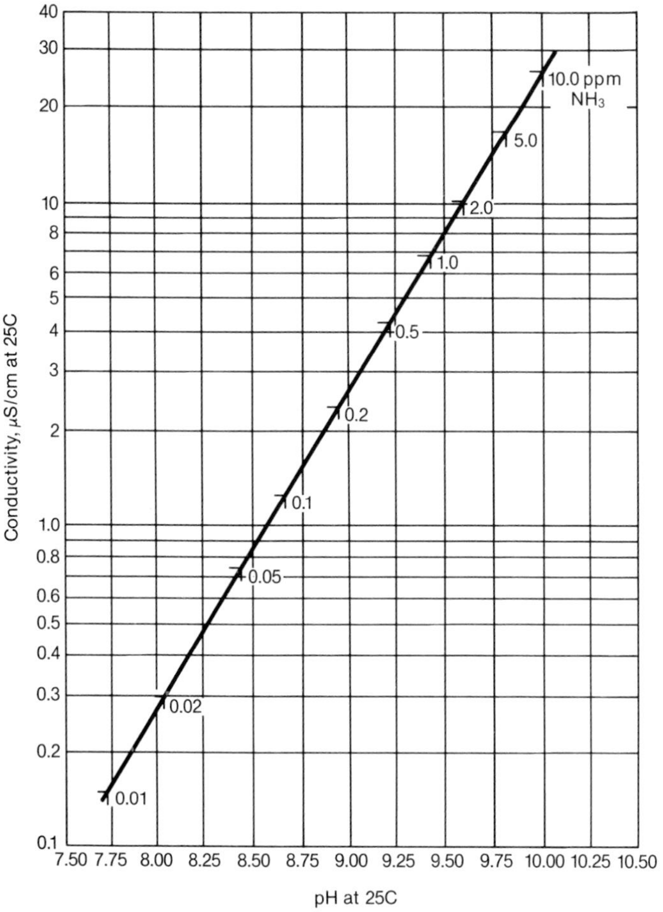

Fig. 15 Theoretical relationship between conductivity and pH for ammonia solutions.

limit of zero means as close as possible to zero. (See discussion on hardness and organics under *Water Treatment for Drum Boilers*.)

Prior to plant startup, water must be circulated through the condensate polishing system to reduce the dissolved material and suspended particles. The cation conductivity of the cycle water must be reduced to 1.0 µS/cm before firing the unit. Temperatures should not exceed 550F (288C) at the convection pass outlet until the iron levels are less than 0.1 ppm at the economizer inlet.

As an alternative to oxygen scavenging to remove the last traces of oxygen from the feedwater, some once-through boilers use oxygen treatment (OT). This is a growing practice in North America. It was developed in Europe and there is also extensive experience in the Commonwealth of Independent States (C.I.S.) (the former Soviet Union). Oxygen treatment can only be used where there is no copper in the preboiler components beyond the condensate polisher and where feedwater is consistently of the highest purity, e.g., cation conductivity ≤ 0.15 µS/cm at 77F (25C). A low concentration of oxygen (0.050 to 0.150 ppm) is added to the condensate. This allows carbon steel to form a more protective oxide layer which decreases iron pickup from erosion/corrosion of carbon steel preboiler components. As a result, the feedwater pH can be reduced, e.g., down to 8.0 to 8.5. The key advantage of oxygen treatment is decreased chemical cleaning frequencies for the boiler. In addition, when oxygen treatment is combined with lower pH water operation, condensate polisher regeneration frequency is also reduced.

Table 2
Recommended Maximum Feedwater Limits for Universal Pressure Boilers

Total solids	0.050 ppm
Silica, SiO_2	0.020 ppm
Iron, Fe	0.010 ppm
Copper, Cu	0.002 ppm
Oxygen, O_2	0.007 ppm
Hardness	0.0 ppm
Organics	0.0 ppm
pH at 77F (25C)	9.5
Cation conductivity	0.5 µS/cm

Condensate treatment

In many cases, condensate does not require treatment prior to reuse. Makeup water is added directly to the condensate to form boiler feedwater. In some cases, however, especially where steam is used in industrial processes, the steam condensate is contaminated by corrosion products or by the in-leakage of cooling water or substances used in the process.

The presence of acidic gases in steam makes the condensate acidic, with consequent corrosion of metal surfaces. In such cases, the corrosion rate can be reduced by adding chemicals (neutralizing or filming amines) that produce alkaline gases or form a protective barrier in the condensate system.

Fig. 16 shows a condensate purification system used in a paper mill boiler cycle. The resin beds remove dissolved impurities by ion exchange and serve as filters to remove suspended solids.

Where significant quantities of magnetic oxide or other magnetic species are present in the condensate feedwater system, an electromagnetic filter (EMF) can effectively remove these suspended solids. The EMF, shown in Fig. 17, consists of a pressure vessel, coil, spheres and a power control unit. Advantages of the EMF are a low pressure drop through the filter, minimal required quantity of flush water, a short backwash process period (minutes) and no chemical requirements.

Condensate polishing systems

Demineralizer systems installed to purify condensate are known as condensate polishing systems. A condensate polishing system is a requisite to maintain the purity required for satisfactory operation of once-through

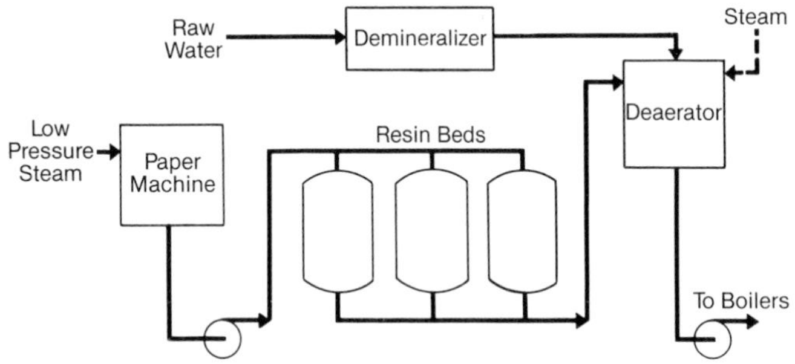

Fig. 16 Condensate purification system (*courtesy of Cochrane Division of Crane Company*).

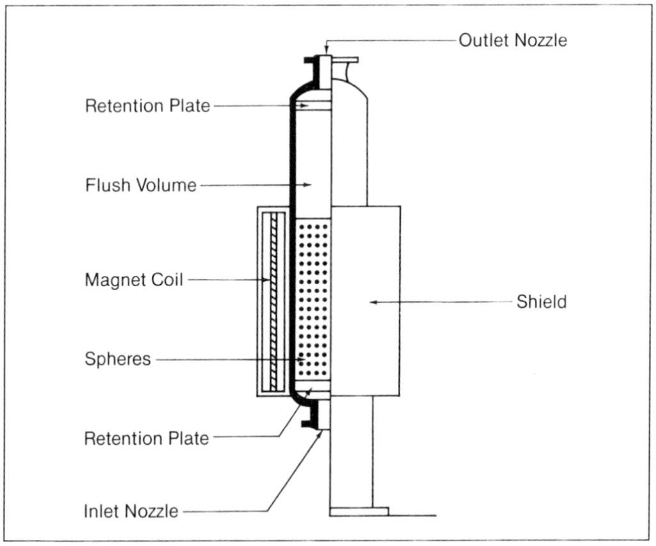

Fig. 17 Sectional view of electromagnetic filter.

boilers. While high pressure drum boilers [greater than 2000 psi (138 bar)] operate satisfactorily without condensate polishing, many utilities recognize the benefits in high pressure plants. These benefits include shorter unit startup time, protection from condenser leakage effects and longer intervals between acid cleanings.

Deep bed demineralizers operate at flow rates of 40 to 60 GPM/ft^2 (27 to 41 l/s m^2) of bed cross-section and require external regeneration facilities. They have a high capacity for removing ionized solids that result from small amounts of condenser leakage and are excellent mechanical filters although not designed as such. The deep bed system is usually operated with the cation resin in the hydrogen form, but the ammonium or morpholinium form can also be used.

The cartridge tubular demineralizer (powdered resin system) uses smaller amounts of disposable, small particle ion exchange resins. There is no regeneration, as the resins are discarded. These systems are less tolerant of condenser leakage.

Chemistry and corrosion control

Two general approaches are used to optimize the chemical environment in power cycles. First, impurities in the water are minimized by purification of makeup water, condensate polishing, deaeration and blowdown. Secondly, chemicals are added to the cycle to alter pH, electrochemical potential and oxygen concentration. Operation for extended periods outside of the specified water chemistry ranges increases the probability of overheat and corrosion related failures. Water chemistry recommendations are given throughout this chapter on boiler water chemistry. Water chemistry control limits/guidelines have been developed and issued by various groups of boiler owners and operators (e.g., ASME,[2] EPRI[11] and VGB[12]), water treatment specialists, utilities and industries. For each boiler, water chemistry limits and treatment practices should be developed and tailored to changing boiler operations and conditions by competent boiler water chemistry specialists. To maintain high quality chemistry control in power cycles, training must be an integral and continual part of operations and should include management, control room operators, chemists and the laboratory staff.

Quality assurance and quality control in water chemistry

Water chemistry specifications are based on the ability of each laboratory to consistently measure the specified parameters. Modern chemistry and radiochemistry laboratories have instituted formal quality assurance programs to quantify and track the precision and bias of the measurements. Detailed procedures cover laboratory structure, training, standardization, calibration, sample collection/storage/analysis, reporting, maintenance records and corrective action procedures. Further discussion of this topic is provided in Reference 13.

Water chemistry computer models

An ever changing array of computer models is available to assist water chemists with data collection, manipulation, trending, interpretation and troubleshooting. Selected tasks performed include: sample tracking, resource management, standard reporting, evaluation of composite mixture properties, evaluation of the corrosion potential for complex chemical systems and specific materials, performance of mass balances, tracking of potential problem areas, and troubleshooting of routine chemistry problems. The most advanced systems include the rudimentary reasoning capability of Expert Systems. These can be invaluable tools in maintaining unit availability.

Chemical injection systems

Chemicals are added to the condensate either downstream of the condensate polishers or at the condensate pump discharge for plants without polishers. Chemicals are also added directly to the boiler for drum boilers. In a few cases, chemicals are added to the steam.

The concentrated solution tank, pump and interconnecting tubing for the aqueous chemical injection systems should be fabricated from corrosion resistant materials. Stainless steel is commonly used, and certain plastics are acceptable for the low pressure components of aqueous chemical injection systems. The high pressure components are preferably stainless steel. The injection nozzle should be stainless steel and should prevent concentrated chemicals from draining onto the surface of the process pipe. Therefore, the nozzle preferably extends into the process stream.

With few exceptions, pH control is achieved using a feedback signal from a specific conductivity monitor.

Hydrazine, aqueous ammonia and possibly other amines present some health hazards which can be minimized with the proper hardware. These chemicals have relatively low vapor pressures and therefore easily become airborne at room temperature. Closed handling systems are available to minimize exposure.

Gaseous chemical injection systems should be fabricated from stainless steel, particularly for oxygen injection into systems that use oxygen treatment (OT). High quality gas regulators and control valves should be used to maintain a consistent flow of oxygen.

Gaseous and liquid injection systems should prevent backflow of process fluid into the supply systems and should shut down upon loss of process flow or chemical supply. In all cases, the material safety data sheets (MSDS) should be in clear view of the injection systems.

Blowdown

Blowdown is the removal of a small fraction of the recirculating water in fossil drum boilers. The primary function of the blowdown system is to control the dissolved solids of the recirculating water. This, in turn, controls carryover and corrosive attack of the units. Blowdown also reduces the concentration of solid corrosion products (metal oxides), but solid particle removal efficiency is very limited. Some utilities have found that corrosion product removal efficiency can be improved by periodic high blowdown rates from lower parts of the boiler. The boiler manufacturer should be consulted before blowing down lower furnace headers when the boiler is in operation.

If the amount of carryover into the steam is small, blowdown can be represented by the following equation:

$$C_b = C_f / b \qquad (3)$$

where

C_b = the concentration of a chemical in the blowdown and recirculating water
C_f = the concentration of the same chemical in the feedwater
b = the fraction of steam flow that is discarded through blowdown

For example, if a 0.1% blowdown were used, the maximum concentration factor (C_b/C_f) would be $1/b = 1/0.001 = 1000$.

To recover the water, blowdown can be purified by one of two methods: 1) it can be routed back to the condenser for purification by the condensate polishers, if available, or 2) a separate blowdown demineralizer. Although both options result in some heat loss, the loss can be minimized by using a regenerative heat exchanger. Also, routing of the blowdown stream back to the condenser is undesirable because the contaminants in the blowdown are diluted by condensate, and the removal efficiency in the condensate polisher is lowered when the impurities are diluted.

Once-through steam generators do not have a defined phase separation or an inventory of recirculating liquid at full power and, therefore, blowdown is not possible. As a result, high purity feedwater is required for these systems. In some once-through units, blowdown can be used at low loads.

Oxygen control and treatment

Cycles that operate with a reducing chemistry limit dissolved oxygen (O_2) to concentrations of ≤ 7 ppb. An oxidizing chemistry [oxygen treatment or Combined Water Treatment (CWT)] is being used in some once-through fossil cycles with sufficiently high water purity [generally <0.15 μS/cm measured at 77F (25C)]. However, even in an oxidizing chemistry, air in-leakage is not acceptable. It complicates the control of oxygen addition, and carbon dioxide (CO_2) and other undesirable atmospheric gases enter with the air. Oxidizing environments are controlled to relatively low concentrations (50 to 150 ppb O_2) preferably by the injection of molecular oxygen.

During operation, oxygen generally enters the cycle with non-deaerated makeup water through leaks in the vacuum portion of the cycle, or from condenser in-leakage. Minimizing blowdown and other water losses minimizes the introduction of air from the makeup water. The total air in-leakage should not be greater than one standard cubic foot per minute per 100 MW of generating capacity (approximately 0.027 Nm^3/100 MW), as measured at the condenser air ejectors.

Mechanical deaerators and chemical oxygen scavengers are used downstream of the condensate polishers to remove the last traces of oxygen and to develop a reducing chemistry. Hydrazine and sodium sulfite are common chemical oxygen scavengers, but sodium sulfite is only used in low pressure boilers. Other oxygen scavengers (erythorbic acid, diethylhydroxylamine, hydroquinone and carbohydrazide) are also used. Chemical oxygen scavengers are not used with oxygen treatment.

pH control

Fossil drum boilers use either a mixture of disodium phosphate (Na_2HPO_4) and trisodium phosphate (Na_3PO_4) or a neutralizing amine, such as ammonium hydroxide, cyclohexylamine or morpholine, for pH control. Amines, other than ammonium hydroxide, are not used in higher pressure drum units because they decompose at the higher steam temperatures.

Fossil (and nuclear) once-through cycles exclusively use neutralizing amines for pH control. The phosphate treatments are not acceptable because there is no recirculating water in which to accumulate these relatively nonvolatile chemicals for eventual blowdown. The high pressure, high temperature once-through fossil cycles are limited to ammonium hydroxide.

Transport of chemicals by steam

Fig. 18 shows the distribution ratio of several solid chemicals in steam and water.[14] In general, the solubility of relatively nonvolatile chemicals in steam decreases with temperature and pressure. Fig. 19 shows this relationship for sodium hydroxide (NaOH) in superheated steam.[15] The solubility in steam of volatile chemicals, such as CO_2 and ammonia (NH_3), generally increases with increasing temperature.

Nonvolatile chemicals that were soluble in the steam at the boiler or steam generator outlet may become insoluble and precipitate as the steam density diminishes as it expands through the turbine. Deposits containing copper and silica have been found in high pressure turbines, and deposits containing sodium hydroxide (NaOH) and sodium chloride (NaCl) have been found in low pressure turbines.

Volatile chemical additives, such as ammonium hydroxide and hydrazine, may minimize the deposition of sodium salts in low pressure turbines. Also, the rapid rate at which the steam expands may limit deposition beyond that indicated by the equilibrium data.

Hideout

Hideout is the accumulation of chemicals on surfaces, in crevices or in deposits within the system during normal, steady-state operation. Traditionally, hideout has been discussed in terms of phosphate deposition from the water

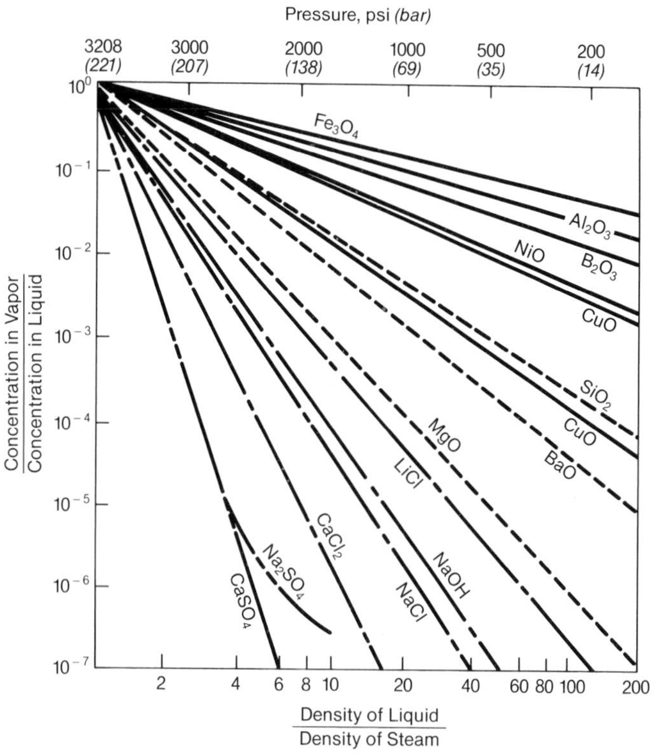

Fig. 18 Impurity carryover coefficients of salts and metal oxides contaminants in boiler water (*adapted from Reference 14*).

in drum boilers or recirculating steam generators (discussed in the *Coordinated and Congruent Phosphate Treatment* and *Equilibrium Phosphate* sections). Hideout return studies are also performed on systems that do not use phosphate treatment, i.e., on systems with high purity feedwater and all volatile treatment (AVT). Because the contaminant concentrations are significantly higher during hideout return, i.e., during load reduction or shutdown, contaminants may be found during this period that were not detectable during steady-state operation. This information can be useful in: 1) determining if deposits are alkaline, neutral or acidic (based on cation/anion ratio), 2) determining if corrosive contaminants are depositing in crevices or under deposits, and 3) locating and correcting water purification or condenser leakage problems. Interpretation of

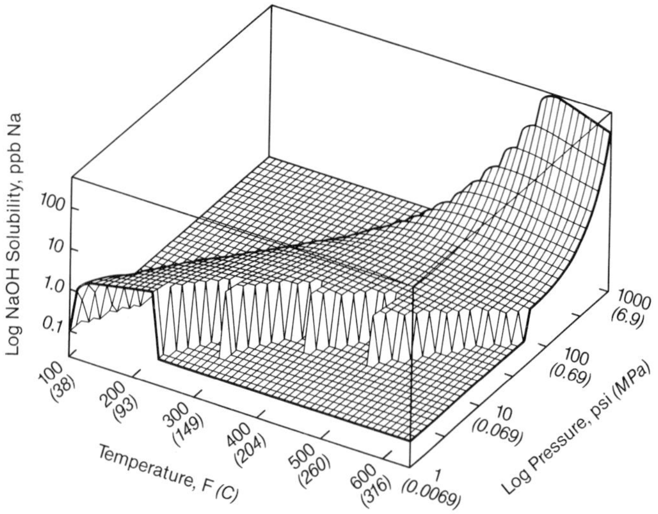

Fig. 19 Sodium hydroxide solubility in superheated steam (*adapted from Reference 15*).

cation/anion ratios may be misleading unless all the ions (including organic anions) are found by analyses.

Resin fragment ingress

Ion exchange resins are used in condensate polishers as well as demineralizers used for makeup water, purification of blowdown, moisture separator drains and feedwater heater drains. Most of the pressure vessels that contain these ion exchange resins have under-drain systems and downstream traps or strainers to prevent leakage of ion exchange resins into the cycle water. This resin can form harmful decomposition products if allowed to enter the high temperature portions of the cycle. Unfortunately, the under-drain system and the traps and strainers are not designed to retain resin fragments that result from resin bead fracture. Also, the resin traps and strainers can fail, resulting in resin bursts.

Resin intrusion can be minimized by controlling flow transients, reducing the strainer's screen size, increasing flow gradually during vessel cut-in, and returning the polisher vessel effluent to the condenser during the first few minutes of cut-in.

Control of organics

Organic contamination can account for the majority of cycle impurities. Organic chemicals can originate from herbicides and pesticides in the makeup water, particulates in the makeup water, lubricating oils, hydraulic fluids, solvents, coatings, plastics and resins. They can also originate from slow oxidative decomposition of the ion exchange resin used in makeup demineralizers and condensate polishers.

The organic chemicals are carbon-based molecules that contribute formic, acetic and organic sulfonic acids as well as CO_2 to the cycle upon decomposition at elevated temperatures. These acids as well as CO_2 can cause undesirable pH suppression, especially in regions of the low pressure turbine where pH additives are not efficient.

Organic molecules can also contain significant amounts of chloride, fluoride and sulfur, which may be released upon decomposition at elevated temperatures and pressures. Most organics can be removed from the makeup water by reverse osmosis (RO) units followed by activated carbon beds and then demineralizers. Some organics are also removed by makeup (ion exchange) demineralizers alone, but the organics may also foul these systems.

Corrosion products and deposits

Products from the corrosion of condensate and feedwater systems are major contributors to boiler and steam generator deposits. These deposits migrate preferentially to the heat transfer surfaces with the highest heat flux according to the following equation:

$$\text{Rate of deposition} \ \alpha \ (\text{heat flux})^n \qquad \textbf{(4)}$$

where α means proportional to and where n is between 1 and 5, depending on the system and other parameters.[16]

Chemical cleaning of fossil fuel equipment

Cleaning frequency

The internal surfaces of boiler water side components (including supply tubes, headers and drums) accumulate deposits even though standard water treatment practices may be followed. These deposits are generally classified as hardness-type scales or soft, porous-type deposits.

Tube samples containing internal water-side deposits should be removed from high heat input zones of the furnace and/or areas where problems have occurred. The deposit weight is first determined by visually selecting a heavily deposited section of the sampled tube. After sectioning the tube (hot and cold sides), the water-formed deposit is removed by scraping from a measured area. The weight of the dry material is reported as weight per unit area: either grams of deposit per square foot of tube surface or mg/cm^2. Procedures for mechanical and chemical methods of deposit removal are provided in ASTM D3483.[10] General guidelines for determining when a boiler should be chemically cleaned are shown in Table 3. The deposit weights given in Table 3 are based on the mechanical scraping method. This removes the porous deposit of external origin and most of the dense, in-site oxide scale. It generates values lower than those obtained from the chemical dissolution method.

Because of the corrosive nature of the fuel and its combustion products, furnace tubes in Kraft recovery and refuse-fired boilers are particularly susceptible to gas-side corrosion which can be aggravated by relatively modest elevated tube metal temperatures. (See Chapters 26 and 27.) Through-wall failures due to external metal corrosion can occur in these tubes at water-side deposit weights much less than 40 g/ft^2 (43 mg/cm^2). In addition, for Kraft recovery boilers there are significant safety concerns for water leakage in the lower furnace. (See Chapters 26 and 43.) For these units, a more conservative water-side condition is recommended for all operating pressures.

Chordal thermocouples

The chordal thermocouple (see Chapter 40) can serve as an effective diagnostic tool for evaluation of deposits on operating boilers. Properly located chordal thermocouples can indicate a tube metal temperature increase caused by internal deposits and can alert the operator to conditions potentially leading to tube failures.

The furnace wall tubes adjacent to the combustion zone are critical boiler components. The heat input is highest in this region and the external tube temperatures are also high. (See Fig. 20.)

Deposition inside tubes can be detected by instrumenting key furnace tubes with chordal thermocouples. These thermocouples compare the surface temperature of the tube exposed to the combustion process with the temperature of saturated water. As deposits grow they insulate the tube from the cooling water and result in tube metal temperature increases.

Beginning with a clean, deposit free boiler, the instrumented tubes are monitored to establish the temperature differential at two or three boiler ratings; this establishes a base curve. At maximum load, with clean tubes, the surface thermocouple typically indicates metal temperatures 25 to 40F (14 to 22C) above saturation in low duty units and 80 to 100F (44 to 56C) in high duty units as shown in Fig. 21. The temperature variation for a typical clean instrumented tube is dependent upon the tube's location in the furnace, tube thickness, inside fluid pressure and the depth of the surface thermocouple. Internal scale buildup is detected by an increase in temperature differential above the base curve. Chemical cleaning should normally be considered if the temperature differential at maximum boiler load reaches 100F (56C).

Initially readings should be taken weekly, preferably using the same equipment and procedure as used for establishing the base curve. Under upset conditions, when deposits are rapidly forming, the checking frequency should be increased.

Table 3
Guidelines for Chemical Cleaning

Unit Operating Pressure, psig (bar)	Water-side Deposit Weight* (g/ft^2)
Below 1000 (69)	20 to 40
1000 to 2000, (69 to 138) including all Kraft recovery and refuse-fired boilers	12 to 20
Above 2000 (138)	10 to 12

* Deposit removed from hot or furnace side of tube using the mechanical scraping method. (1 g/ft^2 = 1.07 mg/cm^2)

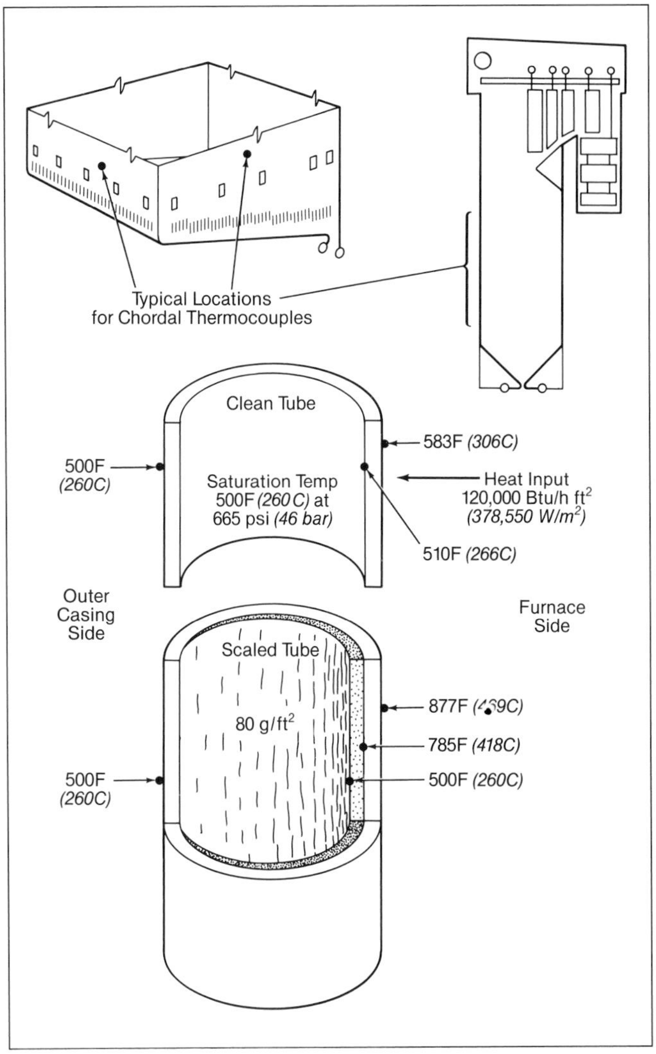

Fig. 20 Typical locations for chordal thermocouples in fossil fuel-fired boiler.

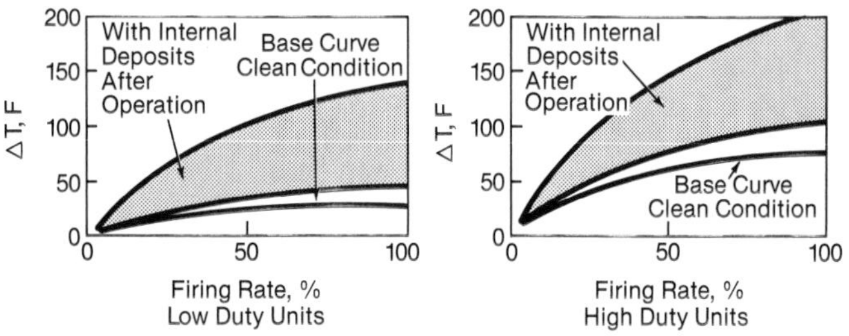

Fig. 21 Temperature difference between surface and saturation thermocouples.

General procedures and methods

In general, four steps are required in a complete chemical cleaning process:

1. The internal heating surfaces are washed with a solvent containing an inhibitor to dissolve or disintegrate the deposits.
2. Clean water is used to flush out loose deposits, solvent adhering to the surface, and soluble iron salts. Corrosive or explosive gases that may have formed in the unit are displaced.
3. The unit is treated to neutralize and *passivate* the heating surfaces. This treatment produces a passive surface, i.e., it forms a very thin protective film on freshly cleaned ferrous surfaces.
4. The unit is flushed with clean water to remove any remaining loose deposits.

The two generally accepted chemical cleaning methods are: 1) continuous circulation of the solvent (Fig. 22), and 2) filling the unit with solvent, allowing it to soak, then flushing the unit (Fig. 23).

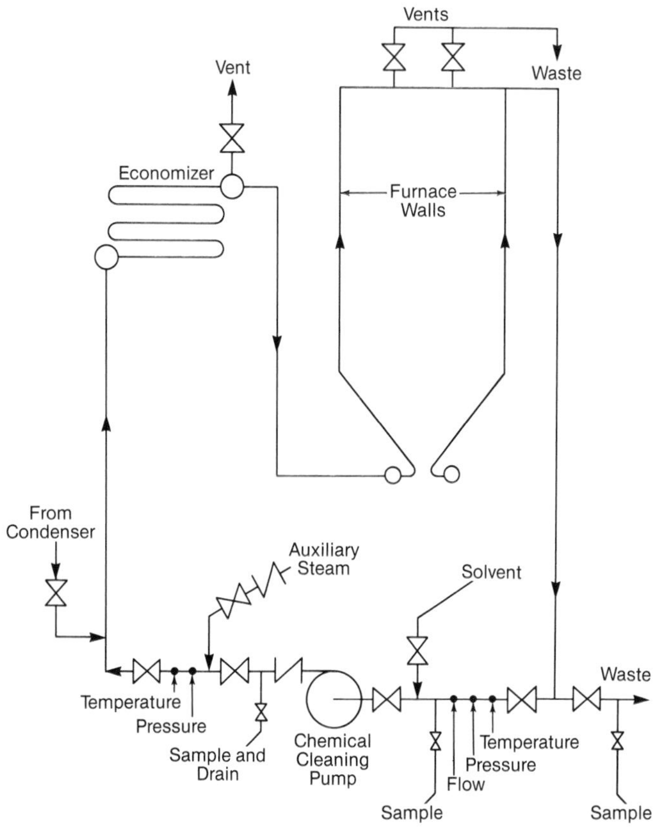

Fig. 22 Chemical cleaning by the circulation method — simplified arrangement of connections for once-through boilers.

Circulation cleaning method

In the circulation method (Fig. 22), after filling the unit, the hot solvent is recirculated until cleaning is completed. Samples of the return solvent are tested periodically during the cleaning. Cleaning is considered complete when the acid strength and the iron content of the returned solvent reach equilibrium (Fig. 24), indicating that no further reaction with the deposits is taking place. In the circulation method, a solvent of lower strength may be used, compared with that of the soaking method, and the solvent is drained from the unit as soon as the reaction is complete. The possibility of damage to the surface cleaned is therefore less in using the circulation method.

The circulation method is particularly suitable for cleaning once-through boilers, superheaters, and economizers with positive liquid flow paths to assure circulation of the solvent through all parts of the unit. Complete cleaning can not be anticipated by this method unless the solvent reaches and passes through every circuit of the unit.

Soaking method

The soaking method (Fig. 23), involves filling the unit with the hot solvent and then allowing the unit to soak for a period of time depending on deposit conditions. To assure complete removal of deposits, the acid strength of the solvent must be somewhat greater than that required by the actual conditions because, unlike the circulation method, control testing during the course of the cleaning is not conclusive, and because samples of solvent drawn from convenient locations may not truly represent conditions in all parts of the unit.

The soaking method is preferable for cleaning units where definite liquid distribution to all circuits by the circulation method is not possible without the use of many chemical inlet connections. The soaking method is also preferred if circulation through all circuits at an appreciable rate can not be assured, except by using a circulating pump of impractical size. These conditions may exist in large natural circulation units that have complex furnace wall cooling systems.

Advantages of the soaking method compared with the circulation method are simplicity of piping connections, assurance that all parts are reached by a solvent of adequate acid strength, and less chemical supervision during the cleaning operation.

Solvents

Many acids and alkaline compounds have been evaluated for removing boiler deposits. Hydrochloric acid (HCl) is the most practical cleaning solvent when using the soaking method on natural circulation boilers. Chelates and other acids have also been used. The chelant based solvents exhibit versatility in that they can be used in both the soak and continuous circulation type cleaning procedures.

An organic acid mixture such as hydroxyacetic-formic is the safest chemical solvent when applying the circulation cleaning method to once-through boilers. These acids decompose into harmless gases in the event of incomplete flushing.

For certain deposits the solvent may require additional reagents, such as ammonium bifluoride, to promote penetration of the deposit. Alloy steel pressure parts, particularly those high in chromium, should generally not be cleaned with acid solvents. Guidelines for solvent selection can be found in References 17, 18 and 19.

Prior to chemically cleaning the unit, it is strongly recommended that a representative tube section be removed and subjected to a laboratory cleaning test to determine the proper solvent chemical and concentrations of that solvent.

Deposits

Scale deposits formed on the internal heating surfaces of a boiler generally come from the water. Most of the constituents belong to one or more of the following groups: iron oxides, metallic copper, carbonates, phosphates, calcium and magnesium sulfates, silica, and silicates. The deposits may also contain various amounts of oil.

Precleaning procedures include analysis of the deposit and tests to determine solvent strength and contact time and temperature. The deposit analyses should include a deposit weight in grams per square foot (or milligrams per square centimeter) and a spectrograph analysis to detect the individual elements. X-ray diffraction identifying the major crystalline constituents is also used.

If the deposit analysis indicates the presence of copper (usually from corrosion in equipment ahead of the boiler), one of three procedures is commonly used: 1) a copper complexing agent is added directly to the acid solvent, 2) a separate cleaning step, featuring a copper solvent, is used followed with an acid solvent, and 3) a chelant based solvent at high temperature is used to remove iron followed by the addition of an oxidizing agent at reduced temperature for copper removal. The decision to use one of these approaches depends on the estimated quantity of copper present in the deposit.

When deposits are dissolved and disintegrated, oil is removed simultaneously, provided it is present only in small amounts. For higher percentages of oil contamination, a wetting agent may be added to the solvent to promote deposit penetration. If the deposit is predominantly oil or grease, boiling out with alkaline compounds must precede the acid cleaning.

Inhibitors

The following equations represent the reactions of hydrochloric acid with constituents of boiler deposits:

$$Fe_3O_4 + 8HCl = 2FeCl_3 + FeCl_2 + 4H_2O \quad (5)$$

$$CaCO_3 + 2HCl = CaCl_2 + H_2O + CO_2 \quad (6)$$

At the same time, however, the acid also reacts with and thins the boiler metal, as represented by the equation:

$$Fe + 2HCl = FeCl_2 + H_2 \quad (7)$$

unless means are provided to slow this reaction without affecting the deposit attack. Today, a number of excellent commercial inhibitors are available to perform this function. The effectiveness of acids in attacking boiler deposits as well as steel increases rapidly with temperature. However, the inhibiting effect decreases as the temperature rises and, at a certain temperature, the inhibitor may decompose. Additionally, all inhibitors are not effective with all acids.

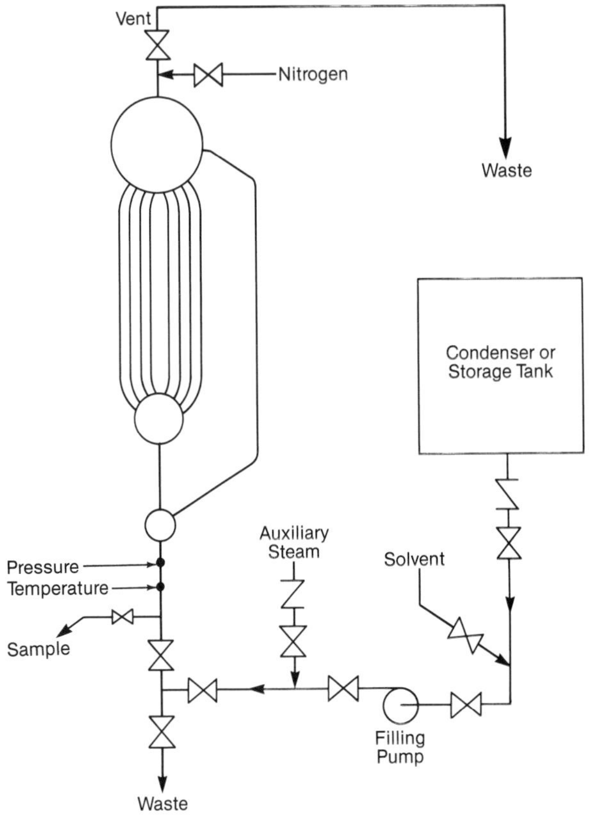

Fig. 23 Chemical cleaning by the soaking method — simplified arrangement of connections for drum-type boilers.

Determination of solvent conditions

Deposit samples The preferred type of deposit sample is a small section of tube with the adhering deposit. In any case, a sample of the deposit representing the worst condition in the unit should be obtained. Selection of the solvent system is made from the deposit analyses. After selection of the solvent system, it is necessary to determine the strength of the solvent, the solvent temperature, and the length of time required for the cleaning process.

Solvent strength The solvent strength should be proportional to the amount of deposit. Commonly used formulations are as follows:

1. Natural circulation boilers (soaking method)
   (a) pre-operational — inhibited 5% hydrochloric acid + 0.25% ammonium bifluoride

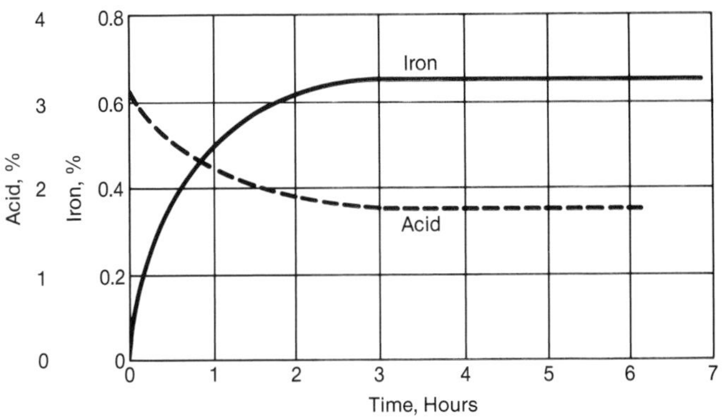

Fig. 24 Solvent conditions during cleaning by the circulation method.

(b) operational — inhibited 5 to 7.5% hydrochloric acid and ammonium bifluoride based on deposit analysis
2 Once-through boilers (circulation method)
(a) pre-operational — inhibited 2% hydroxyacetic-1% formic acids + 0.25% ammonium bifluoride
(b) operational — inhibited 3% hydroxyacetic-1% formic acids + 0.5% ammonium bifluoride

Solvent temperature The temperature of the solvent should be as high as possible without seriously reducing the effectiveness of the inhibitor.

When using hydrochloric acid, commercial inhibitors generally lose their effectiveness above 170F (77C) and corrosion rate increases rapidly. Therefore, the temperature of the solvent, as fed to the unit, should be 160 to 170F (71 to 77C). In using the circulation method with a hydroxyacetic-formic acid mixture, a temperature of 200F (93C) is necessary for adequate cleaning. Chelate-based solvents are generally applied at higher temperatures [about 275F (135C)]. In these cases, the boiler is fired to a specific temperature. The chelate chemicals are introduced and the boiler temperature is cycled by alternately firing and cooling to predetermined limits.

Steam must be supplied from an auxiliary source to heat the acid as it is fed to the unit. When using the circulation method, steam is also used to heat the circulating water to 200F (93C) before injecting the acid solution. Heat should be added by direct contact or closed cycle heat exchangers. The temperature of the solvent should never be raised by firing the unit when using an acid solvent.

Cleaning time When cleaning by the circulation method, process completion is determined by analyzing samples of the return solvent for iron concentration and acid strength. (See Fig. 24.) However, acid circulation for a minimum of four hours is recommended.

In using the soaking method, the cleaning time should be predetermined. A practical guide in determining the solvent retention time for the first few cleanings, using the acid strength previously noted, is as follows:

1. for pre-operational cleaning — four hours minimum,
2. for moderate thickness of relatively soft sludges or for conditions that would otherwise require turbining at not less than yearly intervals — four hours minimum,
3. for thin coatings of relatively hard scale formation — six hours, and
4. for deposits resulting from a severe upset in the feedwater conditioning system — eight hours maximum.

Preparation for cleaning

Heat transfer equipment All parts not to be cleaned should be isolated from the rest of the unit. To exclude acid, appropriate valves should be closed and checked for leaks. Where arrangements permit, parts of the unit such as the superheater can be isolated by filling with demineralized water. Temporary piping should be installed to flush dead legs after cleaning. In addition to filling the superheater with demineralized water, once-through type units should be pressurized with a pump or nitrogen. The pressure should be in excess of the chemical cleaning pump head.

Bronze or brass parts should be removed or temporarily replaced with steel. All valves should be steel or steel alloy. Gauge and meter connections should be closed or removed from the unit.

All parts not otherwise protected by blanking off or by flooding with water should be covered by the inhibited solvent. Vents to a safe discharge should be provided wherever vapors might be pocketed, because acid vapors from the cleaning solution do not retain the inhibitor.

Cleaning equipment The cleaning equipment should be connected as shown in Fig. 22 if the continuous circulation method is used, or as shown in Fig. 23 if the soaking method is used. Continuous circulation requires an inlet connection to assure distribution. It also requires a return line to the chemical cleaning pump from the unit. The soaking method does not require a return line. The pump discharge should be connected to the lowermost unit inlet.

The filling or circulating pump should not be fitted with bronze or brass parts; a standby pump is recommended. A filling pump should have the capacity to deliver a volume of liquid equal to that of the vessel within two hours at 100 psi (6.9 bar). A circulating pump should have sufficient capacity to meet recommended cleaning velocities. With modern once-through boilers, a capacity of 3600 GPM (227 l/s) at 300 psi (20.7 bar) is common. A solvent pump, closed mixing tank and suitable thermometers, pressure gauges, and flow meters are required. An adequate supply of clean water and steam for heating the solvent should be provided. Provision should be made for adding the inhibited solvent to the suction side of the filling or recirculating pump.

Cleaning solutions To determine the amount of solvent required, estimate the content of the vessel and add 10% to allow for losses. Sufficient commercial acid should then be obtained. An inhibitor qualified for use with the solvent also needs to be procured and added to the solvent.

Cleaning procedures

The chemical cleaning of steam generating equipment consists of a series of distinct steps which may include the following:

1. isolation of the system to be cleaned,
2. hydrostatic testing,
3. leak detection during each stage of the process,
4. back flushing of the superheater and forward flushing of the economizer,
5. preheating of the system and temperature control,
6. solvent injection/circulation (if circulation is used),
7. draining and/or displacement of the solvent,
8. neutralization of residual solvent,
9. passivation of cleaned surfaces,
10. flushing and inspection of cleaned surfaces, and
11. layup of the unit.

Every cleaning should be considered as unique and sound engineering judgement should be used throughout the process. The most important design and procedural considerations include reducing system leakage, controlling temperature, maintaining operational flexibility and redundancy, and ensuring personnel safety.

Precautions

Cleaning must not be considered a substitute for proper water treatment. Intervals between cleanings should be extended or reduced as conditions warrant.

Every effort should be used to extend the time between chemical cleanings. Hazards related to chemical cleaning of power plant equipment are fairly well recognized and understood, and appropriate personnel safety steps must be instituted.[17]

Chemical cleaning of superheater, reheater and steam piping

In the past, chemical cleaning of superheaters and reheaters was not performed because it was considered to be unnecessary and expensive. With the use of higher steam temperatures, the development of cleaning procedures for superheaters, reheaters and steam piping has gained acceptance.

When chemically cleaning surfaces that have experienced severe high temperature oxide exfoliation (spalling of hard oxide particles from surfaces), it is important to first remove a tube sample that is representative of the worst condition. Oxidation progresses at about the same rate on the outside of the tubes as on the inside; exfoliation follows a similar pattern.

The tube sample should be tested in a facility capable of producing a flow rate similar to that used in the actual cleaning. This allows development of an appropriate solvent mixture.

To determine the circulating pump size and flows required, it is usually necessary to contact the boiler manufacturer.

Figs. 25 and 26 show possible superheater/reheater chemical cleaning piping schematics for the drum boiler and once-through boiler system, respectively.

If, in the case of a drum boiler, the boiler is to be cleaned along with the superheater and reheater, it is usually necessary to orifice the downcomers to obtain the desired velocities through the furnace walls.

A steam blow to purge all air and to completely fill the system must precede cleaning in all systems containing pendant nondrainable surfaces. Most drainable systems also benefit from such a steam blow.

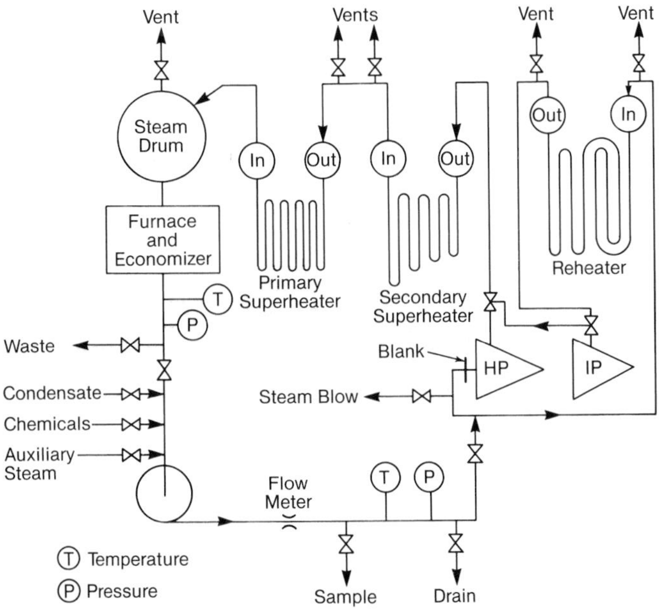

Fig. 25 Typical superheater/reheater chemical cleaning circuit for a drum-type boiler.

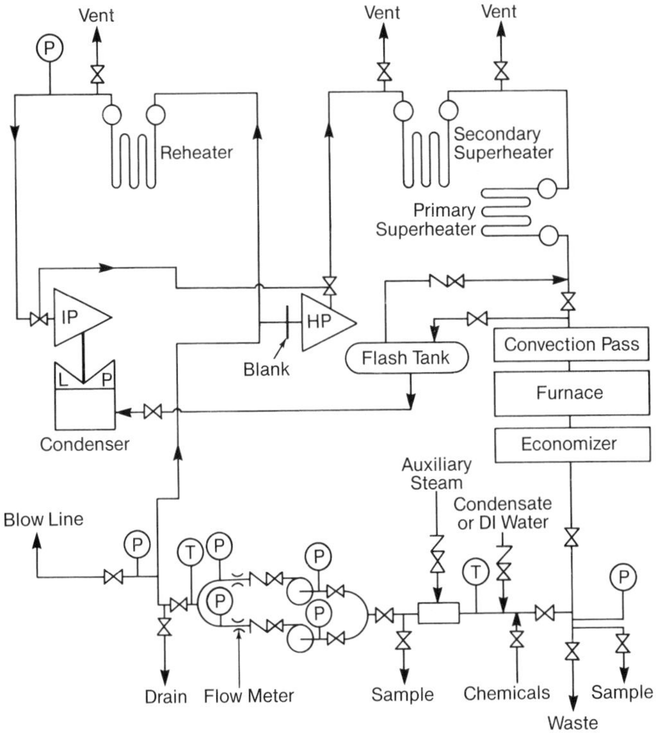

Fig. 26 Typical superheater/reheater chemical cleaning circuit for a once-through boiler.

Presently, two solvent mixtures are available to clean superheater, reheater and steam piping. One is a combination of hydroxyacetic and formic acids containing ammonium bifluoride, and the other is an EDTA (ethylenediaminetetraacetic acid)-based solvent.

Solvent disposal

General considerations A boiler chemical cleaning is not complete until the resultant process waste water stream is disposed of. Selection of handling and disposal methods depends on whether the wastes are classified as hazardous or nonhazardous. Boiler chemical cleaning wastes (BCCW) are different in volume and frequency of generation and have different discharge regulations from other power plant waste streams. Of all power plant discharges BCCWs are most likely to be classified as hazardous. Depending upon the cleaning process, the resultant BCCW may become the driving force in solvent selection.

Under National Pollutant Discharge Elimination System (NPDES) requirements, boiler cleaning wastes are considered chemical metal cleaning wastes. The primary parameters of concern are iron, copper, chromium and pH. In all cases, waste management must be performed in accordance with current regulatory requirements.

Waste management options Table 4 lists the handling practices for BCCW. In *co-ponding*, the BCCW is mixed in an ash pond with other waste streams from the power plant. Acid wastes are neutralized by the alkaline ash, and the metals are precipitated as insoluble metal oxides and hydroxides or absorbed on ash particles. Co-ponding is the least expensive and the easiest disposal option.

Incineration of organic based cleaning wastes by direct injection into the firebox of the utility boiler is another common disposal practice. Potential emissions from the boiler must be carefully monitored to ensure regulatory compliance.

Table 4
Boiler Chemical Cleaning Wastes Practices/Options

Source Reduction
Optimize cleaning frequency
Reduce volume of cleaning solution
Improve boiler water chemistry

Alternate Solutions
Change the cleaning solvent

Disposal
Evaporation
Incineration
Co-ponding
Secured landfill

Treatment
Neutralization
Physical waste treatment
Chemical waste treatment

Recycle and Reuse
Recycle for metal recovery
Reuse acid in alternate applications

Large quantities of BCCW are often disposed in a secure landfill. Evaporation can reduce waste volume and, therefore, reduce overall landfill disposal costs.

HCl cleaning wastes can be treated to NPDES standards using lime or caustic precipitation. It is more difficult to treat the organic cleaning agents (such as EDTA) by current techniques. Treatment methods with permanganate, ultraviolet light, and hydrogen peroxide (wet oxidation) have been used with limited success. Several vendors have proprietary processes which claim to successfully treat chelated wastes.[20,21]

Removing metal ions and reusing chemical cleaning waste are areas receiving increased attention. As the regulatory environment continues to change, more emphasis will be placed on the treatment and reuse of BCCW.

Corrosion in fossil and industrial boilers

Corrosion takes many forms in water and steam systems, and water treatments are specifically designed to prevent corrosion in boilers and steam generating equipment. Most metals form an oxide or hydroxide corrosion film when exposed to water. Under most boiler operating conditions, an oxide layer that acts as a thin coating to protect the metal from further corrosion is formed. This oxide layer must be formed and maintained to avoid most types of corrosion.

Uniform corrosion

Uniform attack is the most common form of boiler corrosion because it reduces the metal thickness by converting metal into an oxide. This oxide layer can be protective or nonprotective. When the layer is protective, as is a layer of magnetite (Fe_3O_4), the corrosion rate typically decreases with time and eventually approaches a low constant rate. If nonprotective scales develop, corrosion rates are significantly higher and are typically linear with time. Boiler water treatments are designed to stabilize the protective oxide films so corrosion decreases with time. Because the metal losses associated with protective oxide films are uniform and occur at a predictable rate, a corrosion allowance can be included in the system design.

Pitting attack

Pitting occurs when the protective oxide film breaks down locally. A schematic of pitting attack compared to uniform corrosion is shown in Fig. 27. Once the protective film is disrupted, an active corrosion site (anode) is created, and it corrodes rapidly. The oxide or metal directly adjacent to the pit is a reduction site (cathode) for the corrosion reaction, and it is not corroded. Numerous pits typically form over the exposed material surface. The distribution, size and number of pits vary with local water conditions as well as material condition.

Galvanic attack

When two dissimilar metals are joined and exposed to corrosive fluids, a galvanic cell is formed. A schematic of galvanic attack is shown in Fig. 28. The galvanic cell creates an electrical potential between the two dissimilar metals. Electric current flows through the metal as a result of the electric potential; this causes accelerated corrosion of one of the metals and protection of the other. The degree to which corrosion is accelerated depends upon a number of factors including the difference in electric potential, the surface area of each metal and the conductivity of the corrosive media. This type of attack is not common in boiler systems.

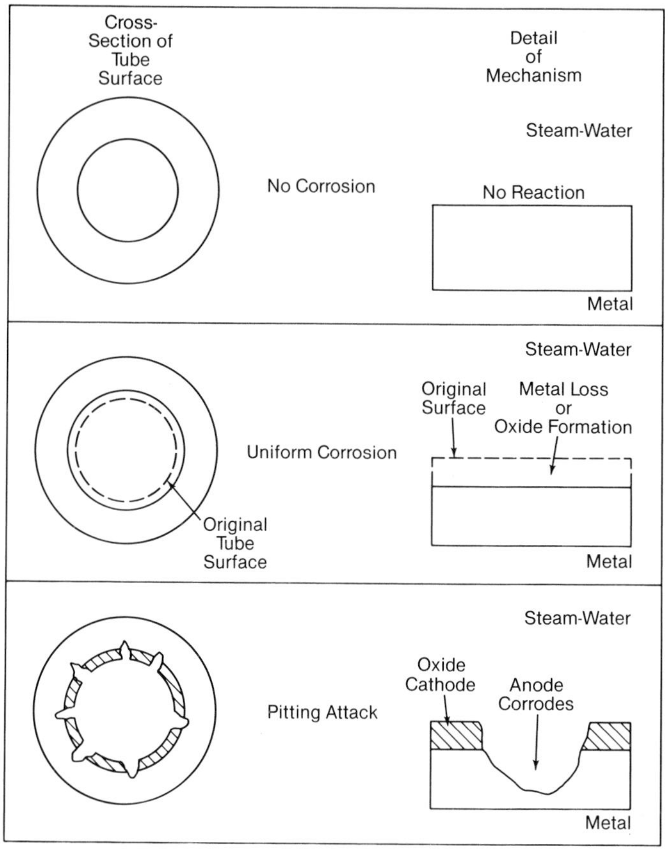

Fig. 27 Schematic showing the differences between pitting attack and uniform corrosion.

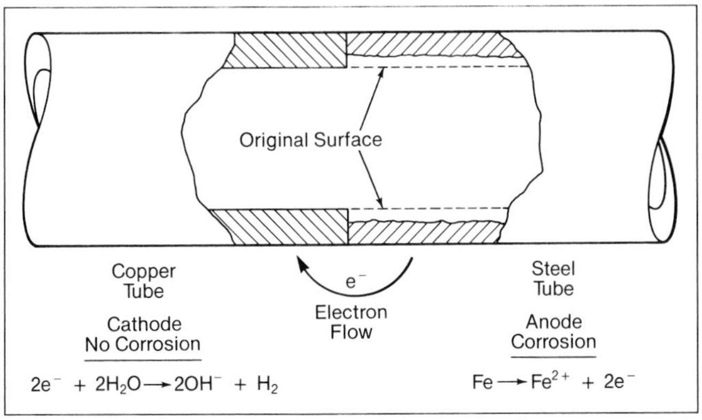

Fig. 28 Schematic of galvanic corrosion occurring between two dissimilar metals.

Hydrogen damage

When corrosion occurs in aqueous systems hydrogen, which is generated as a byproduct, can diffuse into the tube metal. In carbon steels the hydrogen may react with carbon to form methane gas. This reaction causes the metal to be locally decarburized, because the carbon in the steel is being consumed. This decarburization causes the steel to become weakened. The generated methane gas typically collects at grain boundaries and discontinuities within the metal; as the gas pressure builds, small fissures are formed. Eventually these fissures can link and cause a through-wall tube failure. A schematic of hydrogen attack is shown in Fig. 29.

Hydrogen damage is more likely to occur when corrosion is rapid. Tubes have been known to fail within hours from the onset of corrosion. The failures are usually, though not necessarily, associated with a heavy scale on the tube surface. Often, heavy scale is blown off during the failure. The failures are typically described as *window openings* because a small rectangular section is often blown out of the tubing. The prevention of scale on the water side of the tubing as well as tight control of the water chemistry can prevent hydrogen attack.

Stress corrosion cracking

Stress corrosion cracking (SCC) occurs below the material yield strength and is the result of environmental influences. Fig. 30 shows the process of stress corrosion cracking and some typical features. This type of cracking occurs only under specific combinations of materials, environments and stress levels. Stress cracking is time dependent; it may proceed rapidly or many years of trouble free service may be realized before the crack extends to failure. Cracks may be intergranular or transgranular. Stress corrosion cracks are typically branched. Usually there are numerous small secondary cracks associated with the main fracture area. These cracks are mainly seen in austenitic stainless steels.

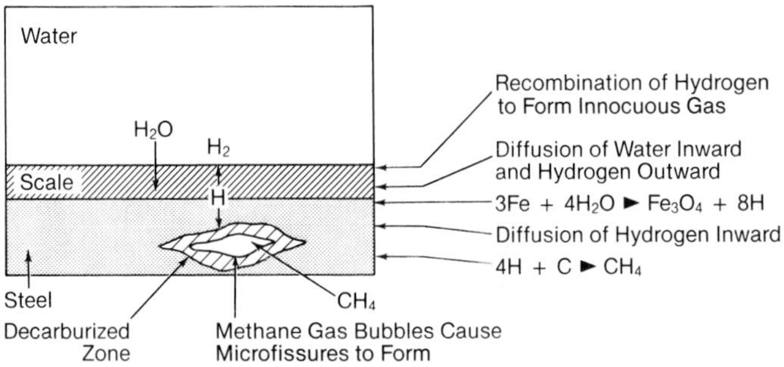

Fig. 29 Schematic of hydrogen attack, showing steps that occur and the final result. Hydrogen attack can occur in both carbon and low alloy steels.

Caustic attack

Caustic attack, also known as grooving or gouging, occurs in steam generating equipment on surfaces with moderate to high heat flux under two distinctly different conditions: 1) vertical tubes where deposits are present on the tube surfaces, and 2) inclined tubes where the heat flux is directed through the upper half of the tube. In vertical tubes, temperature gradients exist through the boiling fluid film and the deposit on the heat transfer surface. The combined fluid film and deposit locally increase the water temperature adjacent to the tube and result in increased boiling within the deposit on the tube surface. This boiling is often described as *wick boiling* (see Fig. 31 from Reference 22) and causes dissolved salts, such as sodium hydroxide (NaOH), to concentrate within the deposit. As a result, local areas of high pH solutions can be formed causing accelerated corrosion. In inclined tubes, steam generated tends to accumulate on the upper surface of the tube whether or not deposits are present. This can potentially result in a steam blanket covering the top surface of the tube in the direction of flow. The temperature gradient through the steam causes the top portion of the tube surface to highly concentrate soluble salts in the boiler water. Because the tubes involved are generally not subject to high heat fluxes, the temperature gradient does not raise tube metal temperatures sufficiently high to cause overheating. Again, local areas of high pH solutions can be formed and cause accelerated corrosion by continuously dissolving and removing the protective magnetite layer. Characteristically, the resulting corrosion is in the form of a wide smooth groove with the groove generally free of deposits and centered on the crown of the tube. The corroded area can sometimes extend for 10 ft (3 m) or more

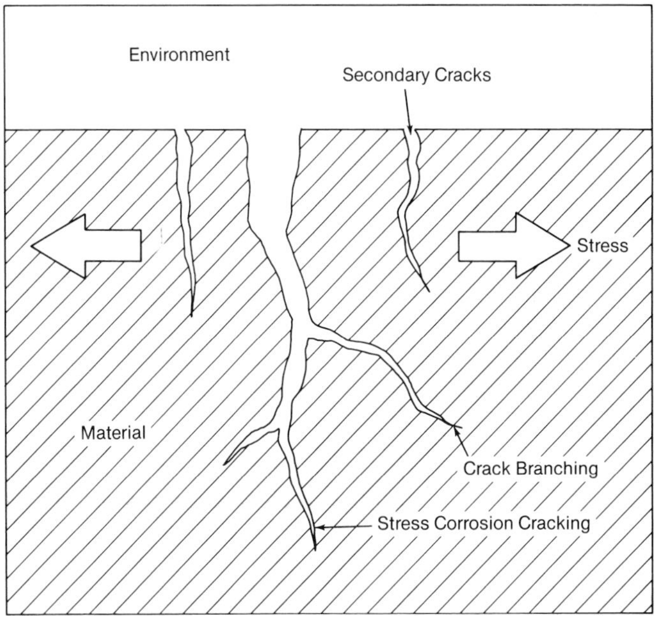

Fig. 30 Schematic of stress corrosion cracking.

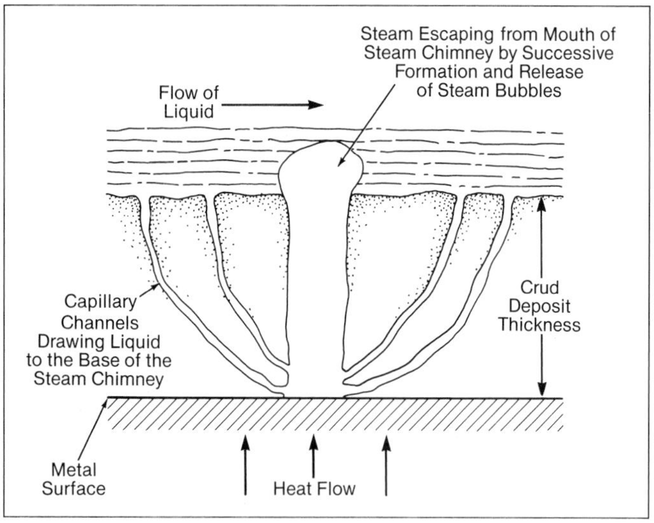

Fig. 31 Schematic of the wick boiling mechanism (*adapted from Reference 22*).

along the length of the tube for as long as the basic physical conditions causing it persist.

Acid attack

Acid attack can occur similarly to the caustic attack in that impurities in the bulk water can become concentrated at the heat transfer surface within deposits. Certain water treatments and impurities can allow a low pH solution to develop under scales during the boiling process. The local areas of low pH solutions under the scale are also prone to rapid corrosion, as shown in Fig. 32.[23] There also appears to be a means of concentrating sodium phosphate within deposits of magnetite on high heat flux surfaces, at least where boiler pressures are in excess of 2600 psig (179 bar gauge) (see References 8 and 9). Under these conditions and where sufficient bulk water phosphate concentrations exist to produce 0.02 molar phosphate concentrations within the boiling film on the deposit surface, the magnetite adsorbs substantial quantities of sodium phosphate in a molar ratio of 2 to 1 (Na:P) or lower.[9] At sufficiently high deposit temperatures and loading of phosphate in the deposit, reactions apparently can take place within the deposit to produce relatively insoluble sodium iron phosphates and rapid corrosion of the underlying metal surface.

Corrosion fatigue

Corrosion fatigue is similar to stress corrosion cracking except that the loads are alternating instead of steady-state. Fatigue in noncorrosive environments can occur at fluctuating loads well below the material yield strength. The addition of a corrosive component speeds the failure.

Corrosion fatigue is the result of a certain combination of stress history, environment and materials. It is time dependent, with a certain number of load cycles needed to initiate cracks. Corrosion fatigue may also occur as the result of thermal cycling in a corrosive environment. In boiler tubing, corrosion fatigue is most common in water-wetted surfaces where there is a mechanical constraint on the tubing. Corrosion fatigue can be found in waterwall tubes adjacent to the windbox floor, buckstay or other welded attachments.

Intergranular corrosion

Intergranular corrosion is accelerated attack along metal grain boundaries. Under most conditions, corrosion at grain boundaries is not a major concern; however, certain heat treating and welding processes can promote this type of corrosion. A classic example is the formation of chromium carbides at the grain boundaries of stainless steels and some nickel base alloys. As a carbide, chromium becomes unavailable to form protective chromium oxide scales; this results in rapid attack along the grain boundaries.

Microbial induced corrosion

Organisms in the water can contribute to corrosion; this is known as *microbial induced corrosion* (MIC). Bacteria that reduce sulfates can lead to sulfidization of a material, with a corresponding increase in corrosion rate. An increasing amount of corrosion is now recognized as being related to biological activity. This type of attack is generally limited to exposed low temperature water such as that used in cooling towers [<200F (93C)].

Erosion-corrosion

When protective scales are removed by flowing water or particles, the corrosion rates increase. There is usually a pattern of corrosion associated with the flow of fluids or particles. The angle of impingement, velocity, nature of fluids or particles, and other factors determine the severity of erosion-corrosion. *Cavitation* is a special form of this corrosion that occurs in fluids. Cavitation causes protective surface scales to be removed by the formation and implosion of gas bubbles in a fluid. Cavitation occurs in flowing fluid systems when the pressure drop causes the local fluid pressure to fall below its vapor pressure at the local temperature. In water systems, erosion-corrosion appears in the form of gullies, grooves or pits, usually with some directional orientation. This type of attack occurs at restrictions and bends in tube circuits where fluid velocity changes.

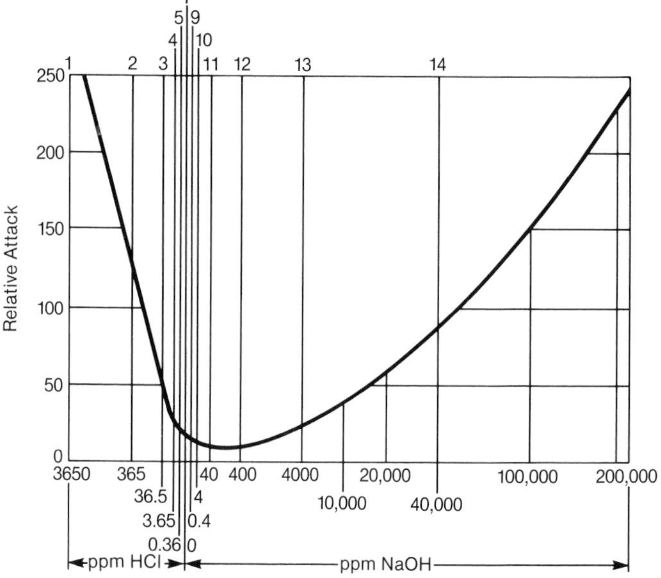

Fig. 32 Relative corrosion attack as a function of caustic (NaOH) or acid HCl) concentration (*adapted from Reference 23*).

Summary of corrosion mechanisms in boilers

Water and steam react with most metals to form oxides or hydroxides. Two example reactions for steel in water and steam systems are:

$$Fe + 2H_2O \rightarrow Fe(OH)_2 + H_2 \quad (8)$$

$$3Fe + 4H_2O \rightarrow Fe_3O_4 + 4H_2 \quad (9)$$

Reactions between steel and water can continue to consume metal at a fairly constant rate as in the first reaction. Conversely, a protective oxide layer can form on the metal surface. This causes reaction rates to slow with time, as is the case with magnetite (Fe_3O_4) formation, described by the second reaction. Most steel surfaces in steam generating equipment are covered with a protective oxide scale due to the tight control of water chemistry. Problems occur when the magnetite oxide scale is perturbed by: upsets in water chemistry, the formation of concentrated boiler chemicals within porous deposits, mechanical stresses or erosion. Various water-side corrosion mechanisms are summarized in Table 5 and typical locations are identified in Figs. 33 and 34. Further discussion of corrosion and failure mechanisms is provided in References 24, 25 and 26.

Table 5
Summary of Water-Side Corrosion in Utility and Industrial Boilers

Type of Corrosion	Examples in Boilers	Remedial Action
Uniform	Oxidation in steam	None required
	Oxidation in water throughout entire boiler	
Pitting	Oxygen pitting of economizer	Lower oxygen in feedwater
		Auxiliary deaerated during startup
	Oxygen pitting of undrained tubes during outages	Use nitrogen blanket, fill with treated water, drain while hot
Hydrogen damage	Waterwall tube window blowouts	Control water chemistry to prevent low pH, limit Cl, etc., remove tube deposits
Stress cracking	Stainless steel superheater/reheater tubes	Avoid water carryover, control hydrotest methods, flush after cleaning
	Ferritic steam tubes	Control attemperator water
Caustic attack	Grooving or gouging of waterwall tubes under deposits fouled surfaces	Chemically clean the boiler, control water chemistry, avoid steam blanketing, reduce hideout, monitor H_2
Acid attack	Waterwall tubes	Control water chemistry, watch for chemcal hideout, chemically clean the boiler, monitor H_2
Corrosion fatigue	Cracking of waterwall tubes near welded attachments	Minimize number of starts/stops, minimize constraints on tubes, lower dissolved oxygen on starts
Erosion-corrosion	Economizer headers and tube inlets	Use resistant materials, minimize flow velocity

Sampling

A key element in chemistry control of steam generators is the effective sampling and transport of boiler water, saturated steam, steam-water mixture and superheated steam. It is critical to:

1. strive to obtain a representative sample,
2. prevent contamination of the sample by sample lines or deposition of sample impurities in the sample lines, and
3. properly store and transport the sample prior to analysis to minimize changes in the sample.

References 27 and 28 provide the most current detailed procedures. In general:

1. Sample lines should be as short as possible and made of stainless steel except where conditions dictate otherwise.
2. Samples should be obtained from continuous blowdown lines where possible following a steam blow.
3. Dilution should be avoided.
4. Samples should be taken during normal steady-state boiler operation.
5. The time between sampling and analysis should be as short as possible.
6. Samples should be cooled to 100F (38C) quickly to avoid loss of contaminants except where high temperature tests are required.
7. Sample nozzles and lines should provide for isokinetic sample velocity and maintain constant high water velocities [minimum of 6 ft/s (1.8 m/s)] to avoid loss of materials.
8. Sample points should be at least 10 diameters downstream of the last bend or disturbance.

Typical errors arise from:

1. deposition in the sample lines,
2. corrosion in the sample lines,
3. air in-leakage,
4. unrepresentative sample acquisition, and
5. errors in the sample collection points.

Sampling of boiler tube deposits occurs through the removal of tube sections, or windows. Particular care must be taken in handling, packaging, labeling and shipping. As discussed earlier, such deposit samples are primarily used as a guide to chemical cleaning frequency and cleaning procedures, as well as in the investigation of the failure mechanisms for boiler tubes. Detailed procedures are provided in Reference 27, ASTM D 887. Advanced techniques such as B&W's FURNIS™ ultrasonic based system can nondestructively determine the relative deposit growth.

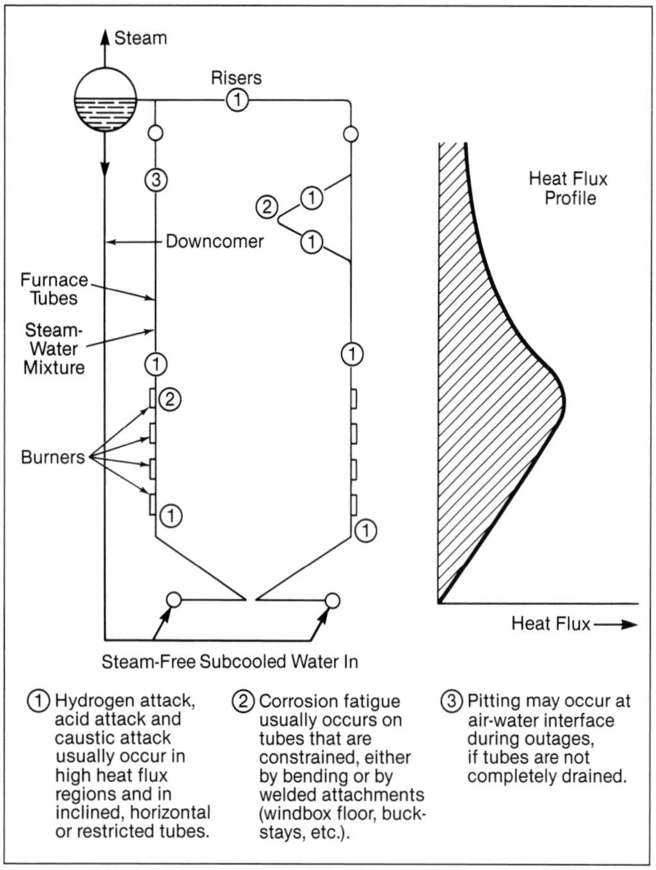

Fig. 33 Typical locations of various types of water-side corrosion in a boiler furnace water circuit.

Measurement techniques

Various measurement techniques can be used to analyze water, water formed deposits and corrosion products. Manual (grab) and on-line measurements are used to control water treatment, determine contaminants, investigate failure mechanisms and aid in chemical cleaning.

B&W has published complete water analysis guides for fossil fuel-fired systems.[29] These guides include discussions of sampling and water treatment control curves.

Boiler water measurement techniques

Manual (grab) samples Guidelines and techniques for measurements in power plant systems are listed in Table 6. The detailed methods are readily available from the American Society for Testing and Materials (ASTM) in Philadelphia, Pennsylvania and the American Society of Mechanical Engineers (ASME) in New York, New York.

The time between sample collection and analysis should be short to minimize absorption of atmospheric gases. Also, contaminant leaching can occur from the walls of sample containers. The measurements listed in Table 6 are for water additives, makeup water contaminants and corrosion products.

Two very useful techniques are atomic absorption and ion chromatography. Atomic absorption (AA) is a means for measuring elemental concentration in sample solutions. This technique involves aspirating the solution into a flame where the elements are atomized. While being atomized, each element is capable of absorbing light at a specific wavelength. Elements are tested one at a time by passing a monochromatic light through the sample. The absorbance of light by the sample is measured and compared to absorbencies of known standards.

Ion chromatography (IC) measures the amount of ions contained in a sample solution. This technique involves: 1) isolating each ion by passing the solution through a polarized column, 2) measuring the peak area generated by each ion as the solution travels through a conductance detector, and 3) comparing this peak area to those generated by known standards. With IC, different types of columns are used to separate different ions; cations and anions can not be analyzed simultaneously. The detection limits are generally below 0.02 ppm for each ion and, in many cases, are close to 0.001 ppm (1 ppb).

Ion chromatography is a multi-ion technique (several ions are determined in a single analysis), requires low volume samples (usually about 20 ml), and is fairly rapid (about 30 minutes). It is used mainly for determining anions, but it can also measure cations, e.g., sodium and ammonium ions. It is also used to determine certain organic acids (formic, acetic, propionic, benzenesulfonic) that are present in cycles. Organic acids can result from decomposition of makeup water organic contaminants, decomposition of certain organic additions, e.g., morpholine, or decomposition of cation exchange resin fragments, e.g., benzenesulfonate.

On-line monitoring and data acquisition On-line monitoring offers many advantages over grab samples. It gives real time data, enables trends to be followed and provides historical data. However, on-line monitors require calibration, maintenance and checks with grab samples or on-line synthesized standard samples to ensure reliability. Table 7 lists important on-line monitoring measurements and ref-

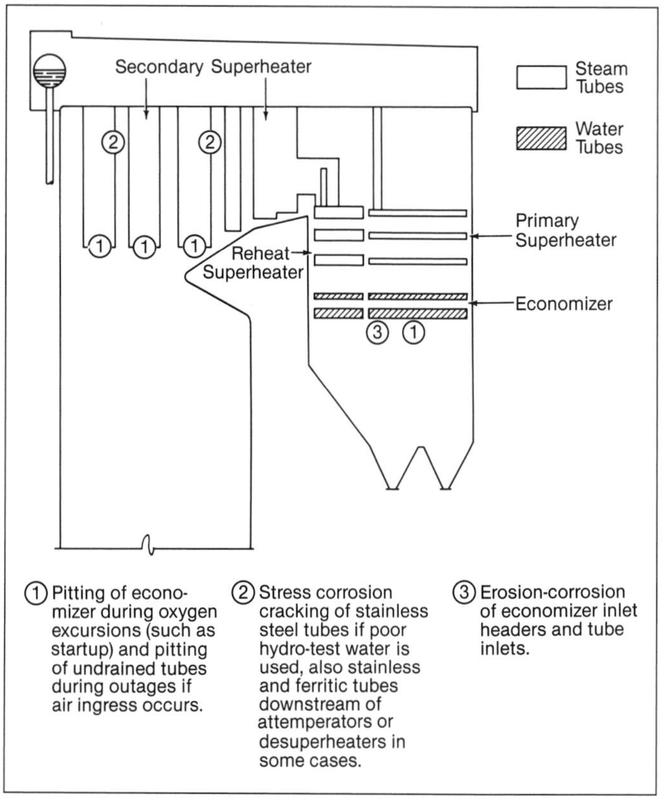

Fig. 34 Boiler convection pass showing typical locations of various types of water-side corrosion.

Table 6
Guidelines for Measurements on Grab Samples

Measurement	Technique(s)	Reference/ Comment (Notes 1 and 2)
pH	Electrometric	ASTM D 1293 Method A
Conductivity	Dip or flow type conductivity cells energized with alternating current at a constant frequency (Wheatstone Bridge)	ASTM D 1125 Methods A or B
Dissolved oxygen	Colorimetric or titrimetric	ASTM D 888 Methods A, B, C
Suspended iron oxides	Membrane comparison charts	ASME PTC 31 Ion exchange equipment
Iron	Photometric (bathophenanthroline) or atomic absorption (graphite furnace)	ASTM D 1068 Method C, D
Copper	Atomic absorption (graphite furnace)	ASTM D 1688 Method C
Sodium	Atomic absorption or flame photometry	ASTM D 4191 or ASTM D 1428
Silica	Colorimetric or atomic absorption	ASTM D 859 or ASTM D 4517
Phosphate	Ion chromatography or photometric	ASTM D 4327 or ASTM D 515 Method A
Ammonia	Colorimetric (nesslerization) or selective ion electrode	ASTM D 1426 Method A or B
Hydrazine	Colorimetric	ASTM D 1385
Chloride	Colorimetric or ion chromatography	ASTM D 512 or ASTM D 4327
Sulfate	Turbidimetric or ion chromatography	ASTM D 516 or ASTM D 4327
Calcium and magnesium	Atomic absorption; gravimetric or titrimetric	ASTM D 511 or ASTM D 1126
Chromium	Photometric or atomic absorption	ASTM D 1687
Fluoride	Selective ion electrode ion chromatography	ASTM D 1179 or ASTM D 4327
Lead	Atomic absorption	ASTM D 3559 Method D
Nickel	Atomic absorption	ASTM D 1886 Method C
Zinc	Atomic absorption	ASTM D 1691
Morpholine	Colorimetric	ASTM D 1942
Hydroxide ion in water	Titrimetric	ASTM D 514
Total organic carbon	Instrumental (oxidation and infrared detection)	ASTM D 4779

Notes:
1. ASME PTC refers to Performance Test Codes of the American Society of Mechanical Engineers, New York, New York.
2. ASTM refers to testing procedures of the American Society for Testing and Materials, Philadelphia, Pennsylvania.

erences to specific methods. In addition to the measurements listed, instrumentation is commercially available to monitor chloride, dissolved oxygen, dissolved hydrogen, silica, phosphate, ammonia and hydrazine. In addition, ion chromatography has been used on-line to measure anions, cations and organic acids.

Increases in the hydrogen content of steam or water may indicate accelerated component corrosion. On-line monitoring of hydrogen has been used to investigate a variety of corrosion mechanisms. On-line instruments based on the principles of thermal conductivity, fuel cell and polarography have been used along with gas chromatographic analysis of grab samples.

Other measurement techniques

There are several other very useful techniques used for the analysis of water, deposits and related materials. The instruments required for these analyses are sophisticated and expensive and require trained operators. Analysis of a single sample may also be expensive. These instruments are not usually found in power stations, but they may be part of well equipped central laboratories such as those of B&W. They may be used more for diagnostic and failure analyses than for routine measurements. Many techniques and instruments are discussed in detail in Reference 30.

Waste water treatment

Industrial plants

Waste water generated by an industrial facility can differ greatly from that of a fossil fuel-fired electric generating facility. The complexity of the generated waste stream depends on the size of the manufacturing facility and the products.

Smaller industrial plants normally generate several waste water streams, including ion exchange regenerate waste, boiler blowdown, cooling system blowdown, process waste water streams and municipal wastes. Most small plants can either send the entire waste water stream to the sanitary sewer or segregate the waste stream. When segregated, the municipal waste goes to the sanitary sewer and industrial waste water is dis-

Table 7
On-line Monitoring Measurements

Measurement	Technique(s)	Reference
pH	Electrometric	ASTM D 5128
Conductivities (general, cation and degassed)	Electrical conductivity measurement before and after hydrogen cation exchanges and at atmospheric boiling water after acidic gas removal	ASTM D 4519
Sodium	Selective ion electrode flame photometry	ASTM D 2791
Total organic carbon	Instrumental (oxidation and measurement of carbon dioxide)	ASTM D 5173

charged to a local waterway after appropriate monitoring and treatment.

Plants that produce large volume waste streams normally have their own treatment facilities. Treatment can be a simple clarification process for suspended solids removal or an integrated treatment process including neutralization, filtrations, oil removal and biological degradation.

Utility plants

All power plants produce waste water streams containing chemicals in various concentrations. Certain waste water streams are generated continuously while others are produced intermittently. Continuous waste water streams arise from ion exchange regeneration, boiler blowdown, cooling tower recirculating system blowdown, once-through cooling water system effluent, chlorination and ash handling sluice water. Examples of intermittent waste water streams include boiler and condenser chemical cleaning solvents and boiler fire side wash water. (See also Chapter 32.)

The common treatment of waste waters in coal-fired plants is the use of ash settling ponds for those plants with a wet ash handling system. Ash is either periodically or continuously sluiced to the pond to allow settling of the suspended matter prior to discharge.

Ion exchange regeneration waste water streams are normally mixed in retaining sumps and are neutralized prior to discharge to the ash pond. The neutralization of regeneration waste streams results in large quantities of precipitating solids. These solids are removed in the ash pond. Ash pond overflow is normally sufficiently low in suspended solids and can be discharged to a local waterway. Settled solids in the pond are periodically removed and shipped to landfills.

References

1. Whirl, S.F., and Purcell, T.E., "Protection against caustic embrittlement by coordinated pH control," Third Annual Meeting of the Water Conference of the Engineers' Society of Western Pennsylvania, Pittsburgh, 1942.

2. Klein, H.A., and Rice, J.K., "A research study on internal corrosion in high pressure boilers," *Journal of Engineering for Power*, Vol. 88, No. 3, pp. 232-242, July 1966.

3. Freier, R.K., "Cover layer formation on steel by oxygen in neutral salt free water," *VGB Speiserwassertagung 1969 Sunderheft*, pp. 11-17 (in German), 1969.

4. Stodola, J., "Review of boiler water alkalinity control," Proceedings of the 47th Annual Meeting of The International Water Conference, Pittsburgh, pp. 235-242, October 27-29, 1986.

5. Cohen, P., ed., *The ASME Handbook on Water Technology in Thermal Power Systems*, The American Society of Mechanical Engineers, New York, 1989.

6. *Consensus on Operating Practices for the Control of Feedwater and Boiler Water Quality in Modern Industrial Boilers*, The American Society of Mechanical Engineers, New York, 1979.

7. *Consensus of Current Practices for Layup of Industrial and Utility Boilers*, The American Society of Mechanical Engineers, New York, 1985.

8. Economy, G., *et al.*, "Sodium phosphate solutions at boiler conditions: solubility, phase equilibrium and interactions with magnetite," Proceedings of the International Water Technology Conference, Pittsburgh, pp. 161-173, 1975.

9. Tremaine, P., *et al.*, "Interactions of sodium phosphate salts with tansitional metal oxides at 360C," Proceedings of the International Conference on Interaction of Iron Based Materials with Water and Steam, Heidelberg, Germany, June 3-5, 1992.

10. *Annual Book of ASTM Standards*, Section 11, Water and Environmental Technology, Vols. 11.01 and 11.02, American Society for Testing and Materials, Philadelphia, 1992.

11. "Interim consensus guidelines on fossil plant cycle chemistry," Report CS-4629, Electric Power Research Institute, Palo Alto, California, June 1986.

12. "VGB guidelines for boiler feedwater, boiler water and steam for water tube boilers with a pressure of 64 bar and higher," *VGB Kraftwerkstechnik*, Vol. 60, No. 10, October 1980.

13. Rice, J., "Quality assurance for continuous cycle chemistry monitoring," Proceedings of the International Conference on Fossil Plant Cycle Chemistry, Report TR-100195, Electric Power Research Institute, Palo Alto, California, December 1991.

14. Martynova, O.I., "Transport and concentration processes of steam and water impurities in steam generating systems," *Water and Steam: Their Properties and Current Industrial Applications*, J. Staub and K. Scheffler, eds., Pergamon Press, Oxford, U.K., pp. 547-562, 1980.

15. Allmon, W.E., "Deposition of corrosive salts from steam,"Report NP-3002, Electric Power Research Institute, Palo Alto, California, April 1983.

16. Cohen, P., "Chemical thermodynamics of steam generating systems," ASME/AIChE National Heat Transfer Conference, Salt Lake City, Utah, 1977.

17. Wackenhuth, E.C., *et al.*, "Manual on chemical cleaning of fossil-fuel steam generating equipment," Report CS-3289, Electric Power Research Institute, Palo Alto, California, 1984.

18. *Industrial Cleaning Manual*, NACE Technical Publication TPC 8, National Association of Corrosion Engineers, Houston, 1982.

19. "Corrosion testing of chemical cleaning solvents," *Material Performance*, Vol. 21, No. 12, p. 48, December 1982.

20. Samuelson, M.L., McConnell, S.B., and Hoy, E.F., "An onsite chemical treatment for removing iron and copper from chelant cleaning wastes," Proceedings of the 49th International Water Conference, Pittsburgh, p. 380, 1988.

21. "Nalmet heavy metal removal program," Nalco Chemical Company, Naperville, Illinois, February 1989.

22. Macbeth, R.V., *et al.*, UKAEA Report No. AAEW-R711, Winfrith, Dorchester, U.K., 1971.

23. Berl, E., and vanTaack, F., "Forschungsarbeiten aus Gebiete des Ingenieurwesens," Heft 330, Berlin, 1930.

24. Uhlig, H.H., *Corrosion and Corrosion Control*, Wiley, New York, 1985.

25. Lamping, G.A., and Arrowood, Jr., R.M., *Manual for Investigation and Correction of Boiler Tube Failures*, Report CS-3945, Electric Power Research Institute, Palo Alto, California, 1985.

26. French, D.N., *Metallurgical Failures in Fossil Fired Boilers*, Wiley, New York, 1983.

27. *Annual Book of ASTM Standards*, Section 11, Water and Environmental Technology, Vol. 11.01 (D 1192 Specification for Equipment for Sampling Water and Steam; D 1066 Practice for Sampling Steam; D 3370 Practice for Sampling Water; and D 4453 Handling of Ultra-pure Water Samples), American Society for Testing and Materials, Philadelphia, 1992.

28. "Water and steam in the power cycle — purity and quality, leak detection, and measurement," ASME Performance Test Code 19.11, The American Society of Mechanical Engineers, New York, 1974 (currently under revision).

29. Upperman, G.T., "Water analysis manual for fossil-fuel fired boilers," Report WCG:91:1078-5, The Babcock & Wilcox Company, Barberton, Ohio, 1991.

30. "Materials characterization," *Metals Handbook*, 9th ed., Vol. 10, American Society of Metals, Metals Park, Ohio, 1986.

Bibliography

Allmon, W. E., "Chemical analysis for fossil power plant water chemistry," Proceedings of EPRI Symposium on Fossil Plant Water Chemistry, Electric Power Research Institute, Palo Alto, California, 1985.

Betz Handbook of Industrial Water Conditioning, 8th Ed., Betz Laboratories, Trevose, Pennsylvania, 1980

Drew Principles of Industry Water Treatment, 3rd ed., Drew Chemical, Boonton, New Jersey, 1979.

Goldstein, P., and Burton, C.L., "A research study on internal corrosion of high-pressure boilers — final report," *Journal of Engineering for Power*, Vol. 91, pp. 75-101, April 1969.

Hall, J.R., and Glysson, G.D., *Monitoring Water in the 1990s: Meeting New Challenges*, STP1102, American Society for Testing and Materials, Philadelphia, 1991.

Jonas, O., and Dooley, B., "International water treatment practices and experiences," Proceedings of the 51st Annual Meeting, International Water Conference, Pittsburgh, pp. 396-404, October 22-24, 1990.

Kemmer, F.N., ed., *The Nalco Water Handbook*, 2nd ed., McGraw-Hill, New York, 1988.

Port, R.D., and Herro, H.M., *The Nalco Guide to Boiler Failure Analysis*, McGraw-Hill, New York, 1991

Babcock & Wilcox built and operates this refuse-to-energy plant in the southern U.S.

Chapter 43
Fossil Fuel Boiler Operations

Effective operation of steam generating equipment is critical to maintaining system efficiency, reliability and availability. The procedures used to run steam generating equipment vary widely depending upon the type of system, fuel and application. Systems can range from simple and fully automated requiring a minimum of attention, such as small gas-fired package boilers, to the very complex requiring constant operator attention and interaction, such as a large process recovery boiler in the pulp and paper industry. There are, however, a set of relatively common fundamental operating guidelines which safeguard personnel and optimize equipment performance and reliability. When combined with equipment specific procedures, these guidelines promote the best possible operations. The following material covers first the fundamentals and then addresses some of the unique elements of Cyclone furnaces, once-through Universal Pressure (UP) boilers, fluidized-bed units and process recovery boilers.

Because of the intimate relationship between equipment design and operation, additional operating guidelines may be found in other chapters of *Steam*. Selected areas of particular interest are listed in Table 1.

Table 1
Index to Additional Operating Guidelines Found in *Steam*

System or Component	Chapter
Air heaters	19
Auxiliary equipment	23
Burners, coal	13
Burners, oil and gas	10
Bypass systems	18
Chemical cleaning	42
Chemical recovery units	26
Cyclone furnaces	14
Fluidized-bed boilers, atmospheric	16
Fluidized-bed boilers, pressurized	29
Particulate removal	33
Pulverizers	12
Scrubbers (flue gas desulfurization)	35
Sootblowers	23
Startup systems	18
Stokers	15
Waste-to-energy systems	27
Water chemistry	42
Water treatment	42

General boiler operations

Fundamental principles

Although boiler design and power production have become sophisticated, basic operating principles still apply. Combustion safety and proper steam/water cooling are essential.

Combustion safety and steam/water cooling requirements Before firing a furnace, there must be no lingering combustible material inside the unit. *Purging*, or removal of this material, assures that the furnace is ready for firing. A standard operating rule is to purge the unit for five minutes at no less than 25% of the maximum continuous rating (MCR) air flow.

Once combustion is established, the correct air/fuel ratio must be maintained. Insufficient air flow may permit the formation of combustible gas pockets and, thereby, provide an explosion potential. Sufficient air flow to match the combustion requirements of the fuel should be maintained and a small amount of excess air should be admitted to cover imperfect mixing and to promote air and fuel distribution.

In addition to these combustion precautions, it is important to verify boiler water levels. Combustion should never be established until adequate cooling water is in the tubes and steam drum.

General safety considerations Pressure part failure remains a major concern in the boiler industry. The American Society of Mechanical Engineers (ASME) has issued extensive codes directed at minimizing these failures. The codes are continuously updated to support quality pressure part design. (See Appendix 2.)

Combustion products have also received considerable safety attention. Carbon dioxide (CO_2), sulfur oxides and chlorine compounds must be considered. While CO_2 is not poisonous, it reduces oxygen availability in the flue gas. Sulfur oxides can form acids when breathed into the lungs. Chlorine compounds can form hydrochloric acids or carcinogens such as dioxin.

Safety of fuel ash must also be considered. While few ashes contain hazardous materials, heavy metals and arsenic can be present in dangerous levels. In addition, residual combustion may continue in ash collected in hoppers.

Finally, safeguards apply to work on out of service boilers. Occasionally a confined space may have insufficient oxygen. In addition, poor lighting in boilers that have been shut down can lead to injuries.

Initial commissioning operations

Operator requirements

Good operation begins before equipment installation is complete. It includes training of the operators as well as preparation of the equipment.

Every operator must be trained to understand and fulfill the responsibility assumed for the successful performance of the equipment and for the safety of all personnel involved. To be prepared for all situations that may arise, the operator must have a complete knowledge of all components — their designs, purposes, limitations and relationships to other components. This includes thoroughly inspecting the equipment and studying the drawings and instructions. The ideal time to become familiar with new equipment is during the pre-operational phase when the equipment is being installed.

Preliminary operation should not be entrusted to inexperienced personnel who are not familiar with the equipment and the correct operating procedures. Considerable damage can result from improper preparation of equipment or its misuse during preliminary checkout.

Operator training takes on special significance during the initial operating period required to prepare the unit for commercial operation. Knowledgeable and experienced operators are valuable during this period when the controls and interlocks are being adjusted, fuel burning equipment is being regulated, operating procedures are being perfected and preliminary tests are being conducted to determine the performance and capabilities of the unit.

Preparations for startup

A systematic approach is required in the preparation for service of a new boiler or any boiler that has undergone major repairs or alterations. The procedure varies with boiler design; however, certain steps are required for all boilers. The steps may be classified as inspection, cleaning, hydrostatic testing, pre-calibration of instruments and controls, auxiliary equipment preparation, refractory conditioning, chemical cleaning, steam line cleaning (blowing), safety valve testing and settings and initial operations for adjustments and testing.

Inspection An inspection of the boiler and auxiliary equipment serves two purposes: 1) it familiarizes the operator with the equipment, and 2) it verifies the condition of the equipment. The inspection should begin some time during the construction phase and continue until all items are completed.

One item frequently overlooked during the inspection is the provision or lack of provision for expansion. The boiler expands as the temperature and pressure are increased and so also do steam lines, flues and ducts, sootblower piping and drain piping. Before pressure is raised in the boiler, temporary braces, hangers or ties used during construction must be removed.

Cleaning Debris and foreign material which accumulate during shipment, storage, erection or repairs must be removed. Debris on the water side can restrict circulation or plug drain lines. Debris on the gas side can alter gas or air flows. Combustible material on the gas side can ignite and burn at uncontrollable rates and cause considerable damage. Glowing embers can be the source of ignition at times when ignition is not desired.

Fuel lines, especially oil and gas lines, should be cleaned to prevent subsequent damage to valves and the plugging of burner parts. Steam cleaning is recommended for all oil and gas lines. Atomizing steam and atomizing air lines should be cleaned.

Hydrostatic test After the pressure parts are assembled, but before the refractory and casing are installed, a hydrostatic test at 1.5 times the boiler design pressure is applied to all new boilers and maintained for a sufficient time to detect any leaks.

The hydrostatic test is normally the first time the new boiler is filled with water; therefore, this is the time to begin using high quality water for the prevention of internal fouling and corrosion. (See Chapter 42.) Demineralized water or condensate treated with 10 ppm of ammonia for pH control and 500 ppm of hydrazine for control of oxygen should be used for all nondrainable superheaters and reheaters. A clear, filtered water, on the other hand, is suitable for components that will be drained immediately after the hydrostatic test.

Temperature plays an important part in hydrostatic testing. First, the metal temperature and, therefore, water temperature should be above the dew point temperature of the surrounding air to prevent the formation of condensate on the parts being tested. Condensation obscures and practically prevents the detection of small leaks. Second, the metal temperature and therefore the water temperature must be at or above applicable Code restrictions for hydrostatic test. For example, these restrictions, as stated in Section I PG99 of the ASME Code, specify that the hydrostatic test temperature will not be below 70F (21C), to take advantage of the inherent toughness of carbon steel materials as related to temperature. Third, the water temperature should be kept low enough so that the pressure parts can be touched and close inspections can be made. It should not be so high that water escaping from small leaks evaporates immediately or flashes to steam. Also, the water temperature should not be more than 100F (56C) above the metal temperature to avoid excessive metal stress transients.

Finally, no air should be trapped in the unit during the hydrostatic test. As the unit is being filled, each available vent should be open until water appears.

Instrumentation and controls Every natural circulation boiler has at least two indicating instruments — a pressure gauge and a gauge glass. If a superheater is involved, some type of steam temperature indicator is also used. Once-through type boilers have pressure gauges, flow meters and temperature indicators. These indicators are important in that the pressure, temperature levels and flows indicated by them must be controlled within design limits. The indicators must, therefore, be correct. Operation should not be attempted until these instruments are calibrated. The calibrations should include corrections for actual operating conditions. Water-leg correction for the pressure gauge is one example. (See Chapter 40.)

Most modern boilers are controlled automatically once

they are fully commissioned. However, before these boilers can be started up, the controls must be operated on manual until the automatic controls have been adjusted for site-specific conditions. Automatic control operations should not be attempted until the automatic functions have been calibrated over the load range.

Auxiliary equipment The auxiliary equipment must also be prepared for operation. This equipment includes fans to supply air for combustion, feedwater pumps to supply water, fuel equipment to prepare and burn the fuel, air heaters to heat the air for combustion, an economizer to heat the water and cool the flue gas, ash removal equipment and a drain system to drain the boiler when required.

Chemical cleaning Water side cleanliness is important for all boilers because water side impurities can lead to boiler tube failures. They can also lead to carryover of solids in the steam, resulting in superheater tube failure or turbine blade deposits.

The flushing of all loose debris from the feedwater system and boiler and the use of high quality water for the hydrostatic test must be supplemented by proper water side cleaning before startup. To remove accumulations of oil, grease and paint, the natural circulation boiler is given a caustic and phosphate boilout after the feedwater system has been given a phosphate flush. The once-through boiler and its associated pre-boiler equipment are given a similar flushing. This boiling out and/or flushing should be accomplished before operation.

After boiling out and flushing are completed, products of corrosion still remain in the feedwater system and boiler in the form of iron oxide and mill scale. It is recommended that acid cleaning for the removal of this mill scale and iron oxide be delayed until operations at fairly high capacities have carried loose scale and oxides from the feedwater system to the boiler. This results in a cleaner boiler for subsequent operations. Chemical cleaning of internal heating surfaces is described in Chapter 42.

Steam line blowing Fine mesh strainers are customarily installed in turbine inlet steam lines to protect turbine blades or valves against damage from scale or other solid material that may be carried by the initial flow of steam. In addition, many operators use high velocity steam to clean the superheater and steam lines of any loose scale or foreign material before coupling the steam line to the turbine. The actual procedure used depends on the design of the unit. Temporary piping to the atmosphere is required with all procedures. This piping must be anchored to resist the high nozzle reaction created during the high velocity blowing period.

Several methods are used for blowing steam lines, including particularly high pressure air and steam blowing. The recommended method and the one most used is steam blowing because temperature shock and high velocities are the most effective means of removing loose scale. Experience has shown that a combination of high velocity and temperature shock is effective. Sufficient shock is obtained with a series of blows where the steam temperature changes during each blow.

Steam line blowing is usually done in two steps:

1. The superheater, main steam piping and cold reheat piping are included in the first step. A temporary crossover is required between the main steam piping and the cold reheat piping. The turbine stop valve is constructed so that the flanged top of the valve can be removed and a temporary piping connection can be bolted in place. Successive steam blows are made through the piping to the atmosphere until it is determined that the targets located at the discharge nozzle of the temporary piping are clean.
2. The second step includes the superheater, main steam piping, reheater, and the cold and hot reheat piping through temporary piping to the atmosphere. Successive blows are made until the targets at the discharge nozzle of the temporary piping are clean.

This two step procedure prevents blowing of a dirty line through the reheater.

There are two basic methods of supplying steam for steam line blowing with a forced circulation boiler. The first method is to use steam returning from the flash tank to the superheater. The second is to use high pressure steam directly from the boiler. The latter method supplies large quantities of steam at higher pressures. Temporary valving is required to control the flow rates during the blowing period.

Boiler pressure and temperature can be maintained during the blowing period by continuous firing. If firing is discontinued during the blowing period, it must be remembered that any change in boiler pressure changes the saturation temperature throughout the system. To avoid excessive thermal shock, changes in boiler pressure should be limited to those corresponding to 75F (42C) in saturation temperature during the relatively short blowing periods.

Safety valves The set point of each safety valve is normally checked, and adjusted if necessary, immediately after reaching full operating pressure for the first time with steam. Safety valve seats are susceptible to damage from wet steam or grit. This is an essential reason for cleaning the boiler and blowing out the superheater and steam line before testing safety valves.

Safety valves on drum-type boilers are normally tested both for set point pressure and for the closing pressures. This requires that the boiler pressure be raised until the valve opens and relieves sufficient pressure for the valves to close.

In order to remain closed without leakage, higher pressure safety valves can not tolerate any damage to the seats. They are normally not tested for closing pressure. Two methods are common for checking high pressure safety valves without permitting them to open fully. One method uses special gags to restrict valve lift and to close the valve as soon as it starts to simmer. Cold water may be used but the set point pressure must be corrected for temperature. A second method uses special calibrated hydraulic jacks or air operated motors to open the valves while the boiler is operating at normal temperature and pressure. This procedure does not require any correction for temperature. The popping pressure is determined by adding the effective jack pressure to the boiler pressure.

The testing of safety valves always requires caution. Safety valve exhaust piping and vent piping should not exert any excessive forces on the safety valve. The lowest set safety valves should be tested first and then gagged to prevent their opening while testing the higher pressure valves. Gags should always be used as a safety measure while making adjustments to the valves.

Startup Operating procedures vary with boiler design. However, certain objectives should be included in the operating procedures of every boiler: 1) protection of pressure parts against corrosion, overheating and thermal stresses, 2) prevention of furnace explosions, and 3) production of steam at the desired temperature, pressure and purity.

Filling In filling the boiler for startup, certain precautions should be taken to protect the pressure parts. First, high quality water should be used to minimize water side corrosion and deposits. Second, the temperature of the water should be regulated to match the temperature of the boiler metal to prevent thermal stresses. High temperature differentials can cause thermal stresses in the pressure parts and, if severe, will limit the life of the pressure parts. High temperature differentials can also distort the pressure parts enough to break studs, lugs and other attachments. Differentials up to 100F (56C) are generally considered acceptable.

A third precaution taken during the filling operation is the use of vents to displace all air with water. This reduces oxygen corrosion and assures that all boiler tubes are filled with water.

A fourth precaution on drum-type boilers is to establish the correct water level before firing begins. The water level rises with temperature; therefore, only 1 in. (25 mm) of water is typically required in the gauge glass except with certain special designs that may require a higher starting level in order to fill all circulating tubes exposed to the hot flue gases.

Circulation Overheating of boiler tubes is prevented by the flow of fluid through the tubes. Flow is produced in the natural circulation-type boiler by the force of gravity acting on fluids of different densities. Flow starts when the density of the water in the heated tubes is less than that in the downcomers. This flow increases as firing rate is increased. (See Chapter 5.) Some drum-type boilers are designed for forced circulation and depend on a circulating pump to assist this flow.

The once-through type boiler depends on the boiler feedwater pump to produce the necessary flow. Whenever a once-through boiler is being fired, a minimum design flow must be maintained through the furnace circuits. With the use of the bypass system, the fluid can bypass the superheater and turbine to maintain this minimum design flow until saturated steam is available for admission to the superheater and until the turbine is using sufficient steam to maintain the design minimum furnace circuit flow. (See Chapter 18.)

Purging Considerable attention has been given to the prevention of furnace explosions, especially on units burning fuel in suspension. Most furnace explosions occur during startup and low load periods. Whenever the possibility exists for the accumulation of combustible gases or combustible dust in any part of the unit no attempt should be made to light the burners until the unit has been thoroughly purged. The accepted practice on multiple burner units is to purge with 25% of full load rated air flow for five minutes.

Protection of economizer Very little water, if any, is added to the drum-type boiler during the pressure raising period; consequently, there is no feedwater flow through the economizer. Economizers are located in relatively low temperature zones. Nevertheless, some economizers generate steam during the pressure raising period. This steam remains trapped until feedwater is fed through the economizer. It not only makes the control of steam drum water level difficult but it causes water hammer. This difficulty is overcome by supplying feedwater continuously by venting the economizer of steam or by recirculating boiler water through the economizer. If a recirculating line is used, the valve in this line must remain open until feedwater is being fed continuously through the economizer to the boiler.

Protection of primary, secondary and reheat superheater During normal operation, every superheater tube must have steam flow sufficient to prevent overheating. During startup, before there is steam flow through every tube, the combustion gas temperature entering the superheater section must be controlled to limit superheater metal temperatures to 900F (482C) for carbon steel tubes and 950 to 1000F (510 to 538C) for various alloy tubes. While firing rate is used primarily to control gas temperatures, other means are useful, e.g., excess air, gas recirculation and burner selection.

Gas temperature entering the superheater during startup is measured with retractable thermoprobes which are removed as soon as steam flow is established in every superheater tube. (See Chapter 40.) These probes are required for the first few startups to establish acceptable firing rates. When fast startups are required with firing rates exceeding those previously established as acceptable, thermocouples are required for every startup. They are always used on the once-through type boiler.

The two prerequisites for steam flow through every superheater tube are: 1) removal of all water from each tube, and 2) a total steam flow equal to or greater than approximately 10% of rated steam flow. Water is removed from drainable superheaters by simply opening the header drains and vents. Nondrainable superheaters are not so simple, because the water must be boiled away. There will be no steam flow through a tube partially filled with water and those portions of the tube not in contact with water will be subjected to excessive temperatures unless the gas temperature is limited.

Thermocouples attached to the outlet legs of nondrainable superheater tubes, where they pass into the unheated vestibule, will read saturation temperature at the existing pressure until that tube is cleared and there is a flow of steam through the tube. The temperature of these outlet legs rises sharply to significant increases above saturation immediately after flow is established. Superheater tubes adjacent to sidewalls and division walls are normally the last to boil clear. These should, therefore, have thermocouples.

Protection of drums and headers In most installations, the time required to place a boiler in service is limited to the time necessary for raising pressure and protecting the superheater and the reheater against overheating. In some cases, however, the time for both starting up and shutting down may be determined by the time required to limit the thermal stresses in the drums and headers.

Various rules, based on thermal stress analysis and supported by operating experience, have been formulated and accepted as general practice. The rules fall into three categories — one for drums and headers with rolled tube joints, a second set for headers with welded tube joints and a third set for steam drums with welded tube joints.

On drums and headers with rolled tube joints, the relatively thin tubes contract and expand at a much faster rate than the thicker tubesheets; therefore, tube seat leaks are likely to occur if heating and cooling rates are not controlled. Heating and cooling rates have, therefore, been established at 100F (56C) change in saturation temperature per hour.

Headers with welded tube connections present no problem with tube seat leaks or with header distortion because the tubes are welded to the headers and the headers are normally filled with fluid at a constant temperature. The concern is mainly with temperature differentials through the header wall and the resulting incipient cracking if excessive thermal stresses occur.

Steam drums with welded tube connections are not subject to tube seat leaks but, because they contain water in the bottom and steam in the top, the heating and cooling rates vary between the top and the bottom. This results in temperature differences between the top and the bottom. Stress analyses show that the principal criterion for reliable rates of heating and cooling should be based on the relationship between the temperature differential through the drum wall and the temperature differential between the top and bottom of the drum, both measured in the same circumferential plane. Stress analysis also shows that the allowable temperature differentials are based on tensile strength, drum diameter, wall thickness and pressure. Therefore, each steam drum has its own allowable temperature differential curves, one set for cooling and one set for heating. Curves for a typical drum are shown in Fig. 1. Actual temperature differentials must be below the allowable differentials shown.

To determine the temperature differentials during the periods of startup and shutdown, it is necessary to continuously and accurately obtain the outside and inside surface temperatures of the drum shell. Temperatures of the outside of the drum shell are best determined by thermocouples. At least six outside thermocouples are required — two on each end, one top and bottom (all outside the internal baffle and scrubber area) and two in the center, one top and one bottom. Temperatures of the inside of the drum shell are best determined by placing thermocouples on one of the riser tubes entering the bottom of the drum, one on each end and one in the center.

There are two periods when the steam in the top of the drum is hotter than the water occupying the lower half. One period is soon after firing begins on boilers with large nondrainable superheaters. Steam forms first in the superheater and flows back to the steam drum where it heats the top of the drum. A second period is during cooling when cold air is pumped through the boiler by the forced draft fans. The boiler water cools, but the steam in the top of the steam drum remains at essentially the same temperature.

The temperatures of the top and bottom can be brought close together by flooding the steam drum with water. Firing must be stopped and water must not be allowed to spill over into the superheater. A high level gauge glass is useful for this operation. However, thermocouples attached to the superheater supply tubes can be used to indicate when the drum is flooded because there is a sharp change in temperature when water enters these supply tubes. Allowable temperature differentials for cooling are stringent, considerably more so than for heating. Fast cooling should, therefore, be supervised with the use of allowable temperature differential curves, thermocouples and a high level gauge glass.

Cooling is particularly important when the boiler is to be drained for short outages. The steam drum must be cooled to permit filling with the available water without exceeding the allowable temperature differentials. (See Fig. 1: each drum requires a unique chart.) A wider latitude is permitted if the water used for filling is hotter than the steam drum.

Before removing the unit from service, a more uniform cooling can be achieved by reducing pressure to approximately two thirds of normal operating pressure. This reduces the temperature of both the water and steam in the steam drum.

Operating techniques for maximum efficiency

Fuel is a major cost in boiler operation. It is therefore important to minimize fuel consumption and maximize steam production. Although a boiler's efficiency is primarily a result of its design, the operator can maintain or significantly improve efficiency by controlling heat losses to the stack and losses to the ash pit.

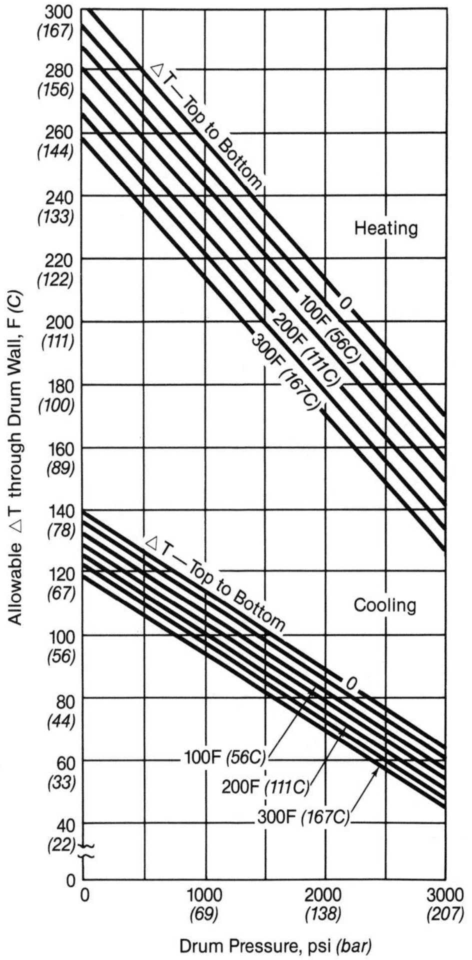

Fig. 1 Allowable temperature differential limits for a typical steam drum.

Stack losses The total heat that exits the stack is controlled by the quantity and the temperature of the flue gas. The quantity of gas is dependent on the fuel being burned, but is also influenced by the amount of excess air supplied to the burners. While sufficient air must be provided to complete the combustion process, excessive quantities of air simply carry extra heat out of the stack.

The temperature of the flue gas is affected by the cleanliness of the boiler heat transfer surfaces. This, in turn, is dependent on sootblower and air heater operation. While high gas exit temperatures waste energy, excessively low temperatures may also be unacceptable. Corrosion can occur at the acid dew point of the gas where corrosive constituents condense on the cooler metal surfaces. Also, ash plugging of heat transfer surfaces can be aggravated by the presence of condensate.

Pressures and temperatures Additional efficiency criteria pertain to the heat cycle of the boiler. Most boilers supply steam to turbines or processes that require heat. These processes rely on design pressures and temperatures. Deviations from the set points may result in lower overall efficiency of the power cycle, loss of production or damage to the product or process.

In units that produce saturated steam, boiler temperature is directly related to the operating pressure. For industrial applications, the process requirements dictate the steam temperature and therefore the operating pressure.

For electric power producing steam systems, steam temperature has a significant impact on turbine efficiency. Modern controls can typically maintain this temperature within 10F (6C) of the desired setting. Each 50F (28C) reduction in steam temperature reduces cycle efficiency by approximately 1% on a supercritical pressure unit.

Variable pressure operation Historically, utility power boilers in the United States (U.S.) have been operated at a constant steam pressure and the turbine load has been controlled by varying the throttle valves at the turbine inlet. This causes an efficiency loss due to the pressure drop across the throttle valves.

In variable pressure operation, boiler pressure is varied to meet turbine requirements. This can significantly improve cycle efficiency when operating at low loads (15 to 40% of maximum continuous rating). However, boiler response to turbine requirements is slower in this mode. When fast response is needed, pressure increments are used. At a given pressure, the throttle valves control the steam to the turbine and provide quick response. As turbine requirements increase and the valves approach full open position, the boiler pressure is increased to the next increment. This arrangement provides high efficiency while retaining quick response.

Key operation functions

Burner adjustments The fuel and combustion air combine and release heat at the burners. To maintain even heat distribution across the width of the furnace, air and fuel flows must be evenly supplied to all burners. While individual burner adjustments affect a particular burner, they also impact adjacent burners. Most boilers have multiple burners in parallel flow paths on the front and/or rear furnace walls. Burners must be similar in design and must be adjusted in the same way to optimize air flow distribution. For a detailed discussion of burner features refer to Chapters 10 and 13.

Adjustments can typically vary the turbulence and flow rates in burners. Increased turbulence increases the air-fuel mixing. It also increases the combustion intensity, provides faster heat release, reduces unburned carbon in the ash, permits operation with less excess air and increases boiler efficiency. However, increased slagging, increased nitrogen oxides (NO_x) emissions and higher fan power consumption can also result. Some burners have adjustments for fuel distribution as well as air flow.

Excess air The total combustion air flow to a boiler is generally controlled by adjusting forced and induced draft fan dampers in relation to the fuel flow. *Excess air* is the amount of additional combustion air over that required to theoretically burn a given amount of fuel. The benefits of increasing excess air include increased combustion intensity, reduced carbon loss and/or CO formation, and reduced slagging conditions. Disadvantages include increased fan power consumption, increased heat loss up the stack, increased tube erosion, and possibly increased NO_x formation.

For most coal ashes, particularly those from Eastern U.S. bituminous coals, the solid to liquid phase changes occur at lower temperatures if free oxygen is not present (reducing conditions) around the ash particles. As a result, more slagging occurs in a boiler operating with insufficient excess air where localized reducing conditions can occur. For some fuels, including Western U.S. subbituminous coals, the oxidizing-reducing temperature differential is much less.

Localized tube metal wastage may also occur in furnace walls under low excess air conditions but the impact is less clearly defined. The absence of free oxygen (a reducing atmosphere) and the presence of sulfur (from the fuel) are known causes of tube metal wastage. The sulfur combines with hydrogen from the fuel to form hydrogen sulfide (H_2S). The H_2S reacts with the iron in the tube metal and forms iron sulfide, which is subsequently swept away with the flue gas. Chlorine also promotes tube wastage. Although most conventional fuels contain very little chlorine, it is a problem in refuse-derived fuels. (See Chapter 27.)

Fuel conditions Fuel conditions are temperature, pressure and, if solid, mean particle size and distribution. Cooler temperatures, lower pressure and larger particle size contribute to less complete combustion and increased unburned carbon in the ash. Conversely, if the fuel is hotter, finer and at a higher pressure, combustion is improved. Unfortunately NO_x emissions and slagging can also increase with these conditions.

Feedwater and boiler water conditioning requirements The main function of a boiler is to transfer heat from combustion gases through tube walls to heat water and produce steam. Clean metal tubes are good conductors, but impurities in the water can collect on the inside surface of the tubes. These deposits reduce heat transfer, elevate tube temperatures and can lead to tube failures. Water conditioning is essential to minimize deposits and maintain unit availability. (See Chapter 42.)

Sootblower operations As discussed in Chapter 23, a *sootblower* is an automated device that uses steam, compressed air, or high pressure water to remove ash deposits from tube surfaces. Sootblowing improves heat transfer by reducing fouling and plugging. However, excessive sootblowing can result in increased operating cost, tube erosion and increased sootblower maintenance. Conversely, infrequent sootblower operation can reduce boiler efficiency and capacity. Optimum sootblowing depends on load conditions, combustion quality and fuel.

Effects of fuel preparation equipment on boiler performance Fuel preparation equipment readies the fuel for combustion. The preparation equipment may be the crushers, pulverizers and drying systems on coal-fired units; fuel oil heaters and pumps on oil-fired units; refuse handling, mixing or drying equipment on refuse-fired boilers; or fuel handling, blending, sizing and delivery equipment on stoker-fired units. If this equipment is not properly operated, the fuel may not be completely burned, leaving unburned carbon in the ash or carbon monoxide (CO) in the flue gas.

The operator must monitor the fuel preparation equipment. Knowledge of the equipment, its maintenance record and operating characteristics is essential.

Personnel safety Operating instructions usually deal primarily with the protection of equipment. Rules and devices for personnel protection are also essential. The items listed here are based on actual operating experience and point out some personnel safety considerations.

1. When viewing flames or furnace conditions, always wear tinted goggles or a tinted shield to protect the eyes from harmful light intensity and flying ash or slag particles.
2. Do not stand directly in front of open ports or doors, especially when they are being opened. Furnace pulsations caused by firing conditions, sootblower operation, or tube failure can blow hot furnace gases out of open doors, even on suction-fired units. Aspirating air is used on inspection doors and ports of pressure-fired units to prevent the escape of hot furnace gases. The aspirating jets can become blocked, or the aspirating air supply can fail. In some cases, the entire observation port or door can be covered with slag, causing the aspirating air to blast slag and ash out into the boiler room.
3. Do not use open ended pipes for rodding observation ports or slag on furnace walls. Hot gases can be discharged through the open ended pipe directly onto its handler. The pipe can also become excessively hot.
4. When handling any type of rod or probe in the furnace — especially in coal-fired furnaces — be prepared for falling slag striking the rod or probe. The fulcrum action can inflict severe injuries.
5. Be prepared for slag leaks. Iron oxides in coal can be reduced to molten iron or iron sulfides in a reducing atmosphere in the furnace resulting from combustion with insufficient air. This molten iron can wash away refractory, seals and tubes and leak out onto equipment or personnel.
6. Never enter a vessel, especially a boiler drum, until all steam and water valves, including drain and blowdown valves, have been closed and locked or tagged. It is possible for steam and hot water to back up through drain and blowdown piping, especially when more than one boiler or vessel is connected to the same drain or blowdown tank.
7. Be prepared for hot water in drums and headers when removing manhole plates and handhole covers.
8. Do not enter a confined space until it has been cooled, purged of combustible and dangerous gases and properly ventilated with precautions taken to keep the entrance open. Station a worker at the entrance, notify a responsible person, or run an extension cord through the entrance.
9. Be prepared for falling slag and dust when entering the boiler setting or ash pit.
10. Use low voltage extension cords, or cords with ground fault interrupters. Bulbs on extension cords and flashlights should be explosion proof.
11. Never step into flyash. It can be cold on the surface yet remain hot and smoldering underneath for weeks.
12. Never use toxic or volatile fluids in confined spaces.
13. Never open or enter rotating equipment until it has come to a complete stop and its circuit breaker is locked open. Some types of rotating equipment can be set into motion with very little force. This type should be locked with a brake or other suitable device to prevent rotation.
14. Always secure the drive mechanism of dampers, gates and doors before passing through them.

Performance tests Many steam generating units are operating day after day with efficiencies at or near design values. Any unit can operate at these efficiencies with the proper instrumentation, a reasonable equipment and instrument maintenance program and proper operating procedures.

Early in the life of the unit when the gas side is clean, the casing is tight, the insulation is new, the fuel burning equipment has been adjusted for optimum performance and the fuel/air ratio has been set correctly, performance tests should be conducted to determine the major controllable heat losses. These losses are the dry gas to the stack and, on coal-fired units, combustible in the ash or slag.

During these tests accurate data should be taken to serve as reference points for future operation. Sampling points which give representative indications of gas temperatures, excess air and combustibles in the ash should be established and data from these points recorded. Items related to the major controllable losses should also be recorded at this time, e.g., draft losses, air flows, burner settings, steam flow, steam and feedwater temperatures, fuel flow and air temperatures.

Procedures for performance tests are provided in the American Society of Mechanical Engineers (ASME) Performance Test Codes — PTC 4.1, *Steam Generating Units*; PTC 4.2, *Coal Pulverizers*; and PTC 4.3, *Air Heaters*.

Abnormal operation

Low water If water level in the drum drops below the minimum required (as determined by the manufacturer), fuel firing should be stopped. Caution should be exercised when adding water to restore the drum level due to the potential of temperature shock from the relatively cooler water coming in contact with hot drum metal. Thermocouples on the top and bottom of the drum will indicate if the bottom of the drum is being rapidly cooled by feedwater addition, which would result in unacceptable top to bottom temperature differentials. If water level indicators show there is still some water remaining in the drum, then feedwater may be slowly added using the thermocouples as a guide. If the drum is completely empty, then water may only be added periodically with soak times provided to allow drum temperature to equalize. (See previous discussion on protection of drums and headers.)

Tube failures Operating a boiler with a known tube leak is not recommended. Steam or water escaping from a small leak can cut other tubes by impingement and set up a chain reaction of tube failures. By the loss of water or steam, a tube failure can alter boiler circulation or flow and result in other circuits being overheated. This is one reason why furnace risers on once-through type boilers should be continuously monitored. A tube failure can also cause loss of ignition and a furnace explosion if re-ignition occurs.

Any unusual increase in furnace riser temperature on the once-through type boiler is an indication of furnace tube leakage. Small leaks can sometimes be detected by the loss of water from the system, the loss of chemicals from a drum-type boiler, or by the noise made by the leak. If a leak is suspected, the boiler should be shut down as soon as normal operating procedures permit. After the leak is then located by hydrostatic testing, it should be repaired.

Several items must be considered when a tube failure occurs. In some cases where the steam drum water level can not be maintained, shut off all fuel flow and completely shut off any output of steam from the boiler. When the fuel has been turned off, purge the furnace of any combustible gases and stop the feedwater flow to the boiler. Reduce the air flow to a minimum as soon as the furnace purge is completed. This procedure reduces the loss of boiler pressure and the corresponding drop in water temperature within the boiler.

The firing rate or the flow of hot gases can not be stopped immediately on some waste heat boilers and some types of stoker-fired boilers. Several factors are involved in the decision to continue the flow of feedwater, even though the steam drum water level can not be maintained. In general, as long as the temperature of the combustion gases is hot enough to damage the unit, the feedwater flow should be continued. (See later discussion for chemical recovery units.) The thermal shock resulting from feeding relatively cold feedwater into an empty steam drum should also be considered. (See previous discussion on protection of drums and headers.) Thermal shock is minimized if the feedwater is hot, the unit has an economizer and the feedwater mixes with the existing boiler water.

After the unit has been cooled, personnel should make a complete inspection for evidence of overheating and for incipient cracks, especially to headers and drums and welded attachments.

An investigation of the tube failure is very important so that the condition(s) causing the tube failure can be eliminated and future failures prevented. This investigation should include a careful visual inspection of the failed tube. In some cases a laboratory analysis or consideration of background information leading up to the tube failure is required. This information should include the location of the failure, the length of time the unit has been in operation, load conditions, startup and shutdown conditions, feedwater treatment and internal deposits.

Shutdown operations

Boiler shutdown is less complicated than startup. The emphasis again is on safety and protection of unit materials. Two shutdown situations may occur, a controlled shutdown or one required in an emergency.

Under controlled conditions the firing rate is gradually reduced. Once the combustion equipment is brought to its minimum capacity, the fuel is shut off and the boiler is purged with fresh air. If some pressure is to be maintained the fans are shut down and the dampers are closed. The drum pressure gradually lowers as heat is lost from the boiler setting; a minimal amount of air drifts out of the stack. If inspection and maintenance are required the draft fans remain on in order to cool the boiler more quickly. If the unit is equipped with a regenerative air heater, it is shut down, allowing the boiler to cool faster. If a tubular air heater is present, this heat trap can only be bypassed to help cool the boiler. The cool down rate should not exceed 100F (55C) per hour of saturation temperature change to prevent damage due to thermal stress.

In an emergency shutdown, the fuel is immediately shut off and the boiler is purged of combustible gases. Additional procedures may apply to the fuel feed equipment. The boiler may be held at a reduced pressure or may be completely cooled as described above.

Operation of Cyclone furnaces

The Cyclone component of a Cyclone furnace is a cylindrical chamber designed to burn crushed coal. The basic design, sizing and general operation of Cyclone furnaces are covered in Chapter 14. The key feature which affects operations is the collection of most of the coal ash as a liquid slag in the Cyclone chamber. This slag is continuously tapped into the furnace through a hole at the discharge end of the Cyclone. The slag collects on the furnace floor and flows through a tap into a water filled tank where the chilling effect of the water leaves the slag in granular form. Unique operational issues center upon maintaining desired furnace slag tapping without excessive maintenance.

Fuels

A key operating parameter of Cyclone units is the selection of a proper coal. Fuels acceptable for Cyclone firing must generally meet the following specifications subject to site-specific conditions:

Bituminous coals:
1. maximum total moisture — 20%
2. ash content — 6 to 25%, dry basis
3. minimum volatile matter — 15%, dry basis
4. ash (slag) viscosity — refer to Table 2
5. ash iron ratio and sulfur content — See Chapter 14, Fig. 4

Subbituminous coals and lignite:
- 1a. maximum total moisture for a direct-fired system but without a pre-dry system — 30%
- 1b. maximum total moisture with pre-dry system — 42%
- 2. minimum high heating value — 6000 Btu/lb (13,956 kJ/kg), as-fired
- 3. ash content — 5 to 25%, dry basis
- 4. ash (slag) viscosity — refer to Table 2.

Table 2
Ash Viscosity Requirements

Coal Rank	Maximum T_{250}* Ash Viscosity	As-Fired Total Moisture %
Bituminous	2450F (1343C)	0 to 20
Subbituminous — direct-fired	2300F (1260C)	21 to 30
Subbituminous/lignite**	2300F (1260C)	31 to 35
Lignite***	2300F (1260C)	36 to 42

* T_{250} is the temperature at which the ash viscosity is 250 poise.
** For lignite firing, a fuel pre-dry system is required.
*** For high moisture lignite firing, pre-dry and moisture separator systems are required.

For proper combustion and efficient unit operation, properly sized crushed coal is required. This is especially important for subbituminous coals, where the higher moisture content requires a finer coal grind to maintain Cyclone temperatures and to minimize unburned carbon in the flyash.

Combustion air

All Cyclones use heated air at a high static pressure. The air temperature ranges from 500 to 750F (260 to 399C), depending on the unit design and the rank of the fuel. The static differential across the Cyclone ranges from 32 to 50 in. wg (8 to 12.5 kPa). This high pressure produces the very high velocities which, in turn, produce the scrubbing action required for complete combustion of the crushed coal.

A key variable in Cyclone operation is excess air. For bituminous coals, which are generally high in sulfur and iron content, 15 to 18% excess air is recommended at full load operation. Operation at lower excess air can promote an oxygen deficient reducing atmosphere, which can significantly increase unit maintenance. Operation with molten slag and deficient air can result in iron sulfide attack (wastage) of the boiler tubes. When operated with insufficient excess air, fuels with high iron content can smelt the iron from the ash. A pig iron, which forms a strong bond to the boiler tubing, can form, resulting in tube damage during deposit removal.

Operation with high levels of excess air leads to lower thermal efficiency and results in Cyclone cooling; this hinders slag tapping. On multi-Cyclone units, it is important to accurately measure the secondary air and coal flow to each Cyclone to ensure it is operating with the proper air/fuel ratio.

Because proper excess air is essential, theoretical air curves should be used to readjust the excess air levels when one unit of a multi-Cyclone furnace is removed from service.

Firing subbituminous coals is generally more difficult due to increased fuel moisture which must be evaporated during combustion. Units designed for subbituminous coal firing have provisions for higher air temperature and generally run with 10 to 12% excess air.

Due to environmental concerns, some bituminous coal-fired Cyclones are being converted to burn low sulfur subbituminous coal or coal blends. Subbituminous coal has a lower ash content than most bituminous coals. As a result, less slag is available to trap the raw coal particles. This, combined with the depressed flame temperature caused by the increased moisture content, can result in less coal being entrapped in the slag where the combustion is completed.

When firing higher moisture subbituminous coal, smaller crushed coal particle size is required, higher transport air temperature is required, and longer lift lines provide the capability of evaporating the moisture before introducing the fuel to the Cyclone burner.

As the moisture content reaches the 20% range at design Cyclone fuel inputs on bin systems, clinkering and unburned carbon in the Cyclone and on the furnace floor increase dramatically. Levels of unburned carbon also increase in the ash in the economizer, air heater and precipitator hoppers. Supplemental firing with fuel oil or gas along with design changes may be required if the unit was not originally designed for this condition.

Several Cyclone units are operating on high moisture North Dakota lignite fuel. This is accomplished by the use of a pre-dry system, as illustrated in Chapter 14. In this system, hot air [750F (399C)] is introduced to the raw coal stream prior to crushing in a pressurized and heated conditioner. The hot air-coal mixture then travels to a cyclone separator, where the saturated air and some coal fines are separated from the coal stream and are injected into the lower furnace. The dried coal is then carried by a lift line to the Cyclone burner.

Operation of once-through boilers

Principles

In once-through operation, high pressure water enters the economizer and high pressure steam leaves the superheater; there is no recirculation of steam or water within the unit. The path is through multiple parallel tube circuits arranged in series. Design and construction details are found in Chapters 18 and 24. This type of boiler can be conceptualized as a long heated pipe with continuous coolant flowing through it. As the fluid progresses through the components, heat is absorbed from the combustion process. The supercritical fluid has no latent heat of vaporization and therefore the fluid temperature constantly increases through the unit. The final steam temperature is dependent on the feedwater inlet temperature and ratio of the fluid flow to the heat available, which have important implications on unit operation.

Operating practice

The once-through steam generator, referred to by Babcock & Wilcox (B&W) as the Universal Pressure (UP) boiler, has operating characteristics not seen in conventional drum units. The UP boiler is capable of operating above the critical pressure point [3208 psi (221.2 bar)] and can deliver steam pressure and temperature conditions without a steam-water separation device.

To start up a UP boiler, a steam separation device (flash tank or vertical steam separator), with appropriate valving and piping to bypass and return fluid to the cycle, is supplied. The modern bypass systems are complex, highly automated and very effective in achieving smooth operation.

The major control functions for fuel, air, water and steam flow are highly interactive placing unique constraints on operation. It is therefore important for the operator to fully understand the relationships between short term (transient) and long term (steady-state) energy transfer effects and to carefully coordinate all control actions. Modern control systems are often referred to as coordinated systems. (See Chapter 41.) The combustion systems are conventional; they are subject to the normal concerns of fuel utilization, efficiency and safety.

The operator has three main controls for operating a UP boiler:

1. firing rate — air and fuel flow controls are integrated for managing the release of combustion energy,
2. feedwater control — the feedwater flow rate is varied by adjusting the boiler feed pump discharge pressure, and
3. steam flow — steam flow is a function of turbine first stage pressure and is controlled by the turbine throttle valves (TTVs).

Firing rate The long term demand for firing rate is directly proportional to load. The short term effects of firing rate impact steam temperature and pressure. During increases in firing rate the steam temperature increases until the feedwater flow is increased to compensate and balance the new firing rate. Steam pressure increases in a similar fashion to that of a drum boiler. In a supercritical pressure application, there is no large rise in specific volume during the transition from liquid to vapor conditions. However, there is still a large increase in specific volume as a supercritical fluid is heated. With this expansion, the turbine control valves are opened and flow is increased to maintain pressure. With the many interactions that are a consequence of firing rate, the operator should strive to position firing rate for the desired electrical load and then manipulate other variables to control temperature and pressure.

Feedwater flow The outlet steam temperature is dependent in steady-state conditions on the ratio of feedwater flow rate to firing rate. Because it is desirable to position firing rate based on electrical demand, feedwater flow control is based on outlet steam temperature. The short term effect of feedwater flow changes is to increase or decrease steam pressure and electrical load. The load change results because energy is placed into or brought out of storage through cooling or heating of the boiler metals. However, this load change is transitory. Pressure is a measure of the balance between steam flow and feedwater flow. The outlet pressure is constant if they are matched. The operator can, therefore, use feedwater to assist in recovery from transients and upsets, but he ultimately must position feedwater flow for the appropriate outlet steam temperature.

Steam flow/steam pressure In steady-state conditions, steam flow is the same as feedwater flow, and boiler pressure is determined by turbine throttle valve position. In the short term, a change in steam flow impacts steam temperature and electrical load. This is because increasing steam flow at a constant firing rate affects the balance of the two and the steam temperature drops initially as a consequence to electrical load increase. This can not be a lasting effect, however, as steam and feedwater flows eventually match each other at a different pressure. Similarly, load may be changed by adjusting the turbine throttle valve which increases or decreases steam flow, but this load change comes from depleting or building stored energy in the form of pressure. As above, steam flow may be manipulated to help restore control of steam temperature and load, but it ultimately must be controlled to achieve the desired operating pressure.

Coordination During normal operation, the plant control system adjusts these highly interactive parameters in a stable system. (See Chapter 41.) Modern systems, referred to as coordinated systems, differentiate long term and short term effects. They apply proportional action to control short term effects and provide integral action to control long term effects. During transients, extreme upset or startup conditions, operator intervention may be required. By knowing the interactions, the operator can stabilize one or two control loops and use the third to mitigate the excursions. Placing a loop in manual control can also separate interactions, improving the stability of the process and the recovery. However, extreme care by experienced operators is critical.

A very useful method for providing coordination is to use steady-state data for reference to conduct open loop response testing. The steady-state data are solid reference tools. Response data are gathered by stepping one control variable and maintaining others at a steady-state. This enhances understanding of long and short term responses. It is very instructive to the operator and provides additional data for control system programming.

Operating skills

Efficiency The supercritical boiler provides a very high cycle efficiency. This inherent efficiency can be maintained throughout boiler life with proper instrumentation and maintenance. The operator must be thoroughly trained to know the optimum operating parameters. Deviations in these parameters may then be addressed with instrument maintenance, direct operator action (such as sootblowing) or an engineering investigation. Performance tests should be conducted on a regular basis. Besides the conventional need to minimize heat losses, these tests can be used to confirm control system calibration. Test results should be reviewed with the operator.

Startup systems evolution Startup systems for UP boilers have evolved as the understanding and experience of the once-through concept have increased. A key requirement for the startup system is to maintain adequate flow in the furnace walls (25% for most supercritical units and 33% for most subcritical units) to protect them from overheating during startup and low load operation. Initial designs (first generation) simply bypassed any excess flow from the furnace, not required for the turbine power generation, directly to the condenser from the high pressure turbine inlet. Second generation startup systems added a steam-water separation device called a flash tank (including steam-water separation equipment) after the convection pass enclosure circuits, but upstream of the primary superheater. This permitted the generation of dry steam for turbine roll and synchronization. The flash tank was effectively a small drum including drum internals. (See Fig. 2.)

Subsequent designs moved the bypass line plus necessary valves (200 and 201) downstream of the primary superheater so that the primary superheater is always in the initial flow circuit. (See Fig. 3 for key elements.)

Steam is produced in the flash tank because the pressure reducing valve (207 in Fig. 3) drops the pressure from supercritical to approximately 1000 psig (69 bar gauge) in current designs. In order for a smooth transition or changeover from startup (flash tank) to once-through operation, the enthalpy of the flash tank steam must be matched with the enthalpy leaving the 201 valve. Because the flash tank generally operates at 1000 psig (69 bar gauge), the throttle pressure to the turbine is only ramped to full pressure after transition from the flash tank. Chapter 18 provides a complete flow schematic of a modern UP bypass system for both constant and variable pressure operating modes.

Startup system operation Because of operating problems and operator difficulties in making the transition from flash tank flow to true once-through operation, the bypass system shown in Fig. 3 (key elements only) has been developed. The 207 bypass valve and piping system to the flash tank and 205 valve and piping system from the flash tank are sized to pass all of the minimum feedwater flow required for furnace protection. Pressure in all the components upstream of the 200/201 valves is maintained at full pressure from initial startup to full load operation. The minimum flow is then initiated by use of the boiler feed pump and the boiler is fired with the 207 valve reducing the pressure to about 1000 psig (69 bar gauge). The steam flow is increased by opening the turbine valves until the primary superheater outlet temperature is 780F (416C). All steam to the secondary superheater is from the flash tank, simulating a drum boiler. At 780F (416C), all fluid entering the flash tank is steam and all the steam returns to the superheater and on to the turbine (Fig. 4). This is referred to as a *dry flash tank*, as there is no drain flow from the flash tank to the condenser. In this situation, the flash tank is no longer a separation device; it acts only as a connecting pipe. Once-through operation is then achieved through swapping flow paths by opening the 201 and closing the 207 valves simultaneously. Because the enthalpy is matched, the conditions are stable and the operator is not required to manipulate firing rate, load or steam flow during the transfer. Once the transfer is made, throttle pressure may be increased by closing the turbine valves until full pressure is achieved. The operator then only needs to adjust for the decreasing secondary superheater outlet enthalpy. As an added feature the contemporary UP boiler may also be equipped with high capacity 201 valves (usually 75% of steam flow rating), so steam pressure is controlled in a variable throttle pressure method (Fig. 5) from the end of transfer up to 75% load. This results in increased cycle efficiency at low loads.

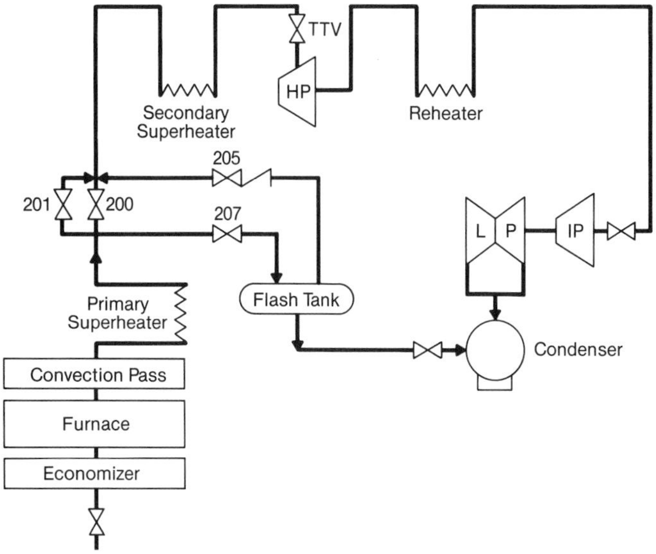

Fig. 3 Startup system with primary superheater in initial flow circuit.

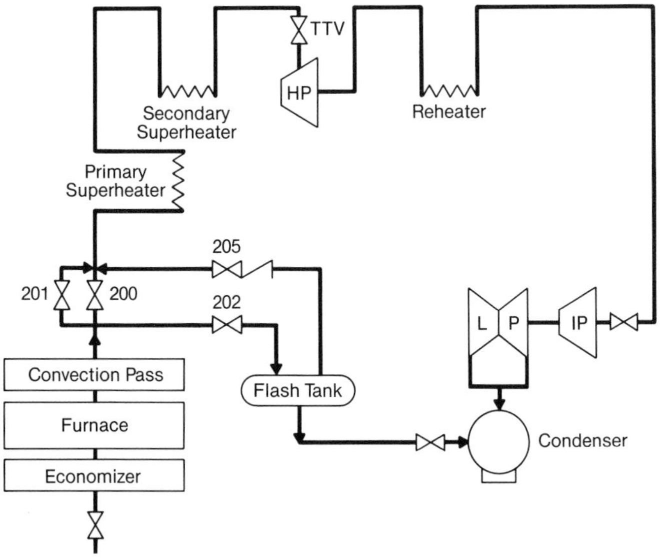

Fig. 2 Second generation UP startup system schematic.

Limits and precautions The once-through nature of UP boilers results in special limits and precautions that must be observed for reliable and dependable operation:

1. Feedwater conductivity — unlike a drum boiler, which can release suspended solids through blowdown, all hardness and other contaminants that enter the boiler in the feedwater are deposited on the water side of the heating surface or in the superheater. Deposition can lead to overheating and tube failures. Boiler firing must be stopped if conductivity exceeds 2.0 micromhos for five minutes or if it exceeds 5.0 micromhos for two minutes. The operator should be trained in water quality control requirements. (See Chapter 42.)
2. Minimum feedwater flow — firing is not permitted unless the boiler feedwater flow is above the specified minimum. The boiler is to be immediately shut down if flow falls below 85% of minimum for twenty seconds or 70% of minimum for one second.
3. Feedwater temperature — the heat input required per unit of water flow is dependent on the difference between the feedwater inlet temperature and the controlled outlet steam temperature. If feedwater temperature is reduced while the outlet temperature is maintained (by overfiring), the heat input to the fur-

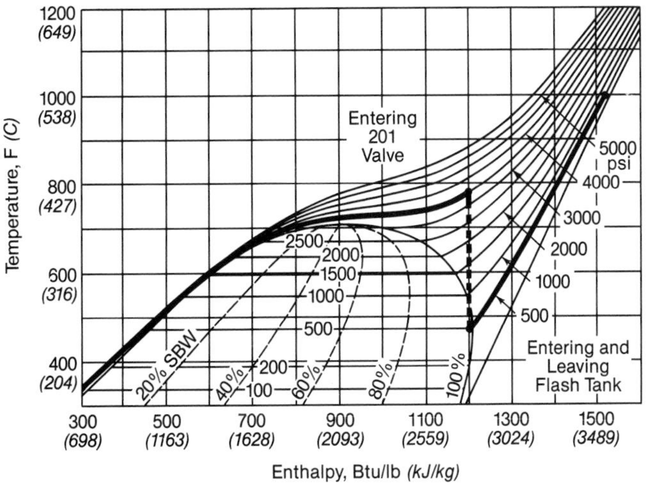

Fig. 4 Temperature-enthalpy diagram for flow through the 201 control valve.

nace can exceed design levels which may result in damage to the tube material. If a unit must be operated with a feedwater heater out of service, the final steam temperature must be reduced to hold furnace heat input rates to within design limits. Unless specific information is provided, the general rule is to reduce steam temperature one degree F for every two degrees F that the feedwater temperature is low.

4. Overfiring — for supercritical units, the fluid in the furnace has no latent heat of vaporization as in a subcritical boiler. Consequently, as heat is absorbed, the fluid temperature is always increasing and the furnace tube metal temperatures increase. Firing in excess of design rates can result in excessive tube metal temperatures in selected furnace locations. The operator must therefore observe a strict limit of transient heat input. A maximum of 1.15 times the steady-state rate is a typical guideline. This rate, at various load conditions, should be clearly established by combustion tests.
5. Tube leaks — operation with a tube leak is not recommended on any boiler. For once-through units, the situation is more critical because the leakage can reduce cooling flow to subsequent tubes or circuits downstream.
6. Low pressure limit — a supercritical pressure boiler must be tripped if the fluid pressure in the furnace falls below 3208 psig (221.2 bar gauge) for fifteen seconds. Failure to do so can result in improper circuit flow, inadequate tube cooling and furnace tube failures.

Operation of fluidized-bed boilers

Chapter 16 provides a detailed discussion of atmospheric pressure fluidized-bed combustion systems. The following sections provide selected remarks on general fluidized-bed boiler operation, both circulating fluidized-bed (CFB) systems and bubbling-bed systems.

CFB systems overview

The CFB is constructed and behaves like a conventional drum boiler in many respects. However, thermal performance, combustion efficiency, furnace absorption pattern and sulfur dioxide (SO_2) control (if used) are strongly influenced by the mass of solids in the bed (inventory), the size distribution of the bed material (bed sizing) and the circulating rate of the bed (flow rate). The bed is classified by two separate sections, the dense bed and dilute bed. Bed density can be described as bulk density which is a measure of the material weight per volume of gas. There is a higher density and inventory in the primary zone, lower density and inventory in the upper zones and a sharp density transition zone (freeboard) between the two. Fig. 6 shows the density of the bed with respect to furnace height and indicates the effect of load.

To integrate these influences into operating procedures, the CFB is equipped with additional instrumentation which provides operator input for bed management. These inputs are as follows:

1. Primary zone ΔP — the gas side differential pressure is measured across the primary furnace zone and is an indication of bed density. The transmitter is typically calibrated for 0 to 50 in. wg (0 to 12.45 kPa) to represent 0 to 100% bed density. Bed density is then a control variable for operation.
2. Secondary zone ΔP — like the primary zone, this ΔP indicates density in the upper zone of the furnace. This is used by the operator in evaluating bed material size, but it is not used as a control variable.
3. Bed temperature — the nature and structure of the fluidized bed allows insertion of multiple thermocouples into the bed. This allows bed temperature to be used as a primary control variable for optimum emission and combustion efficiency controls.
4. Furnace exit gas temperature (FEGT) — thermocouples are inserted at the entrance of the particle separation device to monitor FEGT.

The variables mentioned above are typically not available on conventional boilers. However, the overall operation is similar. Fuel feed is set by steam requirements (pressure) and total air flow is adjusted to follow fuel flow. Bed density is then used to control bed temperature, the most important variable. The bed temperature is usually controlled to 1550F (843C), a temperature resulting in optimal combustion efficiency and SO_2 absorption. The impact of bed temperature on overall combustion is described below:

1500F (816C) or more	Complete carbon combustion
1450 to 1500F (788 to 816C)	Marginal CO burnout
1300 to 1450F (704 to 788C)	Unacceptable CO levels

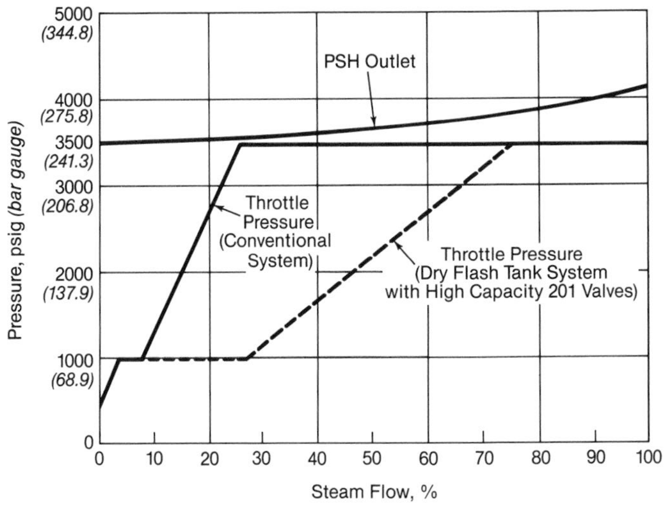

Fig. 5 Variable throttle pressure method control with high capacity 201 valves.

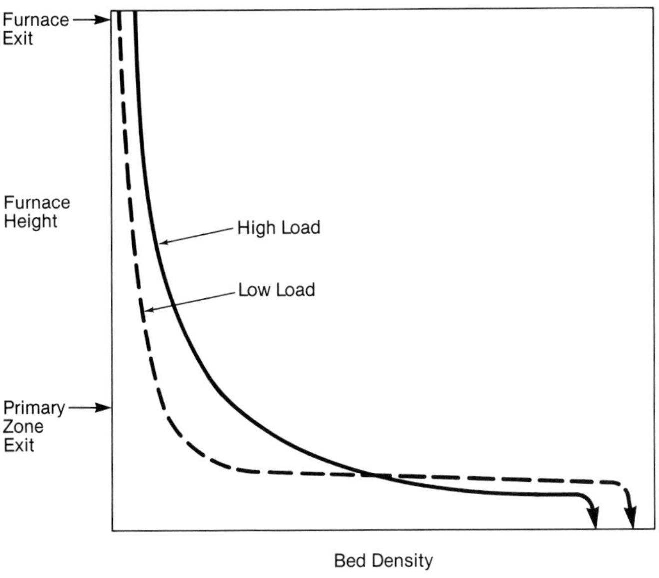

Fig. 6 Typical CFB bed density profile.

The maximum SO_2 absorption occurs at 1550F (843C), but the reaction occurs at a suitable rate from 1450 to 1650F (788 to 899C). The optimum bed operating temperature is from 1500 to 1650F (816 to 899C).

Another variable that is controlled somewhat differently in a CFB is the air admission to the furnace. The total air flow is divided into primary, secondary and tertiary portions. Because total air flow is based on fuel flow, the proportion of the air among the three zones (*air flow split*) becomes another parameter the operator can vary to maintain stability. The split has an impact on bed density because the velocity of primary air determines how much bed material is elutriated and recovered by the particle separation device. A relationship also exists between air flow split, bed density and bed temperature. It can be complex, with changes in one variable impacting several temperatures and flows. Because of these and other interactions, the operator must be trained to control upsets and to establish and maintain stability. These actions are described as follows:

1. Fuel flow control — the operator should always position the fuel feed based upon desired steam flow and steam pressure. Changes in fuel feed can have a short term impact on bed temperature, but this temperature is properly controlled by bed density.
2. Bed density control — bed density is generally controlled by the inventory of bed material within the boiler. To reduce density, material is removed through the bed drain system. To increase density, material is added from the storage or makeup system. An optimum density versus steam flow correlation is established during combustion tests at unit commissioning.
3. Air flow split — the optimum air flow split is also determined from combustion tests and is controlled as a function of steam flow. Staged combustion is desirable for NO_x reduction, but sufficient primary air is needed to establish proper bed conditions. The secondary air should form a penetrating jet for CO burnout and the remainder of the air flow is vented to the secondary ports. As load is increased, excessive secondary air flow to a given set of ports can result in furnace erosion and, therefore, some air flow is diverted to the tertiary ports. Once the optimum splits have been determined and with consistent bed material sizing, there should be no need for the operator to manipulate the splits. An example is shown in Fig. 7.
4. Excess air control — like a conventional solid fuel boiler, the total air flow is best monitored as excess air, with the desired amount determined from combustion tests. The excess air set point increases with decreasing load to maintain steam temperatures, or to maintain primary bed conditions for fluidization and turbulence. Excess air should be closely monitored and controlled by the operator due to its impact on efficiency. An example is shown in Fig. 8.
5. Bed material flows (L-valve) — the bed material particles can be classified as coarse, intermediate, small or fine. Although coarse material often stays in the primary bed or circulates slightly above it in the freeboard area, intermediate to small particles are elutriated, collected and directly returned to the primary bed. Particles may be stored in the collection hopper and the flow may be modulated to the primary bed to influence density. However, the influence is weak and short term, which can interfere with bed temperature control. Therefore, the best approach for the operator is to control the inventory so collected material is returned directly to the primary bed with minimum storage. Dramatic changes in the bed inventory should be avoided, as the storage or depletion of heat in the material can be destabilizing.
6. Bed material flow and makeup — fine material passes through the particle separator and is collected at the boiler outlet in dust collectors. This material is created through attrition and decrepitation of the original bed material. A portion of the small material is usually returned to the primary furnace and the rest of the fines are discarded. This results in a need for makeup material and the operator can supply this material by manipulating the recycle and makeup flows

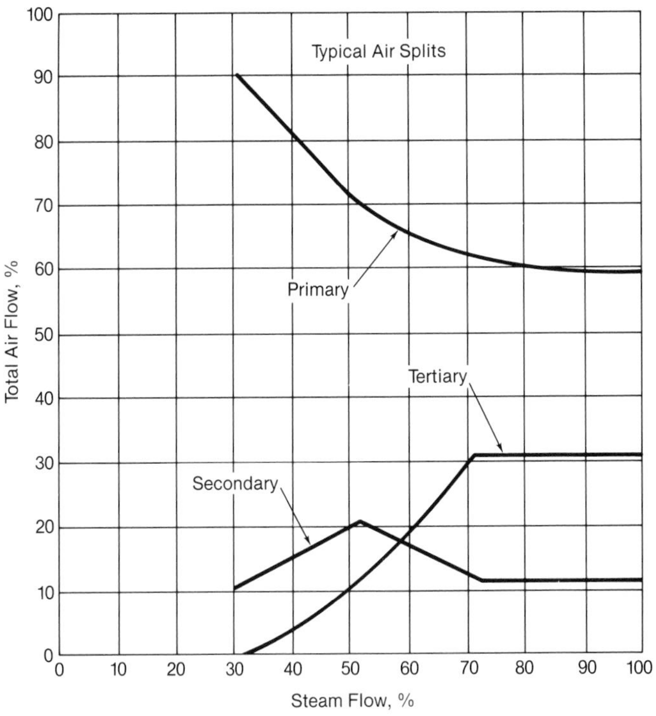

Fig. 7 Typical air splits versus load for a CFB.

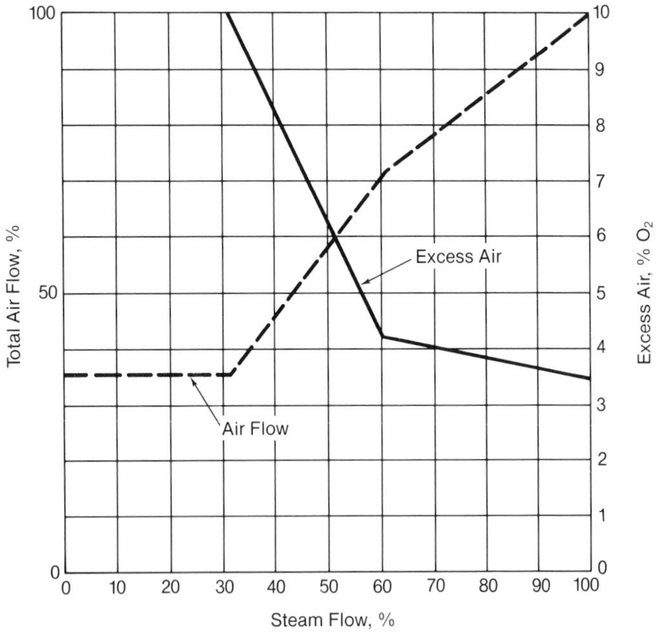

Fig. 8 Excess air diagram for efficiency control.

from storage. Because these flows are modulated through on/off devices, the operator must observe control variable trends and add appropriate material to assure the control variables remain within limits. As load is increased, the need for bed material increases.

7. Bed material flow and drainage — a drain system from the primary bed is provided to remove, cool, screen and recycle bed material from the bottom of the primary zone. Two objectives are to remove coarse particles that may have entered the system as debris and to remove and store bed material for future density control. Load reductions present the most common need for removing this material. As with makeup flow control, this is an intermittent function and is controlled by periodic operation of the drain valves.
8. Stoichiometry — if a bed material such as limestone is being used as a sorbent, the stoichiometry (ratio) is controlled by adding fresh sorbent and discarding spent material. This is accomplished by the makeup and drainage systems, respectively. The operator, by monitoring established SO_2 levels, must then seek an equal balance of fresh sorbent input and spent sorbent removal. (See Chapter 16.)

Fuel sizing and fuel characteristics

Coal firing While difficult to closely control, fuel sizing is very important to successful boiler operation. Fine coal burns in the freeboard area and results in late burnout and an elevated FEGT. To compensate, the operator must bias air flows upward (more secondary and/or tertiary) to increase burnout. At the same time, a compensation in bed density is required. Conversely, if the fuel is coarse, it burns in the primary zone, requiring more air low in the boiler and more density. When fuel variation occurs the operator should be thoroughly trained on the required adjustments. Variations in fuel quality are typically not a major issue with coal-fired CFB units.

Biomass firing Biomass fuels provide a particular challenge in fluidized bed operations because of continuous variations in size, shape, moisture content and heating characteristics. These characteristics require operators to be especially responsive to needs for bed depth changes, air flow splits and fuel feed rate. To minimize operating problems, fuel sources should be continuously blended to achieve as uniform consistency as possible. Modest inventories are needed — large enough to permit blending for uniformity of feed but not large enough to result in excessive inventory storage times which encourage loss of volatiles, fire hazard and general fuel characteristics degradation. Bed material sizing should be checked on a regular basis to ensure that the bed material is not deteriorating. Care must be taken in such sampling procedures because of the high material temperature.

Startup and shutdown considerations Chapter 16 provides general guidelines for the startup of CFBs. An important part of a cold startup is the warming of the bed material. Typically, a CFB is loaded with enough bed material to provide a 10 in. wg (2.5 kPa) pressure differential across the primary zone. During warming and low load operation, the bed behaves like a bubbling bed with high solids elutriation. Solids that leave the bed are collected in the particle separation device and returned to the primary bed. The solids inventory gradually heats up and approaches the FEGT. At about 20% load the solids circulation becomes continuous. It is important for the operator to monitor this heating process. Addition of new (cold) bed material or overfeeding the fuel can result in low bed temperatures and/or excessive CO. Pressure increases must be gradual until the heating is complete. Generally, the allowable rate of drum metal temperature change governs this regime. After bed and drum heating, load may be increased at about 3% per minute. Upon removal of auxiliary fuel, load is raised by adding fuel and air in response to drum pressure. At the same time, the operator must add material to control bed density.

During boiler shutdown, the operator must remove bed material as load is decreased. Above-bed and duct burners must be operated at low loads if the bed temperature drops to 1250F (677C). This also assures complete fuel burnout prior to shutdown and personnel entry. If the boiler is being shut down for maintenance, all bed material should be drained.

Bubbling-bed systems

Overview As is the case with its CFB counterpart, bed management of a bubbling bed is a key operating skill. Bed inventory, bed sizing, makeup and drainage must be integrated into operating procedures from data collected during commissioning tests. The bubbling bed thermal performance is most affected by bed height. Bed height can be minimized by inventory or by reducing the air flow to zero (called *slumping*) at selected areas. The bubbling bed is also equipped with additional instrumentation to assist the operator in developing the unit specific operating guidelines.

1. Primary zone ΔP — the gas side differential pressure is used as an indication of bed height. Because the primary zone is compartmented on the air side, multiple ΔP cells are displayed.
2. Bed temperature — the optimum bed temperature of 1550F (843C) also applies to bubbling beds and is a primary control variable. Because control of this variable is critical, multiple thermocouples are installed in the bed.

The operation of a bubbling bed is similar to that of a conventional drum boiler in that fuel and air demands are set by the required steaming conditions. Bed height is then used to control bed temperature. If an SO_2 sorbent is used, it is usually metered in proportion to the fuel flow. Various interactions can occur when manipulating control variables and the operator must be specifically trained to respond to upsets. These responses include:

1. Fuel flow control — fuel feed is increased or decreased based on steaming requirement. Changes in fuel feed have a short term impact on bed temperature, but this temperature is properly controlled by bed height.
2. Bed height — height is controlled by adding material from the makeup system or by removing material through the drain system. Height is increased to lower bed temperature and decreased to raise bed temperature. This is an operator controlled function and is not automatic. An optimum height versus steam flow relationship is determined during commissioning tests to provide an operator guideline. Because steam temperature is also impacted by the amount of bed material covering the steam-cooled surface, the operator may be required to trim the height to accommodate fuel variations.
3. Excess air control — excess air is primarily controlled based on steam flow. Excess air may be biased by the operator to change bed conditions. Increasing excess air at a given load increases bed turbulence and burnout in the bed while also providing more oxygen to the freeboard area. Reducing excess air at a given load would have the opposite effect. The operator must, therefore, bias the set point based on combustion conditions. Fuel sizing variations and bed material sizing mix are most likely to influence combustion conditions.
4. Bed material flow — material flow to the primary zone is through makeup and bed drains. Because these are generally intermittent devices, the operator must observe control variable trends, particularly height, and make appropriate adjustments.
5. Turndown — conventional methods of air and fuel reduction to achieve turndown have limited application for a bubbling bed. Load can be decreased by up to 20% by reducing bed height and further reduced by slumping one or more of several bed compartments. The operator is responsible for proper selection of these compartments.

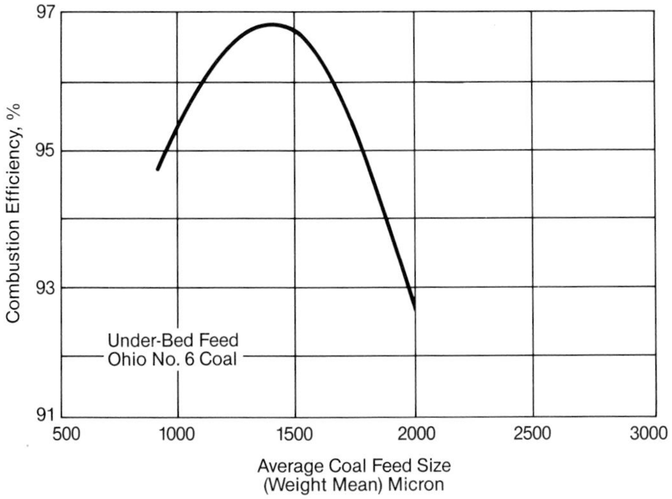

Fig. 9 Typical fuel size diagram for a bubbling bed.

Fuel sizing Careful attention to fuel sizing (minimum and maximum) is critical to successful unit operation for both under-bed and over-bed feed systems. Finer particles are elutriated too quickly and coarser particles tend to take too long to burn completely. In either case, unburned carbon and SO_2 emissions (if being controlled) both increase. An optimum size range exists for each fluidized-bed system and fuel type. An example is provided in Fig. 9. The proper size range is fine tuned through unit operating practice. In addition, for the over-bed feed system fuel size interacts with spreader speed, angle of injection and gas velocity.

Operation of Kraft recovery boilers

Overview

The Kraft recovery boiler has three purposes in today's pulp and paper industry:

1. recovery of sodium and sulfur compounds from the spent pulping liquor in forms suitable for regeneration,
2. efficient heat recovery from burning the liquor to generate steam for process use, and
3. operation in an environmentally responsible manner, cooling the combustion gases to allow back end particulate collection and minimizing the discharge of objectionable gases.

Steam flows can approach 1,000,000 lb/h (126 kg/s) at superheater outlet pressures as high as 1500 psi (103 bar) and final steam temperatures of 1050F (566C). These units burn black liquor with solids contents approaching 75 to 80% and with heating values decreased to about 5800 Btu/lb (13,491 kJ/kg). A detailed review of designs and general operating practice is provided in Chapter 26. Selected operating issues are highlighted below:

1. black liquor combustion process and air systems,
2. emissions,
3. auxiliary burners,
4. operating problems, and
5. smelt-water reactions.

Black liquor combustion process and air flow

Unlike the firing of conventional fossil fuels, recovery boiler combustion goes through several distinct stages. Black liquor firing is composed of drying, volatile burning, char burning and smelt coalescence as discussed in Chapter 26. This has a distinct impact on unit operation and air addition. All modern B&W recovery boilers are equipped with three levels of combustion air. These are known as the primary, secondary and tertiary air streams in order of increasing elevation in the furnace.

Primary air Primary air is in the lowest zone in a recovery boiler furnace. The air ports are located approximately 3 ft (0.9 m) above the furnace floor. Admitted on all four walls of the unit, the primary air provides perimeter air around the bed at low velocity and low penetration, so the boiler does not lose combustion. The primary air is critical to bed stability, bed temperature and reduction efficiency.

Secondary air High pressure secondary air penetrates the unit and is admitted at the top of the bed providing the combustion air for ignition and the heat to dry black liquor droplets and support pyrolysis. This air also helps control SO_2, total reduced sulfur (TRS) and CO emissions. Finally, secondary air controls the shape of the top of the bed. Fig. 10 indicates the proper bed shape with the relative locations of the air ports and liquor nozzles.

Tertiary air Tertiary air is admitted at the upper elevation, above the liquor guns. Like secondary air, it is admitted at high pressure to provide penetration across the width of the furnace to complete the combustion process. This is an oxidizing environment to control CO and TRS emissions.

To maintain combustion in the proper location, the air flow splits and pressures must be properly maintained. Ranges of typical air flow distributions, inlet temperatures and pressure differentials are summarized in Table 3 for the two dominant firing systems, stationary and oscillating.

Successful oscillating or stationary firing depends on the correct liquor droplet size. This size must permit complete in-flight drying before reaching the furnace wall or bed. The variables in black liquor combustion include liquor percent solids, temperature and pressure at the spray nozzles, spray nozzle sizing and splash plate angle, and number and position of black liquor guns in service. Other variables include the vertical and angular movement of the oscillators or the off-horizontal angle of the liquor guns for stationary firing. Other factors on the air side include total excess air; air temperature(s); air splits between the primary, secondary and tertiary ports; and the static air pressure in each of these air zones.

The objective in the manipulation of all of these variables is to control the steps of the drying and combustion of the liquor, and control bed temperatures. The result is stability of operation, with minimum plugging and emissions.

Emissions

The recovery boiler emissions that are covered in this section are SO_2 and TRS. TRS is strictly regulated because of the sulfur's odor. Humans can detect concentrations of less than 1 ppb of hydrogen sulfide, which is the dominant TRS compound. In 1978, the Environmental Protection Agency limited TRS emissions for units with direct contact evaporators to 20 ppm; 5 ppm is the limit for units without these evaporators. Excessive TRS generation results from low bed temperatures. To minimize TRS formation a uniform hot bed is required. Most remaining TRS can be burned to SO_2 with effective distribution and penetration of secondary and tertiary air.

SO_2 is a costly gaseous emission. All of the SO_2 that is generated must be replaced with makeup chemicals. SO_2 is produced when TRS is burned and oxidized in the recovery furnace. It can also be formed as a result of other reactions with sulfur compounds. The SO_2 can be captured by volatilized sodium compounds prior to leaving the boiler and can be recovered as salt cake (sodium sulfate). The release of sodium for sulfur capture is a function of bed temperature with higher temperatures resulting in more sodium release and therefore less SO_2.

SO_2 emissions are also associated with the amount of sulfur in the system. Higher mill sulfidity yields more SO_2 than available sodium and may result in SO_2 emissions.

Table 3
Typical Air Flow Splits and Operating Conditions

	Firing Technique: Stationary Firing	Firing Technique: Oscillating Firing
Primary air	30 to 40% 300F (149C) 3 to 4 in. wg ΔP (0.7 to 1 kPa)	40 to 50% 300F (149C) 1 to 3 in. wg ΔP (0.2 to 0.7 kPa)
Secondary air	40 to 50% 300F (149C) 8 to 18 in. wg ΔP (2 to 4.5 kPa)	20 to 30% 300F (149C) 6 to 10 in. wg ΔP (1.5 to 2.5 kPa)
Tertiary air	10 to 20% 80F (27C) 10 to 20 in. wg ΔP (2.5 to 5 kPa)	20 to 30% 300F (149C) 8 to 12 in. wg ΔP (2 to 3 kPa)
Economizer outlet conditions	< 2.5% O_2 100 to 200 ppm CO	2.5 to 3% O_2 200 to 300 ppm CO

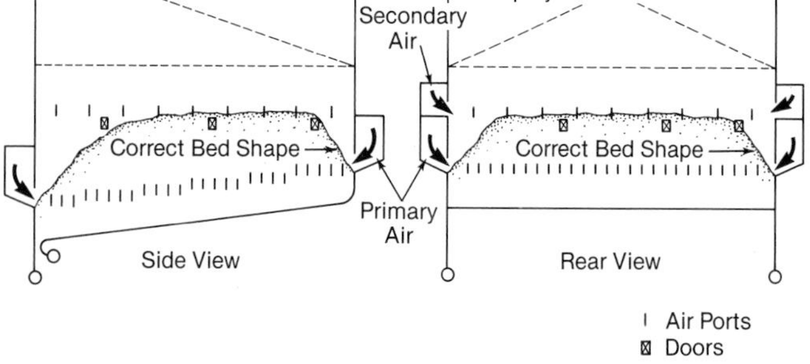

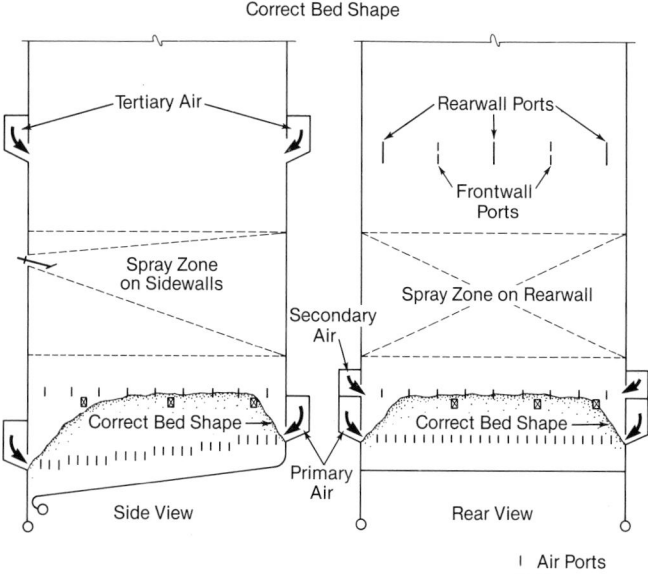

Fig. 10 Proper bed shape for a black liquor recovery boiler.

Auxiliary burners

Recovery boilers are equipped with oil- and/or natural gas-fired auxiliary burners located at the secondary and tertiary air port elevations. The burners in the secondary windboxes are used for unit startup, as black liquor can only be fired into a heated furnace with an auxiliary ignition source. The secondary burners are also used to stabilize the smelt bed during an upset condition and to burn the bed out of the unit on shutdown.

Operating problems

Two problems unique to recovery boiler operations are plugging and aggressive tube corrosion.

Plugging of the superheater, boiler bank or economizer is generally caused by two mechanisms. The first is condensation of the fume or normal gases given off by the black liquor combustion. The hot combustion gases which leave the lower furnace contain vaporized compounds

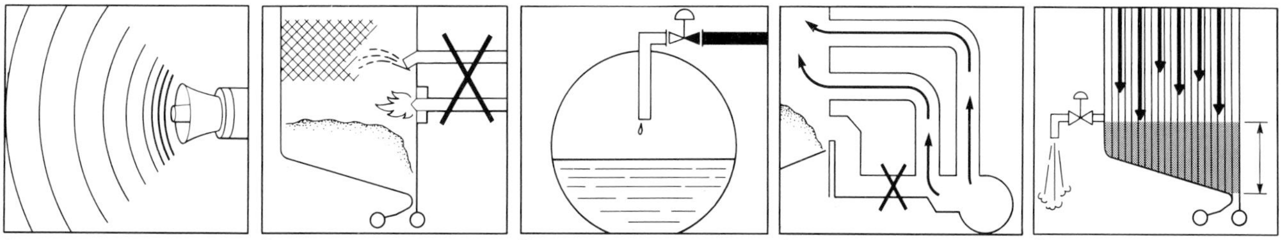

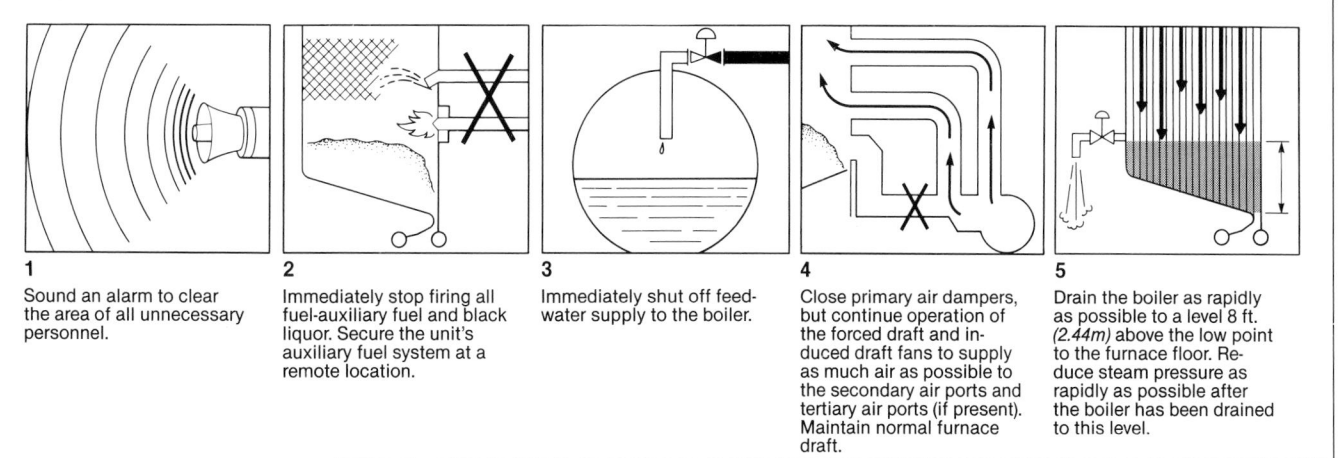

1
Sound an alarm to clear the area of all unnecessary personnel.

2
Immediately stop firing all fuel-auxiliary fuel and black liquor. Secure the unit's auxiliary fuel system at a remote location.

3
Immediately shut off feedwater supply to the boiler.

4
Close primary air dampers, but continue operation of the forced draft and induced draft fans to supply as much air as possible to the secondary air ports and tertiary air ports (if present). Maintain normal furnace draft.

5
Drain the boiler as rapidly as possible to a level 8 ft. *(2.44m)* above the low point to the furnace floor. Reduce steam pressure as rapidly as possible after the boiler has been drained to this level.

Fig. 11 Emergency shutdown procedure for an operating recovery boiler.

that condense when the gas temperature is cooled in the upper zones of the boiler, superheater or boiler bank. This condensation is dependent on gas temperature. The material condenses on the cooled heat transfer surfaces such as furnace screen tubes or primary superheater tubes. It also precipitates out when the gas temperature falls below approximately 1100F (593C). In either case, this material is a source of fouling in convection surfaces. Condensation or precipitation depend upon the temperature regime and chemistry of the fume. The deposit can be in the plastic range and very difficult to remove. On a properly designed and operated unit, this transition of the fume from a dry gas to a sticky substance takes place in the upper furnace, wide spaced screen or superheater. Here the deposits can be controlled by sootblowers. On an overloaded or improperly operated unit, the gas temperature remains elevated farther back into the closer side-spaced boiler bank or economizer, where controlling the plugging with sootblowing equipment becomes difficult. The gas temperature at which the fume turns to a sticky liquid is also liquor chemistry dependent, with higher levels of chlorides or potassium compounds being detrimental to unit cleanability.

The second major cause of plugging is mechanical carryover of smelt, or unburned black liquor, into the convective heat transfer sections of the boiler. This material is predominantly made of sodium carbonate (Na_2CO_3), Na_2SO_4 and sodium sulfide (Na_2S) compounds. The cause of this carryover is related to operation of the liquor nozzles, oscillators and various air port settings. As previously noted, the black liquor firing equipment and air system settings must produce complete in-flight or wall drying of the liquor droplets. If this drying occurs too high above the smelt bed, the less dense droplets are entrained in the gas stream and carry over into the convection sections of the boiler. In practice, most recovery boiler plugging results from a combination of fume condensation and mechanical carryover.

Recovery boilers are also subject to aggressive corrosion compared to conventional fossil fuel-fired units due to the presence of corrosive sulfur, chloride and other trace compounds in an elevated temperature environment. The Kraft recovery boiler also operates in both an oxidizing and reducing atmosphere due to the combustion process. Because of these conditions, the floor and lower furnace walls are constructed with studded carbon steel tubes covered with refractory or bimetallic composite tubing.

Smelt-water reactions

One unique and undesirable feature of a Kraft recovery boiler is the possibility of water entering an operating furnace through a tube leak or external source, resulting in a smelt-water reaction. A smelt-water reaction occurs when water combines with hot or molten smelt; a violent explosion can result. This concern has resulted in years of study and testing and has prompted enhanced industry standards on unit design, operation and maintenance. Modern units are equipped with the provisions for an emergency shutdown procedure. This is initiated if water is suspected to have entered the furnace of an operating recovery boiler. Refer to Fig. 11 for procedural details.

Section VII Maintenance and Life Extension

This section describes the last element of a successful steam generating system life cycle plan — integrated maintenance, condition assessment, and life extension programs.

As owners and operators of steam plants search for optimum performance, efficiency and life cycle for all equipment, issues of maintenance and life extension have become increasingly important. This section of *Steam* has been expanded to address these issues.

The section begins with a discussion of routine maintenance, encountered with all plants both utility and industrial. Maintenance programs are discussed, followed by specifics for both inspection and repair. Condition assessment is then addressed with detailed discussion about examination techniques, assessment of various components, and analysis techniques for determining remaining life. The effects of cycling operation are also addressed.

The section ends with discussions about life extension and upgrades, including specific examples. Aging is discussed and example upgrades are described.

Root pass welding with the tungsten arc process.

Chapter 44
Routine Maintenance

Routine maintenance is a key element in optimizing the production and life of steam generating facilities. When integrated with ongoing unit operations and a program of inspection and upgrades, routine maintenance maximizes safety, reliability, availability, efficiency and environmental protection. While once a relatively straightforward job of restoring damaged or broken components, maintenance has evolved into a sophisticated systematic program of condition assessment, predictive techniques, corrective steps, preventive activities, and ongoing observation and evaluation of plant operations. The goal of a maintenance program is to maximize power production, availability, safety and quality while minimizing costs and the impact on the environment. This goal is achieved in today's power stations through: 1) a comprehensive maintenance program to direct, monitor and control maintenance activities, 2) execution of specific in-service component maintenance functions, 3) performance of specific out-of-service component maintenance activities, and 4) appropriate care of idle equipment.

Maintenance program

Comprehensive approach

The comprehensive maintenance program is based upon a general management maintenance philosophy plus performance standards set for individual steam generating units. These unit performance standards are typically established through a system wide economic evaluation which determines desired unit operating load, availability, cycling requirements and life. Once the overall maintenance philosophy, goals and objectives have been established, a maintenance program is developed around an evaluation of the failure modes and effects of individual components, and the resulting corrective or preventive maintenance options. These options are compiled for the entire plant and then combined with a system of monitoring and control to achieve the plant goals.

Fig. 1 provides a typical decision tree for establishing individual maintenance activities. The options begin with the decision to either wait for an undesirable condition to occur or to initiate some action to avoid a failure.

Corrective maintenance procedures are actions taken to restore a component or system from a failed or deteriorated condition. The decision involved is usually one of repair versus replace.

Preventive maintenance procedures are activities taken to avoid or minimize the probability of failure or deterioration. Preventive actions can be time-, quantity- or condition-directed. Time- and quantity-directed activities are performed independent of equipment condition. They are usually performed when a component has a fairly defined life expectancy or the cost of analysis is higher than the cost of replacement.

Condition-directed activities are performed as needed based on some observed or measured condition. These conditions can be detected in service by inspection or testing, out of service by inspection or testing, or they may be the result of a *predictive maintenance program*. Examples of condition-directed activities include regenerative air heater seal replacement due to an observed oxygen (O_2) increase across the gas side of the air heater, and bearing replacement based on vibration signature analysis. An example of a predictive program is a high pressure, high temperature superheater header replacement program that might be part of an overall life extension program as discussed in Chapter 46.

Plant characterization and option development

A key initial activity is to understand the potential problems of the facility and to develop appropriate responses. First, all potential equipment conditions throughout the plant (such as failures and performance degradations)

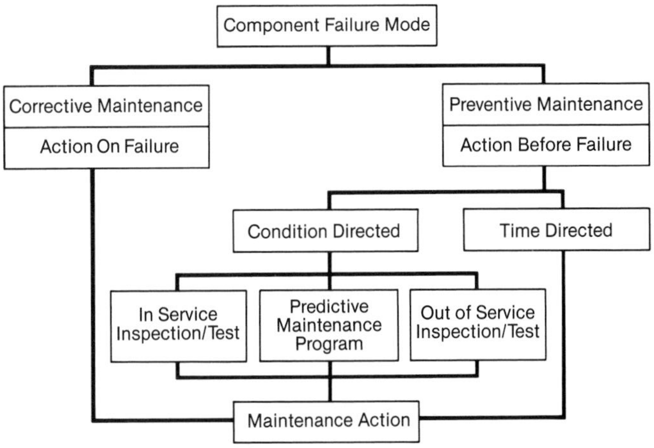

Fig. 1 Maintenance decision tree.

potentially requiring attention must be defined. Second, based on unit-specific performance standards, goals and objectives, the appropriate methods of detection and response for each of the failure conditions must be selected.

One successful method is the reliability centered maintenance (RCM) process which uses a decision logic tree to determine the requirements for equipment maintenance based on safety, operational consequences of possible failures, degradation of production through wear and deposition, and impact on the environment. The process begins by constructing a failure mode evaluation chart (FMEC), which identifies each piece of equipment and its system boundaries. Then for each identified component, all potential modes of failure or degradation are listed along with cause and method of detection. For each individual failure mode, the effects of the failure are listed for the component, system and entire plant. The result is a comprehensive list that identifies each failure mode that can impact system capabilities, jeopardize personnel safety and impact the environment.

Once the FMEC is generated, a family of tables called potential maintenance action tables (PMAT) can be developed for each component. Each table lists possible proactive or reactive responses for each mode of failure (or degenerative condition). The range of responses should cover all of the various options including run-to-failure, time-directed, and condition-directed options. For each response option, a cost benefit value or code is determined.

Action selection

Each potential failure requires the selection of a response consistent with the company's purposes and goals for the equipment and facilities. The result is a set of maintenance programs for each component and failure mode in the facility. When the individual component programs are combined they form the maintenance program for the entire facility. This process is complex and time consuming. Because of the extensive time and effort required for the complete evaluation of a facility, most companies embarking on an RCM program do so in small steps. Generally, the components or sections selected for evaluation are those that cost the most or that have the most impact on availability.

Once detailed maintenance activities are identified, a management program is required to direct, control and measure effectiveness. The three primary elements that make up maintenance actions are *material*, *personnel* and *technology*. A maintenance organization must be able to identify, plan, organize, select and control the correct combination of these three primary elements.

Managed maintenance programs

A managed maintenance program (MMP) consists of well defined procedures that guide the process and flow of maintenance actions and tools, such as computerized database systems, which support and keep the processes operating smoothly.

Fig. 2 illustrates the maintenance process flow. A procedures manual should be developed for each plant defining who is responsible for performing each part of the process, specifying what approvals are needed and who has approval authority, explaining how each block of the process is to be accomplished, describing the transition process between blocks, and compelling closure of the loop to make it self-sustaining. Fig. 3 illustrates the numerous pieces that make up the three elements that must be controlled to have a successful MMP.

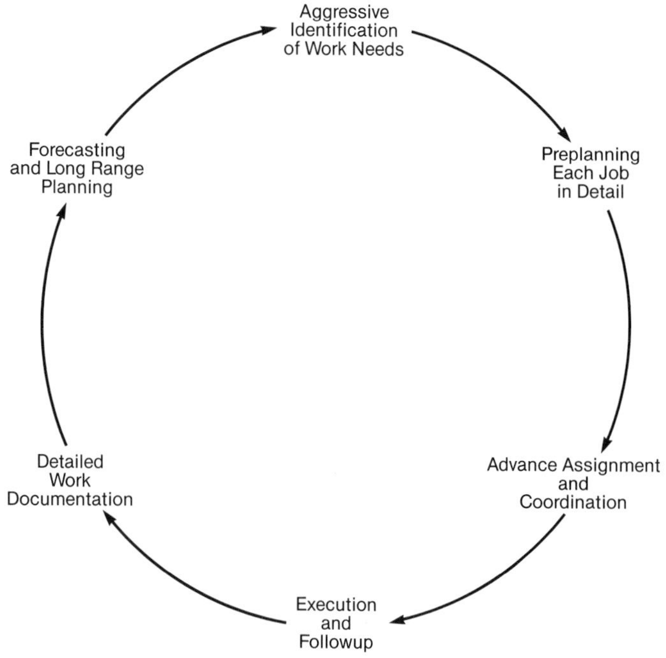

Fig. 2 Maintenance process flow.

An MMP is specifically directed at reliability improvement and cost reduction through the application of the most favorable methods for detecting and correcting abnormal conditions. The systematic approach can more efficiently allocate scarce manpower and equipment resources, avoid unnecessary outages, increase availability, reduce materials consumption and extend normal equipment life spans. The effects of maintenance spending on availability are illustrated in Figs. 4 and 5.

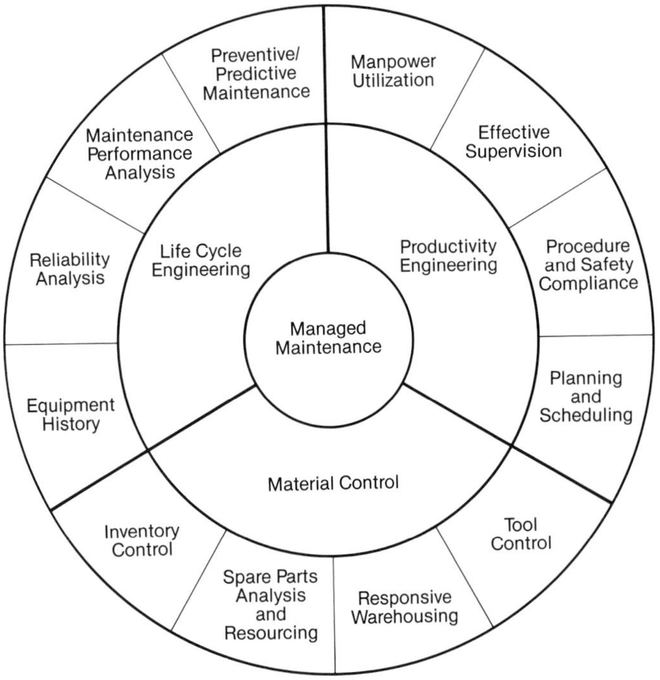

Fig. 3 Elements of the managed maintenance program.

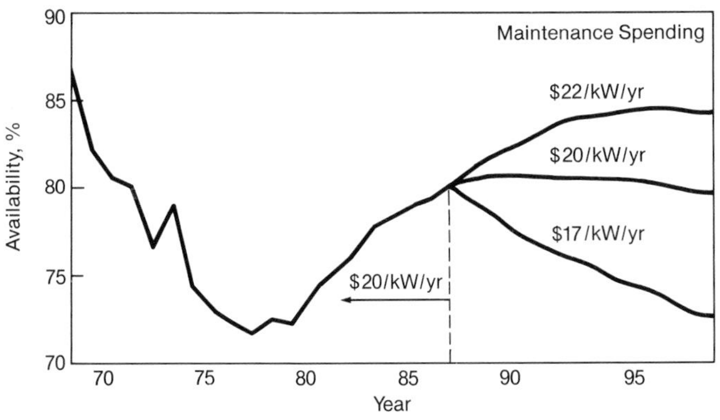

Fig. 4 Effect of maintenance spending: projected coal-fired unit availability.[1]

The results of MMP projects vary based on the initial conditions at a facility, goals set by management and the motivation of the personnel involved. The range of improvements that can be expected, if a moderately successful MMP is implemented, are as follows:

Area of improvement	Range of improvement
Availability	0.5 to 5%
Worker productivity	25 to 100%
Overtime	Reduced to less than 5%
Cost of maintenance	Reduced by 10 to 40%

In-service routine maintenance — fossil fuel equipment

Maintenance begins with the support of safe, efficient in service operation of the steam generator. Maintenance functions should be integrated where possible with ongoing operations to guard against unsafe operating conditions, which can result in injury or unplanned unit down time, and to maintain the highest level of power system efficiency given the constraints of fuel and overall equipment condition.

Safety

Furnace and setting Safety begins by operating the equipment in a manner that avoids any condition that could result in explosive mixtures of fuel in the furnace or other parts of the setting.

Four items of major importance in the prevention of furnace explosions are:

1. up-to-date operating procedures and operating training,
2. frequent checks of combustion monitoring equipment by operations and maintenance personnel to ensure that any loss of ignition is immediately noted,
3. detection of unburned combustibles in the flue gas as a backup to flame observation, and
4. monitoring and maintaining the correct fuel/air ratio.

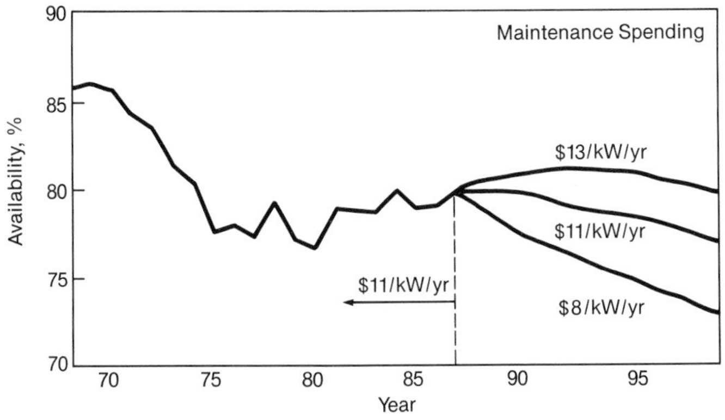

Fig. 5 Effect of maintenance spending: projected oil and gas unit availability.[1]

As these key factors imply, most furnace explosions result from the failure to detect a *loss of ignition* which is followed by the accumulation and subsequent ignition of an explosive mixture in the boiler setting. This may occur even though other indicators such as furnace exit gas temperature, steam temperature and steam pressure only witness moderate reduction, frequently associated with the fuel being incompletely burned.

Precautions outlined in Chapters 10, 12 and 13 provide more specific information concerning the safe handling and combustion of oil, gas and pulverized coal. Means must also be provided to avoid the accumulation of combustible soot or oil on air heater or economizer surfaces in oil-fired units or during oil firing and startup of pulverized coal units.

Pressure parts Protection of pressure parts from excessive thermal stresses or overheating is also critical. Most periodic maintenance of pressure parts is performed during outages; however, there are a number of activities that can and should be performed during operations.

The capacity, relief pressure, and procedures for installation and testing of safety valves have been codified and are law in nearly all jurisdictions worldwide. Modern safety valves are highly dependable when properly installed and checked by inspectors. The operation of these valves should be periodically checked and the set points should be verified.

Adequate water level and flow keep pressure parts from exceeding their safe design limits. Assurance of level and flow are immediate maintenance and safety concerns. In addition they can affect the extended life of the equipment. Routine in-service maintenance and operational checks should be performed to assure that water level and flow systems are functioning properly. Even though some of the systems are redundant and have auctioneering systems to reject false readings, some simple checks to assure proper operation are important to maintain safe operation. Visual checks for drum boilers assure that the sight glass matches the remote transmittal signals. Loose insulation, leakage or impinging leakage from another source could affect sensing devices. For once-through boilers, flow measuring devices can not be physically checked; however, the instruments should be verified by comparison to other readings.

Monitoring and maintaining feedwater and boiler water conditions within prescribed limits, as well as steam purity, are also essential. (See Chapter 42.) To assure that the condition of the water is correct, periodic operational and maintenance checks should be performed on the conditioning equipment and monitoring devices. Chordal thermocouples installed in furnace tubes at strategic locations measure how effectively the water chemistry is controlled. (See Chapter 40.) This serves as an excellent guide to internal cleanliness of the boiler and can be used to indicate the need for chemical cleaning. Tracking chordal thermocouple readings is one of the best predictive maintenance techniques for planning chemical cleaning. Periodic readings should be recorded and maintained by the maintenance engineer.

Monitoring temperatures

Reliable measurement of various temperatures is essential. In general, these temperatures fall into three categories: 1) metal, 2) gas, and 3) steam and water temperatures.

Metal temperatures Proper maintenance of devices for measuring superheater and reheater tube temperatures, chordal thermocouples, and tube temperature thermocouples in various circuits of once-through boilers is essential in preventing excessive metal temperatures. Periodic checks of temperature sensors should be made to assure proper operation. Personnel should look for open circuits, indicating a broken thermocouple, or for readings that do not coincide with historical readings for the same operating conditions. The latter indicates that either the instrument requires calibration or a condition is occurring that requires investigation.

Records of thermocouple readings should be kept for thick wall components as well as steam-cooled tubes. In thick wall, water carrying headers such as those found on economizers, temperature shock records are needed to determine cyclic stress and to predict the onset of cracking. Steam-cooled, thick wall headers such as those found on superheaters and reheaters are subject to cyclic thermal stresses as well as creep fatigue. Thermocouple readings of steam-cooled tubes in the superheater and reheater are needed to calculate creep fatigue and predict remaining life.

Gas temperatures Periodic maintenance and operation checks on air and gas temperature sensors assure safe operation and predict the maintenance condition of the unit. Assuring reliably calibrated equipment to measure gas temperatures can lessen the likelihood of damage. Analysis of gas temperature records is useful in predicting life and/or determining the maintenance needed during upcoming outages.

Steam and water temperatures During startups and restarts, the matching of steam temperatures to allowable turbine temperatures is mandatory. Keeping water and steam temperature measuring devices properly calibrated and in good operating condition is necessary to prolong and enhance the life of all components.

Monitoring pressures

The measurement of pressure and pressure differentials is mandatory. Periodic maintenance and operational checks should be performed to assure that pressure sensors are performing their function.

Water and steam A maintenance file should be kept on internal pressure drops of certain components. Differential pressure serves as an index of internal cleanliness when measured across various portions of the unit. Tracking this differential and relating it to internal inspections can indicate the need for maintenance or cleaning during an outage.

Furnace Differential pressure across the various gas side components of a unit can indicate deterioration in cleanliness and can be used to determine the maintenance needed during an outage. It is also an indicator of the effectiveness of the sootblowing system and can be used to determine operational changes, sootblowing cycle changes or maintenance to the sootblowing system.

Maintaining for efficiency

The best way to analyze efficiency is to examine major efficiency losses, determine the causes and plan the actions to control losses. The process begins by conducting periodic efficiency tests or by installing an on-line unit performance monitoring system which continuously evaluates performance and compares actual performance to expected operation. The results of efficiency tests can be used to monitor trends and make early identification of a particular component contributing to a loss in efficiency. On-line monitoring systems have a built in capability for gathering this information and displaying it in trending graphs for easier analysis and interpretation.

To better understand maintenance for efficiency, the following is a typical heat balance for a selected pulverized coal-fired utility boiler:

Dry gas loss	5.16%
Loss due to hydrogen and moisture in fuel	4.36%
Loss due to unburned combustible	0.50%
Loss due to radiation	0.30%
Loss due to moisture in air	0.13%
Manufacturer's margin and unaccounted loss	1.50%
Overall efficiency	88.05%

Minor losses are usually lumped into one percentage figure and labeled *unaccounted*. These minor losses are quite extensive in number but are small in relation to overall heat loss. Of these losses, the largest is the loss due to sensible heat in ash or slag.

To determine what in-service or out-of-service maintenance actions are needed to correct efficiency losses, it is necessary to understand the parameters affecting these heat losses.

Dry gas loss There are two variables in dry gas loss — stack temperature of the gas and the weight of the gas leaving the unit. Stack temperature varies with the degree of deposition on the heat absorbing surfaces throughout the unit and with the amount of excess combustion air. The effect of excess air increases the gas weight and raises the exit gas temperature; both effects reduce efficiency. A 40F (22C) increase in stack gas temperature on coal-fired installations gives an approximate 1% reduction in efficiency.

When increased losses due to dry gas are indicated, the source of the increase must be determined and appropriate actions can be planned. It must be determined if the dry gas loss is the result of low absorption or increased gas weight. Lower absorption is usually the result of slagging or fouling. This could be the result of inadequate sootblowing which could be operational, maintenance or design related. It could also be the result of improper combustion due to the condition of the fuel supply equipment, fuel burning equipment or fuel/air control system. It is also possible that the problem is the result of a change in the fuel.

If the problem is excessive gas weight due to excess air, investigation should focus on the fuel burning or the fuel/air control system. Once the causes of the loss of efficiency have been narrowed to a relative few, it will take additional off-line and on-line investigation to pinpoint the actual sources.

High exit gas temperatures and high draft losses with normal excess air indicate dirty heat absorbing surfaces and the need for sootblowing or sootblower maintenance. Generally, high excess air increases exit gas draft losses and indicates the need to adjust the fuel/air ratio or perform maintenance on the control devices as discussed above. Even though this may be the most prevalent cause, high excess air can also be caused by excessive casing leaks, excessive cooling or sealing air, or high air heater leakage.

Unburned combustible loss In coal-fired units, unburned combustible loss includes the unburned constituents in the ash pit refuse and in the flyash. These actual carbon constituents may include a variety of compounds; however, in calculating the unburned loss, all constituents are usually considered as pure carbon.

Unburned combustible loss will vary with the volatile matter and the ash content of the coal. These conditions are normally beyond the operator's control. Once again, excess combustion air is the controllable variable affecting this loss.

When high carbon loss is indicated by efficiency testing, maintenance should be concentrated on a few areas. High levels of combustible in the refuse indicate a need for adjustments to or maintenance of fuel preparation and burning equipment. Improperly sized fuel particles, such as excessively large particles coming from a pulverizer or improperly atomized oil, will cause increases in unburned carbon. Secondly, improperly adjusted or damaged fuel-air mixing equipment will result in high carbon loss. Thirdly, improperly adjusted or maintained fuel/air ratios can contribute to unburned carbon loss. In all of these cases, both on-line and off-line inspection and analysis will be necessary to determine the root cause; however, careful off-line inspection will usually narrow the choices to a relative few.

Fig. 6 illustrates the effect on dry gas loss, unburned carbon loss and gross unit efficiency of increasing excess air from the normal operating range to an abnormally high value. The numbers on the curves indicate the three following units from which the data were obtained:

1. a stoker-fired industrial boiler, 35,000 lb/h (4.4 kg/s) capacity,
2. a pulverized coal-fired utility boiler, 600,000 lb/h (75.6 kg/s) capacity, and
3. a pulverized coal-fired utility boiler, 1,170,000 lb/h (147.4 kg/s) capacity.

Fig. 6 shows a net loss of efficiency with increasing excess air in all cases. In each case a small reduction in unburned combustible loss is more than offset by the increased dry gas loss.

Continuous monitoring of flue gas temperatures and flue gas oxygen content by a regularly calibrated recorder or indicator and by periodic checks on unburned combustibles in the refuse will indicate if original efficiencies are being maintained.

Outage routine maintenance

On-line maintenance and investigations are important in the overall maintenance program. However, the major expense maintenance projects and much of the actual physical work can only be done off-line. These are called maintenance outages and vary greatly in length and scope.

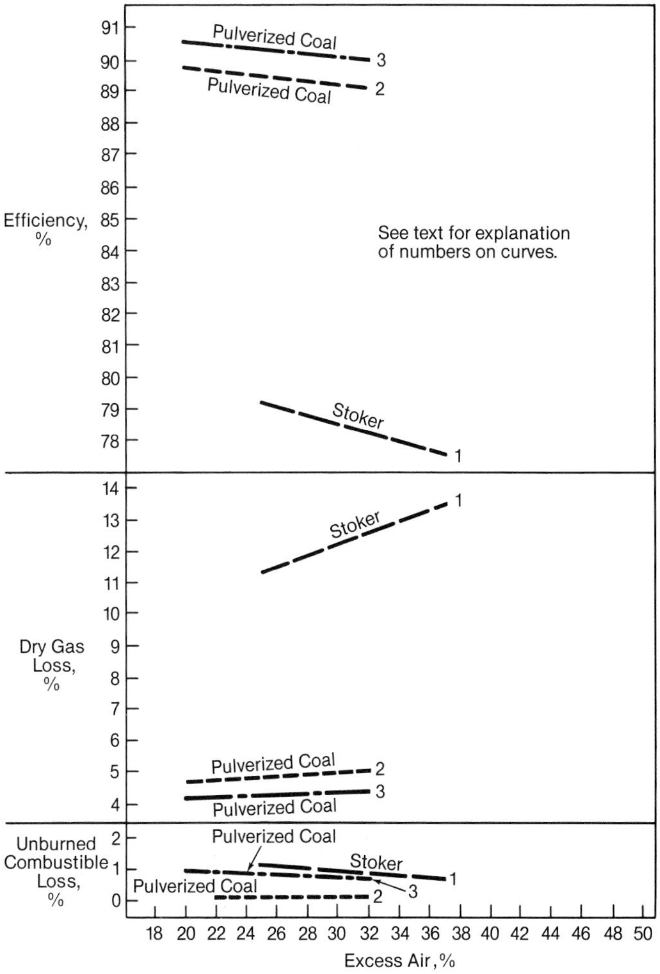

Fig. 6 Efficiency and heat losses versus excess air.

Outages for preventive maintenance are scheduled to prevent equipment failure. In the electric power industry, schedules must take into account system load variations. It is generally preferable not to shut down large electric generating units during the peak load seasons of mid summer or mid winter.

Internal cleanliness and inspection

The cleanliness of internal (water or steam side) surfaces is of prime importance. Internal deposition can lead to chemical attack, corrosion and overheating.

One of the best preventive steps to assure safe, dependable operation is the maintenance of the boiler water to limit internal tube deposits. (See Chapter 42.) Eventually, however, internal tube surface cleaning will be required. There are some on-line methods for determining cleanliness. The chordal type thermocouple (Chapter 40) can be used to measure the extent of internal deposits and to indicate the need for chemical cleaning by tracking the differential temperatures measured by the device. The measurement of progressively increasing water side pressure drop across a once-through boiler can also determine cleanliness.

Even when on-line readings do not indicate a need for cleaning, checking during regularly scheduled out-of-service periods is recommended to assure continued operating capability. This becomes imperative when in-service measurements are not taken.

Low pressure units Low pressure, drum-type boilers lend themselves to careful, visual inspection to establish the need for internal cleaning. Inspection of steam drum internals should be part of any out-of-service inspection, particularly if operating checks of steam leaving the drum show evidence of excessive solids in the steam. All baffling and steam separating equipment should be in place and any joints designed to seal against leakage should be tight. It is usually possible to observe evidence of carryover-producing leaks if the drum is inspected shortly after the boiler is drained and before any internal cleaning is begun.

Cracks may be found in the steam drum internal surface. Most of these are minor, caused by etching of welds during acid cleaning, and can be ground and dye checked as necessary. Other cracks may be the result of thermal shock and are quite serious. If the cracks are not due to etching, the manufacturer's representative should be contacted and a thorough investigation and repair process formulated.

High pressure units Today's high pressure steam generating units, with complex circuitry and all welded construction, make visual examination very difficult and nonconclusive. It is preferable to remove one or more representative tube sections from the high heat input zone for examination and measurement of the internal deposit. Comparison of the internal area and weight of deposit per unit of area, expressed as grams per square foot, with empirically confirmed weight limits and the physical and chemical nature of the deposit establishes the need for internal cleaning.

When removing a tube sample for analysis, certain procedures should be followed to ensure that the sample is removed intact and uncontaminated, and that its exact location is documented.

Chemical or acid cleaning is the quickest and most satisfactory method for the removal of water side deposits. The guidelines and precautions that must be followed for the safe and effective use of chemicals are discussed in Chapters 42 and 45.

Boiler external surface cleanliness and inspection

Two of the most important activities that will occur during a maintenance outage are the cleaning and inspection of the external surfaces of the boiler and its related components. The cleaning will improve heat transfer and reduce the potential for corrosion and erosion. A thorough inspection and testing program will reveal areas needing maintenance.

External surface cleaning begins when the unit is online. Information on boiler and air heater gas temperatures, and gas side pressure drops measured during operation of the unit, are helpful in the planning of the outage. They will show whether external slagging or fouling, not removable by normal sootblower operation, is progressing and will help indicate location.

When ash deposits contain an appreciable amount of sulfur, they should be removed prior to any extended outage. These deposits can absorb ambient moisture and form sulfuric acid that will corrode pressure parts. Preparations for removing these accumulations should be made before the outage.

There are several methods used to clean the external surfaces. Loose or powdery accumulations that have collected in windboxes, flues, ducts and hoppers can usually be cleaned using a large vacuum system specifically designed for this task.

Many ash deposits that result from burning severely fouling solid and liquid fuels are removable by one of various water washing techniques. Units often are designed for water washing, with ample provisions made for drainage of the wash water and ash. Although air heaters may be sectioned so that areas can be isolated and washed without shutdown of the unit, an out of service period is required to clean the remaining boiler surfaces. Because the wash water will reactivate sulfur bearing ash deposits, the wash solution should be acid neutralizing. A good combination is warm water and trisodium phosphate (Na_3PO_4). The amount of Na_3PO_4 needed will be determined by the percent of sulfur in the ash and the total amount of ash accumulation in the unit. The Na_3PO_4 must be diluted into a pumpable solution and metered into the wash water. (See Fig. 7.) It is extremely important to check the pH of the water in the drains to determine whether the deposits are being neutralized. A drain water pH of at least 10.5 to 11.0 should be achieved by regulating the Na_3PO_4 feed rate.

Caution for external cleaning

The wash water first draining from the unit will, in all probability, be acidic. Extreme care must be taken to protect personnel and to dispose of the effluent to avoid pollution. Pumping more water than the drains on the unit can accommodate can overstress the structural parts of the hoppers, and ash accumulation in the hoppers may tend to plug the drains. Observation and rodding can prevent large buildups.

Extremely large slag deposits may require special techniques including blasting with explosives. Care should be taken for some types of ash deposits. The constituents of the ash turn into a concrete-like mass that is extremely hard to remove if allowed to become damp and then dry out. If the ash does become damp the accumulations should remain moist until they can be completely removed. Other types of very hard fouling accumulations are not readily removed by water washing. However, they may soften when exposed to moisture for several hours and can then be removed by high pressure spray.

In many plants the normal sootblower system is suc-

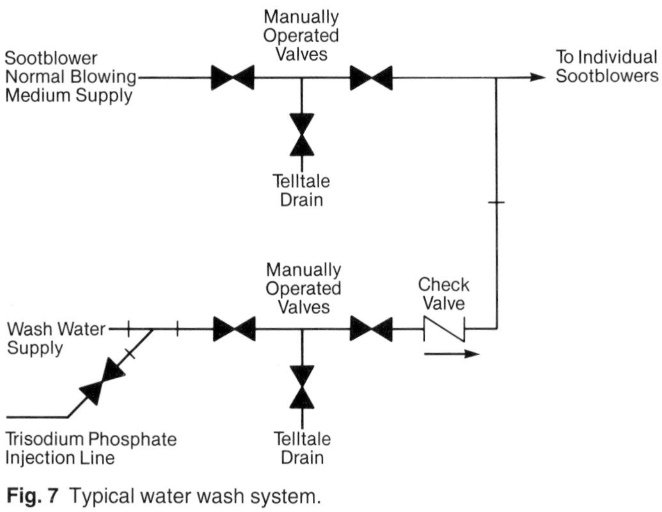

Fig. 7 Typical water wash system.

cessfully used for water washing by connecting a supply of high pressure wash water to the sootblower steam header system. (See Fig. 7.) Drain facilities should be open and ready for discharge of the wash water to catch basins. Wall areas or tubes in wide spaced arrangement lend themselves to cleaning by protracted soaking with the wash water. Areas where surfaces are closely spaced, as in regenerative air heaters, can be cleaned by high pressure water jets. Inspection will indicate areas requiring particular attention.

Out-of-service washing should be scheduled so that surfaces can be dried immediately after cleaning. A unit that must remain out-of-service after washing should be dried at a low firing rate, or by using auxiliary heaters, to prevent corrosion.

There are some important safety and environmental concerns when cleaning a unit. Respiratory and skin protection should always be worn in an uncleaned unit. Head protection should always be worn as protection from falling objects. Units should never be entered if they contain large suspended accumulations. The original equipment manufacturer's operating instructions should be followed.

External surface inspection

Once the unit is brought off-line, it should be thoroughly cooled, vented to remove all harmful gases, external deposits cleaned and lighting and scaffolding put in place for the maintenance activities. The very first activity is an inspection once the unit has been prepared. This may well be the single most important event to occur during the outage. The inspection will:

1. verify scheduled activities for the current outage and confirm the need for preplanned maintenance procedures indicated by information taken on-line,
2. identify additional maintenance activities not indicated during on-line monitoring to correct conditions that would eventually be detrimental to the unit, and
3. provide inspection and testing data to be used in predictive maintenance programs. An example is ultrasonic testing readings of the waterwalls to predict replacement times.

The following areas of the unit should be thoroughly inspected with the indicated objectives.

Tubes — overheating During the inspection, maintenance personnel should check for any possible signs of overheating of tubes, particularly if thermocouples have indicated this possibility. Furnace wall tubes should be examined for swelling, blistering or warping. If signs exist, it may be necessary to gauge tubes either by mechanical or ultrasonic methods to establish the amount of swelling or wall thinning. Superheaters, reheaters and economizers should be inspected in the same way.

If overheating of superheater tubes is suspected to be the result of rapid startup, the startup procedure should be rechecked, and superheater metal temperatures and entering gas temperatures should be closely monitored. If overheating is suspected to be the result of deposits from carryover of solids in steam, the equipment and procedures for maintaining water chemistry should be thoroughly reviewed. Calibration of steam purity measurement equipment should be verified. Also, complete carryover testing may be needed under widely varying conditions to determine the exact cause of the problem.

Tubes — erosion or corrosion To discover any possible signs of erosion or corrosion, a thorough inspection of heat absorbing surfaces should always be part of a planned outage. This inspection should be given added importance if any unusual operating conditions have preceded the outage. In general, erosion appears as a smooth and sometimes shiny loss of metal. It can be caused by impingement of sootblower jets, possibly dictating the need to reduce blowing pressure or realign the blower. Corrosion associated with ash deposits may be hidden and sandblasting or washing of surfaces may be required for detection and measurement.

Additional gas baffling may be required to prevent external erosion caused by channeling of gases between convection surfaces. Readjustment of sootblowing pressures may be required to increase cleaning effectiveness, or to eliminate sootblower erosion in localized areas.

Tubes — misalignment Misalignment of tubes can be caused by warpage or disengagement of support hangers, brackets or spacers. Proper tube alignment maintenance is extremely important in preventing erosion or rubbing from sootblowers and erosion from flue gas particles as a result of unequal gas flow lanes. Misalignment can result in excessive heat absorption and shortening of tube life.

Tubes — slagging or fouling deposits Personnel should inspect for deposits of ash or slag not removed by sootblowing that interfere with heat transfer or free gas flow through the unit. Any heavy deposits that block a major portion of gas lanes can indirectly contribute to erosion problems by creating channels of high velocity in remaining open areas. The presence of deposits indicates a need to check the operation and blowing pressure of sootblowers in that area.

Fuel preparation and burning equipment Any air flow regulating or adjusting equipment, such as dampers or burner register doors, should be free of warpage or overheating and operate freely over the control range.

For coal-fired units, all equipment in contact with the coal should be thoroughly inspected. The abrasive action of coal erodes some metal components faster than others. When a high wear area is found, a periodic record of the wear should be kept and repair or replacement should be made when necessary. For pulverized coal, impellers, coal nozzles and coal pipes, especially in bends, wear faster than straight runs of pipe. One method of reducing wear in coal pipes is to change to ceramic lined piping.

Coal conditioners, pulverizers and feeders should be inspected to determine whether wear parts need replacing, particularly if coal output has been decreasing, or if routine checks of the pulverized coal have shown low fineness.

Impellers and nozzles should be checked for burning, cracking or wear and any accumulations of ash should be removed. Burner throat areas should be checked for accumulations of ash and deterioration of the refractory.

In oil- and gas-fired units, burner gas spuds should be inspected for plugging and overheating, especially after long periods of oil burning. Burner oil atomizer parts should be inspected for wear from debris in oil, especially when oil additives are used. Any coking or clinker deposits should be removed from the burner impellers.

For stoker-fired units, the mechanical operating parts must be thoroughly checked for binding, wear and overheating, and openings should be checked for proper air distribution.

Refractory The condition of any refractory exposed to flue gases, such as burner throats or walls and areas in convection passes, particularly if operating conditions have been unusually severe, should be determined. This protective refractory should be replaced if necessary before returning the unit to service. During unit startup, care should be taken with the firing rate to allow newly installed refractory to properly dry and cure.

Flues, ducts, dampers and windboxes Areas that are used to transport the air and gas should be inspected for erosion, corrosion, cracking and overheating of plates, structural members and welds. Ash hoppers and other areas that may have received heavy ash accumulations should also be thoroughly inspected for possible damage. It is possible to experience a catastrophic failure of these areas through a combination of corrosion damage and excessive loading. While inspecting hoppers, the ash removal systems should be inspected and tested to assure proper operation.

Expansion joints should be thoroughly checked to ensure that the folds of the joints do not contain any ash or other accumulations which can cause corrosion that will lead to leakage or failure.

Dampers should be checked for warpage or other damage. They should be operated (stroked) to assure that they will seal properly and achieve their intended range of travel. Drives and linkages should also be checked.

Unit setting — insulation and casing On boilers with furnaces operating below atmospheric pressure, resealing of the boiler setting may be required to reduce air infiltration evidenced by an increase in excess air and unburned combustible in the ash. On pressurized furnace boilers, leakage is usually accompanied by signs of overheated casing, insulation and lagging. Ash accumulations may also occur around the leaking area. Leaks can usually be located while the unit is on-line, but may require the use of a nontoxic chemical smoke under slightly positive pressure.

Instruments and controls Any instrument or control that is suspect as a result of on-line observation and testing should be thoroughly inspected, tested, repaired or replaced. This should be supplemented by a thorough program of regular periodic inspection, testing and calibration for all of the instruments and control devices on the unit.

Repairs to fossil fuel equipment

Performance monitoring and inspection establish the nature and amount of repair and adjustment required to return a unit to dependable operation. The equipment manufacturer can recommend, guide or carry out the work, but final decisions on what will be acceptable and dependable repairs rest with the operating company and inspectors.

Repairs to pressure parts

Pressure part repairs must meet exacting rules that are included in the American Society of Mechanical Engineers (ASME) Boiler and Pressure Vessel Code, Section I, *Power Boilers*. In selected countries, other codes are also used. The ASME Code was not originally written for repair work but to govern new units. Recently, however, repair codes have been prepared. Whether new construction or repairs, all materials must meet Code requirements.

In the United States (U.S.) repairs to pressure parts are also made under the guidance of the National Board Inspection Code (NBIC), published by the National Board of Boiler and Pressure Vessel Inspectors, Columbus, Ohio. This Code covers problems of inspections and repairs to boilers and auxiliary equipment not otherwise covered in the ASME Code. The NBIC suggests laws and regulations for inspection of pressure vessels and rules for repairs by fusion welding. These rules are acceptable in practically all states in the U.S. In some instances, the ASME Code and the NBIC have been adopted as law by the appropriate regulatory or governing body.

Repairs by welding No repairs by welding are to be made to a boiler without the approval of an authorized inspector. An authorized inspector is licensed by the appropriate governing local jurisdiction in which the boiler is erected. Insurance company inspectors are usually qualified as authorized inspectors.

The welds made in the erection, alteration and repair of a boiler can be divided into three categories: 1) welds joining pressure parts, 2) welds joining nonpressure parts to pressure parts, and 3) welds joining nonpressure parts. Welds joining pressure parts and welds joining nonpressure parts to pressure parts must conform to the requirements of the applicable codes. Codes establish the minimum requirements for base material selection, qualification of welding procedures and welders, preparation of the welding procedure, specification of thermal treatments (such as preheat and postweld heat treatment), and postweld nondestructive examination.

Welds that are made between nonpressure parts such as flues, ducts, casings, hoppers and structural steel must conform to the requirements of engineering detail drawings. In general, the minimum requirements are for thermal treatment, inspection and qualification of procedures. Personnel should comply with the American Welding Society Structural Code, D 1.1.

Where the strength of the structure depends on the strength of the weld, the repair must be made by a qualified operator as defined by the NBIC. The requirements for welder performance qualifications and welding procedure qualifications are detailed in the ASME Code, Section IX, *Welding and Brazing Qualifications*.

For certain repairs where the strength of the structure does not depend on the strength of the weld, as in building up corroded surfaces, the NBIC rules permit the use of an approved operator. The requirements for an approved operator, as outlined in the rules, are somewhat less rigid than those for a qualified operator. Repairs defined by the NBIC include:

1. replacement of tube sections, provided the remaining part of the tube is not less than 75% of its original thickness,
2. seal welding of tubes,
3. building up of certain corroded surfaces, and
4. repairs of cracked ligaments of drums or headers within certain defined limits.

Replacing sections of tubes The replacement of a section of failed tube can be performed by a qualified operator. The length of the replaced section should be a mini-

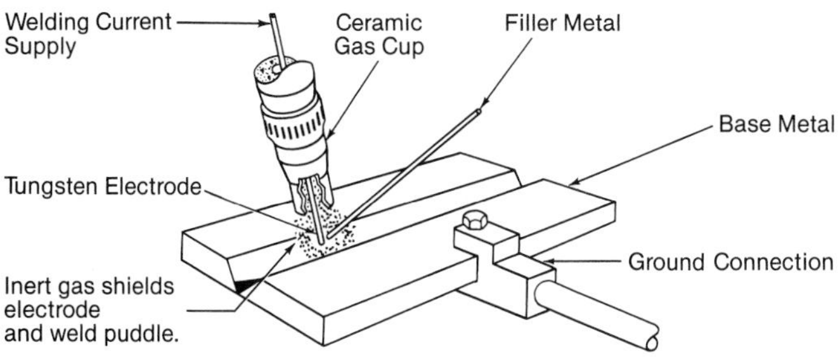

Fig. 8 Essentials of the gas tungsten arc process.

mum of 12 in. (305 mm). The preferred practice for tube removal is with a saw or cutoff tool; however, usual practice is to cut out the defective section with an oxyacetylene torch. When using a torch, care must be taken to prevent slag from entering the tube. This can be accomplished by sawing the lower cut and inserting a dam, then torch cutting the upper cut. The ends should be scarfed by machining with special tools.

Except for the furnaces of high pressure and black liquor boilers, the use of backing rings is generally acceptable. When a backing ring is not used, the tubes should be closely aligned and the space between tube ends should be minimized. The first bead of the weld metal should be run with a 0.09375 in. (2.38 mm) diameter electrode. Use of the gas tungsten arc process is recommended for the root pass (see Fig. 8 and chapter frontispiece). This is a gas arc welding process which uses an inert gas to protect the weld zone from the atmosphere. If the tube to be replaced is of alloy material, the boiler manufacturer should be consulted for procedures and welding techniques. (See also Chapter 38.)

Seal welding Seal welds are used under some conditions for fluid tightness. Structural strength must be achieved by some other means.

In new construction, the ASME Code permits the elimination of the *flare* or *bell* on tubes to be seal welded, provided the throat of the seal weld is not less than 0.1875 in. (4.76 mm) or more than 0.375 in. (9.525 mm). This applies to a seal weld either inside or outside of the drum or header. The NBIC permits seal welding for repair work.

Both Codes require that expanded tubes be re-expanded after seal welding. The re-expansion should be light and the seal weld examined for cracks after the operation is completed.

Building up corroded or eroded surfaces and repairing cracked ligaments Repairs included in these two categories vary considerably and can be extremely complex. It is recommended that each case be reviewed with the boiler manufacturer, operator and authorized inspector to establish acceptable procedures.

Postweld heat treatment Postweld heat treatment has three primary purposes: 1) relief of residual stresses, 2) removal of hydrogen that embrittles the metal, and 3) temper softening of the weld heat affected zone. This zone is the base metal adjacent to the fusion weld that is not melted but is heated to a high temperature.

Following all weld repairs, postweld heat treatment and inspection procedures should be implemented as required by the NBIC and inspection authorities.

Removing tubes and fittings from drums and headers Occasionally tubes must be removed and replaced. There are a variety of ways for removing tubes depending on whether they are to be salvaged, wholly or in part, or scrapped.

With light gauge tubes, it is often possible to cold crimp both ends of the tubes to loosen them in the seats and then drive or jack the tubes out. When the tube wall is too heavy for cold crimping, the two stage heating method may be used. If neither of these methods is applicable, the tubes are usually cut off close to the outside of the shell, and the stubs are removed later in a separate operation. To remove light tube stubs, grooves about 0.75 in. (19 mm) apart should be cut with a round nose chisel. When the tongue (the metal between the two grooves) is knocked free, the stub can be collapsed and removed. (See Fig. 9.)

When a heavy gauge tube must be replaced, but the expanded end joint is still good, it is better to not disturb the joint provided there is outside access to the tube. It is simpler to cut off the tube at a convenient distance from the drum and join the stub end to the new tube with a weld ring. However, if there is no access for such a repair, the cutting can be done from inside the drum by careful use of a cutting torch. The tongues can be prepared without damage to the seat with the grooving tool shown in Fig. 10.

When it is not possible or inadvisable to use a torch, a cutting tool of the type illustrated in Fig. 11 is used.

Seal welded tubes Seal welds are removed from tubes using tools or cutters such as the one shown in Fig. 12. This particular tool will remove seal welds from 2.5 in. (63.5 mm) OD tubes inside headers. It can also be used to cut off excess stock from 2.5, 3 and 3.25 in. (63.5, 76.2 and 82.55 mm) OD tubes. Fig. 13 shows another type of cutter used for removing the seal weld of small diameter (marine superheater) tubes. After removing the seal welds the tubes and tube stubs may be removed as previously described.

Removal of seal welded handhole fittings In the course of repair and maintenance work it is often necessary to remove one or two handhole fittings that have been seal welded. When no alloy material is involved a cutting torch can be used. Considerable care is required to prevent damage to the header in this process. Alternatively, the seal weld can be removed by grinding or machining. While these methods are slower, they do not present the hazard involved in the burning process, and alloy materials can be handled without damage.

If a great number of fittings are to be removed from alloy headers, it may be advantageous to use special tools such as the one shown in Fig. 14. This type of cutting tool is used outside the header to remove seal welds from 3.25, 4 and 4.5 in. (82.55, 101.6 and 114.3 mm) expanded type cup caps, and blind nipples and welded handhole plugs.

Repairs to studded surfaces During the construction, repair and maintenance of boilers it is sometimes necessary to replace or provide additional welded studs on tubes and ductwork to attach insulation, casings, doors or refractories, or for tube protection. In most applications the semi-automatic stud welder may be used. The apparatus consists of a welding machine, a weld timer and a welding gun. The semi-automatic stud welder is not recommended for tubes less than 0.125 in. (3.175 mm) thick. Hand welded studs should be used for these tubes.

Repairs to nonpressure part equipment

Pulverizer maintenance Babcock & Wilcox (B&W) manufactures two types of vertical shaft air-swept pulverizers, the EL which operates on the ball-and-race principle, and

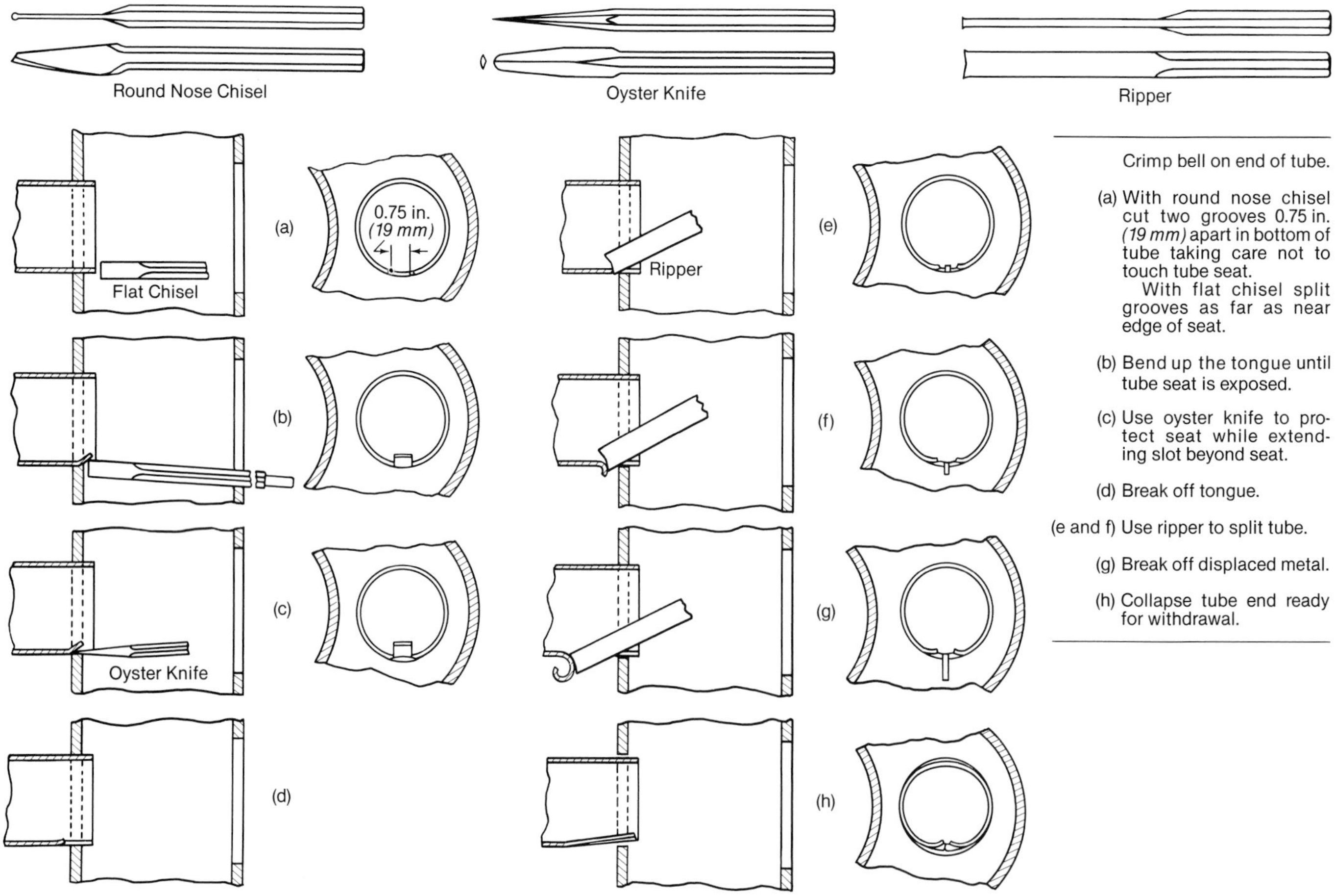

Round Nose Chisel
Oyster Knife
Ripper
Flat Chisel
0.75 in.
(19 mm)
Ripper
Oyster Knife
(a)
(b)
(c)
(d)
(e)
(f)
(g)
(h)
Crimp bell on end of tube.
(a) With round nose chisel cut two grooves 0.75 in. (19 mm) apart in bottom of tube taking care not to touch tube seat.
With flat chisel split grooves as far as near edge of seat.
(b) Bend up the tongue until tube seat is exposed.
(c) Use oyster knife to protect seat while extending slot beyond seat.
(d) Break off tongue.
(e and f) Use ripper to split tube.
(g) Break off displaced metal.
(h) Collapse tube end ready for withdrawal.

Fig. 9 Methods and tools for tube stub removal from headers.

the MPS which operates on the roll-and-race principle. (See Chapter 12.)

In most power station applications, there is a spare pulverizer when firing the design coal. This permits scheduling pulverizer maintenance without unit down time.

Before placing a pulverizer system in service, operators should inspect and record the items noted in the instructions. This information sets the basis by which further maintenance and spare part ordering can be planned. Good records will also show how a change in coal supply can affect maintenance manpower levels or material costs.

The operation and maintenance instructions for each type of pulverizer cover the specific items and procedures to be used. Calibration curves should be provided for the design coal.

The pulverizer has crucial settings such as mounting flatness, throat gap dimension or alignment and ring centering means. After long periods of operation, rework of components may be necessary. In the case of the large roll wheel assemblies in the MPS pulverizer, the condition and tightness of the lip seals, the bearing tolerance, and the condition of the oil should be checked at convenient times during the wear cycle over the life of the equipment.

For many maintenance operations, specific tooling is provided to do the work quickly and safely. Inspection and maintenance of this equipment is also essential to ensure continued safe operation.

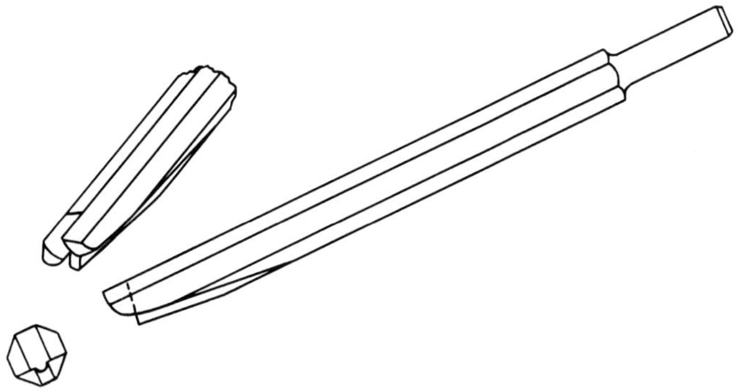

Fig. 10 Grooving tool for tube stub removal.

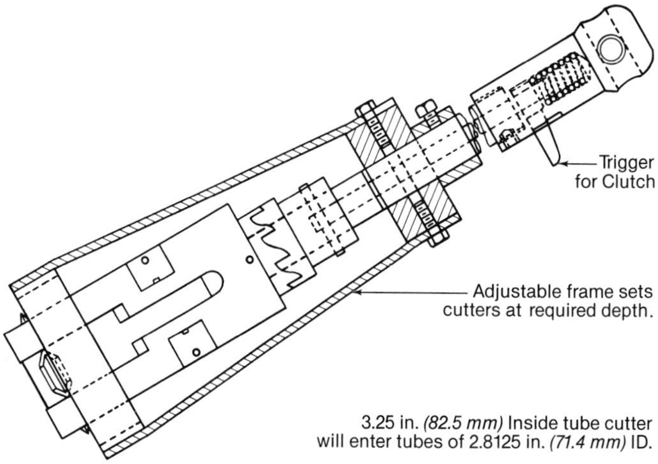

Trigger for Clutch
Adjustable frame sets cutters at required depth.
3.25 in. (82.5 mm) Inside tube cutter will enter tubes of 2.8125 in. (71.4 mm) ID.

Fig. 11 Hand operated inside tube cutter.

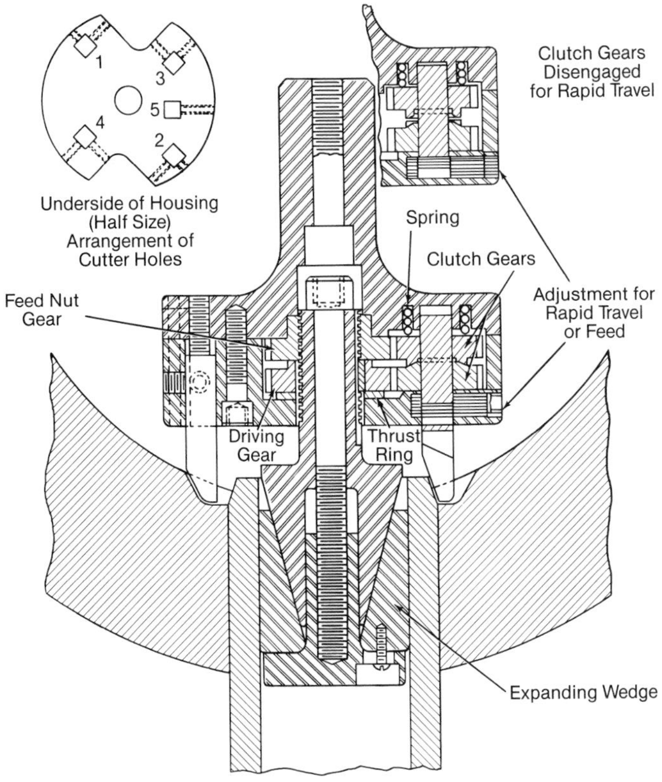

Fig. 12 Mechanism detail for seal weld removal.

Air heater maintenance The two most common types of heat recovery air heaters are the regenerative and tubular air heaters although other types are also used. (See Chapter 19.) The regenerative air heater is made up of baskets that are heated by flue gas then used to heat incoming combustion air. The tubular air heater is a tube and plate arrangement with flue gases on one side of the exchange apparatus and combustion air on the other.

A regenerative air heater should be thoroughly cleaned, then a thorough inspection should be performed checking the following areas:

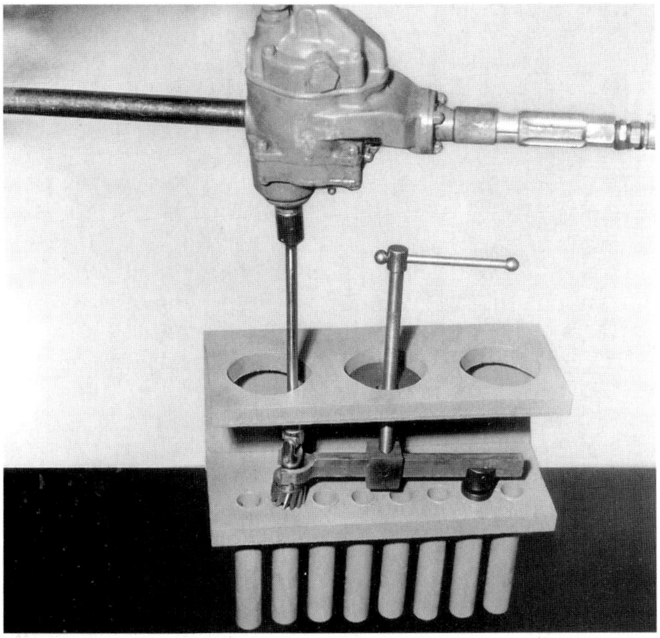

Fig. 13 Cutter for removal of seal welds from small diameter tubes.

1. the heating surface for plugging, erosion or corrosion, broken plates and sootblower damage,
2. the rotor/stator for diaphragm plate warpage or cracks, corroded or eroded members and deteriorated basket supports,
3. all seals,
4. the drive system pinion/pin rack for wear and proper engagement; the gear box for leaks, lubricant contamination and proper level; and the auxiliary air motor,
5. the housing/structure for erosion and corrosion, and expansion joints, and
6. bearings, water wash system, fire detection thermocouples, clear blocked sensing lines and the steam coil air heater.

Tubular air heaters should be thoroughly cleaned, then inspected noting:

1. the sootblowing system for nozzle erosion, travel adjustment and evidence of blowing pressure problems (plugging indicating low pressure and wastage indicating high pressure or improper drainage of the supply system),
2. tubes for erosion, corrosion and plugging,
3. tubesheets and expansion joints for erosion, corrosion, warpage and cracking and shield deterioration, and
4. casing and hoppers for cracked weld joints, expansion joint failures, door gasket deterioration and structural bracing erosion or corrosion.

Sootblowers During the outage, the sootblower system should be thoroughly inspected to assure that it is operating properly. Blowers should be checked for proper

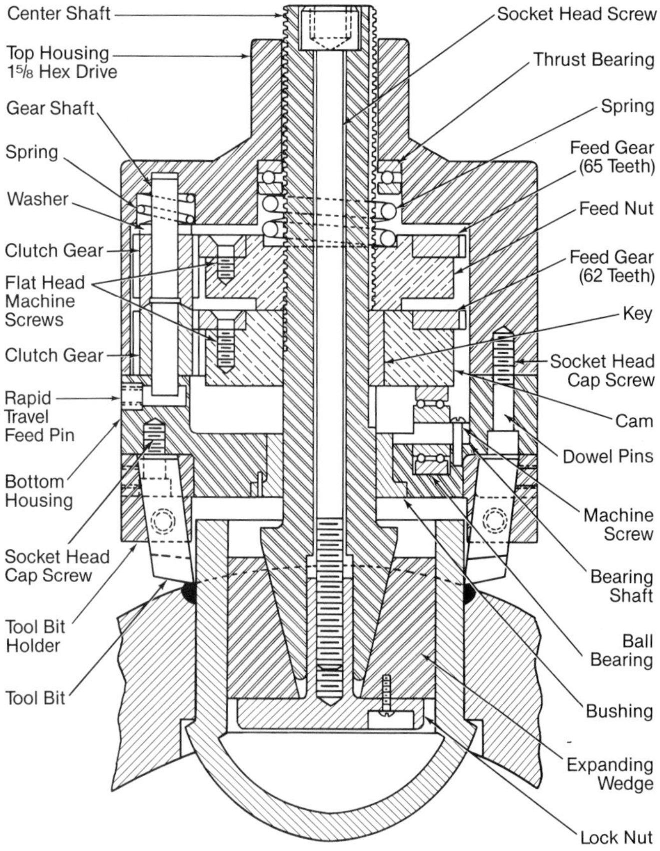

Fig. 14 Tool for removal of seal welds from cup caps or for driving ball bearings on a shaft.

positioning to avoid abrasion damage. The nozzles should be inspected and repaired or replaced if beyond repair. (See Chapter 23.)

An occasional visual inspection of the blower in its rest position is desirable to check whether the nozzle is being fully retracted from the furnace to prevent burnouts of the nozzle block.

Rotating machinery The balancing of rotating machinery is given careful attention by the manufacturer; field balancing is seldom necessary. However, it is sometimes impractical to dynamically balance rotors for field-assembled fans in the manufacturer's shop. These rotors are manufactured with very close tolerances for uniformity of section and are statically balanced. This is usually sufficient. However, when a fan under operating conditions sets up vibrations greater than the manufacturer's limits, it is necessary to balance the rotor in the field.

During the storage and field assembly of ball bearings, care must be taken to prevent the entrance of dirt which is the cause of 90% of failures. Ball bearings are usually assembled on shafts. They can either be pressed on by hand or light power press or driven on with a drift pipe and hammer (Fig. 15).

Overheating must be avoided when the hot oil method is used in preparing a bearing for assembly. The highest permissible temperature is approximately 225F (107C). The shaft size should be checked to determine the tightness of fit. The bearing must be held against the shaft shoulder until it cools enough to grip the shaft tightly or, if there is a lock nut, it should be pulled down before the bearing cools. The bearing should be greased immediately after assembly.

If a bearing is to be used again the same general precautions should be observed in removing it as when it is installed. Correct methods of removal are illustrated in Fig. 16a, b and c. When a gear or some other part on the shaft abuts the inner ring, the bearing can be forced off ahead of the part with a light tap.

When arc welding machinery with ball bearings, welders must be cautioned to attach the ground connection so that the current will not pass through the bearing. If this precaution is not observed, possible arcing between the bearing races and the ball bearings may not only cause pitting but may also reduce the hardness of the bearing surfaces with consequent quicker wear of the bearing.

After installation or repairs and before operation, it is necessary to make certain that all bearings of pulverizers, fans, feeders, pump motors and other rotating equipment are lubricated in accordance with approved practice.

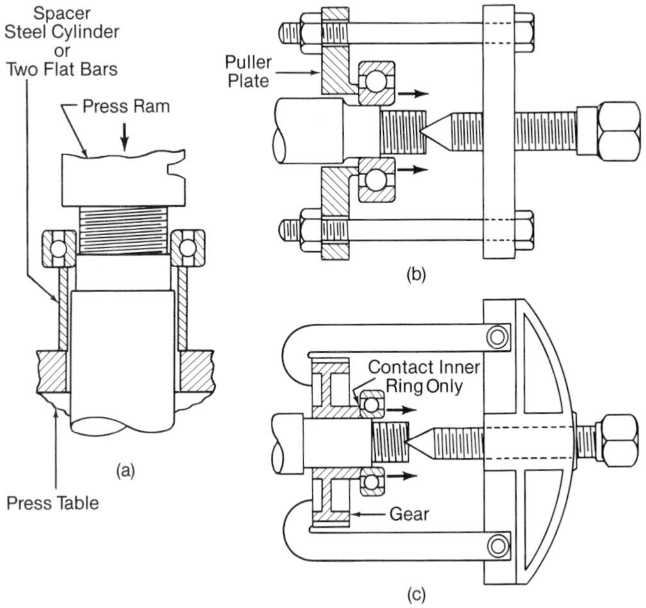

Fig. 16 General methods for safe removal of ball bearings.

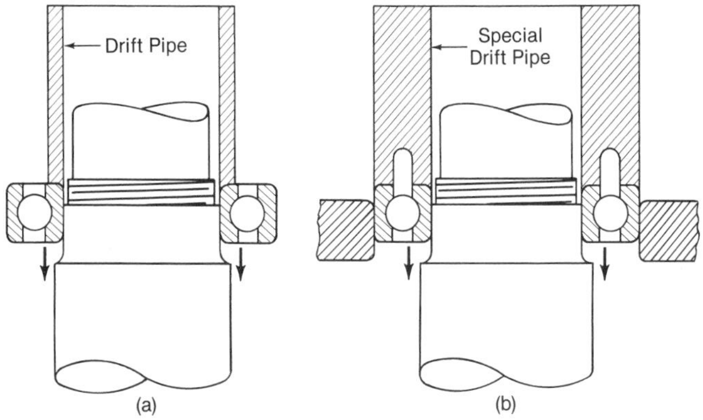

Fig. 15 Typical use of drift pipes in lightly pressing or driving ball bearings on a shaft.

Care of idle equipment

To minimize the possibility of corrosion, boilers to be removed from service must be carefully prepared for the idle period and closely monitored during the outage. Wet and dry storage are two methods available for storing the unit. Although the wet storage procedure is preferred, the availability of good quality water, ambient weather conditions, integrity of the boiler and length of storage period may dictate that the dry storage procedure is more practical.

For both wet and dry storage, the original manufacturer's operating instructions should be followed.

Protection of internal surfaces — dry storage

When a boiler is to be idle for a considerable length of time and a brief period will be allowed for return to service, the dry storage method is recommended. In this method the unit is emptied, thoroughly cleaned internally and externally, dried and then closed to exclude both moisture and air. Trays of lime, silica gel or other moisture absorbent may be placed in the drums to draw off the trapped moisture.

The unit should be properly tagged and appropriate warning signs attached noting that the boiler is stored under nitrogen pressure. Complete exhaustion of the nitrogen must occur before anyone enters the drum otherwise suffocation can result. This procedure is intended to include the economizer and superheater.

Protection of internal surfaces — wet storage

If a boiler is in standby service and must be available for immediate operation, it should be steamed in-service so that boiler water conditions are stabilized and oxygen bubbles cleared from the boiler surfaces. The advantage of using the wet storage procedure is that the unit is stored completely wet with the recommended levels of chemicals to eliminate a wet/dry interface where possible corrosion can occur. It is suggested that volatile chemicals be used to avoid increasing the level of dissolved solids in the boiler water. Solids should be reduced to a minimum prior to the storage period.

The possibility of corrosion must not be underestimated. The boiler should be inspected periodically for possible corrosion damage. Analysis of the boiler water should be conducted periodically and, when necessary, sufficient chemicals added to establish recommended levels. To ensure uniform dispersion of the volatile chemicals, premixing with the water will be necessary, or heat must be applied to induce natural circulation.

Protection of idle superheaters and reheaters

During out of service periods, except when the wet storage method is used, superheaters should always be kept dry and away from the air. Reheaters should be stored in the same manner as superheaters except that the wet storage method is never used.

Protection of external surfaces

The external surfaces of all pressure parts and other metallic surfaces should be completely cleaned of all ash and soot deposits. Other ash, cinder or fuel accumulations on the grates, furnace floors, shelf baffles and in the hoppers should be removed.

The setting should then be closed and kept closed, but periodic inspection should be made to guard against sweating and corrosion of the external surfaces of the pressure parts particularly when the wet storage method is used. It may be necessary to use coke stoves or similar heating devices at convenient points to keep all metal surfaces above the dew point particularly during protracted periods of damp weather accompanied by rising temperatures.

Reference

1. Stoll, H.G., and Szczepanski, R.S., "The influence of upgrading and maintenance spending on power plant performance," Power-Gen '88, Orlando, 1988.

Comprehensive condition assessment programs, including visual inspection, reduce downtime and determine remaining life of components.

Chapter 45
Condition Assessment

The assessment of accumulated damage, or condition assessment, has a long history in the boiler industry. Whenever a component was found to contain damage or had failed, engineers asked what caused the damage and whether other components would fail. These questions typically pertained to tubing and headers, which caused the majority of downtime. As boiler cycling became more common, the need for more routine condition assessment increased to avoid component failure and unscheduled outages.

Condition assessment includes the use of tools or methods in the evaluation of specific components and then the interpretation of the results to identify 1) the remaining life for the component and 2) areas requiring immediate attention.

A boiler's damage assessment, typically compared to its design life, is based on accumulated damage, and can be performed in three phases. In Phase 1 of the assessment, reviews are made of design and overall operating records and interviews are held with operating personnel. In Phase 2, nondestructive examinations, stress analysis, verification of dimensions and operating parameters are undertaken. If required, the more complex Phase 3 includes finite element analysis, operational testing and evaluation and material properties measurement.[1] (See Fig. 1.)

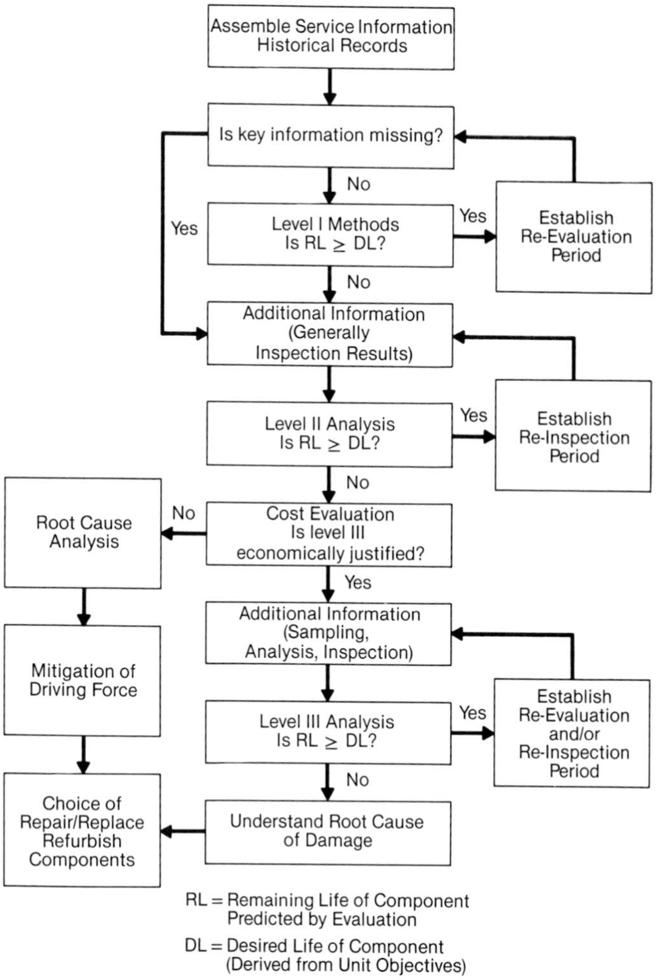

Fig. 1 Three phases (levels) of boiler damage assessment (*courtesy of the Electric Power Research Institute*).[1]

Condition assessment examination methods

The major boiler components must be examined by nondestructive and destructive tests.

Nondestructive examinations

Most nondestructive examination (NDE) methods for fossil fuel-fired plants have been in use for many years, although new methods are being developed for major components. Nondestructive testing does not damage the component.

The NDE methods used in evaluating electric utility power stations and industrial process plants include: visual, magnetic particle, liquid penetrant, ultrasonic, eddy current, radiography, nuclear fluorescence, electromagnetic acoustics, acoustics, metallographic replication and on-line testing.

Visual Whether the inspected component is subject to mechanical wear, chemical attack, or damage from thermal stress, visual examination can detect and identify some of the damage. Visual inspection is enhanced by lighting, magnification, mirrors and optical equipment such as borescopes, fiberscopes and binoculars.

Magnetic particle Magnetic particle testing (MT) and wet fluorescent magnetic particle testing (WFMT) detect surface and near surface flaws. Because a magnetic field must be imparted to the test piece, these tests are only applicable to ferromagnetic materials.[2] The choice between these techniques generally depends on the geometry of the component and the required sensitivity. For typical power plant applications, one of two methods is used — 1) the component is indirectly magnetized using an electromagnetic yoke with alternating current (AC), or 2) the part is directly magnetized by prods driven by AC or direct current (DC). In magnetic particle testing, any discontinuity disrupts the lines of magnetic force passing through the test area creating a leakage field. Iron particles applied to the area accumulate along the lines of magnetic force. Any leakage field created by a discontinuity is easily identified by the pattern of the iron particles. Dry magnetic particle testing is performed using a dry medium composed of colored iron particles that are dusted onto the magnetized area. In areas where a dry medium is ineffective, such as in testing overhead components or the inside surfaces of pressure vessels, the wet fluorescent method is more effective. With this method, fluorescent ferromagnetic particles are suspended in a liquid medium such as kerosene. The liquidborne particles adhere to the test area. Because the particles are fluorescent, they are highly visible when viewed under an ultraviolet light.

Liquid penetrant Liquid penetrant testing (PT) detects surface cracking in a component. PT is not dependent on the magnetic properties of the material and is less dependent on component geometry.[2] It is used by Babcock & Wilcox (B&W) in limited access areas such as tube stub welds on high temperature headers which are generally close spaced. PT detects surface flaws by capillary action of the liquid dye penetrant and is only effective where the discontinuity is open to the component surface. Following proper surface cleaning the liquid dye is applied. The penetrant is left on the test area for about ten minutes to allow it to penetrate the discontinuity. A cleaner is used to remove excess penetrant and the area is allowed to dry. A developer is then sprayed onto the surface. Any dye which has been drawn into the surface at a crack bleeds into the developer by reverse capillary action and becomes highly visible.

Ultrasonic Ultrasonic testing (UT) is the fastest developing technology for nondestructive testing of pressure components. Numerous specialized UT methods have been developed. A piezoelectric transducer is placed in contact with the test material, causing disturbances in the interatomic spacings and inducing an elastic sound wave that moves through the material.[3] The ultrasonic wave is reflected by any discontinuity it encounters as it passes through the material. The reflected wave is received back at the transducer and is displayed on an oscilloscope.

Ultrasonic thickness testing Ultrasonic thickness testing (UTT) is the most basic ultrasonic technology. A common cause of pressure part failure is the loss of material due to oxidation, corrosion or erosion. UTT is relatively fast and is used extensively for measuring wall thicknesses of tubes or piping.

The surface of the component must first be thoroughly cleaned. Because ultrasonic waves do not pass through air, a couplant such as glycerine, a water soluble gel, is brushed onto the surface. The transducer is then positioned onto the component surface within the couplant. A high frequency (2 to 5 MHz) signal is transmitted by the transducer and passes through the metal. UTT is performed using a longitudinal wave which travels perpendicular to the contacted surface. Because the travel time for the reflected wave varies with distance, the metal thickness is determined by the signal displacement, as shown on the oscilloscope screen (Fig. 2).

Ultrasonic oxide measurement In the mid 1980s, B&W developed an ultrasonic technique specifically to evaluate high temperature tubing found in superheaters and reheaters. This NDE method, called the NOTIS® inspection, measures the oxide layer on the internal surfaces of high temperature tubes. The test is generally applicable to low alloy steels because these materials are commonly used in outlet sections of the superheater and reheater.

Low alloy steels grow an oxide layer on their internal surfaces when exposed to high temperatures for long time periods (Fig. 3). The NOTIS® test is not applicable to stainless steels because they do not develop a measurable oxide layer.

The technique used for NOTIS® testing is similar to UTT; the major difference between the two is the frequency range of the ultrasonic signal. A much higher frequency is necessary to differentiate the interface between the oxide layer and inside diameter (ID) surface of the tube. Using data obtained from this NOTIS® testing, tube remaining creep life can also be calculated as discussed later in *Data Analysis Techniques*. NOTIS® and UTT are methods in which the transducer is placed in contact with the tube using a couplant gel. Because of the high sensitivity of the NOTIS® method, it is less tolerant of rough tube surfaces or poor surface preparation.

Ultrasonic measurement of tube deposits Computer technology coupled with ultrasonic instrumentation made possible the development of FURNIS™, a UT method which measures the relative density of deposits inside water-cooled boiler tubes. These deposits are of much lower density than the oxide layer which develops in high temperature superheater or reheater tubes. As a result, a water tube deposit can not be measured by the time delay of an ultrasonic signal.

FURNIS™ is a contact UT method which uses a high frequency transducer to produce longitudinal sound waves similar to UTT. However, instead of measuring the time delay of the reflected sound wave to determine thickness, FURNIS™ measures the energy of the reflected wave to determine deposit density. Multiple reflected waves are displayed on an oscilloscope. The signals are digitized and input into a computer which calculates the waveform energy. From these measurements, values of signal attenuation are calculated. Because the

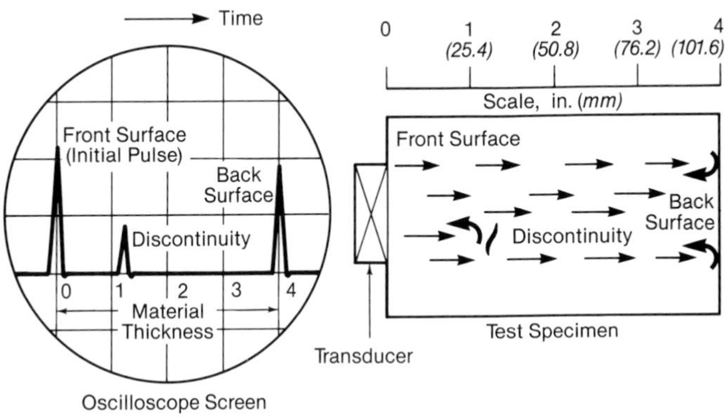

Fig. 2 Typical ultrasonic signal response.[4]

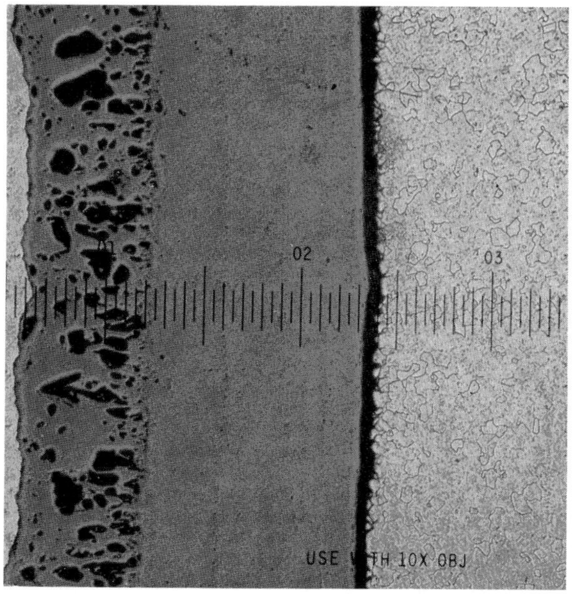

Fig. 3 Steam side oxide scale on ID surface.

energy absorption is proportional to the relative density of the deposit, it is possible to determine the boiler areas with the heaviest deposits.

Ultrasonic measurement of internal tube damage Several ultrasonic methods have been investigated for detecting damage within boiler tubes. All techniques use contact UT where a transducer is placed on the outside diameter (OD) or tube surface using a couplant, and an ultrasonic signal is transmitted through the material. The techniques can be categorized by type of signal evaluation: backscatter, the evaluation of UT wave scatter when reflected by damaged material; attenuation, the evaluation of UT signal loss associated with transmission through damaged material; and velocity, the measurement and comparison of UT wave velocity through the tube material.[5]

When a longitudinal wave passes through a tube, part of the signal is not reflected to the receiver if it encounters damaged material. The damage mechanisms reflect part of the wave at various angles, backscattering the reflected signal. The loss of wave amplitude which is received back at the transducer is then used to evaluate the degree of damage.

Damage in the tube can be assessed by evaluating the loss of signal amplitude (attenuation) as a shear wave is transmitted through the tube wall. The technique uses a fixture with two transducers mounted at angles to each other. One unit transmits a shear wave into the tube and the second transducer, the receiver, picks up the signal as the wave is reflected from the tube ID. A drop in signal amplitude indicates damage in the tube wall. This technology is the basis of the B&W patented FHyNES® test method (Fig. 4).

The velocity test method uses either longitudinal or shear ultrasonic waves. As a wave passes through a chordal section of tube with hydrogen damage, there is a measurable decrease in velocity. Because the signal is not reflected from the tube inside surface, ultrasonic velocity measurement is not affected by damage to the inside of the tube and therefore specifically detects hydrogen damage.

Immersion ultrasonic testing In immersion ultrasonic testing, the part is placed in a water bath which acts as the couplant. B&W uses a form of immersion UT for tube wall thickness measurements. In two-drum industrial power boilers, process recovery boilers and some utility power generation boilers, most of the tubes in the convective bank between the drums are inaccessible for conventional contact UTT measurements. For these applications, an ultrasonic test probe was developed which is inserted into the tubes from the steam drum; it measures the wall thickness from inside the tubes. As the probe is withdrawn in measured increments, the transducers measure the tube wall thicknesses. A limitation of this technique is that the tubes must be relatively clean.

Shear wave ultrasonic testing This is a contact ultrasonic technique in which a shear wave is directed at an angle into the test material. Angles of 45 and 60 deg (0.79 and 1.05 rad) are typically used for defect detection and weld assessment. The entire weld must be inspected for a quality examination.

Eddy current Measuring the effects of induced eddy currents on the primary or driving electromagnetic field is the basis of eddy current testing. The electromagnetic induction needed for eddy current testing is created by using an alternating current. This develops the electromagnetic field necessary to produce eddy currents in a test piece.

Eddy current testing is applicable to any materials which conduct electricity and can be performed on magnetic and nonmagnetic materials. The test is therefore applicable to all metals encountered in power station condition assessment work.

Parameters affecting eddy current testing include the resistivity, conductivity, and magnetic permeability of the test material; the frequency of the current producing the eddy currents; and the geometry and thickness of the component being tested.

Radiography Radiography testing (RT) is the most important NDE method used during field erection of a boiler. Radiography is also valuable in condition assess-

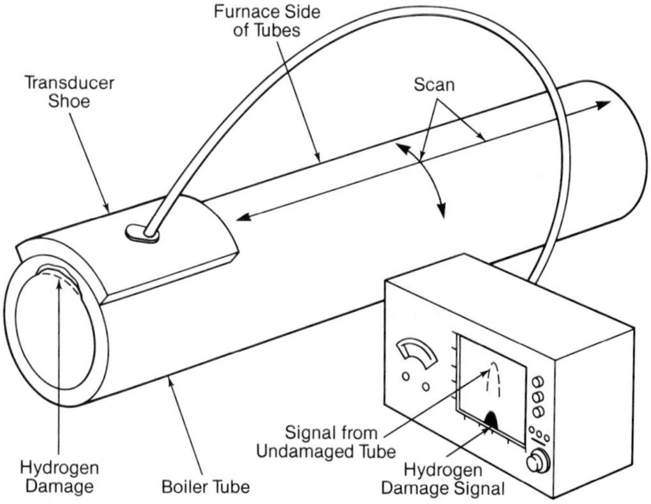

Fig. 4 Shear wave technique for detecting hydrogen damage.

ments of piping. As x-rays and gamma rays pass through a material, some of the rays are absorbed; absorption depends upon material thickness and density. When the rays passing through an object are exposed to a special film, an image of the object is produced due to the partial absorption of the rays.

In practical terms, a radioactive source is placed on one side of a component, such as a pipe, at a weld, and a film is placed on the opposite side. If x-rays are directed through the weld and there is a void within the weld, more rays pass through this void and reach the film, producing a darker image at that point. By examining the radiographic films, the weld integrity can be determined. During the field erection of a boiler and power station, thousands of tube and pipe welds are made and radiographed. (See also Chapters 37 and 38.)

The major disadvantage of radiography is the harmful effect of excessive exposure to the radioactive rays. RT is also limited in its ability to provide the orientation and depth of an indication.

Nuclear fluorescence The primary use of this testing in condition assessment is the verification of alloy materials in high temperature piping systems. When certain elements are exposed to an external source of x-rays they fluoresce (emit) additional x-rays which vary in energy level. This fluorescence is characteristic of the key alloys common to high temperature piping and headers. Chromium and molybdenum are the key elements measured. The nuclear alloy analyzer is a portable instrument that contains a low level source of x-rays. A point on the surface of the pipe is exposed to x-rays emitted from the analyzer. As the source x-rays interact with the atoms of the metal, the alloys emit x-rays back to the analyzer. Within the detector system of the analyzer the fluoresced x-rays are separated into discrete energy regions. By measuring the x-ray intensity in each energy region, the elemental composition is determined.

Electromagnetic acoustics Electromagnetic acoustics combines two nondestructive testing sciences, ultrasonics and electromagnetic induction. Practical application of acoustics to nondestructive boiler testing is only expected to become available in the 1990s. Current ultrasonic tests are contact techniques in which a piezoelectric transducer is coupled to a component surface by a fluid or gel. For electrically conductive materials, ultrasonic waves can be produced by electromagnetic acoustic wave generation.[6] In contrast to conventional contact UT where a mechanical pulse is coupled to the material, the acoustic wave is produced by the interaction of two magnetic sources. The first magnetic source modulates a time dependent magnetic field by electromagnetic induction as in eddy current testing. A second constant magnetic field provided by an AC or DC driven electromagnet or a permanent magnet is positioned near the first field. The interaction of these two fields generates a force, called the Lorentz force, in the direction perpendicular to the two other fields. This Lorentz force interacts with the material to produce a shock wave analogous to an ultrasonic pulse.

Acoustics Acoustics refers to the use of transmitted sound waves for nondestructive testing. It is differentiated from ultrasonics and electromagnetic acoustics in that it features low frequency, audible sound. B&W uses acoustic technology in testing tubular air heaters. Because the sound waves are low frequency, they can only be transmitted through air. A pulse of sound is sent into the air heater tube. As the wave travels along the tube, it is reflected by holes, blockage or partial obstructions. By evaluating the reflected wave on an oscilloscope, the type of flaw and its location along the tube can be determined.

Acoustic emissions Acoustic emissions (AE) detects subsurface crack growth in pressure vessels. When a structure such as a pipe is pressurized and heated, the metal experiences mechanical and thermal stresses. Due to the stress concentration at a defect such as a crack, a small overall stress in the pipe can produce localized yield and fracture stresses resulting in plastic deformation. These localized yields release bursts of energy or stress wave emissions that are commonly called acoustic emissions. AE testing uses acoustic transducers, which are positioned along the vessel being monitored. AE signals are received at various transducers on the vessel. By measuring the time required for the signal to reach each of the transducers, the data can be interpreted to identify the location of the defect.

Metallographic replication Metallographic replication is an in situ test method which enables an image of the metal grain structure to be nondestructively lifted from a component. Replication is important in evaluating high temperature headers and piping because it allows the structure to be examined for creep damage. Prior to the use of replication techniques, it was necessary to remove samples of the material for laboratory analysis. The replication process involves three steps: grinding, polishing and etching, and replicating. In the first step, the surface is rough ground then flapper wheel ground with finer grit paper. In the second step, the surface is polished using increasingly finer grades of diamond paste while intermittently applying an acid solution. The nital acid preferentially attacks the grain boundaries of the metal. In the final step, the replica, a plastic tape, is prepared by coating one face of the tape with acetone for softening. The tape is then firmly pressed onto the prepared surface. Following a suitable drying time, the tape is removed and mounted onto a glass slide for microscopic examination.

Strain measurement Strain measurements are obtained nondestructively by using strain gauges. Gauges used for piping measurements are characterized by an electrical resistance that varies as a function of the applied mechanical strain.[7] For high temperature components, the gauge is made of an alloy, such as platinum-tungsten, which can be used at temperatures up to 1200F (649C). The gauge is welded to the surface of the pipe and the strain is measured as the pipe ramps through a temperature-pressure cycle to operating temperature. Strain gauges used for lower temperature applications such as for analysis of hanger support rods are made of conventional copper-nickel alloy (constantan). These low temperature gauges are made of thin foil bonded to a flexible backing and are attached to the test surface by a special adhesive.

Temperature measurement Most temperature measurements can be obtained with sheathed thermocouples (TC). In special applications where temperature gradients are needed, such as detailed stress analysis of header ligaments, special embedded TCs are used. The embedded unit is constructed by drilling a small hole into the header. A sheathed TC wire is then inserted and peened in place. (See Chapter 40.)

Destructive examinations

Because it is generally more expensive to perform destructive testing, B&W tries to minimize the use of sample analysis. However, for certain components, complete evaluation can only be done by removing and analyzing test samples. Destructive testing is described for two types of specimens, tube and boat samples.

Tube samples Tubes are the most common destructively tested components. Tube samples are generally removed from water- and steam-cooled circuits. A relatively large number of samples may be removed for visual inspections, from which a smaller number are selected for complete laboratory analysis. A tube analysis usually includes the following: 1) as-received sample photo documentation, 2) complete visual, 3) dimensional evaluation of a ring section removed from the sample, 4) material verification by spectrographic analysis, 5) optical metallography, and 6) material hardness measurement. On waterwall tubes removed from the boiler furnace, the analysis includes a measurement of the internal deposit loading [g/ft^2 (g/m^2)] and elemental composition of the deposit. On steam cooled superheater and reheater tubes, the thickness of the high temperature oxide layer is also provided.

Specialized tests are performed as required to provide more in depth information. Failure analysis is a common example. When failures occur in which the root cause is not readily known from standard tests, fractography is performed. Fractography involves examination of the fracture surface using a scanning electron microscope.

Boat samples Boat samples are wedge shaped slices which are removed from larger components such as headers, piping and drums. The shape of the cut allows the material to be replaced by welding. Because the repairs usually require postweld heat treating, the use of boat samples is expensive. In most instances, replication is adequate for metallographic examination of these components and boat sample removal is not required.

Condition assessment of boiler components and auxiliaries

In Phase 1 of a condition assessment program, interviews of plant personnel and review of historical maintenance records help identify problem components. These components are targeted for a closer on site examination during Phase 2 of the program. Nondestructive and destructive examination methods can then be used to evaluate the remaining life of the boiler and its major auxiliaries.

Boiler drums

Steam drum The steam drum is the most expensive boiler component and must be included in any comprehensive condition assessment program. There are two types of steam drums, the all welded design used predominantly in electric utilities where the operating pressures exceed 1800 psi (124 bar), and drums with rolled tubes. The steam drum operates at saturation temperature [less than 700F (371C)]. Because of its relatively low operating temperature, the drum is made of carbon steel and is not subject to significant creep. Creep is defined as increasing strain at a constant stress over time.

Regardless of drum type, damage is primarily due to internal metal loss. The causes of metal loss include: corrosion and oxidation, which can occur during extended outages; acid attack; oxygen pitting; and chelant attack discussed in Chapter 42. Damage can also occur from mechanical and thermal stresses on the drum which concentrate at nozzle and attachment welds. These stresses, most often associated with boilers that are on/off cycled, can result in crack development. Cyclic operation can lead to drum distortion (humping) and can result in concentrated stresses at the major support welds, seam welds, and girth welds. The feedwater penetration area has the greatest thermal differential because incoming feedwater can be several hundred degrees below drum temperature.

A problem unique to steam drums with rolled tube seats is tube seat weepage (slight seeping of water through the rolled joint). If the leak is not stopped, the joint, with its high residual stresses from the tube rolling operation, can experience caustic embrittlement. (See Chapter 42). In addition, the act of eliminating the tube seat leak by repeated tube rolling can overstress the drum shell between tube seats and lead to ligament cracking.

Condition assessment of the steam drum can include visual and fiber optic scope examination, MT, PT, WFMT, UT and replication.

Lower drum The lower or mud drum is found mostly in industrial boilers. (See Chapter 25.) Part of the boiler's water circuit, the lower drum is not subject to large thermal differentials or mechanical stresses. However, as in steam drums with rolled tubes, seat weepage and excessive stresses from tube rolling can occur. In most cases visual inspection, including fiber optic probe examination of selected tube penetrations, is sufficient. Kraft recovery boiler lower drums are subject to corrosion of the tube-drum interface on the OD. This area of the drum is inaccessible, therefore inspections are conducted from the ID using UTT and EMATs to check for cracking and wall thinning.

The downcomers carry water from the steam drum to the mud drum and the various wall circuits. Two areas on the downcomers that should be inspected are termination welds and horizontal runs.

Boiler tubing

Steam-cooled Steam-cooled tubing is found in the superheater (high pressure) and reheat superheater. Both components have tubes subjected to the effects of metal creep. Because most boilers targeted for condition assessment are at least 20 years old, superheaters have become critical assessment components. Creep is a function of temperature, stress and operating time. The creep life of the superheater tubes is reduced by higher operating temperature and by other damage mechanisms, such as erosion and corrosion, causing tube wall thinning and increased stresses. Excessive stresses associated with thermal expansion and mechanical loading can also occur, leading to tube cracks and leaks independent of the predicted creep life.

As discussed in Chapters 18, 20 and 27, superheater tubing can also experience erosion, corrosion and interacting combinations of both.

Condition assessment of the superheater tubes includes visual inspection, NOTIS®, UTT and tube sample analysis. Problems due to erosion, corrosion, expansion, or excessive temperature can generally be located by visual examination.

Water-cooled Water-cooled tubes include those of the economizer, boiler bank and furnace. The convection pass sidewall and screen tubes may also be water-cooled as discussed in Chapter 18. These tubes operate at or below saturation temperature and are not subject to significant creep. Modern boilers in electric utilities and many industrial plants operate at high pressures. Because these boilers are not tolerant of water side deposits, they must be chemically cleaned periodically, which results in some tube material loss. As discussed in Chapter 42, proper water chemistry control will limit tube inside surface material loss due to ongoing operations and cleaning.

With the exception of creep deformation, the factors that reduce steam-cooled tube life can also act upon water-cooled tubes. Erosion is most likely to occur on tube outside surfaces in the boiler or economizer bank from sootblowing or ash particle impingement. Corrosion of the water-cooled tubes is most common on internal tube surfaces and results from excessive water side deposits. Deposit accumulations promote corrosion, caustic gouging or hydrogen damage.

Risers The riser tubes are generally found in the penthouse or over the roof of the boiler. They carry the saturated steam-water mixture exiting the upper waterwall headers to the steam drum. The risers are not generally problem areas on boilers. Condition assessment includes UTT measurements on nondrainable sections and on the extrados of bends. When access is available it is advantageous to perform internal visual inspection with a fiber optic or video probe.

Headers

Headers and their associated problems can be grouped according to operating temperature. High temperature steam-carrying headers are a major concern because they have a finite creep life and their replacement cost is high. Lower temperature water- and steam-cooled headers are not susceptible to creep but may be damaged by corrosion, erosion, or severe thermal stresses.

High temperature The high temperature headers are the superheater and reheater outlets which operate at a bulk temperature of 900F (482C) or higher. Headers operating at high temperature experience creep under normal conditions. The mechanics of creep crack initiation and crack growth are further discussed in the data analysis section of this chapter. Fig. 5 illustrates the locations where cracking is most likely to occur on high temperature headers. In addition to material degradation resulting from creep, high temperature headers can experience thermal and mechanical fatigue. Creep stresses in combination with thermal fatigue stress lead to failure much sooner than those resulting from creep alone.

There are three factors influencing creep fatigue in superheater high temperature headers: combustion, steam flow and boiler load. Most manufacturers design a boiler with burners arranged in the front and/or rear walls. Heat distribution within the boiler is not uniform: burner inputs can vary, air distribution is not uniform; and slagging and fouling can occur. The net effect of these combustion parameters is variations in heat input to individual superheater and reheater tubes. When combined with steam flow differences between tubes within a bank, significant variations in steam temperature entering the header can occur. (See Fig. 6.) Changes in boiler load further aggravate the temperature difference between the individual tube legs and the bulk header. As boiler load increases, the firing rate must increase to maintain pressure. During this transient, the boiler is temporarily over fired to compensate for the increasing steam flow and decreasing pressure. During load decreases, the firing rate decreases slightly faster than steam flow in the superheater with a resulting decrease in tube outlet temperature relative to that of the bulk header (Fig. 7). As a consequence of these temperature gradients, the header experiences localized stresses much greater than those associated with steam pressure and can result in large ligament cracks shown in Fig. 8.

In addition to the effects of temperature variations, the external stresses associated with header expansion and piping loads must be evaluated. Header expansion

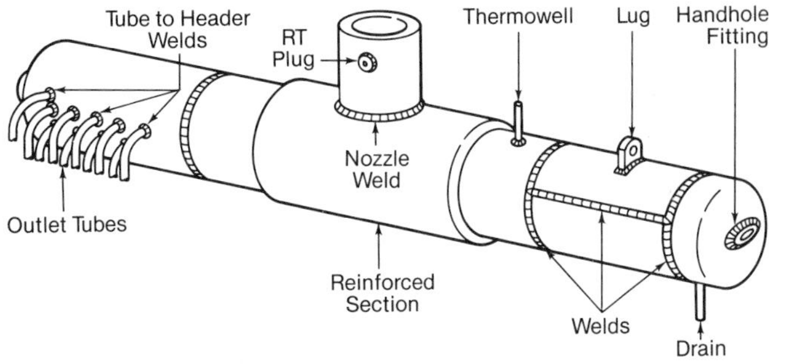

Fig. 5 Header locations susceptible to cracking.

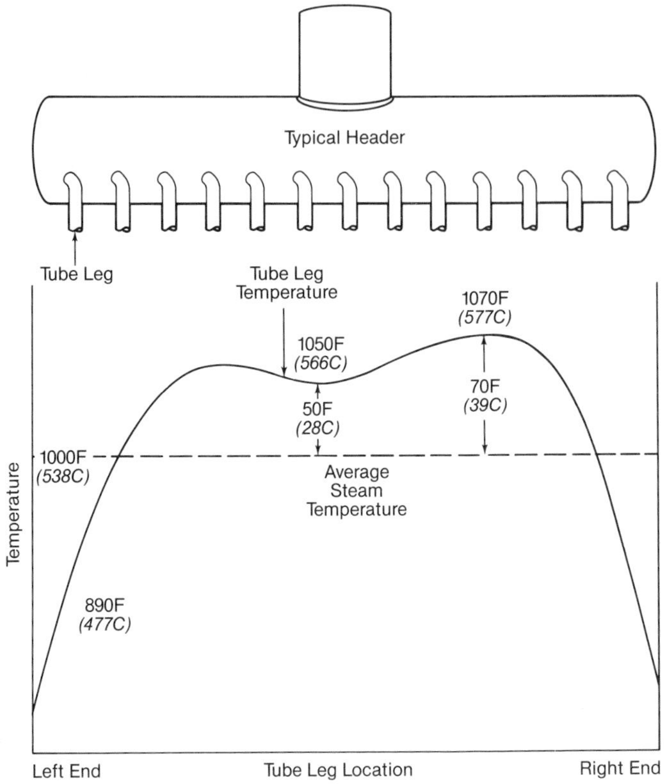

Fig. 6 Steam temperature variation in a header.

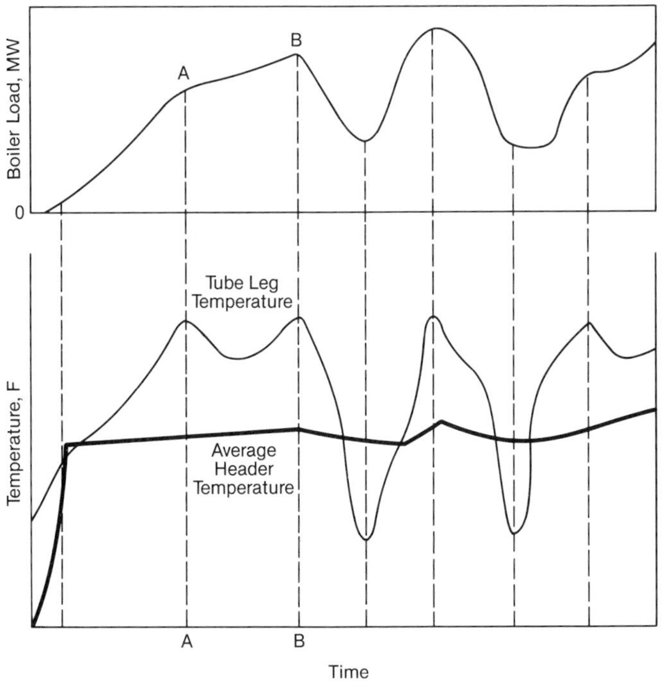

Fig. 7 Superheater tube leg temperatures vary with load.

can cause damage on cycling units resulting in fatigue cracks at support attachments, torque plates, and tube stub to header welds. Steam piping flexibility can cause excessive loads to be transmitted to the header outlet nozzle. These stresses result in externally initiated cracks at the outlet nozzle to header saddle weld.

Condition assessment of high temperature headers should include a combination of NDE techniques which are targeted at the welds where cracks are most likely to develop. Creep of the header causes it to swell; the diameter should be measured at several locations on the header and the outlet nozzle. All major header welds, including the outlet nozzle, torque plates, support lugs, and support plates and circumferential girth welds, should be examined by MT or PT. A percentage of the stub to header welds should be examined by PT. Each section of the header should be examined by eddy current or acid etching to locate the seam if it is not readily apparent. The seam weld is examined for surface indications by MT or PT, and ultrasonic shear wave testing is performed to locate subsurface flaws. To examine the header for creep damage, metallographic replication is performed. The last test, considered the most effective, which should be performed on any high temperature header, is internal examination of at least two tube bore holes. Ideally, the evaluation should correspond to the hottest location along the header.

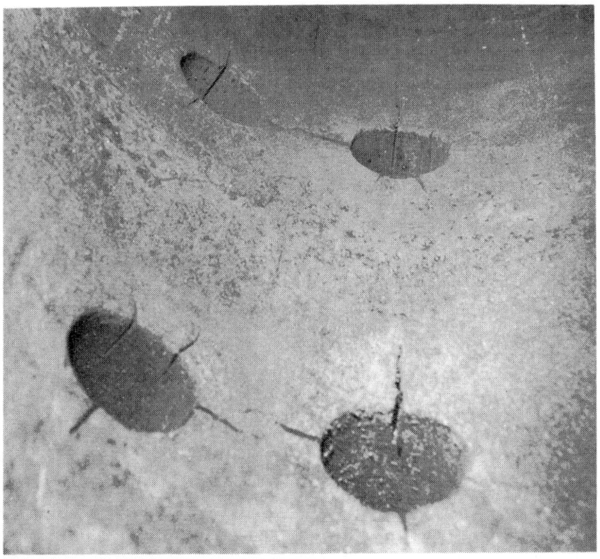

Fig. 8 Large ligament cracks on header ID.

Low temperature The low temperature headers are those operating at temperatures below which creep is a consideration. These include waterwall headers, economizer inlet and outlet headers, and superheater inlet and intermediate headers. Any damage to the low temperature headers is generally caused by corrosion or erosion.

Waterwall headers, found in most electric utility and industrial power generation boilers, are located outside the hostile environment of the combustion zone. One exception is the economizer inlet header; this may be in the gas stream and is subject to unique problems associated with cycling. Boilers that are held overnight in a hot standby condition without firing can experience severe damage to the economizer inlet header in a very short time. This damage is typically caused by thermal shock.

The magnitude of the thermal shock is a function of the temperature differential between the unheated feedwater and the inlet header. It is also a function of water flow, which is usually large because the feedwater piping/valve train is sized for rated boiler capacity. The thermal shock is worst near the header feedwater inlet and rapidly decreases as flow passes into the header and tubes. The primary concern with other low temperature headers is internal and external corrosion during out of service periods. Lower waterwall headers on stoker-fired boilers which burn coal or refuse may experience erosion along the sidewalls adjacent to the stoker grates.

Attemperators

The attemperator, or desuperheater, is located in the piping of the superheater and is used for steam temperature control. (See Chapter 18.) The spray attemperator is the most common type used. (Fig. 9.) In the spray unit, high quality water is sprayed directly into the superheated steam flow where it vaporizes to cool the steam. The attemperator is typically located in the piping between the primary superheater outlet header and the secondary superheater inlet header. Steam exiting the primary header at temperatures of 800 to 900F (427 to

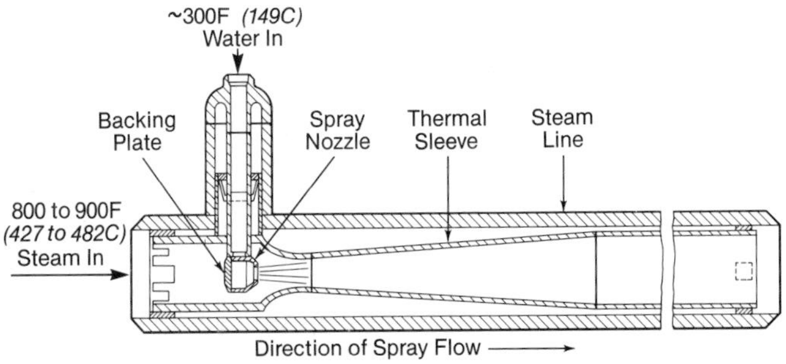

Fig. 9 Typical attemperator assembly.

482C) enters the attemperator, where relatively cool water [300F (149C)] is sprayed into the steam and reduces the temperature to the inlet of the secondary superheater. Because of the large temperature difference between the steam and spray water, parts of the attemperator experience thermal shock each time it is used. Over a period of years this leads to thermal fatigue and eventual failure (Fig. 10).

Condition assessment of the attemperator requires removal of the spray nozzle assembly. The thermal stresses occurring in the attemperator are most damaging at welds, which act as stress concentrators. The spray head and welds on the nozzle assembly are examined visually and by PT to ensure there are no cracks. With the spray head removed the liner can be examined with a video or fiber optic probe. For larger attemperators, it may be necessary to remove radiograph plugs before and after the attemperator to better view the critical liner welds.

High temperature piping

Damage mechanisms Damage to high temperature piping systems operating at more than 800F (427C) arises from creep, cycle fatigue, creep fatigue and erosion-corrosion.

Most modern high temperature piping systems are designed for temperatures ranging from 950 to 1050F (510 to 566C). The American Society of Mechanical Engineers (ASME) allowable material stresses at these temperatures may produce creep rupture in approximately 30 to 40 years. Systems designed from 1950 through 1965 that used 1–1/4 Cr–1/2 Mo alloy steel are underdesigned by today's standards because the ASME has lowered the allowable high temperature stress for this material.

Fatigue damage to a piping system is caused by repeated cyclic loading, which is a result of thermal expansion and contraction, and vibration. Most piping systems are designed with some degree of fatigue resistance. This built in flexibility comes from hangers and supports.

Creep and fatigue can occur together and interact to cause more damage than each mechanism by itself; it is not fully understood which mechanism is the primary cause. This combination of conditions is by far the most prominent because most power piping systems are highly dynamic.

Erosion-corrosion is not as prominent as the creep fatigue failure mechanisms. It is defined as wall thinning that is flow induced and occurs on the fluid side of the piping system. Factors that contribute to erosion-corrosion include bulk fluid velocity, material composition and fluid percent moisture.

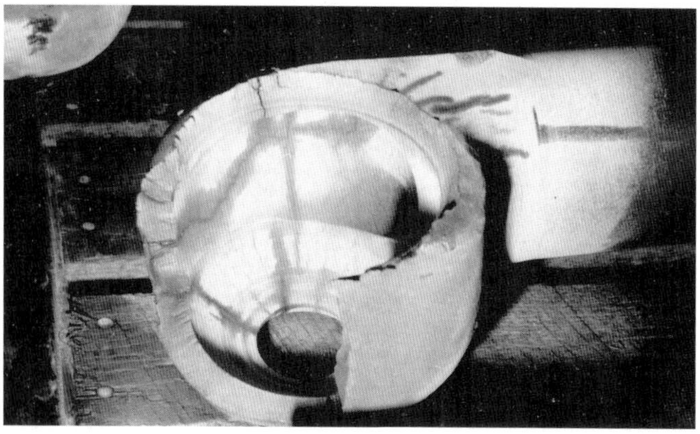

Fig. 10 Failed attemperator spray head.

Overall evaluation program When evaluating high temperature piping, condition assessment may be necessary if the following conditions exist: piping operates above 1000F (538C); was manufactured of SA335-P11 or P22 material, or manufactured of long seam welded material; has had hanger problems; was manufactured with specific weld joint types; has a history of steam leaks; and operates above design conditions. Once a priority list is developed, the evaluation can begin. This evaluation program should be as complete as economically possible.

Detailed evaluation program

Phase I To determine where physical testing is required, the following preliminary steps are part of a phase I evaluation: plant personnel interviews, plant history review, walkdowns, stress analysis and life fraction analysis.

Plant personnel interviews are conducted to gather information that is not readily available from plant records. Significant history may only be found in recollection of experienced personnel.

Operating history reviews complement personnel interviews. They can provide problem histories and design or operating solutions.

Piping system walkdowns serve three major functions: to evaluate pipe supports and hangers, find major bending or warpage, and verify changes.

Pipe hangers and supports should be carefully examined. This can be done by creating a baseline inspection record of all supports.

While the data are being taken on the piping walkdown, the general appearance should also be noted. In particular, inspections may reveal the following damages: necked-down rods or yokes, spring coil fractures, deterioration of the hanger can, and deterioration of tiebacks into building steel.

Many times a walkdown reveals that a modification was performed. If the entire system was not reviewed during the modification, other problems may result.

Stress analysis of the piping system can now be performed. Typically, a computer program is used to perform the stress calculations based on design and any abnormal conditions found during the walkdown. Once a piping system is modeled, the analysis allows the engineer to pinpoint high stress locations. The objective is to limit the nondestructive examination work to these high stress areas.

Life fraction analysis (LFA) of a pipe is done if the primary failure mode is creep due to operating temperatures above 900F (482C). The LFA is based on the unit's operating history and stress levels are calculated using design conditions and minimum wall thicknesses. This analysis is discussed at length later.

Phase II Phase II of the evaluation includes all physical testing of the piping system. The majority of the testing should be nondestructive; however, some destructive testing may be required. The results from Phase I testing provide test location priority. Specific test recommendations are shown in Table 1.

The test data generated from the inspections must be evaluated to determine the remaining component life. This is known as the Phase III evaluation and is covered under *Data Analysis Techniques*.

Table 1

Test Area	Test Type	Optional*
Circumferential welds	A, B, D, E, F, G, H	C, I, J
Longitudinal welds	A, B, D, E, F, G, H	C, I, J
Wye blocks	A, B, E, F, G, H	C, D, I, J
Hanger shear lugs	A, B	C, F, G, J
Hanger bracket and supports	A, B	C, F, G, J
Branch connections	A, B, D, E, F, G, H	C, I, J
RT plugs	A, B	C, F, G, I
Misc. taps and drains	A, B	C, I
Elbows/bend	A, E, F	B, C, G

A - Visual
B - Magnetic particle
C - Liquid penetrant
D - Ultrasonic shear
E - Ultrasonic thickness
F - Replication
G - Material ID
H - Dimensional
I - Radiography
J - Metallography

* Optional tests should be used to gather more detailed information.

Typical failures The most typical steam pipe failure is cracking of attachment welds (support welds or shear lugs). These cracks are caused by thermal fatigue, improper support, or improper welding.

Radiograph plugs usually have cracked seal welds. Although the plug threads are the pressure bearing surfaces, they become disengaged over time due to corrosion, creep swelling or oxidation.

Steam pipe warping is another serious problem. If the pipe has deformed, it has undoubtedly gone through a severe thermal shock. The high strain between the upper and lower sections of pipe can cause permanent deformation.

Two final common failure areas are the boiler outlet headers and turbine stop (throttle) valves. These areas should always be considered in any piping evaluation.

Low temperature piping

Low temperature piping operating at less than 800F (427C) is not damaged by creep. These systems typically fail due to fatigue, erosion or corrosion. The evaluation methods are the same as those for high temperature piping; however, a finite life is not designed. Low temperature pipes, if maintained, last much longer than their high temperature counterparts. Typical systems are reheat inlet steam lines, extraction lines, feedwater lines and general service water lines.

Typical failure Many high temperature failure modes occur in low temperature pipes. Cold reheat lines experience thermal shock because the reheat temperature control is typically an in-line attemperator. The attemperator spray can shock the line if the liner is damaged or the nozzle is broken. Economizer discharge lines, which run from the economizer outlet header to the boiler drum, can be damaged during startup sequences. If the economizer is steaming and flow is initiated as a water slug, the line can experience severe shocks. This can cause line distortion and cracking at the end connections and support brackets. Other low temperature piping can be damaged by oxygen pitting caused by inadequate water treatment. Erosion due to flow cavitation around intrusion points can cause severe wall thinning. If solid particles are entrained in the fluid, erosion of pipe elbows results. General corrosion of the inside pipe surface can be caused during extended outage periods. Proper line draining is recommended unless protective materials are in place.

Tubular air heaters

Tubular air heaters are large heat exchangers that transfer heat from the boiler flue gas to the incoming combustion air, as discussed in Chapter 19. On large utility boilers, tubular air heaters can contain up to 90,000 tubes with lengths of 50 ft (15.2 m) each. These 2 in. (50.8 mm) OD tubes are densely grouped with spacings of 3 to 6 in. (76.2 to 152.4 mm) centers in two directions. Flue gas flow direction is typically opposite that of the combustion air to maximize thermal efficiency. Unfortunately, this promotes corrosion on the gas side cold end.

Condensate formation promotes acid corrosion from the flue gas which causes wall thinning. If left unchecked for several years the tubes eventually corrode through, causing air leakage from the air to gas side.

Because access to air heater tubes is limited, eddy current and acoustic technologies are used to test for blockage, holes and wall thinning. Eddy current technology is used to measure wall thicknesses of thin [< 0.065 in. (< 1.65 mm)] nonferrous heat exchanger tubing. However, standard eddy currents can not penetrate the thick walls [up to 0.150 in. (3.81 mm)] of tubular air heater steel tubes. As a result, B&W developed a specific probe that uses deep penetrating eddy currents at very low frequencies. (See Fig. 11.)

The Deep Eddy™ probe is lowered into a tube by a motor drive attached to an armored cable. As the probe is removed from the tube, areas of general wall wastage from erosion and corrosion are detected. Review of the data, recorded on a computer digitized strip chart, reveals areas of thinning when compared to a calibration sample. The data are represented as a plan view and the thinnest readings are identified.

Holes and partial and complete blockage are located using acoustic technology. When an audible sound is introduced into a tube it travels the length of the tube and exits the open end. If, however, a hole exists in the tube it changes the signal pitch in the same manner as a flutist changes a note pitch. In a like manner, partial or total tube blockage yields a pitch change. B&W uses an inspection probe called The Acoustic Ranger® for this test. (See Fig. 12.)

Boiler settings

The boiler components that are not part of the steam-water pressure boundary are general maintenance items which do not have a significant impact on remaining life of the unit. The nonpressure components include the penthouse, boiler casing, brickwork and refractories, and

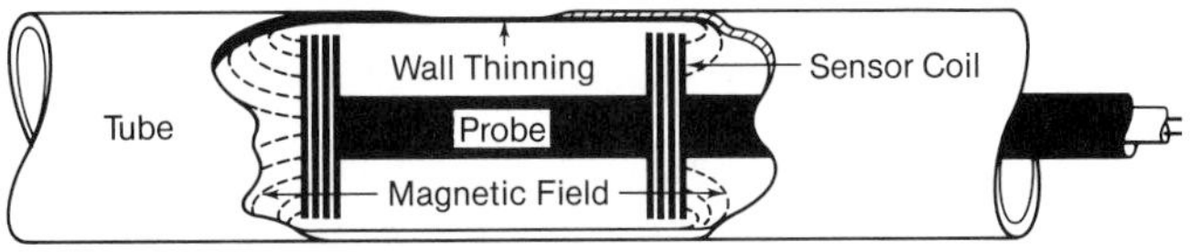

Fig. 11 Deep Eddy™ probe head.

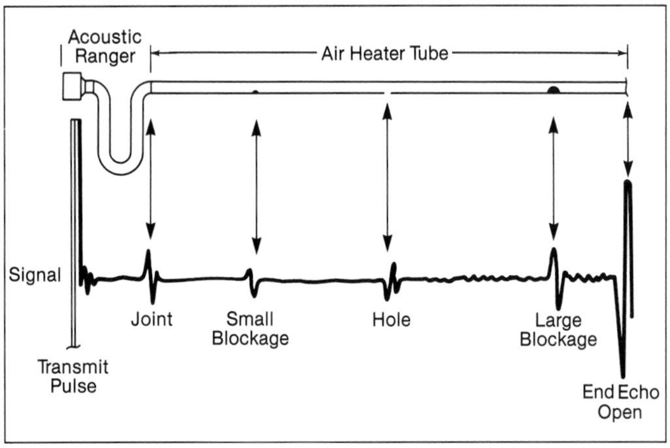

Fig. 12 Acoustic Ranger® schematic.

flues and ducts. Deterioration of these components results from mechanical and thermal fatigue, overheat, erosion and corrosion. In all cases, condition assessment is done by performing a detailed visual inspection. For flues, ducts and casing, it is of value to walk down the in-service boiler to detect hot spots, air leaks and flue gas leaks which can indicate a failed seal.

The structural members of the boiler must be reviewed during a condition assessment inspection. Normally these members along with the support rods above the boiler and auxiliaries last the life of the boiler. However, because nonuniform expansion can lead to boiler load movement, the support system should be examined during the boiler outage inspections. Particular attention should be given to header and drum supports that could be damaged if the vessel is distorted.

Analysis techniques

Once the testing is complete and the data are compiled, the next step in condition assessment is to decide whether to repair, replace or re-inspect certain components. For high temperature components with finite lives, this decision is aided by computers which predict failures by modeling analyses.

Component end of life is defined as the point at which failures occur frequently, the cost of inspection and repair exceed replacement cost or personnel are at risk. Therefore, remaining life can be considered as the interval between the present time (t_p) with accumulated damage and the time at end of life (t_e). This can be written as:

$$R.L. = t_e - t_p \tag{1}$$

For waterwall tubing that is eroding at a linear rate, the remaining life is as follows:

$$R.L. = \frac{(t_c - t_r)}{e.r.} \tag{2}$$

where

$R.L.$ = remaining life
t_c = current wall thickness, in. (mm)
t_r = preset replacement wall thickness, in. (mm)
$e.r.$ = erosion rate, in./yr (mm/yr)

Remaining life of headers is computer calculated using modeling software due to the complexity of crack growth.

Steam-cooled tubes

Steam-carrying superheater and reheater tubes operating at temperatures above 900F (482C) are subject to failure by creep rupture. The creep life of a tube can be estimated from tabulated data, provided the applied (hoop) stress and the operating temperature are known.

When a tube is put in service, the metal contacting the steam begins to form a layer of oxide scale known as magnetite (Fe_3O_4). As the life progresses, the ID oxide layer grows at a rate which is dependent on temperature. This scale acts as a heat transfer barrier and causes an increase in the tube metal temperature as discussed in Chapter 4. The metal temperature, therefore, also gradually increases with time.

Internal oxide thickness measurements are necessary for estimating a tube's operating temperature and remaining creep life as well as for assessing the overall condition of the superheater. In the past, these measurements have been obtained by removing tube samples for laboratory examination. To avoid destructive tests, B&W developed NOTIS®, discussed earlier.

Life prediction methodology The prediction of tube creep life begins with creep rupture data taken in short-term laboratory studies. Creep specimens, similar to cylindrical tensile specimens, are machined from various tube steels. Each specimen is heated to a temperature (T) and is pulled uniaxially at a stress (S) until failure occurs; at this point, a time to failure (t) is measured. A matrix of stress and temperature values has been tested.

The Larson-Miller parameter (LMP) is a function relating S, T, and t. This parameter is defined as:

$$\text{LMP} = (T + 460)\,(20 + \log t)$$

where

T = constant temperature applied to the creep specimen, F
t = time at temperature T, h

Every tube in service has an associated LMP number that increases with time. These LMP data can be related to stress, as is illustrated in Fig. 13. This relationship between stress and LMP can then be used to predict a time to creep rupture for a superheater or reheater tube. Knowing two of the three factors affecting creep rupture, i.e., hours in service and hoop stress, the third factor, temperature history, can be estimated. Because these tubes are in the gas stream, the operating temperature is not known.

There are numerous mathematical models relating steam side oxide thickness, time, and mean metal temperature for low chromium-molybdenum alloys. The following relationship is most widely used:

$$\log x = [0.00022\,(T + 460)\,(20 + \log t)] - 7.25 \tag{3}$$

where x = oxide thickness, mils. Although this formula works well with 1-1/4 Cr alloys, it must be modified for use with higher chromium alloys and carbon steels.

Creep life fraction analysis The life fraction is defined as the ratio of the time a tube withstands a given stress and temperature (t) to the time for creep rupture conditions (t_f).

Robinson's Rule of life fractions states that if the applied stress and temperature conditions are varied, the sum of the life fractions (or damage) associated with each set of conditions equals 1 at failure. It may also be written as follows:

$$(t/t_f)_1 + (t/t_f)_2 + \ldots + (t/t_f)_n = 1 \quad (4)$$

where subscripts 1 through n indicate unique stress temperature conditions.

Example — Part I

Assume a tube operates at a hoop stress of 5 ksi (34,474 kPa) and a temperature of 1050F (566C). What is the predicted time to failure? Using these parameters and the stress-LMP curve presented in Fig. 13, the effective LMP at failure is 38,015.

From the LMP equation, the expected time to failure (t_f) can be calculated:

$$\text{LMP} = [(T+460)(20+\log t_f)]$$

$$t_f = 150,000\ h \quad (5)$$

If this tube has operated for 100,000 hours at these parameters, the creep life fraction expended is:

$$f_{\text{expended}} = t/t_f = \frac{100,000}{150,000} = 0.6667 \quad (6)$$

The creep life fraction remaining is:

$$f_{\text{remaining}} = 1 - t_{\text{expended}} = 1 - 0.6667 = 0.3334$$

$$t_{\text{remaining}} = f_{\text{remaining}} \times t_f$$

$$t_{\text{remaining}} = 0.3334 \times 150,000 \quad (7)$$

$$t_{\text{remaining}} = 50,000\ h$$

Example — Part II

Assume that, after operating at 1050F (566C) for 100,000 hours, tube temperature increases to 1065F (574C).

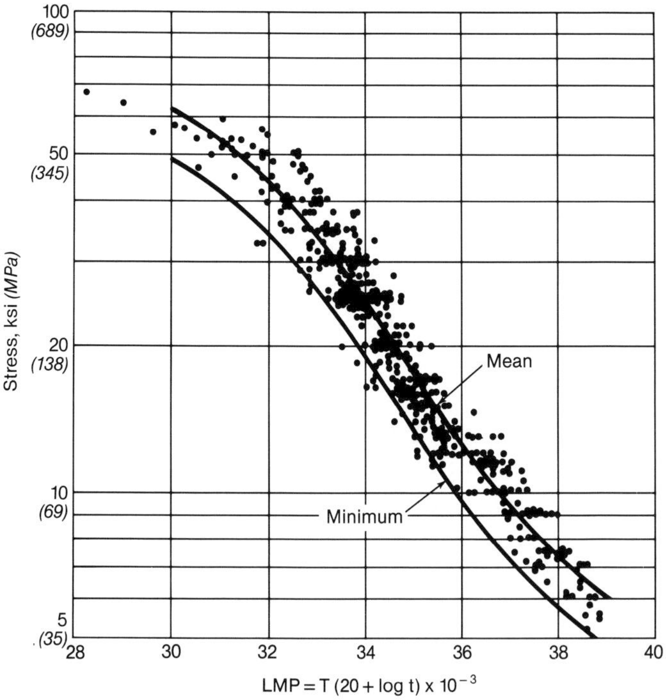

Fig. 13 Stress versus Larsen-Miller parameter (LMP).

Using the same effective LMP, the LMP equation is used to calculate t_f at 1065F (574C) as follows:

$$\text{LMP} = (T+460)(20+\log t_f) \quad (8)$$

$$t_f = 85,000\ h$$

Therefore, a new tube operating at 1065F would have an expected life of 85,000 hours. However, from Part I, this tube has used up two thirds of its life at 1050F and has a remaining life fraction of 0.33.

Robinson's Rule can be applied to determine the higher temperature service life after the tube is exposed to 100,000 hours at 1050F.

Recall Robinson's Rule — the sum of the life fractions is equal to 1:

$$(t/t_f)_{1050} + (t/t_f)_{1065} = 1$$

$$100,000/150,000 + t/85,000 = 1 \quad (9)$$

$$t = 28,000\ h$$

Note that, in this example, the total tube life would be 100,000 + 28,000, or 128,000 hours.

Analysis procedure The following analysis procedure is used with tube wall oxide thickness measurements from the NOTIS® system.

1. The life of the tube, past and future, is broken into time intervals, each of length t:

   An oxide growth rate is determined knowing the present oxide thickness and the time in service and assuming the initial oxide thickness was zero. Once a mathematical function describing oxide thickness with respect to time and temperature is defined, the thickness in each analysis interval can be calculated. The tube metal temperature in each interval, taking into account the insulating property of the oxide, is also calculated.

   A linear wall thinning rate is determined for the tube, knowing the present wall thickness as measured by NOTIS®, the original wall thickness, and the service time of the tube. Once a function describing wall thickness with respect to time is defined, the wall thickness in each analysis interval can be calculated. The hoop stress is calculated using the ASME Boiler Code Section I tube formula in each interval.
2. The creep life fraction used up in each interval is determined:

   Given the stress, the LMP value may be found from the creep rupture database. Given the operating temperature and the LMP, the time a new tube would last at these conditions (t_f) is determined. The interval creep life fraction used up is then t/t_f.
3. Because the life fractions, summed over the analysis intervals, are equal to 1, the remaining life is obtained by subtracting the tube service time from this total life.

Accuracy of creep life prediction analysis Life fraction analysis is the most accurate and widely accepted method for estimating tube life. Although this method is straightforward and well documented, it is not precise.

Result inaccuracies are due to inherent material property variations. Additionally, during service, short excursions to higher temperatures lower the remaining life fraction.

Rather than attempting to pinpoint the time of a creep rupture, the evaluation places each tube into a band of expected remaining lives. These bands take into account the shortcomings of the life fraction analysis as well as the inaccuracy of the operating parameters for the unit being assessed. In effect, these bands are confidence limits.

Capabilities and limitations At elevated temperatures, the external and internal surfaces of boiler tubes slowly oxidize. The external scale is normally removed whereas the internal scale usually remains intact. Typically, the multi-laminated scale that is formed on the inner surface is characterized by an iron-rich inner layer and an oxygen-rich outer layer. The latter generally contains a large number of pores or voids. The ultrasonic response with the NOTIS® system from the inner/outer layer interface is small compared to the signals associated with the metal/oxide and oxide/air interfaces. Therefore, a tightly adhering porous oxide does not affect the accuracy of the NOTIS® system. However, if the two oxide layers become disbanded due to exfoliation processes, the NOTIS® operator may only measure the oxide thickness to the disbanded area and therefore indicate that the oxide is exfoliating. Exfoliation is the flaking of scale particles off of the internal oxide layer.

Water-cooled tubes

Water-carrying tubes operate at or below saturation temperature and are not subject to creep damage. Therefore, these tubes have no defined design life as do high temperature components. However, erosion, corrosion, thermal expansion and mechanical stresses act on water carrying tubes which limits useful practical life.

Outside tube wall thinning Erosion and corrosion are the most common causes of outside diameter wall thinning. Erosion typically occurs on the tube outside diameter in the form of wall thickness loss. A tube's wall thickness is designed to the ASME Boiler and Pressure Vessel Code to withstand a given pressure, temperature, and mechanical load. An example is as follows:

Given:	Design pressure (P)	= 2,400 psi (165 bar)
	Design temperature (T)	= 700F (371C)
	Tube outside diameter (OD)	= 3 in. (76.2 mm)
	Tube material	= SA210 A1

From the ASME Code the wall thickness formula is:

$$t = \frac{PD}{2S + P} + 0.005D \quad \textbf{(10)}$$

where

t = minimum wall thickness
P = design pressure
D = outside diameter
S = allowable stress

By knowing the material and temperature, the allowable stress can be found in the ASME Code. In this case, the allowable stress is 14,400 psi (99,285 kPa). Solving the equation gives t = 0.245 in. (6.22 mm).

In the case of industrial and utility boilers the next higher standard tube thickness would be supplied. In the case of a chemical recovery boiler used in the pulp and paper industry a much higher thickness would be used. This allows for wall thickness reduction when operated in a reducing atmosphere. (See Chapter 26.)

As a guide, the following are B&W recommendations for utility and industrial boiler tubes:

1. Water-cooled tubes should be repaired to original wall thickness or replaced if reduced to 70% of original.
2. Steam-cooled tubes should be repaired to original wall thickness or replaced if reduced to 85% of original.

Inside tube wall corrosion Internal corrosion due to hydrogen damage, caustic gouging and under-deposit corrosion is more difficult to evaluate for remaining life because it is not easily quantified.

In addition to wall thickness, other factors, such as the effect of water quality, microstructure and damaged area size, must also be considered. Internal corrosion data can be mapped similarly to that for ultrasonic wall thickness.

Internal deposits Internal deposits lead to tube failures and provide initiation sites for hydrogen damage and under-deposit corrosion as discussed in Chapter 42. The type of corrosion depends upon the nature of the deposit (dense or porous) and the composition of the chemicals beneath the deposit. Chemical cleaning of the boiler removes these deposits. The cleaning frequency depends on the type of unit, operating pressure, fuel and water treatment program. The hours of operation and data from tube samples also determine the cleaning frequency.

Tube samples should be taken from locations with the heaviest deposits. B&W's FURNIS™ technique can identify these locations nondestructively and data can be collected from several locations in a short period of time.

Headers and piping

High temperature headers and piping typically fail due to creep, fatigue or a combination of the two. Although a header is more complex in geometry, it is essentially a pipe with tubes welded to it. Therefore, many of the remaining life analyses are similar to those for tubes. However, it must be noted that the root cause of damage can be different in headers.

On thick walled sections, where thermal and stress gradients can occur, remaining life is based on crack initiation and propagation. Calculational methods, statistics, dimensional measurements, metallographic methods, and post service creep rupture testing are considered mainly with crack initiation and the events preceding it. When dealing with thick sections, these analytical techniques must be followed by crack growth analyses. Once a component is cracked, the precrack evaluation methods do not apply.

Crack growth analysis (CGA) There are three steps used in predicting crack propagation based on time dependent fracture mechanics. (See also Chapter 7.) The schematic shown in Fig. 14 illustrates the various steps.

Step 1 consists of identifying the creep growth and deformation behavior of the material. Step 2 consists of the expressions for the crack tip driving force for creep, C_t. The basic expression for C_t is:

$$\frac{da}{dt} = bC_t^{\,m} \quad \textbf{(11)}$$

where a is crack depth, t is time, and both b and m are material constants.

The final step is the development of results from steps 1 and 2. These results are the remaining life or crack

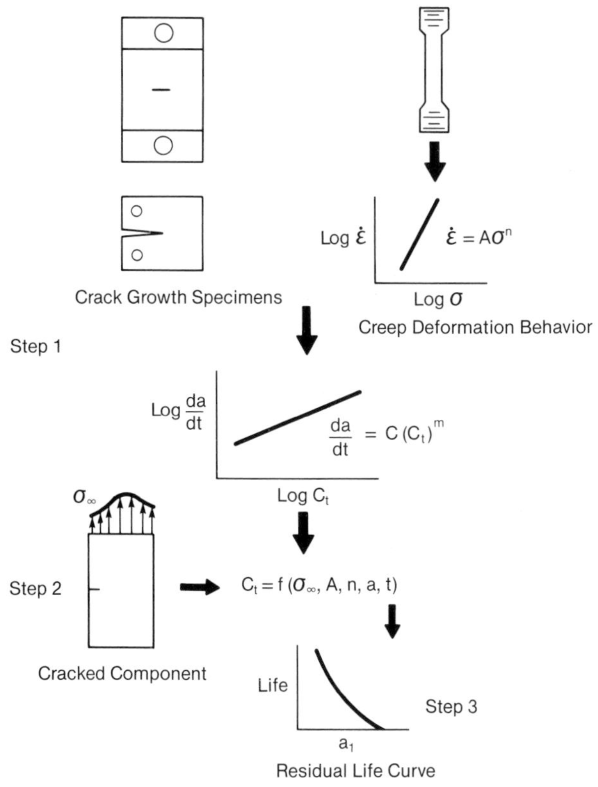

Fig. 14 Three step methodology for crack growth analysis.

growth curves which are presented two ways. The first is a plot of crack versus time. This curve is exponential because the crack grows faster as it becomes larger. The second curve is the inverse of the first, plotting size versus remaining life. Remaining life is based on a critical predetermined through-wall crack size. Once the crack reaches this value, the remaining life is zero. From the previous equation, this can be written:

$$\text{R.L.} = \int_{a_i}^{a_c} \frac{da}{bC_t^{\,m}} \tag{12}$$

where a_c is the critical crack depth and a_i is the initial crack depth.

Leak before break analysis (LBBA) When evaluating cracks that are growing at a stable rate, even though they are through wall, a condition occurs where fluid leaks through the crack but does not cause a catastrophic rupture. If the flaw is characterized by depth a and length $2C$, then the following expression applies to leak before break:

$$\text{R.L.} = \int_{a_i}^{a_c} \frac{da}{bC_t^{\,m}} = \int_{C_i}^{C_{cr}} \frac{dc}{bC_t^{\,m}} \tag{13}$$

where C_{cr} and C_i are the critical crack and initial crack half lengths respectively. Because all variables except C_i are known or determined, the equation can be solved for C_i. B&W has developed a software code, Failure Analysis Diagram (PCFAD), which models many crack scenarios encountered in header and piping systems. The failure assessment procedures uses a safety/failure plane diagram. The plane is defined by the stress intensity factor/fracture toughness ratio (K_r') as the ordinate and the applied stress/net section plastic collapse stress ratio (S_r') as the abscissa. If an assessment point lies within the curve (Fig. 15), the structure is safe. The distance from a point to the curve indicates the margin of safety. Chapter 7 provides further discussion. These two approaches allow the engineer to determine 1) how long a flaw will take to reach a predetermined critical size (crack growth, PC CREEP), and 2) whether the flaw will cause a leak or catastrophic failure.

Life fraction analysis Life fraction analyses (LFA) are performed on piping at temperatures above 900F (482C) in which the primary failure mode is creep. Stress levels are calculated using design conditions. To determine the minimum creep rupture life, published creep rupture data (LMP) are used. The calculation is similar to that for steam-cooled tubes, however piping is not influenced by gas stream heat flux and, therefore, operates at a fairly consistent temperature. The life fraction expended (*LFE*) is expressed as:

$$\text{LFE} = \frac{t}{t_m} \tag{14}$$

By using Minor's sum, an expression can be created to represent the unit's history of operating at different conditions:

$$\text{LFE} = \sum_{i=1}^{z} \frac{t_i}{t_{mi}} \tag{15}$$

where

z = number of different conditions
i = ith temperature and stress level
t_i = time of operation at given conditions
t_{mi} = minimum time to creep rupture (LMP)

When considering only creep rupture stress, a long remaining life results. However, most piping systems are also subjected to fatigue stress. Fatigue damage can be calculated by using standard fatigue curves. By combining the creep rupture and fatigue components, the expression becomes:

$$\text{LFE} = \sum_{i=1}^{z} \frac{t_i}{t_{mi}} + \sum_{j=1}^{x} \frac{c_j}{C_j} \tag{16}$$

where

x = number of different stress modes
j = jth stress level
c_j = number of cycles at stress level
C_j = number of cycles to failure at stress level

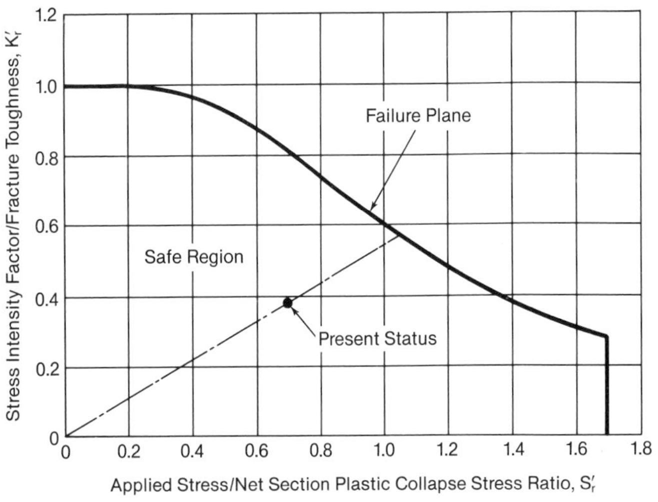

Fig. 15 Sample PCFAD failure plant diagram.

This simple approach requires detailed operating history. In addition, it only provides a gross estimate of expected life.

Stress calculation The calculation of stresses for headers is complex because of tube stub geometry, tube bank loading, differential tube temperatures and piping stresses on the outlet nozzle. Although costly, it is possible to perform a finite element analysis for the entire header. B&W has performed finite element analysis on the tube stub ligament region and found that this area contains very high stresses, especially during temperature transients.

The total stress (S_t) applied to the flaw is equal to the sum of the primary stresses (axial or hoop) (S_p), bending stresses (S_b) and pressure stresses on the crack face (S_c):

$$S_t = S_p + S_b + S_c \quad \textbf{(17)}$$

The primary stress is determined by the orientation of the flaw. If the flaw is located axially along the length of the component, then the hoop stress is primary. If it is located circumferentially, then the axial stress is primary. Bending stresses are caused by dead loads, hanger spacing, thermal differentials and restraints. The pressure stresses applied to the crack face are only considered when the crack is open to the pressure. When this is the case, the applied stress is equal to the internal pressure.

Flaw characterization Flaws found in headers and piping must be characterized prior to crack growth and LBA. This characterization involves accurately determining the flaw's length and depth. Through-wall depth is considered the most critical. The most common characterization methods include standard NDE techniques such as MT, PT, RT and UT.

Eddy current testing has been used by B&W to examine ligament cracks from inside the tube once it has been cleaned of oxide scale. This test method detects the crack depth and location. (See Fig. 16.)

Destructive samples

Destructive sampling is frequently done when data from nondestructive evaluations are inconclusive. Material properties, damage and deposits can be quantified. Tube samples and boat samples are discussed earlier under *Destructive Examinations*.

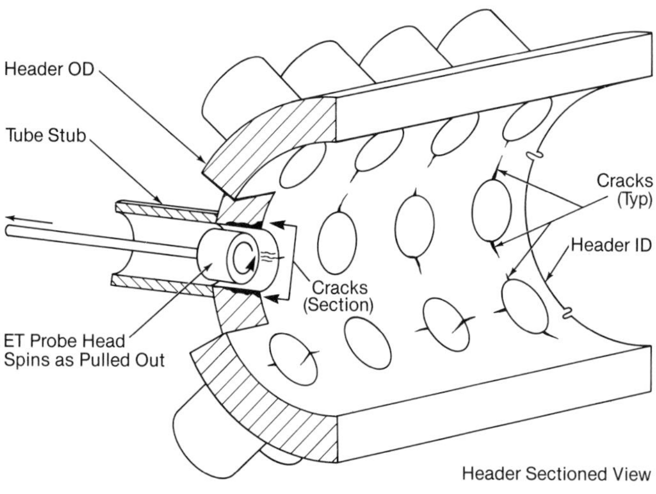

Fig. 16 Eddy current of bore holes.

Leak detection

Leaks in boilers, piping and feedwater heaters are major contributors to power plant unavailability and performance losses. In their early stages, leaks are often undetected because they are inaudible and/or concealed by insulation.

In the early 1980s acoustic monitoring equipment began to be used for leak detection. By using a piezoelectric pressure transducer to detect acoustic energy emitted by a leak, the detection of smaller leaks became possible. Leak noise is transmitted by air and by the structure. Boiler tube leaks are best detected through an airborne sensor because of the large boiler structure volume. Feedwater heaters, headers and piping leaks are best detected through a structural sensor. Because these components are small and self-contained, direct contact monitoring is used.

Leak noise is caused by a fluctuating pressure field associated with turbulence in the fluid. Turbulence is a condition of flow instability in which the inertial effects are highly dominant over the viscous drag effects. Once turbulence is established, the acoustic energy radiated from a leak increases strongly with pressure and flow rates.

Acoustic leak detection technology has been demonstrated through laboratory work, field testing and operational experience. The sensitivity of the sensor depends on three factors: sound radiated from the leak, attenuation of sound between the leak and the sensor, and background noise. Leak noise occurs in a broad band, ranging from below 1 kHz to above 20 kHz. Because of the low frequency background noise and the greater attenuation of high frequencies, most airborne systems operate in the range of 1 to 25 kHz. This is important because the acoustical signal diminishes in amplitude as it travels away from the source. Therefore, in designing a system, there is a tradeoff between sensor spacing and minimum detectable leak.

Table 2 shows typical leak monitor signals for boiler tubes, feedwater heaters, steam piping, and crack detection acoustic emissions.[8]

Boiler leak detection The background noise in a boiler is primarily due to combustion in the furnace. Direct measurement of background noise is needed to determine the spectral characteristics. In addition, the background noise level must be stable. The magnitudes of background noise from different parts of the boiler are similar; a large component is due to sootblower operation.

Using background noise and leak characterization data, full scale tests have been run to optimize sensor listening distances, sensor orientation, and signal processing equipment. A typical 500 MW utility boiler can be monitored using 16 to 24 sensor channels. Fig. 17 shows a typical airborne sensor and waveguide arrangement.

Waveguides are installed through the boiler enclosure at various locations. A typical sensor has a detection range of approximately 50 ft. (15.2 m).

Header and piping leak detection Headers and piping are monitored with structural sensors similar to those of feedwater heaters. The sensors are placed approximately 15 to 20 ft (4.6 to 6.1 m) apart.

These sensors can also be used to detect crack growth. When a crack grows it emits acoustic energy which can be detected. The processing of crack growth signals from leak detection sensors requires different, more powerful com-

Table 2
Typical Acoustic Monitoring Signals[8]

	Leak Detection Boiler Tubes	Leak Detection Feedwater Heaters		Leak Detection Steam Lines	Crack Detection Steam Lines/Headers
Acoustic path	Gasborne	Waterborne	Metalborne	Metalborne	Metalborne
Detection band	0.5 to 25 kHz	5 to 25 kHz	100 to 400 kHz	200 to 500 kHz	100 to 500 kHz
Background noise frequency	Below 2 kHz	Below 8 kHz	Below 125 kHz	Below 200 kHz	Below 200 kHz
Background noise amplitude*	20 to 30 dB	40 to 60 dB	30 to 50 dB	40 to 60 dB	40 to 60 dB
Total system gain	60 dB	50 to 60 dB	40 to 60 dB	40 dB	60 to 80 dB
Number of sensors	12 to 18	1 to 2 per heater	1 per heater	Every 20 ft (6.1 m)	Every 10 to 20 ft (3 to 6.1 m)

* Reference 1 μbar

puting hardware than the systems required for leak detection. With structural leak detection sensors installed, and the proper fitting and signal processing hardware, periodic monitoring for crack growth can be performed.

Cycling effects and solutions

In assessing a boiler's ability to withstand cycling, those components most vulnerable to cycling are reviewed first. These components are discussed from two standpoints: the operating methods which minimize cyclic damage, and design modifications which permit the component to better withstand cycling conditions.

Cycle definition

Two types of cycling service are usually considered: load cycling and on/off cycling. The on/off type has also been called two shifting.

A cycle is considered to start at full load, full temperature steady-state conditions. It goes through a load change, then returns to the initial conditions. A typical load cycle is then composed of three phases: 1) load reduction, 2) low load operation, and 3) reloading. A typical on/off cycle has four phases: 1) load reduction, 2) idle, 3) restart, and 4) reload. The phase that is often ignored, the idle period, can offer the greatest potential for reducing cyclic damage.

Economizer thermal shock

On boilers which are on/off cycled, economizers often show more cyclic damage than the other components. The economizer receives water from the extraction feedwater heater system, and the inner metal surfaces follow the feedwater temperature with practically no time delay. As a result, high rates of metal temperature change can occur with resultingly high local stresses.

Fig. 18 shows economizer inlet temperatures during an overnight shutdown cycle. The first two hours are for load reduction, followed by eight hours of idle or banked condition. Next, the boiler is fired in preparation for restart. The rates of temperature change during the load decrease and increase are usually not excessive, but would represent load cycling conditions for the economizer.

During the banked period, there is some air leakage through the boiler with a resulting decay in boiler pressure. As this happens the drum water level decreases. At the same time, the leaking air passing through the boiler is heated to near saturation temperature, and that air then heats the economizer. An economizer metal temperature can increase at 30 to 50F/h (17 to 28C/h) during this period and can approach saturation temperatures. When the drum level drops, the operator usually refills the boiler so that it is ready for firing. Because there is no extraction steam available, the feedwater temperature is low. This slug of cold water quickly chills the economizer, causing thermal shock as indicated by the solid lines in Fig. 18. The inlet header and tubes receive the greatest shock.

When the boiler is fired in preparation for turbine restart, rollup, and synchronization, the economizer heats up rapidly, often nearing saturation temperature. Feedwater is started when the initial load is applied to the

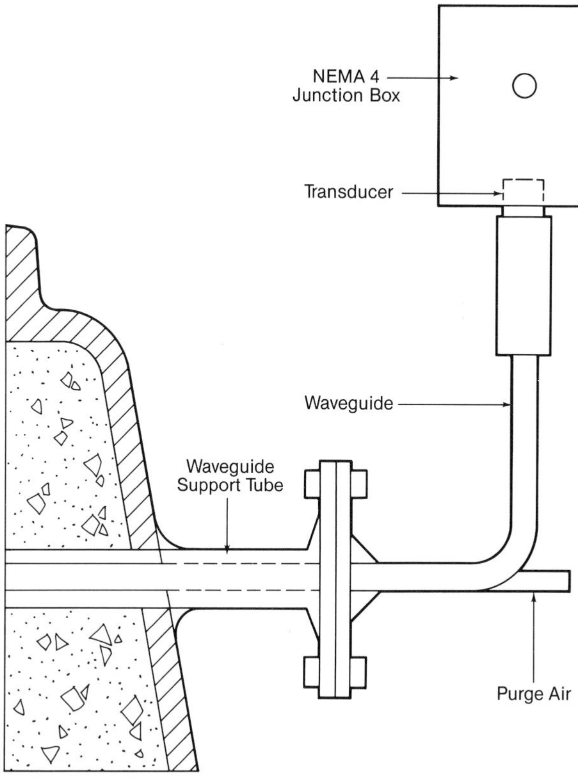

Fig. 17 Typical airborne noise sensors.

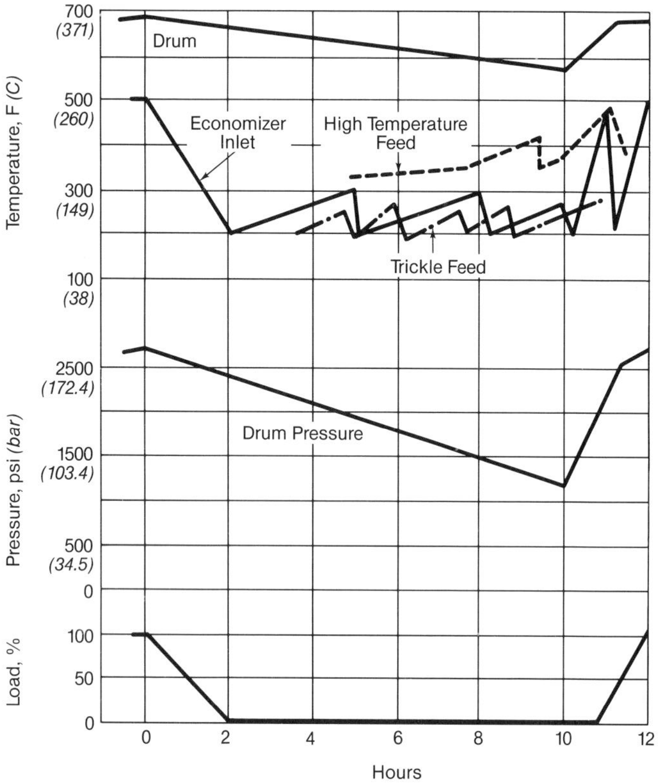

Fig. 18 Economnizer temperatures during overnight shutdown cycles.

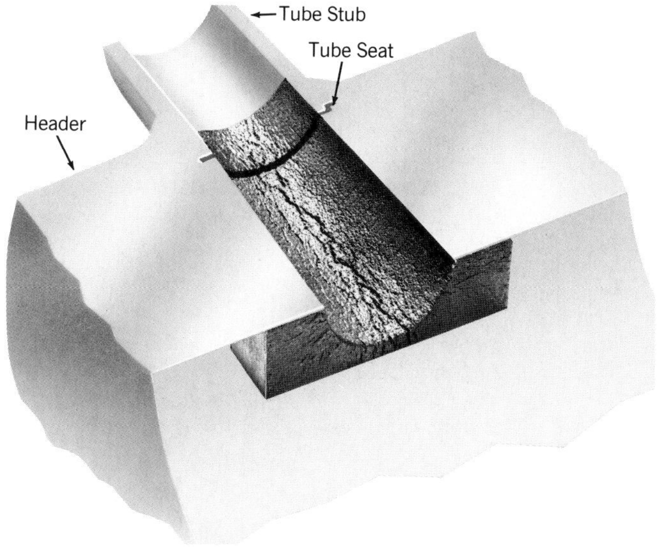

Fig. 19 Cracking in economizer inlet header occurs first in bore holes nearest water inlet.

turbine. Because little extraction heating is available, feedwater temperature is low. A severe shock occurs at this point, as the temperature can increase 300F (167C) in a few minutes.

Typically, early damage consists of cracks initiating in the tube holes of the inlet header which are closest to the feedwater inlet connection (Fig 19).

Other damage has also been seen from this cyclic service. Outlet headers have shown damage similar to inlet headers. Furthermore, some tube bank support systems can not accommodate the high temperature differences between rows.

Solutions are available to reduce the frequency and magnitude of thermal shocks. These have taken two forms and address the out of service and restart conditions.

The first solution is called trickle feed cooling. Very small quantities of feedwater are frequently introduced during the shutdown and restart periods. This prevents the inlet header from reheating and reduces the cooling rate during feedwater introduction. Because feedwater introduction is controlled to limit economizer temperature rise, some drum blowdown may be necessary to prevent a high water level.

A second method of reducing thermal shock is to permit the economizer to reheat during the idle period and then to provide higher temperature feedwater for restart. This can occur by pressurizing a high pressure heater with steam from an auxiliary source or from the drum of the unit. The quantity of steam required is low because it only heats the initial low flow of feedwater.

Furnace subcooling

Several drum boilers which have been subjected to on/off cycling have developed multiple cracks in lower furnace wall tubes where the tubes are restrained from diameter expansion or contraction. Typical cracking areas have been at the lower windbox attachment where filler bars or plates are welded to the tubes (Fig. 20).

Investigation of these failures indicated that, during the shutdown or idle period, relatively cold (cooled below saturation temperature) water settled in the lowest circuits of the furnace bottom. When circulation was started by initial firing or the circulating pump, the cold water interface moved upward through the walls, rapidly cooling the tubes. As the interface moved, its tem-

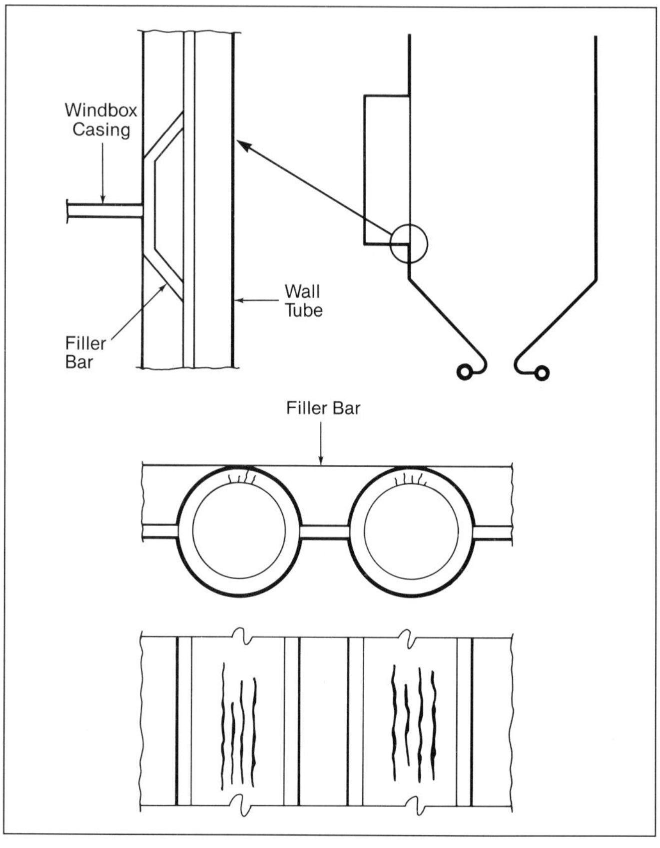

Fig. 20 Lower windbox attachment cracking due to subcooling.

perature gradient decreased and the rate of cooling decreased, therefore reducing damage higher in the furnace. Experience indicates that if the subcooling can be limited to 100F (56C), there is a low probability of damage.

An out of service circulating pump system may be used to limit the subcooling (Fig. 21). This is a low capacity pump which draws from the bottom of the downcomers and discharges water to the drum, therefore preventing the stratification temperatures of water within the unit.

Tube leg flexibility

The enclosure walls of most boilers are water- or steam-cooled. The water-cooled circuits carry boiling water, and the steam-cooled circuits carry steam from the drum. As a result, they operate near the saturation temperature corresponding to the drum pressure. Whether the boiler is being fired or shut down, considerable heat absorption or loss is necessary to change the temperature of the walls. As a result, they change temperature more slowly than the other components.

The economizer, superheater and reheaters penetrate these walls; the penetrations are designed to be gas-tight. At the point of penetration the expansion then follows saturation temperature. However, the header which forms the inlet or outlet of the other circuits expands with the temperature of the steam or water which it is handling. Fig. 22 shows the motions of the header end and the outermost connecting leg for a superheater or reheater outlet. Note that the greatest deflection is when the header temperature is at a maximum.

In the case of economizers and the reheater inlet, the deflection is in the opposite direction because these headers operate below saturation temperature. For these components the greatest temperature difference, and therefore the greatest deflection, is at low loads.

Regardless of the direction, the greatest temperature difference produces the maximum differential expansion

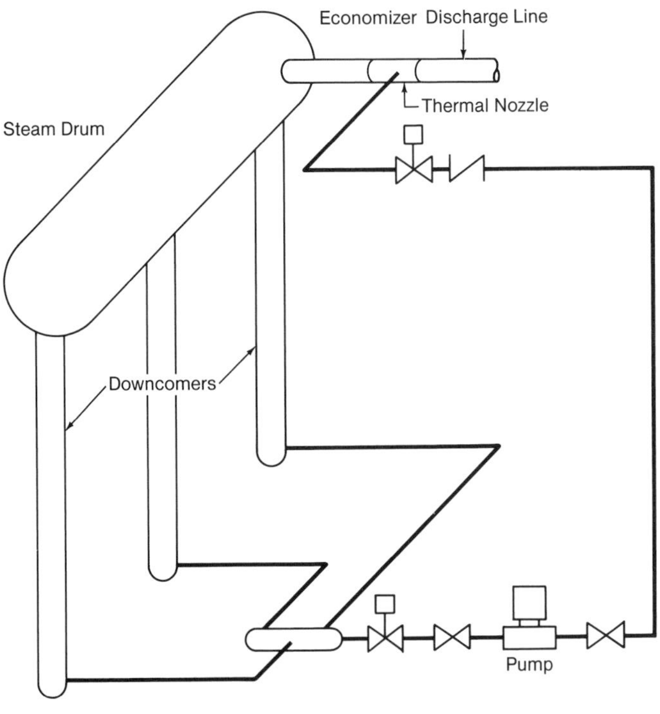

Fig. 21 Out-of-service circulating pump system.

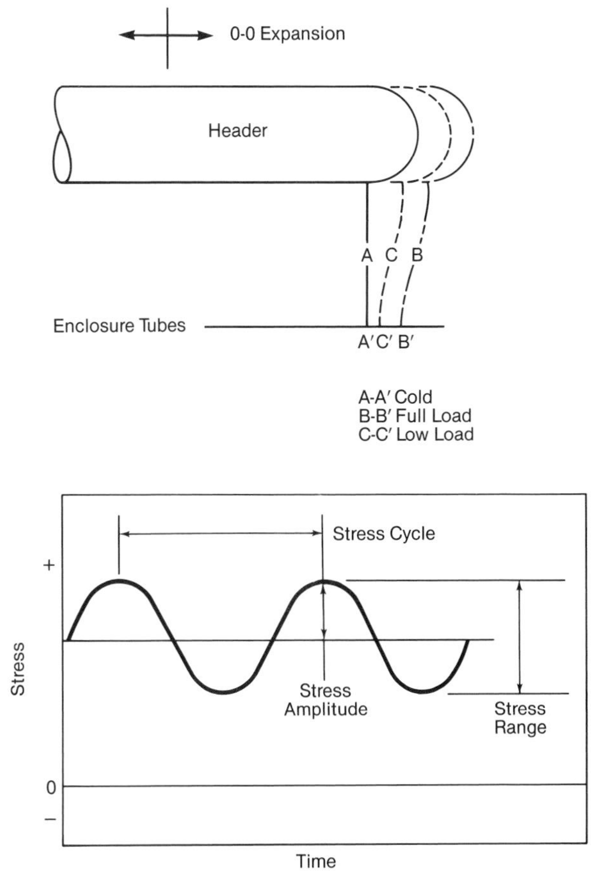

Fig. 22 Superheater tube leg flexibility.

and the maximum bending stresses in the connecting legs. However, the stress range and amplitude dictate component fatigue life.

Consider the superheater outlet and the drum (penetration point) temperature differences at constant and variable drum pressures, as shown in Fig. 23. In this example, for constant drum pressure, the maximum differential occurs at 50% load (hour 1) and is 1000F – 685F = 315F (538C – 363C = 175C). The minimum difference is at the end of the idle period (hour 10) and is 655F – 570F = 85F (346C – 299C = 47C). This is a stress range proportional to 315F – 85F = 230F (157C – 29C = 128C).

For variable drum pressure, the maximum difference occurs at 35% load and is 1000F – 540F = 460F (538C – 282C = 256C). The minimum temperature difference at the end of the idle period is 655F – 400F = 255F (346C – 204C = 142C), producing a stress range proportional to 460F – 255F = 205F (238C – 124C = 114C). The variable drum pressure mode of operation is then slightly less severe for the superheater and reheater outlets and much less severe for the reheater inlet.

In considering operational changes, it should be noted that the steam temperature characteristics shown are the maximum available at a given load. 50 to 75F (28 to 42C) lower temperature differences may be obtained in practice.

The first indications of cyclic damage are external cracks on the tube to header welds or the stub to tube welds of the outermost header legs. This damage is relatively easy to inspect and repair. Successive damage is also usually limited to closely adjacent legs because they

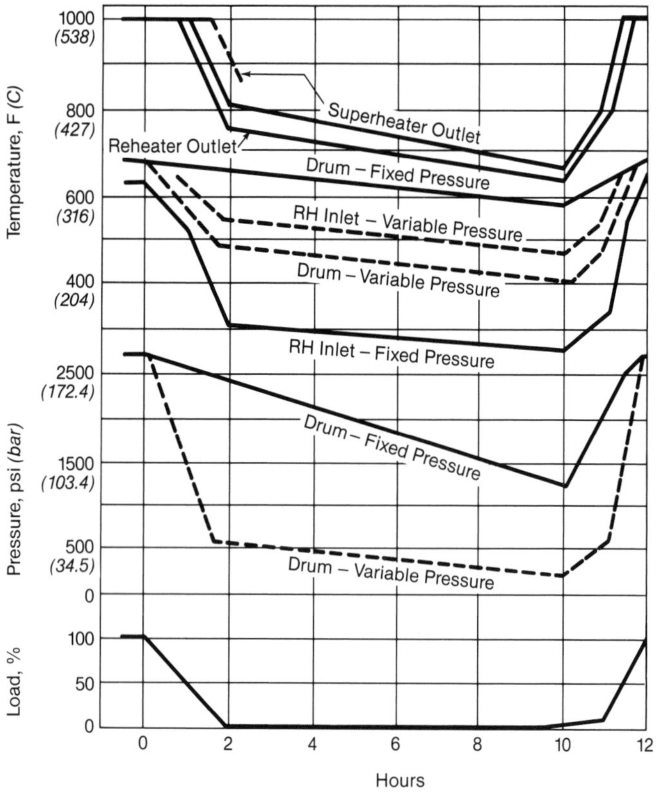

Fig. 23 Overnight shutdown temperatues.

have experienced similar stress levels. Most sensitive are high temperature headers which are a short distance from the penetration seal on wide units.

Drums

It is important to limit the rate of saturation temperature change in a steam drum. When operating in a variable drum pressure mode, considerable overfiring or underfiring is necessary to quickly change the drum pressure. These firing effects on steam temperature control also prevent rapid drum pressure changes.

Top to bottom drum temperature differences must also be limited. Only small differences result when pressure changes are made under load. The greatest differences develop during pressure reduction, when little or no steam is being taken from the drum.

If drum pressure is rapidly lost during the idle period, top to bottom differentials develop and the drum humps. Because most drums have two point supports, the humping is unrestrained and causes little change in the drum or support stress levels. However, the drum acts as a stiff beam, and connected parts move with it (Fig. 24). If those parts do not have sufficient flexibility they can experience unacceptably high stresses.

While most recent units were designed with flexibility in the drum connections, there are some older units where the front furnace wall tubes are routed directly to the drum and are supported by the drum. In such cases, it is difficult to add flexibility by rerouting and the humping must be limited.

New developments

Symptoms of advanced creep and fatigue have been found in some older superheater and reheater outlet headers. Diametral swelling has been observed. Interior cracking around the header ID tube holes and along the length has also been present. To better understand the failure mechanism, finite element stress analyses have been performed. These analyses consider steady-state temperature differences between individual tubes and the bulk header and transient temperature changes of both.

Advances in nondestructive examinations

Innovative techniques are being developed to replace or enhance existing NDE methods. Some are becoming viable due to advancements in microprocessor technology. Others are relatively new and may replace current methods. Advanced techniques include: 1) infrared scanning,[9] 2) automated UT, 3) electromagnetic acoustic transducers (EMATs), a couplant free ultrasonic method,[10] 4) pipe and wall scanners which automatically cover large areas, and 5) through-insulation radiography.

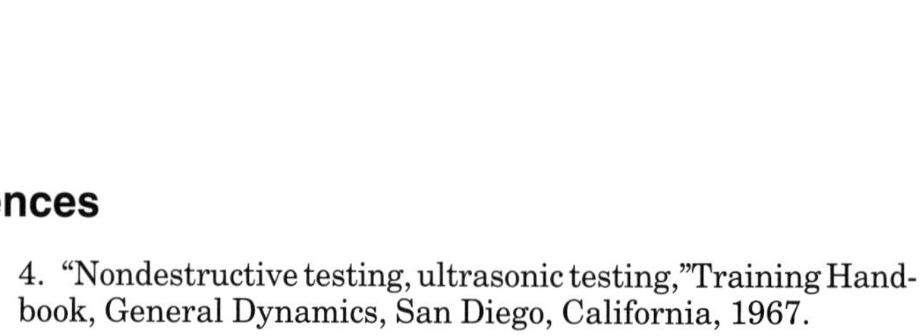

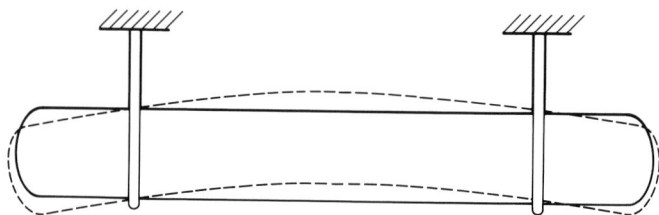

Fig. 24 Drum humping.

References

1. "Condition assessment guidelines for fossil fuel power plant components," Report GS-6724, pp. 1-1 to 1-6, Electric Power Research Institute, Palo Alto, California, March 1990.

2. "Guide for the nondestructive inspection of welds," ANSI/AWS B1.10-86, 2nd ed., pp. 16-17, Americn Welding Society, Miami, Florida, 1984.

3. "Ultrasonic inspection," *Metals Handbook*, 9th ed., Vol. 17, Nondestructive Evaluation and Quality Control, pp. 232-233, 254, ASM International, Metals Park, Ohio, 1989.

4. "Nondestructive testing, ultrasonic testing,"Training Handbook, General Dynamics, San Diego, California, 1967.

5. Alcazar, D.G., Birring, A.S., and Gehl, S., "Ultrasonic detection of hydrogen damage," *Materials Evaluation*, Vol. 47 (3), March 1989.

6. "Ultrasonic inspection," Metals Handbook, 9th ed., Vol. 17, Nondestructive Evaluation and Quality Control, pp. 255-256, ASM International, Metals Park, Ohio, 1989.

7. "Strain measurement for stress analysis," *Metals Handbook*, 9th ed., Vol. 17, Nondestructive Evaluation and Quality Control, pp. 448-449, ASM International, Metals Park, Ohio, 1989.

8. "Acoustic leak detection," Technical Brief TB.CCS.32.9.87, Electric Power Research Institute, Palo Alto, California, September 1987.

9. "Atlantic Electric demonstrates infrared inspection of boiler waterwalls," First Use, Electric Power Research Institute, Palo Alto, California, December 1989.

10. Alers, G.A., and Burns, L.R., "EMATS designs for special applications," *Materials Evaluation*, October 1987.

Upgrade work on the lower hopper of a once-through Universal Pressure boiler.

Chapter 46
Life Extension and Upgrades

Older boilers represent important resources in meeting energy production needs. A strategic approach is required to optimize and extend the life of these units. Initially, routine maintenance is sufficient to maintain high availability. However, as the unit matures and components wear, more significant steps become necessary to extend equipment life. At the same time, boiler upgrades can incorporate technology advances to increase unit efficiency and availability and to address current operating problems or changing operating needs. Design changes can accommodate new fuel sources and permit the boiler to more easily accept cycling service.

Key issues to consider when optimizing a unit include understanding how boilers age, identifying the critical components, and determining the general types of upgrades possible. These elements are then combined with operating experience and equipment condition assessments (see Chapter 45) to develop integrated strategic plans for individual boilers.

Power plant aging

At the beginning of power plant life there is a period in which the operators and maintenance crews learn to work with the new system and minor problems are resolved. This period may be marked by a high forced outage rate, but this quickly declines as the system is broken in.

As the plant matures, the personnel adapt to the new system, and any shortcomings are overcome or better understood. During this phase the forced outage rate remains low, availability is high, and the operating and maintenance costs are minimal. This mature phase normally lasts 25 to 30 years, depending on the design and use of the unit. The power plant is usually operated near rated capacity during this period.

Following this phase, the aging process becomes noticeable. Forced outages and maintenance costs increase, and availability declines. Component end of life usually causes the higher forced outage rate. Occasional operational error and the degradation of boiler components due to erosion, corrosion, creep and fatigue lead to localized failures. The forced outage rate steadily increases during this phase unless major overhauls or component replacements are instituted. A typical variation in forced outage rate with time through this aging process is shown in Fig. 1.

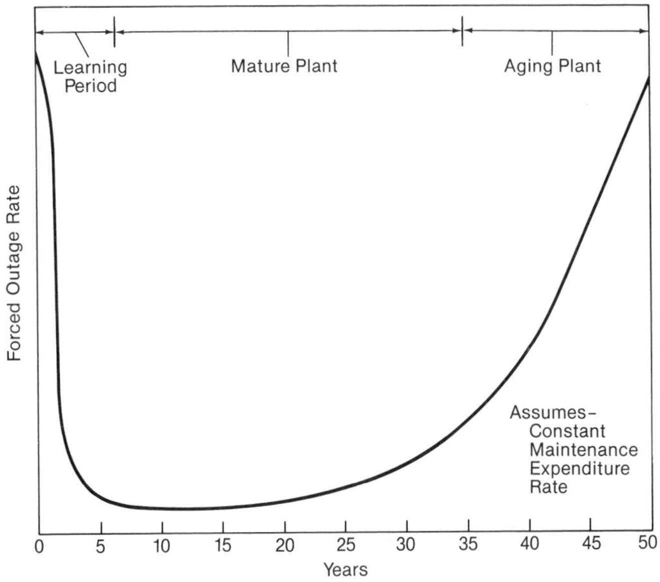

Fig. 1 Theoretical forced outage rate curve with no major action taken.

Traditional roles of the aging plant

As the aging plant becomes less reliable, its role is often changed. Newer, more reliable plants are less costly to maintain and are generally more efficient to handle the base power load. The older plants become auxiliary units or are designated for peaking service. Older plants with higher heat rates, i.e., lower efficiencies, or with low capacity may be retired. Prior to the 1980s, it was assumed that older plants would be torn down to make room for the newer, larger, more efficient units. It was common to retire a plant after 35 to 40 years of service.

This planned obsolescence began to change in the early 1980s. The cost of newer, more efficient plants became more than most boiler operators could readily finance. As a result, new construction was delayed and plans to retire the older plants were put on hold. The need to keep the older units running brought about the new strategy of *life extension*. This is a strategy that delays the plant retirement while maintaining acceptable availability. The strategy requires the replacement of some components to keep the plant running with acceptable forced outage rates and maintenance costs. These replacements or repairs expand upon those traditionally incorporated in a plant maintenance program. Significant capital expenditures are normally required to affect the availability rate. Fig. 2 illustrates the impact of life extension capital on a typical availability curve.

The availability curve drops dramatically after 25 years. Major component failures can cause significant drops in the availability curve due to the resultant long outage times. For example, failures in turbine rotors, steam lines, and steam headers and drums can cause major forced outages. A life extension program that includes capital expenditures to replace this equipment before major forced outages occur can smooth out the availability curve. Higher availabilities usually require higher capital expenditures. The large expenditures needed for high availability in older plants require a strategic plan to yield the best balance of expenditures and availability.

Impact of environmental regulations

The United States (U.S.) federal environmental regulations concerning air pollution limit the release of pollutants from power plants. (See Chapter 32.) The regulations effectively discourage the prolonged life of older units with high emissions. The most recent legislation limits the amount of pollutants released into the atmosphere. As power consumption grows, the resultant increased capacity pushes emissions toward their regulated maximum. There is a strong incentive then to replace the older units with newer, less polluting units or to modify the older units to reduce their pollution contribution. This situation has a strong influence on life extension strategies developed for older units. Power plant owners must consider the investment required to maintain availability and to meet reduced emissions levels.

Another environmental consideration that enters into the life extension strategy is the increase in emissions that may result from capital investments that increase or restore capacity. The increased boiler emissions may push overall system emissions over the regulated limits. Therefore, any life extension program that includes a capacity increase must consider that pollution control equipment may be required.

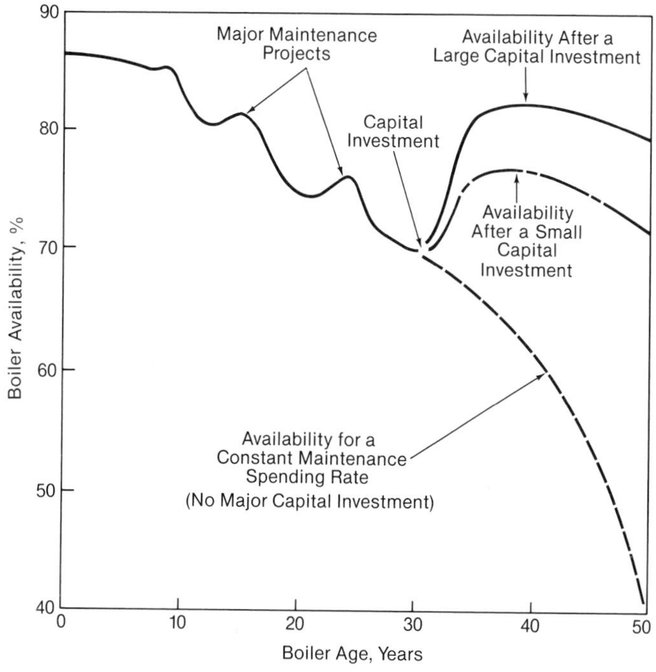

Fig. 2 Typical availability curve for a large, high pressure power boiler with life extension capital expenditures.

Operational improvements included in a life extension program

The capital involved in a life extension program is generally less than half the cost of a newer plant with similar capacity. While the driving force for a life extension program may be to stretch present resources to meet current demands, an extension program can also achieve a return on investment in the older plant. The payback can be realized when the capital improvement results in a lower heat rate, higher capacity, reduced routine maintenance, or lower forced outage rates. There have been numerous life extension programs that have added 30 years to the life of a plant while achieving a payback in as little as one year.

When considering a life extension strategy it is important to first review the boiler and its place in the power system. It is useful to review any operating and maintenance problems or limitations. The long term needs of the system must be understood. Those operating and maintenance characteristics that limit the usefulness or profitability of the plant must be addressed in the life extension program.

Aging of typical boiler components

Various boiler components deteriorate at rates and by mechanisms that are unique to each component. These rates are dependent upon the component design, function and operation. It is useful to first review component aging in a low pressure, low temperature unit operating at less than 1200 psi (82.7 bar) and 900F (482C) final steam temperature. These units are more easily discussed because the component aging mechanisms seldom involve high temperature creep rupture.

Aging of low pressure, low temperature units

Boilers that operate at less than 1200 psi (82.7 bar) pressure and 900F (482C) final steam temperature generally operate for a longer period of time with less maintenance than those units operating above these levels for a given fuel. The major sources of maintenance problems with lower pressure boilers are corrosion, erosion, fatigue and overheating.

Boiler corrosion occurs inside and outside the tubes, pipes, drums and headers. Internal corrosion is usually associated with the boiler water, contaminants in the water, and improper chemical cleaning or poor storage procedures. External corrosion can be caused by corrosive combustion products, a reducing atmosphere in the furnace, moisture between insulation and a component, and the moisture formed on components in the colder flue gas zones when the gas temperature reaches the dew point. Corrosion results in wall metal loss from the boiler component. This wall thinning raises the local stresses of the component and can lead to leaks or structural failure.

Corrosion may also be accelerated by the thermal fatigue stresses associated with startup and shutdown cycles. Furnace wall tubes, in areas of differential and high heat flux, often contain internal longitudinal or external circumferential corrosion fatigue cracks in cycled units.

Corrosion fatigue is a common problem in the drum around rolled tube joints. The residual stresses from the tube rolling process are additive to the operating pressure stresses. Corrosion from chemical cleaning and water chemistry upsets acts on this highly stressed area to produce cracking around the seal weld or the tube hole. Extensive cracking can require drum replacement. Corrosion problems are further discussed in Chapter 42.

Erosion of boiler components is a function of the percent ash in the fuel, ash composition and the local gas velocity. The tube wall loss associated with erosion weakens the component and makes it more likely to fail under normal thermal and pressure stresses. Erosion is common near sootblowers, on the leading edges of superheaters and reheaters, and around eddies in the flue gas caused by closely spaced tube surfaces, slag deposits, or other obstructions including extended surfaces and staggered tube arrangements.

The thermal stresses from temperature differentials that develop between components during boiler startup and shutdown can lead to fatigue cracks. These cracks can develop at tube or pipe bends; at tube-to-header, pipe-to-drum, fitting-to-tube, and support attachment welds; and at other areas of stress concentration. Smaller boilers are less prone to fatigue failures because the thermal differentials operate over small distances in these units. As unit size and steam temperature increase, the potential for thermal stresses and the resulting fatigue cracking rise proportionately.

Overheating is generally a problem that occurs early in the life of the plant and can often result in tube ruptures. The nature of failures attributable to overheating is discussed in Chapters 6, 42 and 45. These problems may often go undetected until a tube failure actually occurs. Overheating attributable to operation is generally resolved during the early stages in the boiler life. Other problems regarding overheating may be difficult to ascertain, and specialized boiler performance testing (see Chapter 40) is generally required to identify the source of the difficulty and to determine the extent of correction needed.

In spite of these aging mechanisms, low pressure and low temperature boilers are normally expected to last more than 50 years without major overhauls unless the unit burns a corrosive fuel, experiences fuel changes, or is improperly operated. When erosion, corrosion, fatigue, or overheating lead to frequent leaking, failures, or the threat of a major safety related failure, then component repair, redesign, or replacement is appropriate.

High pressure, high temperature units

Boilers operating at pressures above 1200 psi (82.7 bar) and 900F (482C) final steam temperature suffer from more complicated aging mechanisms than the lower temperature units. These boilers are generally larger than the low pressure, low temperature units and this increases the likelihood of thermal fatigue resulting from boiler cycling. These high temperature boilers are prone to the same overheating, erosion, and corrosion problems that cause component deterioration in the low temperature units. The higher pressures and associated higher furnace wall temperatures make these units more susceptible to water side corrosion. The high temperatures may promote hydrogen damage of the furnace tubing in areas of high corrosion or heavy internal deposits. Chapter 42 discusses these deterioration mechanisms in more detail. Severe cases of furnace wall hydrogen damage have forced the retirement of older units.

High temperature creep rupture and creep fatigue failure (see Chapters 6 and 7) are the two main aging mechanisms in the high temperature components of high temperature boilers. All components that operate at temperatures above 900F (482C) are subject to some degree of creep. As a result, most of these components have a finite design life and can fail after 20 to 40 years of operation. Chapter 6 discusses creep more fully. Most of the tubes, piping, and headers from the primary superheater to the turbine, including the superheater and reheater, are designed to operate in the creep regime. Replacement and redesign of these components must be considered in any strategy for operating high temperature, high pressure boilers beyond 25 to 35 years. (Fig. 3.)

It is common to replace a superheater after 25 years of service due to creep rupture incidents. Superheater outlet headers have also been replaced after 25 years due to creep fatigue cracking. Chapter 45 discusses methods for predicting creep ruptures in high temperature superheaters and reheaters and for assessing the condition of high temperature piping and headers. Availability of these high temperature units in the later years of operation generally follows the curve shown in Fig. 2, mainly because of creep ruptures and creep fatigue failures.

Fig. 3 Replacement furnace mix panel for Universal Pressure boiler.

The aging process and rate of component degradation vary from unit to unit. Table 1 presents the component replacement sequence for a typical high pressure, high temperature unit. After 40 years of operation most boiler pressure part components have been replaced.

Design limitations addressed in a life extension program

When some portion of the boiler must be replaced, the owner has an opportunity to upgrade this portion. A component upgrade can improve boiler operation and reduce maintenance costs. As previously discussed, these improvements may offer a payback for the upgrade. This section discusses several types of component upgrades. The variety of possible modifications is quite large depending upon the boiler involved, advances in technology, changes in fuels burned and changing operation requirements.

Upgrading boiler components to reduce coal erosion

Most older boiler components do not contain highly wear resistant materials in high erosion areas. These areas, such as coal pulverizers, require frequent and sometimes costly maintenance. Advanced materials have been developed specifically for boiler fuel handling applications. Materials have evolved from basic low carbon steels to ceramic linings. These linings include high density, high alumina ceramics and silicon carbide ceramics. The alumina ceramics with 96% alpha alumina and a density of about 232 lb/ft^3 (3716 kg/m^3) are ideal for components exposed to high velocity coal particles. These alumina linings provide up to ten times the wear resistance of carbon steel. The nitride-bonded silicon carbide ceramics have good erosion and corrosion properties, and they provide high thermal conductivity and superior thermal shock resistance. These ceramics are used in the most severe applications.

Typically, ceramic linings are used in coal transport equipment, pulverizers (see Chapter 12), piping, exhaust fans and burner nozzles. The extent of application depends upon the relative erosive nature of the specific coal and the associated velocities. Table 2 gives some guidelines for the application of ceramic linings to coal piping. Ash composition affects the erosiveness of the coal. Large percentages of clay, alumina, silica, pyrite and quartz also increase coal's abrasiveness. The guidelines given in Table 2 assume a silica composition of 40%. High silica contents increase the abrasiveness of the coal.

Table 1
Component Replacement Schedule for a Typical High Temperature, High Pressure Boiler

Typ. Life (Years)	Component Replaced	Cause for Replacement
20	Miscellaneous tubing	Corrosion,erosion, overheating
	Attemperator	Fatigue
25	Superheater (SH)	Creep
	SH outlet header	Creep fatigue
	Burners and throats	Overheating, corrosion
30	Reheater	Creep
35	Primary economizer	Corrosion
40	Lower furnace	Overheating, corrosion

Note: The actual component life is highly variable depending on the specific design, operation, maintenance and fuel.

Table 2
Recommendations for the Application of High Density Alumina Ceramic Linings in Coal Piping

Ash Content of Coal	Recommendations
Less than 6%	No ceramic lining required.
6% to 9%	Use ceramic lining from the pulverizers to the first major bend plus two pipe diameters beyond the bend. Line all short radius bends and two pipe diameters beyond these bends.
More than 9%	Same as above except, also line all bends and the pipe two diameters beyond these bends.

High temperature outlet header upgrades

Chapter 45 describes the mechanisms for reheater and superheater outlet header failures. The main failure modes are tube-to-header weld failures, ligament cracking and nozzle cracking. These failures are caused by the combined effects of thermal cycling and creep, often called creep fatigue. Advanced design standards for headers have been developed to help reduce these failures. When an outlet header must be replaced, a design upgrade should be considered. An upgraded header can provide long term, reliable service even under cycling conditions.

Nozzle cracking, discussed in Chapter 42, is associated with the weld in a tee nozzle or in the internal sharp corners of a forged design. Any header upgrade could include a forged tee to eliminate the welded connection. The forging would include generous radii to reduce stress concentrations.

Under cycling service, the tube-to-header weld area may develop cracks due to inadequate tube leg flexibility or due to thermal stresses in the header. Cracking due to inadequate tube leg flexibility can be overcome by providing longer or more flexible tube legs between the furnace penetration and the header. However, this modification may require header relocation and/or relocation of the tube penetrations.

Relocating the tube penetrations around the header may also be necessary to avoid header ligament cracking, discussed in Chapter 45. Many headers were designed with closely spaced or nonsymmetrical tube penetrations, as shown in Fig. 4, which are more prone to creep fatigue ligament cracking. Studies have shown that creep fatigue cracking can be decreased by using larger ligaments. Upgraded headers with widely spaced tube penetrations provide larger ligaments and lower ligament temperatures and stresses. As a result, header life is significantly increased.

Redesign of the tube hole penetrations and tube-to-header weld configuration, as shown in Fig. 5, can also increase the life of a high temperature header. Full penetration tube-to-header welds may be used in place of partial penetration socket welds. Elimination of the nonfusion region in the partial penetration weld reduces the stresses around the weld joint and eliminates crack propagation from the end of the gap created by lack of penetration.

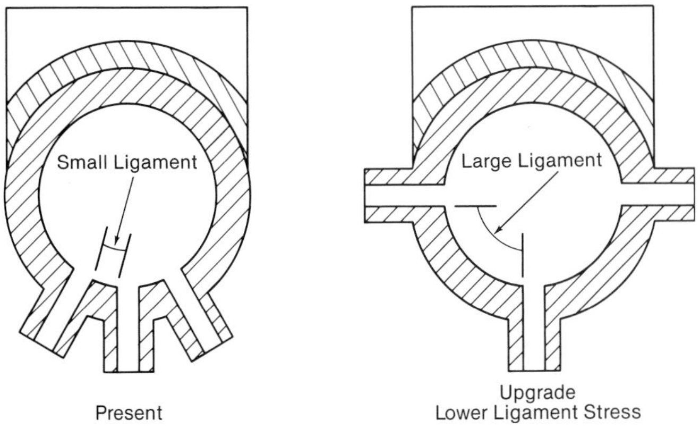

Fig. 4 Increased ligament size for longer design life.

The tube hole in the header creates a sharp discontinuity at the intersection of the hole and header internal surface. Very high stresses associated with this sharp corner cause creep fatigue cracks to initiate from the hole. A redesigned header may also be fabricated with a chamfer at the tube hole penetration as shown in Fig. 5. The chamfer lowers the stress around the hole and reduces creep fatigue cracking.

The easiest upgrade for a high temperature header is an upgrade in material. American Society of Mechanical Engineers (ASME) SA335-P11 headers may be upgraded to SA335-P22 or P91 materials, and SA335-P22 headers may be upgraded to SA335-P91. The material upgrade generally results in a dramatic increase in header life, and in many cases it is anticipated that these upgrades may nearly double the life of certain components.

Cycling boiler upgrades

Many older and larger fossil power boilers were not designed to accommodate frequent on/off cycles. Design criteria for cycling service boilers are limited due to the lack of long term experience with large units. Cycling service can cause fatigue failures in the economizer tubing and inlet header, lower furnace wall tubes and headers, structural components such as buckstays, and some steam drum internals. As discussed in Chapter 45, this fatigue cracking can be caused by the sudden flow of cold water into hot boiler components. Thermal differentials of 200 to 400F (111 to 222C) can be created. Furnace subcooling, boiler forced cooling during a shutdown, and intermittent cold feedwater flow into the boiler during startup are three sources of thermal differentials and cyclic cracking. The thermal stresses produced within the components may be sufficient to produce low cycle fatigue cracks. The solution in most cases is to modify the boiler and/or feedwater system to prevent the sudden entry of cold water into hot components.

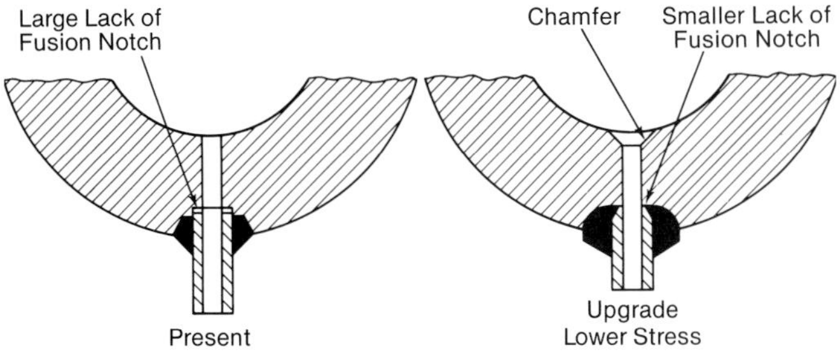

Fig. 5 Header upgrade with redesigned tube penetration.

When a boiler is experiencing cracking due to subcooling and cold feedwater flow upon startup, an off-line pump-assisted circulation system may be installed to reduce the thermal transients. The system, as shown in Fig. 6, consists of an off-line circulation pump, a thermal sleeved tee connection between the off-line pump and the feedwater line, a connection line from the boiler downcomer to the pump, a warming bypass system, various valves and a control system.

The off-line pump is only operated when the boiler is shut down. Its purpose is to provide a small amount of circulation within the furnace circuit and through the economizer to prevent temperature stratification in the water circuits. The tee connection permits the introduction of a small amount of hot water from the furnace into the feedwater stream when feedwater is intermittently supplied to the boiler and before a steady feedwater flow is established. The warm furnace water introduced at the tee connection raises the feedwater temperature enough to prevent thermal shock to the economizer. The connection contains an internal thermal sleeve that protects the tee from a thermal shock when cold feedwater is first fed to the economizer. A control system monitors the feedwater temperature and flow and controls the recirculation pump. When the boiler startup sequence is initiated, the off-line system is shut down and isolated. Warming lines permit natural circulation through the pump when it is shut down and the boiler is off-line.

Experience with off-line recirculation systems has shown that thermal shock differentials can be reduced to less than 100F (56C) from previous levels of 200 to 300F (111 to 167C). Such a reduction may eliminate the fatigue cracking that is associated with frequent unit cycling.

Drum boiler bypass system upgrade

Steam turbine transient stresses associated with on/off operation or load cycling shorten turbine life. Long startup times required to minimize these stresses lead to costly fuel consumption. Drum boilers can be upgraded with a superheater bypass system for improved cycling capability. The bypass system upgrade can be installed on a drum boiler to minimize startup time, provide control of steam

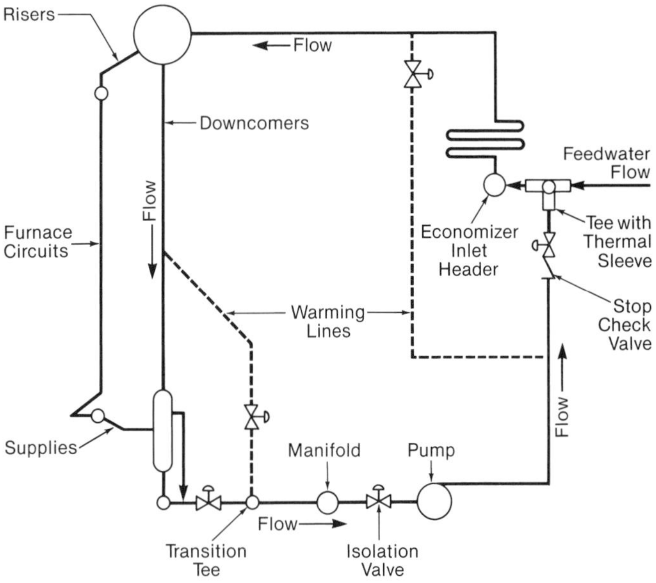

Fig. 6 Off-line recirculation system to reduce thermal shock.

temperature to match turbine metal temperature, and allow dual pressure operation of the boiler and turbine for better load response. These features reduce turbine stresses for improved availability and reduced maintenance costs.

The common means of steam temperature control in a drum boiler (water attemperation in conjunction with parallel convection surface, gas recirculation, excess air, or burner input adjustments) do not permit easy control of turbine temperatures during startup and low loads. This is because there is a large mismatch between flue gas flow and steam flow. The drum boiler bypass system, shown in Fig. 7, provides direct control of the steam temperature by saturated attemperation of the superheat and reheat outlet steam. This arrangement provides the desired steam temperature for the turbine without undue restrictions on the startup firing rate.

The drum boiler bypass system substantially reduces cold startup time because it controls the temperature differences of the saturated boiler surface, superheater surface, and turbine. The bypass system consists of a control system with steam piping and valves (see Fig. 8) and is described in detail in Chapter 18.

The system has a set of superheater stop and bypass control valves downstream of the primary superheater to allow dual pressure operation; the turbine throttle pressure is controlled separately from drum pressure. This control system is designed to maintain constant pressure operation of the drum and boiler furnace circuits and variable pressure operation of the turbine, because the turbine throttle valve is open over most of the load range.

Dual pressure operation minimizes thermal stresses to the boiler and the turbine. A dual pressure shutdown keeps the boiler near full pressure and the turbine metal near full load temperature in preparation for a quick restart. In addition, it allows more rapid load changes than variable boiler pressure operation.

A superheater bypass diverts excess steam from the boiler to the condenser, thereby separating firing rate from drum pressure during shutdown and startup. Gas temperature probes, located near the superheater and reheater outlet tubes, monitor flue gas temperatures.

The superheater bypass system permits cycling the unit rapidly and often without accumulating major turbine damage.

Circulation and capacity upgrades

Water circulation upgrades are often required to overcome operational and maintenance problems. Symptoms of a circulation problem include localized tube failures

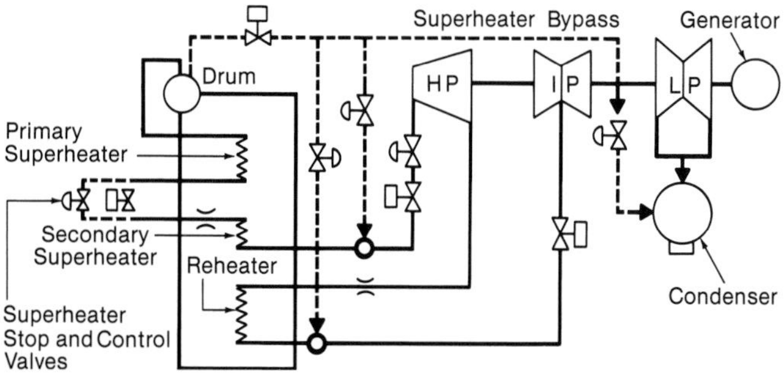

Fig. 7 Typical superheater bypass system for a drum boiler to minimize turbine stress.

Fig. 8 Large diameter piping for bypass system.

in particular boiler circuits, increases in water side and steam side deposits and pressure drops, fluctuations in feedwater control, and drum water level excursions and fluctuations. Occasionally, circulation problems may appear when supply tubes or headers for the furnace waterwall circuits crack and distort. Circulation problems can directly affect boiler performance and, if severe, can reduce capacity and availability.

Determining causes and solutions for circulation problems requires boiler performance testing, heat transfer and circulation computer modeling, and a thorough knowledge of boiler dynamics. Thorough discussions of heat transfer and circulation are covered in Chapters 4 and 5.

Consider an example boiler with circulation problems. This unit was experiencing an abnormally high forced outage rate due to tube failures in the platen wingwalls as well as extensive failures of the furnace rearwall arch tubes. Fig. 9a illustrates the upper furnace configuration and arrangement of the wingwalls.

The majority of tube failures in this unit were found in the lower 17 deg (0.3 rad) inclined portion of the wingwalls. Some time later, the rearwall arch tubes began failing, especially in the lower inclined area. Poor water chemistry and resultant heavy internal tube deposits were originally blamed for the tube failures. If water chemistry is not properly controlled, an internal mineral buildup can occur. These deposits act as insulators. The circulating water can not sufficiently cool the tube which then fails due to short term, localized overheating. However, water chemistry was monitored and found to be acceptable in this case.

The tube failure rate affected availability and the unit had to be derated. The loss of generating revenue was large enough to justify major expenditures to remedy the situation.

A complete engineering study was required to review the plant operating procedures and the original boiler design. The study goals included:

1. analyzing the overall boiler circulation system,
2. reviewing the platen wingwall design,
3. examining the design and performance of steam drum internals,
4. investigating the rearwall arch and supply circuit design, and

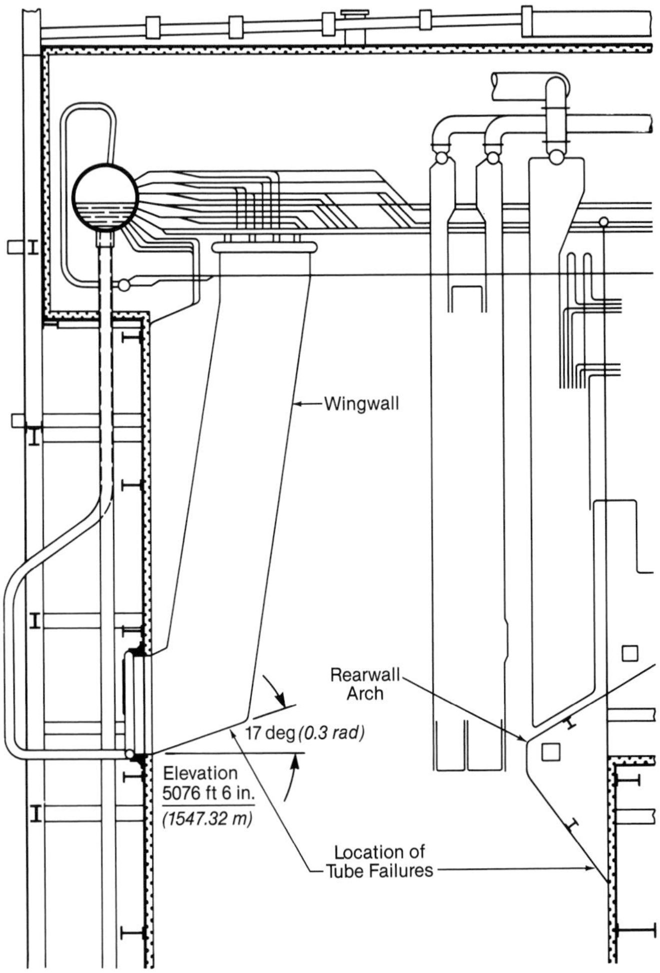

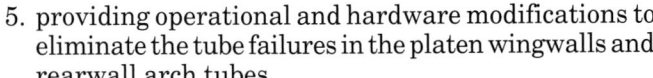

Fig. 9a Original upper furnace arrangement.

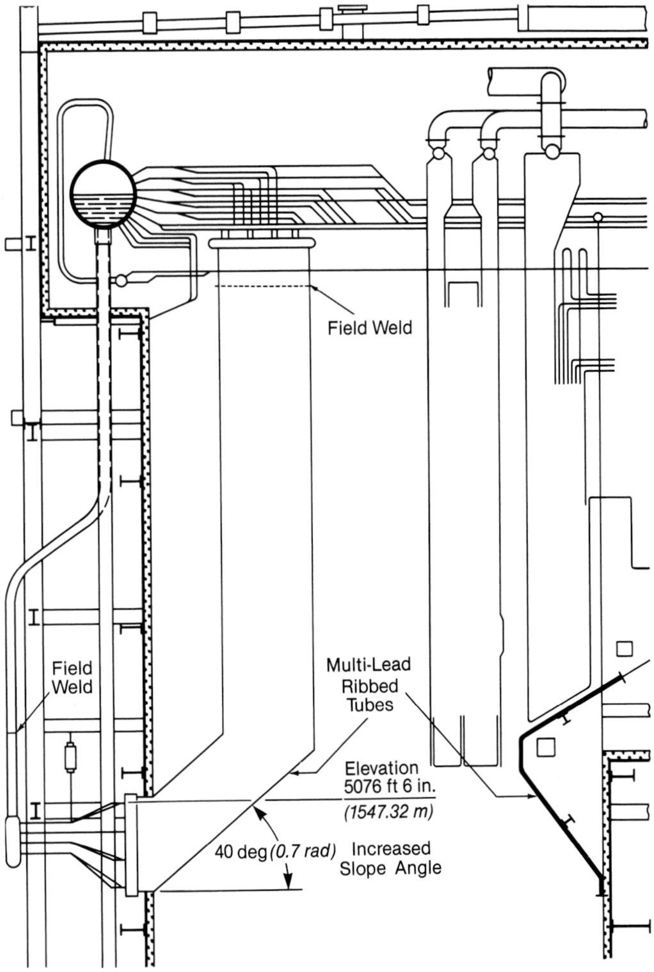

Fig. 9b Upper furnace arrangement for a circulation upgrade.

5. providing operational and hardware modifications to eliminate the tube failures in the platen wingwalls and rearwall arch tubes.

Field tests were conducted to verify the operating parameters. A computer simulation focused on the circulation of the steam drum, wingwall and rearwall arch tube circuits. Mass flow velocities in these tubes were found to be low. These low velocities contributed to flow imbalances between circuits as well as film boiling conditions inside the tubes. The steam film acted as an insulator, producing numerous overheat failures.

The existing drum internals were also found to be significant contributors to the circulation problem. The separators in the steam drum generated a high pressure drop. This high pressure drop reduced the pumping head for the wingwall and rearwall arch circuits, contributing to poor circulation and flow imbalances. The drum water level, 9 in. (230 mm) below the drum centerline, permitted steam to be drawn into the downcomers. This further reduced the effective water density (pumping head) and mass flow velocities in the furnace circuits.

Once the circulation problems were fully identified and analyzed, specific solutions were recommended. The key modifications implemented in the circulation upgrade were as follows (See Fig. 9b):

1. New drum internals. Three rows of highly efficient, low pressure drop cyclone separators replaced the existing two rows of high pressure drop turbo-separators. (See Fig. 10.) The drum water level was increased from −10 in. (−254 mm) to the centerline. These modifications increased the available head for all circuits. However, the wingwall circuits with their extremely flat characteristic predictably received the greatest flow increase.

2. Redesigned wingwalls. The overall length of the wingwalls was increased to raise the heat absorption in these vulnerable circuits, based on the natural circulation principle of more heat, more flow. (Fig. 11.) This also lowered the furnace exit gas temperature by about 35F (19C) which reduced the tendency of slagging beyond the furnace. A third benefit was a slight reduction in superheat and, especially, reheat spray, both being excessive on the existing unit.

   The bottom slope was increased from 17 to 40 deg (0.3 to 0.7 rad) and the tube diameter was decreased. Multi-lead ribbed tubes were used instead of internally smooth tubes in the sloped portion. These measures were aimed at eliminating film boiling and flow instabilities, identified as the root cause of tube failures in these circuits. Film boiling in these circuits promoted formation of internal deposits and corrosion due to stagnant flow conditions. The increased flow resistance was more than offset by the greater pumping head generated by the new design.

Fig. 10 Cyclone steam separator installation.

3. Redesigned water supply system. The existing downcomers were lengthened and closed off by new downcomer bottles. Each bottle was equipped with eight supply pipes, four feeding each vertical wingwall inlet header. To achieve the best flow distribution to all wingwall circuits, each of the inlet headers was divided into four chambers by three diaphragms. These diaphragms received a small, central perforation to allow complete draining of the headers during outages.
4. Redesigned rearwall arch. Multi-lead ribbed tubes were installed in the rearwall arch in the form of shop-membraned tube panels. This eliminated the failures in the lower bend area. The original transition from the 2.5 in. (63.5 mm) OD rearwall tubes to the 3 in. (76.2 mm) arch tubes created a steam-water jet along the unheated side (inside of the extrados) of these bends and a corresponding void on the inside of the tube bends. This had caused the crucial bend sections facing the furnace heat to fail due to insufficient cooling. For safety, the arch tubes were also redesigned as multi-lead ribbed tubes.

Fig. 11 Wingwalls installed to increase heat absorption.

The overall boiler circulation was significantly improved, and a high safety margin prevented the recurrence of failures in the wingwalls or rearwall tubes. The circulation upgrade restored 100% load capability and improved unit reliability and availability.

Superheater and economizer upgrades

Chapters 18 and 19 and earlier portions of this chapter summarize the design, function, operation and aging of superheater and economizer components. The normal aging processes of corrosion, erosion and creep progressively damage these components and they must be replaced from time to time. Any time a replacement is needed, there is an opportunity to upgrade the existing materials, configuration, and/or performance. Upgrades that are performed on older boilers benefit from the development of advanced materials, improved designs, and innovative analytical techniques. (Fig. 12)

Superheater upgrade Consider a utility that had recurring tube failures in the secondary superheater of an oil-fired boiler. Ash corrosion from the fuel oil caused failures in the tubes exposed to the higher operating temperatures. Fig. 13 illustrates the arrangement of the boiler.

The furnace gas temperature at the leading edge of the superheater was nearly 2800F (1538C). High superheater tube temperatures resulted from this unusually high furnace exit gas temperature (FEGT). External tube corrosion was accelerated by this high temperature and the high vanadium and sodium contents in the oil. Tube failures occurred so frequently that the bottom three rows of the horizontal superheater had to be replaced every 18 months.

An upgrade in superheater design usually requires field testing to define the operating conditions followed by a detailed engineering study. To correctly design the upgraded superheater, boiler operating temperatures were obtained. A computerized boiler model was used for the performance testing. This system revealed that the actual FEGT was 125F (69C) higher than the design

Fig. 12 Retrofit economizer redesigned to eliminate plugging and erosion.

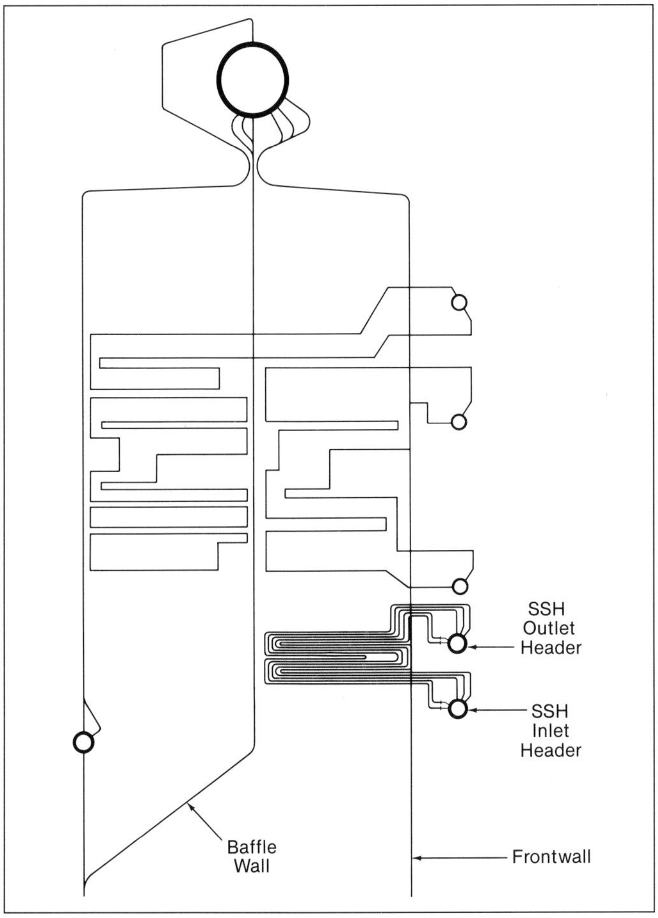

Fig. 13 Boiler arrangement showing secondary superheater (SSH) directly above furnace where gas temperatures reach 2800F (1538C).

value. The high FEGT was a result of furnace design and burner arrangement. Because it was impractical to redesign the furnace or change to a less corrosive fuel, and the owner did not wish to derate the unit, a design study concentrated on a superheater upgrade.

A new superheater was designed using the actual FEGT. This allowed the tubes to continuously function at the higher temperatures without producing overheat or corrosion related failures. In the final arrangement of the superheater upgrade, the three lowest tube rows were replaced with a high chromium nickel alloy material 800H. (See Chapter 6.) The remaining tubes in this bank were replaced with a stainless steel material. A new support method, which allowed greater material expansion at the elevated operating temperatures, was also devised.

Economizer upgrade Another example of an upgrade which improved boiler capacity and availability involved a finned tube economizer. A 600 MW, coal-fired boiler was designed with a staggered, finned tube economizer. The economizer, shown in Fig. 14, was equipped with longitudinal fins. In clean condition these fins increase heat transfer and reduce the amount of tubing required.

Shortly after this unit went into operation, the economizer experienced severe flyash plugging. The utility tried various solutions from extensive sootblowing to forcing the staggered tubes in line. As the flyash plugging spread through both banks, system gas flow resistance more than doubled. The increased flue gas resistance pushed the boiler fan to its limit and limited boiler load. Extremely high ash velocities caused tube erosion and resulted in numerous tube failures. Unit availability and reliability were greatly reduced.

An engineering evaluation determined that the existing finned economizer was not appropriate for the high ash coal used as the main fuel. The staggered tube pattern only increased the severe flyash plugging. An analysis showed that an in-line bare tube economizer with equivalent performance could be fitted into the existing space. Fig. 15 illustrates the upgraded in-line economizer design.

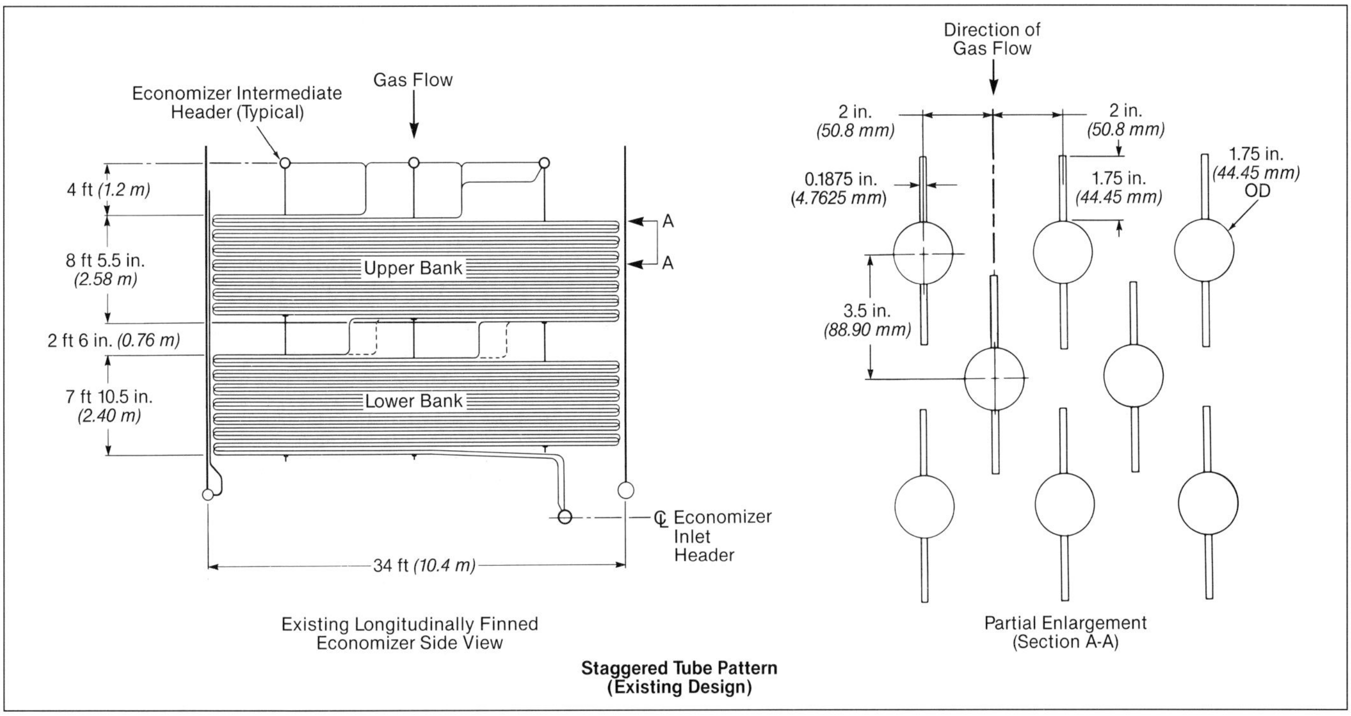

Fig. 14 Original staggered tube, finned economizer that experienced severe flyash plugging.

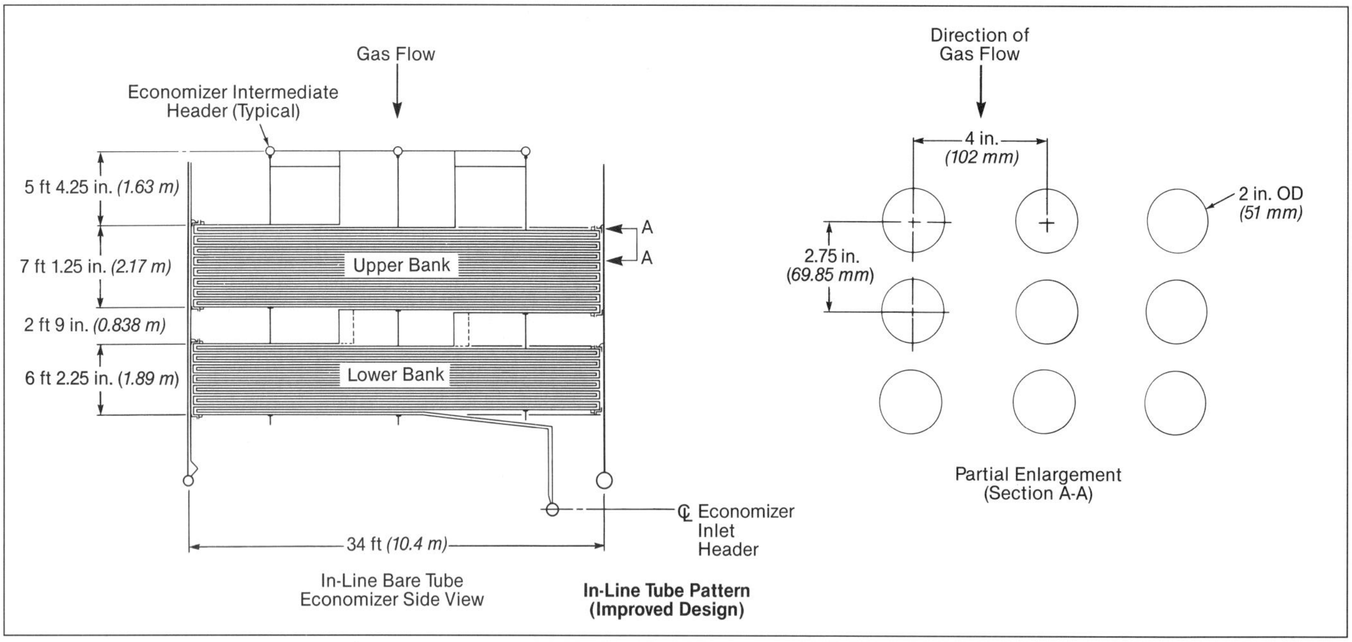

Fig. 15 Upgraded bare tube in-line economizer for high ash coal.

The upgraded economizer included the following features:

1. Vertical spacing between tubes was decreased to allow installation of more tube rows and to minimize gas side resistance.
2. An in-line arrangement, which reduced gas side pressure drop and potential for ash erosion, was used.
3. The economizer tube diameter was increased from 1.75 to 2 in. (44.5 to 50.8 mm) to achieve a higher flue gas velocity and better convection heat transfer without exceeding the allowable velocity limit for the percent ash and percentage of abrasive silica and alumina constituents in the fuel. Bare tubes arranged in line are most conservative in such hostile environments, are least likely to plug, and have the lowest gas side resistance per unit of heat transfer. The weight of the new economizer did not exceed the original weight. Also, the new economizer required no more space than the original design.
4. The tube bends were protected by new erosion barriers on top and bottom of both banks. Properly designed barriers do not noticeably reduce the effective heating surface, but are very effective in throttling the flue gas flow across the tube bends along the enclosure walls. Without barriers, the open space between the return bends and the enclosure walls would be the path of least resistance, and flue gas would stream through these gaps at very high velocities. The effect would be excessive erosion of the bends, reduced overall heat transfer across the banks, and possible damage to uncooled casing. The existing inlet and intermediate headers were reused. Access doors and platforms did not have to be relocated or modified.

The upgraded economizer has experienced no flyash plugging. The recurring tube erosion problem has also been completely eliminated. As a result, the unit's reliability and availability have been significantly restored. The boiler now operates at full load and has not been limited by economizer problems.

Fuel switching and capacity upgrades

Fuel switching is one method that permits utilities to comply with mandated sulfur dioxide (SO_2) emissions limits. The major objective of changing coal type is to reduce the amount of SO_2 produced. Many of the boilers designed and built in the U.S. during the last four decades have fired Eastern bituminous and Midwestern subbituminous coals. These fuels have high heating values and low ash content, but they also usually have an unacceptably high sulfur content. A switch to a low sulfur coal may reduce SO_2 emissions, but it may also have adverse effects on unit performance due to a change in fuel ash content or other fuel characteristics. A thorough study of the change is required. Test burns of a trial fuel are made to evaluate its operational effects. Specific attention is paid to flame stability at different loads, furnace and convection pass slagging and ash loading, pulverizer performance, emissions control equipment performance, and emission changes.

Switching to one of the subbituminous fuels usually reduces SO_2 emissions to acceptable levels but can adversely impact the performance of the boiler island equipment. If the boiler and auxiliaries were designed solely for bituminous firing, switching to a subbituminous fuel could lead to reduced load, increased operating and maintenance expenses, and higher fouling and slagging in the furnace.

While the detailed discussion for rating and analyzing fossil fuels can be found in Chapter 8, the key differences between low and high sulfur coals need to be reviewed. Table 3 highlights the fuel and ash components that most affect boiler performance. An Eastern, high sulfur bituminous coal is used as the reference point. The relative differences between this coal, a Western subbituminous low sulfur coal, and an Eastern bituminous low sulfur coal are shown.

As seen in Table 3, a Western subbituminous fuel generally has a much lower heating value than its high sulfur bituminous counterpart.

Table 3
Comparison of Bituminous and Subbituminous Fuels
Differences in Fuel Properties and Boiler Performance

Fuel Properties	Eastern Bituminous High Sulfur	Western Subbituminous Low Sulfur	Eastern Bituminous Low Sulfur
HHV,* Btu/lb (kJ/kg)	10,570 (24,586)	22% lower	25% higher
Relative fuel flow	1.0	1.27	0.8
Moisture, %	14.25	More than doubled	Less than half
Ash, %	13.41	30% less throughput**	60% less throughput**
Ash slagging potential (furnace)	Medium	High/severe	Low
Ash fouling potential (convection pass)	Low/medium	Medium/high	Low
Volatility		Little change	
Combustion		Little change	
Air preheat	300F (149C)	500F (260C)	300F (149C)
Ash reflectivity	--	Some increase	About same
Coal grindability		Little change	

* Higher heating value
** Throughput based upon constant heat input to boiler

Coal handling equipment considerations If the same electrical generating capacity of the unit is to be maintained, then the fuel flow must increase. The capacities of the coal conveyors, feeders, pulverizers and burners must be checked to determine if these components can accommodate higher fuel flows. One other important factor to be considered when switching to a low sulfur Western fuel is its tendency to be more friable (likely to generate dust). As a result, fugitive dust emissions during unloading, crushing, conveying, storage, and pulverizing increase greatly. This results in an increased risk of fires and explosions. Improved operating procedures can be implemented to manage the fugitive dust. Additional equipment or facilities may also be required. Some of the options that could be required are:

1. enclosed conveyors,
2. improved ventilation systems,
3. conveyor and bunker dust suppression systems,
4. pulverizer fire inerting/fogging equipment,
5. electrical modifications to minimize spark formation,
6. improved housekeeping, and
7. changes in coal receiving facilities (conveyor, barge, truck, rail). (See Chapter 11.)

As the fuel flow is increased to the boiler, higher velocities in the coal pipe can contribute to flame instabilities and increased erosion. Although the total ash content decreases when burning a low sulfur Western fuel, excessive interior erosion of the coal pipe can result due to the higher particle velocities. Flame instabilities can result from increased fuel velocities and a doubling of moisture content. Auxiliary fuel or flame stabilizing lighters could also be required.

While the total amount of required combustion air does not significantly change when switching from a high to low sulfur coal, increased pulverizer drying capacity is required. Additional drying of a high moisture coal can be accomplished by adding air heater surface and by increasing the primary air temperature to the pulverizers. Primary air temperature could increase by as much as 350F (194C) above that required for a high sulfur, low moisture fuel.

Air heater performance is also affected by the increased fouling tendency of Western subbituminous low sulfur coals. Sootblower and water washing protection may be needed to combat heater surface plugging. Air heater adjustments may also be needed to meet the coal drying requirements.

A major limiting factor associated with a coal switch is the capacity of the existing pulverizing equipment. In general, the fuel flow required for the same thermal input to the boiler can not be achieved with existing pulverizers. The coal flow through any pulverizer is affected by the amount of moisture, coal grindability and pulverizer horsepower. In most cases, the existing pulverizer is derated by moisture and decreased grindability.

Solutions for improving pulverizer performance when firing a Western fuel include upgrading the pulverizer internals, increasing the spring tension, or replacing the existing motor. If the remaining plant life is sufficiently long, the installation of larger pulverizers or an increase in the number of mills may be economically justified. Structural and foundation changes would be required. The number, size and routing of coal pipes would also be affected.

When switching coal type, the most significant factors influencing modern boiler operation are the chemical constituents and physical properties of the coal ash. When designing the steam generator, the ash characteristics determine the size, location and spacing of the pressure parts and combustion equipment. Each coal has some tendency to slag or foul various interior boiler surfaces. However, many fuel changes involve replacing bituminous with subbituminous coals in units designed for the former. As a result, any fuel switching is generally preceded by an engineering study and various field tests.

Fuel switching example The following fuel switching example illustrates the types of boiler changes that may be required when firing higher slagging and fouling coals.

Consider a utility with a 365 MW pulverized coal-fired boiler. A cross-sectional drawing of the original unit is shown in Fig. 16.

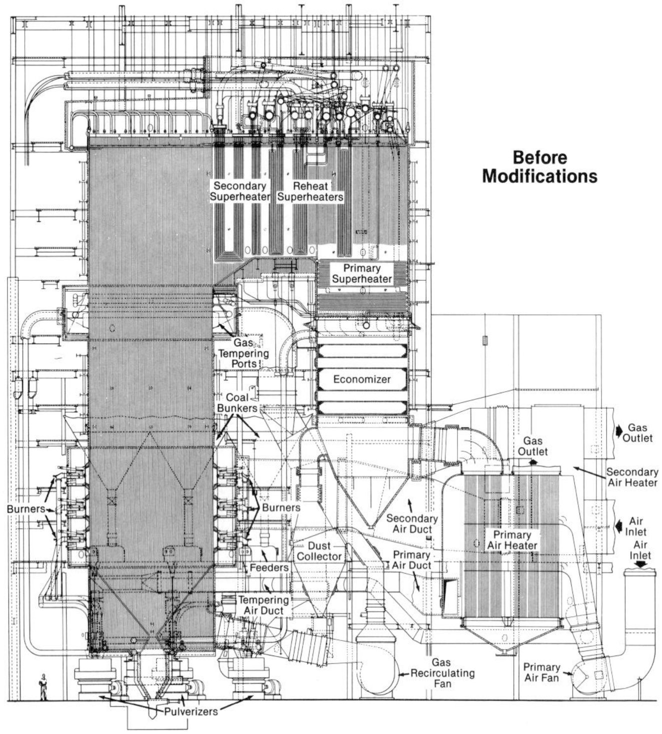

Fig. 16 Original arrangement of coal-fired plant before fuel switching.

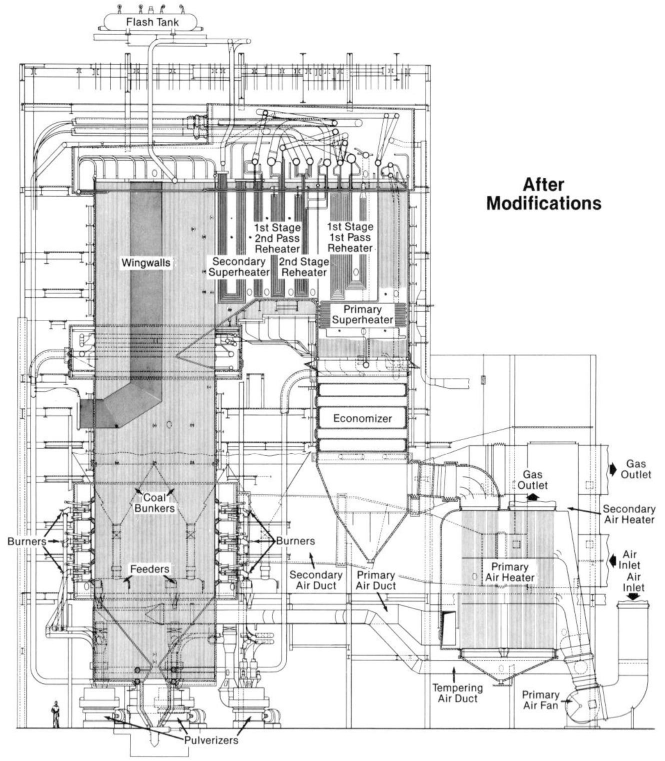

Fig. 17 Upgraded unit design for low sulfur subbituminous Western coal.

Trial burns of a subbituminous coal indicated that a number of functional and operational problems would be encountered. The FEGT increased dramatically, excessive slagging occurred in the furnace and pendant superheater sections, and ash plugging of the horizontal convective sections occurred. As the furnace slagging became severe and furnace absorption decreased, superheater temperatures exceeded design values. This absorption shift added to the slagging of the pendant superheaters. As the tube sections experienced severe plugging, gas velocities increased to the point that tube surface erosion caused extensive leaks. The slagging and fouling increased the draft loss and the maximum fan capacity was reached. Sootblowing was only marginally effective in clearing the fouled sections. The unit was forced to operate at reduced loads and this in turn contributed to reduced sootblower effectiveness.

Extensive field tests were conducted to determine the existing gas temperatures at the furnace exit. The gas temperatures were found to be higher at the bottom of the secondary superheater than at the top. It was determined that the flue gases were vertically stratified through the superheater, with the majority of fouling taking place at the bottom of the superheater. As the lower portion of the bank became plugged, gas flow was biased higher, progressively fouling more of the bank.

A number of modifications were made to restore capacity and to permit burning a Western subbituminous coal. The modifications, illustrated in Fig. 17, included:

1. Furnace wingwalls were installed as primary superheater surfaces to reduce the FEGT from 2500 to 2180F (1371 to 1193C). This eliminated serious fouling of the secondary superheater.
2. A rearwall arch was added to improve vertical distribution of gas flow across the furnace exit and to force the flue gases across the new wingwalls.
3. The horizontal primary superheater surface was reduced to offset the addition of the wingwalls.
4. The gas recirculation system was eliminated. As a result, the first stage reheater was modified to achieve reheat temperature due to reduced gas flows and the lower flue gas temperatures resulting from the addition of the wingwalls.

The modifications shown in Fig. 17 were successful in eliminating the slagging and fouling associated with burning a Western fuel. Lost boiler capacity was restored and the desired SO_2 emission reductions were achieved.

Other system considerations While switching to an alternate low sulfur coal demands careful examination of the steam generator, there are a number of other systems that can also be adversely affected. Flues, ducts and fans, the electrostatic precipitator, and the ash handling system deserve close attention.

Due to increased air and gas flow and increased pressure drops, several options may be considered for flues, ducts and fans. Flues and ducts may be rearranged, the fan wheel may be upgraded, and the fan motor may be upgraded with a higher capacity unit.

Changes in ash loading and resistivity and higher gas volumes may reduce electrostatic precipitator (ESP) performance. In particular, lower SO_2 emissions may adversely affect flyash resistivity. ESP upgrades include adding collecting surface area, adjusting operation, and adding flue gas conditioning. In extreme cases, a larger ESP or baghouse may be retrofitted.

There may also be an increase in furnace ash loading, and additional removal/storage areas may be required. Because fuel ash characteristics, such as agglomerating tendencies, vary, ash handling system upgrades may be necessary.

Switching to oil or natural gas The discussion of fuel switching to reduce SO_2 emissions from utility power plants has been limited to coal, because coal-fired capacity represents the majority of these plants. However, there are circumstances where switching from coal to oil or natural gas may be economically justified. Analyses similar to those for coal type switching must be performed. Additional considerations of fuel price and availability, stability of supplies, handling and storage system changes, and transportation costs affect the decision to switch to oil and/or natural gas. Firing either of these fuels may also require significant changes to pressure part arrangements and tube materials to ensure that overall unit performance is maintained.

Strategies for dealing with the older plant

In developing a long term strategy for a power plant, its current condition must first be evaluated. Performance test and availability data, maintenance records, and operation and condition assessment reports (see Chapter 45) should be compiled into a summary document. This document can outline design and operational deficiencies and can indicate components that need repair or replacement. A schedule for this repair work and a detailed cost breakdown should be included. Recommendations for component upgrades may also be included in this summary.

A second document included in the long term strategy is a load forecast for the plant. The projected load cycling and the anticipated capacity factor are of particular interest. These items are major determinants of plant revenue and operating costs.

The compliance of an older plant with environmental regulations and the associated costs of this compliance are significant factors in this strategy. For plants with high heat rates and emissions levels, retiring the unit is often considered.

When all of the cost and operating data have been summarized, cash flow and return on investment analyses are performed. The required cash flow can be compared to available capital resources, and alternate uses for this capital can be compared to extending unit life.

At this point, a flexible strategy is appropriate. Alternate uses for the older plant can be considered. For example, some plants are candidates for conversion to fluidized-bed units. Any strategy must minimize capital risk and provide contingencies for unexpected load growth. The older plants in a system represent a vital resource that can not be easily replaced; therefore, a careful strategy for life management of these plants is mandatory.

Section VIII
Steam Generation from Nuclear Energy

As a major change from previous editions of *Steam*, all nuclear-related material has been consolidated into one section. This section describes the application of steam generating fundamentals to the design of nuclear steam supply systems, in which steam is generated by heat released from nuclear fuels. The section also describes the manufacture of nuclear equipment and has been expanded to include major operating issues and nuclear services designed to extend the life of operating plants.

The section begins with an overview of nuclear installations, concentrating on the pressurized water reactor. The nuclear fuel cycle is described followed by a discussion about the principles of nuclear reactions. In Chapter 50, in-depth discussions of the steam supply systems are combined with discussions of safety. Chapter 51 then describes the special application of these systems to marine equipment. Manufacturing of nuclear components, primarily steam generators, is discussed in Chapter 52. Major plant operating issues are then described in chapters specific to water treatment and plant operations. The section ends with discussions of life extension, nuclear plant services, and the issues of nuclear waste management.

Oconee Nuclear Power Station, Duke Power Company.

Chapter 47
Nuclear Installations for Electric Power

Since the early 1950s, nuclear fission technology has been explored on a large scale for electric power generation and has evolved into the modern nuclear power plants. (See frontispiece and Fig. 1.) Many advantages of nuclear energy are not well understood by the general public, but this safe, environmentally benign source of electricity is still likely to play a major role in the future world energy picture. Nuclear electric power generation is ideally suited to provide large amounts of power while minimizing the overall environmental impact.

Development of nuclear technology

The concept of an energy generating plant using nuclear fission was first considered by nuclear physicists in the 1930s. However, peaceful use of the atom was delayed until after World War II. The initial peaceful application postulated a controlled nuclear fission chain reaction, generating heat in the fuel and transferring the heat through a fluid medium to a heat exchanger to produce steam. A conventional steam cycle could then be used to generate electricity with a turbine-generator. The proposed nuclear fuels were natural uranium, enriched uranium (containing a higher concentration of the fissionable isotope U-235) or plutonium. It was hoped that these fissile materials would be able to achieve a critical mass for a sustained stable nuclear reaction. The neutrons available for sustaining a chain reaction would be controlled by inserting or removing special neutron-absorbing material. During the late 1940s, calculations showed that nuclear generated electricity could be cost competitive with other sources of generation. Since that time, nuclear power systems have produced competitive power worldwide.

Fundamentals overview

Nuclear energy is derived from the conversion of very small amounts of mass into large amounts of energy, following Albert Einstein's famous relationship:

$$E = mc^2$$

where

E = energy
m = the mass converted into energy
c = the speed of light

As discussed in Chapter 49, the nuclear reactions in today's commercial reactors are based upon a multi-step *fission* or atom splitting process. Neutrons are absorbed by the fissionable atom which is usually U-235. The atom becomes unstable and breaks apart, or fissions, to typically produce two large pieces (new atoms), two or more neutrons and some radiation. The mass of the resulting products is less than that of the initial material by a small amount.

This change in mass produces energy which predominantly takes the form of kinetic energy of the product particles plus some radiation. The kinetic energy eventually appears as internal heating of the fuel which serves as the heat source for the ultimate steam power generation cycle. The neutrons produced serve as the inputs to the next fission reaction resulting in a *chain reaction.* A *critical mass* is achieved when enough nuclear material is brought together so that the fission reactions continue without an external supply of neutrons. In most reactors, the fast moving neutrons are slowed by a material called a *moderator* in order to enhance the chance of a nuclear reaction taking place. Water serves as the moderator as

Fig. 1 Indian Point Station, Unit 1 — Consolidated Edison Company of New York, Inc.

well as fuel coolant in most commercial nuclear reactors today although various combinations of moderator and coolants have also been used (see following discussion).

First generation power plants

The United States (U.S.) had a head start on nuclear technology because of its work in the atomic weapons program. The U.S. Atomic Energy Commission (AEC) took the lead in research and development for a controlled chain reaction application to energy generation. Many concepts were hypothesized and several promising paths were explored, but the real momentum developed when U.S. Navy Captain Hyman G. Rickover established a division in the AEC to develop a nuclear power plant for a submarine. (See Chapter 51.) This program, established in 1949, was to become the forerunner of commercial generating stations in the U.S. and the world. Rickover's design succeeded in 1953. Technology and materials developed by his team became the cornerstone of future U.S. nuclear plants. Concurrently, the AEC established a large testing site in Arco, Idaho where, in 1951, the fast neutron reactor produced the first electricity (100 kW) generated by controlled fission.

The world's first civil nuclear power station became operational in Obninsk in the former Soviet Union in mid 1954 with a generating capability of 5 MW. This was about the same energy level produced in the U.S. submarine design.

In 1953, the Navy canceled Captain Rickover's plans to develop a larger nuclear power plant to be used in an aircraft carrier. However, he subsequently transformed this project into a design for the first U.S. civilian power stations. Duquesne Light Company of Pittsburgh, Pennsylvania agreed to build and operate the conventional portion of the plant and to buy steam from the nuclear facility to offset its cost of operation. On December 2, 1957, the Shippingport, Pennsylvania reactor plant was placed in service with a power output of 60 MW. This event marked the beginning of the first generation U.S. commercial nuclear plants.

Several basic concepts were being explored, developed and demonstrated throughout the world during this period. The U.S. submarine and Shippingport plants were pressurized water reactors (PWR) which maintained subcooled water as the fuel coolant. This water also served as the moderator. The Soviets developed enriched uranium, graphite-moderated, water-cooled designs. English and French engineers explored natural uranium, graphite-moderated, carbon dioxide cooled stations. In all these designs, the coolant (gas, liquid metal or pressurized water) transferred heat to a heat exchanger, where secondary water was vaporized to provide steam to drive a turbine-generator. The other major competing approach in the U.S. was the boiling water reactor (BWR). This was similar to the PWR except that the need for a heat exchanger to transfer heat from the coolant to the secondary steam cycle was eliminated by boiling the water in the reactor core and by using this slightly radioactive steam to drive the turbine. (See also Chapter 1.)

Two main classes of reactor fuel evolved — enriched and natural uranium. In the early stage of nuclear power development only the two major nuclear powers, the U.S. and the Soviet Union, had sufficient fuel production capacity for civilian power production using enriched uranium. Therefore, natural uranium was chosen as the principal fuel in the United Kingdom (U.K.), France, Canada and Sweden. While the enriched uranium-fueled plants could be smaller and therefore required a lower initial investment, higher costs of the enriched fuel over the life of the plant made its use economically equivalent to that of natural uranium. All six countries launched programs to build civil, commercial nuclear power stations. Other countries collaborated with one of these six for construction technology.

In the U.S., the emerging technology focused on the enriched uranium light water reactor (LWR) which used regular water for the coolant and moderator. This had been the technology selected by the Navy programs and, as a result, the civilian sector gained from the experience of naval applications. The AEC also financed construction of the nuclear-powered merchant ship N.S. *Savannah*, which used this technology in a PWR designed and constructed by Babcock & Wilcox (B&W). (See Chapter 51.)

In 1955, the AEC began the Power Reactor Demonstration Program to assist private industry. However, by 1963 only one demonstration plant had started up, and two others had been privately financed. Yankee Rowe, Indian Point Unit 1 in New York, and Dresden 1 in Illinois operated successfully and demonstrated the viability of this technology for larger plants.

Second generation nuclear power plants

From these beginnings, the second generation of nuclear plants began in 1963 when a New Jersey public utility (Jersey Central Power & Light) invited bids for a nuclear power station to be built at Oyster Creek. A comprehensive study showed that it was economically attractive to operate, and the utility was ready to proceed. This bid was won with a turnkey proposal to deliver a plant at a fixed price. The suppliers were expected to absorb initial losses in expectation of profitably building a series of similar plants in the near future. From this study and resulting order came the realization that nuclear power stations could be competitive with other electricity sources. Enthusiasm spread and, in the following two years, more new nuclear generating capacity than conventional fossil fuel capacity was ordered. Orders were placed for more than 50 nuclear plants with outputs from 500 to 1100 MW each for a total of 40,000 MW. Orders averaged 10,000 MW for each of the next three years. A new wave of orders began in 1971 with annual capacities doubling to 20,000 MW; outputs doubled again to more than 40,000 MW in 1972 and 1973. At this time, the total non-nuclear U.S. electrical generation capacity was 430,000 MW. The nuclear program then underway totaled about 200,000 MW from more than 200 plants ordered, under construction, or in-service. The second generation nuclear plants were comparable in size or larger than contemporary fossil plants. Nuclear plant capital costs were generally higher and fuel costs were lower when compared to fossil fuel plants. Therefore, nuclear power costs were lower in the larger plants where fuel cost represented a larger portion of the total cost.

The U.S. designs were also attractive for export. U.S. PWR designs were ordered in Japan, South Korea, the Philippines, Spain, Sweden, Switzerland, Yugoslavia,

Taiwan, Italy, Brazil, Belgium and West Germany, while Italy, Mexico, Spain, India, the Netherlands, Switzerland, Taiwan and Japan ordered BWRs of U.S. design.

In parallel with the rapid U.S. growth in nuclear electric generation capacity, the rest of the economically strong countries also began development and construction. In France, early efforts using natural uranium and gas-cooled reactors could not compete with oil-fired plants and enriched uranium light water reactors. After France developed uranium enrichment capability, the country's national utility, Electricité de France, in conjunction with heavy component designer and manufacturer Framatome, ordered PWR and BWR designs based on U.S. technology. An innovative concept applied by the French was to build a series of nearly identical plants of each design, thereby achieving economies of scale. Subsequently, the PWR design was selected by the French for their standardized series of plants. The basic designs developed were also eventually exported to Belgium, South Korea and South Africa.

In the U.K., initial designs for weapons production reactors were applied to generate commercial power from natural uranium-fueled, graphite-moderated, gas-cooled reactors. In the early 1960s, an advanced gas-cooled reactor (AGR), which substituted enriched for natural uranium, was designed and subsequently built. Italy and Japan also built versions of these British designs.

The Soviet Union developed a graphite-moderated, boiling water-cooled, enriched uranium reactor from its early Obninsk power station design. A major difference in this reactor from similar Western designs was that it did not include a sturdy reactor containment building. The early Soviet goals included the development of naval propulsion reactors based upon the PWR concept which was then applied to civilian use. This PWR design was exported to several Soviet bloc countries and Finland. The Soviet PWR was similar to the U.S. designs except it generally provided significantly lower electrical output.

The excellent moderator characteristics of deuterium oxide (heavy water) became the basis for the very successful Canadian deuterium (CANDU) reactors. These natural uranium-fueled reactors were first operated in 1962 (25 MW Nuclear Power Demonstration reactor at Rolphton, Ontario), and subsequent designs increased output to 800 MW. CANDU designs have been built in India, Pakistan, Argentina, Romania and South Korea.

In Japan, electric utilities were experimenting with several types of imported nuclear generation plant designs. They ordered gas-cooled reactors from the U.K., and BWR and PWR designs from the U.S. The Japanese government also began sponsorship of a breeder reactor design.

The Germans were actively involved in designs primarily associated with high temperature gas reactor concepts. In 1969, the formation of Kraftwerk Union introduced German PWR and BWR designs based on U.S. technology. The intent was to construct a series of these plants to achieve economies of scale. However, unlike the French program, few plants were constructed. Several German PWRs were exported to Brazil, Spain, Switzerland and the Netherlands.

The energy crisis of the mid 1970s caused significant reductions in worldwide electric usage growth rates, and electric utilities began canceling and delaying new nuclear generating capacity construction. Other factors also contributed to this construction decline. The delay of schedules due to slower load growth and/or regulatory hurdles drove unit costs higher due to higher financing costs caused by the longer construction times. The high inflation of the period further increased construction costs. In addition, continually changing safety regulations increased costs as changes in design of plants already under construction were required.

Plant design

By 1978, nine second generation nuclear units designed by B&W were in commercial operation, each generating 850 to 900 MW. The first unit was one of three built at Duke Power Company's Oconee Nuclear Station located near Seneca, South Carolina. The chapter frontispiece shows the three units of this station. The three vertical, cylindrical concrete structures are the reactor containment buildings. Although some equipment is shared among units, each is operationally independent. The rectangular building at the right contains the turbine-generators. The control room, auxiliary systems and fuel handling facilities for Units 1 and 2 are located in a structure between reactor buildings 1 and 2. Corresponding equipment for Unit 3 is located separately. The fol-

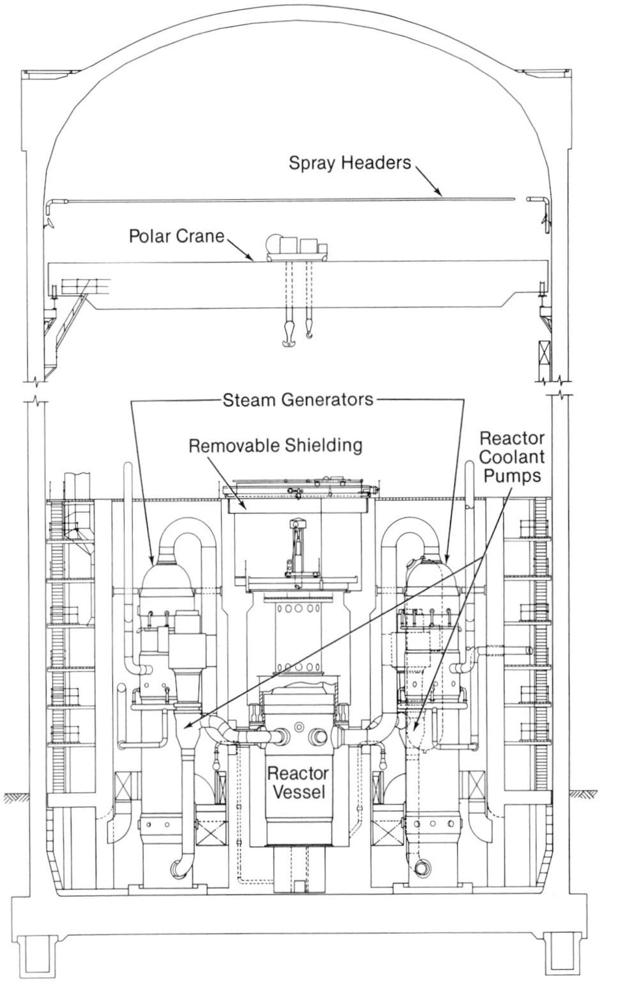

Fig. 2 PWR containment building, sectional view.

lowing general description applies to Oconee Unit 1, although it is generally applicable to all U.S. B&W plants; the concepts are applicable to all PWRs.

Containment building

Fig. 2 is a vertical section of the reactor containment building. The structure is post-tensioned, reinforced concrete with a shallow domed roof and a flat foundation slab. The cylindrical portion is prestressed by a post-tensioning system consisting of horizontal and vertical tendons. The dome has a three way post-tensioning system. The foundation slab is conventionally reinforced with high strength steel. The entire structure is lined with 0.25 in. (6.3 mm) welded steel plate to provide a vapor seal.

The containment building dimensions are: inside diameter 116 ft (35.4 m); inside height 208 ft (63.4 m); wall thickness 3.75 ft (1.143 m); dome thickness 3.25 ft (0.99 m); and foundation slab thickness 8.5 ft (2.59 m). The building encloses the reactor vessel, steam generators, reactor coolant loops, and portions of the auxiliary and safeguard systems. The interior arrangement meets the requirements for all anticipated operating conditions and maintenance, including refueling. The building is designed to sustain all internal and external loading conditions which may occur during its design life.

Reactor vessel and steam generators

Also shown in Fig. 2 are the reactor vessel and the two vertical steam generators, each of which produces approximately 5.6 $\times 10^6$ lb/h (705.6 kg/s) of steam at 910

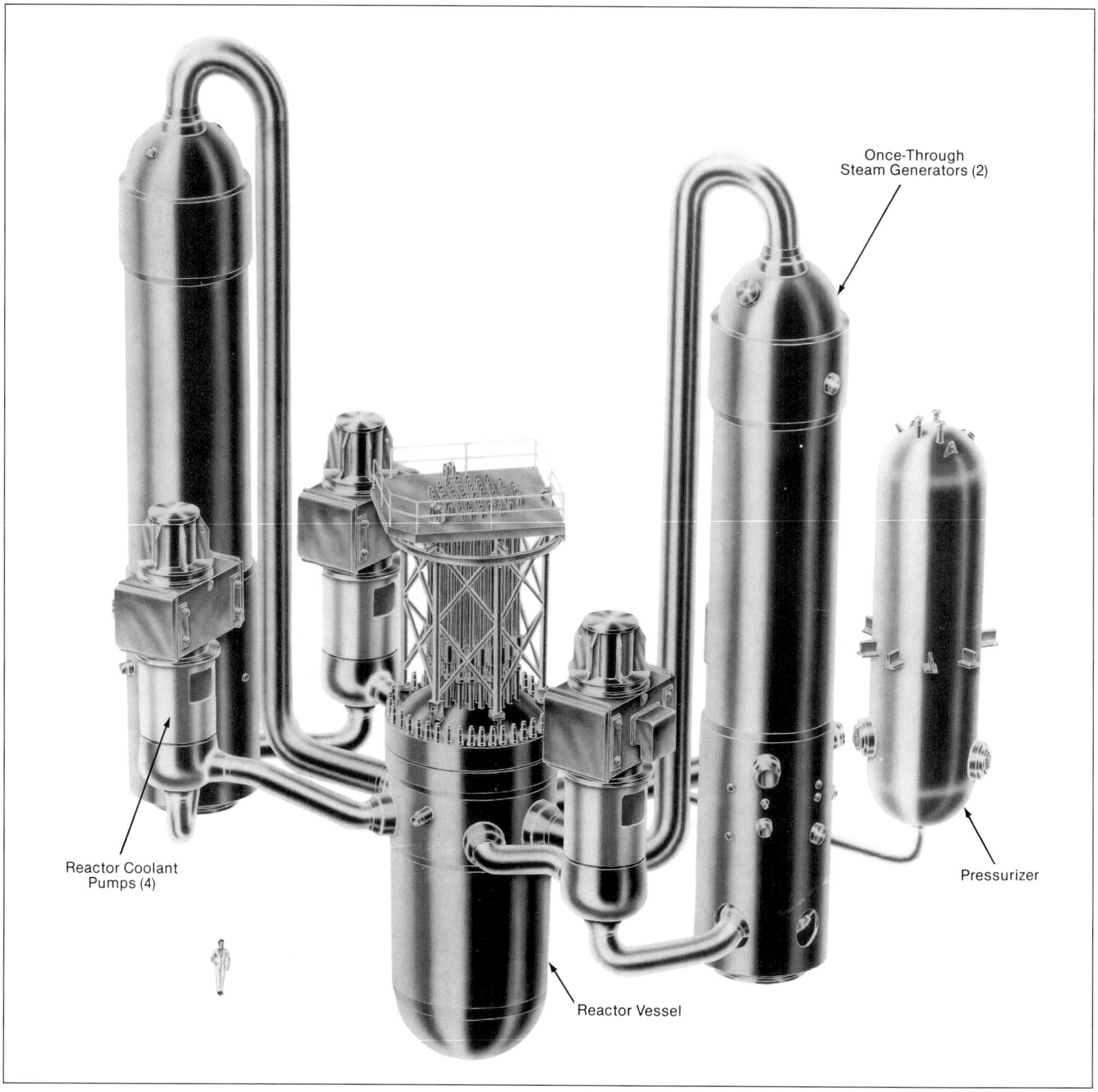

Fig. 3 B&W nuclear steam supply system.

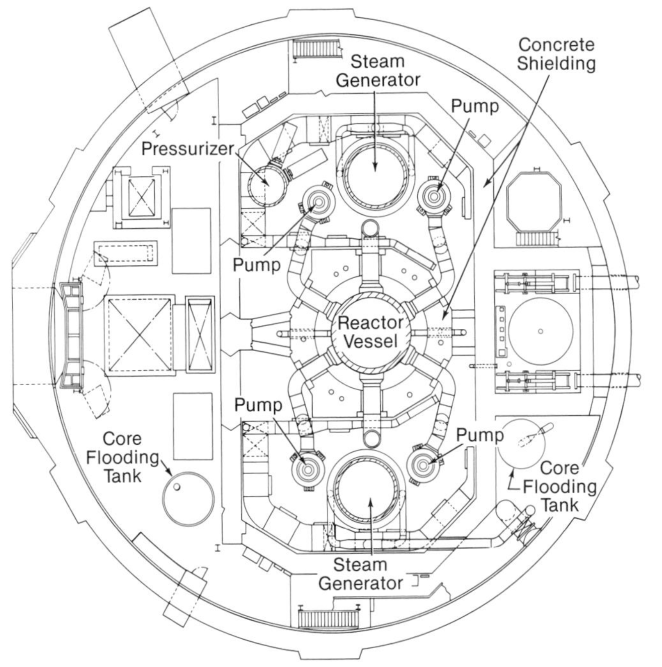

Fig. 4 Reactor containment building, ground floor plan view.

psi (62.7 bar) and 595F (313C). This provides 60F (33C) of superheat. The B&W nuclear steam supply system (NSSS) (Fig. 3) is the only large scale commercial design that provides the added benefit of superheated steam output. A plan view of the containment building is shown in Fig. 4. The concrete shielding around the NSSS components is also shown. Table 1 provides a listing of the important design parameters for the Oconee-type NSSS. Also provided are the design parameters for a larger B&W NSSS. The first of these large 1300 MW designs went into operation in 1987 at the Muelheim Kaerlich power station in West Germany. Three additional units are under construction in the U.S.

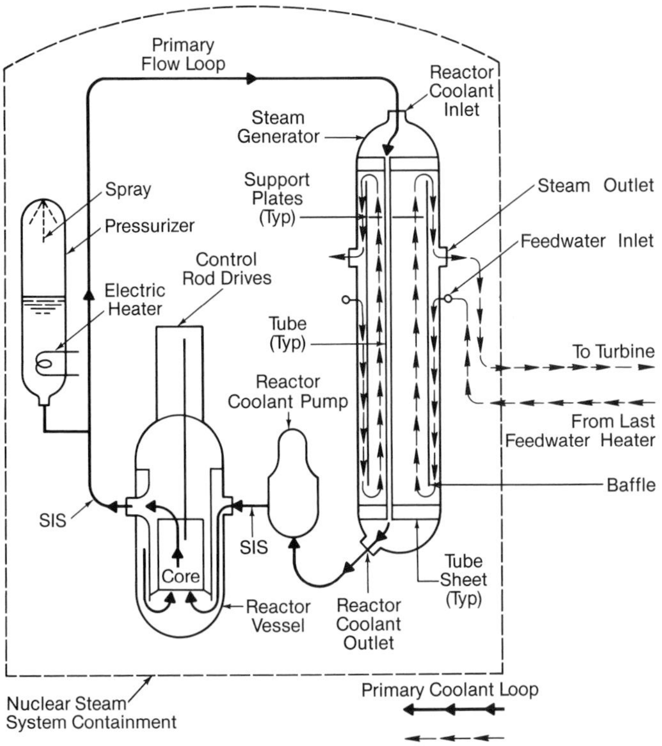

Fig. 5 Simplified schematic of primary and secondary loops.

The NSSS for a typical second generation pressurized water reactor consists of the reactor vessel, two or more steam generators, coolant pumps, a pressurizer and the piping to connect these components. As shown in the very simplified schematic provided in Fig. 5, the NSSS consists of two water flow circuits. The primary flow loop uses pressurized water to transfer heat from the reactor core to the steam generators. Flow is provided by the reactor coolant pump. The pressurizer maintains the primary loop pressure high enough to prevent steam generation in the reactor core during normal operation. The steam generators provide the link between the primary coolant loop and the power producing secondary flow loop. In the B&W design, subcooled secondary side water enters the steam generator and emerges as superheated steam, which is sent to the steam turbine to produce power. The primary side water flow is cooled while supplying the energy to generate the steam. Additional information is included in Table 1 and Chapter 50.

Fig. 6 shows a reactor vessel being placed between two steam generators inside a containment building. The vessel is constructed of low alloy steel with an internal stainless steel cladding. External diameter across the nozzles is 20.75 ft (6.32 m) with an internal height of 37.3 ft (11.38 m) including the closure head. The removable closure head is held in place by sixty 6.5 in. (165 mm) diameter high tensile alloy steel studs. The total dry weight of the vessel and closure head is about 415 t (376 t_m).

The vessel has six major nozzles for reactor coolant flow (two outlet hot leg and four inlet cold leg). The coolant water enters the vessel through the four cold leg nozzles located above the midpoint of the vessel. Water flows downward in the annular space between the reactor vessel and the internal thermal shield, up through the core and upper plenum, down around the inside of the upper portion of the vessel and out through the two hot leg nozzles. Fig. 7 shows a reactor vessel arriving at a nuclear power plant site.

Fig. 6 Reactor vessel placed into position between steam generators.

Table 1
Design Parameters - Nuclear Steam Supply Systems

	Oconee Unit 1	1300 MW Unit*
Reactor hydraulic and thermal design		
Rated heat output (core), MW_t	2,568	3,760
Rated heat output (core), million Btu/h	8,765	12,833
Design overpower, %	114	112
System pressure (nominal), psi	2,185	2,250
Power distribution		
Maximum/average power ratio, radial x local	1.78	1.55
Maximum/average power ratio, axial	1.70	1.67
Overall power ratio	3.03	2.65
Power generated in fuel and cladding, %	97.3	97.3
DNB ratio at rated conditions	2.0	1.81
Minimum DNB ratio at design overpower	1.55	1.40
Coolant flow		
Total flow rate, million lb/h	131.3	158.2
Effective flow rate for heat transfer, million lb/h	124.2	148.5
Effective flow area for heat transfer, ft^2	49.19	56.6
Average velocity along fuel rods, ft/s	15.73	16.85
Coolant temperature, F		
Nominal inlet	554	568.6
Average rise in vessel	50	56.9
Average in vessel	579	597
Average film coefficient, Btu/ft^2 h F	5,000	5,000
Average film temperature difference, F	31	60
Heat transfer at 100% power		
Total heat transfer surface area, ft^2	49,734	63,991
Average heat flux, Btu/ft^2 h	171,470	195,100
Maximum heat flux, Btu/ft^2 h	534,440	517,640
Average thermal output, kW/ft	5.656	5.670
Maximum thermal output, kW/ft	17.63	15.04
Maximum clad surface temperature at nominal pressure, F	654	657
Fuel central temperature, F		
Maximum at 100% power at hot spot	4,250	4,040
Maximum at design overpower	4,650	4,400
Maximum thermal output, kW/ft at design overpower	20.1	16.85
Core mechanical design		
Fuel assemblies		
Number	177	205
Rod pitch, in.	0.568	0.503
Overall dimension, in. (side of square)	8.536	8.536
Total weight, lb	274,350	313,855
Number of spacer grids per assembly	8	8
Fuel assembly pitch spacing, in.	8.587	8.587
Fuel rods		
Number	36,816	54,120
Outside diameter, in.	0.430	0.379
Diametral gap, in.	0.007	0.008
Clad thickness, in. (Zircaloy-4)	0.0265	0.0235
Active fuel length, in.	144	143
Fuel pellets (UO_2 sintered)		
Density, % of theoretical		
First core	93.5	94.0
Subsequent cores	92.5	94.0
Diameter, in.	0.370	0.324
Length, in.	0.7	0.375
Control rod assemblies**		
Neutron absorber	5% Cd-15% In-80% Ag	Ag-In-Cd or boron carbide
Length of poison section, in.	134	134
Cladding material	304SS	304SS
Clad thickness, in.	0.021	0.019
Number of assemblies	61	64
Number of control rods per assembly	16	24
Axial-power-shaping rod assemblies		
Neutron absorber	5% Cd-15% In-80% Ag	Ag-In-Cd or boron carbide
Length of poison section, in.	36	36
Cladding material (poison section)	304SS	304SS
Clad thickness, in.	0.021	0.019
Number of assemblies	8	8
Number of control rods per assembly	16	24
Orifice rod assemblies		
Rod material	304SS	304SS
Number of orifice rods per assembly	16	24
Core structure		
Core barrel ID/OD, in.	141/145	156/161
Thermal shield, ID/OD, in.	147/151	
Nuclear design		
Structural characteristics		
Fuel weight (as UO_2), lb	207,486	233,850
Clad weight (active zone), lb	42,200	48,175
Core diameter, in. (equivalent)	128.9	138.7
Core height, in. (active fuel)	144	143
Reflector thickness and composition		
Top (water plus steel), in.	12	12
Bottom (water plus steel), in.	12	12
Side (water plus steel), in.	18	21.5
Metal/Water (unit cell—volume basis)	0.82	0.83
Fuel rods per fuel assembly	208	264
Performance characteristics		
Loading technique	3 region	modified checkboard
Fuel discharge burnup, MWd/tonne of uranium		
First cycle average	9,600	16,864
Succeeding cycle average	9,700	11,243
Fuel enrichments, wt % U-235		
Core average, first cycle	2.10	2.80
First reload	2.98	3.28
Core average, equilibrium cycle	3.06	3.18
Control characteristics		
Effective multiplication (beginning of life)		
Cold, no power, clean	1.248	1.333
Hot, no power, clean	1.198	1.286
Hot, rated power, Xe and Sm equilibrium	1.132	1.192
Control rod worth ($\Delta k/k$), %	10.9	8.1†
Boron concentrations		
To shut reactor down (k_{eff}=0.99) with all rods inserted (clean), cold/hot ppm	864/587	1178/910

* Approximate net electric power.
** The 1300 MW unit has also 116 burnable poison rod assemblies, having 24 rods per assembly, each rod containing 126 in. length of $Al_2O_3 \bullet B_4C$ contained in Zircaloy-4 cladding tubes 0.032 in. thick.
† Burnable poison rods are worth an additional 5.1%.

Nuclear design (cont'd)	Oconee Unit 1	1300 MW Unit*
Boron concentrations (cont'd)		
Boron worth (hot), % (Δk/k)/ppm	1/85	1/105
Boron worth (cold), % (Δk/k)/ppm	1/64	1/79
Kinetic characteristics (range during life cycle)		
Moderator temperature coefficient, (Δk/k)/F	$+0.5x10^{-4}$ to $-3.0x10^{-4}$	$+0.1x10^{-4}$ to $-3.0x10^{-4}$
Moderator pressure coefficient, (Δk/k)/psi	$-5.0x10^{-7}$ to $+3.0x10^{-6}$	$-0.5x10^{-7}$ to $+3.0x10^{-6}$
Moderator void coefficient, (Δk/k)/% void	$+4.0x10^{-4}$ to $-1.6x10^{-3}$	$+3.0x10^{-4}$ to $-3.0x10^{-3}$
Doppler coefficient, (Δk/k)/F	$-1.1x10^{-5}$ to $-1.7x10^{-5}$	$-1.1x10^{-5}$ to $-1.7x10^{-5}$

Reactor coolant system

System heat output, MW_t	2,584	3,780
System heat output, million Btu/h	8,819	12,901
Operating pressure, psi	2,185	2,250
Reactor inlet temperature, F	554	568.6
Reactor outlet temperature, F	604	625.5
Number of loops	2	2
Design pressure, psi	2,500	2,500
Design temperature, F	650	670
Hydrostatic test pressure (cold), psi	3,125	3,125
Coolant volume, including pressurizer, ft^3	11,478	13,200
Total reactor flow, GPM	352,000	432,800

Reactor coolant system code requirements

Reactor vessel and closure head	ASME III	ASME III
Steam generator and pressurizer	ASME III	ASME III
Reactor coolant piping	ANSI B31.7	ASME III
Reactor coolant pump casing	ASME III**	ASME III

Reactor vessel

Base material	Low alloy steel	
Cladding material	SS	SS and Inconel
Design pressure, psi	2,500	2,500
Design temperature, F	650	670
Operating pressure, psi	2,185	2,250
Inside diameter of shell, in.	171	182
Straight shell thickness, in.	8 7/16	9 1/8
Minimum clad thickness, in.	1/8	1/8
Outside diameter across nozzles, ft-in.	20-9	23-4
Overall height of vessel and closure head (over control rod drive and instrument nozzles), ft-in.	40-8 3/4	42-2

Steam generators

Number of units	2	2
Type	Once-through	
Materials		
Tubes	Inconel	
Shell	Carbon steel	Low alloy steel
Hemispherical heads	Low alloy steel, SS clad	
Tubesheets	Low alloy steel, Inconel clad on reactor coolant side†	
Tube side design pressure, psi	2,500	2,500
Tube side design temperature, F	650	670
Tube side design flow, million lb/h	65.66	79.1

Steam generators (cont'd)	Oconee Unit 1	1300 MW Unit*
Shell side design pressure, psi	1,050	1,235
Shell side design temperature, F	600	631
Operating pressure, tube side, nominal, psi	2,185	2,250
Operating pressure, shell side, nominal, psi	910	1,060
Superheat at outlet at rated load, F (min)	35	35
Hydrostatic test pressure (tube side cold), psi	3,125	3,125
Shell minimum thickness, in.	4 3/16	5 3/4
Maximum outside diameter of straight shell, ft-in.	12-7 1/4	12-6 1/2
Overall height (including supports), ft-in.	73-2 1/2	77-6 1/4

Pressurizer

Material, shell and heads	Carbon steel, SS clad	Low alloy steel, SS clad
Design pressure, psi	2,500	2,500
Design temperature, F	670	670
Steam volume, ft^3	700	1,050
Water volume, ft^3	800	1,200
Electric heater capacity, kW	1,638	1,745
Shell minimum thickness, in.	6 3/16	5 15/16
Shell outside diameter, ft-in.	8-0 3/8	9-11 7/8
Overall height, ft-in.	44-11 3/4	41-4 1/2

Reactor coolant pumps

Number of units	4	4
Type	Vertical, single-stage	
Design pressure, psi	2,500	2,500
Design temperature, F	650	670
Operating pressure, nominal, psi	2,185	2,250
Suction temperature, F	554	568.6
Design capacity, GPM	88,000	108,200
Total developed head, ft	350	370
Hydrostatic test pressure (cold), psi	4,100	3,340
Motor type (single speed)	a-c induction	Squirrel-cage induction
Motor rating (nameplate), hp	9,000	12,500

Reactor coolant piping

Material	Carbon steel, SS clad	
Hot leg (ID), in.	36	38
Cold leg (ID), in.	28	28

Engineered safeguards

Safety injection system		
Number of high pressure injection pumps	3	3
Number of low pressure injection pumps	2‡	2
Reactor building coolers		
Number of units	3	3
Rated capacity, each, million Btu/h	80	140
Core flooding system		
Number of tanks	2	2
Total volume, each, ft^3	1,410	1,800
Reactor building spray		
Number of pumps	2	2

* Approximate net electric power.

** Not code stamped.

† 1300 MW unit lower tubesheet is Inconel-clad on both sides.

‡ Plus one installed spare pump.

Fig. 7 Reactor vessel arriving at plant site.

Fuel assemblies

In the reactor core, 177 individual fuel assemblies rest on a lower grid which is attached to the heavy core barrel. The core barrel and support shield are suspended on a seal ledge near the closure head. Factors considered in structural design of the core are discussed in Chapter 48.

Each of the 177 fuel assemblies contains 208 fuel rods which are made up of uranium dioxide pellets contained in Zircaloy clad tubes, and the gap surrounding the pellets is pressurized with helium. Each fuel assembly is fitted with an instrumentation tube at the center and with 16 guide tubes, each accommodating 16 control rods. (See Fig. 8.) Each group of 16 rods is coupled together to form a control rod assembly as shown in Fig. 9. At any given time, 69 of the 177 fuel assemblies contain a control rod assembly. Because all fuel assemblies are identical and can be used anywhere in the core, the 108 remaining fuel assemblies have their guide tubes partially filled at the top with an orifice rod assembly. This restricts excessive coolant flow through the unused control rod guide tubes and equalizes coolant flow among the fuel assemblies.

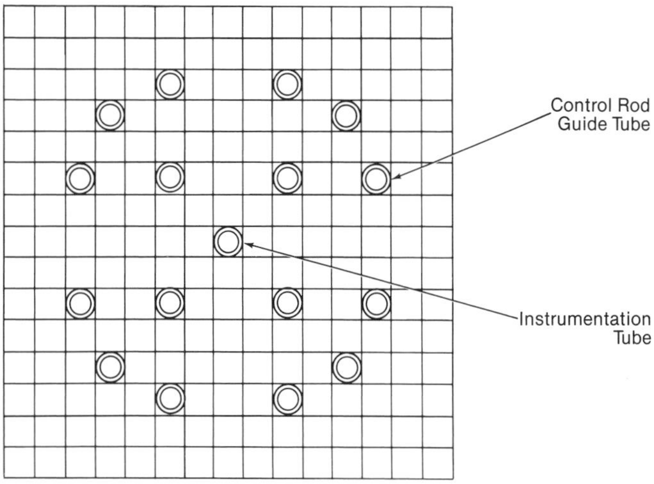

Fig. 8 Diagrammatic cross-section of fuel assemblies showing instrumentation tube and control rod guide tube locations.

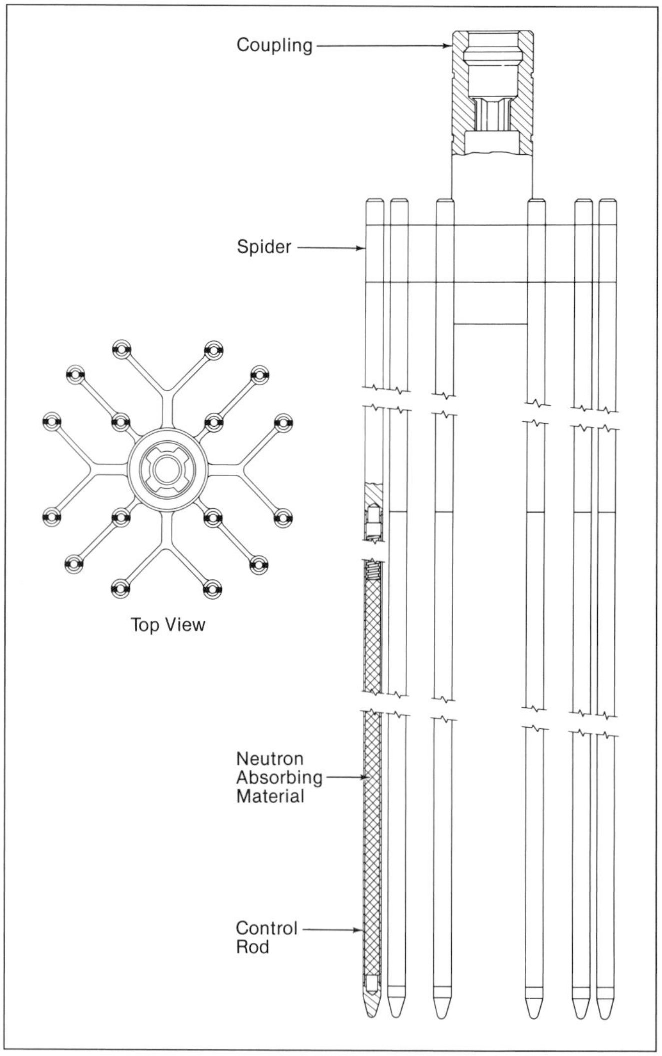

Fig. 9 Control rod assembly.

The control rod assemblies regulate the reactor reactivity and, therefore, the power output. Each control rod consists of an absorber section made from a neutron absorbing material, such as stainless steel with silver-indium-cadmium cladding, and an upper and lower stainless steel end piece. The rods regulate relatively fast reactivity phenomena such as Doppler, xenon buildup and decay, and moderator temperature change effects. The slower reactivity effects, such as fuel burnup, fission product buildup, and hot to cold moderator reactivity deficit, are controlled by a soluble neutron absorbent (usually boron) in the reactor coolant. The concentration of the absorbent is monitored and adjusted by the plant operators. These factors are discussed at length in Chapters 48 and 49.

Controls

The plant power level is maintained by an automated control system. Although this system can be operated manually, the constantly changing parameters are ideally suited for automation with operator supervision and alarm warning features. At B&W designed plants, the control system provides automatic response to signals which originate from the operator or from an automatic load dispatch system. Following a load demand change, the controls change the turbine throttle position and alter feed flow to the steam generators; this maintains a constant steam pressure at the turbine throttle. Simultaneously, the control system adjusts reactor power and maintains the average coolant temperature. The system measures neutron flux and reactor coolant temperature and moves the control rods accordingly. Above 15% power, a constant average coolant temperature is maintained. This requires that the hot leg temperature be raised because, as the turbine steam load increases, the heat transfer in the steam generator increases and the steam generator outlet (cold leg) temperature is reduced. Constant average coolant temperature control results in a relatively constant inventory of water in the coolant system, eliminating the need to add or remove water with changing load. This allows a reduction in waste and soluble absorbent handling for the operator, and control rod movement is reduced.

Coolant loops

The B&W reactor coolant system consists of two loops, each with one steam generator and two coolant pumps. A pressurizer and its associated components are attached to one loop. The loop functions and component characteristics are described in more detail in Chapter 50. The following description is generally applicable to PWRs.

The B&W designed straight tube steam generators are each approximately 68 ft (20.7 m) high and 13 ft (4 m) in diameter. The general arrangement inside the containment building is shown in Figs. 2 and 4. A simplified view (Fig. 10) shows the flow path of the reactor coolant and the secondary water and steam. Coolant enters from the hot leg nozzle at the top of the steam generator and flows downward through more than 15,000 tubes in each steam generator, into the lower hemispherical head, and out one of two nozzles into the cold leg piping. The secondary feedwater enters at the side of the shell, where it is brought to saturation temperature as it flows down the annulus by mixing with steam. The saturated water contacts the heating surface at the bottom and is converted to steam. In its single pass through the lanes between the tubes, the steam is superheated as it contacts the high temperature upper portion of the tubes. At full power, the steam generator produces about 60F (33C) of superheat.

The other major type of steam generator used in nuclear power plants is a vertical recirculating U-tube unit. Various designs are used worldwide and it is by far the most commonly appearing concept. In this type of steam generator, shown in Fig. 11, the reactor coolant enters one side of the lower head from the hot leg and flows up the U-tubes, down into the other side of the lower head, and out through the cold leg nozzle. The feedwater enters midway up the shell and flows down the annulus to the tube region. Feedwater then enters the tube region and begins to absorb heat from the tubes. The top of the U-tubes is always covered with a saturated steam-water mixture. As the feedwater pool continues to absorb heat, saturated steam is produced. This steam is forced through moisture separators at the top of the steam generator before it is sufficiently dried to exit into the main steam headers to the turbine. The moisture extracted from the steam is recirculated to the feedwater pool surrounding the U-tubes. Although the added efficiency of superheat is not achieved by this design, the steam is of sufficient quality for use in turbines designed for this application. B&W has supplied this type of steam generator to virtually all Canadian designed CANDU units worldwide as well as to the replacement steam generator market in the U.S.

Each reactor coolant loop contains one or more pumps in the cold leg. These pumps draw from the outlet of the steam generator and discharge to the inlet nozzle of the

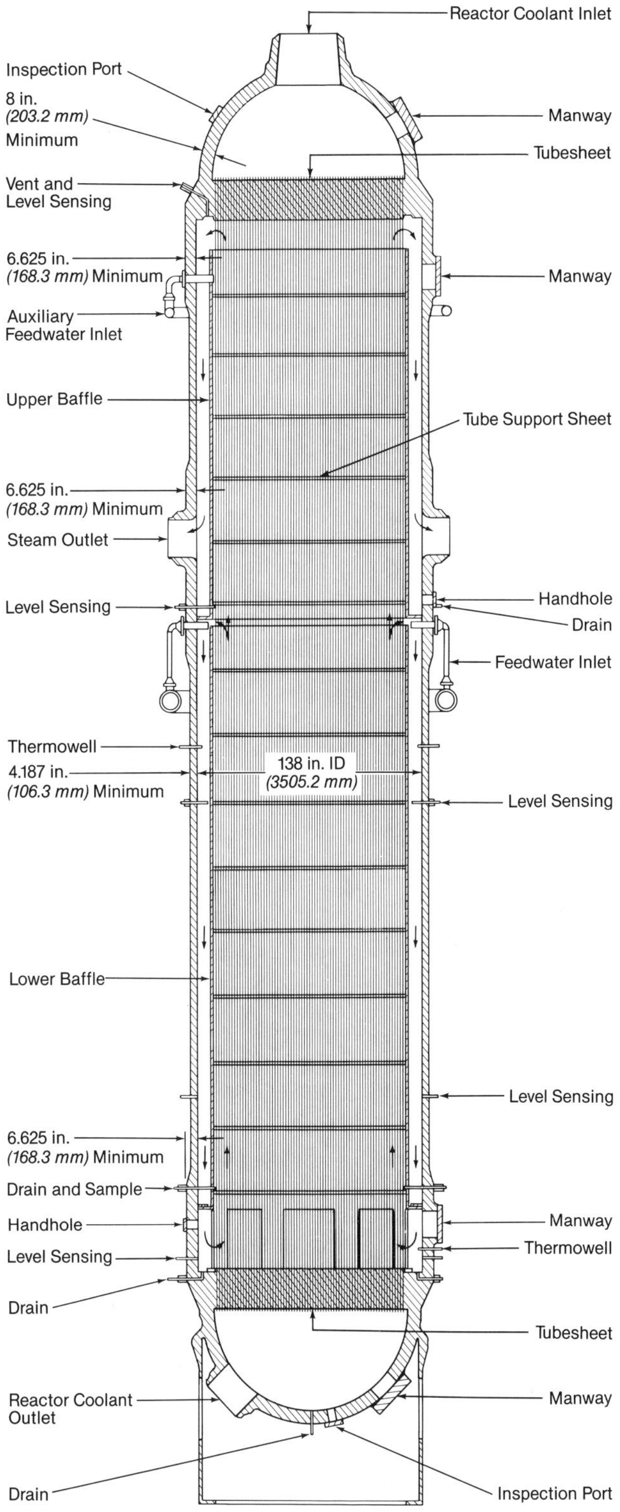

Fig. 10 B&W once-through nuclear steam generator.

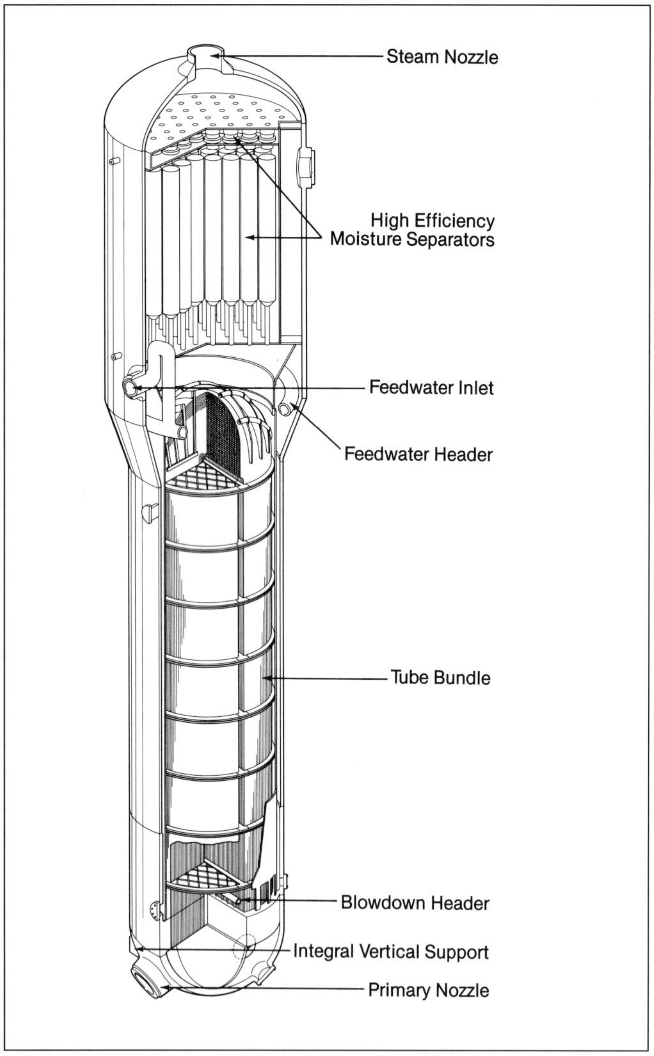

Fig. 11 B&W recirculating nuclear steam generator.

reactor vessel. In the B&W NSSS, there are two pumps in each loop; each one draws from a separate steam generator outlet nozzle. Each pump is a vertical, single stage, centrifugal unit with an inlet pressure of 2185 psi (150.6 bar) and a head of 113 psi (7.8 bar). The four pumps are designed to deliver a total coolant flow of more than 131 $\times 10^6$ lb/h (16,500 kg/s). The pumps are driven by large, vertical, squirrel cage induction motors with air-to-water heat exchangers for motor cooling.

Pressurizer

One reactor coolant loop hot leg is connected to a pressurizer by a surge line pipe. The pressurizer is a large pressure vessel, approximately 42 ft (12.8 m) in internal height with an internal diameter of about 7 ft (2.1 m). The location of the pressurizer is shown in Fig. 4 and a simplified view is shown in Fig. 12. A constant reactor coolant level is maintained in the pressurizer and the pressure is maintained by electric heaters in the lower shell region. The pressurizer provides the overall pressure control for the reactor coolant system. When the coolant system pressure decreases, some of the water in the pressurizer flashes to steam and restores pressure. Similarly, when coolant pressure increases, some of the steam in the pressurizer condenses, thereby reducing pressure. This equilibrium is maintained in a narrow pressure band by controlling the heaters and a condensing spray system.

Later generations of nuclear plants

The technological successes of second generation nuclear power plants around the world have prompted additional advances in design and application. The lessons learned in operating and maintaining the second generation plants have been applied to new designs being developed worldwide. Third generation, or evolutionary, designs have incorporated these improvements and are close to the technical success of the second generation plants. Economic factors will ultimately determine the success of the third generation plants. These economics rely on the ability of constructors to build the plants on schedule and on streamlining the licensing process so that the plants can quickly begin operation. This ability to immediately start producing electric power is an economic key. Other significant economic factors include the accuracy of the electric demand projection, the cost of alternative power sources, the restrictions that may apply to other power sources in the future (such as air quality legislation), and the public's perception of nuclear power.

The design objectives for *third generation evolutionary* plants are focused on plant safety and performance improvements. Increased accident resistance and safety

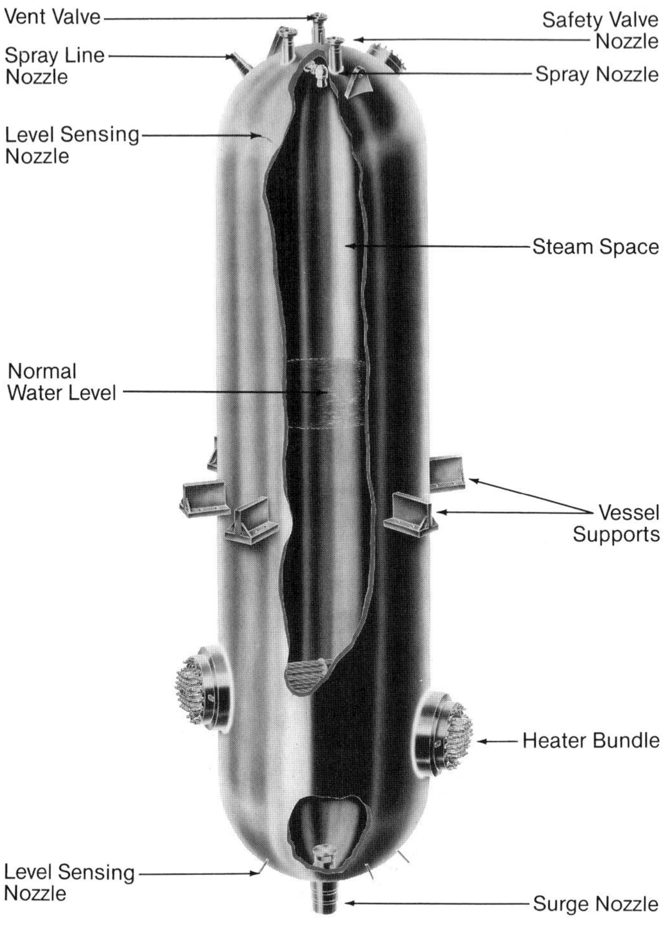

Fig. 12 Nuclear steam system pressurizer.

margins will be required. The safety systems will be simpler and will allow increased operator response time. Instrumentation and controls will feature new technology, and the control room will have enhanced man-machine interfaces. Extension of design life from 40 to 60 years and increased time between refuelings are expected. Additionally, the volume of radioactive waste and the danger of radiation exposure will decrease.

New concepts of plant safety and technology will be introduced into the nuclear industry on a continuing basis. The concept of *passivity* forms the basis for *fourth generation plants*, which take a major step beyond the third generation operationally based designs. The functions of a passive power plant rely on natural forces such as gravity and natural circulation, stored energy from batteries or compressed fluids, check valves and noncycling powered valves rather than on active systems or components. For example, the plant may rely on gravity feed of water from a tank into the reactor coolant system rather than using an electrically driven pump. Further consideration could favor a combination of passive and active design features.

Fuel loading, showing end view of fuel assemblies in place in reactor vessel.

Chapter 48
Nuclear Fuels

The relatively low cost of nuclear fuel per unit of energy is the primary reason that it is sometimes chosen over fossil fuels. Nuclear fuel is a compact source of energy, producing negligible adverse environmental effects and small amounts of residual waste products. In some countries spent nuclear fuel is reprocessed to reclaim the uranium and plutonium in the discharged fuel assemblies. By reprocessing, about 97% of the materials in the spent fuel assembly can be reused.

The use of uranium as an energy source for steam and power generation involves a number of steps collectively referred to as the *nuclear fuel cycle*. These steps include mining, conversion, enrichment, fuel fabrication, power production and disposal. If the spent fuel is not reprocessed, the cycle is referred to as a *once-through* or *open fuel cycle*. When the fuel is reprocessed, it goes through a *closed fuel cycle*. Fig. 1 provides an overview of the closed nuclear fuel cycle from the mining of the uranium ore to disposal. Chapter 56 addresses the important issue of disposal in more depth.

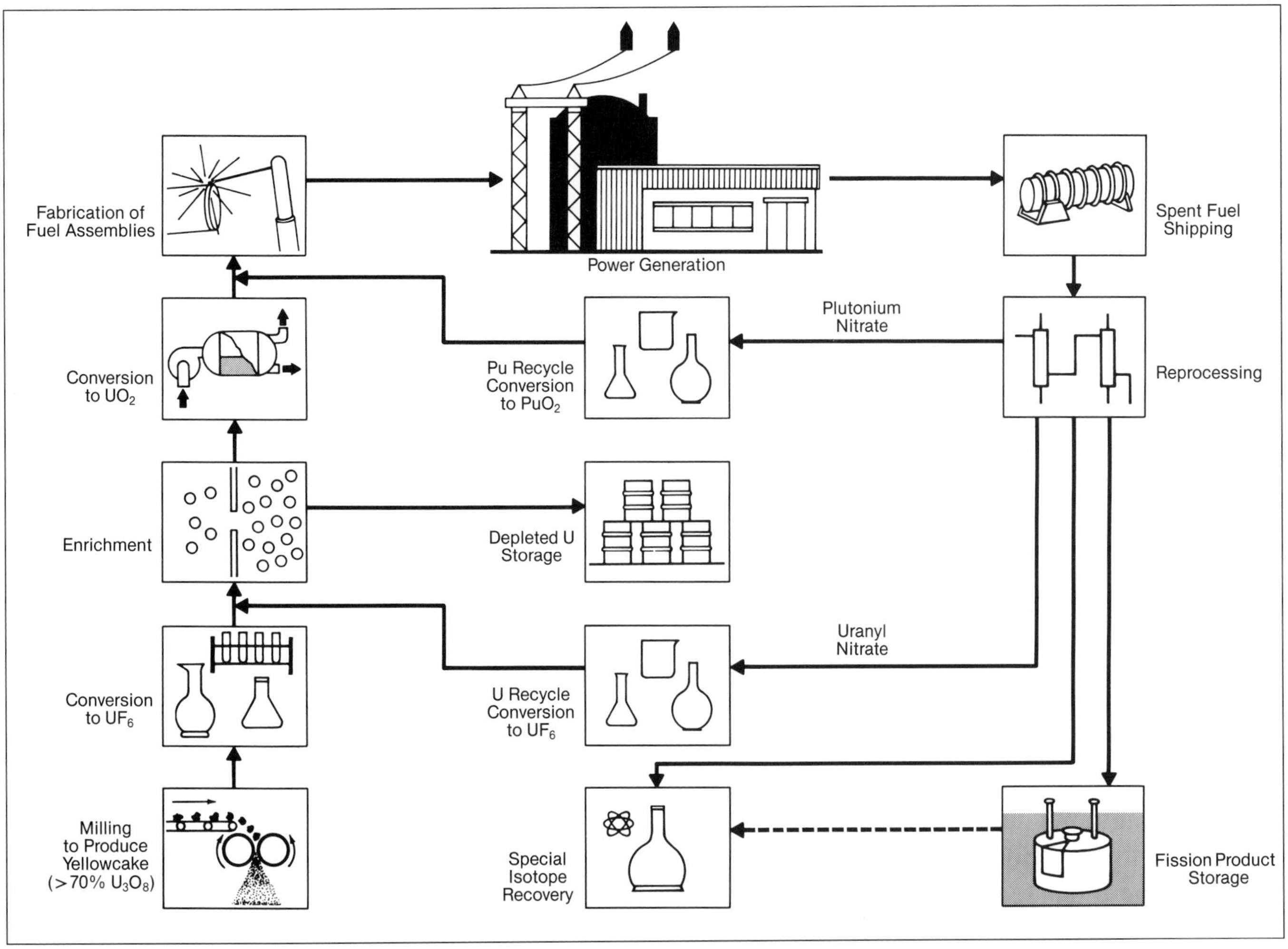

Fig. 1 Closed nuclear fuel cycle.

Nuclear reactor fuel sources

Uranium

Uranium (U) is the basic raw material of the nuclear power industry. It is a heavy, slightly radioactive element with an atomic number of 92. Uranium is the heaviest element that is found naturally in more than trace quantities in the earth's crust.

Uranium ore deposits of current commercial value are located throughout the world and are dominated by the United States (U.S.), Australia, South Africa, Canada and Niger. (See Table 1.) The former Soviet Union (C.I.S.) and the People's Republic of China also have substantial uranium ore deposits and production capabilities. The current reasonably assured resources (RAR, known deposits of such a size, grade and configuration that they are recoverable with proven technology at costs of U.S. $80/kg U or less) are approximately 1,546,000 tonnes U for the World Outside Centrally Planned Economics Areas (WOCA). These deposits have a high probability of commercial recovery. A more speculative category is Estimated Additional Resources (EAR) which covers other estimated discovered and undiscovered resources. Uranium production from these estimated reserves has a lower probability. Uranium production levels are also provided in Table 1 and are expected to range from 40,000 to 50,000 tonnes U annually.

Uranium is a highly reactive metal with three principal valences: +3, +4 and +6. It has three crystalline phases and melts at 2070F (1132C). In the alpha (α) phase, it is reasonably ductile and can be fabricated by standard metalworking techniques. Because small uranium particles or chips are highly pyrophoric (capable of spontaneous combustion when exposed to air), machining operations and scrap storage require special precautions, such as the use of a coolant or an inert atmosphere.

Isotopes and the fundamentals of nuclear physics are discussed in Chapter 49. Natural uranium is a mixture of three isotopes: uranium-234 (U-234) 0.01%, uranium-235 (U-235) 0.71% and uranium-238 (U-238) 99.28%. U-234 is present in small amounts and is not significant. U-235 is a fissionable isotope and U-238 is known as a fertile isotope which can be transmuted to a fissionable material.

These isotopes are slightly radioactive and emit alpha particles. Because natural uranium has a relatively long half life, a minimum of precaution is required when handling it. However, it is chemically toxic and must not be ingested. In handling areas, levels of airborne uranium must be minimized.

U-235 is fissionable because its nucleus can absorb a neutron, become unstable and split into multiple pieces. When 1 gram of U-235 is fissioned, the heat released is equivalent to approximately 1 MWd [24,000 kWh (82×10^6 Btu)]. When one short ton of U-235 is fissioned, the heat released is equivalent to 22×10^9 kWh or 75×10^{12} Btu, which equals the energy contained in approximately 3 million tons of coal.

By itself, U-238 can not sustain a nuclear chain reaction, but it can be fissioned by high energy neutrons. Sufficient U-235 must be present to initiate the fission process and drive the chain reaction. When exposed to neutrons in a nuclear reactor, the U-238 nucleus will capture a neutron and is transformed into plutonium-239 (Pu-239), a fissile isotope. Pu-239 is capable of sustaining a chain reaction and, when fissioned, produces a similar number of free neutrons as U-235. U-238 is therefore considered to be fertile because of this transmutable property. U-238 is 140 times more abundant in nature than U-235 and is the main component of nuclear fuel.

Plutonium

Although trace quantities of plutonium are found in nature, most plutonium is made from U-238 transmutation. All commercial plutonium is made by this transmutation process.

Plutonium is a radioactive element with an atomic number of 94. Although it is classified as a metal, its electrical and thermal conductivities are relatively low making it more like a ceramic material in some ways. Pluto-

Table 1
Selected Uranium Resources and Production Capability Data

Country	Uranium Production (tonnes) 1988 Actual	1989 Expected	Resources (1000 tonnes)* Reasonably Assured** $80/kg U	$130/kg U	Estimated Additional** $80/kg U	$130/kg U
Australia	3,532	3,800	480	58	262	131
Canada	18,400	11,000	139	96	109	95
France	3,394	3,190	47	12	20	16
Gabon	930	950	13	5	1	8
Namibia	3,600	3,600	91	16	30	23
Niger	2,970	3,000	174	2	284	17
South Africa	3,850	2,900	317	102	73	38
Spain	228	216	17	18	0	9
U.S.	5,050	4,600	111	266	515	389

Source—OECD Nuclear Energy Agency and the International Atomic Energy Agency, "Uranium Resources, Production and Demand," 1989.

* Data available January 1, 1989.
** The production cost categories correspond to the mined uranium oxide as follows: U.S. $80/kg U = U.S. $30/lb U_3O_8; U.S. $130/kg U = U.S. $50/lb U_3O_8.

nium has a melting point of 1184F (640C), but it must be melted in a vacuum or inert atmosphere because it is chemically very reactive. It has complicated chemical properties, with five valence states ranging from +2 to +6. Plutonium also has complicated metallurgical properties and is the only element with six allotropes (different physical forms such as charcoal and diamond forms of carbon).

The 15 known isotopes of plutonium range in atomic weight from 232 to 246. All are radioactive, decaying to form stable isotopes of elements such as bismuth and lead. Pu-239 is the most common isotope of plutonium and is the product of fast breeder reactors which are specially designed to produce more fuel than they use. Plutonium generated in commercial power reactors contains substantial amounts of isotopes with mass numbers greater than 239.

Thorium

Although uranium is the basic nuclear power fuel, enormous natural reserves of thorium are also available. Thorium contains no fissionable isotope like U-235; it is primarily a fertile material. The absorption of neutrons within a nuclear reactor converts thorium to U-233. While this is another fissile uranium isotope, it does not occur in natural uranium. A closed fuel cycle that reprocesses the thorium to recover the U-233 is required.

Reserves of thorium, available as thorium oxide (ThO_2), are estimated to exceed half a million tons. A large portion of these resources occurs as monazite sand, which is found in India, Brazil and the U.S. Thorium reserves in the U.S. are estimated at 100,000 tons of thorium oxide equivalent.

Certain nations that lack uranium reserves are developing thorium fuel for nuclear power. However, world nuclear production and development is centered around the uranium-plutonium cycle because of its cost advantage and abundance compared to thorium.

Uranium extraction, enrichment and conversion

Uranium ore mining is conducted in a similar manner to other distributed ore materials. Depending upon the depth, overburden and deposit concentration, surface mining (open pit) and underground mining are common methods used in the U.S. Underground mining accounted for more than half of the U.S. uranium production during the 1980s. In situ solution mining and production resulting as a byproduct from other mining operations and open pit mining account for the balance.

A number of steps are required to extract uranium concentrate from the ore and to prepare the uranium for use in fabricating fuel pellets for commercial reactors. As shown in Fig. 1, these initial steps in the fuel cycle include milling, refining and conversion to uranium hexafluoride (UF_6), enrichment, and conversion to uranium dioxide (UO_2).

Milling

The function of uranium milling is to extract the uranium from the ore, which contains on the average about 5 lb (2.3 kg) of uranium per ton or about 0.25%. The ore is crushed and is then leached with an acid or alkaline reagent to dissolve the uranium which, in turn, is recovered from the solution by solvent extraction or ion exchange techniques. Excess water is removed by calcining, or roasting. The resultant product, known in the industry as *yellowcake*, is a crude uranium concentrate containing from 70 to 90% U_3O_8.

Refining and conversion to UF_6

For the purpose of enrichment, the uranium in yellowcake must be refined and converted into UF_6. Fig. 2 illustrates the process used by the Allied Chemical Corp., in which the essential steps are:

1. The U_3O_8 feed is reduced to raw UO_2 at 550 to 650C with hydrogen or dissociated ammonia in the first converter.
2. The raw UO_2 is converted to UF_4 at 450 to 650C with anhydrous hydrogen fluoride (HF) gas in the second converter.
3. The UF_4 is converted to crude UF_6 at 350 to 500C with elemental fluorine in the last converter.

The final step is fractional distillation for the removal of impurities, principally molybdenum and vanadium. The UF_6 product meets the U.S. Department of Energy (DOE) specifications for impurities allowable in uranium hexafluoride delivered for enrichment in the DOE's separation plants.

Enrichment

Prior to fabrication into fuel assemblies, the uranium must be enriched to the desired concentration of the fissionable isotope, U-235. This is predominantly accomplished in gaseous diffusion plants which are designed to separate isotopes by using the principles of gas kinetics. In a mixture of two gases (U-238 and U-235 hexafluoride), molecules of the lighter gas ($^{235}UF_6$) travel faster and strike the walls of the containing vessels more frequently than do the molecules of the heavier gas ($^{238}UF_6$). If the wall of the vessel is permeable, the lighter molecules will diffuse faster, relative to their concentration, than the heavier molecules. The degree of separation attainable in this manner is very small and the process must be repeated many times before a practical enrichment can be obtained. Therefore, a gaseous diffusion plant consists of hundreds of stages connected in series in an arrangement known as a diffusion *cascade*.

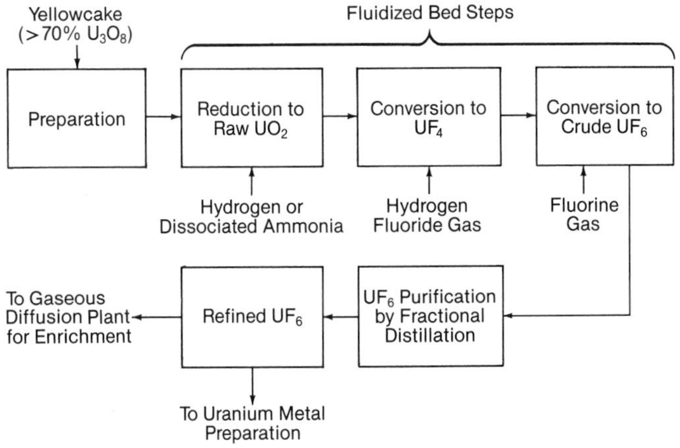

Fig. 2 Flow sheet for uranium refining and conversion to UF_6.

Each diffusion stage consists of a chamber divided into two zones by a finely porous, thin barrier. The gas mixture enters one zone, which is maintained at a higher pressure than the other. Conditions are adjusted so that part of the gas diffuses through the barrier into the lower pressure zone and then is directed to the next stage up the cascade. The remainder flows from the high pressure zone to the next stage down the cascade. The uranium-hexafluoride gas stream diffusing through the barrier is very slightly enriched in the U-235 isotope; the other stream is very slightly depleted in that isotope.

Fig. 3 illustrates the sequence of operations in the cascade. Stages are arranged in a countercurrent cascade similar to a distillation column. At each stage, a pump compresses the gas from the pressure on the downstream side of the diffusion barrier to the pressure on the upstream side and a cooler removes the heat of compression.

The natural uranium feed enters the cascade at an intermediate point. The gas becomes increasingly enriched as it progresses up the cascade. The enriched uranium product is withdrawn at some stage above the feed point, depending on the degree of enrichment desired. Depleted uranium (known as *tails*) is withdrawn from the base of the cascade and held in storage.

The gaseous diffusion process has been used for 30 years in the U.S., the United Kingdom and France. The gas diffusion process is, however, very energy intensive. For each 5.5 kg of UF_6 at a natural circulation of 0.71% U-235, the process produces 1 kg of UF_6 enriched to 4.3% U-235 and 4.5 kg of UF_6 depleted to 0.2% U-235. Alternatively, centrifuge technology has been used by a number of countries including the C.I.S.

Recent technology development efforts have focused upon the development of lasers for use in various enrichment processes.

Conversion to UO_2 powder

Once enriched, UF_6 is converted to UO_2 for use in making the fuel pellets. In one conversion process, enriched UF_6 is first hydrolyzed to UO_2F_6. It is then reacted with ammonia to form ammonium diuranate (ADU). The ADU is calcined to UO_3 and reduced to UO_2 with hydrogen at 800C. A related process is based upon ammonium uranate (AUC) instead.

The UO_2 powder can also be produced by a direct conversion process which involves a solid/gas reaction. In the first step, UF_6 vapor reacts with dry superheated steam to form uranyl fluoride (UO_2F_2) powder and hydrogen fluoride gas. In the second step, pyrohydrolysis takes place in the conversion oven. The UO_2F_2 reacts with a countercurrent of hydrogen and dry superheated steam to form the UO_2 and hydrogen fluoride gas. The ease of operation and constant control of the work parameters provide high quality, consistent powder. Furthermore, the powder produced from direct conversion is more easily sintered.

Fuel fabrication

Fuel assemblies for modern pressurized water reactors (PWRs) consist of enriched UO_2 fuel, Zircaloy-4 fuel cladding, Inconel upper and lower end spacer grids, Zircaloy-4 intermediate spacer grids and stainless steel nozzles as shown in Fig. 4. Manufacturing these components and integrating them into a fuel assembly require special methods and processes.

Pellets

UO_2 powder is the primary constituent of nuclear fuel pellets. The powder is mixed with varying proportions of a second powder, which is prepared by treating U_3O_8 scrap with uranyl nitrate (UNH). This pellet fabrication process is qualified and key process parameters are identified. Certain process changes require a requalification.

In the pelletizing process, the virgin powder is blended and sieved to form a homogeneous lot of about two tons. A pore-forming agent and recycled U_3O_8 are added to this blend. These materials are added in a ratio that attains the desired sintering density while optimizing pellet microstructure.

After blending, the powder is slugged and granulated to produce particles that are easily fed into a press. Before pressing, the granules are spheroidized and mixed with a lubricant. Pressing is performed using a reciprocating hydraulic or rotary unit. The press tonnage is set to provide a green (not yet sintered) pellet with high mechanical strength.

The next step, sintering, involves heating the pressed pellets in a furnace with a dry hydrogen atmosphere. The thermal cycle includes a sintering period of approximately 4.5 hours at 3180F (1749C). To obtain the required diameter, the pellets are machined under water on a centerless grinder equipped with a diamond wheel. They are subsequently speed dried to remove the remaining surface water film. The resulting fuel pellets are shown in Fig. 5. Selected thermophysical properties of the final fuel pellets are summarized in Table 2.

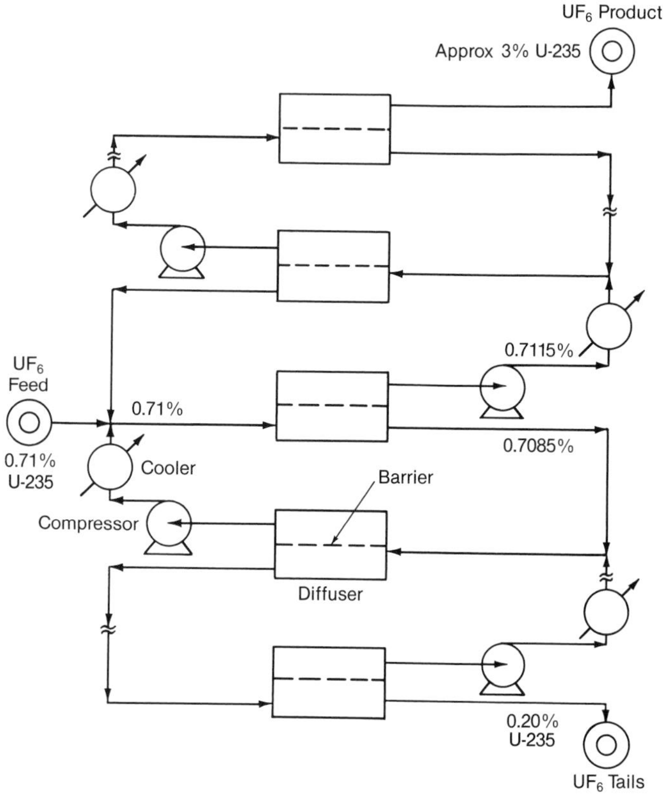

Fig. 3 Uranium enriching cascade.

Cladding

The bar shaped Zircaloy-4 raw material, called *trex*, used to produce the tube shaped cladding is visually inspected by the fabricator; it is then weighed and stored. In production, the qualified material is first reduced in cold pilger (tube drawing) mills. An initial breakdown pass is performed to produce tubing near the required size. This is followed by a second and third pass to achieve the target dimensions. After each pass, the tubing is rinsed and dried.

The cladding is then heat treated in a high vacuum oven, which keeps the temperature uniform along the tube length. Subsequently, the material is cut to length. The outer surfaces are given a final polish and the tubes are rinsed. A final inspection is then performed. This includes an ultrasonic inspection, to verify dimensions and to check for internal defects, and a visual examination with backlighting to identify surface irregularities. Tube length, perpendicularity and straightness are then verified. Several laboratory samples are taken from each production lot to verify chemical composition and mechanical properties.

Fuel rods

Fuel rod manufacture begins with the Zircaloy-4 tubing/cladding previously discussed and includes the components shown in Fig. 6, an enlarged view. In addition to sample dimensional inspection, the tubing is 100% ultrasonically tested for defects before it is released to production.

Table 2
Some Physical Properties of Uranium Dioxide

Temp, F	Thermal Conductivity Btu/h ft F	Coefficient of Linear Expansion* (in./in. F) 10^6**	Specific Heat
300	4.05	5.67	0.056
1000	2.37	5.67	0.067
2000	1.56	5.73	0.072
3000	1.33	6.58	0.075
4000	1.38	7.42	0.081
5000	1.69	—	0.094

* Between 70F and temperature shown.
** Based on 95% density pellets.

The cladding is then prepared for end cap welding. A press fit weld joint configuration is machined on each end and a cleaning plug is passed through the cladding to remove resultant chips. A plenum spring is then loaded and one end cap is welded in place in a chamber pressurized with inert gas. The rod end being welded is inserted into the chamber through a sealing hole and the rod is held stationary while the welding head rotates a tungsten electrode about the weld joint.

Following welding, a unique serial number is stamped on the end cap. Quality data for all components of the fuel rod are traceable by this number. Each end cap weld is automatically tested with ultrasonic equipment within minutes of completing the weld, providing rapid feedback to the welder operator.

At this point, the tubes are transferred to the laser drilling station. Here, the capped end of the cladding is inserted in a small chamber, the chamber is evacuated and a focused laser beam is directed at the center of the end cap. Discharge of the laser generates a small hole in the cap. This hole serves several purposes: it removes

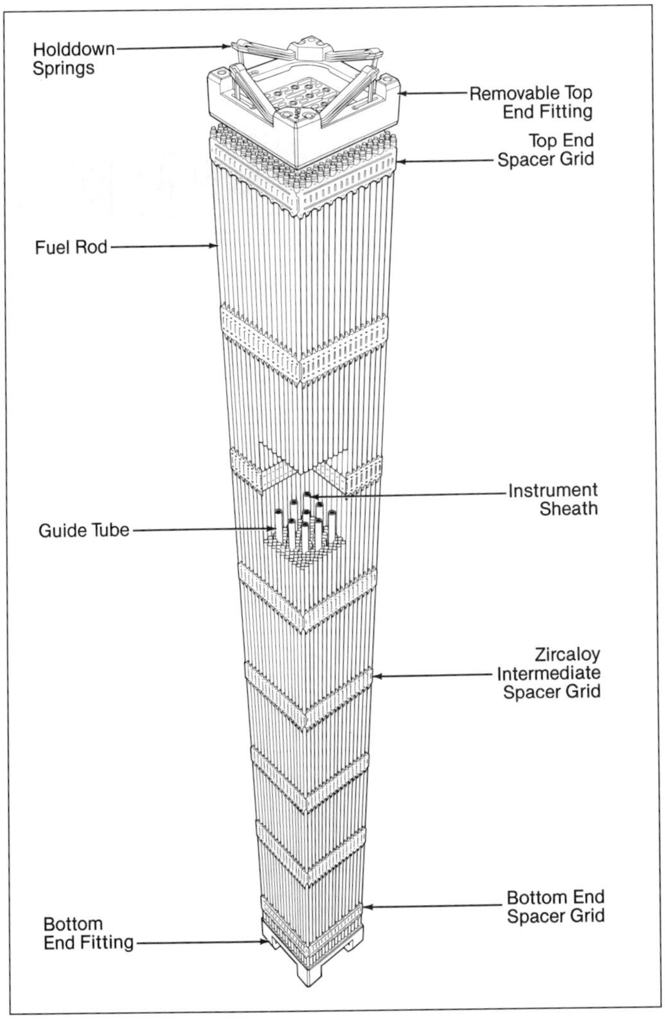

Fig. 4 Nuclear fuel assembly.

Fig. 5 Nuclear fuel pellets.

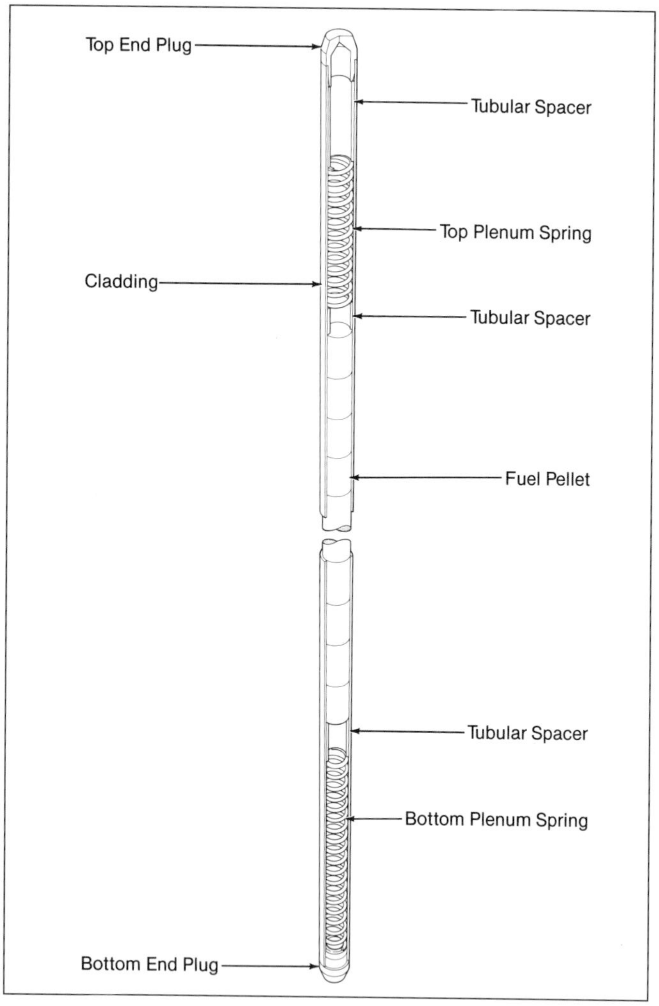

Fig. 6 Typical fuel rod assembly

back-pressure during fuel pellet loading, it provides a path to remove moisture from the fuel pellets and rod internals and it permits access to pressurize the fuel rod. The cladding with one laser drilled end cap in place is then ready for fuel loading.

The fuel pellets containing the proper per rod U-235 content are weighed; this weight is computer documented. The stack of pellets is visually examined for pellet quality and length and the fuel column, with appropriate nonfuel internals, is inserted into the cladding. The plenum at the open rod end is then measured to verify seating and column length and the lower fuel rod spring is inserted. At this point, the second end cap (bearing a code identifying fuel enrichment) is installed in the open end of the rod, causing the plenum spring to preload the fuel column. This end cap is welded and inspected as was the first cap.

Capped fuel rods are accumulated and loaded into a vacuum furnace in lots. Sample rods are included for subsequent hydrogen analysis. The loaded furnace is evacuated, backfilled with inert gas, heated and re-evacuated. This cycle removes remaining moisture from the fuel pellets and internal components. At completion of the cycle, the chamber and rods are refilled with inert gas at atmospheric pressure and the sample rod is evaluated by Quality Control personnel.

Once the correct hydrogen content of the sample pellets is confirmed, the fuel rods are removed from the drying retort. Within an established time limit, the drilled end cap of each rod is inserted into a small chamber for pressurizing and sealing. Once in the chamber, the laser drilled hole is aligned with another laser. The rod is secured and the chamber is sealed and evacuated. The fuel rod is pressurized with dry helium and the laser is discharged to fuse the hole in the end cap. An additional automatic ultrasonic inspection follows the seal weld.

The sealed rods are then transferred to a neutron scanner. Here, the rods are exposed to the neutron beam from a californium source. The resulting activity is measured and analyzed to nondestructively verify the fuel column integrity and length and the enrichment of the pellets. Next, the rods are transported from the scanner to automatic cleaning equipment. Here, the entire fuel rod surface is scrubbed, rinsed with demineralized water and dried in preparation for final inspection.

In the final inspection, the end of the rod through which the pellets were loaded is subjected to alpha counting to ensure that there is no uranium contamination on the outside surface. Following dimensional inspection and a complete visual inspection, helium leak testing is conducted. This is a functional test of the cladding, end caps and welds, verifying rod sealing.

Assembly

The upper and lower grids for fuel assemblies are fabricated from Inconel, an age hardenable nickel alloy. The intermediate spacer grids are made from Zircaloy-4. Strips of the grid alloy are stamped to form the detailed configurations required for contact points, springs and other attributes. After stamping, the strips are inspected and quarantined until released for production.

To assemble the spacer grids, the various types of strips are placed into an array forming a grid. Each grid is numbered and the material's source code is recorded. To accurately position the strips, welding fixture pins are installed into each grid cell. After assembly, the grid is welded at each strip intersection. Inconel and Zircaloy grids are laser welded in an inert atmosphere. For Zircaloy grids, sizing and annealing take place prior to welding; for Inconel grids, age hardening occurs after welding.

Fig. 7 Spacer grid coordinate measuring machine.

Quality Control personnel visually inspect the more than 500 welds, and sample welds from each grid are magnified and dimensionally inspected using a coordinate measuring machine. (See Fig. 7.) A computer printout provides a record of the measurements.

The individual fuel rod cells of the grids are opened for insertion of unique keys. Each cell is opened to a dimension larger than the outside diameter of the fuel rod and is held open by the keys. As a result, stresses are not imposed on the fuel rods during insertion in the grids and the cladding is not scratched. Before fuel bundle assembly begins, a *stacking sheet*, listing all components and assemblies by serial number, is prepared. Additionally, as fuel rods are inserted in the spacer grids, the location of each rod is documented.

The keyed grids are installed into the discs of a horizontal rotary assembly fixture that secures the grids during bundle fabrication. This fixture is mounted on a granite surface plate to assure the assembled bundle is straight. (See Fig. 8.) The fuel rods are inserted into the grids and the grid keys are turned and removed, releasing the cell springs to grip the rods. The rods are seated on lower end fittings.

The control rod guide thimbles are inserted into the grids and sleeves prior to attaching the upper nozzle; the thimbles are rotated to engage the grid restraining system. The upper nozzle is aligned and attached to the sleeves, and the lower nozzle is aligned and attached to the guide thimbles.

Once the fabrication is completed, the fuel assembly is placed on a transfer cart, which is equipped to lift it to the vertical position. The assembly is then removed from the fixture and subsequent handling is done in this position. (See Fig. 9.)

The fuel assembly dimensions are then checked in a vertical envelope gauge. Sensors verify straightness, twist, bow and overall length dimensions in free standing and restrained positions. Strain gauges measure water channels at axial locations along the length of the assembly and signals from these gauges provide rod to rod spacing; the data are compared to specifications. The assembly is then transferred to a pit, where a master control rod gauge is inserted into the fuel assembly guide thimbles. This insertion verifies guide thimble alignment and ensures that the thimbles do not restrict control rod insertion.

Fig. 8 Fuel assembly fabrication.

Fig. 9 Two nuclear fuel assemblies for a large power reactor.

A final visual inspection is performed using high intensity lights. (See Fig. 10.) The assembly is cleaned and sealed in a plastic sheath; it is now ready for storage or shipment to a reactor site.

Fuel assemblies are shipped to the reactor site in *strongbacks* (See Fig. 11.) The fuel bundle is installed by crane into the strongback (which can hold two bundles)

Fig. 10 Final visual inspection of fuel assemblies.

in the vertical position. Next, the strongback is lowered to a horizontal position and secured and the container is closed, sealed and pressurized with dry air. Shock indicators, mounted on the strongback, measure shipment induced shock. The indicators trip if exposed to excessive acceleration forces and these indicators can be viewed at the site prior to opening the container.

Plant operations

Most U.S. power reactors operate for a planned period of between one and two years before the reactivity of the nuclear fuel is depleted and the unit must be reloaded with fresh fuel. Because only part of the fuel is replaced during the shutdown, this is referred to as *batch refueling*. In some reactors, such as the Canadian (CANDU) design, a continuous refueling mode, which does not require reactor shutdown, is used.

Most PWRs were designed for annual refueling, replacing approximately one third of the core at each outage. As operating experience has been gained, the trend has shifted to 18 month refueling cycles. Although fuel costs are higher, the longer refueling cycle is cost effective due to reductions in the number of refueling outages, licensing submittals and replacement power cost. Furthermore, the burnup capability of the fuel has increased, permitting subsequent batch size reductions. (Today, units on 18 month cycles are loaded with batch sizes once used for annual cycles.) In addition to adopting 18 month cycles, most plants have converted from the out/in shuffle scheme to the lumped burnable poison (LBP) scheme, because of its flexibility in accommodating the batch size and enrichment changes. Several utilities have also been successful in adopting two year refueling cycles.

As nuclear fuel is consumed, its characteristics change. Four major issues are burnup dependent: dimensional and structural changes, cladding water-side corrosion and hydriding, fission gas release and pellet cladding interaction. Since the startup of the first Babcock & Wilcox (B&W) commercial nuclear power plant, an ongoing performance monitoring program has collected data to assess these phenomena. To date, more than one million fuel rods have been irradiated and assembly burnups of greater than 50,000 MWd/t_mU have been attained.

Dimensional and structural changes

Dimensional and structural changes primarily include irradiation growth of the fuel rods and fuel assembly, hold down spring relaxation and fuel rod bow. Fuel assembly growth is caused by irradiation induced growth of the guide tubes in the assembly. Fuel rod growth has been comparable to fuel assembly growth, because the fuel rod and assembly guide tubes are made of similar materials.

Other dimensional and structural changes, including changes in the fuel rod diameter and fuel density, can occur in the fuel assembly and its components with continued irradiation. Although the database on these phenomena indicates no life limiting constraints, the phenomena can have combined effects on overall rod performance and susceptibility to other failure mechanisms.

Cladding water-side corrosion and hydriding

Under reactor operating conditions, the fuel rod cladding acquires a thin, adherent oxide layer early in life. After this initial film is established, the extent of further water-side corrosion is a complex function of in-reactor residence time, neutron fluence, local flow conditions and coolant temperature and chemistry. However, increasing data variability has been observed for fuel rods with burnups greater than 35,000 MWd/t_mU. Current data indicate that oxide thickness could become the life limiting factor for fuel rods due to wall thinning and weakening. B&W Fuel Company is currently investigating advanced cladding materials that may slow corrosion and the effects of hydrogen pickup.

The corrosion process also liberates hydrogen from the coolant; less than 20% of it is absorbed by the Zircaloy-4

Fig. 11 Fuel assemblies prepared for shipment.

cladding. Under operating conditions, the hydrogen in the cladding stays in solution for concentrations up to about 200 ppm by weight. The hydrogen precipitates at higher concentrations or lower temperatures, forming zirconium hydride platelets. Because radial orientations of the hydride platelets markedly decrease cladding mechanical properties, the B&W Fuel Company's fuel rod design uses cladding with a crystallographic texture that ensures circumferential hydride platelet orientation. The cladding hydrogen content data through 40,000 MWd/t_mU indicate total hydrogen concentrations below 200 ppm and circumferential hydride platelet orientation at room temperature. Hydrogen concentrations are expected to remain below 400 ppm by weight for fuel rod burnups beyond 50,000 MWd/t_mU, therefore assuring cladding ductility.

Fission gas release

If not accounted for in the fuel rod design, release of fission gas from the fuel at high burnup and high power levels can cause high internal fuel rod pressure. Levels of fission gas release from PWR fuel rods operated at normal linear heat rates to burnups of 50,000 MWd/t_mU have been low, averaging less than 5%. Therefore, the data from high burnup, typical PWR fuel rods do not indicate the significantly increased fission gas release levels that have been observed under abnormal conditions or predicted by some models.

Pellet-cladding-interaction

Pellet-cladding-interaction (PCI) is a broad category of events involving contact between the fuel pellets and the cladding. In this text, however, it refers to significant interactions that cause loss of cladding integrity through mechanical means or by a combination of chemical attack and mechanical interactions. The availability of aggressive fission products and the number and extent of rods experiencing pellet cladding contact increase with burnup. Furthermore, the mechanical properties of the cladding, and therefore its susceptibility to PCI, could change at high burnups. However, observations to date on standard fuel operating under normal reactor conditions do not indicate increased PCI with extended burnups.

Spent fuel and nuclear waste management

Extreme care and detailed handling procedures are required to remove the spent fuel assemblies for disposition. During their removal from the reactor core and near term storage in spent fuel pools, water serves as both shielding and as a cooling medium. The spent fuel assemblies are kept in the spent fuel pool at the reactor site until the radioactive decay and heat generation have declined sufficiently for safe transportation and processing. The fuel can then be disposed of directly (open cycle) or reprocessed to recover 97% of the material. (See Chapter 56.)

Spent fuel storage at a nuclear power plant.

Chapter 49
Principles of Nuclear Reactions

Fundamental particles and structure of the atom[1]

In the fifth century B.C. Greek philosophers postulated that all matter is composed of indivisible particles called *atoms*. Over the next 24 centuries there were many speculations on the basic structure of matter but all theories lacked any experimental basis. It was not until late in the nineteenth and early twentieth centuries that the existence of fundamental particles, the *electron*, *proton* and *neutron*, was confirmed.

The discovery of nuclear fission, credited to Otto Hahn, Lise Meitner and Fritz Strassman, was made possible through an accumulation of knowledge on the structure of matter beginning with Becquerel's 1896 detection of radioactivity.

As shown in Fig. 1, the structure of an atom is pictured as a dense, positively charged nucleus surrounded by an array of negatively charged electrons. Each proton, the equivalent of a hydrogen atom nucleus, carries an elemental positive charge of electricity. The number of protons determines the type of chemical element of the atom. Each neutron is an electrically neutral particle with a mass slightly greater than the proton. Because of their association with the nucleus of the atom, protons and neutrons are also referred to as *nucleons*. Each electron shown orbiting around the nucleus has an elemental negative charge and a mass about 1/2000 that of a proton. The nucleus of an atom with a characteristic number of neutrons and protons is called a *nuclide*.

Despite the minute size of the atom, there is a relatively great distance between the nucleus and the orbiting electrons. This distance, approximately 10^5 times the dimension of the nucleus, accounts for the ability of various radiations to pass through apparently dense materials.

Fig. 1 also indicates the relationship of the positively charged protons in the nucleus and the negatively charged electrons. In the un-ionized state, the number of protons is balanced by an equal number of electrons. An atom becomes ionized by gaining or losing one or more electrons. A gain of electrons yields negative ions and a loss results in positive ions. An atom in an ionized state can interact with other elements to form various compounds.

When an atom has more than two electrons, their orbits are located in a series of separate and distinct groupings of energy levels or *shells*. Each shell is capable of containing a specific number of electrons. In general, an inner shell fills to its maximum number of electrons before electrons begin to form in the next shell. An x-ray is a quantum of electromagnetic energy that is emitted when an electron transitions from an outer shell to an inner shell or between energy levels within the same shell.

The number of electrons in the outermost shell determines certain chemical properties of the elements. The properties are similar for elements which have similar electron distributions in the outer shell regardless of the number of inner shells. This accounts for the repetition of chemical properties in the Periodic Table of the Elements. (See Appendix 1.)

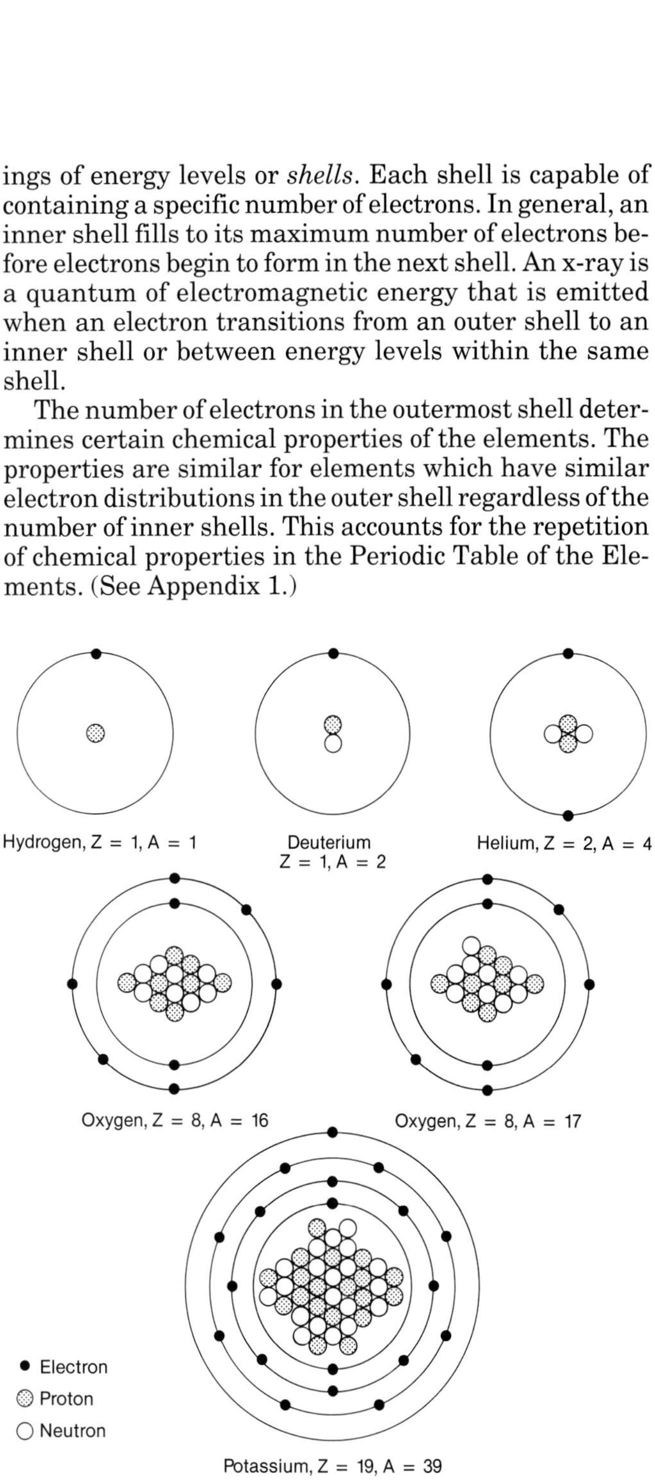

Fig. 1 Structure of the atom (Z = atomic number; A = mass number).

Nuclear radiations

Nuclides that occur in nature are stable in most cases. However, a few are unstable, especially those of atomic number 84 and above. The unstable nuclides undergo spontaneous change at specific rates by radioactive disintegration or decay. Many nuclides decay into other unstable nuclides, resulting in a decay chain that continues until a stable isotope is formed.

There are generally three types of radiation commonly arising out of the decay of specific nuclides: beta (β) particles, gamma (γ) rays and alpha (α) particles.

Beta particle The β particle results from radioactive decay and has the same mass and charge as an electron. It is believed that the nucleus of an atom does not contain electrons and, in radioactive β decay, the β radiation arises from conversion of a neutron into a proton and β particle; therefore:

$$neutron \rightarrow proton + \beta + energy$$

In some instances a positive β particle (called a *positron*) is produced from the conversion of a proton to a neutron.

Gamma ray Gamma rays are electromagnetic radiation resulting from a nuclear reaction. They can be treated like particles in many nuclear reactions and are included among the fundamental particles. Although γ rays have physical characteristics similar to x-rays, their energy is greater and wave length shorter. The only other difference between the two rays is that γ rays originate from within the nucleus while x-rays originate from within the shell structure of the atom.

Alpha particle The α particle is equivalent to the nucleus of a helium atom comprising two neutrons and two protons. It results from radioactive decay of an unstable nuclide and, with very few exceptions, is observed only in the decay of heavy nuclides.

Nuclides and isotopes

A nuclide is characterized by its atomic number Z (number of protons in the nucleus) and the mass number A (total number of protons plus neutrons). When nuclides are described by chemical symbol, the atomic number is the left hand subscript and the mass number the right hand superscript. For example, the α particle can be represented as α, ${}_2\alpha^4$, or ${}_2He^4$. With most chemical elements there are several types of atoms having different mass numbers, i.e., different numbers of neutrons in the nucleus but the same number of protons. An isotope is one of two or more nuclides of the same chemical element having different mass numbers.

Two isotopes of hydrogen and two isotopes of oxygen are depicted in Fig. 1. An ordinary hydrogen atom ${}_1H^1$, contains one proton and no neutrons. Its atomic number Z and mass number A are 1. It combines with oxygen to form H_2O or regular or light water. The deuterium atom, ${}_1H^2$ or ${}_1D^2$, has one proton and one neutron; Z is 1 and A is 2. It combines with oxygen to form D_2O or heavy water. A third isotope of hydrogen, tritium, ${}_1H^3$ or ${}_1T^3$, has one proton and two neutrons.

Heavier nuclides are also identified by chemical name and mass number. Oxygen-16 and oxygen-17 (Fig. 1) have 16 and 17 nucleons respectively, although each has eight protons. For example, symbols for the two isotopes of oxygen described above are ${}_8O^{16}$ and ${}_8O^{17}$.

Although either the atomic number or the chemical symbol could identify the chemical element, subscripts sometimes are useful in accounting for the total number of charges in an equation.

Mass

It is customary to list the mass of atoms and fundamental particles in atomic mass units (AMU). This is a relative scale in which the nuclide ${}_6C^{12}$ (carbon-12) is assigned the exact mass of 12 AMU by agreement at the 1962 International Union of Chemists and Physicists. One AMU is the equivalent of approximately 1.66×10^{-24} grams, or the reciprocal of the presently accepted Avogadro number, 0.602214×10^{24} atoms per gram-atom. A gram-atom of an element is a quantity having a mass in grams numerically equal to the atomic weight of the element.

Table 1 lists the masses of the fundamental particles and the atoms of hydrogen, deuterium and helium in atomic mass units.

Table 1
Masses of Particles and Light Atoms

Isotope or Particle	Mass, AMU
Electron	0.000549
Proton, ${}_1p^1$	1.007277
Neutron, ${}_0n^1$	1.008665
Hydrogen, ${}_1H^1$	1.007825
Deuterium, ${}_1H^2$	2.01410
Helium, ${}_2He^4$	4.00260

Mass defect

The mass of the hydrogen atom ${}_1H^1$ listed in Table 1 is almost but not quite equal to the sum of the masses of its individual particles, one proton and one electron. However, the mass of a deuterium atom ${}_1H^2$ is noticeably less than the sum of its constituents — a neutron, proton and electron. Measurements show that the mass of a nuclide is always less than the sum of the masses of its protons, neutrons and electrons. This difference, the *mass defect* (MD), is customarily calculated as:

$$MD = Zm_h + (A - Z)m_n - m_e \quad \textbf{(1)}$$

where

MD = mass defect, AMU
Z = number of protons in the nucleus of the nuclide
m_h = mass of the hydrogen atom, AMU
$A - Z$ = number of neutrons in the nucleus
m_n = mass of neutron, AMU
m_e = mass of the nuclide including its Z electrons, AMU

Binding energy

Although most nuclei contain a plurality of protons with mutually repulsive positive charges, the nucleus remains tightly bound together, and it takes considerable energy to cause disintegration. This energy, called *binding energy*, is equivalent to the mass defect. From the equivalence of mass and energy, as defined by Einstein's equation $E = mc^2$, one AMU equals 931 million electron volts (*MeV*). An electron volt is the energy gained by a unit electrical charge when it passes, without resistance, through a potential difference of 1 volt. Therefore,

$$Binding\ energy\ (MeV) = \left\{ Zm_h + (A-Z)m_n - m_e \right\} \times 931 \quad \textbf{(2)}$$

This represents the amount of radiant or heat energy released when an atom is formed from neutrons and hydrogen atoms. It also represents the energy which must be added to fission an atom into its basic nucleons, i.e., neutrons and protons.

Dividing Equation 2 by A, the number of nucleons in the nucleus, yields the binding energy per nucleon. This in turn can be plotted as a function of A, the mass number as shown in Fig. 2. The result shows that the binding energy per nucleon rapidly increases for low mass numbers, reaches a maximum for mass numbers in the range of 40 to 80, and then drops off with an increasing slope. On the rising portion of this curve, *fusion* or joining of nucleons to atoms of higher mass number means that there is an increased binding energy per nucleon and consequently a release of energy. On the falling portion of this curve, the *fission* process or splitting of an atom results in nuclides of lesser mass numbers and greater binding energy per nucleon. Again, energy is released because of the increased mass defect.

Radioactivity and decay

Decay rate

The rate at which a radioactive nuclide emits radiation is a characteristic of the nuclide and is unaffected by temperature, pressure or the presence of other elements that may dilute the radioactive substance. Each nucleus of a specific radioactive nuclide has the same probability of decaying in a definite period of time at a rate characterized by its radioactive decay constant. The rate of decay at any time, *t*, always remains proportional to the number of radioactive atoms existing at that particular instant. The decay is calculated by:

$$N(t) = N(0)e^{-\lambda t} \quad \textbf{(3)}$$

where

$N(t)$ = the number of atoms per cm^3 at time t
$N(0)$ = the number of atoms per cm^3 at time zero
λ = radioactive decay constant

Half life

Decay is usually expressed in terms of a unit of time called radioactive half life, $T_{1/2}$. This represents a measurable period of time, the period it takes for a quantity of radioactive material to decay to one half of its original amount. The relationship between half life and decay constant can be determined by substituting ($T_{1/2}$) for *t* and $N(0)/2$ for $N(t)$ in Equation 3 and solving for λ. The result is:

$$\lambda = \frac{0.693}{T_{1/2}} \quad \textbf{(4)}$$

Decay constants for radioactivity isotopes are easily obtainable with a listing of measured half lives.[2]

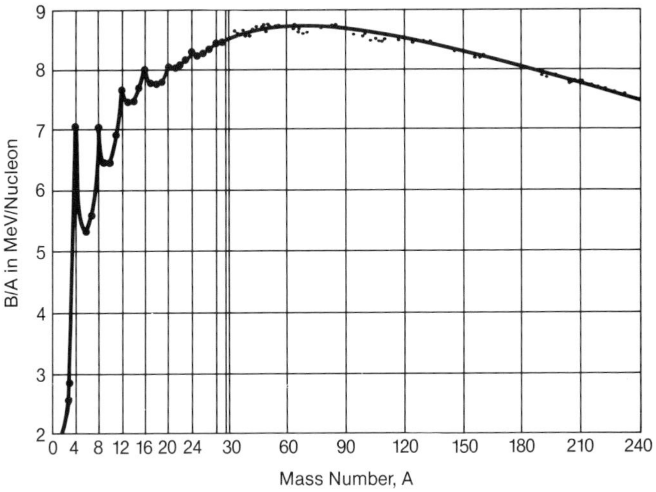

Fig. 2 Binding energy per nucleon versus mass number.

If $N(t)$ represents the number of radioactive atoms present at time *t*, then $\lambda N(t)$ becomes the number of radioactive nuclei that decay per unit of time at time *t*. This is referred to as the *radioactivity*, or more simply the activity, of the atoms and is expressed in *curies*. A curie is 3.70×10^{10} disintegrations per second. $N(t)$ can be converted directly to curies by the relation $N(t)/(3.70 \times 10^{10})$. Fig. 3 shows on a relative scale how the activity decreases during several half lives. This curve applies to all radioactive substances.

It is sometimes difficult to distinguish between stable and radioactive nuclide. All nuclides heavier than bismith-209 (Bi-209) are unstable. However, the specific nuclides thorium-232, uranium-235 (U-235) and U-238 have half lives of 10^8 to 10^{10} years and can be considered stable. In fact, they are generally referred to as the stable isotopes of the heavier chemical elements. Table 2 shows half lives of some nuclides of high atomic number.

Induced nuclear reactions

By providing sufficient energy, all of the fundamental particles can be made to react with various nuclei. In this procedure, the particle strikes or enters an atomic nucleus causing a transfiguration or change in structure of the nucleus and the release of a quantity of energy. Particles which activate these reactions include neutrons, deuterons, α, β, protons, electrons and γ. Many reactions produce artificial radioactive nuclides.

Table 2
Half Lives of Heavy Elements

	Decay Mode	Half Life
Naturally occurring nuclides:		
Thorium-232	α	1.39×10^{10} yr
Uranium-238	α	4.51×10^{9} yr
Uranium-235	α	7.13×10^{8} yr
Artificial nuclides:		
Thorium-233	β	22.1 min
Protactinium-233	β	27.4 d
Uranium-233	α	1.62×10^{5} yr
Uranium-239	β	23.5 min
Neptunium-239	β	2.35 d
Plutonium-239	α	2.44×10^{4} yr

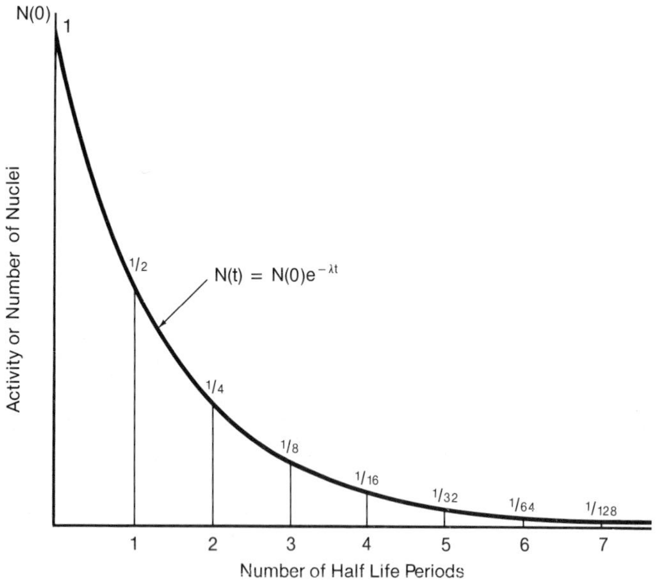

Fig. 3 Exponential decay of radioactive nuclides.

Table 3
Typical Induced Nuclear Reactions

Short Form	Long Form
Alpha:	
$_4Be^9(\alpha,n)_6C^{12}$	$_4Be^9 + {}_2\alpha^4 \rightarrow {}_6C^{12} + {}_0n^1$
$_7N^{14}(\alpha,p)_8O^{17}$	$_7N^{14} + {}_2\alpha^4 \rightarrow {}_9F^{18*} \rightarrow {}_8O^{17} + {}_1p^1$
Deuteron:	
$_{15}P^{31}(d,p)_{15}P^{32}$	$_{15}P^{31} + {}_1d^2 \rightarrow {}_{16}S^{33*} \rightarrow {}_{15}P^{32} + {}_1p^1$
$_4Be^9(d,n)_5B^{10}$	$_4Be^9 + {}_1d^2 \rightarrow {}_5B^{11*} \rightarrow {}_5B^{10} + {}_0n^1$
Gamma:	
$_4Be^9(\gamma,n)_4Be^8$	$_4Be^9 + {}_0\gamma^0 \rightarrow {}_4Be^8 + {}_0n^1$
$_1H^2(\gamma,n)_1H^1$	$_1H^2 + {}_0\gamma^0 \rightarrow {}_1H^{2*} \rightarrow {}_1H^1 + {}_0n$
Neutron:	
$_5B^{10}(n,\alpha)_3Li^7$	$_5B^{10} + {}_0n^1 \rightarrow {}_5B^{11*} \rightarrow {}_3Li^7 + {}_2\alpha^4$
$_{48}Cd^{113}(n,\gamma)_{48}Cd^{114}$	$_{48}Cd^{113} + {}_0n^1 \rightarrow {}_{48}Cd^{114*} \rightarrow {}_{48}Cd^{114} + {}_0\gamma^0$
$_1H^1(n,\gamma)_1H^2$	$_1H^1 + {}_0n^1 \rightarrow {}_1H^{2*} \rightarrow {}_1H^2 + {}_0\gamma^0$
$_8O^{16}(n,p)_7N^{16}$	$_8O^{16} + {}_0n^1 \rightarrow {}_8O^{17*} \rightarrow {}_7N^{16} + {}_1p^1$
Proton:	
$_6C^{12}(p,\gamma)_7N^{13}$	$_6C^{12} + {}_1p^1 \rightarrow {}_7N^{13*} \rightarrow {}_7N^{13} + {}_0\gamma^0$
$_4Be^9(p,d)_4Be^8$	$_4Be^9 + {}_1p^1 \rightarrow {}_5B^{10*} \rightarrow {}_4Be^8 + {}_1d^2$

* Indicates formation of a compound nucleus which is generally unstable.

Charged particles such as alphas, betas and protons do not have great penetration power in matter because the interaction with the existing electrical field within the atoms either slows or stops them. Collisions between like charged particles require exceedingly high kinetic energy. Electrical fields, however, can not deflect electrically neutral neutrons; therefore, they collide with nuclei of the material on a statistical basis. The neutron is therefore the most effective particle for inducing nuclear reactions including fission.

Reactions of principal interest in the design of nuclear reactors are those that involve neutrons and those that involve the interaction of the particles produced by fission and other nuclear processes within the reactor and surrounding materials.

The expression:

$$_{Z1}X^{A1}(P_1P_2)_{Z2}X^{A2} \qquad \textbf{(5)}$$

is generally used to denote a nuclear reaction and is the short form for:

$$_{Z1}X^{A1} + P_1 \rightarrow {}_Z X^{A*} \rightarrow {}_{Z2}X^{A2} + P_2$$

The X^* notation indicates formation of a compound nucleus which is generally unstable. Terms P_1 and P_2 represent incident and resultant particles respectively. The individual sum of the Zs (atomic numbers) and the sum of the As (mass numbers) on the left side of the equation always equals, respectively, the sum of the Zs and the sum of the As on the right side of the equation. (See Table 3).

Nuclear reactions always conserve total mass energy on the two sides of the equation in accordance with Einstein's equation for the equivalence of mass and energy. However, there is usually an energy difference and a mass difference. A good statistical probability for the reaction to occur exists when energy is released as a result of reaction. The more probable reactions generally can be initiated by lower energy particles. The less probable ones, those that require significant energy addition, can be initiated only by high energy particles. Table 3 gives some typical induced nuclear reactions.

In a nuclear reactor, many heavy nuclides with mass numbers greater than 238 are produced by a series of neutron captures accompanied by β and/or α decay. These artificially produced nuclides have many uses — some are readily fissionable, some are fertile because they absorb a neutron and then transmute to a fissionable nuclide, some have high energy α decay modes that can be used in neutron sources, and some are used as sources in medical procedures.

Probability of nuclear reactions — cross-sections

The probability of a nuclear reaction between a neutron, or other fundamental particle, and a particular nuclide is expressed as the *cross-section* for that reaction. The probability is dependent on the energy of the interacting particle. Because neutron interactions are most important in a nuclear reactor, the following discussion is directed toward neutron cross-sections; however, the concepts can be directly applied to all particle interactions.

Two types of nuclear cross-sections are defined:

1. The microscopic cross-section or interaction probability is an intrinsic characteristic of the nuclei of the material. It has dimensions of an area and is normally expressed in *barns* where one barn equals 10^{-24} cm^2.
2. The macroscopic cross-section is a probability of interaction per centimeter of neutron path and takes into account the density of the material. It has the dimensions cm^{-1}.

The symbol σ represents the microscopic cross-section. In the case of a neutron approaching a fissionable atom it becomes possible to consider a total cross-section σ_T in which:

$$\sigma_T = \sigma_c + \sigma_s + \sigma_f = \sigma_a + \sigma_s \qquad \textbf{(6)}$$

where

σ_c = the capture cross-section, a measure of the probability for absorption without fission

σ_s = the scattering cross-section, the probability that the nucleus will scatter the neutron
σ_f = the fission cross-section (present in only a few of the many nuclei), the probability for a neutron to strike and cause a fission to occur
σ_a = the absorption cross-section, the sum of the probabilities for capture and fission

The macroscopic cross-section can be obtained from:

$$\Sigma = N_0 \rho \sigma / M \tag{7}$$

where

Σ = macroscopic cross-section, cm^{-1}
N_0 = Avogadro's number, 0.602214×10^{24} atoms/gram-atom
ρ = density of the material, grams/cm^3
σ = microscopic cross-section, cm^2/atom
M = atomic weight, grams/gram-atom

Experimental measurements determine microscopic cross-sections for each element. Total cross-section measurements are generally made by transmission techniques. For example, placing a material of known density and thickness in front of a neutron source permits measuring the intensity of neutrons at a particular energy on each side of the material. The difference represents the loss or attenuation of neutrons by the material. The cross-section required to obtain this attenuation is derived by calculation.

Isotopes of the same element can have very different cross-sections. For example, the isotope xenon-134 (Xe-134) has a microscopic cross-section for neutron absorption of about 0.2 barn, yet Xe-135 has a microscopic cross-section of 2.7×10^6 barns. U-235 is fissionable at any energy, and U-238 only at high energy.

Cross-sections of some nuclides also contain abrupt peaks called *resonances* at certain energy bands (Fig. 4). Because a cross-section is really a function of relative energy of the neutron and the nucleus, an effective change in the cross-section results when the energies of the neutron and the nucleus increase or decrease as a result of temperature changes.

Where the curve of cross-section versus energy is fairly smooth, the effect of a temperature change of the target nucleus is relatively small. However, the effect of a temperature change is large and important in the vicinity of a resonance. An increase in temperature results in increased vibration of the nucleus with a corresponding increase in the number of probable collisions between the nucleus and neutron occurring at energies in the vicinity of the resonance. Therefore, an increase in the temperature of the nucleus results in an apparent broadening of the energy width of the resonance. This in turn results in a very effective increase in resonance neutron absorption, i.e., capture and fission. Conversely, a decrease in temperature of the nucleus results in narrowing the resonance width and decreasing resonance absorption. The change in resonance energy width with temperature, known as the *Doppler effect*, is important in reactor control.

Fig. 4 illustrates a typical cross-section curve showing the neutron capture cross-section of U-238 as a function of neutron energy. Below 10^2 eV the height of the resonance peaks is about 10^3 barns. Between 10^2 and 4×10^3 eV the great number of resonances makes it impractical to show the cross-section as a curve, although the resonance parameter data are available for use on a computer. Between 4×10^3 and 10^5 eV the curve represents a statistical average of the measurements which have been made.

The fission process

Nuclear fission is the splitting of a nucleus into two or more separate nuclei accompanied by release of a large amount of energy. In the fission of an atom due to a neutron, the mass of the neutron plus its energy must be equal to or greater than the mass defect associated with the two fission products. U-235 is the only naturally occurring nuclide that is capable of undergoing fission by interaction with low energy or slow neutrons. Some of the artificially produced heavy nuclides are also fissionable with slow neutrons, including plutonium-239 (Pu-239) and U-233. Other nuclides such as U-238 and Pu-242 require higher energy neutrons to cause fission.

This difference in fission capability occurs because the binding energy of a nucleus is not only determined by its mass number but also by whether the number of protons and neutrons is even or odd. A nuclide with an even number of neutrons and protons, such as U-238, has the highest binding energy per nucleon and requires the most added energy to fission. Fission only occurs with high energy or

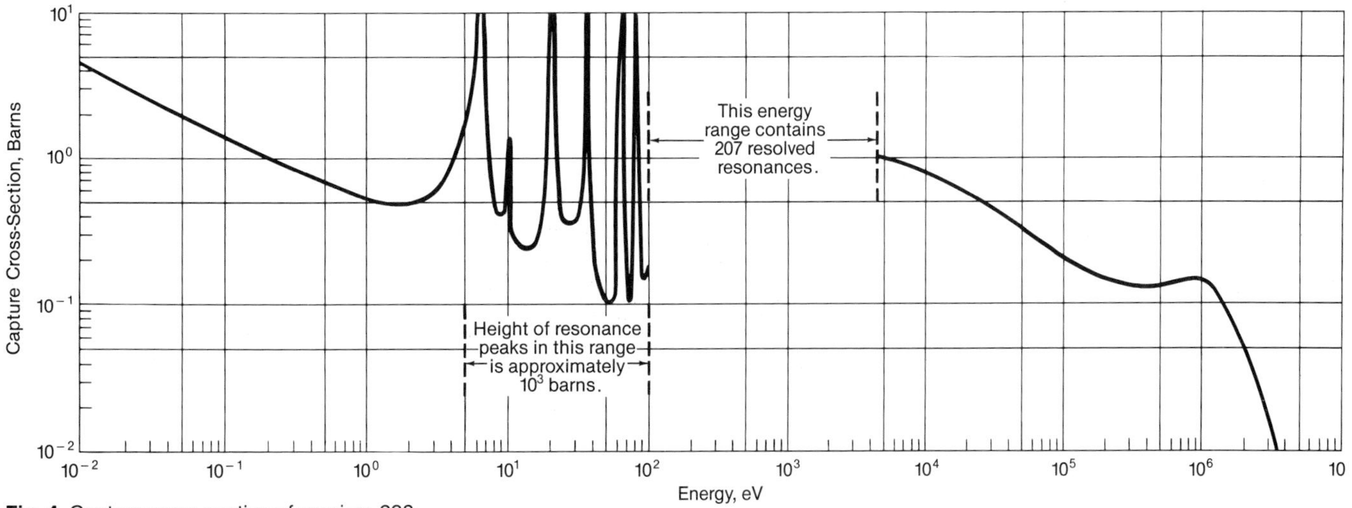

Fig. 4 Capture cross-section of uranium-238.

fast neutrons (> 1MeV). A nuclide with an odd number of protons and an even number of neutrons or an even number of protons and an odd number of neutrons, such as U-235, Pu-239 and U-233, has a lower binding energy per nucleon. Nucleons with both an odd number of protons and neutrons have the lowest binding energy per nucleon for essentially equivalent mass numbers. U-235 and Pu-239 are fissionable with slow or thermal neutrons (~0.025 eV). The term thermal neutron refers to a neutron energy distribution that is in thermal equilibrium with the temperature of the surrounding materials.

Fission occurs when the fissionable nucleus absorbs a neutron. In the case of U-235 the reaction is:

$$_{92}U^{235} + {_0n^1} \rightarrow {_{92}U^{236*}}$$

$$_{92}U^{236*} \rightarrow {_{Z1}X^{A1}} + {_{Z2}Y^{A2}} + 2.43\ {_0n^1} + Energy$$

As previously noted, the asterisk in $_{92}U^{236*}$ indicates an unstable nuclide. The value 2.43 applies to U-235 fission by a thermal neutron and is the statistical average of the number of neutrons produced per reaction. X and Y represent the fission products which are distributed as shown in Fig. 5.

Energy from fission

Fission also produces γ rays, neutrons, β particles and other particles. The energy release per fission amounts to about 204 MeV for U-235 and is distributed as shown in Table 4. Release of approximately this amount of energy per fission can be predicted by examination of Fig. 2 for binding energy, considering that two fission products are formed as shown in Fig. 5. Table 4 also includes data for Pu-239.

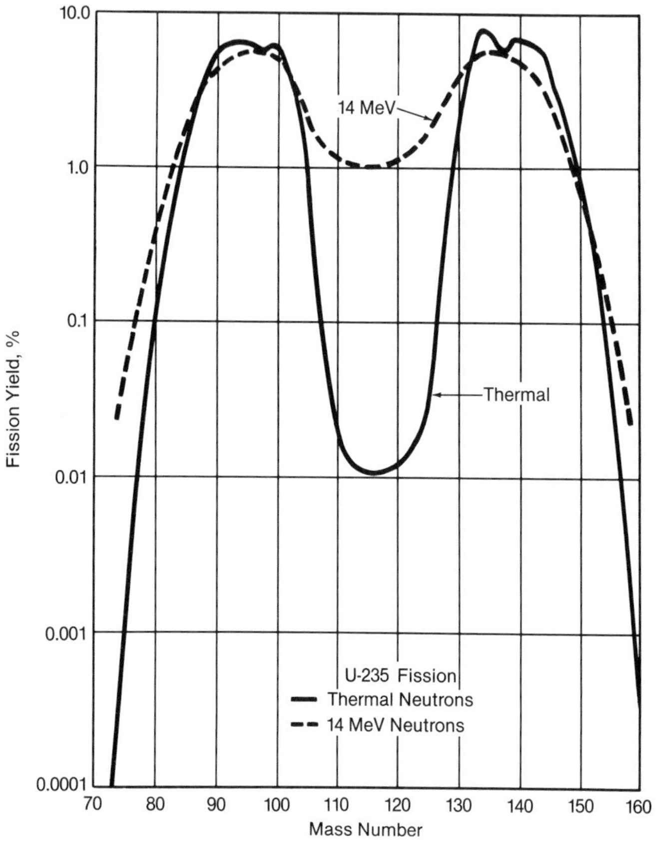

Fig. 5 Mass distribution of fission products from fission of U-235.

Table 4
Energy Produced in Fission (MeV/fission)

	U-235	Pu-239
Instantaneous		
Kinetic energy of fission products	169	175
γ ray energy	8	8
Kinetic energy of fission neutrons	5	6
	182	189
Delayed		
β particles from fission products	8	8
γ rays from fission products	7	6
Neutron-capture γs*	7	10
	22	24
Total	204	213

* Energy produced depends on reactor composition.

Neutrons from fission

The fact that additional neutrons are born or generated by a fission event makes it possible to establish a *chain reaction*, as depicted in Fig. 6, that can sustain itself as long as sufficient fissionable material and neutrons are present. The neutrons produced per fission event, υ, and the average number of neutrons produced per neutron absorbed in the fuel, η, vary with the different fissionable isotopes and with the energy of the neutron producing the fission. Statistical averages for these quantities are given in Table 5. There are individual fission reactions that produce only one neutron, possibly none, or as many as five. Because some of the neutrons are absorbed without producing fission, η becomes a more meaningful quantity in reactor design than υ. The values in the Table are for fission by low energy or thermal neutrons at room temperature [0.025 eV neutron energy or 7218 ft/s (2200 m/s) neutron velocity]. To maintain a chain reaction, the average number of neutrons per absorp-

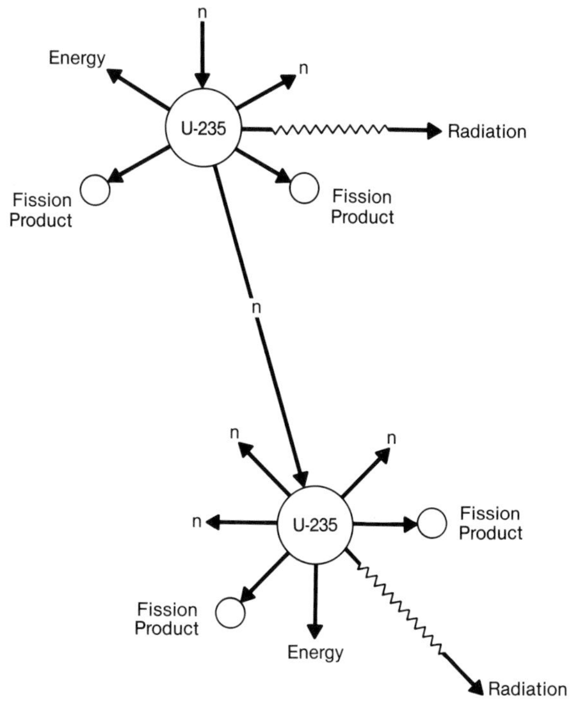

Fig. 6 Chain reaction.

Table 5
Neutrons from Fission

Fuel	Avg. Neutrons per Fission, υ	Avg. Neutrons per Absorption in fuel, η
Uranium-233	2.51	2.28
Uranium-235	2.43	2.07
Natural uranium	2.43	1.34
Plutonium-239	2.90	2.10
Plutonium-241	3.06	2.24

tion must be significantly greater than 1 because some of the neutrons will be lost to absorption in the moderator and structural materials, to absorption in control materials, and to leakage from the core. Control materials such as boron and silver-indium-cadmium are used to maintain a steady-state chain reaction by absorbing any additional neutrons over that needed for steady-state operation.

Fission neutron energy distribution

Neutrons released from fission vary in initial energy over a wide range up to 15 MeV and above. The distribution of neutrons produced by fission as a function of energy has been determined from several experimental measurements, and typical results for U-235 and Pu-239 are shown in Fig. 7. Based on these measurements, the average energy of fission neutrons is about 2 MeV; the peak of the energy distribution occurs at about 0.8 MeV.

Burnup

As a mass of uranium undergoes fission it produces energy. The term *burnup* is used to represent the amount of energy produced per unit mass of the material. The units of burnup are megawatt days per metric ton (MWd/t_m) of initial heavy metal, i.e., uranium. One megawatt day represents about 2.6 x 10^{21} fissions. A nuclear fuel assembly is typically discharged from the reactor when it has achieved a burnup of 50,000 MWd/t_m. In commercial power reactors that are fueled with uranium comprising mainly the isotope U-238, with 4% or less U-235, most of the energy produced comes from the fissioning of U-235. At burnups of 50,000 MWd/t_m only about 5% of the initial uranium content of the fuel assembly has fissioned. Even when the other nuclear reactions that cause loss of the original uranium atoms are considered, there is still more than 90% of the initial uranium remaining in the fuel assembly when it is discharged.

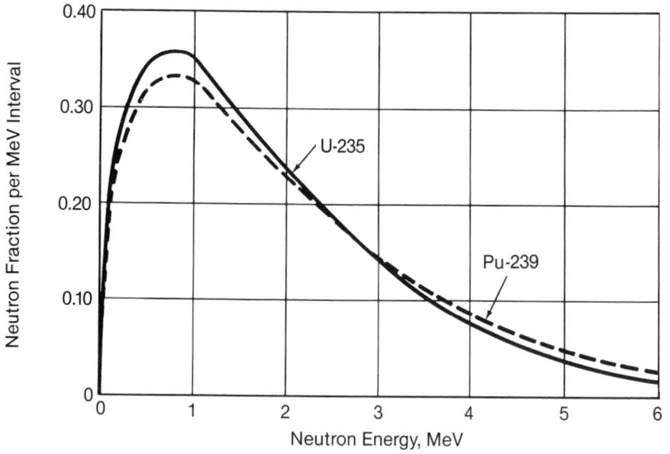

Fig. 7 Fission neutron energy distribution.

Fission products

When a nucleus undergoes fission, experimental measurements have shown that predominantly two fission products or fragments are generated. The distribution of these fission products for U-235 fissions has also been measured and is shown in Fig. 5. In fission, as in any nuclear reaction, there is always conservation of total mass-energy so that one of the two fission products will come from each hump of the distribution.

Examination of Fig. 5 reveals that mass number of the fission products ranges from about 70 to 170 with two plateaus at approximately 95 and 135. Curves are shown for fission caused by thermal energy neutrons and by 14 MeV neutrons. The higher the neutron energy causing fission the more uniform the fission product distribution.

Many fission products interact with neutrons, absorbing them so that they are not available to the chain reaction. As these fission products build up (see Fig. 8), they act as absorbers to retard the chain reaction. All long lived fission products except samarium build up as the core is operated, reaching a maximum effect at the end of core life.

After discharge from the reactor and sitting in storage for a few years, most of the fission products will have decayed due to their relative short half lives. However, some of the long half life fission products will contribute significantly to the γ ray source that must be shielded against in handling the fuel during its ultimate disposal. These isotopes include strontium-90 (Sr-90), ruthenium-106 (Ru-106), cesium-134 (Cs-134) and 137 (Cs-137), cerium-144 (Ce-144) and europium-154 (Eu-154). These isotopes will either be bound in the fuel pellets or on the inside surface of the clad. Other isotopes important to fuel handling are the artificially produced nuclides in the fuel pellet that decay by spontaneously fissioning. These isotopes include Pu-238, americium-241 (Am-241), and curium-242 (Cm-242) and 244 (Cm-244). In addition, the γ emitting isotope cobalt-60 (Co-60) is produced in the structural steels and Inconel through neutron activation and must be shielded against.

Fission product behavior with time

Certain fission products, specifically Xe-135 and samarium-149 (Sm-149), both of which have very high cross-sections for absorption of thermal neutrons, are not only produced directly from fission but also are the decay products of other fission products.

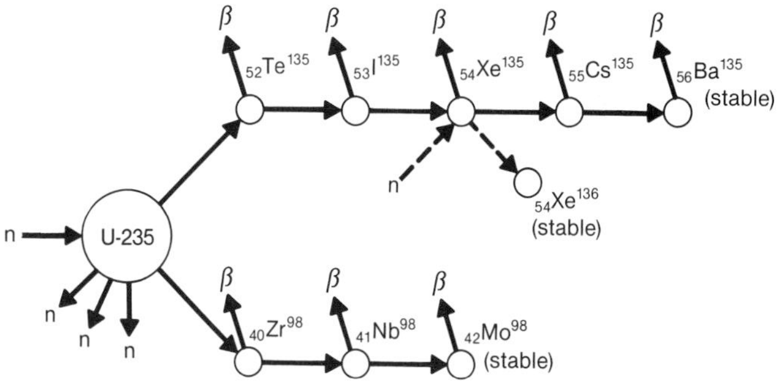

Fig. 8 Fission product chain.

Essentially all initial products of fission are highly radioactive and decay rapidly to less active isotopes with somewhat longer half lives. There are usually several isotopes in the chain before a stable end product is reached. The most significant such decay chain is the following:

$$ {}_{52}Te^{135} \xrightarrow{\beta} {}_{53}I^{135} \xrightarrow{\beta} {}_{54}Xe^{135} \xrightarrow{\beta} {}_{55}Cs^{135} \xrightarrow{\beta} {}_{56}Ba^{135} $$

<1.0 min. 6.7 h 9.2 h 2 x 10^6 yr

In this chain, tellurium-135 (Te-135) with a one minute half life decays to iodine-135 (I-135) with a 6.7 hour half life and then to Xe-135. This xenon isotope, which fortunately has only a 9.2 hour half life, has a macroscopic cross-section for thermal neutron absorption approximately 100,000 cm^{-1}, as great as all the long lived fission products together. Unfortunately, the nuclides of this chain occur abundantly as fission products — a predictable happening, because the mass number 135 occurs at a peak in the fission product distribution (Fig. 5). The Xe-135 absorbs an appreciable fraction of available neutrons as long as the reactor is operating. This changes some of the Xe-135 to Xe-136 (which has negligible neutron absorption). As a result, when a water reactor operates at constant power level, the Xe-135 builds up to its equilibrium value in 36 to 48 hours.

When reactor power lessens and, particularly, when the reactor is shut down, the I-135 formed at the original power level continues for a time to generate Xe-135 at a rate corresponding to the original power level. Therefore, the Xe-135 builds up rapidly after shutdown because fewer neutrons are available for conversion to Xe-136. The buildup reaches a peak 4 to 12 hours after shutdown and then slowly decays. The time behavior of Xe-135 must be addressed in reactor control.

Sm-149, the second important fission product in the reactor core during operation, is generated as follows:

$$ {}_{60}Nd^{149} \rightarrow {}_{61}Pm^{149} \rightarrow {}_{62}Sm^{149} \text{ (stable)} $$

In the chain, neodymium-149 (Nd-149) decays (1.7 hour half life) into promethium-149 (Pm-149), which in turn decays (47 hour half life) into Sm-149. Although Sm-149 is a stable isotope it is destroyed so rapidly by neutron absorption that it reaches an equilibrium value when the reactor operates at constant power. Typically, Sm-149 reaches equilibrium in a pressurized water reactor after 50 to 100 days. Buildup after shutdown is slower and less extensive than for Xe-135.

Decay after shutdown has little consequence for all other fission fragments because of their long lives and comparable capture cross-sections between parent and daughter nuclide.

Nuclear reactor composition

A reactor core is composed of fuel, structural and control materials, and a moderator and/or coolant. Typical pressurized water reactor (PWR) fuel assembly and control component designs are shown in Fig. 9. A depiction of the interactions that take place within a fuel assembly is shown in Fig. 10. (See also Chapter 48.) Neutrons react not only with uranium, but also with the nuclei of most other elements present. Therefore, many materials which have high neutron absorption properties should not be used for structural purposes. Fortunately, a few materials such as aluminum and zirconium have low neutron absorption cross-sections. Inconel, steel and stainless steel also can be used to a limited extent.

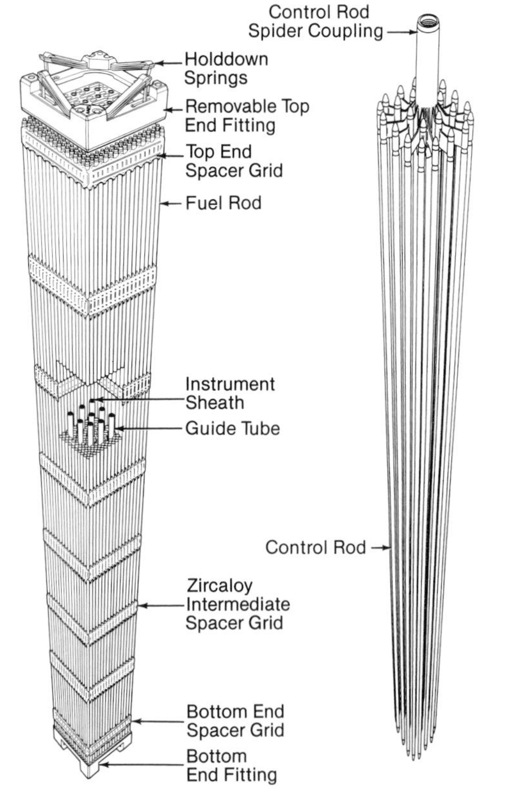

Fig. 9 Typical PWR fuel assembly and control components.

The choice of coolants includes water, helium and liquid sodium. Thermal reactor design, where the neutron energy distribution is essentially in thermal equilibrium with its surroundings, requires a moderator (a material containing atoms of a light element such as hydrogen, deuterium or carbon) in the core to reduce the kinetic energy of the neutrons. Hydrogen is the ideal moderator because its nucleus is as light as a neutron and, therefore, it can absorb the energy of the neutron in a single direct collision. Hydrogen is normally present in the form of water. The heavier the atom the less energy it absorbs per collision, and the less slowing effect on the neutron. Carbon, in the form of graphite, is about the heaviest atom that can be used practically as a moderator. Carbon has the additional advantage of absorbing very few neutrons, even less than hydrogen. Other low weight atoms can be used, but they are less practical because of excess neutron absorption or high material cost. Because the helium nucleus absorbs essentially no neutrons, it would be an ideal moderator except that helium is a gas at reactor operating temperatures. In this state it would be impossible to provide the core with a sufficient number of atoms to be an effective moderator.

All thermal reactors contain some fast (high energy) neutrons because only fast neutrons result from fission. The amount of moderator provided determines the degree of thermalization (energy reduction) or the percentage of thermal neutrons present in the reactor. Perhaps the greatest advantage of the thermal reactor is its compatibility with the water coolant. Water has proven to be the most practical and economical coolant.

The low neutrons per fission available with natural uranium makes it difficult to design a natural uranium (0.71% U-235) reactor. However, it can be accomplished if all materials are of especially high purity and if discrete sections of fuel are placed in a heterogeneous array such that the neutrons can slow down in the moderator and then re-enter the fuel material as slow neutrons. Under these conditions, there is a relative high probability of causing fission. The normal moderators used in natural uranium reactors are either graphite or deuterium oxide (heavy water). The absorption cross-section of normal hydrogen makes it less desirable than the other moderators. Also, minute quantities of any high neutron absorbers can prevent a chain reaction. Nevertheless, all early reactors used natural uranium and many still operate, notably the large plutonium producing reactors and the Canadian CANDU pressurized heavy water reactors. (See Chapter 47.)

Physical description of nuclear chain reactions[3-6]

To use fission as a continuous process for power production, it is necessary to initiate and maintain a fission chain reaction at a controlled rate or level which can be varied with power demand. Obtaining a steady-state chain reaction requires the availability of more than one neutron for each fission produced because nonfission reactions absorb some neutrons and others leak from the reactor. For a reactor as a whole, the neutrons born in each instant of time constitute one generation of neutrons. The term *effective multiplication factor*, k_{eff}, is defined as the ratio of the number of neutrons in one generation to that in the previous generation. With a multiplication factor of less than one, the system is a decaying one and will never be self-sustaining. With a multiplication factor greater than one, a nuclear system produces more neutrons than it uses, and power increases. For a steady-state chain reaction, $k_{eff} = 1$.

The steady-state equation for neutron balance in a chain reaction can also be written as:

$$\textit{Production} = \textit{Absorption} + \textit{Leakage} \qquad \textbf{(8)}$$

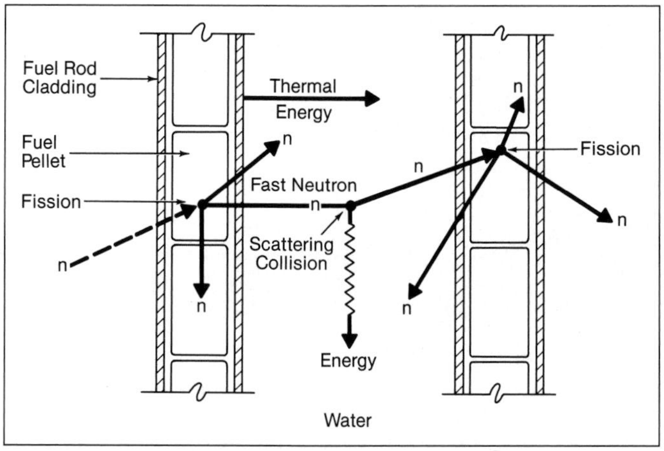

Fig. 10 PWR neutron moderator/coolant-fuel interactions.

When this condition is obtained, the reactor has gone *critical* meaning that the necessary amount of neutron production has been achieved to balance the leakage and the absorption of neutrons. The mass of fissionable material required to achieve this condition is called the *critical mass*.

Production, absorption and leakage all depend heavily on interactions of neutrons with nuclei of the various materials in the reactor. Production depends primarily on those interactions with U-235 nuclei which result in fission. Absorption depends on interactions of neutrons with any nuclei in the core that result in absorption of neutrons with or without fission. Leakage depends on the scattering effect of collisions between neutrons, nuclei and other particles, which results in transport of neutrons toward the boundaries of the chain reacting system and ultimate escape from the system.

Table 5 indicates that each fission of a U-235 atom by a thermal neutron makes available an average of 2.43 neutrons. However, only 2.07 neutrons are produced per thermal neutron absorbed in U-235 as also indicated in Table 5. The other 0.36 neutrons are absorbed in nonfission reactions. If natural uranium is used, the number of neutrons produced per thermal neutron absorbed is reduced to 1.34 because of the neutron absorption in U-238. Most of this absorption results in the ultimate production of fissionable Pu-239. However, neutrons used in this manner are no longer available to help maintain the chain reaction.

A neutron chain reaction is possible because each fission event on the average produces more than one neutron. If the process is controlled such that just one of the neutrons produced from fission causes another fission reaction then a steady state process is maintained and a constant power level is achieved. To best understand the process, a general description of the lifetime of a neutron follows. It must be realized that for a power reactor the number of neutrons crossing a 1 cm^2 surface (neutron flux) near the center of the reactor is on the order of 10^{13} neutrons/cm^2/s. The lifetime of a neutron is on the order of 10^{-9} seconds. Therefore, this description is only representative of what happens in an instant of time.

As described previously, a neutron produced by the fission process has an average energy of about 2 MeV and is termed a fast neutron. At this energy elastic or inelastic scattering reaction with the moderator, the fuel or with the structural materials are most probable. However, about 2% of the fast neutrons cause a fission in U-238 and produce an even higher number of neutrons than produced from thermal fission.

In an elastic scattering reaction some of the neutron energy is transferred to the nucleus with which it collides in the form of kinetic energy; this can cause the nucleus to be displaced from its normal lattice position in the material. In an inelastic collision a compound nucleus is formed in an excited state and it reduces its energy by releasing a lower energy neutron and a γ ray. Some of the original neutron's energy is also deposited as the kinetic energy on the nucleus. The net impact of the scattering reactions is to cause the neutron to lose energy or slow down into the energy range where resonance interactions with various materials (especially U-238) can take place. In the resonance range, neutrons will have a high probability of being captured if their energy coincides with any of the resonances associated with the uranium fuel or the fission products and will not be available to react with the U-235 to help sustain the chain reaction. The resonance escape probability is the fraction of neutrons that escape capture while slowing down to the thermal energy range. The term *resonance escape* refers particularly to U-238 because in the intermediate range of neutron energy there are several resonance peaks of absorption cross-section for this isotope. These resonances are useful in the production of plutonium; however, to maintain criticality in a thermal reactor, sufficient neutrons must escape absorption in the resonance region.

In contrast, collisions with hydrogen atoms in the moderator can slow down the neutron very rapidly, and therefore the resonance escape probability is high for water moderated systems. Once past the resonance range the neutron is said to become thermalized and becomes part of a Maxwellian energy distribution with an average energy of about 0.2 eV. As the fission cross-sections for U-235 and Pu-239 are the highest in the thermal energy range, most fission reactions take place in this range.

The probability of having a fission reaction is referred to as the *thermal utilization* and is the ratio of the fission cross-section to the absorption cross-section (capture plus fission) averaged over the moderator plus the fuel and structural materials.

The other mechanism for loss of neutrons is leakage from the system. This is dependent on the size of the system and the length of the path neutrons travel between source and capture. For a thermal reactor this length is approximately 2.95 in. (75 mm). Today's power reactors are on the order of 13.1 ft (4 m) in diameter and therefore only neutrons born in the fuel assemblies on the periphery of the core have a significant probability of being lost from the system.

Control of the chain reaction

More than 99% of neutrons produced in fission are *prompt* neutrons; that is, they are produced almost instantaneously. About 0.73%, in the case of U-235, are released by the decay of fission products rather than directly from fission. The average half life for these delayed neutrons from fissioning of U-235 is about 13 seconds. This provides time for the reactor operator to respond to small changes in either power demand by the system or reactor system parameters.

The chain reaction can be regulated by placing materials with high neutron absorption capability in the reactor and providing a means of varying their amounts. These materials tend to stop the chain reaction and include boron, cadmium-silver-indium, hafnium and gadolinium. One or more of these materials usually goes into the reactor in the form of control rods that can be withdrawn to start up the reactor or to increase power level and reinserted to reduce power or to shut down the unit. To assure accurate control of power, at least some of the rods must be capable of fine regulation.

Reactivity

The first step in establishing a chain reaction is to bring the reactor to critical condition at essentially no thermal power or zero power level. The reactor power is gradually increased up to the desired level by the removal of a control material and maintained there. The objective in each step of this procedure is to obtain and hold a constant value of $k_{eff} = 1.0$. In view of the changes that continually occur in a nuclear system, it is never possible to keep $k_{eff} = 1.0$ for more than a short period of time without adjustments to compensate for variations. Operating a reactor at any constant power level at an effective multiplication factor of unity corresponds to steering a ship on a compass course. It takes a continual effort, either automatic or manual, to hold the ship on the exact course.

If a reactor operates at a specific power level with $k_{eff} = 1.0$, and if anything changes to increase or decrease the multiplication factor, a reactivity change is said to have occurred. It may be of positive or negative change depending upon the direction of change in k_{eff}. Reactivity, represented by the symbol ρ, is defined as the ratio:

$$\rho = (k_{eff} - 1)/k_{eff} \tag{9}$$

Reactivity is usually expressed in the units of pcm which corresponds to a ρ of 1×10^{-5}. As will be shown most processes in the reactor, except for control rod insertion, generate reactivities of 1 to 10 pcm which constitute a very small change from a k_{eff} of 1.

Calculation of reactor physics parameters[3-6]

The calculation of a chain reaction (or the design of a nuclear reactor) requires solution of the steady-state equation:

$$\textit{Production} = \textit{Absorption} + \textit{Leakage}$$

or its counterpart for nonsteady-state (transient) conditions:

$$\textit{Production} - \textit{Absorption} - \textit{Leakage} = dn/dt \tag{10}$$

where dn/dt is the variation of the neutron density with time.

The production rate is evaluated as:

$$\textit{Production} = \upsilon \Sigma_f \; n \; v \tag{11}$$

where

υ = neutrons per fission
Σ_f = macroscopic fission cross-sections, cm^{-1}
n = neutron density, neutrons/cm^3
v = neutron velocity, cm/s

If the neutron flux is defined as the product of the neutron density and neutron velocity, then the production rate can be defined as:

$$\textit{Production} = \upsilon \Sigma_f \phi \tag{12}$$

where

$\phi = nv$ = neutron flux

Similarly, the absorption rate is defined as:

$$\textit{Absorption} = \Sigma_a \phi \tag{13}$$

where

Σ_a = macroscopic absorption cross-section, cm^{-1}

Leakage is a complicated function of the gradients in the neutron flux at boundaries of the region under consideration.

In performing reactor calculations, the core is divided into discrete spatial regions or nodes, and the continuous neutron energy range is divided into a number of groups. The above equations then become a coupled set of partial differential equations that are solved in both the space and energy domains to find the neutron flux and reaction rates in each node.

In the design of the early reactors, most calculations of reactor physics parameters were done by hand using homogeneous models of the core components. Neutron cross-sections were obtained from experimental plots of total, absorption, scattering and fission cross-sections. Critical experiments were performed mocking up the fuel designs and modeling both the fuel and control rods. From these experiments, modeling adjustments were developed to ensure high accuracy predictions of reactor performance parameters. As with most other disciplines, the advent of fast, large computers has permitted the designer to perform detailed calculations in both space and energy. The following sections describe the current techniques being used in calculations for commercial PWR fuel cycle design and licensing.

Cross-sections

Basic neutron and gamma cross-sections are compiled and verified as part of the industry effort to maintain the cross-section libraries. These libraries provide the basic data needed to calculate the cross-section of each isotope as a function of incident neutron energy. Typically two major energy groups, each containing subgroups, are used by the designer. The first, the fast and epithermal group, spans energies from 1.85 eV to 20 MeV and is composed of 40 or more subgroups. The second, the thermal energy group, covers energies from 0.00001 eV to 1.85 eV and is composed of 50 or more subgroups. In the fast and epithermal energy range, only scattering reactions that decrease neutron energy are used in the calculations. In the thermal energy range, the neutron energies are in thermal equilibrium with the moderator, and energy can be imparted to the neutron during scattering collisions. Therefore, scattering reactions, which can increase or decrease neutron energy, are permitted in the calculations for this group.

When designing a reactor core, specific cross-sections are calculated using the total multigroup cross-section set (40 epithermal plus 50 thermal groups) for each material present. Each cell type in the fuel assembly can then be modeled (Fig. 11). This calculation determines the neutron flux distribution in each energy group and cell. The resulting neutron flux is then used to average the cross-sections over the two major energy groups and over various regions of the fuel assembly. A full core representation, indicating the position of fuel assemblies and control components, is shown in Fig. 12.

The important quantities that describe the reactor core and its performance include critical boron concentration, core power generation distribution, control rod reactivity worths, and the reactivity coefficients associated with changes in reactor conditions.

Critical boron concentration

There is sufficient U-235 in a commercial power reactor to power the unit for 12 to 24 months before refueling. Without additional absorbers present in the system, the k_{eff} would be greater than 1. It is impractical to provide the absorption using control rods. Therefore, for long

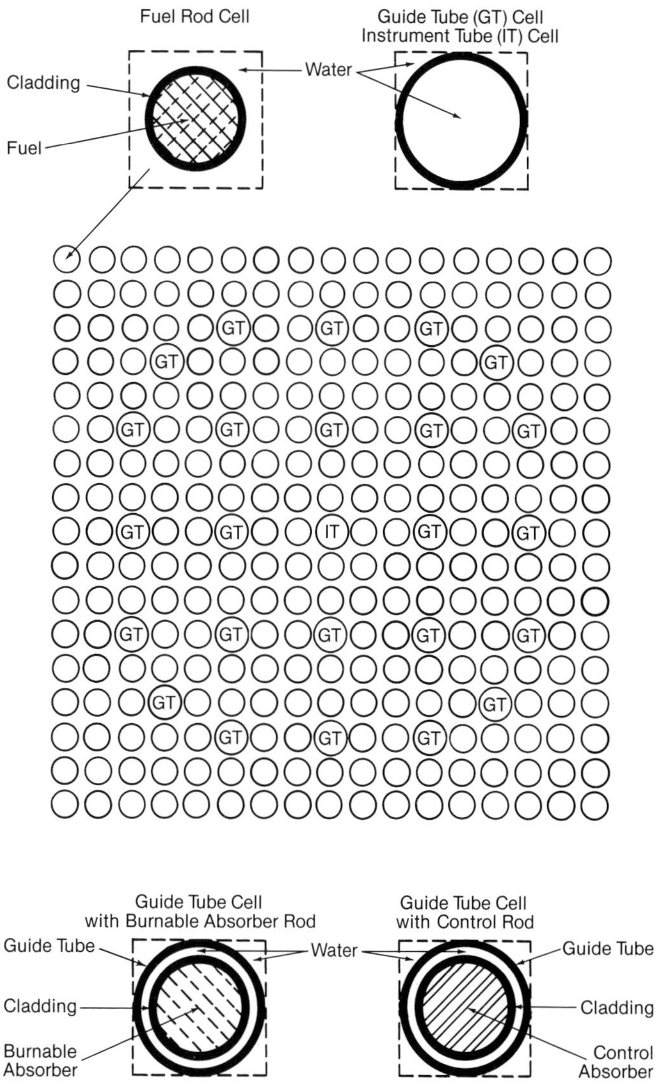

Fig. 11 Cross-sectional view of fuel assembly.

term control $k_{eff} = 1$ is maintained by soluble boron in the moderator (boron dissolved in water) and/or burnable absorbers. Burnable absorbers contain a limited concentration of absorber atoms such that, as neutrons are absorbed, the effectiveness of absorption decreases and essentially burns out with time. Burnable absorbers may be in the fuel rods or in fixed absorber rods that are placed in fuel assembly guide tubes. The boron concentration in the coolant can be near 1800 ppm at the beginning of a fuel cycle. This concentration can decrease to near zero at the end of the cycle, when a significant portion of the uranium is depleted and fission products have built up. The critical boron concentration is the boron level required to maintain steady-state reactor power levels. The burnable absorber limits the amount of soluble boron that is used at the beginning of the cycle.

Power distribution

The primary factor governing acceptable reactor operation is the energy production rate at every point in each fuel rod. If the rate is too high, the fuel can melt and cause rod failure. Excessive energy rates can also cause coolant steaming, which results in poor heat transfer and can lead to clad burnout.

The energy production rate, or power, at each fuel rod location is referred to as the core power distribution. This has typically been calculated by forming a three dimensional model of the reactor and representing each fuel assembly as a homogenized region over an axial interval. A new method for power calculations involves reconstructing the power production in each fuel rod based on detailed fuel assembly calculations. This method is more accurate because it provides a better fuel assembly model that can include water, burnable absorbers or guide tube control rods. With proper fuel assembly placement, the core power distribution can be optimized to limit power peaks in fuel rod segments. As part of this primary analysis, initial enrichments of the fuel and burnable absorbers can be determined, as well as the necessary soluble boron concentration in the coolant over the cycle.

Control rod worths

Control rods are used in a pressurized water reactor to change power levels and to shut down the reactor. Approximately 48 rods, divided into banks of 8, are commonly used and core reactivity is changed by sequentially moving the banks. The control rod banks are grouped into shutdown and control (or regulating) banks. The shutdown banks provide negative reactivity to bring the reactor from hot to cold conditions. During a shutdown, the core temperature decreases and its reactivity increases. The shutdown control rod banks counteract this increase. The controlling banks are used in conjunction with variations in the soluble boron concentration to take the reactor from hot zero power to full power. These banks also provide the reactivity to handle rapid power changes. The negative reactivity caused by the buildup of xenon in the core following a shutdown is offset by a combination of control rod removal and decreases in the coolant boron concentration.

The reactivity *worth* of a control rod is a measure of its ability to reduce core reactivity. Control rod worths are calculated by first calculating the core with the control rod inserted to determine its k_{eff}. The core is then similarly calculated with the rod withdrawn. Control rod worth is defined as:

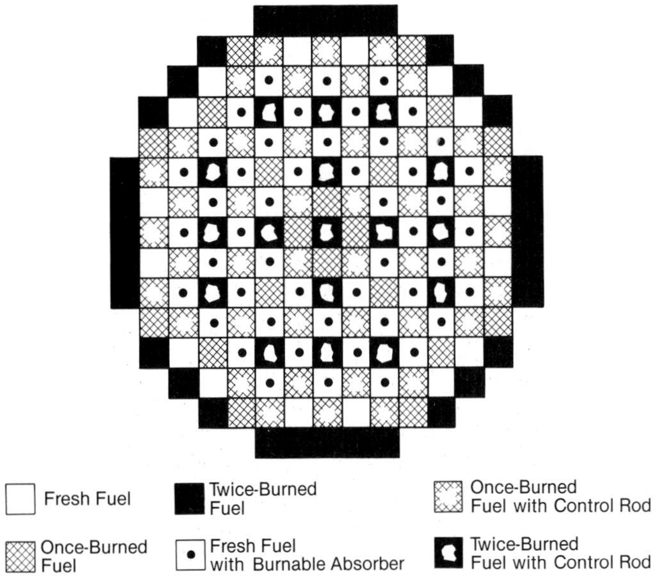

Fig. 12 Cross-sectional computer modeling of PWR core.

$$\frac{k_{eff(i)} - k_{eff(w)}}{k_{eff(w)}\ k_{eff(i)}} \tag{14}$$

where

$k_{eff(w)}$ = multiplication factor, control rod withdrawn
$k_{eff(i)}$ = multiplication factor, control rod inserted

Bank worths are similarly calculated.

Reactivity coefficients

Reactivity coefficients define the rate of core reactivity change associated with the rate of change in reactor conditions. These coefficients indicate the relative sensitivity of the core operation to changes in operating parameters.

In determining a coefficient, the core is first modeled with a given power and/or temperature distribution, and k_{eff} is calculated. The reactor is then similarly modeled at a different power or temperature distribution. The coefficient is defined as the reactivity change per unit of power or temperature. Fig. 13 shows the impact of various core parameter changes on core reactivity.

There are three basic reactivity coefficients that are important to reactor operation: 1) the moderator temperature, 2) Doppler, and 3) power coefficients. The moderator temperature coefficient is defined as the change in reactivity associated with a change in moderator temperature. It includes the effects of moderator density changes and the changes in nuclear cross-sections. The Doppler coefficient is defined as the change in reactivity associated with changes in fuel temperature that occurs primarily because of fuel nuclides with large resonances. The power coefficient is defined as the change in reactivity associated with a change in power level. It is a combination of the moderator and Doppler coefficients and is based on the change in temperature as power changes.

When the coolant temperature increases, the associated density reduction normally decreases its moderating capability. A net decline in core reactivity, i.e., a negative moderator temperature coefficient, occurs. Similarly, when the fuel temperature increases, core reactivity decreases. Both mechanisms provide an inherent safety response for the system. However, if the soluble boron concentration is too high, a positive moderator temperature coefficient can occur, and core reactivity could increase. For this reason boron concentration is limited to ensure a negative moderator coefficient at full power.

Neutron detectors

When operating a nuclear chain reaction system, neutron detectors are used to measure the intensity of the neutron radiation (flux). This radiation is a direct indication of nuclear reaction intensity. Neutron detectors can be counters or ionization chambers. Counters, which detect neutrons by sensing the individual ionizations they produce, are most useful when the neutron flux is low. Ionization chambers are more useful at high neutron fluxes. They measure the electrical current that flows when neutrons ionize gas in a chamber. These two types of detectors, placed on the exterior of the pressure vessel, only measure neutrons leaking from the reactor.

Two other types of detectors, a self-powered neutron detector and a miniature fission chamber, can be used to measure the neutrons inside the instrument tube of operating fuel assemblies. Self-powered detectors are usually arranged in strings and are positioned to continuously measure the neutron reactions at up to seven axial locations in the fuel assembly. The miniature fission chamber contains uranium, and fission events are detected by the electronics. The fission chamber is moved in to and out of the core on a periodic basis, normally monthly, and measures the neutron level along the entire length of the fuel assembly.

Neutron source

During the start of the chain reaction, it is essential that the operator monitor nuclear instrumentation that is counting neutrons and not gamma rays. The mass of fuel in the core is much greater than the critical mass required to sustain a chain reaction. Control rods and soluble boron dissolved in the reactor coolant keep the core subcritical when no power output is required.

When the core is to be brought to criticality, control rods are withdrawn to initiate the chain reaction. However, if the control rods are withdrawn before a measurable neutron flux is available, a reaction could be initiated by a stray neutron and build up to a high power level before the control rods could be reinserted to maintain the desired core output.

To control the rate of buildup of the reaction, it is necessary to have a neutron source present in the core for startup. With a neutron source available, it is possible to measure neutron flux before moving control rods and help ensure a safe reactor startup.

Several types of neutron sources are available. One is an americium-beryllium-curium source rod. The radioactive isotopes americium and curium emit α particles which react with the beryllium to produce neutrons.

Fast reactors

It is possible to design for a chain reaction to occur predominantly with either fast (high energy) neutrons or slow (thermal energy) neutrons. In fast reactors the chain reaction is maintained primarily by fast neutrons. Thermal reactors are those in which the chain reaction occurs primarily from thermal neutrons. Today's commercial water reactors are thermal reactors.

Fertile material such as U-238 can be used as a fuel by first converting it to plutonium in a reactor. Fast reactors and particularly fast breeder reactors accomplish this conversion most effectively. Liquid sodium, which has essentially no moderating effect, is the coolant most often considered for fast breeders. Helium is used as a coolant in high temperature gas-cooled reactor designs.

Conversion and breeding

Most important of the conversion reactions is the capture of a neutron by a U-238 nucleus, resulting in a nucleus of fissionable Pu-239:

$$ {}_{92}U^{238} + {}_{0}n^{1} \rightarrow {}_{92}U^{239} + {}_{0}\gamma^{0} $$

$$ {}_{92}U^{239} \xrightarrow[\text{23 min.}]{\beta} {}_{93}Np^{239} \xrightarrow[\text{2.3 d}]{\beta} {}_{94}Pu^{239} $$

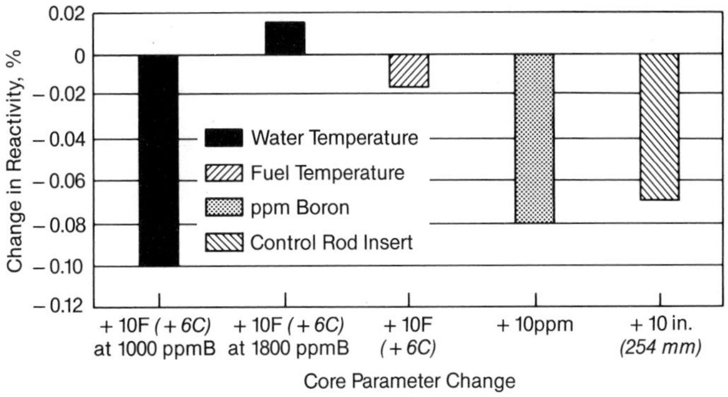

Fig. 13 Core reactivity as a function of operating parameters.

The U-238 nucleus absorbs a neutron and becomes U-239, which decays with a half life of 23 minutes, into neptunium-239 (Np-239). Again this nuclide, with a half life of 2.3 days, is transmuted to Pu-239 by β decay. This is the nuclear process by which the most useful isotope of plutonium is formed.

There are other reactions during operation which form different isotopes of plutonium. Some of these isotopes, including Pu-241, are fissionable, and others, including Pu-240, are converted to fissionable isotopes by neutron absorption. These isotopes can not be separated economically from Pu-239, and therefore must be taken into account when plutonium is used in a reactor.

The large commercial PWRs in the U.S. operate on slightly enriched uranium fuel. These reactors convert U-238 to plutonium and produce about 50% as much plutonium as the U-235 consumed. A reactor is considered to be a converter when the amount of fissionable material produced, e.g., plutonium, is less than the amount of fissionable material consumed, e.g., U-235. A breeder reactor is one in which more fissionable material is produced than consumed.

By definition, a breeder reactor must have the value of η (neutrons produced per neutron absorbed in fissionable fuel, Table 5) greater than 2.0. One neutron is required to maintain the chain reaction, one or more for the breeding, and an additional fraction for absorption in nonfuel materials and leakage.

It is not possible to make a breeder reactor with natural uranium fuel, because η is less than 2.0. Table 5 shows that η does not greatly exceed 2.0 with any common fissionable isotopes for fissions produced by thermal neutrons. Consequently, it becomes difficult and impractical to make a breeder with a thermal reactor.

Fortunately, at high neutron energies, η has a greater value than at thermal energies, particularly with Pu-239. For this reason, and because the absorption cross-sections of most materials are less than at thermal energy, fast reactors breed plutonium more effectively than thermal reactors. In addition, fast reactors operate best with plutonium fuel.

Health physics

Health physics, or radiological physics, is the science of radiation protection. The role of a health physicist in a nuclear plant is similar to that of a safety engineer in an industrial plant. Since the first nuclear reactor went critical in 1942 the potential hazards of radiation have been given very careful attention, and health physics has been an integral part of both public and private U.S. nuclear energy programs. As a direct result, these programs have had an enviable safety record.

The health physicist is concerned with the exposure of individuals to α, β, and γ radiations in all areas of a nuclear facility, and with neutrons in the vicinity of an operating reactor.

Biological effects

The organs of the human body are composed of tissue which is composed of atoms. When electrons are knocked out of or added to atoms (ionized), the chemical bonds which bind atoms together to make molecules are broken. This process is called *ionization*. The recombination of these broken molecules can result in changes in the molecular structure of a cell which may affect the way the cell functions, its growth characteristics and its interaction with other cells. In cases of high radiation exposure cancer may result. The term used to describe the effects of radiation on humans is *biological damage*.

Ionization is a direct result of alpha and beta particles interacting with human tissue. These particles produce a continuous path of ionization as they travel throughout the body. Gamma radiation produces ionization indirectly in matter by the photoelectric effect, the Compton effect and pair production. All three processes yield electrons which in turn produce most of the ionization which occurs within the body.[7]

In the photoelectric effect, the γ ray transfers all its energy to the electron that it strikes. The electron then causes ionization in the medium. In the Compton effect, only part of the γ ray's energy is transferred to the electron as kinetic energy. The remaining energy gives rise to a lower energy γ ray. Pair production occurs when the γ ray has an energy greater than 1.02 MeV. All of its energy is given up and two particles, an electron and a positron, are produced. (All three processes produce electrons which then ionize the absorbing matter.)

Although ionization is what causes biological damage, the seriousness of the damage is determined by many factors such as the types of cells effected (how radiosensitive they are), the age of the person receiving the exposure (young cells are more radiosensitive) and whether the dose is received over a short (acute exposure) or long (chronic exposure) period of time.

Acute exposure is more detrimental because a large amount of damage is incurred rapidly and the cells do not have an adequate amount of time to repair themselves. Under chronic exposure, cells are very effective in repairing injury. The effects of chronic exposure and the amount required to produce damage can only be extrapolated from known cases of acute radiation exposure. Most radiobiology studies have not developed sufficient databases to assess the effects of chronic exposure. It is difficult to clearly define the associated risks from low level exposures. Cancer cases that are induced from exposure usually occur 5 to 25 years after exposure and it is difficult to determine if the cancer is a result of radiation exposure. However, as the nuclear industry ages, the database for low level occupational exposure grows and it becomes more evident that the associated risks are low. The data available seem to suggest that the body can tolerate low doses of radiation received over long periods of time with little risk. This is supported by the exposure and health history of x-ray technicians, physicians and radiation workers.

Radiation protection

Alpha particles, due to their large mass and double positive charge, travel very short distances. Their range in air is only a few centimeters. Fig. 14 illustrates the ranges of various types of radiation. Alpha particles are seldom harmful when the source is located outside the human body because an α particle of the highest energy will barely penetrate the outer layer of skin. If emitted inside the body, however, α radiation can be serious. It is important, therefore, to prevent the ingestion or inhalation of α emitting nuclides. Particular care is required to keep the air in working spaces free of dust containing α emitters. Fabrication of α emitters such as plutonium normally takes place inside gloveboxes that remain under a slight negative pressure. The boxes discharge air effluent through a filter designed to prevent α bearing dust from entering the working spaces and the atmosphere in general. The air is continuously monitored and analyzed.

Beta particles penetrate up to an inch of wood or plastic material and travel several yards in air. The skin and the lenses of the eye are most vulnerable to external β radiation. However, clothing and safety glasses provide adequate protection for external exposure to β radiation. The β particle is not as great an internal hazard as the α particle. The β particle, due to its smaller mass and lower charge, will travel farther than the α particle through tissue and will deposit less energy in a localized area.

Gamma rays penetrate deeply and deposit their energy throughout the entire body. They have great ranges in air and may present a hazard at large distances from the source, however, they do not present as large an internal hazard as α particles.

Neutrons, like gamma radiation, are an external hazard and their damage extends throughout the body. In addition, the effectiveness of neutrons to produce biological damage is 2.5 to 10 times greater than γ rays.

Because radiobiological experiments do not clearly establish a safe level (a threshold), the U.S. federal regulatory exposure limits are very conservative. This concept

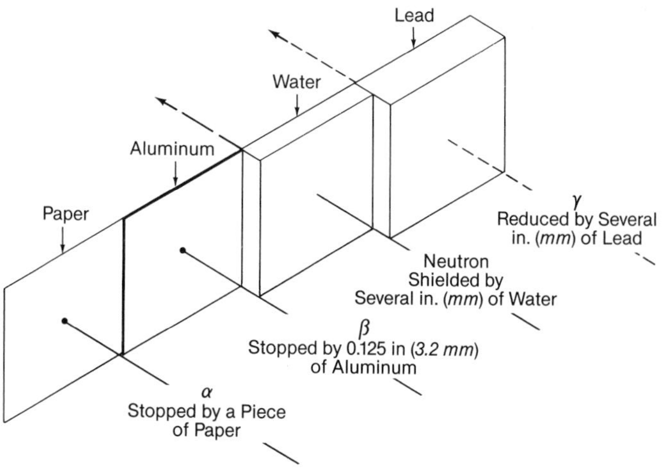

Fig. 14 Ranges of various types of radiations in materials.

is known as *as low as reasonably achievable* (ALARA). ALARA is introduced to balance the risk of radiation injury against the cost of further reduction in radiation exposure and/or the gains of increased productivity.

Shielding

The general public in the U.S. is protected by strict exposure limits established by the federal regulatory agency. Reactor building interiors and exteriors provide shielding to lower exposures incurred by both workers and the public.

Time, distance and shielding are the three fundamental principles of radiation protection. By simply spending less time in a radiation field, one can lower the exposure to radiation. Also, the more distance between a person and the source, the less exposure received. Radiation effects decrease with the square of the distance from the source. Exclusion fences surround nuclear plants to ensure that people can not enter the area without proper precautions.

Pre-job planning helps reduce the amount of time one needs to spend in a radiation field and also informs the workers of the higher radiation areas so they can remove themselves from that area to discuss the job. Radiation work permits (RWPs) are used for various jobs which require activity in a radioactive environment. RWPs describe the job to be completed, the requirements associated with the job, i.e., protective clothing, respirators, etc., the contamination levels and exposure rates that will be incurred and any specific precautions that the worker must take. (See Fig. 15.)

Materials of higher density and high atomic number, such as lead, uranium, thorium and gold, are best suited for γ shields. The use of some of these metals is limited by their high cost and/or weight. Metals such as iron, lead, chromium and nickel are more appropriate from a cost/weight standpoint.

To determine how much shielding is needed, half value and tenth value layers are used. A half value layer (HVL) is defined as that thickness of a material which reduces the radiation intensity to one half its initial value. A tenth value layer (TVL) reduces the intensity to one tenth of its initial value. The following chart illustrates half and tenth value layers for various materials:[8]

Lead (mm)		Concrete (cm)		Iron (cm)	
HVL	TVL	HVL	TVL	HVL	TVL
12	40	6	21	2	7

The shielding that surrounds a nuclear reactor must also moderate and absorb neutrons. For this purpose concrete containing a high volume of water is particularly effective. Fast neutrons escaping from the reactor are slowed down by the water and absorbed by the heavier nuclei in the concrete. Usually a γ ray is emitted in this process and there must be enough additional shielding to stop this secondary γ radiation. Water provides a good shield for research reactors operating at low power.

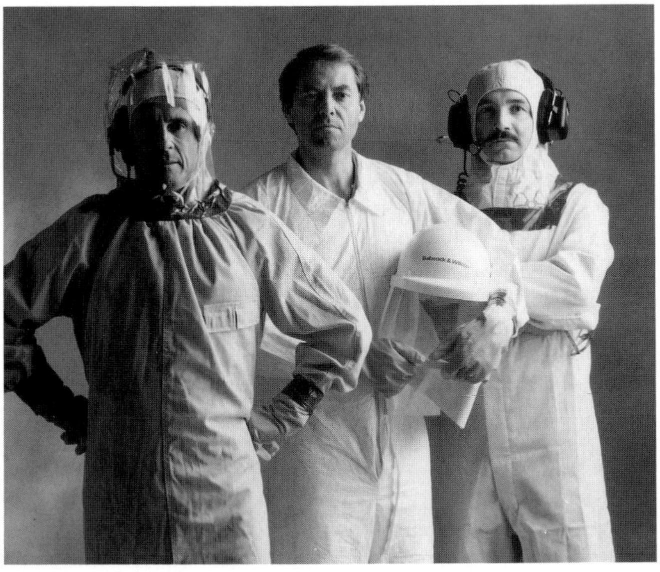

Fig. 15 Technicians in radiation protection garments.

Radiation dose

Health physics is especially concerned with the radiation dose received by workers and the general public. The term *dose* denotes the quantity of radiation absorbed in a specified mass of material.

For γ (or x-) radiation the unit of exposure is the *roentgen* which is defined as that radiation which in dry air at standard conditions of pressure and temperature produces ionization corresponding to 0.33×10^{-9} coulombs per cc. By definition, the roentgen's use extends only to measuring gamma (and x-) radiation, and only in air; therefore, additional measuring units are needed.

For measuring radiation absorbed dose, a unit called the *rad* is used. The rad measures the energy imparted to a specified mass of matter by ionizing radiation (1 rad = 100 ergs/g or 10^{-2} J/kg). It can be used to measure all types of radiation, and has a value approximately equal to the roentgen in air.

Response of biological systems to radiation energy absorbed varies with the type of radiation, and is also influenced by other factors such as the distribution of internally deposited nuclides. For radiation protection purposes, a quantity termed *dose equivalent* is used. To determine the dose equivalent, the dose in rads is multiplied by a quality factor. The quality factor is a proportionality factor that relates the biological effects to the actual incident radiation whether it be from α, β, γ or neutron radiation. The quality factor ranges from 1 for gamma and x-rays and low energy beta rays to 10 for fast neutrons, to 20 for alphas. The special unit of dose equivalent is the roentgen equivalent man (rem).

Federal regulations mandate that a radiation worker may not exceed 1.25 rem/calendar quarter or 5 rem/year. The general public is restricted to 10% of these limits. Effective January 1994, the limit for exposure to the general public will be reduced to 2% of the industry's levels. Exposures actually incurred at nuclear facilities are well below these limits. The industry average for higher dose job categories is 0.65 rem/year. Administrative limits imposed by the employer are typically set at 0.05 rem/day. Table 6 places the risk of such exposure into perspective.

Radiation detection and monitoring

Because no human sense can detect nuclear radiations, the health physicist has had to use special instruments and techniques for detecting and measuring radiation. In general, three types of checks are required as follows:

Table 6
Health Risks Impact on Life Expectancy[9]

Health Risk	Estimated Days of Life Expectancy Lost, Average
Smoking 20 cigarettes/day	2370
Overweight (by 20%)	985
All accidents combined	435
Auto accidents	200
Alcohol consumption (U.S. average)	130
Home accidents	95
Drowning	41
1 rem/year for 30 years	30
Natural background radiation	8
Medical diagnostic x-rays	6
All catastrophes	3.5
1 rem occupational radiation dose	1

1. Measurement of beta and gamma radiations by instruments such as the Geiger counter (Fig. 16) and *ionization chamber* (Fig. 17) provide instantaneous indications of the total level of β and γ. Most instruments are read as dose rate, usually in mr/h (10^{-3} roentgen/h).

   All persons in areas where radiation may be encountered normally wear a film badge or a thermoluminescent dosimeter (TLD). These devices are used for official records for exposure control. However, one's exposure can not be determined without evaluating the device with special processes or instrumentation. These devices measure the amount of β and γ radiation a person absorbs in terms of the total dose of radiation for a day, week or month.
2. Protection against neutrons is provided, primarily, by the shielding around nuclear reactors. When a reactor first goes into service, neutron counters as well as gamma counters are used to make a careful survey of the surrounding area. Should leakage exist, the shielding is modified. Thereafter, periodic checks are made for γ and neutron leakage. In experimental installations, or where neutron surveys indicate the need for continuous surveillance, special film badges and neutron dosimeters may also be used.

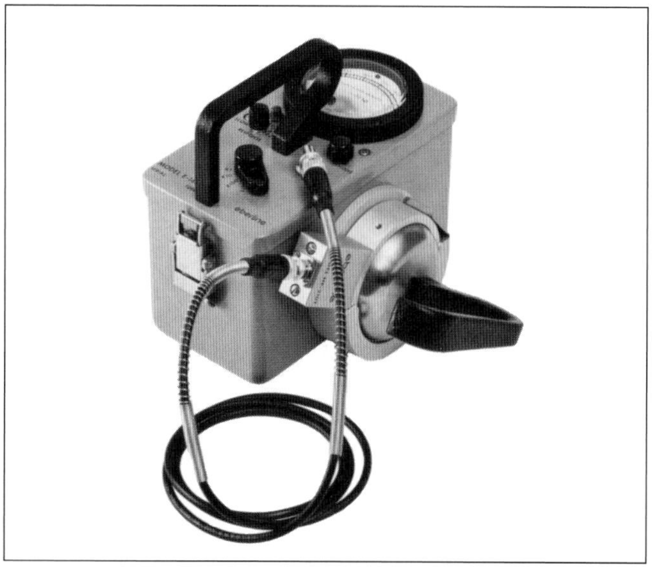

Fig. 16 Geiger counter radiation monitor.

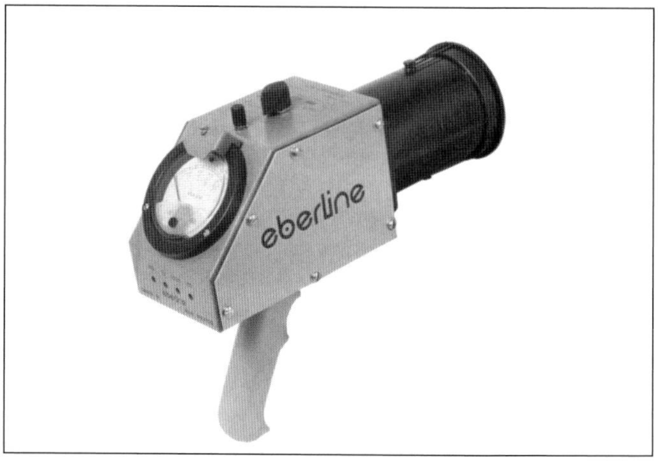

Fig. 17 Cutie Pie radiation monitor.

3. Keeping the concentration of alpha emitters in working spaces and in outgoing air below prescribed tolerances is sufficient to protect against α particles. Alpha counters monitor air quality in working spaces on a regular schedule. Samples of dust accumulating in various areas of a room are monitored periodically for alpha radiation. The air concentrations are then converted into internal exposures for the worker. Whole body counters and urinalysis can also help the health physicist ensure that the workers' internal exposures are well within the regulatory limits. Portable alpha *friskers* are also used to monitor hands, shoes and clothing of personnel leaving risk areas.

When high radioactive air concentrations occur, other precautions such as respirators, self-contained breathing air and glove boxes are used to protect the workers from internal exposures.

Nuclear plant and the environment

U.S. Nuclear Regulatory Commission regulations (Title 10 of the Code of Federal Regulations, Part 20) require the operator of a nuclear power plant to control and monitor radiation exposure of plant employees and to protect the population in the vicinity of the plant. The latter requirement makes it necessary to perform analyses on the air, water and vegetation around the plant. Also, all wastes leaving the plant must be analyzed to make sure that radioactive materials do not exceed maximum permissible concentrations.

Records also must be kept of personnel exposure, test results, radioactive waste discharges, and other items bearing on personnel exposure or environmental contamination. Receipt and shipment of all radioactive materials must be carefully controlled. (See Title 49, Code of Federal Regulations, Part 171-178, for complete requirements.)

All employees who work with radioactive materials or who may be exposed to radiation must be indoctrinated to enable them to protect themselves, other employees, and the general public in the performance of their duties.

The risks associated with nuclear power are much less than for many other industries. The safety regulations are also much more strict for radioactive materials than for other hazardous materials.

References

1. Evans, R.D., *The Atomic Nucleus*, 9th ed., McGraw-Hill, New York, 1955.

2. Lederer, C.M., Hollander, J.M., and Perlman, I., *Table of Isotopes*, 6th ed., Wiley, New York, 1967.

3. Lamarsh, J., *Introduction to Nuclear Reactor Theory*, Addison-Wesley, Reading, Massachusetts, 1977.

4. Duderstadt, J., and Hamilton, L., *Nuclear Reactor Analysis*, Wiley, New York, 1976.

5. Lamarsh, J., *Introduction to Nuclear Engineering*, Addison-Wesley, Reading, Massachusetts, 1977.

6. Bell, G., and Gladstone, S., *Nuclear Reactor Theory*, Van Nostrand, New York, 1970.

7. Moe, H.J., *et al.*, "Radiation Safety Training Technician Training Course," ANL-7291 Rev. 1, Argonne National Laboratory, Argonne, Illinois, 1972.

8. Cember, H., *Introduction to Health Physics*, Pergamon, New York, 1983.

9. "Regulatory Guide 8.29," U.S. Nuclear Regulatory Commission, Washington, D.C., 1981.

10. *Radiation — A Fact of Life*, American Nuclear Society, LaGrange Park, Illinois, 1981.

Reactor vessel in position between two nuclear steam generators during construction.

Chapter 50
Nuclear Steam Supply Systems and Safety

The nuclear steam supply system (NSSS) comprises the reactor core, primary system and steam generating equipment. The majority of United States commercial reactors are cooled and moderated by water and are therefore referred to as light water reactors (LWR). Most of these are pressurized water reactors (PWR); the remainder are boiling water reactors (BWR). The primary coolant is subcooled in a PWR and steam is generated by transferring energy from the primary system to a lower pressure fluid in the steam generators. Steam is produced in the reactor core of a BWR, eliminating the need for separate steam generators but resulting in contact between radioactive primary coolant and the turbine-generators.

The NSSS description that follows is focused on PWRs featuring once-through rather than recirculating steam generators. The once-through steam generator differs from the recirculating unit in that superheated steam is produced without steam separation equipment. The primary coolant traverses the steam generators in straight, vertical tubes rather than in the U-tubes which characterize the recirculating units.

Components

Reactor coolant system

The reactor coolant system (RCS) forms the primary side of the NSSS. The principal RCS components include the reactor vessel, steam generators, pumps, piping and pressurizer. The RCS is shown in Fig. 1.

Reactor vessel

Function and dimensions The reactor vessel, which houses the core, is the central component of the reactor coolant system. The vessel has a cylindrical shell with a spherical bottom head and a ring flange at the top. A closure head is bolted to this flange. (See Chapter 47.) The internals arrangement is shown in Figs. 2 and 3.

The reactor vessel is fabricated from carbon steel and clad with stainless steel. (See Chapter 52.) The vessel with closure head is almost 41 ft (12.5 m) tall and has an inside diameter of 171 in. (4343 mm). The minimum thicknesses of the shell wall and inside cladding are 8.4 and 0.125 in. (213 and 3.18 mm), respectively. Reflective metal (mirror) insulation is installed on the exterior surfaces of the reactor vessel.

Major internal components

The core is the primary component of the reactor vessel. (See Chapter 48.) The remaining major components are the plenum and the core support assemblies.

Plenum assembly The plenum assembly, located directly above the core, includes the plenum cover, upper grid, control rod guide tube assemblies and flanged plenum cylinder. The plenum cylinder has multiple openings for reactor coolant outlet flow. These flow openings are arranged to form the coolant profile leaving the core. The plenum cover is attached to the top flange of the plenum cylinder. It consists of a square lattice of parallel flat plates, a perforated top plate and a flange. The control rod guide tube assemblies are positioned by the upper grid. These assemblies shield the control rods from coolant crossflow and maintain the alignment of the rods.

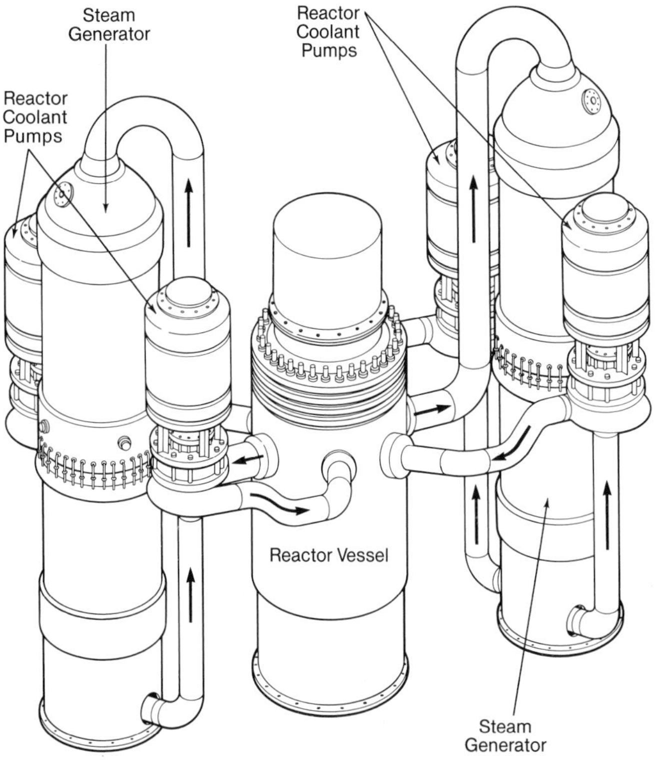

Fig. 1 Reactor coolant system — simplified arrangement.

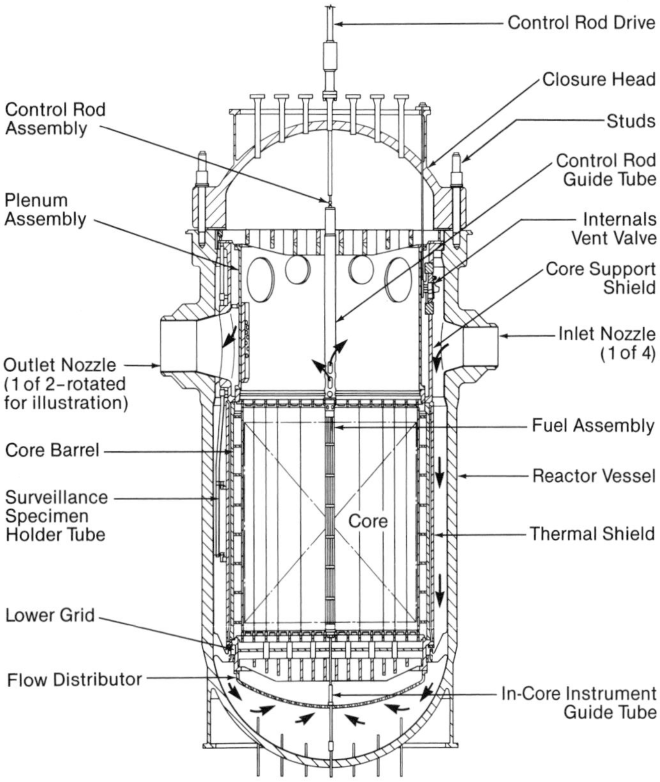

Fig. 2 Cross-sectional view of reactor vessel internals.

Core support assembly The core support assembly consists of the following components:

1. core support shield,
2. core barrel,
3. lower grid assembly,
4. flow distributor,
5. thermal shield,
6. surveillance specimen holder tubes,
7. in-core instrument guide tubes, and
8. internals vent valves.

The core support shield is a flanged cylinder which mates with the reactor vessel opening. The forged top flange rests on the circumferential ledge in the reactor vessel top closure flange, while the core barrel is bolted to the vessel's lower flange. The cylindrical wall of the core support shield has two nozzle openings. These openings form a seal with the reactor vessel outlet nozzles by the differential thermal expansion of stainless and carbon steel. The core support shield also has eight holes in which the internals vent valves are mounted.

The core barrel supports the fuel assemblies, lower grid, flow distributor and in-core instrument guide tubes. This cylinder is flanged at both ends. The upper flange is bolted to the core support shield assembly and the lower flange is bolted to the lower grid assembly. A series of horizontal spacers is bolted to the inside of the cylinder and a series of vertical plates is bolted to the inside surfaces of the horizontal spacers to form a wall enclosing the fuel assemblies. Coolant flows downward along the outside of the core barrel cylinder and upward through the fuel assemblies. A small portion of the coolant also flows upward in the space between the inside of the core barrel cylinder and the inner plate wall.

Core guide lugs are welded to the inside wall of the reactor vessel. In the event of a major internal component failure, these lugs limit the drop of the core barrel to 0.5 in. (12.7 mm) and prevent its rotation about the vertical axis.

The lower grid assembly, consisting of two lattice structures surrounded by a forged flanged cylinder, supports the fuel assemblies, thermal shield and flow distributor. The top flange is bolted to the lower flange of the core barrel. Pads bolted to the top surface of the upper lattice structure provide fuel assembly alignment. A perforated plate midway between the lattice structures aids the uniform distribution of coolant flow to the core.

The flow distributor is a perforated, dished head which is bolted to the bottom flange of the lower grid. The distributor supports the in-core instrument guide tubes and shapes the inlet flow to the core.

The thermal shield is a stainless steel cylinder located in the annulus between the core barrel and the reactor vessel. It is supported and positioned by the top flange of the lower grid. The thermal shield and intervening water annuli reduce the radiation exposure and internal heat generation of the reactor vessel wall by attenuating neutron and gamma radiation.

The surveillance specimen holder tubes are installed on the outer wall of the core support assembly, at approximately mid height of the core.

The in-core instrument guide tube assemblies guide the in-core assemblies between the penetrations in the bottom head of the reactor vessel and the instrument tubes in the fuel assemblies.

Eight co-planar internals vent valves are installed on the core support shield; they are 42 in. (1067 mm) above the centerlines of the reactor vessel inlet and outlet nozzles. The valve seats are inclined 5 deg (0.087 rad) from vertical and the valve discs naturally hang closed. During normal operation, the valves are forced closed by the differential pressure, approximately 43 psi (2.96 bar), between the outer annulus of the reactor vessel and the core outlet region. During a severe accident, such as a large break loss of coolant accident (LBLOCA), however, the differential pressure reverses and the internals vent valves open, permitting steam to be vented directly from the core region to the upper downcomer.

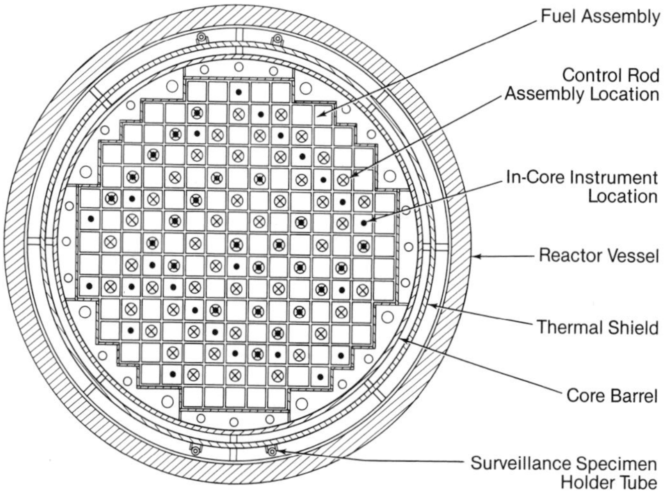

Fig. 3 Plan view of reactor vessel internals.

Flow paths

The four reactor coolant pumps can supply approximately 108% of the design total flow rate of 138 million lb/h (17,388 kg/s).

Approximately 90% of the coolant traverses the following path through the reactor vessel — into the vessel through the four inlet nozzles, one per reactor coolant pump; downward in the outer annulus, between the outside surfaces of the core support shield and thermal shield and the inside surface of the reactor vessel, to the bottom of the reactor vessel; upward through the flow distributor, lower grid and core fuel assemblies to the outlet plenum; and outward through the flow holes in the plenum cylinder to the two outlet nozzles, one per steam generator. Some of the flow traversing the core continues upward within the control rod guide tubes to the upper plenum, above the plenum assembly cover. This fluid rejoins the main flow through holes in the periphery of the plenum cover. Whereas approximately 90% of the coolant follows the paths just described, the remainder bypasses the primary heat transfer surfaces of the core. There are three major components of this bypass flow. First, a portion of the flow at the inlet nozzles proceeds directly to the outlet nozzles through residual clearances between the outlet nozzles of the reactor vessel and the nozzle openings of the core support shield. Second, a portion of the flow in the lower plenum is directed into the gap between the core barrel and the thermal shield, primarily to cool the thermal shield. Finally, bypass flow also occurs in the control rod and instrument guide tubes. The amount of bypass flow through these tubes varies with the configuration of the control components within the core.

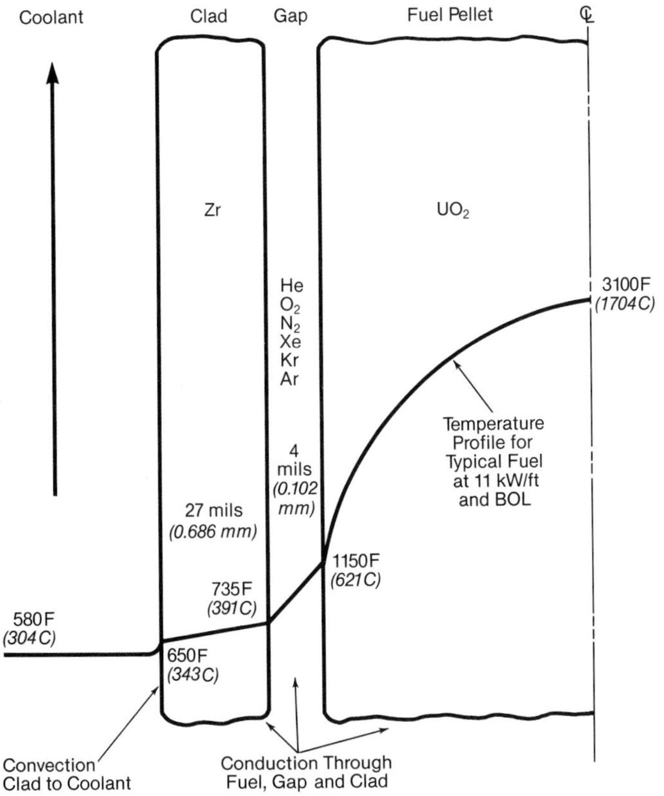

Fig. 4 Radial temperature profile for a fuel rod at 11 kW/ft.

Energy transport

Each fission of uranium-235 (U-235) produces approximately 200 MeV, predominantly in the form of kinetic energy of the fission products. This energy degrades to thermal energy of the fuel. The core heat transfer process involves the fuel, fuel-cladding gap, cladding and coolant. A typical temperature profile is shown in Fig. 4. The mode of heat transfer from cladding to coolant is subcooled forced convection. Film boiling is avoided by maintaining a DNBR above 1.3. DNBR is the ratio of local heat flux at the departure from nucleate boiling (DNB) to actual local heat flux. (See Chapter 5.) The coolant enters the core at approximately 557F (292C) and exits at 607F (319C), thereby establishing the sink temperature for fuel to coolant heat transfer. The cladding to coolant convective and cladding conductive heat transfer rates are relatively high, therefore the fuel temperatures primarily depend on the gap and fuel conductances and the rate of energy deposition. The heat transfer calculation methods are described in Chapters 4 and 5. The melting temperature of unirradiated uranium dioxide (UO_2) is 5080F (2804C). This temperature decreases with burnup at the rate of approximately 58F (32C) per 10,000 MWd/t_mU, due to the accumulation of fission products. A design overpower maximum fuel centerline temperature of 4700F (2593C) has been selected to preclude melting at the fuel centerline.

Instrumentation

The reactor vessel instrumentation consists of the nuclear instrumentation and in-core monitoring system.

Nuclear instrumentation Neutron flux levels are monitored using a combination of source, intermediate and power range detectors. Each of these detectors is located at core mid height, outside the reactor vessel but inside the primary shield. The distribution and types of detectors are:

Range	Detector Type	Number
Source	BF_3 Proportional	2
Intermediate	Compensated ion chamber	2
Power	Uncompensated ion chamber	4

These measurements are supplied to the reactor operator, safety parameter display system, reactor control portion of the integrated control system and reactor protection system. Additionally, the source and intermediate range readings are differentiated to provide startup rate, and the outputs of each pair of power range detectors are differenced to provide top to bottom flux imbalance.

In-core monitoring system Core performance is monitored using 52 in-core detector assemblies. These assemblies are arranged in a spiral fashion outward from the center of the core. Each assembly consists of seven local flux detectors, one background detector and one thermocouple, arranged as shown in Fig. 5. The seven local flux detectors, distributed over the axial core length, indicate the axial flux shape and provide fuel burnup information. The background detector indicates the integrated axial flux. These flux detectors are self-powered rhodium-103 instruments. The thermocouple indicates the exit coolant temperature.

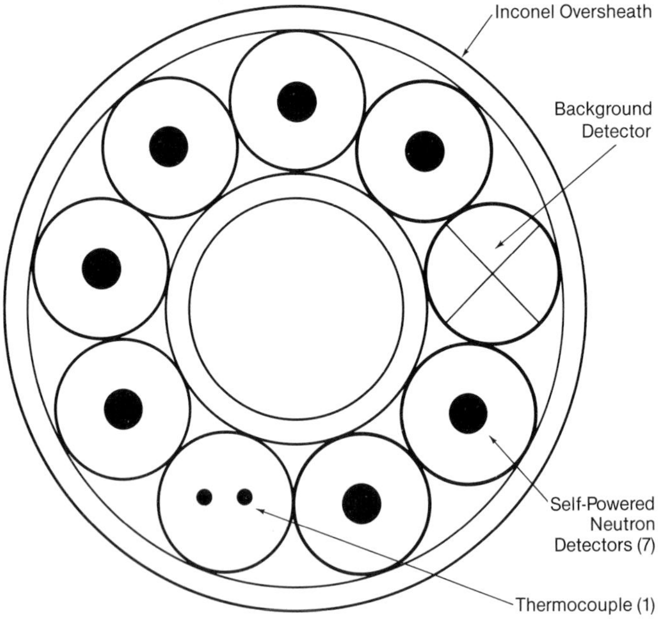

Fig. 5 Instrument tube cross-section.

Steam generators

Function and description

The steam generators transfer heat from the primary system to the secondary system. They function as heat sinks for the reactor core. The decrease of primary coolant temperature through the steam generators is virtually the same as the increase of primary coolant temperature through the core. Any difference is attributable to the fluid energy supplied by the reactor coolant pumps, less the heat losses to ambient. The placement of the steam generators in the reactor coolant system is shown in Fig. 1. The generators are nominally identical and function in parallel. Each is a vertical, straight tube and shell counterflow heat exchanger, approximately 68 ft (20.7 m) tall and 13 ft (3.96 m) in diameter. The more than 15,000 Inconel-600 tubes in each generator are positioned by broached tube support plates; these plates are spaced nonuniformly to minimize tube vibration by avoiding a single harmonic. The distance between tubesheets is 52.1 ft (15.9 m). Main feedwater is introduced into the annular downcomer through thirty-two 3 in. (76 mm) nozzles, 31.9 ft (9.7 m) above the lower tubesheet. Although main feedwater is normally used, an auxiliary feedwater system is also provided. The auxiliary feedwater nozzles are located high in the steam generator, less than 3 ft (0.91 m) below the upper tubesheet. An adjustable orifice is located near the bottom of the downcomer, just above the ports into the tubed region. This device consists of a movable plate overlaying a fixed plate; both plates contain 24 identical flow passages. The movable plate is adjusted to obtain stable hydrodynamic operation by altering the downcomer region pressure drop.

Flow paths

The steam generator internals are shown in Fig. 6. Primary coolant enters at the top, flows downward through the tubes and exits at the bottom through two primary outlets. The generator functions as a counterflow heat exchanger, i.e., the primary coolant flows downward

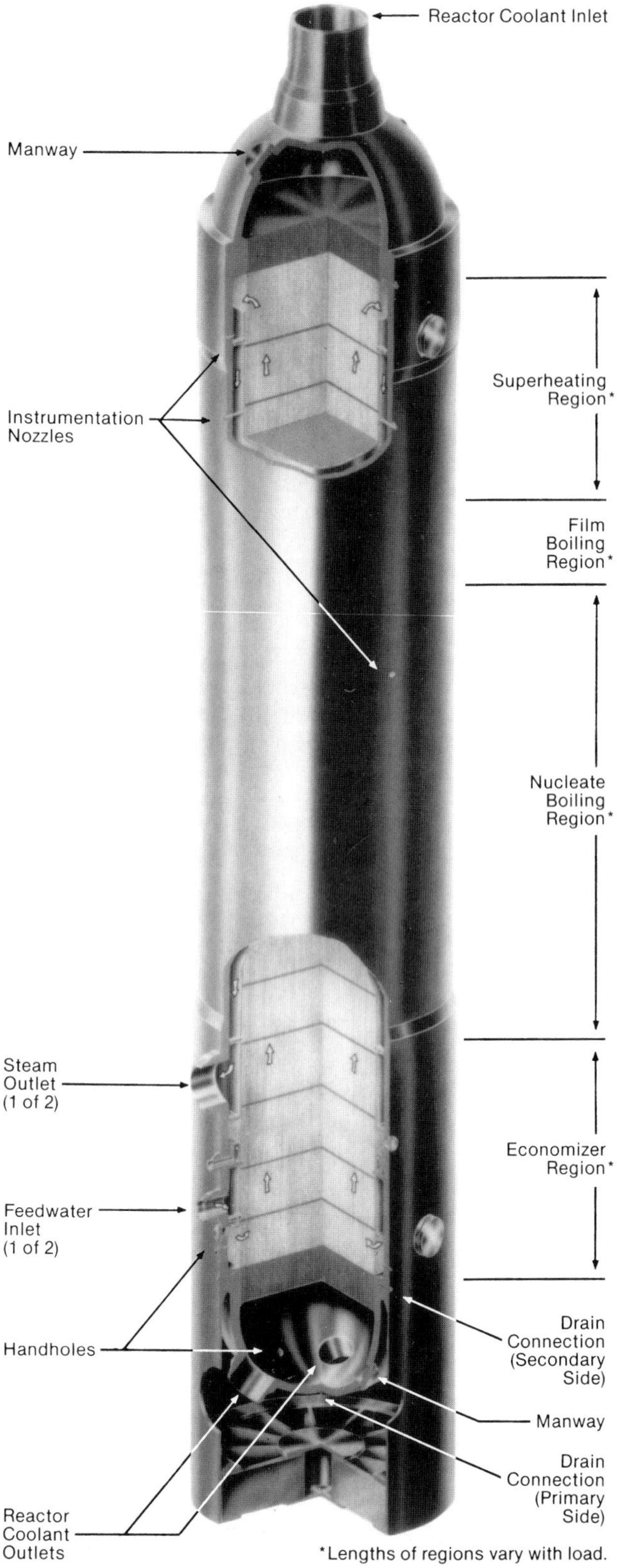

Fig. 6 Steam generator internals.

within the tubes and the secondary fluid flows upward around the tubes. The steam generator is a once-through design in that the secondary fluid makes but one pass through the unit rather than being recirculated as in a U-tube steam generator. (See Chapters 1 and 47.) The once-through steam generator (OTSG) produces superheated steam without mechanical steam separators or dryers. The full range of heat transfer modes is encountered. Feedwater is introduced into the steam generator just above mid height. The subcooled feedwater is heated to saturation by the introduction of tube-region steam. This steam is aspirated through a gap in the baffle surrounding the tubes. This aspiration is self-regulating, i.e., the condensation of steam on the subcooled feed reduces the pressure in the downcomer, drawing in sufficient steam to complete the preheating process. The preheating feed flows downward within the annular region formed by the lower baffle and steam generator inside wall. Near the bottom of the downcomer the feed traverses an adjustable orifice plate, then turns inward to pass through apertures in the bottom of the baffle. The fluid then flows upward around the tubes, passing through broached holes in the 14 tube support plates. Because of aspiration, the steam flow rate below the aspirator is approximately 110% of the rate leaving the steam generator. The steam is well superheated as it exits the top of the tubed region. It turns outward and downward, flowing down the upper annular downcomer to the steam outlet nozzles. (There is no direct flow path from the upper to the lower downcomer.) The preheating of the feed in the lower annular region and the downflow of exiting steam in the upper annular region heat the steam generator shell, minimizing the tube to shell temperature difference.

Modes of secondary side heat transfer

The secondary side feed entering the tubed region is at or near saturation. It is quickly brought to saturation by subcooled (forced) convection heat transfer as it begins to pass up the outside of the tube bundle. The remaining heat transfer modes include nucleate boiling, film boiling and superheating. Nucleate boiling involves the formation of discrete vapor bubbles on the tubes and the expulsion of these bubbles into the secondary coolant. This combination of phase change and bubble motion assists heat transfer; therefore, the nucleate boiling convection heat transfer coefficient is larger than those of the other two modes. (See Chapter 5.) The ratio of steam mass to total fluid mass, i.e., steam quality, increases as the secondary fluid flows upward. As the steam quality approaches 90%, there is insufficient liquid to wet the tubes and nucleate boiling gives way to film boiling, a less efficient heat transfer mechanism. Still farther up the generator, the moisture content of the fluid is depleted, the steam begins to superheat and heat transfer is by forced convection of vapor. This is the least efficient of the three modes. The primary and secondary fluid temperatures and the modes of secondary heat transfer are indicated in Figs. 7 and 8. The secondary temperature remains near saturation through the nucleate and film boiling regions. The counterflowing primary fluid gradually cools as it flows through the steam generator. Therefore, the primary to secondary temperature difference is largest near the top of the film boiling region. However, film boiling heat transfer is less efficient than nucleate boiling heat transfer. Therefore, the largest primary to secondary heat transfer rate occurs near the top of the nucleate boiling region. This is indicated by the rapid decrease of primary fluid temperature at this location. Due to the efficiency of nucleate boiling heat transfer, the temperature of the primary coolant exiting the bottom of the steam generator approaches the secondary saturation temperature. The secondary fluid temperature exceeds the secondary saturation temperature in the superheating region, decreasing the primary to secondary temperature difference. This decrease, coupled with the relatively inefficient secondary side heat transfer in this region, results in a relatively small primary to secondary heat transfer rate, as indicated by the small change of primary fluid temperature with elevation.

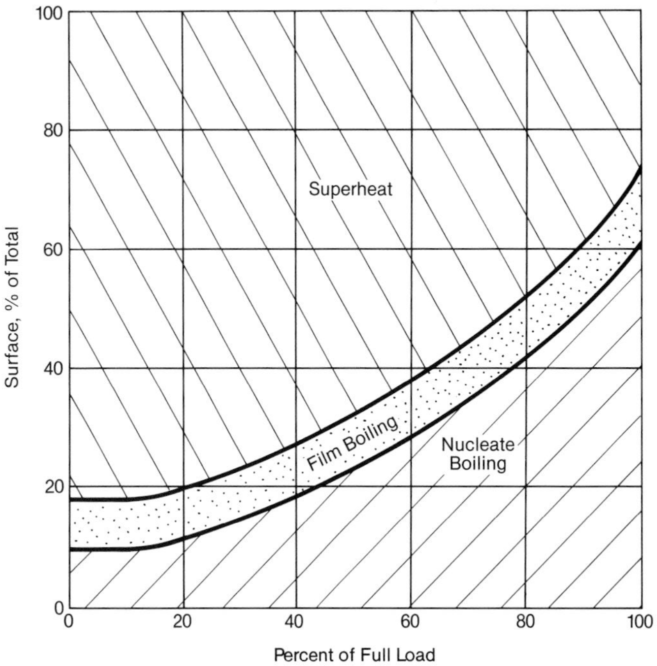

Fig. 7 Steam generator heat transfer.

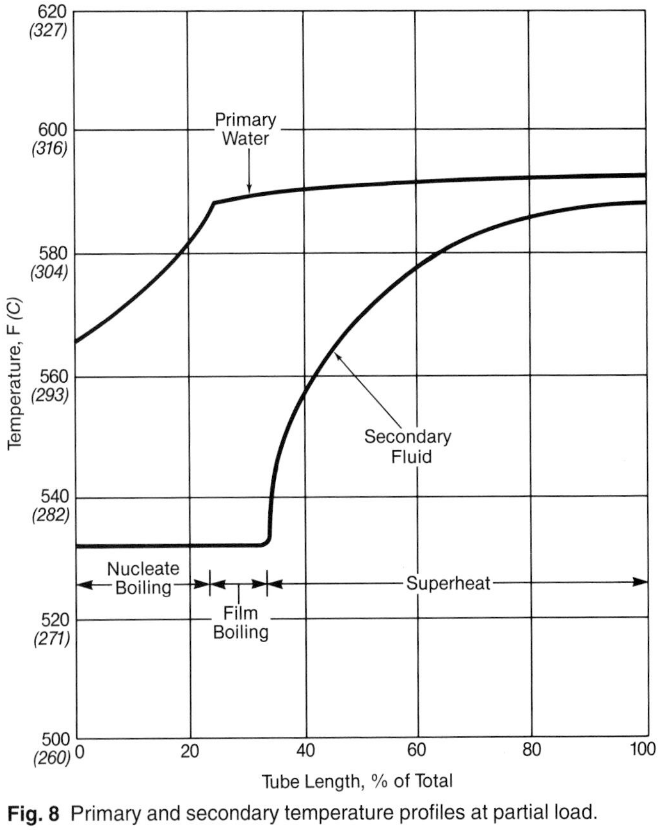

Fig. 8 Primary and secondary temperature profiles at partial load.

Heat transfer versus load

The factors affecting steam generator heat transfer include the primary to secondary temperature difference, the surface areas available for heat transfer and the overall heat transfer coefficient. Although the steam generator total heat transfer area is constant with load (steam demand), the areas influenced by the three modes of secondary heat transfer vary markedly with load. Likewise, the overall primary to secondary heat transfer coefficient versus elevation varies with the modes of secondary heat transfer. The primary to tube convective coefficient is almost constant with power level, as is tube wall conduction. The overall primary to secondary coefficient varies almost solely in response to changes of the secondary side heat transfer coefficient. At steady-state full power, the nucleate boiling mode of heat transfer is predominant. The nucleate boiling, film boiling and superheating lengths are approximately 60, 12 and 28%, respectively, of the total length between tubesheets [52.1 ft (15.9 m)]. At lower power levels, the throughput of steam is reduced and the rate of increase of steam quality increases with elevation, shortening the nucleate boiling length. Also, for a constant steam pressure at the turbine throttle, the requisite steam generator pressure is reduced at lower power levels due to the diminished steam line hydraulic losses. Therefore, steam generator saturation temperature is reduced, resulting in a higher primary to secondary temperature difference (with a constant primary average fluid temperature). For both reasons — lower steam throughput and higher primary to secondary temperature difference — the rate of increase of steam quality with elevation varies inversely with steam demand, therefore decreasing the nucleate boiling length. The film boiling length is nearly constant with steam demand, therefore the superheating length increases with reduced steam demand. The increased superheating length at reduced load results in a net increase of steam superheat. As load is reduced, the steam outlet temperature approaches the primary fluid inlet temperature.

Instrumentation

The steam generator measurements include pressure, temperature and level. Shell temperatures are measured at five equally spaced locations along the length of the generator. These measurements are used to determine tube to shell temperature differences. A steam generator downcomer temperature measurement indicates the temperature of the incoming feed; this temperature is used in the temperature compensation of level. There are three ranges of steam generator level measurement — full, operate and startup. The operate range level is temperature compensated; the others are not. The full range covers almost the entire 52.1 ft (15.9 m) between tubesheets. The startup and operate ranges, in contrast, extend to just above the aspirator port, 32.8 ft (10 m) above the lower tubesheet. The startup range extends from 0.5 ft (0.15 m) above the lower tubesheet, covering 32.3 ft (9.85 m). The operate range begins 8.5 ft (2.59 m) above the lower tubesheet. It is graduated in percents of full (operating) range, from 0% at 102 in. (2.6 m) above the lower tubesheet to 100% at a height of 394 in. (10.0 m). Whereas the common upper tap of the startup and operate range levels extends into the tubed region, their lower taps are in the steam generator lower downcomer. Feed and steam system measurements also indicate steam generator status. These include steam pressure and the pressure drop across the feedwater control valves. The flow rate of main feedwater is indicated in two ranges: startup to 15% of full flow range, and operate range. The majority of these steam generator measurements are supplied to the plant computer and the integrated central system.

Pumps and piping

Description and function Coolant is transported from the reactor vessel to the two steam generators through hot leg piping. The coolant is returned to the reactor vessel from the generators through four cold legs, two per generator. Fluid circulation is provided by four reactor coolant pumps, one per cold leg. The arrangement of the piping and pumps is shown in Fig. 1.

The hot leg and cold leg piping is fabricated from carbon steel and is clad on the inside with type 304 or 316 stainless steel. The inside diameters of the hot leg and cold leg piping are 36 and 28 in. (914 and 711 mm) and their nominal wall thicknesses are 2.75 and 2.5 in. (69.9 and 63.5 mm), respectively. The other piping attached to the RCS is either fabricated from stainless steel or clad with stainless steel.

Thermal sleeves are used to minimize the thermal stresses generated by rapid fluid temperature changes; they are installed on the four high pressure injection (HPI) nozzles and two core flood nozzles.

Pump characteristics The reactor coolant pumps are vertical suction, horizontal discharge, centrifugal units. They are further characterized as diffused flow, single stage, single suction, constant speed, vertical centrifugal, controlled leakage pumps having five-vane impellers. The pumps are driven by constant speed, vertical squirrel cage induction motors. Pump sealing is accomplished using three-stage mechanical seals. The pump and motor are just over 27 ft (8.2 m) tall. The motor develops approximately 9,000 hp (6714 kW) at cold conditions and 1,185 rpm (124 rad/s) using a 6600 volt, three-phase power supply. The locked rotor starting current is approximately 3,800 amperes. At normal operating conditions, the motor develops approximately 8,300 hp (6192 kW), drawing a current of 685 amperes. Each unit's rotational moment of inertia is in excess of 70,000 lbf ft^2 (28,928 Nm^2). This rotational inertia is sufficient to sustain rotation for approximately 45 seconds following a power interruption. Each reactor coolant pump has a nominal capacity of 92,400 GPM (5829 l/s) with a discharge head of 403.5 ft (122.99 m) at operating RCS conditions.

Each pump has a 28 in. (711 mm) bottom suction nozzle, a 41 in. (1041 mm) impeller diameter and a 28 in. (711 mm) side discharge nozzle. The motor is located above the pump and drives it through a vertical, rigidly coupled shaft. The arrangement of a pump is shown in Fig. 9. Each unit is equipped with an anti-rotation device that operates on a wedging principle and prevents reverse rotation of a de-energized pump.

Performance

With one pump inactive, the per pump flow developed by the three active units is larger than that with four pumps running primarily due to the decreased system flow rate and pressure drop. The pump adjacent to the inactive unit

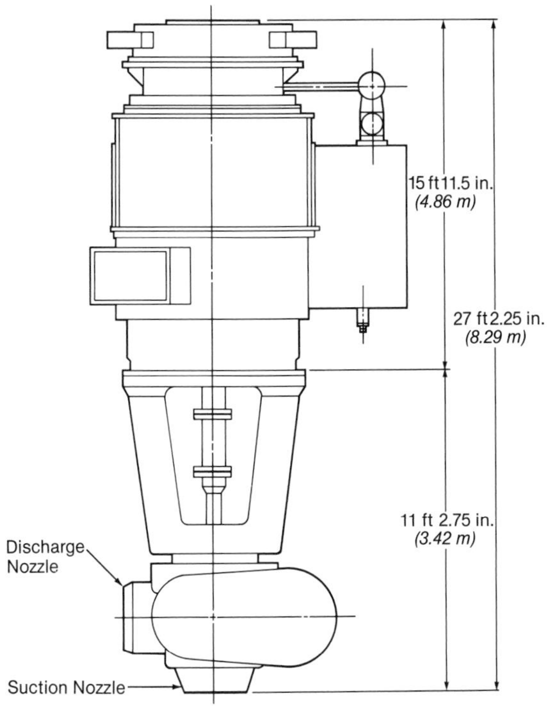

Fig. 9 Reactor coolant pump arrangement.

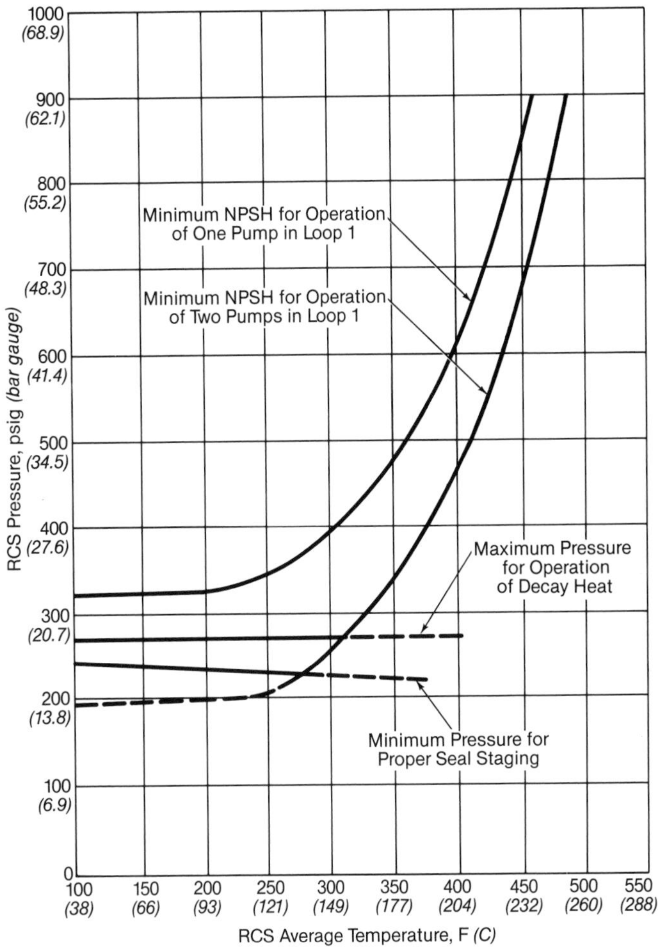

Fig. 10 Pressure and temperature limits for RCS pumps.

is most affected due to the backflow through the inactive pump. The increased flow produced by the active pumps nearly balances the backflow through the inactive unit, and the total core flow with three pumps operating is approximately three fourths of full flow. This approximate balance between increased output and backflow is maintained for various active pump combinations. The net positive suction head (NPSH), the minimum system pressure required for pump operation, is also the minimum pressure necessary to prevent the formation of vapor nuclei in the pump impeller, i.e., cavitation. The NPSH requirements are of greatest concern during reduced pressure operations, such as during plant warmup and cool down. These requirements set the P/T (pressure/temperature) limits for pump operation, as shown in Fig. 10.

Pressurizer

Description

The pressurizer (Fig. 11) is a tall, cylindrical tank connected at the bottom to a hot leg through 10 in. (254 mm) diameter surge line piping. Spray is introduced near the top of the pressurizer through a nozzle and 4 in. (102 mm) diameter line from a cold leg. Three replaceable heater bundles are installed over the lower portion of the pressurizer. Pressure relief devices are mounted at the top of the unit; these include two code safety valves and the power operated relief valve (PORV).

Function

The pressurizer controls primary system pressure. During normal operation, the pressurizer volume of 1500 ft^3 (42.5 m^3) contains equal portions of saturated liquid and saturated steam. Should the primary system liquid expand, such as through an increase of coolant temperature due to a load reduction, the excess volume displaces liquid into the surge line, raising the pressurizer level. This evolution is termed an *insurge*. The pressurizer steam is compressed, raising the primary system pressure and causing some of the steam to be condensed. The specific volume of liquid is much less than that of steam, therefore the steam condensation counteracts the ongoing pressure rise. The opposite occurs during an *outsurge*, the displacement of pressurizer liquid toward the hot leg through the surge line. In this case, the primary system pressure decreases, saturated pressurizer liquid flashes to steam and the net increase of specific volume through the change of phase suppresses the depressurization.

The pressurizer liquid volume is sized to maintain liquid above the pressurizer heaters and to maintain pressure above the HPI actuation point during the outsurge due to a reactor trip. The pressurizer steam volume is sized to retain a steam bubble during the insurge due to a turbine trip. Operation without a pressurizer steam bubble is termed solid plant operation. The pressurizer is sized to avoid this mode in which pressure control is encumbered.

Heaters

The pressurizer heaters maintain the pressurizer liquid at the saturation temperature, replacing the heat losses to ambient. They also raise and/or restore primary system pressure by elevating the saturation temperature of the pressurizer fluid. The three bundles of heaters are divided into five banks. There are two essential banks and three

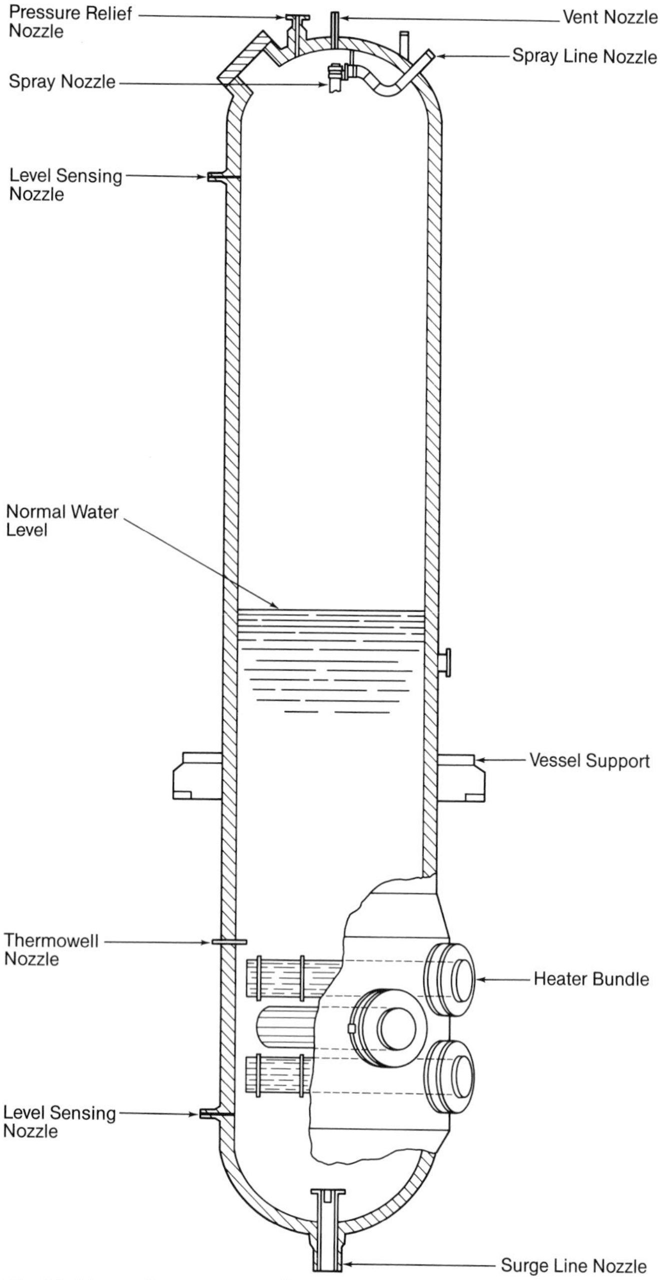

Fig. 11 Pressurizer components.

nonessential banks. The essential banks are energized automatically, based on primary system pressure. Essential bank #1 is energized below 2150 psig (148.2 bar gauge) and is subsequently de-energized above 2160 psig (148.9 bar gauge); the corresponding pressures for essential bank #2 are 2145 and 2155 psig (147.9 and 148.6 bar gauge). Nonessential bank #3 is sized to maintain the pressurizer fluid temperature during normal, steady-state operation. Banks #4 and #5 provide additional heater capacity which can be used during system startup or load changes.

Spray

The pressurizer spray system lowers primary system pressure and counters an increasing pressure transient. The actuation of pressurizer spray introduces 550F (288C) cold leg fluid into the pressurizer steam space. The resulting condensation decreases the pressurizer fluid volume, thereby decreasing pressure.

The spray enters the pressurizer through the spray nozzle. This nozzle imparts rotational motion to the fluid, generating a downward directed, hollow spray cone. This pattern enhances steam condensation efficiency while minimizing contact of the cold spray with the pressurizer vessel walls. Between the pressurizer wall and the spray nozzle, the spray line makes a bend in the vertical plane. This loop seal line configuration is designed to maintain liquid in the spray line, thereby alleviating the temperature changes experienced by the spray nozzle.

The spray line is attached to the cold leg piping just downstream of the reactor coolant pump. The spray is driven by the RCS fluid pressure drop across the reactor vessel, and the full spray flow rate at normal operating conditions is approximately 170 GPM (10.7 l/s). The spray flow control valve is normally operated in a modulated rather than on/off mode. When in automatic mode, the spray flow control valve opens 40% when the primary system pressure increases to 2200 psig (151.7 bar gauge) and closes when pressure decreases to 2150 psig (148.2 bar gauge). The 40% open setting permits a spray flow rate of approximately 90 GPM (5.68 l/s).

A continuous flow rate of approximately 1 GPM (0.063 l/s) is maintained to minimize the thermal transients of the spray nozzle; this is accomplished by a bypass line around the spray flow control valve. An auxiliary system provides pressurizer spray when the reactor coolant pumps are inactive, such as during decay heat removal cooling of the RCS.

Pressure control devices

Pressure increases are countered by pressurizer steam condensation through compression and by pressurizer spray actuation. Should pressure continue to rise, the PORV opens at 2450 psig (168.9 bar gauge). A pressure actuated solenoid operates the pilot valve, which causes the PORV to actuate. The PORV discharges into the pressurizer quench tank and can be isolated by closing the PORV block valve. A hypothetical continuous rod withdrawal accident from low power would cause core power generation to greatly exceed steam generator heat removal. The two code safety valves are sized to prevent RCS pressure from exceeding the 2750 psig (189.6 bar gauge) limit. These valves open at 2500 psig (172.4 bar gauge) and also discharge to the pressurizer quench tank.

Systems

Integrated control system

Function The integrated control system (ICS) simultaneously controls the reactor, steam generators and turbine to obtain a smooth, rapid response to load changes. The ICS operates the control rods to regulate core power, adjusts the feedwater flow to control the rate of steam production and operates the turbine throttle valves to control electrical power output. This control method combines the advantages of two alternative approaches.

The NSSS responds inherently to load changes without an ICS, but this response causes undesirable system variations. Consider the response to an increased load demand without integrated control. The increased electrical load causes the turbine throttle valves to open to maintain turbine speed. The increased steam demand lowers steam pressure at the turbine and within the steam lines and steam generators. Feed flow rate is increased to match the increased steam flow rate. The increased steam generator heat transfer lowers the temperature of the reactor coolant returning to the core. Core power gradually increases due to the reactivity gained from the negative temperature coefficient. This means that the increased reactor coolant density enhances the moderation of neutrons, thereby increasing the thermal neutrons available to cause thermal fissions. Also, the control rods are withdrawn to maintain the average primary fluid temperature. These primary system interactions would be reversed as the core and steam generator stabilized at the new, higher power level. This method of feedback control has the advantage of immediately supplying the required electrical load.

An extremely stable but slow method of NSSS control is termed the turbine-following method. The turbine output is varied only as the steam pressure is adjusted for the revised conditions. Again consider the response to an increased load demand. The control rods are withdrawn to increase reactor power and the feed flow rate is increased to support a higher rate of steam generation. As the turbine throttle pressure begins to increase, the throttle valves are gradually opened, allowing the turbine output to respond to the increased load demand.

The ICS operates all three systems — turbine, feedwater flow and control rods — to combine the rapid response of the unregulated system with the stability of the turbine-following system. The key to ICS operation is the turbine header pressure setpoint. The turbine governor valves maintain turbine speed and turbine header steam pressure. Because of the thermal inertia of the NSSS, the steam production rate can not be changed as rapidly as the steam demand varies in response to a turbine load change. This time delay is handled by temporarily changing the turbine header pressure setpoint. For example, again consider a load increase. The ICS causes the control rods to be withdrawn and the feedwater flow rate to be increased. Simultaneously, the ICS temporarily reduces the turbine header pressure setpoint. The turbine governor valves open to maintain turbine speed and to reduce the header steam pressure to the new setpoint. As core power, primary to secondary heat transfer and steam flow rate gradually increase, the turbine header pressure recovers and its set point is returned to the steady-state value. In this fashion, the ICS obtains a rapid response and a smooth transition between turbine loads.

Subsystems

The ICS contains the following independent subsystems:

1. unit load demand,
2. integrated master control,
3. feedwater control, and
4. reactor control.

The unit load demand (ULD) subsystem establishes the mode of ICS operation. In the manual mode, the operator sets the load at which the system is to operate. In the track mode, the generated power establishes load demand. The ULD subsystem also imposes load and load rate of change limits based on certain restrictive conditions, such as partial pump operation.

The integrated master control (IMC) subsystem processes the actual and set point load demand and turbine header pressures; it generates demand signals for the feedwater and reactor control subsystems. IMC signal processing is based on the steady-state relationship between steam flow rate and load demand. The IMC can also compensate for long term changes of this relationship, such as turbine-generator efficiency or feedwater flow measurement accuracy changes.

The feedwater control (FC) subsystem sets the feedwater flow rate to that demanded by the IMC. Control is applied to a feedwater control valve; the pressure drop across this valve is held constant by variable speed feed pumps. The FC subsystem controls feed to the two steam generators operating in parallel. The system splits the load between the steam generators to equalize their cold leg reactor coolant return temperatures; this control function is particularly significant with partial pump operation. The reactor control (RC) subsystem matches core power to the reactor demand signal supplied by the IMC.

Limits and controls

The ICS recognizes a variety of limiting conditions, such as a reactor coolant pump trip, a feed pump trip or an asymmetric control rod fault. It imposes the appropriate load and load change limits corresponding to these conditions.

Special ICS signals are imposed below 15% of full power. During system startup, the reactor and steam generator power levels are increased to approximately 10% power by steaming through the turbine bypass valves. The turbine-generator is rolled, synchronized with the electrical distribution system and gradually loaded. The turbine control station is placed in automatic mode at approximately 15% turbine load. The reactor is then controlled to maintain a constant average coolant temperature of 582F (306C) above 15% power, using the RC subsystem. From 0 to 15% power, manual control is used to increase the average coolant temperature from 532F to 582F (278 to 306C).

Reactor protection system

Function

The reactor protection system (RPS) trips the reactor when limiting conditions are approached. These limits prevent boiling of the reactor coolant, limit the local core power generation rate (the linear heat rate) and minimize challenges to the PORV.

RPS channels

The RPS operates on the control rod drive power. There are four identical RPS channels; if any two are tripped, the control rod drive breakers are opened, tripping the reactor. The unit can also be tripped manually from the control room. Any single RPS channel can be tested at power without tripping the reactor by actuating the key-operated bypass switch. The RPS then automatically transfers from two out of four to two out of three trip logic.

Trip functions

Each of the RPS channels contains ten trip functions, arranged in series. The trip of any function trips the associated RPS channel. The trip functions are as follows:

1. ARTS main feedwater pump trip,
2. ARTS turbine trip,
3. overpower trip,
4. high outlet temperature trip,
5. high pressure trip,
6. reactor building high pressure trip,
7. pressure/temperature trip,
8. low pressure trip,
9. power/imbalance/flow trip, and
10. power/reactor coolant pumps trip.

ARTS refers to the anticipatory reactor trip system. The two ARTS trip functions trip the reactor before RCS pressure rises to the reactor trip set point, thereby minimizing challenges to the PORV. The main feedwater pump ARTS actuates when both pumps trip, as signalled by their control oil pressure. The turbine ARTS actuates when the main turbine trips, as signalled by its auto-stop oil pressure. Both systems function in parallel with a core power (total flux) bistable. The core power bistable causes the ARTS to be bypassed (disabled) below 18% power and to be reinstated when power increases above 20%.

The power range flux detectors actuate the overpower trip at 104.9% of full power. The hot leg temperature measurements actuate the high reactor outlet temperature trip at 618F (326C). RCS pressure actuates the high pressure trip at 2355 psig (162.4 bar gauge) and the low pressure trip at 1900 psig (131 bar gauge). The reactor building high pressure trip actuates at 4 psig (0.28 bar gauge); each RPS channel is supplied an independent measurement of reactor building pressure. The pressure/temperature trip ensures that RCS pressure is sufficient for the current reactor hot leg outlet temperature. RCS pressure is compared to this minimum pressure. This pressure/temperature requirement, plus the reactor pressure and temperature limits, defines a region of acceptable operation. The power/imbalance/ flow trip uses the upper and lower power range flux detectors and the coolant flow measurements. The two flux measurements are summed to obtain power and differenced to indicate flux imbalance.

The power range flux detectors also supply the power/reactor coolant pump trip. A pump monitor determines the number of operating pumps and the loops in which they are located. The outputs of this monitor and total core power are compared in the power/reactor coolant pump trip bistable.

T_{sat} meters

Also associated with the RPS are two T_{sat} meters which determine the amount of subcooling of the hot leg fluid. The higher of two hot leg temperatures is subtracted from the saturation temperature, which is based on RCS pressure.

Shutdown bypass

A key-operated shutdown bypass switch can disable four of the ten trip functions. This bypass allows control rod drive testing during shutdown and at low RCS pressure, while retaining the remaining trip functions. When the shutdown bypass is actuated, a high RCS pressure trip at 1820 psig (125.5 bar gauge) is inserted into the trip logic; this precludes reactor operation with the shutdown bypass actuated.

Safety features actuation system

The safety features actuation system (SFAS) engages the emergency core cooling system (ECCS) in the event of a breach of the primary system boundary. The two SFAS signals are low RCS pressure and high reactor building pressure. The SFAS operates the following systems:

1. high pressure injection,
2. low pressure injection,
3. reactor building cooling, and
4. reactor building spray.

In addition, the SFAS activates the two emergency diesel generators.

Each of the three independent measurements of RCS pressure is fed to a trip bistable. This input is combined with the output of a reactor building pressure bistable; the trip of either pressure bistable trips the output. This output signal, combined with those of the other two pressure logic signals, engages the SFAS should two out of three trip. Reactor building spray is also initiated using two out of three logic, but is based only on reactor building pressure. In addition, the reactor building spray is delayed five minutes from the time of actuation.

Certain features are bypassed when the system is normally depressurized. These include high pressure injection, low pressure injection, reactor building isolation and reactor building cooling. Bypass must be initiated at an RCS pressure of less than 1850 psig (127.6 bar gauge); the system is automatically reinstated when the pressure exceeds 1850 psig (127.6 bar gauge).

High pressure injection system

The high pressure injection (HPI) system provides water to the RCS during a loss of coolant accident (LOCA). It is designed to prevent core uncovery in the event of a small RCS leak, during which the RCS pressure remains elevated, and to delay core uncovery in the event of an intermediate size break. The HPI system is activated by the SFAS signal when RCS pressure decreases to 1600 psig (110.3 bar gauge) or when reactor building pressure increases to 4 psig (0.28 bar gauge).

The HPI system consists of two pumps, a common discharge header and four lines, which are equipped with motor operated isolation valves leading to the four cold legs. The SFAS signals activate both pumps and their auxiliary equipment. The system also opens the four isolation valves to preset throttled positions to produce 125 GPM (7.89 l/s) per line using either of the two HPI pumps.

The HPI system is designed redundantly. Either pump is sufficient to meet the system requirements. The pumps are located in separate rooms and draw water from the borated water storage tank (BWST) through separate lines. The SFAS channels are physically separated. The HPI pump and valve power supply lines are independent, physically separated and energized from independent power sources. The SFAS also energizes the emergency diesel generators, which provide backup power to the HPI system.

HPI fluid enters each RCS cold leg midway in the down- sloping piping run at the reactor coolant pump discharge. The fluid enters horizontally at the side of the cold leg pipe. The HPI nozzles are equipped with sleeves to minimize the thermal stress caused by the cold water injection.

The HPI system can be used during normal operation as part of the makeup and purification system. Either HPI pump can supply RCS makeup and seal injection flow to the reactor coolant pumps; these functions are normally performed by the makeup pump.

Low pressure injection system

The low pressure injection (LPI) system provides water to the core in the event of a large LOCA which depressurizes the RCS. The two LPI pumps discharge 3000 GPM (189.2 l/s) each at a rated head of 350 ft (106.7 m). [Shutoff head is 435 ft (132.6 m).] The BWST supplies the LPI pumps. The pumps are actuated independently and automatically by the SFAS signal, as are the four motor operated isolation valves. They discharge to the reactor vessel upper downcomer through two core flood nozzles.

The two BWSTs have a capacity of 450,000 gal (1,703,430 l) each. The boron concentration is maintained above 1800 ppm at 80F (27C). The storage tank inventory provides at least 30 minutes of operation of all ECCS pumps. When the BWST inventory is depleted, the LPI pumps are manually transferred to the reactor building emergency sump, which collects water lost from the RCS. In the event that the HPI pumps are still required, such as with RCS pressure higher than the LPI pump shutoff head, the HPI pumps can be realigned to draw from the discharge of the LPI pumps. This is referred to as piggyback operation.

Core flood system

The core flood system provides a passive water supply to the core in the event of a large break loss of coolant accident. The system consists of two tanks, associated piping and valves. Each tank has a volume of 1410 ft^3 (39.9 m^3). Approximately two thirds of this volume, or 7500 gal (28,391 l), is borated water; the remainder is nitrogen gas pressurized to 600 psig (41.4 bar gauge). Should the RCS depressurize to less than 600 psig (41.4 bar gauge) in the event of an LBLOCA, the nitrogen cover gas forces the core flood tank (CFT) water into the RCS. The boron concentration of the CFT water is maintained between 2270 and 3490 ppm.

Each CFT discharges through 14 in. (356 mm) lines, dual check valves and a core flood nozzle into the upper downcomer of the reactor vessel. The two CFTs use separate piping and nozzles. Each tank is equipped with a motor operated isolation valve to prevent tank discharge when the RCS is normally depressurized. These valves are open when the RCS is pressurized.

Reactor building cooling and spray systems

The reactor building emergency cooling and spray systems are actuated in the event of a loss of coolant accident. The spray system also reduces the post accident level of fission products in the reactor building atmosphere through chemical reaction. The systems are designed such that either one can offset the heat released by the escaping reactor coolant.

The systems are activated by the SFAS signals. The emergency cooling system is actuated within 35 seconds after the RCS pressure decreases to 1600 psig (110.3 bar gauge) or the reactor building pressure increases to 4 psig (0.28 bar gauge). The emergency spray system is actuated five minutes after the reactor building pressure reaches 30 psig (2.07 bar gauge). Both systems can also be actuated manually.

The reactor building emergency cooling system includes four units, each consisting of an air circulator and a cooling coil. The 40,000 ft^3/min (18.9 m^3/s) circulators draw air from a point high within the reactor building dome and discharge downward toward the cooling units. Two of the four units are equipped with activated charcoal filters for removing fission products. The cooling units reject heat to the nuclear service cooling water system.

The reactor building emergency spray system includes two spray trains, each consisting of a pump, spray header, spray additive tank and eductor and associated isolation valves, piping and controls. The pumps are supplied from the BWST and, later in an accident, from the reactor building emergency sump.

The 300 hp (224 kW), 1500 GPM (94.6 l/s) spray pumps discharge through the spray additive eductors, drawing in sodium hydroxide. The spray additive concentration is sufficient to bring the entire post-accident inventory of reactor building water to a pH of 9.3. The spray discharge is distributed through 100 spray nozzles per header which are arranged to provide a uniform spray throughout the reactor building, above the operating floor.

Decay heat removal system

The RCS is periodically cooled and depressurized for maintenance and refueling. The first portion of the cooldown is accomplished by circulating the primary coolant using reactor coolant pumps and by removing heat using the steam generators. This method becomes impractical as the RCS pressure is reduced towards the NPSH of the reactor coolant pumps and as the primary to secondary system temperature difference diminishes.

The decay heat removal system (DHRS) is used to complete the RCS cooldown below 225 psig (15.5 bar gauge) and 290F (143C). The DHRS also performs the following functions:

1. purifies the reactor coolant during cold shutdown,
2. refills the RCS following maintenance,
3. cools and adds boron to the spent fuel pool, and
4. transfers water between the borated water storage tank and the fuel transfer canal during refueling.

The DHRS also functions as the low pressure injection system during a loss of coolant accident.

The DHRS is activated approximately six hours after reactor shutdown. It can reduce the RCS temperature from 280F to 140F (138 to 60C) within 14 hours. The DHRS pumps draw reactor coolant from the hot leg piping, just beyond the reactor vessel, through a 12 in. (305 mm) decay heat drop line. The DHRS fluid is discharged through coolers to the reactor vessel upper downcomer through the core flood nozzles. The injected DHRS fluid then follows the usual flow path down the reactor vessel downcomer, up through the core and out the hot leg to the DHRS inlet, thereby completing the flow circuit. The DHRS flow rate and rate of RCS cooling are controlled by bypassing a portion of the DHRS flow around coolers. The flow rate per cooler is limited to approximately 3000 GPM (189.2 l/s). The heat transferred in the coolers is removed by the nuclear service cooling water (NSCW) system, and the temperature difference between the RCS and the DHRS cooler outlet is maintained at approximately 30F (17C). The decay heat removal suction header temperature is measured to monitor RCS cooldown and the return header flow rates are also measured and remotely indicated. The DHRS suction block valve is interlocked to prevent inadvertent actuation at RCS pressures above 225 psig (15.5 bar gauge).

A portion of the DHRS flow rate can be diverted to the makeup and purification system for purifying the reactor coolant. This flow path is from the discharge of a DHRS pump; through the letdown filter, purification demineralizers and makeup filters; and back to a DHRS pump inlet header. The purification demineralizers are limited to a maximum fluid temperature of 135F (57C).

Makeup and purification system

The functions of the makeup and purification system are as follows:

1. control RCS inventory,
2. purify and degasify the reactor coolant,
3. maintain coolant boron concentration,
4. add chemicals to the reactor coolant for pH control,
5. supply seal injection flow to the reactor coolant pumps and handle seal return flow, and
6. add borated water to the core flood tanks.

The makeup and purification system can also function as part of the high pressure injection system.

The makeup and purification system consists of the letdown, purification and makeup portions. The letdown portion draws reactor coolant from the cold leg suction through a 2.5 in. (63.5 mm) line. The effluent is cooled using one of the three letdown coolers. Radioactive nitrogen-16 is allowed to decay in the letdown delay line, a 15 ft (4.6 m) length of 12 in. (305 mm) piping. Depressurization and flow control are provided by the letdown orifice; the nominal flow rate of 45 GPM (2.83 l/s) may be increased to 140 GPM (8.83 l/s) by bypassing the orifice.

The flow rate of 45 GPM (2.83 l/s) processes one RCS volume daily. Letdown flow rate, temperature and pressure are measured. Multiple motor operated valves are used to isolate the system in the event of SFAS operation.

The purification portion of the system consists primarily of the letdown filters, mixed bed demineralizers and makeup filters. The demineralizers are provided temperature protection by temperature actuated isolation valves. The purified effluent may be directed to the makeup tank or diverted to the bleed tanks when reducing the RCS fluid inventory.

The 4400 gal (16,657 l) makeup tank is central to the makeup and purification system. It receives purified letdown flow and seal return flow. It also serves as the point of boron and hydrogen addition to the RCS. Boron is used for reactivity control, lithium hydroxide provides pH control and hydrogen or hydrazine is used for oxygen control. The makeup tank acts as a surge tank to accommodate temporary changes of RCS inventory and provides water for the makeup pump. Finally, it can be used to degasify the reactor coolant.

A 4 in. (101.6 mm) line connects the makeup tank to its pump, with cross connects to the HPI pumps. The pump discharges are recombined and a 2.5 in. (63.5 mm) diameter flow line is routed back to the seal return coolers to cool the operating pumps. The makeup flow rate is measured and remotely indicated and is controlled by the pressurizer level control valve. Makeup is injected into the RCS through the cold leg HPI nozzle at the reactor coolant pump discharge, thereby completing the flow circuit.

During power operation, the makeup and purification system operates continuously to regulate RCS inventory. The makeup flow control valve is adjusted automatically to maintain pressurizer level. The flow rate of the makeup pump is the sum of the makeup flow rate (to the RCS), the seal injection flow rate to the reactor coolant pumps and the makeup pump recirculation flow rate. The net makeup system flow rate to the RCS is the sum of the makeup and the seal injection flow rates, less the seal return flow rate.

Abnormal transient operating guidelines

The abnormal transient operating guidelines (ATOG) represent a symptom-oriented response to plant transients and accidents. These guidelines provide the operator with a clear and effective method of correcting abnormal conditions. ATOG involves the identification and correction of key upset conditions, regardless of their cause. These actions alone are sufficient to ensure core covery and cooling and to ensure plant safety.

The three key symptoms of ATOG are loss of subcooling margin, inadequate heat transfer and excessive heat transfer.

The subcooling margin refers to the difference between the saturation temperature at RCS pressure and an RCS temperature. A positive subcooling margin ensures that the core is covered and therefore cooled. A loss of subcooling margin, on the other hand, indicates steam generation within the RCS and may indicate core uncovery. The operator must take action to restore the subcooling margin, such as by activating HPI pumps or a primary system heat removal process.

Subcooling margin is the key ATOG indicator. However, inadequate or excessive heat transfer indications must also be remedied. Inadequate heat transfer is countered by restoring primary to secondary system cooling and/or by initiating HPI-PORV cooling. In the latter method, core decay heat is removed by actuating full HPI and the PORV. The HPI fluid flows into the cold leg discharge piping leading to the reactor vessel. It flows through the core, out the hot leg, into the pressurizer through the surge line and out the PORV. Excessive heat transfer is remedied by reducing the rate of primary to secondary system heat transfer. This can be done by throttling the flow of auxiliary feedwater to the steam generators, restoring steam generator secondary pressure and reducing the primary to secondary temperature difference.

Exclusive among the events which can give rise to abnormal ATOG symptoms is the steam generator tube rupture (SGTR) accident. This event is readily identified using radiation indications and alarms and indications of steam generator conditions — pressure, temperature and level. Event oriented operator actions are taken to minimize radiation release and to ensure core cooling.

Evolution of reactor safety analyses

The accidents at Three Mile Island Unit 2 (TMI2) in Harrisburg, Pennsylvania and Chernobyl in the former Soviet Union have profoundly affected the perception of reactor safety. Both accidents occurred at reactors and involved extensive operator actions; the similarities end there. The TMI2 accident involved the gradual reduction of RCS fluid inventory after the reactor had been shut down. The core was extensively damaged but the primary coolant system boundary and the containment building held almost all of the radionuclide inventory. Chernobyl, on the other hand, involved an uncontrolled, prompt critical reaction leading to the disruption of the core, primary system and containment, with the resulting extensive contamination. The lessons of Chernobyl have little direct bearing on U.S. commercial PWRs.

The lessons of TMI2 have generated positive steps toward improved reactor safety. The industry has implemented an aggressive program to increase operator and supervisor training. Furthermore, the exchange of technical information among utilities has increased.

More significantly, attention has been focused on system performance and the optimum response to realistic, rather than overly conservative, scenarios. At least in part due to this refocusing, the Nuclear Regulatory Commission (NRC) has authorized best estimate methods of LOCA analysis. Integral systems testing has also been performed to refine the analysis and understanding of the small break LOCA.

The sequence of events at TMI2 and Chernobyl are outlined below. Next, the best estimate method of analysis is discussed. The major system interactions occurring during a small break loss of coolant accident (SBLOCA) are then discussed.

Three Mile Island 2

The TMI2 accident has been extensively documented; comprehensive descriptions are provided in the Kemeny report[1] and Nuclear Safety Analysis Center (NSAC) reports.[2]

On the morning of Wednesday, March 28, 1979, maintenance was being performed on a condensate polisher at TMI2. A mixture of air and water was being used to clear a resin transfer line. On previous occasions, similar maintenance had caused water to leak into the air lines which control the polisher valve positions. The leak may have triggered the condensate pump trip which resulted in the feed pump trip shortly after 0400 (4:00 a.m.).

Unit 2 had been operating at 97% of full power. Within seconds of the feed pump trips, the turbine tripped, the PORV actuated and the reactor tripped due to high pressure. The main turbine tripped automatically on the loss of feedwater. The abrupt reduction of heat removal from the RCS caused it to heat and pressurize, leading to the PORV actuation and reactor trip.

Three emergency feedwater pumps started automatically upon the loss of main feedwater, but they were isolated from the steam generators. This isolation may have occurred during testing of the emergency feedwater pumps two days earlier. This situation was detected and corrected at approximately 0408, several minutes after the steam generators had boiled dry. The eight minute absence of feed was not a major contributor to the accident, but it did complicate plant control.

The RCS rapidly depressurized with the reactor tripped and the PORV open. RCS pressure decreased to the PORV reclosure setpoint within ten seconds of the PORV's actuation. The control room panel light indicating PORV actuation extinguished, but the PORV remained open. It remained open and unblocked for two hours and 22 minutes, during which one third of the RCS fluid volume was discharged through the valve. The PORV drain line temperature remained elevated, but the significance of this indication may have been masked by pre-accident high temperatures due to leaking valves. The continuing PORV discharge filled the drain tank, then overflowed to the reactor building sump. The reactor coolant drain tank relief valve opened at 0403 and the drain tank high temperature alarm was received one minute later. A containment building sump high water level alarm actuated at 0411 and the drain tank rupture disc burst at 0415. The containment building temperature and pressure began to rise and, in response, the operator actuated the fans and coolers. A portion of the sump contents was pumped to the auxiliary building.

Approximately two minutes after the reactor trip and PORV actuation, the rapidly decreasing RCS pressure tripped the SFAS. The HPI pumps actuated, but the operators tripped one of them and throttled the other, apparently to avoid overfilling the RCS. The continuing reduction of RCS pressure, primarily caused by the discharge through the PORV, was lowering the RCS saturation temperature toward the RCS fluid temperature. The RCS fluid attained saturation at approximately 0406. The vapor generation within the RCS expanded the fluid volume which, coupled with the open PORV, caused the pressurizer level to continue to rise. The operator increased the letdown flow rate in an attempt to control pressurizer level and thereby increased the loss rate of primary system inventory.

The generation of voids within the RCS caused the reactor coolant pumps to vibrate while the indicated loop flow rates gradually decreased. Concern for the integrity of the pumps and the loop piping ultimately led to the deactivation of two reactor coolant pumps at 0514; the second two pumps were shut down 27 minutes later.

Voids within the RCS also yielded abnormal indications of core neutron flux; the flux detectors had been calibrated with liquid in the downcomer and core. The two-phase mixture, consisting of an increasing void fraction, resulted in reduced fluid density, reduced flux detector shielding and therefore increased measured neutron flux levels. These indications triggered a manual scram of the reactor at 0423, although the unit had already been tripped. Emergency boration of the RCS was started at 0640.

The termination of forced RCS flow allowed the two-phase mixture to segregate by density, rather than to flow as a mixture. The liquid-vapor interface reached the top of the core near 0600, two hours after the accident began. Fuel cladding began to be breached, releasing gaseous fission products to the RCS and to the reactor building through the PORV. The containment radiation levels began to rise, a site emergency was declared at 0700 and a general emergency was issued at 0724. As core uncovery progressed, the RCS steam temperatures rose. Multiple core outlet temperature alarms sounded and the zirconium cladding reacted with the superheated steam, evolving hydrogen. Some of the hydrogen gas migrated to the containment building, culminating in a hydrogen ignition. This produced a peak pressure of 28 psig (1.93 bar gauge), recorded in the building at 1350.

HPI pumps were started with a high flow rate at 0826, finally recovering the core some two hours later. The core was subsequently uncovered, however, in an attempt to depressurize the RCS; this lasted from 1138 to 1508. Subsequent notable events included intermittent releases of radioactive gases and the several day concern with the RCS gas bubble, which was perceived to be explosive.

Chernobyl

Chernobyl Unit 4 was one of fifteen Soviet reactors of the RBMK-1000 class. It was a heterogeneous, thermal-neutron, pressure-tube reactor and generated 3200 MW_t. It was cooled by boiling light water and moderated using graphite. The 7 m x 11.8 m diameter core contained 1661 pressure tubes.

Three characteristics of an RBMK reactor are noteworthy: complex reactivity control, a positive void coefficient of reactivity, and relatively weak containment. The neutronic coupling of the core required local as well as core-wide reactivity control. To this end, the twenty-four automatic control rods were divided into two groups of twelve rods each — one group for local control, the other for average-power control. The twelve local control rods, located in twelve core zones, were each under automatic control based on twenty-four in-core fission chambers.

A positive void coefficient of reactivity causes the neutron population to increase with the rate of steam production. Moreover, this void coefficient was largest at the lowest power levels. This tendency, plus the inaccuracy of low-power measurements, prohibited reliable continuous operation below 700 MW_t.

A safety test was to be performed during a planned maintenance shutdown. The test involved briefly powering an electrically driven feed pump from the turbogenerator as it coasted down following reactor trip; this test had been tried unsuccessfully in 1982 and 1984. The same procedure was now to be used, except that the operators planned to avoid tripping the reactor so that retests could be performed. Power was to be reduced gradually to the minimum sustainable level of 700 to 1000 MW_t.

The power reduction was begun at 0100 on April 25, 1986. Power was to be gradually reduced to permit xenon equilibration. Xenon is a neutron-absorbing product of iodine decay. A power level stabilization allows the production of iodine by fission; its decay to xenon is then balanced by xenon burnup through neutron absorption. (See Chapter 49.)

At 1305, the electrical dispatcher halted power reduction at 50% power for approximately ten hours. At 1400 the operators blocked the ECCS to avoid spurious triggering. The power reduction was resumed at 2310 but progressed much more rapidly than had been intended. The negative reactivity, aided by the xenon imbalance, caused power to decrease to 30 MW_t before being stabilized at approximately 200 MW_t (still far below the minimum permissible level of 700 MW_t). Stabilization was achieved by withdrawing most of the control rods. Although the flow rates and pump inlet temperatures violated the pump start requirements, two additional circulating pumps were started according to the test plan. The decreased voiding prompted further withdrawal of control rods. Now reactivity was being controlled primarily by coolant voiding, and an increased volume of core coolant was near saturation. Upon initiation of the flow coastdown test, the increased voiding added positive reactivity. As the power began to rise slowly, the reactor was manually scrammed. Shortly thereafter, shocks were felt, the rod insertion was interrupted and there were two successive explosions. The flow coastdown had resulted in a super-prompt-critical power excursion, fuel disintegration, and a steam explosion due to the fuel-coolant interaction which destroyed the primary circuit and the containment.

Best estimate analyses

The prescribed analyses of LOCAs underwent a major revision in September, 1988. The previously mandated Appendix K methods can be used, or a realistic evaluation model may be used. This change was made to CFR 50.46 and Appendix K. The pertinent models, correlations, data and model evaluation procedures of the realistic method are described in the USNRC Regulatory Guide 1.157 of May, 1989 (reference 3). Application of this methodology is demonstrated in NUREG/CR-5249, published in December 1989, as follows:

The criteria set forth in 50.46, Acceptance Criteria for Emergency Core Cooling Systems for Light-Water Nuclear Power Reactors, *and the calculational methods specified in Appendix K, were promulgated in January 1974 after extensive rulemaking hearings and were based on the understanding of ECCS performance available at that time. In the years following the promulgation of those rules the NRC, the nuclear industry and several foreign institutions have conducted an extensive program of research that has greatly improved the understanding of ECCS performance during a postulated loss of coolant accident. The methods specified in Appendix K were found to be highly conservative; that is, the fuel cladding temperatures expected during a loss of coolant accident would be much lower than the temperatures calculated using Appendix K methods. In addition to showing that Appendix K is conservative, the ECCS research provided information that allows for quantification of that conservatism. The results of experiments, computer code development and code assessment allow more accurate calculations, along with reasonable estimates of uncertainty, of ECCS performance during a postulated loss of coolant accident than is possible using the Appendix K procedures.*

It was also found that some plants were being restricted in operating flexibility by limits resulting from conservative Appendix K requirements. Based on the research performed, it was determined that these restrictions could be relaxed through the use of more realistic calculations without adversely affecting safety. The Appendix K requirements tended to divert both NRC and industry re-

sources from matters that are relevant to reactor safety to analyses with assumptions known to be nonphysical . . .

The current revision of 50.46 permits ECCS evaluation models to be fully best estimate *and removes the arbitrary conservatisms contained in the required features of Appendix K for those licensees wishing to use these improved methods. Safety is best served when decisions concerning the limits within which nuclear reactors are permitted to operate are based upon realistic calculations. This approach is currently being used in the resolution of almost all reactor safety issues (e.g., anticipated transients without scram, pressurized thermal shock, and operator guidelines) and is now available for one of the last remaining major issues still treated in a prescriptive manner, the loss of coolant accident . . .*[3]

SBLOCA integral systems tests

SBLOCA events and interactions have been studied extensively using the Multiple Loop Integral Systems Test (MIST) facility (reference 5). MIST evaluations were jointly sponsored by the Babcock & Wilcox Owners Group, the NRC, the Electric Power Research Institute and Babcock & Wilcox. MIST operated at plant fluid conditions and included features of the plant pertinent to SBLOCA. It was extensively instrumented to generate test data for computer code benchmarking and facilitated analysis of the integral system events. The MIST results have been extensively documented in the MIST Final Report. A SBLOCA is characterized by a relatively high primary system pressure compared to that obtained with a larger break size. The challenge of a SBLOCA is therefore to depressurize the RCS while maintaining sufficient primary fluid inventory to guarantee core cooling. The MIST tests have demonstrated several key thermal-hydraulic interactions which determine the course of the post-SBLOCA events. These interactions, which are described below, are as follows:

1. boiler condenser mode (BCM),
2. auxiliary feedwater BCM (AFW-BCM),
3. pool BCM,
4. HPI condensation,
5. HPI-leak cooling,
6. HPI-power operated relief valve (HPI-PORV) cooling,
7. surge line uncovery, and
8. intermediate levels.

Boiler condenser mode

BCM refers to the condensation of primary system vapor within a steam generator. (See Fig. 12.) It is so named because the core acts like a boiler while the steam generator acts like a condenser. BCM is important because it is a method of transferring heat from the primary system to the steam generator secondaries, thereby permitting the depressurization of the primary system when it has been partially voided.

Auxiliary feedwater BCM

In AFW-BCM, the primary system vapor is condensed high in the steam generator due to cooling produced by the injection of highly subcooled auxiliary feedwater. (See Fig. 13.) The primary system level need only be below the elevation of the upper tubesheet, but the auxiliary feedwater system must be active.

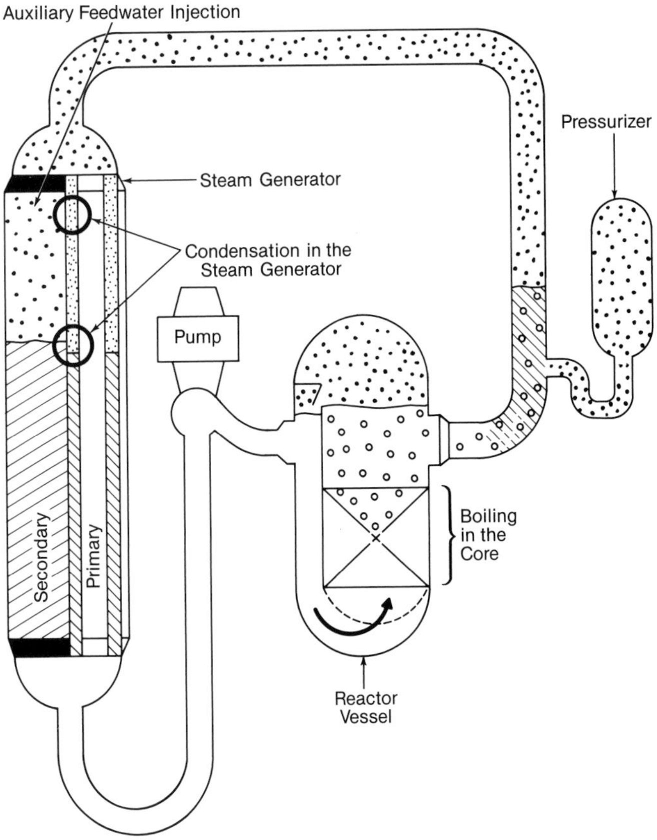

Fig. 12 Boiler condenser mode schematic.

The injection of AFW increases the steam generator secondary side liquid inventory. When the secondary side level has been sufficiently raised, the injection of AFW is halted and the AFW-BCM ceases.

Pool BCM

Pool BCM occurs when the level on the primary side of the steam generator drops below the top of the secondary side pool. (See Fig. 14.) Pool refers to the liquid inventory of the steam generator secondary system. The pool liquid is usually saturated at the secondary side conditions and is quite subcooled compared to the primary side. The primary system vapor condenses in the region where the primary and secondary levels overlap. AFW need not be active to obtain pool BCM.

Pool BCM, like AFW-BCM, is transitory. The primary to secondary heat transfer through pool BCM reduces the local vapor pressure. The steam generator primary side level quickly increases, often at the expense of the primary side level in the alternate steam generator. Also, the reduction of primary system pressure promotes primary system refill. The resulting increase of the steam generator primary side level reduces the level overlap, decreasing the pool BCM activity. However, a pool BCM in one steam generator often causes an AFW-BCM in that unit and/or BCM activity in the alternate generator.

HPI condensation

Steam, produced in the core, flows upward into the outlet plenum, outward through the reactor vessel vent valves and down the reactor vessel downcomer. Subcooled HPI fluid is injected into the down-sloping section of each cold leg and condenses the steam in the down-

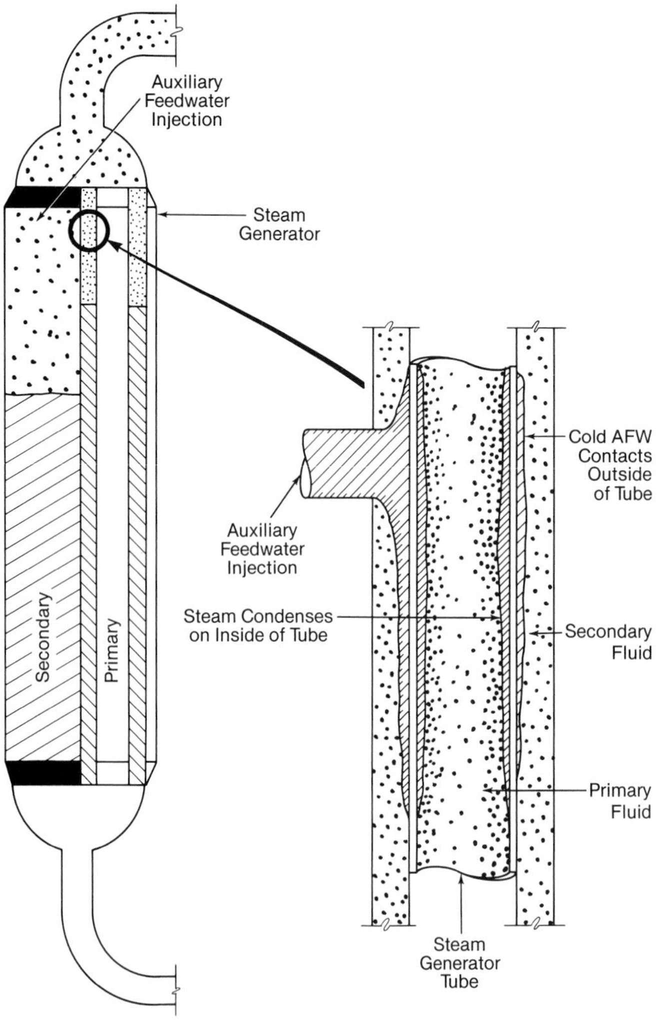

Fig. 13 Auxiliary feedwater boiler condenser mode.

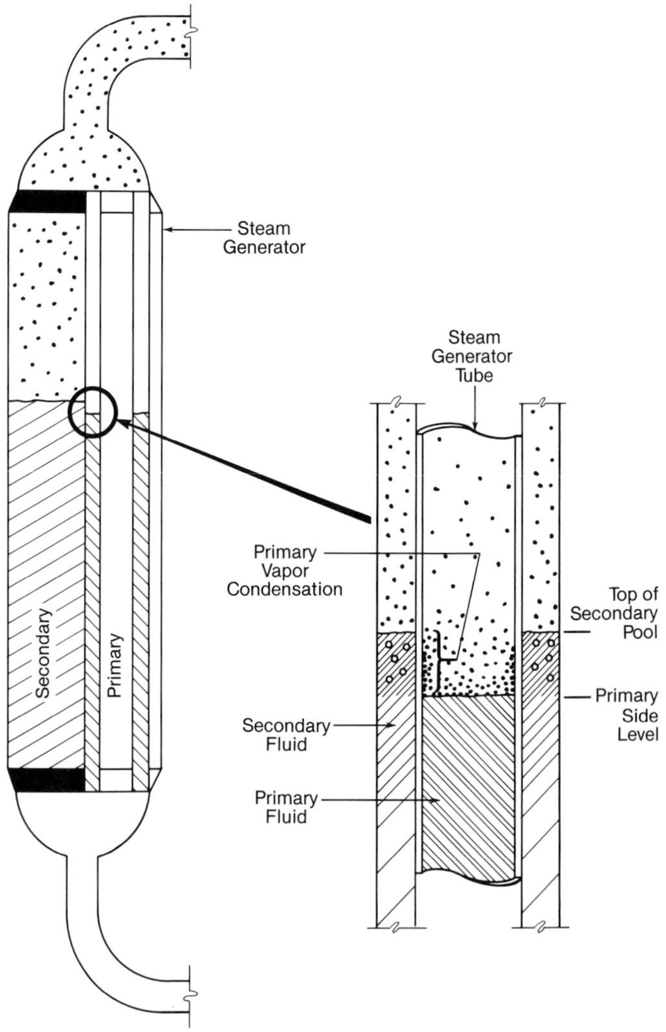

Fig. 14 Pool boiler condenser mode.

comer. This condensation depressurizes the primary system and promotes HPI-leak cooling. (See Fig. 15.)

HPI-leak cooling

HPI-leak cooling is the dominant method of primary system heat removal following an SBLOCA. It is often the sole method of core cooling after loop flow has been interrupted. HPI-leak cooling is commonly associated with HPI condensation of steam in the downcomer. This condensation heats some of the HPI fluid, which then rises. The fluid flows upward along the top of the downsloping cold leg discharge piping. (See Fig. 16.) At the same time, the cooler HPI fluid flows to the downcomer along the bottom of the cold leg discharge piping. This counterflow supplies heated fluid to the leak site, even if it is located in the cold leg suction piping. The injection of cold HPI fluid combined with the discharge of heated leak fluid is an effective method of primary system heat removal.

HPI-PORV cooling

HPI-PORV cooling is an alternate method of cooling the core. The two key components are the high pressure injection pumps and the PORV. HPI water is directed to all four cold legs, and the PORV, which is located at the top of the pressurizer, is kept open. The injection fluid follows the normal path through the downcomer and up to the core, where it is heated. The fluid then flows up the hot leg to the pressurizer surge line connection,

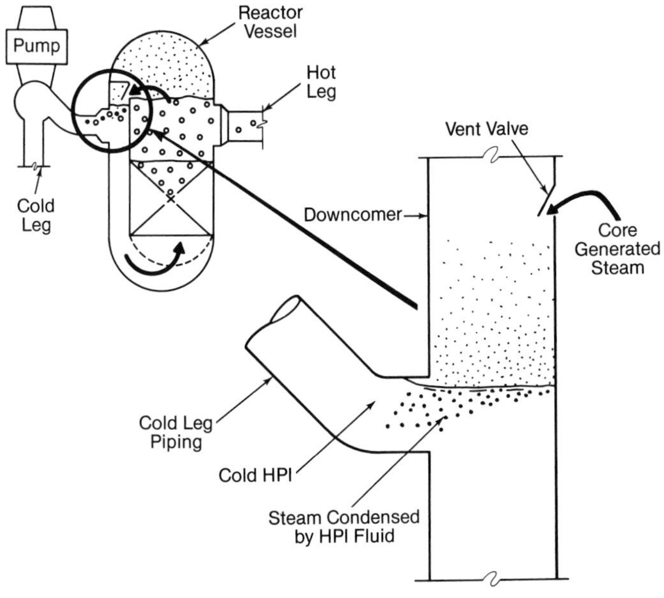

Fig. 15 HPI condensation.

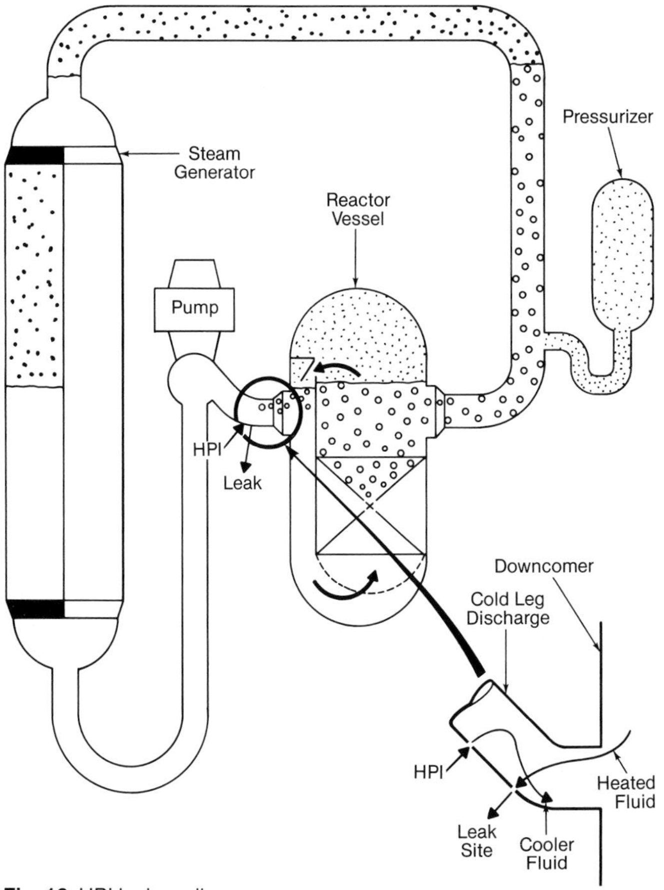

Fig. 16 HPI leak cooling.

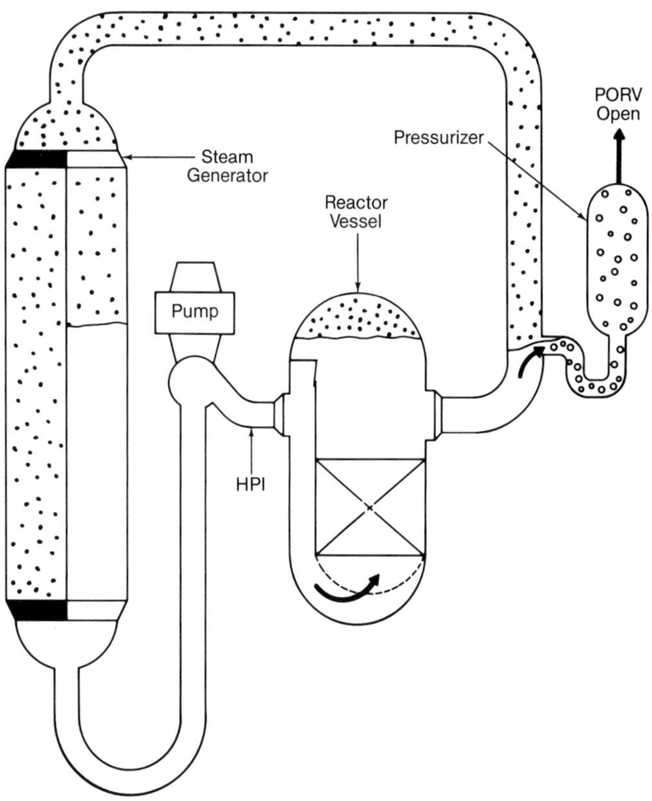

Fig. 17 Surge line uncovery.

through the surge line and pressurizer and out the PORV. HPI-PORV cooling is based on throughput and has commonly been termed feed and bleed cooling.

Surge line uncovery

The PORV discharge rate commonly exceeds the rate of high pressure injection. The hot leg levels slowly descend to the elevation of the pressurizer surge line connection. Then, a steam-water mixture flows through the surge line to the PORV, changing the PORV discharge rates. (See Fig. 17.) The discharge mass flow rate decreases but the volumetric flow rate increases. These flow rate changes enhance the primary system depressurization and stabilize the primary system fluid inventory.

Intermediate levels

Primary to secondary heat transfer is inhibited when the hot leg levels lie within an intermediate range. The two elevations that define this intermediate range are the elevation of the hot leg U-bend spillovers and the elevation of the steam generator upper tubesheets. When a hot leg riser level exceeds the spillover elevation [66.5 ft (20.3 m)], single-phase natural circulation can be maintained. Core outlet fluid is therefore transported to the steam generators, supporting primary to secondary heat transfer.

When the steam generator primary system level drops below the elevation of the upper tubesheets [52.1 ft (15.9 m)], heat can be transferred through AFW-BCM. These two elevations define the boundaries of the intermediate levels. Within this range, the levels are too high for BCM, but too low for single-phase natural circulation.

References

1. "Report of the President's Commission on the accident at Three Mile Island," J.G. Kemeny (Chairman), U.S. Government Printing Office, 1979.

2. "Analysis of Three Mile Island — Unit 2 accident," Nuclear Safety Analysis Center Report NSAC-80-1, Electric Power Research Institute, revised March 1980.

3. "Best-estimate calculations of emergency core cooling system performance," USNRC Regulatory Guide 1.157, May 1989.

4. "Quantifying reactor safety margins," USNRC Report NUREG/CR-5249, December 1989.

5. Multiloop Integral-System Test Final Report, Volume 1 - Summary, USNRC Report NUREG/CR-5395, April 1991.

Launching of the U.S.S. *Nautilus*, the world's first nuclear powered ship.

Chapter 51
Nuclear Ship Propulsion

Nuclear power for propulsion has been applied to both commercial and naval vessels beginning with the U.S.S. *Nautilus*, the world's first nuclear ship. The development of nuclear ship propulsion also formed the basis for commercial pressurized water reactors (PWR) for land-based electric power generation. For submarines, nuclear propulsion has proven to be the greatest single advancement in post-World War II technology. This dramatic achievement virtually abolished range limitations and enabled the submarine to become a true submersible, freed from the need to make regular forays to the surface to recharge batteries.

Nuclear power has also proven extremely valuable for naval surface ships such as aircraft carriers and escort vessels. While the advantages are not as readily apparent as for submarines, the unlimited, sustained power available with nuclear propulsion allows the ship commander to devote his full attention toward effectively executing his mission without the continuous logistical concern for fuel oil supplies.

Unlike naval vessels, commercial nuclear ships have not yet proven their long term viability. Although several demonstration ships, including the Nuclear Ship (N.S.) *Savannah*, have been technological successes, the economic competitiveness of commercial nuclear propulsion is still very much in doubt. Compared with conventionally powered merchant ships, nuclear ships are extremely expensive to construct, require much larger and more highly trained crews, and may be subject to port entry restrictions. Even with fuel cost advantages resulting from higher oil prices, the future of commercial nuclear propulsion remains uncertain at best.

Reactor types

Nuclear reactors for ship propulsion can, in theory, be of any type that can produce sufficient useful power. To date, however, only two types of reactors, the PWR and the sodium-cooled reactor, have actually been applied to operating vessels. The pressurized water reactor, shown in Fig. 1 and described in Chapter 50, has been the predominant reactor type.

Reactor development in the United States (U.S.) began in the 1940s, and by 1948 the U.S. Atomic Energy Commission gave formal project status and high priority to the development of a water-cooled reactor for submarine propulsion. This led to the reactor system for the *Nautilus*. Since the program's beginning, Babcock & Wilcox (B&W) has been the leading supplier of propulsion system components to the U.S. Navy's nuclear fleet.

Also in the 1940s, the U.S. Naval Research Laboratory showed the first major interest in liquid metal coolants for reactors, and B&W was contracted to investigate the use of sodium-potassium alloys as heat transfer media.

Sodium as a reactor coolant for shipboard plants is attractive from the standpoint of temperature and efficiency. The attainable heat transfer coefficients are high and the vapor pressure low. Compared with a PWR, a sodium-cooled reactor permits a tenfold increase in temperature differential across the reactor core and a several hundred degree increase in coolant temperature entering the steam generator. It can produce the same shaft horsepower as a typical pressurized water plant with only 85% of the reactor power. From a theoretical viewpoint, these characteristics would seem to make a sodium-cooled reactor attractive.

From a practical viewpoint, however, there are inherent disadvantages that can outweigh the temperature and efficiency gains possible with sodium coolant. Sodium reacts violently if it comes into contact with water, and it must be kept molten to prevent potential damage to the reactor coolant system. The high temperature sodium installation requires additional shielding and is therefore heavier. It is also more expensive and difficult to

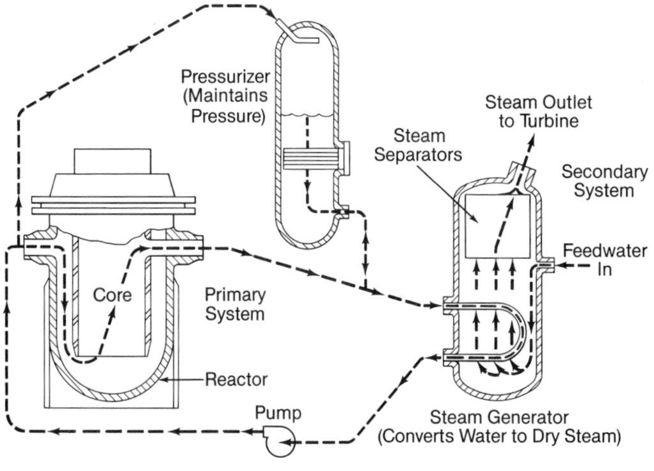

Fig. 1 Pressurized water reactor system — naval type.

maintain and the technology is less mature. As a result, a theoretical advantage of considerable magnitude is outweighed by practical considerations for most marine applications.

Shipboard considerations

There are important differences between shipboard reactors and similar land based installations. These involve weight and space limitations, plant reliability and on-board maintenance, plant safety, and problems inherent with a moving platform.

Weight and space Fig. 2 is a schematic diagram of a typical marine nuclear propulsion plant. The physical size of the shipboard reactor itself does not present a problem, but the size and configuration of the shielded volume around the entire reactor coolant system are of considerable significance.

Weight considerations usually dictate a compact reactor coolant system layout to reduce the size and weight of the secondary shield. However, the need for maintenance access results in a balance such that the permeability (ratio of free space to total volume of space) of the portion of a nuclear plant within the secondary shield typically runs from 65 to 70%. Permeability of other shipboard machinery spaces, exclusive of uptakes, air vents, or cargo access, is approximately 80%.

Reliability and maintenance Because of the exceptionally long time interval between refuelings, plant reliability and the capability to perform required maintenance functions at sea assume increased importance. Many plant components transfer radioactive fluids, and ready access to these components for required maintenance functions is essential. For fluids and other consumables used in the reactor system, such as coolants or auxiliary fluids, provision must be made either for their generation on board or for storage of an adequate reserve. Because of safety and control considerations, exceptionally reliable auxiliary power sources are required.

Plant safety A shipboard nuclear reactor is more subject to external hazards than a similar land based plant. Collision damage protection, secondary containment, special safety devices, and an independent means of propulsion for emergencies are necessary. Vital parts are protected from collision damage to the greatest extent possible by careful location within the ship's machinery spaces. The spreading of radioactivity is the principal hazard from a nuclear accident. A reactor design that uses a coolant with low long term activity and contains its fission product activity in a reasonably permanent form inside solid fuel elements has a great safety advantage. The PWR is such a design.

Another important safety consideration is the fact that a nuclear reactor, even after being shut down, continues to generate heat at low rate from residual radioactivity. Adequate means must be provided for disposing of this residual heat during periods of inoperability or in the event that the vessel is grounded or sunk.

Seakeeping Any nuclear power reactor for shipboard propulsion and its associated equipment must be designed in accordance with all the accepted principles of marine engineering. For example, if large free surfaces are present, special precautions must be taken in the ship's design to assure adequate stability.

Commercial nuclear ships

Nuclear merchant ship *Savannah*

A nuclear merchant ship was first proposed by President Dwight D. Eisenhower in 1955 as evidence of the U.S.'s interest in promoting the peaceful use of atomic energy. After Congress approved the funding in 1956, the President directed the Atomic Energy Commission (AEC) and the Maritime Administration (MARAD) to design and construct the vessel that was subsequently named the *Savannah*.

The program's major objectives were to demonstrate the peaceful use of nuclear energy and resolve the problems of commercial marine reactor operation. Plant requirements included a conservative design with a long

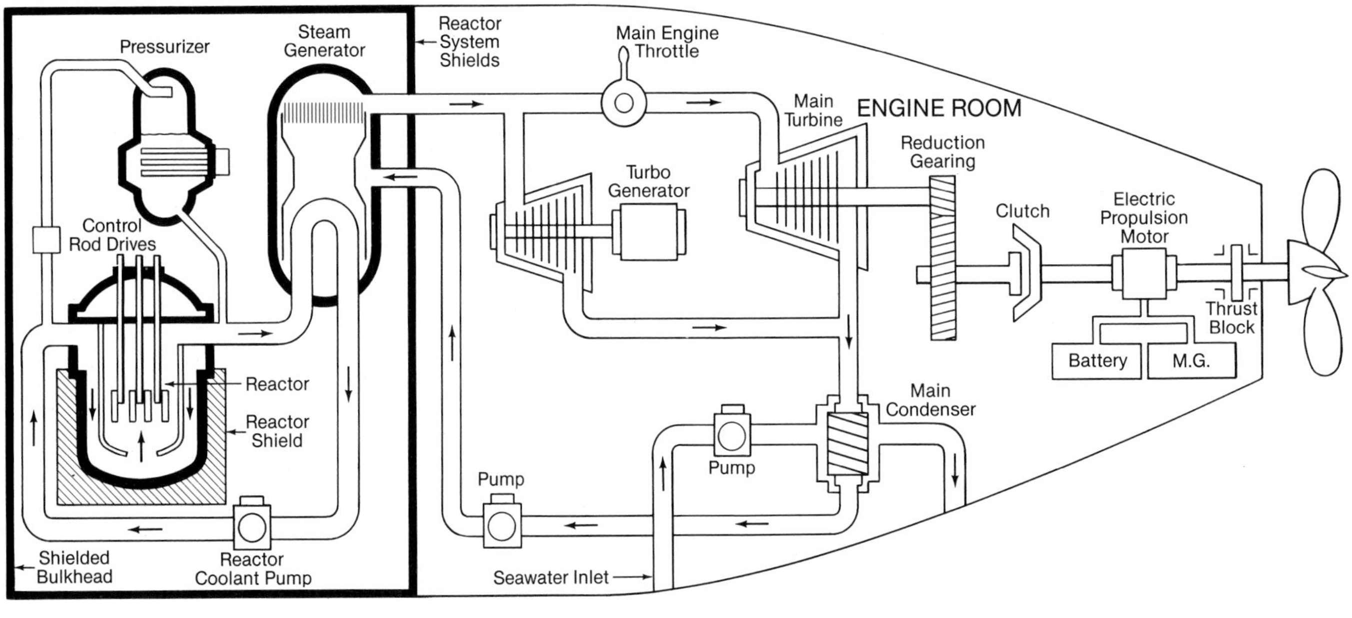

Fig. 2 Shipboard nuclear propulsion system.

Fig. 3 N.S. *Savannah*, first nuclear merchant ship.

core life, use of commercially available materials and equipment wherever practical, and safety of operation.

The *Savannah* (Fig. 3) is a single-screw, geared turbine vessel, 595 ft (181.4 m) long, with a beam of 78 ft (23.8 m), draft of 29.5 ft (9 m), and a displacement of 21,900 t (19,867 t_m). The ship has a design speed of 22 knots (40.7 km/h) at 22,000 shaft hp (16,412 kW), accommodations for 60 passengers and 652,000 ft^3 (18,463 m^3) of cargo space.

B&W supplied the nuclear propulsion plant and auxiliaries for the *Savannah* and trained the operating crew.

Power plant design The *Savannah*'s reactor is of the pressurized water type. The reactor coolant delivers heat through two main coolant loops to generate steam in the two steam generators. This steam expands through the main turbines to the condenser, where pumps return the condensate to the steam generators. There are two 1500 kW turbine generators to provide auxiliary power.

Emergency take-home power is provided by a 750 hp (560 kW) electric motor that can be coupled to the main gearing. Two 750 kW diesel generators, designed to start automatically if the turbine generators fail, provide power for the take-home motor and other essential electrical loads. Pertinent design data for the power plant are given in Table 1.

The *Savannah*'s reactor is fueled by 15,653 lb (7100 kg) of uranium enriched to an average of 4.4% uranium-235. Uranium oxide, in the form of pellets, is contained in stainless steel tubes. Each of the 32 fuel assemblies contains 164 tubes. Individual stainless steel boxes surround each assembly, direct the coolant flow through the fuel assemblies, and form channels for the control rods.

Table 1
N.S. *Savannah*

Power plant design data:	
Maximum shaft power	22,000 hp (16,412 kW)
Reactor power	70 MW
Total steam flow	226,000 lb/h (28.5 kg/s)
Turbine inlet pressure	430 psi (29.6 bar)
Feedwater temperature	340 F (171C)
Reactor coolant data:	
Pressure in reactor	1735 psi (119.6 bar)
Temperature in core	508 F (264C)
Flow	9,400,000 lb/h (1184.4 kg/s)

The reactor is controlled by rods of cruciform section made of stainless steel plates containing boron clad in stainless steel sheets. The rods are actuated by top-mounted electromechanical-hydraulic drives. Zircaloy-2® followers, fastened to the bottom of the absorbing rods, fill the rod channels in the core as the rods are lifted.

The entire primary system is enclosed in a steel containment shell (Fig. 4) which supports the primary, and a portion of the secondary, shielding. The secondary or biological shielding reduces the radiation in crew and passenger spaces to safe levels.

The inherent self-regulating characteristics of a low-enriched uranium-oxide pressurized water reactor, coupled with the centralized instrumentation and control systems of the *Savannah*, permit rapid power transients and maneuvering rates. The reactor power plant was designed to increase load from 20 to 85% in ten seconds and decrease power from 100 to 20% in three seconds.

Construction and operation The keel of the *Savannah* was laid in 1958 and the ship was launched 14 months later. In 1962 the ship was delivered to the operating agent, and port visitations began. On her first fuel core the *Savannah* traveled approximately 330,000 mi (531,089 km) and developed 15,000 full power hours with no shuffling of fuel. By late 1970 the *Savannah* had traveled more than 450,000 mi (724,205 km), visited 32 different U.S. ports in 20 states and 45 different foreign

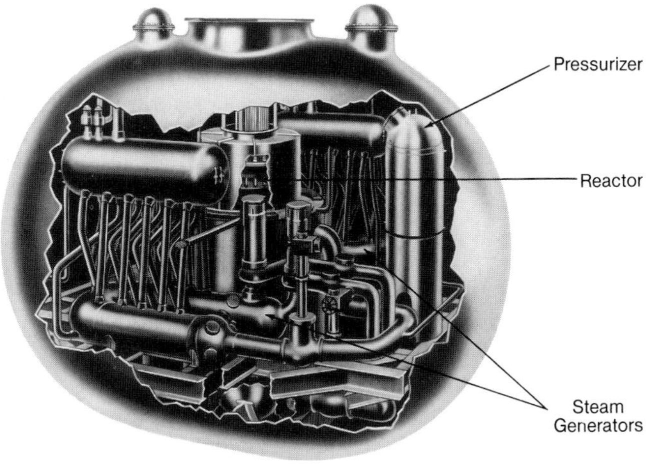

Fig. 4 N.S. *Savannah* nuclear steam supply system arrangement.

ports in 27 countries, and had been visited by more than 1,500,000 people.

The operation of the *Savannah* provided technology for future development of nuclear ships, and established standards for the design of the ship and reactor, operating practices and safety. After completing her mission, the *Savannah* was retired and is now a museum ship located in Charleston, South Carolina.

Nuclear merchant ship *Otto Hahn*

After building the N.S. *Savannah* reactor, B&W developed an improved nuclear marine plant known as the Consolidated Nuclear Steam Generator (CNSG). The CNSG was designed to achieve more economic nuclear propulsion systems for merchant ships, and has potential application for small to medium sized land-based plants.

The CNSG design incorporates the reactor and once-through steam generators within a single pressure vessel, achieving a compact arrangement and eliminating some of the auxiliary equipment.

The CNSG design was used successfully for the nuclear plant of the German N.S. *Otto Hahn* (Fig. 5). This ship is a single-screw geared turbine ore carrier 565 ft (172.2 m) long, with a beam of 77 ft (23.5 m), a draft of 30 ft (9.1 m), dead weight of about 15,000 t (13,608 t_m) and a design speed of 16 knots (30 km/h) at 10,000 shaft hp (7460 kW).

The nuclear plant for this ship was designed and constructed by the German-Babcock Interatom consortium with the assistance of B&W. A cross-section of the pressure vessel is shown in Fig. 6. Pertinent design data for the power plant at normal load are given in Table 2. The *Otto Hahn* began commercial service in 1970 and was decommissioned in 1978.

Nuclear icebreaker *Lenin*

The Soviet icebreaker *Lenin*, launched in 1957, uses nuclear propulsion to operate for long periods in arctic regions without refueling. The *Lenin* is turbo-electric propelled by three screws and has a maximum speed of 18 knots (33 km/h) in clear waters and an operating speed of 2 knots (3.7 km/h) in ice that is 7.875 ft (2.40 m) thick. The ship has a displacement of 16,000 t (14,515 t_m) with a length of 440 ft (134.1 m), beam of 90.5 ft (27.58 m), and draft of 28 ft (8.5 m). Horsepower at the turbine reduction gear is 44,000 (32,824 kW).

Fig. 5 N.S. *Otto Hahn (courtesy of Gesellschaft fur Kernenergieverwertung in Schiffbau und Schiffahrt mbH).*

Table 2
German N. S. *Otto Hahn*

Power plant design data:	
Shaft power	10,000 hp (7460 kW)
Reactor power	38 MW
Total steam flow	141,000 lb/h (17.8 kg/s)
Superheater outlet press.	440 psi (30.3 bar)
Superheater outlet temp.	523 F (273C)
Feedwater temp.	365 F (185C)
Reactor coolant data:	
Pressure in reactor	918 psi (63.3 bar)
Temp. in reactor core	523 F (273C)
Flow	5,280,000 lb/h (665.3 kg/s)

The *Lenin* was originally intended to have two reactors, but it was later decided that three should be provided with any two capable of generating full power. The reactors are arranged in line athwart the ship with the two heat exchangers for each unit located fore and aft of their respective reactors. These are pressurized water reactors, each with 90 MW thermal capacity. A cross-section of the pressure vessel and core is shown in Fig. 7, and pertinent design data for the power plant are given in Table 3. The original core design consisted of fuel chan-

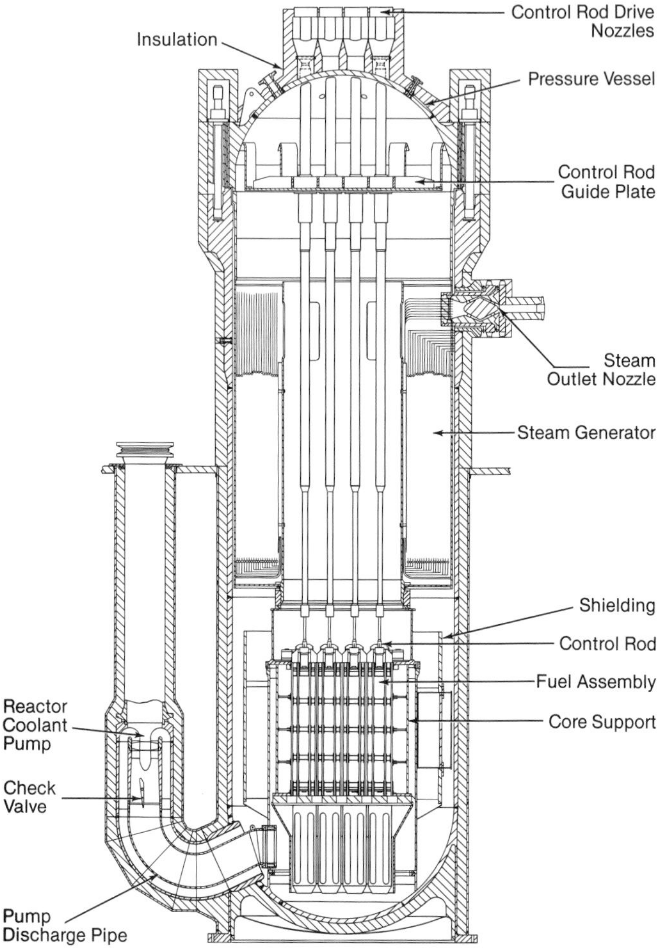

Fig. 6 Reactor vessel cutaway (N.S. *Otto Hahn*) *(courtesy of Gesellschaft fur Kernenergieverwertung in Schiffbau und Schiffahrt mbH).*

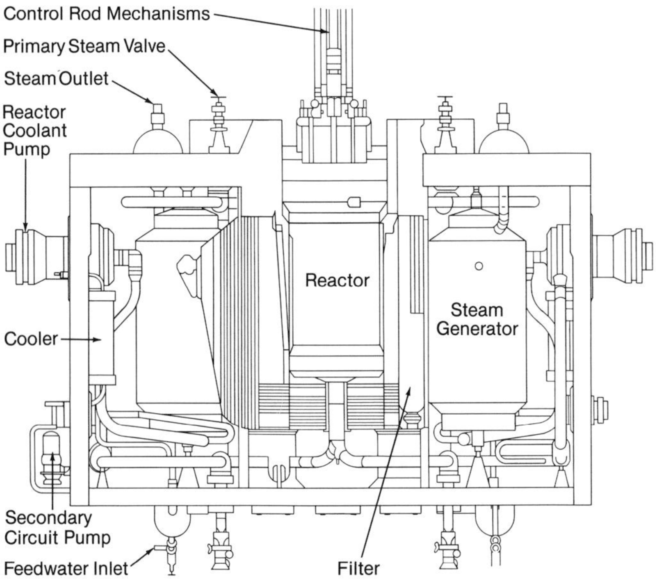

Fig. 7 Reactor unit arrangement (*Lenin*).

nels of stainless steel, approximately 2.17 in. (55 mm) in diameter and 9.8 ft (3 m) in length, each channel containing 36 fuel rods consisting of sintered pellets of uranium oxide, sheathed in zirconium.

Nuclear cargo ship *Mutsu*

In 1967 Japan began building the N.S. *Mutsu* as a nuclear power demonstration ship. The hull was completed in 1970 and fuel was loaded in 1972. The 8213 ton (7451 t_m) cargo ship has an overall length of 426.5 ft (130.0 m) with a beam of 62.3 ft (18.99 m). The *Mutsu* has a design maximum speed of 17 knots (32 km/h) and a crew of 56 (see Table 4).

The *Mutsu*'s pressurized water reactor produces a thermal output of 36 MW. The two-zone core is composed of 32 fuel assemblies consisting of 121 rods arranged in an 11x11 matrix. Nine rods in each assembly contain boron silicate burnable poison. There are 12 cruciform control rods. Core life is expected to be about two years with a 60% load factor.

The reactor coolant system is composed of two loops, each containing a reactor coolant pump and a U-tube steam generator. The main engine is a cross compound saturated steam turbine. An auxiliary boiler produces enough steam to drive the ship at 10 knots.

In 1974, the *Mutsu* sailed from its home port to the open sea for the initial powerup test of the reactor. Despite 2491 t (2260 t_m) of shielding, neutron radiation streaming caused tests to be suspended. Following repairs to the reactor shielding, local opposition kept the ship idle. Testing was finally resumed in 1990, and the *Mutsu* reached a speed of 17.8 knots (33 km/h) at full reactor power during its year long test cruise.

Table 3
Soviet Icebreaker *Lenin*

Power plant design data:	
Power at turbine reduction gear	44,000 hp (32,824 kW)
Number of reactors	3
Reactor power (each)	90 MW
Total steam flow	906,400 lb/h (114.2 kg/s)
Steam pressure	398 psi (27.4 bar)
Steam temperature	590 F (310C)

Table 4
Japanese N.S. *Mutsu*

Power plant design data:	
Shaft power	10,000 hp (7460 kW)
Reactor power	36 MW
Steam pressure	569 psi (39.2 bar)
Steam temperature	482 F (250C)
Reactor coolant data:	
Pressure in reactor	1565 psi (107.9 bar)
Temperature in core	532 F (278C)

Naval nuclear ships

U.S. naval nuclear propulsion program

Background In 1946, the U.S. Navy's Bureau of Ships recognized that atomic energy might ultimately be developed for ship propulsion. Under the direction of Captain (later Admiral) Hyman G. Rickover, a team was established to transform existing theory and concepts into practical engineering designs.

The basic requirements for applying nuclear power to shipboard propulsion were clear. The reactor would have to produce sufficient power so that the ship would have military usefulness, and it would have to produce that power safely and reliably. The reactor plant would have to be rugged enough to meet the stringent requirements of a combatant ship, and be designed for operation by a Navy crew. A suitable reactor would require new corrosion resistant metals which could sustain prolonged periods of intense radiation, effective shielding to protect personnel from radiation, and the development of new components that would operate safely and reliably for prolonged periods.

These problems were even more difficult for submarine applications because the reactor and associated steam plant had to fit within the confines of a small hull, and be able to withstand extreme battle shock. Although the application of nuclear power to submarines was a major challenge, success would revolutionize submarine warfare. No longer limited to submerged operation on battery power, a true submarine was possible — one that could travel submerged at high speed for long periods.

Three reactor concepts were initially considered for naval nuclear propulsion. A study of a gas-cooled reactor showed that this concept was not then suitable. The pressurized water reactor and liquid metal-cooled reactor approaches were found promising and carried through to full scale prototype plants, and thereafter to shipboard applications. B&W became actively involved in both designs.

Design and engineering philosophy Because a warship must be able to perform its mission and return under combat conditions, standards for materials and systems are rigorous. The following are important engineering and design tenets:

- Avoid committing ships and crews to highly developmental and untried systems and concepts.

- Ensure adequate redundancy in design so that the plant can accommodate, without harm to ship or crew, equipment or system failures that will inevitably occur.
- Minimize the need for operator action to accommodate expected transients. With an inherently stable plant, the operator is better able to respond to unusual transients.
- Simplify the system design to rely primarily on direct operator control, rather than on automatic control.
- Select only materials that have been proven by experience for the type of application intended, and insofar as practicable, those that provide the best margin for error in procurement, fabrication and maintenance.
- Confirm reactor and equipment designs through extensive analyses, full scale mockups and testing.
- Conduct extensive accelerated lifetime testing of critical reactor and system components to ensure design adequacy prior to operation.
- Test new reactor designs by use of a land-based prototype of the same design as the shipboard plant. Prototype plants can be subjected to the potential transients that a shipboard plant will experience, so problems can be identified and resolved prior to operation of the shipboard plant.
- Train operators on actual operating prototype reactors. By themselves, simulators do not provide adequate training for naval operators.
- Accept only equipment that meets specification requirements. Verify this by using specially trained inspectors and extensive inspections during manufacturing.
- Concentrate on designing, building and operating the plants to prevent, rather than just cope with, accidents.

U.S.S. *Nautilus*

In 1949, the U.S. Navy's Chief of Naval Operations issued a formal requirement for the development of a nuclear powered submarine. The following year the U.S. Congress authorized funds for a land-based prototype of the pressurized water reactor that would power the world's first nuclear powered ship. B&W was selected to provide major reactor system components. Just three years later, the prototype began operation and, for the first time, a reactor produced sufficient energy to drive power machinery.

On January 17, 1955, the *Nautilus* (Fig. 8) put to sea for the first time and radioed her historic message: "underway on nuclear power." On her shakedown cruise, the *Nautilus* steamed submerged more than 1300 mi (2092 km) in 84 hours — a distance that was 10 times greater than had been traveled continuously by a submerged submarine.

Within months, the *Nautilus* would break virtually every submarine speed and endurance record. Her global odyssey carried her under the seven seas, and on the first voyage in history across the top of the world, passing submerged beneath the North Pole.

On her first fuel core the *Nautilus* steamed more than 62,500 mi (100,584 km), more than half of which were totally submerged. During 25 years of service, she traveled a total of 600,000 mi (965,606 km) on nuclear energy, making the dream of Jules Verne come alive. The *Nautilus* is now a historic museum ship and is located in New London, Connecticut.

Fig. 8 U.S.S. *Nautilus*, world's first nuclear ship.

The *Nautilus* was the first application of a reactor power plant using pressurized water both as the primary coolant and as the heating fluid for converting the secondary water into steam. The recirculating steam generator (Fig. 9) was comprised of a straight tube and shell heat exchanger with riser and downcomer pipes connected to a separate steam drum.

Babcock & Wilcox designed and fabricated the prototype test steam generator, and the reactor vessel and pressurizer for the *Nautilus*.

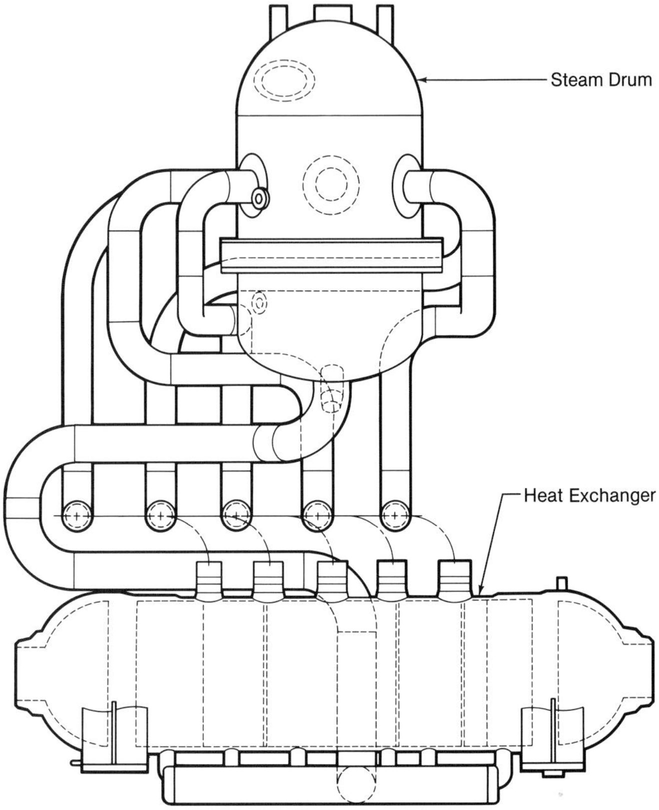

Fig. 9 U.S.S. *Nautilus* steam generator.

U.S.S. *Seawolf*

The U.S.S. *Seawolf*, the second U.S. nuclear submarine, was launched in 1955, and her liquid metal-cooled reactor attained initial criticality in 1956. To help ensure tube integrity, the *Seawolf's* sodium-heated steam generators (Fig. 10) utilized sodium-potassium as a third or *monitoring* fluid in the annulus of the double-tube design. Although the ship operated satisfactorily for almost two years on its sodium-cooled reactor, overriding technical and safety considerations led to the abandonment of this type of reactor for propelling U.S. naval ships.

While liquid sodium is a much more efficient heat transfer medium than water, it can be very troublesome in service. Two problems are particularly noteworthy: the sodium has to be kept molten at all times or it will solidify and can damage primary system piping, and the sodium must be kept isolated from water to prevent a violent reaction.

In 1958, the *Seawolf* entered a shipyard where her sodium-cooled plant was replaced with a pressurized water reactor similar to that in the *Nautilus*. When her sodium plant was shut down for the last time, the *Seawolf* had steamed a total of 71,611 mi (115,247 km) of which 57,118 mi (91,923 km) were fully submerged.

U.S.S. *Skipjack*

The U.S.S. *Skipjack* attack submarine class combined the PWR reactor plant with a streamlined *Albacore* hull shape to provide increased speed and reduced flow noise. In 1956 B&W received a contract for the design and fabrication of steam generators for the *Skipjack*. These were the first vertical recirculating PWR steam generators with integral steam separators and were the forerunner to the recirculating steam generator designs in current commercial PWR plants (Fig. 11).

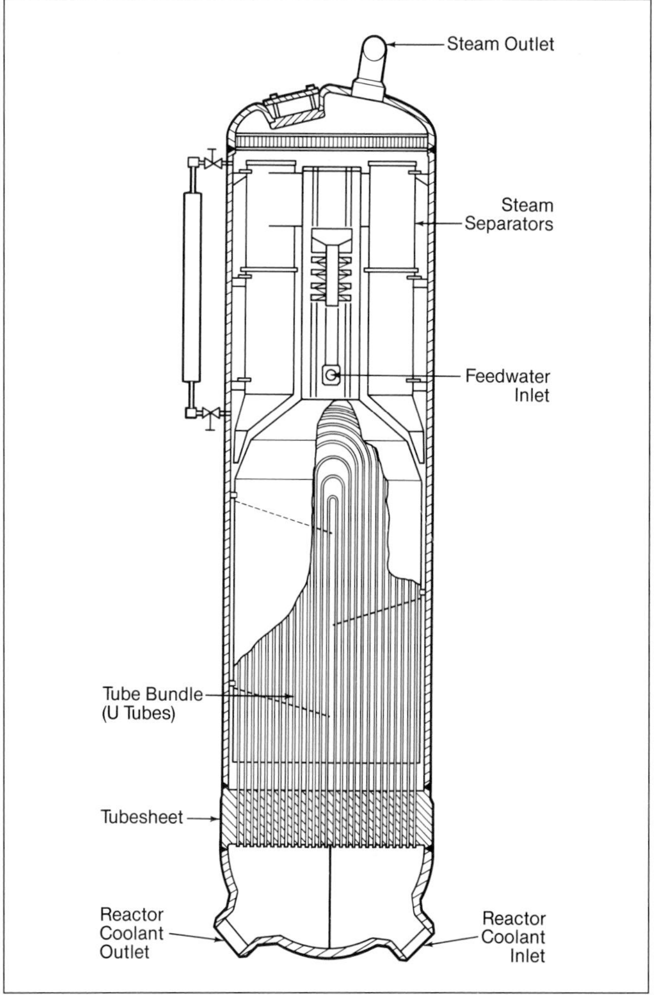

Fig. 11 Vertical recirculating steam generator, patented by B&W.

Technological advancements

Each new naval nuclear propulsion plant is a balance between the desire to make technological advances and a commitment to deliver a reliable fighting ship on schedule. Each new design can incorporate only a portion of all the potential improvements in technology. The ultimate success of any power plant design will depend on how well that balance is achieved.

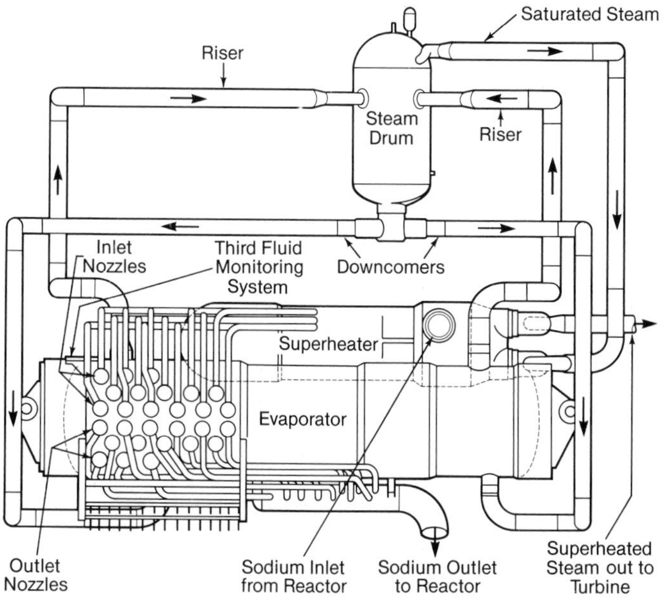

Fig. 10 Sodium-heated steam generator for *Seawolf.*

In steam generators for modern naval vessels (Figs. 12 and 13), improvements have occurred in thermal hydraulic design, materials, structural design, and fabrication. Significant improvements in the art of steam-water separation have contributed greatly to the performance and compactness of naval steam generators. In one of many early performance improvements, a separator (Fig. 14) was developed which afforded a means for secondary separation without increasing the size of the steam drum over that normally required for primary centrifugal separators.

Significant advancements have also been made in reactor fuel technology. Reactor fuel lifetime has been extended from two years for the first *Nautilus* core, to more than 15 years for cores delivered in the 1980s. Efforts are now underway to develop reactor fuel that will last the life of a ship, making expensive and time consuming refueling unnecessary.

MHD thruster research

Recently, the U.S. has begun a series of experiments to evaluate the efficiency of a magnetohydrodynamic (MHD) thruster that may serve as a model for a revolu-

Fig. 12 U.S.S. *Ohio, Trident* ballistic missile submarine.

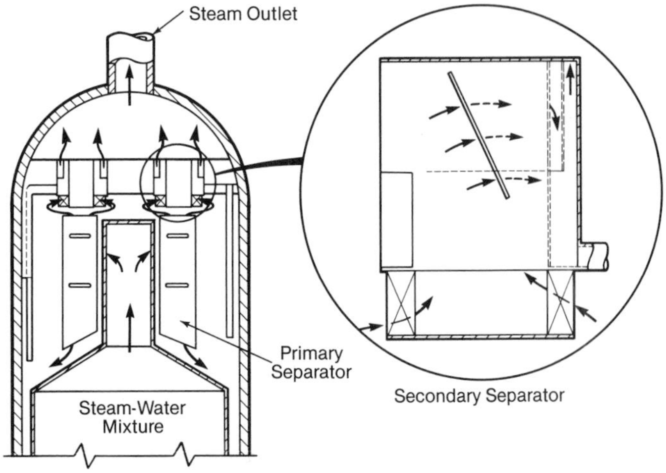

Fig. 14 Secondary steam-water separator patented in 1967 by B&W.

tionary new propulsion system for future Navy nuclear submarines.

MHD propulsion is based on the interaction of a magnetic field on sea water, which is electrically conductive. An electrical charge applied to a magnetic field causes a reaction called a Lorentz force, which forces the water through the field. In an MHD-powered vessel, water would flow through a duct surrounded by a superconducting magnet (one capable of passing electrical current at zero resistance). A current passed through the water precipitates the Lorentz reaction, increasing the velocity of the flow. The movement of the water would then propel the craft either forward or backward, depending on the direction of the applied current.

The use of an MHD system could eventually eliminate the need for a propeller, drive train and gears. Larger

Fig. 13 *Nimitz*-class aircraft carrier.

Fig. 15 Soviet *Typhoon*-class ballistic missile submarine.

electrical generators would be required to provide power to the thruster and to the ship's electrical systems. Eliminating the drive train linked to the propeller would provide greater flexibility in propulsion system location and design, and the resulting system could lead to major improvements in submarine quieting.

International naval nuclear programs

All of the principal maritime nations have studied the application of nuclear power to naval ship propulsion. The U.S., Great Britain, France, the People's Republic of China, and the Soviet Union have built nuclear vessels. These nations all have naval nuclear fleets that rely primarily on pressurized water reactor technology. However, the Soviet Navy reportedly utilizes liquid metal cooled reactors in at least one attack submarine class for higher power output and greater operational speeds.

The Soviet *Typhoon*-class ballistic missile submarine (Fig. 15) is the largest submarine type ever built with a length of 563 ft (171.6 m) and a submerged displacement of 26,500 t (24,041 t_m). The Soviets' titanium-hulled *Alfa*-class attack submarines are reportedly the world's fastest and deepest diving with a speed of 45 knots (83 km/h) and an operational depth of 2500 ft (762 m).

Bibliography

Duncan, F., "Rickover and the nuclear navy," U.S. Naval Institute, Annapolis, Maryland, 1990.

"Economics of defense policy: Adm. H.G. Rickover hearing before Joint Economic Committee, Congress of the United States," January 28, 1982.

Encyclopedia of Science & Technology, Vol. 16, McGraw-Hill, New York.

Janes Fighting Ships, 1990/91 Edition.

Hearing on Nuclear Propulsion for Naval Warships, Joint Committee on Atomic Energy, Congress of the United States, May 5, 1971 to September 30, 1972.

"Lenin — the Russian icebreaker," *Nuclear Engineering*, pp. 432-433, October 1958.

"MHD propulsion for future submarines," *Seapower,* pp. 39-40, February 1991.

Miller, D., and Jordan, J., *Modern Submarine Warfare.*

"Nuclear ship development in Japan," *Nuclear Engineering International*, pp. 558-560, July 1973.

A Review of the United States Naval Nuclear Propulsion Program, U.S. Department of Energy — U.S. Department of Defense, June 1989.

Rickover, Adm. H.G., *et al.*, USN, "Some problems in the application of nuclear propulsion to naval vessels," Presented at Annual Meeting of The Society of Naval Architects of Marine Engineers, Technical Paper, New York, November 14-15, 1957.

Reactor vessel (without closure head) for a B&W nuclear steam supply system being loaded on a barge for shipment from B&W's Mount Vernon works.

Chapter 52
Nuclear Equipment Manufacture

Many manufacturing operations for nuclear steam supply components are similar to those used for fabricating fossil fueled equipment. However, the unique requirements and geometry of nuclear components call for special manufacturing methods, equipment and facilities. Nuclear equipment must operate reliably with minimal maintenance because of its operating environment. For this reason there are rigorous equipment specifications by customers and regulatory authorities. Quality assurance requirements are stringent and special attention must be given to cleanliness and material control.

Selecting the manufacturing sequence and methods requires balancing design requirements with manufacturing capabilities while meeting commercial obligations. As a result, special equipment is usually required.

Typical component description

Commercial nuclear components are typically cylindrical pressure vessels. Some of the components are large and heavy which leads to customized shipping and handling arrangements. Reactor vessels and components for pressurized water reactor (PWR) systems up to 32 ft (9.75 m) in diameter, 125 ft (38.1 m) long and weighing 1000 t (907 t_m) have been built by Babcock & Wilcox (B&W).

Nuclear components must meet stringent cleanliness specifications; therefore, special manufacturing areas are needed. Quality assurance procedures ensure that no material contamination occurs. Some nuclear components also have close assembly tolerances. This requires specialized equipment and extraordinary care during machining and welding operations.

Manufacture of reactor vessels

Different reactor systems have differing reactor vessel designs. B&W has experience in several systems, including the PWR, which is the focus of this chapter.

Each reactor vessel is made up of a cylindrical main section and two hemispherically shaped heads. One of the heads, attached by a bolted flange, has several penetrations for the reactor control rods. Nozzles are welded to the cylindrical section and provide the piping system connections. The inside of the reactor vessel includes many attachments for the fuel support system. (See Chapter 50.)

Fabricating cylindrical shells

Heavy ring forgings or rolled plate may be used as shell *courses* for the cylindrical vessel sections. In forming the plates, they are either hot or cold rolled, depending on thickness, into cylindrical segments or complete cylinders as dictated by the finished vessel size. Fig. 1 shows the equipment used for this operation. Programmed controls regulate the heating rate, temperature, and atmosphere in the furnace and provide a record of each heating cycle.

Where rolled plates are used, the longitudinal seam is joined by submerged arc or electroslag welding. (See Chapter 38.) Electroslag welding requires heating the component to approximately 1600F (871C) then quench-

Fig. 1 Pinch rolls used to form cylindrical courses.

ing below 600F (316C). Finally, a tempering operation provides the mechanical properties of tensile strength and impact strength. (See Chapter 6.) Upon completion of heat treatment, sample pieces are tested to verify that the specifications are met. The shell course is then cleaned of scale by shot blasting.

When heavy ring forgings are used, the material is inspected by ultrasonic (UT) and magnetic particle (MT) techniques. (See Chapter 38.) The forgings are quenched and tempered, then turned and bored on a vertical mill.

Welding the shell courses

The circumferential seams of the individual shell courses are joined by submerged arc welding. The parts being joined are rotated by drum turners under a fixed welding head. (See Fig. 2.) Components are held at a preheat temperature dictated by the material chemistry and thickness. After welding, the shell is stress relieved by slowly heating to a temperature of 1100 to 1350F (593 to 732C), depending on the material. The component is held at temperature for one hour per in. (per 25.4 mm) of thickness, then cooled slowly to 600F (316C).

Fabrication of hemispherical heads

Each reactor vessel requires two hemispherical heads. One head is welded to the cylindrical portion of the shell and the other is bolted on. Both heads are forged or hot pressed from flat plate which is formed into ring or orange peel type segments. The segments are then quenched, tempered, stress relieved and tested. The segments are joined by submerged arc welding to form a hemispherical head.

The closure head is connected to the reactor vessel by bolts and a gasketed joint. This joint is composed of two bolting rings, made from forgings, that are machined, clad (having a layer of dissimilar metal bonded to the parent metal) on the gasket surfaces, then welded to the vessel and head.

Cladding

All reactor components (made of ferritic material) have their internal surfaces clad with austenitic stainless steel or Inconel to provide resistance to corrosion from the reactor coolant. Fig. 3 shows a vessel course being clad using the six wire submerged arc cladding system developed specifically for this purpose. Under normal conditions, a deposition rate of about 80 lb/h (36.3 kg/h) provides a minimum clad layer of 0.1872 in. (4.76 mm) that can be nondestructively examined. Chemistry of the deposit is determined by analyzing material chips. The bond between the cladding and base metal is measured by ultrasonic methods and surface integrity is tested with liquid (dye) penetrant examination (PT).

Nozzles

After completing a cylindrical shell, nozzles are installed using a welding process that suits the application. For larger nozzles up to 48 in. (1219 mm) OD, submerged arc welding is used to weld through the full thickness of the vessel wall. Where two such nozzles are on opposite ends of a vessel, both can be welded simultaneously by welding one from the outside and the second from the inside. Smaller connections for control rods are welded to the inside of hemispherical heads using an integral backing ring.

Fig. 2 Submerged arc welding of circumferential seam in reactor vessel.

Inspection and testing

Throughout the fabrication process, all nuclear pressure vessel welding is examined by radiography. X-ray equipment is used in some cases while radioactive isotopes, such as cobalt-60, are used in others. The choice depends on weld thickness and configuration. Exposure time for a 14 in. (355.6 mm) thick steel shell is about three minutes. Hydrostatic testing is performed on the finished weld.

Shipping

The dock and cranes at B&W's Mount Vernon, Indiana works (see chapter frontispiece) are designed to handle and load heavy vessels. The two cranes, each rated

Fig. 3 Six wire submerged arc welding of stainless steel cladding to inside surface of nuclear component.

at 500 tons, can work in tandem to move vessels from the shop to a barge.

Manufacture of steam generators

B&W manufactures once-through and recirculating steam generators for both PWR and CANDU reactor systems. (See Chapter 47.) Both steam generators have many components and manufacturing methods in common and the following focuses on the recirculating steam generator manufacturing process in particular. Fig. 4 shows a simplified cross section of this design. The steam generator consists of cylindrical and conical shells, hemispherical heads, nozzles, a tubesheet, tubes, lattice grid tube supports, baffle and divider plates, a shroud and steam drum internals. A typical steam generator is about 75 ft (22.9 m) long, 12 to 15 ft (3.7 to 4.6 m) in diameter and weighs up to 400 t (363 t_m).

Cylindrical shells

The shell courses used in fabricating the cylindrical portion of the steam generators are made from forged ring sections or from plate formed by rolls shown in Fig. 5. The rolling method depends upon the size, thickness and material. Cylindrical shells for steam generators are typically less than 4 in. (101.6 mm) thick and are cold rolled. Some designs require thicker sections near the tubesheet area. These shells are usually hot rolled.

The longitudinal seam of each course is submerged arc welded using an automatic guidance system. In preparation for circumferentially welding several courses, each course edge is machined to improve weld quality. The sections are then aligned, tack welded and set up on drum rotators. A stationary submerged arc welder, automatically guided, is positioned over the seam and the circumferential welds are made while accurately rotating the drum.

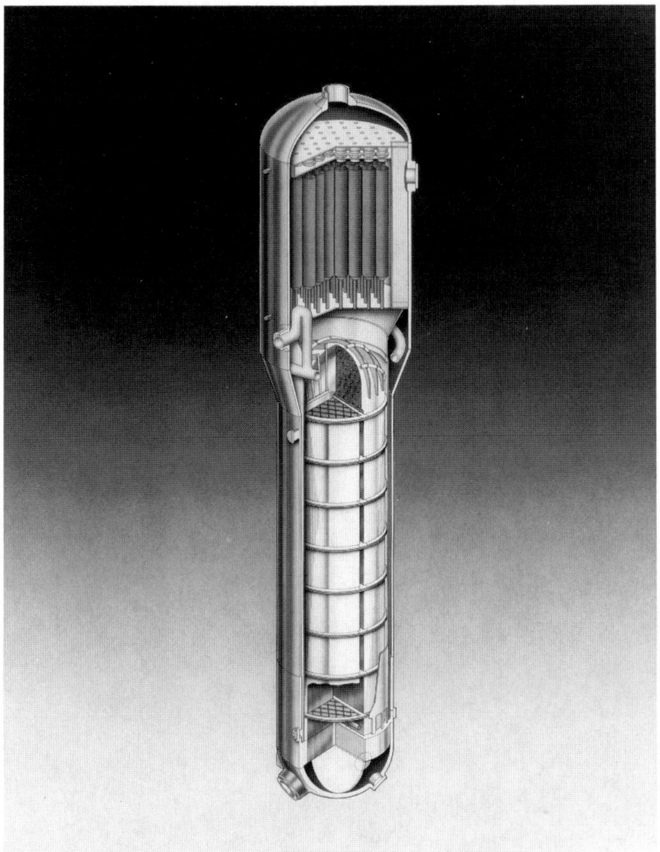

Fig. 4 Advanced series recirculating nuclear steam generator.

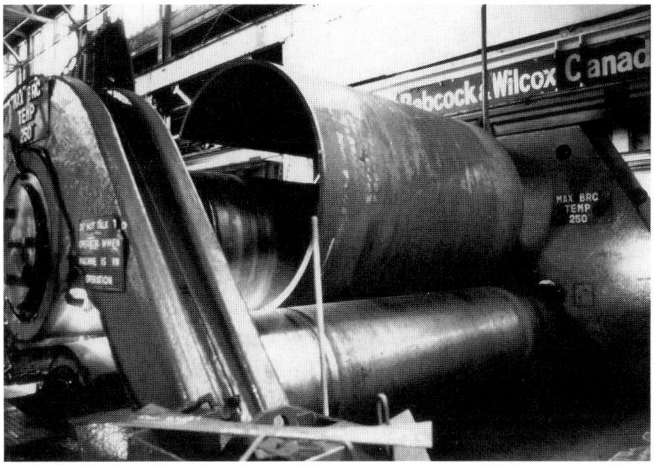

Fig. 5 Cold rolling plate for steam generator cylindrical shell.

Tubesheet manufacture

The tubesheet is made from a forging which can weigh up to 80 t (73 t_m). The first major manufacturing operation is cladding the primary side of the tubesheet with stainless steel or Inconel using metal inert gas (MIG) welding methods, discussed in Chapter 38. The tubesheet is heated before, during and after welding using large electric heaters. Usually the welding head is moved over the stationary tubesheet; however, in some cases the tubesheet geometry requires it to be rotated under a stationary welding head. In these cases, speed control of the rotator is critical to maintaining weld integrity. The clad tubesheet is postweld heat treated to 1100F (593C). The cladding is then machined and examined using ultrasonic testing methods. Cladding thickness after machining is about 0.3125 in. (7.9 mm).

Recirculating steam generators usually have a blowdown system near the tubesheet to permit removal of material from the secondary side water. Some designs feature a hole, typically 3 in. (76.2 mm) OD and 4 ft (1.22 m) deep, that is drilled radially from the outside of the tubesheet toward the center. An ejector drill, consisting of a double walled hollow tool with carbide tips, and driven by a horizontal boring mill (see Fig. 6), makes the hole.

The most critical operation for tubesheet manufacture is drilling the holes through which the tubes are inserted. A typical tubesheet is 24 in. (610 mm) thick and has more than 10,000 holes ranging in diameter from 0.5 to 0.75 in. (12.7 to 19 mm). These holes must be located accurately, be precise in diameter, and be straight and perpendicular to the tubesheet surface. Variations from perpendicularity, often called drill drift, must be limited to 0.015 in. (0.38 mm) through the thickness of the tubesheet. Extreme care is therefore given to gundrilling.

Gundrilling is done on a multi-spindle, computer controlled horizontal drilling machine (Fig. 7). Gundrills are designed to drill holes with large depth to diameter ratios. Coolant travels through the center of the drill and removes cutting chips through a channel along its shank.

Fig. 6 Ejector drilling blowdown hole in tubesheet.

Regular inspections of the drill tools, and of the hole diameters and locations, are conducted. After gundrilling, the tubesheet has its circumferential weld preparations machined on a vertical boring mill.

Steam generator heads

The primary head of a steam generator is typically 10 ft (3 m) in diameter, 6 in. (152.4 mm) thick and weighs up to 35 t (31 t_m). Primary heads used in PWR plant steam generators are clad on the inside surface. Two or three nozzles as large as 41 in. (1041 mm) diameter are attached to the head.

The head is mounted on a large capacity welding positioner so that it can be properly oriented for submerged arc welding of nozzles and attachments. Several attachments for internal components, such as the divider plate seat and manway hinge, are also welded to the inside surface of the head.

The steam drum head of a recirculating steam generator is either semielliptical or hemispherical, is up to 15 ft (4.6 m) in diameter, and is forged or formed from plate. It also contains one large diameter [30 in. (762 mm)] steam outlet nozzle. Special submerged arc machines are mounted on the outlet nozzle and rotate about its centerline to complete the nozzle weld to the head.

Pressure boundary assembly

Pressure boundary components include the primary head, the tubesheet, and vessel and internal supports. The primary head is first assembled by attaching the primary divider plate inside the head. The head is then aligned and tack welded to the primary side of the tubesheet.

This tubesheet and head assembly is aligned and tack welded to the secondary shell assembly. The circumferential seams on the primary and secondary sides of the tubesheet are then submerged arc welded. For steam generators with clad heads, the inside surfaces of the primary head circumferential weld are clad at this stage. Attachments such as vessel and internal supports are positioned and welded on. These attachments may also be added to the secondary shell subassembly to expedite manufacturing.

All of the circumferential and attachment welds are examined by various nondestructive examination methods prior to postweld heat treatment.

Pressure boundary postweld heat treatment An important consideration in planning the assembly of a steam generator is postweld heat treatment (PWHT). All carbon steel pressure welds require PWHT. In the assembly sequence, these welds must be accessible for nondestructive examination (NDE) after PWHT, as specified by American Society of Mechanical Engineers (ASME) Code. (Fig. 8.)

The entire pressure boundary assembly is prepared for PWHT. All openings are sealed. The tubesheet holes are cleaned to ensure the removal of all contaminants which might adhere to the surface during PWHT. Thermocouples are attached in strategic locations to monitor and control the heating and cooling operation. A vacuum is drawn on the vessel and it is backfilled with argon, which is slowly circulated through the inside of the vessel during PWHT to eliminate oxidation. The pressure boundary is then heated in a computer-controlled gas-fired furnace.

PWHT can cause significant temperature differences in the pressure boundary assembly especially if its geometry is complex. Therefore, the pressure boundary is assembled and postweld heat treated prior to installation of the tube bundle to avoid harmful differential thermal expansion effects. Although tube seal welding must be done in a confined area, the primary head is also welded to the tubesheet prior to tube bundle assembly

Fig. 7 Gundrilling a tubesheet.

Fig. 8 Radiographic examination of a steam generator weld.

to avoid PWHT effects. After installing the tube bundle, the final seam is welded and locally heat treated. The welds are then examined according to ASME Code.

Tube bundle components

Tube support assemblies The tube supports of modern steam generators are lattice grids. These grids consist of a ring assembly and a grid of flat bars which are assembled in the ring to form the lattice pattern. The flat bars, intersecting each other on their edges, form a diamond shape around each tube, thereby providing good vibration dampening yet allowing the steam-water mixture to flow through the bundle with minimal pressure drop. (See Fig. 9).

Lattice grids must be accurately manufactured so that each diamond is precisely aligned with the tubesheet drilling pattern. During tube bundle assembly, each tube is inserted through each grid then through the holes in the tubesheet. The accuracy of the lattice grids determines the ease with which the tube bundle is assembled.

The most critical operation in manufacturing lattice grids is cutting slots in the rings and bars where the assembly fits together. Slotting is done using a special milling cutter setup driven by a horizontal boring mill. It is extremely important for the cutter setup to be accurate and the workpiece to be set up true to the machine to control slot depth and angle. The mill must be calibrated to ensure that it is capable of precisely positioning the tool.

Fig. 9 Lattice grid tube bundle support.

CNC machines are mandatory to ensure consistent quality. After final assembly, the completed grid is inspected to verify that the diamonds are properly located.

Baffle plates Baffle plates are installed in the preheater section of steam generators to direct water flow. Baffles are fabricated from machined plates. The plates are set up for gundrilling by stacking them and welding their edges. After the initial holes are drilled, through bolts are then inserted and tightened. This prevents drilling chips from being trapped between the plates. The stack is progressively drilled and bolted until the pattern is complete.

U-bend supports The tube U-bends (see Fig. 10) are protected from flow induced vibration by U-bend supports. These consist of stainless steel flat bars inserted between the tube rows. The flat bars fan out from the center of the U-bend where they are welded to a collector bar. The angular separation of the fan bars depends upon the U-bend size and flow conditions, and the bars are located to minimize unsupported tube lengths. The outer ends of the fan bars are joined after assembly in the bundle with an arch bar which is then connected to the shroud.

Tube bending Tubes for steam generators are typically 0.5 to 0.75 in. (12.7 to 19 mm) diameter with a nominal wall thickness of 0.045 in. (1.14 mm). In the once-through steam generator design, the bundle consists of straight tubes. In the recirculating steam generator, bundles are made up of U-tubes.

Fig. 10 U-bend tube supports.

The U-tubes must have accurate bend radii, straight leg lengths, and must be within a strict ovality and wall thinning tolerance. To achieve this, the tubes are purchased cut to a length tolerance of 0.125 in. (3.18 mm). The tubes are formed on a draw bender over dies to a radius tolerance of ± 0.03 in. (0.8 mm). The dies for the tight radius bends are designed to contain the tube and to minimize ovality during bending. The large radius bends are formed around cone-shaped dies. This allows the tubes to be positioned at different diameters along the cone, permitting fine bend radius adjustments. After the tubes are bent, they are checked to verify that all tolerances have been met.

Shrouds The shrouds are fabricated from steel plates rolled into cylinders and their longitudinal seams are joined by submerged arc welding. The sections of the shroud near the preheater section of the steam generator may have their inside diameters machined round. This ensures that baffle plates can slide axially and can expand during various operating conditions. These shroud sections also contain manifolds to distribute the incoming feedwater and recirculating water.

The inside radius of the shroud is measured by projecting a laser down its centerline and using a radial micrometer. This verifies that sufficient clearance exists to allow lattice grids and baffles to be aligned with the tubesheet pattern.

Tube bundle assembly

The tube bundle is assembled in a *clean room* to ensure that the components are not contaminated. Cleanliness and material control procedures are carefully monitored. All internal components are cleaned just prior to enclosing them.

Installation of shroud and bundle supports The shroud is installed horizontally by using wheels bolted onto one end. This allows the tubesheet end of the shroud to roll into the shell as a crane installs it. The shroud is then aligned using pins located around its circumference.

If the steam generator design includes an integral preheater, the secondary divider plate is installed with a powered cart which adjusts the position of the divider plate. This plate fits into grooves in the shroud. The preheater baffle plates are then installed by mounting them on the end of a boom which is driven by a floor mounted power unit. The holes in the baffle plates must be aligned with those in the tubesheet.

Baffle plates are supported by tie rods, machined to length, which ensure that the plates are parallel to the tubesheet. The baffle plates must have radial clearance to allow for thermal expansion. This clearance, however, creates an installation problem because the baffles must support the tube bundle and remain in alignment with the tubesheet while the tubes are being installed. This problem is overcome by installing a special shim which fits into the clearance between the baffle plates and the shroud. This shim dissolves during vessel operation thereby restoring the proper clearance.

The lattice grids are also installed by mounting them on a boom. (Fig. 11.) The grids are aligned with the tubesheet using scopes which are mounted on the primary side of the tubesheet. A small target is positioned in selected tube locations on the lattice grids, and the grid is adjusted using wedges around its circumference until the target aligns with the scope's line of sight. Alignment rods which are the same diameter as the tube are also used to check the alignment of the lattice grids. The grids are held in place by support blocks which are welded to the inside of the shroud.

Fig. 11 Installation of lattice grid tube bundle support.

Installation of tubes Each tube hole and tube are dry-swabbed clean just prior to assembly. Cleaning operations at this stage are done without liquids. Any moisture trapped between the tube and tube hole can cause seal welding defects.

In a recirculating steam generator, the U-tubes are installed in layers from the inside of the bundle outward (Fig. 12). As each layer is installed, the appropriate U-bend support fan is installed. The partially completed bundle is temporarily supported to prevent sagging due to the weight of the tubes. Each U-tube is inserted until it protrudes about 0.20 in. (5 mm) from the primary face of the tubesheet. This protrusion ensures that a satisfactory fillet (seal) weld can be made on the primary side. The installer also checks for proper clearance between each U-bend. The tubes are tack expanded by inserting an expander inside the tubes about 0.75 in. (19 mm). The expander consists of a split collet actuated by a hydraulically driven tapered mandrel. This yields the tube sufficiently to hold it in place for welding. This expansion method produces less residual stress in the tube than other expansion methods.

Seal welding The tubes and tubesheets are joined with automatic tungsten inert gas (TIG) welding. The seal welding heads are specially designed, close clearance units that clear the primary head when welding the tube ends located at the tubesheet periphery (Fig. 13). During each shift, each operator makes a seal weld test block. The block is inspected with PT, then sectioned for further metallurgical examination.

Each weld is visually compared to a workmanship sample. All welds not meeting the sample quality are machined away, then rewelded. This is followed by PT examination of all seal welds. Finally, a leak test is done by pressurizing the secondary side to 50 psi (344.7 kPa) with a mixture of air and helium. The primary tubesheet face is then monitored for helium leaks at the welds.

Tube expansion Each tube is hydraulically expanded into its hole after seal welding. This expansion closes the crevice between the tube and the hole to avoid a potential corrosion site. The tube may be expanded near the secondary face of the tubesheet or it may be expanded full depth depending upon customer specifications. Hydraulic expansion is the recommended method for nuclear steam generators because it produces less residual stress in the tube and reduces the potential for stress corrosion cracking compared to other expansion methods. The tube is expanded by inserting a probe. The expansion probe has a seal positioned at each end of the expansion zone. Distilled water at 35,000 psi (2413 bar) is pumped through the probe which expands the tube and seals it against the tube hole.

The expansion probe must be carefully positioned with respect to the secondary face of the tubesheet. If the probe is positioned beyond the face, then unacceptable tube deformation could occur. If the probe is too far inside the tube hole, an unacceptably long crevice could possibly result. Therefore, the tubesheet thickness variation is measured and the probe length is adjusted to ensure proper positioning.

Tubes that are located near the tubesheet periphery must be expanded with special probes because the curvature of the head encroaches on the probe insertion area. A flexible extension is attached to the probe which allows it to be inserted in confined areas. In cases where the tube must be expanded full depth, the expansion is done in two steps. Following expansion, each tube is examined with eddy current techniques.

Fig. 12 Installation of tubing.

Fig. 13 Seal welding tubes.

Steam drum internals

Recirculating steam generators require a steam drum and steam separators to remove moisture from the steam (Fig. 14). The steam separators are similar in design to those used on fossil fuel boilers. Generally, the separators in nuclear steam generators are axial flow type, although tangential flow separators have been used. There are typically more than 100 separators in a large recirculating steam generator. Most designs also have smaller secondary separators located above the primary units.

The fabrication and assembly tolerances of steam separators are critical. The shape and assembly clearances of internal components have a significant effect on separator performance. Custom made assembly jigs and fixtures are used to ensure that the required tolerances are met. In addition, all internal welds are carefully ground to avoid discontinuities which affect separator performance.

The steam separators are installed in a deck structure made of steel plate. This deck support structure is welded to the inside of the drum, and the supports are designed to accommodate differential thermal expansion during steam generator operation. The separator and deck structure is usually of modular construction to simplify installation into the steam drum.

Fig. 14 Fabrication of steam separator subassembly.

Final assembly

Closing seam fitting and welding The final assembly operation consists of fitting and welding the closing seam, which is usually one of the circumferential seams on the cone. The steam drum and the cylindrical shell are positioned on rotators and the circumferential seam is aligned. Previous machining of the component mating surfaces assures good alignment.

Once the drum and cylindrical shell are aligned, the seam is tack welded to hold the components in position. The inside of the closing seam is welded first. The tube bundle and the steam separator assembly are sealed off to prevent weld flux and other material from entering these areas. A spider/arm assembly, on which a submerged arc welder is mounted, is then installed inside the vessel. As the vessel is rotated, the inside weld is completed. The root of the weld is then ground back from the outside with a large diameter grinder mounted on a manipulator. (Fig. 15.) The ground area is examined with MT techniques. The outside of the circumferential seam is then submerged arc welded to complete the vessel.

Local postweld heat treatment The closing circumferential seam requires postweld heat treatment. Because temperature differences during this procedure can exceed those experienced during normal operation of the vessel, care must be taken to avoid excessive and harmful differential expansion of internal components. Temperatures can be controlled using an electric furnace which is designed to fit around the closing seam. Insulation is applied to the outside of the vessel shell and to the attachments in the vicinity of the seam. The tube bundle U-bend region is also insulated to reduce radiant heat transfer. Internal heat transfer is further reduced by drawing a vacuum on the secondary side of the vessel. During PWHT, air may be blown through the tube bundle primary side to limit tube temperature. Thermocouples are mounted in each critical area so that the process can be monitored and controlled.

The electric furnace has several zones which can be independently controlled to keep the temperatures within prescribed limits. A computer is used to control the furnace and to provide a record of temperatures.

Final inspection and testing

A hydrostatic test is done on the completed vessel; the primary and secondary sides are tested separately. The test consists of filling the vessel with treated water which is pressurized to 125% of the design pressure. The pressure is held while an inspector examines the vessel for leaks. The metal temperature must be held above 70F (21C) to keep it well above its brittle transition temperature.

After hydrostatic testing, the vessel is drained and dried. A vacuum is drawn on the secondary side to assist drying. Felt plugs are also blown through the tubes to dry them. All accessible pressure boundary welds are then examined by the MT method.

Nozzles and integral support surfaces, called terminal points, connect to mating components in the field. These must be within specified tolerances to simplify field installation and minimize stresses caused by fitup. Typically, terminal points are specified to be within 0.125 in. (3.17 mm) of true position. It is common practice to finish machine nozzle weld preps at this stage to bring them within tolerance.

Fig. 15 Back grinding a circumferential weld.

Just prior to closing the vessel, the insides of the tubes are cleaned by blowing felt plugs through them. The secondary side of the steam generator is examined and cleaned and final drum internal components are installed. All manway covers are bolted on. The primary and secondary sides of the vessel are evacuated and backfilled with nitrogen to reduce the formation of oxides. Steam generators may be coated with a strippable paint to protect them during shipment.

Surface conditioning

Some steam generator customers specify that the inside surfaces of the primary side (excluding the tubing) are to be surface conditioned or electropolished. This process produces a very smooth surface finish, typically 10 microinches or better. Improving component surface finish generally reduces background radiation after operation. A component with low background radiation can be more easily maintained in the field. However, electropolishing is a very expensive and labor intensive operation. For some applications, this added cost may not be offset by savings in future maintenance costs.

The first step in surface conditioning is to grind the surface with abrasive flap wheels. Several passes are required, each one made perpendicular to the previous one and with gradually finer abrasives until a surface finish of 40 microinches is achieved. This operation can be done by machine on an individual component prior to assembly. However, this is usually impractical because the component must then be protected from surface damage for the remainder of the manufacturing operation. Usually, the grinding is done near the end of the assembly sequence with hand tools. This is a time consuming operation, especially if as-welded surfaces are involved.

The second step in surface conditioning, usually done on site, is to use an electrolyte and a shaped cathode to further improve the surface finish. A finish better than 10 microinches can be achieved.

Fig. 16 Steam generator lower section shipped by rail.

Shipping

Steam generators may be shipped by rail (Fig. 16). Special beams are often fabricated so that several cranes can be joined to lift the heavy vessels onto the rail cars. The load must be centered on the car for even weight distribution. If the vessel is long, it can span two railway cars. However, if it is short, a single heavy duty flat car is sufficient. The vessel is oriented to minimize its width and height and is loaded onto bunks fastened to the car. Loads up to 16 ft (4.9 m) in diameter can be shipped along most routes in the United States and Canada. Shipping larger units requires measurement of bridge clearances and other potential obstructions. If the vessel spans two cars, the bunks must be designed to allow movement of the load while negotiating curves. Additional counterweights may be attached to the rail car to lower the center of gravity.

Manufacture of pressurizers

Pressurizers are cylindrical vessels which help stabilize the pressure in the primary heat transport system. (See Chapters 47 and 50.) Each unit is fitted with special penetrations in which electrical heaters are installed.

Pressure boundary assembly

The pressure boundary consists of cylindrical shells and hemispherical heads. The manufacturing methods and quality assurance requirements for these components are the same as those described previously for steam generators.

Heater connections and installation

The heaters are the direct immersion type, sheathed in stainless steel or Inconel and assembled in bundles. Each of the two or three bundles is field assembled through penetrations in the vessel wall and is sealed by means of a bolted closure. The closure is sealed by gaskets or patented mechanical seals. An electrical connection is then made to the end of each heater using a special insulated fitting.

Postweld heat treatment

Postweld heat treatment of pressurizers is simpler than that of steam generators because there are no complex geometries or internals which limit heating and cooling rates. PWHT is done by putting the completed pressurizer in a gas furnace and heating to 1100F (593C).

Some patented mechanical seals used for the heater connections must be installed prior to PWHT because of their required machined surface finish. This machining can not be done after the component is welded onto the vessel because of inadequate access for specialized equipment. As a result, the sealing surface is protected from the furnace environment by coating it with an anti scaling compound. Some minor surface dressing of the sealing surface may be required after PWHT.

Water chemistry monitoring station.

Chapter 53
Water Chemistry Considerations in Nuclear Systems

Water chemistry control and corrosion management are important elements of a comprehensive nuclear steam supply system operating and maintenance program. This is particularly important for nuclear systems because any loss of availability or performance can have a dramatic cost impact.

Many of the elements of water treatment, water chemistry and corrosion discussed in Chapter 42 for fossil fuel systems also apply to nuclear steam systems. Raw water treatment follows similar steps, and many of the Universal Pressure boiler practices such as all volatile treatment (AVT) and condensate polishing have generally been adopted in whole or in part for nuclear plants. The goals remain to minimize corrosion and deposition.

Key differences exist, however. In a pressurized water reactor (PWR) system, two independent coolant circuits are used — the primary or reactor coolant system and the secondary or steam generator condensate feedwater system. Additional differences include: more extensive use of high alloy steel, nickel based alloys and other nonsteel materials; generation of oxygen in the reactor core; the presence of high levels of boric acid in the reactor coolant; and the radiation environment. Because large quantities of thin wall nickel alloy tubing are used, some nuclear steam generators are particularly sensitive to denting, intergranular attack/stress corrosion cracking, crevice corrosion, and pitting, among other localized phenomena.

Although corrosion control systems are very effective, accumulations of corrosion products and deposits are unavoidable during nuclear equipment design life. Proper chemical cleaning processes provide reliable and safe methods for removing and disposing of this accumulated debris.

Water chemistry guidelines and procedures have become more complex as the intricacies of nuclear plants have become better understood with operating experience. Extensive Babcock & Wilcox (B&W) guidelines are presented in Reference 1. The Electric Power Research Institute (EPRI) has also published manuals on both primary[2] and secondary[3] side coolant chemistry, while the Institute of Nuclear Power Operations (INPO) nuclear system chemistry guidelines are presented in Reference 4. The following sections provide brief overview remarks in the following areas:

1. water treatment and conditioning,
2. corrosion, and
3. chemical cleaning.

These sections build upon the general water technology overview presented in Chapter 42.

Water treatment and conditioning

Reactor coolant

In the reactor coolant system (see Chapter 50), the water is conditioned to prevent corrosion by oxygen which is generated by dissociation of the water coolant in the reactor. To minimize oxygen, an excess of hydrogen is maintained. Another important consideration is that the metal enclosure of the primary system consists principally of stainless steel; therefore, halogen impurities (chlorides and fluorides) must be essentially eliminated. Lithium-7, which is not activated by neutrons, is used to control pH and corrosion. An additional consideration is the use of large quantities of boric acid in solution as a reactor control mechanism.

Detailed water chemistry guidelines for the primary side of pressurized water reactor systems have become very complex. Table 1 provides a sample set of selected guidelines for power operation.

Steam generator feedwater

Once-through steam generator The chemistry parameters for B&W once-through steam generator (OTSG) feedwater are summarized in Table 2. These feedwater quality parameters are for normal operation of the plant above 15% full power. The listed chemical parameters are considered for either control or diagnostic use. Control parameters are those which require strict adherence due to steam generator material integrity and performance considerations. Diagnostic parameters assist the plant staff in interpreting steam generator and steam plant chemistry variations.

Sodium is an important parameter for safe turbine operation. The guidelines recommend a 3 ppb maximum for feedwater because sodium is soluble and carried over in the steam. Sodium can be present if there is leakage in the condensate polishing demineralizer effluent. Sodium can also be recycled with the moisture separator drains.

Chloride compounds, such as sodium chloride, can also affect turbine operation. For this reason, the guidelines recommend 5 ppb maximum for the feedwater. The primary source of chloride is leakage into the effluent of the

Table 1
Sample B&W Reactor Coolant Chemistry Guidelines[1] for Power Operations (Note 1)

Parameters	Normal Level	Type of Parameter (Note 2)
Boric acid, ppm	Note 3	D
Lithium as Li-7, ppm	Note 4	C
Conductivity at 77F (25C), μS/cm	Note 4	D
pH at 77F (25C)	Note 4	D
Dissolved oxygen as O_2, ppm	≤ 0.1	C
Chloride as Cl^-, ppm	≤ 0.15	C
Fluoride as F^-, ppm	≤ 0.15	C
Sulfur as $SO_4^=$, ppm	≤ 0.05	D
Hydrogen as H_2, std cc/kg H_2O	25 to 50	C
Suspended solids, ppm	≤ 0.35	D

Notes:
1. Power operations — The reactor is critical and the primary system is >530F and 2155 psig; see Reference 1.
2. D — diagnostic parameter; C — control parameter.
3. Boric acid is used as a chemical shim or supplement for the control rods. Concentration varies with time of core life and operating needs. Typical range: 100 to 13,000 ppm.
4. Lithium-7, conductivity and pH limits are interrelated and are varied in coordination with the boric acid concentration and unit operation (see Reference 1). The typical concentration range for lithium is from 0.2 to 2.5 ppm.

condensate polishing demineralizers (or *breakthrough* of the demineralizers).

Specific or total conductivity can be used to indicate the ammonia concentration or pH of the feedwater. For example, a pH of 9.3, or an ammonia concentration of 0.7 ppm, is equivalent to a pure water conductivity of 5.2 μS/cm (microSiemens per cm) at 77F (25C). However, the presence of ionic impurities can bias these relationships. Therefore, conductivity should be used only as a qualitative indication of ammonia and pH levels.

Although no specification is listed, care should be taken to eliminate, during pretreatment, hardness constituents which can be deposited in the steam generator.

Organics must be strictly limited. Organics with high molecular weights, such as oils and lubricants, can foul ion exchange resins in condensate polishing systems. Organics with low molecular weights can form organic acids and carbon dioxide, both of which can reduce the pH and increase the cation conductivity. These organics can also reduce resin life. Organic compounds with high molecular weights can not be measured at the levels of concern in the steam plant. However, there are techniques for measuring the total organic carbon (TOC) from compounds with low molecular weights in the ppb range. TOC can then be used to monitor some of the organics in the steam plant. In particular, TOC can establish a baseline for low molecular weight organics from which the ingress of these organics can be evaluated.

Use of morpholine for steam cycle pH control One of the disadvantages of using ammonia as the pH control additive is that it is volatilized with the steam from the steam generators through the high and low pressure turbines

Table 2
Steam Generator Water Quality Summary — Normal Operation

Parameter	Once-Through Steam Generator Feedwater (>15% Power)[1]		Recirculating Steam Generator[3] Feedwater		Recirculating Steam Generator[3] Blowdown	
Specific conductivity, μS/cm	Note 1	D (Note 2)	--	--	Note 5	D
Cation conductivity, μS/cm	≤ 0.2	C	≤ 0.2 (Note 3)	D	≤ 0.8	C
Dissolved oxygen, ppb	≤ 5	C	≤ 5	C	--	--
Hydrazine, ppb	> 3x[O_2], 20 ppb min	C	≥ 20	C	--	--
Ammonia, ppb	--	--	--	--	Note 5	D
Total silica, SiO_2, ppb	≤ 20	C	--	--	≤ 300	C
Sodium, Na, ppb	Note 1	D	≤ 3 (Note 3)	D	≤ 20	C
Chloride, Cl, ppb	Note 1	D	--	--	≤ 20	C
Total sulfur, SO_4, ppb	≤ 10	C	--	--	≤ 20	C
Total iron, Fe, ppb	≤ 10	C	≤ 20	C	--	--
Total copper, Cu, ppb	≤ 2	C	≤ 2	C	--	--
Lead as Pb, ppb	< 1	D	--	--	--	--
Suspended solids, ppb	≤ 10	D	--	--	--	--
Total hardness	Note 1	D	--	--	--	--
Organics or TOC	Note 1	D	--	--	--	--
pH at 25C (ferrous)	≥ 9.3	C	9.3 to 9.6 (Note 4)	C	9.0 to 9.5 (Note 4)	C
pH at 25C (ferrous/copper)	--	--	8.8 to 9.2	C	8.5 to 9.0	C

Notes:
1. Refer to the text for further discussion.
2. D — diagnostic parameter; C — control parameter.
3. Addresses turbine purity concerns. Lower limits may be needed to meet blowdown guidelines.
4. Lower value acceptable with condensate polisher in operation.
5. Ammonia, pH and conductivity consistency important.

to the main condenser. This characteristic results in a low pH on the shell side of the feedwater heaters and the steam extraction piping.

One solution to this problem is to use morpholine as the pH control additive. A comparison between the distribution ratio (concentration in vapor/concentration in water) of ammonia and morpholine is shown in Table 3.

The data indicate that about seven times as much morpholine is in the steam condensate at 100 psig (6.9 bar gauge) steam pressure compared to the ammonia treatment. This means that, for equivalent feedwater control ranges, morpholine will provide a more uniform pH throughout the steam cycle. This results in less erosion-corrosion of carbon steel in two-phase environments and less corrosion products being transported to the steam generators.

Morpholine also creates some undesirable effects. At steam generator operating temperatures, it partially decomposes to form other organic compounds (formic and acetic acids) and carbon dioxide (CO_2). The decomposition products can also affect the cation conductivity, which is the chief monitoring parameter for the feedwater. Most plants now use ion chromatographs which can measure these organic acids and other inorganic acids at the ppb levels normally present. EPRI and B&W recommend using a *corrected* cation conductivity from which the contribution of these organic acids has been subtracted.

All operating OTSG plants have converted to secondary system morpholine pH control.

Recirculating steam generator EPRI has also developed a set of standard guidelines for recirculating steam generator (RSG) feedwater and blowdown quality.[3] For reference purposes, selected summary information for normal power operation is provided in Table 2. As for the OTSG system, the guidelines are intended to minimize corrosion and enhance long term availability. The guidelines were formulated to retain operating flexibility while: 1) minimizing ingress of impurities consistent with current best achievable power plant practice, 2) maintaining levels consistent with known corrosion behavior, and 3) defining levels consistent with current commercial detection equipment.

AVT is now used to control water chemistry in most recirculating steam generators (see OTSG discussion). Boric acid feed is being used by some operators to combat various corrosion issues. (See References 3 and 7.)

Corrosion in nuclear systems

A significant portion of the outage time in water-cooled nuclear power generating systems is due to material related problems.[3] The environments causing the degradation include primary and secondary cooling fluids and irradiation produced by the nuclear fission process. Many components are affected.

Table 3
Distribution Ratio for Ammonia and Morpholine

Press., psig (bar gauge)	0 (0)	100 (6.9)	300 (20.7)	500 (34.5)	700 (48.3)	900 (62.1)	1100 (75.9)
Ammonia	10.1	7.1	6.3	5.3	4.2	3.9	3.6
Morpholine	0.4	0.98	1.4	1.2	1.2	1.3	1.3

Pressure vessels of low alloy steels, A533B, A508 and A302B undergo in-service degradation from neutron irradiation resulting in a loss of ductility. Pressure vessels and their welds are also subject to corrosion fatigue and stress corrosion.

Piping for both boiling water reactor (BWR) and PWR systems has a variety of material degradation mechanisms that have caused or could cause serious failures. These include intergranular stress corrosion cracking (IGSCC) of the austenitic stainless steel in both BWR loops and PWR primary piping systems, erosion-corrosion of carbon steel in PWR secondary piping and other components of the steam-water system, fatigue failures of carbon steel piping, and embrittlement of cast stainless steels due to thermal aging.

Four corrosion related phenomena have historically had significant impacts on nuclear steam generator availability: wastage, denting, pitting and fretting. *Wastage* is localized corrosion of the tubes near the tubesheet; this results from enrichment of sludge deposits with phosphate from the treatment chemicals originally used in recirculating steam generators. *Denting* is the physical deformation of the Alloy 600 tubes at the carbon steel support plates where the linear oxide growth rate of the corroding carbon steel causes a collapsing pressure on the tubes. (See Fig. 1.) *Pitting* is discussed in Chapter 42. *Fretting* is primarily the wearing of tubes in their supports due to flow induced vibration and can be aggravated by combined wear/erosion/corrosion mechanisms. These problems have generally been brought under control by switching to all volatile treatment for the secondary side water and by other operating modifications. In some instances, replacement of recirculating steam generators has been required because of excessive tube damage.

Two secondary side water chemistry related problems continue to cause major repairs: 1) secondary side caustic intergranular attack (IGA), and 2) stress corrosion cracking (SCC).

Caustic IGA/SCC occurs in secondary side crevices (open tubesheet and drilled support plate holes) and in sludge piles above the tubesheet due to the concentration of caustic-producing chemicals (including prior phosphate treatments) under heat transfer conditions. Higher temperature differences between the inside and outside of the tubes promote more caustic concentration in the

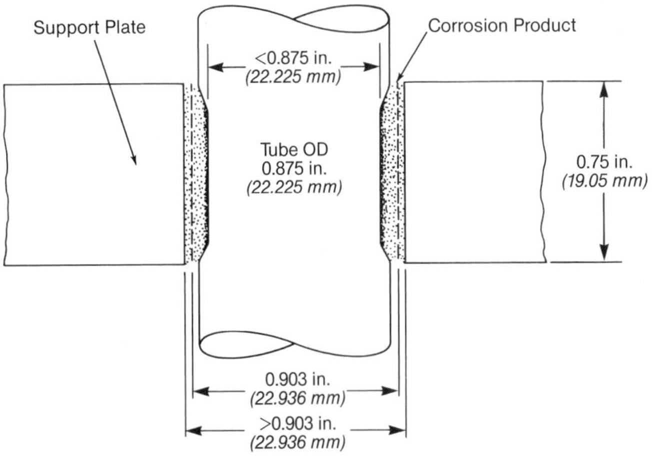

Fig. 1 Nuclear steam generator tube denting damage.

secondary side crevices.[5] Because of the location, attack within the tubesheet is not a safety concern, but tube attack at or above the secondary face of the tubesheet or at a support plate can lead to tube rupture.

Primary side cracking of mill annealed Alloy 600 tubing can occur as early as the first fuel cycle. Failures are located in regions of high residual stresses, e.g., in and near the transitions of tubes roll-expanded into tubesheets and in tight radius U-bends. The mill annealing temperature and carbon concentration of the tubing are critical factors leading to this mode of failure. Low annealing temperatures and low carbon concentrations are detrimental.

Consequently, Alloy 690, which has laboratory-demonstrated resistance to primary side cracking, is now being selected for many new and replacement steam generator tubes. This material, especially when thermally treated, also shows improvement over Alloy 600 in other environments.[6]

Reactor internals, such as bolts, springs, guide pins and other components, are usually made of high strength alloys (Alloy X-750, Alloy 718 or A286 stainless steel) or austenitic stainless steels (types 304 and 316). Irradiation assisted stress corrosion cracking (IASCC), fatigue and corrosion fatigue, and IGSCC are the reported failure modes.

Chemical cleaning of nuclear steam generators

The life and performance of many existing PWR steam generators can be compromised by the accumulation of corrosion products. The products appear as sludge on the tubesheet and they block flow holes in tube support plates.

Corrosion deposits, which consist primarily of iron and copper, can restrict flow, causing increased system pressure drop and loss of steam generator efficiency. Mechanical cleaning methods, such as water lancing, can remove deposits from accessible areas. However, chemical cleaning is the only means to completely remove deposits in a PWR steam generator. Fig. 2 shows how chemical cleaning restored the performance of a steam generator.

While chemical cleaning has a long, successful history in fossil fuel equipment, special considerations are required for nuclear steam generators. The PWR steam generator is essentially a shell and tube heat exchanger. Deposits are found on the shell side of this heat exchanger in contrast to fossil fuel boilers where deposits are on the inside of the tubes.

A PWR steam generator is primarily made with high nickel alloy tubing; most current United States steam generators use Alloy 600. These high nickel alloys do not permit the use of solvents containing halogens, such as hydrochloric acid. In addition, the shell side of a heat exchanger can not be positively flushed of all residual solvent. Therefore, only organic solvents, which decompose at system operating temperature, can be used for cleaning PWR steam generators. Several special solvents have been developed for different steam generator cleaning needs.

Solvents

EPRI/B&W magnetite solvent Based on the results of B&W research and the various EPRI sponsored programs, the solution listed in Table 4 was identified as being capable of dissolving magnetite deposits in a reasonable time without producing excessive corrosion.

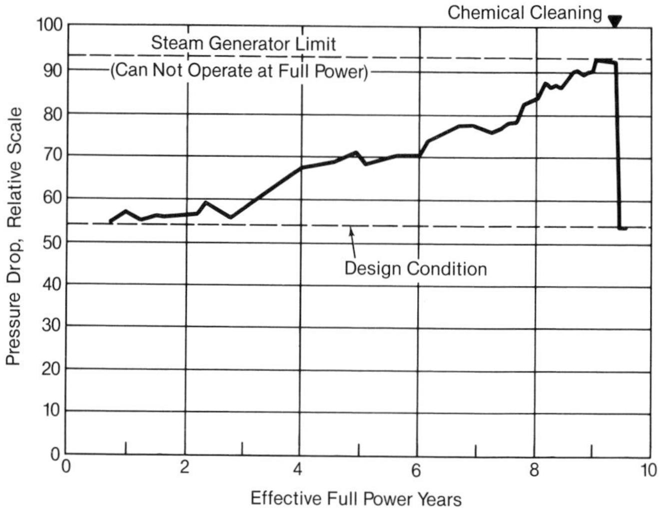

Fig. 2 Effect of chemical cleaning on nuclear steam generator performance.

Copper solvent The solvent developed for dissolving copper is comprised of the constituents listed in Table 5. Optimum copper solvent performance is achieved at the lowest ambient temperature. The solvent reactions are exothermic and can cause a temperature rise. Under no conditions should the initial application temperature exceed 120F (49C).

Alternate solvents In addition to the general purpose processes discussed above, proprietary solvent systems for cleaning PWR steam generators and a high temperature process developed by EPRI are also available. These alternate systems have been applied successfully to units in Europe. Because of differences in steam generator design and materials of construction, they have not been fully qualified for U.S. application.

Process qualification

Before a chemical cleaning process can be safely applied to a steam generator, site specific qualification testing is required. Details include:

1. solvent optimization testing,
2. determination of expected corrosion during solvent application,
3. identification of a corrosion monitoring scheme, and
4. definition of corrosion allowances.

Deposit samples and loading Solvent contact time is based on deposit dissolution studies. Plant deposit samples from the steam generator are required. The deposits in a PWR steam generator generally come from

Table 4
EPRI/B&W Magnetite Solvent for Cleaning Nuclear Steam Generators

100 to 200 g/l of EDTA*
10 g/l of hydrazine (N_2H_4)
5 to 10 ml/l CCI-801 corrosion inhibitor
Initial pH** adjusted to 7.0 with NH_4OH
Application temperature of 93C

* EDTA — ethylenediaminetetraacetic acid
** Room temperature pH of the solvent

Table 5
Copper Solvent for Cleaning Nuclear Steam Generators

50 to 100 g/l of EDTA
20 to 30 g/l of hydrogen peroxide (H_2O_2)
pH adjusted to 7.0 with NH_4OH followed by
pH adjusted to 9.2 to 10.0 with ethylenediamine (EDA)

the water and consist of corrosion products from unit materials. These deposits consist primarily of iron oxides (primarily magnetite) and metallic copper.

In addition to obtaining deposit samples, the total inventory of deposits in the steam generator must be estimated. The solvent capacity for dissolved species depends upon the initial chelate concentration and the volume of solvent. There is also a volumetric limit on the amount of dissolved species a solvent can hold.

Preparatory procedures include analysis of the deposit and cleaning tests to determine solvent strength and process conditions. The deposit analyses should include elemental analysis to detect the individual elements and x-ray diffraction to identify the major crystalline constituents present in the deposit.

Process optimization testing Deposit dissolution testing is required to obtain the most efficient chemical cleaning process for a specific steam generator. The tests should provide the basic solubility and kinetics data required to define the final cleaning procedures. Data from these tests include:

1. effect of contaminants and deposit morphology on dissolution rate,
2. amount of insoluble material present in the deposits,
3. effective solvent capacity, and
4. estimate of the total and cycle cleaning times.

Corrosion considerations An evaluation of the expected corrosion is essential in any nuclear steam generator chemical cleaning program. The objectives of this evaluation are: 1) to determine the effects of the solvents and cleaning processes on the steam generator materials, and 2) to define a chemical process operating envelope, which ensures that the corrosion allowances are not exceeded.

Solvent delivery system design

The components for a nuclear steam generator solvent delivery system are similar to those for the fossil fuel systems discussed in Chapter 42. Therefore, fossil chemical cleaning experience is a foundation for nuclear units.

The basic off-line cleaning process is a variation of the soak method of cleaning. This method allows the solvent to be mixed, heated and replenished outside the steam generator. The fill, soak and drain method also ensures that all steam generator areas periodically receive fresh solvent during the cleaning process.

As with fossil chemical cleanings the most important design considerations include reducing system leakage, controlling temperature and steam generator level, maintaining operational flexibility and redundancy, and ensuring personnel safety. Also, the design must permit installing the cleaning system's major components outside the reactor building prior to unit outage. Installation and removal of piping and equipment inside the reactor building can occur during a normal refueling outage.

Waste water

Nuclear steam generating plants produce waste water streams that are similar to those produced by fossil fuel plants. These wastes include makeup demineralizer regeneration effluents, floor drain flows, chlorinated cooling water and laboratory chemical discharges. Nuclear plants also generate radioactive liquid wastes.

The current trend in most nuclear plants is to recycle most process liquids. Normally, liquid waste water is segregated then processed to produce high quality recycled water. Examples of treatment processes include evaporators for concentrating primary coolant, ion exchange for cleanup of primary letdown, and chemical waste tanks for holding solutions to permit radioactive decay.

References

1. *Water Chemistry Manual, 177 Fuel Assembly Plants*, Report BAW-1385, B&W Nuclear Service Company, Lynchburg, Virginia, February 7, 1990.

2. *PWR Primary Water Chemistry Guidelines*, Revision 1, Report NP-7077-SR, Electric Power Research Institute, Palo Alto, California, September 1988.

3. *PWR Secondary Water Chemistry Guidelines*, Revision 2, Change No. 1 (7-17-91), Report EPRI NP-6239, Electric Power Research Institute, Palo Alto, California, December 1988.

4. *Guideline for Chemistry at Nuclear Power Stations*, Report INPO 88-021, Institute of Nuclear Power Operations, Atlanta, Georgia, December 1988.

5. Fletcher, W.D., and Picone, L.F., "Secondary water treatment for PWR steam generators," International Water Conference, Pittsburgh, October 24-26, 1972.

6. Theus, G. J., and Nakaysu, F., "Corrosion of steam generating tubing," appearing in *Proceedings of the 2nd International Topical Meeting on Nuclear Power Plant Thermal Hydraulics and Operation,* pp. 10-56 and 10-64, J. Nakabayashi and H. Nariai, eds., Atomic Energy Society of Japan, Tokyo, April 1986.

7. Cohen, P., ed., *The ASME Handbook on Water Technology for Thermal Power Systems*, American Society of Mechanical Engineers, New York, 1989.

Nuclear power plant control room (*courtesy of Duke Power Company*).

Chapter 54
Nuclear Operations

Nuclear power plant operations described in this chapter are generally applicable to light water moderated and cooled pressurized water reactor (PWR) systems, as depicted in Fig. 1, and are modeled on plants with Babcock & Wilcox (B&W) designed reactors. As design details vary among PWR types, operational differences are noted.

This chapter provides an overview of PWR operations including licensing, operator training, testing, criticality and startup, power escalation, routine power operation and scheduled and unscheduled outages. (See Chapters 47 and 50 for additional details on nuclear systems and equipment.)

Major licensing requirements

In the United States (U.S.), civilian use of nuclear energy was authorized by the Atomic Energy Act of 1954. The act also established an Atomic Energy Commission (AEC) for developing and regulating nuclear energy. A major objective of the AEC was to protect the public health and safety. The Energy Reorganization Act of 1974 replaced the AEC with the Nuclear Regulatory Commission (NRC), established to regulate the construction and operation of nuclear power plants as well as research facilities. These regulations are described in Section 10 of the Code of Federal Regulations. Public health and safety continue to be a major goal of the NRC.

The safety of nuclear power plants is based on:

1. *Defense in depth* This PWR feature is defined as three barriers between the nuclear fuel and the environment: 1) the fuel cladding, 2) the reactor coolant system (RC or RCS) pressure boundary, i.e., the reactor vessel, piping, pressurizer, steam generators and pumps, and 3) the reactor containment building, which is typically a pre-stressed concrete structure with an inner steel liner.
2. *Design approval* This includes NRC review and approval of the site and the design of safety systems and safety related equipment.

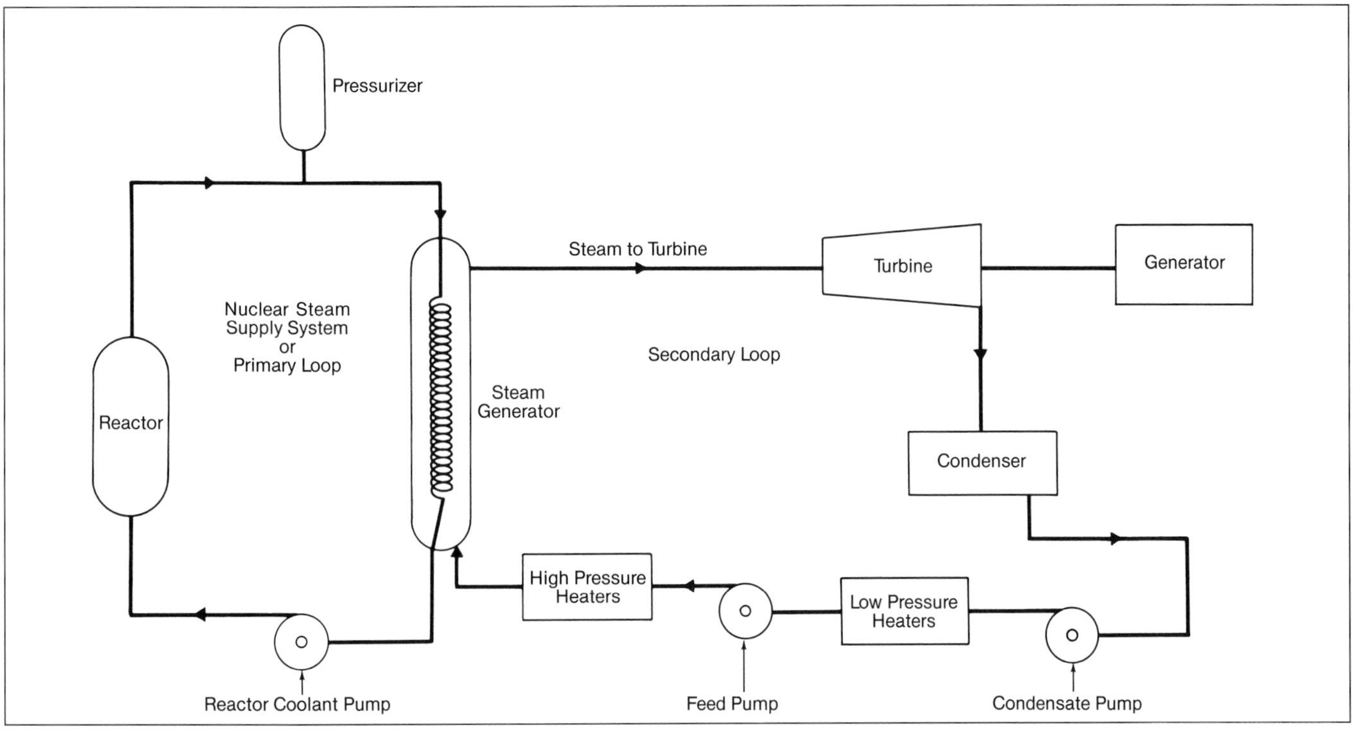

Fig. 1 Diagram of a pressurized water reactor (PWR) system.

3. *Quality assurance program* Construction of the station and erection of the equipment are guided by NRC approved quality assurance programs.
4. *Operating license review* Following construction, the NRC grants the utility (owner) an operating license. The license and the technical specifications define the operating bases of the plant.

The nuclear licensing steps, described as follows, are primarily designed to ensure public safety.

Plant licensing

In addition to licensing by the NRC, approvals for the plant site, construction and operating licenses must be obtained. Additional regulating agencies include the Federal Energy Regulatory Commission, the Environmental Protection Agency and the Corps of Engineers. Public utilities commission (PUC), state environmental control board and local governmental approvals are also often required for nuclear power plant siting, construction and operation.

NRC power plant licensing takes place in two major steps: the construction permit and the operating license. Completion of each step entails hearings conducted by an advisory safety licensing board.

In the construction permit stage, certain documentation must be prepared for the NRC. The proposed plant site is characterized, the preliminary design features are outlined, and conformance to the Code of Federal Regulations standards is presented.

A preliminary safety analysis report, which contains design and safety information, is prepared by the utility and its contractors, and is submitted to the NRC. The report includes the following topics:

1. general description of the station,
2. site characteristics,
3. design criteria for structures, components, equipment and systems,
4. reactor and fuel design,
5. RCS and its components,
6. engineered safety features,
7. instrumentation and controls,
8. electric power output,
9. auxiliary systems,
10. steam and power conversion system,
11. radioactive waste management,
12. radiation protection,
13. conduct of operations,
14. initial tests and operation,
15. accident analyses,
16. technical specifications, and
17. quality assurance measures.

The review includes questions by the NRC and subsequent responses from the plant owner. Along with the review, these become part of the public record and licensing process. The NRC staff prepares a safety evaluation report as a result of their review. This report leads to an Atomic Safety and Licensing Board hearing, which is open to the public. Upon approval by the Board a construction permit is issued by the NRC and plant construction can begin.

As the plant design becomes final and construction proceeds, the information contained in the review is revised and expanded. A complete description of the site characteristics, safety systems, quality assurance programs and technical specifications is included. This revised information becomes the final safety analysis report. This report is also submitted to the NRC for review and approval. As with the preliminary report, review includes NRC questions and owner responses. Following review, the NRC again prepares a safety evaluation report of the site and plant. This provides the basis for a second public hearing to approve an operating license for the utility. During construction, on-site NRC inspectors are assigned to assure conformance to the final report.

Upon receipt of the operating license, the utility performs system pre-operational and hot functional testing and prepares to load fuel. Approval by the NRC is required for fuel loading and low power operation at about 5% of full power. Following successful fuel loading and low power operation, the NRC grants permission for power escalation.

In 1989, the nuclear power plant approval procedure was revised to allow one step licensing. The objectives of this simplified procedure are to streamline the licensing process, to promote standardized vendor designs and to minimize NRC required design changes during and after construction.

The simplified rule differs from the original licensing process in that: 1) a whole plant design certification by the NRC is required, 2) following design certification, a combined construction and operating permit may be obtained, and 3) site permits may be obtained prior to design certification. The nuclear industry expects the simplified licensing process to reduce licensing and construction time from about 12 to 5 years.

Technical specifications

Technical specifications are, in the broadest sense, the licensed limits for operation. These specifications are based on the analyses and evaluations of the final safety analysis report and include the following:

Safety limits Limits are imposed on important process variables to prevent an uncontrolled release of radioactivity. If any limit is exceeded the reactor must be shut down. The licensee must notify the commission, review the matter and record the results of the review. The review includes the cause of the condition and the corrective action taken to prevent reoccurrence. Operation shall not be resumed until authorized by the commission.

Limiting conditions for operation These conditions are the lowest permissible equipment performance levels for safe operation of the facility. When a limiting condition for a nuclear reactor is exceeded, the licensee shall shut down the reactor or follow any remedial action permitted by the technical specifications until the condition can be met. In most cases, the licensee must notify the commission, review the incident, record the results of the review, and submit corrective actions.

Surveillance requirements Safety related nuclear power plant equipment must be tested periodically; frequency depends on the specific equipment. Typical equipment in the surveillance program includes the following systems:

1. reactor protection system,
2. safety feature actuation system,
3. control rods,
4. high pressure injection system,
5. low pressure injection system,

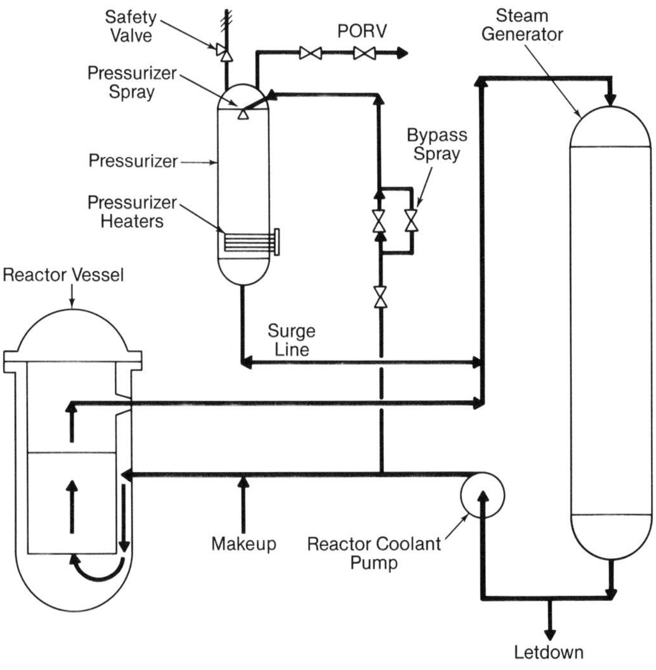

Fig. 2 Diagram of the reactor coolant system.

6. emergency feedwater system,
7. emergency diesel generators,
8. core flood tank isolation and check valves,
9. containment leak rate, and
10. all isolation valves.

If safety equipment fails a surveillance test, corrective action must be implemented within a specified time. If corrections can not be made within this time period, the plant must be shut down.

Design features Certain plant design features are included in the technical specifications. These features, such as construction materials and layout, would affect plant safety if altered.

Administrative controls Administrative controls are the provisions relating to organization, management, procedures, record keeping, review and audit, and reporting that are necessary to assure safe operation of the facility.

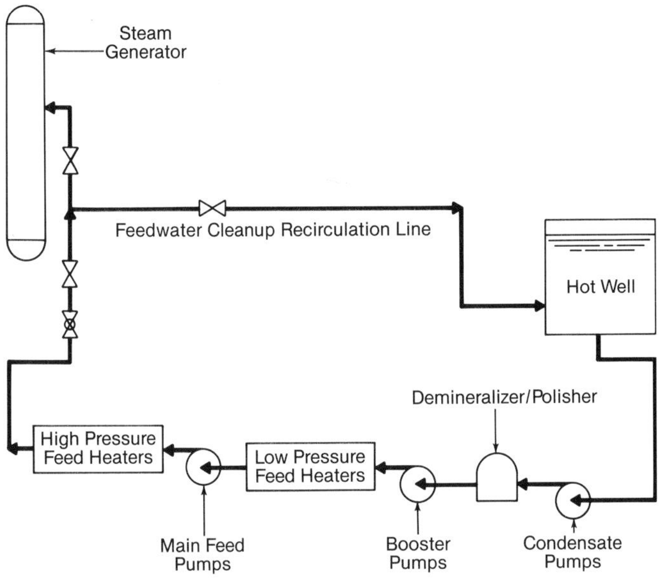

Fig. 3 Typical feedwater system.

Pre-operational testing

System checkout During nuclear power plant construction, it is desirable to test each system (electrical, hydraulic, pneumatic, etc.) as soon as possible. Pre-operational testing is integrated with the plant erection schedule so that system testing can begin as soon as support systems are available. Test procedures are prepared for each component and system. The tests are performed and witnessed by designated engineers and reviewers.

Pre-critical testing is performed on all systems and components. These tests are conducted on the reactor coolant system (Fig. 2), the feedwater system (Fig. 3), the steam system (Fig. 4), the reactor makeup and purification system (Fig. 5), the emergency feedwater system (Fig. 6), and the control and protective systems (Figs. 7, 9 and 10). In general, pre-critical testing consists of the following sequence of events:

1. *Cleaning and inspection* Fluid systems are flushed to remove construction debris and are inspected for cleanliness and proper installation of piping and components.
2. *Electrical tests* AC and DC power wiring and distributions are inspected and tested for proper hookup and circuit continuity. Rotating and moving equipment, such as pumps and valves, are given no-load tests to determine proper direction of motion.
3. *Instrumentation tests* Instrumentation and control wiring is inspected and tested for proper hookup and electrical continuity. The instrumentation (transmitters, temperature sensors, flow meters and level indicators) is calibrated and the control systems are tested. A control simulator is used by some utilities for cold system testing.
4. *Hydrostatic tests* All fluid systems and components except the RCS are cold tested hydrostatically to greater than design pressure. The RCS, because of the relatively high nil ductility temperatures of the reactor vessel welds, is hydrotested at a temperature slightly above ambient.
5. *Functional tests* Functional tests which demonstrate component performance are conducted. Typically, system flow rates, pump head capacity relationships, valve stroke times and controls are tested.

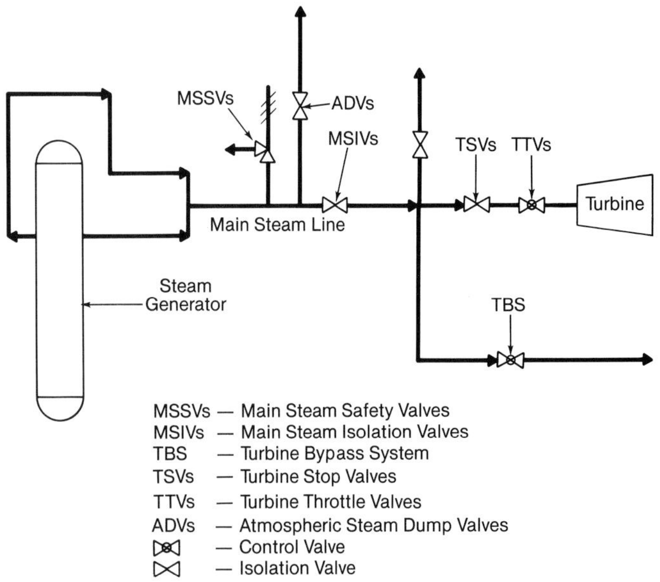

Fig. 4 Simplified steam system diagram.

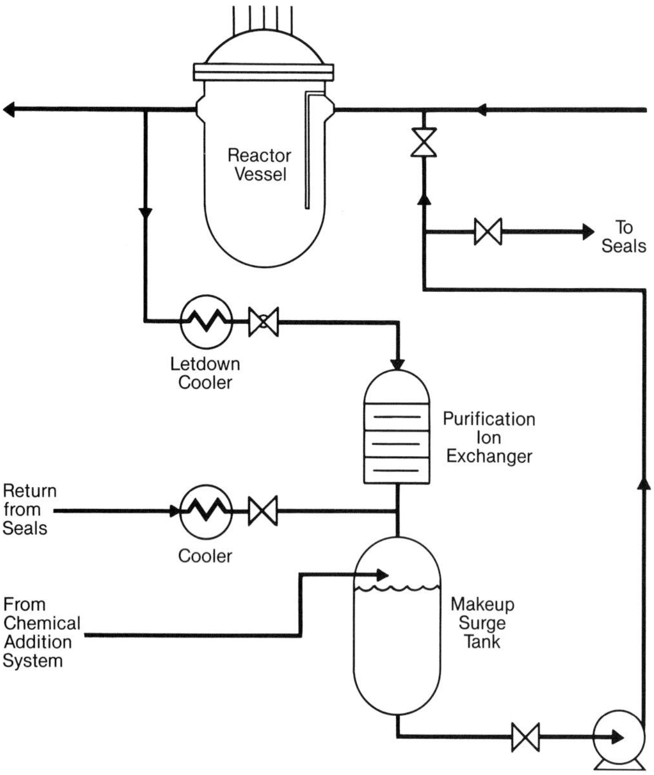

Fig. 5 Typical RCS makeup and purification system.

6. *Operational tests* Additional tests at near normal operating conditions are performed for low temperature and pressure systems. Instrument air systems, the spent fuel pool cooling system, component cooling water systems and the decay heat cooling systems are typical systems tested.

Performance of the high temperature and pressure systems, such as the RCS, the makeup and letdown system, the feedwater system, the steam system and the emergency feedwater system is verified during hot functional testing.

Hot functional testing Hot functional testing (HFT) occurs after all nuclear plant systems have been erected and have completed pre-operational testing. This testing provides warmup and operation at zero power operating conditions. The fuel is not loaded in the core for hot functional testing. In B&W plants, a plate is placed in the core region to simulate the core pressure drop for the RCS pump and flow testing.

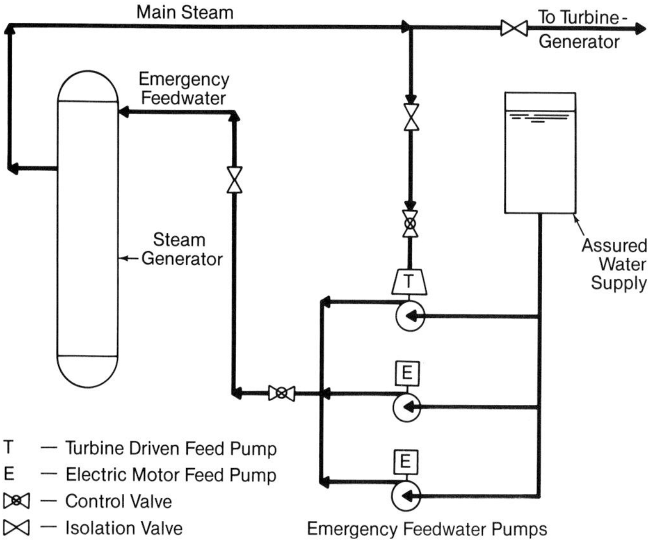

Fig. 6 Typical emergency feedwater system.

Before hot functional testing, the water quality of the RCS, the feedwater system and the steam generators is brought within specification. The RCS water requirements are shown in Table 1. The feedwater and steam generator water quality requirements are given in Table 2. Additional details concerning water quality are described in Chapter 53.

The RCS water quality is obtained by thorough cleaning of the system, use of demineralized borated water (about 2000 ppm boric acid) and by use of the makeup and purification system (Fig. 5). The purification system circulates the RCS volume about once per day through the letdown demineralizers.

The steam generator and feedwater system water quality is obtained by thoroughly cleaning and flushing the system, by raising the feedwater temperature to about 180F (83C) and by circulating the feedwater through a cleanup loop. The steam generator water level is then established with clean feedwater and the warmup can begin.

The hot functional test begins by establishing water levels in the RCS, as indicated by the pressurizer level, in the steam generators and in the condenser hot well.

The pressurizer heaters (Fig. 2) are energized to raise the RCS to a pressure which provides a net positive suction head of about 250 psig (17 bar gauge). This permits operation of the reactor coolant pumps. Upon reaching this pressure, the RC pumps are energized to heat the reactor coolant and steam systems at a rate of about 30F/h (17C/h) from about 70 to 530F (21 to 277C).

There are several temperature hold points during the HFT warmup at which specific tests are performed. Typical testing includes the pressurizer spray bypass flow cali-

Table 1
PWR Coolant Water Quality

Total solids (dissolved and undissolved, excluding lithium-7 hydroxide and boric acid), ppm maximum	1.0
Boric acid, ppm	<13,000
Lithium-7, ppm	0.2 to 2.0
pH at 77F (25C)	4.8 to 8.5
Dissolved oxygen as O_2, ppm	0.1 (controlled by excess hydrogen)
Chlorides as Cl^-, ppm maximum	0.1
Fluorides as F^-, ppm maximum	0.1
Hydrogen as H_2, standard cc/kg H_2O	15 to 40
Total gas, standard cc/kg H_2O, maximum	100

Table 2
PWR Steam Generator Feedwater Quality

Maximum total solids (dissolved and suspended), ppb	50
Maximum dissolved oxygen, ppb	7
Maximum total silica (as SiO_2), ppb	20
Maximum total iron (as Fe), ppb	10
Maximum total copper (as Cu), ppb	2
pH at 77F (25C) (adjusted with ammonia)	9.3 to 9.5
Organics:	
Lead	0
Maximum cation conductivity, ohm/cm	0.5

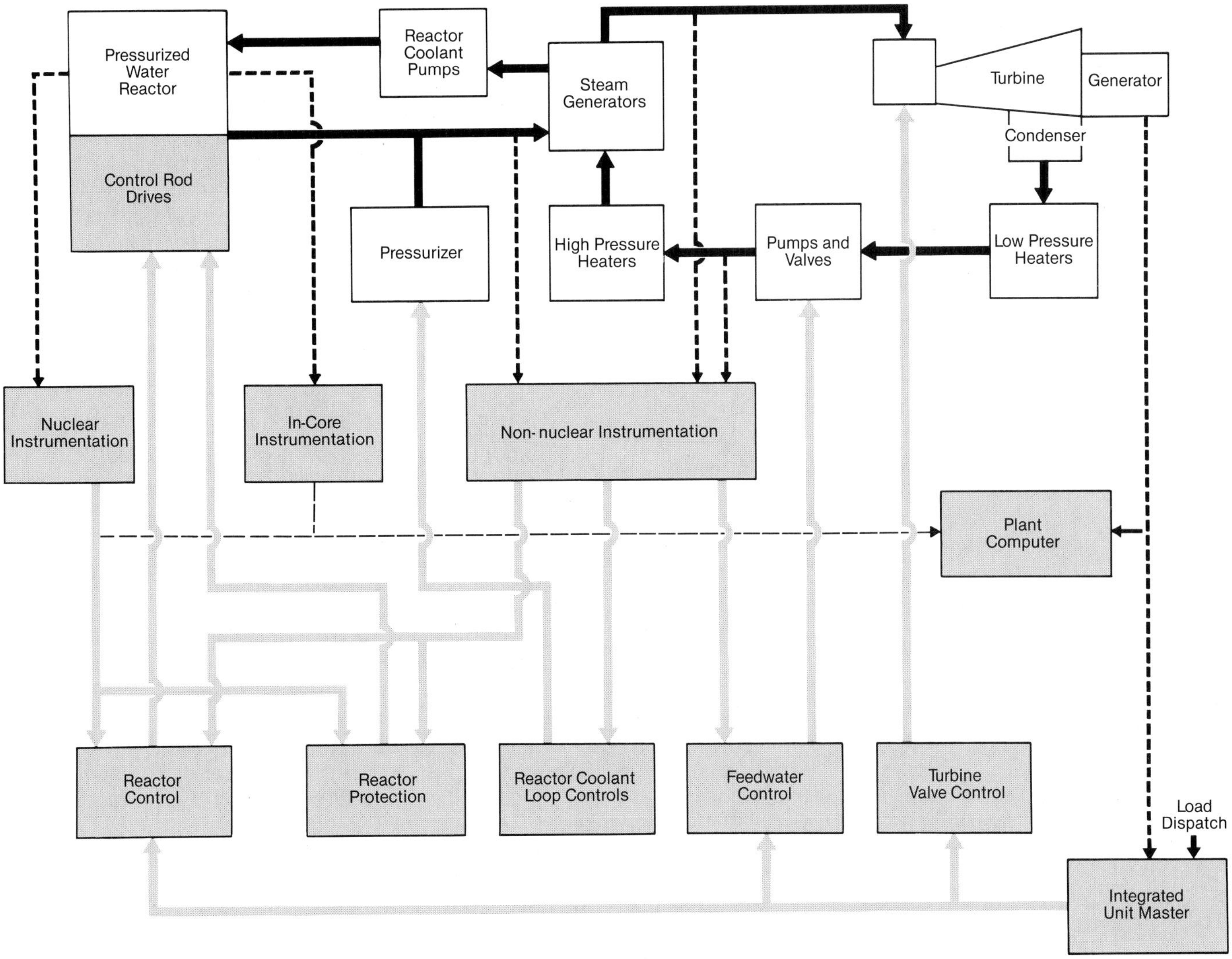

Fig. 7 Instrumentation, control and protection functions for a nuclear PWR station.

bration. This bypass has a trickle flow of about 1 GPM (0.06 l/s), which keeps the pressurizer spray line and the spray nozzle at a relatively uniform temperature.

At about 600 psig (42 bar gauge), the RCS high pressure safety injection pumps (Fig. 8) are tested to demonstrate flow capability and to adjust the safety injection flow distribution.

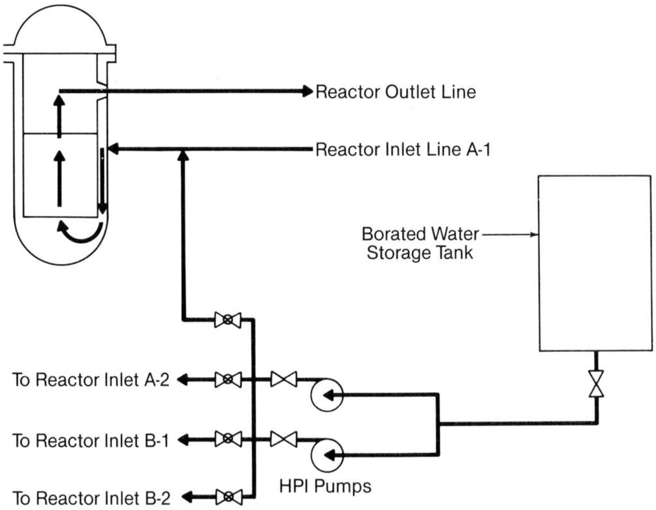

Fig. 8 Diagram of the high pressure injection (HPI) system.

Following testing at the hold points, system heating continues to the hot zero power (HZP) level. Typically, in pressurized water reactors, the hot zero power temperature and pressure ranges are:

RC temperature — 500 to 550F (260 to 288C)
RC pressure — 2100 to 2200 psig (145 to 152 bar gauge)
Steam temperature — 525 to 570F (274 to 299C)
Steam pressure — 800 to 1200 psig (55 to 83 bar gauge)

Before fuel loading, the reactor protection system (RPS) and the control systems (Figs. 10a, 10b and 11) are calibrated and checked.

In addition, all protective system instrumentation, including the out of core neutron detectors (Fig. 10b), RCS pressure transmitters and temperature sensors, nuclear instrumentation and transmitters, RC pump monitors and the feedwater pump turbine hydraulic oil pressure switches, are calibrated and tested.

The reactor coolant system pressure and inventory, feedwater, steam pressure and electrical generation are controlled by the integrated control system (ICS) (Figs. 10a and 10b) and the non-nuclear instrumentation (NNI) (Fig. 7). The NNI provides most of the instrumentation for the reactor, feed and steam systems, including signal inputs to the ICS, the control console indicators and recorders, and the plant computer as well as many of the

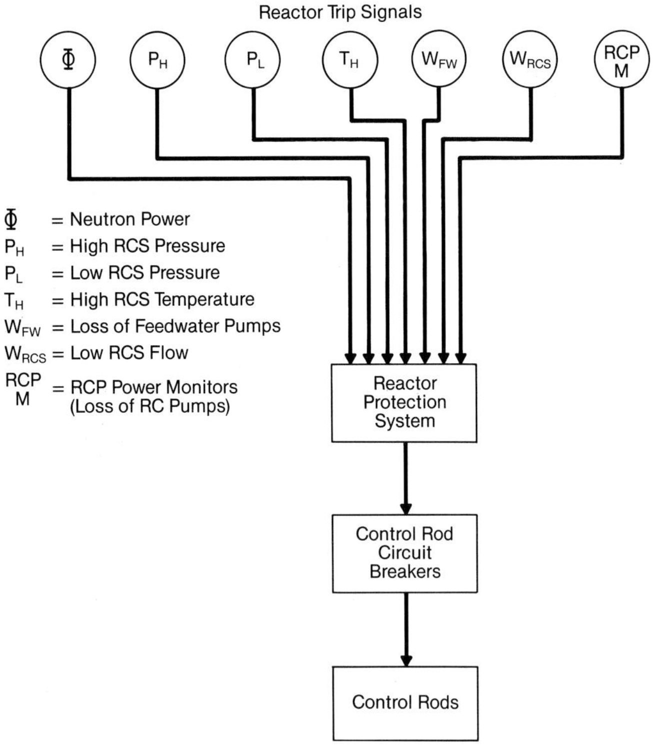

Fig. 9 Diagram of the reactor protection system (RPS).

test inputs to the reactimeter. The ICS is made up of unit load demand controls, integrated master controls, turbine and turbine bypass system controls, feedwater controls and reactor control subsystems. The ICS also controls the reactor, feedwater flow and the turbine-generator output to meet the electrical power demand.

The NNI also controls the RC system makeup flow, the pressurizer heaters, the pressurizer power operated relief valve (PORV) and the reactor coolant pump seal flow.

The feedwater and steam control subsections of the ICS provide automatic control over the full power range; manual reactor control is used up to 15% power and automatic control is available from 15% to full power. Manual control can be used in each subsection of the ICS or its actuated components such as control rods, feedwater valves, feed pump turbine and turbine bypass valves. Alternately, the turbine can be manually operated from the control room.

Below 15% power, the ICS maintains the steam generator water level at a constant set point and the turbine bypass system similarly controls the steam pressure. The RCS temperature rises with power. The temperature is typically 532F (278C) at zero load and increases to about 580F (304C) at 15% power. As the load increases to full power, the temperature remains at about 580F (304C). Above 15% power, the RCS temperature and steam pressure are controlled by the ICS.

PWRs designed and built by other suppliers use recirculating steam generators (RSG), which have different operating characteristics than the once-through steam generators (OTSG). (See Chapter 47.) In the RSG plants, the RCS average temperature rises with the power level, the steam generator operating pressure drops with increasing power (Fig. 12), and the steam generator water level is controlled to a constant RSG secondary inventory or to a relatively narrow variable RSG band.

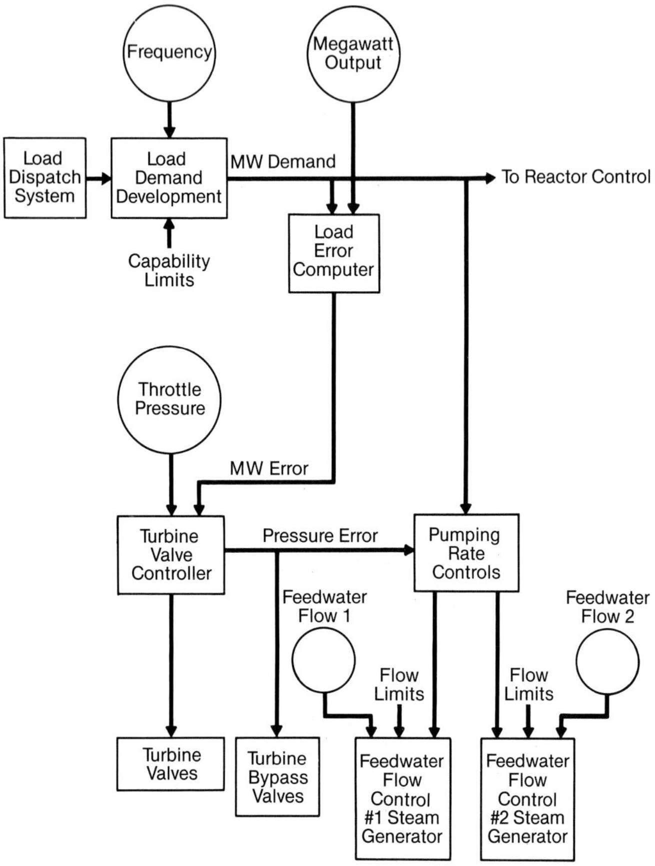

Fig. 10a Integrated control system.

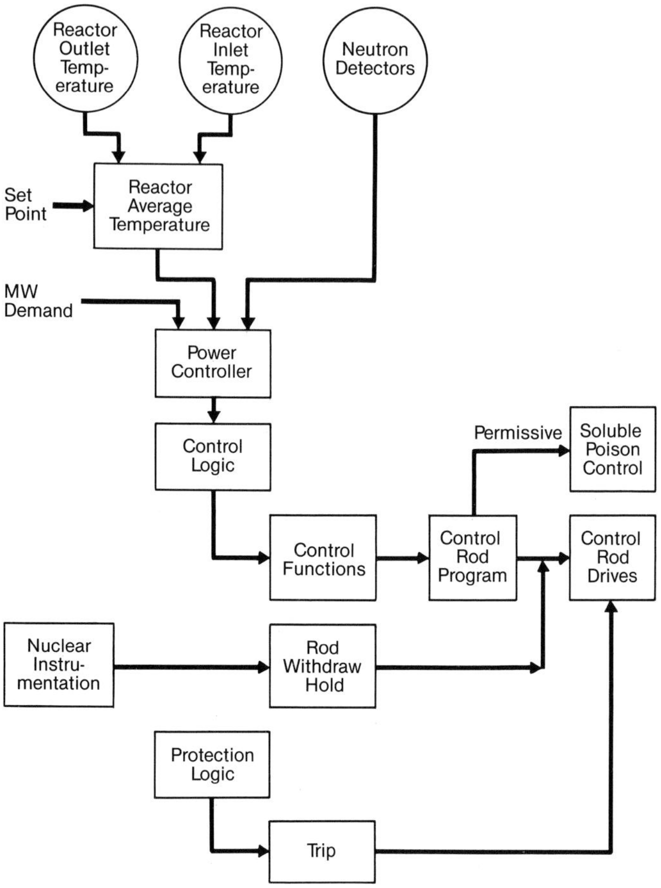

Fig. 10b Integrated control system — reactor controls.

An RSG plant's average RCS temperature is controlled to a variable set point; the feedwater flow is controlled by a multi-element unit which monitors steam flow, feedwater flow, steam generator water level and, for some plants, RC temperature and steam pressure.

The RSG units typically control electrical output by controlling the turbine to a variable first stage pressure signal. This control allows the steam pressure to change with power level.

Typical functional testing performed at HZP includes the following system tests:

1. pressurizer heaters,
2. pressurizer spray,
3. PORV,
4. RC pumps,
5. RC system flow,
6. makeup and purification,
7. RC pump seal flow and flow control,
8. feedwater and condensate pumps,
9. turbine bypass valves and atmospheric dump valves,
10. emergency feedwater pumps and system,
11. high pressure injection pumps,
12. low pressure injection pumps,
13. decay heat removal system,
14. reactor building cooling,
15. turbine roll,
16. loose parts monitoring system,
17. instrumentation and control system, and
18. baseline characteristics of operating equipment.

The duration of an HFT is usually 500 hours for a first of a kind unit and about 250 hours for all other units.

A significant portion of the HFT period is allocated to operator training because the plant conditions are near those of normal operation. In performing the hot tests, the operators gain valuable experience in managing the systems under normal and abnormal conditions. This operating experience constitutes one of the final stages of operator training prior to the NRC operator's license examination.

After completion of the test, the RCS is cooled and depressurized and the reactor vessel head is removed. The steam, feedwater and fluid safety systems are inspected and modified as necessary. Following successful HFT, permission for fuel loading is granted by the NRC.

Operator training

The Code of Federal Regulations requires that the reactor operators and supervisory personnel be licensed. An individual who has the basic operational qualifications is called a *reactor operator*; a *senior reactor operator* has completed more intensive training. The utilities develop, manage and conduct NRC approved operator training and requalification programs to satisfy regulatory requirements and to maintain the required operating staff. The training programs cover all facets of nuclear operations, including equipment malfunctions, design faults and severe transients.

Operator trainees develop an understanding of the plant by studying the equipment descriptions, drawings and instruction manuals. Plant walkdowns of all accessible equipment are performed as part of the training program. Oral and written examinations are given by the NRC. In addition, trainees receive hands on experience at the station, at an operating nuclear plant or at plant simulators.

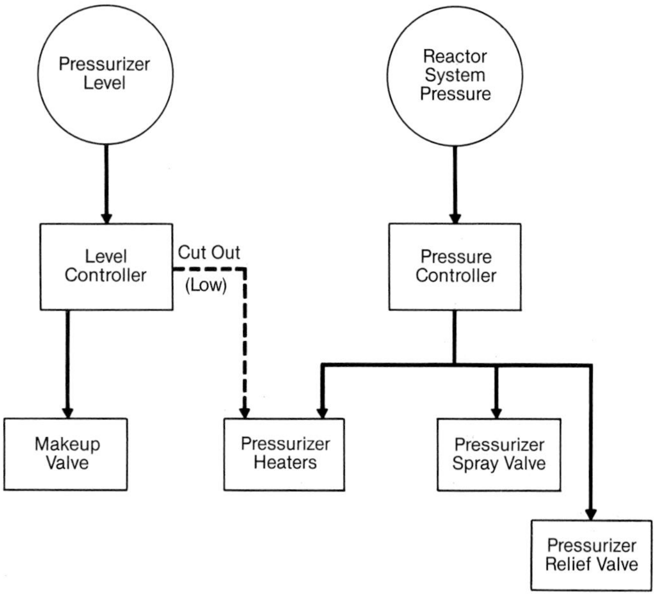

Fig. 11 Reactor coolant loop controls.

An operator's license is issued for a period of two years and is renewable by requalification. Requalification training is scheduled during the operator's normal shift and is conducted by the station training department. This training is similar to the previously described instruction and includes lectures, on the job training, review of facility or procedural changes, and additional testing by the NRC.

Fuel loading and zero power testing

The nuclear fuel (see Chapter 48) is first loaded into the reactor vessel only after pre-loading tests and practice fuel handling operations are completed. Neutron detectors and sources, placed in the reactor vessel, are used to monitor neutron multiplication during the core loading process.

Fuel loading takes place with the reactor vessel (Chapter 47) and the fuel handling pool filled with ambient temperature demineralized water. The borated water is a neutron poison and assures a multiplication factor of 0.95 or less with a fully loaded core.

With the system prepared fuel loading begins. Neutron multiplication measurements, slow fuel assembly inser-

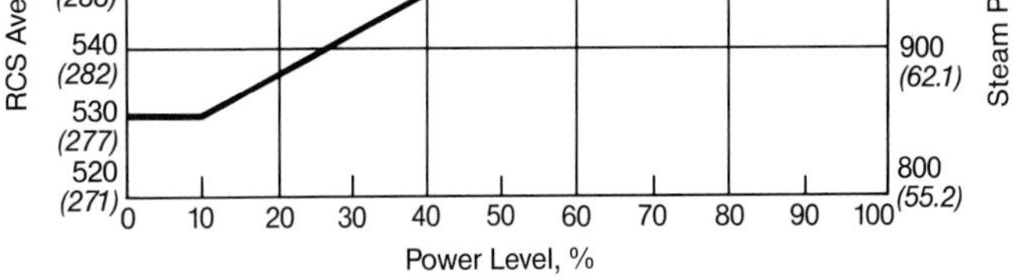

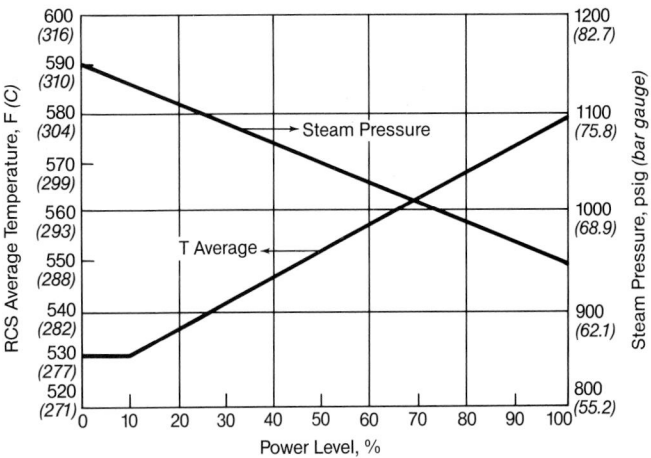

Fig. 12 Typical recirculating steam generator plant RCS average temperature and steam pressure.

tion and frequent pauses to relocate sources and neutron monitors take place until all fuel assemblies are installed. After loading, the system is closed and prepared for low power. Testing following fuel loading includes calibrating neutron source range detectors, assuring sufficient source range signal for warmup to hot zero power and verification that the reactor is sufficiently subcritical for the warmup.

Following completion of this testing, safety control rod groups are withdrawn to provide a subcritical shutdown margin should a reactor trip be required during the warmup.

After reaching about 250 psig (17 bar gauge) in the RCS and with one or more RCS pumps running, the system high points and control rod drives are vented to release noncondensable gases. Venting of the drives is particularly important because gas in the control drive path can interfere with the hydraulic snubbers which prevent impact of the rods upon trip.

Upon reaching HZP conditions, excess heat from the RCS pumps is removed by the turbine bypass system (Fig. 4). This maintains constant RCS temperature and steam pressure conditions.

Initial reactor criticality is achieved in one of two ways. One method is to dilute the boric acid concentration in the RCS until a subcritical margin is achieved. Criticality is then achieved by slowly withdrawing the control rods. A second method is preferred by most B&W plant owners. All control rods are withdrawn while maintaining subcriticality by boric acid concentration. The concentration is then diluted until criticality is achieved. After initial criticality is attained, a series of zero power physics tests are performed:

1. *Control rod reactivity worth* This verifies that the individual rod worth, control rod group (banks) rod worth, and total rod worth are within acceptance criteria. The tests must show that the control rod worth is capable of power operation. Performance of the safety function, which brings the reactor to a subcritical condition following a trip, is also verified. Typical control rod reactivity worth for each of seven banks is shown in Table 3.
2. *Core reactivity coefficients* The reactivity coefficients are the moderator deficit (or coefficient), the fuel temperature (Doppler) coefficient, the RCS pressure coefficient and the power (overall) coefficient. This testing verifies that the reactivity coefficients are within acceptance criteria and that the power coefficient is negative at levels above about 20% power.
3. *Soluble poison (boric acid) reactivity worth* Tests are conducted by exchanging control rod position with soluble poison concentration to determine that the reactivity worth of the poison is within core acceptance criteria.

   Boric acid dissolved in the reactor coolant is used to limit the core's reactivity throughout the fuel cycle. The boron concentration at HZP is about 1500 ppm. As xenon and other reactivity poisons build into the fuel and the fuel is depleted, the boron concentration is reduced to maintain full power (see Fig. 13).
4. *Nuclear instrumentation range and overlap tests* At criticality, the three neutron detector ranges (startup, intermediate and power) must be on scale and must have the proper overlap (Fig. 14). Testing is conducted at initial criticality to assure proper neutron detection signal strength.

Table 3
Control Rod Group Reactivity Worths at Hot Zero Power, Beginning of Fuel Cycle

Group	Purpose	Reactivity Worth % Δ K/K
1	Safety	0.95
2	Safety	3.05
3	Safety	.79
4	Safety	1.75
5	Control	1.15
6	Control	1.26
7	Control	1.17
Total		10.12

5. *Pseudo-control rod ejection tests* In addition to the standard control rod testing, pseudo-control rod tests are performed. This testing assures that the control rod with the most reactivity worth does not exceed the licensed limits.

The zero power physics tests as well as testing of the protection systems are performed to:

1. measure operational parameters,
2. measure performance against design criteria,
3. provide compliance with NRC requirements, and
4. obtain data for design improvements and potential power upgrades.

Most initial operation parameters are recorded by special equipment as well as by the plant computer. The special equipment is often a data acquisition system, called a *reactimeter*, which uses portable digital computers.

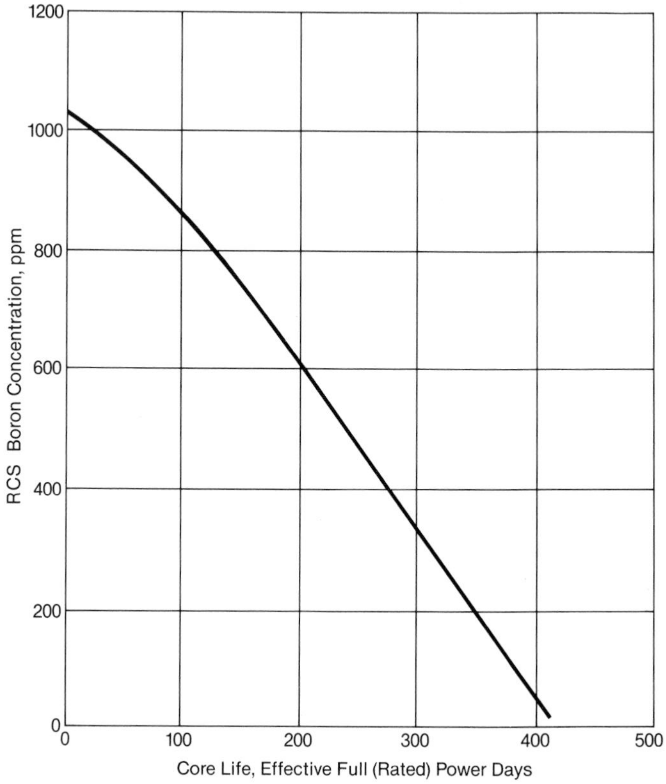

Fig. 13 Boron concentration in the RCS during a fuel cycle.

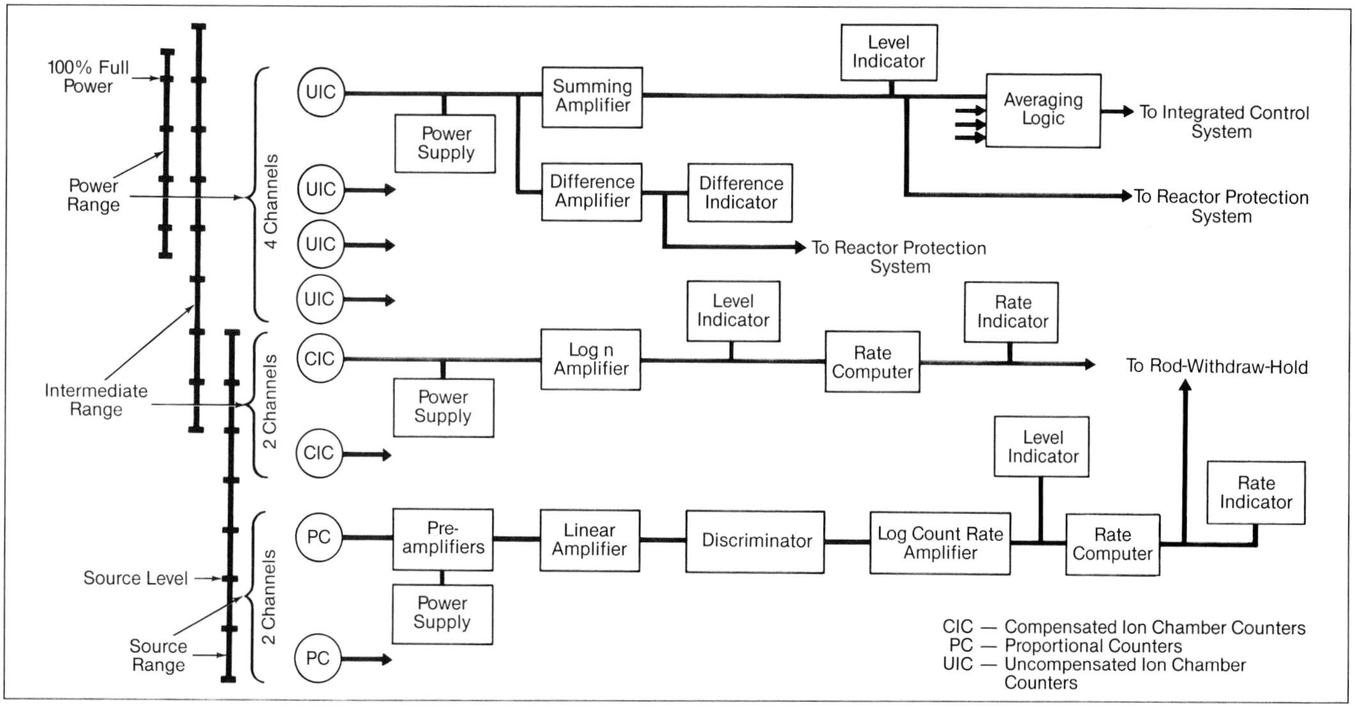

Fig. 14 Nuclear instrumentation and neutron overlap.

Power ascension tests

After completion of the zero power physics tests and low power operation (5% or less), the power is raised in stages. Typically, the power plateaus are 15, 45, and 75% and full power.

The objectives of power escalation include:

1. determining the plant operating characteristics,
2. familiarizing the operators with the plant,
3. verifying that the plant is capable of operating safely and within the design and technical specifications,
4. verifying that the operating procedures provide reliable operation, and
5. bringing the unit to rated capacity.

Testing at each stage includes operating the reactor systems and the entire power plant. The test time at each plateau is sufficient to provide adequate operator familiarization and training.

During the power escalation sequence, numerous tests are performed. The testing includes:

1. radiochemistry,
2. RCS chemistry,
3. steam generator chemistry,
4. plant shutdown from outside the control room,
5. RCS hot leakage,
6. vibration and loose parts baseline,
7. effluent monitoring,
8. nuclear steam system (NSS) heat balance,
9. ICS steady-state,
10. ICS transient,
11. loss of off-site power,
12. in-core monitoring system,
13. nuclear instrumentation calibration,
14. biological shield survey,
15. plant computer,
16. core power distribution,
17. ejected rod,
18. dropped rod,
19. reactivity coefficients at power,
20. power imbalance detector correlation,
21. reactor trip,
22. turbine trip, and
23. natural circulation (first of a kind units).

Operation to the first test plateau is with manual control of the reactor and with the turbine-generator off-line. Plant heat removal is by the turbine bypass system, which controls steam flow from the steam generators to the condenser. In this low power range, the operating performance of the turbine bypass system, feedwater system, and the feedwater and steam pressure controls is tested for the first time. Initial heat balances are performed by calculating the primary heat balances (RCS flow and temperature change across the reactor) and the secondary heat balances (feedwater flow and enthalpy rise across the steam generators).

The heat balances, in addition to RCS flow measurements, permit calibrating the nuclear instrumentation using the plant thermal power as a baseline. The secondary heat balances also provide an initial cross check of the RCS flow measurements.

Near the completion of 15% power level testing, a reactor shutdown from outside the control room is performed. Reactor control is from a remote panel and the reactor is brought to and maintained at HZP conditions. This test demonstrates safe control and shutdown of the unit should a fire or toxic gas require the operators to evacuate the main control room.

The performance at the next power plateau is extrapolated from the previous plateau. For example, operation at the 45% power plateau is estimated from the operating experience at the 15% plateau. Following satisfactory operation at 15% power, the system may be placed in automatic control and the power level raised to the 45% level. Power escalation, however, is typically performed with the reactor in manual control and the feedwater and steam systems in automatic control. As the power is raised above 15%, the turbine-generator is brought up to speed, loaded and synchronized to the electrical grid. At the 45% power level, heat balances are again calculated and the neutron power and RCS flow are calibrated. Control system tests, consisting of decreasing and increasing power ramps, are also conducted. At the 45% power level, the in-core instrumentation (rhodium detectors) is on-line, and the core performance indicators can be investigated by mapping the core power distribution. Fig. 15 shows the significant reactor and steam system parameters during power ramp control testing at the 45% power plateau.

Following successful testing at the 45% power level, the power is raised to about 75%. During this increase, the second main feed pump is typically placed in operation at about 45 to 50% power. Heat balances at 75% power are again calculated for the reactor and the secondary systems, and the neutron detectors and RCS flow are calibrated.

Testing with three of the four reactor coolant pumps is conducted at 75% power. This provides the core and RCS temperature distribution characteristics. It also verifies proper feedwater flow to the steam generators. Fig. 16 shows several important parameters during a reactor coolant pump trip test with a reactor runback.

Ramp power changes at rates up to 10% per minute are also conducted at the 75% power level, similar to that shown in Fig. 15. This tests the control system operation and permits control setting changes.

As testing is completed at the 75% power level, the unit's performance is reviewed and its full power performance is extrapolated. The reactor is then gradually raised to full power and further testing preparations are made.

At full power, the RCS and the secondary system heat balances are again performed, the neutron instrumentation is calibrated and final reactor coolant flow calibrations are performed. Typical heat balance equations are as follows.

RCS heat balance

The heat balance equation for the RCS is:

$$Q_{RCS} = W_{RCA}(H_{OA} - H_{INA}) + W_{RCB}(H_{OB} - H_{INB}) + Q_{RCP} + W_{MU}H_{MU} - W_{LD}H_{LD} - Q_{HL} \quad (1)$$

where

Q_{RCS} = RCS power, Btu/h

W_{RCA} and W_{RCB} = RCS flow rates in loop A and loop B, respectively, taken from flow meters in each hot leg, lb/h

H_{OA} and H_{OB} = reactor outlet enthalpy of each loop, obtained from temperature and pressure measurements in each outlet line, Btu/lb

H_{INA} and H_{INB} = reactor inlet enthalpy of each loop, obtained from averaged temperatures of each inlet line and from the calculated pressures at the temperature sensors, Btu/lb

Q_{RCP} = hydraulic power input from the four RCS pumps, Btu/h

W_{MU} = makeup flow rate, lb/h

H_{MU} = enthalpy of makeup flow, obtained from makeup temperature and RCS pressure, Btu/lb

W_{LD} = letdown flow rate, lb/h

H_{LD} = enthalpy of letdown flow, from reactor inlet line from which letdown is taken, Btu/lb

Q_{HL} = estimated heat loss from the RCS, Btu/h

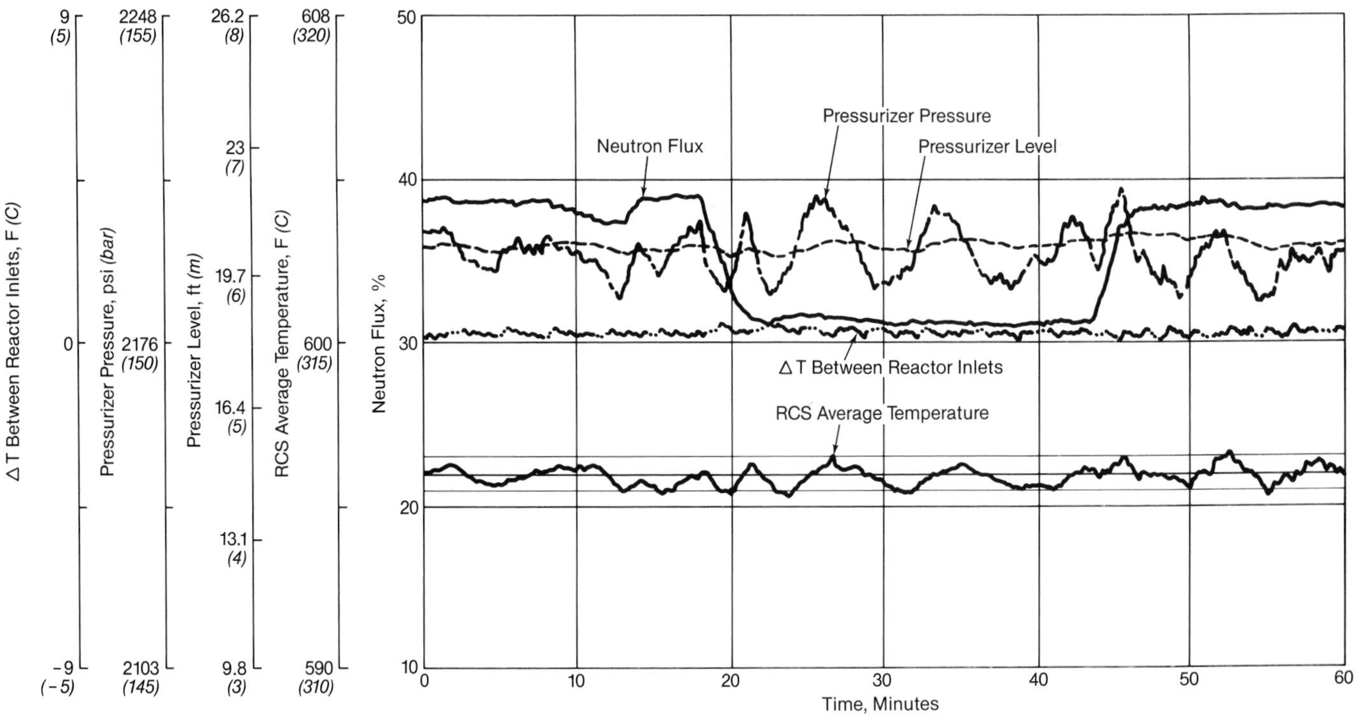

Fig. 15 Power escalation testing: control system ramp power changes.

At full power, the RCS heat losses are small compared to the RCS power and are usually neglected.

Secondary system heat balance The heat balance equation for the secondary system is:

$$Q_{SEC} = W_{FWA}(H_{STMA} - H_{FWA}) + W_{FWB}(H_{STMB} - H_{FWB}) - Q_{SHL} \quad (2)$$

where

Q_{SEC} = energy removed from the RCS, Btu/h

W_{FWA} and W_{FWB} = feedwater flow rates to the A and B steam generators, respectively, lb/h

H_{STMA} and H_{STMB} = enthalpy of superheated steam exiting each steam generator, obtained from temperature and pressure measurements at each generator's outlet piping, Btu/lb

H_{FWA} and H_{FWB} = enthalpy of feedwater flowing to each steam generator, obtained from temperature and pressure measurements in the feedwater lines, Btu/lb

Q_{SHL} = estimated heat losses from the secondary systems, Btu/h

RC flow calibration The secondary heat balance is usually more accurate than the primary balance because of the large difference between the feedwater temperature (and enthalpy) and the steam temperature (and enthalpy). Additionally, the feedwater flow meters are laboratory calibrated and special feedwater flow measurement equipment is often also used to obtain accurate feedwater flow measurement. The RC flow rate is calibrated to the secondary heat balance by the following relationship:

$$W_{RCA} = Q_{SECA} / (H_{OA} - H_{INA}) \quad (3)$$

$$W_{RCB} = Q_{SECB} / (H_{OB} - H_{INB}) \quad (4)$$

$$W_{RCS} = W_{RCA} + W_{RCB} \quad (5)$$

where

W_{RCS} = total RCS flow rate, lb/h

Q_{SECA} and Q_{SECB} = secondary heat balances for loops A and B, Btu/h

During steady-state, full power operation, power maps are obtained from the in-core neutron monitors and the core neutron flux distribution is evaluated against the acceptance criteria. Final tests of the reactivity coefficients, control rod worth and boron reactivity worth are also performed and compared with design values.

Several important safety tests are conducted at the 100% power level, including:

1. *RC flow coast down testing* For first of a kind plants, the RCS pumps are tripped in various combinations to characterize the flow coast down performance and to test the trip functions related to loss of RCS flow or pumps. A typical B&W RCS flow coast down during the first seconds after pump trip is shown in Fig. 17.
2. *Reactor trip testing* A reactor trip test is performed from full power to verify that the control rod insertion rates, the pressurizer performance and the control of steam pressure and feedwater flow are within acceptance criteria.
3. *Loss of off-site power* Testing of the unit's response to separation from the electrical grid is conducted to verify that performance during the subsequent reactor trip, startup of the emergency diesel generators, operation of the turbine driven emergency feedwater pumps and the natural circulation of the RCS are within acceptance criteria.

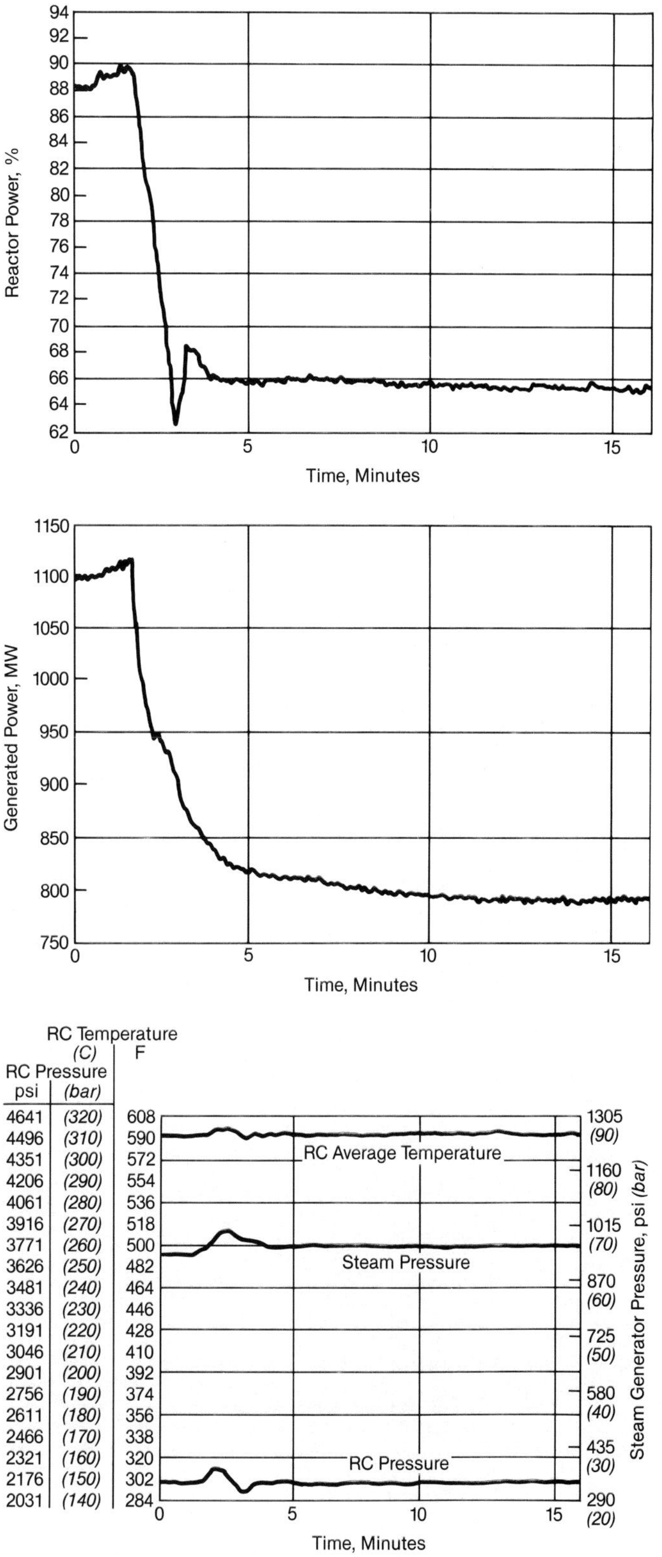

Fig. 16 Power escalation testing: reactor coolant flow response to trip of one RC pump.

4. *Control system tests* Steady-state and transient control systems are tested at full power to verify their performance.
5. *Turbine trip testing* The turbine-generator is tripped at full power to verify that the turbine, the main steam system, the RCS and the feedwater system perform as designed to bring the unit to reduced power. The response of several RCS and steam system parameters during a turbine trip with reactor runback at the B&W designed Muelheim Kaerlich (Germany) plant is shown in Fig. 18.

During the warmup and power escalation test periods, the water in the RCS and in the feedwater and steam generators is monitored to assure that it meets the chemistry requirements given in Tables 1 and 2. A key to proper operation and long equipment life is strict adherence to the primary and secondary water chemistry requirements. (See Chapter 53.)

Acceptance testing

After testing at full power is complete, final acceptance testing is performed before turning the unit over to the utility. Acceptance testing demonstrates that the steam pressure, flow and temperatures meet specifications and that the electrical output of the turbine-generator is as designed. Additional tests, such as transient capability, op-

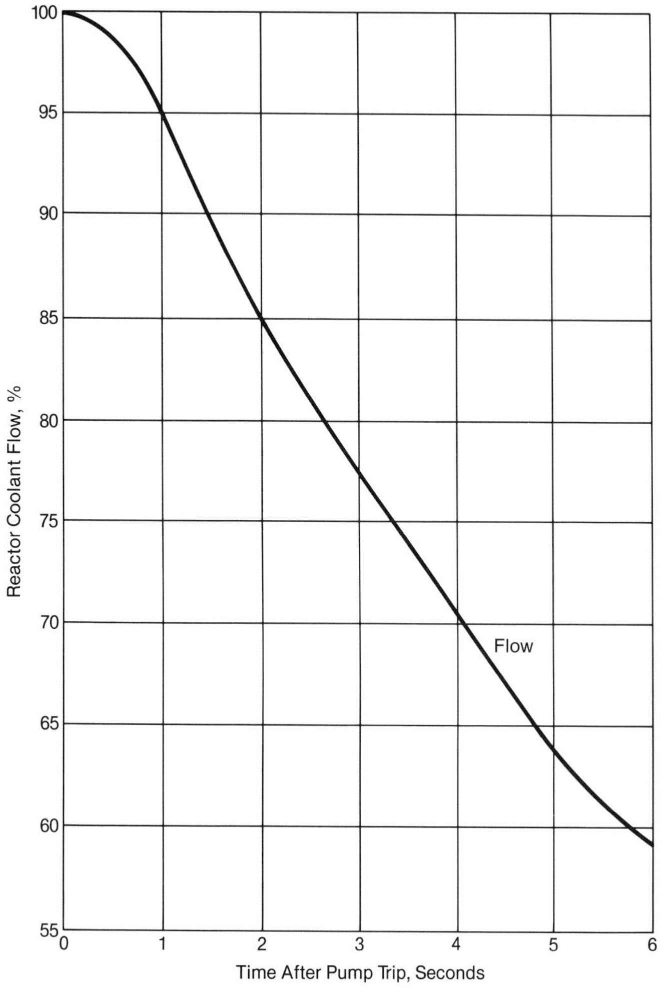

Fig. 17 Power escalation testing: reactor coolant flow response to trip of all RC pumps.

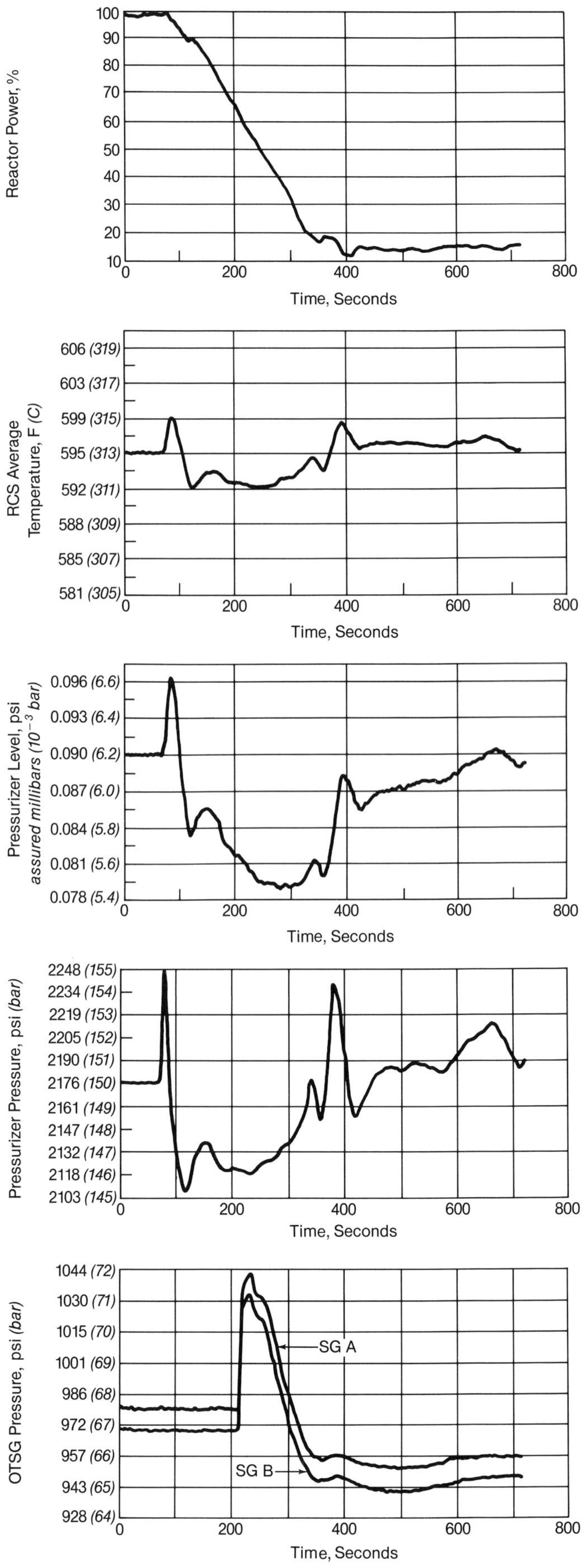

Fig. 18 Acceptance testing: turbine trip from full power with reactor runback.

eration with three of four RC pumps, and load rejection capability, are often included in the acceptance testing.

Typically, steady-state acceptance testing to demonstrate that steam conditions and electric generation meet the warranties is carried out for about 100 hours. Parameters such as reactor power, RCS temperatures, feed-water flow and temperatures, steam temperature and pressure, and electrical generation are monitored to maintain a running account of the performance.

Transient tests are scheduled and are typically of short duration. A typical transient acceptance test is as follows. Power is reduced from 100 to 60% at the design rate. 60% power is held until peak xenon poison builds into the core, which is normally six to eight hours after power reduction. The unit is then returned to full power at the design rate. This test demonstrates the transient capability of the control systems as well as the boron handling system's capability to dilute and then increase the RCS boric acid concentration.

Certain plant parameters, and those necessary to demonstrate performance, are recorded by the reactimeter and the plant computer. The data are used for performance evaluation and to demonstrate that the warranties are met.

Upon satisfactory completion of the acceptance tests, the plant is turned over to the utility and begins commercial operation.

Commercial operation

Although the B&W nuclear plants are designed for electrical load following, most U.S. nuclear power plants are base loaded, primarily because of associated lower fuel costs. During base load operation, the plant operates at or near full power, regardless of demand.

Electrical power production

The main control room and its adjacent control equipment, offices and work areas form the center of nuclear plant operations. The operations personnel, including the reactor operators, auxiliary operators and supervisory personnel, perform their functions in and around the control room. All control stations, switches, controllers and indicators necessary to start up, operate and shut down the nuclear units are located here, as shown in the chapter frontispiece. Operations which maintain safe conditions after upsets are also coordinated from this central control room. Controls for certain auxiliary systems are located at remote stations.

The information necessary for routine monitoring of the reactor, safety systems and the station is displayed on the control room console, on panel boards and on CRTs clearly visible to the operator. Control equipment frequently used or protective equipment needed to support an emergency is mounted on the operating console. Recorder and radiation monitoring readouts are mounted on panels. Indicators and controllers needed primarily during startup or shutdown are mounted on adjacent side panels.

A computer is available in the control room for alarm and performance monitoring and for data entry. On-demand printouts are available to the operator in addition to the periodic printouts of the station variables. Visible and audible alarms are also placed in the control room to warn of an unsafe condition.

The station operator has many duties including monitoring plant performance and making adjustments as necessary. The operator must periodically reduce the RCS boron concentration, as indicated in Fig. 13. This compensates for reactivity loss due to fuel burnup and maintains full power with the control rods within a specified operating band. The operator must also maintain a log of operations, review and approve maintenance activities, and supervise and perform surveillance tests.

One of the most time consuming functions of the plant operating staff is scheduling and performing mandatory surveillance testing of the safety equipment. There are about 400 surveillance tests required for each nuclear power plant. The emergency diesel generators, for example, are tested monthly. Most fluid systems, such as high pressure injection, emergency feedwater and decay heat removal, are also tested monthly. Finally, the reactor protection system and related safety systems are tested on a similar schedule.

During power production the staff plans and conducts plant maintenance. Preventive maintenance and, more recently, reliability centered maintenance programs have been instituted and have contributed to improved station reliability.

Nuclear power plant operations are audited periodically by teams from the Institute for Nuclear Power Operations and the NRC.

Forced outages

Unscheduled or forced outages occur when upsets, such as turbine trips, a loss of off-site power, or an excessive unidentified RCS leak force a reactor shutdown. Forced outages are hopefully of short duration to minimize purchased power and the loss of plant availability.

Maintenance and repair of equipment outside the reactor building are conducted during power operation to support the system and to avoid forced outages.

Scheduled outages

Each utility uses outage planners to schedule inspections, major maintenance, plant modifications and refueling operations for planned outages. The refueling frequency normally ranges from 12 to 24 months. Because of today's longer refueling cycles, midcycle planned maintenance outages may be necessary. (See Chapter 55.)

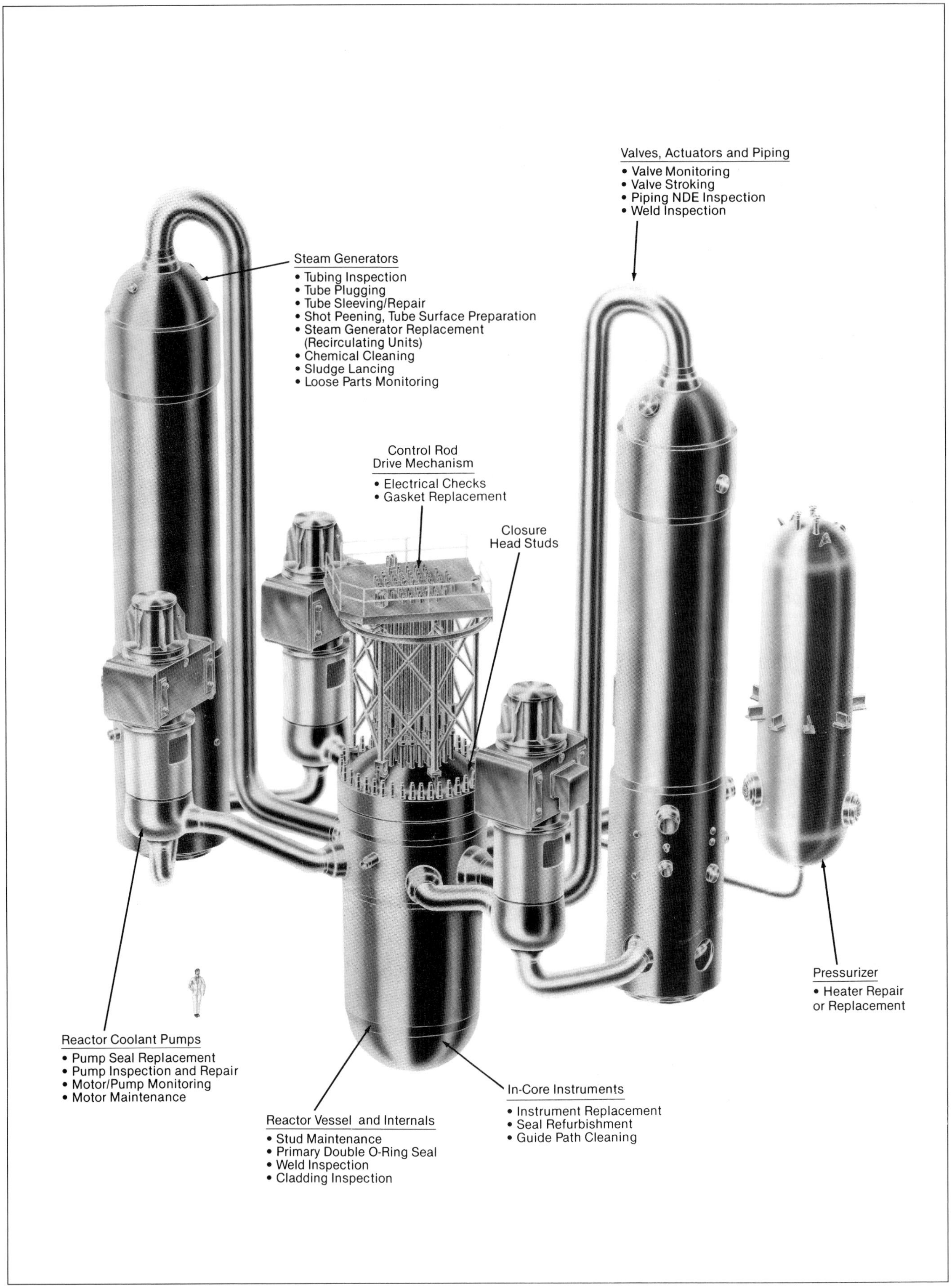
Valves, Actuators and Piping
• Valve Monitoring
• Valve Stroking
• Piping NDE Inspection
• Weld Inspection
Steam Generators
• Tubing Inspection
• Tube Plugging
• Tube Sleeving/Repair
• Shot Peening, Tube Surface Preparation
• Steam Generator Replacement (Recirculating Units)
• Chemical Cleaning
• Sludge Lancing
• Loose Parts Monitoring
Control Rod Drive Mechanism
• Electrical Checks
• Gasket Replacement
Closure Head Studs
Pressurizer
• Heater Repair or Replacement
Reactor Coolant Pumps
• Pump Seal Replacement
• Pump Inspection and Repair
• Motor/Pump Monitoring
• Motor Maintenance
In-Core Instruments
• Instrument Replacement
• Seal Refurbishment
• Guide Path Cleaning
Reactor Vessel and Internals
• Stud Maintenance
• Primary Double O-Ring Seal
• Weld Inspection
• Cladding Inspection

Fig. 1 Key areas of service and life extension in a pressurized water reactor.

Chapter 55
Service, Life Extension and Enhancements

Maintenance in nuclear plants, as in other industrial and fossil utility installations, is important to proper operation. Equipment efficiency can be maintained and forced outages can be minimized with a well executed maintenance program.

Nuclear plant routine maintenance is done during scheduled refueling outages. Because of the extended time between these outages (12 to 24 months), long range maintenance planning is essential.

As plant equipment ages or new technology becomes available, plant enhancements can be implemented to improve efficiency, provide more reliable operation, or simplify maintenance. If replacement of a component is not practical or economical, life extension is considered. This may include analyzing the component's history, or it may involve more detailed inspections. Extension past the design life is becoming more common as the world's nuclear plants age.

The following focuses on maintenance procedures as well as the unique aspects of working in a nuclear facility. Key enhancements are discussed to highlight recent technological advances. Life extension, including the required licensing, is also covered. Although most of the material concentrates on the primary reactor coolant system, many of the issues also pertain to the remainder of the plant.

Examples and figures primarily describe current methods implemented in pressurized water reactor (PWR) systems in the United States (U.S.). (See Fig. 1, facing page.)

Maintenance — key issues

Radiological considerations

Although precautions are taken, nuclear plant maintenance workers are exposed to low levels of radiation, as discussed in Chapters 49 and 50.

When a worker enters a radioactive area, he must wear protective clothing. This typically consists of cotton overalls, paper and rubber shoe covers, cotton and rubber gloves and a cotton hood or paper hat. Respiratory protection may also be required.

Limited accessibility

A major restraint to performing nuclear plant maintenance is the limited accessibility of much of the equipment. The primary reactor system is not accessible during power operation. As a result, this equipment must be maintained during outages. Because fuel cycles range from 12 to 24 months, reactor system maintenance must be thorough. Parts are often replaced based on past data even if deterioration is not apparent.

Most of the plant equipment is required by license to be operable during normal operation. If a safety-related system fails, the operator typically has a maximum of 72 hours to repair it, or the entire unit must be shut down. Examples of this critical equipment include the high pressure injection system, low pressure/decay heat removal system, core flood system, building spray system, reactor protection system and the nuclear instrumentation.

As fuel cycles have increased in length, midcycle planned shutdowns are often acceptable alternatives to possible unplanned forced outages.

Waste disposal

Waste from nuclear plant maintenance includes rags, lubricants and replaced parts. The protective clothing worn by the technicians must be laundered. Furthermore, certain tools and equipment used for repairs can not be decontaminated and must be disposed of or stored at a licensed facility. As a result, care is taken to minimize the waste. For example, hand tools can be covered with plastic which can later be discarded.

Application of remote technology

As utilities strive to reduce personnel radiation exposure, remote technology is advancing. A principle known as *time, distance and shielding* is the cornerstone of performing maintenance in a dose-efficient manner. In applying remote technology, emphasis is placed on time and distance.

When considering whether remote technology is applicable, the level of radiation, the time required for the maintenance and the work place environment are evaluated. Key areas in which this technology has been applied are the reactor vessel and its internals and the steam generators. Although the reactor vessel is highly radioactive, it is submerged in water. While the water shields the workers from radiation, remote equipment facilitates the underwater repairs.

Although remote technology often encompasses the use of robots, it can also be simply an extension of a hand tool to increase distance from the workpiece. Each maintenance task requires a different degree of sophistication based on the environment and the complexity of the task. As remote technology advances, the potential for radiation exposure declines.

Manpower planning

Because extensive maintenance of nuclear power plants takes place during relatively short unit outages, manpower planning is critical. Technicians are regulated by Nuclear Regulatory Commission (NRC) dose limits and some companies further curtail exposure. As a result, work loads for highly skilled technicians must be carefully planned.

The most effective way to increase utilization of manpower in a nuclear unit is to reduce the level of radiation. This can be done by decontaminating the area, by removing the source of radiation and by shielding existing radiation sources.

As previously indicated, refueling cycles are approaching 24 months. In addition, these outages are usually planned for the spring and fall months, when electricity demands are at their minimum. As a result, requirements for skilled maintenance technicians are also the greatest during these periods and manpower plans must take this into account.

Component maintenance

Much of the maintenance performed in nuclear power plants is similar to that for conventional systems. This includes equipment lubrication, fluid level checks and adjustments. Because most of the active systems are fluid (water, steam or air) systems, most of the work is performed on pumps, valves, fans and filters. In addition, the electrical distribution systems and the instrument and control (IC) systems require regularly scheduled maintenance. As previously discussed, nuclear systems are unique in that many components are inaccessible.

Reactor vessel and internals

The reactor vessel and internals are static components requiring little maintenance. Activities that are performed during each refueling outage focus on the integrity of the reactor vessel/head joint. During refueling, the reactor head must be removed from the vessel to gain access to the core. Each of the studs that retain the head is also removed from the vessel for cleaning. Special precautions are taken to contain the contaminated particulate that is released by stud cleaning equipment. Following cleaning, each stud is lubricated for reinsertion into the vessel flange.

The primary seal between the reactor vessel flange and head is a double O-ring design. Both mating surfaces must be thoroughly cleaned before installing the O-rings, which are replaced each time the head is removed. The seals are installed on the head and, just prior to setting the head, the vessel flange is cleaned and inspected for scratches that could cause leakage. Each stud hole is cleaned following head installation.

There is little routine maintenance required for the reactor internals because they are passive components. However, because the internals are highly radioactive, any repairs must be done under water. The bolted joints of the reactor internals have required the most repair activity. Stress corrosion cracking of high strength steel or Inconel bolts has been the most common cause for repair. Although the performance of A-286 stainless and X-750 Inconel bolts was expected to be good, pure water stress corrosion cracking exceeded early design parameters.

In replacing these bolts, standard wrenches, drills and sockets have been adapted to be operated from an extension or underwater platform, as shown in Fig. 2. Underwater cameras facilitate these remote repairs. Replacing the bolts in a typical single joint can take up to 16 days working around the clock with a crew of about 24 technicians.

In addition to using improved bolt materials, bolt torquing operations have become more sophisticated. Bolts were previously tightened to a specified torque with the equivalent of a common torque wrench. Actual bolt torques varied considerably, due to variations in surface finish and lubrication and because the tightening was done under water. Ultrasonic load measurement now permits a more consistent bolt torque. When torqued, a bolt elongates and the elongation is proportional to the tightening load applied. Ultrasonics are used to measure the length of the bolt before tightening. As the bolt is tightened, its length is continuously monitored. Once it has elongated a specified amount, tightening is stopped and the predetermined torque has then been applied.

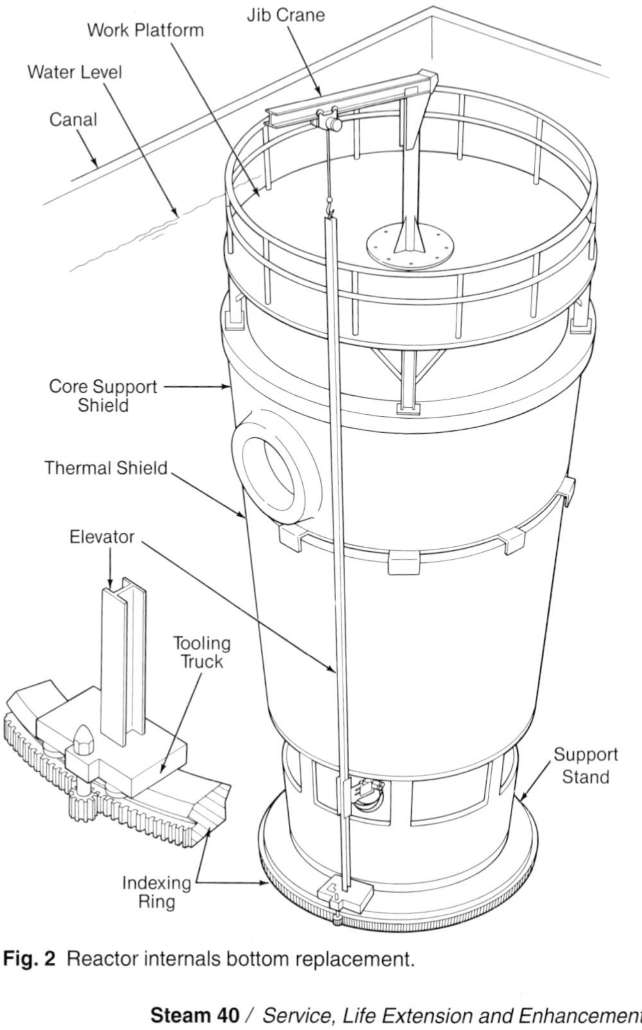

Fig. 2 Reactor internals bottom replacement.

Chapter 57
Balance of Plant Equipment

The production of power from steam involves the interaction of several major components and subsystems. As shown in Fig. 1 the steam supply system includes fuel handling and preparation equipment, the boiler system, and emissions control equipment. To generate electrical power, a *prime mover*, such as a turbine-generator, is required to convert the thermal energy of steam into mechanical and ultimately electrical power. A number of heat exchanger systems are also used to improve cycle efficiency and reduce power generation costs. The water-cooled condenser and cooling tower reject unused energy to the environment while minimizing back-pressure on the steam turbine. In selected cases, these can be replaced by air-cooled steam condensers. Finally, as discussed in Chapter 2, regenerative feedwater heaters can be used to improve overall cycle efficiency.

In addition to this more traditional power production approach, cogeneration plants combine power and process steam generation. Where combustion or gas turbines are used for primary power production, exhaust gas from the turbines can be used to generate steam for further process or power production applications.

Steam turbines

A steam turbine takes the thermal energy of the steam from a boiler and, through expansion, converts it into useful mechanical work. Steam is first introduced through small stationary nozzles in the turbine. The working fluid expands and attains high velocity, converting thermal to kinetic energy as it passes through these openings. The steam kinetic energy is then converted into mechanical energy by interaction with the moving turbine blades. The thermodynamics of this process are discussed in Chapter 2. References 1, 2 and 3 address performance evaluation and design. (See also References 4, 5 and 6.)

Types of steam turbines

Steam turbines are made in a variety of sizes ranging from 1 hp (0.75 kW) units used as mechanical drives for pumps, compressors and other process equipment to 1500

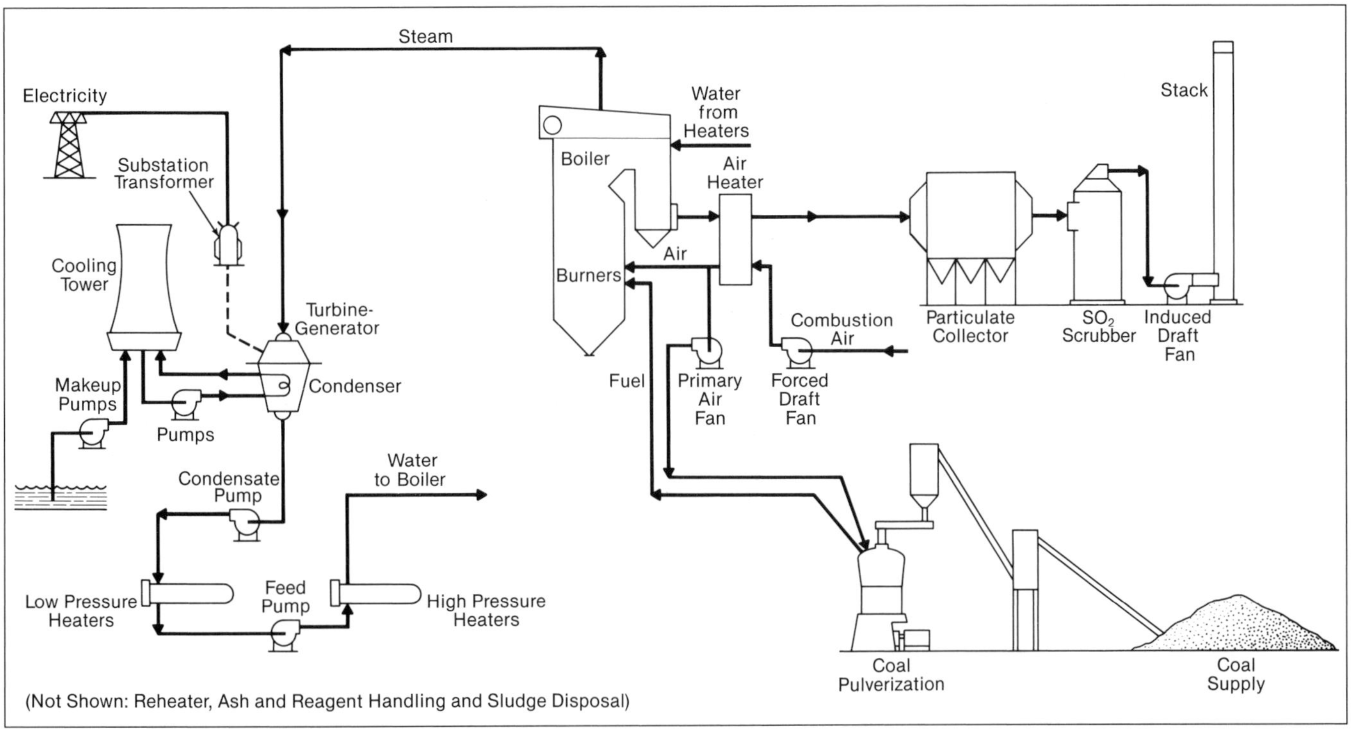

Fig. 1 Coal-fired utility power plant schematic.

MW capacity units for electric generator drives in the largest power plants. The chapter frontispiece and Fig. 2 provide views of turbines at a large electric power plant.

Turbines can be classified in several ways.

Steam supply and exhaust conditions These conditions include condensing, noncondensing, reheat, extraction and induction. The type of steam turbine most often used as electrical generator drives in electric power plants is the *condensing turbine*. These units exhaust steam at less than atmospheric pressure to a condenser.

Noncondensing or *back-pressure turbines* are most widely used in process plants and can fulfill a variety of output requirements. The exhaust pressure is controlled by a regulating station set to maintain the desired process steam pressure.

Reheat, condensing turbines are used almost exclusively in electric power plants. In these units the main steam flow exhausts from the high pressure section and is returned to the boiler where its temperature is increased. The steam is then returned to the turbine at the next intermediate turbine section inlet for further expansion.

Both controlled *extraction* and *induction turbines* are normally found in process plants. Extraction turbines direct steam from the turbine by takeoffs to other parts of the process requiring steam. In induction turbines, low pressure steam is introduced into the unit at an intermediate stage to produce additional power.

Casing or shaft arrangement These arrangements include single casing, tandem compound and cross compound turbines. *Single casing* units represent the basic arrangement where a single casing and shaft are used for smaller units. In the *tandem compound* arrangement there are two or more casings with the shafts coupled in line driving a single generator. *Cross compound* designs feature two or more shafts that are not in line; these shafts drive two generators and often operate at different rotational speeds.

Fig. 2 Large electric power turbine with casing removed (*courtesy of* Power *magazine*).

Details of stage design

To maximize turbine efficiency, the steam is expanded and does work in a number of steps or stages. These stages are then characterized by how the energy is extracted from the steam. Turbine stages are known as impulse or reaction, although most combine elements of the two. In an *impulse turbine*, often compared to a water wheel, impulse nozzles orient the steam so it flows in well formed high speed jets (Fig. 3a). These jets contain kinetic energy, which the moving blades, or *buckets*, convert into shaft rotation (mechanical work) as the steam jet changes direction. The pressure drop occurs across the stationary blades only.

Reaction turbines make use of the reaction force produced as the steam accelerates through the nozzles created by the blades themselves. Steam enters a fixed vane or nozzle passage and leaves as a jet that fills the entire circumference of the rotor (Fig. 3b). The steam then flows through moving blades which are shaped to act like moving nozzles. Here the steam changes direction and drops in pressure, and its speed increases relative to that of the blades.

Modern steam turbines normally combine impulse and reaction. For example many of the impulse turbines use from 10 to 50% reaction in their designs. All control stages are the impulse type while most low pressure section blades are the reaction type.

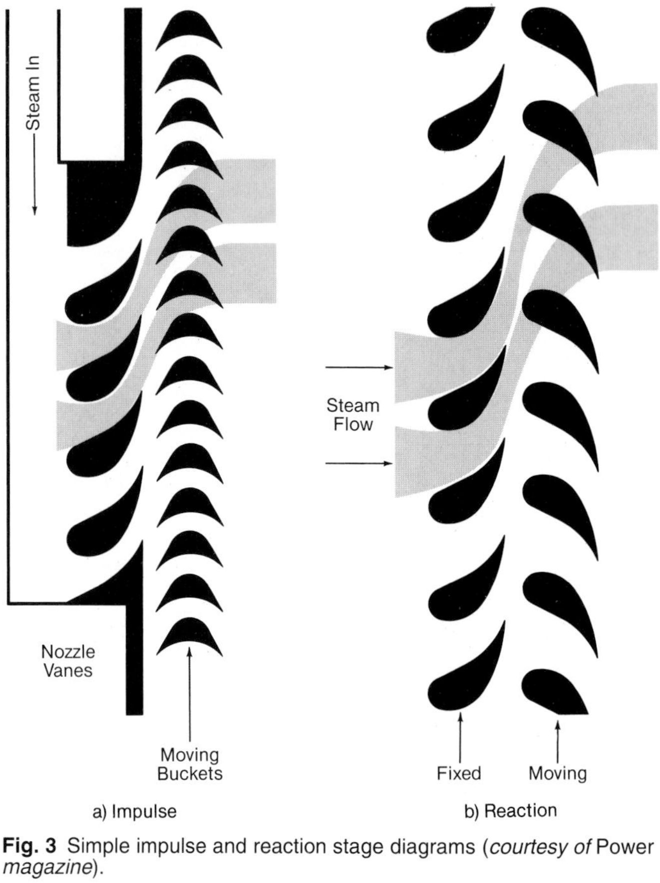

Fig. 3 Simple impulse and reaction stage diagrams (*courtesy of* Power *magazine*).

Pressure staging and velocity compounding

When the steam first enters the turbine, it is directed through the stationary impulse nozzles at high velocity. These nozzles organize the steam flow into high speed jets, which are directed against the moving blades (Fig. 4). The blades absorb the kinetic energy of the jets and convert it into shaft rotation.

When the buckets (or turbine blades) are stationary, the jets enter and leave with equal speed, developing the maximum force but producing no mechanical work. As the buckets increase speed, the jets become slower relative to the buckets and the force then becomes smaller. Under ideal conditions the steam jets do the most work when the moving blades are traveling at approximately one half the velocity of these jets. However, to meet this one half velocity guideline in a single step, the rotating elements would have to rotate at excessive speeds. Therefore, pressure staging or velocity compounding is used to reduce the required velocity of the rotating buckets. *Pressure staging* consists of restricting the pressure drop in each set of nozzles. After the steam passes through the rotating blades (one stage), it is expanded in the next set of nozzles.

In *velocity compounding*, steam energy is absorbed in a series of constant pressure steps by using a set of stationary blades to reverse the steam flow between sets of rotating blades. The high speed steam jet gives up part of its kinetic energy in the first row of moving blades. Then the fixed reversing blades redirect the slower steam into a second row of moving blades, where most of the remaining kinetic energy is absorbed. The steam then continues on to a series of *reaction stages*.

Turbine components

As shown in Fig. 5, the steam turbine basically consists of a shaft containing one or more disks to which the moving blades are attached, plus a casing in which the stationary blades and nozzles are mounted. The shaft is mounted within the casing by means of support bearings to carry the vertical and circumferential load and axial thrust bearings to resist the axial load induced by the steam flow. Seals are provided where the shaft penetrates the casing to prevent steam loss and air in-leakage. Seals are also provided within the casing to minimize steam bypassing the turbine stages.

Blades The blades are located on the circumference of the disks, where they receive the steam and convert it into work. There are many blades attached in each turbine stage; larger turbines contain more stages. Turbine blades are usually made from low carbon stainless steel. For high

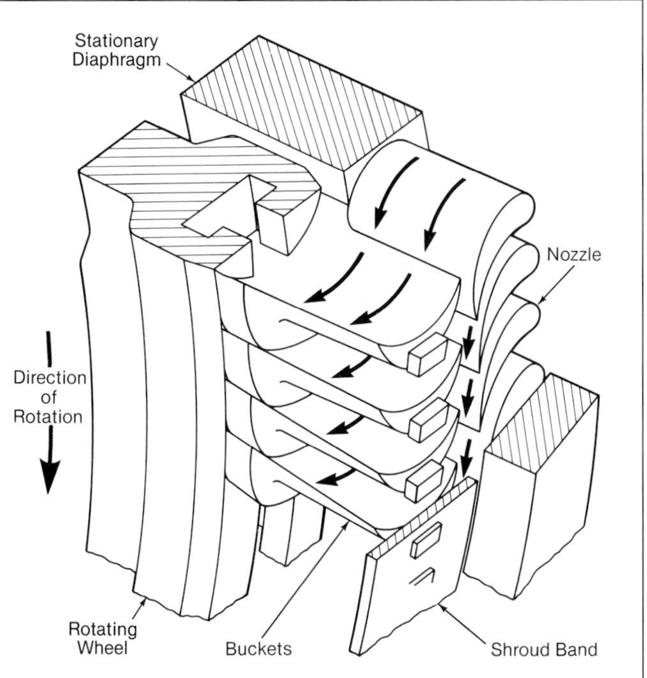

Fig. 4 Turbine impulse stage schematic (*adapted from TPC Training Systems, a division of Telemedia, Inc.*).

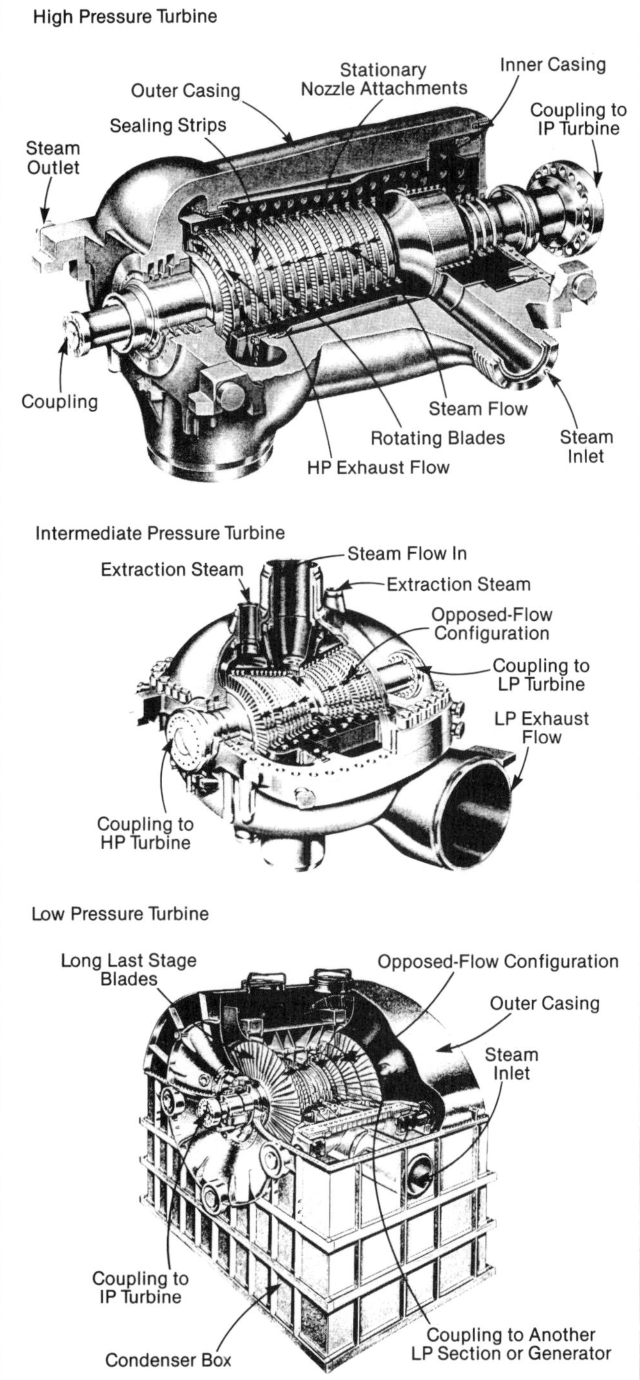

Fig. 5 Typical high, intermediate and low pressure turbine schematics (*courtesy of* Power *magazine*).

temperature [750F (399C) and above] applications, particularly in large units, alloyed chrome steel provides the required strength and erosion resistance.

As the steam expands through the stages, its volume increases. Taking this into account, the blade length and stage diameter must increase in each succeeding stage. (See Fig. 6.) In general, longer blades increase efficiency.

When the blade length becomes a significant part of the stage diameter, the ratio of steam speed to blade speed changes over its length. To accommodate this, contoured blades, which provide impulse flow at the root (bottom) and reaction flow at the tip, are used (Fig. 7). Also, because the steam approaches the blade from a direction nearly opposite to its motion, the blade section must be twisted to receive the steam smoothly.

Ideally steam enters the nozzle in an axial direction and leaves circumferentially, forming a vortex flow that is confined by the casing before the steam enters the moving blades.

The economy and performance of the turbine depend upon the design and construction of the blades. They must withstand the action of the steam and the centrifugal force caused by the high speeds at which the turbine operates. Therefore, blade design must be a compromise between strength and efficiency.

There is very little clearance between the stationary and moving elements in the turbine. Because of this, any vibration of the moving blades causes them to contact stationary components. As a result extreme care must be taken to eliminate vibration.

Rotors, shafts and bearings A turbine rotor consists of a shaft on which one or more disks are mounted. Turbine disks are either heat shrunk and keyed to the machined shaft or welded to the rotor shaft. For larger turbines, or those operated in excess of 750F (399C), the wheel disks are integral with the shaft forging. These forgings usually contain small amounts of alloying elements such as nickel, chromium, vanadium and molybdenum.

The rotors and blades are among the most critical and highly stressed components in the power plant. As the turbine ages, the rotors become susceptible to fatigue cracking.

The rotor shaft is supported at each end by bearings. These are normally ball bearings on the smaller turbines,

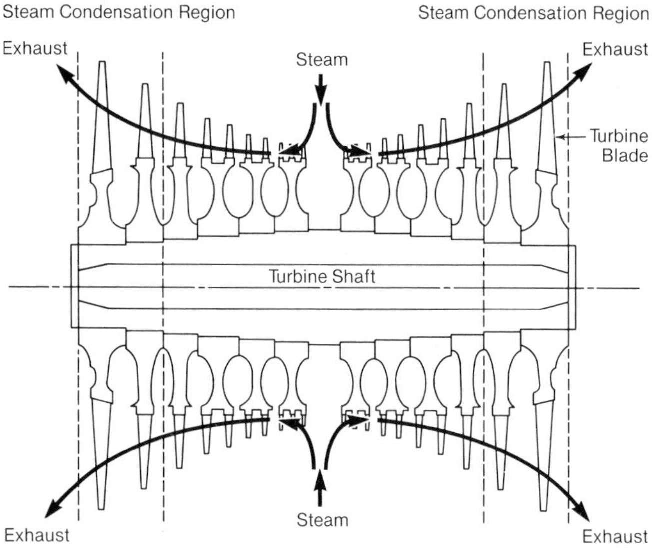

Fig. 6 Low pressure, double flow turbine showing blade size variation (*courtesy of* Power *magazine*).

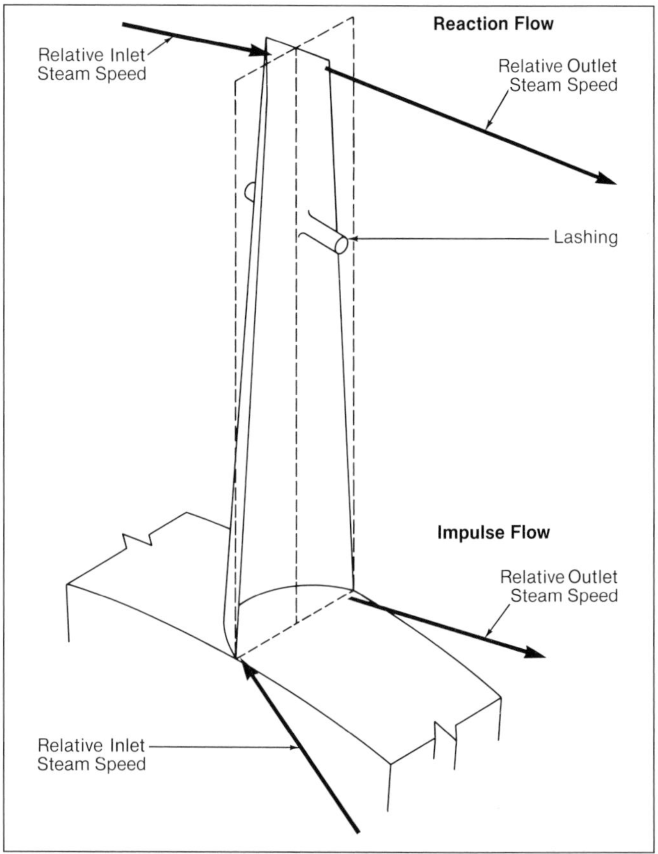

Fig. 7 Contoured turbine blades accommodate flow and speed variations along their length (*adapted from* Power *magazine*).

while larger turbines use a pressure lubricated journal bearing. The journal bearings have oil grooves in their top halves, creating an oil wedge that presses down on the journal. Tilting pad designs which create the oil wedge are also used.

Bearings must also be designed to counterbalance the net thrust along the shaft resulting from the steam pressure differential across the turbine stages. To maintain the very small clearances between the stationary and moving elements in each stage of the turbine, a thrust bearing is used.

Casings and seals The casing is an integral component of the turbine system. One function of the casing is to support the rotor bearings. Its internal surfaces must be designed to efficiently move the steam through the nozzles and blades. The casing must also support the nozzles in all turbine stages. Casings are steel castings, which permit the complicated shapes often required. The alloy compositions provide good weldability, castability and physical properties.

At the turbine inlet, steam enters through a stop valve and steam chest. In high temperature units these elements are separate from the main turbine structure. However, in smaller units the steam chest is usually mounted directly on the casing. At the outlet of the turbine, an exhaust hood guides the flowing steam from the last stage to the condenser.

Besides carrying the dead weight of various turbine components, the casing must also resist the mechanical stresses caused by the reaction forces on the stationary nozzles and blades, plus the thermal stresses caused by the temperature swings present during operation.

Designing the front high pressure end of the casing is especially difficult, because it must be allowed to move axially while retaining radial alignment to maintain the minute clearances between the moving and stationary parts. On very high pressure, high temperature turbines, the front end is built using two casings (inner and outer) with the steam flowing in the annulus region. This design allows for a thinner casing which can better handle the expansion and contraction caused by the large temperature differentials.

Because the shaft passes through the casing, seals are needed to minimize steam leakage. Carbon packing ring seals are often used on smaller, low temperature applications for this purpose (Fig. 8). In these seals the carbon packing rides directly on the shaft and is held in place by top anchored springs. Any steam that passes the seals is either channeled to a lower pressure stage or feedwater heater or is vented to the atmosphere.

Stepped labyrinth gland seals are also used to control steam leakage on larger, higher temperature applications (Fig. 8). In this type of seal, leaks are either channeled into an intermediate chamber and then directed to lower turbine stages or feedwater heaters, or sent to the condenser. Some turbines use a combination of the two types of seals, with a series of labyrinth seals flanked by a set of carbon ring seals at the shaft ends.

Steam turbine sections

Large high pressure turbines usually use two or three separate casings and the turbines are typically broken down into three sections: high pressure (HP), intermediate pressure (IP) and low pressure (LP). The typical turbines for these applications are shown in Fig. 5. The high pressure (high temperature) turbine typically uses a double casing for the reasons discussed. Exhaust from the high pressure turbine in fossil fuel-fired applications is typically returned to the boiler for reheating before being sent to the intermediate turbine. Frequently the high pressure and intermediate pressure turbines are combined into one casing with opposed flow directions to balance axial forces. In nuclear units the high pressure turbine exhaust is usually passed through a separate moisture separator, or moisture separator reheater (see Reference 6 for a discussion).

The intermediate pressure turbine is used with steam generators which use a reheat cycle. Depending upon the pressure and volumetric flow, the intermediate pressure turbine may use single flow or double flow (as shown in Fig. 5) designs. A double reheat cycle uses two intermediate pressure turbines.

The low pressure turbine receives flow from a *crossover* from either the high pressure turbine (nonreheat units) or intermediate pressure turbine. While a single flow design may be used, a double flow design shown in Fig. 5 is more common. The turbine blades in low pressure turbines are quite long to accommodate the very high volumetric flows. It is not uncommon to have multiple (two to four) low pressure casings each with double flow in larger power stations configured for tandem compounding and/or cross compounding.

Turbine flow control

A key factor in determining the part load performance and responsiveness of the steam turbine power system is the method used to control the steam flow to the first turbine stage. Two basic methods are used: adjusting the active nozzle area (partial arc admission) or changing the inlet steam pressure (full arc admission). With partial arc admission, the first stage nozzles are divided into groups and controlled separately with throttling valves. To increase load, the throttling valves for each group of nozzles are successively opened to full throttle position until the desired load is reached. In full arc admission, the steam pressure entering all of the first stage nozzles is adjusted by either operating the boiler at constant pressure and throttling the steam flow (all valves open together until the desired load is reached) or by varying the boiler operating pressure with the throttle valves basically full open except at the lowest loads.

Constant throttle pressure, full arc admission operation is simple but wastes pumping power at partial loads. Throttling the steam flow also results in a turbine inlet steam temperature drop. Constant throttle pressure, partial arc admission reduces the throttling energy losses but is typically slightly less efficient at full load because of the turbine design considerations. Variable boiler pressure operation tends to result in a less responsive system.

Each system therefore has a different impact on part load system efficiency, turbine inlet steam temperature and system responsiveness. Hybrid systems combine two or three of these systems to provide more optimized solutions for specific application requirements.

Maintenance and operation

On the large turbines used in electric power plants, many of the maintenance and operational problems can be attributed to one of three factors:

1. increased cycling imposed on the turbine,
2. operation of the turbine beyond its design life, and
3. the operational need for higher efficiency and output from the turbine.

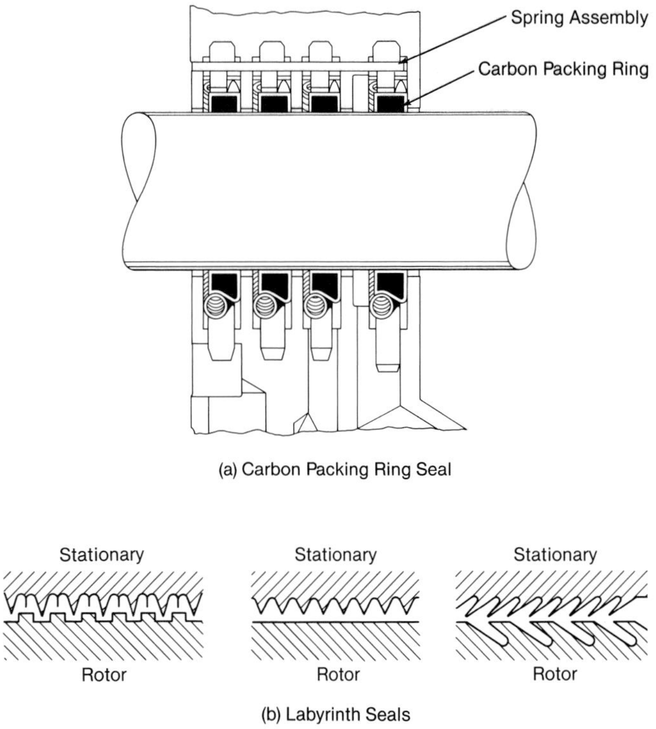

Fig. 8 Steam turbine shaft seals.

While the latter two factors increase turbine fatigue stresses, cycling operation produces the widest range of problems. Cycling can cause solid particle erosion in the high pressure and intermediate pressure internals while causing corrosion in the low pressure stages.

Solid particle erosion occurs when the high temperature oxide scale exfoliates from the superheater and reheater tubes in the boiler or from the steam outlet piping. The exfoliated scale becomes entrained in the steam flowing to the turbine and erodes components in its path, especially the blades and nozzles. Solid particle erosion can cause efficiency losses and crack initiation and propagation, possibly resulting in turbine failure. The first stages of high pressure and intermediate pressure turbines are particularly sensitive to this problem. To reduce this risk, nondestructive testing is used to detect the presence and thickness of oxide on the high temperature boiler components. Chemical cleaning of the boiler and steam lines can minimize these deposits. (See Chapter 42.) Improved blading and chromizing of the blades can also reduce damage.

The severity of solid particle erosion in any turbine is directly related to the amount and type of cycling duty, the type of boiler producing the steam, the presence of a steam bypass system, and the turbine blade material used.

While the high and intermediate pressure turbines are affected by erosion-induced problems, cycling and low load operation intensify corrosion related attack in low pressure turbine stages. Low load operation can significantly add to the concentration of chemicals in low pressure turbine stages, especially near the wet/dry interface. These corrosive elements can then contribute to stress corrosion cracking (SCC) and corrosion fatigue which are two of the most prevalent failure mechanisms in low pressure turbine stages.

Additional causes of operating performance deterioration and maintenance include increasing steam leakage due to wear, foreign material damage, deposits and general corrosion. Wear can result from the rubbing of seal packings as well as contact between moving and stationary parts. Surface contact can arise from vibrations, thermal distortion, bearing failures and solid particle erosion among others.

Damage from foreign material is distinguished from solid particle erosion because it occurs much more quickly. Identification of potential damage can be made from routine inspections of blades and nozzles as well as by monitoring vibration characteristics of the turbine.

Deposits and corrosion arise predominantly from vaporous, soluble and insoluble material transport through the preboiler and boiler systems. Deposits arise mainly when the solubility limits of the entrained material are reached. Because each compound has a different solubility limit as a function of temperature and different initial concentrations are present, contaminants will deposit at different locations throughout the turbine. Deposits generally reduce the efficiency, capacity and reliability of turbines by changing the shape and smoothness of the nozzles and turbine blades. During maintenance periods, it is common practice to attempt to remove these deposits. Because mechanical carryover from modern high pressure fossil boilers is usually less than 0.1%, the majority of deposits must arise from vaporous carryover. Therefore, it is critical to maintain strict boiler feedwater quality limits so that the vaporous carryover can be controlled within acceptable limits. (See Chapter 42.) The major impurities causing turbine corrosion are sodium hydroxide, sodium chloride, and various organic and inorganic acids.[7]

Power plant applications

Fossil fuel plants Steam turbines used in fossil fuel-fired power plants have varied widely in size and cycle application (References 4, 5 and 6). Sizes typically have ranged from 100 to 1300 MW. In the U.S., the steam cycle conditions reached a maximum of 5000 psig/1200F/1050F/1050F (344.8 bar gauge/649C/566C/566C) double reheat. However, operating problems and higher costs with this design have focused most plant designs since the early 1970s on one of two system cycles:

1. subcritical drum boiler systems: 2400 psig/1000F/1000F (165.5 bar gauge/538C/538C), and
2. supercritical once-through boiler systems: 3500 psig/1000F/1000F (241.3 bar gauge/538C/538C).

Typical turbine heat balances for both of these cycles are presented in Chapter 2. Rotational speeds for the turbines are typically 3600 revolutions per minute (rpm) for 60 Hz applications and 3000 rpm for 50 Hz applications. The maximum exhaust steam quality in the turbine is generally limited to 10 to 15% steam by weight. Reheat extraction pressures are typically designed for 20 to 30% of the maximum continuous rated superheater outlet pressure. Corresponding typical *net turbine heat rates* for the 2400 and 3500 psig cycles are 7911 Btu/kWh and 7703 Btu/kWh respectively.

Nuclear plants Turbines for nuclear power applications generally fall in the range of 500 to 1300 MW, although units as large as 1500 MW have been installed outside of the U.S.[4] The steam cycle is normally based upon 1000 psig (69.0 bar gauge) with either saturated or slightly superheated steam temperatures. Rotational speed is 1800 rpm for 60 Hz and 1500 rpm for 50 Hz generators. Reheat and nonreheat cycles are used.

Steam system interfaces

The steam turbine and steam generating system must be well integrated to reliably achieve the specified plant performance. A number of critical interface items have been identified above. In summary, these include:

1. Steam cycle:
   - maximum continuous pressure and temperature
   - maximum flow rate
   - reheat requirements
2. Steam temperature:
   - specified superheat
   - specified reheat
   - permissible variations
   - rate of change
3. Control method:
   - constant or variable boiler outlet pressure
4. Water chemistry:
   - boiler outlet/turbine inlet requirements
   - upstream materials of construction
5. Transient conditions:
   - startup
   - rate of load change
   - minimum flow rate

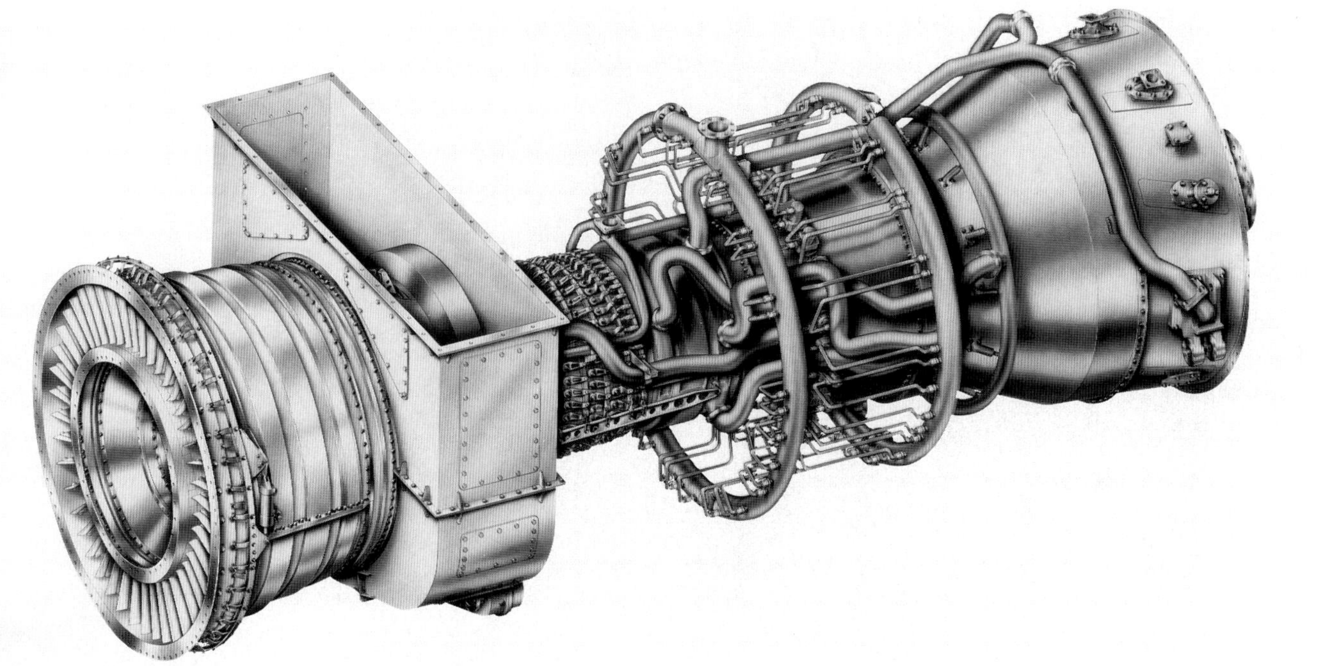

Fig. 9 Aeroderivative gas turbine, model GE LM6000 (*courtesy of General Electric Company*).

Gas turbines

Gas or combustion turbines have continuously evolved into highly reliable, low capital cost means of producing power (Figs. 9 and 10). In electrical power applications, they have historically been used for peaking duty, where rapid startup and short runs are needed. Advances in efficiency and reliability have made these machines available for intermediate and, in some cases, base load capability, depending upon the cost and availability of fuel.

In addition to simple cycle power production, gas turbines are now frequently integrated with steam producing systems to increase power production efficiency and capture more of the available energy in cogeneration applications.

Basic components

All gas turbines rely on three basic components: a compressor, combustor and turbine.[8-13] The simplest gas turbine is the open cycle unit, shown in Fig. 11. As the compressor rotates, it draws in ambient air, pressurizes it and steadily forces it into the combustor. The temperature of the air rises due to the compression.

Compressors are grouped into two major categories: axial flow and centrifugal. In an *axial flow compressor*, rotor blades shaped like airfoils take in air, increase its velocity

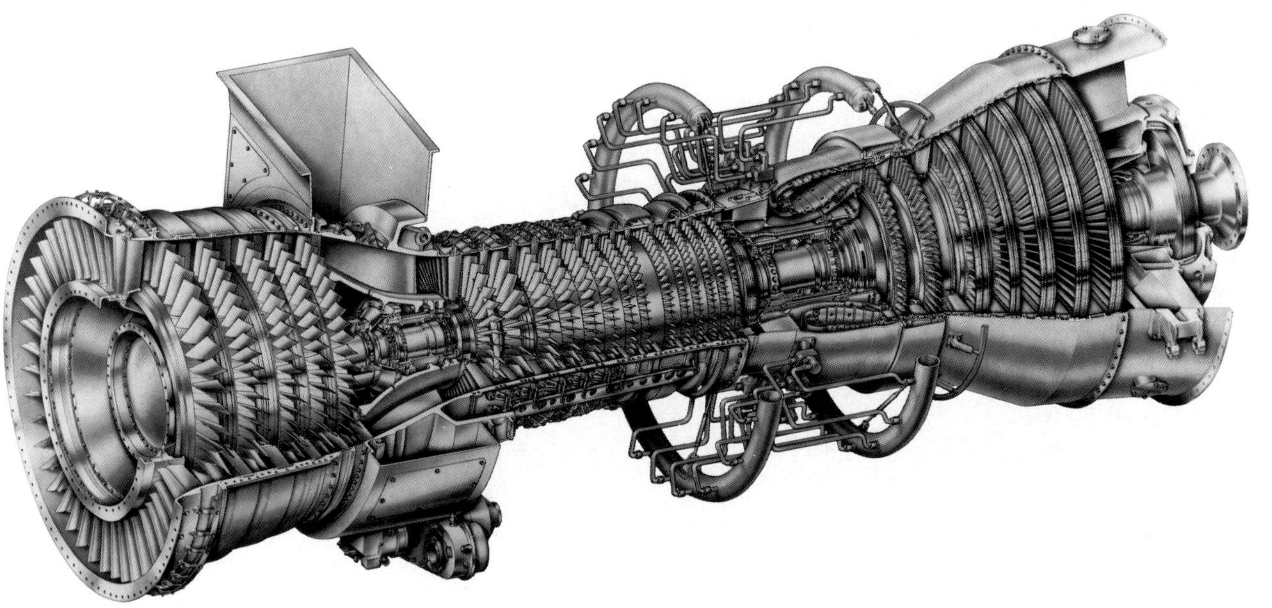

Fig. 10 Sectional view of gas turbine shown in Fig. 9 (*courtesy of General Electric Company*).

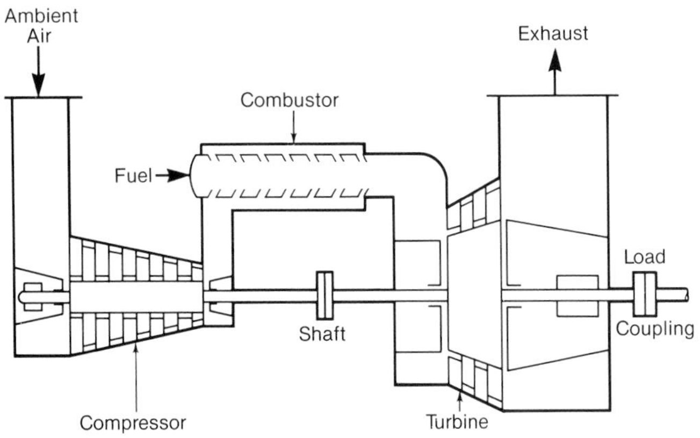

Fig. 11 Simple open cycle gas turbine schematic.

and force it into stationary blade passages. These passages are shaped to form diffusers, which then slow and pressurize the incoming air.

Air enters the *centrifugal compressor* at the center of a rotating bladed impeller. Blade rotation imparts a centrifugal force on the air, driving it radially outward at very high speed into a stationary diffuser, which is wrapped around the rotating impeller. As in the axial flow design, the stationary diffuser section slows and pressurizes the air.

The blades in a compressor must be precisely shaped and clean to achieve efficiencies of about 90%. Any blade surface fouling quickly reduces compressor performance.

Fuel is burned with the compressed air in the combustor. This fuel is most often natural gas, but it can also be high quality fuel oil, synthetic gas (produced by gasifying fuel oil, coal or refuse), or liquefied coal. As the fuel burns it causes the gas temperature to rise substantially in an essentially constant pressure process. Although very high local temperatures occur in the combustor, the unit is designed to provide mixing, dilution and cooling so that the gases leave at a mixed average temperature.

Combustor lengths are designed to provide adequate time and space to completely burn residual fuel and to allow lighter fuels to burn easily. Properly designed combustor lengths also provide an acceptable temperature profile of the combustion gases as they enter the turbine nozzles and buckets. Advances in combustor design have focused on reducing the formation of nitrogen oxides (NO_x). NO_x emission levels range from 70 ppmv in older machines to less than 9 ppmv in advanced designs at 15% O_2.[9]

The high energy combustion gases expand in the nozzle and bucket (turbine) section just as steam does in a steam turbine. Here the temperature and pressure of the gases drop as their energy is converted into mechanical work. This conversion takes place in two steps. In the nozzle section the hot gases expand and a portion of the thermal energy is converted into kinetic energy. A portion of this kinetic energy is then transferred to the rotating buckets and converted to work.

Approximately two thirds of this work is needed to drive the compressor. Therefore, the turbine's useful output, which drives an external mechanical load through a shaft coupling, is represented by only one third of this work. Chapter 2 discusses the thermodynamics of this Brayton cycle process.

The turbine section, especially the first stage moving buckets and the disks carrying them, must withstand extremely high [2200F (1204C) or more] inlet gas temperatures. These high temperatures, combined with contaminants in the air and fuel, expose the nozzles and buckets to accelerated oxidation, or hot corrosion. Because these contaminants are costly to eliminate and difficult to control, high grade alloys or coatings are used to provide corrosion resistance and maximize component life.

Types of gas turbines

Gas turbines can be categorized as single-, two- or three-shaft designs. The single-shaft turbine (Fig. 12a) is configured with one continuous shaft and all stages operate at the same speed. These units are typically used for generator drive applications.

In the simple cycle, two-shaft gas turbine (Fig. 12b), the low pressure or power turbine rotor is mechanically separated from the high pressure turbine and compressor rotor. In a two-shaft turbine all of the work developed by the power turbine is available to drive the load equipment, because the work developed by the high pressure turbine supplies the energy necessary to drive the compressor.

There is also a three-shaft gas turbine design (Fig. 12c), which consists of two compressor turbine shafts and a turbine output shaft. In the three-shaft unit the low pressure turbine drives through the high pressure turbine and high pressure compressor to power the low pressure compressor.

Aeroderivative gas turbines In addition to the heavyweight industrial gas turbines, there are the aircraft derivative or aeroderivative gas turbines. Since their introduction in the 1960s these turbines, modeled after the familiar jet engine, have been at the forefront of simple cycle efficiency.

Aeroderivative gas turbines have benefited tremendously from the resources spent on aircraft engine research and development. Other gas turbines have also gained from this technology but to a lesser degree. However, the aeroderivative engine must be modified for industrial use, increasing the initial cost of the unit. Unique hardware must be added to adapt the high efficiency, high volume aircraft components to drive a generator or other industrial load.

When choosing between a heavy duty industrial gas turbine and an aeroderivative unit, several factors should be considered, including the intended use (generation or process), efficiency, heat rate and fuel(s). An economic analysis of the project should be performed considering initial cost, expected life and future costs of fuel, operation and maintenance.

Regenerators and combined cycles

A typical simple cycle gas turbine can generally convert 25 to 32% of the fuel input into shaft output. Most of the remainder is released as exhaust heat. The exhaust gases reach the stack at atmospheric pressure and high temperature. One way to improve the thermal efficiency of a simple open cycle gas turbine is to recover some of this energy-rich exhaust heat and return it to the cycle. This can be accomplished by adding a regenerator or by using a combined cycle.

A *regenerator* is basically a compact, counterflow, one pass heat exchanger that transfers exhaust heat to the compressor discharge air before it enters the combustor. The design objective of the regenerator is to bring the tem-

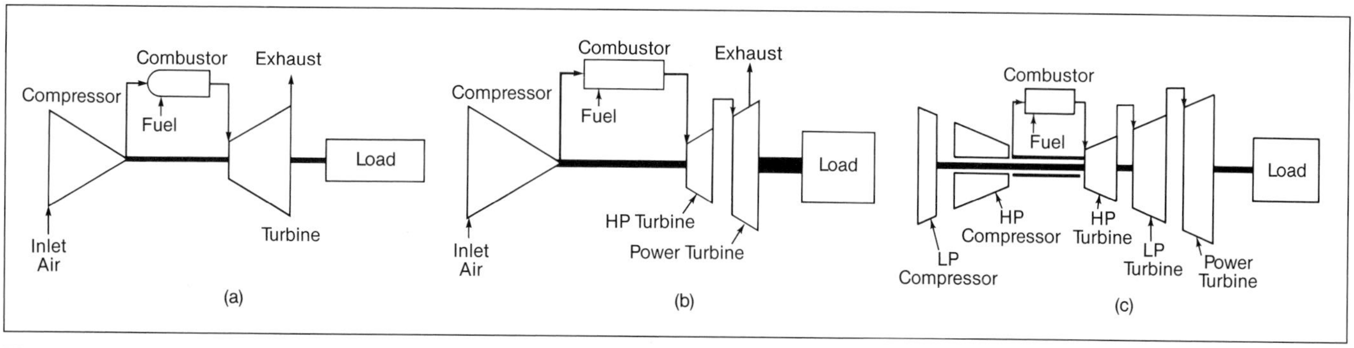

Fig. 12 Alternate gas turbine shaft configurations: (a) single-shaft, (b) two-shaft and (c) three-shaft designs.

perature of the compressor discharge air close to the temperature of the exhaust gas. This objective is measured in terms of regenerator *effectiveness*. (See Chapter 4.) A regenerator can reduce the amount of fuel needed to heat the air to combustion temperature by 25 to 30%. A regenerative cycle, two-shaft gas turbine schematic is shown in Fig. 13.

A *combined cycle* also uses this exhaust heat, attaining much higher utilization of the fuel input. Typical combined cycles feature one or more gas turbines with heat recovery steam generators located in their exhaust path. (See Chapter 31.)

Many options are available in combined cycles, depending on whether the steam's thermal energy is used for power production, process or a combination of the two. By application of the more complex heat recovery cycles, very high utilization of the fuel input to the gas turbine can be attained. It is common to achieve more than 80% fuel input utilization by producing a combination of electric power and process heat. Combined cycles that produce electric power alone potentially reach thermal efficiencies approaching 50% when advanced gas turbine designs are used.

Application considerations

Gas turbines are available in a broad range of sizes, generally from 0.5 MW to 150 MW. Single units producing up to 200 MW are in their early demonstration phases. The larger advanced machines are characterized by turbine inlet temperatures approaching 2300F (1260C) and increased cycle efficiency.

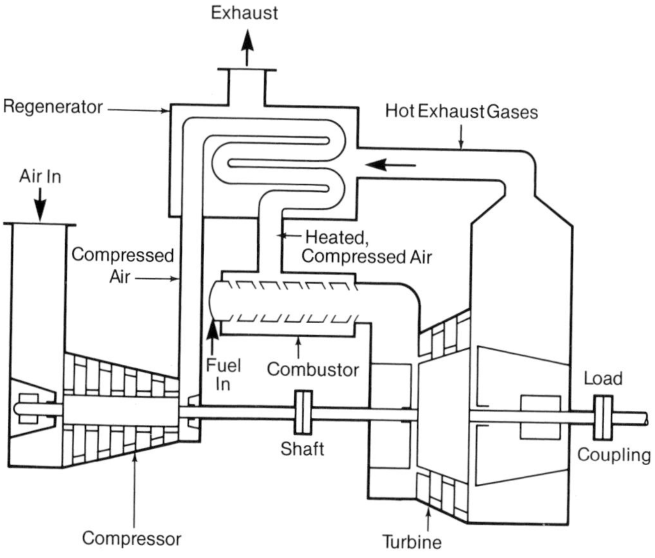

Fig. 13 Gas turbine with regenerator — schematic.

Gas turbine-generator systems include a number of auxiliaries. These include the lube oil system, filters, fuel supply systems, sound dampers, controls and, if required by local regulation, selective catalytic reduction (SCR) deNO_x systems. (See Chapter 34.) The lube oil system is particularly important. Lube oil pumps are positive displacement designs and are typically powered directly from the gas turbine shaft. This feature permits the turbine to *run down* following a power failure without a serious risk of damage.

Gas turbine systems are relatively simple to operate and maintain compared to other power producing systems. Fully automated operation with remote control through power dispatchers is used in many applications. Special attention is given to preventing and controlling compressor and turbine fouling to maintain high efficiency and availability. Fouling of the turbine may be particularly severe when firing crude and heavy oils because they contain high levels of particulate and corrosive compounds.

In addition to the effect of deposits, gas turbine efficiency and power output decline as the unit ages.[9] Gas turbine blades experience erosion and oxidation in the high temperature, high flow environment. Components such as vane carriers and housings can change shape, resulting in flow losses. Mechanical wear occurs due to surface contact in vanes, blades and seals. General corrosion of the rotors, housing and piping also occurs.

Repowering aging power plants

New gas turbines can be used in conjunction with existing steam generation equipment to increase plant efficiency and reduce plant heat rate. There are five systems commonly considered in utility repowering applications:

1. steam turbine or substitute repowering,
2. supplemental repowering,
3. process repowering,
4. boiler repowering, and
5. station repowering.

In *steam turbine repowering*, gas turbines generate power and exhaust to waste heat boilers, which generally replace the plant's steam generators. Steam output from these boilers drives the existing steam turbine. The waste heat boilers may be unfired, with the gas turbine exhaust providing the heat input, or they may be supplementally fired to increase steam production.

Steam turbine repowering is by far the most common application. It is especially adaptable to small, low pressure plants, where the gas turbines and waste heat boilers can be easily matched to the existing steam cycle. The benefits of steam turbine repowering are as follows:

1. additional generating capacity can be quickly added; the plant's output can be increased by as much as 200%,
2. the initial cost of the gas turbine and waste heat boiler compares favorably to the installed cost of a simple cycle gas turbine,
3. operating costs are reduced as plant efficiency can be increased by 5 to 30%, and
4. the combined cycle offers considerable generating flexibility.

In *supplemental repowering* the exhaust from the gas turbine is used to produce superheated steam in a waste heat boiler. This steam is then directed to the superheater outlet of the existing boiler at the plant's original steam pressure and temperature. The partially cooled combustion exhaust is then used in preheating the boiler combustion air.

The *process repowering* concept is essentially the same as steam turbine repowering, except the steam produced in the waste heat boiler is used in the process. In electric utility plants the steam is used for feedwater heating. Although process repowering has been used extensively for process requirements, utility feedwater heating applications are limited.

In *boiler repowering*, gas turbines generate power and act as forced draft fans for the existing boiler as they supply hot exhaust gas for combustion air. However this approach is highly site specific because of the need to match combustion air requirements and the turbine exhaust flow. Boiler modifications in areas such as the windbox are usually necessary to withstand the hot exhaust gases. Another limitation is that many older plants do not have the space to accommodate the required hot gas ductwork.

When a power plant would otherwise be retired, *station repowering* can be considered. In this case a gas turbine or combined cycle power plant is installed; much of the existing power generation equipment is abandoned or demolished. Some existing systems, such as cooling towers, transmission lines and water and fuel systems, may be retained.

Condensers

The power plant condenser receives exhaust steam from the low pressure turbine and condenses it to liquid for reuse. While direct contact condensers which spray cooling water directly into the steam flow have been used in some applications, the water-cooled surface condenser is the dominant technology used in modern central power stations. In selected cases, air-cooled steam condensers provide an attractive alternative to surface condensers where cooling water is not readily available or the plume from a wet cooling tower is not acceptable.

The surface condenser serves three important functions:

1. provides a low back-pressure at the turbine exhaust to maximize plant thermal efficiency and reduce the heat rate (see Chapter 2),
2. conserves the high purity water for reuse in the boiler-turbine system to minimize water treatment costs,
3. deaerates the condensate to minimize corrosion potential in the balance of the power system components, and
4. serves as a collection point for all condensate drains.

To serve these functions well, modern power condensers must be very tight with effectively no leakage of ambient air or cooling water into the condensing steam space. Large volumes of cooling water are provided to absorb the heat given off by the condensing steam — typically 50 to 80 times the steam mass flow.[14] Condenser back-pressures usually range from 1.0 to 4.5 in. Hg absolute (3.4 to 15.2 kPa) with higher pressures possible where the cooling water temperature is elevated or an air-cooled steam condenser is used. Under normal operation, the residual oxygen (O_2) level is generally less than 42 ppb, with levels down to 7 ppb possible under certain conditions.[15,16]

Arrangement

Surface condensers are basically shell and tube heat exchangers with the exhaust steam condensing on the shell side and the cooling water flowing in one, two or four passes inside the tubes. While small condensers are frequently cylindrical, larger power condensers are generally rectangular in shape as shown in Fig. 14. The core of the equipment consists of the tube bundles where the tube size selected typically ranges from 0.625 to 2.0 in. (15.9 to 50.8 mm) OD. Tube bundle configurations are typically unique to each supplier with key features including low steam side pressure drop, good steam flow distribution to all bundle areas, high heat transfer rates, and the effective collection and discharge of noncondensable gases including oxygen and carbon dioxide (CO_2). A number of baffles and plates support the tubes, distribute the flow, guide the noncondensable gas flow towards its outlet, and minimize flow induced vibrations which could result from the high steam velocities. *Water boxes* or open plenums on both ends of the tube bundles distribute the cooling water flow. In larger units, the water boxes are generally made with curved surfaces to better withstand the internal pressure [up to 80 psig (0.55 MPa)].

A variety of condenser configurations are used depending upon low pressure turbine arrangement and specific site requirements. Most condensers are configured to receive steam from the top in downflow from the low pressure turbine. (See Fig. 14.) Side and axial flow entry designs are also occasionally used. The condenser may be designed with a single cooling water pass in a single shell as shown in Fig. 14. Other configurations include twin shell single pass, single shell double pass, twin shell double pass designs and various single pass, multi-pressure geometries.

Condensate is collected in the bottom of the condenser in the *hot well*. (See Fig. 14.) From here the condensate feed pump supplies the subcooled water to the feedwater heaters.

The condenser tube bundle is arranged so that the steam flow sweeps the O_2, CO_2 and other noncondensable gases to an *air removal zone* where they are removed by an *exhauster* connection at the end of the condenser. The air removal zone may be shrouded or unshrouded. Ineffective designs can permit the buildup of noncondensable gases which can result in reduced heat transfer performance as the gases migrate to and effectively block the condenser tube surfaces.

Movement due to temperature differences between the turbine and condenser is usually accommodated by a stainless steel bellows or rubber belt type expansion joint. For smaller units the condenser may be supported on springs and rigidly connected to the turbine.

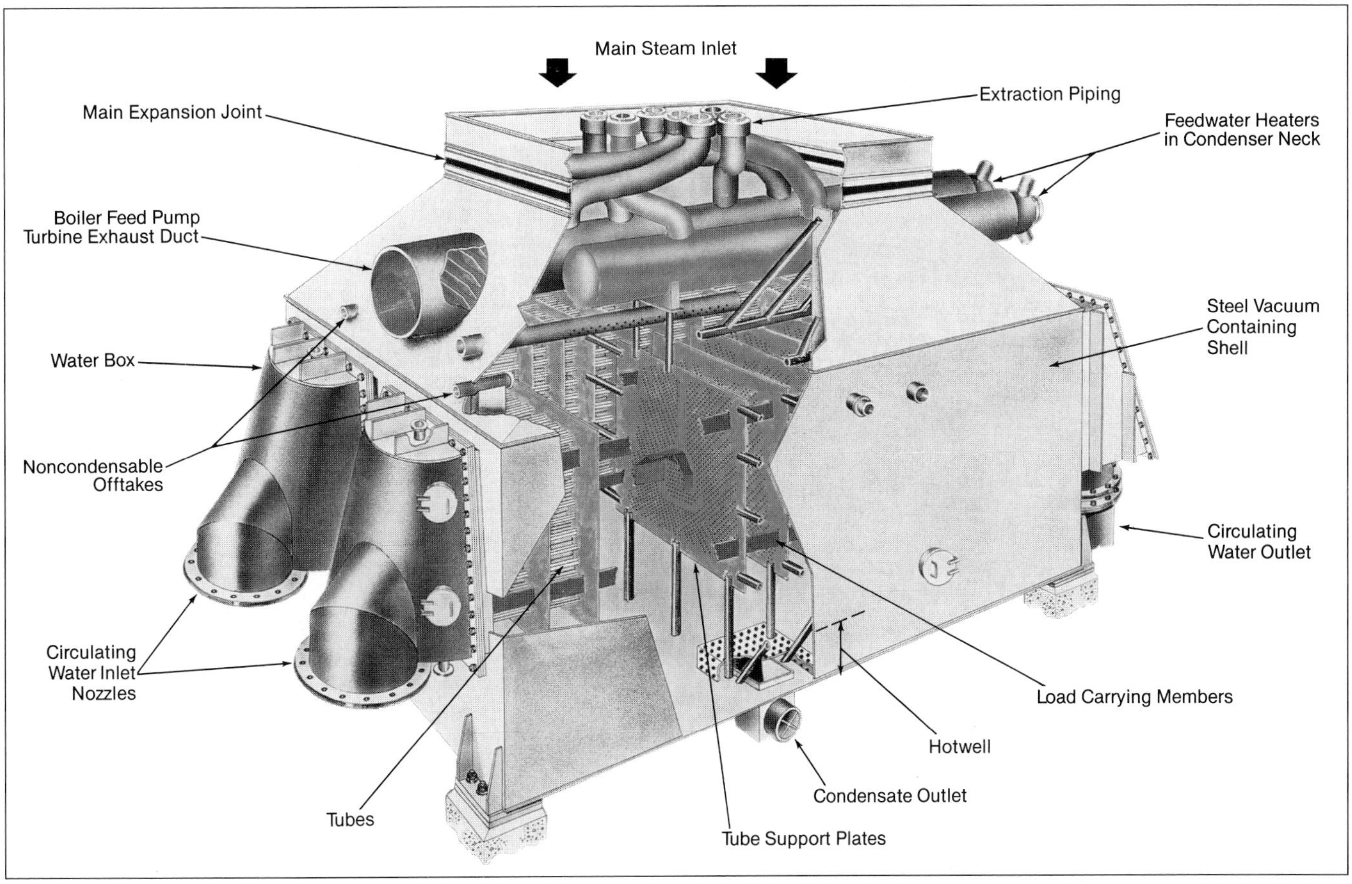

Fig. 14 Schematic view of single shell, single pass steam surface condenser (*adapted from Senior Engineering Company*).

Design parameters

The base thermal load to be handled by the condenser is set by the turbine heat balance at maximum load conditions.[14-18] Unless otherwise specified, the condenser duty for turbine service is assumed to be the product of the steam flow in lb/h or kg/s and either 950 Btu/lb or 2210 kJ/kg.[15] Condensers are also frequently sized to handle transient thermal load conditions which can exceed the maximum load conditions. A typical example would be part or full load turbine steam bypass to the condenser from the boiler.

Cooling water velocities are usually established by balancing a number of factors including economics, fouling, erosion potential, water quality and tubing material, among others. Typical cooling water velocities[15] include: 6 ft/s (1.8 m/s) for sea water with entrained sand, 7 ft/s (2.1 m/s) for clean water and 8 ft/s (2.4 m/s) for clean cooling tower water. Design velocities are also adjusted for tube material: 6.5 ft/s (2.0 m/s) for aluminum-brass tubes, 7 ft/s (2.1 m/s) for Admiralty metal and 8 ft/s (2.4 m/s) or more for stainless steel and titanium.[38] Velocities below 3 ft/s (0.9 m/s) are not recommended because of the potential for poor tube side flow distribution.[15]

The cooling water temperature increase is frequently between 15 and 25F (8 and 14C). The minimum terminal temperature difference is not less than 5F (3C). Obviously the cooling water temperature significantly affects the minimum available turbine back-pressure. In general, a 55F (13C) inlet water temperature will provide a condenser back-pressure of about 1.5 in. Hg (5.1 kPa) while a 95F (35C) temperature generally corresponds to a 3.5 in. Hg (11.9 kPa) back-pressure.

A variety of materials have been used for thin wall condenser tubing, depending upon the cooling water service, steam chemistry requirements and economics. Biofouling resistance and high thermal conductivity, as well as resistance to erosion-corrosion, pitting, crevice corrosion, microbiologically induced corrosion and general corrosion, are all considerations in the material selection. Copper alloys present particular problems in high pressure boiler and nuclear reactor systems because of the potential for copper transport into the boiler and turbine where deposition can occur. Typical materials include Admiralty metal (70 to 73% Cu, 0.9 to 1.2% Sn, 0.07% maximum Fe and balance Zn), arsenical copper, aluminum brass, aluminum bronze, 90-10 Cu-Ni, 70-30 Cu-Ni, cold rolled carbon steel, stainless steels (304/316), corrosion resistant stainless steels and titanium. Though more expensive, titanium and corrosion resistant stainless steels are being used more often because of their high corrosion resistance and reduced risk of tube leaks. However, care must be exercised in applying these thin wall tubes to avoid flow induced vibrations and the potential for galvanic corrosion with dissimilar metal tubesheets.

Most condensers have expanded tube-to-tubesheet connections. Welded joints are also used.

Thermal and hydraulic performance of surface condensers can be treated in general terms using the information provided in Chapters 3 and 4, while detailed discussions are provided in References 17 and 18. Estimated overall heat transfer surface requirements can also be evaluated using the correlations and correction factors provided in Reference 15. Key parameters affecting overall performance and turbine back-pressure include turbine and cooling system configurations, cooling water source temperature, initial steam to cooling water temperature difference, increase in cooling water temperature, terminal temperature difference, fouling, tube material, tube thickness and cooling water velocity. To account for the inevitable reduction in performance which occurs due to fouling between condenser cleanings and design uncertainty, a standard cleanliness factor of 85% is frequently used although this may not properly account for the actual unit behavior.[16] The condenser fouling is usually attributed to sedimentation, scaling, corrosion and biological growth. Reference 15 also provides guidelines for tube bundle supports used to minimize problems due to tube vibrations.

Key issues in the design and operation of steam condensers include:[19,20]

1. avoiding flow induced vibration and its associated tubing wear and failure,
2. optimizing steam side flow distribution to use all areas of the condenser surface effectively,
3. providing internal axial bundle restraints to maintain the integrity of the water boxes and peripheral areas of the tube bundle,
4. distributing steam and water with baffle and connection designs which prevent harm to tubes,
5. bracing the shell against large vacuum loads,
6. preventing air and circulating water in-leakage, and
7. maintaining clean condenser tubes to permit satisfactory unit performance.

When tubes become plugged with leaves, marine life and other debris, valving arrangements can provide backwashing and flushing to remove this debris. For cleaning of internal tube surfaces, rubber plugs or wire brushes are manually inserted, or sponge balls or brushes can be automatically recirculated through the condenser. Chemical water treatment can also mitigate biofouling.

The performance of surface condensers can be evaluated using the American Society of Mechanical Engineers (ASME) Performance Test Code 12.2.

Steam system interface

The condenser performance may have a dramatic effect on the steam producing systems. Air and cooling water leakage may result in excessive boiler corrosion and deposition. Poor overall condenser performance results in high back-pressures and lower overall power plant electricity production and efficiency. Higher steam system firing rates needed to make up for the power loss due to condenser fouling or air in-leakage increase stress on the boiler and other components.

Special provisions during the design phase may also be required to match the condenser capacity with the boiler and turbine steam bypass requirements. Cycling steam power systems frequently incorporate steam bypass systems around the turbine (see Chapter 18) which can pass full load steam flow directly to the condenser. The resulting thermal load on the condenser may exceed the thermal load for normal full load system operation. This may be accompanied by high steam temperatures which can potentially cause undesirable thermal expansion of the condenser tubes or low pressure turbine casing.

Cooling towers

Heat rejection systems

The amount of heat rejected from modern thermal power stations is significant, representing 60 to 68% of the total heat input. 10 to 15% is typically rejected up the stack with the flue gas. Most of the balance must be absorbed from condensing the steam in the Rankine cycle. (See Chapter 2.) Ultimately all of this energy is absorbed in the atmosphere although it may first be rejected to a body of water.

The heat rejection systems are generally classified as either once-through (open) or closed. In *once-through systems*, water is withdrawn from a lake, river or ocean and then pumped through the condenser where its temperature is increased by about 15 to 20F (8.3 to 11.1C). The warmer water is then discharged back to its source. In *closed loop systems*, heat is rejected to the atmosphere through the use of either a cooling tower (wet or dry) or some form of outdoor body of water such as a spray pond or cooling lake.

Once-through cooling systems are generally no longer acceptable in most developed regions because withdrawal sites are either already fully developed or because of increasingly stringent environmental regulations. As a result, virtually all new power systems use closed loop heat rejection systems. Cooling ponds or lakes are primarily found at existing older plant sites. Most newer systems use either the wet cooling towers discussed below or dry systems discussed in the next section.

In wet cooling tower systems, water is circulated through the condenser to absorb the heat from the steam. It is then circulated through the cooling tower where the heat is rejected to the atmosphere by evaporation of some of the circulating water and by heating the air. The cool water is then returned to the condenser by a circulating pump. (See the left side of Fig. 1.) A river or other water source must then only provide sufficient makeup water flow to compensate for the water lost through evaporation and blowdown. The cooling tower performs three functions:

1. dissipates heat from the steam condenser,
2. limits water use, and
3. provides low temperature water to the condenser to maintain plant thermal efficiency.

General arrangement and types of cooling towers

Wet cooling towers are specialized direct contact heat exchangers where warm water is brought into contact with relatively dry air, and energy is transferred by means of sensible heat loss to the air and by latent heat loss with the evaporated water. As with any direct contact heat exchanger, the heat transfer rate is dependent upon maximizing the contact area and residence time between the air and water. All cooling tower designs have five common elements:

1. air circulation system,
2. water distribution/spray system,
3. packing or fill to maximize the contact between the air and water,
4. cooling water collection/discharge basin, and
5. mist eliminators to minimize droplet carryover and water loss.

Wet cooling towers are generally configured with the air and water in counterflow or crossflow. In *counterflow*, water falls down through the fill while air is drawn upward. In *crossflow* units, water cascades downward while the air moves horizontally, perpendicular to the water flow. Such cooling towers are generally classified by the means used to move the air: mechanical forced draft, mechanical induced draft and natural draft (hyperbolic) towers.

In *mechanical forced draft* cooling towers, fans are typically located upstream of the air/water interface, usually close to ground level. This permits easy inspection and maintenance while minimizing the water pumping head. Limitations associated with this design typically include nonuniform air inlet flow distribution due to ground effects, fan icing, recirculation of the moist air from the tower discharge to the inlet, and fan blade diameter.

In *mechanical induced draft* cooling towers, the fan is located downstream of the air/water interface. (See Fig. 15.) In this case air is drawn through the cooling tower. Air flow distribution is typically more uniform and less prone to ground interference or recirculation. The fans are, however, less accessible.

In the *natural draft* cooling tower (see Figs. 16 and 17), air flow is drawn through the packing or fill by means of the small density difference between the warm moist air inside the tower and the cooler ambient air outside the tower. Because of the relatively small temperature and density difference, such cooling towers tend to be very tall [typically 300 to 500 ft (91 to 152 m)]. These towers are relatively open cylindrical structures with the packing and water distribution at their base. Modern units are sometimes referred to as *hyperbolic* towers because of their shape. The hyperbolic shape has little to do with the efficient production of the natural draft, but rather offers superior strength and resistance to wind loadings. As a result, less material is needed.

The key tradeoff between mechanical and natural draft systems basically balances the higher capital cost of the natural draft tower versus the higher operating (power) and maintenance costs of the forced draft system. *Hybrid* tower designs are also available which combine shorter stacks with fan assistance to minimize overall costs.

Design considerations[21-29]

Fill *Tower fill* or *packing* is one of the most important elements in the cooling tower. Its purpose is to increase the contact area between the air and water as well as to increase the water residence time. The tower fill is generally classified as film type or splash type.

When *film type fill* is used, a thin water layer is directed along a plate or sheet, and air is forced past the water. Several sheets are generally placed together at fixed angles to maximize the air-water contact area and water residence time. These fills are generally denser than the splash types. Therefore a smaller volume of film type fill is required to remove a given thermal load. However, the increased density per unit depth induces a larger pressure loss and also requires higher fan power. Film type fills are usually made from polystyrene, polypropylene, high density polyethylene or polyvinyl chloride (PVC).

Splash type fills increase the air-water contact area and residence time by a splashing action. In these units water enters the tower and falls on splash bars. These bars divide the large water droplets into smaller ones, increasing their surface area. The bars also slow the fall of the droplets, increasing residence time. Splash bars are normally made from fiberglass or PVC and previously have been redwood. These splash fills are usually found in crossflow cooling towers.

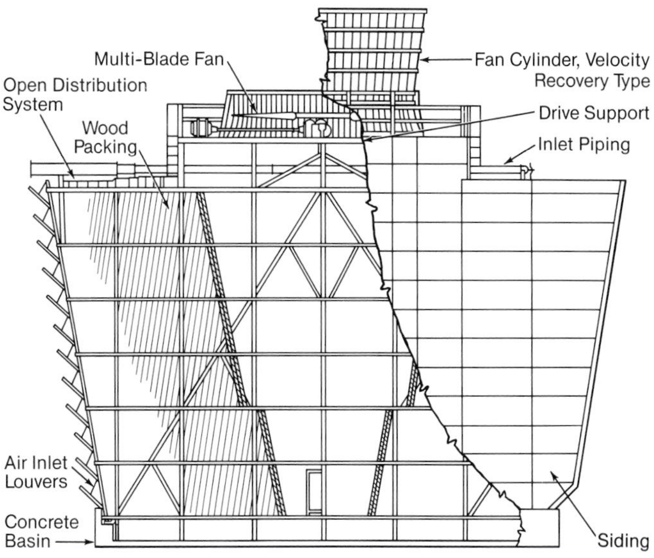

Fig. 15 Crossflow induced draft cooling tower (*courtesy of* Power *magazine*).

In a typical splash type configuration the fills are horizontally spaced and nailed to a stringer. This comprises one deck and the decks are vertically spaced. Water cascades over the decks and air is introduced in a crossflow configuration.

Construction materials A variety of materials are used in modern cooling towers:

1. frame — redwood, Douglas fir, galvanized steel, stainless steel, fiberglass or concrete,
2. fill — redwood, plastics, fiberglass and PVC,
3. casing — fiberglass, aluminum, stainless steel, PVC or concrete,

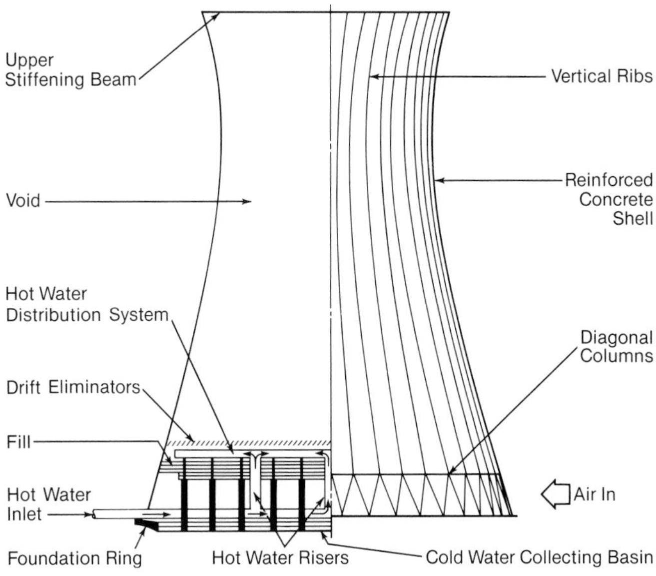

Fig. 16 Natural draft, hyperbolic cooling tower — counterflow design (*courtesy of* Power *magazine*).

Fig. 17 Natural draft, hyperbolic cooling tower — crossflow design.

4. fan blades — aluminum, fiber reinforced plastics, stainless steel or Monel™,
5. fan hubs — cast iron, galvanized steel or stainless steel,
6. fan stack — fiberglass reinforced plastic, steel or concrete,
7. drift eliminators — PVC or fiber reinforced plastic,
8. water spray nozzles — PVC, polypropylene or ceramic,
9. louvers — PVC, Douglas fir or fiber reinforced plastic, and
10. natural draft cooling tower shell — high density, type II reinforced concrete.

Fire protection Although concrete or steel cooling towers do not burn, they can be severely damaged by exposure to fire. If certain components, such as the fan deck, are combustible, there is potential for fire damage. To provide protection, typical wood structure mechanical draft towers have fire extinguishing systems. However, with the increasing use of noncombustible materials such as PVC and concrete, constructions without fire extinguishing systems have been approved for use.

When fire protection is required, simple wet down systems or intricate sprinkler systems may be installed.[30] *Wet down* systems are simple piping and nozzle arrangements that provide a small, continuous water flow at the top of the tower. This flow only maintains dampness in primary wood components. *Sprinkler* systems are complex arrangements of piping, nozzles, valves, and sensors or fusible heads that deluge the tower with water at the first sign of a fire.

Performance The overall size and performance of a cooling tower are affected by the heat dissipation or load, cooling water temperature change (or *range*), *approach to saturation* temperature and the *wet bulb* temperature at the site. Natural draft cooling towers are also affected by relative humidity and atmospheric conditions. The active area of the tower increases almost proportionately with the total heat dissipation load. Tower size can be reduced by increasing: 1) the range, 2) the approach to saturation temperature, and 3) the wet bulb temperature. Provision must also be made to avoid exhaust air recirculation and ground interference. An approach to sizing wet cooling towers is presented in References 21, 22, and 26 to 29.

Operating issues

An operator must be aware of and make provision for several potential operating problems:[23] packing or fill performance degradation, fill silt buildup, icing in freezing weather and wind effects on performance. Operators must also ensure that the water loading is balanced and that fans and motors are running properly.

While tower fill performance is affected by the type of fill, it is also related to water and air loadings. A low water loading results in poor water distribution and air short circuiting, while excessive loading causes the tower to flood, producing excessive air pressure losses. In both cases, a deterioration of performance occurs.

When operating cooling towers in cold climates, it is important to guard against ice formation. Of particular concern are the air intakes and fill, where a buildup can jeopardize thermal performance. Severe ice formation may be reduced by temporarily reversing the direction of the draft fans, shifting relatively warm water outward over the ice buildup. Several natural draft towers have been fitted with a lattice in the air intake for protection against icing. This lattice is deliberately iced during cold weather to reduce air flow and prevent fill freezing.[23] Hot water is used to melt the ice buildup and restore air flow. Counterflow cooling tower designs tend to be less sensitive to icing problems than crossflow towers.

It is necessary to provide a blowdown stream from the cooling water systems to control the water solids buildup and concentration. Means are usually required to treat this stream before discharge (see Chapter 32 for limits). Multiple-effect evaporators can be used to concentrate the blowdown stream for disposal.

Thermal performance testing procedures are provided in ASME Performance Test Code 23.

Air-cooled steam condensers

Air-cooled steam condenser systems serve the same functions as the water-cooled condenser and wet cooling tower. They provide a low turbine back-pressure, conserve high purity water and deaerate the condensate.

However, in place of the evaporation in the wet tower, all of the heat rejected from the steam is absorbed in the form of sensible heat gain in the ambient air. The air-cooled steam condenser system may be mechanical or natural draft although most modern installations are mechanical draft.

Air-cooled steam condenser systems can be attractive for a number of reasons:

1. makeup water is not required, thereby permitting plant siting without regard for large cooling water supplies,
2. maintenance is less expensive,
3. large amounts of treatment chemicals are not required, and there are no associated problems of blowdown disposal, and
4. fogging, misting and local icing resulting from wet cooling tower plume are not encountered.

The main disadvantage of air-cooled systems is that the minimum temperature and corresponding steam condenser pressure are not as low as with wet cooling tower systems. This results in a higher turbine back-pressure, lower plant efficiency and increased heat rejection for the same power production. Turbine back-pressures of 3.7 to 9 in. Hg (13 to 30 kPa) are encountered in air-cooled steam condensers while once-through and wet cooling tower systems under similar conditions provide back-pressures of 1.0 to 4.5 in. Hg (3.4 to 15.2 kPa).[31]

Two basic systems are available: 1) direct steam condenser system, and 2) indirect system with intermediate coolant loop. In the *direct system* steam condenser, the turbine exhaust steam is directed to the air-cooled steam condenser where the heat is absorbed directly by the air, effectively combining the conventional condenser and cooling tower in one device. This system has the thermodynamic advantage of requiring only one heat exchanger temperature drop versus two in the indirect system. A key disadvantage is the need for large ducts to carry the steam from the low pressure turbine exhaust to the steam condenser. In the *indirect system*, a surface or spray steam condenser absorbs heat from the steam, and a secondary water loop transfers the energy to a separate air cooler which then rejects the heat to the atmosphere. The expensive, large diameter steam ducts are replaced with the water piping and an additional heat exchanger. In the indirect system, which uses a surface condenser, an additional thermal loss also arises from the condenser temperature drop.

Direct air-cooled steam condensers — arrangement

Air-cooled steam condensers are effectively simple air-cooled heat exchangers where steam condenses on the inside of the tubes while ambient air flows over and is heated on the outside by convective heat transfer.[31-34] Fins on the tube outside surface increase contact area and overall heat transfer rate. (See Chapter 4.) These tubes are assembled into large staggered tube array bundles connected to a steam inlet header and an outlet condensate header. These are generally arranged into an A-frame as shown in Figs. 18 and 19 to conserve space. High air flow rates are provided by large horizontally mounted fans located below the A-frame. Sufficient space below the fan is required to admit the necessary air volume. Steam flows down the tubes and condensate flows by gravity through a condensate drain pot to a condensate storage tank. Provision is made to remove noncondensable gases which could cause tube blockage. Condenser tubes are generally 1 to 2 in. (25.4 to 51.8 mm) outside diameter and a variety of fin types are used — extruded, extruded serrated, embedded or wrap-on. Besides the A-frame configuration, horizontal and vertical bundle orientations are also used. The thermal performance and pressure drop can generally be evaluated using the procedures outlined in Chapters 3 and 4 and those discussed in References 31 to 34. Overall turbine exhaust temperature is controlled by the ambient air *dry bulb* temperature, heat exchanger approach temperature and temperature range.

Fig. 18 Air-cooled steam condenser installation.

Indirect air-cooled steam condenser systems

In this case, a conventional surface condenser is used to condense the steam, and a secondary water flow loop provides the cooling water to the condenser inlet and supplies warm water to an air-cooled heat exchanger (air cooler) to reject the heat. The air cooler is conventional in nature and can be evaluated using standard techniques.

New indirect systems being tested replace the secondary single phase cooling water loop with a two-phase sys-

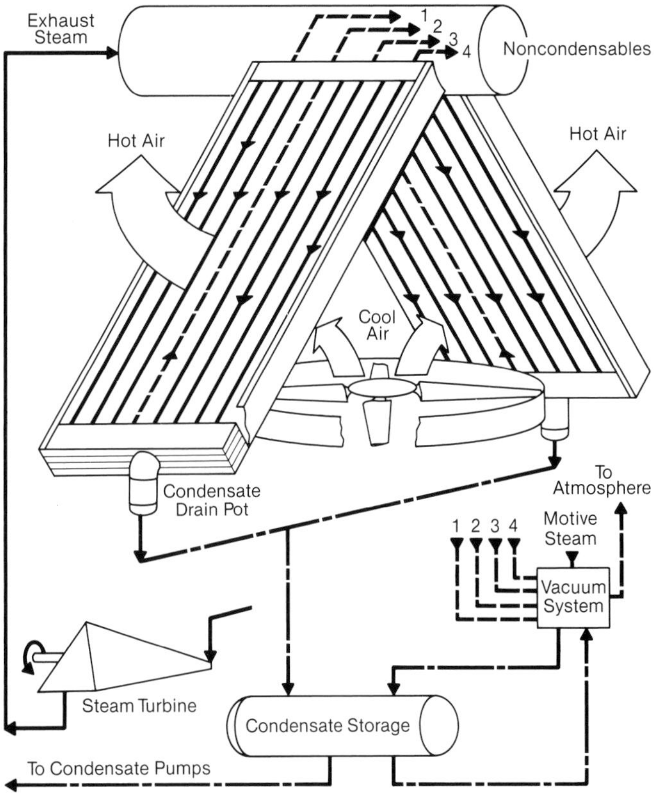

Fig. 19 Air-cooled steam condenser system schematic.

tem which uses a more volatile and lower freezing point heat transfer fluid such as ammonia or isopentane. The secondary loop pressure is controlled so that the heat transfer fluid will evaporate in the steam condenser as it absorbs the heat given up by the steam and then condense again to a liquid in the air cooler as the heat transfer fluid gives up its latent heat. Such a system gains thermodynamically because of two factors:

1. it is effectively isothermal, and
2. smaller temperature losses will occur as the heat transfer fluid absorbs, transports and discharges the energy because of the high evaporation and condensation heat transfer coefficients. This has the added benefit of reducing equipment size.

Operation

Air-cooled steam condensers have been successfully used on power plants as large as 360 MW in the U.S. and 665 MW internationally. By 1991 over 7000 MW of plant capacity worldwide operated using direct air-cooled steam condensers with an additional 8400 MW of capacity using indirect air-cooled systems.[34] During the summer months, when the average dry bulb temperature exceeds 90F (32C), the condensing temperature can typically range from 150 to 160F (66 to 71C), corresponding to a saturated steam pressure of 3.7 to 4.7 in. Hg (12.5 to 15.9 kPa). Such high back-pressure operation requires special turbine modifications. Several schemes are available to help compensate for the potential high turbine back-pressures.

If the air-cooled steam condenser is installed in a cold environment, provision must be made in hardware design and operation to avoid freezing of the condensate in the tubes. If permitted to occur, such freezing cycles will likely result in tube wall failures.

Feedwater heaters

Feedwater heaters have two main functions in modern power stations. They provide thermodynamic gains of the regenerative steam cycle discussed in Chapter 2 and they raise the feedwater temperature to minimize thermal effects in the boiler. Feedwater heaters use superheated or saturated steam from selected turbine extraction points to preheat feedwater from the condenser prior to feeding the water to the boiler economizer or drum. These heaters are classified as closed or open designs. In addition, they are designed for low or high pressure operation.[35-38]

Closed feedwater heaters are specialized shell and tube heat exchangers with superheated or saturated steam condensing on the shell side and the feedwater flowing on the tube side. Fig. 20 provides a schematic of a high pressure closed feedwater heater. Most modern feedwater heaters are installed in the horizontal position shown although vertical units have been used. While vertical units minimize floor space and building volume, higher maintenance costs and increased operational problems have led to a general preference for the horizontal design. Horizontal feedwater heaters have generally had fewer operating problems because a greater liquid (*drains*) volume is available for level control during plant transients.

Low pressure heaters, located upstream of the boiler feed pump, are generally designed for tube side pressures of less than 900 psig (62.1 bar gauge). Low pressure heaters are usually furnished with full access channel covers to facilitate maintenance. Units built with a hemispherical or elliptical head include a bolted and gasketed manway for access. Low pressure heaters generally have the tubes only expanded into the tubesheet although units for boiling water reactor service are also welded to the tubesheet. *High pressure heaters*, which are downstream of the feed pump, withstand pressures up to 1500 psig (103.4 bar gauge) in nuclear units and up to 5000 psig (344.8 bar gauge) in fossil fuel plants. Generally, high pressure heaters have been furnished with hemispherical head designs and manway access. When manufacturing limits permit, full diameter access designs have also been supplied; these increase the ease of routine maintenance but at a higher capital cost. Welded tube-to-tubesheet joints are provided.

Open feedwater heaters (deaerators) in fossil power plants serve the dual purpose of regenerative feedwater heating to increase plant efficiency and deaeration to remove gases which may cause equipment corrosion. They can also provide high purity feedwater storage for the boiler feed pump. Heat transfer is by direct contact between the

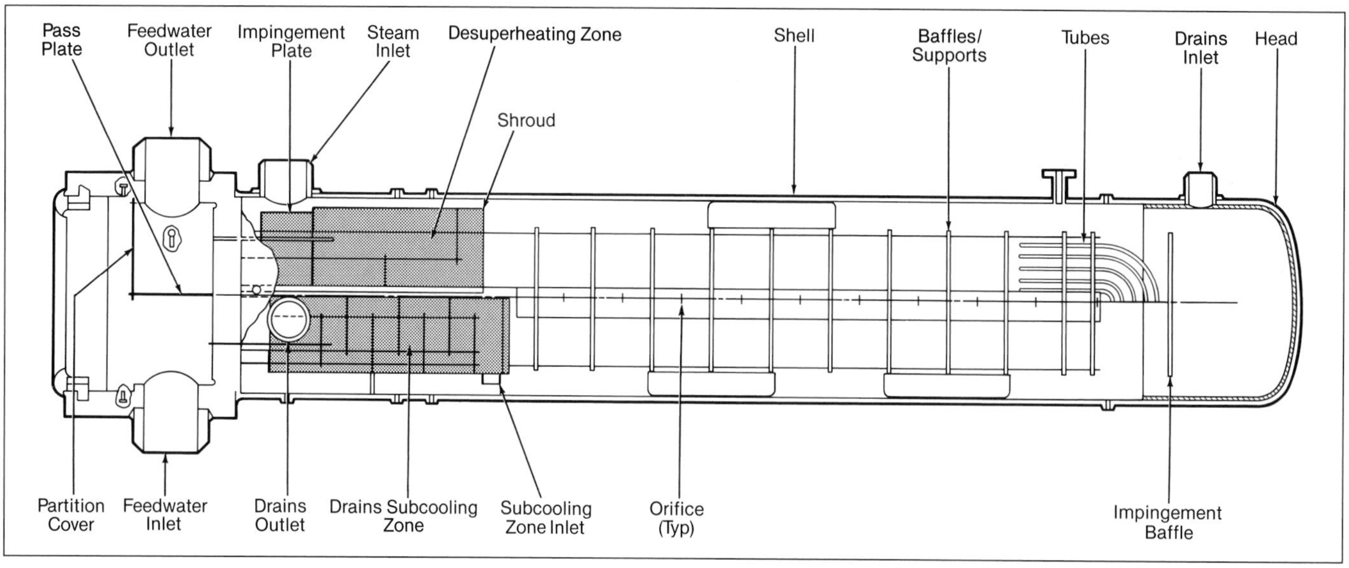

Fig. 20 Typical three zone, high pressure closed feedwater heater (*adapted from Reference 44*).

feedwater and extraction steam in bubbling, tray, spray or combination systems. The drains from the high pressure feedwater heaters usually flow into the deaerator. Noncondensable gases are vented to the atmosphere, with the O_2 content typically maintained below 7 ppb and free CO_2 reduced to negligible amounts in large power plants.[36] Further discussion of open feedwater heaters and deaerators is provided in Reference 38.

The number and type of feedwater heaters used depend upon the steam cycle, pressure and economics with smaller plants having fewer units. In utility and industrial plants, typically five to seven feedwater heater stages are used with multiple parallel feedwater heaters per stage in some cases. A general rule is to divide the feedwater heating range into reasonable increments of temperature rise, typically 50 to 80F (28 to 44C) depending upon the extraction pressures available and the heat rate benefit derived from having fewer or more feedwater heating stages. Parallel feedwater heater trains have typically been used where the flow rate exceeds 3,000,000 lb/h (378 kg/s) in high pressure heaters, while single strings of low pressure heaters have been used for flow rates up to a range of 4,000,000 to 5,000,000 lb/h (504 to 630 kg/s).

Closed feedwater heaters

Arrangement Most modern closed feedwater heaters are composed of bundles of a large number of U-tubes which are expanded and/or welded into stationary tubesheets at one end of the shell. (See Fig. 20.) As shown, a variety of baffles and tube support plates are used to direct flow, minimize tube vibration, reduce erosion and promote high heat transfer rates. Some designs replace the tubesheet with inlet and outlet headers within the heat exchanger shell. Tubes are usually laid out in a triangular pitch with a pitch to diameter ratio of 1.25 and a minimum clearance between tubes of 0.1875 in. (4.763 mm).

The lowest cost feedwater heaters are typically long, horizontal, two-pass designs with high water velocities. The length of the heat exchanger bundle is typically limited by the available tube lengths, with the longest heat exchanger typically ranging from 40 to 48 ft (12.2 to 14.6 m) for U-tube units. Water side pressure drop restrictions may result in large numbers of tubes and larger heater shells. When flow rates are low [250,000 to 300,000 lb/h (3.15 to 37.8 kg/s)] and heater length is a limitation, a four-pass feedwater heater design may be used with its associated larger shell and higher equipment costs.

The closed feedwater heater can be divided into one, two or three functional zones depending upon whether the extraction steam is superheated or saturated and the desired degree of condensate subcooling. In the simplest case, the heat exchanger serves as a one zone *condensing heater*, where the entire shell side basically operates at the saturated steam temperature corresponding to the shell side pressure. To this design a *drains subcooling zone* can be added by the installation of an appropriate enclosure. (See Fig. 20.) This permits the cold feedwater to extract additional energy from the shell side by subcooling the condensate before it is drained from the unit (typically within 10F (6C) of the inlet feedwater temperature).

When superheated extraction steam is supplied to the feedwater heater, additional improvement in performance can be obtained by creating a *desuperheating zone* on the shell side of the bundle at the extraction steam inlet. (See Fig. 20.) The enclosure baffles partially isolate the superheated steam from the rest of the bundle. This arrangement allows the outlet feedwater temperature to approach and sometimes exceed the saturation temperature corresponding to the shell side pressure.

Heat transfer rates The heat transfer rates for each feedwater heater zone must be evaluated separately using the techniques discussed in Chapter 4. The rates are primarily affected by feedwater velocity, tube metal, tube wall thickness, mean temperature differences, water temperatures, shell side pressure and extraction steam flow as well as drain flow in the subcooling zone. In the condensing zone the overall heat transfer coefficient typically falls within 500 to 900 Btu/h ft^2 F (2.8 to 5.1 kW/m^2 K) while the desuperheating zone range is 80 to 110 Btu/h ft^2 F (0.45 to 0.62 kW/m^2 K) and the drain subcooling zone range is 300 to 450 Btu/h ft^2 F (1.7 to 2.6 kW/m^2 K).[38] Typically 60 to 80% of the total heat transfer takes place in the condensing zone. The Heat Exchange Institute recommends adding a fouling resistance of 0.0002 h ft^2 F/Btu (35×10^{-6} m^2 K/W) on the entire tube side as well as an additional 0.0003 h ft^2 F/Btu (53×10^{-6} m^2 K/W) on the shell side in any desuperheating or drain subcooling zones used.[37]

Design elements

Industry standards for the design of closed feedwater heaters are provided in References 37 and 39. Beyond mechanical design, these standards address installation, cleaning, maintenance and inspection. General mechanical design is discussed in References 39 and 40. Feedwater heater mechanical design generally takes place in accordance with the rules of the ASME Boiler and Pressure Vessel Code, Section VIII or other applicable codes or regulations. In order for the feedwater heater to achieve the desired performance goals, the optimum arrangement of tubes and baffles must be selected. Among the other key issues are:

1. proper venting to remove corrosive noncondensable gases,
2. impingement protection to prevent tube erosion,
3. tube support to prevent excessive tube vibration,
4. proper location of level connections to prevent improper level control,
5. proper tube-to-tubesheet attachment,
6. mechanical design to avoid low cycle fatigue in cycling service, and
7. access to the tubesheet.

The Heat Exchange Institute specifies the following maximum velocities of feedwater passing through the tubes, evaluated at the average of the inlet and outlet temperatures:[38]

stainless steel, Monel™, Inconel™	10.0 ft/s (3.0 m/s)
copper-nickel (70-30, 80-20 and 90-10)	9.0 ft/s (2.7 m/s)
Admiralty and copper	8.5 ft/s (2.6 m/s)
carbon steel	8.0 ft/s (2.4 m/s)

Construction materials

Because of concerns about potential transport of copper from the feedwater heaters into the boiler and turbine, stainless steel and carbon steel are generally used for feedwater heater tubing. When chlorides are a concern, as with a plant using sea water cooling, 90-10 Cu-Ni alloy is used for low pressure heater tubing, and 70-30 Cu-Ni alloy is used in high pressure units. Additional newer corrosion resistant alloys are under development.

Operating issues

Feedwater heater damage has predominantly been found to involve tube failures from a number of causes:[41-43]

1. steam impingement erosion,
2. cavitation type erosion of the drains subcooling zone entrance,
3. tube vibration,
4. inlet tube end erosion/corrosion,
5. oxygen pitting,
6. exfoliation,
7. stress corrosion cracking,
8. laminated corrosion,
9. dealloying, and
10. seal weld cracks due to embrittlement.

Tube joint failures, improper plugging and poor maintenance practices also contribute to feedwater heater downtime and shorter heater life.

Many problems frequently reported for high pressure feedwater heaters result from impingement erosion, cavitation type erosion, erosion/corrosion and flow induced tube vibrations. Reliable designs to prevent these destructive tube vibrations must consider normal operation and the effects of abnormal operating conditions.

Inlet end erosion/corrosion can be significant with carbon steels. Lower design velocities, inlet water nozzles and proper pH control typically minimize this issue.

Tube leakage (and consequent plugging) causes considerable heater downtime and the need to plug tubes can cause a partial or full forced outage. Furthermore, improper plugging can damage tubesheets and tube joints.

Overall operating thermal performance can be evaluated using the ASME Performance Test Code 12.1 for feedwater heaters.

Maintenance

Feedwater heaters require virtually no day to day maintenance; however certain components, such as valves, need constant surveillance. Nondestructive examination (NDE) is a powerful inspection technique for feedwater heaters; it is useful in tube side and shell side inspections during routine outages. Because tube condition is critical, sonic pulse, eddy current and ultrasonic testing are used to assess tube integrity. Eddy current testing is particularly valuable for determining wall thinning and identifying cracks, although it is slower than sonic pulse testing.

When leaks are detected, tubes are typically plugged using one of a variety of tube plugging techniques. Tapered or mechanical plugs are used in tubes which are expanded into the tubesheet while explosive plugs are used in heaters with welded tube joints. From 10 to 30% of the tubes may be plugged in some units. This results in a significantly higher pressure drop, but has less of an impact on thermal performance because the higher feedwater velocities tend to increase the heat transfer rates in the remaining tubes.

Steam system interface

The feedwater heater system can have a significant impact on the steam system performance. Experience has shown that much of the deposits found in steam generating systems are derived from corrosion products and contaminants picked up in the feedwater heating system. Copper based tubing is being replaced with other materials to reduce dissolved copper carryover. Proper chemistry control and specified deaerator system operation can minimize corrosion in the feedwater system and reduce corrosion product carryover into the steam generator. Proper oxygen control is especially important to minimize unacceptable steam generator system corrosion.

Feedwater heaters may sometimes be taken out of service (bypassed) because of maintenance problems or because of the need for more power on a short term basis. The feedwater system must be designed for this option. The extraction steam passes through the rest of the steam turbine resulting in more overall power production. However, this increased output comes at the expense of cycle efficiency (less regenerative heating). In addition, if one of the last heaters is removed while the steam generator is still on-line, the temperature entering the boiler will fall, reducing the average feedwater inlet temperature. This results in more heat pickup in the economizer and potentially less in the air heater. A major impact of this is the increase in enthalpy needed to heat water to the desired steam temperature. A higher firing rate is needed to achieve the desired steam flow. This has a significant impact on boiler operation, especially the primary superheater. Lower economizer temperatures may also lead to increased economizer corrosion.

References

Steam turbines

1. Spencer, R.C., Cotton, K.C., and Cannon, C.N., "A method for predicting the performance of steam turbine-generators... 16,500 kW and larger," Technical Paper 62-WA-209, American Society of Mechanical Engineers, New York, 1962.

2. Salisbury, J.K., *Steam Turbines and Their Cycles*, Wiley, New York, 1950 (Reprint R.E. Krieger, Huntington, New York, 1974).

3. Kearton, W.J., *Steam Turbine Theory and Practice*, Pitman, London, 1958.

4. "Steam turbines and auxiliaries," *Power*, Vol. 133, No. 6, pp. 5.1 to 5.63, June 1989.

5. "Steam turbine fundamentals" and "Steam turbines," Chapters 2.1 and 2.2, appearing in appearing in *Standard Handbook of Powerplant Engineering,* T.C. Elliot, ed., McGraw-Hill, New York, pp. 2.3 to 2.62, 1989.

6. Schofield, P., "Steam turbines," Chapter 3, appearing in *The ASME Handbook on Water Technology for Thermal Power Systems*, P. Cohen, ed., American Society of Mechanical Engineers, New York, pp. 117 to 177, 1989.

7. Schleithoff, K., and Termuchlea, H., "Steam purity in German power plants: standards, operational data and the effects on turbine components," Technical Paper 78-JPGC-PWR-20, American Society of Mechanical Engineers, New York, 1978.

Gas turbines

8. Makansi, J., "Gas turbines," *Power*, Vol. 132, No. 3, pp. 15 to 24, March 1988.

9. "Gas turbine fundamentals" and "Gas turbines," Chapters 2.4 and 2.5, appearing in *Standard Handbook of Powerplant Engineering*, T.C. Elliot, ed., McGraw-Hill, New York, pp. 2.109 to 2.147, 1989.

10. *1992 International Gas Turbine and Aeroengine Technology Report*, International Gas Turbine Institute, ASME, Atlanta, Georgia, 1992.

11. Cohen, H., Rogers, G.F.C., and Saravanamuttoo, H. I. S., *Gas Turbine Theory*, Longman Group Ltd., London, 1974.

12. Ushiyama, L., "Theoretically estimating the performance of gas turbines under varying conditions," *Journal of Engineering for Power*, Vol. 98, No. 1, pp. 69 to 78, January 1976.

13. Hesse, W.J., and Mumford, N.V.S., *Jet Propulsion for Aerospace Applications*, Pitman, London, 1964.

Condensers

14. Aschner, F.S., *Planning Fundamentals of Thermal Power Plants*, pp. 102 to 129, Wiley, New York, 1978.

15. *Standards for Steam Surface Condensers*, 8th ed., Heat Exchange Institute, Cleveland, Ohio, 1984 (also Addendum 1, 1989).

16. Coit, R., "Condensers," Chapter 5, appearing in *The ASME Handbook on Water Chemistry in Thermal Power Systems*, P. Cohen, ed., American Society of Mechanical Engineers, New York, pp. 197 to 227, 1989.

17. Marto, P.J., and Nunn, R.H., eds., *Power Condenser Heat Transfer Technology*, Hemisphere, Washington, D.C., 1981.

18. Butterworth, D., "Steam power plant and process condensers," Chapter 11, appearing in *Boilers, Evaporators and Condensers*, S. Kakac, ed., Wiley, New York, pp. 571 to 633, 1991.

19. Syrett, B.C., and Coit, R., "Causes and prevention of condenser tube failures," American Nuclear Society Meeting, Portland, Oregon, July 25-28, 1982.

20. Bell, R.J., and Conley, E.F., "Recommended practices for operating and maintaining steam surface condensers," Report CS-5235, Electric Power Research Institute, Palo Alto, California, 1987.

Cooling towers

21. Hensley, J.C., "Cooling towers," Chapter 1.9, appearing in *The Standard Handbook of Powerplant Engineering,* T.C. Elliot, ed., McGraw Hill, New York, pp. 1.207 to 1.239, 1989.

22. Baker, D.R., and Shryock, H.A., "A comprehensive approach to the analysis of cooling tower performance," *Journal of Heat Transfer*, Vol. 83, No. 3, pp. 339 to 350, 1961.

23. Bartz, J.A., "Improvements in cooling tower performance," *2nd International Conference on Improved Coal-Fired Power Plants*, A. Armor, *et al.*, eds., Electric Power Research Institute, Palo Alto, California, pp. 72-3 to 72-18, 1989.

24. "Cooling towers — A special report," *Power*, Vol. 117, No. 3, pp. S-1 to S-24, March 1973.

25. Marks, R.H., "Cooling towers — A special report," *Power*, Vol. 107, No. 3, pp. S-1 to S-16, March 1963.

26. Cheremisinoff, N.P., and Cheremisinoff, P.N., *Cooling Towers*, Ann Arbor Science Publishing, Ann Arbor, Michigan, 1981.

27. *Performance Curves*, Cooling Tower Institute, Houston, Texas, 1967.

28. Hallett, G.F., "Performance curves for mechanical draft cooling towers," *Journal of Engineering for Power,* Vol. 97, No. 3., pp. 503 to 508, October 1975.

29. "Cooling systems," Section 4.8 appearing in *Handbook of Energy Systems Engineering*, L. Wilbur, ed., Wiley, New York, pp. 494 to 522, 1985.

30. "Standards on water cooling towers," Standard NPFA 214, National Fire Protection Association, Quincy, Massachusetts, 1990.

Air-cooled steam condensers

31. Li, K.W., and Priddy, A.P., *Power Plant System Design*, Wiley, New York, pp. 474 to 521, 1985.

32. Larinoff, M.W., "Dry cooling tower plant design specifications and performance characteristics," *Dry and Wet/Dry Cooling Towers for Power Plants,* American Society of Mechanical Engineers, New York, 1973.

33. Miliares, E.S., *Power Plants and Air-Cooled Condensing Systems*, MIT Press, Cambridge, Massachussets, 1974.

34. Elliot, T.C., and Kals, W., "Air-cooled condensers," *Power*, Vol. 134, No. 1, pp. 13 to 21, January 1990.

Feedwater heaters

35. *Feedwater Heaters*, The Griscom-Russell Company, Massillon, Ohio, 1958.

36. Coit, R., "Feedwater heaters," and Oliker, I.I., "Deaeration," Chapters 4 and 15, appearing in *The ASME Handbook on Water Technology for Thermal Power Systems*, P. Cohen, ed., American Society of Mechanical Engineers, New York, pp. 181 to 196 and 1291 to 1320, 1989.

37. *Standards for Closed Feedwater Heaters*, 4th ed., Heat Exchange Institute, Cleveland, Ohio, 1984.

38. Spencer, J.R., and Stephani, R.M., "Power plant heat exchangers," Section 9.5, appearing in *Mark's Standard Handbook for Mechanical Engineers*, 9th ed., E.A. Avallone and T. Baumeister III, eds., McGraw-Hill, New York, pp. 9-60 to 9-79, 1987.

39. *Standards of Tubular Exchanger Manufacturers Association*, 7th ed., TEMA, Tarrytown, New York, 1988.

40. Singh, K.P., and Soler, A.I., *Mechanical Design of Heat Exchangers and Pressure Vessel Components*, Arcturus Publishers, Cherry Hill, New Jersey, 1984.

41. Bell, R.J., Rose, D.M., and Diaz-Tous, I.A., "High pressure feedwater heaters for advanced fossil-fired power plants," *1st Int'l. Conf. on Improved Coal-Fired Power Plants*, A. Armor, *et al.*, eds., Electric Power Research Institute, Palo Alto, California, pp. 4-93 to 4-118, 1988.

42. Bart, J.A., *et al.*, "Improved feedwater systems and components for coal-fired power plants," *2nd Int'l. Conf. on Improved Coal-Fired Power Plants*, A.F. Armor, *et al.*, eds., Electric Power Research Institute, Palo Alto, California, pp. 75-3 to 75-31, 1989.

43. Yokell, S., *A Working Guide to Shell-and-Tube Heat Exchangers*, McGraw-Hill, New York, 1990.

44. Andreone, C.F., and Rose, D.M., "Closed feedwater heaters: New emphasis on efficiency, reliability and maintainability," *Power Engineering*, Vol. 87, No. 11, pp. 62 to 64, November 1983.

Appendices

Appendix 1
Conversion Factors, SI Steam Properties and Useful Tables

This appendix provides tables, charts and figures that can be useful in evaluating the generation and use of steam. They supplement the material presented throughout *Steam* and are divided into four subsections: selected conversion factors; SI versions of the Steam Tables and graphical data; useful charts and figures; and typical operating parameters for a 500 MW pulverized coal-fired power plant.

Table 1.1
Length

Starting With → Multiply By To Obtain ↓	Inch	Foot	Yard	Statute Mile	Millimeter	Centimeter	Meter	Kilometer
Inch	1	12	36	63.36×10^{3}	3.937×10^{-2}	0.3937	39.37	3.937×10^{4}
Foot	0.08333	1	3	5,280	3.281×10^{-3}	3.281×10^{-2}	3.281	3,281
Yard	0.02778	0.3333	1	1,760	1.094×10^{-3}	1.094×10^{-2}	1.093	1,094
Statute Mile	1.578×10^{-5}	1.894×10^{-4}	5.682×10^{-4}	1	6.214×10^{-7}	6.214×10^{-6}	6.214×10^{-4}	0.6214
Millimeter	25.40	304.8	914.4	$1{,}609 \times 10^{6}$	1	10.00	1,000	1.000×10^{6}
Centimeter	2.540	30.48	91.44	1.609×10^{5}	0.1000	1	100	1.000×10^{5}
Meter	0.0254	0.3048	0.9144	1,609	1.000×10^{-3}	0.010	1	1,000
Kilometer	2.540×10^{-5}	3.048×10^{-4}	9.144×10^{-4}	1.609	1.000×10^{-6}	1×10^{-5}	0.0010	1

Table 1.2
Length, Small Units

Starting With → Multiply By To Obtain ↓	Mil	Inch	Angstrom	Nanometer	Micron	Millimeter	Centimeter
Mils	1	1,000	3.937×10^{-6}	3.937×10^{-5}	3.937×10^{-2}	39.37	393.7
Inch	0.0010	1	3.937×10^{-9}	3.937×10^{-8}	3.937×10^{-5}	3.937×10^{-2}	0.3937
Angstrom	2.540×10^{5}	2.540×10^{8}	1	10.00	1.000×10^{4}	1.000×10^{7}	1.000×10^{8}
Nanometer	2.540×10^{4}	2.540×10^{7}	0.100	1	1,000	1.000×10^{6}	1.000×10^{7}
Micron	25.40	2.540×10^{4}	1.000×10^{-4}	1.000×10^{-3}	1	1,000	1.000×10^{4}
Millimeter	2.540×10^{-2}	25.40	1.000×10^{-7}	1.000×10^{-6}	1.000×10^{-3}	1	10
Centimeter	2.540×10^{-3}	2.54	1.000×10^{-8}	1.000×10^{-7}	1.000×10^{-4}	0.100	1

Table 1.3
Area

Starting With → Multiply By To Obtain ↓	Square Inch	Square Foot	Square Yard	Acre	Square Centimeter	Square Meter	Square Kilometer
Square Inch	1	144.0	1,296	6.273×10^{6}	0.1550	1550	1.550×10^{9}
Square Foot	6.944×10^{-3}	1	9.000	4.356×10^{4}	1.076×10^{-3}	10.76	1.076×10^{7}
Square Yard	7.716×10^{-4}	0.1111	1	4840	1.196×10^{-4}	1.196	1.196×10^{6}
Acre	1.594×10^{-7}	2.296×10^{-5}	2.066×10^{-4}	1	2.471×10^{-8}	2.471×10^{-4}	247.1
Square Mile	2.491×10^{-10}	3.587×10^{-8}	3.228×10^{-7}	1.563×10^{-3}	3.861×10^{-11}	3.861×10^{-7}	0.3861
Square Centimeter	6.452	929.0	8,361	4.047×10^{7}	1	1.000×10^{4}	1.000×10^{10}
Square Meter	6.452×10^{-4}	0.09290	0.8361	4,047	1.000×10^{-4}	1	1.000×10^{6}
Square Kilometer	6.452×10^{-10}	9.290×10^{-8}	8.361×10^{-7}	4.047×10^{-3}	1.000×10^{-10}	1.000×10^{-6}	1

Table 1.4
Volume

Starting With → Multiply By To Obtain ↓	Fluid Ounce	Quart	Gallon	Barrel (Petroleum)	Cubic Foot	Cubic Yard	Cubic Centi-Meter (milliliter)	Liter	Cubic Meter
Fluid Ounce	1	32.00	128.0	5376	957.5	2.585×10^{4}	3.381×10^{-2}	33.82	3.382×10^{4}
Quart	0.03125	1	4.000	168.0	29.92	807.9	1.057×10^{-3}	1.057	1,057
Gallon	7.813×10^{-3}	0.2500	1	42.00	7.481	202.0	2.642×10^{-4}	0.2642	264.2
Barrel (Petroleum)	1.860×10^{-4}	5.952×10^{-3}	0.0238	1	0.1781	4.809	6.290×10^{-6}	6.290×10^{-3}	6.290
Cubic Foot	1.044×10^{-3}	3.342×10^{-2}	0.1336	5.615	1	27.00	3.532×10^{-5}	3.532×10^{-2}	35.32
Cubic Yard	3.868×10^{-5}	1.238×10^{-3}	4.951×10^{-3}	0.2079	3.704×10^{-2}	1	1.308×10^{-6}	1.308×10^{-3}	1.308
Cubic Centimeter (Milliliter)	29.57	946.4	3785	1.590×10^{5}	2.832×10^{4}	7.646×10^{5}	1	1,000	1.000×10^{6}
Liter	2.957×10^{-2}	0.9464	3.785	159.0	28.32	764.6	1.000×10^{-3}	1	1,000
Cubic Meter	2.957×10^{-5}	9.464×10^{-4}	3.785×10^{-3}	0.1590	2.832×10^{-2}	0.7646	1.000×10^{-6}	1.000×10^{-3}	1

Table 1.5
Mass

Starting With → Multiply By To Obtain ↓	Pound	Short Ton	Long Ton	Gram	Kilogram	Metric Ton
Pound	1	2,000	2240	2.205×10^{-3}	2.205	2,205
Short Ton	5.000×10^{-4}	1	1.12	1.102×10^{-6}	1.102×10^{-3}	1.102
Long Ton	4.464×10^{-4}	0.8929	1	9.842×10^{-7}	9.842×10^{-4}	0.9842
Gram	453.6	9.072×10^{5}	1.016×10^{6}	1	1,000	1.000×10^{6}
Kilogram	0.4536	907.2	1016	1.000×10^{-3}	1	1,000
Metric Ton	4.536×10^{-4}	0.9072	1.016	1.000×10^{-6}	0.0010	1

Table 1.6
Force

Starting With → Multiply By To Obtain ↓	Newton	Grams	Kilograms	Pounds
Newton	1	9.807×10^{-3}	9.807	4.448
Grams (f)	102.0	1	1000	453.6
Kilograms (f)	0.1020	0.001	1	0.4536
Pounds (f)	0.2248	2.205×10^{-3}	2.205	1

Table 1.7
Pressure or Force per Unit Area

Starting With → Multiply By / To Obtain ↓	Atmospheres	Cm of mercury at 0C	Inch of mercury at 0C	Inch of water at 4C	Feet of water at 4C	Kilograms per square meter	Pounds per square inch (abs)	Pascal	Bar
Atmospheres	1	1.316×10^{-2}	3.342×10^{-2}	2.458×10^{-3}	2.950×10^{-2}	9.678×10^{-5}	6.804×10^{-2}	9.869×10^{-6}	0.9869
Cm of mercury at 0C	76.00	1	2.540	0.1868	2.242	7.356×10^{-3}	5.171	7.501×10^{-4}	75.01
Inch of mercury at 0C	29.92	0.3937	1	7.355×10^{-2}	0.8826	2.896×10^{-3}	2.036	2.953×10^{-4}	29.53
Inch of water at 4C	406.8	5.354	13.6	1	12	3.937×10^{-2}	27.68	4.015×10^{-3}	401.9
Feet of water at 4C	33.90	0.4460	1.133	8.333×10^{-2}	1	3.281×10^{-3}	2.307	3.345×10^{-4}	33.49
Kilograms per square meter	1.033×10^{4}	136.0	345.3	25.40	304.8	1	703.1	1.0197×10^{-1}	1.0197×10^{4}
Pounds per square inch (abs)	14.70	0.1934	0.4912	3.613×10^{-2}	0.4335	1.422×10^{-3}	1	1.450×10^{-4}	14.50
Pascal	1.01325×10^{5}	1.333	3,387.4	249.0	2,989	9.80665	6,895	1	1.000×10^{5}
Bar	1.013	1.333×10^{-2}	3.387×10^{-2}	2.49×10^{-3}	2.989×10^{-2}	9.80665×10^{-5}	6.895×10^{-2}	1×10^{-5}	1

PRESSURE is the force per unit area exerted on or by a fluid.
Absolute pressure is the sum of atmospheric and gauge pressure.
Standard atmospheric pressure is 14.696 lbf/in or 29.92 mm Hg.

Table 1.8
Energy, Work and Heat

Starting With → Multiply By / To Obtain ↓	British thermal units	Centimeter-grams	Foot-pounds	Horsepower-hours	Joules watt-second	Kilowatt-hours	Kilogram meters	Watt-hours
British thermal units	1	9.297×10^{-8}	1.285×10^{-3}	2545	9.480×10^{-4}	3,413	9.297×10^{-3}	3.413
Centimeter-grams	1.076×10^{7}	1	1.383×10^{4}	2.737×10^{10}	1.020×10^{4}	3.671×10^{10}	10^{5}	3.671×10^{7}
Foot-pounds	778	7.233×10^{-5}	1	1.98×10^{6}	0.7376	2.655×10^{6}	7.233	2655
Horsepower-hours	3.929×10^{-4}	3.654×10^{-11}	5.050×10^{-7}	1	3.725×10^{-7}	1.341	3.653×10^{-6}	1.341×10^{-3}
Joules watt-second	1054.8	9.807×10^{-5}	1.356	2.684×10^{6}	1	3.6×10^{6}	9.807	3600
Kilowatt-hours	2.93×10^{-4}	2.724×10^{-11}	3.766×10^{-7}	0.7457	2.778×10^{-7}	1	2.724×10^{-6}	0.001
Kilogram-meters	107.6	10^{-5}	0.1383	2.737×10^{5}	0.1020	3.671×10^{5}	1	367.1
Watt-hours	0.293	2.724×10^{-8}	3.766×10^{-4}	745.7	2.778×10^{-4}	1000	2.724×10^{-3}	1

Table 1.9
Power or Rate of Doing Work

Starting With → Multiply By / To Obtain ↓	Btu per minute	Foot-pounds per minute	Foot-pounds per second	Horsepower	Kilowatts	Watts
Btu per minute	1	1.285×10^{-3}	7.716×10^{-2}	42.44	56.91	5.691×10^{-2}
Foot-pounds per minute	778	1	60	3.3×10^{4}	4.426×10^{4}	44.26
Foot-pounds per second	12.97	1.667×10^{-2}	1	550	737.6	0.7376
Horsepower	2.357×10^{-2}	3.030×10^{-5}	1.818×10^{-3}	1	1.341	1.341×10^{-3}
Kilowatts	1.757×10^{-2}	2.260×10^{-5}	1.356×10^{-3}	0.7457	1	10^{-3}
Watts	17.5725	2.260×10^{-2}	1.356	745.7	1000	1

Table 1.10
Density (Mass/Volume)

Starting With → Multiply By / To Obtain ↓	Pounds per Cubic Foot	Pounds per Gallon (U.S. Liquid)	Grams per Cubic Centimeter	Kilograms per Liter	Kilograms per Cubic Meter
Pounds per Cubic Foot	1	7.481	62.43	62.43	6.243×10^{-2}
Pounds per Gallon (U.S. Liquid)	0.1337	1	8.345	8.345	8.345×10^{-3}
Grams per Cubic Centimeter	1.602×10^{-2}	0.1198	1	1.000	1.000×10^{-3}
Kilograms per Liter	1.602×10^{-2}	0.1198	1.000	1	1.000×10^{-3}
Kilograms per Cubic Meter	16.02	119.8	1,000	1,000	1

Table 1.11
Thermal, Heat Transfer and Flow Parameters

Thermal Conductivity

1 Btu/h ft F = 1 Btu/h ft R
= 12 Btu in./h ft^2 F
= 1.7307 W/m K

Heat Content

1 Btu/lb = 2.326 kJ/kg
1 Btu/Standard cubic foot = 0.039338 MJ/Nm^3

Heat Flux, or Heat Transfer Rate

1 Btu/h ft^2 = 3.1546 W/m^2

Volumetric Heat Release Rate

1 Btu/h ft^3 = 10.3497 W/m^3

Heat Transfer Coefficient or Conductance

1 Btu/h ft^2 F = 5.6784 W/m^{2}K

Mass Flow Rate

1 lb/h = 0.000126 kg/s

Mass Flux

1 lb/h ft^2 = 0.001356 kg/m^{2}s

Volumetric Flow Rate

1 ft^3/min (ACFM) = 0.000472 m^3/s
1 GPM = 0.06308 l/s

Table 1.12
Multiplying Prefixes

Multiplication Factors	Prefix	SI Symbol
1 000 000 000 000 = 10^{12}	tera	T
1 000 000 000 = 10^{9}	giga	G
1 000 000 = 10^{6}	mega	M
1 000 = 10^{3}	kilo	k
100 = 10^{2}	hecto*	h
10 = 10^{1}	deka*	da
0.1 = 10^{-1}	deci*	d
0.01 = 10^{-2}	centi*	c
0.001 = 10^{-3}	milli	m
0.000 001 = 10^{-6}	micro	μ
0.000 000 001 = 10^{-9}	nano	n
0.000 000 000 001 = 10^{-12}	pico	p
0.000 000 000 000 001 = 10^{-15}	femto	f
0.000 000 000 000 000 001 = 10^{-18}	atto	a

* To be avoided where possible.

Table 1.13
Selected Additional Conversion Factors

Length and area

100 feet (ft)/minute (min)	= 0.508 meter (m)/ second(s)
1 square mile (mi^2)	= 640 acres
	= 259 hectares (ha)
1 m/s	= 196.9 ft/min
1 square kilometer (km^2)	= 100 ha
	= 0.3861 mi^2
1 ha	= 10,000 m^2
	= 2.471 acres
1 nautical mile	= 6080 ft
	= 1.853 km
1 nautical mile/hour (h)	= 1 knot

Weight

1 pound (lb)	= 16 ounces (oz)
	= 7000 grains (gr)
	= 0.454 kilogram (kg)
1 oz	= 0.0625 lb
	= 28.35 grams (g)
1 grain (gr)	= 64.8 milligrams (mg)
	= 0.0023 oz
1 lb/ft	= 1.488 kg/m
1 g	= 1000 mg
	= 0.03527 oz
	= 15.43 grains
1 kg/m	= 0.672 lb/ft

Volume

1 cubic foot (ft^3)	= 1728 cubic inches (in.^{3})
1 in.^{3}	= 16,390 cubic millimeters (mm^3)
1 imperial gallon	= 277.4 in.^{3}
	= 4.55 liters (l)
1 U.S. gallon (gal)	= 0.833 imperial gallon
	= 3.785 l
	= 231 in.^{3}
1 U.S. barrel (bbl) (petroleum)	= 42 U.S. gallons
	= 35 imperial gallons
1 liter (1)	= 10^6 mm^3
	= 0.220 imperial gallon
	= 0.2642 U.S. gallon
	= 61.0 in.^{3}
1 board ft	= 12 in. x 12 in. x 1 in. thick
	= 144 in.^{3}
1 ft^3/min	= 1.699 m^3/h
1 m^3/h	= 0.589 ft^3/min

Temperature, measured

F	= (C x 9/5) + 32
	= R − 459.67
C	= (F − 32) x 5/9
	= K − 273.15

Density

1 ft^3/lb	= 0.0624 m^3/kg
1 grain/ft^3	= 2.288 g/m^3
1 grain/U.S. gallon	= 17.11 g/m^3
	= 17.11 mg/l
1 m^3/kg	= 16.02 ft^3/lb
1 g/m^3	= 0.437 grain/ft^3
	= 0.0584 grain/ U.S. gallon
1 g/l	= 58.4 grain/U.S. gallon

Water at 62F (16.7C)

1 ft^3	= 62.3 lb
1 lb	= 0.01604 ft^3
1 U.S. gallon	= 8.33 lb

Water at 39.2F (4C), maximum density

1 ft^3	= 62.4 lb
1 m^3	= 1000 kg
1 lb	= 0.01602 ft^3
1 l	= 1.0 kg
1 kg/m^3	= 1 g/l
	= 1 part per thousand
1 g/m^3	= 1 mg/l
	= 1 part per million (ppm)

Pressure

1 atmosphere (metric)	= 98,066.5 pascals (Pa)
	= 1 kg force (kgf)/square centimeter (cm^2)
	= 10 m head of water
	= 14.22 lb/in.^{2}
1 lb/ft^2	= 47.88 Pa
	= 0.1924 in. of water
	= 4.88 kg/m^2
1 ton (t)/in.^{2}	= 13,789 kilopascals (kPa)
	= 1.406 kg/mm^2
1 in. head of water	= 5.20 lb/ft^2
1 m head of water	= 9806 Pa
	= 0.1 kg/cm^2
1 m head of mercury (Hg)	= 133.3 kPa
	= 1.360 kg/cm^2
	= 1333 millibars
1 kg/m^2	= 9.806 Pa
	= 1 mm head of water
	= 0.2048 lb/ft^2
1 kg/cm^2	= 98.066 kPa
	= 735.5 mm Hg
	= 0.981 bar
	= 14.22 lb/in.^{2}
1 kg/mm^2	= 9.8066 megapascals (MPa)
	= 0.711 t/in^2

In these conversions, inches and feet of water are measured at 62F (16.7C), millimeters and meters of water at 39.2F (4C), and inches, millimeters, and meters of mercury at 32F (0C).

Table 2.1
SI Properties of Saturated Steam and Saturated Water (Temperature)

Temp K	Press. MPa	Volume, m^3/kg Water v_f	Volume, m^3/kg Steam v_g	Enthalpy, kJ/kg Water H_f	Enthalpy, kJ/kg Evap H_{fg}	Enthalpy, kJ/kg Steam H_g	Entropy, kJ/(kg K) Water s_f	Entropy, kJ/(kg K) Evap s_{fg}	Entropy, kJ/(kg K) Steam s_g
273.16	0.0006113	0.001000	206.1	0.0	2500.9	2500.9	0.0	9.1555	9.1555
275	0.0006980	0.001000	181.7	7.5	2496.8	2504.3	0.0274	9.0792	9.1066
280	0.0009912	0.001000	130.3	28.1	2485.4	2513.5	0.1015	8.8765	8.9780
285	0.001388	0.001001	94.67	48.8	2473.9	2522.7	0.1749	8.6803	8.8552
290	0.001919	0.001001	69.67	69.7	2462.2	2531.9	0.2475	8.4903	8.7378
295	0.002620	0.001002	51.90	90.7	2450.3	2541.0	0.3193	8.3061	8.6254
300	0.003536	0.001004	39.10	111.7	2438.4	2550.1	0.3900	8.1279	8.5179
305	0.004718	0.001005	29.78	132.8	2426.3	2559.1	0.4598	7.9551	8.4149
310	0.006230	0.001007	22.91	153.9	2414.3	2568.2	0.5285	7.7878	8.3163
315	0.008143	0.001009	17.80	175.1	2402.0	2577.1	0.5961	7.6255	8.2216
320	0.01054	0.001011	13.96	196.2	2389.8	2586.0	0.6626	7.4682	8.1308
325	0.01353	0.001013	11.04	217.3	2377.6	2594.9	0.7280	7.3156	8.0436
330	0.01721	0.001015	8.809	238.4	2365.3	2603.7	0.7924	7.1675	7.9599
335	0.02171	0.001018	7.083	259.4	2353.0	2612.4	0.8557	7.0236	7.8793
340	0.02718	0.001021	5.737	280.5	2340.5	2621.0	0.9180	6.8838	7.8018
345	0.03377	0.001024	4.680	301.5	2328.0	2629.5	0.9793	6.7479	7.7272
350	0.04166	0.001027	3.844	322.5	2315.4	2637.9	1.0397	6.6156	7.6553
355	0.05105	0.001030	3.178	343.4	2302.9	2646.3	1.0991	6.4869	7.5860
360	0.06215	0.001034	2.643	364.4	2290.1	2654.5	1.1577	6.3615	7.5192
365	0.07521	0.001037	2.211	385.3	2277.3	2662.6	1.2155	6.2391	7.4546
370	0.09047	0.001041	1.860	406.3	2264.3	2670.6	1.2725	6.1198	7.3923
375	0.1082	0.001045	1.573	427.3	2251.2	2678.5	1.3288	6.0032	7.3320
380	0.1288	0.001049	1.337	448.3	2237.9	2686.2	1.3843	5.8894	7.2737
385	0.1524	0.001053	1.142	469.3	2224.5	2693.8	1.4393	5.7779	7.2172
390	0.1795	0.001058	0.9800	490.4	2210.9	2701.3	1.4936	5.6688	7.1624
395	0.2104	0.001062	0.8445	511.5	2197.0	2708.5	1.5473	5.5621	7.1094
400	0.2456	0.001067	0.7308	532.7	2182.9	2715.6	1.6005	5.4573	7.0578
405	0.2854	0.001072	0.6349	554.0	2168.6	2722.6	1.6532	5.3546	7.0078
410	0.3302	0.001077	0.5537	575.3	2154.0	2729.3	1.7054	5.2537	6.9591
415	0.3806	0.001082	0.4846	596.7	2139.1	2735.8	1.7572	5.1545	6.9117
420	0.4370	0.001087	0.4256	618.2	2123.9	2742.1	1.8085	5.0570	6.8655
425	0.4999	0.001093	0.3750	639.8	2108.4	2748.2	1.8594	4.9611	6.8205
430	0.5699	0.001099	0.3314	661.4	2092.7	2754.1	1.9099	4.8667	6.7766
435	0.6474	0.001104	0.2938	683.1	2076.6	2759.7	1.9599	4.7737	6.7336
440	0.7332	0.001110	0.2612	705.0	2060.0	2765.0	2.0096	4.6820	6.6916
445	0.8277	0.001117	0.2328	726.9	2043.2	2770.1	2.0590	4.5914	6.6504
450	0.9315	0.001123	0.2080	749.0	2025.9	2774.9	2.1080	4.5020	6.6100
455	1.045	0.001130	0.1864	771.1	2008.2	2779.3	2.1567	4.4136	6.5703
460	1.170	0.001137	0.1673	793.4	1990.1	2783.5	2.2050	4.3263	6.5313
465	1.306	0.001144	0.1506	815.7	1971.6	2787.3	2.2530	4.2399	6.4929
470	1.454	0.001152	0.1358	838.2	1952.6	2790.8	2.3007	4.1544	6.4551
475	1.615	0.001159	0.1227	860.8	1933.0	2793.8	2.3482	4.0695	6.4177
480	1.789	0.001167	0.1111	883.5	1913.0	2796.5	2.3953	3.9855	6.3808
490	2.181	0.001184	0.09150	929.3	1871.4	2800.7	2.4887	3.8193	6.3080
500	2.637	0.001202	0.07585	975.6	1827.5	2803.1	2.5813	3.6550	6.2363
510	3.163	0.001222	0.06323	1022.6	1781.0	2803.6	2.6731	3.4921	6.1652
520	3.766	0.001244	0.05296	1070.4	1731.7	2802.1	2.7644	3.3301	6.0945
530	4.453	0.001267	0.04454	1119.1	1679.1	2798.2	2.8555	3.1681	6.0236
540	5.233	0.001293	0.03758	1168.9	1622.9	2791.8	2.9466	3.0055	5.9521
550	6.112	0.001322	0.03179	1219.9	1562.7	2782.6	3.0382	2.8413	5.8795
560	7.100	0.001355	0.02694	1272.5	1497.8	2770.3	3.1306	2.6746	5.8052
570	8.206	0.001391	0.02284	1326.9	1427.5	2754.4	3.2241	2.5044	5.7285
580	9.439	0.001433	0.01934	1383.3	1350.9	2734.2	3.3193	2.3291	5.6484
590	10.81	0.001482	0.01635	1442.3	1266.6	2708.9	3.4167	2.1468	5.5635
600	12.33	0.001540	0.01375	1504.6	1172.5	2677.1	3.5174	1.9543	5.4717
610	14.02	0.001611	0.01146	1571.1	1065.6	2636.7	3.6231	1.7468	5.3699
620	15.88	0.001704	0.009422	1644.3	939.6	2583.9	3.7370	1.5154	5.2524
630	17.95	0.001837	0.007532	1729.3	781.4	2510.7	3.8671	1.2404	5.1075
640	20.25	0.002076	0.005626	1842.9	550.5	2393.4	4.0389	0.8602	4.8991
647.29	22.089	0.003155	0.003155	2098.8	0.0	2098.8	4.4289	0.0	4.4289

Courtesy of Department of Mechanical Engineering, Stanford University

Table 2.2
SI Properties of Saturated Steam and Saturated Water (Pressure)

Press. MPa	Temp K	Volume, m³/kg Water v_f	Steam v_g	Enthalpy, kJ/kg Water H_f	Evap H_{fg}	Steam H_g	Entropy, kJ/(kg K) Water s_f	Evap s_{fg}	Steam s_g
0.00080	276.92	0.001000	159.7	15.4	2492.4	2507.8	0.0559	9.0007	9.0566
0.0010	280.13	0.001000	129.2	28.6	2485.1	2513.7	0.1034	8.8714	8.9748
0.0012	282.81	0.001000	108.7	39.7	2479.0	2518.7	0.1429	8.7654	8.9083
0.0014	285.13	0.001001	93.92	49.3	2473.6	2522.9	0.1768	8.6754	8.8522
0.0016	287.17	0.001001	82.76	57.8	2468.9	2526.7	0.2065	8.5972	8.8037
0.0018	288.99	0.001001	74.03	65.5	2464.5	2530.0	0.2330	8.5280	8.7610
0.0020	290.65	0.001002	67.00	72.4	2460.6	2533.0	0.2569	8.4659	8.7228
0.0025	294.23	0.001002	54.25	87.5	2452.1	2539.6	0.3083	8.3340	8.6423
0.0030	297.23	0.001003	45.67	100.1	2445.0	2545.1	0.3510	8.2258	8.5768
0.0040	302.12	0.001004	34.80	120.7	2433.2	2553.9	0.4197	8.0541	8.4738
0.0050	306.03	0.001005	28.19	137.2	2423.8	2561.0	0.4740	7.9203	8.3943
0.0060	309.31	0.001007	23.74	151.0	2415.9	2566.9	0.5191	7.8105	8.3296
0.0080	314.66	0.001009	18.10	173.7	2402.8	2576.5	0.5915	7.6364	8.2279
0.010	318.96	0.001010	14.67	191.8	2392.4	2584.2	0.6488	7.5006	8.1494
0.012	322.57	0.001012	12.36	207.1	2383.5	2590.6	0.6964	7.3891	8.0855
0.014	325.70	0.001013	10.69	220.3	2375.8	2596.1	0.7371	7.2946	8.0317
0.016	328.47	0.001015	9.433	231.9	2369.1	2601.0	0.7728	7.2124	7.9852
0.018	330.96	0.001016	8.445	242.4	2362.9	2605.3	0.8045	7.1397	7.9442
0.020	333.22	0.001017	7.649	251.9	2357.4	2609.3	0.8332	7.0745	7.9077
0.025	338.12	0.001020	6.204	272.6	2345.1	2617.7	0.8947	6.9359	7.8306
0.030	342.26	0.001022	5.229	289.9	2334.9	2624.8	0.9458	6.8220	7.7678
0.040	349.02	0.001026	3.993	318.3	2318.0	2636.3	1.0279	6.6413	7.6692
0.050	354.48	0.001030	3.240	341.3	2304.1	2645.4	1.0930	6.5001	7.5931
0.060	359.09	0.001033	2.732	360.6	2292.4	2653.0	1.1471	6.3841	7.5312
0.080	366.65	0.001038	2.087	392.3	2273.0	2665.3	1.2344	6.1994	7.4338
0.10	372.78	0.001043	1.694	418.0	2257.0	2675.0	1.3038	6.0548	7.3586
0.101325	373.14	0.001043	1.673	419.5	2256.1	2675.6	1.3079	6.0462	7.3541
0.12	377.96	0.001047	1.428	439.7	2243.4	2683.1	1.3617	5.9356	7.2973
0.14	382.46	0.001051	1.237	458.6	2231.4	2690.0	1.4115	5.8341	7.2456
0.16	386.47	0.001054	1.091	475.5	2220.5	2696.0	1.4553	5.7456	7.2009
0.18	390.09	0.001058	0.9775	490.8	2210.6	2701.4	1.4945	5.6670	7.1615
0.20	393.38	0.001061	0.8857	504.7	2201.5	2706.2	1.5300	5.5963	7.1263
0.25	400.59	0.001067	0.7187	535.2	2181.3	2716.5	1.6068	5.4451	7.0519
0.30	406.70	0.001073	0.6058	561.2	2163.7	2724.9	1.6710	5.3201	6.9911
0.40	416.78	0.001084	0.4625	604.3	2133.8	2738.1	1.7755	5.1196	6.8951
0.50	425.01	0.001093	0.3749	639.8	2108.4	2748.2	1.8594	4.9611	6.8205
0.60	432.00	0.001101	0.3157	670.1	2086.3	2756.4	1.9299	4.8293	6.7592
0.80	443.59	0.001115	0.2404	720.7	2048.0	2768.7	2.0451	4.6169	6.6620
1.0	453.06	0.001127	0.1944	762.5	2015.1	2777.6	2.1378	4.4479	6.5857
1.2	461.14	0.001139	0.1633	798.5	1985.9	2784.4	2.2160	4.3065	6.5225
1.4	468.22	0.001149	0.1408	830.2	1959.4	2789.6	2.2838	4.1847	6.4685
1.6	474.56	0.001159	0.1238	858.8	1934.8	2793.6	2.3440	4.0770	6.4210
1.8	480.30	0.001168	0.1104	884.9	1911.8	2796.7	2.3981	3.9805	6.3786
2.0	485.57	0.001176	0.09963	908.9	1890.2	2799.1	2.4474	3.8927	6.3401
2.5	497.15	0.001197	0.07998	962.4	1840.2	2802.6	2.5549	3.7018	6.2567
3.0	507.05	0.001216	0.06668	1008.7	1795.0	2803.7	2.6461	3.5400	6.1861
4.0	523.55	0.001252	0.04978	1087.6	1713.4	2801.0	2.7968	3.2725	6.0693
5.0	537.14	0.001286	0.03944	1154.5	1639.4	2793.9	2.9206	3.0520	5.9726
6.0	548.79	0.001319	0.03244	1213.7	1570.2	2783.9	3.0271	2.8613	5.8884
7.0	559.03	0.001352	0.02737	1267.4	1504.3	2771.7	3.1216	2.6909	5.8125
8.0	568.22	0.001385	0.02352	1317.0	1440.5	2757.5	3.2073	2.5351	5.7424
10.	584.22	0.001453	0.01803	1407.9	1316.4	2724.3	3.3600	2.2533	5.6133
11.	591.30	0.001489	0.01599	1450.2	1255.0	2705.2	3.4296	2.1224	5.5520
12.	597.90	0.001527	0.01426	1491.2	1193.2	2684.4	3.4960	1.9956	5.4916
13.	604.09	0.001567	0.01278	1531.1	1130.7	2661.8	3.5599	1.8717	5.4316
14.	609.90	0.001610	0.01149	1570.4	1066.8	2637.2	3.6220	1.7490	5.3710
16.	620.59	0.001710	0.009307	1648.9	931.3	2580.2	3.7441	1.5007	5.2448
18.	630.22	0.001840	0.007492	1731.4	777.4	2508.8	3.8703	1.2336	5.1039
20.	638.96	0.002041	0.005836	1828.5	581.0	2409.5	4.0172	0.9093	4.9265
22.089	647.29	0.003155	0.003155	2098.8	0.0	2098.8	4.4289	0.0	4.4289

Courtesy of Department of Mechanical Engineering, Stanford University

Table 2.3
SI Properties of Compressed Water (Temperature and Pressure)

Press. MPa		Temperature, K (Sat. Pressure, MPa) 400 (0.2456)	425 (0.4999)	450 (0.9315)	475 (1.615)	500 (2.637)	525 (4.098)	550 (6.112)	575 (8.806)	600 (12.33)
sat	ρ,kg/m³	937.35	915.08	890.25	862.64	831.71	796.64	756.18	708.38	649.40
	H,kJ/kg	532.69	639.71	748.98	860.80	975.65	1094.63	1219.93	1354.82	1504.56
	s,kJ/(kg K)	1.60049	1.85933	2.10801	2.34815	2.58128	2.80995	3.03821	3.27144	3.51742
0.50	ρ,kg/m³	937.51	915.08							
	H,kJ/kg	532.82	639.71							
	s,kJ/(kg K)	1.60020	1.85933							
0.70	ρ,kg/m³	937.62	915.22							
	H,kJ/kg	532.94	639.84							
	s,kJ/(kg K)	1.59999	1.85914							
1.00	ρ,kg/m³	937.79	915.41	890.30						
	H,kJ/kg	533.12	640.02	749.01						
	s,kJ/(kg K)	1.59968	1.85884	2.10793						
1.40	ρ,kg/m³	938.01	915.66	890.58						
	H,kJ/kg	533.37	640.27	749.20						
	s,kJ/(kg K)	1.59928	1.85843	2.10740						
2.00	ρ,kg/m³	938.33	916.03	890.99	862.93					
	H,kJ/kg	533.76	640.64	749.50	860.92					
	s,kJ/(kg K)	1.59868	1.85779	2.10661	2.34748					
3.00	ρ,kg/m³	938.86	916.63	891.66	863.70	832.04				
	H,kJ/kg	534.42	641.26	750.01	861.24	975.68				
	s,kJ/(kg K)	1.59771	1.85672	2.10529	2.34578	2.58049				
5.00	ρ,kg/m³	939.90	917.80	892.99	865.23	833.88	797.71			
	H,kJ/kg	535.77	642.49	751.04	861.95	975.97	1094.50			
	s,kJ/(kg K)	1.59579	1.85454	2.10267	2.34248	2.57633	2.80757			
7.00	ρ,kg/m³	940.93	918.96	894.30	866.74	835.70	800.06	757.63		
	H,kJ/kg	537.13	643.73	752.08	862.71	976.34	1094.30	1219.41		
	s,kJ/(kg K)	1.59390	1.85237	2.10006	2.33927	2.57233	2.80247	3.03517		
10.00	ρ,kg/m³	942.45	920.66	896.23	868.97	838.39	803.48	762.36	711.24	
	H,kJ/kg	539.18	645.59	753.67	863.91	976.99	1094.16	1217.91	1353.16	
	s,kJ/(kg K)	1.59110	1.84912	2.09618	2.33455	2.56651	2.79513	3.02530	3.26566	
14.00	ρ,kg/m³	944.45	922.89	898.75	871.89	841.87	807.86	768.28	720.13	656.02
	H,kJ/kg	541.93	648.10	755.82	865.57	977.99	1094.20	1216.32	1348.37	1499.43
	s,kJ/(kg K)	1.58742	1.84484	2.09110	2.32843	2.55903	2.78578	3.01295	3.24764	3.50461
20.00	ρ,kg/m³	947.40	926.15	902.43	876.12	846.90	814.10	776.50	731.86	675.64
	H,kJ/kg	546.09	651.88	759.11	868.19	979.70	1094.61	1214.62	1342.80	1485.18
	s,kJ/(kg K)	1.58199	1.83852	2.08365	2.31954	2.54829	2.77251	2.99579	3.22363	3.46589
30.00	ρ,kg/m³	952.15	931.39	908.31	882.86	854.83	823.75	788.76	748.41	700.23
	H,kJ/kg	553.10	658.28	764.75	872.87	983.09	1096.17	1213.36	1336.77	1469.95
	s,kJ/(kg K)	1.57321	1.82824	2.07165	2.30546	2.53159	2.75225	2.97029	3.18966	3.41631
50.00	ρ,kg/m³	961.23	941.29	919.30	895.30	869.22	840.82	809.63	774.92	735.63
	H,kJ/kg	567.28	671.32	776.46	882.95	991.08	1101.29	1214.33	1331.34	1454.04
	s,kJ/(kg K)	1.55640	1.80868	2.04904	2.27933	2.50117	2.71624	2.92655	3.13458	3.34342
100.00	ρ,kg/m³	982.01	963.53	943.46	921.97	899.14	875.01	849.47	822.32	793.27
	H,kJ/kg	603.30	705.01	807.50	910.95	1015.43	1121.02	1227.92	1336.47	1447.18
	s,kJ/(kg K)	1.51781	1.76443	1.99876	2.22248	2.43683	2.64289	2.84180	3.03479	3.22326

Courtesy of Department of Mechanical Engineering, Stanford University

Table 2.4 (Part 1 of 3)
SI Properties of Superheated Steam (Temperature and Pressure)
(Pressure from 0.0010 to 0.0400 MPa)

Press. MPa (Sat Temp, K)		Temperature, K Sat.	350	400	450	500	550	600	650	700
0.0010	v,m³/kg	129.2	161.5	184.6	207.7	230.7	253.8	276.9	300.0	323.1
(280.1)	H,kJ/kg	2513.7	2644.4	2739.0	2834.7	2931.8	3030.2	3130.1	3231.7	3334.8
	s,kJ/(kg K)	8.9748	9.3913	9.6439	9.8693	10.0737	10.2614	10.4353	10.5978	10.7506
0.0020	v,m³/kg	67.00	80.73	92.28	103.8	115.4	126.9	138.4	150.0	161.5
(290.7)	H,kJ/kg	2533.0	2644.3	2738.9	2834.7	2931.7	3030.2	3130.1	3231.6	3334.8
	s,kJ/(kg K)	8.7228	9.0710	9.3238	9.5493	9.7538	9.9414	10.1153	10.2779	10.4307
0.0040	v,m³/kg	34.80	40.35	46.13	51.91	57.68	63.45	69.22	74.99	80.76
(302.1)	H,kJ/kg	2553.9	2644.0	2738.8	2834.6	2931.7	3030.1	3130.1	3231.6	3334.8
	s,kJ/(kg K)	8.4738	8.7504	9.0035	9.2292	9.4338	9.6215	9.7954	9.9579	10.1108
0.0070	v,m³/kg	20.53	23.04	26.35	29.66	32.96	36.25	39.55	42.85	46.15
(312.2)	H,kJ/kg	2572.0	2643.5	2738.5	2834.4	2931.5	3030.0	3130.0	3231.6	3334.7
	s,kJ/(kg K)	8.2750	8.4911	8.7447	8.9707	9.1753	9.3631	9.5371	9.6996	9.8525
0.0100	v,m³/kg	14.67	16.12	18.44	20.75	23.07	25.38	27.69	29.99	32.30
(319.0)	H,kJ/kg	2584.2	2643.0	2738.2	2834.2	2931.4	3030.0	3130.0	3231.5	3334.7
	s,kJ/(kg K)	8.1494	8.3254	8.5796	8.8058	9.0106	9.1984	9.3724	9.5349	9.6878
0.0200	v,m³/kg	7.649	8.044	9.210	10.37	11.53	12.68	13.84	14.99	16.15
(333.2)	H,kJ/kg	2609.3	2641.4	2737.3	2833.7	2931.0	3029.7	3129.7	3231.3	3334.5
	s,kJ/(kg K)	7.9077	8.0019	8.2579	8.4849	8.6901	8.8781	9.0522	9.2148	9.3678
0.0400	v,m³/kg	3.993	4.005	4.595	5.179	5.759	6.339	6.917	7.495	8.073
(349.0)	H,kJ/kg	2636.3	2638.2	2735.5	2832.5	2930.3	3029.1	3129.3	3231.0	3334.3
	s,kJ/(kg K)	7.6692	7.6747	7.9344	8.1631	8.3690	8.5574	8.7318	8.8945	9.0476

Press. MPa (Sat Temp, K)		Temperature, K 750	800	850	900	950	1000	1050	1100	1150
0.0010	v,m³/kg	346.1	369.2	392.3	415.4	438.4	461.5	484.6	507.7	530.7
(280.1)	H,kJ/kg	3439.6	3546.1	3654.3	3764.3	3876.0	3989.5	4104.7	4221.7	4340.5
	s,kJ/(kg K)	10.8952	11.0327	11.1639	11.2896	11.4104	11.5268	11.6392	11.7481	11.8536
0.0020	v,m³/kg	173.1	184.6	196.1	207.7	219.2	230.8	242.3	253.8	265.4
(290.7)	H,kJ/kg	3439.6	3546.1	3654.3	3764.3	3876.0	3989.5	4104.7	4221.7	4340.5
	s,kJ/(kg K)	10.5753	10.7128	10.8440	10.9697	11.0905	11.2069	11.3193	11.4282	11.5337
0.0040	v,m³/kg	86.53	92.30	98.07	103.8	109.6	115.4	121.1	126.9	132.7
(302.1)	H,kJ/kg	3439.6	3546.1	3654.3	3764.3	3876.0	3989.5	4104.7	4221.7	4340.5
	s,kJ/(kg K)	10.2554	10.3929	10.5241	10.6498	10.7706	10.8870	10.9994	11.1083	11.2138
0.0070	v,m³/kg	49.45	52.74	56.04	59.34	62.63	65.93	69.23	72.52	75.82
(312.2)	H,kJ/kg	3439.5	3546.0	3654.3	3764.3	3876.0	3989.5	4104.7	4221.7	4340.5
	s,kJ/(kg K)	9.9971	10.1346	10.2658	10.3915	10.5123	10.6287	10.7412	10.8500	10.9556
0.0100	v,m³/kg	34.61	36.92	39.23	41.54	43.84	46.15	48.46	50.77	53.07
(319.0)	H,kJ/kg	3439.5	3546.0	3654.3	3764.2	3876.0	3989.5	4104.7	4221.7	4340.4
	s,kJ/(kg K)	9.8324	9.9699	10.1011	10.2269	10.3477	10.4641	10.5765	10.6854	10.7910
0.0200	v,m³/kg	17.30	18.46	19.61	20.77	21.92	23.07	24.23	25.38	26.54
(333.2)	H,kJ/kg	3439.4	3545.9	3654.2	3764.2	3875.9	3989.4	4104.7	4221.7	4340.4
	s,kJ/(kg K)	9.5124	9.6499	9.7812	9.9069	10.0277	10.1442	10.2566	10.3655	10.4710
0.0400	v,m³/kg	8.650	9.228	9.805	10.38	10.96	11.54	12.11	12.69	13.27
(349.0)	H,kJ/kg	3439.2	3545.7	3654.0	3764.0	3875.8	3989.3	4104.6	4221.6	4340.3
	s,kJ/(kg K)	9.1923	9.3299	9.4611	9.5869	9.7077	9.8242	9.9366	10.0455	10.1511

Courtesy of Department of Mechanical Engineering, Stanford University

Table 2.4 (Part 2 of 3)
SI Properties of Superheated Steam (Temperature and Pressure)
(Pressure from 0.070 to 2.000 MPa)

Press. MPa (Sat. Temp, K)		Temperature, K Sat.	400	450	500	550	600	650	700	750
0.070	v,m³/kg	2.365	2.617	2.953	3.287	3.619	3.950	4.281	4.612	4.942
(363.1)	H,kJ/kg	2659.6	2732.7	2830.8	2929.1	3028.3	3128.7	3230.5	3333.8	3438.8
	s,kJ/(kg K)	7.4789	7.6707	7.9019	8.1090	8.2980	8.4727	8.6357	8.7889	8.9337
0.101325	v,m³/kg	1.673	1.802	2.036	2.268	2.498	2.727	2.956	3.185	3.413
(373.1)	H,kJ/kg	2675.6	2729.7	2829.0	2927.9	3027.4	3128.0	3229.9	3333.4	3438.4
	s,kJ/(kg K)	7.3541	7.4942	7.7281	7.9365	8.1261	8.3012	8.4644	8.6177	8.7627
0.200	v,m³/kg	0.8857	0.9024	1.025	1.144	1.262	1.379	1.495	1.612	1.728
(393.4)	H,kJ/kg	2706.2	2720.2	2823.2	2924.0	3024.6	3125.8	3228.2	3332.0	3437.3
	s,kJ/(kg K)	7.1263	7.1616	7.4044	7.6168	7.8085	7.9847	8.1486	8.3024	8.4477
0.400	v,m³/kg	0.4625		0.5053	0.5671	0.6273	0.6866	0.7455	0.8040	0.8624
(416.8)	H,kJ/kg	2738.1		2811.0	2916.0	3018.8	3121.5	3224.8	3329.2	3435.0
	s,kJ/(kg K)	6.8951		7.0634	7.2848	7.4808	7.6594	7.8248	7.9795	8.1255
0.700	v,m³/kg	0.2729		0.2822	0.3197	0.3552	0.3899	0.4240	0.4579	0.4915
(438.1)	H,kJ/kg	2763.1		2791.3	2903.4	3010.0	3114.8	3219.5	3325.0	3431.5
	s,kJ/(kg K)	6.7072		6.7709	7.0073	7.2104	7.3928	7.5605	7.7168	7.8637
1.000	v,m³/kg	0.1944			0.2206	0.2464	0.2712	0.2955	0.3194	0.3432
(453.1)	H,kJ/kg	2777.6			2890.2	3000.9	3108.0	3214.2	3320.7	3428.0
	s,kJ/(kg K)	6.5857			6.8223	7.0333	7.2198	7.3898	7.5476	7.6956
2.000	v,m³/kg	0.09963			0.1044	0.1191	0.1326	0.1454	0.1578	0.1701
(485.6)	H,kJ/kg	2799.1			2840.5	2968.4	3084.5	3196.1	3306.2	3416.1
	s,kJ/(kg K)	6.3401			6.4242	6.6683	6.8704	7.0491	7.2122	7.3639

Press. MPa (Sat. Temp, K)		Temperature, K 800	850	900	950	1000	1050	1100	1150	1200
0.070	v,m³/kg	5.272	5.602	5.932	6.262	6.592	6.922	7.251	7.581	7.911
(363.1)	H,kJ/kg	3545.4	3653.8	3763.8	3875.6	3989.1	4104.4	4221.5	4340.2	4460.7
	s,kJ/(kg K)	9.0713	9.2027	9.3284	9.4493	9.5658	9.6783	9.7872	9.8927	9.9953
0.101325	v,m³/kg	3.641	3.870	4.098	4.326	4.554	4.781	5.009	5.237	5.465
(373.1)	H,kJ/kg	3545.1	3653.5	3763.6	3875.4	3989.0	4104.3	4221.3	4340.1	4460.6
	s,kJ/(kg K)	8.9004	9.0317	9.1576	9.2785	9.3950	9.5075	9.6164	9.7220	9.8245
0.200	v,m³/kg	1.844	1.959	2.075	2.191	2.306	2.422	2.537	2.653	2.768
(393.4)	H,kJ/kg	3544.2	3652.7	3762.9	3874.8	3988.5	4103.8	4220.9	4339.7	4460.3
	s,kJ/(kg K)	8.5856	8.7172	8.8432	8.9642	9.0808	9.1933	9.3023	9.4079	9.5105
0.400	v,m³/kg	0.9206	0.9787	1.037	1.095	1.153	1.210	1.268	1.326	1.384
(416.8)	H,kJ/kg	3542.2	3651.1	3761.5	3873.6	3987.4	4102.9	4220.1	4339.0	4459.6
	s,kJ/(kg K)	8.2639	8.3959	8.5221	8.6433	8.7601	8.8728	8.9818	9.0875	9.1901
0.700	v,m³/kg	0.5250	0.5584	0.5917	0.6249	0.6581	0.6912	0.7244	0.7575	0.7905
(438.1)	H,kJ/kg	3539.3	3648.6	3759.4	3871.8	3985.8	4101.5	4218.9	4337.9	4458.6
	s,kJ/(kg K)	8.0029	8.1354	8.2621	8.3836	8.5006	8.6135	8.7226	8.8284	8.9312
1.000	v,m³/kg	0.3668	0.3902	0.4137	0.4370	0.4603	0.4836	0.5068	0.5300	0.5532
(453.1)	H,kJ/kg	3536.4	3646.1	3757.3	3870.0	3984.3	4100.1	4217.6	4336.8	4457.6
	s,kJ/(kg K)	7.8356	7.9686	8.0957	8.2175	8.3348	8.4478	8.5571	8.6631	8.7659
2.000	v,m³/kg	0.1821	0.1941	0.2060	0.2178	0.2295	0.2413	0.2530	0.2646	0.2763
(485.6)	H,kJ/kg	3526.5	3637.8	3750.2	3863.9	3979.0	4095.5	4213.5	4333.1	4454.2
	s,kJ/(kg K)	7.5064	7.6413	7.7698	7.8928	8.0108	8.1245	8.2343	8.3406	8.4437

Table 2.4 (Part 3 of 3)
SI Properties of Superheated Steam (Temperature and Pressure)
(Pressure from 3.0 to 100 MPa)

Press. MPa (Sat. Temp, K)		Temperature, K Sat.	600	650	700	750	800	900	1000	1100
3.0	v,m³/kg	0.06668	0.08628	0.09532	0.1040	0.1124	0.1206	0.1367	0.1526	0.1684
(507.1)	H,kJ/kg	2803.7	3059.6	3177.3	3291.3	3404.0	3516.5	3743.1	3973.7	4209.4
	s,kJ/(kg K)	6.1861	6.6516	6.8402	7.0091	7.1646	7.3098	7.5767	7.8196	8.0442
4.0	v,m³/kg	0.04978	0.06304	0.07024	0.07699	0.08849	0.08981	0.1021	0.1142	0.1261
(523.6)	H,kJ/kg	2801.0	3033.0	3157.8	3276.1	3391.6	3506.3	3735.8	3968.3	4205.3
	s,kJ/(kg K)	6.0693	6.4847	6.6846	6.8600	7.0194	7.1674	7.4378	7.6827	7.9085
6.0	v,m³/kg	0.03244	0.03958	0.04507	0.04998	0.05459	0.05901	0.06750	0.07571	0.08375
(548.8)	H,kJ/kg	2783.9	2973.8	3116.3	3244.4	3366.3	3485.5	3721.2	3957.5	4197.0
	s,kJ/(kg K)	5.8884	6.2201	6.4486	6.6385	6.8067	6.9605	7.2382	7.4872	7.7154
8.0	v,m³/kg	0.02352	0.02759	0.03239	0.03643	0.04011	0.04359	0.05018	0.05648	0.06260
(568.2)	H,kJ/kg	2757.5	2904.1	3071.1	3210.9	3340.0	3464.0	3706.2	3946.6	4188.7
	s,kJ/(kg K)	5.7424	5.9938	6.2617	6.4691	6.6472	6.8073	7.0927	7.3459	7.5766
10.0	v,m³/kg	0.01803	0.02008	0.02468	0.02825	0.03141	0.03434	0.03979	0.04494	0.04992
(584.2)	H,kJ/kg	2724.3	2818.3	3021.4	3175.6	3312.6	3442.0	3691.0	3935.6	4180.3
	s,kJ/(kg K)	5.6133	5.7722	6.0983	6.3270	6.5162	6.6833	6.9767	7.2344	7.4676
15.0	v,m³/kg	0.01034		0.01404	0.01723	0.01975	0.02196	0.02593	0.02956	0.03301
(615.4)	H,kJ/kg	2610.1		2868.5	3077.4	3239.6	3384.3	3652.0	3907.7	4159.4
	s,kJ/(kg K)	5.3090		5.7192	6.0295	6.2536	6.4404	6.7559	7.0254	7.2653
20.0	v,m³/kg	0.00584		0.00790	0.01156	0.01386	0.01575	0.01900	0.02188	0.02456
(639.0)	H,kJ/kg	2409.5		2625.1	2961.0	3159.4	3323.0	3611.7	3879.3	4138.5
	s,kJ/(kg K)	4.9265		5.2616	5.7622	6.0363	6.2477	6.5881	6.8702	7.1172
30.0	v,m³/kg	0.00543	0.00786	0.00951	0.01087	0.01208	0.01318	0.01421	0.01518	0.01612
	H,kJ/kg	2633.3	2973.5	3189.4	3367.6	3528.0	3678.1	3821.6	3960.6	4096.5
	s,kJ/(kg K)	5.1776	5.6490	5.9280	6.1442	6.3276	6.4900	6.6373	6.7729	6.8993
40.0	v,m³/kg	0.00260	0.00483	0.00640	0.00760	0.00863	0.00954	0.01039	0.01117	0.01192
	H,kJ/kg	2221.6	2752.8	3043.7	3258.4	3441.6	3607.8	3763.3	3911.6	4054.6
	s,kJ/(kg K)	4.5363	5.2720	5.6483	5.9089	6.1185	6.2982	6.4578	6.6025	6.7357
50.0	v,m³/kg	0.00203	0.00323	0.00459	0.00567	0.00658	0.00738	0.00811	0.00878	0.00941
	H,kJ/kg	2074.8	2535.1	2893.2	3147.2	3354.5	3537.3	3705.1	3862.8	4013.2
	s,kJ/(kg K)	4.2943	4.9293	5.3926	5.7009	5.9380	6.1358	6.3080	6.4619	6.6019
60.0	v,m³/kg	0.00183	0.00251	0.00350	0.00444	0.00526	0.00597	0.00661	0.00720	0.00775
	H,kJ/kg	2013.6	2387.9	2754.9	3039.5	3269.1	3468.0	3647.8	3814.7	3972.3
	s,kJ/(kg K)	4.1795	4.6954	5.1697	5.5152	5.7779	5.9930	6.1776	6.3405	6.4872
70.0	v,m³/kg	0.00172	0.00217	0.00287	0.00364	0.00435	0.00498	0.00556	0.00609	0.00658
	H,kJ/kg	1977.8	2301.9	2644.3	2941.8	3188.5	3401.4	3592.2	3767.9	3932.5
	s,kJ/(kg K)	4.1031	4.5498	4.9920	5.3530	5.6353	5.8656	6.0615	6.2329	6.3861
80.0	v,m³/kg	0.00164	0.00197	0.00248	0.00310	0.00371	0.00427	0.00479	0.00527	0.00571
	H,kJ/kg	1953.8	2248.4	2563.0	2858.7	3115.3	3339.0	3539.4	3722.9	3893.9
	s,kJ/(kg K)	4.0448	4.4510	4.8571	5.2159	5.5094	5.7514	5.9571	6.1362	6.2953
100.0	v,m³/kg	0.00153	0.00176	0.00207	0.00247	0.00291	0.00335	0.00377	0.00416	0.00453
	H,kJ/kg	1923.7	2186.1	2461.2	2736.7	2995.5	3230.7	3444.2	3640.0	3821.8
	s,kJ/(kg K)	3.9567	4.3186	4.6735	5.0077	5.3037	5.5581	5.7772	5.9684	6.1375

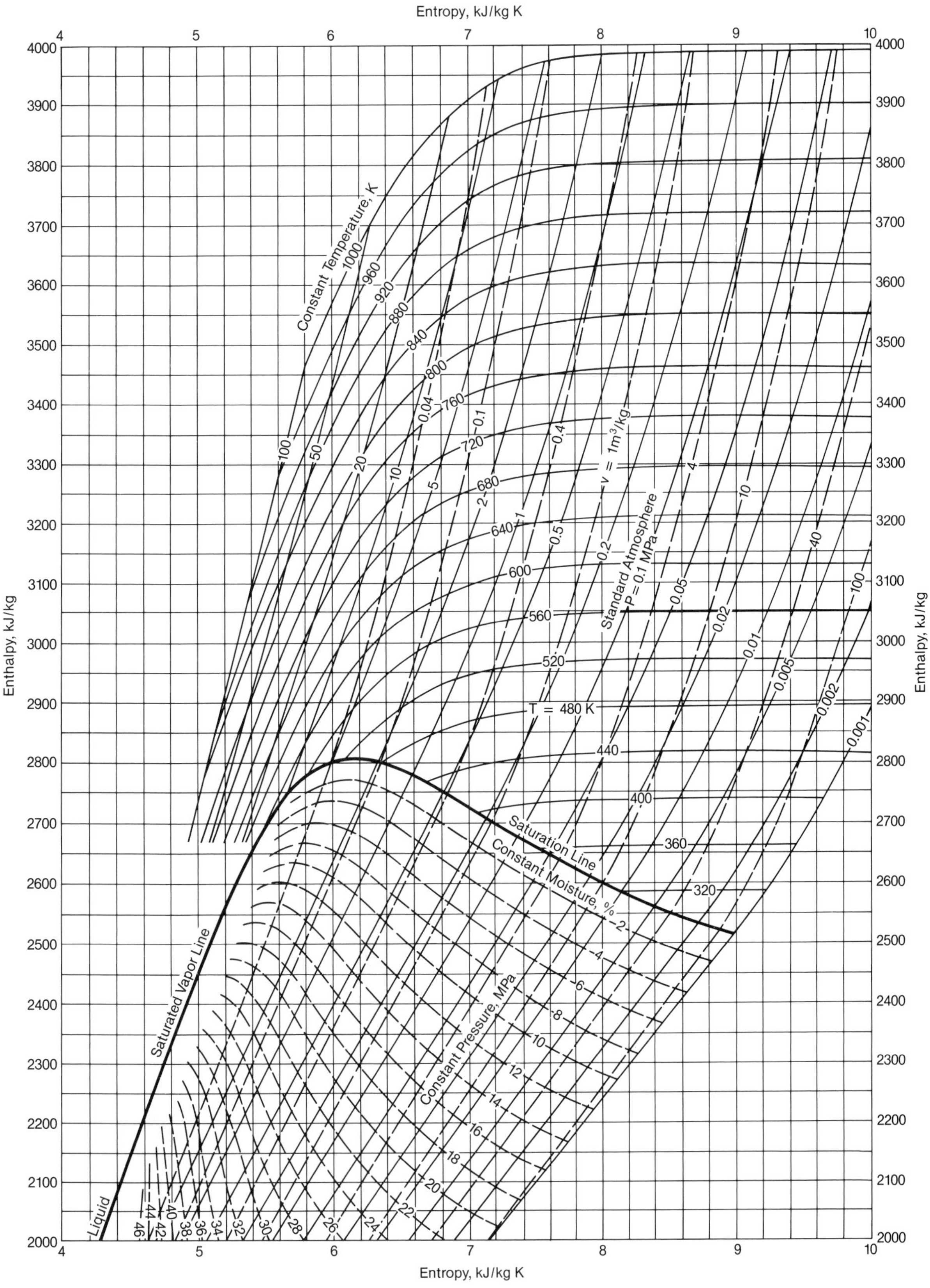

Fig. 2.1 Mollier diagram (*H-s*) for steam — SI units.

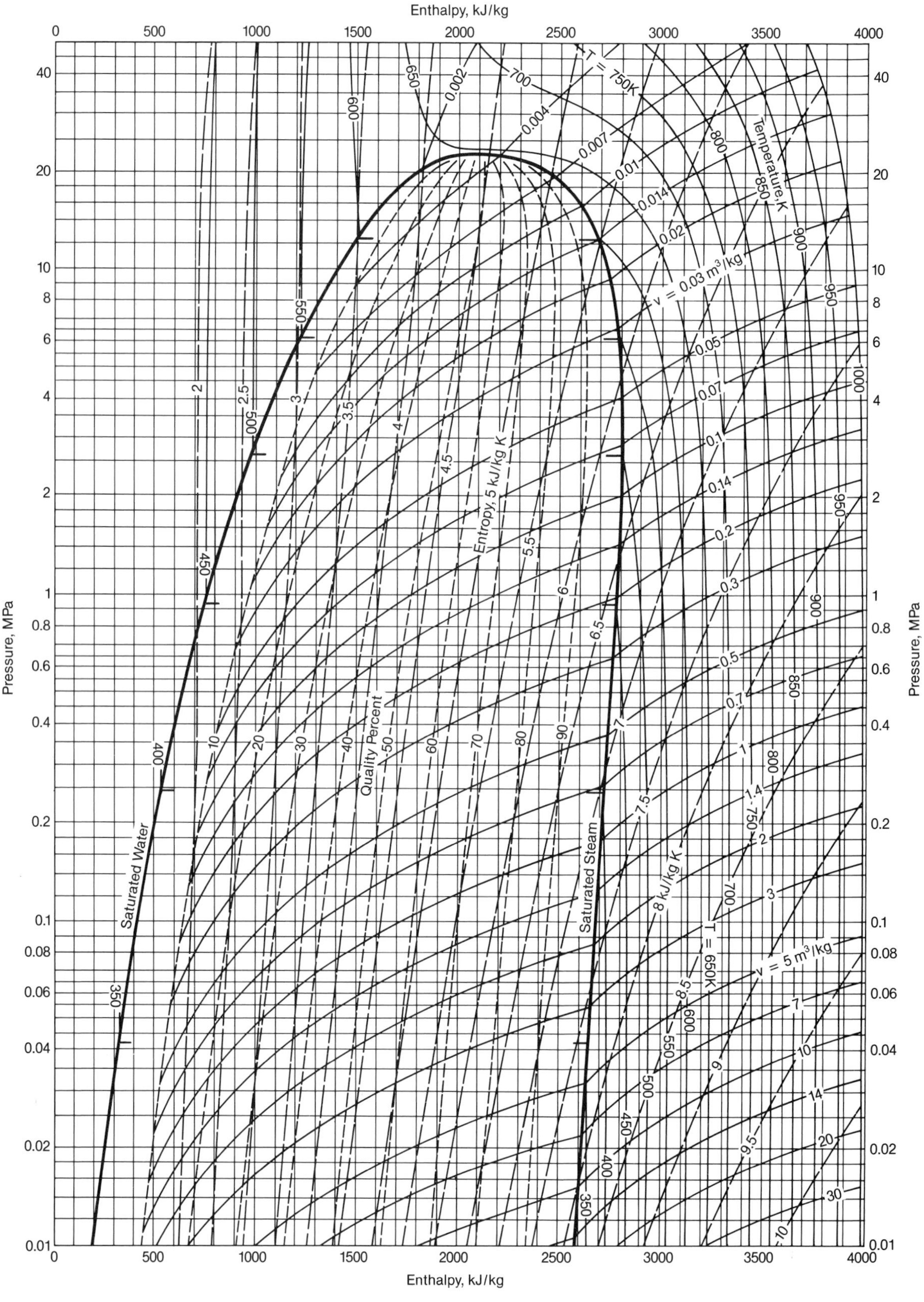

Fig. 2.2 Pressure-enthalpy diagram for steam — SI units.

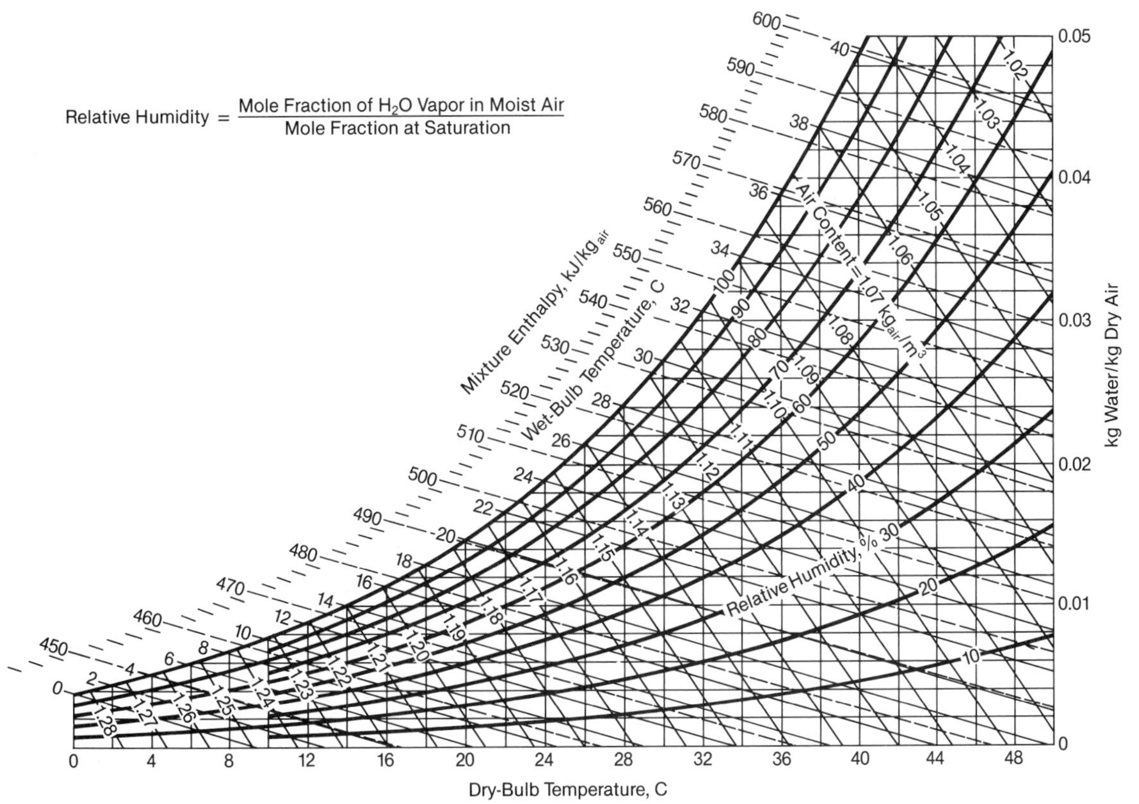

Fig. 2.3 Psychrometric chart — SI units.

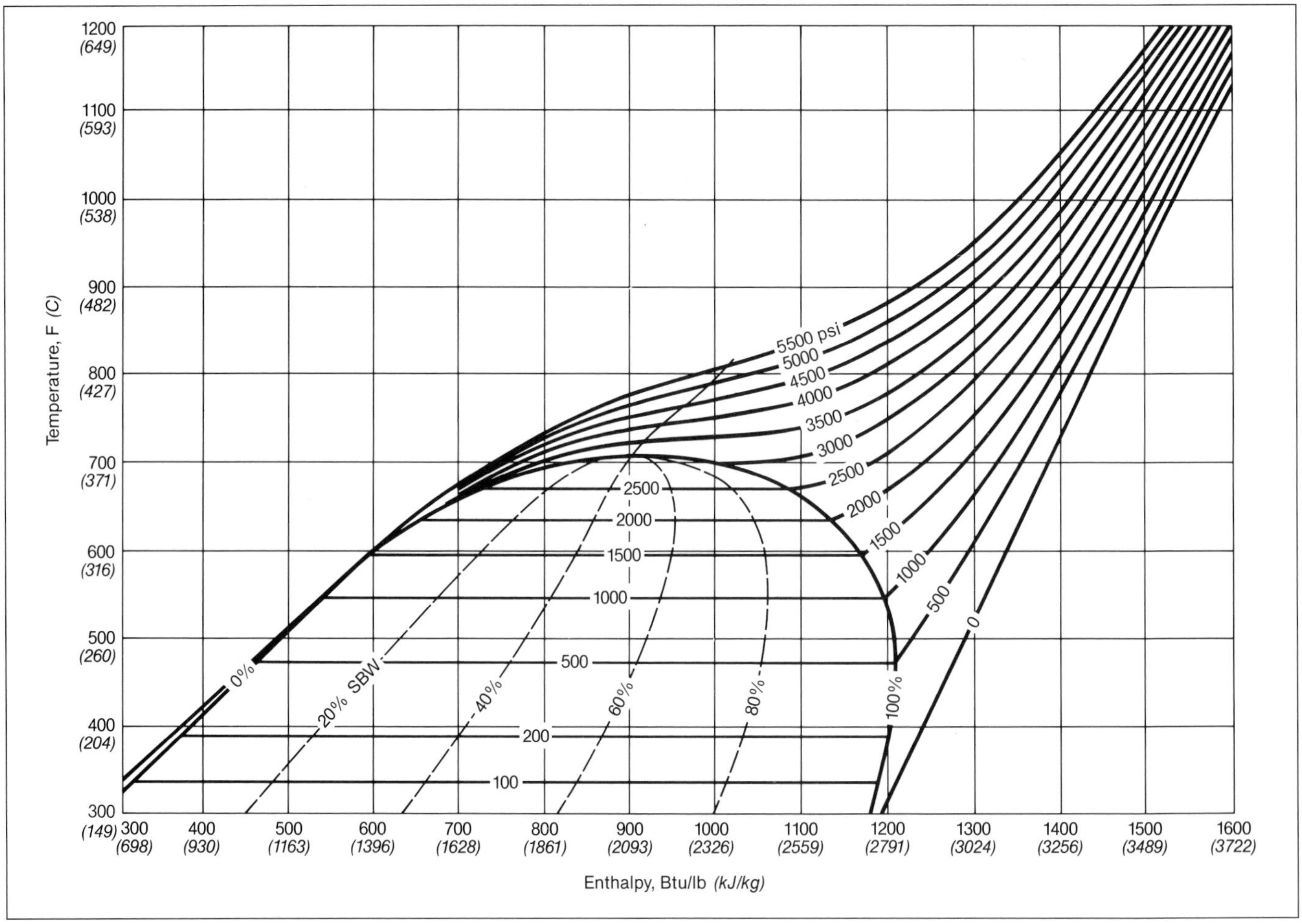

Fig. 3.1 Temperature-enthalpy diagram (bar = 0.06896 × psi).

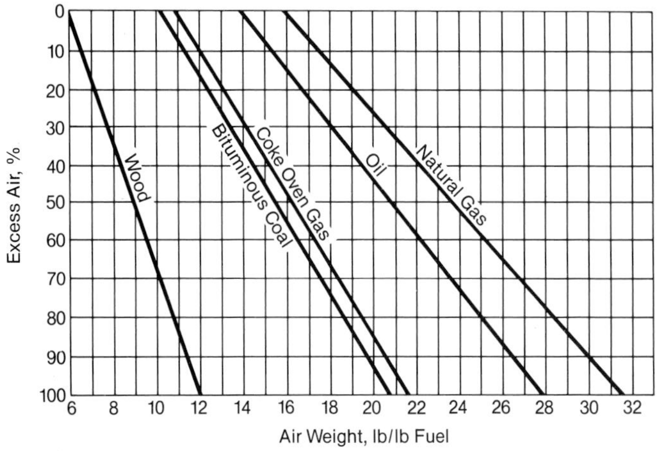

Fig. 3.2 Estimates of the weight of air required for selected fuels and excess air levels.

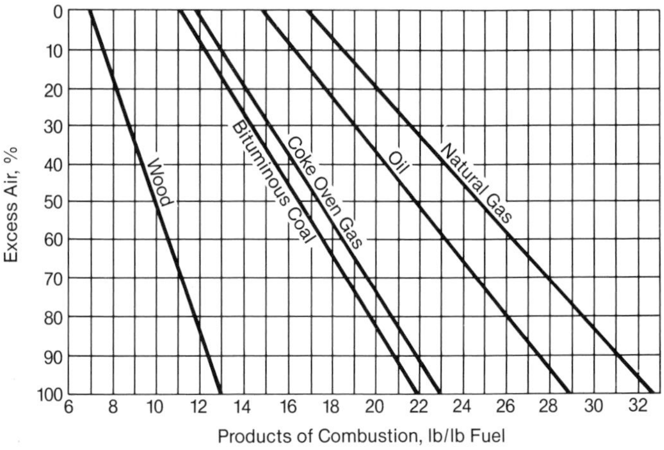

Fig. 3.3 Estimates of the weight of the products of combustion for selected fuels and excess air levels.

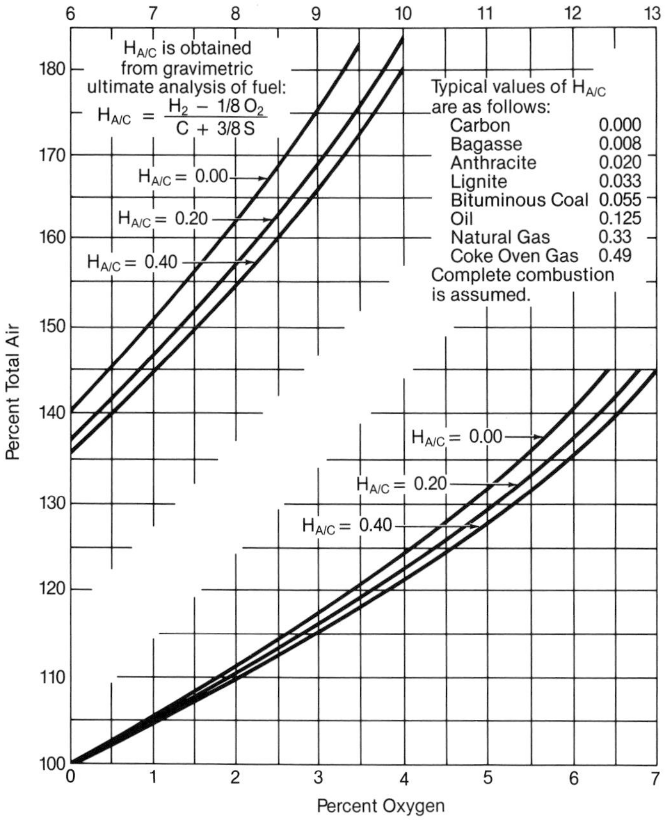

Fig. 3.4 Estimates of the percent total air for flue gas percent oxygen for selected fuels.

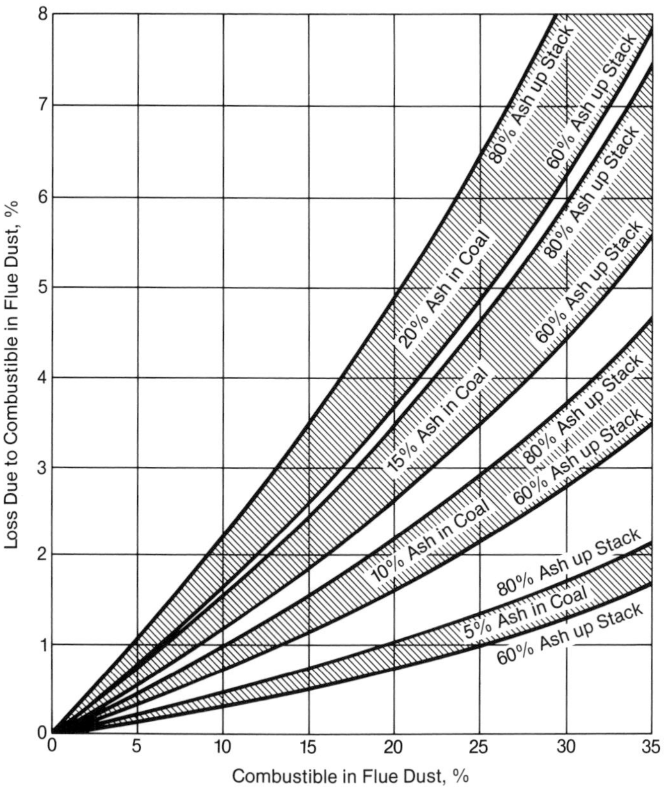

Fig. 3.5 Estimates of efficiency loss due to unburned combustibles for selected fuels and ash contents.

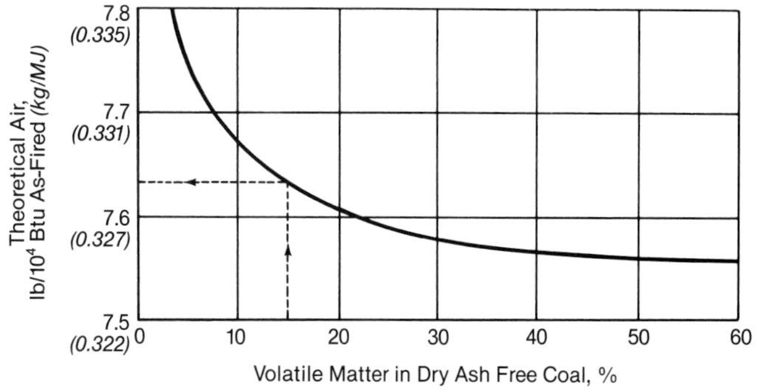

Fig. 3.6 Estimates of theoretical air required for the combustion of coal based upon volatile matter content.

Table 3.1
SO_2, SO_3 Estimated Emissions from Boilers Firing Coal or Fuel Oil

SO_2:

$$\text{lb } SO_2 \text{ produced/lb fuel} = K \left(\frac{\text{lb sulfur}}{\text{lb fuel}}\right) \left(\frac{64 \text{ lb } SO_2\text{/mole}}{32 \text{ lb sulfur/mole}}\right)$$

$$\text{lb } SO_2 \text{ produced/h} = K \left(\frac{\text{lb fuel}}{\text{h}}\right) \left(\frac{\% \text{ sulfur in fuel} \times 2}{100}\right)$$

$$\% \, SO_2 \text{ by weight} = \frac{\text{lb } SO_2\text{/h} \times 10^2}{\text{lb flue gas/h}}$$

$$\text{ppm } SO_2 \text{ by weight} = (\% \, SO_2 \text{ by weight}) \times 10^4$$

$$\% \, SO_2 \text{ by volume} = \frac{100\,(\% \, SO_2 \text{ by weight}/64)}{\frac{(\% \, SO_2 \text{ by weight})}{64} + \frac{100 - (\% \, SO_2 \text{ by weight})}{M}}$$

$$\text{ppm } SO_2 \text{ by volume} = \% \, SO_2 \text{ by volume} \times 10^4$$

where K = 0.95 Cyclone furnace firing
0.97 dry bottom PC firing
0.99 oil firing

M = molecular weight of flue gas
30.2 coal firing
29.0 oil firing

SO_3:
SO_3 quantity = 1% of SO_2 quantity (coal firing)
= 2% of SO_2 quantity (oil firing)

Table 3.2
Periodic Table of the Elements

Group	Atomic Number	Symbol	Name	Atomic Weight (1)	Oxidation States	Boiling Point, K	Melting Point, K	Density at 300 K (2) (g/cm³)	Electron Configuration
IA	1	H	Hydrogen	1.0079	1	20.268	14.025	0.0899*	$1s^1$
VIII	2	He	Helium	4.00260		4.215	0.95 (at 26 atm)	0.1787*	$1s^2$
IA	3	Li	Lithium	6.941	1	1615	453.7	0.53	$1s^2 2s^1$
IIA	4	Be	Beryllium	9.01218	2	2745	1560	1.85	$1s^2 2s^2$
IIIB	5	B	Boron	10.81	3	4275	2300	2.34	$1s^2 2s^2 p^1$
IVB	6	C	Carbon	12.011	±4,2	4470*	4100*	2.62	$1s^2 2s^2 p^2$
VB	7	N	Nitrogen	14.0067	±3,5,4,2	77.35	63.14	1.251*	$1s^2 2s^2 p^3$
VIB	8	O	Oxygen	15.9994	−2	90.18	50.35	1.429*	$1s^2 2s^2 p^4$
VIIB	9	F	Fluorine	18.998403	−1	84.95	53.48	1.696*	$1s^2 2s^2 p^5$
VIII	10	Ne	Neon	20.179		27.096	24.553	0.901*	$1s^2 2s^2 p^6$
IA	11	Na	Sodium	22.98977	1	1156	371.0	0.97	$[Ne]3s^1$
IIA	12	Mg	Magnesium	24.305	2	1363	922	1.74	$[Ne]3s^2$
IIIB	13	Al	Aluminum	26.98154	3	2793	933.25	2.70	$[Ne]3s^2 p^1$
IVB	14	Si	Silicon	28.0855	4	3540	1685	2.33	$[Ne]3s^2 p^2$
VB	15	P	Phosphorus	30.97376	±3,5,4	550	317.30	1.82	$[Ne]3s^2 p^3$
VIB	16	S	Sulfur	32.06	±2,4,6	717.75	388.36	2.07	$[Ne]3s^2 p^4$
VIIB	17	Cl	Chlorine	35.453	±1,3,5,7	239.1	172.16	3.17*	$[Ne]3s^2 p^5$
VIII	18	Ar	Argon	39.948		87.30	83.81	1.784*	$[Ne]3s^2 p^6$
IA	19	K	Potassium	39.0983	1	1032	336.35	0.86	$[Ar]4s^1$
IIA	20	Ca	Calcium	40.08	2	1757	1112	1.55	$[Ar]4s^2$
IIIA	21	Sc	Scandium	44.9559	3	3104	1812	3.0	$[Ar]3d^1 4s^2$
IVA	22	Ti	Titanium	47.90	4,3	3562	1943	4.50	$[Ar]3d^2 4s^2$
VA	23	V	Vanadium	50.9415	5,4,3,2	3682	2175	5.8	$[Ar]3d^3 4s^2$
VIA	24	Cr	Chromium	51.996	6,3,2	2945	2130	7.19	$[Ar]3d^5 4s^1$
VIIA	25	Mn	Manganese	54.9380	7,6,4,2,3	2335	1517	7.43	$[Ar]3d^5 4s^2$
VIIIA	26	Fe	Iron	55.847	2,3	3135	1809	7.86	$[Ar]3d^6 4s^2$
VIIIA	27	Co	Cobalt	58.9332	2,3	3201	1768	8.90	$[Ar]3d^7 4s^2$
VIIIA	28	Ni	Nickel	58.70	2,3	3187	1726	8.90	$[Ar]3d^8 4s^2$
IB	29	Cu	Copper	63.546	2,1	2836	1357.6	8.96	$[Ar]3d^{10} 4s^1$
IIB	30	Zn	Zinc	65.38	2	1180	692.73	7.14	$[Ar]3d^{10} 4s^2$
IIIB	31	Ga	Gallium	69.72	3	2478	302.90	5.91	$[Ar]3d^{10} 4s^2 p^1$
IVB	32	Ge	Germanium	72.59	4	3107	1210.4	5.32	$[Ar]3d^{10} 4s^2 p^2$
VB	33	As	Arsenic	74.9216	±3,5	876 (subl.)	1081 (28 atm)	5.72	$[Ar]3d^{10} 4s^2 p^3$
VIB	34	Se	Selenium	78.96	−2,4,6	958	494	4.80	$[Ar]3d^{10} 4s^2 p^4$
VIIB	35	Br	Bromine	79.904	±1,5	332.25	265.90	3.12	$[Ar]3d^{10} 4s^2 p^5$
VIII	36	Kr	Krypton	83.80		119.80	115.78	3.74*	$[Ar]3d^{10} 4s^2 p^6$
IA	37	Rb	Rubidium	85.4678	1	961	312.64	1.53	$[Kr]5s^1$
IIA	38	Sr	Strontium	87.62	2	1650	1041	2.6	$[Kr]5s^2$
IIIA	39	Y	Yttrium	88.9059	3	3611	1799	4.5	$[Kr]4d^1 5s^2$
IVA	40	Zr	Zirconium	91.22	4	4682	2125	6.49	$[Kr]4d^2 5s^2$
VA	41	Nb	Niobium	92.9064	5,3	5017	2740	8.55	$[Kr]4d^4 5s^1$
VIA	42	Mo	Molybdenum	95.94	6,5,4,3,2	4912	2890	10.2	$[Kr]4d^5 5s^1$
VIIA	43	Tc	Technetium	(98)	7	4538	2473	11.5	$[Kr]4d^5 5s^2$
VIIIA	44	Ru	Ruthenium	101.07	2,3,4,6,8	4423	2523	12.2	$[Kr]4d^7 5s^1$
VIIIA	45	Rh	Rhodium	102.9055	2,3,4	3970	2236	12.4	$[Kr]4d^8 5s^1$
VIIIA	46	Pd	Palladium	106.4	2,4	3237	1825	12.0	$[Kr]4d^{10}$
IB	47	Ag	Silver	107.868	1	2436	1234	10.5	$[Kr]4d^{10} 5s^1$
IIB	48	Cd	Cadmium	112.41	2	1040	594.18	8.65	$[Kr]4d^{10} 5s^2$
IIIB	49	In	Indium	114.82	3	2346	429.76	7.31	$[Kr]4d^{10} 5s^2 p^1$
IVB	50	Sn	Tin	118.69	4,2	2876	505.06	7.30	$[Kr]4d^{10} 5s^2 p^2$
VB	51	Sb	Antimony	121.75	±3,5	1860	904	6.68	$[Kr]4d^{10} 5s^2 p^3$
VIB	52	Te	Tellurium	127.60	−2,4,6	1261	722.65	6.24	$[Kr]4d^{10} 5s^2 p^4$
VIIB	53	I	Iodine	126.9045	±1,5,7	458.4	386.7	4.92	$[Kr]4d^{10} 5s^2 p^5$
VIII	54	Xe	Xenon	131.30		165.03	161.36	5.89*	$[Kr]4d^{10} 5s^2 p^6$
IA	55	Cs	Cesium	132.9054	1	944	301.55	1.87	$[Xe]6s^1$
IIA	56	Ba	Barium	137.33	2	2171	1002	3.5	$[Xe]6s^2$
IIIA	57	La ★	Lanthanum	138.9055	3	3730	1193	6.7	$[Xe]5d^1 6s^2$
IVA	72	Hf	Hafnium	178.49	4	4876	2500	13.1	$[Xe]4f^{14} 5d^2 6s^2$
VA	73	Ta	Tantalum	180.9479	5	5731	3287	16.6	$[Xe]4f^{14} 5d^3 6s^2$
VIA	74	W	Tungsten	183.85	6,5,4,3,2	5828	3680	19.3	$[Xe]4f^{14} 5d^4 6s^2$
VIIA	75	Re	Rhenium	186.207	7,6,4,2,−1	5869	3453	21.0	$[Xe]4f^{14} 5d^5 6s^2$
VIIIA	76	Os	Osmium	190.2	2,3,4,6,8	5285	3300	22.4	$[Xe]4f^{14} 5d^6 6s^2$
VIIIA	77	Ir	Iridium	192.22	2,3,4,6	4701	2716	22.5	$[Xe]4f^{14} 5d^7 6s^2$
VIIIA	78	Pt	Platinum	195.09	2,4	4100	2045	21.4	$[Xe]4f^{14} 5d^9 6s^1$
IB	79	Au	Gold	196.9665	3,1	3130	1337.58	19.3	$[Xe]4f^{14} 5d^{10} 6s^1$
IIB	80	Hg	Mercury	200.59	2,1	630	234.28	13.53	$[Xe]4f^{14} 5d^{10} 6s^2$
IIIB	81	Tl	Thallium	204.37	3,1	1746	577	11.85	$[Xe]4f^{14} 5d^{10} 6s^2 p^1$
IVB	82	Pb	Lead	207.2	4,2	2023	600.6	11.4	$[Xe]4f^{14} 5d^{10} 6s^2 p^2$
VB	83	Bi	Bismuth	208.9804	3,5	1837	544.52	9.8	$[Xe]4f^{14} 5d^{10} 6s^2 p^3$
VIB	84	Po	Polonium	(209)	4,2	1235	527	9.4	$[Xe]4f^{14} 5d^{10} 6s^2 p^4$
VIIB	85	At	Astatine	(210)	±1,3,5,7	610	575	—	$[Xe]4f^{14} 5d^{10} 6s^2 p^5$
VIII	86	Rn	Radon	(222)		211	202	9.91*	$[Xe]4f^{14} 5d^{10} 6s^2 p^6$
IA	87	Fr	Francium	(223)	1	950	300	—	$[Rn]7s^1$
IIA	88	Ra	Radium	226.0254	2	1809	973	5	$[Rn]7s^2$
IIIA	89	Ac ★★	Actinium	227.0278	3	3473	1323	10.07	$[Rn]6d^1 7s^2$
IVA	104	Unq	(Unnilquadium)	(261)	—	—	—	—	$[Rn]5f^{14} 6d^2 7s^2$ †
VA	105	Unp	(Unnilpentium)	(262)	—	—	—	—	$[Rn]5f^{14} 6d^3 7s^2$ †
VIA	106	Unh	(Unnilhexium)	(263)	—	—	—	—	$[Rn]5f^{14} 6d^4 7s^2$ †
★	58	Ce	Cerium	140.12	3,4	3699	1071	6.78	$[Xe]4f^1 5d^1 6s^2$
★	59	Pr	Praseodymium	140.9077	3,4	3785	1204	6.77	$[Xe]4f^3 6s^2$
★	60	Nd	Neodymium	144.24	3	3341	1289	7.00	$[Xe]4f^4 6s^2$
★	61	Pm	Promethium	(145)	3	3785	1204	6.475	$[Xe]4f^5 6s^2$
★	62	Sm	Samarium	150.4	3,2	2064	1345	7.54	$[Xe]4f^6 6s^2$
★	63	Eu	Europium	151.96	3,2	1870	1090	5.26	$[Xe]4f^7 6s^2$
★	64	Gd	Gadolinium	157.25	3	3539	1585	7.89	$[Xe]4f^7 5d^1 6s^2$
★	65	Tb	Terbium	158.9254	3,4	3496	1630	8.27	$[Xe]4f^9 6s^2$
★	66	Dy	Dysprosium	162.50	3	2835	1682	8.54	$[Xe]4f^{10} 6s^2$
★	67	Ho	Holmium	164.9304	3	2968	1743	8.80	$[Xe]4f^{11} 6s^2$
★	68	Er	Erbium	167.26	3	3136	1795	9.05	$[Xe]4f^{12} 6s^2$
★	69	Tm	Thulium	168.9342	3,2	2220	1818	9.33	$[Xe]4f^{13} 6s^2$
★	70	Yb	Ytterbium	173.04	3,2	1467	1097	6.98	$[Xe]4f^{14} 6s^2$
★	71	Lu	Lutetium	174.967	3	3668	1936	9.84	$[Xe]4f^{14} 5d^1 6s^2$
★★	90	Th	Thorium	232.0381	4	5061	2028	11.7	$[Rn]6d^2 7s^2$
★★	91	Pa	Protactinium	231.0359	5,4	—	—	15.4	$[Rn]5f^2 6d^1 7s^2$
★★	92	U	Uranium	238.029	6,5,4,3	4407	1405	18.90	$[Rn]5f^3 6d^1 7s^2$
★★	93	Np	Neptunium	237.0482	6,5,4,3	—	910	20.4	$[Rn]5f^4 6d^1 7s^2$
★★	94	Pu	Plutonium	(244)	6,5,4,3	3503	913	19.8	$[Rn]5f^6 7s^2$
★★	95	Am	Americium	(243)	6,5,4,3	2880	1268	13.6	$[Rn]5f^7 7s^2$
★★	96	Cm	Curium	(247)	3	—	1340	13.511	$[Rn]5f^7 6d^1 7s^2$
★★	97	Bk	Berkelium	(247)	4,3	—	—	—	$[Rn]5f^9 7s^2$
★★	98	Cf	Californium	(251)	3	—	900	—	$[Rn]5f^{10} 7s^2$
★★	99	Es	Einsteinium	(252)	—	—	—	—	$[Rn]5f^{11} 7s^2$
★★	100	Fm	Fermium	(257)	—	—	—	—	$[Rn]5f^{12} 7s^2$
★★	101	Md	Mendelevium	(258)	—	—	—	—	$[Rn]5f^{13} 7s^2$
★★	102	No	Nobelium	(259)	—	—	—	—	$[Rn]5f^{14} 7s^2$
★★	103	Lr	Lawrencium	(260)	—	—	—	—	$[Rn]5f^{14} 6d^1 7s^2$

KEY (example: 30, Zn, Zinc, 65.38, 2, 1180, 692.73, 7.14, $[Ar]3d^{10}4s^2$): ATOMIC NUMBER; ATOMIC WEIGHT (1); OXIDATION STATES (Bold most stable); SYMBOL; BOILING POINT, K; MELTING POINT, K; DENSITY at 300 K (2) (g/cm³); ELECTRON CONFIGURATION; NAME

† The names and symbols of elements 104 - 106 are those recommended by IUPAC as systematic alternatives to those suggested by the purported discoverers. Berkeley (USA) researchers have proposed Rutherfordium, Rf, for element 104 and Hahnium, Ha, for element 105. Dubna (USSR) researchers, who also claim the discovery of these elements have proposed different names (and symbols).

The A & B subgroup designations, applicable to elements in rows 4, 5, 6, and 7, are those recommended by the International Union of Pure and Applied Chemistry. It should be noted that some authors and organizations use the opposite convention in distinguishing these subgroups.

* Estimated Values

NOTES:

(1) Based upon carbon-12. ( ) indicates most stable or best known isotope.

(2) Entries marked with asterisks refer to the gaseous state at 273 K and 1 atm and are given in units of g/l.

NSRDS

SARGENT-WELCH SCIENTIFIC COMPANY
7300 NORTH LINDER AVENUE, SKOKIE, ILLINOIS 60077

Catalog Number S-18806

Table 4.1
Coal and Ash Analysis, An Example of a High Sulfur Bituminous Coal

Proximate Analysis, %		Ultimate Analysis, %		Sulfur Forms, %	
Moisture	12.0	Moisture	12.0	Pyritic	2.0
VM	35.7	Carbon	60.6	Organic	2.0
FC	40.3	Hydrogen	4.1	Sulfate	0.1
Ash	12.0	Nitrogen	1.2		4.1
		Sulfur	4.1		
		Oxygen	6.0		
		Ash	12.0		

Ash Fusion Temperatures, F (Reducing Atmosphere)	
Initial Deformation	1950
Softening	2030
Hemispherical	2100
Fluid	2150

Other

Btu/lb = 10,950 as received
Grindability = 56 HGI
Theoretical air = 7.6 lb/10,000 Btu

Ash Analysis, %	
Si as SiO_2	45.0
Al as Al_2O_3	20.0
Fe as Fe_2O_3	18.0
Ti as TiO_2	1.0
Ca as CaO	7.0
Mg as MgO	1.0
Na as Na_2O	0.6
K as K_2O	1.9
S as SO_3	3.5
P as P_2O_5	0.2

Table 4.2
Mechanical Equipment 500 MW PC-Fired Plant

Steam Generator
Balanced draft, direct-fired, pulverized coal

Pulverizers
Six, roller-mill, 50 t/h each

Main Steam		Reheat Steam	
Flow	3.625 x 10^6 lb/h	Flow	3.336 x 10^6 lb/h
Temp	1005F	Temp	1000F
Pressure	2415 psi	Pressure	532 psi

Turbine/Generator
Tandem-compound, four-flow, 3600 rpm, 640,000 kVa at 24 kV, 3 phase, 60 Hz.

Fans
FD: 2 motor-driven, total 4,000,000 lb/h
PA: 2 motor-driven, total 710,000 lb/h
ID: 4 motor-driven, total 5,174,000 lb/h

Fan Horsepower Requirements
FD: 2400 HP based on static rise from -0.5 to 14 in. wg
PA: 1700 HP based on static rise from -0.5 to 16 in. wg
ID: 3800 HP based on static rise from -16 to -1 in. wg and 200F gas inlet.

Fan data assumes 8% air heater leakage and 87% fan/motor efficiency.

Feedwater Pumps
Main: 2 turbine-driven, 50% capacity, 5000 GPM each, TDH: 7200 ft
Booster: 2 turbine-driven, 50% capacity, 4800 GPM each, TDH: 1280 ft
Condensate: 2 motor-driven, 100% capacity, 1400 hp, 7000 GPM each, TDH: 630 ft

Table 4.3
Environmental Equipment 500 MW PC-Fired Plant

Wet Limestone Scrubber (full load)

Limestone usage rate:	30.5 t/h
Fixed sludge waste:	106 t/h
SO_2 at scrubber inlet:	36,568 lb/h
SO_2 at scrubber outlet:	3,657 lb/h
Required SO_2 removal:	90%

Particulate Control

Type:	Electrostatic precipitator
Emissions:	0.03 lb/10^6 Btu
Specific collector area:	350 ft^2/1000 SCFM

Table 4.4
Plant Performance Summary 500 MW PC-Fired Plant

Turbine Steam Conditions	
Pressure	2400psig
Temperature	1000F
Reheat temperature	1000F

Fuel Input
Coal feed 223 t/h
BTU input 4.88 x 10^9 Btu/h

Ash Flow Rates	t/h
Furnace bottom ash	5.1
Economizer ash	2.2
Fly ash	20.7
Total through unit	28.0
Mill rejects (pyrites)	0.5

Heat Rates at Full Load	
Steam cycle	7880 Btu/kWh
Gross	8985 Btu/kWh
Net	9725 Btu/kWh

Rated Output Power Production	
Total gross	536 MW
Total Net	502 MW

Thermal Discharge	10^6 Btu/h
Cooling towers	2500
Stack	587

Controlled Emissions	lb/h
SO_2	3700
NO_x as NO_2	2900*
Particulate	146

Efficiencies	
Boiler (coal to steam)	88%
Plant (coal to busbar)	35%

Untreated Flue Gas Characteristics

Component	Volume % Dry	Wet
Nitrogen	81.07	73.47
Carbon dioxide	15.28	13.84
Oxygen	3.26	2.95
Water	---	9.39
Sulfur dioxide	0.39	0.35
	100.00	100.00

Flue Gas Molecular Weight
30.85 (dry basis)
29.64 (wet basis)

Excess air leaving economizer 17.9%

Mass Flows	lb/10,000 Btu
Actual dry air	8.888
Wet gas from fuel	0.800
Moisture in air	0.116
Wet gas weight	9.804
Water from fuel	0.444
Water in wet gas	0.560
Dry gas weight	9.244

Water in wet gas (as%) 5.711%

Losses	%
Dry gas loss	4.659
Water from fuel	5.056
Moisture in air	0.116
Unburned Combustibles	0.500
Radiation	0.170
Unaccounted for Manufacturer's Margin	1.500
	12.001

*Federal Standard. Current technology can attain lower.

Appendix 2
Codes and Standards

There are many codes and specifications which control the design and construction of boilers, steam generators, pressure vessels and piping. Two of the most important are the American Society of Mechanical Engineers (ASME) Boiler and Pressure Vessel Code (BPVC) and the ASME B31 Pressure Piping Code. Pressure-containing components, collectively called pressure parts, may be required by law to meet the requirements of these codes. In the United States (U.S.) and Canada (and recently in some other countries), state, city and provincial laws enforced by local jurisdictions require new pressure parts to comply with the requirements of the ASME BPVC. The design and construction of pressure parts for repair or replacement are established by the local jurisdiction (see Note below) and/or insurance carrier generally following the National Board Inspection Code (NBIC) in the U.S.

In addition to the BPVC and B31 Pressure Piping Codes, the ASME Performance Test Code provides uniform procedures for the testing and performance evaluation of power plant equipment. This code is frequently used to establish contract compliance of a particular component.

The ASME Boiler and Pressure Vessel Code

The ASME BPVC establishes rules of safety governing the design, fabrication and inspection during construction of boilers, pressure vessels and nuclear power plant components. The objectives of the rules are to assure reasonably certain protection of life and property and to provide a margin for deterioration in service. These rules do not provide criteria for thermal performance, but rather set minimum necessary guidelines for structural integrity to ensure safe operation during the expected component life. The BPVC provides a systematic approach to evaluating the stresses and applying material properties in a way which provides a safe pressure vessel design. The rules are established through a structured voluntary consensus code writing system and are implemented through specific contract terms and/or adoption by various jurisdictions (state, local, city, etc.).

Note: *Local jurisdiction* refers to the municipal, state, provincial or federal authorities enforcing boiler and pressure vessel laws or regulations.

ASME BPVC history

To understand why the ASME Boiler and Pressure Vessel Committee was formed and how it operates, some historical background is necessary.

During the mid 1880s, explosions in fire tube boilers were common. Design knowledge and construction expertise generally were held by large companies, such as Babcock & Wilcox (B&W), and were based essentially on experience.

In 1887, the American Boiler Manufacturers Association (ABMA) was organized to develop general rules for design and construction of boilers and pressure vessels. However, participants were reluctant to give up trade secrets and the effort to develop rules or standards failed. There was, however, some exchange of technical information such as materials data, riveting methods, safety factors, head and flange design rules and hydrostatic testing which helped manufacturers improve the safety of their products.

From 1898 to 1903, more than 1200 people were killed in the U.S. in 1900 separate boiler explosions. The catastrophic explosion of a fire tube boiler in a factory in Brockton, Massachusetts, in 1905 killed 58 people. This tragedy led the governor of Massachusetts to request and obtain the first legal set of rules for the design and construction of boilers. The next year a similar set of boiler design and construction rules was issued by the State of Ohio. Other jurisdictions followed with their own rules. This wide array of rules and regulations caused problems for manufacturers and users because equipment which was acceptable and met the rules in one state was frequently not acceptable in another.

Colonel E.D. Meier, a founder of the ABMA, campaigned to get acceptance of uniform boiler laws. In 1911, he was elected president of the ASME. Finally, on February 13, 1915, the first ASME Boiler and Pressure Vessel Code of 147 pages was issued. It was entitled *Boiler Construction Code, 1914 Edition*.

ASME BPVC additions

After the 1914 Edition was issued, the following sections were issued:

1921 Section III — Boilers for Locomotives. (With the 1962 Edition, this Section was integrated into Section I and the Section III designation was later reassigned.)

1922 Section V — Miniature Boilers. (With the 1962 Edition, this Section was integrated into Section I and the Section V designation was later reassigned.)
1923 Section IV — Low Pressure Heating Boilers.
1924 Section II — Material specifications. (Until 1924, materials were included as part of Section I.)
1925 Section VIII — Unfired Pressure Vessels. (With the 1968 Edition, this title was changed to Pressure Vessels, Division 1 and a new division was issued.)
1926 Section VI — Rules for Inspection (of Power Boilers). (With the 1970 Addenda, this section was reassigned with a new title.)
1926 Section VII — Suggested Rules for Care of Power Boilers.
1937 Section IX — Welding Qualifications. (This section was originally a supplement to Section VIII, but with the 1941 Edition the section was published separately.)
1963 Section III — Nuclear Vessels. (With the 1971 Edition, this section became Nuclear Power Plant Components.)
1968 Section VIII, Division 2 — Alternative Rules for Pressure Vessels. (With the 1968 Edition, Section VIII became Section VIII, Division 1 Pressure Vessels.)
1968 Section X — Fiber-Reinforced Plastic Pressure Vessels.
1970 Section XI — Rules for Inservice Inspection of Nuclear Power Plant Components.
1970 Section VI — Recommended Rules for Care and Operation of Heating Boilers.
1971 Section V — Nondestructive Examination.
1975 Section III, Division 2 — Code for Concrete Reactor Vessels and Containments.

As of 1992, the current sections of the BPVC are summarized in Table 1.

Table 1
ASME Boiler and Pressure Vessel Code Sections

I	Power Boilers
II	Materials
III-1	Division 1, Rules for Nuclear Power Plant Components
III-2	Division 2, Code for Concrete Reactor Vessels and Containments
IV	Heating Boilers
V	Nondestructive Examination
VI	Recommended Rules for the Care and Operation of Heating Boilers
VII	Recommended Guidelines for the Care of Power Boilers
VIII-1	Pressure Vessels, Division 1
VIII-2	Pressure Vessels, Division 2 — Alternative Rules
IX	Welding and Brazing Qualifications
X	Fiber-Reinforced Plastic Pressure Vessels
XI	Rules for Inservice Inspection of Nuclear Power Plant Components

Organization of the ASME Boiler and Pressure Vessel Committee

The first ASME Boiler and Pressure Vessel Committee in 1911 had seven members. By the early 1990s, the ASME Boiler and Pressure Vessel Committee had grown to about 800 people involved at all levels from working groups to the Boiler and Pressure Vessel Main Committee. The BPVC tries to maintain balance on the committees with representatives from manufacturers, users, insurance carriers, jurisdictions and other areas. However, the representatives from these groups may maintain their position within the Boiler and Pressure Vessel Committee even if they change their affiliation.

The administrative structure of the ASME Boiler and Pressure Vessel Committee consists of the Main Committee, the Executive Committee, the Conference Committee and the Headquarters Staff. Subcommittees report to the Main Committee.

In general, there are three different types of subcommittees reporting to the main committee. First, there are the component or *book* committees which are responsible for all the rules of design and construction for a particular type of component and for that Section of the BPVC in which the rules are given. Presently, the book subcommittees are:

Section I — Subcommittee on Power Boilers (SC I)
Section III — Subcommittee on Nuclear Power (SC III)
Section IV — Subcommittee on Heating Boilers (SC IV)
Section VIII — Subcommittee on Pressure Vessels (SC VIII)
Section X — Subcommittee on Reinforced Plastic Pressure Vessels (SC X)

Second, there are the service book committees. These subcommittees are responsible for developing a specific set of rules which may be used by the component committee. Currently, these are:

Section II — Subcommittee on Materials (SC II)
Section V — Subcommittee on Nondestructive Examination (SC V)
Section IX — Subcommittee on Welding (SC IX)
Section XI — Subcommittee on Nuclear Inservice Inspection (SC XI)

Finally, there are the service committees which are responsible for a specific area of technology or administration of the BPVC that may be applied to the book sections. These currently are:

Subcommittee on Design (SC D)
Subcommittee on Safety Valve Requirements (SC SVR)
Subcommittee on Boiler and Pressure Vessel Accreditation (SC BPVA)
Subcommittee on Nuclear Accreditation (SC NA)

Code Editions, Code Addenda and Code Cases

A new edition of the BPVC is issued every three years (. . ., 1986, 1989, 1992, . . .). The code requirements become mandatory on the date of issue. No items are included in the new edition of the BPVC which have not been covered by the previous edition or its three Addenda.

Additions, revisions and deletions to the BPVC are contained in the Code Addenda, issued every year on December 31. The Code Addenda becomes mandatory six months after the date of issue although its contents are optional up to that date. This permits a thorough review by the public without having to comply with mandatory rules. It also gives a transition period for changing any affected company standards.

Except for Section III, the date of the appropriate edition and addenda of the BPVC for a particular contract is the date the contract is negotiated or signed. Most sections indicate that it is the date of the contract, although the actual signing of the contract may come very much later. Section III permits, for some applications, the use of an edition and addenda which are not the latest versions. In general, this is used for construction of a component which is used for repair or replacement. The other sections of the BPVC do not control the repair or replacement of components after the product has been installed; consequently, the decision of the date of the edition and addenda is the responsibility of the local jurisdiction and/or the insurance carrier.

Code Cases are contained in two special volumes independent from the BPVC Sections. Code Cases are issued four times a year after each Main Committee meeting and are subject to the same approval procedures as other changes and additions. Code Cases permit the use of a material or construction for which there are no current BPVC rules. The Code Cases are kept very specific to limit their usage. Each is reviewed for incorporation into the text of the appropriate BPVC Section. Code Cases are numbered sequentially, and nuclear Code Cases carry an N prefix. Code Case use is optional, not mandatory. When one is chosen, all parts are mandatory and the Code Case number shall be listed on the Manufacturer's Data Report form.

Technical Inquiries

When any user of the BPVC is unable to interpret a specific rule, a Technical Inquiry may be sent to the Boiler and Pressure Vessel Committee for an interpretation. Each Section of the BPVC includes an appendix that describes the preparation of Technical Inquiries. It is important that the Inquiry meet all of the requirements so that action can be taken without delay. Inquiries are approved by two methods. The first is consideration by a special group of five people who must vote unanimously on the answer. If that method fails, the inquiry must be considered and approved by the entire subcommittee. Once the inquiry and reply are approved, a reply is sent to the inquirer. The question and reply are also published in the *Interpretations*, issued twice each year to BPVC subscribers.

There are several reasons that an inquiry will not be handled by the Committee:

1. indefinite question — no reference to BPVC paragraph or rule,
2. semi-commercial question — doubt as to whether question is related to BPVC requirements or is asking for design approval,
3. approval of specific design — inquirer wants approval of specific design or construction as meeting BPVC requirements, and
4. basis or background of BPVC rules — inquirer wants rationale or basis of BPVC requirements. These are not given. Suggested revisions and additions are accepted for consideration.

General design philosophy and safety factors

The basic design philosophy of the BPVC from its beginning was one in which the primary membrane stress, or maximum direct stress, does not exceed the allowable stress. (See Chapter 7 for an in depth discussion of stress analysis and stress categorization, e.g., primary membrane stress.) The calculated stress is based on the Maximum Stress Theory. This design philosophy is still used in Section I; Section VIII, Division 1; and Section IV. In most instances, these sections do not call for a detailed analysis to determine the local and secondary stresses, instead they rely on a design-by-formula approach.

The BPVC makes a significant distinction between allowable stress limits and the physical properties of the material such as ultimate tensile strength. The published allowable stresses incorporate a safety factor which reduces measured properties to account for: 1) the degree of complexity of the stress evaluation method, 2) certain levels and types of stress concentration, 3) certain nonuniformity in the materials, and 4) geometric factors. The safety factors are based upon experience, experimental evidence and theoretical evaluations. In general, where the BPVC allows a more detailed exact stress evaluation, a smaller safety factor is permitted in determining the allowable stress. The BPVC therefore specifies the evaluation procedures and the applicable allowable stresses (including safety factors) to provide a uniform, safe design procedure.

In the original BPVC in 1914, the allowable stress limits for these three sections were based on a safety factor of five applied to the minimum specified ultimate tensile strength of the material. However, in December 1931, a joint American Petroleum Institute (API)-ASME Committee on Unfired Pressure Vessels was formed to develop a special code for Petroleum Liquids and Gases. The first edition was issued in September 1934. The API-ASME Code permitted a safety factor of four with only minor changes in the design equations used for establishing the minimum wall thickness of cylinders and heads, for reinforced openings, and for other parts. At that time, the ASME BPVC maintained a safety factor of five with some special provisions which permitted a safety factor of four. One of the reasons for the special provisions was to conserve metal during World War II.

The safety factor of five, except for the special provisions in Section VIII, was used by all sections of the BPVC through the 1949 Edition. Based upon the satisfactory experience in using a safety factor of four with the API-ASME Code and the special paragraphs of Section VIII, the 1950 editions of Section I and Section VIII were issued with a safety factor of four. The safety factor of Section IV remains at five.

During the 1950s, Admiral Hyman Rickover led the development of the first nuclear submarine, the *Nautilus*. Westinghouse's Bettis Atomic Power Laboratory was the prime contractor, and B&W designed and built the reactor vessel, pressurizer and steam generators. Bettis, General Electric's Knolls Atomic Power Laboratory and the U.S. Atomic Energy Commission developed and published rules for the stress analysis and design basis. This was entitled *Tentative Structural Design Basis for Reactor Vessels*. These rules permitted increased allowable stresses because a thorough and accurate stress analysis was required.

During the late 1950s, the ASME Code Committee and the U.S. government began to see the need for an ASME code which would contain design rules similar to the tentative basis. The first consideration was to do a major upgrading of Section VIII either by a supplemental document or major revision of Section VIII. After much discussion in various committees, it was decided to develop a new code section.

In 1958, the ASME established a Special Committee to Review Code Stress Basis. This special group met many times to develop and review all facets of nuclear vessel requirements — not only design rules, but also fabrication, inspection, materials and other kinds of rules. Fatigue analysis methods were introduced for explicit use for the first time.

In 1963, a public hearing was held in Baltimore, Maryland. As a result of the meeting, Section III, *Nuclear Vessels*, was issued which permitted a safety factor of three (on tensile strength). However, in order to use this factor, many requirements had to be met. A very thorough stress analysis procedure, called design by analysis, and the maximum shear stress theory had to be applied. Also, more rigid fabrication, examination and testing procedures were required to allow a reduction in the safety factor.

While the development of the Nuclear Code was being concluded in late 1961, a task group was formed to develop the original planned extension of the Section VIII. The original title was *Task Group on Code for Industrial Pressure Vessels of Superior Quality*. Later, this name was changed to *Alternative Rules for Pressure Vessels* and ultimately it became Section VIII, Division 2.

In 1968, the new Division was issued as an alternative method for constructing pressure vessels: Section VIII, Division 2. It required a much more extensive structural investigation to reduce the safety factor to three.

Allowable stress values

For Section I, Section VIII-1 and Section III-1 Classes 2 and 3:

The BPVC provides the basis for establishing allowable stress values for Section I, Section VIII-1 and Section III-1 Classes 2 and 3. (See Chapter 7.) For wrought or cast ferrous and nonferrous materials, the allowable stress values are based on the following criteria:

1. At temperatures below the creep range, the lowest of the following:

   1/4 of the specified minimum tensile strength at room temperature,
   (1.1)/4 of the tensile strength at temperature,
   2/3 of the specified minimum yield strength at room temperature, and
   2/3 of the yield strength at temperature.

   In addition, for austenitic stainless steels and certain nonferrous nickel alloys, excluding bolting, flanges and other strain sensitive usage where slightly greater deformation is not objectionable, the factor on yield strength at temperature may be increased from 2/3 to 0.90.

2. At temperatures in the creep range, the lowest of the following:

   100% of the average stress to produce a creep rate of 0.01% in 1000 hours,
   80% of the minimum stress to cause rupture at the end of 100,000 hours, and
   67% of the average stress to cause rupture at the end 100,000 hours.

For structural grades, the following special limits shall be applied at temperatures below the creep range with no allowable stresses in the creep range:

0.92 additional multiplying factor applied to all allowable stresses and the tensile strength value not to exceed 55 ksi regardless of the actual value in the specification.

Method for development of allowable stress values

Allowable stress values are developed by the Subcommittee on Materials (SC II) by applying the various safety factors to data which the committee has obtained from tests conducted for the committee or various other industrial sources. The published allowable stress values are then the lowest stresses evaluated using the criteria discussed above. (See also Chapter 7.)

For Section VIII-2 and Section III-1 Class 1:

The BPVC also provides the basis for establishing allowable stress values, called *design stress intensity values*, for Section VIII-2 and Section III-1 Class 1. No design stress intensity values have been established in the creep range for Section VIII-2. For Section III-1 Class 1 applications in the elevated temperature range, rules for some materials are given in Code Case N-47. For the various materials permitted and except for bolting materials, the design stress intensity values, i.e., allowable stress values, are based on the lowest value of the following:

1/3 of the specified minimum tensile strength at room temperature,
(1.1)/3 of the tensile strength at temperature,
2/3 of the specified minimum yield strength at room temperature, and
2/3 of the yield strength at temperature except for most austenitic stainless steels and certain nonferrous materials where the factor may reach as high as 90% of the yield strength at temperature. This may result in a permanent strain of as much as 0.1%. When this amount of deformation is not acceptable, the designer must reduce the allowable stress to obtain an acceptable amount of deformation.

There are tables in Section VIII-2 which list factors for limiting permanent strain in high alloy steels and in nickel and high-nickel alloys. In effect, the criteria permit the designer to choose a factor between 67 and 90% of the yield strength depending on the tolerable deformation. For example, suppose that by some limiting clearances the tolerable strain is limited to 0.05% in a shell. The designer can then refer to the table of limiting permanent strain for 18-8 stainless steel and see that the factor on yield strength should be limited to 0.80. The designer then goes to the yield strength values to establish his own allowable stress values.

Bolting materials have their design stress intensity values, i.e., allowable stresses, established by similar, but more restrictive, criteria. (See Appendix 3 of Section VIII-2.)

Strength theories

In the design of boilers and pressure vessels, two theories of failure are considered: failure through excessive elastic deformation, including elastic instability based upon the theory of elasticity, and failure through excessive plastic deformation and plastic instability (incremental collapse). (See Chapter 7.) For the design of boiler components and pressure vessels, elastic failure is usually assumed. Elastic failure is assumed to occur when the elastic limit of the material is reached. Beyond this limit, excessive deformation or rupture is expected. For metallic materials, the elastic limit is measured in terms of tensile strength, yield strength and rupture strength. The latter term, however, is also used in plastic failure.

The three commonly used theories to predict elastic failure are the maximum principal stress theory, the maximum shear stress theory and the distortion energy theory as discussed in Chapter 7.

Test results show that for ductile materials, such as those ferrous and nonferrous materials permitted by the BPVC, either the maximum shear stress theory or the distortion energy theory predicts, yield and fatigue failure better than the maximum stress theory. However, the maximum stress theory is easy to apply in boiler components and pressure vessels where the circumferential and the longitudinal stresses are calculated using simple formulas with an adequate safety factor to set the allowable stress values. Where a more exact analysis is desired and the safety factor may be less, the BPVC uses the maximum shear stress theory. Section I, Section IV, Section VIII-1, and Section III-1 Subsections NC, ND and NE use the maximum stress theory while Section III-1 Subsection NB and Section VIII-2 use the maximum shear stress theory. The former use a safety factor of four or five on tensile strength; the latter use a safety factor of three.

Design criteria

The design criteria for different sections of the BPVC are related to the theory of failure and safety factors which are applied to the allowable stress values. For both Section I, *Power Boilers,* and Section IV, *Heating Boilers*, the requirements call for the calculation of the required minimum wall thickness which will keep the basic circumferential stress at or below the allowable stress limit for that section. Additional rules, equations and charts are given to determine the minimum required thicknesses of components other than those with circular cross-sections. A more detailed stress analysis may be required for special designs and configurations for which there are no rules. Both Section I and Section IV recognize that high local stresses (discussed below) and secondary stresses may exist within a component, but those sections have established design rules that keep stresses at a safe level with a minimum of additional analysis.

Section VIII-1, *Pressure Vessels,* and Section III-1 Class 2 and 3, *Nuclear Components*, have similar design criteria to Sections I and IV except Sections VIII-1 and III-1 Class 2 and 3 require the minimum thickness of cylindrical shells to be calculated in both the circumferential and the longitudinal directions. The actual minimum required thickness of the cylinder may be set by stress in either direction. In addition, Sections VIII-1 and III-1 Class 2 and 3 have rules which permit the combination of primary membrane stress and primary bending stress to be as high as 1.5 S where S is the basic allowable stress value. At elevated temperatures, consideration must be given to inelastic strain due to creep.

The design criteria for Section VIII-2 provide rules for the more common shell and head geometries for temperatures where the design stress intensity values have been set by either tensile strength or yield strength. A detailed stress analysis is required for most complex geometries and loadings in addition to internal pressure loading. For Section III-1 Class 2 and 3, the requirements are similar; however, there are fewer formulas and rules which require analysis most of the time.

Both Sections VIII-2 and III-1 Class 1 have adopted the maximum shear stress criteria. This method requires calculated stresses to be assigned to various categories and subcategories that permit different allowable stresses for various combinations of calculated stresses. The various categories and subcategories of stresses are:

1. primary stresses
   a. general primary membrane stress
   b. local primary membrane stress
   c. primary bending stress
2. secondary stresses
3. peak stresses

Primary stresses are caused by loadings which are necessary to satisfy the laws of equilibrium between applied forces and moments, and they are not self-limiting by deformation and redistribution. Secondary stresses are developed by self-constraint in the structure. A basic characteristic of secondary stresses is that they are self-limiting. That is, rotation, deflection or deformation takes place until the forces and moments are in balance even though some permanent geometric change may have occurred. Peak stresses are highly localized stresses in a structure usually caused by an abrupt change in the geometry such as a notch or a nozzle fillet weld. Peak stresses are considered only during fatigue analysis of cyclic loadings.

Failure modes, stress limits and stress categories are related. One of the most important aspects of a stress analysis is to make sure that various stresses are assigned to the proper stress categories. Limits on primary stress are set to prevent plastic deformation and burst. Secondary stress limits are set to prevent excessive plastic deformation which may lead to incremental collapse and to ensure the validity of the use of an elastic analysis for making a fatigue analysis. Peak stress limits are set to prevent fatigue failure due to excessive cyclic loadings. For analysis according to Sections VIII-2 and III-1 Class 1, thermal stresses are considered only in the secondary and peak categories. Those thermal stresses which cause a distortion of the structure are categorized as secondary stresses; those thermal stresses which are caused by suppression of expansion without distortion are categorized as peak stresses.

The ASME Code for Pressure Piping, B31

The ASME Code for Pressure Piping is divided into many sections relative to different applications. Currently, these sections are:

B31.1	Power Piping
B31.2	Fuel Gas Piping
B31.3	Chemical Plant and Petroleum Refinery Piping
B31.4	Liquid Transportation Systems for Hydrocarbons, Liquid Petroleum Gas, Anhydrous Ammonia and Alcohols
B31.5	Refrigeration Piping

B31.8 Gas Transmission and Distribution Piping Systems
B31.11 Slurry Transportation Piping Systems

The scope of jurisdiction of the ASME BPVC Section I applies to the boiler proper and to the boiler external piping. The boiler external piping is that piping connecting the boiler proper with the nonboiler external piping as defined in PG-58 of Section I. For boiler external piping, the materials, design, fabrication, installation and testing shall be done according to ASME B31.1, *Power Piping*. In addition, most of the connecting piping outside of the Section I jurisdiction is done according to ASME B31.1.

Editions, Addenda, Cases and Interpretations

With the responsibility of the B31 Pressure Piping Code Committee resting with ASME, all of the issuance dates and procedures of releasing editions, addenda, cases and interpretations are now the same for the Piping Code as for the BPVC.

Design philosophy, safety factors and strength theory

The design philosophy of the Piping Code is one where the primary membrane stress is not permitted to exceed the allowable stress. In the Piping Code, the longitudinal pressure stress is a major consideration because it combines with longitudinal bending stress developed from the supporting of dead loads and from flexibility stress due to the thermal expansion of the piping system. Although circumferential pressure stress is calculated and sets the minimum required thickness for internal pressure, the longitudinal stresses usually control.

Two sections of the ASME B31 Code for Pressure Piping are primarily used in combination with the ASME BPVC sections. The two are: ASME B31.1 Code for Power Piping, and ASME B31.3 Code for Chemical Plant and Petroleum Refinery Piping. B31.1 is closely associated with Section I while B31.3 is associated with Section VIII. However, there are no ASME requirements assigning one or the other. This is essentially done by the jurisdictions and the laws they enforce.

Both sections of the Piping Code use the maximum stress theory for combining and evaluating stresses. The allowable stress values for Section I and B31.1 are the same as are the safety factors. However, B31.3 is unique in that it uses the maximum stress theory but permits higher allowable stresses which are based on a safety factor of three on ultimate tensile strength similar to Section VIII-2 and Section III-1 Class 1.

Symbols and Abbreviations

Symbols

Symbols used in *Steam* are defined within the chapters. The following is a summary of those that are commonly used. Specialized symbols are not repeated here. Symbols for chemical elements appear in Appendix 1, Table 3.2. The frequently used units are listed for each parameter, for illustration purposes. Other units used are defined in the chapter text.

a	crack size, ft or in. (m or mm)
a	inside radius, ft or in. (m or mm)
a, b, c	general coefficients
a	major axis of an ellipsoid, ft or in. (m or mm)
A	area, cross-sectional area or surface area, ft^2 (m^2)
A	mass number (total number of protons plus neutrons)
A_n	cross-sectional area of nozzles, ft^2 (m^2)
b	minor axis of an ellipsoid, ft or in. (m or mm)
b	outside radius, ft or in. (m or mm)
B	barometric pressure, in. Hg
B	ratio of throat diameter to pipe diameter, dimensionless
B_w	moisture fraction, lb/lb (kg/kg)
c	distance, ft (m)
c	speed of light
c_p	specific heat at constant pressure, Btu/lb F (J/kg K)
c_{pa}	specific heat of air, Btu/lb F (J/kg K)
c_{pg}	specific heat of gas, Btu/lb F (J/kg K)
c_v	specific heat at constant volume, Btu/lb F (J/kg K)
C	crack geometry factor, dimensionless
C or c	coefficient
C_a	concentration of specie at duct conditions, lb/lb (kg/kg)
C_{eff}	collection/removal efficiency, dimensionless or percent
C_f	cleanliness factor, dimensionless
C_i	concentration of specie in the impinger, lb/lb (kg/kg)
C_{in}	concentration at inlet, lb/lb (kg/kg)
C_m	mass flow of specie, lb/h (kg/s)
C_{out}	concentration at outlet, lb/lb (kg/kg)
C_p	pitot tube correction factor, dimensionless
$C_{ppm,d}$	concentration of specie on a dry basis, ppm
$C_{ppm,w}$	concentration of specie on a wet basis, ppm
C_q	coefficient of discharge, dimensionless
C_s	concentration of specie at standard conditions, lb/lb (kg/kg)
C_t	energy release rate (power) parameter, in. lb/in.^{2} h (J/m^2 s)
C_t	thermal capacitance, Btu/ft^3 F (J/m^3 K)
C_V	valve flow coefficient
d	diameter, ft or in. (m or mm)
d_a	percent reduction in area
D	diameter, ft or in. (m or mm)
D or d	displacement, ft (m)
D_e or D_H	equivalent or hydraulic diameter = 4A/Per, ft or in. (m or mm)
DL	dead load, psi (Pa)
e	available energy, Btu/lb (J/kg)
E	emission rate, dry, at 0% O_2, lb/h (kg/s)
E	energy, Btu (J)
E	weld joint efficiency or ligament efficiency, dimensionless
E	Young's modulus of elasticity, psi (Pa)
E_b	blackbody emissive power, Btu/h ft^2 (W/m^2)
EQ	seismic loading, psi (Pa)
Eu	Euler number = $\rho V^2/P$, dimensionless
f	friction factor, dimensionless
f_x	body force
F	geometry correction factor, dimensionless
F	heat exchanger arrangement factor, dimensionless
F	radiation configuration factor, dimensionless
F_a	crossflow arrangement factor, dimensionless
F_A	allowable axial compressive stress, psi (Pa)
F_d	F-factor, dry, at 0% O_2, dimensionless
F_d	tube bundle depth factor, dimensionless

F_e	effectiveness factor, dimensionless
F_{pp}	fluid property factor
FS	factor of safety, dimensionless
F_T	fluid temperature factor, dimensionless
$\mathcal{F}$	total radiation exchange factor, dimensionless
g	acceleration of gravity, 32.17 ft/s^2 (9.8 m/s^2)
g	Gibbs free energy, Btu/lb (J/kg)
g_c	proportionality factor, 32.17 lbm ft/lbf s^2 (1 kg m/N s^2)
G	incident thermal radiation, Btu/h ft^2 (W/m^2)
G	mass flux or mass velocity, lb/h ft^2 (kg/m^2 s)
G_{max}	critical flow mass flux, lb/h ft^2 (kg/m^2 s)
Gr	Grashof number $= q\,\beta(T_s - T_f)\,\rho^3 L/\mu$, dimensionless
h	heat transfer coefficient, Btu/h ft^2 F (W/m^2 K)
h	thickness, ft or in. (m or mm)
h_{ct}	contact coefficient, Btu/h ft^2 F (W/m^2 K)
h_c	crossflow heat transfer coefficient, Btu/h ft^2 F (W/m^2 K)
h_c'	crossflow velocity and geometry factor
h_l	longitudinal heat transfer coefficient, Btu/h ft^2 F (W/m^2 K)
h_l'	longitudinal flow velocity and geometry factor
H	enthalpy, Btu/lb (J/kg)
H_f	enthalpy of saturated water, Btu/lb (J/kg)
H_{fg}	enthalpy of vaporization, Btu/lb (J/kg)
H_{fw}	enthalpy of feedwater, Btu/lb (J/kg)
H_g	enthalpy of saturated steam, Btu/lb (J/kg)
I	electrical current, amperes
I	intensity
I	moment of inertia, in.^{4} (m^4)
J	mechanical equivalent of heat, 778.17 ft lb/Btu (1 N m/J)
J	radiosity, Btu/h ft^2 (W/m^2)
J_I	J-integral, in. lb/in.^{2} (J/m^2)
J_{IC}	J-integral, critical, in. lb/in.^{2} (J/m^2)
J_R	material crack growth resistance, in. lb/in.^{2} (J/m^2)
k	coefficient or parameter
k	specific heat ratio, dimensionless
k	thermal conductivity, Btu/h ft F (W/m K)
k_c	insulation thermal conductivity, Btu/h ft F (W/m K)
k_T	isothermal compressibility
K	discharge coefficient, dimensionless
K	factor or coefficient
K	thermal conductance, Btu/h F (W/K)
K_{eq}	equilibrium constant, dimensionless
K_I	stress intensity factor, ksi in.$^{1/2}$ (MPa m$^{1/2}$)
K_{IC}	critical stress intensity factor or fracture toughness, ksi in.$^{1/2}$ (MPa m$^{1/2}$)
K_r	stress intensity factor/toughness ratio, dimensionless
K_y	mass transfer coefficient, lb/ft^2 s (kg/m^2 s)
$l_{\parallel}$	tube centerline spacing parallel to flow, ft (m)
$l_{\perp}$	tube centerline spacing transverse to flow, ft (m)
l_e	equivalent plate thickness, ft or in. (m or mm)
L or l	length or dimension, ft or in. (m or mm)
L_h	fin height, ft or in. (m or mm)
$LMTD$	log mean temperature difference, F (C)
LS	limestone consumption, lb/h (kg/s)
L_t	fin spacing, ft or in. (m or mm)
m	mass, lb (kg)
$\dot{m}$	mass flow rate, lb/h (kg/s)
m_s	mass of steam, lb (kg)
m_w	mass of water, lb (kg)
M	moisture in steam sample, percent
M	molecular weight, lb/lb-mole (kg/kg-mole)
M_d	molecular weight of gas, dry basis
M_o	redundant bending moment, lb ft (N m)
M_w	molecular weight of gas, wet basis
n	parameter, coefficient or number of cycles
N	coefficient, number of cycles or number of concentric thermocouple shields
Nu	Nusselt number $= h\,L/k$, dimensionless
N_v or N	number of velocity heads, dimensionless
p	particle size, in. (mm)
p_r	partial pressure, atm
P	parameter
P	pressure or partial pressure, psi (Pa)
P	temperature ratio for surface arrangement factor, dimensionless
Pe	Peclet number = RePr, dimensionless
Per	perimeter, ft or in. (m or mm)
PL	furnace gas pressure loading, psi (Pa)
P_m	pressure at meter, psi (Pa)
Pr	Prandtl number $= c_p\mu/k$, dimensionless
P_s	absolute pressure at duct, psi (Pa)
ΔP	pressure differential, psi (Pa)
q	heat flow rate, Btu/h (W)
q''	heat flux, Btu/h ft^2 (W/m^2)
q''_{CHF}	heat flux at CHF conditions, Btu/h ft^2 (W/m^2)
q'''	volumetric heat generation rate, Btu/h ft^3 (W/m^3)
q_{rel}	heat release, Btu/h ft^3 (W/m^3)
q_{rev}	reversible heat input per unit mass, Btu/lb (J/kg)
Q or q	heat input, Btu (J), or heat input per unit mass, Btu/lb (J/kg)
Q	gas flow rate, lb/h (kg/s)
Q	heat absorption or heat added to system, Btu (J)
Q_{ac}	actual volumetric flow rate, ft^3/h (m^3/s)
Q_{sd}	dry volumetric flow rate at standard conditions, ft^3/h (m^3/s)
r	radius or inside radius, ft or in. (m or mm)
R	constant or coefficient
R	forcing function
R	radius or outside radius, ft or in. (m or mm)
R	specific gas constant = **R**/M, lbf ft/lbm R (J/kg K)
R	temperature ratio for surface arrangement factor, dimensionless

R	thermal resistance, h ft^2 F/Btu (m^2 K/W)
R	universal gas constant, 1545 ft lb/lb-mole R (8.3143 kJ/kg-mole K)
Ra	Rayleigh number = GrPr, dimensionless
Re	Reynolds number = $\rho VL / \mu = GL/\mu$, dimensionless
R_f	fouling index — bituminous ash
R_s	slagging index — bituminous ash
R_s*	slagging index — lignitic ash
R_{vs}	slagging index — viscosity
s	specific entropy or entropy, Btu/lb F (J/kg K)
S	area or total exposed surface area for a finned surface, ft^2 (m^2)
S	general source term
S	quantity of entropy, Btu/F (J/K)
S	steam-water slip ratio, dimensionless
SE	stack effect, in. wg/ft
S_f	fin surface area; sides plus peripheral area, ft^2 (m^2)
S_L	longitudinal (parallel) spacing, ft or in. (m or mm)
S_m	allowable stress intensity, psi (Pa)
S_r	applied stress/net section plastic collapse stress ratio, dimensionless
S_T	transverse (side) spacing, ft (m)
St	Stanton number = Nu/RePr, dimensionless
S_u	material tensile strength, psi (Pa)
S_y	material yield strength, psi (Pa)
t	thickness, ft or in. (m or mm)
t	time, h or s
T	temperature, F or R (C or K)
T_{cv}	temperature at critical viscosity, F (C)
T_{sat}	saturation temperature, F (C)
T_w	wall temperature, F (C)
T_{250}	temperature at 250 poise viscosity, F
T°	temperature at initial time, F (C)
T_o	sink temperature, F or R (C or K)
TD	terminal temperature difference, F (C)
TS	tensile strength at temperature, psi (Pa)
Δt	time interval, h or s
ΔT	temperature difference, F (C)
ΔT_{LMTD}	log mean temperature difference, F (C)
ΔT_{sat}	$T_{sat}-T_w$, F (C)
u	internal stored energy, Btu/lb (J/kg)
u, v, w	velocity in x, y, z coordinates respectively, ft/s (m/s)
U	overall heat transfer coefficient, Btu/h ft^2 F (W/m^2 K)
v	specific volume, ft^3/lb (m^3/kg)
V	electrical voltage, volts
V	velocity, ft/s (m/s)
V	volume, ft^3 (m^3)
V_d	dry gas volume, ft^3 (m^3)
V_m	dry gas volume at meter conditions, ft^3 (m^3)
V_n	sample gas velocity through the nozzle, ft/s (m/s)
V_o	redundant shear force, psi (Pa)
V_s	velocity of the flue gas, ft/s (m/s)
w	work per unit mass, ft-lb/lbm (J/kg)
w_{rev}	reversible work per unit mass, ft-lb/lbm (J/kg)
W	width, ft or in. (m or mm)
W	work, ft-lb (J)
W_1	leakage flow rate, lb/h (kg/s)
WL	wind loading, psi (Pa)
x	dimension, ft or in. (m or mm)
x	steam quality, mass fraction steam
x_{CHF}	steam quality at CHF conditions, mass fraction steam
x, y, z	dimensions in Cartesian coordinate system, ft or in. (m or mm)
Δx	change in length, ft or in. (m or mm)
X	weight fraction, dimensionless
Y	compressibility factor, dimensionless
Y	Schmidt fin geometry factor, dimensionless
Y_g	concentration in bulk fluid, lb/lb (kg/kg)
Y_i	concentration at condensate interface, lb/lb (kg/kg)
Z or z	elevation or length, ft or in. (m or mm)
Z	atomic number (number of protons in the nucleus)
Z	Schmidt fin geometry factor, dimensionless

Greek symbols

α	absorptivity, dimensionless
α	angle, deg (rad)
α	coefficient of thermal expansion, ft/ft (m/m)
α	thermal diffusivity = $k / \rho c_p$, ft^2/s (m^2/s)
α	void fraction, dimensionless
β	deflection or rotation influence coefficients
β	volume coefficient of expansion, 1/R (1/K)
Γ	effective diffusion coefficient
δ	diameter or film thickness, ft or in. (m or mm)
ε	emissivity, dimensionless
ε	roughness, in. (mm)
ε	strain, ft/ft (m/m)
η	efficiency as a fraction or percentage
η	Schmidt fin efficiency, dimensionless
η_{net}	net efficiency, dimensionless
η_{th}	thermal efficiency, dimensionless
θ	angle, deg (rad)
μ	dynamic viscosity, lbm/ft s (kg/m s)
μ	Poisson's ratio, dimensionless
ν	kinetic viscosity, ft^2/s (m^2/s)
π	3.1415926
ρ	density, lb/ft^3 (kg/m^3)
ρ	reflectivity, dimensionless
$\bar{\rho}_d$	average downcomer density, lb/ft^3 (kg/m^3)

ρ_{hom}	homogenous density, lb/ft^3 (kg/m^3)
$\rho(z)$	local density, lb/ft^3 (kg/m^3)
σ	atomic cross-section, barns
σ	standard deviation
σ	Stefan-Boltzmann constant, 0.1713×10^{-8} Btu/h ft^2 R^4 (5.669×10^{-8} W/m^2 K^4)
σ	stress, psi (Pa)
σ_a	allowable alternating stress, psi (Pa)
σ_r, σ_t, σ_z	radial, tangential and axial stress, psi (Pa)
σ_t	thermal stress, psi (Pa)
$\sigma_{y.p.}$	yield point stress, psi (Pa)
σ_1	longitudinal stress, psi (Pa)
σ_2	hoop stress, psi (Pa)
τ	shear stress, psi (Pa)
τ	transmissivity, dimensionless
ϕ^2_{LO}	two-phase friction multiplier, all fluid as a liquid, dimensionless
ϕ	general dependent variable
Φ	two-phase multiplier for a steam-water separator, dimensionless

Subscripts

1	initial or inlet condition
2	final or outlet condition
a	air
amb	ambient
aux	auxliary component or system
b	bulk fluid
c	contact or conduction
cg	convection, gas-side
cv	convection
$clean$	unfouled or clean condition
e	node point east
$econ$	economizer
eq or e	equivalent
f	fluid, fin or saturated water
g	gas or saturated steam
i	inside or ith parameter or initial condition
in	entering a system
j	jth parameter
l	liquid
max	maximum
min	minimum
o	outside or final condition
out	leaving a system
p	node point under evaluation
P	products
r	radiation
rg	radiation, gas-side
R	reactants
s	steam
sg	surface-to-gas
SE	stack effect
t	thermal
w	node point west
w	wall
δ	liquid film surface (gas liquid interface)

Abbreviations

AA	atomic absorption
ABMA	American Boiler Manufacturers Association
abs	absolute
ABS	American Bureau of Shipping
AC	alternating current
ACF	actual cubic feet
ACFM	actual cubic feet per minute
ADS	air density separator
AE	acoustic emissions
AEC	Atomic Energy Commission, U.S.
AFB	atmospheric pressure fluidized bed
AFW	auxiliary feedwater
AGR	advanced gas-cooled reactor
AH	air heater
AIB	auxiliary input burner
AISC	American Institute of Steel Construction
ALARA	as low as reasonably achievable
amb	ambient
AMU	atomic mass units
ANSI	American National Standards Institute
API	American Petroleum Institute
approx	approximate
ARIS	Automated Reactor Inspection System
ARTS	Anticipatory Reactor Trip System
ASD	allowable stress design
ASME	American Society of Mechanical Engineers
ASTM	American Society for Testing and Materials
attemp	attemperator
atm	atmosphere
ATOG	abnormal transient operating guidelines
avg	average
AVT	all-volatile treatment
B/A	base to acid ratio
B&W	Babcock & Wilcox
BACT	best available control technology
BAT	best available technology
bbl	barrel
BCC	body-centered cubic
BCCW	boiler chemical cleaning waste
BCM	boiler condenser mode
BCT	body-centered tetragonal
BCTMP	bleached chemical thermomechanical pulp
BFB	bubbling fluidized bed
BFW	boiler feed water
BGC	British Gas Corporation
bit or bitum	bituminous
BOF	basic oxygen furnace
BOL	beginning of life
BOOS	burners out-of-service
BPVC	Boiler and Pressure Vessel Code (ASME)
BRC	below regulatory concern
BSW	bottom sediment and water
Btu	British thermal unit
BWR	boiling water reactor
BWST	borated water storage tank

C	Celsius or centigrade degrees
C.I.S.	Confederation of Independent States
CAA	Clean Air Act, U.S.
CANDU	Canadian Deuterium Uranium
cc	cubic centimeter
CC	combined cycle
CCZ	Controlled Combustion Zone
CEMS	continuous emissions monitoring systems
CFB	circulating fluidized bed
cfm	cubic feet per minute
CFR	Code of Federal Regulations
CFT	core flood tank
CGA	crack growth analysis
CHF	critical heat flux
CH-TRU	contact-handled transuranic waste
cm	centimeter
CNC	computer numerical control
CNSG	Consolidated Nuclear Steam Generator
coef	coefficient
col	column
cond	condenser
COG	coke oven gas
cP	centipoise
CRDM	control rod drive mechanism
CVD	chemical vapor deposition
CWA	Clean Water Act
CWP	coal water paste
d	day
dB	decibels
DC	direct current
DC	drain cooler
deg or °	degree
DESAH	dual economizer steam air heater
DGV	sampled dry gas volume
DHRS	decay heat removal system
DNB	departure from nucleate boiling
DNBR	DNB ratio
DNC	distributed numerical control
DOE	Department of Energy, U.S.
DPFAD	deformation plasticity failure assessment diagram
DRB	dual register burner
DRE	destruction and removal efficiency
DSCF(M)	dry standard cubic feet (per minute)
EAR	estimated additional resources
EBW	electron beam welding
EC	European Community
ECCS	emergency core cooling system
ECON	economizer
ECS	environmental control systems
ECT	eddy current testing
EDM	electrical discharge machining
EDTA	ethylenediaminetetraacetic acid
eff	efficiency
EI	enhanced ignition (burner)
EMAT	electromagnetic acoustic transducer
emf	electromotive force
EMF	electromagnetic filter
EOR	enhanced oil recovery
EPA	Environmental Protection Agency, U.S.
EPFM	elastic-plastic fracture mechanics
EPRI	Electric Power Research Institute
ERW	electric-resistance welded
ESP	electrostatic precipitator
ESW	electroslag welding
evap	evaporation
F	Fahrenheit degrees
FB	firebrick
FBC	fluidized-bed combustion
FC	fixed carbon
FC/VM	fixed carbon/volatile matter ratio
FC	feedwater control subsystem
FCC	face-centered cubic
FCC	fluid catalytic cracking
FD	forced draft
FEA	finite element analysis
FEGT	furnace exit gas temperature
FES	finite element system
FGD	flue gas desulfurization
FGR	flue gas recirculation
FHyNES	furnace hydrogen damage nondestructive examination system
FMB	marine package boilers
FMEC	failure mode evaluation chart
FRP	fiberglass reinforced plate
FS	factor of safety
FSS	flame safety system
ft	foot/feet
FT	fluid temperature
FURNIS	furnace inspection system
FW	field weld
g	gram
Ga	gauge
gal	gallon
gen	generator
GJ	gigajoule
GMAW	gas metal arc welding
GPM	gallons per minute
gr	grains
GR	gas recirculation
GTAW	gas tungsten arc welding
GTCC	greater than class C
h	hour
HAZ	heat-affected zone
HFRW	high frequency resistance welding
HGI	Hardgrove Grindability Index
HHV	higher heating value
HLW	high level waste
hp	horsepower
HP	high pressure (turbine)

HPI	high pressure injection
HRSG	heat recovery steam generator
HT	hemispherical temperature
HTW	high temperature Winkler
HV	high volatile
HVL	half value layer
HVT	high velocity thermocouple
IAPS	International Association for the Properties of Steam
IASCC	irradiation assisted stress corrosion cracking
IC	ion chromatography
ICS	integrated control system
ID	induced draft, or inside diameter
IDT	initial deformation temperature
IGA	intergranular attack
IGCC	integrated gasification combined cycle
IGSCC	intergranular stress corrosion cracking
IMC	integrated master control subsystem
in.	inch
in. Hg	inches of mercury
in. wg	inches water gauge
INPO	Institute of Nuclear Power Operations
IP	intermediate pressure (turbine)
IPP	independent power producer
IR	infrared radiation
ISO	International Standards Organization
IT	initial deformation temperature
ITS-90	International Temperature Scale 1990
J	joule
K	Kelvin degrees
kcal	kilocalories
kg	kilogram
kHz	kilohertz
kJ	kilojoule
km	kilometer
kmole or kg-mole	kilogram-mole
kPa	kilopascal
ksi	1000 pounds per square inch
kV	kilovolt
kW	kilowatt
kWh	kilowatt hour
l	liter
L/G	liquid to gas ratio
LAER	lowest achievable emission rate
LAF	laboratory ashing furnace
lb	pound
LBBA	leak before break analysis
lbf	pound force
LBLOCA	large break loss of coolant accident
lbm	pound mass
lb-mole	pound-mole
LBP	lumped burnable poison
LEA	low excess air
LEFM	linear elastic fracture mechanics
LFA	life fraction analysis
LFE	life fraction expended
LHV	lower heating value
LLW	low level waste
LMP	Larson-Miller parameter
LMTD	log mean temperature difference
LNCB	low NO_x Cell burner
LNG	liquefied natural gas
LOCA	loss of coolant accident
LP	low pressure (turbine)
LPI	low pressure injection
LPMS	loose parts monitoring system
LSA	low specific activity
LVS	limited vertical sweep
LWR	light water reactor
m	meter
MACT	maximum achievable control technology
MAF	moisture and ash free
MARAD	Maritime Administration, U.S.
max	maximum
MCR	maximum continuous rating
MD	mass defect
med	medium
MeV	million electron volts
MFR	mass flow rate
mg	milligram
MHD	magnetohydrodynamics
MHVT	multiple shield, high velocity thermocouple
mi	mile
MIC	microbial induced corrosion
MIG	metal inert gas
min	minimum
MIST	multiple loop integral systems test
MJ	megajoule
ml	milliliter
mm	millimeter
MMP	managed maintenance program
MMT	minimum metal temperature
mol	molecular
MOV	motor operated valve
mPa	millipascal
MPa	megapascal
mph	miles per hour
MRS	monitored retrievable storage
MSDS	material safety data sheets
MSW	municipal solid waste
MT	magnetic particle testing
MTD	mean temperature difference
mV	millivolt
MV	motor vessel
MW	megawatt, megawatts-electric
MWd	megawatt-day
MW_t	megawatts-thermal
μS	microSiemens

N	Newton
N_{2a}	atmospheric nitrogen
NAPAP	National Acid Precipitation Assessment Program
NBIC	National Boiler Inspection Code
NC	numerical control
NDE	nondestructive examination
NDTT	nil-ductility transition temperature
NFPA	National Fire Protection Association
Nm^3	normal cubic meter
NMVOC	nonmethane volatile organic compounds
NOTIS	nondestructive oxide thickness inspection system
NPDES	national pollutant discharge elimination system
NPHR	net plant heat rate
NPSH	net positive suction head
NRC	Nuclear Regulatory Commission, U.S.
N.S.	nuclear ship
NSAC	Nuclear Safety Analysis Center
NSCW	nuclear service cooling water
NSFO	Navy special fuel oil
NSPS	New Source Performance Standards
NSSS	nuclear steam supply system
NUG	nonutility generator
NWMS	nuclear waste management system
NWPA	Nuclear Waste Policy Act
OBW	oversize bulky waste
OCRWM	Office of Civilian Radioactive Waste Management, U.S.
OD	outside diameter
OFA	overfire air
OT	oxygen treatment
OTSG	once-through steam generator
oz	ounce
Pa	pascal
PA	primary air
PA/PC	primary air and pulverized coal mixture
PAW	plasma arc welding
PAX	primary air exchange (burner)
PC	pulverized coal
PCB	polychlorinated biphenyls
PCDD	polychlorinated dibenzoparadioxin
PCDF	polychlorinated dibenzofurans
PCFAD	computer-based failure analysis diagram
PFBC	pressurized fluidized-bed combustion
pH	negative log of hydrogen ion concentration
PHWR	pressurized heavy water reactor
PIP	peak impact pressure
PMAT	potential maintenance action table
POHC	principal organic hazardous constituent
PORV	power (pilot) operated relief valve
ppb	parts per billion (mass)
ppm	parts per million (mass)
ppmv	parts per million by volume
PR	process recovery
press.	pressure
pri	primary
prox	proximate
PRV	pressure reducing valve
PSD	prevention of significant deterioration
psi	pounds per square inch (absolute or difference)
psia	pounds per square inch absolute
psig	pounds per square inch gauge
P/T	pressure/temperature
PT	liquid/dye penetrant examination
PTC	Performance Test Code
PUC	Public Utilities Commission
pulv	pulverizer
PURPA	Public Utiity Regulatory Policies Act of 1978
PVC	polyvinyl chloride
PWHT	postweld heat treatment
PWR	pressurized water reactor
R	Rankine degrees
rad	radian
RAR	reasonably assured resources
RB	Radiant boiler
RBC	Radiant boiler Carolina type
RBE	Radiant boiler El Paso type
RBT	Radiant boiler Tower type
RC	reactor control subsystem
RC	reactor coolant
RCM	reliability centered maintenance
RCRA	Resource Conservation and Recovery Act
RCS	reactor coolant system
RDF	refuse-derived fuel
recirc	recirculation
red.	reducing atmosphere
rem	roentgen equivalent man
rev	revolutions
RH	reheater
RHAS	radiant heat absorbing surface
RHSH	reheat superheater
RH-TRU	remote handled transuranic waste
RO	reverse osmosis
rpm	revolutions per minute
RPS	reactor protection system
RSG	recirculating steam generator
RT	radiography testing
RWP	radiation work permit
s	second(s)
SAE	Society of Automotive Engineers
sat	saturated conditions
SAW	submerged arc welding
SBLOCA	small break loss of coolant accident
SBW	steam by weight
SCC	submerged chain conveyor
SCC	stress corrosion cracking
SCE	submerged combustion evaporator
SCF(M)	standard cubic feet (per minute)
SCR	selective catalytic reduction

SD	stack draft
SE	stack effect
sec	second(s)
SFAS	safety features actuation system
SG	steam generator
SGTR	steam generator tube rupture
SH	superheater
SI	international system of units
SIP	state implementation plans
SMAW	shielded metal arc welding
SNAME	Society of Naval Architects and Marine Engineers
SNCR	selective noncatalytic reduction
SNG	synthetic natural gas
SOA	state-of-the-art
SPB	Stirling power boiler
sp gr	specific gravity
SP-TRU	special transuranic waste
SR	refractory surface
SR	stoichiometric ratio
SS	stainless steel
S.S.	steam ship
SSU	seconds Saybolt universal
ST	softening temperature
subbitum	subbituminous
SW	water-cooled surface
SWUP	spiral wound Universal Pressure (boiler)
t	ton
T_{250}	viscosity at 250 centipoise
TBS	turbine bypass system
TC	thermocouple
TD	terminal temperature difference
TDF	tire-derived fuel
TDFM	time dependent fracture mechanics
TEG	turbine exhaust gas
temp	temperature
theo	theoretical
TIG	tungsten inert gas
thk	thick
t_m	metric ton
TMI2	Three Mile Island Unit 2
TOC	total organic carbon
TR set	transformer rectifier set
TRC	total residual chlorine
TRS	total reduced sulfur
TRU	transuranic
TS	tensile strength
TSC	two-stage combustion
TSP	total suspended particulate
TSS	total suspended solids
TSV	turbine stop valve
TTT	time-temperature-transformation
TTV	turbine throttle valve
TVL	tenth value layer
UBC	unburned carbon
UCL	unburned carbon loss
UIC	uncompensated ion chamber counter
ULD	unit load demand subsystem
UP	Universal Pressure (boiler)
UPC	Universal Pressure coal-fired (boiler)
U.S.	United States
USCG	United States Coast Guard
USNRC	United States Nuclear Regulatory Commission
U.S.S.	United States ship
UT	ultrasonic
UTT	ultrasonic thickness testing
UV	ultraviolet absorption
VAC	volts alternating current
V	volts
VM	volatile matter
VOC	volatile organic compounds
vol	volume
vs	versus
W	watt
WFMT	wet magnetic particle testing
wg	water gauge
WHRB	waste heat recovery boiler
WIPP	Waste Isolation Pilot Plant
WOCA	world outside centrally planned economies areas
WRC	Welding Research Council
wt	weight
YP	yield point
yr	year
YS	yield strength

Index